Definite Integrals

1 $\displaystyle\int_0^{\pi/2} \sin^2 nx\,dx = \int_0^{\pi/2} \cos^2 nx\,dx = \frac{\pi}{4} \qquad (n \geq 1)$

2 $\displaystyle\int_0^{2\pi} \sin mx \sin nx\,dx = \int_0^{2\pi} \cos mx \cos nx\,dx = 0 \qquad (m \neq n)$

3 $\displaystyle\int_0^{1} (1 - x^2)^{(2n+1)/2}\,dx = \frac{1 \cdot 3 \cdot 5 \cdots (2n+1)}{2 \cdot 4 \cdot 6 \cdots (2n+2)}\,\frac{\pi}{2}$

4 $\displaystyle\int_0^{\pi/2} \sin^{2n} x\,dx = \int_0^{\pi/2} \cos^{2n} x\,dx = \frac{(2n)!}{2^{2n}(n!)^2}\,\frac{\pi}{2}$

5 $\displaystyle\int_0^{\pi/2} \sin^{2n+1} x\,dx = \int_0^{\pi/2} \cos^{2n+1} x\,dx = \frac{2^{2n}(n!)^2}{(2n+1)!}$

6 $\displaystyle\int_0^{\infty} e^{-x^2}\,dx = \frac{1}{2}\sqrt{\pi}$

7 $\displaystyle\int_0^{1} x^m (1 - x)^n\,dx = \frac{m!n!}{(m+n+1)!}$

8 $\displaystyle\iint\limits_{\substack{x \geq 0,\, y \geq 0,\\ x+y \leq 1}} x^m y^n (1 - x - y)^p\,dx\,dy = \frac{m!n!p!}{(m+n+p+2)!}$

9 $\displaystyle\iiint\limits_{\substack{x \geq 0,\, y \geq 0,\, z \geq 0\\ x+y+z \leq 1}} x^p y^q z^r (1 - x - y - z)^s\,dx\,dy\,dz = \frac{p!q!r!s!}{(p+q+r+s+3)!}$

Polar Coordinates

$$\begin{cases} x = r\cos\theta \\ y = r\sin\theta \end{cases}$$

$dA = r\,dr\,d\theta$

Cylindrical Coordinates

$x = r\cos\theta$

$y =$

$z =$

$dV =$

Spherical Coordinates

$$\begin{cases} x = \rho\sin\phi\cos\theta \\ y = \rho\sin\phi\sin\theta \\ z = \rho\cos\phi \end{cases}$$

$dV = \rho^2 \sin\phi\,d\rho\,d\phi\,d\theta$

Vectors

$\mathbf{a} \cdot \mathbf{b} = a_1 b_1 + a_2 b_2 + a_3 b_3$

$\mathbf{a} \times \mathbf{b} = \begin{vmatrix} \mathbf{i} & \mathbf{j} & \mathbf{k} \\ a_1 & a_2 & a_3 \\ b_1 & b_2 & b_3 \end{vmatrix}$

$\nabla = \mathbf{i}\dfrac{\partial}{\partial x} + \mathbf{j}\dfrac{\partial}{\partial y} + \mathbf{k}\dfrac{\partial}{\partial z}$

$\nabla f = \text{grad}\, f = (f_x, f_y, f_z)$

$\nabla \cdot \mathbf{v} = \text{div}\,\mathbf{v} = u_x + v_y + w_z$

$\nabla \times \mathbf{v} = \text{curl}\,\mathbf{v} = \begin{vmatrix} \mathbf{i} & \mathbf{j} & \mathbf{k} \\ \partial/\partial x & \partial/\partial y & \partial/\partial z \\ u & v & w \end{vmatrix}$

Length, Area, and Volume

Triangle

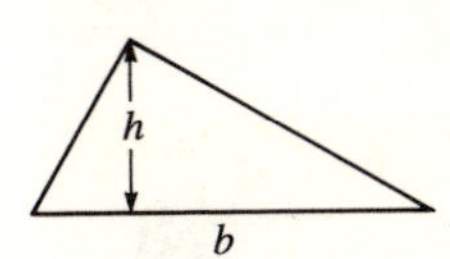

$A = \frac{1}{2}bh$

Parallelogram

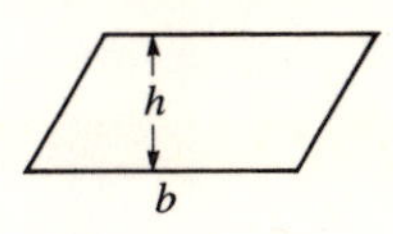

$A = bh$

Circle

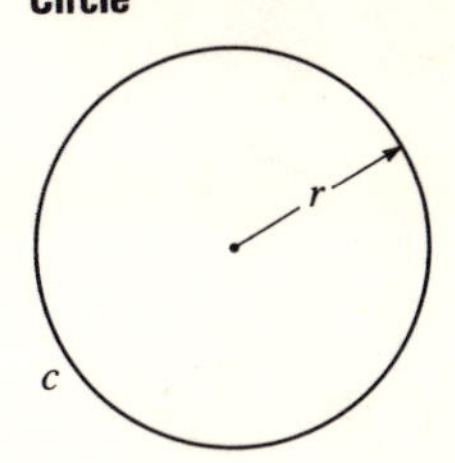

$c = 2\pi r$
$A = \pi r^2$

Right circular cylinder

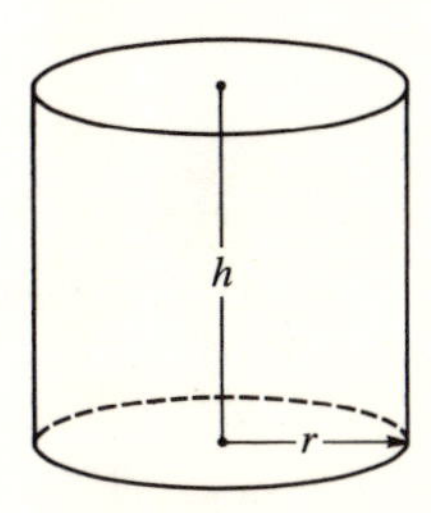

$A = 2\pi rh$
$V = \pi r^2 h$

Right circular cone

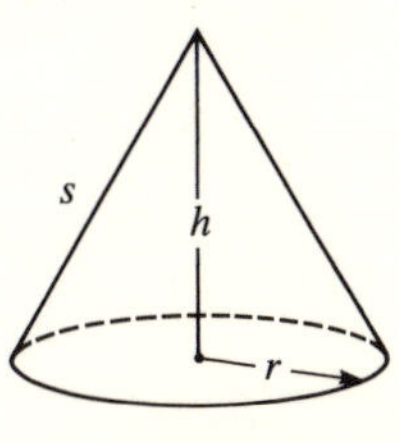

$s = \sqrt{r^2 + h^2}$
$A = \pi rs$
$V = \frac{1}{3}\pi r^2 h$

Sphere

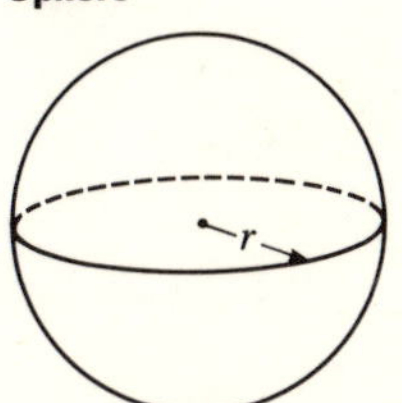

$A = 4\pi r^2$
$V = \frac{4}{3}\pi r^3$

Calculus

Calculus Harley Flanders

Florida Atlantic University

W. H. Freeman and Company
New York

Library of Congress Cataloging in Publication Data

Flanders, Harley.
Calculus.

Includes index.
1. Calculus. I. Title.
QA303.F582 1985 515 84-18773
ISBN 0-7167-1643-7

Book and cover design: Valerie Pettis Design

Printed in the United States of America

1 2 3 4 5 6 7 8 9 0 KP 4 3 2 1 0 8 9 8 7 6 5

Table of Contents

Preface

Calculus can be exciting; no other college subject offers a student so much new scope and power. Teaching the student how to set up and solve calculus problems—how to *use* calculus—is the main objective of this text. In support of this objective, my aim is to provide enough intuition and theory so that the student will understand and have confidence in the techniques of calculus. For each topic, this requires a small amount of introductory theory (with technical details postponed), worked examples to illustrate techniques and applications, and many exercises. Harder theory and unusual topics are clearly marked as optional.

How calculus should be taught is a much debated and dynamic topic. Many years ago, college students generally took a year of precalculus mathematics; the following one-year calculus course was aimed mainly at engineering and physics students. The textbooks were terse, and most applications were to geometry and physics. Things began to change in the fifties, when many first-year students went directly into calculus and the textbooks began to expand, adding analytic geometry and differential equations; the one-year calculus course grew to three—even four—semesters. In the sixties, much more theoretical material was added to introductory calculus as the traditional advanced calculus course disappeared and the level of the upper-division real analysis course rose. The next decade saw two main trends. First, many diverse applications were added to the introductory calculus course in order to service its broadly expanding clientele, which began to include many students outside the fields of mathematics, physics, and engineering. Second, there was a shift from real variable theory to intuition, reflecting both unsatisfactory results in teaching concentrated theory and a growing number of less prepared students. The influence of the computer began to show, particularly in the availability of scientific calculators, now so inexpensive.

This book, for the mid-eighties, is aimed at mainstream calculus students and strives for an optimal balance of intuition and rigor. While a wide range of applications is presented, I have intentionally avoided those that require lengthy digressions into peripheral topics, which can cause a beginning student to lose sight of the main path. Because drill is so important, the exercise sets concentrate on easy and middle level problems; the more challenging exercises are starred. The exercises strike a balance between straightforward drill, computation with newly learned techniques, and applications; they are graded in difficulty, and I hope that I have succeeded in making them accessible and interesting.

Organization Chapter 1 is a review of precalculus mathematics, including the trigonometric and exponential functions. Some instructors will omit this chapter entirely; most will spend only two or three lectures on it, depending on their students' backgrounds. Chapters 2 and 3 cover the derivative and its usual applications; Chapter 2 also includes inverse and implicit functions. Chapter 4 completes the introduction of the elementary transcendental functions.

Chapters 5 to 7 cover integration. The introduction is through easily understood step functions, but Riemann sums are discussed also and used later in applications. The fundamental theorem is introduced as early as possible.

Chapter 8 covers the standard topics of plane analytic geometry, using calculus tools when appropriate. Chapters 9 to 11 form a unit on numerical methods, series, and power series. Chapters 12 and 13 cover vector geometry in three-space, vector functions, and curves. Chapters 14 to 16 cover the differential calculus of several variables, and Chapters 17 and 18 the integral calculus. Finally, Chapter 19 is a brief introduction to differential equations. The answer section contains answers to all odd-numbered exercises. The index, which I prepared with the help of a computer, is unusually thorough, and it includes many synonyms not actually used in the text.

Design Mathematics is hard for students to read, partly because each mathematical expression contains a great deal of information packed into a small space. In the course of much expository and text writing and journal editing, I have given a lot of thought to the issue. The design of this book reflects my conclusions. First, the layout of the text clearly identifies and separates the various elements that make it up: sections, examples, definitions, remarks, etc., so that the student may always know precisely what he is reading and why. Second, there is extra space around mathematical expressions in English sentences. Third, I omit most punctuation marks around mathematical expressions because I feel that commas and periods can interfere with a student's reading of mathematics. Fourth, the length of fraction bars and the spaces within nested parentheses and elsewhere have been adjusted so the parsing is clear. Other such design features will be apparent to those familiar with calculus books, and I hope that all of them will make this book easier for students to read.

The text is lavishly illustrated. Each figure is designed to convey a maximum of information with clarity and a minimum of clutter. In particular, space objects have been drawn with careful attention to the principles of projection, and they are simple line drawings rather than airbrushed or computer-generated drawings, so that students can reproduce them. With rare exceptions, figures appear next to the text that refers to them.

Study Guide The Study Guide was prepared by Marshall Fraser and me. For each section of the book, it contains a summary of objectives, new notation, new vocabulary, warnings, hints on solution strategies, and detailed solutions to selected exercises of each type.

Acknowledgments Most of this book is new material, but I acknowledge the many ideas that have filtered into it from contributions by my coauthors, Robert R. Korfhage and Justin J. Price, of the following texts: *Calculus* (1970), *A First Course in Calculus* (1973), *A Second Course in Calculus* (1974), and *Calculus with Analytic Geometry* (1978), all published by Academic Press, Inc.

In many respects this book represents a team effort. The following mathematicians analyzed my previous calculus books or criticized drafts of parts of this book: Don Albers (Menlo College), Dean Arganbright (Whitworth College), J. E. Benson (Fairleigh Dickinson University), Daniel Drucker (Wayne State University), Underwood Dudley (De Pauw University), Bruce Edwards (University of Florida, Gainesville), David Ellis (San Francisco State University), William R. Fuller (Purdue University), Stuart Goldenberg (California Polytechnic State University, San Luis Obispo), Otis Kenny (Boise State University), Michael Martin (University of Denver), John Neff (Georgia Instutite of Technology), Jack Schiller (Temple University), Robert G. Stein (California State College, San Bernardino), Jim Vick (University of Texas, Austin). In addition, Marshall Fraser, Fred Greenleaf (New York University), Eugene Krause (University of Michigan), and Keith Phillips (New Mexico State University, Las Cruces) patiently reviewed draft after draft of the whole book, contributing greatly to its improvement.

All calculations in the text were checked by Dean R. Hickerson, Anthony Barcellos, and Andrew Kudlacik. Dean Hickerson and Tony Barcellos solved all the exercises; I also did so independently. After such intense cross-checking, we hope that the book and the answer section will be quite accurate, but I accept the ultimate responsibility for errors and will appreciate hearing of them.

The publisher, W. H. Freeman and Company, has done everything humanly and corporately possible to develop and produce this book. The Mathematics Editor, Peter Renz (now back in college teaching), analyzed every draft and every review, as did the Consulting Editor, Victor Klee, and they added more than I can ever thank them for. The Developmental Editor, Carol Pritchard-Martinez, and the Production Editor, Andrew Kudlacik, did absolutely first-rate and very critical jobs with every conceivable detail. The manuscript was typed on a computer by Dawn Schwartz, typesetting was by Progressive Typographers, proofreading by me and by Mary Ann Rosenberg and Elizabeth Marraffino, and page makeup by Ken Ekkens of Publishing Synthesis, Ltd. The book was designed by Valerie Pettis; the illustrations for the text were drawn by J & R Services and for the answer section by Keithley and Associates, Inc. All did outstanding jobs. Numerous others at Freeman contributed to the project: Linda Chaput, Patrick Cunningham, Lisa Douglis, Jill Feldheim, Jerry Lyons, Megan Newman, Neil Patterson, and Lynne Sheehan. Finally, the Mathematics and Computer Science Department of Florida Atlantic University generously supported me with facilities and services.

Harley Flanders
Boca Raton, Florida
October 1984

Calculus

1 Functions and Graphs

This chapter is a review of important material that you studied in college algebra, trigonometry, and precalculus courses, and that is an essential prerequisite for studying calculus. If some of these topics were not covered deeply enough in your previous courses, you will find some new material here, or at least find a different approach to familiar topics.

Some instructors may omit this review chapter and start with Chapter 2, where the subject of calculus begins. Even if your instructor does so, you will probably refer from time to time to topics covered here. You should look this chapter over and make sure that you can work the exercises. If you get stuck on an exercise, study the relevant text carefully. If you are still stuck, ask your instructor for help.

This chapter cannot serve as a substitute for precalculus courses, but it is a source for the things you need to review. Some proofs are included here, but it is perfectly all right to use a mathematical statement (theorem) without knowing how to prove it. What you must know precisely are its hypotheses and its conclusion, and how to apply the theorem to problems.

The first section reviews the real number system, with special emphasis on the number line, order, and inequalities. The following three sections cover functions, coordinates, graphs of functions, and polynomial and rational functions. Section 1-5 contains the distance formula and some of its applications. The next two sections review the trigonometric functions, and Section 1-8 reviews exponents and exponential functions.

1-1 The Real Number System

The real number system, denoted **R**, is the habitat of scientific measurement and of calculus. Everyone is familiar with its arithmetic. In more advanced courses, the system **R** is defined and developed rigorously. However, that deep and lengthy project is not appropriate for a beginning calculus text. Instead, we shall assume the main properties of the real number system as needed. To start with, we shall take for granted the arithmetic operations, addition and multiplication, and their inverses, subtraction and division. We shall also assume their familiar properties, such as the associative, commutative, and distributive laws. See Table 1-1-1 for examples of these laws.

Table 1-1-1
Examples of the laws of arithmetic

Commutative laws

$4.7 + 8.2 = 8.2 + 4.7 = 12.9$

$(4.7)(8.2) = (8.2)(4.7) = 38.54$

Associative laws

$[3.2 + (-7.1)] + 0.5$
$= 3.2 + [(-7.1) + 0.5]$
$= -3.4$

$[(3.2)(-7.1)](0.5)$
$= (3.2)[(-7.1)(0.5)]$
$= -11.36$

Distributive law

$(-9.3)[(6.1) + (-2.9)]$
$= (-9.3)(6.1) + (-9.3)(-2.9)$
$= -29.76$

The Real Line

Real numbers can be used to label points on a line. First we choose a point on the line and mark it 0. Then we choose a point to the right of 0 and mark it 1. In other words, we choose a starting point, a unit length, and a positive direction (the direction from 0 toward 1). Then we mark the points $2, 3, 4, \ldots$ to the right and the points $-1, -2, -3, \ldots$ to the left. (See Figure 1-1-1.) To complete the labeling, we must make a fundamental assumption. We take it as an axiom that *there is a one-to-one correspondence between the points on the line and the system* **R** *of real numbers.* More precisely:

- To each real number a there corresponds a point P_a on the line.
- Each point P on the line is the correspondent of a single real number a.
- If P_a is left of P_b, then $b - a$ is the distance from P_a to P_b.

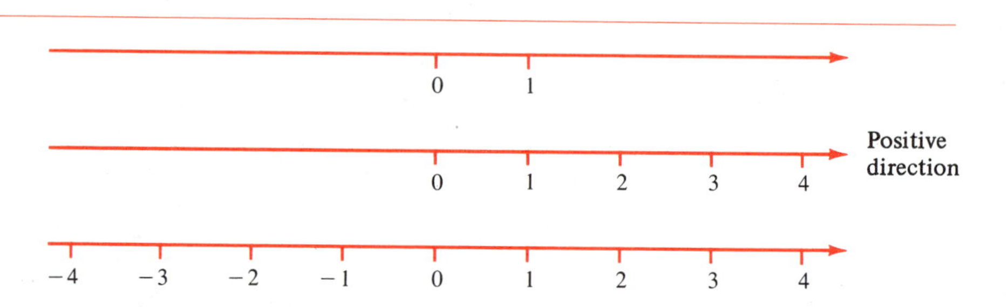

Figure 1-1-1
Labeling the points on a line

Because of the close association of the real number system **R** with the set of points on a line, it is common to refer to a line as the real number system and to the real number system as a line—the **real line** or the **number line.** For instance, in a mathematical discussion, the real number 5.2 and the point labeled 5.2 might be considered the same. We say "3.7 lies to the left of 5.2." Although not correct logically, such language almost never causes confusion; in fact it often sharpens our feeling for a problem.

Completeness

Our geometric intuition about a line tells us that a line has no holes in it. There is no way to move in a plane from one side of a line to the other side without meeting the line. More precisely, suppose that L is a line in the plane and that A and B are two points, one on each side of the line L.

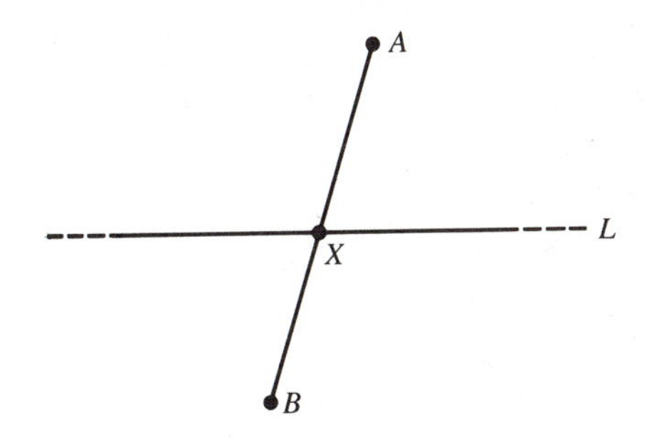

Figure 1-1-2
Completeness: There is no way that AB can slip through L.

See Figure 1-1-2. Then the segment AB cannot somehow slip through L; it must meet L in some point X.

This property, applied to the real line, says something very important about the real number system **R**. It says that in a certain sense **R** has no holes; it is **complete.** We shall make this idea of the **completeness** of the real number system more precise in Chapter 3, and use it for several decisive arguments in calculus.

Order

If number a lies to the left of number b, then we say that "a is **less than** b" and we write

$$a < b$$

Alternatively, we say that "b is **greater than** a" and we write

$$b > a$$

The notation $a \leq b$, which is read "a is less than or equal to b," means that either $a < b$ or $a = b$. The same condition can be written $b \geq a$; then it is read "b is greater than or equal to a."

The relations "less than," "greater than," and so on impose what is called **order** on the real number system. The most basic properties of order are the following:

Properties of Order

- Reflexivity: $a \leq a$.
- Anti-symmetry: If $a \leq b$ and $b \leq a$, then $a = b$.
- Transitivity: If $a \leq b$ and $b \leq c$, then $a \leq c$.
- Trichotomy: If a and b are real numbers, then exactly one of these three relations holds:

$$a < b \qquad a = b \qquad a > b$$

The arithmetic operations $+$, $-$, $\times$, and $\div$ are closely linked to the order relations. The following rules describe this connection:

Rules Connecting Arithmetic and Order

1 If $a < b$ and if c is any real number, then $a + c < b + c$ and $a - c < b - c$.

2 If $a < b$ and $c > 0$, then $ac < bc$ and $a/c < b/c$.

3 If $a < b$ and $c < 0$, then $ac > bc$ and $a/c > b/c$; that is, $bc < ac$ and $b/c < a/c$. In particular, $-1 < 0$, so $-b < -a$.

4 If $0 < a < b$ or $a < b < 0$, then $1/b < 1/a$.

5 The statement $a < b$ is equivalent to the statement $b - a > 0$.

The rules apply just as well with $\leq$ in place of $<$, except that in rule 4 you must have $0 < a \leq b$ or $a \leq b < 0$ because division by 0 is not allowed.

An important and frequently used consequence of these rules is the fact that the square of a real number is always non-negative. Precisely:

6 If x is any real number, then $x^2 \geq 0$. If $x^2 = 0$, than $x = 0$.

Example 1

Start with the relation $4 < 7$.

a Rule 1 with $a = 4$, $b = 7$, and $c = \frac{1}{3}$ implies

$$4 + \tfrac{1}{3} < 7 + \tfrac{1}{3} \quad \text{and} \quad 4 - \tfrac{1}{3} < 7 - \tfrac{1}{3}$$

b Rule 2 with $a = 4$, $b = 7$, and $c = \frac{1}{7}$ implies $\frac{4}{7} < 1$.
c Rule 3 with $a = 4$, $b = 7$, and $c = -1$ implies $-7 < -4$.
d Rule 4 with $a = \frac{2}{3}$ and $b = 1$ implies $1 < \frac{3}{2}$.

Rules 2, 3, and 4 imply the so-called rules of signs:

Rules of signs	Examples
• positive × positive = positive	$5 \times 7 = 35$
• negative × negative = positive	$(-5) \times (-7) = 35$
• positive × negative = negative	$5 \times (-7) = -35$
• $(\text{positive})^{-1}$ = positive	$5^{-1} = 1/5 = \frac{1}{5}$
• $(\text{negative})^{-1}$ = negative	$(-7)^{-1} = 1/(-7) = -\frac{1}{7}$

For instance, to prove that negative × negative = positive, let $a < 0$ and $c < 0$. Then apply rule 3 with $b = 0$ to deduce $ac > 0$. The other rules of signs can be proved similarly.

The following three natural extensions of the rules connecting arithmetic and order are useful in computations:

Additional Rules Concerning Arithmetic and Order

7 If $a < A$ and $b < B$, then $a + b < A + B$.
8 If $0 \leq a < A$ and $0 \leq b < B$, then $ab < AB$.
9 Suppose $a \geq 0$ and $b \geq 0$. Then $a < b$ if and only if $a^2 < b^2$.

Example 2

a $(\sqrt{2})^2 = 2 < \frac{9}{4} = (\frac{3}{2})^2$; hence $\sqrt{2} < \frac{3}{2}$ by rule 9.
b $3 < \frac{49}{16} = (\frac{7}{4})^2$; hence $\sqrt{3} < \frac{7}{4}$ by rule 9.
c $\sqrt{2} + \sqrt{3} < \frac{3}{2} + \frac{7}{4} = \frac{13}{4}$ by rule 7.
d $\sqrt{2} \cdot \sqrt{3} < \frac{3}{2} \cdot \frac{7}{4}$; that is, $\sqrt{6} < \frac{21}{8}$ by rule 8.

Rules 7 and 8 apply also to three or more inequalities. For instance, if $a < A$, $b < B$, and $c < C$, then $a + b + c < A + B + C$.

Absolute Value

The **absolute value** of a real number a, written $|a|$, is a measure of the size of a, regardless of its sign.

Definition Absolute Value

$$|a| = \begin{cases} a & \text{if } a \geq 0 \\ -a & \text{if } a < 0 \end{cases}$$

For example, $|7| = 7$ and $|-7| = 7$. Again, $|-0.035| = 0.035$ and $|0.035| = 0.035$. Also $|0| = 0$. Thus, each real number except 0 has positive absolute value. Even though it is not true that $-6 > 4$, it is correct that $|-6| > |4|$.

By its very definition, $|a|$ is the distance on the number line from a to 0. Thus, the "size" of a is measured by its *distance* from 0, and this distance is a non-negative quantity no matter whether a is to the right or to the left of 0. For any real numbers a and b, we say that b **is larger in magnitude than** a if $|b| > |a|$. Thus 4 is greater than -6, but -6 is larger in magnitude than 4.

Rules for Absolute Value

1 $|-a| = |a|$

2 $|ab| = |a| \cdot |b|$

3 $\left|\dfrac{a}{b}\right| = \dfrac{|a|}{|b|} \qquad (b \neq 0)$

4 $|a + b| \leq |a| + |b| \qquad$ (**triangle inequality**)

Example 3

a $|(-6)(-5)| = |30| = 30 = |-6| \cdot |-5|$

b $|-\frac{3}{4}| = |\frac{3}{4}| = \frac{3}{4} = |-3|/|4|$

c $|3 + 4| = |7| = 7 \leq |3| + |4|$

d $|-3 + (-4)| = |-7| = 7 \leq |-3| + |-4|$

e $|5 - 2| = |3| = 3 < |5| + |-2|$

Rule 4, the triangle inequality, is an equality if a and b have the same sign and is a strict inequality if they have opposite signs. Rules 2 and 4 extend to three or more numbers. For instance

$$|abc| = |a| \cdot |b| \cdot |c| \quad \text{and} \quad |a + b + c| \leq |a| + |b| + |c|$$

Rules for Extending the Triangle Inequality

5 $|a - b| \leq |a| + |b|$

6 $|a| - |b| \leq |a - b|$

To obtain rule 5, replace b by $-b$ in rule 4. To obtain rule 6, apply rule 4 to the sum $(a - b) + b = a$:

$$|a| = |(a - b) + b| \leq |a - b| + |b| \quad \text{so} \quad |a| - |b| \leq |a - b|$$

Geometric Considerations

Once real numbers have been identified with points on the line, many arithmetic statements can be translated into geometric statements, and vice versa. Here are a few examples:

Arithmetic	Geometric
a is positive	a lies to the right of 0.
$a > b$	a lies to the right of b.
$a - b = c > 0$	a lies c units to the right of b.
$a < b < c$	a lies to the left of c, and b lies between a and c.
$\|a - 3\| < \frac{1}{2}$	a is less than $\frac{1}{2}$ unit from 3.
$\|a\| < \|b\|$	a is closer to the origin than b is.

Figure 1-1-3
The distance between points a and b, that is, the length of the line segment connecting them, is

$$L = |a - b| = |b - a|$$

For any two points a and b on the number line, the distance between them is $|a - b|$. See Figure 1-1-3. Thus the distance between 5 and 9 is $|5 - 9| = |-4| = 4$. Since $|a - b| = |b - a|$, it doesn't matter which you subtract from which; the answers come out the same.

The close relation between arithmetic and geometry is extremely important; often we can use arithmetical reasoning to solve geometrical problems or geometrical reasoning to solve arithmetical problems. Finding two different ways of looking at a problem increases the chances of solving it.

Intervals

The set of all numbers between two fixed numbers is called an **interval** on the number line. An interval may include one or both of its end points, or neither. An interval that includes both end points is called a **closed interval.** Let $a < b$; then we use the notation

$$[a, b]$$

for the *closed* interval of all points x satisfying $a \leq x \leq b$.

An interval that excludes both end points is called an **open interval.** We use the notation

$$(a, b)$$

for the *open* interval of all points x satisfying $a < x < b$.

For example, $[-2, 1]$ describes the *closed* interval of all numbers between -2 and 1, *including* the end points, whereas $(3, 7)$ describes the *open* interval of all numbers strictly between 3 and 7, that is, *excluding* the end points (Figure 1-1-4).

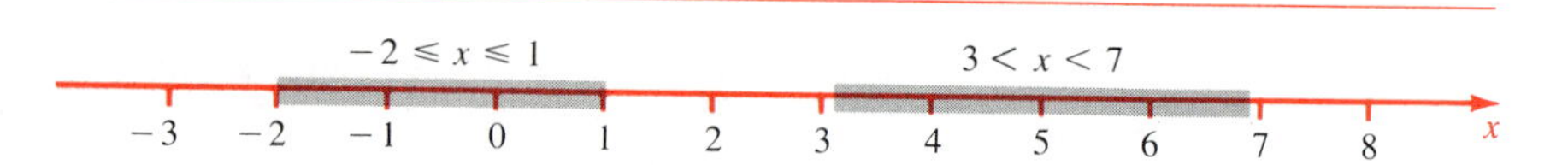

Figure 1-1-4
The closed interval $[-2, 1]$ and the open interval $(3, 7)$

We also need notation for hybrid intervals called **half-open** (or **half-closed**) **intervals.** See Figure 1-1-5. The suggestive notation for these intervals is

$(a, b]$ for the set of x satisfying $a < x \leq b$

$[a, b)$ for the set of x satisfying $a \leq x < b$

The intervals that we have defined so far are *bounded.* We also deal with *unbounded* intervals, that is, intervals that go off indefinitely in one direction or the other. (See Figure 1-1-6 for examples.)

Figure 1-1-5
The half-open intervals $(-3, 0]$ and $[2, 5)$

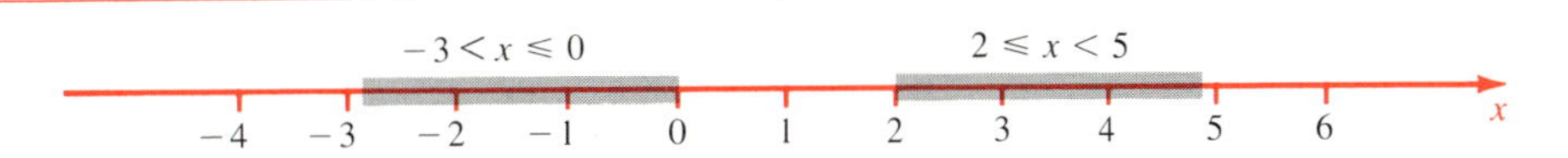

Figure 1-1-6
Unbounded intervals. Note the notation:

$$(-\infty, 3) = \{x \mid x < 3\}$$

$$[7, +\infty) = \{x \mid 7 \leq x\}$$

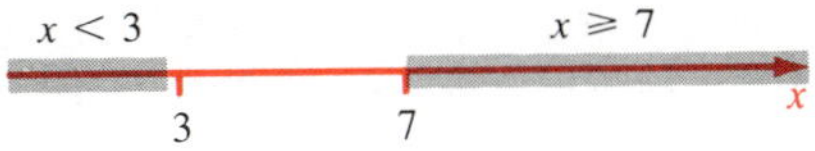

$x < 3$ $x \geq 7$

3 7 x

Using absolute values and inequalities, we can develop a nice shorthand to express geometrical facts about distances and intervals. For example, $|x - 4| < 1$ describes the set of points x whose distance from the point 4 is less than 1. In other words, $|x - 4| < 1$ describes the open interval $(3, 5)$. Similarly, $|x - 4| \leq 1$ describes the closed interval $[3, 5]$.

Centered Intervals

- The inequality $|x - a| < r$ describes the *open* interval $(a - r, a + r)$.
- The inequality $|x - a| \leq r$ describes the *closed* interval $[a - r, a + r]$.

We can think of $|x - a| < r$ as representing the open interval with center at a and "radius" equal to r. See Figure 1-1-7. Then $|x| < r$ represents the open interval with center 0 and radius r, because $|x| = |x - 0|$. In calculus, the Greek letter ε (epsilon) generally denotes a small positive number, so $|x - a| < \varepsilon$ describes a small interval centered at a. These terms can be used to express a simple, yet important principle:

- If $|x| < \varepsilon$ for every $\varepsilon > 0$, then $x = 0$.

Figure 1-1-7
The open interval $(a - r, a + r)$

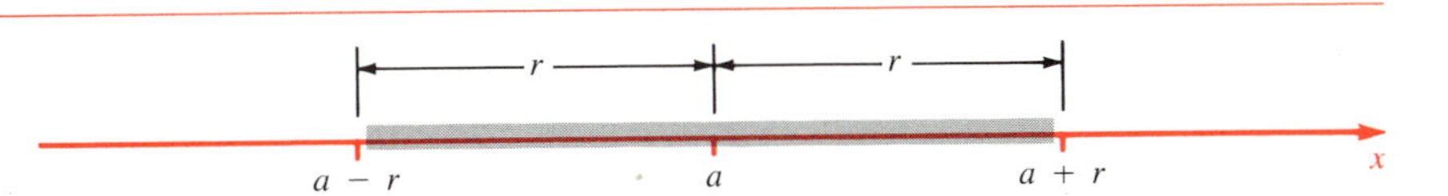

In other words, if $|x|$ is smaller than every positive number, then $x = 0$. This is clear geometrically; the only point contained in *every* open interval centered at 0, no matter how small the interval, is the point 0.

Inequalities

When you are asked to solve an inequality, the required answer is the interval or the union of intervals that consists of the set of all numbers satisfying the given inequality.

Example 4

Solve

a $5x + 7 \geq \frac{1}{2}x$ **b** $x^2 \leq \frac{1}{4}$ **c** $x^2 > \frac{1}{4}$

Solution

a By rule 1 on page 3, you can transpose terms in inequalities, that is, add or subtract the same quantity from both sides of the inequality. Thus:

$$5x + 7 \geq \tfrac{1}{2}x \quad \text{is equivalent to} \quad 5x - \tfrac{1}{2}x \geq -7$$

That is, $\tfrac{9}{2}x \geq -7$. By rule 2, this is equivalent to

$$x \geq \tfrac{2}{9}(-7) = -\tfrac{14}{9}$$

Thus the answer is $x \geq -\tfrac{14}{9}$, an unbounded interval.

b $x^2 = |x|^2$ and $|x| \geq 0$. By rule 9, page 4,

$$|x|^2 \leq \tfrac{1}{4} \quad \text{if and only if} \quad |x| \leq \tfrac{1}{2}$$

so the answer is the closed interval $[-\tfrac{1}{2}, \tfrac{1}{2}]$. See Figure 1-1-8.

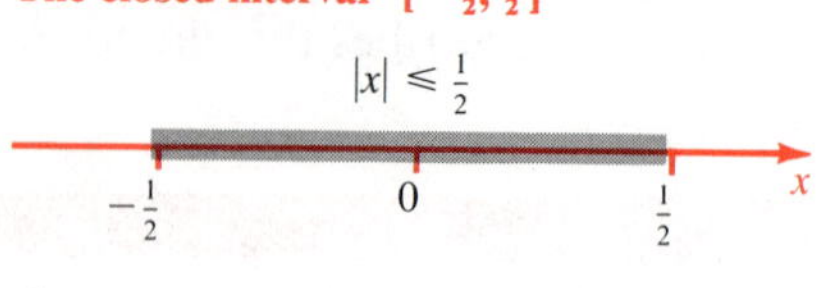

Figure 1-1-8
The closed interval $[-\tfrac{1}{2}, \tfrac{1}{2}]$

c Similarly

$$|x|^2 > \tfrac{1}{4} \quad \text{if and only if} \quad |x| > \tfrac{1}{2}$$

so the answer is either $x < -\tfrac{1}{2}$ or $x > \tfrac{1}{2}$. Thus the answer is the union of two unbounded intervals (Figure 1-1-9). ●

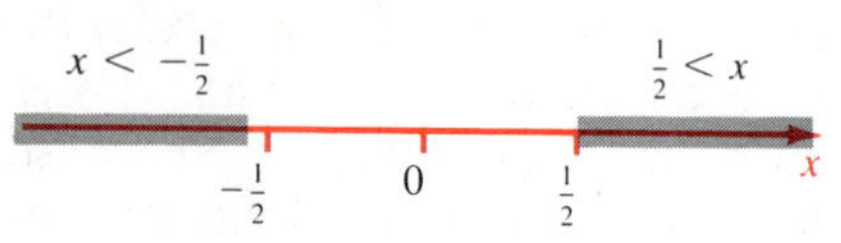

Figure 1-1-9
The unbounded open intervals described by $|x| > \tfrac{1}{2}$

Example 5

Solve

a $\dfrac{1}{x+1} > 2$ **b** $\dfrac{x+1}{2x+1} \geq -2$

Solution

a Clearly $x = -1$ is ruled out because the left side is then undefined. If $x < -1$, then $x + 1 < 0$, so $1/(x+1) < 0$, whereas we want $1/(x+1) > 2$. Therefore, $x \leq -1$ is excluded. If $x > -1$, then $x + 1 > 0$, so rule 4 on page 3 applies:

$$\frac{1}{x+1} > 2 > 0 \quad \text{so} \quad x + 1 < \tfrac{1}{2} \qquad \text{hence} \qquad x < \tfrac{1}{2} - 1 = -\tfrac{1}{2}$$

Thus the answer is the set of all x satisfying $-1 < x < -\tfrac{1}{2}$, that is, the open interval $(-1, -\tfrac{1}{2})$.

b There are two cases, depending on the sign of the denominator. In the first case, $2x + 1 > 0$. Multiplying both sides of the inequality by $2x + 1$ preserves the direction of the inequality:

$$\frac{x+1}{2x+1}(2x+1) \geq -2(2x+1)$$

that is, $x + 1 \geq -4x - 2$. Hence

$$5x \geq -3 \quad \text{so} \quad x \geq -\tfrac{3}{5}$$

But $2x + 1 > 0$ is equivalent to $x > -\tfrac{1}{2}$. Since $-\tfrac{1}{2} > -\tfrac{3}{5}$, the solution set in this case is the unbounded interval $(-\tfrac{1}{2}, +\infty)$.

In the other case, $2x + 1 < 0$. This time, multiplying by $2x + 1$ reverses the inequality:

$$\frac{x+1}{2x+1}(2x+1) \le -2(2x+1)$$

that is, $x + 1 \le -4x - 2$. Hence

$$5x \le -3 \quad \text{so} \quad x \le -\tfrac{3}{5}$$

The inequality $2x + 1 < 0$ is equivalent to $x < -\frac{1}{2}$. But $-\frac{3}{5} < -\frac{1}{2}$, so the solution set in this case is the unbounded interval $(-\infty, -\frac{3}{5}]$. Thus the answer consists of the union of two unbounded intervals, $(-\frac{1}{2}, +\infty)$ and $(-\infty, -\frac{3}{5}]$. See Figure 1-1-10. ●

Figure 1-1-10
The solution of $\dfrac{x+1}{2x+1} \ge -2$

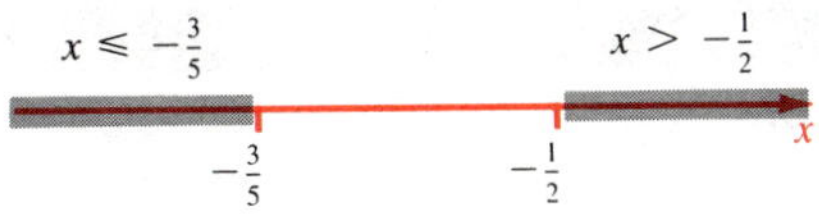

$x \le -\frac{3}{5}$ $x > -\frac{1}{2}$ $-\frac{3}{5}$ $-\frac{1}{2}$ x

Example 6

Solve $|2x + 1| < 3$.

Solution This inequality with an absolute value can be rewritten as a pair of inequalities without absolute value:

$$-3 < 2x + 1 < 3$$

Add -1, then divide by 2:

$$-4 < 2x < 2 \qquad -2 < x < 1$$

Thus the answer is the open interval $(-2, 1)$.

Alternative Solution Divide the given inequality by 2:

$$|x + \tfrac{1}{2}| < \tfrac{3}{2} \qquad \text{that is} \qquad |x - (-\tfrac{1}{2})| < \tfrac{3}{2}$$

Thus the inequality requires x to be within $\frac{3}{2}$ of $-\frac{1}{2}$, so the answer is

$$-\tfrac{1}{2} - \tfrac{3}{2} < x < -\tfrac{1}{2} + \tfrac{3}{2} \qquad \text{that is} \qquad -2 < x < 1$$ ●

Example 7

Solve $|x| < |x + 4|$.

Solution We can divide the problem into three cases (Figure 1-1-11). First case: $x \le -4$. Then $x < 0$ and $x + 4 \le 0$, so

$$|x| = -x \quad \text{and} \quad |x + 4| = -(x + 4) = -x - 4$$

Thus $|x| < |x + 4|$ means

$$-x < -x - 4 \qquad \text{that is} \qquad 0 < -4$$

which is never true. Hence no $x \le -4$ is in the solution set.

Second case: $-4 < x < 0$. Then

$$|x| = -x \quad \text{and} \quad |x + 4| = x + 4$$

so $|x| < |x + 4|$ means

$$-x < x + 4 \qquad \text{that is} \qquad 2x > -4 \quad \text{so} \quad x > -2$$

Figure 1-1-11
Three cases to consider in solving the inequality $|x| < |x + 4|$

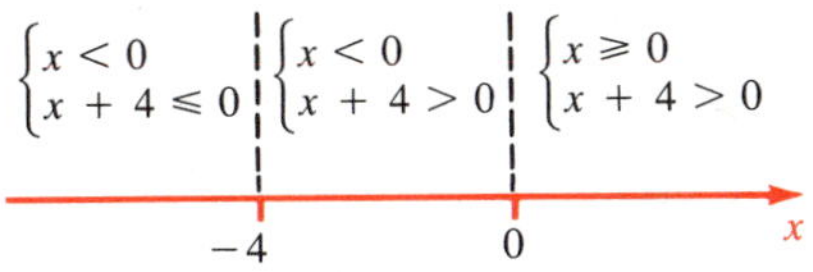

$\begin{cases} x < 0 \\ x + 4 \le 0 \end{cases}$ $\begin{cases} x < 0 \\ x + 4 > 0 \end{cases}$ $\begin{cases} x \ge 0 \\ x + 4 > 0 \end{cases}$

-4 0 x

Final case: $x \geq 0$. Then

$$|x| = x \quad \text{and} \quad |x + 4| = x + 4$$

Thus $|x| < |x + 4|$ means

$$x < x + 4$$

which is always true, so the given inequality is true for all $x \geq 0$. Thus the complete answer is the unbounded interval $(-2, +\infty)$. ●

Approximation

Much of calculus concerns approximating numbers by other numbers, and the triangle inequality often is needed to make approximations. For instance, suppose x is close to a and y is close to b. Then $x + y$ should be close to $a + b$. How close?

Since $|x - a|$ measures how close x and a are, the statement that x is close to a can be written

$$|x - a| < \varepsilon_1$$

where ε_1 is some small positive number. Similarly

$$|y - b| < \varepsilon_2$$

where ε_2 is another small positive number. To measure how close $x + y$ is to $a + b$, we use the triangle inequality:

$$\begin{aligned} |(x + y) - (a + b)| &= |(x - a) + (y - b)| \\ &\leq |x - a| + |y - b| < \varepsilon_1 + \varepsilon_2 \end{aligned}$$

In words, $x + y$ approximates $a + b$ to within $\varepsilon_1 + \varepsilon_2$.

Example 8

Suppose $|x - 3| < 0.001$ and $|y - 5| < 0.002$. Estimate

a $|(x + y) - 8|$ **b** $|(2x + 4y) - 26|$ **c** $|xy - 15|$

Solution

a We must measure how close $x + y$ is to $3 + 5 = 8$. By the triangle inequality,

$$\begin{aligned} |(x + y) - 8| &= |(x - 3) + (y - 5)| \leq |x - 3| + |y - 5| \\ &< 0.001 + 0.002 = 0.003 \end{aligned}$$

b We must measure how close $2x + 4y$ is to $2 \cdot 3 + 4 \cdot 5 = 26$. By the triangle inequality,

$$\begin{aligned} |(2x + 4y) - 26| &= |2(x - 3) + 4(y - 5)| \\ &\leq |2(x - 3)| + |4(y - 5)| \\ &= 2|x - 3| + 4|y - 5| \\ &< 2(0.001) + 4(0.002) = 0.01 \end{aligned}$$

c Since x is close to 3 and y is close to 5, we suspect that xy is close to $3 \cdot 5 = 15$. How close? We must estimate the size of $|xy - 15|$. In order to apply the triangle inequality, we must first manipulate the expression $xy - 15$. This is done by a frequently used technique. We write

$$\begin{aligned} xy - 15 &= xy - 3y + 3y - 15 \\ &= (x-3)y + 3(y-5) \end{aligned}$$

By subtracting and adding $3y$, we have converted $xy - 15$ into an expression to which we can apply the given estimates for $|x - 3|$ and $|y - 5|$:

$$\begin{aligned} |xy - 15| &= |(x-3)y + 3(y-5)| \\ &\le |(x-3)y| + |3(y-5)| \qquad \text{(triangle inequality)} \\ &< (0.001)|y| + 3(0.002) \end{aligned}$$

Since y is so close to 5, surely $|y| < 6$. Hence

$$|xy - 15| < 6(0.001) + 3(0.002) = 0.012 < 0.02$$

(More precisely, $|y| < 5.002$, but using that number does not significantly sharpen the estimate of $|xy - 15|$.) ●

Rational and Irrational Numbers

The real number system includes the rational number system. Recall that a **rational number** is a real number that is the quotient of integers. For instance

$$\tfrac{2}{3} = 2 \div 3 \qquad -\tfrac{7}{5} = (-7) \div 5 \qquad 3.1416 = 31416 \div 10000$$

$$0.11111 \cdots = 1 \div 9 \qquad 0.27272727 \cdots = 3 \div 11$$

Each rational number (also called **fraction** in school arithmetic) can be written in lowest terms, that is, as a quotient of integers with no common factor larger than 1. For instance

$$\frac{2}{4} = \frac{1}{2} \qquad \frac{10}{4} = \frac{5}{2} \qquad \frac{31416}{10000} = \frac{3927}{1250}$$

A real number that is not rational is called **irrational.** The familiar number $\sqrt{2}$ is irrational. Why? Briefly, if $\sqrt{2} = a/b$, where a and b are integers; then $a^2 = 2b^2$. If this equation is true, then 2 divides a^2 an even number of times (twice as many times as 2 divides a) but divides $2b^2$ an odd number of times (one more than twice as many times as 2 divides b). That is impossible.

Most real numbers are irrational. This statement is difficult to explain precisely, but roughly it means that if a machine could choose real numbers at random, it would hardly ever come up with a rational choice. Although rationals are rare among the reals, they are the numbers always used in machine calculations. Every number that goes into or comes out of a computer or calculator is a rational number (a finite binary or decimal expansion).

Rational approximations to irrational numbers have been used for ages. The familiar number π is irrational, and

$$\frac{22}{7} = 3.142857\cdots \qquad \frac{256}{81} = 3.16049\cdots$$

are ancient approximations to π. For $\sqrt{2}$ we have

$$\sqrt{2} = 1.41421\ 356\cdots$$

$$\frac{577}{408} = 1.41421\ 568\cdots$$

$$\frac{577001}{408001} = 1.41421\ 467\cdots$$

It is convenient to round off unending decimal expansions like this. Then we use $\approx$ for approximate equality:

$$\sqrt{2} \approx 1.41421\ 36 \qquad \frac{577}{408} \approx 1.41421\ 57$$

$$\frac{577001}{408001} \approx 1.41421\ 47$$

How close is $\frac{577}{408}$ to $\sqrt{2}$? Certainly

$$0 < \frac{577}{408} - \sqrt{2} \approx 1.41421\ 57 - 1.41421\ 36$$

$$= 0.00000\ 21 < 0.000003 = 3 \times 10^{-6}$$

Exercises

Find

1 $|6 - 13|$

2 $|(-3)(5)|$

3 $|-7/5|$

4 $|3/(-4)|$

5 $|3 \cdot 8 - 5 \cdot 9|$

6 $|(5 - 17)/(-4)|$

7 Prove that if $0 < x < y$, then $0 < x/y < 1$.

8 Prove that if $x \neq y$, then $x^2 + y^2 > 2xy$.
[Hint Consider $(x - y)^2$.]

Prove

9 if $a \le b$ and $-a \le b$, then $|a| \le b$

10 if a and b have the same sign or if either is zero, then $|a + b| = |a| + |b|$

11 if a and b have opposite signs, then $|a + b| < |a| + |b|$

12 if a and b are any real numbers, then $||a| - |b|| \le |a - b|$

Solve the inequality

13 $2x - 3 < 5$

14 $3x + 7 \le -1$

15 $4x > x - 2$

16 $6x - 1 < 3$

17 $2x + 10 \ge 70$

18 $4x - 5 \le 8x + 1$

19 $3(x - 4) > \frac{1}{2}x - 6$

20 $-4 < 2x + 6 < 16$

21 $6 \le \frac{1}{2}(x + 3) \le 10$

22 $5(x - 2) \le 2x + 7$

23 $\frac{2}{3}x \ge x + 7$

24 $x < x - 3 \le 2x$

25 $\dfrac{1}{x + 5} > \frac{1}{8}$

26 $\frac{1}{2} > \dfrac{4}{3 - x}$

27 $\dfrac{2x + 1}{4x + 1} < 1$

28 $\dfrac{1}{x + 1} > \dfrac{2}{x}$

29 $\dfrac{x}{x - 3} < 0$

30 $\dfrac{x}{8x - 3} > 0$

Express using absolute values

31 x is either 2 or -2

32 x is farther from a than from b

33 x is at least as close to a as to b

34 x is either to the left of -3 or to the right of 3

35 x is between 16 and 18

36 x is within distance 2 of 7

37 Find all points that are 3 times as far from 5 as from 1.

38 Find all points that are 10 times as far from 5 as from 1.

39 Explain how you can tell at a glance that there is no x for which both $|x-1|<2$ and $|x-12|<3$.

40 Explain why $|a|+|b|+|c|>0$ is algebraic shorthand for "at least one of the numbers a, b, c is different from 0."

Solve the inequality; express your answer in terms of intervals, without using absolute values

41 $|3x| \le 12$

42 $|\frac{1}{2}x| \le 5$

43 $|x| \le 0.01$

44 $0<|x-3|<10^{-4}$

45 $|x-3| \le 1$

46 $|x+4| \le 2$

47 $|-5x|<10$

48 $0<|x+5|<5\times 10^{-3}$

49 $|x-4|<5$

50 $|2x-1|<4$

51 $|3x+1|>1$

52 $|7-x|<6$

53 $|x^2-1|<8$

54 $|x|<|x+5|$

55 $0<|x+2|<0.1$

56 $0<|x-9|<2$

57 $|x|+|x-4| \le 4$

58 $|x|+|x-2|<3$

59 Suppose $|x-a|<10^{-6}$. Show that $|7x-7a|<10^{-5}$.

60 Suppose $|x-7|<10^{-6}$ and $|y-5|<10^{-6}$. Show that $|(x+y)-12|<10^{-5}$.

61 Suppose $|x+3|<10^{-5}$ and $|y-4|<10^{-5}$. Show that $|(x-y)+7|<2\times 10^{5}$.

62 Suppose $|x+2|<10^{-6}$ and $|y+3|<10^{-6}$. Show that $|(x-3y)-7|<4\times 10^{-6}$.

63 Suppose $|x-5|<10^{-6}$ and $|y-7|<10^{-6}$. Show that $|xy-35|<2\times 10^{-5}$.

64 Suppose $|x+1|<10^{-5}$ and $|y-1|<10^{-5}$. Show that $|xy+1|<3\times 10^{-5}$.

65 Suppose $|x+2|<10^{-6}$ and $|y+3|<10^{-6}$. Show that $|xy-6|<6\times 10^{-6}$.

66 Suppose $|x-3|<10^{-6}$. Prove that $|x^2-9|<10^{-5}$. [Hint Factor x^2-9.]

67 (cont.) Prove also that $|x^3-27|<5\times 10^{-5}$.

68 Suppose $|x-1|<10^{-5}$. Prove that $|1/x-1|<1.1\times 10^{-5}$.

1-2 Functions and Their Graphs

Let the symbol x represent any real number belonging to a certain set D of real numbers. Suppose there is a rule that associates with each such x a real number y. Then this rule is called a **function** whose **domain** is D.

For instance, suppose that to each real x a number y is assigned by the rule $y=x^2$. Then this assignment is a function whose domain D is the set of all real numbers.

For another example, let D be the set of all $x \ge 0$. To each x in D assign $y=\sqrt{x}$. (In this book, $\sqrt{x}$ always denotes the *non-negative* square root.)

Figure 1-2-1
A function as a "black box"

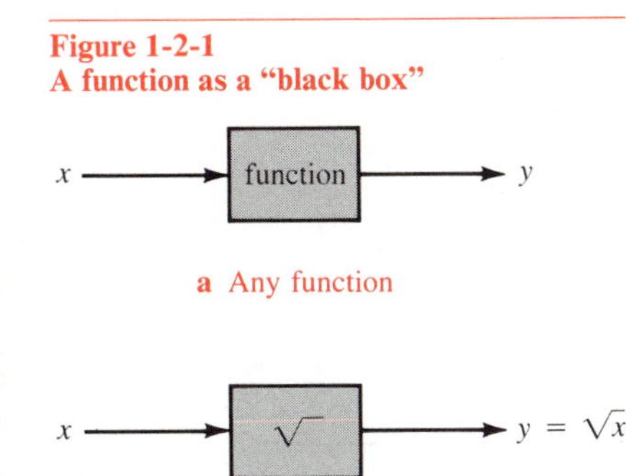

a Any function

b The square root function

We can think of a function as a "black box," that is, a machine whose inner workings are hidden from us. We input an x from D; the black box outputs a y. See Figure 1-2-1a. For the previous example, we draw a square root machine, Figure 1-2-1b.

The set of all numbers y that a function assigns to the numbers x in its domain is called the **range** of the function. For example, the range of the function given by the rule $y=2x$ is the set of all real numbers; the range of the function given by $y=x^2$ is the set of all *non-negative* real numbers. We sometimes say that a function **maps** its domain onto its range.

Notation

The symbol used to denote a typical real number in the domain of a function is sometimes called the **independent variable.** The symbol used to denote the typical real number in the range is called the **dependent variable.**

Generally, but not always, variables are denoted by lowercase letters such as t, x, y, and z. Functions generally are denoted by f, g, h, and capital letters.

If f denotes a function, x the independent variable, and y the dependent variable, then it is common practice to write $y = f(x)$, read "y equals f of x" or "y equals f at x." This means that the function f assigns to each x in its domain a number $f(x)$, which is also written y.

There are several common variations of this notation. For instance, if f is the function that assigns to each real number its square, then we write $f(x) = x^2$ or $y = x^2$. For this example, $f(3) = 3^2 = 9$ and $f(-5) = (-5)^2 = 25$.

Remark 1 It is logically incorrect to say "the function $f(x)$" or "the function x^2" or the function "$y = f(x)$." The symbols $f(x)$, y, and x^2 represent *numbers,* the numbers assigned by the function f to the numbers x. A function is not a number, but a rule that assigns a number $f(x)$ to each number x in a certain domain. Nevertheless, these slight inaccuracies are so universal that we do not try to avoid them.

Remark 2 A function is not a formula, and need not be specified by a formula. It is true that in practice most functions are *evaluated* by means of formulas. For instance, f may assign to each real number x the real number y that is computed by formulas such as $y = x^2$ or $y = (\sqrt{x^2 + 1})/(1 + 7x^4)$, and so on. Yet there are perfectly good functions not readily given by formulas:

Example 1

a $f(x) = \begin{cases} 1 & \text{if } x > 0 \\ 0 & \text{if } x = 0 \\ -1 & \text{if } x < 0 \end{cases}$ $\qquad D = \mathbf{R}$

$f(3.14) = 1 \qquad f(-5) = -1$

b $f(x) = \begin{cases} 1 & \text{if } x \text{ is rational} \\ 0 & \text{if } x \text{ is irrational} \end{cases}$ $\qquad D = \mathbf{R}$

$f(0.5) = 1 \qquad f(-\frac{4}{7}) = 1 \qquad f(1 + \sqrt{3}) = 0$

c $f(x) = \begin{cases} \sqrt{x} & \text{if } 0 < x < 1 \\ x & \text{if } x \geq 1 \end{cases}$ $\qquad D = (0, +\infty)$

$f(\frac{1}{2}) = \frac{1}{2}\sqrt{2} \qquad f(17) = 17$ ●

Example 2

Let $f(x)$ denote the largest integer $y \leq x$. This function is often written $[x]$ and called the **greatest integer function.** For each x, the value $[x]$ is the unique integer satisfying

$$[x] \leq x < [x] + 1$$

(A modern notation is $\lfloor x \rfloor$, called the "floor" of x.)

$[3] = 3 \qquad [3.6] = 3 \qquad [-3.6] = -4$ ●

Keep in mind that $f(x)$ is the *number* assigned to x by the function f. If, for instance, $f(x) = x^2 + 3$, then $f(1) = 4$, $f(2) = 7$, and $f(3) = 12$. Similarly

$$f(x+1) = (x+1)^2 + 3 \quad \text{and} \quad f(x^2) = (x^2)^2 + 3 = x^4 + 3$$

and so on. For this particular function, you must boldly square whatever appears in the parentheses (or brackets), no matter what is is called, then add 3:

$$f(x+y) = (x+y)^2 + 3 \qquad f(1/x) = (1/x)^2 + 3$$

$$f[f(x)] = [f(x)]^2 + 3 = (x^2+3)^2 + 3 = x^4 + 6x^2 + 12$$

Most functions arising in practice have simple domains. The most common domains are the whole line **R**, an interval, a half-line such as $[0, +\infty)$ or $(-\infty, 2)$ and some simple combinations of these. The following table lists some functions and their domains.

Function	Domain
$f(x) = 2x + 1$	**R**
$f(x) = \sqrt{x+2}$	$[-2, +\infty)$
$f(x) = \sqrt{1 - x^2}$	$[-1, 1]$
$f(x) = 1/x$	$(-\infty, 0) \cup (0, +\infty)$

Note Frequently a function is given by a formula, but its domain is not stated. Then the domain of the function is automatically taken to be the set of all real numbers for which the formula can be evaluated.

Example 3

Find the domain D of the given function:

a $f(x) = x^3 + 2x + 5$ **c** $f(x) = \sqrt{x+1}$

b $f(x) = \dfrac{1}{x+2}$ **d** $f(x) = \dfrac{x-1}{x^2-1}$

Solution

a $D =$ **R**.

b The denominator is 0 if $x = -2$; hence $D = \{x \mid x \neq -2\} = (-\infty, -2) \cup (-2, +\infty)$.

c $x + 1 < 0$ if $x < -1$, and negative numbers do not have real square roots; hence $D = [-1, +\infty)$.

d The denominator is 0 if $x = -1$ or $x = 1$; hence $D = \{x \mid x \neq -1 \text{ and } x \neq 1\}$
$= (-\infty, -1) \cup (-1, 1) \cup (1, +\infty)$. ●

The Construction of Functions

There are several standard methods for building new functions out of old ones. We consider the most common of these constructions.

1 Addition of Functions If f and g are functions defined on the same domain, then their **sum** $f+g$ is a function defined on that domain by

$$(f+g)(x) = f(x) + g(x)$$

For example, let $f(x) = 2x - 3$ and $g(x) = x^2 - x - 1$. Then

$$(f+g)(x) = (2x-3) + (x^2 - x - 1) = x^2 + x - 4$$

The **difference** $f - g$ is defined similarly.

It is generally agreed that if f and g do not have the same domain, then $f+g$ is defined *only* on the common part of the domains of f and g. The same remark applies to the difference and to the product, defined below. For instance, let $f(x) = x - \sqrt{x}$ with domain $[0, +\infty)$ and let $g(x) = \sqrt{x} + \sqrt{1-x}$ with domain $[0, 1]$. Then $h = f + g$ has domain $[0, 1]$, and for x in this domain, $h(x) = x + \sqrt{1-x}$. Note that the expression $x + \sqrt{1-x}$ is defined on the interval $(-\infty, 1]$, but that interval is *not* the domain of $f+g$.

2 Multiplication of a Function by a Constant If c is a real number and f is a function, the function cf is defined by

$$(cf)(x) = cf(x)$$

For example, if $f(x) = x^2 - 2x - 1$, then

$$(-5f)(x) = (-5)(x^2 - 2x - 1) = -5x^2 + 10x + 5$$

3 Multiplication of Functions If f and g are functions defined on the same domain, then their **product** fg is defined by

$$(fg)(x) = f(x)g(x)$$

For example, if $f(x) = 2x - 1$ and $g(x) = 3x + 4$, then

$$(fg)(x) = (2x-1)(3x+4) = 6x^2 + 5x - 4$$

4 Composition of Functions If g is a function whose range is a subset of the domain of a second function f, then the **composite** $f \circ g$ of f and g is defined by the formula

$$(f \circ g)(x) = f[g(x)]$$

Think of inserting one function into the other, or replacing the independent variable of f by the function value $g(x)$. See Figure 1-2-2 for a "black box" interpretation. (The electrical analogy is a series or cascade connection of functions.) The domain of $f \circ g$ is exactly the domain of g.

Figure 1-2-2
Composition as a black box: The output $g(x)$ of g is the input of f.

input x

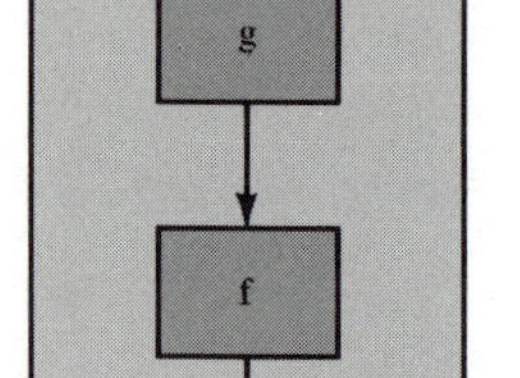

output $(f \circ g)(x)$

Example 4

a $f(y) = y^2 + 2y$ and $g(x) = -3x$

The domain of $f(y)$ is $\{\text{all } y\}$ and the domain of $g(x)$ is $\{\text{all } x\}$.

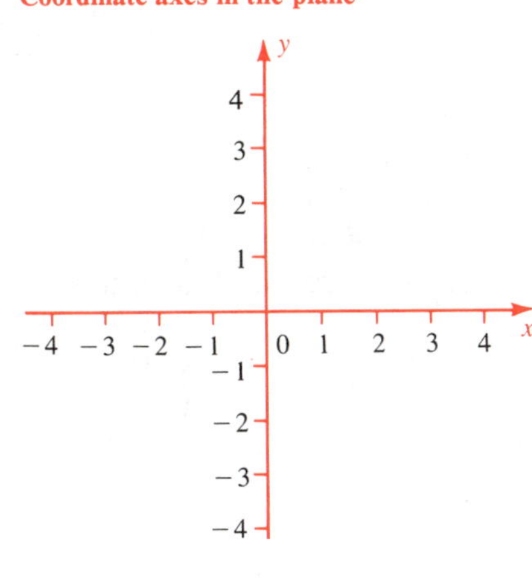

Figure 1-2-3
Coordinate axes in the plane

Certainly the range of $g(x)$ is a subset of the domain of $f(y)$, and

$$(f \circ g)(x) = f[g(x)] = [g(x)]^2 + 2[g(x)] = (-3x)^2 + 2(-3x) = 9x^2 - 6x$$

b $f(y) = 3y - 4$ and $g(x) = 2x^2 - x + 1$

Again, the domain of $f(y)$ is $\{\text{all } y\}$, the domain of $g(x)$ is $\{\text{all } x\}$, and the range of $g(x)$ lies in the domain of $f(y)$. We have

$$(f \circ g)(x) = f[g(x)] = 3g(x) - 4 = 3(2x^2 - x + 1) - 4 = 6x^2 - 3x - 1$$

c $f(y) = \sqrt{y-1}$ and $g(x) = 4x^2$ with domain $\{x \mid |x| > 1\} = (-\infty, -1) \cup (1, +\infty)$.

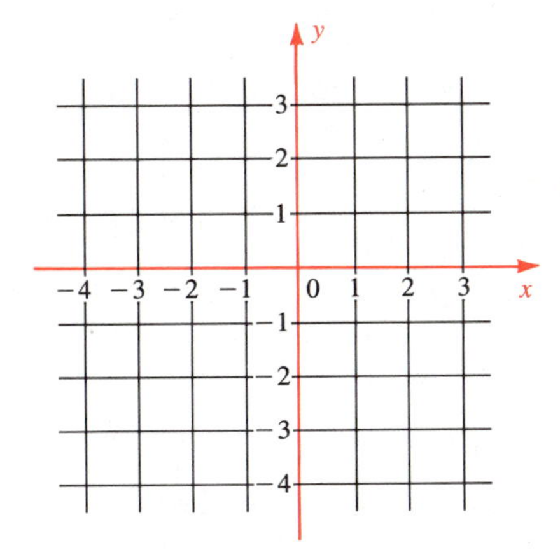

Figure 1-2-4
Rectangular coordinate grid

The domain of f is $[1, +\infty)$. The range of g is $(4, +\infty)$, which lies in the domain of f. We have

$$(f \circ g)(x) = f[g(x)] = \sqrt{g(x) - 1} = \sqrt{4x^2 - 1}$$

and the domain of $f \circ g$ is the domain of g.

d $f(y) = y^2 + 1$ and $g(x) = \sqrt{x}$ with domain $[0, +\infty)$

In this case

$$(f \circ g)(x) = f[g(x)] = [g(x)]^2 + 1 = x + 1$$

and the domain of $f \circ g$ is $[0, +\infty)$, the domain of $g(x)$. It doesn't matter that the expression $x + 1$ can be evaluated for any real x. The domain of $(f \circ g)(x)$ is limited to precisely the domain of $g(x)$, by definition. ●

The Coordinate Plane

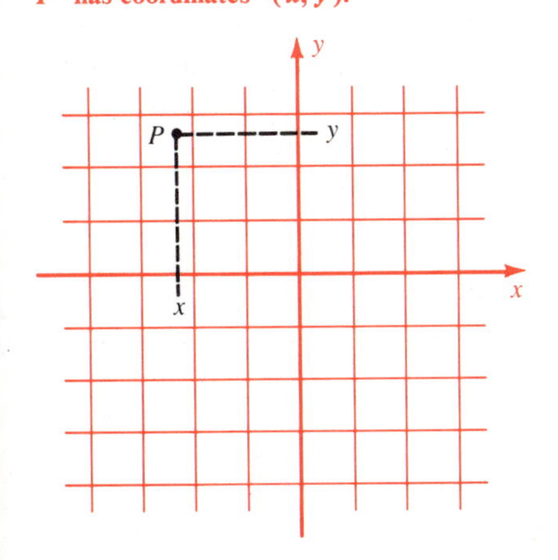

Figure 1-2-5
P has coordinates (x, y).

If the points of a line are labeled by real numbers (as in Section 1-1), we say that the line is **coordinatized:** each point has a label, or **coordinate.**

In order to graph functions, we label (coordinatize) the points of a plane. We draw two perpendicular lines in the plane, mark their intersection 0, and coordinatize each line as shown in Figure 1-2-3. The horizontal line is called the ***x*-axis** and the vertical line the ***y*-axis.**

Consider all lines parallel to the x-axis and all lines parallel to the y-axis (Figure 1-2-4). These two systems of parallel lines impose a rectangular grid on the whole plane. We use this grid to coordinatize the points of the plane.

If P is any point of the plane, then one vertical line and one horizontal line pass through P. See Figure 1-2-5. They meet the axes in points x and y, respectively. We associate with P the ordered pair (x, y), which completely describes the location of P.

Conversely, if (x, y) is any ordered pair of real numbers, then the vertical line through x on the x-axis and the horizontal line through y

on the y-axis meet in a point P whose coordinates are precisely (x, y). Thus there is a one-to-one correspondence

$$P \longleftrightarrow (x, y)$$

between the set of points of the plane and the set of all ordered pairs of real numbers. The numbers x and y are the **coordinates** of P. Thus x is the **x-coordinate** of P, and y is the **y-coordinate** of P. The point $(0, 0)$ is called the **origin.** The coordinate axes divide the plane into four **quadrants** numbered as in Figure 1-2-6.

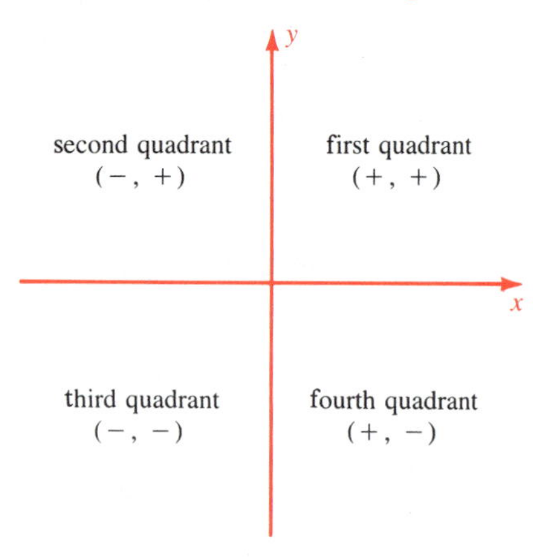

Figure 1-2-6
The four quadrants and the signs of the coordinates found in each quadrant

The coordinate system we have introduced is called a **rectangular** or **Cartesian** coordinate system. Some writers refer to the horizontal coordinate of a point as its **abscissa** and the vertical coordinate as its **ordinate.** Sometimes the pair (x, y) is called (ungrammatically) the "coordinates" of the corresponding point.

Warning Unfortunately, the same notation is used for a point (a, b) in the plane and for the open interval (a, b). You can always tell from the context which is meant.

In Figure 1-2-3, the same unit length is used on both axes, that is, both axes have the same scale. If we coordinatize the plane in order to study its geometry, then usually we use the same scale on both axes. But if our purpose is to graph functions, or to represent data, there is no need to scale the axes the same way. For instance, if we plot barrels of oil versus dollars, the unit distance on the y-axis may represent 10^6 barrels and the unit distance on the x-axis may represent $\$3.5 \times 10^7$. If we plot distance versus time, one unit on the y-axis may represent 100 kilometers and one unit on the x-axis may represent 1 hour. There is no reason for the units on different axes to have the same geometric length.

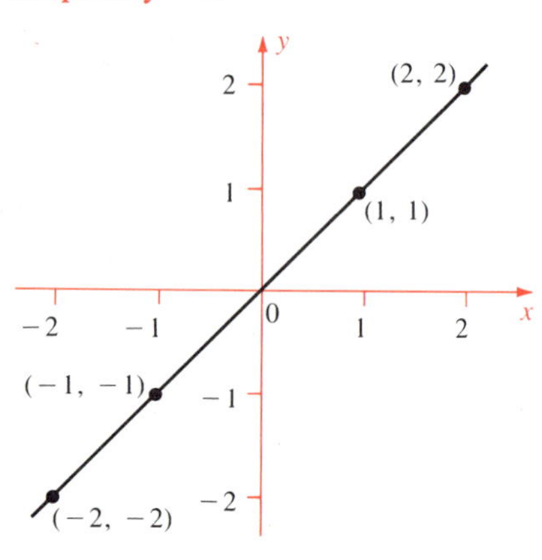

Figure 1-2-7
Graph of $y = x$

The Graph of a Function

Given a function f, we can construct a geometric picture of the function. For each number x in the domain of f, we find the corresponding number $y = f(x)$ in the range and then we plot the point (x, y). The set of all such points is called the **graph** of $f(x)$. For example, if $f(x) = x$, then for each real number x, the corresponding number is $y = x$, so we plot all points of the form (x, x). The set of all these points (Figure 1-2-7) is a straight line. If $f(x) = x + 1$, we plot all points of the form $(x, x + 1)$. The resulting set of points is again a straight line, parallel to the graph of $y = x$ and one unit above it (Figure 1-2-8).

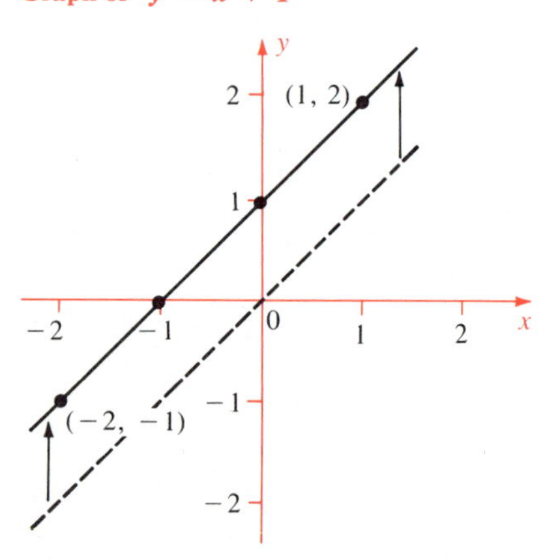

Figure 1-2-8
Graph of $y = x + 1$

Terminology Instead of referring to the "graph of $f(x) = x + 1$," we often say "the graph of $y = x + 1$," or when we are especially lazy, "the graph of $x + 1$."

Example 5

Graph the function $f(x) = |x|$.

Solution By definition, $|x| = x$ when $x \geq 0$, and $|x| = -x$ when $x \leq 0$. Consider the two cases separately. If $x \geq 0$, then $f(x) = x$; hence for $x \geq 0$ the graph is identical with the one shown in

Figure 1-2-9
Graph of $y = |x|$

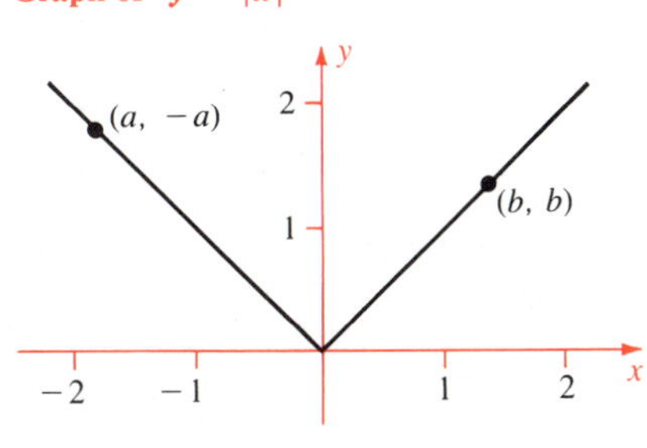

Figure 1-2-7. However, if $x \le 0$, then $f(x) = -x$. Therefore this portion of the graph consists of all points $(x, -x)$, such as $(-1, 1)$, $(-2, 2)$, and $(-3, 3)$. Plot a few points; you will see that the graph for $x \le 0$ is a half line (ray) at angle $135°$ to the positive x-axis (Figure 1-2-9). ●

Example 6

Graph $y = 1/x$.

Solution The domain of $f(x) = 1/x$ consists of all real numbers except $x = 0$, so the graph is not defined at $x = 0$. Plot a few points to get a general picture: $(1, 1)$, $(2, \frac{1}{2})$, $(3, \frac{1}{3})$, and so on. As x increases, y decreases, so the graph approaches the x-axis from above (Figure 1-2-10a). As x decreases from 1 toward 0, the curve rises steeply, as is seen from plotting $(\frac{1}{2}, 2)$, $(\frac{1}{3}, 3)$, $(\frac{1}{4}, 4)$, and so on. Hence the graph approaches the positive y-axis (Figure 1-2-10b).

Figure 1-2-10
Graph of $y = 1/x$

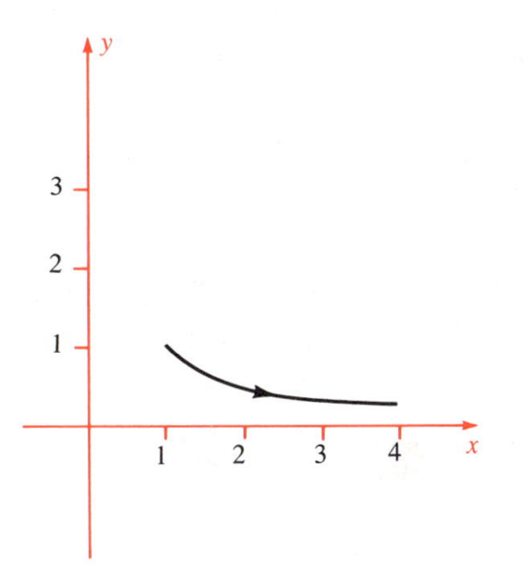

a y decreases as x increases.

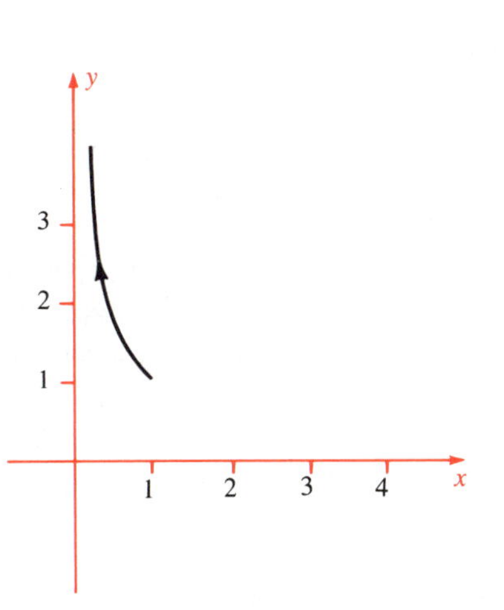

b y increases rapidly as $x \to 0$ from the right.

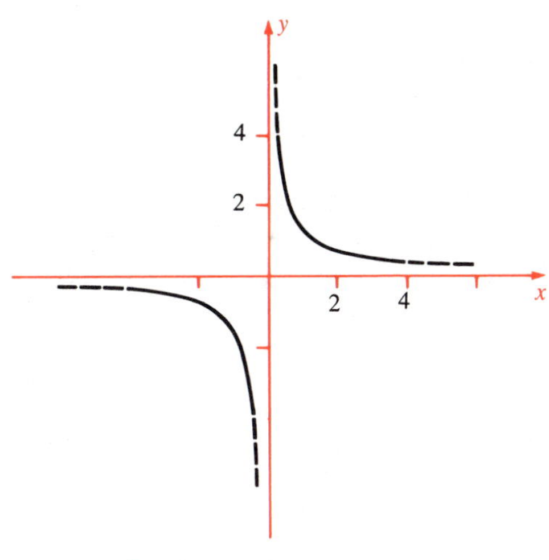

c Complete graph. Note the symmetry of the two branches.

Finally, plot a few points for $x < 0$; for instance, $(-\frac{1}{2}, -2)$, $(-1, -1)$, $(-2, -\frac{1}{2})$, and $(-3, -\frac{1}{3})$. This part of the curve is obviously symmetric to the part for $x > 0$, but lies *below* the x-axis. Combine all this information to sketch the complete graph (Figure 1-2-10c). ●

Each function has a graph. The graph of a function is a subset of the coordinate plane with the special property that it meets each vertical line at most once. Conversely any (non-empty) subset G of the plane that meets each vertical line in at most one point defines a function $f(x)$ whose graph is G. To each x that occurs as a first coordinate of a point (x, y) of G, it assigns the number y. In other words, $f(x)$ is the "height" of the set G above the x-axis.

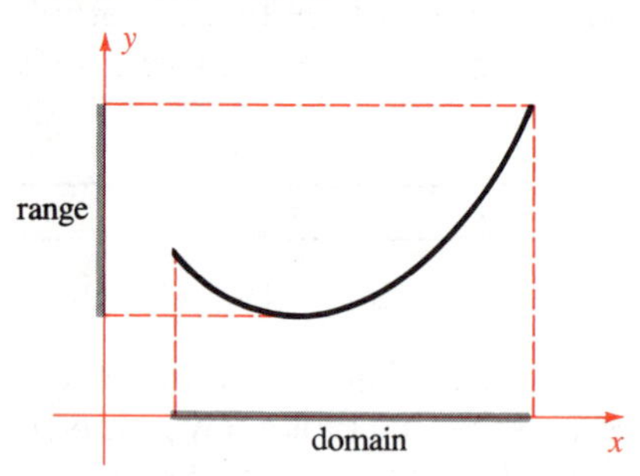

Figure 1-2-11
Reading the domain and range of a function from its graph

Graphical definition of functions is standard procedure in science. For instance, a scientific instrument recording temperature or blood pressure on graph paper is defining a function of time. There is hardly ever an explicit formula for such a function.

The graph of a function provides a picture of the function's domain and range. The projection of the graph onto the x-axis is the domain; the projection onto the y-axis is the range (Figure 1-2-11).

Exercises

1 Let $f(x) = 2x + 5$. Compute
a $f(0)$ **c** $f(\frac{1}{2})$ **e** $f(x-3)$
b $f(2)$ **d** $f(1/x)$

2 Let $f(x) = x^2 + x + 1$. Compute
a $f(0)$ **c** $f(x^2)$ **e** $f(x+h) - f(x)$
b $f(-x)$ **d** $f(\sqrt{x})$

Find the domain and the range of $f(x)$.

3 $f(x) = 3x - 2$
4 $f(x) = -7x + 6$
5 $f(x) = 4x - 5$
6 $f(x) = 7 - x$
7 $f(x) = 1/(2x-3)$
8 $f(x) = x/(x+2)$
9 $f(x) = x/(3x-5)$
10 $f(x) = 1/\sqrt{1-x}$
11 $f(x) = \sqrt{x-6}$
12 $f(x) = \sqrt{5-2x}$
13 $f(x) = \sqrt{4-9x^2}$
14 $f(x) = \sqrt{15x^2+11}$
15 $f(x) = \sqrt{2x-3}$
16 $f(x) = \dfrac{1}{\sqrt{x+4}}$
17 $f(x) = \sqrt{\frac{1}{4} - x^2}$
18 $f(x) = \sqrt{x^2-1}$
19 $f(x) = \sqrt{(x-1)(x-4)}$
20 $f(x) = \sqrt{x^3+1}$

Find $(f+g)(x)$ and $(fg)(x)$ for

21 $f(x) = 3x + 1$ $\quad g(x) = -2$
22 $f(x) = 2x - 1$ $\quad g(x) = 2x + 3$
23 $f(x) = x^2$ $\quad g(x) = -2x + 1$
24 $f(x) = x^2 + 1$ $\quad g(x) = -x^2 + x$

Find $f \circ g$ and $g \circ f$ for

25 $f(x) = 3x + 1$ $\quad g(x) = x - 2$
26 $f(x) = 2x - 1$ $\quad g(x) = -x^2 + 3x$
27 $f(x) = 2x^2$ $\quad g(x) = -x - 1$
28 $f(x) = x + 1$ $\quad g(x) = -x + 1$
29 $f(x) = 2x$ $\quad g(x) = -2x$
30 $f(x) = x + 3$ $\quad g(x) = -x + 1$
31 $f(x) = x^2$ $\quad g(x) = 3$
32 $f(x) = \pi x^2$ $\quad g(x) = 2x + 5$

Indicate on the coordinate plane the set of all points (x, y) for which

33 $x = -3$
34 $y = 2$
35 x and y are positive
36 either x or y is zero (or both are)
37 $1 \le x \le 3$
38 $-1 \le y \le 2$
39 $-2 \le x \le 2$ and $-2 \le y \le 2$
40 $x > 2$ and $y < 3$
41 both x and y are integers
42 $x^2 > 4$
43 $|x| \ge 1$ and $|y| \le 2$
44 $|x| \ge 2$ and $|y| \ge 2$
45 $xy > 0$ and $|x| \le 3$
46 $|x| + |y| > 0$

Graph

47 $f(x) = x + 2$
48 $f(x) = x - 1$
49 $f(x) = -x$
50 $f(x) = -x + 1$
51 $f(x) = -17$
52 $f(x) = 0.03$
53 $f(x) = x + 0.01$
54 $f(x) = -x - 2.5$
55 $f(x) = |x|$
56 $f(x) = |x - 1|$
57 $f(x) = \begin{cases} 0 & \text{if } x \le 0 \\ 2x & \text{if } x > 0 \end{cases}$
58 $f(x) = \begin{cases} x - 1 & \text{if } x \le 3 \\ 2 & \text{if } x > 3 \end{cases}$
59 $f(x) = \begin{cases} 1 & \text{if } x > 0 \\ 0 & \text{if } x = 0 \\ -1 & \text{if } x < 0 \end{cases}$
60 $f(x) = 1 - |x|$

61 Does it make sense to add the functions $y = \sqrt{1-x}$ and $y = \sqrt{x-2}$?

62 Does it make sense to form $f \circ g$ if $f(x) = \sqrt{2x-5}$ and $g(x) = 1 - x^2$?

63 If $f(x) = x$, and $g(x)$ is any function, find $f \circ g$.

64 If $g(x) = x$ and $f(x)$ is any function, find $f \circ g$.

65 Let $f(x) = 1 - x$. Compute $(f \circ f)(x)$.

66 Let $f(x) = 1/x$ for $x \neq 0$. Compute $(f \circ f)(x)$.

67 Prove that if $f(x) = 3x - 5$, then $f[\frac{1}{2}(x_0 + x_1)] = \frac{1}{2}[f(x_0) + f(x_1)]$.

68 If $f(x) = 1/x$, show that $f[\frac{1}{2}(x_0 + x_1)] = 2f(x_0 + x_1)$.

69 If $f(x) = 1/x^2$, show that $f(x_0x_1) = f(x_0)f(x_1)$.

70 A function f is called **strictly increasing** if $f(x_1) < f(x_2)$ whenever $x_1 < x_2$. Show that the sum of two strictly increasing functions is strictly increasing.

The graph of $y = f(x)$ is given. Graph $y = (f \circ f)(x)$.

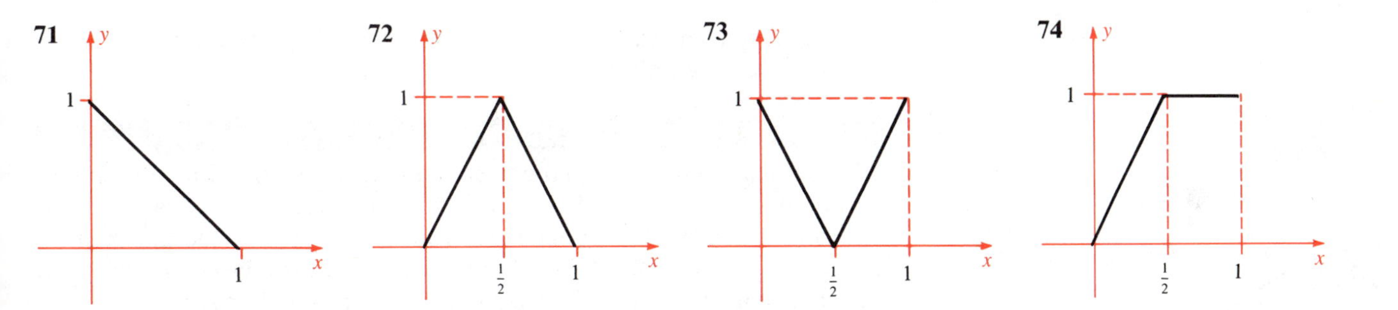

1-3 Linear and Quadratic Functions

A function $f(x)$ is called **linear** if there are constants a and b such that

$$f(x) = ax + b$$

for all real values of x. In the special case that $a = 0$, then $f(x) = b$, a **constant** function. It assigns the same value b to each x.

The graph of $y = ax + b$ is a non-vertical straight line (which accounts for the name *linear*). We shall prove this for one case: $a > 0$ and $b > 0$. First, we note that the points on the graph corresponding to $x = 0$ and to $x = 1$ are

$$(0, b) \quad \text{and} \quad (1, a + b)$$

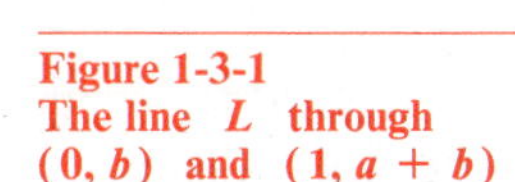
Figure 1-3-1
The line L through $(0, b)$ and $(1, a + b)$

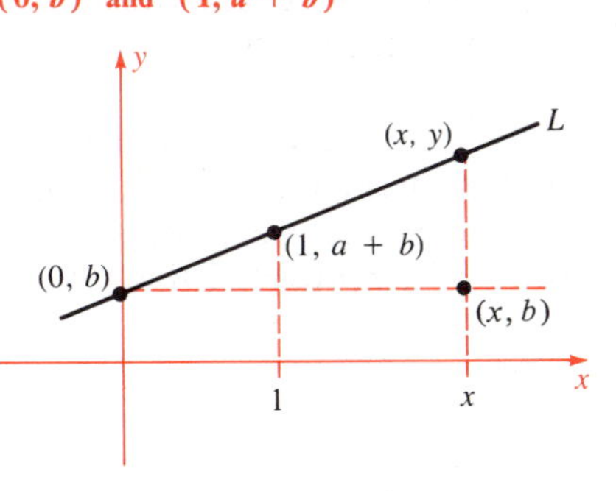

We shall show that the line L through these two points is the graph of $y = ax + b$.

To do so, we choose any point (x, y) on L with $x > 0$. See Figure 1-3-1. Then we pick out two similar right triangles, shown separately in Figure 1-3-2. (The case $x < 0$ requires slightly modified figures.) Similarity (all corresponding angles equal) implies that corresponding sides have the same ratio. Therefore

Figure 1-3-2
Similar triangles from Figure 1-3-1

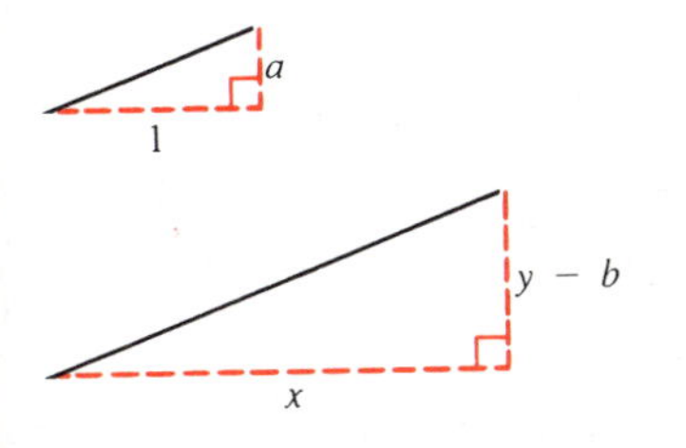

$$\frac{y - b}{a} = \frac{x}{1} \qquad \text{hence} \qquad y - b = ax$$

It follows that $y = ax + b$ for each point (x, y) on L. The reasoning can be reversed to show that if $x > 0$ and $y = ax + b$, then (x, y) is on L.

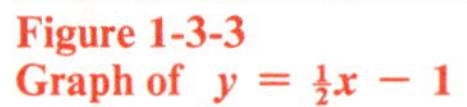
Figure 1-3-3
Graph of $y = \frac{1}{2}x - 1$

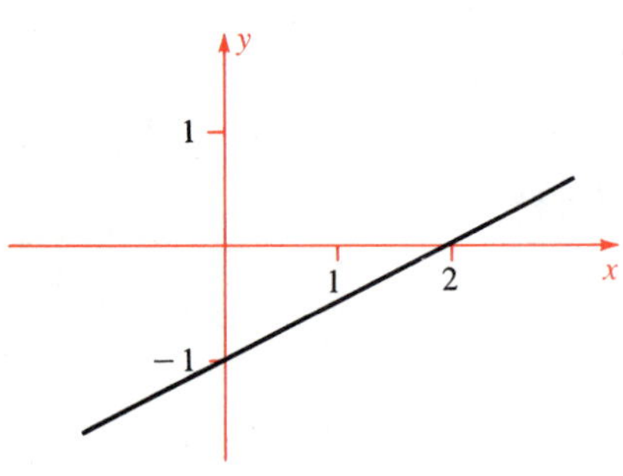

Conversely, if L is a non-vertical straight line, then it intersects the y-axis in some point $(0, b)$ and it intersects the vertical line $x = 1$ in a point we can write as $(1, a + b)$. The graph of $y = ax + b$, as we have just seen, is also a line through these two points; hence it is L. Therefore L is the graph of $y = ax + b$.

See Figures 1-3-3 and 1-3-4 for examples.

The Graph of a Linear Function

The graph of a linear function is a non-vertical straight line. Conversely, each non-vertical straight line is the graph of a linear function. ●

Note that a vertical straight line is represented by an equation $x = c$, but it is *not* the graph of a function.

Figure 1-3-4
Graph of $y = -\frac{2}{3}x - 2$

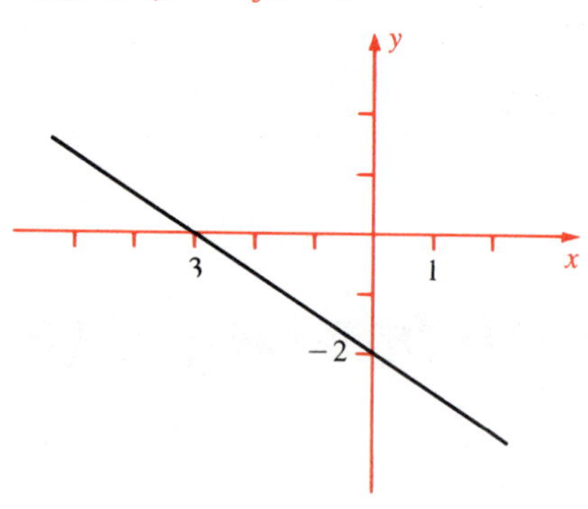

Slope

We now define a measure of the steepness of a non-vertical line. We choose two points (x_0, y_0) and (x_1, y_1) on the line. As x advances from x_0 to x_1, the variable y changes from y_0 to y_1, so the change in y is $y_1 - y_0$. See Figure 1-3-5. The slope is the ratio of the change in y to the change in x, that is, the change in y per unit change in x.

Slope

The **slope** of the line through (x_0, y_0) and (x_1, y_1) is

$$\frac{y_1 - y_0}{x_1 - x_0}$$ ●

If the line rises as x increases, then both $x_1 - x_0$ and $y_1 - y_0$ have the same sign; hence the slope is positive. If the line falls as x increases, then $x_1 - x_0$ and $y_1 - y_0$ have opposite signs; hence the slope is negative.

The more steeply a line rises, the greater is the change in y compared to the change in x; hence the greater is the slope. For a line making a $45°$ angle with the positive x-axis, $y_1 - y_0 = x_1 - x_0$, so its slope is 1. (This assumes equal scales on the axes.) For a horizontal line, $y_1 - y_0 = 0$, so its slope is 0.

Figure 1-3-5
$$\text{slope} = \frac{\text{change in } y}{\text{change in } x} = \frac{y_1 - y_0}{x_1 - x_0}$$

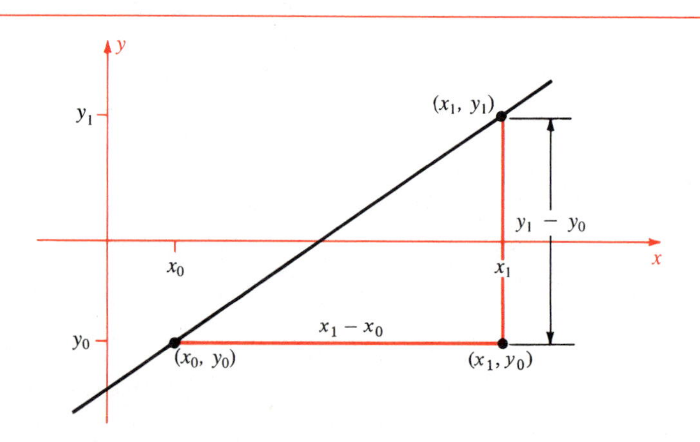

The slope of a line depends only on the line, not on the choice of the two points (x_0, y_0) and (x_1, y_1). In fact, to compute the slope of the line $y = ax + b$, we choose *any* two distinct points on the line and have

$$\text{slope} = \frac{y_1 - y_0}{x_1 - x_0} = \frac{(ax_1 + b) - (ax_0 + b)}{x_1 - x_0}$$
$$= \frac{a(x_1 - x_0)}{x_1 - x_0} = a$$

The number b is called the **y-intercept** of the line $y = ax + b$ because the line cuts the y-axis at $(0, b)$.

Slope and *y*-intercept

The graph of $y = ax + b$ is the line of slope a and y-intercept b.

Example 1

a $y = 3x - 5$ slope $= 3$ y-intercept $= -5$
b $y = -3x + 5$ slope $= -3$ y-intercept $= 5$
c $y = -2$ slope $= 0$ y-intercept $= -2$

Equations of Lines

If a line has slope a and passes through $(0, b)$, then the line is the graph of $f(x) = ax + b$, that is, $y = ax + b$ is an equation of the line.

Suppose, however, that a line has slope a and passes through a point (x_0, y_0), not necessarily on the y-axis. How can we find its equation? By the definition of slope, a point $(x, y) \neq (x_0, y_0)$ lies on the line provided

$$\frac{y - y_0}{x - x_0} = a$$

This is the desired equation. It is usually written in the form $y - y_0 = a(x - x_0)$, in which we may even substitute $x = x_0$ (without dividing by 0).

What if we are given two points on a line, (x_0, y_0) and (x_1, y_1), where $x_0 \neq x_1$. How do we find an equation of the line? We use the preceding result with a computed by

$$a = \frac{y_1 - y_0}{x_1 - x_0}$$

Often the resulting equation is written in the form

$$\frac{y - y_0}{x - x_0} = \frac{y_1 - y_0}{x_1 - x_0}$$

A special case is sometimes handy. If the x-intercept is c and the y-intercept is d, that is, the line passes through $(c, 0)$ and $(0, d)$, then this equation can be written in the form

$$\frac{x}{c} + \frac{y}{d} = 1$$

We leave the derivation as an exercise.

Finally, an equation of the form

$$ax + by = c$$

with $b \neq 0$ can be solved for y:

$$y = -\frac{a}{b}x + \frac{c}{b}$$

so it is another form of the equation of a line.

Equations of Lines

Slope-intercept form	$y = ax + b$
Point-slope form	$y - y_0 = a(x - x_0)$
Two-point form	$\dfrac{y - y_0}{x - x_0} = \dfrac{y_1 - y_0}{x_1 - x_0}$
Two-intercept form	$\dfrac{x}{c} + \dfrac{y}{d} = 1$
Standard form	$ax + by = c \quad (b \neq 0)$
Vertical line	$x = c$

Example 2

Find the slope-intercept form for the line

a through $(2, 3)$ and $(-1, 1)$ (Figure 1-3-6)

b through $(-1, 2)$ with slope -5 (Figure 1-3-7)

c given by the equation $3x + 4y = 5$ (Figure 1-3-8)

Figure 1-3-6
Graph of $y = \frac{2}{3}x + \frac{5}{3}$

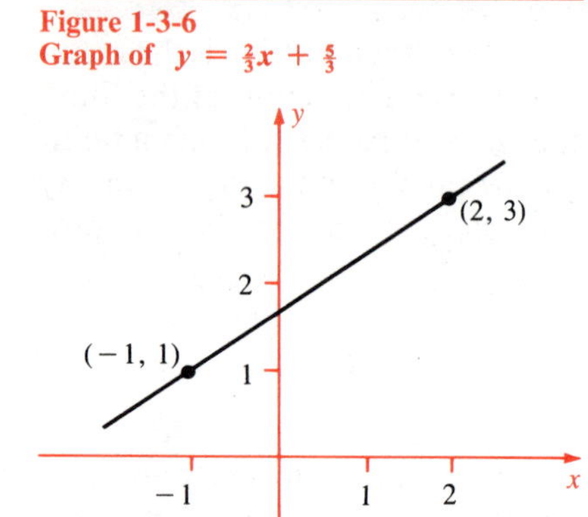

Solution

a Two points are given, so we start with the two-point form

$$\frac{y - 3}{x - 2} = \frac{1 - 3}{-1 - 2} = \tfrac{2}{3}$$

Now we change this into slope-intercept form by solving for y:

$$y - 3 = \tfrac{2}{3}(x - 2) = \tfrac{2}{3}x - \tfrac{4}{3} \qquad y = \tfrac{2}{3}x + \tfrac{5}{3}$$

Figure 1-3-7
Graph of $y = -5x - 3$

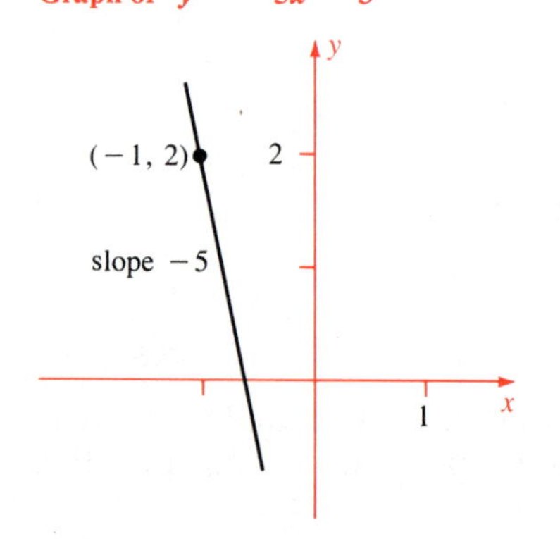

b We must start with the point-slope form:

$$y - 2 = -5(x + 1) = -5x - 5$$

Now we solve for y to obtain $y = -5x - 3$.

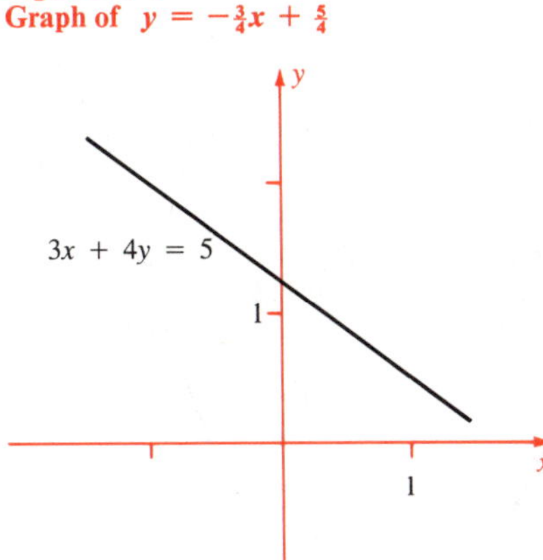

Figure 1-3-8
Graph of $y = -\frac{3}{4}x + \frac{5}{4}$

c We solve for y:

$$4y = -3x + 5 \qquad y = -\tfrac{3}{4}x + \tfrac{5}{4}$$

To close this discussion of linear functions, we give the conditions for lines to be parallel or perpendicular.

Parallel and Perpendicular Lines

Suppose two lines are given:

$$L_1:\ y = a_1x + b_1 \qquad \text{and} \qquad L_2:\ y = a_2x + b_2$$

Then:

- L_1 and L_2 are parallel if and only if $a_1 = a_2$.
- L_1 and L_2 are perpendicular if and only if $a_1a_2 = -1$.

To prove the first statement, we solve a little puzzle. Suppose we start with a line $L_1: y = ax + b_1$. See Figure 1-3-9. Suppose also, for simplicity, that $b_2 > b_1$. If we move each point of L_1 up $b_2 - b_1$ units, then the result is a line L_2, parallel to L_1. What is its equation?

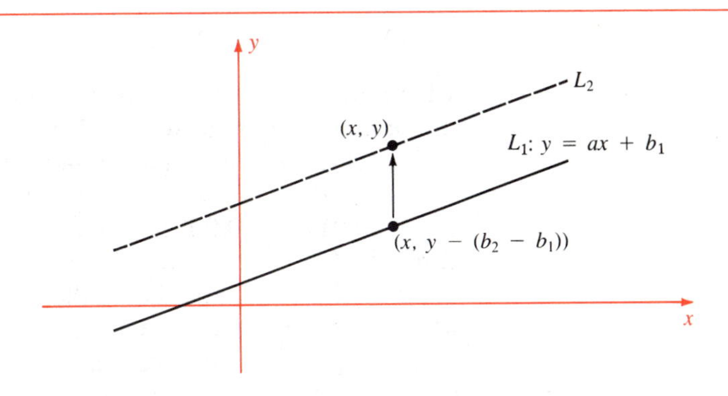

Figure 1-3-9
The line L_1 is shifted up $b_2 - b_1$ units.

If (x, y) is any point on L_2, then $(x, y - (b_2 - b_1))$ is a point of L_1; hence

$$y - (b_2 - b_1) = ax + b_1$$

which implies

$$y = ax + b_2$$

Thus the equation of L_2 is $y = ax + b_2$.

What this means is that if a is fixed, the graphs of the linear functions

$$y = ax + b \qquad (b \text{ any real number})$$

form the family of parallel lines that all have slope a. See Figure 1-3-10.

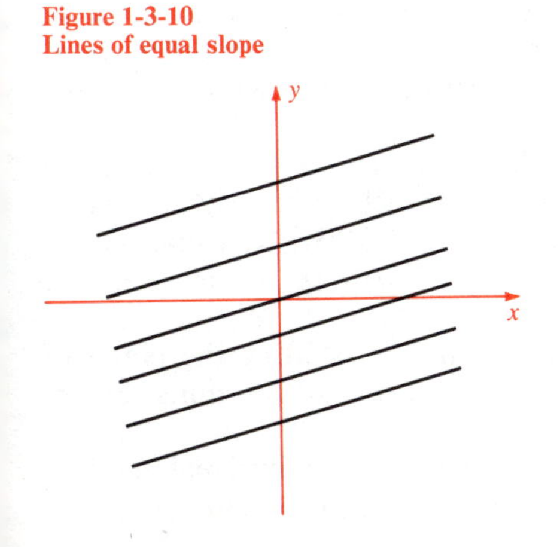

Figure 1-3-10
Lines of equal slope

We postpone the proof of the second statement until page 47.

Figure 1-3-11
Graph of $y = x^2$

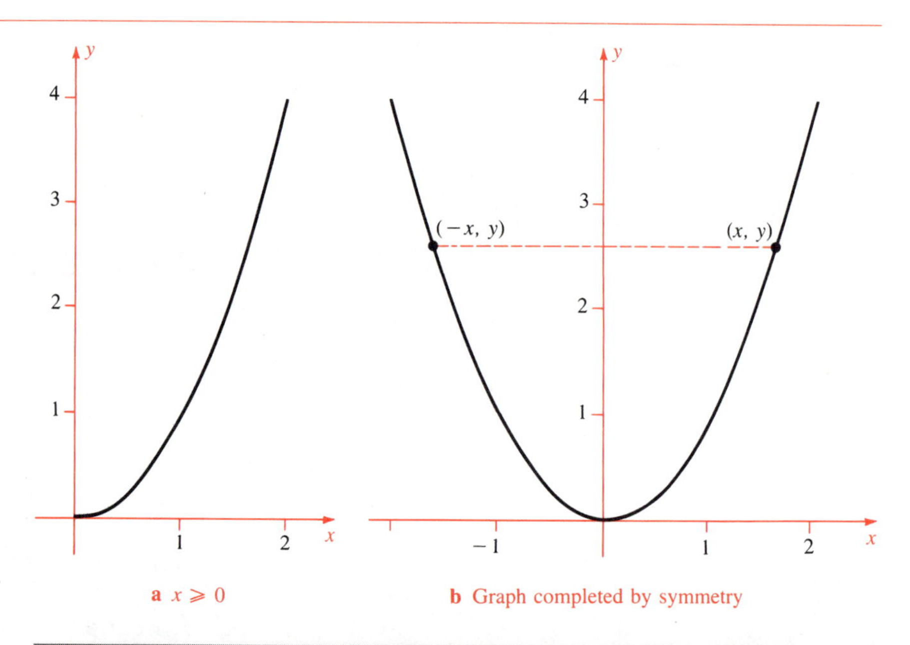

Quadratic Functions

A function $f(x)$ is called **quadratic** if

$$f(x) = ax^2 + bx + c$$

where a, b, and c are constants and $a \neq 0$. The domain of a quadratic function is $\mathbf{R} = \{\text{all } x\}$.

Let us graph the simplest quadratic, $y = x^2$. We consider first only $x \geq 0$. As x increases, y increases, but very slowly at first. For example $(0.01)^2 = 0.0001$, and $(0.1)^2 = 0.01$. The curve passes through $(1, 1)$ and then begins to rise rapidly, passing through $(2, 4)$, $(3, 9)$, $(4, 16)$, and so on. Plotting some of these points gives a rough idea of the graph for $x \geq 0$. See Figure 1-3-11a.

For $x < 0$, we note that $(-x)^2 = x^2$. Hence for each point (x, y) on the curve, the point $(-x, y)$ is also on the curve. It follows that the graph is symmetric about the y-axis (Figure 1-3-11b). The curve is called a **parabola.**

Next let us graph $y = ax^2$, assuming first that $a > 0$. The graph of $y = ax^2$ can be obtained from the graph of $y = x^2$ in a simple way: Each point (x, y) on $y = x^2$ is changed to (x, ay). In other words, the graph $y = x^2$ is stretched (or shrunk) by the factor a in the y-direction only (Figure 1-3-12).

If $a < 0$, then $-a > 0$, and the graph of $y = ax^2$ is obtained from the graph of $y = (-a)x^2$ by changing each y to $-y$, that is, by forming a mirror image in the x-axis (Figure 1-3-13). Note that $(0, 0)$ is the lowest point on the graph of $y = ax^2$ if $a > 0$, and is the highest point on the graph if $a < 0$.

The graph of $y = ax^2 + c$ is obtained by shifting the graph of $y = ax^2$ up (if $c > 0$) or down (if $c < 0$) by $|c|$ units (Figure 1-3-14).

Next let us graph $y = a(x + h)^2$, where h is a fixed real number. First let us assume that $h > 0$. For each point (x, y) on this curve, the

Figure 1-3-12
Graphs of $y = ax^2$ for $a > 0$

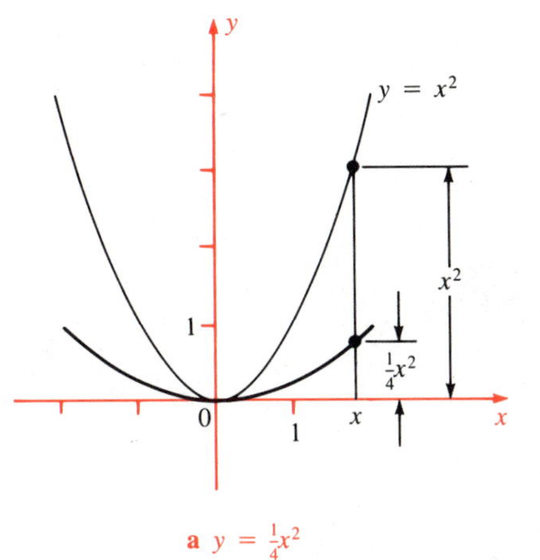

a $y = \frac{1}{4}x^2$

b $y = \frac{3}{2}x^2$

Figure 1-3-13
Graphs of $y = ax^2$ for $a < 0$

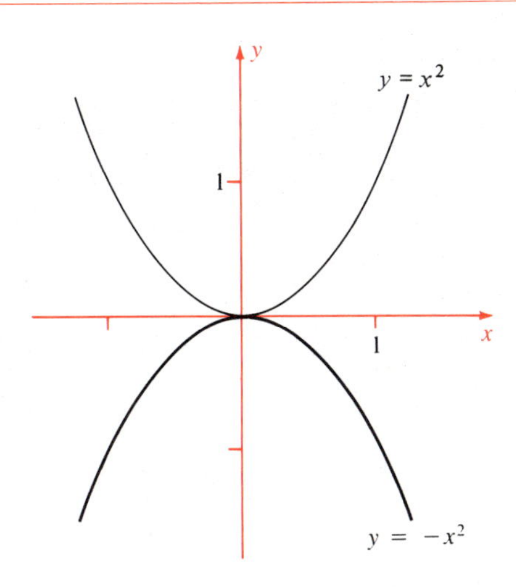

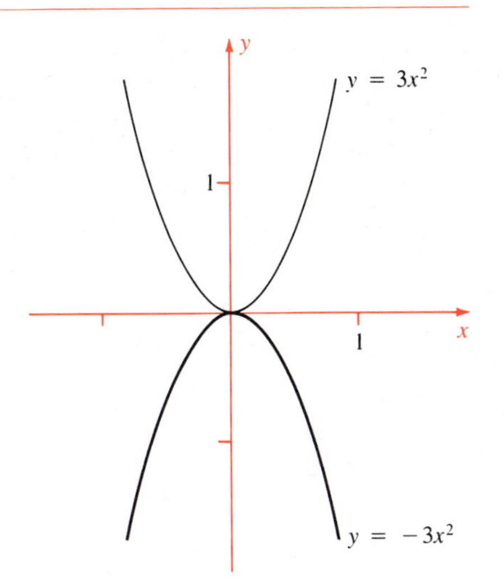

Figure 1-3-14
Graphs of $y = ax^2 + c$

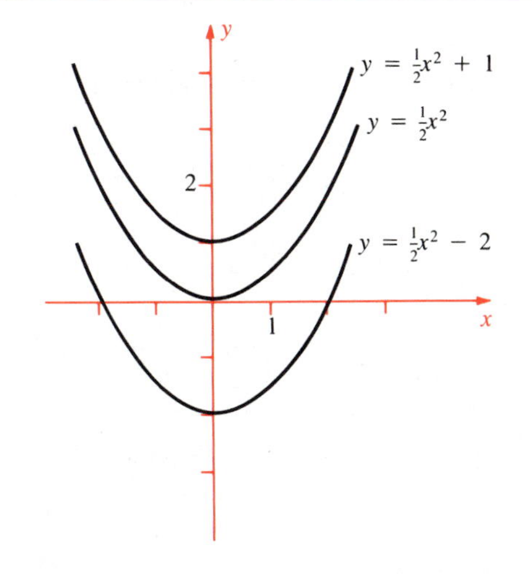

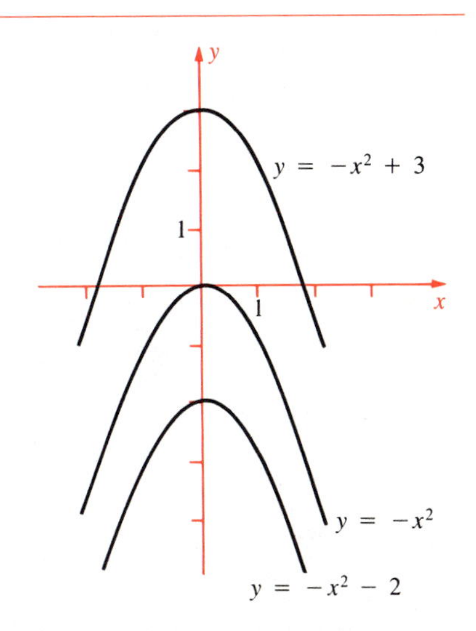

Figure 1-3-15
Graphs of $y = a(x + 5)^2$ and $y = a(x - 5)^2$

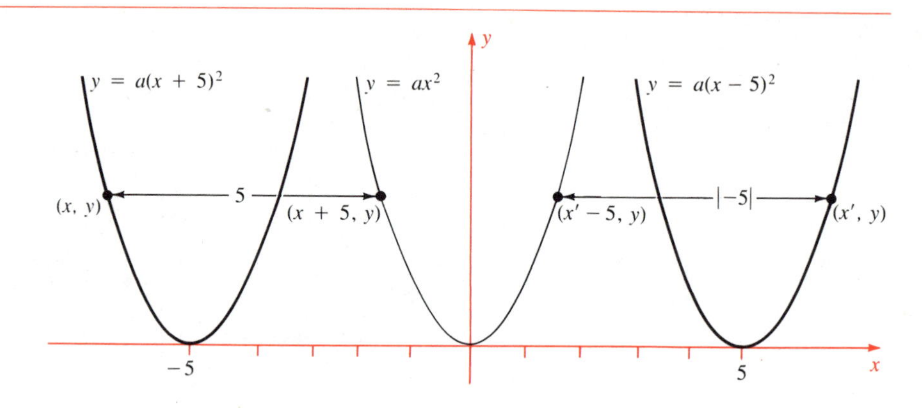

point $(x + h, y)$ is on the curve $y = ax^2$. The point $(x + h, y)$ lies h units to the right of (x, y) because $h > 0$. Thus, if we start with the graph of $y = a(x + h)^2$ and move each point h units to the right, we get the curve $y = ax^2$. In other words, the curve $y = a(x + h)^2$ is the curve $y = ax^2$ shifted h units to the left. If $k < 0$, similar reasoning shows that $y = a(x + k)^2$ is the curve $y = ax^2$ shifted $|k|$ units to the right (Figure 1-3-15).

To graph the general quadratic function we complete the square. For instance, to graph

$$y = x^2 - 6x + 5$$

we write

$$y = x^2 - 6x + 9 - 9 + 5 = (x - 3)^2 - 4$$

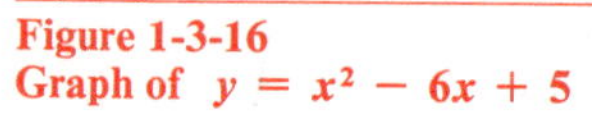
Figure 1-3-16
Graph of $y = x^2 - 6x + 5$

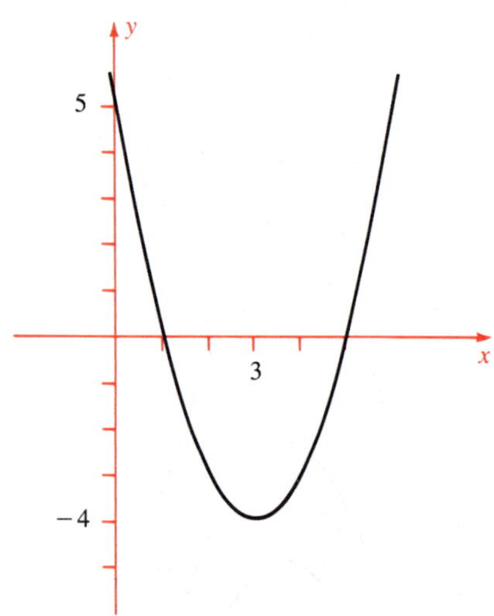

so the graph is obtained from that of $y = x^2$ by horizontal and vertical shifts (Figure 1-3-16).

In general, to graph

$$y = ax^2 + bx + c \qquad (a \neq 0)$$

we first complete the square:

$$\begin{aligned} y &= a\left(x^2 + \frac{b}{a}x + \frac{c}{a}\right) \\ &= a\left(x^2 + \frac{b}{a}x + \frac{b^2}{4a^2} + \frac{c}{a} - \frac{b^2}{4a^2}\right) \\ &= a\left(x^2 + \frac{b}{a}x + \frac{b^2}{4a^2}\right) + a\left(\frac{c}{a} - \frac{b^2}{4a^2}\right) \\ &= a\left(x + \frac{b}{2a}\right)^2 + \left(c - \frac{b^2}{4a}\right) \end{aligned}$$

Hence for $a \neq 0$, the graph of $y = ax^2 + bx + c$ is the graph of

$$y = a\left(x + \frac{b}{2a}\right)^2 + \frac{4ac - b^2}{4a} = a(x + h)^2 + c'$$

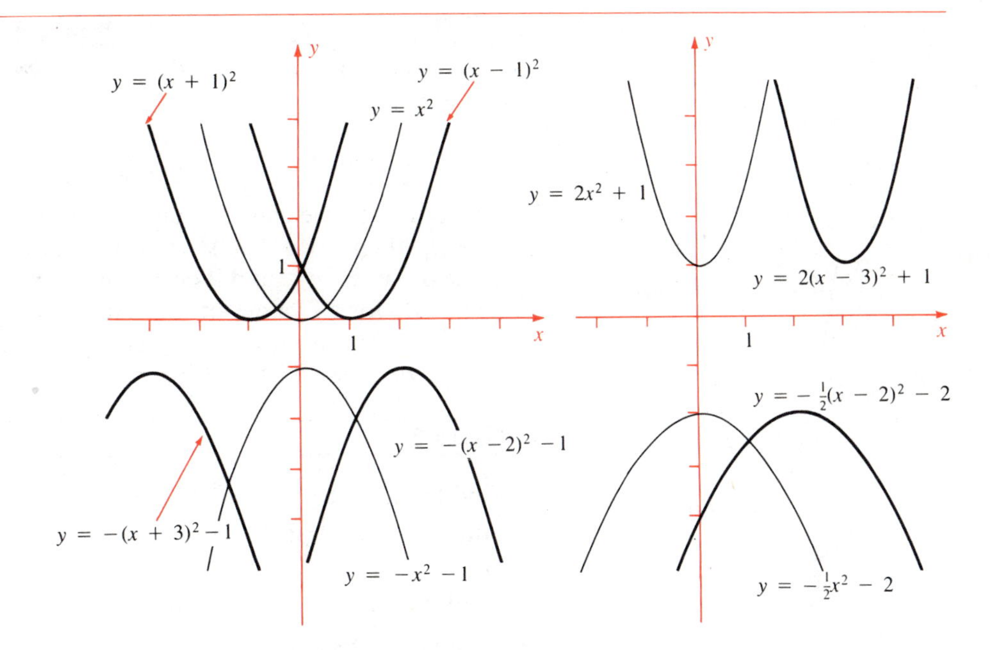

Figure 1-3-17
Graphs of $y = a(x + h)^2 + c'$

where $h = b/2a$ and $c' = (4ac - b^2)/4a$. Therefore the graph of $y = ax^2 + bx + c$ is the graph of $y = ax^2$ shifted horizontally $|b/2a|$ units and vertically $|c'|$ units. The horizontal shift is left if $b/2a > 0$, right if $b/2a < 0$. The vertical shift is up if $c' > 0$, down if $c' < 0$. See Figure 1-3-17.

The Quadratic Formula

This is a good place to recall that a **zero** of a function $f(x)$ is a root of the equation $f(x) = 0$. Thus r is a zero of $f(x)$ provided r is in the domain of $f(x)$ and $f(r) = 0$.

If $f(x) = ax^2 + bx + c \quad (a \neq 0)$ is a quadratic function, then $f(x)$ has a real zero if and only if its **discriminant,** $b^2 - 4ac$, is non-negative. Then its zeros are given by the **quadratic formula:**

$$r = \frac{-b \pm \sqrt{b^2 - 4ac}}{2a}$$

If $b^2 - 4ac = 0$, then there is only one zero, $r = -b/2a$, and $f(x) = a(x - r)^2$. If $b^2 - 4ac > 0$, then $f(x)$ has two distinct zeros r_1 and r_2, given by the quadratic formula, and

$$f(x) = a(x - r_1)(x - r_2)$$

For instance, if $f(x) = 3x^2 + 5x + 1$, then $b^2 - 4ac = 25 - 12 = 13$, so

$$r_1 = \frac{-5 + \sqrt{13}}{6} \qquad r_2 = \frac{-5 - \sqrt{13}}{6}$$

and $f(x) = 3(x - r_1)(x - r_2)$.

Figure 1-3-18
Graph of $y = x^2 - 6x$

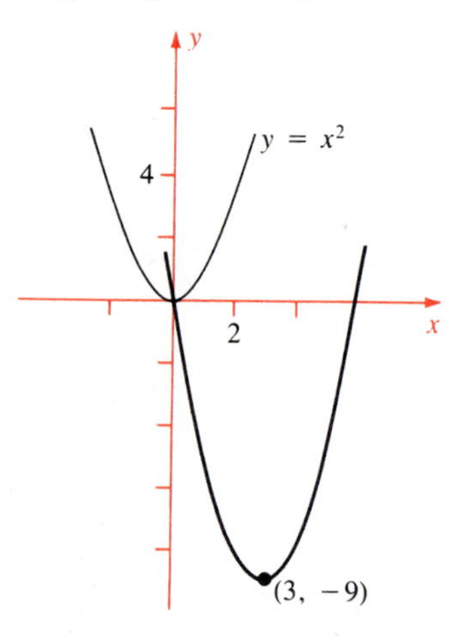

Figure 1-3-19
Graph of $y = -2x^2 - 4x + 1$

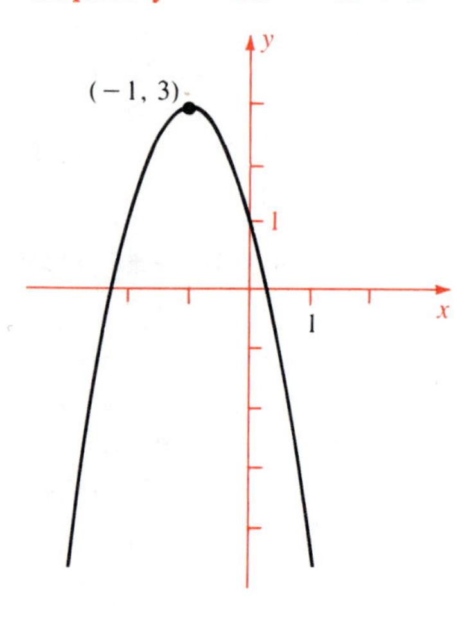

Figure 1-3-20
perimeter $= 2x + 2w = 60$
Hence $w = 30 - x$,
so area $= xw = x(30 - x)$.

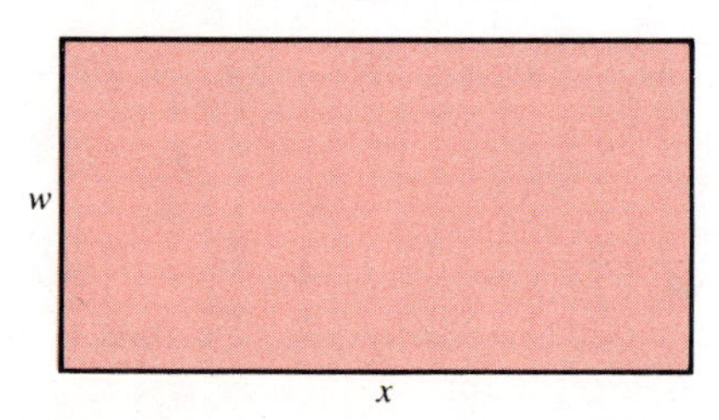

Maxima and Minima of Quadratic Functions

Completing the square enables us to write the quadratic $y = ax^2 + bx + c$ as

$$y = a(x + h)^2 + c'$$

From just this we can read off some valuable information. We see that if $a > 0$, then $a(x + h)^2 \geq 0$ for all values of x. Hence the *minimum* value of y is c', and it occurs at $x = -h$. If $a < 0$, then $a(x + h)^2 \leq 0$ for all values of x. Hence the *maximum* value of y is c', and it occurs at $x = -h$. Thus the range of the function $f(x) = ax^2 + bx + c$ is $[c', +\infty)$ if $a > 0$, and $(-\infty, c']$ if $a < 0$.

Example 3

a Find the lowest point on the curve $y = x^2 - 6x$.
b Find the highest point on the curve $y = -2x^2 - 4x + 1$.

Solution

a Complete the square:

$$y = x^2 - 6x = x^2 - 6x + 9 - 9 = (x - 3)^2 - 9$$

The value of y is least at $x = 3$. The lowest point is $(3, -9)$. See Figure 1-3-18.

b Complete the square:

$$y = -2x^2 - 4x + 1 = -2(x^2 + 2x) + 1$$
$$= -2(x^2 + 2x + 1) + 2 + 1 = -2(x + 1)^2 + 3$$

The value of y is greatest at $x = -1$. The highest point is $(-1, 3)$. See Figure 1-3-19. ●

Example 4

What is the largest possible area of a rectangular rug whose perimeter is 60 feet?

Solution Let x be the length of the rug. By Figure 1-3-20, the area is $A = x(30 - x)$. Since x is a length, $x > 0$. Because x is less than half the perimeter, $x < 30$. The problem, therefore, is to find the largest value of A for $0 < x < 30$. Complete the square:

$$A = 30x - x^2 = -(x^2 - 30x) = -(x^2 - 30x + 15^2) + 15^2$$
$$= -(x - 15)^2 + 225$$

Clearly the largest value of A is 225, which occurs at $x = 15$, an acceptable value of x. Therefore the largest possible area is 225 ft^2. ●

In Chapter 3 we develop tools of calculus for finding maximum and minimum values of general functions. But for *quadratic* functions, calculus is not needed; completing the square does the job.

Exercises

Graph

1 $y = 2x - 3$ for $0 \le x \le 4$

2 $y = 2x - 3$ for $-2 \le x \le 0$

3 $y = 2x + 9$ for $1 \le x \le 2$

4 $y = -3x + 1$ for $0 \le x \le 1$

5 $y = -3x + 1$ for $-5 \le x \le 5$

6 $y = -2x + 1$ for $-20 \le x \le -10$

7 $y = 3x + 40$ for $25 \le x \le 50$

8 $y = 9x - 50$ for $100 \le x \le 200$

9 $y = 0.1x + 1.5$ for $2 \le x \le 3$

10 $y = -0.3x + 0.2$ for $-1 \le x \le 1$

Graph (t in seconds, x in feet)

11 $x = 0.2t - 1$ for $0 \le t \le 5$

12 $x = 25t + 15$ for $50 \le t \le 100$

13 $x = 9t - 9$ for $1 \le t \le 2$

14 $x = -100t + 20$ for $-1 \le t \le 1$

15 $x = -t + 10$ for $25 \le t \le 50$

16 $x = 40t + 40$ for $0 \le t \le 100$

Find the slope of the line through the given points

17 $(0, 0)$ $(3, 4)$

18 $(0, 0)$ $(2, 6)$

19 $(-1, 2)$ $(1, 2)$

20 $(-1, 2)$ $(1, 0)$

21 $(0, 1)$ $(1, 2)$

22 $(0, -1)$ $(1, 2)$

23 $(-1, -1)$ $(1, 2)$

24 $(-1, 2)$ $(2, -1)$

25 $(-3, 1)$ $(-2, 2)$

26 $(-2, -2)$ $(3, -4)$

Find the equation of the line with given slope a and passing through the given point

27 $a = 1$ $(1, 2)$

28 $a = -1$ $(2, -1)$

29 $a = 0$ $(4, 3)$

30 $a = 2$ $(1, 3)$

31 $a = \frac{1}{2}$ $(2, -2)$

32 $a = \frac{2}{3}$ $(-1, 1)$

Find the equation of the line through the two given points

33 $(0, 0)$ $(1, 2)$

34 $(1, 0)$ $(3, 0)$

35 $(-1, 0)$ $(2, 4)$

36 $(-1, -1)$ $(2, 6)$

37 $(\frac{1}{2}, 1)$ $(\frac{3}{2}, 2)$

38 $(-2, 0)$ $(-\frac{1}{2}, -1)$

39 $(0.1, 3.0)$ $(0.3, 2.0)$

40 $(-2.01, 4.10)$ $(-2.00, 4.00)$

Find the slope and y-intercept

41 $3x - y - 7 = 0$

42 $x + 2y + 6 = 0$

43 $3(x - 2) + y + 5 = 2(x + 3)$

44 $2(x + y + 1) = 3x - 5$

Find both intercepts

45 $x/2 + y/3 = 1$

46 $x/a + y/b = 1$

47 $2x + 3y = 1$

48 $ax + by = 1$

Find the equation of the line

49 through $(0, 0)$ and parallel to $y = -5x + 2$

50 through $(-1, 4)$ and parallel to $3x - 2y = 4$

51 through $(2, -1)$ and parallel to the line through $(1, 2)$ and $(2, 3)$

52 through $(-1, 3)$ and parallel to the line through $(2, 5)$ and $(-3, 1)$

53 through $(4, 5)$ and perpendicular to $x - 3y - 4 = 0$

54 through $(2, 7)$ and perpendicular to $y = -2x + 1$

55 through $(0, 0)$ and perpendicular to the line through $(1, 1)$ and $(-2, -4)$

56 through $(-2, -1)$ and perpendicular to the line through $(1, 0)$ and $(-1, 0)$

Graph on the indicated domain (use different scales on the axes if necessary)

57 $y = 0.1x^2$ $[0, 100]$

58 $y = -x^2$ $[-0.1, 0]$

Graph

59 $y = -\frac{1}{2}x^2$

60 $y = \frac{1}{2}x^2$

61 $y = x^2 + 3$

62 $y = -x^2 - 3$

63 $y = 2x^2 - 1$

64 $y = -2x^2 - 1$

65 $y = -\frac{1}{4}x^2 + 2$

66 $y = -\frac{1}{4}x^2 - 2$

Graph and locate the highest or lowest point

67 $y = x^2 - 4x + 1$

68 $y = x^2 + 2x - 5$

69 $y = x^2 - x + 1$

70 $y = -x^2 - 2x$

71 $y = -x^2 - 4x - 3$

72 $y = -x^2 + 4x + 1$

73 $y = 2x^2 + 4x$

74 $y = 3x^2 + 12x - 8$

75 $y = -2x^2 + 8x - 10$

76 $y = -2x^2 + 12x$

77 $y = 2x^2 + 2x + 2$

78 $y = 2x^2 - 3x$

79 $y = x^2 + x - 4$

80 $y = 3x^2 + 3x$

81 $y = -x^2 - 2x$

82 $y = -2x^2 + x$

83 Show that the graph of $y = ax^2 + bx$ passes through the origin for all choices of a and b.

84 For what value of c does the lowest point of the graph of $y = x^2 + 6x + c$ fall on the x-axis?

85 Under what conditions is the lowest point of the graph of $y = x^2 + bx + c$ on the y-axis?

86 A farmer will make a rectangular pen with 100 ft of fencing, using part of a wall of the barn for one side of the pen. What is the largest area that can be enclosed?

87 A 4-ft line is drawn across a corner of a rectangular room, cutting off a triangular region. Show that the area A of the triangular region cannot exceed 4 ft^2. [Hint Use the Pythagorean theorem and work with A^2.]

88 A rectangular solid has a square base, and the sum of its 12 edges is 4 ft. Show that its total surface area (sum of the areas of its 6 faces) is largest if the solid is a cube.

1-4 Graphing Techniques

This section features techniques and shortcuts that can reduce work in graphing. Some of the ideas showed up in Section 1-3 when we graphed quadratic functions. It is a good idea to spell them out.

Figure 1-4-1
Symmetry about the y-axis:
If (x, y) is on the graph, so is its mirror image $(-x, y)$ in the y-axis.

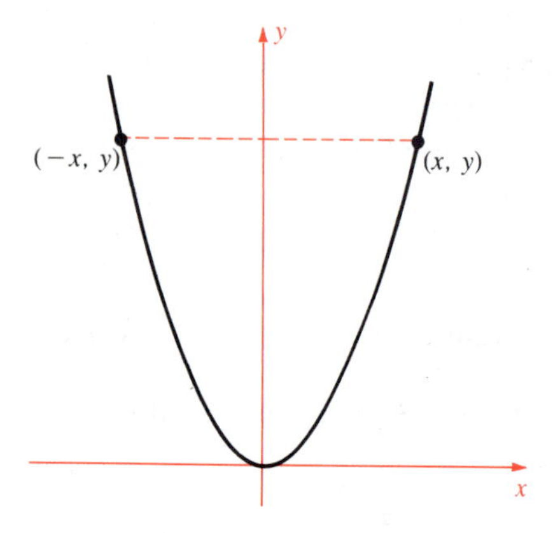

Symmetry

Consider the graphs of $y = x^2$ and $y = x^3$ in Figures 1-4-1 and 1-4-2. These graphs possess certain symmetries. The one in Figure 1-4-1 is symmetric about the y-axis. The one in Figure 1-4-2 is symmetric about the origin; that is, to each point of the graph corresponds an equidistant opposite point as seen through a peephole in the origin. In either case we need plot the curve only for $x \geq 0$, then obtain the rest by symmetry. The work is thus cut in half.

When we plot $y = f(x)$, how can we recognize symmetry in advance? Look at Figure 1-4-1. The curve $y = f(x)$ is **symmetric about the y-axis** if for each x, the value of y at $-x$ is the same as at x, that is, $f(-x) = f(x)$. If $f(x)$ satisfies this condition, it is called an *even* function.

Look at Figure 1-4-2. The curve $y = f(x)$ is **symmetric about the origin** if for each x, the value of y at $-x$ is the negative of the value at x, that is, $f(-x) = -f(x)$. If $f(x)$ satisfies this condition, it is called an *odd* function.

Figure 1-4-2
Symmetry about the origin:
If (x, y) is on the graph,
so is the opposite point $(-x, y)$.

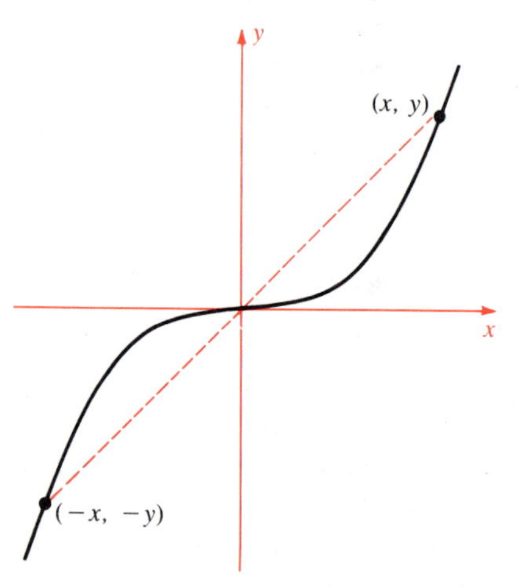

Even and Odd Functions

- An **even** function $f(x)$ is one for which $f(-x) = f(x)$. The graph of an even function is symmetric about the y-axis.
- An **odd** function $f(x)$ is one for which $f(-x) = -f(x)$. The graph of an odd function is symmetric about the origin.

Vertical and Horizontal Shifts

We know that if a positive constant c is added to or subtracted from $f(x)$, the graph of $y = f(x)$ is shifted up or down c units. Now let us consider *horizontal* shifts. How can we shift the graph of $y = f(x)$ three units to the right? More generally how can we find a function $g(x)$ for which the graph of $y = g(x)$ is precisely that of $y = f(x)$ shifted c units to the right?

Consider Figure 1-4-3. For each point (x, y) on the curve $y = g(x)$, there corresponds a point $(x - c, y)$ on the curve $y = f(x)$. The values of y are the same. But on the first curve, $y = g(x)$, and on the second curve, $y = f(x - c)$. Conclusion: $g(x) = f(x - c)$ for all x. This makes sense: If x represents time, then the value of g "now" is the same as the value that f had c time units ago.

The same reasoning shows that the graph of $y = f(x + c)$ is the graph of $y = f(x)$ shifted c units to the left. See Figure 1-3-15.

Shifted Graphs

$$\left.\begin{array}{l} y = f(x) + c \\ y = f(x) - c \\ y = f(x - c) \\ y = f(x + c) \end{array}\right\} \text{ is the graph of } y = f(x) \text{ shifted } c \text{ units } \left\{\begin{array}{l} \text{up} \\ \text{down} \\ \text{right} \\ \text{left} \end{array}\right.$$

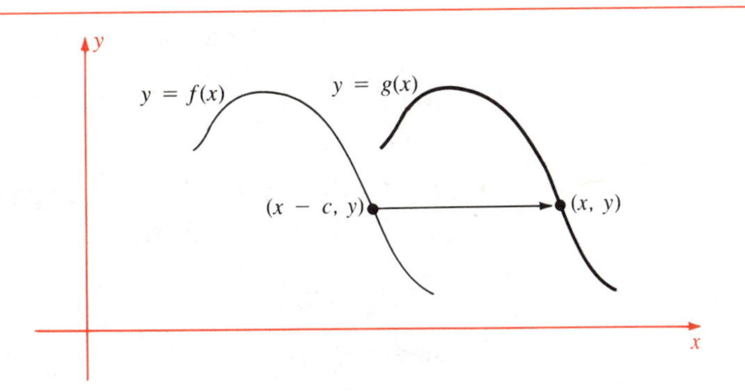

Figure 1-4-3
Horizontal shift $(c > 0)$.
The graph of $g(x) = f(x - c)$ is the graph of $f(x)$ shifted c units to the right.

Stretching and Reflecting

If $c > 0$, then the graph of $y = cf(x)$ is obtained from that of $y = f(x)$ by stretching by a factor of c in the y-direction. Each point (x, y) is replaced by (x, cy). Note: "stretching" by a factor less than one is interpreted as shrinking (Figure 1-4-4).

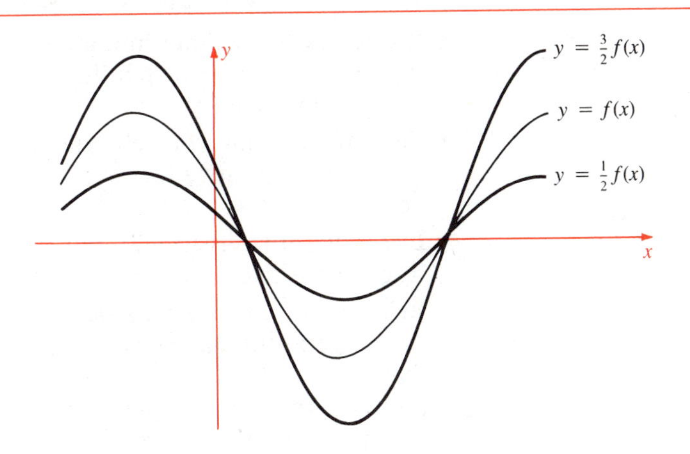

Figure 1-4-4
Stretching in the y-direction

The graph of $y = -f(x)$ is obtained by reflecting the graph of $y = f(x)$ in the x-axis (turning it upside down). That is because each point (x, y) is replaced by the point $(x, -y)$. See Figure 1-4-5.

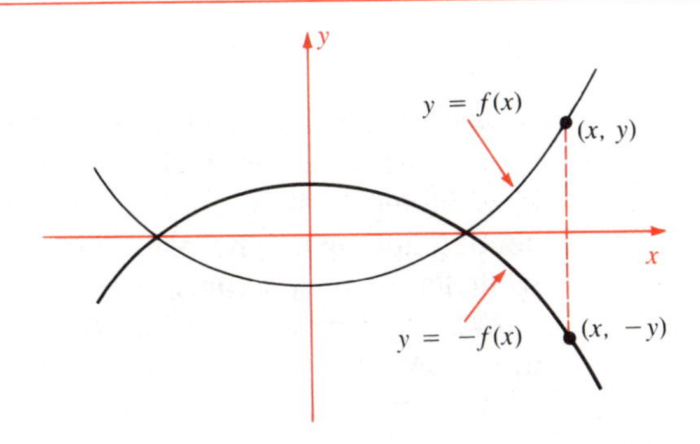

Figure 1-4-5
Reflection in the x-axis

If $c < 0$, the graph of $y = cf(x)$ is obtained from that of $y = f(x)$ by reflecting in the x-axis, then stretching in the y-direction. For instance, the graph of $y = -3f(x)$ is obtained in two steps from the graph of $y = f(x)$:

- **Step 1** Reflect to obtain the graph of $y = -f(x)$.
- **Step 2** Stretch by 3 to obtain the graph of

$$y = 3[-f(x)] = -3f(x)$$

These steps also could be done in reverse order.

Table 1-4-1

x	$f(x) = x^3$
0.0	0.000
0.1	0.001
0.2	0.008
0.3	0.027
0.4	0.064
0.5	0.125
0.6	0.216
0.7	0.343
0.8	0.512
0.9	0.729
1.0	1.000

Table 1-4-2

x	$f(x) = x^3$
0	0
1	1
2	8
3	27
4	64
5	125
6	216
7	343
8	512
9	729
10	1000

Free Information about Graphs

Very often you can get valuable information about a graph just by looking at its equation. It is good practice not to start right off plotting points, but to take a minute to think. Look for symmetry, shifts, and stretching and reflecting. Here are some further points you can check.

- **Domain** Is y defined for all real x or is there some restriction on x? For example, $y = \sqrt{1 - x^2}$ is defined only for $|x| \le 1$, and $y = 1/[(x-1)(x-4)]$ is not defined for $x = 1$ and $x = 4$.
- **Range** Is there some limitation on y? For example, if we consider $y = 1/(1 + x^2)$, then by inspection, $0 < y \le 1$. The graph does not extend above the level $y = 1$ or below the level $y = 0$.
- **Sign of y** Can you tell where $y > 0$ or $y < 0$? For example, $y = 1/x$ is positive for $x > 0$ and negative for $x < 0$. Also y is never 0.
- **Increasing or Decreasing?** For example, $1/x$ decreases as x increases through positive values.
- **Behavior for $|x|$ Very Large** Is y very large or very small when x is very large positive or very large negative? For example, we see by inspection that $y = (x+1)/x^2$ is very small (positive) when x is very large positive. Another example: $y = (x^2 + 1)/(x + 1)$ is very large negative when x is very large negative.

You will not always be able to check all these points. At least see what you can find easily.

Polynomial Functions

A function of the form

$$f(x) = a_n x^n + a_{n-1}x^{n-1} + \cdots + a_1 x + a_0$$

is called a **polynomial function.** Unless stated otherwise, the domain of any polynomial function is $\mathbf{R} = \{\text{all } x\}$. Provided $a_n \ne 0$, the integer n is the **degree** of the polynomial function. Later we shall learn a great deal about graphing polynomial functions by using tools from calculus. Meanwhile let us study some special cases.

We begin with **power functions** $y = x^n$, where n is any positive integer. We are already familiar with $y = x$ and $y = x^2$, so we start with $y = x^3$.

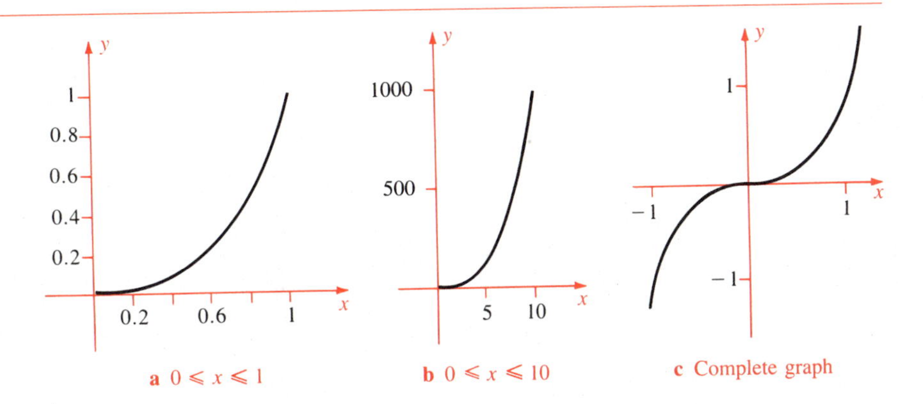

Figure 1-4-6
Graph of $y = x^3$

Because $f(x) = x^3$ is an odd function, its graph is symmetric about the origin. Therefore we concentrate on the right half of the graph, where $x \geq 0$. As x starts from 0 and increases, x^3 also starts from 0 and increases. Let us compute some values (Table 1-4-1) for $0 \leq x \leq 1$. The table gives a pretty good idea of the graph from 0 to 1. See Figure 1-4-6a. The curve is quite flat near $x = 0$, much flatter than the graph of $y = x^2$. See Figure 1-4-8 (next page). Now let us consider some larger values of x in Table 1-4-2. The graph rises very fast as x increases (Figure 1-4-6b). We now have a good idea of the graph for $x \geq 0$. By symmetry, we sketch the complete graph (Figure 1-4-6c).

We obtain the graph of $y = x^4$ in a similar way. This time, $f(x)$ is an even function, so its graph is symmetric about the y-axis. See Figure 1-4-7, on the next page.

It is interesting to compare the graphs of x^2, x^3, and x^4 for small x and for large x. When x is small, x^2 is very small, x^3 is even smaller, and x^4 is even smaller yet. But when x is large, x^2 is very large, x^3 is even larger, and x^4 is larger still. Tables 1-4-3 and 1-4-4 and the graphs in Figure 1-4-8 (next page) show this clearly.

The graphs of $y = x^5$, $y = x^7$, $y = x^9$, $\cdots$ in which the exponent is odd, are all more or less like the graph of $y = x^3$. They are increasingly flat near $x = 0$ and grow increasingly rapidly for x large.

Table 1-4-3
Powers of small x

x	x^2	x^3	x^4
0.0	0.00	0.000	0.0000
0.1	0.01	0.001	0.0001
0.2	0.04	0.008	0.0016
0.3	0.09	0.027	0.0081
0.4	0.16	0.064	0.0256
0.5	0.25	0.125	0.0625
0.6	0.36	0.216	0.1296
0.7	0.49	0.343	0.2401
0.8	0.64	0.512	0.4096
0.9	0.81	0.729	0.6561
1.0	1.00	1.000	1.0000

Table 1-4-4
Powers of $x \geq 1$

x	x^2	x^3	x^4
1	1	1	1
2	4	8	16
3	9	27	81
4	16	64	256
5	25	125	625
6	36	216	1296
7	49	343	2401
8	64	512	4096
9	81	729	6561
10	100	1000	10000

Figure 1-4-7
Graph of $y = x^4$

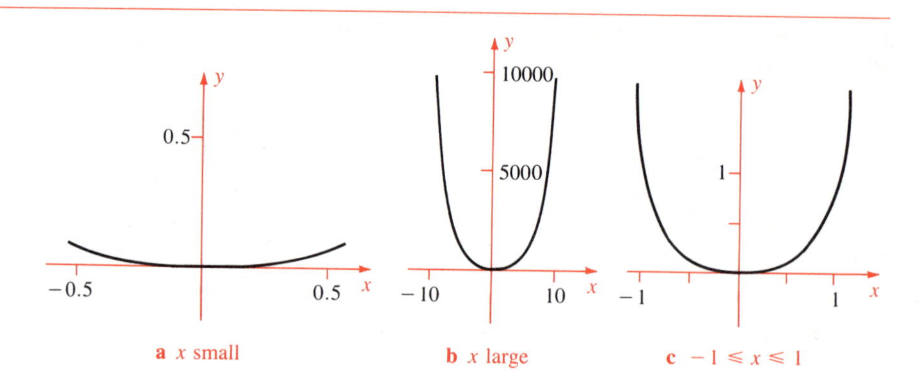

Figure 1-4-8
Graphs of $y = x^2 \quad y = x^3 \quad y = x^4$

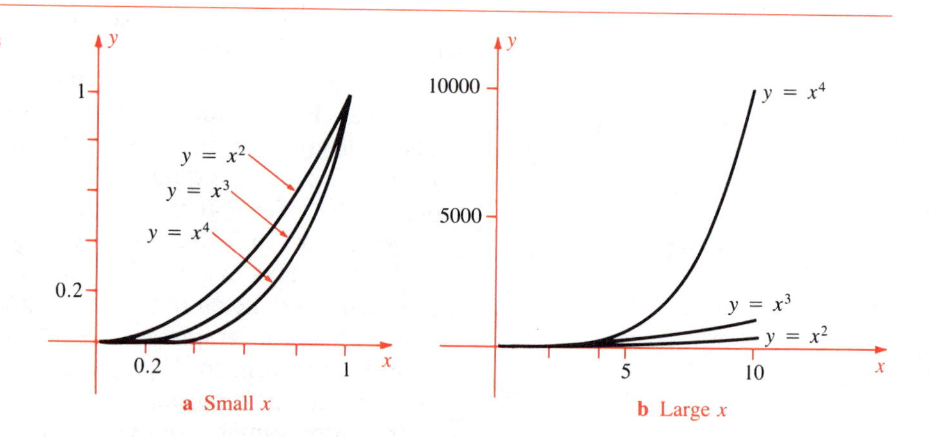

Figure 1-4-9
Graph of $y = x^5$

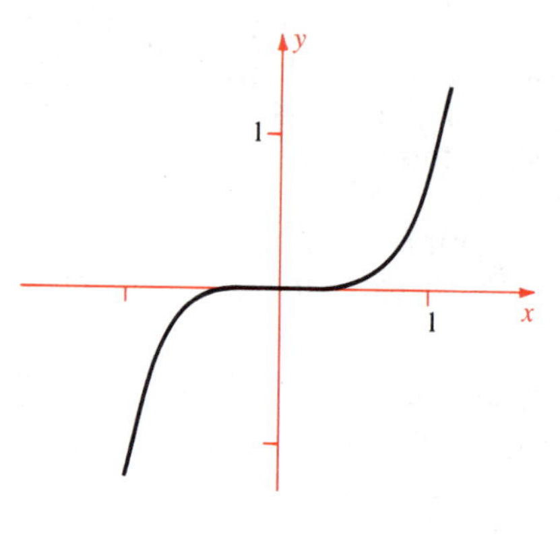

They are all symmetric through the origin (Figure 1-4-9). Similar remarks apply to the graphs of $y = x^4$, $y = x^6$, $y = x^8$, $y = x^{10}$, $\cdots$, $y = x^{2n}$, $\cdots$ in which the exponent is even (Figure 1-4-10), except that these graphs are symmetric about the x-axis.

It is important to remember that the graph of $y = x^n$ crosses the x-axis at $x = 0$ if n is odd; it touches the x-axis but does not cross it if n is even. Algebraically this is obvious: If n is odd, x^n changes sign as x changes from negative to positive; if n is even, x^n is positive everywhere except at $x = 0$. Note that for $n = 1$, the graph of $y = x^n = x$ crosses the axis sharply, but for $n = 3, 5, 7, \cdots$ the graph of $y = x^n$ slithers across.

The graph of $y = c(x - r)^n$ is similar to that of $y = x^n$, except that it touches the x-axis at r instead of 0, and is stretched or contracted by a factor $|c|$ in the y-direction (and reflected in the x-axis if $c < 0$). See Figure 1-4-11.

Figure 1-4-10
Graph of $y = x^6$

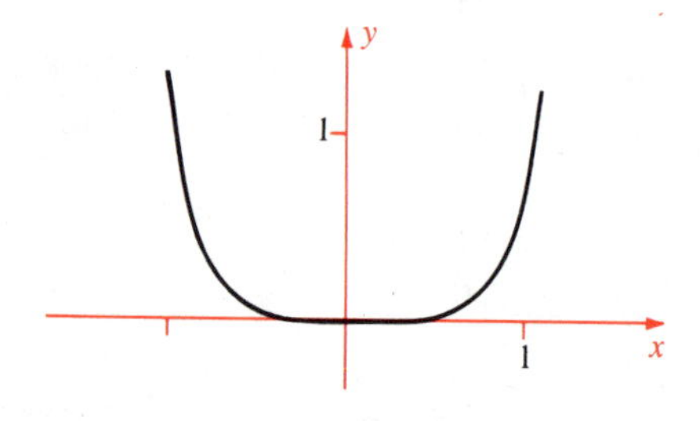

Limit Notation

Now we introduce some useful notation that will be made precise later. This notation helps us handle very large values of variables.

Let $y = x^n$, where $n \geq 1$. Then intuitively speaking the graph of y rises without bound as x increases without bound. We write either

$$x^n \to +\infty \quad \text{as} \quad x \to +\infty \qquad \text{or} \qquad \lim_{x \to +\infty} x^n = +\infty$$

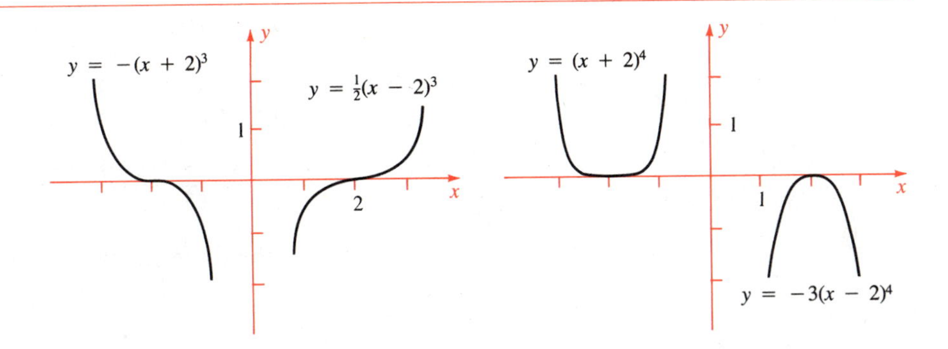

Figure 1-4-11
Graphs of $y = c(x - r)^n$ for various c and r

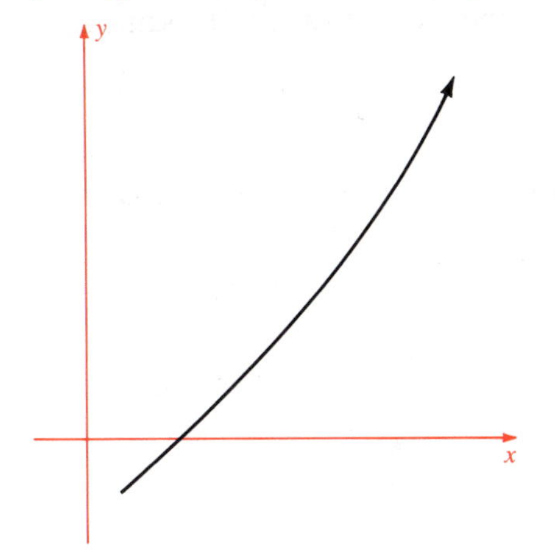

Figure 1-4-12
"y increases without bound as x increases without bound." This means that y eventually exceeds any preassigned number, no matter how large.

We read the first expression "x^n approaches plus infinity as x approaches plus infinity," and we read the second expression "the limit of x^n as x approaches plus infinity is plus infinity." This is strictly a suggestive notation; we are not creating a new *number* called "infinity." The notation is shorthand for "x^n increases without bound as x increases without bound." See Figure 1-4-12.

If n is even, the graph of $y = x^n$ also rises without bound as x decreases without bound. We write either

$$x^n \to +\infty \quad \text{as} \quad x \to -\infty \qquad \text{or} \qquad \lim_{x\to-\infty} x^n = +\infty$$

If n is odd, the graph falls without bound as x decreases (moves to the left) without bound. We then write either

$$x^n \to -\infty \quad \text{as} \quad x \to -\infty \qquad \text{or} \qquad \lim_{x\to-\infty} x^n = -\infty$$

Remark What does the expression "increases without bound" mean? The sequence of numbers

$$1 \quad 4 \quad 9 \quad 16 \quad 25 \quad \cdots \quad n^2 \quad \cdots$$

increases without bound. Give me any large number L and I'll give you one of these numbers that's larger. For instance $(1001)^2 > 1{,}000{,}000$. On the other hand, the sequence of numbers

$$0.9 \quad 0.99 \quad 0.999 \quad 0.9999 \quad 0.99999 \quad \cdots$$

increases all right, but *with* bound; the numbers never get past 1.0.

Similarly, the function $f(x) = (x - 1)/x$ for $x > 0$ increases as x increases without bound, but its value never gets past $y = 1$. See Figure 1-4-13.

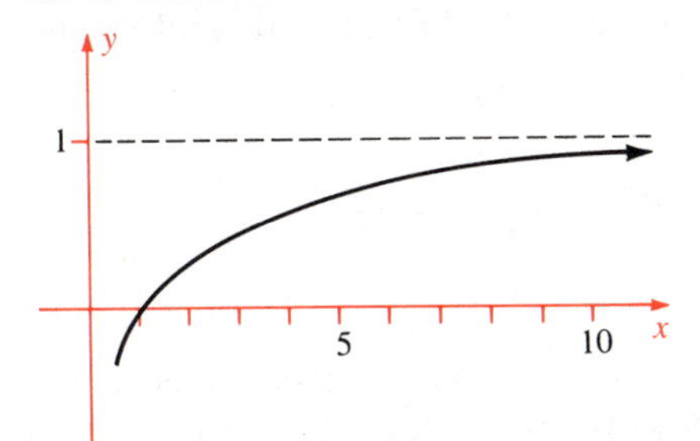

Figure 1-4-13
$y = \dfrac{x-1}{x}$ increases *with* bound as x increases without bound.

We know that $x^3 \to +\infty$ as $x \to +\infty$. What can we say about the behavior of a cubic polynomial like

$$y = x^3 - 5x^2 - 10x + 7$$

as $x \to +\infty$? Our numerical evidence (Table 1-4-4) is that x^3 overwhelms $5x^2$, $-10x$, and 7 as $x \to +\infty$ so that

$$y = x^3 - 5x^2 - 10x + 7 \to +\infty \quad \text{as} \quad x \to +\infty$$

must be the case. We can convince ourselves of this assertion by examining the ratio of y to x^3:

$$\frac{y}{x^3} = \frac{x^3 - 5x^2 - 10x + 7}{x^3} = 1 - \frac{5}{x} - \frac{10}{x^2} + \frac{7}{x^3}$$

As $x \to +\infty$, the three terms

$$-\frac{5}{x} \qquad -\frac{10}{x^2} \qquad \frac{7}{x^3}$$

get closer and closer to 0, and so does their sum. We conclude that $y/x^3 \approx 1$ for x large. We write this in the suggestive notation

$$\frac{y}{x^3} \to 1 \quad \text{as} \quad x \to +\infty$$

Thus, for large x, the functions y and x^3 are about the same size. Another useful abbreviation is

$$y \sim x^3 \quad \text{as} \quad x \to +\infty$$

meaning simply that $y/x^3 \to 1$. Since $x^3 \to +\infty$, it seems clear that $y \to +\infty$ also.

In general, when

$$\frac{f(x)}{g(x)} \to 1 \quad \text{as} \quad x \to +\infty$$

we write

$$f(x) \sim g(x) \quad \text{as} \quad x \to +\infty$$

(also for $x \to -\infty$). We read this expression as "$f(x)$ is **asymptotic** to $g(x)$ as $x \to +\infty$."

Figure 1-4-14
Graph of $y = (x - 1)^2(x - 2)$

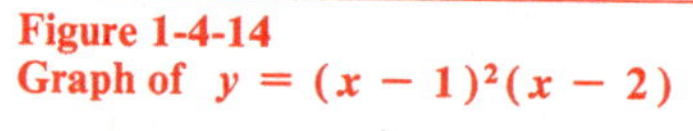

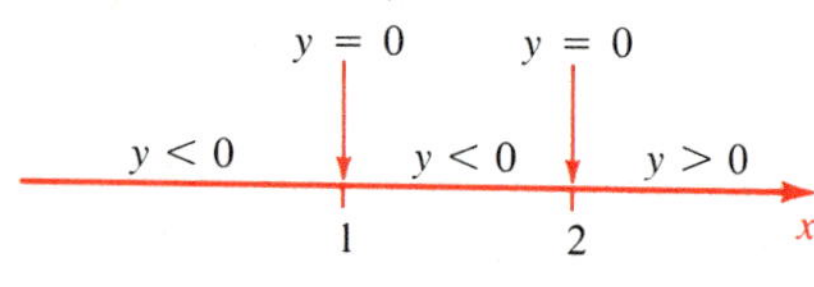

a Sign diagram

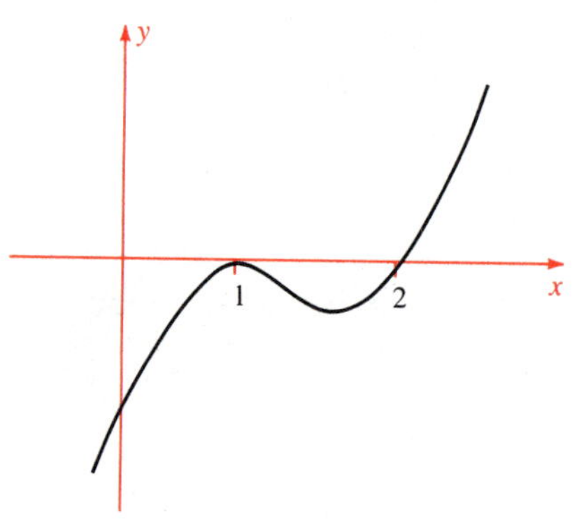

b Rough graph

Example 1

Graph $y = (x - 1)^2(x - 2)$.

Solution First of all, $y = 0$ at $x = 1$ and at $x = 2$, so the graph meets the x-axis at the points $(1, 0)$ and $(2, 0)$. If $x \neq 1$, then $(x - 1)^2 > 0$, so the sign of y is the same as the sign of $x - 2$. We conclude that y changes sign as x passes through 2. However, y does not change sign as x passes through 1. We show this information in a sign diagram (Figure 1-4-14a).

Next, we expand y:

$$\begin{aligned} y &= (x - 1)^2(x - 2) = (x^2 - 2x + 1)(x - 2) \\ &= x^3 - 4x^2 + 5x - 2 \end{aligned}$$

Thus y is a cubic polynomial in x, that is, a polynomial of degree 3. As shown above, the highest term x^3 dominates y for $|x|$ large:

$$y \sim x^3 \quad \text{as} \quad x \to +\infty$$

Therefore

$$y \to +\infty \quad \text{as} \quad x \to +\infty \qquad \text{and} \qquad y \to -\infty \quad \text{as} \quad x \to -\infty$$

This gives us enough information for a rough graph (Figure 1-4-14b). ●

Example 2

Graph $y = (x + 3)(x - 1)(x - 6)$.

Solution We see that $y = 0$ at $x = -3$, at $x = 1$, and at $x = 6$. As x passes through -3, the factor $x + 3$ changes sign while $(x - 1)(x - 6)$ has constant sign. Hence y changes sign as x passes through -3. Similarly, y changes sign as x passes through 1 and through 6. Also, for $|x|$ large,

$$y = x^3 + \cdot \cdot \cdot + 18 \sim x^3$$

so

$$y \to +\infty \quad \text{as} \quad x \to +\infty \qquad \text{and} \qquad y \to -\infty \quad \text{as} \quad x \to -\infty$$

We summarize the sign information in Figure 1-4-15a and then sketch the graph in Figure 1-4-15b. ●

Figure 1-4-15
Graph of
$y = (x + 3)(x - 1)(x - 6)$

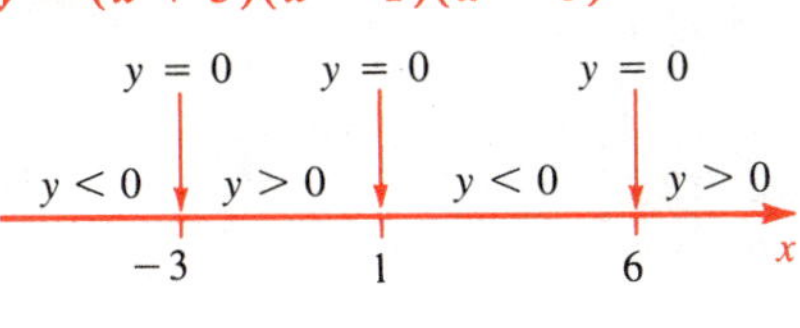

a Sign diagram

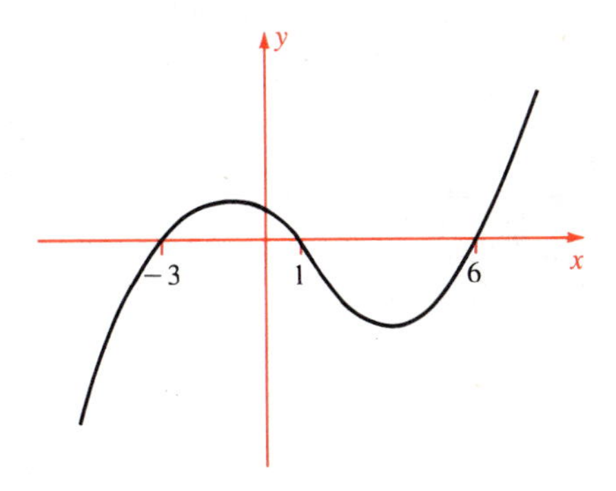

b Rough graph

The Graph of $1/x^n$

The power function $f(x) = 1/x^n$ is an even function if n is an even integer and an odd function if n is an odd integer. Hence in either case, if we sketch the graph for $x > 0$, the graph for $x < 0$ can be obtained by symmetry.

All these graphs pass through $(1, 1)$. Since $x^n \to +\infty$ as $x \to +\infty$, the reciprocal $1/x^n \to 0$ as $x \to +\infty$. The larger n is, the faster $1/x^n \to 0$. As $x \to 0$ through positive values, $x^n \to 0$, hence $1/x^n \to +\infty$. This gives the general shape of the graphs for $x > 0$. We complete the graphs using even or odd symmetry. See Figure 1-4-16 (below left) and Figure 1-4-17 (next page).

The graph of $y = 1/(x - r)^n$ is simply a horizontal shift of one of the graphs of Figures 1-4-16 and 1-4-17.

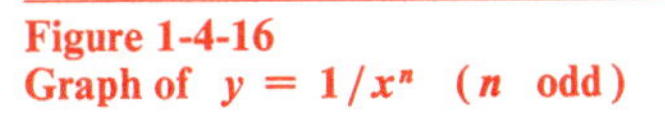

Figure 1-4-16
Graph of $y = 1/x^n$ (n odd)

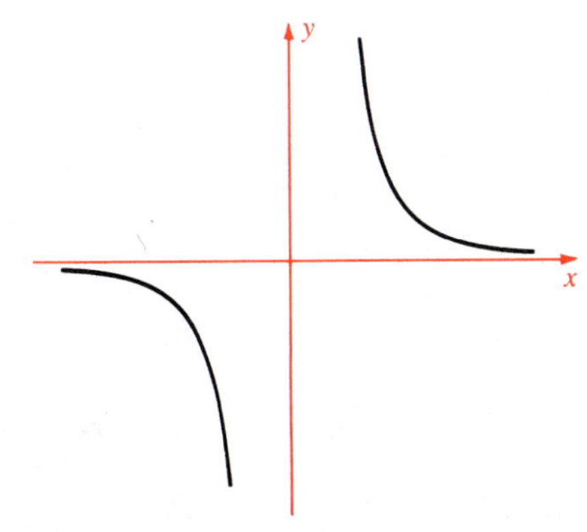

Rational Functions

A **rational function** is the quotient of two polynomials:

$$r(x) = \frac{p(x)}{q(x)}$$

Let us agree that all rational functions will be expressed in lowest terms; that is, the numerator and denominator will have no common polynomial factors. For example, we write

$$r(x) = \frac{1}{x + 1} \qquad \text{not} \quad r(x) = \frac{x - 1}{x^2 - 1}$$

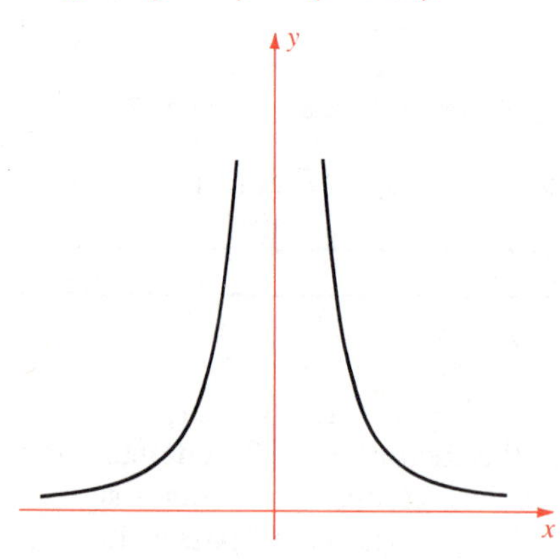

Figure 1-4-17
Graph of $y = 1/x^n$ (n even)

The domain of a rational function in lowest terms consists of all real numbers except the zeros of the denominator.

It will be convenient to introduce the notation

$$x \to a$$

(read "x approaches a") when x takes values arbitrarily close to the real number a. If we wish to restrict x to values on the right of a, then we modify the notation to

$$x \to a+$$

which is shorthand for "x approaches a through values greater than a"; that is, "x approaches a from the right." Similarly, $x \to a-$ is shorthand for "x approaches a through values less than a"; that is, "x approaches a from the left."

Graphs of rational functions can be more complicated than those of polynomials. Nevertheless, many rational functions can be graphed without much trouble. One important source of information is that the behavior of $f(x)$ as $x \to \pm\infty$ is predictable. Suppose

$$r(x) = \frac{a_m x^m + a_{m-1}x^{m-1} + \cdots + a_1 x + a_0}{b_n x^n + b_{n-1}x^{n-1} + \cdots + b_1 x + b_0}$$

$$(a_m \neq 0, \quad b_n \neq 0)$$

Then as $|x| \to +\infty$,

$$\begin{cases} |r(x)| \to +\infty & \text{if } m > n \\ r(x) \to 0 & \text{if } m < n \\ r(x) \to \dfrac{a_m}{b_n} & \text{if } m = n \end{cases}$$

Thus, if the degree of the numerator exceeds the degree of the denominator (the top-heavy case), then $|r(x)| \to +\infty$ as $|x| \to +\infty$. In the opposite (bottom-heavy) case, $r(x) \to 0$ as $|x| \to +\infty$. If the degrees of the numerator and denominator are equal, then $r(x)$ tends to a finite non-zero number, the quotient of the leading coefficients.

Let us sketch a proof of these assertions. The idea is that for $|x|$ large, the leading terms of both numerator and denominator dominate their other terms. Hence for $|x|$ large,

$$r(x) \sim \frac{a_m x^m}{b_n x^n} = \frac{a_m}{b_n} x^{m-n}$$

and the various assertions follow.

Example 3

Discuss the behavior as $x \to \pm\infty$ of the given rational function:

a $r(x) = \dfrac{x^5 + 3x}{x^2 + 12} \sim \dfrac{x^5}{x^2} = x^3$

$$r(x) \to -\infty \text{ as } x \to -\infty \quad \text{and} \quad r(x) \to +\infty \text{ as } x \to +\infty$$

b $r(x) = \dfrac{6x^2 + 7x - 3}{2x^3 + x^2 + 1} \sim \dfrac{6x^2}{2x^3} = \dfrac{3}{x}$

$r(x) \to 0+$ as $x \to +\infty$ and $r(x) \to 0-$ as $x \to -\infty$

c $r(x) = \dfrac{2x^3 + 1}{5x^3 - 4x^2 - x - 7} \sim \dfrac{2x^3}{5x^3} = \dfrac{2}{5}$

$r(x) \to \frac{2}{5}$ as $x \to +\infty$ ●

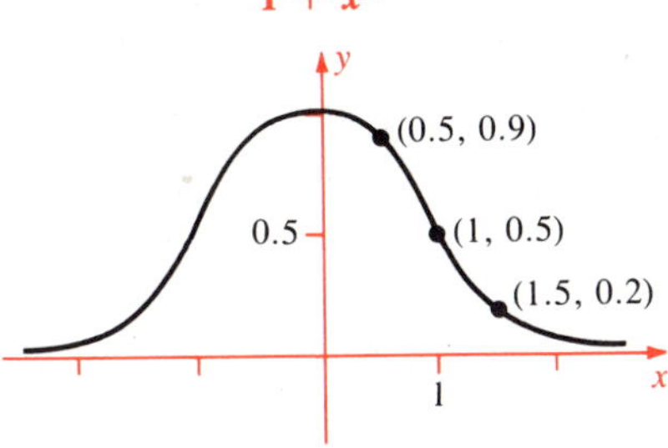

Figure 1-4-18
Graph of $y = \dfrac{1}{1 + x^4}$

Example 4

Graph $y = \dfrac{1}{1 + x^4}$

Solution The graph is defined for all x. Note that $1 + x^4 \geq 1$ for all x, so $0 < y \leq 1$. In fact, $y = 1$ only at $x = 0$, so the highest point on the curve is $(0, 1)$. The function is even, so the graph is symmetric about the y-axis. Clearly $y \to 0+$ as $x \to \pm\infty$. This free information alone is enough for a fairly good idea of the curve. Plotting a few points helps fix the shape (Figure 1-4-18). ●

Example 5

Graph $y = \dfrac{x - 1}{x^2}$

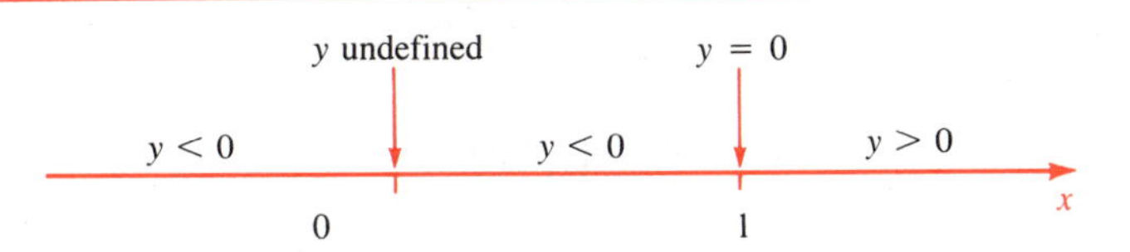

Figure 1-4-19
Sign diagram for $y = (x - 1)/x^2$

Solution First collect some free information. The graph is undefined at $x = 0$. If $x \neq 0$, then $x^2 > 0$ so the sign of y is the same as that of $x - 1$; this information is summarized in a sign diagram (Figure 1-4-19). Further immediate information: the rational function is bottom-heavy; hence $y \to 0$ as $x \to +\infty$.

Now let's look at how the function behaves near the special point $x = 0$ where y is undefined. Write

$$y = \frac{x - 1}{x^2} \sim -\frac{1}{x^2} \quad \text{near} \quad x = 0$$

Combine all the information (Figure 1-4-20); it suggests the desired graph (Figure 1-4-21, next page). ●

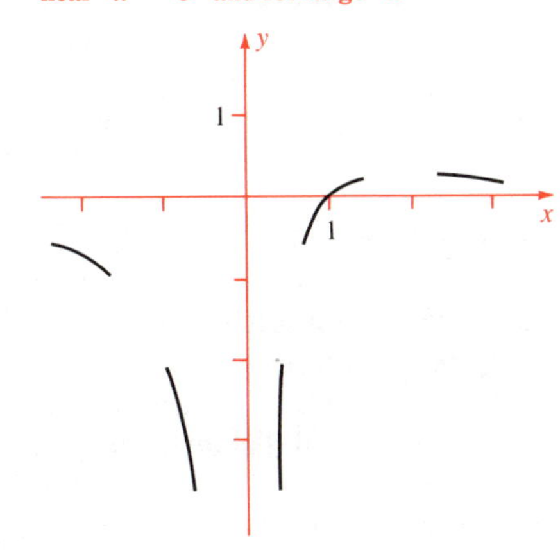

Figure 1-4-20
Behavior of $y = (x - 1)/x^2$ near $x = 0$ and for large x

Example 6

Graph $y = \dfrac{x^2}{(x + 2)(x - 1)}$

Figure 1-4-21
Graph of $y = (x - 1)/x^2$

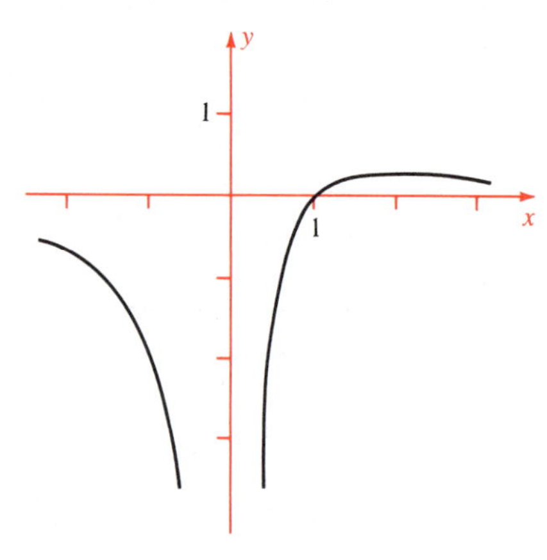

Solution The graph is undefined at $x = -2$ and $x = 1$. It meets the x-axis at $x = 0$ and nowhere else. The function is positive for large values of x and changes sign at $x = 1$ and $x = -2$, but not at $x = 0$. See Figure 1-4-22.

Next look at the behavior of the function near the three special points $x = 0$, -2, and 1 where either $y = 0$ or y is undefined. Near the point $x = 0$,

$$y \sim \frac{1}{(0 + 2)(0 - 1)} x^2 = -\tfrac{1}{2}x^2$$

Near $x = -2$,

$$y \sim \frac{(-2)^2}{(-2 - 1)} \cdot \frac{1}{x + 2} = \frac{-\frac{4}{3}}{x + 2}$$

Near $x = 1$,

$$y \sim \frac{(1)^2}{(1 + 2)} \cdot \frac{1}{x - 1} = \frac{\frac{1}{3}}{x - 1}$$

Figure 1-4-22
Sign diagram for
$y = x^2/[(x + 2)(x - 1)]$

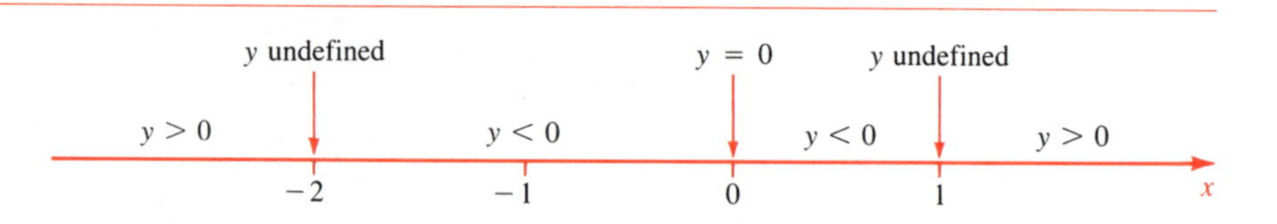

Finally, because

$$y = \frac{x^2}{(x + 2)(x - 1)} = \frac{x^2}{x^2 + x - 2}$$

our criterion for the behavior of rational functions shows that $y \to 1$ as $x \to \pm\infty$. One thing is not quite obvious, however: Does $y \to 1$ from above or from below? For large positive values of x, the denominator is greater than the numerator, so y is below 1. Hence $y \to 1-$ as $x \to +\infty$. For large negative values of x, it is the other way around: $y \to 1+$ as $x \to -\infty$.

We sketch this information (Figure 1-4-23). A rough graph is suggested (Figure 1-4-24). ●

Figure 1-4-23
Behavior of $y = x^2/[(x + 2)(x - 1)]$
near special points

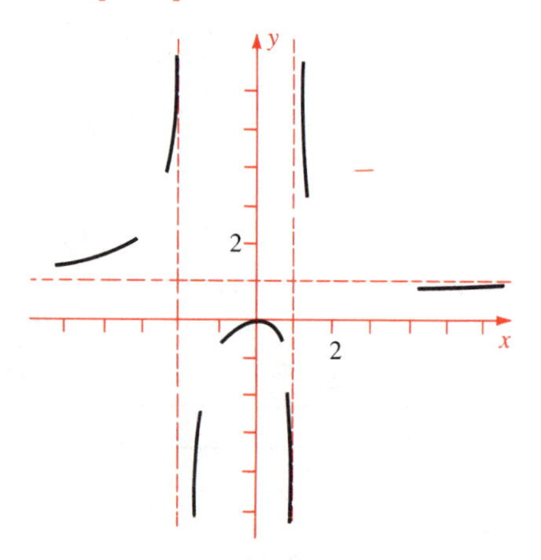

Two more rational functions are graphed in Figures 1-4-25 and 1-4-26. A glance at Figure 1-4-25 shows that the lines $x = 1$, $x = -1$, and $y = 1$ play a special role. These lines are called asymptotes of the graph. A line $x = a$ is called a **vertical asymptote** of the graph $y = f(x)$ if either $f(x) \to +\infty$ or $f(x) \to -\infty$ as $x \to a$. A non-vertical line L is called an **asymptote** of a graph if the vertical distance between the line and the graph approaches 0 as $x \to +\infty$ or $x \to -\infty$ (or both), for example, the line $y = 1$ in Figure 1-4-26.

Figure 1-4-24
Graph of $y = \dfrac{x^2}{(x+2)(x-1)}$

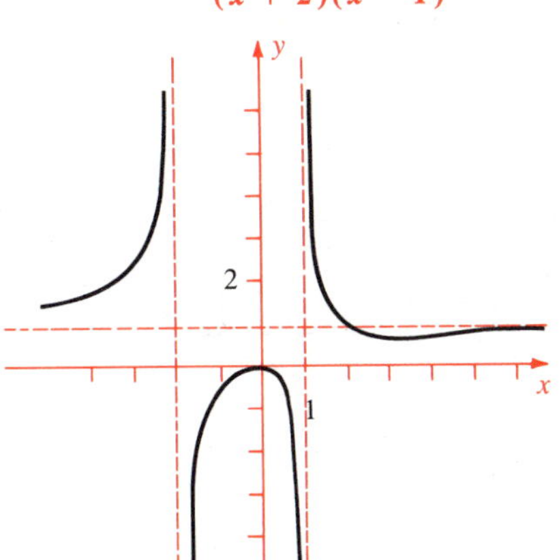

Figure 1-4-25
Graph of $y = \dfrac{x(x+2)}{(x+1)(x-1)}$

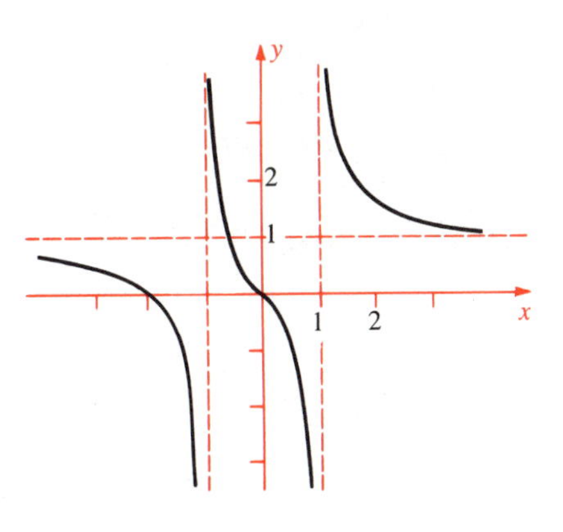

Figure 1-4-26
Graph of $y = \dfrac{x^2(x-2)}{(x-1)^3}$

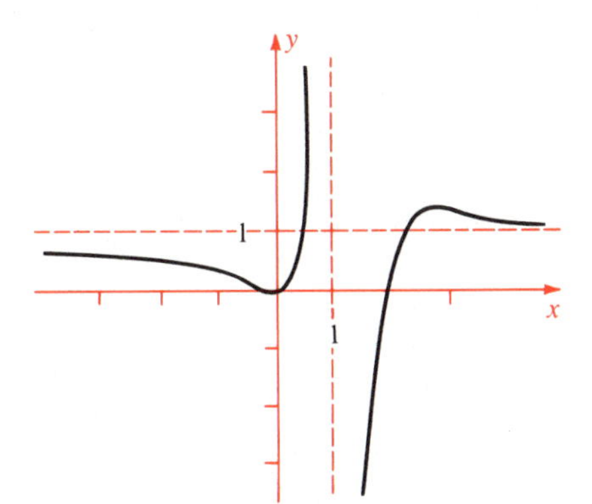

Exercises

1 Which of the functions are even?

$$x^2 \qquad x^3 \qquad x^4 \qquad \frac{1}{x^2+1} \qquad \frac{1}{x^3+1}$$

$$\frac{x}{x^2+1} \qquad \frac{1}{x^3+x} \qquad x^3+x^2+1$$

2 (cont.) Which of the functions are odd?

Graph

3 $y = \frac{1}{4}x^4$
4 $y = -x^4$
5 $y = \frac{1}{8}x^3$
6 $y = -x^3 + 1$
7 $y = -x^6$
8 $y = x^3 - 2$
9 $y = x^4 - 1$
10 $y = x^5 + 1$
11 $y = -x^5$
12 $y = \frac{1}{16}x^5$
13 $y = (x+1)^3$
14 $y = (x-1)^4$
15 $y = -(x+1)^3$
16 $y = -(x-1)^4$
17 $y = (x-1)^3 + 1$
18 $y = -(x+1)^4 - 1$
19 $y = \frac{1}{3}(x+2)^4$
20 $y = -\frac{1}{2}(x+2)^4 - 4$
21 $y = (x - \frac{1}{2})^5 - 1$
22 $y = (x + \frac{1}{2})^6 - 2$

Draw a sign diagram; then use it to sketch the graph of $y = f(x)$

23 $f(x) = x(x-1)(x-2)$
24 $f(x) = (x+2)x(x-2)$
25 $f(x) = (x+2)(x+1)(x-1)$
26 $f(x) = (x+2)(x-1)(x-2)$
27 $f(x) = x^2(x-1)$
28 $f(x) = (x-1)(x-2)^2$
29 $f(x) = -(x+1)(x-1)^2$
30 $f(x) = -x^2(x-1)$
31 $f(x) = \frac{1}{6}(x-1)(x-2)(x-3)(x-4)$
32 $f(x) = \frac{1}{24}x(x-2)(x-3)(x-4)$
33 $f(x) = x^2(x^2-1)$
34 $f(x) = \frac{1}{4}(x-1)^2(x^2-4)$
35 $f(x) = -\frac{1}{4}(x+1)^2(x^2-4)$
36 $f(x) = -\frac{1}{12}x^2(x-2)(x-3)$
37 $f(x) = x(x-1)^3$
38 $f(x) = \frac{1}{8}(x^2-1)^2$

Graph

39 $y = \dfrac{1}{x-4}$
40 $y = \dfrac{-1}{(x+2)^2}$
41 $y = \dfrac{1}{(x+1)^2} - 2$
42 $y = \dfrac{1}{4x-3}$
43 $y = \dfrac{1}{1+x^2}$
44 $y = \dfrac{1}{x} + 3$
45 $y = -\dfrac{1}{x} + 1$
46 $y = -\dfrac{1}{x^2} - 2$
47 $y = \dfrac{x}{5x-3}$
48 $y = \dfrac{2}{1+x^4}$
49 $y = \dfrac{-1}{4+x^2}$
50 $y = \dfrac{-x}{3x+7}$
51 $y = \dfrac{x^2}{x+1}$
52 $y = \dfrac{x-1}{x^2}$

Describe the behavior of $r(x)$ as $x \to +\infty$

53 $r(x) = \dfrac{1}{x+1} - \dfrac{1}{x-1}$
54 $r(x) = x^2 - \dfrac{1}{x^2}$

Graph

55 $y = \dfrac{(x+1)(x-1)}{x^3}$

56 $y = \dfrac{x^3}{(x+1)(x-1)}$

57 $y = \dfrac{(x+2)^2}{x^3}$

58 $y = \dfrac{x^3}{(x-1)^2}$

59 $y = \dfrac{-x^2}{(x+1)^2}$

60 $y = \dfrac{x^3}{(x+1)^3}$

Fill in the table for the given functions

x	10	100	1000	10^6	-10	-100	-1000	-10^6
$f(x)$								
$g(x)$								
$f(x)/g(x)$								

61 $f(x) = \dfrac{3x^5 - 7x^4 + 2}{x^3 + 2x + 5}$ $\quad g(x) = 3x^2$

62 $f(x) = \dfrac{x^4 - 2x^3 + x^2 + x + 7}{8x^3 + x^2 - 3}$ $\quad g(x) = \frac{1}{8}x$

63 $f(x) = \dfrac{x^2 + 3}{x^3 - x + 9}$ $\quad g(x) = 1/x$

64 $f(x) = \dfrac{x + 10}{3x^4 + 9x^3 + x}$ $\quad g(x) = \dfrac{1}{3x^3}$

65 $f(x) = \dfrac{2x^5 - x + 1}{3x^5 + x^4}$ $\quad g(x) = \frac{2}{3}$

66 $f(x) = \dfrac{10x^4 + x^3 + 9x^2}{2x^4 - 11x + 1}$ $\quad g(x) = 5$

1-5 The Distance Formula

The distance formula is a coordinate plane version of the most famous result of plane geometry, the Pythagorean theorem. Let us first recall the theorem: If a and b are the legs of a right triangle and c the hypotenuse (Figure 1-5-1), then

$$c^2 = a^2 + b^2$$

Figure 1-5-1
Pythagorean theorem: $c^2 = a^2 + b^2$

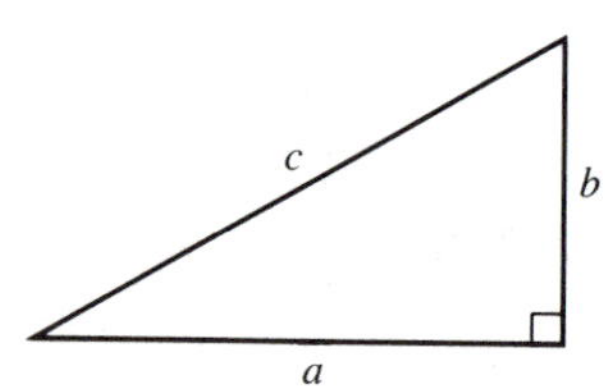

There are many different proofs. One of them is sketched in Figure 1-5-2.

The Pythagorean theorem has a converse: If a, b, and c are the sides of a triangle and if $c^2 = a^2 + b^2$, then the triangle is a right triangle with hypotenuse c. This is intuitively clear mechanically. Imagine rods of lengths a and b hinged at one end (Figure 1-5-3). When the angle γ between them is a right angle, then $c = (a^2 + b^2)^{1/2}$. When γ increases from a right angle, then c increases, so $c > (a^2 + b^2)^{1/2}$.

Figure 1-5-2
Cut 4 triangles from a square of side $a + b$ as shown: the result is a square of side c. Do it another way: the result is two squares, of side a and of side b. Hence $c^2 = a^2 + b^2$.

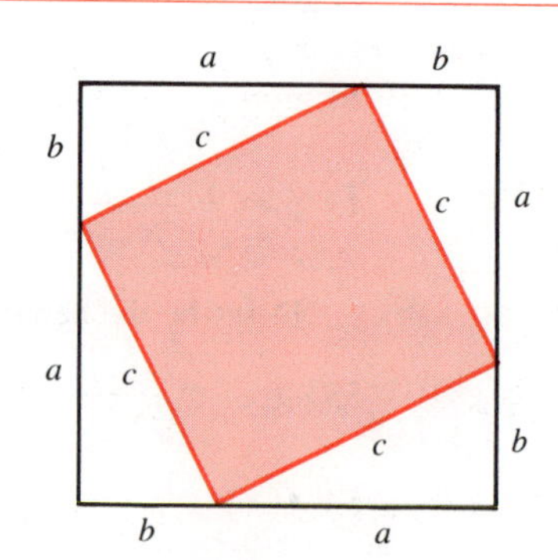

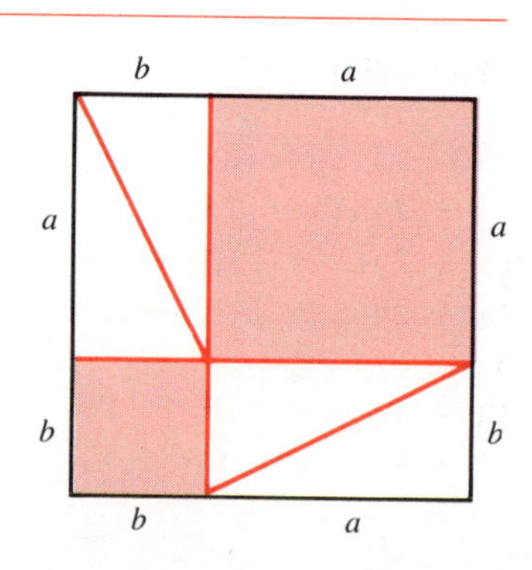

Figure 1-5-3

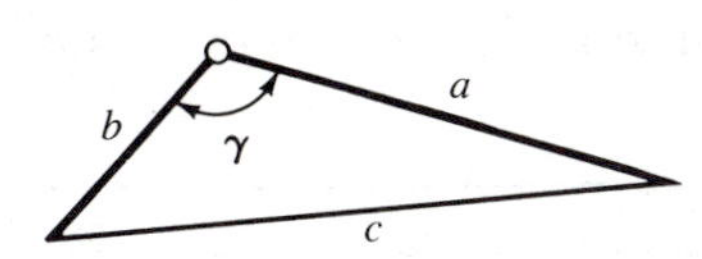

When γ decreases from a right angle, then side c decreases, so $c < (a^2 + b^2)^{1/2}$. See the *law of cosines,* page 61, for a convincing mathematical reason why $c^2 = a^2 + b^2$ forces γ to be a right angle. Let us restate the Pythagorean theorem including its converse:

Pythagorean Theorem

Let a, b, and c be the sides of a triangle T. Then T is a right triangle with hypotenuse c if and only if

$$c^2 = a^2 + b^2$$

Now we apply the Pythagorean theorem to the distance formula. The problem is to find the distance between two points whose coordinates are (x_1, y_1) and (x_2, y_2). To do so, we introduce the auxiliary point (x_2, y_1), forming a right triangle as shown in Figure 1-5-4. The legs have lengths $|x_2 - x_1|$ and $|y_2 - y_1|$, so by the Pythagorean theorem,

$$d^2 = |x_2 - x_1|^2 + |y_2 - y_1|^2 = (x_2 - x_1)^2 + (y_2 - y_1)^2$$

Figure 1-5-4
Derivation of the distance formula

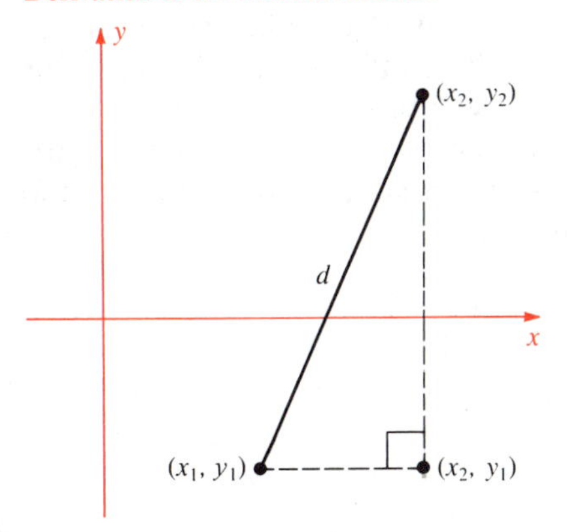

Distance Formula

The distance between (x_1, y_1) and (x_2, y_2) is

$$\sqrt{(x_2 - x_1)^2 + (y_2 - y_1)^2}$$

Warning The distance formula, a geometric result, is correct only if the same unit of length is used on both axes.

Circles

The set of all points in the plane one unit from the origin is a circle of radius 1, called the **unit circle.** By the distance formula, a point (x, y) is on the unit circle if and only if $(x - 0)^2 + (y - 0)^2 = 1^2$, that is, $x^2 + y^2 = 1$. This formula is the equation of the unit circle (Figure 1-5-5).

Figure 1-5-5
The unit circle: $x^2 + y^2 = 1$

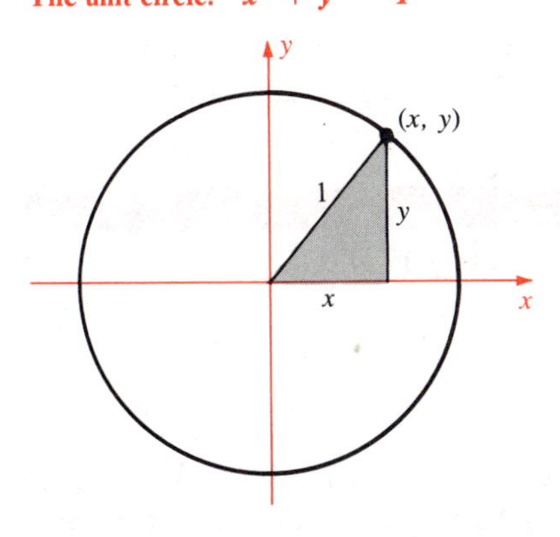

The Unit Circle

The unit circle consists of all points (x, y) in the plane that satisfy the condition

$$x^2 + y^2 = 1$$

In general, the set of all points at a fixed distance r from a point (a, b) is the circle with center (a, b) and radius r. See Figure 1-5-6 (next page). By the distance formula, the distance from (x, y) to (a, b) is r if and only if

$$(x - a)^2 + (y - b)^2 = r^2$$

Equation of a Circle

The circle with center (a, b) and radius r is the set of all points (x, y) satisfying the equation

$$(x - a)^2 + (y - b)^2 = r^2$$

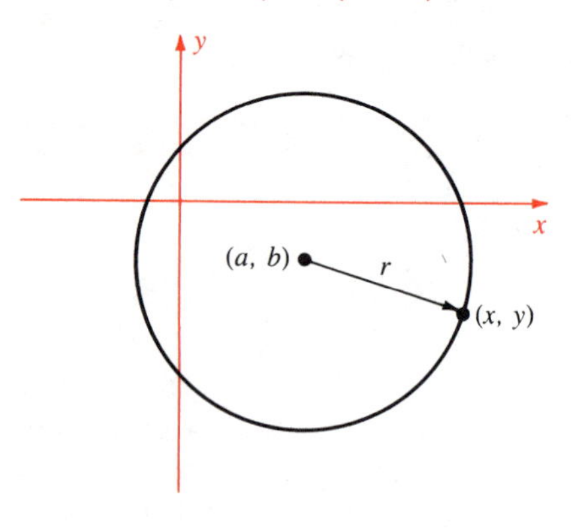

Figure 1-5-6
The circle with center (a, b) and radius r: $(x - a)^2 + (x - b)^2 = r^2$

To use this equation for a particular circle, we must find the center (a, b) and the radius r of the circle, then substitute these quantities into the equation.

Example 1

Find the equation of the circle with center $(2, 2)$ and tangent to the line $x = 3$.

Solution From Figure 1-5-7, the point of tangency is $(3, 2)$, at distance 1 from the center $(2, 2)$. Hence $r = 1$ and the equation is

$$(x - 2)^2 + (y - 2)^2 = 1$$

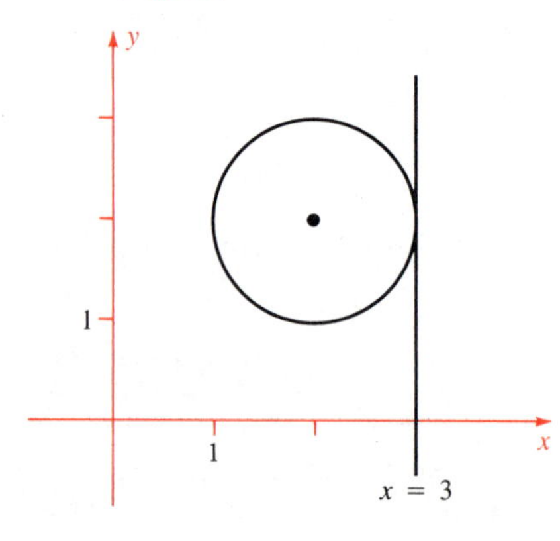

Figure 1-5-7
See Example 1.

Parabolas

A **parabola** is defined as the set of all points equidistant from a fixed line D and a fixed point P not on D. We call D the **directrix** and P the **focus** of the parabola. To find an equation for a parabola, we set up the coordinate system (Figure 1-5-8) so that $P = (0, p)$ and D is the line $y = -p$. (This guarantees that $(0, 0)$ will be on the parabola.)

By definition, a point (x, y) is on the parabola if and only if the distances d_1 and d_2 in Figure 1-5-9 are equal, or equivalently, if and only if $d_1^2 = d_2^2$. By the distance formula,

$$d_1^2 = (x - 0)^2 + (y - p)^2 \quad \text{and} \quad d_2^2 = (y + p)^2$$

Hence

$$(x - 0)^2 + (y - p)^2 = (y + p)^2$$

$$x^2 + y^2 - 2py + p^2 = y^2 + 2py + p^2$$

We cancel y^2 and p^2 and combine the two terms $2py$:

$$x^2 = 4py$$

The steps can be read backwards to show that if $x^2 = 4py$, then (x, y) is a point of the parabola with focus $P = (0, p)$ and directrix D: $y = -p$.

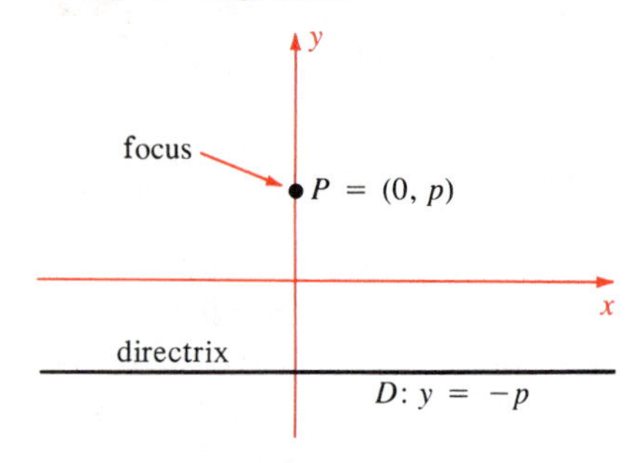

Figure 1-5-8
Set-up for the parabola

Parabola

The equation of the parabola with focus $(0, p)$ and directrix $y = -p$ is

$$y = \frac{1}{4p} x^2$$

With our choice of axes, the parabola is the graph of a quadratic function of the form $y = ax^2$. Conversely, given any quadratic function $y = ax^2$, its graph is the parabola with focus $(0, p)$ and directrix $y = -p$, where $p = 1/4a$. It follows that the graph of any quadratic function $y = ax^2 + bx + c$ is a parabola. For by completing the square, we can write the quadratic in the form $y = a(x + h)^2 + c'$. Hence its graph is the parabola $y = ax^2$ shifted horizontally $|h|$ units and vertically $|c'|$ units, still a parabola.

Figure 1-5-9
The parabola:
$d_1 = d_2$ implies the equation $x^2 = 4py$

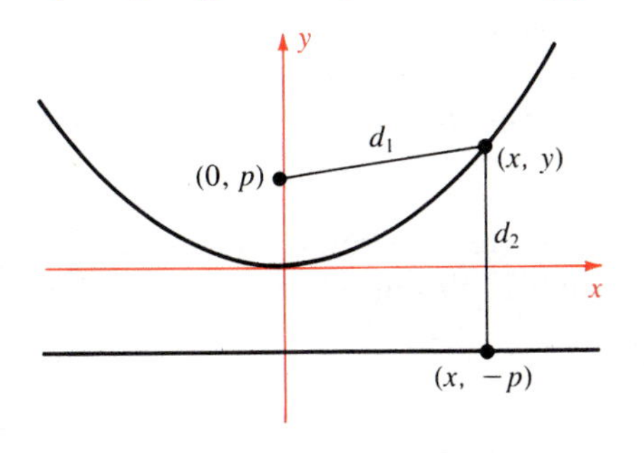

Figure 1-5-10
Perpendicular lines: $\overline{PQ}^2 = \overline{OP}^2 + \overline{OQ}^2$

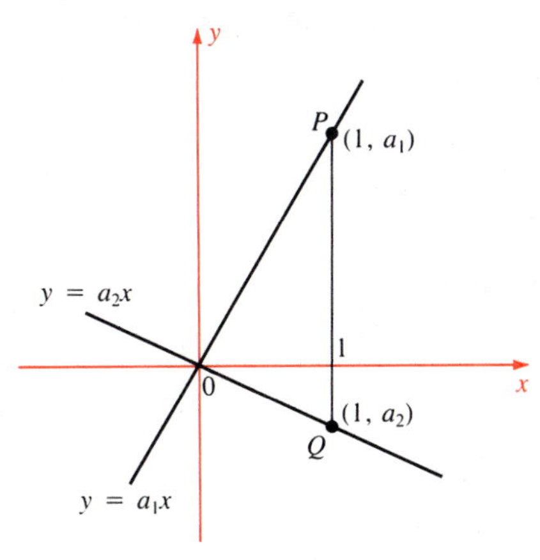

Perpendicular Lines and Midpoints

On page 25 we stated the condition $a_1a_2 = -1$ for two lines

$$y = a_1x + b_1 \quad \text{and} \quad y_2 = a_2x + b_2$$

to be perpendicular (provided neither is horizontal or vertical). Now we prove that this condition holds if and only if the lines are perpendicular.

The line $y = a_1x$ is parallel to the line $y = a_1x + b_1$ so we may assume, without loss of generality, that $b_1 = 0$. Similarly we may assume $b_2 = 0$. Now we see in Figure 1-5-10 that the lines are perpendicular if and only if the segment from $P = (1, a_1)$ to $Q = (1, a_2)$ is the hypotenuse of a right triangle with the right angle at O. By the Pythagorean theorem, this is so if and only if

$$\overline{PQ}^2 = \overline{OP}^2 + \overline{OQ}^2$$

But by the distance formula

$$\overline{PQ}^2 = (a_2 - a_1)^2 \qquad \overline{OP}^2 = 1 + a_1^2 \qquad \text{and} \qquad \overline{OQ}^2 = 1 + a_2^2$$

so the two lines are perpendicular if and only if

$$(a_2 - a_1)^2 = (1 + a_1^2) + (1 + a_2^2)$$

which simplifies to $a_1a_2 = -1$. ●

We often remember this result in the form $a_2 = -1/a_1$. In words: Two non-vertical lines are perpendicular if and only if the slope of one is the negative reciprocal of the slope of the other.

We pass on to a further application of the distance formula.

Midpoint Formula

The midpoint of the line segment joining (x_0, y_0) and (x_1, y_1) is the point

$$(\bar{x}, \bar{y}) = (\tfrac{1}{2}(x_0 + x_1), \tfrac{1}{2}(y_0 + y_1))$$

In words, each coordinate of the midpoint is the average of the corresponding coordinates of the two given points. ●

To prove this, let d denote the distance from (x_0, y_0) to (x_1, y_1), so that

$$d^2 = (x_1 - x_0)^2 + (y_1 - y_0)^2$$

Let d_0 denote the distance from $(\bar{x}, \bar{y})$ to (x_0, y_0). Then

$$\begin{aligned} d_0^2 &= (\bar{x} - x_0)^2 + (\bar{y} - y_0)^2 \\ &= [\tfrac{1}{2}(x_0 + x_1) - x_0]^2 + [\tfrac{1}{2}(y_0 + y_1) - y_0]^2 \\ &= [\tfrac{1}{2}(x_1 - x_0)]^2 + [\tfrac{1}{2}(y_1 - y_0)]^2 \\ &= \tfrac{1}{4}[(x_1 - x_0)^2 + (y_1 - y_0)^2] = \tfrac{1}{4}d^2 = (\tfrac{1}{2}d)^2 \end{aligned}$$

Hence $d_0 = \tfrac{1}{2}d$. Similarly, if d_1 denotes the distance from $(\bar{x}, \bar{y})$ to

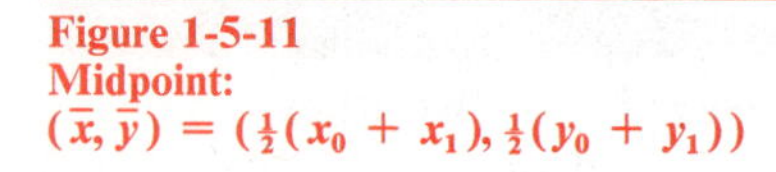
Figure 1-5-11
Midpoint:
$(\bar{x}, \bar{y}) = (\frac{1}{2}(x_0 + x_1), \frac{1}{2}(y_0 + y_1))$

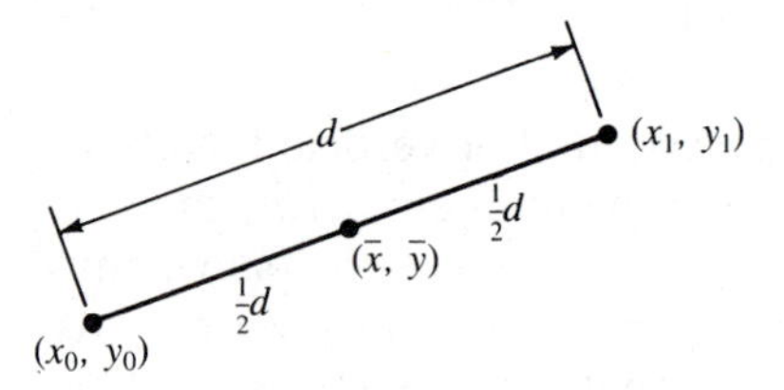

(x_1, y_1), then $d_1 = \frac{1}{2}d$. It follows (Figure 1-5-11) that $(\bar{x}, \bar{y})$ is the midpoint. ●

Example 2

Find the perpendicular bisector of the segment joining $(1, 2)$ and $(3, -5)$.

Solution The bisector (Figure 1-5-12) is the line perpendicular to the segment and passing through its midpoint. The midpoint is

$$(\tfrac{1}{2}(1 + 3), \tfrac{1}{2}(2 - 5)) = (2, -\tfrac{3}{2})$$

The slope a of the bisector is the negative reciprocal of the slope of the segment:

$$a = -\frac{3 - 1}{-5 - 2} = \tfrac{2}{7}$$

Therefore, by the point-slope form of a line, the equation of the bisector is

$$y - (-\tfrac{3}{2}) = \tfrac{2}{7}(x - 2) \qquad \text{that is} \qquad y = \tfrac{2}{7}x - \tfrac{4}{7} - \tfrac{3}{2}$$

which simplifies to $y = \frac{2}{7}x - \frac{29}{14}$. ●

Figure 1-5-12
See Example 2.

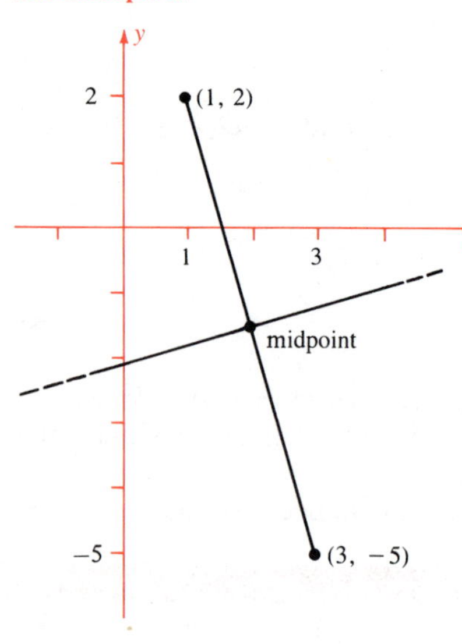

Reflection in the Line $y = x$

The following result will be particularly useful when we discuss inverse functions.

• Any pair of points (x, y) and (y, x), with $x \neq y$, are mirror images of each other with respect to the line $y = x$. ●

Proof "Mirror images" means that $y = x$ is the perpendicular bisector of the segment joining the two points (Figure 1-5-13). Check that it is. First of all, the midpoint of the segment joining (x, y) and (y, x) is

$$(\tfrac{1}{2}(x + y), \tfrac{1}{2}(y + x))$$

which lies on the line $y = x$ because the coordinates are equal. Second, the slope of $y = x$ is 1 and the slope of the segment is

$$\frac{x - y}{y - x} = -1$$

the negative reciprocal of 1. Hence $y = x$ is perpendicular to the segment and passes through its midpoint. ●

The symmetry of the points (x, y) and (y, x) relative to the line $y = x$ is often useful. For example, consider the graph of $y = 1/x$. For each point (x, y) on the graph, the reversed point (y, x) is also on the graph because the variables x and y can be interchanged in the equation $y = 1/x$. That is, $x = 1/y$ is just $y = 1/x$ expressed differently. Therefore the graph is symmetric with respect to the line $y = x$. See Figure 1-5-14.

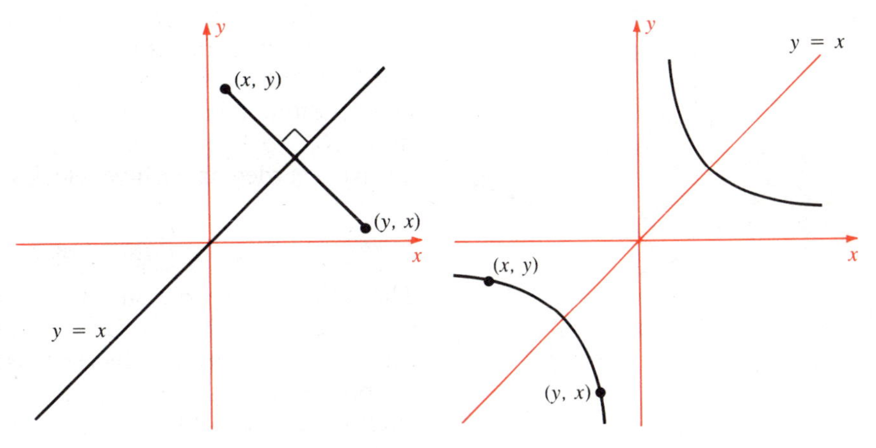

Figure 1-5-13
(x, y) and (y, x) are mirror images in the line $y = x$.

Figure 1-5-14
If x and y are interchangeable in the equation $y = f(x)$, then the graph is symmetric about the line $y = x$.

Exercises

Write the equation of the circle

1 center $(1, 3)$, radius 6

2 center $(5, 12)$, radius 13

3 center $(-4, 3)$, radius 5

4 center $(-2, -1)$, radius 1

5 center $(1, 5)$, through $(0, 0)$

6 center $(3, 3)$, through $(-2, -4)$

7 center $(-5, 2)$, tangent to y-axis

8 center $(1, 2)$, tangent to x-axis

9 diameter from $(0, 1)$ to $(3, 3)$

10 diameter from $(-2, -3)$ to $(4, 1)$

Write the equation for the most general circle

11 radius 3, tangent to x-axis

12 center in first quadrant, tangent to both axes

13 passing through $(0, 0)$

14 tangent to the line $y = 3$

Do the circles intersect?

15 $x^2 + y^2 = 4$ and $(x - 3)^2 + (y + 2)^2 = 1$

16 $(x - 1)^2 + (y - 5)^2 = 9$ and $(x - 4)^2 + (y - 3)^2 = 4$

17 Show that each point of the circle $(x - 4)^2 + y^2 = 4$ is twice as far from $(0, 0)$ as from $(3, 0)$.

18 Prove that the circle $(x - 1)^2 + (x + 2)^2 = 9$ lies inside the circle $x^2 + y^2 = 36$.

Use a calculator to determine whether the given point is inside, on, or outside the circle of radius 4 with center $(3, -1)$

19 $(5.10729, 2.39993)$

20 $(-0.21046, -3.38597)$

21 $(4 + \sqrt{6}, -19 + \sqrt{401})$

22 $(\sqrt[3]{3}, \sqrt[3]{19})$

Find the midpoint of

23 $(0, 1)$ and $(2, 3)$

24 $(-1, 7)$ and $(2, 0)$

25 $(3, -4)$ and $(5, 4)$

26 $(3, -5)$ and $(-5, 3)$

Find the perpendicular bisector of the segment through

27 $(0, 0)$ and $(-3, 5)$

28 $(0, 2)$ and $(4, 0)$

29 $(3, 5)$ and $(6, -2)$

30 $(1, -1)$ and $(-3, 6)$

Graph the parabola; find its focus and its directrix

31 $x = y^2$

32 $y = -8x^2$

33 $y = x^2 - 4x + 2$

34 $x = 3y^2 - 3y + 1$

35 Find the set of centers of all circles that are tangent to the x-axis and pass through the point $(0, 2)$.

36 Show that (x, y) and $(-y, -x)$ are symmetric about the line $x + y = 0$.

37 Show that $(5x, 5y)$ and $(3x + 4y, 4x - 3y)$ are symmetric about the line $2y = x$.

38 Find the mirror image of $(2, 0)$ in the line $y = 3x$.

1-6 Sine and Cosine Functions

In this and the next section we review the trigonometric functions, stressing their features most important for calculus.

In plane geometry, an angle is defined as one of the two regions bounded by two rays with the same vertex (Figure 1-6-1).

Figure 1-6-1
Static angle

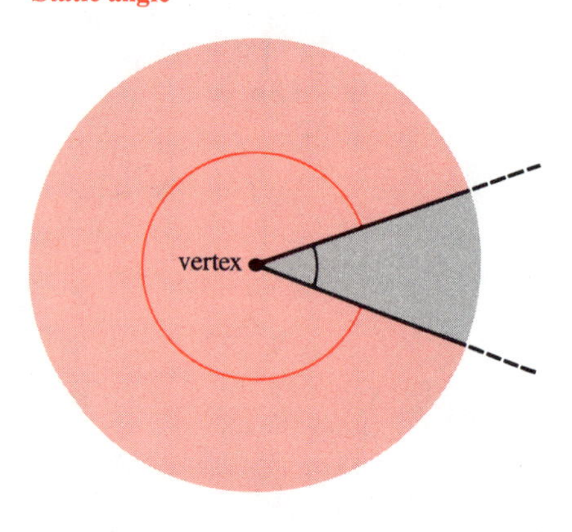

In calculus we need a dynamic approach to angle. An angle is the result of a rotation (Figure 1-6-2) of a ray about its origin, from an initial position R_0 to its terminal position R_1. Counterclockwise rotation is always regarded as positive, clockwise rotation as negative.

Radian Measure

The unit of angle measure that is most convenient in calculus and for scientific work is the *radian.* To measure an angle we place it in what is called *standard position:* Its vertex is at the origin and its initial side is along the positive x-axis. The angle then subtends (cuts out) an arc on the unit circle (Figure 1-6-3). The **radian measure** of the angle is defined to be the *length of the subtended arc.*

Figure 1-6-2
Dynamic angle

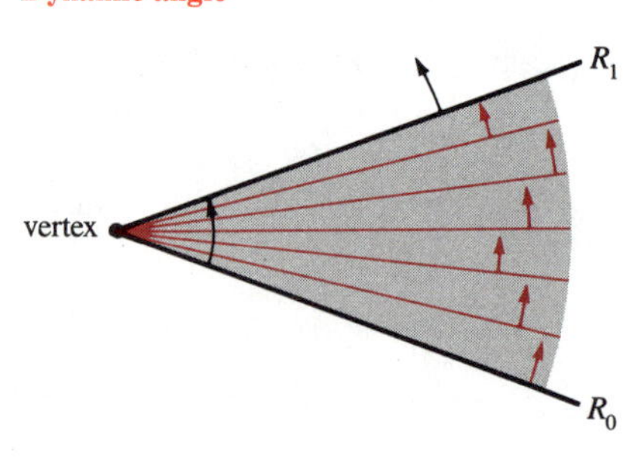

Remark One of the achievements of calculus is to prove that each smooth curve joining two points has a definite length. In particular, each arc of a circle has a length. The theory of arc length will be discussed only much later in this text, but we shall use its results for circular arcs freely. The same applies to area of circles and sectors of circles. In particular, we accept for now that there is a real number π such that any circle of radius r has circumference $2\pi r$ and encloses area πr^2.

In a unit circle, a central angle θ subtends an arc of length θ. A central angle of $360°$ subtends the entire circle, whose circumference is 2π. Hence we have a conversion relation between radians and degrees: 2π radians $= 360°$.

Figure 1-6-3
Radian measure of angle
= length of subtended arc on the unit circle

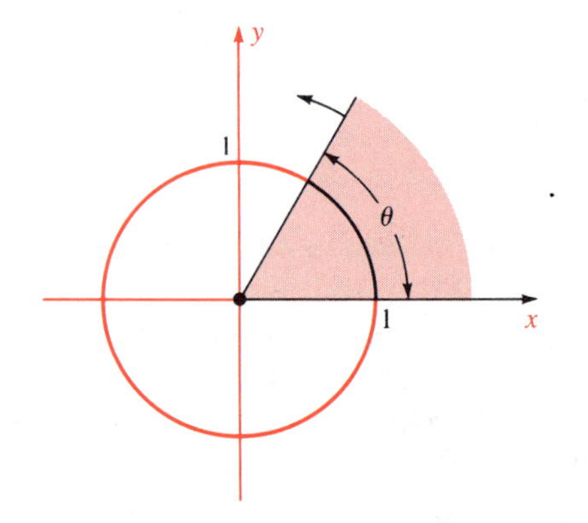

Conversion Radians to Degrees, Degrees to Radians

π rad $= 180°$ 1 rad $\approx 57.2958°$ $1° \approx 0.0174533$ rad

We follow the convention of omitting rad in radian measure. It is accepted procedure to write $\pi = 180°$ and to speak of angles such as $\frac{1}{2}\pi$, $\frac{2}{3}\pi$, 0.752, and so on.

Now consider Figure 1-6-4. If the segment OP starts at position OA and swings counterclockwise, then angle θ starts at 0 and increases. After a complete revolution, $\theta = 2\pi$. After another revolution, $\theta = 4\pi$, and so on. In general, starting from a given position, one counterclockwise revolution of OP increases θ by 2π. If OP starts at OA and swings clockwise, then θ is considered to be a negative angle. For example, after a quarter revolution clockwise, $\theta = -\frac{1}{2}\pi$; after three full clockwise revolutions, $\theta = -3 \cdot 2\pi = -6\pi$.

According to this scheme, each real number θ determines a unique position of OP and a unique angle. However, each position of OP corresponds to infinitely many angles, differing from each other by integer multiples of 2π. For example, if OP points straight up, then the corresponding angles are $\frac{1}{2}\pi$, $\frac{1}{2}\pi \pm 2\pi$, $\frac{1}{2}\pi \pm 4\pi$, $\frac{1}{2}\pi \pm 6\pi$, and so on.

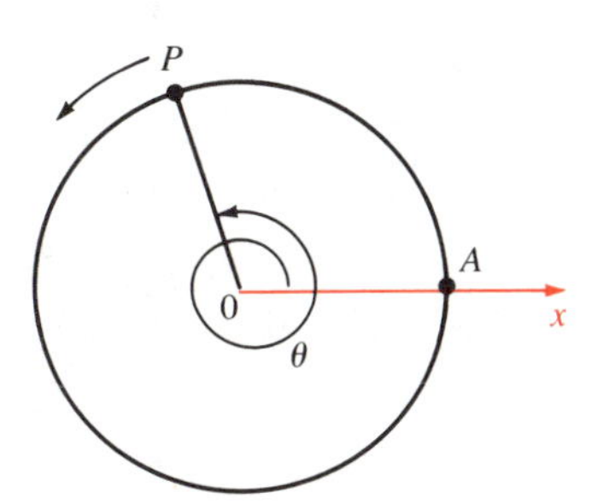

Figure 1-6-4
Multiple rotations

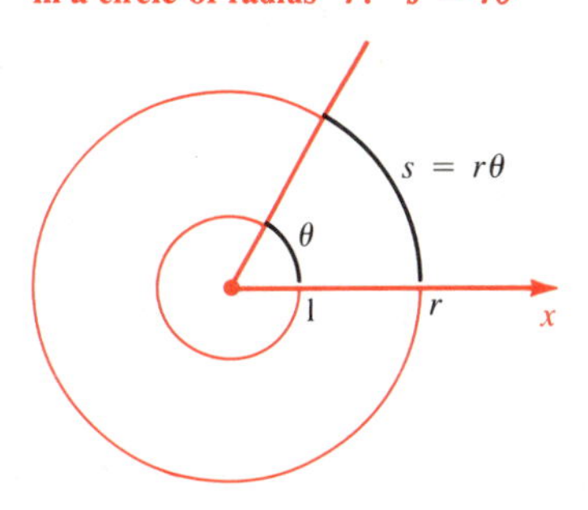

Figure 1-6-5
Arc s subtended by central angle θ in a circle of radius r: $s = r\theta$

Remark Suppose a central angle θ in a circle of radius r subtends an arc of length s. From the similarity of the two circular sectors in Figure 1-6-5 follows the proportion $s/\theta = r/1$. Hence $s = r\theta$.

Sine and Cosine

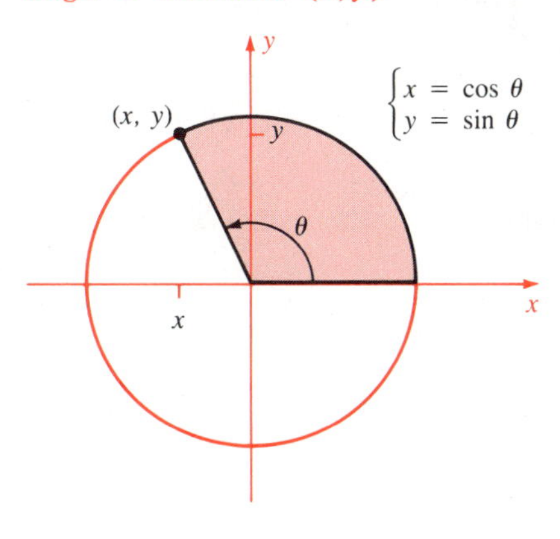

Figure 1-6-6
Angle θ determines (x, y)

Each real number θ determines an angle in standard position, which in turn determines a point (x, y) on the unit circle. In this way, the real number x is determined by θ; in other words, to each θ corresponds an x. This means we have a function whose domain is $\{\text{all } \theta\}$. It is called the **cosine** function, and we write $x = \cos\theta$. Similarly, to each real θ corresponds a real y. This correspondence determines the **sine** function and we write $y = \sin\theta$. See Figure 1-6-6.

Since (x, y) lies on the unit circle, $-1 \le x \le 1$ and $-1 \le y \le 1$. It is clear geometrically that $x = \cos\theta$ takes all values between -1 and 1, so the *range* of $x = \cos\theta$ is the interval $[-1, 1]$ of the x-axis. Similarly the range of $y = \sin\theta$ is the interval $[-1, 1]$ of the y-axis. Table 1-6-1 lists some frequently used exact values of the functions $\sin\theta$ and $\cos\theta$.

Table 1-6-1

θ	$\sin\theta$	$\cos\theta$
0	0	1
$\frac{1}{6}\pi$	$\frac{1}{2}$	$\frac{1}{2}\sqrt{3}$
$\frac{1}{4}\pi$	$\frac{1}{2}\sqrt{2}$	$\frac{1}{2}\sqrt{2}$
$\frac{1}{3}\pi$	$\frac{1}{2}\sqrt{3}$	$\frac{1}{2}$
$\frac{1}{2}\pi$	1	0
$\frac{2}{3}\pi$	$\frac{1}{2}\sqrt{3}$	$-\frac{1}{2}$
$\frac{3}{4}\pi$	$\frac{1}{2}\sqrt{2}$	$-\frac{1}{2}\sqrt{2}$
$\frac{5}{6}\pi$	$\frac{1}{2}$	$-\frac{1}{2}\sqrt{3}$
π	0	-1
$\frac{3}{2}\pi$	-1	0

To obtain approximate values of $\sin\theta$ and $\cos\theta$ from your calculator, set the deg/rad switch to rad.

2 . 5 8 sin yields $\sin 2.58 \approx 0.53253$

1 . 7 2 +/− cos yields $\cos(-1.72) \approx -0.14865$

Basic Identities

The numbers θ and $\theta + 2\pi$ determine the same point (x, y) on the unit circle. Therefore we have the following relation:

Periodicity of Cosine and Sine

$$\cos(\theta + 2\pi) = \cos\theta \qquad \sin(\theta + 2\pi) = \sin\theta$$

Thus the values of $\cos\theta$ and $\sin\theta$ repeat when θ increases by 2π. We say that these functions are **periodic** with **period** 2π.

By definition, $(\cos\theta, \sin\theta)$ is a point on the unit circle $x^2 + y^2 = 1$. Therefore the sine and cosine functions satisfy the following relation:

Figure 1-6-7
Reflection in the x-axis:
$(x, y) \to (x, -y) \qquad \theta \to -\theta$

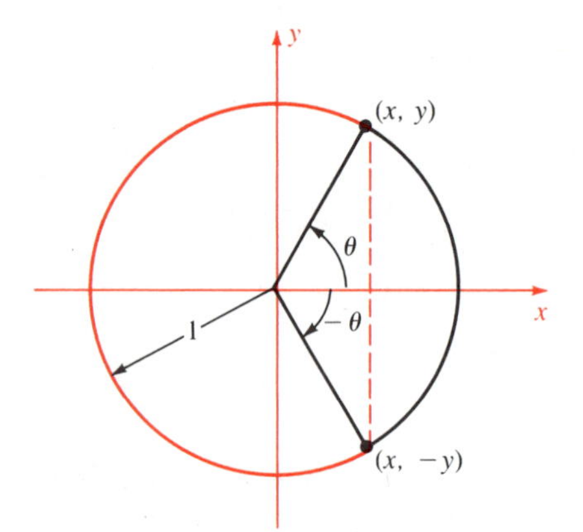

Figure 1-6-8
Reflection in $(0, 0)$:
$(x, y) \to (-x, -y) \qquad \theta \to \theta + \pi$

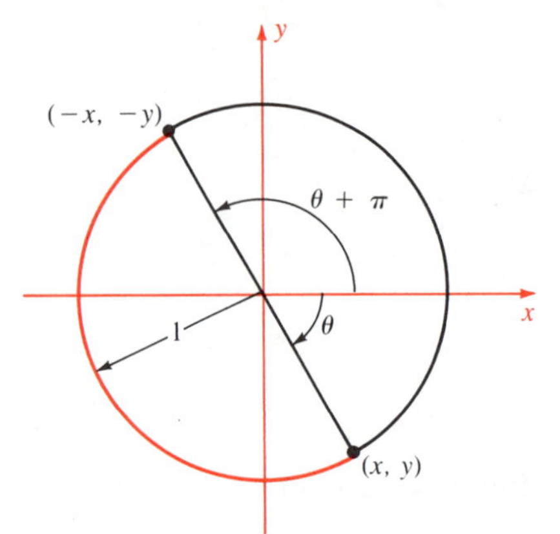

Figure 1-6-9
Reflection in $y = x$:
$(x, y) \to (y, x) \qquad \theta \to \frac{1}{2}\pi - \theta$

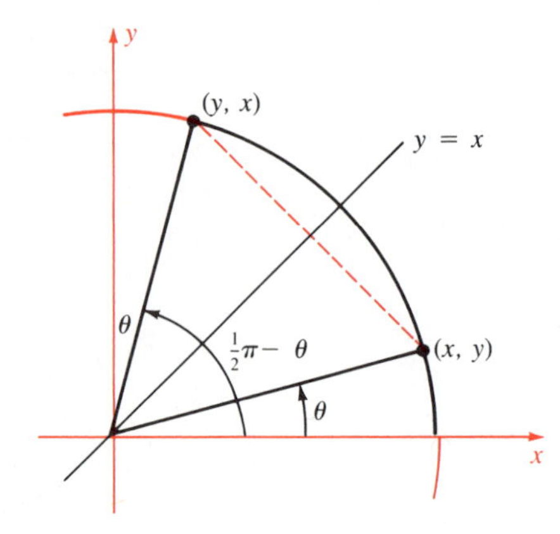

A Fundamental Identity

$$\sin^2\theta + \cos^2\theta = 1$$

Other basic properties of $\cos\theta$ and $\sin\theta$ are shown in Figure 1-6-7.

Evenness and Oddness of Sine and Cosine

$$\cos(-\theta) = \cos\theta \qquad \sin(-\theta) = -\sin\theta$$

Thus $\cos\theta$ is an even function and $\sin\theta$ is an odd function.

From Figure 1-6-8,

- $\cos(\theta + \pi) = -\cos\theta \qquad \sin(\theta + \pi) = -\sin\theta$

Now examine Figure 1-6-9. It shows that if angle θ determines (x, y) on the unit circle, then (y, x) is determined by angle $\frac{1}{2}\pi - \theta$. Therefore $\cos(\frac{1}{2}\pi - \theta) = y = \sin\theta$ and $\sin(\frac{1}{2}\pi - \theta) = x = \cos\theta$.

- $\cos(\frac{1}{2}\pi - \theta) = \sin\theta \qquad \sin(\frac{1}{2}\pi - \theta) = \cos\theta$

Similar arguments yield the further relations

- $\cos(\pi - \theta) = -\cos\theta \qquad \sin(\pi - \theta) = \sin\theta$
- $\cos(\frac{1}{2}\pi + \theta) = -\sin\theta \qquad \sin(\frac{1}{2}\pi + \theta) = \cos\theta$

Among the most basic formulas in mathematics are the *addition laws* for sine and cosine.

Addition Laws for Sine and Cosine

$$\sin(\alpha + \beta) = \sin\alpha\cos\beta + \cos\alpha\sin\beta$$

$$\cos(\alpha + \beta) = \cos\alpha\cos\beta - \sin\alpha\sin\beta$$

Figure 1-6-10
The shaded angle is $\alpha + \beta$.

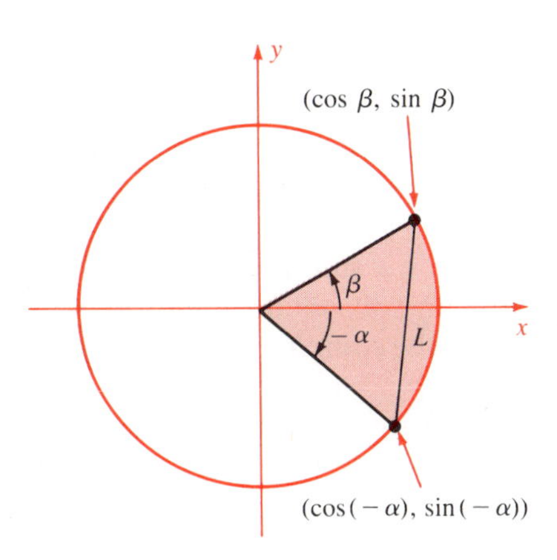

Each of these formulas implies the other. For instance, if you replace α by $-\frac{1}{2}\pi + \alpha$ in the cosine formula, the sine formula follows. Thus if we prove one of the formulas, say the cosine formula, then the other follows. To do so, we draw angles β and $-\alpha$ in standard position in a unit circle (Figure 1-6-10). By the distance formula, the chord length L satisfies

$$\begin{aligned} L^2 &= [\cos\beta - \cos(-\alpha)]^2 + [\sin\beta - \sin(-\alpha)]^2 \\ &= (\cos\beta - \cos\alpha)^2 + (\sin\beta + \sin\alpha)^2 \\ &= (\cos^2\beta - 2\cos\beta\cos\alpha + \cos^2\alpha) \\ &\quad + (\sin^2\beta + 2\sin\beta\sin\alpha + \sin^2\alpha) \\ &= (\cos^2\beta + \sin^2\beta) + (\cos^2\alpha + \sin^2\alpha) \\ &\quad - 2\cos\beta\cos\alpha + 2\sin\beta\sin\alpha \\ &= 2 - 2(\cos\alpha\cos\beta - \sin\alpha\sin\beta) \end{aligned}$$

Now we rotate the triangle about the origin through angle α. This does not change the chord length L; by Figure 1-6-11,

Figure 1-6-11
Rotate the shaded angle in Figure 1-6-10 counterclockwise α radians.

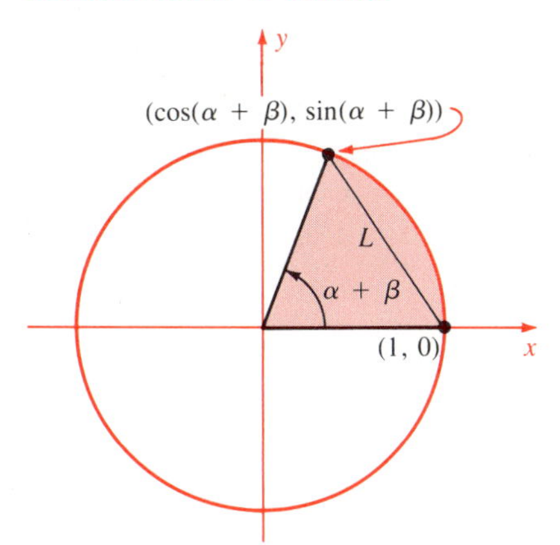

$$\begin{aligned} L^2 &= [\cos(\alpha+\beta) - 1]^2 + [\sin(\alpha+\beta) - 0]^2 \\ &= [\cos^2(\alpha+\beta) - 2\cos(\alpha+\beta) + 1] + \sin^2(\alpha+\beta) \\ &= 2 - 2\cos(\alpha+\beta) \end{aligned}$$

We equate the two expressions for L^2:

$$2 - 2(\cos\alpha\cos\beta - \sin\alpha\sin\beta) = 2 - 2\cos(\alpha+\beta)$$

The formula $\cos(\alpha+\beta) = \cos\alpha\cos\beta - \sin\alpha\sin\beta$ follows.

We obtain alternative versions of the addition formulas by substituting $-\beta$ for β, then using $\cos(-\beta) = \cos\beta$ and $\sin(-\beta) = -\sin\beta$:

Alternative Addition Laws for Sine and Cosine

$$\sin(\alpha - \beta) = \sin\alpha\cos\beta - \cos\alpha\sin\beta$$

$$\cos(\alpha - \beta) = \cos\alpha\cos\beta + \sin\alpha\sin\beta$$

By setting $\alpha = \beta = \theta$ in the addition laws, we obtain the *double angle* formulas for $\sin 2\theta$ and $\cos 2\theta$ in terms of $\sin\theta$ and $\cos\theta$:

Double Angle Formulas for Sine and Cosine

$$\sin 2\theta = 2\sin\theta\cos\theta \qquad \cos 2\theta = \cos^2\theta - \sin^2\theta$$

The second formula has two alternative forms, both derived from $\cos^2\theta + \sin^2\theta = 1$:

$$\cos 2\theta = 1 - 2\sin^2\theta = 2\cos^2\theta - 1$$

Graphs

Let us graph $y = \sin\theta$. Since $\sin\theta$ has period 2π, we need only plot $\sin\theta$ on the interval $[-\pi, \pi]$. We can then extend the graph indefinitely to the right and to the left, making it repeat every 2π. Actu-

Table 1-6-2

θ	$\sin\theta$
0.0	0.00
0.1π	0.31
0.2π	0.59
0.3π	0.81
0.4π	0.95
0.5π	1.00
0.6π	0.95
0.7π	0.81
0.8π	0.59
0.9π	0.31
π	0.00

ally, because $\sin\theta$ is an odd function, we need only plot $\sin\theta$ for $0 \le \theta \le \pi$. The portion of the graph where $-\pi \le \theta \le 0$ is the reflection of this portion in the origin.

To graph $y = \sin\theta$ for $0 \le \theta \le \pi$, we first make Table 1-6-2, values of $\sin\theta$ rounded to two places for 11 equally spaced points: 0, $0.1\pi, \cdots, \pi$. (It may be easier on your calculator to use degrees and calculate $\sin 18°$, $\sin 36°$, $\cdots$ than to use radians and calculate $\sin 0.1\pi$, $\sin 0.2\pi$, $\cdots$.)

We plot this data in Figure 1-6-12 and fill in a smooth curve. Then we extend this curve to $[-\pi, 0]$ by symmetry, to obtain Figure 1-6-13. Finally we use the periodicity of $\sin\theta$ to get the complete graph, shown in Figure 1-6-14.

From the graph of $y = \sin\theta$, we obtain the graph of $y = \cos\theta$ free of charge. For the relation $\cos\theta = \sin(\theta + \frac{1}{2}\pi)$ shows that the graph of $y = \cos\theta$ is just the graph of $y = \sin\theta$ shifted $\frac{1}{2}\pi$ units to the left (Figure 1-6-15).

Figure 1-6-12
Graph of $y = \sin\theta$ for $0 \le \theta \le \pi$

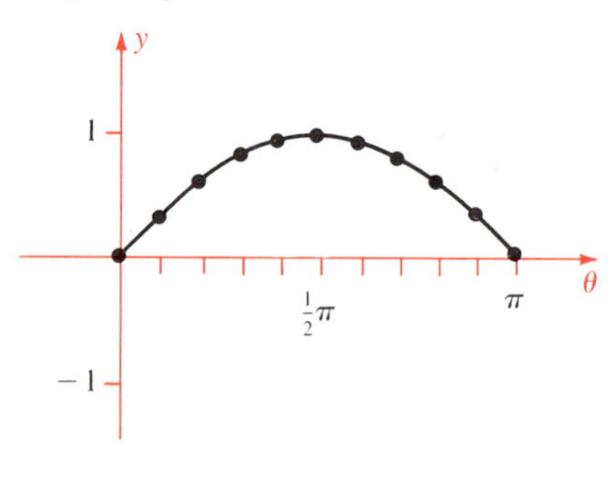

Multiple Angles

The graph of $y = \sin\theta$ makes one complete cycle on the interval $[0, 2\pi]$. The graph of $y = \sin 2\theta$ makes one complete cycle on the interval $[0, \pi]$ because 2θ runs from 0 to 2π as θ runs from 0 to π. See Figure 1-6-16. Therefore $\sin 2\theta$ oscillates twice as fast as $\sin\theta$. Similarly, $\sin\frac{1}{2}\theta$ oscillates half as fast as $\sin\theta$. See Figure 1-6-17.

In general, $\sin k\theta$ and $\cos k\theta$ make k cycles on $0 \le \theta \le 2\pi$, or equivalently, one full cycle on $0 < \theta \le 2\pi/k$. In other words, $\sin k\theta$ and $\cos k\theta$ are periodic functions with period $2\pi/k$. To confirm this assertion analytically, we note that

$$\sin k(\theta + 2\pi/k) = \sin(k\theta + 2\pi) = \sin k\theta$$

The assertion about $\cos k\theta$ can be confirmed similarly.

Figure 1-6-13
Graph of $y = \sin\theta$ for $-\pi \le \theta \le \pi$

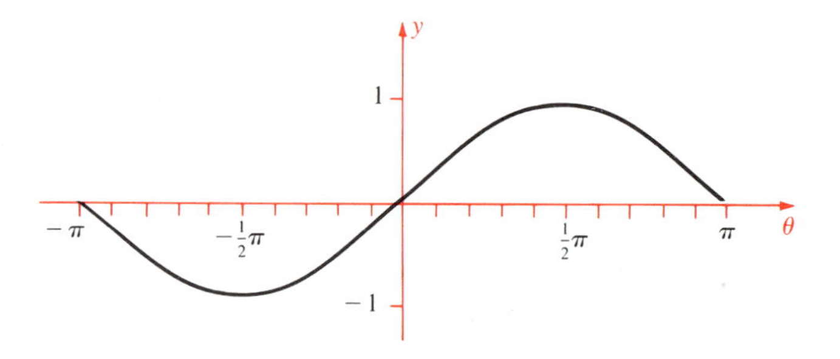

Figure 1-6-14
Graph of $y = \sin\theta$

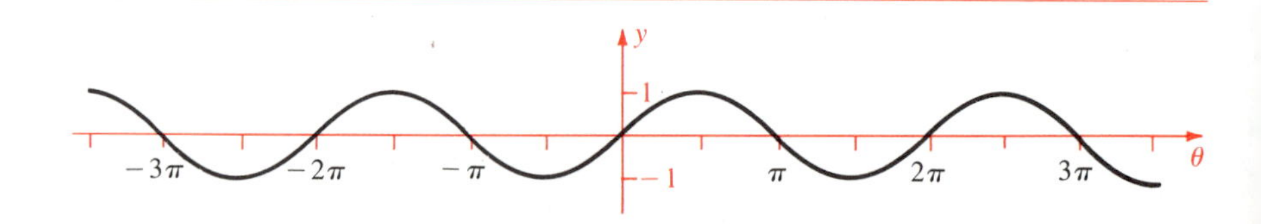

Figure 1-6-15
Graph of $y = \cos\theta$

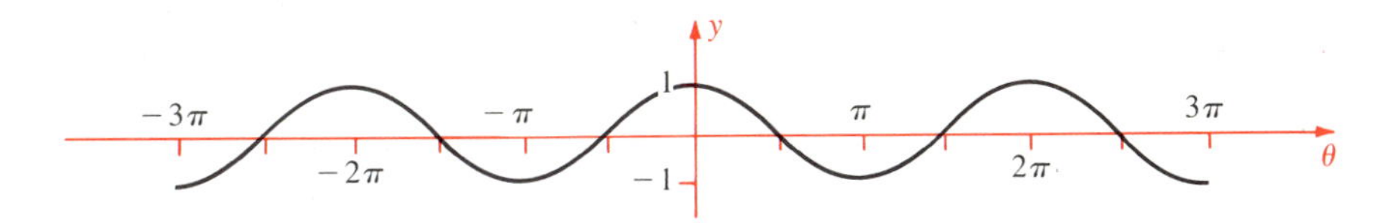

Figure 1-6-16
The graph of $y = \sin 2\theta$ oscillates twice as fast as the graph of $y = \sin\theta$.

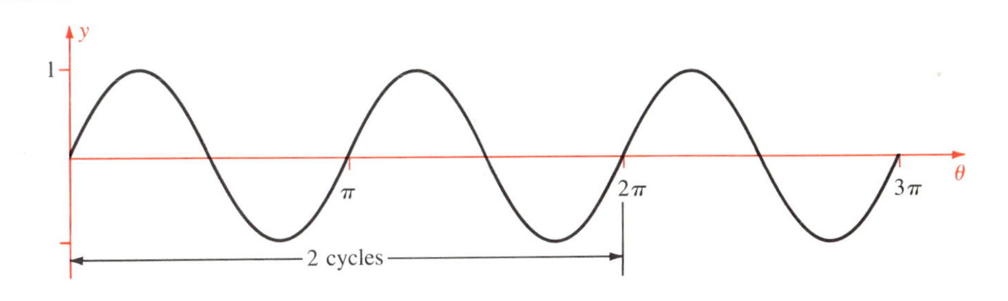

Figure 1-6-17
The graph of $y = \sin\frac{1}{2}\theta$ oscillates half as fast as the graph of $y = \sin\theta$.

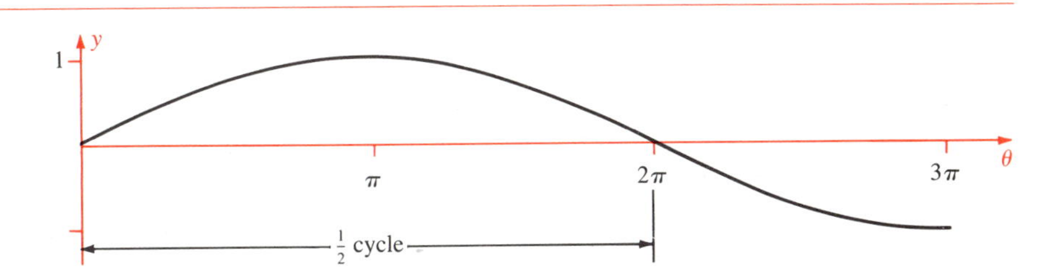

Exercises

Convert to radians (for example, $30° = \frac{1}{6}\pi$)

1 $60°$ $150°$ $-240°$ $390°$ $-450°$ $900°$

2 $90°$ $-120°$ $210°$ $420°$ $-330°$ $480°$

3 $45°$ $-180°$ $270°$ $630°$ $-135°$ $495°$

4 $-90°$ $135°$ $-315°$ $405°$ $-270°$ $15°$

Convert to degrees

5 $\frac{1}{2}\pi$ $-\frac{2}{3}\pi$ $\frac{5}{3}\pi$ 3π $-\frac{8}{3}\pi$ $\frac{50}{3}\pi$

6 $-\pi$ $-\frac{3}{2}\pi$ $\frac{7}{6}\pi$ 5π $-\frac{13}{6}\pi$ $\frac{17}{3}\pi$

7 $\frac{1}{4}\pi$ $\frac{2}{15}\pi$ $\frac{5}{12}\pi$ $\frac{1}{36}\pi$ $-\frac{11}{12}\pi$ $-\frac{1}{15}\pi$

8 $\frac{7}{2}\pi$ $-\frac{3}{4}\pi$ $\frac{7}{15}\pi$ $\frac{3}{8}\pi$ $\frac{1}{9}\pi$ $-\frac{1}{18}\pi$

Find the coordinates of the point on the unit circle with central angle $\theta =$

9 $\frac{1}{4}\pi$ $-\frac{3}{4}\pi$ $\frac{11}{4}\pi$

10 $\frac{3}{2}\pi$ $-\frac{1}{4}\pi$ $\frac{13}{4}\pi$

11 π $-\frac{2}{3}\pi$ $\frac{13}{6}\pi$

12 $\frac{1}{6}\pi$ $\frac{5}{6}\pi$ $-\frac{2}{3}\pi$

Find all θ in radians, $0 \le \theta < 2\pi$, such that

13 $\sin\theta = \frac{1}{2}\sqrt{2}$

14 $\sin\theta = -\frac{1}{2}$

15 $\cos\theta = -\frac{1}{2}\sqrt{3}$

16 $\cos\theta = \frac{1}{2}\sqrt{2}$

17 $\sin(\theta + \pi) = 1$

18 $\cos(\theta + \pi) = 1$

19 $\sin\theta = \cos\theta$

20 $\sin(\theta - \pi) = -1$

Prove that the function is periodic and find the smallest period that you can

21 $\sin 2\pi x$

22 $\cos\frac{1}{3}x$

23 $\sin x \sin 3x$

24 $\sin 3x + \cos 4x$

25 $\cos^2 x$

26 $\cos x \cos 2x$

Show how these formulas (given on page 52) are consequences of those given previously

27 $\cos(\pi - \theta) = -\cos\theta$ and $\sin(\pi - \theta) = \sin\theta$

28 $\cos(\frac{1}{2}\pi + \theta) = -\sin\theta$ and $\sin(\frac{1}{2}\pi + \theta) = \cos\theta$

Write the addition laws in these special cases

29 $\beta = \frac{1}{4}\pi$

30 $\beta = \frac{1}{6}\pi$

Express in terms of $\sin\theta$ and $\cos\theta$

31 $\cos(\theta - \frac{1}{3}\pi)$

32 $\sin(\theta + \frac{1}{6}\pi)$

33 $\sin 3\theta$

34 $\cos 3\theta$

35 $\sin 4\theta$

36 $\cos 4\theta$

37 Prove $\cos\alpha\cos\beta = \frac{1}{2}[\cos(\alpha + \beta) + \cos(\alpha - \beta)]$. [Hint Start with the two formulas for $\cos(\alpha \pm \beta)$.]

38 (cont.) Find similar formulas for $\sin\alpha\cos\beta$ and $\sin\alpha\sin\beta$.

***39** Prove $\cos x \cos 2x \cos 4x \cos 8x = \frac{1}{8}(\cos x + \cos 3x + \cos 5x + \cdots + \cos 13x + \cos 15x)$

40 Prove $\cos \frac{1}{2}\theta = \pm\sqrt{\frac{1}{2}(1 + \cos\theta)}$. If $-2\pi \le \theta \le 2\pi$, when is the sign $+$ and when $-$?

41 Prove $\sin \frac{1}{2}\theta = \pm\sqrt{\frac{1}{2}(1 - \cos\theta)}$. If $-2\pi \le \theta \le 2\pi$, when is the sign $+$ and when $-$?

42 (cont.) Compute $\sin 15°$ in two ways, using $15° = 45° - 30° = \frac{1}{2}(30°)$, to prove

$$\tfrac{1}{4}(\sqrt{6} - \sqrt{2}) = \tfrac{1}{2}\sqrt{2 - \sqrt{3}}$$

Graph

43 $y = \cos 3\theta$

44 $y = \sin 4\theta$

45 $y = 2 - \sin\theta$

46 $y = 1 + \cos\theta$

47 $y = \cos 2\pi t$

48 $x = \cos(\theta + \frac{1}{4}\pi)$

Convert to radians (4-place accuracy)

49 $38.26°$

50 $172.91°$

51 $-35.81°$

52 $-111.44°$

Convert to degrees (2-place accuracy)

53 0.2863

54 2.9919

55 -0.2342

56 -1.8520

Estimate to 5 places

57 $\sin 21.34°$

58 $\cos 103.41°$

59 $\sin 0.2110$

60 $\cos 2.8146$

61 $\cos(-0.3042)$

63 $\sin(-2.0013)$

1-7 Additional Trigonometric Functions

In the previous section we reviewed the sine and cosine. Now let us recall the definitions of the other trignometric functions:

The Tangent, Cotangent, Secant, and Cosecant Functions

$$\tan x = \frac{\sin x}{\cos x} \qquad \cot x = \frac{\cos x}{\sin x}$$

$$\sec x = \frac{1}{\cos x} \qquad \csc x = \frac{1}{\sin x}$$

These four functions are defined in terms of $\sin x$ and $\cos x$. We can easily see some of their basic properties.

1 Domains Because of zeros in the denominators, $\tan x$ and $\sec x$ are not defined where $\cos x = 0$: at $\pm\frac{1}{2}\pi$, $\pm\frac{3}{2}\pi$, $\pm\frac{5}{2}\pi$, and so on. Similarly, $\cot x$ and $\csc x$ are not defined where $\sin x = 0$: at 0, $\pm\pi$, $\pm 2\pi$, and so on.

2 Periods The functions $\sin x$ and $\cos x$ are periodic with period 2π; the other four trigonometric functions inherit this period. For example,

$$\sec(x + 2\pi) = \frac{1}{\cos(x + 2\pi)} = \frac{1}{\cos x} = \sec x$$

Similarly, $\tan(x + 2\pi) = \tan x$, $\cot(x + 2\pi) = \cot x$, and $\csc(x + 2\pi) = \csc x$. More can be said about $\tan x$ and $\cot x$. Recall that

$$\sin(x + \pi) = -\sin x \quad \text{and} \quad \cos(x + \pi) = -\cos x$$

Therefore

$$\tan(x + \pi) = \frac{\sin(x + \pi)}{\cos(x + \pi)} = \frac{-\sin x}{-\cos x} = \tan x$$

and similarly for $\cot x$. Thus $\tan x$ and $\cot x$ have period π.

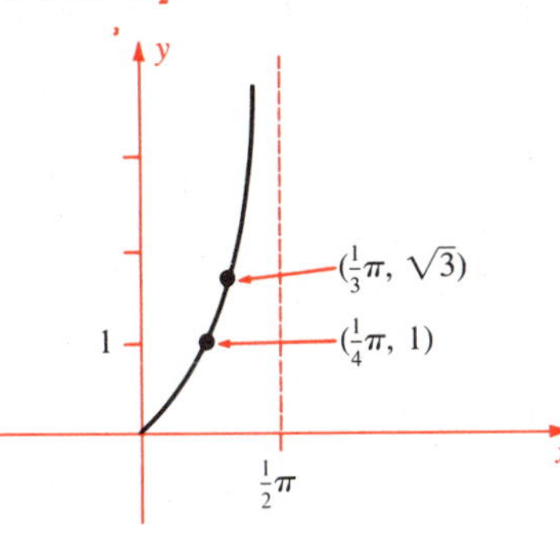

Figure 1-7-1
Graph of $y = \tan x$ for $0 \leq x < \frac{1}{2}\pi$

Periodicity

- The functions $\tan x$ and $\cot x$ have period π.
- The functions $\sec x$ and $\csc x$ have period 2π.

3 Evenness and Oddness Recall that $\sin(-x) = -\sin x$ and $\cos(-x) = \cos x$. Hence

$$\tan(-x) = \frac{\sin(-x)}{\cos(-x)} = \frac{-\sin x}{\cos x} = -\tan x$$

$$\sec(-x) = \frac{1}{\cos(-x)} = \frac{1}{\cos x} = \sec x$$

Similarly,

$$\cot(-x) = -\cot x \quad \text{and} \quad \csc(-x) = -\csc x$$

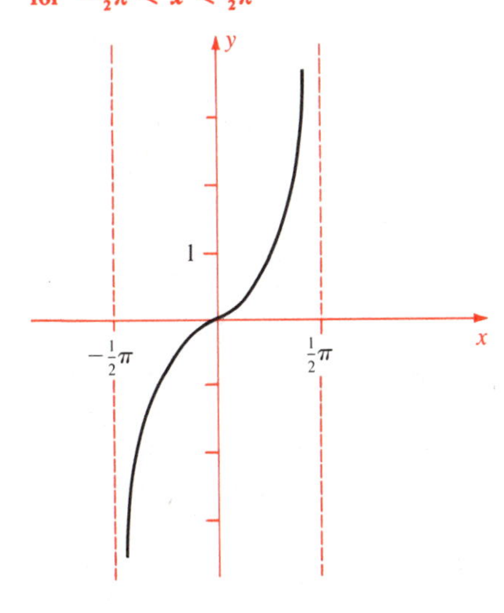

Figure 1-7-2
Graph of $y = \tan x$ for $-\frac{1}{2}\pi < x < \frac{1}{2}\pi$

Oddness and Evenness

- The functions $\sin x$, $\tan x$, $\cot x$, and $\csc x$ are odd functions.
- The functions $\cos x$ and $\sec x$ are even functions.

Graphs

Let us graph $y = \tan x$. Since $\tan x$ is an odd function and has period π, it suffices to sketch the part of the curve from 0 to $\frac{1}{2}\pi$. Then the complete graph can be obtained by symmetry and periodicity.

Since $\tan x = (\sin x)/(\cos x)$, we see that $\tan 0 = 0$. Furthermore, as x increases from 0 towards $\frac{1}{2}\pi$, the numerator increases from 0 towards 1, while the denominator decreases from 1 towards 0. Therefore $\tan x$ increases, and $\tan x \to +\infty$ as $x \to \frac{1}{2}\pi$. See Figure 1-7-1. Since $\tan x$ is an odd function, we can extend the graph (Figure 1-7-2) from 0 to $-\frac{1}{2}\pi$ by symmetry. Then we obtain the complete graph (Figure 1-7-3) of $y = \tan x$ by periodicity (period π).

Figure 1-7-3
Graph of $y = \tan x$

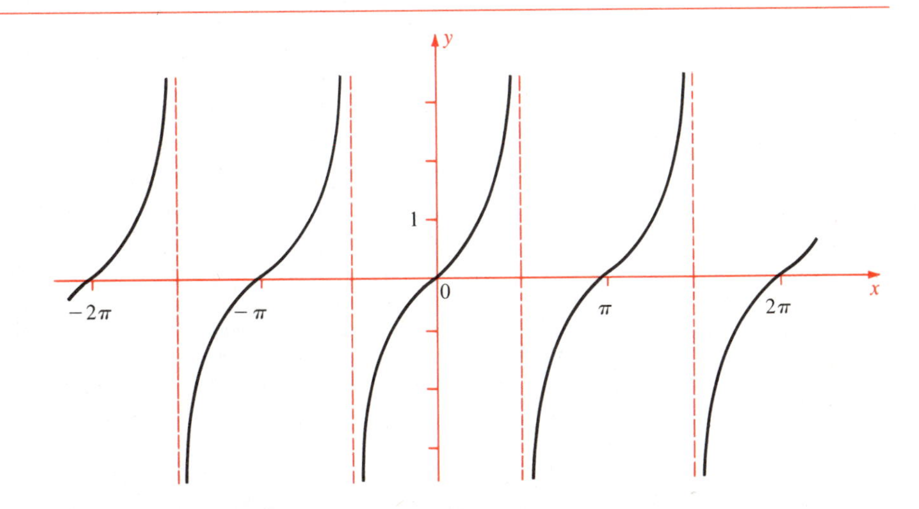

By similar reasoning, we obtain the graph (Figure 1-7-4) of $y = \cot x$. Note that on each interval of length π, both $\tan x$ and $\cot x$ take every real value once. Thus the range of each function is **R**.

Let us sketch $y = \csc x$. This function is odd and has period 2π, so by our usual argument, we need only plot the graph on the interval $0 < x < \pi$. The points $x = 0$ and $x = \pi$ are excluded because $\csc x$ is not defined there.

Figure 1-7-4
Graph of $y = \cot x$

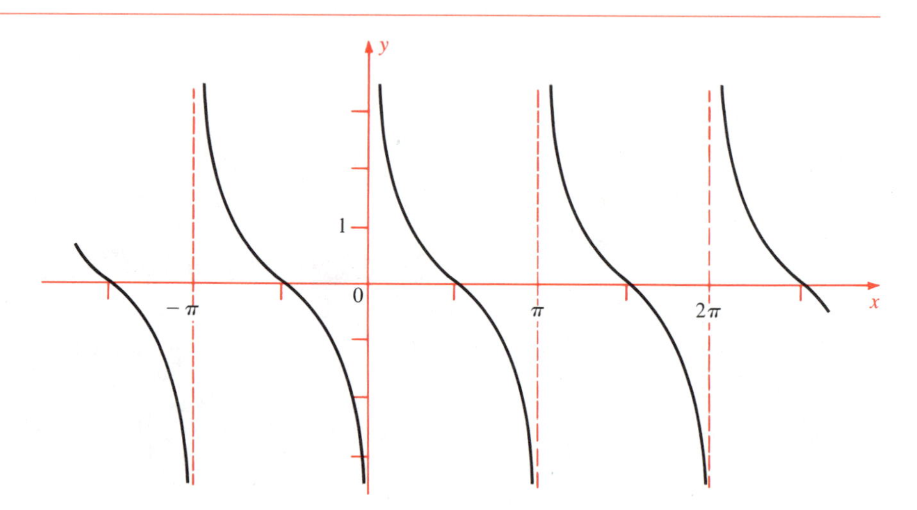

Assume $0 < x < \pi$. Then $0 < \sin x \leq 1$; so it follows that $\csc x = 1/\sin x \geq 1$. Actually, $\csc x = 1$ only at $x = \frac{1}{2}\pi$, where $\sin x = 1$, so the graph has a minimum point at $(\frac{1}{2}\pi, 1)$. If $x \to 0+$ or $x \to \frac{1}{2}\pi-$, then $1/\sin x \to +\infty$ because the denominator approaches 0. One other useful piece of information: the graph is symmetric about the line $x = \frac{1}{2}\pi$, a property $y = \csc x$ inherits from $y = \sin x$. (You should check that $\sin(\frac{1}{2}\pi - x) = \sin(\frac{1}{2}\pi + x)$.) We now have enough data for a reasonable sketch of $y = \csc x$ on $(0, \pi)$. We extend the curve to $(-\pi, 0)$ by oddness, then obtain the complete graph (Figure 1-7-5) by periodicity. (There is another approach to graphing $y = \csc x$: First sketch $y = \sin x$, then sketch its reciprocal.)

We obtain the graph of $y = \sec x$ free of charge, just as we obtained the graph of $y = \cos x$ from that of $y = \sin x$. From the relation $\cos x = \sin(x + \frac{1}{2}\pi)$ follows $\sec x = \csc(x + \frac{1}{2}\pi)$. Hence the graph of $y = \sec x$ is just the graph of $y = \csc x$ shifted $\frac{1}{2}\pi$ units to the left (Figure 1-7-6).

Identities

We start with the identity

$$\sin^2 x + \cos^2 x = 1$$

and divide both sides first by $\cos^2 x$, then by $\sin^2 x$:

Figure 1-7-5
Graph of $y = \csc x$

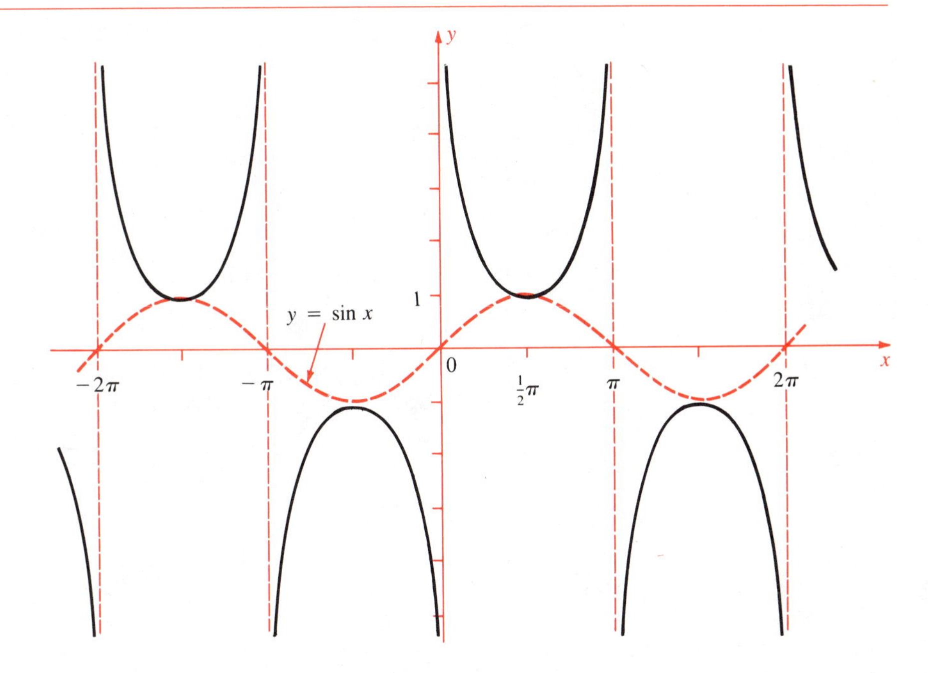

Figure 1-7-6
Graph of $y = \sec x$

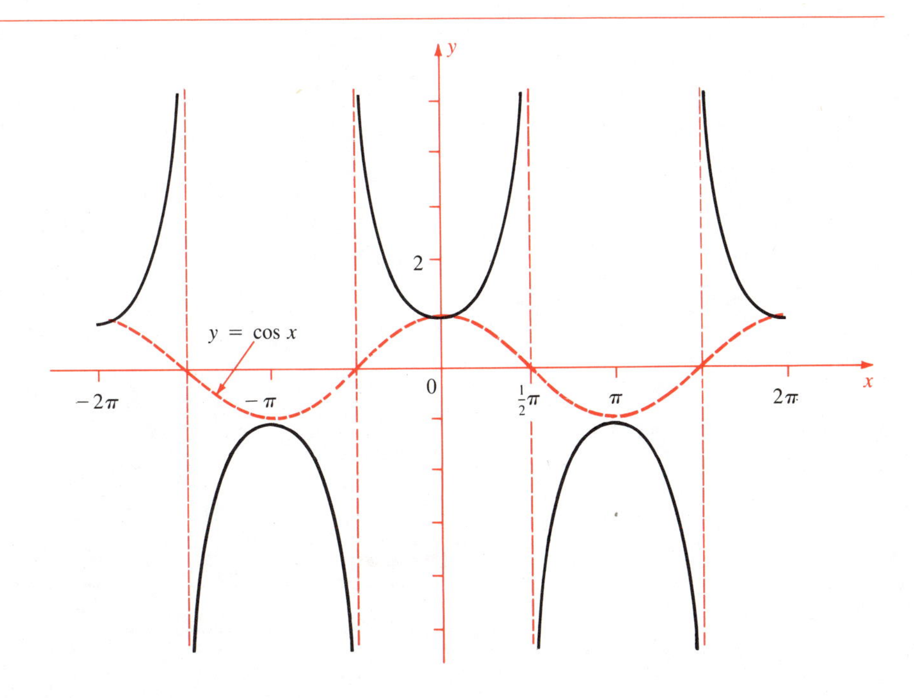

Basic Identities

$$\tan^2 x + 1 = \sec^2 x \qquad \cot^2 x + 1 = \csc^2 x$$

These identities are helpful in expressing one trigonometric function in terms of another. For example,

$$\sin x = \pm\sqrt{1 - \cos^2 x} \qquad \sec x = \pm\sqrt{\tan^2 x + 1}$$

$$\cot x = \pm\sqrt{\csc^2 x - 1} = \pm\frac{\sqrt{1 - \sin^2 x}}{\sin x}$$

In each case, we need information about the quadrant of x in order to choose the sign correctly.

Example 1

Express $\cos x$ and $\sin x$ in terms of $\tan x$ for $0 < x < \frac{1}{2}\pi$.

Solution

$$\cos x = \frac{1}{\sec x} = \frac{1}{\pm\sqrt{\tan^2 x + 1}}$$

$$\sin x = \tan x \cos x = \frac{\tan x}{\pm\sqrt{\tan^2 x + 1}}$$

Since $\sin x$, $\cos x$, and $\tan x$ are positive for $0 < x < \frac{1}{2}\pi$, choose the positive square root.

The addition laws for the sine and cosine yield an addition law for the tangent. Write

$$\tan(\alpha + \beta) = \frac{\sin(\alpha + \beta)}{\cos(\alpha + \beta)} = \frac{\sin\alpha\cos\beta + \cos\alpha\sin\beta}{\cos\alpha\cos\beta - \sin\alpha\sin\beta}$$

Divide numerator and denominator by $\cos\alpha\cos\beta$:

$$\tan(\alpha + \beta) = \frac{\dfrac{\sin\alpha}{\cos\alpha} + \dfrac{\sin\beta}{\cos\beta}}{1 - \dfrac{\sin\alpha}{\cos\alpha}\cdot\dfrac{\sin\beta}{\cos\beta}} = \frac{\tan\alpha + \tan\beta}{1 - \tan\alpha\tan\beta}$$

Addition Law for Tangent

$$\tan(\alpha + \beta) = \frac{\tan\alpha + \tan\beta}{1 - \tan\alpha\tan\beta}$$

In particular, for $\alpha = \beta = \theta$, we have the double angle formula:

Double Angle Formula for Tangent

$$\tan 2\theta = \frac{2\tan\theta}{1 - \tan^2\theta}$$

Triangles

We list here the formulas used to solve triangles (Figures 1-7-7 and 1-7-8). To review their proofs, refer to any trigonometry or precalculus textbook.

Figure 1-7-7
Right triangle

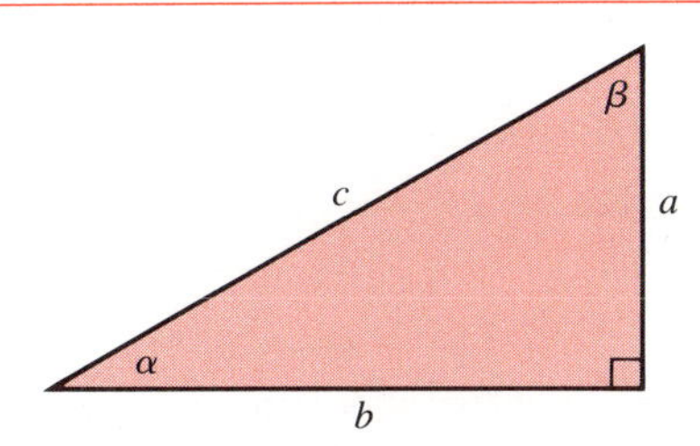

$0 < \alpha < \frac{1}{2}\pi \qquad 0 < \beta < \frac{1}{2}\pi \qquad \alpha + \beta = \frac{1}{2}\pi$

$a = c\cos\beta \qquad b = c\sin\beta \qquad b = a\tan\beta$

$\text{area} = \frac{1}{2}ab$

Figure 1-7-8
Oblique triangle

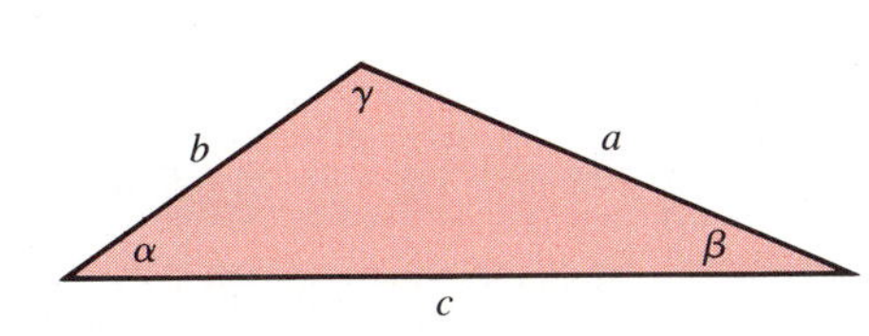

$0 < \alpha < \pi \qquad 0 < \beta < \pi \qquad 0 < \gamma < \pi$

$\alpha + \beta + \gamma = \pi$

$\text{area} = \frac{1}{2}ab\sin\gamma$

Law of sines:

$$\frac{a}{\sin\alpha} = \frac{b}{\sin\beta} = \frac{c}{\sin\gamma}$$

Law of cosines:

$$c^2 = a^2 + b^2 - 2ab\cos\gamma$$

Heron's formula:

$$\text{area} = \sqrt{s(s-a)(s-b)(s-c)}$$

where $s = \frac{1}{2}(a + b + c)$

Exercises

1 Express $\sin x$ in terms of $\cot x$.
2 Express $\sin x$ in terms of $\sec x$.
3 Express $\cot^2 x$ in terms of $\cos^2 x$.
4 Express $\cot x$ in terms of $\sec x$.

Prove

5 $\cot(\alpha - \beta) = \dfrac{\cot\alpha\cot\beta + 1}{\cot\beta - \cot\alpha}$

6 $\cot 2\theta = \dfrac{\cot^2\theta - 1}{2\cot\theta}$

7 $\tan\theta = \dfrac{\sin 2\theta}{1 + \cos 2\theta}$

8 $\sin 2\theta = \dfrac{2\tan\theta}{1 + \tan^2\theta}$

Find the smallest period

9 $\tan x - \cot x$

10 $\sec\theta - \csc\theta$

Graph

11 $y = \cot 2x$
12 $y = 1 - \csc x$
13 $y = \tan^2 x$
14 $y = \sec^2 x$
15 $y = \tan(x - \frac{1}{4}\pi)$
16 $y = -\sec x$

Suppose $0 \le \theta \le \pi$. Prove

17 $\sin\frac{1}{2}\theta = \sqrt{(1 - \cos\theta)/2}$
18 $\cos\frac{1}{2}\theta = \sqrt{(1 + \cos\theta)/2}$
19 Prove $\tan\frac{1}{2}\theta = (\sin\theta)/(1 + \cos\theta) = (1 - \cos\theta)/\sin\theta$

20 Prove $\sin\theta = \dfrac{2\tan\frac{1}{2}\theta}{1 + \tan^2\frac{1}{2}\theta}$

21 Prove $\sin\alpha + \sin\beta = 2\sin\frac{1}{2}(\alpha + \beta)\cos\frac{1}{2}(\alpha - \beta)$.
22 Prove $\cos\alpha + \cos\beta = 2\cos\frac{1}{2}(\alpha + \beta)\cos\frac{1}{2}(\alpha - \beta)$.

Find the angles of a triangle whose sides a, b, c are

23 1.7423 2.1106 1.4191
24 7.051 6.988 6.356

Find the remaining sides and angles, given

25 $a = 4.908 \quad b = 3.157 \quad \gamma = 0.8031$
26 $a = 2.114 \quad b = 8.356 \quad \gamma = 2.312$
27 $a = 4.000 \quad \beta = 1.612 \quad \gamma = 0.4410$
28 $a = 3.002 \quad \beta = 1.001 \quad \gamma = 1.395$

1-8 Exponents and Exponential Functions

In this section a, b, and c denote positive real numbers. We begin our review with integer exponents.

If n is any positive integer, then

$$a^n = a \cdot a \cdot a \cdot \cdot \cdot a \qquad (n \text{ factors})$$

This definition is extended to all integer exponents by

$$a^0 = 1 \qquad a^{-n} = \frac{1}{a^n}$$

The *rules of exponents* for integer exponents are the following:

- If m and n are integers, then

$$a^{m+n} = a^m a^n \qquad (a^m)^n = a^{mn}$$

$$(ab)^n = a^n b^n \qquad a^{m-n} = a^m / a^n$$

These rules can be proved by mathematical induction.

Figure 1-8-1
It is reasonable that $x^n = a$ has a solution $x = \sqrt[n]{a}$.

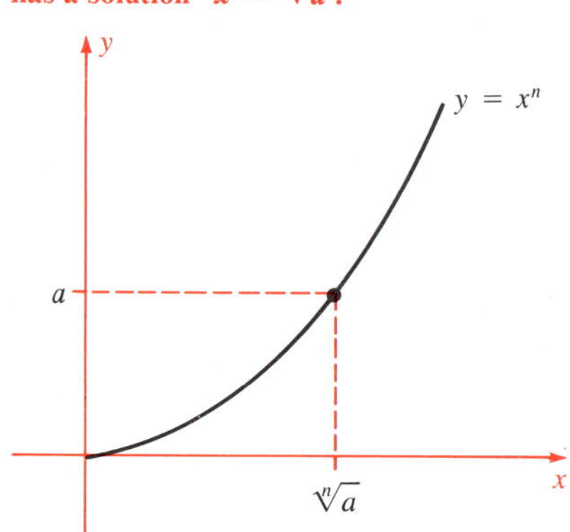

Roots and Rational Exponents

Let n be a whole number, and consider the equation

$$x^n = a$$

Experience leads us to assume that this equation has a unique positive solution. (Remember that a is positive.) See Figure 1-8-1. The solution is written

$$\sqrt[n]{a} \quad \text{or} \quad a^{1/n}$$

and read "the n-th root of a" or "a to the 1 over n." In Chapter 4, we shall see how the existence of n-th roots can be proved.

Example 1

Steps on a calculator with algebraic logic:

7 [√‾]	$\sqrt{7} \approx 2.64575\ 131$
3 . 1 4 1 5 9 [y^x] 3 [$1/x$] [=]	$(3.14159)^{1/3} \approx 1.46459\ 148$
5 . 7 [y^x] 9 [±] [$1/x$] [=]	$(5.7)^{-1/9} \approx 0.82416\ 450$

Now let r be any positive rational number. Then $r = m/n$ where m and n are positive integers. We define

$$a^r = (a^m)^{1/n}$$

An alternative definition is

$$a^r = (a^{1/n})^m$$

which turns out to be the same thing. Also, if we represent the rational number r in two different ways, we get the same result. That is, if

$$r = \frac{m}{n} = \frac{p}{q} \qquad (m, n, p, q \text{ positive integers})$$

then

$$(a^m)^{1/n} = (a^p)^{1/q}$$

so a^r really is defined unambiguously. The definition is extended to all rational exponents (not just positive ones) by

$$a^0 = 1 \qquad a^{-r} = 1/a^r$$

We can then prove the *rules of exponents* for rational exponents:

- If r and s are rational numbers, then

$$a^{r+s} = a^r a^s \qquad (a^r)^s = a^{rs} \qquad (ab)^r = a^r b^r \qquad a^{r-s} = a^r/a^s$$

We omit the rather tedious "algebra" proofs of these and other facts in this section. In Chapter 4, calculus will provide tools both for defining a^x and for proving its properties. Those tools are easier and more satisfactory than those of elementary mathematics.

Scientific Notation The familiar scientific, or exponential, representation of numbers used in calculators and programming languages is a convenient use of exponents. Thus for instance $3{,}500{,}000 = 3.5 \times 10^6$ and $0.00035 = 3.5 \times 10^{-4}$.

Inequalities

Suppose first that $a > 1$. Then, considered as a *function* of the rational number r, the quantity a^r is a strictly increasing function (that is, if $r < s$, then $a^r < a^s$). In the case $0 < a < 1$, the function a^r is a strictly decreasing function of r. The case $a = 1$ is trivial: then $a^r = 1$ for all r. (Peek ahead at Figure 1-8-3.)

Exponential Inequalities

Let $a > 0$ and let r and s be rational numbers such that $r < s$. Then the following hold:

- If $a > 1$ then $a^r < a^s$.
- If $0 < a < 1$ then $a^r > a^s$.

Real Exponents

It is possible to define a^x for any $a > 0$ and *any real* x. The definition is somewhat technical, and we shall only indicate how it works. This definition leads to a function $f(x) = a^x$ *for all real* x that satisfies the rules of exponents and inequalities (previously stated only for rational exponents).

For example, suppose we want to compute $y = (3.5)^x$ where $x = \sqrt{2}$, a known irrational. We can estimate y by rational powers of 3.5, which we already know how to compute. Since

$$1.414213 < \sqrt{2} < 1.414214$$

the first exponential inequality implies

$$(3.5)^{1.414213} < (3.5)^{\sqrt{2}} < (3.5)^{1.414214}$$

By calculator we obtain

$$5.880687 < (3.5)^{1.414213} \quad \text{and} \quad (3.5)^{1.414214} < 5.880695$$

so we conclude that

$$(3.5)^{\sqrt{2}} \approx 5.8807$$

A more accurate approximation is 5.8806916. An (algebraic logic) calculator sequence for $(3.5)^{\sqrt{2}}$ is

3 . 5 y^x 2 $\sqrt{\ }$ =

Exponential Functions

Fix $a > 0$. Then there exists a unique function $f(x)$ with domain **R** such that

1 $f(x_1 + x_2) = f(x_1)f(x_2)$

2 If $a > 1$ then $f(x)$ is positive and strictly increasing.
That is, if $x_1 < x_2$, then $0 < f(x_1) < f(x_2)$.
If $0 < a < 1$ then $f(x)$ is positive and strictly decreasing.
That is, if $x_1 < x_2$, then $f(x_1) > f(x_2) > 0$.
If $a = 1$ then $f(x) = 1$ for all x.

3 If r is rational, then $f(r)$ has the value a^r given by the definition of rational exponents.

Remark Property 3 can be replaced by the simple relation $f(1) = a$. It is then a tedious algebra exercise to prove that property **3** follows. For starters,

$$f(2) = f(1 + 1) = f(1)f(1) = a \cdot a = a^2$$

Similarly,

$$f(3) = f(2 + 1) = f(2)f(1) = a^2 \cdot a = a^3 \qquad \text{and so on}$$

The common notation for this function $f(x)$ is a^x. Exponential functions are among the most frequently used functions in mathematics and its applications. They obey the following rules:

Rules of Exponents

Let $a > 0$ and $b > 0$. Then for any real numbers x and y,

1 $a^{x+y} = a^x a^y$ **3** $(ab)^x = a^x b^x$

2 $(a^x)^y = a^{xy}$ **4** $a^{x-y} = a^x / a^y$

Rule 4 is a direct consequence of rule 1:

$$1 = a^0 = a^{x-x} = a^x a^{-x}$$

so $a^{-x} = 1/a^x$ and $a^{x-y} = a^x a^{-y} = a^x/a^y$. Rules 2 and 3 require more effort to prove.

Graphs

In Figure 1-8-2 we show some graphs $y = a^x$ for $a > 1$. As the graphs suggest:

$$\text{If } a > 1 \text{ then } \lim_{x\to+\infty} a^x = +\infty \text{ and } \lim_{x\to-\infty} a^x = 0$$

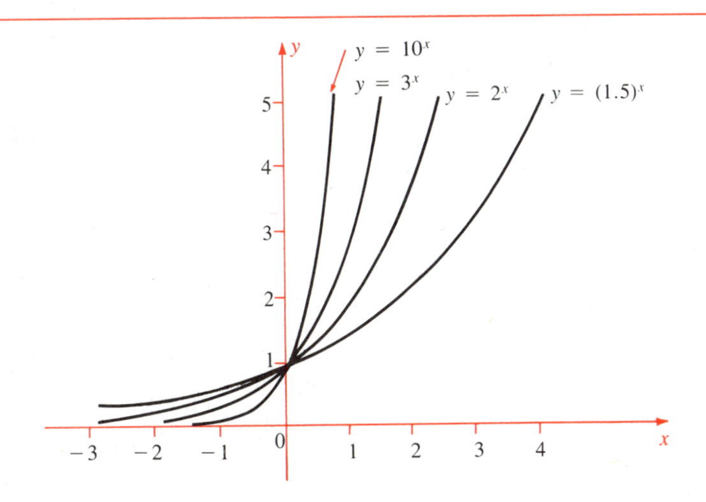

Figure 1-8-2
Graphs of $y = a^x$ for $a > 1$

If $0 < a < 1$ then $1/a > 1$. Since $a^x a^{-x} = 1$, it follows that

$$a^x = \frac{1}{a^{-x}} = \left(\frac{1}{a}\right)^{-x}$$

Therefore the graph of $y = a^x$ is obtained by replacing each point (x, y) on the graph of $y = (1/a)^x$ by $(-x, y)$. In other words, the graph of $y = a^x$ is the reflection in the y-axis of the graph of $y = (1/a)^x$. See Figure 1-8-3.

A more detailed discussion of exponential functions will be given at the beginning of Chapter 4.

Figure 1-8-3
The graph of $y = (0.5)^x$ is obtained by reflecting the graph of $y = 2^x$ in the y-axis.

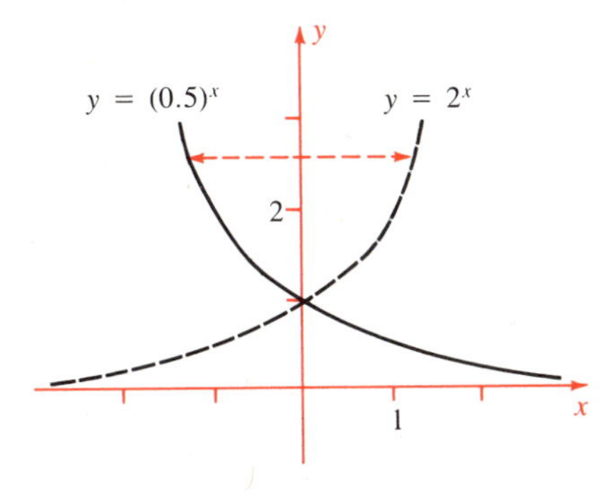

Exercises

Graph

1 $y = 5^x$

2 $y = (1.2)^x$

3 $y = (0.8)^x$

4 $y = 2^{x-1}$

5 $y = 2^{x+1}$

6 $y = 2^{x^2}$

7 $y = 3^{|x|}$

8 $y = 3^{-|x|}$

9 $y = 1 - 2^x$

10 $y = 2^x - x$

11 Suppose $0 < a < b$ and $x > 0$. Prove $a^x < b^x$.

12 Show that $36.461 < \pi^\pi < 36.463$.

Graph, with particular attention to large values of x

13 $y = 2^x/x \quad (x > 0)$

14 $y = 2^x/(1 + x^2)$

15 $y = 3^x - 100 \cdot 2^x$

16 $y = 2^x - 10 \cdot (1.8)^x$

1-9 Review Exercises

1 Given $f(x) = 3x + 1$
a Compute $f(0)$ $f(-2)$ $f[f(x)]$
b Show that $f(a+b) = f(a) + f(b) - 1$

2 Find a linear function whose graph passes through $(0, 3)$ and $(1, 5)$.

3 Graph and find the lowest point of $y = 2x^2 - 12x + 14$.

4 If $f(x) = \sqrt{x}$ and $g(x) = 3 - x$, compute
a $(f \circ g)(x)$ and b $(g \circ f)(x)$
In each case, state the domain of the function.

5 Find a linear function $f(x) = ax + b$ whose graph passes through $(0, 6)$ and is parallel to the graph of $y = -x$.

6 For what numbers b is the value of $x^2 + bx + 1$ positive regardless of the choice of x?

7 Plot on the same set of coordinate axes:
a $y = (x-2)^2$ b $y = -3(x-2)^2$

8 Find a cubic polynomial whose graph crosses the x-axis at $x = -1$, is tangent to the x-axis at $x = 2$, and crosses the y-axis at $y = 8$.

9 Construct a rational function with vertical asymptotes $x = 0$ and $x = 4$, and horizontal asymptote $y = 3$.

10 What is the relation between the graph of $y = ax^2 + bx + c$ and that of $y = ax^2 - bx + c$?

Graph

11 $y = x^3 - 3x$
12 $y = \frac{1}{4}(x+2)(x-1)^3$
13 $y = x^3 - 2x^2$
14 $y = \frac{1}{4}x^2(x+2)(x-3)^2$
15 $y = x + 1/x$
16 $y = x^2/(x-3)^2$
17 $y = x(x-1)/(x^2-4)$
18 $y = (x-2)^2/(x+1)$

Prove

19 $\cos 2x = \dfrac{1 - \tan^2 x}{1 + \tan^2 x}$
20 $\cot x - \tan x = 2 \cot 2x$
21 $\sin\alpha - \sin\beta = 2\sin\frac{1}{2}(\alpha - \beta)\cos\frac{1}{2}(\alpha + \beta)$
22 (cont.)

$$\cos\theta + \cos 3\theta + \cdots + \cos(2n-1)\theta = \frac{\sin 2n\theta}{2\sin\theta}$$

[Hint Find $2\sin\theta\cos(2j-1)\theta$ by Exercise 21.]

Graph

23 $y = \cos(x - \frac{1}{4}\pi)$
24 $y = \sin(\frac{1}{2}\pi + 1/x)$ $x > 0$
25 $y = \sin(2\pi x^2)$
26 $y = \cos\left(\dfrac{2\pi}{x^2+1}\right)$
27 $y = 2^{-1.5x}$
28 $y = 2^{1/x}$ $x > 0$
29 $y = 3^{1/(1+x^2)}$
30 $y = 2^{x/(1+x^2)}$

2 The Derivative

2-1 The Slope Problem

The graph of a linear function is a straight line; its steepness is the same at every point. The graph of a general function $y = f(x)$ is a curve; we must expect its steepness to vary from point to point. We measure how steep the graph of a linear function is by its slope. Another way to put it is that the slope a of $y = ax + b$ measures how fast y grows relative to x. In this chapter we attack the problem: What is the slope of a more general function $y = f(x)$? This problem is far more important than it appears at first glance; its solution leads to the *derivative*, one of the most useful ideas in mathematics. The derivative, or slope, of a function is the rate of increase of the height of the function's graph as x increases. It turns out that velocity, marginal cost, and many other rates of change can be interpreted as slopes of functions, so they are derivatives.

Let us consider the graph in Figure 2-1-1. At A the graph is increasing, so we expect the slope to be positive. At B the graph is increasing faster, so we expect the slope to be larger than at A. Both at C and at D the graph is decreasing, so we expect negative slopes. Because the graph is steeper at D than at C, we expect the slope at D to be less than that at C (hence larger in absolute value).

Figure 2-1-1
The graph of $y = f(x)$
$0 <$ (slope at A) $<$ (slope at B)
$0 >$ (slope at C) $>$ (slope at D)

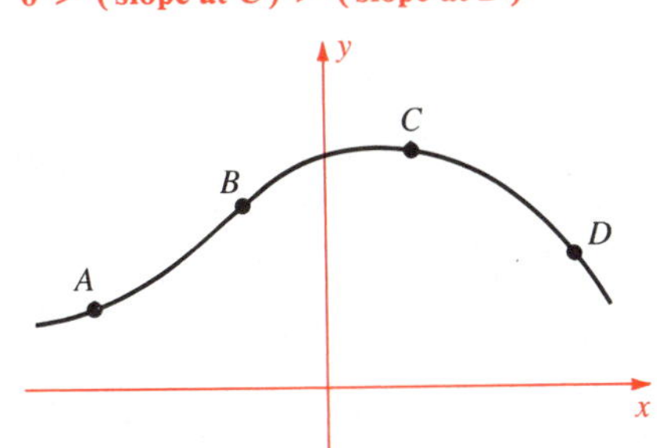

The problem is to define the slope of a function $y = f(x)$ at a point $P = (a, f(a))$ of its graph (Figure 2-1-2, next page). Our method for solving this problem is the key idea of differential calculus. First we choose a point Q of the graph, very close to P. Then we compute the slope of the secant through P and the nearby point Q. (The word **secant** simply means a line through two or more distinct points of a curve.)

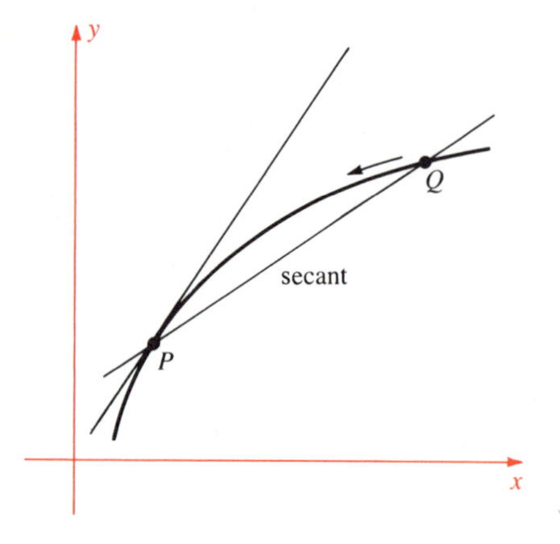

Figure 2-1-2
The geometry of slope: P is fixed on the graph of $y = f(x)$ and $Q \to P$. Then slope of secant $PQ \to$ slope of f at P.

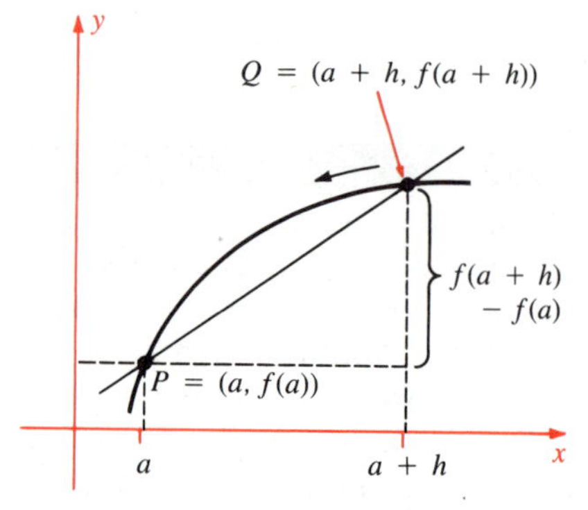

Figure 2-1-3
The algebra of slope:

$$\text{slope}(PQ) = \frac{f(a+h) - f(a)}{h}$$

Generally, as Q is chosen closer and closer to P, the slope of PQ will become closer and closer to some number, and this "limiting value" will be the slope of the graph at P.

To get a point Q that is near $P = (a, f(a))$, we let Q be the point of the graph corresponding to $x = a + h$, where h is a positive or negative number close to 0. Then $Q = (a + h, f(a + h))$. From Figure 2-1-3, the slope of the secant through P and Q is

$$\text{slope}(PQ) = \frac{f(a+h) - f(a)}{(a+h) - a} = \frac{f(a+h) - f(a)}{h}$$

The ratio $[f(a+h) - f(a)]/h$ is called the **difference quotient** of $f(x)$ at $x = a$. The figure suggests that the difference quotient is close to the slope of the graph at P provided that Q is close to P, in other words, provided that h is small (but not 0, because division by 0 is not allowed). Consequently, we expect to find the slope at P by examining the difference quotient for smaller and smaller values of h.

Example 1

Find the slope of $y = x^2$ at $x = 1$.

Solution At $x = 1$ we have $P = (1, 1^2) = (1, 1)$. A nearby point is $Q = Q_h = (1 + h, (1 + h)^2)$, and the difference quotient is

$$\text{slope}(PQ_h) = \frac{(1+h)^2 - 1}{(1+h) - 1} = \frac{(1+h)^2 - 1}{h}$$

This is shown in Figure 2-1-4. Next we experiment with some small positive and negative values of h to see what this slope looks like (a calculator helps). For instance, if $h = 0.1$, then

$$\text{slope}(PQ_h) = \frac{(1.1)^2 - 1}{0.1} = \frac{1.21 - 1}{0.1} = \frac{0.21}{0.1} = 2.1$$

Let us also try $h = 0.001$ and $h = -0.001$:

$$\frac{(1.001)^2 - 1}{0.001} = 2.001 \qquad \frac{(1 - 0.001)^2 - 1}{-0.001} = 1.999$$

Try h even smaller, $h = 0.00001$ and $h = -0.00001$:

$$\frac{(1.00001)^2 - 1}{0.00001} = 2.00001 \qquad \frac{(0.99999)^2 - 1}{-0.00001} = 1.99999$$

The message is clear: As h gets smaller and smaller, Q_h moves closer and closer to P, and the slope of the secant gets closer and closer to 2. This experimental evidence strongly suggests that the slope of $y = x^2$ at $P = (1, 1)$ equals 2. ●

The difference quotient in the solution of Example 1 can be simplified by algebra:

$$\text{slope}(PQ_h) = \frac{(1+h)^2 - 1}{h} = \frac{(1 + 2h + h^2) - 1}{h} = 2 + h$$

What we have observed numerically is no accident. From the simplified expression for the slope, it is clear that as h becomes smaller and smaller, the slope becomes closer and closer to 2.

Example 2

Find the slope of $y = x^2$ at

a $x = 2$ and **b** $x = a$

Solution

a The point on the graph corresponding to $x = 2$ is

$$P = (2, 2^2) = (2, 4)$$

The typical nearby point is $Q_h = (2 + h, (2 + h)^2)$, and the difference quotient is

$$\text{slope}(PQ_h) = \frac{(2+h)^2 - 4}{h}$$

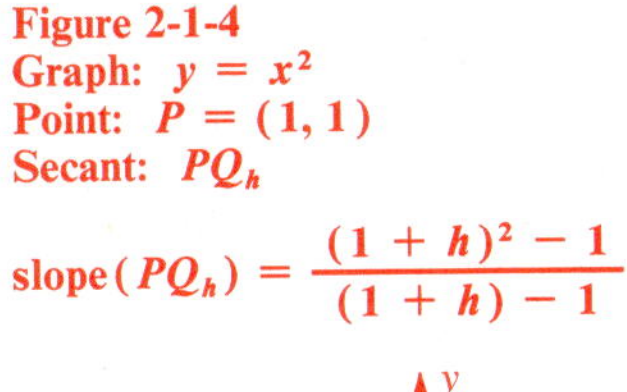

Figure 2-1-4
Graph: $y = x^2$
Point: $P = (1, 1)$
Secant: PQ_h

$$\text{slope}(PQ_h) = \frac{(1+h)^2 - 1}{(1+h) - 1}$$

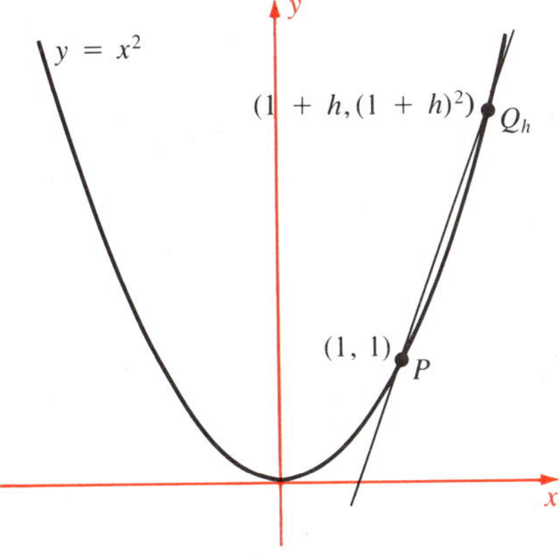

Let us simplify this expression by algebra:

$$\text{slope}(PQ_h) = \frac{(4 + 4h + h^2) - 4}{h} = \frac{4h + h^2}{h} = 4 + h$$

With or without a calculator, we easily find

h	0.1	−0.01	0.001	−0.000001
slope(PQ_h)	4.1	3.99	4.001	3.999999

The numerical evidence strongly suggests that the slope of $y = x^2$ at $(2, 4)$ equals 4.

b The point P on the graph of $y = x^2$ corresponding to $x = a$ is $P = (a, a^2)$. The typical nearby point is $Q_h = (a + h, (a + h)^2)$, where $|h|$ is small but not 0. The corresponding difference quotient is

$$\frac{(a+h)^2 - a^2}{(a+h) + a} = \frac{(a+h)^2 - a^2}{h}$$

We cannot use numerical tests directly because we do not know the value of a. However, we can use algebra to simplify the expression:

$$\text{slope}(PQ_h) = \frac{(a+h)^2 - a^2}{h} = \frac{(a^2 + 2ah + h^2) - a^2}{h}$$

$$= \frac{2ah + h^2}{h} = 2a + h$$

If h is very small, then $2a + h$ is very close to $2a$, as close as we please if we choose h small enough. We conclude that the slope of

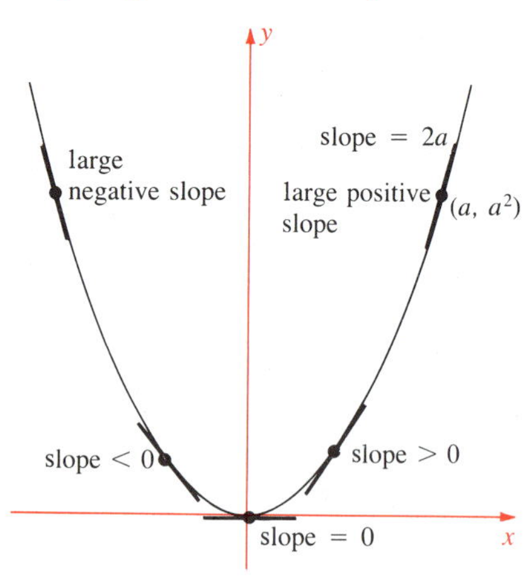

Figure 2-1-5
Slope of $y = x^2$ at various points

$y = x^2$ at (a, a^2) is $2a$. A glance at Figure 2-1-5 shows that we are in the right ball park: The slope at (a, a^2) is positive where $a > 0$, negative where $a < 0$, and zero where $a = 0$. Furthermore, $2a$ increases as a increases, which agrees with the graph's increasing steepness. (Note also that the slope $2a$ obtained by algebra agrees with the answer in part **a** based on experimental evidence.) ●

Example 3

Find the slope of $y = x^3$ at

a $x = -2$ and **b** $x = a$

Solution

a The point on the graph of $y = x^3$ where $x = -2$ is

$$P = (-2, (-2)^3) = (-2, -8)$$

Write down the difference quotient and simplify it:

$$\text{slope}(PQ_h) = \frac{(-2+h)^3 - (-2)^3}{h}$$

$$= \frac{(-8 + 12h - 6h^2 + h^3) + 8}{h} = 12 - 6h + h^2$$

If h is small, then $-6h$ and h^2 are both small. Therefore $12 - 6h + h^2$ is close to 12. The smaller h is, the closer $12 - 6h + h^2$ is to 12. We conclude that the slope of $y = x^3$ at $(-2, -8)$ is 12.

b Now $P = (a, a^3)$ and $Q_h = (a + h, (a + h)^3)$, so the difference quotient is

$$\text{slope}(PQ_h) = \frac{(a+h)^3 - a^3}{h}$$

$$= \frac{(a^3 + 3a^2h + 3ah^2 + h^3) - a^3}{h}$$

$$= 3a^2 + 3ah + h^2$$

As h approaches 0, both $3ah$ and h^2 approach 0 while $3a^2$ remains fixed. We conclude that the slope of $y = x^3$ at (a, a^3) is $3a^2$.

This answer makes good sense geometrically. The slope $3a^2$ is always positive, except that it is zero at $a = 0$. As $|a|$ increases, $3a^2$ increases rapidly, so the curve becomes very steep (Figure 2-1-6). ●

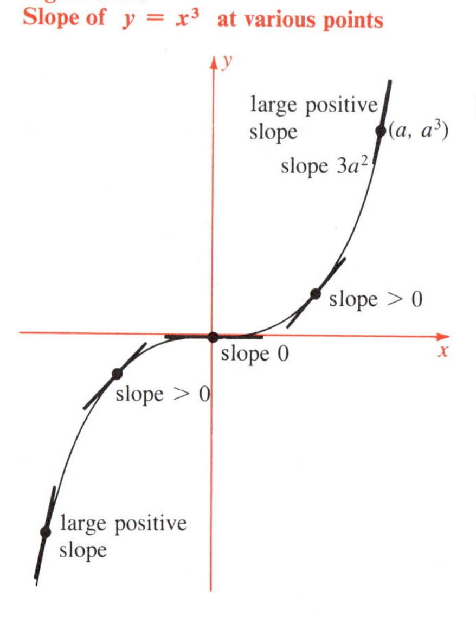

Figure 2-1-6
Slope of $y = x^3$ at various points

Exercises

Given $y = f(x)$ and $x = a$, compute $P = (a, f(a))$ and find the slope of the graph of $y = f(x)$ at P by experimenting with the difference quotient. Use your calculator to compute $[f(a+h) - f(a)]/h$ for $h = \pm 0.01$, ± 0.001, and ± 0.0001 and guess the slope from your numerical work

1 $f(x) = 3x^2$
$a = 2$

2 $f(x) = 3x^2$
$a = -5$

3 $f(x) = x^2 - x$
$a = 1$

4 $f(x) = x^2 - x$
$a = -2$

5 $f(x) = x^2 + 3x$
$a = 2$

6 $f(x) = x^2 + 3x$
$a = -1$

7 $f(x) = 2x^3 + x$
$a = 3$

8 $f(x) = 2x^3 + x$
$a = 0$

9 $f(x) = x^3 - x^2$
$a = -3$

10 $f(x) = x^3 - x^2$
$a = 2$

11 $f(x) = x^3 - 5x + 4$
$a = -1$

12 $f(x) = x^3 - 5x + 4$
$a = 2$

Find the slope of the following curves, using the algebraic method of Example 3b

13 $y = 3x^2$ at $(a, 3a^2)$

14 $y = x^2 - x$ at $(a, a^2 - a)$

15 $y = x^2 + 3x$ at $(a, a^2 + 3a)$

16 $y = 2x^3 + x$ at $(a, 2a^3 + a)$

17 $y = x^3 - x^2$ at $(a, a^3 - a^2)$

18 $y = x^3 - 5x + 4$ at $(a, a^3 - 5a + 4)$

19 Let $y = 1/x$ and $a = 0.5$. Use a calculator to find the difference quotient for $h = 0.01,\ -0.01,\ 0.00001$, and -0.00001. Then estimate the slope of $y = 1/x$ at $(0.5, 2)$.

20 Let $y = 1/x^2$ and $a = 2$. Use a calculator to find the difference quotient for $h = \pm 0.01$ and ± 0.001. Then estimate the slope of $y = 1/x^2$ at $(2, 0.25)$.

21 Let $y = \sin x$ and $a = 0$. Use a calculator to find the difference quotient for $h = \pm 0.01$ and ± 0.001. (Be sure your calculator is in *radian mode.*) What is your conclusion?

22 Let $y = 10^x$ and $a = 0$. Use your calculator to find the difference quotient for $h = \pm 0.001$ and ± 0.00001. What is your conclusion?

2-2 Limits of Functions

In the previous section, we found the slope of a function $f(x)$ at $x = a$ by an informal study of the difference quotient $[f(a + h) - f(a)]/h$ for values of h close to 0. We found the slope of $f(x)$ at $x = a$ as a sort of "limiting value" of the difference quotient as h approached 0. Differential calculus was developed by this sort of informal, intuitive consideration for about two centuries, and this approach usually serves today. But as calculus was applied in more and more ways, doubts arose about its validity when used in complex problems remote from intuition. It then became necessary to find a precise, rigorous basis for limits, so that practical rules for evaluating limits could be proved.

The cornerstone of calculus is the limit concept. In this section we give its precise definition, in the following section the most important and applicable properties of limits, and in Section 2-4 their proofs, which are optional. Once you have grasped the idea and mastered the basic properties of limits, you can use the concept without always returning to the definition or to the calculations in this section. Therefore, while the details of this section may be considered optional, you should go through them because they will help you to understand the limit concept. The definition and Example 1 may be adequate background for the pragmatic rules of Section 2-3.

The limit idea is applicable in many situations, not just finding the limit of a difference quotient. We shall define

$$\lim_{x \to a} f(x) = L$$

in general, not merely for the special case of difference quotients. In the discussion of limits that follows, we shall use the traditional Greek letters ε (epsilon) and δ (delta). We think of ε and δ as small positive real numbers.

We want these symbols to say that the values of $f(x)$ are as close to L as desired, *provided* x is close enough (but not equal) to a. By "as close to L as desired" we mean that for each positive number ε no matter how small, $|f(x) - L| < \varepsilon$, provided x is in the domain of $f(x)$ and x is sufficiently close to a. We think of ε as *very* small: 10^{-10}, 10^{-100}, and even smaller.

The phrase "x is sufficiently close to a" means that x lies in a small interval centered at a, but $x \neq a$. Such a "punctured" interval (Figure 2-2-1) is described by

$$0 < |x - a| < \delta \quad \text{where} \quad \delta > 0$$

Let us summarize this discussion with a formal definition:

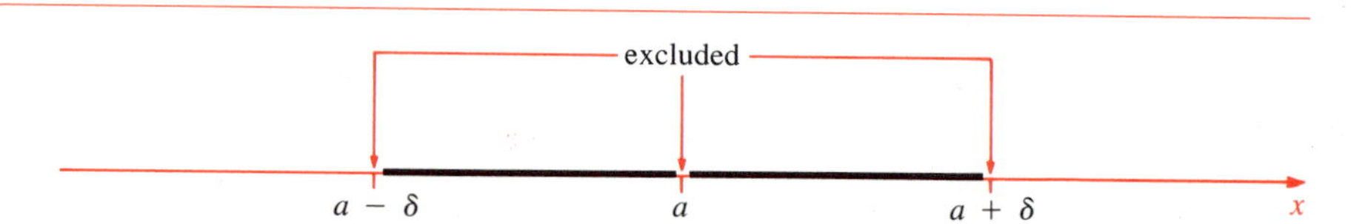

Figure 2-2-1
The "punctured" interval

$$(a - \delta, a) \cup (a, a + \delta)$$

consists of all x for which $0 < |x - a| < \delta$.

Definition Limit

Let $f(x)$ be a function with domain D. Let a be a point of the real axis such that D contains points different from a that are arbitrarily close to a. Let L be a real number. Then

$$\lim_{x \to a} f(x) = L$$

if for each number $\varepsilon > 0$ there exists a number $\delta > 0$ such that

$$|f(x) - L| < \varepsilon$$

for all x in D that satisfy the inequality

$$0 < |x - a| < \delta$$

There are two important points to note about this basic definition: First, it does not matter whether a is in the domain D of $f(x)$ or not. Even if a is in D, the value of $f(x)$ at a itself does not enter into the definition, because we consider only points x for which $0 < |x - a|$. The definition depends only on the values of $f(x)$ for x *near* a.

Second, D must contain points different from a but arbitarily close to a. The most important cases are when D is an interval and a is a point of D, and when D is an interval and a is an end point of D but not necessarily in D. The expression "arbitrarily close to a" means that, given any positive δ, *no matter how small,* there is a point x in D such that $0 < |x - a| < \delta$.

Terminology The expression

$$\lim_{x \to a} f(x) = L$$

is read, "The limit as x approaches a of $f(x)$ is L." We also say that "$f(x)$ approaches L as x approaches a," and write

$$f(x) \to L \quad \text{as} \quad x \to a$$

In practice, the definition works this way. If challenged with an ε, then you must produce a suitable δ. You must be able to do so for *every* $\varepsilon > 0$, not just a particular ε. Note that the δ you produce will depend on the ε you are challenged with. A δ that works for one particular ε generally will not work for a much smaller ε.

Example 1

Use the definition of limit to prove

a $\lim_{x\to 3} 2x = 6$ **b** $\lim_{x\to 7} x^2 = 49$

Solution

a In this case $f(x) = 2x$ with domain $\mathbf{R}$. Here $a = 3$ and the domain surely contains points different from a that are arbitrarily close to a. The value of the limit to be established is $L = 6$. Let $\varepsilon > 0$. Our problem is to find a $\delta > 0$ so that $|f(x) - L| < \varepsilon$ whenever $0 < |x - a| < \delta$. Now

$$|f(x) - L| = |2x - 6| = |2(x - 3)| = 2|x - 3|$$

Therefore $|f(x) - L| < \varepsilon$ is equivalent to

$$2|x - 3| < \varepsilon$$

which in turn is equivalent to

$$|x - 3| < \tfrac{1}{2}\varepsilon$$

This inequality shows us how we can choose δ: We take $\delta = \frac{1}{2}\varepsilon$. If, as in the definition of limit, we require x to be so close to 3 that $0 < |x - 3| < \delta$, then

$$|f(x) - 6| = 2|x - 3| < 2\delta = 2(\tfrac{1}{2}\varepsilon) = \varepsilon$$

Therefore $\lim_{x\to 3} 2x = 6$ by the definition of limit.

b In this case $f(x) = x^2$ with domain $\mathbf{R}$. Also, $a = 7$ and $L = 49$. The domain $\mathbf{R}$ contains points different from a that are arbitrarily close to a. Let us analyze what else we must do. We want to make $|f(x) - 49| = |x^2 - 49|$ small by choosing x sufficiently close to 7. Now

$$|x^2 - 49| = |(x + 7)(x - 7)| = |x + 7| \cdot |x - 7|$$

The factor $|x - 7|$ will certainly be small when x is close to 7. What about the other factor? Surely, if x is very close to 7 then $x + 7$ will be very close to 14. Indeed, we can guarantee $|x + 7| < 15$ by keeping x between 6 and 8, that is, by insisting that $|x-7|<1$. Thus we have

$$|x^2-49|=|x+7|\cdot|x-7|<15\cdot|x-7|<15\delta$$

provided $|x-7|<\delta$ and $\delta<1$.

Now let ε be given. Our problem is to find δ both smaller than 1 and so small that

$$|x^2-49|<\varepsilon$$

provided $|x-7|<\delta$. This will be so if $\delta=\frac{1}{15}\varepsilon$ or $\delta=1$, whichever is smaller. In either case

$$|x-7|<\tfrac{1}{15}\varepsilon \quad\text{and}\quad |x+7|<15$$

so $|x^2-49|=|x+7|\cdot|x-7|<15\cdot\frac{1}{15}\varepsilon=\varepsilon$. Thus, we simply choose δ to be the smaller of 1 and $\frac{1}{15}\varepsilon$. Then if $|x-7|<\delta$, we have $|x^2-49|<\varepsilon$. Therefore $\lim_{x\to7}x^2=49$ by the definition of limit. ●

Example 1 shows how the ε-δ method is used. There are many tricks of the trade and refinements, some of which are shown in the examples that follow. Working out limits using ε's and δ's is a skill that is learned only with much practice. Fortunately, we shall soon develop rules that allow us to do most of calculus without looking at ε-δ proofs!

Example 2

Prove $\displaystyle\lim_{x\to0}5\sqrt{x}=0$.

Solution The function $f(x)=5\sqrt{x}$ has domain $D=[0,+\infty)$, and $a=0$ is the left end point of the domain. Notice that the definition of limit applies perfectly well if a is an end point of the function's domain.

To prove $\lim_{x\to0}5\sqrt{x}=0$, we must prove: Given any $\varepsilon>0$, there is a $\delta>0$ such that $5\sqrt{x}<\varepsilon$ whenever $0<x<\delta$. In this case, we can omit the absolute values of the definition because $x>0$, so

$$|f(x)-0|=|5\sqrt{x}-0|=|5\sqrt{x}|=5\sqrt{x}$$

and $|x-0|=x$. Since $\sqrt{x}>0$, the inequality

$$5\sqrt{x}<\varepsilon \quad\text{is equivalent to}\quad 25x<\varepsilon^2$$

(by squaring both sides). Therefore $5\sqrt{x}<\varepsilon$ will be the case if

$$0<x<\tfrac{1}{25}\varepsilon^2$$

This tells us how to choose δ. Given $\varepsilon>0$, we choose $\delta=\frac{1}{25}\varepsilon^2$. If $0<x<\delta$, then we indeed have

$$|f(x)-0|=5\sqrt{x}<5\sqrt{\delta}=5(\tfrac{1}{5}\varepsilon)=\varepsilon$$

This proves that $\displaystyle\lim_{x\to0}5\sqrt{x}=0$ by the definition of limit. ●

Example 3

Prove $\lim_{x\to 1} 1/x = 1$.

Solution The function $f(x) = 1/x$ is defined for all $x \neq 0$. This domain splits into two intervals, $(-\infty, 0)$ and $(0, +\infty)$. Because we are interested only in values of x very near 1, let us stick to the right half, $(0, +\infty)$. We can force x to be positive by considering only δ's that satisfy $\delta < 1$. Then $|x - 1| < \delta < 1$ forces $x > 0$.

We next try to make $|1/x - 1|$ small by taking x sufficiently close to 1. Now

$$\left|\frac{1}{x} - 1\right| = \left|\frac{1-x}{x}\right| = \frac{1}{x}|1 - x| = \frac{1}{x}|x - 1|$$

The factor $1/x$ is a problem. How can we make it small? Note that it is large only when x is very small. But we want only x's near 1, so certainly we can require $\delta < \frac{1}{2}$. Then $|x - 1| < \delta$ forces $x > \frac{1}{2}$ so that $1/x < 2$. See Figure 2-2-2. We then have

$$\left|\frac{1}{x} - 1\right| = \frac{1}{x}|x - 1| < 2|x - 1|$$

provided $|x - 1| < \frac{1}{2}$.

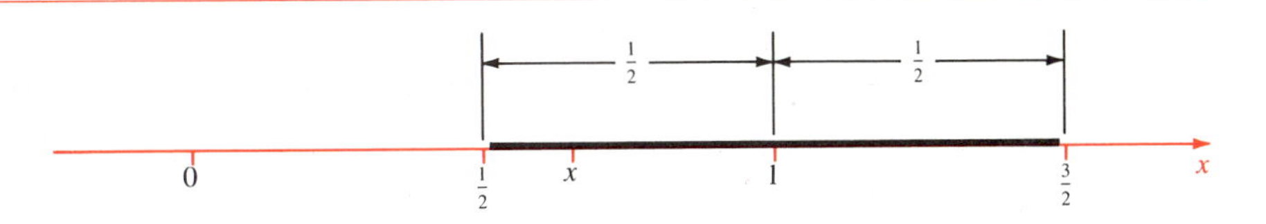

Figure 2-2-2
By requiring $\delta < 1/2$, we guarantee that $|x - 1| < \delta$ forces $x > 1/2$, that is, $1/x < 2$.

Now suppose we are given $\varepsilon > 0$. We want $|1/x - 1| < \varepsilon$. Certainly this will be true provided both $|x - 1| < \frac{1}{2}$ *and*

$$2|x - 1| < \varepsilon \qquad \text{that is} \qquad |x - 1| < \tfrac{1}{2}\varepsilon$$

Consequently we choose δ to be the smaller of $\frac{1}{2}$ and $\frac{1}{2}\varepsilon$. Then if $|x - 1| < \delta$, we have

$$\left|\frac{1}{x} - 1\right| = \frac{1}{x}|x - 1| < 2|x - 1| < 2\delta \leq 2(\tfrac{1}{2}\varepsilon) = \varepsilon$$

Therefore $\lim_{x\to 1} 1/x = 1$ by the definition of limit. ●

Intuition and Proof

In each of the previous examples it would have been pretty clear just by looking at the function what its limit was at the point in question. For instance, one can look at the graph of $y = f(x) = 1/x$ for $x > 0$ (Figure 2-2-3, next page) and say: Of course $\lim_{x\to 1} 1/x = f(1) = 1$; what else could it be?

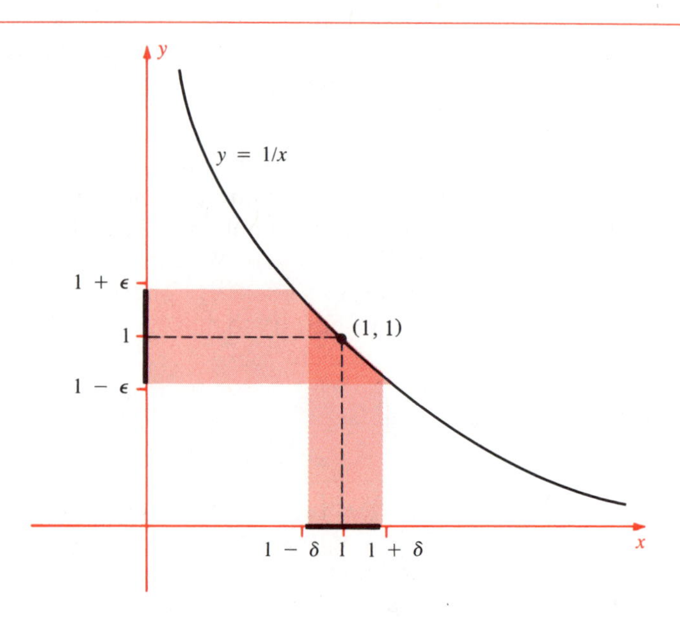

Figure 2-2-3
Graphical evidence that $\lim_{x\to 1} 1/x = 1$. $f(x) = 1/x$ takes each point of the x-interval $(1-\delta, 1+\delta)$ into the y-interval $(1-\varepsilon, 1+\varepsilon)$ provided δ is small enough.

The definition of limit bears out our intuition. When it is obvious graphically what a limit is, that is usually the right answer, even if there are technical difficulties in proving it so. There is usually no mystery about limits; the limit language expresses in precise terms what we know intuitively from our experience. The precise definition and the previous proofs of various limits makes us 100% convinced of their correctness. However, limits are not always obvious, and when they are not, you must be careful and very precise. Would you guess any of the following limits at first glance?

$$\lim_{x\to 0} \frac{1-\cos x}{x^2} = \tfrac{1}{2} \qquad \lim_{x\to 0} x^x = 1 \qquad \lim_{x\to 0} \frac{x}{\sin x} = 1$$

Proving these statements is beyond us right now, but we shall get to them in later chapters. However, you may test them on your calculator.

Example 4

a Guess the value of $\displaystyle\lim_{x\to 1} \frac{x^3-1}{x^2-1}$.

b Prove that your guess is correct.

Solution Set $f(x) = (x^3-1)/(x^2-1)$, a rational function with domain all the real numbers except for $x = \pm 1$. The values $+1$ and -1 are excluded because the denominator of $f(x)$ equals 0 for these values of x.

a It is a great help to simplify the expression for $f(x)$ as much as possible. If we factor numerator and denominator, terms may cancel, resulting in a simpler formula for $f(x)$:

$$f(x) = \frac{x^3-1}{x^2-1} = \frac{(x-1)(x^2+x+1)}{(x-1)(x+1)} = \frac{x^2+x+1}{x+1}$$

The expression on the right, unlike $(x^3 - 1)/(x^2 - 1)$, has a well-defined value for $x = 1$, leading us to *guess* that

$$\lim_{x \to 1} f(x) = \frac{1^2 + 1 + 1}{1 + 1} = \frac{3}{2} = \tfrac{3}{2}$$

b For x in the domain of $f(x)$, we have

$$\begin{aligned} f(x) - \tfrac{3}{2} &= \frac{x^2 + x + 1}{x + 1} - \tfrac{3}{2} \\ &= \frac{2(x^2 + x + 1) - 3(x + 1)}{2(x + 1)} \\ &= \frac{2x^2 - x - 1}{2(x + 1)} = \frac{(2x + 1)(x - 1)}{2(x + 1)} \end{aligned}$$

Therefore

$$|f(x) - \tfrac{3}{2}| = \tfrac{1}{2}|2x + 1| \cdot \frac{1}{|x + 1|} \cdot |x - 1|$$

Given $\varepsilon > 0$, we want to choose δ so that $|f(x) - \frac{3}{2}| < \varepsilon$ for all x satisfying $0 < |x - 1| < \delta$. Clearly we can make the final factor $|x - 1|$ as small as we please, so our first task is to control the size of the remaining factors. We do so by insisting that δ be at most 1. Then $|x - 1| < \delta$ forces $0 < x < 2$, so

$$|2x + 1| \le 2|x| + 1 < 5$$

and

$$|x + 1| = x + 1 > 1 \qquad \text{so that} \qquad \frac{1}{|x + 1|} < 1$$

Consequently

$$\begin{aligned} |f(x) - \tfrac{3}{2}| &= \tfrac{1}{2}|2x + 1| \cdot \frac{1}{|x + 1|} \cdot |x - 1| \\ &< \tfrac{1}{2} \cdot 5 \cdot 1 \cdot |x - 1| = \tfrac{5}{2}|x - 1| \end{aligned}$$

Clearly, $\frac{5}{2}|x - 1| < \varepsilon$ provided that we have $|x - 1| < \frac{2}{5}\varepsilon$. Hence, given $\varepsilon > 0$, we choose δ to be the smaller of the numbers 1 and $\frac{2}{5}\varepsilon$. Then if $|x - 1| < \delta$, we have

$$|f(x) - \tfrac{3}{2}| < \tfrac{5}{2}|x - 1| < \tfrac{5}{2}\delta < \tfrac{5}{2} \cdot \tfrac{2}{5}\varepsilon = \varepsilon$$

From the definition of limit, $\lim_{x \to 1} f(x) = \frac{3}{2}$ follows. ●

We have seen how to find a limit from numerical evidence and by algebraic simplification. We also have seen how to prove that our answer is correct by using the definition of limit. The definition of limit allows us to certify that a given L is indeed the limit of $f(x)$ as x approaches a.

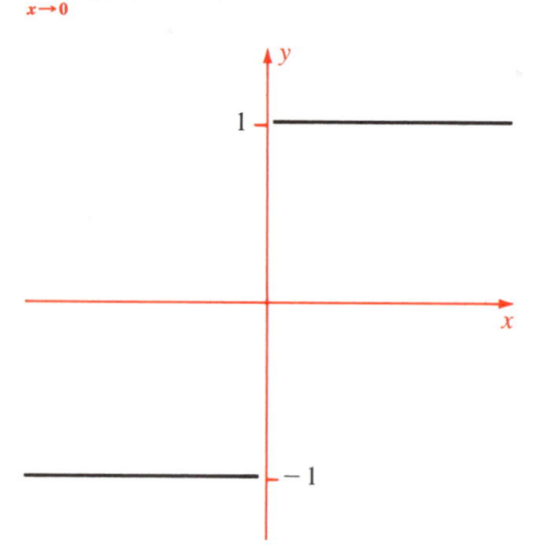

Figure 2-2-4
$\lim_{x \to 0} x/|x|$ does not exist.

However, a function may not necessarily have a limit at a given point. For instance, none of the following "limits" exists:

$$\lim_{x \to 0} \frac{x}{|x|} \qquad \lim_{x \to 0} \frac{1}{x} \qquad \lim_{x \to 0} \sin(1/x)$$

The graphs in Figures 2-2-4, 2-2-5, and 2-2-6 are evidence that these limits do not exist, and it can be proved that this is so.

One-Sided Limits

In Section 1-4, we introduced the notations

$$x \to a+ \quad \text{and} \quad x \to a-$$

which we now clarify. We write

$$\lim_{x \to a+} f(x) = L$$

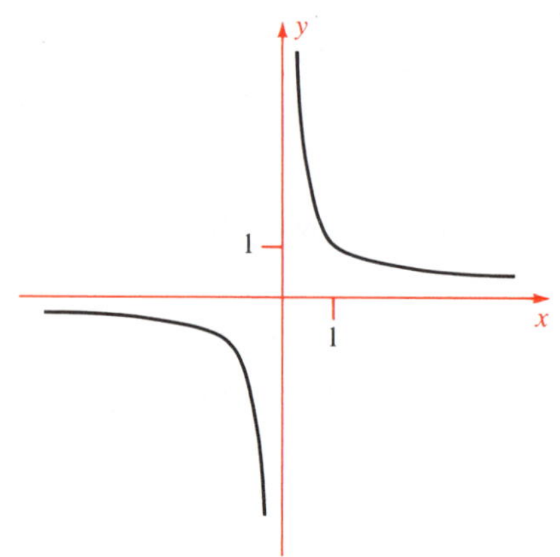

Figure 2-2-5
$\lim_{x \to 0} 1/x$ does not exist.

to mean we consider *only* values of x that are both in the domain of $f(x)$ and to the right of a, that is, $x > a$. Thus

$$\lim_{x \to a+} f(x) = L$$

means

1 The domain D of $f(x)$ contains points $x > a$ arbitrarily close to a.
2 For each $\varepsilon > 0$ there exists a $\delta > 0$ such that

$$|f(x) - L| < \varepsilon$$

for all x in D such that $a < x < a + \delta$.

The expression $\lim_{x \to a+} f(x)$ is read "the limit of $f(x)$ as x approaches a from the right." The expression

Figure 2-2-6
$\lim_{x \to 0} \sin(1/x)$ does not exist.

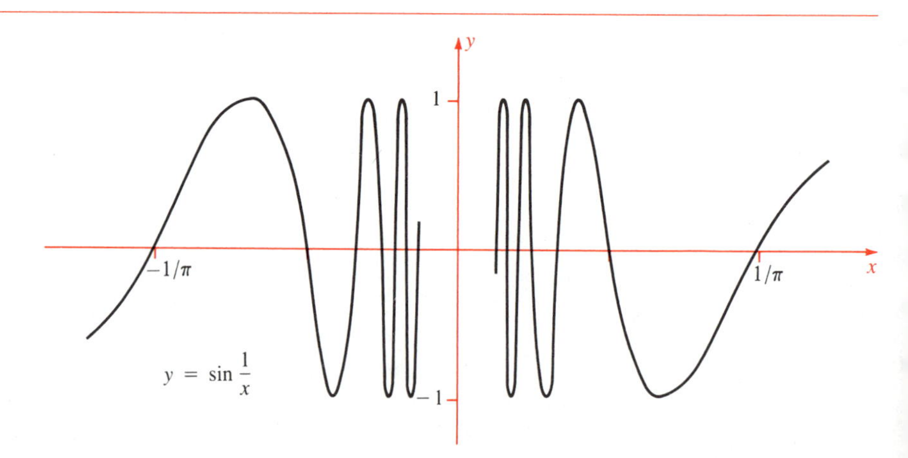

$$\lim_{x \to a-} f(x) = L$$

is defined similarly. In this case, "x approaches a from the left."

Example 5

a $\lim_{x \to 1-} \sqrt{1 - x} = 0$ **b** $\lim_{x \to 0+} \dfrac{x}{|x|} = 1$ **c** $\lim_{x \to 0-} \dfrac{x}{|x|} = -1$ ●

Note One-sided limits can be included in the definition of limit on page 72. We simply define a new domain E of a function to be

$$E = \{x \mid x > a \text{ and } x \text{ is in } D\}$$

Then

$$\lim_{x \to a+} f(x) \qquad [f(x) \text{ with domain } D]$$

is the same as

$$\lim_{x \to a} f(x) \qquad [f(x) \text{ with domain } E]$$

Therefore any properties we later give for limits hold automatically for one-sided limits as well.

Exercises

Guess the limit and prove your answer is correct by using the definition of limit.

1 $\lim_{x \to -1} 4x$

2 $\lim_{x \to 2} (-3x)$

3 $\lim_{x \to 2} (3x - 1)$

4 $\lim_{x \to -1} (5x - 4)$

5 $\lim_{x \to 2} 3x^2$

6 $\lim_{x \to -2} (-3x^2)$

7 $\lim_{x \to 0} (x^2 + 2x)$

8 $\lim_{x \to 10} (x^2 - x)$

9 $\lim_{x \to 1} 1/x$

0 $\lim_{x \to -1} 3/x$

1 $\lim_{x \to 2} [-1/(2x)]$

12 $\lim_{x \to -3} [-3/(2x)]$

13 $\lim_{x \to 3} \dfrac{1}{1 + x}$

14 $\lim_{x \to 1} \dfrac{x^2}{2 - x}$

15 $\lim_{x \to -1} \dfrac{x + 1}{x^2 - x - 2}$

16 $\lim_{x \to -2} \dfrac{x + 1}{x^2 - x - 2}$

17 $\lim_{x \to 0} \dfrac{x + 2}{x + 1}$

18 $\lim_{x \to 1} \dfrac{x + 1}{x + 2}$

19 $\lim_{x \to -2} \dfrac{x^2 + 5x + 6}{x^2 + 3x + 2}$

20 $\lim_{x \to 1} \dfrac{x^2 - 2x + 1}{x^2 + 2x - 3}$

Use a calculator to fill in the table for each given function $f(x)$. Then guess the value of $\lim_{x \to 0} f(x)$.

x	-0.01	-0.0001	0.0001	0.01
$f(x)$				

21 $f(x) = \dfrac{x^2 - x}{x^3 - x}$

22 $f(x) = \dfrac{x^3 - x}{x^5 - x}$

23 $f(x) = \dfrac{x^2 + \sin x}{x}$ (x in radians)

24 $f(x) = \dfrac{\sin(3 \sin x)}{2x}$ (x in radians)

25 $f(x) = (1 + x)^{1/x}$

26 $f(x) = (1 + 1/x)^x$

2-3 Rules for Limits

The examples in Section 2-2 show that the ε-δ game can be tedious and time-consuming. We need systematic techniques for evaluating limits, techniques that do not require searching for a δ in each case. The following rules are just that. In stating these rules, we understand that $f(x)$ and $g(x)$ have a common domain, and that a is a number suitable for applying $\lim\limits_{x\to a}$ both to $f(x)$ and to $g(x)$.

Theorem 1 Basic Limit Rules

Suppose both of the limits

$$\lim_{x\to a} f(x) = L \quad \text{and} \quad \lim_{x\to a} g(x) = M$$

exist. Suppose c is a constant. Then the following four limits exist and have the values stated:

1 $\lim\limits_{x\to a} [cf(x)] = cL$

2 $\lim\limits_{x\to a} [f(x) \pm g(x)] = L \pm M$

3 $\lim\limits_{x\to a} [f(x)g(x)] = LM$

4 If $M \neq 0$ then $\lim\limits_{x\to a} \dfrac{f(x)}{g(x)} = \dfrac{L}{M}$

In words, rule 2 says that the limit of a sum is the sum of the limits—likewise for the difference. Rule 3 says that the limit of a product is the product of the limits, and rule 4 says the same for quotients. Note that rule 1 is the special case $g(x) = c$ of rule 3. ●

Rules 2, 3, and 4 also hold for three or more summands, factors, or divisors. For instance, if we also assume $\lim\limits_{x\to 0} h(x) = N$, then

$$\lim_{x\to a} [f(x) + g(x) - h(x)] = L + M - N$$

$$\lim_{x\to a} \frac{f(x)}{g(x)h(x)} = \frac{L}{MN} \qquad (\text{provided } MN \neq 0)$$

We shall give proofs in the next section. More important for us now is to learn the use of the limit rules. They are rules for finding unknown new limits from known old ones. We shall use two basic building blocks. First is a constant function $f(x) = c$. For it we have

$$\lim_{x\to a} c = c$$

Indeed, given any $\varepsilon > 0$, take *any* $\delta > 0$. Then

$$|f(x) - c| = |c - c| = 0 < \varepsilon$$

for all x satisfying $|x - a| < \delta$.

The next building block is the so-called **identity function,** $f(x) = x$. For this linear function we have

$$\lim_{x \to a} x = a$$

To prove it, note that $|f(x) - a| = |x - a|$, so if $|x - a| < \varepsilon$, then $|f(x) - a| < \varepsilon$. Hence, given ε, we simply choose $\delta = \varepsilon$. Pretty easy!

Now we can use our limit rules to make easy work of some harder results. By rule 3,

$$\lim_{x \to a} x^2 = \lim_{x \to a} x \cdot x = (\lim_{x \to a} x)(\lim_{x \to a} x) = a \cdot a = a^2$$

Now by rule 2 we can find limits such as

$$\lim_{x \to a} (x + 1) = \lim_{x \to a} x + \lim_{x \to a} 1 = a + 1$$

$$\lim_{x \to a} (x^2 + x + 1) = \lim_{x \to a} x^2 + \lim_{x \to a} (x + 1) = a^2 + a + 1$$

After a preliminary simplification and by rule 4, we can find the limit in Example 4 of Section 2-2

$$\lim_{x \to 1} \frac{x^3 - 1}{x^2 - 1} = \lim_{x \to 1} \frac{x^2 + x + 1}{x + 1} = \frac{\lim_{x \to 1} (x^2 + x + 1)}{\lim_{x \to 1} (x + 1)}$$

$$= \frac{3}{2} = \tfrac{3}{2}$$

Note how difficulty at $x = 1$ has been avoided: The limit $\lim_{x \to 1} f(x)$ does not depend on $f(1)$. However, the first two expressions for $f(x)$ are equal for $x \neq 1$. This is a lot easier than the previous solution.

Let us redo Example 3 of Section 2-2 the same way. By rule 4,

$$\lim_{x \to 1} \frac{1}{x} = \frac{\lim_{x \to 1} 1}{\lim_{x \to 1} x} = \frac{1}{1} = 1$$

Example 1

a $\lim_{x \to -3} (4x - 5x^2) = 4 \lim_{x \to -3} x - 5 \lim_{x \to -3} x^2$

$$= 4 \cdot (-3) - 5 \cdot 9 = -57$$

b $\lim_{x \to 2} \frac{x^2 - 4}{x - 2} = \lim_{x \to 2} \frac{(x - 2)(x + 2)}{x - 2}$

$$= \lim_{x \to 2} (x + 2) = 2 + 2 = 4$$

c $\lim_{x \to -1} \frac{x + 2}{x^2} = \frac{\lim_{x \to -1} (x + 2)}{\lim_{x \to -1} x^2} = \frac{-1 + 2}{1} = 1$

Note that in part **b** a preliminary simplification is made before taking the limit. This avoids a zero denominator. Such simplifications are often made when evaluating limits. ●

Theorem 2 Further Limit Rules

Suppose both limits

$$\lim_{x \to a} f(x) = L \quad \text{and} \quad \lim_{x \to a} g(x) = M$$

exist. Then:

5 If $f(x) \geq 0$ for all x in the domain of $f(x)$, then $L \geq 0$.
If $L > 0$, then there is a $\delta > 0$ such that $f(x) > 0$ for all x in the domain of $f(x)$ such that $0 < |x - a| < \delta$.

6 If $f(x) \geq g(x)$ for all x in the common domain of $f(x)$ and $g(x)$, then $L \geq M$.

7 If $f(x) \geq 0$ for all x in the domain of $f(x)$, then

$$\lim_{x \to a} \sqrt{f(x)} = \sqrt{L}$$

Rules 1–7 apply to the one-sided limits $\lim_{x \to a+}$ and $\lim_{x \to a-}$ also. As noted on page 79, all properties of limits hold equally well for regular limits and for one-sided limits.

Notice also that the conclusion given in rule 7 can also be written $\lim_{x \to a} f(x)^{1/2} = L^{1/2}$. Later we shall show more generally that we have $\lim_{x \to a} f(x)^r = L^r$ for any rational r. (Please do not use this fact yet to solve exercises.)

Example 2

Consider $f(x) = 1 - x^2$ with domain the closed interval $[-1, 1]$. On this domain, $f(x) \geq 0$.

a $\lim_{x \to 0} \sqrt{1 - x^2} = \sqrt{1 - 0} = 1$

b $\lim_{x \to 1-} \sqrt{1 - x^2} = \sqrt{1 - 1} = 0$

c $\lim_{x \to -1+} \sqrt{1 - x^2} = \sqrt{1 - 1} = 0$

The limit notation in **b** and **c** makes explicit the inherent restrictions on the domain of the function $y = \sqrt{1 - x^2}$. However, we could have written **b** in the form

$$\lim_{x \to 1} \sqrt{1 - x^2}$$

because the given domain of $f(x)$ contains *only* numbers to the left of $x = 1$. In **b**, the notation $x \to 1-$ in the limit provides emphasis and reminds us of the restriction on x.

Example 3

$$\lim_{x \to a} |x| = \lim_{x \to a} \sqrt{x^2} = \sqrt{a^2} = |a|$$

Limits as $x \to +\infty$ and $x \to -\infty$

Suppose $f(x)$ is defined on a domain that includes arbitrarily large real numbers, and suppose L is a real number. We define

$$\lim_{x\to+\infty} f(x) = L$$

to mean that whenever $\varepsilon > 0$, there is a real N such that

$$|f(x) - L| < \varepsilon$$

for all x in the domain of $f(x)$ such that $x > N$. See Figure 2-3-1.

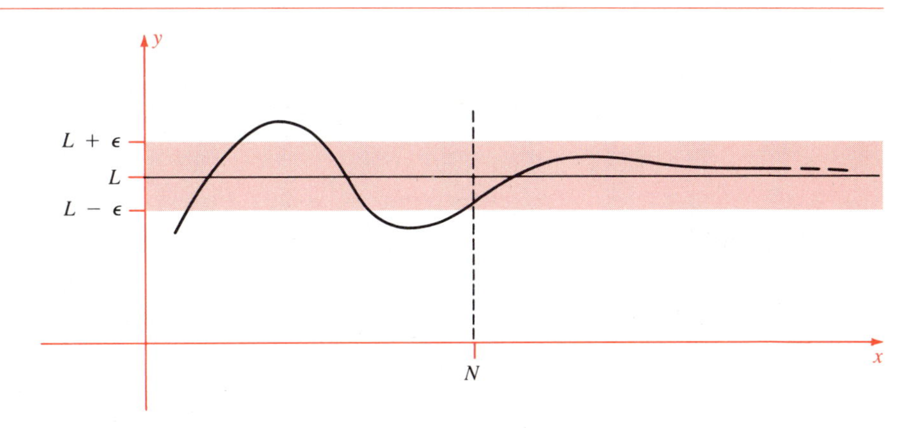

Figure 2-3-1
$\lim_{x\to+\infty} f(x) = L$ means that $f(x)$ can be restricted to any given strip

$$L - \varepsilon < f(x) < L + \varepsilon$$

by making $x > N$ for a suitable N that depends on ε.

We shall use this definition later. Right now it is worth mentioning that the rules for limits in Theorems 1 and 2 are also valid for $\lim_{x\to+\infty}$. The proofs in Section 2-4 require only slight rewordings to be valid for $\lim_{x\to+\infty}$. Finally, a similar definition applies to $x \to -\infty$. (The only change is to restrict x to $x < N$ rather than $x > N$.)

Example 4

Show that

a $\displaystyle\lim_{x\to+\infty} \frac{1}{x} = 0$ **b** $\displaystyle\lim_{x\to+\infty} \frac{3x+1}{x-2} = 3$

Solution

a Let $f(x) = 1/x$ and suppose $\varepsilon > 0$. We want to make

$$|f(x) - 0| < \varepsilon$$

by making x sufficiently large. For convenience, we take $N > 0$ so that $x > 0$ and

$$|f(x) - 0| = \left|\frac{1}{x}\right| = \frac{1}{x}$$

To make $1/x < \varepsilon$, we need only require that $x > 1/\varepsilon$, so we choose $N = 1/\varepsilon$. Then $|f(x) - 0| < \varepsilon$ provided $x > N$. This proves that $\lim_{x\to+\infty} 1/x = 0$.

b We first do a little algebraic manipulation, then apply the rules of limits and $\lim_{x\to+\infty} 1/x = 0$:

$$\lim_{x\to+\infty} \frac{3x+1}{x-2} = \lim_{x\to+\infty} \frac{3+(1/x)}{1-2(1/x)} = \frac{\lim_{x\to+\infty} [3+(1/x)]}{\lim_{x\to+\infty} [1-2(1/x)]} = \frac{3}{1} = 3$$

●

Exercises

Use the theorems on limits to find the required limit

1 $\lim_{x\to a} x^3$

2 $\lim_{x\to 0} (3+x)^3$

3 $\lim_{x\to 0} \dfrac{1}{1+x}$

4 $\lim_{x\to 0} \dfrac{1+x}{1-x}$

5 $\lim_{x\to 0} \dfrac{1}{(x-1)^3}$

6 $\lim_{x\to -1} \dfrac{1+2x}{1-x}$

7 $\lim_{t\to 2} (t^2-1)$

8 $\lim_{t\to 3} (t^3+3)$

9 $\lim_{h\to 1} \dfrac{1-2h}{1+h}$

10 $\lim_{y\to -1} \dfrac{1+2y}{1-2y}$

11 $\lim_{y\to 0} \dfrac{y}{(1+2y)^2-1}$

12 $\lim_{h\to 0} \dfrac{h}{(2+h)^3-8}$

13 $\lim_{z\to 1} \dfrac{z-1}{z^4-1}$

14 $\lim_{x\to -2} \dfrac{x^6-64}{x^3+8}$

15 $\lim_{x\to -2} \dfrac{\sqrt{x+3}}{x+1}$

16 $\lim_{t\to 2} \dfrac{1-\sqrt{t}}{\sqrt{t^2-2}-t}$

17 $\lim_{x\to 1} \dfrac{1}{1+\dfrac{1}{1+x}}$

18 $\lim_{x\to 1} \dfrac{1}{1+\dfrac{1}{1-x}}$

Find

19 $\lim_{x\to 0+} x^{5/4}$ [Hint $x^{1/4} = \sqrt{\sqrt{x}}$]

20 $\lim_{x\to 0-} \dfrac{2x+|x|}{x-|x|}$ [Hint $x < 0$ implies $|x| = ?$]

21 $\lim_{x\to 1+} \dfrac{|x-1|}{x-1}$

22 $\lim_{x\to +\infty} \dfrac{x^2}{x^2+2}$

23 $\lim_{x\to +\infty} \dfrac{x^2}{x^3+1}$

24 $\lim_{x\to -\infty} \dfrac{x+1}{x}$

25 Suppose $\lim_{x\to 0} [f(x)/x] = L$. Find $\lim_{x\to 0} f(x)$.

26 Suppose $f(x) \ge 0$ and $f(x)^2 + 6f(x) = x^2$. Find $\lim_{x\to 1} f(x)$.

2-4 More on Limits [Optional]

Let us look at the definition of limit from a slightly different point of view. If $f(x) \to L$ as $x \to a$, then the difference between $f(x)$ and L becomes smaller and smaller as $x \to a$. In other words, the difference approaches 0. Think of L as an approximation to $f(x)$ for x near a. The error in this approximation is

$$E(x) = f(x) - L$$

so $f(x)$ is the sum of the approximation L and the error $E(x)$:

$$f(x) = L + E(x)$$

Because $|f(x) - L| = |E(x)|$, we see that

$$\lim_{x \to a} f(x) = L \quad \text{if and only if} \quad \lim_{x \to a} E(x) = 0$$

Proofs of the Limit Rules

Our proofs of the limit rules will be based on several lemmas that pave the way for us. In order to avoid cluttering the proofs with distracting details we shall take two short cuts. First, we shall not mention domains. Second, when we are interested in $\lim_{x \to a} f(x)$, it doesn't matter how $f(x)$ behaves far away from a. Once we have $f(x)$ behaving properly for x very close to a, we can ignore what $f(x)$ does outside of this very small vicinity (neighborhood) of a.

Lemma 1 Suppose $\lim_{x \to a} f(x) = 0$ and $\lim_{x \to a} g(x) = 0$. Then

$$\lim_{x \to a} [f(x) + g(x)] = 0 \quad \text{and} \quad \lim_{x \to a} [f(x) - g(x)] = 0$$ ●

Lemma 2 Suppose B is a constant,

$$\lim_{x \to a} f(x) = 0 \quad \text{and} \quad |g(x)| \le B$$

Then $\lim_{x \to a} [f(x)g(x)] = 0$. ●

Proof of Lemma 1 Suppose $\varepsilon > 0$; we are challenged to make $|[f(x) \pm g(x)] - 0|$ less than ε. Since $\lim_{x \to a} f(x) = 0$, we can choose $\delta_1 > 0$ so that $|f(x)| = |f(x) - 0| < \frac{1}{2}\varepsilon$ whenever $0 < |x - a| < \delta_1$. Similarly we can choose δ_2 so that $|g(x)| < \frac{1}{2}\varepsilon$ whenever $0 < |x - a| < \delta_2$. Now let δ be the *smaller* of δ_1 and δ_2. Then if $0 < |x - a| < \delta$, *both*

$$|f(x)| < \tfrac{1}{2}\varepsilon \quad \text{and} \quad |g(x)| < \tfrac{1}{2}\varepsilon$$

By the triangle inequality (page 5)

$$\begin{aligned} |[f(x) \pm g(x)] - 0| &= |f(x) \pm g(x)| \\ &\le |f(x)| + |g(x)| < \tfrac{1}{2}\varepsilon + \tfrac{1}{2}\varepsilon = \varepsilon \end{aligned}$$

for all such x. This completes the proof of Lemma 1. ●

Proof of Lemma 2 If $B = 0$, then $g(x) = 0$ and the conclusion is obvious. Assume $B > 0$ and suppose $\varepsilon > 0$. We are challenged to make $|f(x)g(x) - 0|$ less than ε. Since $\lim f(x) = 0$, we can choose $\delta > 0$ so that $|f(x)| < \varepsilon/B$ whenever $0 < |x - a| < \delta$. Then

$$|f(x)g(x) - 0| = |f(x)g(x)| = |f(x)||g(x)| < (\varepsilon/B)B = \varepsilon$$

This completes the proof of Lemma 2. ●

The proofs that follow all have the same basic structure. First we compute the error in estimating a function by the constant that we think is its limit as $x \to a$. We must prove that this error approaches 0. The error will turn out to be bounded by a sum of terms, and we prove that each separately approaches 0, usually by using Lemma 2. Then Lemma 1 completes the job.

Rule 1 If $\lim_{x\to a} f(x) = L$, then $\lim_{x\to a} [cf(x)] = cL$

Proof This is the easiest of the proofs. Write

$$f(x) = L + F(x)$$

where $F(x)$ is the error term for $f(x)$, so $\lim_{x\to a} F(x) = 0$. Then

$$cf(x) = cL + E(x)$$

where the error term for $cf(x)$ is $E(x) = cF(x)$. By Lemma 2, $\lim E(x) = \lim [cF(x)] = 0$; hence $\lim [cf(x)] = cL$. ●

Rule 2 If $\lim_{x\to a} f(x) = L$ and $\lim_{x\to a} g(x) = M$, then

$$\lim_{x\to a} [f(x) \pm g(x)] = L \pm M$$

Proof Write

$$f(x) = L + F(x) \qquad g(x) = M + G(x)$$

where $\lim_{x\to a} F(x) = 0$ and $\lim_{x\to a} G(x) = 0$. Add (or subtract):

$$f(x) \pm g(x) = (L \pm M) + E(x)$$

with the new error term $E(x) = F(x) \pm G(x)$. By Lemma 1, $\lim_{x\to a} E(x) = 0$. Therefore

$$\lim_{x\to a} [f(x) \pm g(x)] = L \pm M$$ ●

Rule 3 If $\lim_{x\to a} f(x) = L$ and $\lim_{x\to a} g(x) = M$, then

$$\lim_{x\to a} [f(x)g(x)] = LM$$

Proof Write

$$f(x) = L + F(x) \qquad g(x) = M + G(x)$$

where $\lim F(x) = 0$ and $\lim G(x) = 0$. Multiply these expressions:

$$f(x)g(x) = LM + E(x)$$

with the new error term $E(x) = LG(x) + F(x)M + F(x)G(x)$.

By Lemma 2,

$$\lim_{x\to a}[LG(x)] = 0 \quad \text{and} \quad \lim_{x\to a}[F(x)M] = 0$$

Since $F(x) \to 0$, we have $|F(x)| < 1$ if $|x - a|$ is sufficiently small. Hence by Lemma 2 again,

$$\lim_{x\to a}[F(x)G(x)] = 0$$

Thus $E(x)$ is the sum of three terms, each of which has limit 0 as $x \to a$. By Lemma 1, $\lim_{x\to a} E(x) = 0$. But $E(x)$ is the error in the approximation of $f(x)g(x)$ by LM. Thus $f(x)g(x) \to LM$. ●

To prove rule 4 we have to control the size of a denominator—to make sure it isn't too small. We need another lemma for this.

Lemma 3 Suppose $\lim_{x\to a} g(x) = M$ and $M \neq 0$. Then there is a $\delta > 0$ such that

$$\left|\frac{1}{g(x)}\right| < \frac{2}{|M|} \quad \text{whenever} \quad 0 < |x - a| < \delta$$ ●

Proof To make $|1/g(x)| < 2/|M|$, we need to make $|g(x)| > \frac{1}{2}|M|$. To do so, we define $G(x)$ by $g(x) = M + G(x)$, so that $\lim_{x\to a} G(x) = 0$. Now we have been challenged with $\varepsilon = \frac{1}{2}|M| > 0$, so by the definition of $\lim_{x\to a} G(x) = 0$, there exists a $\delta > 0$ such that $|G(x)| < \varepsilon = \frac{1}{2}|M|$ whenever $0 < |x - a| < \delta$. This δ does the trick. Indeed, by the triangle inequality,

$$|M| = |g(x) - G(x)| \le |g(x)| + |G(x)| < |g(x)| + \tfrac{1}{2}|M|$$

Consequently $|g(x)| > |M| - \frac{1}{2}|M| = \frac{1}{2}|M|$ as required. This completes the proof of Lemma 3. ●

Rule 4 If $\lim_{x\to a} f(x) = L$ and $\lim_{x\to a} g(x) = M$, where $M \neq 0$, then $\lim_{x\to a}[f(x)/g(x)] = L/M$.

Proof Write $f(x) = L + F(x)$ and $g(x) = M + G(x)$, where $\lim_{x\to a} F(x) = 0$ and $\lim_{x\to a} G(x) = 0$. Then

$$\begin{aligned}
\frac{f(x)}{g(x)} - \frac{L}{M} &= \frac{L + F(x)}{M + G(x)} - \frac{L}{M} \\
&= \frac{[L + F(x)]M - L[M + G(x)]}{M[M + G(x)]} \\
&= \frac{MF(x) - LG(x)}{Mg(x)} \\
&= \frac{1}{g(x)}F(x) - \frac{L}{Mg(x)}G(x)
\end{aligned}$$

Therefore

$$\frac{f(x)}{g(x)} = \frac{L}{M} + E(x)$$

where our new error term is

$$E(x) = \frac{1}{g(x)}\,F(x) - \frac{L}{Mg(x)}\,G(x)$$

Since $M \neq 0$, Lemma 3 applies: $|1/g(x)| < 2/|M|$ for all x sufficiently close to a. It follows by Lemma 2 that

$$\lim_{x \to a}\left[\frac{1}{g(x)}\,F(x)\right] = 0$$

Similarly for all x near enough to a,

$$\left|\frac{L}{Mg(x)}\right| < \left|\frac{L}{M}\right|\left|\frac{M}{2}\right| \qquad \text{so} \qquad \lim_{x \to a}\left[\frac{L}{Mg(x)}\,G(x)\right] = 0$$

By Lemma 1, $\lim_{x \to a} E(x) = 0$. This means

$$\lim_{x \to a} \frac{f(x)}{g(x)} = \frac{L}{M}$$

which completes the proof. ●

Rule 5 There are two statements in rule 5; the first is

If $f(x) \geq 0$ and $\lim_{x \to a} f(x) = L$, then $L \geq 0$.

Proof This proof has a different flavor from the previous proofs because we are proving a property of a limit rather than finding its value.

Take any $\varepsilon > 0$. Then $|f(x) - L| < \varepsilon$ if x is sufficiently close to a, that is,

$$L - \varepsilon < f(x) < L + \varepsilon$$

We need only half of this: $f(x) < L + \varepsilon$. Since $f(x) \geq 0$, we deduce that $0 < L + \varepsilon$ for all $\varepsilon > 0$. To eliminate the possibility that $L < 0$, we then simply take $\varepsilon = -L$ to obtain the contradiction $0 < 0$. Therefore $L \geq 0$. This completes the proof of the first part of rule 5. ●

The second part of rule 5 says:

If $L > 0$, then there is a $\delta > 0$ such that $f(x) > 0$ for all x in the domain of $f(x)$ satisfying $0 < |x - a| < \delta$.

Proof For any $\varepsilon > 0$, there is a $\delta > 0$ such that

$$L - \varepsilon < f(x) < L + \varepsilon$$

for $0 < |x - a| < \delta$. But if we choose $\varepsilon = L$, then $L - \varepsilon = 0$. Therefore $f(x) > 0$ for $0 < |x - a| < \delta$ and x in the domain of $f(x)$. This completes the proof of the second part of rule 5. ●

Rule 6 If $f(x) \geq g(x)$, $\lim_{x \to a} f(x) = L$, and $\lim_{x \to a} g(x) = M$, then $L \geq M$.

Proof By rule 2, $\lim_{x \to a}[f(x) - g(x)] = L - M$. But we have $f(x) - g(x) \geq 0$, so $L - M \geq 0$ by rule 5. ●

Rule 7 If $f(x) \geq 0$ and $\lim_{x \to a} f(x) = L$, then

$$\lim_{x \to a} \sqrt{f(x)} = \sqrt{L}$$

Proof Note that $L \geq 0$ by rule 5, so $\sqrt{L}$ exists. We must split the proof into two cases, so we shall take the easier case first. Note in both cases that ingenuity is required.

Case 1 $L = 0$ We have $\lim_{x \to a} f(x) = 0$. Choose any $\varepsilon > 0$. We slyly apply the definition of $\lim_{x \to a} f(x) = 0$, but with the positive number ε^2 in place of the usual ε. There exists a $\delta > 0$ such that $f(x) < \varepsilon^2$ whenever $0 < |x - a| < \delta$. Then we have

$$0 \leq \sqrt{f(x)} < \sqrt{\varepsilon^2} = \varepsilon$$

Therefore $\lim_{x \to a} \sqrt{f(x)} = 0 = \sqrt{L}$.

Case 2 $L > 0$ As usual, $f(x) = L + F(x)$ where $\lim_{x \to a} F(x) = 0$. We write

$$\sqrt{f(x)} = \sqrt{L} + E(x)$$

and our problem is to show that $E(x) \to 0$ as $x \to a$. Our approach is to express $E(x)$ in such a way that $F(x)$ appears as a factor. We write

$$F(x) = f(x) - L = (\sqrt{f(x)} + \sqrt{L})(\sqrt{f(x)} - \sqrt{L})$$
$$= (\sqrt{f(x)} + \sqrt{L})E(x)$$

Thus the error in approximating $\sqrt{f(x)}$ by $\sqrt{L}$ is

$$E(x) = \frac{1}{\sqrt{f(x)} + \sqrt{L}} F(x)$$

Now $\sqrt{f(x)} + \sqrt{L} \geq \sqrt{L}$, so

$$0 < \frac{1}{\sqrt{f(x)} + \sqrt{L}} \leq \frac{1}{\sqrt{L}}$$

By Lemma 2, $\lim_{x \to a} E(x) = 0$. This completes the proof. ●

Exercises

Find the limit and give a reason for your conclusion

1 $\lim_{x\to 2} \frac{1}{x}$

2 $\lim_{x\to 11} \sqrt{2x+3}$

3 $\lim_{x\to -2} \frac{1}{x^2+1}$

4 $\lim_{x\to 0} |x|$

5 $\lim_{x\to 0} \frac{x^2}{x^2}$

6 $\lim_{x\to 1} \frac{x^2-1}{x-1}$

7 $\lim_{x\to 0} \frac{x^2}{x}$

8 $\lim_{x\to 1}(3x^5 - 7x^3 + 2x^2 + x - 5)$

9 $\lim_{x\to 0} \sqrt{1+x^3}$

10 $\lim_{x\to 0+} (\sqrt{x} - x)$

11 $\lim_{h\to 0} \frac{3+h+h^2}{5-4h+h^3}$

12 $\lim_{h\to 0} \sqrt{1+h^2}$

13 Let $f(x) = 1$ for $x \geq 0$ and $f(x) = 0$ for $x < 0$. Show that $\lim_{x\to 0+} f(x)$ and $\lim_{x\to 0-} f(x)$ both exist. Does $\lim_{x\to 0} f(x)$ exist? If not, why not?

14 Suppose the domain of $f(x)$ is $(-1, 1)$ and $\lim_{x\to 0} f(x) = L$. Prove $\lim_{x\to 0+} f(x) = L$.

15 Suppose both one-sided limits exist and $\lim_{x\to a-} f(x) = \lim_{x\to a+} f(x)$. Prove that $\lim_{x\to a} f(x)$ exists and equals the one-sided limits.

16 Define $f(x) = \sqrt{x}$ for $x > 0$ and $f(x) = x$ for $x < 0$. Prove $\lim_{x\to 0} f(x)$ exists and find it. [Hint Use Exercise 15.]

17 Prove $x \sin(1/x) \to 0$ as $x \to 0$.

18 Suppose $f(x) \to L$ as $x \to a$. Prove $f(x)^3 \to L^3$ as $x \to a$.

19 Find $\delta > 0$ so $\left|\frac{h}{3+h}\right| < 10^{-5}$ provided $|h| < \delta$.

20 Find $\delta > 0$ so $\left|\frac{1-h}{2+h} - \frac{1}{2}\right| < 10^{-4}$ provided $|h| < \delta$.

21 Suppose $\lim_{x\to a} f(x) > 10$. Prove $f(x) > 10$ for x sufficiently near a, but different from a.

22 Prove $\lim_{x\to 0} \frac{x^2}{x + \sin x} = 0$.

2-5 Continuous Functions

In this section we shall study a type of function that is extremely important in calculus, the *continuous* function. In our work, the domain of a continuous function will always be an interval, or a union of intervals. We introduced finite closed intervals $[a, b] = \{x \mid a \leq x \leq b\}$ and open intervals $(a, b) = \{x \mid a < x < b\}$ in Chapter 1. Now we make precise what we mean in general by an interval. Intuitively, an interval is either the whole real line, or what you can draw on the real line by starting someplace and either continuing indefinitely (a ray) or stopping (a finite interval). The end points may or may not be included. The following formal definition covers all cases.

Definition Interval

An **interval** is a set I of real numbers such that:

1 I contains at least two points.

2 If a and b are points of I, then every point between a and b is a point of I.

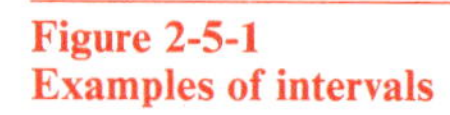

Figure 2-5-1
Examples of intervals

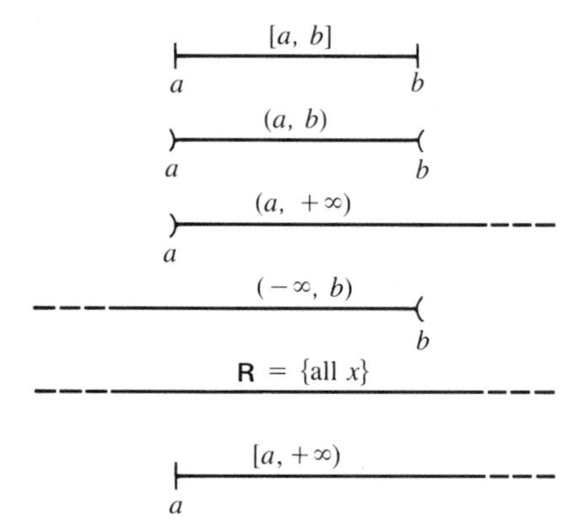

The real line **R** itself is an interval. Typical intervals are finite open intervals like $(-3, 12)$, finite closed intervals like $[-3, 12]$, finite half-open intervals like $(-3, 12] = \{x \mid -3 < x \leq 12\}$, open semi-infinite intervals like $(-3, +\infty) = \{x \mid -3 < x\}$, and so on. Figure 2-5-1 shows some of the possibilities.

In this part of calculus, the calculus of functions of a single real variable, each function that we work with has as its domain either an interval, or the union of finitely many intervals, or even the union of an infinite sequence of intervals.

Example 1

a $f(x) = \sqrt{1 - x^2}$ $\quad D = [-1, 1]$
b $f(x) = \sqrt{x^2 - 1}$ $\quad D = (-\infty, -1] \cup [1, +\infty)$
c $f(x) = 1/x^2$ $\quad D = (-\infty, 0) \cup (0, +\infty)$
d $f(x) = \csc x$ $\quad D = \{x \mid x \neq n\pi \text{ for any interger } n\}$

(See Figure 1-7-5, page 59, for a graph of $y = \csc x$.) ●

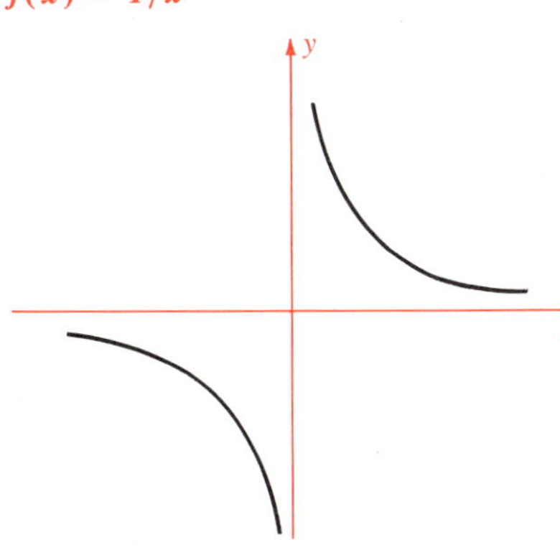
Figure 2-5-2
$f(x) = 1/x$

Continuity

The following is a fundamental definition in calculus:

Definition Continuous Function

Let $f(x)$ be a function whose domain D is an interval, or a union of intervals.

1 Let a be a point of D. Then $f(x)$ is **continuous at** a if

$$\lim_{x \to a} f(x) = f(a)$$

2 $f(x)$ is **continuous on** D if $f(x)$ is continuous at each point D. ●

To say that $f(x)$ is continuous at a means that the value $f(a)$ is "predictable" from the values of $f(x)$ near $x = a$. Namely, $\lim_{x \to a} f(x)$ must exist, and it must equal the "correct" value, $f(a)$.

To say that $f(x)$ is continuous on an *interval* I means, intuitively, that you can draw the graph of $y = f(x)$ without lifting your pencil. The word for continuous in some languages (Russian, for instance) translates literally as "unbroken," a term that expresses the concept well.

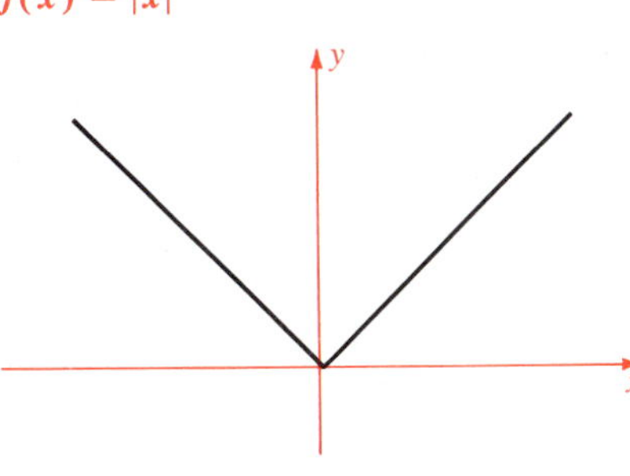
Figure 2-5-3
$f(x) = |x|$

Example 2

a $f(x) = 1/x \quad D = \{x \mid x \neq 0\} = (-\infty, 0) \cup (0, +\infty)$
$f(x)$ is continuous on D (Figure 2-5-2).
Note that 0 is not in the domain of $f(x)$.

b $f(x) = |x| \quad D = \mathbf{R}$
$f(x)$ is continuous (Figure 2-5-3).
Note that $\lim_{x \to 0} f(x) = \lim_{x \to 0} |x| = 0 = f(0)$.

c $f(x) = \begin{cases} 0 & \text{for } x < 0 \\ \frac{1}{2} & \text{for } x = 0 \\ 1 & \text{for } x > 0 \end{cases} \quad D = \mathbf{R}$

The graph (Figure 2-5-4) suggests that $f(x)$ is *not* continuous because $f(x)$ is not continuous at $x = 0$. This is the case because $\lim_{x \to 0} f(x)$ does not even exist.

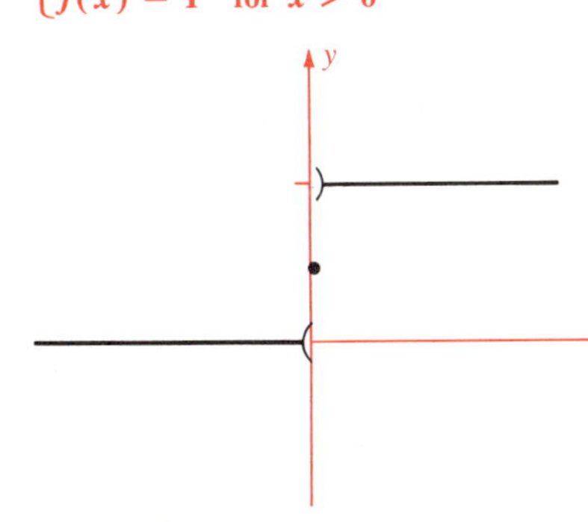
Figure 2-5-4
$\begin{cases} f(x) = 0 & \text{for } x < 0 \\ f(0) = \frac{1}{2} \\ f(x) = 1 & \text{for } x > 0 \end{cases}$

d $f(x) = \begin{cases} 2x & \text{for } x \neq 1 \\ -1 & \text{for } x = 1 \end{cases} \quad D = \mathbf{R}$

The graph (Figure 2-5-5, next page) suggests that $f(x)$ is *not* continuous. This time the limit exists at every point of the function's domain, but $\lim_{x \to 1} f(x) = \lim_{x \to 1} 2x = 2$, which differs from $f(1) = -1$. Thus

Figure 2-5-5
$\begin{cases} f(x) = 2x & \text{for } x \neq 1 \\ f(1) = -1 \end{cases}$

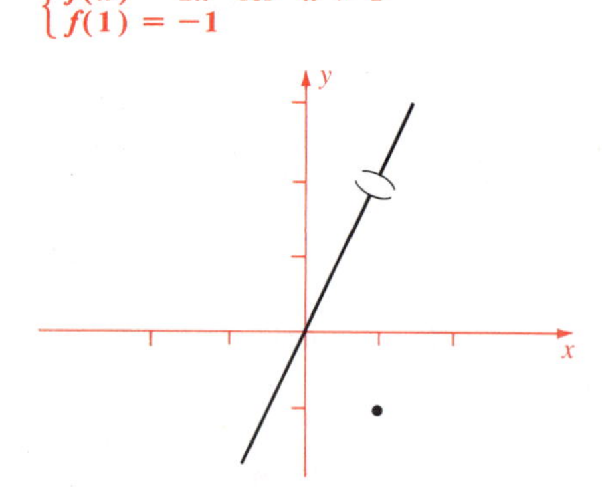

$\lim_{x\to 1} f(x) \neq f(1)$. The condition for continuity at $x = 1$ is not satisfied, because $f(1)$ is not predictable from values of $f(x)$ arbitrarily close to 1. ●

Terminology In a situation like Example 2d we say "$f(x)$ is continuous except for $x = 1$." What this means is that if the domain of $f(x)$ is restricted by chopping out the single point $x = 1$, then the function on the resulting smaller domain is continuous. Similarly, in Example 2c we say "$f(x)$ is continuous for $x \neq 0$."

Continuity is a really important new idea, which you *must* master. It is important to grasp the precise definitions. A careful reading of this section with attention to the definitions and examples will give you a good start. Then go back to Section 2-2, on limits, and check the similar points there, making sure you understand how they relate to continuity.

In what follows, the domains are always intervals or unions of intervals; that is understood. We shall not spell this out each time.

Theorem 1

Let $f(x)$ and $g(x)$ be continuous on the same domain D and let c be a constant. Then:

1 $cf(x)$ is continuous.
2 $f(x) \pm g(x)$ is continuous.
3 $f(x)g(x)$ is continuous.
4 If $g(x)$ is never 0 on D, then $\dfrac{f(x)}{g(x)}$ is continuous. ●

Each of these statements is a direct consequence of the corresponding fact about limits in Theorem 1 in Section 2-3. For instance, the proof of 3 is

$$\lim_{x\to a}[f(x)g(x)] = [\lim_{x\to a} f(x)][\lim_{x\to a} g(x)] = f(a)g(a) = (fg)(a)$$

We can use Theorem 1 of this section to construct new continuous functions out of known ones. Our initial building blocks will be constant functions $f(x) = c$ and the identity function $f(x) = x$. A constant function $f(x) = c$ is continuous because

$$\lim_{x\to a} f(x) = \lim_{x\to a} c = c = f(a)$$

The identity function $f(x) = x$ is continuous because

$$\lim_{x\to a} f(x) = \lim_{x\to a} x = a = f(a)$$

From these facts and repeated applications of Theorem 1, part 3, it follows that each monomial function

$$a_0 \qquad a_1x \qquad a_2x^2 \qquad a_3x^3 \qquad \ldots$$

is continuous. For instance, $x^2 = x \cdot x$ is the product of two continuous functions (both the identity function), hence is continuous; a_2x^2 is the product of two continuous functions (the constant a_2 and x^2), hence is continuous; and so on. By the continuity of monomial functions and by Theorem 1, part 2, the sum of several monomial functions is continuous. Therefore we have the following result:

Theorem 2

Each polynomial function

$$f(x) = a_n x^n + a_{n-1} x^{n-1} + \cdot \cdot \cdot + a_0$$

is continuous. ●

The next theorem concerns rational functions.

Theorem 3

Each rational function is continuous on its domain. ●

Proof A rational function (page 39) is a quotient of polynomials, and its domain is the set of real numbers where the denominator is not 0. Theorem 3 now follows from Theorem 2 and Theorem 1, part 4. ●

Example 3

Each of the functions

$$\frac{2x^3 - x + 1}{x^5 - 1} \qquad \frac{x^4 + 4}{x^2 + 4} \qquad \frac{x^3 - 3x^2 + 2x + 1}{x^2 - 7x + 9}$$

is continuous on its domain (where its denominator is not 0). ●

Functions like the following are also continuous:

$$\sqrt{x} \quad \text{on} \quad [0, +\infty) \qquad \sqrt{x^2 - 1} \quad \text{on} \quad (-\infty, -1] \cup [1, +\infty)$$

Theorem 4

If $f(x)$ is continuous and $f(x) \geq 0$, then $g(x) = \sqrt{f(x)}$ is continuous. ●

Proof We use Theorem 2, part 7, on page 82.

$$\lim_{x \to a} g(x) = \lim_{x \to a} \sqrt{f(x)} = \sqrt{\lim_{x \to a} f(x)} = \sqrt{f(a)} = g(a)$$

Therefore $g(x)$ is continuous. ●

Example 4

The following functions are continuous:

a $\sqrt{x^2 + 1}$ on $\mathbf{R}$

b $\sqrt{\dfrac{x}{x^4 + 6x^2 + 1}}$ on $[0, +\infty)$

c $\dfrac{3x}{x - 1} + \sqrt{\dfrac{x^2 + 1}{x + 1}}$ on $(-1, 1) \cup (1, +\infty)$ ●

Remark Theorem 4 asserts that $f(x)^{1/2}$ is continuous if $f(x) \geq 0$ and $f(x)$ is continuous. More generally, we shall show later that $f(x)^r$ is continuous under the same conditions. For instance, $(x^2 + 1)^{1/3}$ is continuous for all x, and $x^{-2/5}$ is continuous for $x > 0$.

Trigonometric Functions

Let us, for a moment, take for granted the relations

$$\lim_{h\to 0} \sin h = 0 \qquad \lim_{h\to 0} \cos h = 1$$

Given these facts, we can prove that $\sin x$ and $\cos x$ are continuous functions. To do so, we must use the addition laws for sine and cosine (page 52). For $\sin x$ we have

$$\begin{aligned}
\lim_{x\to a} \sin x &= \lim_{h\to 0} \sin(a + h) \\
&= \lim_{h\to 0} (\sin a \cos h + \cos a \sin h) \\
&= (\sin a)(\lim_{h\to 0} \cos h) + (\cos a)(\lim_{h\to 0} \sin h) \\
&= (\sin a) \cdot 1 + (\cos a) \cdot 0 = \sin a
\end{aligned}$$

Hence $\sin x$ is continuous. The proof for the continuity of $\cos x$ is similar.

It follows from part 4 of Theorem 1 that

$$\tan x = \frac{\sin x}{\cos x} \qquad \csc x = \frac{1}{\sin x} \qquad \text{and so on}$$

are all continuous functions on their domains.

Theorem 5

Each of the functions

$$\sin x \qquad \cos x \qquad \tan x \qquad \csc x \qquad \sec x \qquad \cot x$$

is continuous on its domain.

Example 5

By Theorem 5, the following functions are continuous:

a $4 \sin x + \cos x$
b $x^3 \sin^2 x - 3 \cos^3 x$
c $\tan^3 x$ for $x \neq (n + \frac{1}{2})\pi$, where n is any integer
d $\sqrt{\cos x}$ on $[-\frac{1}{2}\pi, \frac{1}{2}\pi]$ (and wherever $\cos x \geq 0$)
e $\sqrt{x + \sin x}$ on $[0, +\infty)$

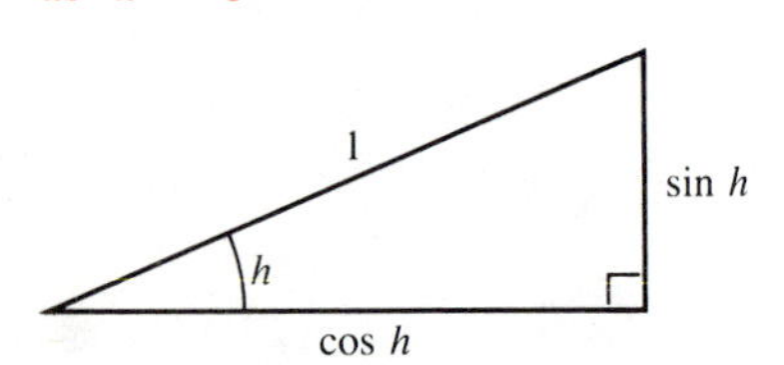

Figure 2-5-6
$\sin h \to 0$ as $h \to 0$
$\cos h = \sqrt{1 - \sin^2 h} \to \sqrt{1 - 0} = 1$ as $h \to 0$

Limits of Sine and Cosine

The limits $\lim_{t\to 0} \sin t = 0$ and $\lim_{t\to 0} \cos t = 1$ are intuitively clear (Figure 2-5-6). They are also strongly suggested by calculator evidence. Set the mode to *radians.* Then

Figure 2-5-7
$|\text{chord}| < |\text{arc}|$
$2 \sin h < 2h$
$\sin h < h$

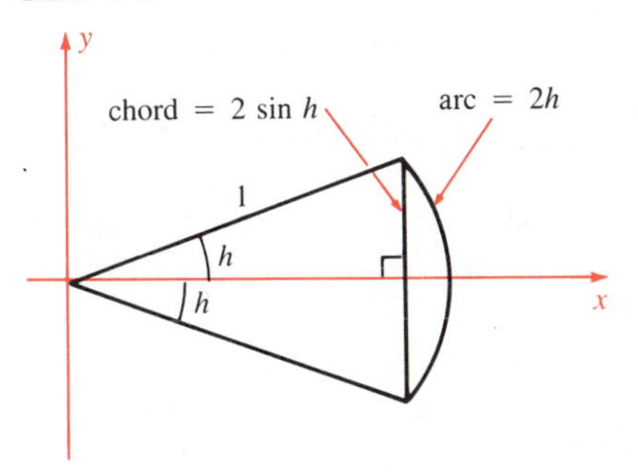

$\sin 0.10 \approx 0.998$ $\qquad$ $\cos 0.10 \approx 0.995$

$\sin 0.05 \approx 0.04998$ $\qquad$ $\cos 0.05 \approx 0.9988$

$\sin 0.01 \approx 0.00999\,98$ $\qquad$ $\cos 0.01 \approx 0.99995$

For a proof, we start with Figure 2-5-7, from which the inequality $0 < \sin h < h$ follows for $0 < h < \frac{1}{2}\pi$. (You have to believe that the shortest path joining two points is the straight path.) From this and $\sin(-h) = -\sin h$ we obtain $|\sin h| < |h|$ for $|h| < \frac{1}{2}\pi$. It follows that $\lim_{h\to 0} \sin h = 0$.

Next,

$$1 - \cos h = \frac{(1 - \cos h)(1 + \cos h)}{1 + \cos h} = \frac{\sin^2 h}{1 + \cos h}$$

Because

$$1 + \cos h > 1 \quad \text{for} \quad -\tfrac{1}{2}\pi < h < \tfrac{1}{2}\pi$$

we have

$$|1 - \cos h| = \frac{1}{|1 + \cos h|}|\sin^2 h| < |\sin h|^2 < |h|^2$$

for $0 < |h| < \frac{1}{2}\pi$. It follows from this inequality that $\lim_{h\to 0} \cos h = 1$.

Continuity of Composite Functions

The preceding theorems allow us to spot many continuous functions at a glance. For example

$$f(x) = \frac{\tan x}{\sqrt{x^2 + \sin^2 x}} \quad \text{and} \quad g(x) = \frac{\sin x + 5\cos x}{x^3 + \sqrt{\sec x}}$$

are continuous. Imagine trying from scratch to prove such functions continuous! Still, we can easily think of combinations not covered by these theorems, for instance

$$\sin(x^2) \quad \text{and} \quad \cos(2 - 3\sin x)$$

Each is a "continuous function of a continuous function." In other words, it is a composite of continuous functions. The next theorem covers all cases like this one. Before reading it, you may wish to review the definitions of *range* (page 13) and *composite function* (page 16) and Figure 1-2-11.

Theorem 6

Let $y = g(x)$ and $z = f(y)$ be continuous functions for which the range of $g(x)$ is contained in the domain of $f(y)$. Then the composite function

$$z = (f \circ g)(x) = f[g(x)]$$

is continuous. More precisely, if $g(x)$ is continuous at a point a and $f(y)$ is continuous at $b = g(a)$, then $(f \circ g)(x)$ is continuous at $x = a$.

Proof We must prove that

$$\lim_{x \to a} f[g(x)] = f[g(a)]$$

for each a in the domain D of $g(x)$. The previous limit theorems do not apply, so we go right back to the definition of continuity. (We shall use the Greek letter eta η; it rhymes with beta β and theta θ.)

In a nutshell, the idea of the proof is this: if y is sufficiently close to $g(a)$, then $f(y)$ is as close as we please to $f[g(a)]$, by the continuity of $f(y)$. But if x is sufficiently close to a, then $y = g(x)$ is indeed quite close to $g(a)$, by the continuity of $g(x)$. Therefore $f[g(x)]$ is as close as we please to $f[g(a)]$ provided x is sufficiently close to a. This is the essential idea; now we fill in the technical details.

Given a, we set $b = g(a)$. Now we let $\varepsilon > 0$. We want to make

$$|f[g(x)] - f[g(a)]| = |f[g(x)] - f(b)| < \varepsilon$$

by restricting x to be close to a. Because $f(y)$ is continuous at $y = b$, there exists an $\eta > 0$ such that

$$|f(y) - f(b)| < \varepsilon \qquad (*)$$

whenever $0 < |y - b| < \eta$. Actually, this is so whenever $|y - b| < \eta$, because $y = b$ is in the domain of $f(y)$.

Now a is in the domain of the continuous function $g(x)$, and $\eta > 0$, so there is a $\delta > 0$ such that

$$|g(x) - b| = |g(x) - g(a)| < \eta$$

whenever $0 < |x - a| < \delta$. Therefore if $|x - a| < \delta$, we can take $y = g(x)$ in relation $(*)$. We conclude that

$$|f[g(x)] - f(b)| < \varepsilon$$

whenever $0 < |x - a| < \delta$.

For each ε we have found a δ as required by the definition of limit, so we have proved

$$\lim_{x \to a} f[g(x)] = f(b) = f[g(a)]$$

Consequently $f[g(x)]$ is continuous at each point a of its domain. This completes the proof. ●

Example 6

The following functions are continuous:

a $h(x) = \sin[x/(1 + \cos x)]$ on $(-\pi, \pi)$

Here $h(x) = (f \circ g)(x)$ is the composite of $f(y) = \sin y$ and $g(x) = x/(1 + \cos x)$. Note that $g(x)$ is defined on $(-\pi, \pi)$ and $f(y)$ is defined on **R**.

b $h(x) = \tan\sqrt{x}$ on $[0, (\frac{1}{2}\pi)^2)$

Here $h(x) = (f \circ g)(x)$ with $f(y) = \tan y$ and $g(x) = \sqrt{x}$ with domain $[0, (\frac{1}{2}\pi)^2)$. The function $g(x)$ is continuous, and its range is $[0, \frac{1}{2}\pi)$, which is contained in the domain of $f(y)$. ●

Exponential Functions

We reviewed exponential functions in Section 1-8. We now show that they are continuous.

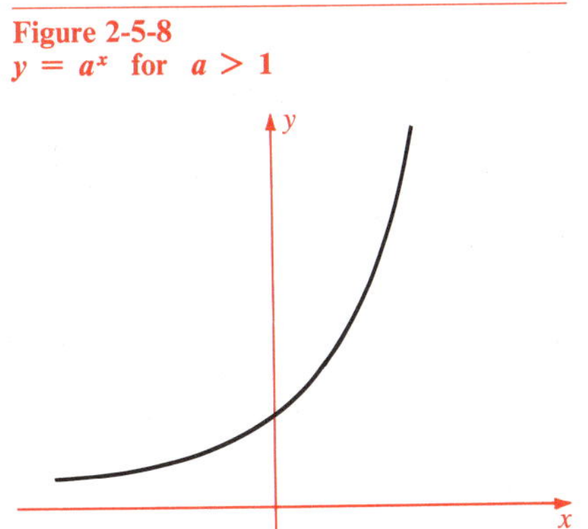

Figure 2-5-8
$y = a^x$ for $a > 1$

Theorem 7

Each exponential function $f(x) = a^x$ with $a > 0$ is continuous on the real line **R**. ●

Intuitively this statement seems correct because the graph of $y = a^x$ appears smooth (Figure 2-5-8). To prove it so we must show that

$$\lim_{x \to b} a^x = a^b$$

We use the rules of exponents (see page 64) to reduce the general case of continuity at $x = b$ to the special case of continuity at $x = 0$:

$$\lim_{x \to b} a^x = \lim_{h \to 0} a^{b+h} = \lim_{h \to 0} (a^b a^h) = a^b \lim_{h \to 0} a^h$$

Therefore to complete this proof, we require the relation $\lim_{h \to 0} a^h = 1$. At this stage, a complete proof of this limit formula would be technical and tedious, so we omit it. In Section 4-1, the limit formula will follow easily from other results. However, a few calculator experiments give convincing numerical evidence that this limit is correct. A partial proof will be sketched after the next example.

Example 7

Each of the functions

$$2^{\sin x} \qquad \frac{1}{x + 10^x} \qquad \cos(x^2 - 3x \cdot 5^{x^2 - x})$$

is continuous. ●

Partial Proof That a^x Is Continuous at $x = 0$ [Optional] To give a hint of how one might prove that $a^x \to 1$ as $x \to 0$, we'll sketch a special case: $a = 2$. Actually we'll prove only the one-sided limit

$$\lim_{x \to 0+} 2^x = 1$$

Because $2^{-x} = 1/2^x$, it is a short step from here to the two-sided limit. We need an inequality: If $x > 0$ and $n \geq 2$ is an integer, then

$$1 + nx < (1 + x)^n$$

This can be proved either by mathematical induction or by using the binomial theorem, which implies that $(1+x)^n = 1 + nx +$ (positive terms). We substitute $x = 1/n$ in the inequality to obtain

$$2 = 1 + n\frac{1}{n} < \left(1 + \frac{1}{n}\right)^n$$

Also $1 < 2$. Now we take n-th roots: $1 < 2^{1/n} < 1 + 1/n$. From here the proof is a routine ε-δ proof, based on the fact that 2^x is a strictly increasing function.

Suppose $\varepsilon > 0$. Choose an integer $n \geq 2$ such that $n > 1/\varepsilon$. Then $1/n < \varepsilon$. If $0 < x < 1/n$, then

$$1 < 2^x < 2^{1/n} < 1 + 1/n$$

Hence $0 < 2^x - 1 < 2^{1/n} - 1 < 1/n < \varepsilon$. This does the trick, with $\delta = 1/n$. ●

Exercises

Use only the continuity of constant functions and the identity function, and Theorems 1 and 4 to prove, step by step, the continuity of

1 $f(x) = x^2 - 4x + 6$

2 $f(x) = x^3 + 3x$

3 $f(x) = \dfrac{x}{x^2 + 3}$

4 $f(x) = \dfrac{x^2 + 1}{x^4 + 1}$

5 $f(x) = \sqrt{1 + x^2}$

6 $f(x) = \sqrt{x^6 + 3x^2 + 7}$

7 Use Theorems 2 and 4 to prove that $f(x) = |x|$ is continuous.

8 (cont.) Prove this directly from the definition of continuity.

9 Suppose $f(x)$ is continuous and $f(a) > 0$. Prove that there exists a $\delta > 0$ such that $f(x) > 0$ whenever $|x - a| < \delta$. [Hint Try $\varepsilon = f(a)$.]

10 Suppose there is a triangle T with sides a, b, c. Let $f(x)$ be the area of the triangle T_x with sides $a + x$, $b + x$, $c + x$. Prove $f(x)$ is continuous for $x \geq 0$. [Hint See Heron's formula in Figure 1-7-8, page 61.]

11 Let $f(x)$ denote the larger of the two roots of the quadratic equation $y^2 + xy - 3 = 0$. Prove that $f(x)$ is continuous on **R**.

12 (Splicing functions) Suppose $g(x)$ has domain $(-\infty, 0]$ and $h(x)$ has domain $[0, +\infty)$, and both $g(x)$ and $h(x)$ are continuous. Suppose also that $g(0) = h(0)$. Define $f(x)$ on **R** by $f(x) = g(x)$ if $x < 0$ and $f(x) = h(x)$ if $x \geq 0$. Prove $f(x)$ is continuous.

13 For $a > 0$, prove $\lim_{x \to a} \sqrt[3]{x} = \sqrt[3]{a}$. [Hint $x - a = (\sqrt[3]{x} - \sqrt[3]{a})(\sqrt[3]{x^2} + \sqrt[3]{a}\sqrt[3]{x} + \sqrt[3]{a^2})$]

14 (cont.) Prove $f(x) = \sqrt[3]{x}$ is continuous on **R**.

Numerical Exercises

15 $f(x) = x^2$ is continuous at $x = 3$, and $f(3) = 9$. Tabulate $f(x)$ for $x = 3 \pm h$, where

$h = 0.005 \quad 0.001 \quad 0.0005 \quad 0.0001$

16 (cont.) The evidence should indicate that

$$|(3 + h)^2 - 9| \leq 7|h|$$

if $|h|$ is small. Prove this by algebra.

17 $f(x) = \sqrt{x}$ is continuous at $x = 4$, and $f(4) = 2$. Tabulate $f(x)$ for $x = 4 \pm h$, where h has the values in Exercise 15.

18 (cont.) The evidence should indicate that

$$\sqrt{4 + h} \approx 2 + \tfrac{1}{4}h$$

if $|h|$ is small. Prove this by algebra.

19 The function $f(x) = \sin x$ is continuous at $x = \frac{1}{6}\pi$, and $f(\frac{1}{6}\pi) = \frac{1}{2}$. Set your calculator on *radians* and tabulate $f(\frac{1}{6}\pi \pm h)$ where h has the values in Exercise 15. Any conclusion on the relative sizes of $|\sin(\frac{1}{6}\pi + h) - \frac{1}{2}|$ and $|h|$ for $|h|$ small?

20 By Example 4, $f(x) = \sqrt{1 + x^2}$ is continuous for $x = 0$, and $f(0) = 1$. Tabulate $f(x)$ for $x = 0.05$, 0.01, 0.0005, 0.0001. Any conclusion on the relative sizes of $\sqrt{1 + x^2} - 1$ and $|x|^2$ for $|x|$ small?

2-6 Derivative of a Function

We are finally ready to make precise the idea of slope discussed in Section 2-1. It will be understood from now on that each function $f(x)$ that we consider has for its domain either an interval or a union of intervals. The slope of the graph of $y = f(x)$ at a point $(a, f(a))$ will be a number $f'(a)$, called the derivative of $f(x)$ at $x = a$, defined precisely as a limit.

Definition Derivative

Let a be a point of the domain of $f(x)$. The **derivative** of $f(x)$ at $x = a$ is the limit

$$f'(a) = \lim_{h \to 0} \frac{f(a+h) - f(a)}{h}$$

provided this limit exists. If it does exist, we say $f(x)$ is **differentiable** at $x = a$; otherwise $f(x)$ is **not differentiable** at $x = a$.

There is a useful alternative form of the limit defining the derivative. Replace $a + h$ by x, and note that $x \to a$ is equivalent to $h \to 0$:

Alternative Definition Derivative

$$f'(a) = \lim_{x \to a} \frac{f(x) - f(a)}{x - a}$$

Example 1

Find $f'(a)$ for

a $f(x) = x^2$ **b** $f(x) = x^3$ **c** $f(x) = 5x^3 - 7x^2$

Solution

a We use the definition of the derivative with results on limits proved earlier in this chapter.

$$f'(a) = \lim_{h \to 0} \frac{f(a+h) - f(a)}{h}$$

$$= \lim_{h \to 0} \frac{(a+h)^2 - a^2}{h} = \lim_{h \to 0} \frac{(a^2 + 2ah + h^2) - a^2}{h}$$

$$= \lim_{h \to 0} \frac{2ah + h^2}{h} = \lim_{h \to 0} (2a + h) = 2a$$

[Dividing out h is valid because, in general, $\lim_{h \to 0} g(h)$ does not depend on the value of g at $h = 0$.]

Another approach is to use the alternative definition of the derivative:

$$\lim_{x \to a} \frac{f(x) - f(a)}{x - a} = \lim_{x \to a} \frac{x^2 - a^2}{x - a} = \lim_{x \to a} \frac{(x+a)(x-a)}{x - a}$$

$$= \lim_{x \to a} (x + a) = a + a = 2a$$

Therefore for any real number a, we have $f'(a) = 2a$.

b We use the alternative definition of derivative:

$$f'(a) = \lim_{x \to a} \frac{f(x) - f(a)}{x - a} = \lim_{x \to a} \frac{x^3 - a^3}{x - a}$$

But

$$\frac{x^3 - a^3}{x - a} = \frac{(x^2 + ax + a^2)(x - a)}{x - a} = x^2 + ax + a^2$$

so

$$f'(a) = \lim_{x \to a} (x^2 + ax + a^2) = a^2 + a \cdot a + a^2 = 3a^2$$

Therefore $f'(a) = 3a^2$ for each real number a.

c We use the alternative definition of derivative:

$$\begin{aligned} \frac{f(x) - f(a)}{x - a} &= \frac{(5x^3 - 7x^2) - (5a^3 - 7a^2)}{x - a} \\ &= \frac{(5x^3 - 5a^3) - (7x^2 - 7a^2)}{x - a} \\ &= 5\frac{x^3 - a^3}{x - a} - 7\frac{x^2 - a^2}{x - a} \end{aligned}$$

To take the limit, we can combine what we already know from **a** and **b** with our limit rules (page 80). See how useful these rules are!

$$\begin{aligned} f'(a) &= \lim_{x \to a} \frac{f(x) - f(a)}{x - a} \\ &= 5 \cdot \lim_{x \to a} \frac{x^3 - a^3}{x - a} - 7 \cdot \lim_{x \to a} \frac{x^2 - a^2}{x - a} \\ &= 5 \cdot 3a^2 - 7 \cdot 2a = 15a^2 - 14a \end{aligned}$$

●

The derivative $f'(a)$ is a number defined at each point a of the domain of $f(x)$ for which a certain limit exists. Therefore $f'(a)$ can be considered as the value at $x = a$ of a new *function*, defined on part of the domain of $f(x)$, maybe on the whole domain. This point of view suggests writing $f'(x)$ for the slope at $(x, f(x))$. Thus the function $f'(x)$ is defined by

$$f'(x) = \lim_{h \to 0} \frac{f(x + h) - f(x)}{h}$$

for all x at which the limit exists.

With this notation, the results of **a** and **b** in Example 1 can be written

a If $f(x) = x^2$, then $f'(x) = 2x$.

b If $f(x) = x^3$, then $f'(x) = 3x^2$.

Notation and Terminology

A function whose derivative exists is called a **differentiable** function. You **differentiate** a function when you take its derivative; the process is called **differentiation.**

Many different notations are commonly used for the derivative, and it is important to recognize them. If $y = f(x)$, then some of the ways the derivative may be denoted are

$$f'(x) \qquad f' \qquad y'(x) \qquad y' \qquad \frac{dy}{dx} \qquad \frac{df}{dx} \qquad \frac{df(x)}{dx}$$

You read the first one "f prime of x"; the fifth, "dee-*y*-dee-*x*." Note that the second, fourth, fifth, and sixth do not show the point where the derivative is to be evaluated. To show the value of the derivative at $x = a$, write

$$y'(a) \quad \text{or} \quad \left.\frac{dy}{dx}\right|_{x=a} \quad \text{or} \quad \left.\frac{df}{dx}\right|_{a} \quad \text{or} \quad \frac{df}{dx}(a) \quad \text{or} \quad \left(\frac{dy}{dx}\right)_a$$

Sometimes it pays to think of the derivative as an operator or machine (black box) into which you feed $f(x)$ and out of which comes $f'(x)$. The operator d/dx is applied to the function $y = f(x)$ to produce its derivative dy/dx.

The variable may be denoted by a letter other than x; that does not change the process. For instance,

$$\frac{d}{du}(u^2) = 2u \qquad \frac{d}{dy}(y^3) = 3y^2$$

The letter t is usually used for time. There is a special and commonly used notation for time derivatives, a dot instead of a prime:

$$\frac{d}{dt}[x(t)] = \dot{x}(t)$$

Thus if $x(t) = t^2$, then $\dot{x} = 2t$.

The derivative $f'(x)$ is also called the *first derivative* of $f(x)$—a hint at more derivatives to come.

Remark The notation dy/dx is probably the one most widely used in books and articles where calculus is applied. It is a historical notation from the early days of calculus, before the limit concept was fully understood. Then the derivative was thought of as the quotient $(dy)/(dx)$. Here dy was the "infinitesimal" change in y resulting from dx, an "infinitesimal" change in x. In those days, people were comfortable with "infinitesimals," thought of as some sort of non-zero numbers smaller in magnitude than all positive real numbers. We shall use dy/dx *only* as a formal notation and not think of it as a quotient.

Some Basic Derivatives

Derivatives are so important that sooner or later we must learn how to differentiate every function in sight. Let us first attack some simple, but basic ones.

The simplest is a constant function $f(x) = c$. Its graph is a horizontal line, with slope 0, so $f'(x) = 0$. More formally

$$f'(x) = \lim_{h\to 0} \frac{f(x+h) - f(x)}{h} = \lim_{h\to 0} \frac{c - c}{h}$$

$$= \lim_{h\to 0} \frac{0}{h} = \lim_{h\to 0} 0 = 0$$

so $f'(x) = 0$.

Now consider a linear function $f(x) = mx + b$. The graph of $y = mx + b$ is a straight line of slope m, so $f'(x) = m$. Formal verification:

$$f'(x) = \lim_{h\to 0} \frac{f(x+h) - f(x)}{h}$$

$$= \lim_{h\to 0} \frac{[m(x+h) + b] - [mx + b]}{h}$$

$$= \lim_{h\to 0} \frac{mh}{h} = \lim_{h\to 0} m = m$$

so $f'(x) = m$.

Next we differentiate $f(x) = x^n$, where n is any positive integer. The case $n = 1$ is $f(x) = x$, a linear function, which has been taken care of already. We can therefore suppose $n \geq 2$. By a factoring formula,

$$f(x) - f(a) = x^n - a^n$$

$$= (x - a)(x^{n-1} + ax^{n-2} + a^2x^{n-3} + \cdots + a^{n-2}x + a^{n-1})$$

Hence

$$f'(a) = \lim_{x\to a} \frac{f(x) - f(a)}{x - a}$$

$$= \lim_{x\to a} (x^{n-1} + ax^{n-2} + a^2x^{n-3} + \cdots + a^{n-1})$$

As $x \to a$,

$$\lim_{x\to a} ax^{n-2} = a \lim_{x\to a} x^{n-2} = a \cdot a^{n-2} = a^{n-1}$$

$$\lim_{x\to a} a^2x^{n-3} = a^2 \lim_{x\to a} x^{n-3} = a^2 \cdot a^{n-3} = a^{n-1} \quad \text{and so on}$$

Thus each term has the same limit, a^{n-1}. Now the limit of the sum equals the sum of the limits. Since there are n terms in the sum, we obtain $f'(a) = na^{n-1}$. Therefore

$$\frac{d}{dx} x^n = nx^{n-1}$$

(See Exercise 46 for an alternative derivation.)

This is also true for $n = 0$ and $n = 1$ if interpreted with a grain of salt (that is, if the special point $x = 0$ is ignored). Here we think of x^0 as 1, so that

$$\frac{d}{dx}1 = \frac{d}{dx}x^0 = 0x^{0-1} = 0 \quad \text{and}$$

$$\frac{d}{dx}x = \frac{d}{dx}x^1 = 1 \cdot x^{1-1} = 1 \cdot x^0 = 1$$

Finally, let us differentiate $f(x) = 1/x$, defined for $x \neq 0$. First, for $a \neq 0$,

$$f(x) - f(a) = \frac{1}{x} - \frac{1}{a} = \frac{a - x}{ax}$$

Hence

$$f'(a) = \lim_{x \to a} \frac{f(x) - f(a)}{x - a} = \lim_{x \to a} \frac{(a - x)}{ax(x - a)}$$

$$= \lim_{x \to a}\left(-\frac{1}{ax}\right) = -\frac{1}{a^2}$$

Therefore $f'(x) = -1/x^2$. We have established the following formulas:

Basic Differentiation Formulas

$$\frac{d}{dx}(c) = 0 \qquad \frac{d}{dx}(mx + b) = m$$

$$\frac{d}{dx}(x^n) = nx^{n-1} \quad (n \geq 1) \qquad \frac{d}{dx}\left(\frac{1}{x}\right) = -\frac{1}{x^2} \quad (x \neq 0)$$

Example 2

Find **a** $y'(-2)$ for $y = x^4$ **b** $y'(\frac{1}{3})$ for $y = 1/x$

Solution

a $y = x^4$ $\quad \left.\frac{dy}{dx}\right|_{-2} = \left.4x^3\right|_{-2} = 4 \cdot (-2)^3 = -32$

b $y = \frac{1}{x}$ $\quad \left.\frac{dy}{dx}\right|_{1/3} = -\left.\frac{1}{x^2}\right|_{1/3} = -\frac{1}{(\frac{1}{3})^2} = -9$

Example 3

Find all a such that the slope of $y = x^2$ at (a, a^2) equals the slope of $y = x^3$ at (a, a^3).

Solution The two required slopes are the derivatives

$$\left.\frac{d}{dx}x^2\right|_a = 2a \quad \text{and} \quad \left.\frac{d}{dx}x^3\right|_a = 3a^2$$

so we require $2a = 3a^2$. The solutions of this equation are

$$a = 0 \quad \text{and} \quad a = \tfrac{2}{3}$$

Rate of Change

The derivative has an important interpretation besides slope of a curve. Suppose, for instance, that $f'(a) = 3$. This means that

$$\frac{f(a+h) - f(a)}{h} \longrightarrow 3 \quad \text{as} \quad h \longrightarrow 0$$

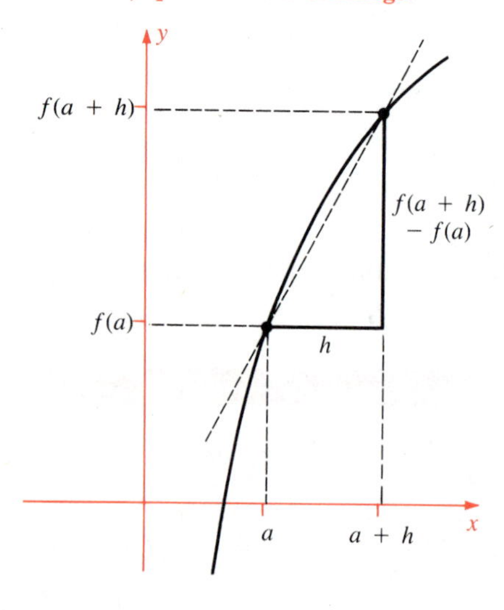

Figure 2-6-1
The difference quotient (average rate of change) $[f(a+h) - f(a)]/h$ approximates the rate of change of y with respect to x. Its limit, the derivative, *equals* the rate of change.

Now $f(a+h) - f(a)$ is the change in $y = f(x)$ as x changes by amount h from a to $a + h$. Therefore the difference quotient represents the *average rate of change* of $f(x)$ over the interval from a to $a + h$. When h is very small, the difference quotient, or average rate of change, is very close to 3. Thus, a change in x causes a change in y of about 3 times as much. We say that the rate of change of y with respect to x is 3 at $x = a$. Geometrically, the curve $y = f(x)$ is rising at $(a, f(a))$ three times as fast as it is moving to the right (Figure 2-6-1).

In general, $f'(a)$ is the **instantaneous rate of change** of y with respect to x at $x = a$. For a linear function $f(x) = mx + b$, this interpetation of the derivative is natural. The graph is a straight line, and since $f'(a) = m$ at all points, the instantaneous rate of change of y with respect to x is everywhere equal to the slope m. For other functions, the graph $y = f(x)$ is curved, and the instantaneous rate of change of y with respect to x varies from point to point. In the next two chapters, we examine some applications in detail. Meanwhile, let us look briefly at some of the ways these ideas are applied.

Velocity Suppose $x = f(t)$ is the distance in feet that a vehicle travels in t seconds. If, for instance, $f(t_0 + 5) - f(t_0) = 210$ ft, then the average speed of the vehicle in the 5-second time interval $t_0 \le t \le t_0 + 5$ is $210/5 = 42$ ft/sec.

The derivative $f'(t_0)$ is the rate of change of distance with respect to time at time t_0, commonly known as the **velocity** (or **instantaneous velocity**) of the vehicle.

Marginal Cost In economics, anything that is supplied by a supplier and purchased in a market is called a **good**. The cost C of a good depends on the quantity Q supplied: $C = f(Q)$. Assuming that $f(Q)$ is a differentiable function, economists define the **marginal cost** as the derivative dC/dQ. Similarly, if $P = g(Q)$ is the profit function, defined as the profit when the quantity Q is sold, then dP/dQ is the **marginal profit**. The economist usually uses the word "marginal" to mean derivative.

Heat Treatment In the heat treatment of a metal object, the rate at which the object cools is critical. If at any instant it cools too slowly, then the metal will fail to develop its required characteristics, such as ductility and strength. But if it cools too rapidly, the object might develop internal stresses and surface flaws. If the temperature T is known as a function of time, $T = f(t)$, then the derivative dT/dt is the critical quantity that must be controlled. By analyzing this derivative function, one finds which cooling processes are acceptable.

Exercises

Find

1 $\dfrac{dy}{dx}$ for $y = x^2$

2 $\dfrac{dV}{dP}$ for $V = P^2$

3 $\dfrac{dy}{dx}$ for $y = -4(x - 5)$

4 $\dfrac{ds}{dt}$ for $s = -3(4 - 5t)$

5 $\dfrac{d}{dx}(x)$

6 $\dfrac{d}{dx}(x^2)$

7 $\dfrac{df}{dx}$ for $f(x) = 3x + 2$

8 $\dfrac{df}{dx}$ for $f(x) = 12x - 7$

9 $f'(x)$ for $f(x) = 8x$

10 $F'(z)$ for $F(z) = z^2$

11 $\dfrac{dx}{dy}$ for $x = y^3$

12 $\dfrac{dF}{dx}$ for $F(x) = x^3$

13 $\left.\dfrac{d}{dt}(5)\right|_{t=1}$

14 $\dfrac{d}{dx}(4^3)$

15 $\dfrac{dy}{dx}$ for $y = x^3$ at $x = 0, 3, -3$

16 $\dfrac{ds}{dt}$ for $s = t^3$ at $t = 0, 1, 2, 3$

17 $\left.\dfrac{dG}{dz}\right|_6$ for $G(z) = z^3$

18 $\left.\dfrac{dy}{dx}\right|_{-4}$ for $y = x^3$

19 $V'(4)$ and $V'(a)$ for $V(P) = P^3$

20 $s'(5)$ and $s'(t_0)$ for $s(t) = t^3$

Calculate

21 $f'(-6)$, $f'(12)$, $f'(1)$, for $f(x) = x^2$

22 $G'(-1)$, $G'(0)$, $G'(1)$, for $G(x) = x^2$

23 $f'(-1)$, $f'(1)$, $f'(a)$, $f'(-a)$, for $f(x) = 1/x$

24 $f'(-\frac{1}{2})$, $f'(\frac{1}{2})$, $f'(2)$, $f'(3)$, for $f(x) = 1/x$

25 $\left.\dfrac{dy}{dx}\right|_{x=a}$ and $\left.\dfrac{dy}{dx}\right|_{x=1/a}$ if $y = \dfrac{1}{x}$

26 $\left.\dfrac{dV}{dP}\right|_{P=1/4}$ and $\left.\dfrac{dV}{dP}\right|_{P=4}$ if $V = \dfrac{1}{P}$

27 $\left.\dfrac{d}{ds}\left(\dfrac{1}{s}\right)\right|_b$

28 $\left.\dfrac{d}{dx}\left(\dfrac{1}{x}\right)\right|_t$

29 $\dfrac{d}{dx}(x^2)$ at $x = 3, 7, 11$

30 $\dfrac{d}{dP}(P^2)$ at $P = -1, 1, 0$

31 $\dfrac{d}{dx}(13x + 5)$ at $x = 2$

32 $\dfrac{d}{dt}(-13t + v_0)$ at $t = 0$

33 $\left.\dfrac{dy}{dx}\right|_{x=3}$ for $y = x^2$

34 $\left.\dfrac{dR}{dI}\right|_{I=-2}$ for $R = I^2$

35 $\left.\dfrac{dv}{dt}\right|_9$ for $v = -32t + 200$

36 $\left.\dfrac{ds}{dt}\right|_{10}$ for $s = t^2$

37 Find all points where $y = x^2$ has slope 6.

38 Find a point on $y = x^2$ where the slope equals that of the line $x + 2y + 7 = 0$.

39 Find all points on the curve $y = x^3$ where the slope is 12.

40 Find where the curve $y = 1/x$ has slope $-\frac{1}{2}$.

41 Does the graph of $y = x^3$ ever have slope 0? If so, where?

42 Do the curves $y = 1/x$ and $y = x^3$ ever have the same slope? If so, where?

43 Find all positive values of x where $y = x^2$ is steeper (has greater slope) than $y = x^3$.

44 Find all positive values of x where $y = x^3$ is steeper than $y = x^2$.

*45 Give an interpretation for $x = 1$ of the identity

$$\frac{x^n - 1}{x - 1} = 1 + x + x^2 + \cdots + x^{n-1}$$

*46 Use the first definition of the derivative and the binomial theorem to give another derivation of the formula $d(x^n)/dx = nx^{n-1}$.

2-7 Sums, Products, and Quotients

The definition of derivative given in the previous section is a nuisance to apply each time we differentiate a function. In practice, we use the formulas that we have already developed and rules that we develop now to express derivatives of new functions in terms of derivatives we already know. The two simplest of these new rules are the following:

Constant Factor Rule

Let $u(x)$ be differentiable and let c be a constant. Then $cu(x)$ is differentiable and

$$\frac{d}{dx}[cu(x)] = c\frac{d}{dx}u(x) \qquad \text{that is} \qquad (cu)' = cu'$$

In words, the derivative of a constant times a function equals the constant times the derivative of the function. [Note that cu' means $(c)(u')$, not $(cu)'$.]

Sum Rule

Let $u(x)$ and $v(x)$ be differentiable functions. Then $u(x) + v(x)$ and $u(x) - v(x)$ are differentiable and

$$\frac{d}{dx}[u(x) \pm v(x)] = \frac{d}{dx}u(x) \pm \frac{d}{dx}v(x)$$

that is,

$$(u \pm v)' = u' \pm v'$$

In words, the derivative of a sum equals the sum of the derivatives.

Proofs These rules hold because they reflect corresponding properties of limits. For the first, use the definition of derivative and limit rule 1, on page 80:

$$\begin{aligned}\frac{d}{dx}[cu(x)] &= \lim_{h\to 0}\frac{cu(x+h) - cu(x)}{h}\\ &= \lim_{h\to 0} c\,\frac{u(x+h) - u(x)}{h}\\ &= c\lim_{h\to 0}\frac{u(x+h) - u(x)}{h} = c\frac{d}{dx}u(x)\end{aligned}$$

The sum rule is proved by applying limit rule 2. We shall do it for sums; the proof for differences is practically the same:

$$\begin{aligned}&\frac{d}{dx}[u(x) + v(x)]\\ &\quad= \lim_{h\to 0}\frac{[u(x+h) + v(x+h)] - [u(x) + v(x)]}{h}\\ &\quad= \lim_{h\to 0}\left[\frac{u(x+h) - u(x)}{h} + \frac{v(x+h) - v(x)}{h}\right]\\ &\quad= \lim_{h\to 0}\frac{u(x+h) - u(x)}{h} + \lim_{h\to 0}\frac{v(x+h) - v(x)}{h}\\ &\quad= \frac{d}{dx}u(x) + \frac{d}{dx}v(x)\end{aligned}$$

The sum rules holds for any number of summands, not just two. For instance

$$(u + v - w)' = u' + v' - w' \quad \text{and so on}$$

Example 1

Differentiate **a** $5x^3 - 8x^2 - 7$ **b** $\frac{1}{2}x^5 + 3/x$

Solution

a We know the derivatives of x^3, of x^2, and of the constant 7. Apply the sum rule for three terms, then the constant factor rule:

$$\begin{aligned}\frac{d}{dx}(5x^3 - 8x^2 - 7) &= \frac{d}{dx}(5x^3) - \frac{d}{dx}(8x^2) - \frac{d}{dx}(7)\\ &= 5\frac{d}{dx}(x^3) - 8\frac{d}{dx}(x^2) - 0\\ &= 5 \cdot 3x^2 - 8 \cdot 2x\\ &= 15x^2 - 16x\end{aligned}$$

b Same procedure:

$$\begin{aligned}\frac{d}{dx}\left(\tfrac{1}{2}x^5 + \frac{3}{x}\right) &= \frac{d}{dx}(\tfrac{1}{2}x^5) + \frac{d}{dx}\left(\frac{3}{x}\right)\\ &= \tfrac{1}{2}\frac{d}{dx}(x^5) + 3\frac{d}{dx}\left(\frac{1}{x}\right)\\ &= \tfrac{1}{2}(5x^4) + 3\left(\frac{-1}{x^2}\right)\\ &= \tfrac{5}{2}x^4 - \frac{3}{x^2}\end{aligned}$$

●

Remark In practice, problems like these are usually done in one step, by inspection. It's just a matter of knowing the rules cold, so you are confident of making no errors.

Products

For the derivative of a product $u(x)v(x)$, a natural first guess is $u'(x)v'(x)$. Intuition fails; the natural guess is just plain wrong! For instance, if $u(x) = x$ and $v(x) = x^2$, then $u'v' = 1 \cdot 2x = 2x$. But $uv = x^3$, so $(uv)' = 3x^2$, certainly not $2x$.

To determine the correct relation, consider a variable rectangle (Figure 2-7-1a, next page) whose sides are $u(x)$ and $v(x)$. Its area is $A(x) = u(x)v(x)$. We want to find how fast $A(x)$ changes relative to x. Suppose x changes to $x + h$. From Figure 2-7-1b,

$$\begin{aligned}A(x + h) = A(x) &+ [u(x + h) - u(x)]v(x + h)\\ &+ u(x)[v(x + h) - v(x)]\end{aligned}$$

Figure 2-7-1

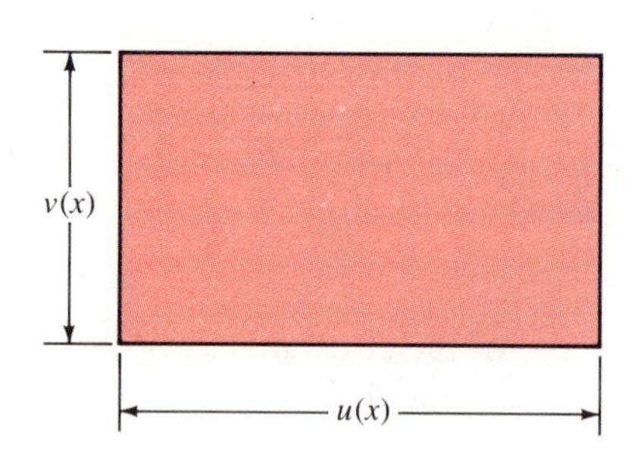

a Rectangle with sides $u(x), v(x)$ that depend on x.
area $= A(x) = u(x)V(x)$

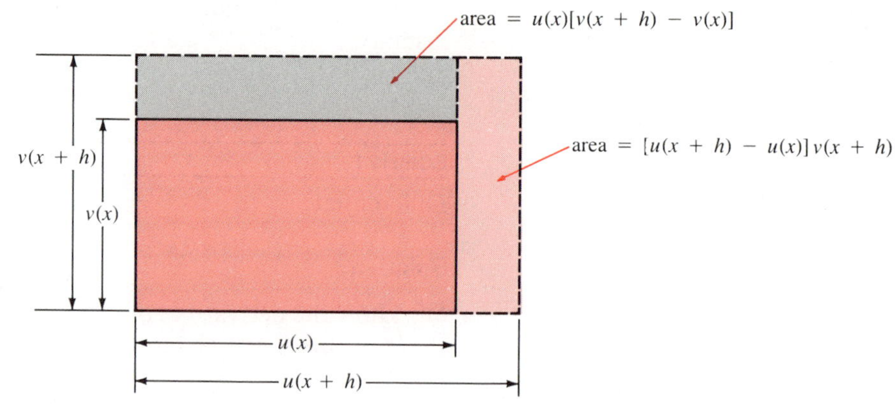

b Change in $A(x)$ as x changes to $x + h$
$=$ added area
$= [u(x+h) - u(x)]v(x+h) + u(x)[v(x+h) - v(x)]$

Consequently the difference quotient for the area function is

$$\frac{A(x+h) - A(x)}{h} = \left[\frac{u(x+h) - u(x)}{h}\right] v(x+h) + u(x)\left[\frac{v(x+h) - v(x)}{h}\right]$$

If we knew that $v(x)$ were continuous, then we would know that $\lim_{h\to 0} v(x+h) = v(x)$. By letting $h \to 0$, we would conclude that

$$(uv)' = A'(x) = u'(x)v(x) + u(x)v'(x)$$

This is the right track; now we state the result formally, but postpone the details of its proof until the end of this section.

Product Rule

If $u(x)$ and $v(x)$ are differentiable functions, then $u(x)v(x)$ is differentiable and

$$\frac{d}{dx}[u(x)v(x)] = \left[\frac{d}{dx}u(x)\right] \cdot v(x) + u(x) \cdot \left[\frac{d}{dx}v(x)\right]$$

Briefly, $(uv)' = u'v + uv'$. [Here the term uv' means $(u)(v')$; like exponents, primes take priority over products.]

Example 2

Let $u(x) = x$ and $v(x) = x^2$ so that $u(x)v(x) = x^3$. By the product rule,

$$\frac{d}{dx}(x \cdot x^2) = \left(\frac{d}{dx}x\right) \cdot x^2 + x \cdot \left(\frac{d}{dx}x^2\right)$$
$$= 1 \cdot x^2 + x \cdot 2x = 3x^2$$

This is indeed the derivative of x^3.

Example 3

Apply the product rule to differentiate

$$x^{m+n} = x^m x^n$$

and show that you get the expected derivative.

Solution We expect

$$\frac{d}{dx} x^{m+n} = (m+n)x^{m+n-1}$$

The product rule gives us

$$\begin{aligned}\frac{d}{dx}(x^m x^n) &= \left(\frac{d}{dx} x^m\right) x^n + x^m \left(\frac{d}{dx} x^n\right) \\ &= (mx^{m-1})x^n + x^m(nx^{n-1}) \\ &= mx^{m+n-1} + nx^{m+n-1} = (m+n)x^{m+n-1}\end{aligned}$$

This agrees with the derivative obtained directly. ●

Example 4

Differentiate

a $y = (3x-2)(x^2+5x+1)$ **b** $y = \dfrac{1}{x^2}$

Solution

a Apply the product rule with $u(x) = 3x - 2$ and $v(x) = x^2 + 5x + 1$:

$$\begin{aligned}\frac{dy}{dx} &= \left[\frac{d}{dx}(3x-2)\right](x^2+5x+1) \\ &\quad + (3x-2)\frac{d}{dx}(x^2+5x+1) \\ &= 3(x^2+5x+1) + (3x-2)(2x+5) \\ &= (3x^2+15x+3) + (6x^2+11x-10) \\ &= 9x^2+26x-7\end{aligned}$$

b Write $y = (1/x)(1/x)$ and apply the product rule:

$$\begin{aligned}\frac{dy}{dx} &= \left(\frac{d}{dx}\frac{1}{x}\right)\frac{1}{x} + \frac{1}{x}\left(\frac{d}{dx}\frac{1}{x}\right) \\ &= \frac{-1}{x^2}\cdot\frac{1}{x} + \frac{1}{x}\cdot\frac{-1}{x^2} \\ &= \frac{-2}{x^3}\end{aligned}$$

●

Quotients

Next on the agenda is a rule for the derivative of a quotient. The natural guess,

$$\frac{d}{dx}\left[\frac{u(x)}{v(x)}\right] = \frac{u'(x)}{v'(x)}$$

is again wrong. (Find some examples showing that this formula is false.) To guess the correct formula, we suppose that u/v has a derivative and apply the product rule to $u = (u/v)v$ in order to find it:

$$u = \left(\frac{u}{v}\right)v \qquad u' = \left[\left(\frac{u}{v}\right)v\right]' = \left(\frac{u}{v}\right)'v + \left(\frac{u}{v}\right)v'$$

Now solve for $(u/v)'$:

$$\left(\frac{u}{v}\right)'v = u' - \left(\frac{u}{v}\right)v' = \frac{u'v - uv'}{v}$$

$$\left(\frac{u}{v}\right)' = \frac{u'v - uv'}{v^2}$$

What this shows is that *if* u/v *is differentiable,* then its derivative can be computed as indicated. The following quotient rule says more than this; it will be proved at the end of this section.

Quotient Rule

If $u(x)$ and $v(x)$ are differentiable and $v(x) \neq 0$, then the quotient $u(x)/v(x)$ is differentiable and

$$\frac{d}{dx}\left[\frac{u(x)}{v(x)}\right] = \frac{u'(x)v(x) - u(x)v'(x)}{v(x)^2}$$

Briefly,

$$\left(\frac{u}{v}\right)' = \frac{u'v - uv'}{v^2}$$

Example 5

Differentiate

a $\dfrac{x}{x^2 + 1}$ **b** $\dfrac{x^2 - 7}{x^3}$

Solution

a Take $u = x$ and $v = x^2 + 1$. Then

$$\frac{d}{dx}\left(\frac{x}{x^2 + 1}\right) = \frac{(x)'(x^2 + 1) - x(x^2 + 1)'}{(x^2 + 1)^2}$$

$$= \frac{1 \cdot (x^2 + 1) - x \cdot 2x}{(x^2 + 1)^2} = \frac{1 - x^2}{(x^2 + 1)^2}$$

b Take $u = x^2 - 7$ and $v = x^3$. Then

$$\frac{d}{dx}\left(\frac{x^2-7}{x^3}\right) = \frac{(x^2-7)'x^3 - (x^2-7)(x^3)'}{(x^3)^2}$$

$$= \frac{2x \cdot x^3 - (x^2-7)\cdot 3x^2}{x^6}$$

$$= \frac{2x^2 - 3(x^2-7)}{x^4} = \frac{-x^2+21}{x^4}$$

The special case $u = 1$ in the quotient rule yields a useful formula for the derivative of a reciprocal.

Reciprocal Rule

If $v(x) \neq 0$, then $\dfrac{d}{dx}\left(\dfrac{1}{v}\right) = -\dfrac{v'}{v^2}$

This follows directly from the quotient rule, because if $u = 1$, then $u' = 0$. Therefore

$$\frac{d}{dx}\left(\frac{1}{v}\right) = \frac{0\cdot v - 1\cdot v'}{v^2} = \frac{v'}{v^2}$$

Example 6

$$\frac{d}{dx}\left(\frac{1}{x^4-3x+1}\right)_{x=1} = -\left.\frac{(x^4-3x+1)'}{(x^4-3x+1)^2}\right|_{x=1}$$

$$= \left.\frac{-4x^3+3}{(x^4-3x+1)^2}\right|_{x=1}$$

$$= \frac{-4+3}{(1-3+1)^2} = -1$$

Example 7

Use the product and reciprocal rules to derive a formula for $[1/(uv)]'$.

Solution 1

$$\left(\frac{1}{uv}\right)' = -\frac{(uv)'}{(uv)^2} = -\frac{u'v+uv'}{(uv)^2}$$

Solution 2

$$\left(\frac{1}{uv}\right)' = \left[\left(\frac{1}{u}\right)\left(\frac{1}{v}\right)\right]' = \left(\frac{1}{u}\right)'\left(\frac{1}{v}\right) + \left(\frac{1}{u}\right)\left(\frac{1}{v}\right)'$$

$$= \left(-\frac{u'}{u^2}\right)\left(\frac{1}{v}\right) + \left(\frac{1}{u}\right)\left(-\frac{v'}{v^2}\right)$$

$$= -\frac{u'}{u^2v} - \frac{v'}{uv^2} = -\frac{u'v+uv'}{u^2v^2}$$

The Derivative of x^{-n}

In Section 2-6 (page 102) we derived the formula

$$\frac{d}{dx}(x^n) = nx^{n-1}$$

for $n = 1, 2, 3, \ldots$ and also for $n = -1$. The formula is actually true for all exponents n, whether integers or not. While we cannot yet prove this in such generality, we can prove that the formula is correct for all negative integers n, not just $n = -1$. To do so, we apply the reciprocal rule to $x^{-m} = 1/x^m$, where m is a *positive* integer:

$$\frac{d}{dx}(x^{-m}) = \frac{d}{dx}\left(\frac{1}{x^m}\right) = \frac{-1}{(x^m)^2}\frac{d}{dx}(x^m)$$

$$= -\left(\frac{1}{x^{2m}}\right)(mx^{m-1}) = -mx^{-m-1}$$

Now let $n = -m < 0$. We have just proved $d(x^n)/dx = nx^{n-1}$ for any negative integer n.

Derivative of x^n

For any integer n, positive, negative, or zero,

$$\frac{d}{dx}(x^n) = nx^{n-1}$$

(Recall that for $n = 0$ this is interpreted as $d1/dx = 0$ and for $n = 1$ as $dx/dx = 1$.) ●

Example 8

a $\dfrac{d}{dx}(x^{-7}) = -7x^{-8}$

b $\dfrac{d}{dx}\left(\dfrac{1}{x^{13}}\right) = \dfrac{d}{dx}(x^{-13}) = -13x^{-14} = -\dfrac{13}{x^{14}}$

c $\dfrac{d}{dx}\left(x^5 + \dfrac{1}{x^5}\right) = \dfrac{d}{dx}(x^5) + \dfrac{d}{dx}\left(\dfrac{1}{x^5}\right)$

$$= 5x^4 - \frac{5}{x^6} = \frac{5(x^{10} - 1)}{x^6}$$

●

Proof of the Product Rule [Optional]

We saw on page 107 how to approach this proof by geometry. The analytic proof uses the device of subtracting and adding a suitable term in the numerator of the difference quotient; then the rules for limits are applied. However, as we have already seen, at a certain point in the proof we must know that any differentiable function is continuous, so we prove that first as a lemma.

Lemma If $f(x)$ is differentiable at $x = a$, then $f(x)$ is continuous at $x = a$. ●

Proof of the Lemma We must prove that $\lim_{x\to a} f(x) = f(a)$, or, what is the same thing, $\lim_{h\to 0} f(a+h) = f(a)$. Another way to put it is

$$\lim_{h\to 0} [f(a+h) - f(a)] = 0$$

But

$$\begin{aligned}\lim_{h\to 0} [f(a+h) - f(a)] &= \lim_{h\to 0} \left[\frac{f(a+h) - f(a)}{h} h\right] \\ &= \left[\lim_{h\to 0} \frac{f(a+h) - f(a)}{h}\right](\lim_{h\to 0} h) \\ &= f'(a)\cdot 0 = 0\end{aligned}$$

●

Proof of the Product Rule The numerator of the difference quotient for the product $u(x)v(x)$ is

$$\begin{aligned}&u(x+h)v(x+h) - u(x)v(x) \\ &\quad = u(x+h)v(x+h) - u(x)v(x+h) \\ &\qquad\qquad + u(x)v(x+h) - u(x)v(x) \\ &\quad = [u(x+h) - u(x)]v(x+h) + u(x)[v(x+h) - v(x)]\end{aligned}$$

The term $u(x)v(x+h)$ is subtracted and added, with no net change. Consequently

$$\begin{aligned}(uv)' &= \lim_{h\to 0} \frac{u(x+h)v(x+h) - u(x)v(x)}{h} \\ &= \lim_{h\to 0} \left[\frac{u(x+h) - u(x)}{h} v(x+h) + u(x)\frac{v(x+h) - v(x)}{h}\right] \\ &= \left[\lim_{h\to 0} \frac{u(x+h) - u(x)}{h}\right][\lim_{h\to 0} v(x+h)] + u(x)\left[\lim_{h\to 0} \frac{v(x+h) - v(x)}{h}\right] \\ &= u'(x)v(x) + u(x)v'(x)\end{aligned}$$

In the final step we used the lemma for $\lim_{h\to 0} v(x+h) = v(x)$. ●

Proof of the Quotient Rule [Optional]

Note first that the quotient rule is an easy consequence of the product rule and the reciprocal rule:

$$\left(\frac{u}{v}\right)' = \left(u\frac{1}{v}\right)' = u'\left(\frac{1}{v}\right) + u\left(\frac{1}{v}\right)' = \frac{u'}{v} + u\left(-\frac{v'}{v^2}\right)$$

$$= \frac{u'v - uv'}{v^2}$$

Therefore, to prove the quotient rule, it suffices to prove the reciprocal rule, which we do by a limit argument, using the lemma once again:

$$\left(\frac{1}{v(x)}\right)' = \lim_{h\to 0}\frac{1}{h}\left[\frac{1}{v(x+h)} - \frac{1}{v(x)}\right]$$
$$= \lim_{h\to 0}\frac{1}{h}\left[\frac{v(x)-v(x+h)}{v(x)v(x+h)}\right]$$
$$= \left[\lim_{h\to 0}\frac{1}{v(x)v(x+h)}\right]\left[\lim_{h\to 0}\frac{v(x)-v(x+h)}{h}\right]$$
$$= \frac{-1}{v(x)}\left[\lim_{h\to 0}\frac{1}{v(x+h)}\right]\left[\lim_{h\to 0}\frac{v(x+h)-v(x)}{h}\right]$$
$$= \frac{-1}{v(x)}\cdot\frac{1}{v(x)}v'(x) = -\frac{v'(x)}{v(x)^2}$$

Exercises

Find dy/dx for $y =$

1 $x^2 + 3x$
2 $5x^3 - x$
3 $-x^4 + 2x^2 - 1$
4 $3x^2 + 2x + 1$
5 $x + 1/x$
6 $5x^5 - 7/x$
7 $x^3 - 3x + 1$ at $x = -1$
8 $x^2 + 2/x$ at $x = 3$
9 $x^3 + x - 2/x$ at $x = 2$
10 $5x^4 - 4x^3 + x$ at $x = 1$
11 $(x + 1)(x^2 + 1)$
12 $(x^2 - 1)(x^2 + 3)$
13 $(3x + 4) \times (x^2 - 2x - 3)$
14 $(2x - 7)(x^3 + 1)$
15 $(x^5 - 2) \times (x^3 + x - 3)$
16 $(x^4 - 2x)(2x^3 - 3x)$

Differentiate with respect to x by applying the product rule twice

17 $(x + 1)(x + 2)(x + 3)$
18 $(x + 1)(x^2 + 1)(x^3 + 1)$
19 $(2x + 1)(x - 3)(3x + 4)$
20 $(3x^2 - 1)(x - 1)(2x + 3)$

Differentiate by the rules for sums and powers, then by the product rule. Verify that you get the same result

21 $x^2 - 1 = (x + 1)(x - 1)$
22 $x^3 + 1 = (x + 1)(x^2 - x + 1)$
23 $x^4 - 1 = (x^2 + 1)(x^2 - 1)$
24 $x^6 - 1 = (x^2 - 1)(x^4 + x^2 + 1)$
25 $x^5 - 1 = (x - 1)(x^4 + x^3 + x^2 + x + 1)$
26 $x^4 + 4 = (x^2 + 2x + 2)(x^2 - 2x + 2)$

Test the "formula" $(u/v)' = u'/v'$ by computing both sides

27 $u = x \quad v = x$
28 $u = x^2 \quad v = x$
29 $u = x \quad v = x^2$
30 $u = x^5 \quad v = x^3$

Differentiate with respect to x

31 $x/(x + 1)$
32 $x/(1 - x)$
33 $\dfrac{3x + 1}{x + 5}$
34 $\dfrac{x - 2}{x + 2}$
35 $\dfrac{x^2 + x + 3}{x + 4}$
36 $\dfrac{x^2}{x^3 + 1}$
37 $\dfrac{x^2 + 1}{x^2 - 1}$
38 $\dfrac{x^2 + x}{x^2 + 1}$
39 $\dfrac{2x}{x^2 + x + 1}$
40 $\dfrac{x^3 - x - 1}{x + 2}$
41 $\dfrac{2x + 7}{x^2 + 3}$
42 $\dfrac{-x + 5}{3x^3 - 1}$
43 $\dfrac{(x + 1)(x^2 + 1)}{x^3 + 2}$
44 $\dfrac{x^3 + 2}{(x + 1)(x^2 + 1)}$
45 $\dfrac{(x + 1)(x + 2)}{(x + 3)(x + 4)}$
46 $\dfrac{(x + 2)(x + 4)}{(x + 1)(x + 3)}$
47 $x^2 - x^{-2}$
48 $\dfrac{1}{x} + \dfrac{1}{x^3} + \dfrac{1}{x^5}$
49 $\frac{1}{3}x^{-3} - \frac{1}{7}x^{-7}$
50 $x^{-7} - \dfrac{1}{x^7}$
51 $\frac{1}{2}x^{-6} + \frac{1}{3}\sqrt[9]{x}$
52 $\dfrac{3}{x^3} + \dfrac{5}{x^5} - \dfrac{7}{x^7}$

53 Prove $\left(\dfrac{fg}{h}\right)' = \dfrac{(fgh)' - 2fgh'}{h^2}$

54 Prove $\left(\dfrac{u}{v^n}\right)' = \dfrac{u'v - nuv'}{v^{n+1}}$

2-8 The Chain Rule

Suppose we must differentiate

$$y = (x^3 - 4x + 2)^9$$

How do we do it? One way is to multiply together nine factors $x^3 - 4x + 2$, write down the resulting polynomial of degree 27, and differentiate it term by term according to the methods we have studied. This procedure is a great deal of work and is prone to error because it involves so much calculation. There must be a better method.

We can analyze the matter as follows: We write

$$y = u^9 \quad \text{where} \quad u = x^3 - 4x + 2$$

Thus we are dealing with a composite function

$$y = f[g(x)]$$

where

$$y = f(u) = u^9 \quad \text{and} \quad u = g(x) = x^3 - 4x + 2$$

Certainly we can find dy/du and du/dx, rather easily in this case:

$$\frac{dy}{du} = 9u^8 \qquad \frac{du}{dx} = 3x^2 - 4$$

Our problem is to fit the pieces together and come up with dy/dx. The derivative notation dy/dx suggests that $dy/dx = (dy/du)(du/dx)$. Does this make any sense?

Suppose we have a composite function $y = f[g(x)]$, where we know $f'(u)$ and $g'(x)$. Now $g'(x)$ is the rate of change of $g(x)$ with respect to x, that is, how fast $u = g(x)$ changes compared with how fast x changes, and $f'(u)$ represents how fast $y = f(u)$ changes compared with how fast u changes. Consider some concrete numbers. Suppose, say, that u is increasing 5 times as fast as x is increasing and y is increasing 7 times as fast as u is increasing. Can we guess how many times faster y is increasing than x is increasing? Sure we can; common sense says that y is increasing $7 \cdot 5 = 35$ times as fast as x is increasing.

If this guess is on the right track, then the derivative of the composite function $f[g(x)]$ with respect to x must be $f'[g(x)] \cdot g'(x)$. For the example we started with,

$$\frac{d}{dx}(x^3 - 4x + 2)^9 = 9(x^3 - 4x + 2)^8(3x^2 - 4)$$

This is much easier than doing it the hard way by expansion!

Once again, if a child runs 5 times as fast as a turtle, and the turtle runs 7 times as fast as a snail, then the child runs 35 times as fast as the snail. The general idea is stated in the all-important *chain rule,* whose proof is given at the end of this section.

Chain Rule

If $y = f(u)$ and $u = g(x)$ are differentiable, then so is the composite function $y = f[g(x)]$ where it is defined, and

$$(f \circ g)'(x) = f'[g(x)]g'(x)$$

that is,

$$\frac{d}{dx} f[g(x)] = \left(\frac{df}{du}\bigg|_{u=g(x)}\right)\left(\frac{dg}{dx}\bigg|_{x}\right)$$
$$= \left(\frac{df}{du}[g(x)]\right)\left(\frac{dg}{dx}(x)\right)$$

Briefly, $\dfrac{dy}{dx} = \dfrac{dy}{du} \cdot \dfrac{du}{dx}$

Note that the brief version in the "dee-y-dee-x" notation contains a nice memory aid: "Cancel du." (However, do not think of du/dx as a fraction. So far, we have not given the separate quantities du, dx, and dy meanings.)

Example 1

Differentiate

a $y = (x^2 + 1)^{13}$

b $y = 5(x^2 + 1)^6 - 8(x^2 + 1)^2 + 9$

c $y = \dfrac{1}{(x^2 + 5x + 1)^3}$ at $x = 0$

Solution

a The function $y(x)$ is the composite of $y = u^{13}$ and $u = x^2 + 1$. By the chain rule,

$$\frac{dy}{dx} = \frac{dy}{du} \cdot \frac{du}{dx} = (13u^{12})(2x)$$
$$= 13(x^2 + 1)^{12} \cdot 2x = 26x(x^2 + 1)^{12}$$

b The function $y(x)$ is the composite of $y = 5u^6 - 8u^2 + 9$ and $u = x^2 + 1$. By the chain rule,

$$\frac{dy}{dx} = \frac{dy}{du} \cdot \frac{du}{dx} = (30u^5 - 16u)(2x)$$
$$= (2x)[30(x^2 + 1)^5 - 16(x^2 + 1)]$$
$$= 4x(x^2 + 1)[15(x^2 + 1)^4 - 8]$$

c The function $y(x)$ is the composite of $y = u^{-3}$ and $u = x^2 + 5x + 1$. By the chain rule,

$$\frac{dy}{dx} = \frac{dy}{du} \cdot \frac{du}{dx} = (-3u^{-4})(2x + 5)$$

If $x = 0$, then $u = 1$ and $2x + 5 = 5$. Hence

$$\left.\frac{dy}{dx}\right|_{x=0} = (-3)(5) = -15$$

Example 2

Express F' in terms of f' where

a $F(x) = f(x + c)$ **b** $F(x) = f(kx)$

and c and k are constants.

Solution

a By the chain rule with $u(x) = x + c$, we have

$$F'(x) = f'(u)(x + c)' = f'(u) = f'(x + c)$$

b By the chain rule with $u(x) = kx$, we have

$$F'(x) = f'(u)(kx)' = kf'(u) = kf'(kx)$$

Let r be any rational number and consider the corresponding power function $y = x^r$ with domain $(0, +\infty)$. The following useful result greatly extends what we already know if r is an integer, namely, that $(x^r)' = rx^{r-1}$.

Power Rule

If r is a rational number and $y(x) = x^r$ for $x > 0$, then $y(x)$ is differentiable and

$$\frac{dy}{dx} = rx^{r-1} \quad \text{that is} \quad \frac{d}{dx}x^r = rx^{r-1}$$

In the particular case $r = \frac{1}{2}$, we have the frequently used square-root rule:

$$\frac{d}{dx}\sqrt{x} = \frac{1}{2\sqrt{x}}$$

Proof We shall postpone the proof that $y(x)$ is differentiable until the end of this section. However, if we *assume* this first, then the formula for dy/dx is an easy consequence of the chain rule.

Write the rational exponent r in the form $r = m/n$, where m and n are integers. Then $y(x) = x^{m/n}$, so take the n-th power of both sides:

$$[y(x)]^n = x^m$$

Differentiate both sides by the rule for *integer* powers, using the chain rule on the left side:

$$n[y(x)]^{n-1}\frac{dy}{dx} = mx^{m-1}$$

Multiply both sides by y and replace y^n by x^m:

$$ny^n \frac{dy}{dx} = myx^{m-1} \qquad nx^m \frac{dy}{dx} = myx^{m-1}$$

Cancel x^m, divide by n, replace m/n with r, and finally replace y with x^r:

$$n \frac{dy}{dx} = myx^{-1} \qquad \frac{dy}{dx} = ryx^{-1} = rx^r x^{-1} = rx^{r-1}$$

This completes the proof. ●

Remark The power rule is also valid for $x < 0$ provided $r = m/n$ in lowest terms with n an *odd* integer, so that the function $y = x^r$ is defined. For instance,

$$\frac{d}{dx} x^{1/3} = \tfrac{1}{3} x^{-2/3}$$

is valid for all $x \neq 0$, not just for $x > 0$.

Example 3

Differentiate $(x^2 + 7)^{5/4}$

Solution By the chain rule and the power rule,

$$\begin{aligned} \frac{d}{dx}(x^2 + 7)^{5/4} &= \tfrac{5}{4}(x^2 + 7)^{1/4} \frac{d}{dx}(x^2 + 7) \\ &= \tfrac{5}{4}(x^2 + 7)^{1/4}(2x) \end{aligned}$$

●

Example 4

Differentiate **a** $y = \dfrac{1}{\sqrt[4]{x^2 + 1}}$ **b** $y = (x^{-2/3} + x^{4/3})^{3/5}$

Solution

a Apply the chain rule with $y = u^{-1/4}$ and $u = x^2 + 1$:

$$\frac{dy}{dx} = \frac{dy}{du} \cdot \frac{du}{dx} = -\tfrac{1}{4} u^{-5/4} \cdot 2x = \frac{-x}{2(x^2 + 1)^{5/4}}$$

b Apply the chain rule with $y = u^{3/5}$ and $u = x^{-2/3} + x^{4/3}$:

$$\begin{aligned} \frac{dy}{dx} = \frac{dy}{du} \cdot \frac{du}{dx} &= (\tfrac{3}{5} u^{-2/5})(-\tfrac{2}{3} x^{-5/3} + \tfrac{4}{3} x^{1/3}) \\ &= \tfrac{3}{5}(x^{-2/3} + x^{4/3})^{-2/5}(-\tfrac{2}{3} x^{-5/3} + \tfrac{4}{3} x^{1/3}) \end{aligned}$$

This formula can be simplified to

$$\frac{dy}{dx} = \frac{2(2x^2 - 1)}{5x^{7/5}(1 + x^2)^{2/5}}$$

●

Composition of Three or More Functions

In applications one frequently meets composite functions involving two or more successive compositions, such as

$$[(3x^2+1)^{1/3}+1]^{1/4}$$

Such functions are differentiated by applying the chain rule repeatedly. For instance, if $y = y(u)$, $u = u(v)$, and $v = v(x)$, then

$$\frac{dy}{dx} = \frac{dy}{du}\cdot\frac{du}{dx} \quad \text{but} \quad \frac{du}{dx} = \frac{du}{dv}\cdot\frac{dv}{dx}$$

Hence $\dfrac{dy}{dx} = \dfrac{dy}{du}\cdot\dfrac{du}{dv}\cdot\dfrac{dv}{dx}.$

Example 5

Differentiate

a $y = [(3x^2+1)^{1/3}+1]^{1/4}$
b $y = [(3x^2+1)^{1/3}+x]^{1/4}$

Solution

a Write $y = u^{1/4}$, $u = v^{1/3}+1$, and $v = 3x^2+1$. Then

$$\frac{dy}{dx} = \frac{dy}{du}\cdot\frac{du}{dv}\cdot\frac{dv}{dx} = (\tfrac{1}{4}u^{-3/4})(\tfrac{1}{3}v^{-2/3})(6x)$$

$$= \frac{x}{2[(3x^2+1)^{1/3}+1]^{3/4}(3x^2+1)^{2/3}}$$

b This is different because of the x term inside the square brackets. The chain rule still applies, but you must use the sum rule also. Write $y = u^{1/4}$, $u = v^{1/3}+x$, and $v = 3x^2+1$. Then

$$\frac{dy}{dx} = \frac{dy}{du}\cdot\frac{du}{dx} = \tfrac{1}{4}u^{-3/4}\frac{du}{dx}$$

Continue:

$$\frac{du}{dx} = \frac{d}{dx}v^{1/3} + \frac{d}{dx}x = \frac{d}{dx}v^{1/3} + 1 \quad \text{and}$$

$$\frac{d}{dx}v^{1/3} = \frac{d}{dv}v^{1/3}\frac{dv}{dx} = \tfrac{1}{3}v^{-2/3}(6x)$$

Now put the pieces together:

$$\frac{dy}{dx} = \tfrac{1}{4}u^{-3/4}[\tfrac{1}{3}v^{-2/3}(6x)+1]$$

$$= \frac{1}{4[(3x^2+1)^{1/3}+x]^{3/4}}\left[\frac{2x}{(3x^2+1)^{2/3}}+1\right]$$

●

Remark After some practice, you will soon do problems like these by inspection, without using the intermediate variables u, v, and so on. In numerical work, however, the extra variables can be useful because they split the calculation into several stages, for which you may have subroutines. For instance, in Example 5a, suppose you had to compute dy/dx for 25 values of x. Then it would make good sense to organize the computation in the form

$$v = 3x^2 + 1 \qquad u = v^{1/3} + 1 \qquad w = u^{3/4}v^{2/3} \qquad y' = x/2w$$

This sort of organization is particularly helpful when you are using a programmable calculator or computer.

A Friendly Tip Make a special effort to get the chain rule straight, because misuse of the chain rule is the cause of most mistakes in differentiation. Answers can end up rather long and messy, as Example 5 shows. Therefore keep cool, work systematically, and especially take care not to forget the innermost differentiation.

Example 6

Typical mistake	Correct answer
$\dfrac{d}{dx}(5x)^3 = 3(5x)^2$	$3(5x)^2 \cdot 5$
$\dfrac{d}{dx}\dfrac{1}{2x+7} = \dfrac{-1}{(2x+7)^2}$	$\dfrac{-2}{(2x+7)^2}$
$\dfrac{d}{dx}\dfrac{1}{7-2x} = \dfrac{-2}{(7-2x)^2}$	$\dfrac{2}{(7-2x)^2}$
$\dfrac{d}{dx}\sqrt{3+4x^2} = \dfrac{1}{2\sqrt{3+4x^2}} \cdot 4$	$\dfrac{1}{2\sqrt{3+4x^2}} \cdot 8x$
$\dfrac{d}{dx}[1+5(1+4x)^2]^3 = 3[1+5(1+4x)^2]^2[10(1+4x)]$	$3[1+5(1+4x)^2]^2 \times [10(1+4x)] \cdot 4$

Differentiability of x^r [Optional]

We tackle the proof that $y(x) = x^r$ is differentiable before the chain rule proof because it's easier.

Let $r = m/n$ with integers m and n. It suffices to prove that $x^{1/n}$ is differentiable because then $x^r = (x^{1/n})^m$ will be differentiable by the chain rule. Thus let $y = \sqrt[n]{x}$ and $b = \sqrt[n]{a}$, where $a > 0$. First we prove $\sqrt[n]{x}$ is continuous at $x = a$, then pull ourselves up by our bootstraps to prove $\sqrt[n]{x}$ is differentiable at $x = a$.

We start with the algebraic identity

$$(y - b)(y^{n-1} + by^{n-2} + b^2y^{n-3} + \cdots + b^{n-1})$$
$$= y^n - b^n = x - a$$

We solve for $y - b$:

$$\sqrt[n]{x} - \sqrt[n]{a} = y - b = F(x)(x - a)$$

where

$$F(x) = \frac{1}{y^{n-1} + by^{n-2} + \cdots + b^{n-1}}$$

Remember that $a > 0$, so that $b > 0$ also. We are trying to prove continuity at $x = a$, so we may assume $x > 0$. Then $y > 0$, and we have

$$y^{n-1} + by^{n-2} + \cdots + b^{n-1} > b^{n-1}$$

It follows that

$$0 < F(x) < \frac{1}{b^{n-1}}$$

By Lemma 2 on page 85, applied to $\sqrt[n]{x} - \sqrt[n]{a} = F(x)(x - a)$, we have

$$\lim_{x \to a} (\sqrt[n]{x} - \sqrt[n]{a}) = 0 \qquad \text{that is} \qquad \lim_{x \to a} \sqrt[n]{x} = \sqrt[n]{a}$$

This proves $\sqrt[n]{x}$ is continuous for $x > 0$.

By the formula above for $F(x)$, the difference quotient may be written

$$\frac{\sqrt[n]{x} - \sqrt[n]{a}}{x - a} = \frac{1}{y^{n-1} + by^{n-2} + \cdots + b^{n-1}}$$

Now by continuity, $\lim_{x \to a} y = b$; hence

$$\lim_{x \to a} (y^{n-1} + by^{n-2} + b^2y^{n-3} + \cdots + b^{n-1})$$
$$= b^{n-1} + b^{n-1} + b^{n-1} + \cdots + b^{n-1} = nb^{n-1}$$

Therefore

$$\lim_{x \to a} \frac{\sqrt[n]{x} - \sqrt[n]{a}}{x - a} = \frac{1}{nb^{n-1}} = \frac{1}{na^{(n-1)/n}} = \frac{1}{n} a^{1/n - 1}$$

which completes the proof that $x^{1/n}$ is differentiable for $x > 0$. In fact, it shows that

$$\frac{d}{dx} x^{1/n} = \frac{1}{n} x^{1/n - 1} \quad \text{for} \quad x > 0$$

in agreement with the power rule. ●

Proof of the Chain Rule [Optional]

Let us recall what is given and what is to be proved. First, we assume $y = f(u)$ is differentiable at $u = b$. Next, we assume that the range of $g(x)$ is a subset of the domain of $f(u)$ and that $b = g(a)$. Finally, we assume that $u = g(x)$ is differentiable at $x = a$. Thus we have

$$\lim_{u \to b} \frac{f(u) - f(b)}{u - b} = f'(b)$$

and [remember that $g(a) = b$]

$$\lim_{x \to a} \frac{g(x) - b}{x - a} = g'(a)$$

We must prove

$$\lim_{x \to a} \frac{f[g(x)] - f(b)}{x - a} = f'(b)g'(a)$$

If we are not too fussy, we can simply write

$$\lim_{x \to a} \frac{f[g(x)] - f(b)}{x - a} = \lim_{x \to a} \left(\frac{f[g(x)] - f(b)}{g(x) - b} \right) \left(\frac{g(x) - b}{x - a} \right)$$
$$= \left(\lim_{u \to b} \frac{f(u) - f(b)}{u - b} \right) \left(\lim_{x \to a} \frac{g(x) - g(a)}{x - a} \right) = f'(b)g'(a)$$

This argument is flawed in two ways. First of all, $g(x) - b$ may equal 0 for many values of x near a, so that we are possibly dividing by 0. Second, we would have to justify the switch from $x \to a$ to $u \to b$ in the first factor.

Both of these difficulties can be overcome. However, we prefer a different approach that proves the chain rule without these problems, and that gives more insight into differentiability.

Lemma Suppose $f(u)$ has domain D and is differentiable at $u = b$. Then there is a continuous function $F(u)$ with domain D such that $F(b) = 0$ and

$$f(u) = f(b) + (u - b)[f'(b) + F(u)]$$ ●

Proof We simply define $F(u)$ by $F(b) = 0$ and

$$F(u) = \frac{f(u) - f(b)}{u - b} - f'(b) \quad \text{for} \quad u \neq b$$

Because $f(u)$ is continuous on D, it follows that $F(u)$ is continuous at all points of D different from b. It is also continuous at b. That is because the differentiability of $f(u)$ at $u = b$ implies

$$\lim_{u \to b} F(u) = f'(b) - f'(b) = 0 = F(b)$$

What is more, $f(u)$ and $F(u)$ are related as asserted. For $u = b$ this is obvious, and for $u \neq b$, it follows from the formula that defines $F(u)$ in that case. This completes the proof of the lemma. ●

To prove the chain rule, we apply the lemma both to $f(u)$ at $u = b$ and to $g(x)$ at $x = a$:

$$y = f(u) = f(b) + (u - b)[f'(b) + F(u)]$$

where $\lim_{u\to b} F(u) = F(b) = 0$ and

$$u = g(x) = b + (x - a)[g'(a) + G(x)]$$

where $\lim_{x\to a} G(x) = G(a) = 0$.

We substitute $g(x)$ for u in the first formula, using the second formula to eliminate the factor $u - b = g(x) - b$:

$$\begin{aligned} y &= f[g(x)] \\ &= f(b) + (x - a)[g'(a) + G(x)] \cdot [f'(b) + F(g(x))] \end{aligned}$$

Consequently, the difference quotient for the composite function is

$$\frac{f[g(x)] - f(b)}{x - a} = [g'(a) + G(x)] \cdot [f'(b) + F(g(x))]$$

To find its limits as $x \to a$, we compute limits of its two factors. First,

$$\begin{aligned} \lim_{x\to a} [g'(a) + G(x)] &= g'(a) + \lim_{x\to a} G(x) \\ &= g'(a) + 0 = g'(a) \end{aligned}$$

and second (because F is continuous),

$$\begin{aligned} \lim_{x\to a} [f'(b) + F(g(x))] &= f'(b) + \lim_{x\to a} F[g(x)] \\ &= f'(b) + F[\lim_{x\to a} g(x)] = f'(b) + F(b) = f'(b) \end{aligned}$$

It follows that the difference quotient for $f[g(x)]$ has limit $f'(b)g'(a)$ as $x \to a$. This completes the proof of the chain rule. ●

Exercises

Differentiate with respect to x

1 $(x^3 + 1)^4 - 3(x^3 + 1)^2$

2 $(x^2 - x)^9 + 4(x^2 - x)^7$

3 $(x + 1/x)^5 - (x + 1/x)^3$

4 $5(1 - 2x)^7 - 7(1 - 2x)^5$

5 $-3(2x - 1/x)^{10}$

6 $\dfrac{3x^6 + 1}{3x^6 - 1}$

7 $(1 - x^2)^5$

8 $(x^3 - 2)^4$

9 $(x^2 - 2x + 1)^3$

10 $(\frac{1}{3}x - 1)^6$

11 $(x^2 + 1)^3(x - 1)^2$

12 $(x^2 + 1)^2(x - 1)^3$

13 $(x^3 + 6)^2/x$

14 $(x^2 + 2)^4/x$

15 $1/x^9$

16 $\dfrac{1}{(4x + 3)^5}$

17 $\dfrac{1}{(x^2 + x)^4}$

18 $\dfrac{1}{(2x - 3)^6}$

19 $\left(\dfrac{x + 1}{x + 3}\right)^2$

20 $\dfrac{x^3 + 1}{x^3 - 1}$

21 $\dfrac{1}{(x^2 + 1)^6}$

22 $\dfrac{(x^2 - 1)^2}{(x^2 + 3)^7}$

23 $\dfrac{3x^2 + 1}{2x^2 - 1}$

24 $\dfrac{x^3}{(x^3 + 1)^2}$

25 $\sqrt{x + 3}$

26 $\sqrt{x^2 + x}$

27 $\sqrt{\dfrac{x}{x + 1}}$

28 $(1 + 2\sqrt{x})^3$

29 $\dfrac{x}{\sqrt{1 - x^2}}$

30 $\sqrt{\dfrac{x + a}{x - a}}$

31 $x^2\sqrt{x^2 - a^2}$

32 $\sqrt{x + \sqrt{x}}$

33 $\dfrac{4 - x}{\sqrt{8x - x^2}}$

34 $\dfrac{1 - \sqrt{3x}}{1 + \sqrt{3x}}$

35 $\left(\dfrac{x}{\sqrt{x}+1}\right)^3$

36 $\left(\dfrac{2}{x^2}+\dfrac{3}{x^5}\right)^{-2}$

37 $\dfrac{1}{\sqrt{1+2x+3x^2}}$

38 $\dfrac{1}{\sqrt{2x+1}}$

39 $[(3x^2+1)^2+1]^{1/5}$

40 $[(x^2+2x)^{1/3}-1]^{1/2}$

41 $[(x^2+2x)^{1/3}-3x]^{1/2}$

***42** $\dfrac{\left(\dfrac{x+1}{x-1}\right)^{1/2}+1}{\left(\dfrac{x+1}{x-1}\right)^{1/2}-1}$

43 $(x^3+2x)^3 + 2(x^3+2x)$

44 $\dfrac{1}{2+\sqrt[3]{x}}$

45 $\dfrac{1}{x\sqrt[3]{x^2+a^2}}$

46 $\sqrt[5]{x+\sqrt[3]{x+1}}$

Find dy/dx for $y =$

47 $\dfrac{1}{1-2u}$ where $u = \dfrac{x-1}{2x}$

48 $\dfrac{-u+1}{u-2}$ where $u = \dfrac{2x+1}{x+1}$

49 u^2+1 where $u = v^2+1$ and $v = x^2+1$

50 u^2+1 where $u = v^3+1$ and $v = x^4+1$

51 $\sqrt{u+1}$ where $u = x^{2/3}+1$

52 $u^{1/5}$ where $u = v^2+1$ and $v = 1/x$

53 u^2+u+1 where $u = \sqrt[3]{v}$ and $v = x^2+x$

54 u^{-3} where $u = (v+1)/(v+3)$ and $v = \sqrt[3]{x}$

2-9 Trigonometric Functions

If your familiarity with the trigonometric functions has faded, now is the time to renew it by reading Sections 1-6 and 1-7. In this section, we shall find the derivatives of $y = \sin x$ and $y = \cos x$ where, as usual in calculus, the variable x is measured in radians. You should recall the graphs of these functions (Figure 2-9-1). We now state the formulas for $(\sin x)'$ and $(\cos x)'$, and shall give their proofs after working some examples.

Figure 2-9-1

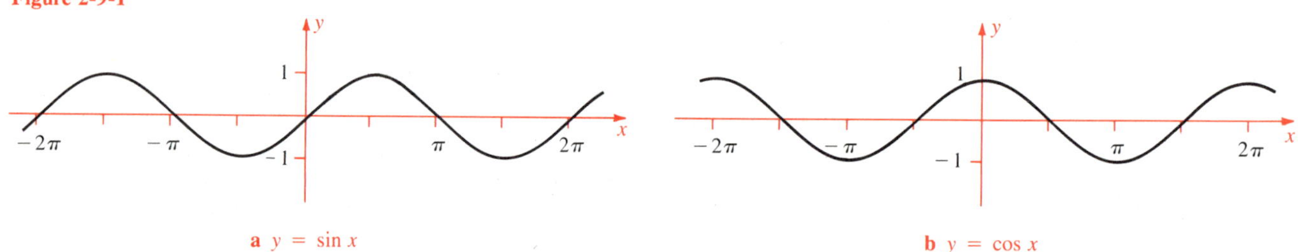

a $y = \sin x$ b $y = \cos x$

Derivatives of Sine and Cosine

$$\frac{d}{dx}(\sin x) = \cos x \qquad \frac{d}{dx}(\cos x) = -\sin x$$

For the geometric meaning of these formulas, see Figure 2-9-2.

We can easily differentiate many other functions by combining these formulas with our previous rules and derivative formulas.

Example 1

Differentiate **a** $y = x\sin x + \cos x$ **b** $y = \frac{1}{2}x^2\cos x$

Solution

a $y' = (x\sin x)' + (\cos x)' = [(x)'(\sin x) + x(\sin x)'] - \sin x$
$= (\sin x + x\cos x) - \sin x = x\cos x$

b $y' = \frac{1}{2}(x^2)'(\cos x) + \frac{1}{2}x^2(\cos x)' = x\cos x - \frac{1}{2}x^2\sin x$

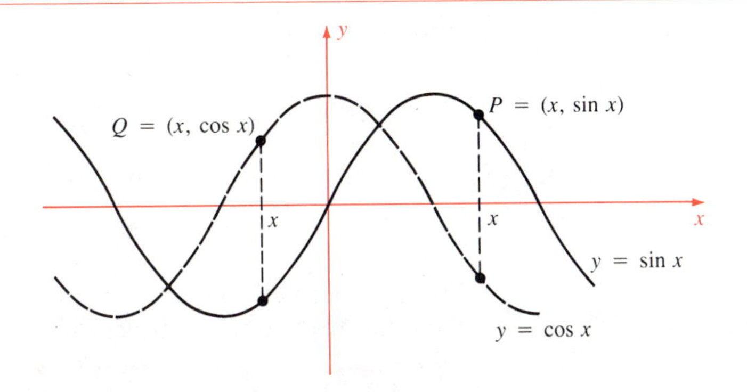

Figure 2-9-2
The graph of $y = \cos x$ is just the graph of $y = \sin x$, displaced horizontally. The slope of $y = \sin x$ at $P = (x, \sin x)$ equals $\cos x$. The slope of $y = \cos x$ at $Q = (x, \cos x)$ equals $-\sin x$.

Example 2

Differentiate

a $\sin \frac{1}{3}x$ **b** $\sqrt{8 - 7\cos 2x}$ **c** $\cos^2 x$

Solution

a Let $u = \frac{1}{3}x$. By the chain rule,

$$\frac{d}{dx}\sin u = (\cos u)\frac{du}{dx} = (\cos \tfrac{1}{3}x)\cdot\tfrac{1}{3} = \tfrac{1}{3}\cos\tfrac{1}{3}x$$

b Let $u = 8 - 7\cos 2x$. Then

$$\frac{d}{dx}\sqrt{u} = \frac{1}{2\sqrt{u}}\cdot\frac{du}{dx} = \frac{1}{2\sqrt{u}}\cdot\frac{d}{dx}(8 - 7\cos 2x)$$
$$= \frac{1}{2\sqrt{u}}\left[-7\frac{d}{dx}\cos 2x\right]$$
$$= \frac{1}{2\sqrt{u}}[-7(-\sin 2x)\cdot 2] = \frac{7\sin 2x}{\sqrt{8 - 7\cos 2x}}$$

Note that two applications of the chain rule are needed. Don't forget the second.

c Let $u = \cos x$. Then

$$\frac{d}{dx}u^2 = 2u\frac{du}{dx} = (2\cos x)(-\sin x) = -2\cos x\sin x$$

The answer can also be written $-\sin 2x$. However, note the relation $\cos^2 x = \frac{1}{2}(\cos 2x + 1)$. ●

Example 3

Show that

$$\frac{d}{dx}(\sin x - \tfrac{1}{3}\sin^3 x) = \cos^3 x$$

Solution

$$\frac{d}{dx}(\sin x - \tfrac{1}{3}\sin^3 x) = \frac{d}{dx}\sin x - \tfrac{1}{3}\frac{d}{dx}\sin^3 x$$
$$= \cos x - \tfrac{1}{3}(3\sin^2 x)(\cos x)$$
$$= (\cos x)(1 - \sin^2 x) = \cos^3 x \qquad \bullet$$

From the derivatives of $\sin x$ and $\cos x$, we easily find the derivatives of the other trigonometric functions:

Derivatives of Tangent, Cotangent, Secant, and Cosecant

$$\frac{d}{dx}\tan x = \sec^2 x \qquad \frac{d}{dx}\sec x = \sec x \tan x$$

$$\frac{d}{dx}\cot x = -\csc^2 x \qquad \frac{d}{dx}\csc x = -\csc x \cot x \qquad \bullet$$

For example, to find the derivative of $\tan x$, we have

$$\frac{d}{dx}\tan x = \frac{d}{dx}\left(\frac{\sin x}{\cos x}\right)$$
$$= \frac{\left(\frac{d}{dx}\sin x\right)\cos x - \sin x \frac{d}{dx}(\cos x)}{\cos^2 x}$$
$$= \frac{(\cos x)(\cos x) - (\sin x)(-\sin x)}{\cos^2 x}$$
$$= \frac{\cos^2 x + \sin^2 x}{\cos^2 x}$$
$$= \frac{1}{\cos^2 x} = \sec^2 x$$

Another example:

$$\frac{d}{dx}\sec x = \frac{d}{dx}\left(\frac{1}{\cos x}\right) = -\frac{\frac{d}{dx}\cos x}{\cos^2 x} = -\frac{-\sin x}{\cos^2 x}$$
$$= \frac{1}{\cos x} \cdot \frac{\sin x}{\cos x} = \sec x \tan x$$

The derivatives of $\cot x$ and $\csc x$ are found similarly.

By using trigonometric identities, we can express these formulas in many different forms. For instance

$$\frac{d}{dx}(\tan x) = \sec^2 x = 1 + \tan^2 x$$

$$\frac{d}{dx}(\sec x) = \sec x \tan x = \frac{\sin x}{\cos^2 x} = \frac{\sin x}{1 - \sin^2 x}$$

Example 4

Differentiate

a $\tan^4 3x$ **b** $\sec x \cot x$

Solution

a $$\frac{d}{dx}(\tan 3x)^4 = 4(\tan 3x)^3 \frac{d}{dx}\tan 3x$$
$$= 4(\tan^3 3x)(3\sec^2 3x) = 12\tan^3 3x \sec^2 3x$$

b $$\frac{d}{dx}(\sec x \cot x) = \left(\frac{d}{dx}\sec x\right)\cot x + (\sec x)\left(\frac{d}{dx}\cot x\right)$$
$$= (\sec x \tan x)\cot x + (\sec x)(-\csc^2 x)$$
$$= \sec x - \sec x \csc^2 x$$
$$= (\sec x)(1 - \csc^2 x) = (\sec x)(-\cot^2 x)$$
$$= \left(\frac{1}{\cos x}\right)\left(-\frac{\cos^2 x}{\sin^2 x}\right) = -\frac{\cos x}{\sin^2 x}$$

Alternative Solution of b It often pays to use identities *before* differentiating:

$$\sec x \cot x = \frac{1}{\cos x}\cdot\frac{\cos x}{\sin x} = \frac{1}{\sin x} = \csc x$$

Therefore

$$\frac{d}{dx}(\sec x \cot x) = \frac{d}{dx}\csc x = -\csc x \cot x$$

●

Proofs of the Differentiation Formulas

It suffices to prove

a $\left(\frac{d}{dx}\sin x\right)_{x=0} = 1$ and **b** $\left(\frac{d}{dx}\cos x\right)_{x=0} = 0$

Suppose these two relations are indeed true. By the addition law for sine (page 52),

$$\sin(a + x) = \sin a \cos x + \cos a \sin x$$

By the chain rule (review Example 2a, page 117)

$$\left(\frac{d}{dx}\sin x\right)_{x=a} = \left(\frac{d}{dx}\sin(a+x)\right)_{x=0}$$
$$= (\sin a)\left(\frac{d}{dx}\cos x\right)_{x=0} + (\cos a)\left(\frac{d}{dx}\sin x\right)_{x=0}$$
$$= (\sin a)\cdot 0 + (\cos a)\cdot 1 = \cos a$$

Therefore $(\sin x)' = \cos x$. Similarly, by the addition law for cosines

$$\cos(a + x) = \cos a \cos x - \sin a \sin x$$

it follows that $(\cos x)' = -\sin x$.

Thus it only remains to prove **a** and **b**. To do so, we need some inequalities. Now look back at page 95. There we proved

$$\begin{cases} |\sin h| < |h| \\ |\cos h - 1| < |h|^2 \end{cases} \quad \text{for} \quad 0 < |h| < \tfrac{1}{2}\pi$$

From the second of these inequalities we have

$$\left| \frac{\cos h - 1}{h} \right| < |h|$$

The left side is the difference quotient for $\cos x$ at $x = 0$. Consequently

$$\left(\frac{d}{dx} \cos x \right)_{x=0} = \lim_{h \to 0} \frac{\cos h - \cos 0}{h - 0} = \lim_{h \to 0} \frac{\cos h - 1}{h} = 0$$

This takes care of **b**. For the sine derivative we need a further geometric inequality. The shaded sector in Figure 2-9-3 has area $\frac{1}{2}h$ because

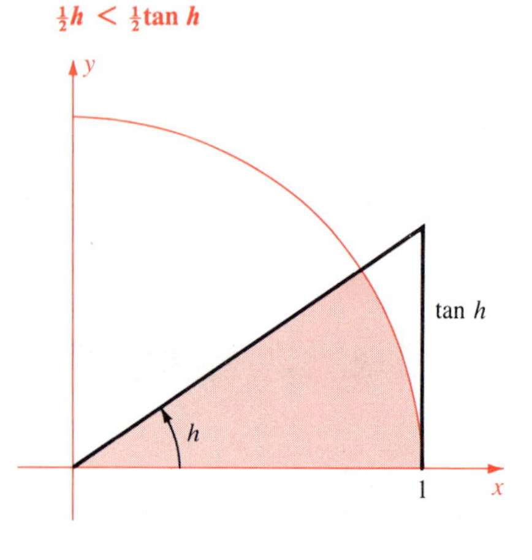

Figure 2-9-3
The area of the shaded sector is less than the area of the triangle:

$\frac{1}{2}h < \frac{1}{2}\tan h$

$$\frac{\text{shaded area}}{\pi \cdot 1^2} = \frac{\text{shaded area}}{\text{area of circle}} = \frac{h}{2\pi}$$

The triangle has area $= \frac{1}{2}(\text{base})(\text{height}) = \frac{1}{2}(1)(\tan h) = \frac{1}{2}\tan h$. Since the area of the shaded sector is less than the area of the triangle,

$$\frac{h}{2} < \tfrac{1}{2}\tan h = \tfrac{1}{2}\,\frac{\sin h}{\cos h}$$

which implies $\cos h < (\sin h)/h$ for $0 < h < \frac{1}{2}\pi$. But we also have $|\sin h| < |h|$ so that $\sin h < h$ for $h > 0$. Putting this all together we have

$$\cos h < \frac{\sin h}{h} < 1 \quad \text{for} \quad 0 < h < \tfrac{1}{2}\pi$$

The three terms in this inequality are all even functions, so the inequality is true for $0 < |h| < \frac{1}{2}\pi$. Since $\lim_{h\to 0} \cos h = \cos 0 = 1$, it is clear from the inequality that $\lim_{h\to 0} (\sin h)/h = 1$. From this,

$$\left(\frac{d}{dx} \sin x \right)_{x=0} = \lim_{h \to 0} \frac{\sin h - \sin 0}{h - 0} = \lim_{h \to 0} \frac{\sin h}{h} = 1$$

which completes the proof. ●

Squeezing a Limit

The proof above used a certain "obvious" principle for squeezing a limit between two other limits. We have not yet stated it explicitly; it is contained in the lemmas below.

Lemma 1 Squeezing to 0 If $|f(x)| \le |g(x)|$ and $\lim_{x\to a} g(x) = 0$, then $\lim_{x\to a} f(x) = 0$. ●

Proof Let $\varepsilon > 0$. There is a δ such that $|g(x)| < \varepsilon$ whenever $0 < |x - a| < \delta$. Consequently $|f(x)| \le |g(x)| < \varepsilon$ whenever $0 < |x - a| < \delta$. Therefore $\lim_{x\to a} f(x) = 0$ by the definition of limit. ●

Lemma 2 "Squeezing" Principle If $f(x) \le g(x) \le h(x)$, if $\lim_{x\to a} f(x) = L$, and if $\lim_{x\to a} h(x) = L$, then $\lim_{x\to a} g(x) = L$. ●

Proof We have $0 \le g(x) - f(x) \le h(x) - f(x)$ and

$$\lim_{x\to a} [h(x) - f(x)] = \lim_{x\to a} h(x) - \lim_{x\to a} f(x) = L - L = 0$$

By Lemma 1, we have $\lim_{x\to a} [g(x) - f(x)] = 0$. But we may write $g(x) = [g(x) - f(x)] + f(x)$. Therefore

$$\lim_{x\to a} g(x) = \lim_{x\to a} [g(x) - f(x)] + \lim_{x\to a} f(x) = 0 + L = L \quad ●$$

Remark With little extra effort we can actually prove

$$\lim_{h\to 0} \frac{1 - \cos h}{h^2} = \tfrac{1}{2}$$

Because $(1 - \cos h)(1 + \cos h) = 1 - \cos^2 h = \sin^2 h$, we have

$$1 - \cos h = (\sin^2 h)\frac{1}{1 + \cos h}$$

$$\frac{1 - \cos h}{h^2} = \left(\frac{\sin h}{h}\right)^2 \frac{1}{1 + \cos h}$$

$$\lim_{h\to 0} \frac{1 - \cos h}{h^2} = \left(\lim_{h\to 0} \frac{\sin h}{h}\right)^2 \frac{1}{\lim_{h\to 0}(1 + \cos h)} = 1^2 \cdot \tfrac{1}{2} = \tfrac{1}{2}$$

See Table 2-9-1 for numerical evidence.

Table 2-9-1

h	$\dfrac{1 - \cos h}{h^2}$
0.50	0.49
0.10	0.4996
0.05	0.49989 6
0.01	0.49999 58

Angles in Degrees

Our differentation formulas for the trigonometric functions require angles to be in radians. We can also differentiate these functions when angles are expressed in degrees or in some other system, but the formulas do not turn out as simple. For example, suppose we want $d(\sin\theta)/d\theta$ when θ is in degrees. We convert to radians, $\theta \text{ deg} = (\pi/180)\theta \text{ rad}$, then differentiate using our standard formula and the chain rule:

$$\frac{d}{d\theta} \sin\left(\frac{\pi}{180}\theta\right) = \frac{\pi}{180}\left(\cos\frac{\pi}{180}\theta\right)$$

Therefore

$$\frac{d}{d\theta}\sin\theta° = \frac{\pi}{180}\cos\theta°$$

The awkward constant $\pi/180$ will appear in the derivatives of all trigonometric functions if they are expressed in degrees.

Exercises

Differentiate

1 $\sin^4 x$
2 $\cos^3 x$
3 $x\cos x - \sin x$
4 $x^2 \sin x$
5 $\sqrt{\cos x}$
6 $\sin(x + a)$
7 $(\sin x)/x$
8 $\sin(1/x)$
9 $x\tan x$
10 $(\tan x)\sqrt{x}$
11 $\sec^2 x$
12 $(\sin x)(1 + \cos x)$
13 $\dfrac{\sin x + \cos x}{\sin x - \cos x}$
14 $\dfrac{\sec x}{1 + \tan x}$
15 $\sec x - \csc x$
16 $(\cot x - 1)^3$
17 $\tan x + \cot x$
18 $\tan^5 x$
19 $\sec^3 x - 3\tan^2 x$
20 $\csc(1/x^2)$
21 $\sec^4 x - \tan^4 x$
22 $x + \sec x - \tan x$
23 $\sqrt{1 + \sec^2 x}$
24 $x^3\sqrt{1 + \cot^2 x}$

2-10 Implicit and Inverse Functions

In this section we shall learn two useful methods for computing derivatives. For the time being we shall postpone the theory behind the first method, implicit differentiation, but we shall return to it in Chapter 14.

Sometimes a function $y = f(x)$ is not known explicitly, but a relation between it and the variable x is known. Then y is called an **implicit function** of x. In such a case $f'(x)$ can be found by differentiating both sides of the given relation, using the chain rule as is appropriate.

Example 1

Suppose $y = y(x)$ is the function with domain $(-1, 1)$ such that

$$x^2 + y^2 = 1 \quad \text{and} \quad y > 0$$

Express dy/dx in terms of x and y.

Figure 2-10-1
Graph of $x^2 + y^2 = 1$. It defines two functions of x, and we choose the top one.

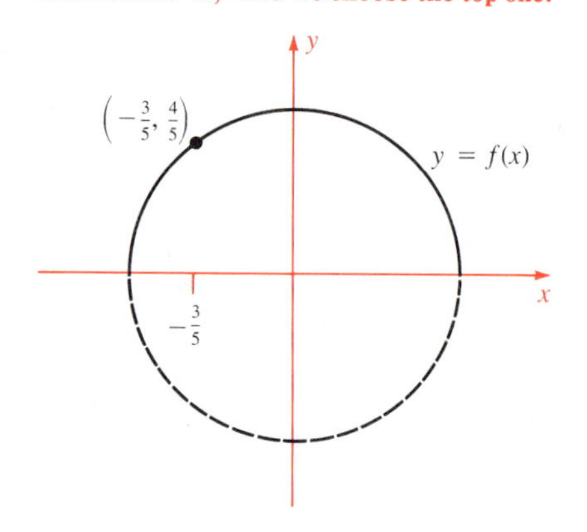

Solution First we graph the relation $x^2 + y^2 = 1$; the graph is the unit circle (Figure 2-10-1). This graph defines *two* functions of x. The top half of the graph defines one function, the bottom half the other. [We have chosen the top (solid) half, so that $y = y(x) > 0$ for $-1 < x < 1$.] To find dy/dx for (either) function, we differentiate both sides of $x^2 + y^2 = 1$ with respect to x:

$$\frac{d}{dx}(x^2 + y^2) = \frac{d}{dx}(1) \qquad 2x + \left(\frac{d}{dy}y^2\right)\left(\frac{dy}{dx}\right) = 0$$

$$2x + 2y\frac{dy}{dx} = 0 \quad \text{so} \quad \frac{dy}{dx} = -\frac{x}{y}$$

For instance, $y(-\frac{3}{5}) = \frac{4}{5}$, and the corresponding slope is

$$\left.\frac{dy}{dx}\right|_{-3/5} = -\frac{-\frac{3}{5}}{\frac{4}{5}} = \frac{3}{4}$$

Check In this case we can solve for y explicitly:

$$y = \sqrt{1 - x^2}$$

We differentiate directly, not implicitly, by the chain rule:

$$\frac{dy}{dx} = \frac{d}{dx}\sqrt{1 - x^2} = \frac{1}{2\sqrt{1 - x^2}}\frac{d}{dx}(1 - x^2)$$

$$= \frac{1}{2\sqrt{1 - x^2}}(-2x) = \frac{-x}{\sqrt{1 - x^2}} = -\frac{x}{y}$$ ●

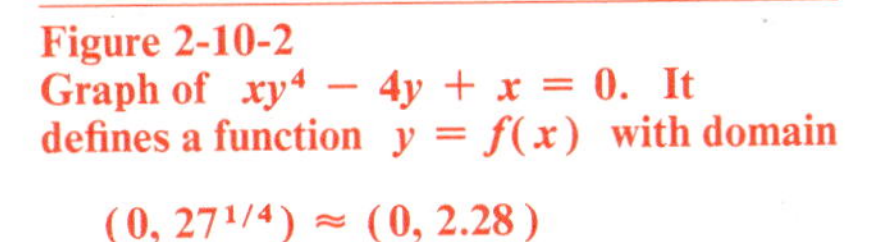

Figure 2-10-2
Graph of $xy^4 - 4y + x = 0$. It defines a function $y = f(x)$ with domain $(0, 27^{1/4}) \approx (0, 2.28)$

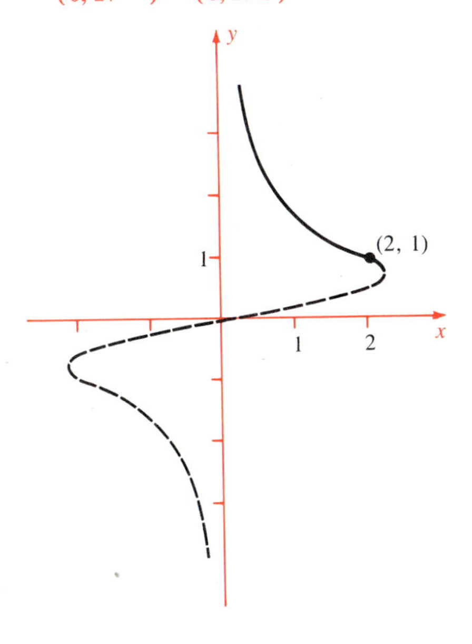

Example 2

The relation

$$xy^4 - 4y + x = 0$$

is graphed in Figure 2-10-2. Let $y = y(x)$ denote the function defined by the solid portion of the graph. Find its slope at $P = (2, 1)$.

Solution First check that the point P is indeed on the graph:

$$xy^4 - 4y + x = (2)(1^4) - 4(1) + 2 = 0$$

Now differentiate the relation $xy^4 - 4y + x = 0$ with respect to x:

$$y^4 + 4xy^3\frac{dy}{dx} - 4\frac{dy}{dx} + 1 = 0$$

Solve for dy/dx:

$$4(xy^3 - 1)\frac{dy}{dx} + (y^4 + 1) = 0 \qquad \frac{dy}{dx} = \frac{y^4 + 1}{4(1 - xy^3)}$$

Therefore

$$\left.\frac{dy}{dx}\right|_{(2,1)} = \frac{1^4 + 1}{4[1 - (2)(1^3)]} = \frac{2}{-4} = -\frac{1}{2}$$ ●

Figure 2-10-3
Graph of $y^6 + y + xy - x = 0$

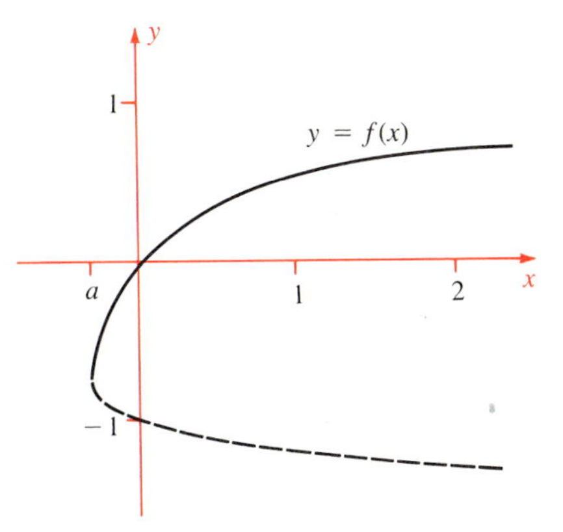

Example 3

Part of the graph of

$$y^6 + y + xy - x = 0$$

is shown in Figure 2-10-3. The given relation defines y as an implicit function of x on an interval $(a, +\infty)$ including $x = 0$. Find a formula for dy/dx in terms of x and y and find $y'(0)$.

Solution Differentiate the defining relation with respect to x:

$$\frac{d}{dx}(y^6 + y + xy - x) = \frac{d}{dx}0 = 0$$

There are four terms on the left-hand side:

$$\frac{d}{dx}y^6 = 6y^5\frac{dy}{dx} \qquad \frac{d}{dx}y = \frac{dy}{dx}$$

$$\frac{d}{dx}(xy) = 1 \cdot y + x\frac{dy}{dx} \qquad \frac{d}{dx}(-x) = -1$$

Therefore

$$6y^5\frac{dy}{dx} + \frac{dy}{dx} + y + x\frac{dy}{dx} - 1 = 0$$

Solve for dy/dx:

$$(6y^5 + x + 1)\frac{dy}{dx} + (y - 1) = 0 \qquad \frac{dy}{dx} = \frac{1 - y}{6y^5 + x + 1}$$

Since $(0, 0)$ lies on the graph,

$$\left.\frac{dy}{dx}\right|_{x=0} = \left.\frac{1 - y}{6y^5 + x + 1}\right|_{(x,y)=(0,0)} = \frac{1}{1} = 1$$

●

Inverse Functions

Inverse functions come in pairs. Each function of the pair undoes what the other does.

Example 4

The functions

$$f(x) = 2x - 5 \quad \text{and} \quad g(y) = \tfrac{1}{2}(y + 5)$$

are inverses of each other. To see that each undoes what the other does, we compute the two composite functions $g \circ f$ and $f \circ g$:

$$\begin{aligned}(g \circ f)(x) = g[f(x)] &= \tfrac{1}{2}[f(x) + 5] \\ &= \tfrac{1}{2}[(2x - 5) + 5] = x \\ (f \circ g)(y) = f[g(y)] &= 2g(y) - 5 \\ &= 2[\tfrac{1}{2}(y + 5)] - 5 = y\end{aligned}$$

Starting with x, applying f to it and applying g to the result gets us back to x, where we started. So g undoes f. Similarly, starting with y, applying g to it and applying f to the result gets us back to y, where we started. So f undoes g. ●

Example 5

$$f(x) = x^3 \qquad g(y) = y^{1/3}$$

This time

$$(g \circ f)(x) = g[f(x)] = g(x^3) = (x^3)^{1/3} = x$$
$$(f \circ g)(y) = f[g(y)] = f(y^{1/3}) = (y^{1/3})^3 = y$$

So again f and g undo each other. ●

Now we formally define inverse functions.

Definition Inverse Functions

Let $f(x)$ have domain D and $g(y)$ have domain E. The functions $f(x)$ and $g(y)$ are called **inverse functions** provided

- E is the range of $f(x)$ and D is the range of $g(y)$.
- $g[f(x)] = x$ for all x in D and $f[g(y)] = y$ for all y in E. In other words, $g \circ f$ is the identity function on D and $f \circ g$ is the identity function on E.

Each of the functions $f(x)$ and $g(y)$ is the **inverse function** of the other. ●

If $f(x)$ has an inverse function $g(y)$, then $f(x)$ is a one-to-one function on its domain D. That is $x_1 \neq x_2$ implies that $f(x_1) \neq f(x_2)$. For if $f(x_1) = f(x_2)$, then

$$g[f(x_1)] = g[f(x_2)] \qquad \text{that is} \qquad x_1 = x_2$$

Conversely, if $f(x)$ is a one-on-one function on its domain D, then $f(x)$ has an inverse function $g(y)$. For let E denote the range of $f(x)$. Then each y in E is the image under f of a *unique* x in D. That is, for each y in E, there exists a unique x in D such that $f(x) = y$. Set $x = g(y)$, defining g on E. Then $g[f(x)] = x$ by the definition of g, and also $f[g(y)] = f(x) = y$. That is, $f(x)$ and $g(y)$ are inverse functions.

The following theorem establishes the existence of inverse functions in most cases.

Theorem Existence of Inverse Functions

A continuous function $f(x)$ on an interval has an inverse function if and only if $f(x)$ is strictly increasing or strictly decreasing. In either case, the range of $f(x)$, which is the domain of its inverse function, is also an interval, and the inverse function $g(y)$ is continuous. ●

The complete proof of the statement is beyond the scope of this course. Observe, however, that if $f(x)$ is strictly increasing, then $x_1 < x_2$ implies $f(x_1) < f(x_2)$, so surely $f(x_1) \neq f(x_2)$. That is, $f(x)$ is a one-to-one function. The hard parts of the theorem are proving that the range of $f(x)$ is an interval and that the inverse function $g(y)$ is continuous. These are basic properties of continuous functions, and we'll have more to say on the subject in the next chapter.

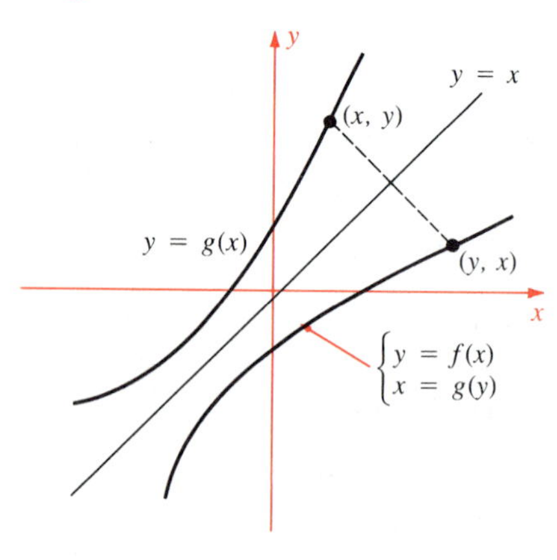

Figure 2-10-4
Graphs of inverse functions

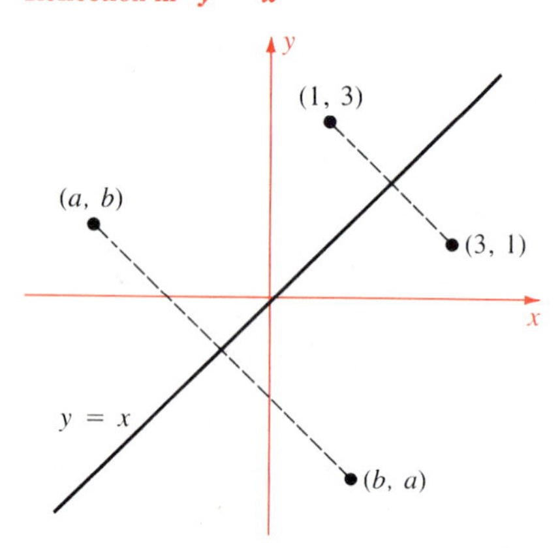

Figure 2-10-5
Reflection in $y = x$

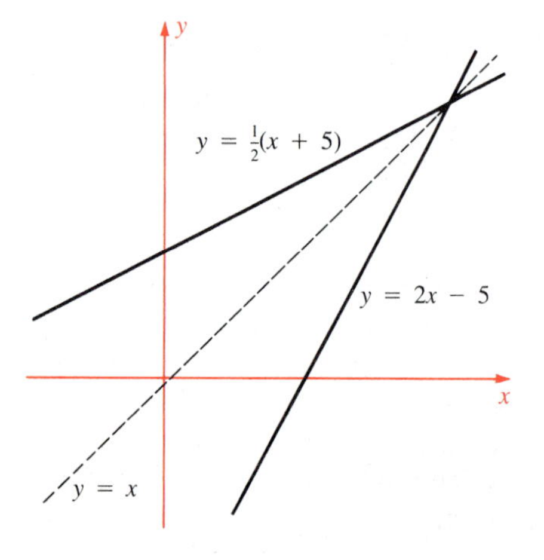

Figure 2-10-6
$f(x) = 2x - 5 \quad g(x) = \frac{1}{2}(x + 5)$

Graphs of Inverse Functions

Suppose $f(x)$ has an inverse function $g(y)$. What is the relation between the graphs of these two functions? The points on the graph of $y = f(x)$ can be written in two ways, either as

$$(x, f(x)) \quad \text{or as} \quad (g(y), y)$$

Hence the graph of $x = g(y)$ is the same as the graph of $y = f(x)$, except that the vertical y-axis is the axis of the independent variable for $x = g(y)$.

That should be all there is to it. Yet the notation $x = g(y)$ seems awkward, as we are accustomed to x as the *independent* variable and y as the *dependent* variable. So we reverse the roles of x and y and graph $y = g(x)$ instead. Then each point (x, y) on the graph of $y = g(x)$ corresponds to a point (y, x) on the graph of $y = f(x)$. See Figure 2-10-4.

The correspondence between points (x, y) and (y, x) is that each is the reflection of the other in the diagonal line $y = x$. See Figure 2-10-5. Therefore the relation between the graphs of $y = f(x)$ and $y = g(x)$, where f and g are a pair of inverse functions, is that each is the mirror image of the other in the line $y = x$. Examples are shown in Figures 2-10-6 and 2-10-7.

Remark If two non-horizontal lines are the reflections of each other in the line $y = x$, then their slopes are reciprocals. In Figure 2-10-6, the slopes are 2 and $\frac{1}{2}$. In general, suppose one of the lines is $y = ax + b$ with $a \neq 0$. Then $x = a^{-1}(y - b)$, so the other (reflected) line is $y = a^{-1}(x - b)$ with slope a^{-1}.

Derivatives of Inverse Functions

One way to look at inverse functions is in terms of solving equations. Given a one-to-one function $x = g(y)$, if we solve the relation $g(y) = x$ for y in terms of x, then the solution is the inverse function $y = f(x)$. For instance, if we solve $x = y^3 - 1$ for y, we find the inverse function $y = (x + 1)^{1/3}$.

Another example: $x = y^5 + y$ is a strictly increasing continuous function of y, so it has an inverse function. But it is not possible to solve for y explicitly in the form $y = f(x)$. Yet the inverse function $y = f(x)$ exists, even though we cannot write down a formula for it. (However, given any *particular* x, we can solve *approximately* for the corresponding y to any required accuracy, but not exactly.)

The function $g(y) = y^5 + y$ is a differentiable function, and its derivative is $g'(y) = 5y^4 + 1$. We ought to be able to differentiate the inverse function $f(x)$, assuming it is also a differentiable function. The following theorem takes care of this and gives a formula for the derivative.

Derivatives of Inverse Functions

Suppose $x = g(y)$ is a differentiable function with domain an open interval of the y-axis. Suppose $g'(y) > 0$ for all y in this domain. Then $g(y)$ is strictly increasing, and the range of $g(y)$ is an open interval on the x-axis. Also $g(y)$ has an inverse function $y = f(x)$, where $f(x)$ is a differentiable function with domain the range of $g(y)$.

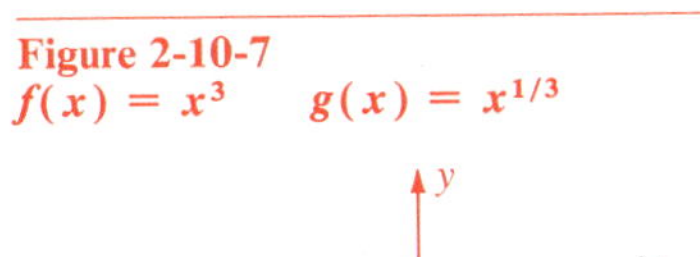

Figure 2-10-7
$f(x) = x^3 \qquad g(x) = x^{1/3}$

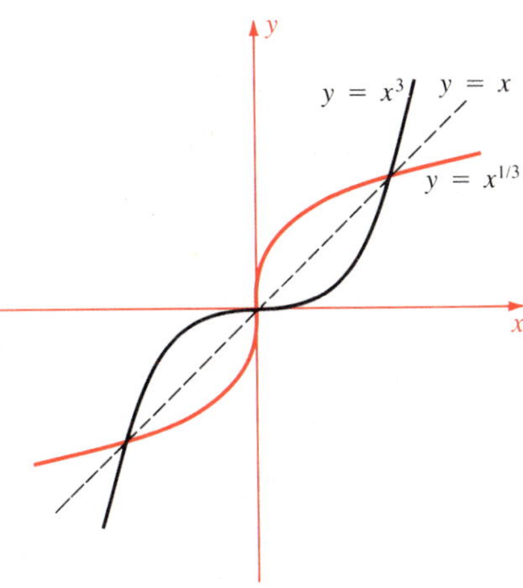

Finally,

$$\frac{d}{dx}f(x) = \frac{1}{\left.\dfrac{d}{dy}g(y)\right|_{y=f(x)}}$$

The conclusion that $g(y)$ is a strictly increasing function should be intuitively clear: if $g'(y)$, representing the rate of change of $g(y)$, is always positive, then $g(y)$ is always increasing. But that is not a proof, and we shall return to this point in the next chapter.

The final statement of the theorem is the important one for us here. If everything else is granted, then it is a consequence of the chain rule, applied to compute d/dx of both sides of $g(y) = x$, where $y = f(x)$:

$$\frac{d}{dx}g(y) = \frac{d}{dx}x \qquad \text{hence} \qquad \left(\left.\frac{d}{dy}g(y)\right|_{y=f(x)}\right)\left(\frac{dy}{dx}\right) = 1$$

The differentiation formula has a geometric interpretation (Figure 2-10-8). The graphs of $y = f(x)$ and $y = g(x)$ are reflections of each other in the line $y = x$. At corresponding points (a, b) and (b, a), their tangents F and G are also reflections of each other, so their slopes are reciprocal:

$$(\text{slope of } G)|_{x=b} = 1/(\text{slope of } F)|_{x=a}$$

However, the slope of F is the derivative $f'(a)$ and the slope of G is the derivative $g'(b)$. Therefore $g'(b) = 1/f'(a)$. (If F happens to be horizontal at a, then G is vertical at b. The slope of G is undefined, and $g(x)$ is not differentiable at the corresponding $x = b$.)

Figure 2-10-8
If $y = f(x)$ and $y = g(x)$ represent inverse functions, then the tangents at corresponding points have reciprocal slopes.

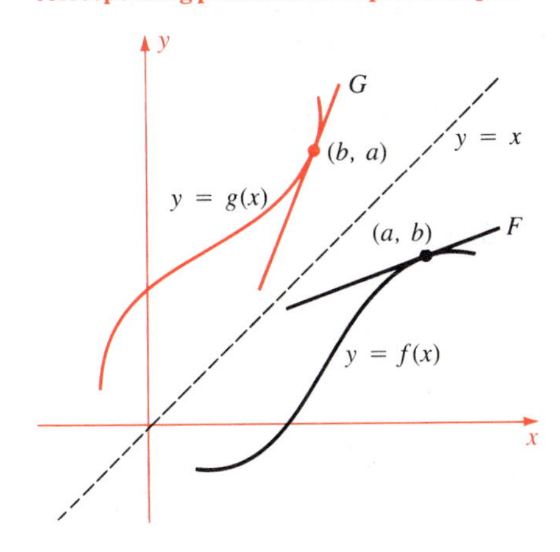

Example 6

Suppose

$$x = g(y) = y + \cos y \qquad (-\tfrac{3}{2}\pi < y < \tfrac{1}{2}\pi)$$

Show that there is an inverse function $y = f(x)$ with domain the open interval $(-\frac{3}{2}\pi, \frac{1}{2}\pi)$. Find

$$\left.\frac{dy}{dx}\right|_{x=1} \quad \text{and} \quad \left.\frac{dy}{dx}\right|_{x=-\pi/2}$$

Solution A graph (Figure 2-10-9, next page) helps us understand this example. First we must compute $g'(y)$:

$$g'(y) = \frac{d}{dy}g(y) = \frac{d}{dy}(y + \cos y) = 1 - \sin y$$

Now $\sin y \le 1$ with equality only at $y = \frac{1}{2}\pi + 2\pi n$ so $g'(y) > 0$ on the open interval $(-\frac{3}{2}\pi, \frac{1}{2}\pi)$. It is precisely at the end points that $g'(y) = 0$; that is why we stay inside the open interval $(-\frac{3}{2}\pi, \frac{1}{2}\pi)$. The function $x = g(y)$ increases in its domain $(-\frac{3}{2}\pi, \frac{1}{2}\pi)$. At the end points of this interval, $g(-\frac{3}{2}\pi) = -\frac{3}{2}\pi$ and $g(\frac{1}{2}\pi) = \frac{1}{2}\pi$. Therefore

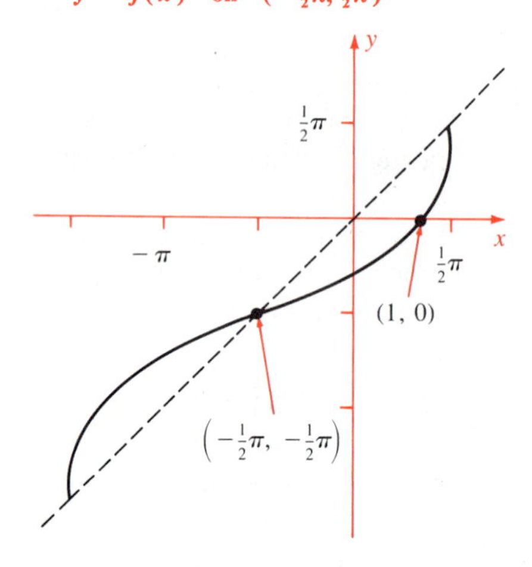

Figure 2-10-9
The graph of the function
$x = y + \cos y$ on $(-\frac{3}{2}\pi, \frac{1}{2}\pi)$
is also the graph of the inverse function
$y = f(x)$ on $(-\frac{3}{2}\pi, \frac{1}{2}\pi)$

the inverse function $y = f(x)$ also has domain $(-\frac{3}{2}\pi, \frac{1}{2}\pi)$. However, there is no way we can hope to solve the equation

$$y + \cos y = x$$

for $y = f(x)$ explicitly. In other words, there is no formula for $f(x)$. Still, we can compute dy/dx. In fact, the derivative formula yields

$$\frac{dy}{dx} = \frac{1}{g'(y)} = \frac{1}{1 - \sin y}$$

Since $g(0) = 1$ and $g(-\frac{1}{2}\pi) = -\frac{1}{2}\pi$ we have

$$\left.\frac{dy}{dx}\right|_{x=1} = \left.\frac{1}{1 - \sin y}\right|_{y=0} = 1$$

$$\left.\frac{dy}{dx}\right|_{x=-\pi/2} = \left.\frac{1}{1 - \sin y}\right|_{y=-\pi/2} = 2$$

●

Remark There is an intuitive explanation for the key relation $f'(x) = 1/g'(y)$ connecting the derivatives of inverse functions. Suppose x increases 3 times as fast as y. This means that a very small increase in y results in an increase in x of about 3 times as much. Now then, how much faster than x is y increasing? Well, a very small increase in x will result in an increase in y of about $\frac{1}{3}$ as much, so y is increasing $\frac{1}{3}$ as fast as x is at the instant in question. That's really all there is to the relation $dy/dx = 1/(dx/dy)$.

The method of differentiating an inverse function also applies if $x = g(y)$ has $g'(y) < 0$ on its domain. Then $g(y)$ is a strictly decreasing function so that its inverse function $y = f(x)$ exists, and $f'(x) = 1/g'(y)$.

Example 7

Suppose $x = g(y) = y^4$ with domain $(-\infty, 0)$. Find the derivative of the inverse function $y = f(x)$.

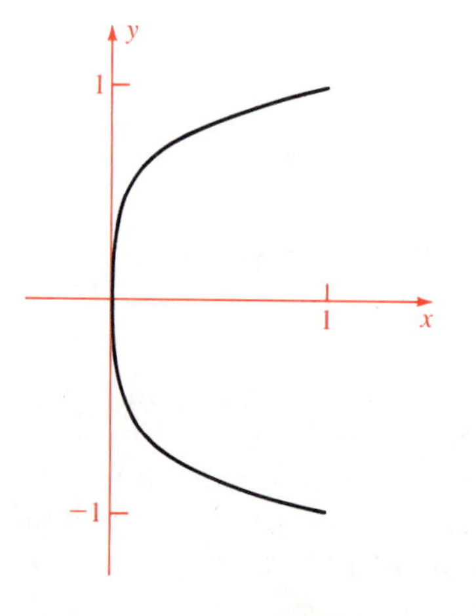

Figure 2-10-10
Graph of $x = y^4$

Solution As usual, a graph (Figure 2-10-10) helps us see what is happening. Evidently the domain of the inverse function $y = f(x)$ is $(0, +\infty)$. We can verify that $g(y)$ is strictly decreasing by computing its derivative:

$$\frac{dx}{dy} = g'(y) = 4y^3 \qquad \text{so that} \qquad g'(y) < 0 \quad \text{for} \quad y < 0$$

For the inverse function $y = f(x)$, we have

$$\frac{dy}{dx} = f'(x) = \frac{1}{g'(y)} = \frac{1}{4y^3}$$

In terms of x,

$$\frac{dy}{dx} = \frac{1}{4(-x^{1/4})^3} = -\tfrac{1}{4}x^{-3/4}$$

Check In this case we can carry out the solution explicitly for $f(x)$; namely $y = -x^{1/4}$, so

$$\frac{dy}{dx} = -\frac{d}{dx}x^{1/4} = -\tfrac{1}{4}x^{-3/4}$$

●

Remark The connection between implicit functions and inverse functions should not escape our notice. If $x = g(y)$, then we can write the *relation*

$$x - g(y) = 0$$

and differentiate with respect to x by the chain rule:

$$1 - g'(y)(dy/dx) = 0 \quad \text{so} \quad dy/dx = 1/g'(y)$$

We are back to the formula for the derivative of an inverse function.

Exercises

Find explicitly the function $y = f(x)$ defined implicitly by the given relation

1 $2x - y - 6 = 0$
2 $x^2y - 2x + 3y = 0$
3 $y^2 - 2xy - 4 = 0 \quad f(0) = -2$
4 $\sqrt{x} + \sqrt{y} = 3 \quad (x > 0,\ y > 0)$

Find a formula for $y'(x)$ in terms of x and y, where

5 $x^3 + y^3 = 1$
6 $x^2 - y^3 = -1$
7 $y^3 + xy + x^2 = 0$
8 $y^5 - x^3 = 0$
9 $x^2 + y^3 = xy$
10 $x^4 - y^4 = 3x^2y^3$
11 $x + y = \sin y$
12 $\sin y + xy = x$
13 $x^4 + 3y^4 = x + y$
14 $y + x \tan y = x$
15 $y = x^4 + x^3y^5$
16 $(x + 1)y = (y + 1)^3x$

Give the inverse function in the form $y = f(x)$

17 $x = 3y - 7$
18 $x = -2y + 5$
19 $x = -1/y \quad (y \neq 0)$
20 $x = 3/(10y - 7) \quad (y \neq \frac{7}{10})$
21 $x = (2y - 7)/(y + 4) \quad (y \neq -4)$
22 $x = (y + 1)/(y - 1) \quad (y \neq 1)$
23 $x = (y + 2)/(y + 3) \quad (y \neq -3)$
24 $x = (3y + 4)/(2y + 3) \quad (y \neq -\frac{3}{2})$
25 $x = (y^3 + 2)/(y^3 + 3) \quad (y^3 \neq -3)$
26 $x = (3y + 4)^3/(2y + 3)^3 \quad (y \neq -\frac{3}{2})$
27 $x = \sqrt{2y - 8} \quad (y > 4)$
28 $x = y^5 + 1$
29 $x = 9/(y + 7) - 7 \quad (y \neq -7)$
***30** $x = y + 1/y \quad (y > 1)$

Find $f'(c)$, where $y = f(x)$ is the inverse function of $x = g(y)$

31 $g(y) = y^3 + y \quad c = 30$
32 $g(y) = y^5 + y \quad c = 2$
33 $g(y) = y^3 + y^2 + y \quad c = -1$
34 $g(y) = y - 1/y \quad c = 0$
35 $g(y) = y^3/(1 + y^2) \quad c = -\frac{8}{5}$
36 $g(y) = y^7 + 2y^5 + y \quad c = 4$
37 $g(y) = y^{-3} + y^{-9} \quad c = -2$
38 $g(y) = 2y^3 - 9y^2 + 18y \quad c = 0$
39 $g(y) = \sin y$ domain $(-\frac{1}{2}\pi, \frac{1}{2}\pi) \quad c = \frac{1}{2}$
40 $g(y) = \cos y$ domain $(0, \pi) \quad c = 0$
41 $g(y) = 2y + \sin y \quad c = 0$
42 $g(y) = 3y - 2\cos y \quad c = -2$
43 $g(y) = \tan y$ domain $(-\frac{1}{2}\pi, \frac{1}{2}\pi) \quad c = 1$
44 $g(y) = \csc y$ domain $(0, \frac{1}{2}\pi) \quad c = \sqrt{2}$
45 $g(y) = y \sin y$ domain $(0, \frac{1}{2}\pi) \quad c = \frac{1}{12}\pi$
46 $g(y) = y \tan y$ domain $(0, \frac{1}{2}\pi) \quad c = \frac{1}{4}\pi$

2-11 Higher Derivatives

Start with a differentiable function $f(x)$. Its derivative $f'(x)$ is itself a function. If $f'(x)$ in turn is differentiable, then

$$\frac{d}{dx}[f'(x)]$$

is called the **second derivative** of $y = f(x)$ and is written

$$\frac{d^2y}{dx^2} \quad \text{or} \quad f''(x) \quad \text{or} \quad y''$$

A function that has a second derivative is called **twice differentiable.**

Remark The symbol d^2y/dx^2 is read "dee-two-y-dee-x-squared." The notation is natural in terms of *operators.* If we think of the second derivative operator as the composite of the operator d/dx with itself, then

$$\frac{d^2}{dx^2} = \left(\frac{d}{dx}\right) \circ \left(\frac{d}{dx}\right) = \left(\frac{d}{dx}\right)^2$$

Example 1

a $f(x) = x^3 - 4x \qquad f'(x) = 3x^2 - 4 \qquad f''(x) = 6x$

b $H(t) = 2(t-1)^4 \qquad \dfrac{dH}{dt} = 8(t-1)^3 \qquad \dfrac{d^2H}{dt^2} = 24(t-1)^2$ ●

Example 2

Compute the second derivative of **a** $y = \sqrt{x}$ **b** $y = (x^3 + 6)^5$

Solution

a $$\frac{dy}{dx} = \frac{d}{dx}\sqrt{x} = \frac{1}{2\sqrt{x}} = \tfrac{1}{2}x^{-1/2}$$

$$\frac{d^2y}{dx^2} = \frac{d}{dx}\left(\frac{dy}{dx}\right) = \frac{d}{dx}\left(\tfrac{1}{2}x^{-1/2}\right) = \tfrac{1}{2}\frac{d}{dx}\left(x^{-1/2}\right) = -\tfrac{1}{4}x^{-3/2}$$

b $$\frac{dy}{dx} = \frac{d}{dx}(x^3+6)^5 = 5(x^3+6)^4(3x^2) = 15x^2(x^3+6)^4$$

By the product and chain rules,

$$\begin{aligned}\frac{d^2y}{dx^2} &= \frac{d}{dx}\left(\frac{dy}{dx}\right) = \frac{d}{dx}[15x^2(x^3+6)^4] \\ &= 15\left[x^2\frac{d}{dx}(x^3+6)^4 + (x^3+6)^4\frac{d}{dx}x^2\right] \\ &= 15[x^2 \cdot 4(x^3+6)^3 \cdot 3x^2 + (x^3+6)^4 \cdot 2x] \\ &= 30x(x^3+6)^3[6x^3 + (x^3+6)] \\ &= 30x(x^3+6)^3(7x^3+6)\end{aligned}$$ ●

The second derivative $f''(x)$ is the rate of change with respect to x of the first derivative $f'(x)$; hence $f''(x)$ measures how fast the slope of the curve $y = f(x)$ is changing. This useful fact will be applied in the next chapter.

If the second derivative $f''(x)$ of $y = f(x)$ is differentiable, then its derivative $[f''(x)]'$ is called the **third derivative** of $f(x)$ and is written

$$\frac{d^3y}{dx^3} \quad \text{or} \quad f'''(x) \quad \text{or} \quad y''' \quad \text{or} \quad y^{(3)}$$

Example 3

a $f(x) = x^{10} \qquad f'(x) = 10x^9 \qquad f''(x) = 90x^8 \qquad f'''(x) = 720x^7$

b $g(x) = x^4 - 5x^3 + 3x + 7 \qquad g'(x) = 4x^3 - 15x^2 + 3$

$g''(x) = 12x^2 - 30x \qquad g'''(x) = 24x - 30$ ●

Fourth, fifth, and **higher derivatives** are defined similarly by repeated differentiation. The n-th derivative of $y = f(x)$ is written

$$\frac{d^ny}{dx^n} \quad \text{or} \quad f^{(n)}(x) \quad \text{or} \quad y^{(n)}$$

Example 4 (continuing Example 3)

a $f^{(4)}(x) = 720(7x^6) = 5040x^6 \qquad f^{(5)}(x) = 5040(6x^5) = 30{,}240x^5$

$f^{(6)}(x) = 30{,}240(5x^4) = 151{,}200x^4$ and so on

b $g^{(4)}(x) = 24 \qquad g^{(5)}(x) = 0 \qquad g^{(6)}(x) = g^{(7)}(x) = \cdots = 0$ ●

Example 5

Compute

a $y^{(4)}$ for $y = (\frac{1}{2}x - 3)^8$

b $y^{(5)}$ for $y = 19x^4 + 36x^3 + 499x^2 - 125x - 62$

Solution

a Differentiate four times, using the chain rule. Don't forget the factor $\frac{1}{2}$ at each step.

$$y' = 8(\tfrac{1}{2}x - 3)^7(\tfrac{1}{2}) \qquad y'' = 8 \cdot 7(\tfrac{1}{2}x - 3)^6(\tfrac{1}{2})^2$$

$$y''' = 8 \cdot 7 \cdot 6(\tfrac{1}{2}x - 3)^5(\tfrac{1}{2})^3$$

$$y^{(4)} = 8 \cdot 7 \cdot 6 \cdot 5(\tfrac{1}{2}x - 3)^4(\tfrac{1}{2})^4 = 105(\tfrac{1}{2}x - 3)^4$$

b By inspection $y^{(5)} = 0$, because the derivative of a polynomial is a polynomial of one lower degree. Hence y, y', y'', y''', $y^{(4)}$ have degrees 4, 3, 2, 1, and 0. Thus $y^{(4)}$ is a constant, so $y^{(5)} = 0$. ●

Example 6

Compute the first through fourth derivatives of

a $y = \sin x \qquad$ **b** $z = \cos x$

Solution

a $y' = \cos x \qquad y'' = (\cos x)' = -\sin x$

$y''' = (-\sin x)' = -\cos x \qquad y^{(4)} = (-\cos x)' = \sin x$

b $z' = -\sin x$

We can use the results of part **a**, because $-\sin x = -y$:

$$z'' = -y' = -\cos x \qquad z''' = -y'' = \sin x$$

$$z^{(4)} = -y^{(3)} = \cos x$$

●

For the next example we need to recall the **factorial function** on non-negative integers:

$$0! = 1 \qquad 1! = 1 \qquad 2! = 1 \cdot 2 = 2 \qquad 3! = 1 \cdot 2 \cdot 3 = 6$$

$$4! = 1 \cdot 2 \cdot 3 \cdot 4 = 24$$

In general

$$n! = (n-1)! \cdot n = 1 \cdot 2 \cdot 3 \cdots (n-1) \cdot n$$

Your calculator may have a factorial key n! that gives estimates up to $69! \approx 1.71 \times 10^{98}$. (The next one $70! \approx 2.20 \times 10^{100}$ is too large for most calculators.)

Example 7

Given $y = 1/x$, find

a $y^{(25)}$ **b** $y^{(n)}$

Solution

a Compute a few derivatives:

$$y' = -\frac{1}{x^2} \qquad y'' = \frac{2}{x^3} \qquad y''' = -\frac{2 \cdot 3}{x^4}$$

$$y^{(4)} = \frac{2 \cdot 3 \cdot 4}{x^5}$$

A clear pattern emerges: The numerators are $1!$, $2!$, $3!$, $4!$, and so on, and the denominators are x^2, x^3, x^4, x^5, and so on. The signs alternate, minus for odd-order derivatives, plus for even. According to this pattern,

$$y^{(25)} = -\frac{25!}{x^{26}}$$

b By the same reasoning, $y^{(n)} = \pm \dfrac{(\text{factorial})}{(\text{power of } x)}$.

The factorial must be $n!$ with the same n as in $y^{(n)}$. The exponent of x is one larger, that is, $n + 1$. For the sign, introduce the factor

$(-1)^n$, an automatic sign changer. Its value is $+1$ for n even and -1 for n odd.

Answer $\quad y^{(n)} = (-1)^n \dfrac{n!}{x^{n+1}}$ ●

Remark 1 It is easy to make mistakes when finding general formulas as in Example 7b. Be especially careful in case of an alternating sign; note that $(-1)^{n+1}$ is the sign changer that is $+1$ for *n odd* and -1 for *n even.* Check your answer for a few small integers n and, if in doubt, prove the formula by induction.

Remark 2 It may be difficult or impossible to find a formula for the n-th derivative of a given function. Just try computing four or five derivatives of $y = (x^2 - 2)/(x^3 + 1)$ and you will soon be convinced that formulas for n-th derivatives are not always easy to find.

Derivatives of Products

Suppose you want the second or third derivative of $y = x^3(2x - 1)^{10}$. You can expand $(2x - 1)^{10}$ by the binomial theorem, multiply by x^3, then differentiate. But that's a lot of work; it's much easier to treat y as the product of x^3 and $(2x - 1)^{10}$.

In general, if $y = uv$, then $y' = u'v + uv'$. Differentiate again, using the product rule on each term:

$$\begin{aligned} y'' &= (u'v)' + (uv')' = (u''v + u'v') + (u'v' + uv'') \\ &= u''v + 2u'v' + uv'' \end{aligned}$$

For the third derivative, differentiate again, using the product rule on each term. The result is

$$y''' = u'''v + 3u''v' + 3u'v'' + uv'''$$

Similarly, the fourth derivative is

$$y^{(4)} = u^{(4)}v + 4u'''v' + 6u''v'' + 4u'v''' + uv^{(4)}$$

Note the similarity of these formulas to the binomial expansions

$$\begin{aligned} (u+v)^2 &= u^2 + 2uv + v^2 \\ (u+v)^3 &= u^3 + 3u^2v + 3uv^2 + v^3 \\ (u+v)^4 &= u^4 + 4u^3v + 6u^2v^2 + 4uv^3 + v^4 \end{aligned}$$

Consider the special case

$$y(x) = xv(x)$$

which occurs frequently in practice. Then

$$y' = v + xv' \qquad y'' = v' + (v' + xv'') = 2v' + xv''$$

$$y''' = 2v'' + (xv'')' = 2v'' + (v'' + xv''') = 3v'' + xv'''$$

$$y^{(4)} = 3v''' + (xv''')' = 3v''' + (v''' + xv^{(4)}) = 4v''' + xv^{(4)}$$

In general,

$$y^{(n)} = nv^{(n-1)} + xv^{(n)}$$

Example 8

$$(x \sin x)^{(9)} = 9(\sin x)^{(8)} + x(\sin x)^{(9)}$$
$$= 9 \sin x + x \cos x$$

Example 9

Compute y'' where $y = x^3(2x - 1)^{10}$.

Solution Let $u = x^3$ and $v = (2x - 1)^{10}$. Then $y = uv$ so

$$\begin{aligned} y'' &= u''v + 2u'v' + uv'' \\ &= 6x(2x - 1)^{10} + 2 \cdot 3x^2[10 \cdot 2(2x - 1)^9] \\ &\quad + x^3[10 \cdot 9 \cdot 2^2(2x - 1)^8] \\ &= 6x(2x - 1)^8[(2x - 1)^2 + 20x(2x - 1) + 60x^2] \\ &= 6x(2x - 1)^8(104x^2 - 24x + 1) \end{aligned}$$

Remark The general case is handled by the **Leibniz formula:**

$$(uv)^{(n)} = u^{(n)}v + \binom{n}{1} u^{(n-1)}v' + \binom{n}{2} u^{(n-2)}v'' + \cdots + \binom{n}{n-1} u'v^{(n-1)} + uv^{(n)}$$

which parallels the *binomial theorem* for powers:

$$(u + v)^n = u^n + \binom{n}{1} u^{n-1}v + \binom{n}{2} u^{n-2}v^2 + \cdots + \binom{n}{n-1} uv^{n-1} + v^n$$

The general term in the Leibniz formula is

$$\binom{n}{k} u^{(n-k)}v^{(k)} \qquad \text{where} \qquad \binom{n}{k} = \frac{n!}{k!(n-k)!}$$

Exercises

Find d^2y/dx^2 for $y =$

1 $3x^2 - 2x + 1$

2 $x^3 - 7x$

3 $x^2(1 - x)$

4 $(x^2 - 1)(x^2 - 2)$

5 $x^4(x + 1)^2$

6 $x^9 - 8x^7$

7 $\dfrac{1}{1 + x^2}$

8 $\dfrac{x}{x^2 + 4}$

9 $\dfrac{x}{x - 2}$

10 $\dfrac{x - 1}{x + 1}$

11 $\dfrac{x^2 + a}{x^2 - a}$

$\dfrac{x^2}{x^3+2}$

$\dfrac{1}{\sqrt{1-3x}}$

$\dfrac{1}{1+\sqrt{x}}$

$\sqrt{x^2+9}$

$\dfrac{x}{\sqrt{1-x^2}}$

17 $\sqrt[3]{x} - 5\sqrt[3]{1+x}$

18 $(x^{1/5}+2)^6$

19 $\cot x$

20 $\sec x$

21 $\csc x$

22 $\sqrt{\sin x}$

23 $\tan^3 x$

24 $(\sin x)/x$

Find $\left.\dfrac{d^2y}{dt^2}\right|_{t=c}$

5 $y = \frac{4}{3}\pi t^3 \qquad c = 2$

$y = \dfrac{1}{3t-1} \qquad c = 1$

7 $y = \frac{1}{2}t^{5/4} \qquad c = 81$

3 $y = \dfrac{3}{t} - 8t^{10} \qquad c = -1$

$y = t^2(t-3)^2 \qquad c = 0$

$y = t^8(4t-5)^2 \qquad c = 1$

$y = t^2\cos t \qquad c = \frac{1}{2}\pi$

2 $y = \sin^3 t \qquad c = \frac{1}{2}\pi$

3 $y = (1-\cos t)/t \qquad c = \frac{1}{2}\pi$

$y = \tan t + \cot t \qquad c = -\frac{1}{4}\pi$

5 $y = \sqrt{1-t^2} \qquad c = 0$

6 (cont.) $y = t\sqrt{1-t^2} \qquad c = 0$

37 Find a formula for $\dfrac{d^n}{dx^n}(x^{1/2})$

38 (cont.) The numerator of your answer should involve the product $1\cdot3\cdot5\cdot7\cdots(2n-3)$. Express this product in terms of factorials and a power of 2.

39 Find a formula for $\dfrac{d^n}{dx^n}(x^{1/3})$

40 Find a formula for $\dfrac{d^n}{dx^n}\left(\dfrac{x+1}{x-1}\right)$

Use the formula of Leibniz to find

41 $\dfrac{d^2}{dx^2}[x^4(2x+1)^6]$

42 $\dfrac{d^2}{dx^2}[\sqrt{x}\,(x-10)^5]$

43 $\dfrac{d^5}{dx^5}[x(x-3)^8]$

44 $\dfrac{d^3}{dx^3}[x^5(x+3)^6]$

45 $\dfrac{d^{10}}{dx^{10}}[(x+1)\sqrt{x}\,]$

46 $\dfrac{d^2}{dx^2}[\sqrt{x-1}\,\sqrt[3]{x-2}\,]$

47 $\dfrac{d^5}{dx^5}[x^3(x-1)^{10}]$ at $x = 1$

48 $\dfrac{d^5}{dx^5}[x^5\sqrt{x^3+16}\,]$ at $x = 0$

Compute

49 $\dfrac{d}{dx}\left[x - \dfrac{f(x)}{f'(x)}\right]$ **50** $f''[f(x)] - \dfrac{d}{dx}\left[\dfrac{f'[f(x)]}{f'(x)}\right]$

2-12 Review

We started this chapter with the intuitive idea of the slope of a curve. In order to make that notion precise we came to the idea of the derivative of a function, but only through a lengthy study of limits and continuous functions. We shall come back to slope as the first order of business in the next chapter. We shall do so armed with the technique of differentiation, the main tool for dealing with slope and many other quantities.

It really is important to know how to differentiate functions without hesitation. Let us review the main concepts and results.

Definition of Limit $\lim_{x\to a} f(x) = L$ means $f(x)$ can be made arbitrarily close to L by taking x sufficiently close to a. Precisely:

For each $\varepsilon > 0$ there exists a $\delta > 0$ such that

$$0 < |x-a| < \delta \text{ implies } |f(x) - L| < \varepsilon$$

[Shorthand notation: $f(x) \to L$ as $x \to a$.]

Properties of Limits With suitable qualifications,

$$\lim_{x\to a}[f(x)+g(x)] = \lim_{x\to a} f(x) + \lim_{x\to a} g(x)$$

$$\lim_{x\to a}[f(x)g(x)] = [\lim_{x\to a} f(x)][\lim_{x\to a} g(x)]$$

$$\lim_{x\to a}\frac{f(x)}{g(x)} = \frac{\lim_{x\to a} f(x)}{\lim_{x\to a} g(x)} \qquad \lim_{x\to a}\sqrt{f(x)} = \sqrt{\lim_{x\to a} f(x)}$$

$$f(x) \geq g(x) \text{ implies } \lim_{x\to a} f(x) \geq \lim_{x\to a} g(x)$$

Definition of Continuous Function A function $f(x)$ is continuous at a point a if $\lim_{x\to a} f(x) = f(a)$.

Properties of Continuous Functions If $f(x)$ and $g(x)$ are continuous, then (with suitable qualifications on domains) so are

$$af(x) + bg(x) \qquad f(x)g(x) \qquad \frac{f(x)}{g(x)} \qquad \sqrt{f(x)} \qquad (f \circ g)(x)$$

Polynomials are continuous; so are rational functions, the trigonometric functions, and a^x.

Definition of the Derivative The derivative of $f(x)$ at $x = a$ is

$$f'(a) = \lim_{h\to 0}\frac{f(a+h)-f(a)}{h} = \lim_{x\to a}\frac{f(x)-f(a)}{x-a}$$

provided that the limit exists.

Rules for Differentiation

$$[af(x) + bg(x)]' = af'(x) + bg'(x) \qquad \text{(linearity)}$$

$$[f(x)g(x)]' = f'(x)g(x) + f(x)g'(x) \qquad \text{(product rule)}$$

$$\left[\frac{f(x)}{g(x)}\right]' = \frac{f'(x)g(x) - f(x)g'(x)}{[g(x)]^2} \qquad \text{(quotient rule)}$$

The Chain Rule $\{f[g(x)]\}' = f'[g(x)]g'(x)$

Special Functions

$$(x^r)' = rx^{r-1}$$

$$(\sin x)' = \cos x \qquad (\cos x)' = -\sin x$$

$$(\tan x)' = \sec^2 x \qquad (\cot x)' = -\csc^2 x$$

$$(\sec x)' = \sec x \tan x \qquad (\csc x)' = -\csc x \cot x$$

Implicit Functions If a function $y = f(x)$ is defined implicitly by an equation connecting x and y, then that equational relation can be differentiated term by term and the result solved for $f'(x)$.

Inverse Functions If $y = f(x)$ and $x = g(y)$ are inverse functions, then

$$f'(x) = \frac{1}{g'[f(x)]}$$

Higher Derivatives

$$f''(x) = \frac{d}{dx}f'(x) = \frac{d^2f}{dx^2} \qquad f'''(x) = \frac{d}{dx}f''(x) \qquad \cdots$$

Exercises

Find the following limits

1 $\lim_{x\to 0+} \frac{1+\sqrt{x}}{3x^3 - 2}$

2 $\lim_{x\to 2} \frac{x-2}{x^3-8}$

3 $\lim_{x\to 0} \frac{1-\sqrt{1+x}}{x}$

4 $\lim_{x\to 1} (1 + \sqrt[3]{1+7x})(3-x)$

5 $\lim_{h\to 0} \frac{h^2}{\sin h}$

6 $\lim_{h\to 0} \left(\frac{\sin h}{3h}\right)^2$

7 Find $\lim_{x\to 0} \frac{\sqrt{2+x} - \sqrt{2}}{x}$

[Hint Rationalize the numerator; that is, multiply and divide by $\sqrt{2+x} + \sqrt{2}$.]

8 Find $\lim_{x\to 0} \frac{(2+x)^{1/3} - (2)^{1/3}}{x}$

[Hint Proceed as in Exercise 7, using the identity $y^3 - z^3 = (y-z)(y^2+yz+z^2)$.]

9 Suppose $\lim_{x\to 4} f(x) = 10^{-6}$. Prove that there is a small segment centered at $x = 4$ on which $f(x) > 0$ provided $x \neq 4$.

10 Where is $f(x) = |x+1| + |x+2| + |x+3|$ continuous? differentiable?

11 Where is $f(x) = x|x|$ continuous? differentiable?

12 Let $f(x) = x^2$ for $x \le 1$ and $f(x) = 2x - 1$ for $x > 1$. Show that $f(x)$ is continuous and differentiable for all x.

13 Show that

$$\frac{1}{a+h} = \frac{1}{a} - \frac{h}{a^2} + \frac{h^2}{a^2(a+h)}$$

Use this identity with the definition of the derivative to calculate the derivative of $1/x$ at $x = a$.

14 Show that

$$\sqrt{a+h} = \sqrt{a} + \frac{h}{2\sqrt{a}} - \frac{h^2}{2\sqrt{a}\,(\sqrt{a+h} + \sqrt{a})^2}$$

Use this identity with the definition of the derivative to calculate the derivative of $\sqrt{x}$ at $x = a$.

Differentiate

15 $x^9 \sin^2 x$

16 $x^9/(x + \sin x)$

17 $(2 + \sqrt{3x})^5$

18 $\frac{1}{x\sqrt{9x^2+4}}$

19 $(x+1)(2x+1)^2 \times (3x+1)^3$

20 $\left[1 + \left(\frac{3x}{1+x^2}\right)^{1/3}\right]^{1/2}$

21 $x^2 \tan x$

22 $\cos 2x - x \sec x$

Find dy/dx in terms of x and y where

23 $x^4 + y^3 = 1$

24 $x + \sin(xy) = y + 1$

25 $x = y^5 + y$

26 $x = y^2 + \cos y$

Compute

27 $\frac{d^2}{dx^2}\left(\frac{3x+7}{x+2}\right)$

28 $\frac{d^3}{dx^3}\sqrt{3x-2}\,\Big|_{x=1}$

29 $\frac{d^6}{dx^6}(x \cos x)$

30 $\frac{d^n}{dx^n}\left(\frac{x^2}{1-x}\right)$

31 Find $\lim_{x\to\infty} [ax]/x$ where $[x]$ is the greatest integer function (page 14).

32 Let $h(x) = f[g(x)] - g[f(x)]$, where $f(x)$ and $g(x)$ are differentiable. Suppose $f(c) = g(c) = c$, a constant. Prove that $h'(c) = 0$.

33 Set $y = \cfrac{1}{1 + \cfrac{1}{1 - \cfrac{1}{x}}}$ and find dy/dx.

34 (cont.) Find $\lim_{x\to -\infty} y$.

3 Applications of Differentiation

In the previous chapter we learned what the derivative of a function is and techniques for computing derivatives. In this chapter we apply derivatives to a variety of problems: graphing functions, approximation, motion, the shapes of graphs, and finding the largest and smallest values of functions. We also cover some basic theory that is needed for these applications.

3-1 Tangents, Graphs, and Approximation

We sometimes refer to a differentiable function as a **smooth** function and to its graph as a smooth graph or smooth curve. Let $y = f(x)$ be a smooth function, and let $P = (a, f(a))$ be a point on its graph. By the **slope** of the graph at P we mean simply the derivative $f'(a)$.

The **tangent** to the graph at P is the line passing through P whose slope equals the slope $f'(a)$ of the graph at P. By the point-slope form, the equation of this line is

$$y - f(a) = f'(a)(x - a)$$

that is

$$y = f(a) + f'(a)(x - a)$$

Example 1

Find the tangent to

a $y = x^2$ at $(-2, 4)$ **b** $y = 1/x$ at $(2, \frac{1}{2})$

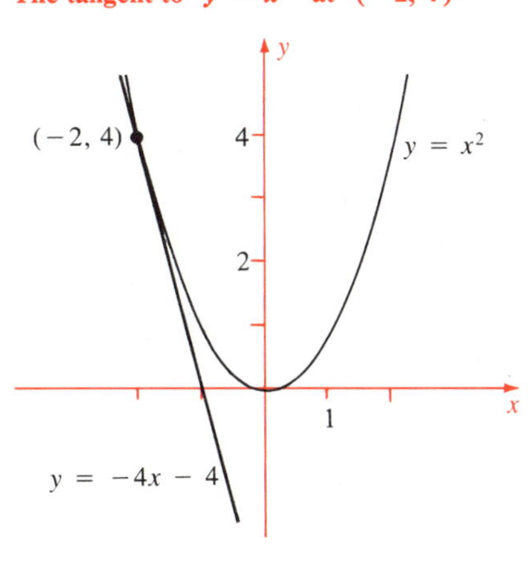

Figure 3-1-1
The tangent to $y = x^2$ at $(-2, 4)$

Solution

a The slope of $y = x^2$ at $(-2, 4)$ is

$$\left.\frac{dy}{dx}\right|_{-2} = 2x\Big|_{-2} = -4$$

The line through $(-2, 4)$ with slope -4 is

$$y - 4 = -4[x - (-2)] \qquad \text{that is} \qquad y = -4x - 4$$

See Figure 3-1-1.

b The slope of $y = 1/x$ at $(2, \frac{1}{2})$ is

$$\left.\frac{dy}{dx}\right|_{2} = -\left.\frac{1}{x^2}\right|_{2} = -\tfrac{1}{4}$$

The tangent is

$$y - \tfrac{1}{2} = -\tfrac{1}{4}(x - 2) \qquad \text{that is} \qquad y = -\tfrac{1}{4}x + 1$$

See Figure 3-1-2 ●

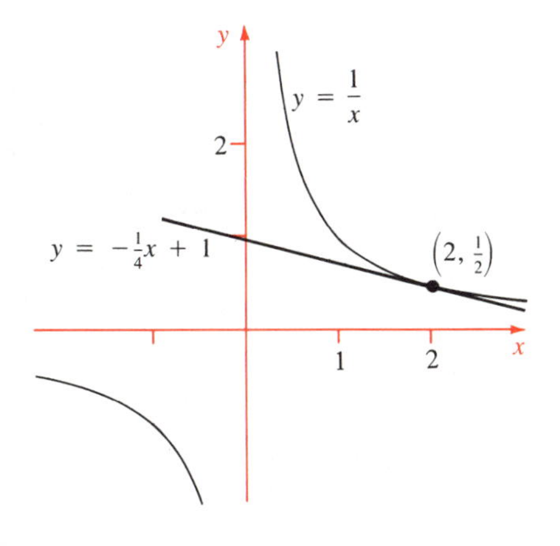

Figure 3-1-2
The tangent to $y = 1/x$ at $(2, \frac{1}{2})$

At each point of the graph of a function, the derivative is the slope of the graph. Where the derivative is positive, the graph slopes upward; that is, the function increases. Where the derivative is negative, the graph slopes downward; that is, the function decreases. Where the derivative is zero, the graph is horizontal. This information is of great help in sketching curves.

Example 2

Sketch $y = x^3 - 3x + 1$.

Solution The derivative is

$$y' = 3x^2 - 3 = 3(x^2 - 1)$$

The sign of $y'(x)$ depends on whether $x^2 > 1$ or $x^2 < 1$:

$$\begin{cases} y' > 0 & \text{if } x^2 > 1 & \text{that is} & \text{if } x < -1 \text{ or } x > 1 \\ y' < 0 & \text{if } x^2 < 1 & \text{that is} & \text{if } -1 < x < 1 \\ y' = 0 & \text{if } x^2 = 1 & \text{that is} & \text{if } x = +1 \text{ or } x = -1 \end{cases}$$

The function increases and decreases as indicated in Figure 3-1-3. This information indicates a high point where $x = -1$ and a low point where $x = 1$. These are important features of the graph, so we find the corresponding values of y:

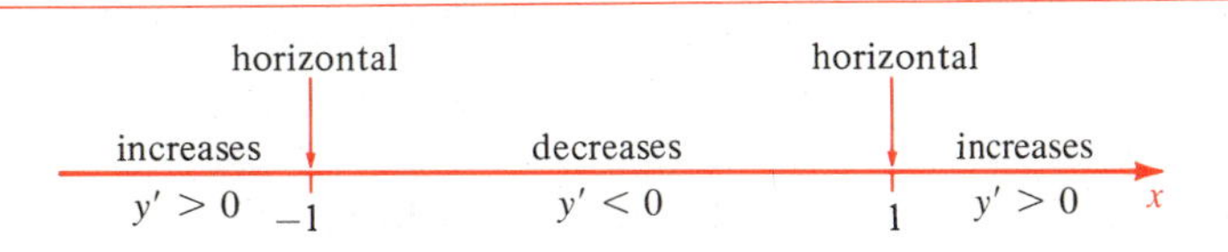

Figure 3-1-3
The sign of the derivative of $y = x^3 - 3x + 1$

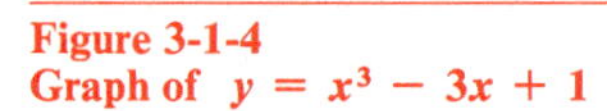
Figure 3-1-4
Graph of $y = x^3 - 3x + 1$

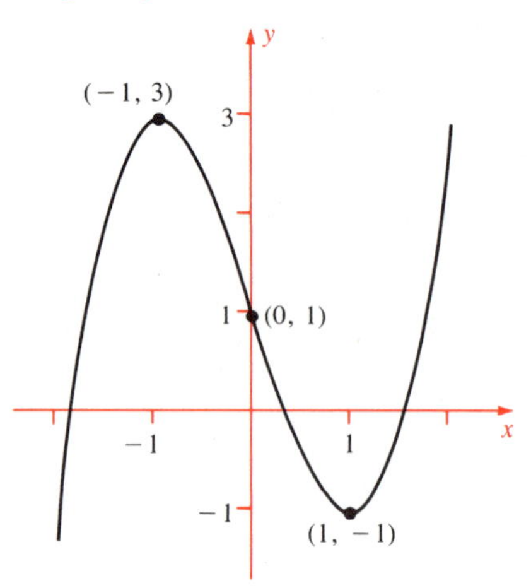

$$y(-1) = (-1)^3 - 3(-1) + 1 = 3$$
$$y(1) = 1^3 - 3 + 1 = -1$$

Then we plot the high point $(-1, 3)$ and the low point $(1, -1)$. Also y is easy to compute for $x = 0$, so we find that $(0, 1)$ is on the graph. Next, if $x \to +\infty$ or $x \to -\infty$, then

$$y' = 3(x^2 - 1) \to +\infty$$

so the graph is increasingly steep to the right and to the left. This is enough information for a reasonable sketch (Figure 3-1-4). ●

Suppose we can plot a few points of a graph, including all of the points where a graph is horizontal. In many cases, this is enough information to make a rough sketch of the graph. That is, we may not have to worry about the signs of $f'(x)$ between consecutive horizontal points. The graph must rise from one horizontal point to the next if the next horizontal point is higher, and the graph must fall if that next horizontal point is lower. The following example illustrates this method.

Figure 3-1-5
A few plotted points and all horizontal tangents of the graph of $y = \cos 2x + 2\cos x$ on $[0, 2\pi]$

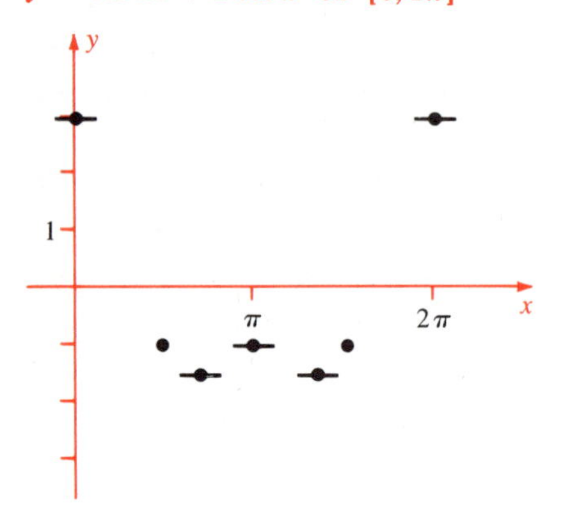

Example 3

Sketch the curve $y = \cos 2x + 2\cos x$

Solution The curve is periodic with period 2π, so it is enough to sketch it for $0 \le x \le 2\pi$, then extend the curve by its periodicity. For a reasonable sketch without too much work, we first find a few points on the curve that are easy to compute:

$$(0, 3) \quad (\tfrac{1}{2}\pi, -1) \quad (\pi, -1) \quad (\tfrac{3}{2}\pi, -1) \quad (2\pi, 3)$$

Next we locate the horizontal points, the points where $y' = 0$. To do so, we compute

$$\frac{dy}{dx} = \frac{d}{dx}(\cos 2x + 2\cos x) = -2\sin 2x - 2\sin x$$
$$= -4\sin x\cos x - 2\sin x = -2(\sin x)(2\cos x + 1)$$

Therefore $dy/dx = 0$ where either $\sin x = 0$ or $\cos x = -\frac{1}{2}$. The corresponding horizontal points are

$$(0, 3) \quad (\pi, -1) \quad (2\pi, 3) \qquad \text{and} \qquad (\tfrac{2}{3}\pi, -\tfrac{3}{2}) \quad (\tfrac{4}{3}\pi, -\tfrac{3}{2})$$

All of this information is shown in Figure 3-1-5, and the data is interpolated by a smooth curve in Figure 3-1-6. Finally the curve is extended in Figure 3-1-7, by means of the periodicity of y. ●

Figure 3-1-6
The curve suggested on one period by the points in Figure 3-1-5

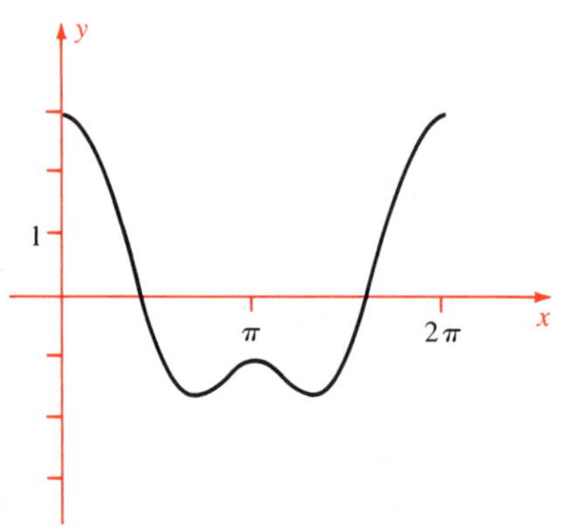

Approximating Functions

Approximating complicated functions by simpler functions is one of the important processes of calculus. Let us consider the simplest type of approximation, by linear functions.

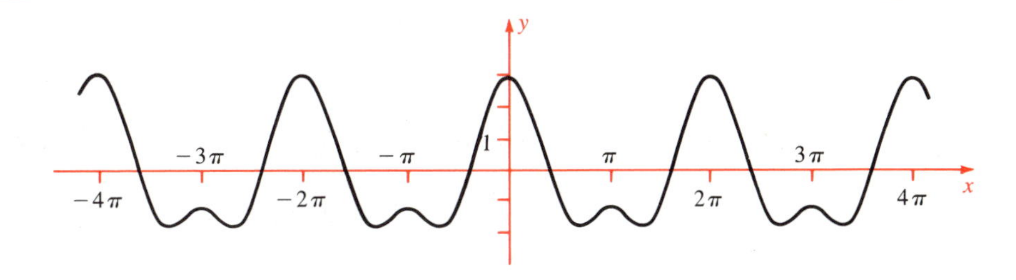

Figure 3-1-7
Graph of $y = \cos 2x + 2 \cos x$

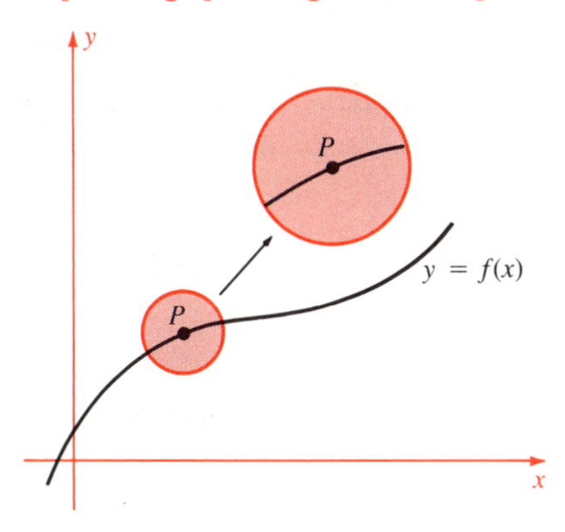

Figure 3-1-8
Inspect the graph through a microscope.

Under a high-powered microscope, a smooth graph $y = f(x)$ appears nearly straight (Figure 3-1-8). The tangent at $P = (a, f(a))$ is almost indistinguishable from the curve, at least very near P. See Figure 3-1-9. Since the equation of the tangent is $y = f(a) + f'(a)(x - a)$, the linear function

$$y = g(x) = f(a) + f'(a)(x - a)$$

ought to be a good approximation to $f(x)$, provided x is near a.

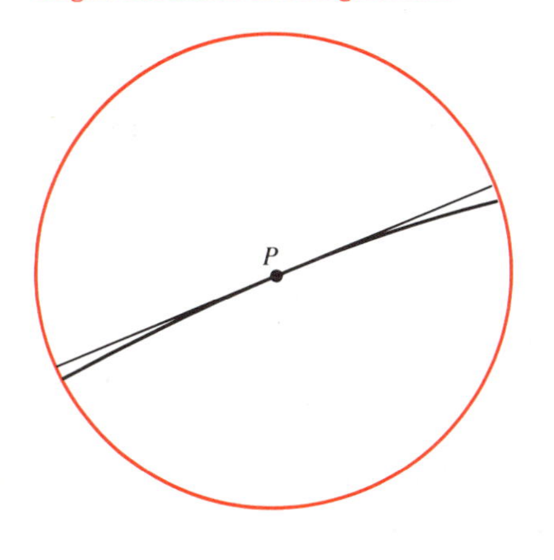

Figure 3-1-9
Higher magnification: The curve and its tangent are almost indistinguishable.

Example 4

Approximate $f(x) = \sqrt{x}$ at $a = 100$, and check the closeness of the approximation at $x = 98, 99, \cdots, 102$.

Solution Since

$$f(100) = 10 \quad \text{and} \quad f'(100) = \frac{1}{2\sqrt{100}} = \frac{1}{20} = 0.05$$

the tangent at $a = 100$ is

$$y = g(x) = 10 + 0.05(x - 100) = 0.05x + 5$$

This is the linear approximation. Values are tabulated in Table 3-1-1. Note that for $x = 102$, which differs from 100 by 2%, the error is about 5.0×10^{-4}, only about 0.005% of $f(102)$. Not bad! ●

Table 3-1-1

x	$\sqrt{x}$	$0.05x + 5$
98	9.89949	9.90000
99	9.94987	9.95000
100	10.00000	10.00000
101	10.04988	10.05000
102	10.09950	10.10000

In general, to see how closely the linear approximation $g(x)$ approximates $f(x)$, we must measure the error in the approximation, that is, the difference between $f(x)$, the true value, and $g(x)$, the approximate value. The error is

$$E(x) = f(x) - g(x) = f(x) - f(a) - f'(a)(x - a)$$

In many cases, we can show by direct calculation that $|E(x)|$ is smaller than a constant times $(x - a)^2$. That square is good news because $(x - a)^2$ is very small when x is near a, much smaller than $|x - a|$ itself. We see this in the next example.

Example 5

Approximate $f(x) = x^2$ at $a = -2$.

Solution We found the tangent to $y = f(x)$ in Example 1a:

$$y = g(x) = -4x - 4$$

Hence if we approximate the curve by its tangent, the error is

$$E(x) = f(x) - g(x) = x^2 - (-4x - 4) = (x + 2)^2$$
$$= [x - (-2)]^2$$

Accordingly, if x is within 0.1 of -2, that is, if $|x + 2| < 0.1$, then $E(x) < (0.1)^2 = 0.01$. If $|x + 2| < 0.01$, then $E(x) < 0.0001$. In general, the error is the square of the distance from x to -2. ●

Example 6

Approximate $f(x) = 1/x$ at $a = 2$.

Solution We found the tangent to $y = f(x)$ in Example 1b:

$$y = g(x) = -\tfrac{1}{4}x + 1$$

Hence if we approximate the curve by its tangent, then the error is

$$E(x) = f(x) - g(x) = 1/x - (-\tfrac{1}{4}x + 1)$$

Let us tabulate to six places a few values of $E(x)$ for x near 2. See Table 3-1-2. The approximation is extremely accurate. This is so because $E(x)$ is divisible by $(x - 2)^2$:

$$E(x) = \frac{1}{x} + \tfrac{1}{4}x - 1 = \frac{x^2 - 4x + 4}{4x} = \frac{(x - 2)^2}{4x}$$

If $x \approx 2$ then $4x \approx 8$, so $E(x) \approx \tfrac{1}{8}(x - 2)^2$. For instance, if $|x - 2| < 0.1$ then $x > 1.9$. Consequently

$$0 < E(x) < \frac{|x - 2|^2}{4 \times 1.9} < 0.002$$

Table 3-1-2

x	$f(x)$	$g(x)$	$E(x)$
1.900	0.526316	0.525000	0.001316
1.950	0.512820	0.512500	0.000320
1.990	0.502513	0.502500	0.000013
1.995	0.501253	0.501250	0.000003
1.999	0.500250	0.500250	0.000000
2.100	0.476191	0.475000	0.001191
2.050	0.487805	0.487500	0.000305
2.010	0.497512	0.497500	0.000012
2.005	0.498753	0.498750	0.000003
2.001	0.499750	0.499750	0.000000

Similarly,

If $|x - 2| < 0.01$ then $0 \le E(x) < 0.00002$

If $|x - 2| < 0.001$ then $0 \le E(x) < 2 \times 10^{-7}$

These estimates compare well with the more precise numbers in Table 3-1-2. ●

We shall return to this subject in Chapter 9.

Exercises

Find the equation of the tangent to the curve

1 $y = x^2$ through $(2, 4)$

2 $y = x^3$ through $(2, 8)$

3 $y = \cos x$ through $(\frac{1}{3}\pi, \frac{1}{2})$

4 $y = \sin x$ through $(-\frac{1}{6}\pi, -\frac{1}{2})$

5 $y = x - 1/x$ at $x = -1$

6 $y = x/(x^2 + 4)$ at $x = 0$

7 $y = 1/x^2$ at $x = -3$

8 $y = (\frac{1}{2}x - 1)^5$ at $x = 4$

Find the equation(s) of the tangent line(s) to the curve and determine where each tangent crosses the coordinate axes

9 $y = x^2 + 3x - 1$ through $(2, 9)$

10 $y = x^3 - 8x^2$ through $(10, 200)$

11 $y = 1/x$ and the tangent has slope -81

12 $y = 1/x$ and the tangent has slope -6

Find all points on the graph with the given property; then find the equation(s) of the tangent(s) to the curve at these points

13 $y = x^2$ all points where the slope is 10

14 $y = x^3$ all points where the slope is 27

15 $y = x^2$ all points for which the tangent crosses the y-axis at $y = -16$

16 $y = x^3$ all points for which the tangent crosses the y-axis at $y = -128$

Sketch the curve

17 $y = x^3 - x$

18 $y = 3x^3 - 4x$

19 $y = x^3 - 3x$

20 $y = 12x - x^3$

21 $y = 4x^3 - 2x^2$

22 $y = 6x^2 - x^3$

23 $y = -x^3 + 3x - 4$

24 $y = x^3 - 3x + 5$

25 $y = \frac{1}{3}x - x^3$

26 $y = x^3 - 3x^2$

27 $y = x^3 - x^2 - 8x + 4$

28 $y = 7 - 3x^2 - 2x^3$

29 $y = \sin x + \cos 2x$

30 $y = \sin 2x - \sin x$

31 $y = \tan x + \cot x$

32 $y = \sec x - \csc x$

Find the equation of the line tangent to the curve at the specified point; also find the error $E(x)$ made in approximating the curve by its tangent

33 $y = 1 - x^2$ at $(0, 1)$

34 $y = 2x^2 + 3$ at $(1, 5)$

35 $y = x^3$ at $(3, 27)$

36 $y = x^2 + x + 1$ at $(-1, 1)$

37 $y = 1/(3x + 4)$ at $(-1, 1)$

38 $y = x^2 - x^3$ at $(1, 0)$

Find the error $E(x)$ in approximating $y = f(x)$ at $x = a$ and show that $|E(x)| \le K|x - a|^2$ for a suitable constant K if x is sufficiently close to a

39 $f(x) = x^3$ $a = -1$

40 $f(x) = 1/x^2$ $a = 1$

41 Let $y = g(x)$ be the tangent to $y = \sqrt{x}$ at $(1, 1)$. Compute $g(x)$ and tabulate x, $\sqrt{x}$, $g(x)$, and $E(x) = \sqrt{x} - g(x)$ for $x = 0.98$, 0.99, 1.02, and 1.01.

42 Let $y = g(x)$ be the tangent to $y = x^5 + 2x$ at $(0, 0)$. Compute $g(x)$ and tabulate $x^5 + 2x$, $g(x)$, and $E(x) = (x^5 + 2x) - g(x)$ for $x = -0.2$, -0.1, 0.1, and 0.2.

Find the area of the triangle bounded by the coordinate axes and the line tangent to $y = 1/x$

43 at $x = 2$

44 at $x = a$ where $a \ne 0$

45 Show that the tangents to the parabola $y = x^2$ at $(\pm 3, 9)$ cross on the y-axis. Where?

46 Exactly one of the lines $y = 3x + b$ is tangent to the parabola $y = x^2$. Which line?

***47** Let $y = g(x)$ be the tangent to $y = x^3$ at $(2, 8)$. Show $|x^3 - g(x)| \le 7|x - 2|^2$ for $|x - 2| < 1$.

48 Let $y = g(x)$ be the tangent to $y = 1/x^2$ at $(-1, 1)$. Show that $|1/x^2 - g(x)| \le 16|x + 1|^2$ for $-\frac{3}{2} < x < -\frac{1}{2}$.

3-2 Rectilinear Motion

Rectilinear motion is motion along a straight line. Its study leads to two important concepts: velocity and acceleration.

Suppose a particle is moving along the x-axis. Its position x depends on the time t. Therefore we write x as a function of t, namely, $x = x(t)$. Thus the position of the particle (its past history, its present position, and its future) is described by a function whose domain is an interval of the time axis.

The variables t and x have physical dimensions: time for t, length for x. **Average velocity** over a time interval $[a, b]$ is defined as

$$\frac{x(b) - x(a)}{b - a}$$

(This is often referred to as the average speed, but we want to use the word "speed" in a different way.) For instance, if a car starts on a trip and arrives at its destination 90 kilometers down the road 1.2 hours later, then its average velocity is $90/1.2 = 75$ km/hr.

We now define the *instantaneous* velocity of a particle as a derivative:

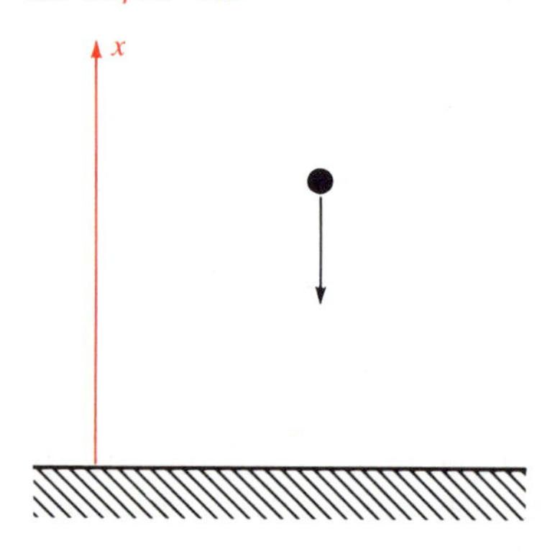

Figure 3-2-1
Falling body: height decreases, so $dx/dt < 0$

Velocity and Speed

Let $x = x(t)$ be the position at time t of a particle moving on the x-axis.

- Its **velocity** is $v(t) = dx/dt$.
- Its **speed** is $|v| = |dx/dt|$.

Notice that dx/dt may be negative. This typically happens for a falling body (Figure 3-2-1) whose position is measured up from ground level. Its position $x(t)$ is a decreasing function of t, so $dx/dt < 0$. Speed is the magnitude of the rate of motion, without regard to its direction, whereas velocity also takes into account its direction.

Notation The use of a dot to indicate derivative with respect to time is common in physics and engineering. Instead of dx/dt, you may see $\dot{x}(t)$ or just $\dot{x}$.

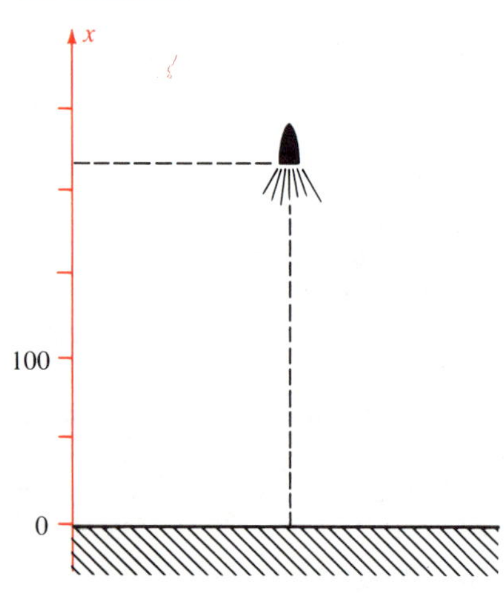

Figure 3-2-2
See Example 1

Example 1

During its initial 60 sec of flight, a certain rocket fired vertically reaches an elevation of $50t^2$ ft above the ground in t sec. How fast is the rocket rising 12 sec after it is fired?

Solution Let the x-axis be vertical with its origin at ground level, and calibrate it in feet (Figure 3-2-2). Measure time in seconds, starting at the instant of blast-off. Then during the rocket's initial 60 sec of flight, its distance above ground is

$$x(t) = 50t^2$$

Differentiate to find the velocity:

$$v(t) = \frac{dx}{dt} = 100t \qquad \text{hence} \qquad v(12) = 1200 \text{ ft/sec}$$

Remark Note that the physical dimension of velocity (and of speed) is length/time. If the units of length and time are feet and seconds, the unit of velocity (speed) is feet per second, abbreviated ft/sec. In the metric system, the unit is meter/second, abbreviated m/sec. When we use metric units, we generally stick with the MKS system—meters, kilograms, seconds. This is also referred to as the system of SI units—the Système Internationale, or International System.

Example 2

A ball is thrown up from a support 180 ft above the ground. Assume the ball's height above ground after t sec to be

$$x = 180 + 64t - 16t^2 \text{ ft}$$

Find **a** its velocity after 1 sec, **b** its maximum height, and **c** its speed as it hits the ground.

Solution Take the x-axis as in Figure 3-2-2 and measure time from the instant the ball is thrown. Notice that the relation

$$x(t) = 180 + 64t - 16t^2$$

is consistent with the initial height $x = 180$ ft at the initial time $t = 0$ sec. Differentiate to find the velocity:

$$v(t) = \frac{dx}{dt} = 64 - 32t = 32(2 - t)$$

a After 1 sec, the velocity is $v(1) = 32(2 - 1) = 32$ ft/sec. (Since the velocity is positive, the ball is rising.)

b To find the maximum of $x(t)$, determine the sign of $v(t) = dx/dt$:

$$\begin{cases} v(t) > 0 & \text{for } t < 2 \\ v(t) = 0 & \text{for } t = 2 \\ v(t) < 0 & \text{for } t > 2 \end{cases}$$

Therefore $x(t)$ increases for $t < 2$, stops instantaneously at $t = 0$, and decreases for $t > 2$. Therefore the maximum height is reached when $t = 2$. This maximum height is $x(2) = 244$ ft.

c The ball hits the ground when $x(t) = 0$. We solve for t:

$$x(t) = 180 + 64t - 16t^2 = 0$$

$$4t^2 - 16t - 45 = 0$$

$$t = \frac{16 \pm \sqrt{16^2 + 4 \cdot 4 \cdot 45}}{8} = 2 \pm \frac{\sqrt{16 + 45}}{2}$$

There is only one *positive* time t for which $x(t) = 0$; it is $t = 2 + \frac{1}{2}\sqrt{61}$. At this instant, the velocity is

$$v(2 + \tfrac{1}{2}\sqrt{61}) = 32[2 - (2 + \tfrac{1}{2}\sqrt{61})] = -16\sqrt{61}$$

$$\approx -125.0 \text{ ft/sec}$$

The velocity is negative because the ball is falling. The corresponding speed is the magnitude of the velocity: $|v| = 125.0$ ft/sec. ●

Acceleration

During takeoff, an airplane moves faster and faster; its velocity increases. A car with its brakes applied moves slower and slower; its velocity decreases. In many applications, it is important to know how velocity changes during motion.

Definition Acceleration

Let $v(t)$ be the velocity of a moving particle at time t. Then the **acceleration** of the particle is defined as $a(t) = dv/dt$. ●

Acceleration is the derivative of the velocity function. It measures the rate of change of velocity with respect to time. Positive acceleration indicates increasing velocity; negative acceleration, decreasing velocity; zero acceleration, constant velocity.

Remember that velocity itself is a derivative:

$$v(t) = \frac{dx}{dt} = \dot{x}(t)$$

where $x = x(t)$ is the position at time t. Thus acceleration is a *second derivative,* being the derivative of a derivative:

$$\text{acceleration} = \frac{d}{dt}\left(\frac{dx}{dt}\right) = \frac{d^2x}{dt^2} = \ddot{x}(t)$$

●

Remark Acceleration is

$$\frac{\text{change in velocity}}{\text{change in time}} = \frac{\text{distance/time}}{\text{time}} = \frac{\text{distance}}{(\text{time})^2}$$

If the units of distance and time are feet and seconds, then the unit of acceleration is $\text{ft/sec/sec} = \text{ft/sec}^2$. In the metric (MKS) system, the unit of acceleration is m/sec^2.

Example 3 (Continuation of Example 2)

A ball is $x(t) = 180 + 64t - 16t^2$ ft above the ground at time t sec. Find its acceleration at time t.

Solution Differentiate twice:

$$v(t) = \frac{dx}{dt} = 64 - 32t \text{ ft/sec} \quad \text{so} \quad a(t) = \frac{dv}{dt} = -32 \text{ ft/sec}^2$$ ●

Remark The *negative* acceleration means that the velocity is decreasing (from positive to negative to more negative). If you are driving on a level road and you take your foot off the accelerator, then engine drag and air drag create negative acceleration, so your car slows down.

Derivative Zero

Suppose a car is on the road and its speedometer registers (correctly) 0 for one hour. How far does the car move? Obviously it is parked and goes nowhere. Mathematically this says that if a function $x(t)$ is differentiable on an interval and its derivative equals zero on the interval, then $x(t)$ is a constant function on that interval. Put this way, the fact is less obvious, and requires a proof, which will be given in Section 3-4. Furthermore, it has many important consequences that are not at all obvious. We state it now as a formal theorem about functions in general, not just $x = x(t)$.

Theorem Zero Derivative

Suppose $y = f(x)$ is a differentiable function on an interval I. Suppose $f'(x) = 0$ for each x in I. Then $f(x)$ is a constant function.

Figure 3-2-3
Two cars with the same velocity always remain a constant distance apart.

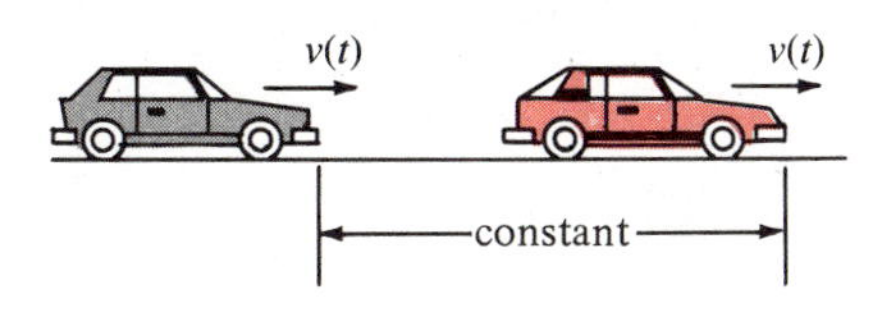

Next, we state and prove some useful consequences of this theorem. The first, in our context of rectilinear motion, says that if two moving cars have the same velocity at each instant of time, then one car stays a fixed distance in front of the other (Figure 3-2-3).

Corollary 1 Equal Derivatives

Suppose $f(x)$ and $g(x)$ are differentiable functions. Suppose $f'(x) = g'(x)$ for each x in an interval. Then $f(x) = g(x) + c$ on that interval, where c is a constant.

Proof

Set $h(x) = f(x) - g(x)$. Then $h(x)$ is a differentiable function on the interval and

$$h'(x) = f'(x) - g'(x) = 0$$

by the sum rule, page 106. By the theorem, $h(x) = c$. It follows that $f(x) - g(x) = c$, that is, $f(x) = g(x) + c$.

The next corollary says that a curve of constant slope is necessarily a straight line, and a curve of linear slope is necessarily a parabola (graph of a quadratic function). In the context of motion, the first statement says that if a car moves on a straight road with constant velocity, then its position is a linear function of time.

Corollary 2 Constant Derivative

Suppose $f(x)$ is a differentiable function on an interval I.

a If $f'(x) = a$ for each x in I, where a is a constant, then $f(x) = ax + b$ on I, where b is a constant.

b If $f'(x) = ax + b$ is a linear function, then $f(x) = \frac{1}{2}ax^2 + bx + c$ on I, where c is a constant.

Proof

a Set $g(x) = ax$. Then $f'(x) = a = g'(x)$. Hence Corollary 1 applies: $f(x) = g(x) + b = ax + b$.

b Set $g(x) = \frac{1}{2}ax^2 + bx$. Then $g'(x) = ax + b = f'(x)$ so Corollary 1 applies: $f(x) = g(x) + c = \frac{1}{2}ax^2 + bx + c$. ●

We have assumed, without proving it, that if the derivative of a function is positive on an interval, then the function is increasing. In our present context, if $x(t)$ is the height above ground of a particle at time t and if $v(t) > 0$, then the particle is rising. We take this for granted now, and shall justify it later, in Section 3-4.

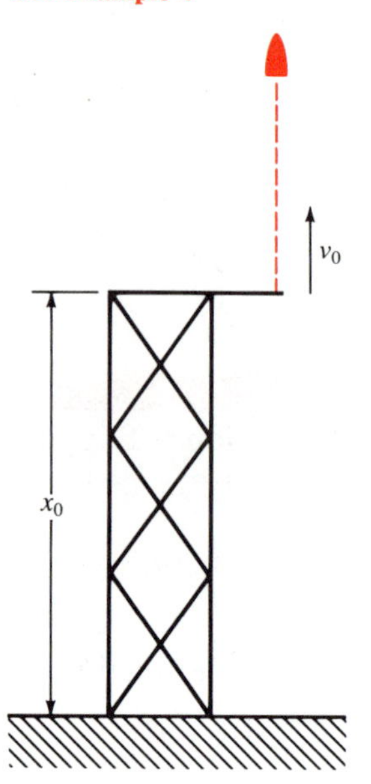

Figure 3-2-4
See Example 4

Example 4

A bullet is shot straight up from a platform x_0 meters above ground (Figure 3-2-4) with an initial velocity of v_0 m/sec. Gravity causes a constant negative acceleration of $-g$ m/sec^2. After t sec, what are **a** the velocity of the bullet and **b** its height above ground level?

Solution

a First find a formula for the velocity $v(t)$. Since acceleration is dv/dt, we have $dv/dt = -g$. By Corollary 2a applied to $v(t)$,

$$v(t) = -gt + b$$

To find the constant b that fits this problem, remember that the value $v(0) = v_0$ is given. Set $t = 0$:

$$v_0 = -g \cdot 0 + b \qquad \text{that is} \qquad b = v_0$$

Hence

$$v(t) = -gt + v_0 \quad \text{m/sec}$$

is the required formula.

b Now Corollary 2b applies to $x(t)$ because

$$\frac{dx}{dt} = v(t) = -gt + v_0$$

and both g and v_0 are constants. Therefore

$$x(t) = -\tfrac{1}{2}gt^2 + v_0 t + c$$

for some appropriate constant c. To find the value of c, remember that the value $x(0) = x_0$ is given. Set $t = 0$:

$$x_0 = 0 + 0 + c \qquad \text{hence} \qquad c = x_0$$

Consequently

$$x(t) = -\tfrac{1}{2}gt^2 + v_0 t + x_0 \quad \text{meters}$$ ●

In Example 4 we solved a **differential equation,** that is, an equation involving the derivatives of a function in which the function itself is the unknown. The data of Example 4 can be written:

$$\frac{d^2x}{dt^2} = -g \quad \text{where} \quad x(0) = x_0 \quad \text{and} \quad \frac{dx}{dt}(0) = v_0$$

The first equation is the differential equation; the other equations are **initial conditions.** The subject of differential equations is one of the most important in applied mathematics. Some say that the whole purpose of learning calculus is to prepare for differential equations!

Example 5

An alpha particle enters a linear accelerator (Figure 3-2-5). It is immediately subject to a constant acceleration that increases its velocity from 1000 m/sec to 5000 m/sec in 0.001 sec, when it strikes its target. Compute its acceleration. How long is the accelerator path?

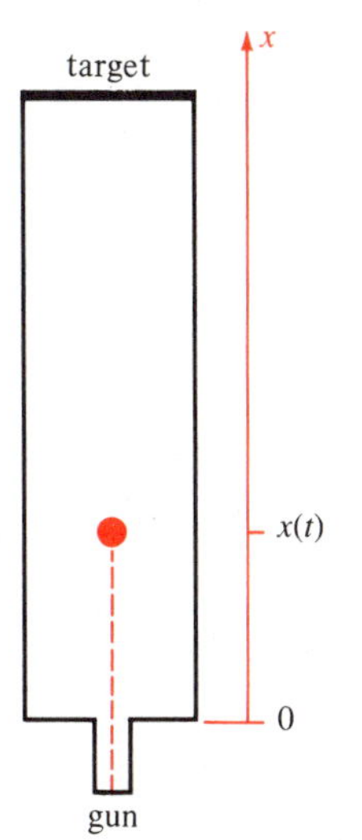

Figure 3-2-5
Alpha particle in accelerator (top view)

Solution Choose the x-axis as shown in the figure and also assume the particle enters the accelerator when $t = 0$. Its position t sec later is $x(t)$. We are given $d^2x/dt^2 = a$, where $x(0) = 0$, $v(0) = 1000$, and $v(0.001) = 5000$.

The constant a is the unknown constant acceleration, and the velocity $v(t) = dx/dt$ as usual. Now $d^2x/dt^2 = dv/dt$, so $dv/dt = a$. Corollary 2a applies: $v(t) = at + b$. To find the constant b, we substitute $t = 0$:

$$b = v(0) = 1000$$

Therefore $dx/dt = v(t) = at + 1000$.

We were given $v(0.001) = 5000$; hence $5000 = 0.001a + 1000$ so $0.001a = 4000$ and it follows that

$$a = 4 \times 10^6 \text{ m/sec}^2$$

This solves the first part of the problem. The length of the accelerator path is the distance the particle travels in 0.001 sec. Now

$$\frac{dx}{dt} = v(t) = (4 \times 10^6)t + 1000$$

is a linear function, so we apply Corollary 2b:

$$x(t) = \tfrac{1}{2}(4 \times 10^6)t^2 + 1000t + c$$

We have $c = 0$ because $x(0) = 0$. The required distance is

$$x(10^{-3}) = \tfrac{1}{2}(4 \times 10^6)(10^{-3})^2 + (1000)(10^{-3})$$
$$= 2 + 1 = 3 \text{ meters}$$

●

In all of our examples so far, the acceleration has been constant. Constant acceleration is common in nature. For instance, near the earth's surface, the force of gravity on a (rising or falling) body is effectively constant, hence causes constant acceleration. An example of non-constant acceleration is a body falling into a star from a long distance away. Then the acceleration of the body due to the gravitational attraction of the star is

$$\frac{d^2x}{dt^2} = \frac{k}{x^2}$$

where k is a constant. This differential equation is difficult to solve at this stage. However, see Chapter 19.

Figure 3-2-6
Locomotive pulling train

Example 6

The distance that its locomotive pulls a certain train from rest in t seconds is $x(t) = At^{3/2}$, where A is constant. Show that $v(t)a(t)$ is a constant (Figure 3-2-6).

Solution Compute v and a:

$$v(t) = \frac{d}{dt}x(t) = \tfrac{3}{2}At^{1/2} \qquad a(t) = \frac{d}{dt}v(t) = \tfrac{3}{4}At^{-1/2}$$

Now multiply:

$$v(t)a(t) = (\tfrac{3}{2}At^{1/2})(\tfrac{3}{4}At^{-1/2}) = \tfrac{9}{8}A^2$$

●

Remark If m is the mass of the train, then $P = mva$ is the *power* that the locomotive generates. Thus in this case the power is constant, which is natural for an engine under load. Note that $a(t) \to +\infty$ as $t \to 0+$. This explains why a train seems to start with a sudden jolt.

It can be shown that functions $x = x(t)$ for which $v(t)a(t)$ is constant and $v(0) = 0$ are always of the form $x = At^{3/2}$. See Chapter 19.

Exercises

1 A projectile shot straight up has height $x = -16t^2 + 900t$ ft after t sec. Compute its average velocity between $t = 2$ and $t = 3$, also between $t = 2$ and $t = 2.1$. Compute its instantaneous velocity when $t = 2$.

2 During the initial stages of flight, a rocket reaches a height of $50t^2 + 500t$ ft in t sec. What is the average velocity between $t = 2$ and $t = 3$? Find the instantaneous velocity when $t = 2$ and when $t = 3$. What is the average of the two instantaneous velocities?

3 A projectile launched vertically from a plane has height $x = -5t^2 + 100t + 1500$ m after t sec. Find its maximum height. Find its velocity after 15 sec, and upon striking the ground.

4 An object shot upward has height $x = -5t^2 + 30t$ m after t sec. Compute its velocity after 1.5 sec, its maximum height, and the speed with which it strikes the ground.

5 A ball is thrown straight up from the top of a 600-foot tower. After t sec, it is $x = -16t^2 + 24t + 600$ ft above ground. When does the ball begin to descend? What is its speed when 605 ft above ground while going up? while coming down? (What general rule does this suggest?)

6 A shell fired at angle of elevation $45°$ with initial speed 300 m/sec has height $y = 150t\sqrt{2} - 5t^2$ m and horizontal distance $x = 150t\sqrt{2}$ m from its initial position t sec after firing. How far from its initial point does it strike the ground? What is its maximum elevation?

7 A body moves along a horizontal line according to the law $x = t^3 - 9t^2 + 24t$ ft. **a** When is x increasing and when decreasing? **b** When is the velocity increasing and when decreasing? **c** Find the total distance traveled between $t = 0$ and $t = 6$ sec.

8 Solve Exercise 7 if the law of motion is $x = t^3 - 3t^2 - 9t$ ft.

9 A ball is thrown straight up with an initial velocity of 48 ft/sec. Gravity causes a constant negative acceleration, -32 ft/sec². How high will the ball go if it is thrown from a height of 4 ft?

10 A car coasts down a 200-foot-long hill with acceleration 8 ft/sec^2. If the car starts from rest (zero velocity) at the top of the hill, when does it reach the bottom? How fast is it going then?

11 Starting from rest, what constant acceleration must a car undergo to move 75 ft in 5 sec?

12 The makers of a certain automobile advertise that it will accelerate from 0 to 100 mph in 1 min. If the acceleration is constant, how far will the car go in this time?

13 During the initial stages of flight after blast-off, a rocket shot straight up has acceleration 6 m/sec^2. The engine cuts out at $t = 10$ sec, after which only the gravitational acceleration, -10 m/sec^2, retards its motion. How high will the rocket go? How long does it take to reach its maximum height?

14 An airplane taking off from a landing field makes a run of 1000 m. If it starts its run with speed 7 m/sec, moves with constant acceleration, and completes the takeoff in 40 sec, with what speed does it take off?

15 A subway train starts from rest at a station and accelerates at the rate of 2 m/sec^2 for 10 sec. It then runs at constant speed for 60 sec, after which it decelerates at the rate of 3 m/sec^2 until it stops at the next station. Find the total distance it travels between the stations.

16 Gravitational acceleration on the moon is 0.165 times that on the earth. If a bullet shot straight up from the earth will rise 1 km, how far would it rise if shot on the moon?

Solve the differential equations subject to the given initial conditions

17 $dy/dx = -16x \quad y(0) = 12$

18 $dy/dt = 3t^2 + 4 \quad y(1) = -3$

19 $d^2y/dt^2 = -32 \quad y(1) = 48 \quad (dy/dt)|_1 = 64$

20 $d^2y/dx^2 = 8 \quad y(0) = 2 \quad y'(0) = 1$

21 $d^2y/dt^2 = 2t - 1 \quad y(0) = 5 \quad (dy/dt)|_0 = 3$

22 $d^2y/dx^2 = 3 - 4x \quad y(1) = 2 \quad y'(1) = 6$

3-3 Related Rates

If two changing quantities are related, then their rates of change are also related. Many problems can be solved by finding the relation between rates of change.

Example 1

A 15-foot ladder leans against a vertical wall. If the top slides downward at the rate of 2 ft/sec, find the speed of the lower end when it is 12 ft from the wall.

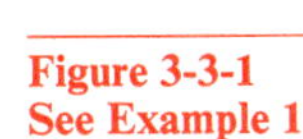
Figure 3-3-1
See Example 1

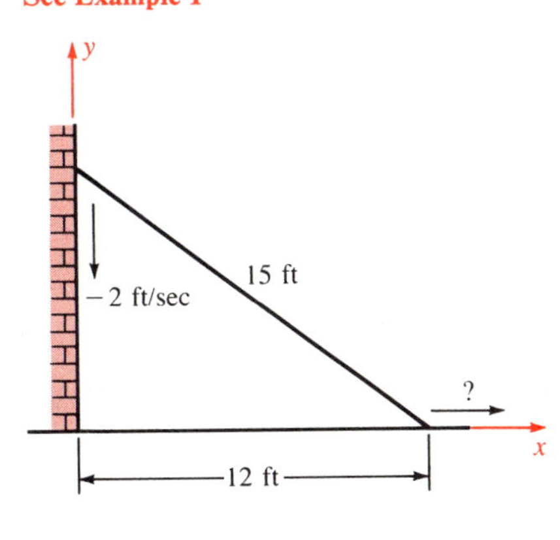

Solution We make a sketch placing axes as in Figure 3-3-1. Then $x = x(t)$ and $y = y(t)$ are related functions. We are given $dy/dt = -2$ and asked to find dx/dt at the instant when $x = 12$. The figure shows that the geometric relation between x and y is

$$x^2 + y^2 = 15^2$$

We differentiate with respect to time by the chain rule; then solve for dx/dt:

$$2x\frac{dx}{dt} + 2y\frac{dy}{dt} = 0 \qquad \text{hence} \qquad \frac{dx}{dt} = -\frac{y}{x}\cdot\frac{dy}{dt}$$

To find dx/dt at the instant in question, we need the values of y, dy/dt, and x at that instant. We are given $dy/dt = -2$ and $x = 12$. From the relation $x^2 + y^2 = 15^2$, we find $y = 9$. Therefore

$$\frac{dx}{dt} = -\left(\frac{9}{12}\right)(-2) = \frac{3}{2} \text{ ft/sec}$$

Example 1 is a typical related-rate problem. We are given the time derivative of one quantity and asked for the time derivative of a related quantity at a certain instant. We must find a relation between the two quantities, then differentiate it with respect to time to get a relation between their derivatives. Finally we substitute the data at the instant in question. Finding this data may require some side computations. Note that we do not need formulas for $x = x(t)$ or $y = y(t)$ as *explicit* functions of t.

Example 2

A large spherical rubber balloon (Figure 3-3-2) is inflated by a pump that injects helium at the rate of $10\ \text{ft}^3/\text{sec}$. At the instant when the balloon contains $972\pi\ \text{ft}^3$ of gas, how fast is its radius increasing? (Ignore changes in pressure.)

Figure 3-3-2
Sphere
Radius: r
Volume: $V = \frac{4}{3}\pi r^3$
Surface area: $A = 4\pi r^2$

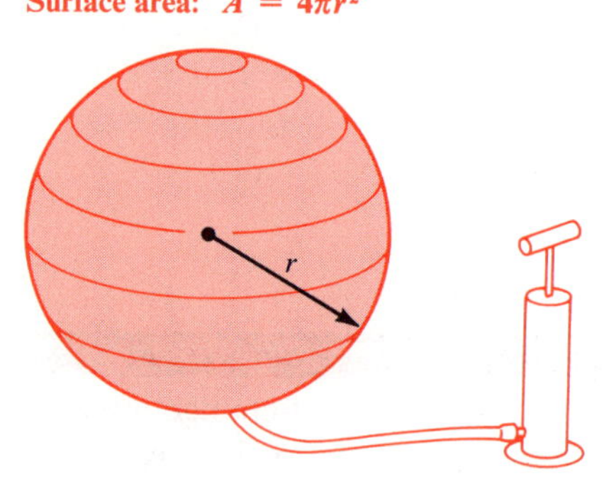

Solution We denote the radius and volume of the balloon at time t by $r(t)$ and $V(t)$. The derivative $dV/dt = 10\ \text{ft}^3/\text{sec}$ is given. The derivative dr/dt is required at a specific time. The formula for the volume V of a sphere of radius r is $V = \frac{4}{3}\pi r^3$. Hence

$$V(t) = \tfrac{4}{3}\pi[r(t)]^3$$

To find a relation between dr/dt and dV/dt, we differentiate V with respect to t, using the chain rule:

$$\frac{dV}{dt} = \frac{d}{dt}\left(\tfrac{4}{3}\pi r^3\right) = \frac{d}{dr}\left(\tfrac{4}{3}\pi r^3\right)\frac{dr}{dt} = 4\pi r^2\frac{dr}{dt}$$

Now we solve for dr/dt:

$$\frac{dr}{dt} = \frac{1}{4\pi r^2}\cdot\frac{dV}{dt} = \frac{10}{4\pi r^2}$$

This formula tells us the rate of change of the radius at any instant, in terms of the radius. At the instant in question, the volume is $972\pi\ \text{ft}^3$, so we can find the radius from the formula $V = \frac{4}{3}\pi r^3$:

$$\frac{4}{3}\pi r^3 = 972\pi \qquad r^3 = \frac{3}{4\pi}972\pi = 729 \qquad r = 9$$

But when $r = 9$,

$$\frac{dr}{dt} = \frac{10}{4\pi\cdot 9^2} = \frac{10}{324\pi} \approx 9.82\times 10^{-3}\ \text{ft/sec}$$

●

Example 3 (Continuation of Example 2)

How fast is the surface area of the sphere increasing at the instant in question?

Solution The formula for the surface area A of a sphere of radius r is $A = 4\pi r^2$. We differentiate with respect to t and substitute the expression we found in Example 2 for dr/dt:

$$\frac{dA}{dt} = \frac{d}{dt}(4\pi r^2) = \frac{d}{dr}(4\pi r^2)\frac{dr}{dt} = 8\pi r\frac{dr}{dt}$$

$$= (8\pi r)\left(\frac{10}{4\pi r^2}\right) = \frac{20}{r}$$

At the instant in question, $r = 9$, as we found, so

$$\frac{dA}{dt} = \frac{20}{9} \approx 2.22 \text{ ft}^2/\text{sec}$$

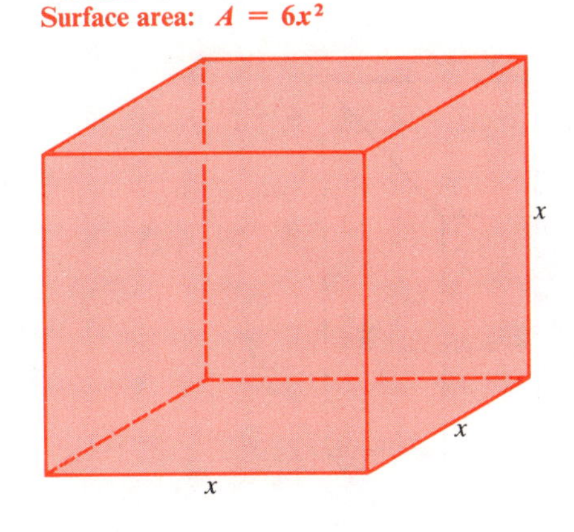

Figure 3-3-3
Cube
Side: x
Volume: $V = x^3$
Surface area: $A = 6x^2$

Example 4

The volume of an evaporating cube of dry ice (Figure 3-3-3) is decreasing at the rate of $4 \text{ cm}^3/\text{sec}$, but always keeping its shape. How fast is its surface area changing when the surface area is 24 cm^2?

Solution Let x be the edge of the cube. Then its area and volume are

$$A = 6x^2 \quad \text{and} \quad V = x^3$$

Express A in terms of V. Since $x = V^{1/3}$ and $A = 6x^2$,

$$A = 6V^{2/3}$$

Differentiate with respect to time, using the chain rule:

$$\frac{dA}{dt} = 6\frac{d}{dt}(V^{2/3}) = 6\frac{d}{dV}(V^{2/3})\frac{dV}{dt}$$

$$= 6 \cdot \tfrac{2}{3}V^{-1/3}\frac{dV}{dt} = 4V^{-1/3}\frac{dV}{dt}$$

Now find the value of V at the given instant, that is, when $A = 24$. From the relation $A = 6V^{2/3}$ it follows that

$$24 = 6V^{2/3} \qquad \text{hence} \qquad V^{2/3} = 4 \quad \text{and} \quad V = 8$$

Use the value $V = 8$ and the given value of the derivative $dV/dt = -4$:

$$\frac{dA}{dt} = (4)(8)^{-1/3}(-4) = -8 \text{ cm}^2/\text{sec}$$

Alternative Solution Instead of eliminating x, differentiate both relations $A = 6x^2$ and $V = x^3$ with respect to t:

$$\frac{dA}{dt} = 12x\frac{dx}{dt} \quad \text{and} \quad \frac{dV}{dt} = 3x^2\frac{dx}{dt}$$

When $A = 24$, then $6x^2 = 24$, so $x = 2$. But we were given $dV/dt = -4$; consequently

$$3x^2\frac{dx}{dt} = -4 \quad \text{so that} \quad 3 \cdot 4\frac{dx}{dt} = -4 \qquad \text{hence} \qquad \frac{dx}{dt} = -\tfrac{1}{3}$$

Finally,

$$dA/dt = 12x(dx/dt) = 12(2)(-\tfrac{1}{3}) = -8\ \text{cm}^2/\text{sec}$$

Thus the surface area is *decreasing* at the rate of $8\ \text{cm}^2/\text{sec}$. ●

Example 5

A rectangular tank has a sliding panel that divides it into two tanks of adjustable length and of width 3 ft. See Figure 3-3-4. Water is poured into the left compartment at the rate of $5\ \text{ft}^3/\text{min}$. At the same time, the sliding panel is moved to the right at the rate of $3\ \text{ft}/\text{min}$. When the left compartment is 10 ft long it contains $70\ \text{ft}^3$ of water. Is the water level rising or falling? How fast?

Figure 3-3-4
See Example 5

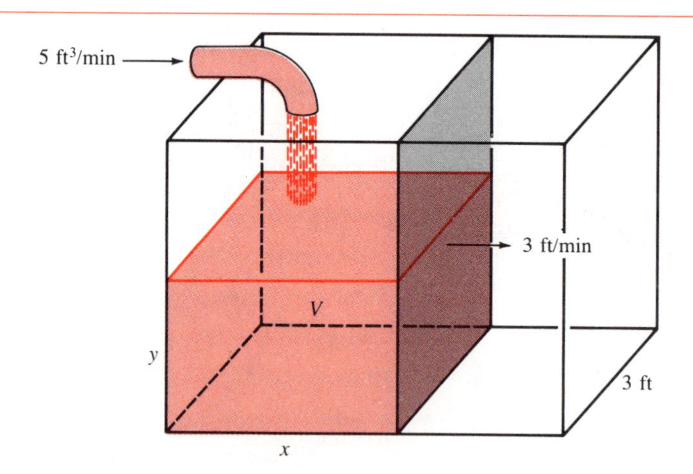

Solution Let x be the length of the left compartment; let y and V be the depth and the volume of the water in the left compartment. Then x, y, and V are all functions of time. Given:

$$\frac{dx}{dt} = 3\ \text{ft}/\text{min} \quad \text{and} \quad \frac{dV}{dt} = 5\ \text{ft}^3/\text{min}$$

The problem is to compute dy/dt when $x = 10$ and $V = 70$. To do so, find the relation between x, y, and V. By the figure, $V = 3xy$. Instead of differentiating the equation $V = 3xy$, we can solve for y first, then differentiate:

$$y = \tfrac{1}{3}\frac{V}{x} \qquad \text{hence} \qquad \frac{dy}{dt} = \tfrac{1}{3}\frac{x(dV/dt) - V(dx/dt)}{x^2}$$

At the given instant,

$$\frac{dy}{dt} = \tfrac{1}{3}\frac{10 \cdot 5 - 70 \cdot 3}{10^2} = -\frac{16}{30} = -\tfrac{8}{15}$$

The water level is falling at the rate of $\tfrac{8}{15}\ \text{ft}/\text{min}$. ●

Example 6

A point P moves counterclockwise at a constant speed of one revolution per minute (1 rpm) around a circle of radius 50 ft. As P moves, the tangent at P crosses the line OA at a moving point T. See Figure 3-3-5. Compute the speed of T when $\theta = \frac{1}{4}\pi$

Figure 3-3-5
See Example 6

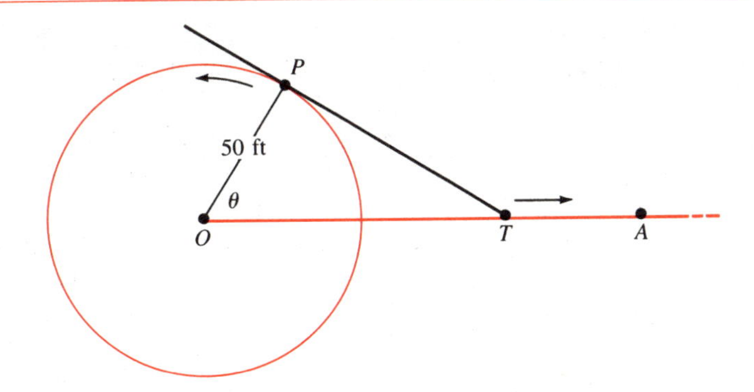

Solution We are given $d\theta/dt = 2\pi$ rad/min and are asked to find dx/dt, where $x = \overline{OT}$. The right triangle OPT suggests a relation between x and θ, namely, $x = 50 \sec \theta$. We differentiate with respect to t:

$$\frac{dx}{dt} = 50 \sec \theta \tan \theta \frac{d\theta}{dt}$$

When $\theta = \frac{1}{4}\pi$, then $\sec \theta = \sqrt{2}$ and $\tan \theta = 1$. At that instant, $dx/dt = 50\sqrt{2} \cdot 2\pi$. Answer: $100\pi\sqrt{2} \approx 444.3$ ft/min ●

Figure 3-3-6
See Example 7

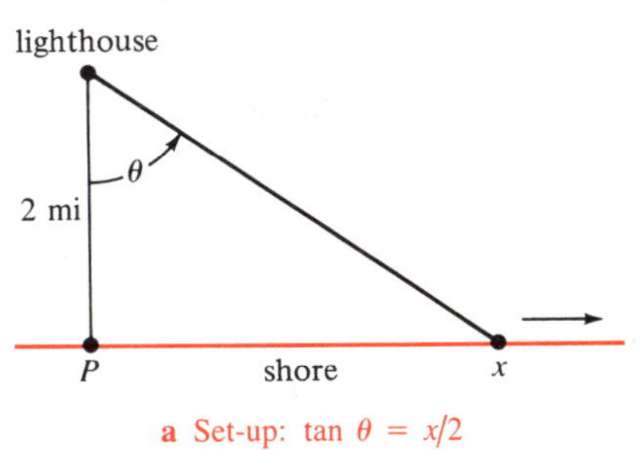

a Set-up: $\tan \theta = x/2$

b $\text{Sec}\ \theta = \sqrt{13}/2$

Example 7

A lighthouse stands 2 miles off a long straight shore, opposite a point P. Its light rotates counterclockwise at the constant rate of 1.5 rpm. How fast is the beam moving along the shore as it passes a point 3 miles to right of P?

Solution Set up coordinate axes with P at the origin and with x-axis along the shore (Figure 3-3-6a). The beam hits the shore at x. The rate of change of the angle θ is given as 1.5 rpm. Since 1 rev $= 2\pi$ rad, this translates to $d\theta/dt = 3\pi$ rad/min. The problem is to compute dx/dt at the instant when $x = 3$. The figure suggests a relation between x and θ:

$$\frac{x}{2} = \tan \theta \qquad \text{that is} \qquad x = 2 \tan \theta$$

Now differentiate with respect to t:

$$\frac{dx}{dt} = \frac{dx}{d\theta} \cdot \frac{d\theta}{dt} = (2 \sec^2\theta)(3\pi) = 6\pi \sec^2\theta$$

When $x = 3$, Figure 3-3-6b shows that $\sec\theta = \sqrt{13}/2$. Therefore at this instant,

$$\frac{dx}{dt} = 6\pi(\tfrac{1}{2}\sqrt{13})^2 = \frac{39\pi}{2} \approx 61.26 \text{ mi/min}$$

Exercises

Two functions $x = x(t)$ and $y = y(t)$ satisfy the given relation. Find dy/dt for the given data (and be sure to test the data for consistency).

1 $x^2 + 2 = y^3$
$x = 5 \quad y = 3 \quad dx/dt = -1$

2 $y + 1/y = x + 2$
$x = \frac{4}{3} \quad y = 3 \quad dx/dt = 10$

3 $x = \dfrac{y+1}{y-1}$
$x = -1 \quad y = 0 \quad dx/dt = 2$

4 $x^2 + xy + y^2 = 3$
$x = 1 \quad y = -2 \quad dx/dt = -5$

5 $y + \sqrt{1 + y^2} = x$
$x = 3 \quad y = \frac{4}{3} \quad dx/dt = 1$

6 $x^3(1 + y^2) = 2$
$x = 1 \quad y = 1 \quad dx/dt = 2$

7 A stone thrown into a pond produces a circular ripple, which expands from the point of impact. When the radius is 8 ft, it is observed that the radius is increasing at the rate 1.5 ft/sec. How fast is the area enclosed by the ripple increasing at that instant?

8 Two cars leave an intersection P. After 60 sec, the car traveling north has speed 50 ft/sec and distance 2000 ft from P, and the car traveling west has speed 75 ft/sec and distance 2500 ft from P. At that instant, how fast are the cars separating from each other?

9 A conical tank with its vertex pointing down has height 4 m and radius 1 m at the top. Oil flows in at the rate 0.05 m^3/min. When the depth is 2 m, how fast is the level rising? [Hint The volume of a right circular cone of base radius r and height h is $V = \frac{1}{3}\pi r^2 h$.]

10 (cont.) After the tank is filled, it is emptied through a tap at the very bottom. It is known that the rate of discharge is $dV/dt = -k\sqrt{h}$ where k is a constant and h is the oil depth. Suppose when the tank is full that $dV/dt = -0.2$ m^3/min. Find how fast the level is falling when the depth is 3 m.

11 A point P moves along the curve $y = x^3 - 3x^2$. When P is at $(1, -2)$, its x-coordinate is increasing at rate 3 cm/sec. Find the rate of increase of r, the distance from the origin to P. [Hint Work with r^2.]

12 A point P moves along the curve $y = x^4 + x + 1$. When P is at $(1, 3)$ its y-coordinate increases at rate 1. Find the rate of increase of the distance from the origin to P.

13 Yarn of radius 2 mm is being wound on a ball at the rate of 60 cm/sec. Assume that we can reasonably approximate the ball by a perfect sphere at each instant, and that it consists entirely of thread with no empty space. Find the rate of increase of the radius when the radius is 5 cm.

14 Two concentric circles are expanding. At a certain instant the outer radius is 10 ft and it is expanding at rate 2 ft/sec, while the inner radius is 3 ft and it is expanding at rate 5 ft/sec. Find the rate of change of the area between the circles at that instant.

15 Solve Exercise 14 for spheres and volume.

16 A train on a track 30 ft above the ground crosses a (perpendicular) street at the rate of 50 ft/sec at the instant that an automobile, approaching at the rate of 30 ft/sec, is 40 ft up the street. Find how fast the train and the auto are separating 2 sec later. [Hint In space, $\text{dist}^2 = (\text{EW-dist})^2 + (\text{NS-dist})^2 + (\text{vert dist})^2$.]

17 If a vertical chord sweeps across a circle of radius 10 ft at the rate of 6 ft/sec, how fast is the length of the chord decreasing when it is $\frac{3}{4}$ of the way across?

18 The power P in watts dissipated by an R-ohm resistor with V volts across it is $P = V^2/R$. At a certain instant, V is 112 volts, R is 10,000 ohms, and V and R are changing at the rate of 3 volts/min and -200 ohms/min, respectively. Find the rate of change of P in watts/min.

19 The volume V and pressure P of a gas in a constant-temperature engine cylinder are related by $PV = k$, a constant. Express dP/dt in terms of P and dV/dt.

20 Ship A sails due south toward a port P at 5 mph. Ship B sails due east away from P at 10 mph. At a given instant, A is a miles from P, and B is b miles from P. Show that the ships are getting closer together if $a > 2b$ and farther apart if $a < 2b$.

21 Express the rate at which the chord x in Figure 3-3-7 is lengthening in terms of the radius a, the central angle θ, and the angular speed $\omega = d\theta/dt$.

22 Express the rate at which the segment x in Figure 3-3-8 is lengthening in terms of a, b, θ, and $\omega = d\theta/dt$.

Figure 3-3-7
See Exercise 21

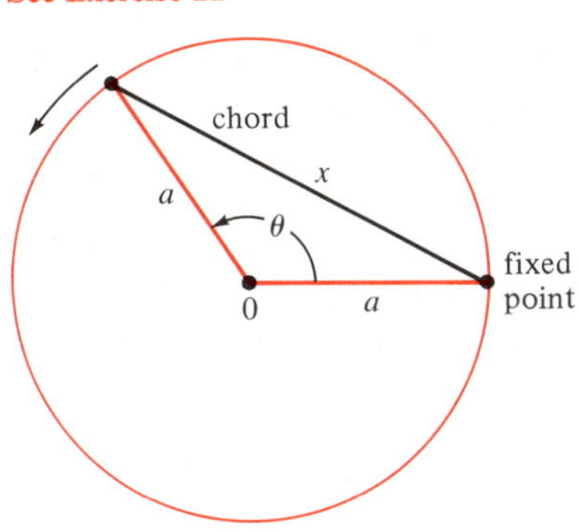

Figure 3-3-8
See Exercise 22

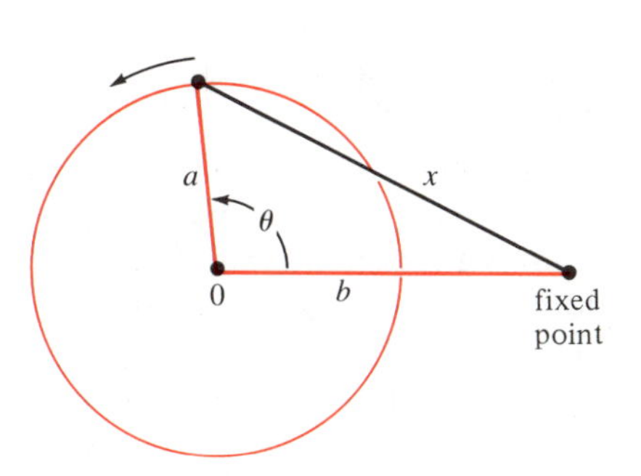

Figure 3-3-9
See Exercise 23

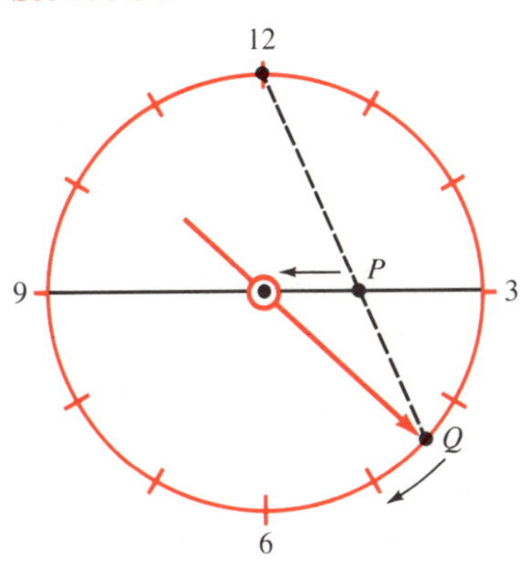

23 Suppose Figure 3-3-9 represents an electric clock of radius 3 inches and that Q is the tip of its second hand. How fast is the point P moving at 20 sec past 3:47 PM?

24 The hour hand of an electric clock has length a cm and the minute hand has length b cm. How fast are the tips of the hands separating at 3:00? How fast at 8:00?

25 A low-flying jet in level flight passes 450 ft directly over an observer on the ground. Shortly afterward its angle of elevation is $\frac{1}{6}\pi$ and decreasing at the rate of $\frac{1}{9}\pi/\text{sec}$. Compute the plane's speed, assumed constant.

26 If the target in Figure 3-3-10 is 90 m from the range finder and sailing away at 11 m/sec, how fast is θ increasing in deg/sec?

27 In a tangent galvanometer (ammeter), a current of I amperes produces a deflection angle θ, where I and θ are related by $I = k \tan \theta$. Compute the instrument's **sensitivity,** defined as

$$S = \frac{I/\theta}{dI/d\theta}$$

Show that $S \approx 1$ for $\theta \approx 0$.

28 To measure the diameter D of a shallow circular hole (Figure 3-3-11a), a needle gauge, of length L less than D, is rocked back and forth with end E fixed (Figure 3-3-11b). Let $2x$ denote the "rock." Express D in terms of x and show that the **sensitivity** of the gauge, defined by $S = (x/D)(dD/dx)$, is given by

$$S = \frac{x^2}{L^2 - x^2}$$

(Therefore for x small, D is relatively insensitive to a relative change in x, so the gauge is highly accurate.)

Figure 3-3-10
See Exercise 26

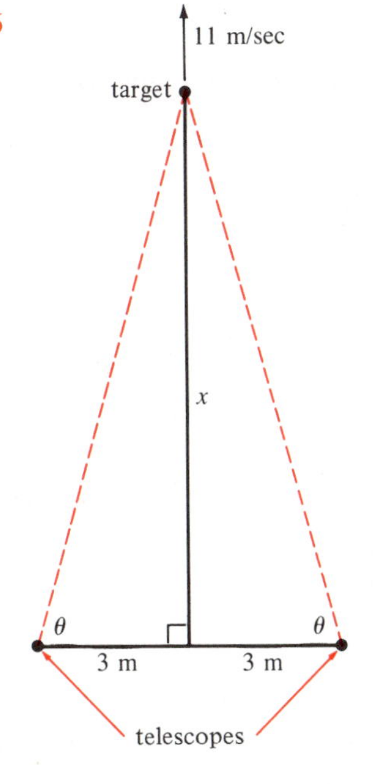

Figure 3-3-11
See Exercise 28

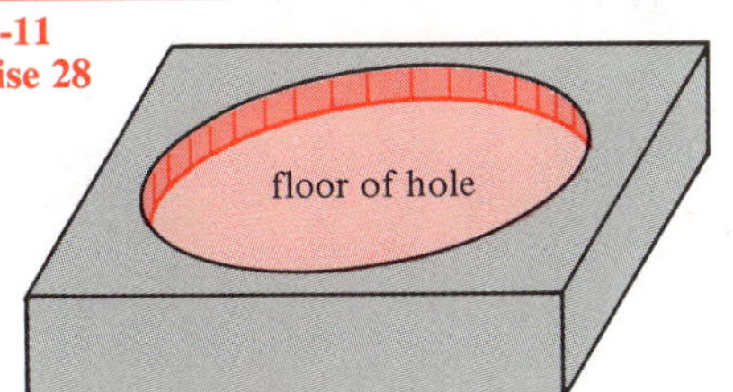

a Hole of diameter D

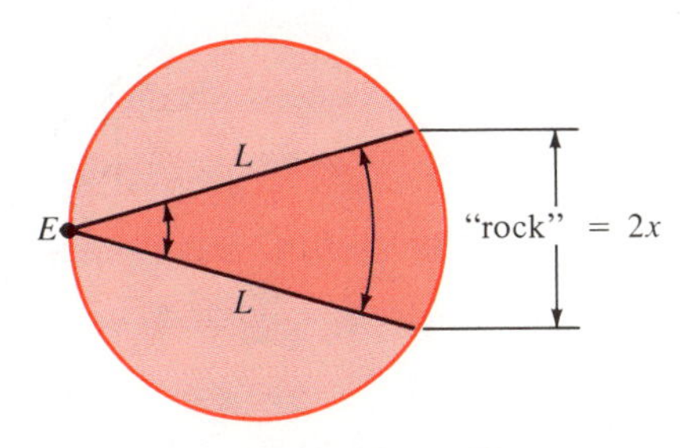

b Top view of rocking needle gauge

3-4 Properties of Continuous and Differentiable Functions

In this section we cover very important facts about continuous and differentiable functions, facts essential for understanding the subsequent sections. We shall postpone the theoretical discussion of these matters until Section 3-10.

The following theorem states a very basic property of continuous functions. It is concerned with the range of a continuous function whose domain is a closed interval. Recall that the **range** of a function $f(x)$ whose domain is a set D is the set of all real numbers $y = f(x)$ where x is in D. The theorem says that the range of a continuous function whose domain is a closed interval cannot be any old set; it must be a closed interval itself (or a single point). See Figure 3-4-1.

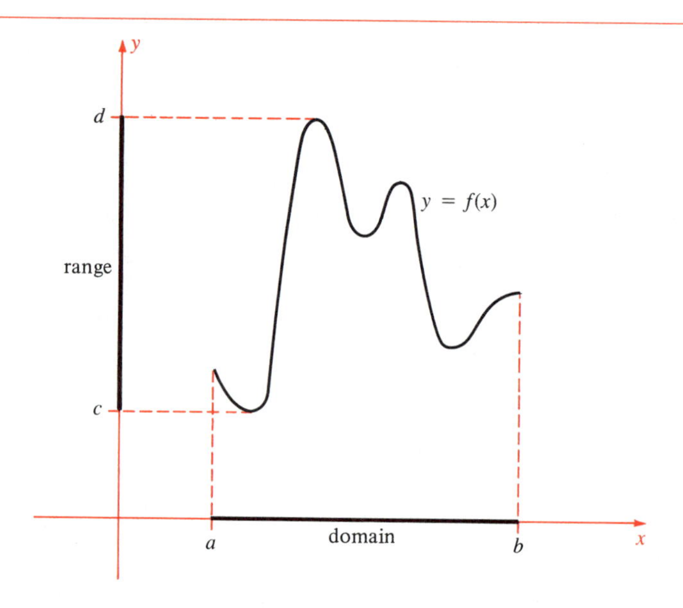

Figure 3-4-1
$f(x)$ continuous
Domain: closed interval $[a, b]$
Range: closed interval $[c, d]$
(or a single point)

Theorem 1 Range of a Continuous Function

Let $y = f(x)$ be a continuous function whose domain is a closed interval $[a, b]$ of the x-axis. Then the range of $f(x)$ is a closed interval $[c, d]$ of the y-axis, or a single point if $f(x)$ is a constant function. ●

The proof of this theorem is beyond the scope of a first calculus course. It depends on the real number system's having no holes—the *completeness* property introduced on page 2 and discussed in Section 3-10. The theorem has striking consequences, some of which we now give as corollaries.

Corollary 1 Boundedness

Let $y = f(x)$ be a continuous function on closed interval $[a, b]$. Then $f(x)$ is **bounded.** That is, there are constants c and d such that $c \le f(x) \le d$ for all x in $[a, b]$. ●

The corollary says that the values of a continuous function on a closed interval cannot go off the map; they are confined to a finite interval.

In fact, let the *range* of $f(x)$ be the closed interval $[c, d]$, as given by Theorem 1 (or $c = d$ if $f(x) = c$ is constant). Then $c \le f(x) \le d$ for all x in $[a, b]$ by the very definition of *range*. The number c, the left end point of the closed interval $[c, d]$, is in the range of $f(x)$. That is, $c = f(x_0)$ for some x_0 in $[a, b]$. Thus

$$f(x_0) \le f(x) \quad \text{for each } x \text{ in } [a, b]$$

This means that $f(x_0)$ is the *minimum* of $f(x)$ on $[a, b]$. Similarly, there is an x_1 in $[a, b]$ such that $f(x_1) = d$, the *maximum* of $f(x)$ on $[a, b]$. See Figure 3-4-2. We have proved the following corollary:

Figure 3-4-2
$f(x)$ continuous on $[a, b]$
A maximum $f(x_1)$ exists.
A minimum $f(x_0)$ exists.

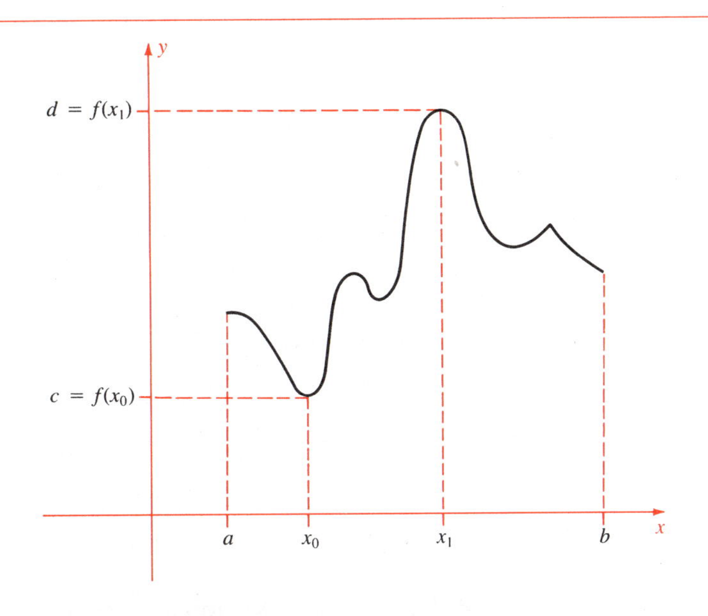

Corollary 2 Existence of Maxima and Minima

Let $f(x)$ be a continuous function on a closed interval $[a, b]$. Then $f(x)$ has both a maximum and a minimum on the interval. That is, there exist x_0 and x_1 in $[a, b]$ such that $f(x_0) \le f(x) \le f(x_1)$ for all x in $[a, b]$. ●

Figure 3-4-3
$y = f(x) = x$ domain $(0, 1)$
No matter where x is chosen in the domain, there are x_1 and x_2 in the domain such that
$f(x_1) < f(x)$ and $f(x_2) > f(x)$

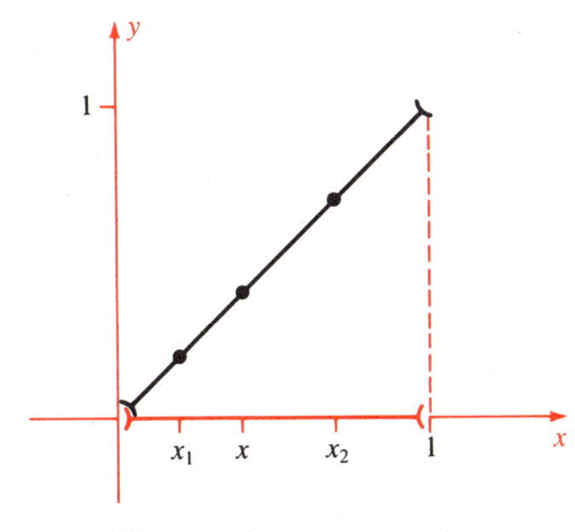

The following example shows functions that do not satisfy the hypotheses of Theorem 1 or its corollaries, so they cannot be expected to satisfy its conclusions. Note that their domains are not *closed* intervals.

Example 1

a $f(x) = x$ domain $(0, 1)$, an open interval

The range is $(0, 1)$ on the y-axis. The function is bounded: $0 < f(x) < 1$, but it does not have a minimum or a maximum on its domain. That is because if x is in the domain, then there are points x_1 and x_2 in the domain such that $f(x_1) < f(x)$ and $f(x_2) > f(x)$. For instance, choose $x_1 = \frac{1}{2}x$ and $x_2 = \frac{1}{2}(x + 1)$. See Figure 3-4-3.

Figure 3-4-4
$y = f(x) = 1/x$
Domain $(0, 1]$

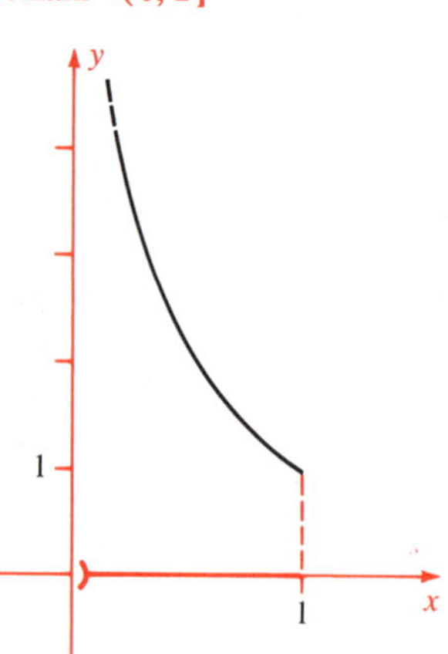

Figure 3-4-5
$y = f(x) = 1/x$
Domain $(0, +\infty)$

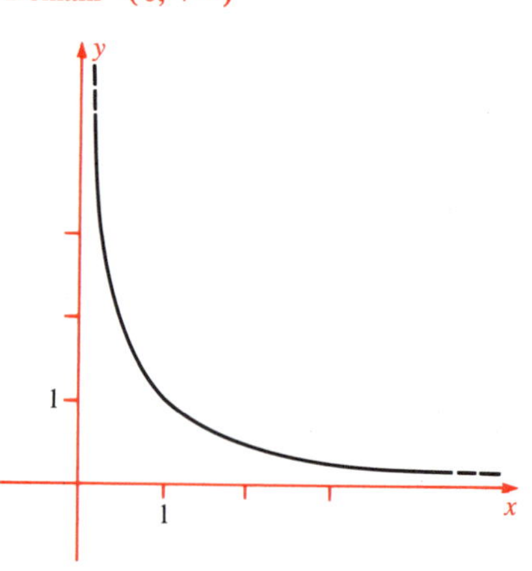

b $f(x) = 1/x$ domain $(0, 1]$

The range is the interval $[1, +\infty)$ of the y-axis. The function is bounded below, $1 \le f(x)$, but unbounded above. It has a minimum $f(1)$, because $f(x) \ge f(1) = 1$, but no maximum (Figure 3-4-4).

c $f(x) = 1/x$ domain $(0, +\infty)$

The range is the interval $(0, +\infty)$ of the y-axis. The function is bounded below: $f(x) > 0$, but unbounded above. It has no minimum and no maximum (Figure 3-4-5). You might think its minimum is 0, but that is wrong because there is no x_0 in its domain such that $f(x_0) = 0$.

Figure 3-4-6
$y = \tan x$
Domain $(-\frac{1}{2}\pi, \frac{1}{2}\pi)$

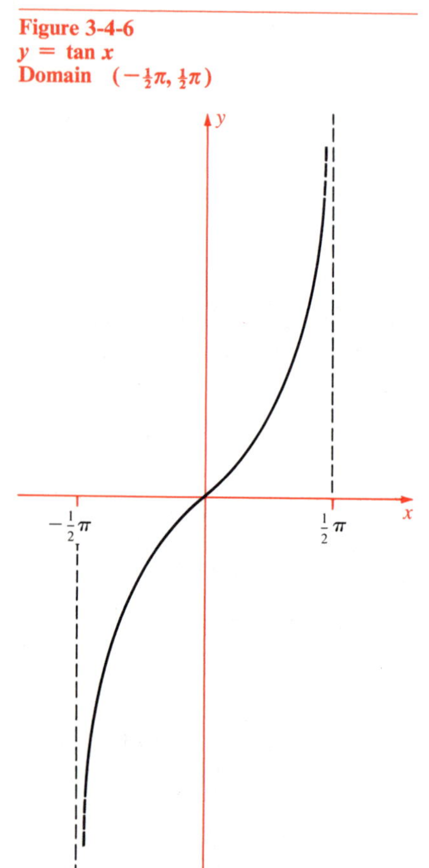

d $f(x) = \tan x$ domain $(-\frac{1}{2}\pi, \frac{1}{2}\pi)$

The range is the whole y-axis. The function is unbounded above and below, so it certainly has no minimum and no maximum (Figure 3-4-6). ●

Theorem 1 can be interpreted intuitively by saying that you can draw the graph of a continuous function without lifting your pen. The following corollary expresses this idea more strongly: It says that the graph of a continuous function on an interval never jumps over a horizontal line (Figure 3-4-7).

Corollary 3 Intermediate Value Theorem

Let $f(x)$ be continuous on an interval I. Suppose a and b are two points of I and k is a real number between the values $f(a)$ and $f(b)$. That is, either

$$f(a) < k < f(b) \quad \text{or} \quad f(a) > k > f(b)$$

Then there is a number x_0 between a and b such that $f(x_0) = k$. ●

Figure 3-4-7
$f(x)$ is continuous on $[a, b]$.
k is between $f(a)$ and $f(b)$.
The graph of $y = f(x)$ *must* intersect the line $y = k$.

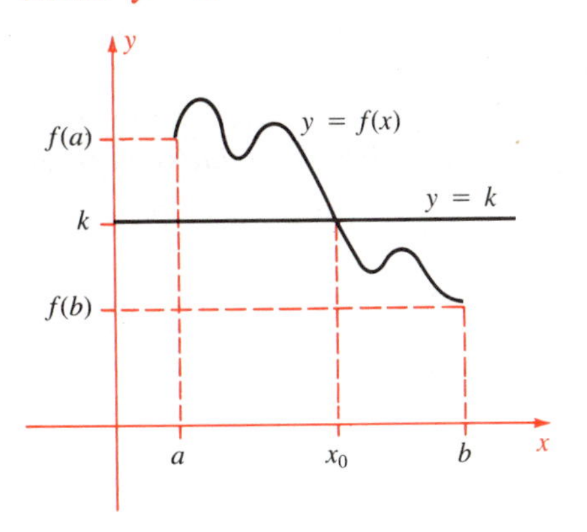

Figure 3-4-8
If $f'(c) > 0$, then for x near c,
$\begin{cases} f(x) > f(c) & \text{for } x > c \\ f(x) < f(c) & \text{for } x < c \end{cases}$

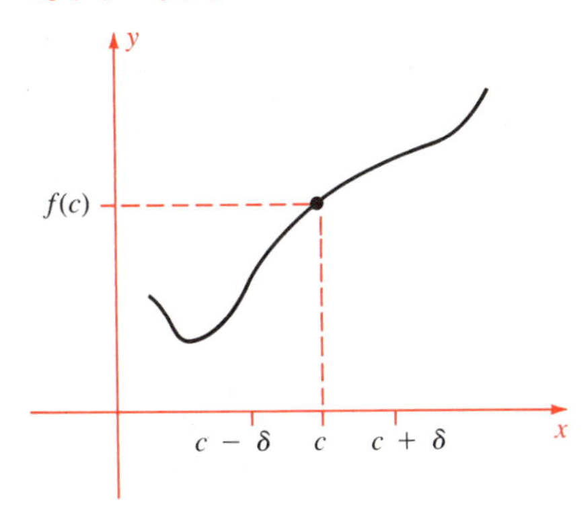

Figure 3-4-9
If $f'(c) < 0$, then for x near c,
$\begin{cases} f(x) < f(c) & \text{for } x > c \\ f(x) > f(c) & \text{for } x < c \end{cases}$

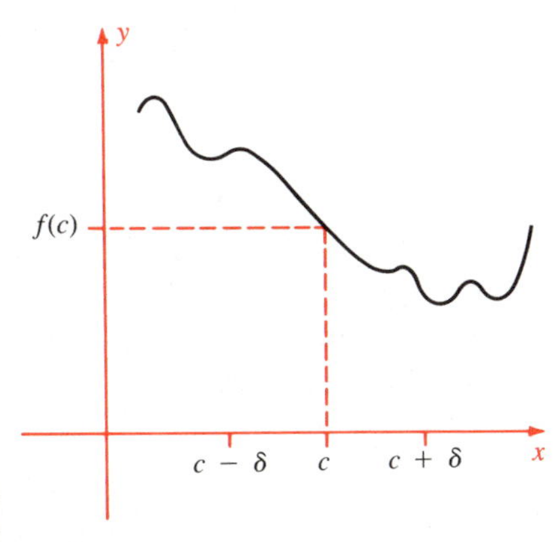

Non-zero Derivative

We move on to functions that are not merely continuous, but are differentiable. That gives us more to work with, so we expect more results. Before we look at important consequences of being differentiable on an interval, we take up what is called a "local" result. We assume something about a derivative at a single point and prove that something happens very close to that point. Precisely, if $f'(c) > 0$, then $f(x)$ is larger than $f(c)$ slightly to the right of $x = c$ and smaller slightly to the left. A corresponding result holds for $f'(c) < 0$. Intuitively the theorem means that $f'(c)$ really measures the *rate of change* of $f(x)$ at c: A positive derivative at c means that $f(x)$ is truly increasing at c: that is, as x moves from left to right through c, the function increases in value. See Figure 3-4-8.

Theorem 2 Function Increasing Locally

Let the domain of $f(x)$ be an open interval (a, b), and let $f(x)$ be differentiable at $x = c$, where $a < c < b$. Suppose $f'(c) > 0$. Then there exists a $\delta > 0$ such that

$$\begin{cases} f(x) > f(c) & \text{on } (c, c+\delta) \\ f(x) < f(c) & \text{on } (c-\delta, c) \end{cases}$$ ●

Proof

From the very definition of the derivative,

$$\lim_{x\to c} \frac{f(x) - f(c)}{x - c} = f'(c) > 0$$

This implies, by one of our basic results on limits (Theorem 2, part 5, in Section 2-3), that there is a $\delta > 0$ such that

$$\frac{f(x) - f(c)}{x - c} > 0$$

for $0 < |x - c| < \delta$, that is, for x in $(c - \delta, c)$ or in $(c, c + \delta)$. Suppose that x is in the right-hand interval, $(c, c + \delta)$. Then $c < x < c + \delta$, so that $0 < x - c < \delta$. Hence

$$f(x) - f(c) = (x - c)\frac{f(x) - f(c)}{x - c} > 0 \quad \text{so} \quad f(x) > f(c)$$

Similarly, if x is in the left-hand interval $(c - \delta, c)$, then $f(x) - f(c) < 0$ so $f(x) < f(c)$. This completes the proof. ●

Remark Suppose $f'(c) < 0$. Then $-f'(c) > 0$, so the theorem applies to $-f(x)$. The conclusion is that $f(x) > f(c)$ for $c - \delta < x < c$ and $f(x) < f(c)$ for $c < x < c + \delta$. See Figure 3-4-9.

Also, do not read into the conclusion of Theorem 2 more than it says. It does not follow from $f'(c) > 0$ that $f(x)$ is strictly increasing in some open interval around c.

Figure 3-4-10
Maximum: $f(x_0)$
Local maxima: $f(x_0)$ and $f(x_2)$
Minimum: $f(x_1)$
Local minima: $f(x_1)$ and $f(x_3)$

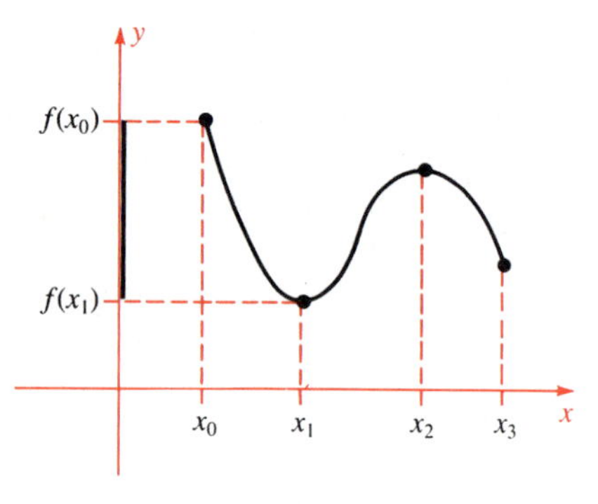

Local Maxima and Minima

We now come to an important concept: local maxima and minima. Consider the function graphed in Figure 3-4-10. Its range is the interval $[f(x_1), f(x_0)]$ of the y-axis. Its minimum is $f(x_1)$; its maximum is $f(x_0)$. But $f(x_2)$ is also a maximum of $f(x)$, at least for values of x that are close to x_2. Also $f(x_3)$ is a minimum of $f(x)$ if you don't go too far from x_3. We call $f(x_0)$ and $f(x_2)$ local maxima of $f(x)$, and $f(x_1)$ and $f(x_3)$ local minima. (The adjective *relative* is often used instead of *local.*) In general we make the following definition.

Definition Local Maxima and Minima

Let $f(x)$ be a continuous function and let c be a point of its domain. The function $f(x)$ has a **local minimum** at $x = c$ if there is a $\delta > 0$ such that

$$f(x) \geq f(c)$$

for all x in the domain of f satisfying $|x - c| < \delta$. The function $f(x)$ has a **local maximum** at $x = c$ if there is a $\delta > 0$ such that

$$f(x) \leq f(c)$$

for all x in the domain of f satisfying $|x - c| < \delta$. ●

This definition paves the way for one of the main tools used in applying differential calculus. It says that local extrema (maxima and minima) of a differentiable function that occur strictly inside its domain occur only where the tangent is horizontal (Figure 3-4-11).

Figure 3-4-11
Inside the domain of a differentiable function $f(x)$, extrema occur only where $f'(x) = 0$.

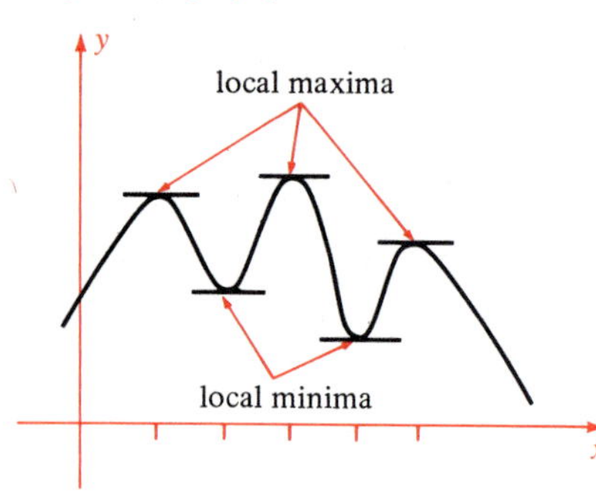

Theorem 3

Let $f(x)$ be a differentiable function on the closed interval $[a, b]$. Suppose that $a < c < b$ and that $f(c)$ is either a local maximum or a local minimum of $f(x)$. Then $f'(c) = 0$. ●

Proof There are just three possibilities:

$$f'(c) > 0 \qquad f'(c) < 0 \qquad f'(c) = 0$$

Theorem 2 rules out the first two. For if $f'(c) > 0$, then $f(x) > f(c)$ just to the right of $x = c$ and $f(x) < f(c)$ just to the left of $x = c$, so $f(c)$ is neither a local maximum nor local minimum. Similarly, $f'(c) < 0$ is impossible. The only possibility is $f'(c) = 0$. ●

Rolle's Theorem and the Mean Value Theorem

We are preparing for a very fundamental theorem about differentiable functions called the mean value theorem. It is precisely what we need to prove that a function whose derivative is everywhere zero is constant, that a function whose derivative is everywhere positive is increasing, and many other results. Its proof depends on the following preliminary result, actually a special case of the mean value theorem.

Theorem 4 Rolle's Theorem

Let $f(x)$ be a continuous function whose domain is a closed interval $[a, b]$, and suppose $f(x)$ is differentiable on the open subinterval (a, b). Also assume $f(a) = f(b)$. Then there exists a point c such that $a < c < b$ and $f'(c) = 0$. ●

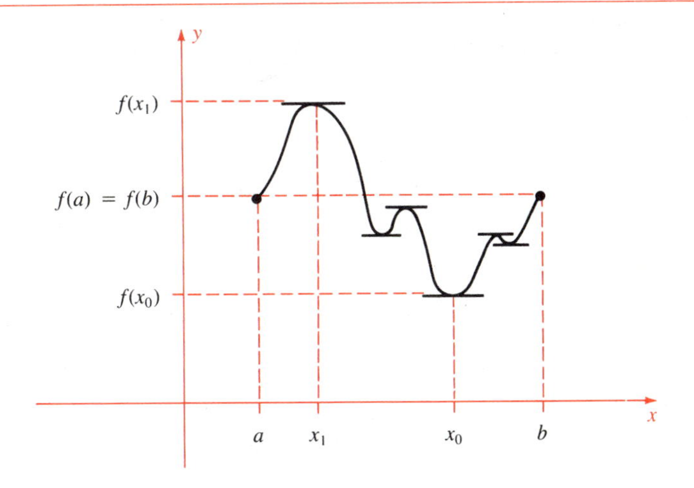

Figure 3-4-12
Rolle's theorem: $f(x)$ is continuous on $[a, b]$, $f(a) = f(b)$, and $f(x)$ is differentiable on (a, b). Then $f'(x) = 0$ someplace on (a, b).

Proof See Figure 3-4-12 for a geometric explanation of Rolle's theorem and of the proof. First we apply Corollary 2 of Theorem 1: There exist x_0 and x_1 in the closed interval $[a, b]$ such that

$$f(x_0) \le f(x) \le f(x_1)$$

for all x in the domain $[a, b]$ of f. If $a < x_0 < b$ then $f'(x_0) = 0$ by the preceding theorem. If $a < x_1 < b$, then $f'(x_1) = 0$ for the same reason. If neither, then x_0 and x_1 must both be end points of the interval $[a, b]$. But $f(a) = f(b)$, so no matter whether x_0 and x_1 are a or b, we have

$$f(x_0) = f(x_1) = f(a) = f(b)$$

But every value of $f(x)$ lies between its (equal) maximum and minimum values, so $f(x)$ is a constant on the interval. This implies $f'(c) = 0$ for *every* c in $[a, b]$, more than we wanted! ●

The next result, known as the mean value theorem or the law of the mean, can be interpreted as an "oblique" form of Rolle's theorem. Rolle's theorem guarantees the existence of a horizontal tangent provided that $f(a) = f(b)$, that is, a tangent parallel to the (horizontal) chord joining $(a, f(a))$ and $(b, f(b))$. The mean value theorem guarantees the existence of a tangent parallel to the chord joining $(a, f(a))$ and $(b, f(b))$, whether that chord is horizontal or oblique. Thus it says that the *average slope* of $f(x)$ over the interval is equal to its *instantaneous slope* somewhere on the interval. See Figure 3-4-13 (next page) for the geometry of the mean value theorem, which we shall refer to as the MVT.

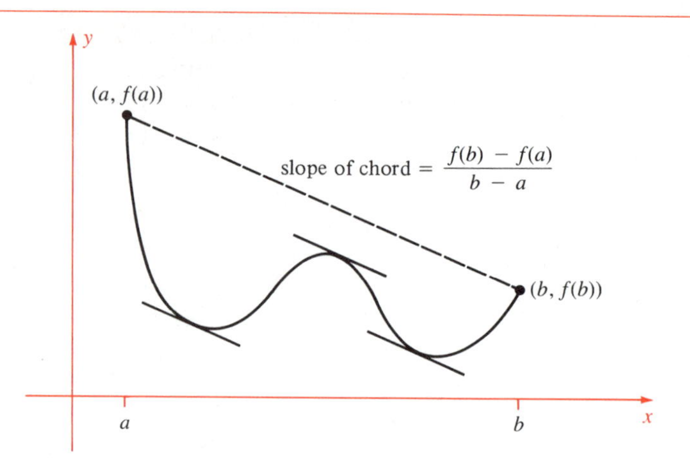

Figure 3-4-13
Mean value theorem: There exist tangent(s) parallel to the chord.

Theorem 5 Mean Value Theorem (MVT)

Let $f(x)$ be a continuous function whose domain is a closed interval $[a, b]$. Suppose $f(x)$ is differentiable on the open subinterval (a, b). Then there exists a point c such that $a < c < b$ and

$$f'(c) = \frac{f(b) - f(a)}{b - a}$$

Proof First, we seek the equation of the line through $(a, f(a))$ and $(b, f(b))$. Its slope is

$$m = \frac{f(b) - f(a)}{b - a}$$

so by the point-slope form (page 24), the equation of the line is

$$y = g(x) = m(x - a) + f(a)$$

Thus $g(x)$ is the linear function such that $g(a) = f(a)$ and $g(b) = f(b)$. The vertical distance from the graph of $g(x)$ · to that of $f(x)$ is

$$h(x) = f(x) - g(x)$$

This difference $h(x)$ is a continuous function on $[a, b]$ and is differentiable on (a, b). What is more,

$$h(a) = f(a) - g(a) = 0 \quad \text{and} \quad h(b) = f(b) - g(b) = 0$$

so $h(a) = h(b)$. Hence Rolle's theorem applies to $h(x)$: For some c with $a < c < b$ we have $h'(c) = 0$. But

$$h'(x) = f'(x) - g'(x) = f'(x) - m$$

Therefore the relation $h'(c) = 0$ implies that

$$f'(c) = m = \frac{f(b) - f(a)}{b - a}$$

Our first application of the MVT will be to prove that a function with zero derivative on its entire domain is constant. This is unfinished business from page 155.

Corollary 1 Zero Derivative

Let $f(x)$ be continuous on the closed interval $[a, b]$ and differentiable on the open subinterval (a, b). Suppose $f'(x) = 0$ everywhere on (a, b). Then $f(x)$ is constant on $[a, b]$.

Proof Let $a < x_1 \leq b$. By the MVT, applied to the interval $[a, x_1]$, there exists a c such that $a < c < x_1$ and

$$\frac{f(x_1) - f(a)}{x_1 - a} = f'(c)$$

But $f'(c) = 0$; hence $f(x_1) - f(a) = 0$. That is, $f(x_1) = f(a)$. This is true for every x_1 in the interval $(a, b]$; hence f is constant. This completes the proof and confirms the obvious: If the speed of a car is 0 during a time interval, then it really doesn't move!

Our final result in this section again confirms what is intuitively obvious: if the derivative of a function is positive on an interval, then the function increases on that interval. Note that this is a "global" statement in that it deals with properties of a function throughout an entire interval. In contrast, Theorem 2 is a "local" statement in that it refers only to an immediate neighborhood of a single point.

Corollary 2 Increasing and Decreasing Functions

Let $f(x)$ be continuous on the closed interval $[a, b]$ and differentiable on the open subinterval (a, b).

1. If $f'(x) \geq 0$ on (a, b), then $f(x)$ is an **increasing** function. That is, $f(x_0) \leq f(x_1)$ whenever $a \leq x_0 < x_1 \leq b$.
2. If $f'(x) > 0$ on (a, b), then $f(x)$ is a **strictly increasing** function. That is, $f(x_0) < f(x_1)$ whenever $a \leq x_0 < x_1 \leq b$.
3. Corresponding results hold for $f' \leq 0$ (**decreasing**) and $f' < 0$ (**strictly decreasing**).

Proof Let $a \leq x_0 < x_1 \leq b$. Apply the MVT to the interval $[x_0, x_1]$: There exists a point c such that $x_0 < c < x_1$ and

$$f'(c) = \frac{f(x_1) - f(x_0)}{x_1 - x_0}$$

that is

$$f(x_1) - f(x_0) = f'(c)(x_1 - x_0)$$

But $x_1 - x_0 > 0$, so if $f'(c) \geq 0$, then $f(x_1) - f(x_0) \geq 0$, that is, $f(x_1) \geq f(x_0)$. And if $f'(c) > 0$, then $f(x_1) - f(x_0) > 0$, that is, $f(x_1) > f(x_0)$. This completes the proof.

Summary

This has been a big slug of theory in the middle of a chapter on applications. Why? Because this mathematics is essential to the applications on curve sketching and on maxima and minima problems that follow. No one will expect you to master all of this theory right away, and this is a section to which you will frequently refer. For now, you should leave the section believing the following facts:

1 A continuous function $f(x)$ on $[a, b]$ has a maximum and a minimum (Corollary 2 of Theorem 1).

2 If a differentiable function $f(x)$ has a local maximum or minimum at $x = c$ in (a, b), then $f'(c) = 0$ (Theorem 3).

3 If $f'(x) = 0$ on an interval, then $f(x)$ is constant (Corollary 1 of Theorem 5).

4 If $f'(x) \geq 0$ on an interval, then $f(x)$ is increasing. If $f'(x) > 0$ on an interval, then $f(x)$ is strictly increasing (Corollary 2 of Theorem 5). Similar statements hold for decreasing functions.

5 The mean value theorem (Theorem 5). Try to memorize Figure 3-4-13; it says all you need to understand the MVT.

Exercises

Show that the hypotheses of Rolle's theorem are satisfied by $y = f(x)$ and find *all* c on the open interval (a, b) such that $f'(c) = 0$

1 $y = 1 - x^2$ domain $[-1, 1]$

2 $y = x(1 - x)$ domain $[0, 1]$

3 $y = 3x - x^2$ domain $[0, 3]$

4 $y = x^2(1 - x)$ domain $[0, 1]$

5 $y = \sin x$ domain $[0, 4\pi]$

6 $y = \cos x$ domain $[0, 2\pi]$

7 $y = x^2(x - 1)(x - 2)$ domain $[0, 1]$

8 $y = x^2(x^2 - 1)$ domain $[-1, 1]$

9 $y = \sqrt{1 - x^2}$ domain $[-1, 1]$

10 $y = \sqrt{x} + \sqrt{1 - x}$ domain $[0, 1]$

Show that the hypotheses of the MVT are satisfied by $y = f(x)$. Sketch the graph and find *all* c on the open subinterval (end points removed) such that $f'(c)$ equals the slope of the chord. Show the chord and the parallel tangents on your graph.

11 $y = x^2$ domain $[0, 1]$

12 $y = x^2$ domain $[1, 4]$

13 $y = x^3$ domain $[0, 1]$

14 $y = x^3$ domain $[-1, 1]$

15 $y = x^3$ domain $[-1, 2]$

16 $y = x^3 - x$ domain $[-2, 1]$

17 $y = \sqrt{1 - x^2}$ domain $[-1, 0]$

18 $y = -\sqrt{1 - x^2}$ domain $[-\frac{3}{5}, 1]$

19 Use Corollary 2 of Theorem 1 to prove that $f(x) = x(1 - x)(x^{10} + x^8 + 1)$ has a positive maximum on $[0, 1]$.

20 Is the function $f(x) = x^3/(1 + x^4)$ bounded for all x? If so, find a bound C such that $|f(x)| \leq C$ for all x.

21 Is $f(x) = 1/x - 1/x^2$ bounded for $x > 0$?

22 Is the function in Exercise 21 bounded for $x < -1$?

3-5 Curve Sketching

This section is concerned with the finer points of graphing functions. So far, in graphing a function $y = f(x)$, we have used values of the function to plot points and to determine behavior at $+\infty$ and $-\infty$. We have used values of the first derivative $f'(x)$ to find intervals where $f(x)$ is increasing or decreasing and to locate horizontal tangents. In this section we use additional techniques for graphing based on the first derivative and, in addition, introduce further graphing techniques based on the second derivative. The following example shows the need for such further techniques.

Example 1

Graph $y = 16x^2(1-x)^2$

Solution The points $(0,0)$ and $(1,0)$ are on the graph. Also $y \to +\infty$ as $x \to +\infty$ and as $x \to -\infty$ because y is a polynomial of even degree. Next we seek horizontal tangents. Now

$$y' = 16[2x(1-x)^2 - 2x^2(1-x)] = 32x(1-x)(1-2x)$$

so $y' = 0$ at $x = 0$, $\frac{1}{2}$, and 1. Also $y' < 0$ for $x < 0$, so y is decreasing on $(-\infty, 0]$. Similarly, y is increasing on $[0, \frac{1}{2}]$, decreasing on $[\frac{1}{2}, 1]$, and increasing on $[1, +\infty)$. The value $x = \frac{1}{2}$ corresponds to the point $(\frac{1}{2}, 1)$ on the graph. This information is plotted in Figure 3-5-1 and a smooth curve is interpolated in Figure 3-5-2. ●

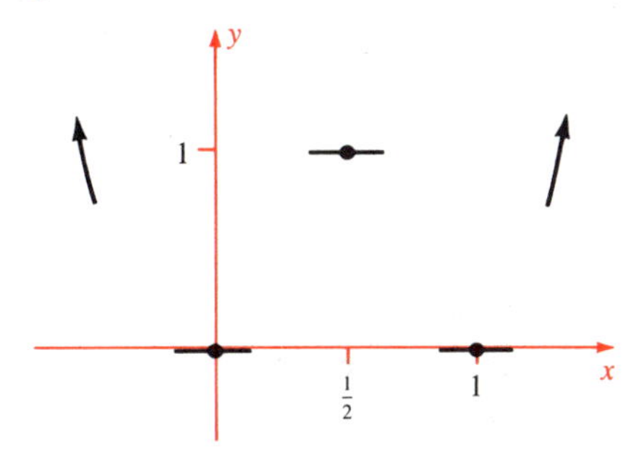

Figure 3-5-1
$y = 16x^2(1-x)^2$ has horizontal tangents at $(0,0)$, $(\frac{1}{2}, 1)$, and $(1,0)$. $y \to +\infty$ as $x \to \pm\infty$

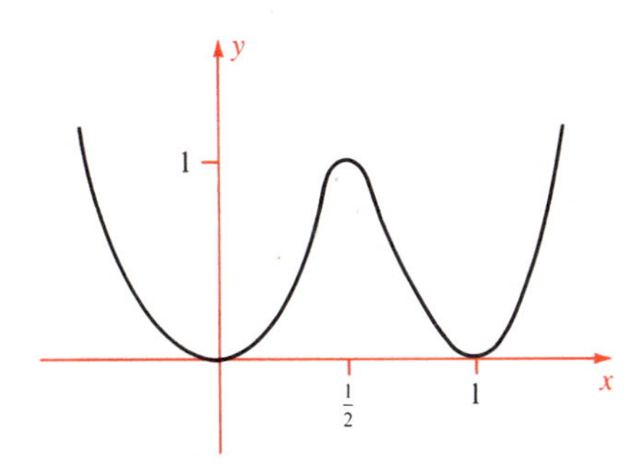

Figure 3-5-2
Graph of $y = 16x^2(1-x)^2$

Figure 3-5-2 actually assumes a lot. For instance, why can't the piece of curve joining the horizontal tangents at $(0,0)$ and $(\frac{1}{2}, 1)$ be wiggly like Figure 3-5-3? It cannot be, and the methods we develop in this section will show why.

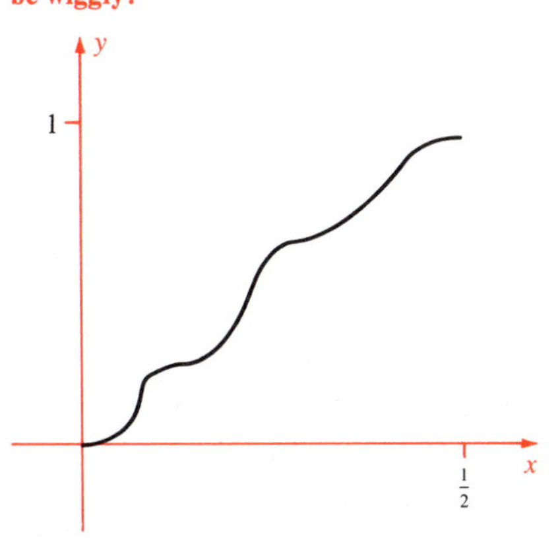

Figure 3-5-3
Can the graph of $y = 16x^2(1-x)^2$ be wiggly?

Concave Functions

Suppose $f(x)$ is a differentiable function whose first derivative $f'(x)$ is a strictly increasing function of x. Then as x increases, the slope of the graph of $y = f(x)$ increases, that is, the tangent line to $y = f(x)$ turns counterclockwise (Figure 3-5-4, next page).

Look carefully at Figure 3-5-4. It seems to imply that the graph is always curving in the same direction. Now that means if you draw any chord to the graph, then that chord is entirely above the graph (Figure 3-5-5, next page). This property will be shown in Section 3-10 to follow from the assumption of increasing slope. We should mention that this chord property is sometimes taken as the *definition* of concave.

Figure 3-5-4
f' strictly increasing:
The tangent turns counterclockwise as x increases.

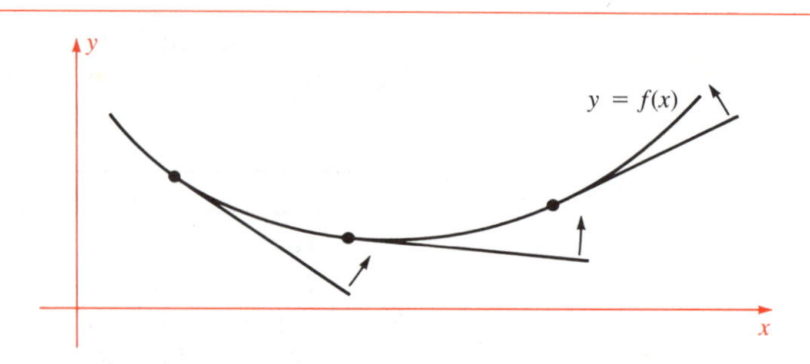

Figure 3-5-5
Chord property:
Each chord is above the graph.

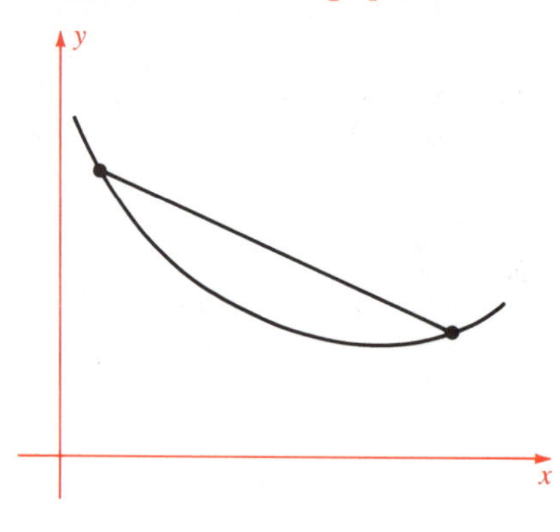

Definition Concave Functions

Let $f(x)$ be continuous on an interval, and differentiable except possibly at the end points.

- $f(x)$ is **concave up** if $f'(x)$ is a strictly increasing function.
- $f(x)$ is **concave down** if $f'(x)$ is a strictly decreasing function.

Example 2

Graph $y = f(x) = x^3 - 3x$ and discuss its concavity.

Solution The point $(0, 0)$ is on the graph. Now

$$y' = 3x^2 - 3 = 3(x^2 - 1) = 3(x + 1)(x - 1)$$

so $y' = 0$ for $x = -1$ and for $x = 1$. Thus $(-1, 2)$ and $(1, -2)$ are points where the tangent is horizontal. All of this is old hat for us now and leads easily to the graph in Figure 3-5-6. What is new is the question of concavity. As x increases from $-\infty$ to 0, the derivative $y'(x)$ decreases strictly from $+\infty$ to -3. Thus $f(x) = x^3 - 3x$ is concave down on $(-\infty, 0]$. As x increases from 0 to $+\infty$, the derivative $y'(x)$ increases strictly from -3 to $+\infty$, so $f(x)$ is concave up on $[0, +\infty)$. See Figure 3-5-7.

Figure 3-5-6
Graph of $y = f(x) = x^3 - 3x$

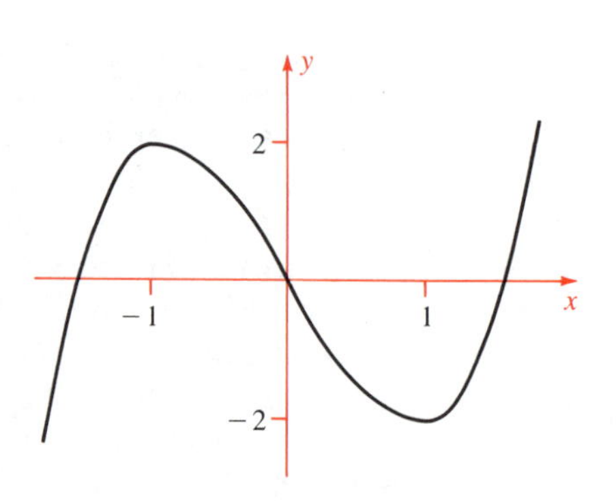

Figure 3-5-7
Concavity of $f(x) = x^3 - 3x$

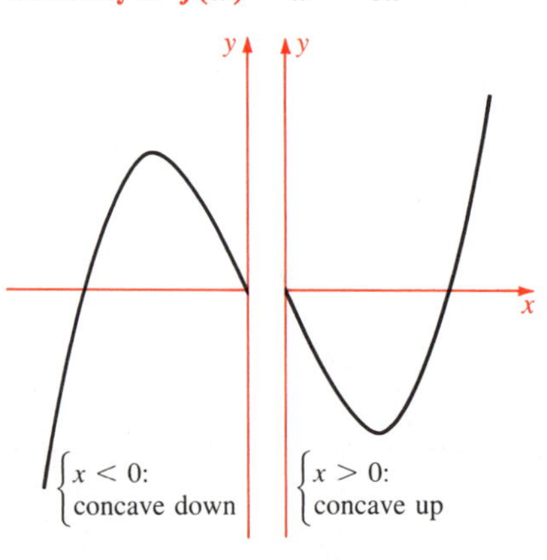

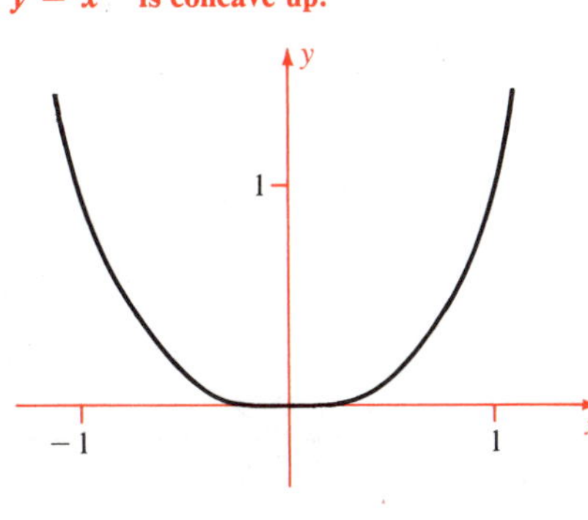

Figure 3-5-8
$y = x^4$ is concave up.

Example 3

a $f(x) = x^4$ domain $\mathbf{R} = (-\infty, +\infty)$

$f'(x) = 4x^3$ is strictly increasing; hence $f(x)$ is concave up (Figure 3-5-8).

b $f(x) = 1/x$ domain $(0, +\infty)$

$f'(x) = -1/x^2$ is strictly increasing; hence $f(x)$ is concave up (Figure 3-5-9).

c $f(x) = 1/x$ domain $(-\infty, 0)$

$f'(x) = -1/x^2$ is strictly decreasing; hence $f(x)$ is concave down (Figure 3-5-10). ●

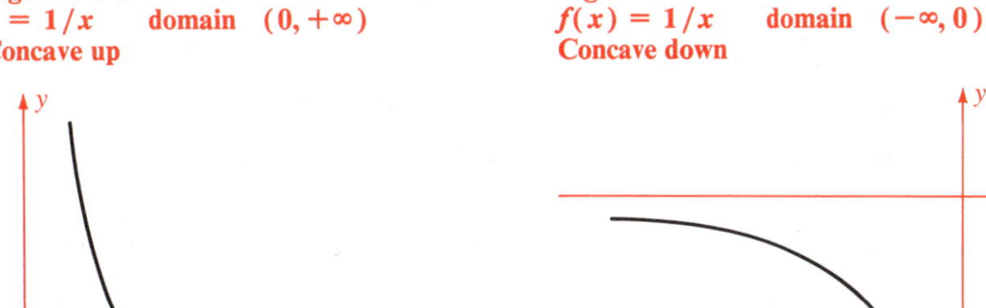

Figure 3-5-9
$y = 1/x$ domain $(0, +\infty)$
Concave up

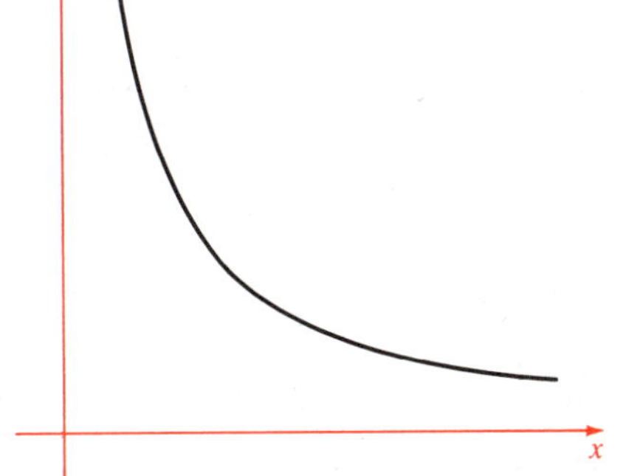

Figure 3-5-10
$f(x) = 1/x$ domain $(-\infty, 0)$
Concave down

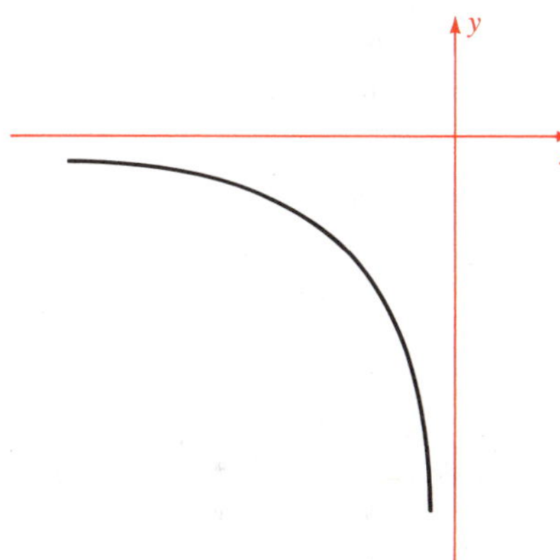

We know that if the derivative of a function is positive on an interval, then the function is strictly increasing, and if the derivative is negative, then the function is strictly decreasing. Suppose we apply these statements to $g(x) = f'(x)$. Then we find the following tests for concavity in terms of the second derivative:

Theorem 1 Second Derivative Test for Concavity

Let $f(x)$ be a continuous function whose domain is an interval. Suppose $f(x)$ is twice differentiable on the interval, except possibly at the end points.

- If $f''(x) > 0$, then $f(x)$ is concave up.
- If $f''(x) < 0$, then $f(x)$ is concave down. ●

Example 4

a $f(x) = x^3$

Here $f''(x) = 6x$, so $f'' < 0$ for $x < 0$, and $f'' > 0$ for

Figure 3-5-11
$y = x^3$
Concave up on $[0, +\infty)$
Concave down on $(-\infty, 0]$

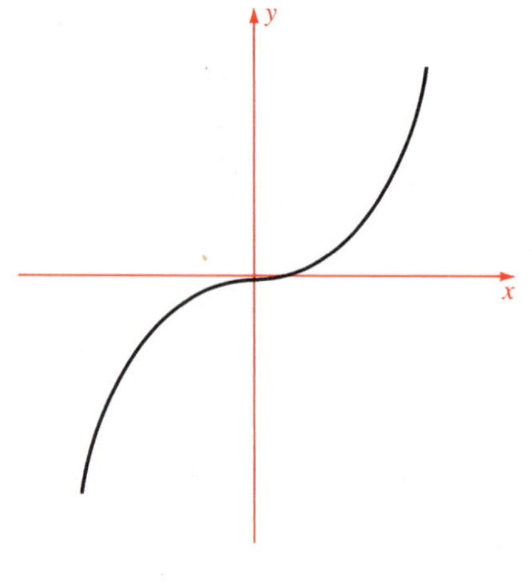

Figure 3-5-12
$y = \cos x$ domain $[\frac{1}{2}\pi, \frac{3}{2}\pi]$
Concave up

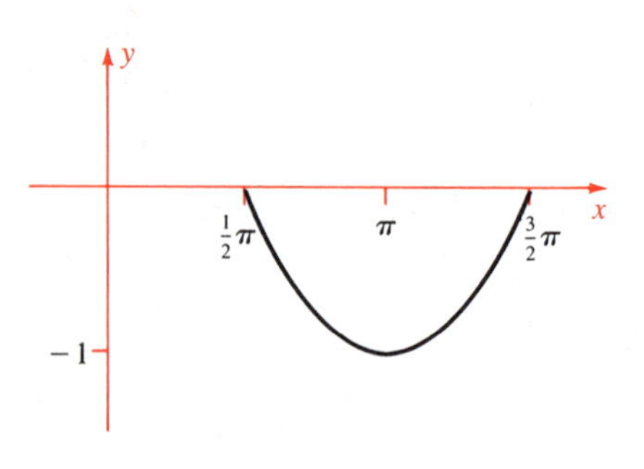

Figure 3-5-13
$y = 2 - \dfrac{6}{3 + x^2}$ domain $[-1, 1]$
Concave up

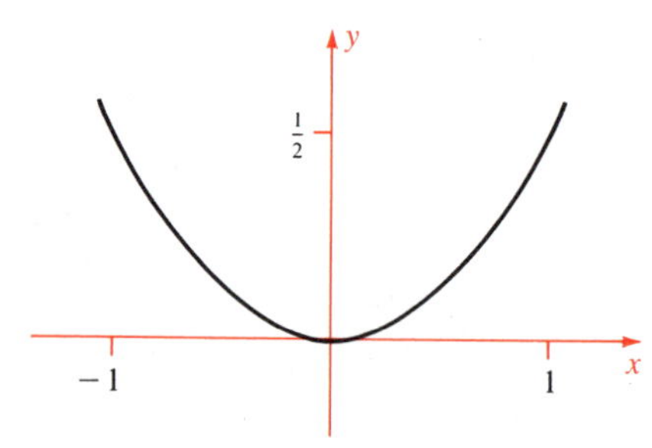

$x > 0$. Therefore $f(x)$ is concave down on $(-\infty, 0]$ and concave up on $[0, +\infty)$. See Figure 3-5-11.

b $f(x) = \cos x$ domain $[\frac{1}{2}\pi, \frac{3}{2}\pi]$

Here $f''(x) = -\cos x = -f(x) > 0$ on $(\frac{1}{2}\pi, \frac{3}{2}\pi)$; hence $f(x)$ is concave up (Figure 3-5-12).

c $f(x) = 2 - \dfrac{6}{3 + x^2}$ domain $[-1, 1]$

We compute

$$f'(x) = \frac{12x}{(3 + x^2)^2} \quad \text{and} \quad f''(x) = \frac{36(1 - x^2)}{(3 + x^2)^3} > 0$$

on $(-1, 1)$; hence $f(x)$ is concave up (Figure 3-5-13). ●

Example 5

a $y = \sqrt{x}$ domain $[0, +\infty)$

We find

$$y' = \tfrac{1}{2}x^{-1/2} \quad \text{and} \quad y'' = -\tfrac{1}{4}x^{-3/2} < 0$$

on $(0, +\infty)$; hence $y(x)$ is concave down (Figure 3-5-14).

Figure 3-5-14

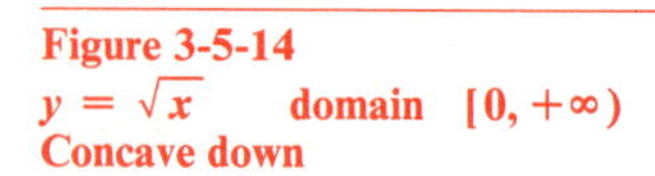

$y = \sqrt{x}$ domain $[0, +\infty)$
Concave down

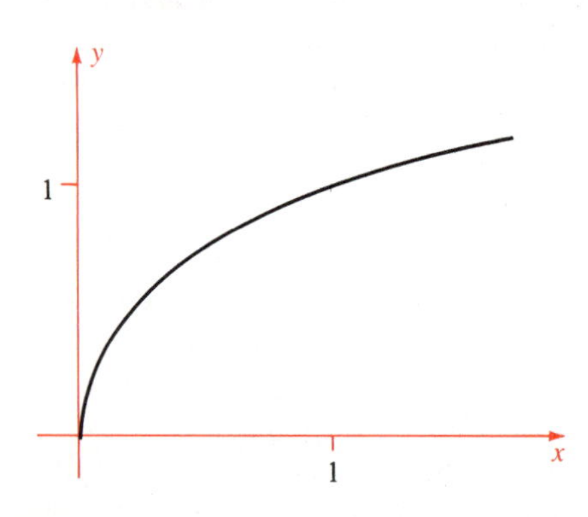

b $y = \sqrt{1 - x^2}$ domain $[-1, 1]$

We'll use the second derivative test; it is easiest to differentiate y implicitly:

$$y^2 = 1 - x^2 \quad \text{hence} \quad 2yy' = -2x \quad \text{so} \quad y' = -x/y$$

Now we have

$$y'' = -\frac{y - xy'}{y^2} = -\frac{y - x(-x/y)}{y^2} = -\frac{1}{y^3}$$

Because y is positive, $y'' < 0$ for $-1 < x < 1$, so $y(x)$ is concave down (Figure 3-5-15). ●

Figure 3-5-15
$y = \sqrt{1 - x^2}$ domain $[-1, 1]$
Concave down

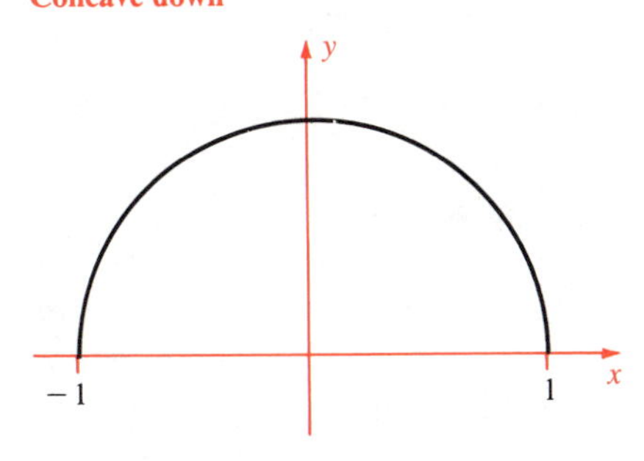

It seems plausible that if $y = f(x)$ is concave up, then its tangent at any point is below its graph. This is the content of the following theorem, which will be proved in Section 3-10.

Theorem 2 Tangent Theorem

Let $f(x)$ be a function that is concave up on an interval. Let c be a point of this interval at which $f(x)$ is differentiable. Then the graph of $y = f(x)$ is above the tangent line to the graph at $(c, f(c))$, at each point of the interval except $(c, f(c))$ itself.

If the graph is concave down instead, then the graph is below the tangent.

Example 6

a $y = x^2 \qquad c = -1$

The function is concave up since $y'' = 2 > 0$. Its graph (Figure 3-5-16) is entirely above the tangent line at $(-1, 1)$ except where they touch, at $(-1, 1)$.

Figure 3-5-16
$f(x) = x^2$
Concave up
Graph *above* tangent

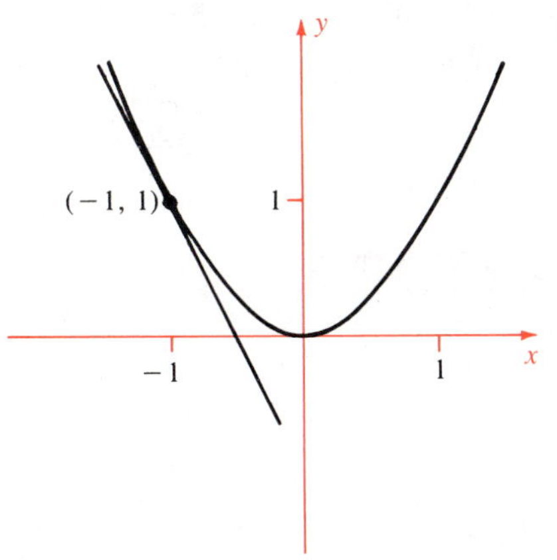

b $y = x^{3/2} \qquad c = 0$

The function $f(x) = x^{3/2}$ is continuous on the interval $[0, +\infty)$. It is differentiable at $x = 0$ because

$$f'(0) = \lim_{x \to 0+} \frac{f(x) - f(0)}{x - 0} = \lim_{x \to 0+} \frac{x^{3/2}}{x} = \lim_{x \to 0+} x^{1/2} = 0$$

(Note that $x^{3/2}$ cannot be defined for $x < 0$; hence the one-sided limit.) Thus the tangent at $(0, 0)$ is horizontal. For $x > 0$, we have $f'(x) = \frac{3}{2}x^{1/2}$, a strictly increasing function; hence $f(x)$ is concave up. Its graph (Figure 3-5-17) lies above the tangent at $(0, 0)$, as predicted by Theorem 2.

The next theorem states in another way that concavity makes a graph look like it is always curving in the same direction. From below, a graph that is concave up looks like the outside of a magnifying lens. The theorem will be proved in Section 3-10.

Figure 3-5-17
$f(x) = x^{3/2}$ domain $[0, +\infty)$
Concave up
Graph above tangent at $(0, 0)$

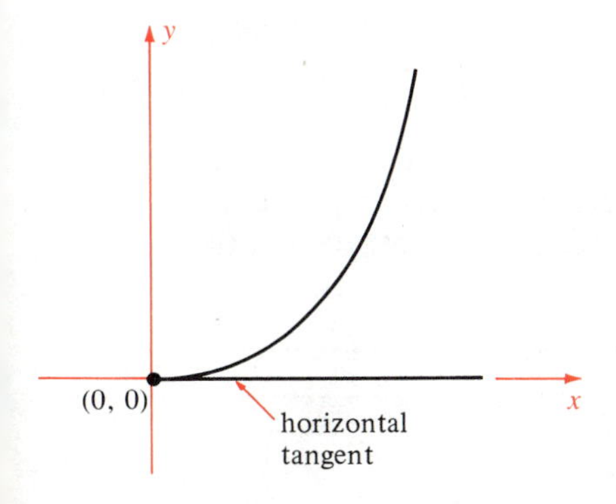

Theorem 3 Chord Theorem

Let $f(x)$ be a function that is concave up on an interval and suppose $x_0 < x < x_1$ are three points of that interval. Then the point $(x, f(x))$ of the graph is below the chord joining

$$(x_0, f(x_0)) \quad \text{to} \quad (x_1, f(x_1))$$

Remark Some writers use the terms *convex* and *concave* instead of concave up and concave down, and some vice versa.

Inflection Points

If $f(x)$ is concave up or concave down, and if L is its tangent line at $x = c$, then the graph of $y = f(x)$ is entirely above L or entirely below L, except at the point of tangency $(c, f(c))$. In particular, the graph does not *cross* the tangent line L at $(c, f(c))$. An inflection point on a graph is a point where the sense of concavity changes.

Definition Inflection Point

Let $f(x)$ be a differentiable function on an open interval (a, b) and let $a < c < b$. Then $(c, f(c))$ is an **inflection point** of the graph of $y = f(x)$ if $f(x)$ is concave up on one side of $x = c$ and concave down on the other, provided attention is restricted to points sufficiently close to $x = c$.

Figure 3-5-18
Graph of $f(x) = x^3$

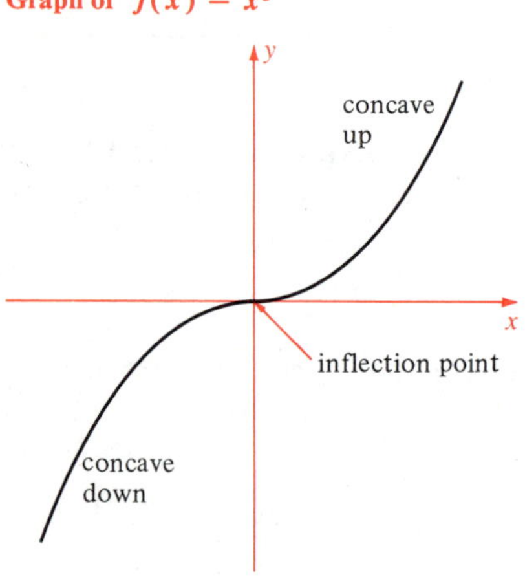

Example 7

$$f(x) = x^3 \qquad c = 0$$

The graph of $y = x^3$ is concave up for $x \geq 0$ and concave down for $x \leq 0$. Therefore $(0, 0)$ is an inflection point of the graph (Figure 3-5-18).

How can we find inflection points? Look carefully at Figure 3-5-18. The graph is concave down on $(-\infty, 0]$ because $f''(x) < 0$ for $x < 0$. The graph is concave up on $[0, +\infty)$ because $f''(x) > 0$ for $x > 0$. Therefore the sign of $f''(x)$ changes at $x = 0$. This is the key, and leads us to a practical test for inflection points.

Theorem 4 Inflection Point Test

Let $f(x)$ be twice differentiable on an open interval (a, b) and let $a < c < b$. Then

1 If $(c, f(c))$ is an inflection point of the graph of $y = f(x)$, then $f''(c) = 0$.

2 If $f''(x)$ changes sign at $x = c$, then $(c, f(c))$ is an inflection point of the graph of $y = f(x)$.

Figure 3-5-19
Graph of $f(x) = x^4$.
$f''(0) = 0$, but $(0, 0)$ is *not* an inflection point.

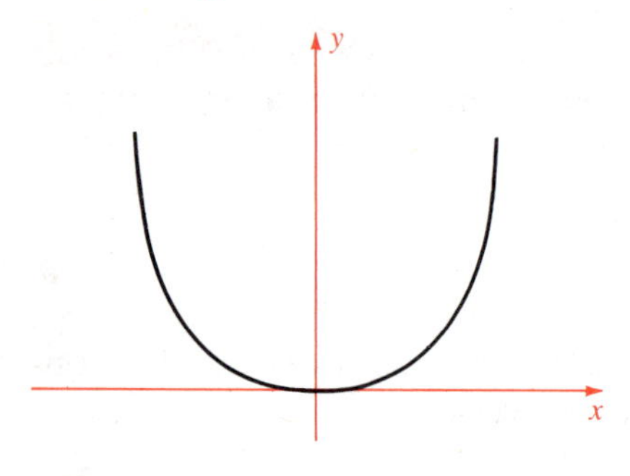

The proof of part 1 involves some technicalities, so we'll postpone it until Section 3-10. However, if we are willing to assume not only that $f(x)$ is twice differentiable, but also that $f''(x)$ is continuous, then $f''(c) = 0$ follows from $f''(x)$ having opposite signs on either side of $x = c$, by the intermediate value theorem.

The proof of part 2 is straightforward. If $f''(x)$ changes sign at $x = c$, then $f''(x)$ is positive on one side of $x = c$ and negative on the other side. Hence $f(x)$ is concave up on one side of $x = c$ and concave down on the other. Done.

The converse of part 1 is not true in general. Just the assumption $f''(c) = 0$ by itself does not guarantee an inflection point at $(c, f(c))$. For example, $f(x) = x^4$ does not have an inflection point at $(0, 0)$ even though $f''(0) = 0$. See Figure 3-5-19.

Example 8

$$y = (x - 1)^3 + \tfrac{1}{2}(x - 1) + 1$$

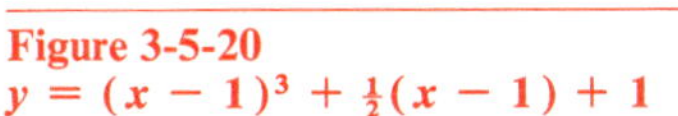

Figure 3-5-20
$y = (x-1)^3 + \frac{1}{2}(x-1) + 1$

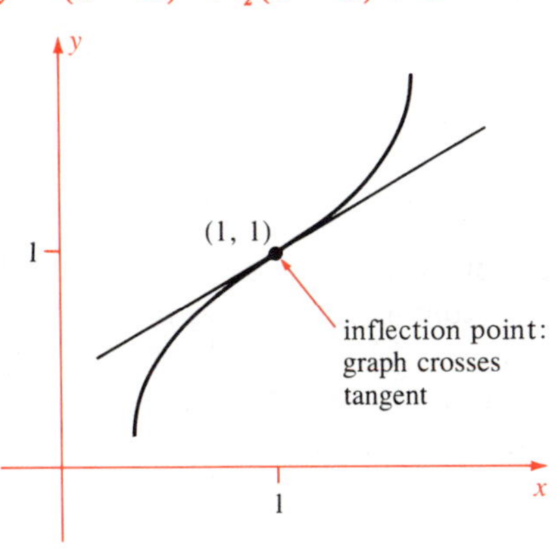

We have

$$y' = 3(x-1)^2 + \tfrac{1}{2} \quad \text{and} \quad y'' = 6(x-1)$$

Since $y'' = 0$ only for $x = 1$ and $y(1) = 1$, the only possible inflection point is $(1, 1)$. Clearly $y'' > 0$ for $x > 1$ and $y'' < 0$ for $x < 1$, so y'' changes sign at $x = 1$. Therefore $(1, 1)$ is an inflection point. The slope of the graph at $(1, 1)$ is $y'(1) = \frac{1}{2}$. See Figure 3-5-20. ●

A property of inflection points that follows easily from the tangent theorem is that *at an inflection point, a graph always crosses its tangent.* For suppose, say, that $f(x)$ is concave up for $x < c$ and concave down for $x > c$. Then the graph of $y = f(x)$ lies above the tangent at $x = c$ for $x < c$ and lies below for $x > c$.

Figure 3-5-21
$y = x^{1/3}$
Vertical inflection at $(0, 0)$

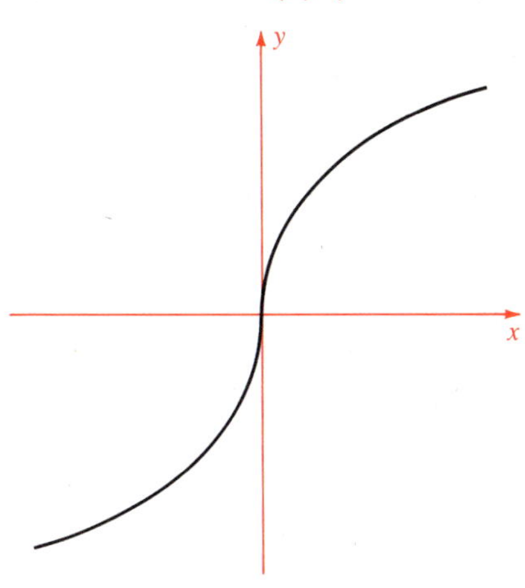

Remark Sometimes we speak of a *vertical inflection,* with a function such as $y = x^{1/3}$ at $(0, 0)$ in mind (Figure 3-5-21). Certainly this graph is concave up for $x \leq 0$ and concave down for $x \geq 0$, so the sense of concavity changes at $x = 0$. But the function $f(x) = x^{1/3}$ is not differentiable at $x = 0$, so the definition of inflection point does not apply. However, in some sense the graph has a vertical tangent at $(0, 0)$. In what sense? Simply this: The *inverse function* $x = y^3$ has a horizontal tangent at $(0, 0)$ in the y, x-plane, and $(0, 0)$ is an inflection point of that function.

In general, we say that $y = f(x)$ has a vertical inflection at $(c, f(c))$ if there is an inverse function $x = g(y)$ near that point with an inflection point at $(f(c), c)$ and whose tangent there is horizontal.

Graphing

Let us apply our new tools to the problem of graphing $y = f(x)$. In addition to information from our old methods, we now get information from the second derivative. We seek all intervals on which $f''(x) > 0$; there the graph is concave up. Similarly, the graph is concave down on any interval where $f''(x) < 0$. Any point on the graph separating these two types of intervals is an inflection point.

Example 9

Graph the function

$$y = x^3 - 3x^2 + 2$$

Indicate all intervals of concavity and all inflection points.

Solution Just by looking at the function, we can see that $y \to -\infty$ as $x \to -\infty$ and $y \to +\infty$ as $x \to +\infty$, and that $(0, 2)$ is on the graph. This is "free" information. Now we differentiate the function $y = x^3 - 3x^2 + 2$ twice:

$$y' = 3x^2 - 6x = 3x(x-2)$$

$$y'' = 6x - 6 = 6(x-1)$$

The graph is horizontal where $y' = 0$: at $x = 0$ and at $x = 2$. These values correspond to $(0, 2)$ and $(2, -2)$ on the curve. Clearly

$$\begin{cases} y'' > 0 & \text{for} \quad x > 1 \\ y'' = 0 & \text{for} \quad x = 1 \\ y'' < 0 & \text{for} \quad x < 1 \end{cases}$$

so the graph is concave up for $x > 1$, concave down for $x < 1$, and has an inflection point where $x = 1$, that is, at the point $(1, 0)$, since $y(1) = 0$. At this point the slope is $y'(1) = -3$. It will be helpful to find all crossings of the x-axis and the slopes at these points. Now $(1, 0)$ is one crossing, corresponding to the solution $x = 1$ of

$$y = x^3 - 3x^2 + 2 = 0$$

To find the others, we factor out $x - 1$ and find the zeros of the resulting quadratic:

$$x^3 - 3x^2 + 2 = (x - 1)(x^2 - 2x - 2)$$

$$x^2 - 2x - 2 = 0 \qquad x = \frac{2 \pm \sqrt{12}}{2} = 1 \pm \sqrt{3}$$

Thus the graph crosses the x-axis at approximately $(-0.73, 0)$ and $(2.73, 0)$. The corresponding slopes (values of y') are

$$\begin{aligned} y'(1 \pm \sqrt{3}) &= 3(1 \pm \sqrt{3})[(1 \pm \sqrt{3})-2] \\ &= 3(1 \pm \sqrt{3})(-1 \pm \sqrt{3}) \\ &= 3(-1 + 3) = 6 \end{aligned}$$

See Figure 3-5-22. ●

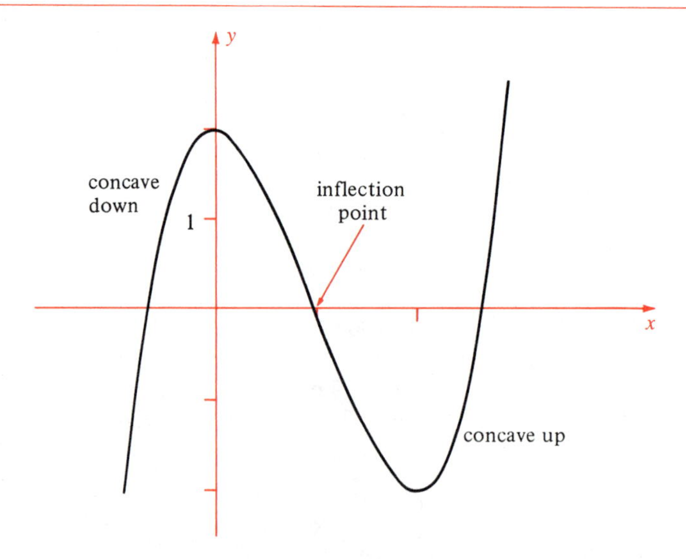

Figure 3-5-22
Graph of $y = x^3 - 3x^2 + 2$

Example 10

Graph $y = 16x^2(1 - x)^2$. Find all intervals of concavity and all inflection points. (We started the section with this example.)

Solution Differentiate:

$$y' = 32x(1 - x)(1 - 2x) = 32(2x^3 - 3x^2 + x)$$
$$y'' = 32(6x^2 - 6x + 1)$$

The solutions of $y'' = 0$ are the solutions of the quadratic equation $6x^2 - 6x + 1 = 0$:

$$x = \frac{6 \pm \sqrt{36 - 24}}{12} = \frac{3 \pm \sqrt{3}}{6}$$

Set

$$x_1 = \tfrac{1}{6}(3 - \sqrt{3}) \approx 0.21 \quad \text{and} \quad x_2 = \tfrac{1}{6}(3 + \sqrt{3}) \approx 0.79$$

Since the graph of the *quadratic* $6x^2 - 6x + 1$ is a parabola opening up, it is clear that $y'' > 0$ for $x < x_1$ and for $x > x_2$, and that $y'' < 0$ for $x_1 < x < x_2$. Therefore the original curve is concave up for $x < x_1$ and for $x > x_2$. It is concave down for $x_1 < x < x_2$, and there are inflection points at $x = x_1$ and $x = x_2$. By calculator,

$$y'(x_1) \approx 3.1 \quad \text{and} \quad y'(x_2) \approx -3.1$$

See Figure 3-5-23. ●

Figure 3-5-23
Graph of $y = 16x^2(1 - x)^2$

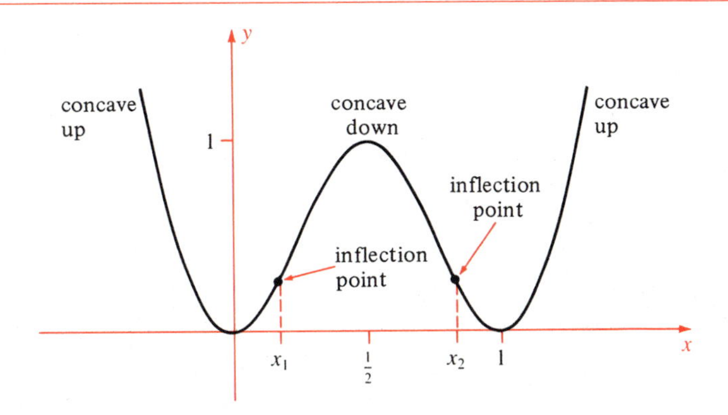

Remark Earlier we raised the possibility that this graph might be "wiggly" from $x = 0$ to $x = \frac{1}{2}$ (Figure 3-5-3). We now know this is impossible because each wiggle means another inflection point, but we know that there is only one inflection point for x between 0 and $\frac{1}{2}$.

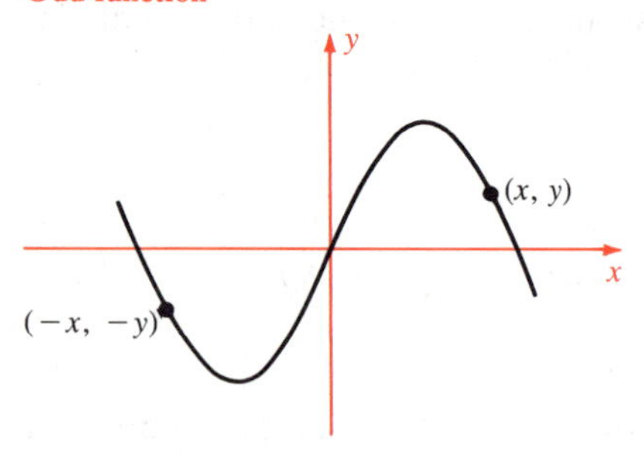

Figure 3-5-24
Odd function

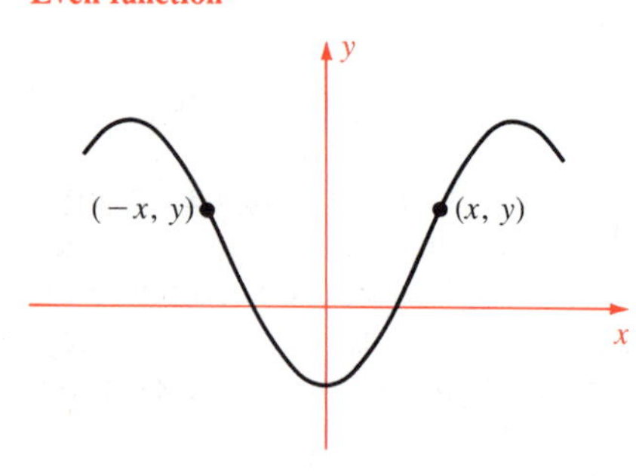

Figure 3-5-25
Even function

Hints for Sketching $y = f(x)$

1 Get as much "free" information as you can by inspection:

- If easily done, find where y is positive, negative, or zero.
- Look for symmetry. If $f(x)$ is an odd function, that is, if $f(-x) = -f(x)$, then the graph of $y = f(x)$ is symmetric about the origin (Figure 3-5-24). If $f(x)$ is an even function, that is, if $f(-x) = f(x)$, then the graph of $y = f(x)$ is symmetric about the y-axis (Figure 3-5-25).
- Find the behavior of y as $x \to +\infty$ and as $x \to -\infty$.
- Find values of x, if any, for which the curve is not defined. Try to see by inspection whether $y \to +\infty$ or $y \to -\infty$ as x approaches such a point from the left and from the right (vertical asymptote; see page 42).

2 Take the derivative. Its sign will tell you where the curve is rising, falling, or horizontal. Plot all points where the tangent is horizontal, that is, where $y' = 0$. Check for local maxima and minima at these points, and for horizontal inflections. If you can, find the behavior of y' as $x \to +\infty$ and as $x \to -\infty$.

3 Take the second derivative. Where not zero, its sign will indicate the sense of concavity. Locate and plot all inflection points.

4 For accuracy, plot a few points. Look for points that are easy to compute. Try $x = 0$ and, if it is not too hard, see where $y = 0$.

These hints are suggestions, not sacred rules. Be flexible; there is no substitute for common sense.

Exercises

Locate where $y = f(x)$ is concave up, concave down, and where its graph has inflection points

1 $f(x) = 4(x-3)^2$

2 $f(x) = -3(x+7)^2$

3 $f(x) = (x-2)^3 + 2x$

4 $f(x) = x^3 + 3x^2 + 7x + 1$

5 $f(x) = -x^3 + 4x - 5$

6 $f(x) = x^3 - 6x^2 + 3x + 1$

7 $f(x) = \dfrac{1}{\sqrt{x^2+5}}$

8 $f(x) = x^4 - 12x^3 + 6x^2 + 4$

9 $f(x) = \sin x$

10 $f(x) = \cos x$

11 $f(x) = \tan x$

12 $f(x) = \cot x$

13 $f(x) = \sec x$

14 $f(x) = \csc x$

Sketch the graph. Show all intervals of concavity and inflection points

15 $y = 3x^3 + x$

16 $y = x^3 + \frac{1}{2}x + 1$

17 $y = x^2(x-3)$

18 $y = x(x+1)(x+2)$

19 $y = x^3(x+2)$

20 $y = x^4 - 6x^2$

21 $y = \dfrac{1}{x^2+1}$

22 $y = \dfrac{x}{x^2+1}$

23 $y = \dfrac{x}{x^2-1}$ $\quad (x^2 \neq 1)$

24 $y = \dfrac{x-1}{x^2}$ $\quad (x \neq 0)$

25 $y = x^2 + x^{1/2}$ $\quad (x \geq 0)$

26 $y = \frac{1}{6}x^3 + \frac{9}{2}x^{1/3}$

27 $y = \dfrac{x^2}{1+x^3}$ $\quad (x \neq -1)$

28 $y = \dfrac{x^3}{1+x^2}$

29 $y = 4\sin x + \sin 2x$ $\quad (-\frac{1}{2}\pi \leq x \leq \frac{1}{2}\pi)$

30 $y = 8\sin x - \tan x$ $\quad (-\frac{1}{2}\pi < x < \frac{1}{2}\pi)$

3-6 Optimization

The title of this section means finding the largest value or the smallest value a continuous function takes on its domain. Each such value is called an **extremum** of the function (plural: **extrema**). The solution of the problem of finding extrema is one of the most striking, satisfactory, and important applications of differential calculus.

Let us begin with the maximization problem. Suppose we are given $f(x)$, continuous on a closed interval $[a, b]$ and differentiable on (a, b). See Figure 3-6-1. We seek the maximum of $f(x)$. A basic theoretical result (Corollary 2, page 167) says that there is a maximum $f(c)$ of $f(x)$ for at least one c in the interval. If c is not an end point, that is, if $a < c < b$, then another basic result (Theorem 3, page 170) says that $f'(c) = 0$. Geometrically, the graph $y = f(x)$ is horizontal at $x = c$.

Figure 3-6-1
$f(x)$ is differentiable.
At a maximum point $(c, f(c))$, where $a < c < b$, the tangent is horizontal.

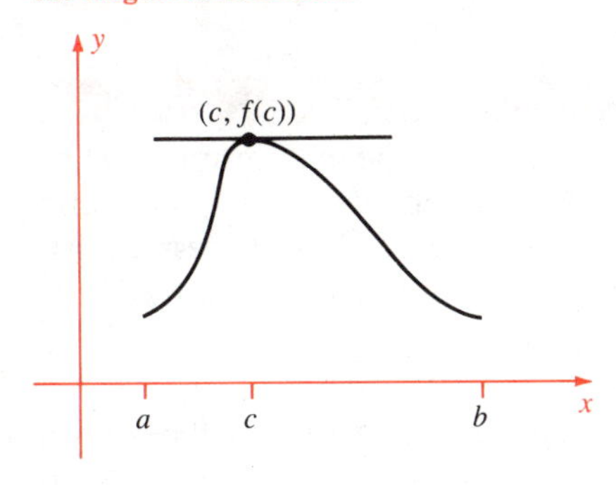

Therefore, the maximum value of $f(x)$ must occur at one (or both) of the end points of the interval $[a, b]$ and/or at one or more of the inside points of the interval where $f'(x) = 0$. The same applies to the minimum of $f(x)$.

So far we have assumed that $f(x)$ is differentiable in (a, b), that is, that $f'(x)$ exists whenever $a < x < b$. We also have to deal with continuous functions whose derivatives do not exist at certain points. Such points have to be tested for maxima or minima on a case-by-case basis. Some possibilities are shown in Figure 3-6-2.

Figure 3-6-2
Functions differentiable on R except at $x = c$

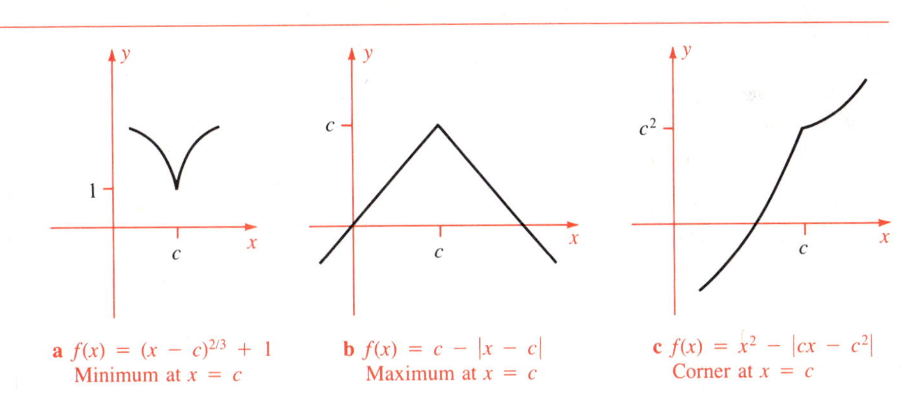

a $f(x) = (x - c)^{2/3} + 1$
Minimum at $x = c$

b $f(x) = c - |x - c|$
Maximum at $x = c$

c $f(x) = x^2 - |cx - c^2|$
Corner at $x = c$

In most cases that we shall encounter, particularly in applications of optimization, there will rarely be points where the derivative does not exist, and there will be only a few points where the derivative equals zero. Anyhow, there are exactly three possible types of points where a continuous function on an interval has a maximum or a minimum: points where the derivative is zero, end points, and points where the derivative does not exist. These points are called critical points of the function.

Definition Critical Point

Let $f(x)$ be continuous on an interval I. The **critical points** of $f(x)$ are the following:

1 Points of I where $f(x)$ is differentiable and $f'(x) = 0$. These are called **stationary points** of $f(x)$.
2 End points of I.
3 Points of I where $f'(x)$ does not exist.

We summarize the discussion of maxima and minima in the following basic theorem:

Theorem 1

Let $f(x)$ be continuous on an interval I. Then the maximum of $f(x)$ occurs at a critical point, that is, either a point where $f'(x)$ exists and equals zero, or an end point of I, or a point where $f'(x)$ doesn't exist. (The maximum may be attained at more than one critical point.)
The same holds for the minimum of $f(x)$.

For most functions $f(x)$ that come up in practice, there will be only a small number of points c_i with $a < c_1 < c_2 < \cdots < c_n < b$ such that $f'(c_i) = 0$ or $f'(x)$ does not exist. Then the values $f(a)$, $f(c_1)$, $f(c_2)$, . . . , $f(c_n)$, $f(b)$ are *all possible candidates* for the maximum and for the minimum of $f(x)$. The maximum of $f(x)$ is the largest of these values; the minimum of $f(x)$ is the smallest (Figure 3-6-3).

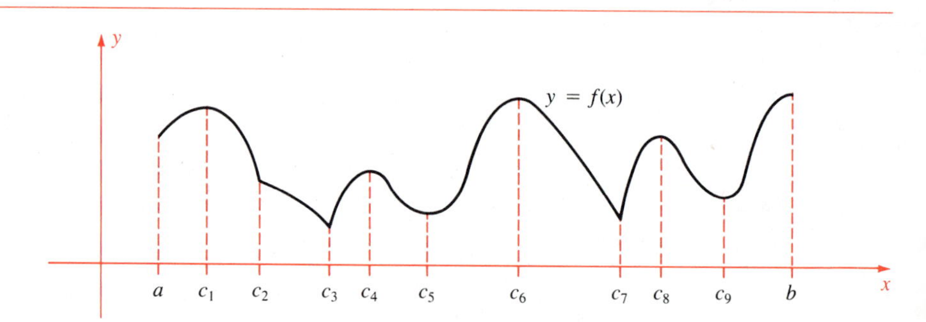

Figure 3-6-3
$f(x)$ is continuous on $[a, b]$.
Its critical points are
End points: a, b
Stationary points: c_1, c_4, c_5, c_6, c_8, c_9
Points of no derivative: c_2, c_3, c_7

Optimization Procedure

To find the maximum and minimum of $f(x)$ in the closed interval $[a, b]$, locate all points where $f'(x) = 0$ or $f'(x)$ doesn't exist. Call them $c_1, c_2, \cdots, c_n$. The maximum is the largest of the numbers

$$f(a) \qquad f(c_1) \qquad f(c_2) \qquad \cdots \qquad f(c_n) \qquad f(b)$$

The minimum is the smallest.

Example 1

Find the maximum and minimum on $[-2, 2]$ of $f(x) = x^3 - 3x + 1$.

Solution The function is differentiable on $(-2, 2)$ and

$$f'(x) = 3x^2 - 3 = 3(x^2 - 1) = 3(x + 1)(x - 1)$$

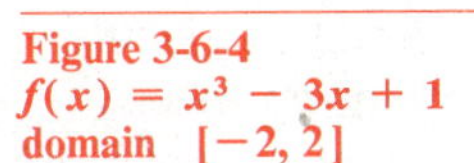

Figure 3-6-4
$f(x) = x^3 - 3x + 1$
domain $[-2, 2]$

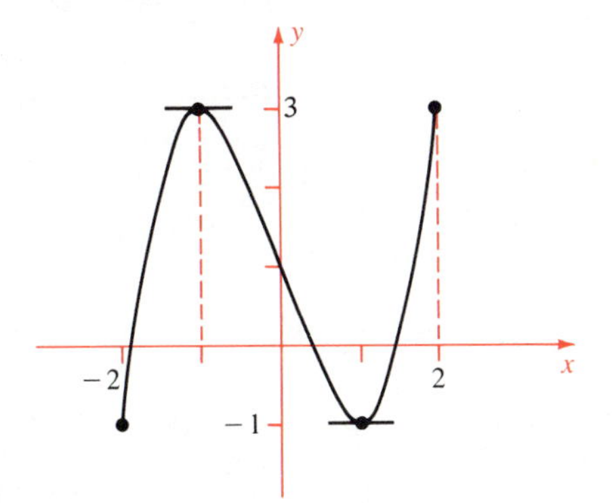

The stationary points of $f(x)$, that is, the zeros of $f'(x)$ are $c_1 = -1$ and $c_2 = 1$. The list of candidates for extrema consists of the values of $f(x)$ at these two stationary points and the values of $f(x)$ at the two end points (the only other critical points):

$$f(a) = f(-2) = -1 \qquad f(c_1) = f(-1) = 3$$
$$f(c_2) = f(1) = -1 \qquad f(b) = f(2) = 3$$

Conclusions (Figure 3-6-4):

$$\max(f) = f(-1) = f(2) = 3$$
$$\min(f) = f(-2) = f(1) = -1$$

●

Example 2

Find the maximum and minimum on $[-2, 1]$ of $f(x) = x^4 + 2x^3$.

Solution The function is differentiable on $(-2, 1)$ and

$$f'(x) = 4x^3 + 6x^2 = 2x^2(2x + 3)$$

Thus $f'(x) = 0$ for $x = -\frac{3}{2}$ and $x = 0$. Therefore the candidates for $\max(f)$ and $\min(f)$ are

$$f(-2) = 0 \qquad f(-\tfrac{3}{2}) = -\tfrac{27}{16} \qquad f(0) = 0 \qquad f(1) = 3$$

Figure 3-6-5
$y = x^4 + 2x^3$ domain $[-2, 1]$

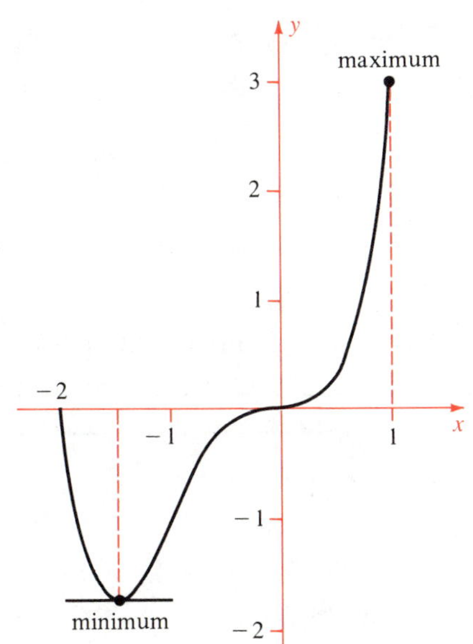

The largest is 3; the smallest is $-\frac{27}{16}$. Therefore (Figure 3-6-5)

$$\max(f) = f(1) = 3 \quad \text{and} \quad \min(f) = f(-\tfrac{3}{2}) = -\tfrac{27}{16}$$

●

Non-Closed Intervals

Some problems require the maximum or minimum of a function over an interval that is not closed. In such a case, nothing guarantees a maximum or minimum. But geometrical or physical considerations may imply that an extremum exists. We can often see that the function changes in such a way that there must be a maximum or a minimum.

Example 3

Find the maximum of $f(x) = -x^2 + 6x - 5$.

Solution The interval in this case is the whole real line $\mathbf{R}$ and $f(x)$ is differentiable. As $x \to +\infty$ or $x \to -\infty$, the quadratic term $-x^2$ dominates the function. Hence

$$f(x) \to -\infty \quad \text{as} \quad x \to +\infty \quad \text{and as} \quad x \to -\infty$$

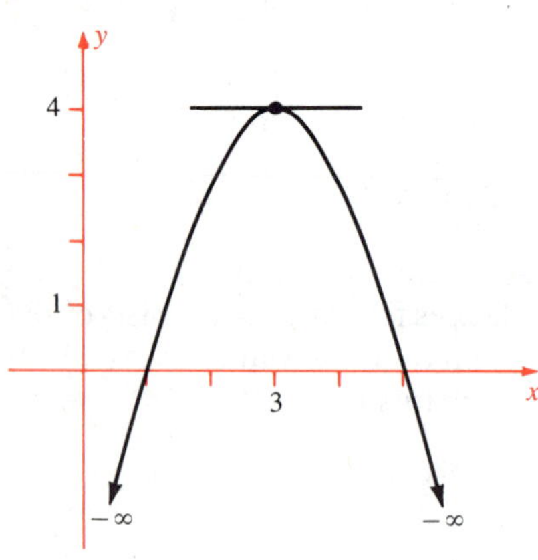

Figure 3-6-6
$y = -x^2 + 6x - 5$ domain **R**

Therefore $f(x)$ must have a maximum at some critical point, and the only possibility is a stationary point, that is, a zero of $f'(x)$. But

$$f'(x) = -2x + 6 = -2(x - 3)$$

so $f'(x) = 0$ only for $x = 3$. Therefore $f(3)$ *must be* the maximum (Figure 3-6-6). Conclusion: $\max(f) = f(3) = 4$. ●

Note that this result can also be obtained by algebra (without calculus) from $f(x) = -(x - 3)^2 + 4$. However, this is something special for quadratic functions. Calculus solves the extremum problem for far more general functions.

Example 4

Find the minimum of $f(x) = x^4 - 4x$.

Solution The function is differentiable on **R** and

$$f(x) \to +\infty \quad \text{as} \quad x \to +\infty \quad \text{or} \quad x \to -\infty$$

so $f(x)$ *must* have a minimum someplace on **R** and the only possible critical points are zeros of $f'(x)$. Consequently, we differentiate and solve the equation $f'(x) = 0$:

$$f'(x) = 4(x^3 - 1) \qquad \text{hence} \qquad x^3 - 1 = 0 \quad \text{so} \quad x = 1$$

There is only one stationary point (Figure 3-6-7), so it must give the minimum: $\min(f) = f(1) = 1 - 4 = -3$. ●

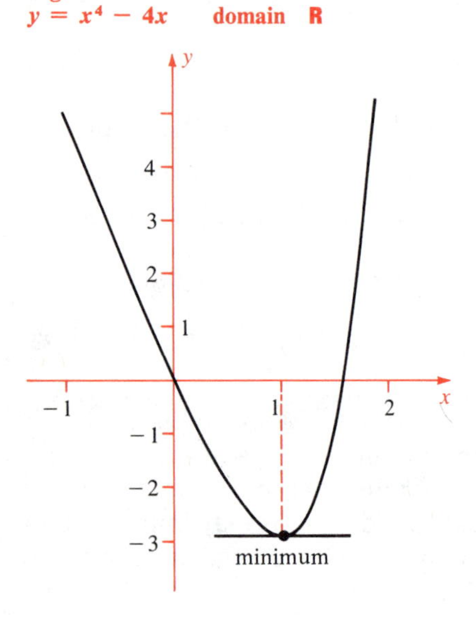

Figure 3-6-7
$y = x^4 - 4x$ domain **R**

We can formalize the reasoning used in these examples as a theorem. It doesn't cover all cases, but does indicate how you reason in similar circumstances.

Theorem 2

Suppose $f(x)$ is continuous on (a, b) (where possibly $a = -\infty$ or $b = +\infty$). Suppose

$$\lim_{x \to a+} f(x) = +\infty \quad \text{and} \quad \lim_{x \to b-} f(x) = +\infty$$

Then $f(x)$ has a minimum. ●

What this means is shown in Figure 3-6-8; a proof is given in Section 3-10.

Example 5

Find the maximum of $f(x) = \dfrac{x}{2x^3 + 1}$ on $[0, +\infty)$.

Solution We have $f(0) = 0$ and $f(x) > 0$ for $x > 0$. Also the rational function $f(x)$ is bottom-heavy (page 40), so

$$f(x) \to 0+ \quad \text{as} \quad x \to +\infty$$

Figure 3-6-8
$f(x)$ continuous on (a, b)
$f(x) \to +\infty$ as $\begin{cases} x \to a+ \\ x \to b- \end{cases}$
Then $f(x)$ has a minimum.

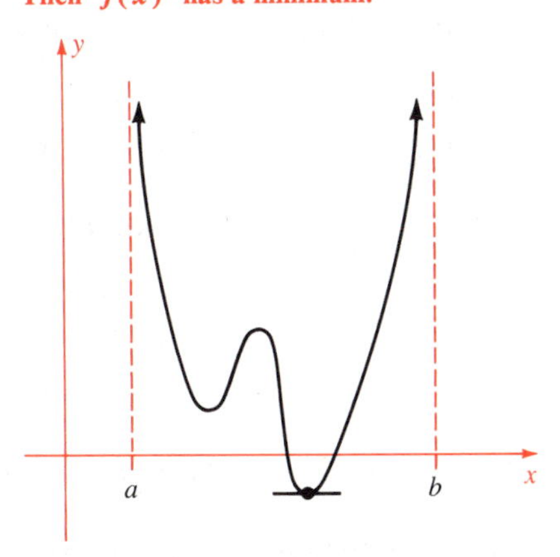

Figure 3-6-9
$y = \dfrac{x}{2x^3 + 1}$ domain $[0, +\infty)$

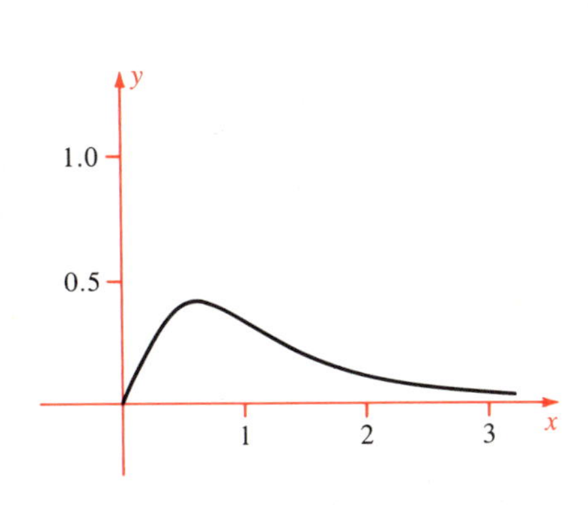

The function increases at first, starting from $f(0) = 0$, but dies out to 0 as x increases indefinitely. Since $f(x)$ is continuous, it must have a maximum someplace (Figure 3-6-9). Since $f(x)$ is differentiable in $(0, +\infty)$ and the value $f(0)$ at the critical point $x = 0$ is not a maximum, the only possibility is a stationary point. We test $f'(x) = 0$:

$$f'(x) = \frac{(2x^3 + 1) - 6x^3}{(2x^3 + 1)^2} = \frac{1 - 4x^3}{(2x^3 + 1)^2}$$

This quotient can equal zero only if its numerator is zero. Thus to solve $f'(x) = 0$, we solve

$$1 - 4x^3 = 0 \qquad \text{that is} \qquad x = \sqrt[3]{\tfrac{1}{4}} = \tfrac{1}{2}\sqrt[3]{2} \approx 0.629961$$

There is only one horizontal tangent for $x > 0$; it must yield the maximum:

$$\text{maximum} = f(\tfrac{1}{2}\sqrt[3]{2}) \approx 0.419974$$

●

Figure 3-6-10
$f(x) = \dfrac{1}{1 + x^2}$ domain **R**
Maximum, no minimum

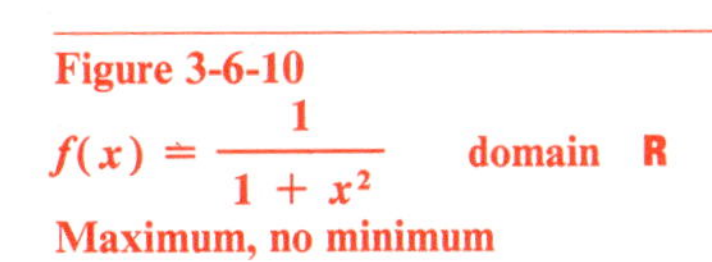
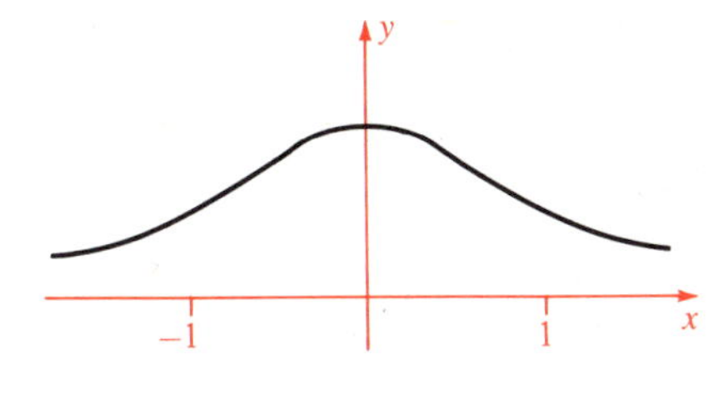

In general, a continuous function $f(x)$ need not have a maximum or a minimum unless x is restricted to a closed interval. For example, take $f(x) = 1/(1 + x^2)$ where x is unrestricted (Figure 3-6-10). This function has a maximum value 1, but no minimum value (it takes all values between 0 and 1, excluding zero). For another example, take $f(x) = 1/x$ for $x > 0$. This function has neither a maximum nor a minimum (Figure 3-6-11).

Figure 3-6-11
$f(x) = 1/x$ domain $(0, +\infty)$
No maximum, no minimum

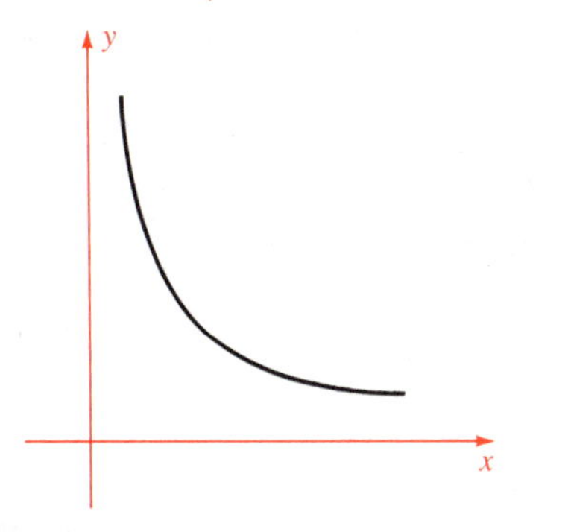

First Derivative Test

The following theorem is a test for local extrema of differentiable functions. Part of it we know already: The derivative is always 0 at a local maximum or minimum that is not an end point. The new part is a test that tells in many cases whether a point of zero derivative is a local maximum or a local minimum. Intuitively, the theorem says that if a function increases, pauses, and then decreases, then it has a local maximum.

Theorem 3 First Derivative Test

Suppose $f(x)$ is differentiable on an open interval (a, b). Then:

1 If $f(x)$ has a local extremum at $x = c$, then $f'(c) = 0$.

2 Suppose $f'(c) = 0$ and suppose there is a $\delta > 0$ such that

$$\begin{cases} f'(x) > 0 & \text{on} \quad (c - \delta, c) \\ f'(x) < 0 & \text{on} \quad (c, c + \delta) \end{cases}$$

Then $f(x)$ has a local *maximum* at $x = c$.

3 Similarly, if

$$\begin{cases} f'(x) < 0 & \text{on} \quad (c - \delta, c) \\ f'(x) > 0 & \text{on} \quad (c, c + \delta) \end{cases}$$

then $f(x)$ has a local *minimum* at $x = c$.

Proof Statement 1 is just a restatement of Theorem 3, page 170. Statement 2 follows easily from Corollary 2, page 173. Indeed, $f(x)$ is strictly increasing on $(c - \delta, c]$ and strictly decreasing on $[c, c + \delta)$, so $f(c)$ is the one and only maximum of $f(x)$ on $(c - \delta, c + \delta)$. Statement 3 follows similarly. This completes the proof. See Figure 3-6-12.

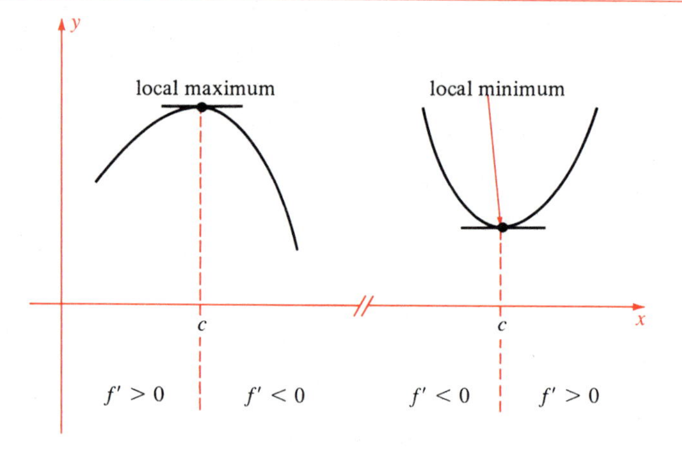

Figure 3-6-12
Sufficient conditions for local extrema

Example 6

Classify the points of

$$y = f(x) = x^4 - 4x^3 + 4x^2 - 1$$

at which $f'(x) = 0$, and graph the function.

Solution

$$\begin{aligned} f'(x) &= 4x^3 - 12x^2 + 8x \\ &= 4x(x^2 - 3x + 2) = 4x(x - 1)(x - 2) \end{aligned}$$

Figure 3-6-13
Sign diagram for $y = f'(x)$
[*not for* $f(x)$] in Example 6

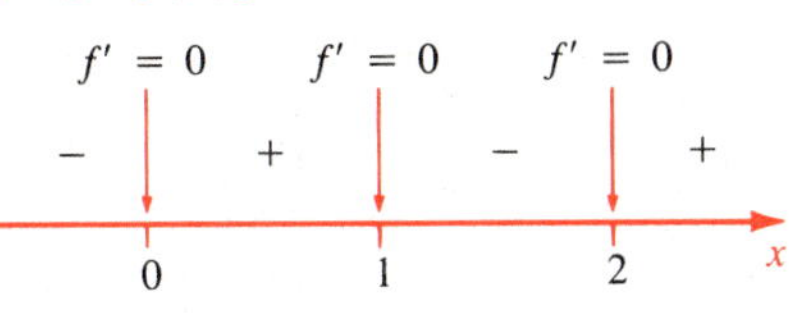

Make a sign diagram (Figure 3-6-13) for $f'(x)$. By the first derivative test, $f(x)$ has local minima at $x = 0$ and at $x = 2$, and a local maximum at $x = 1$. This with the additional information

$$f(0) = -1 \qquad f(1) = 0 \qquad f(2) = -1$$
$$f(x) \to +\infty \quad \text{as} \quad x \to \pm\infty$$

is enough for a rough graph (Figure 3-6-14). ●

Figure 3-6-14
Graph of $y = x^4 - 4x^3 + 4x^2 - 1$

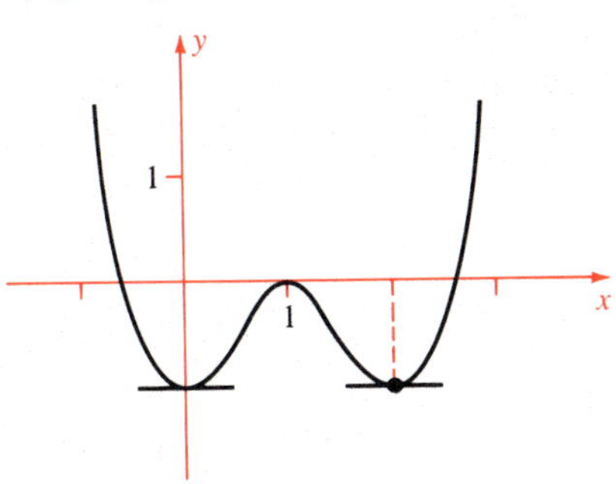

Theorem 3 takes care of the cases in which $f'(x)$ changes sign at $x = c$. What happens if $f'(c) = 0$ but $f'(x)$ does not change sign at $x = c$? The typical example is $f(x) = x^3$ at $x = 0$. See Figure 3-6-15. The derivative $f'(x) = 3x^2 = 0$ at $x = 0$, but it is positive otherwise. In other words, $y = x^3$ has an inflection point at $(0, 0)$, and has neither a local minimum nor a local maximum there.

This is typical. If the derivative is positive on both sides of $x = c$, where $f'(c) = 0$, then $f(x)$ increases, pauses, then increases some more, so $f(x)$ has neither a maximum nor a minimum at $x = c$.

Figure 3-6-15
Graph of $y = x^3$

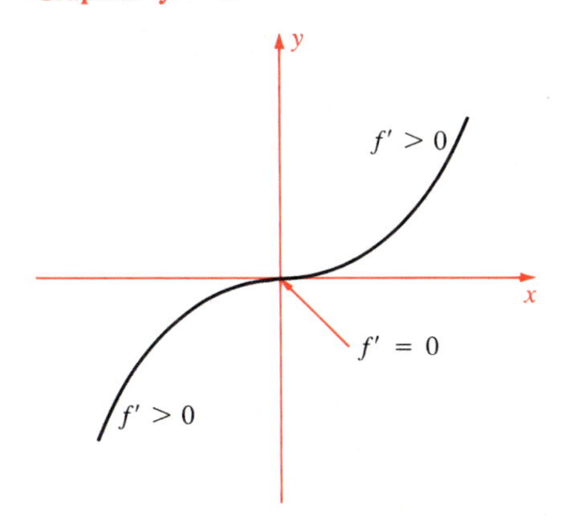

Example 7

Locate all local extrema of $f(x) = x^4(x-1)^3(x-2)^2$.

Solution The function is a polynomial (hence differentiable) with leading term x^9. Therefore

$$f(x) \to +\infty \quad \text{as} \quad x \to +\infty$$
$$f(x) \to -\infty \quad \text{as} \quad x \to -\infty$$

The function cannot have a maximum or a minimum, but it can have local extrema. At each local extremum, $f'(x) = 0$. So we first differentiate:

$$\begin{aligned} f'(x) &= 4x^3(x-1)^3(x-2)^2 + 3x^4(x-1)^2(x-2)^2 \\ &\qquad + 2x^4(x-1)^3(x-2) \\ &= x^3(x-1)^2(x-2)[4(x-1)(x-2) \\ &\qquad + 3x(x-2) + 2x(x-1)] \\ &= x^3(x-1)^2(x-2)(9x^2 - 20x + 8) \end{aligned}$$

The zeros of the last factor are, by the quadratic formula,

$$x = \tfrac{1}{9}(10 \pm \sqrt{28}) \approx 0.52, \quad 1.70$$

Therefore, the points where $f'(x) = 0$ are, in increasing order,

$$c_1 = 0 \qquad c_2 = \tfrac{1}{9}(10 - \sqrt{28}) \qquad c_3 = 1$$
$$c_4 = \tfrac{1}{9}(10 + \sqrt{28}) \qquad c_5 = 2$$

Near $x = 0$,

$$(x-1)^2(x-2)(9x^2 - 20x + 8) \approx -16$$

Hence

$$f'(x) \approx -16x^3 \quad \text{for} \quad x \approx 0$$

Therefore $f'(x)$ changes sign from $+$ to $-$ as x passes through 0 (from left to right), so there is a local maximum at $x = 0$.

Near $x = 1$,

$$x^3(x-2)(9x^2 - 20x + 8) \approx 3$$

Hence

$$f'(x) \approx 3(x-1)^2 \quad \text{for} \quad x \approx 1$$

It follows that $f'(x) > 0$ for $x \approx 1$ except at $x = 1$ itself. Thus $f(x)$ is strictly increasing on *both* sides of $x = 1$, provided x is close to 1. Therefore there is an inflection point with horizontal tangent at $x = 1$.

Similarly, near $x = 2$,

$$f'(x) \approx 32(x-2)$$

so $f'(x)$ changes sign from $-$ to $+$ as x passes through 2. Hence there is a local minimum at $x = 2$.

Since c_2 and c_4 are simple zeros of its quadratic factor $9x^2 - 20x + 8$, we know that $f'(x)$ changes sign at $x = c_2$. But which way? We can settle it because $f'(x) < 0$ for x immediately to the right of 0 and $f'(x) > 0$ for x immediately to the left of 1. Therefore $f'(x)$ must change sign from $-$ to $+$ as x passes through c_2, so $f(x)$ has a local minimum at $x = c_2$. Similarly $f(x)$ has a local maximum at $x = c_4$. This information is shown in a sign diagram (Figure 3-6-16), and the graph is shown in Figure 3-6-17. ●

Figure 3-6-16
Sign diagram for $f'(x)$ in Example 7

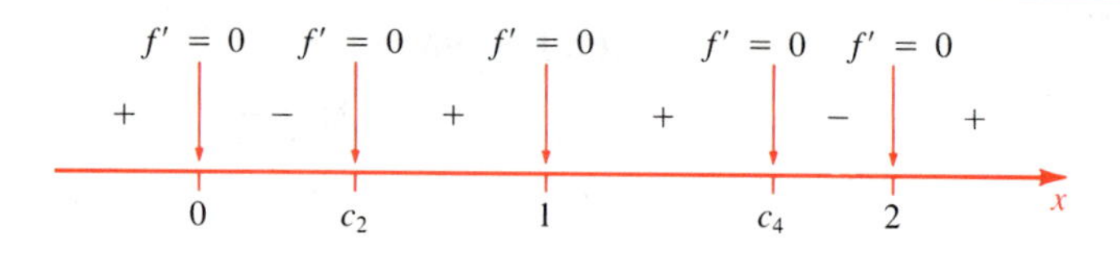

Figure 3-6-17
Graph of $y = x^4(x-1)^3(x-2)^2$
$$\begin{cases} c_2 = \frac{1}{9}(10 - \sqrt{28}) \approx 0.52 \\ c_4 = \frac{1}{9}(10 + \sqrt{28}) \approx 1.70 \end{cases}$$

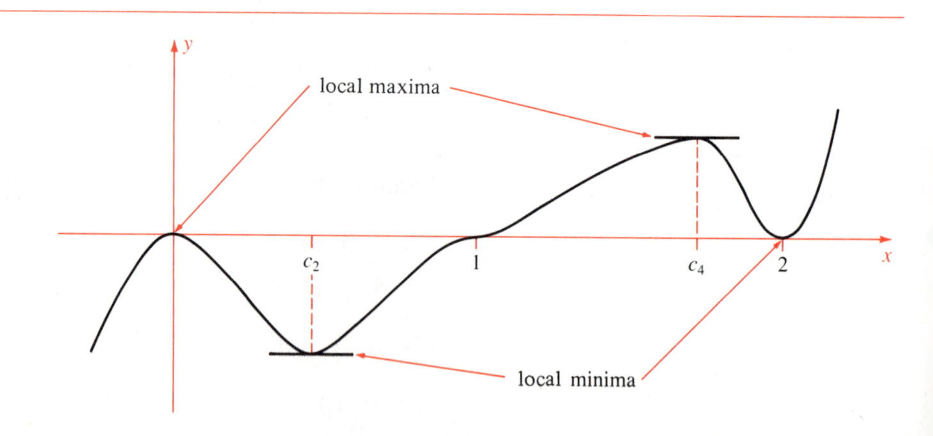

Example 8

Locate all local extrema of $f(x) = x + 2\cos x$ and give the corresponding values of $f(x)$.

Solution The domain of $f(x)$ is **R**, with no end points. Consequently, at each local extremum, $f'(x) = 0$, so we differentiate and solve the resulting equation

$$f'(x) = 1 - 2\sin x = 0$$

Thus $f'(x) = 0$ for $\sin x = \frac{1}{2}$, that is, for

$$x = \tfrac{1}{6}\pi + 2\pi n \quad \text{and} \quad x = \tfrac{5}{6}\pi + 2\pi n$$

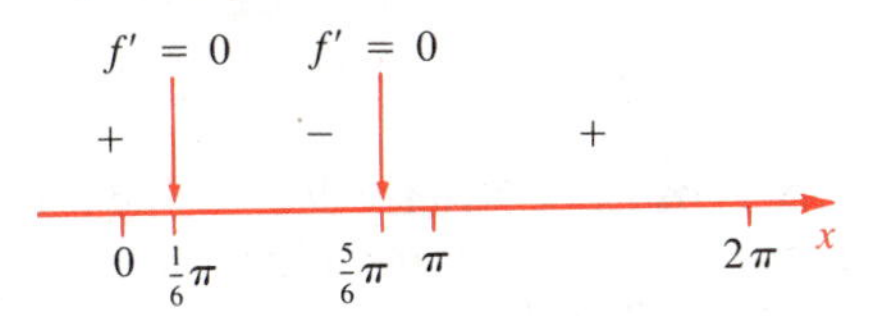

Figure 3-6-18
Sign diagram for $f'(x)$ in Example 8

The function $f'(x)$ is periodic of period 2π, so it suffices to make a sign diagram (Figure 3-6-18) for an interval of length 2π, say $[0, 2\pi]$.

We conclude that $f(x)$ has a local maximum at $x = \frac{1}{6}\pi$ and a local minimum at $x = \frac{5}{6}\pi$. By periodicity, the same holds for $\frac{1}{6}\pi + 2\pi n$ and $\frac{5}{6}\pi + 2\pi n$. Since $\cos\frac{1}{6}\pi = \frac{1}{2}\sqrt{3}$ and $\cos\frac{5}{6}\pi = -\frac{1}{2}\sqrt{3}$, our conclusions are

$$\text{Local maximum: } x = \tfrac{1}{6}\pi + 2\pi n \qquad f(x) = \tfrac{1}{6}\pi + 2\pi n + \sqrt{3}$$
$$\text{Local minimum: } x = \tfrac{5}{6}\pi + 2\pi n \qquad f(x) = \tfrac{5}{6}\pi + 2\pi n - \sqrt{3}$$

where n is any integer. See Figure 3-6-19. ●

Some Non-Differentiable Examples

On rare occasions we meet functions that have critical points where the derivative does not exist. This doesn't happen often, but we should be prepared for it.

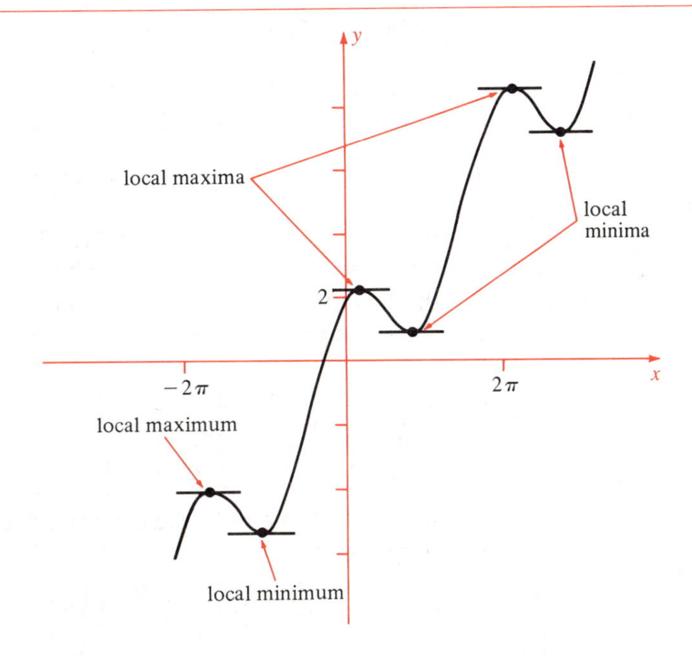

Figure 3-6-19
Graph of $y = x + 2\cos x$

Example 9

Find all local extrema of $f(x) = x + \frac{3}{2}x^{2/3}$ and graph $y = f(x)$.

Solution The function is continuous on **R** and differentiable except at $x = 0$. There is a critical point at $x = 0$ because $f'(0)$ does not exist, so $f(0)$ is a candidate for a local extremum. Since $0 < \frac{2}{3} < 1$, the (positive) term $\frac{3}{2}x^{2/3}$ dominates x near zero; hence $f(x) > 0$ for x near zero. We conclude that $f(0)$ is a local minimum. Next

$$f'(x) = 1 + x^{-1/3}$$

and we locate all stationary points by solving $f'(x) = 0$:

$$x^{-1/3} = -1$$

It follows that $x = -1$ is the only zero of $f'(x)$. We compute

$$f(-1) = -1 + \tfrac{3}{2}(-1)^{2/3} = -1 + \tfrac{3}{2} = \tfrac{1}{2}$$

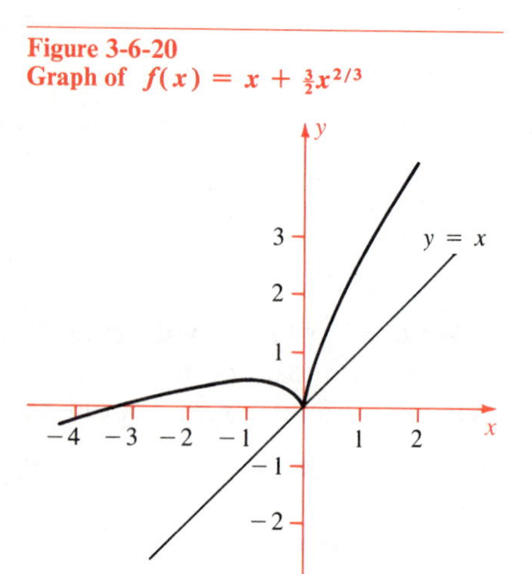

Figure 3-6-20
Graph of $f(x) = x + \frac{3}{2}x^{2/3}$

Since $f(0) = 0 < \frac{1}{2}$, the function $f(x)$ must be strictly decreasing on the interval $(-1, 0)$. Since $f'(x) > 0$ for x large negative, the function is strictly increasing on $(-\infty, -1)$. Therefore $f(-1) = \frac{1}{2}$ is a local maximum of $f(x)$. We check that the graph of $y = f(x) = x + \frac{3}{2}x^{2/3}$ crosses the x-axis only if $x = 0$ or $x^{1/3} + \frac{3}{2} = 0$, that is $x = (-\frac{3}{2})^3 = -\frac{27}{8}$. Finally, the graph is above $y = x$, but not much relative to the size of x for $|x|$ large, because $f(x)/x = 1 + \frac{3}{2}x^{-1/3}$. All this information leads to Figure 3-6-20. ●

Example 10

Find all local maxima and minima of $f(x) = x^{2/3}(1 - x)^{2/3}$ and graph $y = f(x)$.

Solution The derivative is

$$\begin{aligned} f'(x) &= \tfrac{2}{3}[x^{-1/3}(1 - x)^{2/3} - x^{2/3}(1 - x)^{-1/3}] \\ &= \tfrac{2}{3}x^{-1/3}(1 - x)^{-1/3}[(1 - x) - x] \\ &= \tfrac{2}{3}x^{-1/3}(1 - x)^{-1/3}(1 - 2x) \end{aligned}$$

provided $x \neq 0$ and $x \neq 1$. The derivative of $f(x)$ does not exist for $x = 0$ or for $x = 1$. The derivative is 0 only at $x = \frac{1}{2}$. Thus 0, $\frac{1}{2}$, and 1 are all critical points of $f(x)$. From the factored form of $f'(x)$ it follows that

$$\begin{cases} f'(x) < 0 & \text{for } x < 0 \\ f'(x) > 0 & \text{for } 0 < x < \frac{1}{2} \\ f'(x) < 0 & \text{for } \frac{1}{2} < x < 1 \end{cases}$$

so $f(x)$ has a local maximum at $x = \frac{1}{2}$. The corresponding value of $f(x)$ is

$$f(\tfrac{1}{2}) = (\tfrac{1}{2})^{2/3}(\tfrac{1}{2})^{2/3} = (\tfrac{1}{2})^{4/3} \approx 0.397$$

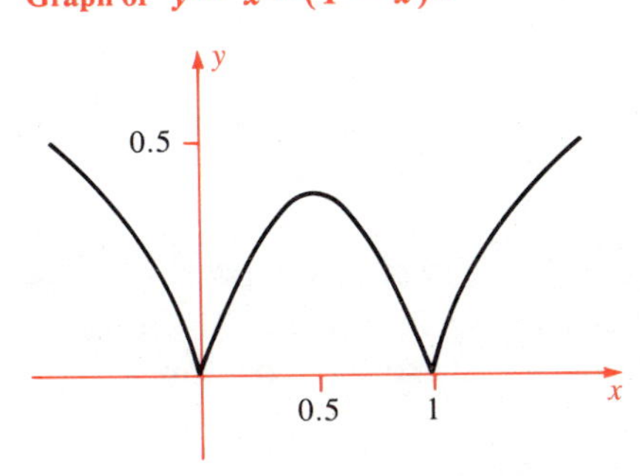

Figure 3-6-21
Graph of $y = x^{2/3}(1 - x)^{2/3}$

At the other two critical points we have

$$f(0) = 0 \quad \text{and} \quad f(1) = 0$$

Since $f(x) = [x^{1/3}(1 - x)^{1/3}]^2$, it is clear that $f(x) > 0$ if $x \neq 0$ or $x \neq 1$, so $f(x)$ has local minima at $x = 0$ and at $x = 1$. See Figure 3-6-21.

Alternative Solution For this particular function, a special device can be used. The function $z = y^3$ is strictly increasing; hence $f(x)$ has a local maximum or minimum precisely where the composite function $g(x) = f(x)^3$ has a local maximum or minimum. But

$$g(x) = x^2(1 - x)^2$$

There are no problems with this function; it has a derivative everywhere, and

$$g'(x) = 2x(1 - x)(1 - 2x)$$

The zeros of $g'(x)$ are 0, $\frac{1}{2}$, and 1. We easily find that $g(x)$ has a local maximum at $x = \frac{1}{2}$ and local minima at $x = 0$ and $x = 1$. The same is true of $f(x)$. ●

Remark We shall use this device in some of our applications later when we have to find extrema of $\sqrt{f(x)}$ where $f(x) \geq 0$, and it is easier to deal with $f(x)$.

Exercises

Find the maximum and minimum of $f(x)$ on the given domain (**R** if not stated otherwise) and where they occur; also sketch $y = f(x)$

1 $f(x) = x^2 - 4x + 6$
2 $f(x) = 2x^2 - 9x + 12$
3 $f(x) = -x^2 + 6x + 4$
4 $f(x) = -3x^2 + 3x + 1$
5 $f(x) = 12x + 1/3x \quad (0, +\infty)$
6 $f(x) = 12x + 1/3x \quad (-\infty, 0)$
7 $f(x) = 2x^3 - 3x^2 - 12x + 1 \quad [0, 3]$
8 $f(x) = 2x^3 - 3x^2 - 12x + 1 \quad [-2, 2]$
9 $f(x) = \dfrac{1}{(x - 1)(2 - x)} \quad (1, 2)$
10 $f(x) = \dfrac{1}{x^2 - x + 1}$
11 $f(x) = \frac{1}{3}x + 1/x \quad [1, 3]$
12 $f(x) = 1/x - 1/x^2 - 1/x^3 \quad [1, +\infty)$
13 $f(x) = 1/x - 1/x^2 - 1/x^3 \quad [-3, -1]$
14 $f(x) = x - x^4 \quad [0, 1]$
15 $f(x) = 4x^2 + 1/x \quad [\frac{1}{4}, 1]$
16 $f(x) = x/(2 + x^3) \quad [-1, +\infty)$
17 $f(x) = x^4 - 2x^2 + 1 \quad [-1, 2]$
18 $f(x) = 3x^4 - 16x^3 + 18x^2 + 12 \quad [0, 4]$
19 $f(x) = -3x^4 - 16x^3 - 18x^2 - 12 \quad [-3, 0]$
20 $f(x) = x^3 + x + 2/x \quad (0, +\infty)$
21 $f(x) = \sqrt{1 - x^4} \quad [-1, 1]$
22 $f(x) = \sqrt[3]{1 - x^2} \quad [-1, 1]$
23 $f(x) = x\sqrt{1 - x^2} \quad [-1, 1]$
24 $f(x) = x^2\sqrt{1 - x^2} \quad [-1, 1]$
25 $f(x) = \dfrac{x}{1 + x^2}$
26 $f(x) = \dfrac{x}{1 + x^3} \quad (-1, +\infty)$

Find the coordinates of all local maxima and minima, and then sketch the graph.

27 $f(x) = \dfrac{x^2}{1 + x^3} \quad (-1, +\infty)$
28 $f(x) = \dfrac{x^3}{1 + x^4}$
29 $f(x) = \dfrac{\cos^2 x}{1 + \cos^2 x}$
30 $f(x) = \dfrac{\sin x}{1 + \cos^2 x}$

31 $f(x) = \sin x - \cos 2x$

32 $f(x) = 4x - \tan x$ $(-\frac{1}{2}\pi, \frac{1}{2}\pi)$

33 $f(x) = x^{2/3} - x$

34 $f(x) = x^{2/3} - x^{2/5}$

35 $f(x) = (\sin x)^{2/3}$

36 $f(x) = (1 + \sin x)^{2/3}$

37 $f(x) = x^{2/3} + (1 - x)^{2/3}$

38 $f(x) = \sqrt{1 - x^2}$ on $[-1, 1]$ $f(x)$ periodic of period 2 on **R**

39 $f(x) = |x^2 - 1|$

*40 $f(x) = x^{2/3} - (1 - x)^{2/3}$

3-7 On Problem Solving

A major part of your time in calculus and other courses is devoted to solving problems. It is worth your while to develop sound techniques. Here are a few suggestions.

- **Think** Before plunging into a problem, take a moment to think. Read the problem again and think about it. What are its essential features? Have you seen a problem like it before? What techniques are needed?

  Try to make a rough estimate of the answer. It will help you understand the problem and will serve as a check against unreasonable answers. A car will *not* go 1800 miles in 3 hours; a weight dropped from 10,000 feet will *not* hit the earth at 5 miles per hour; the volume of a tank is *not* -275 gallons.

- **Examine the data** Be sure you understand what is given. Translate the data into mathematical language. Whenever possible, make a clear diagram and label it accurately. Place the axes so as to simplify computations. If you get stuck, check to see that you are using *all* the data.

- **Avoid sloppiness**

a Avoid sloppiness in language. Mathematics is written in English sentences. A typical mathematical sentence is "$y = 4x + 1$." The equal sign is the verb in this sentence; it means "equals" or "is equal to." The equal sign is not to be used in place of "and" or as a punctuation mark. *Quantities on opposite sides of an equal sign must be equal.*

Use short simple sentences. Avoid pronouns such as "it" and "which." Give names to quantities and use them. Otherwise you may write gibberish like the following:

"To find the minimum of it, differentiate it and set it equal to zero, then solve it which if you substitute it, is the minimum."

Better: "To find the minimum of $f(x)$, set its derivative $f'(x)$ equal to zero. Let x_0 be a solution of the resulting equation. Then $f(x_0)$ is a possible minimum value of $f(x)$."

b Avoid sloppiness in computation. Do calculations in sequences of neat, orderly steps. Include all steps except utterly trivial ones. This will help eliminate errors, or at least make errors easier to find. Check any numbers used; be sure that you have not dropped a minus sign or transposed digits. Lots of test points are lost because of $-$ instead of $+$ and $+$ instead of $-$.

c Avoid sloppiness in units. If you start out measuring in feet, all lengths must be in feet, all areas in square feet, and all volumes in cubic feet. Do not mix feet and acres, seconds and years.

d Avoid sloppiness in the answer. Be sure to answer the question that is asked. If the problem asks for the *maximum value* of $f(x)$, the answer is not the *point* where the maximum occurs. If the problem asks for a *formula*, the answer is not a *number*.

Example 1

Find the minimum of $f(x) = x^2 - 2x + 1$.

Solution 1 $\quad 2x - 2 \quad x = 1 \quad 1^2 - 2 \cdot 1 + 1 \quad 0$

Unbearable. This is just a collection of marks on the paper. There is absolutely no indication of what these marks mean or of what they have to do with the problem. When you write, it is your responsibility to inform the reader of what you are doing. Assume he is intelligent, but not a mind reader.

Solution 2

$$\frac{df}{dx} = 2x - 2 = 0 = 2x = 2 = x = 1 = f(x)$$
$$= 1^2 - 2 \cdot 1 + 1 = 0$$

Poor. The equal sign is badly mauled. This solution contains such enlightening statements as "$0 = 2 = 1$," and it does not explain what the writer is doing.

Solution 3 $\quad \dfrac{df}{dx} = 2x - 2 = 0 \quad 2x = 2 \quad x = 1$

This is better than Solution 2, but contains three errors. Error 1: The first statement, "$df/dx = 2x - 2 = 0$," muddles two separate steps. First the derivative is computed, then the derivative is equated to zero. Error 2: The solution is incomplete because it does not give what the problem asks for, the minimum value of f. Instead, it gives the point x at which the minimum is assumed. Error 3: The solution doesn't show that $f(1)$ is actually the minimum.

Solution 4 The derivative of $f(x)$ is $\quad f'(x) = 2x - 2$

At a minimum, $f'(x) = 0$. Hence $\quad 2x - 2 = 0 \quad x = 1$

The corresponding value of f is $\quad f(1) = 1^2 - 2 \cdot 1 + 1 = 0$

If $x > 1$, then $f'(x) = 2(x - 1) > 0$, so f is increasing.
If $x < 1$, then $f'(x) = 2(x - 1) < 0$, so f is decreasing.
Hence f is minimal at $x = 1$, and the minimum value of f is 0.

This solution is absolutely correct, but long. For homework assignments the following may be satisfactory (check with your instructor):

Solution 5 $\quad f'(x) = 2x - 2$

At min, $f' = 0 \quad 2x - 2 = 0 \quad x = 1$.
For $x > 1 \quad f'(x) = 2(x - 1) > 0 \quad f$ increases.
For $x < 1 \quad f'(x) = 2(x - 1) < 0 \quad f$ decreases.
Hence $x = 1$ yields min: $f_{\min} = f(1) = 1^2 - 2 \cdot 1 + 1 = 0$.

The next solution was submitted by a student who took a minute to think.

Solution 6 $f(x) = x^2 - 2x + 1 = (x - 1)^2 \geq 0$

But $f(1) = (1 - 1)^2 = 0$. Hence the minimum value of $f(x)$ is 0. ●

The Wyatt Earp Principle That legendary gunfighter Wyatt Earp survived many a shootout in the old West. Yet Earp carried only one gun and used no fancy tricks. His secret? He took an extra split second to *aim.* While the bad guy blazed away wildly with two guns, Earp got his man on the first shot.

Try to face a calculus problem the way Wyatt Earp faced a gunfight. Instead of differentiating wildly with both hands, take a minute to think. You may find the problem simpler than it looks at first. Certainly you will have a better chance of winning the showdown.

Example 2

A gardener wishes to fence off a rectangular plot as large in area as possible, using part of a long straight wall as one side of the boundary and 200 ft of fencing for the other three sides. Find the dimensions and area of the largest plot.

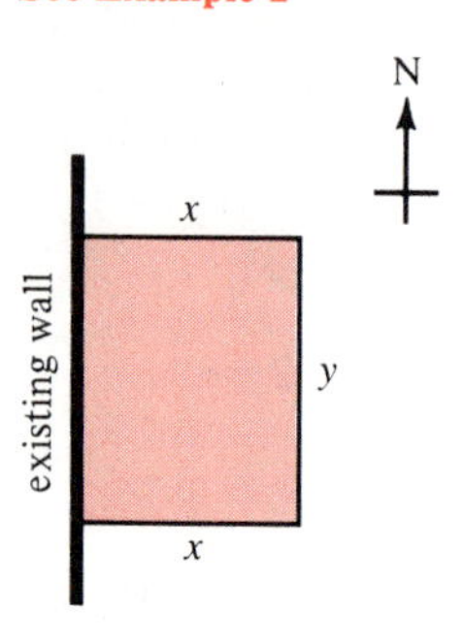

Figure 3-7-1
See Example 2

Solution Label the unknown sides as in Figure 3-7-1. The area is $A = xy$, and this must be maximized. Now the two east-west sides and the north-south side must add up to 200 ft. Hence

$$2x + y = 200 \quad \text{so} \quad y = 200 - 2x$$

By substituting, express A as a function of x:

$$A = A(x) = x(200 - 2x) = 200x - 2x^2$$

The possibilities $x < 0$ and $x > 100$ are ruled out because there $A < 0$. Therefore the domain of $A(x)$ is the interval $[0, 100]$. Since $A(x) > 0$ inside this domain and $A(0) = A(100) = 0$, there is a maximum on $(0, 100)$. To find it, differentiate and solve $A'(x) = 0$:

$$A'(x) = 200 - 4x$$

$$A'(x) = 0 \quad \text{if} \quad 200 - 4x = 0 \qquad \text{that is} \qquad x = 50$$

The corresponding $y = 200 - 2x = 200 - 100 = 100$ and

$$A_{\text{max}} = xy = 50 \cdot 100 = 5000 \text{ ft}^2$$ ●

Example 3

An open box is constructed by removing a small square from each corner of a square tin sheet and then folding up the sides. If the sheet is L cm on each side, what is the largest possible volume of the box?

Figure 3-7-2
See Example 3

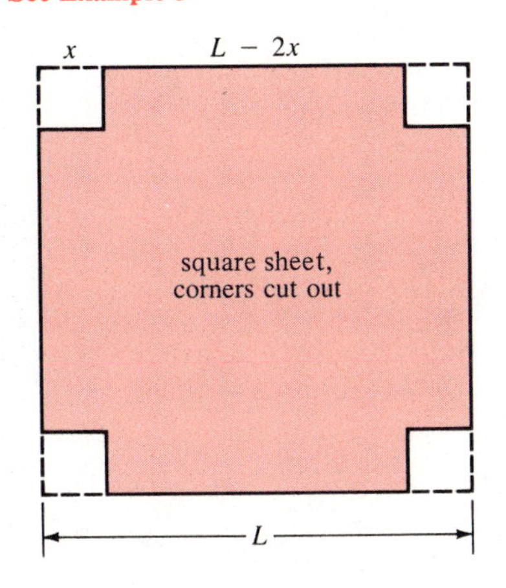

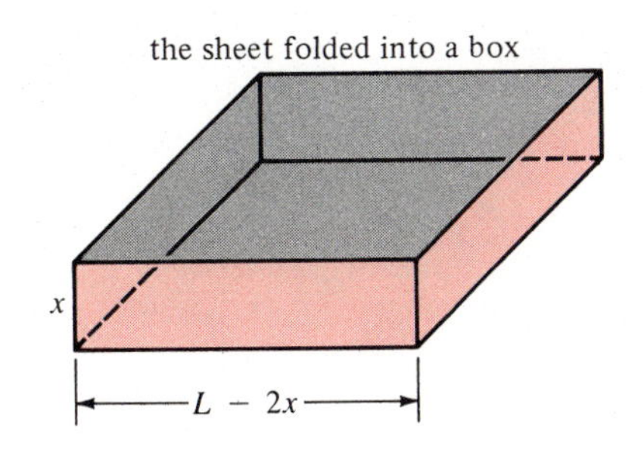

Solution Let each cutout square have side x. See Figure 3-7-2. Express the volume of the box as a function of x:

$$\text{volume} = (\text{area of base}) \cdot (\text{height})$$

The base is a square of side $L - 2x$, and the height is x. So the volume of the box is

$$V(x) = (L - 2x)^2 x = (L^2 - 4Lx + 4x^2)x$$
$$= L^2 x - 4Lx^2 + 4x^3$$

By the nature of the problem, x must be positive but less that $\frac{1}{2}L$, half the side of the sheet. The problem can now be stated in mathematical terms: Find the largest value of $V(x)$ in the domain $(0, \frac{1}{2}L)$. Differentiate:

$$V'(x) = L^2 - 8Lx + 12x^2 = (L - 2x)(L - 6x)$$

The factor $L - 2x$ is positive because $x < \frac{1}{2}L$. Therefore the sign of $V'(x)$ is the same as the sign of $L - 6x$:

$$\begin{cases} V'(x) > 0 & \text{for} \quad 0 < x < \frac{1}{6}L \\ V'(\frac{1}{6}L) = 0 \\ V'(x) < 0 & \text{for} \quad \frac{1}{6}L < x < \frac{1}{2}L \end{cases}$$

Therefore $V(x)$ has a maximum at $x = \frac{1}{6}L$, and

$$V_{\max} = V(\tfrac{1}{6}L) = \left(L - \frac{2L}{6}\right)^2\left(\frac{L}{6}\right) = \left(\frac{2L}{3}\right)^2\left(\frac{L}{6}\right)$$
$$= \tfrac{2}{27}L^3 \text{ cm}^3$$

Example 4

The illumination of an object by a light source is directly proportional to the strength of the source and inversely proportional to the square of the distance between the source and the object. Two light sources, one five times as strong as the other, are 1 m apart (Figure 3-7-3). At what point on the line between the sources should a screen be placed so that the illumination it receives is minimal?

Figure 3-7-3
See Example 4

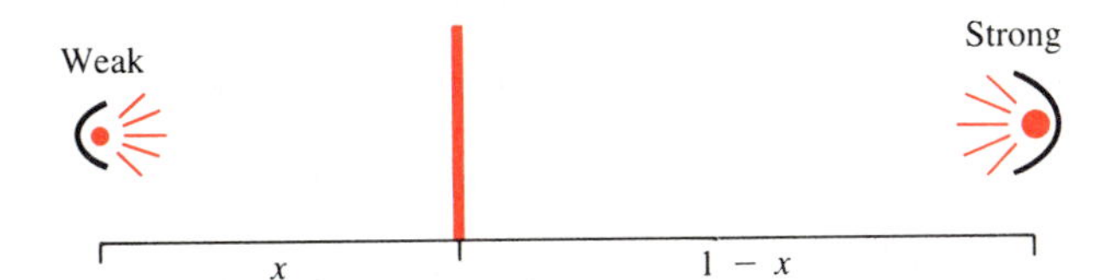

Solution Apparently the screen should be closer to the weaker source, so $x < \frac{1}{2}$. Even though one source is five times as strong as the other, the screen cannot be too close to the weaker source because of the inverse square rule. A reasonable guess: x is between 0.2 and 0.4.

The illumination from the weaker source is $I_1 = k/x^2$, where the constant k depends on the units of measurement. The illumination from the stronger source is

$$I_2 = \frac{5k}{(1-x)^2}$$

The problem is to minimize

$$I = I_1 + I_2 = \frac{k}{x^2} + \frac{5k}{(1-x)^2}$$

on the domain $(0, 1)$. Differentiate:

$$\frac{dI}{dx} = -\frac{2k}{x^3} + \frac{10k}{(1-x)^3}$$

The derivative equals 0 provided

$$\frac{2k}{x^3} = \frac{10k}{(1-x)^3} \qquad \text{that is} \qquad 5x^3 = (1-x)^3$$

Take cube roots: $\quad (\sqrt[3]{5})x = 1 - x \qquad (1 + \sqrt[3]{5})x = 1$

$$x = \frac{1}{1+\sqrt[3]{5}} \approx 0.369 \text{ m}$$

Since $I \to +\infty$ as $x \to 0+$ or $x \to 1-$, this value of x must give the minimal I. ●

Example 5

Ship A leaves a port at noon and sails due north at 10 mph. Ship B is 100 mi east of the port at noon, sailing due west at 6 mph. When will the ships be nearest each other?

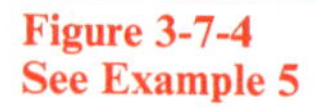
Figure 3-7-4
See Example 5

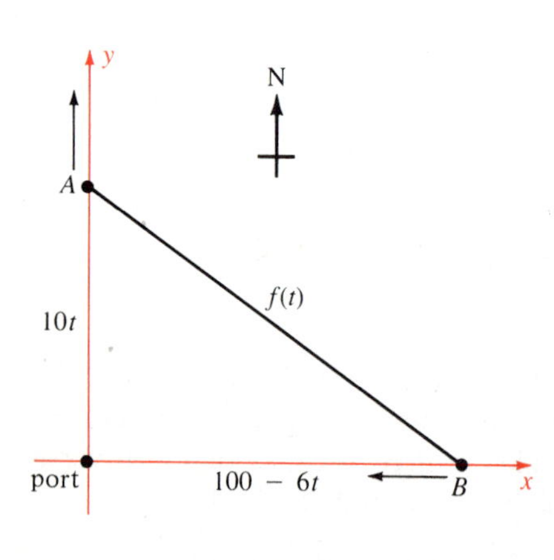

Solution Set up axes with the port at the origin and the y-axis pointing north. The relative position of the ships at t hr past noon is shown in Figure 3-7-4. Since ship A is in port before noon, $t \geq 0$ is required. Since ship B is in port at noon $+ \frac{100}{6} =$ 4:40 AM the next morning, $t \leq \frac{100}{6}$ is required. Thus the domain in question is the closed interval $[0, \frac{100}{6}]$ of the t-axis.

The distance between the ships is

$$f(t) = \sqrt{(100-6t)^2 + (10t)^2}$$

The square root is annoying because when we differentiate, the resulting expression will be messy. A simple device for getting rid of the square root is squaring $f(t)$. Set

$$g(t) = (100-6t)^2 + (10t)^2 = f(t)^2$$

Because $g(t) > 0$ and because the square root function $\sqrt{z}$ is strictly increasing for $z > 0$, the function $g(t)$ has its minimum precisely where $f(t) = \sqrt{g(t)}$ has its minimum. It is easier to minimize the quadratic function $g(t)$ than to minimize $f(t)$. We have

$$\begin{aligned} g(t) &= (100 - 6t)^2 + (10t)^2 \\ &= 10{,}000 - 1200t + 36t^2 + 100t^2 \\ &= 10{,}000 - 1200t + 136t^2 \end{aligned}$$

Hence

$$g'(t) = -1200 + 272t$$

$$g'(t) = 0 \quad \text{for} \quad t = \frac{1200}{272} \approx 4.41 \approx 4 \text{ hr } 25 \text{ min}$$

Since $g(t)$ is quadratic, so its graph is a parabola, and $0 < \frac{1200}{272} < \frac{100}{6}$, we know that we have found the minimum. Thus the ships are closest at about 4:25 PM. ●

Remark Ask yourself this: Must the problem have a solution strictly between the two end points? Suppose ship B moves a *lot* slower than A; would that matter? Or suppose B moves a lot *faster* than A; then what?

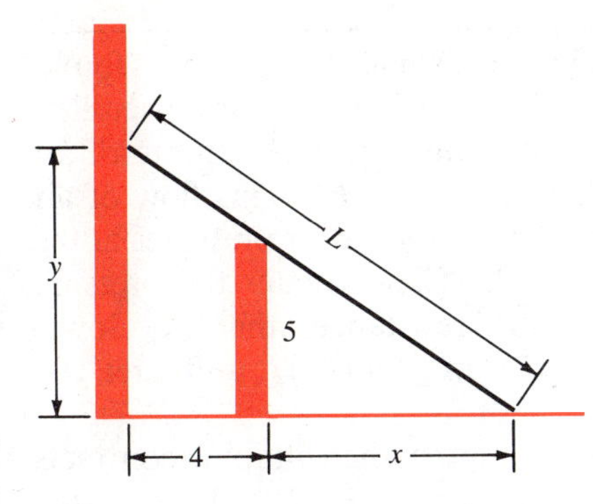

Figure 3-7-5
See Example 6

Example 6

A 5-foot fence stands 4 ft from a high wall (Figure 3-7-5). How long is the shortest ladder than can reach from the ground outside the fence to the wall?

Solution Take a moment to think. Note that if x is very small and positive, the ladder will be nearly vertical, certainly longer than is necessary. If x is large, the ladder will be nearly horizontal, again too long. The best choice of x might be somewhere around 5 or 6, surely between 2 and 10. In fact, as x increases from near 0, it seems that L should decrease, reach a minimum, then increase thereafter.

To start the computation, note that

$$L^2 = (x + 4)^2 + y^2$$

Eliminate y by finding a relation between x and y. By similar triangles,

$$\frac{y}{x + 4} = \frac{5}{x} \qquad \text{that is} \qquad y = \frac{5(x + 4)}{x}$$

Hence

$$L^2 = (x + 4)^2 + \frac{25(x + 4)^2}{x^2} = (x + 4)^2\left(1 + \frac{25}{x^2}\right)$$

Rather than take the square root, minimize L^2 over the domain $(x > 0)$. To do so, differentiate L^2:

$$\frac{d}{dx}(L^2) = 2(x+4)\left(1+\frac{25}{x^2}\right) + (x+4)^2\left(\frac{-50}{x^3}\right)$$

$$= 2(x+4)\left[1+\frac{25}{x^2} - \frac{25(x+4)}{x^3}\right]$$

$$= 2(x+4)\left[\frac{x^3+25x-25(x+4)}{x^3}\right]$$

$$= \frac{2(x+4)(x^3-100)}{x^3}$$

There is only one positive value of x for which the derivative is zero: $x = \sqrt[3]{100} \approx 4.64$. The derivative is negative for $x < \sqrt[3]{100}$, positive for $x > \sqrt[3]{100}$. Thus our physical intuition was correct: L^2 decreases, reaches a minimum near $x = 5$, then increases.

From the formula for L^2,

$$L = (x+4)\sqrt{1+\frac{25}{x^2}} = \left(1+\frac{4}{x}\right)\sqrt{x^2+25}$$

Therefore

$$L_{\min} = L(\sqrt[3]{100}) = \left(1+\frac{4}{\sqrt[3]{100}}\right)\sqrt{100^{2/3}+25} \approx 12.7 \text{ ft}$$ ●

Example 7

The **resistance** R to air flow through a tube of small radius r was found experimentally by J. L. M. Poiseuille in 1846, and is known as Poiseuille's law: $R = k/r^4$, where k is a positive constant. The volume of air moving out of the tube per unit time is the **flow** F. This flow of air through the tube depends on the difference P of the air pressures at the ends of the tube, according to $F = P/R$. (The greater the pressure difference, the more flow; the more the resistance, the less flow.) We measure F in cm^3/sec and P in $(\text{force})/(\text{unit area}) = \text{dyne}/\text{cm}^2$.

When a person draws air through his windpipe, the windpipe contracts from its initial radius r_0 when $P = 0$ to $r = r_0 - aP$, where a is a positive constant. This is known experimentally to hold for $r_0 \geq r \geq \frac{1}{2}r_0$. Find the radius at which the *speed* of the air stream in the windpipe is as large as possible

Solution F is the volume of air moving through a fixed cross-section of the tube in a unit of time. Therefore the airspeed is $v = F/(\text{area}) = F/\pi r^2$. Thus

$$v = \frac{F}{\pi r^2} = \frac{P}{\pi r^2 R} = \frac{(r_0 - r)/a}{(\pi r^2)(k/r^4)} = \frac{1}{\pi a k}(r_0 - r)r^2$$

We seek the maximum of v on the interval $[\frac{1}{2}r_0, r_0]$. It will be simpler to drop the constants and work with $f(r) = (r_0 - r)r^2$. The candidates for maximum are the values of $f(r)$ at the end points, $f(\frac{1}{2}r_0) = \frac{1}{8}r_0^3$

and $f(r_0) = 0$, as well as all values of $f(r)$ where r is inside the domain and $f'(r) = 0$. Now

$$f(r) = r_0 r^2 - r^3 \quad \text{so} \quad f'(r) = r(2r_0 - 3r)$$

Therefore $f'(r) = 0$ for $r = \frac{2}{3}r_0$, which is in the domain, and for $r = 0$, which isn't. Now

$$f(\tfrac{2}{3}r_0) = (\tfrac{1}{3}r_0)(\tfrac{2}{3}r_0)^2 = \tfrac{4}{27}r_0^3$$

Since $\frac{4}{27} > \frac{1}{8}$, the maximum is attained at radius $r = \frac{2}{3}r_0$. ●

Exercises

1 A ball thrown straight up reaches a height of $3 + 40t - 16t^2$ ft in t sec. How high will it go?

2 The power output P of a battery is given by $P = EI - RI^2$, where E and R are positive constants and I is current. Find the current for which the power output is a maximum and find the maximum power. (Here E is the voltage the battery delivers and R is its internal resistance.)

3 A farmer with 300 m of fencing wishes to enclose a rectangular area and divide it into 5 pens with fences parallel to the short end of the rectangle. What dimensions of the enclosure make its area a maximum?

4 Show that the rectangle of largest possible area, for a given perimeter, is a square.

5 An open rectangular box has volume $15\ \text{ft}^3$. The length of its base is 3 times its width. Materials for the sides and base cost 60¢ and 40¢ per ft^2, respectively. Find the dimensions of the cheapest such box.

6 Find the dimensions of the rectangle of largest area inscribed in a right triangle with legs of length a and b, if two sides of the rectangle lie along the legs of the triangle.

7 A length of wire 28 ft long is cut into two pieces. One piece is bent into a 3:4:5 right triangle and the other piece is bent into a square. Minimize the sum of the areas of the two figures.

8 (cont.) Maximize the area sum.

9 What points on the curve $xy^2 = 1$ are nearest the origin?

10 Find the point on the graph of the equation $y = \sqrt{x}$ nearest to the point $(1, 0)$.

11 Two particles moving in the plane have coordinates $(2t, 8t^3 - 24t + 10)$ and $(2t + 1, 8t^3 + 6t + 1)$ at time t. How close do the particles come to each other?

12 Find the minimum vertical distance between the curves $y = 27x^3$ and $y = -1/x$ if $x \neq 0$.

13 Find the maximum slope of the curve $y = 6x^2 - x^3$.

14 Find the minimum of $\overline{AX} + \overline{XB}$ in Figure 3-7-6, where X moves along segment CD.

Figure 3-7-6
See Exercise 14

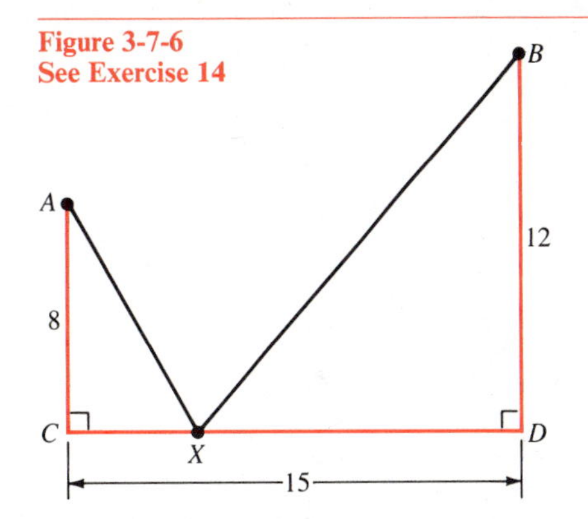

Figure 3-7-7
See Exercise 15

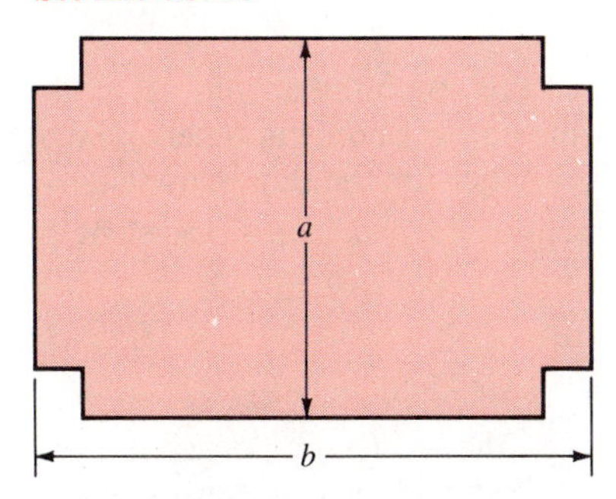

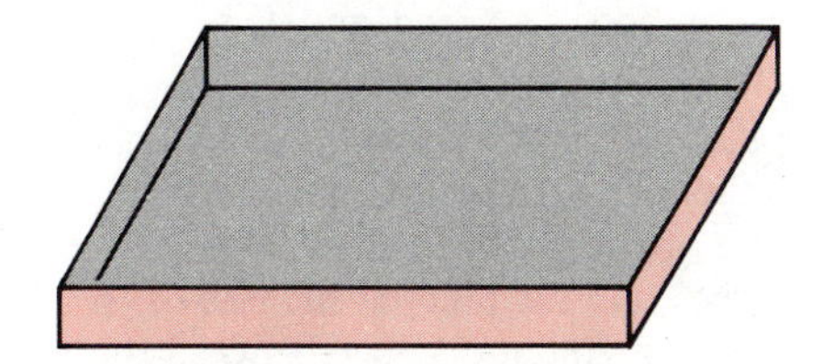

15 Equal squares are cut out of the four corners of an $a \times b$ rectangle of sheet metal and the result is folded into an open rectangular box (Figure 3-7-7). Find the maximum volume of the box, assuming $b = 2a$.

16 Squares and rectangles are cut from an $a \times b$ rectangle, as indicated in Figure 3-7-8. The result is folded into a closed rectangular box. Assuming $b = 2a$, find the maximum volume of the box.

Figure 3-7-8
See Exercise 16

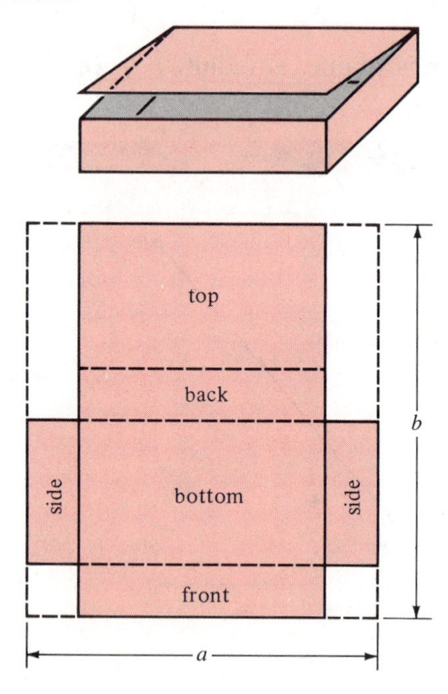

17 Suppose, in Example 5 of this section, that ship A is already 70 mi north of the port at noon, but otherwise the problem is the same. Now when are the ships closest?

18 The speed v of a surface wave depends on its wavelength λ according to

$$v = \sqrt{\frac{g}{2\pi}\lambda + \frac{2\pi\sigma}{\rho\lambda}}$$

where the constants are g, the acceleration of gravity, σ, the surface tension of the fluid, and ρ, the density of the fluid. Find the minimum possible speed and the corresponding wavelength.

19 As a hiker starts across a 200-foot bridge, a small boat passes directly beneath the center of the bridge on a straight path perpendicular to the bridge. Suppose the boat is moving at the rate of 8 ft/sec and the hiker at the rate of 6 ft/sec. Find the shortest horizontal distance between the boat and the hiker.

20 Suppose, in Exercise 19, that the bridge is 50 ft high. Find the shortest *distance* between the hiker and the boat. [Hint In space, $\text{dist}^2 = (\text{EW-dist})^2 + (\text{NS-dist})^2 + (\text{vert dist})^2$.]

Suppose the cost of producing x units is $f(x)$ dollars, and the price of x units is $h(x)$ dollars. Calculate the maximum net revenue (receipts less cost) possible for a manufacturer if

21 $f(x) = 7.5x + 400 \qquad h(x) = 10x - 0.0005x^2$

22 $f(x) = 50x + 1200 \qquad h(x) = 65x - 0.001x^2$

23 The cost per hour in dollars for fuel to operate a certain airliner is $0.012v^2$, where v is the speed in mph. If fixed charges amount to \$4000/hr, find the most economical speed for a 1500 mi trip.

24 During a typical eight-hour work day the quantity of gravel produced in a plant is $60t + 12t^2 - t^3$ tons, where t represents hours worked from the start of the work day. When in the day is the rate of production at a maximum?

25 Suppose a liter of water (at 25 °C) contains x moles of hydronium ions (H_3O^+) and y moles of hydroxide ions (OH^-). Experiments show that $xy = 10^{-14}$. Minimize the sum $x + y$.

26 In Example 7 of this section, find the radius for which the flow F is maximum and find the corresponding pressure (for optimal coughing).

27 The energy of a certain diatomic molecule is

$$U = \frac{a}{x^{12}} - \frac{b}{x^6}$$

where a and b are positive constants and x is the distance between the atoms. Find the **dissociation energy,** the maximum of $-U$.

28 A one-port network (Figure 3-7-9) at fixed frequency ω is to be terminated in a resistance x so the power

$$P = \frac{E^2x}{(2\pi\omega L)^2 + R^2 + 2Rx + x^2}$$

dissipated by x (in heat) is maximal. Find x and P_{max}.

Figure 3-7-9
See Exercise 28

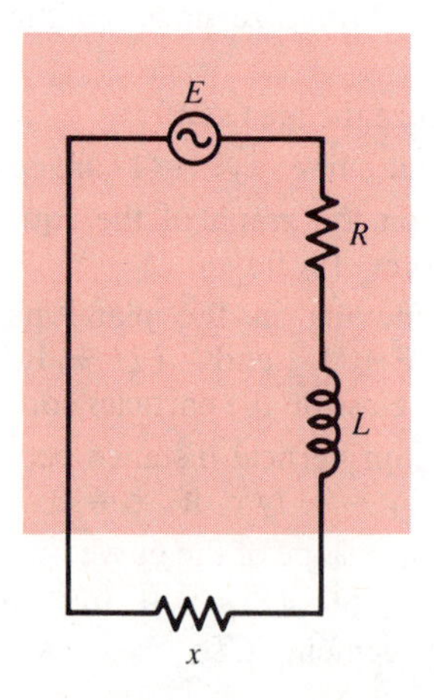

3-8 Further Applications

Many of our applications are geometric and require formulas for the perimeters, areas, and volumes of common shapes. These formulas will be mentioned as needed. However, note the table of geometric formulas inside the front cover.

We now consider some extremum problems in the plane.

Example 1

Find the largest possible area of an isosceles triangle inscribed in a circle of radius r.

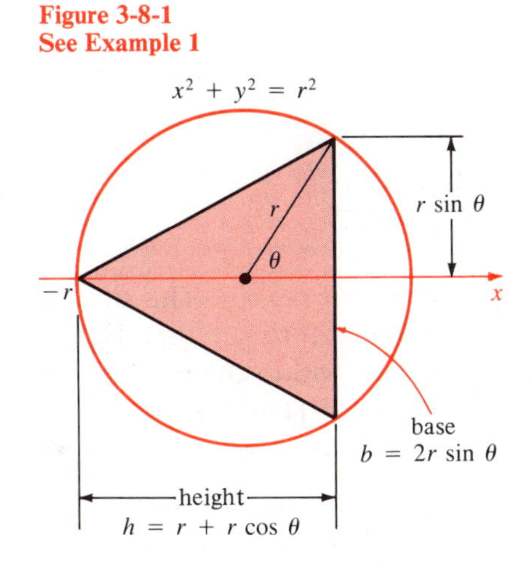

Figure 3-8-1
See Example 1

Solution The first thing is to draw a figure (Figure 3-8-1), choose a variable, and set up the area function. This is usually the hardest part of a problem like this, and worth some time. Choose the apex of the isosceles triangle at $(-r, 0)$ and the central angle θ as the independent variable. Then the base and height of the triangle are

$$b = 2r \sin \theta \quad \text{and} \quad h = r(1 + \cos \theta)$$

The area is

$$A = A(\theta) = \tfrac{1}{2}bh = r^2 \sin \theta (1 + \cos \theta)$$

From the figure it is clear that we need consider only θ such that $0 < \theta < \pi$, and on this domain $A(\theta) > 0$. Also we observe that $A(0) = A(\pi) = 0$, so $A(\theta)$ has no maximum at those points, but at some point where $A'(\theta) = 0$. Now

$$A'(\theta) = r^2[\cos \theta (1 + \cos \theta) - \sin^2\theta]$$

Consequently $A'(\theta) = 0$ if

$$\cos \theta (1 + \cos \theta) = \sin^2\theta = 1 - \cos^2\theta$$

$$2 \cos^2\theta + \cos \theta - 1 = 0$$

This quadratic equation for $\cos \theta$ has solutions

$$\cos \theta = -1 \quad \text{and} \quad \cos \theta = \tfrac{1}{2}$$

Since $0 < \theta < \pi$, only the second value is admissible, and it implies $\theta = \frac{1}{3}\pi$. The maximum area is

$$\begin{aligned} A_{\max} &= A(\tfrac{1}{3}\pi) = r^2(\sin \tfrac{1}{3}\pi)(1 + \cos \tfrac{1}{3}\pi) \\ &= r^2(\tfrac{1}{2}\sqrt{3})(\tfrac{3}{2}) = \tfrac{3}{4}r^2\sqrt{3} \end{aligned}$$

It is easily seen that the corresponding triangle is equilateral. ●

Example 2

A man is in a rowboat 1 mi offshore. His home is 5 mi farther along the straight shoreline. If he can walk twice as fast as he can row, what is his quickest way home?

Figure 3-8-2
See Example 2

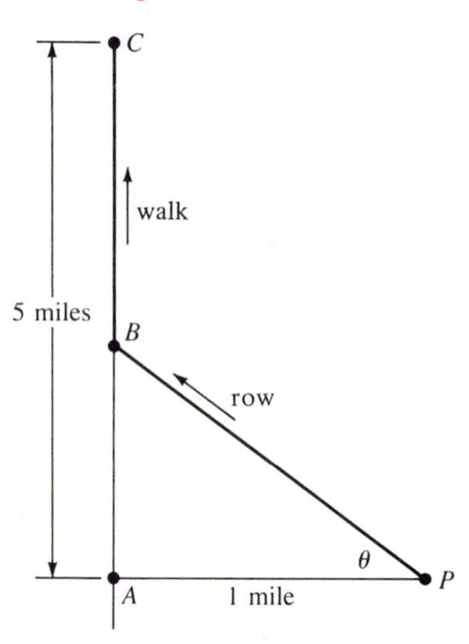

Solution Draw a diagram (Figure 3-8-2). He should row from point P to a point B on the shore, then walk to his home at C. Express everything in terms of the angle θ:

$$\overline{PB} = \sec\theta \quad \text{and} \quad \overline{BC} = 5 - \overline{AB} = 5 - \tan\theta$$

Let v be his rowing speed and $2v$ his walking speed. The time required to reach home is

$$T = T(\theta) = \frac{\overline{PB}}{v} + \frac{\overline{BC}}{2v} = \frac{\sec\theta}{v} + \frac{5 - \tan\theta}{2v}$$

Since B must be between A and C, angle θ is at least 0 and at most the angle whose tangent is 5. Differentiate:

$$\frac{dT}{d\theta} = \frac{\sec\theta\tan\theta}{v} - \frac{\sec^2\theta}{2v}$$

$$= \frac{\sin\theta}{v\cos^2\theta} - \frac{1}{2v\cos^2\theta} = \frac{2\sin\theta - 1}{2v\cos^2\theta}$$

The derivative is 0 if $\sin\theta = \frac{1}{2}$, that is, $\theta = \frac{1}{6}\pi$. The sign of $dT/d\theta$ changes from minus to plus as θ increases through $\frac{1}{6}\pi$. Hence T has its minimum there. Notice that $\frac{1}{6}\pi$ falls within the permissible domain of θ because $\tan\frac{1}{6}\pi = 1/\sqrt{3} < 5$. Therefore the man should row to shore at an angle of $\frac{1}{6}\pi$, then walk the rest of the way. ●

Example 3

At a corner inside a rectangular building, a hall of width a meets a hall of width b. How long a pole can be pushed around the corner while remaining on the floor?

Solution Draw a figure (Figure 3-8-3) and choose a variable θ as indicated. The pole will be as long as possible and just fit around the corner provided it just touches the inside corner P at some point of turning. Therefore, what we really want is the *shortest* segment that has its ends on the outside walls and touches the inside corner.

The length L of the segment is

$$L = L(\theta) = \overline{AP} + \overline{BP} = \frac{a}{\cos\theta} + \frac{b}{\sin\theta}$$

Figure 3-8-3
See Example 3

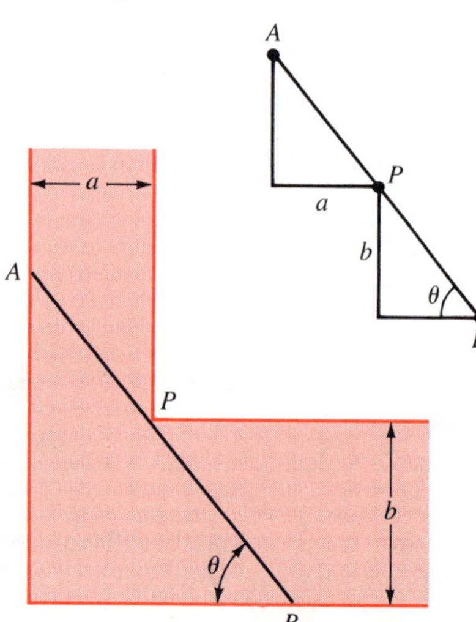

The pole will turn through no more than a right angle, so $0 < \theta < \frac{1}{2}\pi$. Also $L(\theta) \to +\infty$ if $\theta \to 0+$ or $\theta \to \frac{1}{2}\pi-$. Therefore $L(\theta)$ has its minimum somewhere in its domain where $L'(\theta) = 0$. Now

$$L'(\theta) = \frac{a\sin\theta}{\cos^2\theta} - \frac{b\cos\theta}{\sin^2\theta}$$

so $L'(\theta) = 0$ provided that

$$\frac{a\sin\theta}{\cos^2\theta} = \frac{b\cos\theta}{\sin^2\theta}$$

Figure 3-8-4
See Example 4

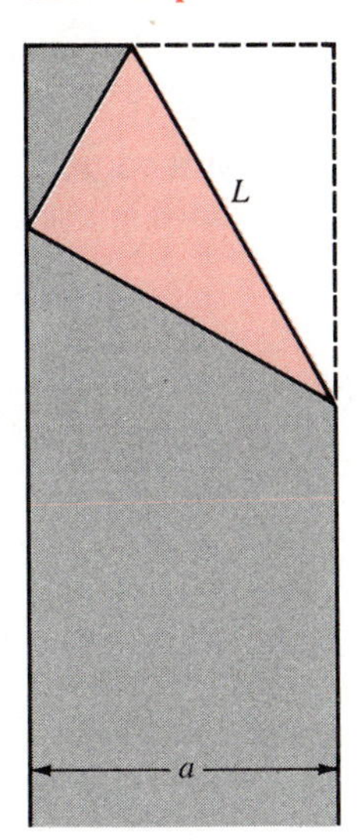

Multiply by $\sin^2\theta/(a\cos\theta)$:

$$\frac{\sin^3\theta}{\cos^3\theta} = \frac{b}{a} \qquad \text{hence} \qquad \tan^3\theta = \frac{b}{a}$$

so that $\tan\theta = (b/a)^{1/3}$.

The function $\tan\theta$ is strictly increasing for $0 < \theta < \frac{1}{2}\pi$, so there is precisely one value of θ with $\tan\theta = (b/a)^{1/3}$. To find the corresponding value of L, we need $\sin\theta$ and $\cos\theta$ expressed in terms of $\tan\theta$. (See Example 1 in Section 1-7.)

$$\begin{cases} \sin\theta = \dfrac{\tan\theta}{\sqrt{\tan^2\theta + 1}} = \dfrac{(b/a)^{1/3}}{\sqrt{(b/a)^{2/3} + 1}} = \dfrac{b^{1/3}}{\sqrt{a^{2/3} + b^{2/3}}} \\ \cos\theta = \dfrac{1}{\sqrt{\tan^2\theta + 1}} = \dfrac{1}{\sqrt{(b/a)^{2/3} + 1}} = \dfrac{a^{1/3}}{\sqrt{a^{2/3} + b^{2/3}}} \end{cases}$$

Consequently

$$L_{\min} = \frac{a}{\cos\theta} + \frac{b}{\sin\theta} = \frac{a\sqrt{a^{2/3} + b^{2/3}}}{a^{1/3}} + \frac{b\sqrt{a^{2/3} + b^{2/3}}}{b^{1/3}}$$
$$= (a^{2/3} + b^{2/3})\sqrt{a^{2/3} + b^{2/3}}$$
$$= (a^{2/3} + b^{2/3})^{3/2}$$

This is the answer we want: the length of the longest pole that can be pushed through the corner without getting stuck. ●

Figure 3-8-5
See Example 4

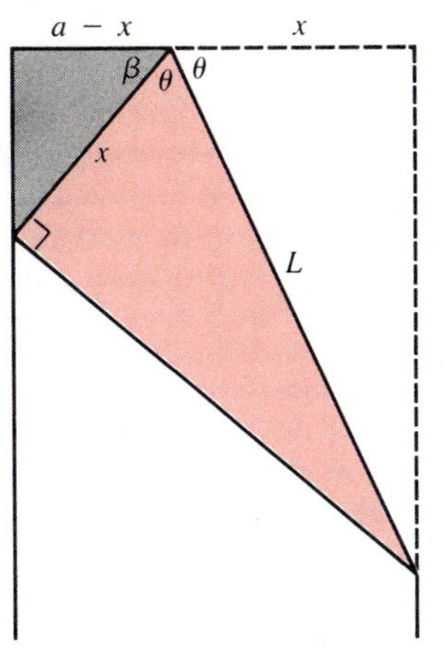

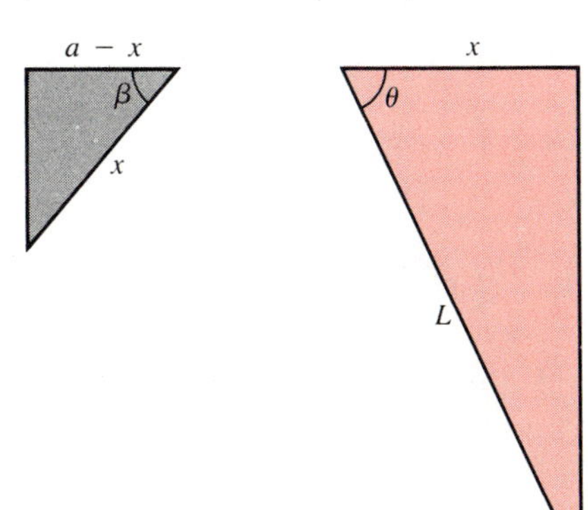

Example 4

One corner of a long rectangular strip of width a is folded over and just reaches the opposite edge (Figure 3-8-4). Find the minimum length L of the crease.

Solution We choose x in Figure 3-8-5 for the independent variable. Our first problem is to express L in terms of x alone. The angles θ and β will be useful, but we'll get rid of them eventually. From the figure we have

$$2\theta + \beta = \pi \qquad \cos\beta = \frac{a - x}{x} \quad \text{and} \quad L = \frac{x}{\cos\theta}$$

From here we use trigonometric identities to eliminate β and θ. First $\beta = \pi - 2\theta$ so

$$\cos\beta = \cos(\pi - 2\theta) = -\cos 2\theta = 1 - 2\cos^2\theta$$

Hence

$$\cos^2\theta = \tfrac{1}{2}(1 - \cos\beta) = \tfrac{1}{2}\left(1 - \frac{a - x}{x}\right) = \frac{2x - a}{2x}$$

We want to avoid taking square roots if possible, so we work with L^2 instead of L. We set

$$f(x) = L^2 = \left(\frac{x}{\cos\theta}\right)^2 = \frac{x^2}{\cos^2\theta} = \frac{2x^3}{2x - a}$$

The domain of $f(x)$ is $(\frac{1}{2}a, a]$; think why. Now $f(x) > 0$ and $f(x) \to +\infty$ as $x \to \frac{1}{2}a+$. Also $f(a) = 2a^2$; this end point value is one candidate for the minimum of $f(x)$. For other candidates we set $f'(x) = 0$ as usual. First

$$\frac{df}{dx} = \frac{6x^2(2x - a) - 4x^3}{(2x - a)^2} = \frac{2x^2}{(2x - a)^2}[3(2x - a) - 2x]$$
$$= \frac{2x^2(4x - 3a)}{(2x - a)^2}$$

Thus $df/dx = 0$ only for $4x - 3a = 0$, that is, for $x = \frac{3}{4}a$. The corresponding value of $f(x)$ is

$$f(\tfrac{3}{4}a) = \frac{2(\frac{3}{4}a)^3}{2(\frac{3}{4}a) - a} = \frac{\frac{27}{32}a^3}{\frac{1}{2}a} = \tfrac{27}{16}a^2$$

Thus the only candidates for the minimum of $f(x)$ are $f(a) = 2a^2$ and $f(\frac{3}{4}a) = \frac{27}{16}a^2$. The smaller one is $f(\frac{3}{4}a)$. Therefore

$$f_{\min} = \tfrac{27}{16}a^2 \quad \text{and} \quad L_{\min} = \sqrt{f_{\min}} = \tfrac{3}{4}a\sqrt{3}$$ ●

Now we consider some extremum problems in space.

Figure 3-8-6
See Example 5

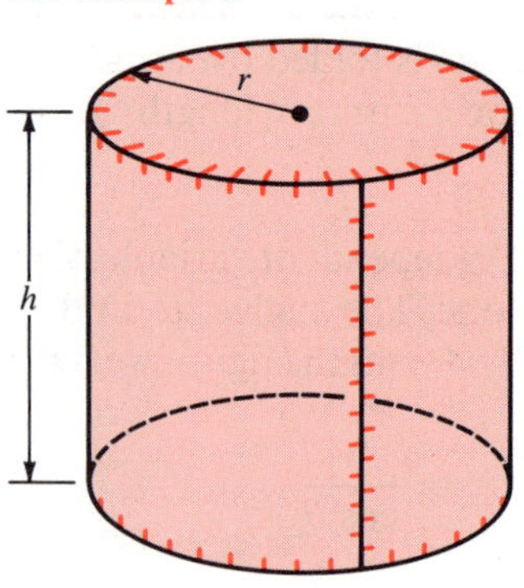

Example 5

A closed tank is a right circular cylinder in shape. It is made from three sheets with three welded seams (Figure 3-8-6). Because the seams may leak eventually, the total length of the seams should be made as short as possible. Assuming the tank will have volume 1000 liters, find the radius and height in meters.

Solution First of all, 1000 liters $= 1\ \text{m}^3$, and the volume of a right circular cylinder is $V = \pi r^2 h$, hence

$$\pi r^2 h = 1 \qquad \text{so that} \qquad h = \frac{1}{\pi r^2}$$

The total seam length (top, bottom, and side welds) is

$$L = 2\pi r + 2\pi r + h = 4\pi r + h = 4\pi r + \frac{1}{\pi r^2}$$

The domain of $L = L(r)$ is $(0, +\infty)$; thus the only critical points are stationary points. Set $dL/dr = 0$ to find the extrema:

$$\frac{dL}{dr} = 4\pi - \frac{2}{\pi r^3}$$

This derivative is a strictly increasing function of r, and is zero only if

$$\frac{2}{\pi r^3} = 4\pi \qquad \text{that is} \qquad r^3 = \frac{1}{2\pi^2} \quad \text{so} \quad r = \frac{1}{(2\pi^2)^{1/3}}$$

Let us give this value of r the name c. Then $dL/dr < 0$ for $r < c$ and $dL/dr > 0$ for $r > c$. Hence L is strictly decreasing for $r < c$ and strictly increasing for $r > c$. Consequently $L(c)$ is the minimum of $L(r)$. The corresponding h is

$$h = \frac{1}{\pi c^2} = \frac{(2\pi^2)^{2/3}}{\pi} = (4\pi)^{1/3}$$

The dimensions of the tank are

$$\text{radius} = \frac{1}{(2\pi^2)^{1/3}} \approx 0.370 \text{ m} \qquad \text{height} = (4\pi)^{1/3} \approx 2.325 \text{ m}$$ ●

Remark It is a better practice to carry along the constant V until the end, and then substitute $V = 1 \text{ m}^3$. Doing so gives a check by physical dimension. In this case we would have found

$$r = \frac{1}{(2\pi^2)^{1/3}} V^{1/3} \quad \text{and} \quad h = (4\pi)^{1/3} V^{1/3}$$

These relations check physically because V, a volume, has physical dimension $(\text{length})^3$, so $V^{1/3}$ has dimension (length).

Example 6

A conical drinking cup is made from a circular sheet of radius r by cutting out a central sector and bringing the edges together (Figure 3-8-7). Find the maximum volume of the cup.

Solution First we need to know the volume of a right circular cone of height h and base radius x: It is $V = \frac{1}{3}\pi x^2 h$. The cup has fixed slant height r. Its base radius x and its height h are related by $x^2 + h^2 = r^2$. Hence

$$V = V(x) = \tfrac{1}{3}\pi x^2 h = \tfrac{1}{3}\pi x^2 \sqrt{r^2 - x^2}$$

Figure 3-8-7
See Example 6

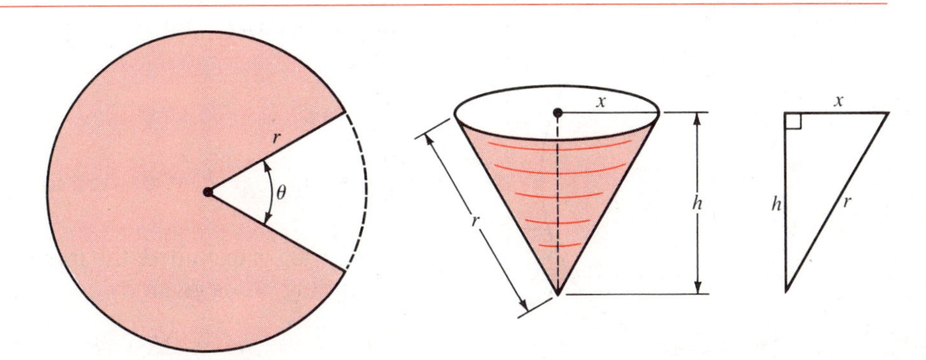

We seek x in the domain $[0, r]$. Now $V(x)$ is continuous on $[0, r]$, and $V(x) > 0$ on $(0, r)$ but $V(0) = V(r) = 0$. Therefore the function $V(x)$ must have its maximum in $(0, r)$ and $V'(x) = 0$ at the maximum. Now

$$V'(x) = \tfrac{1}{3}\pi\left(2x\sqrt{r^2 - x^2} - \frac{x^3}{\sqrt{r^2 - x^2}}\right)$$

$$= \tfrac{1}{3}\pi\left(\frac{2x(r^2 - x^2) - x^3}{\sqrt{r^2 - x^2}}\right) = \tfrac{1}{3}\pi\frac{x(2r^2 - 3x^2)}{\sqrt{r^2 - x^2}}$$

We see that $V'(x) = 0$ with $0 < x < r$ only for $3x^2 = 2r^2$, that is, $x^2 = \frac{2}{3}r^2$. Therefore

$$V_{\max} = \tfrac{1}{3}\pi x^2\sqrt{r^2 - x^2} = \tfrac{1}{3}\pi(\tfrac{2}{3}r^2)\sqrt{r^2 - \tfrac{2}{3}r^2}$$

$$= \tfrac{2}{9}\pi r^2\sqrt{\tfrac{1}{3}r^2} = \tfrac{2}{27}\pi r^3\sqrt{3}$$ ●

Figure 3-8-8
Cone inscribed in sphere

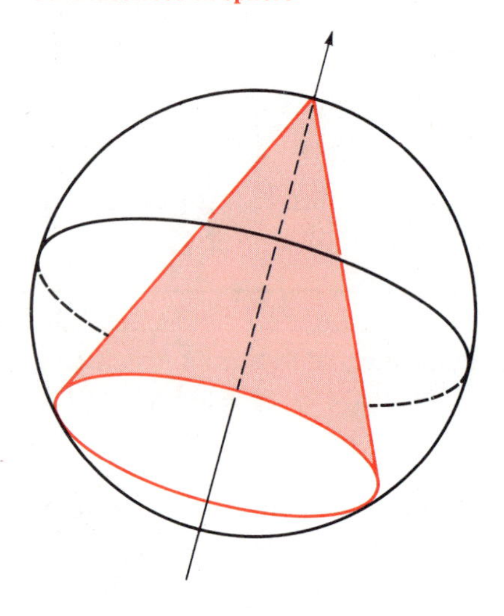

Example 7

Find the right circular cone of largest volume inscribed in a sphere of radius R. See Figure 3-8-8.

Solution The volume V of a cone is

$$V = \tfrac{1}{3}\pi r^2 h$$

where r is the radius of its base and h is its height. Because the cone is inscribed in a sphere, there must be a relation between r, h, and R. Make a careful drawing of a cross-section (Figure 3-8-9). From the drawing, $r^2 + (h - R)^2 = R^2$; hence

$$r^2 = R^2 - (h - R)^2 = 2Rh - h^2$$

Substitute this expression in V to eliminate r:

$$V = \tfrac{1}{3}\pi r^2 h = \tfrac{1}{3}\pi(2Rh - h^2)h = \tfrac{1}{3}\pi(2Rh^2 - h^3)$$

Since $V = 0$ for $h = 0$ and for $h = 2R$, you must maximize the positive function $V(h) = \frac{1}{3}\pi(2Rh^2 - h^3)$ in the interval $0 < h < 2R$. The maximum occurs at a zero of the derivative:

Figure 3-8-9
Cross section of Figure 3-8-8

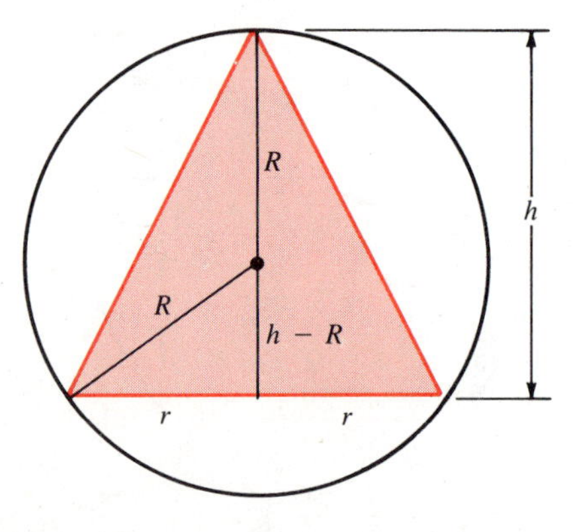

$$\frac{dV}{dh} = \tfrac{1}{3}\pi(4Rh - 3h^2) = \tfrac{1}{3}\pi h(4R - 3h)$$

Hence

$$\frac{dV}{dh} = 0 \quad \text{for} \quad h = 0 \quad \text{and for} \quad h = \tfrac{4}{3}R$$

Since $h = 0$ is excluded, the maximum must occur for $h = \frac{4}{3}R$. The corresponding r is given by

$$r^2 = 2Rh - h^2 = 2R \cdot \tfrac{4}{3}R - (\tfrac{4}{3}R)^2 = \tfrac{8}{9}R^2$$

Consequently $r = \frac{2}{3}R\sqrt{2}$ and

$$V_{\max} = \tfrac{1}{3}\pi r^2 h = \tfrac{1}{3}\pi(\tfrac{8}{9}R^2)(\tfrac{4}{3}R) = \tfrac{32}{81}\pi R^3$$ ●

Remark The answer checks dimensionally: (volume) = (length)^3. Since the sphere has volume $\frac{4}{3}\pi R^3$, it follows that the largest cone inscribed in a sphere contains $\frac{8}{27}$ of the volume of the sphere.

Snell's Law

Fermat's principle of least time states that light traveling between two points in a transparent substance will take the path that requires the least time.

Example 8

Assume the upper half of the x, y-plane is a substance in which the speed of light is v_1 and the lower half is another substance in which the speed of light is v_2. Describe the path of a light ray traveling between two points in opposite halves of the plane.

Figure 3-8-10
See Example 8

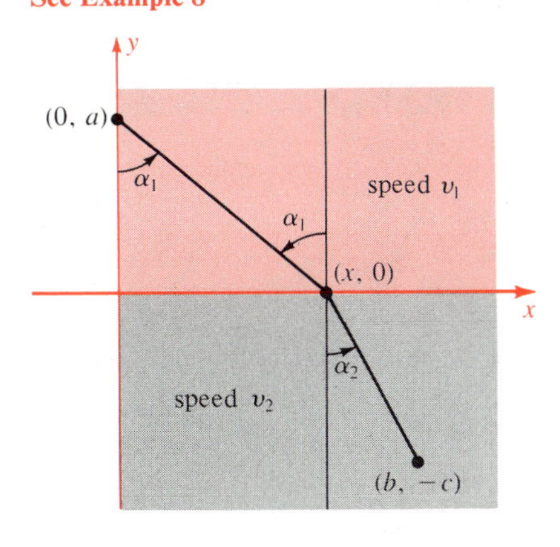

Solution First note that within each substance, the least-time path is the shortest-distance path, that is, a straight line. Let the two points be $(0, a)$ and $(b, -c)$ in Figure 3-8-10. A ray will travel from $(0, a)$ along a straight line to some point $(x, 0)$ and then along another straight line to $(b, -c)$. A value x must be found so that the time of travel is a minimum.

The time required for a ray to travel from $(0, a)$ to $(x, 0)$ is

$$T_1 = \frac{\text{distance}}{\text{speed}} = \frac{\sqrt{x^2 + a^2}}{v_1}$$

The time required from $(x, 0)$ to $(b, -c)$ is

$$T_2 = \frac{\sqrt{(b - x)^2 + c^2}}{v_2}$$

We must minimize

$$T = T_1 + T_2 = \frac{\sqrt{x^2 + a^2}}{v_1} + \frac{\sqrt{(b - x)^2 + c^2}}{v_2}$$

Figure 3-8-11
See Example 8

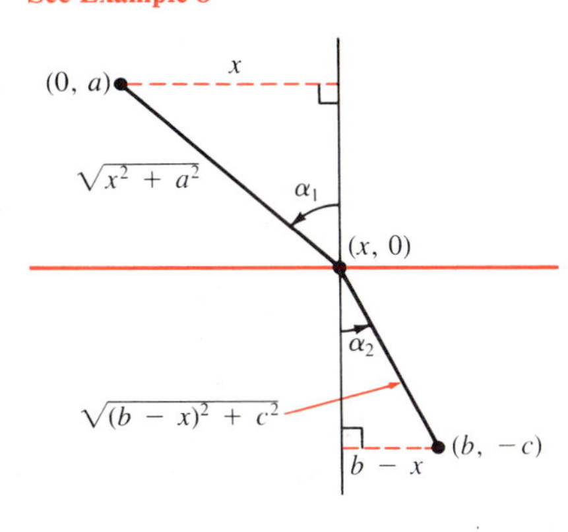

The function $T = T(x)$ is differentiable for all x. We compute dT/dx:

$$\frac{dT}{dx} = \frac{x}{v_1\sqrt{x^2 + a^2}} - \frac{b - x}{v_2\sqrt{(b - x)^2 + c^2}}$$

This derivative looks complicated, but from Figure 3-8-11,

$$\frac{x}{\sqrt{x^2 + a^2}} = \sin\alpha_1 \quad \text{and} \quad \frac{b - x}{\sqrt{(b - x)^2 + c^2}} = \sin\alpha_2$$

Hence the derivative has the simple form

$$\frac{dT}{dx} = \frac{\sin \alpha_1}{v_1} - \frac{\sin \alpha_2}{v_2}$$

The derivative is zero provided that x is chosen to satisfy

$$\frac{\sin \alpha_1}{v_1} = \frac{\sin \alpha_2}{v_2}$$

This equation is known as **Snell's law of refraction.** To see that it describes the path of *least* time, note that dT/dx is the difference of two terms. As x increases from $-\infty$ to $+\infty$, the first term $(\sin \alpha_1)/v_1$ increases steadily and the second term $(\sin \alpha_2)/v_2$ decreases steadily. Consequently, dT/dx steadily increases. Therefore, the unique minimum T occurs at the x for which $dT/dx = 0$.

Answer The path is the broken line for which

$$\frac{\sin \alpha_1}{v_1} = \frac{\sin \alpha_2}{v_2}$$

Exercises

1 A triangle has two of its vertices on a fixed circle of radius r and its third vertex at the center of the circle. Find the largest possible area that the triangle can have.

2 A page is to contain 27 in^2 of print. The margins at the top and bottom are 1.5 in, at the sides 1 in. Find the most economical dimensions of the page, that is, minimize its area.

3 Find the largest possible area of an isosceles triangle whose equal legs have length L ft.

4 Find the dimensions of the rectangle of maximum area that can be inscribed in the region bounded by the parabola $y = -8x^2 + 16$ and the x-axis, provided that one side of the rectangle is on the x-axis.

5 An athletic field of 400-meter perimeter consists of a rectangle with a semicircle at each end. Find the dimensions of the field so that the area of the rectangular portion is the largest possible.

6 Find the dimensions of the rectangle of largest area that can be inscribed in an equilateral triangle of side s, if one side of the rectangle lies on the base of the triangle.

7 Find the dimensions of the rectangle of largest area that is inscribed in a right triangle with legs a and b and has one corner at the right angle of the triangle.

8 (cont.) The same, except one edge of the rectangle lies on the hypotenuse of the triangle. Which rectangle has the larger area, the one in Exercise 7 or this one?

9 Describe the isosceles triangle of smallest area that circumscribes a circle of radius r. [Hint Use the angle θ in Figure 3-8-12 as the independent variable.]

Figure 3-8-12
See Exercise 9

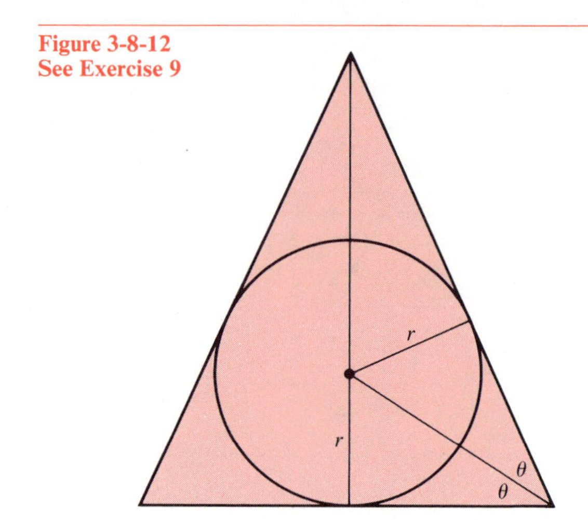

10 Find the largest area of a rectangle inscribed in a semicircle of radius r, with one side of the rectangle on the diameter.

11 Find the largest area of a trapezoid (Figure 3-8-13) with slant sides a and one base b.

Figure 3-8-13
See Exercise 11

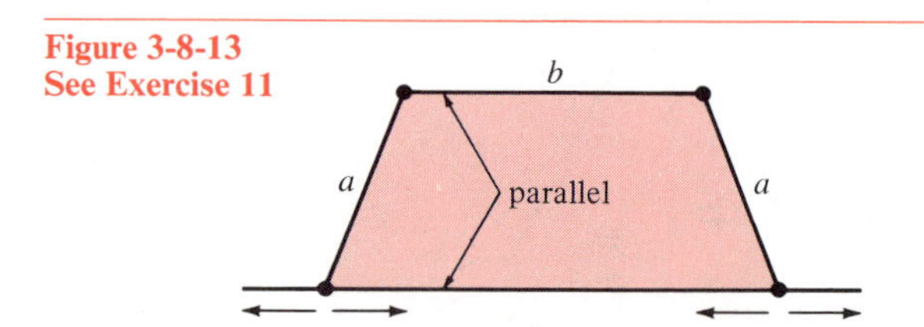

2 Among all isosceles triangles of fixed perimeter p, which has the largest area?

3 A window of perimeter 16 ft has the form of a rectangle topped by a semicircle. For what radius of the semicircle is the window area greatest?

4 (cont.) For what radius of the semicircle is the most light admitted, if the semicircle admits half as much light per unit area as the rectangle admits per unit area?

5 In Figure 3-8-14, the vertical lines are parallel. Find X on or below C so the shaded area is as small as possible. Express $\overline{CX}$ in terms of a and b.
[Hint Look for similar triangles.]

Figure 3-8-14
See Exercise 15.
A, B, and C are fixed.

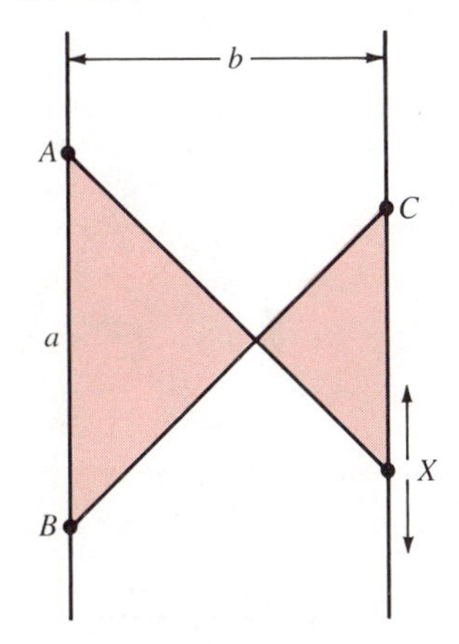

16 Through each point P of the parabola $y = x^2$ a normal line is drawn (a line perpendicular to the tangent at P). It meets the parabola in another point Q. Minimize $\overline{PQ}$.

17 The strength of a beam of fixed length and rectangular cross section is proportional to the width and to the square of the depth of the cross section. Find the proportions of the beam of greatest strength that can be cut from a circular log of radius r. See Figure 3-8-15.

18 (cont.) The *stiffness* of the beam (resistance to sagging) is proportional to the width and to the cube of the depth. Find the proportions for greatest stiffness.

Figure 3-8-15
See Exercise 17

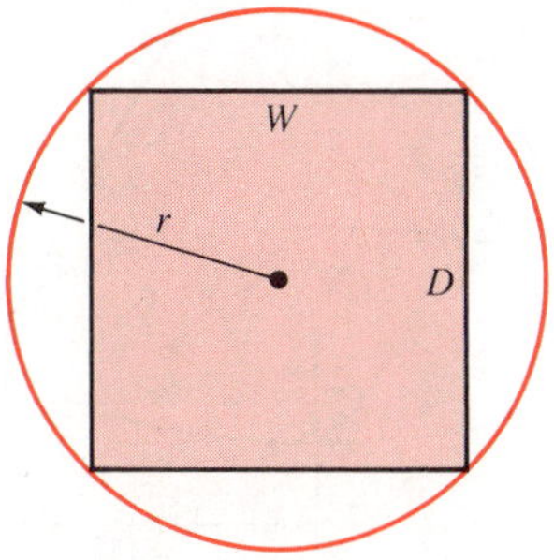

19 An 8-foot ladder leans against the top of a 4-foot fence. Find the largest horizontal distance the ladder can reach beyond the fence.

20 Suppose in Example 3 that the pole can be tilted so its ends touch the floor and the ceiling, which is c ft from the floor. Now how long can the pole be?

21 A woman in a rowboat 3 mi off a long straight shore wants to reach a point 5 mi up the shore. If she can row 2 mph and walk 4 mph, describe her fastest route.

22 (cont.) Suppose the boat has a motor. How fast must the boat be able to go so that the fastest route is entirely by boat?

23 A tinsmith makes a pyramid with a square base, starting with a square of sheet metal of side a and cutting away the corners as indicated in Figure 3-8-16. Maximize the volume of the pyramid. (A pyramid of base area B and height h has volume $V = \frac{1}{3}Bh$.)

Figure 3-8-16
See Exercise 23

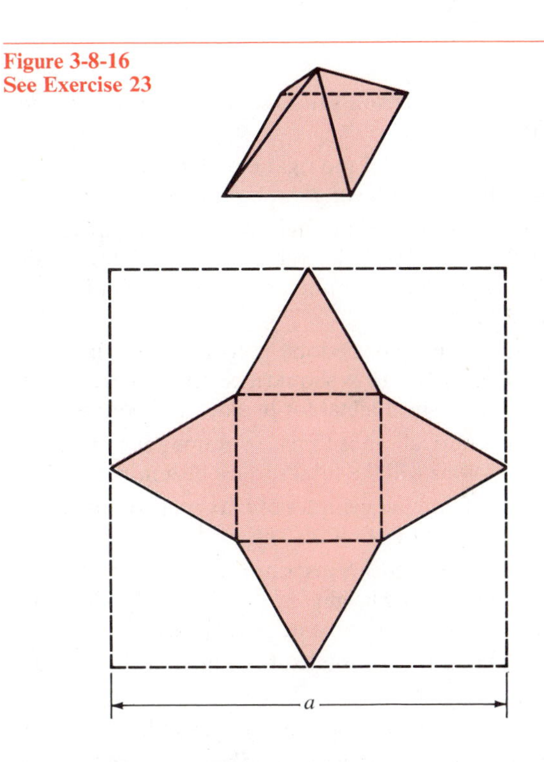

24 One end of a rod of length a is hinged at the origin (Figure 3-8-17). The other end is hinged to one end of a rod of length b. Finally, the free end of the second rod is free to slide along the x-axis. Maximize the shaded area.

Figure 3-8-17
See Exercise 24

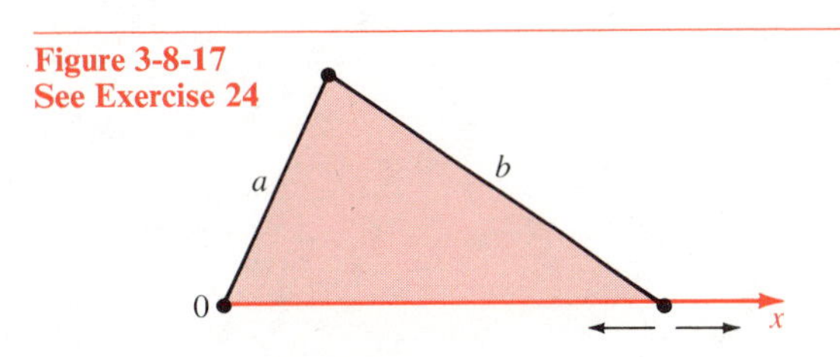

25 A light bulb of fixed strength is free to move on the vertical z-axis (Figure 3-8-18). Because of the inverse square law and the tilt of the beam of light relative to the vertical, the light intensity per unit area on the floor at fixed distance a from the z-axis is

$$I = \frac{kz}{(a^2 + z^2)^{3/2}}$$

where k is a constant. Maximize I and find where the maximum occurs.

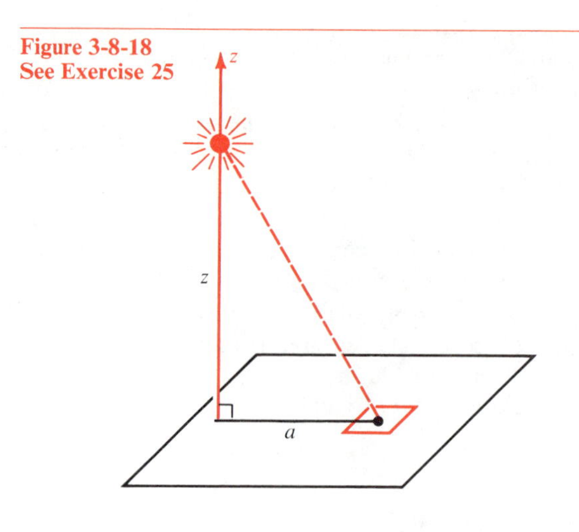

Figure 3-8-18
See Exercise 25

26 A steady electric current flows in a circular wire loop (Figure 3-8-19) of radius r. Through the center of the circle, a z-axis is drawn perpendicular to the plane of the current loop. According to Ampere's law, the current generates a magnetic field. The force exerted by this field on a unit magnetic dipole along the z-axis at distance z from the plane is

$$F = k\frac{z}{(a^2 + z^2)^{5/2}}$$

where k is a constant. Find the maximum of F and where it occurs.

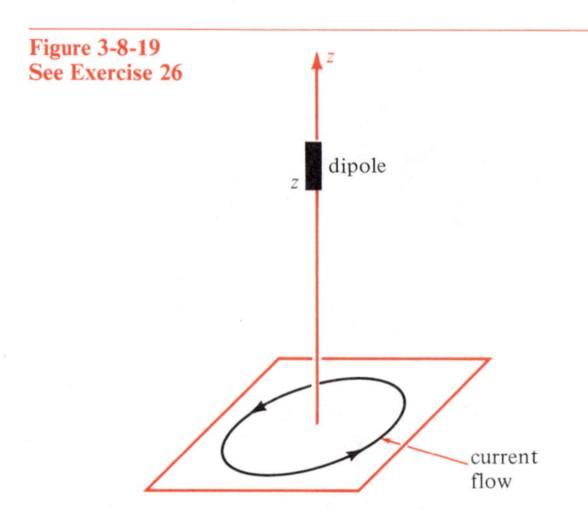

Figure 3-8-19
See Exercise 26

27 Find the dimensions of the right circular cone having the greatest volume for a given slant height a.

28 Find the dimensions of the right circular cone of volume V with the least possible lateral surface area. (If the radius of the base is r and the *slant* height is s, then the lateral area is $A = \pi rs$.)

29 A cylindrical tank (open top) is to hold V liters (1 liter = 1000 cm^3). How should it be made so as to use the least amount of sheet metal for its sides and bottom?

30 A closed cylindrical can is to have volume 500 cm^3. For what dimensions will the total surface be a minimum?

31 Find the right circular cylinder of maximal volume that is inscribed in a sphere of radius r.

32 A right circular cylinder is inscribed in a right circular cone of base radius r and height h. See Figure 3-8-20. Maximize the curved surface area of the cylinder.

33 (cont.) Maximize the total surface area of the cylinder, assuming $h > 2r$.

34 (cont.) Maximize the volume of the cylinder.

35 An isosceles triangle is circumscribed about a circle of radius a. Minimize its area.

36 A right circular cone is circumscribed about a sphere of radius a. Minimize its volume.

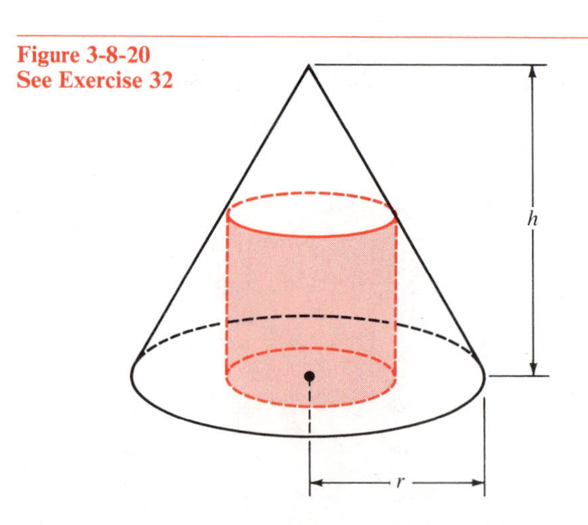

Figure 3-8-20
See Exercise 32

3-9 Second Derivative Test

If a derivative $f'(x)$ changes from positive to negative as x passes through $x = c$, then $f(x)$ has a local maximum at $x = c$; that is what the first derivative test says. Now one way to be sure that $f'(x)$ changes sign from $+$ to $-$ as x passes over a point where $f'(x) = 0$ is to have the second derivative $f''(c) < 0$. For then we know that $f'(x) > f'(c) = 0$ slightly to the left of $x = c$ and $f'(x) < f'(c) = 0$ slightly to the right. That is a consequence of Theorem 2, page 169, applied to $f'(x)$. This reasoning proves the following useful test.

Figure 3-9-1

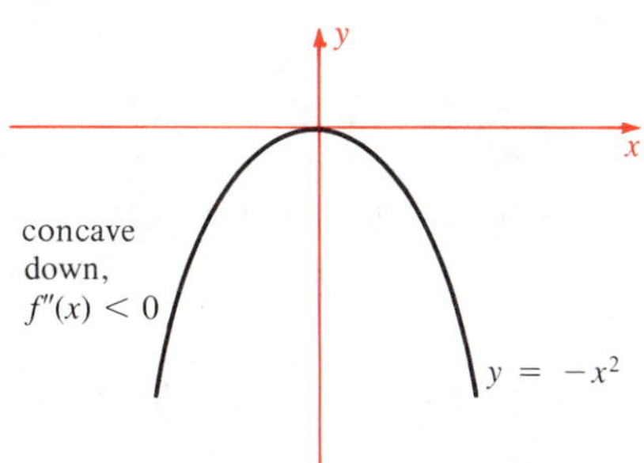

Second Derivative Test

Suppose that $f(x)$ is a differentiable function in the open interval (a, b) and that $f'(c) = 0$, where $a < c < b$. Suppose $f''(c)$ exists.

- If $f''(c) < 0$, then $f(x)$ has a local maximum at $x = c$.
- If $f''(c) > 0$, then $f(x)$ has a local minimum at $x = c$.

Hint To keep the signs straight in your mind, remember that $f''(x) < 0$ goes with concave down (like $y = -x^2$) and $f''(x) > 0$ goes with concave up (like $y = x^2$). See Figure 3-9-1.

Remark In the case $f'(c) = 0$ and $f''(c) = 0$, the theorem tells us nothing. For instance look at $y = x^3$, $y = x^4$, and $y = -x^4$. In all three cases $f'(0) = f''(0) = 0$. The first has a horizontal inflection, the second a local minimum, the third a local maximum at $x = 0$. See Figure 3-9-2.

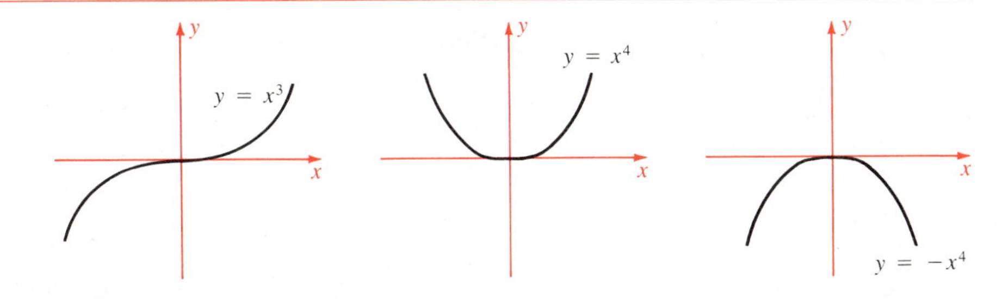

Figure 3-9-2
For all of these functions, $f''(0) = 0$.

Example 1

Use the second derivative test to locate all local maxima and minima of

$$f(x) = 3x^5 - 50x^3 + 135x + 15$$

Solution Because the first term dominates the polynomial as $x \to \pm\infty$, it follows that $f(x) \to +\infty$ as $x \to +\infty$ and $f(x) \to -\infty$ as $x \to -\infty$. Therefore $f(x)$ does not have an absolute maximum or minimum on $\mathbf{R}$. To find local extrema, differentiate $f(x)$ and solve $f'(x) = 0$:

$$\begin{aligned} f'(x) &= 15x^4 - 150x^2 + 135 \\ &= 15(x^4 - 10x^2 + 9) \\ &= 15(x^2 - 1)(x^2 - 9) \\ &= 15(x + 1)(x - 1)(x + 3)(x - 3) \end{aligned}$$

Consequently

$$f'(x) = 0 \quad \text{for} \quad x = -1, \quad x = 1, \quad x = -3, \quad \text{and} \quad x = 3$$

Next, compute $f''(x)$ and note its sign at the four critical points:

$$f''(x) = 15(4x^3 - 20x) = 60x(x^2 - 5)$$

$$f''(-1) = (60)(-1)(-4) > 0 \qquad f''(1) = (60)(1)(-4) < 0$$

$$f''(-3) = (60)(-3)(4) < 0 \qquad f''(3) = (60)(3)(4) > 0$$

By the second derivative test, $f(x)$ has local maxima at $x = 1$ and $x = -3$ and local minima at $x = -1$ and $x = 3$. ●

Example 2

Use the second derivative test to find all local maxima and minima of

$$f(x) = 2 \sin x + x$$

Solution Differentiate twice:

$$f'(x) = 2 \cos x + 1 \qquad f''(x) = -2 \sin x$$

Solve $f'(x) = 0$:

$$2 \cos x + 1 = 0 \qquad \cos x = -\tfrac{1}{2}$$

Hence

$$\begin{aligned} x &= \tfrac{2}{3}\pi + 2\pi n \quad \text{(2nd quadrant)} \\ x &= -\tfrac{2}{3}\pi + 2\pi n \quad \text{(3rd quadrant)} \end{aligned}$$

are all solutions of $f'(x) = 0$. Now test the second derivative:

$$\begin{aligned} f''(\tfrac{2}{3}\pi) &= -2 \sin(\tfrac{2}{3}\pi) < 0 \\ f''(-\tfrac{2}{3}\pi) &= -2 \sin(-\tfrac{2}{3}\pi) > 0 \end{aligned}$$

Conclusion

$$f(x) \quad \text{has a} \quad \begin{cases} \text{local max} & \text{at} \quad x = \tfrac{2}{3}\pi + 2\pi n \\ \text{local min} & \text{at} \quad x = -\tfrac{2}{3}\pi + 2\pi n \end{cases}$$

A rough graph (Figure 3-9-3) confirms this. ●

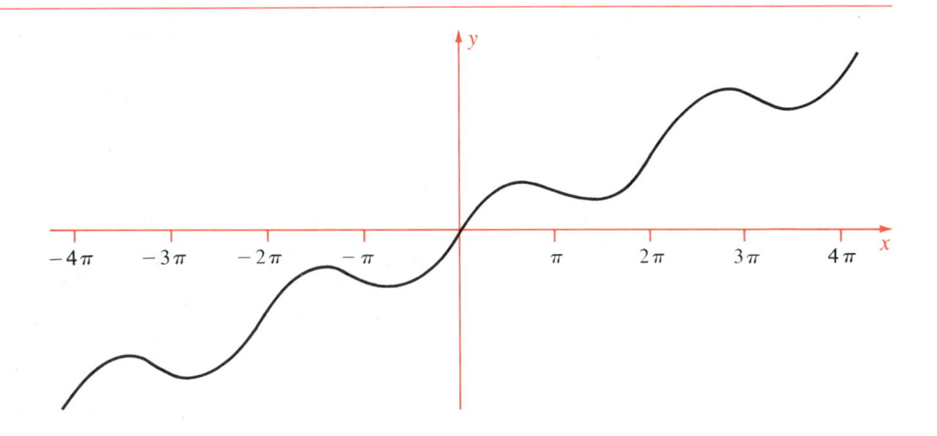

Figure 3-9-3
Local maxima and minima of $y = 2 \sin x + x$

Remarks on Finding Maxima and Minima In most maximum and minimum problems involving a differentiable function, there are only one or two zeros of the derivative to consider, and possibly the end points. Often you can rule out end points by the meaning of the problem. Then you have to decide which zero of the derivative gives the maximum or the minimum. If it is easy to compute the second derivative, do so. If not, or if the second derivative is zero, try observing the sign of the derivative near the point in question. Better yet, graph the function if that is easy. Be flexible.

Exercises

Use the second derivative test to locate all local maxima and minima

1 $y = x^4 - 2x^2$

2 $y = x^5 - 20x - 3$

3 $y = 3x^4 - 4x^3 - 12x^2 + 2$

4 $y = x^3 + 4x^2 + 5x + 2$

5 $y = 2x - 27/x^2$

6 $y = 2x + 27/x^2$

7 $y = 1/x^2 - 1/x^3$

8 $y = x^2 - 1/x$

9 $y = \dfrac{x^2}{1 + x^4}$

10 $y = \dfrac{x^3}{1 + x^4}$

11 $y = \dfrac{x}{\sqrt{x^4 + 16}}$

12 $y = \dfrac{\sqrt{x}}{5x + 4}$

Let $n \geq 2$. Find the maximum for $x \geq 0$

3 $y = \dfrac{x}{(x + 2)^n}$

14 $y = \dfrac{x}{(1 + x^2)^n}$

5 Assume $a > 0$ and $b > 0$. Find all local maxima and minima of $y = (a + x)\sqrt{b^2 + (a - x)^2}$. Consider various cases depending on a and b.

6 Prove that the maximum of

$$y = \frac{(1 + x)^2}{1 + \sqrt{1 + x^4}}$$

is taken at $x = c$ where $c^3 - c - 2 = 0$. Then show that $c \approx 1.5214$ and $y_{\max} = (1 + c)^2/(2 + c) \approx 1.8054$.

17 Given that $f''(c) = 0$ and $f'''(c) > 0$, draw a conclusion about the graph $y = f(x)$ near $x = c$.

18 Find the two positive numbers x and y for which $x + y = 1$ and such that x^3y^4 is maximum.

19 Find the maximum for $x \geq 0$ of $x/(1 + x)^2$.

20 Find the maximum and minimum of $(1 + x^2)/(1 + x^4)$.

21 Given n numbers $a_1, a_2, \cdots, a_n$, show that

$$(x - a_1)^2 + (x - a_2)^2 + \cdots + (x - a_n)^2$$

is least when $x = \bar{a}$, the average of the numbers.

***22** Given n numbers $a_1, \cdots, a_n$ such that

$$\sum_1^n a_i = 0 \quad \text{and} \quad \sum_1^n a_i^3 = 0$$

set $f(x) = \sum_1^n (x - a_i)^4$. Find the minimum value of $f(x)$ and prove it is so.

3-10 Basic Theory and Proofs [Optional]

From time to time in this calculus course we must use certain fundamental properties of continuous functions, such as Theorem 1, page 166. Mostly, we have to omit their proofs, which are hard and belong in theory courses, like Real Analysis. These proofs depend on a property of the real number system called *completeness.* This property can be stated in several equivalent ways, each more or less easy to believe. In a particular situation, one form of completeness may seem easier to apply than the others. What completeness says is that there are no holes in the real number system, as there are in the rational number system.

For instance, define two "intervals" of rational numbers by

$$\begin{cases} S = \{x \mid x \text{ is rational and } x < \sqrt{2}\} \\ T = \{x \mid x \text{ is rational and } x > \sqrt{2}\} \end{cases}$$

Then S is the set of all rationals to the left of $\sqrt{2}$ and T is the set of all rationals to the right of $\sqrt{2}$. Each rational number is either in S or in T. The irrational number $\sqrt{2}$ is between S and T; it represents a "hole" in the rational number system (Figure 3-10-1).

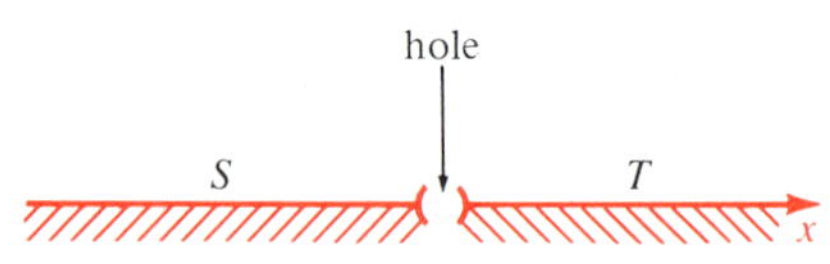

Figure 3-10-1 The rational line has a hole at $\sqrt{2}$.

There are ways to *construct* the real number system from the rational number system. These involve much work, but as a result, the completeness property of the real number system can be *proved.* Instead, we simply consider the completeness property as an axiom—a "self-evident truth," which we accept on faith and use. Our first version is exactly the "no holes" idea mentioned already.

Completeness Axiom (First Version)

Let S and T be non-empty sets of real numbers such that:

- Each real number either belongs to S or belongs to T.
- If s is in S and t is in T then $s < t$.

Then there exists a real number c such that

$$s \le c \le t \quad \text{for all } s \text{ in } S \text{ and } t \text{ in } T$$

As we said, there are equivalent forms of this completeness axiom; we state one other version which, perhaps, is more "self-evident" than the first version:

Completeness Axiom (Second Version)

Let S be a non-empty set of real numbers that is bounded above. That is, assume there is a real number b such that $s \le b$ for all s in S. Then there exists a real number c, called the **least upper bound** of S such that:

- $s \le c$ for all s in S.
- If x is any real number such that $s \le x$ for all s in S, then $c \le x$.

Let us rephrase this version in terms of intervals. If S is a set of real numbers, an **upper bound** for S is any real number b such that $s \le b$

for all s in S. A set may not have any upper bound at all. The real number system **R** itself is such a set; the set **Z** of integers is another example. However, if a set S does have at least one upper bound, the completeness property says that the set of *all* of its upper bounds is an interval of the form $[c, +\infty)$, *including* the left-hand end point.

Let us review quickly the fundamental property of continuous functions in Section 3-4. It is Theorem 1, which says: *The range of a continuous function with domain a closed interval is itself a closed interval or a single point.* This theorem is a consequence of the completeness axiom, but its proof is too hard for a first calculus course.

Concave Functions

We postponed the proofs of two theorems in Section 3-5, the tangent theorem and the chord theorem. Let us start with the **tangent theorem.** We are given a function $f(x)$ that is concave up on an interval, and a point c of the interval such that $f'(c)$ exists. The conclusion is that the graph of $y = f(x)$ is above the tangent line to the graph at $(c, f(c))$. We shall prove this conclusion in two steps.

Figure 3-10-2
Horizontal tangent: $f'(c) = 0$

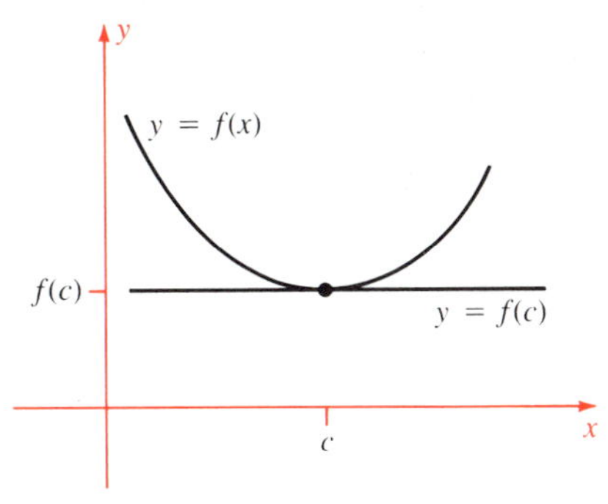

We start with the special case $f'(c) = 0$ as shown in Figure 3-10-2. In this case, the tangent line at $(c, f(c))$ is the horizontal line $y = f(c)$. Consequently we must prove that $f(x) > f(c)$ for $x \neq c$. Now *concave up* means that $f'(c)$ is strictly increasing. But $f'(c) = 0$, so we must have

$$f'(x) < 0 \quad \text{for} \quad x < c \quad \text{and} \quad f'(x) > 0 \quad \text{for} \quad x > c$$

Therefore $f(x)$ is strictly decreasing for $x < c$ and $f(x)$ is strictly increasing for $x > c$ by Corollary 2, page 173. Accordingly, we have $f(x) > f(c)$ for $x < c$ and $f(x) > f(c)$ for $x > c$. This completes the proof in the special case under consideration.

In the general case (where $f'(c)$ is not necessarily 0) we use a trick: we subtract a linear function from $f(x)$. We set

$$g(x) = f(x) - [f'(c)]x$$

Then $g'(x) = f'(x) - f'(c)$ so $g'(x)$ is a strictly increasing function, just as $f'(x)$ is. This means that $g(x)$ is concave up. But also $g'(c) = 0$, so we are back in the special case, and we conclude that $g(x) > g(c)$ if $x \neq c$. However, this says that

$$f(x) - [f'(c)]x > f(c) - [f'(c)]c$$

that is

$$f(x) > f'(c)(x - c) + f(c) \quad \text{for} \quad x \neq c$$

But the equation of the tangent to the graph at $(c, f(c))$ is

$$y = f'(c)(x - c) + f(c)$$

so the proof is complete; the graph is indeed above the tangent. ●

Figure 3-10-3

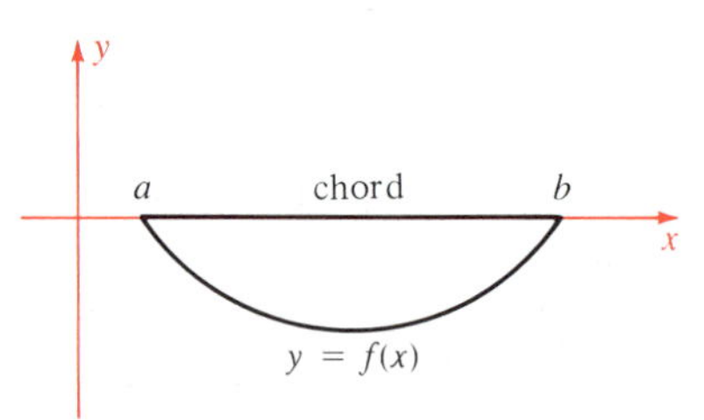

The **chord theorem** starts with a function $f(x)$ that is concave up on an interval, and with three points $a < c < b$ of this interval. The conclusion is that $(c, f(c))$ lies below the chord joining $(a, f(a))$ to $(b, f(b))$.

Again we start the proof with a special case. We suppose that $f(a) = f(b) = 0$. Then the chord is simply the segment $[a, b]$ of the x-axis, and we have to prove that $f(c) < 0$. See Figure 3-10-3. Suppose, on the contrary, that $f(c) \geq 0$. By the MVT, there is a point x_1 in the interval (a, c) such that

$$f'(x_1) = \frac{f(c) - f(a)}{c - a} = \frac{f(c)}{c - a} \geq 0$$

Similarly, there is a point x_2 in the interval (c, b) such that

$$f'(x_2) = \frac{f(b) - f(c)}{b - c} = \frac{-f(c)}{b - c} \leq 0$$

Therefore both $x_1 < x_2$ and $f'(x_1) \geq 0 \geq f'(x_2)$. But $f(x)$ is concave up, which means that $f'(x)$ is a *strictly increasing* function, so we have a contradiction. This finishes off the special case.

In the general case, we subtract a suitable linear function (the chord) from $f(x)$. We let

$$y = h(x) = mx + p$$

be the equation of the chord. Then $h(a) = f(a)$ and $h(b) = f(b)$. (We could express the constants m and p in terms of a, b, $f(a)$, and $f(b)$, but these details really do not matter.) We define

$$\begin{aligned} g(x) &= f(x) - h(x) \\ &= f(x) - mx - p \end{aligned}$$

Then $g'(x) = f'(x) - m$ is a strictly increasing function because $f'(x)$ is so. Also $g(a) = g(b) = 0$, so we are in the special case, and we conclude that $g(c) < 0$, that is,

$$f(c) < h(c)$$

This completes the proof that the graph is below the chord. ●

Inflection-Point Test

We postponed the proof of the first part of Theorem 4, page 180. We are given a twice differentiable function $f(x)$ on an open interval with an inflection point at $x = c$. Our problem is to prove that $f''(c) = 0$.

As already noted, if $f''(x)$ is continuous, then there is no difficulty with this proof. However, even if $f''(x)$ is not continuous, the assertion can still be proved. For definiteness, we assume $f(x)$ is concave down for $x < c$ and concave up for $x > c$. Thus $f''(x) \leq 0$ for $x < c$ and $f''(x) \geq 0$ for $x > c$. We wish to conclude from these assumptions that $f''(c) = 0$. This is implied by the following lemma, applied to $g(x) = f'(x)$.

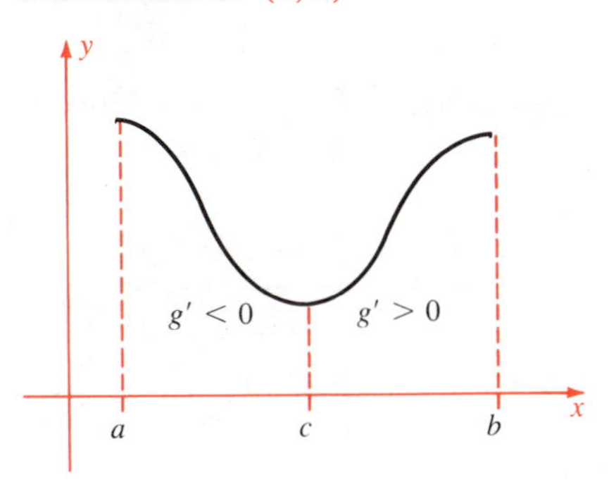

Figure 3-10-4
$y = g(x)$
Differentiable on (a, b)

Lemma Let $g(x)$ be differentiable on an open interval (a, b) and let $a < c < b$. Assume $g'(x) \le 0$ for $x < c$ and $g'(x) \ge 0$ for $x > c$. Then $g'(c) = 0$. ●

Proof Since $g'(x) \le 0$ for $x < c$, it follows that $g(x)$ is a decreasing function on $(a, c]$. Since $g'(x) \ge 0$ for $x > c$, it follows that $g(x)$ is an increasing function on $[c, b)$. See Figure 3-10-4. Therefore $g(c)$ is the *minimum* of $g(x)$ on the interval (a, b). Since $g(x)$ is differentiable on this interval, we must have $g'(c) = 0$. ●

Existence of Extrema

In Theorem 2, page 188, we claimed that if $f(x)$ is continuous on an *open* interval (a, b) and if $f(x) \to +\infty$ both as $x \to a+$ and as $x \to b-$, then $f(x)$ has a minimum. (The cases $a = -\infty$ and/or $b = +\infty$ are included.)

Proof Choose any c on the interval, so $a < c < b$. Then choose a_1 and b_1 with $a < a_1 < c < b_1 < b$ such that

$$\begin{cases} f(x) > f(c) & \text{for} \quad a < x < a_1 \\ f(x) > f(c) & \text{for} \quad b_1 < x < b \end{cases}$$

These inequalities rule out a minimum of $f(x)$ on (a, a_1) or on (b_1, b) because all values of $f(x)$ are too large on either of these intervals. See Figure 3-10-5. That leaves us with $[a_1, b_1]$.

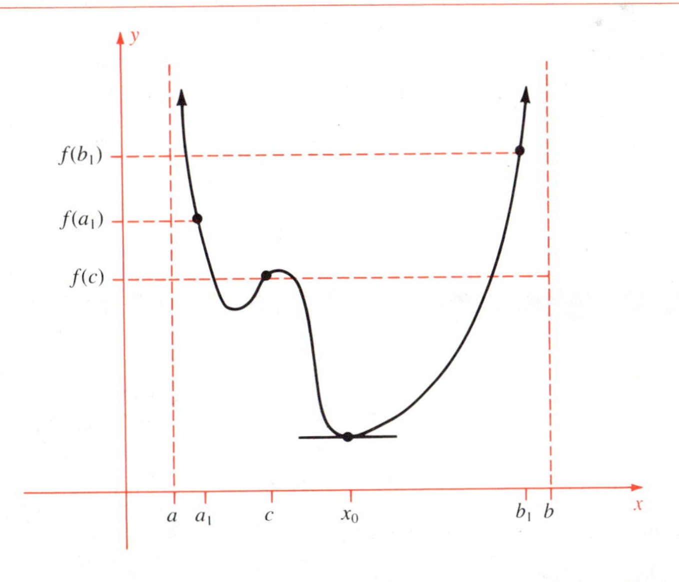

Figure 3-10-5
$f(x) > f(c)$ on $(a, a_1]$ and on $[b_1, b)$, so $f_{\min}$ must occur on $[a_1, b_1]$.

Therefore we now consider $f(x)$ with the *restricted* domain $[a_1, b_1]$, a continuous function on a closed interval. It has a minimum there, $f(x_0)$. Since values of $f(x)$ are too large anywhere but on $[a_1, b_1]$, it is clear that $f(x_0)$ is the minimum of $f(x)$ on the whole domain (a, b). This completes the proof. ●

Exercises

1 Let $f(x)$ be concave up, $x < y$, and $0 < t < 1$. Prove $f[tx + (1-t)y] < tf(x) + (1-t)f(y)$.

2 (cont.) Let $x < z < y$. Prove that

$$(y-x)f(z) < (y-z)f(x) + (z-x)f(y)$$

In Exercises 3–8 you may assume that second derivatives exist if you wish.

3 Prove that the sum of two concave-up functions is concave up.

4 If $f(x) > 0$ and $f(x)$ is concave up, prove that $f(x)^2$ is concave up.

5 (cont.) Give an example of a positive concave-up function $g(x)$ such that $\sqrt{g(x)}$ is not concave up.

6 Let $f(x)$ and $g(x)$ be positive concave-up functions, both increasing or both decreasing. Prove that $f(x)g(x)$ is concave up.

7 Let $f(y)$ and $g(x)$ be concave-up functions, $f(y)$ increasing. Prove that the composite function $f[g(x)]$ is concave up.

8 (cont.) Find and prove a similar statement about concave-down functions.

9 Let $y = x^4 + a_1x^3 + a_2x^2 + a_3x + a_4$. Assume $y'(c) = 0$ and $y''(c) < 0$. Prove that y has exactly two local minima.

10 (cont.) Give an explicit example of such a polynomial.

11 Prove $1 + x < e^x < 1 + (e-1)x$ for $0 < x < 1$.

12 Prove $(2/\pi)x < \sin x < x$ for $0 < x < \frac{1}{2}\pi$.

13 Let $f(x)$ be a twice differentiable function on an open interval. For each x on the interval define $\theta(x)$ by $f'(x) = \tan\theta(x)$ and $-\frac{1}{2}\pi < \theta(x) < \frac{1}{2}\pi$. Interpret $\theta(x)$ geometrically, and show that it is a differentiable function.

14 (cont.) Now suppose that $f(x)$ is concave up. What do you conclude about $\theta'(x)$? Interpret geometrically.

The object of the next two exercises is to prove an important inequality: if $a_1, a_2, \ldots, a_n$ are any positive numbers, then

$$\sqrt[n]{a_1a_2\cdots a_n} \le \frac{a_1 + a_2 + \cdots + a_n}{n}$$

In words, the **geometric mean** of a set of numbers does not exceed the **arithmetic mean** (average). We abbreviate the inequality by the notation $G_n \le A_n$.

***15** Show that the maximum value of the ratio

$$f(x) = \frac{\sqrt[n+1]{a_1a_2\cdots a_nx}}{\dfrac{1}{n+1}(a_1 + a_2 + \cdots + a_n + x)}$$

occurs for $x = A_n$, and compute the maximum. Conclude that

$$\frac{G_{n+1}}{A_{n+1}} \le \left(\frac{G_n}{A_n}\right)^{n/(n+1)}$$

***16** By repeated applications of Exercise 15, show that

$$\frac{G_n}{A_n} \le \left(\frac{G_1}{A_1}\right)^{1/n}$$

and therefore $G_n \le A_n$. Explain why $G_n = A_n$ if and only if $a_1 = a_2 = \cdots = a_n$.

3-11 Review Exercises

1 Graph $y = \dfrac{2x+1}{x^2+4}$

2 Where is $y = (1 + x^3)^{-1}$ concave up and where concave down for $x > 0$?

3 A 30-foot ladder leans against a vertical wall. Its foot is pushed towards the wall at the rate of 2 ft/sec. Find the speed of the upper end when the lower end is 10 ft from the wall.

4 A body moves along a horizontal line according to the law $s = 12t - t^3$, where $t \ge 0$.

a When is s increasing and when decreasing?
b What is the maximum velocity?
c What is the acceleration at $t = 3$?
The units are feet and seconds.

5 A train moving at 90 ft/sec slows up with a constant negative acceleration of 6 ft/sec². How long is it until the train stops? How far does it go?

6 A car is sitting 25 feet from a marker on a highway. At 12:00 noon it starts to move away from the marker. Its velocity during the next 10 seconds is given by

$$v = 5t + t^{1.08} \text{ ft/sec} \quad (0 \le t \le 10)$$

a Find an expression for its acceleration at time t for $0 \le t \le 10$.
b How far from the marker is it at 10 seconds past noon? (Give a numerical answer.)

7 A balloon rises straight up from the ground at a constant rate of 5 ft/sec. At the instant it reaches an altitude of

100 ft, how fast is its angle of inclination changing as seen from the ground 100 ft from the point of release?

8 A spring of length 1 ft is hung vertically. A weight attached to the free end stretches the spring 4 ft. If the weight is displaced 2 ft lower and released, then its distance (measured down from the ceiling) after t sec is $y = 5 + 2\cos\omega t$, where $\omega^2 = g/4$ and $g = 32.2$ ft/sec^2. Describe the motion of the weight; give its velocity and acceleration.

9 A certain pendulum swings out a circular arc when set in motion. Suppose that at time t the tip of the pendulum has horizontal position $x = A \sin 2\pi t$. Describe this horizontal motion, giving velocity, acceleration, and position at critical values of t. (It is the motion of the shadow of the pendulum bob cast by the sun directly overhead.)

10 Graph $y = 2x^3 - 9x^2 + 12x - 5$. Plot all points where the tangent is horizontal, and plot all points of inflection.

11 A rectangular box with a square bottom is to have volume 648 cubic inches. If the material on the top and bottom costs three times as much as that on the sides, find the most economical dimensions.

12 Compute the largest area of all rectangles inscribed in the ellipse $x^2/9 + y^2/4 = 1$, if the sides of the rectangles are parallel to the axes of the ellipse.

13 Compute the largest volume of all cones that can be generated by rotating a right triangle with hypotenuse c about one of its legs. [Hint $V = \frac{1}{3}\pi r^2 h$.]

14 A child M in a rowboat 1 km offshore from point A wants to reach a point B 4 km up the coast (Figure 3-11-1). The child rows at 4 km/hr and walks at w km/hr. It turns out that the fastest journey consists of rowing to the midpoint P of AB and then walking to B. Find w.

15 If x is a number between 0 and 1, then $x > x^3$. Find the positive number x that exceeds its cube by the greatest possible amount.

16 Find the largest value of x^2y if x and y are positive numbers whose sum is 15.

Figure 3-11-1
See Exercise 14

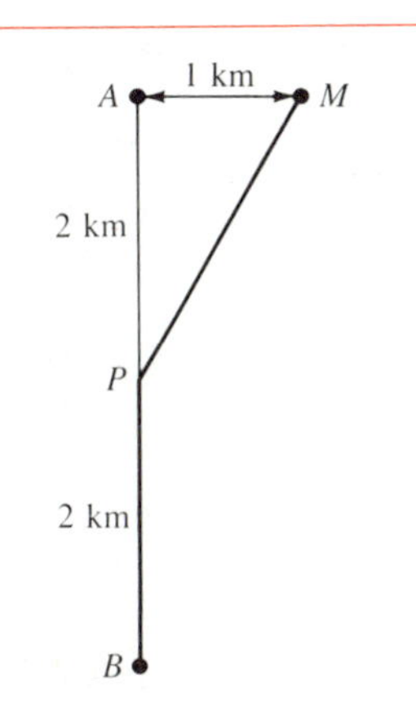

17 A farmer has 100 m of fencing and a long brick wall. He wants to fence in three identical-size rectangular pens against the wall as shown in Figure 3-11-2. What is the largest area (total area of the three pens) that he can enclose?

Figure 3-11-2
See Exercise 17

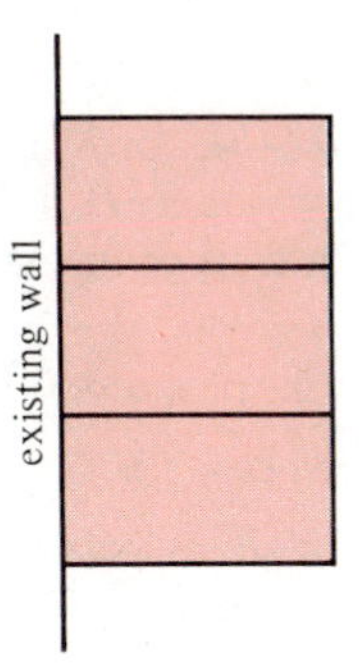

18 Find the line tangent to the curve $y = 4 - x^2$ at a point of the first quadrant that cuts from the first quadrant a triangle of minimum area.

19 A line tangent to the unit circle at a point in the first quadrant meets the coordinate axes in points $(x, 0)$ and $(0, y)$. Minimize $x + y$.

20 The power radiated in direction θ by an accelerating relativistic charged particle is

$$P = \frac{k \sin^2\theta}{(1 - \beta\cos\theta)^5} \qquad 0 < \beta < 1$$

Find θ so P is maximal.

21 Find the maximum of $y = \frac{1}{2}x - \sin x$ for $0 \le x \le 4\pi$.

22 An open-top cylindrical can is to hold one liter. The bottom is made of material that costs 10% more per cm^2 than the side. Find the measurements of the can that costs least to make.

23 A railroad will run a special train if at least 200 people subscribe. The fare will be \$8 per person if 200 people go but will decrease 1¢ for each additional person who goes. (For example, if 250 people go, the fare will be \$7.50.) What number of passengers will bring the railroad maximum revenue?

24 Of all lines of negative slope through the point (a, b) in the first quadrant, find the one that cuts from the first quadrant a triangle of least area.

25 A wire 30 inches long is cut into two parts, one of which is bent into a circle, and the other into a square. How should the wire be cut so that the sum of the areas of the circle and the square is a minimum?

26 What is the maximum volume of the cylinder generated by rotating a rectangle of perimeter 48 cm about one of its sides?

27 Locate all local extrema of $f(x) = \sin x \sin 2x$ on $[0, 2\pi]$.

28 Set $f(x) = x^4 - 2x^3 + 2x^2 - x + 1$. By algebraic magic, $f(x) = (x^2 - x + \frac{1}{2})^2 + \frac{3}{4} \geq \frac{3}{4}$. By using calculus, find a larger lower bound for $f(x)$.
[Hint Try $f'(\frac{1}{2})$.]

29 The drag on an airplane at subsonic speed v is

$$D = av^2 + b/v^2$$

where $a > 0$ and $b > 0$, provided the speed is not too small. Find the speed at which D is least. (At this speed the range is greatest for a given fuel supply.)

30 (cont.) At a steady speed v, the thrust of the engine just balances the drag, and the power developed by the engine is

$$P = (\text{thrust})(\text{speed}) = Dv = av^3 + b/v$$

Find the speed at which P is least. (Since P is proportional to the rate of consumption of fuel, the time airborne will be greatest at this speed.)

31 The distances x and y from a thin convex lens to an object on its axis and to the object's image, respectively, are related by $1/x + 1/y = 1/f$, where f is the (constant) focal length. Minimize $x + y$.

32 The magnetic force at a point on the axis of a conducting loop at distance x from the loop's center is given by $F = kx/(x^2 + a^2)^{5/2}$, where k and a are positive constants. Find the intervals of concavity of F and where F has inflection points.

33 A projectile is fired towards a hill from the foot of the hill, which rises at angle α with level ground. If the gun's elevation from the horizon is θ, where $\alpha < \theta < \frac{1}{2}\pi$, and v_0 is the initial speed of the projectile, then the shell hits the hill at horizontal distance

$$x = \frac{v_0{}^2}{g}[\sin 2\theta - (\tan \alpha)(1 + \cos 2\theta)]$$

Find $x_{\max}$, and interpret the maximizing angle θ geometrically.

34 Prove that light from A to B, reflecting off the mirror M in Figure 3-11-3 will take the path so that $\alpha = \beta$. [Review Fermat's principle of least time, page 211.]

Figure 3-11-3
See Exercise 33

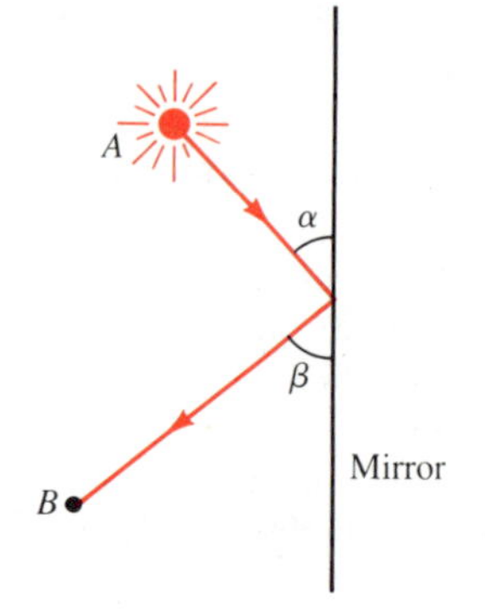

35 The surface magnetic field strength $B = B(t)$ of a pulsar is related to its period $P = P(t)$ by the formula $B^2 = kP \cdot (dP/dt)$. Express dB/dt in terms of P, dP/dt, and d^2P/dt^2.

36 A swampy region shares a long straight border with a region of farm land. A telephone cable is to be constructed connecting two locations, one in each region. Its cost is d_1 dollars per mile in the swampy region and d_2 dollars per mile in the farm land. What path should the cable take for its cost to be least?

37 One corner of a long rectangular strip of width a is folded over and just reaches the opposite edge, as shown in Figure 3-11-4. Find the largest possible area A.

Figure 3-11-4
See Exercises 37 and 38

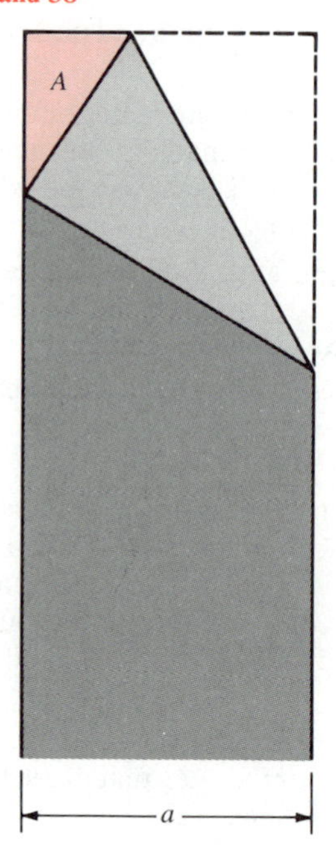

***38** (cont.) Find the minimum area of the light gray triangle. (See Example 4, page 207.)

***39** Prove $\left(\dfrac{x^4 + 1}{2}\right)^{1/4} \geq \left(\dfrac{x^3 + 1}{2}\right)^{1/3}$ for all $x \geq 0$.

***40** An isosceles triangle is circumscribed about a circle of radius r. Find its altitude if its perimeter is to be as small as possible.

4 Special Functions

We have developed and applied algebraic and trigonometric functions and their derivatives. We now continue this work with exponential, logarithmic, inverse trigonometric, and related functions. Each new family of functions gives us tools for further applications. The functions introduced in this chapter complete the list of what are sometimes called "elementary functions," the building blocks from which the functions of calculus are constructed. We begin with exponential functions, which are used in problems of natural growth and decay.

4-1 Exponential Functions

In Chapter 1 we reviewed exponential functions $f(x) = a^x$ from the point of view of college algebra. In Chapter 2 we noted that such functions are continuous, and we even sketched how this fact might be proved. The point of these early discussions was to gain familiarity with an important family of functions so as to have more concrete examples to work with. However, the discussions were incomplete as rigorous mathematics. Now we reexamine the matter from the point of view of calculus, rather than college algebra. Since we now know quite a bit of calculus, let's use it.

The following theorem covers precisely what must be said about the *existence* of exponential functions.

Theorem 1 Existence of Exponential Functions

For each positive real number a there exists a unique continuous function $\exp_a x$ with domain $\mathbf{R}$ such that

1 $\exp_a 1 = a$ and **2** $\exp_a(x + y) = (\exp_a x)(\exp_a y)$ ●

Note carefully the word *unique* in the theorem. It says that there is *exactly one* continuous function with domain $\mathbf{R}$ that satisfies statements 1 and 2.

At this point we accept Theorem 1 as true. It does require a proof, and there are two approaches to proving it. We shall sketch one proof at the end of this section. The other approach establishes the existence of $\exp_a x$ from the existence of its inverse function, $\log_a x$. This approach, which begins by proving that the inverse function exists, usually makes use of integration. We shall discuss it in the next chapter.

The proof of Theorem 1 at the end of this section will show that $\exp_a n = a^n$ for integers n in the usual algebraic sense, and then that $\exp_a r = a^r$ for rational numbers r. These facts justify using the conventional, standard notation a^x instead of $\exp_a x$ for exponential functions.

We shall usually restrict ourselves to $a > 1$. When we have to deal with b^x for $0 < b < 1$, we shall simply use instead

$$a^{-x} = b^x \quad \text{where} \quad a = b^{-1} > 1$$

For instance $(\frac{1}{2})^x = 2^{-x}$.

It will be useful to restate Theorem 1 in a^x notation. This time we also state properties of a^x that can be proved readily from Theorem 1.

Theorem 2 Properties of Exponential Functions

Let $a > 1$. Then there exists a unique continuous function $f(x) = a^x$ with domain $\mathbf{R}$ such that

1 $a^1 = a$ and **2** $a^{x+y} = a^x a^y$

This function has the following properties:

3 $a^0 = 1$ **4** $a^{-x} = 1/a^x$ **5** $(a^x)^y = a^{xy}$ **6** $(ab)^x = a^x b^x$

7 a^x is strictly increasing with range $(0, +\infty)$

8 $\lim_{x\to+\infty} a^x = +\infty$ and $\lim_{x\to-\infty} a^x = 0$

9 For r rational, a^r has the usual algebraic meaning. In particular, if m is a positive integer, then

$$a^m = a \cdot a \cdot a \cdots a \quad (m \text{ factors})$$

and $a^{1/m} = \sqrt[m]{a}$. ●

Properties 1–6 are the *laws of exponents* of algebra. In property 7, the phrase **strictly increasing** means that if $x < y$, then $a^x < a^y$.

The first limit relation in property 8 can be proved in many ways. Since a^x is strictly increasing, it suffices to prove that

$$\lim_{n\to+\infty} a^n = +\infty \quad \text{where } n \text{ varies over integers}$$

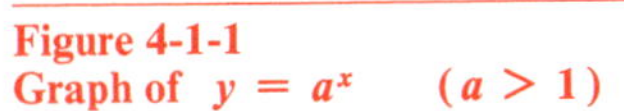

Figure 4-1-1
Graph of $y = a^x$ $(a > 1)$

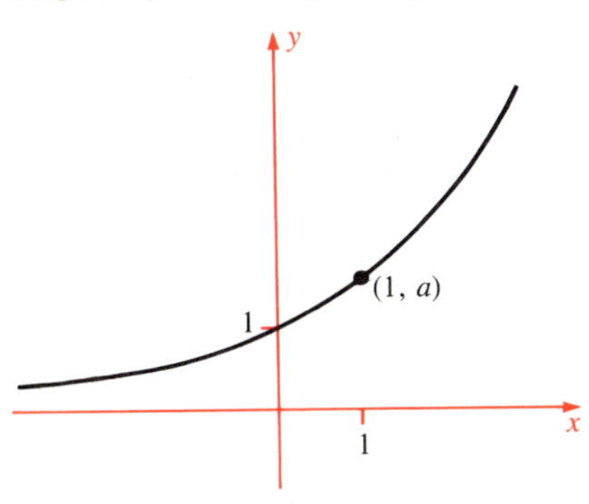

Figure 4-1-2
Graph of $y = a^x$ $(0 < a < 1)$
Note that $y = b^{-x}$ where $b = a^{-1}$

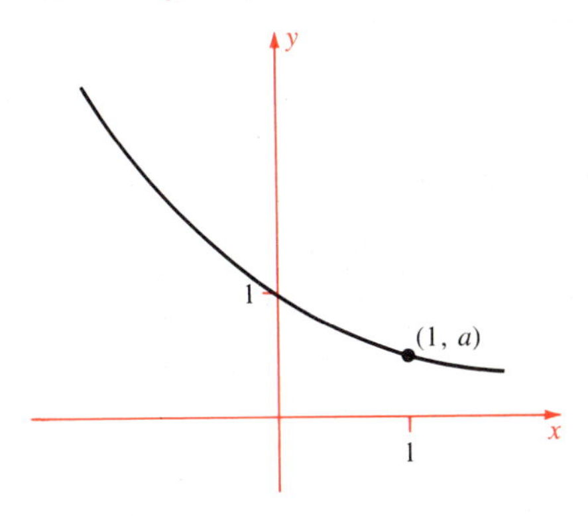

To do so, write $a = 1 + b$ where $b > 0$. By an easy induction (or by the binomial theorem), for $n > 1$,

$$a^n = (1 + b)^n > 1 + nb + \binom{n}{2} b^2 + \cdots + b^n > 1 + nb$$

Since $1 + nb \to +\infty$, it follows that $a^n \to +\infty$ as $n \to +\infty$. The second limit relation now follows from $a^x = 1/a^{-x}$.

From the limit relations 8, the continuity of a^x, and the intermediate value theorem, it follows that the range of a^x is $(0, +\infty)$.

The graph of $y = a^x$ for $a > 1$ is drawn in Figure 4-1-1. We shall see later that the function is concave up as indicated by the graph. Figure 4-1-2 shows what the graph looks like if $0 < a < 1$. As already noted, when $0 < a < 1$, we shall work with b^{-x} where $b = 1/a > 1$ instead of with a^x.

Derivatives

We begin with some numerical evidence that each exponential function $f(x) = a^x$ is differentiable at $x = 0$. The difference quotient at $x = 0$ is

$$\frac{f(0 + h) - f(0)}{h} = \frac{a^h - 1}{h}$$

Let us start with $f(x) = 2^x$ and test various h. See Table 4-1-1. The numerical evidence strongly suggests that

$$\left.\frac{d}{dx} 2^x\right|_{x=0} = \lim_{h \to 0} \frac{2^h - 1}{h}$$

exists and its value is about 0.69315. Actually, to six places, this derivative is 0.693147. Again, let's try $f(x) = 3^x$. Difference quotients are tabulated in Table 4-1-2. This time the numerical evidence points to

$$\left.\frac{d}{dx} 3^x\right|_{x=0} \approx 1.0986$$

Table 4-1-1

h	$\dfrac{2^h - 1}{h}$	$\dfrac{2^{-h} - 1}{-h}$
.0009	.69336	.69293
.0008	.69334	.69296
.0007	.69332	.69298
.0006	.69329	.69300
.0005	.69327	.69303
.0004	.69324	.69305
.0003	.69322	.69308
.0002	.69320	.69310
.0001	.69317	.69312

Table 4-1-2

h	$\dfrac{3^h - 1}{h}$	$\dfrac{3^{-h} - 1}{-h}$
.0009	1.09916	1.09807
.0008	1.09910	1.09813
.0007	1.09903	1.09819
.0006	1.09897	1.09825
.0005	1.09891	1.09831
.0004	1.09885	1.09837
.0003	1.09879	1.09843
.0002	1.09873	1.09849
.0001	1.09867	1.09855

Actually, to six places this derivative is 1.098612. Note in both tables that as h decreases toward 0 from above, the difference quotient decreases, and as h increases towards 0 from below, the difference quotient increases. That behavior is consistent with Figure 4-1-1, which implies that the graph of $y = a^x$ is concave up—a fact we shall prove soon. Thus we get the idea that for any $a > 1$, the function a^x is differentiable at $x = 0$.

Lemma 1 If $a > 1$, then $\left.\dfrac{d}{dx} a^x\right|_{x=0}$ exists. ●

We shall take this for granted now—a proof is sketched at the end of this section. The law of exponents, $a^{x+h} = a^x a^h$, now enables us to differentiate $f(x) = a^x$ in general:

$$\frac{d}{dx} a^x = \lim_{h \to 0} \frac{a^{x+h} - a^x}{h} = \lim_{h \to 0} \frac{a^x(a^h - 1)}{h}$$

$$= (a^x)\left(\lim_{h \to 0} \frac{a^h - 1}{h}\right) = (a^x)\left(\left.\frac{d}{dx} a^x\right|_{x=0}\right)$$

The quantity

$$\left.\frac{d}{dx} a^x\right|_{x=0}$$

is a number $k = k(a)$ that depends only on a. Geometrically, $k(a)$ is the slope at $x = 0$ of $y = a^x$. If we change the notation slightly, we have established the following theorem:

Theorem 3 Derivative of a^x

For each $a > 1$,

$$\frac{d}{dx} a^x = k \cdot a^x$$

where

$$k = k(a) = \left.\frac{d}{dx} a^x\right|_{x=0} > 0$$ ●

Example 1

a $\dfrac{d}{dx} 2^x = k \cdot 2^x$ where $k = k(2) \approx 0.69315$ to five significant figures.

b $\dfrac{d}{dx} 3^x = k \cdot 3^x$ where $k = k(3) \approx 1.0986$ to five significant figures. ●

Figure 4-1-3
$1 < a < b$

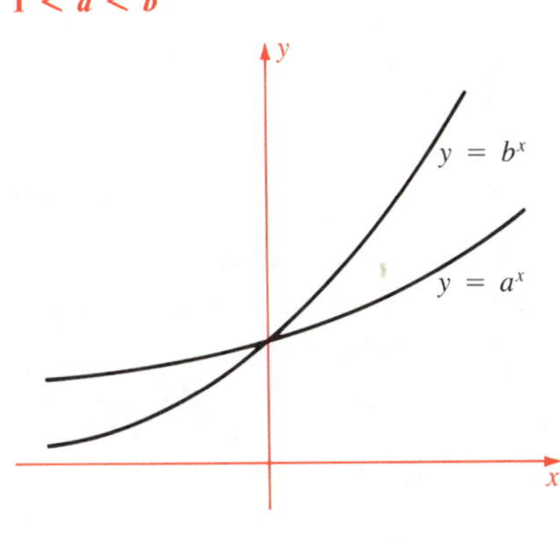

If we compare $y = a^x$ with $y = b^x$, where $1 < a < b$, it is clear from the graphs that the slope of $y = a^x$ is smaller than the slope of $y = b^x$ at $x = 0$. See Figure 4-1-3. Therefore in the relation

$$\frac{d}{dx} a^x = ka^x \qquad \text{where} \quad k = k(a)$$

the larger a is, the larger $k(a)$ is. We know that $k(2) < 1$ and $k(3) > 1$, which suggests that somewhere between 2 and 3 there is an a for which $k(a) = 1$.

This particular a, whose existence will next be established, is always denoted e. Let us try to find e in the form $e = 2^c$, where c is to be determined. Then $e^x = 2^{cx}$. We set $k = k(2)$ so that

$$\frac{d}{dx}(2^x)\bigg|_{x=0} = k \approx 0.693147$$

By the chain rule, with $u = cx$,

$$\frac{d}{dx} e^x\bigg|_{x=0} = \frac{d}{dx} 2^{cx}\bigg|_{x=0} = c\frac{d}{du} 2^u\bigg|_{u=0} = ck$$

If we set $c = 1/k$, then $ck = 1$ and $e = 2^{1/k}$. Hence

$$\frac{d}{dx} e^x\bigg|_{x=0} = 1$$

which is the desired property of the number e. Numerically, we have $e = 2^{1/k} \approx 2^{1/0.693147} \approx 2.71828$. We now state the main conclusion of this section, one of the most important results in calculus.

Theorem 4 Derivative of e^x

There exists a unique real number e such that the exponential function e^x satisifes

$$\frac{d}{dx} e^x = e^x$$

Numerically, $e \approx 2.71828$.

Figure 4-1-4
Graph of $y = e^x$
The slope at (x, e^x) is e^x.

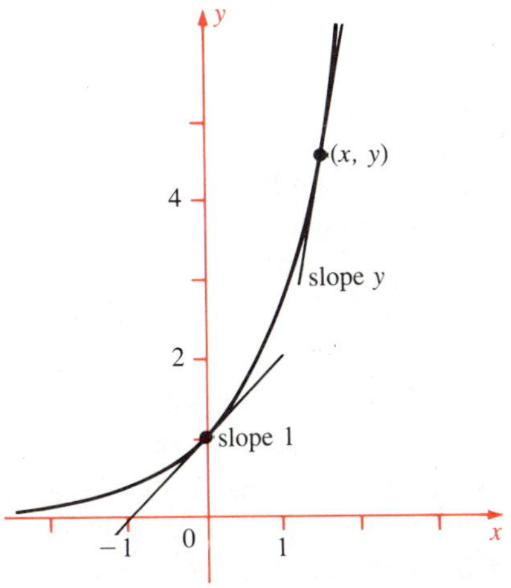

The function $f(x) = e^x$ stands out among all other exponential functions because of its strikingly simple derivative formula. It is honored with the name **the exponential function.** Its graph is shown in Figure 4-1-4. Note in particular that the graph passes through $(0, 1)$ with slope 1. The exponential function is also written $\exp x$. This is particularly useful notation when a complicated expression is substituted for x.

A composite function of the form $y = e^{u(x)} = \exp u(x)$ can be differentiated by the chain rule:

$$\frac{d}{dx} e^{u(x)} = \left(\frac{d}{du} e^u\bigg|_{u=u(x)}\right)\left(\frac{du}{dx}\right) = e^{u(x)}\frac{du}{dx}$$

Example 2

a $\dfrac{d}{dx} e^{3x} = 3e^{3x}$ **b** $\dfrac{d}{dx} \exp x^2 = 2x \exp x^2$

c $\dfrac{d}{dx} e^{ax}\cos bx = ae^{ax}\cos bx - be^{ax}\sin bx$

Example 3

Minimize $f(x) = e^x/x$ on the open interval $(0, +\infty)$.

Solution $f'(x) = \dfrac{(e^x)'x - e^x}{x^2} = \dfrac{e^x}{x^2}(x-1)$

Since e^x is never 0, the only positive x for which $f'(x) = 0$ is $x = 1$. Because e^x/x^2 is always positive,

$$f'(x) > 0 \quad \text{for} \quad x > 1 \qquad \text{and} \qquad f'(x) < 0 \quad \text{for} \quad 0 < x < 1$$

Therefore $f_{\min} = f(1) = e$.

Properties of $y = e^x$

We know from its defining properties that e^x is a strictly increasing function. But now calculus gives us another reason, because

$$\frac{d}{dx} e^x = e^x > 0$$

The second derivative is also positive:

$$\frac{d^2}{dx^2} e^x = \frac{d}{dx}\left(\frac{d}{dx} e^x\right) = \frac{d}{dx} e^x = e^x > 0$$

Therefore e^x is concave up. Thus the way we have been drawing the graph is quite correct.

Properties of *e* The number e has many remarkable features; we first mention the following limit relation:

$$e = \lim_{x \to +\infty} \left(1 + \frac{1}{x}\right)^x$$

We shall prove this relation in Section 9-2. Meanwhile, let us test it on a calculator; some numerical evidence is shown in Table 4-1-3.

Like the number π, the number e is a fundamental constant of nature, independent of units of measurement. It has been computed to great accuracy. To 15 places

$$e \approx 2.71828\ 18284\ 59045$$

Many interesting properties of e have been discovered; for example, e is an irrational number; indeed, e is not a root of any polynomial equation $x^n + a_{n-1}x^{n-1} + \cdots + a_0 = 0$ with *rational* coefficients.

Table 4-1-3

x	$\left(1 + \dfrac{1}{x}\right)^x$
10	2.59374
10^2	2.70481
10^3	2.71692
10^4	2.71815
10^5	2.71827
10^6	2.71828

The Function a^x

The strictly increasing continuous function e^x has range $(0, +\infty)$. Therefore it has an inverse function, written $\ln x$, whose domain is $(0, +\infty)$ and whose range is **R**. We shall study $\ln x$ in detail in Sections 4-3 and 4-4; for the moment we need it only to write any given positive a as a value of e^x:

$$a = e^{\ln a} \quad \text{if} \quad a > 0$$

Now we can express each exponential function a^x in terms of e^x in a simple way. We set $k = \ln a$ so that $a = e^k$. Then

$$a^x = (e^k)^x = e^{kx} \quad \text{where} \quad k = \ln a$$

Clearly, if $a > 1$, then $k = \ln a > 0$. Thus once we have *the* exponential function $y = e^x$ we essentially know all exponential functions a^x.

We differentiate by the chain rule:

$$\frac{d}{dx}a^x = \frac{d}{dx}e^{kx} = ke^{kx} = ka^x$$

Therefore the scale factor k can be characterized by the relation

$$k = \frac{d}{dx}a^x \bigg|_{x=0}$$

We summarize this discussion in a theorem:

Theorem 5

Let $a > 1$. Then

$$a^x = e^{kx} \qquad \text{where} \quad k = \ln a = \frac{d}{dx}a^x \bigg|_{x=0}$$

Rate of Growth of e^x

We know that $e^x \to +\infty$ as $x \to +\infty$, but how rapidly? As we shall see in the next section, applications of exponential functions arise frequently so it is important to know how rapidly e^x grows and how rapidly e^{-x} decays. We tabulate some approximate values in Table 4-1-4. Evidently e^x increases *very* rapidly, much faster than x increases. It is a safe bet that

$$\frac{e^x}{x} \to +\infty \quad \text{as} \quad x \to +\infty$$

In fact much more is true: e^x increases faster than x^2, than x^3, and even faster than *any* positive power of x. For example, let us compare e^x with x^{10}, itself a rapidly increasing function. To do so, we tabulate the ratio e^x/x^{10} for some large x. See Table 4-1-5. After a slow start, e^x completely overwhelms x^{10}. The same holds for e^x versus any power x^n.

Table 4-1-4

x	e^x
5	1.5×10^2
10	2.2×10^4
50	5.2×10^{21}
100	2.7×10^{43}
500	1.4×10^{217}
1000	2.0×10^{434}

Table 4-1-5

x	e^x/x^{10}
5	1.5×10^{-5}
10	2.2×10^{-6}
50	5.3×10^4
100	2.7×10^{23}
500	1.4×10^{190}
1000	2.0×10^{404}

- **Rapid Growth of e^x** For any positive n

$$\lim_{x \to +\infty} \frac{e^x}{x^n} = +\infty$$

A proof is sketched at the end of the section. For now, we can get a good idea why the statement is correct if we consider the case $n = 2$. We set

$$f(x) = \frac{e^x}{x^2}$$

Suppose x increases by 1. Then e^x increases to

$$e^{x+1} = e^x e$$

so that e^x more than doubles. On the other hand, x^2 increases to $(x + 1)^2$. This is not much of an increase *relative* to the size of x^2. In fact

$$\frac{(x+1)^2}{x^2} = (1 + 1/x)^2 \approx 1 \quad \text{for } x \text{ large}$$

Therefore for x large, $f(x + 1) > 2f(x)$ because

$$\frac{f(x+1)}{f(x)} = \frac{e^{x+1}/(x+1)^2}{e^x/x^2} = e\frac{x^2}{(x+1)^2} > 2$$

By induction, this means $f(x + n) > 2^n f(x)$ for x sufficiently large, which is convincing evidence that $f(x) \to +\infty$ as $x \to +\infty$.

The function $e^x \to 0$ as $x \to -\infty$. In other words, $e^{-x} \to 0$ as $x \to +\infty$. Just as e^x increases more rapidly than x^n as $x \to +\infty$, so the function e^{-x} decreases to 0 more rapidly than x^{-n} as $x \to +\infty$.

- **Rapid Decay of e^{-x}** For any positive n,

$$\lim_{x \to +\infty} x^n e^{-x} = 0$$

This is so because

$$\frac{1}{x^n e^{-x}} = \frac{e^x}{x^n} \to +\infty$$

Proofs [Optional]

Theorem 1 asserts that if $a > 0$, then there exists a unique continuous function $\exp_a x$ such that $\exp_a 1 = a$ and

$$\exp_a(x + y) = (\exp_a x)(\exp_a y)$$

Let us assume $a > 1$ for definiteness. The existence of $\exp_a x$ can be proved by first defining $\exp_a n$ for integers n and then $\exp_a r$ for r rational. Then we use a limit argument based on approximating real numbers by rationals. Of course there are some technical details, but the proof is straightforward, although long, and no special tricks are involved.

For a positive integer n, we define $\exp_a n = a^n$ as usual by algebra. Then we define $\exp_a(-n) = a^{-n} = 1/a^n$ and $\exp_a 0 = 1$. To define $\exp_a(m/n)$ where m and n are integers and $n > 0$, we use n-th roots. Specifically, we solve $x^n = a^m$. To do so, we note that $f(x) = x^n$ is continuous and strictly increasing on the domain $[0, +\infty)$, that $f(0) = 0$, and that $f(x) \to +\infty$ as $x \to +\infty$. It follows that the range of $f(x) = x^n$ is $[0, +\infty)$, so there is a unique $x > 0$ such that $x^n = a^m$. We then define

$$\exp_a(m/n) = x$$

Note that this last step, proving the existence of n-th roots, uses methods of calculus. In school algebra, the existence of such roots is assumed without proof.

Thus $\exp_a r$ is a real-valued function defined on the set $\mathbb{Q}$ of rational numbers. By essentially algebraic arguments one can show that this function is strictly increasing (recall that $a > 1$) and that for r restricted to rational numbers,

$$\lim_{r\to-\infty} \exp_a r = 0 \quad \text{and} \quad \lim_{r\to+\infty} \exp_a r = +\infty$$

Finally, we can prove continuity in the following sense:

- If $\varepsilon > 0$, then there exists a $\delta > 0$ such that whenever r is rational and $|r| < \delta$, then $|\exp_a r - 1| < \varepsilon$.
- Suppose $B > 0$ and $\varepsilon > 0$. Then there exists a $\delta > 0$ such that whenever r and s are rational, $|r| < B$ and $|s| < B$, and $|r - s| < \delta$, then $|\exp_a r - \exp_a s| < \varepsilon$.

The main technical step in defining $\exp_a x$ for irrational x is proving that the following limits exist and are equal:

$$\lim_{r\to x-} \exp_a r = \lim_{r\to x+} \exp_a r \qquad (r \text{ rational})$$

Once done, $\exp_a x$ is defined as the common value of these limits. From this point it is routine to complete the proof of Theorem 1, including the uniqueness part. ●

Next we sketch a proof of Lemma 1, that a^x is differentiable at $x = 0$, in other words, that

$$\lim_{x\to 0} \frac{a^x - 1}{x}$$

exists. The proof depends on the following inequality, which more or less says that $f(x) = a^x$ is concave up (for $a > 1$). See Figure 4-1-5.

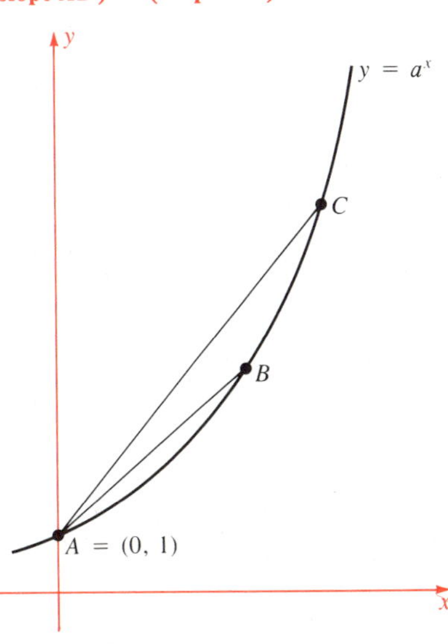

Figure 4-1-5
If B is right of A and C is right of B, then (slope AB) < (slope AC)

Lemma 2 Let $a > 1$ and $0 < x < y$. Then

$$\frac{a^x - 1}{x} < \frac{a^y - 1}{y}$$ ●

Proof If we clear the denominators, what we want to prove is

$$xa^y - ya^x - x + y > 0 \quad \text{for} \quad a > 1$$

This suggests that we hold x and y fixed and consider the function

$$g(t) = xt^y - yt^x - x + y \quad \text{for} \quad t \geq 1$$

Then $g(1) = x - y - x + y = 0$, and for $t > 1$,

$$\frac{dg}{dt} = xyt^{y-1} - xyt^{x-1} = xyt^{-1}(t^y - t^x)$$

Since $y > x$ and $t > 1$, we have $t^y > t^x$; hence $dg/dt > 0$. Therefore $g(t)$ is strictly increasing. Since $a > 1$ we have

$$g(a) > g(1) = 0$$

Consequently

$$xa^y - ya^x - x + y > 0$$

which completes the proof of Lemma 2. ●

Remark We have cheated a little here because we really know power functions $f(t) = t^r$ only for r rational. However, once the inequality is proved for x and y rational, it is a routine limit argument to establish it for x and y irrational also.

Lemma 2 says that the difference quotient

$$F(x) = \frac{a^x - 1}{x}$$

is a (strictly) increasing function on $(0, +\infty)$. Also $F(x) > 0$ since $a^x > 1$ for $x > 0$. An argument using the completeness of the real number system (page 166) implies that

$$k = \lim_{x \to 0+} \frac{a^x - 1}{x}$$

exists. The essential idea is to consider the set S of all values of $(a^x - 1)/x$ for $x > 0$. Then k equals the greatest lower bound of S. (In Chapter 3, we stated completeness in terms of least upper bound, but of course there is an equivalent statement in terms of greatest lower bound).

On the negative side of 0 we have

$$\lim_{x \to 0+} \frac{a^{-x} - 1}{-x} = \lim_{x \to 0+} \left(\frac{1}{a^x}\right)\left(\frac{a^x - 1}{x}\right)$$

$$= \left(\lim_{x \to 0+} \frac{1}{a^x}\right)\left(\lim_{x \to 0+} \frac{a^x - 1}{x}\right) = 1 \cdot k = k$$

This completes the proof that a^x is differentiable at $x = 0$. ●

Finally, we prove that

$$\lim_{x \to +\infty} \frac{e^x}{x^n} = +\infty$$

It suffices to prove this fact when n is an integer. For instance $x^{9.3} < x^{10}$ for $x > 1$; hence $e^x/x^{9.3} > e^x/x^{10}$. If we prove $e^x/x^{10} \to +\infty$, then certainly $e^x/x^{9.3} \to +\infty$ also. The proof depends on another inequality.

Lemma 3 For each integer $n \geq 0$ and each positive real x,

$$e^x > 1 + \frac{x}{1!} + \frac{x^2}{2!} + \frac{x^3}{3!} + \cdot \cdot \cdot + \frac{x^n}{n!}$$

Proof First recall the factorial notation:

$$n! = 1 \cdot 2 \cdot 3 \cdot \cdot \cdot n$$

Thus $3! = 1 \cdot 2 \cdot 3 = 6$, $7! = 1 \cdot 2 \cdot 3 \cdot 4 \cdot 5 \cdot 6 \cdot 7 = 5040$, and so on. Now set

$$p_n(x) = 1 + \frac{x}{1!} + \cdot \cdot \cdot + \frac{x^n}{n!}$$

for each integer n. The interesting thing about this sequence of polynomials is that

$$\frac{d}{dx} p_n(x) = p_{n-1}(x)$$

which is clear by inspection. Now we prove by induction on n that $e^x > p_n(x)$ for $x > 0$. This is true for $n = 0$ because $e^x > 1 = p_0(x)$ for $x > 0$. Suppose it is true for a certain n. Set

$$f(x) = e^x - p_{n+1}(x)$$

Then

$$f'(x) = e^x - p_n(x) > 0$$

for $x > 0$. Therefore $f(x)$ is a strictly increasing function. Consequently if $x > 0$, then

$$f(x) > f(0) = e^0 - p_{n+1}(0) = 1 - 1 = 0$$

Thus $f(x) > 0$ for all $x > 0$. This means $e^x > p_{n+1}(x)$ for all $x > 0$, which completes the proof of the lemma. ●

Now, to prove $e^x/x^n \to +\infty$ we use Lemma 3 with n replaced by $n + 1$:

$$\frac{e^x}{x^n} > \frac{p_{n+1}(x)}{x^n} > \frac{1}{x^n} \cdot \frac{x^{n+1}}{(n+1)!} = \frac{1}{(n+1)!} x$$

Figure 4-1-6
Polynomial approximation to $y = e^x$

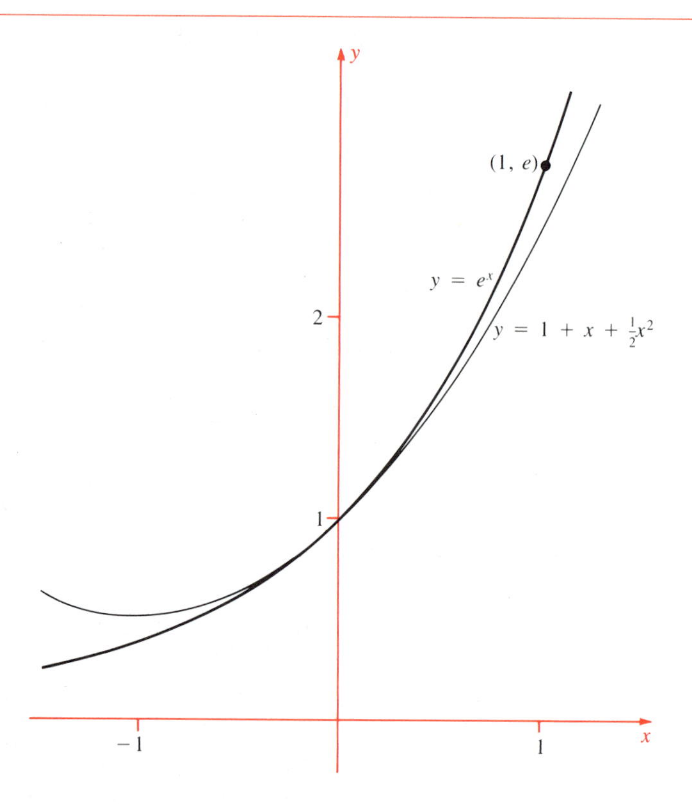

Because $(n + 1)!$ is a constant, $x/(n + 1)! \to +\infty$ as $x \to +\infty$. Therefore $e^x/x^n \to +\infty$ also. ●

Remark The choice of the polynomials $p_n(x)$ is no accident. It happens that these polynomials are approximations to the exponential function, each better than the previous one. See Figure 4-1-6 for $n = 2$ and Table 4-1-6 for various $p_n(x)$. For $x = 1$, the numbers $p_n(1)$ give excellent approximations to e. We can write $e = \lim p_n(1)$, that is

$$e = \lim_{n \to +\infty} 1 + \frac{1}{1!} + \frac{1}{2!} + \cdots + \frac{1}{n!}$$

Some numerical evidence is given in Table 4-1-7 to be compared with $e \approx 2.71828\ 18285$. With $n = 10$ we already get seven-place accuracy, something that requires $n \approx 2 \times 10^7$ when we use the formula $e \approx (1 + 1/n)^n$.

Table 4-1-6

x	$p_5(x)$	$p_{10}(x)$	e^x
0.1	1.10517	1.10517	1.10517
0.5	1.64870	1.64872	1.64872
1.0	2.71667	2.71828	2.71828
2.0	7.26667	7.38899	7.38906
−1.0	0.36667	0.36788	0.36788
−2.0	0.06667	0.13538	0.13534

Table 4-1-7

n	$p_n(1)$
5	2.7167
6	2.71806
8	2.718279
10	2.71828 1801
15	2.71828 18285

Exercises

Graph

1 $y = e^{x-1}$

2 $y = e^{1-x}$

3 $y = e^x/x \quad (x > 0)$

4 $y = 1/(1 + e^x)$

5 $y = \exp(\sin x)$

6 $y = \exp(\tan x)$

7 $y = e^{-x}\sin 10x \quad (x \geq 0)$

8 $y = e^{-x}\cos 20x \quad (x \geq 0)$

Graph; show all inflection points and intervals of concavity

9 $y = \exp(-x^2)$

0 $y = \exp(-1/x) \quad (x > 0)$

11 $y = \exp(1/x) \quad (x > 0)$

12 $y = \exp[1/(1 + x^2)]$

Differentiate with respect to x

3 $\exp(3x^2)$

4 $e^{1/x}$

5 $\exp(x^4)$

6 $e^x/(1 + e^{-x})$

7 e^x/x

8 x^2e^{2x}

9 $\dfrac{1}{1 + e^{-x}}$

0 $\sqrt{x}\exp(-\sqrt{x})$

1 $\left(\dfrac{x}{e^{2x} + 1}\right)^3$

22 $\dfrac{xe^x}{x^2 + 1}$

23 $\dfrac{e^x + e^{-x}}{e^x - e^{-x}}$

24 $e^{5x}(x^2 - 3x + 6)$

25 $\exp(\sin x)$

26 $\sin e^x$

27 $e^{u(x)}$ where $u(x) = ae^{bx}$

28 $\exp u(x)$ where $u(x) = 3\cos 2\pi x$

Find

29 $\dfrac{d^6}{dx^6} e^{-2x}$

30 $\dfrac{d^{27}}{dx^{27}} e^{-x}$

Find the maximum of $f(x)$ for $x \geq 0$ and where it is located

31 $f(x) = xe^{-x}$

32 $f(x) = x^2e^{-x}$

33 $f(x) = 3e^x - e^{3x}$

34 $f(x) = e^{-x} - e^x$

35 Construct Table 4-1-1 for $y = 5^x$. Conclude that

$$\frac{d}{dx} 5^x = k \cdot 5^x \quad \text{where} \quad k \approx 1.6094$$

36 Construct Table 4-1-1 for $y = 10^x$. Conclude that

$$\frac{d}{dx} 10^x = k \cdot 10^x \quad \text{where} \quad k \approx 2.3026$$

37 Justify $2^x \approx 1 + (0.6931)x$ for $|x|$ small.

38 (cont.) Test this estimate on $\sqrt{2} = 2^{1/2}$.

39 Justify $e^x \approx 1 + x$ for $|x|$ small.

40 (cont.) Show that this estimate is consistent with $e^x \cdot e^{-x} = 1$.

4-2 Applications of Exponential Functions

The function $y = e^t$ is equal to its own derivative. That is also true of $y = ce^t$:

$$\frac{d}{dt}(ce^t) = c\frac{d}{dt}e^t = ce^t$$

More generally, the derivative of $y = ce^{kt}$ is k times y:

$$\frac{d}{dt}y = \frac{d}{dt}(ce^{kt}) = c\frac{d}{dt}e^{kt} = cke^{kt} = ky$$

For $k > 0$, the differential equation

$$dy/dt = ky$$

describes a quantity *growing* at a rate proportional to its own size: the larger y is, the faster it increases. For $k < 0$, the equation describes a quantity *decaying* at a rate proportional to its own size: the smaller y is, the more slowly it decays. For applications, it is important to know that the functions $y = ce^{kt}$ are the *only* solutions of $dy/dt = ky$.

Theorem

If $y = y(t)$ is differentiable on an interval and $dy/dt = ky$, then $y = ce^{kt}$ for some constant c.

Proof It suffices to prove that $y/e^{kt} = ye^{-kt}$ is constant. We have

$$\frac{d}{dt}(ye^{-kt}) = \frac{dy}{dt}e^{-kt} - kye^{-kt} = kye^{-kt} - kye^{-kt} = 0$$

so ye^{-kt} is a constant function: $ye^{-kt} = c$. Thus $y = ce^{kt}$.

Many time-dependent processes obey the **natural growth law** $dy/dt = ky$. We now examine several cases.

Bacterial Growth

A colony of bacteria with unlimited food and no enemies grows at a rate proportional to its own size. We want a formula for $n(t)$, the number of bacteria in the colony at time t.

To attack the problem we make an approximation. The function $n(t)$ is not continuous since it jumps by one each time a new bacterium is produced. However, $n(t)$ is generally very large, and bacteria are produced at tiny time intervals, so we smooth out the problem by treating $n(t)$ as a continuous, even differentiable, function. In practice, this leads to satisfactory results. The growth law of $n(t)$ is

$$\frac{dn}{dt} = kn \qquad (k > 0)$$

Therefore

$$n(t) = ce^{kt}$$

To evaluate c, we set $t = 0$:

$$n(0) = ce^0 = c$$

Thus $c = n(0) = n_0$ is the number of bacteria at time 0, and

$$n(t) = n_0e^{kt}$$

In practice, k is found from additional data.

Example 1

There are 10^5 bacteria in a culture at the start of an experiment and 10^6 after 5 hours. Find a formula for $n(t)$.

Solution The initial number of bacteria $n_0 = 10^5$ is given. Therefore for any t,

$$n(t) = n_0e^{kt} = 10^5e^{kt}$$

To use the other given information, $n(5) = 10^6$, set $t = 5$:

$$n(5) = 10^6 = 10^5e^{5k}$$

Therefore $e^{5k} = 10$. Now we have two possible ways to continue. First, since $e^{kt} = (e^k)^t$, it is enough to find e^k. But

$$e^k = 10^{1/5} \approx 1.58$$

Therefore $n(t) = 10^5(10^{1/5})^t$.

The other route is to use the inverse function $\ln x$ of e^x in order to find k:

$$\ln e^{5k} = \ln 10 \quad \text{so} \quad 5k = \ln 10$$

Using the ln key on a hand calculator, we find $k = \frac{1}{5}\ln 10 \approx 0.46$. Therefore $n(t) = 10^5 \exp[(\frac{1}{5}\ln 10)t]$. Numerically,

$$n(t) \approx 10^5 \times 1.58^t \approx 10^5 e^{0.46t}$$

●

Radioactive Decay

A radioactive element decays (into other products) at a rate proportional to the amount present. Its **half-life** is the time required for half of the original material to decay.

Example 2

Carbon-14 has a half-life of 5568 years. Find its decay law.

Solution Let $m(t)$ kilograms be the mass of a quantity of carbon-14 at time t years. Then

$$\frac{dm}{dt} = -\lambda m$$

where the **decay constant** λ is positive. The solution of this differential equation is

$$m(t) = m_0 e^{-\lambda t}$$

where $m_0 = m(0)$, the initial mass. We are given

$$m(5568) = \tfrac{1}{2}m_0 \qquad \text{that is} \qquad m_0 e^{-5568\lambda} = \tfrac{1}{2}m_0$$

Therefore $e^{-5568\lambda} = \frac{1}{2}$; that is,

$$e^{5568\lambda} = 2 \qquad \text{hence} \qquad e^{\lambda} = 2^{1/5568} \approx 1.000124$$

Consequently

$$m(t) = m_0 e^{-\lambda t} = m_0 \cdot 2^{-t/5568}$$

Alternatively, $5568\lambda = \ln e^{5568\lambda} = \ln 2$, so

$$\lambda = \frac{\ln 2}{5568} \approx 1.24 \times 10^{-4}$$

Consequently $m(t) = m_0 e^{-\lambda t}$ where $\lambda \approx 1.24 \times 10^{-4}$. ●

Example 3

A waste product in a certain nuclear reactor has a half-life of 84 years. How long before the mass of undecayed waste drops to $\frac{1}{10}$ of 1% of its original mass?

Solution As in Example 2, the mass $m(t)$ of the undecayed waste after t years is given by

$$m(t) = m_0 e^{-\lambda t} \quad \text{where} \quad e^{84\lambda} = 2$$

To find λ, apply ln to both sides of the second relation:

$$84\lambda = \ln 2 \qquad \text{hence} \qquad \lambda = \tfrac{1}{84} \ln 2$$

Now set $m(t) = (0.001)m_0$ and solve for t:

$$(0.001)m_0 = m_0 e^{-\lambda t}$$

$$e^{-\lambda t} = 0.001 \qquad \text{hence} \qquad e^{\lambda t} = 1000$$

Now apply ln: $\lambda t = \ln 1000$. Hence

$$t = \frac{\ln 1000}{\lambda} = \frac{84 \ln 1000}{\ln 2} \approx 837 \text{ years}$$

●

Compound Interest

Money is not a natural substance even though precious metals and gems are found in the ground. However, money does have a natural growth law called *compound interest.*

First comes **simple interest.** A dollar invested for one year at 20% interest per year is worth $1.20 at the end of the year. But suppose the money is allowed to ride for another year. Then what is it worth? If the contract says *simple* annual interest, then it earns another $0.20, so is worth $1.40 at the end of the second year. But that is not fair because the investor has left the $0.20 earned the first year invested for another year. It should also earn interest at the annual rate of 20% for that second year, that is, it should earn $0.04, so the original dollar should be worth $1.44 at the end of two years. This would be the case if the contract specified **compound interest,** which means that interest is also paid on accumulated interest.

In business, interest is usually specified by an *annual* percentage rate and a payment period. An example is 18% annual interest compounded monthly. What this means is that the monthly interest rate is $\frac{1}{12}(18\%) = \frac{1}{12}(0.18) = 0.015$, and that interest is paid on accumulated interest as well as on principal (the initial amount).

In general, if the principal amount P is invested at the (decimal) interest rate i per payment period for n payment periods, then it grows to

$$A_n = P(1 + i)^n$$

This is the **law of compound interest.** It is easy enough to prove. Initially the amount is P. During the first payment period it earns Pi, so the amount after one payment period is $P + Pi = P(1 + i)$. The rule is this: Whatever amount A_{k-1} the account is worth at the beginning of the k-th payment period, it is worth $A_k = A_{k-1}(1 + i)$ at the end of the k-th payment period. Thus the successive values of the account are

$$P \quad P(1 + i) \quad P(1 + i)^2 \quad P(1 + i)^3 \quad \ldots \quad P(1 + i)^n = A_n$$

Now suppose that the annual rate is I and that interest is compounded m times per year. Suppose also that we invest for N years. Then the payment period is $1/m$ year, there are mN payment periods, and the interest rate per payment period is $i = I/m$. Therefore P grows to

$$A_{mN} = P(1 + i)^{mN} = P\left(1 + \frac{I}{m}\right)^{mN}$$

Since we can use calculus, we naturally ask what is the limit of A_{mN} as $m \to +\infty$, that is, what happens if interest is compounded more and more times per year. A similar limit that we know is

$$\lim_{x \to +\infty} \left(1 + \frac{1}{x}\right)^x = e$$

If we set $x = m/I$ in this limit, then $mN = xIN$, and we have

$$\lim_{m \to +\infty} \left(1 + \frac{I}{m}\right)^{mN} = \left[\lim_{x \to +\infty} \left(1 + \frac{1}{x}\right)^x\right]^{IN} = e^{IN}$$

Thus the value of the money **compounded continuously** for N years is

$$V_N = \lim_{m \to +\infty} A_{mN} = Pe^{IN}$$

This is the natural growth law for money. We write the law in our usual form for growth laws:

$$V(t) = Pe^{It} \qquad \text{so that} \qquad \frac{dV}{dt} = IV$$

where t is measured in years, P is the initial amount, and I is the annual interest rate, to be compounded continuously.

Example 4

A thousand dollars is deposited in a 7-year CD (certificate of deposit) paying $11\frac{3}{4}\%$ annual interest. Estimate its value upon maturity

a if interest is compounded daily, and

b if interest is compounded continuously.

Solution

a The annual rate is $11\frac{3}{4}\% = 0.1175$, so the daily interest rate is $i = 0.1175/365$. The number of payments is $7 \times 365 = 2555$. Therefore the maturity value of the CD is

$$A = 1000(1+i)^{2555} = 1000\left(1 + \frac{0.1175}{365}\right)^{2555} \approx \$2275.88$$

(We have taken the year as 365 days; however, some financial institutions use a 360-day year for interest calculations.)

b If $V(t)$ denotes the value of the CD after t years, then

$$\frac{dV}{dt} = 0.1175V \qquad \text{so that} \qquad V(t) = 1000e^{0.1175t}$$

Therefore

$$V(7) = 1000e^{(0.1175)(7)} \approx \$2276.18$$

This is not much different from the answer in part **a**. ●

Exercises

Numerical Computation To solve $a^x = b$ on most calculators, use the sequence of keystrokes

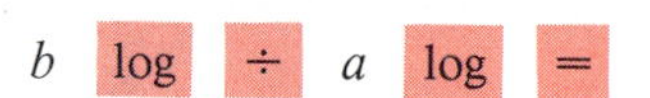

b log ÷ a log =

You can substitute ln for log in both places if that is convenient. It works; don't worry why. We'll explain this in the next section.

1 A culture of bacteria has a population of 3×10^6 initially, and 9×10^6 two hours later. What is the growth law? How long does it take the colony to double?

2 A bacteria colony has a population of 5.2×10^6 after 2 days and 8.7×10^6 after 3 more days. How many bacteria were there initially?

3 Assume that population grows at a rate proportional to the population itself. In 1960 the U.S. population was 179 million, and in 1980 it was 227 million. Make predictions for the years, 2000, 2025, and 2050.

4 (cont.) Estimated world population was 1.6 billion in 1900 and 4.5 billion in 1980. Make predictions for 2000, 2050, and 2100.

5 Radium-224 has a half-life of 3.64 days. Find its decay law. How long will it take for $\frac{2}{3}$ of a quantity to disintegrate?

6 After 3 days exactly 0.5 kg of a 2-kilogram sample of a certain radioactive substance remains undecayed. At what rate in kg/day is the substance decaying at 4 days?

7 A 5-gram sample of radioactive material contains 2 grams of polonium-210, which has a half-life of 138.3 days, and 3 grams of radium-224, which has a half-life of 3.64 days. When will the sample contain equal amounts of polonium-210 and radium-224?

8 A salt in solution decomposes into other substances at a rate proportional to the amount still unchanged. If 10 kg of a salt reduces to 5 kg in $\frac{1}{2}$ hr, how much is left after 15 hr?

9 In a certain calculus course, the number of students dropping out each class day was proportional to the number still enrolled. If 2000 started out and 10% dropped after 12 classes, estimate the number left after 36 classes.

10 Under certain ideal conditions the rate of change of pressure above sea level is proportional to the pressure. If the barometer reads 762 mm at sea level, and 635 mm at 1200 m, find the barometric pressure at 6500 m.

11 At what annual interest rate will money compounded continuously double in a year?

12 How long will it take a sum of money compounded continuously at 12.5% per annum to show a 100% return.

13 Suppose a quantity of hot fluid is stirred so that at any time t its temperature $u(t)$ is uniform (the same) throughout the fluid. Let a denote the constant outside temperature. **Newton's law of cooling** says that the rate of decrease of u, due to heat loss at the surface, is proportional to $u - a$, that is

$$\frac{du}{dt} = -k(u-a) \quad \text{where} \quad k > 0$$

Solve for u. Show that $u(t) \to a$ as $t \to +\infty$. [Hint Set $v(t) = u(t) - a$.]

14 (cont.) Suppose $a = 0°\text{C}$, and the fluid cools from $100°\text{C}$ to $50°\text{C}$ in 5 minutes. How much longer will it take to cool to $5°\text{C}$?

15 The electric energy of a certain diatomic molecule is

$$U = K(e^{-2a(x-c)} - 2e^{-a(x-c)})$$

where k, a, and c are positive constants and x is the distance between the atoms. Find the **dissociation energy,** the maximum of $-U$.

16 The rate of growth of the mass m of a falling raindrop is λm, where $\lambda > 0$. Show that $m = m_0 e^{\lambda t}$.

17 Living wood absorbs atmospheric radiocarbon (^{14}C, half-life 5568 yr), and at equilibrium the rate of absorption exactly balances the rate of loss by radioactive decay. Therefore the amount of radiocarbon per gram of total carbon in all living wood can be considered constant. When wood dies, it no longer absorbs fresh ^{14}C, and its ^{14}C content decays at the rate of 6.68 disintegrations per minute (dpm) per gram. Wood excavated by archeologists in 1950 from a Babylonian city built in the time of King Hammurabi, the law-giver, measured 4.09 dpm. Show that Hammurabi lived about 1990 B.C.

18 (cont.) Wood left by prehistoric cave dwellers was found in 1950 in the Lascaux Caves in France, and measured 0.97 dpm. Estimate the age of the famous cave paintings in 1950.

4-3 The Logarithm Function

The logarithm function is the inverse function of the exponential function. Let us review what this means. First, the exponential function $y = e^x = \exp x$ has domain **R**, the x-axis, and range the interval $(0, +\infty)$ of the y-axis. See Figure 4-3-1. Also, $\exp x$ is a strictly increasing function on **R**. In fact,

Figure 4-3-1
Graph of $y = e^x$

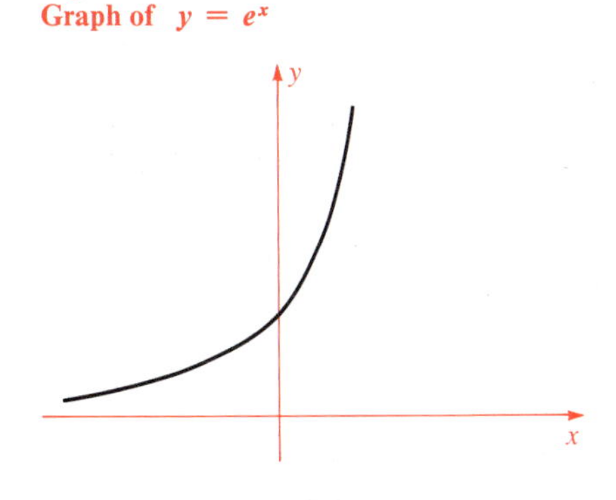

$$\frac{d}{dx}\exp x = \exp x > 0$$

According to our discussion of inverse functions in general (page 133), the relation $y = \exp x$ can be "solved" for x. That is, the function $y = \exp x$ has an inverse function $x = \ln y$, called the **natural logarithm** function. Its domain is $(0, +\infty)$, the range of $\exp x$. Its range is **R**, the domain of $\exp x$. As usual, we graph $y = \ln x$ rather than $x = \ln y$. According to our discussion (page 134) of graphs of inverse functions, the graph of $y = \ln x$ is the reflection in the diagonal line $y = x$ of the graph of $y = e^x$. See Figure 4-3-2.

Figure 4-3-2
The graph of $y = \ln x$ is the reflection in $y = x$ of the graph of $y = e^x$.

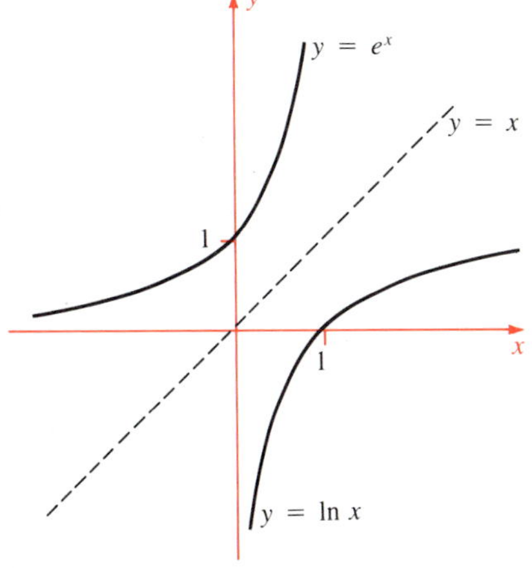

Logarithm Function

The function $y = \ln x$ is the inverse function of $y = \exp x$. Thus it satisfies the relations

$$\exp(\ln x) = x \quad \text{for } x > 0 \qquad \text{and} \qquad \ln(\exp x) = x \quad \text{for all } x$$

The function $\ln x$ is strictly increasing with domain $(0, +\infty)$ and range **R**. What is more,

$$\begin{cases} \ln x < 0 & \text{for } 0 < x < 1 \\ \ln 1 = 0 & \\ \ln x > 0 & \text{for } x > 1 \end{cases}$$

Example 1

a $\ln e^4 = 4$
b $\ln[\exp(-2.3)] = -2.3$
c $\ln e^\pi = \pi$
d $e^{\ln 7} = 7$
e $e^{\ln 10.6} = 10.6$
f $\exp(\ln \frac{1}{2}) = \frac{1}{2}$

From Figure 4-3-2 it is not clear whether the graph of $y = \ln x$ rises arbitrarily high or not. Actually, the graph reaches all levels because the range of $\ln x$ is $(0, +\infty)$; that is, each $y > 0$ is the natural logarithm of something, namely of e^y. For example,

$$10 = \ln e^{10} \qquad 100 = \ln e^{100} \qquad 1000 = \ln e^{1000}$$

Therefore $y = \ln x$ approaches $+\infty$ as x approaches $+\infty$.

Similarly, each real number $y < 0$ is the natural logarithm of something: $y = \ln e^y$. For example,

$$-10 = \ln e^{-10} \qquad -100 = \ln e^{-100} \qquad -1000 = \ln e^{-1000}$$

Therefore $y = \ln x$ approaches $-\infty$ as $x \to 0+$.

$$\lim_{x \to +\infty} (\ln x) = +\infty \qquad \lim_{x \to 0+} (\ln x) = -\infty$$

Algebraic Properties of $\ln x$

Recall the rules of exponents (Theorem 2, page 226) as applied to the exponential function e^x:

- $e^a e^b = e^{a+b}$
- $e^a / e^b = e^{a-b}$
- $e^{-a} = 1/e^a$
- $(e^a)^b = e^{ab}$

Somehow, these properties ought to rub off onto the inverse function $\ln x$. They do—each of the four statements can be translated into a corresponding statement about natural logarithms. Set $e^a = x$ and $e^b = y$; that means $a = \ln x$ and $b = \ln y$. The first rule

$$e^a e^b = e^{a+b} \quad \text{translates to} \quad xy = e^{\ln x + \ln y}$$

Hence, $\ln(xy) = \ln x + \ln y$. Similarly, the other three rules translate into properties of $\ln x$.

Rules of Logarithms

Let x and y be positive. Then

- $\ln xy = \ln x + \ln y$
- $\ln x/y = \ln x - \ln y$
- $\ln 1/x = -\ln x$
- $\ln x^b = b \ln x$

Warning There are no nice formulas for $\ln(x + y)$ or for $(\ln x)(\ln y)$.

Example 2

Simplify

a $e^{(\ln 25)/2}$ **b** $\ln 2 + \ln 4 + \ln 8 + \cdots + \ln 128$

Solution

a $e^{(\ln 25)/2} = e^{\ln \sqrt{25}} = e^{\ln 5} = 5$

Alternative: $e^{(\ln 25)/2} = (e^{\ln 25})^{1/2} = 25^{1/2} = 5$

b $\ln 2 + \ln 4 + \cdots + \ln 128 = \ln 2 + \ln 2^2 + \cdots + \ln 2^7$

$$= \ln 2 + 2\ln 2 + \cdots + 7\ln 2$$

$$= (1 + 2 + 3 + \cdots + 7)\ln 2 = 28\ln 2$$

Derivative of $\ln x$

The natural logarithm function inherits useful algebraic properties from e^x. It also inherits a simple derivative:

$$\frac{d}{dx}(\ln x) = \frac{1}{x} \qquad (x > 0)$$

This follows from the rule for differentiating inverse functions (page 134). In this case, if $y = \ln x$ then $x = e^y$, so by the chain rule,

$$1 = \frac{dx}{dx} = \frac{de^y}{dy} \cdot \frac{dy}{dx} = e^y \frac{dy}{dx} = x\frac{dy}{dx}$$

Therefore

$$\frac{d(\ln x)}{dx} = \frac{dy}{dx} = \frac{1}{x}$$

The derivative formula agrees with the graph of $y = \ln x$ in Figure 4-3-2. The slope $1/x$ is always positive, it becomes very large as $x \to 0+$, and it dies out as $x \to +\infty$.

Remark The formula for $(\ln x)'$ fills a certain gap in our repertory of derivatives. From $(x^{\alpha+1})' = (\alpha + 1)x^\alpha$ we have

$$x^\alpha = \frac{d}{dx}\left(\frac{x^{\alpha+1}}{\alpha + 1}\right) \quad \text{for} \quad \alpha \neq -1$$

which gives each power of x, except x^{-1}, as a derivative. The additional formula

$$x^{-1} = \frac{d}{dx}\ln x$$

fills this gap.

If $u(x)$ is differentiable and $u(x) > 0$, then the composite function $\ln u(x)$ is differentiable. By the chain rule,

$$\frac{d}{dx}\ln u(x) = \left(\frac{d}{du}\ln u\right)\left(\frac{du}{dx}\right) = \frac{1}{u(x)} \cdot \frac{du}{dx}$$

Therefore

$$\frac{d}{dx}\ln u(x) = \frac{u'(x)}{u(x)} \quad \text{for} \quad u(x) > 0$$

Example 3

Differentiate

a $y = \ln(x^2 + 1)$ **b** $y = \ln\left(\dfrac{1 + x}{1 - x}\right)$ for $-1 < x < 1$

Solution

a $\dfrac{dy}{dx} = \dfrac{u'}{u} = \dfrac{(x^2 + 1)'}{x^2 + 1} = \dfrac{2x}{x^2 + 1}$

b Clearly, $1 + x > 0$ and $1 - x > 0$ for $-1 < x < 1$. By one of the rules of logarithms, $y = \ln(1 + x) - \ln(1 - x)$. Hence

$$\frac{dy}{dx} = \frac{(1 + x)'}{1 + x} - \frac{(1 - x)'}{1 - x}$$

$$= \frac{1}{1 + x} - \frac{-1}{1 - x} = \frac{2}{1 - x^2}$$

●

Logs to Other Bases

The natural logarithm function is also called the **log to the base** e, because it is the inverse of the exponential function $x = e^y$. Now if $a > 1$, then $x = a^y$ is also an exponential function with positive derivative, so it has an inverse function called the **log to the base** a and written

$$y = \log_a x$$

Thus

$$a^{\log_a x} = x \quad \text{for } x > 0 \qquad \text{and} \qquad \log_a a^y = y \quad \text{for all } y$$

We seem to have a whole slew of new logarithm functions. Actually not; they are only constant multiples of $\ln x$. This is clear if we apply the natural logarithm to the relation $a^y = x$. Indeed, we set $y = \log_a x$ so that $a^y = x$. Then we apply ln:

$$\ln a^y = \ln x \quad \text{so} \quad y \ln a = \ln x \qquad \text{that is} \qquad y = \frac{\ln x}{\ln a}$$

Since $y = \log_a x$, this proves the following:

For each $a > 1$ we have $\log_a x = \dfrac{\ln x}{\ln a}$ ●

The rules for $\ln x$ also apply to $\log_a x$:

- $\log_a xy = \log_a x + \log_a y$
- $\log_a x/y = \log_a x - \log_a y$
- $\log_a 1/x = -\log_a x$
- $\log_a x^b = b \log_a x$ ●

The particular case $a = 10$ is useful in calculations, and $\log_{10} x$ is called the **common logarithm** function. It is also written simply log without a subscript. Before electronic computers and calculators were invented, calculations were done using log tables. An important feature of log is its relation to scientific notation:

$$\log(a \times 10^n) = \log a + \log 10^n = \log a + n$$

Logarithms to any base can be expressed in terms of log. Indeed, if $\log_a x = c$, then $a^c = x$. Apply log to this equation:

$$\log a^c = \log x \qquad c \log a = \log x \qquad c = \frac{\log x}{\log a}$$

Therefore

$$\log_a x = \frac{\log x}{\log a}$$

Example 4

Use the "log tables" in your calculator to estimate to five places

a $\ln 2$ **b** $\log_7 23$

Solution

a $\ln 2 = \dfrac{\log 2}{\log e} \approx \dfrac{0.30103}{0.43429} \approx 0.69315$

Check By calculator $\ln 2 \approx 0.693147$

b $\log_7 23 = \dfrac{\log 23}{\log 7} \approx \dfrac{1.36173}{0.84510} \approx 1.61132$

Check On an algebraic calculator,

7 [y^x] 1 . 6 1 1 3 2 [=] 22.99976 · · ·

Rate of Growth of $\ln x$

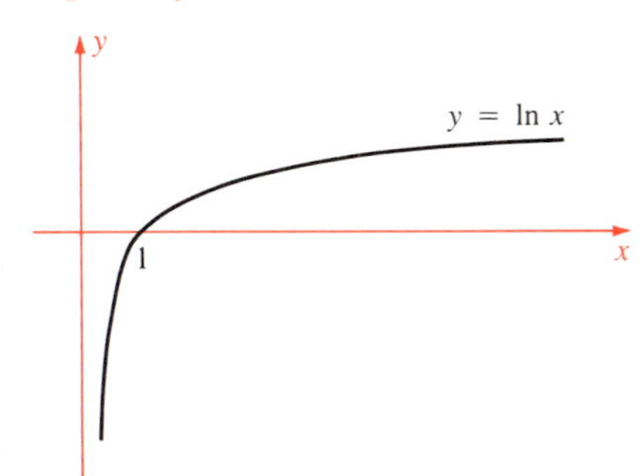

Figure 4-3-3
The growth of $y = \ln x$ is painfully slow as $x \to +\infty$.

We know that $\ln x$ is strictly increasing and that $\ln x \to +\infty$ as $x \to +\infty$. Its graph (Figure 4-3-3) increases slowly, being the mirror image in $y = x$ of the rapidly increasing graph of $y = e^x$.

Actually, the rate of increase of $y = \ln x$ is agonizingly slow. The curve does not reach the level $y = 10$ until $x = e^{10} \approx 22{,}000$; it does not reach the level $y = 100$ until $x = e^{100} \approx 2.7 \times 10^{43}$. Obviously, the larger x is, the smaller $\ln x$ is relative to x. Precisely,

$$\lim_{x \to +\infty} \frac{\ln x}{x} = 0$$

This assertion will be proved if we can show that for any positive integer n, no matter how large,

$$\frac{\ln x}{x} < \frac{1}{n}$$

for x sufficiently large. We use the corresponding fact (page 232) about the rapid growth of e^x compared to x^n: Given any positive integer n we have

$$\lim_{x \to +\infty} \frac{e^x}{x^n} = +\infty$$

This implies $e^x > x^n$ when x is sufficiently large. Hence

$$\ln e^x > \ln x^n \qquad x > n \ln x \qquad \frac{\ln x}{x} < \frac{1}{n}$$

which is what we wanted to prove.

Perhaps comparing $\ln x$ with x is unrealistic. Maybe we should compare $\ln x$ with a smaller function, for example with $x^{1/2}$, which is much smaller than x. How dows $\ln x$ compare with $x^{1/2}$ or with $x^{1/3}$, which is smaller yet? The answer is that *any* positive power of x overwhelms $\ln x$.

If $p > 0$ then $\lim_{x \to +\infty} \dfrac{\ln x}{x^p} = 0$

This follows from the behavior of $(\ln x)/x$. For if we set $x^p = y$, then $x = y^{1/p}$ and $y \to +\infty$ as $x \to +\infty$. Therefore

$$\frac{\ln x}{x^p} = \frac{\ln y^{1/p}}{y} = \frac{(1/p) \ln y}{y} = \frac{1}{p} \cdot \frac{\ln y}{y} \to 0$$

as $x \to +\infty$. We list some values of $(\ln x)/x^{1/3}$ for large x in Table 4-3-1.

Behavior of $\ln x$ as $x \to 0+$

We have seen that $\ln x \to -\infty$ as $x \to 0+$. We ask how fast $\ln x$ approaches $-\infty$, fast enough that $x \ln x$ also approaches $-\infty$ as $x \to 0+$, or less fast? Here is the precise answer:

- If $p > 0$, then $\lim_{x \to 0+} (x^p \ln x) = 0$
- In particular, $\lim_{x \to 0+} (x \ln x) = 0$

To prove this assertion, replace x by $1/y$, so $y \to +\infty$ as $x \to 0+$. Then

$$\lim_{x \to 0+} (x^p \ln x) = \lim_{y \to +\infty} \frac{\ln(1/y)}{y^p} = \lim_{y \to +\infty} \frac{-\ln y}{y^p} = 0$$

We list some values for the function $x^{1/4} \ln x$ in Table 4-3-2.

Table 4-3-1

x	$(\ln x)/x^{1/3}$
10^3	0.691
10^6	0.138
10^9	0.0207
10^{30}	6.91×10^{-9}
10^{300}	6.91×10^{-98}

Table 4-3-2

x	$x^{1/4} \ln x$
10^{-1}	-1.29
10^{-2}	-1.46
10^{-6}	-0.437
10^{-20}	-4.61×10^{-4}
10^{-100}	-2.30×10^{-23}

Exercises

Find the inverse function

1 $x = e^{-y}$

2 $x = \exp(-y^3)$

3 $x = 1/\ln y \quad (y > 1)$

4 $x = 1/\ln y$ $(0 < y < 1)$

5 $x = \ln(y + 5)$ $(y > -5)$

6 $x = \ln(\ln y)$ $(y > 1)$

Simplify

7 $\ln e^{a+2}$

8 $\exp(\ln x^2)$

9 $e^{-\ln x}$

10 $\exp(2 \ln x + 5 \ln y)$

11 $\ln \sqrt{e}$

12 $\ln(1/e^{3/2})$

Differentiate with respect to x

13 $\ln 5x$

14 $3 \ln 4x$

15 $2 \ln x^2$

16 $\ln x^4$

17 $\ln(1/x)$

18 $\ln(x^2 + x)$

19 $\ln(\sin x)$

20 $\ln(\cos x)$

21 $\ln\left(\dfrac{x+1}{x-1}\right)$

22 $\ln(\ln x)$

23 $\sqrt{\ln x}$

24 $(\ln x)^2$

25 $\dfrac{1}{\ln x}$

26 $\ln\left(\dfrac{x+2}{2x+1}\right)$

27 $\ln(x + \sqrt{x^2 + 1})$

28 $\ln(\sec x + \tan x)$

29 $-\frac{1}{2} \ln\left(\dfrac{2 + \sqrt{x^2 + 4}}{x}\right)$

30 $\dfrac{1}{b} \ln(a + bx)$

31 $(\ln x)^n/n \quad n \neq 0$

32 $x^n(\ln x - 1/n)/n \quad n \neq 0$

Find the tangent line to $y = \ln x$

33 at $x = 1$

34 at $x = e$

35 Show that $y = \ln x$ is concave down.

36 Show that $y = \ln(\ln x)$ is concave down on its domain.

37 Minimize $y = x \ln x$ for $x > 0$.

38 Minimize $y = x^2 \ln x$ for $x > 0$.

39 Minimize $y = x^3 \ln x$ for $x > 0$.

40 Let $p > 0$ and $q > 0$. Maximize $y = \dfrac{(\ln x)^p}{x^q}$ for $x \geq 1$.

Graph

41 $y = x \ln x$

42 $y = \dfrac{\ln x}{x}$

Find

43 $\displaystyle\lim_{x \to +\infty} \frac{x}{10^6 \ln x}$

44 $\displaystyle\lim_{x \to +\infty} \frac{(x+1)\ln x}{x^2}$

45 $\displaystyle\lim_{x \to +\infty} \frac{(\ln x)^3}{x^{1/4}}$

46 $\displaystyle\lim_{x \to +\infty} \frac{\log_{10} x}{x}$

47 $\displaystyle\lim_{x \to +\infty} \frac{x^2 + 1}{x(\ln x)^2}$

48 $\displaystyle\lim_{x \to 0+} \sqrt{x} \ln x$

49 $\displaystyle\lim_{x \to 0+} x(\ln x)^{20}$

50 $\displaystyle\lim_{x \to +\infty} \frac{x \ln x}{x^p + 1} \quad (p > 1)$

51 $\displaystyle\lim_{x \to +\infty} \frac{3(\ln x)^2 + 1}{x}$

52 $\displaystyle\lim_{x \to +\infty} \frac{\ln(\ln x)}{\ln x}$

53 Show that $x = (\ln y)/y$ for $y > e$ has an inverse function $y = g(x)$, and compute $g'(3/e^3)$.

54 Let $p > 0$ and $q > 0$. Prove $\displaystyle\lim_{x \to +\infty} \frac{(\ln x)^p}{x^q} = 0$.

Find the integer n for which

55 $e^n < 1000 < e^{n+1}$

56 $10^n < e^{100} < 10^{n+1}$

Harder Exercises

57 Prove that $\ln x \leq x - 1$ for $0 < x$, with equality only if $x = 1$. [Hint Draw a graph.]

58 (cont.) Prove that $(x - 1)/x \leq \ln x$ for $0 < x$, with equality only if $x = 1$.

59 Use Exercises 57 and 58 to prove Napier's inequality: If $0 < x < y$, then

$$\frac{1}{y} < \frac{\ln y - \ln x}{y - x} < \frac{1}{x}$$

60 Prove $\ln x \leq n(x^{1/n} - 1)$ for $n \geq 1$ and $0 < x$, with equality only if $x = 1$. [Hint Use Exercise 57.]

61 (cont.) Prove $\left(1 + \dfrac{x}{n}\right)^n < e^x$ for $x > 0$ and $n \geq 1$.

62 Prove $\ln x \geq n(1 - x^{-1/n})$ for $n \geq 1$ and $0 < x$, with equality only if $x = 1$. [Hint Use Exercise 57.]

63 (cont.) Prove $e^x < \left(1 - \dfrac{x}{n}\right)^{-n}$ for $x > 0$ and $n \geq 1$.

64 Use the results of Exercises 61 and 63 to prove

$$\left(1 + \frac{1}{n}\right)^n < e < \left(1 + \frac{1}{n}\right)^{n+1} \quad \text{for all } n \geq 1$$

[Hint Set $x = n/(n + 1)$.]

65 Prove $\ln x < \frac{1}{2}(x - 1/x)$ for $x > 1$. [Hint Set $f(x) = \frac{1}{2}(x - 1/x) - \ln x$ and differentiate.]

66 (cont.) Prove $\ln x < \sqrt{x} - 1/\sqrt{x}$ for $x > 1$.

67 Compare the inequalities in Exercises 57, 60 (for $n = 2$), and 66 by completing the table to five places

x	$\ln x$	$x - 1$	$2(\sqrt{x} - 1)$	$\sqrt{x} - 1/\sqrt{x}$
1.2				
1.5				
2.0				
e				
3.0				

68 From Exercise 66, deduce Kepler's inequality

$$\frac{\ln y - \ln x}{y - x} < \frac{1}{\sqrt{xy}} \quad \text{for } 0 < x < y$$

69 Set $f(x) = \ln x - \frac{1}{2}(x - 2) + \frac{1}{8}(x - 2)^2$. Show that $f(x)$ is strictly increasing for $1 \le x \le 2$. Deduce that $\ln 2 > \frac{5}{8}$.

70 Set $f(x) = \ln x - \frac{1}{3}(x - 3) + \frac{1}{18}(x - 3)^2$. Show that $f(x)$ is strictly increasing for $2 \le x \le 3$. Deduce that $\ln 3 > \ln 2 + \frac{7}{18}$.

71 Combine Exercises 69 and 70 to prove $\ln 3 > 1$. Conclude that $e < 3$.

72 Generalize Exercise 66 for $n \ge 2$ and $x > 1$:

$$\ln x < \frac{n}{2(n-1)}\left(x^{(n-1)/n} - \frac{1}{x^{(n-1)/n}}\right)$$

[Hint Replace x in Exercise 65 by a suitable expression.]

73 Prove that $\ln x < a + (x - a)/a$ for $0 < a$, $0 < x$, $x \ne a$.

74 (cont.) Prove $\ln(n + 1) - \ln n < 1/n$ for $n = 1, 2, 3, \cdots$.

75 (cont.) Prove $\ln(n + 1) - \ln n > 1/(n + 1)$ for $n = 1, 2, 3, \cdots$.

76 (cont.) Finally, prove

$$\tfrac{1}{2} + \tfrac{1}{3} + \cdots + 1/n < \ln n < 1 + \tfrac{1}{2} + \tfrac{1}{3} + \cdots + 1/n$$

for $n = 2, 3, \cdots$. Test this numerically for $n = 5$ and $n = 10$, and for $n = 100$ if you can.

4-4 Applications of Logarithms

Logarithms are sometimes used to prove assertions about limits. Suppose $f(x) > 0$ and $L > 0$. If

$$\lim_{x\to a} f(x) = L \qquad \text{then} \qquad \lim_{x\to a} [\ln f(x)] = \ln L$$

because $\ln y$ is continuous. Conversely, if

$$\lim_{x\to a} \ln f(x) = \ln L \qquad \text{then} \qquad \lim_{x\to a} \exp[\ln f(x)] = \exp(\ln L)$$

because $\exp y$ is continuous. However, $\ln$ and $\exp$ are inverse functions, so this says that $\lim_{x\to a} f(x) = L$. To summarize:

- Suppose $f(x) > 0$ and $L > 0$. Then

$$\lim_{x\to a} f(x) = L \qquad \text{if and only if} \qquad \lim_{x\to a} \ln[f(x)] = \ln L$$

Example 1

Let $a > 0$. Prove $\lim_{x\to+\infty} a^{1/x} = 1$.

Solution It is equivalent to prove $\lim_{x\to+\infty} \ln a^{1/x} = \ln 1 = 0$. But

$$\ln a^{1/x} = \frac{\ln a}{x} \to 0 \quad \text{as} \quad x \to +\infty$$

Remark You can test this result on a calculator. Start with any number $a > 0$. Take repeated square roots, obtaining

$$a^{1/2} \qquad a^{1/4} \qquad a^{1/8} \qquad a^{1/16} \qquad \cdots$$

and watch the numbers approach 1. Starting with $a = 500$ for example, you obtain

22.36068	4.72871	2.17456	1.47464	1.21435
1.10197	1.04975	1.02457	1.01221	1.00609
1.00304	1.00152	$\cdots$.		

Example 2

Prove $\lim_{x\to 0+} x^x = 1$.

Solution It is equivalent to prove $\lim_{x\to 0} \ln x^x = \ln 1 = 0$. But $\ln x^x = x \ln x$, and according to a result on page 248,

$$\lim_{x\to 0+} x \ln x = 0$$

so we are done. A quick calculator check:

$$(0.1)^{0.1} \approx 0.794 \qquad (0.01)^{0.01} \approx 0.955$$

$$(0.001)^{0.001} \approx 0.993 \qquad (0.0001)^{0.0001} \approx 0.99908$$

●

Exponential Equations

An equation of the form

$$b^x = c \qquad (b > 0,\ c > 0)$$

can be solved by logs. Use any base a and apply $\log_a$ to the equation:

$$\log_a b^x = \log_a c \qquad \text{that is} \qquad x \log_a b = \log_a c$$

Hence

$$x = \frac{\log_a c}{\log_a b}$$

In particular

$$x = \frac{\ln c}{\ln b} = \frac{\log c}{\log b}$$

Either of these two expressions for x can be used in practice because the functions ln and log are available on all scientific calculators. (The function ln is available in most high-level computing languages, but log not necessarily.)

Example 3

How long will it take for money to triple if it is invested at 14% annual interest rate, compounded daily?

Solution Let n be the length of the investment in years. The set-up is

$$\left(1 + \frac{0.14}{365}\right)^{365n} = 3$$

Therefore

$$365n = \frac{\ln 3}{\ln\left(1 + \frac{0.14}{365}\right)} \approx \frac{1.098612}{3.834881 \times 10^{-4}} \approx 2.86479 \times 10^3$$

Divide by 365 to find $n \approx 7.849$ yr (0.849 yr is about 10 months and 6 days). ●

Logarithms can be used to estimate very large numbers.

Example 4

The fast fission reactions in a nuclear bomb double every 10^{-8} sec, a unit of time called a *shake*. At this rate, a single fission gives rise to 2^{1000} fissions after 10^{-5} sec. Estimate 2^{1000}.

Solution If you simply key in

2 [y^x] 1 0 0 0 [=]

on an algebraic calculator, you get an overflow error, since a calculator only goes up to $9.999 \cdots \times 10^{99}$. The trick is to use the keys [log] and [10^x] cleverly. If

$$K = 2^{1000} \qquad \text{then} \qquad \log K = 1000 \log 2$$

Begin by computing $1000 \log 2$:

2 [log] [×] 1 0 0 0 [=] 301.02999 · · ·

Now subtract the integer part and apply 10^x to the difference:

[−] 3 0 1 [=] [10^x] 1.0715 · · ·

Therefore the answer is $2^{1000} \approx 1.07 \times 10^{301}$. This is a rather large number, explaining why a nuclear blast takes considerably less time than 10^{-5} sec. ●

Power Functions

In Chapter 2 we proved the power rule (page 117)

$$\frac{d}{dx} x^r = r x^{r-1}$$

for r rational. What about functions like $f(x) = x^{\sqrt{2}}$ or $f(x) = x^e$ in which the exponent is an irrational number?

General Power Rule

If p is any real number, then

$$\frac{d}{dx}x^p = px^{p-1} \qquad (x > 0)$$

Proof Set $y = x^p$. Then $\ln y = p \ln x$; hence

$$\frac{d}{dx}\ln y = \frac{d}{dx}(p \ln x) = \frac{p}{x}$$

But by the chain rule, $(\ln y)' = y'/y$; hence

$$\frac{y'}{y} = \frac{p}{x} \qquad \text{so that} \qquad y' = p\frac{y}{x} = p\frac{x^p}{x} = px^{p-1}$$

Example 5

a $\displaystyle\frac{d}{dx}x^{\sqrt{2}} = (\sqrt{2})x^{\sqrt{2}-1}$ **b** $\displaystyle\frac{d}{dx}x^e\Big|_{x=1} = ex^{e-1}\Big|_{x=1} = e$

The next example uses the logarithm to derive an interesting inequality.

Example 6

a Maximize $y = (\ln x)/x$ for $x > 0$.

b Show that $x^e \le e^x$ for all $x > 0$, with equality only for $x = e$.

Solution

a Differentiate $y = (\ln x)/x$:

$$\frac{dy}{dx} = \frac{x(1/x) - \ln x}{x^2} = \frac{1 - \ln x}{x^2}$$

It follows that $y' = 0$ if $x = e$,

$$y' > 0 \quad \text{if} \quad 0 < x < e \qquad \text{and} \qquad y' < 0 \quad \text{if} \quad e < x$$

Consequently, y is strictly increasing as x goes from 0 to e and strictly decreasing as x continues beyond e. Therefore the maximum of y is $y(e) = (\ln e)/e = 1/e$.

b By **a** we have

$$\frac{\ln x}{x} \le \frac{1}{e}$$

for all $x > 0$, with equality only at $x = e$. Consequently, for $x > 0$ and $x \ne e$,

$$e \ln x < x \qquad \ln x^e < x = \ln e^x$$

It follows that $x^e < e^x$. ●

Remark A puzzle asks which is larger, π^e or e^π? This example shows that e^π is larger. Numerically, $e^\pi \approx 23.1407$ and $\pi^e \approx 22.4592$.

Logarithmic Differentiation

Differentiation of products, quotients, and powers can often be simplified by taking logs before differentiating. Be careful, however, to take logs of positive functions only.

Example 7

Find the derivative of

$$y = \left(\frac{x^2}{x^4+1}\right)^{1/3} \quad \text{for} \quad x \neq 0$$

Solution Take logs:

$$\ln y = \tfrac{1}{3}[\ln x^2 - \ln(x^4+1)] = \tfrac{1}{3}[2\ln x - \ln(x^4+1)]$$

Therefore

$$\frac{y'}{y} = (\ln y)' = \tfrac{1}{3}\left(\frac{2}{x} - \frac{4x^3}{x^4+1}\right)$$

$$y' = \frac{y}{3}\left(\frac{2}{x} - \frac{4x^3}{x^4+1}\right) = \tfrac{1}{3}\left(\frac{x^2}{x^4+1}\right)^{1/3}\left(\frac{2}{x} - \frac{4x^3}{x^4+1}\right)$$ ●

For the next example, we need a slight extension of the formula for the derivative of $\ln u(x)$. If $u(x) < 0$, then $|u(x)| > 0$, so $\ln|u(x)| = \ln[-u(x)]$ is defined. By the chain rule, its derivative is

$$\frac{d}{dx}\ln|u| = \frac{d}{dx}\ln(-u) = \frac{(-u)'}{-u} = \frac{u'}{u}$$

We combine the cases $u(x) > 0$ and $u(x) < 0$ into a single formula:

Logarithmic Derivative

If $u(x) \neq 0$, then the **logarithmic derivative** of $u(x)$ is defined to be

$$\frac{d}{dx}\ln|u| = \frac{u'}{u}$$ ●

The logarithmic derivative of $u(x)$ is the *relative* rate of change of $u(x)$ with respect to x, and it is often useful.

The next example concerns derivatives of factored polynomials and rational functions. If $f(x) = (x-a)(x-b)$, then

$$\frac{f'}{f} = \frac{(x-b)+(x-a)}{(x-a)(x-b)} = \frac{1}{x-a} + \frac{1}{x-b}$$

If f has three or more factors, a similar formula holds, but it is clumsy to derive by direct differentiation. Logarithmic differentiation comes to the rescue.

Example 8

Prove

a If $f(x) = (x - a)(x - b) \cdots (x - d)$, then

$$\frac{f'(x)}{f(x)} = \frac{1}{x - a} + \frac{1}{x - b} + \cdots + \frac{1}{x - d} \qquad (x \neq a, b, \cdots, d)$$

b If $f(x) = (x - a)^m (x - b)^n \cdots (x - d)^p$, where $m, n, \cdots, p$ are positive or negative integers, then

$$\frac{f'(x)}{f(x)} = \frac{m}{x - a} + \frac{n}{x - b} + \cdots + \frac{p}{x - d} \qquad (x \neq a, b, \cdots, d)$$

Solution It suffices to derive **b** since **a** is the special case $m = n = \cdots = p = 1$. The natural impulse is to take logs first, but to be safe, take the log of $|f(x)|$:

$$|f(x)| = |x - a|^m |x - b|^n \cdots |x - d|^p$$

$$\ln|f(x)| = m \ln|x - a| + n \ln|x - b| + \cdots + p \ln|x - d|$$

Now differentiate:

$$\frac{f'(x)}{f(x)} = m \frac{d}{dx} \ln|x - a| + \cdots + p \frac{d}{dx} \ln|x - d|$$

$$= \frac{m}{x - a} + \cdots + \frac{p}{x - d}$$

●

Example 9

Differentiate $y = x^x$.

Solution The function $y = x^x$ is defined for $x > 0$. Take logs:

$$\ln y = \ln(x^x) = x \ln x$$

Now take d/dx:

$$\frac{y'}{y} = \frac{d}{dx}(x \ln x) = \ln x + x \cdot \frac{1}{x} = \ln x + 1$$

$$y' = (\ln x + 1) y$$

$$\frac{d}{dx} x^x = (\ln x + 1) x^x$$

●

We noted earlier that the exponential function occurs in many diverse natural phenomena. It is not surprising that its inverse, the logarithm function, also occurs in nature as do the related power functions. For example, in the subject of psychophysics, Fechner's law says that a stimulus of magnitude x causes a sensation of magnitude $y = K \ln x$.

Further examples: When a mole of an ideal gas undergoes a reversible isothermal expansion from volume V_0 to V_1, its entropy increase is $S_1 - S_0 = R \ln(V_1/V_0)$, where R is the universal gas constant. If $\pi(n)$ denotes the number of prime numbers from 2 to n, then $\pi(n) \approx n/\ln n$ for n large. (A prime number is an integer like 19 or 67 that has no factors other than 1 and itself. Thus $77 = 7 \times 11$ is not a prime.)

Applications to Economics

We now consider an important concept in economics. Suppose that x and y are positive variables and that y depends on x. The **elasticity** of y with respect to x is defined as

$$\frac{Ey}{Ex} = \frac{d(\ln y)}{d(\ln x)} = \frac{x}{y} \cdot \frac{dy}{dx}$$

Its meaning is this: Imagine that both x and y depend on time. In a time interval from t to $t + h$, the variable y changes by the amount $y(t+h) - y(t)$, so its *relative* change is the quotient $[y(t+h) - y(t)]/y(t)$. The elasticity of y with respect to x is the limit as $h \to 0$ of the ratio of the relative change of y to that of x:

$$\begin{aligned}\frac{Ey}{Ex} &= \lim_{h\to 0} \frac{[y(t+h) - y(t)]/y(t)}{[x(t+h) - x(t)]/x(t)} \\ &= \frac{x(t)}{y(t)} \lim_{h\to 0} \frac{[y(t+h) - y(t)]/h}{[x(t+h) - x(t)]/h} \\ &= \frac{x}{y} \cdot \frac{dy/dt}{dx/dt} \\ &= \frac{x}{y} \cdot \frac{dy}{dx}\end{aligned}$$

Example 10

For what functions $y = y(x)$ is Ey/Ex constant?

Solution Suppose $Ey/Ex = k$. Set $v = \ln y$ and $u = \ln x$. Then $dv/du = k$ so that $v = ku + c$, where c is a constant. That is,

$$\ln y = k \ln x + c = \ln x^k + c$$

Set $c = \ln a$. Then

$$\ln y = \ln x^k + \ln a = \ln ax^k$$

Therefore $y = ax^k$. ●

In economics, a **good** is anything that is supplied and purchased, like bread, furnace repair, soybean futures, or life insurance. To each possible price p of the good corresponds a **demand** x, the amount of the good that would be purchased at price p. Thus we can write $x = x(p)$. We usually assume that x is a strictly decreasing differentiable function of p, so we can write the inverse $p = p(x)$. The **revenue** is $R = xp$, the total money taken in by the sale of x units of the good at price p. The rate of increase of the revenue per unit change in the price is called the **marginal revenue with respect to price.** It is simply the derivative dR/dp.

Example 11

Express the marginal revenue dR/dp in terms of the demand x and the elasticity η of demand with respect to price.

Solution By definition of elasticity,

$$\eta = \frac{Ex}{Ep} = \frac{p}{x} \cdot \frac{dx}{dp}$$

The marginal revenue is

$$\frac{dR}{dp} = \frac{d}{dp}(px) = x + p\frac{dx}{dp}$$

$$= x + x\frac{p}{x} \cdot \frac{dx}{dp} = x + x\eta$$

Hence

$$\frac{dR}{dp} = x(1 + \eta)$$

This result is important in economics. ●

Poisson's Gas Equation

Suppose a fixed quantity of a diatomic gas (such as O_2, N_2, or CO) is subject to a reversible adiabatic process. Here **reversible** means that the process can go forward and then back to its original state and that no energy is dissipated. **Adiabatic** means that no heat is transferred to or from the gas. You can think of the process as taking place in a Thermos bottle.

Then the pressure p and the volume v of the gas satisfy

$$\frac{dp/dt}{p} + \gamma\frac{dv/dt}{v} = 0 \quad \text{where} \quad \gamma = \tfrac{7}{5} = 1.4$$

This relation says that $(1/p)(dp/dt)$, the relative rate of change of pressure, is proportional to $-(1/v)(dv/dt)$, the relative rate of decrease of volume.

Example 12

How are p and v related?

Solution The given relation between p and v can be written

$$\frac{d(\ln p)}{dt} + \gamma \frac{d(\ln v)}{dt} = 0$$

that is,

$$\frac{d}{dt}(\ln pv^{\gamma}) = 0$$

Hence $\ln pv^{\gamma} = a$, a constant, so that

$$pv^{\gamma} = c$$

This result, known as *Poisson's equation,* is used in thermodynamics and meteorology. ●

Exercises

Find

1 $\lim_{x\to 0+} x^{\sqrt{x}}$

2 $\lim_{x\to 0+} x^{(x^2)}$

3 $\lim_{x\to +\infty} x^{1/x}$

4 $\lim_{x\to 0+} x^{1/x}$

5 $\lim_{x\to 0+} x^{(x^x)}$

6 $\lim_{x\to 0+} x^{-1/\ln x}$

7 $\lim_{x\to 0+} x^{\sin x}$

8 $\lim_{x\to +\infty} x(a^{1/x} - 1)$ $(a > 0)$

9 $\left.\frac{d}{dx} 5^x\right|_{x=2}$

10 $\frac{d}{dx} 2^{\sqrt{x}}$

11 $\left.\frac{d}{dx} 3^{\sin x}\right|_{x=\pi/4}$

12 $\frac{d}{dx} 2^{\tan x}$

Solve to five-place accuracy

13 $3^x = 10$

14 $7^x = 4$

15 $4^x - 2 \cdot 2^x - 3 = 0$

16 $25^x - 7 \cdot 5^x + 12 = 0$

17 The largest known prime number in June 1979 was $2^{44497} - 1$. Estimate it to three significant figures.

18 The largest known prime number in October 1983 was $2^{132,049} - 1$. How many decimal digits does it contain?

Maximize

19 $\frac{\ln x}{x^2}$

20 $\frac{x^2}{2^x}$ $(x \geq 0)$

21 $\frac{\ln(3x)}{x}$

22 $x^{-1/\ln x}$

Graph

23 $y = x^x$

24 $y = x^{1/x}$

Differentiate with respect to x

25 $y = x^{x-1}$

26 $y = (x + 2)^{x+3}$

27 $y = x^{1/x}$

28 $y = 2^{2x}$

29 $y = 3^{\ln x}$

30 $y = (1 + 1/x)^x$

31 $y = 10^{(x^2)}$

32 $y = x^{(x^x)}$

33 $y = (\ln x)^x$

34 $y = (2 + \sin x)^x$

35 $y = \log_x b$

36 $y = \log_x x$

37 $y = x^{\ln x}$

38 $y = x^{\sin x}$

Differentiate using logarithmic differentiation

39 $y = (x^3 + 2)^{1/2}$

40 $y = \left(\frac{x + 1}{(x + 2)^4}\right)^{1/3}$

41 $y = \left(\frac{x^2 + 4}{x + 7}\right)^6$

42 $y = \frac{(x + 1)(x + 2)}{(x + 4)(x + 5)}$

43 $y = (2x + 3)^{1/3}e^{-x^2}$

44 $y = \frac{x^2 e^{3x^3}}{(x + 3)^2}$

45 $y = \frac{e^x(x^3 - 1)}{\sqrt{2x + 1}}$

46 $y = \sqrt{\frac{x^2 - 1}{x^2 + 1}}$

Compute y'/y, the relative rate of change of y with respect to x

47 $y = x^{5/6}$

48 $y = (2x^4 + 1)^{-2/7}$

49 $y = (x - 3)(x - 5)$

50 $y = (x - 1)(x - 3)(x - 5)$

51 $y = (x + 2)(x + 7)^3$

52 $y = (x + 2)^2(x + 3)^3/(x + 4)^4$

Prove the following formulas for elasticity

53 $\dfrac{E(uv)}{Ex} = \dfrac{Eu}{Ex} + \dfrac{Ev}{Ex}$

54 $\dfrac{E}{Ex}\left(\dfrac{u}{v}\right) = \dfrac{Eu}{Ex} - \dfrac{Ev}{Ex}$

55 $\dfrac{Ey}{Ex} = \dfrac{Ey}{Eu} \cdot \dfrac{Eu}{Ex}$ (chain rule)

56 $\dfrac{E}{Ex}(y + c) = \dfrac{y}{y + c} \cdot \dfrac{Ey}{Ex}$

Find

57 $\dfrac{E}{Ex}(cx^p)$

58 $\dfrac{E}{Ex}(ax + b)$

59 $\dfrac{E}{Ex}(ce^{kx})$

60 $\dfrac{E}{Ex}(xy)$

The functions in the next two exercises are called **Einstein functions** in radiation theory.

61 Show that $y = \ln(1 - e^{-x})$ is strictly increasing.

62 Show that $y = \dfrac{x}{e^x - 1} - \ln(1 - e^{-x})$ is strictly decreasing for $x > 0$.

63 Find the equation of the general straight line plotted on semi-log paper (Figure 4-4-1: the scale on one axis is linear; on the other axis, logarithmic).

64 Find the equation of the general straight line plotted on log-log paper (Figure 4-4-2: the scales on both axes are logarithmic).

Harder Exercises

65 Let $e < a < b$. Prove $a^b > b^a$.
[Hint Consider $y = (\ln x)/x$.]

66 Find all integers m and n such that $1 \le m < n$ and $m^n = n^m$.
[Hint Consider $y = (\ln x)/x$.]

67 Arrange these functions according to increasing size as $x \to +\infty$:

$$y_1 = 2^x \qquad y_2 = x^{(\ln x)^3} \qquad y_3 = (\sqrt{x})^x$$
$$y_4 = x^{\sqrt{x}} \qquad y_5 = e^{x^2}$$

68 Let $f(x)$ and $g(x)$ be differentiable, with $f(x) > 0$. Prove $f(x)^{g(x)}$ is differentiable.

69 A measure of the brightness of a visible star S is its **flux** $f(S)$. This is the amount of radiant energy from the star falling on a unit area of Earth (perpendicular to the light from the star) per second. Another measure is the **magnitude** $m(S)$ of the star. This is defined so that if $f(S_2) = Kf(S_1)$, then $m(S_1) - m(S_2)$ depends only on K. Guess this dependence, the relation between $m(S_1) - m(S_2)$ and $K = f(S_2)/f(S_1)$ if a flux ratio $K = 100$ yields a magnitude difference of 5. Assume $m(S)$ decreases as $f(S)$ increases.

70 Let $a > 1$. Prove the inequality of James Bernoulli:

$$(1 + x)^a \ge 1 + ax \quad \text{for} \quad x > -1$$

with equality only for $x = 0$.

Figure 4-4-1
See Exercise 63

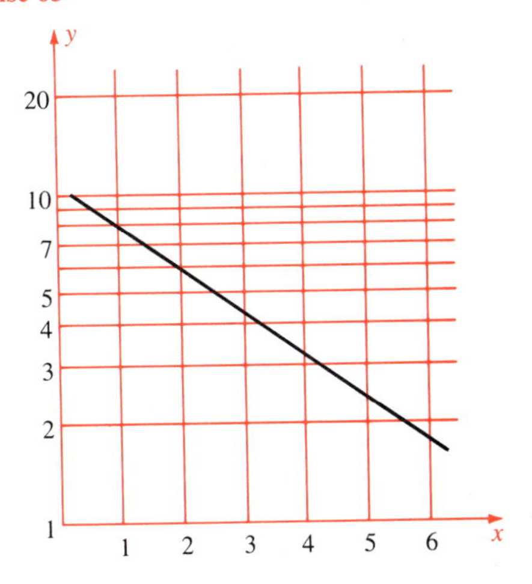

Figure 4-4-2
See Exercise 64

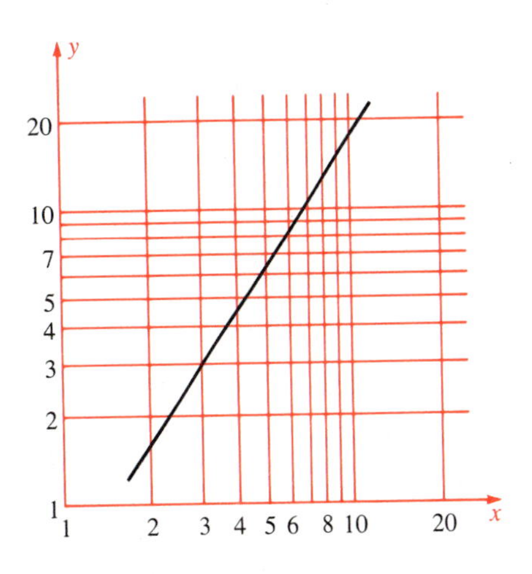

4-5 Inverse Trigonometric Functions

Consider the sine function $y = \sin x$, but only on the interval $[-\frac{1}{2}\pi, \frac{1}{2}\pi]$ of the x-axis. On this restricted domain, $\sin x$ is a strictly increasing function (Figure 4-5-1). We know this from the definition of the sine function; we also know it from

$$\frac{d}{dx}\sin x = \cos x > 0 \quad \text{for} \quad -\tfrac{1}{2}\pi < x < \tfrac{1}{2}\pi$$

Figure 4-5-1
Graph of $y = \sin x$
for $-\frac{1}{2}\pi \le x \le \frac{1}{2}\pi$

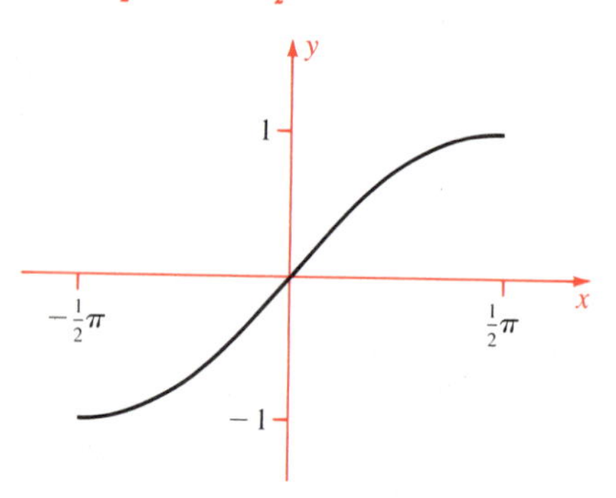

The range of $y = \sin x$ with x restricted to $[-\frac{1}{2}\pi, \frac{1}{2}\pi]$ is the interval $[-1, 1]$ on the y-axis. Its inverse function is called the *arc sine* function and written $y = \arc \sin x$. The domain of $y = \text{arc sin } x$ is the interval $[-1, 1]$ and its graph is the reflection in $y = x$ of the graph of $y = \sin x$. It is drawn in Figure 4-5-2. The graph has vertical tangents at $(1, \frac{1}{2}\pi)$ and $(-1, -\frac{1}{2}\pi)$, reflecting the horizontal tangents of $y = \sin x$ at $(\frac{1}{2}\pi, 1)$ and $(-\frac{1}{2}\pi, -1)$.

Figure 4-5-2
Graph of $y = \text{arc sin } x$
Domain $[-1, 1]$
Range $[-\frac{1}{2}\pi, \frac{1}{2}\pi]$

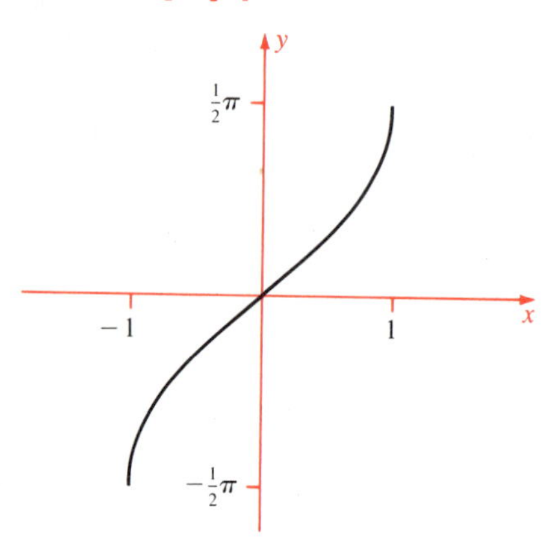

Arc Sine Function

The inverse function of the function $y = \sin x$ with x-domain $[-\frac{1}{2}\pi, \frac{1}{2}\pi]$ is the **arc sine** function

$$y = \text{arc sin } x$$

Arc sine is a strictly increasing function with domain $[-1, 1]$ and range $[-\frac{1}{2}\pi, \frac{1}{2}\pi]$.

Example 1

Some special values of $\text{arc sin } x$, seen by inspection:

$$\text{arc sin } 1 = \tfrac{1}{2}\pi \qquad \text{arc sin}(-1) = -\tfrac{1}{2}\pi \qquad \text{arc sin } 0 = 0$$
$$\text{arc sin } \tfrac{1}{2} = \tfrac{1}{6}\pi \qquad \text{arc sin}(-\tfrac{1}{2}\sqrt{2}) = -\tfrac{1}{4}\pi \qquad \text{arc sin } \tfrac{1}{2}\sqrt{3} = \tfrac{1}{3}\pi$$

For instance, $\sin(-\frac{1}{4}\pi) = -\frac{1}{2}\sqrt{2}$; hence the fifth of these relations follows.

Since sine and arc sine are inverse functions, we have

$$\begin{cases} \text{arc sin}(\sin\theta) = \theta & \text{for} \quad -\frac{1}{2}\pi \le \theta \le \frac{1}{2}\pi \\ \sin(\text{arc sin } x) = x & \text{for} \quad -1 \le x \le 1 \end{cases}$$

Because $\sin x$ is an odd function, it follows that $\text{arc sin } x$ is also odd:

$$\text{arc sin}(-x) = -\text{arc sin } x$$

Example 2

Suppose $\theta = \text{arc sin } 0.3$. Find $\sin\theta$ and $\cos\theta$.

Solution Since sine and arc sine are inverses,

$$\sin\theta = \sin(\text{arc sin } 0.3) = 0.3$$

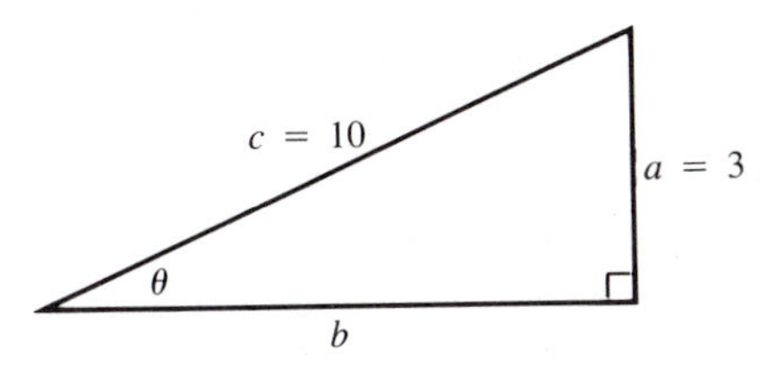

Figure 4-5-3
See Example 2

Since θ is in the first quadrant,

$$\cos\theta = \sqrt{1 - \sin^2\theta} = \sqrt{1 - 0.09} = \sqrt{0.91}$$

An alternative is to draw a right triangle (Figure 4-5-3) one of whose angles is $\theta = \text{arc sin } 0.3$. The side b adjacent to θ is given by

$$b = \sqrt{c^2 - a^2} = \sqrt{100 - 9} = \sqrt{91}$$

Therefore $\cos\theta = b/c = \frac{1}{10}\sqrt{91}$. ●

Our next chore is to differentiate $\theta = \text{arc sin } x$. As usual, we differentiate the inverse relation $x = \sin\theta$ with respect to x by the chain rule:

$$\frac{d}{dx}x = \frac{d}{dx}\sin\theta = \left(\frac{d}{d\theta}\sin\theta\right)\frac{d\theta}{dx}$$

$$1 = (\cos\theta)\frac{d\theta}{dx} \qquad \text{hence} \qquad \frac{d\theta}{dx} = \frac{1}{\cos\theta}$$

This is valid for $-1 < x < 1$, that is, $-\frac{1}{2}\pi < \theta < \frac{1}{2}\pi$. There $\cos\theta > 0$, so

$$\cos\theta = \sqrt{1 - \sin^2\theta} = \sqrt{1 - x^2}$$

Hence

$$\frac{d\theta}{dx} = \frac{1}{\cos\theta} = \frac{1}{\sqrt{1 - x^2}}$$

But $\theta = \text{arc sin } x$, so we have derived the important formula

$$\frac{d}{dx}\text{arc sin } x = \frac{1}{\sqrt{1 - x^2}} \qquad \text{for} \quad -1 < x < 1$$ ●

Example 3

Find $\dfrac{d}{dx}\text{arc sin }\sqrt{x}$.

Solution Write $y = \text{arc sin } u$, where $u = \sqrt{x}$. Then

$$\frac{dy}{dx} = \frac{dy}{du}\cdot\frac{du}{dx} = \frac{1}{\sqrt{1 - u^2}}\cdot\frac{du}{dx}$$

$$= \frac{1}{\sqrt{1 - (\sqrt{x})^2}}\cdot\frac{1}{2\sqrt{x}}$$

$$= \frac{1}{\sqrt{1 - x}}\cdot\frac{1}{2\sqrt{x}}$$

$$= \frac{1}{2\sqrt{x - x^2}}$$ ●

Figure 4-5-4
Graph of $y = \cos x$ for $0 \le x \le \pi$

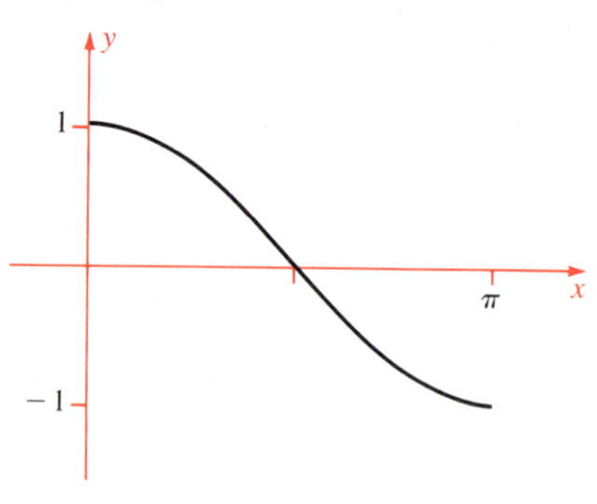

Inverse Cosine Function

Just as with $\sin x$, the inverse of $y = \cos x$ can be defined only if the domain of $\cos x$ is suitably restricted. A logical choice is to restrict x to the interval $[0, \pi]$. See Figure 4-5-4. On this domain, $\cos x$ is a strictly decreasing function. We know this from the definition of the cosine function; we also know it from

$$\frac{d}{dx}\cos x = -\sin x < 0 \quad \text{for} \quad 0 < x < \pi$$

The range of $y = \cos x$ for $0 \le x \le \pi$ is the interval $[-1, 1]$ of the y-axis. Its inverse function is called the *arc cosine* function, and its domain is $[-1, 1]$. The graph of $y = \text{arc}\cos x$ is the reflection in $y = x$ of the graph of $y = \cos x$. See Figure 4-5-5. It has vertical tangents at $(-1, \pi)$ and at $(1, 0)$.

Figure 4-5-5
Graph of $y = \text{arc}\cos x$
Domain $[-1, 1]$ Range $[0, \pi]$

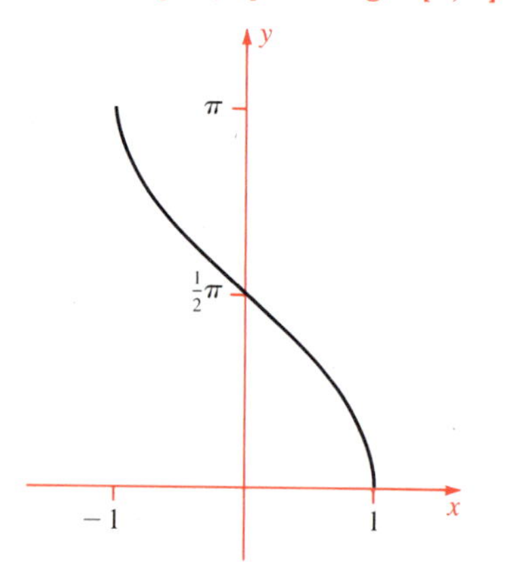

Arc Cosine Function

The inverse function $y = \cos x$ with x-domain $[0, \pi]$ is the **arc cosine** function

$$y = \text{arc}\cos x$$

Arc cosine is a strictly decreasing function with domain $[-1, 1]$ and range $[0, \pi]$.

Example 4

Some special values of $\text{arc}\cos x$, seen by inspection:

$$\text{arc}\cos 1 = 0 \qquad \text{arc}\cos(-1) = \pi \qquad \text{arc}\cos 0 = \tfrac{1}{2}\pi$$

$$\text{arc}\cos \tfrac{1}{2} = \tfrac{1}{3}\pi \qquad \text{arc}\cos \tfrac{1}{2}\sqrt{2} = \tfrac{1}{4}\pi \qquad \text{arc}\cos(-\tfrac{1}{2}\sqrt{2}) = \tfrac{3}{4}\pi$$

For instance, $\cos \frac{3}{4}\pi = -\frac{1}{2}\sqrt{2}$, which explains the last of these formulas.

Figure 4-5-6
$y = \text{arc}\cos x$ is symmetric about $(0, \frac{1}{2}\pi)$; hence
$\text{arc}\cos(-x) - \frac{1}{2}\pi = -(\text{arc}\cos x - \frac{1}{2}\pi)$
That is, $\text{arc}\cos(-x) = \pi - \text{arc}\cos x$.

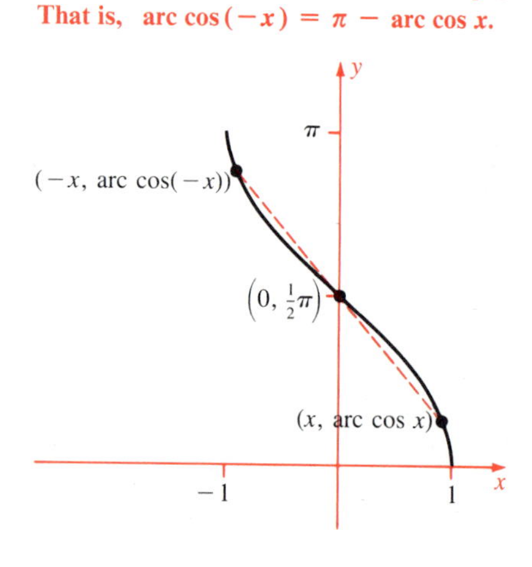

Since cosine and arc cosine are inverse functions, we have

$$\begin{cases} \text{arc}\cos(\cos\theta) = \theta & \text{for} \quad 0 \le \theta \le \pi \\ \cos(\text{arc}\cos x) = x & \text{for} \quad -1 \le x \le 1 \end{cases}$$

We note two further relations involving $\text{arc}\cos x$:

$$\text{arc}\cos x + \text{arc}\cos(-x) = \pi \qquad \text{arc}\sin x + \text{arc}\cos x = \tfrac{1}{2}\pi$$

The first expresses analytically the geometric statement that the graph of $y = \text{arc}\cos x$ is symmetric about the point $(x, y) = (0, \frac{1}{2}\pi)$. See Figure 4-5-6. The second statement says that if we flip the arc sine graph across the x-axis and raise the result $\frac{1}{2}\pi$, we get the arc cosine graph. See Figure 4-5-7.

From the second of the relations,

$$\text{arc}\cos x = \tfrac{1}{2}\pi - \text{arc}\sin x$$

Figure 4-5-7
The curve $y = \text{arc sin } x$ reflected in the x-axis is the curve $y = \text{arc cos } x - \frac{1}{2}\pi$. Therefore

$$\text{arc cos } x - \tfrac{1}{2}\pi = \text{arc sin}(-x) = -\text{arc sin } x$$

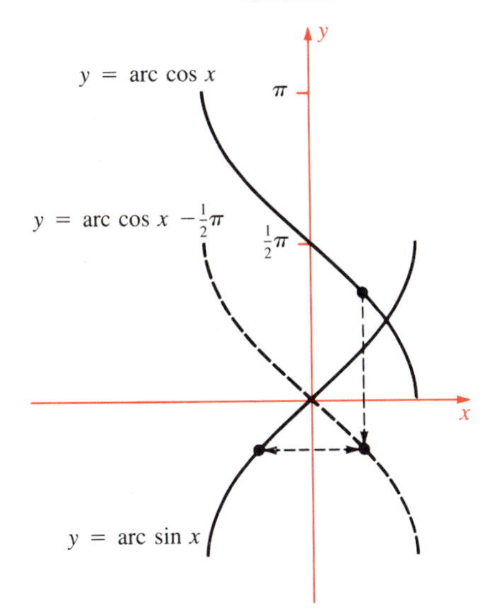

Therefore arc cos x is not really a new function once we have arc sin x. From this relation,

$$\frac{d}{dx}\text{arc cos } x = \frac{d}{dx}\left(\tfrac{1}{2}\pi - \text{arc sin } x\right)$$

$$= -\frac{d}{dx}\text{arc sin } x = \frac{-1}{\sqrt{1-x^2}}$$

Therefore $\quad \dfrac{d}{dx}\text{arc cos } x = \dfrac{-1}{\sqrt{1-x^2}}$

Inverse Tangent Function

Recall the graph (Figure 4-5-8) of $y = \tan x$. To obtain an inverse function, it is natural to consider just one branch of the graph. Therefore we restrict the domain of $y = \tan x$ to the x-interval $(-\frac{1}{2}\pi, \frac{1}{2}\pi)$. On this domain the function is strictly increasing because

$$\frac{d}{dx}\tan x = \sec^2 x > 0$$

and its range is **R**. Its inverse function is called the *arc tangent* function. The graph of $y = \text{arc tan } x$ is the reflection in $y = x$ of the graph of $y = \tan x$. See Figure 4-5-9. Note the horizontal asymptotes $y = -\frac{1}{2}\pi$ and $y = \frac{1}{2}\pi$.

Arc Tangent Function

The inverse function of the function $y = \tan x$ with x-domain $(-\frac{1}{2}\pi, \frac{1}{2}\pi)$ is the **arc tangent** function

$$y = \text{arc tan } x$$

Arc tangent is a strictly increasing function with domain **R** and range $(-\frac{1}{2}\pi, \frac{1}{2}\pi)$. What is more

$$\lim_{x \to +\infty} \text{arc tan } x = \tfrac{1}{2}\pi \qquad \text{and} \qquad \lim_{x \to -\infty} \text{arc tan } x = -\tfrac{1}{2}\pi$$

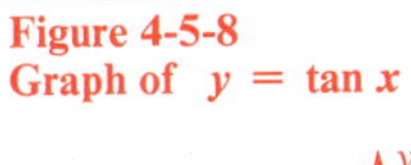

Figure 4-5-8
Graph of $y = \tan x$

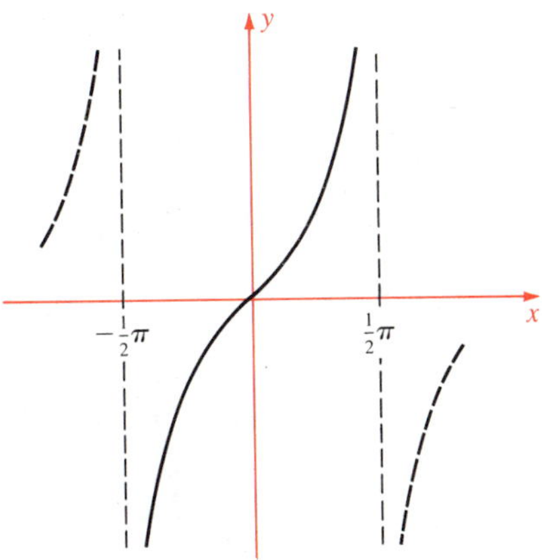

Figure 4-5-9
$y = \text{arc tan } x$
Domain **R**
Range $(-\frac{1}{2}\pi, \frac{1}{2}\pi)$

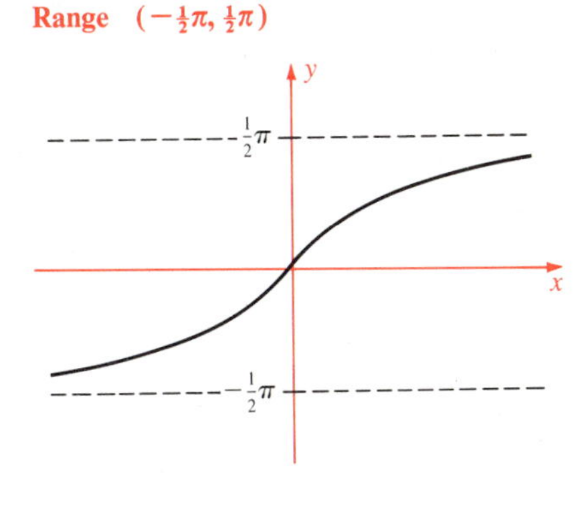

To illustrate the behavior of arc tan x for large x, note that

$$\tfrac{1}{2}\pi - \text{arc tan } 100 \approx 1.5708 - 1.5608 \approx 0.01$$

$$\tfrac{1}{2}\pi - \text{arc tan } 1000 \approx 1.5708 - 1.5698 \approx 0.001$$

Example 5

Some special values of arc tan x:

$$\text{arc tan } 0 = 0 \qquad \text{arc tan } 1 = \tfrac{1}{4}\pi \qquad \text{arc tan } \sqrt{3} = \tfrac{1}{3}\pi$$

$$\text{arc tan}(-\sqrt{3}) = -\tfrac{1}{3}\pi \qquad \text{arc tan } \tfrac{1}{3}\sqrt{3} = \tfrac{1}{6}\pi$$

Since tangent and arc tangent are inverse functions, we have

$$\begin{cases} \text{arc tan}(\tan\theta) = \theta & \text{for } -\tfrac{1}{2}\pi < \theta < \tfrac{1}{2}\pi \\ \tan(\text{arc tan } x) = x & \text{for all } x \end{cases}$$

Since $\tan\theta$ is an odd function, arc tan x is also odd:

$$\text{arc tan}(-x) = -\text{arc tan } x$$

To differentiate $\theta = \text{arc tan } x$, we write $x = \tan\theta$ and differentiate by the chain rule:

$$\frac{d}{dx}x = \frac{d}{dx}\tan\theta = \left(\frac{d}{d\theta}\tan\theta\right)\frac{d\theta}{dx}$$

$$1 = (\sec^2\theta)\frac{d\theta}{dx}$$

Hence

$$\frac{d\theta}{dx} = \frac{1}{\sec^2\theta}$$

But $\sec^2\theta = 1 + \tan^2\theta = 1 + x^2$, so we have obtained another important formula:

$$\frac{d}{dx}\text{arc tan } x = \frac{1}{1 + x^2}$$

Example 6

Differentiate arc tan$(5x^2 + 1)$.

Solution Write $y = \text{arc tan } u$, where $u = 5x^2 + 1$. Then

$$\frac{dy}{dx} = \frac{1}{1 + u^2}\cdot\frac{du}{dx} = \frac{1}{1 + u^2}\cdot 10x$$

$$= \frac{10x}{1 + (5x^2 + 1)^2}$$

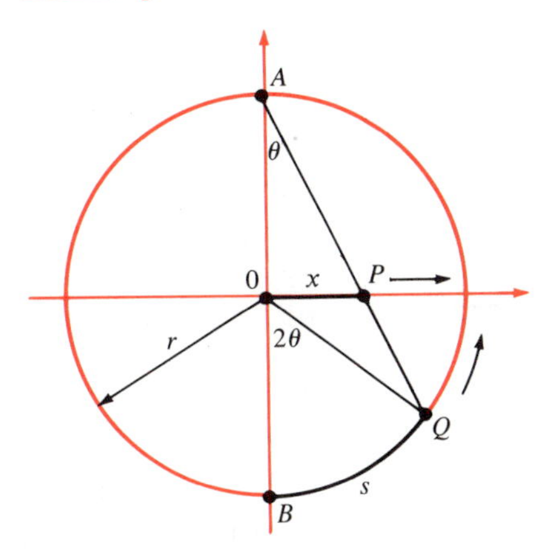

Figure 4-5-10
See Example 7

Example 7

The circle shown in Figure 4-5-10 has radius r meters. As the point P moves to the right at the rate of v m/sec, how fast is the length of the arc BQ increasing?

Solution The central angle is 2θ, twice the inscribed angle θ. Now we can express the arc length s in terms of the angle θ. In fact, by the definition of radian measure, $s = 2r\theta$. Next we express θ in terms of x. From the triangle $A0P$,

$$\theta = \arc\tan\frac{x}{r} \qquad \text{hence} \qquad s = 2r\arc\tan\frac{x}{r}$$

We have found a relation between s and x, so we now have a straightforward related-rates problem. We differentiate both sides of the relation with respect to time:

$$\frac{ds}{dt} = 2r\frac{d}{dt}\left(\arc\tan\frac{x}{r}\right) = 2r\frac{1}{1+(x/r)^2}\left(\frac{1}{r}\cdot\frac{dx}{dt}\right)$$
$$= \frac{2r^2}{x^2+r^2}\cdot v = \frac{2r^2v}{x^2+r^2} \quad \text{m/sec}$$

As a rough check, we see from Figure 4-5-10 that BQ should increase most rapidly when $x = 0$, then less rapidly as x increases. According to the answer, it does so. ●

Notation The following alternative notation is common for inverse trigonometric functions:

$$\arc\sin x = \sin^{-1}x \qquad \arc\tan x = \tan^{-1}x$$

and so on. Do not confuse

$$\sin^{-1}x \quad \text{with} \quad \frac{1}{\sin x}$$

The notation $\sin^{-1}x$ is awkward because we do write $\sin^n x$ for $(\sin x)^n$ when $n > 0$.

Other Inverse Trigonometric Functions

Inverses of the functions $\cot\theta$, $\sec\theta$, and $\csc\theta$ can be defined in a similar manner. Rather than discuss them in detail, we show their graphs (Figures 4-5-11, 4-5-12, and 4-5-13, on the next page) and list a few basic relations:

- $\arc\tan x + \arc\cot x = \frac{1}{2}\pi$ for all x
- $\arc\sec x + \arc\csc x = \frac{1}{2}\pi$ for $|x| \geq 1$
- $\arc\sec x = \arc\cos 1/x$ for $|x| \geq 1$
- $\arc\csc x = \arc\sin 1/x$ for $|x| \geq 1$ ●

Figure 4-5-11
Graph of $y = \text{arc cot } x$
Domain R
Range $(0, \pi)$

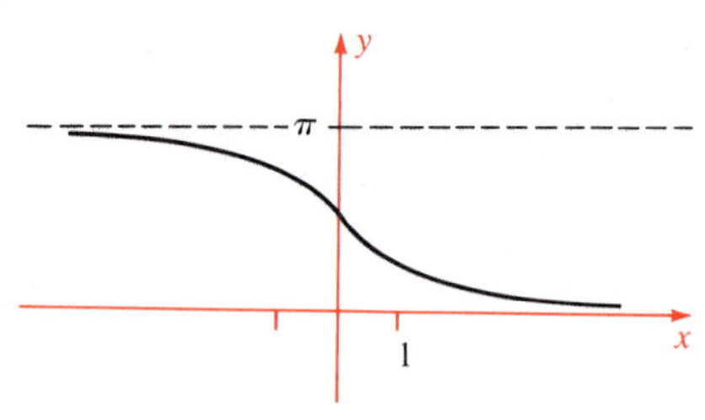

Figure 4-5-12
Graph of $y = \text{arc sec } x$
Domain $(-\infty, -1] \cup [1, +\infty)$
Range $(\frac{1}{2}\pi, \pi] \cup [0, \frac{1}{2}\pi)$

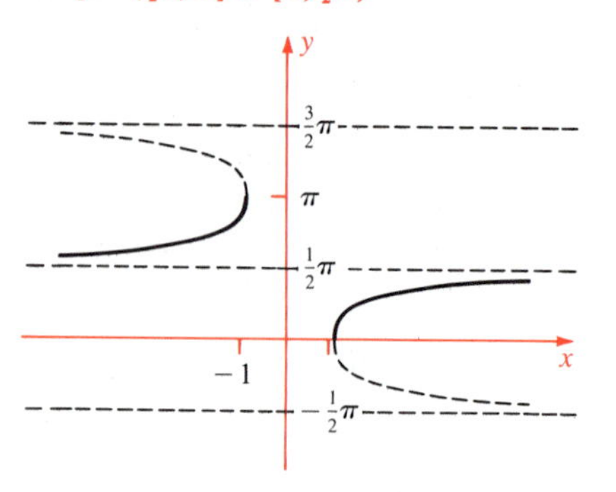

Figure 4-5-13
Graph of $y = \text{arc csc } x$
Domain $(-\infty, -1] \cup [1, +\infty)$
Range $[-\frac{1}{2}\pi, 0) \cup (0, \frac{1}{2}\pi]$

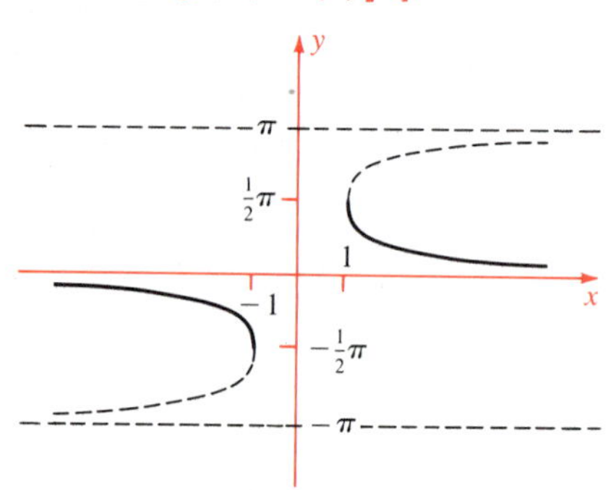

The remaining derivative formulas are given next. They easily follow from the relations just noted and the previous derivative formulas.

- $$\frac{d}{dx}(\text{arc tan } x) = \frac{1}{1 + x^2}$$
- $$\frac{d}{dx}(\text{arc cot } x) = \frac{-1}{1 + x^2}$$
- $$\frac{d}{dx}(\text{arc sec } x) = -\frac{d}{dx}(\text{arc csc } x) = \begin{cases} \dfrac{1}{x\sqrt{x^2 - 1}} & \text{for } x > 1 \\ \dfrac{-1}{x\sqrt{x^2 - 1}} & \text{for } x < -1 \end{cases}$$

Example 8

The Statue of Liberty is 150 ft tall and stands on a 150-foot pedestal. How far from the base should you stand so you can photograph only the statue (not its base) with largest possible angle? Assume camera level is 5 ft.

Solution Draw a diagram, labeling the various distances and angles as indicated (Figure 4-5-14). The problem is to choose x in such a way that the angle θ is greatest. If x is very small or very large, θ will be small. Certainly the optimal value of x is, say, between 50 and 1000 ft.

Express θ as a function of x. From Figure 4-5-14,

$$\theta = \beta - \alpha \qquad \cot\alpha = \frac{x}{145} \quad \text{and} \quad \cot\beta = \frac{x}{295}$$

Hence

$$\theta = \text{arc cot}\frac{x}{295} - \text{arc cot}\frac{x}{145}$$

This is the function of x to be maximized. Its domain is $(0, +\infty)$; there are no end points. Differentiate:

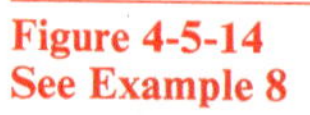

Figure 4-5-14
See Example 8

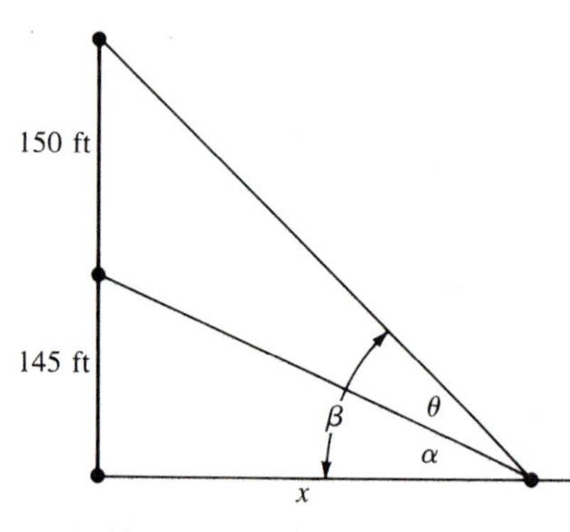

$$\frac{d\theta}{dx} = \frac{-1/295}{1 + (x/295)^2} + \frac{1/145}{1 + (x/145)^2}$$
$$= \frac{-295}{(295)^2 + x^2} + \frac{145}{(145)^2 + x^2}$$

Therefore

$$\frac{d\theta}{dx} = 0 \quad \text{if} \quad \frac{145}{(145)^2 + x^2} = \frac{295}{(295)^2 + x^2}$$

Solve for x^2:

$$(145)(295)^2 + 145x^2 = (295)(145)^2 + 295x^2$$
$$x^2(295 - 145) = (145)(295)(295 - 145)$$
$$x^2 = (145)(295)$$

The positive root of this equation is the answer:

$$x = \sqrt{(145)(295)} = 5\sqrt{1711} \approx 206.8 \text{ ft}$$ ●

Example 9

Maximize θ in Figure 4-5-15 if x is free to move on the positive x-axis.

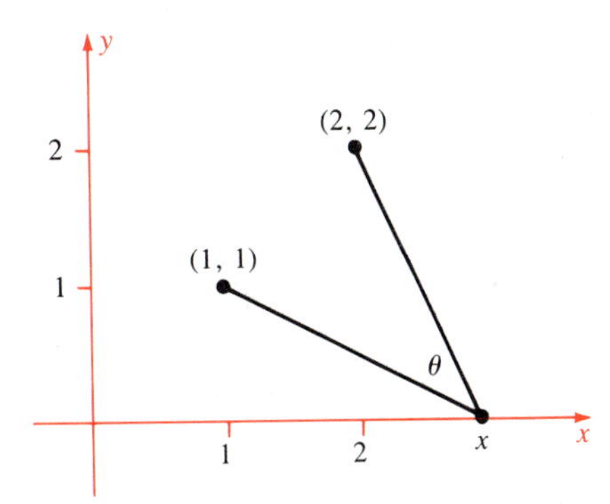

Figure 4-5-15
Maximize θ for $x > 0$. See Example 9

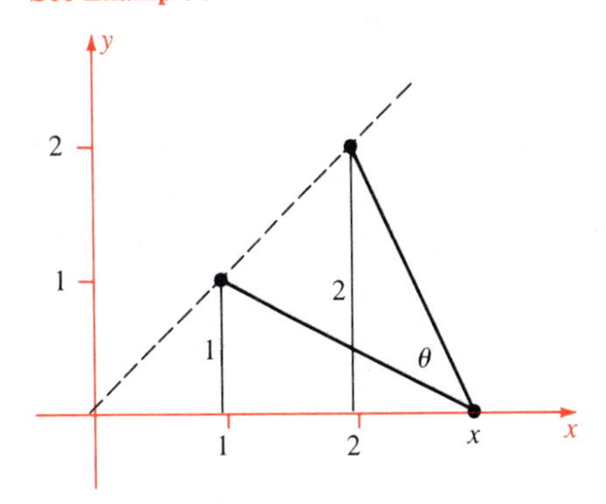

Figure 4-5-16
$\theta = \text{arc cot}\,\frac{1}{2}(x - 2) - \text{arc cot}(x - 1)$
See Example 9

Solution By Figure 4-5-16, we have

$$\theta = \text{arc cot}\,\tfrac{1}{2}(x - 2) - \text{arc cot}(x - 1)$$

Hence

$$\frac{d\theta}{dx} = \frac{-\frac{1}{2}}{1 + [\frac{1}{2}(x - 2)]^2} - \frac{-1}{1 + (x - 1)^2}$$

Therefore $d\theta/dx = 0$ for $1 + \frac{1}{4}(x - 2)^2 = \frac{1}{2} + \frac{1}{2}(x - 1)^2$, that is,

$$4 + (x - 2)^2 = 2 + 2(x - 1)^2$$

This equation simplifies to $x^2 = 4$; hence $x = 2$ and

$$\theta = \text{arc cot } 0 - \text{arc cot } 1 = \tfrac{1}{2}\pi - \tfrac{1}{4}\pi = \tfrac{1}{4}\pi$$

Since $\theta \to 0$ if $x \to 0+$ or $x \to +\infty$, the value $\frac{1}{4}\pi$ is the maximum. ●

Query What does the solution $x = -2$ of $x^2 = 4$ mean geometrically in this problem?

Exercises

Evaluate

1 $\text{arc sin}\,(\frac{1}{2}\sqrt{2}\,)$
2 $\text{arc cos}\,(-\frac{1}{2}\sqrt{2}\,)$
3 $\text{arc tan}\,\sqrt{3}$
4 $\text{arc sec}\,1$
5 $\text{arc csc}\,(\frac{2}{3}\sqrt{3}\,)$
6 $\text{arc cot}\,(-1)$
7 $\text{arc sin}\,\frac{1}{2} - \text{arc sin}\,(-\frac{1}{2}\sqrt{3}\,)$
8 $\text{arc cos}\,\frac{1}{2} - \text{arc cos}\,(-\frac{1}{2})$
9 $\text{arc tan}\,(\tan \frac{1}{7}\pi)$
10 $\cos(\text{arc cos}\,0.35)$
11 $\cos(\text{arc sin}\,\frac{3}{5})$
12 $\cot(\text{arc tan}\,2)$
13 $\text{arc sin}\,\frac{2}{9} + \text{arc cos}\,\frac{2}{9}$
14 $\text{arc tan}\,6.2 + \text{arc cot}\,6.2$

Find the inverse function $\theta = g(x)$ and its domain

15 $x = \ln \sin \theta$, $0 < \theta \leq \frac{1}{2}\pi$
16 $x = (\text{arc tan}\,\theta)^3$

Differentiate

17 $\text{arc sin}\,\frac{1}{3}x$
18 $\text{arc cos}\,2x$
19 $\text{arc tan}\,(x^2)$
20 $x\,\text{arc sin}\,(2x + 1)$
21 $(\text{arc sin}\,3x)^2$
22 $\text{arc tan}\,\sqrt{x}$
23 $\text{arc cot}\,\dfrac{1}{x}$
24 $\text{arc sin}\,\dfrac{x}{x+3}$
25 $\text{arc tan}\,\dfrac{x-1}{x+1}$
26 $\text{arc tan}\,\dfrac{1}{x} + \text{arc cot}\,x$
27 $2x\,\text{arc tan}\,2x - \ln\sqrt{1 + 4x^2}$
28 $x\,\text{arc cot}\,x + \ln\sqrt{1 + x^2}$
29 $x\,\text{arc sin}\,\frac{1}{4}x + \sqrt{16 - x^2}$
30 $\frac{1}{2}(x^2 + 1)\,\text{arc tan}\,x - \frac{1}{2}x$

In Exercises 31–35, express $d\theta/dx$ in terms of the variable x and the constant lengths a and b

31

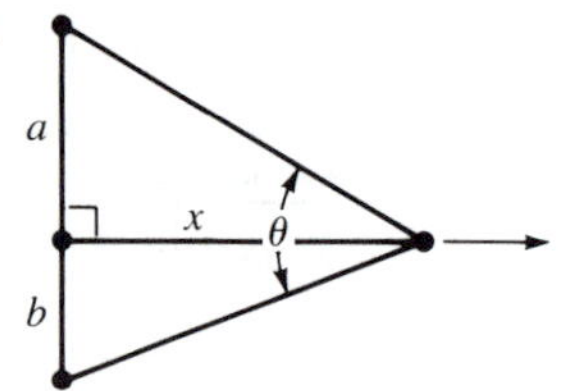

34

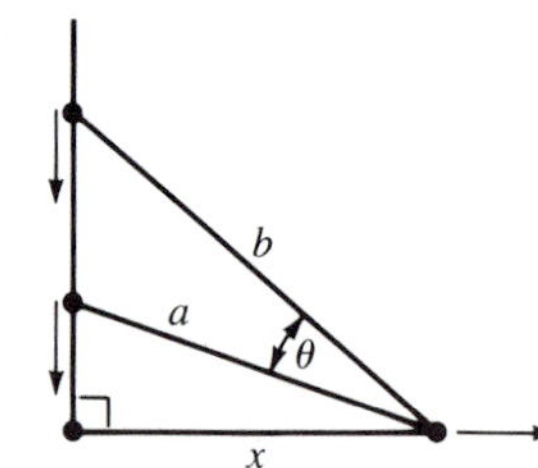

32

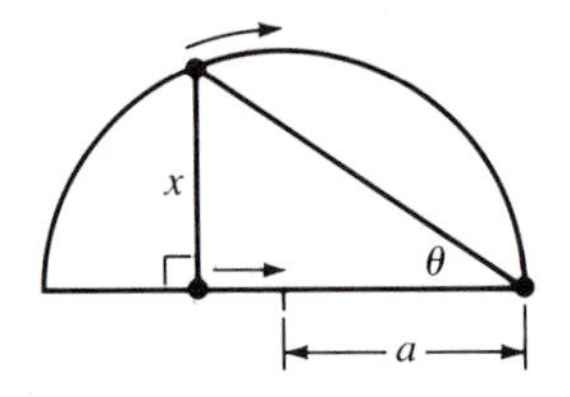

35

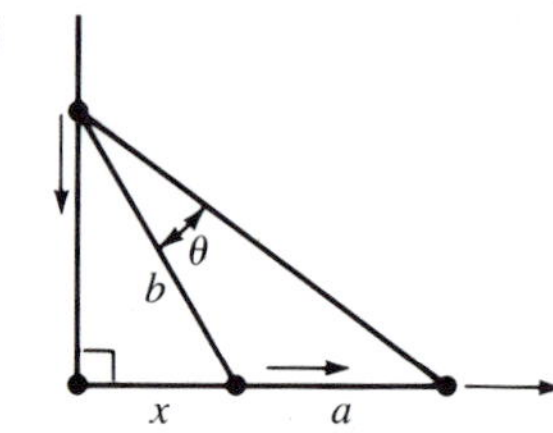

33

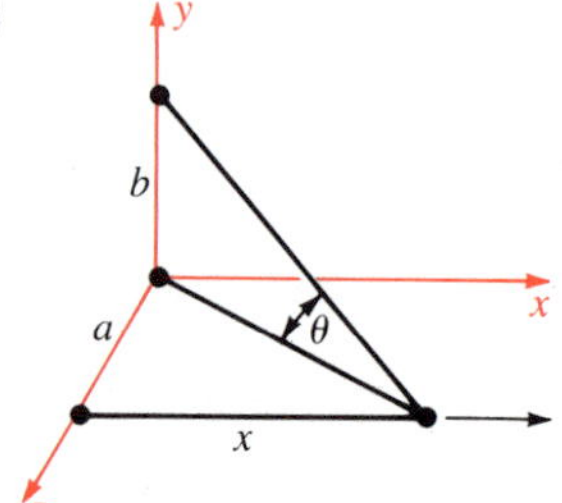

36 Show in Exercise 34 that $d\theta/dx > 0$
37 Maximize θ in Figure 4-5-17, where a and b are constant lengths and $x > 0$
38 (cont.) Evaluate $\theta + 2\alpha$ when θ is maximal.

Figure 4-5-17
See Exercises 37 and 38

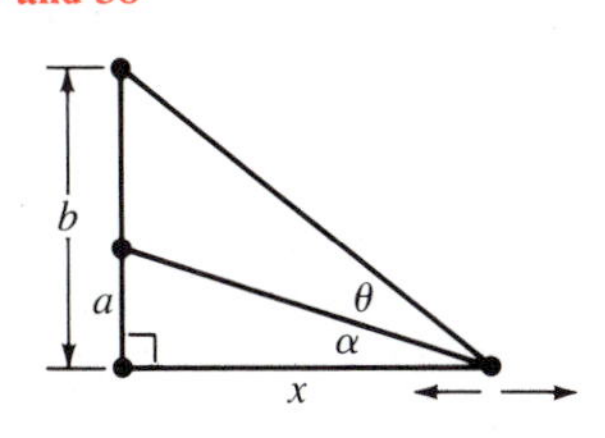

Figure 4-5-18
See Exercises 39, 40, and 41

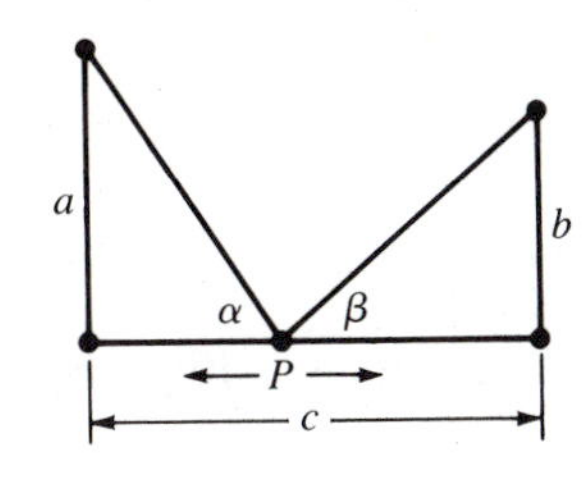

39 In Figure 4-5-18 the lengths a, b, and c are constant and P moves over the segment of length c. Prove that if $\alpha + \beta$ is minimal, then $b \sin^2\alpha = a \sin^2\beta$.

40 (cont.) Suppose $a = b$. Find $\min(\alpha + \beta)$.

41 A point P is on the positive y-axis moving towards the origin at speed 5 cm/sec. When the point is at $(0, 2)$, find how fast the angle APB is changing, where $A = (-1, 0)$ and $B = (1, 0)$. (Lengths in cm.)

42 A balloon is released from eye level and rises 10 ft/sec. According to an observer 100 ft from the point of release, how fast is the balloon's elevation angle increasing 4 sec later?

43 Prove $\arctan x + \operatorname{arc\,cot} x = \frac{1}{2}\pi$. [Hint See the corresponding relation for arc sine and arc cosine.]

44 Prove $\operatorname{arc\,sec} x = \arccos(1/x)$ for $|x| \ge 1$.

45 Prove $\operatorname{arc\,sec} x + \operatorname{arc\,csc} x = \frac{1}{2}\pi$ for $|x| \ge 1$.

46 Prove $\arcsin x + \arccos x = \frac{1}{2}\pi$ for $0 < x < 1$ by a right-triangle argument.

47 Prove $\arctan(1/x) = \operatorname{arc\,cot} x$ for $x > 0$.

48 (cont.) Find and prove the corresponding formula for $x < 0$.

49 Prove $\sin(\arccos x) = \sqrt{1 - x^2}$ for $-1 \le x \le 1$.

50 For which x is $y = \arctan(\cot x)$ defined? For these x, express y in terms of x without using trigonometric functions.

51 Prove $\arccos(2x^2 - 1) = 2 \arccos x$ for $0 \le x \le 1$. [Hint What trigonometric identity does this suggest?]

52 (cont.) Find the corresponding formula for $-1 \le x \le 0$.

53 Prove $\arctan x + \arctan y = \arctan \dfrac{x + y}{1 - xy}$ provided $xy < 1$. You may assume that $xy < 1$ implies that $|\arctan x + \arctan y| < \frac{1}{2}\pi$.

Use the result of Exercise 53 to prove

54 $\arctan \frac{1}{2} + \arctan \frac{1}{3} = \frac{1}{4}\pi$

55 $2 \arctan \frac{1}{3} + \arctan \frac{1}{7} = \frac{1}{4}\pi$

56 $2 \arctan \frac{1}{5} + \arctan \frac{1}{7} + 2 \arctan \frac{1}{8} = \frac{1}{4}\pi$

57 $5 \arctan \frac{1}{7} + 2 \arctan \frac{3}{79} = \frac{1}{4}\pi$

58 $4 \arctan \frac{1}{5} - \arctan \frac{1}{239} = \frac{1}{4}\pi$

59 $6 \arctan \frac{1}{8} + 2 \arctan \frac{1}{57} + \arctan \frac{1}{239} = \frac{1}{4}\pi$

60 $12 \arctan \frac{1}{18} + 8 \arctan \frac{1}{57} - 5 \arctan \frac{1}{239} = \frac{1}{4}\pi$
(This is particularly useful for computing π to high accuracy because $\arctan x$ for x small can be approximated by Taylor polynomials. See Section 9-5.)

4-6 Hyperbolic Functions

The hyperbolic functions are certain combinations of exponential functions, with properties similar to those of the trigonometric functions. They are useful for solving differential equations and evaluating integrals, and have other uses in mathematics and physics. (Hyperbolic functions are closely related to the curve called the hyperbola. See the exercises at the end of Section 8-6 and the material on hyperbolas in Section 13-3 for this relation.)

The three basic hyperbolic functions are the **hyperbolic sine,** the **hyperbolic cosine,** and the **hyperbolic tangent:**

- $\sinh x = \frac{1}{2}(e^x - e^{-x}) \qquad \cosh x = \frac{1}{2}(e^x + e^{-x})$
- $\tanh x = \dfrac{\sinh x}{\cosh x} = \dfrac{e^x - e^{-x}}{e^x + e^{-x}}$

These functions are defined for all x. Tables of their values are available, and they are found on many scientific calculators. From the definitions,

- $\sinh(-x) = -\sinh x$ • $\cosh(-x) = \cosh x$
- $\tanh(-x) = -\tanh x$

Thus $\sinh x$ and $\tanh x$ are odd functions, but $\cosh x$ is an even function. Other immediate consequences of the definitions are that $\cosh x > 0$ for all x and that $|\tanh x| < 1$ for all x. (The numerator of $\tanh x$ is always a bit less in absolute value than the denominator.) Furthermore, since $e^{-x} \to 0$ as $x \to +\infty$ and $e^{x} \to 0$ as $x \to -\infty$, we have

$$\sinh x \approx \tfrac{1}{2}e^{x} \qquad \cosh x \approx \tfrac{1}{2}e^{x} \qquad \tanh x \approx 1 \quad \text{as} \quad x \to +\infty$$

$$\sinh x \approx -\tfrac{1}{2}e^{-x} \qquad \cosh x \approx \tfrac{1}{2}e^{-x} \qquad \tanh x \approx -1 \quad \text{as} \quad x \to -\infty$$

The less commonly used **hyperbolic cotangent, hyperbolic secant,** and **hyperbolic cosecant** are defined by

- $\coth x = \dfrac{1}{\tanh x}$ • $\operatorname{sech} x = \dfrac{1}{\cosh x}$ • $\operatorname{csch} x = \dfrac{1}{\sinh x}$

Here are the derivatives of the hyperbolic functions:

- $\dfrac{d}{dx}\sinh x = \cosh x$ • $\dfrac{d}{dx}\coth x = -\operatorname{csch}^2 x$
- $\dfrac{d}{dx}\cosh x = \sinh x$ • $\dfrac{d}{dx}\operatorname{sech} x = -\operatorname{sech} x \tanh x$
- $\dfrac{d}{dx}\tanh x = \operatorname{sech}^2 x$ • $\dfrac{d}{dx}\operatorname{csch} x = -\operatorname{csch} x \coth x$

These formulas are very easily checked. For instance,

$$\frac{d}{dx}\cosh x = \frac{d}{dx}\tfrac{1}{2}(e^{x} + e^{-x}) = \tfrac{1}{2}(e^{x} - e^{-x}) = \sinh x$$

The formulas are analogous to the derivative formulas for the corresponding trigonometric functions, but not *exactly*. There are sign differences!

Example 1

Show that $y = \sinh 3x$ satisfies the differential equation $y'' - 9y = 0$.

Solution Use the differentiation formulas for $\sinh x$ and $\cosh x$ with the chain rule. If $y = \sinh 3x$, then

$$y' = 3\cosh 3x \qquad y'' = (3\cosh 3x)' = 9\sinh 3x$$

Therefore $y'' - 9y = 9\sinh 3x - 9\sinh 3x = 0$. ●

Graphs and Identities

Let us graph the three basic hyperbolic functions, using our knowledge of their derivatives. For $\sinh x$ and $\tanh x$, the derivatives are positive; these functions are strictly increasing. For $y(x) = \cosh x$, we observe that

$$y'(0) = \sinh 0 = 0 \quad \text{and} \quad y''(x) = \cosh x > 0$$

Hence the graph of $y = \cosh x$ is concave up, with a minimum at $x = 0$. We now have plenty of information to sketch $\sinh x$, $\cosh x$, and $\tanh x$. See Figures 4-6-1, 4-6-2, and 4-6-3.

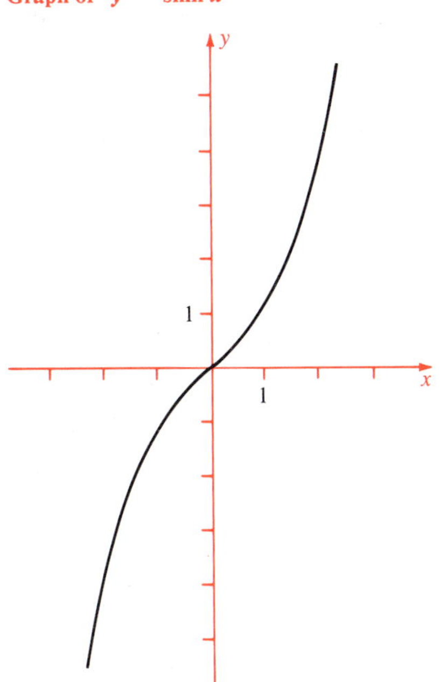

Figure 4-6-1
Graph of $y = \sinh x$

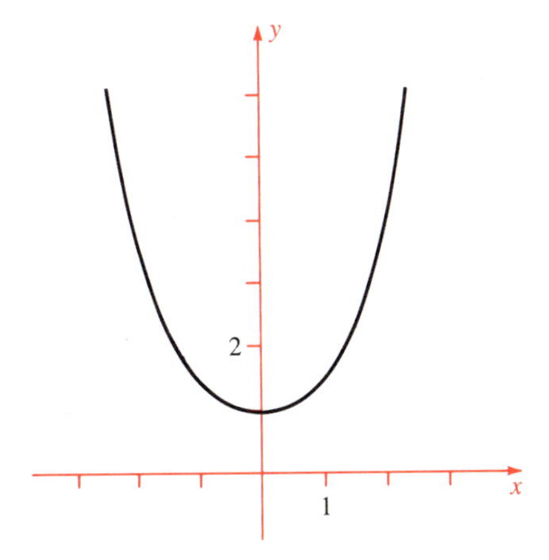

Figure 4-6-2
Graph of $y = \cosh x$

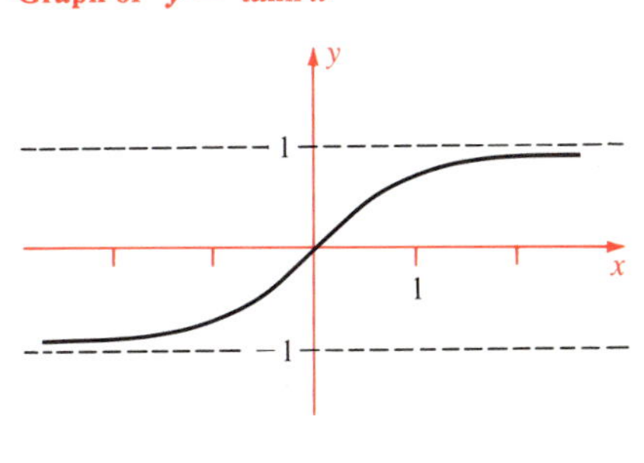

Figure 4-6-3
Graph of $y = \tanh x$

The hyperbolic functions are related to each other by identities similar to trigonometric identities. For example, since

$$\cosh^2 x = [\tfrac{1}{2}(e^x + e^{-x})]^2 = \tfrac{1}{4}(e^{2x} + 2 + e^{-2x})$$

and

$$\sinh^2 x = [\tfrac{1}{2}(e^x - e^{-x})]^2 = \tfrac{1}{4}(e^{2x} - 2 + e^{-2x})$$

it follows that

$$\cosh^2 x - \sinh^2 x = 1$$

Easy consequences are the identities

$$\tanh^2 x + \operatorname{sech}^2 x = 1 \qquad \coth^2 x - \operatorname{csch}^2 x = 1$$

Note the similarity to trigonometric identities, except for signs. Virtually every trigonometric identity has a hyperbolic analogue. But you must be careful with signs!

Example 2

Prove the identity $\cosh(u+v) = \cosh u \cosh v + \sinh u \sinh v$.

Solution Express the right-hand side in terms of exponentials and simplify algebraically:

$$\begin{aligned}
&\cosh u \cosh v + \sinh u \sinh v \\
&\quad = \tfrac{1}{4}(e^u + e^{-u})(e^v + e^{-v}) + \tfrac{1}{4}(e^u - e^{-u})(e^v - e^{-v}) \\
&\quad = \tfrac{1}{4}(e^{u+v} + e^{u-v} + e^{-u+v} + e^{-u-v}) \\
&\qquad + \tfrac{1}{4}(e^{u+v} - e^{u-v} - e^{-u+v} + e^{-u-v}) \\
&\quad = \tfrac{1}{2}(e^{u+v} + e^{-(u+v)}) = \cosh(u+v)
\end{aligned}$$

●

Inverse Hyperbolic Functions

The function $y = \sinh x$ is strictly increasing, with domain **R** and range **R**. Hence $\sinh x$ has an inverse, written* either $\sinh^{-1}x$ or $\operatorname{arg\,sinh} x$. Thus the statements

$$y = \sinh^{-1}x \quad \text{and} \quad x = \sinh y$$

are equivalent. The graph of $y = \sinh^{-1}x$, shown in Figure 4-6-4, is the reflection in the line $y = x$ of the graph of $y = \sinh x$.

Figure 4-6-4
Graph of $y = \sinh^{-1}x$

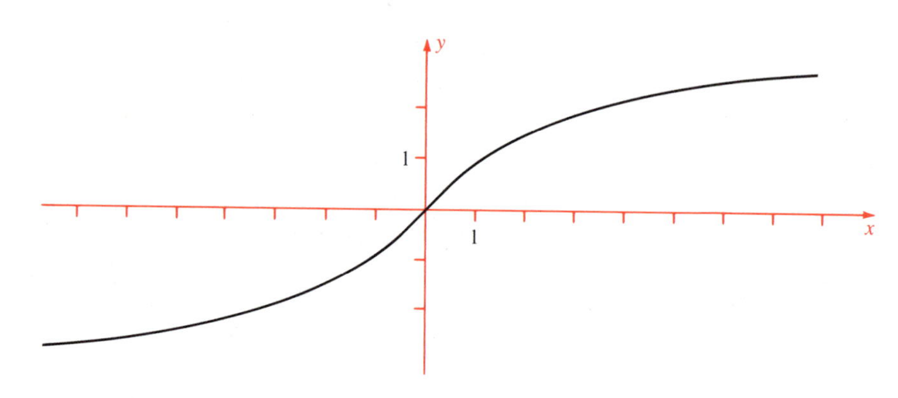

Warning Do not confuse $\sinh^{-1}x$ with $1/\sinh x$.

Since $\sinh x$ is defined in terms of exponentials, it seems reasonable that $\sinh^{-1}x$ should be expressible in terms of logarithms. Indeed, if $y = \sinh^{-1}x$, then

$$x = \sinh y = \tfrac{1}{2}(e^y - e^{-y})$$

Hence

$$e^y - 2x - e^{-y} = 0 \qquad \text{so that} \qquad e^{2y} - 2xe^y - 1 = 0$$

* When $y = \sinh x$, usually x is *not* an angle or arc, and x is called the **argument** of $\sinh x$. Hence the name arg sinh.

Figure 4-6-5
Graph of $y = \cosh^{-1}x$

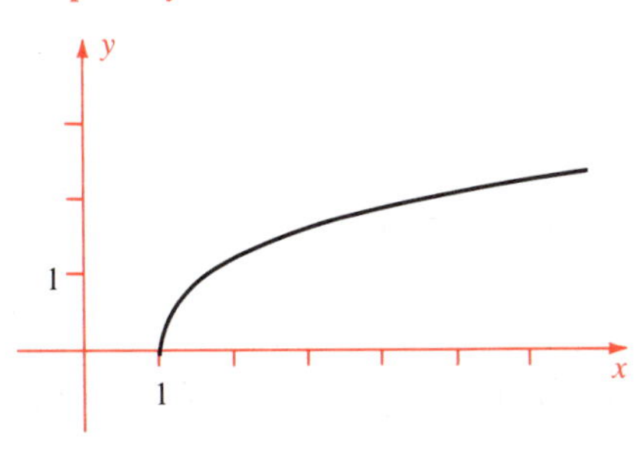

This is a quadratic equation for e^y. By the quadratic formula, we have $e^y = x \pm \sqrt{x^2+1}$. Since $e^y > 0$, the correct choice of sign is plus:

$$e^y = x + \sqrt{x^2+1} \qquad \text{therefore} \qquad y = \ln(x + \sqrt{x^2+1})$$

Consequently:

$$\sinh^{-1}x = \ln(x + \sqrt{x^2+1})$$

To obtain an inverse for $y = \cosh x$, we restrict its domain to the x-interval $[0, +\infty)$ on which $\cosh x$ is a strictly increasing function. On this domain its range is $[1, +\infty)$. Hence its inverse function $y = \cosh^{-1}x$ has domain $[1, +\infty)$. It can be expressed in terms of logarithms by a derivation similar to the one given above for $\sinh^{-1}$:

$$\cosh^{-1}x = \ln(x + \sqrt{x^2-1}) \qquad (x \geq 1)$$

Figure 4-6-6
Graph of $y = \tanh^{-1}x$

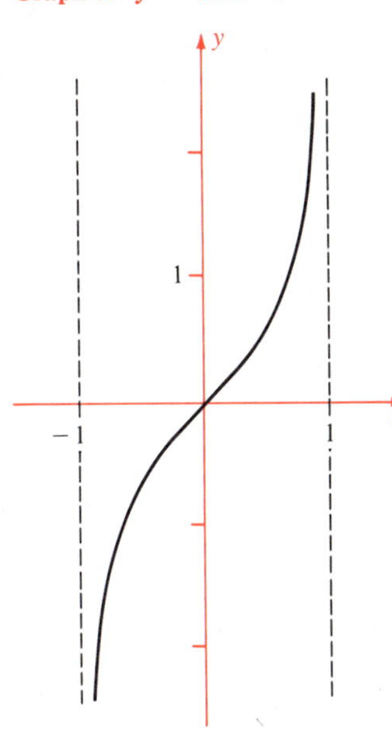

The inverse hyperbolic tangent is defined on $(-1, 1)$ and can be expressed by the formula

$$\tanh^{-1}x = \tfrac{1}{2}\ln\left(\frac{1+x}{1-x}\right) \qquad (-1 < x < 1)$$

Graphs of $\cosh^{-1}x$ and $\tanh^{-1}x$ are shown in Figures 4-6-5 and 4-6-6.

Now we consider derivatives. If $y = \sinh^{-1}x$, then $x = \sinh y$. Therefore,

$$\frac{dy}{dx} = \frac{1}{dx/dy} = \frac{1}{\cosh y} = \frac{1}{\sqrt{1+\sinh^2 y}} = \frac{1}{\sqrt{1+x^2}}$$

In this way, we obtain the formulas

- $$\frac{d}{dx}\sinh^{-1}x = \frac{1}{\sqrt{1+x^2}}$$
- $$\frac{d}{dx}\cosh^{-1}x = \frac{1}{\sqrt{x^2-1}}$$
- $$\frac{d}{dx}\tanh^{-1}x = \frac{1}{1-x^2}$$

Exercises

Prove

1 $\sinh(u+v) = \sinh u \cosh v + \cosh u \sinh v$.

2 $\tanh(u+v) = \dfrac{\tanh u + \tanh v}{1 + \tanh u \tanh v}$

3 $\cosh 2x = \cosh^2 x + \sinh^2 x$

4 $\sinh(u+v) - \sinh(u-v) = 2\cosh u \sinh v$

5 $\sinh 3x = 3\sinh x + 4\sinh^3 x$

6 $1 + 2(\cosh x + \cosh 2x + \cdots + \cosh nx) = [\sinh(n+\frac{1}{2})x]/\sinh\frac{1}{2}x$

Differentiate with respect to x

7 $\sinh 5x$

8 $\cosh\sqrt{x}$

9 $\tanh(x^2+1)^{1/2}$

10 $\tanh^3 x$

11 $\frac{1}{3}e^{2x}(2\cosh x - \sinh x)$

12 $x\cosh x - \sinh x$

13 $\ln\cosh x$

14 $\sqrt{\cosh 4x}$

15 $x^2\sinh x - 2x\cosh x + 2\sinh x$

16 $\frac{1}{4}\sinh 2x - \frac{1}{2}x$

Find $\lim_{x\to+\infty}$ of

17 $e^{-x}\sinh x$

18 $e^{-x}\cosh x$

19 $e^{x}\operatorname{sech} x$

20 $e^{2x}(1 - \tanh x)$

21 $\dfrac{\cosh x}{\cosh 2x}$

22 $\dfrac{\ln \sinh x}{x}$

Prove

23 $\dfrac{d}{dx}\tanh x = \operatorname{sech}^2 x$

24 $\dfrac{d}{dx}\operatorname{csch} x = -\operatorname{csch} x \coth x$

25 $\cosh^{-1}x = \ln(x + \sqrt{x^2 - 1})$

26 $\tanh^{-1}x = \frac{1}{2}\ln\left(\dfrac{1 + x}{1 - x}\right)$

27 $\dfrac{d}{dx}\cosh^{-1}x = \dfrac{1}{\sqrt{x^2 - 1}}$

28 $\dfrac{d}{dx}\tanh^{-1}x = \dfrac{1}{1 - x^2}$

29 $\sinh^{-1}|\tan\theta| = \cosh^{-1}(\sec\theta) \qquad (-\frac{1}{2}\pi < \theta < \frac{1}{2}\pi)$

30 $\dfrac{d}{d\theta}\tanh^{-1}(\sin\theta) = \sec\theta$

4-7 Review Exercises

Differentiate with respect to x

1 $x^2 e^{-3/x}$

2 $x^3 e^{-x}$

3 x/e^x

4 e^{xu} where $u = e^x$

5 $(x \ln x)^{4/3}$

6 $x^{(x^3)}$

7 $2^{\sqrt{x}}$

8 $\arcsin\left(\dfrac{x + a}{x - a}\right)$

9 $\sqrt{1 + x^2}\arctan x$

10 $\sinh^3 2x$

11 $\frac{1}{6}\ln\left(\dfrac{1 + x + x^2}{(1 - x)^2}\right) + \dfrac{1}{\sqrt{3}}\arctan\left(\dfrac{2x + 1}{\sqrt{3}}\right)$

12 $\frac{1}{4}\ln\left(\dfrac{1 + x}{1 - x}\right) + \frac{1}{2}\arctan x$

Find

13 $\dfrac{d}{dx}e^{-ex}\Big|_{x=1/e}$

14 $\lim_{t\to 0}\dfrac{e^{kt} - 1}{t}$

15 $\lim_{x\to 0}\left(\dfrac{x}{e^x - 1}\right)^2$

16 $\lim_{x\to 1}\dfrac{1 - x}{e^x - e}$

17 $\lim_{x\to+\infty}\dfrac{5x + 1}{2x + \ln x}$

18 $\lim_{x\to+\infty}\dfrac{\sinh x}{\sinh(x + 1)}$

19 Find the minimum of e^{-x/e^x} for $x \geq 0$.

20 Find the maximum of $xe^{-\sqrt{x}}$.

21 Show that $e^x \geq 1 + x$ for all x.

22 Let $\alpha > 1$ and $y = \exp(-x^\alpha)$ for $x > 0$. Find all x where the graph has an inflection point.

23 Assume that the population of a certain city grows at a rate proportional to the population itself. If the population was 100,000 in 1940 and 150,000 in 1980, predict what it will be in 2000.

24 The graph of $y = f(x)$ has the following property. The tangent line at each point (x, y) meets the x-axis at $(x - 1, 0)$. Find $f(x)$.

25 When the switch is closed in the circuit of Figure 4-7-1, the current $I = I(t)$ satisfies the conditions

$$L\frac{dI}{dt} + RI = E \qquad I_0 = I(0) = 0$$

Compute $\dfrac{d}{dt}(Ie^{Rt/L})$, and use the result to find a formula for I.

Figure 4-7-1
See Exercise 25

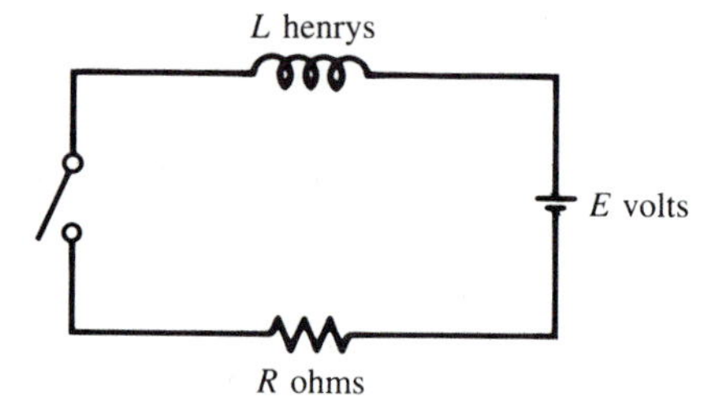

26 When the line from the boat (Figure 4-7-2) is wrapped around the rough mooring post, friction causes the tension T in the held end of the rope, as a function of the length L of rope in contact with the post, to decrease at a rate proportional to T. Precisely,

$$\frac{dT}{dL} = -\frac{\mu}{a}T \qquad (\mu > 0)$$

Show that if n turns are taken, then a force of merely $T_0 e^{-2\pi n\mu}$ will hold the boat against the gale, where T_0 is the tension in the boat end of the rope.

Figure 4-7-2
See Exercise 26

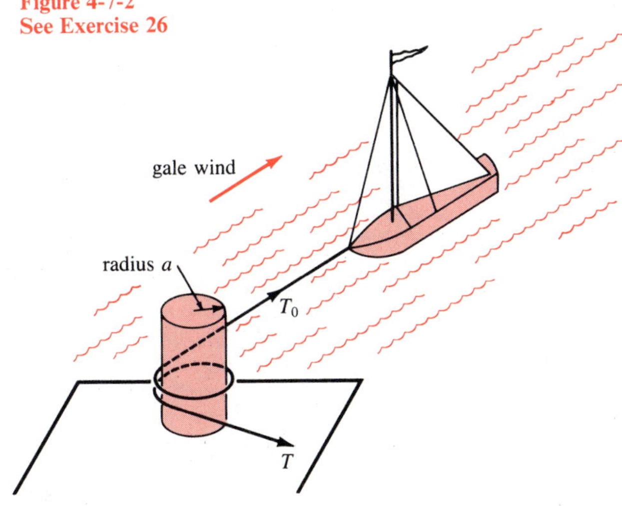

Solve for x to three significant figures

27 $\ln \ln x = 3$

28 $\ln \ln x = 6$

29 Find the point of intersection of the tangents to the graph $y = \ln x$ at $(e, 1)$ and $(1/e, -1)$.

30 Suppose $\theta = (\text{arc sin } x)^\circ$ in *degrees*. Find $d\theta/dx$.

31 The "doomsday" equation of von Foerster et al. predicts the population $N(t)$ in year t as

$$N(t) = \frac{1.79 \times 10^{11}}{(2026.87 - t)^{0.99}} \quad \text{for} \quad t < 2026.87$$

Find the inverse function $t = t(N)$.

32 Prove that $e^{-x} = x^3$ has exactly one solution.

33 Suppose $f(x) > 0$ and $y = f(x)$ is twice differentiable and concave down for $a < x < b$. Show that $z = \ln f(x)$ is concave down.

34 Find $\lim_{x \to +\infty} x(a^{1/x} - 1)$ where $a > 0$.

Prove

35 $\cosh^4 x - \sinh^4 x = \cosh 2x$

36 $2 \cosh x \sinh x = \dfrac{2 \tanh x}{1 - \tanh^2 x}$

37 $\tanh x = 1 - \dfrac{2}{\sinh 2x + \cosh 2x + 1}$

38 $\text{arc tan } x - \text{arc tan } \dfrac{x - 1}{x + 1} = \frac{1}{4}\pi$ for $x > -1$

39 $\frac{1}{2}\pi + \text{arc tan } x > 2 \text{ arc tan } 2x$

40 $\ln x > 2(x - 1)/(x + 1)$ for $x > 1$

41 Some programming languages (ALGOL 60, Pascal, and others) contain only sin, cos, and arc tan as standard trig functions, so it is necessary to express other trig functions in terms of these. Find a formula of the form $\text{arc cos } x = \text{arc tan}(?)$, and state where it is valid.

42 (cont.) Find a similar formula for arc sin x.

43 In radiation theory, $f(x) = x^2 e^x/(e^x - 1)^2$ is known as an **Einstein function.** Show that $f(x) \to 1$ as $x \to 0+$, and $f(x) \to 0+$ as $x \to +\infty$.

*44 (cont.) Show that $f(x)$ is strictly decreasing for $x > 0$.

5 Integration

5-1 The Area Problem

The ancient Greek mathematician Archimedes used ingenious methods to compute the area bounded by a parabola and a chord (Figure 5-1-1). In this chapter we shall develop tools of calculus that make the solution of this and similar problems routine. Although their invention was motivated by such area problems, these tools now have a wide range of scientific and technical applications.

Figure 5-1-1
The problem of Archimedes:
Find the area bounded by the chord and the parabola.

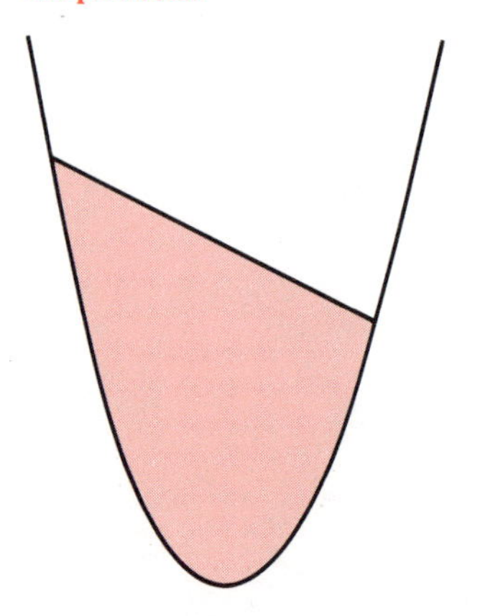

We start with a general problem: Find the area under the graph of a non-negative function $f(x)$ defined on a closed interval $[a, b]$. See Figure 5-1-2. Just as the slope problem led to the first major concept in calculus, the derivative, this area problem leads to the second major concept, the integral.

Part of the problem is the meaning of area. In elementary geometry we learned the areas of rectangles and figures derived from rectangles: triangles and other polygons. We need a reasonable definition of area for more general regions and a method for computing such areas.

For a few paragraphs, we shelve functions and their graphs and concentrate instead on plane regions, like that in Figure 5-1-3. If a region is not too complicated, it ought to have area. Let us suppose in advance that "uncomplicated" regions, the regions that we can draw with a pencil, do indeed have area, and that the area of each such region is a positive real number. What do we feel intuitively about the areas of such regions?

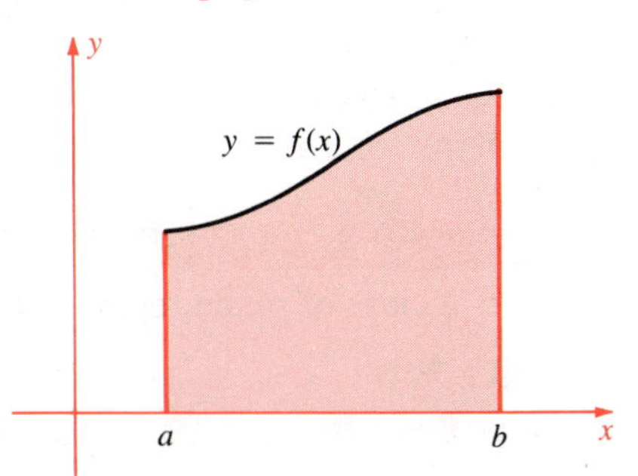

Figure 5-1-2
Area under a graph

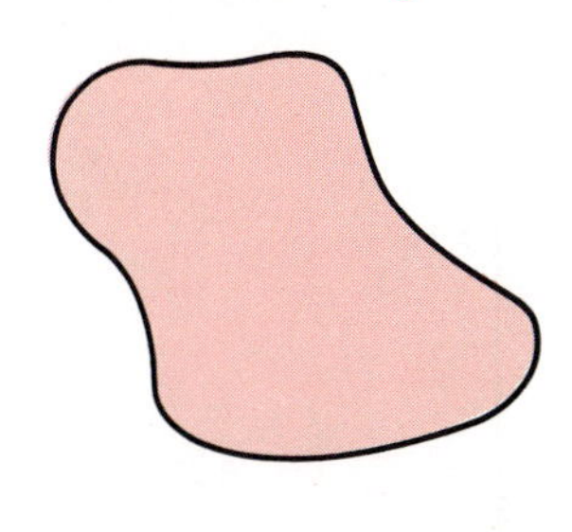

Figure 5-1-3
An uncomplicated plane region. What is its area in square units?

One intuitive feeling is that areas "add up." If the shaded region in Figure 5-1-3 were a piece of thin sheet metal, its weight would be proportional to its area. If we snipped it into two pieces, the weight of the whole would be the sum of the weights of the pieces, so the area of the whole must be the sum of the areas of the pieces. Similarly, the area of an assembled jigsaw puzzle is the sum of the areas of the pieces (Figure 5-1-4).

Figure 5-1-4
Areas add up.

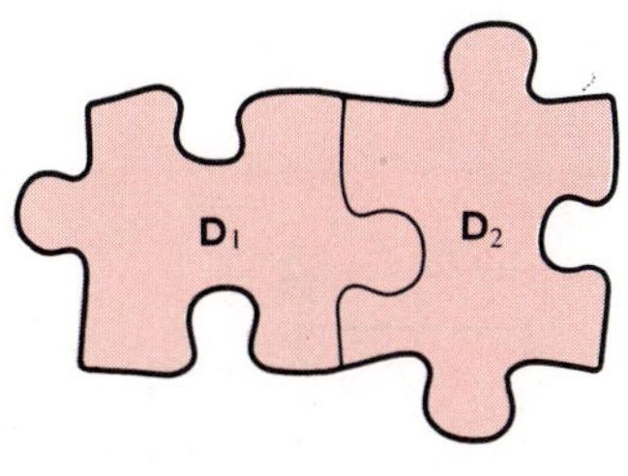

a Split **D** into two subregions

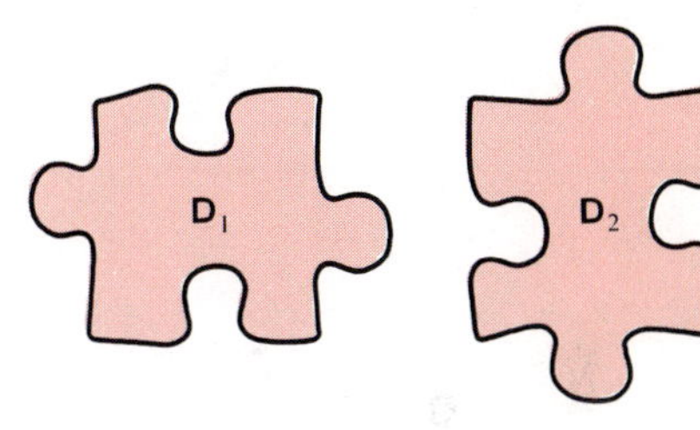

b area (**D**) = area ($\mathbf{D}_1$) + area ($\mathbf{D}_2$)

A second intuitive feeling for area is that if one region is contained in another (Figure 5-1-5), then the contained region has less area than the containing region. This is in fact a logical consequence of areas adding up (Figure 5-1-6).

Figure 5-1-5
$\mathbf{D}_1$ is contained in **D**, so

$$\text{area}(\mathbf{D}_1) < \text{area}(\mathbf{D})$$

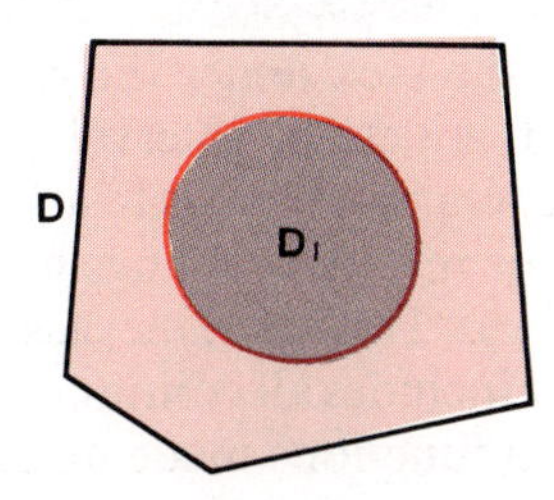

Figure 5-1-6
D is split into $\mathbf{D}_1$ and $\mathbf{D}_2$; hence

$$\text{area}(\mathbf{D}) = \text{area}(\mathbf{D}_1) + \text{area}(\mathbf{D}_2) > \text{area}(\mathbf{D}_1)$$

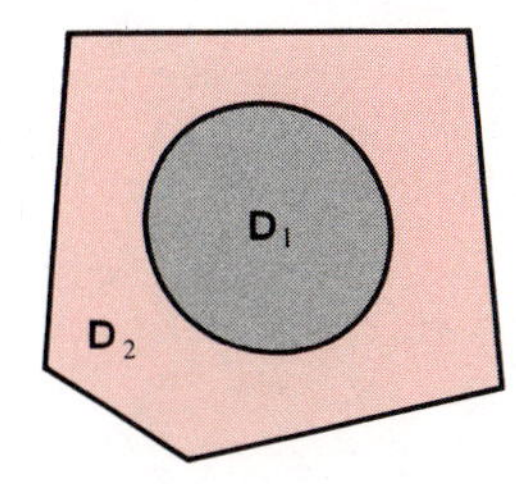

Figure 5-1-7
Rectangle: area (**D**) = hw

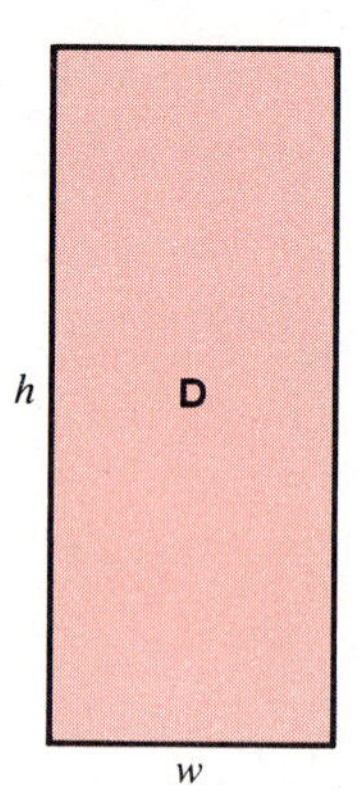

Finally, our basic building blocks are rectangles (Figure 5-1-7), and we know the area of a rectangle, height times width, from elementary geometry.

These self-evident physical notions lead us to four mathematical requirements for area:

Attributes of Area

There is a class of plane regions **D** that have area.

1 area (**D**) > 0

2 If $\mathbf{D} = \mathbf{D}_1 \cup \mathbf{D}_2$ and $\mathbf{D}_1$ and $\mathbf{D}_2$ do not overlap, then

$$\text{area}(\mathbf{D}) = \text{area}(\mathbf{D}_1) + \text{area}(\mathbf{D}_2)$$

3 If $\mathbf{D}_1 \subseteq \mathbf{D}$, then $\text{area}(\mathbf{D}_1) \leq \text{area}(\mathbf{D})$

4 The area of any rectangle equals the product (height) × (width).

Our method for finding the area of a region **D** is a two-pronged attack, as shown in Figures 5-1-8 and 5-1-9. One attack is to squeeze down from outside; the other attack is to squeeze up from inside. If this pincers maneuver succeeds, then the outer and inner approximations will converge to a common value. That value is then defined as the area of **D**.

Figure 5-1-8
Attack from the outside:
Cover **D** with non-overlapping rectangles.
The sum of their areas
is an *upper* bound for area (**D**).

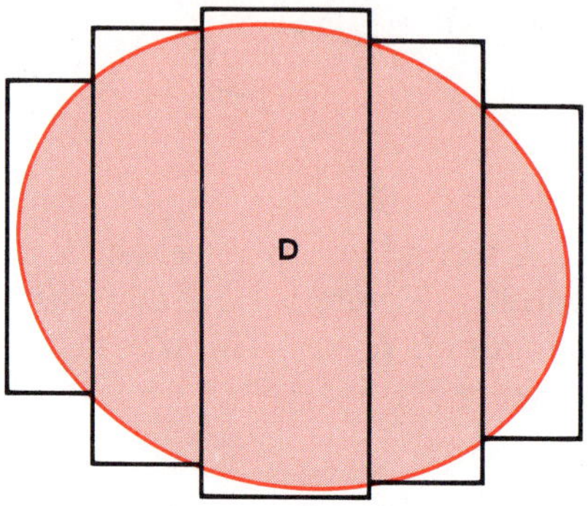

Figure 5-1-9
Attack from the inside:
Pack non-overlapping rectangles into **D**.
The sum of their areas
is a *lower* bound for area (**D**).

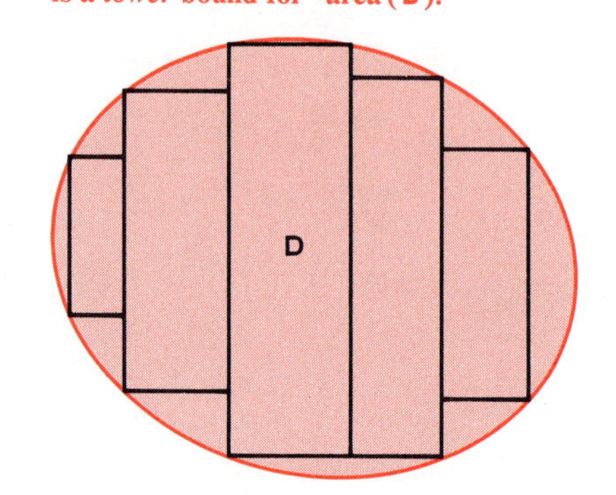

Area under a Graph

We return to the area under the graph of a non-negative function $f(x)$ whose domain is a closed interval $[a, b]$. This is the situation shown in Figure 5-1-10. Our problem is to assign to the shaded region a number that agrees with our intuitive feelings for area. We cannot do this for all such functions $f(x)$. However, the class of functions for which we can is large enough that it includes all continuous functions and all piecewise continuous functions (functions made up of several continuous pieces).

Figure 5-1-10
Area under the graph of a non-negative function $f(x)$ with domain a closed interval $[a, b]$

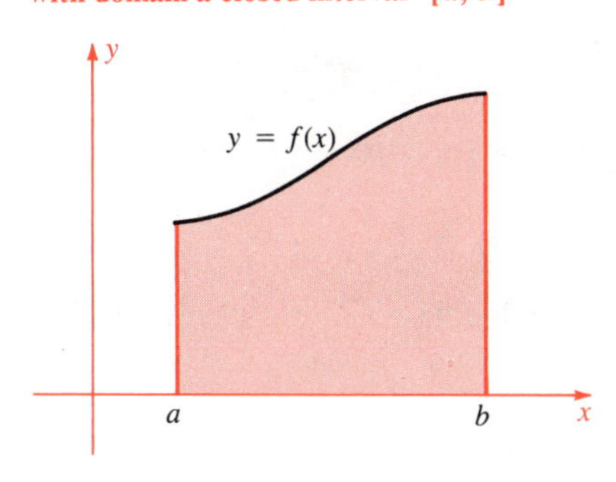

When we can assign an area to the region under the graph of $f(x)$, we shall denote it by the usual notation

$$\int_a^b f(x)\,dx$$

read "the integral of $f(x)$ with respect to x from a to b."

Although Figure 5-1-10 shows $f(x) \geq 0$ on the interval $[a, b]$, we shall define the integral without that restriction. Our intuitive attributes of area translate into properties we want the integral to have:

Properties of the Integral

Suppose $f(x)$ and $g(x)$ have integrals on a closed interval $[a, b]$.

1 If $a < c < b$, then

$$\int_a^b f(x)\,dx = \int_a^c f(x)\,dx + \int_c^b f(x)\,dx \qquad \text{(Figure 5-1-11)}$$

2 If $f(x) \leq g(x)$, then

$$\int_a^b f(x)\,dx \leq \int_a^b g(x)\,dx \qquad \text{(Figure 5-1-12)}$$

3 If $f(x) = K$, a constant function, then

$$\int_a^b K\,dx = K(b - a) \qquad \text{(Figure 5-1-13)}$$

Figure 5-1-11
Total area = area($\mathbf{D}_1$) + area($\mathbf{D}_2$)

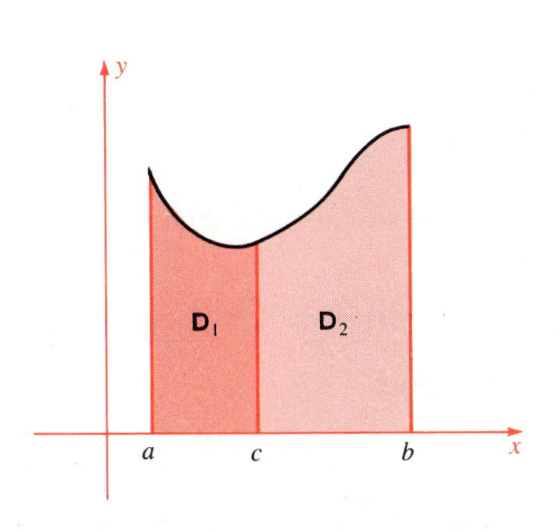

Figure 5-1-12
$f(x) \leq g(x)$ implies area(f) $\leq$ area(g)

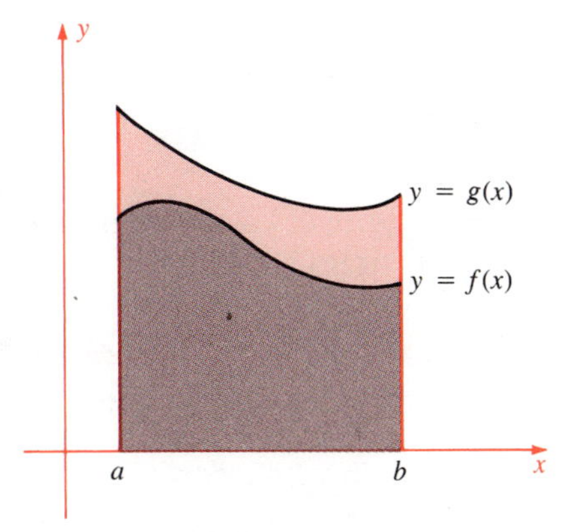

Figure 5-1-13
$f(x) = K$, a constant function, implies $\int_a^b K\,dx = K(b - a)$

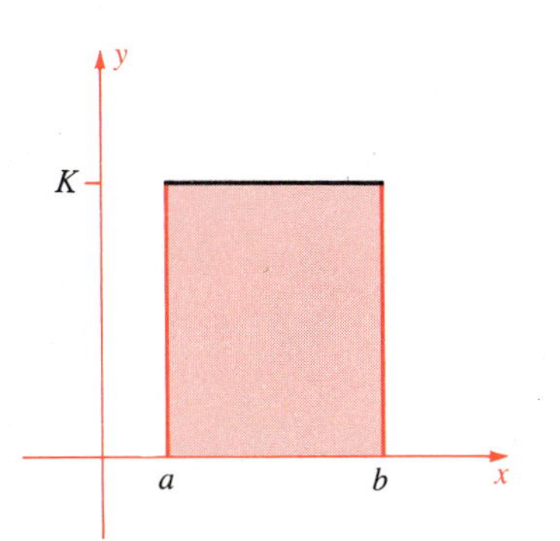

Step Functions

Our strategy for defining the integral of a function is to squeeze the function between functions whose integrals we know. The squeezing, or approximating, functions we use are the *step functions,* as shown in Figure 5-1-14.

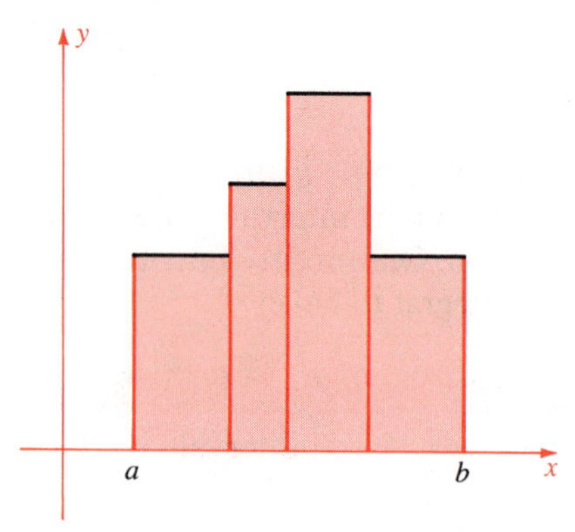

Figure 5-1-14
A positive step function and the region under its graph

Step functions are functions that are piecewise constant. The region under the graph of a step function is a finite union of non-overlapping rectangles, so we can easily calculate its area.

Definitions Partitions and Step Functions

1 A **partition** of a closed interval $[a, b]$ is an increasing finite sequence

$$a = x_0 < x_1 < x_2 < \cdots < x_{n-1} < x_n = b$$

of points of the interval, including the end points. These points divide $[a, b]$ into non-overlapping closed **subintervals**

$$[x_0, x_1] \quad [x_1, x_2] \quad \cdots \quad [x_{n-1}, x_n]$$

2 A **step function** on $[a, b]$ is a function that is piecewise constant. That is, there is a partition

$$a = x_0 < x_1 < \cdots < x_n = b$$

such that $s(x)$ is constant on each *open* interval (x_{j-1}, x_j) for $j = 1, 2, \ldots, n$. [The values $s(x_j)$ of $s(x)$ at the partition points need not be related to the values of $s(x)$ on either adjoining subinterval.] A step function is shown in Figure 5-1-15. ●

Figure 5-1-15
A step function

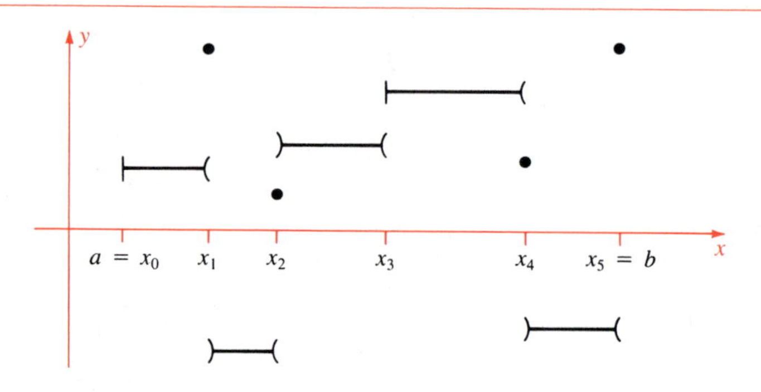

The integral of a step function is exactly what we expect it to be: the sum of heights times widths.

Definition Integral of a Step Function

Let $s(x)$ be a step function with partition points

$$a = x_0 < x_1 < x_2 < \cdots < x_n = b$$

Suppose $s(x) = K_j$ on (x_{j-1}, x_j). The **integral** of $s(x)$ on $[a, b]$ is defined as the sum

$$\int_a^b s(x)\,dx$$
$$= K_1(x_1 - x_0) + K_2(x_2 - x_1) + \cdots + K_n(x_n - x_{n-1})$$ ●

Figure 5-1-16

$\int_1^5 s(x)\,dx = \text{area}(\mathbf{D}) = 14$

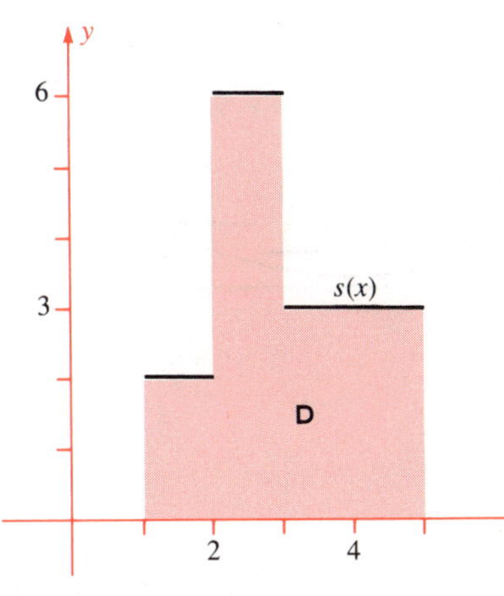

Note that the values $s(x_j)$ of $s(x)$ at the partition points do not enter the definition.

Example 1

Suppose $s(x) = 2$ for $1 < x < 2$ and $s(x) = 6$ for $2 < x < 3$ and $s(x) = 3$ for $3 < x < 5$ as in Figure 5-1-16. Then

$$\int_1^5 s(x)\,dx = 2(2-1) + 6(3-2) + 3(5-3)$$

$$= 2 + 6 + 6 = 14 \qquad \bullet$$

The properties that we require for the integrals of fairly general functions do indeed hold for step functions.

Figure 5-1-17
The areas add up:

$$\int_a^b S(x)\,dx = \int_a^c S(x)\,dx + \int_c^b S(x)\,dx$$

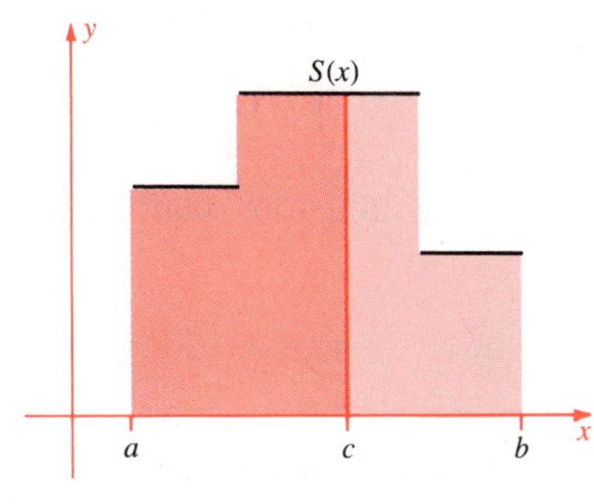

Theorem 1 Properties of Integrals of Step Functions

Let $s(x)$ and $S(x)$ be step functions on a closed interval $[a, b]$.

1 If $a < c < b$, then

$$\int_a^b S(x)\,dx = \int_a^c S(x)\,dx + \int_c^b S(x)\,dx \qquad \text{(Figure 5-1-17)}$$

2 If $s(x) \le S(x)$, then

$$\int_a^b s(x)\,dx \le \int_a^b S(x)\,dx \qquad \text{(Figure 5-1-18)}$$

3 If $S(x) = K$, a constant function, then

$$\int_a^b S(x)\,dx = \int_a^b K\,dx = K(b-a) \qquad \text{(Figure 5-1-13)} \quad \bullet$$

Statement 1 is obvious, and statements 2 and 3 are clear geometrically. Their proofs are not hard, but somewhat tedious, so we omit them.

Figure 5-1-18
$s(x) \le S(x)$ implies

$$\int_a^b S(x)\,dx - \int_a^b s(x)\,dx = \text{lightly shaded area} \ge 0$$

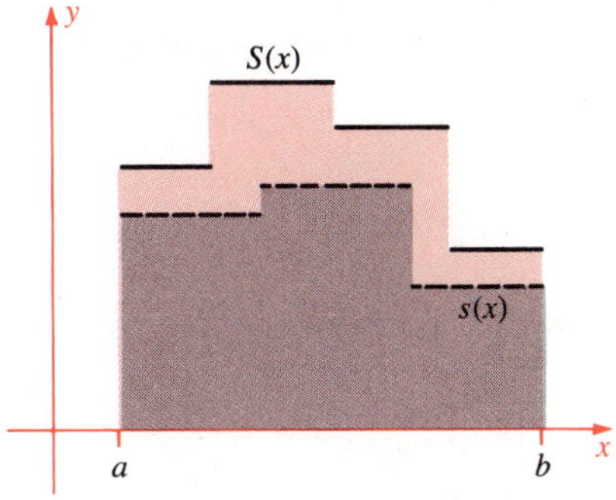

The Approximation Process

Given a function $f(x)$ on a closed interval $[a, b]$, we want to trap it between two step functions as shown in Figure 5-1-19. We seek step functions $s(x)$ and $S(x)$ such that

- $s(x) \le f(x) \le S(x)$
- The difference between the area under $y = S(x)$ and the area under $y = s(x)$ is small. (The difference equals the area of the lightly shaded region in the figure.) In symbols the difference equals

$$\int_a^b S(x)\,dx - \int_a^b s(x)\,dx = \text{“small”}$$

The following theorem makes this idea precise.

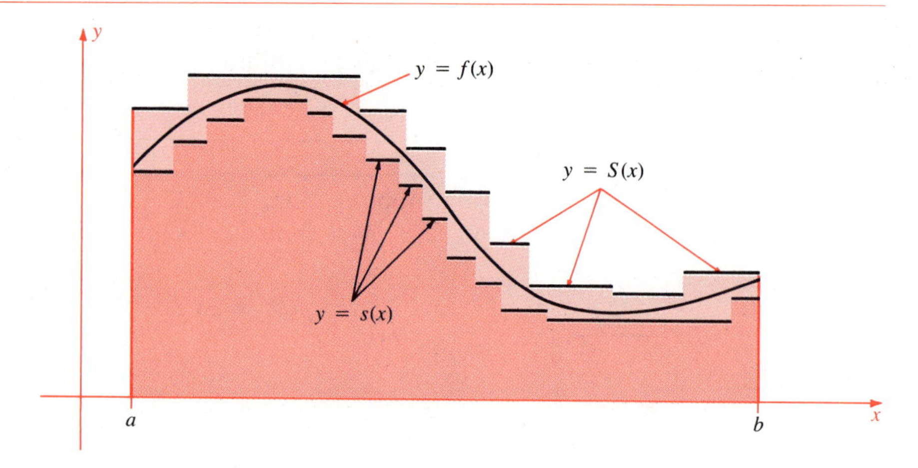

Figure 5-1-19
Trapping a function between two step functions:
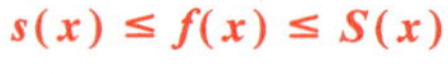
$s(x) \leq f(x) \leq S(x)$

Theorem 2 Existence of the Integral

Assume $f(x)$ is a function whose domain is a closed interval $[a, b]$. Suppose for each $\varepsilon > 0$ there exist step functions $S_\varepsilon(x)$ and $s_\varepsilon(x)$ on $[a, b]$ such that

$$s_\varepsilon(x) \leq f(x) \leq S_\varepsilon(x) \quad \text{and} \quad \int_a^b S_\varepsilon(x)\,dx - \int_a^b s_\varepsilon(x)\,dx < \varepsilon$$

Then there exists a unique real number, denoted $\int_a^b f(x)\,dx$, such that if $s(x)$ and $S(x)$ are *any* two step functions that satisfy $s(x) \leq f(x) \leq S(x)$, then

$$\int_a^b s(x)\,dx \leq \int_a^b f(x)\,dx \leq \int_a^b S(x)\,dx$$

When this is the case, $f(x)$ is called **integrable** and the number

$$\int_a^b f(x)\,dx$$

is called the **integral** of $f(x)$ on $[a, b]$. ●

We shall sketch the proof of Theorem 2 without going into technical details. The key word is *unique.* There is one and only one real number trapped between all integrals of step functions

$$\int_a^b s(x)\,dx \quad \text{and} \quad \int_a^b S(x)\,dx$$

where $s(x) \leq f(x) \leq S(x)$. We know by Theorem 1 that

$$\int_a^b s(x)\,dx \leq \int_a^b S(x)\,dx$$

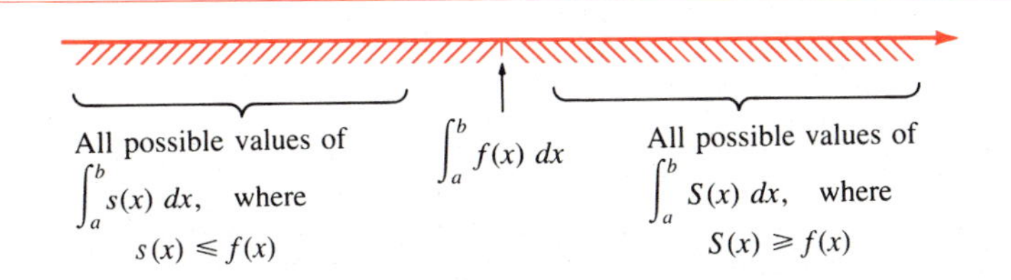

Figure 5-1-20
$\int_a^b f(x)\,dx$ is trapped between all integrals of step functions below $f(x)$ and all above $f(x)$.

so all integrals of such $s(x)$ are less than all integrals of such $S(x)$, as in Figure 5-1-20. The hypothesis of Theorem 2, shown graphically in Figure 5-1-21, is that there are such integrals arbitrarily close together. Since the real number system has no holes (completeness property), there must be a unique real number, the integral of $f(x)$, as claimed.

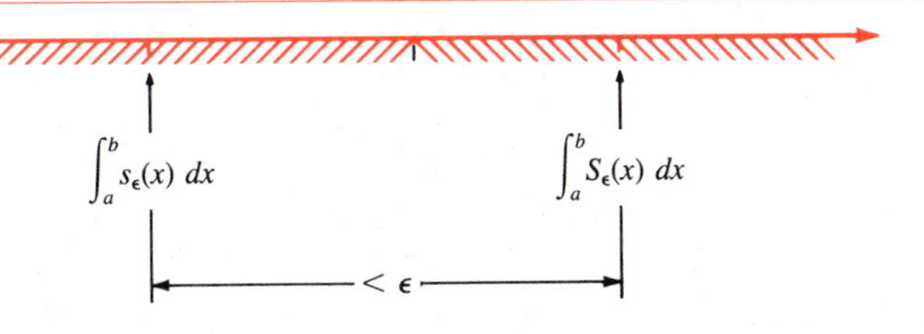

Figure 5-1-21
Hypothesis of Theorem 2:
For each $\varepsilon > 0$, we can find integrals of step functions within ε of each other, as drawn above.

Remark The combination of symbols

$$\int_a^b f(x)\,dx$$

should be regarded as a single entity, a single number. Even though the shorter notation

$$\int_a^b f$$

is sometimes convenient, the extended notation will prove very useful. Sometimes x is referred to as a "dummy" variable.

It is pretty obvious that step functions are integrable. So are all continuous functions; this discovery is a major achievement of calculus.

Theorem 3 Integrability of Continuous Functions

Let $f(x)$ be a continuous function on a closed interval $[a, b]$. Then $f(x)$ is integrable on $[a, b]$.

Remark The complete proof of Theorem 3 is given in courses on real analysis. The basic step is to prove that given any $\varepsilon > 0$, there exists a step function $g(x)$ such that

$$|f(x) - g(x)| < \varepsilon$$

for all x in $[a, b]$. This assertion uses what is called *uniform continuity,* a deep property of continuous functions on closed intervals. Uniform continuity is one more consequence of the completeness of the real number system. If the existence of $g(x)$ is accepted, then

$$s(x) = g(x) - \varepsilon \quad \text{and} \quad S(x) = g(x) + \varepsilon$$

are step functions that satisfy $s(x) \le f(x) \le S(x)$ and

$$\int_a^b S(x)\,dx - \int_a^b s(x)\,dx < 2\varepsilon(b - a)$$

Clearly $2\varepsilon(b - a)$ can be made as small as we please by choosing ε appropriately.

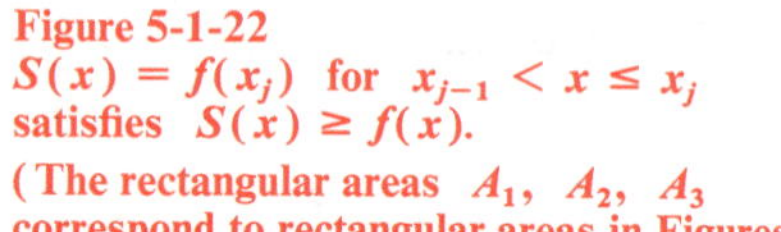

Figure 5-1-22
$S(x) = f(x_j)$ for $x_{j-1} < x \le x_j$ satisfies $S(x) \ge f(x)$.
(The rectangular areas A_1, A_2, A_3 correspond to rectangular areas in Figures 5-1-23.)

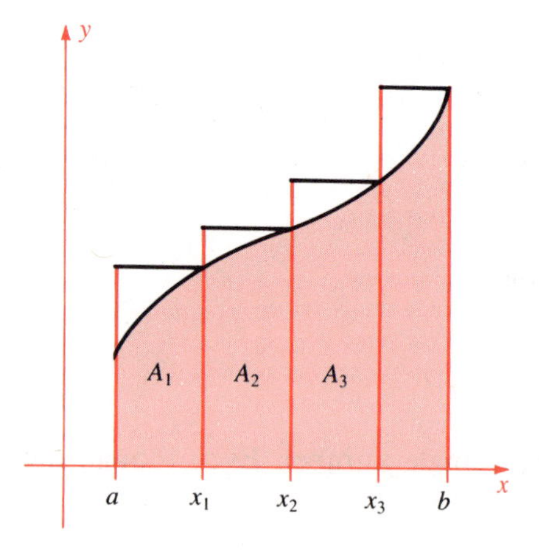

Most of the functions that we meet in practice have a finite number of ups and downs. It is comparatively easy to prove Theorem 3 for such functions. The following theorem, which we prove completely, shows the idea for **monotone** functions, that is, functions that are either increasing everywhere or decreasing everywhere. We state and prove the theorem for the increasing case; the decreasing case is handled similarly.

Theorem 4 Integrability of Monotone Functions

Suppose $f(x)$ is an increasing function on a closed interval $[a, b]$. Then $f(x)$ is integrable on $[a, b]$.

Proof Partition $[a, b]$ into n equal parts by

$$a = x_0 < x_1 < \cdots < x_n = b \qquad x_j - x_{j-1} = h = \frac{b - a}{n}$$

Define a step function $S(x)$ by

$$S(x) = f(x_j) \quad \text{on} \quad (x_{j-1}, x_j] \qquad \text{and} \qquad S(a) = f(a)$$

See Figure 5-1-22. Define another step function $s(x)$ by

$$s(x) = f(x_{j-1}) \quad \text{on} \quad [x_{j-1}, x_j) \qquad \text{and} \qquad s(b) = f(b)$$

Figure 5-1-23
$s(x) = f(x_{j-1})$ for $x_{j-1} \le x < x_j$ satisfies $s(x) \le f(x)$.
(The rectangular areas A_1, A_2, A_3 correspond to rectangular areas in Figure 5-1-22.)

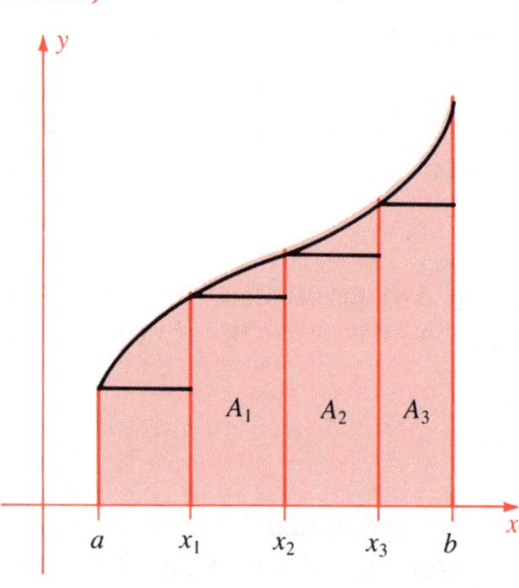

See Figure 5-1-23. Because $f(x)$ is increasing we have

$$s(x) \le f(x) \le S(x)$$

Let us compute the integrals of $s(x)$ and $S(x)$, first for $n = 4$, as shown in Figures 5-1-22 and 5-1-23:

$$\int_a^b S(x)\,dx = \frac{b - a}{4}[f(x_1) + f(x_2) + f(x_3) + f(b)]$$

$$\int_a^b s(x)\,dx = \frac{b - a}{4}[f(a) + f(x_1) + f(x_2) + f(x_3)]$$

When we subtract, all the intermediate terms cancel, and we have

$$\int_a^b S(x)\,dx - \int_a^b s(x)\,dx = \frac{b - a}{4}[f(b) - f(a)]$$

The general case is the same, except that 4 is replaced by n, the number of subintervals:

$$\int_a^b S(x)\,dx - \int_a^b s(x)\,dx = \frac{b-a}{n}[f(b) - f(a)]$$

This difference can be made as small as we please by choosing n sufficiently large. Therefore $f(x)$ is integrable. ●

Once we have Theorem 4 and the analogous statement for decreasing functions, the integrability of functions with a finite number of ups and downs can be proved by using part 1 of Theorem 1 several times.

Example 2

Estimate $\int_0^1 x^2\,dx$ by using a partition of 10 equal subintervals.

Solution The function $f(x) = x^2$ on the closed interval $[0, 1]$ is increasing, hence integrable by Theorem 4. We take $n = 10$ and construct step functions $S(x)$ and $s(x)$ as in the proof of the theorem. Then $b - a = 1 - 0 = 1$, so

$$\begin{aligned}\int_0^1 S(x)\,dx &= \tfrac{1}{10}(0.1^2 + 0.2^2 + \cdots + 0.9^2 + 1.0^2)\\ &= (0.1)(0.01 + 0.04 + \cdots + 0.81 + 1.00)\\ &= (0.1)(3.85) = 0.385\end{aligned}$$

and

$$\begin{aligned}\int_0^1 s(x)\,dx &= \tfrac{1}{10}(0.0^2 + 0.1^2 + \cdots + 0.8^2 + 0.9^2)\\ &= (0.1)(0.00 + 0.01 + \cdots + 0.64 + 0.81)\\ &= (0.1)(2.85) = 0.285\end{aligned}$$

We conclude that

$$0.285 \le \int_0^1 x^2\,dx \le 0.385$$ ●

Riemann Sums

There is another way to look at the integral. Suppose that $f(x)$ is integrable on $[a, b]$, and let $x_0, \cdots, x_n$ be any partition of the interval $[a, b]$. In each subinterval $[x_{j-1}, x_j]$ of the partition choose a point z_j. The choice of the z_j is completely arbitrary (Figure 5-1-24).

Figure 5-1-24

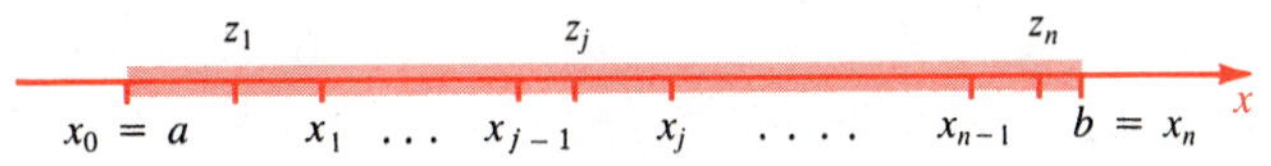

z_1 z_j z_n x

$x_0 = a$ x_1 $\ldots$ x_{j-1} x_j $\ldots$ x_{n-1} $b = x_n$

Figure 5-1-25
Riemann sum

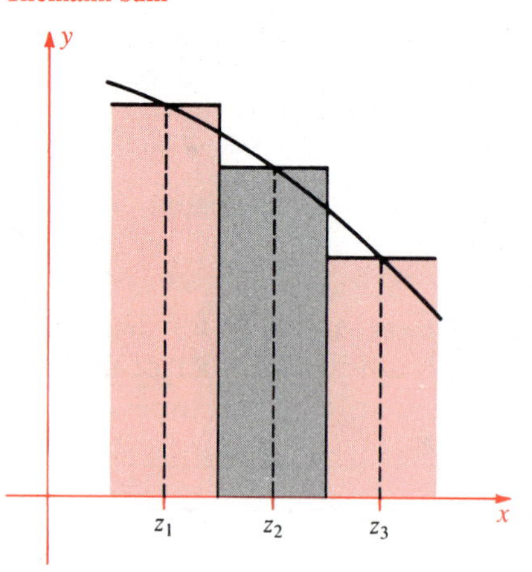

We define a step function $s(x)$ by

$$s(x) = f(z_j) \quad \text{for} \quad x_{j-1} < x < x_j$$

(Figure 5-1-25). The values $s(x_j)$ at the partition points can be anything, but let us make $s(x_j) = f(x_j)$. It is clear that $s(x)$ is a kind of approximation to $f(x)$. Indeed, we expect that if n is very large and the successive partition points are close together, then the area under $y = s(x)$ is very close to the area under $y = f(x)$. The area under $y = s(x)$ is

$$\int_a^b s(x)\,dx = f(z_1)(x_1 - x_0) + f(z_2)(x_2 - x_1) + \cdots + f(z_n)(x_n - x_{n-1})$$

The sum on the right is called a **Riemann sum** for $f(x)$. Many Riemann sums correspond to a given partition, because each sum depends on the choice of the points z_j. The precise relation of Riemann sums to integrals is expressed in the following theorem:

Theorem 5

Let $f(x)$ be integrable on $[a, b]$ and let $\varepsilon > 0$. Then there exists a $\delta > 0$ with the following property:

Suppose $x_0, \cdots, x_n$ is any partition of $[a, b]$ whose subintervals all have length less than δ. Suppose R is any Riemann sum corresponding to the partition. Then

$$\left| \int_a^b f(x)\,dx - R \right| < \varepsilon$$

The striking feature of Theorem 5 is this: Once you choose any partition, all of whose subintervals are sufficiently short, then it doesn't matter how you choose the z_j; the corresponding Riemann sum is within ε of the integral of $f(x)$. We omit the proof of the theorem.

Riemann sums are very important in applications of integration. We usually set up an integral for some physical or geometrical quantity by forming Riemann sums and then taking their limit as the partition lengths approach zero. This technique will be explained in Chapter 7.

Exercises

The step function $s(x)$ is defined on $[a, b]$ with the partition points

$$x_0 = a \quad x_1 \quad x_2 \quad x_3 \quad x_4 = b$$

and $s(x) = K_j$ on (x_{j-1}, x_j). In each exercise, the data is given in the form

$$\begin{matrix} x_0 & & x_1 & & x_2 & & x_3 & & x_4 \\ & K_1 & & K_2 & & K_3 & & K_4 & \end{matrix}$$

Graph $y = s(x)$ and find $\int_a^b s(x)\,dx$

1 $\begin{matrix} 0 & & 1 & & 2 & & 3 & & 4 \\ & 2 & & 1 & & 1 & & 2 & \end{matrix}$

2 $\begin{matrix} 0 & & 1 & & 2 & & 3 & & 4 \\ & -1 & & -1 & & 2 & & -2 & \end{matrix}$

3 $\begin{matrix} 0 & & \frac{1}{4} & & \frac{1}{2} & & \frac{3}{4} & & 1 \\ & 1 & & 0 & & 1 & & 1 & \end{matrix}$

4 $0 \quad \frac{1}{4} \quad \frac{1}{2} \quad \frac{3}{4} \quad 1$
$\quad -2 \quad 6 \quad -6 \quad 2$

5 $0 \quad \frac{1}{3} \quad \frac{1}{2} \quad \frac{2}{3} \quad 1$
$\quad 3 \quad 6 \quad 6 \quad 3$

6 $0 \quad \frac{1}{3} \quad \frac{1}{2} \quad \frac{2}{3} \quad 1$
$\quad 5 \quad 4 \quad 3 \quad 2$

7 $-2 \quad -1 \quad 0 \quad 1 \quad 2$
$\quad 1 \quad 1 \quad 1 \quad 0$

8 $-2 \quad -1 \quad 0 \quad 1 \quad 2$
$\quad \frac{1}{3} \quad \frac{1}{2} \quad -\frac{1}{2} \quad \frac{1}{3}$

9 Divide the interval $[0, 1]$ into five parts by $x_j = j/5$ for $j = 0, 1, \cdots, 5$. Let z_j be the midpoint of $[x_{j-1}, x_j]$ for $j = 1, \cdots, 5$. Compute the corresponding Riemann sum for

$$\int_0^1 x^2\, dx$$

10 In Exercise 9, change z_j to the point that is $\frac{2}{3}$ of the way from x_{j-1} to x_j. Thus $z_1 = \frac{2}{15}$, and so on. Now compute the corresponding Riemann sum. (You should get $0.36444 \cdots$, which is an approximation to the exact answer, $\frac{1}{3}$.)

5-2 Sums and Examples

In later sections we shall develop practical and effective indirect ways of computing integrals. In this section, however, we shall compute several integrals directly from the definition. In each case, we approximate the given function by suitable step functions and find their integrals. Now the integral of a step function is a finite sum, so we need to do some preliminary work on computing sums. (Our results will have applications to other than integrals later.)

Summation Notation

It is common to denote a sum by a capital Greek sigma. We write

$$\sum_{i=m}^{n} a_i = a_m + a_{m+1} + a_{m+2} + \cdots + a_n$$

The expression on the left is read "the sum of a_i for i from m to n." The letter i is the *index of summation* and can be any letter. It is sometimes referred to as a "dummy" index. Thus

$$\sum_{i=m}^{n} a_i \qquad \text{and} \qquad \sum_{j=m}^{n} a_j$$

are symbols that stand for the same sum

$$a_m + a_{m+1} + a_{m+2} + \cdots + a_n$$

The symbol

$$\sum_{i=m}^{n}$$

does not represent a number or a function. It stands for a *process* to be performed, as does the (partial) symbol

$$\int_a^b \cdots dx$$

You may already know summation in another language. For instance

$$S = \sum_{I=M}^{N} A_I$$

can be written in two common computer languages in the following way:

BASIC	Pascal
10 LET S = 0	**begin**
20 FOR I = M TO N	S := 0;
30 LET S = S + A(I)	**for** I := M **to** N **do**
40 NEXT I	S := S + A[I]
	end;

Example 1

a $\sum_{i=1}^{5} c_i = c_1 + c_2 + c_3 + c_4 + c_5$

b $\sum_{j=0}^{3} b_j = b_0 + b_1 + b_2 + b_3$

c $\sum_{k=-2}^{2} 3k^2 = 3(-2)^2 + 3(-1)^2 + 3(0)^2 + 3(1)^2 + 3(2)^2 = 30$

d $\sum_{k=7}^{10} \frac{1}{k+1} = \frac{1}{8} + \frac{1}{9} + \frac{1}{10} + \frac{1}{11}$ ●

Summation Formulas

The first sum we need is

$$S_n = \sum_{j=1}^{n} j = 1 + 2 + 3 + \cdots + (n-1) + n$$

To find this sum, we write S_n twice, once in each direction, and add:

$$\begin{aligned} S_n &= 1 + 2 + \cdots + (n-1) + n \\ S_n &= n + (n-1) + \cdots + 2 + 1 \\ \hline 2S_n &= (n+1) + (n+1) + \cdots + (n+1) + (n+1) \\ &= n(n+1) \end{aligned}$$

Hence

$$S_n = \tfrac{1}{2}n(n+1)$$

Next we need the sum of the first n squares. The formula is

$$\sum_{j=1}^{n} j^2 = 1^2 + 2^2 + 3^2 + \cdots + n^2 = \tfrac{1}{6}n(n+1)(2n+1)$$

To verify this formula, set $S_n = \tfrac{1}{6}n(n+1)(2n+1)$. Then $S_1 = 1$, and you can check by a little algebra that $S_k - S_{k-1} = k^2$. Consequently

$$1^2 + 2^2 + 3^2 + \cdots + n^2$$
$$= 1 + (S_2 - S_1) + (S_3 - S_2) + \cdots + (S_n - S_{n-1}) = S_n$$

because the terms cancel in pairs.

There is also a useful formula for the sum of the first n cubes:

$$\sum_{j=1}^{n} j^3 = \tfrac{1}{4}n^2(n+1)^2$$

The proof is similar to the preceding one, so we omit it. However, do test it for a few small n.

Next is the formula for the sum of a geometric progression:

$$1 + r + r^2 + \cdots + r^{n+1} = \frac{r^n - 1}{r - 1} \qquad (r \neq 1)$$

It is true because

$$(r-1)(r^{n-1} + r^{n-2} + \cdots + r + 1)$$
$$= (r^n - r^{n-1}) + (r^{n-1} - r^{n-2}) + \cdots$$
$$+ (r^2 - r) + (r - 1)$$
$$= r^n - 1$$

Again, terms cancel in pairs. Let us summarize:

Elementary Sum Formulas

- $\displaystyle\sum_{i=1}^{n} c = nc$
- $\displaystyle\sum_{i=1}^{n} i = \tfrac{1}{2}n(n+1)$
- $\displaystyle\sum_{i=1}^{n} i^2 = \tfrac{1}{6}n(n+1)(2n+1)$
- $\displaystyle\sum_{i=1}^{n} i^3 = \tfrac{1}{4}n^2(n+1)^2$
- $\displaystyle\sum_{k=0}^{n-1} r^k = \frac{r^n - 1}{r - 1} \qquad (r \neq 1)$

Sums have a linearity property that is a natural consequence of the commutative, associative, and distributive laws of arithmetic. For instance,

$$(ha_1 + kb_1) + (ha_2 + kb_2) + (ha_3 + kb_3)$$
$$= h(a_1 + a_2 + a_3) + k(b_1 + b_2 + b_3)$$

In general:

Linearity of Sums

$$\sum_{j=m}^{n} (ha_j + kb_j) = h\sum_{j=m}^{n} a_j + k\sum_{j=m}^{n} b_j$$

Figure 5-2-1
The integral of $y = x$ on $[0, b]$

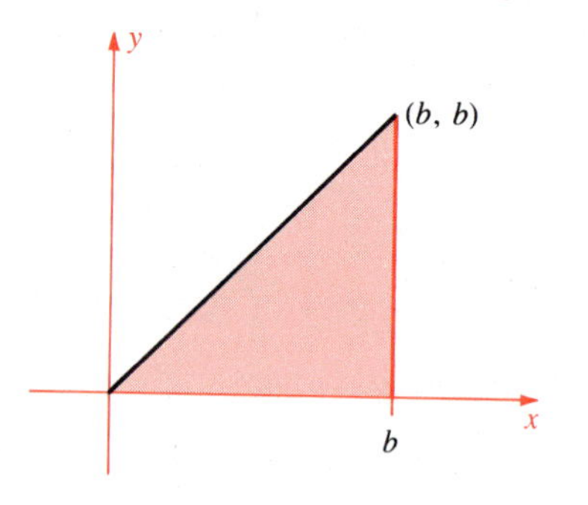

Figure 5-2-2
A step function satisfying $S(x) \geq x$. Here $n = 4$ and $h = \frac{1}{4}b = x_j - x_{j-1}$.

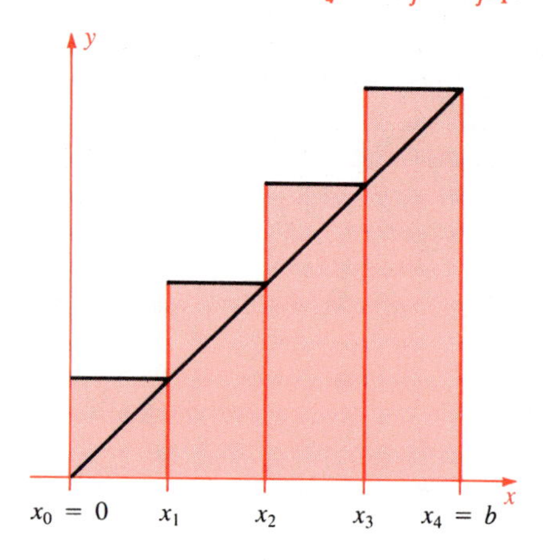

Figure 5-2-3
A step function satisfying $s(x) \leq x$. Here $n = 4$ and $s(x) = S(x) - \frac{1}{4}b$.

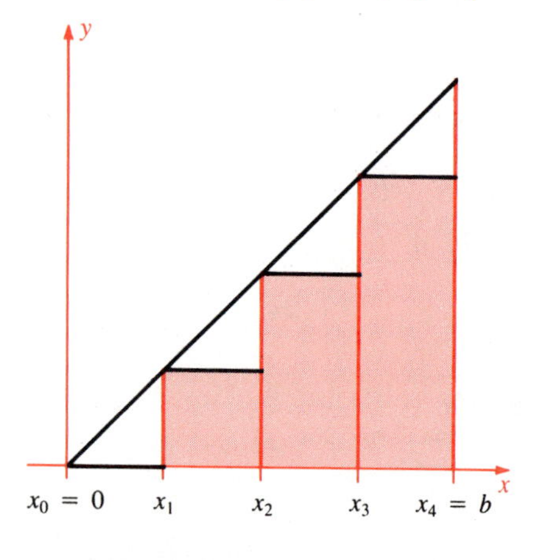

Some Integrals

Example 2

Find $\displaystyle\int_0^b x\,dx$

Solution It is obvious that the shaded region in Figure 5-2-1 is a right triangle with equal legs of length b. Hence its area is $\frac{1}{2}b^2$. The problem is to show that the definition of the integral leads to the same answer.

The function $f(x) = x$ is increasing, so we can use the same approach as we used to prove Theorem 4 in the previous section.

Partition $[0, b]$ into n equal parts by the points

$$x_j = hj \qquad \text{where} \quad h = \frac{b}{n} \quad \text{and} \quad 0 \leq j \leq n$$

Define a step function $S(x)$ by

$$S(0) = 0 \quad \text{and} \quad S(x) = x_j \quad \text{for} \quad x_{j-1} < x \leq x_j$$

See Figure 5-2-2. Clearly $x \leq S(x)$ for $0 \leq x \leq b$, so

$$\int_0^b x\,dx \leq \int_0^b S(x)\,dx$$

Now compute:

$$\int_0^b S(x)\,dx = x_1h + x_2h + \cdots + x_nh = h(x_1 + \cdots + x_n)$$

$$= h\sum_{j=1}^{n} x_j = h\sum_{j=1}^{n} hj = h^2\sum_{j=1}^{n} j$$

$$= h^2[\tfrac{1}{2}n(n+1)]$$

Substitute $h = b/n$ and simplify:

$$\int_a^b S(x)\,dx = \frac{b^2}{n^2}[\tfrac{1}{2}n(n+1)] = \tfrac{1}{2}b^2\frac{n+1}{n}$$

Now we need a step function below $f(x)$. If you simply shift $S(x)$ down by h, the resulting function is below $f(x)$. Thus set $s(x) = S(x) - h$. Then $s(x) \leq f(x)$ for $0 \leq x \leq b$ so

$$\int_0^b s(x)\,dx \leq \int_0^b f(x)\,dx$$

See Figure 5-2-3. Consequently

$$\int_0^b s(x)\,dx = \int_0^b [S(x) - h]\,dx = \left[\int_0^b S(x)\,dx\right] - hb$$

$$= \tfrac{1}{2}b^2\frac{n+1}{n} - \frac{b^2}{n}$$

Now assemble the inequalities:

$$\tfrac{1}{2}b^2\,\frac{n+1}{n} - \frac{b^2}{n} \le \int_0^b x\,dx \le \tfrac{1}{2}b^2\,\frac{n+1}{n}$$

Now *squeeze;* that is, let n increase indefinitely. Since

$$\lim_{x\to+\infty}\frac{1}{n} = 0 \quad\text{and}\quad \lim_{n\to+\infty}\frac{n+1}{n} = \lim_{n\to+\infty}\left(1+\frac{1}{n}\right) = 1$$

the conclusion is

$$\tfrac{1}{2}b^2 \le \int_0^b x\,dx \le \tfrac{1}{2}b^2$$

The squeezing process has succeeded completely; the integral is squeezed between two equal numbers! Therefore

$$\int_0^b x\,dx = \tfrac{1}{2}b^2$$

●

Example 3

Find $\int_a^b x\,dx$ where $0 < a < b$.

Solution The integral equals the area of the trapezoidal region in Figure 5-2-4. We know that

$$\int_0^b x\,dx = \int_0^a x\,dx + \int_a^b x\,dx$$

But by Example 2,

$$\int_0^b x\,dx = \tfrac{1}{2}b^2 \qquad\text{and}\qquad \int_0^a x\,dx = \tfrac{1}{2}a^2$$

Therefore

$$\int_a^b x\,dx = \tfrac{1}{2}b^2 - \tfrac{1}{2}a^2 = \tfrac{1}{2}(b^2 - a^2)$$

Note that the answer can be written $\frac{1}{2}(b+a)(b-a)$, that $\frac{1}{2}(b+a)$ is the average of the (vertical) bases of the trapezoid, and that $b-a$ is its height. Therefore the answer agrees with the formula from plane geometry for the area of a trapezoid. ●

Figure 5-2-4
$\int_a^b x\,dx$

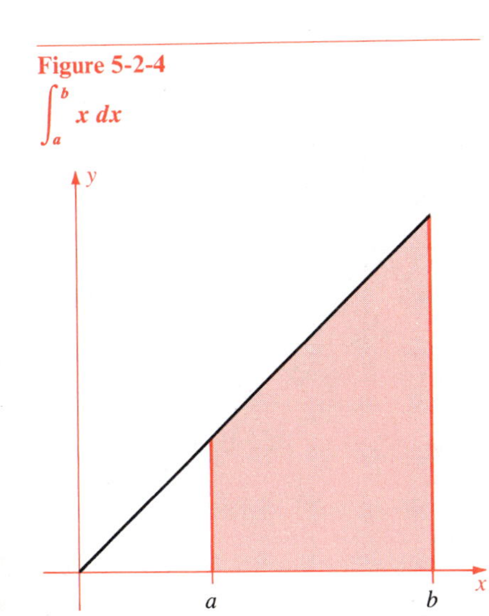

Example 4

Find $\int_0^b x^2\,dx$

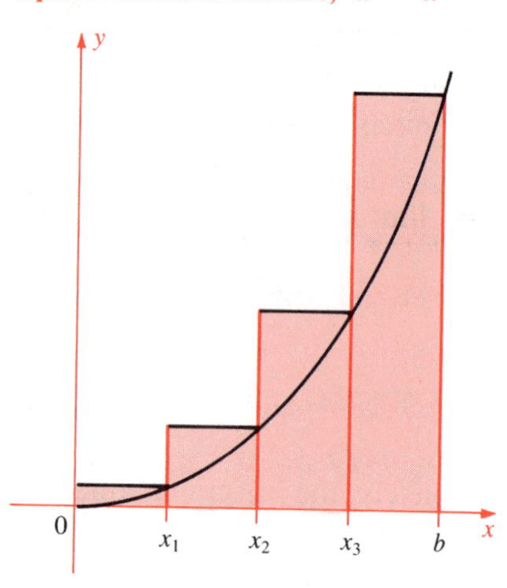

Figure 5-2-5
Step function above $y = x^2$, equal divisions. In this case, $n = 4$.

Solution We'll do only the upper half of this one, and leave the lower half for the exercises. As in the solution of Example 2, we partition the interval $[0, b]$ into n equal parts by

$$x_j = hj \quad \text{where} \quad h = b/n \quad \text{and} \quad 0 \le j \le n$$

We define a step function (Figure 5-2-5) by

$$S(0) = 0 \quad \text{and} \quad S(x) = x_j^2 \quad \text{for} \quad x_{j-1} < x \le x_j$$

Then $x^2 \le S(x)$ on $[0, b]$, so we have

$$\int_0^b x^2\,dx \le \int_0^b S(x)\,dx$$

We compute the integral of the step function $S(x)$ by using the formula for the sum of the first n squares:

$$\int_0^b S(x)\,dx = \sum_{j=1}^n x_j^2 h = h \sum_{j=1}^n (hj)^2 = h^3 \sum_{j=1}^n j^2$$
$$= \tfrac{1}{6}h^3 n(n+1)(2n+1)$$

Since $h = b/n$, we have

$$\int_0^b x^2\,dx \le \int_0^b S(x)\,dx = \tfrac{1}{6}b^3 \frac{(n+1)(2n+1)}{n^2}$$

We squeeze down by increasing n. Now

$$\lim_{n\to+\infty} \frac{(n+1)(2n+1)}{n^2} = \lim_{n\to+\infty}\left(1 + \frac{1}{n}\right)\left(2 + \frac{1}{n}\right)$$
$$= 1 \cdot 2 = 2$$

so we conclude that

$$\int_0^b x^2\,dx \le \tfrac{1}{3}b^3$$

Similarly (Exercise 16), $\int_0^b x^2 \ge \tfrac{1}{3}b^3$, so we conclude that

$$\int_0^b x^2\,dx = \tfrac{1}{3}b^3$$

●

Example 5

Find $\int_0^b x^3\,dx$

Solution As before, we partition $[0, b]$ into n equal subintervals by the points

$$hj \quad \text{where} \quad h = \frac{b}{n} \quad \text{and} \quad 0 \le j \le n$$

The step function $S(x)$ defined by

$$S(0) = 0 \quad \text{and} \quad S(x) = x_j^3 \quad \text{for} \quad x_{j-1} < x \le x_j$$

satisfies $x^3 \le S(x)$ on $[0, b]$, so we have

$$\int_0^b x^3\,dx \le \int_0^b S(x)\,dx$$

By using the formula for the sum of the first n cubes, we obtain

$$\begin{aligned}\int_0^b S(x)\,dx &= \sum_{j=1}^{n} (hj)^3 h = h^4 \sum_{j=1}^{n} j^3 \\ &= h^4[\tfrac{1}{4}n^2(n+1)^2] = \frac{b^4}{4n^4}n^2(n+1)^2 \\ &= \frac{b^4}{4}\left(\frac{n+1}{n}\right)^2\end{aligned}$$

We squeeze down by increasing n:

$$\int_0^b x^3\,dx \le \frac{b^4}{4} \lim_{n\to+\infty} \left(\frac{n+1}{n}\right)^2 = \tfrac{1}{4}b^4$$

Similarly (Exercise 17),

$$\int_0^b x^3\,dx \ge \tfrac{1}{4}b^4 \qquad \text{so finally} \qquad \int_0^b x^3\,dx = \tfrac{1}{4}b^4$$

●

Example 6

Find $\displaystyle\int_0^b e^x\,dx$

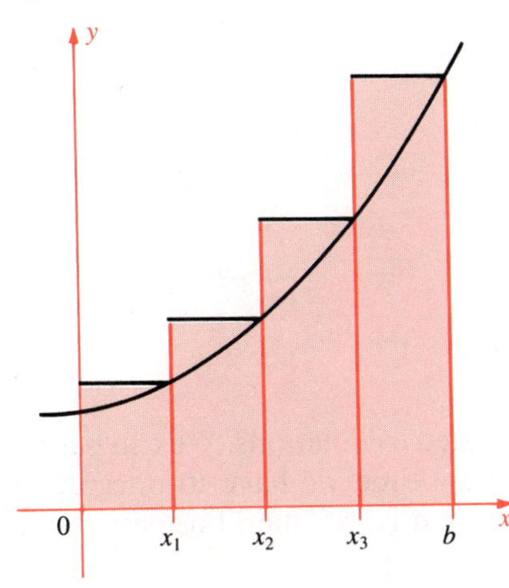

Figure 5-2-6
Step function above $y = e^x$, equal divisions.

Solution Partition $[0, b]$ into n equal parts. The partition points are

$$x_j = hj \qquad \text{where} \quad h = b/n \quad \text{and} \quad 0 \le j \le n$$

as in Examples 2, 4, and 5. Define a step function $S(x)$ by

$$S(0) = 1 \quad \text{and} \quad S(x) = \exp(x_j) \quad \text{for} \quad x_{j-1} < x \le x_j$$

See Figure 5-2-6. Since e^x is an increasing function, $e^x \le S(x)$ for $0 \le x \le b$. Now compute:

$$\int_0^b S(x)\,dx = \sum_{j=1}^{n} \exp(x_j)h = h\sum_{j=1}^{n} e^{hj}$$

We must sum a geometric progression with ratio e^h:

$$\sum_{j=1}^{n} e^{hj} = e^h \sum_{j=0}^{n-1} e^{hj} = e^h \frac{e^{hn}-1}{e^h - 1}$$

Plug this into the last factor of the previous formula, then replace hn by b:

$$\int_0^b S(x)\,dx = he^h \frac{e^b - 1}{e^h - 1} = (e^b - 1)e^h \frac{h}{e^h - 1}$$

Conclusion:

$$\int_0^b e^x\,dx \le (e^b - 1)e^h \frac{h}{e^h - 1}$$

Now let $n \to +\infty$ so that $h \to 0$. The factor $e^h \to 1$, so we obtain

$$\int_0^b e^x\,dx \le (e^b - 1)\lim_{h\to 0} \frac{h}{e^h - 1}$$

The quantity whose limit must be found is the reciprocal of a difference quotient, whose limit we know as a derivative:

$$\lim_{h\to 0} \frac{e^h - 1}{h} = \frac{d}{dx} e^x \bigg|_{x=0} = 1 \qquad \text{so} \qquad \lim_{h\to 0} \frac{h}{e^h - 1} = \frac{1}{1} = 1$$

Consequently

$$\int_0^b e^x\,dx \le e^b - 1$$

We can derive the reverse inequality similarly (Exercise 18). Hence

$$\int_0^b e^x\,dx = e^b - 1$$

a surprisingly simple answer indeed. ●

Let us summarize the results of the examples; they are all useful formulas:

- $$\int_a^b x\,dx = \tfrac{1}{2}(b^2 - a^2) \qquad (0 < a < b)$$
- $$\int_0^b x^2\,dx = \tfrac{1}{3}b^3 \qquad (0 < b)$$
- $$\int_0^b x^3\,dx = \tfrac{1}{4}b^4 \qquad (0 < b)$$
- $$\int_0^b e^x\,dx = e^b - 1 \qquad (0 < b)$$

●

Their derivations involved some complicated calculations. We can hardly go through life doing so much work every time we have to integrate a function. We need a systematic method; that is the main business of the next section.

Exercises

Find

1 $\sum_{j=1}^{50} 0$

2 $\sum_{j=1}^{100} 1$

3 $\sum_{j=0}^{100} 1$

4 $\sum_{j=0}^{13} 3$

5 $\sum_{j=1}^{100} j$

6 $\sum_{j=11}^{100} j$

7 $\sum_{j=1}^{10} j^2$

8 $\sum_{j=11}^{10} j^2$

Derive the formula

9 $\sum_{j=0}^{n} a_j = \sum_{i=0}^{n} a_{n-i}$

0 $\sum_{j=1}^{n} a_j = \sum_{j=0}^{n-1} a_{j+1}$

11 $\sum_{j=1}^{n} a_j = \sum_{j=11}^{n+10} a_{j-10}$

12 $\sum_{j=1}^{n} a_j = \sum_{j=-n}^{-1} a_{-j}$

3 Set $S_n = \frac{1}{2}n(n+1)$. Show that $S_n - S_{n-1} = n$ and use this to prove that $1 + 2 + \cdots + n = S_n$.

4 Set $S_n = \frac{1}{4}n^2(n+1)^2$. Show that $S_n - S_{n-1} = n^3$. Use this to derive the formula for $1^3 + 2^3 + \cdots + n^3$.

5 Derive a formula for

$$\int_a^b x^2\,dx \qquad (0 < a < b)$$

by the method of Example 3.

6 Complete the solution of Example 4. Define $s(x)$ by $s(b) = b^2$ and $s(x) = x_{j-1}^2$ for $x_{j-1} \le x < x_j$.

7 Complete the solution of Example 5. Define $s(x)$ by $s(b) = b^3$ and $s(x) = x_{j-1}^3$ for $x_{j-1} \le x < x_j$.

8 Complete the solution of Example 6. Define $s(x)$ by $s(b) = e^b$ and $s(x) = \exp(x_{j-1})$ for $x_{j-1} \le x < x_j$.

In the following exercises we derive the formula

$$\int_1^b \frac{1}{x}\,dx = \ln b$$

where $b > 1$. We partition $[1, b]$ by the points of a *geometric progression*

$$r^0 = 1 \quad r \quad r^2 \quad r^3 \quad \cdots \quad r^n = b$$

Thus $r = b^{1/n}$.

19 Define a corresponding $S(x)$ by $S(b) = 1/b$ and $S(x) = r^{-j}$ where $r^j \le x < r^{j+1}$ and $0 \le j \le n-1$. Find

$$\int_1^b S(x)\,dx$$

*20 (cont.) By using a difference-quotient argument, show that

$$\int_1^b \frac{dx}{x} \le \ln b$$

*21 (cont.) Prove similarly that

$$\ln b \le \int_1^b \frac{dx}{x}$$

and conclude that equality holds.

*22 (cont.) Apply the same technique to rederive the formula

$$\int_1^b x\,dx = \tfrac{1}{2}(b^2 - 1)$$

5-3 The Fundamental Theorem

Let us introduce some standard terminology. When we write

$$A = \int_a^b f(x)\,dx$$

the function $f(x)$ is called the **integrand** of the integral. The number A is called the **definite integral** of $f(x)$ from $x = a$ to $x = b$, also called "the integral of f from a to b." The numbers a and b are the **limits of integration;** a is the **lower limit** and b is the **upper limit.** (The word *limit* is traditional here although *bound* or *end point* would be better terms. This term *limit* has nothing to do with our usual technical use of the word.)

The following theorem summarizes four major properties of the definite integral, properties that we use time and time again. They are consequences of corresponding properties of step functions, but we omit their proofs.

Figure 5-3-1
The area between $y = f(x) + 1$ and $y = f(x)$

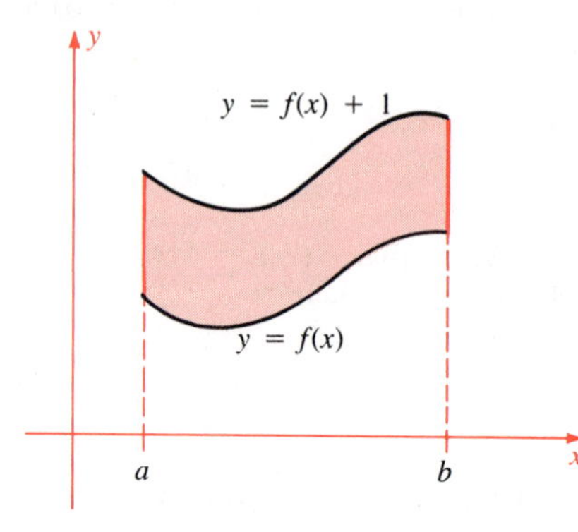

Figure 5-3-2

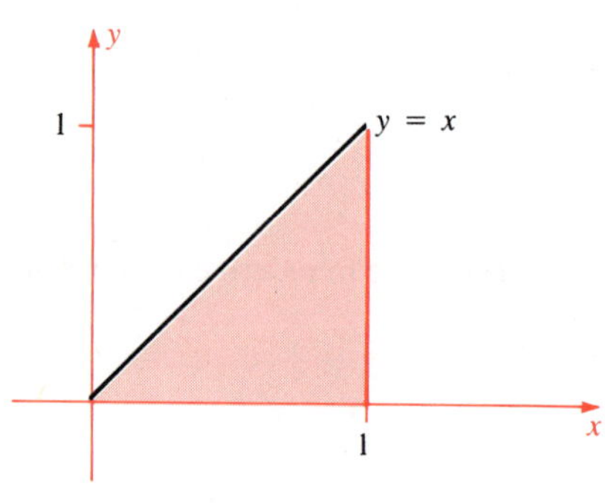

a Area under $y = x$

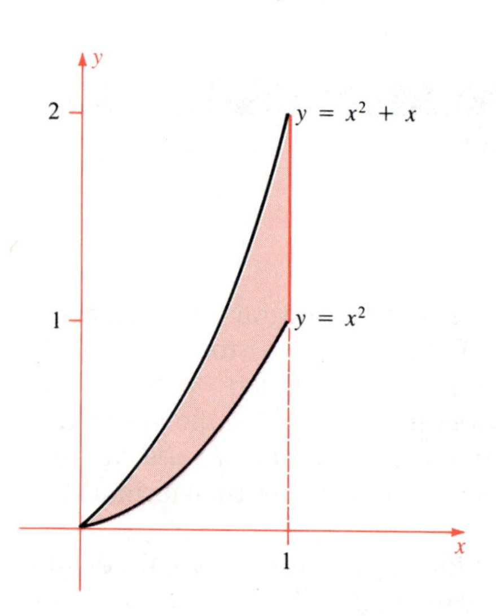

b Area between $y = x^2 + x$ and $y = x^2$

Theorem 1 Properties of the Definite Integral

Let $f(x)$ and $g(x)$ be integrable functions on $[a, c]$.

1 If $a < b < c$, then

$$\int_a^c f(x)\,dx = \int_a^b f(x)\,dx + \int_b^c f(x)\,dx$$

2 If $f(x) \le g(x)$, then

$$\int_a^c f(x)\,dx \le \int_a^c g(x)\,dx$$

3 If K is a constant, then

$$\int_a^c K\,dx = K(c - a)$$

4 Linearity of the Integral If h and k are constants, then

$$\int_a^c [hf(x) + kg(x)]\,dx = h\int_a^c f(x)\,dx + k\int_a^c g(x)\,dx$$

Example 1

By linearity,

$$\int_0^2 (5x - 2x^2)\,dx = 5\int_0^2 x\,dx - 2\int_0^2 x^2\,dx$$

$$= 5 \cdot \tfrac{1}{2} \cdot 2^2 - 2 \cdot \tfrac{1}{3} \cdot 2^3 = \tfrac{14}{3}$$

Theorem 1 has geometric consequences that are not at all obvious at first sight.

Example 2

Find the shaded area in Figure 5-3-1.

Solution The area is

$$A = \int_a^b [f(x) + 1]\,dx - \int_a^b f(x)\,dx$$

$$= \int_a^b f(x)\,dx + \int_a^b 1\,dx - \int_a^b f(x)\,dx$$

$$= \int_a^b 1\,dx = b - a$$

Think about it.

Example 3

Show, without actually computing them, that the shaded areas in Figures 5-3-2a and 5-3-2b are equal.

Solution The area in Figure 5-3-2a is $\int_0^1 x\,dx$.

The area in Figure 5-3-2b is

$$\int_0^1 (x^2 + x)\,dx - \int_0^1 x^2\,dx = \int_0^1 x^2\,dx + \int_0^1 x\,dx - \int_0^1 x^2\,dx$$
$$= \int_0^1 x\,dx$$

They are the same. ●

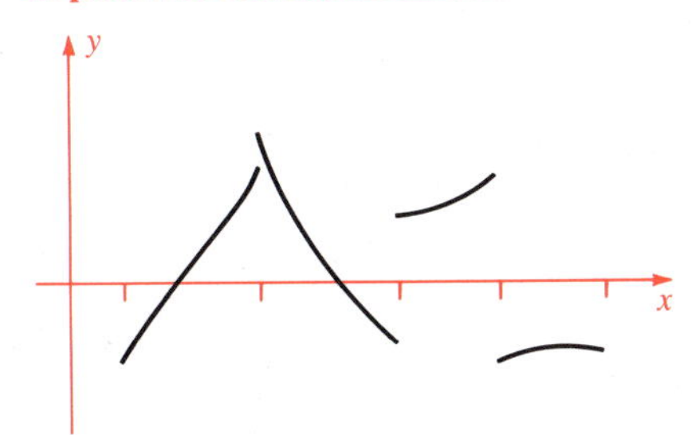

Figure 5-3-3
A piecewise continuous function

Piecewise Continuous Functions

The integral applies to a large class of functions, much larger than the classes of step functions and continuous functions lumped together. For practical applications of calculus we do not need much more than step and continuous functions, but we shall introduce a useful slightly larger class of functions that are integrable. Figure 5-3-3 is typical of this type of function, which we now define formally.

Definition

A function $f(x)$ whose domain is a closed interval $[a, b]$ is **piecewise continuous** provided there is a partition of the domain and there is a continuous function, $g_j(x)$ on each *closed* subinterval $[x_{j-1}, x_j]$ of the partition such that $f(x) = g_j(x)$ for all x in the *open* interval (x_{j-1}, x_j). ●

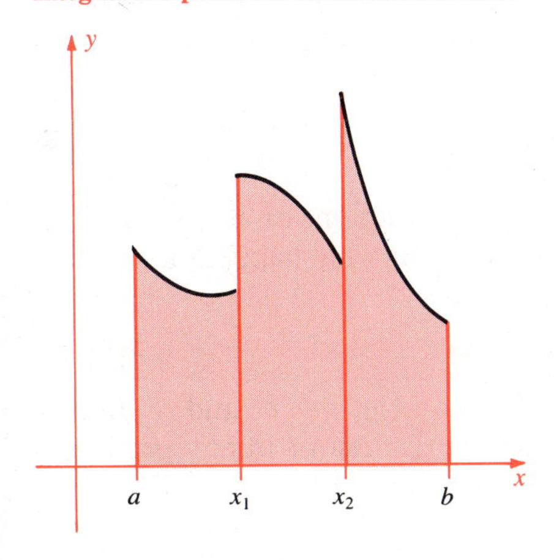

Figure 5-3-4
Integral of a piecewise continuous function

Note that (like step functions) piecewise continuous functions need not be specified at the partition points x_j. Continuous functions are piecewise continuous, and step functions are piecewise continuous. If $f(x)$ is piecewise continuous, then $f(x)$ is integrable, and its integral is given by

$$\int_a^b f(x)\,dx = \sum_{j=1}^{n} \int_{x_{j-1}}^{x_j} g_j(x)\,dx$$

where the notation follows the preceding definition. This integral equals the shaded area in Figure 5-3-4.

Integrals with $a \geq b$

We have defined

$$\int_a^b f(x)\,dx$$

for $a < b$. Now we extend the definition to the cases $a = b$ and $a > b$. From Theorem 1 we know that

$$\int_a^c f(x)\,dx = \int_a^b f(x)\,dx + \int_b^c f(x)\,dx$$

provided that $a < b < c$. We try to devise an extended definition so that this property holds regardless of how a, b, and c are ordered. Before we give the official definition, let us assume the previous relation holds without the restrictions $a < b < c$ and see what happens. First let us try $b = a$:

$$\int_a^c f(x)\,dx = \int_a^a f(x)\,dx + \int_a^c f(x)\,dx$$

This forces $\int_a^a f(x)\,dx = 0$.

Now we start again and take $c = a$. Then

$$\int_a^a f(x)\,dx = \int_a^b f(x)\,dx + \int_b^a f(x)\,dx$$

Since the left-hand side is 0, we are forced to

$$\int_b^a f(x)\,dx = -\int_a^b f(x)\,dx$$

Thus we are led naturally to a definition.

Definition

We define

$$\int_a^a f(x)\,dx = 0 \quad \text{and} \quad \int_a^b f(x)\,dx = -\int_b^a f(x)\,dx \quad \text{for} \quad a > b$$

Example 4

a $\int_5^5 x^2\,dx = 0$ **b** $\int_1^0 x\,dx = -\frac{1}{2}$ **c** $\int_3^2 K\,dx = -K$

Now part 1 of Theorem 1 holds without restrictions on the order of a, b, and c. The proof is just a matter of checking what happens for each of the various arrangements of a, b, and c.

Theorem 2

Let $f(x)$ be integrable on a closed interval and let a, b and c be any three points of this domain, not necessarily distinct or in any particular order. Then

$$\int_a^c f(x)\,dx = \int_a^b f(x)\,dx + \int_b^c f(x)\,dx$$

The Fundamental Theorem of Calculus

So far in this chapter we have developed the definition of the integral and its formal properties. The definition can be used to calculate integrals, but it does not give an easy method for integrating functions and getting answers.

In Section 5-2, we needed a whole bag of tricks when working with step functions, so such direct procedures seem hopelessly hard to use regularly. Our problem is where to go from here?

If you do not see how to proceed, that is not surprising because it took mathematicians about 2000 years to find the right techniques. The ancient Greeks, Archimedes in particular, solved the area problem in a few special cases, always by ingenious geometric tricks. Nothing much happened from then until after the Renaissance, when the tricks used in Section 5-2 were discovered.

The breakthrough came with the idea of changing the problem from a static one to a dynamic one. Instead of computing the area under a curve between two fixed vertical lines, compute it between a fixed line and a second *moving* line. That is the key idea. At first, suppose $f(t)$ is continuous and $f(t) > 0$. Denote by $A(x)$ the area under the curve $y = f(t)$ between $t = a$ and $t = x$. See Figure 5-3-5. Then

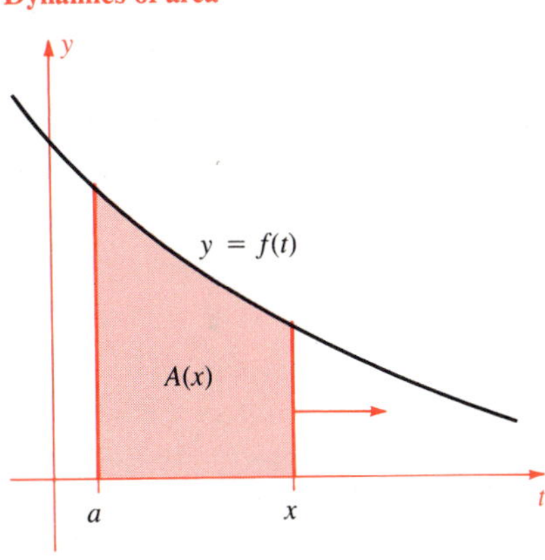

Figure 5-3-5
Dynamics of area

$$A(x) = \int_a^x f(t)\,dt$$

Clearly $A(x)$ is a function of x. For instance, if $f(t) = t^3$ and $a = 0$, then

$$A(x) = \int_0^x t^3\,dt = \tfrac{1}{4}x^4$$

Our aim is to find an explicit formula for $A(x)$. For this purpose, we study how $A(x)$ varies as x varies. Figure 5-3-5 shows that as x increases, the area builds up, so $A(x)$ increases. But how fast? What change in $A(x)$ results from a small change h in x? Figure 5-3-6 provides a clue. When $f(x)$ is large, the corresponding change in $A(x)$ is large; when $f(x)$ is small, the change is small. Apparently the rate of change of $A(x)$ relative to x is very much like the value of $f(x)$. In the language of calculus, we suspect that the derivative dA/dx is proportional to $f(x)$. Let us test our hunch with $f(t) = t^3$ and $a = 0$.

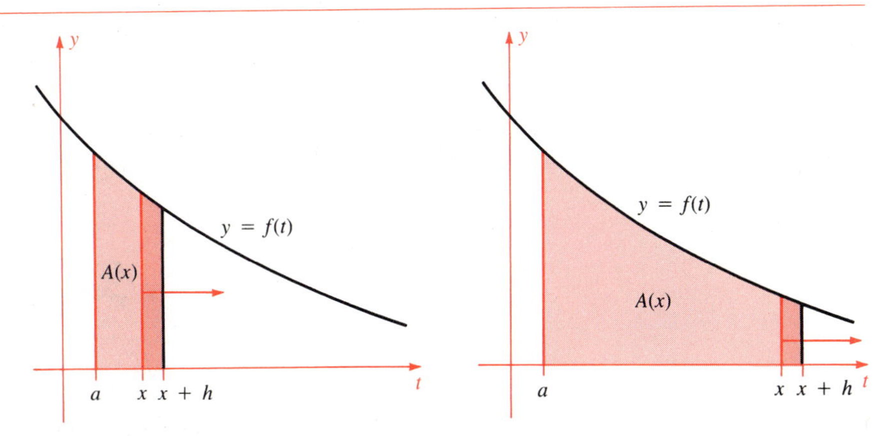

Figure 5-3-6
The rate of growth of $A(x)$ seems proportional to $f(x)$.

a $f(x)$ large: $A(x)$ increases rapidly

b $f(x)$ small: $A(x)$ increases slowly

Then

$$A(x) = \tfrac{1}{4}x^4 \qquad \text{hence} \qquad \frac{dA}{dx} = x^3$$

Result: dA/dx is not only *proportional* to $f(x)$, but dA/dx is actually *equal* to $f(x)$. We have hit upon the crucial fact: that in general $dA/dx = f(x)$.

Fundamental Theorem of Calculus (Derivative of an Integral)

Let $f(t)$ be continuous on the closed interval $[a, b]$. For each x in the domain, let

$$A(x) = \int_a^x f(t)\,dt$$

Then $A(x)$ is differentiable and

$$\frac{dA}{dx} = \frac{d}{dx}\int_a^x f(t)\,dt = f(x)$$

A proof will be given at the end of this section. Note that the fundamental theorem applies no matter what the upper limit is called. For instance

$$\frac{d}{dv}\left[\int_a^v f(x)\,dx\right] = f(v)$$

Example 5

Find

a $\displaystyle \frac{d}{du}\int_0^u te^{4t}\,dt$ **b** $\displaystyle \frac{d}{dx}\int_0^{3x^2} te^{4t}\,dt$

Solution

a Set $\displaystyle F(u) = \int_0^u te^{4t}\,dt$

By the fundamental theorem, $\displaystyle \frac{d}{du}F(u) = ue^{4u}$.

b We seek the derivative with respect to x of $F(3x^2)$. By the chain rule

$$\frac{d}{dx}F(u) = \frac{d}{du}F(u)\cdot\frac{du}{dx} = \frac{d}{du}F(u)\cdot\frac{d}{dx}(3x^2)$$
$$= (ue^{4u})(6x) = (3x^2e^{12x^2})(6x) = 18x^3e^{12x^2}$$

Terminology Given $f(x)$, any function $F(x)$ whose derivative equals $f(x)$ is called an **antiderivative** of $f(x)$. [It is sometimes also called a **primitive** or **indefinite integral** of $f(x)$.] If the domain of $f(x)$ is an interval, then any two antiderivatives of $f(x)$ differ by a constant. This statement is the content of Corollary 1 on page 155.

We have almost achieved our goal of finding a formula for $A(x)$. We know that $A(x)$ is one of the antiderivatives of $f(x)$, but which one? Well, since they all differ by constants, it shouldn't be hard to sort out. Let $F(x)$ be *any* antiderivative of $f(x)$. Then $A(x) = F(x) + c$ because $A'(x) = F'(x)$. To find c, note that the area is 0 where we start, that is, at $x = a$:

$$0 = A(a) = F(a) + c \qquad c = -F(a)$$

Therefore $A(x) = F(x) - F(a)$, for *every* antiderivative $F(x)$ of $f(x)$.

In particular, $A(b) = F(b) - F(a)$. This gives us a supremely practical alternative version of the fundamental theorem:

Fundamental Theorem of Calculus (Evaluation of the Definite Integral)

Let $f(x)$ be continuous on $[a, b]$ and let $F(x)$ be any antiderivative of $f(x)$. Then

$$\int_a^b f(x)\,dx = F(b) - F(a) = F(x)\Big|_a^b$$

We have reached the end of a chain of thought that began by thinking of the definite integral as a function of its upper limit. The evaluation rule transforms a difficult problem, evaluating integrals, into a much easier one, finding antiderivatives. To compute $\int_a^b f(x)\,dx$, just recall, work out, or look up an antiderivative $F(x)$; then evaluate $F(b) - F(a)$.

For instance, in Section 5-2 we laboriously churned out the formulas

$$\int_a^b x\,dx = \tfrac{1}{2}(b^2 - a^2) \qquad \text{and} \qquad \int_0^b e^x\,dx = e^b - 1$$

Now they are obvious consequences of

$$x = \frac{d}{dx}(\tfrac{1}{2}x^2) \qquad \text{and} \qquad e^x = \frac{d}{dx}e^x$$

Because we know many derivatives, we know many antiderivatives, so many integrals are now within our grasp.

Example 6

Evaluate

a $\displaystyle\int_1^2 x^5\,dx$ **b** $\displaystyle\int_3^5 \frac{dx}{x^2}$ **c** $\displaystyle\int_{\pi/4}^{\pi/3} \sec^2 x\,dx$

Solution

a $x^5 = \dfrac{d}{dx}(\tfrac{1}{6}x^6)$ hence

$$\int_1^2 x^5\,dx = \tfrac{1}{6}x^6\Big|_1^2 = \tfrac{1}{6}(2^6 - 1^6) = \tfrac{63}{6} = \tfrac{21}{2}$$

b $\dfrac{1}{x^2} = \dfrac{d}{dx}\left(-\dfrac{1}{x}\right)$ hence

$$\int_3^5 \frac{dx}{x^2} = -\frac{1}{x}\bigg|_3^5 = -\tfrac{1}{5} - (-\tfrac{1}{3}) = \tfrac{2}{15}$$

c $\sec^2 x = \dfrac{d}{dx}\tan x$ hence

$$\int_{\pi/4}^{\pi/3} \sec^2 x\,dx = \tan x\bigg|_{\pi/4}^{\pi/3} = \tan\tfrac{1}{3}\pi - \tan\tfrac{1}{4}\pi = \sqrt{3} - 1$$

●

Example 7

Find $\displaystyle\int_{-2}^{2} |x+1|\,dx$

Solution

$$|x+1| = \begin{cases} -(x+1) & \text{for } x \le -1 \\ x+1 & \text{for } x \ge -1 \end{cases}$$

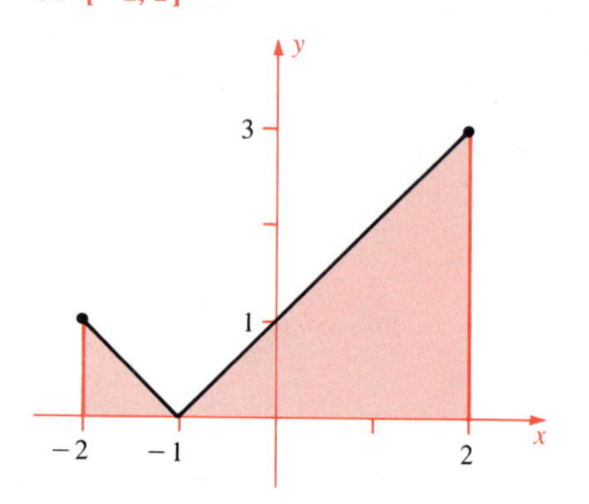

Figure 5-3-7
The integral of $y = |x+1|$ on $[-2, 2]$

so we decompose the integral into two parts and deal with each separately (Figure 5-3-7). We have

$$x + 1 = \frac{d}{dx}\tfrac{1}{2}(x+1)^2 \quad\text{and}\quad -(x+1) = \frac{d}{dx}[-\tfrac{1}{2}(x+1)^2]$$

Hence

$$\begin{aligned}\int_{-2}^{2}|x+1|\,dx &= \int_{-2}^{-1}|x+1|\,dx + \int_{-1}^{2}|x+1|\,dx \\ &= \int_{-2}^{-1} -(x+1)\,dx + \int_{-1}^{2}(x+1)\,dx \\ &= -\tfrac{1}{2}(x+1)^2\bigg|_{-2}^{-1} + \tfrac{1}{2}(x+1)^2\bigg|_{-1}^{2} \\ &= -\tfrac{1}{2}(0-1) + \tfrac{1}{2}(9-0) = 5\end{aligned}$$

●

Dummy Variables

The integral

$$\int_a^b f(x)\,dx$$

depends on three things: its lower limit a, its upper limit b, and its integrand f. We might write

$$\int_a^b f(x)\,dx = \int_a^b f = I(a, b; f)$$

to display this dependence. The integral does *not* depend on x; for that reason x is sometimes called a **dummy variable.** Thus

$$\int_a^b f(x)\,dx = \int_a^b f(y)\,dy = \int_a^b f(t)\,dt = \cdots$$

There are very good reasons for using the full notation complete with the (x) and the dx. When we study integration by substitution and integration by parts in the next chapter, the reasons will become apparent.

Proof of the Fundamental Theorem

We are given a continuous function $f(t)$ in the closed interval $[a, b]$, and we define

$$A(x) = \int_a^x f(t)\,dt \qquad (a \le x \le b)$$

We must prove $A'(x) = f(x)$. A preliminary lemma will help us do so.

Figure 5-3-8
If $|g(x)| \le K$, then

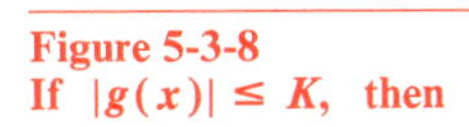

$$\left|\int_a^b g(x)\,dx\right| \le K|b - a|$$

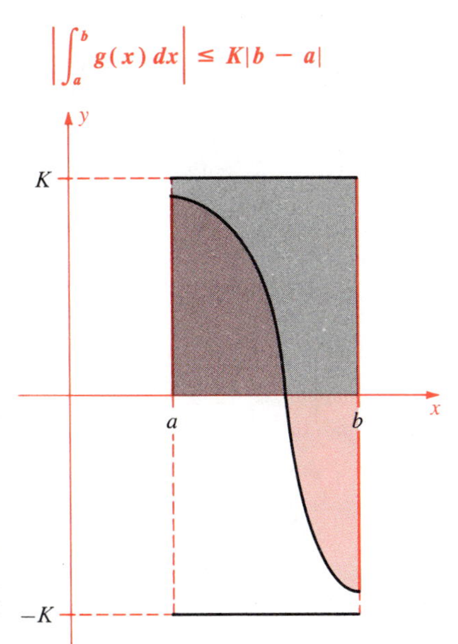

Lemma Suppose $g(x)$ is continuous on an interval and $|g(x)| \le K$, a constant. Suppose a and b are in the domain of $g(x)$ and $a \ne b$. See Figure 5-3-8. Then

$$\left|\int_a^b g(x)\,dx\right| \le K|b - a| \qquad \bullet$$

Proof We may assume $a < b$, because if $b < a$, we can interchange a and b without changing either side of the inequality. Since $|g(x)| \le K$ we have by Theorem 1,

$$\int_a^b g(x)\,dx \le \int_a^b K\,dx = K(b - a)$$

Since $-g(x) \le K$ we also have

$$-\int_a^b g(x)\,dx = \int_a^b -g(x)\,dx \le \int_a^b K\,dx = K(b - a)$$

Therefore

$$\left|\int_a^b g(x)\,dx\right| \le K(b - a) = K|b - a| \qquad \bullet$$

Proof of the Fundamental Theorem We fix x in the interval and form the difference

$$\begin{aligned} A(x + h) - A(x) &= \int_a^{x+h} f(t)\,dt - \int_a^x f(t)\,dt \\ &= \int_x^{x+h} f(t)\,dt \end{aligned}$$

Next,

$$\int_x^{x+h} f(t)\,dt = \int_x^{x+h} \{f(x) + [f(t) - f(x)]\}\,dt$$
$$= \int_x^{x+h} f(x)\,dt + \int_x^{x+h} [f(t) - f(x)]\,dt$$

The first term on the right is an integral with respect to t of $f(x)$, where x is fixed. This means that $f(x)$ is a constant as far as that integration process is concerned. Hence

$$\int_x^{x+h} f(x)\,dt = f(x)\int_x^{x+h} dt = f(x)h$$

Therefore

$$\int_x^{x+h} f(t)\,dt = hf(x) + \int_x^{x+h} [f(t) - f(x)]\,dt$$

Combine these relations:

$$A(x+h) - A(x) - hf(x) = \int_x^{x+h} [f(t) - f(x)]\,dt$$

Since $f(t)$ is continuous, given $\varepsilon > 0$, there exists a $\delta > 0$ such that we have $|f(t) - f(x)| < \varepsilon$ whenever $|t - x| < \delta$. Therefore if $0 < |h| < \varepsilon$ we have

$$\left|\int_x^{x+h} [f(t) - f(x)]\,dt\right| \le \varepsilon|h|$$

by the lemma. Consequently

$$|A(x+h) - A(x) - hf(x)| \le \varepsilon|h|$$

so we have

$$\left|\frac{A(x+h) - A(x)}{h} - f(x)\right| \le \varepsilon$$

We have proved that for each $\varepsilon > 0$ there is a $\delta > 0$ so this inequality holds whenever $0 < |h| < \delta$. This means

$$\lim_{h\to 0} \frac{A(x+h) - A(x)}{h} = f(x)$$

hence $A'(x) = f(x)$. (Note that if $x = a$ we must restrict h to positive values to keep $x + h$ within $[a, b]$. Likewise, if $x = b$ we must restrict h to negative values.) ●

Exercises

Evaluate

1 $\int_{-1}^{1} dx$

2 $\int_{-2}^{0} dx$

3 $\int_{-1}^{2} x\,dx$

4 $\int_{0}^{2} -4\,dx$

5 $\int_{-1}^{0} -3\,dx$

6 $\int_{-3}^{-1} x\,dx$

7 $\int_{0}^{2} (1 - x)\,dx$

8 $\int_{-1}^{1} (2 + 3x)\,dx$

9 $\int_{0}^{2} 6x^2\,dx$

10 $\int_{0}^{1} (2x - 3x^2)\,dx$

11 $\int_{1}^{2} 12t^3\,dt$

12 $\int_{-2}^{-1} 7u^2\,du$

13 $\int_{-2}^{1} 8u\,du$

14 $\int_{-3}^{-2} 20y^3\,dy$

15 $\int_{-2}^{-1} \frac{1}{x^2}\,dx$

16 $\int_{-4}^{-2} \left(\frac{1}{x^2} + x\right) dx$

17 $\int_{-2}^{2} (3 + 2x - x^2)\,dx$

18 $\int_{-1}^{2} (x - 1) \times (3x + 1)\,dx$

19 $\int_{0}^{4} x(2 - x)\,dx$

20 $\int_{1}^{3} (x^2 - 4x + 4)\,dx$

21 $\int_{-a}^{a} (x - a)^2\,dx$

22 $\int_{a}^{b} (b - x)(x - a)\,dx$

23 $\int_{1}^{2} e^{1-x}\,dx$

24 $\int_{-1}^{1} e^{3x}\,dx$

25 $\int_{0}^{2} (2x - 4e^{-x})\,dx$

26 $\int_{-1}^{0} (2e^x - e^{2x})\,dx$

27 $\int_{0}^{\pi/2} \cos\theta\,d\theta$

28 $\int_{\pi/2}^{3\pi/2} \cos 3\theta\,d\theta$

29 $\int_{0}^{\pi} \sin 2\theta\,d\theta$

30 $\int_{-\pi/2}^{0} -2\cos 2\theta\,d\theta$

31 $\int_{0}^{3\pi} (\cos t + \sin t)\,dt$

32 $\int_{0}^{\pi} (3\sin 3t + 2\cos 2t)\,dt$

33 $\int_{3\pi}^{3\pi} \sin\theta\cos\theta\,d\theta$

34 $\int_{\pi/2}^{\pi/2} \sin 4\theta\,d\theta$

35 $\int_{0}^{\pi/4} \sec^2\theta\,d\theta$

36 $\int_{\pi/6}^{3\pi/4} \csc\theta\cot\theta\,d\theta$

37 $\int_{2}^{0} 4t^3\,dt$

38 $\int_{\pi}^{0} \sin\theta\,d\theta$

39 $\int_{\pi/2}^{-\pi/2} 3\cos\theta\,d\theta$

40 $\int_{0}^{-1} t^4\,dt$

Find dy/dx

41 $y = \int_{0}^{x} \sec\theta\,d\theta$

42 $y = \int_{1}^{x+1} \frac{1}{t}\,dt$

43 $y = \int_{0}^{\sin x} \sqrt{t}\,dt$

44 $y = \int_{3x}^{10} u^4\,du$

45 The point x moves to the right at the rate of 4 cm/sec. Let $A(x)$ be the area under the curve $y = \sin^2 u$ between $u = 0$ and $u = x$. Find dA/dt when $x = \frac{1}{3}\pi$. The unit of length on both the u-axis and the y-axis is the centimeter.

46 Criticize: $\int_{-1}^{1} \frac{1}{x^2}\,dx = -\frac{1}{x}\Big|_{-1}^{1} = -1 - 1 = -2$

5-4 Applications

For our first application of the definite integral, let us compute the areas of some regions.

Example 1

Find the area of the region bounded by the curve $y = 1 + x^3$, the two axes, and the line $x = 1$.

Solution The first step is to make a drawing (Figure 5-4-1, next page). The region in question is simply the region under the graph of $y = 1 + x^3$ between the vertical lines $x = 0$ and $x = 1$. Its area is

$$\int_{0}^{1} (1 + x^3)\,dx = \left(x + \tfrac{1}{4}x^4\right)\Big|_{0}^{1} = \left(1 + \tfrac{1}{4}\right) - 0 = \tfrac{5}{4}$$

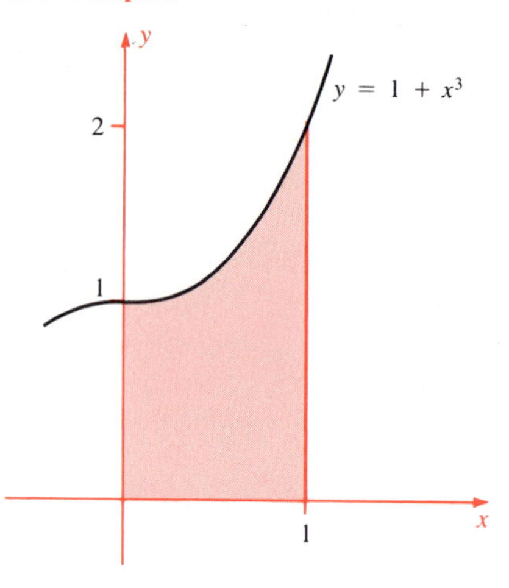

Figure 5-4-1
See Example 1

Example 2

Find the area of the parabolic segment bounded by the curve $y = x^2$ and the line $y = 2$.

Solution Sketch the region (Figure 5-4-2). The line and the curve meet where $x^2 = y = 2$, that is, where $x = \pm\sqrt{2}$. One way to find the shaded area is to compute the area under the curve and subtract it from the area of the rectangle:

$$\begin{aligned}\text{area} &= (2\sqrt{2})(2) - \int_{-\sqrt{2}}^{\sqrt{2}} x^2\,dx = 4\sqrt{2} - \tfrac{1}{3}x^3\Big|_{-\sqrt{2}}^{\sqrt{2}} \\ &= 4\sqrt{2} - \tfrac{1}{3}[(\sqrt{2})^3 - (-\sqrt{2})^3] = 4\sqrt{2} - \tfrac{4}{3}\sqrt{2} = \tfrac{8}{3}\sqrt{2}\end{aligned}$$

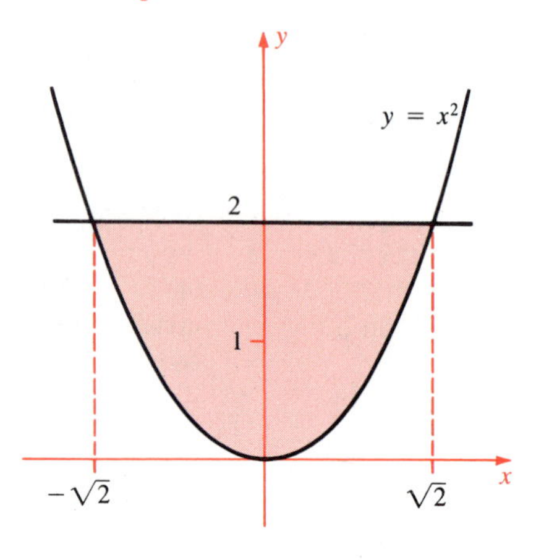

Figure 5-4-2
See Example 2

Algebraic Area

When an integrand is positive, its definite integral is an area. How can we interpret an integral when the integrand is not always positive? Suppose $s(x)$ is a step function and $s(x) \le 0$ on $[a, b]$. An example is shown in Figure 5-4-3. By definition

$$\begin{aligned}\int_a^b s(x)\,dx &= -A(x_1 - a) - B(x_2 - x_1) - C(b - x_2) \\ &= -[A(x_1 - a) + B(x_2 - x_1) + C(b - x_2)] \\ &= -(\text{shaded area})\end{aligned}$$

Clearly this is true for any negative step function. Since the integral of a continuous function is obtained by squeezing it between step functions, the integral of a negative function is the negative of an area, and the same holds for piecewise continuous negative functions. See Figure 5-4-4.

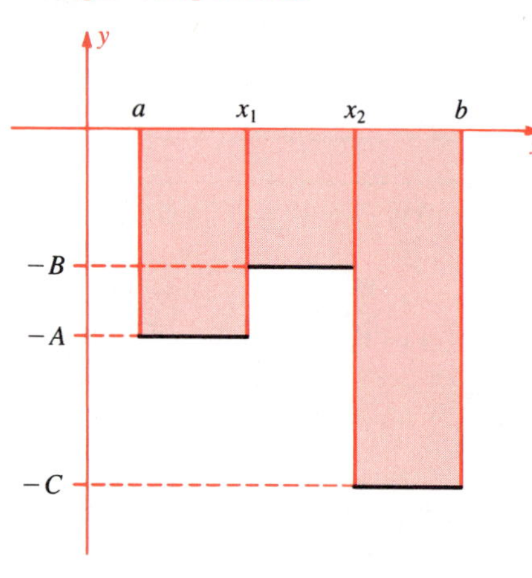

Figure 5-4-3
A negative step function

Integral of a Negative Function

Suppose $f(x)$ is piecewise continuous on $[a, b]$ and $f(x) \le 0$. Then

$$\int_a^b f(x)\,dx$$

is the negative of the area of the region bounded by $y = f(x)$, $x = a$, $x = b$, and the x-axis.

In general, a piecewise continuous function may change sign several times, so its graph and the x-axis bound several regions, some above the x-axis, some below. We see that its integral equals the *algebraic sum* of these areas—each area above the x-axis is added, each area below is subtracted.

For instance consider the function in Figure 5-4-5. We have

$$\int_a^b f(x)\,dx = \int_a^{c_1} f(x)\,dx + \int_{c_1}^{c_2} f(x)\,dx + \int_{c_2}^b f(x)\,dx$$

Figure 5-4-4
If $f(x) \le 0$, then

$$\text{area} = -\int_a^b f(x)\,dx$$

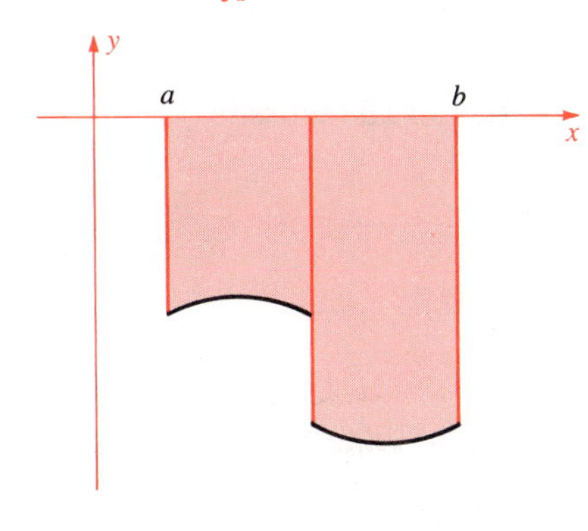

The middle term equals the area A_2, but each of the two end terms equals the negative of an area, so

$$\int_a^b f(x)\,dx = -A_1 + A_2 - A_3$$

So it goes in general. For example (Figure 5-4-6), we can see without computing that

$$\int_0^{2\pi} \sin x\,dx = 0$$

since the areas above and below the x-axis cancel each other.

Example 3

One can of paint will cover $5\ \text{m}^2$. How much paint is needed to cover the region bounded by $y = \sin x$ and the x-axis between $x = 0$ and $x = 2\pi$? Assume the unit on each axis is 1 meter.

Figure 5-4-5
The integral of a function that crosses the x-axis is the algebraic sum of areas.

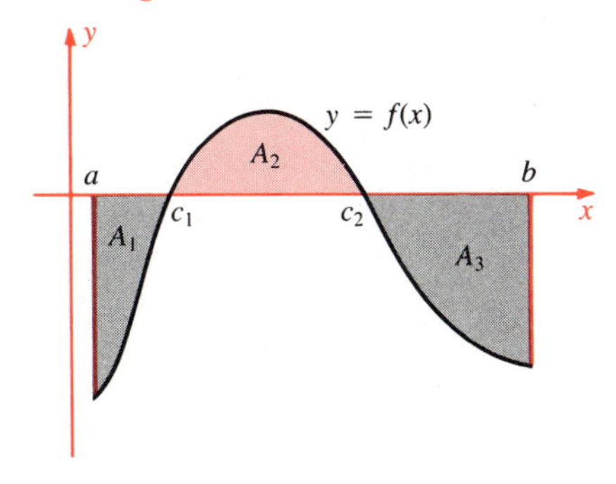

Solution If we treat the problem carelessly and integrate $\sin x$ between 0 and 2π, we reach a ridiculous conclusion: it takes no paint at all to cover the region! But remember, the integral

$$\int_a^b f(x)\,dx$$

gives the actual area under $y = f(x)$ only if $f(x) \ge 0$ between a and b.

Look at Figure 5-4-6. The two humps of the curve have equal area: compute the area of the hump between 0 and π, then double the result. Since $(-\cos x)' = \sin x$ we have

$$\int_0^{\pi} \sin x\,dx = -\cos x\Big|_0^{\pi} = (-\cos \pi) - (-\cos 0)$$
$$= 1 - (-1) = 2$$

Thus the area of one hump is $2\ \text{m}^2$; the total area is $4\ \text{m}^2$. This will require $4/5 = 0.8$ cans of paint. ●

Figure 5-4-6

$$\int_0^{2\pi} \sin x\,dx = 0$$

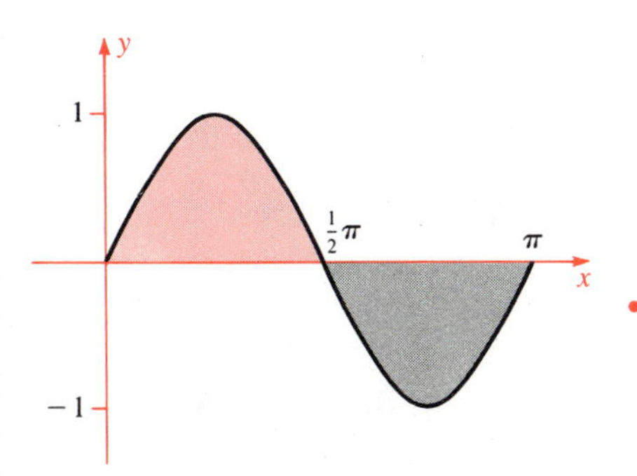

Antiderivative of $1/x$

The formula $1/x = d(\ln x)/dx$ is valid for $x > 0$. Thus $\ln x$ is an antiderivative of $1/x$, at least for $x > 0$. But what about x negative? The trouble is that $\ln x$ is not defined for $x < 0$; however, $\ln|x| = \ln(-x)$ is, because $-x > 0$. By the chain rule,

$$\frac{d}{dx}\ln(-x) = \frac{(-x)'}{-x} = \frac{-1}{-x} = \frac{1}{x} \quad \text{for} \quad x < 0$$

• An antiderivative of $\dfrac{1}{x}$ is $\begin{cases} \ln x & \text{for } x > 0 \\ \ln(-x) & \text{for } x < 0 \end{cases}$ ●

In other words: $\ln|x|$ is an antiderivative of $1/x$ for $x \ne 0$. Compare the discussion on page 245.

Example 4

Compute

a $\displaystyle\int_1^3 \frac{1}{x}\,dx$ **b** $\displaystyle\int_{-3}^{-1} \frac{1}{x}\,dx$

Solution Use the result above and the fundamental theorem:

a $$\int_1^3 \frac{1}{x}\,dx = \ln x\Big|_1^3 = \ln 3 - \ln 1 = \ln 3$$

b $$\int_{-3}^{-1} \frac{1}{x}\,dx = \ln|x|\Big|_{-3}^{-1} = \ln 1 - \ln 3 = \ln \tfrac{1}{3}$$

Remark The two answers are negatives of each other. A glance at Figure 5-4-7 shows why; the integrals represent areas that are of equal magnitude but opposite sign.

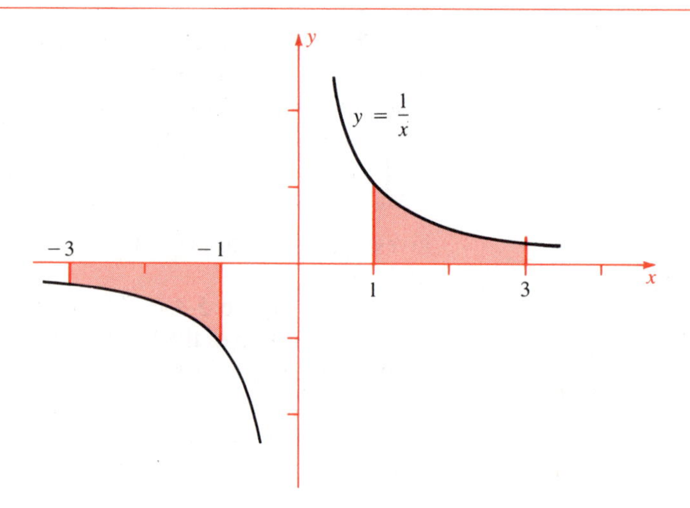

Figure 5-4-7
The areas are equal;
the integrals have opposite signs.

Warning If $a < 0$ and $b > 0$, then

$$\int_a^b \frac{1}{x}\,dx$$

is undefined, because $1/x$ is unbounded in any interval that includes $x = 0$. Therefore $1/x$ cannot be squeezed between any step functions at all; it is not an integrable function on such an interval. Actually we shall learn in Chapter 10 how to integrate some unbounded functions, but the process fails for this particular case because $\lim_{x\to 0+}|\ln x| = +\infty$.

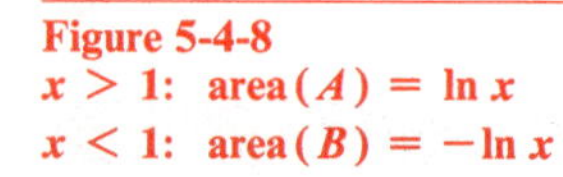
Figure 5-4-8
$x > 1$: area$(A) = \ln x$
$x < 1$: area$(B) = -\ln x$

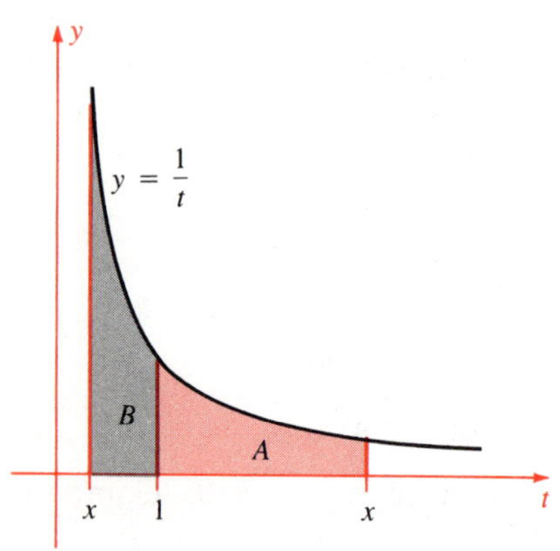

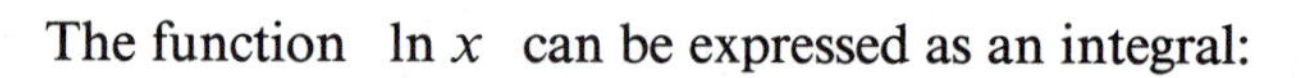
The function $\ln x$ can be expressed as an integral:

$$\ln x = \int_1^x \frac{1}{t}\,dt \quad \text{for} \quad x > 0$$

Therefore $\ln x$ represents an area if $x > 1$, or the negative of an area if $0 < x < 1$. See Figure 5-4-8.

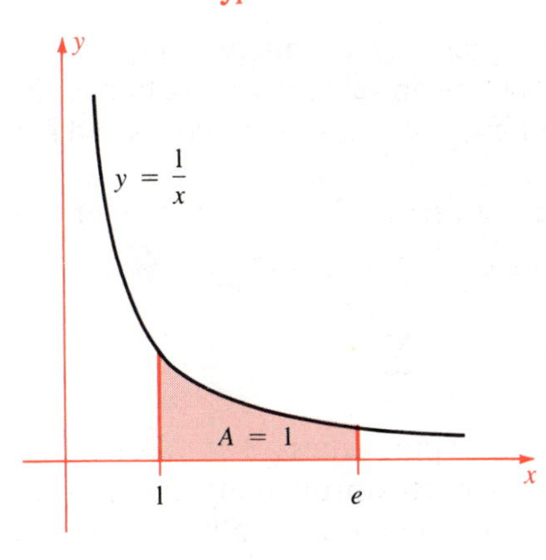

Figure 5-4-9
e is defined by $\int_1^e \frac{dx}{x} = 1$

From $\ln e = 1$, we obtain a geometric interpretation of the number e. It is the unique number greater than 1 such that the area under $y = 1/x$ from 1 to e equals 1 (Figure 5-4-9).

From this definition, we can give a geometric proof that $2.5 < e < 3$. See Figure 5-4-10.

Remark In Chapter 4 we introduced the exponential function $y = e^x$ and then defined the logarithm function as its inverse function. An alternative approach is to *define* the logarithm function first by

$$\ln x = \int_1^x \frac{dt}{t} \quad \text{for} \quad x > 0$$

then to define the exponential function as its inverse. Thus

$$x = \int_1^{e^x} \frac{dt}{t}$$

It is now a big job to prove all of the properties of e^x. See Exercises 33–36 for a start.

Figure 5-4-10
Proof that $2.5 < e < 3$

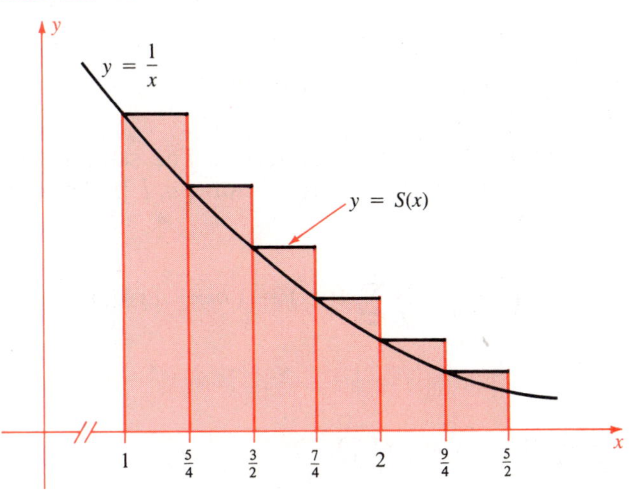

a $S(x) = 4/n$ for $n/4 < x < (n+1)/4$

$$\int_1^{5/2} dx/x < \int_1^{5/2} S(x)\,dx$$

$$= \tfrac{1}{4}\left(1 + \tfrac{4}{5} + \tfrac{2}{3} + \tfrac{4}{7} + \tfrac{1}{2} + \tfrac{4}{9}\right) = \tfrac{2509}{2520} < 1$$

Therefore $e > \frac{5}{2}$

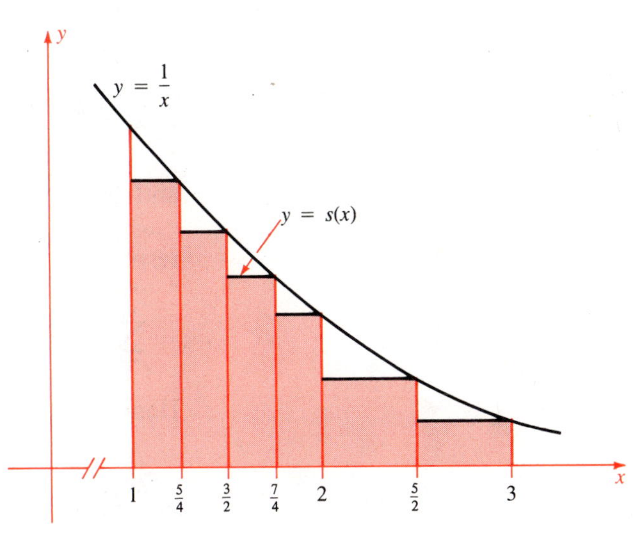

b $s(x) = 4/(n+1)$ for $n/4 < x < (n+1)/4 \leq 2$
$s(x) = 2/(n+1)$ for $2 \leq n/2 < x < (n+1)/2$

$$\int_1^3 dx/x > \int_1^3 s(x)\,dx$$

$$= \tfrac{1}{4}\left(\tfrac{4}{5} + \tfrac{2}{3} + \tfrac{4}{7} + \tfrac{1}{2}\right) + \tfrac{1}{2}\left(\tfrac{2}{5} + \tfrac{1}{3}\right) = \tfrac{841}{840} > 1$$

Therefore $e < 3$

Mean Value of a Function

Consider this problem: Find the mean (average) value of a function $f(x)$ on an interval $[a, b]$. Now the (arithmetic) mean, or average, of *numbers* $x_1, \cdots, x_n$ is defined by

$$\bar{x} = \frac{x_1 + \cdots + x_n}{n}$$

This formula won't work here because there are infinitely many values $f(x)$. Indeed, it is not at all clear what the "mean value" of a function is. So part of the problem is to define the problem, just as it was for slopes and for area.

One possible approach is this. Take the mean of a large sample of the values of $f(x)$:

$$M_n(f) = \frac{f(z_1) + \cdots + f(z_n)}{n} = \frac{1}{n}\sum_{j=1}^{n} f(z_j)$$

and see what happens to $M_n(f)$ as n approaches $+\infty$. For a fair sample, choose $z_1, \ldots, z_n$ well distributed throughout the interval $[a, b]$, just as a political pollster samples opinions in all parts of the country.

This approach seems reasonable but hard to carry out. What saves the day is that $M_n(f)$ can be interpreted as a Riemann sum. Partition $[a, b]$ into n *equal* subintervals

$$a_0 = x_0 < x_1 < \cdots < x_n = b$$

with

$$x_j - x_{j-1} = \frac{b - a}{n} \qquad (j = 1, 2, \cdots, n)$$

and let z_j be any point in the j-th subinterval. The sample $z_1, z_2, \cdots, z_n$ is then fairly well distributed throughout $[a, b]$. The corresponding Riemann sum is

$$\sum_{j=1}^{n} f(z_j)(x_j - x_{j-1}) = \left(\frac{b - a}{n}\right)\sum_{j=1}^{n} f(z_j) = (b - a)M_n(f)$$

Hence

$$M_n(f) = \frac{1}{b - a}\sum_{j=1}^{n} f(z_j)(x_j - x_{j-1})$$

so

$$\lim_{n\to+\infty} M_n(f) = \frac{1}{b - a}\int_a^b f(x)\,dx$$

This reasoning motivates the following definition:

Definition

Let $f(x)$ be an integrable function on $[a, b]$. Its **mean value** is

$$M(f) = \frac{1}{b - a}\int_a^b f(x)\,dx$$

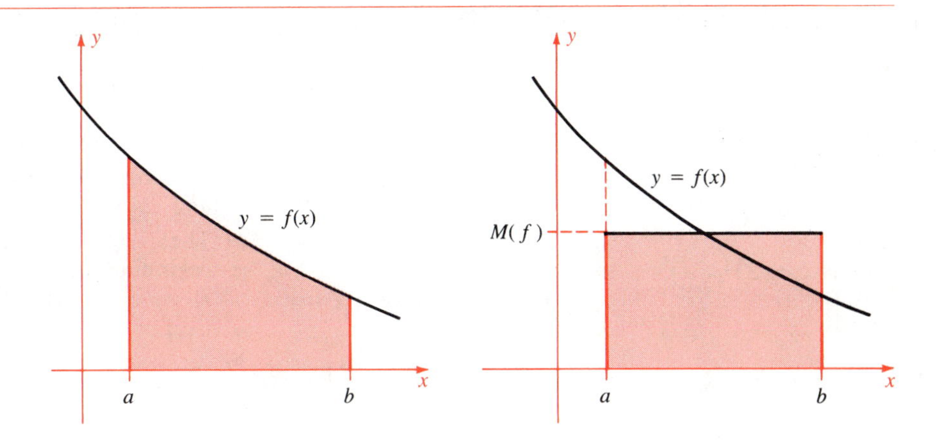

Figure 5-4-11
Geometric interpretation of the mean $M(f)$:
The shaded areas are equal.

The definition has a simple geometric interpretation: The algebraic area under $f(x)$ between $x = a$ and $x = b$ equals the (algebraic) area of a rectangle with base $b - a$ and height $M(f)$. See Figure 5-4-11.

Example 5

A 35-mg (milligram) sample of a certain radioactive substance decays in t days to $35e^{-t/10}$ mg. Compute the average mass $\overline{m}$ of undecayed material present during the first three days.

Solution The mean value of $f(t) = 35e^{-t/10}$ from $t = 0$ to $t = 3$ is

$$\overline{m} = M(f) = \frac{1}{3-0}\int_0^3 f(t)\,dt = \tfrac{35}{3}\int_0^3 e^{-t/10}\,dt$$

The problem now is to integrate a function of the form e^{kt}, where k is a non-zero constant. Since $(e^{kt})' = ke^{kt}$, we see that e^{kt}/k is an antiderivative of e^{kt}. In our case, $k = -\frac{1}{10}$, so an antiderivative of $e^{-t/10}$ is $-10e^{-t/10}$. Therefore the average mass is

$$\overline{m} = \tfrac{35}{3}(-10e^{-t/10})\Big|_0^3 = -\tfrac{350}{3}(e^{-t/10})\Big|_0^3$$
$$= -\tfrac{350}{3}(e^{-0.3} - 1) \approx -\tfrac{350}{3}(0.7408 - 1) \approx 30.24 \text{ mg}$$

The average rate of change of a function $f(x)$ on $[a, b]$ was defined in Chapter 2 as

$$\frac{f(b) - f(a)}{b - a}$$

But now we know better: what we should define as the average rate of change is the mean value $M(f')$ of the *instantaneous* rate of change $f'(x)$ over $[a, b]$. But do we really know better? Think this through carefully; your understanding of calculus will increase.

Exercises

Sketch the graph of $y = f(x)$ and compute the area under the graph over the indicated closed interval

1 $y = (x-1)^2$ $[1, 4]$

2 $y = 4 - x^2$ $[-2, 2]$

3 $y = 4(x-3)^2$ $[0, 6]$

4 $y = 4x - x^2$ $[0, 4]$

5 $y = \cos x$ $[-\frac{1}{2}\pi, \frac{1}{2}\pi]$

6 $y = \sin 2x$ $[0, \frac{1}{2}\pi]$

7 $y = \sin x + 4\cos x$ $[0, \frac{1}{2}\pi]$

8 $y = x + \sin x$ $[\pi, 2\pi]$

9 $y = (x^2 - 1)^2$ $[-1, 1]$

10 $y = e^{-2x}$ $[-1, 2]$

Sketch the graph and find the area of the region determined by

11 the y-axis $y = x^3$ and $y = 8$

12 $x \geq 0$ $y \geq 0$ and $y = 1 - x^2$

Sketch and find the *geometric* area of the region bounded by

13 the x-axis $y = \cos x$ $x = 0$ and $x = \pi$

14 the x-axis $y = x - \dfrac{2}{x^2}$ $x = 1$ and $x = 2$

15 the x-axis $y = x^2 - 5x + 6$ $x = 0$ and $x = 3$

16 the x-axis $y = x^2 - 7x + 12$ $x = 1$ and $x = 5$

17 The graph $y = x^2$ for $x \geq 0$ may be looked at as the graph of $x = \sqrt{y}$ for $y \geq 0$. Show that this implies geometrically the relation

$$\int_0^b x^2\,dx + \int_0^{b^2} \sqrt{y}\,dy = b^3 \qquad (b > 0)$$

Now verify the relation by computing the integrals.

18 (cont.) Find a similar formula for $y = x^3$.

19 Use your knowledge of the area of a circle to compute

$$\int_{-a}^{a} \sqrt{a^2 - x^2}\,dx$$

***20** (cont.) Find the area enclosed by the ellipse

$$\frac{x^2}{a^2} + \frac{y^2}{b^2} = 1 \qquad (a > 0,\ b > 0)$$

Find the mean value $M(f)$ on $[a, b]$

21 $f(x) = x^5$ $[-2, 2]$

22 $f(x) = \sin x$ $[-\frac{1}{3}\pi, \frac{1}{3}\pi]$

23 $f(x) = x^4$ $[0, 3]$

24 $f(x) = 1/x^3$ $[1, 2]$

25 $f(x) = e^x$ $[-1, 1]$

26 $f(x) = x^n$ $[0, b]$

27 If x shares of a certain stock are sold, the price in dollars per share is $P = 37 + [2.5 \times 10^6/(x + 500)^2]$. Find the average price per share on sales of 0 to 2000 shares.

28 Find the average area of circles with radius between 1 and 2 cm.

29 The rainfall per day in Erewhon, x days after the beginning of the year, is expressed by the formula $R = (5.1 \times 10^{-5})(6511 + 366x - x^2)$ cm/day. Estimate average daily rainfall for the first 100 days.

30 A certain car, starting from rest, accelerates at the constant rate of 11 ft/sec^2. Find its average speed during the first 10 sec.

Figure 5-4-10b contained a proof that $e < 3$. The next two exercises suggest another approach.

31 Find the tangent line $y = t(x)$ to $y = f(x) = 1/x$ at $(2, \frac{1}{2})$. Prove that $t(x) < 1/x$ for $x > 0$, except that $t(2) = \frac{1}{2}$.

32 (cont.) Integrate this inequality on $[1, 3]$ to prove $\ln 3 > 1$, and hence $e < 3$.

The purpose of the remaining exercises is to start the development of the logarithm and the exponential function from an integral definition. In total, this is a long story, and these exercises are just the beginning, included here to show an alternative approach to those functions. Temporarily forget that you ever saw the functions $y = \ln x$ and $y = e^x$. Define

$$f(x) = \int_1^x \frac{dt}{t} \quad \text{for } x > 0$$

33 Prove that $f(x)$ is strictly increasing, that its range is an interval, and that it has a strictly increasing inverse function $y = F(x)$.

34 Fix $a > 0$. Prove that

$$f(ax) - f(x)$$

is a constant. Evaluate the constant and hence establish the functional equation

$$f(xy) = f(x) + f(y) \quad \text{for } x > 0 \text{ and } y > 0$$

[Hint Differentiate.]

35 Conclude that if x and y are in the domain of $F(x)$, then so are $x + y$ and $x - y$, and

$$\begin{cases} F(x+y) = F(x)F(y) \\ F(x-y) = F(x)/F(y) \end{cases}$$

36 Deduce that the domain of $F(x)$ is $\mathbf{R}$.

5-5 Inequalities and Estimates

The definite integral satisfies inequalities that have useful numerical applications.

Inequalities

Let $f(x)$ and $g(x)$ be integrable on $[a, b]$.

1 If $f(x) \le g(x)$, then $\int_a^b f(x)\,dx \le \int_a^b g(x)\,dx$

See Figure 5-5-1.

2 If $m \le f(x) \le M$, then $m(b-a) \le \int_a^b f(x)\,dx \le M(b-a)$

See Figure 5-5-2.

3 $\left|\int_a^b f(x)\,dx\right| \le \int_a^b |f(x)|\,dx$

See Figure 5-5-3.

4 If $|f(x)| \le M$, then $\left|\int_a^b f(x)g(x)\,dx\right| \le M\int_a^b |g(x)|\,dx$

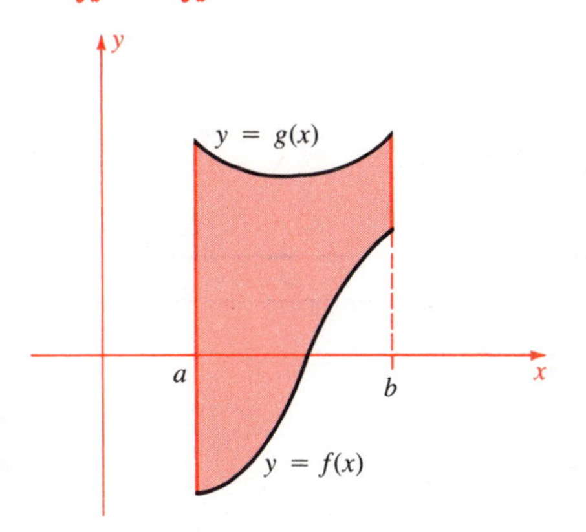

Figure 5-5-1
If $f(x) \le g(x)$, then $\int_a^b g - \int_a^b f = \text{area} \ge 0$

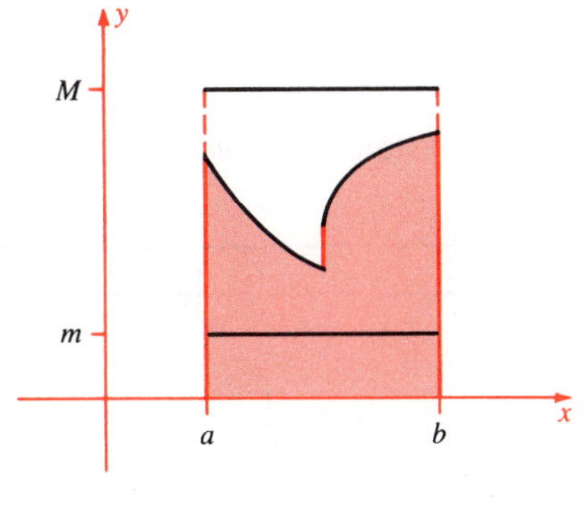

Figure 5-5-2
If $m \le f(x) \le M$, then $m(b-a) \le \int_a^b f \le M(b-a)$

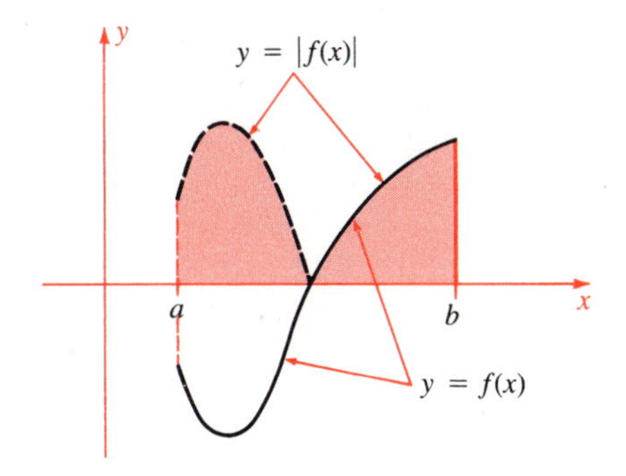

Figure 5-5-3
$f \le |f|$ so $\int_a^b f \le \int_a^b |f|$. Similarly $-f \le |f|$ so $-\int_a^b f \le \int_a^b |f|$.

Inequality 1 was given on page 279. Inequality 2 is a direct consequence of inequality 1. For $m \le f(x) \le M$ implies

$$\int_a^b m\,dx \le \int_a^b f(x)\,dx \le \int_a^b M\,dx$$

(Inequality 2 also follows from basics—constant functions *are* step functions.)

Inequality 3 also follows from inequality 1 because $f(x) \le |f(x)|$ and $-f(x) \le |f(x)|$. Consequently we have both

$$\int_a^b f(x)\,dx \le \int_a^b |f(x)|\,dx \quad \text{and} \quad -\int_a^b f(x)\,dx \le \int_a^b |f(x)|\,dx$$

Hence

$$\left|\int_a^b f(x)\,dx\right| \le \int_a^b |f(x)|\,dx$$

Inequality 4 holds because $|f(x)g(x)| = |f(x)||g(x)| \le M|g(x)|$ so

$$\left|\int_a^b f(x)g(x)\,dx\right| \le \int_a^b |f(x)g(x)|\,dx$$

$$= \int_a^b |f(x)||g(x)|\,dx \le \int_a^b M|g(x)|\,dx = M\int_a^b |g(x)|\,dx$$

Remark Suppose in inequality 1 that $f(x)$ and $g(x)$ are *continuous* functions on $[a, b]$ and that not only $f(x) \le g(x)$ on $[a, b]$, but for some c in the interval, $f(c) < g(c)$. Then it can be shown (Exercises 19–20) that not only

$$\int_a^b f(x)\,dx \le \int_a^b g(x)\,dx$$

but actually

$$\int_a^b f(x)\,dx < \int_a^b g(x)\,dx$$

Similar statements hold for inequalities 2–4 also.

Example 1

Prove

a $\displaystyle\int_0^{\pi/3} \sin^5 x\,dx \le \int_0^{\pi/3} \sin^4 x\,dx$

b $\displaystyle\sqrt{3}\int_1^4 x^2\,dx \le \int_1^4 x^2\sqrt{3 + e^{-x}}\,dx \le 2\int_1^4 x^2\,dx$

Solution

a If $0 \le x \le \frac{1}{3}\pi$, then

$$0 \le \sin x < 1 \qquad \text{hence} \qquad \sin^5 x \le \sin^4 x$$

Now integrate this inequality, that is, apply inequality 1.

b If $x \geq 0$, then $0 < e^{-x} \leq 1$. Therefore

$$\sqrt{3+0} < \sqrt{3+e^{-x}} \leq \sqrt{3+1} = 2$$

so $x^2\sqrt{3} < x^2\sqrt{3+e^{-x}} \leq 2x^2$. Now apply inequality 1. ●

Example 2

Prove $2\int_a^b f(x)g(x)\,dx \leq \int_a^b f^2(x)\,dx + \int_a^b g^2(x)\,dx$

Solution From $[f(x) - g(x)]^2 \leq 0$ follows

$$f^2(x) + g^2(x) \geq 2f(x)g(x)$$

Now apply inequality 1. ●

Estimates

Sometimes we need the exact value of an integral. At other times a highly accurate approximation to its value is satisfactory. But sometimes a rough order of magnitude is adequate. In such cases, quick estimates are usually possible by means of simple techniques based on inequalities.

Example 3

Show that $\int_1^3 \frac{dx}{1+x^4} < \frac{1}{3}$

Solution $1/(1+x^4) < 1/x^4$ for $x > 0$. Hence

$$\int_1^3 \frac{dx}{1+x^4} < \int_1^3 \frac{dx}{x^4} = \frac{-1}{3x^3}\bigg|_1^3 = \frac{1}{3} - \frac{1}{81} < \frac{1}{3}$$ ●

Remark Example 3 illustrates an important technique: replacing the integrand by a slightly larger one that is easily integrated. Generally, this technique yields better results than the use of inequality 2 does. For instance, all we can deduce from inequality 2 is

$$\int_1^3 \frac{dx}{1+x^4} < 1$$

Here $M = \frac{1}{2}$, the largest value of $1/(1+x^4)$ in the interval $[1, 3]$. At times, however, inequality 2 yields adequate results.

Example 4

Estimate $\int_6^8 \frac{dx}{x^3 + x + \sin 2x}$

Solution Find bounds for the integrand. Since $-1 \leq \sin 2x \leq 1$ and $x^3 + x$ is increasing (positive derivative),

$$221 = 6^3 + 6 - 1 \leq x^3 + x + \sin 2x \leq 8^3 + 8 + 1 = 521$$

on the domain $[6, 8]$. Take reciprocals (and reverse the inequalities):

$$1.9 \times 10^{-3} < \tfrac{1}{521} \leq \frac{1}{x^3 + x + \sin 2x} \leq \tfrac{1}{221} < 4.6 \times 10^{-3}$$

for $6 \leq x \leq 8$. By inequality 2,

$$3.8 \times 10^{-3} < \int_6^8 \frac{dx}{x^3 + x + \sin 2x} < 9.2 \times 10^{-3}$$

●

Example 5

Show that $\left| \int_2^5 \frac{\sin x}{(1 + x)^2}\, dx \right| \leq \frac{1}{6}$

Solution Since $|\sin x| \leq 1$, we use inequality 4:

$$\left| \int_2^5 \frac{\sin x}{(1 + x)^2}\, dx \right| \leq \int_2^5 \frac{dx}{|1 + x|^2}$$

$$= \int_2^5 \frac{dx}{(1 + x)^2} = \left. \frac{-1}{1 + x} \right|_2^5 = \tfrac{1}{3} - \tfrac{1}{6} = \tfrac{1}{6}$$

●

Example 6

Show that $\int_0^3 \exp(-x^2)\, dx$ is a good approximation for $\int_0^{100} \exp(-x^2)\, dx$. How good?

Solution The error in this approximation is

$$\int_0^{100} \exp(-x^2)\, dx - \int_0^3 \exp(-x^2)\, dx = \int_3^{100} \exp(-x^2)\, dx$$

Since $\exp(-x^2)$ is extremely small even for moderate values of x, the integral on the right is small. Estimate the integral by inequality 2. Because $\exp(-x^2)$ is a decreasing function, its largest value in the interval $[3, 100]$ occurs at the left end point; this value is e^{-9}. By inequality 2,

$$\int_3^{100} \exp(-x^2)\, dx < (100 - 3)e^{-9} < (97)(1.3 \times 10^{-4})$$

$$< 1.3 \times 10^{-2}$$

A better estimate can be obtained using inequality 1. Note that $x^2 \geq 3x$ when $x \geq 3$. It follows that $\exp(-x^2) \leq e^{-3x}$ for $x \geq 3$. Therefore

$$\int_3^{100} \exp(-x^2)\, dx \leq \int_3^{100} e^{-3x}\, dx = \left. \frac{e^{-3x}}{-3} \right|_3^{100} = \frac{e^{-9}}{3} - \frac{e^{-300}}{3}$$

$$< \frac{e^{-9}}{3} < \frac{1.3 \times 10^{-4}}{3} < 5 \times 10^{-5}$$

We conclude that

$$\int_0^3 \exp(-x^2)\,dx \quad \text{approximates} \quad \int_0^{100} \exp(-x^2)\,dx$$

to within 5×10^{-5}. ●

Remark This example illustrates an important labor-saving device. If you intend to estimate

$$\int_0^{100} \exp(-x^2)\,dx$$

by an approximate integration method (see Section 5-6) you can save a great deal of work by applying the method to

$$\int_0^3 \exp(-x^2)\,dx$$

Ignoring the rest of the integral introduces an error less than 5×10^{-5}. If that is not precise enough, you might apply approximate integration to

$$\int_0^4 \exp(-x^2)\,dx$$

Then by the same argument as in the example, ignoring the rest of the integral introduces an error less than $\frac{1}{4}e^{-16} < 3 \times 10^{-8}$.

Exercises

In the following exercises you are asked to find bounds for certain integrals. There are generally several ways to estimate each integral. Try to obtain the bound given or to improve on it. If you cannot, at least find *some* bound. Show

1 $1 < \int_0^1 \exp(x^2)\,dx < e - 1$

2 $\int_0^1 \frac{dx}{4 + x} < \int_0^1 \frac{dx}{4 + x^3} < \frac{1}{4}$

3 $3 \int_3^5 \frac{dx}{x} < \int_3^5 \frac{\sqrt{3 + 2x}}{x}\,dx < 4 \int_3^5 \frac{dx}{x}$

4 $\frac{1}{15} < \int_1^2 \frac{dx}{x^3 + 3x + 1} < \frac{1}{5}$

5 $\int_0^{100} e^{-x}\sin^2 x\,dx < 1$

6 $\frac{3}{10} < \int_1^4 \frac{dx}{x^2 + x + 1} < \frac{3}{4}$

7 $6 < \int_5^{11} \frac{x}{\sqrt{5 + 4x}}\,dx < 10$

8 $2 < \int_0^4 \frac{dx}{1 + \sin^2 x} < 4$

9 $\frac{1}{2}\pi < \int_0^{\pi/2} \frac{d\theta}{\sqrt{1 - k^2\sin^2\theta}} < \frac{\pi}{2\sqrt{1 - k^2}}$ $(0 < k < 1)$

10 $\int_0^{\pi/3} \sin 2\theta \cos\theta\,d\theta < \frac{3}{4}$

11 $\frac{15}{16} < \int_1^2 x^3 2^{-x}\,dx < \frac{15}{8}$

12 $\frac{9}{10}(1 - e^{-1}) < \int_1^{10} \frac{1 - e^{-x}}{x^2}\,dx < \frac{9}{10}$

13 $\int_0^1 \frac{\sin x + \cos x}{(1 + x)^2}\,dx < \frac{1}{2}\sqrt{2}$

[Hint $\sin x + \cos x = \sqrt{2}\cos(x - \frac{1}{4}\pi)$.]

***14** $\frac{2 - \sqrt{3}}{\sqrt{2}} < \int_0^{\pi/6} \sqrt{\sin x}\,dx < \frac{2}{3}\left(\frac{\pi}{6}\right)^{3/2}$

[Hint Show that $\sqrt{2}\sin x < \sqrt{\sin x} < \sqrt{x}$ on the interval.]

15 $\dfrac{99\pi}{400} < \displaystyle\int_1^{100} \frac{\arctan x}{x^2}\,dx < \frac{99\pi}{200}$

16 $(1 - e^{-1})\ln 10 < \displaystyle\int_1^{10} \frac{1 - e^{-x}}{x}\,dx < \ln 10$

17 Show that the error in estimating

$$\int_0^3 \frac{1}{(x^2+1)^2}\,dx \quad \text{by} \quad \int_0^2 \frac{1}{(x^2+1)^2}\,dx$$

is less than 0.04.

18 Show that $2 < \displaystyle\int_3^6 \frac{dx}{\sqrt{1+\cos^2 x}} < 3.$

***19** Suppose that $f(x)$ is continuous on $[a, b]$, that $f(x) \geq 0$ on $[a, b]$, and that $f(c) > 0$ for some c in $[a, b]$. Prove that

$$\int_a^b f(x)\,dx > 0$$

[Hint Use continuity to find an interval including c and an $\varepsilon > 0$ such that $f(x) > \varepsilon$ on that interval. Then construct a step function $s(x) \leq f(x)$ with positive integral.]

***20** (cont.) Suppose that $f(x)$ and $g(x)$ are continuous on $[a, b]$, that $f(x) \leq g(x)$, and that $f(c) < g(c)$ for some c in $[a, b]$. Prove that

$$\int_a^b f(x)\,dx < \int_a^b g(x)\,dx$$

5-6 Approximate Integration

We must frequently approximate a definite integral rather than compute its exact value. This happens in two cases:

Case 1 The function to be integrated is given experimentally, say by instrument readings. There is no formula for it.

Case 2 We have a formula for the function, but we do not know an antiderivative.

Let us consider these cases. In Case 1, we are given values

$$f(x_0) \qquad f(x_1) \qquad \cdots \qquad f(x_n)$$

at $n + 1$ points x_j. Often the points are equally spaced on an interval $[a, b]$:

$$x_0 = a \qquad x_1 = a + h \qquad x_2 = a + 2h \qquad \cdots \qquad x_n = a + nh = b$$

Thus the interval is divided into n equal subintervals, each of length

$$h = \frac{b - a}{n}$$

and we know the values of $f(x)$ only at the division points $x_0, \cdots, x_n$. We *assume* that there is a smooth function $f(x)$ on the interval taking these values. For instance we might be given readings

x_j	1.0	1.5	2.0	2.5	3.0
$f(x_j)$	1.00	0.67	0.50	0.40	0.33

The problem is to find a reasonable estimate for $\int_1^3 f(x)\,dx$.

In Case 2, the problem is to estimate the integral of $f(x)$ without knowing an antiderivative of $f(x)$. An example is

$$\int_1^2 \frac{e^x}{x}\,dx$$

We have not seen an antiderivative of e^x/x, so we cannot compute this integral exactly. (In fact, it can be proved that the antiderivative of e^x/x cannot be expressed in terms of the elementary functions of calculus.) Still we should somehow be able to estimate it with reasonable accuracy.

We shall handle Case 2 by computing values of $f(x)$ at equally spaced points and using these values to estimate the integral. Thus we shall treat both cases by the same method. It seems reasonable to expect better results in Case 2, however. Since we have a formula for $f(x)$, we can compute more and more data points for greater and greater accuracy. In addition, we can use further information about $f(x)$ to find limits on the possible error in estimating the integral.

Remark It is standard to use the letter h for $(b-a)/n$. The symbol Δx is another standard notation; we prefer h.

Rectangular and Trapezoidal Approximations

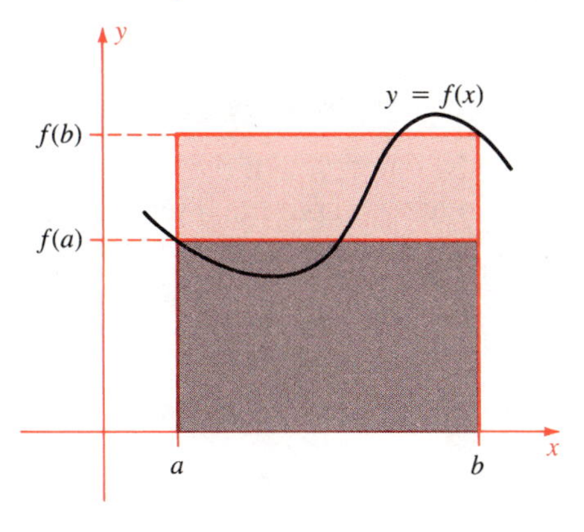

Figure 5-6-1
Two rectangular approximations

The simplest case of the problem is this: Given $f(a)$ and $f(b)$—nothing else—estimate

$$\int_a^b f(x)\,dx$$

Geometry (Figure 5-6-1) suggests two possible estimates by (algebraic) areas of rectangles:

$$\int_a^b f(x)\,dx \approx (b-a)f(a) \quad \text{and} \quad \int_a^b f(x)\,dx \approx (b-a)f(b)$$

Notice that each uses only part of the data. Which is the better estimate? That depends on $f(x)$. If $f(x)$ is an increasing function, at least we can say the following:

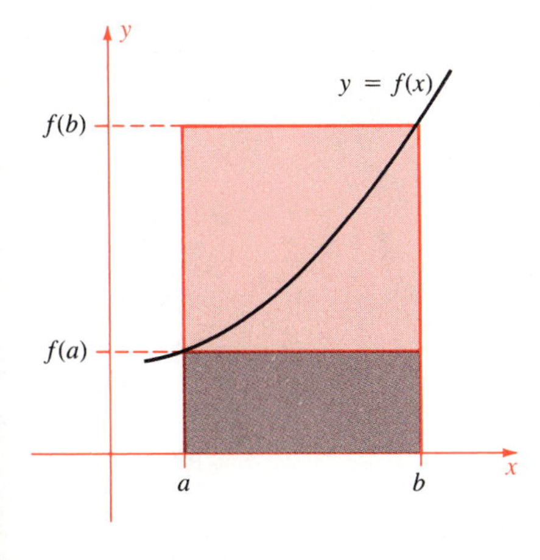

Figure 5-6-2
Over and under rectangular approximations, curve increasing

$$(b-a)f(a) \le \int_a^b f(x)\,dx \le (b-a)f(b)$$

This assertion is obvious geometrically (Figure 5-6-2). If $f(x)$ is decreasing, the same is true with the sense of the inequalities reversed. Thus of the two rectangular approximations, $(b-a)f(a)$ and $(b-a)f(b)$, one is an overestimate and the other is an underestimate, at least when $f(x)$ is increasing or decreasing. Common sense suggests *averaging* these estimates:

$$\int_a^b f(x)\,dx \approx (b-a)\frac{f(a)+f(b)}{2}$$

Now we have used all the data—that is a good sign. Furthermore, the average $\frac{1}{2}(b-a)[f(a)+f(b)]$ of the two rectangular areas is itself an area, the (algebraic) area of a trapezoid (Figure 5-6-3a, next page).

Figure 5-6-3
Trapezoidal approximation to $\int_a^b f(x)\,dx$

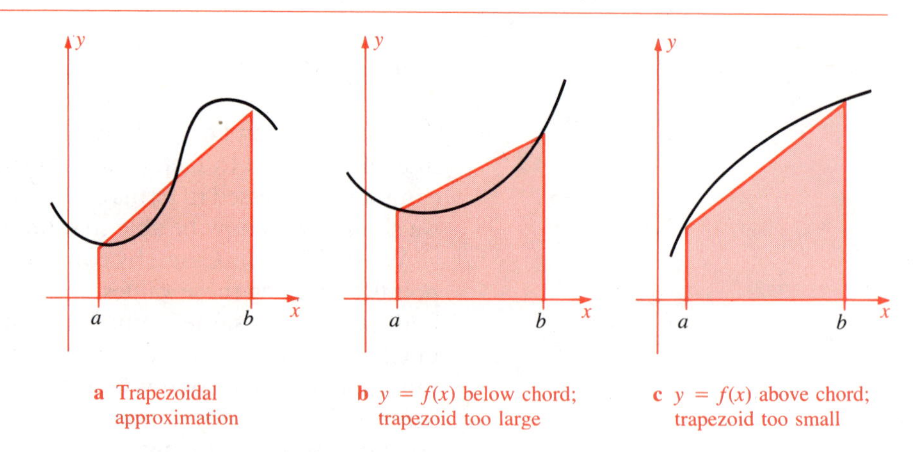

a Trapezoidal approximation

b $y = f(x)$ below chord; trapezoid too large

c $y = f(x)$ above chord; trapezoid too small

The figure suggests that the trapezoidal area may be a reasonable approximation to the area under the curve.

Can we draw any conclusion about the trapezoidal approximation, knowing that $f(x)$ increases? No. If $f(x)$ lies below the chord (Figure 5-6-3b), the trapezoidal estimate is too large; if $f(x)$ lies above, it is too small (Figure 5-6-3c). But if neither is the case (Figure 5-6-3a), what happens is anybody's guess.

The Trapezoidal Rule

The trapezoidal estimate of an integral may be inaccurate if the interval is long (Figure 5-6-4). Hoping for a better estimate, we split the interval into n equal parts. On each subinterval we estimate the integral of $f(x)$ by the area of a trapezoid, then add up these areas (Figure 5-6-5). The result is an approximation called the *trapezoidal rule.*

Figure 5-6-4
One big trapezoid: inaccurate

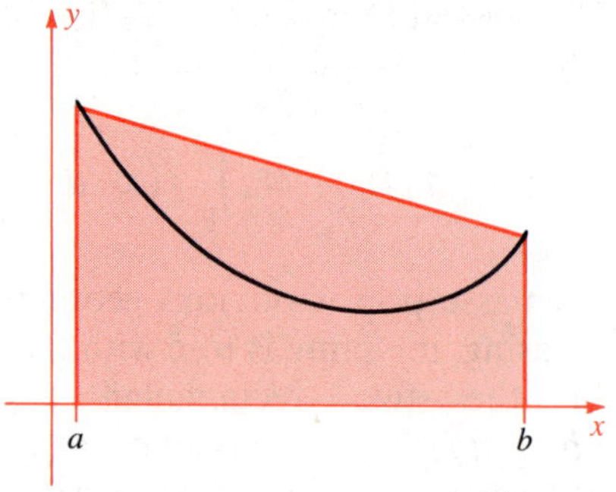

Figure 5-6-5
Many trapezoids: improved accuracy

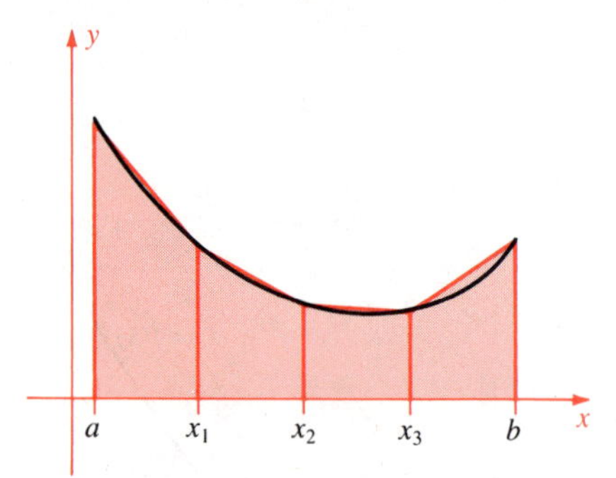

To derive a working formula, we write

$$a = x_0 < x_1 < \cdots < x_n = b$$

where

$$x_j - x_{j-1} = h = \frac{b-a}{n}$$

We apply the trapezoidal estimate on each subinterval:

$$\int_a^b f(x)\,dx = \int_{x_0}^{x_1} + \int_{x_1}^{x_2} + \cdots + \int_{x_{n-1}}^{x_n} f(x)\,dx$$

$$\approx (x_1 - x_0)\frac{f(x_0) + f(x_1)}{2}$$

$$+ \cdots + (x_n - x_{n-1})\frac{f(x_{n-1}) + f(x_n)}{2}$$

$$= \tfrac{1}{2}h[f(x_0) + f(x_1)] + \tfrac{1}{2}h[f(x_1) + f(x_2)] + \tfrac{1}{2}h[f(x_2) + f(x_3)] + \cdots + \tfrac{1}{2}h[f(x_{n-1}) + f(x_n)]$$

$$= \tfrac{1}{2}h[f(x_0) + 2f(x_1) + 2f(x_2) + \cdots + 2f(x_{n-1}) + f(x_n)]$$

We have derived the following rule:

Trapezoidal Rule

Let $f(x)$ be integrable on $[a, b]$. Set $h = (b - a)/n$, $x_j = a + jh$, and $f_j = f(x_j)$. Then

$$\int_a^b f(x)\,dx \approx \tfrac{1}{2}h(f_0 + 2f_1 + 2f_2 + \cdots + 2f_{n-1} + f_n)$$

Example 1

Estimate $\displaystyle\int_1^3 \frac{dx}{x}$ with $n = 2,\ 4,\ 10,\ 20,\ 100.$

Solution For $n = 2$, we have $x_0 = 1$, $x_1 = 2$, $x_2 = 3$, and $h = 1$, so

$$\int_1^3 \frac{dx}{x} \approx \tfrac{1}{2}[f_0 + 2f_1 + f_2] = \tfrac{1}{2}[\tfrac{1}{1} + \tfrac{2}{2} + \tfrac{1}{3}] = \tfrac{7}{6} \approx 1.17$$

For $n = 4$, we have $x_0 = 1$, $x_1 = \tfrac{3}{2}$, $x_2 = 2$, $x_3 = \tfrac{5}{2}$, $x_4 = 3$, and $h = \tfrac{1}{2}$, so

$$\int_1^3 \frac{dx}{x} \approx \tfrac{1}{4}[1 + 2(\tfrac{2}{3} + \tfrac{1}{2} + \tfrac{2}{5}) + \tfrac{1}{3}] = \tfrac{67}{60} \approx 1.117$$

For $n = 10$, we have $h = 0.2$ and we use a calculator:

$$\int_1^3 \frac{dx}{x} \approx \frac{1}{2}(0.2)\left[1 + 2\left(\frac{1}{1.2} + \frac{1}{1.4} + \cdots + \frac{1}{2.8}\right) + \frac{1}{3}\right]$$

$$\approx 1.1016$$

For $n = 20$ and $n = 100$, we use a programmable calculator or computer:

$$\int_1^3 \frac{dx}{x} \approx \tfrac{1}{2}(0.1)\left[1 + 2\left(\frac{1}{1.1} + \frac{1}{1.2} + \cdots + \frac{1}{2.9}\right) + \frac{1}{3}\right]$$

$$\approx 1.09935$$

$$\int_1^3 \frac{dx}{x} \approx \tfrac{1}{2}(0.02)\left[1 + 2\left(\frac{1}{1.02} + \cdots + \frac{1}{2.98}\right) + \frac{1}{3}\right]$$

$$\approx 1.098642$$

We summarize our results in the following table:

n	2	4	10	20	100
approx.	1.17	1.117	1.1016	1.09935	1.098642

Because $f(x) = 1/x$ is concave up for $x > 0$, each of the approximations is too large. Our computations yield approximations that decrease as n increases, a good sign. The actual value of the integral to 10 places is

$$\int_1^3 \frac{dx}{x} = \ln x \Big|_1^3 = \ln 3 \approx 1.09861\ 22887$$

If $f(x)$ is a linear function, then

$$\int_a^b f(x)\,dx = \frac{b-a}{2}[f(a) + f(b)]$$

as is easily checked. Thus the trapezoidal approximation is exact for linear functions. Consequently the error in the trapezoidal rule for a general function $f(x)$ should depend on how much $f(x)$ deviates from being linear. Now the second derivative of a linear function is zero, so the second derivative of a general function measures its deviation from linearity. Therefore we expect an error estimate in the trapezoidal rule to involve the second derivative of $f(x)$. The following is such an estimate, but its proof is too hard to include here.

Error in the Trapezoidal Rule

Suppose $|f''(x)| \leq M$ on $[a, b]$. Set $h = (b-a)/n$, $x_j = a + jh$, and $f_j = f(x_j)$. Then

$$\int_a^b f(x)\,dx = \frac{h}{2}(f_0 + 2f_1 + \cdots + 2f_{n-1} + f_n) + \text{error}$$

where

$$|\text{error}| \leq \frac{Mnh^3}{12} = \frac{M(b-a)}{12}h^2$$

Example 2

Find an upper bound for the error in the trapezoidal rule for

$$\int_1^3 \frac{dx}{x} \quad (=\ln 3) \quad \text{with} \quad n = 100$$

Solution Set $f(x) = 1/x$. Then $f''(x) = 2/x^3$, which has its largest value on $[1, 3]$ at $x = 1$. Hence take $M = f''(1) = 2$. Since $h = 0.02$, we have

$$|\text{error}| \le \frac{Mnh^3}{12} = \frac{2(100)(0.02)^3}{12} = \tfrac{4}{3} \times 10^{-4} < 1.4 \times 10^{-4}$$

This confirms again that the estimate computed in Example 1 is correct to at least three places. ●

The Midpoint Rule

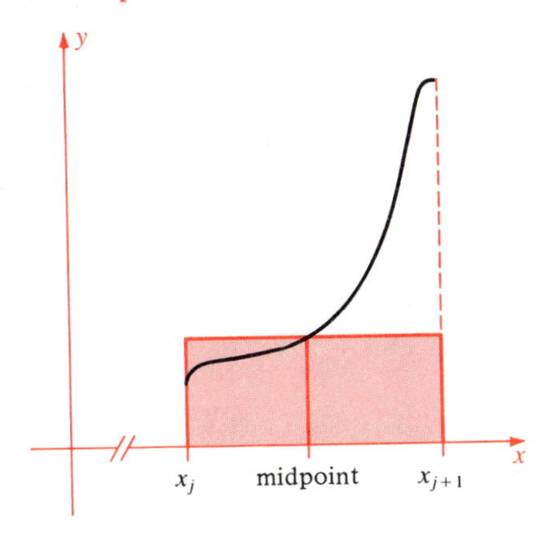

Figure 5-6-6
The midpoint rule

Again divide the interval $[a, b]$ into n equal parts, each of length $h = (b - a)/n$. The j-th subinterval is

$$[a + jh, \quad a + (j + 1)h]$$

with midpoint $x_j = a + (j + \frac{1}{2})h$. The *midpoint approximation* is the Riemann sum based on choosing function values at three midpoints (Figure 5-6-6):

Midpoint Rule

Let $f(x)$ be integrable on $[a, b]$. Set $h = (b - a)/n$ and set $f_j = f(x_j)$, where $x_j = a + (j + \frac{1}{2})h$. Then

$$\int_a^b f(x)\,dx \approx h(f_0 + f_1 + \cdots + f_{n-1})$$

Like the trapezoidal rule, the midpoint rule is exact for linear functions. Hence we expect the error in the midpoint rule to be controlled by the magnitude of the second derivative. This turns out to be so, and in fact the midpoint approximation is slightly more accurate than the trapezoidal approximation.

Error in the Midpoint Rule

Suppose $|f''(x)| \le M$ on $[a, b]$. Set $h = (b - a)/n$, $x_j = a + (j + \frac{1}{2})h$, and $f_j = f(x_j)$. Then

$$\int_a^b f(x)\,dx = h(f_0 + f_1 + \cdots + f_{n-1}) + \text{error}$$

where

$$|\text{error}| \le \frac{Mnh^3}{24} = \frac{M(b - a)}{24} h^2$$

Which should we use, the trapezoidal rule or the midpoint rule? If the function is known by experimental readings, then the data usually fits the trapezoidal rule, not the midpoint rule, so there is no choice. But if function values are computable, then the expected error for the midpoint rule is about half that for the trapezoidal rule. Also the midpoint rule actually costs slightly less in amount of computing (n function evaluations instead of $n + 1$). Therefore the midpoint rule has a slight edge, particularly if automatic computing is available. In Chapter 9 we shall discuss more accurate methods of approximate integration.

Terminology Formulas such as the trapezoidal and midpoint rules for approximate integration are sometimes called **numerical quadrature** formulas. The word *quadrature* by itself is an old term for integration.

Exercises

Approximate the integral by the trapezoidal rule with $n = 4$ (three significant figures), $n = 10$ (four significant figures), and if a programmable calculator or computer is available, $n = 50$ (six significant figures)

1 $\displaystyle\int_1^2 \frac{dx}{x}$

2 $\displaystyle\int_0^3 \frac{dx}{x+1}$

3 $\displaystyle\int_1^3 \frac{x-1}{x+1}\,dx$

4 $\displaystyle\int_{-1}^1 \frac{dx}{1+x^2}$

5 $\displaystyle\int_1^2 \frac{x}{1+x^2}\,dx$

6 $\displaystyle\int_0^1 \sqrt{1-x^2}\,dx$

7 $\displaystyle\int_0^4 \sqrt{1+x^2}\,dx$

8 $\displaystyle\int_0^2 xe^x\,dx$

9 $\displaystyle\int_0^{10} xe^{-x}\,dx$

10 $\displaystyle\int_0^4 \exp(-x^2)\,dx$

11 $\displaystyle\int_1^2 e^{-1/x}\,dx$

12 $\displaystyle\int_0^\pi e^x \sin x\,dx$

13 $\displaystyle\int_1^2 \log_{10} x\,dx$

14 $\displaystyle\int_1^2 x\log_{10} x\,dx$

15 $\displaystyle\int_0^{\pi/4} \tan x\,dx$

16 $\displaystyle\int_0^\pi x\sin x\,dx$

17 $\displaystyle\int_0^{\pi/2} x^2\sin x\,dx$

18 $\displaystyle\int_{-\pi/4}^{\pi/4} \sec x\,dx$

19 $\displaystyle\int_0^{\pi/2} \sin^2 x\,dx$

20 $\displaystyle\int_0^{\pi/2} x\cos x\,dx$

21–40 For these exercises follow the instructions for the corresponding integrals in Exercises 1 to 20, except use the midpoint rule. For instance, Exercise 27 is the integral of Exercise 7 approximated by the midpoint rule ($n = 4, 10, 50$).

41 An automobile starting from rest accelerates for 15 sec. Velocity readings (in feet per second) taken at 1-second intervals are:

t	v	t	v	t	v
0	0.0				
1	0.5	6	18.7	11	63.9
2	2.1	7	25.5	12	75.9
3	4.7	8	33.3	13	88.9
4	8.3	9	43.1	14	102.9
5	13.0	10	53.0	15	118.0

Estimate the distance the car traveled during this time.

42 Depth soundings in feet are taken across a 90-foot-wide river at 5-foot intervals, resulting in these readings:

x	y	x	y	x	y
0	0.0				
5	1.0	35	12.0	65	14.0
10	2.5	40	13.0	70	10.5
15	5.0	45	12.5	75	9.0
20	8.0	50	13.5	80	6.5
25	10.5	55	16.0	85	4.0
30	11.5	60	16.0	90	0.0

If the flow in the river is 5 ft/sec, estimate to the nearest million cubic feet the daily flow of water.

Compare the exact value, trapezoidal rule with $n = 2$, and midpoint rule with $n = 2$

43 $\displaystyle\int_0^2 x^2\,dx$

44 $\displaystyle\int_1^2 x^3\,dx$

45 $\displaystyle\int_1^3 \frac{1}{x}\,dx$

46 $\displaystyle\int_1^3 \frac{1}{x^2}\,dx$

47 Show that the trapezoidal rule with $n = 100$ approximates

$$\int_4^7 e^{-x} \sin 2x \, dx \qquad \text{to at least four places}$$

48 Show that the trapezoidal rule with $n = 100$ approximates

$$\int_0^1 \frac{dx}{x^3 + 10} \qquad \text{to within } 10^{-6}$$

49 Suppose the trapezoidal rule is used to estimate

$$\int_0^2 xe^x \, dx$$

Use the error estimate in the text to find the least n so that you are sure the error is at most 2×10^{-5}.

50 Suppose the midpoint rule is used to estimate

$$\int_0^4 x^2 e^x \, dx$$

Use the error estimate in the text to find the least n so that you are sure the error is at most 5×10^{-6}.

51 Use the trapezoidal rule to prove

$$\tfrac{1}{2}[\ln 1 + 2\ln 2 + 2\ln 3 + \cdots + 2\ln(n-1) + \ln n] < \int_1^n \ln x \, dx$$

52 (cont.) Prove $n! < \sqrt{n}\,(n^n)e^{-n+1}$.
[Hint $x \ln x - x$ is an antiderivative of $\ln x$.]

5-7 Review Exercises

1 Define $f(x)$ by $f(x) = 1$ if $2n \le x < 2n + 1$ for some integer n and $f(x) = 0$ otherwise. Sketch $y = f(x)$ and evaluate

$$\int_0^{10} f(x) \, dx$$

2 Define $f(x)$ by $f(x) = n$ for $n \le x < n + 1$, where n is any integer. Sketch $y = f(x)$ and evaluate

$$\int_0^{100} f(x) \, dx$$

Evaluate

3 $\displaystyle\int_{-2}^{2} (x^2 - 4)^2 \, dx$

4 $\displaystyle\int_0^{-a} 14(x - a)^6 \, dx$

5 $\displaystyle\int_1^2 \left(x - \frac{1}{x^2}\right) dx$

6 $\displaystyle\int_{5\pi/4}^{7\pi/4} \cos x \, dx$

7 $\displaystyle\left(\frac{d}{dx}\int_0^x \frac{dt}{\sqrt{t^4 + 3}}\right)\bigg|_{x=1}$

8 $\displaystyle\left(\frac{d}{dx}\int_0^{x^2} \frac{dt}{\sqrt{t^2 + 9}}\right)\bigg|_{x=2}$

9 $\displaystyle\frac{d}{dx}\int_x^1 \exp(-t^2) \, dt$

10 $\displaystyle\frac{d}{dx}\int_x^{x+1} t^3 e^{-t} \, dt$

11 Find the area bounded by the x-axis, $x = -2$, $x = 0$, and the curve $y = x^3 + 4x + 5$.

12 Find the area bounded by $x = 1$, $x = 2$, $y = x^2$, and $y = x^4$.

Find the mean value of

13 $f(x) = \sin x$ on $[0, \pi]$

14 $f(x) = e^{x/k}$ on $[-k, k]$ where $k > 0$

15 $f(x) = |x|$ on $[-1, 3]$

16 $f(x) = \begin{cases} -1 & \text{for } -\frac{3}{2} \le x < 0 \\ 1 & \text{for } 0 \le x \le 3 \end{cases}$ on $[-\frac{3}{2}, 3]$

17 Find $f(x) = \displaystyle\int_0^x |t| \, dt$

18 Graph $y = \displaystyle\int_0^x f(t) \, dt$ where

$$f(t) = \begin{cases} 1 & \text{for } 4n - 1 \le t < 4n + 1 \\ -1 & \text{for } 4n + 1 \le t < 4n + 3 \end{cases}$$

Define $f(x) = \displaystyle\int_0^x \frac{dt}{1 + t^4}$ on **R**

***19** Prove that $|f(x)| < \frac{4}{3}$.

20 Prove that $f(x)$ has an inverse function $g(y)$ and that $g'(y) = 1 + g(y)^4$.

Estimate $\displaystyle\int_0^\pi \theta \sin\theta \, d\theta$ to four significant figures by the

21 trapezoidal rule with $n = 10$

22 midpoint rule with $n = 10$

6 Techniques of Integration

6-1 Introduction

Let us begin with some notation and terminology. The symbol

$$\int f(x)\,dx$$

which is called the **indefinite integral** of $f(x)$, denotes the most general antiderivative of $f(x)$. For example,

$$\int x^2\,dx = \tfrac{1}{3}x^3 + C \qquad \int \cos x\,dx = \sin x + C$$

The letter C always means an arbitrary constant. It is there because any two antiderivatives of a function (with domain an interval) differ by a constant. To each *differentiation* formula, there corresponds an *indefinite integral* formula. For instance, to

$$\frac{d}{dx}(\tan x) = \sec^2 x \quad \text{corresponds} \quad \int \sec^2 x\,dx = \tan x + C$$

We have developed rules for differentiating most functions that arise in practice. The reverse process, antidifferentiating a function, that is, finding its indefinite integral is harder. There is no systematic procedure for antidifferentiation. However, there are important techniques and a large bag of miscellaneous tricks. Success is not guaranteed; there are simple-looking functions that are not derivatives of any of the functions we encounter in calculus, for example

$$\exp(-x^2) \qquad \frac{\sin x}{x} \qquad \text{and} \qquad \sqrt{(1-x^2)(1-4x^2)}$$

It is important to know how to find antiderivatives. Many applications of integration and of differential equations lead to relations involving definite integrals; the most satisfactory and useful form of the answer often requires carrying out the integration. The formulations of many laws of nature are the result of evaluating integrals. For instance, the law of conservation of energy in mechanics is the result of evaluating an integral that arises from Newton's equations of motion.

This chapter covers indispensable methods for finding antiderivatives. However, the single most important practical method of finding an integral is to look it up in a table of integrals. You should start to familiarize yourself with the table just inside the covers of this book. You should also look at a more extensive table, like the one in the *CRC Standard Mathematical Tables,* to see how it is organized.

One of the main reasons for learning various techniques of integration is to learn the use of integral tables. Tables cannot possibly list every conceivable integral, but they do list integrals to which you can reduce many other integrals by some preliminary, routine work. For example, every integral table includes

$$\int \sqrt{x^2 + a^2}\, dx$$

but probably no integral table includes

$$\int \sqrt{3x^2 + 2}\, dx \qquad \text{or} \qquad \int e^{-x}\sqrt{e^{-2x} + 1}\, dx$$

It is your job to recognize that each of these two integrals can be reduced to the first one, and to carry out the necessary preliminary manipulations.

Symbolic manipulation, a recent development in computer science, has resulted in programs that can antidifferentiate many functions. These include languages like MACSYMA and SETL, made for symbolic manipulation, and software packages, like muMATH, for microcomputers. In time these programs will be more powerful and more available, and will take much of the work out of evaluating integrals. Eventually you will not need integral tables any more than you now need logarithm tables—they also will be built into small computers.

One good feature of finding integrals is that you can easily check your work. You start off with a function $f(x)$. You do manipulations, guesswork, and table searching, ending up with a function $F(x)$ that is supposed to be an antiderivative of $f(x)$. To check, you simply differentiate $F(x)$ to see whether you indeed get $f(x)$.

Table 6-1-1
Essential derivative formulas

$(x^p)' = px^{p-1}$

$(\sin x)' = \cos x$
$(\cos x)' = -\sin x$
$(\tan x)' = \sec^2 x$

$(\sinh x)' = \cosh x$
$(\cosh x)' = \sinh x$
$(\tanh x)' = \operatorname{sech}^2 x$

$(\arcsin x)' = \dfrac{1}{\sqrt{1 - x^2}}$

$(\arctan x)' = \dfrac{1}{1 + x^2}$

$(e^x)' = e^x \qquad (\ln|x|)' = 1/x$

Table 6-1-2
Derivative formulas sometimes used

$(\sec x)' = \sec x \tan x$
$(\csc x)' = -\cot x \csc x$
$(\cot x)' = -\csc^2 x$

Table 6-1-3
Derivative formulas seldom used

$(\operatorname{sech} x)' = -\operatorname{sech} x \tanh x$
$(\operatorname{csch} x)' = -\operatorname{csch} x \coth x$
$(\coth x)' = -\operatorname{csch}^2 x$

$$(\sinh^{-1} x)' = [\ln(x + \sqrt{x^2 + 1})]' = \frac{1}{\sqrt{x^2 + 1}}$$

$$(\cosh^{-1} x)' = [\ln(x + \sqrt{x^2 - 1})]' = \frac{1}{\sqrt{x^2 - 1}} \qquad (x > 1)$$

$$(\tanh^{-1} x)' = \left(\tfrac{1}{2}\ln\frac{1 + x}{1 - x}\right)' = \frac{1}{1 - x^2} \qquad (|x| < 1)$$

What to Memorize

Table 6-1-1 is a list of derivative formulas that are an absolute must; you must know these forwards and backwards. Table 6-1-2 is a secondary list: formulas very good to know, but not as absolutely essential as the formulas in Table 6-1-1. Finally, Table 6-1-3 includes other derivative formulas we derived earlier, but which hardly anyone remembers. We can forget about them for now, but they will reappear later as part of our work.

Throughout this chapter and many places in this text where integrals are evaluated, formulas such as

$$\int \sqrt{a^2 - x^2}\, dx = \tfrac{1}{2}x\sqrt{a^2 - x^2} + \tfrac{1}{2}a^2 \arcsin x/a + C$$

$$\int \sin^2 x\, dx = \tfrac{1}{2}x - \tfrac{1}{4}\sin 2x + C$$

will occur, and possibly you may memorize some of them temporarily. But probably you will forget them soon afterwards because they are complicated and because they are less fundamental than the basic building blocks in Table 6-1-1. You must know how to find such formulas, and how to apply them to integrals that are almost, but not quite, the same, like

$$\int \sqrt{a^2 - b^2x^2} \quad \text{and} \quad \int \sin^2 3x\, dx$$

Constant of Integration

Each integration formula we derive has the form

$$\int f(x)\, dx = F(x) + C$$

The right side means *all* functions with derivative $f(x)$. The symbol C represents an arbitrary constant. Sometimes it is convenient to lump together several such constants of integration. For instance, from

$$\int e^x\, dx = e^x + C_1 \quad \text{and} \quad \int \cos x\, dx = \sin x + C_2$$

we deduce that

$$\int (e^x - 3\cos x)\, dx = e^x - 3\sin x + C$$

The constant of integration may conceal different forms of an answer. For instance

$$\frac{d}{dx}\sin^2 x = 2\sin x\cos x \quad \text{and} \quad \frac{d}{dx}(-\cos^2 x) = 2\sin x \cos x$$

so we can write either

$$\int 2\sin x\cos x\, dx = \sin^2 x + C$$

or

$$\int 2 \sin x \cos x \, dx = -\cos^2 x + C$$

Because of the identity $-\cos^2 x = \sin^2 x - 1$, these are different forms of the same answer. If you work an integration problem in two different ways and get two seemingly different answers, both answers may be correct! That is because some identity may imply that the two answers differ only by a constant.

Finally, note that integral tables usually omit the constant of integration altogether; the user is supposed to supply it.

Organization of the Chapter

In the next section, we introduce the concept of *differentials,* a tool of great power and simplicity. It is also a topic of importance apart from its application to integration.

The *change-of-variable formula* is one of the two most important techniques for indefinite integration. We take it up in Sections 6-2 and 6-3. The other of the most important techniques is *integration by parts,* which we cover in Section 6-4. Integration by parts, like the change-of-variable formula, is important in its own right and has applications beyond antidifferentiation.

Sections 6-5 to 6-7 are devoted to various techniques, the bag of tricks, in other words. Section 6-8 is important because it teaches the use of integral tables. Finally, Section 6-9 covers iterative methods, a most interesting and useful subject.

6-2 Differentials and Change of Variable

Figure 6-2-1
Differentials

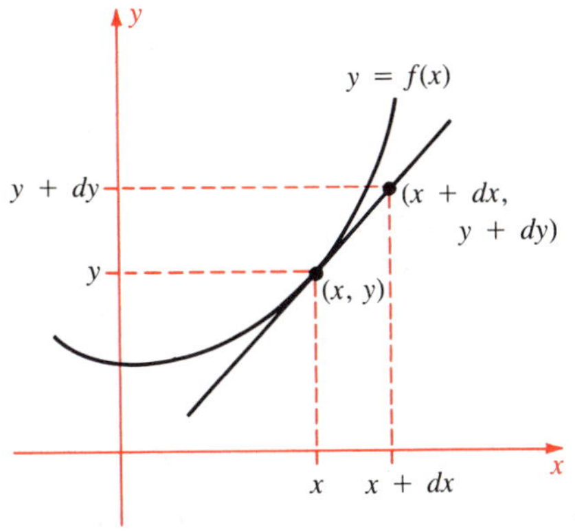

We start with a differentiable function $y = f(x)$ and a point (x, y) of its graph (Figure 6-2-1). We let $(x + dx, y + dy)$ be any point on the tangent to the graph at (x, y). Here for the first time, the quantities dx and dy are real numbers, not the result of operations on x and y. Since the slope of the tangent line is $f'(x)$, the relation between dy and dx is

$$dy = f'(x)\, dx$$

Thus for a given x, we think of dx as an independent variable and dy as a dependent variable that depends linearly on dx. Of course dy really depends both on dx *and* on x because $f'(x)$ itself depends on x. For instance, if $y = f(x) = x^3$, then $dy = 3x^2\, dx$.

We can build such variables from any differentiable function. For instance

$$x = \sin \theta \quad \text{and} \quad dx = \cos \theta \, d\theta$$

$$z = e^t \quad \text{and} \quad dz = e^t\, dt = z\, dt$$

In the particular case $y = x$ we have $dy = dx$.

Variables such as dx, dy, and dt are called **differentials.** They are precisely the expressions that appear under the integral sign. In fact, when we write an integral like

$$\int g(x)\,dx$$

we are posing a problem: Find $y = f(x)$ so that $dy = g(x)\,dx$, that is, so that $f'(x) = g(x)$.

Differentials have an important inner consistency that is a consequence of the chain rule. For example, suppose

$$w = \sqrt{1 - v^2} \quad \text{and} \quad v = \sin u$$

Then

$$dw = \frac{-v\,dv}{\sqrt{1 - v^2}} \quad \text{and} \quad dv = \cos u\,du$$

But we can substitute the expression for v into that for w, so that w is now a function of u, and then compute dw:

$$w = \sqrt{1 - \sin^2 u} = \cos u \qquad \text{hence} \qquad dw = -\sin u\,du$$

We have reached *two* apparently different dw's. They are actually the same. That is the consistency:

$$\begin{aligned} dw &= \frac{-v\,dv}{\sqrt{1 - v^2}} = \frac{-(\sin u)(\cos u\,du)}{\sqrt{1 - \sin^2 u}} \\ &= \frac{-\sin u \cos u\,du}{\cos u} = -\sin u\,du \end{aligned}$$

In general, suppose

$$w = G(v) \quad \text{and} \quad v = F(u)$$

On the one hand we have

$$dw = \frac{dG}{dv}\,dv \quad \text{and} \quad dv = \frac{dF}{du}\,du$$

(The fractional expressions are derivatives in the usual notation, *not* quotients of differentials.) By substituting the expression for dv into the expression for dw, we obtain

$$dw = \frac{dG}{dv} \cdot \frac{dF}{du}\,du$$

On the other hand, $w = G[F(u)] = H(u)$, where H denotes the composite function $H = G \circ F$, so

$$dw = \frac{dH}{du}\,du$$

However, by the chain rule,

$$\frac{dH}{du} = \frac{dG}{dv} \cdot \frac{dF}{du}$$

so we obtain the same relation between dw and du by whichever method it is computed.

Change of Variable

Before we state a formal rule, let us look at one example.

Example 1

Find $\int \sin^3 x \cos x\, dx$

Solution Write the integrand in the form

$$(\sin^3 x)(\cos x\, dx)$$

Since $d(\sin x) = \cos x\, dx$, this suggests setting $u = \sin x$. Then $du = \cos x\, dx$. Hence

$$\int \sin^3 x \cos x\, dx = \int u^3\, du = \tfrac{1}{4}u^4 + C = \tfrac{1}{4}\sin^4 x + C$$

Check: $\dfrac{d}{dx}(\tfrac{1}{4}\sin^4 x) = (\sin^3 x)\dfrac{d}{dx}(\sin x) = \sin^3 x \cos x$ ●

Now we state the general rule.

Change of Variable (Substitution)

$$\int f[u(x)]\frac{du}{dx}\, dx = \int f(u)\, du \quad \text{where} \quad u = u(x)$$

Suppose $F(u)$ is an antiderivative of $f(u)$. Then in addition

$$\int f[u(x)]\frac{du}{dx}\, dx = F[u(x)] + C$$

Use of this formula requires spotting an integrand in the form $f[u(x)]u'(x)$, a skill that comes with practice.

Example 2

Find **a** $\int \exp(x^2)\, x\, dx$ **b** $\int (e^x + x)^2(e^x + 1)\, dx$

c $\int \cos(\sqrt{x})\frac{1}{\sqrt{x}}\, dx$

Solution

a The function $\exp(x^2)$ can be simplified by setting $u = x^2$. Then $du = 2x\,dx$, hence

$$\int \exp(x^2)x\,dx = \int e^u \cdot \tfrac{1}{2}\,du = \tfrac{1}{2}\int e^u\,du$$
$$= \tfrac{1}{2}e^u + C = \tfrac{1}{2}\exp(x^2) + C$$

Note that the final *answer* must be given in terms of x, not u, because the *problem* is given in terms of x.

b In this problem the function $e^x + x$ is the villain. But $e^x + 1$ is its derivative, which suggests setting $u = e^x + x$. Then the differential is $du = (e^x + 1)\,dx$. Hence

$$\int (e^x + x)^2(e^x + 1)\,dx = \int u^2\,du = \tfrac{1}{3}u^3 + C$$
$$= \tfrac{1}{3}(e^x + x)^3 + C$$

c Clearly we should try $u = \sqrt{x}$. Then $du = \tfrac{1}{2}\,dx/\sqrt{x}$ so that we have $dx/\sqrt{x} = 2\,du$, and

$$\int \cos(\sqrt{x})\frac{1}{\sqrt{x}}\,dx = \int (\cos u)(2\,du) = 2\sin u + C$$
$$= 2\sin(\sqrt{x}) + C$$

●

Example 3

Find **a** $\displaystyle\int \frac{\arctan x}{1 + x^2}\,dx$ **b** $\displaystyle\int \frac{\ln x}{x}\,dx$

Solution

a Set $u = \arctan x$. Then $du = \dfrac{dx}{1 + x^2}$, so

$$\int \frac{\arctan x}{1 + x^2}\,dx = \int u\,du = \tfrac{1}{2}u^2 + C = \tfrac{1}{2}(\arctan x)^2 + C$$

b Set $u = \ln x$. Then $du = dx/x$, so

$$\int \frac{\ln x}{x}\,dx = \int u\,du = \tfrac{1}{2}u^2 + C = \tfrac{1}{2}(\ln x)^2 + C$$

●

Example 4

Find **a** $\displaystyle\int \frac{dx}{\sqrt{3x + 5}}$ **b** $\displaystyle\int \frac{4x - 5}{x^2 + 1}\,dx$

Solution

a Set $u = 3x + 5$. Then $du = 3\,dx$ so that $dx = \tfrac{1}{3}\,du$. Hence

$$\int \frac{dx}{\sqrt{3x+5}} = \int \frac{\frac{1}{3}\,du}{\sqrt{u}} = \frac{1}{3}\int \frac{du}{\sqrt{u}} = \frac{1}{3}\int u^{-1/2}\,du$$
$$= \tfrac{1}{3}(2u^{1/2}) + C = \tfrac{2}{3}\sqrt{3x+5} + C$$

Alternative Solution To eliminate the radical, set $u^2 = 3x + 5$. Then $2u\,du = 3\,dx$ so that $dx = \frac{2}{3}u\,du$ and

$$\int \frac{dx}{\sqrt{3x+5}} = \frac{2}{3}\int \frac{u\,du}{u} = \tfrac{2}{3}u + C = \tfrac{2}{3}\sqrt{3x+5} + C$$

b Use a little algebra to split the integral into two parts:

$$\int \frac{4x-5}{x^2+1}\,dx = 4\int \frac{x\,dx}{x^2+1} - 5\int \frac{dx}{x^2+1}$$

In the first integral on the right, the numerator is nearly the differential of the denominator. Set $u = x^2 + 1$, so $du = 2x\,dx$. Then throw in the needed factor 2:

$$4\int \frac{x\,dx}{x^2+1} = 2\int \frac{2x\,dx}{x^2+1} = 2\int \frac{du}{u}$$
$$= 2\ln|u| + C_1 = 2\ln(x^2+1) + C_1$$
$$= \ln(x^2+1)^2 + C_1$$

The second integral is $\arctan x + C_2$. Therefore

$$\int \frac{4x-5}{x^2+1}\,dx = \ln(x^2+1)^2 - 5\arctan x + C$$

where the constants are lumped together: $C = C_1 - 5C_2$. ●

Definite Integrals

The method of substitution, or change of variables, applies to definite integrals as well as indefinite integrals. As we shall now show, the limits of integration must be changed the same way the variable is changed to get the correct value.

By the chain rule, if $F(u)$ is an antiderivative of $f(u)$, then $F[u(x)]$ is an antiderivative of $f[u(x)]u'(x)$. Therefore, by two applications of the fundamental theorem we have

$$\int_a^b f[u(x)]\frac{du}{dx}\,dx = F[u(x)]\Big|_{x=a}^{x=b} = F[u(b)] - F[u(a)]$$
$$= F(u)\Big|_{u=u(a)}^{u=u(b)} = \int_{u(a)}^{u(b)} f(u)\,du$$

Thus once the integral is expressed in terms of u, the computation can be done entirely in terms of u, *provided* the limits of the integrals are changed from $x = a$ and $x = b$ to $u = u(a)$ and $u = u(b)$. We have proved the following result.

Change of Variable (Definite Integrals)

$$\int_a^b f[u(x)]\frac{du}{dx}\,dx = \int_c^d f(u)\,du$$

where $c = u(a)$ and $d = u(b)$.

Example 5

Compute $\int_0^4 2\sqrt{x^2+9}\,x\,dx$

Solution (old way) First evaluate the indefinite integral

$$\int 2\sqrt{x^2+9}\,x\,dx$$

Make the substitution $u = x^2 + 9$, $du = 2x\,dx$:

$$\int 2\sqrt{x^2+9}\,x\,dx = \int u^{1/2}\,du = \tfrac{2}{3}u^{3/2} + C$$
$$= \tfrac{2}{3}(x^2+9)^{3/2} + C$$

Finally, put in the limits of integration:

$$\int_0^4 2\sqrt{x^2+9}\,x\,dx = \tfrac{2}{3}(x^2+9)^{3/2}\Big|_0^4 = \tfrac{2}{3}(5^3 - 3^3) = \tfrac{196}{3}$$

Solution (new way) Again substitute $u = u(x) = x^2 + 9$. Note that $u(4) = 25$ and $u(0) = 9$. Therefore

$$\int_0^4 2\sqrt{x^2+9}\,x\,dx = \int_9^{25} u^{1/2}\,du = \tfrac{2}{3}u^{3/2}\Big|_9^{25} = \tfrac{2}{3}(5^3 - 3^3) = \tfrac{196}{3}$$

Alternative Solution (avoiding fractional exponents) Set $u^2 = u(x)^2 = x^2 + 9$. Then $2u\,du = 2x\,dx$. Now $u(4) = 5$ and $u(0) = 3$, consequently

$$\int_0^4 2\sqrt{x^2+9}\,x\,dx = \int_3^5 (2u)(u\,du) = \tfrac{2}{3}u^3\Big|_3^5$$
$$= \tfrac{2}{3}(5^3 - 3^3) = \tfrac{196}{3}$$

Example 6

Evaluate $\int_{1/2}^{1/\sqrt{2}} \frac{\arcsin x}{\sqrt{1-x^2}}\,dx$

Solution Since the derivative of $\arcsin x$ is $1/\sqrt{1-x^2}$, it is reasonable to substitute

$$u = u(x) = \arcsin x \quad \text{so that} \quad du = \frac{dx}{\sqrt{1-x^2}}$$

Then $u(\frac{1}{2}) = \frac{1}{6}\pi$ and $u(1/\sqrt{2}) = \frac{1}{4}\pi$ so we have

$$\int_{1/2}^{1/\sqrt{2}} \frac{\text{arc sin } x}{\sqrt{1-x^2}}\,dx = \int_{\pi/6}^{\pi/4} u\,du = \tfrac{1}{2}u^2\Big|_{\pi/6}^{\pi/4}$$
$$= \tfrac{1}{2}(\tfrac{1}{16}\pi^2 - \tfrac{1}{36}\pi^2) = 5\pi^2/288$$ ●

Exercises

Find the indefinite integral. Make free use of Tables 6-1-1 and 6-1-2 on page 328

1 $\int \sin x \cos x\,dx$

2 $\int \sin^4 x \cos x\,dx$

3 $\int \cos^2 x \sin x\,dx$

4 $\int \exp(x^3)x^2\,dx$

5 $\int 5e^{5x}\,dx$

6 $\int (e^x + 3x)\times(e^x + 3)\,dx$

7 $\int \sin x \exp(\cos x)\,dx$

8 $\int \frac{\sin x}{\cos^2 x}\,dx$

9 $\int (1 + \sin x)^3 \cos x\,dx$

10 $\int \tan^3 x \sec^2 x\,dx$

11 $\int \cos 3x\,dx$

12 $\int \sec^4 x \tan x\,dx$

13 $\int x^2 \sec^2(x^3)\,dx$

14 $\int \frac{3x^2}{x^3 + 4}\,dx$

15 $\int \frac{e^x + 2x}{e^x + x^2 + 1}\,dx$

16 $\int \frac{\exp(\sqrt{x})}{\sqrt{x}}\,dx$

17 $\int \frac{x\,dx}{\sqrt{1 - x^2}}$

18 $\int \frac{dx}{\sqrt{1 + 5x}}$

19 $\int \frac{x\,dx}{(1 + x^2)^3}$

20 $\int 8x(1 + 4x^2)^5\,dx$

21 $\int (3x + 1)^4\,dx$

22 $\int x(x^2 + 5)^6\,dx$

23 $\int \frac{e^x\,dx}{1 + e^{2x}}$

24 $\int x^2 \exp(-x^3)\,dx$

25 $\int \frac{dx}{(5 - 3x)^2}$

26 $\int \frac{4x - 3}{1 + x^2}\,dx$

27 $\int \frac{e^x - e^{-x}}{e^x + e^{-x}}\,dx$

28 $\int \frac{\ln(2x + 7)}{2x + 7}\,dx$

29 $\int \frac{x^3\,dx}{1 + x^4}$

30 $\int \frac{x\,dx}{1 + x^4}$

31 $\int \frac{dx}{x \ln x}$

32 $\int (5x - 7)^{10}\,dx$

33 $\int x\sqrt{5x^2 + 3}\,dx$

34 $\int (4x^3 + 1)\times\sqrt{x^4 + x}\,dx$

35 $\int \tan^2\tfrac{1}{3}x \sec^2\tfrac{1}{3}x\,dx$

36 $\int \frac{\sin x}{\cos x}\,dx$

37 $\int e^x \exp(e^x)\,dx$

38 $\int \exp(e^x + x)\,dx$

39 $\int \frac{x^3 - x}{x^4 - 2x^2}\,dx$

40 $\int \frac{dx}{x(\ln x)^{5/2}}$

Compute the definite integral by making an approximate substitution and changing the limits of integration

41 $\int_0^1 x(x^2 + 1)^7\,dx$

42 $\int_0^1 x(x^2 - 2)^3\,dx$

43 $\int_1^3 x \exp(x^2)\,dx$

44 $\int_1^2 \frac{x^2}{(x^3 + 1)^4}\,dx$

45 $\int_0^{\sqrt{\pi}} x \sin x^2\,dx$

46 $\int_{\pi/6}^{\pi/4} \frac{\cos x}{\sin^2 x}\,dx$

47 $\int_0^{\pi/4} (1 + \tan x)^3\times\sec^2 x\,dx$

48 $\int_0^1 \frac{\text{arc tan } x}{1 + x^2}\,dx$

49 $\int_0^{\pi/4} \cos^3 x \sin x\,dx$

50 $\int_0^1 \frac{e^x\,dx}{(e^x + 1)^2}$

51 $\int_0^4 \frac{x\,dx}{\sqrt{x^2 + 9}}$

52 $\int_{-1}^1 (3x - 2)^6\,dx$

53 $\int_{-1}^0 \frac{x}{(x^2 - 4)^3}\,dx$

54 $\int_0^1 \frac{4x^3 + 2x}{(x^4 + x^2 + 1)^3}\,dx$

55 $\int_1^2 \frac{(\ln x)^3}{x}\,dx$

56 $\int_{\pi/6}^{\pi/2} \frac{\cos x \ln \sin x}{\sin x}\,dx$

57 $\int_0^1 \frac{x^2\,dx}{(x^3 + 1)^3}$

58 $\int_{-1}^1 (x^5 + 5x + 1)^2\times(x^4 + 1)\,dx$

59 $\int_0^{\pi/3} \sec^2\theta \tan^5\theta\,d\theta$

60 $\int_{-2}^2 x(x^2 + 1)^{7/2}\,dx$

A particle moving on the x-axis has velocity $v = t/(1 + t^2)$ m/sec

61 How far does it go from time $t = 0$ to $t = 10$ sec?

62 How many more seconds will elapse before it goes twice as far?

6-3 Examples of Substitutions

Frequently an integral can be simplified by an appropriate substitution.

Example 1

Find $\displaystyle\int x\sqrt{x+1}\,dx$

Solution The radical $\sqrt{x+1}$ is a problem. To eliminate it, set $u^2 = x+1$, so that $\sqrt{x+1} = u$. Now express x and dx in terms of u and du:

$$x = u^2 - 1 \quad \text{and} \quad dx = 2u\,du$$

Therefore

$$\int x\sqrt{x+1}\,dx = \int (u^2-1)(u)(2u\,du) = 2\int (u^4 - u^2)\,du$$
$$= 2(\tfrac{1}{5}u^5 - \tfrac{1}{3}u^3) + C = \tfrac{2}{15}u^3(3u^2-5) + C$$

Finally substitute $u = \sqrt{x+1}$ to express the answer in terms of x:

$$\int x\sqrt{x+1}\,dx = \tfrac{2}{15}(x+1)^{3/2}[3(x+1)-5] + C$$
$$= \tfrac{2}{15}(x+1)^{3/2}(3x-2) + C$$

●

Example 2

Find $\displaystyle\int \frac{x\,dx}{(x-a)^3}$

Solution Complicated expressions in the denominator are usually harder to handle than in the numerator. This suggests that we try $u = x - a$. Then $x = u + a$ and $dx = du$. Hence

$$\int \frac{x\,dx}{(x-a)^3} = \int \frac{(u+a)\,du}{u^3} = \int \left(\frac{1}{u^2} + \frac{a}{u^3}\right) du$$
$$= -\frac{1}{u} - \frac{a}{2u^2} + C$$
$$= -\frac{1}{(x-a)} - \frac{a}{2(x-a)^2} + C$$
$$= \frac{-2x+a}{2(x-a)^2} + C$$

●

Example 3

Find $\displaystyle\int \frac{dx}{1+\sqrt[3]{x}}$

Solution The cube root is awkward, so we set $x = u^3$ in order to eliminate it. Then $dx = 3u^2\,du$ and

$$\int \frac{dx}{1 + \sqrt[3]{x}} = 3\int \frac{u^2\,du}{1 + u}$$

The fraction $u^2/(1 + u)$ is top-heavy; we always carry out a long division in such situations so that the resulting remainder is bottom-heavy (higher degree denominator than numerator):

$$\frac{u^2}{1 + u} = u - 1 + \frac{1}{1 + u}$$

Now we integrate:

$$\begin{aligned}\int \frac{dx}{1 + \sqrt[3]{x}} &= 3\int \left(u - 1 + \frac{1}{1 + u}\right) du \\ &= 3(\tfrac{1}{2}u^2 - u + \ln|1 + u|) + C \\ &= 3(\tfrac{1}{2}x^{2/3} - x^{1/3} + \ln|1 + x^{1/3}|) + C \\ &= \tfrac{3}{2}x^{2/3} - 3x^{1/3} + \ln|1 + x^{1/3}|^3 + C\end{aligned}$$

●

Often a suitable substitution can change an integral into one that is already known.

Example 4

Find **a** $\displaystyle\int \frac{dx}{a^2 + x^2}$ **b** $\displaystyle\int \frac{dx}{\sqrt{a^2 - x^2}}$

where $a > 0$.

Solution

a We already have the basic formula

$$\int \frac{dx}{1 + x^2} = \arctan x + C$$

from Table 6-1-1. Since this integral is so like the one in our problem, we try to change that integral into the known form. To do so, we set $x = au$ so that a^2 can be factored out of the denominator, leaving the desired form $1 + u^2$. Then $dx = a\,du$ and

$$\begin{aligned}\int \frac{dx}{a^2 + x^2} = \int \frac{a\,du}{a^2 + a^2u^2} &= \frac{1}{a}\int \frac{du}{1 + u^2} \\ &= \frac{1}{a}\arctan u + C \\ &= \frac{1}{a}\arctan \frac{x}{a} + C\end{aligned}$$

b This integral recalls the basic formula

$$\int \frac{dx}{\sqrt{1-x^2}} = \text{arc sin } x + C$$

Again we set $x = au$ so that $a^2 - x^2$ reduces to a^2 times $1 - u^2$:

$$\int \frac{dx}{\sqrt{a^2-x^2}} = \int \frac{a\,du}{\sqrt{a^2-a^2u^2}} = \int \frac{du}{\sqrt{1-u^2}}$$

$$= \text{arc sin } u + C = \text{arc sin } \frac{x}{a} + C$$

●

Example 5

Find **a** $\displaystyle\int \frac{dx}{\sqrt{x^2+a^2}}$ **b** $\displaystyle\int \frac{dx}{\sqrt{x^2-a^2}}$

where $a > 0$.

Solution

a We already have the formula

$$\int \frac{dx}{\sqrt{x^2+1}} = \ln(x + \sqrt{x^2+1}) + C$$

from Table 6-1-3. To deal with a^2, we again set $x = au$. Then

$$\begin{aligned}
\int \frac{dx}{\sqrt{x^2+a^2}} &= \int \frac{a\,du}{\sqrt{a^2u^2+a^2}} \\
&= \int \frac{du}{\sqrt{u^2+1}} \\
&= \ln(u + \sqrt{u^2+1}) + C_1 \\
&= \ln\left[\frac{x}{a} + \sqrt{\left(\frac{x}{a}\right)^2 + 1}\right] + C_1 \\
&= \ln\left[\frac{1}{a}(x + \sqrt{x^2+a^2})\right] + C_1 \\
&= \ln(x + \sqrt{x^2+a^2}) - \ln a + C_1 \\
&= \ln(x + \sqrt{x^2+a^2}) + C
\end{aligned}$$

Note that $-\ln a$ is absorbed into the final constant of integration.

b The substitution $x = au$ reduces this integral also to one of the formulas in Table 6-1-3, page 328. We obtain

$$\int \frac{dx}{\sqrt{x^2-a^2}} = \ln|x + \sqrt{x^2-a^2}| + C \qquad (|x| > |a|)$$

This result can be checked by differentiation.

●

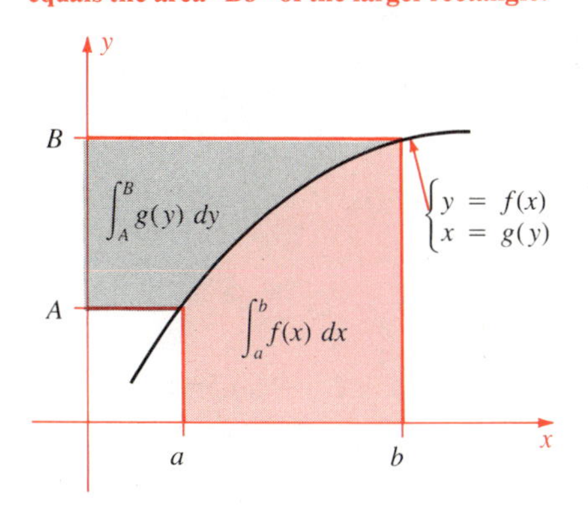

Figure 6-3-1
The sum of the two shaded areas plus the area Aa of the smaller rectangle equals the area Bb of the larger rectangle.

There is a useful relation between the integral of a function and the integral of its inverse function.

Integrals of Inverse Functions

Let $f(x)$ be a strictly increasing continuous function with domain $[a, b]$. Set $A = f(a)$ and $B = f(b)$, and let $g(y)$ be the inverse function of $f(x)$ so that the domain of $g(y)$ is $[A, B]$. Then

$$\int_a^b f(x)\,dx + \int_A^B g(y)\,dy = Bb - Aa$$

The proof is entirely geometrical (Figure 6-3-1). We illustrate it in the simplest case, $0 < a < b$ and $0 < A < B$. Other cases can be reduced to this one by shifting.

Example 6

Show that $\displaystyle\int_1^b \ln x\,dx = b\ln b - b + 1.$

Solution Apply the formula with $f(x) = \ln x$, $g(y) = e^y$, $a = 1$, $A = 0$, and $B = \ln b$:

$$\int_1^b \ln x\,dx + \int_0^{\ln b} e^y\,dy = b\ln b$$

But $\displaystyle\int_0^{\ln b} e^y\,dy = e^y\Big|_0^{\ln b} = b - 1,$ so the result follows.

Exercises

Find the indefinite integral

1. $\int x\sqrt{x+3}\,dx$
2. $\int (\sin x)e^{\cos x}\,dx$
3. $\int \frac{x}{\sqrt{2x+5}}\,dx$
4. $\int \frac{dx}{1+b^2x^2}$
5. $\int \frac{x^2\,dx}{(x-1)^3}$
6. $\int \frac{dx}{\sqrt{1-4x^2}}$
7. $\int \frac{dx}{1+\sqrt{x}}$
8. $\int e^{2x}\sqrt{1+e^x}\,dx$
9. $\int \frac{dx}{x+\sqrt[4]{x}}$
10. $\int \frac{(x-1)\,dx}{\sqrt{1-x^2}}$
11. $\int \frac{e^{2x}}{1+e^{4x}}\,dx$
12. $\int x^3\sqrt{x^2+1}\,dx$
13. $\int \frac{dx}{1+(5x+2)^2}$
14. $\int \frac{x+1}{a^2+b^2x^2}\,dx$
15. $\int \frac{x^3\,dx}{\sqrt{1-x^2}}$
16. $\int \frac{x}{1+\sqrt{x}}\,dx$
17. $\int \frac{\sin 2x\,dx}{3+\cos 2x}$
18. $\int (x^2+x+1)\times \sqrt{x+1}\,dx$
19. $\int x\sqrt[3]{2x+1}\,dx$
20. $\int \frac{x^3-5}{(x+2)^2}\,dx$
21. $\int \frac{x^3}{x^2+1}\,dx$
22. $\int x^2\sqrt{x+3}\,dx$
23. $\int \frac{dx}{\sqrt{9x^2+1}}$
*24. $\int \frac{\sqrt{x}+1}{x+1}\,dx$

Compute the definite integral by making an appropriate substitution and changing the limits of integration

25 $\int_0^2 x^3\sqrt{x^4+9}\,dx$

26 $\int_1^2 (x-1)^3 \times (x+2)\,dx$

27 $\int_4^5 \frac{x}{(x-2)^3}\,dx$

28 $\int_0^5 x\sqrt{x+4}\,dx$

29 $\int_0^1 x(2x-1)^5\,dx$

30 $\int_{-\ln 2}^{-(\ln 2)/2} \frac{e^x\,dx}{\sqrt{1-e^{2x}}}$

31 $\int_0^3 \frac{dx}{1+\sqrt{1+x}}$

32 $\int_0^{12} \frac{x^2}{\sqrt{2x+1}}\,dx$

33 $\int_0^1 \frac{x^3\,dx}{\sqrt{4-x^2}}$

34 $\int_0^2 \frac{dx}{(x+2)\sqrt{x+1}}$

35 $\int_0^4 (x-1)(x-2) \times (x-3)\,dx$

36 $\int_0^1 [x^n-(1-x)^n]\,dx$

37 Prove $\int_a^b f(c+x)\,dx = \int_{c+a}^{c+b} f(x)\,dx.$

38 Prove $\int_a^b f(-x)\,dx = \int_{-b}^{-a} f(x)\,dx.$

39 Suppose $0 < a < b$ and $c > 0$. Use the change of variable formula to prove

$$\int_{ca}^{cb} \frac{dt}{t} = \int_a^b \frac{dt}{t}$$

Explain this result in terms of logarithms.

40 Find $\int_x^1 \text{arc}\cos t\,dt.$ [Hint Draw a figure.]

6-4 Integration by Parts

This important technique converts one integration problem into a different one, hopefully an easier one. Integration by parts comes from the product rule:

$$\frac{d}{dx}(uv) = u\frac{dv}{dx} + v\frac{du}{dx}$$

This rule can be expressed in terms of differentials. We multiply by the differential dx:

$$\frac{d}{dx}(uv)\,dx = u\frac{dv}{dx}\,dx + v\frac{du}{dx}\,dx$$

The term on the left is $d(uv)$. On the right we have

$u\,dv$ and $v\,du$

We have obtained the following:

Product Rule in Differential Form

$$d(uv) = u\,dv + v\,du$$

Rearranging the terms, we have $u\,dv = d(uv) - v\,du$ and consequently

$$\int u\,dv = \int d(uv) - \int v\,du$$

But $\int d(uv) = uv + C$, so we have the following formula:

Integration by Parts (Indefinite Integrals)

$$\int u\,dv = uv - \int v\,du$$

(The constant of integration is absorbed into the second integral.)

This formula converts the problem of integrating $u\,dv$ into that of integrating $v\,du$. In many cases, $v\,du$ is easy to integrate. The trick is to choose u and v appropriately. As we shall see in examples, it is often straightforward to do so.

Example 1

Find **a** $\int xe^x\,dx$ **b** $\int x\cos x\,dx$

Solution

a Write $xe^x\,dx = u\,dv$. There are two obvious choices: $u = x$ and $dv = e^x\,dx$, or $u = e^x$ and $dv = x\,dx$. Try the first:

$$\int xe^x\,dx = \int u\,dv \quad \text{where} \quad u = x \quad \text{and} \quad dv = e^x\,dx$$

Then $du = dx$ and one choice of v is $v = e^x$. By the formula for integration by parts,

$$\int xe^x\,dx = uv - \int v\,du = xe^x - \int e^x\,dx$$

$$= xe^x - e^x + C$$

The choice $u = x$ and $dv = e^x\,dx$ succeeds because $u' = 1$ is *less* complicated than u whereas $v = e^x$ is *no more* complicated than the e^x in $dv = e^x\,dx$.

We also try the other possibility:

$$\int xe^x\,dx = \int u\,dv \quad \text{where} \quad u = e^x \quad \text{and} \quad dv = x\,dx$$

Then $du = e^x\,dx$ and we may take $v = \frac{1}{2}x^2$:

$$\int xe^x\,dx = uv - \int v\,du = \tfrac{1}{2}x^2e^x - \tfrac{1}{2}\int x^2e^x\,dx$$

This is unsatisfactory since we have "reduced" the given integral to a more complicated integral. It fails because $u' = e^x$ is just as complicated as $u = e^x$ whereas $v = \frac{1}{2}x^2$ is *more* complicated than the function x in $dv = x\,dx$.

b Interpret the integral as

$$\int x\,d(\sin x)$$

Apply the formula for integration by parts with

$$u = x \qquad dv = \cos x\,dx$$
$$du = dx \qquad v = \sin x$$

to obtain

$$\int x\cos x\,dx = uv - \int v\,du = x\sin x - \int \sin x\,dx$$
$$= x\sin x + \cos x + C$$

The following example requires two techniques, integration by parts and change of variables.

Example 2

Find **a** $\int \text{arc sin } x\,dx$ **b** $\int x^3\exp(x^2)\,dx$

Solution

a The whole problem here is arc sin x. Because its derivative is algebraic rather than trigonometric, our first guess is to try $u = \text{arc sin } x$ and $dv = dx$:

$$u = \text{arc sin } x \qquad dv = dx$$
$$du = \frac{dx}{\sqrt{1-x^2}} \qquad v = x$$

Then

$$\int \text{arc sin } x\,dx = uv - \int v\,du = x\,\text{arc sin } x - \int \frac{x\,dx}{\sqrt{1-x^2}}$$

There are several ways to do the integral on the right. For example, substitute $t^2 = 1 - x^2$ so $t\,dt = -x\,dx$:

$$\int \frac{x\,dx}{\sqrt{1-x^2}} = -\int \frac{t\,dt}{t} = -\int dt = -t + C_1$$
$$= -\sqrt{1-x^2} + C_1$$

Therefore

$$\int \text{arc sin } x\,dx = x\,\text{arc sin } x + \sqrt{1-x^2} + C$$

b Because $\exp(y)$ is a simpler function than $\exp(x^2)$, we first try $y = x^2$ so that $dy = 2x\,dx$. Then

$$\int x^3\exp(x^2)\,dx = \int x^2\exp(x^2)(x\,dx) = \int ye^y(\tfrac{1}{2}\,dy)$$

Now we can integrate by parts as in Example 1a to obtain

$$\int ye^y\,dy = ye^y - \int e^y\,dy = ye^y - e^y + C_1 = (y-1)e^y + C_1$$

Therefore

$$\int x^3 \exp(x^2)\,dx = \tfrac{1}{2}(x^2-1)\exp(x^2) + C$$

Note The variable t was used in the substitution step in Example 2a because u was used in the first step. Always take care not to confuse variables.

By putting in limits of integration, we can express the formula for integration by parts in terms of definite integrals:

Integration by Parts (Definite Integrals)

If $u(x)$, $v(x)$, $u'(x)$, and $v'(x)$ are continuous on $[a, b]$, then

$$\int_a^b u(x)\frac{dv}{dx}\,dx = u(x)v(x)\Big|_a^b - \int_a^b v(x)\frac{du}{dx}\,dx$$

Example 3

Compute **a** $\displaystyle\int_0^\pi x\sin x\,dx$ **b** $\displaystyle\int_1^2 \ln x\,dx$

Solution

a We take $u = x$ because $u' = 1$ is simpler than u. The rest of the integral is $dv = \sin x\,dx$:

$$u = x \qquad dv = \sin x\,dx$$
$$du = dx \qquad v = -\cos x$$

Therefore

$$\int_0^\pi x\sin x\,dx = uv\Big|_0^\pi - \int_0^\pi v\,du$$
$$= -x\cos x\Big|_0^\pi - \int_0^\pi (-\cos x)\,dx$$
$$= \pi + \int_0^\pi \cos x\,dx = \pi + \sin x\Big|_0^\pi = \pi$$

b The derivative of $\ln x$ is $1/x$, which is simpler than $\ln x$, so we try

$$u = \ln x \qquad dv = dx$$
$$du = \frac{dx}{x} \qquad v = x$$

Then

$$\int_1^2 \ln x\,dx = uv\Big|_1^2 - \int_1^2 v\,du = x\ln x\Big|_1^2 - \int_1^2 x\cdot\frac{dx}{x}$$

$$= (2\ln 2 - \ln 1) - x\Big|_1^2 = 2\ln 2 - 1$$

Repeated Integration by Parts

Some problems require two or more integrations by parts.

Example 4

Find $\int x(\ln x)^2\,dx$

Solution To simplify the integral, try

$$u = (\ln x)^2 \qquad dv = x\,dx$$

$$du = \frac{2\ln x}{x}\,dx \qquad v = \tfrac{1}{2}x^2$$

Then

$$\int x(\ln x)^2\,dx = \tfrac{1}{2}x^2(\ln x)^2 - \int \tfrac{1}{2}x^2\,\frac{2\ln x}{x}\,dx$$

$$= \tfrac{1}{2}x^2(\ln x)^2 - \int x\ln x\,dx$$

The problem now is to evaluate

$$\int x\ln x\,dx$$

which is simpler than the original integral because $\ln x$ appears only to the first power. Therefore another integration by parts should reduce the integral to

$$\int x\,dx$$

Try it. Integrate by parts again with

$$u = \ln x \qquad dv = x\,dx$$

$$du = \frac{dx}{x} \qquad v = \tfrac{1}{2}x^2$$

Then

$$\int x\ln x\,dx = \tfrac{1}{2}x^2\ln x - \int \tfrac{1}{2}x^2\,\frac{dx}{x} = \tfrac{1}{2}x^2\ln x - \tfrac{1}{2}\int x\,dx$$

$$= \tfrac{1}{2}x^2\ln x - \tfrac{1}{4}x^2 + C_1$$

Combine the results:

$$\int x(\ln x)^2\,dx = \tfrac{1}{2}x^2(\ln x)^2 - (\tfrac{1}{2}x^2 \ln x - \tfrac{1}{4}x^2 + C_1)$$
$$= \tfrac{1}{4}x^2[2(\ln x)^2 - 2\ln x + 1] + C$$

●

Example 5

Find $\int x^3 e^x\,dx$

Solution Repeated differentiation of x^3 leads to ever simpler functions, whereas repeated integration of e^x leads to the same function. Thus we integrate by parts three times, always with $dv = e^x\,dx$:

$$\int x^3 e^x\,dx = x^3 e^x - 3\int x^2 e^x\,dx$$
$$\int x^2 e^x\,dx = x^2 e^x - 2\int x e^x\,dx$$
$$\int x e^x\,dx = x e^x - \int e^x\,dx = x e^x - e^x + C_1$$

We combine the results:

$$\int x^3 e^x\,dx = x^3 e^x - 3x^2 e^x + 6x e^x - 6e^x + C$$
$$= e^x(x^3 - 3x^2 + 6x - 6) + C$$

●

Example 6

Find $\int e^x \cos x\,dx$

Solution The functions e^x and $\cos x$ both differentiate and integrate into equally complicated functions. We first try

$$u = e^x \qquad dv = \cos x\,dx$$
$$du = e^x\,dx \qquad v = \sin x$$

Then

$$\int e^x \cos x\,dx = e^x \sin x - \int e^x \sin x\,dx$$

Not much progress, so let us try again with the same u:

$$u = e^x \qquad dv = \sin x\,dx$$
$$du = e^x\,dx \qquad v = -\cos x$$

Then

$$\int e^x \sin x\,dx = -e^x \cos x + \int e^x \cos x\,dx$$

The integral on the right is what we started with! Have we gone in a circle? No, because substitution of this expression into the result of the first integration by parts yields

$$\int e^x \cos x\,dx = e^x \sin x + e^x \cos x - \int e^x \cos x\,dx$$

The minus sign on the right saves us from disaster. Solve:

$$\int e^x \cos x\,dx = \tfrac{1}{2}(e^x \sin x + e^x \cos x)$$

We have found one antiderivative; to find all, we add the constant of integration:

$$\int e^x \cos x\,dx = \tfrac{1}{2}e^x(\sin x + \cos x) + C$$

●

Exercises

Find

1 $\int x \sin 2x\,dx$

2 $\int x \cos 3x\,dx$

3 $\int xe^{2x}\,dx$

4 $\int x \sec^2 x\,dx$

5 $\int \sqrt{x} \ln x\,dx$

6 $\int \text{arc cos } x\,dx$

7 $\int \text{arc tan } x\,dx$

8 $\int \ln(x^2 + 1)\,dx$

9 $\int x \text{ arc tan } x\,dx$

10 $\int x^5 \exp(x^2)\,dx$

11 $\int e^{2x} \sin 3x\,dx$

12 $\int x^2 e^{-x}\,dx$

13 $\int x \cosh x\,dx$

14 $\int x^2 \sinh 3x\,dx$

15 $\int x^2 \cos ax\,dx$

16 $\int x^2 \ln x\,dx$

17 $\int (\ln x)^2\,dx$

18 $\int x^2(\ln x)^2\,dx$

19 $\int \left(\ln \ln x + \dfrac{1}{\ln x}\right) dx$

20 $\int \dfrac{\ln(\ln x)}{x}\,dx$

***21** $\int \sin(\ln x)\,dx$

***22** $\int (2 \sec^3 x - \sec x)\,dx$

Compute

23 $\int_{\pi}^{2\pi} x \cos x\,dx$

24 $\int_0^{\pi/4} x \sin 2x\,dx$

25 $\int_0^1 xe^{3x}\,dx$

26 $\int_{-1}^1 xe^{-2x}\,dx$

27 $\int_1^e (\ln x)^2\,dx$

28 $\int_1^2 x^2 \ln x\,dx$

29 $\int_0^{1/2} \text{arc sin } 2x\,dx$

30 $\int_0^{\pi/2} x^2 \sin x\,dx$

31 $\int_0^1 e^x(x + 3)^2\,dx$

32 $\int_0^{\pi/2} e^{2x} \sin x\,dx$

33 $\int_0^{\pi/3} x \sin^2 x \cos x\,dx$

***34** $\int_0^{\pi/2} x \sin^2 x\,dx$

35 Prove that

$$\int_{-1}^1 x^2 \exp(x^2)\,dx = e - \tfrac{1}{2}\int_{-1}^1 \exp(x^2)\,dx$$

36 (cont.) Find a relation between

$$\int_{-1}^1 x^4 \exp(x^2)\,dx \quad \text{and} \quad \int_{-1}^1 \exp(x^2)\,dx$$

7 Show that

$$\int_0^{2\pi} f(x)\cos x\,dx = -\int_0^{2\pi} f'(x)\sin x\,dx$$

8 If $f(a) = f(b) = 0$, show that

$$\int_a^b f(x)\,dx = -\tfrac{1}{2}\int_a^b (x-a)(x-b)f''(x)\,dx$$

9 Show that

$$\int_0^{2\pi} \cos x \cos 2x\,dx = 0$$

by integrating twice by parts.

40 Let $P(x)$ be a polynomial. Show that

$$\int P(x)e^x\,dx = [P(x) - P'(x) + P''(x) - P^{(3)}(x) + \cdots]e^x + C$$

***41** Prove $\displaystyle\int_0^x \frac{\sin t}{t}\,dt > 0$ for all $x > 0$.
[Hint Try $v = 1 - \cos t$. There is no real problem at $t = 0$.]

***42** Suppose $f(x)$ has a continuous derivative on $[0, 1]$. Prove

$$\lim_{x\to+\infty}\int_0^1 f(t)\sin(xt)\,dt = 0$$

6-5 Use of Algebra

When working integration problems, keep in mind the possibility of simplifying the integrand by algebraic manipulation. Such tactics as long division, factoring, combining fractions, and using trigonometric identities may convert an unfamiliar integrand into a combination of familiar ones.

Example 1

Find $\displaystyle\int \frac{x^4}{1+x^2}\,dx$

Solution The rational function is top-heavy, so, as usual, we use long division:

$$\frac{x^4}{1+x^2} = x^2 - 1 + \frac{1}{1+x^2}$$

The result is a polynomial $x^2 - 1$ plus the familiar integrand $1/(1+x^2)$, so we can integrate:

$$\int \frac{x^4}{1+x^2}\,dx = \int\left(x^2 - 1 + \frac{1}{1+x^2}\right)dx$$
$$= \tfrac{1}{3}x^3 - x + \arctan x + C$$

●

Example 2

Find $\displaystyle\int \sqrt{\frac{1+x}{1-x}}\,dx$

Solution We multiply numerator and denominator by $\sqrt{1+x}$:

$$\sqrt{\frac{1+x}{1-x}} = \sqrt{\frac{1+x}{1-x}}\cdot\frac{\sqrt{1+x}}{\sqrt{1+x}} = \frac{1+x}{\sqrt{1-x^2}}$$

We recognize the two summands and integrate:

$$\int \sqrt{\frac{1+x}{1-x}}\,dx = \int \frac{dx}{\sqrt{1-x^2}} + \int \frac{x\,dx}{\sqrt{1-x^2}}$$
$$= \arc\sin x - \sqrt{1-x^2} + C$$

●

Use of Trigonometric Identities

Integrals involving the direct trigonometric functions can often be evaluated by using appropriate relations between the functions. Facility with trigonometric functions is a great asset here.

Example 3

Find **a** $\int \cos^3 x\,dx$ **b** $\int \cos^3 x \sin^2 x\,dx$

Solution

a Save $\cos x\,dx = d(\sin x)$ and convert the rest of the integrand into powers of $\sin x$:

$$\cos^3 x\,dx = \cos^2 x(\cos x\,dx) = (1-\sin^2 x)(\cos x\,dx)$$

Now integrate:

$$\int \cos^3 x\,dx = \int \cos x\,dx - \int \sin^2 x \cos x\,dx$$
$$= \sin x - \tfrac{1}{3}\sin^3 x + C$$

b Same technique; save $\cos x\,dx$ and convert the rest to sines:

$$\cos^3 x \sin^2 x\,dx = \cos^2 x \sin^2 x(\cos x\,dx)$$
$$= (1-\sin^2 x)(\sin^2 x)(\cos x\,dx)$$

$$\int \cos^3 x \sin^2 x\,dx = \int \sin^2 x \cos x\,dx - \int \sin^4 x \cos x\,dx$$
$$= \tfrac{1}{3}\sin^3 x - \tfrac{1}{5}\sin^5 x + C$$

●

The method used in Example 3 can be applied to any problem of the form

$$\int \cos^{2m+1} x \sin^{2n} x\,dx \quad \text{or} \quad \int \cos^{2m} x \sin^{2n+1} x\,dx$$

In the first case, save $\cos x\,dx$ and convert the rest to sines; in the second case, save $\sin x\,dx$ and convert the rest to cosines. An integral of the form

$$\int \cos^{2m+1} x \sin^{2n+1} x\,dx$$

can be handled either way, for instance by saving $\cos x\,dx$ and converting $\cos^{2m}x$ to $(1 - \sin^2 x)^m$. Integrands of the form $\cos^{2m}x \sin^{2n}x$ require a different technique—use of double-angle formulas. Recall that

$$\cos 2x = \cos^2 x - \sin^2 x = 2\cos^2 x - 1 = 1 - 2\sin^2 x$$

Hence

$$\cos^2 x = \tfrac{1}{2}(1 + \cos 2x) \quad \text{and} \quad \sin^2 x = \tfrac{1}{2}(1 - \cos 2x)$$

Example 4

Find **a** $\int \sin^2 x\,dx$ **b** $\int \cos^2 x\,dx$ **c** $\int \sin^4 x\,dx$

Solution

a Use the double-angle formula $\sin^2 x = \frac{1}{2}(1 - \cos 2x)$:

$$\int \sin^2 x\,dx = \tfrac{1}{2}\int dx - \tfrac{1}{2}\int \cos 2x\,dx = \tfrac{1}{2}x - \tfrac{1}{4}\sin 2x + C$$

b Use the result of part **a** and $\cos^2 x = 1 - \sin^2 x$:

$$\begin{aligned}\int \cos^2 x\,dx &= \int (1 - \sin^2 x)\,dx \\ &= \int dx - \int \sin^2 x\,dx \\ &= x - (\tfrac{1}{2}x - \tfrac{1}{4}\sin 2x) + C = \tfrac{1}{2}x + \tfrac{1}{4}\sin 2x + C\end{aligned}$$

Alternatively, start with $\cos^2 x = \frac{1}{2}(1 + \cos 2x)$ and proceed as in part **a**.

c First express $\sin^4 x$ in terms of $\cos 2x$ by means of a double-angle formula:

$$\begin{aligned}\sin^4 x &= (\sin^2 x)^2 = [\tfrac{1}{2}(1 - \cos 2x)]^2 \\ &= \tfrac{1}{4}(1 - 2\cos 2x + \cos^2 2x)\end{aligned}$$

Now integrate:

$$\begin{aligned}\int \sin^4 x\,dx &= \tfrac{1}{4}\int dx - \tfrac{1}{2}\int \cos 2x\,dx + \tfrac{1}{4}\int \cos^2 2x\,dx \\ &= \tfrac{1}{4}x - \tfrac{1}{4}\sin 2x + \tfrac{1}{4}\int \cos^2 2x\,dx\end{aligned}$$

To find the remaining integral in this expression, use the double-angle formula $\cos^2 x = \frac{1}{2}(1 + \cos 2x)$ with x replaced by $2x$, namely, $\cos^2 2x = \frac{1}{2}(1 + \cos 4x)$:

$$\int \cos^2 2x\,dx = \tfrac{1}{2}\int dx + \tfrac{1}{2}\int \cos 4x\,dx = \tfrac{1}{2}x + \tfrac{1}{8}\sin 4x + C_1$$

Combine results:

$$\int \sin^4 x\, dx = (\tfrac{1}{4}x - \tfrac{1}{4}\sin 2x) + \tfrac{1}{4}(\tfrac{1}{2}x + \tfrac{1}{8}\sin 4x) + C$$

$$= \tfrac{3}{8}x - \tfrac{1}{4}\sin 2x + \tfrac{1}{32}\sin 4x + C$$

●

Example 5

Find **a** $\int \tan x\, dx$ **b** $\int \tan^2 x\, dx$ **c** $\int \tan^3 x\, dx$

Solution

a By the identity $\tan x = (\sin x)/(\cos x)$:

$$\int \tan x\, dx = \int \frac{\sin x}{\cos x}\, dx = \int \frac{-du}{u} = -\ln|u| + C$$

where $u = \cos x$. Hence

$$\int \tan x\, dx = -\ln|\cos x| + C = \ln|\sec x| + C$$

b Use the identity $\tan^2 x = \sec^2 x - 1$:

$$\int \tan^2 x\, dx = \int (\sec^2 x - 1)\, dx = \int \sec^2 x\, dx - \int dx$$

$$= \tan x - x + C$$

c $$\int \tan^3 x\, dx = \int \tan x(\sec^2 x - 1)\, dx$$

$$= \int \tan x \sec^2 x\, dx - \int \tan x\, dx$$

The first integral is of the form $\int u\, du$, where $u = \tan x$; the second integral was done in part **a**. Hence

$$\int \tan^3 x\, dx = \tfrac{1}{2}\tan^2 x + \ln|\cos x| + C$$

●

Integrals of $\sec x$ and $\csc x$

Look closely at two basic differentiation formulas that involve $\sec x$:

$$(\tan x)' = \sec^2 x \qquad (\sec x)' = \tan x \sec x$$

The ingenious idea of *adding* the formulas probably does not occur to everyone!

$$(\tan x + \sec x)' = \sec^2 x + \tan x \sec x$$

$$= (\tan x + \sec x)\sec x$$

Table 6-5-1
Trigonometric integrals (constants omitted)

$$\int \sin x\,dx = -\cos x$$
$$\int \cos x\,dx = \sin x$$
$$\int \tan x\,dx = -\ln|\cos x|$$
$$\int \cot x\,dx = \ln|\sin x|$$
$$\int \sec x\,dx = \ln|\sec x + \tan x|$$
$$\int \csc x\,dx = -\ln|\csc x + \cot x|$$

From this follows

$$\int \sec x\,dx = \int \frac{d(\tan x + \sec x)}{\tan x + \sec x} = \ln|\tan x + \sec x| + C$$

Similarly

$$\int \csc x\,dx = -\ln|\csc x + \cot x| + C$$

We can now list (Table 6-5-1) the integrals of the six basic trigonometric functions.

Completing the Square

In Examples 4 and 5 in Section 6-3, we derived the following formulas:

$$\int \frac{dx}{x^2 + a^2} = \frac{1}{a}\operatorname{arc\,tan}\frac{x}{a} + C \qquad (a > 0)$$

$$\int \frac{dx}{\sqrt{a^2 - x^2}} = \operatorname{arc\,sin}\frac{x}{a} + C \qquad (|x| < a \text{ and } a > 0)$$

$$\int \frac{dx}{\sqrt{x^2 + a^2}} = \ln(x + \sqrt{x^2 + a^2}) + C \qquad (a > 0)$$

$$\int \frac{dx}{\sqrt{x^2 - a^2}} = \ln|x + \sqrt{x^2 - a^2}| + C \qquad (|x| > a > 0)$$

One other integral of this type will be derived in Section 6-7:

$$\int \frac{dx}{a^2 - x^2} = \frac{1}{2a}\ln\left|\frac{a + x}{a - x}\right| + C \qquad (a > 0)$$

These are formulas that occur in all integral tables. It is important to know how to reduce related integrals to these. This can often be done if the integrand involves a quadratic polynomial or the square root of a quadratic polynomial. The method is our old friend completing the square.

Example 6

Find $\displaystyle\int \frac{dx}{x^2 - 10x + 29}$

Solution Complete the square:

$$x^2 - 10x + 29 = (x^2 - 10x + 25) + 4 = u^2 + a^2$$

where $u = x - 5$ and $a = 2$. Therefore $du = dx$ and

$$\int \frac{dx}{x^2 - 10x + 29} = \int \frac{du}{u^2 + a^2} = \frac{1}{a}\operatorname{arc\,tan}\frac{u}{a} + C$$

$$= \tfrac{1}{2}\operatorname{arc\,tan}\frac{x - 5}{2} + C$$

●

Example 7

Find $\int \frac{dx}{\sqrt{3 - x - x^2}}$

Solution Complete the square:

$$3 - x - x^2 = 3 - (x^2 + x) = 3 - (x^2 + x + \tfrac{1}{4}) + \tfrac{1}{4}$$
$$= \tfrac{13}{4} - (x + \tfrac{1}{2})^2 = a^2 - u^2$$

where $u = x + \frac{1}{2}$ and $a = \frac{1}{2}\sqrt{13}$. Therefore

$$\int \frac{dx}{\sqrt{3 - x - x^2}} = \int \frac{dx}{\sqrt{\frac{13}{4} - (x + \frac{1}{2})^2}} = \int \frac{du}{\sqrt{a^2 - u^2}}$$
$$= \text{arc sin } \frac{u}{a} + C = \text{arc sin}\left(\frac{2x + 1}{\sqrt{13}}\right) + C$$

●

Example 8

Find $\int \frac{dx}{\sqrt{5x^2 - 2x}}$

Solution Complete the square:

$$5x^2 - 2x = 5(x^2 - \tfrac{2}{5}x + \tfrac{1}{25} - \tfrac{1}{25}) = 5[(x - \tfrac{1}{5})^2 - \tfrac{1}{25}]$$
$$= 5(u^2 - a^2)$$

where $u = x - \frac{1}{5}$ and $a = \frac{1}{5}$. Therefore

$$\int \frac{dx}{\sqrt{5x^2 - 2x}}$$
$$= \int \frac{dx}{\sqrt{5}\sqrt{(x - \frac{1}{5})^2 - \frac{1}{25}}}$$
$$= \frac{1}{\sqrt{5}} \int \frac{du}{\sqrt{u^2 - a^2}}$$
$$= \frac{1}{\sqrt{5}} \ln|u + \sqrt{u^2 - a^2}| + C_1$$
$$= \frac{1}{\sqrt{5}} (\ln|5x - 1 + \sqrt{25x^2 - 10x}| - \ln 5) + C_1$$

Consequently

$$\int \frac{dx}{\sqrt{5x^2 - 2x}} = \frac{1}{\sqrt{5}} \ln|5x - 1 + \sqrt{25x^2 - 10x}| + C$$

●

Exercises

Compute

1 $\displaystyle\int \frac{1 + x^4}{9 + x^2}\,dx$

2 $\displaystyle\int \frac{x^2\,dx}{x^2 + 3}$

3 $\displaystyle\int \frac{2x + 1}{x - 4}\,dx$

4 $\displaystyle\int \frac{x^2}{x - 1}\,dx$

5 $\displaystyle\int \frac{(x + 1)^3}{x^2}\,dx$

6 $\displaystyle\int \frac{x^8 + 1}{x^2 + 1}\,dx$

7 $\displaystyle\int \sqrt{\frac{1 + ax}{1 - ax}}\,dx$

8 $\displaystyle\int \frac{dx}{\sqrt{x + 5} - \sqrt{x}}$

9 $\displaystyle\int \sec x \csc x\,dx$

10 $\displaystyle\int \cos x \csc x\,dx$

11 $\displaystyle\int \cos^3 x \sin^4 x\,dx$

12 $\displaystyle\int (\cos x - \sin x)^2\,dx$

13 $\displaystyle\int \sin^3 x \cos^2 x\,dx$

14 $\displaystyle\int \sin^3 ax\,dx$

15 $\displaystyle\int \cos^5 3x\,dx$

16 $\displaystyle\int \cot^3 x\,dx$

17 $\displaystyle\int \cos^4 x\,dx$

18 $\displaystyle\int \sin^2 ax \cos^2 ax\,dx$

19 $\displaystyle\int \tan^4 x\,dx$

20 $\displaystyle\int \sec^4 x\,dx$

21 $\displaystyle\int x \tan x^2\,dx$

22 $\displaystyle\int \tan^2 x \sec^4 x\,dx$

23 $\displaystyle\int \frac{dx}{1 - \sin x}$

24 $\displaystyle\int (\sin x) \times \left(\frac{\sec x + 1}{\sec x - 1}\right)^{1/2} dx$ $\quad (-\tfrac{1}{2}\pi < x < \tfrac{1}{2}\pi)$

Evaluate the definite integral

25 $\displaystyle\int_{3\pi/4}^{\pi} \tan x\,dx$

26 $\displaystyle\int_{-2}^{-1} \frac{dx}{\sqrt{4x^2 - 1}}$

27 $\displaystyle\int_0^{\pi/2} \sin 2x\,dx$

28 $\displaystyle\int_0^1 \cos^2 \pi x\,dx$

29 $\displaystyle\int_0^{2\pi} \sin x \cos 3x\,dx$

[Hint $\sin A \cos B = \frac{1}{2}(\sin ? - \sin ?)$]

30 $\displaystyle\int_0^{2\pi} \cos 3x \cos 4x\,dx$

[Hint $\cos A \cos B = \frac{1}{2}(\cos ? + \cos ?)$]

Compute

31 $\displaystyle\int \frac{dx}{x^2 + 2x + 5}$

32 $\displaystyle\int \frac{dx}{2x^2 + x + 6}$

33 $\displaystyle\int \frac{dx}{\sqrt{6x - x^2}}$

34 $\displaystyle\int \frac{3x + 10}{\sqrt{x^2 + 2x + 5}}\,dx$

35 $\displaystyle\int \frac{x\,dx}{\sqrt{4x - x^2}}$

36 $\displaystyle\int \frac{x^2\,dx}{x^2 - 4x + 9}$

37 $\displaystyle\int \frac{x\,dx}{\sqrt{3x^4 - 4x^2 + 1}}$

38 $\displaystyle\int \frac{2x\,dx}{1 - x^2 - x^4}$

39 $\displaystyle\int \frac{dx}{bx - ax^2}$ $\quad (a > 0$ and $b > 0)$

40 $\displaystyle\int \frac{dx}{a^2x^2 + x}$ $\quad (a > 0)$

6-6 Change of Independent Variable

There are two ways to use the change-of-variable formula. The first way, which we have used a lot, is shown by the example

$$\int f(\sin x) \cos x\,dx = \int f(u)\,du$$

in which we introduce a new variable u for the function $\sin x$. Sometimes, however, we go in the other direction and replace x itself by a function of a new variable. For instance, to evaluate

$$\int \frac{dx}{x^2 + 25}$$

we can set $x = 5 \tan \theta$ and rewrite the integral in terms of θ:

$$\int \frac{dx}{x^2 + 25} = \int \frac{5 \sec^2 \theta\,d\theta}{25(\tan^2 \theta + 1)} = \frac{1}{5} \int d\theta$$

$$= \frac{1}{5}\theta + C = \frac{1}{5} \arctan \frac{1}{5}x + C$$

Integrals involving $a^2 - x^2$ or $a^2 + x^2$ are sometimes simplified by trigonometric substitutions. The substitution $x = a \sin \theta$ changes $a^2 - x^2$ into $a^2\cos^2\theta$. The substitution $x = a \tan \theta$ changes $a^2 + x^2$ into $a^2\sec^2\theta$. These substitutions are particularly useful where a square root is involved.

Example 1

Find $\displaystyle\int \frac{dx}{x^2\sqrt{4 - x^2}}$ for $-2 < x < 2$

Solution Set $x = 2 \sin \theta$, where $-\frac{1}{2}\pi < \theta < \frac{1}{2}\pi$. Then

$$\sqrt{4 - x^2} = \sqrt{4 - 4\sin^2\theta} = 2\sqrt{1 - \sin^2\theta}$$
$$= 2\sqrt{\cos^2\theta} = 2\cos\theta > 0$$

Now

$$\int \frac{dx}{x^2\sqrt{4 - x^2}} = \int \frac{2\cos\theta\, d\theta}{(2\sin\theta)^2\sqrt{4 - 4\sin^2\theta}}$$
$$= \int \frac{2\cos\theta\, d\theta}{4\sin^2\theta \cdot 2\cos\theta} = \frac{1}{4}\int \frac{d\theta}{\sin^2\theta}$$

Hence

$$\int \frac{dx}{x^2\sqrt{4 - x^2}} = \frac{1}{4}\int \csc^2\theta\, d\theta = -\frac{1}{4}\cot\theta + C$$

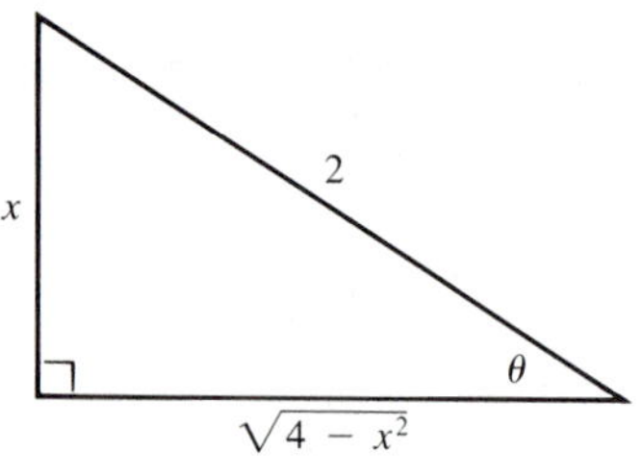

Figure 6-6-1
$\sin\theta = \dfrac{x}{2}$ $\cot\theta = \dfrac{\sqrt{4 - x^2}}{x}$

As a final step, express $\cot\theta$ in terms of x. This can be done quickly by the practical technique of drawing an appropriate right triangle (Figure 6-6-1). Set the hypotenuse equal to 2 and the side opposite θ equal to x. Then $(\text{opp})/(\text{hyp}) = x/2 = \sin\theta$, so $x = 2\sin\theta$ as above. By the Pythagorean theorem, the side adjacent to θ equals

$$\sqrt{2^2 - x^2} = \sqrt{4 - x^2}$$

Since all three sides are now expressed in terms of x, we can express any trigonometric function of θ in terms of x. In particular,

$$\cot\theta = \frac{(\text{adj})}{(\text{opp})} = \frac{\sqrt{4 - x^2}}{x}$$

Therefore

$$\int \frac{dx}{x^2\sqrt{4 - x^2}} = -\frac{\sqrt{4 - x^2}}{4x} + C$$

●

Example 2

Find $\displaystyle\int \frac{dx}{\sqrt{a^2 + x^2}}$ $(a > 0)$

Solution Set $x = a\tan\theta$ with $-\frac{1}{2}\pi < \theta < \frac{1}{2}\pi$. Then

$$\int \frac{dx}{\sqrt{a^2+x^2}} = \int \frac{a\sec^2\theta\, d\theta}{\sqrt{a^2(1+\tan^2\theta)}} = \int \frac{a\sec^2\theta\, d\theta}{a\sec\theta}$$

$$= \int \sec\theta\, d\theta = \ln|\sec\theta + \tan\theta| + C$$

Figure 6-6-2

$\tan\theta = \dfrac{x}{a} \qquad \sec\theta = \dfrac{\sqrt{a^2+x^2}}{a}$

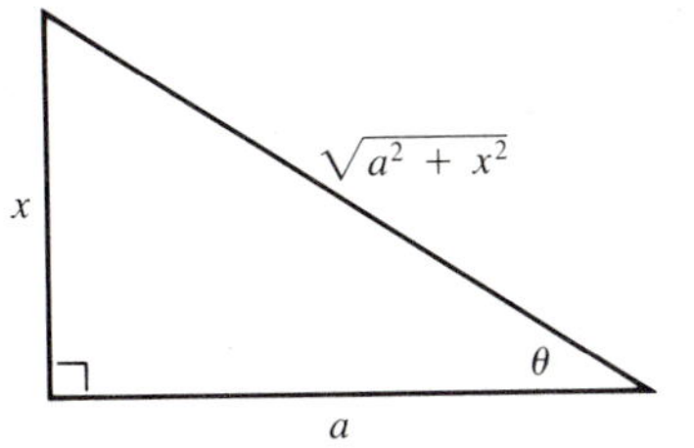

Draw a right triangle (Figure 6-6-2) with angle θ, opposite side x and adjacent side a. Then $(\text{opp})/(\text{adj}) = x/a = \tan\theta$, so it follows that $x = a\tan\theta$. The hypotenuse is $\sqrt{a^2+x^2}$. Hence

$$\sec\theta = \frac{(\text{hyp})}{(\text{adj})} = \frac{\sqrt{a^2+x^2}}{a}$$

Therefore

$$\int \frac{dx}{\sqrt{a^2+x^2}} = \ln\left|\frac{\sqrt{a^2+x^2}}{a} + \frac{x}{a}\right| + C_1$$

$$= \ln|\sqrt{a^2+x^2} + x| - \ln|a| + C_1$$

$$= \ln|\sqrt{a^2+x^2} + x| + C$$

$$= \ln(\sqrt{a^2+x^2} + x) + C$$

Note that $\sqrt{a^2+x^2} + x > 0$ for any x. That is why the absolute value is not needed. ●

Example 3

Compute the definite integral $\displaystyle\int_0^1 \frac{dx}{(1+x^2)^2}$

Solution The form of the denominator suggests that we try $x = \tan\theta$. Then $\theta = 0$ for $x = 0$, and $\theta = \frac{1}{4}\pi$ for $x = 1$. Hence

$$\int_0^1 \frac{dx}{(1+x^2)^2} = \int_0^{\pi/4} \frac{\sec^2\theta\, d\theta}{(1+\tan^2\theta)^2} = \int_0^{\pi/4} \frac{d\theta}{\sec^2\theta}$$

$$= \int_0^{\pi/4} \cos^2\theta\, d\theta$$

We know the indefinite integral of $\cos^2\theta$ from Example 4b in Section 6-5. Consequently

$$\int_0^1 \frac{dx}{(1+x^2)^2} = \int_0^{\pi/4} \cos^2\theta\, d\theta = \left(\tfrac{1}{2}\theta + \tfrac{1}{4}\sin 2\theta\right)\Big|_0^{\pi/4} = \tfrac{1}{8}\pi + \tfrac{1}{4}$$

●

Example 4

Find the indefinite integral $\displaystyle\int \frac{dx}{(1+x^2)^2}$

Figure 6-6-3

$\tan\theta = \dfrac{x}{1} = x$

$\sin 2\theta = 2\sin\theta\cos\theta$

$= 2\left(\dfrac{x}{\sqrt{1+x^2}}\right)\left(\dfrac{1}{\sqrt{1+x^2}}\right)$

$= \dfrac{2x}{1+x^2}$

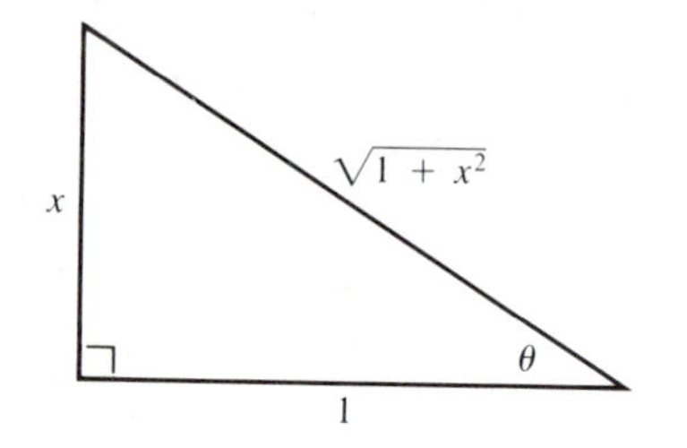

Solution From the solution of the last example,

$$\int \frac{dx}{(1+x^2)^2} = \tfrac{1}{2}\theta + \tfrac{1}{4}\sin 2\theta + C \quad \text{where} \quad x = \tan\theta$$

It remains to express this function of θ in terms of x. Obviously, $\theta = \arctan x$. To express $\sin 2\theta$ in terms of x, draw an appropriate right triangle (Figure 6-6-3) showing $x = (\text{opp})/(\text{adj}) = \tan\theta$. It follows that

$$\sin 2\theta = 2\sin\theta\cos\theta$$
$$= 2\left(\frac{x}{\sqrt{1+x^2}}\right)\left(\frac{1}{\sqrt{1+x^2}}\right) = \frac{2x}{1+x^2}$$

Therefore

$$\int \frac{dx}{(1+x^2)^2} = \tfrac{1}{2}\arctan x + \frac{\tfrac{1}{2}x}{1+x^2} + C$$

Integrals involving $x^2 - a^2$ can sometimes be simplified by the substitution $x = a\sec\theta$ because

$$x^2 - a^2 = a^2(\sec^2\theta - 1) = a^2\tan^2\theta$$

If $x \geq a > 0$, then $x/a \geq 1$, so we can take $0 \leq \theta < \tfrac{1}{2}\pi$. Then $\tan\theta > 0$, so $\sqrt{x^2 - a^2} = +a\tan\theta$.

Example 5

Find $\displaystyle\int \frac{dx}{x\sqrt{x^2 - a^2}} \qquad (x > a > 0)$

Solution Set $x = a\sec\theta$. Then

$$\int \frac{dx}{x\sqrt{x^2-a^2}} = \int \frac{a\sec\theta\tan\theta\,d\theta}{(a\sec\theta)(a\tan\theta)}$$
$$= \frac{1}{a}\int d\theta = \frac{1}{a}\theta + C$$
$$= \frac{1}{a}\,\text{arc sec}\,\frac{x}{a} + C$$
$$= \frac{1}{a}\arccos\frac{a}{x} + C$$

Hyperbolic Substitutions

For integrals involving $\sqrt{x^2 \pm a^2}$, a hyperbolic substitution is often useful. Because $\cosh^2 t - \sinh^2 t = 1$, the substitution $x = a\sinh t$ transforms $x^2 + a^2$ into $a^2\cosh^2 t$. Likewise, setting $x = a\cosh t$ transforms $x^2 - a^2$ into $a^2\sinh^2 t$.

Example 6

Find $\displaystyle\int \frac{dx}{\sqrt{x^2 - a^2}} \qquad (x > a > 0)$

Solution Substitute $x = a\cosh t$. Then $dx = a\sinh t\,dt$ and

$$\int \frac{dx}{\sqrt{x^2 - a^2}} = \int \frac{a\sinh t\,dt}{\sqrt{a^2\cosh^2 t - a^2}} = \int \frac{a\sinh t\,dt}{\sqrt{a^2\sinh^2 t}}$$
$$= \int dt = t + C_1 = \cosh^{-1}\frac{x}{a} + C_1$$
$$= \ln\left[\frac{x}{a} + \sqrt{\left(\frac{x}{a}\right)^2 - 1}\,\right] + C_1$$
$$= \ln(x + \sqrt{x^2 - a^2}\,) + C$$

●

Manipulations with Differentials [Optional]

Integrals involving $\sqrt{x^2 \pm a^2}$ can also be handled by working directly with differentials.

Example 7

Find $\displaystyle\int \frac{dx}{\sqrt{x^2 \pm a^2}}$

Solution Substitute $y^2 = x^2 \pm a^2$. Then differentiate:

$$y\,dy = x\,dx \qquad \text{hence} \qquad \frac{dx}{y} = \frac{dy}{x}$$

By elementary algebra, if

$$\frac{a}{b} = \frac{c}{d} \qquad \text{then} \qquad \frac{a}{b} = \frac{a + c}{b + d}$$

Apply this to $dx/y = dy/x$:

$$\frac{dx}{y} = \frac{dx + dy}{y + x} = \frac{d(x + y)}{x + y}$$

so that $\displaystyle\int \frac{dx}{y} = \int \frac{d(x + y)}{x + y} = \ln|x + y| + C$. Therefore

$$\int \frac{dx}{\sqrt{x^2 \pm a^2}} = \ln|x + \sqrt{x^2 \pm a^2}\,| + C$$

●

Example 8

Find $\displaystyle\int \sqrt{x^2 \pm a^2}\,dx$

Solution We want the integral of $y\,dx$, where y is the quantity introduced in the solution of Example 7:

$$y\,dx = \frac{y^2\,dx}{y} = \frac{x^2 \pm a^2}{y}\,dx = \frac{x^2\,dx}{y} \pm a^2\frac{dx}{y}$$
$$= \frac{x}{y}(y\,dy) \pm a^2\frac{dx}{y} = x\,dy \pm a^2\frac{dx}{y}$$

But also $y\,dx = d(xy) - x\,dy$. Add to eliminate $x\,dy$:

$$2y\,dx = d(xy) \pm a^2\frac{dx}{y}$$

Integrate, using the result of Example 7:

$$\int y\,dx = \tfrac{1}{2}\int d(xy) \pm \tfrac{1}{2}a^2\int\frac{dx}{y} = \tfrac{1}{2}xy \pm \tfrac{1}{2}a^2\ln|x+y| + C$$

Therefore

$$\int\sqrt{x^2 \pm a^2}\,dx = \tfrac{1}{2}x\sqrt{x^2 \pm a^2} \pm \tfrac{1}{2}a^2\ln|x + \sqrt{x^2 \pm a^2}| + C$$

●

Similarly one can evaluate

$$\int\frac{\sqrt{x^2 \pm a^2}}{x}\,dx \quad \text{and} \quad \int\sqrt{a^2 - x^2}\,dx$$

See Exercises 24 and 30 respectively.

Exercises

Use a trigonometric substitution to find

1 $\displaystyle\int\frac{dx}{\sqrt{1+x^2}}$

2 $\displaystyle\int\frac{dx}{1-x^2}$ $(|x| < 1)$

3 $\displaystyle\int\frac{x^3\,dx}{\sqrt{4-x^2}}$

4 $\displaystyle\int\frac{\sqrt{9-x^2}}{x^2}\,dx$

5 $\displaystyle\int\frac{dx}{x^2\sqrt{16-x^2}}$

6 $\displaystyle\int\frac{x^2\,dx}{\sqrt{a^2-x^2}}$ $(|x| < a)$

7 $\displaystyle\int\frac{\sqrt{x^2+a^2}}{x^2}\,dx$ $(a > 0)$

8 $\displaystyle\int\frac{x^2\,dx}{(1+x^2)^{3/2}}$

9 $\displaystyle\int\frac{x^2\,dx}{(1+x^2)^2}$

10 $\displaystyle\int\frac{dx}{x^2\sqrt{1+x^2}}$

11 $\displaystyle\int\frac{dx}{\sqrt{x^2-a^2}}$ $(x > a > 0)$

12 $\displaystyle\int\frac{\sqrt{x^2-a^2}}{x}\,dx$ $(x \geq a > 0)$

Let $a > 0$. Use a hyperbolic substitution to find

13 $\displaystyle\int\frac{dx}{\sqrt{x^2+a^2}}$

14 $\displaystyle\int\sqrt{x^2+a^2}\,dx$

15 $\displaystyle\int\frac{x^2\,dx}{\sqrt{x^2+a^2}}$

16 $\displaystyle\int\frac{\sqrt{x^2+a^2}}{x^2}\,dx$

17 $\displaystyle\int\frac{dx}{a^2-x^2}$

18 $\displaystyle\int\frac{dx}{x\sqrt{a^2-x^2}}$

Express in terms of logarithms

19 $\displaystyle\int_0^1\frac{dx}{\sqrt{1+x^2}}$

20 $\displaystyle\int_{\sqrt{2}}^2\frac{dx}{\sqrt{x^2-1}}$

21 $\displaystyle\int_0^{1/2}\frac{dx}{1-x^2}$

22 $\displaystyle\int_0^{\sqrt{2}/2}\frac{x^2\,dx}{1-x^2}$

Reduce each integral to one that does not involve a radical, and evaluate it. Use the "manipulations with differentials" method with $y^2 = x^2 \pm a^2$, so $y\,dy = x\,dx$, and the results of Examples 7 and 8. Assume $a > 0$ and $x > 0$ and also $x > a$ if $y^2 = x^2 - a^2$.

3 $\displaystyle\int \frac{dx}{xy}$

25 $\displaystyle\int \frac{y\,dx}{x^2}$

4 $\displaystyle\int \frac{y\,dx}{x}$

*26 $\displaystyle\int \frac{dx}{x^2 y}$

Reduce the integral to one without a radical and evaluate it. Assume $0 < x < a$. Set $y^2 = a^2 - x^2$ so that $y\,dy = -x\,dx$ and

$$\int \frac{dx}{y} = -\int \frac{dy}{x} = \arcsin \frac{x}{a} + C$$

27 $\displaystyle\int \frac{dx}{xy}$

29 $\displaystyle\int \frac{dx}{x^2 y}$

28 $\displaystyle\int \frac{y\,dx}{x^2}$

*30 $\displaystyle\int y\,dx$

6-7 Rational Functions

Recall that a **rational function** is the quotient of two polynomials. This section contains systematic methods for integrating rational functions.

Our first step is a preliminary long division (if necessary) to make the numerator of lower degree than the denominator. If $f(x)/g(x)$ is a rational function, where $f(x)$ and $g(x)$ are polynomials, and if it happens that $\deg f(x) \geq \deg g(x)$, then by long division we obtain

$$\frac{f(x)}{g(x)} = q(x) + \frac{r(x)}{g(x)}$$

where $g(x)$ and $r(x)$ are polynomials and $r(x)$ has degree less than the degree of $g(x)$ [or $r(x) = 0$]. Therefore

$$\int \frac{f(x)}{g(x)}\,dx = \int q(x)\,dx + \int \frac{r(x)}{g(x)}\,dx$$

Since we can integrate the polynomial $q(x)$ by inspection, this reduces the problem to integrating a bottom-heavy rational function.

Example 1

Find $\displaystyle\int \frac{x^4 + 1}{x^2 + 1}\,dx$

Solution Divide:

$$\frac{x^4 + 1}{x^2 + 1} = x^2 - 1 + \frac{2}{x^2 + 1}$$

Now integrate:

$$\int \frac{x^4 + 1}{x^2 + 1}\,dx = \int \left(x^2 - 1 + \frac{2}{x^2 + 1}\right) dx$$

$$= \tfrac{1}{3}x^3 - x + 2 \arctan x + C$$

●

Partial Fractions

One of the basic theorems of algebra says that if

$$g(x) = x^n + a_1 x^{n-1} + \cdots + a_n$$

is a polynomial with real coefficients, then $g(x)$ can be factored into a product of linear factors $x - a$ and irreducible quadratic factors $x^2 + bx + c$, all with real coefficients. **Irreducible** means that $x^2 + bx + c$ does not have real zeros, so that the quadratic cannot be factored into real linear factors. (Recall the test for irreducibility of $x^2 + bx + c$: negative discriminant, $b^2 - 4c < 0$.) We shall assume that the denominator $g(x)$ of $f(x)/g(x)$ has been factored in this way, and study some special cases.

Any rational function of the form

$$\frac{cx + d}{(x - a)(x - b)} \qquad (a \neq b)$$

can be split into the sum of two simpler fractions:

$$\frac{A}{x - a} + \frac{B}{x - b}$$

This decomposition into **partial fractions** simplifies integration since each term is easy to integrate.

Example 2

Find $\displaystyle\int \frac{2x + 1}{(x - 3)(x - 4)}\, dx$

Solution First decompose the integrand into partial fractions:

$$\frac{2x + 1}{(x - 3)(x - 4)} = \frac{A}{x - 3} + \frac{B}{x - 4}$$

where A and B are constants to be determined. Multiply through by $(x - 3)(x - 4)$ to clear fractions:

$$2x + 1 = A(x - 4) + B(x - 3) = (A + B)x - (4A + 3B)$$

The coefficients of x on both sides of this identity must be equal, and so must the constant terms. Hence

$$A + B = 2 \qquad \text{and} \qquad -4A - 3B = 1$$

The unknowns A and B must satisfy these two equations simultaneously. Solve the linear system to obtain $A = -7$ and $B = 9$. Now integrate:

$$\int \frac{2x + 1}{(x - 3)(x - 4)}\, dx = \int \left(\frac{-7}{x - 3} + \frac{9}{x - 4} \right) dx$$

$$= \ln \left| \frac{(x - 4)^9}{(x - 3)^7} \right| + C$$

●

Remark There is a different way to compute A and B. Return to the equation

$$2x + 1 = A(x - 4) + B(x - 3)$$

This must hold for *every* value of x, in particular for $x = 3$ and $x = 4$:

$$x = 3: \quad 6 + 1 = A(3 - 4) + 0 \quad \text{hence} \quad A = -7$$

$$x = 4: \quad 8 + 1 = 0 + B(4 - 3) \quad \text{hence} \quad B = 9$$

Example 3

Find $\displaystyle\int \frac{dx}{a^2 - x^2} \quad (a > 0)$

Solution Decompose the integrand into partial fractions:

$$\frac{1}{a^2 - x^2} = \frac{1}{(a - x)(a + x)} = \frac{A}{a - x} + \frac{B}{a + x}$$

To find the constants A and B, first clear of fractions:

$$1 = A(a + x) + B(a - x)$$

Set $x = a$ to obtain $A = 1/2a$. Then set $x = -a$ to obtain $B = 1/2a$. Hence

$$\frac{1}{a^2 - x^2} = \frac{1}{2a}\left(\frac{1}{a - x} + \frac{1}{a + x}\right)$$

Now integrate:

$$\int \frac{dx}{a^2 - x^2} = \frac{1}{2a}\int\left(\frac{1}{a - x} + \frac{1}{a + x}\right) dx$$

$$= \frac{1}{2a}(-\ln|a - x| + \ln|a + x|) + C$$

$$= \frac{1}{2a}\ln\left|\frac{a + x}{a - x}\right| + C$$

Query The substitution $x = a \sin\theta$ in this result leads to

$$\int \sec\theta \, d\theta = \tfrac{1}{2}\ln\left|\frac{1 + \sin\theta}{1 - \sin\theta}\right| + C$$

Can you explain why this is the same as

$$\int \sec\theta \, d\theta = \ln|\sec\theta + \tan\theta| + C?$$

Example 4

Find $\displaystyle\int \frac{x^3 + 4}{x^2 + x}\,dx$

Solution First divide $x^3 + 4$ by $x^2 + x$:

$$\frac{x^3 + 4}{x^2 + x} = x - 1 + \frac{x + 4}{x^2 + 4}$$

Next decompose the remainder into partial fractions:

$$\frac{x + 4}{x^2 + x} = \frac{x + 4}{x(x + 1)} = \frac{A}{x} + \frac{B}{x + 1}$$

To determine the constants A and B, clear fractions:

$$x + 4 = A(x + 1) + Bx$$

Then set $x = 0$ and $x = -1$ to obtain $A = 4$ and $B = -3$. Now assemble the integrand and integrate:

$$\begin{aligned}\int \frac{x^3 + 4}{x^2 + x}\,dx &= \int \left(x - 1 + \frac{4}{x} - \frac{3}{x + 1}\right) dx \\ &= \tfrac{1}{2}x^2 - x + 4\ln|x| - 3\ln|x + 1| + C \\ &= \tfrac{1}{2}x^2 - x + \ln\left|\frac{x^4}{(x + 1)^3}\right| + C\end{aligned}$$

●

The method of partial fractions can be used to integrate any rational function $f(x)/g(x)$, provided that the denominator can be completely factored into linear and quadratic factors. In practice, this is hard to do for polynomials of degree 3 or more, except in special cases.

Assume the degree of $g(x)$ exceeds that of $f(x)$, and assume that $g(x)$ is factored into linear and quadratic factors. The algebraic theory of partial fractions, which is really beyond the scope of this course, tells us the following about the decomposition of $f(x)/g(x)$ into partial fractions:

- For each factor $x - a$ there is a term: $\dfrac{A}{x - a}$
- If $(x - a)^2$ occurs, there are two terms: $\dfrac{A}{x - a} + \dfrac{B}{(x - a)^2}$
- If $(x - a)^3$ occurs, there are three terms:

$$\frac{A}{x - a} + \frac{B}{(x - a)^2} + \frac{C}{(x - a)^3}$$

- For each irreducible quadratic factor $x^2 + ax + b$ there is a term:

$$\frac{Ax + B}{x^2 + ax + b}$$

- If $(x^2 + ax + b)^2$ occurs, there are two terms:

$$\frac{Ax + B}{x^2 + ax + b} + \frac{Cx + D}{(x^2 + ax + b)^2}$$

and so on.

For instance:

$$\frac{1}{(x-a)(x-b)(x-c)} = \frac{A}{x-a} + \frac{B}{x-b} + \frac{C}{x-c} \qquad (a,\ b,\ c \text{ distinct})$$

$$\frac{1}{(x-a)^2(x-b)} = \frac{A}{x-a} + \frac{B}{(x-a)^2} + \frac{C}{x-b} \qquad (a \neq b)$$

$$\frac{1}{(x-a)(x^2+bx+c)} = \frac{A}{x-a} + \frac{Bx + C}{x^2 + bx + c} \qquad (b^2 - 4c < 0)$$

$$\frac{1}{(x-a)(x^2+b^2)^2} = \frac{A}{x-a} + \frac{Bx + C}{x^2 + b^2} + \frac{Dx + E}{(x^2 + b^2)^2} \qquad (b \neq 0)$$

The constants can always be determined by clearing fractions and then equating coefficients of like powers of x. Some preliminary substitution of special values for x often shortens the work.

Example 5

Find $\displaystyle\int \frac{dx}{x^4 - 1}$

Solution First decompose the integrand:

$$\frac{1}{x^4 - 1} = \frac{1}{(x-1)(x+1)(x^2+1)} = \frac{A}{x-1} + \frac{B}{x+1} + \frac{Cx + D}{x^2 + 1}$$

Multiply through by $(x-1)(x+1)(x^2+1)$:

$$1 = A(x+1)(x^2+1) + B(x-1)(x^2+1) + Cx(x-1)(x+1) + D(x-1)(x+1)$$

Set $x = 1$ and $x = -1$ to obtain $A = -B = \frac{1}{4}$. Set $x = 0$ to obtain $1 = A - B - D = \frac{1}{4} + \frac{1}{4} - D$. Hence $D = -\frac{1}{2}$. Choose any other value of x to find C. Try $x = 2$ for example:

$$1 = 15A + 5B + 6C + 3D = \frac{15}{4} - \frac{5}{4} + 6C - \frac{3}{2}$$

from which $C = 0$. Now integrate:

$$\int \frac{dx}{x^4 - 1} = \int \left(\frac{\frac{1}{4}}{x - 1} - \frac{\frac{1}{4}}{x + 1} - \frac{\frac{1}{2}}{x^2 + 1} \right) dx$$

$$= \tfrac{1}{4} \ln|x - 1| - \tfrac{1}{4} \ln|x + 1| - \tfrac{1}{2}\text{arc tan } x + C$$

$$= \ln \left| \frac{x - 1}{x + 1} \right|^{1/4} - \tfrac{1}{2}\text{arc tan } x + C$$

Example 6

Find $\displaystyle\int \frac{2x + 5}{(x - 1)(x + 3)^2}\, dx$

Solution Decompose:

$$\frac{2x + 5}{(x - 1)(x + 3)^2} = \frac{A}{x - 1} + \frac{B}{x + 3} + \frac{C}{(x + 3)^2}$$

Multiply through by $(x - 1)(x + 3)^2$:

$$2x + 5 = A(x + 3)^2 + B(x - 1)(x + 3) + C(x - 1)$$

Set $x = 1$ to obtain $A = \frac{7}{16}$: Set $x = -3$ to obtain $C = \frac{1}{4}$. Choose any other value of x to find B, for example $x = 0$:

$$5 = 9A - 3B - C = \tfrac{63}{16} - 3B - \tfrac{1}{4}$$

from which $B = -\frac{7}{16}$. Now integrate:

$$\int \frac{2x + 5}{(x - 1)(x + 3)^2}\, dx$$

$$= \tfrac{7}{16} \int \frac{dx}{x - 1} - \tfrac{7}{16} \int \frac{dx}{x + 3} + \tfrac{1}{4} \int \frac{dx}{(x + 3)^2}$$

$$= \tfrac{7}{16} \ln|x - 1| - \tfrac{7}{16} \ln|x + 3| - \frac{\frac{1}{4}}{x + 3} + C$$

$$= \tfrac{7}{16} \ln \left| \frac{x - 1}{x + 3} \right| - \frac{1}{4(x + 3)} + C$$

Rational Functions of Sine and Cosine [Optional]

In theory, partial fraction decompositions allow the integration of any rational function $r(t)$. An interesting substitution, based on the half-angle formulas of trigonometry, reduces any rational function of $\sin \theta$ and $\cos \theta$ to a rational function $r(t)$. Thus, at least in theory, integrals such as

$$\int \frac{d\theta}{3 + \cos \theta} \quad \text{and} \quad \int \frac{1 + \sin^3\theta \cos \theta}{4 \cos^5\theta - 3 \sin \theta}\, d\theta$$

can be computed explicitly. However, the computations may be quite formidable. When a *definite* integral is required, approximation methods like the trapezoidal rule and the rules to be discussed in Section 9-6 usually involve less work.

We state the half-angle formulas in the form needed for computing integrals.

Half-Angle Formulas

Set $t = \tan \frac{1}{2}\theta$. Then

$$\sin \theta = \frac{2t}{1+t^2} \qquad \cos \theta = \frac{1-t^2}{1+t^2} \qquad t = \frac{\sin \theta}{1+\cos \theta}$$

$$d\theta = \frac{2\,dt}{1+t^2}$$

The first three formulas are usually given in trigonometry courses, and they follow easily from the standard double-angle formulas. We shall give a geometric derivation in Section 13-3.

The fourth formula, the relation between $d\theta$ and dt, is obtained by differentiating $t = \tan \frac{1}{2}\theta$:

$$dt = \tfrac{1}{2}(\sec^2 \tfrac{1}{2}\theta)\,d\theta = \tfrac{1}{2}(1 + \tan^2 \tfrac{1}{2}\theta)\,d\theta = \tfrac{1}{2}(1+t^2)\,d\theta$$

With this preparation we can state a rule for transforming integrals of functions of $\sin \theta$ and $\cos \theta$. Suppose $f(x, y)$ is a polynomial in two variables or a quotient of two polynomials. Then

$$\int f(\sin \theta, \cos \theta)\,d\theta = 2 \int f\left(\frac{2t}{1+t^2}, \frac{1-t^2}{1+t^2}\right) \frac{dt}{1+t^2}$$

where $t = \tan \frac{1}{2}\theta$.

Note that this is a formal procedure. When you apply it to a definite integral, you must be careful that the θ-domain does not include a zero of the denominator of the integrand.

Example 7

Compute $\displaystyle\int \frac{d\theta}{3 + \cos \theta}$

Solution Use the half-angle substitution $t = \tan \frac{1}{2}\theta$. Then

$$\int \frac{d\theta}{3+\cos\theta} = 2\int \frac{dt/(1+t^2)}{3 + [(1-t^2)/(1+t^2)]}$$

$$= 2\int \frac{dt}{3(1+t^2) + (1-t^2)} = 2\int \frac{dt}{4+2t^2}$$

$$= \int \frac{dt}{2+t^2} = \frac{1}{\sqrt{2}} \arctan \frac{t}{\sqrt{2}} + C$$

$$= \frac{1}{\sqrt{2}} \arctan \left(\frac{1}{\sqrt{2}} \tan \tfrac{1}{2}\theta\right) + C$$

Exercises

Decompose the integrand into partial fractions and integrate

1. $\displaystyle\int \frac{dx}{(x+1)(x-1)}$

2. $\displaystyle\int \frac{x\,dx}{(x+2)(x+3)}$

3. $\displaystyle\int \frac{x^2\,dx}{(x+1)(x-2)}$

4. $\displaystyle\int \frac{dx}{(x+1)(x+2)(x+3)}$

5. $\displaystyle\int \frac{x\,dx}{(x+1)(x+2)(x+3)}$

6. $\displaystyle\int \frac{dx}{(x+1)(x^2+4)}$

7. $\displaystyle\int \frac{x^4}{(x^2+1)^2}\,dx$

8. $\displaystyle\int \frac{x^3-1}{x(x^2+1)}\,dx$

9. $\displaystyle\int \frac{x+1}{(x-1)(x^2+4)}\,dx$

10. $\displaystyle\int \frac{dx}{x(x+1)^2}$

11. $\displaystyle\int \frac{dx}{x^2-3x+2}$

12. $\displaystyle\int \frac{dx}{(x-2)(x+4)}$

13. $\displaystyle\int \frac{x+3}{x^2+x}\,dx$

14. $\displaystyle\int \frac{x^2+1}{x^2-5x+6}\,dx$

15. $\displaystyle\int \frac{2x+3}{x^3+x}\,dx$

16. $\displaystyle\int \frac{x\,dx}{(x+1)^2(x-3)}$

17. $\displaystyle\int \frac{dx}{(x-2)^2(x+9)}$

18. $\displaystyle\int \frac{dx}{3x^2-13x+4}$

19. $\displaystyle\int \frac{x^4\,dx}{x^3-1}$

20. $\displaystyle\int \frac{x\,dx}{x^4-1}$

21. $\displaystyle\int \frac{x^3\,dx}{x^2+3x+2}$

22. $\displaystyle\int \frac{dx}{(x-1)(x-2)(x-3)}$

23. $\displaystyle\int \frac{x^2+x+1}{(x-3)(x^2+2x+2)}\,dx$

24. $\displaystyle\int \frac{dx}{x(x-3)^2}$

Evaluate the definite integral

25. $\displaystyle\int_0^1 \frac{dx}{x^2-5x+6}$

26. $\displaystyle\int_1^2 \frac{dx}{x^2(x+1)}$

27. $\displaystyle\int_1^4 \frac{dx}{x^3+8}$

28. $\displaystyle\int_0^3 \frac{x\,dx}{(x+1)(x^2+9)}$

29. $\displaystyle\int_0^{\pi/2} \frac{\cos\theta\,d\theta}{\sin^2\theta+7\sin\theta+10}$

30. $\displaystyle\int_4^9 \frac{dx}{\sqrt{x}\,(1+\sqrt{x})(2+\sqrt{x})}$

Compute

31. $\displaystyle\int \frac{d\theta}{\sin\theta+\cos\theta}$

32. $\displaystyle\int \frac{4-\sin\theta}{3-\cos\theta}\,d\theta$

6-8 Integral Tables and Loose Ends

This book contains a short table of indefinite integrals inside the covers. Much longer tables are available, for example those in the *C.R.C. Standard Mathematical Tables.* We suggest that you get one of the more complete integral tables and spend some time browsing through it. Become familiar with the type of integral you can expect to find there. Not every integral is listed in a table, but many can be transformed into integrals that are listed.

Example 1

Use integral tables to find $\displaystyle\int x^3\sqrt{3-4x^2}\,dx$

Solution Most tables include a section on integrals involving $\sqrt{a^2 - x^2}$. A formula in the C.R.C. tables states that

$$\int x^3\sqrt{a^2 - x^2}\,dx = -(\tfrac{1}{5}x^2 + \tfrac{2}{15}a^2)(a^2 - x^2)^{3/2} + C$$

This is very close to what is wanted, except that $\sqrt{3 - 4x^2}$ appears instead of $\sqrt{a^2 - x^2}$. There are two ways of changing this integral into the formula from the table. The first method is to divide out the 4:

$$\sqrt{3 - 4x^2} = \sqrt{4(\tfrac{3}{4} - x^2)} = 2\sqrt{\tfrac{3}{4} - x^2} = 2\sqrt{a^2 - x^2}$$

where $a^2 = \frac{3}{4}$. Now use the table formula:

$$\int x^3\sqrt{3 - 4x^2}\,dx = 2\int x^3\sqrt{\tfrac{3}{4} - x^2}\,dx$$
$$= -2(\tfrac{1}{5}x^2 + \tfrac{2}{15}\cdot\tfrac{3}{4})(\tfrac{3}{4} - x^2)^{3/2} + C$$

The second method is to make the substitution $u = 2x$, so $x = \frac{1}{2}u$ and $dx = \frac{1}{2}\,du$. Then

$$\sqrt{3 - 4x^2} = \sqrt{3 - (2x)^2} = \sqrt{a^2 - u^2}$$

Now express everything in terms of u and use the table formula with $a^2 = 3$:

$$\int x^3\sqrt{3 - 4x^2}\,dx = \int (\tfrac{1}{2}u)^3\sqrt{3 - u^2}\,(\tfrac{1}{2}\,du)$$
$$= \tfrac{1}{16}\int u^3\sqrt{3 - u^2}\,du$$
$$= -\tfrac{1}{16}(\tfrac{1}{5}u^2 + \tfrac{2}{15}\cdot 3)(3 - u^2)^{3/2} + C$$
$$= -\tfrac{1}{16}(\tfrac{4}{5}x^2 + \tfrac{2}{5})(3 - 4x^2)^{3/2} + C$$

A little algebra shows that the two answers agree. ●

Example 2

Use integral tables to find $\int e^{2x}\sin^3 x\,dx$

Solution In the C.R.C. tables under *Exponential Forms* is the formula

$$\int e^{ax}\sin^n bx\,dx$$
$$= \frac{1}{a^2 + n^2b^2}\left[(a\sin bx - nb\cos bx)e^{ax}\sin^{n-1}bx + n(n-1)b^2\int e^{ax}\sin^{n-2}bx\,dx\right] + C$$

This is a formula that lowers the power of $\sin bx$ by two. Apply it with $a = 2$, $b = 1$, $n = 3$:

$$\int e^{2x} \sin^3 x \, dx$$

$$= \frac{1}{4+9}\left[(2 \sin x - 3 \cos x) e^{2x} \sin^2 x + 6 \int e^{2x} \sin x \, dx\right] + C$$

The integral on the right is also given in the C.R.C. tables. Its value is

$$\tfrac{1}{5} e^{2x}(2 \sin x - \cos x)$$

It follows that

$$\int e^{2x} \sin^3 x \, dx$$

$$= \tfrac{1}{13} e^{2x}[(2 \sin x - 3 \cos x) \sin^2 x + \tfrac{6}{5}(2 \sin x - \cos x)] + C \quad \bullet$$

Integral tables use abbreviations for common expressions. For instance, one section of the C.R.C. tables contains formulas involving X and $\sqrt{X}$, where $X = a + bx + cx^2$.

Example 3

Use integral tables to find

$$\int \frac{\sqrt{5x^2 + 2x + 3}}{x} \, dx$$

Solution According to one of the formulas in the C.R.C. tables,

$$\int \frac{\sqrt{X}}{x} \, dx = \sqrt{X} + \tfrac{1}{2} b \int \frac{dx}{\sqrt{X}} + a \int \frac{dx}{x\sqrt{X}}$$

The integrals on the right are also given in the tables:

$$\int \frac{dx}{\sqrt{X}} = \frac{1}{\sqrt{c}} \ln\left|\sqrt{X} + x\sqrt{c} + \frac{b}{2\sqrt{c}}\right| + C \qquad (c > 0)$$

$$\int \frac{dx}{x\sqrt{X}} = -\frac{1}{\sqrt{a}} \ln\left|\frac{\sqrt{X} + \sqrt{a}}{x} + \frac{b}{2\sqrt{a}}\right| + C \qquad (a > 0)$$

Set

$$X = 3 + 2x + 5x^2 \qquad a = 3 \qquad b = 2 \qquad c = 5$$

The result is

$$\int \frac{\sqrt{5x^2 + 2x + 3}}{x}\,dx$$

$$= \sqrt{X} + \frac{1}{\sqrt{5}} \ln\left|\sqrt{X} + x\sqrt{5} + \frac{1}{\sqrt{5}}\right|$$

$$- \sqrt{3} \ln\left|\frac{\sqrt{X} + \sqrt{3}}{x} + \frac{1}{\sqrt{3}}\right| + C$$

where $X = 5x^2 + 2x + 3.$ ●

Even and Odd Functions

Sometimes a definite integral

$$\int_a^b f(x)\,dx$$

can be simplified because the integrand $f(x)$ has certain symmetry. We now discuss some labor-saving devices for evaluating integrals with certain symmetries.

Integrals of Even Functions

If $f(x)$ is an even function, that is, $f(-x) = f(x)$, then

$$\int_{-a}^{a} f(x)\,dx = 2\int_0^a f(x)\,dx$$ ●

Proof Since

$$\int_{-a}^{a} f(x)\,dx = \int_{-a}^{0} f(x)\,dx + \int_0^a f(x)\,dx$$

we must prove that

$$\int_{-a}^{0} f(x)\,dx = \int_0^a f(x)\,dx$$

This is obvious from the graph (Figure 6-8-1), and follows easily by means

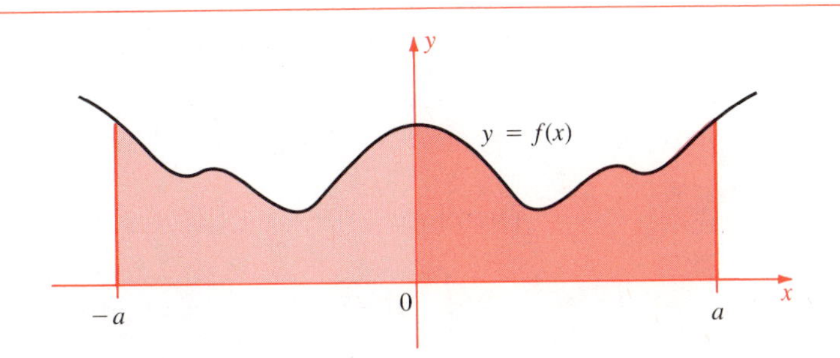

Figure 6-8-1
Even function: left area = right area.

of the change of variable $x = -u$:

$$\int_{-a}^{0} f(x)\,dx = \int_{a}^{0} f(-u)(-du) = -\int_{a}^{0} f(u)\,du$$

$$= \int_{0}^{a} f(u)\,du = \int_{0}^{a} f(x)\,dx$$

Example 4

Find $\displaystyle\int_{-\pi/2}^{\pi/2} \cos \tfrac{1}{3}x\,dx$

Solution The integrand is an even function. Therefore

$$\int_{-\pi/2}^{\pi/2} \cos \tfrac{1}{3}x\,dx = 2\int_{0}^{\pi/2} \cos \tfrac{1}{3}x\,dx$$

$$= 6 \sin \tfrac{1}{3}x\Big|_{0}^{\pi/2} = 6 \sin \tfrac{1}{6}\pi = 3$$

Integrals of Odd Functions

If $f(x)$ is an odd function, that is, $f(-x) = -f(x)$, then

$$\int_{-a}^{a} f(x)\,dx = 0$$

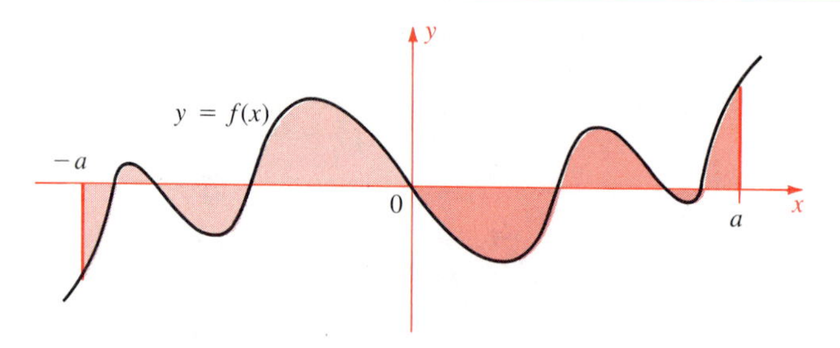

Figure 6-8-2
Odd function: left area cancels right area.

This is obvious from a graph (Figure 6-8-2), and follows easily by the change of variable $x = -u$:

$$\int_{-a}^{a} f(x)\,dx = \int_{a}^{-a} f(-u)(-du) = \int_{a}^{-a} f(u)\,du$$

$$= -\int_{-a}^{a} f(u)\,du = -\int_{-a}^{a} f(x)\,dx$$

Therefore

$$2\int_{-a}^{a} f(x)\,dx = 0 \quad \text{so that} \quad \int_{-a}^{a} f(x)\,dx = 0$$

Example 5

Integrals of odd functions over intervals $[-a, a]$:

a $\displaystyle\int_{-\pi/4}^{\pi/4} \sin^9 x\, dx = 0$ **b** $\displaystyle\int_{-6}^{6} \frac{x^5}{x^4+1}\, dx = 0$ ●

Example 6

Show that

a $\displaystyle\int_{-3}^{3} (x\sin^2 x - 14x^3 + 1)\, dx = 6$

b $\displaystyle\int_{-8}^{10} \frac{x}{x^2+4}\, dx = \int_{8}^{10} \frac{x}{x^2+4}\, dx$

Solution

a Write the integral as

$$\int_{-3}^{3} (x\sin^2 x - 14x^3)\, dx + \int_{-3}^{3} dx$$

The first integral is 0 because the integrand is odd; the second integral equals 6 by inspection.

b The integrand is odd. Therefore

$$\int_{-8}^{10} \frac{x}{x^2+4}\, dx = \int_{-8}^{8} + \int_{8}^{10} = 0 + \int_{8}^{10} \frac{x}{x^2+4}\, dx$$ ●

Periodic Functions

Recall that a function $f(x)$ is **periodic** of **period** p if

$$f(x+p) = f(x)$$

Trigonometric functions are the most familiar periodic functions. We note two basic properties of integrals of periodic functions:

- Suppose $f(x)$ has period p. Then

1 $\displaystyle\int_{a+p}^{b+p} f(x)\, dx = \int_{a}^{b} f(x)\, dx$ **2** $\displaystyle\int_{a}^{a+p} f(x)\, dx = \int_{0}^{p} f(x)\, dx$ ●

The geometric content of relation 1 is clear from Figure 6-8-3 (next page). It is easily proved by the change of variable $x = p + u$:

$$\int_{a+p}^{b+p} f(x)\, dx = \int_{a}^{b} f(p+u)\, du = \int_{a}^{b} f(u)\, du = \int_{a}^{b} f(x)\, dx$$

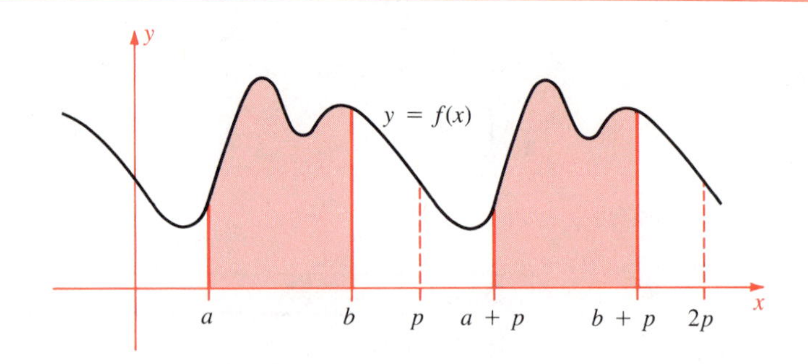

Figure 6-8-3
Translation for a periodic function:
$$\int_{a+p}^{b+p} f(x)\,dx = \int_a^b f(x)\,dx$$

Relation 2 is illustrated in Figure 6-8-4. It says that all integrals of $f(x)$ over intervals of length p are equal. To prove it analytically, write

$$\int_a^{a+p} = \int_0^{a+p} - \int_0^a \qquad \text{and} \qquad \int_0^{a+p} = \int_0^p + \int_p^{a+p}$$

Therefore

$$\int_a^{a+p} f(x)\,dx = \int_0^p f(x)\,dx + \int_p^{a+p} f(x)\,dx - \int_0^a f(x)\,dx$$

But by relation 1,

$$\int_p^{a+p} f(x)\,dx = \int_{0+p}^{a+p} f(x)\,dx = \int_0^a f(x)\,dx$$

Therefore relation 2 follows. ●

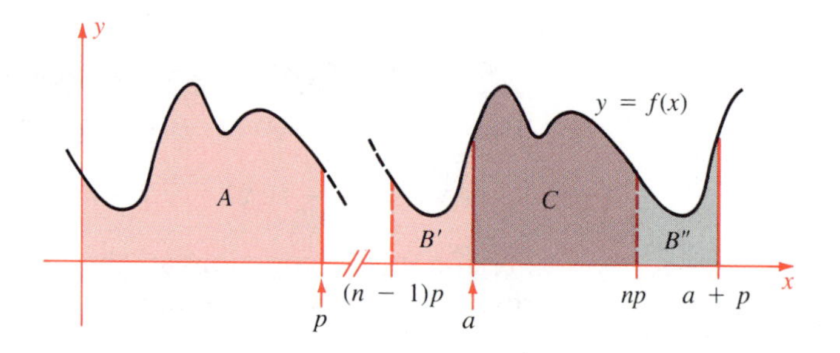

Figure 6-8-4
area B'' = area B'. Hence
$$C + B'' = C + B' = A$$
That is,
$$\int_a^{a+p} f(x)\,dx = \int_0^p f(x)\,dx$$
for any a.

Example 7

Let $f(y)$ be integrable on $[-1, 1]$, and let n be a positive integer. Prove that

$$\int_0^{2\pi} f(\sin nx)\,dx = \int_0^{2\pi} f(\cos nx)\,dx$$

Solution The integrand has period 2π (indeed, $2\pi/n$, but 2π is sufficient). Make the change of variable $x = u + \pi/2n$, so that

$$\sin nx = \sin(nu + \tfrac{1}{2}\pi) = \cos nu$$

Hence, $f(\sin nx) = f(\cos nu)$. Then

$$\int_0^{2\pi} f(\sin nx)\,dx = \int_{-\pi/2n}^{2\pi-\pi/2n} f(\cos nu)\,du$$

Now apply relation 2 with $a = -\pi/2n$:

$$\int_{-\pi/2n}^{2\pi-\pi/2n} f(\cos nu)\,du = \int_0^{2\pi} f(\cos nu)\,du = \int_0^{2\pi} f(\cos nx)\,dx$$

For example, if $f(y) = \sqrt{1 + y^3}$ then

$$\int_0^{2\pi} \sqrt{1 + \sin^3 x}\,dx = \int_0^{2\pi} \sqrt{1 + \cos^3 x}\,dx$$

●

Example 8

Use Example 7 to find $\displaystyle\int_0^{2\pi} \sin^2 nx\,dx$

Solution Set

$$A = \int_0^{2\pi} \cos^2 nx\,dx \quad \text{and} \quad B = \int_0^{2\pi} \sin^2 nx\,dx$$

By Example 7, we have $A = B$, so $A + B = 2B$. But

$$A + B = \int_0^{2\pi} (\cos^2 nx + \sin^2 nx)\,dx = \int_0^{2\pi} dx = 2\pi$$

Therefore, $A = B = \pi$.

●

Exercises

Find, using tables

1 $\displaystyle\int e^{-2x} \sin 5x\,dx$

2 $\displaystyle\int \sqrt{4 - x^2}\,dx$

3 $\displaystyle\int x^2\sqrt{1 - 4x^2}\,dx$

4 $\displaystyle\int \frac{x^2\,dx}{\sqrt{4 - 3x^2}}$

5 $\displaystyle\int \frac{x^2\,dx}{2 + 5x^2}$

6 $\displaystyle\int (4 - x^2)^{3/2}\,dx$

7 $\displaystyle\int x^3 \sin \tfrac{1}{2}x\,dx$

8 $\displaystyle\int (\ln x)^4\,dx$

9 $\displaystyle\int \frac{x^2 - 6x - 2}{x\sqrt{10x^2 + 7}}\,dx$

10 $\displaystyle\int (x + 3)^2 \times \sqrt{x^2 + 2x + 5}\,dx$

Compute

11 $\displaystyle\int_0^1 \frac{x^4 + 2x^2 - 3}{x^4 + 2x^2 + 1}\,dx$

12 $\displaystyle\int_0^1 \frac{x^2\,dx}{1 + 3x^2}$

13 $\displaystyle\int_0^{\pi} e^{3x} \cos^6 x\,dx$

14 $\displaystyle\int_0^{2\pi} \cos^2 x \sin^8 x\,dx$

15 $\displaystyle\int_0^1 \frac{dx}{(1 + 3x)^2(2 + 5x)}$

16 $\displaystyle\int_1^2 (x \ln x)^3\,dx$

Reduce by symmetry (and periodicity) to an integral over as short an interval as you can, but do not evaluate

17 $\displaystyle\int_{-1}^{2} (x^3 - 5x)\,dx$

18 $\displaystyle\int_{-4}^{4} \sin(x^2)\,dx$

19 $\displaystyle\int_{-2}^{3} \frac{x}{x^4 + x^2 + 1}\,dx$

20 $\displaystyle\int_{-4}^{4} \sqrt{x^6 + 4x^2 + 6}\,dx$

21 $\displaystyle\int_{0}^{3\pi} (\sin x + \cos x)\,dx$

22 $\displaystyle\int_{-\pi/2}^{5\pi/2} (\sin x - \cos x)\,dx$

23 $\displaystyle\int_{0}^{4\pi} \sin^2 x\,dx$

24 $\displaystyle\int_{0}^{3\pi} \frac{\sin x}{2 + \cos x}\,dx$

25 $\displaystyle\int_{-100\pi}^{100\pi} \sin(\tfrac{1}{12}x - 4)\,dx$

26 $\displaystyle\int_{1}^{4} \sin(\pi x + 3)\,dx$

27 A vertical line $x = c$ is called an **axis of symmetry** of $f(x)$ if $f(c - x) = f(c + x)$. If this is the case, prove that

$$\int_{c-a}^{c+a} f(x)\,dx = 2\int_{c}^{c+a} f(x)\,dx$$

28 A point $(c, 0)$ of the x-axis is called a **point of symmetry** of $f(x)$ if $f(c - x) = -f(c + x)$. If this is the case, prove that

$$\int_{c-a}^{c+a} f(x)\,dx = 0$$

Determine all axes of symmetry $x = c$ and all points $(c, 0)$ of symmetry (see Exercises 27 and 28)

29 $y = (\cos x)\exp(x^2)$

30 $y = x^3 - 5x$

31 $y = (x^2 - 1)(x + 3)$

32 $y = \exp(\sin x)$

33 $y = \sin x + \cos x$

34 $y = (x^2 + x + 1)^{-1}$

35 $y = [1 + (x + 1)^2]^{1/2}$

36 $y = (x^2 + 2x)(x + 3)(x + 5)$

37 Find all functions $f(x)$ defined on $[-a, a]$ that are both even and odd.

38 Suppose $f(x)$ is defined on $(-\infty, \infty)$ and has both $x = 0$ and $x = c$ as axes of symmetry (Exercise 27), where $c > 0$. Show that $f(x)$ is periodic.

A function $f(x)$ with domain $[a - h, a + h]$ is **symmetric** in (a, b), that is,

$$\tfrac{1}{2}[f(a - x) + f(a + x)] = b \qquad (0 \le x \le h)$$

39 Find $f(a)$

40 Find $\displaystyle\int_{a-h}^{a+h} f(x)\,dx$

41 Suppose $f(x)$ is differentiable on $\mathbf{R}$ and periodic of period p. Suppose $f'(x)$ is continuous. Prove

$$\int_{a}^{a+p} f'(x)\,dx = 0$$

42 Suppose that $f(x)$ has a continuous derivative and is periodic of period $2P$. Show that

$$\int_{-P}^{P} x^2 f'(x)\,dx + 2\int_{-P}^{P} x f(x)\,dx = 0$$

43 Set $f(x) = (\sin x)/\sin(x + \frac{1}{4}\pi)$. Show that $f(x) + f(\frac{1}{2}\pi - x) = \sqrt{2}$, and then find $\displaystyle\int_{0}^{\pi/2} f(x)\,dx$. [Hint See Exercise 40.]

44 Set $f(x) = (x - \frac{1}{4}\pi)^2(\sin x)/(\sin x + \cos x)$. Show that $f(x) + f(\frac{1}{2}\pi - x) = (x - \frac{1}{4}\pi)^2$, and then find $\displaystyle\int_{0}^{\pi/2} f(x)\,dx$.

6-9 Iteration Formulas

The integral

$$\int x^2 e^x\,dx$$

requires two integrations by parts. Each integration lowers the power of x by one until x disappears. In the same way

$$\int x^3 e^x\,dx \quad \text{and} \quad \int x^4 e^x\,dx$$

require three and four integrations by parts, respectively. It is convenient to have an **iteration formula** (also called **reduction formula**), a formula that reduces

$$\int x^n e^x \, dx \quad \text{to} \quad \int x^{n-1} e^x \, dx$$

Repeated use of such a formula reduces

$$\int x^n e^x \, dx \quad \text{to} \quad \int e^x \, dx$$

Example 1

Derive a reduction formula for $\int x^n e^x \, dx$

Solution Integrate by parts with

$$u = x^n \qquad dv = e^x \, dx$$
$$du = nx^{n-1} \, dx \qquad v = e^x$$

The result is

$$\int x^n e^x \, dx = x^n e^x - n \int x^{n-1} e^x \, dx$$

For abbreviation, write

$$J_n = \int x^n e^x \, dx$$

Then the reduction formula is $J_n = x^n e^x - nJ_{n-1}$. ●

Example 2

Find $\int x^5 e^x \, dx$

Solution Use the reduction formula just derived to find J_5. With $n = 5$, the reduction formula yields

$$J_5 = x^5 e^x - 5J_4$$

With $n = 4$, it yields $J_4 = x^4 e^x - 4J_3$. Hence

$$J_5 = x^5 e^x - 5(x^4 e^x - 4J_3) = x^5 e^x - 5x^4 e^x + 20J_3$$

By repeated further use of the reduction formula,

$$J_3 = x^3 e^x - 3J_2 = x^3 e^x - 3(x^2 e^x - 2J_1)$$
$$= x^3 e^x - 3x^2 e^x + 6(xe^x - J_0)$$

The integral J_0 is easy:

$$J_0 = \int x^0 e^x \, dx = e^x + C$$

Hence

$$J_3 = e^x(x^3 - 3x^2 + 6x - 6) + C$$

and consequently

$$\begin{aligned} J_5 &= x^5e^x - 5x^4e^x + 20e^x(x^3 - 3x^2 + 6x - 6) + C \\ &= e^x(x^5 - 5x^4 + 20x^3 - 60x^2 + 120x - 120) + C \end{aligned}$$

●

Question Study the polynomial in the answer. How does each term follow from the preceding term? Can you write down the value of $\int x^6e^x\,dx$ by inspection?

Example 3

Derive a reduction formula for $\displaystyle\int \cos^n x\,dx$

Solution Write

$$J_n = \int \cos^n x\,dx = \int \cos^{n-1}x \cos x\,dx$$

and integrate by parts with

$$u = \cos^{n-1}x \qquad dv = \cos x\,dx$$

$$du = -(n-1)\cos^{n-2}x \sin x\,dx \qquad v = \sin x$$

The result is

$$\int \cos^n x\,dx = \cos^{n-1}x \sin x + (n-1)\int \cos^{n-2}x(1 - \cos^2 x)\,dx$$

Therefore $J_n = \cos^{n-1}x \sin x + (n-1)J_{n-2} - (n-1)J_n$. Combine the terms in J_n:

$$nJ_n = \cos^{n-1}x \sin x + (n-1)J_{n-2}$$

Now dividing by n gives the desired reduction formula:

$$\int \cos^n x\,dx = \frac{\cos^{n-1}x \sin x}{n} + \frac{n-1}{n}\int \cos^{n-2}x\,dx$$

●

Remark This reduction formula lowers the power of $\cos x$ by two. Therefore, repeated application will ultimately reduce J_n to J_0 or J_1, according to whether n is even or odd. But both of these are easy:

$$J_0 = \int \cos^0 x\,dx = \int dx = x + C$$

$$J_1 = \int \cos x\,dx = \sin x + C$$

Example 4

Compute $\displaystyle\int_0^{\pi/2} \cos^6 x\,dx$

Solution Set

$$K_n = \int_0^{\pi/2} \cos^n x\,dx$$

(This notation distinguishes from the J_n we have used previously for indefinite integrals.) Then by the reduction formula of Example 3,

$$K_n = \frac{\cos^{n-1}x \sin x}{n}\Bigg|_0^{\pi/2} + \frac{n-1}{n}\int_0^{\pi/2} \cos^{n-2}x\,dx$$

Hence

$$K_n = 0 + \frac{n-1}{n}K_{n-2} = \frac{n-1}{n}K_{n-2}$$

Apply this formula with $n = 6$, then repeat with $n = 4$ and $n = 2$:

$$K_6 = \tfrac{5}{6}K_4 = \tfrac{5}{6}\cdot\tfrac{3}{4}K_2 = \tfrac{5}{6}\cdot\tfrac{3}{4}\cdot\tfrac{1}{2}K_0$$

Therefore

$$\int_0^{\pi/2} \cos^6 x\,dx = \frac{5\cdot 3\cdot 1}{6\cdot 4\cdot 2}\int_0^{\pi/2} dx = \frac{5\cdot 3\cdot 1}{6\cdot 4\cdot 2}\cdot\frac{\pi}{2} = \tfrac{5}{32}\pi \qquad \bullet$$

Integral tables usually include a number of useful reduction formulas.

Exercises

Find a formula reducing J_n to J_{n-1}

1 $J_n = \int (\ln x)^n\,dx$

2 $J_n = \int x^n e^{-2x}\,dx$

3 $J_n = \int x^2(\ln x)^n\,dx$

4 $J_n = \int \dfrac{dx}{(a^2 + x^2)^n}$

5 $J_n = \int \dfrac{dx}{(x^2 - a^2)^n}$

6 $J_n = \int \dfrac{dx}{(a^3 + x^3)^n}$

Find a formula reducing J_n to J_{n-2}

7 $J_n = \int \tan^n x\,dx$

8 $J_n = \int \dfrac{dx}{x^n\sqrt{x^2 + a^2}}$

9 $J_n = \int \sec^n x\,dx$

***10** $J_n = \int e^{ax}\sin^n bx\,dx$

11 $J_n = \int \sin^n x\,dx$

***12** $J_n = \int (\arcsin x)^n\,dx.$

Compute by means of an appropriate reduction formula

13 $\int_0^{\pi/2} \sin^7 x\,dx$

14 $\int_0^{\pi/2} \sin^8 x\,dx$

15 $\int_0^{\pi/4} \tan^{10} x\,dx$

16 $\int_0^1 \dfrac{dx}{(1 + x^2)^3}$

17 $\int_1^2 (\ln x)^4\,dx$

***18** $\int_{\pi/2}^{\pi} x^4 \sin x\,dx$

19 Set $J_n = \int e^{ax}\tan^n x\,dx$. Express J_n in terms of J_{n-1} and J_{n-2}.

***20** Prove $\displaystyle\int_a^b (x-a)^m(b-x)^n\,dx = \frac{m!n!}{(m+n+1)!}(b-a)^{m+n+1}$

6-10 Review Exercises

Compute

1 $\displaystyle\int \frac{dx}{(x-a)(x-b)}$

2 $\displaystyle\int \frac{x^2\,dx}{\sqrt{x+1}}$

3 $\displaystyle\int \frac{\sin 2x\,dx}{\sqrt{5+\cos 2x}}$

4 $\displaystyle\int \frac{dx}{(2-5x)^2}$

5 $\displaystyle\int \sec^4 3x \tan^3 3x\,dx$

6 $\displaystyle\int \tan^5 x\,dx$

7 $\displaystyle\int \sin^6 x \cos^3 x\,dx$

8 $\displaystyle\int \sin^4 x\,dx$

9 $\displaystyle\int x^5\sqrt{1+x^2}\,dx$

10 $\displaystyle\int \frac{x^3\,dx}{x^2+4x+13}$

11 $\displaystyle\int \sin\sqrt{x}\,dx$

12 $\displaystyle\int \frac{dx}{e^x+1}$

13 $\displaystyle\int \frac{x\,dx}{\sqrt{2x-x^2}}$

14 $\displaystyle\int \frac{dx}{(1-x^2)^{3/2}}$

15 $\displaystyle\int x^2 \arctan x\,dx$

16 $\displaystyle\int \frac{(x+1)e^x}{xe^x+1}\,dx$

17 $\displaystyle\int \frac{dx}{a^4x^2+b^2x^4}$

18 $\displaystyle\int \frac{x^4+a^4}{x^4-a^4}\,dx$

19 $\displaystyle\int \frac{dx}{e^x+5+4e^{-x}}$

20 $\displaystyle\int \frac{\ln(2+\sqrt{x})}{\sqrt{x}}\,dx$

21 $\displaystyle\int \frac{dx}{x(\ln x)(\ln\ln x)}$

22 $\displaystyle\int \frac{x\,dx}{\sqrt{1-9x^4}}$

23 $\displaystyle\int x^2 \sin x\,dx$

24 $\displaystyle\int \frac{e^x\,dx}{\sqrt{1+e^{2x}}}$

25 $\displaystyle\int \frac{dx}{x\sqrt{1+x^2}}$

26 $\displaystyle\int \frac{\sqrt{a^2x^2-1}}{x}\,dx$

27 Evaluate $\displaystyle\int_{-a}^{a} \sqrt{a^2-x^2}\,dx$ by inspection.
[Hint Interpret as an area.]

28 According to Table 6-5-1

$$\int_{-\alpha}^{\alpha} \sec x\,dx = \ln|\sec x + \tan x|\Big|_{-\alpha}^{\alpha} = \ln|\sec\alpha + \tan\alpha| - \ln|\sec\alpha - \tan\alpha|$$

But secant is an even function, so

$$\int_{-\alpha}^{\alpha} \sec x\,dx = 2\int_{0}^{\alpha} \sec x\,dx = 2\ln|\sec x + \tan x|\Big|_{0}^{\alpha} = 2\ln|\sec\alpha + \tan\alpha|$$

The answers appear different. Explain.

29 Find $\displaystyle\int_0^b x(b-x)^n\,dx$

30 Find $\displaystyle\int \frac{dx}{\sqrt{e^x-1}}$

7 Applications of Integration

7-1 Introduction

The definite integral

$$\int_a^b f(x)\,dx$$

introduced in order to compute areas, turns out to be a powerful tool not only in area problems, but in a surprisingly large number of other applications. The reason for its great versatility is that the integral can "sum" lots of tiny quantities, whatever their nature.

For example, take the problem of computing the area of a region under the curve $y = f(x)$. The region can be sliced vertically into a large number of thin pieces, each approximately a rectangle of area $f(x)\,\Delta x$, where x falls somewhere in the slice and Δx is the width of the slice. The integration process "adds up" all these tiny areas into a Riemann sum, an approximation to the area of the region, and then passes to its limit. Even the notation is suggestive: $f(x)\,dx$ represents $f(x)\,\Delta x$, and the symbol

$$\int_a^b f(x)\,dx$$

means "sum" the quantities $f(x)\,dx$ between $x = a$ and $x = b$. (The $\int$ sign was originally an S for sum.) After this introductory section, we shall use dx rather than Δx when setting up integrals.

Figure 7-1-1
Radius of slice $\approx f(x)$
Area of slice $\approx \pi f(x)^2$
Volume of slice $\approx \pi f(x)^2 \Delta x$
Volume of vase: $V = \int_a^b \pi f(x)^2\, dx$

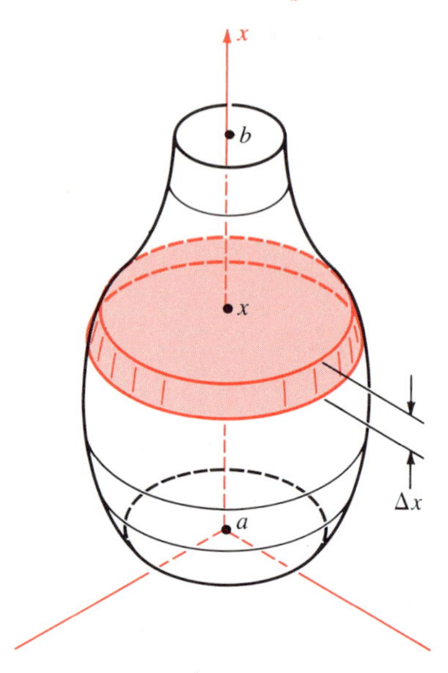

It happens in many applications that a quantity can be divided into a large number of small parts, each given by an expression of the type $f(x)\,\Delta x$. The integration process first adds up these parts into a Riemann sum and then takes its limit, just as with area. We consider three typical examples.

Volume of Revolution

A vase is being shaped on a potter's wheel (Figure 7-1-1). At each x between a and b, its cross section is a circle of radius $f(x)$. What is the volume of the vase?

We slice the vase into thin slabs by cuts perpendicular to the x-axis. The slab at height x is nearly a thin cylindrical disk of volume

$$(\text{area of base}) \cdot (\text{thickness}) = [\pi f(x)^2]\,\Delta x$$

The integration process

$$\int_a^b \pi f(x)^2\, dx$$

first adds up these small volumes into a Riemann sum, an approximation to the volume of the vase. Then it takes the limit of the sum, the total volume of the vase.

Figure 7-1-2
A particle at x pulled by a force $f(x)$

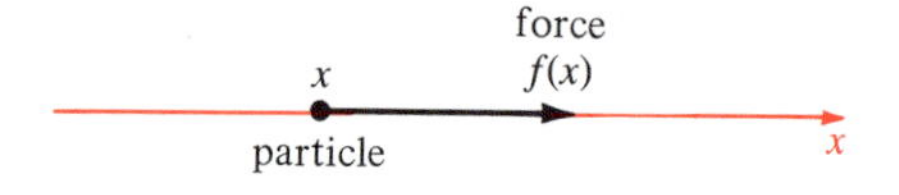

Work

Suppose at each point of the x-axis there is a force of magnitude $f(x)$ pulling a particle (Figure 7-1-2). How much work is done by the force in moving the particle from $x = a$ to $x = b$?

We partition the interval from a to b into a large number of small pieces of length Δx. In the piece at x, the force is nearly constant, so the work it does there is approximately

$$(\text{force}) \cdot (\text{distance}) = f(x)\,\Delta x$$

The integral

$$\int_a^b f(x)\, dx$$

first adds up these little bits of work into a Riemann sum, an approximation to the total work. Then it gives the limit of the sum, the total work done.

Figure 7-1-3
Motion of a particle with velocity $v(t)$

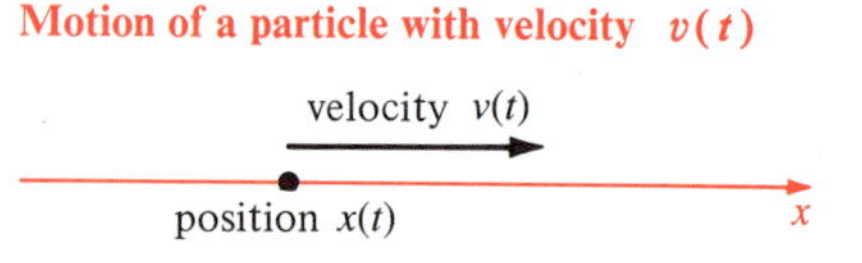

Distance

If a particle (Figure 7-1-3) moves to the right along the x-axis with velocity $v(t)$ at time t, how far does it move between $t = a$ and $t = b$?

We partition the time interval into a large number of very short equal time intervals, each of duration Δt. In the short interval around time t, the velocity is practically constant, so the distance traveled in this short period of time is approximately

$$(\text{velocity}) \cdot (\text{time}) = v(t)\,\Delta t$$

The integral

$$\int_a^b v(t)\,dt$$

first adds up all these little distances into a Riemann sum, an approximation to the distance traveled. Then it gives the limit of the sum, the overall distance traveled. See Figure 7-1-4.

Figure 7-1-4
Velocity in the Δt time interval: $\approx v(t)$
Distance traveled in this time interval: $\approx v(t)\,\Delta t$
Distance traveled in the time interval $[a, b]$:

$$D = \int_a^b v(t)\,dt$$

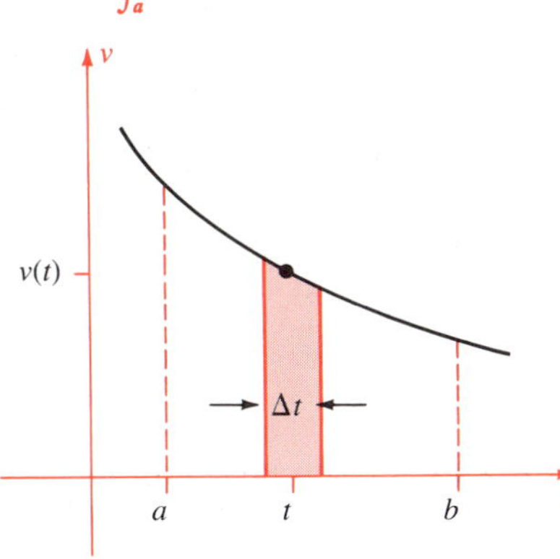

Summary

The integral "adds up" many small quantities:

Area = limit of sum of thin rectangles of area $f(x)\,\Delta x$

$$= \int_a^b f(x)\,dx$$

Volume of revolution

$$= \text{limit of sum of thin cylindrical disks of volume } \pi f(x)^2\,\Delta x$$

$$= \pi \int_a^b f(x)^2\,dx$$

Work = limit of sum of small amounts of work $f(x)\,\Delta x$

$$= \int_a^b f(x)\,dx$$

Distance = limit of sum of short distances $v(t)\,\Delta t$

$$= \int_a^b v(t)\,dt$$

7-2 Area

Suppose we want to find the area of the region (Figure 7-2-1, next page) bounded by the curves $y = f(x)$ and $y = g(x)$, where $g(x) \le f(x)$, and the lines $x = a$ and $x = b$. We think of the region as split into a large number of thin rectangles. A typical one, shown in Figure 7-2-1 (next page), has height $[f(x) - g(x)]$, width dx, and area $[f(x) - g(x)]\,dx$. The integral

$$A = \int_a^b [f(x) - g(x)]\,dx$$

sums these areas and gives the limit, the required area of the region.

We shall refer to the differential

$$dA = [f(x) - g(x)]\,dx$$

as the **element of area.** It is a suggestive terminology for setting up applications of integration.

Figure 7-2-1
Length of rectangle: $\approx f(x) - g(x)$
Area of rectangle: $\approx [f(x) - g(x)]\,dx$
Area of region:

$$A = \int_a^b [f(x) - g(x)]\,dx$$

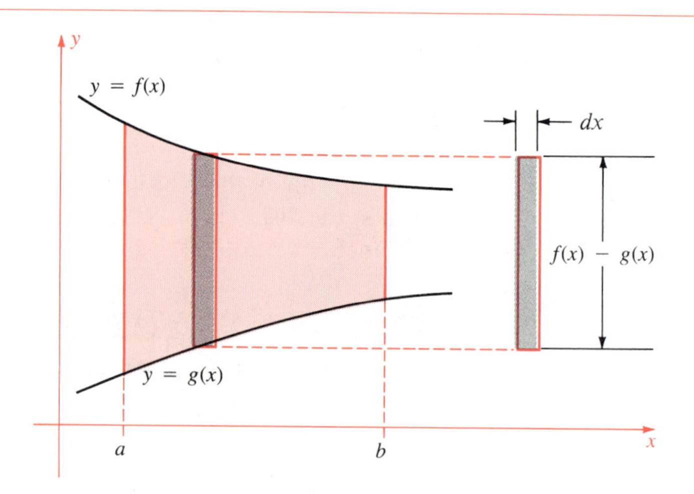

The following steps should be followed in solving area problems by integration:

1. Draw a figure.
2. Choose an appropriate variable and determine its domain.
3. Express the element of area in terms of the variable and its differential.
4. Integrate.

Sometimes the region must be split into several regions. Then apply steps 2–4 to each subregion.

Example 1

Compute the area of the region bounded by the curves $y = e^{x/2}$ and $y = 1/x^2$, and the lines $x = 2$ and $x = 3$.

Solution Sketch the region (Figure 7-2-2). Think of it as being composed of thin rectangular slabs. The area of the typical slab is

$$dA = \left(e^{x/2} - \frac{1}{x^2}\right) dx$$

Therefore, in the limit,

$$A = \int_2^3 \left(e^{x/2} - \frac{1}{x^2}\right) dx = 2e^{x/2} + \frac{1}{x}\bigg|_2^3$$

$$= (2e^{3/2} + \tfrac{1}{3}) - (2e + \tfrac{1}{2}) = 2(e^{3/2} - e) - \tfrac{1}{6}$$ ●

Figure 7-2-2
See Example 1

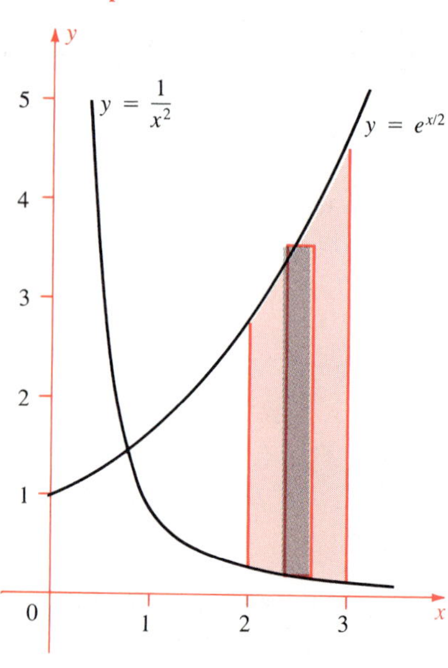

It is important to remember which is the upper boundary and which is the lower boundary. If we got the upper and lower boundaries reversed, then we would compute

$$\int_a^b [g(x) - f(x)]\,dx$$

which is not the area but the negative of the area.

If the curves cross, say at $x = b$, then the upper and lower boundaries reverse (Figure 7-2-3). In that case, we must compute the shaded area by

$$\int_a^b [f(x) - g(x)]\,dx + \int_b^c [g(x) - f(x)]\,dx$$

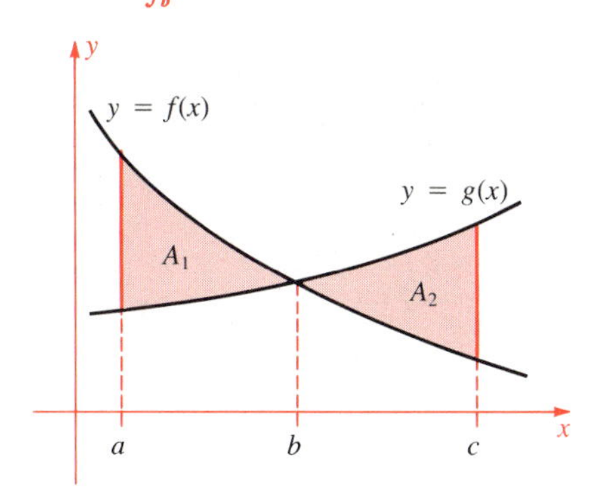

Figure 7-2-3
The *geometric* area is $A = A_1 + A_2$
$= \int_a^b [f(x) - g(x)]\,dx + \int_b^c [g(x) - f(x)]\,dx$

Under each integral sign, the upper curve comes first. If we computed just

$$\int_a^c [f(x) - g(x)]\,dx$$

then the two areas would be counted with opposite signs, and the result would not be the geometric area. However, $|f(x) - g(x)|$, the absolute value of the difference, *always* has the correct sign for measuring geometric area, so we can state a principle:

- The *geometric area* of the region bounded by the curves $y = f(x)$ and $y = g(x)$, and the lines $x = a$ and $x = b$, is

$$\int_a^b |f(x) - g(x)|\,dx$$

In particular, if $g(x) \le f(x)$, then the geometric area is

$$\int_a^b [f(x) - g(x)]\,dx$$

Example 2

Find the area bounded by the curves $y = x$ and $y = 4/x$, and the lines $x = 1$ and $x = 4$.

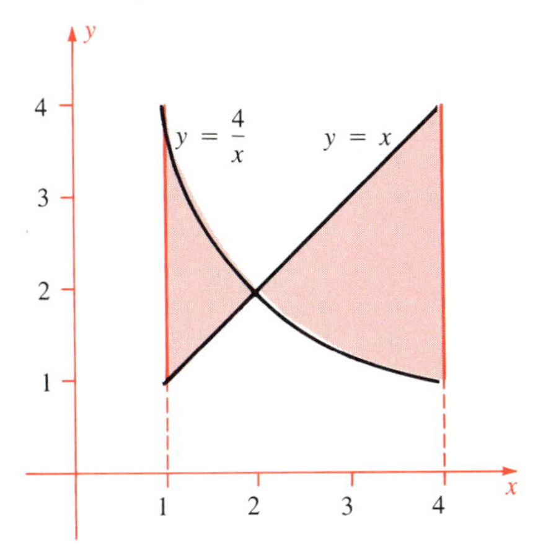

Figure 7-2-4
See Example 2

Solution First we need a figure (Figure 7-2-4). The curves evidently cross at $(2, 2)$. For $1 \le x \le 2$, the curve $y = 4/x$ is the upper boundary of the region. For $2 \le x \le 4$, the curve $y = x$ is the upper boundary. Therefore

$$\begin{aligned} A &= \int_1^2 \left(\frac{4}{x} - x\right) dx + \int_2^4 \left(x - \frac{4}{x}\right) dx \\ &= \left(4 \ln x - \tfrac{1}{2}x^2\right)\Big|_1^2 + \left(\tfrac{1}{2}x^2 - 4 \ln x\right)\Big|_2^4 \\ &= [(4 \ln 2 - 2) - (4 \ln 1 - \tfrac{1}{2})] \\ &\quad + [(8 - 4 \ln 4) - (2 - 4 \ln 2)] \\ &= [4 \ln 2 - \tfrac{3}{2}] + [6 - 4 \ln 2] = \tfrac{9}{2} \end{aligned}$$

Example 3

Compute the area of the region bounded by the curves $y = x^2$ and $y = x + 2$.

Figure 7-2-5
By algebra,
$P = (-1, 1)$ and $Q = (2, 4)$
See Example 3

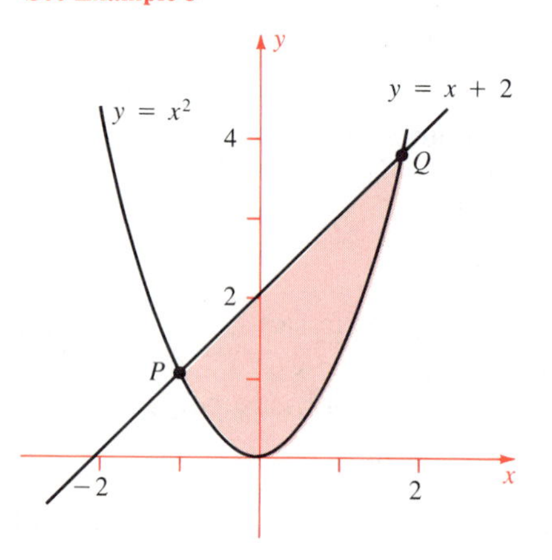

Solution It is not clear from the statement of the problem what the region is; a graph is most helpful (Figure 7-2-5). It shows that the region is a parabolic segment. Clearly, our first task is to find the intersections P and Q, that is, the solutions of the system

$$\begin{cases} y = x^2 \\ y = x + 2 \end{cases}$$

To solve the system, we eliminate y. The resulting equation for x is

$$x^2 - x - 2 = 0$$

with zeros $x = 2$ and $x = -1$. Thus, we have either $x = 2$ and $y = x + 2 = 4$ or $x = -1$ and $y = x + 2 = 1$, so the two intersections are

$$P = (-1, 1) \quad \text{and} \quad Q = (2, 4)$$

Consequently

$$A = \int_{-1}^{2} [(x + 2) - x^2]\, dx = \int_{-1}^{2} (-x^2 + x + 2)\, dx$$

$$= (-\tfrac{1}{3}x^3 + \tfrac{1}{2}x^2 + 2x)\Big|_{-1}^{2}$$

$$= -\tfrac{1}{3}(8 + 1) + \tfrac{1}{2}(4 - 1) + 2(2 + 1) = \tfrac{9}{2}$$ ●

Example 4

Compute the geometric area of the region (Figure 7-2-6) bounded by the curves

$$y = f(x) = x^3 - 2x^2 - 5x + 6$$

and

$$y = g(x) = -x^3 + 8x^2 - 9x - 10$$

Figure 7-2-6
$f(x) = x^3 - 2x^2 - 5x + 6$
$g(x) = -x^3 + 8x^2 - 9x - 10$
(The scales are distorted.)

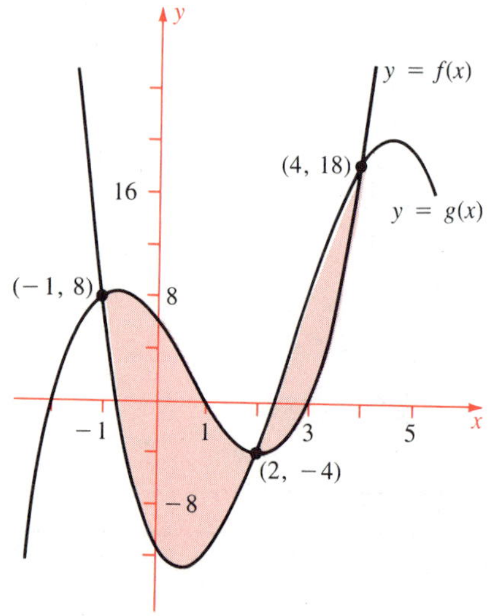

Solution To sketch the region we need the intersections of the graphs. Thus we must solve $f(x) - g(x) = 0$. But

$$f(x) - g(x) = 2x^3 - 10x^2 + 4x + 16$$

so we must solve the cubic equation

$$x^3 - 5x^2 + 2x + 8 = 0$$

By trial and error, its roots are $x = -1$, $x = 2$, and $x = 4$, and the corresponding values of y are $y = 8$, $y = -4$, and $y = 18$. Hence the intersections are $(-1, 8)$, $(2, -4)$, and $(4, 18)$. This information together with our general knowledge of the shape of cubics is enough for a graph (Figure 7-2-6). Clearly $g(x) \le f(x)$ for $-1 \le x \le 2$, and $f(x) \le g(x)$ for $2 \le x \le 4$. Therefore

$$A = \int_{-1}^{4} |f(x) - g(x)|\,dx$$
$$= \int_{-1}^{2} [f(x) - g(x)]\,dx + \int_{2}^{4} [g(x) - f(x)]\,dx$$

Since $f(x) - g(x) = 2x^3 - 10x^2 + 4x + 16$, the first integral is

$$\int_{-1}^{2} [f(x) - g(x)]\,dx = (\tfrac{1}{2}x^4 - \tfrac{10}{3}x^3 + 2x^2 + 16x)\Big|_{-1}^{2} = \tfrac{63}{2}$$

Similarly,

$$\int_{2}^{4} [g(x) - f(x)]\,dx = \tfrac{32}{3}$$

Consequently $A = \frac{63}{2} + \frac{32}{3} = \frac{253}{6}$. ●

Remark In this example, the *algebraic area* is

$$\int_{-1}^{4} [f(x) - g(x)]\,dx = \tfrac{63}{2} - \tfrac{32}{3} = \tfrac{125}{6}$$

which is quite a different answer.

The methods of this section apply also to regions bounded by curves $x = f(y)$ and $x = g(y)$ and lines $y = c$ and $y = d$. The area of such a region is given by the integral

$$\int_{c}^{d} |f(y) - g(y)|\,dy$$

It is the limit of the sum of horizontal—rather than vertical—slices.

Example 5

Compute the area of the region bounded by the curves $y = x$, $y = 1/x^2$, and the line $y = 2$.

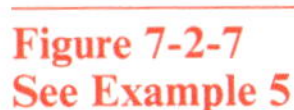
Figure 7-2-7
See Example 5

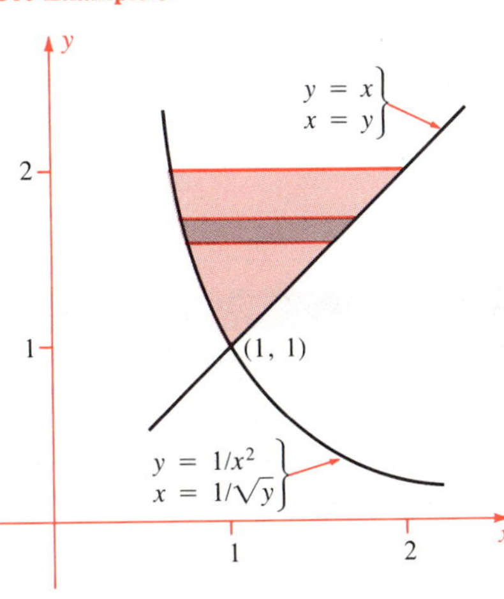

Solution Sketch the region (Figure 7-2-7). The curves obviously intersect at $(1, 1)$. If you compute the area by vertical slicing, you will need two integrals, because the lower boundary is made up of two different curves. So slice horizontally instead. Then the right-hand boundary is $x = y$ and the left-hand boundary is $y = 1/x^2$, that is, $x = 1/\sqrt{y}$. The typical horizontal slice has element of area

$$dA = \left(y - \frac{1}{\sqrt{y}}\right)dy$$

Therefore the area of the region is

$$\int_{1}^{2} \left(y - \frac{1}{\sqrt{y}}\right)dy = (\tfrac{1}{2}y^2 - 2\sqrt{y})\Big|_{1}^{2} = \tfrac{7}{2} - 2\sqrt{2}$$ ●

Note When you slice horizontally, you obtain $dA = (\text{something})\,dy$. That "something" must be expressed in terms of y before you can integrate. That is why in Example 5 we wrote the boundaries as $x = y$ and $x = 1/\sqrt{y}$.

Exercises

The region is bounded as indicated. Sketch the region and compute its area

1 $y = e^x \quad y = 1 + x \quad x = 0 \quad x = 1$

2 $y = x^2 \quad y = -x^2 \quad x = -1 \quad x = 1$

3 $y = x^3 \quad y = x + 1 \quad x = -1 \quad x = 0$

4 $y = e^x \quad y = \ln x \quad x = 1 \quad x = 2$

5 $y = 1/x \quad y = 1/x^2 \quad x = 1 \quad x = 3$

6 $y = 1/x \quad y = x \quad x = \frac{1}{2} \quad x = 1$

7 $y = 1 - x^2 \quad y = x^2 - 1$

8 $y = |x| \quad y = -|x| \quad x = -1 \quad x = 2$

9 $y = \sin x \quad y = \cos x \quad x = \frac{1}{2}\pi \quad x = \pi$

10 $y = 2 + \sin 2x \quad y = \sin x \quad x = 0 \quad x = 2\pi$

Compute the area of the region bounded by

11 $y = 8 - x^2 \quad y = -2x$

12 $y = x^2 + 5 \quad y = 6x$

13 $y = 3x^2 \quad y = -3x^2 \quad x = -1 \quad x = 1$

14 $y = 1 - x^3 \quad y = x^2 - 1 \quad x = -1$

15 $y = x^3 - 5x^2 + 6x \quad y = x^3$

16 $y = x^2 - 8x \quad y = x$

17 $y = \cos x \quad y = \sin x \quad x = \frac{1}{4}\pi \quad x = \frac{5}{4}\pi$

18 $y = 2\sin(\frac{1}{4}\pi x) \quad y = (x - 3)(x - 4)$
$x = 2 \quad x = 4$

19 $y = e^x \quad y = e^{-x} \quad y = e^2$

20 $y = \cos x - 1 \quad y = 1 - \cos 2x$
$x = 0 \quad x = 2\pi$

21 $y = 1/x^2 \quad y = x \quad y = 8x$

22 $y = 1/x^2 \quad y = 0 \quad y = x^2 \quad x = 3$

23 $y^2 = 2x \quad y = \frac{1}{2}x - 3$

24 $x = y^2 \quad x = 6 - y^4$

25 Find a so that the area bounded by $y = x^2 - a^2$ and $y = a^2 - x^2$ is 9.

26 Find the fraction of the area of one hump of the curve $y = \sin x$ that lies above $y = \frac{1}{2}$.

27 Let $P = (a, ka^2)$ and $Q = (b, kb^2)$ be two points of the parabola $y = kx^2$. Find the area bounded by the parabola and the segment PQ, assuming $k > 0$ and $a < b$.

28 (cont.) Prove that this area is $\frac{4}{3}$ the area of the triangle PTQ, where T is the point of the parabola whose tangent is parallel to PQ. (This result is Archimedes' quadrature of the parabolic segment.)

29 The larger square in Figure 7-2-8 is fixed; the smaller square slides on the x-axis, the center of its base at x. Graph the function $A(x)$, the area of the intersection of the squares.

30 (cont.) Find the average value of $A(x)$ for $-3 \le x \le 3$.

Figure 7-2-8
See Exercise 29

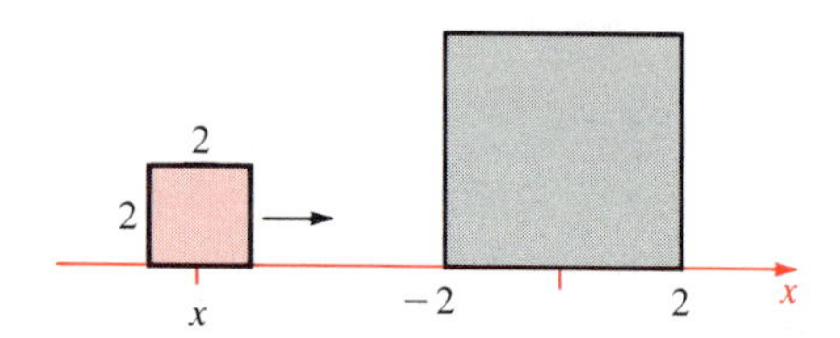

7-3 Volume

Before computing volumes, let us look at a simple area problem whose solution involves some useful ideas.

Example 1

Find the area of a circle of radius r. (Assume the formula $c = 2\pi r$ for its circumference.)

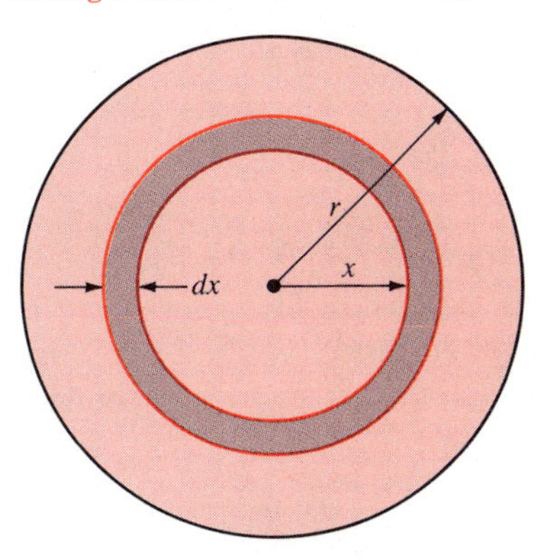

Figure 7-3-1
Slicing a circle into concentric rings

Solution Cut the circle into thin *concentric rings* (Figure 7-3-1). Let x denote the radial distance of a ring from the center, so $0 \le x \le r$. The typical ring has length $2\pi x$ and width dx. Hence the element of area is $dA = 2\pi x\, dx$. Therefore the area of the circle is

$$A = \int_0^r 2\pi x\, dx = \pi x^2 \Big|_0^r = \pi r^2$$

The strategy used in solving Example 1 is important and worth reviewing. First we slice the circle into thin pieces, each having area approximately $2\pi x\, dx$. Then we find the limit of the sum of these small areas by integrating.

Similar strategy applies to finding the volume of a solid. We slice the solid into many thin pieces, each of which is approximately a familiar shape of known volume. The element of volume is $dV = A(x)\, dx$, where the base area $A(x)$ of the typical slice is multiplied by its thickness dx. Then we find the limit of the sum of these little volumes by integrating. Thus the general plan of attack on a particular solid consists of four steps:

1 Sketch the solid.
2 Choose a method of slicing the solid.
3 Choose a variable x that locates the typical slice, find the domain $[a, b]$ that applies to the problem, and work out an expression $A(x)$ for the area of the slice. The x-axis must be perpendicular to the slice. Then the element of volume is $dV = A(x)\, dx$.
4 Evaluate

$$V = \int_a^b A(x)\, dx$$

Remark In practice, not all volume problems can be solved in this way. In later chapters we develop the double and the triple integral as additional tools.

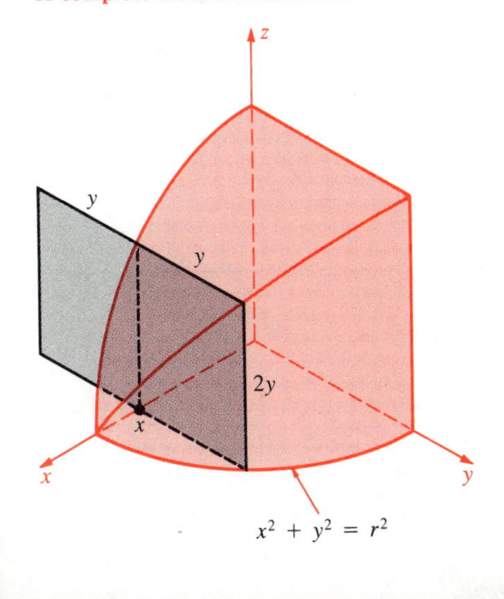

Figure 7-3-2
One quarter of the solid in Example 2. A complete cross section is shown.

Example 2

A solid resting on the x, y-plane has a circular base bounded by the circle $x^2 + y^2 = r^2$. Each slice by a plane perpendicular to the x-axis is a square. Find the volume of the solid.

Solution A figure showing one-quarter of the solid helps us to visualize it (Figure 7-3-2). A cross section is shown passing through $(x, 0)$ on the x-axis. It is a square of side $2y$, so its area as a function of x is

$$A(x) = (2y)^2 = 4y^2 = 4(r^2 - x^2)$$

The element of volume is $dV = A(x)\, dx$. Because the base of the solid is bounded by $x^2 + y^2 = r^2$, we have $-r \le x \le r$, so the set-up is

$$V = \int_{-r}^{r} A(x)\, dx = \int_{-r}^{r} 4(r^2 - x^2)\, dx$$

This is easy to evaluate:

$$V = 4\int_{-r}^{r} (r^2 - x^2)\,dx = 8\int_{0}^{r} (r^2 - x^2)\,dx$$
$$= 8(r^2x - \tfrac{1}{3}x^3)\Big|_0^r = 8(r^3 - \tfrac{1}{3}r^3) = \tfrac{16}{3}r^3$$

●

Remark Note that the answer is correct dimensionally. The radius r and the coordinates x and y have the dimension (length), and the volume has the dimension $(\text{length})^3$. If, for instance, we had come up with $V = 8(r^3 - \frac{1}{3}r^2)$, we would have known automatically that we had made a mistake.

Volume of Revolution

Let R be the region in the x, y-plane under the curve $y = f(x)$, where $a \le x \le b$. If R is revolved about the x-axis, a solid of revolution is swept out (Figure 7-3-3). What is its volume?

Figure 7-3-3
The solid swept out when R revolves about the x-axis

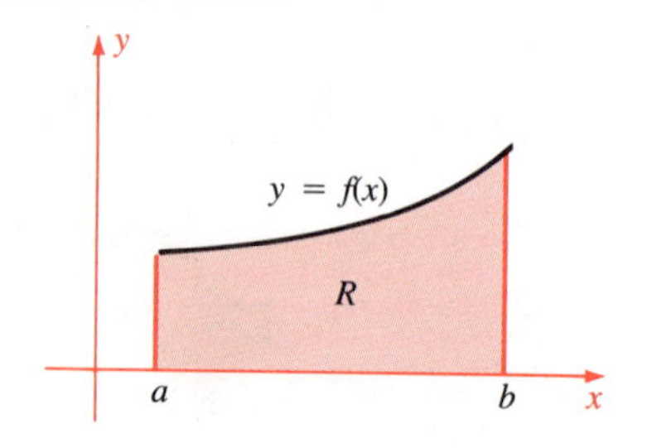

Figure 7-3-4
Thin "rectangle," dimensions $f(x) \times dx$

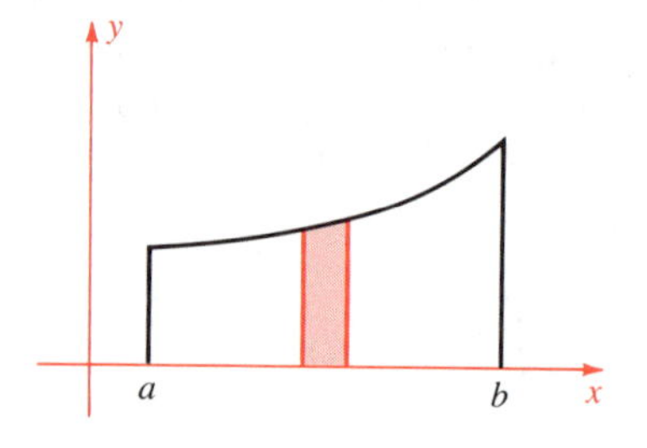

Figure 7-3-5
Resulting slab has base area $\pi f(x)^2$ and thickness dx. Hence its volume is $dV = \pi f(x)^2\,dx$.

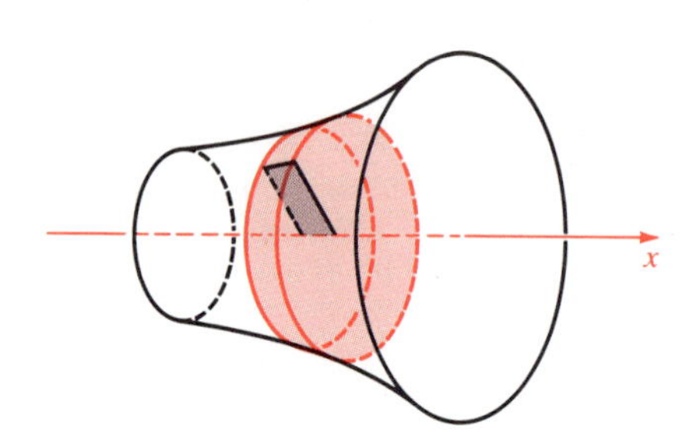

We follow our general strategy. First we slice R into thin rectangles (Figure 7-3-4). Each sweeps out a circular slab (Figure 7-3-5), so we have sliced the solid into slabs. The typical slab located at x has radius $f(x)$, base area $\pi f(x)^2$, and thickness dx, and hence its element of volume is $dV = \pi f(x)^2\,dx$. Consequently, the volume of the solid of revolution is

$$V = \int_a^b \pi f(x)^2\,dx$$

Example 3

Find the volume of a sphere of radius r.

Solution The sphere is a solid of revolution obtained by rotating a semicircle (Figure 7-3-6) about its diameter. Take the semicircle bounded by the x-axis and $y = \sqrt{r^2 - x^2}$. Each slab corresponding to a rectangle has radius $y = \sqrt{r^2 - x^2}$ and thickness dx, so the element of volume

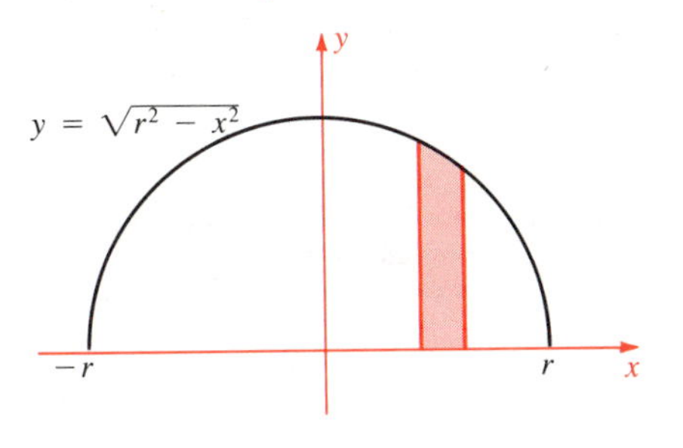

Figure 7-3-6
Rotating the semicircle about its diameter generates a sphere. The thin "rectangle" has dimensions $y \times dx$.

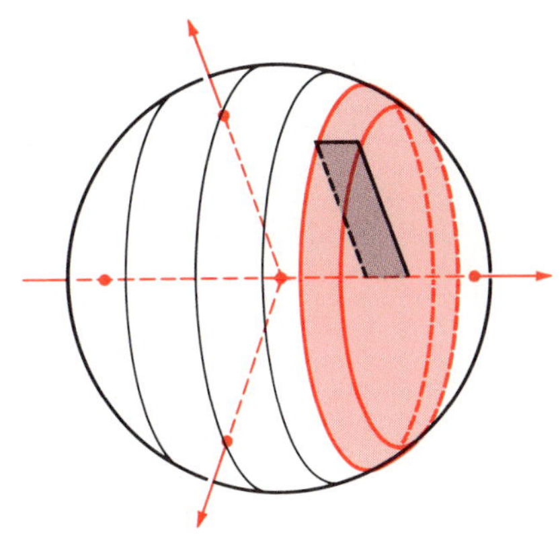

Figure 7-3-7
Resulting slab has volume
$dV = \pi y^2 dx = \pi(r^2 - x^2)\,dx$

(shown in Figure 7-3-7) is

$$dV = \pi y^2\,dx = \pi(r^2 - x^2)\,dx$$

Integrate from $x = -r$ to $x = r$:

$$V = \int_{-r}^{r} \pi(r^2 - x^2)\,dx = 2\pi \int_0^r (r^2 - x^2)\,dx$$

$$= 2\pi\left(r^2 x - \tfrac{1}{3}x^3\right)\Big|_0^r = 2\pi\left(r^3 - \tfrac{1}{3}r^3\right) = \tfrac{4}{3}\pi r^3$$

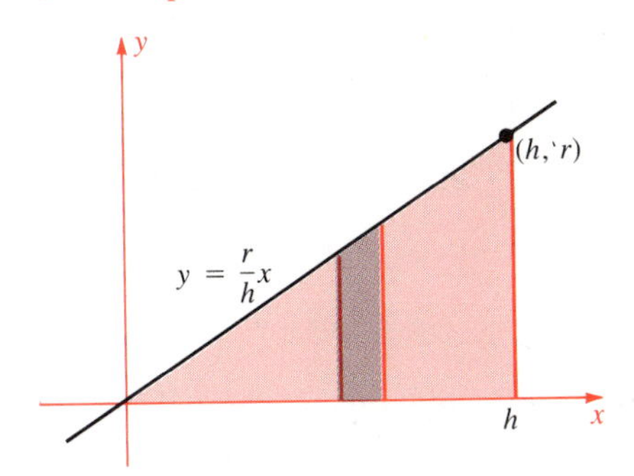

Figure 7-3-8
Rotate the triangle about the x-axis. See Example 4.

Example 4

A right circular cone of height h is constructed over a base of radius r. Compute its volume.

Solution The cone is a solid of revolution, obtained by revolving a right triangle about one leg. Choose the triangle indicated in Figure 7-3-8, and rotate it about the x-axis to generate the cone (Figure 7-3-9). Slice the cone into thin slabs, each of width dx. Note that the triangle is the region under the curve $y = (r/h)x$, where $0 \le x \le h$. Thus each slab has radius $(r/h)x$ and volume

$$dV = \pi\left(\frac{r}{h}x\right)^2 dx$$

Therefore the volume of the cone is

$$V = \int_0^h \pi \frac{r^2}{h^2} x^2\,dx = \pi \frac{r^2}{h^2}\left(\tfrac{1}{3}x^3\right)\Big|_0^h = \pi \frac{r^2}{h^2}\left(\tfrac{1}{3}h^3\right) = \tfrac{1}{3}\pi r^2 h$$

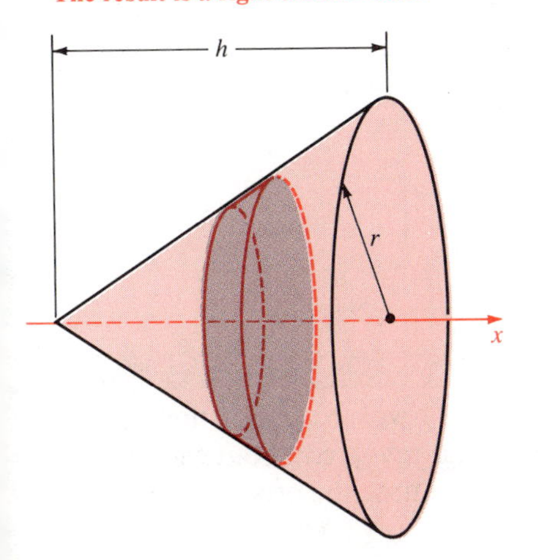

Figure 7-3-9
The result is a right circular cone.

If the region we revolve does not touch the axis of the revolution, then the strips generate washer-shaped slabs instead of cylindrical slabs. This situation is no harder to handle.

Example 5

The region in the x, y-plane bounded by the parabola $y = x^2$, the line $x = a$, and the x-axis is revolved about the y-axis. (Assume $a > 0$.) Find the volume of the resulting solid.

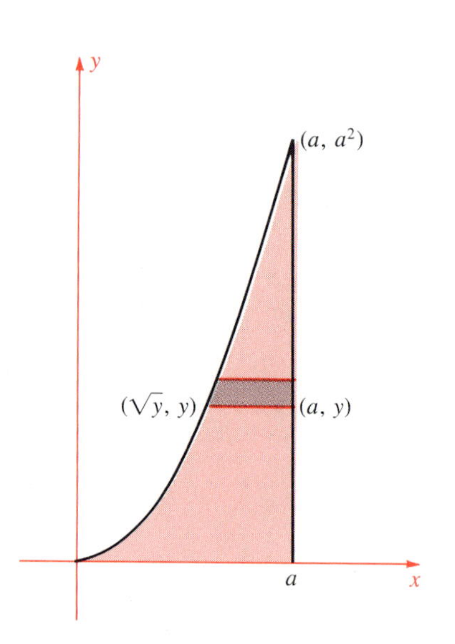

Figure 7-3-10
Slice the region into strips parallel to the x-axis. See Example 5.

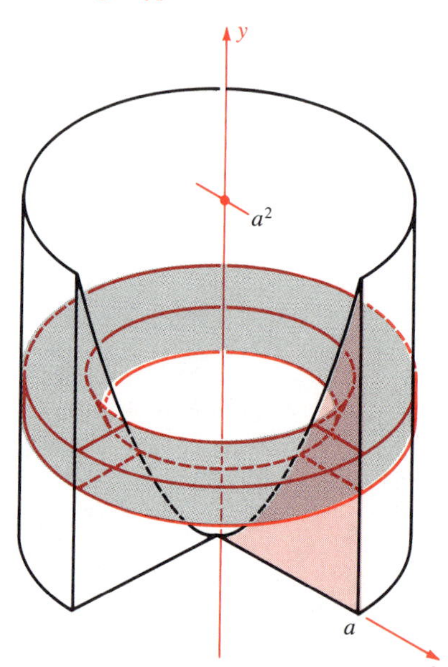

Figure 7-3-11
Cutaway view of the solid of revolution, showing a typical washer

Solution Slice the plane region into thin rectangles parallel to the x-axis (Figure 7-3-10). When the typical resulting strip is revolved about the y-axis, it sweeps out a thin circular washer (Figure 7-3-11). The base of the washer is the region between two concentric circles of radii a and $\sqrt{y}$. Hence its area is

$$\pi a^2 - \pi(\sqrt{y})^2 = \pi(a^2 - y)$$

Since its thickness is dy, the corresponding element of volume is $dV = \pi(a^2 - y)\,dy$. Integrate these small volumes from level $y = 0$ to level $y = a^2$:

$$V = \int_0^{a^2} \pi(a^2 - y)\,dy = \pi(a^2 y - \tfrac{1}{2}y^2)\Big|_0^{a^2}$$
$$= \pi(a^4 - \tfrac{1}{2}a^4) = \tfrac{1}{2}\pi a^4$$

●

Cylindrical Shells

Suppose the region R in Figure 7-3-12 is revolved about the x-axis. Slice it into *horizontal* slabs. Each slab, when revolved about the x-axis, generates a cylindrical shell with element of volume the product of its three dimensions:

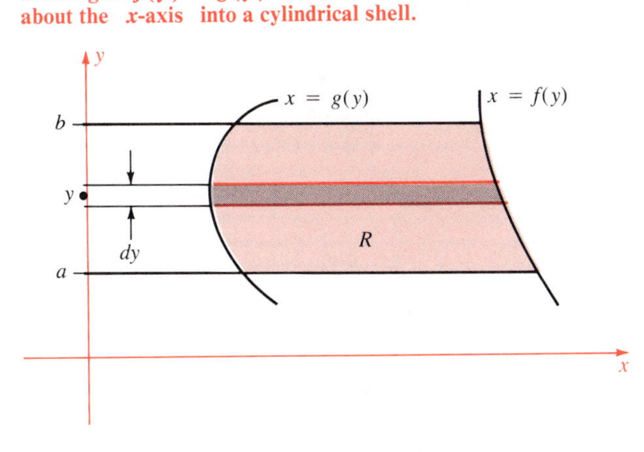

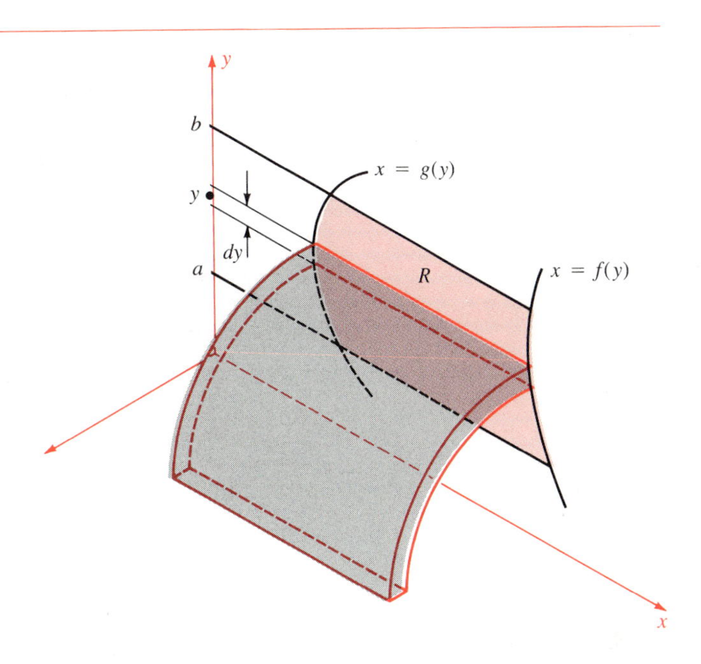

Figure 7-3-12
The region R is bounded by $x = f(y)$, $x = g(y)$, $y = b$, and $y = a$. The horizontal strip at y of width dy and length $f(y) - g(y)$ revolves about the x-axis into a cylindrical shell.

$$\begin{aligned} dV &= (\text{circumference})(\text{height})(\text{thickness}) \\ &= 2\pi y[f(y) - g(y)]\,dy \end{aligned}$$

Therefore the volume of the solid of revolution is

$$V = 2\pi \int_a^b y[f(y) - g(y)]\,dy$$

We shall use this approach for alternative solutions of the previous two examples.

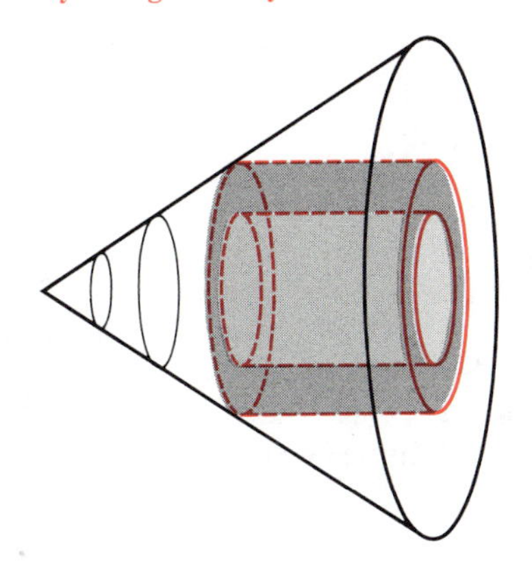

Figure 7-3-13
Finding the volume of a cone by slicing it into cylindrical shells

Example 4a

A right circular cone of height h is constructed over a base of radius r. Compute its volume.

Solution Slice the cone into cylindrical shells rather than slabs (Figure 7-3-13), and choose y as the variable. This corresponds to slicing the triangle into thin strips parallel to the x-axis (Figure 7-3-14). Each strip sweeps out a thin cylindrical shell with radius y, height $h - (h/r)y$, and thickness dy. The volume of this shell is

$$dV = (\text{circumference})(\text{height})(\text{thickness}) = (2\pi y)\left(h - \frac{hy}{r}\right) dy$$

Therefore

$$\begin{aligned} V &= \int_0^r (2\pi y)\left(h - \frac{h}{r}y\right) dy = 2\pi h \int_0^r \left(y - \frac{y^2}{r}\right) dy \\ &= 2\pi h\left(\tfrac{1}{2}y^2 - \tfrac{1}{3}y^3/r\right)\Big|_0^r = 2\pi h\left(\tfrac{1}{2}r^2 - \tfrac{1}{3}r^2\right) = \tfrac{1}{3}\pi r^2 h \end{aligned}$$

●

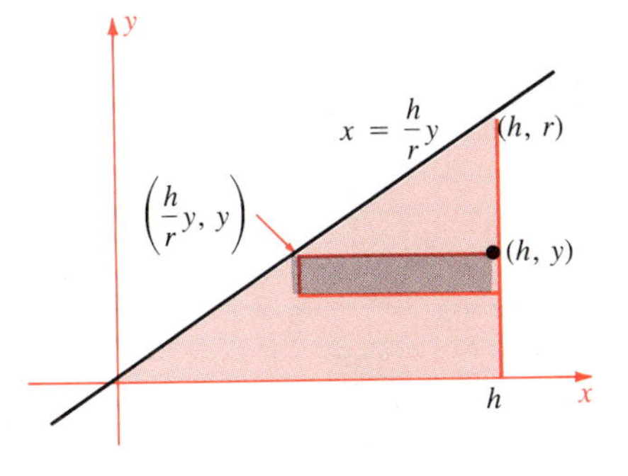

Figure 7-3-14
Rotating the horizontal strip about the x-axis results in the cylindrical shell shown in Figure 7-3-13.

Example 5a

The region in the x, y-plane bounded by the parabola $y = x^2$, the line $x = a$, and the x-axis is revolved about the y-axis. (Assume $a > 0$.) Find the volume of the resulting solid.

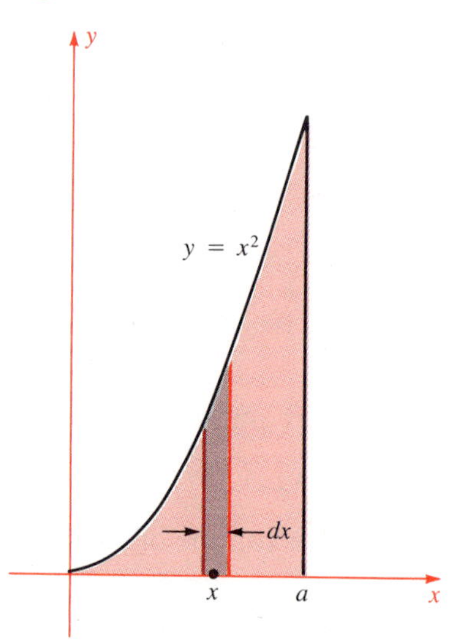

Figure 7-3-15
Rotating the vertical strip about the y-axis results in the cylindrical shell shown in Figure 7-3-16.

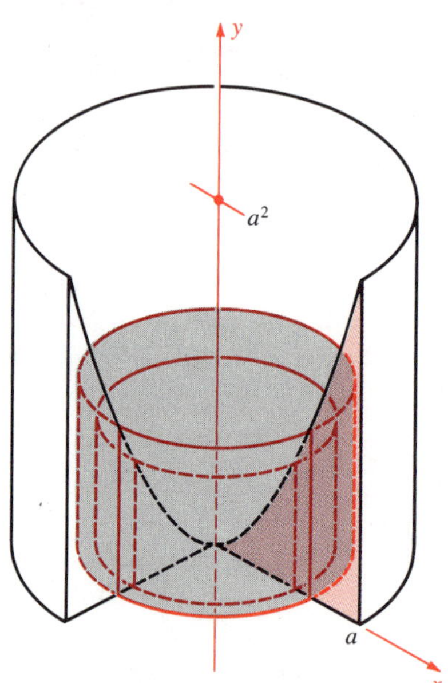

Figure 7-3-16
Solution of Example 5a by cylindrical shells

Solution The region under the parabola in Figure 7-3-15 is split into thin *vertical* strips parallel to the y-axis. Now x is the variable, and the solid of revolution is sliced into thin cylindrical shells (Figure 7-3-16). The typical shell has radius x, height x^2, and thickness dx. Hence element of volume is $dV = (2\pi x)x^2\,dx$. The volume of the solid is

$$V = \int_0^a (2\pi x)x^2\,dx$$

$$= \int_0^a 2\pi x^3\,dx$$

$$= 2\pi\left(\tfrac{1}{4}x^4\right)\Big|_0^a = \tfrac{1}{2}\pi a^4$$

Other Volumes

The volume of certain figures other than solids of revolution can also be found with the tools at our disposal.

Example 6

A cone of height h has an irregular base of area B. Find the volume of the cone.

Figure 7-3-17
Cone with irregular base

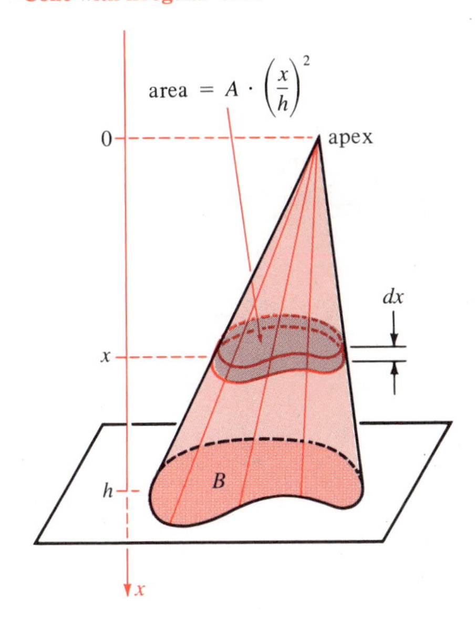

Solution Let x denote distance measured from the apex towards the plane of the base (Figure 7-3-17). The typical cross section of the cone by a plane parallel to the base, distance x from the apex, is a plane region similar to the base. This cross section has linear dimensions proportional to x, hence area proportional to x^2.

Let $A(x)$ denote this area. Then $A(x) = cx^2$. To find the constant c, note that $A(h) = B$. Therefore $ch^2 = B$ so that $c = B/h^2$ and

$$A(x) = \frac{B}{h^2}x^2$$

Slice the cone into slabs by planes parallel to the base. A typical slab has base area $A(x)$, thickness dx, and volume element

$$dV = A(x)\,dx = \frac{B}{h^2}x^2\,dx$$

Hence the volume of the cone is

$$V = \int_0^h \frac{B}{h^2}x^2\,dx = \frac{B}{h^2}\left(\tfrac{1}{3}x^3\right)\Big|_0^h = \tfrac{1}{3}Bh$$ ●

Area of a Sphere

The following discussion shows that sometimes you can put the cart before the horse! We know that the *volume* of a sphere of radius r is $\frac{4}{3}\pi r^3$. Let us use this information to find the *surface area* of the sphere. We don't know yet "officially" what surface area is. Nevertheless, let's assume that it exists.

Example 7

Find the surface area A of a sphere of radius r.

Figure 7-3-18
Cutaway view of sphere and a concentric spherical shell

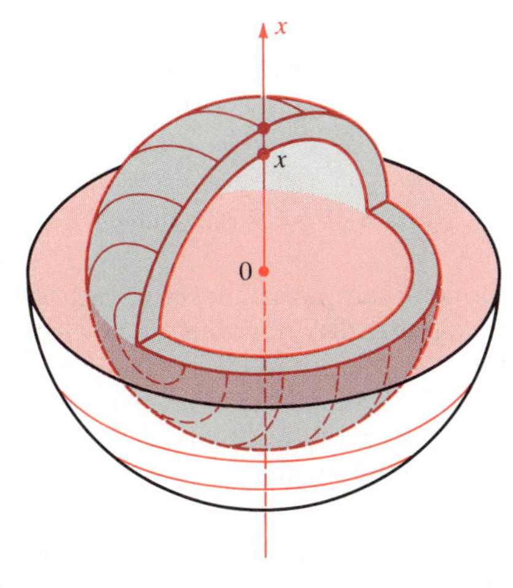

Solution Find the volume of the sphere by slicing it into concentric spherical shells (Figure 7-3-18). Let $S(x)$ be the surface area of the typical shell at distance x from the center $(0 \le x \le r)$. Then the element of volume of the sphere is the volume of the shell:

$$dV = S(x)\,dx$$

It follows that

$$V(r) = \int_0^r S(x)\,dx$$

By the fundamental theorem,

$$S(r) = \frac{d}{dr}V(r) = \frac{d}{dr}\left(\tfrac{4}{3}\pi r^3\right) = 4\pi r^2$$

Hence $A = 4\pi r^2$. ●

Exercises

1 A solid resting on the x, y-plane has as its base the semicircle bounded by the x-axis and the part $y \ge 0$ of $x^2 + y^2 = r^2$. Each cross section of the solid by a plane perpendicular to the x-axis is a semicircular disk. Find the volume of the solid.

2 A solid resting on the x, y-plane has as its base the first-quadrant region bounded by the y-axis, $y = b > 0$, and $y = x^2$. Each cross section of the solid by a plane perpendicular to the y-axis is a semicircular disk. Find the volume of the solid.

3 The base of a solid is the same as in Exercise 2. Its cross sections by planes perpendicular to the x-axis are semicircular disks. Find the volume of the solid.

4 The answer to Exercise 2 is b^2 times a constant factor. The answer to Exercise 3 is $b^{5/2}$ times a constant factor. Both answers seem dimensionally impossible. Think this matter through carefully and explain the apparent paradoxes.

The region of the x, y-plane whose boundary curves are given is revolved about the x-axis. Find the volume of the resulting solid of revolution

5 x-axis $\quad y = 2x + 3 \quad x = 0 \quad x = 3$

6 x-axis $\quad 2y + x = 3 \quad x = -1 \quad x = 1$

7 x-axis $\quad y = 1/(x + 1) \quad x = 0 \quad x = 4$

8 x-axis $\quad y = -3/(x + 4) \quad x = -1 \quad x = 1$

9 x-axis $\quad y = 1/x^2 \quad x = 1 \quad x = 3$

10 x-axis $\quad y = \sin x \quad x = 0 \quad x = \pi$

11 x-axis $\quad y = e^x \quad x = a \quad x = b \quad (a < b)$

12 x-axis $\quad y = \ln x \quad x = 1 \quad x = b \quad (1 < b)$

13 x-axis $\quad y = 3\sqrt{x} \quad x = 0 \quad x = 4$

14 $x = 16y^2 \quad x = 2 \quad x = 4 \quad (y \ge 0)$

15 x-axis $\quad y = \sinh x \quad x = a \quad (a > 0)$

16 $y = \cosh x \quad y = b \quad (b > 1)$

17 The region of the x, y-plane bounded by the x-axis, $y = x + 1$, $x = 1$, and $x = 4$ is revolved about the line $y = -3$. Find the volume of the resulting solid.

18 The region of the x, y-plane bounded by the lines $x = 1$, $y = x$, and $x = -2y + 6$ is revolved about the line $y = -2$. Find the volume of the resulting solid.

19 The region of the x, y-plane bounded by $y = -1$, $y = e^{2x}$, $x = 0$, and $x = 2$ is revolved about the line $y = -1$. Find the volume of the resulting solid.

20 The region of the x, y-plane bounded by the y-axis, $x^2 = \sin y$, $y = 0$, and $y = \pi$ is revolved about the y-axis. Find the volume of the resulting solid.

21 Find the volume of a frustum of a right circular cone with lower radius b, upper radius a, and height h. See Figure 7-3-19.

Figure 7-3-19
See Exercise 21

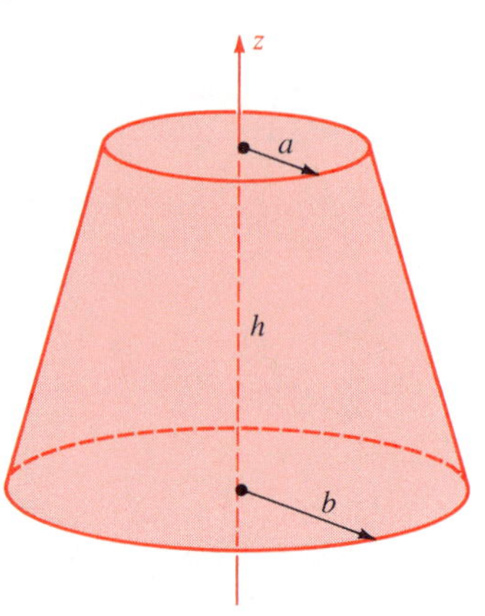

22 Find the remaining volume in a sphere of radius r if a hole of radius b is drilled through its center.

23 A plane at distance h from the center of a sphere of radius r cuts off a spherical cap of height $r - h$. Find the volume of the cap.

24 Find the volume of the solid formed by revolving the triangle in the x, y-plane with vertices $(1, 1)$, $(0, 2)$, and $(2, 2)$ about the x-axis.

25 A circular hole is cut on center vertically through a sphere, leaving a ring of height h. Calculate the volume of the ring.

26 A circle of radius a is revolved about a line in the same plane at distance b from the center of the circle. Assume $b > a$. Show that the resulting torus has volume $2\pi^2 a^2 b$. [Hint Use washers.]

27 The rectangle $-1 \le x \le 1$, $-2 \le y \le 2$, $z = 0$ moves upwards, always perpendicular to the z-axis, and its center always on the z-axis. It rotates counterclockwise at a uniform rate as it rises, and has turned $90°$ when it reaches $z = 1$. Find the volume swept out.

28 The region bounded by the x-axis, $y = f(x) = k/x$, $x = a$, and $x = b$ is revolved about the x-axis. Assume $k > 0$ and $0 < a < b$. Let $g(a)$ denote the limit of the resulting volume as $b \to +\infty$. Find k so $g(a) = f(a)$ for all $a > 0$.

29 For a certain type of tornado, each horizontal cross section of the funnel cloud is a circular disk of radius $x = ke^{az}$, where z is the altitude. In one of these storms it was observed that the radius at the ground was 50 ft while the radius at the top of the tornado, altitude 1000 ft, was 100 ft. Find the volume of the funnel.

30 Each horizontal cross section of a certain kitchen mixing bowl is a circle. When filled to depth h with batter, the volume of the batter is $V = ah^2 + 2bh$, where $a = 10.0$ cm and $b = 100.0$ cm^2. Express the radius of the top of the batter in terms of h. Estimate the radius when $h = 10.0$ cm.

A region on the surface of a sphere of radius r has area S. Each point of the region is connected to the center of the sphere by a line segment. Find the volume of the resulting solid.

32 (cont.) Suppose the region has hair growing radially outward. That is, from each point of the region, a segment of length a is constructed outward along the radial direction. Find the volume of the resulting solid mass of hair.

7-4 Work

In elementary physics, we are taught that the work W done by a constant force F in moving an object through distance D on a line is $W = FD$; that is, work equals force times distance.

When the force is variable, work is defined by an integral. The definition is based on the following reasoning. Suppose a continuous force $f(x)$ acts over an interval $[a, b]$ of the x-axis, and suppose an object is moved from a to b by the force. We partition the interval into many small subintervals. On the typical subinterval at x of length dx, the force is almost a constant $f(x)$. Therefore the work over this interval must be approximately $dW = f(x)\,dx$. We sum these approximations and take the limit to *define* work. Thus the **element of work** is

$$dW = f(x)\,dx$$

and the **work** done by the force in moving the object from $x = a$ to $x = b$ is

$$W = \int_a^b f(x)\,dx$$

Units In the English system, work is measured in foot-pounds (ft-lbs). In the CGS metric system, the unit of work is one **erg** = one dyne-centimeter; and in the MKS system, it is one **joule** (J) = one newton-meter. It follows that one joule = 10^7 ergs and one foot-pound $\approx$ 1.356 joules.

The MKS force unit, the **newton** (N) is the force that applied to a one-kilogram mass will impart an acceleration of one m/sec^2. A kilogram weight exerts a force of g newtons, where $g \approx 9.807$.

To solve a work problem, first sketch the situation (if it is not clear), and determine the distance interval. Next, express the force in terms of distance. Then express the element of work in terms of the distance and its differential. Finally, integrate.

Example 1

At each point of the x-axis (marked off in feet) there is a force of $5x^2 - x + 2$ pounds pulling an object. Compute the work done in moving it from $x = 1$ to $x = 4$.

Solution The element of work is $dW = (5x^2 - x + 2)\,dx$, hence

$$W = \int_1^4 (5x^2 - x + 2)\,dx = \left(\tfrac{5}{3}x^3 - \tfrac{1}{2}x^2 + 2x\right)\Big|_1^4$$

$$= \tfrac{5}{3}(64 - 1) - \tfrac{1}{2}(16 - 1) + 2(4 - 1) = 103.5 \text{ ft-lb}$$ ●

Figure 7-4-1
See Example 2

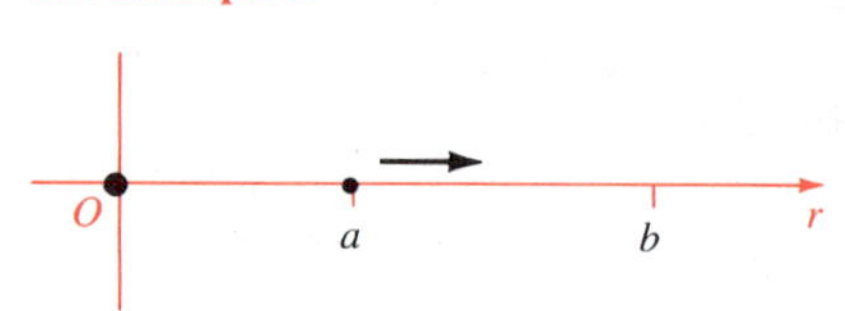

Figure 7-4-2
Hooke's Law. See Example 3

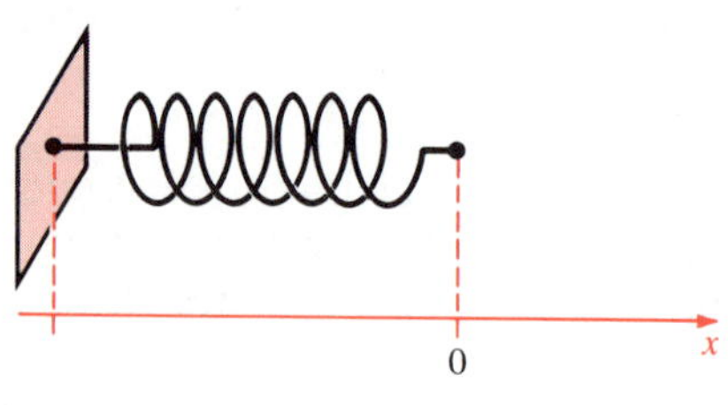

a Equilibrium position of spring

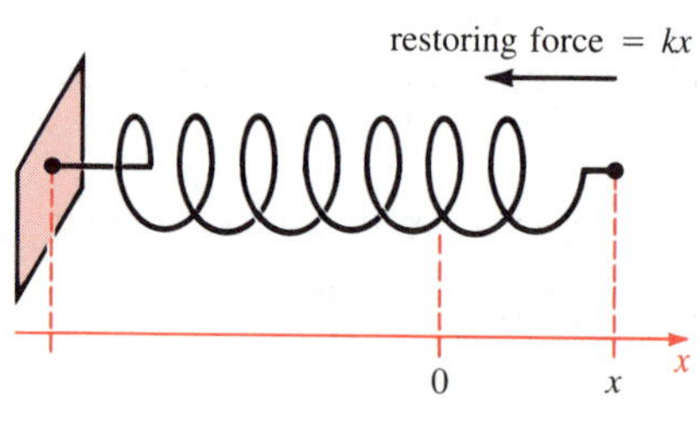

b Stretched position

Example 2

According to Newton's law of gravitation, two bodies attract each other with a force F proportional to the product m_1m_2 of their masses and inversely proportional to the square of the distance r between them:

$$F = G\frac{m_1m_2}{r^2}$$

where G is the **gravitational constant.** If one of the bodies is fixed at the origin (Figure 7-4-1), how much work is needed to move the other body from $r = a$ to $r = b$, where a and b are positive?

Solution The element of work is

$$dW = F\,dr = G\frac{m_1m_2}{r^2}\,dr$$

hence

$$W = \int_a^b G\frac{m_1m_2}{r^2}\,dr = Gm_1m_2\left(-\frac{1}{r}\right)\Bigg|_a^b$$
$$= Gm_1m_2\left(\frac{1}{a} - \frac{1}{b}\right)$$

●

Remark 1 When $a < b$ the work is positive because you do work against the gravitational force. But when $a > b$ the free mass moves towards the fixed mass, *opposite* to your direction of pull. Hence this counts as negative work. Imagine moving the free body from a to b and then back to a. The total work is zero. Why?

Remark 2 In the MKS system, m_1 and m_2 are in kilograms, r in meters, F in newtons, W in joules, and

$$G \approx 6.670 \times 10^{-11}\ \text{N-m}^2/\text{kg}^2$$

Example 3

When a spring is stretched a small amount, there is a restoring force proportional to the distance of the moving end from its equilibrium point (Hooke's Law). Suppose 2 joules work are needed to stretch a certain spring 10 cm. How much work is needed to stretch it 25 cm?

Solution Let x denote the displacement in centimeters of the free end from equilibrium (Figure 7-4-2). The force needed at x to oppose the restoring force is kx, so the work in stretching the spring from $x = 0$ to $x = b$ is

$$W = \int_0^b kx\,dx = \tfrac{1}{2}kb^2 \text{ joules}$$

When $b = 10$, then $W = 2$, hence $2 = \frac{1}{2}k(10)^2$, so $k = \frac{1}{25}$. Therefore when $b = 25$,

$$W = \tfrac{1}{2}(\tfrac{1}{25})(25)^2 = 12.5 \text{ joules}$$

●

Example 4

A sunken tank (Figure 7-4-3) has the shape of a right circular cone with its apex down. Compute the work done in pumping a tankful of water to ground level.

Figure 7-4-3
See Example 4

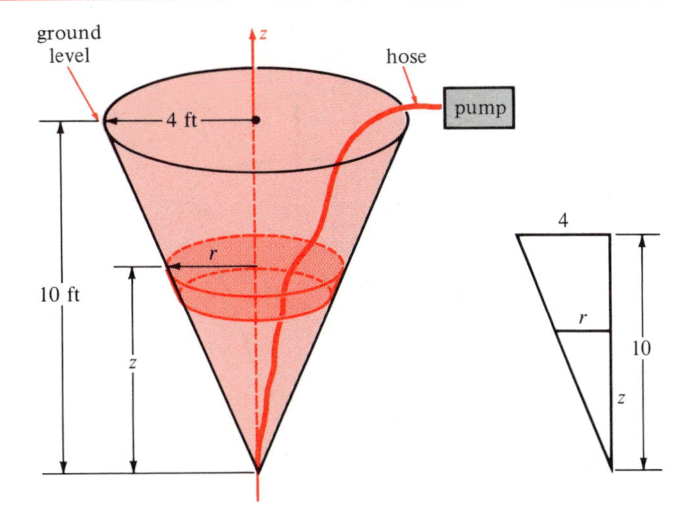

Solution Set up axes so the vertex of the cone is the origin and the axis of the cone is the z-axis. Imagine the tank sliced into thin slabs perpendicular to its axis. The idea is to compute the work done in raising each slab of water to the level $z = 10$, then to add up these elements of work by integrating. For convenience, set $\delta = 62.4 \text{ lb/ft}^3$, the density of water.

The first problem is to find the volume of a slab at level z. By similar triangles, its radius r satisfies

$$\frac{r}{4} = \frac{z}{10} \qquad \text{hence} \qquad r = \tfrac{2}{5}x$$

Therefore the volume of the slab is

$$\pi r^2\, dz = \tfrac{4}{25}\pi z^2\, dz$$

The upward force required to lift the slab equals its weight (to overcome the downward force of gravity). Since weight in pounds equals density δ times volume, the force is $\frac{4}{25}\pi\delta z^2\, dz$. The slab must be raised a distance $10 - z$. Therefore the element of work is

$$dW = \tfrac{4}{25}\pi\delta(10 - z)z^2\, dz$$

Consequently the total work is

$$W = \tfrac{4}{25}\pi\delta \int_0^{10} (10 - z)z^2\, dz = \tfrac{4}{25}\pi\delta\left(\tfrac{10}{3}z^3 - \tfrac{1}{4}z^4\right)\Big|_0^{10}$$

$$= \tfrac{400}{3}\delta\pi = \tfrac{400}{3}(62.4)\pi \approx 26{,}138 \text{ ft-lb}$$

●

Example 5

A heavy buoy of weight w in the shape of a cone of revolution (Figure 7-4-4a) floats in a lake with its lowest point at depth h. The buoy is raised by a winch until it just clears the water. How much work is done?

Figure 7-4-4
See Example 5

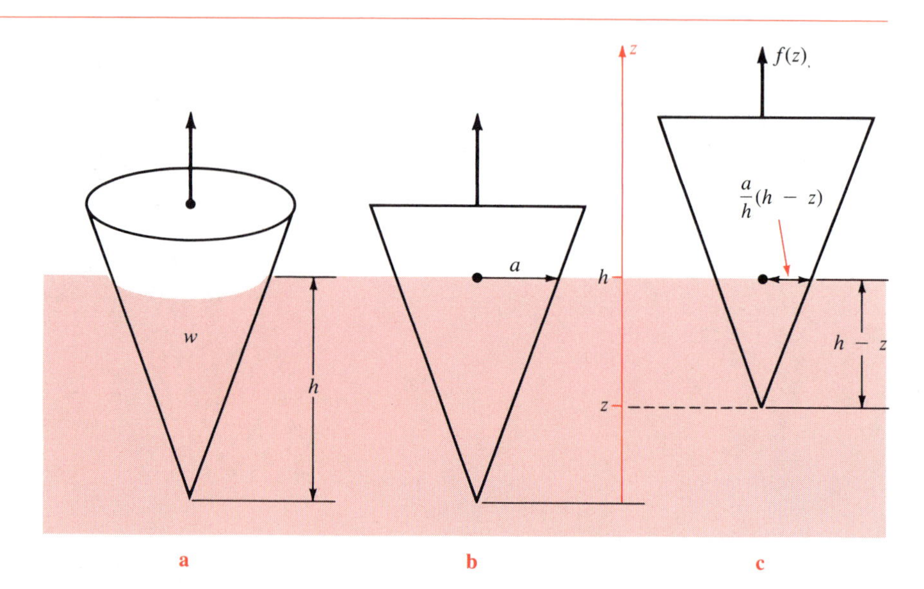

Solution Fix the z-axis as in Figure 7-4-4b, and denote by a the radius of the buoy *at the water level.* By Archimedes' principle for floating bodies, the buoy is acted on by an upward buoyant force of the water equal to the weight of the water displaced. When the buoy floats, this buoyant force exactly balances the downward force of gravity w. Thus

$$w = \tfrac{1}{3}\pi a^2 h\delta$$

where the right-hand side is the product of the volume $\frac{1}{3}\pi a^2 h$ of the submerged cone, and the density δ of the lake water.

Suppose the buoy is hoisted z units (Figure 7-4-4c). The force $f(z)$ required to hold it in this position is the weight of the buoy minus the buoyant force of the water. But now the part submerged has radius $(a/h)(h - z)$ and height $h - z$. Hence

$$f(z) = w - \tfrac{1}{3}\pi\left[\frac{a}{h}(h - z)\right]^2 (h - z)\delta$$

$$= w - \frac{\pi a^2 \delta}{3h^2}(h - z)^3 = w - \frac{w}{h^3}(h - z)^3$$

Therefore the work done lifting the buoy out of the water is

$$W = \int_0^h f(z)dz = w\int_0^h dz - \frac{w}{h^3}\int_0^h (h - z)^3\,dz$$

Now

$$\int_0^h dz = h \qquad \text{and} \qquad \int_0^h (h - z)^3\, dz = -\tfrac{1}{4}(h - z)^4 \Big|_0^h = \tfrac{1}{4}h^4$$

so

$$W = wh - \frac{w}{h^3} \cdot \tfrac{1}{4}h^4 = wh - \tfrac{1}{4}wh = \tfrac{3}{4}wh$$

Exercises

1 Find the work done by a force $f(x) = 3x + 2$ N in moving an object from $x = 1$ m to $x = 7$ m.

2 At each point of the x-axis (marked off in feet) there is a force of $x^2 - 5x + 6$ lb pushing to the right against an object. Compute the work done in moving the object from $x = 1$ to $x = 5$.

3 A 50-foot chain weighing 2 lb/ft is attached to a cylindrical drum hung from the ceiling (Figure 7-4-5). The ceiling is high enough so that the free end of the chain does not touch the floor. How much work is required to wind the chain around the drum? Assume that the radius of the drum is negligibly small.

Figure 7-4-5
See Exercises 3 and 4

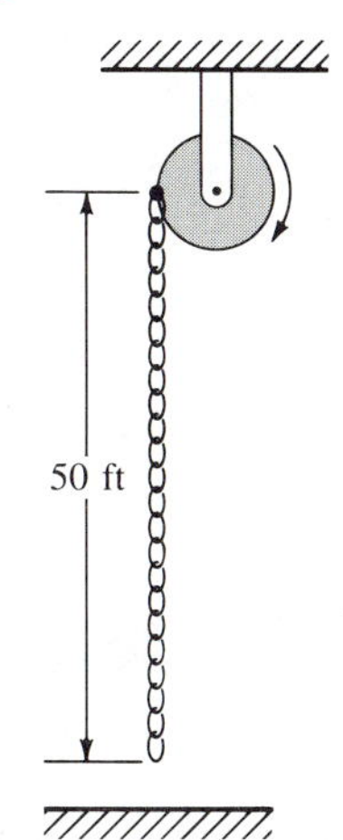

4 In the previous exercise, suppose that a 200-pound weight is attached to the free end. Now how much work is required to wind up the chain?

5 The force in pounds required to stretch a certain spring x ft is $F = 8x$. How much work is required to stretch the spring 6 in.? 1 ft? 2 ft?

6 A 3-pound force will stretch a spring 0.5 ft. How much work is required to stretch the spring 2 ft?

7 A 100-pound bag of sand is hoisted 50 ft at a rate of 5 ft/sec. Because of a hole in the bag, 2 lb of sand is lost each second. Compute the work done.

8 A 5-pound monkey (Figure 7-4-6) is attached to the free end of a 20-foot hanging chain that weighs 0.25 lb/ft. The monkey climbs the chain to the top. How much work does he do?

Figure 7-4-6
See Exercise 8

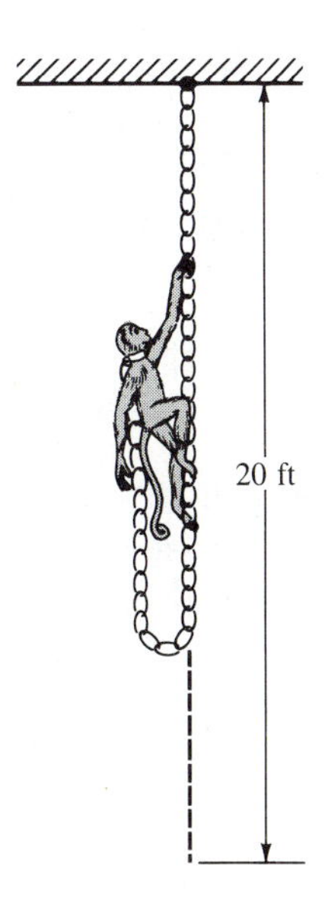

9 How much work is required to lift a 500-kilogram payload from the surface of Earth to height 500 km? height 1000 km? You may assume Earth has radius $r_E \approx 6.37 \times 10^6$ m and mass $m_E \approx 5.98 \times 10^{24}$kg. The gravitational force on a payload of mass m at height x from the center of Earth is $F = Gm_Em/x^2$, where

$$G \approx 6.67 \times 10^{-11}\ \text{N-m}^2/\text{kg}^2$$

Figure 7-4-7
See Exercise 10

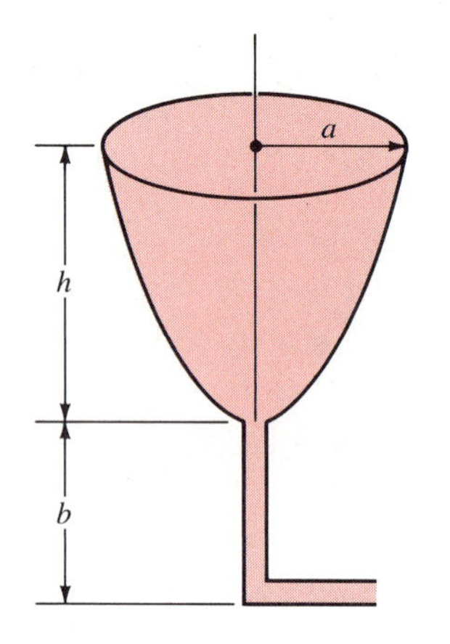

Figure 7-4-8
See Exercise 11

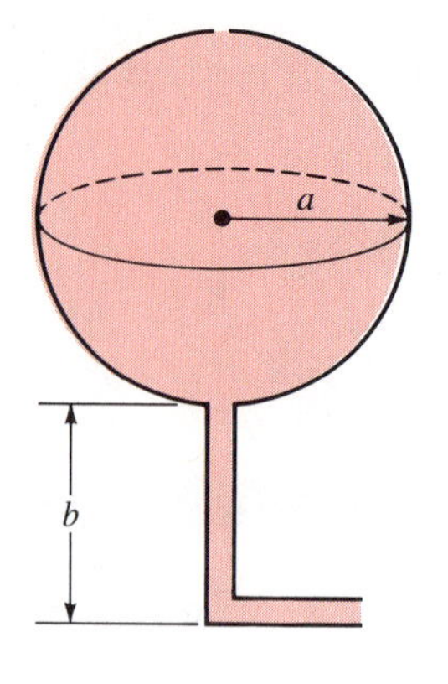

Figure 7-4-9
See Exercise 12

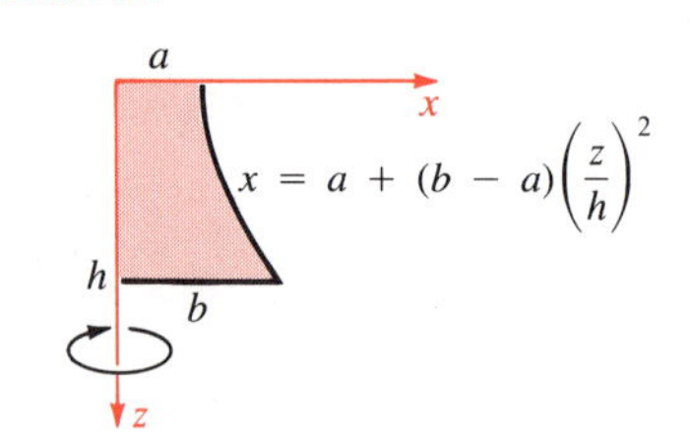

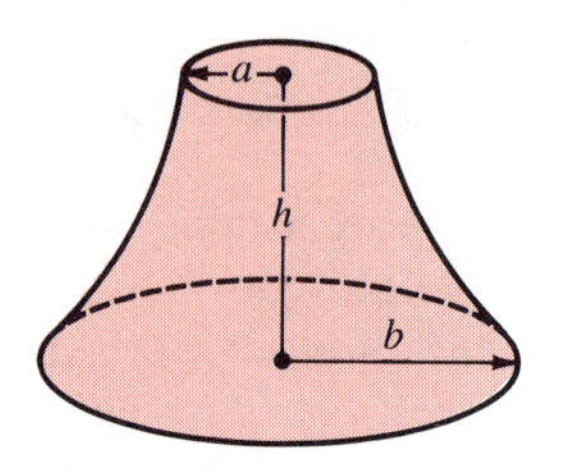

10 How much work is required to fill the tank in Figure 7-4-7 with water pumped from the bottom of the pipe. The tank is in the shape of a paraboloid of revolution, obtained by revolving a parabola about its axis. Assume that the diameter of the pipe is negligible.

11 How much work is required to pump water from the bottom of the pipe and fill the spherical tank in Figure 7-4-8? Assume that the diameter of the pipe is negligible.

12 A tank is obtained by revolving a parabolic segment as indicated in Figure 7-4-9. How much work is required to pump a tankful of fluid of density δ to the level of its top?

13 Suppose in Example 5 that the top of the buoy is $\frac{1}{4}h$ from the water's surface. How much work is required to push the buoy down until its top is at water level?

***14** A spherical mine has enough flotation so its average density equals 1030 kg/m^3, that of the seawater it floats in—just touching the surface. Suppose its radius is 0.60 m. How much work is required to lift it so it just clears the water?

15 A particle of mass m is constrained to move on a vertical circle of radius a. (Think of a pendulum bob.) Suppose it is pushed from its downward rest position through a central angle ϕ How much work is done? [Hint Resolve the gravitational force into components parallel and perpendicular to the circle; only the former must be opposed.]

***16** An open-top cylindrical tank of radius R and height H is filled with water. A cylindrical buoy of radius r and weight w floats on end in the tank, its base at depth h, where $r < R$ and $h < H$. How much work is required to raise the buoy until its base just clears the water. (Note that the water level goes down as the buoy comes up.)

17 A 50-meter chain weighs 3 kg/m. One end is attached to a drum 25 m above the floor; otherwise the chain is coiled on the floor. Find the work required to wind the chain around the drum. Assume that the radius of the drum is negligible.

18 A tank of chemical waste initially weighs 700 kg. It is hoisted 30 m by a crane at the rate of 0.25 m/sec. While being hoisted, the tank leaks out waste at the rate of 0.5 kg/sec. Find the work done in hoisting the tank.

7-5 Fluid Pressure

Fluid pressure is measured by the force it exerts on any piece of surface immersed in the fluid. At each point of the surface this force is exerted against the surface in the direction perpendicular to the surface.

Technically, **pressure** is magnitude of force per unit area. In Figure 7-5-1, the pressure p exerts a force dF against the immersed element dA. Because of Pascal's law (pressure at a point within a fluid is equal in all directions) pressure exerts an equal but oppositely directed force against the other side of the element dA (so nothing happens). However, if dA is part of the boundary of the fluid, that is, part of the container wall, then the force is not opposed by fluid pressure on the other side (Figure 7-5-2). The wall's own strength must hold it up.

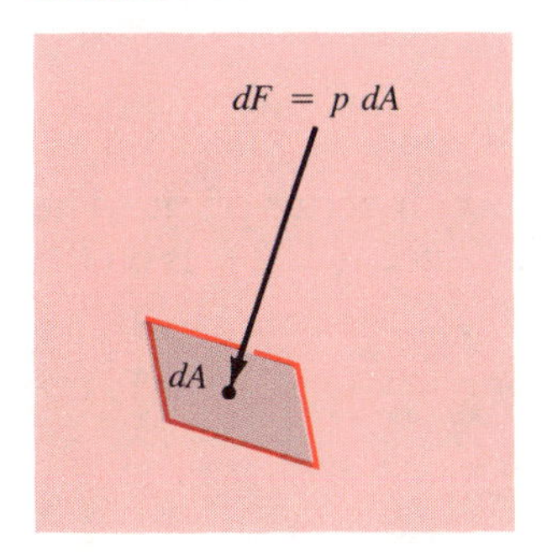

Figure 7-5-1
Pressure inside the fluid

The pressure at depth y in a fluid is $p = \delta g y$, where δ is the density of the fluid and g the acceleration of gravity. (We often take $g = 1$ by measuring mass and force in the same units.) The total force due to pressure against a *plane* surface submerged in the fluid is obtained by "summing" the elements of force $dF = p\,dA$ over the surface. Since p is constant at depth y, this can usually be done by integration. The result is (the magnitude of) a force that is directed perpendicular to the plane surface. This force is important in the design of tanks, dams, large buildings (wind pressure), and airfoils (lift).

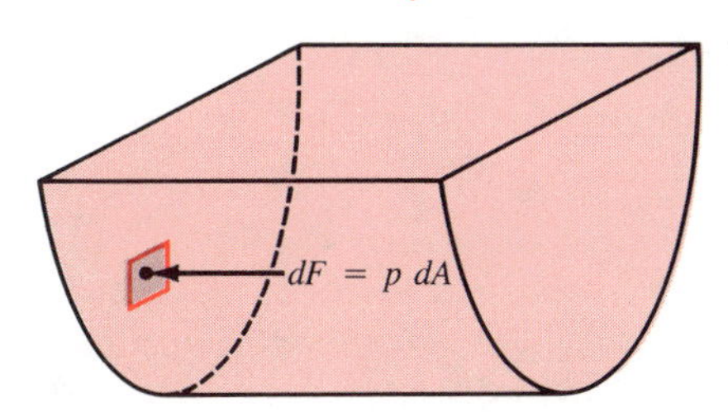

Figure 7-5-2
Pressure at the boundary of a fluid

Remarks on Units Pressure units are a bit of a jungle. The English system units are the lb/in^2 and the lb/ft^2. The MKS (SI) unit is the N/m^2 (newton/meter squared, also called the Pascal, Pa), which is usually too small to be practical. The unit kg/cm^2 ($\approx 9.807 \times 10^4\ \text{N/m}^2$) is common. Here kg means one kg *force,* not mass. One kg force is the force that gravity at the surface of Earth exerts on one kg of mass. A tire gauge in Europe might read $2.1\ \text{kg/cm}^2$, equivalent to about $30\ \text{lb/in}^2$.

Other common units:

- the **atmosphere,** where

$$1\ \text{atm} = 1.01325 \times 10^5\ \text{N/m}^2 \approx 1.033\ \text{kg/cm}^2 \approx 14.70\ \text{lb/in}^2$$

- The **bar** $= 10^5\ \text{N/m}^2$ (common in meteorology, where atmospheric pressure is often reported in millibars)
- the **torr** $= \frac{1}{760}$ atm (used in low-pressure work, and also called mmHg = millimeters of mercury).

Most pressure measurements at the surface of the Earth are "gauge" pressures, so atmospheric pressure is the reference point for zero. For instance $25\ \text{lb/in}^2$ pressure in a tire means $25\ \text{lb/in}^2$ *above* atmospheric pressure, which itself is about $14.7\ \text{lb/in}^2$ above vacuum. Thus the tire pressure is about $39.7\ \text{lb/in}^2$ above vacuum.

But for low-pressure work, the reference point for zero is usually the true zero pressure of a vacuum. For instance $3\ \text{lb/in}^2$ pressure in a space ship en route to Mars means $3\ \text{lb/in}^2$ above the outside pressure.

The density (mass/volume) of water is

$$1000\ \text{kg/m}^3 = 1\ \text{kg/liter} = 1\ \text{g/cm}^3 \approx 62.4\ \text{lb/ft}^3$$

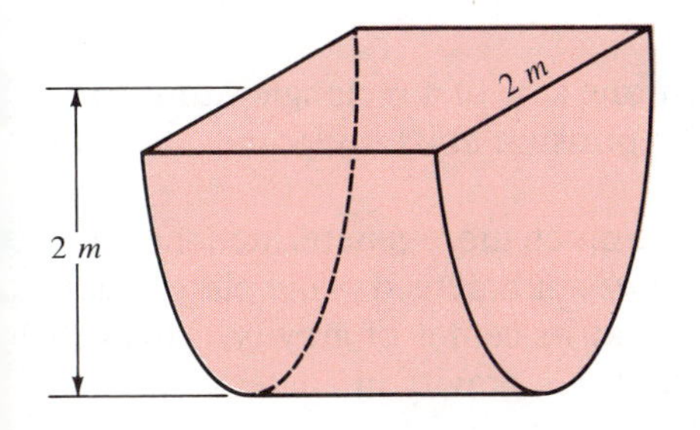

Figure 7-5-3
Parabolic tank. See Example 1

Example 1

The ends of the cylindrical tank of water in Figure 7-5-3 are parabolas with vertical axes. Find the force in kilograms due to fluid pressure that the water exerts on each end of the tank.

Solution Choose axes for one end as in Figure 7-5-4 (next page), the scale in meters. The strip shown, at height y from the bottom of the tank, has area $dA = 2x\,dy$ and lies at depth $2 - y$. The average pressure against the strip is $p = \delta(2 - y) = 1000(2 - y)\ \text{kg/m}^2$, so the element of force against the strip is

$$dF = 1000(2 - y)(2x\,dy) = 2000(2 - y)x\,dy$$

Figure 7-5-4
See Example 1

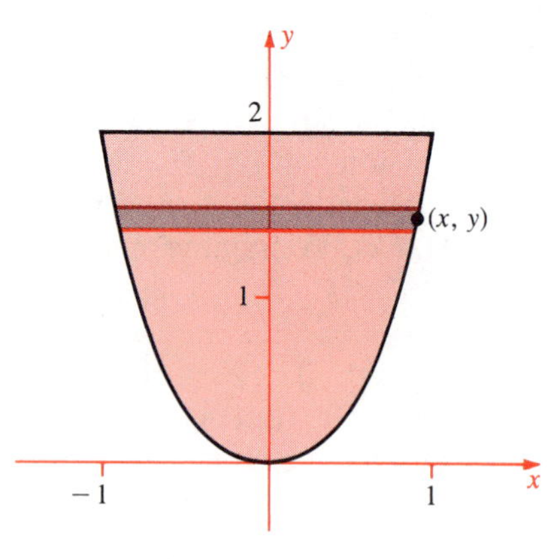

The force in kg due to fluid pressure against the end of the tank is

$$F = 2000\int_0^2 (2 - y)x\,dy$$

To evaluate the integral, first we must express x in terms of y. The equation of the parabola has the form $y = kx^2$. Since $(1, 2)$ is on the parabola, $k = 2$. Therefore the equation is

$$y = 2x^2 \qquad \text{that is} \qquad x = \sqrt{y/2} = \tfrac{1}{2}\sqrt{2}\,\sqrt{y}$$

Therefore

$$\begin{aligned} F &= 1000\sqrt{2}\int_0^2 (2 - y)\sqrt{y}\,dy \\ &= 1000\sqrt{2}\int_0^2 (2y^{1/2} - y^{3/2})\,dy \\ &= 1000\sqrt{2}\left(\tfrac{4}{3}y^{3/2} - \tfrac{2}{5}y^{5/2}\right)\Big|_0^2 \\ &= 1000\sqrt{2}\left(\tfrac{8}{3}\sqrt{2} - \tfrac{8}{5}\sqrt{2}\right) \\ &= 16{,}000\left(\tfrac{1}{3} - \tfrac{1}{5}\right) = \tfrac{1}{3}(6400) = 2133\tfrac{1}{3}\text{ kg} \end{aligned}$$

Vertical Slicing

Sometimes it is convenient to compute force due to fluid pressure by slicing a surface into vertical strips. To do so effectively we need a property of force on rectangles.

Suppose a rectangle stands vertically in a fluid (Figure 7-5-5). Choose axes as indicated. Then the element of force against the horizontal strip at depth y is

$$dF = (\delta y)(a\,dy) = \delta a y\,dy \qquad (\delta = \text{density})$$

Therefore

$$\begin{aligned} F &= \int_L^{L+b} \delta a y\,dy = \tfrac{1}{2}\delta a y^2\Big|_L^{L+b} \\ &= \tfrac{1}{2}\delta a(2Lb + b^2) = (ab)[\delta(L + \tfrac{1}{2}b)] \end{aligned}$$

Figure 7-5-5
Submerged rectangular plate, standing vertically

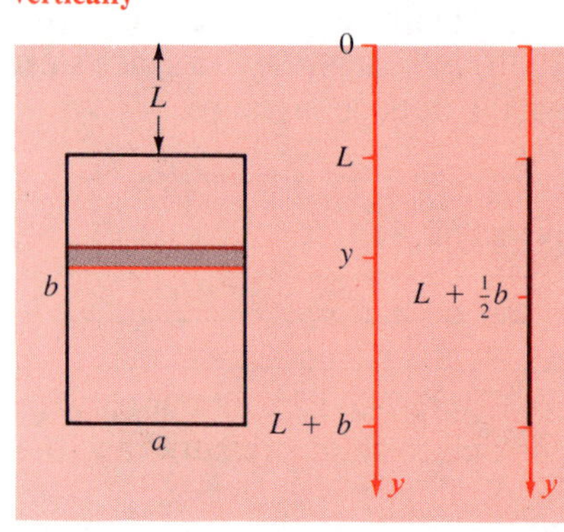

But ab is the area of the rectangle and $L + \frac{1}{2}b$ is the depth of the midpoint of the rectangle.

- The force due to fluid pressure against one side of a rectangle submerged vertically in a fluid is its area times the pressure at its midpoint.

Remark This result is a special case of a much more general fact: the force due to fluid pressure against one side of *any* submerged plane plate equals the area of the plate times the pressure at its center of gravity. This will become clear after our discussion of center of gravity in a later chapter.

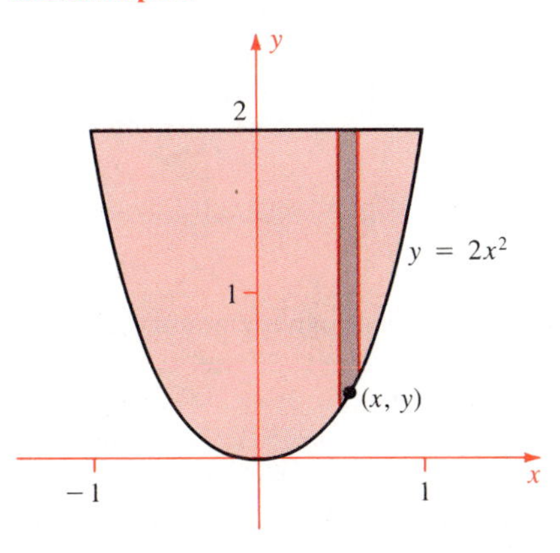

Figure 7-5-6
See Example 2

Example 2

Solve Example 1 by vertical slicing.

Solution The vertical strip in Figure 7-5-6 has area $(2 - y)\,dx$ and its midpoint lies at depth $\frac{1}{2}(2 - y)$, so the pressure at the midpoint is

$$p = \tfrac{1}{2}(2 - y)\delta = 500(2 - y)\,\text{kg}$$

Therefore the element of force against this vertical strip is

$$dF = [500(2 - y)][(2 - y)\,dx] = 500(2 - y)^2\,dx$$

But $y = 2x^2$ so

$$dF = 500(2 - 2x^2)^2\,dx = 2000(1 - x^2)^2\,dx$$

Therefore

$$F = \int_{-1}^{1} 2000(1 - x^2)^2\,dx = 4000\int_0^1 (1 - 2x^2 + x^4)\,dx$$

$$= 4000(x - \tfrac{2}{3}x^3 + \tfrac{1}{5}x^5)\Big|_0^1 = 4000(1 - \tfrac{2}{3} + \tfrac{1}{5})$$

$$= (4000)(\tfrac{8}{15}) = 2133\tfrac{1}{3}\,\text{kg}$$

Exercises

Each figure in Exercises 1–10 is the end of a tank filled with a fluid of density δ. Express the force due to fluid pressure on the end of the tank in terms of the data.

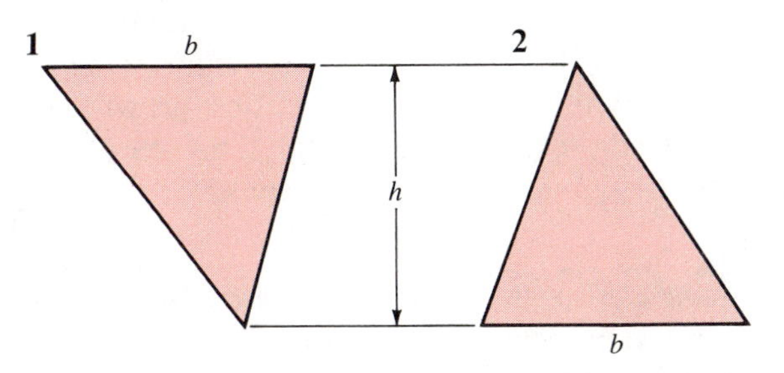

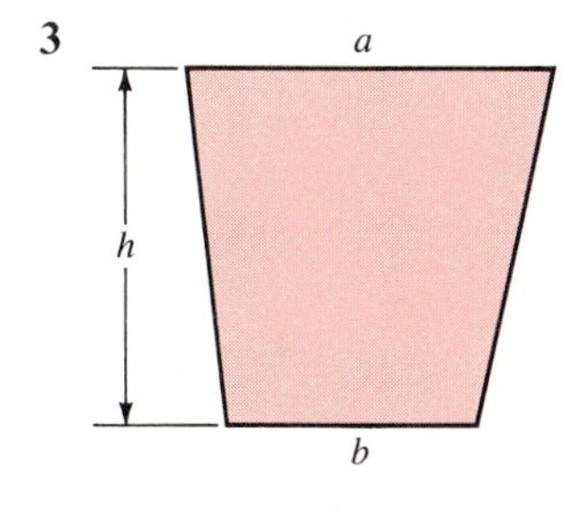

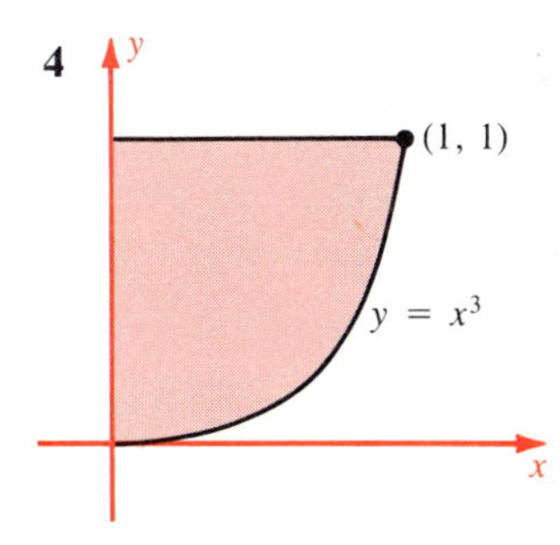

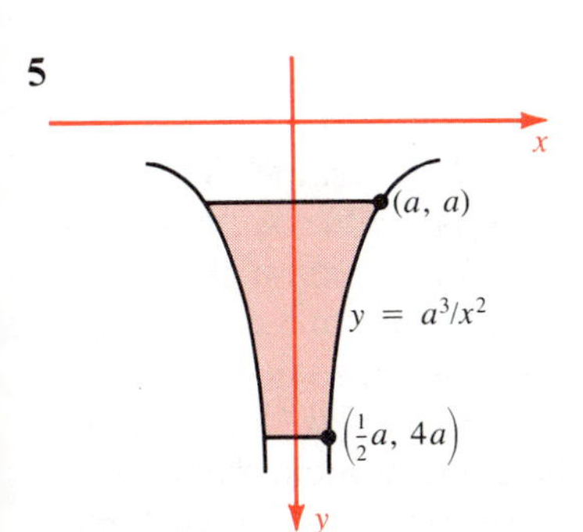

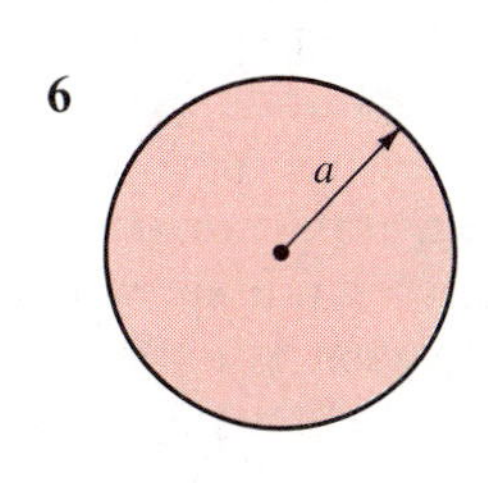

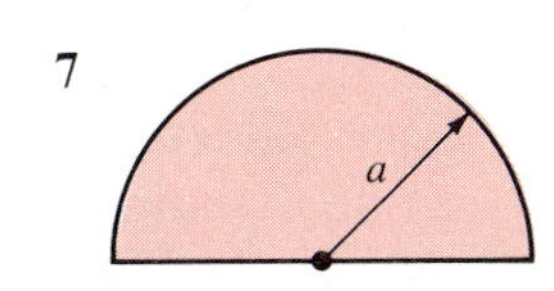

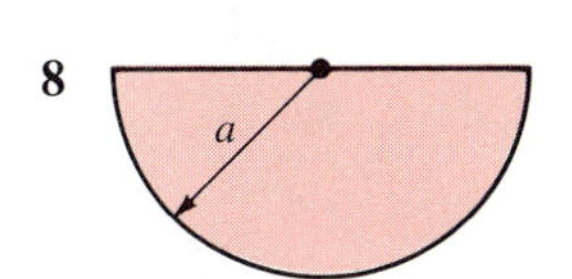

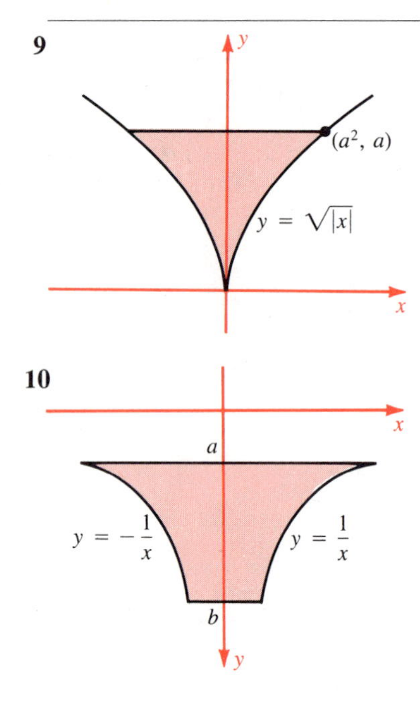

Air pressure at height h meters above sea level is given by $p = p_0 e^{-ah}$ kg/m^2, where $p_0 = 1.03 \times 10^4$ and $a = 1.25 \times 10^{-4}$.

11 Find the force due to air pressure against a rectangular wall 40 m wide and 120 m high whose base is 200 m above sea level.

12 Find the force due to air pressure against a triangular wall with base 10 m at sea level and height 10 m.

13 A tank of water has the shape of a high rectangular column with square base and height h meters, and its volume is 1 m^3. Let F be the force due to fluid pressure against one of the vertical sides and let A be the area of the side. Find c so FA^c is independent of h.

14 Let F be the force due to fluid pressure against one end of a tank. Suppose a second tank (filled with the same fluid) is exactly the same shape as the first tank except that each linear dimension is enlarged by the factor k. Find the force due to fluid pressure against the corresponding end of the second tank.

15 One end of a tank has area A and height h. Suppose the end is symmetric about the horizontal line at level $\frac{1}{2}h$. Express the force due to fluid pressure against the end in terms of A, h, and the fluid density δ.

16 One end of a tank has area A and height h, and the force due to fluid pressure against it is F_1. A second tank is the same shape turned upside down, and the force due to fluid pressure against the corresponding face is F_2. Find the relation between F_1, F_2, A, h, and the fluid density δ.

7-6 Miscellaneous Applications

The previous topics—area, volume, work, and pressure—should convince anyone of the enormous applicability of integration. As further evidence, we include some topics from a variety of directions: growth of money, probability, suspension bridges, rotating fluids, and the shape of ice cubes. Additional applications are contained in the exercises.

Present Value of Future Income

We discussed compound interest in Section 4-2. Let us recall that if amount A is invested at annual interest rate r, compounded continuously, then its value after t years is Ae^{rt}.

Now we consider compounding of funds invested not in one lump sum, but over a period of time. Suppose money is deposited continuously into an initially empty account, at the rate of $f(t)$ dollars per year at time t. If the constant annual interest rate is r, what will be the value of the account after T years? Also, what is its fair present value, that is, the largest amount you would be willing to pay today for ownership of the account in T years?

To answer these questions, we divide the time interval $[0, T]$ into many small pieces of duration dt. In the typical short interval at time t, the element of money deposited is $f(t)\,dt$. This remains in the account for $T - t$ more years; hence it grows to the element of amount $dA = e^{r(T-t)}f(t)\,dt$. The total amount $A(t)$ of the account at time T is the integral of these elements:

$$A(T) = \int_0^T e^{r(T-t)}f(t)\,dt = e^{rT}\int_0^T e^{-rt}f(t)\,dt$$

It is called the **value** of the account at time T. The **present value** of the account is defined as the amount V which, if deposited today, will grow to amount $A(T)$ in T years. Therefore

$$Ve^{rT} = A(T) = e^{rT}\int_0^T e^{-rt}f(t)\,dt \qquad V = \int_0^T e^{-rt}f(t)\,dt$$

Let us summarize the discussion:

- Suppose that money is deposited continuously into an account at the rate of $f(t)$ dollars per year and that r is the constant annual interest rate. Then the **value** of the account after T years will be

$$A(T) = e^{rT}\int_0^T e^{-rt}f(t)\,dt$$

Its **present value** is

$$V(T) = \int_0^T e^{-rt}f(t)\,dt$$

Example 1

Funds are deposited at a continuous rate of $10,000 per year for 10 years, and the interest rate is 11%. Find the value of the account after 10 years and its present value.

Solution The value after 10 years is

$$\begin{aligned} A(10) &= e^{10(0.11)}V(10) \\ &= e^{1.1}\int_0^{10} e^{-(0.11)t}\,10{,}000\,dt \\ &= e^{1.1}\left(10{,}000\,\frac{e^{-(0.11)t}}{-0.11}\right)\Bigg|_0^{10} \\ &\approx (3.00417)[(10{,}000)(1/0.11)(1 - 0.332871)] \\ &\approx (3.00417)(60{,}648.1) \approx \$182{,}197 \end{aligned}$$

The present value is $V(10) \approx \$60{,}648$. This is the lump sum which, deposited today at 11%, will grow to $182,197 in 10 years.

Remark In economics the function $V = V(T)$ is called the **present capital value** of the **income stream** $f(t)$.

Expectation and Variance

Let X be a random variable on a sample space Ω. This means that Ω is the set of all possible outcomes of a certain experiment and that X is a real-valued function on Ω. We assume that the values of X fall into a closed interval $[a, b]$. For each subinterval $[c, d]$ of $[a, b]$ there is a (non-negative) probability

$$P(c \leq X \leq d)$$

that X falls into $[c, d]$. Since it is certain that $a \leq X \leq b$, we have

$$P(a \leq X \leq b) = 1$$

Also, if $[c, e]$ is a subinterval of $[a, b]$ and $c \leq d \leq e$, then

$$P(c \leq X \leq e) = P(c \leq X \leq d) + P(d \leq X \leq e)$$

Thus the probability function is like an integral of a non-negative function so let us assume that it really is an integral.

Thus we assume that there exists an integrable function $f(t)$ on $[a, b]$, called the **density,** or **probability density,** of the random variable X, such that

$$P(c \leq X \leq d) = \int_c^d f(t)\, dt$$

This density function $f(t)$ on $[a, b]$ is assumed to have three properties:

1 $f(t)$ is integrable on $[a, b]$

2 $f(t) \geq 0$

3 $\displaystyle\int_a^b f(t)\, dt = 1$

Sometimes we write $f(t) = f_X(t)$ to denote the dependence of f on X.

Now we are on familiar ground, even though the preceding paragraph has not told you what a sample space is, what a probability function is, or what a random variable is. Indeed, these are deep matters, studied in courses on probability. However, we do understand what a density function is in terms of properties 1–3 above. We now consider two important concepts in probability: expectation and variance. The first, **expectation,** also called **expected value,** is defined to be

$$E(X) = \int_a^b t f(t)\, dt$$

Note that $t \geq a$ in the integrand, so

$$E(X) = \int_a^b t f(t)\, dt \geq \int_a^b a f(t)\, dt = a \int_a^b f(t)\, dt = a$$

Similarly $E(X) \le b$, so we have

$$a \le E(X) \le b$$

Example 2

Find $E(X)$ for the (uniform) density $f(t) = 1/(b-a)$.

Solution The constant function $f(t)$ is a probability density on $[a, b]$. For it is integrable and non-negative, and

$$\int_a^b f(t)\,dt = \int_a^b \frac{dt}{b-a} = \frac{b-a}{b-a} = 1$$

The expectation is

$$E(X) = \int_a^b tf(t)\,dt = \frac{1}{b-a}\int_a^b t\,dt$$

$$= \frac{1}{b-a}\cdot \tfrac{1}{2}(b^2-a^2) = \tfrac{1}{2}(a+b)$$

This means roughly that if all points of $[a, b]$ are equally likely values of X, then the expected value of X is the midpoint of $[a, b]$. ●

Example 3

Given $\lambda > 0$, find c so that $f(t) = ce^{-\lambda t}$ is a density on $[0, 1]$. Then find the expectation of the corresponding random variable X.

Solution Clearly $f(t)$ is integrable and non-negative, and

$$\int_0^1 f(t)\,dt = c\int_0^1 e^{-\lambda t}\,dt = -\frac{c}{\lambda}e^{-\lambda t}\Big|_0^1 = \frac{c}{\lambda}(1-e^{-\lambda})$$

Therefore $c = \lambda/(1-e^{-\lambda})$ makes the integral 1.

Next,

$$E(X) = \int_0^1 tf(t)\,dt$$

$$= c\int_0^1 te^{-\lambda t}\,dt = c\left(-\frac{1}{\lambda}te^{-\lambda t} - \frac{1}{\lambda^2}e^{-\lambda t}\right)\Big|_0^1$$

$$= c\left[-\frac{e^{-\lambda}}{\lambda} + \frac{1}{\lambda^2}(1-e^{-\lambda})\right] = \frac{-e^{-\lambda}}{1-e^{-\lambda}} + \frac{1}{\lambda}$$ ●

In a long series of experiments, you would expect the values X of the outcomes to average to $E(X)$. The **variance** of X measures the tendency of the experimental results to disperse from $E(X)$. It is defined by

$$\operatorname{var}(X) = \int_a^b [t - E(X)]^2 f(t)\,dt$$

Example 4

Find the variance in Example 2.

Solution We found $E(X) = \frac{1}{2}(a+b)$. Therefore

$$\begin{aligned}
\text{var}(X) &= \int_a^b [t - E(X)]^2 f(t)\,dt = \frac{1}{b-a}\int_a^b [t - E(X)]^2\,dt \\
&= \frac{1}{b-a}\cdot\frac{1}{3}[t - E(X)]^3\Big|_a^b \\
&= \frac{1}{3(b-a)}\{[b - E(X)]^3 - [a - E(X)]^3\} \\
&= \frac{1}{3(b-a)}\{[\tfrac{1}{2}(b-a)]^3 - [-\tfrac{1}{2}(b-a)]^3\} \\
&= \tfrac{1}{12}(b-a)^2
\end{aligned}$$

The Suspension Bridge

What is the shape of the cable that supports the roadway of a suspension bridge (Figure 7-6-1)? Our model is based on the following simplifying assumptions:

- The weight of the cable and of the suspension rods is negligible.
- The weight of the roadway is uniform, δ per unit length.
- The suspension rods are so close together that the horizontal loading of the cable is the uniform weight of the roadway: δ per unit length.

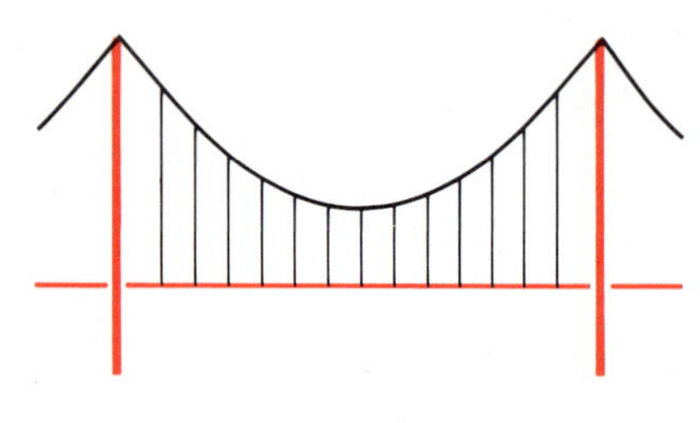

Figure 7-6-1
Suspension bridge

We choose axes as in Figure 7-6-2 and consider the portion of the cable for $0 \le x \le a$. Three forces act to hold it in equilibrium: the horizontal tension at the left end of magnitude T_0, the tension at the right end in the tangent direction with magnitude $T(a)$, and the downward weight $a\delta$.

We suppose the cable has the shape of a curve $y = f(x)$. If θ denotes the angle of the tangent at a, then clearly

$$\tan\theta = \frac{dy}{dx}\Big|_a$$

Since the three forces balance, their horizontal components balance, as do their vertical components. This gives us two relations:

$$T(a)\cos\theta = T_0 \qquad T(a)\sin\theta = a\delta$$

Therefore

$$\tan\theta = \frac{T(a)\sin\theta}{T(a)\cos\theta} = \frac{\delta}{T_0}a$$

Since this is true for any a, we deduce that

$$\frac{dy}{dx} = \frac{\delta}{T_0}x$$

Figure 7-6-2

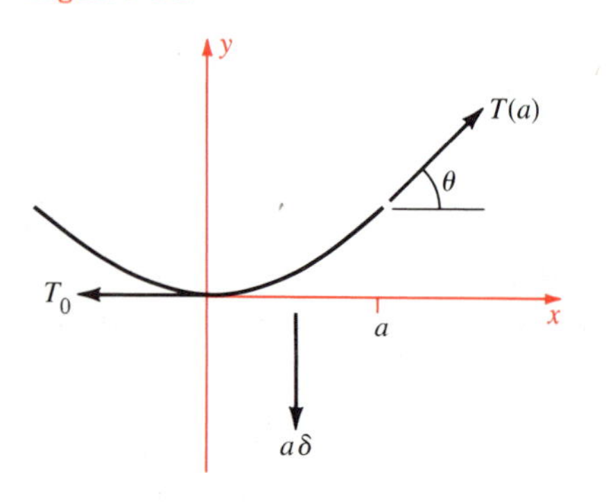

Figure 7-6-3
Rotating fluid

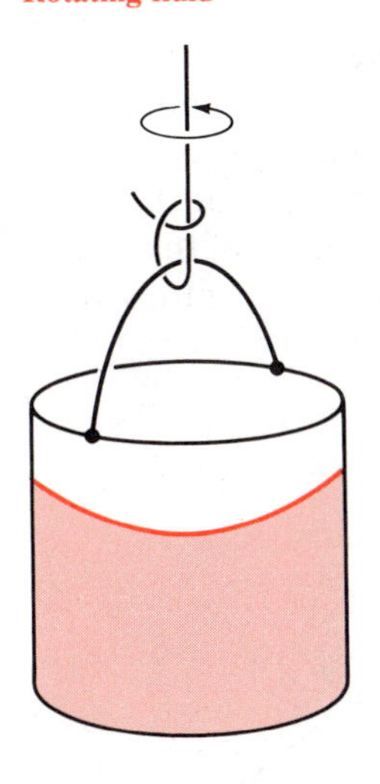

Therefore

$$y = \int_0^x \frac{\delta}{T_0} u\,du = \frac{\delta}{2T_0} x^2$$

We conclude that the cable has the shape of a parabola.

Free Surface of a Rotating Fluid

A partly filled bucket of water (Figure 7-6-3) is rotating at a steady speed. It was brought to this state gradually, so the water inside rotates with the bucket. The surface of the water is a surface of revolution, and our problem is to describe it.

Imagine that this surface is obtained by revolving $y = f(x)$ about the y-axis (Figure 7-6-4). Now consider a small particle of mass dm at the water's surface. As it rotates with the fluid, it keeps its relative position on the surface because of three forces: (1) a buoyant force, due to fluid pressure, acting perpendicular to the surface, (2) its weight acting downwards, (3) centrifugal force acting horizontally outwards. These forces are shown in Figure 7-6-5. The centrifugal force has magnitude $x\omega^2\,dm$, where ω is the angular speed in rad/sec.

Figure 7-6-4
Cross section of the surface

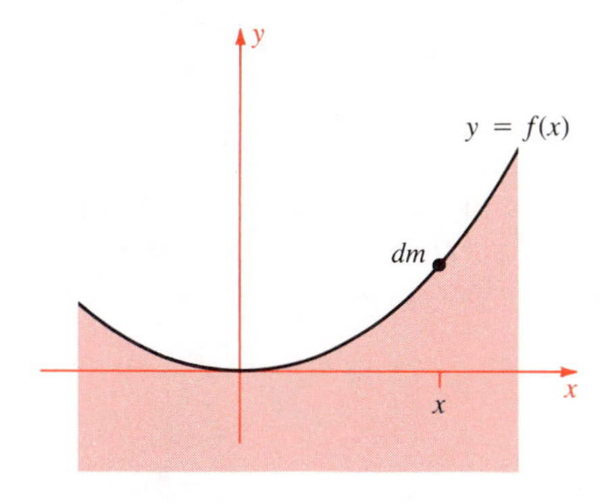

These forces must balance; in particular their components in the direction tangent to the curve must balance. Hence

$$(g\,dm)\cos(\tfrac{1}{2}\pi - \theta) = (x\omega^2\,dm)\cos\theta$$

that is

$$g\sin\theta = x\omega^2\cos\theta \qquad \text{so that} \qquad \tan\theta = \frac{\omega^2}{g}x$$

(The buoyant force acts perpendicularly to the curve; hence it makes no contribution to this tangential relation.) But $\tan\theta = dy/dx$, so we have obtained the relation

$$\frac{dy}{dx} = \frac{\omega^2}{g}x$$

where ω and g are constants. Consequently

$$y = \int_0^x \frac{\omega^2}{g} u\,du = \frac{\omega^2}{2g}x^2$$

Therefore the rotating surface is a paraboloid of revolution.

Figure 7-6-5
Detail of forces in balance at the fluid surface. ($x\omega^2\,dm$ is centrifugal force.)

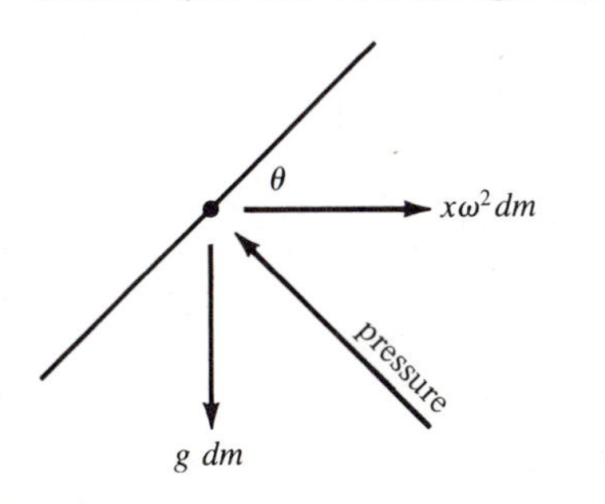

Freezing of Liquids

The top of an ice cube is never flat; it is always curved a bit and often has a sharp peak. The following is a simplified model to explain this surface shape.

At the moment water freezes to ice, its volume expands by a factor $1 + \beta$, where $\beta > 0$. In fact, $\beta = \frac{1}{8}$ for water, but we shall retain β for generality, since the model applies to other fluids as well.

Suppose that water is frozen in a cylindrical container having perfect radial symmetry, and that freezing takes place from the outside curved wall inwards. At an intermediate stage of the freezing, there is a flat cylinder of unfrozen water surrounded by a cylinder of ice (Figure 7-6-6). The water is higher than its initial level h because the expanding ice is squeezing it, forcing it up. Therefore, when freezing is complete, the top of the ice is peaked (Figure 7-6-7). The problem is to describe the shape of the frozen surface, that is, to identify the profile curve $y = f(x)$ in Figure 7-6-8.

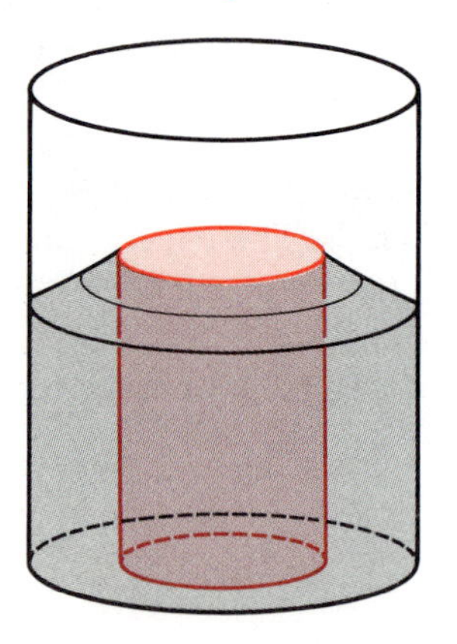

Figure 7-6-6
Partly frozen water in a can.
As freezing proceeds from the outside, water builds up towards the center.

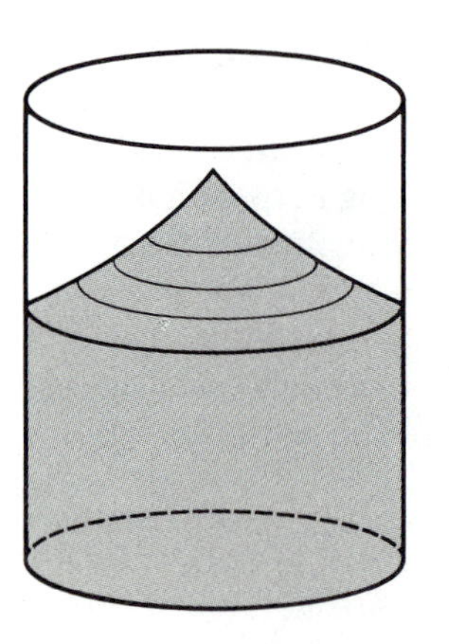

Figure 7-6-7
Can of frozen water

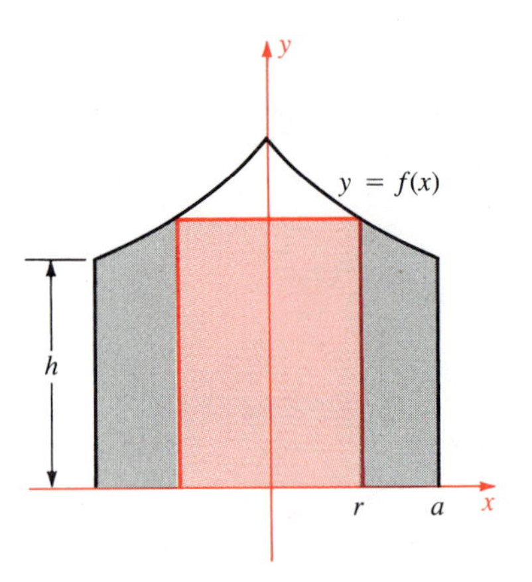

Figure 7-6-8
Cross section

At an intermediate stage, the central core of water is a cylinder of radius r, where $0 < r < a$, having volume $\pi r^2 f(r)$. Eventually this core will freeze into a solid of revolution whose upper surface is swept out by the curve $y = f(x)$. Its volume will be

$$\int_0^r 2\pi x f(x)\, dx$$

But this volume of ice is $1 + \beta$ times the volume of the water it came from, hence

$$\int_0^r 2\pi x f(x)\, dx = (1 + \beta)\pi r^2 f(r) \qquad (0 < r < a)$$

The desired function $f(x)$ must satisfy this equation subject to the initial condition $f(a) = h$.

To find $f(x)$, we first differentiate with respect to r:

$$\frac{d}{dr}\int_0^r 2\pi x f(x)\, dx = (1 + \beta)\pi \frac{d}{dr}[r^2 f(r)]$$

that is,

$$2\pi r f(r) = (1 + \beta)\pi[2r f(r) + r^2 f'(r)]$$

Then we cancel πr:

$$2f(r) = (1 + \beta)[2f(r) + rf'(r)]$$

Hence

$$(1 + \beta)rf'(r) = -2\beta f(r)$$

Remember that r was an intermediate value of x with $0 < r < a$. So let us call it x again since we are seeking $y = f(x)$. Thus

$$(1 + \beta)xf'(x) = -2\beta f(x)$$

so

$$xf'(x) = pf(x)$$

where

$$p = \frac{-2\beta}{1 + \beta}$$

We can rewrite the equation as

$$\frac{f'}{f} = \frac{p}{x} \qquad \text{that is} \qquad \frac{d}{dx}\ln f = \frac{d}{dx}\ln x^p$$

Therefore $\ln f(x) = \ln x^p + C$, that is, $f(x) = kx^p$. We choose k so that $f(a) = h$, that is, $ka^p = h$. Thus we arrive at the surface shape given by

$$y = h\left(\frac{x}{a}\right)^p$$

where

$$p = \frac{-2\beta}{1 + \beta}$$

Figure 7-6-9
Fluid expands on freezing.

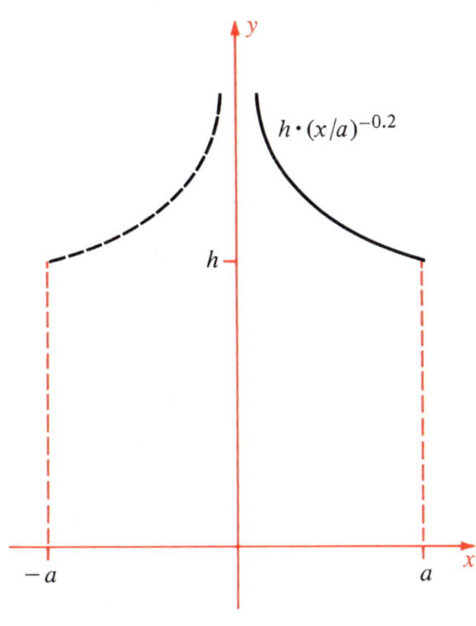

For $0 < \beta < 1$, this curve is shown in Figure 7-6-9. Note that $p < 0$. Look for this shape when you remove the top (not the bottom) from a can of frozen juice.

If the liquid is a molten metal, then it contracts upon solidifying instead of expanding. Hence β is slightly less than 0. The same analysis applies, only this time $0 < p < 1$, yielding the shape shown in Figure 7-6-10.

Figure 7-6-10
Fluid contracts on freezing.

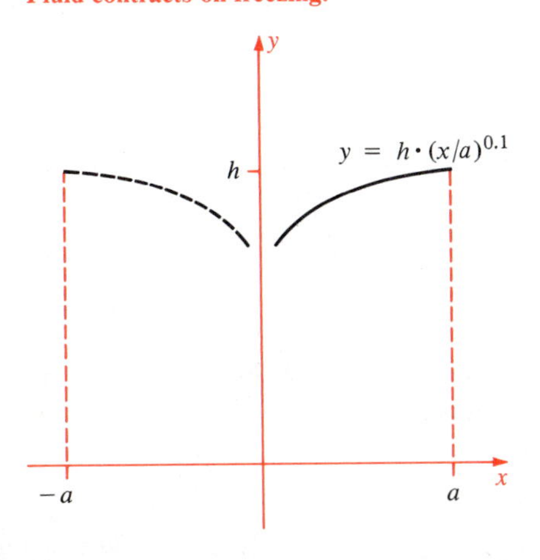

Remark 1 The model fails when the cylinder is very thin because factors like surface tension, which we have ignored, become significant.

Remark 2 The mathematics in this example consisted of deriving a differential equation by differentiating an equation involving an integral with a variable (upper) limit. This is an important technique in physics.

Exercises

1 Find the present value $V(T)$ of the constant income stream $f(t) = c$, interest rate r, time T years.

2 (cont.) A company deposits funds continuously at the constant rate of Y dollars per year in an account paying 8% interest, compounded continuously. Find Y so that the investment will be worth 10^7 dollars in 10 years.

3 Find the present value $V(T)$ of the income stream $f(t) = bt$, interest rate r, time T years.

4 (cont.) A company deposits funds continuously in an 8% account, starting initially at the rate of $\$10^6/\text{yr}$ and increasing the rate steadily to $\$2 \times 10^6/\text{yr}$ at the end of 5 years. What is the value of the account at that time?

In Exercises 5–10, the income stream $f(t)$ at annual interest rate r for t years has present value $V = V(T)$.

5 Suppose money is steadily invested for one year, steadily withdrawn at the same rate the next year, then again steadily invested, and so on (cyclic behavior). Thus $f(t) = c$ for $2n \le t < 2n + 1$ and $f(t) = -c$ for $2n + 1 \le t < 2n + 2$. Find $V(T)$ for T an even integer, $T = 2N$.

6 (cont.) Find the ultimate value, $\lim V(2N)$ as $N \to +\infty$.

7 Suppose $f(t) = c \sin \pi t$, the simplest possible smooth periodic function in certain respects. Find $V(T)$ for T an even integer, $T = 2N$.

8 (cont.) Find $\lim V(2N)$ as $N \to +\infty$.

9 Find $f(T)$ if $V(T) = cT$.

10 Find $f(T)$ if $V(T) = cTe^{-kT}$.

11 Suppose that the interest rate $r = r(t)$ varies for $t \ge 0$. Show that an initial amount A_0 will grow in time t to

$$A(t) = A_0 e^{\phi(t)} \quad \text{where} \quad \phi(t) = \int_0^t r(u)\, du$$

12 (cont.) Find the present value $V(T)$ of the income stream $f(t)$.

In Exercises 13–20, show that the given function is a probability density, find $E(X)$, and find $\operatorname{var}(X)$

13 $f(t) = |t|$ on $[-1, 1]$

14 $f(t) = 1 - |t|$ on $[-1, 1]$

15 $f(t) = 3t^2$ on $[0, 1]$

16 $f(t) = \frac{5}{2}t^4$ on $[-1, 1]$

17 $f(t) = (2\pi)^{-1}(1 - \cos t)$ on $[-\pi, \pi]$

18 $f(t) = (2\pi)^{-1}(1 - \cos t)$ on $[0, 2\pi]$

19 $f(t) = 0$ if $\frac{1}{3} < t < \frac{2}{3}$, otherwise $f(t) = \frac{3}{2}$ on $[0, 1]$

20 $f(t) = 0$ if $\frac{1}{4} < t < \frac{3}{4}$, otherwise $f(t) = 2$ on $[0, 1]$

Let $f(t)$ be the probability density function on $[a, b]$ of a random variable X. The **distribution function** of X is defined by

$$F(t) = \int_a^t f(u)\, du \qquad (a \le t \le b)$$

The value $F(t)$ is the probability that X lies in $[a, t]$.

21 Find the distribution function for $f(t)$ of Exercise 13.

22 Find the distribution function for $f(t)$ of Exercise 18.

23 Suppose the density function $f(t)$ on $[a, b]$ is continuous and $F(t)$ is the corresponding distribution function. Find $F'(t)$.

24 (cont.) Show that

$$E(X) = b - \int_a^b F(t)\, dt$$

25 Express the tension $T(x)$ in a suspension-bridge cable in terms of T_0, δ, and x.

26 Suppose the roadway of the suspension bridge is thicker at the ends than in the middle, say its linear density (weight per unit length) is $\delta(x) = A + Bx^2$, where x is the horizontal distance from the middle of the bridge. Find the shape of the cable.

27 Where does a teeter-totter (seesaw) balance? A thin, stiff, horizontal rod (Figure 7-6-11) of varying linear density $\delta(x)$ is to be balanced on a fulcrum at $\bar{x}$. Each element of mass is acted on by gravity, causing a turning moment (torque) about x equal to the product of the mass by its distance from the fulcrum. Set up integrals for the moments on either side of the fulcrum. Equate these to find $\bar{x}$.

Figure 7-6-11
See Exercise 27

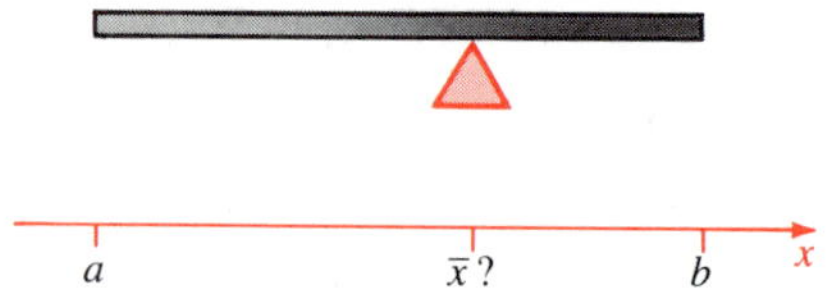

a x̄? b x

28 Solve the ice cube problem when the container is a thin rectangular slab (Figure 7-6-12) and cooling is applied only at the ends (symmetrically). The flat faces and bottom are insulators.

29 In a reactor, the uranium oxide fuel generates heat energy at the rate of $Q = 362\ \text{watts}/\text{cm}^3$. This heat is extracted by a coolant passing over the surface of the fuel. Suppose a long cylindrical fuel element has radius a cm. It is known that

$$Q = \frac{4}{a^2} \int_{T_1}^{T_0} \mu(T)\, dT$$

Figure 7-6-12
See Exercise 28

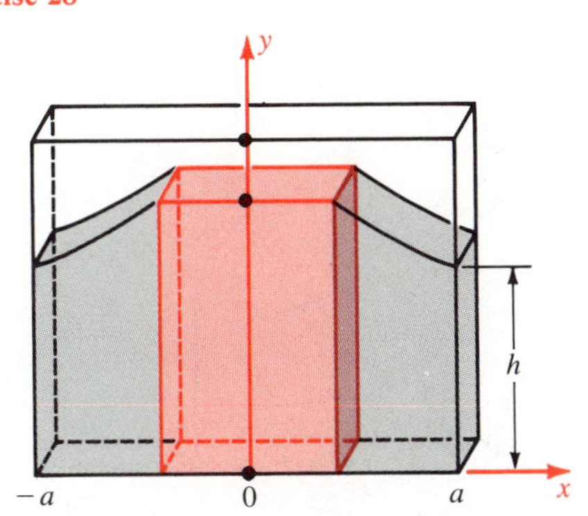

where T_0 is the temperature on the axis of the cylinder, T_1 is the surface temperature of the element or temperature of the coolant, and μ is the thermal conductivity of uranium oxide. At high temperatures, $\mu(T) = 31.7/T$ watts per centimeter per degree kelvin, where the temperature T is measured in K. Given that $a = 0.610$ cm and $T_1 = 820$ K, compute T_0.

30 Two point masses m_0 and m_1 at distance d exert an attractive force on each other of magnitude

$$F = G\frac{m_0 m_1}{d^2}$$

where G is the **constant of gravitation.** Find the gravitational attraction of the uniform rod of linear density δ in Figure 7-6-13 on the point mass m_0.

Figure 7-6-13
See Exercise 30

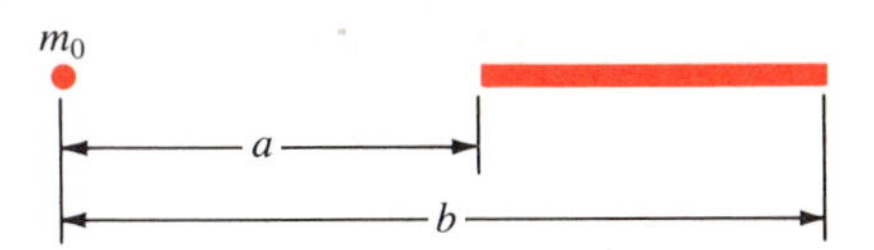

7-7 Review Exercises

1 Compute the area of the region bounded by the curves $y = x^3$ and $y = 2x^2$.

2 Compute the area of the region bounded by the curves $y = e^{2x}$, $y = -x^2$, $x = 0$, and $x = 3$.

3 Let R be the region under the curve $y = 1/x^2$ from $x = 1$ to $x = 10$. Find a vertical line that divides R into two regions of equal area.

4 In Figure 7-7-1, we have $a < x < b$. Find the maximal area of the triangle.

Figure 7-7-1
See Exercise 4

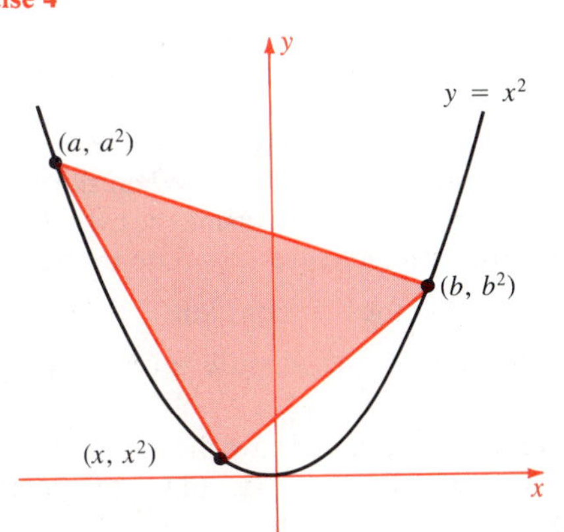

5 Find the area of the first-quadrant region bounded by $y = x^2$, $y = 1/x^2$, and $y = 3$.

6 Find the area bounded by $y = x^2$, $y = 1/x^2$, and $x = 3$.

7 Find the area bounded by $y = x^3$, $y = x^5$, and the lines $x = \frac{1}{2}$ and $x = 2$.

8 Find the area bounded by $y = x^3$ and $y = x^5$.

9 The portion of the curve $y = x^3 + 1$ between $x = 0$ and $x = 1$ is rotated about the line $y = \frac{1}{2}$. Find the volume of the resulting solid.

10 The area bounded by the curve $y = x - x^2$ and the x-axis is revolved around the y-axis. Find the volume of the resulting solid.

11 Compute the volume of a pinch-waist tank that is in the shape of the solid generated by rotating the plane region $-1 \le x \le \frac{1}{9}y^2$, $-3 \le y \le 3$ about the line $x = -1$. Take feet as the units.

12 (cont.) How much work is required to fill the tank with water from the level of its bottom? Assume water weighs 62.4 lb/ft^3.

The tank in Figure 7-7-2 has its rectangular base on the horizontal x, y-plane. Its sides are trapezoids; the back sides are vertical, the two forward sides are oblique to the horizontal. The tank is filled with fluid of density δ.

13 Find the volume V of the tank.
[Hint Slice by horizontal planes.]

14 How much work is required to fill the tank with fluid from the level of its base?

15 Compute the total pressure F_L on the (shaded) left end of the tank.

16 Find the total pressure F_R on the (shaded) right end of the tank. (Remember that pressure acts perpendicular to any surface, even an oblique one.)

Figure 7-7-2
See Exercises 13–16

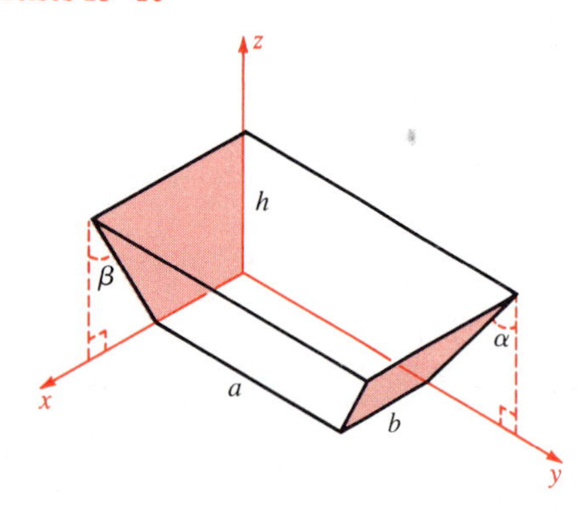

17 At time $t = 0$, a point is at the origin. It moves to the right along the x-axis, its speed at time t given by $1/\sqrt{3t+1}$. Find its position at $t = 4$.

18 One particle is at $x = 0$ and another at $x = 5$. Each attracts a third particle p with a force $k/(\text{distance})^2$. Compute the work done moving p from $x = 10$ to $x = 15$.

The tank in Figure 7-7-3 consists of a standing right circular cylinder cut off by an oblique plane that cuts the horizontal x, y-plane at angle α. It is filled with fluid of density δ.

19 Use slabs perpendicular to the y-axis to set up the volume as an integral.

20 Evaluate this integral. Can you interpret the result geometrically?

21 Set up an integral for the work required to fill the tank with fluid from the level of its base.

22 The wedge in Figure 7-7-4 has been cut from a right circular cylinder by a plane oblique to the horizontal base. Set up the volume as an integral in two ways and compute it both ways.

Figure 7-7-3
See Exercises 19–21

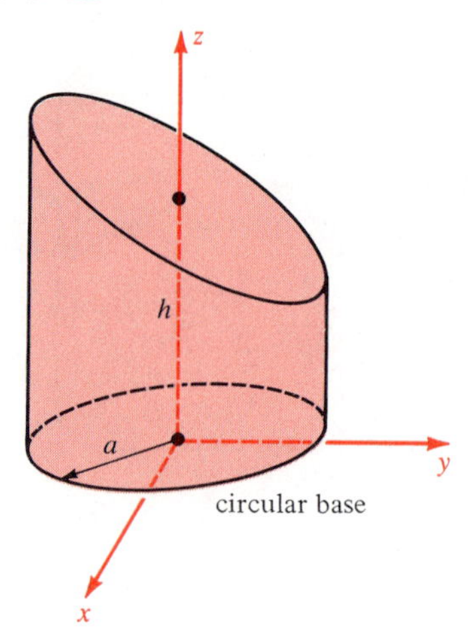

Figure 7-7-4
See Exercise 22

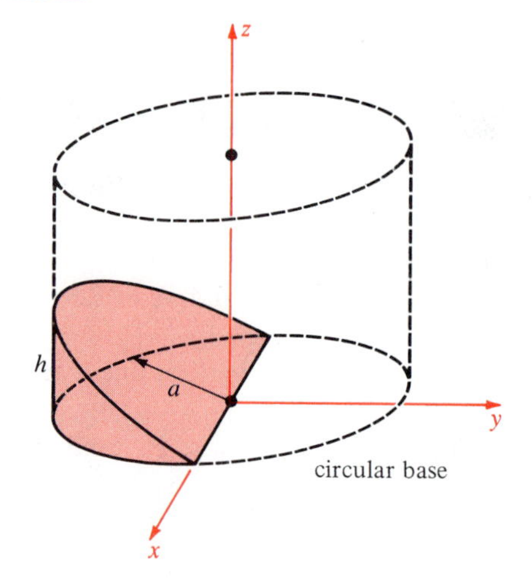

23 The piston of Figure 7-7-5 has cross-sectional area A, and it is kept at constant temperature. When the piston is initially at x_0, the gas pressure inside the cylinder equals the outside atmospheric pressure p_0. Use the gas law

$$(\text{pressure})(\text{volume}) = \text{constant}$$

to derive a formula for the pressure $p(x)$ when the piston is at x.

24 (cont.) Find the work done by the external force F to move the piston from x_0 to x. (Don't forget that atmospheric pressure is helping you.)

Figure 7-7-5
See Exercises 23 and 24

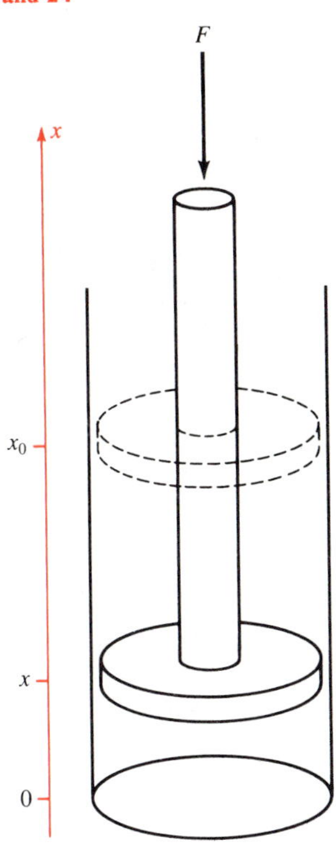

25 A metal weight of density 5 g/cm^3 has the shape of a truncated pyramid with square cross section. Its top measures 10×10 cm, its bottom 20×20 cm, and its height is 10 cm. The weight is submerged in a lake of density 1 g/cm^3 with its top just at the water's surface. How much work (in g-cm) is needed to lift the weight just clear of the water?

26 Find the (shaded) area between the three semicircles in Figure 7-7-6.

Figure 7-7-6
See Exercise 26

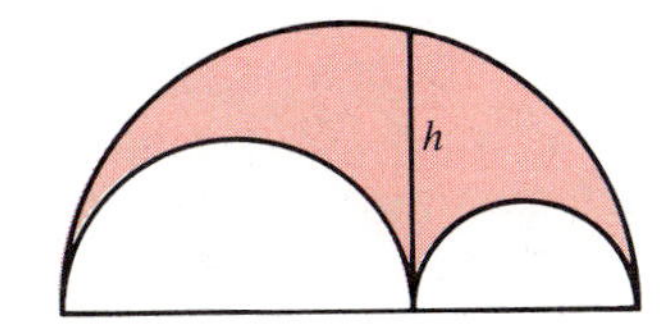

27 The office space part of the Transamerica Building in San Francisco (Figure 7-7-7) consists of a truncated pyramid. Its height is 621 ft. Its base is a square of side 152 ft, and its top is a square of side 40 ft. Find its volume by integration, then use the trapezoidal rule to estimate the total floor space of its 46 floors. (The ground floor is floor 1; the roof doesn't count.)

Figure 7-7-7
See Exercise 27

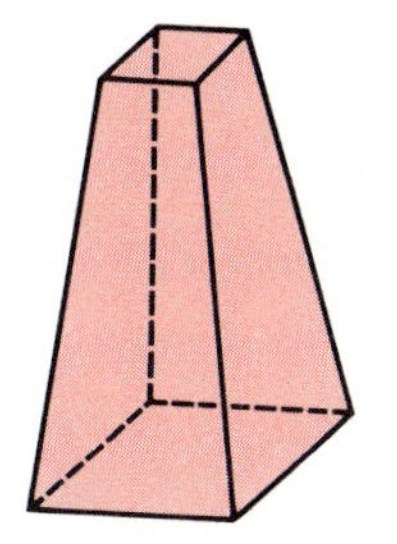

28 Find the gravitational attraction of the uniform rod of linear density δ in Figure 7-7-8 on the point mass m_0. (See Exercise 30 in Section 7-6.) By symmetry, the horizontal component of the force is 0, so only the vertical component is required. [Hint The integration is easy if you use an angle as the variable.]

Figure 7-7-8
See Exercise 28

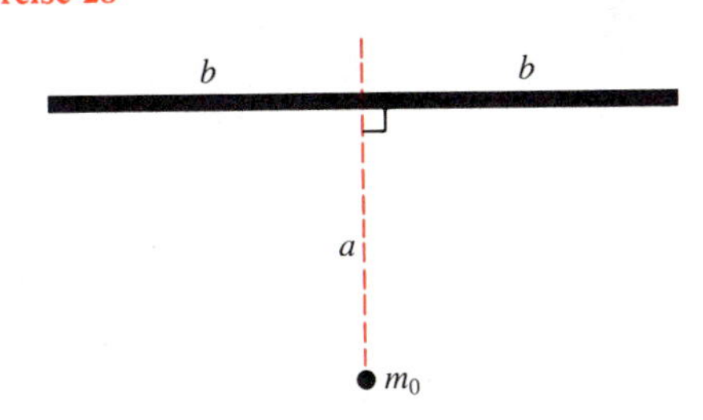

29 Find k so that $f(t) = kt^3$ is a probability density function on $[0, 2]$. Find the expectation $E(X)$ of the corresponding random variable X.

30 (cont.) Find $\operatorname{var}(X)$.

8 Analytic Geometry

The philosopher and mathematician René Descartes (1596–1650) slept till noon every day and invented analytic geometry. Analytic geometry is the study of geometry by means of algebra and coordinate systems. In honor of Descartes, rectangular coordinate systems are called *Cartesian.*

8-1 Translation and Circles

A useful skill in working with coordinate systems is knowing where to place the axes. While doing a problem in a given coordinate system, it may be convenient to introduce new coordinate axes parallel to the given ones. This operation is called **shifting** or **translating** the axes.

After a translation of axes, a point with coordinates (x, y) relative to the original axes acquires new coordinates $(\bar{x}, \bar{y})$ relative to the new axes. It is easy to express the relation between the old coordinates and the new ones. The translation sets up a new origin $\bar{\mathbf{0}}$ at a point (h, k) in the old coordinates (Figure 8-1-1). From Figure 8-1-2, we have $x = \bar{x} + h$ and $y = \bar{y} + k$, no matter what quadrant $\bar{\mathbf{0}}$ lies in. Conversely, $\bar{x} = x - h$ and $\bar{y} = y - k$.

Translation of Axes

If the coordinate axes are translated so that the origin is moved to (h, k), then the new coordinates $(\bar{x}, \bar{y})$ and the old coordinates (x, y) of a point are related by the equations

$$\begin{cases} \bar{x} = x - h \\ \bar{y} = y - k \end{cases} \qquad \begin{cases} x = \bar{x} + h \\ y = \bar{y} + k \end{cases}$$

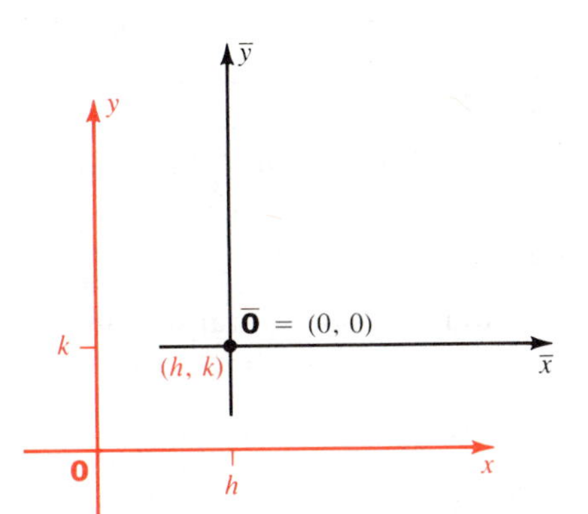

Figure 8-1-1
Choice of new origin and axes

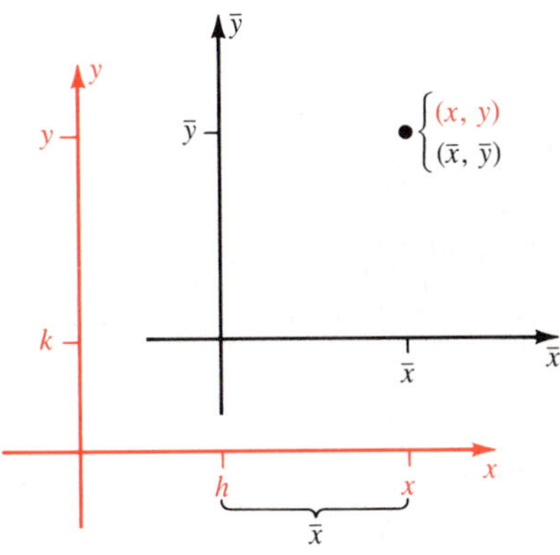

Figure 8-1-2
By comparing horizontal distances (with due regard for sign), $x = \bar{x} + h$. Similarly, $y = \bar{y} + k$.

To keep the signs right, remember that $x = h$ and $y = k$ must imply $\bar{x} = 0$ and $\bar{y} = 0$, and conversely.

Translation of axes is helpful in simplifying equations and computations. Suppose some geometric set, a line, a circle, a parabola, or anything is described by a relation between x and y in the x, y-coordinate system. If we translate coordinates, so that each point of the geometric set has new coordinates $(\bar{x}, \bar{y})$, then the set is described by a relation between its new coordinates $\bar{x}$ and $\bar{y}$. If the origin of the new coordinate system is carefully chosen, this new relation might be simpler than the original one between x and y.

Example 1

$$y + 3 = (x - 2)^2$$

Take for the new origin $(h, k) = (2, -3)$. Then

$$\bar{x} = x - 2 \qquad \bar{y} = y + 3$$

so the curve described by $y + 3 = (x - 2)^2$ has the simpler equation

$$\bar{y} = \bar{x}^2$$

in the new coordinate system (Figure 8-1-3). ●

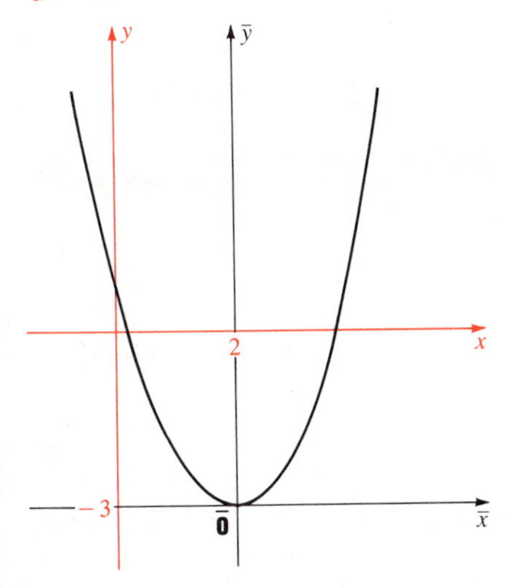

Figure 8-1-3
$\begin{cases} y + 3 = (x - 2)^2 \\ \bar{y} = \bar{x}^2 \end{cases}$

Lines

Suppose we translate coordinates by

$$x = \bar{x} + h \qquad y = \bar{y} + k$$

The equation of a straight line,

$$ax + by = c$$

in the old coordinates becomes $a(\bar{x} + h) + b(\bar{y} + k) = c$, that is,

$$a\bar{x} + b\bar{y} = c - (ah + bk)$$

in the new coordinates. Thus the equation of a line is the same in either coordinate system except for the constant on the right side.

Example 2

Find all translations that convert

$$x + 2y = 1 \quad \text{into} \quad \bar{x} + 2\bar{y} = 3$$

Solution Under a translation $x = \bar{x} + h$, $y = \bar{y} + k$, the equation $x + 2y = 1$ becomes $(\bar{x} + h) + 2(\bar{y} + k) = 1$, that is,

$$\bar{x} + 2\bar{y} = 1 - (h + 2k)$$

Therefore, the condition on (h, k) is

$$1 - (h + 2k) = 3 \qquad \text{that is} \qquad h + 2k = -2$$

See Figure 8-1-4 for a geometric interpretation. ●

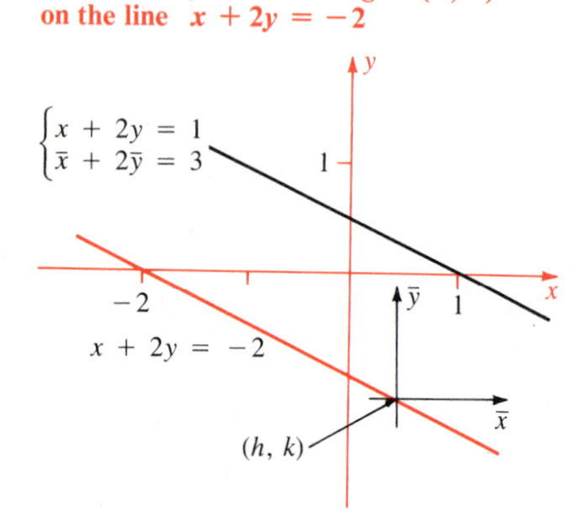

Figure 8-1-4
The equation of the line $x + 2y = 1$ becomes $\bar{x} + 2\bar{y} = 3$ in any new coordinate system with origin (h, k) on the line $x + 2y = -2$

Division of a Segment

Suppose we are given two points of the plane,

$$P = (a, b) \quad \text{and} \quad Q = (c, d)$$

We know how to find the midpoint of the segment PQ; it is

$$M = (\tfrac{1}{2}(a + c), \tfrac{1}{2}(b + d))$$

See page 47. The midpoint M is the point on the segment PQ that is $\frac{1}{2}$ of the way from P to Q.

More generally, suppose $0 < t < 1$. We seek the point R that is t of the way from P to Q. (For instance, if $t = \frac{2}{5}$, then we want the point that is $\frac{2}{5}$ of the way from P to Q.) To find the x-coordinate of R, start from a and go t of the way from a to c:

$$x = a + t(c - a) = (1 - t)a + tc$$

See Figure 8-1-5. A similar formula holds for the y-coordinate.

Division of a Segment

Let

$$P = (a, b) \quad Q = (c, d) \quad \text{and} \quad 0 < t < 1$$

The point R that is on the segment PQ and t of the way from P to Q has coordinates

$$x = (1 - t)a + tc \qquad y = (1 - t)b + td$$ ●

Remark In case $P = (0, 0)$, this says that the point t of the way from the origin to (c, d) is (tc, td), which should be clear.

Figure 8-1-5
$a+t(c-a)$ is t of the way from a to c.

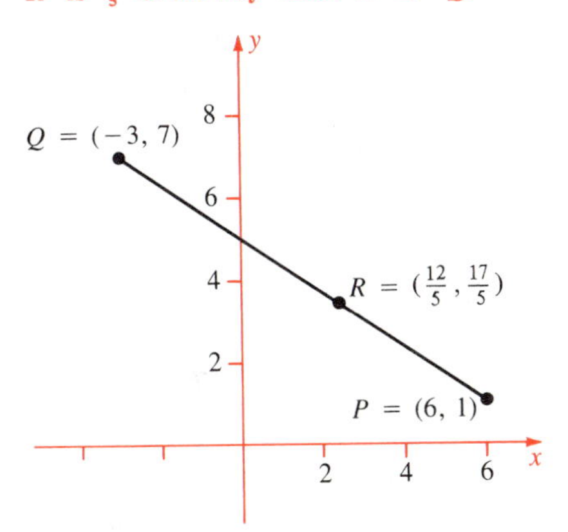

Figure 8-1-6
R is $\frac{2}{5}$ of the way from P to Q.

Example 3

The point that is $\frac{2}{5}$ of the way from $(6, 1)$ to $(-3, 7)$ is

$$R = \left(\tfrac{3}{5}\cdot 6 + \tfrac{2}{5}\cdot(-3),\ \ \tfrac{3}{5}\cdot 1 + \tfrac{2}{5}\cdot 7\right)$$
$$= \left(\tfrac{12}{5}, \tfrac{17}{5}\right)$$

See Figure 8-1-6. ●

The Circle

The equation of the circle with center (a, b) and radius r is

$$(x-a)^2 + (y-b)^2 = r^2$$

(See pages 45 and 46.) This equation is the algebraic statement that the distance from each point (x, y) on the circle to its center equals the radius. If new axes are centered at (a, b), the equation becomes simply $\bar{x}^2 + \bar{y}^2 = r^2$. Let us examine the general equation of a circle. Expanded, the equation above becomes

$$(x^2 - 2ax + a^2) + (y^2 - 2by + b^2) = r^2$$

that is,

$$x^2 + y^2 - 2ax - 2by + (a^2 + b^2 - r^2) = 0$$

Now suppose we start with any equation of the form

$$x^2 + y^2 - 2ax - 2by + c = 0$$

What is the set of all points (x, y) that satisfy it? (Of course we suspect a circle.) We complete the squares for both the x and the y terms:

$$(x^2 - 2ax + a^2) + (y^2 - 2by + b^2) + c = a^2 + b^2$$
$$(x-a)^2 + (y-b)^2 = a^2 + b^2 - c$$

Now here is a hitch: the left side is a sum of squares, hence non-negative. But the right side is not guaranteed to be non-negative! We are forced to consider three cases:

- $a^2 + b^2 - c < 0$. Then no point (x, y) satisfies the equation; the set is empty.
- $a^2 + b^2 - c = 0$. Then only $(x, y) = (a, b)$ satisfies the equation because the left side is positive for any other point; the set consists of the single point (a, b).

Figure 8-1-7
$(x - a)^2 + (y - b)^2 = r^2 \quad (r > 0)$

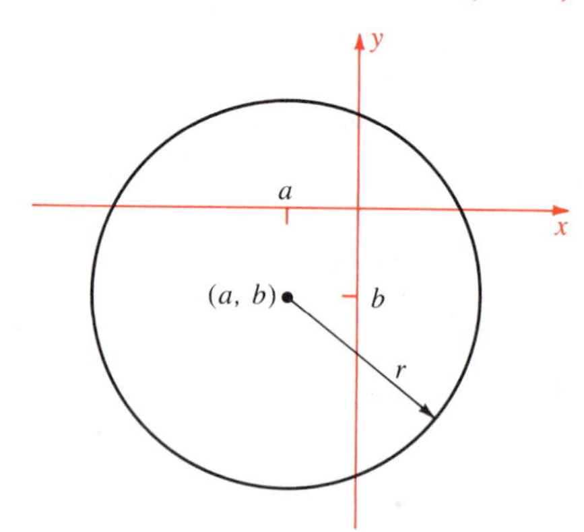

- $a^2 + b^2 - c > 0$. Then we set $r = \sqrt{a^2 + b^2 - c}$ and the equation becomes

$$(x - a)^2 + (y - b)^2 = r^2$$

The set in this case is an honest circle (Figure 8-1-7).

Example 4

Describe the set of points (x, y) that satisfy

a $x^2 + y^2 - 2x + 4y - 4 = 0$ **c** $x^2 + y^2 - 2x + 4y + 6 = 0$
b $x^2 + y^2 - 2x + 4y + 5 = 0$

Solution Complete the squares in x and y:

$$x^2 - 2x + y^2 + 4y = (x - 1)^2 + (y + 2)^2 - 5$$

Hence the three cases become

a $(x - 1)^2 + (y + 2)^2 = 9 = 3^2$ circle: center $(1, -2)$ radius 3
b $(x - 1)^2 + (y + 2)^2 = 0$ single point: $(1, -2)$
c $(x - 1)^2 + (y + 2)^2 = -1$ empty set ●

Figure 8-1-8

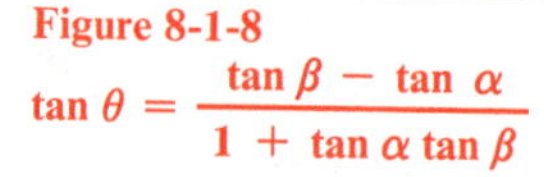

$\tan\theta = \dfrac{\tan\beta - \tan\alpha}{1 + \tan\alpha\tan\beta}$

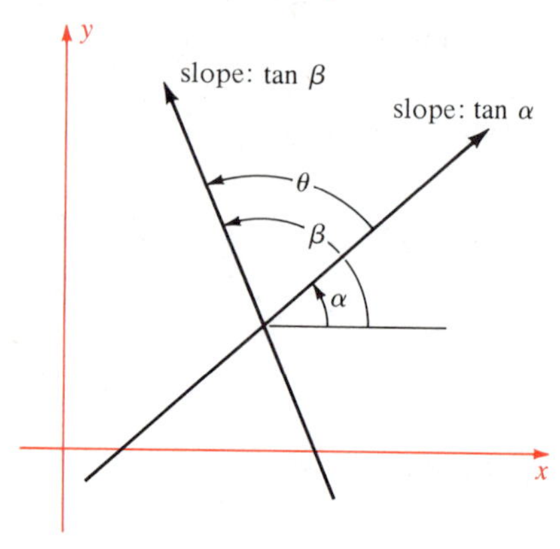

Angle between Lines

Later we shall need a formula for the angle, measured counterclockwise, between two directed lines in terms of the slopes of the lines. With the notation chosen in Figure 8-1-8 we have $\theta = \beta - \alpha$. By the addition formula for tangent (page 60) with α replaced by $-\alpha$,

$$\tan\theta = \tan(\beta - \alpha) = \frac{\tan\beta - \tan\alpha}{1 + \tan\alpha\tan\beta}$$

Since $\tan\alpha$ and $\tan\beta$ are the respective slopes of the given lines, this is the required formula.

Exercises

An $\bar{x}, \bar{y}$-coordinate system is introduced with its origin at $(-7, 3)$. Find the $\bar{x}, \bar{y}$-coordinates of the point with x, y-coordinates

1 $(1, 6)$
2 $(5, 4)$
3 $(0, 0)$
4 $(-8, 1)$
5 $(-5, -6)$
6 $(0, 2)$

With the $\bar{x}, \bar{y}$-coordinate system as above, find the x, y-coordinates for the point

7 $(\bar{x}, \bar{y}) = (1, 3)$
8 $(\bar{x}, \bar{y}) = (4, -2)$
9 $(\bar{x}, \bar{y}) = (-6, -7)$
10 $(\bar{x}, \bar{y}) = (0, 3)$
11 $(\bar{x}, \bar{y}) = (-7, 3)$
12 $(\bar{x}, \bar{y}) = (1, 6)$

Describe a translation of axes that converts the first equation into the second

13 $3x - 2y = 1 \qquad 3\bar{x} - 2\bar{y} = 0$
14 $x + 2y = 5 \qquad \bar{x} + 2\bar{y} = 0$
15 $y = x^2 + 2x + 2 \qquad \bar{y} = \bar{x}^2 + 1$
16 $y = 3 + \dfrac{x - 1}{1 + (x - 1)^2} \qquad \bar{y} = \dfrac{\bar{x}}{1 + \bar{x}^2}$
17 $y = \sin(x - \frac{1}{6}\pi) - 1 \qquad \bar{y} = \sin\bar{x}$
18 $y = \cos x + \cos(x + \frac{1}{3}\pi) + \cos(x + \frac{2}{3}\pi)$
$\bar{y} = \cos(\bar{x} - \frac{1}{3}\pi) + \cos\bar{x} + \cos(\bar{x} + \frac{1}{3}\pi)$

Find the point

19 $\frac{2}{3}$ of the way from $(1, 1)$ to $(4, -5)$

20 $\frac{1}{5}$ of the way from $(4, 0)$ to $(0, 4)$

21 Interpret

$$(x, y) = ((1 - r)x_1 + rx_2,\ (1 - r)y_1 + ry_2)$$

when $r > 1$.

22 (cont.) Do the same when $r < 0$.

Describe the set of points that satisfy

23 $x^2 + y^2 - 4x - 4y = 0$

24 $x^2 + y^2 - 6x = 0$

25 $x^2 + y^2 + 2x + 6y = 26$

26 $x^2 + y^2 - x + y = \frac{33}{2}$

27 $x^2 + y^2 - x + 2y = 0$

28 $x^2 + y^2 + 6x - 8y = 25$

29 $2x^2 + 2y^2 - 3x - 5y + 1 = 0$

30 $3x^2 + 3y^2 - x - y = 0$

31 Let $(x - a_1)^2 + (y - b_1)^2 = r_1^2$ and $(x - a_2)^2 + (y - b_2)^2 = r_2^2$ be two non-concentric circles. The **radical axis** of the two circles is the graph of

$$(x - a_1)^2 + (y - b_1)^2 - (x - a_2)^2 - (y - b_2)^2 = r_1^2 - r_2^2$$

Show that the radical axis is a line perpendicular to the line of centers.

32 (cont.) Let P be any point on the radical axis, outside of both circles. Suppose A is a point of the first circle such that PA is a tangent to that circle, and suppose B is a similarly chosen point of the second circle. Prove that $\overline{PA} = \overline{PB}$.

8-2 Intersections of Lines and Circles

Given a circle and a line, there are three possibilities:

- They do not intersect.
- They have exactly one common point—the line is **tangent** to the circle.
- They intersect in two distinct points.

From the equations of the circle and line, we find the points of intersection (if any). The algebra is simplest if the circle has its center at the origin. Otherwise, we make a preliminary translation to move the origin to the center.

Suppose the equations are

$$\begin{cases} x^2 + y^2 = r^2 \\ ax + by = c \end{cases} \qquad \text{where} \qquad a^2 + b^2 > 0$$

The points of intersection are the common solutions (x, y) of the two equations.

Either $a \neq 0$ or $b \neq 0$. We assume $a \neq 0$ and multiply the first equation by a^2:

$$a^2x^2 + a^2y^2 = a^2r^2$$

We substitute $ax = c - by$, to eliminate x:

$$(c - by)^2 + a^2y^2 = a^2r^2$$

We expand and collect terms:

$$(a^2 + b^2)y^2 - 2bcy + (c^2 - a^2r^2) = 0$$

This is a quadratic equation for y; it has 0, 1, or 2 solutions, depending on its discriminant D.

If $D < 0$, then there are no real solutions, hence no points of intersection. If $D \geq 0$, we solve for y by the quadratic formula, then find x from the relation $ax = c - by$. Note that if $D = 0$, there is one point of intersection (tangency), and if $D > 0$, there are two points of intersection.

In case $a = 0$, then $b \neq 0$ and a similar argument applies.

Example 1

Find the intersections of $x^2 + y^2 = 4$ and $x + 2y = 1$.

Solution Replace x^2 by $(1 - 2y)^2$:

$$(1 - 2y)^2 + y^2 = 4 \qquad 1 - 4y + 5y^2 = 4$$

$$5y^2 - 4y - 3 = 0$$

The discriminant of the quadratic equation is

$$D = (-4)^2 - 4(5)(-3) = 16 + 60 = 76 > 0$$

The quadratic has two solutions:

$$y = \frac{4 \pm \sqrt{76}}{10} = \frac{4 \pm 2\sqrt{19}}{10} = \tfrac{2}{5} \pm \tfrac{1}{5}\sqrt{19}$$

The corresponding values of x are

$$x = 1 - 2y = 1 - \tfrac{2}{5}(2 \pm \sqrt{19}) = \tfrac{1}{5} \mp \tfrac{2}{5}\sqrt{19}$$

See Figure 8-2-1. Hence the two intersection points are

$$(\tfrac{1}{5} - \tfrac{2}{5}\sqrt{19}, \tfrac{2}{5} + \tfrac{1}{5}\sqrt{19}) \quad \text{and} \quad (\tfrac{1}{5} + \tfrac{2}{5}\sqrt{19}, \tfrac{2}{5} - \tfrac{1}{5}\sqrt{19})$$

●

Figure 8-2-1

$$\begin{cases} P = (\tfrac{1}{5} - \tfrac{2}{5}\sqrt{19}, \tfrac{2}{5} + \tfrac{1}{5}\sqrt{19}) \\ Q = (\tfrac{1}{5} + \tfrac{2}{5}\sqrt{19}, \tfrac{2}{5} - \tfrac{1}{5}\sqrt{19}) \end{cases}$$

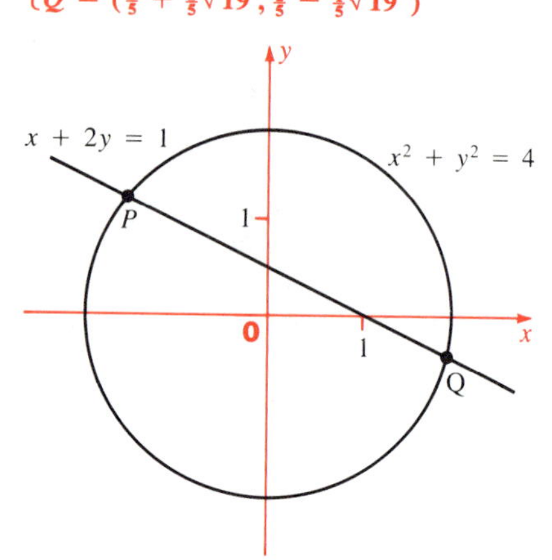

Example 2

Find the intersection of $(x - 1)^2 + (y - 4)^2 = 25$ and $x - y = -2$.

Solution Translate the origin to $(1, 4)$. Then $\bar{x} = x - 1$ and $\bar{y} = y - 4$. The new equation of the circle is $\bar{x}^2 + \bar{y}^2 = 25$, and the new equation of the line is

$$(\bar{x} + 1) - (\bar{y} + 4) = -2 \qquad \text{that is} \qquad \bar{x} - \bar{y} = 1$$

Now solve the system of equations

$$\bar{x}^2 + \bar{y}^2 = 25 \qquad \bar{x} - \bar{y} = 1$$

Eliminate $\bar{x}$:

$$(\bar{y} + 1)^2 + \bar{y}^2 = 25 \qquad 2\bar{y}^2 + 2\bar{y} - 24 = 0$$

$$\bar{y}^2 + \bar{y} - 12 = 0 \qquad (\bar{y} + 4)(\bar{y} - 3) = 0$$

Figure 8-2-2
$P = (-2, 0)$ and $Q = (5, 7)$

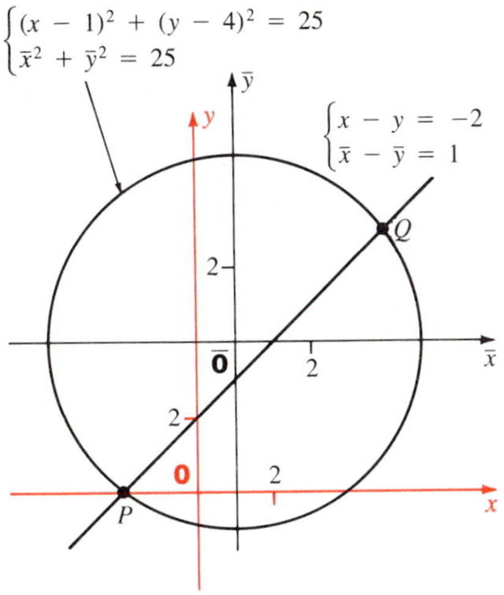

$\begin{cases} (x-1)^2 + (y-4)^2 = 25 \\ \bar{x}^2 + \bar{y}^2 = 25 \end{cases}$ $\begin{cases} x - y = -2 \\ \bar{x} - \bar{y} = 1 \end{cases}$

Hence

$$\bar{y} = -4 \quad \text{or} \quad \bar{y} = 3$$

The corresponding values of $\bar{x} = \bar{y} + 1$ are $\bar{x} = -3$ and $\bar{x} = 4$, so the two points of intersection are $(\bar{x}, \bar{y}) = (-3, -4)$ and $(\bar{x}, \bar{y}) = (4, 3)$. The (x, y) coordinates of these points are

$$(x, y) = (\bar{x} + 1, \bar{y} + 4) = (-2, 0) \quad \text{and} \quad (5, 7)$$ ●

See Figure 8-2-2.

Remark Example 2 could also be solved without shifting axes. For instance x could be eliminated from the original pair of equations and the resulting quadratic solved for y.

Intersection of Two Circles

Given two non-concentric circles, find their intersection. This problem can be reduced to the previous problem of a circle and a line. We suppose the equations of the circles are

$$\textbf{1} \quad x^2 + y^2 - 2a_1x - 2b_1y = c_1$$

$$\textbf{2} \quad x^2 + y^2 - 2a_2x - 2b_2y = c_2$$

The centers are (a_1, b_1) and (a_2, b_2) respectively, as can be seen by completing the squares. We subtract equation 2 from equation 1 to eliminate $x^2 + y^2$:

$$\textbf{3} \quad 2(a_2 - a_1)x + 2(b_2 - b_1)y = c_1 - c_2$$

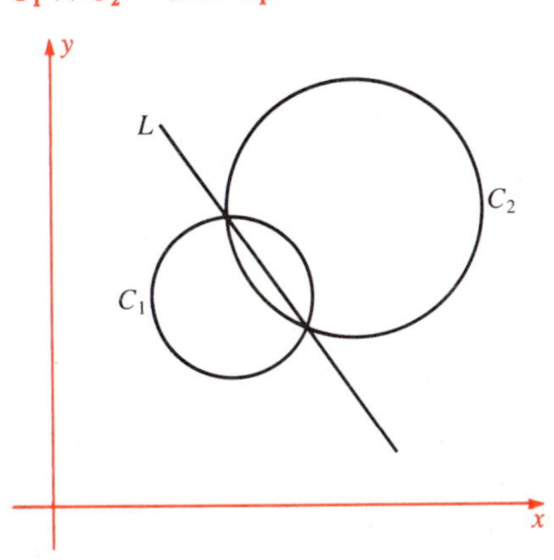

Figure 8-2-3
$C_1 \cap C_2 = L \cap C_1$

Because the circles are not concentric, $(a_1, b_1) \neq (a_2, b_2)$. Hence $a_2 - a_1$ and $b_2 - b_1$ are not both 0, so 3 is the equation of a line L. See Figure 8-2-3.

This line L intersects either circle in precisely the same points where the circles intersect each other. Why? Because if (x, y) satisfies both equations 1 and 2, then (x, y) also satisfies equation 3, the difference of 1 and 2. Therefore (x, y) satisfies 1 and 3. Conversely if (x, y) satisfies both equations 1 and 3, then (x, y) also satisfies equation 2, the difference of 1 and 3. Hence (x, y) satisfies 1 and 2. (Similarly, the intersections of L and the second circle are again the same points as the intersections of the two circles.)

Thus finding the intersection of two circles is equivalent to finding the intersection of a circle and a line. But we know how to do that!

Example 3

Find the intersection of

$$x^2 + y^2 = 25 \quad \text{and} \quad x^2 + y^2 + 2x - 14y = -25$$

Solution Subtract the first equation from the second:

$$2x - 14y = -50 \qquad x - 7y = -25$$

This is the equation of a line. To obtain its intersection with the first circle, solve the system

$$x^2 + y^2 = 25 \qquad x - 7y = -25$$

Eliminate x:

$$(7y - 25)^2 + y^2 = 25$$
$$50y^2 - 350y + 625 = 25$$
$$y^2 - 7y + 12 = 0$$

The solutions are $y = 3$ and $y = 4$. The corresponding values of $x = 7y - 25$ are $x = -4$ and $x = 3$. Therefore there are two points of intersection,

$$(-4, 3) \text{ and } (3, 4)$$

See Figure 8-2-4. ●

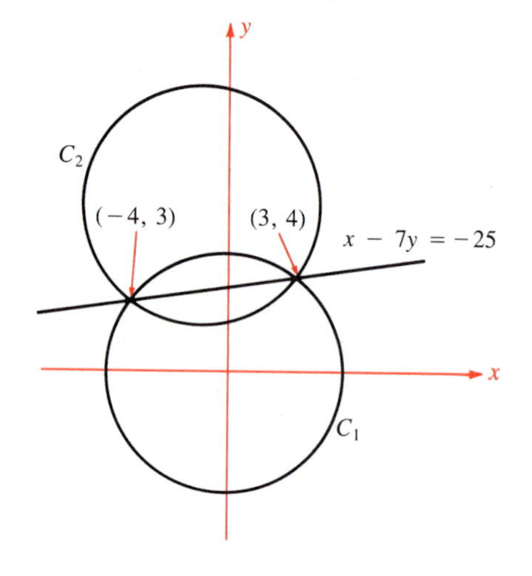

Figure 8-2-4
C_1: $x^2 + y^2 = 25$
C_2: $x^2 + y^2 + 2x - 14y = -25$

Tangents to Circles

A line tangent to a circle is perpendicular to the radius at the point of contact. Thus, if (u, v) is a point on the circle $x^2 + y^2 = r^2$, the tangent at (u, v) is perpendicular to the radial line through (u, v). Recall that the slopes of two perpendicular lines are negative reciprocals of each other. Since the radial line has slope v/u, it follows that the tangent has slope $-u/v$. Therefore the equation of the tangent is

$$y - v = -\frac{u}{v}(x - u) \qquad \text{that is} \qquad ux + vy = u^2 + v^2 = r^2$$

See Figure 8-2-5.

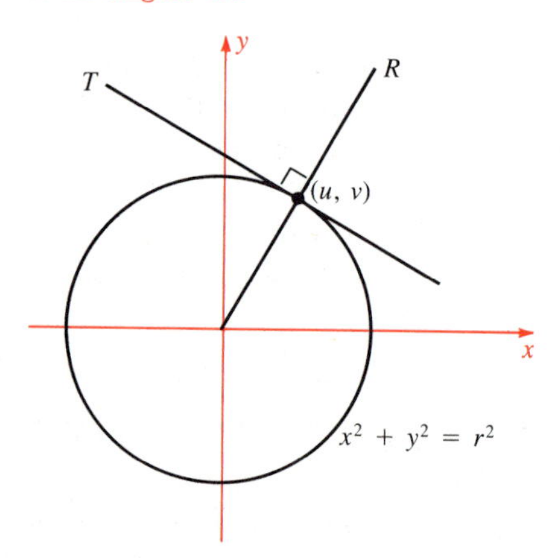

Figure 8-2-5
The radial line R is perpendicular to the tangent T.

Remark The circle is such a simple curve that we do not need calculus at all to find its slope at (u, v). However, it is reassuring to verify the result by calculus. Except at $(u, v) = (\pm r, 0)$, where the tangent is vertical, we can find dy/dx by differentiating $x^2 + y^2 = r^2$ implicitly:

$$2x + 2y\left.\frac{dy}{dx}\right|_{(u,v)} = 0 \qquad \text{hence} \qquad \left.\frac{dy}{dx}\right|_{(u,v)} = -\frac{u}{v}$$

Example 4

The tangent to $x^2 + y^2 = 25$ at $(4, 3)$ is

$$4x + 3y = 25$$ ●

The problem of finding the two tangent lines to a circle from an external point is harder, but the basic idea is the same.

Example 5

Find the two tangents to the circle $x^2 + y^2 = 4$ from $(3, 0)$.

Solution Suppose a tangent meets the circle at (u, v). Then the radial line through (u, v) is perpendicular to the tangent line, the line through $(3, 0)$ and (u, v). The radial line has slope v/u, and the tangent line has slope $v/(u - 3)$. But the product of the slopes of two perpendicular lines equals -1, so

$$\frac{v}{u} \cdot \frac{v}{u - 3} = -1$$

which can be written $u^2 + v^2 = 3u$. Therefore (u, v) satisfies the system

$$u^2 + v^2 = 4 \qquad u^2 + v^2 = 3u$$

It follows that $3u = 4$, so $u = \frac{4}{3}$ and

$$v^2 = 4 - u^2 = 4 - \frac{16}{9} = \frac{20}{9} \qquad v = \pm\frac{2}{3}\sqrt{5}$$

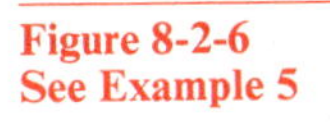
Figure 8-2-6
See Example 5

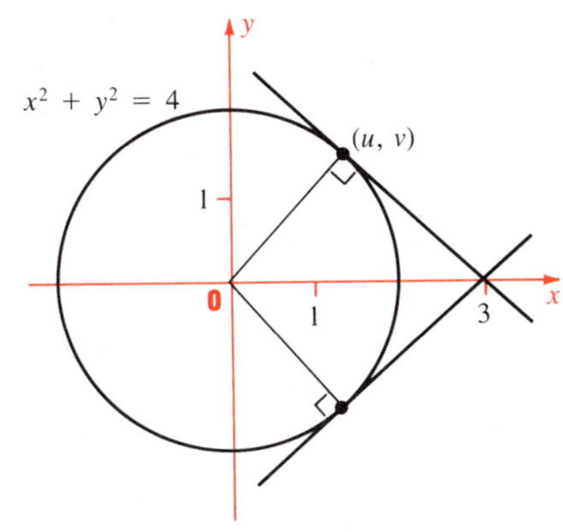

Therefore $(u, v) = (\frac{4}{3}, \pm\frac{2}{3}\sqrt{5})$ are the two points of tangency (Figure 8-2-6). The equations of the corresponding tangents are, by the two-point form of the equation of a line,

$$\frac{y - 0}{x - 3} = \frac{\pm\frac{2}{3}\sqrt{5}}{\frac{4}{3} - 3} = \frac{\pm\frac{2}{3}\sqrt{5}}{-\frac{5}{3}} = \mp\frac{2}{5}\sqrt{5}$$

that is, $y = \pm\frac{2}{5}\sqrt{5}\,(x - 3)$.

Alternative Solution Let (u, v) be any point on the circle $x^2 + y^2 = 4$. Then the tangent to the circle at (u, v) has the equation $ux + vy = 4$. We seek those points (u, v) for which this tangent line passes through $(3, 0)$. This is the case provided that

$$3u + 0 \cdot v = 4 \qquad \text{that is} \qquad u = \frac{4}{3}$$

But $u^2 + v^2 = 4$, hence

$$v^2 = 4 - (\frac{4}{3})^2 = \frac{20}{9} \qquad v = \pm\frac{2}{3}\sqrt{5}$$

Therefore the tangent lines are

$$\frac{4}{3}x \pm \frac{2}{3}\sqrt{5}\,y = 4 \qquad \text{that is} \qquad \frac{2}{3}\sqrt{5}\,y = \pm(4 - \frac{4}{3}x)$$

Divide out $\frac{4}{3}$ on the right side, and then divide the whole equation by the coefficient of y:

$$\frac{2}{3}\sqrt{5}\,y = \pm\frac{4}{3}(3 - x)$$

$$y = [(\pm\frac{4}{3})/(\frac{2}{3}\sqrt{5})](3 - x) = \mp\frac{2}{5}\sqrt{5}\,(x - 3)$$

●

Remark The algebra in the solution of Example 5 turns out to be fairly simple because $(3, 0)$ is on the x-axis. For an external point off the axes, the algebra will be more complicated.

Exercises

Find the intersection of the circle and the line

1 $x^2 + y^2 = 9 \qquad y = x + 1$

2 $x^2 + y^2 = 10 \qquad x + y = 1$

3 $x^2 + y^2 = 5 \qquad x + 2y = 5$

4 $x^2 + y^2 = 6 \qquad y = 2x - 7$

5 $x^2 + y^2 - 4x - 2y + 4 = 0 \qquad 2x - 5y = 6$

6 $x^2 + y^2 + 4x + 5y = 0 \qquad x - 4y = 1$

7 circle with center $(1, 1)$ and radius 5, line through $(0, 2)$ and $(4, 0)$

8 circle with center at $(3, 4)$ and passing through $(0, 0)$, line through $(1, 1)$ parallel to $y = 3x$

Find the point or points of intersection (if any) of the circles

9 $x^2 + y^2 = 9 \qquad x^2 + y^2 + 8x + 12 = 0$

10 $x^2 + y^2 - 2x - 2y = 0$
$(x - 2)^2 + (y - 3)^2 = 4$

11 $x^2 + y^2 - 10x + 6y + 33 = 0$
$x^2 + y^2 + 2x + 4y - 4 = 0$

12 $(x - 1)^2 + (y - 2)^2 = \frac{5}{4}$
$(x - 3)^2 + (y - 6)^2 = \frac{45}{4}$

Find the tangent to $x^2 + y^2 = 1$ at

13 $(-1, 0)$

14 $(0, -1)$

15 $(\frac{1}{2}\sqrt{2}, -\frac{1}{2}\sqrt{2})$

16 $(-\frac{1}{2}\sqrt{2}, -\frac{1}{2}\sqrt{2})$

17 $(\frac{1}{2}\sqrt{3}, \frac{1}{2})$

18 $(-\frac{1}{2}, \frac{1}{2}\sqrt{3})$

19 Find the tangents to $(x - 1)^2 + (y - 2)^2 = 1$ that pass through $(0, 0)$.

20 Find the tangents to $x^2 + y^2 = 1$ that pass through $(2, 2)$.

21 Find the tangents to $x^2 + y^2 = 13$ that pass through $(-5, 1)$.

22 Find the tangents to $(x - 1)^2 + y^2 = 1$ that pass through $(0, -2)$.

23 Show that the circles $x^2 + y^2 - 2x - 4y - 6\frac{1}{4} = 0$ and $x^2 + y^2 - 6x - 12y + 43\frac{3}{4} = 0$ are tangent. [Hint Compute the distance between their centers.]

***24** Find all common tangents to the circles $x^2 - 2x + y^2 = 0$ and $x^2 + 4x + y^2 = 0$.

25 Show that the circles $x^2 + y^2 = 1$ and $25x^2 + 25y^2 - 8x - 6y = 15$ are tangent.

26 Show that the tangent to $(x - a)^2 + (y - b)^2 = r^2$ at a point (u, v) on the circle is $(u - a)(x - a) + (v - b)(y - b) = r^2$.

27 Find the tangents to $x^2 + y^2 = 25$ with slope $\frac{3}{4}$.

28 Find the circle with center $(2, -1)$ tangent to $3x + y = 0$.

29 Find the circles passing through $(1, 8)$ and tangent to both axes.

30 Find the circle tangent to the x-axis at $(a, 0)$ and passing through (b, c). Assume $c \neq 0$.

31 Prove that the circles $x^2 + y^2 - 8x + 2y + 8 = 0$ and $x^2 + y^2 - 2x + 10y + 22 = 0$ are tangent, and find their point of tangency.

32 Find the circle through $(2, 2)$, $(4, 1)$, and $(3, -1)$.

33 Show that $y = mx$ is tangent to the circle

$$x^2 + y^2 + 2ax + 2by + c = 0$$

if and only if $(a + mb)^2 = c(1 + m^2)$.

34 Show that if the two circles

$$x^2 + y^2 - 2a_1x - 2b_1y + c_1 = 0$$
$$x^2 + y^2 - 2a_2x - 2b_2y + c_2 = 0$$

intersect at right angles, then

$$2(a_1a_2 + b_1b_2) = c_1 + c_2$$

and conversely. [Hint Find the two radii and the distance between centers.]

8-3 Curves Defined Geometrically

In this short section we look at some sets in the plane that are defined by geometric conditions. The sets that concern us here turn out to be curves. Now a curve may be defined *analytically*, that is, in terms of a relation between the coordinates of each point on the curve. For example,

$$\{(x, y) \mid y = x^2\}$$

This is the set of all points (x, y) for which $y = x^2$. A curve may also be described *geometrically*, that is in terms that make no reference to a coordinate system. For example,

$$\{X \mid \overline{PX} = 3\}$$

is the set of all points X whose distance from a fixed point P equals 3, obviously a circle of center P and radius 3.

Remark The older term **locus** is also used for a set of points in the plane (usually a curve).

Suppose a curve is specified by a geometric condition, and we want to know what the curve looks like. For instance, can we recognize it as a familiar curve like a line or a circle. Analytic geometry provides a very effective mechanism for solving this problem. The steps are

1. Choose the coordinate system so the data is represented simply.
2. Write the geometric condition as an equation.
3. Simplify this equation as much as possible.
4. Either recognize the graph of the equation as a familiar curve, or at least use it to plot enough points so the shape is clear.
5. If possible, translate the analytic description of the curve into a standard geometric description.

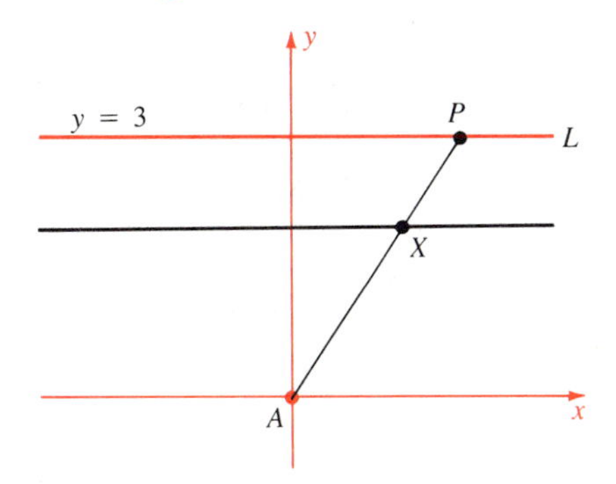

Figure 8-3-1
See Example 1

Example 1

Let L be a line and A a point not on L. For each point P of L, let X be the point that is $\frac{2}{3}$ of the way from A to P. Find the set of all such points X.

Solution Choose the coordinate system so that $A = (0, 0)$ and L is the horizontal line $y = 3$. See Figure 8-3-1. A point P is on L provided $P = (t, 3)$, where t is a real number. The point $\frac{2}{3}$ of the way from A to P is $X = (\frac{2}{3}t, 2)$. We conclude that the set of all such X is the horizontal line $y = 2$. In geometric terms, the curve in question is the line parallel to L that is $\frac{2}{3}$ of the way from A to L. ●

Example 2

Let A and B be distinct points in the plane. Find the set of all points X that are three times as far from A as from B.

Solution First choose a coordinate system so that $A = (0, 0)$ and $B = (1, 0)$. By the distance formula, the condition on $X = (x, y)$ is

$$\sqrt{x^2 + y^2} = 3\sqrt{(x - 1)^2 + y^2}$$

Square and simplify:

$$\begin{aligned} x^2 + y^2 &= 9[(x - 1)^2 + y^2] \\ &= 9(x^2 + y^2 - 2x + 1) \end{aligned}$$

that is,

$$8x^2 + 8y^2 - 18x + 9 = 0$$

Divide by 8 and complete the square:

$$x^2 + y^2 - \tfrac{9}{4}x + \tfrac{9}{8} = 0$$
$$(x - \tfrac{9}{8})^2 + y^2 = \tfrac{81}{64} - \tfrac{9}{8} = \tfrac{9}{64} = (\tfrac{3}{8})^2$$

Figure 8-3-2
See Example 2

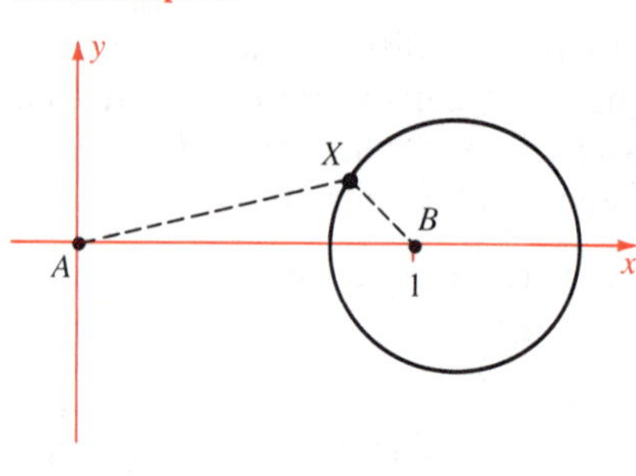

The set of all such X is a circle with center $(\tfrac{9}{8}, 0)$ and radius $\tfrac{3}{8}$. In geometric terms, the curve in question is a circle whose radius is $\tfrac{3}{8}\overline{AB}$ and whose center is on the ray from A through B, $\tfrac{1}{8}\overline{AB}$ past B. See Figure 8-3-2. ●

Hindsight If we had chosen $B = (8, 0)$ instead of $(1, 0)$, there would have been no fractions, and the final equation would have been $(x - 9)^2 + y^2 = 3^2$.

In Examples 1 and 2, we could choose the unit of distance at our convenience, so we could specify numerical values for coordinates of points and coefficients of equations. This is not always possible. In the next example there is no way to choose the coordinate system so that both the equation of the given circle and the coordinates of the given point involve only specific numbers, but still all cases are included. Therefore we choose letters for the radius of the circle and for one coordinate of the point.

Example 3

Let A be a point and C a circle. For each point P on C, let X be the midpoint of A and P. Find the set of all such points X.

Solution The ratio of the radius of the circle to the distance of A from the center of the circle is not given, so choose a coordinate system so that C is the circle $x^2 + y^2 = r^2$ and $A = (a, 0)$. The typical point of C is $P = (u, v)$, where $u^2 + v^2 = r^2$. The midpoint of AP is

$$X = (x, y) \quad \text{where} \quad x = \tfrac{1}{2}(a + u) \quad \text{and} \quad y = \tfrac{1}{2}v$$

To find how x and y are related, we must solve for u and v in terms of x and y, and substitute into $u^2 + v^2 = r^2$:

Figure 8-3-3
See Example 3

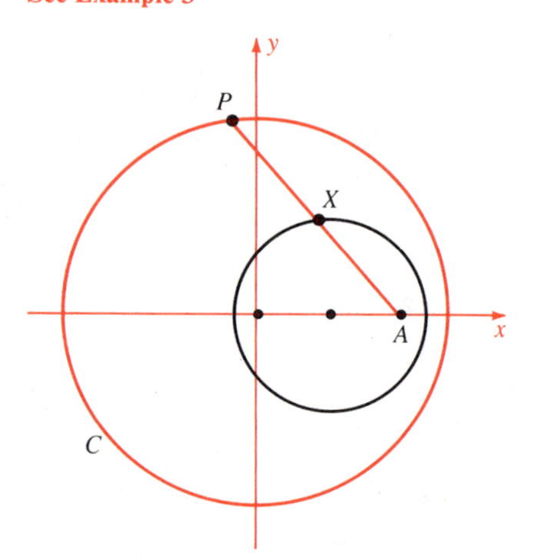

$$u = 2x - a \quad \text{and} \quad v = 2y$$
$$(2x - a)^2 + (2y)^2 = r^2$$
$$(x - \tfrac{1}{2}a)^2 + y^2 = (\tfrac{1}{2}r)^2$$

Therefore the set of all such points X is the circle with center $(\tfrac{1}{2}a, 0)$ and radius $\tfrac{1}{2}r$. In geometric terms, the curve in question is the circle whose center is the midpoint of A and the center of C, and whose radius is half the radius of C. It doesn't matter whether A is inside C, on C, or outside C. See Figure 8-3-3 for the case in which A is inside C. ●

Example 4

Suppose point A is inside of a circle C. Find the set of all midpoints X of chords of C through A.

Solution Choose a coordinate system so that $A = (0, 0)$ and C is the circle $(x - c)^2 + y^2 = r^2$, where $0 < c < r$. Then the center of C is $B = (c, 0)$. The typical chord through A is a segment of a line $y = mx$. If $X = (x, y)$ is the midpoint of this chord, then the line through X and B is perpendicular to the chord, so its slope is $-1/m$. Therefore

$$\frac{y - 0}{x - c} = -\frac{1}{m} \qquad \text{that is} \qquad -my = x - c$$

so (x, y) satisfies the system of equations

$$y = mx \qquad -my = x - c$$

We easily find

$$x = \frac{c}{1 + m^2} \quad \text{and} \quad y = \frac{cm}{1 + m^2}$$

To find how x and y are related, we must eliminate m. The quickest way is

$$x^2 + y^2 = \frac{c^2 + (cm)^2}{(1 + m^2)^2} = \frac{c^2(1 + m^2)}{(1 + m^2)^2} = c\frac{c}{1 + m^2} = cx$$

$$x^2 + y^2 - cx = 0$$

$$(x - \tfrac{1}{2}c)^2 + y^2 = (\tfrac{1}{2}c)^2$$

This is a circle with center $(\frac{1}{2}c, 0)$, the midpoint of AB, and radius $\frac{1}{2}c$. Therefore the curve in question is the circle with AB as a diameter, where B is the center of the given circle C. See Figure 8-3-4. ●

Figure 8-3-4
See Example 4

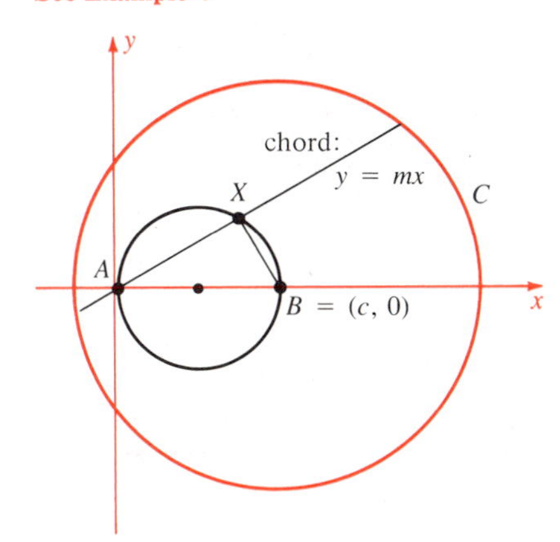

Exercises

1 Let L be a line and A a point not on L. Find the set of all midpoints of AP, where P varies over L.

2 A 10-foot ladder leans against a wall, and its foot slides along the floor as its top slides down the wall. Find the path of its midpoint.

3 Find the set of midpoints of all chords of length 6 of a circle of radius 5.

4 At each point of a circle of radius 1 is drawn a segment of length 3, tangent at one end to the circle. Find the path traced by the other end of the segment.

5 Find the set of all points whose distance from $(0, 0)$ is five times the distance from $(3, 4)$.

6 Do Example 3 by choosing the origin at A.

7 Solve Example 4, assuming that A is on the circle C.

8 Solve Example 4, assuming that A is outside of the circle C.

9 Let $P = (u, v) \neq (0, 0)$ trace a circle C passing through $(0, 0)$. Find the set of all points

$$X = (x, y) = \left(\frac{u}{u^2 + v^2}, \frac{v}{u^2 + v^2}\right)$$

10 (cont.) Solve the same problem where P traces a line L not passing through $(0, 0)$.

11 Let C_1 and C_2 be two circles that are external to each other, and let $a > 0$. Describe the set of all points X such that the length of the tangents from X to C_1 is a times the length of the tangents from X to C_2.

12 Let X and Y be fixed perpendicular lines and A a point on neither. A variable line L through A meets X in P; also the line M through A perpendicular to L meets Y in Q. Find the locus of the midpoint of PQ.

13 Find the foot of the perpendicular dropped from the point (u, v) to the line $ax + by = c$.

***14** (cont.) Given lines L_1, L_2, and M, for each Q on M, let P_i be the foot of the perpendicular from Q to L_i. Describe the set of the midpoints of P_1P_2.

8-4 Conics — The Parabola

The ancient Greek geometers discovered that on intersecting a right circular cone by various planes, they obtained three remarkable curves called **conic sections,** or **conics.** If the plane is parallel to a generator of the cone, the curve is called a *parabola* (Figure 8-4-1a). If the plane is not parallel to a generator, but intersects only one nappe of the cone, the curve is called an *ellipse* (Figure 8-4-1b). A special case is when the plane is parallel to the base of the cone; then the ellipse is just a circle. Finally, if the plane intersects both nappes, the curve has two branches and is called a *hyperbola.* (In all cases, we assume the plane does not pass through the apex of the cone.

Figure 8-4-1
Conic sections (conics)

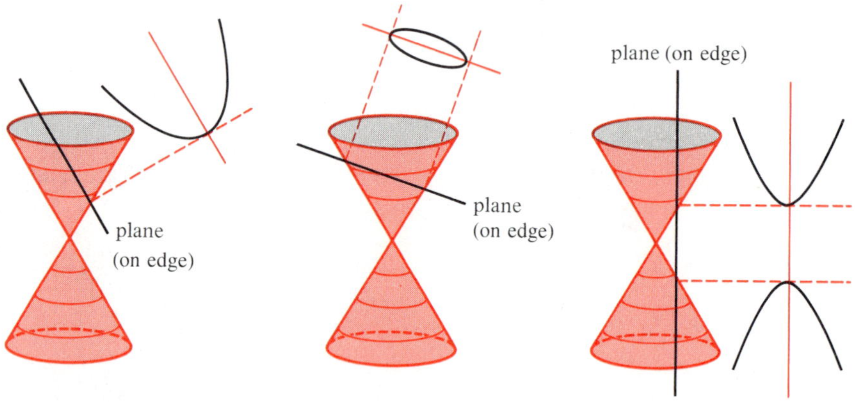

a Parabola: plane parallel to generator

b Ellipse: plane cuts one nappe, not parallel to generator

c Hyperbola: plane cuts both nappes

By definition, a *nappe* of a cone is the part of the cone on one side of any plane that passes through the apex but doesn't meet the cone in any other points. Thus each cone has two nappes.)

We shall not use this three-dimensional approach of intersecting a cone with a plane. Instead, we shall give alternative plane-geometry definitions of the conics.

Figure 8-4-2
Definition of the parabola

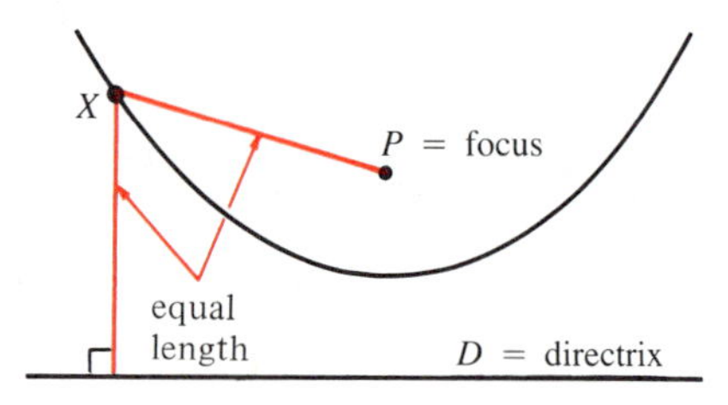

The Parabola

A **parabola** is defined as the set of all points X equidistant from a fixed line D and a fixed point P not on D. See Figure 8-4-2. We call D the **directrix** and P the **focus** of the parabola. For convenience let us take the focus to be $P = (0, p)$ with $p > 0$, and the directrix D to be the line $y = -p$. Then the equation of the parabola follows easily, as was shown on p. 46. The curve is shown in Figure 8-4-3.

Figure 8-4-3
Equation of the parabola:
$y = \dfrac{1}{4p}x^2 \qquad (p > 0)$

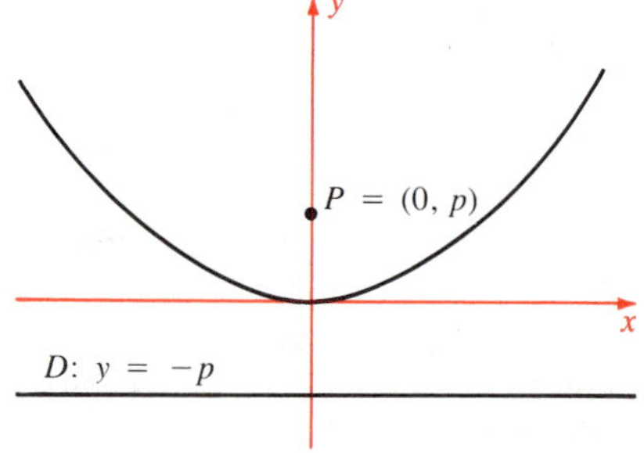

The Parabola (Standard Position)

The equation of the parabola with focus $(0, p)$ and directrix $y = -p$ is

$$y = \frac{1}{4p}x^2$$

Figure 8-4-4
Geometry of the parabola

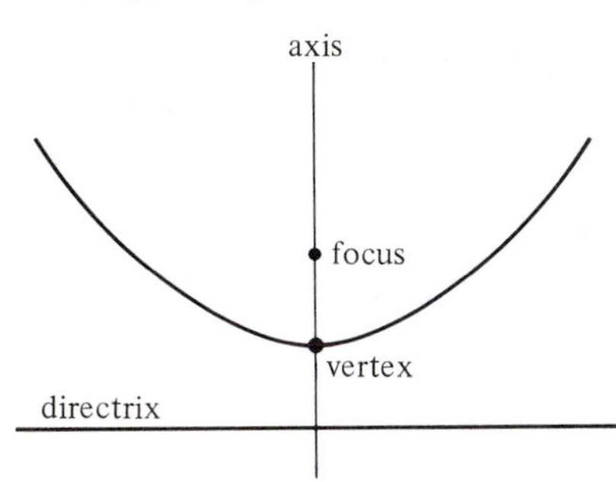

For our choice of axes, the parabola is the graph of a quadratic polynomial. Conversely, the graph of any quadratic polynomial $y = ax^2 + bx + c$ (with $a \neq 0$) is a parabola. This is just a matter of completing the square:

$$y = a(x - h)^2 + k \qquad y - k = a(x - h)^2$$

where h and k are easily determined. Now translate the origin to (h, k). Then the new coordinates are related to the old by $\bar{x} = x - h$ and $\bar{y} = y - k$, so the curve in the *new* coordinates is $\bar{y} = a\bar{x}^2$. This is a parabola with $4p = 1/a$. If $a > 0$, the curve opens upward; if $a < 0$, it opens downward. Its focus is $(\bar{x}, \bar{y}) = (0, p) = (0, 1/4a)$ in the *new* coordinates, that is, $(x, y) = (h, p + k) = (h, 1/4a + k)$ in the *old* coordinates. Its directrix is $\bar{y} = -p$ in the *new* coordinates, that is, $y = -p + k = -1/4a + k$ in the *old* coordinates.

The line through the focus perpendicular to the directrix is called the **axis** of the parabola. The point of intersection of the axis with the parabola is the **vertex** of the parabola (Figure 8-4-4).

By interchanging the roles of x and y, we see that a parabola whose axis is parallel to the x-axis is the locus of an equation $x = ay^2 + by + c$. Some examples are shown in Figure 8-4-5.

Figure 8-4-5
Parabolas with horizontal axes

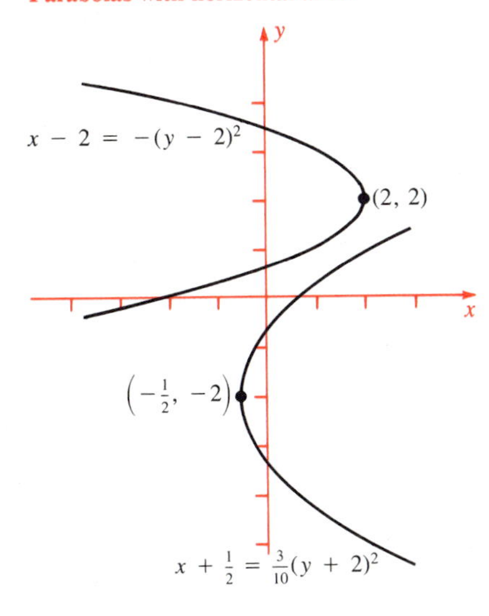

Equation of the Parabola

The equation

$$y - k = a(x - h)^2$$

which is obtained from $y = ax^2$ by translating the origin to (h, k), represents a parabola with vertex at (h, k) and axis vertical. The parabola opens upward if $a > 0$, downward if $a < 0$. Its focus is $(x, y) = (h, 1/4a + k)$ and its directrix is $y = -1/4a + k$.

The equation

$$x - h = a(y - k)^2$$

represents a parabola with vertex at (h, k) and axis horizontal. The parabola opens to the right if $a > 0$, to the left if $a < 0$. Its focus is $(x, y) = (1/4a + h, k)$ and its directrix is $x = -1/4a + h$. ●

Figure 8-4-6
See Example 1

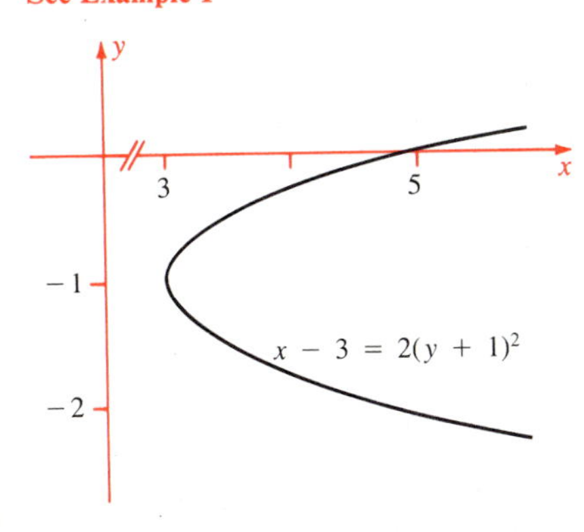

Example 1

Show that $x = 2y^2 + 4y + 5$ is the equation of a parabola and sketch the curve. Find its vertex, axis, focus, and directrix.

Solution Complete the square:

$$x = 2(y^2 + 2y + 1) - 2 + 5 \qquad \text{hence} \qquad x - 3 = 2(y + 1)^2$$

The curve is a parabola with vertex $(3, -1)$, axis $y = -1$, and opening to the right. To find its focus and its directrix, write the equation in the form $x - 3 = (1/4p)(y + 1)^2$, where $p = \frac{1}{8}$. The focus is $(3 + \frac{1}{8}, -1)$; the directrix is $x = 3 - \frac{1}{8} = \frac{23}{8}$. See Figure 8-4-6. ●

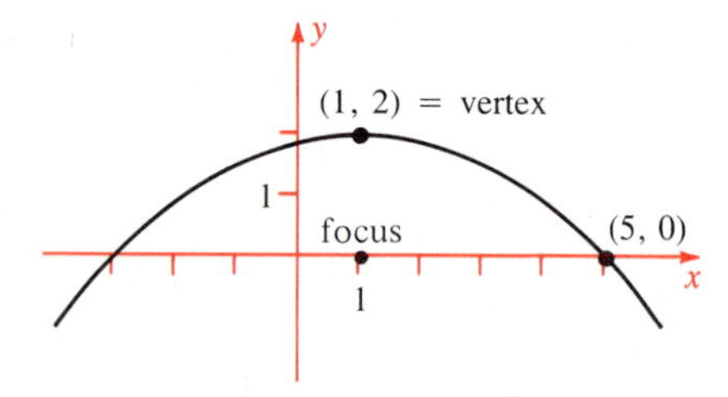

Figure 8-4-7
See Example 2

Example 2

Find the equation of the vertical parabola with vertex $(1, 2)$ and passing through $(5, 0)$. Then find its focus.

Solution The equation is of the form

$$y - 2 = a(x - 1)^2$$

To find a, set $x = 5$ and $y = 0$:

$$-2 = a(5 - 1)^2 \qquad a = -\tfrac{1}{8}$$

Hence the desired parabola is $y - 2 = -\frac{1}{8}(x - 1)^2$. See Figure 8-4-7. The focus is $|1/4a|$ below the vertex, hence at $(1, 2 - 2) = (1, 0)$. ●

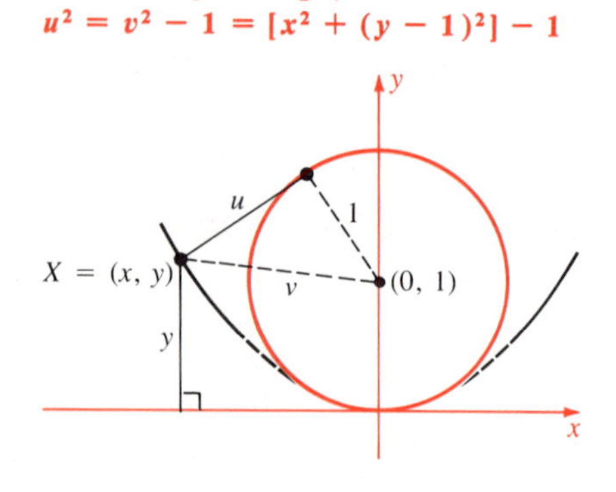

Figure 8-4-8
From the right triangle, we have $u^2 = v^2 - 1 = [x^2 + (y - 1)^2] - 1$

Example 3

Find the set of all points X whose distance to the x-axis equals the length of the tangent(s) from X to the circle $x^2 + (y - 1)^2 = 1$.

Solution Draw an accurate figure (Figure 8-4-8) and write $X = (x, y)$. The condition on the distance is $y = u$, so

$$y^2 = u^2 = v^2 - 1 = [x^2 + (y - 1)^2] - 1$$

Simplify:

$$x^2 - 2y = 0 \qquad y = \tfrac{1}{2}x^2$$

Therefore the set of points X is a parabola. ●

Example 4

For each point P, let X be the point that is $\frac{1}{3}$ of the way from $(-6, 0)$ to P. Find the curve traced by X as P varies over the parabola $y = x^2$.

Solution Write $P = (u, v)$ where $v = u^2$ and set $X = (x, y)$. Then

$$x = \tfrac{2}{3}(-6) + \tfrac{1}{3}u \quad \text{and} \quad y = \tfrac{2}{3}(0) + \tfrac{1}{3}v$$

by the formula for division of a segment (page 418). Solve for u and v:

$$-12 + u = 3x \qquad u = 3x + 12 \qquad v = 3y$$

Now substitute into $v = u^2$:

$$3y = (3x + 12)^2 = 9(x + 4)^2$$

$$y = 3(x + 4)^2$$

Therefore the required curve is another parabola (Figure 8-4-9). ●

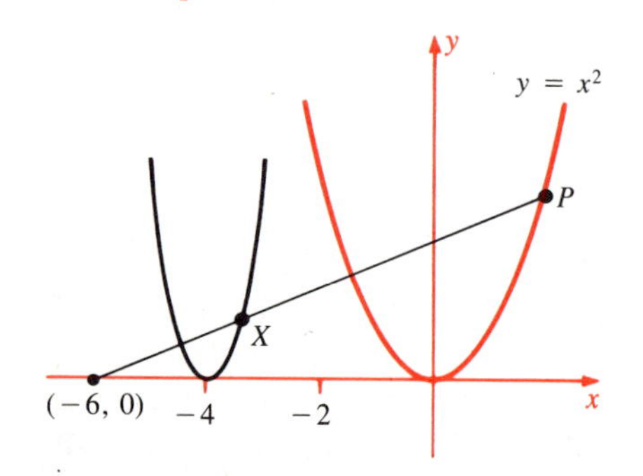

Figure 8-4-9
See Example 4

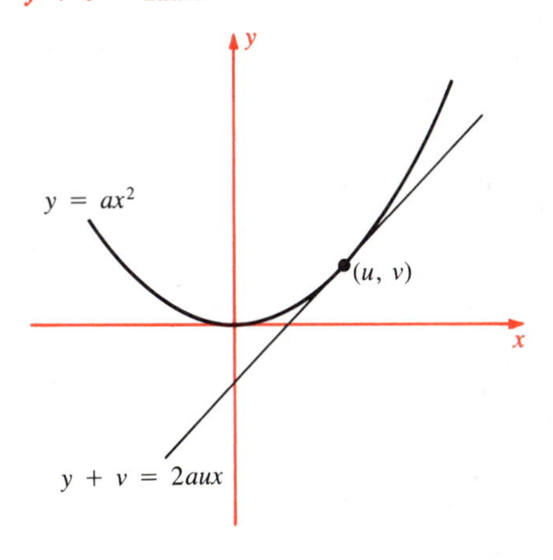

Figure 8-4-10
The tangent to $y = ax^2$ at (u, v): $y + v = 2aux$

Let us find the tangent to the parabola $y = ax^2$, where $a > 0$, at one of its points (u, v). The slope of the tangent is

$$\left.\frac{dy}{dx}\right|_u = 2ax\Big|_u = 2au$$

hence the equation of the tangent is $y - v = 2au(x - u)$, that is, $y - v + 2au^2 = 2aux$. But $au^2 = v$ because the point (u, v) is on the parabola, so the tangent line can be written as $y + v = 2aux$. See Figure 8-4-10.

Tangent to a Parabola

The tangent to $y = ax^2$ at (u, v) is

$$y + v = 2aux$$

Remark The equation of the tangent has a remarkable symmetry property: the roles of (x, y) and (u, v) are interchangeable! That is, if we interchange x and u and simultaneously interchange y and v in the equation $y + v = 2aux$, then the resulting equation is the same. This has the following interpretation: Suppose (x, y) and (u, v) are two points of the plane such that $y + v = 2aux$. Then if either is on the parabola $y = ax^2$, then the line joining the two points is tangent to the parabola.

We now have a tool for finding the tangents to the parabola that pass through an outside point (u, v). Indeed, each point of tangency (x, y) must satisfy $y + v = 2aux$, so it is a solution of the system

$$\begin{cases} y = ax^2 \\ y + v = 2aux \end{cases}$$

See Figure 8-4-11. To solve the system, we eliminate y:

$$ax^2 + v = 2aux \qquad \text{hence} \qquad ax^2 - 2aux + v = 0$$

so that

$$x = \frac{au \pm \sqrt{a(au^2 - v)}}{a}$$

The solution is valid if $a(au^2 - v) \geq 0$. If equality holds, $au^2 = v$, then (u, v) is on the parabola and $x = u$; nothing new. If $a(au^2 - v) > 0$, then (u, v) is outside of the parabola and there are two values of x (with two corresponding values of $y = ax^2$), hence two tangents. In the remaining case, $a(au^2 - v) < 0$; then (u, v) is inside of the parabola and there are no (real) solutions, hence no tangents.

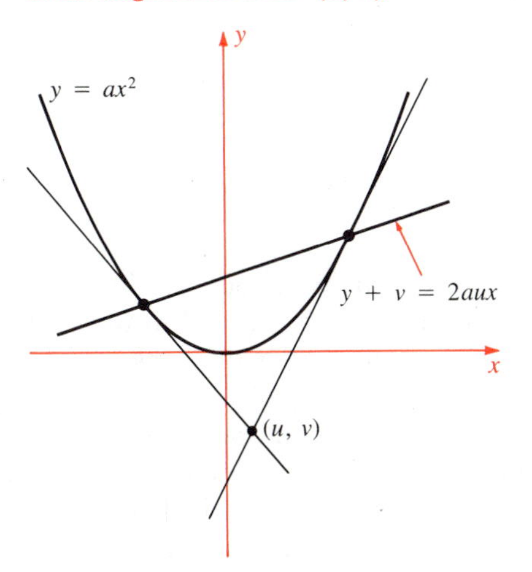

Figure 8-4-11
(u, v) is outside the parabola: Then $y + v = 2aux$ is the line through the points of tangency of the tangent lines from (u, v).

Remark 1 The line $y + v = 2aux$ is called the **polar** of the point (u, v) with respect to the parabola $y = ax^2$.

Remark 2 This discussion can be worked out for any parabola, not just $y = ax^2$.

Example 5

Find the tangents to $y = x^2$ from $(-1, -3)$.

Solution Here $a = 1$ and the external point is $(-1, -3)$. By the formulas, the tangents through $(-1, -3)$ touch the parabola at (x, y), where

$$y = x^2 \quad \text{and} \quad y - 3 = -2x$$

Eliminate y:

$$x^2 - 3 = -2x \qquad \text{hence} \qquad x^2 + 2x - 3 = 0$$

There are two solutions, $x_1 = 1$ and $x_2 = -3$. The corresponding values of y are $y_1 = 1$ and $y_2 = 9$, so the two points of tangency are $(1, 1)$ and $(-3, 9)$. Since the tangent to $y = ax^2$ at (u, v) is $y + v = 2aux$, the two tangent lines in this case are

$$y = 2x - 1 \quad \text{and} \quad y = -6x - 9$$

Reflection Property of the Parabola

The parabola has a remarkable geometric property with practical applications. Think of the inside of the parabola as a mirror. If a point source of light is placed at the focus, the light rays striking the mirror at various places will all be reflected parallel to the axis of the parabola, forming a beam (Figure 8-4-12a). This is the principle of the parabolic searchlight.

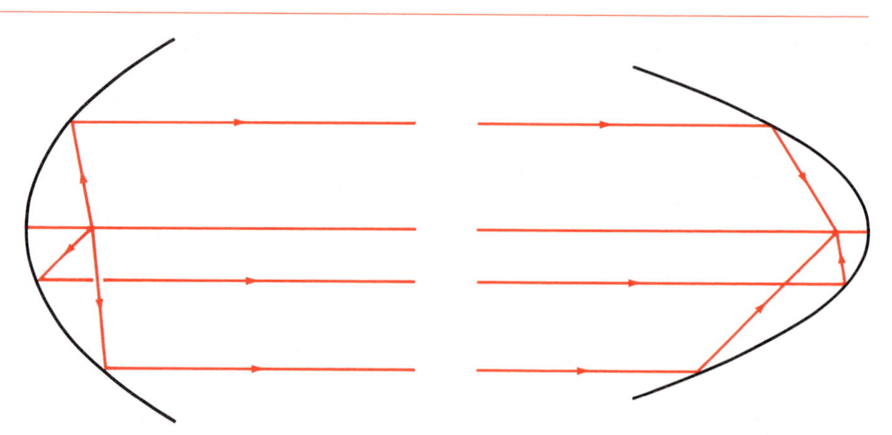

Figure 8-4-12
Reflection property of the parabola

a Light from the focus is reflected in a parallel beam.

b Light from "infinity" is reflected to the focus.

Conversely, light rays from infinity entering the parabola parallel to its axis will bounce off the mirror to the focus (Figure 8-4-12b). Thus they are concentrated (focused) at this single point. This is the principle of the parabolic receiving antenna and the reflecting telescope.

Let us prove this property of the parabola. We take the standard parabola $4py = x^2$ with $p > 0$. A light ray from the focus $(0, p)$, striking the parabola at (u, v), is reflected so that "the angle of incidence equals the angle of reflection," in other words, so that $\alpha = \beta$ in Figure 8-4-13.

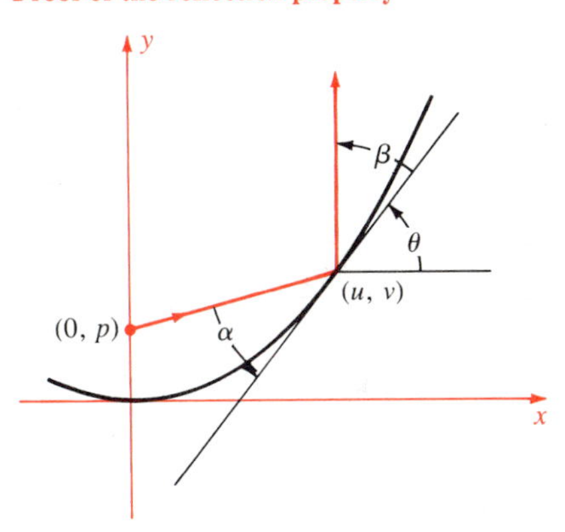

Figure 8-4-13
Proof of the reflection property

Since $y = (1/4p)x^2$, we have

$$\tan\theta = \text{slope of the tangent at } (u, v) = \left.\frac{dy}{dx}\right|_u = \frac{u}{2p}$$

The light ray from $(0, p)$ to (u, v) has slope $(v - p)/u$. We can apply the formula for the angle between lines (page 420) to find $\tan\alpha$:

$$\tan\alpha = \frac{\dfrac{u}{2p} - \dfrac{v-p}{u}}{1 + \left(\dfrac{v-p}{u}\right)\left(\dfrac{u}{2p}\right)} = \frac{u^2 - 2p(v-p)}{2up + u(v-p)}$$

$$= \frac{2pv + 2p^2}{up + uv} = \frac{2p(v+p)}{u(v+p)} = \frac{2p}{u}$$

But $\beta = \alpha$, so $\tan\beta = 2p/u$. If we now consider

$$\tan\beta = \frac{2p}{u} \quad\text{and}\quad \tan\theta = \frac{u}{2p} = \frac{1}{\tan\beta} = \cot\beta$$

we are forced to the conclusion that $\beta + \theta = \frac{1}{2}\pi$. Hence the reflected light ray is vertical, that is, parallel to the axis of the parabola. This is what we set out to prove.

Exercises

Give the focus, directrix, axis, and vertex

1 $x - 3 = 3y^2$

2 $x + 2 = -y^2$

3 $x = 2(y+1)^2$

4 $2y = -3(x-2)^2$

5 $x^2 + 4x - 6y = 0$

6 $2y^2 - 4y + x + 2 = 0$

Find the equation of the parabola

7 vertex $(0, 0)$ through the point $(1, 3)$ axis vertical

8 vertex $(0, 0)$ through the point $(6, -1)$ axis vertical

9 vertex $(1, 2)$ through the point $(-3, 4)$ axis horizontal

10 vertex $(-5, 0)$ through the point $(0, 8)$ axis horizontal

11 vertex $(2, -3)$ focus $(2, 1)$

12 vertex $(2, -3)$ focus $(10, -3)$

Find the set of centers of all circles that are simultaneously tangent to the y-axis and the circle

13 $x^2 + y^2 = 2x$

14 $x^2 + y^2 = 1$

15 $(x-1)^2 + y^2 = 4$

16 $(x-3)^2 + y^2 = 1$

17 Let P be a parabola and L a line. Find the set of midpoints of all chords of P parallel to L.

18 Find the set of all points X whose distance to the x-axis equals the length of the tangents from X to the circle $x^2 + (y-b)^2 = r^2$, where $b \geq 0$ and $r > 0$.

19 Let P be the focus of the parabola $y = \frac{1}{4}x^2$ and let X be any point on the parabola. Show that the circle with diameter PX is tangent to the x-axis.

20 Suppose a circle C intersects the parabola $y = x^2$ in four points (x_i, y_i) where $i = 1, \cdots, 4$. Prove that $x_1 + x_2 + x_3 + x_4 = 0$.

Find the tangent to

21 $4y = x^2$ at $(-2, 1)$

22 $x = 2y^2$ at $(2, 1)$

23 $y = (x-1)^2$ at $(3, 4)$

24 $x = -3y^2$ at $(-3, -1)$

Find the tangents to

25 $4y = x^2$ from $(-1, -1)$

26 $x = 2y^2$ from $(0, 2)$

27 $y = (x-1)^2$ from $(-1, 3)$

28 $x = -3y^2$ from $(3, 0)$

29 Let (u, v) lie inside the parabola $y = ax^2$, $a > 0$, and let (x, y) lie on the polar of (u, v). Show that (x, y) lies outside the parabola.

30 Let (u, v) be any point not on the parabola $y = ax^2$, and let L denote its polar. How are the polars of all points (x, y) of L related?

31 Find all parabolas with focus at the origin and axis the y-axis.

32 (cont.) Show that any two that open in opposite directions intersect at right angles.

33 Find the (acute) angle of intersection of the two tangents to the parabola $y = x^2$ from the point $(2, 1)$.

34 Let C be a parabola, either horizontal or vertical, in the x, y-plane. Suppose there is a change of scale on each axis according to $x = a\bar{x}$ and $y = b\bar{y}$ where a and b are positive. Show that with respect to the barred variables C is still a parabola.

***35** Consider all chords AB of the parabola $y = x^2$ such that AOB is a right angle, where O is the origin. Show that these chords all have a point in common.

***36** (cont.) Find the set of midpoints of all these chords AB.

***37** Let P be the focus of the parabola $y = \frac{1}{4}x^2$ and consider all possible chords AB of the parabola that pass through P. Find the the set of midpoints of AB.

***38** (cont.) Show for each such chord AB, that the circle with diameter AB is tangent to the directrix of the parabola.

8-5 Conics—The Ellipse

An **ellipse** is defined as the set of all points X such that

$$\overline{XP} + \overline{XQ} = 2a$$

where P and Q are two fixed points such that $\overline{PQ} < 2a$, and a is a positive constant. In words, the sum of the distances of X from two fixed points is a constant. The points P and Q are called the **foci** (plural of **focus**) of the ellipse.

Figure 8-5-1
The ellipse: $\overline{XP} + \overline{XQ} = 2a$

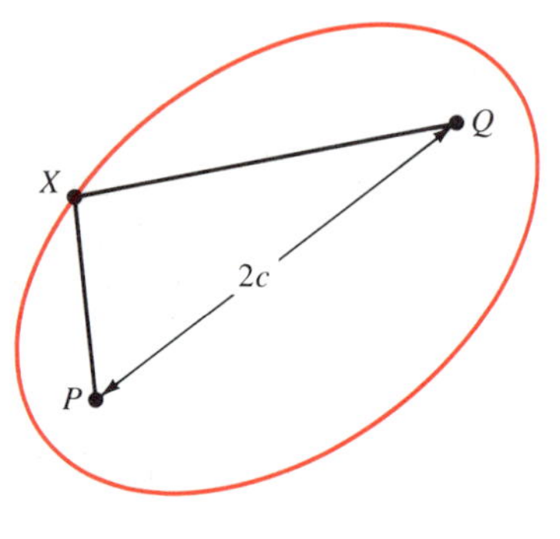

Let the distance between the foci be denoted $2c$. Thus $\overline{PQ} = 2c$ and $0 < c < a$. See Figure 8-5-1. To find an equation for the ellipse, we choose coordinates so that the x-axis goes through the foci and the origin is their midpoint (Figure 8-5-2). A point $X = (x, y)$ is on the ellipse if and only if its distance sum is $2a$, that is, if and only if

$$\overline{XP} + \overline{XQ} = 2a$$

that is,

$$\sqrt{(x - c)^2 + y^2} + \sqrt{(x + c)^2 + y^2} = 2a$$

This unpleasant expression is an equation for the ellipse. To derive an equivalent equation without radicals requires squaring twice and doing some algebra. It also requires a tedious check (which we omit) that no undesired points creep in. (Squaring is tricky. For instance, squaring $x = 2$ leads to $x^2 = 4$, which has a solution that is not a solution of $x = 2$. In general, when you square an equation and then find all solutions of the squared equation, then you must *check* which of these actually satisfy the original equation.)

Figure 8-5-2
Standard position: foci $(\pm c, 0)$

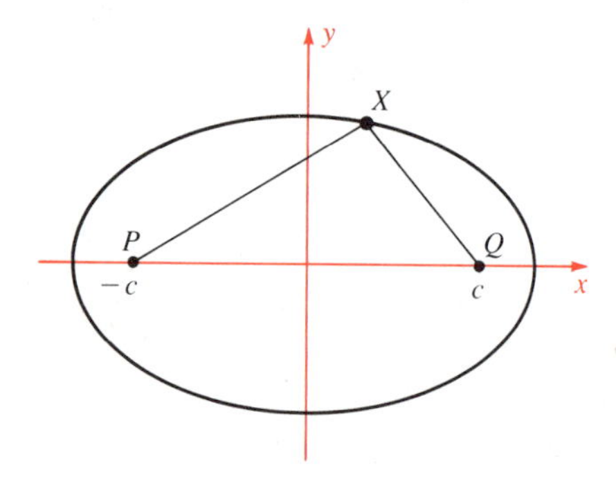

To eliminate the radicals, we transpose the second radical and square:

$$\sqrt{(x - c)^2 + y^2} = 2a - \sqrt{(x + c)^2 + y^2}$$

$$(x - c)^2 + y^2 = 4a^2 - 4a\sqrt{(x + c)^2 + y^2} + (x + c)^2 + y^2$$

We simplify this relation to

$$a^2 + cx = a\sqrt{(x + c)^2 + y^2}$$

Now we square again:

$$a^4 + 2a^2cx + c^2x^2 = a^2[(x + c)^2 + y^2]$$

and we simplify to

$$(a^2 - c^2)x^2 + a^2y^2 = a^2(a^2 - c^2)$$

Since $0 < c < a$ we have $a^2 > c^2$, so we may define $b > 0$ by $b^2 = a^2 - c^2$. Then

$$b^2x^2 + a^2y^2 = a^2b^2 \qquad \text{that is} \qquad \frac{x^2}{a^2} + \frac{y^2}{b^2} = 1$$

The Ellipse (Standard Position)

The equation of the ellipse with foci $(-c, 0)$ and $(c, 0)$ and length sum $2a$, where $a > c$, is

$$\frac{x^2}{a^2} + \frac{y^2}{b^2} = 1$$

where $b^2 = a^2 - c^2$. (Note that $a > b$.)

Conversely, $x^2/a^2 + y^2/b^2 = 1$, where $a > b > 0$, describes an ellipse whose foci are $(\pm c, 0)$, where $c^2 = a^2 - b^2$, and whose distance sum is $2a$.

Remark If $a = b$, the equation of the ellipse becomes

$$\frac{x^2}{a^2} + \frac{y^2}{a^2} = 1 \qquad \text{that is} \qquad x^2 + y^2 = a^2$$

the equation of a circle of radius a. Thus a circle can be considered as an extreme case of an ellipse, where $c = 0$. (The foci come together at one point, the center.)

Let us sketch this ellipse. Because both terms on the left side of the equation are non-negative, it follows that

$$\frac{x^2}{a^2} \le 1 \quad \text{so} \quad x^2 \le a^2 \qquad \text{and} \qquad \frac{y^2}{b^2} \le 1 \quad \text{so} \quad y^2 \le b^2$$

Therefore the curve lies in the box $-a \le x \le a$, $-b \le y \le b$.

Next, we note symmetry: if (x, y) satisfies the equation, then so do $(-x, y)$, $(x, -y)$, and $(-x, -y)$. Therefore the curve is symmetric about both axes and about the origin. We need plot it only in the first quadrant, then extend the curve to the other quadrants by symmetry.

We solve for y:

$$y = \frac{b}{a}\sqrt{a^2 - x^2}$$

(The positive square root applies in the first quadrant.) If x starts at 0 and increases to a, then y starts at b and decreases to 0. The curve has a horizontal tangent at $(0, b)$, a vertical tangent at $(a, 0)$, and is concave down. We postpone the proofs of these assertions until the exercises. We now have enough information for a reasonable sketch (Figure 8-5-3).

The points $(\pm a, 0)$ and $(0, \pm b)$ are called the **vertices** of the ellipse. The numbers a and b are known (historically) by the peculiar names **semi-major axis** and **semi-minor axis.** The point halfway between the foci is called the **center** of the ellipse. Actually, the **major axis** of the ellipse is the line through its foci, and the **minor axis** is the perpendicular line through its center.

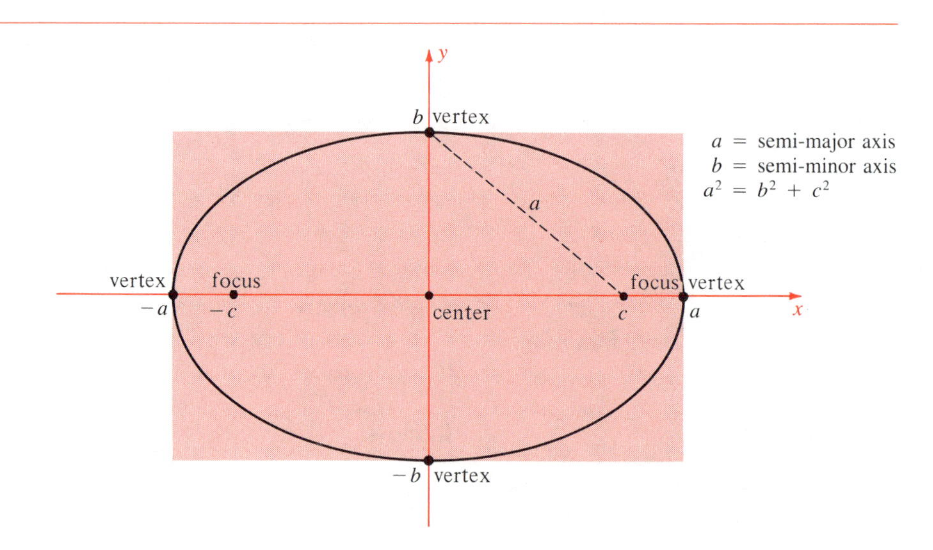

Figure 8-5-3
Geometry of the ellipse
$(a > b \quad c = \sqrt{a^2 - b^2})$

As Figure 8-5-3 shows, a is the distance from a focus to one of the vertices $(0, \pm b)$. That is because $a^2 = b^2 + c^2$.

If $a > b$, then

$$\frac{x^2}{b^2} + \frac{y^2}{a^2} = 1$$

is an ellipse, but with its major axis along the y-axis instead of the x-axis, and its foci at $(0, \pm c)$, where $a^2 = b^2 + c^2$. See Figure 8-5-4.

By translation of axes, we get

$$\frac{(x - h)^2}{a^2} + \frac{(y - k)^2}{b^2} = 1$$

the equation of an ellipse centered at (h, k).

Figure 8-5-4
$\frac{x^2}{b^2} + \frac{y^2}{a^2} = 1$ with $0 < b < a$
$(c = \sqrt{a^2 - b^2})$

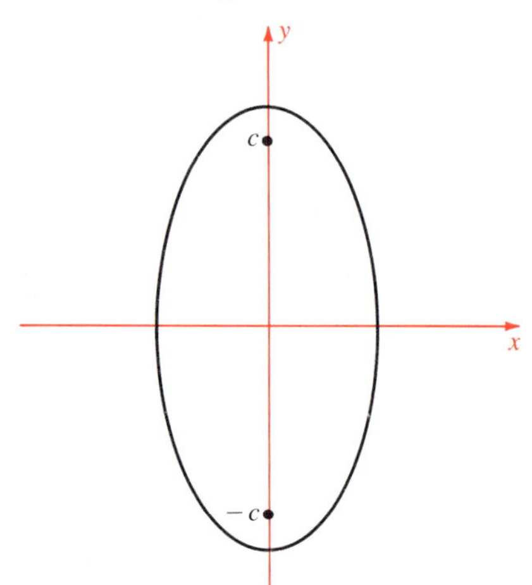

Remark To construct an ellipse, tie a string of length $2a$ to two fixed pins $2c$ units apart $(a > c)$. Place your pencil against the string and move it so the string is taut. The curve generated is an ellipse. Why? If you move the pins closer and closer together $(c \to 0)$, the ellipse becomes more and more like a circle.

Example 1

Show that

$$4x^2 + 9y^2 - 8x + 54y + 49 = 0$$

defines an ellipse. Locate its center, vertices, and foci.

Solution Complete the squares in x and y:

$$4(x^2 - 2x + 1) + 9(y^2 + 6y + 9) + 49 - 4 - 81 = 0$$

$$4(x - 1)^2 + 9(y + 3)^2 = 36$$

$$\frac{(x-1)^2}{9} + \frac{(y+3)^2}{4} = 1$$

This is an ellipse with horizontal major axis, center at $(1, -3)$, semi-major axis $a = 3$, semi-minor axis $b = 2$. Hence its vertices are

$$(1 \pm a, -3) \quad \text{and} \quad (1, -3 \pm b)$$

that is,

$$(4, -3) \qquad (-2, -3) \qquad \text{and} \qquad (1, -1) \qquad (1, -5)$$

Since $a > b$, its foci are on the horizontal axis $y = -3$ and are

$$(1 \pm c, -3) \quad \text{where} \quad c^2 = a^2 - b^2 = 5 \quad \text{so} \quad c = \sqrt{5}$$

Therefore the foci are the points $(1 \pm \sqrt{5}, -3)$. ●

Example 2

Find the equation of the ellipse having foci at $(\pm 3, 1)$ and vertices at $(\pm 4, 1)$.

Solution The center is $(0, 1)$, halfway between the foci. Therefore, the equation has the form

$$\frac{(x-0)^2}{a^2} + \frac{(y-1)^2}{b^2} = 1$$

The distance from the center to either vertex on the major axis is $a = 4$. To find b, use $b^2 = a^2 - c^2$, where c is the distance from the center to either focus. Clearly $c = 3$, so $b^2 = 4^2 - 3^2 = 7$. Hence the equation of the ellipse is

$$\frac{x^2}{16} + \frac{(y-1)^2}{7} = 1$$

●

Parametrization of the Ellipse

Consider again the ellipse

$$\frac{x^2}{a^2} + \frac{y^2}{b^2} = 1$$

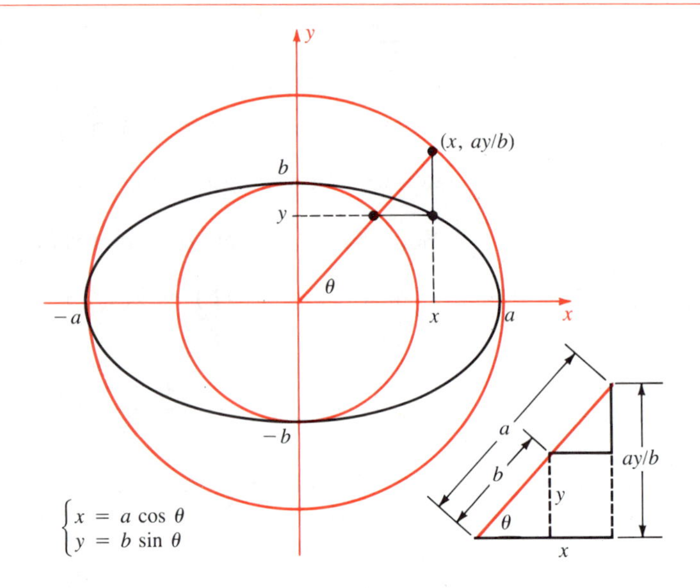

Figure 8-5-5
Parametrization of the ellipse
$\dfrac{x^2}{a^2} + \dfrac{y^2}{b^2} = 1$

Graph it (Figure 8-5-5) and draw the two circles, center $(0, 0)$, radii a and b. Let (x, y) be any point on the ellipse. Then

$$x^2 + \left(\frac{ay}{b}\right)^2 = a^2$$

so the point $(x, ay/b)$ is on the larger circle. If θ denotes the corresponding central angle, then

$$x = a\cos\theta \quad \text{and} \quad \frac{ay}{b} = a\sin\theta$$

that is,

$$x = a\cos\theta \qquad y = b\sin\theta$$

Similarly $(bx/a, y)$ lies on the smaller circle, and this leads to precisely the same relations, so the figure is correct.

As θ runs from 0 to 2π, the point $(x, y) = (a\cos\theta, b\sin\theta)$ traverses the ellipse once in the counterclockwise sense, starting at $(a, 0)$. Note in Figure 8-5-5 that the parameter θ is not the central angle of (x, y); rather θ is the central angle of the two circles related to the ellipse.

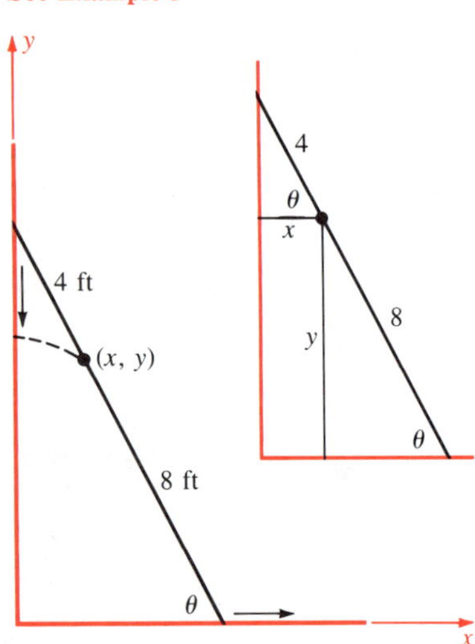

Figure 8-5-6
See Example 3

Example 3

A 12-foot ladder leans against a wall. Its bottom is pulled along the floor away from the wall and its top slides down the wall. Find the path of the point of the ladder 4 ft from its top.

Solution Choose axes as indicated in Figure 8-5-6, and let θ be the angle the ladder makes with the floor. The result is that $x = 4\cos\theta$ and $y = 8\sin\theta$; hence $(x/4)^2 + (y/8)^2 = 1$ so the point moves on an ellipse. As the ladder slides, θ runs from $\frac{1}{2}\pi$ to 0, so the path of the point is a quarter of the ellipse, traversed clockwise. The standard equation of the ellipse is

$$\frac{x^2}{4^2} + \frac{y^2}{8^2} = 1$$

Note that its major axis is vertical.

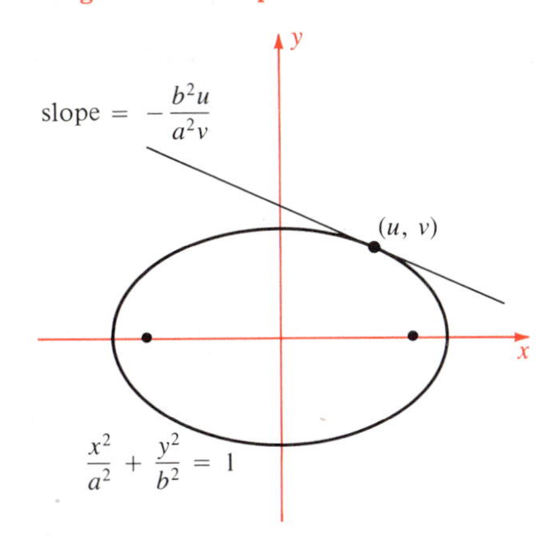

Figure 8-5-7
Tangents to the ellipse

Tangents to the Ellipse

Let us find the tangent to the ellipse

$$\frac{x^2}{a^2} + \frac{y^2}{b^2} = 1 \qquad (a > 0, b > 0)$$

at one of its points (u, v). See Figure 8-5-7.

For the slope, we need dy/dx at $x = u$. The equation of the ellipse actually defines two functions of x, corresponding to $y > 0$ and $y < 0$. However, both satisfy

$$\frac{x^2}{a^2} + \frac{y^2}{b^2} = 1$$

so we can simply differentiate this relation implicitly and set $(x, y) = (u, v)$:

$$\frac{2x}{a^2} + \frac{2y}{b^2} \cdot \frac{dy}{dx} = 0$$

Hence

$$\left.\frac{dy}{dx}\right|_{(u,v)} = -\left.\frac{b^2x}{a^2y}\right|_{(u,v)} = -\frac{b^2u}{a^2v}$$

(The sign takes care of itself.) Now we can write down the equation of the tangent:

$$y - v = -\frac{b^2u}{a^2v}(x - u)$$

If we transpose everything to the left side and multiply by v/b^2, the result is

$$\frac{u}{a^2}(x - u) + \frac{v}{b^2}(y - v) = 0$$

Hence

$$\frac{ux}{a^2} + \frac{vy}{b^2} = \frac{u^2}{a^2} + \frac{v^2}{b^2} = 1$$

Tangent to the Ellipse

The tangent to $\dfrac{x^2}{a^2} + \dfrac{y^2}{b^2} = 1$ at (u, v) is $\dfrac{ux}{a^2} + \dfrac{vy}{b^2} = 1$

Remark The reasoning fails if $v = 0$, but the result is still correct—just reverse the roles of x and y to prove it.

As in the case of the parabola, (u, v) and (x, y) are interchangeable in the tangent formula. It follows, as for the parabola, that if (u, v) lies outside of the ellipse and (x, y) is the point of contact with the ellipse of one of the two tangents through (u, v), then $ux/a^2 + vy/b^2 = 1$.

Remark The line $ux/a^2 + vy/b^2 = 1$ is called the **polar** of (u, v) with respect to the ellipse $x^2/a^2 + y^2/b^2 = 1$.

Example 4

Find the two tangents from $(-2, 3)$ to the ellipse $\frac{1}{8}x^2 + \frac{1}{2}y^2 = 1$.

Solution Set $(u, v) = (-2, 3)$. If (x, y) is the point of tangency of either of the tangents through (u, v), then

$$\frac{ux}{8} + \frac{vy}{2} = 1 \qquad \text{that is} \qquad -\tfrac{1}{4}x + \tfrac{3}{2}y = 1$$

To find (x, y) we must solve the system

$$\tfrac{1}{8}x^2 + \tfrac{1}{2}y^2 = 1 \qquad -\tfrac{1}{4}x + \tfrac{3}{2}y = 1$$

Clear both equations of fractions and solve the second equation for x in terms of y:

$$x^2 + 4y^2 = 8$$

$$-x + 6y = 4 \qquad \text{that is} \qquad x = 6y - 4 = 2(3y - 2)$$

Now substitute this expression for x into the first equation:

$$4(3y - 2)^2 + 4y^2 = 8$$

that is,

$$(9y^2 - 12y + 4) + y^2 = 2$$

Collect terms:

$$10y^2 - 12y + 2 = 0 \qquad \text{hence} \qquad 5y^2 - 6y + 1 = 0$$

This quadratic equation factors: $(5y - 1)(y - 1) = 0$. The solutions are

$$y = \tfrac{1}{5} \qquad x = 6y - 4 = -\tfrac{14}{5} \qquad \text{and} \qquad y = 1 \qquad x = 2$$

so the two points of tangency are $(-\frac{14}{5}, \frac{1}{5})$ and $(2, 1)$. The corresponding tangents are

$$\frac{-\frac{14}{5}x}{8} + \frac{\frac{1}{5}y}{2} = 1 \qquad \text{and} \qquad \frac{2x}{8} + \frac{y}{2} = 1$$

that is,

$$-7x + 2y = 20 \qquad \text{and} \qquad x + 2y = 4$$

Remark If (u, v) is inside the ellipse, that is, if $u^2/a^2 + v^2/b^2 < 1$, there are no tangents through (u, v). When you eliminate y (or x) from the system $x^2/a^2 + y^2/b^2 = 1$, $ux/a^2 + vy/b^2 = 1$, the resulting quadratic will have negative discriminant, hence no (real) roots.

Reflection Property of the Ellipse

Like the parabola, the ellipse has a remarkable reflection property. Think of an elliptical pool table. Then a ball cued from one focus will always pass through the other focus after one rebound off the side. This is the principle of whispering galleries. Sound rays broadcast from one focus of an elliptical chamber will bounce off the walls and pass through the other focus. Hence a listener at one focus hears clearly a whisper from the other focus.

Figure 8-5-8
Reflection property: $\alpha = \beta$

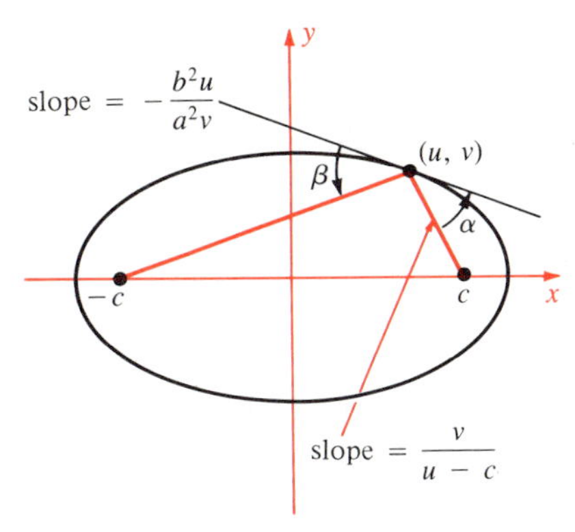

Let us prove this property of the ellipse. We take the standard ellipse $x^2/a^2 + y^2/b^2 = 1$ with foci at $(\pm c, 0)$. See Figure 8-5-8. We must verify that $\alpha = \beta$. For then, the ray from $(c, 0)$ will strike the ellipse and be reflected to the focus at $(-c, 0)$.

By the formula for the angle between two lines (page 420),

$$\tan\alpha = \frac{-\dfrac{b^2u}{a^2v} - \dfrac{v}{u-c}}{1 + \left(\dfrac{v}{u-c}\right)\left(-\dfrac{b^2u}{a^2v}\right)} = \frac{-b^2u(u-c) - a^2v^2}{a^2v(u-c) - b^2uv}$$

To simplify this expression, we use the relation $c^2 = a^2 - b^2$ and the relation $u^2/a^2 + v^2/b^2 = 1$ in the form $b^2u^2 + a^2v^2 = a^2b^2$:

$$\begin{aligned}\tan\alpha &= \frac{b^2cu - (b^2u^2 + a^2v^2)}{(a^2 - b^2)uv - a^2cv}\\ &= \frac{b^2cu - a^2b^2}{c^2uv - a^2cv}\\ &= \frac{b^2(cu - a^2)}{cv(cu - a^2)} = \frac{b^2}{cv}\end{aligned}$$

Note that $cu - a^2 \neq 0$ because $c < a$ and $|u| \le a$, so $|cu| < a^2$.

Now we look at the corresponding angle β with respect to the other focus $(-c, 0)$. We can use the preceding formula for $\tan\alpha$ with c replaced by $-c$. By Figure 8-5-8, however, the angle *from* the ray *to* the tangent is $-\beta$, so the formula yields

$$\tan(-\beta) = \frac{b^2}{(-c)v} \qquad \text{that is} \qquad \tan\beta = \frac{b^2}{cv} = \tan\alpha$$

Therefore $\beta = \alpha$.

Exercises

Give the center, major and minor semi-axes (a and b), vertices, and foci

1 $\frac{1}{25}x^2 + \frac{1}{9}y^2 = 1$

2 $x^2 + 4y^2 = 4$

3 $2(x+1)^2 + (y-2)^2 = 2$

4 $4x^2 + y^2 - 2y = 0$

5 $2x^2 + y^2 - 12x - 4y = -21$

6 $x^2 + 2y^2 + 8y = 0$

Write the equation of the ellipse

7 center at $(1, 4)$ vertices at $(10, 4)$ and $(1, 2)$

8 center at $(-2, -3)$ vertices at $(7, -3)$ and $(-2, -7)$

9 foci at $(2, 0)$ and $(8, 0)$ vertex at $(0, 0)$

10 foci at $(0, 3)$ and $(0, -3)$ semi-major axis $= 10$

11 foci at $(-1, 0)$ and $(3, 0)$ $c/a = \frac{1}{2}$

12 vertices at $(0, 2)$ and $(0, 6)$ $c/a = \frac{3}{4}$ major axis vertical

13 Show that $x^2/a^2 + y^2/b^2 = 1$ has horizontal tangents at $(0, \pm b)$ and vertical tangents at $(\pm a, 0)$.

14 Show that the upper half of the ellipse $x^2/a^2 + y^2/b^2 = 1$ is concave down.

15 Prove that the points on an ellipse farthest from its center are the two vertices on the major axis.

16 Prove that the points on an ellipse nearest to its center are the two vertices on the minor axis.

17 A rod moves with one end on the x-axis and the other on the y-axis. Describe the path of any other point of the rod.

***18** Find the set of midpoints of all chords of slope m of the ellipse $x^2/a^2 + y^2/b^2 = 1$.

19 Let D be a fixed diameter of a circle of radius r. At each point A of the circle drop a perpendicular to D. Find the set of midpoints M of these perpendiculars.

20 (cont.) Extend each perpendicular beyond the circle so that its length increases by a factor $k > 1$. Find the set of end points E of these segments.

***21** Find the set of centers of all circles that are simultaneously tangent at different points to the circles $x^2 + y^2 = 2ax$ and $x^2 + y^2 = 2bx$, where $0 < a < b$.

***22** A gadget (Figure 8-5-9) consists of two disks glued together and a point P on their line of centers. The small disk can move only in the vertical tracks, and the larger disk can move only in the horizontal tracks. Describe the curve traced by P.

23 Show by integrating that the area of the shaded region in Figure 8-5-10 is $\frac{1}{2}ab\theta - \frac{1}{2}xy$, where the ellipse is parameterized by $x = a\cos\theta$, $y = b\sin\theta$.

24 (cont.) Find in the figure a region whose area is $\frac{1}{2}ab\theta$.

Find the tangent to

25 $x^2 + 2y^2 = 1$ at $(1, 0)$

26 $\frac{1}{8}x^2 + \frac{1}{18}y^2 = 1$ at $(2, -3)$

27 $\frac{1}{15}x^2 + \frac{1}{40}y^2 = 1$ at $(3, 4)$

28 $\frac{1}{6}x^2 + \frac{1}{3}y^2 = 1$ at $(-2, 1)$

Find the tangents to

29 $x^2 + 2y^2 = 1$ from $(1, 3)$

30 $x^2 + 2y^2 = 1$ from $(2, 0)$

31 $3x^2 + 2y^2 = 3$ from $(-1, 1)$

32 $3x^2 + y^2 = 1$ from $(-2, -1)$

Figure 8-5-9
See Exercise 22

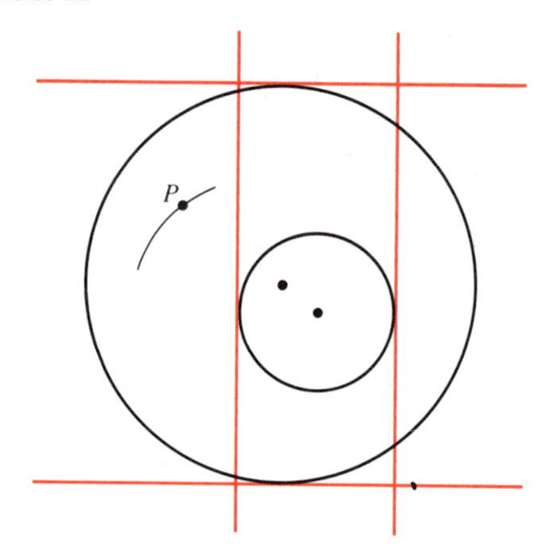

Figure 8-5-10
See Exercise 23

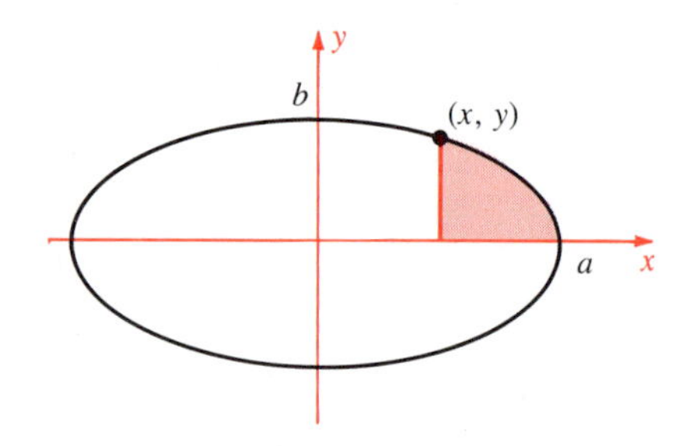

8-6 Conics — The Hyperbola

A **hyperbola** is defined as the set of all points X such that the difference of the distances of X from two fixed points P and Q has constant absolute value $2a$. Thus

$$|\overline{XP} - \overline{XQ}| = 2a \qquad (a > 0)$$

is the defining relation. The points P and Q are the **foci** of the hyperbola. (Note that the definition of hyperbola is the same as that of ellipse except for *difference* instead of *sum.*)

Figure 8-6-1
Hyperbola: $|\overline{XP} - \overline{XQ}| = 2a$

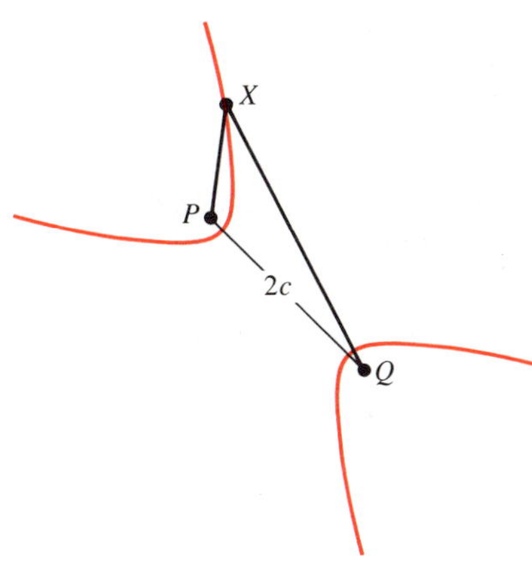

Suppose $\overline{PQ} = 2c.$ In Figure 8-6-1 we have $\overline{XP} + \overline{PQ} > \overline{XQ}.$ Hence $2c = \overline{PQ} > \overline{XQ} - \overline{XP} = 2a.$ Therefore if the hyperbola is to have any points on it that are not collinear with P and Q we must have $2c > 2a,$ that is, $c > a.$ In our discussion of hyperbolas, we always assume this.

To obtain an equation for the hyperbola, we set up coordinate axes so the foci are $(\pm c, 0)$. See Figure 8-6-2. Then a point (x, y) is on the hyperbola if and only if

$$\left|\sqrt{(x-c)^2 + y^2} - \sqrt{(x+c)^2 + y^2}\right| = 2a$$

Just as we did to derive the ellipse, we can eliminate the radicals by some algebraic manipulations. The end result is the following:

The Hyperbola (Standard Position)

The equation of the hyperbola with foci $(-c, 0)$ and $(c, 0)$ and absolute distance difference $2a$ is

$$\frac{x^2}{a^2} - \frac{y^2}{b^2} = 1$$

where $b^2 = c^2 - a^2.$

Conversely, $x^2/a^2 - y^2/b^2 = 1,$ where $a > 0$ and $b > 0,$ describes a hyperbola with foci $(\pm c, 0),$ where $c^2 = a^2 + b^2,$ and with absolute distance difference $2a.$

Figure 8-6-2
Standard position: $\dfrac{x^2}{a^2} - \dfrac{y^2}{b^2} = 1$
where $b^2 = c^2 - a^2$

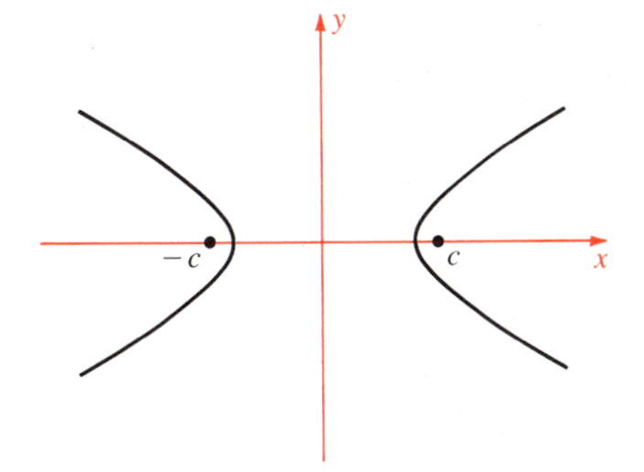

The size of b depends on the relative sizes of c and $a.$ Both $b \le a$ and $b > a$ are possible for the hyperbola. (For the *ellipse* with foci on the x-axis, only $b \le a$ is possible since $b^2 = a^2 - c^2.$)

Let us sketch the hyperbola

$$\frac{x^2}{a^2} - \frac{y^2}{b^2} = 1$$

The curve is symmetric about both axes and about the origin; we need plot it only in the first quadrant, then extend the curve to the other quadrants by symmetry. We solve for y:

$$y = \frac{b}{a}\sqrt{x^2 - a^2}$$

(The positive square root applies for the first quadrant.) Since the quantity under the radical must be non-negative, the locus is defined only for

$x \geq a$. Now when x starts at a and increases, y starts at 0 and increases. When x is very large, we suspect that y is slightly less than bx/a. To confirm this suspicion, we rationalize the numerator of the difference:

$$\begin{aligned}
\frac{b}{a}x - y &= \frac{b}{a}(x - \sqrt{x^2 - a^2}\,) \\
&= \frac{b}{a}(x - \sqrt{x^2 - a^2}\,)\,\frac{x + \sqrt{x^2 - a^2}}{x + \sqrt{x^2 - a^2}} \\
&= \left(\frac{b}{a}\right)\frac{x^2 - (x^2 - a^2)}{x + \sqrt{x^2 - a^2}} \\
&= \frac{ab}{x + \sqrt{x^2 - a^2}} < \frac{ab}{x}
\end{aligned}$$

It follows that $(b/a)x - y$ is positive, but becomes smaller and smaller as x becomes larger and larger. This means the curve approaches the line $y = bx/a$ (from below) as x increases.

Further information: the hyperbola has a vertical tangent at $(a, 0)$ and is concave down in the first quadrant. (See Exercises 19–20.) We can now make a reasonable sketch of the curve (Figure 8-6-3).

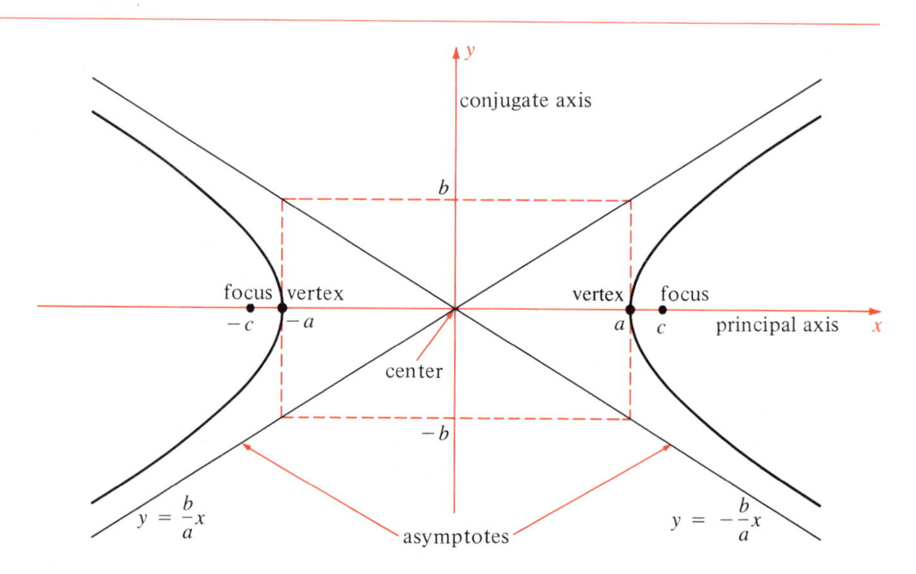

Figure 8-6-3
Geometry of the hyperbola $\dfrac{x^2}{a^2} - \dfrac{y^2}{b^2} = 1$

The lines $y = \pm bx/a$ are the asymptotes of the hyperbola. A neat way to remember this fact is to write

$$\frac{x^2}{a^2} - \frac{y^2}{b^2} = \left(\frac{x}{a} + \frac{y}{b}\right)\left(\frac{x}{a} - \frac{y}{b}\right)$$

The expression on the left is zero if and only if one of the factors on the right is zero, that is, if and only if $y = \pm bx/a$.

Asymptotes

The **asymptotes** of the hyperbola $\dfrac{x^2}{a^2} - \dfrac{y^2}{b^2} = 1$ are the lines

$$y = \frac{b}{a}x \quad \text{and} \quad y = -\frac{b}{a}x$$

or equivalently, the set of all points (x, y) that satisfy the equation

$$\frac{x^2}{a^2} - \frac{y^2}{b^2} = 0$$

Here is some of the official terminology for hyperbolas. A hyperbola consists of two **branches:** one where $\overline{XP} - \overline{XQ} = 2a,$ and the other one where $\overline{XQ} - \overline{XP} = 2a.$ The point halfway between the foci is the **center** of the hyperbola. The line through the foci is the **principal axis** and the line through the center perpendicular to the principal axis is the **conjugate axis.** The points where the hyperbola meets its principal axis are its **vertices.**

A hyperbola is **rectangular** if its asymptotes are perpendicular. This happens when the slopes of the two asymptotes are negative reciprocals of each other:

$$\left(\frac{b}{a}\right)\left(-\frac{b}{a}\right) = -1 \qquad \text{that is} \qquad b^2 = a^2 \quad \text{so} \quad b = a$$

Thus the locus of $x^2 - y^2 = a^2$ is a rectangular hyperbola.

By translation,

$$\frac{(x-h)^2}{a^2} - \frac{(y-k)^2}{b^2} = 1$$

is the equation of a hyperbola with center at (h, k) and horizontal principal axis. Its asymptotes are the lines

$$y - k = \pm\frac{b}{a}(x - h)$$

By interchanging the roles of x and $y,$ we see that the equation

$$-\frac{x^2}{b^2} + \frac{y^2}{a^2} = 1$$

describes a hyperbola with center at the origin, vertical principal axis, foci at $(0, \pm c),$ where $c^2 = a^2 + b^2,$ and absolute distance difference $2a.$ See Figure 8-6-4.

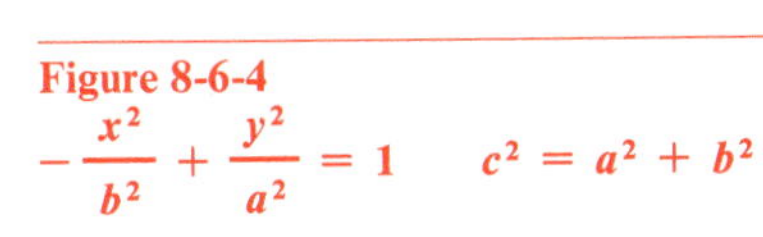

Figure 8-6-4
$-\dfrac{x^2}{b^2} + \dfrac{y^2}{a^2} = 1 \qquad c^2 = a^2 + b^2$

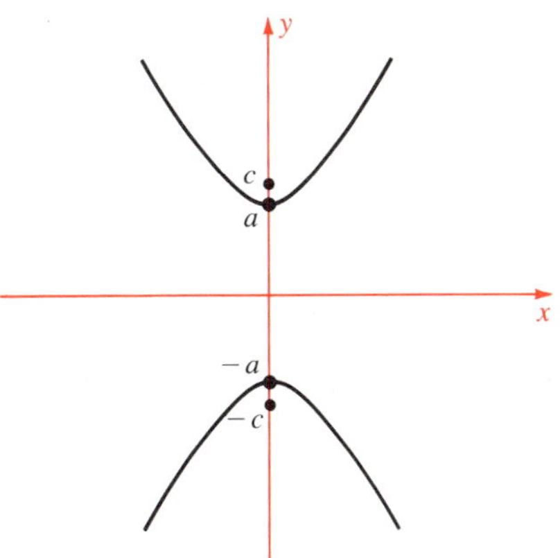

Remark The equation $x^2/a^2 + y^2/b^2 = 1$ and the equation $x^2/a^2 - y^2/b^2 = 1$ differ by a little minus sign, but that makes all the difference in the world. The first equation, where the sign is plus, requires $x^2 \le a^2$ and $y^2 \le b^2$; the locus is confined. The second imposes no such restriction; x^2/a^2 and y^2/b^2 can both be enormous, yet differ by 1.

Example 1

Show that $3x^2 - 12x - 8y^2 = 12$ is the equation of a hyperbola. Locate its center, axes, foci, and asymptotes, and sketch the curve.

Solution Complete the square in x:

$$3(x-2)^2 - 8y^2 = 12 + 12 = 24$$

Hence

$$\frac{(x-2)^2}{8} - \frac{y^2}{3} = 1$$

The curve is a hyperbola with center $(2, 0)$ and $a^2 = 8$ and $b^2 = 3$. Its principal axis is the x-axis and its foci are

$$(2 \pm c, 0) \quad \text{where} \quad c = \sqrt{a^2 + b^2} = \sqrt{11}$$

Its asymptotes are

$$y = \pm\frac{b}{a}(x-2) = \pm\sqrt{\tfrac{3}{8}}\,(x-2)$$

See Figure 8-6-5. ●

Figure 8-6-5
$3x^2 - 12x - 8y^2 = 12$

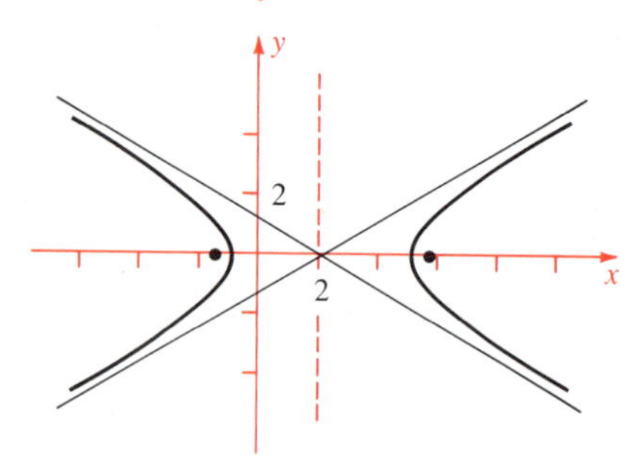

Example 2

Find the equation of the hyperbola with

a vertices $(-1, 1)$, $(-1, 5)$ and foci $(-1, 0)$, $(-1, 6)$.
b foci $(0, 2)$, $(10, 2)$ and asymptotes having slopes ± 3.

Solution

a The principal axis is vertical and the center is $(-1, 3)$. Therefore the equation has the form

$$\frac{(y-3)^2}{a^2} - \frac{(x+1)^2}{b^2} = 1$$

Now $2a$ is the distance between the vertices; hence $2a = 4$, $a = 2$. Similarly, $2c$ is the distance between the foci; hence $2c = 6$, $c = 3$. Finally, $b^2 = c^2 - a^2$, so $b^2 = 9 - 4 = 5$. Hence the desired equation is

$$\frac{(y-3)^2}{4} - \frac{(x+1)^2}{5} = 1$$

b The principal axis is horizontal and the center is $(5, 2)$. Therefore the equation has the form

$$\frac{(x-5)^2}{a^2} - \frac{(y-2)^2}{b^2} = 1$$

The distance between the foci is $10 = 2c$; hence $c = 5$. It follows that $a^2 + b^2 = c^2 = 25$. Furthermore, the slopes of the asymptotes are $\pm 3 = \pm b/a$, so $b^2 = 9a^2$. Therefore

$$25 = a^2 + b^2 = 10a^2 \qquad a^2 = \tfrac{5}{2} \qquad b^2 = \tfrac{45}{2}$$

Hence the equation of the hyperbola is

$$\frac{(x-5)^2}{\frac{5}{2}} - \frac{(y-2)^2}{\frac{45}{2}} = 1$$

●

Example 3

Find the set of centers of all circles that are tangent to the x-axis and cut off a (variable) segment of length $2k$ on the y-axis.

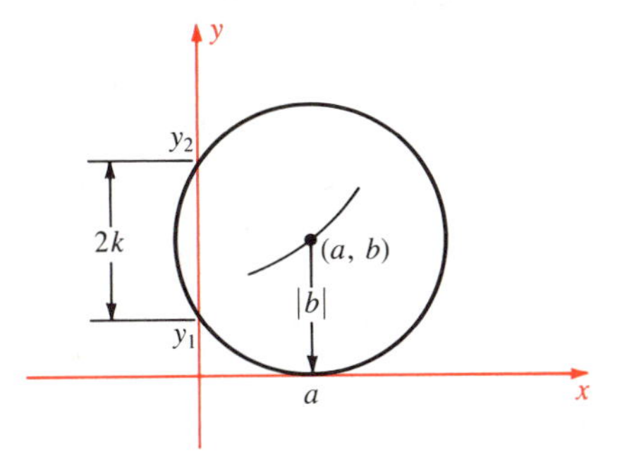

Figure 8-6-6
$y_2 - y_1 = 2k$. See Example 3.

Solution Draw a picture (Figure 8-6-6). If the center of a circle is (a, b) and the circle is tangent to the x-axis, then its radius is $|b|$ (absolute value in case $b < 0$). The equation of the circle is

$$(x-a)^2 + (y-b)^2 = b^2$$

that is,

$$x^2 + y^2 - 2ax - 2by + a^2 = 0$$

Suppose the circle intersects the y-axis at $(0, y_1)$ and at $(0, y_2)$ with $y_1 < y_2$. Then the segment condition is $y_2 - y_1 = 2k$.

To find y_1 and y_2 set $x = 0$ and solve for y:

$$y^2 - 2by + a^2 = 0 \qquad \text{hence} \qquad y = b \pm \sqrt{b^2 - a^2}$$

Assume $b^2 > a^2$; otherwise the circle either does not intersect the y-axis or is tangent to it. Then

$$y_2 - y_1 = 2\sqrt{b^2 - a^2} = 2k \quad \text{so that} \quad b^2 - a^2 = k^2$$

Hence the centers (a, b) all lie on the rectangular hyperbola

$$y^2 - x^2 = k^2$$

Conversely, each point on this hyperbola is a point of the required set, as is easily checked. ●

Example 4

Find all points that are 10 units from the origin and two units closer to $(3, 0)$ than to $(-3, 0)$.

Solution The first condition means that the point (x, y) lies on the circle $x^2 + y^2 = 10^2$. The second condition means that it lies on one branch of the hyperbola with foci $(\pm 3, 0)$ and difference of the distances 2. This hyperbola is in standard position with principal axis horizontal, so its equation has the standard form

$$\frac{x^2}{a^2} - \frac{y^2}{b^2} = 1$$

The absolute distance difference is $2a$, and we are given that this is 2. Hence $2a = 2$ so $a = 1$. The distance between the foci is $2c$ and this is the distance between $(3, 0)$ and $(-3, 0)$. Hence $2c = 6$ so $c = 3$. Therefore $b^2 = c^2 - a^2 = 8$ so $1/b^2 = \frac{1}{8}$.

The desired points are the intersections of the circle and the right-hand branch of the hyperbola [where the points are closer to $(3, 0)$ than to $(-3, 0)$]. Thus the pair of equations

$$x^2 + y^2 = 100 \qquad x^2 - \tfrac{1}{8}y^2 = 1$$

must be solved simultaneously. Subtract to eliminate x and solve for y:

$$\tfrac{9}{8}y^2 = 99 \qquad \text{hence} \qquad y^2 = 88 \quad \text{so that} \quad y = \pm\sqrt{88}$$

from which

$$x^2 = 100 - y^2 = 100 - 88 = 12 \qquad \text{hence} \qquad x = \pm\sqrt{12}$$

On the right-hand branch of the hyperbola, $x > 0$, so only the value $x = \sqrt{12}$ is acceptable. Therefore the desired points are $(\sqrt{12}, \pm\sqrt{88})$. ●

Exercises

Find the principal axis, center, foci, and asymptotes

1 $\frac{1}{4}x^2 - \frac{1}{9}y^2 = 1$

2 $\frac{1}{9}x^2 - \frac{1}{4}y^2 = 1$

3 $-\frac{1}{9}x^2 + \frac{1}{4}y^2 = 1$

4 $-\frac{1}{4}x^2 + \frac{1}{9}y^2 = 1$

5 $(x + 1)^2 - (y - 1)^2 = 1$

6 $-(x - 2)^2 + 4(y + 1)^2 = 4$

7 $x^2 - 5y^2 + 4x - 20y = 0$

8 $-x^2 + 2y^2 - 6x - 20y + 47 = 0$

9 $4x^2 - y^2 - 24x - 2y + 31 = 0$

10 $3x^2 - 3y^2 - 3x - 2y = \frac{31}{12}$

Write the equation of the hyperbola with

11 foci $(0, \pm 5)$ vertex $(0, -4)$

12 vertices $(\pm 3, 0)$ focus $(-5, 0)$

13 asymptote $y = -2x$ vertices $(\pm 2, 0)$

14 foci $(1, 7)$ and $(1, -3)$ vertex $(1, 6)$

15 asymptotes $y = \pm(x - 1)$ curve passes through $(3, 1)$

16 asymptotes $y = \pm 2x$ curve passes through $(1, 1)$

17 Show that $x^2 - y^2 + 2ax + 2by + c = 0$ represents a rectangular hyperbola. Assume $a^2 \neq b^2 + c$.

18 Show that $3x^2 - y^2 + 6ax + 2by + c = 0$ represents a hyperbola whose asymptotes form a 60° angle. Assume $3a^2 \neq b^2 + c$.

19 Show that $x^2/a^2 - y^2/b^2 = 1$ has a vertical tangent at each vertex.

20 (cont.) Show that the right branch is concave up with respect to the y-axis.

21 Let $0 < s < r$ and $r + s < 2a$. Describe the set of centers of all circles that are simultaneously tangent externally to the circles

$$(x + a)^2 + y^2 = r^2 \quad \text{and} \quad (x - a)^2 + y^2 = s^2$$

22 (cont.) Identify the other branch geometrically.

***23** (cont.) Find another hyperbola lurking in this configuration.

24 A rifle at point A on level ground is shot at a target at point B. Find the curve of all observers X who hear the shot and the impact of the slug simultaneously.

25 Three listening posts A, B, and C record an explosion. Post A is 10 km west of post B, and post C is 8 km south of B. Posts B and C hear the explosion simultaneously; A hears it 6 sec later. Assuming the speed of sound in air is $\frac{1}{3}$ km/sec, locate the explosion.

26 Two firms are $2c$ km apart, and they both sell the same item. Assume their shipping costs per item per km are equal, but one firm's price at the factory is k times that of the other, where $k > 1$. Find the curve of equal prices.

The next three exercises explain why the hyperbolic functions are so named.

27 Show that the right-hand branch of $x^2/a^2 - y^2/b^2 = 1$ is parameterized by $x = a\cosh t$, $y = b\sinh t$.

28 (cont.) Show that the shaded region in Figure 8-6-7 has area $\frac{1}{2}xy - \frac{1}{2}abt$.

29 (cont.) Find in the figure a region whose area is $\frac{1}{2}abt$.

30 For what values of m does $y = mx$ intersect the hyperbola $x^2/a^2 - y^2/b^2 = 1$?

31 Let (u, v) be a point of $x^2/a^2 - y^2/b^2 = 1$. Show that the tangent at (u, v) is $xu/a^2 - yv/b^2 = 1$.

32 (cont.) Find the tangents to $2x^2 - y^2 = 1$ from $(2, 3)$.

***33** Prove that an ellipse and a hyperbola with the same foci always intersect at right angles.

***34** Prove in Figure 8-6-8 that $\alpha = \beta$, the "reflection property" of the hyperbola.

Figure 8-6-7
See Exercise 28.

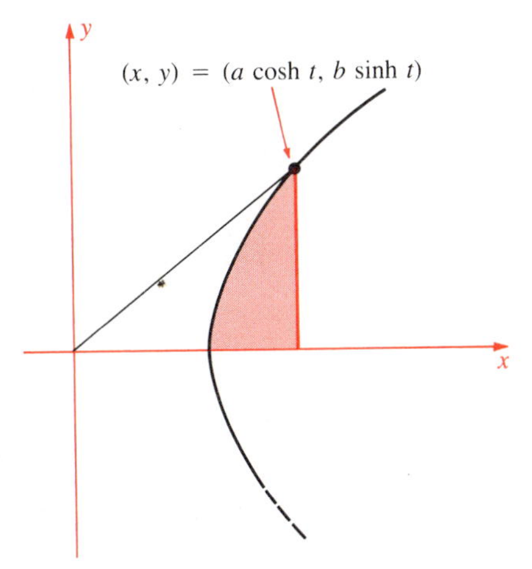

Figure 8-6-8
Foci P and Q. See Exercise 34.

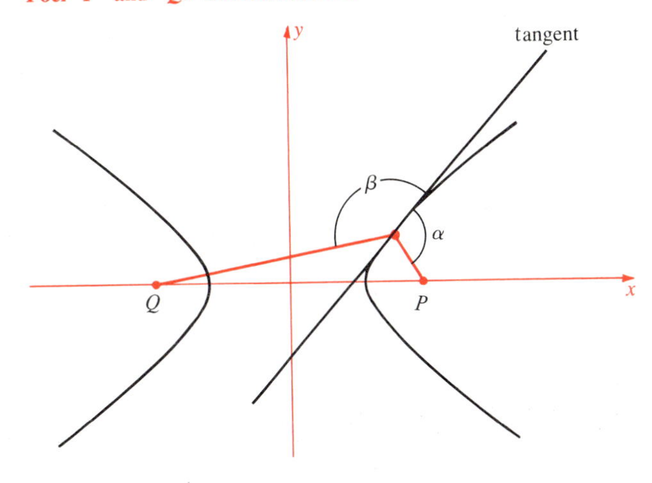

8-7 Polar Coordinates

The usual rectangular coordinate system is convenient in some situations but clumsy in others. Sometimes another coordinate system, called the **polar coordinate** system, fits a problem much more naturally. The idea of polar coordinates is that you identify a point by telling how far it is from a given reference point **0** and in what direction. (This is the principle of the radar screen.)

In a rectangular coordinate system, two families of grid lines, $x =$ constant and $y =$ constant, fill the plane. Each point X is the intersection of two of these lines, $x = a$ and $y = b$, and receives the coordinates (a, b). See Figure 8-7-1a.

Polar coordinates work on a similar principle. There are two families of grid lines: all circles centered at **0**, and all rays from **0**. See Figure 8-7-1b. Each point X different from **0** is the intersection of one circle and one ray. The circle is identified by a positive number r, its radius, and the ray is identified by a real number θ, its angle in radians from the positive x-axis. Thus X is assigned the **polar coordinates** $[r, \theta]$. Since θ is determined only up to a multiple of 2π, we agree that the polar coordinates

$$[r, \theta + 2\pi n]$$

Figure 8-7-1
Rectangular and polar coordinates

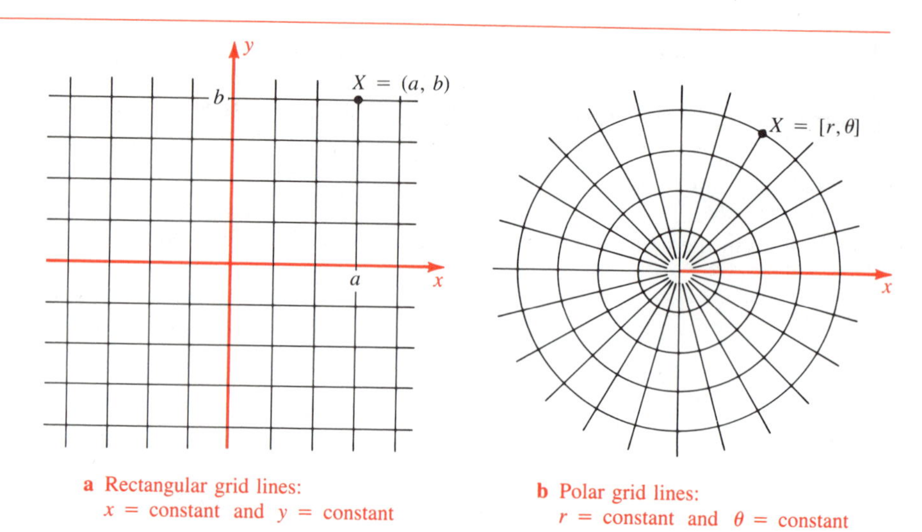

a Rectangular grid lines: $x =$ constant and $y =$ constant

b Polar grid lines: $r =$ constant and $\theta =$ constant

where n is any integer, all represent the same point. The point **0** does not determine an angle θ. Nonetheless, it is customary to say that any pair $[0, \theta]$ represents **0**. The positive x-axis is sometimes called the **polar axis** or ***r*-axis.** There is no "θ-axis" in any sense.

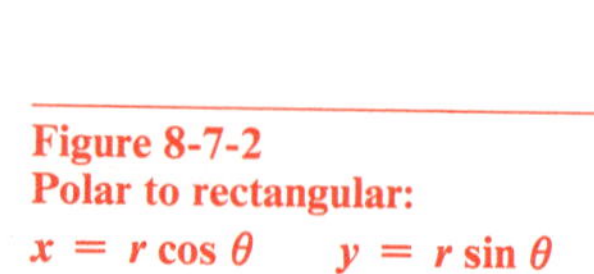

Figure 8-7-2
Polar to rectangular:
$x = r\cos\theta \qquad y = r\sin\theta$

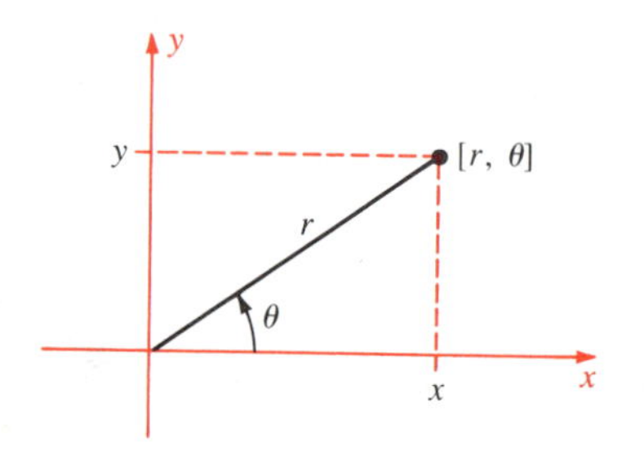

Notation We write $[r, \theta]$ with square brackets to denote the polar coordinates of a point, to distinguish from (x, y), its rectangular coordinates. Some authors write (r, θ) and (x, y), and expect you to know from the context which is which. This is particularly confusing when actual numbers are involved. What point is meant by $(5, 0.6435)$? Is it $(5, 0.6435)$ or $[5, 0.6435] \approx (4, 3)$? Another notation is $(a, b)_{\text{pol}}$, which is pretty cumbersome.

Given the polar coordinates of a point, what are its rectangular coordinates? If its polar coordinates are $[r, \theta]$, then the point is r units from **0** in the direction θ. Hence $x = r\cos\theta$, $y = r\sin\theta$. See Figure 8-7-2.

Conversely, given the rectangular coordinates (x, y), what are the polar coordinates? Figure 8-7-3 shows that $r = \sqrt{x^2 + y^2}$, and that $\cos\theta = x/r$ and $\sin\theta = y/r$.

Figure 8-7-3
Rectangular to polar:
$r = \sqrt{x^2 + y^2}$

$\cos\theta = \dfrac{x}{r} \qquad \sin\theta = \dfrac{y}{r}$

Polar to Rectangular	**Rectangular to Polar**
$\begin{cases} x = r\cos\theta \\ y = r\sin\theta \end{cases}$	$\begin{cases} r = \sqrt{x^2 + y^2} \\ \cos\theta = \dfrac{x}{\sqrt{x^2+y^2}} = \dfrac{x}{r} \\ \sin\theta = \dfrac{y}{\sqrt{x^2+y^2}} = \dfrac{y}{r} \end{cases}$

Remark We cannot just say $\theta = \arcsin y/r$ or $\theta = \arccos x/r$ without some consideration of the ranges of these functions. Remember that

$$-\tfrac{1}{2}\pi \le \arcsin u \le \tfrac{1}{2}\pi \quad \text{and} \quad 0 \le \arccos u \le \pi$$

We must take the *quadrant* of (x, y) into account before mechanically punching the arc sin or the arc cos button on a calculator. For example, suppose that $(x, y) = (-3, -4)$, a point in the third quadrant. Then $r = 5$ and $\arcsin(-\frac{4}{5}) \approx -53.13°$ (fourth quadrant) and $\arccos(-\frac{3}{5}) \approx 126.87°$ (second quadrant), both wrong! The correct answer is either

$$180° - \arcsin(-\tfrac{4}{5}) \approx 180° + 53.13° = 233.13°$$

or

$$360° - \arccos(-\tfrac{3}{5}) \approx 360° - 126.87° = 233.13°$$

If (x, y) is in the first or fourth quadrant, then $\theta = \arcsin y/r$. If (x, y) is in the first or second quadrant, then $\theta = \arccos x/r$. But if θ is in the third quadrant, then (in radians)

$$\theta = 2\pi - \arccos y/r = \pi - \arcsin x/r$$

Example 1

a Convert $(2, -2\sqrt{3})$ to polar coordinates.
b Convert $[3, \frac{1}{6}\pi]$ to rectangular coordinates.

Solution

a $r^2 = 4 + 12 = 16$ so that $r = 4$. Also, we have $\cos\theta = \frac{2}{4} = \frac{1}{2}$ and $\sin\theta = \frac{1}{4}(-2\sqrt{3}) = -\frac{1}{2}\sqrt{3}$, so $\theta = \frac{5}{3}\pi$. Therefore the polar coordinates are $[4, \frac{5}{3}\pi]$.

b $x = r\cos\theta = 3\cos\frac{1}{6}\pi = \frac{3}{2}\sqrt{3}$ $\quad y = r\sin\theta = 3\sin\frac{1}{6}\pi = \frac{3}{2}$

Therefore the rectangular coordinates are $(\frac{3}{2}\sqrt{3}, \frac{3}{2})$. ●

Negative r

In some applications it is convenient to allow points $[r, \theta]$ with $r < 0$. Consider a ray and a point $[r, \theta]$ on the ray (Figure 8-7-4a). Suppose the point moves towards **0**, through **0**, and keeps on going! Then r

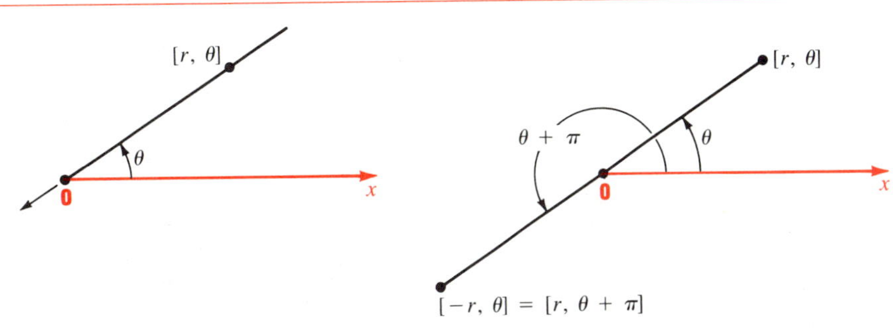

Figure 8-7-4
The meaning of negative r

a Hold θ fixed and move $[r, \theta]$ through the origin.

b Identify $[-r, \theta]$ with $[r, \theta + \pi]$.

decreases, becomes 0, but then what? So that θ won't jump abruptly to $\theta + \pi$, we agree that θ remains constant, but r becomes negative. This amounts to agreeing that the polar coordinates

$$[-r, \theta] \quad \text{and} \quad [r, \theta + \pi]$$

represent the same point (Figure 8-7-4b). For example, the point $(-1, -1)$ has polar coordinates $[-\sqrt{2}, \frac{1}{4}\pi]$ as well as $[\sqrt{2}, \frac{5}{4}\pi]$.

Lines and Circles

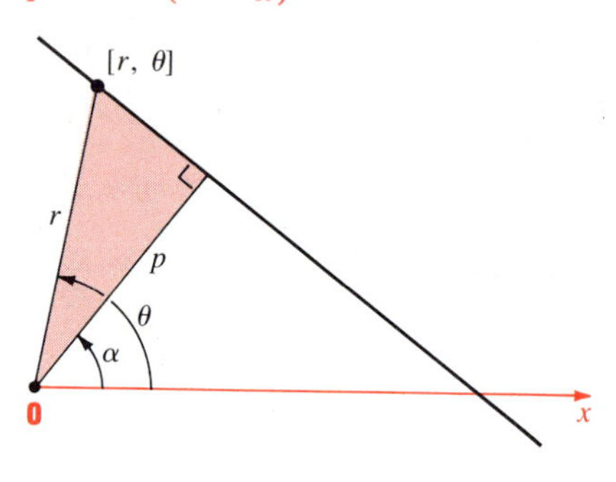

Figure 8-7-5
From the right triangle,
$p = r\cos(\theta - \alpha)$

In polar coordinates, the equation of a line through **0** is $\theta = \theta_0$, where θ_0 is the angle the line makes with the positive x-axis. What is the equation of a line L not through the origin? Drop a perpendicular to L from **0**. It has length $p > 0$ and polar angle α. See Figure 8-7-5.

The figure shows that for each point $[r, \theta]$ on the line, we have $r\cos(\theta - \alpha) = p$.

Polar Equation of a Line

The polar equation of a line not passing through **0** is

$$r\cos(\theta - \alpha) = p \qquad (p > 0)$$

Here p is the distance from **0** to the line, and the point $[p, \alpha]$ is the foot of the perpendicular from **0** to the line.

The equation $r\cos(\theta - \alpha) = p$ has an analogue in rectangular coordinates. Use the trigonometric addition law for $\cos(\theta - \alpha)$, then replace $r\cos\theta$ by x and $r\sin\theta$ by y:

$$r\cos\theta\cos\alpha + r\sin\theta\sin\alpha = p \qquad x\cos\alpha + y\sin\alpha = p$$

The latter equation is called the **normal form** of the line (in rectangular coordinates). Note that even lines through the origin satisfy such an equation with $p = 0$. (However, α is then determined up to a multiple of π rather than a multiple of 2π, as when $p > 0$.)

Normal Form of a Line

Each line in the plane has an equation

$$x\cos\alpha + y\sin\alpha = p \qquad (p \geq 0)$$

If $p > 0$, then $(p\cos\alpha, p\sin\alpha)$ is the foot of the perpendicular from **0** to the line.

To convert an equation $ax + by = c$ into normal form, just divide by $\pm\sqrt{a^2 + b^2}$, taking the sign to be the same as that of c:

$$\frac{a}{\pm\sqrt{a^2 + b^2}}x + \frac{b}{\pm\sqrt{a^2 + b^2}}y = \frac{c}{\pm\sqrt{a^2 + b^2}}$$

Since

$$\left(\frac{a}{\pm\sqrt{a^2+b^2}}\right)^2 + \left(\frac{b}{\pm\sqrt{a^2+b^2}}\right)^2 = 1$$

there is an angle α such that the equation can be written

$$(\cos\alpha)x + (\sin\alpha)y = p \qquad (p \geq 0)$$

Example 2

Find the point of $x - 3y = 7$ closest to the origin.

Solution Divide both sides by $\sqrt{1^2 + 3^2} = \sqrt{10}$ to obtain the normal form

$$\frac{x}{\sqrt{10}} - \frac{3y}{\sqrt{10}} = \frac{7}{\sqrt{10}}$$

Then

$$\cos\alpha = \frac{1}{\sqrt{10}} \qquad \sin\alpha = \frac{-3}{\sqrt{10}} \qquad \text{and} \qquad p = \frac{7}{\sqrt{10}}$$

The point on the line closest to **0** is

$$(p\cos\alpha, p\sin\alpha) = (\tfrac{7}{10}, -\tfrac{21}{10})$$

●

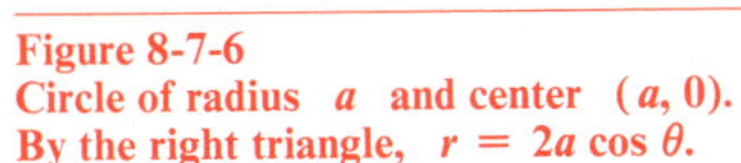

Figure 8-7-6
Circle of radius a and center $(a, 0)$. By the right triangle, $r = 2a\cos\theta$.

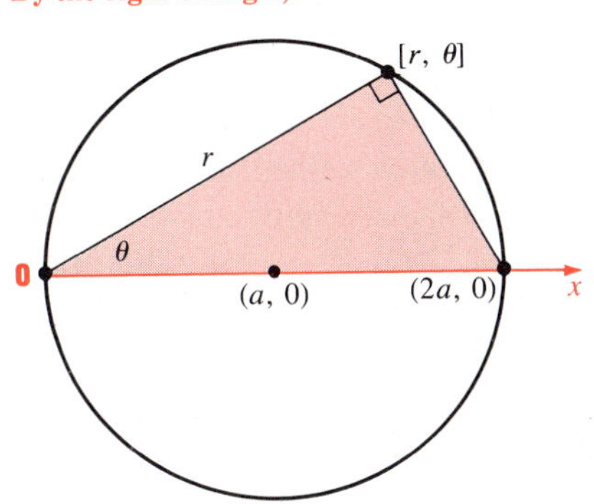

In polar coordinates, the equation of a circle of radius a, center **0**, is simply $r = a$. Consider next the circle in Figure 8-7-6 of radius a and center $(a, 0)$. Its Cartesian equation is

$$(x - a)^2 + y^2 = a^2 \qquad \text{that is} \qquad x^2 + y^2 = 2ax$$

Substitute $x = r\cos\theta$ and $y = r\sin\theta$:

$$r^2 = 2ar\cos\theta$$

If $r \neq 0$, then $r = 2a\cos\theta$. But $r = 0$ represents only the point **0**, which is already on the curve $r = 2a\cos\theta$ for $\theta = \frac{1}{2}\pi$. Hence canceling r does not change the curve, so

$$r = 2a\cos\theta$$

is the polar equation of the given circle. The right triangle in Figure 8-7-6 shows that the relation $r = 2a\cos\theta$ is satisfied by every point on the circle.

Let us see how $[r, \theta]$ moves on the circle $r = 2a\cos\theta$ as θ makes a complete revolution. If θ starts at 0, then r starts at $2a$. If θ increases through the first quadrant, from 0 to $\frac{1}{2}\pi$, then r decreases to 0. (Think of a rod turning counterclockwise and shrinking.) Hence $[r, \theta]$ traces the upper half of the circle (Figure 8-7-7a, next page).

If θ then increases through the second quadrant, from $\frac{1}{2}\pi$ to π, then r decreases from 0 to $-2a$. Since r is negative, the point

Figure 8-7-7
The circle $r = 2a \cos\theta$ traced as $0 \le \theta \le \pi$

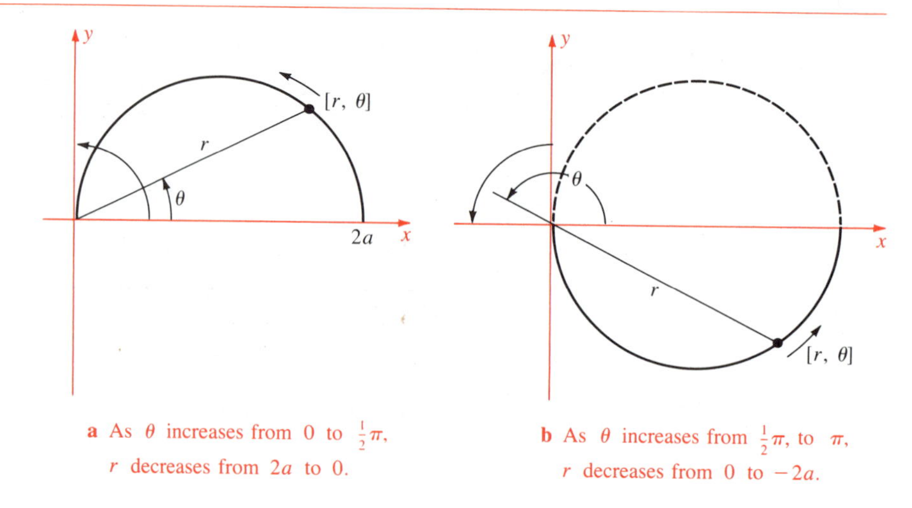

a As θ increases from 0 to $\frac{1}{2}\pi$, r decreases from $2a$ to 0.

b As θ increases from $\frac{1}{2}\pi$, to π, r decreases from 0 to $-2a$.

$[r, \theta]$ is measured "backward" and moves through the *fourth* quadrant, tracing the lower half of the circle (Figure 8-7-7b).

Thus the full circle is described as θ runs through the first and second quadrants, from 0 to π. As θ runs through the third and fourth quadrants, from π to 2π, the same circle is traced again. For when θ is in the third quadrant, $r < 0$, so $[r, \theta]$ describes the semicircle in the first quadrant; when θ is in the fourth quadrant, $r > 0$, so $[r, \theta]$ describes the semicircle in the fourth quadrant. Therefore, in one complete revolution of θ, the circle is traced twice.

Polar Equation of a Circle Tangent to the y-axis at $(0, 0)$

The graph of the equation $r = 2a \cos\theta$ is a circle of radius a and center $[a, 0]$. The circle is traced twice as θ makes a complete revolution.

Figure 8-7-8
Apply the law of cosines to express d in terms of the polar coordinates.

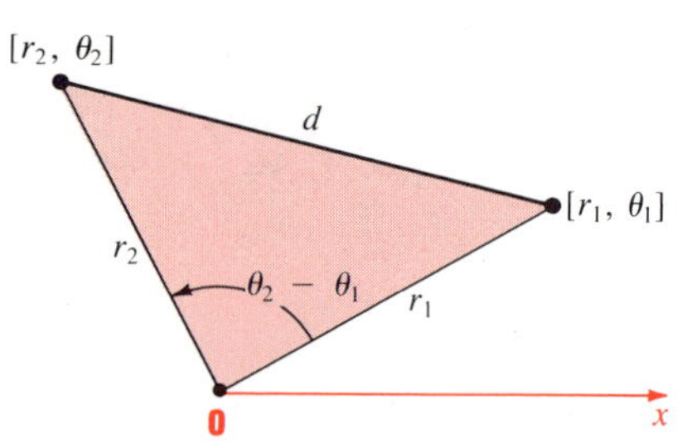

In rectangular coordinates, the distance formula follows from the Pythagorean theorem; in polar coordinates it follows from the law of cosines. Take two points $[r_1, \theta_1]$ and $[r_2, \theta_2]$ in Figure 8-7-8. In the shaded triangle, the angle at **0** is $\theta_2 - \theta_1$. The side opposite this angle has length d, and the two adjacent sides have lengths r_1 and r_2. According to the law of cosines (Figure 1-7-8), we have:

Distance Formula (Polar Form)

If d is the distance between the points $[r_1, \theta_1]$ and $[r_2, \theta_2]$ then

$$d^2 = r_1{}^2 - 2r_1r_2\cos(\theta_2 - \theta_1) + r_2{}^2$$

An immediate corollary is the polar equation of the circle of radius a, center $[p, \alpha]$.

Polar Equation of a Circle

The graph of the equation

$$r^2 - 2rp\cos(\theta - \alpha) + p^2 = a^2$$

is a circle of radius a and center $[p, \alpha]$.

Note two special cases: If the center is **0**, then $p = 0$, and the equation boils down to $r^2 = a^2$, that is, $r = a$ or $r = -a$. If the center is at the point $(a, 0)$, then $p = a$, $\alpha = 0$ and the equation becomes

$$r^2 - 2ra\cos\theta + a^2 = a^2 \qquad \text{that is} \qquad r = 2a\cos\theta$$

Exercises

Give the rectangular coordinates

1 $[1, \frac{1}{2}\pi]$

2 $[1, -\frac{1}{2}\pi]$

3 $[1, -\frac{1}{6}\pi]$

4 $[1, \frac{1}{3}\pi]$

5 $[2, -\frac{3}{4}\pi]$

6 $[2, \frac{5}{4}\pi]$

Give the polar coordinates

7 $(1, 1)$

8 $(0, -1)$

9 $(-1, 1)$

10 $(-\frac{1}{2}, \frac{1}{2}\sqrt{3})$

11 $(\sqrt{3}, -1)$

12 $(\sqrt{2}, -\sqrt{2})$

Find an equation in polar coordinates

13 line through **0** and $[3, \frac{1}{4}\pi]$

14 circle with center **0**, radius 5

15 line through $[1, 0]$ and $[1, \frac{1}{2}\pi]$

16 line perpendicular to $\theta = \frac{3}{4}\pi$, tangent to circle $r = 1$

17 circle with center $[a, \pi]$, radius a

18 circle with center $[a, \frac{1}{2}\pi]$, radius a

19 circle through **0**, center $[1, -\frac{3}{4}\pi]$

20 circle through **0**, center $[2, \frac{1}{4}\pi]$

21 line through $(1, 2)$, slope $-\frac{1}{2}$

22 line through $(-2, 1)$, slope 3

23 circle with center $[5, \frac{1}{6}\pi]$, radius 4

24 circle with center $[5, \frac{1}{6}\pi]$, through $[3, -\frac{1}{2}\pi]$

Give the line in normal form

25 $3x - 4y = 5$

26 $-5x + 12y = -26$

27 $7x + 24y = -25$

28 $5x - 3y = 2$

29 through $(1, 2)$ and $(2, 1)$

30 through $(-1, -3)$, slope 2

Suppose the normal form of a line is $x\cos\alpha + y\sin\alpha = p$.

31 If the origin is translated to (h, k), find the normal form in the new coordinates.

32 (cont.) Find all translations of coordinates that convert the equation to $x\cos\alpha + y\sin\alpha = 0$.

8-8 Polar Graphs

Graphing a function $r = f(\theta)$ in polar coordinates is tricky at first, because you must change your point of view. For $y = f(x)$, in rectangular coordinates, you think of x running along the x-axis, with the corresponding point (x, y) measured above or below. Basically your mental set is "left-right" and "up-down."

In polar coordinates, however, you must think of the angle θ swinging around (like a radar scope) and repeating after 2π. For each θ, you must measure forward from the origin a distance $f(\theta)$, or backward if $f(\theta) < 0$. Your mental set must be "round and round" and "in and out."

Because of the special nature of points $[r, \theta]$ where $r < 0$, pay close attention to the sign of $f(\theta)$ and be sure to plot points "backwards" if $f(\theta) < 0$.

Look for symmetries and periodicity. For example if $f(\theta + 2\pi) = f(\theta)$, the polar graph $r = f(\theta)$ will repeat after 2π. There are many symmetries possible; we mention only two, $f(\theta)$ even and $f(\theta)$ odd. If $f(\theta)$ is even, that is, $f(-\theta) = f(\theta)$, then the point $[r, -\theta]$ is on the graph whenever $[r, \theta]$ is; that is, the curve is symmetric about the

Figure 8-8-1
Graph of $r = f(\theta)$ where $f(\theta)$ is *even:* $f(-\theta) = f(\theta)$

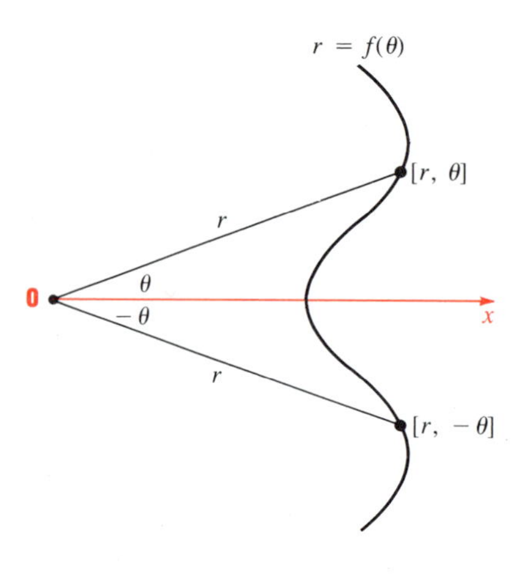

Figure 8-8-2
Graph of $r = f(\theta)$ where $f(\theta)$ is *odd:* $f(-\theta) = -f(\theta)$

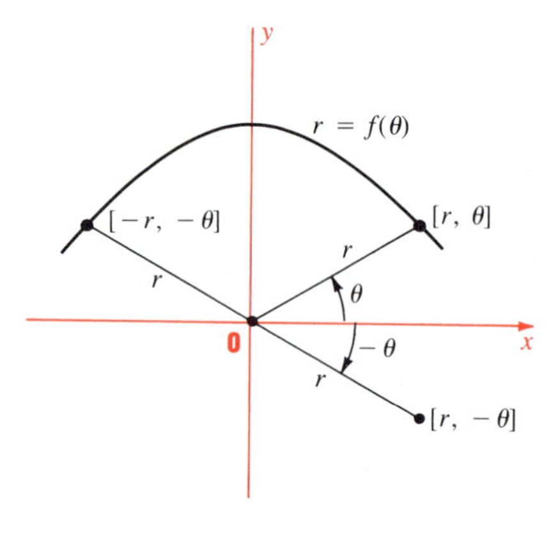

x-axis (Figure 8-8-1). If $f(\theta)$ is odd, $f(-\theta) = -f(\theta)$, then the point $[-r, -\theta]$ is on the graph whenever $[r, \theta]$ is; that is, the curve is symmetric about the y-axis (Figure 8-8-2).

Example 1

Graph the **spiral of Archimedes,** $r = \theta$.

Solution If θ increases, starting from 0, then r increases steadily, also starting from 0. Hence the locus goes round and round, its distance from 0 becoming greater and greater. The result is a spiral. Since $f(\theta) = \theta$ is an odd function, we obtain the locus for $\theta < 0$ by reflection in the y-axis (Figure 8-8-3). ●

Figure 8-8-3
Spiral of Archimedes: $r = \theta$
(arrows indicate θ increasing)

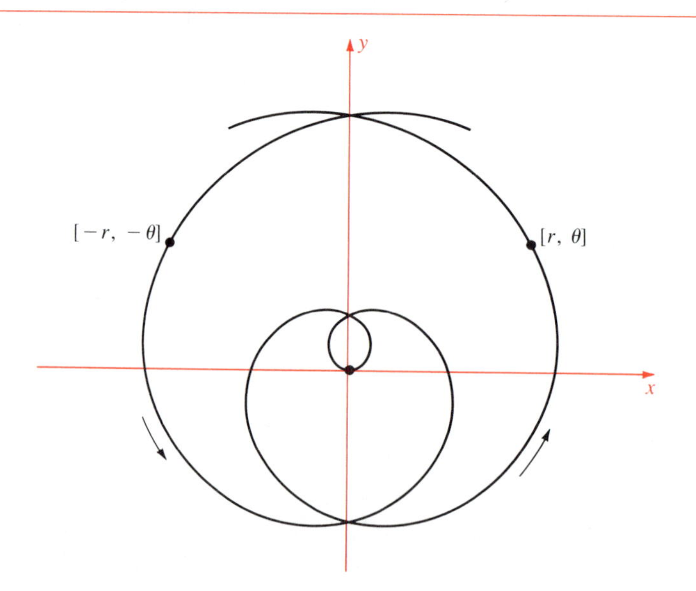

Example 2

Graph the **rose** $r = a\cos 2\theta$, where $a > 0$.

Solution Since $\cos 2(\theta + 2\pi) = \cos 2\theta$, the curve repeats every 2π, so we need plot it only for a single interval of length 2π.

A preliminary sketch showing the sign of $\cos 2\theta$ is helpful (Figure 8-8-4). If θ starts at 0 and increases to $\frac{1}{4}\pi$, then $\cos 2\theta$ starts at 1 and decreases to 0. Since $\cos 2\theta$ is an even function, this part of the graph is repeated below the x-axis forming a loop (Figure 8-8-5).

Figure 8-8-4
Signs of $r = a\cos 2\theta \quad (a > 0)$

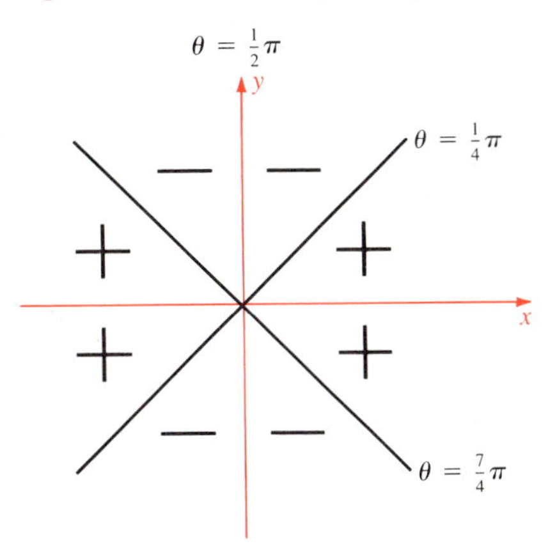

Figure 8-8-5
Partial graph of $r = a\cos 2\theta$

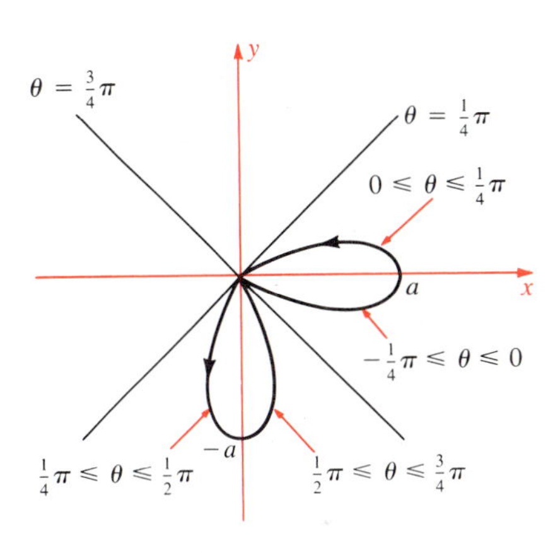

If θ increases from $\frac{1}{4}\pi$ to $\frac{1}{2}\pi$ to $\frac{3}{4}\pi$, then $\cos 2\theta$ is negative and goes from 0 to -1 and back to 0. Thus we get another loop, but between $\frac{5}{4}\pi$ and $\frac{7}{4}\pi$. For θ going from $\frac{3}{4}\pi$ to $\frac{5}{4}\pi$, we get a third loop plotted forward, and from $\frac{5}{4}\pi$ to $\frac{7}{4}\pi$ a fourth loop plotted backwards. Figure 8-8-6 is the complete graph. ●

Figure 8-8-6
Complete graph of $r = a\cos 2\theta$
$(a > 0)$

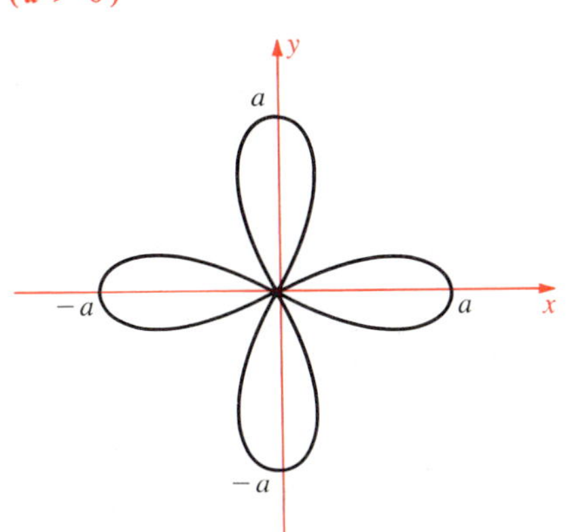

Hindsight It is necessary to plot only one of the petals in Figure 8-8-5. From $\cos(\theta + \pi) = -\cos\theta$ follows $\cos 2(\theta + \frac{1}{2}\pi) = -\cos 2\theta$. Hence $\cos 2\theta$ repeats itself (but with opposite sign) every $\frac{1}{2}\pi$. Therefore, rotate the first loop backward by $\frac{1}{2}\pi$; the result is the next loop of the curve. Rotate again, and once again, and you have generated the whole curve.

For an accurate picture of the petals, plot some points. One thing can be said without plotting; the petals are rounded at their ends, not pointed. That stems from a property of the cosine: $\cos 2\theta \approx 1 - 2\theta^2$ for small angles. Hence for θ small (near the tip of the first petal to the right), the curve $r = a\cos 2\theta$ looks like the circle $r = a$.

Polar Equation of the Ellipse

The orbit of a planet around a fixed star is an ellipse with the star at one focus. In astronomy one measures angles rather than distances, so it is natural to study the polar equation of an ellipse with one focus at the origin.

Figure 8-8-7
Polar form of ellipse:
$r(a - c\cos\theta) = a^2 - c^2 = b^2$

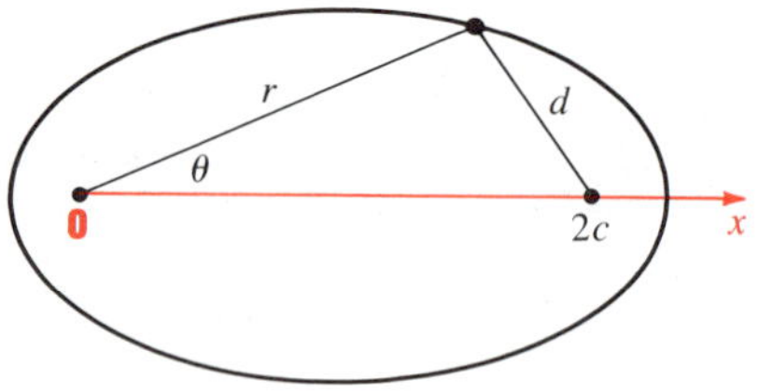

Place the origin at one focus and take the polar axis through the other focus, $[2c, 0]$. See Figure 8-8-7. By definition of the ellipse, we have $r + d = 2a$ so that $d = 2a - r$. Hence

$$d + r = 2a \quad \text{and} \quad d - r = 2a - r - r = 2(a - r)$$

But by the distance formula (page 456)

$$d^2 = r^2 + (2c)^2 - 2r(2c)\cos\theta = r^2 + 4c^2 - 4rc\cos\theta$$

Therefore

$$(d + r)(d - r) = d^2 - r^2 = 4(c^2 - rc\cos\theta)$$

Replace $d + r$ by $2a$ and $d - r$ by $2(a - r)$ in this relation:

$$4a(a - r) = 4(c^2 - rc\cos\theta)$$

Hence

$$a(a - r) = c^2 - rc\cos\theta$$

Transpose all terms with r to the left and everything else to the right, and change signs:

$$r(a - c\cos\theta) = a^2 - c^2$$

Finally replace $a^2 - c^2$ by b^2.

Polar Equation of the Ellipse

The ellipse with foci at **0** and $[2c, 0]$ and with distance sum $2a$ has the equation

$$r(a - c\cos\theta) = b^2 = a^2 - c^2$$

Remark It is an interesting exercise to derive the rectangular equation of the ellipse from the polar equation. Of course you use both of the relations $r^2 = x^2 + y^2$ and $x = r\cos\theta$. You should end with the equation $(x - c)^2/a^2 + y^2/b^2 = 1$ because the origin is at a focus, not at the center of the ellipse.

Eccentricity

Define the **eccentricity** of the ellipse to be the number $e = c/a$. Since $c < a$, it follows that $0 < e < 1$. Also define the number p to be $p = b^2/2ae = b^2/2c$. In terms of this notation, the polar equation $r(a - c\cos\theta) = b^2$ of the ellipse becomes the following:

Polar Equation of the Ellipse (Eccentricity Form)

$$r(1 - e\cos\theta) = 2ep$$

The eccentricity determines the shape of the ellipse. If e is near zero, then c is small compared to a. That means the foci are close together relative to the semi-major axis; hence the ellipse is circle-like. If e is close

Figure 8-8-8
Ellipses of various eccentricities

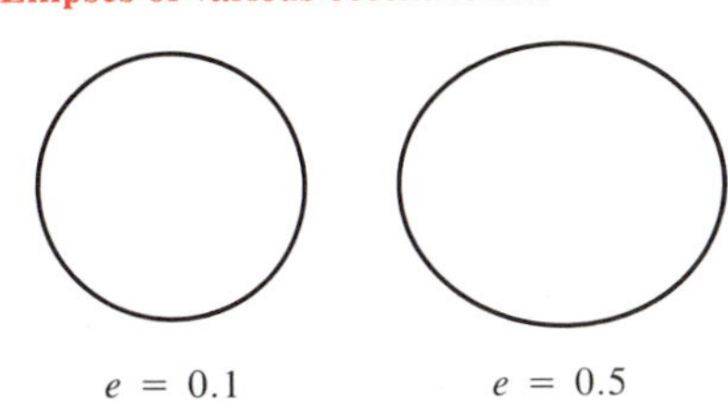

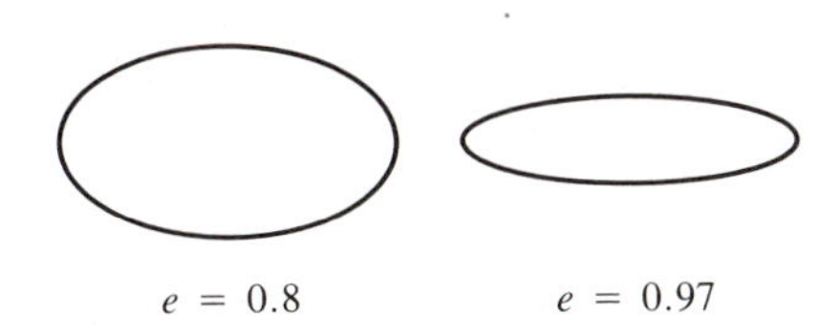

to 1, the foci are relatively far apart and the ellipse is long and thin (Figure 8-8-8). Once e is given, the scale factor p determines the size of the ellipse (as the radius does for a circle). Note the limiting case $e = 0$. Then $c = 0$ and $b = a$, so the equation becomes $r = b$, a circle with center at the origin.

Example 3

Find the equation of the ellipse having eccentricity $\frac{3}{4}$ and foci at $(-5, -1)$ and $(1, -1)$.

Solution The center is $(-2, -1)$ so the equation is of the form

$$\frac{(x+2)^2}{a^2} + \frac{(y+1)^2}{b^2} = 1$$

where $a > b$ because the major axis is horizontal.

The distance between the foci is $6 = 2c$; hence $c = 3$. By the definition of eccentricity, $c = ae$; hence $a = c/e = 3/(\frac{3}{4}) = 4$. Finally, $b^2 = a^2 - c^2 = 16 - 9 = 7$. Therefore, the equation of the ellipse is

$$\frac{(x+2)^2}{16} + \frac{(y+1)^2}{7} = 1$$

●

Example 4

An ellipse has semi-major axis a and eccentricity e. Find in terms of a and e the length of the chord through one of the foci and perpendicular to the major axis.

Solution Let L denote the required chord length. Choose the ellipse as in Figure 8-8-7, so that its polar equation is

$$r(1 - e\cos\theta) = 2pe$$

The chord in question meets the ellipse in two points. The upper point is $[r, \frac{1}{2}\pi]$, where, by the equation, $r = 2pe$, which is half the chord length L. Therefore $L = 4pe$, and our problem is to express L in terms of a and e. Now

$$c = ae \quad \text{and} \quad b^2 = a^2 - c^2 = a^2(1 - e^2)$$

By the definition of p, we have $2pe = b^2/a = a(1 - e^2)$. Therefore

$$L = 4pe = 2a(1 - e^2)$$

●

Polar Equation of the Hyperbola

Fix the polar coordinate system so the foci of a hyperbola are the origin and $[2c, \pi]$, and let the absolute distance difference be $2a$, so $c > a$. See Figure 8-8-9, on the next page. Let us insist that d and r are genuine distances, that is, greater than zero.

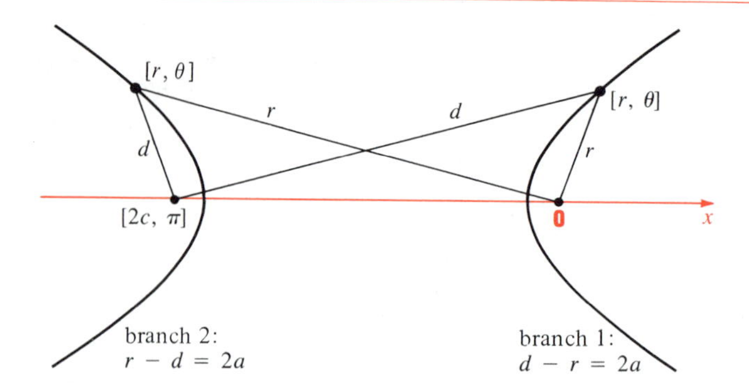

Figure 8-8-9
Set-up for the hyperbola in polar coordinates

Branch 2 is defined by $r - d = 2a$, that is, $d = r - 2a$, and branch 1 by $d - r = 2a$, that is $d = r + 2a$. Hence both branches are defined by $d = r \pm 2a$. Therefore

$$d - r = \pm 2a \quad \text{and} \quad d + r = r \pm 2a + r = 2(r \pm a)$$

By the distance formula (page 456)

$$d^2 = r^2 + 4c^2 - 4cr\cos(\pi - \theta) = r^2 + 4(c^2 + cr\cos\theta)$$

Therefore

$$(d - r)(d + r) = d^2 - r^2 = 4(c^2 + cr\cos\theta)$$

Replace $d - r$ by $\pm 2a$ and $d + r$ by $2(r \pm a)$ and simplify:

$$\pm 4a(r \pm a) = 4(c^2 + cr\cos\theta)$$
$$\pm ar + a^2 = c^2 + cr\cos\theta$$

Now rearrange terms:

$$\pm ar - cr\cos\theta = c^2 - a^2 \qquad \text{that is} \qquad r(\pm a - c\cos\theta) = b^2$$

Thus the polar equations of branch 1 and branch 2 respectively are

$$r(a - c\cos\theta) = b^2 \quad \text{and} \quad r(a + c\cos\theta) = -b^2$$

We have insisted that $r > 0$, so there are restrictions on θ. Since $c > a > 0$, we must have:

Branch 1: $\arccos(a/c) < \theta < 2\pi - \arccos(a/c)$

Branch 2: $\pi - \arccos(a/c) < \theta < \pi + \arccos(a/c)$

See Figure 8-8-10.

The **eccentricity** of the hyperbola is defined to be $e = c/a$. Clearly $e > 1$. We also define p by $p = b^2/2ae = b^2/2c$. Then the polar equation can be expressed as follows:

Figure 8-8-10
Restrictions on θ in the polar equation of the hyperbola

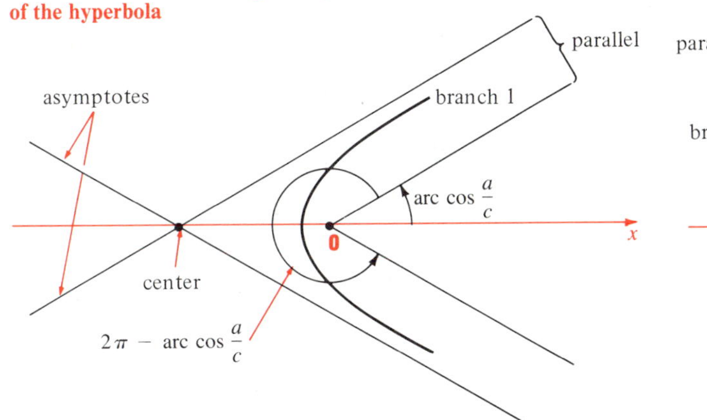

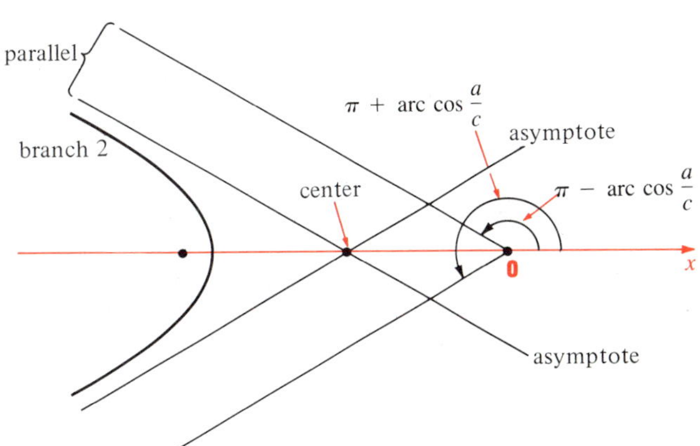

Polar Equation of the Hyperbola

The hyperbola with foci **0** and $[2c, \pi]$ and absolute distance difference $2a$ has the equation

$$r(\pm 1 - e\cos\theta) = 2ep$$

Its eccentricity determines the shape of the hyperbola. Suppose

$$\frac{x^2}{a^2} - \frac{y^2}{b^2} = 1$$

is a hyperbola in standard position with foci at $(\pm c, 0)$. Its asymptotes have slopes

$$\pm\frac{b}{a} = \pm\frac{\sqrt{c^2 - a^2}}{a} = \pm\frac{\sqrt{a^2e^2 - a^2}}{a} = \pm\sqrt{e^2 - 1}$$

Hence if e is near 1, the asymptotes have small slopes and the hyperbola is squeezed into a narrow angle. The larger the eccentricity, the broader the hyperbola (Figure 8-8-11, next page).

Uniform Definition of Conics

Surprisingly enough, a slight change in the focus-directrix definition of the parabola results in one definition for all of the conics. Let D be a fixed line and P a fixed point at distance $2p$ from D. See Figure 8-8-12, next page. Fix $e > 0$. The set of all points X such that $\overline{PX}$ equals e times the distance from X to D defines a conic.

To prove this, choose P at the origin and let D be the line $x = -2p$. Assuming $X = (x, y) = [r, \theta]$ lies to the right of D, the defining relation is

$$r = e(x + 2p) = e(r\cos\theta + 2p)$$

Figure 8-8-11
Eccentricities near 1 and far from 1; note the relative positions of the foci.

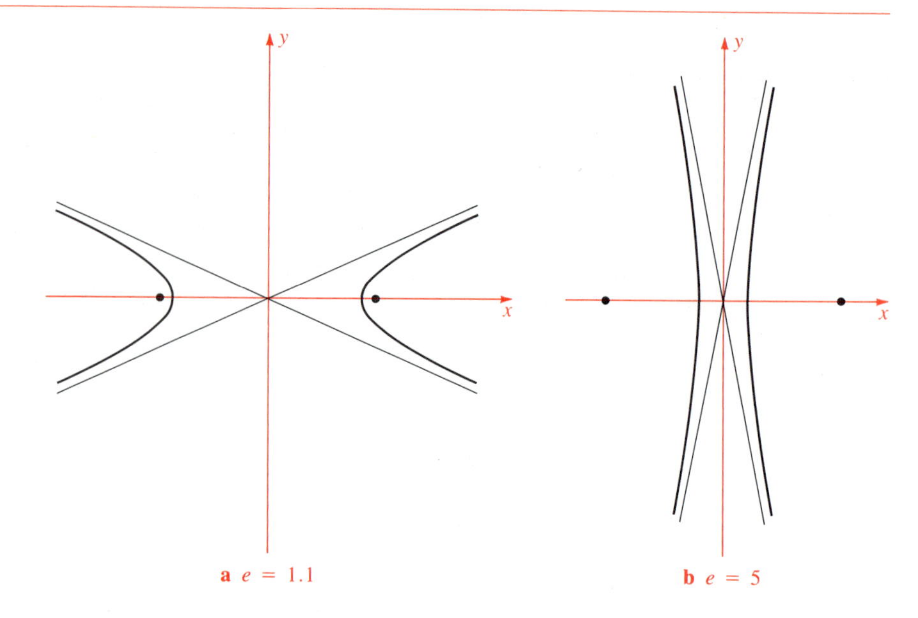

Figure 8-8-12
The general conic: $\overline{PX} = e\overline{XD}$

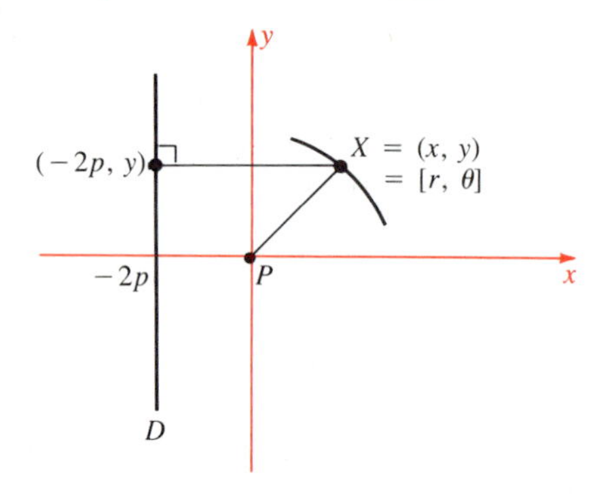

Therefore

$$r(1 - e\cos\theta) = 2pe \qquad (*)$$

If $0 < e < 1$, we recognize the relation $(*)$ as the polar equation of an ellipse with eccentricity e and one focus at the origin, the other at $(2c, 0)$, where the relations

$$e = c/a \qquad 2pe = b^2/a \qquad \text{and} \qquad c^2 = a^2 - b^2$$

define a, b, and c.

If $e > 1$, we recognize the relation $(*)$ as the equation of the right branch of a hyperbola, as in Figure 8-8-9. The left branch comes from the relation $r = e(-x - 2p) = -e(r\cos\theta + 2p)$, where X is to the *left* of D. That is,

$$r(-1 - e\cos\theta) = 2pe$$

If $e = 1$, then the relation $(*)$ is the equation of the parabola with focus P and directrix D. (Prove it.)

In general, D is called the **directrix** of the conic and P is one of its **foci.** For an ellipse or a hyperbola, there is a second directrix corresponding to the other focus. Note that $2p$ is the distance from P to its associated directrix D—a geometric interpretation of p that we lacked until now.

Exercises

Graph

1 $r = 2\sin 2\theta$

2 the **rose** $r = \sin 5\theta$

3 $r = \cos 3\theta$

4 $r = -\cos 4\theta$

5 $r = \theta^2$

6 the **lemniscate** $r^2 = \cos 2\theta$

7 the **cissoid** $r = \sec\theta - \cos\theta = \sin\theta\tan\theta$

8 the **strophoid** $r = \cos 2\theta \sec \theta$
[Hint for 7 and 8 Use $x = r\cos\theta$ to find the vertical asymptote.]

9 the **cardioid** $r = 1 - \cos\theta$

0 the **limaçon** $r = 2 + \cos\theta$

1 the **limaçon** $r = 1 + 2\cos\theta$

2 the **bifolium** $r = \sin\theta\cos^2\theta$

3 the **conchoid** $r = \csc\theta - 2$

4 Graph $r = a + b\cos\theta$ $(a > 0,\ b > 0,\ a \le 2b)$ in general. [Hint Use Exercises 10 and 11.]

5 If an ellipse has eccentricity e and we write $e = \cos\alpha$ where $0 < \alpha < 90°$, then we call the ellipse an "α degree ellipse." Interpret α geometrically and express b/a in terms of α. (Templates for drawing ellipses go by degree.)

6 (cont.) Draw ellipses of $15°$, $30°$, $45°$, and $60°$.

17 The orbit of the earth is approximately an ellipse with the sun at one focus and semi-major and semi-minor axes 9.3×10^7 and 9.1×10^7 miles. Compute the eccentricity of the orbit.

18 (cont.) Find the distance from the sun to the other focus of the ellipse.

19 What is the eccentricity of a rectangular hyperbola?

20 Show that the distances of any point (x, y) of the ellipse $x^2/a^2 + y^2/b^2 = 1$, where $a > b$, to its foci are $a \pm ex$.

21 Let P be a focus of an ellipse and XY a chord through P. Show that $1/\overline{XP} + 1/\overline{YP} = 1/ep$.

22 (cont.) Find a corresponding result for hyperbolas.

23 Find the polar equation of the parabola whose focus is at the origin and whose directrix is $x = -2p$.

24 (cont.) Find a result like that of Exercise 21 for parabolas.

8-9 Rotation of Axes

Suppose we start with a polar coordinate system and create a new system by rotating the polar axis counterclockwise through an angle α. See Figure 8-9-1. A point with coordinates $[r, \theta]$ acquires new coordinates $[\bar{r}, \bar{\theta}]$. From the figure it is clear that $\bar{r} = r$ and $\bar{\theta} = \theta - \alpha$

Figure 8-9-1
Rotation of polar coordinates

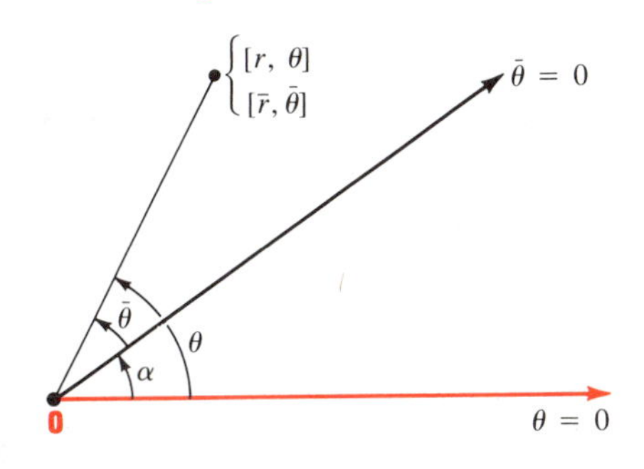

Rotation of Axes (Polar Coordinates)

If the polar axis is rotated by an angle α, a point $[r, \theta]$ acquires the new coordinates $[\bar{r}, \bar{\theta}]$, where

$$\bar{r} = r \quad \text{and} \quad \bar{\theta} = \theta - \alpha$$

As an application, let us find the polar equation of the line L that is p units from the origin and perpendicular to the ray $\theta = \alpha$. See Figure 8-9-2. Relative to the tilted axis, the line has equation $\bar{r}\cos\bar{\theta} = p$. Its r, θ-equation therefore is $r\cos(\theta - \alpha) = p$. This is a quick derivation of the equation given on page 454.

By the same reasoning the relation $r = f(\theta - \alpha)$ represents the curve $r = f(\theta)$ rotated counterclockwise through angle α. For example, knowing that $r = 2a\cos\theta$ represents the circle of radius a and center $[a, 0]$, we can instantly write down the equation of the circle of radius a and center $[a, \frac{1}{2}\pi]$. The equation is

$$r = 2a\cos(\theta - \tfrac{1}{2}\pi) = 2a\cos(\tfrac{1}{2}\pi - \theta) = 2a\sin\theta$$

Figure 8-9-2
Polar equation of line L:
$\bar{r}\cos\bar{\theta} = p$ or $r\cos(\theta - \alpha) = p$

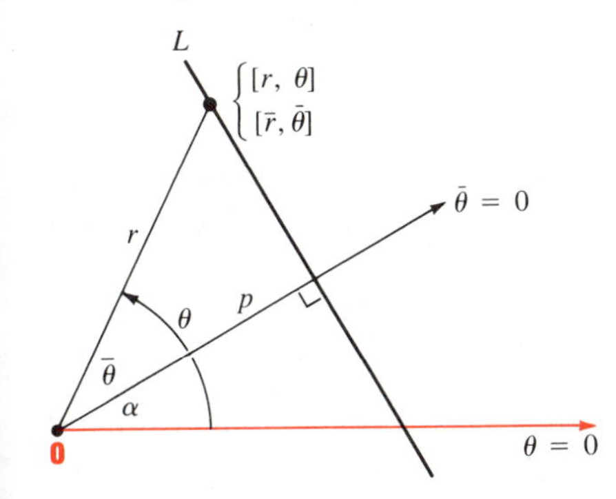

Now let us look at the effect of rotation of axes on rectangular coordinates. Suppose we rotate the x- and y-axes through an angle α, obtaining new axes that we call the $\bar{x}$- and $\bar{y}$-axes (Figure 8-9-3, next page). They define a new rectangular coordinate system.

A point in the plane with coordinates (x, y) now acquires new coordinates $(\bar{x}, \bar{y})$. From Figure 8-9-3 it is not obvious what relation exists

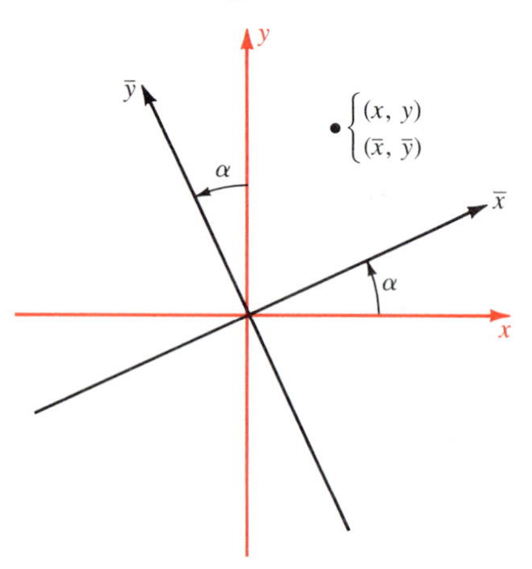

Figure 8-9-3
Rotation of rectangular coordinates

between x, y and $\bar{x}, \bar{y}$. Still, knowing α, we should be able to express $\bar{x}, \bar{y}$ in terms of x, y and vice versa. The trick is to pass through polar coordinates, for which the rotation rule is so easy: $\bar{r} = r$, $\bar{\theta} = \theta - \alpha$. Indeed,

$$\begin{aligned} \bar{x} = \bar{r}\cos\bar{\theta} &= r\cos(\theta - \alpha) = r\cos\theta\cos\alpha + r\sin\theta\sin\alpha \\ &= x\cos\alpha + y\sin\alpha \\ \bar{y} = \bar{r}\sin\bar{\theta} &= r\sin(\theta - \alpha) = -r\cos\theta\sin\alpha + r\sin\theta\cos\alpha \\ &= -x\sin\alpha + y\cos\alpha \end{aligned}$$

This gives expressions for $\bar{x}$ and $\bar{y}$ in terms of x and y. To obtain x and y in terms of $\bar{x}$ and $\bar{y}$ we can solve the pair of linear equations for x and y. Another method is to argue that the relation of (x, y) to $(\bar{x}, \bar{y})$ is the same as the relation of $(\bar{x}, \bar{y})$ to (x, y) except that α is replaced by $-\alpha$ since $\theta = \bar{\theta} + \alpha = \bar{\theta} - (-\alpha)$. Thus

$$\begin{aligned} x &= \bar{x}\cos(-\alpha) + \bar{y}\sin(-\alpha) = \bar{x}\cos\alpha - \bar{y}\sin\alpha \\ y &= -\bar{x}\sin(-\alpha) + \bar{y}\cos(-\alpha) = \bar{x}\sin\alpha + \bar{y}\cos\alpha \end{aligned}$$

Rotation of Axes (Rectangular Coordinates)

Suppose the plane is rotated through angle α so that the x- and y-axes, under this rotation, become the $\bar{x}$- and $\bar{y}$-axes. Then the x, y-coordinates and $\bar{x}, \bar{y}$-coordinates of any point are related by

$$\begin{cases} x = \bar{x}\cos\alpha - \bar{y}\sin\alpha \\ y = \bar{x}\sin\alpha + \bar{y}\cos\alpha \end{cases} \qquad \begin{cases} \bar{x} = x\cos\alpha + y\sin\alpha \\ \bar{y} = -x\sin\alpha + y\cos\alpha \end{cases}$$

For example, if $\alpha = 45°$, then

$$x = \tfrac{1}{2}\sqrt{2}\,(\bar{x} - \bar{y}) \qquad y = \tfrac{1}{2}\sqrt{2}\,(\bar{x} + \bar{y})$$

For another example, if $\alpha = -30°$, then we have $\cos\alpha = \frac{1}{2}$ and $\sin\alpha = -\frac{1}{2}\sqrt{3}$. Hence

$$x = \tfrac{1}{2}(\bar{x} + \sqrt{3}\,\bar{y}) \qquad y = \tfrac{1}{2}(-\sqrt{3}\,\bar{x} + \bar{y})$$

Conics

We have learned how to graph quadratic equations of the form

$$ax^2 + cy^2 + dx + ey + f = 0 \qquad (a^2 + c^2 > 0)$$

By completing squares, we generally obtain one of the conic sections (sometimes a degenerate conic, or no locus at all). Now we tackle the most general quadratic equation

$$ax^2 + bxy + cy^2 + dx + ey + f = 0$$

It is the term bxy that makes life difficult. Where does it come from and how can we get rid of it? We can learn a good deal from an experiment.

Figure 8-9-4
Problem: To express $\frac{1}{9}x^2 + \frac{1}{4}y^2 = 1$ as a relation between $\bar{x}$ and $\bar{y}$

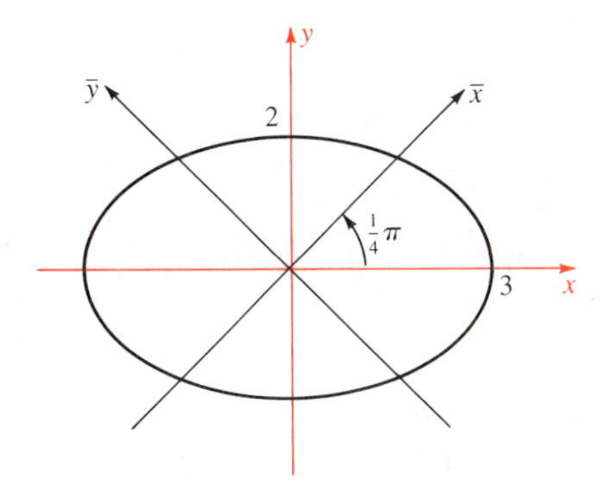

Figure 8-9-5
$\frac{13}{72}\bar{x}^2 + \frac{10}{72}\bar{x}\bar{y} + \frac{13}{72}\bar{y}^2 = 1$

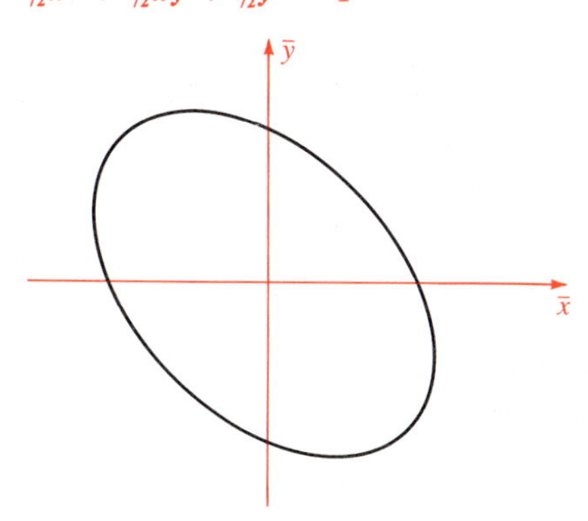

Example 1

Find the equation of the ellipse $\frac{1}{9}x^2 + \frac{1}{4}y^2 = 1$ in the $\bar{x}, \bar{y}$-coordinate system that results from a $\frac{1}{4}\pi$ rotation of the x, y-coordinate system (Figure 8-9-4).

Solution The rotation formulas are

$$x = \tfrac{1}{2}\sqrt{2}\,(\bar{x} - \bar{y}) \qquad y = \tfrac{1}{2}\sqrt{2}\,(\bar{x} + \bar{y})$$

Substitute:

$$\frac{x^2}{9} + \frac{y^2}{4} = \frac{1}{2}\left[\frac{(\bar{x} - \bar{y})^2}{9} + \frac{(\bar{x} + \bar{y})^2}{4}\right]$$
$$= \tfrac{1}{2}\left[\tfrac{13}{36}\bar{x}^2 + \tfrac{10}{36}\bar{x}\bar{y} + \tfrac{13}{36}\bar{y}^2\right]$$

Therefore the $\bar{x}, \bar{y}$-equation of the ellipse (Figure 8-9-5) is

$$\tfrac{13}{72}\bar{x}^2 + \tfrac{10}{72}\bar{x}\bar{y} + \tfrac{13}{72}\bar{y}^2 = 1$$ ●

The experiment suggests that the $\bar{x}\bar{y}$ term is due to the tilt of the coordinate axes relative to the axes of the ellipse. If so, then the same should be true for hyperbolas and parabolas. Example 1 suggests that an xy term occurs when the axes are "incorrectly" placed. Perhaps it can be eliminated by rotating the axes through a cleverly chosen angle. Let us look for a suitable angle. Now a rotation of coordinates,

$$\begin{cases} x = \bar{x}\cos\alpha - \bar{y}\sin\alpha \\ y = \bar{x}\sin\alpha + \bar{y}\cos\alpha \end{cases}$$

changes a linear polynomial $dx + ey + f$ in x and y into a linear polynomial in $\bar{x}$ and $\bar{y}$. Of more interest to us is what it does to the quadratic polynomial $ax^2 + bxy + cy^2$. Substitute:

$$\begin{aligned} ax^2 + bxy + cy^2 = {} & a(\bar{x}\cos\alpha - \bar{y}\sin\alpha)^2 \\ & + b(\bar{x}\cos\alpha - \bar{y}\sin\alpha)(\bar{x}\sin\alpha + \bar{y}\cos\alpha) \\ & + c(\bar{x}\sin\alpha + \bar{y}\cos\alpha)^2 \end{aligned}$$

Multiply out and collect terms in $\bar{x}^2$, $\bar{x}\bar{y}$, and $\bar{y}^2$. Result:

Effect of Rotation on a Quadratic Polynomial

Under a rotation through an angle α, the quadratic polynomial

$$ax^2 + bxy + cy^2 + dx + ey + f$$

is changed to

$$\bar{a}\bar{x}^2 + \bar{b}\bar{x}\bar{y} + \bar{c}\bar{y}^2 + \bar{d}\bar{x} + \bar{e}\bar{y} + \bar{f}$$

$$\text{where} \quad \begin{cases} \bar{a} = a\cos^2\alpha + b\cos\alpha\sin\alpha + c\sin^2\alpha \\ \bar{b} = 2(c - a)\sin\alpha\cos\alpha + b(\cos^2\alpha - \sin^2\alpha) \\ \bar{c} = a\sin^2\alpha - b\sin\alpha\cos\alpha + c\cos^2\alpha \end{cases}$$ ●

We are most concerned with the formula for $\bar{b}$, which we can write as

$$\bar{b} = (c - a)\sin 2\alpha + b\cos 2a$$

It is always possible to choose the rotation angle α so that $\bar{b} = 0$, that is, so that $(c - a)\sin 2a + b\cos 2a = 0$. For if $c = a$, we take $\alpha = \pm\frac{1}{4}\pi$; if $c \neq a$, we choose α so $\tan 2\alpha = b/(a - c)$.

Rotation to Eliminate the *xy*-Term

A quadratic relation

$$ax^2 + bxy + cy^2 + dx + ey + f = 0$$

is changed into a quadratic relation

$$\bar{a}\bar{x}^2 + \bar{c}\bar{y}^2 + \bar{d}\bar{x} + \bar{e}\bar{y} + \bar{f} = 0$$

without an $\bar{x}\bar{y}$ term by rotating the axes through angle α, where

$$\tan 2\alpha = \frac{b}{a - c} \quad \text{if} \quad a \neq c; \qquad \alpha = \pm\tfrac{1}{4}\pi \quad \text{if} \quad a = c$$

Because the tangent has period π, the angle 2α is determined up to a multiple of π. Hence α is determined only up to a multiple of $\frac{1}{2}\pi$. Therefore we can always choose α in the first quadrant.

In numerical examples, we must compute $\bar{a}$ and $\bar{c}$ from a, b, c, and $\tan 2\alpha$. We write the formulas for $\bar{a}$ and $\bar{c}$ in the form

$$\begin{cases} \bar{a} = a\cos^2\alpha + \frac{1}{2}b\sin 2\alpha + c\sin^2\alpha \\ \bar{c} = a\sin^2\alpha - \frac{1}{2}b\sin 2\alpha + c\cos^2\alpha \end{cases}$$

From $\tan 2\alpha$ we can find $\sin 2\alpha$ and $\cos 2\alpha$:

$$\sin 2\alpha = \frac{\pm\tan 2\alpha}{\sqrt{1 + \tan^2 2\alpha}} \qquad \cos 2\alpha = \frac{\pm 1}{\sqrt{1 + \tan^2 2\alpha}}$$

From $\cos 2\alpha$ we can find $\cos^2\alpha$ and $\sin^2\alpha$:

$$\cos^2\alpha = \tfrac{1}{2}(1 + \cos 2\alpha) \qquad \sin^2\alpha = \tfrac{1}{2}(1 - \cos 2\alpha)$$

Everything ties together neatly.

Example 2

Describe the conic $xy = 1$.

Solution In this case $a = c = 0$ and $b = 1$. Therefore we choose $\alpha = \frac{1}{4}\pi$ to make $\bar{b} = 0$. The rotation is

$$x = \tfrac{1}{2}\sqrt{2}\,(\bar{x} - \bar{y}) \qquad y = \tfrac{1}{2}\sqrt{2}\,(\bar{x} + \bar{y})$$

so by direct computation,

$$xy = \tfrac{1}{2}(\bar{x} - \bar{y})(\bar{x} + \bar{y}) = \tfrac{1}{2}(\bar{x}^2 - \bar{y}^2)$$

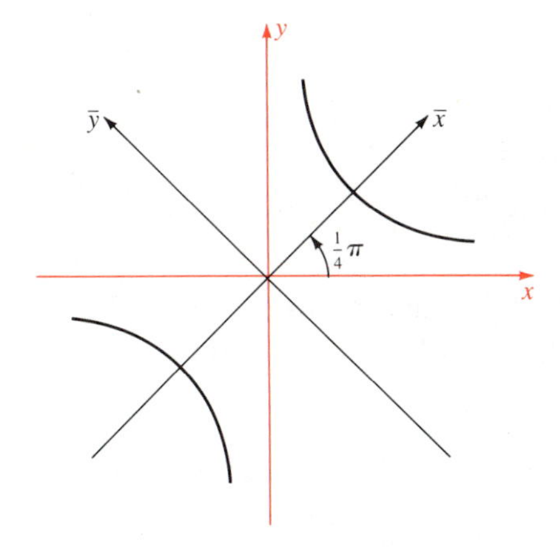

Figure 8-9-6
Before rotation: $xy = 1$
After rotation: $\frac{1}{2}\bar{x}^2 - \frac{1}{2}\bar{y}^2 = 1$

Therefore in the new coordinate system the equation of the conic is

$$\tfrac{1}{2}(\bar{x}^2 - \bar{y}^2) = 1$$

We recognize this as a rectangular hyperbola in standard position (Figure 8-9-6). ●

Example 3

Describe the conic $x^2 - 2xy + 3y^2 = 1$.

Solution Rotate the axes through angle α where

$$\tan 2\alpha = \frac{b}{a-c} = \frac{-2}{1-3} = 1 \quad \text{so} \quad 2\alpha = \tfrac{1}{4}\pi \quad \alpha = \tfrac{1}{8}\pi$$

Hence $\sin 2\alpha = \cos 2\alpha = \frac{1}{2}\sqrt{2}$ and

$$\cos^2\alpha = \tfrac{1}{2}(1 + \tfrac{1}{2}\sqrt{2}) \qquad \sin^2\alpha = \tfrac{1}{2}(1 - \tfrac{1}{2}\sqrt{2})$$

Substitute these values with $a = 1$, $b = -2$, $c = 3$ into the formulas for $\bar{a}$ and $\bar{c}$:

$$\begin{cases} \bar{a} = \frac{1}{2}(1 + \frac{1}{2}\sqrt{2}) - \frac{1}{2}\sqrt{2} + \frac{3}{2}(1 - \frac{1}{2}\sqrt{2}) = 2 - \sqrt{2} \\ \bar{c} = \frac{1}{2}(1 - \frac{1}{2}\sqrt{2}) + \frac{1}{2}\sqrt{2} + \frac{3}{2}(1 + \frac{1}{2}\sqrt{2}) = 2 + \sqrt{2} \end{cases}$$

Therefore, in the $\bar{x}, \bar{y}$-coordinate system, the equation is

$$(2 - \sqrt{2})\bar{x}^2 + (2 + \sqrt{2})\bar{y}^2 = 1$$

Because $2 - \sqrt{2}$ and $2 + \sqrt{2}$ are both positive, this is an ellipse in standard form:

$$\frac{\bar{x}^2}{A^2} + \frac{\bar{y}^2}{B^2} = 1$$

where

$$A^2 = \frac{1}{2 - \sqrt{2}} = \tfrac{1}{2}(2 + \sqrt{2})$$

$$B^2 = \frac{1}{2 + \sqrt{2}} = \tfrac{1}{2}(2 - \sqrt{2})$$

See Figure 8-9-7. ●

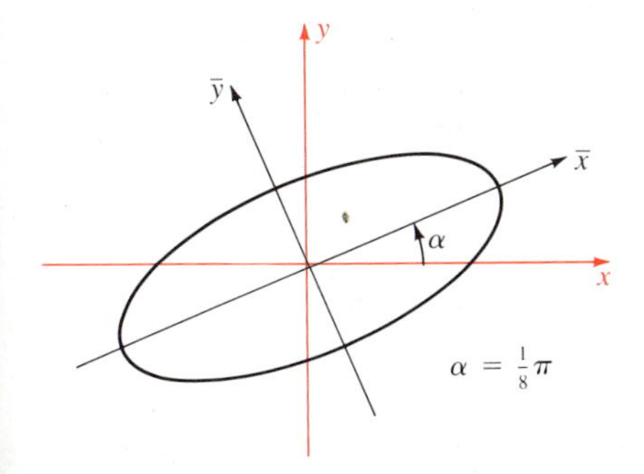

Figure 8-9-7
Before rotation: $x^2 - 2xy + 3y^2 = 1$
After rotation:
$$\frac{\bar{x}^2}{\frac{1}{2}(2 + \sqrt{2})} + \frac{\bar{y}^2}{\frac{1}{2}(2 - \sqrt{2})} = 1$$

The Discriminant Test

Given the most general quadratic relation

$$ax^2 + bxy + cy^2 + dx + ey + f = 0$$

is there any way to tell at a glance what type of conic its graph is? The answer is almost, but not quite, yes. If the relation can be reduced by a preliminary translation to the form

$$ax^2 + bxy + cy^2 = 1$$

then one can describe the type of graph by inspection! We define the **discriminant** by

$$\Delta = 4ac - b^2$$

If $\Delta > 0$, then the graph is an ellipse. If $\Delta < 0$, then the graph is a hyperbola. If $\Delta = 0$, then the graph is a pair of straight lines (possibly coincident) or the empty set in the silly case $a = b = c = 0$. We shall omit the proof; however, see Exercise 22.

Note Few people can remember the formulas of this section. What is important is to understand what they do and how to apply them.

Exercises

1 Solve the system of linear equations

$$x = \bar{x}\cos\alpha - \bar{y}\sin\alpha \qquad y = \bar{x}\sin\alpha + \bar{y}\cos\alpha$$

for $\bar{x}$ and $\bar{y}$. Explain your answer.

2 Let $x = \bar{x}\cos\alpha - \bar{y}\sin\alpha$ and $y = \bar{x}\sin\alpha + \bar{y}\cos\alpha$. Compute $x^2 + y^2$. Explain your answer.

3 Let (x_1, y_1) and (x_2, y_2) be two points in the x, y-coordinate system. Let $(\bar{x}_1, \bar{y}_1)$ and $(\bar{x}_2, \bar{y}_2)$ be their coordinates in the $\bar{x}, \bar{y}$-coordinate system obtained by a rotation. Compute $x_1x_2 + y_1y_2$ in terms of $\bar{x}_1$, $\bar{x}_2$, $\bar{y}_1$, $\bar{y}_2$, and α, the angle of rotation.

4 Follow a rotation through angle α by a rotation through angle β. The result is obviously a rotation through angle $\alpha + \beta$. Use this observation and the rotation formulas to verify the addition laws for sine and cosine.

5 Let $(x, y) = (\cos\theta, \sin\theta)$ be a point on the unit circle. Rotate the axes by an angle α and show geometrically that $(\bar{x}, \bar{y}) = (\cos(\theta - \alpha), \sin(\theta - \alpha))$.

6 (cont.) Combine this result with the rotation formulas to get a new verification of the addition laws for the sine and cosine.

Make a suitable rotation and write the $\bar{x}, \bar{y}$-equation without an xy term

7 $x^2 - xy = 1$

8 $xy - y^2 = 1$

9 $xy + y^2 = 1$

10 $2xy + y^2 = 1$

Determine the type of the conic and the directions of its principal axes

11 $x^2 + xy + y^2 = 1$

12 $x^2 - xy + y^2 = 1$

13 $x^2 + xy - y^2 = 1$

14 $x^2 - xy - y^2 = 1$

15 $x^2 + xy + 2y^2 = 1$

16 $x^2 - xy + 2y^2 = 1$

17 $x^2 - 2xy + y^2 = 2y$

18 $x^2 - 4xy + 4y^2 = x$

19 $2x^2 - 6xy + y^2 = 1$

20 $x^2 + 3xy - y^2 = 1$

Suppose a rotation converts $ax^2 + bxy + cy^2$ into $\bar{a}\bar{x}^2 + \bar{b}\bar{x}\bar{y} + \bar{c}\bar{y}^2$. Prove

21 $a + c = \bar{a} + \bar{c}$

***22** $4ac - b^2 = 4\bar{a}\bar{c} - \bar{b}^2$

8-10 Review Exercises

Identify the curve, graph it, and find its main features—foci, asymptotes, and so on

1 $x^2 + y^2 - 8x + 12y = -43$

2 $y^2 + 4y - 2x = 0$

3 $x^2 + 25y^2 - 6x + 50y = 66$

4 $x^2 - 25y^2 - 6x - 50y = 116$

Identify the graph and find its polar equation

5 $x - 2y = 3$

6 $x^2 + (y + 1)^2 = 1$

7 Fix A in the plane. Find the curve traced by the midpoint of AP, where P traces the parabola $y = x^2$.

8 Find the set of centers of all circles that are simultaneously tangent (not at the origin) to the x-axis and to the circle $x^2 + y^2 = 2y$.

Graph

9 The **hyperbolic spiral** $r\theta = 1 \quad (\theta > 0)$

10 The **lituus** $r^2\theta = 1$

11 The **logarithmic spiral** $\ln r = \theta$

12 The **parabolic spiral** $(r - 1)^2 = 4\theta$

Figure 8-10-1
See Exercise 13

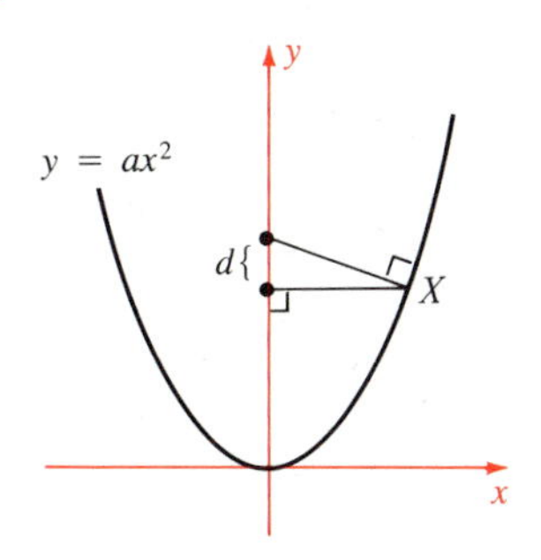

13 In Figure 8-10-1, show that d is independent of X.

14 Find the equation of the rectangular hyperbola with foci (c, c) and $(-c, -c)$.

15 Show for the ellipse in Figure 8-10-2 that uv/w^2 is constant. Here u and v are measured along the major axis.

16 (cont.) Give a similar result for the hyperbola.

Figure 8-10-2
See Exercise 15

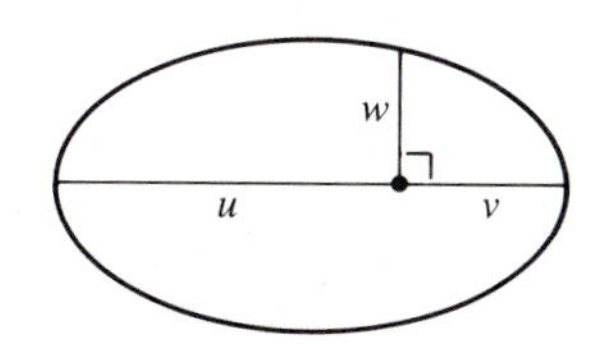

17 Find the polar equation for the set of all points the product of whose distances from $(-a, 0)$ and $(a, 0)$ is a^2.

18 Two firms are $2c$ km apart, and they both sell an item at the same price. However, the shipping cost per km (as the crow flies) for one firm is k times that for the other, where $k > 1$. Find the curve of equal cost.

Here are two drawing board and string constructions for curves. The string ends are attached at P and A (and the pencil point really touches the rule). Find the curves.

19

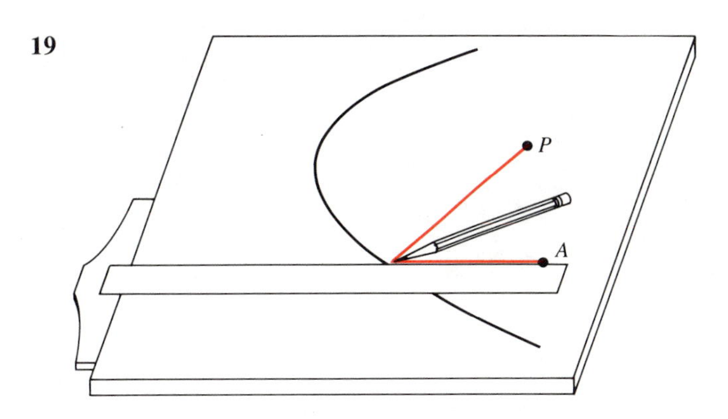

20

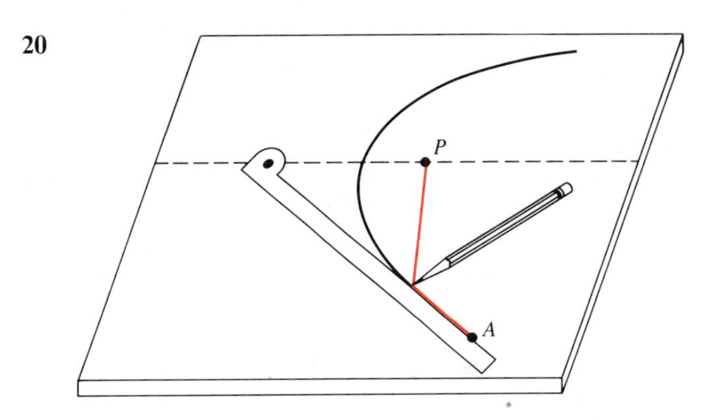

9 Numerical Calculus

This chapter is an introduction to a vast subject called numerical analysis. The main topics of the chapter are approximations of functions by polynomials, approximate integration, and approximations to zeros of functions.

An approximation is useful only if we are able to measure how accurately it approximates, that is, to estimate its error. Error estimation requires a deeper study of Rolle's theorem and the mean value theorem than we made in Chapter 3. This is where we begin our work. The first application will be an important method of finding limits, called l'Hospital's rule.

9-1 Mean Value Theorem

In Chapter 3, we used Rolle's theorem and the mean value theorem to establish some basic properties of differentiable functions. See page 173. Let us recall the statement of Rolle's theorem.

Rolle's Theorem

Let $f(x)$ be a continuous function on the closed interval $[a, b]$ and suppose $f'(x)$ exists on the open subinterval (a, b). Assume $f(a) = f(b)$. Then $f'(c) = 0$ for some c between a and b. ●

Sometimes we won't bother with the fussy "continuous on $[a, b]$ and differentiable on the open interval (a, b)," but merely say "differentiable on $[a, b]$," which assumes slightly more. You should keep in mind the example $f(x) = \sqrt{1 - x^2}$ on $[-1, 1]$ to remember the difference between being differentiable on the interval *including* its end points and just being continuous.

Example 1

Apply Rolle's theorem to $f(x) = e^{kx}\cos x$ on the closed interval $[-\frac{1}{2}\pi, \frac{1}{2}\pi]$. What is your conclusion?

Solution The derivative is

$$f'(x) = ke^{kx}\cos x - e^{kx}\sin x$$

Since $f(-\frac{1}{2}\pi) = 0$ and $f(\frac{1}{2}\pi) = 0$, the hypotheses of Rolle's theorem are satisfied, so there is a number c between $-\frac{1}{2}\pi$ and $\frac{1}{2}\pi$ such that $f'(c) = 0$. Since $e^{kc} \neq 0$, we have

$$k\cos c - \sin c = 0 \qquad \text{that is} \qquad \tan c = k$$

since $\cos c \neq 0$. But k can be any real number, so we have proved that the range of $\tan x$ is $\mathbf{R}$. ●

A number of interesting results are obtained by applying Rolle's theorem to cleverly chosen auxiliary functions. The mean value theorem itself is one case in point. Let us recall the mean value theorem of differential calculus. (As we shall see, there is also a mean value theorem of integral calculus.)

Figure 9-1-1
The MVT is a skew, or sheared, version of Rolle's theorem.

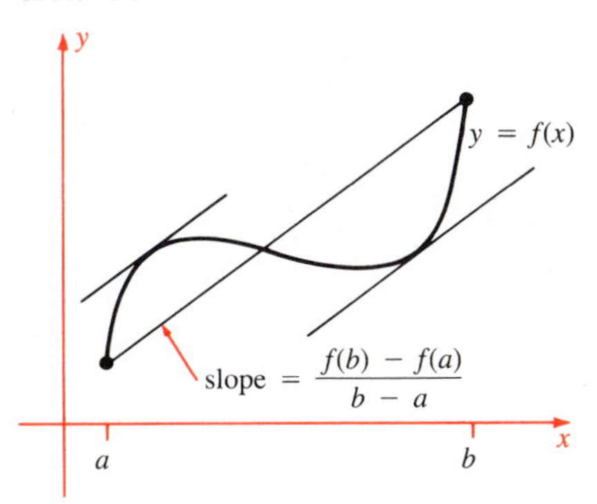

a Mean value theorem: Somewhere the tangent is parallel to the chord.

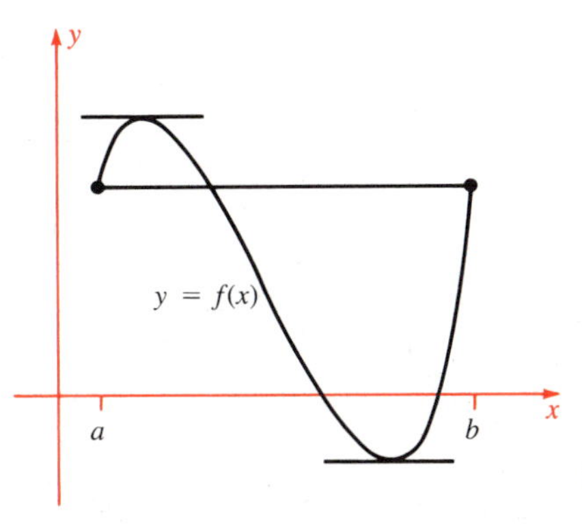

b Rolle's theorem: If $f(a) = f(b)$, then somewhere the tangent is horizontal.

Mean Value Theorem (MVT)

Let $f(x)$ be a continuous function on the closed interval $[a, b]$, and suppose $f'(x)$ exists on the open interval (a, b). Then

$$\frac{f(b) - f(a)}{b - a} = f'(c)$$

for some c between a and b. ●

Note carefully where the MVT differs from Rolle's theorem. In the latter, we add the hypothesis $f(a) = f(b)$ and draw the conclusion that $f'(c) = 0$ for some c between a and b. Geometrically, the mean value theorem says that somewhere between $x = a$ and $x = b$, the tangent is parallel to the chord (Figure 9-1-1a). Rolle's theorem is the special case in which the chord is horizontal (Figure 9-1-1b). You might say that the MVT is a skew, or sheared, version of Rolle's theorem.

In most applications, the MVT is expressed in the form

$$f(b) - f(a) = f'(c)(b - a) \qquad (a < c < b)$$

and is used to estimate the difference between the values of $f(x)$ at two points.

Example 2

a Show that $|\sin b - \sin a| \leq |b - a|$ for any a and b. To prove this, set $f(x) = \sin x$. Then $|f'(x)| = |\cos x| \leq 1$, so

$$|\sin b - \sin a| = |\cos c| \cdot |b - a| \leq |b - a|$$

b Show that $|\sqrt{102} - \sqrt{101}| < 0.05$. To prove this, set $f(x) = \sqrt{x}$. Then

$$|\sqrt{102} - \sqrt{101}| = |f'(c)||102 - 101|$$
$$= \frac{1}{2\sqrt{c}} < \frac{1}{2\sqrt{100}} = 0.05$$

●

The MVT compares the change in $f(x)$ with the change in the function $g(x) = x$:

$$\frac{f(b) - f(a)}{b - a} = \frac{f(b) - f(a)}{g(b) - g(a)} = f'(c)$$

The following generalized MVT compares the change in $f(x)$ to the change in another function $g(x)$. The only restriction is that the derivative $g'(x)$ must be non-zero.

Generalized Mean Value Theorem

Let $f(x)$ and $g(x)$ be differentiable functions on $[a, b]$ and suppose $g'(x) \neq 0$ on (a, b). Then

$$\frac{f(b) - f(a)}{g(b) - g(a)} = \frac{f'(c)}{g'(c)}$$

for some c between a and b.

●

Proof Set

$$h(x) = [g(b) - g(a)][f(x) - f(a)] - [f(b) - f(a)][g(x) - g(a)]$$

Then $h(x)$ is differentiable and $h(a) = h(b) = 0$. By Rolle's theorem, $h'(c) = 0$ for some c such that $a < c < b$. But

$$h'(x) = [g(b) - g(a)]f'(x) - [f(b) - f(a)]g'(x)$$

Therefore

$$[g(b) - g(a)]f'(c) = [f(b) - f(a)]g'(c)$$

Since $g'(c) \neq 0$, it follows that $g(b) - g(a) \neq 0$ by Rolle's theorem. Hence we can divide to obtain the desired formula. [The only reason we assume $g'(x) \neq 0$ is so we can divide out.] ●

Example 3

Show that the generalized MVT applies to $f(x) = x^2$ and $g(x) = x^4$ on the interval $[1, 2]$, and find a number c that satisfies the conclusion.

Solution We have

$$f'(x) = 2x \quad \text{and} \quad g'(x) = 4x^3$$

Clearly $g'(x) \neq 0$ on $[1, 2]$. Therefore the generalized MVT applies and says there is a number c such that $1 < c < 2$ and

$$\frac{f(2) - f(1)}{g(2) - g(1)} = \frac{f'(c)}{g'(c)} \quad \text{that is} \quad \frac{4 - 1}{16 - 1} = \frac{2c}{4c^3}$$

This equation for c can be simplified to $2c^2 = 5$, which has the solution $c = \sqrt{\frac{5}{2}} \approx 1.6$ in the interval $[1, 2]$. ●

The MVT of differential calculus starts with a function on an interval and gives information about its derivative at an intermediate point. The MVT of integral calculus starts with a definite integral on an interval and gives information about the integrand at an intermediate point. Our first form of the MVT for integrals simply says that the average of a continuous function over a closed interval equals a value actually assumed by the function.

MVT of Integral Calculus

Let $f(x)$ be a continuous function on the interval $[a, b]$. Then

$$\frac{1}{b - a} \int_a^b f(x)\, dx = f(c)$$

for some c on the interval. ●

We can skip the proof since it is included in the proof that is coming up. We pass immediately to a more useful generalized mean value theorem. This deals with two functions whose product is integrated. The result has a lot of applications to estimating errors.

Generalized MVT of Integral Calculus

Let $f(x)$ and $g(x)$ be continuous functions on the interval $[a, b]$ with $g(x) \geq 0$. Then

$$\int_a^b f(x) g(x)\, dx = f(c) \int_a^b g(x)\, dx$$

for some c in the interval $[a, b]$. ●

Proof We may assume

$$\int_a^b g(x)\, dx > 0$$

For if not, then $g(x) = 0$ and both sides of the desired relation equal 0 so any c will do.

Since $f(x)$ is continuous on the closed interval $[a, b]$ it takes on its maximum and minimum (Theorem 1, Corollary 2, on page 167):

$$f(c_0) \leq f(x) \leq f(c_1)$$

where c_0 and c_1 lie in the interval. Since $g(x) \ge 0$, we have

$$f(c_0)g(x) \le f(x)g(x) \le f(c_1)g(x)$$

Therefore

$$f(c_0)\int_a^b g(x)\,dx \le \int_a^b f(x)g(x)\,dx \le f(c_1)\int_a^b g(x)\,dx$$

that is,

$$f(c_0) \le \frac{\displaystyle\int_a^b f(x)g(x)\,dx}{\displaystyle\int_a^b g(x)\,dx} \le f(c_1)$$

Thus the quotient in the middle falls between two values of $f(x)$. Therefore this quotient is a value $f(c)$ of $f(x)$ for some c between c_0 and c_1. That is because the range of the continuous function $f(x)$ on the closed interval $[c_0, c_1]$ is a closed interval (or single point) by Theorem 1 of Section 3-4 (page 166). ●

Remark The previous MVT is the case $g(x) = 1$ of the generalized MVT. For then the integral of g equals $b - a$.

Example 4

a $f(x) = x^3 \qquad g(x) = e^{-x} \qquad$ on $\quad [-1, 1]$

$$\int_{-1}^{1} x^3e^{-x}\,dx = c^3\int_{-1}^{1} e^{-x}\,dx = c^3(e - e^{-1})$$

for some c in $[-1, 1]$.

b $f(x) = \sin x \qquad g(x) = x^2 \qquad$ on $\quad [0, \pi]$

$$\int_0^{\pi} x^2\sin x\,dx = (\sin c)\int_0^{\pi} x^2\,dx = \tfrac{1}{3}\pi^3\sin c$$

for some c in $[0, \pi]$. ●

Exercises

1 Apply Rolle's theorem to $f(x) = e^{-kx}\sin x$ on $[0, \pi]$. What do you conclude?

2 The same question for $f(x) = e^{kx}\sqrt{1 - x^2}$ on $[-1, 1]$.

Use the MVT to show that

3 $\ln 51 - \ln 50 < 0.02$

4 $\sqrt[3]{1001} < 10.0034$

5 $\arctan 6 - \arctan 5 < 0.04$

6 $\tan \frac{2}{9}\pi > 0.82$ [Hint Start at $\frac{1}{4}\pi$.]

7 $\frac{1}{6}\pi + \frac{1}{15}\sqrt{3} < \arcsin \frac{3}{5} < \frac{1}{6}\pi + \frac{1}{8}$

8 $\frac{1}{6}\pi - \frac{1}{15}\sqrt{3} < \arcsin \frac{2}{5} < \frac{1}{6}\pi - \frac{1}{42}\sqrt{21}$

Show that the hypotheses of the generalized MVT are satisfied and find a number c that works

9 $f(x) = x^3 \qquad g(x) = x^2 \qquad [0, 1]$

10 $f(x) = \sin x \qquad g(x) = \cos x \qquad [0, \frac{1}{2}\pi]$

11 $f(x) = x^3 \quad g(x) = (x-1)^2 \quad [0, 1]$

12 $f(x) = x^3 \quad g(x) = (x-1)^2 \quad [1, 2]$

Apply the generalized MVT of integral calculus to the given functions and state the conclusion

13 $f(x) = e^x \quad g(x) = x^3 \quad [0, 1]$

14 $f(x) = \tan x \quad g(x) = \cos x \quad [0, \frac{1}{2}\pi]$

15 Show that $g(x) = 10x^9 - 16x^7 + 6x^2 - 1$ has a zero on the open interval $(0, 1)$.
[Hint Apply Rolle's theorem to $f(x) = \int_0^x g(t)\,dt$.]

16 (cont.) Suppose

$$\frac{a_0}{n+1} + \frac{a_1}{n} + \cdots + \frac{a_{n-1}}{2} + a_n = 0$$

Show that $g(x) = a_0x^n + a_1x^{n-1} + \cdots + a_n$ has a zero on $(0, 1)$.

17 Suppose $f(x)$ is differentiable on $[a, b]$ and $f(a) = f(b) = 0$. Prove there exists a number c such that $a < c < b$ and $f(c) + cf'(c) = 0$.
[Hint Try to guess a function $g(x)$ such that $g'(x) = f(x) + xf'(x)$ and use Rolle's theorem.]

18 (cont.) What does Exercise 17 imply if $f(x) = \sin x$?

19 (cont.) What does Exercise 17 imply if $f(x) = x^2 - 1 + \sin \pi x$?

20 Suppose $0 < a < b$ and $f(x)$ is differentiable on $[a, b]$. Assume $f(a) = f(b) = 0$. Prove there exists a number c in (a, b) such that $f(c) - cf'(c) = 0$.
[Hint See the hint for Exercise 17.]

21 (cont.) Apply Exercise 20 to $\sin x$ on $[\pi, 2\pi]$. What is your conclusion?

22 (cont.) Apply Exercise 20 to $x^4 - 11x^2 + 30$ on $[\sqrt{5}, \sqrt{6}]$. Conclusion?

23 Let $f(x)$ and $g(x)$ be differentiable on $[a, b]$ and $f(a) = f(b) = 0$. Suppose $f'g - fg'$ has no zero in this interval. Prove that g has a zero in (a, b).
[Hint Apply Rolle's theorem to a suitable combination of f and g. What combination does the expression $f'g - fg'$ suggest?]

24 (cont.) Find an example of Exercise 23.

25 Let $f(x)$ be differentiable on $[a, b]$. Prove there exists a number c in (a, b) such that

$$f'(c) = \frac{f(b) - f(c)}{c - a}$$

[Hint Look at $g(x) = (x - a)[f(b) - f(x)]$ and use Rolle's theorem.]

26 Apply Exercise 25 to $f(x) = \sin^2 x$ on $[0, \frac{1}{2}\pi]$. Conclusion?

27 Suppose $f'(x)$ is continuous for $|x - a| < \delta$. Use the MVT to find $\lim_{h\to 0}[f(a+h) - f(a-h)]/2h$.

28 Suppose $f(x)$ is continuous for $|x - a| < \delta$ and differentiable near $x = a$ except possibly at $x = a$. Suppose $\lim_{x\to a} f'(x) = L$. Prove $f(x)$ is differentiable at $x = a$ and $f'(a) = L$.

29 Criticize the following "proof" of the generalized MVT: By the MVT, $f(b) - f(a) = f'(c)(b - a)$. Similarly $g(b) - g(a) = g'(c)(b - a)$. Divide:

$$\frac{f(b) - f(a)}{g(b) - g(a)} = \frac{f'(c)(b-a)}{g'(c)(b-a)} = \frac{f'(c)}{g'(c)}$$

[Hint If you can't see anything wrong, try it for $f(x) = x^3$ and $g(x) = x^2$ on $[0, 1]$.]

*30 Let $f(x)$, $g(x)$, $h(x)$ be continuous on $[a, b]$ and differentiable on (a, b). Prove there exists a c in (a, b) such that the determinant

$$\begin{vmatrix} f(a) & g(a) & h(a) \\ f(b) & g(b) & h(b) \\ f'(c) & g'(c) & h'(c) \end{vmatrix} = 0$$

9-2 L'Hospital's Rule

We sometimes have to find limits of the form

$$\lim_{x\to a} \frac{f(x)}{g(x)}$$

where $f(a) = g(a) = 0$, what we might loosely call "limits of the form $0/0$." Each of the following limits is a case in point:

$$\lim_{x\to 0} \frac{\tan x}{e^x - 1} \qquad \lim_{x\to \pi} \frac{x - \pi}{\sin x} \qquad \lim_{x\to\infty} \frac{\frac{1}{2}\pi - \arctan x}{x^{-1}}$$

L'Hospital's rule helps evaluate such limits. We shall look at several forms of the rule, including also limits of the form ∞/∞. (Incidentally, l'Hospital is pronounced low'-pea-tal, with the long a as in father. It is also spelled l'Hôpital and Lhospital.)

L'Hospital's Rule I

Let $f(x)$ and $g(x)$ be differentiable functions near $x = a$ such that $f(a) = g(a) = 0$. Suppose that $g(x) \neq 0$ and $g'(x) \neq 0$ for $x \neq a$. If the limit

$$\lim_{x\to a} \frac{f'(x)}{g'(x)} = L$$

exists, where L is a real number or $+\infty$ or $-\infty$, then also

$$\lim_{x\to a} \frac{f(x)}{g(x)} = L$$

Briefly, if the limit on the right side exists, then

$$\lim_{x\to a} \frac{f(x)}{g(x)} = \lim_{x\to a} \frac{f'(x)}{g'(x)}$$

This rule is an easy consequence of the generalized MVT. If $x \neq a$, then there exists a number c between a and x such that

$$\frac{f(x)}{g(x)} = \frac{f(x) - f(a)}{g(x) - g(a)} = \frac{f'(c)}{g'(c)}$$

If $x \to a$, then $c \to a$, so

$$\frac{f(x)}{g(x)} = \frac{f'(c)}{g'(c)} \to L$$

Example 1

Find **a** $\lim_{x\to 0} \frac{\tan x}{e^x - 1}$ **b** $\lim_{x\to 0} \frac{1 - \cos x}{x^2}$

Solution

a $\tan 0 = 0 = e^0 - 1$, and both $e^x - 1 \neq 0$ and $e^x \neq 0$ for $x \neq 0$, so l'Hospital's rule applies:

$$\lim_{x\to 0} \frac{\tan x}{e^x - 1} = \lim_{x\to 0} \frac{(\tan x)'}{(e^x - 1)'} = \lim_{x\to 0} \frac{\sec^2 x}{e^x} = \frac{\sec^2 0}{e^0} = 1$$

b $1 - \cos 0 = 0^2 = 0$, and both $x^2 \neq 0$ and $2x \neq 0$ for $x \neq 0$, so l'Hospital's rule applies:

$$\lim_{x\to 0} \frac{1 - \cos x}{x^2} = \lim_{x\to 0} \frac{(1 - \cos x)'}{(x^2)'} = \lim_{x\to 0} \frac{\sin x}{2x}$$

This is another limit of the form $0/0$, so l'Hospital's rule applies a second time:

$$\lim_{x \to 0} \frac{1 - \cos x}{x^2} = \lim_{x \to 0} \frac{\sin x}{2x} = \lim_{x \to 0} \frac{\cos x}{2} = \tfrac{1}{2}$$

Remark It is important to check the assumptions $f(a) = 0$ and $g(a) = 0$ before you apply l'Hospital's rule. For example, suppose $f(x) = x^3 + 1$ and $g(x) = x^2 + 1$. Clearly

$$\lim_{x \to 0} \frac{f(x)}{g(x)} = \tfrac{1}{1} = 1$$

But by "l'Hospital's rule,"

$$\lim_{x \to 0} \frac{f(x)}{g(x)} = \lim_{x \to 0} \frac{f'(x)}{g'(x)} = \lim_{x \to 0} \frac{3x^2}{2x} = \lim_{x \to 0} \tfrac{3}{2}x = 0$$

This is incorrect because the hypotheses $f(0) = 0$ and $g(0) = 0$ of l'Hospital's rule are not satisfied.

We next consider limits of the form $\pm\infty/\infty$.

L'Hospital's Rule II

Let $f(x)$ and $g(x)$ be differentiable functions near $x = a$ such that

$$f(x) \to \pm\infty \quad \text{and} \quad g(x) \to \pm\infty$$

as $x \to a$. Suppose that $g(x) \neq 0$ and $g'(x) \neq 0$ for $x \neq a$. If

$$\lim_{x \to a} \frac{f'(x)}{g'(x)} = L$$

where L is finite or $+\infty$ or $-\infty$, then also

$$\lim_{x \to a} \frac{f(x)}{g(x)} = L$$

If the first limit is a one-sided limit (say $x \to a+$), then so is the second.

Remark The proof of this rule is delicate. Let us just sketch the idea for the case in which L is finite.

Let $a < x < z$. Then

$$\frac{f(z) - f(x)}{g(z) - g(x)} = \frac{f'(c)}{g'(c)}$$

for some c in (x, z). Choose z very close to a. Then c is very close to a; hence $f'(c)/g'(c)$ is close to L. Therefore

$$\frac{f(z) - f(x)}{g(z) - g(x)} \approx L$$

and this is true for any x such that $a < x < z$.

Now hold z fixed and let $x \to a$. By hypothesis, $f(x)$ and $g(x)$ are unbounded. Hence

$$\frac{f(z) - f(x)}{g(z) - g(x)} = \frac{f(x)}{g(x)}\left(\frac{1 - f(z)/f(x)}{1 - g(z)/g(x)}\right) \approx \frac{f(x)}{g(x)}$$

Therefore $f(x)/g(x)$ is close to L when x is close to a.

Example 2

Find $\lim_{x \to 0+} x \ln x$.

Solution Write

$$x \ln x = \frac{\ln x}{1/x}$$

which is of the form $f(x)/g(x)$, where

$$f(x) = \ln x \to -\infty \quad \text{and} \quad g(x) = \frac{1}{x} \to +\infty$$

as $x \to 0+$. By l'Hospital's rule II,

$$\lim_{x \to 0+} x \ln x = \lim_{x \to 0+} \frac{\ln x}{1/x} = \lim_{x \to 0+} \frac{1/x}{-1/x^2} = \lim_{x \to 0+} (-x) = 0$$

●

The final version of l'Hospital's rule deals with limits of the form $0/0$ or ∞/∞, but where x approaches infinity rather than a real number. We shall state the result for $x \to +\infty$, but the same result holds for $x \to -\infty$.

L'Hospital's Rule III

Let $f(x)$ and $g(x)$ be differentiable functions for all large x such that either

$$\begin{cases} f(x) \to 0 \\ g(x) \to 0 \end{cases} \quad \text{or} \quad \begin{cases} f(x) \to \pm\infty \\ g(x) \to \pm\infty \end{cases}$$

as $x \to +\infty$. Suppose that $g(x) \neq 0$ and $g'(x) \neq 0$ for large x. If

$$\lim_{x \to +\infty} \frac{f'(x)}{g'(x)} = L$$

where L is finite or $+\infty$ or $-\infty$, then also

$$\lim_{x \to +\infty} \frac{f(x)}{g(x)} = L$$

●

We omit the rather complicated proof of this rule.

Example 3

Find **a** $\lim_{x\to+\infty} x \operatorname{arc\,cot} x$ **b** $\lim_{x\to+\infty} \dfrac{(\ln x)^2}{x}$

Solution

a Write

$$x \operatorname{arc\,cot} x = \frac{\operatorname{arc\,cot} x}{1/x}$$

which is of the form $f(x)/g(x)$, where $f(x) = \operatorname{arc\,cot} x \to 0$ and $g(x) = 1/x \to 0$ as $x \to +\infty$. By l'Hospital's rule,

$$\lim_{x\to+\infty} x \operatorname{arc\,cot} x = \lim_{x\to+\infty} \frac{\operatorname{arc\,cot} x}{1/x} = \lim_{x\to+\infty} \frac{-1/(1+x^2)}{-1/x^2}$$

$$= \lim_{x\to+\infty} \frac{x^2}{1+x^2} = \lim_{x\to+\infty} \frac{1}{x^2+1} = 1$$

b This is of the form ∞/∞, so l'Hospital's rule applies:

$$\lim_{x\to+\infty} \frac{(\ln x)^2}{x} = \lim_{x\to+\infty} \frac{2(\ln x)/x}{1} = 2 \lim_{x\to+\infty} \frac{\ln x}{x}$$

Once again

$$\lim_{x\to+\infty} \frac{(\ln x)^2}{x} = 2 \lim_{x\to+\infty} \frac{\ln x}{x}$$

$$= 2 \lim_{x\to+\infty} \frac{1/x}{1} = 2 \lim_{x\to+\infty} \frac{1}{x} = 0$$ ●

Many limits of the form $\lim f(x)^{g(x)}$ can be handled by l'Hospital's rule. The trick is to take logs first, then use the continuity of $\ln x$ and its inverse function e^x.

Example 4

Find $\lim_{x\to 0+} x^{x^2}$

Solution Set

$$y = x^{x^2} \quad \text{so that} \quad \ln y = \ln x^{x^2} = x^2 \ln x$$

By l'Hospital's rule (or what we know about the rate of growth of $\ln x$),

$$\lim_{x\to 0+} \ln y = \lim_{x\to 0+} x^2 \ln x = 0$$

Hence

$$\lim_{x\to 0+} y = \lim_{x\to 0+} \exp(\ln y) = \exp(\lim_{x\to 0+} \ln y) = e^0 = 1$$ ●

Example 5

Find $\lim_{x\to+\infty}\left(1+\frac{1}{x}\right)^x$

Solution Set

$$y=\left(1+\frac{1}{x}\right)^x \quad \text{so that} \quad \ln y = x\ln\left(1+\frac{1}{x}\right)$$

Now set $t = 1/x$. Then

$$\ln y = \frac{\ln(1+t)}{t}$$

Now

$$\lim_{x\to+\infty}\ln y = \lim_{t\to0+}\ln y = \lim_{t\to0+}\frac{\ln(1+t)}{t}$$

By l'Hospital's rule,

$$\lim_{t\to0}\frac{\ln(1+t)}{t} = \lim_{t\to0}\frac{1}{1+t} = 1$$

Hence $\lim_{x\to+\infty} y = \exp(\lim_{x\to+\infty}\ln y) = \exp(1) = e$. ●

The l'Hospital Habit

Try to resist the habit of applying l'Hospital's rule automatically to every limit problem you face. Often there is a simpler way, for instance, use of limits already known or algebraic simplifications.

Example 6

Find

a $\lim_{x\to+\infty}\frac{\sqrt{2+5x^2}}{x}$ **b** $\lim_{x\to0}\frac{\sin^2 x}{x^2}$

Solution

a This has the form ∞/∞, so by l'Hospital,

$$\lim_{x\to+\infty}\frac{\sqrt{2+5x^2}}{x} = \lim_{x\to+\infty}\frac{5x/\sqrt{2+5x^2}}{1}$$

$$= \lim_{x\to+\infty}\frac{5x}{\sqrt{2+5x^2}}$$

$$= \frac{5}{\lim_{x\to+\infty}\frac{\sqrt{2+5x^2}}{x}} = \frac{5}{?}$$

back to where we started. Ugh!

Alternative Solution for Part a

$$\frac{\sqrt{2 + 5x^2}}{x} = \sqrt{\frac{2}{x^2} + 5} \rightarrow \sqrt{0 + 5} = \sqrt{5}$$

b This has the form $0/0$, so by l'Hospital, applied twice

$$\lim_{x\to 0} \frac{\sin^2 x}{x^2} = \lim_{x\to 0} \frac{2 \sin x \cos x}{2x} = \lim_{x\to 0} \frac{\cos^2 x - \sin^2 x}{1}$$
$$= \lim_{x\to 0} \cos^2 x - \lim_{x\to 0} \sin^2 x = 1 - 0 = 1$$

This solution is correct, but long.

Alternative Solution for Part b As we have seen, $(\sin x)/x \rightarrow 1$ as $x \rightarrow 0$. Therefore

$$\frac{\sin^2 x}{x^2} = \left(\frac{\sin x}{x}\right)^2 \rightarrow 1^2 = 1$$

Be alert to the possibility of adapting known limits to new situations. For example, from $(\sin x)/x \rightarrow 1$ as $x \rightarrow 0$ follow (without l'Hospital)

$$\frac{x}{\tan x} = \frac{x \cos x}{\sin x} = \frac{\cos x}{(\sin x)/x} \rightarrow \frac{1}{1} = 1 \quad \text{as } x \rightarrow 0$$
$$\frac{\sin 3x}{2x} = \frac{\sin 3x}{3x} \cdot \frac{3x}{2x} \rightarrow 1 \cdot \tfrac{3}{2} = \tfrac{3}{2} \quad \text{as } x \rightarrow 0$$
$$\frac{\sin(x^2)}{x} = \frac{\sin(x^2)}{x^2} \cdot x \rightarrow 1 \cdot 0 = 0 \quad \text{as } x \rightarrow 0$$

Exercises

Find

1 $\lim_{x\to 0} \dfrac{x}{e^x - 1}$

2 $\lim_{x\to 1} \dfrac{\ln x}{1 - x}$

3 $\lim_{x\to 0} \dfrac{x - \sin x}{x^3}$

4 $\lim_{x\to 0} \dfrac{\cos x - 1 + \frac{1}{2}x^2}{x^4}$

5 $\lim_{x\to 0} \dfrac{\cosh x - 1}{x^2}$

6 $\lim_{x\to 0} \dfrac{\tan x^2}{\sin^2 x}$

7 $\lim_{x\to \pi/2-} (x - \frac{1}{2}\pi) \tan x$

8 $\lim_{x\to \pi+} (x - \pi) \csc x$

9 $\lim_{x\to 0} \dfrac{x \tan x}{\cos x - 1}$

10 $\lim_{x\to 0} \dfrac{\sinh x}{\sin x}$

11 $\lim_{x\to \pi} \dfrac{\cos x + 1}{x(\tan \frac{1}{4}x - 1)}$

12 $\lim_{x\to 1} \dfrac{(\ln x)^2}{\sin^2 \pi x}$

13 $\lim_{x\to 0+} x^{\tan x}$

14 $\lim_{x\to 0+} x^{\ln(1+x)}$

15 $\lim_{x\to 0+} [\ln(1 + x)]^x$

16 $\lim_{x\to +\infty} \left(1 + \dfrac{1}{x^2}\right)^x$

17 $\lim_{x\to +\infty} \dfrac{(\ln x)^3}{x}$

18 $\lim_{x\to +\infty} \dfrac{e^x}{x^{10}}$

19 $\lim_{x\to +\infty} x(\frac{1}{2}\pi - \operatorname{arc\,tan} x)$

20 $\lim_{x\to +\infty} x^2(\operatorname{arc\,cot} x)^3$
[Hint Don't overdo the l'Hospital habit.]

21 $\lim_{x\to +\infty} (\sqrt{x^2 + 1} - x)$

22 $\lim_{x\to +\infty} \dfrac{\sqrt{x^2 + 1}}{x}$

23 $\lim_{x\to 0} \left(\dfrac{1}{x} - \dfrac{1}{\sin x}\right)$

24 $\lim_{x\to +\infty} x^3(1 - e^{-1/x^2})$

25 $\lim_{x\to +\infty} x \operatorname{arc\,tan} x$

26 $\lim_{x\to 0} \dfrac{1/x - 1/\sin x}{e^x - 1}$

27 $\lim_{x\to 1+} (x-1)\ln[-\ln(x-1)]$

28 $\lim_{x\to 1+} (\ln x)\ln[-\ln(x-1)]$

29 $\lim_{x\to 0} \left(\dfrac{a^{1+x}-1}{a-1}\right)^{1/x} \quad (a>1)$

30 $\lim_{x\to +\infty} \dfrac{\ln\ln x}{\ln x}$

31 Assume $f(x)$ is continuously differentiable near $x = c$. Find

$$\lim_{x\to 0} \frac{f(c+2x)-f(c+x)}{x}$$

32 Let $f(x) = a_0x^n + a_1x^{n-1} + \cdots + a_n$, where $a_0 > 0$. Find $\lim_{x\to+\infty} f(x)^{1/x}$.

*33 Suppose

$$\lim_{x\to+\infty} f(x) = L \quad \text{and} \quad \lim_{x\to+\infty} f'(x) = M$$

where L is a real number. Prove that $M = 0$.

*34 (cont.) Suppose $\lim_{x\to+\infty} f'(x) = +\infty$. Prove that $\lim_{x\to+\infty} f(x)/x = +\infty$.

35 Find $\displaystyle\lim_{x\to+\infty} \frac{1}{\exp x^2}\int_0^x \exp u^2\, du.$

*36 Prove $\displaystyle\int_0^x \exp u^2\, du \sim \frac{1}{x}\exp x^2$ as $x \to +\infty$.

37 Find $\displaystyle\lim_{x\to+\infty} (\exp x^2)\int_x^\infty \exp(-u^2)\, du.$

*38 Prove $\displaystyle\int_x^\infty \exp(-u^2)\, du \sim \frac{1}{2x}\exp(-x^2)$ as $x \to +\infty$.

9-3 Polynomial Approximation

The approximation of functions by polynomials is very important because values of polynomials can be calculated by arithmetic. It is by means of such approximations that function values are usually calculated. For instance, did you ever wonder how your calculator computes $\tan x$? As we shall show eventually,

$$\tan x \approx x + \tfrac{1}{3}x^3 + \tfrac{2}{15}x^5 + \tfrac{17}{315}x^7$$

for x small. For instance, if $x = 0.25$, then the value of the polynomial to 8 places is 0.25534 184. To the same accuracy, $\tan 0.25 \approx$ 0.25534 192. For $x = 0.2$,

$$\tan 0.2 \approx 0.20271\ 004 \qquad \text{poly} \approx 0.20271\ 002$$

This should give some indication of the importance of polynomial approximations and arouse your curiosity about how they are discovered.

We begin with approximations by linear polynomials. We touched this ground much earlier, at the beginning of Chapter 3. There we claimed that the best linear approximation to a (differentiable) function $f(x)$ at $x = a$ is the tangent line through $(a, f(a))$. Now we make this claim more precise.

The tangent line is quite close to the curve, at least near the point of tangency (Figure 9-3-1). The slope of the tangent is $f'(a)$, so by the point-slope formula the equation of the tangent line is

$$y - f(a) = f'(a)(x-a)$$

that is,

$$y = f(a) + f'(a)(x-a)$$

Since the tangent is close to the curve, it is reasonable to expect that the linear function $f(a) + f'(a)(x-a)$ is a good approximation to

Figure 9-3-1

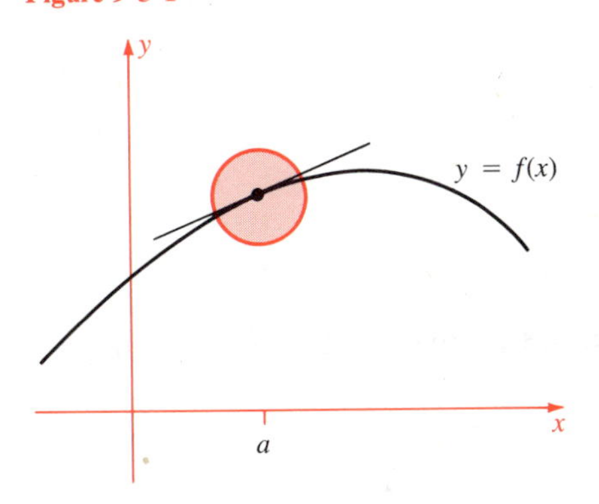

a Graph of a function and its tangent at $f(a)$

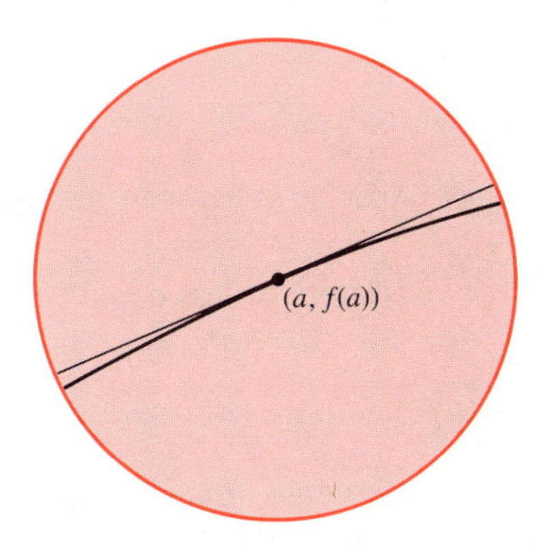

b Enlargement of a small neighborhood

$f(x)$, at least for x near a. Just how good the approximation is depends on the *error*

$$E(x) = f(x) - [f(a) + f'(a)(x - a)]$$

In fact $E(x)$ is smaller than $x - a$. Precisely:

Linear Approximation

Suppose $f(x)$ is differentiable at $x = a$, and set

$$E(x) = f(x) - [f(a) + f'(a)(x - a)]$$

Then

$$\lim_{x \to a} \frac{E(x)}{x - a} = 0$$

This is nothing deeper than the definition of the derivative:

$$\lim_{x \to a} \frac{E(x)}{x - a} = \lim_{x \to a} \left(\frac{f(x) - f(a)}{x - a} - f'(a) \right)$$
$$= f'(a) - f'(a) = 0$$

We can write the definition of $E(x)$ in the form

$$f(x) = f(a) + f'(a)(x - a) + E(x)$$

that is,

$$f(x) = (\text{linear approximation}) + (\text{error})$$

Although the conclusion $E(x)/(x - a) \to 0$ is good, usually the error is even smaller, as we shall soon see.

Remark No other linear function approximates $f(x)$ as well as $f(a) + f'(a)(x - a)$. For suppose

$$f(x) = A + B(x - a) + F(x) \quad \text{and} \quad F(x)/(x - a) \to 0$$

as $x \to a$. Then obviously we have $F(x) \to 0$ so it follows that $A = \lim_{x \to a} f(x) = f(a)$. But now

$$\frac{f(x) - f(a)}{x - a} = \frac{f(x) - A}{x - a} = B + \frac{F(x)}{x - a}$$

As x approaches a, the left side approaches $f'(a)$ and the right side approaches $B + 0 = B$. Therefore $B = f'(a)$ so

$$A + B(x - a) = f(a) + f'(a)(x - a)$$

Example 1

How closely does the tangent to $y = x^3$ at $(1, 1)$ approximate the function?

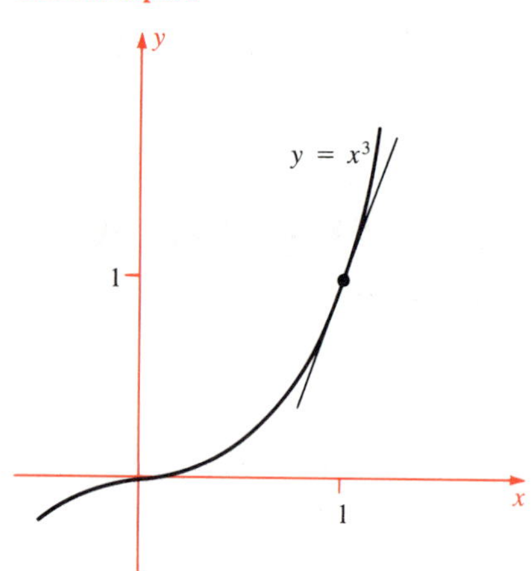

Figure 9-3-2
See Example 1

Solution Clearly $y(1) = 1$, so the given point lies on the graph (Figure 9-3-2). Next,

$$y' = 3x^2 \qquad \text{hence} \qquad y'(1) = 3$$

The tangent line at $(1, 1)$ is

$$y = 1 + 3(x - 1)$$

Therefore, the error made in approximating the curve by its tangent is

$$\begin{aligned} E(x) &= x^3 - 1 - 3(x - 1) \\ &= (x - 1)(x^2 + x + 1) - 3(x - 1) \\ &= (x - 1)(x^2 + x - 2) = (x - 1)^2(x + 2) \end{aligned}$$

Suppose we limit attention to $|x - 1| < 1$, that is, to $0 < x < 2$. Then $|x + 2| < 4$, so

$$|E(x)| < 4|x - 1|^2$$

For instance, if $|x - 1| < 10^{-4}$, then $|E(x)| < 4 \times 10^{-8}$. ●

Note in the example that the error $E(x)$ is more or less equal to a constant times $(x - a)^2$:

$$E(x) = (x + 2)(x - 1)^2 \approx 3(x - 1)^2$$

near $x = 1$. This is a stronger statement than merely $E(x)/(x - a) \to 0$ as $x \to a$. Later we shall prove that this is typical behavior provided $f(x)$ is smooth enough.

Second Degree Approximation

The linear or **first degree approximation** to $f(x)$ at $x = a$ is

$$P_1(x) = f(a) + f'(a)(x - a)$$

Near $x = a$, the function $P_1(x)$ appears to be a good approximation to $f(x)$. This is not surprising since

$$P_1(a) = f(a) \quad \text{and} \quad P_1'(a) = f'(a)$$

Hence both $y = P_1(x)$ and $y = f(x)$ pass through the point $(a, f(a))$ with the same slope.

For greater accuracy, we approximate $f(x)$ by a polynomial whose graph passes through $(a, f(a))$ with the same slope as $y = f(x)$ and whose second derivative also agrees with that of $f(x)$. Therefore we seek a polynomial $P_2(x)$ such that

$$P_2(a) = f(a) \qquad P_2'(a) = f'(a) \quad \text{and} \quad P_2''(a) = f''(a)$$

Let us try a quadratic

$$P_2(x) = A + B(x - a) + C(x - a)^2$$

Table 9-3-1
e^x and its first two polynomial approximations

x	e^x	$P_1(x)$	$P_2(x)$
Small values of x			
-0.4	0.6703	0.6000	0.6800
-0.3	0.7408	0.7000	0.7450
-0.2	0.8187	0.8000	0.8200
-0.1	0.9048	0.9000	0.9050
0.0	1.0000	1.0000	1.0000
0.1	1.1052	1.1000	1.1050
0.2	1.2214	1.2000	1.2200
0.3	1.3499	1.3000	1.3450
0.4	1.4918	1.4000	1.4800
Larger values of x			
-2.0	0.1353	-1.0000	1.0000
-1.5	0.2231	-0.5000	0.6250
-1.0	0.3679	0.0000	0.5000
-0.5	0.6065	0.5000	0.6250
0.0	1.0000	1.0000	1.0000
0.5	1.6487	1.5000	1.6250
1.0	2.7183	2.0000	2.5000
1.5	4.4817	2.5000	3.6250
2.0	7.3891	3.0000	5.0000

We note that

$$P_2(a) = A \qquad P_2'(a) = B \quad \text{and} \quad P_2''(a) = 2C$$

(Verify these statements.) Since we want

$$P_2(a) = f(a) \qquad P_2'(a) = f'(a) \quad \text{and} \quad P_2''(a) = f''(a)$$

the only possibility is

$$A = f(a) \qquad B = f'(a) \quad \text{and} \quad C = \tfrac{1}{2}f''(a)$$

- The **second degree approximation** to $f(x)$ at $x = a$ is

$$P_2(x) = f(a) + f'(a)(x - a) + \tfrac{1}{2}f''(a)(x - a)^2$$

The polynomial $P_2(x)$ agrees with $f(x)$ at $x = a$, and its derivative and second derivative agree with those of $f(x)$ at $x = a$.

The first two terms of $P_2(x)$ are $f(a) + f'(a)(x - a)$, which is $P_1(x)$. Thus $P_2(x)$ consists of the linear approximation to $f(x)$ plus another term, $\frac{1}{2}f''(a)(x - a)^2$, which corrects some of the error in linear approximation.

Example 2

Approximate e^x near $x = 0$ by the first and second degree polynomials $P_1(x)$ and $P_2(x)$. Test their accuracy.

Solution Use the formula

$$P_2(x) = f(a) + f'(a)(x - a) + \tfrac{1}{2}f''(a)(x - a)^2$$

where $f(x) = e^x$ and $a = 0$. In this case $f(x) = f'(x) = f''(x) = e^x$, so $f(0) = f'(0) = f''(0) = 1$. Therefore

$$P_1(x) = 1 + x \quad \text{and} \quad P_2(x) = 1 + x + \tfrac{1}{2}x^2$$

Table 9-3-1 compares e^x with $P_1(x)$ and $P_2(x)$ for various values of x. From the table we see that $P_1(x)$ and $P_2(x)$ are good estimates of e^x for x near 0, but that $P_2(x)$ is considerably better than $P_1(x)$. Both estimates become poor as x moves away from 0, but $P_2(x)$ is accurate over a wider range because its graph is curved like that of e^x near $x = 0$. See Figure 9-3-3.

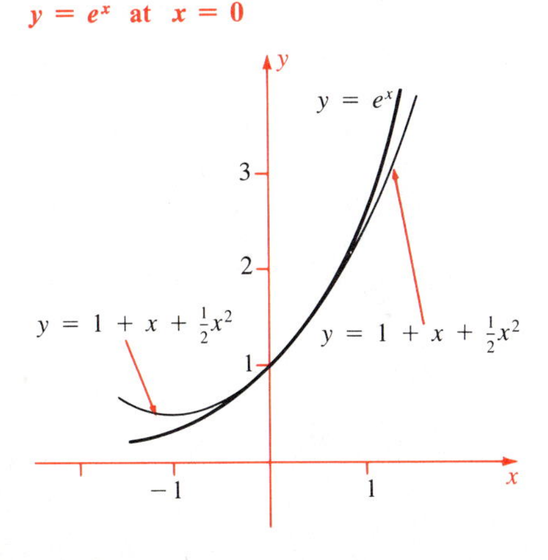

Figure 9-3-3
Second degree approximation to $y = e^x$ at $x = 0$

Example 3

Estimate $y = 1/x$ near $x = 1$ by its first and second degree approximations. Test their accuracy.

Solution

$$y = \frac{1}{x} \qquad y' = -\frac{1}{x^2} \qquad \text{and} \qquad y'' = \frac{2}{x^3}$$

Table 9-3-2
$1/x$ and its first two polynomial approximations

x	$1/x$	$P_1(x)$	$P_2(x)$
x near 1			
0.85	1.1765	1.1500	1.1725
0.90	1.1111	1.1000	1.1100
0.95	1.0526	1.0500	1.0525
1.00	1.0000	1.0000	1.0000
1.05	0.9524	0.9500	0.9525
1.10	0.9091	0.9000	0.9100
1.15	0.8696	0.8500	0.8725
Other values of x			
0.25	4.0000	1.7500	2.3125
0.50	2.0000	1.5000	1.7500
0.75	1.3333	1.2500	1.3125
1.00	1.0000	1.0000	1.0000
1.25	0.8000	0.7500	0.8125
1.50	0.6667	0.5000	0.7500
1.75	0.5714	0.2500	0.8125

Hence $y(1) = 1$, $y'(1) = -1$, and $y''(1) = 2$. Therefore

$$P_1(x) = 1 - (x - 1)$$

and

$$P_2(x) = 1 - (x - 1) + (x - 1)^2$$

Since we are dealing with numbers near 1, it is convenient to leave $P_1(x)$ and $P_2(x)$ in terms of $x - 1$.

Table 9-3-2 compares $1/x$ with $P_1(x)$ and $P_2(x)$. From the table, we see that $P_1(x)$ is a good approximation to $1/x$ provided x is near 1, but that $P_2(x)$ is much better. Both estimates become poor as x moves away from 1, but $P_2(x)$ is accurate over a wider range. ●

Error Estimates

In our example of a linear approximation, the error had order of magnitude $|x - a|^2$. By analogy, we might expect the error in a second degree approximation

$$E(x) = f(x) - P_2(x)$$

to have order of magnitude $|x - a|^3$. By that we mean

$$E(x) \approx c(x - a)^3 \quad \text{for } x \text{ near } a$$

where c is a suitable constant. We shall prove this conjecture in Section 9-5. Meanwhile, let us verify it in an example.

Example 4

How closely does $P_2(x)$ approximate the function $y = 1/x$ near $x = 1$?

Solution By Example 3,

$$E(x) = \frac{1}{x} - P_2(x) = \frac{1}{x} - 1 + (x - 1) - (x - 1)^2$$

Hence

$$\begin{aligned} E(x) &= \frac{1 - x}{x} + (x - 1) - (x - 1)^2 \\ &= (x - 1)\left[-\frac{1}{x} + 1 - (x - 1)\right] \\ &= (x - 1)\left[\frac{x - 1}{x} - (x - 1)\right] \\ &= (x - 1)^2\left(\frac{1}{x} - 1\right) = -\frac{1}{x}(x - 1)^3 \end{aligned}$$

For x near 1,

$$E(x) \approx -(x-1)^3$$

so the order of magnitude of the error is as expected.

For a more precise estimate, not involving the vague symbol $\approx$, let us restrict our attention to a specific interval around $x = 1$, say $(\frac{1}{2}, \frac{3}{2})$. On this interval $|1/x| < 2$. Hence we get the specific estimate

$$|E(x)| = \left|-\frac{1}{x}\right| |x-1|^3 < 2|x-1|^3$$

For instance, if $|x - 1| < 10^{-3}$, then $|E(x)| < 2 \times 10^{-9}$ so the approximation is very close indeed. ●

Exercises

Given $f(x)$ and a, find $P_1(x)$ at $x = a$

1 $f(x) = 1 - x^2$ $\quad a = 0$

2 $f(x) = 2x^2 + 3$ $\quad a = 1$

3 $f(x) = x^3$ $\quad a = -1$

4 $f(x) = x^3$ $\quad a = 2$

5 $f(x) = 1/x$ $\quad a = \frac{1}{2}$

6 $f(x) = x^2 - x^3$ $\quad a = 1$

7 $f(x) = x^4$ $\quad a = 1$

8 $f(x) = x^5$ $\quad a = -1$

9 $f(x) = \cos x$ $\quad a = 0$

10 $f(x) = \sin x$ $\quad a = \frac{1}{4}\pi$

11 $f(x) = xe^x$ $\quad a = 0$

12 $f(x) = 1/(1 + e^x)$ $\quad a = 0$

Given $f(x)$ and a, find $P_1(x)$ at $x = a$ and express the error in the form $E(x) = (x - a)^2 g(x)$

13 $f(x) = 1/x^2$ $\quad a = -1$

14 $f(x) = 1/(x^3 + 1)$ $\quad a = 1$

Given $f(x)$ and a, find $P_2(x)$ at $x = a$ and express the error in the form $E(x) = (x - a)^3 g(x)$

15 $f(x) = x + 1$ $\quad a = 0$

16 $f(x) = (x - 2)^3 + 3(x - 2)^2 - 4(x - 2) + 1$ $\quad a = 2$

17 $f(x) = 1/x$ $\quad a = -2$

18 $f(x) = 1/(x^2 + 1)$ $\quad a = 0$

19 $f(x) = x^2/(x + 1)$ $\quad a = 0$

20 $f(x) = x^3/(x + 1)$ $\quad a = 1$

Given $f(x)$ and a, find $P_2(x)$ at $x = a$

21 $f(x) = e^{2x}$ $\quad a = 0$

22 $f(x) = \cos x$ $\quad a = 0$

23 $f(x) = \tan x$ $\quad a = 0$

24 $f(x) = \sinh x$ $\quad a = 0$

Complete the table (to four-place accuracy—this means with $|\text{error}| \le 5 \times 10^{-5}$) at $x = 0$

25

x	xe^x	$P_1(x)$	$P_2(x)$
0.1			
0.2			
0.3			
0.4			
0.5			

26

x	$\dfrac{1}{1-x^2}$	$P_1(x)$	$P_2(x)$
0.1			
0.2			
0.3			
0.4			
0.5			

27 Justify the approximation $\ln x \approx 2(x - 1)/(x + 1)$ near $x = 1$. Test it numerically for $x = 0.5, 0.8, 1.2, 1.5$, and 2.0, and compare $2(x - 1)/(x + 1)$ to $P_2(x)$ for $\ln x$.

28 Find a, b, c so that $e^{-x} \approx (ax + b)/(cx + 1)$ near $x = 0$ in the sense that both sides have the same $P_2(x)$. (This is called Padé approximation.)

9-4 Taylor Approximations

Let us extend the ideas of Section 9-3 to the approximation of functions by polynomials of arbitrary degree. We start with polynomials themselves and observe the following algebraic fact: Each polynomial can be expressed not only in powers of x, but also in powers of $x - a$, where a is any number. This form of the polynomial is convenient for computations near $x = a$.

Example 1

a Express $p(x) = x^2 + x + 2$ in terms of $x - 1$.
b Express $p(x) = x^4$ in terms of $x + 1$.

Solution

a Set $u = x - 1$ so $x = u + 1$:

$$\begin{aligned} p(x) &= x^2 + x + 2 \\ &= (u+1)^2 + (u+1) + 2 \\ &= (u^2 + 2u + 1) + (u + 1) + 2 \\ &= u^2 + 3u + 4 \\ &= (x-1)^2 + 3(x-1) + 4 \end{aligned}$$

b $$\begin{aligned} p(x) = x^4 &= [(x+1) - 1]^4 \\ &= (x+1)^4 - 4(x+1)^3 + 6(x+1)^2 - 4(x+1) + 1 \end{aligned}$$ ●

In general, to express

$$p(x) = B_0 + B_1 x + B_2 x^2 + B_3 x^3 + \cdots + B_n x^n$$

in powers of $x - a$, write $u = x - a$, then substitute $u + a$ for x:

$$p(x) = B_0 + B_1(u + a) + B_2(u + a)^2 + \cdots + B_n(u + a)^n$$

Now expand each power by the binomial formula and collect like powers of u. The result is a polynomial in $u = x - a$, as desired.

This method is laborious when the degree of $p(x)$ exceeds three or four. If

$$p(x) = A_0 + A_1(x - a) + A_2(x - a)^2 + \cdots + A_n(x - a)^n$$

is the desired expansion of $p(x)$ in terms of $x - a$, we can compute the coefficients $A_0, A_1, \cdots$ directly, without a lot of algebra. First $p(a) = A_0$, so finding A_0 is no problem at all. But how do we get A_1? The trick is to *differentiate* $p(x)$, then set $x = a$:

$$p'(x) = A_1 + 2A_2(x - a) + \cdots + nA_n(x - a)^{n-1}$$

$$\text{so} \quad p'(a) = A_1$$

Differentiation can be repeated:

$$p''(x) = 2A_2 + 3 \cdot 2A_3(x-a) + \cdots + n(n-1)A_n(x-a)^{n-2} \quad \text{so} \quad p''(a) = 2A_2$$

$$p'''(x) = 3 \cdot 2A_3 + \cdots + n(n-1)(n-2)A_n(x-a)^{n-3} \quad \text{so} \quad p'''(a) = 3 \cdot 2A_3 = 3!\,A_3$$

Continuing in this way, we find ($p^{(k)}$ denotes k-th derivative)

$$p^{(4)}(a) = 4!\,A_4 \qquad p^{(5)}(a) = 5!\,A_5 \qquad \cdots \qquad p^{(n)}(a) = n!\,A_n$$

• If $p(x)$ is a polynomial of degree n and if a is any number, then

$$p(x) = p(a) + p'(a)(x-a) + \frac{1}{2!}p''(a)(x-a)^2 + \frac{1}{3!}p'''(a)(x-a)^3 + \cdots + \frac{1}{n!}p^{(n)}(a)(x-a)^n$$ ●

We now solve Example 1 again, but by this technique.

Example 2

a Express $p(x) = x^2 + x + 2$ in terms of $x - 1$.
b Express $p(x) = x^4$ in terms of $x + 1$.

Solution

a $p(x) = x^2 + x + 2 \qquad p'(x) = 2x + 1 \qquad p''(x) = 2$

$p(1) = 4 \qquad p'(1) = 3 \qquad p''(1) = 2$

Therefore

$$p(x) = 4 + 3(x-1) + \tfrac{1}{2} \cdot 2(x-1)^2 = 4 + 3(x-1) + (x-1)^2$$

b $p(x) = x^4 \qquad p'(x) = 4x^3 \qquad p''(x) = 12x^2$

$p'''(x) = 24x \qquad p^{(4)}(x) = 24$

$p(-1) = 1 \qquad p'(-1) = -4 \qquad p''(-1) = 12$

$p'''(-1) = -24 \qquad p^{(4)}(-1) = 24$

Therefore

$$p(x) = 1 - 4(x+1) + \frac{12}{2!}(x+1)^2 - \frac{24}{3!}(x+1)^3 + \frac{24}{4!}(x+1)^4$$

$$= 1 - 4(x+1) + 6(x+1)^2 - 4(x+1)^3 + (x+1)^4$$

●

Taylor Polynomials

Let us return to the main problem: given a function $f(x)$ and a number a, to find a polynomial $P(x)$ of degree n that approximates $f(x)$ for values of x near a. In view of the way that we found $P_2(x)$ in Section 9-3, it now seems reasonable to construct a polynomial $P_n(x)$ so that

$$P_n(a) = f(a) \qquad P_n'(a) = f'(a)$$

$$P_n''(a) = f''(a) \qquad \cdots \qquad P_n^{(n)}(a) = f^{(n)}(a)$$

Thus $P_n(x)$ mimics $f(x)$ and its first n derivatives at $x = a$.

Let us find $P_n(x)$ explicitly. We write

$$P_n(x) = A_0 + A_1(x - a) + A_2(x - a)^2 + \cdots + A_n(x - a)^n$$

and determine the coefficients A_k appropriately. Now we know that $A_k = P_n^{(k)}(a)/k!$. Since we want $P_n^{(k)}(a) = f^{(k)}(a)$, we must choose $A_k = f^{(k)}(a)/k!$.

Taylor Polynomials

Suppose $f(x)$ is n times differentiable near $x = a$. Then the n-th **degree Taylor polynomial** of $f(x)$ at $x = a$ is

$$\begin{aligned} P_n(x) &= f(a) + f'(a)(x - a) + \frac{1}{2!}f''(a)(x - a)^2 \\ &\quad + \cdots + \frac{1}{n!}f^{(n)}(a)(x - a)^n \\ &= f(a) + \sum_{k=1}^{n} \frac{f^{(k)}(a)}{k!}(x - a)^k \end{aligned}$$

Remark When $f(x)$ is itself a polynomial of degree n, then $P_n(x) = f(x)$ is precisely the expression for $f(x)$ in powers of $x - a$. Furthermore, in this case,

$$P_n(x) = P_{n+1}(x) = P_{n+2}(x) = \cdots$$

(Why?) Thus for an n-th degree polynomial $f(x)$, its n-th and all higher degree Taylor polynomials equal $f(x)$.

The first three Taylor polynomials explicitly are

$$P_1(x) = f(a) + f'(a)(x - a)$$

$$P_2(x) = f(a) + f'(a)(x - a) + \tfrac{1}{2}f''(a)(x - a)^2$$

$$\begin{aligned} P_3(x) &= f(a) + f'(a)(x - a) + \tfrac{1}{2}f''(a)(x - a)^2 \\ &\quad + \tfrac{1}{6}f'''(a)(x - a)^3 \end{aligned}$$

The first two are old friends from Section 9-3. In general, each Taylor polynomial is derived from the preceding one by addition of a single term:

$$P_{n+1}(x) = P_n(x) + \frac{1}{(n+1)!} f^{(n+1)}(a)(x-a)^{n+1}$$

We expect that $P_n(x)$ is a good approximation to $f(x)$; the error is $f(x) - P_n(x)$. We try to reduce this error by adding an additional term

$$\frac{1}{(n+1)!} f^{(n+1)}(a)(x-a)^{n+1}$$

to $P_n(x)$, thereby obtaining $P_{n+1}(x)$, an even better approximation.

Example 3

Find the n-th degree Taylor polynomial of

a $f(x) = e^x$ at $x = 0$ **b** $f(x) = \ln x$ at $x = 1$

Test the accuracy of P_2, P_3, and P_4 in part **a** by tabulating (calculator) values.

Solution

a Compute successive derivatives of $f(x) = e^x$ and evaluate them at $x = 0$:

$$f(x) = e^x \quad f'(x) = e^x \quad f''(x) = e^x \quad \cdots$$
$$f(0) = 1 \quad f'(0) = 1 \quad f''(0) = 1 \quad \cdots$$

Hence, according to the recipe,

$$\begin{aligned} P_n(x) &= f(0) + f'(0)x + \frac{1}{2!}f''(0)x^2 \\ &\quad + \cdots + \frac{1}{n!}f^{(n)}(0)x^n \\ &= 1 + x + \frac{x^2}{2!} + \frac{x^3}{3!} + \cdots + \frac{x^n}{n!} = \sum_{k=0}^{n} \frac{x^k}{k!} \end{aligned}$$

Table 9-4-1 is evidence of the accuracy of these approximations.

Table 9-4-1

x	e^x	$P_2(x)$	$P_3(x)$	$P_4(x)$
0.1	1.10517	1.10500	1.10517	1.10517
0.2	1.22140	1.22000	1.22133	1.22140
0.3	1.34986	1.34500	1.34950	1.34984
0.4	1.49182	1.48000	1.49067	1.49173
0.5	1.64872	1.62500	1.64583	1.64844
0.75	2.11700	2.03125	2.10156	2.11475
1.0	2.71828	2.50000	2.66667	2.70833

b Compute successive derivatives of $f(x) = \ln x$ and evaluate them at $x = 1$:

$$f' = \frac{1}{x} \qquad f'' = -\frac{1}{x^2} \qquad f''' = \frac{2!}{x^3}$$

$$f^{(4)} = -\frac{3!}{x^4} \qquad \cdots \qquad f^{(k)} = (-1)^{k-1}\frac{(k-1)!}{x^k}$$

Hence

$$f(1) = 0 \qquad f'(1) = 1 \qquad f''(1) = -1 \qquad f'''(1) = 2$$

$$f^{(4)}(1) = -3! \qquad \cdots \qquad f^{(k)}(1) = (-1)^{k-1}(k-1)!$$

Therefore

$$\frac{f^{(k)}(1)}{k!} = \frac{(-1)^{k-1}(k-1)!}{k!} = \frac{(-1)^{k-1}}{k}$$

and

$$P_n(x) = \sum_{k=1}^{n} (-1)^{k-1}\frac{(x-1)^k}{k}$$

$$= (x-1) - \frac{(x-1)^2}{2} + \frac{(x-1)^3}{3} - \cdots$$

$$+ (-1)^{n-1}\frac{(x-1)^n}{n}$$

●

Example 4

Compute the Taylor polynomials at $x = 0$ for

a $f(x) = \sin x$ **b** $f(x) = \cos x$

Test the accuracy of P_1, P_3, and P_5 in part **a** by tabulating (calculator) values.

Solution

a Compute derivatives:

$$f(x) = \sin x \qquad f'(x) = \cos x \qquad f''(x) = -\sin x$$

$$f'''(x) = -\cos x \qquad f^{(4)}(x) = \sin x \qquad \cdots$$

repeating in cycles of four. At $x = 0$, the values are

$$0 \quad 1 \quad 0 \quad -1 \quad 0 \quad 1 \quad 0 \quad -1 \quad 0 \quad \cdots$$

Hence the n-th degree Taylor polynomial of $\sin x$ is

$$P_n(x) = x - \frac{x^3}{3!} + \frac{x^5}{5!} - \frac{x^7}{7!} + \frac{x^9}{9!} - \cdots$$

For example,

$$P_3(x) = P_4(x) = x - \frac{x^3}{3!}$$

$$P_5(x) = P_6(x) = x - \frac{x^3}{3!} + \frac{x^5}{5!}$$

$$P_7(x) = P_8(x) = x - \frac{x^3}{3!} + \frac{x^5}{5!} - \frac{x^7}{7!}$$

In general, $P_{2m-1}(x) = P_{2m}(x)$, and the sign of the last term is plus if m is odd, minus if m is even:

$$\begin{aligned} P_{2m-1}(x) = P_{2m}(x) &= x - \frac{x^3}{3!} + \frac{x^5}{5!} - \cdots \\ &\qquad + (-1)^{m-1}\frac{x^{2m-1}}{(2m-1)!} \\ &= \sum_{i=1}^{m} (-1)^{i-1} \frac{x^{2i-1}}{(2i-1)!} \end{aligned}$$

Table 9-4-2 is evidence of the accuracy of these Taylor polynomials. Figure 9-4-1 compares the graph of $\sin x$ with the graphs of $P_1(x)$, $P_3(x)$, and $P_5(x)$.

Table 9-4-2

x	$\sin x$	$P_1(x) = P_2(x)$	$P_3(x) = P_4(x)$	$P_5(x) = P_6(x)$
0.1	0.09983	0.10000	0.09983	0.09983
0.2	0.19867	0.20000	0.19867	0.19867
0.3	0.29552	0.30000	0.29550	0.29552
0.4	0.38942	0.40000	0.38933	0.38942
0.5	0.47943	0.50000	0.47917	0.47943
1.0	0.84147	1.00000	0.83333	0.84167

Figure 9-4-1
Taylor approximations to $y = \sin x$ at $x = 0$

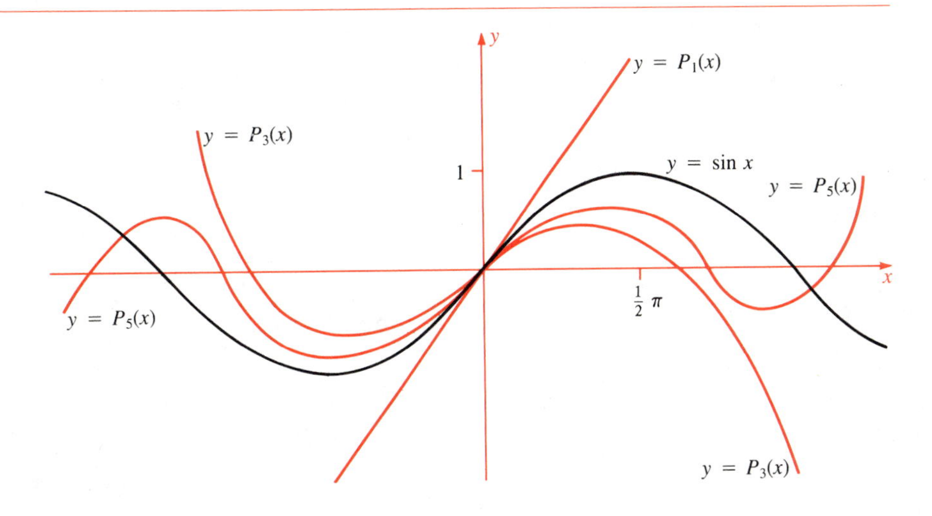

b Since $\cos x$ is the derivative of $\sin x$, we can read off its derivatives at $x = 0$ from those of $\sin x$ found in part **a**. Beginning with the "zeroth derivative" $f^{(0)}(0) = f(0)$, they are

$$1 \quad 0 \quad -1 \quad 0 \quad 1 \quad 0 \quad -1 \quad 0 \quad \cdots$$

repeating in cycles of four. Using these values, we obtain

$$\begin{aligned} P_{2m}(x) &= P_{2m+1}(x) \\ &= \sum_{i=0}^{m} (-1)^i \frac{x^{2i}}{(2i)!} \\ &= 1 - \frac{x^2}{2!} + \frac{x^4}{4!} - \cdots + (-1)^m \frac{x^{2m}}{(2m)!} \end{aligned}$$

Summary of Taylor Approximations

$$e^x \approx 1 + \frac{x}{1!} + \frac{x^2}{2!} + \frac{x^3}{3!} + \cdots + \frac{x^n}{n!}$$

$$\ln x \approx (x-1) - \frac{(x-1)^2}{2} + \frac{(x-1)^3}{3} - \cdots + (-1)^{n-1}\frac{(x-1)^n}{n}$$

$$\sin x \approx x - \frac{x^3}{3!} + \frac{x^5}{5!} - \frac{x^7}{7!} + \cdots + (-1)^{n-1}\frac{x^{2n-1}}{(2n-1)!}$$

$$\cos x \approx 1 - \frac{x^2}{2!} + \frac{x^4}{4!} - \frac{x^6}{6!} + \cdots + (-1)^n \frac{x^{2n}}{(2n)!}$$

Shortcuts

If you know the Taylor polynomials of several functions, you can easily find those of certain related functions. Suppose $f(x)$ has the n-th degree Taylor polynomial $P_n(x)$ at $x = a$, and that c is a constant. Then $cf(x)$ has the n-th degree Taylor polynomial $cP_n(x)$ at $x = a$.

If also $g(x)$ has the n-th degree Taylor polynomial $Q_n(x)$ at $x = a$, then $f(x) + g(x)$ has the n-th degree Taylor polynomial $P_n(x) + Q_n(x)$ at $x = a$.

Suppose now that $f(x)$ has the n-th degree Taylor polynomial $P_n(x)$ at $x = 0$ and that c is a constant. Then $f(cx)$ has the n-th degree Taylor polynomial $P_n(cx)$ at $x = 0$. All of these results are routine to prove.

Example 5

Find the fourth degree Taylor polynomial at $x = 0$ of

a $f(x) = 3e^x$ **b** $f(x) = 3e^x + \sin x$ **c** $f(x) = e^{3x}$

Solution

a Use the result of Example 3a:

$$P_4(x) = 3\left(1 + \frac{1}{1!}x + \frac{1}{2!}x^2 + \frac{1}{3!}x^3 + \frac{1}{4!}x^4\right)$$
$$= 3 + 3x + \tfrac{3}{2}x^2 + \tfrac{1}{2}x^3 + \tfrac{1}{8}x^4$$

b Use what we have just found for $3e^x$ and the result of Example 4a:

$$P_4(x) = (3 + 3x + \tfrac{3}{2}x^2 + \tfrac{1}{2}x^3 + \tfrac{1}{8}x^4) + (x - \tfrac{1}{6}x^3)$$
$$= 3 + 4x + \tfrac{3}{2}x^2 + \tfrac{1}{3}x^3 + \tfrac{1}{8}x^4$$

c Again from the fourth-degree Taylor polynomial of e^x we have

$$P_4(x) = 1 + \frac{1}{1!}(3x) + \frac{1}{2!}(3x)^2 + \frac{1}{3!}(3x)^3 + \frac{1}{4!}(3x)^4$$
$$= 1 + 3x + \tfrac{9}{2}x^2 + \tfrac{9}{2}x^3 + \tfrac{27}{8}x^4$$

Exercises

Express $p(x)$ in terms of $u = x - a$, where $p(x) =$

1 $x^2 + 5x + 2 \quad a = 1$
2 $x^3 - 3x^2 + 4x \quad a = 2$
3 $2x^3 + 5x^2 + 13x + 10 \quad a = -1$
4 $3x^3 - 2x^2 - 2x + 1 \quad a = 1$
5 $2x^4 + 5x^3 + 4x + 16 \quad a = -2$
6 $x^4 - 5x^2 + x + 2 \quad a = 2$
7 $5x^5 + 4x^4 - 3x^3 - 2x^2 + x + 1 \quad a = -1$
8 $x^5 + 2x^4 + 3x^2 + 4x + 5 \quad a = -2$

Compute the fifth-degree Taylor polynomial of

9 xe^{-x} at $x = 0$
10 $x^2 \cos x$ at $x = 0$
11 $(1 + x^3)\sin x$ at $x = 0$
12 $x^3/(1 + x^2)$ at $x = 0$
13 $\sin x$ at $x = \frac{1}{4}\pi$
14 $\cos x$ at $x = 0$
15 $1/x^3$ at $x = -1$
16 $\exp(x^2)$ at $x = 0$
17 $\sin x$ at $x = \frac{1}{2}\pi$
18 $\cos x$ at $x = -\frac{1}{2}\pi$
19 $\ln x$ at $x = 2$
20 $\ln x$ at $x = \frac{1}{2}$
21 e^x at $x = 1$
22 e^{-x} at $x = -1$
23 $\sin x + 2\cos x$ at $x = 0$
24 $e^x - e^{-x}$ at $x = 0$
25 $\sin 3x$ at $x = 0$
26 $\cos \pi x$ at $x = 0$

Compute $P_4(x)$ at $x = 0$, and tabulate $f(x)$ and $P_4(x)$ to five places for $-0.5 \le x \le 0.5$ by increments of 0.1

27 $f(x) = xe^x$
28 $f(x) = x\sin x - x^2$
29 $f(x) = 1/(1 - x)$
30 $f(x) = 1/(1 - x)^2$

9-5 Taylor's Formula

We introduced the Taylor polynomials $P_n(x)$ in the preceding section with the hope that $P_n(x)$ would be a good approximation to a given function $f(x)$. Our optimism will be justified if the error

$$R_n(x) = f(x) - P_n(x)$$

is very small compared to $x - a$. The next two statements assure us that it is. The first gives an explicit formula for the error, the second an estimate of its size.

In this subject the error is usually called the **remainder.** That's why we write $R_n(x)$ rather than our usual $E(x)$.

Taylor's Formula with Remainder

Suppose $f(x)$ has continuous derivatives up to and including $f^{(n+1)}(x)$ near $x = a$. Write

$$f(x) = P_n(x) + R_n(x)$$

where $P_n(x)$ is the n-th degree Taylor polynomial at $x = a$ and $R_n(x)$ is the remainder (or error). Then

$$R_n(x) = \frac{1}{n!}\int_a^x (x - t)^n f^{(n+1)}(t)\,dt$$

Usually the integral expressing $R_n(x)$ cannot be computed exactly. Nevertheless, the integral can be estimated. One important estimate depends on a bound for the $(n + 1)$-th derivative, $f^{(n+1)}(x)$.

Estimate of Remainder

Let $f(x) = P_n(x) + R_n(x)$, where $P_n(x)$ is the n-th Taylor polynomial of $f(x)$ at $x = a$. Suppose that

$$|f^{(n+1)}(x)| \le M$$

in some interval including a, say $b \le a \le c$. Then

$$|R_n(x)| \le \frac{M}{(n+1)!}|x - a|^{n+1} \quad \text{for} \quad b \le x \le c$$

Let us assume Taylor's formula temporarily. (It will be derived at the end of this section.) The remainder estimate then follows easily:

$$\begin{aligned}
|R_n(x)| &= \frac{1}{n!}\left|\int_a^x (x - t)^n f^{(n+1)}(t)\,dt\right| \\
&\le \frac{1}{n!}\left|\int_a^x |(x - t)^n f^{(n+1)}(t)|\,dt\right| \\
&\le \frac{M}{n!}\left|\int_a^x |(x - t)^n|\,dt\right| = \frac{M}{(n+1)!}|x - a|^{n+1}
\end{aligned}$$

Note We have used inequality 4, page 313. The absolute value signs outside of the integral take care of the possibility that $x < a$.

Example 1

Let e^x be approximated by its n-th degree Taylor polynomial at $x = 0$. Estimate the remainder, assuming $|x| \le B$.

Solution The $(n + 1)$-th derivative is $f^{(n+1)}(x) = e^x$. If $x \ge 0$, then the largest value of $f^{(n+1)}(x)$ between 0 and B is $e^B < 3^B$. By the remainder estimate with $M = 3^B$,

$$|R_n(x)| \le \frac{3^B}{(n+1)!}x^{n+1} \quad \text{for} \quad 0 \le x \le B$$

If $x \le 0$, then the largest value of $f^{(n+1)}(t)$ between 0 and $-B$ is $e^0 = 1$. By the remainder estimate with $M = 1$,

$$|R_n(x)| \le \frac{|x|^{n+1}}{(n+1)!} \quad \text{for} \quad x \le 0$$

●

How can we compute a five-place table of e^x for a certain range of x? One method is to approximate e^x by a Taylor polynomial $P_n(x)$ with n large enough that $|R_n(x)| < 5 \times 10^{-6}$ on the given range, then use values of P_n. (For the sake of economy we should use the smallest n that does the trick.) This is a practical method because computing many values of a polynomial is relatively easy, especially with a computer. (There are other ways of approaching this problem, for instance by interpolation.)

Remark When we say that 1.12750 is a five-place approximation to $e^{0.12}$, we mean that $|e^{0.12} - 1.12750| \le 5 \times 10^{-6}$. Similarly 1.127497 is a six-place approximation to $e^{0.12}$.

Example 2

Use the conclusion of Example 1 to estimate the smallest n such that the remainder in Taylor's formula for e^x at $x = 0$ satisfies the inequality $|R_n(x)| < 5 \times 10^{-6}$ for

a $|x| < 0.5$ **b** $|x| < 1$

Solution By Example 1

$$|R_n(x)| < \frac{3^B}{(n+1)!} B^{n+1} \qquad \text{for} \quad |x| \le B$$

The problem is to choose n as small as possible so that

$$\frac{3^B}{(n+1)!} B^{n+1} < 5 \times 10^{-6}$$

a Here $B = \frac{1}{2}$, so we want

$$\frac{3^{1/2}}{(n+1)!} \left(\frac{1}{2}\right)^{n+1} < 5 \times 10^{-6}$$

that is,

$$2^{n+1}(n+1)! > \frac{\sqrt{3}}{5 \times 10^{-6}}$$

Now

$$\frac{\sqrt{3}}{5} \times 10^6 < \frac{2}{5} \times 10^6 = 4 \times 10^5$$

so it suffices to choose n to satisfy the inequality

$$2^{n+1}(n+1)! \ge 4 \times 10^5$$

By trial and error,

$$2^7 \times 7! > 6 \times 10^5 \qquad \text{but} \qquad 2^6 \times 6! < 5 \times 10^4$$

The correct choice is $n + 1 = 7$, that is, $n = 6$.

b This time $B = 1$. By similar reasoning, n must satisfy

$$\frac{3}{(n+1)!} < 5 \times 10^{-6}$$

so it suffices to choose n such that

$$(n+1)! > \frac{3}{5} \times 10^6 = 6 \times 10^5$$

Since $9! < 4 \times 10^5$ and $10! > 3 \times 10^6$, the correct choice now is $n + 1 = 10$, that is, $n = 9$. ●

Remark By calculation, $|R_5(0.5)| \approx 2.3 \times 10^{-5}$, so $n = 5$ doesn't work; 6 is the smallest possible n for part **a**. For part **b**, $n = 9$ is not the smallest value that works. It can be shown (with some difficulty) that $|R_8(x)| < 4 \times 10^{-6}$ for $|x| < 1$. Since $|R_7(1)| \approx 2.8 \times 10^{-5}$, we see that $n = 8$ is the smallest possible. Can you see where we lost ground in our estimates? We often are closer to target than the error bound shows.

Example 3

Estimate the remainders in Taylor's formula for

a $\sin x$ **b** $\cos x$ (Refer to Example 4, page 494.)

Solution

a A bound for the $(n+1)$-th derivative is easy: $|f^{(n+1)}(x)| \le 1$ because $f^{(n+1)}(x) = \pm\sin x$ or $\pm\cos x$. Hence

$$|R_n(x)| \le \frac{|x|^{n+1}}{(n+1)!}$$

For $f(x) = \sin x$, we showed that $P_{2m-1}(x) = P_{2m}(x)$; it follows that

$$|R_{2m-1}(x)| = |R_{2m}(x)| \le \frac{|x|^{2m+1}}{(2m+1)!}$$

b The estimate for $|R_n(x)|$ is exactly the same as in part **a**. However, for $f(x) = \cos x$ we showed that $P_{2m}(x) = P_{2m+1}(x)$. It follows that

$$|R_{2m}(x)| = |R_{2m+1}(x)| \le \frac{|x|^{2m+2}}{(2m+2)!}$$ ●

Example 4

Find the lowest degree Taylor polynomial for $\sin x$ at $x = 0$ such that the estimate of Example 3a implies

$$|R_n(x)| < 5 \times 10^{-6} \quad \text{for} \quad |x| \le \tfrac{1}{4}\pi$$

Solution By Example 3a,

$$|R_{2m-1}(x)| \le \frac{|x|^{2m+1}}{(2m+1)!}$$

Hence we want to choose m so that

$$\frac{(\frac{1}{4}\pi)^{2m+1}}{(2m+1)!} < 5 \times 10^{-6}$$

By trial and error we find

$$\frac{(\frac{1}{4}\pi)^7}{7!} > \frac{(\frac{3}{4})^7}{7!} > 2 \times 10^{-5} \quad \text{and} \quad \frac{(\frac{1}{4}\pi)^9}{9!} < \frac{1}{9!} < 3 \times 10^{-6}$$

Therefore the correct choice is $2m + 1 = 9$, that is, $m = 4$. The corresponding Taylor polynomial is

$$P_{2m-1}(x) = P_7(x) = x - \frac{x^3}{3!} + \frac{x^5}{5!} - \frac{x^7}{7!}$$ ●

Remark Since $|R_5(\frac{1}{4}\pi)| \approx 3.6 \times 10^{-5}$, the smallest possible n is 7. Compare the remark after Example 2.

Example 5

If $f(x) = \ln x$ is approximated by its n-th degree Taylor polynomial at $x = 1$, estimate the remainder for $|x - 1| < 0.5$. (Refer to Example 3b, page 493.)

Solution As was shown in that example, $f^{(n+1)}(x) = \pm n!/x^{n+1}$. If $|x - 1| < 0.5$, then $x > 0.5$. Hence

$$|f^{(n+1)}(x)| < \frac{n!}{(0.5)^{n+1}} = 2^{n+1}n!$$

Therefore

$$|R_n(x)| \le \frac{2^{n+1}n!}{(n+1)!}|x-1|^{n+1} = \frac{|2(x-1)|^{n+1}}{n+1} < \frac{1}{n+1}$$

since

$$|2(x-1)| < 1 \quad \text{for} \quad |x-1| < 0.5$$ ●

Lagrange Form of the Remainder

There are several useful ways to write the remainder in Taylor's formula. The integral formula we have used is often referred to as the **Cauchy form.** Another popular remainder formula uses a derivative instead of an integral; this **Lagrange form of the remainder** is

$$R_n(x) = \frac{f^{(n+1)}(c)}{(n+1)!}(x-a)^{n+1}$$

where c is between a and x. It can be proved with a slightly weaker assumption than that needed for the Cauchy form, namely that $f^{(n+1)}(x)$ exists, but is not necessarily continuous. We omit the rather technical proof.

Proof of Taylor's Formula [Optional]

Recall that $P_n(x)$ is constructed so that

$$P_n(a) = f(a)$$
$$P_n'(a) = f'(a)$$
$$P_n''(a) = f''(a)$$
$$\cdots$$
$$P_n^{(n)}(a) = f^{(n)}(a)$$

Consequently the remainder $R_n(x) = f(x) - P_n(x)$ satisfies

$$R_n(a) = R_n'(a) = R_n''(a) = \cdots = R_n^{(n)}(a) = 0$$

The following lemma shows that such a function can be expressed as an integral.

Lemma Let $g(x)$ have continuous derivatives near $x = a$ up to and including $g^{(n+1)}(x)$ and suppose

$$g(a) = g'(a) = g''(a) = \cdots = g^{(n)}(a) = 0$$

Then

$$g(x) = \frac{1}{n!}\int_a^x (x-t)^n g^{(n+1)}(t)\,dt$$

●

We shall prove the lemma in a moment. To derive Taylor's formula from the lemma, we take $g(x) = R_n(x) = f(x) - P_n(x)$. Then $g(x)$ satisfies the hypotheses of the lemma and $g^{(n+1)}(x) = f^{(n+1)}(x)$ since $P_n(x)$ is a polynomial of degree n or less. Hence

$$R_n(x) = \frac{1}{n!}\int_a^x (x-t)^n f^{(n+1)}(t)\,dt$$

Proof of the Lemma The lemma is proved by repeated integration by parts. Fix a and x, set

$$u(t) = \frac{(x-t)^n}{n!} \quad \text{and} \quad v(t) = g^{(n)}(t)$$

Then

$$du = \frac{-(x-t)^{n-1}}{(n-1)!}\,dt \quad \text{and} \quad dv = g^{(n+1)}(t)\,dt$$

The integral in the lemma is

$$\frac{1}{n!}\int_a^x (x-t)^n g^{(n+1)}(t)\,dt = \int_a^x u(t)\,dv(t)$$

$$= u(t)v(t)\Big|_a^x - \int_a^x v(t)\,du(t)$$

$$= 0 - 0 + \frac{1}{(n-1)!}\int_a^x (x-t)^{n-1}g^{(n)}(t)\,dt$$

The two zeros come from

$$u(x)v(x) = 0 \cdot v(x) = 0$$

and

$$u(a)v(a) = u(a)g^{(n)}(a) = 0$$

Therefore

$$\frac{1}{n!}\int_a^x (x-t)^n g^{(n+1)}(t)\,dt$$

$$= \frac{1}{(n-1)!}\int_a^x (x-t)^{n-1}g^{(n)}(t)\,dt$$

Thus n has been decreased by one. Repeat the process (mathematical induction) until the exponent of $x - t$ reaches 0:

$$\frac{1}{n!}\int_a^x (x-t)^n g^{(n+1)}(t)\,dt = \cdots = \frac{1}{0!}\int_a^x g'(t)\,dt$$

$$= g(x) - g(a) = g(x)$$

This completes the proof of the lemma. ●

Exercises

Find $P_n(x)$ and some upper bound for $|R_n(x)|$ in terms of $x - a$

1 $\sin 2x \quad a = 0$

2 $\sin 2x \quad a = \frac{1}{2}\pi$

3 $xe^x \quad a = 0$

4 $xe^x \quad a = 1$

5 $x^2 \ln x \quad a = 1$

6 $x^2 \ln x \quad a = e$

7 $x^2e^{-x} \quad a = 0$

8 $x^2e^{-x} \quad a = 1$

9 $x \sin x \quad a = 0$

10 $x \cos x \quad a = 0$

11 $\sin x + \cos x \quad a = 0$

12 $\cosh x \quad a = 0$

13 $\sinh x \quad a = 0$

14 $e^x + e^{2x} \quad a = 0$

15 $1/(1 + x) \quad a = 1$

***16** $x/(1 + x^4) \quad a = 0$

17 Estimate the error in approximating $f(x) = \ln(1 + x)$ for $-\frac{1}{3} \le x \le \frac{1}{3}$ by its tenth degree Taylor polynomial near $x = 0$.

18 Approximate $f(x) = 1/(1 - x)^2$ for $-\frac{1}{4} \le x \le \frac{1}{4}$ to three decimal places by a Taylor polynomial about $x = 0$.

19 Approximate $\sin^2 x$ by its fourth degree Taylor polynomial near $x = 0$. Estimate the error if $|x| \le 0.1$. [Hint $\sin^2 x = \frac{1}{2}(1 - \cos 2x)$.]

20 Show that for $100 \le x \le 101$, the approximation $\sqrt{x} \approx 10 + \frac{1}{20}(x - 100)$ is correct to within 0.0002.

Let $R_n(x)$ be the remainder for $\sin x$ at $a = 0$. Show that

21 $|R_3(x)| < 5 \times 10^{-6}$ for $|x| < 0.22 \approx 12.6°$

22 $|R_5(x)| < 5 \times 10^{-6}$ for $|x| < 0.59 \approx 33.8°$

23 $|R_7(x)| < 5 \times 10^{-6}$ for $|x| < 1.06 \approx 60.7°$

24 $|R_9(x)| < 5 \times 10^{-6}$ for $|x| < 1.61 \approx 92.2°$

25 Find k so that $R_2(x)$ for $\cos x$ at $a = 0$ satisfies $|R_2(x)| < 5 \times 10^{-6}$ for $|x| < k\pi$.

26 (cont.) How can you estimate quite simply $\sin(\frac{19}{40}\pi)$ to five places?

27 How many terms of the Taylor polynomial for $\ln x$ at $a = 1$ are needed to compute $\ln 1.25$ to five places?

28 (cont.) The same for $\ln 0.75$.

29 Show that $P_3(x)$ yields five-place accuracy in the approximation of $\sin x$ at $a = \frac{1}{4}\pi$ provided $|x - \frac{1}{4}\pi| < 0.1 \approx 5.7°$.

30 (cont.) Show that $P_3(x)$ yields eight-place accuracy for $44° < x < 46°$. Write out $P_3(x)$.

31 Find $P_3(x)$ for $f(x) = \sqrt{1-x}$ at $a = 0$, and estimate the error.

32 In Exercise 28 in Section 3-3, we found that the diameter D of a hole in which a needle gauge of length L has "rock" x is $D = L^2/\sqrt{L^2 - x^2}$. Justify the approximation

$$D \approx L + \frac{x^2}{2L} + \frac{3x^4}{8L^3}$$

*33 If the interest rate is i and interest is compounded monthly, then money will double in n years, where

$$\left(1 + \frac{i}{12}\right)^{12n} = 2 \qquad \text{hence} \qquad n = \frac{\frac{1}{12}\ln 2}{\ln(1 + i/12)}$$

Prove the "rule of 72": $n \approx 72/I$ where $I = 100i$ is the interest rate in %. [Hint Find $P_1(I)$ at $I = 50$.]

*34 (cont.) Prove the more accurate estimate

$$n \approx \frac{69.3}{I} + 0.003$$

Test these estimates for $I =$ 6%, 10%, 16%, and 20%.

Use Taylor's formula (not l'Hospital's rule) to find

35 $\displaystyle\lim_{x\to 0} \frac{e^x - 1 - x - \frac{1}{2}x^2}{x^3}$

36 $\displaystyle\lim_{x\to 0} \frac{\cos x - 1 + \frac{1}{2}x^2}{x^4}$

37 $\displaystyle\lim_{x\to 0} \frac{\sin x - x + \frac{1}{6}x^3}{x^5}$

38 $\displaystyle\lim_{x\to 0} \frac{\tan x - x}{x^3}$

39 Suppose $f(0) = g(0) = 0$, and suppose $f'(x)$ and $g'(x)$ are continuous near 0 and $g'(0) \neq 0$. Use Taylor approximations to prove

$$\lim_{x\to 0} \frac{f(x)}{g(x)} = \frac{f'(0)}{g'(0)}$$

[Hint Use the Lagrange form of the remainder R_0.]

*40 (cont.) More generally, suppose

$$f(0) = f'(0) = \cdots = f^{(n)}(0)$$
$$= g(0) = g'(0) = \cdots = g^{(n)}(0) = 0$$

and suppose $f^{(n+1)}(x)$ and $g^{(n+1)}(x)$ are continuous near 0 and $g^{(n+1)}(0) \neq 0$. Prove $g(x) \neq 0$ for $0 < |x| < \delta$ with δ sufficiently small and

$$\lim_{x\to 0} \frac{f(x)}{g(x)} = \frac{f^{(n+1)}(0)}{g^{(n+1)}(0)}$$

*41 Suppose $f^{(n+1)}(x)$ is continuous near $x = a$ and $f'(a) = f''(a) = \cdots = f^{(n)}(a) = 0$. Suppose n is odd. Prove if $f^{(n+1)}(a) > 0$, then $f(a)$ is a local minimum; if $f^{(n+1)}(a) < 0$, then $f(a)$ is a local maximum. (This generalizes the second derivative test.) [Hint Use the Lagrange form of R_n.]

*42 (cont.) Suppose n is even and $f^{(n+1)}(a) \neq 0$. What conclusion can you draw?

*43 Suppose in the integral for $R_{n-1}(x)$ in Taylor's formula you make the change of variable from t to u given by $t = a + u(x - a)$. Give the resulting formula for $R_n(x)$.

*44 (cont.) Use this formula to prove that if $f(x)$ has $n + 1$ continuous derivatives near $x = a$ and

$$f(a) = f'(a) = f''(a) = \cdots = f^{(n-1)}(a) = 0$$

then $f(x) = (x - a)^n g(x)$ where $g(x)$ is continuous near $x = a$, including at $x = a$ itself.

9-6 Approximate Integration

In Section 5-6 we learned two methods of approximate integration, the trapezoidal rule and the midpoint rule. Let us recall the trapezoidal rule:

$$\int_a^b f(x)\,dx \approx \tfrac{1}{2}h[f_0 + 2f_1 + \cdots + 2f_{n-1} + f_n]$$

where $h = (b - a)/n$ and $f_j = f(a + jh)$. We also stated an error estimate for the trapezoidal rule: If $|f''(x)| \leq M$ for $[a, b]$, then

$$|\text{error}| \le \frac{M(b-a)}{12} h^2$$

When $n = 1$, the trapezoidal rule says

$$\int_a^b f(x)\,dx \approx \frac{b-a}{2}[f(a) + f(b)] = \int_a^b p_1(x)\,dx$$

where $p_1(x)$ is an approximation to $f(x)$, namely, the *linear* function that passes through $(a, f(a))$ and $(b, f(b))$.

Simpson's Three-Point Rule

For greater accuracy, we approximate $f(x)$ by the *quadratic* function $p(x)$ that passes through $(a, f(a))$ and $(b, f(b))$, and also through

$$(c, f(c)) \quad \text{where} \quad c = \tfrac{1}{2}(a + b)$$

(Note that c is the midpoint of the interval of integration $[a, b]$.) Then we approximate the integral of $f(x)$ by the integral of $p(x)$.

Let us use the notation

$$h = \tfrac{1}{2}(b - a) \qquad x_0 = a \qquad x_1 = a + h \qquad x_2 = b = a + 2h$$

Given a quadratic polynomial $p(x)$, we need a formula for

$$\int_a^b p(x)\,dx$$

in terms of $p_0 = p(x_0)$, $p_1 = p(x_1)$, and $p_2 = p(x_2)$:

Lemma If $p(x)$ is a quadratic polynomial, then

$$\int_a^b p(x)\,dx = \tfrac{1}{3}h[p_0 + 4p_1 + p_2]$$

where $h = \tfrac{1}{2}(b - a)$. ●

Proof Write $p(x) = A + B(x - x_1) + C(x - x_1)^2$. The left side of the equation to be proved is

$$\int_a^b p(x)\,dx = \int_{x_1-h}^{x_1+h} [A + B(x - x_1) + C(x - x_1)^2]\,dx$$
$$= 2Ah + \tfrac{2}{3}Ch^3$$

The right side is

$$\tfrac{1}{3}h(p_0 + 4p_1 + p_2) = \tfrac{1}{3}h[(A - Bh + Ch^2) + 4A + (A + Bh + Ch^2)]$$
$$= \tfrac{1}{3}h(6A + 2Ch^2) = 2Ah + \tfrac{2}{3}Ch^3$$

The two sides agree. ●

Remark The formula also holds for the special cubic polynomial $p(x) = (x - x_1)^3$. In fact, the left side equals 0 because $y = p(x)$ is symmetric about the point $(x_1, 0)$, and the right side equals 0 because $p_1 = 0$ and $p_0 = -p_2$. But any cubic can be written as the sum of a quadratic and $D(x - x_1)^3$, so we conclude that the formula is correct not only for quadratics, but also for cubic polynomials—an important bonus.

Given a continuous function $f(x)$ on the interval $[a, b]$, let $p(x)$ be the quadratic polynomial with the same values as $f(x)$ at the three points $x = x_0$, $x = x_1$, and $x = x_2$. Then $p_0 = f_0$, $p_1 = f_1$, and $p_2 = f_2$. Since $p(x)$ is an approximation to $f(x)$, its integral is an approximation to the integral of $f(x)$. Therefore

$$\int_a^b f(x)\,dx \approx \int_a^b p(x)\,dx = \tfrac{1}{3}h[p_0 + 4p_1 + p_2]$$

$$= \tfrac{1}{3}h[f_0 + 4f_1 + f_2]$$

Simpson's Three-Point Rule

$$\int_a^b f(x)\,dx \approx \tfrac{1}{3}h[f_0 + 4f_1 + f_2]$$

where $h = \tfrac{1}{2}(b - a)$ and

$$f_0 = f(a) \qquad f_1 = f(a - h) \qquad f_2 = f(b) = f(a + 2h)$$

Remark As noted above, Simpson's three-point rule is exact, not just an approximation, provided $f(x)$ is a polynomial of degree three or less.

Example 1

Use Simpson's three-point rule to estimate

a $\displaystyle\int_0^\pi \sin x\,dx$ **b** $\displaystyle\int_0^1 e^x\,dx$

Bound the error in each case.

Solution

a Here $h = \tfrac{1}{2}\pi$, so

$$\int_0^\pi \sin x\,dx \approx \tfrac{1}{3}(\tfrac{1}{2}\pi)(\sin 0 + 4\sin\tfrac{1}{2}\pi + \sin\pi)$$

$$= \tfrac{2}{3}\pi \approx 2.0944$$

The exact value is

$$\int_0^\pi \sin x\,dx = -\cos x\Big|_0^\pi = 2$$

Therefore $|\text{error}| < 0.095 = 9.5 \times 10^{-2}$.

Figure 9-6-1
Good and poor approximations to different functions $f(x)$ by a quadratic $p(x)$

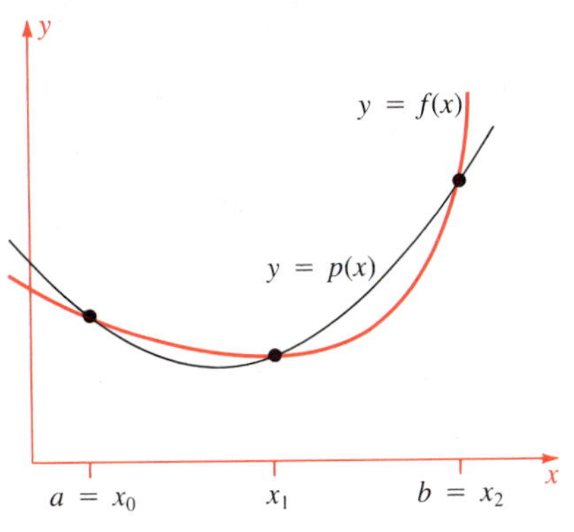

a Good approximation

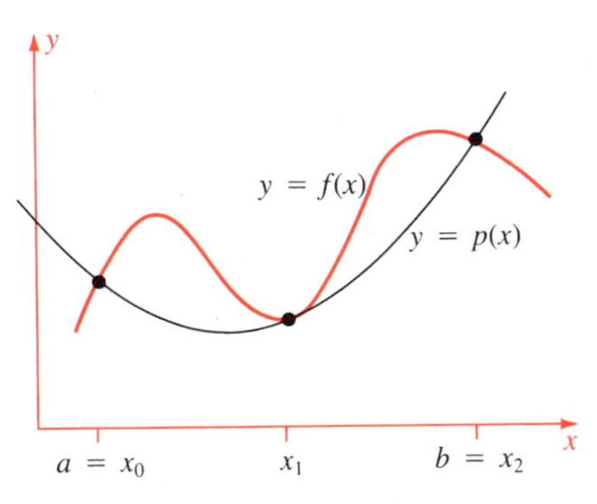

b Poor approximation

b In this case $h = \frac{1}{2}$, and

$$\int_0^1 e^x\,dx \approx \tfrac{1}{3}(\tfrac{1}{2})(1 + 4e^{1/2} + e) \approx 1.7189$$

But to four places,

$$\int_0^1 e^x\,dx = e^x\Big|_0^1 = e - 1 \approx 1.7183$$

Hence $|\text{error}| < 0.001 = 10^{-3}$. ●

General Simpson's Rule

Does Simpson's three-point rule always provide a good approximation to the integral? That depends on the graph of $f(x)$. See Figure 9-6-1. The same interpolating quadratic $p(x)$ appears in Figure 9-6-1a and Figure 9-6-1b since the three points that determine $p(x)$ are the same. Yet the approximation

$$\int_a^b f(x)\,dx \approx \int_a^b p(x)\,dx$$

is good in Figure 9-6-1a and poor in Figure 9-6-1b.

In the first case, $p(x)$ approximates $f(x)$ closely. Furthermore,

$$\int_{x_0}^{x_1} p(x)\,dx < \int_{x_0}^{x_1} f(x)\,dx$$

whereas

$$\int_{x_1}^{x_2} p(x)\,dx > \int_{x_1}^{x_2} f(x)\,dx$$

so the errors partly cancel.

In Figure 9-6-1b, however, $p(x)$ is a poor approximation to $f(x)$. To make matters worse, the errors do not cancel; they accumulate. The trouble is that the points x_0, x_1, x_2 are too widely spaced. It would be much better to apply the three-point rule twice, once from x_0 to x_1 and once from x_1 to x_2 (Figure 9-6-2).

In general, to improve the approximation we divide $[a, b]$ into an even number of equal pieces and apply Simpson's three-point rule serially, taking three division points at a time.

Figure 9-6-2
Three-point rule applied twice: on $[x_0, x_1]$ and on $[x_1, x_2]$

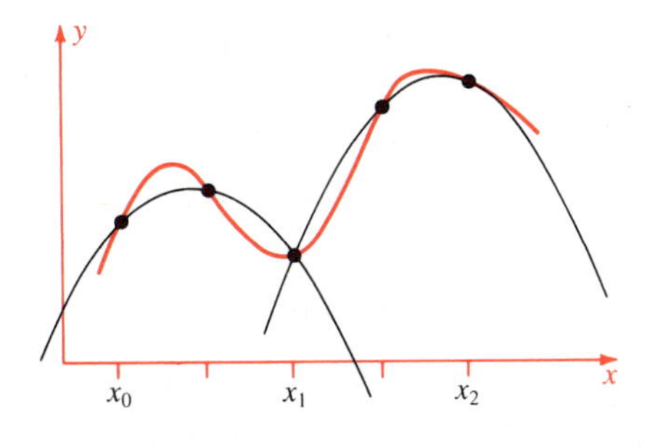

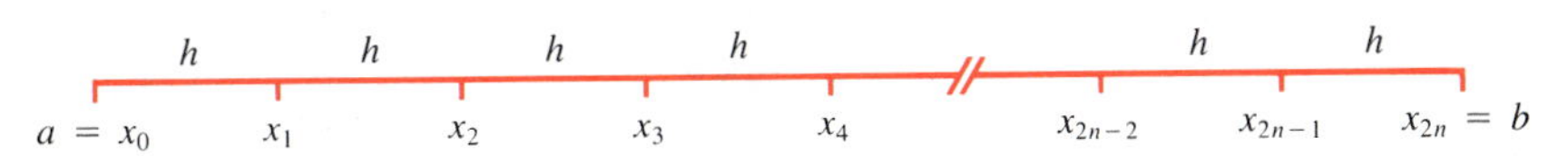

Let us divide $[a, b]$ into $2n$ subintervals, each of the same length $h = (b - a)/2n$. We write

$$\int_a^b f(x)\,dx = \int_{x_0}^{x_2} + \int_{x_2}^{x_4} + \int_{x_4}^{x_6} + \cdots + \int_{x_{2n-2}}^{x_{2n}}$$

and apply three-point approximation to each integral on the right side:

$$\int_{x_0}^{x_2} f(x)\,dx \approx \tfrac{1}{3}h[f_0 + 4f_1 + f_2]$$

$$\int_{x_2}^{x_4} f(x)\,dx \approx \tfrac{1}{3}h[f_2 + 4f_3 + f_4]$$

$$\int_{x_4}^{x_6} f(x)\,dx \approx \tfrac{1}{3}h[f_4 + 4f_5 + f_6]\ \ldots$$

$$\int_{x_{2n-2}}^{x_{2n}} f(x)\,dx \approx \tfrac{1}{3}h[f_{2n-2} + 4f_{2n-1} + f_{2n}]$$

Adding these estimates, we obtain Simpson's rule:

Simpson's Rule

$$\int_a^b f(x)\,dx \approx \tfrac{1}{3}h[f_0 + 4f_1 + 2f_2 + 4f_3 + 2f_4 + 4f_5 + \cdots + 2f_{2n-2} + 4f_{2n-1} + f_{2n}]$$

where $h = \frac{1}{2}(b - a)/n$ and $f_0, f_1, \cdots, f_{2n}$ are the values of f at the successive points of division of $[a, b]$ into $2n$ equal parts.

Note the sequence of coefficients:

1 4 2 4 2 4 $\cdots$ 4 2 4 1

The number of subintervals of $[a, b]$ is $2n$ (even). The number of points of division, counting the end points, is $2n + 1$ (odd). These odd numbers are used to describe various versions of Simpson's rule. Thus a seven-point Simpson approximation refers to the following division:

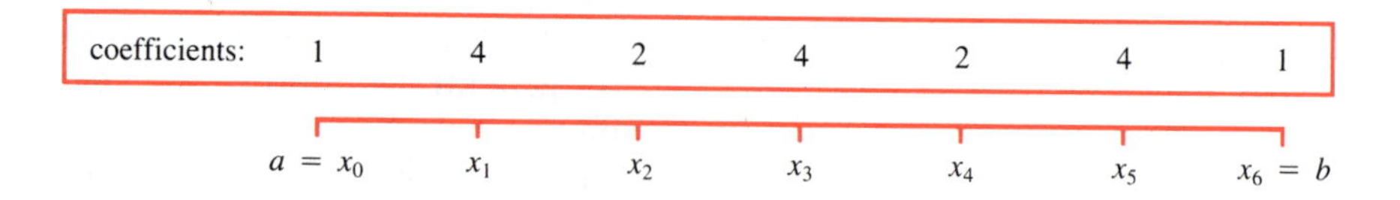

In this case $n = 3$, $2n = 6$, $2n + 1 = 7$, and Simpson's rule states that

$$\int_a^b f(x)\,dx \approx \tfrac{1}{3}h[f_0 + 4f_1 + 2f_2 + 4f_3 + 2f_4 + 4f_5 + f_6]$$

Remark Simpson's rule is exact if $f(x)$ is a cubic polynomial. That is the case because the same is true of Simpson's three-point rule.

We shall discuss the error in Simpson's rule after some examples.

Example 2

Use Simpson's rule with five and with eleven points to estimate

$$\int_{-1}^{1} x^4\,dx$$

to four places. Compare the estimates with the exact value.

Solution The exact value is

$$\int_{-1}^{1} x^4\,dx = \tfrac{1}{5}x^5\Big|_{-1}^{1} = \tfrac{2}{5} = 0.4000$$

If $2n + 1 = 5$, then $n = 2$ and $h = (b - a)/2n = \frac{1}{2}$. The approximation is

$$\begin{aligned}\int_{-1}^{1} x^4\,dx &\approx \tfrac{1}{3}h[f_0 + 4f_1 + 2f_2 + 4f_3 + f_4]\\ &= \tfrac{1}{6}[1 + 4(-\tfrac{1}{2})^4 + 0 + 4(\tfrac{1}{2})^4 + 1]\\ &= \tfrac{1}{6}\cdot\tfrac{5}{2} = \tfrac{5}{12} \approx 0.4167\end{aligned}$$

so $|\text{error}| < 0.02$.

Similarly, if $2n + 1 = 11$, then $n = 5$, $h = \frac{2}{10}$, and

$$\begin{aligned}\int_{-1}^{1} x^4\,dx &\approx \tfrac{1}{15}[1 + 4(-\tfrac{8}{10})^4 + 2(-\tfrac{6}{10})^4 + 4(-\tfrac{4}{10})^4 + 2(-\tfrac{2}{10})^4\\ &\qquad + 0 + 2(\tfrac{2}{10})^4 + 4(\tfrac{4}{10})^4 + 2(\tfrac{6}{10})^4 + 4(\tfrac{8}{10})^4 + 1]\\ &= \tfrac{2}{15}\times 10^{-4}(10^4 + 4\times 8^4 + 2\times 6^4\\ &\qquad + 4\times 4^4 + 2\times 2^4)\\ &= \tfrac{2}{15}\times 10^{-4}\times 30{,}032 \approx 0.4004\end{aligned}$$

This time $|\text{error}| < 5\times 10^{-4}$. ●

It pays to exploit symmetry when applying Simpson's rule. In this way, you can generally improve accuracy with no additional computation. Compare the following example with the last one.

Example 3

Use Simpson's rule with five points to estimate $\int_{-1}^{1} x^4\,dx$ to four places. Exploit symmetry.

Solution The integrand is an even function; hence

$$\int_{-1}^{1} x^4\,dx = 2\int_{0}^{1} x^4\,dx$$

Now use Simpson with five points on $[0, 1]$, which is as good as using Simpson with nine points on $[-1, 1]$. The result is

$$\int_{-1}^{1} x^4\, dx \approx \tfrac{2}{12}[0 + 4(\tfrac{1}{4})^4 + 2(\tfrac{1}{2})^4 + 4(\tfrac{3}{4})^4 + 1] \approx 0.4010$$

Clearly $|\text{error}| < 0.0015$, which is better than what we got in Example 2 with the five-point approximation: $|\text{error}| < 0.02$. ●

Example 4

The *exact* value of

$$\int_0^2 \exp(-x^2)\, dx$$

cannot be found by methods of calculus. To eleven places it equals 0.88208 10350 6. Table 9-6-1 gives the results of several trapezoidal and Simpson approximations to this integral. The table shows that Simpson's rule is much more precise than the trapezoidal rule. For instance, with five points of subdivision, the errors are approximately 0.0015 (trapezoidal) versus 0.00027 (Simpson); with eleven points of subdivision, 0.00024 (trapezoidal) versus 0.000008 (Simpson). ●

Table 9-6-1

Approximations to $\int_0^2 \exp(-x^2)\, dx$

Trapezoidal		*Simpson*	
$n+1$	$\int$	$2n+1$	$\int$
3	0.877037	3	0.829944
5	0.880618	5	0.881812
7	0.881415	7	0.882031
9	0.881704	9	0.882066
11	0.881837	11	0.882073
21	0.882020	21	0.882081

An Application to Volume [Optional]

A **prismoid** is a region in space bounded by two parallel planes and one or more surfaces joining the planes (Figure 9-6-3). There is a nice formula for the volume V of a prismoid whose cross-sectional area $f(x)$ is a cubic function of the height x, measured along an axis perpendicular to the base planes:

$$V = \tfrac{1}{6}h(A_0 + 4M + A_1)$$

where h is the distance between base planes, A_0 and A_1 are the areas of the bases, and M is the cross-sectional area halfway between the bases. This is sometimes called the **prismoidal formula.** The proof is a direct application of Simpson's rule (Exercise 29). Note that "cubic" includes linear and quadratic polynomials, the usual cases in applications of the prismoidal formula.

Figure 9-6-3
Prismoid: $A(x)$ = a cubic in x

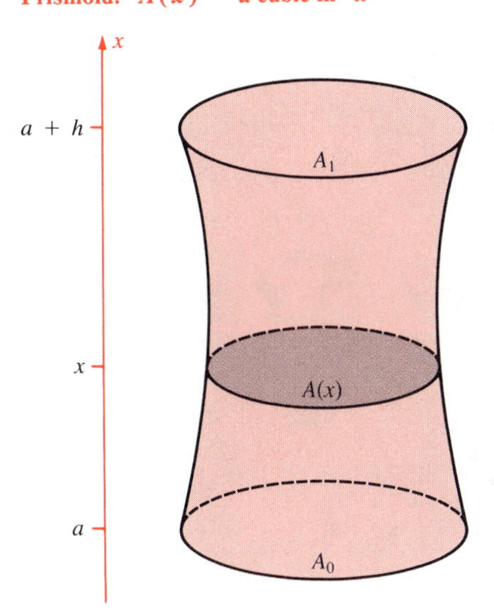

Example 5

Find the volume of a sphere of radius a by the prismoidal formula.

Solution The cross-sectional area is a quadratic function of the height. Hence the formula applies with

$$h = 2a \qquad A_0 = A_1 = 0 \qquad M = \pi a^2$$

Conclusion: $V = \frac{1}{6}(2a)(0 + 4\pi a^2 + 0) = \frac{4}{3}\pi a^3$ ●

Error in Simpson's Rule [Optional]

Simpson's rule is exact for cubic polynomials. Now $f(x)$ is a cubic if and only if $f^{(4)}(x) = 0$. Hence $f^{(4)}(x)$ measures how far $f(x)$ differs from being a cubic, so it is reasonable that $|f^{(4)}(x)|$ should appear in error estimates from Simpson's rule. The following estimate for the three-point rule is proved in courses in numerical analysis.

Error in Simpson's Rule

Suppose $f(x)$ has a continuous fourth derivative on $[a, b]$, and $|f^{(4)}(x)| \le M$. Then

$$\int_a^b f(x)\,dx = \tfrac{1}{3}h(f_0 + 4f_1 + f_2) + E \quad \text{where} \quad |E| \le \frac{Mh^5}{90}$$ ●

To obtain an error estimate for the general Simpson's rule, we divide the interval into $2n$ parts, and apply the preceding estimate n times, taking $h = (b-a)/2n$. This yields

$$|\text{error}| \le \frac{Mnh^5}{90} = \frac{M}{90 \times 2^5} \cdot \frac{(b-a)^5}{n^4} = \frac{M}{2880} \cdot \frac{(b-a)^5}{n^4}$$

Thus the maximum error is proportional to $1/n^4$. That is why a large n yields high accuracy.

The corresponding error estimate for the trapezoidal rule, where the interval is divided into n parts and $|f''(x)| \le M$ on the interval, is

$$|\text{error}| \le \frac{M}{12} \cdot \frac{(b-a)^3}{n^2}$$

It should be clear that even for a modest-size n, Simpson's rule is much more accurate. Note, however, that the n in Simpson's rule means $2n$ subintervals, whereas in the trapezoidal rule it means n subintervals. Thus the correct comparison is $2n + 1$ Simpson with $2n$ trapezoidal:

$$|\text{error(Simpson)}| \le \frac{1}{2880} M_4 \frac{(b-a)^5}{n^4} \qquad [|f^{(4)}(x)| \le M_4]$$

$$|\text{error(trapezoidal)}| \le \frac{1}{48} M_2 \frac{(b-a)^3}{n^2} \qquad [|f''(x)| \le M_2]$$

In both cases we are evaluating the function at exactly the same $2n + 1$ points.

Example 6

Estimate $\ln 2$ by Simpson's rule with $h = 0.1$. Find a bound for the error.

Solution Set $f(x) = 1/x$. Then

$$\ln 2 = \int_1^2 f(x)\,dx$$

To have $h = 0.1$ we take $n = 5$, dividing the interval $[1, 2]$ into 10 parts. Simpson's rule yields

$$\ln 2 \approx \tfrac{1}{3}h(f_0 + 4f_1 + 2f_2 + \cdots + 4f_9 + f_{10})$$

$$= \tfrac{1}{3}(0.1)\left(\frac{1}{1} + \frac{4}{1.1} + \frac{2}{1.2} + \frac{4}{1.3} + \frac{2}{1.4} + \frac{4}{1.5} + \frac{2}{1.6} + \frac{4}{1.7} + \frac{2}{1.8} + \frac{4}{1.9} + \frac{1}{2}\right)$$

$$\approx 0.693150$$

For the error we have the bound

$$|\text{error}| \le \frac{M}{2880} \cdot \frac{1}{5^4} \qquad \text{where} \qquad |f^{(4)}(x)| \le M$$

But $f^{(4)}(x) = 2 \cdot 3 \cdot 4/x^5 = 24/x^5$ with its largest value $f^{(4)}(1) = 24$. Therefore

$$|\text{error}| \le \frac{24}{2880 \times 5^4} < 1.4 \times 10^{-5}$$

This is merely an upper bound for the error. Actually, to six places, $\ln 2 \approx 0.693147$, so the true error is less than 4×10^{-6}. ●

Exercises

Estimate to four places by Simpson's three-point rule, and give an estimate of $|\text{error}|$

1 $\int_0^{\pi/2} \sin x\,dx$

2 $\int_{\pi/2}^{\pi} \sin x\,dx$

3 $\int_{-\pi/4}^{\pi/4} \cos x\,dx$

4 $\int_{-1}^{0} e^x\,dx$

5 $\int_1^2 e^x\,dx$

6 $\int_0^1 \frac{dx}{1 + x^2}$

7 $\int_1^2 \frac{dx}{x^2}$

8 $\int_{-1}^{1} x^5\,dx$

In Exercises 9–20, remember that Simpson's rule with n implies $2n$ subintervals. Estimate to five places by Simpson's rule with $n = 2$ and $n = 4$

9 $\int_0^1 e^x\,dx$

10 $\int_0^{\pi} \sin x\,dx$

11 $\int_0^1 x^9\,dx$

12 $\int_0^2 \sqrt{1 + x^3}\,dx$

13 $\int_1^3 \sqrt[4]{1 + x^2}\,dx$

14 $\int_1^2 \frac{x^2\,dx}{1 + x^4}$

15 $\int_{-1}^{1} \sin x^2\,dx$

16 $\int_{-\pi/2}^{\pi/2} \frac{\sin x}{x}\,dx$

17 $\int_1^4 e^{1/x}\,dx$

18 $\int_{\pi}^{2\pi} \frac{\cos x}{x}\,dx$

19 $\int_1^3 \sqrt{1 + e^x}\,dx$

20 $\int_1^3 e^{\sqrt{x}}\,dx$

Estimate to seven places with $n = 10$

21 $\displaystyle\int_0^1 e^x\,dx$ **22** $\displaystyle\int_0^{\pi} \sin x\,dx$

23 Compute exactly the Simpson approximation to

$$\frac{1}{\pi}\int_0^{4\pi} \sqrt{4 + \sin x}\,dx \quad \text{with} \quad n = 2$$

24 (cont.) Exploit symmetry in the same problem.

Show that the prismoidal formula applies to the solid of revolution and compute the volume

25 the region bounded by $x = y^2$ and $x = a$ revolved about the x-axis

26 the region bounded by the x-axis, $y = x^{3/2}$, and $x = a$ revolved about the x-axis

27 the ellipse $x^2/a^2 + y^2/b^2 = 1$ revolved about the x-axis

28 the region bounded by the hyperbola $-x^2/a^2 + y^2/b^2 = 1$, $x = -c$, and $x = c$, revolved about the x-axis.

29 Prove the prismoidal formula.

30 Use the error estimate to find the least n for which it is safe to estimate

$$\tfrac{1}{4}\pi = \int_0^1 \frac{dx}{1 + x^2}$$

to five places by Simpson's rule.

31 Estimate $\displaystyle\int_0^1 x \ln x\,dx$ to four places.

32 The **sine integral** function is defined by

$$\operatorname{Si}(x) = \int_0^x \frac{\sin t}{t}\,dt$$

Show that $\operatorname{Si}(1) \approx 0.94608$.

9-7 Root Finding; Newton's Method

In this section we shall study methods for approximating the roots of equations. We begin with equations of the special form $\phi(x) = x$. See Figure 9-7-1. It will turn out that each equation of the form $f(x) = 0$ can be reduced to this special form in a variety of ways. For instance, if we set $\phi(x) = x - f(x)$, then the equations $\phi(x) = x$ and $f(x) = 0$ are equivalent. [However, this is not necessarily the best way to deal with $f(x) = 0$.] The first method we shall study is called *iteration*. An example will show the idea.

Figure 9-7-1
The solution of $\phi(x) = x$ is given by the intersection of $y = \phi(x)$ and $y = x$.

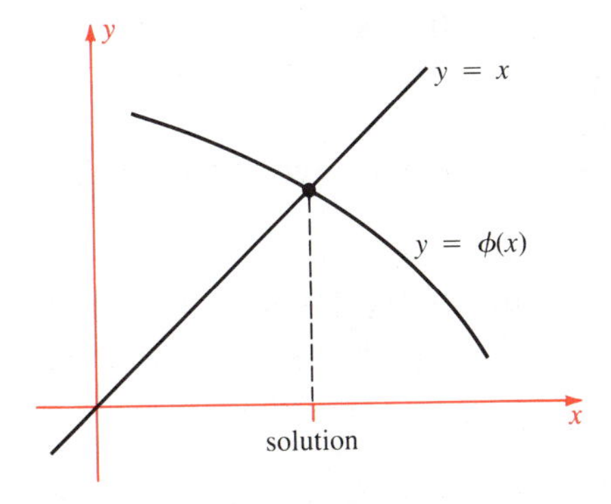

Example 1

Estimate the solution of $\cos x = x$ by iteration.

Solution First draw $y = \cos x$ and $y = x$ on the same graph (Figure 9-7-2, next page). The curves meet at one point, so $\cos x = x$ has a unique solution, roughly between $x = 0.5$ and $x = 1.0$.

Now choose any real number x_0. Set $x_1 = \cos x_0$, then $x_2 = \cos x_1$, then $x_3 = \cos x_2$, and so on. You will notice that the successive x_n tend to stabilize as n grows (Figure 9-7-3, next page). For instance, starting with $x_0 = 1$, a four-place computation yields successive values of x_n as shown in Table 9-7-1, next page. (Note how easy iteration is on a calculator!)

Conclusion: To four-place accuracy,

$$\cos(0.7391) \approx 0.7391$$

Table 9-7-1
Iteration: $x_0 = 1,\ x_{n+1} = \cos x_n$
(done on a four-place computer that rounds)

n	x_n	n	x_n
1	0.5403	13	0.7375
2	0.8576	14	0.7402
3	0.6543	15	0.7383
4	0.7935	16	0.7396
5	0.7014	17	0.7387
6	0.7639	18	0.7393
7	0.7221	19	0.7389
8	0.7504	20	0.7392
9	0.7314	21	0.7390
10	0.7442	22	0.7391
11	0.7356	23	0.7391
12	0.7414		

Figure 9-7-2
Solution of $\cos x = x$

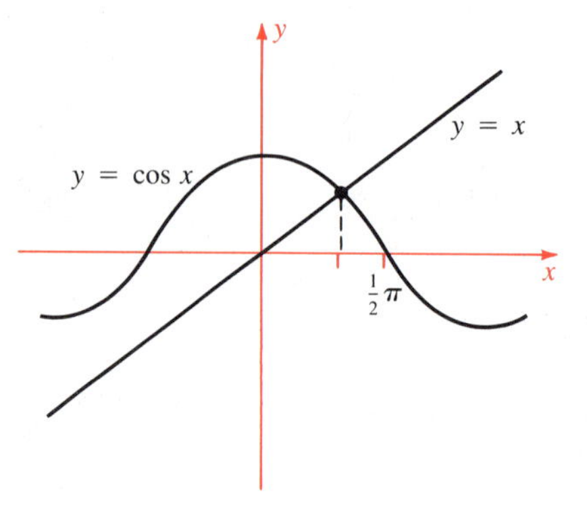

Figure 9-7-3
The iteration process: Up to (x_0, x_1), across to (x_1, x_1), down to (x_1, x_2), across to (x_2, x_2), and so on.

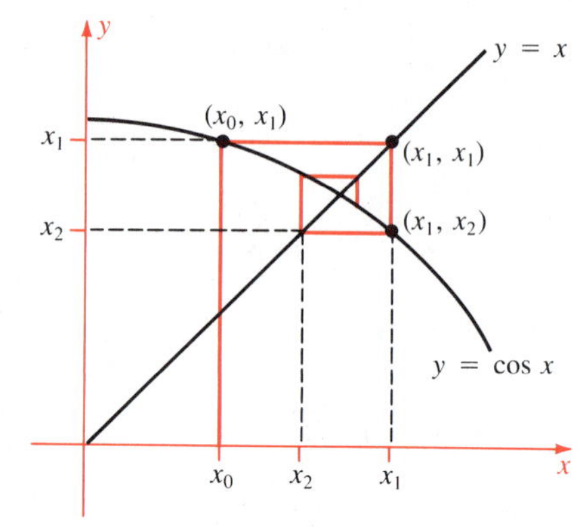

If we start afresh with $x_0 = 0.739100$ and work to six-place accuracy, we find

$$x_9 = x_{10} = 0.739085$$

Hence $\cos(0.739085) \approx 0.739085$ to six-place accuracy. ●

Under certain conditions, an equation $x = \phi(x)$ can be solved (approximately) by **iteration.** We start with an initial value x_0 and define successively

$$x_1 = \phi(x_0) \quad x_2 = \phi(x_1) \quad x_3 = \phi(x_2) \quad \cdots \quad x_{n+1} = \phi(x_n)$$

We hope these values stabilize, and do so rapidly; then the process yields an approximate solution of $x = \phi(x)$. For $x_n \approx x_{n+1}$ is the same as $x_n \approx \phi(x_n)$; hence x_n is an approximate solution.

The process does not always work. For instance, if $\phi(x) = 2\cos x$, then $x = 1.02986\ 653$ is a solution of $\phi(x) = x$ to eight places. However, set $x_0 = 1$ and iterate to five-place accuracy. This means compute, round to five places, compute, round, and so on:

$$x_1 = 1.08060 \qquad x_2 = 0.94160 \qquad x_3 = 1.17699$$
$$x_4 = 0.76741 \qquad x_5 = 1.43942 \qquad x_6 = 0.26200$$
$$x_7 = 1.93175 \qquad x_8 = -0.70633$$

Thus even though we start within 0.03 of the answer, the process is unstable (Figure 9-7-4).

Figure 9-7-4
Iteration fails.

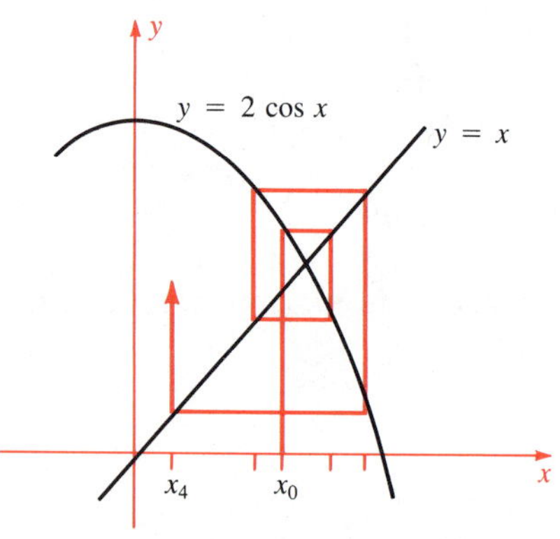

Remark We can salvage this iteration by modifying it slightly. Instead of defining x_{n+1} as $2\cos x_n$, let us try the average of x_n and $2\cos x_n$. Therefore, we define a new function.

$$\psi(x) = \tfrac{1}{2}(x + 2\cos x)$$

Table 9-7-2
$\psi(x) = \frac{1}{2}(x + 2\cos x)$, $x_0 = 1$

n	x_n	n	x_n
1	1.040302	9	1.029869
2	1.026111	10	1.029866
3	1.031204	11	1.029867
4	1.029388	12	1.029866
5	1.030037	13	1.0298667
6	1.029806	14	1.0298665
7	1.029888	15	1.0298665
8	1.029859		

If we can solve $x = \psi(x)$, then $x = 2\cos x$, so we shall have solved the original equation $\phi(x) = x$. Let us again start with $x_0 = 1$ and iterate (Table 9-7-2).

We conclude that $x = 1.02986\ 65$ approximates the solution of $x = 2\cos x$ to seven places. This shows that a clever choice of $\phi(x)$ may pay off.

The method of iteration has its ups and downs; sometimes it works, sometimes it fails. We need a way of testing $\phi(x)$ in advance to be sure the method will work. The key is the size of $\phi'(x)$ near the solution. If $|\phi'(x)| < 1$, the method works; otherwise, it generally fails.

Example 2

a $\phi(x) = \cos x$ so that $|\phi'(x)| = |-\sin x| < 1$ for $x \approx 0.74$. Iteration works.

b $\phi(x) = 2\cos x$ so that $|\phi'(x)| = |-2\sin x| > 1$ for $x \approx 1.0$. Iteration fails.

c $\phi(x) = \frac{1}{2}(x + 2\cos x)$ so that $|\phi'(x)| = |\frac{1}{2} - \sin x| < 1$ for $x \approx 1.0$. Iteration works. ●

Let us gather our observations into a precise statement:

Iteration

Let $\phi(x)$ be a differentiable function that maps a closed interval $[a, b]$ into itself, that is, $a \le \phi(x) \le b$ whenever $a \le x \le b$. Assume

$$|\phi'(x)| \le M < 1$$

on the interval, where M is a constant. Finally, assume there exists a point c in the interval such that $c = \phi(c)$. Choose any x_0 on the interval and define

$$x_1 = \phi(x_0) \qquad x_2 = \phi(x_1) \qquad x_3 = \phi(x_2) \qquad \cdots$$

As n increases, x_n becomes as close as we please to c, so x_n approximates c to any required degree of accuracy. ●

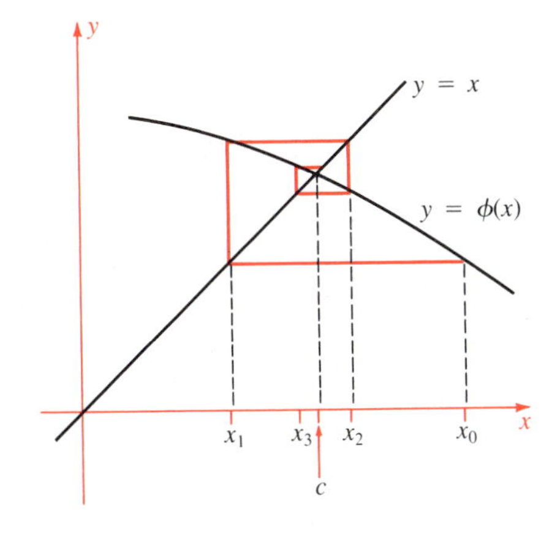

Figure 9-7-5
The case $-1 < \phi'(x) < 0$: iteration succeeds.

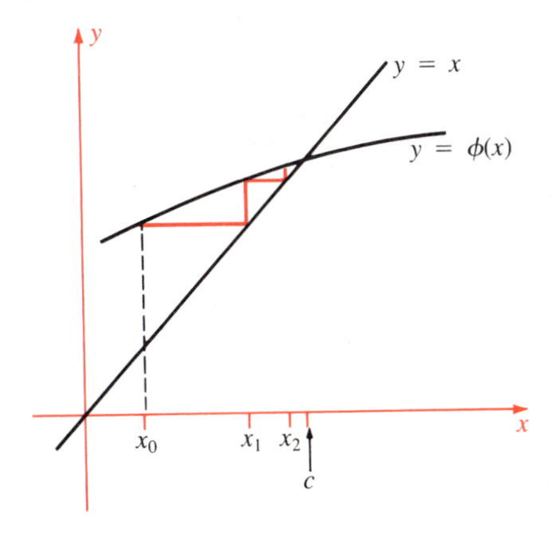

Figure 9-7-6
The case $0 < \phi'(x) < 1$: iteration succeeds.

Instead of proving this statement formally, let us try to see geometrically why it makes sense. Figure 9-7-5 shows what to expect in the case $-1 < \phi'(x) < 0$. Then the numbers x_n alternate from one side of $x = c$ to the other, always getting closer. The process seems to spiral into c. Figure 9-7-6 shows what happens if $0 < \phi'(x) < 1$. Then the x_n stay on one side of $x = c$ but get closer and closer to c. The process seems to zigzag into c.

Remark 1 By one use of the MVT and mathematical induction, we can prove that $|x_n - c| \le M^n|x_0 - c|$. Therefore the powers of M control how fast x_n approaches c. If $M \le 0.1$, then each x_n is less than one-tenth the distance from c of x_{n-1}, so we gain at least one more decimal-place accuracy with each iteration. But if $M = 0.9$, then it takes about 22 more x's to be sure of one more decimal-place accuracy, because $(0.9)^{21} > 0.1$ and $(0.9)^{22} < 0.1$. In any case, if $M < 1$, then the successive iterations steadily move towards c.

Remark 2 The first step in the formal proof is to show that

$$|\phi(x) - c| \leq M|x - c|$$

This is easy by the MVT because

$$\phi(x) - c = \phi'(z)(x - c)$$

with z between c and x, and $|\phi'(z)| \leq M$ by hypothesis. Thus the distance from $\phi(x)$ to c is less than the distance from x to c by a factor of M or less. This is fine, but we would much prefer an inequality like

$$|\phi(x) - c| \leq K|x - c|^2$$

Then the distance from $\phi(x)$ to c is at most proportional to the square of the distance from x to c. No matter what the constant K is, once x is close to c, then $\phi(x)$ is *very* close to c.

Remark 3 It is known that the assumption $c = \phi(c)$ for some c is not necessary; the other assumptions guarantee the existence of a unique solution.

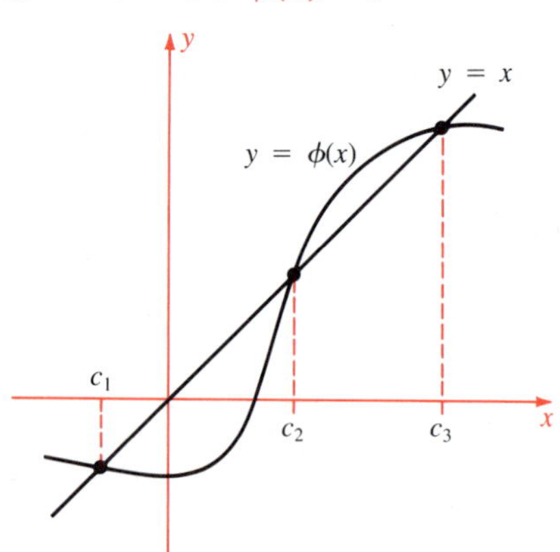

Figure 9-7-7
Fixed points of $\phi(x)$: points c where $\phi(c) = c$

Roots and Fixed Points

A point c such that $\phi(c) = c$ is called a **fixed point** of the function $\phi(x)$. It is a value of x where the graphs of $y = x$ and $y = \phi(x)$ intersect (Figure 9-7-7). It is also a root of the equation $\phi(x) - x = 0$.

The idea of fixed points is important in solving equations. For given an equation $f(x) = 0$, we can set $\phi(x) = x + f(x)g(x)$, where $g(x)$ is any convenient function. Then each root of the equation $f(x) = 0$ is a fixed point of the function $\phi(x)$. Conversely, if $\phi(c) = c$ and $g(c) \neq 0$, then c is a root of $f(x) = 0$.

Newton-Raphson Method

Now we shall learn a powerful method for estimating roots of $f(x) = 0$. We associate with $f(x)$ a function $\phi(x)$ whose fixed points are the roots of $f(x)$, and whose iterates estimate the fixed points very rapidly.

The **Newton-Raphson method** (also called **Newton's method**) is based on linear approximation. Suppose $x = c$ is an (unknown) zero of $f(x)$. We make an initial guess of x for c. We then take the tangent to the graph $y = f(x)$ at $(x, f(x))$, and let $\phi(x)$ be its x-intercept (Figure 9-7-8). It appears that $\phi(x)$ should be a lot closer to c than x is.

The equation of the tangent is

$$\frac{Y - f(x)}{X - x} = f'(x)$$

Figure 9-7-8
Newton-Raphson method: $\phi(x)$ is the x-intercept of the tangent to $y = f(x)$ at $(x, f(x))$.

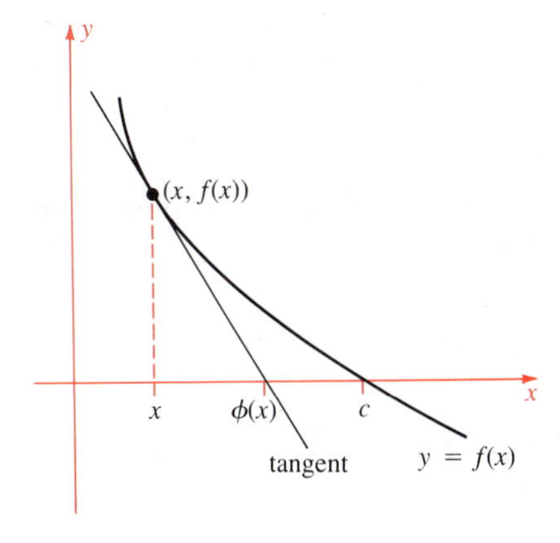

For the x-intercept of the tangent, we set $Y = 0$ and solve for X:

$$\frac{-f(x)}{X - x} = f'(x) \qquad \text{so} \qquad X - x = -\frac{f(x)}{f'(x)}$$

and

$$X = x - \frac{f(x)}{f'(x)}$$

Therefore, given $f(x)$, we *define*

$$\phi(x) = x - \frac{f(x)}{f'(x)}$$

assuming $f'(x) \neq 0$. Obviously, if $f(c) = 0$, then $\phi(c) = c$. Conversely if $\phi(c) = c$, then $f(c) = 0$. Thus the root problem for $f(x) = 0$ is transformed into the fixed-point problem for $\phi(x)$.

Still assuming that c is a root of $f(x) = 0$, we want to approximate c by iteration. We know that the smaller $|\phi'(x)|$ is near $x = c$, the faster the iterates x_0, $x_1 = \phi(x_0)$, $x_2 = \phi(x_1)$, $\cdots$ will tend towards c. So let us compute the derivative:

$$\begin{aligned}\phi'(x) &= \frac{d}{dx}\left[x - \frac{f(x)}{f'(x)}\right] \\ &= 1 - \frac{[f'(x)]^2 - f(x)f''(x)}{[f'(x)]^2}\end{aligned}$$

Hence

$$\phi'(x) = \frac{f(x)f''(x)}{[f'(x)]^2} \qquad \text{and} \qquad \phi'(c) = 0$$

Great! That means $|\phi'(x)|$ is *very* small near $x = c$. Therefore, if x_0 is chosen sufficiently close to c in the first place, the iterates x_n will tend to $x = c$ rapidly (Figure 9-7-9).

Let us summarize this discussion.

Figure 9-7-9
Typical functions for which $x_n \to c$ rapidly

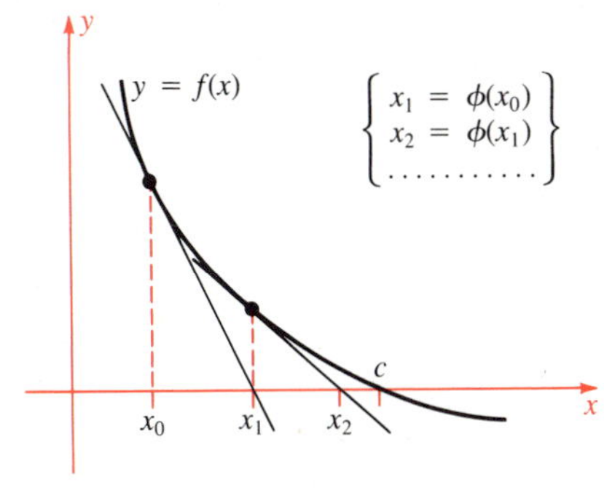

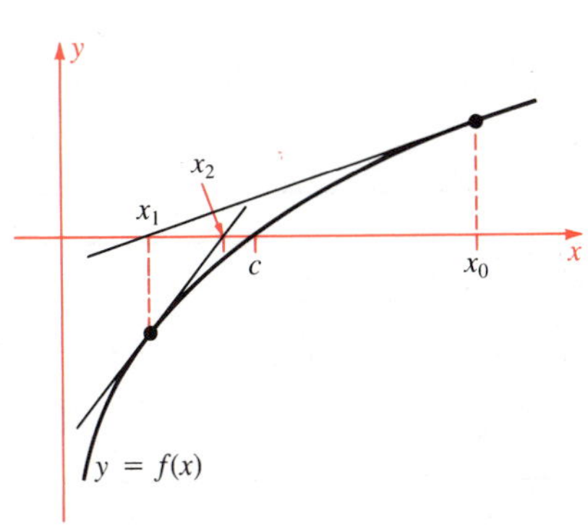

Newton-Raphson Method

Suppose $f(c) = 0$, $f'(c) \neq 0$, and $f''(x)$ is continuous near $x = c$. Set

$$\phi(x) = x - \frac{f(x)}{f'(x)}$$

If x_0 is sufficiently close to c, then the iterates

$$x_1 = \phi(x_0) \qquad x_2 = \phi(x_1) \qquad \cdots \qquad x_n = \phi(x_{n-1}) \qquad \cdots$$

approach as close as we please to c. Also $\phi'(c) = 0$, so convergence is rapid.

Example 3

Estimate to six places the roots of $x^5 - x + 1 = 0$.

Solution A graph of $y = f(x) = x^5 - x + 1$ will show how many roots there are and their approximate locations. The derivative

$$f'(x) = 5x^4 - 1$$

is 0 at $x = \pm\sqrt[4]{\frac{1}{5}} \approx \pm 0.67$ and changes sign at these two points. This gives useful information about the graph:

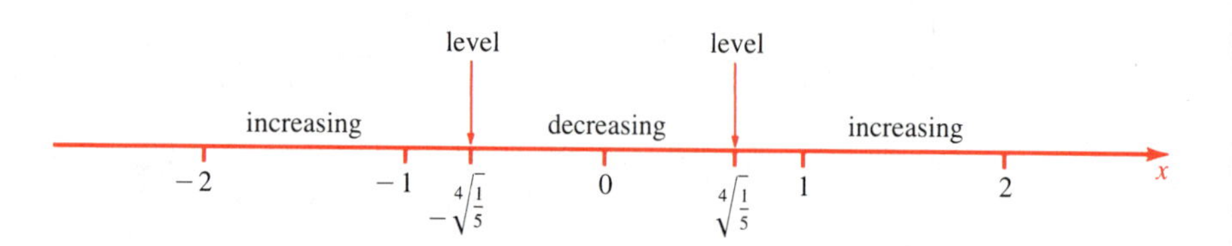

Clearly $f(x)$ has a local maximum at $-(\frac{1}{5})^{1/4}$ and a local minimum at $(\frac{1}{5})^{1/4}$. Their values are

$$f_{\text{max}} = f[-(\tfrac{1}{5})^{1/4}] \approx 1.53 \quad \text{and} \quad f_{\text{min}} = f[(\tfrac{1}{5})^{1/4}] \approx 0.47$$

This is enough for a sketch (Figure 9-7-10).

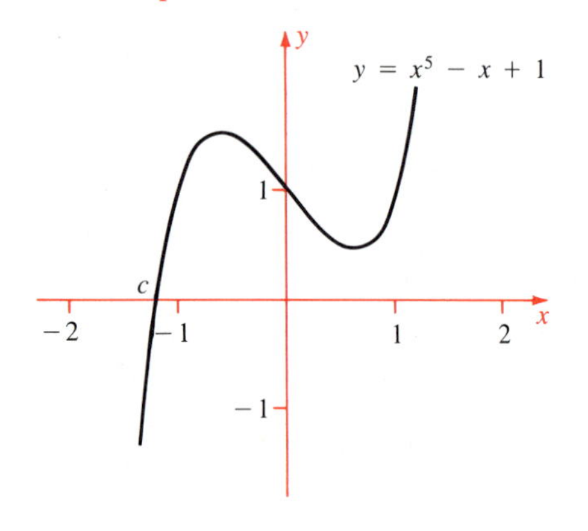

Figure 9-7-10
See Example 3

The graph shows that there is exactly one root, and it is negative. By some trial and error, $f(-2) = -29$ and $f(-1) = 1$, hence the root is between -2 and -1, probably closer to -1.

Now the work starts. Set

$$\phi(x) = x - \frac{f(x)}{f'(x)} = x - \frac{x^5 - x + 1}{5x^4 - 1} = \frac{4x^5 - 1}{5x^4 - 1}$$

and $x_0 = -1$. Then

$$x_1 = \phi(-1) = -\tfrac{5}{4} = -1.25$$
$$x_2 = \phi(-1.25) \approx -1.178459$$
$$x_3 = \phi(x_2) \approx -1.167537$$
$$x_4 = \phi(x_3) \approx -1.167304$$
$$x_5 = \phi(x_4) \approx -1.167304$$

The process has stabilized to at least six places, so to six-place accuracy, $x \approx -1.167304$ is the one root of $f(x) = 0$. ●

Example 4

Estimate the root of $\cos x = x$ to six places.

Solution Set $f(x) = x - \cos x$ and

$$\phi(x) = x - \frac{f(x)}{f'(x)} = x - \frac{x - \cos x}{1 + \sin x} = \frac{x \sin x + \cos x}{1 + \sin x}$$

Start with $x_0 = 1$. Iteration yields

$$x_1 = \phi(x_0) \approx 0.750364 \qquad x_2 = \phi(x_1) \approx 0.739113$$
$$x_3 = \phi(x_2) \approx 0.739085 \qquad x_4 = \phi(x_3) \approx 0.739085$$

Therefore 0.739085 is a root of $\cos x = x$ to six places. ●

Remark It took only four iterations for six-place accuracy. Compare the solution of Example 1 where it took 23 iterations for only four-place accuracy. Clearly Newton-Raphson is a superior method.

Convergence

It is important to know how well the Newton-Raphson method works in a particular problem, that is, how rapidly the iterates x_n approach their limiting value c. The following is a useful practical result in this connection.

Theorem

Let $f(x)$ have continuous first and second derivatives on the interval $[a, b]$, satisfying

$$|f'(x)| \geq m > 0 \quad \text{and} \quad |f''(x)| \leq M$$

where m and M are constants. Suppose $f(c) = 0$ for some c on the interval. Set

$$\phi(x) = x - \frac{f(x)}{f'(x)}$$

Then

$$|\phi(x) - c| \leq \frac{M}{2m}|x - c|^2$$ ●

We omit the proof, which requires a sophisticated application of Taylor's theorem.

Example 5

Suppose Newton's method is used to estimate $\sqrt{2}$, starting with $x_0 = 1$. How far should the process be continued to guarantee 20-place accuracy?

Solution Set $f(x) = x^2 - 2$ and

$$\phi(x) = x - \frac{f(x)}{f'(x)} = x - \frac{x^2 - 2}{2x} = \frac{x^2 + 2}{2x}$$

Since $f(1) = -1$ and $f(2) = 2$, we should seek our solution on the interval $[1, 2]$. There $|f'(x)| = |2x| \geq 2 > 0$ and $|f''(x)| = |2| = 2$, so we choose $m = 2$ and $M = 2$. Consequently, by the estimate in the theorem,

$$|\phi(x) - \sqrt{2}\,| \le \tfrac{1}{2}|x - \sqrt{2}\,|^2$$

Since $\sqrt{2} < 1.5$,

$$|x_0 - \sqrt{2}\,| = |1 - \sqrt{2}\,| < |1.5 - 1| = \tfrac{1}{2}$$

Now pull the bootstrap:

$$|x_1 - \sqrt{2}\,| \le \tfrac{1}{2}|x_0 - \sqrt{2}\,|^2 < \tfrac{1}{2}(\tfrac{1}{2})^2 = 1/2^3$$
$$|x_2 - \sqrt{2}\,| \le \tfrac{1}{2}|x_1 - \sqrt{2}\,|^2 < \tfrac{1}{2}(1/2^3)^2 = 1/2^7$$
$$|x_3 - \sqrt{2}\,| < \tfrac{1}{2}(1/2^7)^2 = 1/2^{15}$$
$$|x_4 - \sqrt{2}\,| < \tfrac{1}{2}(1/2^{15})^2 = 1/2^{31}$$

and in general,

$$|x_n - \sqrt{2}\,| < 1/2^N \qquad \text{where} \qquad N = 2^{n+1} - 1$$

We want $|x_n - \sqrt{2}\,| < 5 \times 10^{-21}$. This will be so provided that

$$1/2^N < 5 \times 10^{-21} \qquad \text{where} \qquad N = 2^{n+1} - 1$$

that is,

$$2^N > \tfrac{1}{5} \times 10^{21} \qquad \text{so that} \qquad 2^{N-1} > 10^{20}$$

But $2^{67} \approx 1.5 \times 10^{20}$ (by calculator), so we choose $N \ge 68$. The least n for which $2^{n+1} - 1 \ge 68$ is $n = 6$. Thus x_6 provides 20-place accuracy. ●

Remark 1 Note that $\phi(x) = (x^2 + 2)/(2x) = \frac{1}{2}(x + 2/x)$ so $\phi(x)$ is the average of x and $2/x$. This is reasonable since $\sqrt{2}$ is that number c for which $c = 2/c$.

Remark 2 The first six x_n are

$$x_0 = 1 \qquad x_1 = \tfrac{3}{2} \qquad x_2 = \tfrac{17}{12} \qquad x_3 = \tfrac{577}{408}$$
$$x_4 = \tfrac{665857}{470832} \qquad x_5 = \tfrac{88\,67310\,88897}{62\,70135\,66048}$$

We find by a computer that

$$\sqrt{2} \approx 1.41421\;35623\;73095\;04880\;169$$
$$x_4 \approx 1.41421\;35623\;747$$
$$x_5 \approx 1.41421\;35623\;73095\;04880\;1690$$

The theory predicts that $|x_4 - \sqrt{2}\,| < 2^{-31} < 5 \times 10^{-10}$, hence that x_4 is accurate to at least nine places. But error estimates usually lose something; actually x_4 is accurate to 11 places. Similarly the theory predicts that $|x_5 - \sqrt{2}\,| < 2^{-63} < 1.1 \times 10^{-19}$, assuring 18-place accuracy for x_5; actually x_5 is accurate to at least 23 places. (See Exercise 42.)

Exercises

Use iteration to estimate the solution of $\phi(x) = x$; work to four-place accuracy

1 $\phi(x) = e^{-x/2} \quad x_0 = 1$

2 $\phi(x) = e^{-x} \quad x_0 = 1$

3 $\phi(x) = \sqrt{3 + x} \quad x_0 = 3$

4 $\phi(x) = \sqrt{3 + x} \quad x_0 = 100$

5 $\phi(x) = 2 \arctan x \quad (x > 0) \quad x_0 = 2$

6 $\phi(x) = \pi + \arctan x \quad x_0 = 0$

7 $\phi(x) = \sqrt{\frac{1}{2}(1 + x)} \quad x_0 = 0$

8 $\phi(x) = \sqrt{\frac{1}{2}(1 + x)} \quad x_0 = 1000$

9 $\phi(x) = x(\frac{5}{4} - x^2) \quad x_0 = 1$

0 $\phi(x) = x(\frac{5}{4} - x^2) \quad x_0 = 1.25$

1 Set $\phi(x) = 0.6316x + 0.7368 \cos x$ and $x_0 = 1$. Find x_3 and x_4 to eight places.

2 (cont.) $\phi(x) = x$ is equivalent to what simple problem previously considered?

Use the Newton-Raphson method to estimate all roots to five places (ten places)

3 $x^2 = 10$

4 $x^3 = 10$

5 $5x^4 = 1$

6 $x \ln x = 1$

7 $x^3 - x + 1 = 0$

8 $e^{-5x} = x$

19 $x^3 + x^2 - 4 = 0$

20 $x^3 + x + 5 = 0$

21 $e^x + x = 0$

22 $x^3 - 2x^2 + x + 3 = 0$

23 $5x^2 + 4x - 3 = 0$

24 $x^3 - 3x + 1 = 0$

25 $x^3 - 2x^2 - x + 3 = 0$

26 $x^3 - 6x^2 + 9x - 1 = 0$

27 $e^x = 2 \cos x \quad (x > -\pi)$

28 $e^x = 3x$

29 $x = 2 \sin x$

30 $x = 2 \cos 2x$

31 $x^5 - 5x + 1 = 0$

32 $x^3 + 1.5x^2 - 5.75x + 3.37 = 0$

33 $10 \cos x - 8 + x^2 = 0$

34 $x^4 - 10^4 x + 1 = 0$

35 $e^{-x^2} = \dfrac{1}{1 + 4x^2}$

*36 $x^4 - 1.73x^2 + 0.46x + 1.275 = 0$

37 Find to five places the largest c such that $y = cx$ is tangent to $y = \cos x$.

38 Find to five places the largest c such that $\cosh x \geq cx$ for all x.

39 Find to five places the minimum vertical distance between the graphs $y = 1.5e^x$ and $y = x^3$.

*40 Estimate to three-place accuracy the solution of the system $x^2 + y^4 = 100$ and $e^x - e^y = 1$.

*41 Show in Example 5 that $|x_6 - \sqrt{2}| < 10^{-43}$.

*42 (cont.) Show that $x_1 > x_2 > x_3 > \cdots > \sqrt{2}$. Conclude that $|x_5 - \sqrt{2}| < 5 \times 10^{-23}$.

43 Let $x_0 > 0$ and $\phi(x) = \sqrt{\frac{1}{2}(1 + x)}$. Show that $x_n \to 1$ by estimating $\phi'(x)$.

44 Set $\phi(x) = x - f(x)/f'(x)$ and assume $f(c) = 0$ and $f'(c) \neq 0$. Find $\lim [\phi(x) - c]/(x - c)^2$ as $x \to c$ by l'Hospital's rule.

9-8 Review Exercises

Use the MVT to show that

1 $\sqrt[3]{28} < 3\frac{1}{27}$

2 $\cos x \geq 1 - \frac{1}{2}x$ for $0 \leq x \leq \frac{1}{6}\pi$

Find

3 $\displaystyle\lim_{x \to \pi/2-} \frac{\cos x}{1 - \sin x}$

4 $\displaystyle\lim_{x \to +\infty} \frac{e^{3x}}{e^{2x} + x^2}$

5 $\displaystyle\lim_{\theta \to \pi/2} (\tfrac{1}{2}\pi - \theta) \tan \theta$

6 $\displaystyle\lim_{x \to +\infty} [1 + a(b^{1/x} - 1)]^x$

7 Find the fifth degree Taylor polynomial of $\cos x$ at $x = \frac{1}{3}\pi$.

8 Find the 15-th degree Taylor polynomial of $\exp(-x^3)$ at $x = 0$.

9 Examine the graphs $y = \cos x$ and $y = \sqrt{1 - x^2}$ near $(0, 1)$. Is the cosine curve inside or outside of the circle? [Hint Use Taylor polynomials.]

10 Let $P_2(x)$ be the second degree Taylor polynomial of $\sqrt{x}$ at $x = 100$. Show that

$$|\sqrt{x} - P_2(x)| < 5 \times 10^{-6} \quad \text{if} \quad 100 \leq x < 102$$

Estimate to four places by Simpson's rule

11 $\displaystyle\int_1^3 \frac{e^x}{x}\, dx$

12 $\displaystyle\int_2^5 \frac{dx}{\ln x}$

13 $\displaystyle\int_{0.6}^{1.4} \ln \tan x\, dx$

14 $\displaystyle\int_0^{\pi/2} \sqrt{1 + \cos^2 x}\, dx$

15 $\displaystyle\int_0^3 x \sin^2(\tfrac{1}{4}\pi x)\, dx$

16 $\displaystyle\int_1^4 (\ln x)^2\, dx$

Estimate to five places the roots of

17 $e^{-x/2} = x + 3$

18 $2 \arctan x = x$

19 $(2\pi + x) \tan x = 1 \qquad (0 < x < \frac{1}{2}\pi)$

20 $0.4x = \tan x \qquad (\frac{1}{2}\pi < x < \frac{3}{2}\pi)$

21 For what choice of the constant c does the weighted average

$$\phi(x) = cx + (1 - c)\frac{a}{x^2}$$

iterate as rapidly as possible towards $\sqrt[3]{a}$?

22 Suppose $f(0) = f'(0) = f''(0) = \cdots = f^{(n)}(0) = 0$ and $f^{(n+1)}(x) \geq 0$ for all $x \geq 0$. Prove that $f(x) \geq 0$ for all $x > 0$ and that if $f(c) = 0$ for some $c > 0$, then $f(x) = 0$ for all $0 \leq x \leq c$.
[Hint Taylor's formula]

23 Suppose $f(x)$ is differentiable for $x \geq a$ and $f'(x) \geq m > 0$. Prove $f(x) \to +\infty$ as $x \to +\infty$.

24 Let $P_4(x)$ be the fourth degree Taylor polynomial of e^x at $x = 0$. Which of the following gives the best approximation to e?

$$P_4(1) \qquad [P_4(0.5)]^2 \qquad [P_4(0.1)]^{10}$$

25 Show that

$$\sin x \approx \tfrac{1}{2}(e^x - e^{-x}) - \tfrac{1}{3}x^3[1 + \tfrac{1}{840}(x^4 + \tfrac{1}{7920}x^8)]$$

with error less than 5×10^{-9} for $|x| < \frac{1}{2}\pi$. (Estimates like this and the one in Exercise 26 are sometimes used in calculators.)

26 How good is the estimate
$\sin x \approx \frac{1}{3}x[10(\frac{1}{20}x^2 + 1)^{-1} - 7]$ for x small?
[Hint Use Taylor polynomials.]

10 Sequences and Series

In this chapter we study sequences and their limits. We also develop the properties of infinite series, the sums of infinite sequences. Series are used to express numbers such as e and π as infinite sums of rational numbers. In the next chapter we shall express transcendental functions such as $\exp x$ and $\sin x$ as infinite sums. In this chapter, we also cover improper integrals, which are closely related to infinite series.

10-1 Sequences and Limits

We begin with infinite sequences of real numbers, like

$$1 \quad 1/2 \quad 1/3 \quad 1/4 \quad 1/5 \quad \cdots \quad 1/n \quad \cdots$$

$$\frac{1}{1+1} \quad \frac{2}{1+2} \quad \frac{3}{1+3} \quad \frac{4}{1+4} \quad \frac{5}{1+5} \quad \cdots \quad \frac{n}{1+n} \quad \cdots$$

$$1 \quad 2 \quad 4 \quad 8 \quad 16 \quad 32 \quad \cdots \quad 2^n \quad \cdots$$

In mathematics, a **sequence** is a real-valued function whose domain is a set of integers. Our concern is with limits of *infinite* sequences. Unless otherwise stated, each of the sequences we deal with will be assumed infinite, that is, its domain will be an *infinite* set of integers. What is more, we shall limit attention to sequences whose domains are sets of non-negative integers. The domain is sometimes referred to as the **index set** of the sequence.

Notation

For convenience and by tradition, sequence notation differs from standard function notation. The value of a sequence a at n is written a_n rather than $a(n)$. The sequence is "written out" as $a_1,\ a_2,\ a_3,\ \cdots,$ and compact notations for the sequence are

$$\{a_n\} \quad \text{or} \quad \{a_n\}_{n=1}^{\infty} \quad \text{or} \quad \{a_n\}_1^{\infty}$$

The first element of a sequence need not be a_1; it may be a_0, or a_3, or a_{10}, for example. We write

$$\{a_n\}_0^{\infty} \quad \text{for} \quad a_0 \quad a_1 \quad a_2 \quad \cdots$$

$$\{a_n\}_{10}^{\infty} \quad \text{for} \quad a_{10} \quad a_{11} \quad a_{12} \quad \cdots$$

Other enumerations are possible, too. For example,

$$\{a_{2n}\}_1^{\infty} = a_2 \quad a_4 \quad a_6 \quad \cdots$$

$$\{b_{4n+1}\}_0^{\infty} = b_1 \quad b_5 \quad b_9 \quad \cdots$$

Many sequences we encounter have some mathematical rule of formation, sometimes an explicit formula for the n-th element, sometimes a formula for the n-th element in terms of one or several previous elements. In the latter case, the sequence is defined *iteratively.*

Example 1

a $a_n = \dfrac{1}{n} \quad (n \geq 1)$: $\quad 1 \quad 1/2 \quad 1/3 \quad 1/4 \quad \cdots$

b $a_n = \dfrac{n}{1+n} \quad (n \geq 1)$: $\quad \dfrac{1}{1+1} \quad \dfrac{2}{1+2} \quad \dfrac{3}{1+3} \quad \cdots$

c $a_n = \dfrac{n}{\ln n} \quad (n \geq 2)$: $\quad \dfrac{2}{\ln 2} \quad \dfrac{3}{\ln 3} \quad \dfrac{4}{\ln 4} \quad \cdots$

d $a_0 = 1$ and $a_n = a_{n-1}/2$ for $n \geq 1$:

$$1 \quad 1/2 \quad 1/4 \quad 1/8 \quad 1/16 \quad 1/32 \quad \cdots$$

e $a_1 = 1 \quad a_2 = 1 \quad a_{n+2} = a_{n+1} + a_n$:

$$1 \quad 1 \quad 2 \quad 3 \quad 5 \quad 8 \quad 13 \quad 21 \quad \cdots$$

●

Not every sequence has a simple formula, or even any formula at all, for example $\{d_n\}$, where d_n is the n-th digit in the decimal representation of π, or $\{p_n\}$, where p_n is a patient's blood pressure n minutes after an operation.

Limit of a Sequence

We wish to know whether a sequence $\{a_n\}$ converges to a limit. For instance, consider the sequence $\{1/n\}$:

$$1 \quad 1/2 \quad 1/3 \quad 1/4 \quad \cdots$$

It seems reasonable in this case to write

$$\lim_{n\to+\infty} \frac{1}{n} = 0$$

Now consider the example $\{n^2/(n^2+1)\}$. Some terms are

$$\frac{1}{2} \quad \frac{4}{5} \quad \frac{9}{10} \quad \cdots \quad \frac{10{,}000}{10{,}001} \quad \cdots$$

In this case it seems reasonable to write

$$\lim_{n\to+\infty} \frac{n^2}{n^2+1} = 1$$

Next consider $\{(-1)^n\}$ for $n \geq 0$. Its terms are

$$1 \quad -1 \quad 1 \quad -1 \quad 1 \quad -1 \quad \cdots$$

which do not approach any single real number. There are two candidates for the limit, $+1$ and -1, and that is one too many. It seems reasonable to say that the sequence $\{(-1)^n\}$ does not converge (has no limit).

Finally consider $\{n\}$ for $n \geq 1$. Its terms are

$$1 \quad 2 \quad 3 \quad 4 \quad \cdots$$

There is no candidate for the limit, so the sequence does not converge.

These examples prepare us for a formal definition.

Limit of a Sequence

A sequence $\{a_n\}$ **converges** to limit L if for each $\varepsilon > 0$ there exists an integer N such that

$$|a_n - L| < \varepsilon$$

for all n in the index set of the sequence such that $n \geq N$. If so, we write

$$\lim_{n\to+\infty} a_n = L \quad \text{or} \quad a_n \to L$$

If $\{a_n\}$ does not converge to a limit, we say that $\{a_n\}$ **diverges.**

Remark This definition is closely related to our previous definition of

$$\lim_{x\to+\infty} f(x)$$

This is not surprising if we remember that $\{a_n\}$ is a real-valued *function* whose domain is an infinite set of positive integers.

Let us try to visualize geometrically what convergence means. Suppose $\lim_{n\to+\infty} a_n = L$. Let $\varepsilon > 0$ be given. The definition requires that eventually all terms be within ε of L. In other words, the inequality $|a_n - L| < \varepsilon$ must hold eventually. In geometric terms, given any inter-

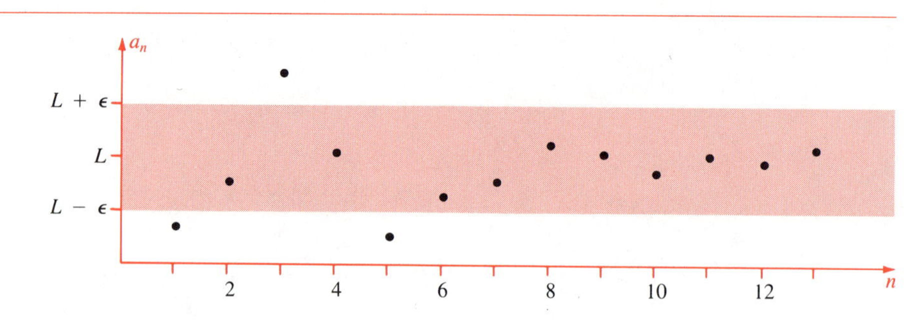

Figure 10-1-1
Eventually all a_n fall into the target interval $(L - \varepsilon, L + \varepsilon)$. (In this case, "eventually" means starting with a_6.)

val centered at L, no matter how small, the elements of the sequence eventually get into that interval *and stay there* (Figure 10-1-1).

Remark A sequence like $\{n^2\}$ that eventually remains larger that any number B **diverges to** $+\infty$, and we write $\lim_{n\to+\infty} n^2 = +\infty$. A sequence like $\{-3n\}$ that eventually remains less than any number $-B$ **diverges to** $-\infty$, and we write $\lim_{n\to\infty}(-3n) = -\infty$. These conventions fit into the notation we use for limits of convergent sequences.

A sequence with a finite limit is called a **convergent** sequence, and a sequence without a finite limit is called a **divergent** sequence. If a sequence is convergent, then it has only one limit since its elements cannot eventually be arbitrarily close to two different numbers.

Uniqueness of the Limit

If $a_n \to L_1$ and $a_n \to L_2$, then $L_1 = L_2$.

We leave the formal proof as an exercise. Because of this theorem, we may speak of *the* limit of a convergent sequence.

Proving that $a_n \to L$ is a kind of game: I challenge you with an ε, and you must find some appropriate N. It is not necessary to find the *smallest* N such that $|a_n - L| < \varepsilon$ for all $n \geq N$; that may be hard. But for each ε, you must find *some* N. This is much like the ε-δ game played to prove that a function tends to a given limit.

Example 2

If $a_n = 1/n$, prove that $a_n \to 0$.

Solution Let $\varepsilon > 0$ be given. We must find an N such that

$$|a_n - 0| = |a_n| < \varepsilon$$

for all $n \geq N$. Thus we want $|a_n| = 1/n < \varepsilon$. It is not hard to see how large N must be. We simply choose N larger than $1/\varepsilon$. Then $1/N < \varepsilon$, hence

$$|a_n| = \frac{1}{n} < \frac{1}{N} < \varepsilon \quad \text{for all} \quad n \geq N$$

This completes the proof.

Remark The proof is valid for *each* choice of ε. If you produce an N that works only for $\varepsilon = 0.1$ or $\varepsilon = 0.001$, that is not sufficient. In general the smaller ε is, the larger N must be. That makes sense: the closer you want to approximate the limit, the farther out in the sequence you must go.

Example 3

If $a_n = \dfrac{n^2}{2n^2 + 1}$, prove that $a_n \to \frac{1}{2}$.

Solution Write out a few terms:

$$1/3 \quad 4/9 \quad 9/19 \quad 16/33 \quad 25/51 \cdots$$

By algebra, $a_n = 1/(2 + 1/n^2)$. A good guess is that $1/n^2 \to 0$ as $n \to +\infty$, so $a_n \to \frac{1}{2}$ is plausible. How do we prove this is so? Let $\varepsilon > 0$ be given. We must find an N such that $|a_n - \frac{1}{2}| < \varepsilon$ for all $n \geq N$. Now

$$|a_n - \tfrac{1}{2}| = \left| \frac{n^2}{2n^2 + 1} - \frac{1}{2} \right| = \frac{1}{2(2n^2 + 1)} < \frac{1}{4n^2}$$

We want to make $|a_n - \frac{1}{2}| < \varepsilon$. Clearly it suffices to make $1/4n^2 < \varepsilon$, that is, $n^2 > 1/4\varepsilon$. This suggests that we choose N so large that

$$N^2 > 1/4\varepsilon \qquad \text{that is} \qquad N > \sqrt{1/4\varepsilon}$$

Then for all $n \geq N$,

$$|a_n - \tfrac{1}{2}| < \frac{1}{4n^2} \leq \frac{1}{4N^2} < \varepsilon$$

This completes the proof. ●

The idea of the limit of a sequence is fundamental to our decimal representation of the real number system. For what does it really mean to express a number as an infinite decimal, for instance,

$$\pi = 3.14159\ 2653 \cdots ?$$

It means that π is the limit of the sequence $\{a_n\}$, where

$$a_1 = 3.1 \qquad a_2 = 3.14 \qquad a_3 = 3.141 \qquad a_4 = 3.1415 \qquad \cdots$$

Note that $|\pi - a_n|$ is the error when π is approximated by n places of its decimal expansion, and $|\pi - a_n| < 10^{-n}$

Before continuing, we wish to dispel two old-wives' tales concerning limits of sequences:

Tale 1 If $a_n \to L$, then the numbers a_n get closer and closer to L without ever reaching L. Not necessarily! For example,

$$3 \quad 3 \quad 3 \quad 3 \quad \cdots \qquad \text{and} \qquad 0 \quad 1/2 \quad 0 \quad 1/3 \quad 0 \quad 1/4 \quad \cdots$$

The first sequence converges to 3, the second to 0. Each "reaches its limit" infinitely often.

Tale 2 If $a_n \to L$, the numbers increase toward L or decrease toward L. Not necessarily! For example,

$$1 \quad -1/2 \quad 1/3 \quad -1/4 \quad \cdots$$

converges to 0 but jumps around. In fact, take the sequence

$$1 \quad 1/2 \quad 1/3 \quad 1/4 \quad \cdots$$

and sprinkle in minus signs at random. The resulting irregular sequence still converges to zero.

An even more striking example is this one:

$$1/10 \quad 1/10^2 \quad 2/10^2 \quad 1/10^3 \quad 2/10^3 \quad 3/10^3 \quad 1/10^4 \quad 2/10^4 \quad 3/10^4 \quad 4/10^4 \quad \cdots$$

This sequence converges to zero, yet it has longer and longer strings of terms that move away from zero!

A **subsequence** of $\{a_n\}$ is a sequence $\{a_{n_j}\}$ whose index set $\{n_j\}$ is an (infinite) subset of $\{n\}$ the index set of $\{a_n\}$. For example,

$$a_1 \quad a_3 \quad a_5 \quad a_7 \quad \cdots$$

$$a_4 \quad a_5 \quad a_6 \quad a_7 \quad \cdots$$

$$a_1 \quad a_4 \quad a_9 \quad a_{16} \quad \cdots$$

are subsequences of $\{a_n\}_1^\infty$.

The idea is to thin out the original sequence and then relabel what remains with the new index j. Clearly the j-th term of the subsequence is at least as far out as the j-th term of the original.

Just as in finding limits of functions, the formal definition of the limit of a sequence is important in making the concept precise, but tedious to apply. In practice, we use properties of limits that allow us to determine the convergence of many sequences from that of known convergent sequences. A very simple property of this type concerns subsequences of convergent sequences.

Convergence of Subsequences

If $a_n \to L$, then any subsequence of $\{a_n\}$ also converges to L. ●

The proof is not hard; we leave it for an exercise. As an application of this principle, we easily deduce that the following sequences $\{b_n\}$ all converge to zero because each is a subsequence of $\{1/n\}$, which we know converges to zero:

$1 \quad \frac{1}{3} \quad \frac{1}{5} \quad \frac{1}{7} \quad \cdots \qquad b_n = \frac{1}{2n-1}$

$\frac{1}{2} \quad \frac{1}{4} \quad \frac{1}{8} \quad \frac{1}{16} \quad \cdots \qquad b_n = \frac{1}{2^n}$

$1 \quad \frac{1}{2^2} \quad \frac{1}{3^2} \quad \frac{1}{4^2} \quad \cdots \qquad b_n = \frac{1}{n^2}$

$1 \quad \frac{1}{2^3} \quad \frac{1}{3^3} \quad \frac{1}{4^3} \quad \cdots \qquad b_n = \frac{1}{n^3}$

The next property says that a sequence trapped between two sequences that converge to the same limit is itself squeezed to that limit. Figure 10-1-2 shows why it works.

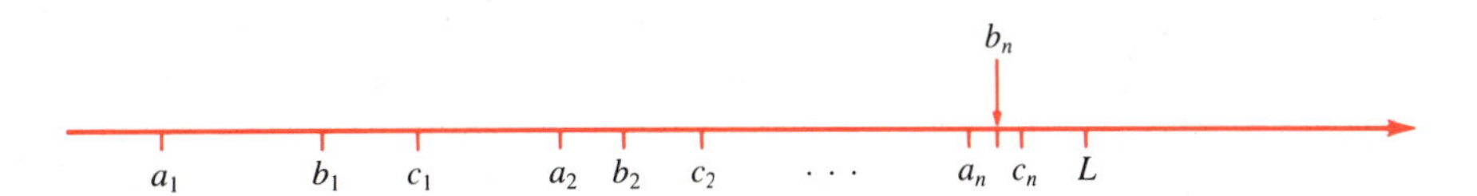

Figure 10-1-2
$a_n \to L$, $c_n \to L$, and $a_n < b_n < c_n$. Then $b_n \to L$ also.

Theorem (Trapped Sequences)

Suppose $\{a_n\}$ and $\{c_n\}$ are convergent sequences with the same limit L, and suppose $\{b_n\}$ is a sequence such that $a_n \le b_n \le c_n$ for $n \ge 1$. Then $\{b_n\}$ also converges to L. ●

For example, suppose $a_n = 1 - 1/n$ and $c_n = 1 + 1/n$, so that $a_n \to 1$ and $c_n \to 1$ as $n \to +\infty$. Let $b_n = 1 + (\sin n)/n$. Then b_n is trapped between a_n and c_n so we conclude that $b_n \to 1$ also.

The special case $a_n = 0$ of this theorem is particularly useful:

Corollary (Squeezing to Zero)

Suppose $0 \le b_n \le c_n$ and $c_n \to 0$. Then $b_n \to 0$. ●

As applications of the corollary, we deduce immediately that the following sequences $\{b_n\}$ all converge to zero:

$$\{1/n^{3/2}\} \qquad \{1/n^2\} \quad \{(\sin^2 n)/n\}$$

In each case, $0 \le b_n \le 1/n$. Since $1/n \to 0$, the corollary shows that $b_n \to 0$.

Remark The result on trapped sequences has an obvious analogue for functions: If $f(x) \le g(x) \le h(x)$ and

$$\lim_{x \to a} f(x) = L \quad \text{and} \quad \lim_{x \to a} h(x) = L$$

then $\lim_{x \to a} g(x) = L$. The trapped-sequence theorem is essentially the case $a = +\infty$ of this result for functions.

Arithmetic of Convergent Sequences

Given a sequence $\{a_n\}$ we can form a new sequence $\{|a_n|\}$ by taking absolute values, and another new sequence $\{ca_n\}$ by multiplying each term by a number c. Given two sequences $\{a_n\}$ and $\{b_n\}$ we can form new sequences $\{a_n + b_n\}$, $\{a_nb_n\}$, and $\{a_n/b_n\}$ by adding, multiplying, or dividing termwise. The following basic rules concerning convergence of sequences are analogous to the rules for limits of functions.

Rules for Limits

Suppose

$$\lim_{n\to+\infty} a_n = A \quad \text{and} \quad \lim_{n\to+\infty} b_n = B$$

Then

- **Absolute Value Rule** $\lim|a_n| = |A|$
- **Constant-Multiple Rule** $\lim ca_n = cA$
- **Sum Rule** $\lim(a_n \pm b_n) = A \pm B$
- **Product Rule** $\lim(a_nb_n) = AB$
- **Quotient Rule** $\lim \dfrac{a_n}{b_n} = \dfrac{A}{B}$ (provided $b_n \neq 0$ and $B \neq 0$)

Each of these rules can be stated informally in words. For instance, the quotient rule says that the limit of a quotient is the quotient of the limits, provided the denominator does not tend towards 0. We omit the proofs, which are close in spirit to those of the corresponding results on limits of functions. These rules will help us evaluate very complex limits. To show how they work, we begin with easy examples.

Example 4

Find $\displaystyle\lim_{n\to+\infty}\left(2+\frac{1}{n}\right)\left(3-\frac{4}{n^2}\right)$

Solution We know that $1/n \to 0$ and $1/n^2 \to 0$. By the constant-multiple rule with $c = 4$, we have $4/n^2 \to 0$. By the sum rule

$$2+\frac{1}{n} \to 2 \quad \text{and} \quad 3-\frac{4}{n^2} \to 3$$

Finally, by the product rule

$$\left(2+\frac{1}{n}\right)\left(3-\frac{4}{n^2}\right) \to 2 \cdot 3 = 6$$

Example 5

Find $\displaystyle\lim_{n\to+\infty}\frac{n^2-2n-5}{n^3}$

Solution Write

$$a_n = \frac{n^2 - 2n - 5}{n^3} = \frac{1}{n} - 2\left(\frac{1}{n^2}\right) - 5\left(\frac{1}{n^3}\right)$$

By the various rules, we have

$$\lim a_n = 0 - 2 \cdot 0 - 5 \cdot 0 = 0$$

●

Remark In practice, we skip many of the steps illustrated in the solutions of Examples 4 and 5. For example, to find

$$\lim \frac{3n^2 + n - 7}{4n^2 + 3n + 6}$$

we divide all terms by n^2, then use the various rules, combining some obvious steps:

$$\lim \frac{3n^2 + n - 7}{4n^2 + 3n + 6} = \lim \frac{3 + \dfrac{1}{n} - \dfrac{7}{n^2}}{4 + \dfrac{3}{n} + \dfrac{6}{n^2}} = \frac{3 + 0 - 0}{4 + 0 + 0} = \frac{3}{4}$$

Example 6

Let a be a constant. Prove that $a^n \to 0$ if $0 < a < 1$ and $a^n \to +\infty$ if $a > 1$.

Solution First assume $0 < a < 1$. Then $1/a > 1$, so it follows that $1/a = 1 + p$ with $p > 0$. By the binomial theorem, for $n > 1$,

$$\frac{1}{a^n} = (1 + p)^n = 1 + pn + (\text{positive terms}) > pn$$

Therefore

$$0 < a^n < \frac{1}{pn} = \frac{1}{p} \cdot \frac{1}{n}$$

Since $1/n \to 0$ and $1/p$ is fixed, $a^n \to 0$ by the corollary to the trapped sequence theorem.

Now assume $a > 1$. Then $a = 1 + b$ where $b > 0$. By the binomial theorem, for $n > 1$,

$$a^n = (1 + b)^n = 1 + nb + (\text{positive terms}) > 1 + nb \to +\infty$$

Therefore $a^n \to +\infty$. ●

Exercises

From the given formula for a_n write the first five terms of the sequence, starting with $n = 1$, and guess its limit

1 $1 - \dfrac{1}{n}$

2 $(n - 1)^2$

3 $\dfrac{n - 1}{n + 1}$

4 $\dfrac{1}{n} - \dfrac{1}{n + 1}$

5 $\dfrac{n}{n^2 + n + 1}$

6 $\dfrac{(n + 1)^2 - n^2}{n}$

7 $(-\frac{1}{2})^n$

8 $(0.1)^n$

9 $(1.01)^n$

10 $(1 + 2/n)^n$

Find N so that for $n \geq N$,

11 $\left| \dfrac{3n + 1}{4n + 5} - \frac{3}{4} \right| < 0.001$

12 $\dfrac{2^n}{(2n)!} < 10^{-5}$

Using the definition of limit, show that

13 $\lim\limits_{n \to +\infty} \dfrac{1}{\sqrt{n}} = 0$

14 $\lim\limits_{n \to +\infty} \dfrac{n^2}{(n + 1)(n + 2)} = 1$

Find $\lim a_n$ if $a_n =$

15 $\dfrac{n + 1}{2n + 3}$

16 $\dfrac{4n}{n^2 + 1}$

17 $\dfrac{n^2 + 2n - 8}{n^3 + 7n + 9}$

18 $\dfrac{n^2 + 5}{3n^2 + 12n + 5}$

19 $\dfrac{n}{\sqrt{n^2 + 1}}$

20 $\dfrac{2n}{\sqrt{n^3 + 6}}$

21 $\dfrac{\sqrt{n}}{3 + 2\sqrt{n}}$

22 $\dfrac{1}{n^2 + \sin n}$

23 $\left(1 + \dfrac{2}{n}\right)^3$

24 $\left(1 + \dfrac{3n}{n^2 + 5}\right)^4$

25 $\dfrac{10^n}{n!}$

26 $\dfrac{(n + 1)(n + 2)(n + 3)}{(n + 4)(n + 5)(n + 6)}$

27 $\left(\dfrac{3n}{n + 7}\right)[1 - (\frac{1}{2})^n]$

28 $\dfrac{7^n + 1}{8^n}$

29 $\dfrac{\sin n}{n}$

30 $\dfrac{1}{1 + \ln n}$

31 $\dfrac{n(2 - e^{-n})}{3n + 1}$

32 $\dfrac{e^n - 1}{e^n + 1}$

33 If $\{a_n\}$ converges and $a_n \geq 0$ for $n \geq 1$, show that $\lim a_n \geq 0$.

34 (cont.) If $\{a_n\}$ and $\{b_n\}$ converge and if $a_n \geq b_n$ for $n \geq 1$, show that $\lim a_n \geq \lim b_n$.

35 Give an example of two divergent sequences $\{a_n\}$ and $\{b_n\}$ such that $\{a_n + b_n\}$ diverges.

36 (cont.) Give an example such that $\{a_n + b_n\}$ converges.

37 Give an example of two divergent sequences $\{a_n\}$ and $\{b_n\}$ such that $\{a_n b_n\}$ diverges.

38 (cont.) Give an example such that $\{a_n b_n\}$ converges.

39 If $|a_n| \to |a|$, does it follow that $a_n \to a$? Give a proof or a counterexample.

40 If $\lim a_n = 0$, prove from the definition that $\lim a_n{}^2 = 0$.

41 Prove that inserting or deleting a finite number of terms does not affect the convergence of a sequence.

42 If $a_n \to L$ and $b_n \to L$, show that the sequence

$$a_1 \quad b_1 \quad a_2 \quad b_2 \quad a_3 \quad b_3 \quad \cdots$$

also converges to L.

43 Under what circumstances can a sequence of *integers* have a finite limit?

44 Show by an example that a sequence may diverge yet contain subsequences that converge.

45 Suppose $a_n \to 0$ and $|b_n| \leq B$, a positive constant. Show that $a_n b_n \to 0$.

46 (cont.) Prove that

$$\frac{3 + e^{-n}}{n(2 + \sin n)} \to 0$$

47 Prove that if $a_n \to L$, then any subsequence of $\{a_n\}$ also converges to L.

48 Prove that if $a_n \to L_1$ and $a_n \to L_2$, then $L_1 = L_2$. [Hint Write $L_1 - L_2 = (L_1 - a_n) - (L_2 - a_n)$.]

Find a formula for a_n for $n \geq 1$

49 $a_n = f^{(n)}(1)$ where $f(x) = 1/\sqrt{x}$

50 $a_n = f^{(n)}(0)$ where $f(x) = e^{x/2}$

10-2 Properties of Limits

There is a useful principle relating limits of sequences to continuous functions. It allows us to deduce limits such as

$$\sin a_n \to \sin L \quad \text{and} \quad \exp(a_n) \to e^L$$

from $a_n \to L$.

Continuous Function of Limits

Let $f(x)$ be continuous at $x = L$, and suppose $\lim_{n\to+\infty} a_n = L$. Then

$$\lim_{n\to\infty} f(a_n) = f(L)$$

Informal Proof If n is large, then a_n is near L because $a_n \to L$. If x is near L, then $f(x)$ is near $f(L)$ because $f(x)$ is continuous. Putting these statements together, we deduce that if n is large, then $f(a_n)$ is near $f(L)$, that is, $f(a_n) \to f(L)$.

Example 1

Let $a > 0$. Prove $\lim_{n\to+\infty} \sqrt[n]{a} = 1$.

Solution The idea is to use

$$\sqrt[n]{a} = a^{1/n} = \exp(\ln a^{1/n}) = \exp[(\ln a)/n] = e^{(\ln a)/n}$$

Now $(\ln a)/n \to 0$ as $n \to +\infty$. Hence

$$\sqrt[n]{a} = e^{(\ln a)/n} \to e^0 = 1$$

since $f(x) = e^x$ is continuous.

Monotone Sequences

Each time we have proved that a sequence converges, we have actually found its limit. Sometimes it is difficult or impossible to find the exact limit. However, in some applications the limit itself is not needed, only the knowledge that the sequence has a limit, that is, it converges.

In such situations we need "intrinsic" criteria for convergence. These are tests for convergence that use the nature of the sequence itself and do not require knowledge of the limit. We shall begin with such a test for monotone sequences.

Monotone Sequences

A sequence $\{a_n\}$ is called **increasing** if

$$a_1 \le a_2 \le a_3 \le \cdots$$

A sequence $\{a_n\}$ is called **decreasing** if

$$a_1 \ge a_2 \ge a_3 \ge \cdots$$

A sequence that is either increasing or decreasing is called **monotone** (moving in one direction).

We are particularly interested in monotone increasing sequences whose terms do not increase without bound and in monotone decreasing sequences whose terms do not decrease without bound. We need still another formal definition to clarify this idea.

Bounded Sequences

A sequence $\{a_n\}$ is **bounded above** if there is a number B such that $a_n \le B$ for each element a_n.

A sequence is **bounded below** if there is a number A such that $a_n \ge A$ for each a_n.

Now we can state an important intrinsic criterion for convergence. The criterion is a consequence of the completeness property of the real number system (page 218), but we leave its proof to more advanced courses.

Convergence of Monotone Sequences

An increasing sequence that is bounded above converges:

$$\text{If } a_1 \le a_2 \le a_3 \le \cdots \le B \text{ then } \lim_{n\to+\infty} a_n \le B.$$

A decreasing sequence that is bounded below converges:

$$\text{If } a_1 \ge a_2 \ge a_3 \ge \cdots \ge A \text{ then } \lim_{n\to+\infty} a_n \ge A.$$

Thus an increasing sequence is either bounded above and sneaks up to a limit, or is unbounded above and marches off the map to $+\infty$. An analogous statement holds for decreasing sequences.

Example 2

Prove $\{a_n\}$ converges, where

$$a_n = \left(1 - \frac{1}{2^2}\right)\left(1 - \frac{1}{3^2}\right)\left(1 - \frac{1}{4^2}\right)\cdots\left(1 - \frac{1}{n^2}\right)$$

Solution Clearly $a_n > 0$ and

$$a_{n+1} = a_n\left(1 - \frac{1}{(n+1)^2}\right) < a_n$$

so $a_1 > a_2 > a_3 > \cdots > 0$. In other words, $\{a_n\}$ is a decreasing sequence that is bounded below by 0. Therefore $\{a_n\}$ converges.

Example 3

Define a sequence $\{a_n\}$ by

$$a_n = 1 + \frac{1}{1!} + \frac{1}{2!} + \cdots + \frac{1}{n!}$$

Show that $\{a_n\}$ converges.

Solution Since $a_{n+1} = a_n + 1/(n+1)! > a_n$, the sequence is monotone increasing. To prove convergence, we show the sequence is bounded above. We use an important technique: replacing the terms making up a_n by slightly larger terms that can be summed easily. Now one thing we can sum for sure is a geometric progression. This suggests the comparison (for $n \geq 3$)

$$n! = 1 \cdot 2 \cdot 3 \cdot \cdot \cdot n > 1 \cdot 2 \cdot 2 \cdot 2 \cdot \cdot \cdot = 2^{n-1}$$

$$\frac{1}{n!} < \frac{1}{2^{n-1}}$$

It follows that

$$a_n = 1 + 1 + \frac{1}{2} + \frac{1}{6} + \frac{1}{24} + \cdot \cdot \cdot + \frac{1}{n!}$$
$$< 1 + \left(1 + \frac{1}{2} + \frac{1}{2^2} + \frac{1}{2^3} + \cdot \cdot \cdot + \frac{1}{2^{n-1}}\right)$$
$$= 1 + \frac{1 - (\frac{1}{2})^n}{1 - \frac{1}{2}} = 1 + 2\left(1 - \frac{1}{2^n}\right) < 3$$

Therefore $\{a_n\}$ increases and is bounded above by 3, so $\{a_n\}$ converges. We can say (from this proof) that $\lim a_n \leq 3$, no more. ●

Remark In the next chapter we shall show that $a_n \rightarrow e \approx 2.71828$.

Sometimes a sequence is defined by an iteration formula. Instead of a formula for a_n in terms of n, we are given a formula for a_n in terms of one or more previous a's and possibly n too. The following example contains a sequence in which the first a is given and each succeeding a is expressed in terms of the previous a.

Example 4

Define a sequence $\{a_n\}$ by

$$a_1 = 2 \qquad a_{n+1} = \frac{1}{2}\left(a_n + \frac{2}{a_n}\right)$$

Prove $\{a_n\}$ converges and then find its limit.

Solution Compute a few elements:

$$a_1 = 2 \qquad a_2 = \frac{1}{2}\left(2 + \frac{2}{2}\right) = \frac{3}{2} = 1.5$$

$$a_3 = \frac{1}{2}\left(\frac{3}{2} + \frac{2}{\frac{3}{2}}\right) = \frac{17}{12} \approx 1.4167$$

$$a_4 = \frac{1}{2}\left(\frac{17}{12} + \frac{2}{\frac{17}{12}}\right) = \frac{577}{408} \approx 1.4142$$

The sequence appears to be decreasing. Since its elements are bounded below by 0, we are optimistic about the chances of convergence. Probably the sequence converges to a positive limit L, somewhere near 1.4.

Assuming the sequence does converge, what is its limit? We exploit the defining relation

$$a_{n+1} = \frac{1}{2}\left(a_n + \frac{2}{a_n}\right)$$

If $a_n \to L$, then the right side approaches $\frac{1}{2}(L + 2/L)$. The left side approaches L because $\{a_{n+1}\}$ is a subsequence of $\{a_n\}$. Therefore, the limit L must satisfy

$$L = \frac{1}{2}\left(L + \frac{2}{L}\right) \qquad \text{that is} \qquad 2L^2 = L^2 + 2 \qquad L^2 = 2$$

Since $L \geq 0$, the only possibility is $L = \sqrt{2} \approx 1.414214$.

The reasoning is correct *provided* $\{a_n\}$ converges. To prove it does, it suffices to show that $\{a_n\}$ is decreasing. We would like to argue as follows:

$$a_{n+1} = \frac{1}{2}\left(a_n + \frac{2}{a_n}\right) \leq \frac{1}{2}(a_n + a_n) = a_n$$

so $a_{n+1} \leq a_n$.

This argument is valid provided $2/a_n < a_n$, that is, provided $2 \leq a_n^2$. So the final step is to prove this inequality for $n \geq 2$. Now

$$a_n^2 = \left[\frac{1}{2}\left(a_{n-1} + \frac{2}{a_{n-1}}\right)\right]^2 = \frac{1}{4}\left(a_{n-1} - \frac{2}{a_{n-1}}\right)^2 + 2 \geq 2$$

This ties up the last loose end. ●

Remark Note the connection with the Newton-Raphson method (page 517): if $f(x) = 2 - x^2$, then

$$\phi(x) = x - \frac{f(x)}{f'(x)} = x - \frac{2 - x^2}{-2x} = \frac{x^2 + 2}{2x}$$
$$= \frac{1}{2}\left(x + \frac{2}{x}\right)$$

Sometimes a sequence $\{a_n\}$ jumps around a lot at first, then eventually settles down to a nice behavior. This doesn't matter, because convergence depends on what happens in the long run, not at first. In general, convergence tests are unaffected by a finite number of exceptions. This is an important practical point, and should be remembered. For instance, if a sequence is bounded above and *eventually* increasing, then it converges.

Comparison Tests

In certain situations we need only compare *successive* terms of the sequence. The following two tests for convergence guarantee that if successive terms are always sufficiently close together, then the sequence converges.

First Comparison Test

Suppose $0 < c < 1$ and $0 < B$, and suppose $\{a_n\}$ is a sequence for which

$$|a_n - a_{n+1}| \le Bc^n \qquad \text{for all } n$$

Then $\{a_n\}$ converges. ●

We omit the technical proof of this test. The following application of it justifies the use of unending decimals.

Example 5

Suppose $\{d_n\}$ is a sequence of integers, where $0 \le d_n \le 9$ and

$$a_n = 0.d_1d_2 \cdots d_n = \frac{d_1}{10} + \frac{d_2}{10^2} + \cdots + \frac{d_n}{10^n}$$

Prove that the sequence $\{a_n\}$ converges.

Solution

$$|a_n - a_{n+1}| = \frac{d_{n+1}}{10^{n+1}} \le \frac{9}{10^{n+1}} = \frac{9}{10}\left(\frac{1}{10}\right)^n$$

The comparison test applies with $B = \frac{9}{10}$ and $c = \frac{1}{10}$. ●

Second Comparison Test

Suppose $0 < c < 1$, and suppose $\{a_n\}$ is a sequence for which

$$|a_{n+1} - a_{n+2}| \le c|a_n - a_{n+1}| \qquad \text{for all } n$$

Then $\{a_n\}$ converges. ●

Proof We work iteratively from $|a_n - a_{n+1}|$ down to $|a_1 - a_2|$:

$$\begin{aligned} |a_n - a_{n+1}| &\le c|a_{n-1} - a_n| \le c^2|a_{n-2} - a_{n-1}| \\ &\le \cdots \le c^{n-1}|a_1 - a_2| = Bc^n \end{aligned}$$

where $B = |a_1 - a_2|/c$. Now the previous test applies. ●

Remark Note carefully the distinction between the two comparison tests. In the first we must show that the difference $|a_n - a_{n+1}|$ decreases at least geometrically. In the second we must show that each difference is at most a fixed proportion of the previous difference.

Example 6

Set $a_1 = 1$ and $a_{n+1} = 1/(1 + a_n)$. Prove that $\{a_n\}$ converges and find its limit.

Solution The first four terms are 1, $\frac{1}{2}$, $\frac{2}{3}$, $\frac{3}{5}$, $\frac{5}{8}$, $\frac{8}{13}$. They seem to increase and decrease alternately. Apparently 1 is the largest, $\frac{1}{2}$ is the smallest. We can easily prove this last guess by induction. For if $\frac{1}{2} \le a_n \le 1$, then $\frac{3}{2} \le 1 + a_n \le 2$, so

$$\tfrac{2}{3} \ge a_{n+1} = 1/(1 + a_n) \ge 1/2$$

Hence certainly $\frac{1}{2} \le a_{n+1} \le 1$.

Next, we use this rough estimate on the size of a_n to compare successive terms:

$$\begin{aligned} |a_{n+1} - a_{n+2}| &= \left| \frac{1}{1 + a_n} - \frac{1}{1 + a_{n+1}} \right| \\ &= \left| \frac{a_{n+1} - a_n}{(1 + a_n)(1 + a_{n+1})} \right| \end{aligned}$$

But

$$(1 + a_n)(1 + a_{n+1}) \ge (1 + \tfrac{1}{2})^2 = \tfrac{9}{4}$$

Hence

$$|a_{n+1} - a_{n+2}| \le \tfrac{4}{9}|a_n - a_{n+1}| \quad \text{for all } n$$

Therefore $\{a_n\}$ converges by the second comparison test.

Knowing that the sequence converges, we can now find its limit. Let $a_n \to L$. Then $L \ge \frac{1}{2} > 0$ because $a_n \ge \frac{1}{2}$ for all n. From $a_{n+1} = 1/(1 + a_n)$ follows $L = 1/(1 + L)$ by taking the limit of both sides. Hence (since $L \ge 0$)

$$L^2 + L - 1 = 0 \qquad L = \frac{\sqrt{5} - 1}{2} \approx 0.61803$$

●

Note that $a_{15} \approx 0.61803$ is already quite close to the limit L, so the sequence converges rapidly.

Cauchy Sequences [Optional]

If $a_n \to L$, then the numbers a_n get close to L, hence close to each other. More precisely, suppose $\varepsilon > 0$ is given. There exists an N such that if $m \ge N$ and $n \ge N$, then both $|a_m - L| < \frac{1}{2}\varepsilon$ and $|a_n - L| < \frac{1}{2}\varepsilon$. Consequently

$$\begin{aligned} |a_m - a_n| = |(a_m - L) - (a_n - L)| &< |a_m - L| + |a_n - L| \\ &< \tfrac{1}{2}\varepsilon + \tfrac{1}{2}\varepsilon = \varepsilon \end{aligned}$$

for all m and n greater than or equal to N.

Cauchy Sequences

A sequence $\{a_n\}$ is called a **Cauchy sequence** if for each $\varepsilon > 0$, there is a positive integer N such that

$$|a_m - a_n| < \varepsilon \quad \text{for all } m \ge N \text{ and all } n \ge N$$

(Cauchy is pronounced koh′-shee. A Cauchy sequence is sometimes called a **fundamental sequence.**)

●

For an example of a Cauchy sequence, take $\{a_n\}$ where $a_n = 1/n$. Then

$$|a_m - a_n| = \left|\frac{1}{m} - \frac{1}{n}\right| < \frac{1}{m} + \frac{1}{n}$$

Therefore $|a_m - a_n| < \varepsilon$ if

$$\frac{1}{m} < \frac{\varepsilon}{2} \quad \text{and} \quad \frac{1}{n} < \frac{\varepsilon}{2}$$

that is, if

$$m > \frac{2}{\varepsilon} \quad \text{and} \quad n > \frac{2}{\varepsilon}$$

For example, if $\varepsilon = 0.001$, then $|a_m - a_n| < \varepsilon$ if $m > 2000$ and $n > 2000$. For instance, $|a_{2177} - a_{14508}| < 0.001$.

Not only is every convergent sequence a Cauchy sequence, but conversely, every Cauchy sequence is a convergent sequence.

Cauchy Criterion

A sequence converges if and only if it is a Cauchy sequence.

We have proved the "only if" part of this assertion: each convergent sequence is a Cauchy sequence. The converse, the "if" part, says that if a sequence is a Cauchy sequence, then it is a convergent sequence. This is a deep result which goes right back to the very fundamental completeness of the real number system. Its proof is beyond the scope of this course.

The Cauchy criterion is intrinsic in that it depends on the sequence itself and on nothing else. Once it is proved, the other intrinsic criteria we have been discussing all can be proved by routine techniques.

It is a very common misconception that if $|a_{n+1} - a_n| \to 0$, then the sequence $\{a_n\}$ is a Cauchy sequence and hence converges. This is very wrong, and to wipe it out of your mind, you only have to remember one good example, the sequence $\{a_n\}$ where $a_n = \ln n$. This sequence certainly does not converge because $\ln n \to +\infty$ as $n \to +\infty$, so also it is not a Cauchy sequence. But

$$\begin{aligned} |a_{n+1} - a_n| &= \ln(n+1) - \ln n \\ &= \ln \frac{n+1}{n} \end{aligned}$$

Clearly $(n+1)/n \to 1$ as $n \to +\infty$, so

$$\ln \frac{n+1}{n} \to \ln 1 = 0$$

Therefore $|a_{n+1} - a_n| \to 0$.

Exercises

Find $\lim a_n$, where $a_n =$

1 $e^{-1/n}$

2 $\cos \dfrac{2\pi}{\sqrt{n}}$

3 $\arc\tan\left(\dfrac{n}{n+3}\right)$

4 $\left(\dfrac{n^2+5n+1}{4n^2+3}\right)^{1/2}$

Prove convergence of the sequence $\{a_n\}$ by showing that it is monotone and bounded

5 $a_n = \dfrac{1}{2}\cdot\dfrac{4}{5}\cdot\dfrac{7}{8}\cdots\dfrac{3n-2}{3n-1}$

6 $a_n = \left(1-\dfrac{1}{8}\right)\left(1-\dfrac{1}{27}\right)\cdots\left(1-\dfrac{1}{n^3}\right)$

7 $a_n = 1 + \dfrac{1}{(2!)^2} + \dfrac{1}{(3!)^2} + \cdots + \dfrac{1}{(n!)^2}$

***8** $a_n = \dfrac{1}{n+1} + \dfrac{1}{n+2} + \cdots + \dfrac{1}{2n}$

9 $a_n = c^n \quad (0 < c < 1)$

10 $a_0 = 1,\ a_1 = \sqrt{5},\ \cdots,\ a_{n+1} = \sqrt{5a_n}$

11 $a_0 = 1,\ a_1 = \frac{3}{2},\ \cdots,\ a_{n+1} = 1 + \frac{1}{2}a_n$

12 $a_0 = \frac{1}{2}\pi,\ a_1 = \sin\frac{1}{2}\pi,\ \cdots,\ a_{n+1} = \sin a_n$

Use the method of Example 4 to evaluate the limit

13 in Exercise 10

14 in Exercise 12

15 Define $\{x_n\}$ by $x_0 = 0$, $x_1 = 1$, and $x_{n+2} = \frac{1}{2}(x_n + x_{n+1})$. Prove $\{x_n\}$ is convergent.

16 (cont.) Find $\lim x_n$.

17 Define $a_1 = \sqrt{2}$ and $a_{n+1} = \sqrt{2 + a_n}$ for $n = 1, 2, \cdots$. Prove that $\{a_n\}$ is convergent.

18 (cont.) Find $\lim a_n$.

19 Suppose $x > 0$. Define $a_1 = \sqrt{x}$ and $a_{n+1} = \sqrt{x + a_n}$ for $n = 1, 2, \cdots$. Prove $\{a_n\}$ is bounded.

20 (cont.) Prove $\{a_n\}$ is increasing, hence has a limit. Find the limit.

21 Suppose $0 < x < y$. Define two sequences $\{a_n\}$ (harmonic means) and $\{b_n\}$ (arithmetic means) by $a_0 = x$, $b_0 = y$,

$$a_{n+1} = \frac{2}{(1/a_n) + (1/b_n)} \quad\text{and}\quad b_{n+1} = \frac{a_n + b_n}{2}$$

Prove $a_0 < a_1 < a_2 < \cdots < b_2 < b_1 < b_0$. [Hint If $0 < x < y$, show that $x < 2xy/(y+x) < \frac{1}{2}(x+y) < y$.]

22 (cont.) Find $\lim a_n$ and $\lim b_n$.

23 Suppose $0 < x < 2$. Define $\{b_n\}$ by $b_0 = 1 - x$ and $b_{n+1} = b_n^2$. Show that $b_n \to 0$.

24 (cont.) Define $\{a_n\}$ by $a_0 = 1$, $a_{n+1} = a_n(1 + b_n)$. Prove that $a_n \to 1/x$. [Hint Show by induction that $a_n = (1 - b_n)/x$.] This provides an algorithm for division on a calculator that has only $+$, $-$, and $\times$.

25 Suppose $0 < x < 2$. Define $\{b_n\}$ by $b_0 = 1 - x$ and $b_{n+1} = \frac{1}{4}b_n^2(3 + b_n)$. Prove $b_n \to 0$.

26 (cont.) Define $\{a_n\}$ by $a_0 = x$ and $a_{n+1} = a_n(1 + \frac{1}{2}b_n)$. Prove $a_n \to \sqrt{x}$. [Hint Show that $a_n = \sqrt{x(1 - b_n)}$.] This provides an algorithm for $\sqrt{\ }$ in terms of $+$, $-$, and $\times$.

27 Suppose $0 < a < 1$ and $p > 0$. Find $\lim n^p a^n$. [Hint $(\ln n)/n \to 0$]

***28** Let $a_1 = \sqrt{12}$ and $a_{n+1} = \sqrt{12 - a_n}$ for $n = 1, 2, \cdots$. Analyze $\{a_n\}$ for possible convergence.

Use the First Comparison Test to prove the convergence of $\{a_n\}$, where

29 $a_n = 1 - \dfrac{1}{1!} + \dfrac{1}{2!} - \dfrac{1}{3!} + \cdots + \dfrac{(-1)^n}{n!}$

30 $a_n = \dfrac{1}{2} - \dfrac{2}{2^2} + \dfrac{3}{2^3} + \cdots + (-1)^{n-1}\dfrac{n}{2^n}$

31 Suppose $0 < x$. Set $b_0 = \frac{1}{2}(x + 1/x)$ and $b_{n+1} = \sqrt{\frac{1}{2}(1 + b_n)}$. Show that $b_n \to 1$.

***32** (cont.) Set $a_0 = \frac{1}{2}(x - 1/x)$ and $a_{n+1} = a_n/b_{n+1}$. Prove that $a_n \to \ln x$. This provides an algorithm for computing logs on a calculator with $+$, $-$, $\times$, and $\sqrt{\ }$. [Hint Express b_n and a_n in terms of x.]

33 Let $1 \leq x < 2$. Define two sequences recursively as follows: $a_0 = x$ and $c_0 = 0$. If $a_n^2 < 2$, then $a_{n+1} = a_n^2$ and $c_{n+1} = c_n$. If $a_n^2 \geq 2$, then $a_{n+1} = \frac{1}{2}a_n^2$, and $c_{n+1} = c_n + (\frac{1}{2})^{n+1}$. Prove that

$$c_n + \frac{\log_2 a_n}{2^n} = \log_2 x$$

34 (cont.) Prove that $1 \leq a_n < 2$, that $\{c_n\}$ is increasing, and that $\lim_{n\to\infty} c_n = \log_2 x$. (This provides an algorithm for logs on a computer that only adds, squares, and divides by 2.)

***35** Let $0 < x < y$. Prove (by the MVT)

$$(n+1)x^n < \frac{y^{n+1} - x^{n+1}}{y - x} < (n+1)y^n$$

Hence $x^n[(n+1)y - nx] < y^{n+1}$ and $y^n[(n+1)x - ny] < x^{n+1}$.

***36** (cont.) Set $a_n = (1 + 1/n)^n$. Prove $\{a_n\}$ is increasing and $a_n < 4$.

***37** (cont.) Set $b_n = (1 + 1/n)^{n+1}$. Prove $\{b_n\}$ is decreasing.

***38** (cont.) Conclude that $\{a_n\}$ and $\{b_n\}$ converge to the same limit. (The limit, of course, is e.)

10-3 Infinite Series

One of the most important topics in mathematical analysis, both in theory and applications, is infinite series. There are many problems whose solution is in the form of a sum

$$a_1 + a_2 + a_3 + \cdots + a_n + \cdots$$

of all of the terms of an infinite sequence. Our problem is to attach a meaning to such an infinite sum. Let us consider for instance

$$1 + \frac{1}{2} + \frac{1}{4} + \cdots + \frac{1}{2^n} + \cdots$$

and start adding up terms. We find $1, \frac{3}{2}, \frac{7}{4}, \frac{15}{8}, \frac{31}{16}, \cdots,$ numbers getting closer and closer to 2. The message is clear: these finite sums should have limit 2. Therefore in some sense, the infinite sum equals 2. In contrast, if we consider the sum

$$1 + 1 + 1 + \cdots + 1 + \cdots$$

and start adding up terms, we find the sums $1, 2, 3, 4, \cdots$. These numbers become larger and larger, eventually exceeding any given real number. In this case there is no reasonable sum of the series.

Before we write formal definitions, let us consider in some detail two important infinite sums.

Geometric Series

A **geometric series** is an infinite sum in which the ratio of any two consecutive terms is always the same:

$$a + ar + ar^2 + \cdots + ar^n + \cdots \qquad (a \neq 0, \quad r \neq 0)$$

Let s_n denote the sum of all terms up to ar^n,

$$s_n = a + ar + ar^2 + \cdots + ar^n$$

If $r \neq 1$, there is a simple formula for s_n, the sum of a geometric progression (page 289):

$$s_n = a(1 + r + r^2 + \cdots + r^n) = a\left(\frac{1 - r^{n+1}}{1 - r}\right)$$

If $|r| < 1$, then $r^{n+1} \to 0$ as n increases. Hence

$$\lim s_n = \frac{a}{1 - r} \qquad (|r| < 1)$$

so a logical choice for the sum of the series is $a/(1 - r)$. But if $|r| > 1$, then $|r^{n+1}| \to +\infty$ as $n \to +\infty$ in the sense that $|r^{n+1}|$ becomes arbitrarily large. Clearly also $|s_n| \to +\infty$ as $n \to +\infty$ in this case. There is no sensible choice for the sum of the series. If $r = 1$, then

$$s_n = a + a + \cdots + a = (n + 1)a$$

so $s_n \to \pm\infty$. In the remaining case, $r = -1$, the s_n alternate between a and 0. Again there is no sensible choice for the sum.

- An infinite geometric series

$$a + ar + ar^2 + \cdots + ar^n + \cdots \qquad (a \neq 0, \quad r \neq 0)$$

converges to $a/(1-r)$ if $|r| < 1$, but diverges if $|r| \geq 1$.

Harmonic Series

The series

$$1 + \tfrac{1}{2} + \tfrac{1}{3} + \cdots + 1/n + \cdots$$

is known as the **harmonic series.** Although it is not at all obvious, the sums $s_n = 1 + \frac{1}{2} + \frac{1}{3} + \cdots + 1/n$ are unbounded, so the series has no sum. To see why, we observe that

$$s_1 = 1 > \tfrac{1}{2} \qquad s_2 = s_1 + \tfrac{1}{2} > \tfrac{1}{2} + \tfrac{1}{2} = \tfrac{2}{2}$$

$$s_4 = s_2 + (\tfrac{1}{3} + \tfrac{1}{4}) > s_2 + (\tfrac{1}{4} + \tfrac{1}{4}) > \tfrac{2}{2} + \tfrac{1}{2} = \tfrac{3}{2}$$

$$s_8 = s_4 + (\tfrac{1}{5} + \tfrac{1}{6} + \tfrac{1}{7} + \tfrac{1}{8}) > s_4 + (\tfrac{1}{8} + \tfrac{1}{8} + \tfrac{1}{8} + \tfrac{1}{8})$$

$$> \tfrac{3}{2} + \tfrac{1}{2} = \tfrac{4}{2}$$

Similarly

$$s_{16} > \tfrac{5}{2} \qquad s_{32} > \tfrac{6}{2} \qquad \cdots \qquad s_{2^n} > \tfrac{1}{2}(n+1)$$

Thus the sequence of sums s_n increases, and our estimates show s_n eventually passes any given positive number. (The sums grow very slowly. About 2^{15} terms are needed before s_n exceeds 10 and about 2^{29} terms before it exceeds 20.)

Both the geometric series for $0 < r < 1$ and the harmonic series have positive terms that decrease toward zero, yet one series has a sum and the other does not. This indicates the subtlety we must expect in our further study of infinite series.

Convergence and Divergence

It is time to formulate our ideas more precisely.

- An **infinite series** is a formal sum

$$\sum a_n = a_1 + a_2 + a_3 + \cdots$$

of the terms of a sequence $\{a_n\}_1^\infty$. Associated with each infinite series is its sequence $\{s_n\}$ of **partial sums** defined by

$$s_1 = a_1 \qquad s_2 = s_1 + a_2 \quad \text{and in general} \quad s_n = s_{n-1} + a_n$$

A series **converges** to the real number S, or has **sum** S, if $\lim_{n\to+\infty} s_n = S$. A series **diverges,** or has no finite sum, if $\lim s_n$ does not exist. A series that converges is called **convergent**; a series that diverges is called **divergent.** If a series converges to S, we write

$$a_1 + a_2 + a_3 + \cdots = S \quad \text{or equivalently} \quad \sum_{n=1}^{\infty} a_n = S$$

We have a precise definition for $\lim s_n = S$. Let us rephrase the definition of convergence of series accordingly:

- The infinite series $a_1 + a_2 + a_3 + \cdots$ converges to S if for each $\varepsilon > 0$, there is a positive integer N such that

$$|(a_1 + a_2 + \cdots + a_n) - S| < \varepsilon$$

whenever $n \geq N$.

Thus, no matter how small ε, you will get within ε of S by adding up enough terms. For each ε, the N tells how many terms are "enough." Naturally the smaller ε is, the larger N may have to be.

Remark If $s_n \to +\infty$ as $n \to +\infty$, we say that $\Sigma_{n=1}^{\infty} a_n$ **diverges to** $+\infty$, and write $\Sigma_{n=1}^{\infty} a_n = +\infty$. Similarly if $s_n \to -\infty$, we say $\Sigma_{n=1}^{\infty} a_n$ **diverges to** $-\infty$, and write $\Sigma_{n=1}^{\infty} a_n = -\infty$.

Example 1

Consider the series $3 + \frac{3}{2} + \frac{3}{4} + \frac{3}{8} + \cdots$

a Show that its sum is 6.

Find N so that

b $|s_n - 6| < 10^{-4}$ for $n \geq N$

c $|s_n - 6| < 10^{-8}$ for $n \geq N$

Solution

a This is a geometric series with first term $a = 3$ and common ratio $r = \frac{1}{2}$. Since $|r| < 1$, the series converges to

$$\frac{a}{1 - r} = \frac{3}{1 - \frac{1}{2}} = 6$$

b The n-th partial sum is

$$s_n = 3\left(1 + \frac{1}{2} + \frac{1}{4} + \cdots + \frac{1}{2^{n-1}}\right)$$

$$= 3 \cdot \frac{1 - (\frac{1}{2})^n}{1 - \frac{1}{2}} = 6[1 - (\tfrac{1}{2})^n]$$

Hence

$$|s_n - 6| = 6(\tfrac{1}{2})^n$$

Therefore $|s_n - 6| < 10^{-4}$ provided $6(\frac{1}{2})^n < 10^{-4}$, that is,

$$(\tfrac{1}{2})^n < \tfrac{1}{6} \times 10^{-4} \quad \text{so} \quad 2^n > 6 \times 10^4$$

Now $2^{15} < 33{,}000$ and $2^{16} > 65{,}000$, so $N = 16$ works.

c As in part **b** we must have $2^n > 6 \times 10^8$, that is,

$$n \log 2 > \log(6 \times 10^8) = 8 + \log 6$$

so

$$n > \frac{8 + \log 6}{\log 2} \approx 29.2$$

Therefore $N = 30$ works. ●

Remark We extend the notion of infinite series to series that do not necessarily start at $n = 1$:

$$\sum_{n=k}^{\infty} a_n = a_k + a_{k+1} + a_{k+2} + \cdots$$

is the sum of the sequence $\{a_n\}_k^\infty$. It should be clear how to modify the discussion to include this slight generalization. For example, the sum of the geometric progression $\{(\frac{1}{2})^n\}_0^\infty$ is

$$\sum_{n=0}^{\infty} (\tfrac{1}{2})^n = 1 + \tfrac{1}{2} + \tfrac{1}{4} + \tfrac{1}{8} + \cdots = 2$$

Another example: $\displaystyle\sum_{n=2}^{\infty} \frac{(-1)^n}{\ln n}$. [Note that $1/(\ln 1)$ is undefined.]

Convergence Tests

When we study the convergence of an infinite series $\Sigma\, a_n$, we really study the convergence of the sequence $\{s_n\}$ of partial sums. Thus we actually have *two* sequences to consider, the sequence $\{a_n\}$ that defines the series and the derived sequence $\{s_n\}$ of partial sums. The definition of convergence concentrates on $\{s_n\}$, and we can apply everything we know about the convergence of sequences to $\{s_n\}$. However, what we really want are tests for the convergence of $\Sigma\, a_n$ *in terms of the sequence* $\{a_n\}$.

For example, we know that adding the same constant to all sufficiently large terms of $\{s_n\}$ does not affect its convergence or divergence. Consequently inserting, deleting, or altering a finite number of terms of an infinite series only adds a constant to each s_n beyond a certain point, hence does not affect the convergence or divergence of the series. For instance, if we delete the first ten terms of the series $a_1 + a_2 + a_3 + \cdots$, then the effect is that we decrease each partial sum s_n (for $n > 10$) by the amount s_{10}. If the original series diverges, then so does the modified series. If the original series converges to S, then the modified series converges to $S - s_{10}$.

Note In problems where we must decide whether a given infinite series converges or diverges, we may, without prior notice, ignore or change a (finite) batch of terms at the beginning. This does not affect convergence.

We now state, without proof, an important test for whether a given series is convergent or divergent.

Cauchy Test

An infinite series $\Sigma\, a_n$ converges if and only if for each $\varepsilon > 0$, there is a positive integer N such that

$$|a_{n+1} + a_{n+2} + \cdot \cdot \cdot + a_m| < \varepsilon$$

whenever $m > n \geq N$.

Thus beyond a certain point in a convergent series, any block of consecutive terms, *no matter how long,* must have a very small sum.

An important corollary of the "only if" part of the Cauchy test is a necessary condition for convergence of an infinite series. If a series does converge, then the Cauchy test is satisfied. In particular, it is satisfied in the special case $n = m - 1$. Then the block consists of just one term, a_m, so $|a_m| < \varepsilon$ when $m \geq N$. In other words, $a_m \to 0$.

n-th Term Test for Divergence

If $\lim a_n$ does not exist, or if $\lim a_n$ exists but $\lim a_n \neq 0$, then $\Sigma\, a_n$ diverges. Equivalently, if $\Sigma\, a_n$ converges, then $\lim_{n\to+\infty} a_n = 0$.

For emphasis we have stated the test in two equivalent forms. We remark that it can easily be proved from scratch, independently of the Cauchy test. For, if $\Sigma\, a_n = S$, then $s_n \to S$. Hence

$$a_n = s_n - s_{n-1} \to S - S = 0$$

Warning Pay attention here, because misuse of this test causes lots of errors. It is really a test for *divergence,* not a test for convergence. It says that if $\lim_{n\to+\infty} a_n \neq 0$, then $\Sigma\, a_n$ diverges. It says nothing if $\lim_{n\to+\infty} a_n = 0$. In that case the series may diverge—or it may converge. Keep in mind the harmonic series, in which $a_n = 1/n$; then $\lim_{n\to+\infty} a_n = 0$ but $\Sigma\, a_n$ diverges.

Exercises

Compute the finite sum

1 $1 + \dfrac{1}{3} + \dfrac{1}{3^2} + \cdot \cdot \cdot + \dfrac{1}{3^9}$

2 $1 - \dfrac{1}{3} + \dfrac{1}{3^2} - + \cdot \cdot \cdot - \dfrac{1}{3^9}$

3 $\dfrac{1}{2} + \dfrac{1}{4} + \dfrac{1}{8} + \cdot \cdot \cdot + \dfrac{1}{256}$

4 $(\frac{2}{3})^2 + (\frac{2}{3})^3 + \cdot \cdot \cdot + (\frac{2}{3})^6$

5 $3 + \dfrac{3^2}{x} + \dfrac{3^3}{x^2} + \cdot \cdot \cdot + \dfrac{3^{n+1}}{x^n}$

6 $1 - y^2 + y^4 - + \cdot \cdot \cdot + y^{20}$

7 $r^{1/2} + r + r^{3/2} + \cdot \cdot \cdot + r^4$

8 $(x + 1) + (x + 1)^2 + \cdot \cdot \cdot + (x + 1)^5$

Sum the infinite series

9 $1 - \frac{2}{5} + (\frac{2}{5})^2 - (\frac{2}{5})^3 + - \cdots$

10 $\frac{1}{2} - \frac{1}{4} + \frac{1}{8} - \frac{1}{16} + - \cdots$

11 $\dfrac{1}{2^{10}} + \dfrac{1}{2^{11}} + \dfrac{1}{2^{12}} + \cdots$

12 $\dfrac{1}{3} + \dfrac{1}{27} + \dfrac{1}{243} + \cdots$

13 $\dfrac{4+1}{9} + \dfrac{8+1}{27} + \dfrac{16+1}{81} + \cdots$

14 $12 - 6 + 3 - \frac{3}{2} + \frac{3}{4} - + \cdots$

15 $\dfrac{1}{2+x^2} + \dfrac{1}{(2+x^2)^2} + \dfrac{1}{(2+x^2)^3} + \cdots$

16 $\dfrac{\cos\theta}{2} + \dfrac{\cos^2\theta}{4} + \dfrac{\cos^3\theta}{8} + \cdots$

17 A certain rubber ball when dropped will bounce back to half the height from which it is released. If the ball is dropped from 3 ft and continues to bounce, find the total distance through which it moves.

18 Trains A and B are 60 miles apart on the same track and start moving toward each other, each traveling at the rate of 30 mph. At the same time, a fly starts at train A and flies to train B at 60 mph. Then it returns to train A, then to B, and so on. Use a geometric series to compute the total distance it flies until the trains meet.

19 (cont.) Do Exercise 18 without geometric series.

20 A line segment of length L is drawn and (step 1) its middle third is erased. Then (step 2) the middle third of each of the two remaining segments is erased. Then (step 3) the middle third of each of the four remaining segments is erased, and so on. After step n, what is the total length of all the segments deleted?

Interpret the repeating decimal as a geometric series and express its sum as a fraction

21 0.11111 · · ·

22 0.101010 · · ·

23 0.434343 · · ·

24 0.185185185 · · ·

Show that the series diverge

25 $\frac{1}{2} + \frac{1}{4} + \frac{1}{6} + \frac{1}{8} + \cdots$

26 $1 + \frac{1}{3} + \frac{1}{5} + \frac{1}{7} + \cdots$

27 Find n so large that $\dfrac{1}{101} + \dfrac{1}{102} + \cdots + \dfrac{1}{n} > 2.$

28 Aristotle summarized Zeno's paradoxes as follows:

I can't go from here to the wall. For to do so, I must first cover half the distance, then half the remaining distance, then again half of what still remains. This process can always be continued and can never be completed.

Explain what is going on here.

29 Use partial fractions (page 360) to write $1/n(n+1)$ in the form $A/n + B/(n+1)$. Use the resulting expression to sum

$$\frac{1}{1\cdot 2} + \frac{1}{2\cdot 3} + \frac{1}{3\cdot 4} + \cdots$$

30 Use a method similar to that in Exercise 29 to sum

$$\sum_{n=1}^{\infty} \frac{2n+1}{[n(n+1)]^2}$$

31 Describe all convergent series of *integers*.

32 Show how the harmonic series fails the Cauchy test for $\varepsilon = \frac{1}{2}$.

***33** Let $\{a_n\}$ be a convergent sequence with limit 0. Define the averages

$$b_n = \frac{a_1 + \cdots + a_n}{n}$$

Prove $\lim b_n = 0$. [Hint Choose N so that $|a_n| < \frac{1}{2}\varepsilon$ for $n > N$ and write

$$b_{N+j} = \frac{a_1 + a_2 + \cdots + a_N}{N+j} + \frac{a_{N+1} + \cdots + a_{N+j}}{N+j}$$

34 (cont.) If $\lim a_n = L$, prove that $\lim b_n = L$. [Hint Write $a_n = L + c_n$.]

10-4 Series of Positive Terms

In this section we deal with infinite series having only non-negative terms. (Everything we do could be restated for series having only non-positive terms by the simple expedient of changing all signs.) The partial sums of a series of non-negative numbers form an increasing sequence, $s_1 \le s_2 \le s_3 \le s_4 \le \cdots$. Recall that an increasing sequence must be one of two types: either (a) the sequence is bounded above, in which case it converges; or (b) it is not bounded above, and it marches off the map to $+\infty$, that is, diverges to $+\infty$.

We deduce corresponding statements about series:

- A series $a_1 + a_2 + a_3 + \cdots$ with $a_n \geq 0$ converges if and only if there exists a positive number M such that

$$a_1 + a_2 + \cdots + a_n \leq M \qquad \text{for all} \quad n \geq 1$$

Using this fact, we can often establish the convergence or divergence of a given series by comparing it with a familiar series.

First Comparison Test

Suppose $\Sigma\, a_n$ and $\Sigma\, b_n$ are series with non-negative terms.

1 If $\Sigma\, a_n$ converges and $b_n \leq a_n$ for all $n \geq 1$, then $\Sigma\, b_n$ also converges,

2 If $\Sigma\, a_n$ diverges and $b_n \geq a_n$ for all $n \geq 1$, then $\Sigma\, b_n$ also diverges.

Proof Let s_n and t_n denote the partial sums of $\Sigma\, a_n$ and $\Sigma\, b_n$ respectively. Then $\{s_n\}$ and $\{t_n\}$ are increasing sequences.

1 Since $\Sigma\, a_n$ converges, $s_n \leq \Sigma\, a_n = M$ for all $n \geq 1$. Since $b_k \leq a_k$ for all k, we have $t_n \leq s_n$ for all n. Hence $t_n \leq s_n \leq M$ for all $n \geq 1$, so $\Sigma\, b_n$ converges.

2 Since $\Sigma\, a_n$ diverges, the sequence $\{s_n\}$ is unbounded. Since $b_k \geq a_k$, we have $t_n \geq s_n$. Hence $\{t_n\}$ is also unbounded, so $\Sigma\, b_n$ diverges.

It is important to apply the comparison test correctly. Roughly speaking, part 1 says that "smaller than small is small" and part 2 says that "bigger than big is big." However the phrases "smaller than big" and "bigger than small" contain little useful information.

The first comparison test, as well as the other tests we shall derive, applies if the given condition holds from some point on, not necessarily starting at $n = 1$. For example, if $\Sigma\, a_n$ converges, and if $b_n \leq a_n$ only for $n \geq 500$, then the series $\Sigma\, b_n$ also converges. The initial finite sum $b_1 + b_2 + \cdots + b_{499}$ can be anything; only the ultimate behavior of a series counts toward convergence or divergence.

Example 1

Test for convergence or divergence:

a $\displaystyle\sum \frac{\sin^2 n}{3^n}$ **b** $\displaystyle\sum \frac{1}{\sqrt{n}}$ **c** $\displaystyle\sum \frac{n}{2n+1}$

Solution

a $(\sin^2 n)/3^n \leq 1/3^n$, and $\Sigma\, 1/3^n$ converges, so the given series converges.

b $1/\sqrt{n} \geq 1/n$, and $\Sigma\, 1/n$ diverges, so the given series diverges.

c $a_n = n/(2n+1) \geq \frac{1}{3}$, and $\Sigma\, \frac{1}{3}$ diverges, so the given series diverges.

The comparison test is useful provided you have a good supply of known series. An excellent class of series for comparisons are those of the form $\Sigma\, 1/n^p$.

p-Series

The series $\displaystyle\sum \frac{1}{n^p}$ diverges if $p \le 1$ and converges if $p > 1$.

Proof If $0 < p \le 1$, then $1/n^p \ge 1/n$ and the series diverges by comparison with the divergent harmonic series $\Sigma\, 1/n$.

If $p > 1$, we shall show that the partial sums of the series are bounded. We use an important trick: we interpret s_n as an area and compare it with a region below the curve $y = 1/x^p$. See Figure 10-4-1.

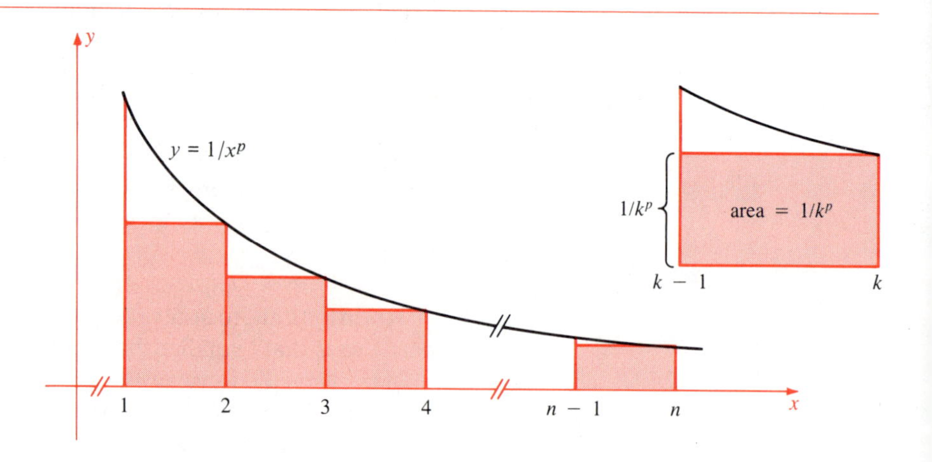

Figure 10-4-1
The rectangular sum is less than the area under the curve.

The combined area of the rectangles shown is less than the area under the decreasing curve between $x = 1$ and $x = n$. Therefore

$$\frac{1}{2^p} + \frac{1}{3^p} + \cdots + \frac{1}{n^p} < \int_1^n \frac{dx}{x^p} = \frac{-1}{p-1} \cdot \frac{1}{x^{p-1}}\bigg|_1^n$$

$$= \frac{1}{p-1}\left(1 - \frac{1}{n^{p-1}}\right)$$

Since $p - 1 > 0$, the right side is a positive number, a little less than $1/(p-1)$ for all values of n. Hence

$$s_n = 1 + \left(\frac{1}{2^p} + \frac{1}{3^p} \cdots + \frac{1}{n^p}\right) < 1 + \frac{1}{p-1}$$

for $n \ge 1$. Thus the partial sums are bounded if $p > 1$, so the series converges.

Suppose $\Sigma\, a_n$ is a given series, and $c \neq 0$. Then the two series $\Sigma\, a_n$ and $\Sigma\, ca_n$ either both converge or both diverge. For the partial sums of the series are $\{s_n\}$ and $\{cs_n\}$, sequences that converge or diverge together.

We can extend these remarks to a pair of series $\Sigma\, a_n$ and $\Sigma\, b_n$ where the ratios b_n/a_n are not constant, but restricted to a suitable range.

Second Comparison Test

Let $\Sigma\, a_n$ and $\Sigma\, b_n$ be given series with positive terms. Suppose there exist positive numbers c and d such that

$$c \leq \frac{b_n}{a_n} \leq d$$

for all sufficiently large n. Then the series both converge or both diverge. ●

Proof If $\Sigma\, a_n$ converges, then so does $\Sigma\, da_n$. But $b_n \leq da_n$, so $\Sigma\, b_n$ converges. If $\Sigma\, a_n$ diverges, then so does $\Sigma\, ca_n$. But $b_n \geq ca_n$, so $\Sigma\, b_n$ diverges. ●

The conditions of the preceding test are automatically satisfied if the ratios b_n/a_n actually approach a positive limit L. Then, by the definition of limit with $\varepsilon = \frac{1}{2}L$, the ratios eventually satisfy $\frac{1}{2}L < b_n/a_n < \frac{3}{2}L$.

Third Comparison Test

Let $\Sigma\, a_n$ and $\Sigma\, b_n$ have positive terms. If $\lim b_n/a_n = L$ exists and if $L > 0$, then either both series converge or both series diverge. ●

Example 2

Test for convergence or divergence:

a $\displaystyle\sum \frac{1}{n + \sqrt{n}}$ **b** $\displaystyle\sum \frac{4n + 1}{3n^3 - n^2 - 1}$

Solution

a When n is very large, n is much larger than $\sqrt{n}$. This suggests that the terms behave roughly like $1/n$, so we begin to suspect divergence. Let $a_n = 1/n$ and $b_n = 1/(n + \sqrt{n}\,)$. We could establish divergence right away if it were true that $b_n > a_n$. Unfortunately that is false. [It would be true if b_n were $1/(n - \sqrt{n}\,)$.] Instead we examine the ratio of b_n to a_n:

$$\frac{b_n}{a_n} = \frac{n}{n + \sqrt{n}} = \frac{1}{1 + 1/\sqrt{n}} \longrightarrow \frac{1}{1 + 0} = 1 \quad \text{as} \quad n \longrightarrow \infty$$

The ratios have a positive limit. Therefore $\Sigma\, b_n$ diverges since $\Sigma\, 1/n$ diverges.

b When n is very large, the terms appear to behave like the quantities $4n/3n^3 = 4/3n^2$. This suggests comparison with the convergent series $\Sigma\, 1/n^2$. Let

$$a_n = 1/n^2 \quad \text{and} \quad b_n = (4n + 1)/(3n^3 - n^2 - 1)$$

Then

$$\frac{b_n}{a_n} = \frac{(4n + 1)n^2}{3n^3 - n^2 - 1} = \frac{4 + 1/n}{3 - 1/n - 1/n^3} \longrightarrow \frac{4}{3}$$

The ratios have a positive limit. Therefore $\Sigma\, b_n$ converges because $\Sigma\, 1/n^2$ converges. ●

The Ratio Tests

In a geometric series, the ratio a_{n+1}/a_n is a constant r. If $|r| < 1$, the series converges, basically because its terms decrease rapidly. By analogy with the two preceding tests, we should expect convergence in general if the ratios are small but not necessarily constant.

First Ratio Test

Let $\Sigma\, a_n$ be a series of positive terms.

1 The series converges if there is a number $r < 1$ such that $a_{n+1}/a_n \le r$ eventually.

2 The series diverges if $a_{n+1}/a_n \ge 1$ eventually. ●

Proof

1 Suppose $a_{n+1}/a_n \le r < 1$ starting with $n = N$. Then

$$a_{N+1} \le a_N r$$

$$a_{N+2} \le a_{N+1} r \le a_N r^2$$

and by induction, $a_{N+k} \le a_N r^k$. That is, $a_n \le a_N r^{n-N} = (a_N r^{-N}) r^n$ for all $n \ge N$. It follows that the series $\Sigma\, a_n$ converges by comparison with the convergent geometric series $\Sigma\, r^n$.

2 From some point on, $a_{n+1} \ge a_n$. The terms are non-decreasing, hence the series diverges. ●

Remark The ratio test is a sort of "self-comparison" test. Instead of comparing $\Sigma\, a_n$ with $\Sigma\, b_n$, it compares $\Sigma\, a_n$ with itself! (The ratio test is also known as d'Alembert's test.)

Warning Note that the test for convergence requires the condition $a_{n+1}/a_n \le r < 1$, not just $a_{n+1}/a_n < 1$. The ratios must eventually *stay away* from 1. If $a_{n+1}/a_n < 1$ but $a_{n+1}/a_n \to 1$, we may have divergence. For example, we take $a_n = 1/n$. Then

$$a_{n+1}/a_n = n/(n + 1) < 1$$

but $\Sigma\, 1/n$ diverges.

It often happens that the ratios a_{n+1}/a_n approach a limit. Then we can cast the ratio test into a different form.

Second Ratio Test

Let $\Sigma\, a_n$ be a series of positive terms. Suppose

$$\lim_{n\to\infty} \frac{a_{n+1}}{a_n} = r$$

1 The series converges if $r < 1$.
2 The series diverges if $r > 1$.
3 If $r = 1$, the test is inconclusive; the series may either converge or diverge.

Proof

1 If $r < 1$, choose ε so small that $r + \varepsilon < 1$. By definition of the statement $a_{n+1}/a_n \to r$, there exists a sufficiently large positive integer N such that $a_{n+1}/a_n < r + \varepsilon < 1$ for all $n \geq N$. Therefore the series converges by the preceding test.
2 Similarly, if $r > 1$, then $a_{n+1}/a_n > r - \varepsilon > 1$ from some point on. The series diverges.
3 If $r = 1$, this test cannot distinguish between convergent and divergent series. For example, take $a_n = 1/n^p$. The series converges for $p > 1$, diverges for $p \leq 1$. But for all values of p,

$$\frac{a_{n+1}}{a_n} = \frac{n^p}{(n+1)^p} = \left(\frac{n}{n+1}\right)^p = \left(1 - \frac{1}{n+1}\right)^p \to (1-0)^p = 1$$

Example 3

Test for convergence or divergence

a $\displaystyle\sum \frac{n}{2^n}$ **b** $\displaystyle\sum \frac{10^n}{n!}$

Solution

a Set $a_n = n/2^n$. Then

$$\frac{a_{n+1}}{a_n} = \frac{(n+1)/2^{n+1}}{n/2^n} = \frac{n+1}{2n} = \tfrac{1}{2}\left(1 + \frac{1}{n}\right) \to \tfrac{1}{2}$$

Since $\frac{1}{2} < 1$, the series converges by the ratio test.

b Set $a_n = 10^n/n!$. Then

$$\frac{a_{n+1}}{a_n} = \frac{10^{n+1}/[1 \cdot 2 \cdots n(n+1)]}{10^n/(1 \cdot 2 \cdots n)} = \frac{10}{n+1} \to 0$$

Since $0 < 1$, the series converges by the ratio test.

The Root Test

The ratio test works because if eventually $a_{n+1}/a_n \le r < 1$, then $\Sigma\, a_n$ can be compared with a geometric series. The following *root test* also works because it allows comparison with a geometric series. Yet it is of rather a different nature from the ratio test. As we shall see, it applies sometimes when the ratio test does not. (The root test is also known as Cauchy's test.)

Root Test

Let $\Sigma\, a_n$ be a series of positive terms. Suppose $r < 1$ and eventually $\sqrt[n]{a_n} \le r$. Then $\Sigma\, a_n$ converges.

We leave the proof for an exercise.

Example 4

Suppose $a_n = \dfrac{1}{[3 + (-1)^n]^n}$

Show that $\Sigma\, a_n$ is convergent, but that the ratio test does not apply.

Solution If n is odd, then $a_n = 1/2^n$, but if n is even, then $a_n = 1/4^n$. Therefore the ratio a_{n+1}/a_n is equal to $1/2^{n+2}$ if n is odd and to 2^{n-N} if n is even. Clearly we learn nothing from the ultimate behavior of the ratios. The n-th roots tell us more. If n is odd, then $\sqrt[n]{a_n} = \frac{1}{2}$ and if n is even, then $\sqrt[n]{a_n} = \frac{1}{4}$, so certainly $\sqrt[n]{a_n} \le \frac{1}{2}$ in all cases. Therefore $\Sigma\, a_n$ converges.

Exercises

Determine whether the series converges or diverges

1. $\sum \dfrac{1}{n^2 + 1}$
2. $\sum \dfrac{1}{2^n\sqrt{n}}$
3. $\sum \dfrac{n}{4n + 3}$
4. $\sum \dfrac{1}{(2n - 1)^3}$
5. $\sum \dfrac{1}{n\sqrt{n + 3}}$
6. $\sum \dfrac{1 + \sqrt[3]{n}}{n}$
7. $\sum \dfrac{n^2}{2n^4 + 7}$
8. $\sum \dfrac{1}{\ln n}$
9. $\sum \dfrac{1}{n^n}$
10. $\sum \dfrac{n}{(n + 1)(n + 3)(n + 5)}$
11. Prove that if $\Sigma\, a_n$ and $\Sigma\, b_n$ converge, then so does $\Sigma\,(a_n + b_n)$, and find the sum.
12. If $\Sigma\, a_n$ and $\Sigma\, b_n$ diverge, show by examples that $\Sigma\,(a_n + b_n)$ may either converge or diverge.

Let $\Sigma\, a_n$ be a convergent series of positive terms

13. Prove that $\Sigma\, a_n{}^2$ converges.
14. Show by examples that $\Sigma \sqrt{a_n}$ may either converge or diverge.

Test for convergence or divergence

15. $\sum \dfrac{1}{n^2 - 3}$
16. $\sum \dfrac{1}{\sqrt{2n^3 - n}}$
17. $\sum \dfrac{1}{4n - 1}$
18. $\sum \dfrac{5 + \sqrt{n}}{1 + n}$
19. $\sum \dfrac{n^3}{n!}$
20. $\sum n^3(\frac{3}{4})^n$
21. $\sum ne^{-n}$
22. $\sum \dfrac{3^n + 1}{5e^n + n}$
23. $\sum \dfrac{2^n + n}{3^n - n}$
24. $\sum \dfrac{1}{(\ln n)^n}$
25. $\sum \dfrac{n!}{1 \cdot 3 \cdot 5 \cdots (2n - 1)}$
26. $\sum \dfrac{(n!)^2}{(2n)!}$

27 Prove the root test.

28 Let $\Sigma\, a_n$ and $\Sigma\, b_n$ be series with positive terms. Suppose $b_n/a_n \to 0$. Find an example where $\Sigma\, b_n$ converges while $\Sigma\, a_n$ diverges. Does this contradict the text?

An **infinite product** is an expression of the form

$$\prod_{n=1}^{\infty}(1 + a_n) = (1 + a_1)(1 + a_2)(1 + a_3)\cdots$$

Its sequence of **partial products** is $\{p_n\}$, where $p_n = (1 + a_1)(1 + a_2)\cdots(1 + a_n)$. In the following exercises we shall assume $a_n \geq 0$.

29 Show that $p_n \geq 1 + a_1 + \cdots + a_n$.

30 (cont.) Suppose $\{p_n\}$ converges. Prove that $\Sigma\, a_n$ converges.

31 (cont.) Show that $\ln p_n \leq a_1 + \cdots + a_n = s_n$.

32 (cont.) Suppose $\Sigma\, a_n$ converges. Prove that $\{p_n\}$ converges.

33 Let $\Sigma\, a_n$ and $\Sigma\, b_n$ be series with positive terms. Suppose $\Sigma\, b_n$ converges and $a_{n+1}/a_n \leq b_{n+1}/b_n$ for all n. Prove $\Sigma\, a_n$ converges.

34 Define $\{a_n\}$ by $a_0 = 1$ and $a_{n+1} = \dfrac{a_n + 1}{a_n + 2}$. Prove $\{a_n\}$ is convergent and find its limit.

35 Given $1 + \dfrac{1}{2^2} + \dfrac{1}{3^2} + \dfrac{1}{4^2} + \cdots = \dfrac{\pi^2}{6}$, show that

$$1 + \frac{1}{3^2} + \frac{1}{5^2} + \frac{1}{7^2} + \cdots = \frac{\pi^2}{8}$$

36 (cont.) Find $1 + \dfrac{1}{5^2} + \dfrac{1}{7^2} + \dfrac{1}{11^2} + \dfrac{1}{13^2} + \dfrac{1}{17^2} + \dfrac{1}{19^2} + \cdots$

10-5 Series with Positive and Negative Terms

Infinite series with both positive and negative terms are generally more complicated than series with terms all of the same sign. In this section, we discuss two common types of mixed series, alternating series and absolutely convergent series.

Alternating Series

An **alternating series** is one whose terms are alternately positive and negative. Examples:

$$1 - \tfrac{1}{2} + \tfrac{1}{3} - \tfrac{1}{4} + - + - \cdots$$

$$1 - x^2 + x^4 - x^6 + - + - \cdots \qquad (\text{alternating for all } x \neq 0)$$

$$x - \frac{x^2}{4} + \frac{x^3}{9} - \frac{x^4}{16} + - + - \cdots$$

$$(\text{alternating only for } x > 0)$$

Such series have some extremely useful properties, two of which we now state.

Alternating Series Test

Suppose the terms of an alternating series $\Sigma\, a_n$ decrease in absolute value towards zero, that is,

$$|a_1| > |a_2| > |a_3| > \cdots > |a_n| > \cdots$$

and $\lim_{n\to+\infty} a_n = 0$. Then the series converges.

Remainder Estimate

If such a series is broken off at the n-th term, then the remainder (in absolute value) is less than the absolute value of the $(n+1)$-th term.

Formal proofs of these two assertions will be given at the end of the section. They provide a very simple convergence criterion and an immediate remainder estimate for *alternating* series. Let us show geometrically that they make good sense. Suppose Σa_n is an alternating series whose terms decrease in absolute value towards zero. (To be definite, assume $a_1 > 0$.) The partial sums $s_n = a_1 + a_2 + \cdots + a_n$ oscillate back and forth, as shown in Figure 10-5-1. But since the terms decrease towards zero, the oscillations become shorter and shorter. The odd partial sums decrease and the even ones increase, squeezing down on some number S.

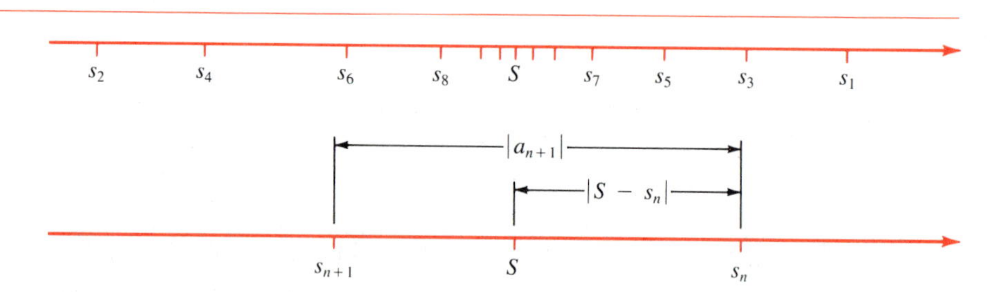

Figure 10-5-1
The partial sums of an alternating series

If the series is broken off after n terms, the remainder is $|S - s_n|$. But from Figure 10-5-1,

$$|S - s_n| < |s_{n+1} - s_n| = |a_{n+1}|$$

Thus, the remainder is less than the absolute value of the $(n+1)$-th term.

Example 1

Prove convergent

a $1 - \dfrac{1}{\sqrt{2}} + \dfrac{1}{\sqrt{3}} - \dfrac{1}{\sqrt{4}} + - \cdots$

b $\dfrac{1}{2} - \dfrac{\sqrt{2}}{3} + \dfrac{\sqrt{3}}{4} - \dfrac{\sqrt{4}}{5} + - \cdots$

Solution

a If Σa_n denotes the series, then $|a_n| = 1/\sqrt{n}$. The numbers $|a_n|$ clearly decrease toward 0. Hence the series converges by the alternating-series test.

b This time $|a_n| = \sqrt{n}/(n+1)$. The series alternates, and $|a_n| \to 0$. It will be enough to show that the sequence $\{\sqrt{n}/(n+1)\}$ decreases. To do so show that the *function* $y(x) = \sqrt{x}/(x+1)$ decreases. Compute its derivative:

$$y'(x) = \frac{(x+1)/2\sqrt{x} - \sqrt{x}}{(x+1)^2} = \frac{1-x}{2\sqrt{x}\,(x+1)^2}$$

Clearly $y'(x) < 0$ for $x > 1$, therefore $y(x)$ decreases. In particular,

$$y(n+1) < y(n) \quad \text{or equivalently } |a_{n+1}| < |a_n|$$

Hence the series converges by the alternating series test. ●

Example 2

Find all values of x for which the series converges:

$$x + \frac{x^2}{2} + \frac{x^3}{3} + \frac{x^4}{4} + \cdots$$

Solution First take $x \geq 0$. Then all terms are positive, and

$$\frac{a_{n+1}}{a_n} = \frac{x^{n+1}/(n+1)}{x^n/n} = x\frac{n}{n+1} \longrightarrow x$$

By the ratio test, the series converges for $0 \leq x < 1$ and diverges for $x > 1$. The test is inconclusive for $x = 1$. However, in that case the series is the harmonic series,

$$1 + \frac{1}{2} + \frac{1}{3} + \frac{1}{4} \cdots$$

which diverges.

Now take $x < 0$. Then the series alternates. If $x < -1$, then $|x|^n/n \longrightarrow \infty$; hence the series diverges. If $-1 \leq x < 0$, then $|x|^n/n$ decreases to 0; hence the series converges by the alternating series test.

Answer: The series converges precisely for $-1 \leq x < 1$. ●

Example 3

In the next chapter we shall show that the series $\Sigma_{n=0}^{\infty} x^n/n!$ converges to e^x for all x. Use this series to estimate $1/e$ to three-place accuracy.

Solution Set $x = -1$. Then $e^{-1} = \Sigma_{n=0}^{\infty} (-1)^n/n!$ The signs alternate and the terms decrease in absolute value to 0. Therefore,

$$e^{-1} = 1 - \frac{1}{1!} + \frac{1}{2!} + \cdots + \frac{(-1)^n}{n!} + \text{remainder}$$

where

$$|\text{remainder}| < \frac{1}{(n+1)!}$$

For three-place accuracy, we need $|\text{remainder}| \le 5 \times 10^{-4}$, so we want an n for which

$$\frac{1}{(n+1)!} \le 5 \times 10^{-4}$$

that is,

$$(n+1)! \ge \frac{1}{5 \times 10^{-4}} = 2000$$

Now $6! = 720$ and $7! = 5040$. So we choose $n + 1 = 7$, that is, $n = 6$. Then

$$e^{-1} \approx 1 - 1 + \tfrac{1}{2} - \tfrac{1}{6} + \tfrac{1}{24} - \tfrac{1}{120} + \tfrac{1}{720} \approx 0.368$$

Absolutely Convergent Series

How is it that the harmonic series $\Sigma\, 1/n$ diverges but the alternating harmonic series $\Sigma\,(-1)^{n-1}/n$ converges? Essentially, the harmonic series diverges because its terms don't decrease quite fast enough, like $1/n^2$ or $1/2^n$ for example. Its partial sums consist of many small terms that have a large total. The terms of $\Sigma\, 1/2^n$, however, decrease so fast that the total of any large number of them is bounded.

The alternating harmonic series converges, not by smallness of its terms alone, but also because strategically placed minus signs cause lots of cancellation. Just look at two consecutive terms:

$$+\frac{1}{n} - \frac{1}{n+1} = \frac{1}{n(n+1)}$$

Cancellation produces a term of a convergent series! Thus $\Sigma\,(-1)^{n-1}/n$ converges because its terms get small *and* because a delicate balance of positive and negative terms produces important cancellations.

Some series with mixed terms converge by the smallness of their terms alone; they would converge even if all the signs were $+$. Such a series has a special name: an absolutely convergent series.

Absolute Convergence

A series $\Sigma\, a_n$ is called **absolutely convergent** provided

$$\sum |a_n|$$

is convergent.

As we might expect, absolute convergence implies (is even stronger than) convergence. The following statement states this precisely; its proof will be discussed at the end of this section.

- If a series $\Sigma\, a_n$ converges absolutely, then it converges.

In studying series with mixed terms, it is a good idea to check first for absolute convergence. Just change all signs to $+$, then test for convergence of the positive series.

Example 4

Test for convergence and absolute convergence the series

a $1 + \dfrac{1}{2^2} - \dfrac{1}{3^2} + \dfrac{1}{4^2} + \dfrac{1}{5^2} - \dfrac{1}{6^2} + + - \cdots$

b $1 - \dfrac{1}{\sqrt{2}} + \dfrac{1}{\sqrt{3}} - \dfrac{1}{\sqrt{4}} + - \cdots$

Solution

a The series of absolute values is $\Sigma\, 1/n^2$, which converges. The series is absolutely convergent, hence convergent.

b The series of absolute values is $\Sigma\, 1/\sqrt{n}$, which diverges. Hence the given series does not converge absolutely. It does converge, nevertheless, because it satisfies the test for alternating series: its terms decrease in absolute value towards zero. ●

Proofs [Optional]

First we prove that an alternating series whose terms decrease in absolute value towards zero converges. Suppose

$$a_0 \geq a_1 \geq a_2 \geq \cdots \geq 0 \quad \text{and} \quad \lim_{n\to+\infty} a_n = 0$$

Set

$$s_n = a_0 - a_1 + a_2 - \cdots + (-1)^n a_n$$

Then s_n is the n-th partial sum of the alternating series

$$\sum_{i=0}^{\infty} (-1)^i a_i$$

whose convergence is to be proved.

We first prove that the subsequence $\{s_{2n}\}$ of even-numbered terms converges. First of all,

$$s_{2n} = (a_0 - a_1) + (a_2 - a_3) + \cdots + (a_{2n-2} - a_{2n-1}) + a_{2n}$$

is a sum of non-negative numbers. Hence $s_{2n} \geq 0$ for all n. Next,

$$s_{2n} = s_{2n-2} - (a_{2n-1} - a_{2n}) \leq s_{2n-2}$$

Therefore $\{s_{2n}\}$ is a monotone decreasing sequence of non-negative numbers:

$$s_0 \geq s_2 \geq s_4 \geq \cdots \geq s_{2n} \geq \cdots \geq 0$$

Therefore $\{s_{2n}\}$ converges to a limit S:

$$\lim_{n\to+\infty} s_{2n} = S$$

and clearly $S \ge 0$. Since $s_{2n+1} = s_{2n} - a_{2n+1}$ we have

$$\lim_{n\to+\infty} s_{2n+1} = \lim_{n\to+\infty} s_{2n} - \lim_{n\to+\infty} a_{2n+1} = S - 0 = S$$

so the odd-numbered subsequence $\{s_{2n+1}\}$ also converges, to the same limit. It follows easily that the whole sequence converges to S, that is, $\lim_{n\to+\infty} s_n = S$. ●

To prove the remainder estimate we write

$$|S - s_n| = a_{n+1} - a_{n+2} + a_{n+3} - \cdots$$
$$= a_{n+1} - (a_{n+2} - a_{n+3} + a_{n+4} - \cdots)$$

But

$$T = a_{n+2} - a_{n+3} + a_{n+4} - \cdots$$

is a sum just like S: the sum of an alternating series whose terms decrease in absolute value. Therefore $T \ge 0$ and

$$|S - s_n| = a_{n+1} - T \le a_{n+1}$$ ●

Next we prove that if $\Sigma\, a_n$ converges absolutely, then it converges. The proof uses the Cauchy test (page 545). By the Cauchy test, for each $\varepsilon > 0$ there is an N such that

$$|a_{n+1}| + |a_{n+2}| + \cdots + |a_m| < \varepsilon \quad \text{provided} \quad m > n \ge N$$

But

$$|a_{n+1} + a_{n+2} + \cdots + a_m| \le |a_{n+1}| + \cdots + |a_m| < \varepsilon$$

by the triangle inequality. Therefore $\Sigma\, a_n$ converges by the Cauchy test. ●

Exercises

Test for convergence and absolute convergence

1 $\sum (-1)^n \dfrac{1}{\ln n}$

2 $\sum (-1)^n \dfrac{n}{2^n}$

3 $\sum (-1)^n \dfrac{n}{3n+1}$

4 $\sum (-1)^n \dfrac{\ln n}{n}$

5 $\sum (-1)^n \dfrac{n^2}{(1.01)^n}$

6 $\sum (-1)^n \dfrac{5n^2}{2n^3 - 1}$

7 $\sum (-1)^n \dfrac{1}{n + \ln n}$

8 $\sum (-1)^n \sin \dfrac{2\pi}{n}$

9 $\sum \dfrac{\sin n}{n^2}$

10 $\sum_1^\infty (-1)^n \dfrac{1}{|n - 100.5|}$

11 $\sum \dfrac{\cos n\pi}{\sqrt{n+3}}$

12 $\sum \dfrac{(-1)^n}{1 + (n - 100)^2}$

13 $\sum \dfrac{(-1)^n}{500 + (1/n)^2}$

14 $\sum (-1)^{2n-1} \dfrac{n-1}{(n+1)^3}$

5 $1 - \frac{1}{10} + \frac{1}{2} - \frac{1}{11} + \frac{1}{3} - \frac{1}{12} + - \cdots$

6 $(1 - \frac{1}{10}) + (\frac{1}{2} - \frac{1}{11}) + (\frac{1}{3} - \frac{1}{12}) + \cdots$

7 $1 + \frac{1}{2} - \frac{1}{3} + \frac{1}{4} + \frac{1}{5} - \frac{1}{6} + + - \cdots$

8 $1 + \frac{1}{2} - \frac{1}{3} - \frac{1}{4} + \frac{1}{5} + \frac{1}{6} - - + + \cdots$

9 $\frac{1}{2} - 1 + \frac{1}{4} - \frac{1}{3} + \frac{1}{6} - \frac{1}{5} + - \cdots$

0 $(\frac{1}{2} - 1) + (\frac{1}{4} - \frac{1}{3}) + (\frac{1}{6} - \frac{1}{5}) + \cdots$

Estimate to four places by the method of Example 3

1 $1/\sqrt{e}$ 22 $1/\sqrt[3]{e}$

3 Suppose $\Sigma\, a_n{}^2$ and $\Sigma\, b_n{}^2$ both converge. Show that $\Sigma a_n b_n$ converges absolutely. [Hint $2xy \le x^2 + y^2$.]

*24 Suppose $\Sigma\, a_n$ converges, but not absolutely. Let $\Sigma\, b_j$ and $\Sigma\, c_k$ be the series made of the positive a_n's and negative a_n's respectively. Prove that both $\Sigma\, b_j$ and $\Sigma\, c_k$ diverge.

Find all real numbers x for which the series converges absolutely

25 $\sum \dfrac{x^{2n}}{n!}$

26 $\sum \dfrac{\sin nx}{n^2}$

27 $\sum (3x)^{2n}$

28 $\sum nx^{2n}$

10-6 Improper Integrals

In scientific problems, one frequently meets definite integrals in which one (or both) of the limits is infinite. A definite integral whose upper limit is $+\infty$, whose lower limit is $-\infty$, or both, is called an **improper integral.**

In order to give a precise definition of improper integrals, we must recall the meaning of the statement $\lim_{x\to+\infty} F(x) = L$.

Limit at Infinity

Let $F(x)$ be defined for $x \ge a$, where a is some real number. Then

$$\lim_{x\to+\infty} F(x) = L$$

if for each $\varepsilon > 0$, there is a number b such that

$$|F(x) - L| < \varepsilon \qquad \text{for all} \quad x \ge b$$

For increasing and decreasing functions the basic fact about limits is analogous to the one about sequences. Its proof depends on the completeness of the real number system and is left for a more advanced course.

- Let F be an increasing function. Then $\lim_{x\to+\infty} F(x)$ exists if and only if $F(x)$ is bounded above, that is, if and only if there exists a number M such that

$$F(x) \le M$$

for all x in the domain of F.

Similarly if $F(x)$ is a decreasing function, then $\lim_{x\to+\infty} F(x)$ exists if and only if $F(x)$ is bounded below.

Suppose $f(x)$ is continuous for $x \ge a$. Set

$$F(b) = \int_a^b f(x)\,dx$$

Now $\lim_{b\to+\infty} F(b)$ may or may not exist.

Improper Integral

Define

$$\int_a^\infty f(x)\,dx = \lim_{b\to+\infty} \int_a^b f(x)\,dx$$

provided the limit exists. If it does, the integral is said to **converge,** otherwise to **diverge.** If the limit is $+\infty$, we say that the integral **diverges to** $+\infty$ and write $\int_a^\infty f(x)\,dx = +\infty$.

Similarly, define

$$\int_{-\infty}^b f(x)\,dx = \lim_{a\to-\infty} \int_a^b f(x)\,dx$$

provided the limit exists.

Finally, define

$$\int_{-\infty}^\infty f(x)\,dx = \int_{-\infty}^0 f(x)\,dx + \int_0^\infty f(x)\,dx$$

provided both integrals on the right converge.

Integrals of these three kinds are called **improper integrals.**

Remarks An integral from $-\infty$ to $+\infty$ may be split at any convenient finite point just as well as at 0.

An improper integral need not converge. For example, the integral

$$\int_1^\infty \frac{dx}{x} \quad \text{diverges because} \quad \lim_{b\to+\infty} \int_1^b \frac{dx}{x} = \lim_{b\to+\infty} \ln b = +\infty$$

Example 1

Evaluate

a $\displaystyle\int_0^\infty \frac{dx}{1+x^2}$ **b** $\displaystyle\int_{-\infty}^3 e^x\,dx$

Solution

a $\displaystyle\int_0^b \frac{dx}{1+x^2} = \arctan x\Big|_0^b = \arctan b$

Let $b \to +\infty$. Then $\arctan b \to \frac{1}{2}\pi$. Hence

$$\int_0^\infty \frac{dx}{1+x^2} = \lim_{b\to+\infty} \int_0^b \frac{dx}{1+x^2} = \lim_{b\to+\infty} \arctan b = \tfrac{1}{2}\pi$$

b $\displaystyle\int_a^3 e^x\,dx = e^x\Big|_a^3 = e^3 - e^a$

Let $a \to -\infty$. Then $e^a \to 0$. Hence

$$\int_{-\infty}^3 e^x\,dx = \lim_{a\to-\infty} \int_a^3 e^x\,dx = \lim_{a\to-\infty} (e^3 - e^a) = e^3$$

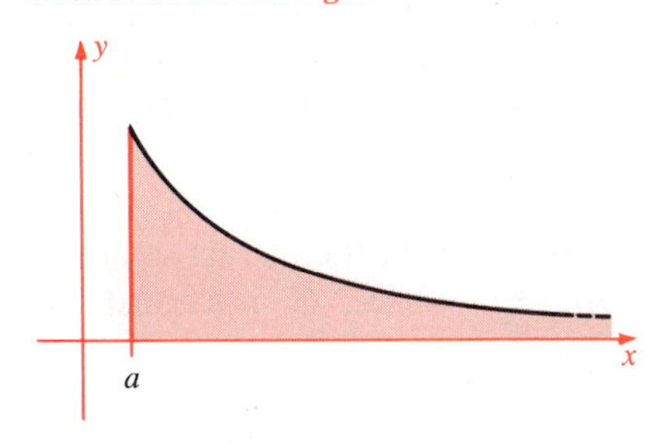

Figure 10-6-1
Area of an infinite region

Remember that a definite integral of a positive function represents the area under a curve. We interpret the improper integral

$$\int_a^\infty f(x)\,dx \quad \text{where} \quad f(x) \geq 0$$

as the area of the infinite region in Figure 10-6-1. If the integral converges, the area is finite; if the integral diverges, the area is infinite.

At first it may seem unbelievable that a region of infinite extent can have finite area. But it can, and here is another, especially simple example. Take the region under the curve $y = 2^{-x}$ to the right of the y-axis (Figure 10-6-2). The rectangles shown in Figure 10-6-2 have base 1 and heights $1,\ 1/2,\ 1/4,\ 1/8,\ \ldots$ Their total area is

$$1 + 1/2 + 1/4 + 1/8 + \cdots = 2$$

Therefore, the shaded infinite region has finite area less than 2.

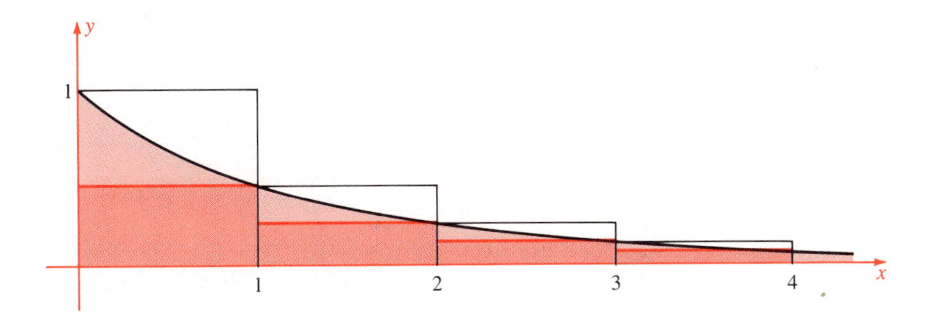

Figure 10-6-2
Area under $y = 2^{-x}$ from 0 to $+\infty$

Example 2

Compute the exact area of the shaded region in Figure 10-6-2.

Solution The area is given by the improper integral

$$\int_0^\infty 2^{-x}\,dx = \lim_{b\to+\infty} \int_0^b 2^{-x}\,dx$$

An antiderivative of 2^{-x} is $-2^{-x}/\ln 2$ (because $2^{-x} = e^{-x\ln 2}$). Hence

$$\int_0^\infty 2^{-x}\,dx = \lim_{b\to+\infty} \frac{1}{\ln 2}(1 - 2^{-b}) = \frac{1}{\ln 2} \approx 1.44270$$

●

Remark The answer is reasonable. Since the darker shaded region in Figure 10-6-2 has area

$$\tfrac{1}{2} + \tfrac{1}{4} + \tfrac{1}{8} + \cdots = 1$$

the answer is between 1 and 2. A closer look shows it is slightly less than 1.5. Why?

Example 3

A particle P of mass m is at the origin. Find the work done against the gravitational attraction of P in moving a unit mass from $x = a$ to $x = +\infty$. (This work is called the **gravitational potential** due to P. You may wish to review the topic of *work* on page 395.)

Solution According to Newton's law of gravitation, the attractive force of P on the unit mass at x is $F = Gm/x^2$, where G is a constant. Therefore the work in moving the unit mass to $+\infty$ is

$$W = \int_a^\infty \frac{Gm}{x^2}\,dx = \frac{-Gm}{x}\bigg|_a^\infty = \frac{Gm}{a}$$

Both Limits Infinite

Now let us try an integral where both limits are infinite.

Example 4

Evaluate $\displaystyle\int_{-\infty}^{\infty} \frac{(\arctan x)^2}{1+x^2}\,dx$

Solution By definition, the value of this integral is

$$\int_{-\infty}^{0} \frac{(\arctan x)^2}{1+x^2}\,dx + \int_{0}^{\infty} \frac{(\arctan x)^2}{1+x^2}\,dx$$

provided *both* improper integrals converge. By inspection,

$$\int \frac{(\arctan x)^2}{1+x^2}\,dx = \tfrac{1}{3}(\arctan x)^3 + C$$

Now $\arctan b \to \frac{1}{2}\pi$ as $b \to +\infty$, and $\arctan a \to -\frac{1}{2}\pi$ as $a \to -\infty$. Therefore

$$\int_0^\infty \frac{(\arctan x)^2}{1+x^2}\,dx = \lim_{b\to+\infty} \tfrac{1}{3}(\arctan x)^3\bigg|_0^b = \tfrac{1}{3}(\tfrac{1}{2}\pi)^3 = \tfrac{1}{24}\pi^3$$

and similarly $\displaystyle\int_{-\infty}^{0} \frac{(\arctan x)^2}{1+x^2}\,dx = \tfrac{1}{24}\pi^3.$

Both improper integrals converge; the answer is the sum of their values:

$$\int_{-\infty}^{\infty} \frac{(\arctan x)^2}{1+x^2}\,dx = \tfrac{1}{24}\pi^3 + \tfrac{1}{24}\pi^3 = \tfrac{1}{12}\pi^3$$

Remark Do you prefer this snappy calculation?

$$\int_{-\infty}^{\infty} \frac{(\arctan x)^2}{1+x^2}\,dx = \tfrac{1}{3}(\arctan x)^3\bigg|_{-\infty}^{\infty}$$

$$= \tfrac{1}{3}(\tfrac{1}{2}\pi)^3 - \tfrac{1}{3}(-\tfrac{1}{2}\pi)^3 = \tfrac{1}{12}\pi^3$$

Warning Try the same method on $\int_{-\infty}^{\infty} \frac{dx}{x^2}$. It fails! Why?

Exercises

Evaluate

1 $\int_2^{\infty} \frac{dx}{x^3}$

2 $\int_5^{\infty} e^{-x}\,dx$

3 $\int_0^{\infty} xe^{-x}\,dx$

4 $\int_{-\infty}^{-1} \frac{dx}{x^2}$

5 $\int_{-\infty}^{-1} \frac{dx}{1+x^2}$

6 $\int_4^{\infty} \frac{dx}{x\sqrt{x}}$

7 $\int_{-\infty}^{\infty} e^{-|x|}\,dx$

8 $\int_{-\infty}^{\infty} x\exp(-x^2)\,dx$

9 $\int_4^{\infty} \frac{dx}{x\sqrt{9+x^2}}$

10 $\int_1^{\infty} \frac{dx}{x(x+3)}$

11 $\int_0^{\infty} \frac{x\,dx}{x^4+1}$ (let $u = x^2$)

12 $\int_1^{\infty} \frac{dx}{(x^2+1)^2}$

13 $\int_{-\infty}^{\infty} \frac{dx}{1+x^2}$

14 $\int_{-\infty}^{\infty} e^{-|x-2|}\,dx$

$F(s) = \int_0^{\infty} e^{-sx}f(x)\,dx$ is called the **Laplace transform** of $f(x)$. Find

15 $\int_0^{\infty} xe^{-sx}\,dx \quad (s > 0)$

16 $\int_0^{\infty} x^2e^{-sx}\,dx \quad (s > 0)$

*17 $\int_0^{\infty} x^ne^{-sx}\,dx \quad (s > 0)$

18 $\int_0^{\infty} e^{ax}e^{-sx}\,dx \quad (s > a)$

19 $\int_0^{\infty} e^{-sx}\sin x\,dx \quad (s > 0)$

20 $\int_0^{\infty} e^{-sx}\cosh x\,dx \quad (s > 1)$

21 $\int_0^{\infty} xe^{ax}e^{-sx}\,dx \quad (s > a)$

22 $\int_0^{\infty} xe^{-sx}\sin x\,dx \quad (s > 0)$

Is the area under the curve finite or infinite?

23 $y = 1/x$ from $x = 5$ to $x = +\infty$

24 $y = 1/x^2$ from $x = 1$ to $x = +\infty$

25 $y = \sin^2 x$ from $x = 0$ to $x = +\infty$

26 $y = (1.001)^{-x}$ from $x = 0$ to $x = +\infty$

27 $y = 1/(3x+50)$ from $x = 2$ to $x = +\infty$

28 $y = x/(x^2+10)$ from $x = 0$ to $x = +\infty$

29 Evaluate $\int_{-\infty}^{\infty} x^5\exp(-x^2)\,dx$

30 Without evaluating, show that

$$\int_1^{\infty} \frac{dx}{1+x^2} = \int_0^1 \frac{dx}{1+x^2}$$

31 Suppose $a > 0$ and $\int_0^{\infty} f(x)\,dx = L.$ Find $\int_0^{\infty} f(ax)\,dx.$

32 Suppose $\int_{-\infty}^{\infty} f(x)\,dx = L.$ Find $\int_{-\infty}^{\infty} f(x+b)\,dx.$

33 Show for $a > 0$ that

$$\int_a^{\infty} \frac{x^2+1}{x^2}e^{-x^2/2}\,dx = \frac{e^{-a^2/2}}{a}$$

*34 (cont.) Show for $a > 0$ that

$$\frac{a}{a^2+1}e^{-a^2/2} < \int_a^{\infty} e^{-x^2/2}\,dx < \frac{e^{-a^2/2}}{a}$$

(This integral is important in statistics.)

35 Prove for each non-negative integer k the convergence of

$$I_k = \int_{-\infty}^{\infty} x^{2k}e^{-x^2/2}\,dx$$

36 (cont.) Prove that

$$I_k = 1\cdot 3\cdot 5\cdot 7\cdot\cdot\cdot(2k-1)I_0$$

[We shall prove $I_0 = (2\pi)^{1/2}$ when we study double integrals.]

10-7 Convergence and Divergence Tests

Whether an improper integral converges or diverges may be a subtle matter. Let us illustrate this point with a useful class of improper integrals. They are analogous to p-series, discussed in Section 10-4.

p-Integrals

The improper integral

$$\int_a^\infty \frac{dx}{x^p} \qquad (a > 0)$$

diverges to $+\infty$ if $p \le 1$ and converges if $p > 1$.

Proof Suppose $p \ne 1$. Then

$$\int_a^b \frac{dx}{x^p} = -\frac{1}{p-1}\cdot\frac{1}{x^{p-1}}\bigg|_a^b$$
$$= \frac{1}{p-1}\left(\frac{1}{a^{p-1}} - \frac{1}{b^{p-1}}\right)$$

Now it makes a big difference whether $p - 1$ is positive or negative. For as $b \to +\infty$,

$$\frac{1}{b^{p-1}} \to \begin{cases} 0 & \text{if } p-1>0 \\ +\infty & \text{if } p-1<0 \end{cases}$$

Hence if $p > 1$, then

$$\lim_{b\to+\infty}\int_a^b \frac{dx}{x^p} = \frac{1}{p-1}\cdot\frac{1}{a^{p-1}}$$

If $p < 1$, then

$$\lim_{b\to+\infty}\int_a^b \frac{dx}{x^p} = \lim_{b\to+\infty}\frac{1}{1-p}(b^{1-p} - a^{1-p}) = +\infty$$

These conclusions mean the given integral converges if $p > 1$ and diverges to $+\infty$ if $p < 1$. If $p = 1$, then

$$\lim_{b\to+\infty}\int_a^b \frac{dx}{x} = \lim_{b\to+\infty}(\ln b - \ln a) = +\infty$$

so the integral also diverges to $+\infty$.

For a graphical interpretation of these results, see Figure 10-7-1. For $p > 0$, the curves $y = 1/x^p$ all decrease as x increases. The key is in their *rate* of decrease. If $p \le 1$, the curve decreases slowly enough that the shaded area (Figure 10-7-1a) increases without bound as $b \to +\infty$. If $p > 1$, the curve decreases fast enough that the shaded area (Figure 10-7-1b) is bounded by a fixed number, $1/(p-1)a^{p-1}$, no matter how large b is.

Figure 10-7-1

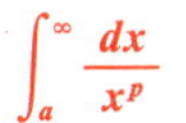

$\int_a^\infty \frac{dx}{x^p}$

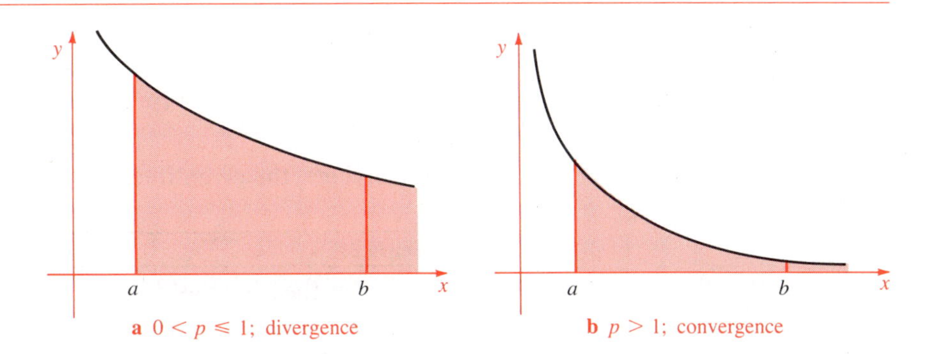

a $0 < p \leq 1$; divergence **b** $p > 1$; convergence

The figure suggests that for a positive function $f(x)$ the convergence or divergence of

$$\int_a^\infty f(x)\,dx$$

depends on how rapidly $f(x) \to 0$ as $x \to +\infty$. For the function $f(x) = 1/x^{10}$, which decreases very fast, the integral converges with plenty to spare. But for $f(x) = 1/x^{1.01}$, which decreases much more slowly, the integral barely makes it.

Comparison Tests

We shall state comparison tests for convergence and divergence of improper integrals. In spirit, the discussion parallels that for infinite series on page 547, and the proofs, which we omit, are similar.

First Comparison Test for Integrals

Suppose $f(x) \geq 0$ and $g(x) \geq 0$ for $x \geq a$.

- If $g(x) \leq f(x)$, then the convergence of $\int_a^\infty f(x)\,dx$ implies the convergence of $\int_a^\infty g(x)\,dx$.
- If $g(x) \geq f(x)$, then the divergence of $\int_a^\infty f(x)\,dx$ implies the divergence of $\int_a^\infty g(x)\,dx$.

Example 1

Show that the integrals converge:

a $\displaystyle\int_1^\infty \frac{dx}{x^2 + \sqrt{x}}$ **b** $\displaystyle\int_0^\infty \frac{e^{-x}}{x+1}\,dx$ **c** $\displaystyle\int_3^\infty \frac{\sin^2 x}{x^3}\,dx$

Solution We have the inequalities

$$\frac{1}{x^2 + \sqrt{x}} < \frac{1}{x^2} \qquad (1 \leq x < \infty)$$

$$\frac{e^{-x}}{x+1} < e^{-x} \qquad (0 \leq x < \infty)$$

$$\frac{\sin^2 x}{x^3} \leq \frac{1}{x^3} \qquad (3 \leq x < \infty)$$

Since the integrals

$$\int_1^\infty \frac{dx}{x^2} \qquad \int_0^\infty e^{-x}\,dx \qquad \int_3^\infty \frac{dx}{x^3}$$

all converge, the given integrals converge by the first comparison test. ●

Example 2

Show that the integrals diverge:

a $\displaystyle\int_1^\infty \frac{\sqrt{x}}{1+x}\,dx$ **b** $\displaystyle\int_2^\infty \frac{dx}{\sqrt{x}-\sqrt[3]{x}}$ **c** $\displaystyle\int_3^\infty \frac{\ln x}{x}\,dx$

Solution We have the inequalities

$$\frac{\sqrt{x}}{1+x} \ge \frac{1}{1+x} \qquad (1 \le x < \infty)$$

$$\frac{1}{\sqrt{x}-\sqrt[3]{x}} > \frac{1}{\sqrt{x}} \qquad (2 \le x < \infty)$$

$$\frac{\ln x}{x} \ge \frac{\ln 3}{x} > \frac{1}{x} \qquad (3 \le x < \infty)$$

Since the integrals

$$\int_1^\infty \frac{dx}{1+x} \qquad \int_2^\infty \frac{dx}{\sqrt{x}} \qquad \text{and} \qquad \int_3^\infty \frac{dx}{x}$$

all diverge, the given integrals diverge by the first comparison test. ●

The following test does not require the integrand to be non-negative.

Second Comparison Test for Integrals

Suppose $f(x) \ge 0$ and $g(x)$ is bounded, that is $|g(x)| \le M$ for some constant M. Then the convergence of

$$\int_a^\infty f(x)\,dx \quad \text{implies the convergence of} \quad \int_a^\infty f(x)g(x)\,dx$$

●

Example 3

Show that the integrals converge:

a $\displaystyle\int_0^\infty e^{-x}\sin^3 x\,dx$ **b** $\displaystyle\int_1^\infty \frac{\ln x}{x^3}\,dx$ **c** $\displaystyle\int_{-1/2}^\infty \frac{\arctan x}{1+x^3}\,dx$

Solution Apply the second comparison test:

a Since $\int_0^\infty e^{-x}\,dx$ converges and $|\sin^3 x| \le 1$, the given integral converges.

b Write $\dfrac{\ln x}{x^3} = \dfrac{1}{x^2} \cdot \dfrac{\ln x}{x}$

The integral $\int_1^\infty dx/x^2$ converges and $(\ln x)/x$ is bounded. (Its maximum value is $1/e$.) Hence the given integral converges.

c Break the integral into a sum:

$$\int_{-1/2}^{\infty} \frac{\arctan x}{1 + x^3}\,dx = \int_{-1/2}^{1} + \int_{1}^{\infty}$$

The first integral on the right is an integral of a continuous function on a finite interval, so it exists. Thus the problem is equivalent to showing that

$$\int_{1}^{\infty} \frac{\arctan x}{1 + x^3}\,dx$$

converges. But $|\arctan x| < \frac{1}{2}\pi$, and

$$\int_{1}^{\infty} \frac{dx}{1 + x^3} < \int_{1}^{\infty} \frac{dx}{x^3}$$

which converges. Hence the given integral converges. ●

Remark The splitting of the integral in part **c** was necessary. You cannot prove the convergence of

$$\int_{-1/2}^{\infty} \frac{dx}{1 + x^3} \quad \text{by comparison with} \quad \int_{-1/2}^{\infty} \frac{dx}{x^3}$$

because the latter integral is undefined. (There is a zero in the denominator at $x = 0$.) Always keep in mind that the convergence of an integral $\int_a^\infty f(x)\,dx$ depends only on the ultimate behavior of $f(x)$, that is, its behavior for large x.

Absolute Convergence

An improper integral $\int_a^\infty f(x)\,dx$ is called **absolutely convergent** if $\int_a^\infty |f(x)|\,dx$ is convergent. Just as with infinite series, an absolutely convergent integral is convergent. This statement will follow easily from the next comparison test.

Third Comparison Test for Integrals

If $\int_a^\infty f(x)\,dx$ is convergent and if $|g(x)| \le f(x)$ for $a \le x < +\infty$, then $\int_a^\infty g(x)\,dx$ is convergent and absolutely convergent.

In particular, if $\int_a^\infty g(x)\,dx$ converges absolutely, then it converges. ●

A series may converge, but not converge absolutely. Similarly, an integral may converge, but not converge absolutely. An example is $\int_1^\infty f(x)\,dx$ where $f(x) = (-1)^n/n$ for $n \le x < n + 1$. Is this cheating? Not really; $\int_0^\infty [(\sin x)/x]\,dx$ is another example (with a continuous integrand), as we shall see in the next section.

Exercises

Test for convergence

1 $\displaystyle\int_0^\infty \frac{dx}{x+1}$

2 $\displaystyle\int_1^\infty \frac{dx}{x^2+x}$

3 $\displaystyle\int_0^\infty \frac{x^2e^{-x}}{1+x^2}\,dx$

4 $\displaystyle\int_2^\infty e^{-x^3}\,dx$

5 $\displaystyle\int_0^\infty \cosh x\,dx$

6 $\displaystyle\int_0^\infty \frac{x\,dx}{\sqrt{x^2+3}}$

7 $\displaystyle\int_{-\infty}^\infty \frac{\sin x}{1+x^2}\,dx$

8 $\displaystyle\int_0^\infty \sin x\,dx$

9 $\displaystyle\int_1^\infty \frac{\cos x}{\sqrt{x}\,(x+4)}\,dx$

10 $\displaystyle\int_2^\infty \frac{dx}{\ln x}$

11 $\displaystyle\int_3^\infty \frac{x^3}{x^4-1}\,dx$

12 $\displaystyle\int_0^\infty \frac{dx}{1+x+e^x}$

13 $\displaystyle\int_{-\infty}^0 x^2e^x\,dx$

14 $\displaystyle\int_{-5}^\infty \frac{dx}{\sqrt{(x+6)(x+7)(x+8)}}$

15 Show that $\displaystyle\int_2^\infty \frac{dx}{x(\ln x)^p}$ converges if $p>1$, diverges if $p\le 1$. [Hint Use the substitution $u=\ln x$.]

16 Show that $\displaystyle\int_3^\infty \frac{dx}{x\ln x[\ln(\ln x)]^p}$ converges if $p>1$, diverges if $p\le 1$.

17 Show that $\displaystyle\int_1^\infty \frac{\ln x}{x^p}\,dx$ converges if $p>1$, diverges if $p\le 1$.
[Hint Recall that $(\ln x)/x^r\to 0$ as $x\to+\infty$ if $r>0$.]

18 Denote by R the infinite region under $y=1/x$ to the right of $x=1$. Suppose R is rotated around the x-axis, forming an infinitely long horn. Show that the volume of this horn is finite. Its surface area, however, is infinite. (The surface area is certainly larger than the area of R; show that the area of R is infinite.) Here is an apparent paradox: You can fill the horn with paint, but you cannot paint it. Where is the fallacy?

Find all values of s for which the integral converges

19 $\displaystyle\int_0^\infty \frac{x\,dx}{\sqrt{1+x^s}}$

20 $\displaystyle\int_{-1}^\infty (x+3)^s\,dx$

21 $\displaystyle\int_{-\infty}^\infty \frac{dx}{(x^2+1)^s}$

22 $\displaystyle\int_{-\infty}^\infty e^{-sx}\,dx$

23 $\displaystyle\int_0^\infty e^{-sx}e^x\,dx$

24 $\displaystyle\int_0^\infty \frac{e^{-sx}}{1+x^2}\,dx$

25 $\displaystyle\int_0^\infty e^{-sx}e^{-x^2}\,dx$

26 $\displaystyle\int_1^\infty \frac{x^s}{(1+x^3)^s}\,dx$

The **exponential integral** for $x<0$ is $\mathrm{Ei}(x)=\displaystyle\int_{-\infty}^x \frac{e^t}{t}\,dt$. Set

$$E(x)=-\mathrm{Ei}(-x)=\int_x^\infty \frac{e^{-t}}{t}\,dt \quad\text{for } x>0$$

27 Clearly $E(x)$ is very large for x near 0. How large? Prove the **asymptotic expansions**

$$\begin{aligned}E(x)&=-e^{-x}\ln x+\int_x^\infty e^{-t}\ln t\,dt\\ &=-e^{-x}(\ln x+x\ln x-x)\\ &\quad+\int_x^\infty te^{-t}(\ln t-1)\,dt\end{aligned}$$

[Hint Integration by parts]

28 (cont.) Clearly $E(x)$ is very small for x large. How small? Prove

$$\begin{aligned}E(x)&=e^{-x}\left(\frac{1}{x}-\frac{1}{x^2}+\frac{2!}{x^3}-\cdots\right.\\ &\quad\left.+(-1)^{n-1}\frac{(n-1)!}{x^n}\right)\\ &\quad+(-1)^n n!\int_x^\infty \frac{e^{-t}}{t^{n+1}}\,dt\end{aligned}$$

[Hint Integration by parts]

29 Let $K_n=\displaystyle\int_0^\infty x^ne^{-x^2}\,dx$. Prove $K_{n+2}=\frac{1}{2}(n+1)K_n$.
Conclude that $K_{2n+1}=\frac{1}{2}(n!)$ and $K_{2n}=\dfrac{(2n)!}{2^{2n}n!}K_0$.
(It will be proved later that $K_0=\frac{1}{2}\sqrt{\pi}$.)

30 (cont.) Prove $K_n^2<K_{n-1}K_{n+1}$.
[Hint Show that $K_{n+1}+2tK_n+t^2K_{n-1}>0$.]

*31 Suppose $\displaystyle\int_0^\infty \frac{f(x)}{x}\,dx$ converges. Find $\displaystyle\int_0^\infty \frac{f(ax)}{x}\,dx$ for $a>0$.

*32 (cont.) Suppose $f(x)$ is continuous for $x\ge 0$ and for each $\varepsilon>0$, the integral

$$\int_\varepsilon^\infty \frac{f(x)}{x}\,dx$$

converges. Let $0<a<b$. Prove the existence and value given of

$$\int_0^\infty \frac{f(ax)-f(bx)}{x}\,dx=f(0)\ln\frac{b}{a}$$

33 A certain model of a freezing ice cube involves solving the integral equation

$$\int_0^x tf(t)\,dt = \tfrac{1}{2}(1+\beta)x^2 f(x)$$

for $f(x)$. We claim that $f(x) = x^p$ is a solution. What condition does this impose on β?

34 Set

$$F(x) = \int_0^\infty \frac{\arctan(xt)}{1+t^2}\,dt$$

Find $F(x) + F(1/x)$ for $x > 0$.
[Hint Use the change of variable $t = 1/u$.]

10-8 Relation between Integrals and Series

We have already seen similarities between infinite series and infinite integrals. In this section we discuss a useful test for convergence or divergence of a series in terms of a related integral. This is important, for usually it is easier to find the value of an integral than the sum of a series.

Consider the relation between the series

$$\frac{1}{2^2} + \frac{1}{3^2} + \cdots + \frac{1}{n^2} + \cdots$$

and the convergent integral

$$\int_1^\infty \frac{dx}{x^2}$$

(See Figure 10-8-1.) The rectangles shown in Figure 10-8-1 have areas $1/2^2,\ 1/3^2,\ \cdots$. Obviously the sum of these areas is finite, being less than the finite area under the curve. Hence, the series converges. This illustrates a general principle:

Figure 10-8-1
$\int_1^\infty f(x)\,dx > f(2) + f(3) + f(4)\cdots$

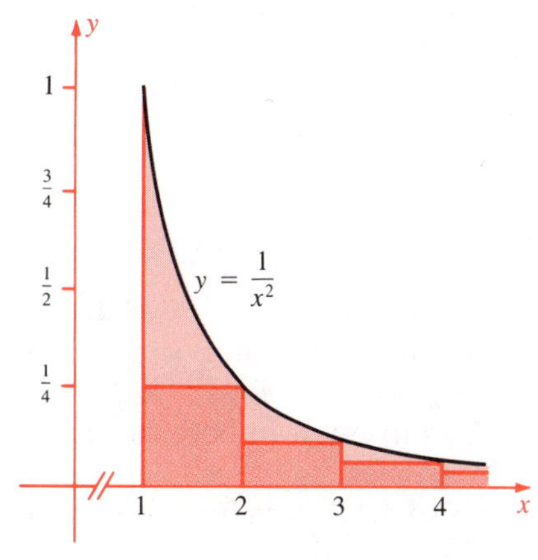

Integral Test

Suppose $f(x)$ is a positive decreasing function. Then the series

$$f(1) + f(2) + \cdots + f(n) + \cdots$$

converges if the integral

$$\int_1^\infty f(x)\,dx$$

converges, and diverges if the integral diverges.

Proof The argument given above for $f(x) = 1/x^2$ holds for any positive decreasing function $f(x)$. Figure 10-8-1 indicates that

$$f(2) + f(3) + \cdots + f(n) \le \int_1^n f(x)\,dx$$

Figure 10-8-2
$f(1) + f(2) + f(3) + \cdots > \int_1^\infty f(x)\,dx$

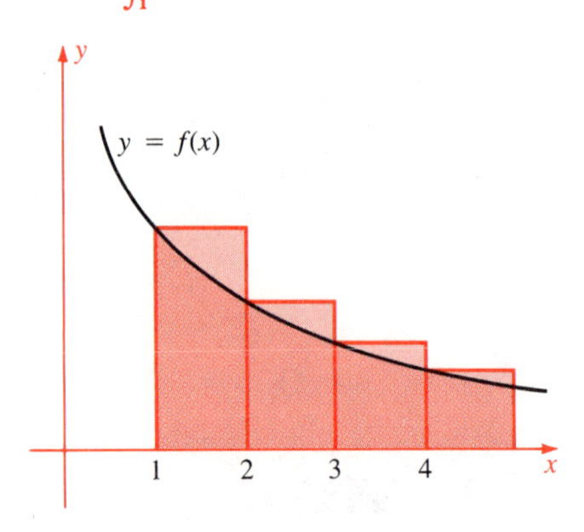

If the infinite integral converges, then

$$s_n = f(1) + f(2) + \cdots + f(n) \le f(1) + \int_1^n f(x)\,dx$$
$$\le f(1) + \int_1^\infty f(x)\,dx$$

Hence the increasing partial sums are bounded; the series converges.

If the infinite integral diverges, the rectangles are drawn above the curve (Figure 10-8-2). Their areas are $f(1)$, $f(2)$, ... This time

$$s_n = f(1) + f(2) + \cdots + f(n) \ge \int_1^{n-1} f(x)\,dx.$$

But the integrals on the right are unbounded. Hence the increasing sequence $\{s_n\}$ is unbounded; the series diverges. ●

Remark As usual we may ignore a finite number of terms or a finite part of the integral. The series can start at $f(k)$ and the integral can just as well be

$$\int_a^\infty f(x)\,dx \qquad a > 0$$

Example 1

Does the series

$$\frac{1}{2\ln 2} + \frac{1}{3\ln 3} + \frac{1}{4\ln 4} + \cdots$$

converge or diverge?

Solution Let $a_n = 1/(n\ln n)$. On the one hand, $a_n < 1/n$, but that doesn't help because $\Sigma\, 1/n$ diverges. On the other hand, since $\ln n$ increases so slowly, $a_n > 1/n^2$, or even $a_n > 1/n^p$ for $p > 1$. But that again does not help because $\Sigma\, 1/n^p$ converges. In both cases, the inequalities go the wrong way.

However, $f(x) = 1/(x\ln x)$ is a positive decreasing function, so we can use the preceding test. Substituting $u = \ln x$, we have

$$\int_2^\infty \frac{dx}{x\ln x} = \int_2^\infty \frac{1}{\ln x}\left(\frac{1}{x}\,dx\right) = \int_{\ln 2}^\infty \frac{1}{u}\,du$$

This integral diverges. Hence the series diverges. ●

Example 2

Show that

$$\int_0^\infty \frac{\sin x}{x}\,dx$$

converges, but not absolutely. See Figure 10-8-3.

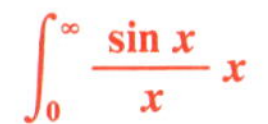

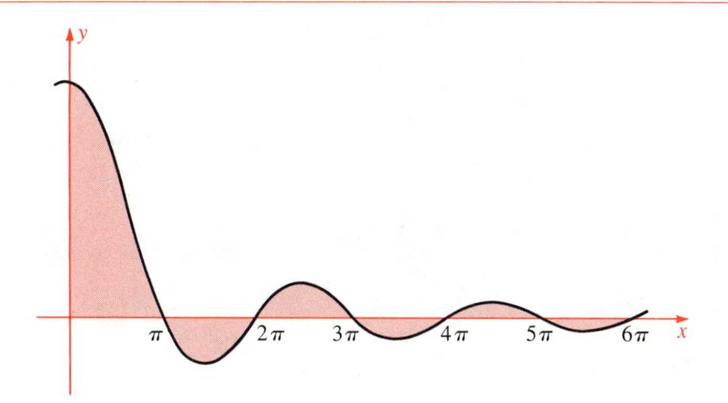

Figure 10-8-3
$\int_0^\infty \frac{\sin x}{x}\, x$

Solution First we get away from 0; then integrate by parts:

$$\int_0^b \frac{\sin x}{x}\,dx = K + \int_\pi^b \frac{\sin x}{x}\,dx \qquad \text{where} \quad K = \int_0^\pi \frac{\sin x}{x}\,dx$$

$$\int_\pi^b \frac{\sin x}{x}\,dx = -\int_\pi^b \left(\frac{1}{x}\right) d(\cos x)$$

$$= -\frac{\cos x}{x}\bigg|_\pi^b - \int_\pi^b \frac{\cos x}{x^2}\,dx$$

As $b \to +\infty$,

$$-\frac{\cos x}{x}\bigg|_\pi^b = \frac{\cos \pi}{\pi} - \frac{\cos b}{b} \to \frac{\cos \pi}{\pi} = -\frac{1}{\pi}$$

and

$$\int_\pi^b \frac{\cos x}{x^2}\,dx \to \int_\pi^\infty \frac{\cos x}{x^2}\,dx$$

a convergent integral by the comparison $|(\cos x)/x^2| \le 1/x^2$. Therefore the given integral converges:

$$\int_0^\infty \frac{\sin x}{x}\,dx = \lim_{b\to+\infty} \int_0^b \frac{\sin x}{x}\,dx$$

$$= \int_0^\pi \frac{\sin x}{x}\,dx - \frac{1}{\pi} + \int_\pi^\infty \frac{\cos x}{x^2}\,dx$$

Now we consider absolute convergence. We have

$$\int_{n\pi}^{(n+1)\pi} \frac{|\sin x|}{x}\,dx \ge \frac{1}{(n+1)\pi} \int_{n\pi}^{(n+1)\pi} |\sin x|\,dx = \frac{2}{(n+1)\pi}$$

so

$$\int_0^\infty \frac{|\sin x|}{x}\,dx = \sum_{n=0}^\infty \int_{n\pi}^{(n+1)\pi} \frac{|\sin x|}{x}\,dx \ge \sum_{n=0}^\infty \frac{2}{(n+1)\pi}$$

diverges by comparison with the harmonic series. Thus the given integral converges, but not absolutely. ●

Remark The example is interesting because the indefinite integral of $(\sin x)/x$ cannot be expressed in terms of the functions of calculus.

Exercises

Use the integral test to test for convergence

1 $1 + \dfrac{1}{\sqrt{2}} + \dfrac{1}{\sqrt{3}} + \dfrac{1}{\sqrt{4}} + \cdots$

2 $\dfrac{1}{e} + \dfrac{2}{e^2} + \dfrac{3}{e^3} + \cdots$

3 $1 + \dfrac{1}{2^3} + \dfrac{1}{3^3} + \cdots$

4 $\dfrac{1}{1+1^2} + \dfrac{1}{1+2^2} + \dfrac{1}{1+3^2} + \cdots$

5 $\sum \dfrac{2n-1}{n^2}$

6 $\sum \dfrac{1}{n(\ln n)^3}$

7 $\sum \dfrac{1 + \ln n}{n^n}$

8 $\sum \dfrac{n}{(n+1)^3}$

9 $\sum (1 - \tanh n)$

10 $\sum (\frac{1}{2}\pi - \arctan n)$

11 Show geometrically that the sum of

$$1 + \frac{1}{2^2} + \frac{1}{3^2} + \frac{1}{4^2} + \cdots$$

is less than 2. See Figure 10-8-1. (It is a remarkable fact that the exact sum is $\frac{1}{6}\pi^2$.)

12 Use the method of inscribing and circumscribing rectangles to show that

$$\ln(n+1) < 1 + \frac{1}{2} + \frac{1}{3} + \cdots + \frac{1}{n} < 1 + \ln n$$

Is $1 + \dfrac{1}{2} + \dfrac{1}{3} + \cdots + \dfrac{1}{1000}$ more or less than 10?

13 Estimate how many terms of the series $1 + \frac{1}{2} + \frac{1}{3} + \frac{1}{4} + \cdots$ must be added before the sum exceeds 1000.

***14** (cont.) The same for

$$\frac{1}{2\ln 2} + \frac{1}{3\ln 3} + \cdots + \frac{1}{n\ln n} > 10$$

15 Show geometrically that $\ln(n!) > \int_1^n \ln x\,dx$. Conclude that $n! > e(n/e)^n$.

16 (cont.) Show that $100! > 10^{157}$.

17 Prove

$$\int_0^\infty \frac{\sin x}{x}\,dx = \int_0^\infty \frac{1-\cos x}{x^2}\,dx$$

***18** Let $f(x)$ be the saw-toothed function (Figure 10-8-4). Prove $\displaystyle\int_0^\infty \frac{f(x)}{x}\,dx$ converges.

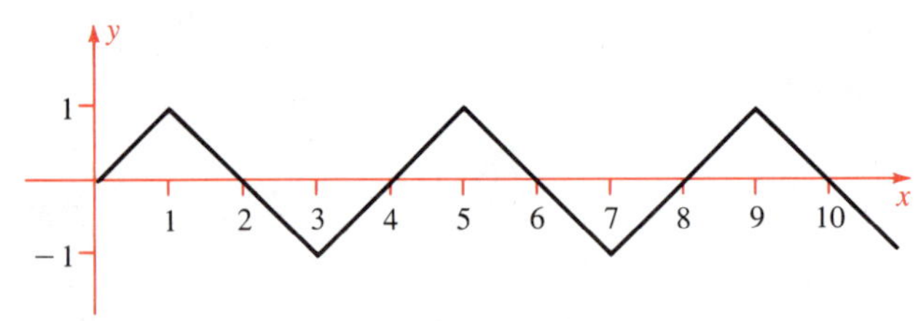

Figure 10-8-4
The saw-toothed function

19 Prove that $\displaystyle\int_1^\infty \frac{\sin x}{\sqrt{x}}\,dx$ converges.

[Hint Integrate by parts.]

20 (cont.) Prove that $\displaystyle\int_1^\infty \sin(x^2)\,dx$ converges.

[Hint Set $x^2 = u$.]

***21** We know that if $\alpha > 0$, then

$$\sum \frac{1}{n^{1+\alpha}}$$

converges. We also know that $1/n \to 0$ as $n \to +\infty$. Does the series

$$\sum \frac{1}{n^{1+1/n}}$$

converge or diverge?

***22** More generally, determine the convergence or divergence of

$$\sum \frac{1}{n^{1+1/n^\varepsilon}} \qquad (\varepsilon > 0)$$

10-9 Review Exercises

Find

1 $\lim_{n\to+\infty}(e^{-n} + e^{-2/n})$

2 $\lim_{n\to+\infty}(n^3 - n^2 + 1)/(n^3 + n^2 + 1)$

3 Suppose $\lim_{n\to+\infty} a_n = L$ and $\lim_{n\to+\infty} b_n = M$. Prove that $\lim_{n\to+\infty}(a_{2n} - b_{2n+1})$ exists and find its value.

4 Let $a_0 = 1$ and $a_n = a_{n-1}[1 - (n+1)/n^3]$. Prove that $\{a_n\}$ converges.

5 Suppose $\{a_n\}$ converges. Show that

$$a_1,\ a_2,\ a_2,\ a_3,\ a_3,\ a_3,\ a_4,\ a_4,\ a_4,\ a_4,\ \cdots$$

also converges.

6 Each real x between 0 and 1 has a *binary* expansion $x = 0.b_1b_2b_3\ldots$ where $b_i = 0$ or 1. This means

$$x = \sum_{n=1}^{\infty} \frac{b_n}{2^n}$$

Show that the series converges. When do two different expansions yield the same x?

Test for convergence

7 $\dfrac{1}{5} + \dfrac{1}{15} + \dfrac{1}{25} + \dfrac{1}{35} + \cdots$

8 $\dfrac{1+2}{1+3} + \dfrac{1+2+4}{1+3+9} + \dfrac{1+2+4+8}{1+3+9+27} + \cdots$

9 $\sum \dfrac{n(n+1)}{(n+2)(n+3)(n+4)}$

10 $\sum \dfrac{n^3}{2^n}$

11 $\sum_{n=1}^{\infty} \dfrac{n}{4+n^3}$

12 $\sum_{n=2}^{\infty} \dfrac{\ln n}{1+n^2}$

13 $\sum_{n=1}^{\infty} \dfrac{n^2e^{-n}}{1+n^2}$

14 $\sum_{n=2}^{\infty} \dfrac{1}{\sqrt{n}\,\ln n}$

Test for convergence and absolute convergence

15 $\sum_{n=1}^{\infty} \dfrac{(-1)^{n-1}}{n+\sqrt{n}}$

16 $\sum_{n=1}^{\infty} \dfrac{(-1)^{n-1}}{1-1/n^2}$

17 $\sum_{n=0}^{\infty} \dfrac{(-1)^n}{\ln(1+n)}$

18 $\sum_{n=2}^{\infty} \dfrac{(-1)^n}{n(\ln n)^2}$

Test for convergence

19 $\displaystyle\int_0^{\infty} \frac{\sin^2 x}{1+x^2}\,dx$

20 $\displaystyle\int_1^{\infty} \frac{x^5\,dx}{1+x^6}$

21 $\displaystyle\int_1^{\infty} \frac{dx}{x^{3/4}}$

22 $\displaystyle\int_{-\infty}^{\infty} \frac{dx}{1+\exp(x^2)}$

23 Let

$$a_n = \frac{(1+1)(1+4)(1+9)\cdots(1+n^2)}{(2+1)(2+4)(2+9)\cdots(2+n^2)}$$

Prove that $\{a_n\}$ converges.

24 (cont.) Prove that $\sum a_n$ diverges. [Hint Prove $a_n \geq 2/3n$.]

25 If a series of non-negative terms $\sum a_n$ converges, show that $\sum a_{2n}$ also converges.

26 Show by example that the statement in Exercise 25 is false without the assumption $a_n \geq 0$.

11 Power Series

In this chapter we study power series

$$\sum_{n=0}^{\infty} a_n x^n = a_0 + a_1 x + a_2 x^2 + \cdots + a_n x^n + \cdots$$

and their applications. Power series serve a number of important purposes in both theoretical and applied mathematics. First, they are particularly suitable for computation and are indispensable in many numerical problems. Second, they provide an alternative way of expressing many familiar functions; hence they aid our understanding of these functions. Finally, with power series we can define functions that are hard or impossible to specify otherwise. Certainly nobody objects to defining a function by a polynomial,

$$f(x) = a_0 + a_1 x + a_2 x^2 + \cdots + a_n x^n$$

So why not define a function by a power series,

$$f(x) = a_0 + a_1 x + a_2 x^2 + \cdots + a_n x^n + \cdots$$

provided that the series converges?

11-1 Convergence and Divergence

Given a power series, we first ask: Does it converge? If so, for which values of x does it converge? Now for each fixed x, a power series is an infinite series of numbers. So whatever we know about infinite series applies. In particular, convergence and divergence are defined in terms of the sequence of partial sums.

Convergence of Power Series

A power series $\Sigma_{k=0}^{\infty} a_k x^k$ **converges** at a point x if the sequence of its partial sums

$$p_n(x) = \sum_{k=0}^{n} a_k x^k$$

converges. The power series **diverges** at x if the sequence $\{p_n(x)\}$ diverges.

If $\lim_{n \to +\infty} p_n(x) = f(x)$ for each x in a set D, we write

$$f(x) = \sum_{n=0}^{\infty} a_n x^n \quad \text{on } D$$

For example, take the geometric series

$$1 + x + x^2 + x^3 + \cdots$$

Its n-th partial sum is

$$p_n(x) = 1 + x + x^2 + \cdots + x^n = \begin{cases} \dfrac{1 - x^{n+1}}{1 - x} & \text{for } x \neq 1 \\ n + 1 & \text{for } x = 1 \end{cases}$$

If $|x| < 1$, then $x^{n+1} \to 0$. Hence $p_n(x) \to 1/(1 - x)$. If $|x| \geq 1$, the sequence $\{p_n(x)\}$ diverges. Therefore

$$1 + x + x^2 + x^3 + \cdots = \frac{1}{1 - x} \quad \text{on} \quad (-1, 1)$$

and the series diverges for all other values of x.

The geometric series is especially nice because there is a formula for its partial sums. This not only helps the discussion of convergence and divergence, but also leads to a neat formula for the sum of the series where it converges. Generally there is no nice formula for the partial sums of a given power series. Still we can use the techniques of the last chapter to investigate convergence and divergence.

For example, consider the power series

$$1 + x + \frac{x^2}{2^2} + \frac{x^3}{3^3} + \cdots + \frac{x^n}{n^n} + \cdots$$

For any fixed x, eventually $n > 2|x|$ so that $|x/n| < \frac{1}{2}$. This implies

$$\left| \frac{x^n}{n^n} \right| < \frac{1}{2^n}$$

for all sufficiently large n. Hence the series converges (absolutely) by comparison with the geometric series $\Sigma\, 1/2^n$. There is no obvious formula for the sum; we consider the power series as *defining* a new function $f(x)$ whose domain is all real x.

One further example:

$$1 + x + 2^2x^2 + 3^3x^3 + \cdots + n^nx^n + \cdots$$

Obviously, this power series converges at $x = 0$. However, it diverges everywhere else: if $x \neq 0$, then

$$|n^nx^n| = |nx|^n \to +\infty \quad \text{as} \quad n \to +\infty$$

Interval of Convergence

What is the domain of convergence of a power series? The preceding examples show that there are at least three possibilities: the domain may consist of a single point, a bounded interval, or the entire real axis. In fact, these are the *only* possibilities. This is stated formally in the following theorem. Note that we generalize from powers of x to powers of $x - c$, where c is fixed.

Theorem Interval of Convergence

Given a power series $\Sigma\, a_n(x - c)^n$, precisely one of the following three cases holds:

1 The series converges only for $x = c$.

2 The series converges for all values of x.

3 There is a positive number R such that the series converges for each x satisfying $|x - c| < R$ and diverges for each x satisfying $|x - c| > R$.

The complete proof of this theorem is somewhat long. The main idea is to show that if $\Sigma\, a_nx_0{}^n$ converges for a particular $x_0 \neq 0$, then $\Sigma\, a_nx^n$ converges provided $|x| < |x_0|$. (In fact, it converges absolutely.) Here is why. Since $\Sigma\, a_nx_0{}^n$ converges, its terms are bounded: $|a_nx_0{}^n| \leq B$ for some B and all n. Consequently

$$\sum |a_nx^n| = \sum |a_nx_0{}^n| \left|\frac{x}{x_0}\right|^n \leq \sum B \left|\frac{x}{x_0}\right|^n$$

But $|x/x_0| < 1$, so $\Sigma\, |a_nx^n|$ converges by comparison with a geometric series.

Case 1 of the theorem is an extreme case. It occurs when the coefficients a_n increase so rapidly that the power series can converge only if all terms after a_0 vanish. An example is $\Sigma\, n^nx^n$. Power series of this type are of no use.

Case 2 occurs when the coefficients become small very rapidly. An example is $\Sigma\, x^n/n!$, where for any x the general term $x^n/n!$ eventually tends to zero quickly because the coefficients $a_n = 1/n!$ become small so fast. Power series of this type are the nicest kind since they never cause any problems with convergence.

Case 3 lies between. The coefficients do not increase so rapidly that the series never converges except for $x = c$, nor do they decrease so rapidly that the series always converges. A typical example is the geometric series $\Sigma\, x^n$, where each $a_n = 1$. This series converges for $|x| < 1$ and diverges for $|x| > 1$, hence $R = 1$.

In Case 3 the set of all points x for which the series $\Sigma\, a_n(x - c)^n$ converges is called its **interval of convergence** (or **domain of convergence.**) This set consists of the interval $(c - R, c + R)$ and possibly one or both of its end points. The number R is called the **radius of convergence** of the power series (Figure 11-1-1).

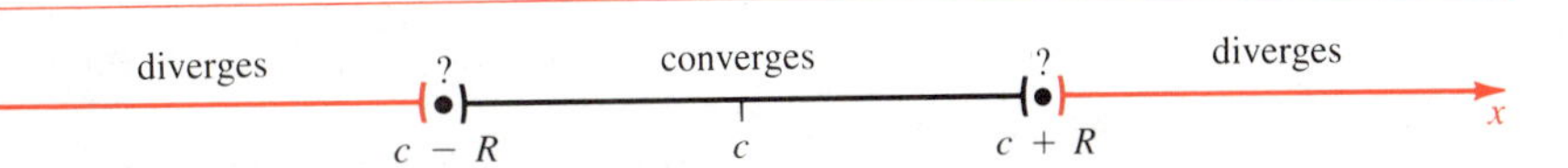

Figure 11-1-1
Interval of convergence

By convention, $R = 0$ in Case 1 (convergence for $x = c$ only); and $R = +\infty$ in Case 2 (convergence for all x). The interval of convergence in Case 1 is the single point c; in Case 2 it is $\mathbf{R}$, the entire x-axis.

Example 1

Find the radius of convergence, interval of convergence, and sum of the power series

a $1 + 4x + 4^2x^2 + 4^3x^3 + \cdots$

b $1 - \dfrac{1}{2}(x + 3) + \dfrac{1}{2^2}(x + 3)^2 - \dfrac{1}{2^3}(x + 3)^3 + \cdots$

Solution Each series is of the form

$$1 + y + y^2 + y^3 + \cdots$$

This geometric series converges only for $|y| < 1$, and there its sum is $1/(1 - y)$.

a Here $y = 4x$. The sum is $1/(1 - 4x)$ and the series converges if and only if $|4x| < 1$, that is, $|x| < \frac{1}{4}$. Hence $R = \frac{1}{4}$ and the interval of convergence is $(-\frac{1}{4}, \frac{1}{4})$.

b Here $y = -\frac{1}{2}(x + 3)$. Hence the sum of the geometric series is $1/[1 + \frac{1}{2}(x + 3)] = 2/(x + 5)$ and the series converges if and only if $|\frac{1}{2}(x + 3)| < 1$, that is, $|x + 3| < 2$. Hence $R = 2$ and the interval of convergence is $(-5, -1)$. ●

Example 2

Find the interval of convergence

a $x + \dfrac{x^2}{2} + \dfrac{x^3}{3} + \cdots + \dfrac{x^n}{n} + \cdots$

b $\dfrac{x}{10} + \dfrac{x^2}{10^4} + \dfrac{x^3}{10^9} + \cdots + \dfrac{x^n}{10^{n^2}} + \cdots$

Solution

a Since $|x^n/n| < |x|^n$, the series converges for $|x| < 1$ by comparison with the geometric series. Hence its interval of convergence includes the interval $|x| < 1$. However, the series diverges at $x = 1$ (harmonic series) and converges at $x = -1$ (alternating harmonic series). By part 3 of the previous theorem, the series diverges for $|x| > 1$ because it diverges for $x = 1$. Therefore its interval of convergence is $[-1, 1)$.

b For any fixed x, choose a positive integer p such that $|x| < 10^p$. Then if $n > p$,

$$\left|\frac{x^n}{10^{n^2}}\right| < \frac{10^{np}}{10^{n^2}} = \frac{1}{10^{n(n-p)}} \leq \frac{1}{10^n}$$

Hence the series converges by comparison with the geometric series $\Sigma\, 1/10^n$. The interval of convergence is **R**. ●

Remark Anything is possible at the end points of the interval of convergence. For instance, the geometric series diverges at both end points of its interval of convergence. The series in Example 2a converges at the end point $x = -1$ of its interval of convergence (alternating harmonic series) and diverges (harmonic series) at the other end point, $x = 1$. There exist power series that converge at both end points. See Exercise 32.

The Ratio Test

Often the radius of convergence of a given power series can be found by the following ratio test, a consequence of the ratio test for series of constants.

Ratio Test for Power Series

Suppose the power series

$$a_0 + a_1(x - c) + a_2(x - c)^2 + \cdots + a_n(x - c)^n + \cdots$$

has non-zero coefficients. If

$$\left|\frac{a_n}{a_{n+1}}\right| \to R \quad \text{as} \quad n \to +\infty$$

where R is 0, positive, or $+\infty$, then R is the radius of convergence. ●

Proof For simplicity of notation assume $c = 0$. First suppose that $0 < R < +\infty$. If $|x| < R$, then

$$\frac{|a_{n+1}x^{n+1}|}{|a_n x^n|} = \left|\frac{a_{n+1}}{a_n}\right| |x| \to \frac{|x|}{R} < 1$$

Hence the series $\Sigma\, a_n x^n$ converges (absolutely) by the first ratio test for a series of constants (page 550). If $|x| > R$, then

$$\frac{|a_{n+1}x^{n+1}|}{|a_n x^n|} \to \frac{|x|}{R} > 1$$

Hence for n sufficiently large, $\{|a_n x^n|\}$ is an increasing sequence. Therefore $\Sigma\, a_n x^n$ diverges because its terms do not approach 0. Thus the series converges for $|x| < R$ and diverges for $|x| > R$. In other words, its radius of convergence is R.

Suppose $|a_n/a_{n+1}| \to R = 0$. If $x \neq 0$, then

$$\frac{|a_{n+1}x^{n+1}|}{|a_n x^n|} = \left|\frac{a_{n+1}}{a_n}\right| |x| \to +\infty$$

Hence the series diverges for all $x \neq 0$; its radius of convergence is 0.

Finally, suppose $|a_n/a_{n+1}| \to R = +\infty$. Then for any x,

$$\frac{|a_{n+1}x^{n+1}|}{|a_n x^n|} = \left|\frac{a_{n+1}}{a_n}\right| |x| \to 0$$

Hence the series converges for all x; its radius of convergence is $+\infty$. This completes the proof in all cases. ●

Example 3

Find the radius of convergence

a $1 + x/4 + x^2/7 + x^3/10 + \cdots + x^n/(3n+1) + \cdots$

b $(x-1) - 4(x-1)^2 + 9(x-1)^3 - \cdots + (-1)^{n-1}n^2(x-1)^n + \cdots$

c $1 + \dfrac{x}{2+1} + \dfrac{x^2}{2^2+2} + \dfrac{x^3}{2^3+3} + \cdots + \dfrac{x^n}{2^n+n} + \cdots$

d $\dfrac{x^3}{1\cdot 3\cdot 5} + \dfrac{x^4}{1\cdot 3\cdot 5\cdot 7} + \cdots + \dfrac{x^n}{1\cdot 3\cdot 5\cdots(2n-1)} + \cdots$

e $x^3/\sqrt{3} + x^6/\sqrt{6} + x^9/\sqrt{9} + \cdots + x^{3n}/\sqrt{3n} + \cdots$

Solution In each case apply the ratio test for power series:

a Here $a_n = 1/(3n+1)$ and $a_{n+1} = 1/(3n+4)$, so

$$\left|\frac{a_n}{a_{n+1}}\right| = \frac{1/(3n+1)}{1/(3n+4)} = \frac{3n+4}{3n+1} \to 1 \quad \text{as} \quad n \to +\infty$$

Hence $R = 1$.

b Here $a_n = (-1)^{n-1}n^2$ and $a_{n+1} = (-1)^n(n+1)^2$, so

$$\left|\frac{a_n}{a_{n+1}}\right| = \frac{n^2}{(n+1)^2} = \left(\frac{n}{n+1}\right)^2 \to 1 \quad \text{as} \quad n \to +\infty$$

Hence $R = 1$.

c Here, $a_n = 1/(2^n + n)$ and $a_{n+1} = 1/(2^{n+1} + n + 1)$, so

$$\left|\frac{a_n}{a_{n+1}}\right| = \frac{1/(2^n+n)}{1/(2^{n+1}+n+1)} = \frac{2^{n+1}+n+1}{2^n+n}$$

$$= \frac{2 + (n+1)\cdot 2^{-n}}{1 + n\cdot 2^{-n}} \to \frac{2+0}{1+0} = 2 \quad \text{as} \quad n \to +\infty$$

Hence $R = 2$.

d Here $a_n = 1/[1 \cdot 3 \cdot 5 \cdot \cdot \cdot (2n - 1)]$. As $n \to +\infty$,

$$\left|\frac{a_n}{a_{n+1}}\right| = \frac{1 \cdot 3 \cdot 5 \cdot \cdot \cdot (2n-1)(2n+1)}{1 \cdot 3 \cdot 5 \cdot \cdot \cdot (2n-1)} = 2n + 1 \to +\infty$$

so $R = +\infty$. Notice that the ratios make sense only for $n \geq 3$ because $a_0 = a_1 = a_2 = 0$ in the given series. However, it is perfectly correct to apply the ratio test from some point on since we may ignore a finite number of terms when studying convergence and divergence.

e The ratio test does not apply directly because "two-thirds" of the coefficients in this power series are zero. Nevertheless, the series may be written

$$\frac{y}{\sqrt{3}} + \frac{y^2}{\sqrt{6}} + \frac{y^3}{\sqrt{9}} + \cdot \cdot \cdot + \frac{y^n}{\sqrt{3n}} + \cdot \cdot \cdot$$

where $y = x^3$. The ratio test does apply to the series in this form:

$$\left|\frac{a_n}{a_{n+1}}\right| = \frac{\sqrt{3(n+1)}}{\sqrt{3n}}$$

$$= \sqrt{\frac{n+1}{n}} \to 1 \quad \text{as} \quad n \to +\infty$$

Hence the y-series converges for $|y| < 1$ and diverges for $|y| > 1$. Therefore, the original series converges for $|x^3| < 1$ and diverges for $|x^3| > 1$, that is, for $|x| < 1$ and $|x| > 1$, respectively. Hence $R = 1$. ●

The following root test for the radius of convergence of a power series may work when the ratio test fails. We omit its rather technical proof.

Root Test for Power Series

Suppose $a_n \neq 0$ for all $n \geq 0$ and

$$\lim_{n \to +\infty} \frac{1}{|a_n|^{1/n}} = R$$

Then

$$\sum_{n=0}^{\infty} a_n x^n$$

has radius of convergence R. ●

Example 4

Find the radius of convergence of

$$\frac{x}{10^2} + \frac{x^2}{10} + \frac{x^3}{10^4} + \frac{x^4}{10^3} + \frac{x^5}{10^6} + \frac{x^6}{10^5} + \frac{x^7}{10^8} + \frac{x^8}{10^7} + \cdot \cdot \cdot$$

Solution For any n,

$$a_n = \frac{1}{10^{n\pm 1}}$$

with $+$ if n is odd, $-$ if n is even. Therefore

$$|a_n|^{1/n} = \frac{1}{10^{1\pm 1/n}} \to \frac{1}{10}$$

hence $R = 10$ by the root test. ●

The ratio test fails for this example because alternately $a_n/a_{n+1} = \frac{1}{10}$ or 1000.

Exercises

Find the radius of convergence

1 $1 + x + 2x^2 + 3x^3 + \cdots$

2 $x + \frac{x^2}{4} + \frac{x^3}{9} + \frac{x^4}{16} + \cdots$

3 $x - \frac{x^3}{3} + \frac{x^5}{5} - \frac{x^7}{7} + - \cdots$

4 $1 + \frac{x}{1\cdot 2} + \frac{x^2}{2\cdot 2^2} + \frac{x^3}{3\cdot 2^3} + \cdots$

5 $\frac{x}{3+2} + \frac{x^2}{3^2+2^2} + \frac{x^3}{3^3+2^3} + \cdots$

6 $\frac{x-1}{1\cdot 2\cdot 3} + \frac{(x-1)^2}{2\cdot 3\cdot 4} + \frac{(x-1)^3}{3\cdot 4\cdot 5} + \cdots$

7 $x + \sqrt{2}\,x^2 + \sqrt{3}\,x^3 + \sqrt{4}\,x^4 + \cdots$

8 $\frac{1}{2} - \frac{x}{2^2} + \frac{x^2}{2^4} - \frac{x^3}{2^8} + \frac{x^4}{2^{16}} - + \cdots$

9 $(e^2 - 2)x^2 + (e^3 - 3)x^3 + (e^4 - 4)x^4 + \cdots$

10 $\frac{2x}{1^3} + \frac{3x^2}{2^3} + \frac{4x^3}{3^3} + \frac{5x^4}{4^3} + \cdots$

11 $\frac{x^2}{(\ln 2)^2} + \frac{x^3}{(\ln 3)^3} + \frac{x^4}{(\ln 4)^4} + \cdots$

12 $1 + x + 2!x^2 + 3!x^3 + \cdots$

13 $\frac{x^2}{2+\ln 2} + \frac{x^3}{3+\ln 3} + \frac{x^4}{4+\ln 4} + \cdots$

14 $\frac{ab}{1\cdot c}x + \frac{a(a+1)b(b+1)}{2!c(c+1)}x^2 + \frac{a(a+1)(a+2)b(b+1)(b+2)}{3!c(c+1)(c+2)}x^3 + \cdots \quad (c > 0)$

15 $4\cdot 5x^4 + 8\cdot 9x^8 + 16\cdot 13x^{12} + 32\cdot 17x^{16} + 64\cdot 21x^{20} + \cdots$

16 $(1+2)x + (1+2+4)x^2 + (1+2+4+8)x^3 + \cdots$

17 $\frac{1\cdot 4}{2\cdot 5}x^2 + \frac{1\cdot 4\cdot 7}{2\cdot 5\cdot 8}x^3 + \frac{1\cdot 4\cdot 7\cdot 10}{2\cdot 5\cdot 8\cdot 11}x^4 + \cdots$

18 $\frac{(2!)^3}{6!}x^2 + \frac{(3!)^3}{9!}x^3 + \frac{(4!)^3}{12!}x^4 + \cdots$

Find the sum of the series and its domain of convergence

19 $1 + (x-3) + (x-3)^2 + (x-3)^3 + \cdots$

20 $\frac{x}{5} + \left(\frac{x}{5}\right)^2 + \left(\frac{x}{5}\right)^3 + \left(\frac{x}{5}\right)^4 + \cdots$

21 $1 - e^x + e^{2x} - e^{3x} + \cdots$

22 $\cos^2 x + \cos^4 x + \cos^6 x + \cos^8 x + \cdots$

23 $\ln x + \ln(\sqrt{x}) + \ln(\sqrt[4]{x}) + \ln(\sqrt[8]{x}) + \cdots$

24 $\frac{1}{x} + \frac{1}{x^2} + \frac{1}{x^3} + \frac{1}{x^4} + \cdots$

Give the interval of convergence of the following modifications of the series $\Sigma\, x^n/n$, with particular attention to the end points.

25 $x + \frac{x^2}{2} - \frac{x^3}{3} - \frac{x^4}{4} + \frac{x^5}{5} + \frac{x^6}{6} - \frac{x^7}{7} - \frac{x^8}{8} + \cdots$

26 $x + \frac{x^2}{2} - \frac{x^3}{3} + \frac{x^4}{4} + \frac{x^5}{5} - \frac{x^6}{6} + \frac{x^7}{7} + \frac{x^8}{8} - \frac{x^9}{9} + \cdots$

27 $x + \frac{x^2}{2} + \frac{x^3}{3} - \frac{x^4}{4} + \frac{x^5}{5} + \frac{x^6}{6} + \frac{x^7}{7} - \frac{x^8}{8} + \cdots$

28 The same idea, but with these signs

$+ + + - - - + + + - - - \cdots$

Find the radius of convergence by the root test

29 $\Sigma_{n=1}^{\infty} 2^{n+(-1)^n} x^n$

30 $\Sigma_{n=1}^{\infty} [3 + (-1)^n] x^n$

31 Give an example of a power series with radius of convergence π.

32 Give an example of a power series that converges at both end points of a finite interval of convergence.

33 Suppose that infinitely many coefficients of a power series are non-zero integers. Show that the radius of convergence is at most 1.

34 Suppose $\Sigma a_n x^n$ has radius of convergence R. If $|b_n| \leq |a_n|$ for each n, what can be said about the radius of convergence of $\Sigma b_n x^n$?

Find the domain of convergence

35 $\displaystyle\sum_{n=1}^{\infty} \frac{(\sin x)^n}{n^2}$

36 $\displaystyle\sum_{n=1}^{\infty} \frac{(x^2 + 1)^n}{n(n + 1)}$

37 $\displaystyle\sum_{n=1}^{\infty} \frac{e^{nx}}{n}$

38 $\displaystyle\sum_{n=1}^{\infty} n^2 (\cos \pi x)^n$

11-2 Taylor Series

The function $f(x) = 1/(1 - x)$ can be expressed as the sum of the power series $\Sigma_0^{\infty} x^n$ for $|x| < 1$. Is this unusual or can other functions be expressed as power series? If they can, how do we find a power series for a given function?

A systematic approach to these questions is based on Taylor polynomials. Let us recall their definition: if $f(x)$ has n derivatives at $x = c$, we associate with $f(x)$ its n-th degree Taylor polynomial at $x = c$,

$$P_n(x) = f(c) + \frac{f'(c)}{1!}(x - c) + \frac{f''(c)}{2!}(x - c)^2 + \cdots + \frac{f^{(n)}(c)}{n!}(x - c)^n$$

We know that $P_n(x)$ approximates $f(x)$ near $x = c$, and we hope that the approximations improve as n increases. This suggests associating with $f(x)$ the infinite series whose partial sums are the Taylor polynomials $P_n(x)$ at $x = c$.

Taylor Series

Suppose $f(x)$ has derivatives of all orders at $x = c$. We associate with $f(x)$ its **Taylor series** at $x = c$:

$$f(c) + \frac{f'(c)}{1!}(x - c) + \frac{f''(c)}{2!}(x - c)^2 + \cdots = \sum_{n=0}^{\infty} \frac{f^{(n)}(c)}{n!}(x - c)^n$$

[We are using the conventions that $f^{(0)}(c) = f(c)$ and $0! = 1$.] ●

Terminology The Taylor series of $f(x)$ at $x = c$ is often called the **Taylor expansion** of $f(x)$ at $x = c$. We speak of **expanding** or **representing** $f(x)$ in a Taylor series. A Taylor series at $x = 0$ is sometimes called a **Maclaurin series.**

The familiar functions e^x, $\sin x$, and $\cos x$ have derivatives of all orders, so they have Taylor series. We worked out their Taylor polynomials at $x = 0$ in Section 9-4. Let us now restate those results in terms of Taylor series.

Function	Taylor series at $x = 0$	Source
e^x	$\sum_{n=0}^{\infty} \frac{x^n}{n!}$	Example 5a, page 493
$\sin x$	$\sum_{n=1}^{\infty} (-1)^{n-1} \frac{x^{2n-1}}{(2n-1)!}$	Example 6a, page 494
$\cos x$	$\sum_{n=0}^{\infty} (-1)^{n} \frac{x^{2n}}{(2n)!}$	Example 6b, page 496

These three Taylor series converge for all x. Consider, for instance, the series for e^x:

$$\sum_{n=0}^{\infty} \frac{x^n}{n!}$$

If $x \neq 0$, then the ratio test applies:

$$\left| \frac{a_n}{a_{n+1}} \right| = \frac{1/n!}{1/(n+1)!} = n + 1 \rightarrow +\infty \quad \text{as} \quad n \rightarrow +\infty$$

Thus the radius of convergence is $R = +\infty$, that is, the series converges for all x. Similarly, the Taylor series for $\sin x$ and $\cos x$ converge for all x.

Example 1

Find the Taylor series of $f(x) = \ln(1 + x)$ at

a $x = 0$ **b** $x = 2$

and determine the interval of convergence of each.

Solution Compute successive derivatives:

$$f'(x) = \frac{1}{1+x}$$

$$f''(x) = \frac{-1}{(1+x)^2}$$

$$f'''(x) = \frac{2}{(1+x)^3}$$

$$f^{(4)}(x) = \frac{-2 \cdot 3}{(1+x)^4} \quad \cdots \quad f^{(n)}(x) = \frac{(-1)^{n-1}(n-1)!}{(1+x)^n}$$

Hence

$$\frac{f^{(n)}(0)}{n!} = \frac{(-1)^{n-1}}{n} \quad \text{and} \quad \frac{f^{(n)}(2)}{n!} = \frac{(-1)^{n-1}}{n \cdot 3^n}$$

Substitute these values into the formula with $c = 0$ in case **a** and $c = 2$ in case **b**. The constant terms in the two series are respectively $\ln(1 + 0) = 0$ and $\ln(1 + 2) = \ln 3$. The resulting Taylor series are

a $$x - \frac{x^2}{2} + \frac{x^3}{3} - \frac{x^4}{4} + \cdots = \sum_{n=1}^{\infty} (-1)^{n-1} \frac{x^n}{n}$$

(Compare Example 3b, page 494)

b $$\ln 3 + \frac{x - 2}{1 \cdot 3} - \frac{(x - 2)^2}{2 \cdot 3^2} + \cdots$$
$$= \ln 3 + \sum_{n=1}^{\infty} \frac{(-1)^{n-1}}{n \cdot 3^n}(x - 2)^n$$

Now find the radius of convergence R by the ratio test. In case **a**, the n-th coefficient is $a_n = (-1)^{n-1}/n$, so

$$\left|\frac{a_n}{a_{n+1}}\right| = \frac{n + 1}{n} \to 1$$

Hence $R = 1$. Since the series diverges at $x = -1$ and converges at $x = 1$, the interval of convergence is $(-1, 1]$. (Compare Example 2a, page 577.)

In case **b**, the n-th coefficient is $a_n = (-1)^{n-1}/3^n n$, so

$$\left|\frac{a_n}{a_{n+1}}\right| = \frac{(n + 1)3^{n+1}}{3^n n} = \frac{3(n + 1)}{n} \to 3$$

Hence $R = 3$. As in case **a**, the series diverges at the left end point, $x = -1$, and converges at the right end point, $x = 5$. Hence the interval of convergence is $(2 - 3, 2 + 3] = (-1, 5]$. ●

Example 2

Find the Taylor series of $f(x) = 1/\sqrt{x}$ at $x = 9$ and determine its radius of convergence.

Solution Compute successive derivatives:

$$f(x) = \frac{1}{x^{1/2}} \qquad f'(x) = \frac{-1}{2x^{3/2}} \qquad f''(x) = \frac{3}{2^2 x^{5/2}}$$

$$f'''(x) = \frac{-3 \cdot 5}{2^3 x^{7/2}} \qquad \cdots$$

$$f^{(n)}(x) = \frac{(-1)^n 1 \cdot 3 \cdot 5 \cdots (2n - 1)}{2^n x^{(2n+1)/2}}$$

It is convenient to express the product of successive odd integers in the numerator of $f^{(n)}(x)$ in terms of factorials, particularly because most scientific calculators have a factorial key. We use the device of multiplying and dividing by the product of successive even integers. For example,

$$1 \cdot 3 \cdot 5 \cdot 7 = \frac{1 \cdot 2 \cdot 3 \cdot 4 \cdot 5 \cdot 6 \cdot 7 \cdot 8}{2 \cdot 4 \cdot 6 \cdot 8} = \frac{8!}{2 \cdot 1 \cdot 2 \cdot 2 \cdot 2 \cdot 3 \cdot 2 \cdot 4} = \frac{8!}{(2 \cdot 2 \cdot 2 \cdot 2)(1 \cdot 2 \cdot 3 \cdot 4)} = \frac{8!}{2^4 \cdot 4!}$$

In general

$$1 \cdot 3 \cdot 5 \cdot \cdots (2n-1) = \frac{1 \cdot 2 \cdot 3 \cdot 4 \cdot 5 \cdot \cdots \cdot (2n-1)(2n)}{2 \cdot 4 \cdot 6 \cdots (2n)} = \frac{(2n)!}{2^n n!}$$

It follows that

$$f^{(n)}(x) = \frac{(-1)^n}{2^n x^{(2n+1)/2}} \cdot \frac{(2n)!}{2^n n!} = \frac{(-1)^n (2n)!}{2^{2n} n! x^{(2n+1)/2}}$$

In particular, since $9^{(2n+1)/2} = 3^{2n+1} = 3 \cdot 3^{2n}$, we have

$$f^{(n)}(9) = \frac{(-1)^n (2n)!}{(n!) 2^{2n} 3^{2n+1}} = \frac{(-1)^n (2n)!}{3(n!) 2^{2n} 3^{2n}} = \frac{(-1)^n (2n)!}{3(n!) 6^{2n}}$$

(Note that this formula is correct for $n = 0$.) Therefore the desired Taylor series is

$$\sum_{n=0}^{\infty} \frac{f^{(n)}(9)}{n!}(x-9)^n = \sum_{n=0}^{\infty} \frac{(-1)^n (2n)!}{3(n!)^2 6^{2n}}(x-9)^n$$

To find its radius of convergence, apply the ratio test:

$$\left|\frac{a_n}{a_{n+1}}\right| = \frac{(2n)!}{3(n!)^2 6^{2n}} \cdot \frac{3[(n+1)!]^2 6^{2n+2}}{(2n+2)!} = \frac{36(n+1)^2}{(2n+1)(2n+2)} \to 9 \quad \text{as} \quad n \to +\infty$$

Hence $R = 9$. ●

Validity of Taylor Series

In the preceding examples, we formally wrote down the Taylor series associated with various functions; we did not prove that the functions actually equal the sums of their Taylor series. There is no guarantee that they do. In fact, there exist functions that are not the sums of their Taylor series. (See Exercises 25–28.)

Suppose $f(x)$ has derivatives of all orders at $x = c$. Write

$$f(x) = f(c) + \frac{f'(c)}{1!}(x-c) + \cdots + \frac{f^{(n)}(c)}{n!}(x-c)^n + R_n(x)$$

Then $f(x)$ is the sum of its Taylor series if and only if the remainders $R_n(x) \to 0$ as $n \to +\infty$. Our trump card in dealing with $R_n(x)$ is Taylor's formula, page 498. It says that

$$R_n(x) = \frac{1}{n!} \int_c^x (x - t)^n f^{(n+1)}(t)\, dt$$

From this follows, as on page 498, a useful estimate for the remainder. Suppose that

$$|f^{(n+1)}(x)| \leq M$$

in some closed interval including c, say $[a, b]$. Then

$$|R_n(x)| \leq \frac{M}{(n+1)!} |x - c|^{n+1}$$

for $a \leq x \leq b$.

Let us now prove that some familiar functions are actually equal to the sums of their Taylor series.

- $$e^x = 1 + x + \frac{x^2}{2!} + \cdots = \sum_{n=0}^{\infty} \frac{x^n}{n!} \quad \text{on } \mathbf{R}$$

- $$\sin x = x - \frac{x^3}{3!} + \frac{x^5}{5!} - \cdots = \sum_{n=1}^{\infty} (-1)^{n-1} \frac{x^{2n-1}}{(2n-1)!} \quad \text{on } \mathbf{R}$$

- $$\cos x = 1 - \frac{x^2}{2!} + \frac{x^4}{4!} - \cdots = \sum_{n=0}^{\infty} (-1)^n \frac{x^{2n}}{(2n)!} \quad \text{on } \mathbf{R}$$

Proof By examples 3 and 4 in Section 9-4, each series is the Taylor series associated with the given function. In each case, we must write the function as $f(x) = P_n(x) + R_n(x)$ and prove that $R_n(x) \to 0$ for all values of x. We shall carry out the proof only for $f(x) = e^x$. The details for $\sin x$ and $\cos x$ are similar.

Thus let $f(x) = e^x$. Assume at first that $-B \leq x \leq B$. In this interval, $|f^{(n+1)}(x)| = e^x \leq e^B$. Apply the estimate for the remainder, with $M = e^B$:

$$|R_n(x)| < e^B \frac{|x|^{n+1}}{(n+1)!} \leq e^B \frac{B^{n+1}}{(n+1)!} \quad \text{for } |x| \leq B$$

Since e^B is fixed, to prove that $R_n(x) \to 0$ as $n \to +\infty$ it is enough to show that $B^n/n! \to 0$. But $B^n/n!$ is the n-th term of the *convergent* series $\Sigma\, B^n/n!$. Therefore it indeed has limit 0 as $n \to +\infty$. Therefore $|R_n(x)| \to 0$ for all $|x| \leq B$. The argument is valid for each positive B, so $R_n(x) \to 0$ for all x. ●

Example 3

Find the sum of the series and its domain of convergence

a $1 - \dfrac{(2x)^2}{2!} + \dfrac{(2x)^4}{4!} - \dfrac{(2x)^6}{6!} + \cdots$

b $\dfrac{x^4}{2!} - \dfrac{x^6}{3!} + \dfrac{x^8}{4!} - \dfrac{x^{10}}{5!} + \cdots$

c $1 + \ln x + \dfrac{(\ln x)^2}{2!} + \dfrac{(\ln x)^3}{3!} + \cdots \qquad (x > 0)$

Solution

a The series has the form

$$1 - \frac{y^2}{2!} + \frac{y^4}{4!} - \frac{y^6}{6!} + \cdots$$

with $y = 2x$. This series converges to $\cos y$ for all real y. Therefore, given any real x, it is legitimate to substitute $y = 2x$. Thus the sum of the given series is $\cos 2x$, for all x.

b The series has the form

$$\left(1 + \frac{y}{1!} + \frac{y^2}{2!} + \frac{y^3}{3!} + \cdots\right) - 1 - y$$

with $y = -x^2$. The series in parentheses converges to e^y for all real values y, in particular for $y = -x^2$. Thus the given series converges to $e^{-x^2} - 1 + x^2$ for all x.

c The series has the form $\Sigma_0^\infty y^n/n!$, which converges to e^y for all real y. In particular, for $y = \ln x$, it converges to $\exp(\ln x) = x$. ●

In the next two sections we discuss ways of establishing the validity of various Taylor series without investigating their remainders. Such methods will be of great use since estimates of $R_n(x)$ can be tricky. For example, proving the validity of the relation

$$\ln(1 + x) = x - \frac{x^2}{2} + \frac{x^3}{3} - \cdots \qquad \text{for } |x| < 1$$

(Example 1) is difficult using remainders, but comes out easily in Section 11-4. Let us accept it for the time being.

Exercises

Find the Taylor series of the function at $x = c$

1 e^{3x} $\quad c = 0$

2 e^x $\quad c = 2$

3 $\cos x$ $\quad c = \frac{1}{4}\pi$

4 $\sin x$ $\quad c = \frac{1}{6}\pi$

5 $\sinh x$ $\quad c = 0$

6 $\sinh x$ $\quad c = 1$

7 $\dfrac{1}{1 - x}$ $\quad c = \frac{1}{3}$

8 $\dfrac{1}{1 - x}$ $\quad c = 3$

9 $\ln(a + x)$ $\quad c = 0 \quad a > 0$

10 $\dfrac{1}{a + bx}$ $\quad c = 0 \quad a, b > 0$

11 $(1 + x)e^x$ $\quad c = 0$

12 $e^{ax} + e^{bx}$ $\quad c = 0$

13 $\sqrt{x}$ $\quad c = 4$

14 $1/x^2$ $\quad c = 1$

15 $5x^3 + 12x - 7$ $\quad c = 0$

16 $x^4 - 6x^2$ $\quad c = -1$

Find the terms through x^5 of the Taylor series at $x = 0$ of

17 $\tan x$

18 $\exp(-x^2)$

Find the sum of the series

19 $1 - \dfrac{(x-1)^2}{1!} + \dfrac{(x-1)^4}{2!} - \dfrac{(x-1)^6}{3!} + \cdots$

20 $\dfrac{\pi^2}{2!}x^2 - \dfrac{\pi^4}{4!}x^4 + \dfrac{\pi^6}{6!}x^6 - + \cdots$

21 Let $\Sigma_0^\infty a_n x^n$ be the Taylor series associated with $f(x)$ at $x = 0$. Find the Taylor series associated with $F(x) = \int_0^x f(t)\,dt$ at $x = 0$.

22 Suppose the derivatives of $f(x)$ have this property: for each interval $|x - c| \le B$, there exist positive constants a and K such that $|f^{(n)}(x)| \le aK^n$. Prove that the Taylor series of $f(x)$ at $x = c$ converges to $f(x)$ for all x. (The constants a and K may depend on B.)

23 Compute the Taylor series of $\cosh x$ at $x = 0$ and prove that it converges to $\cosh x$ for all x. [Hint Use Exercise 22.]

24 Show that the Taylor series of $e^x \cos x$ at $x = 0$ converges to $e^x \cos x$ for all x. Do not compute the series. [Hint Use Exercise 22.]

The following four exercises establish the existence of a function that is not equal to the sum of its Taylor series.

***25** Define

$$f(x) = \begin{cases} e^{-1/x^2} & x \ne 0 \\ 0 & x = 0 \end{cases}$$

Show that $f(x)$ is differentiable at $x = 0$ and that $f'(0) = 0$.

26 (cont.) Let $x \ne 0$. Prove by induction that $f^{(n)}(x)$ is a sum of terms of the form $ae^{-1/x^2}/x^k$ for $n \ge 1$.

27 (cont.) Prove by induction that $f(x)$ has derivatives of all orders at $x = 0$ and that $f^{(n)}(0) = 0$ for $n \ge 1$.

28 (cont.) Conclude that the Taylor series of $f(x)$ at $x = 0$ converges for all x, but not to $f(x)$.

29 Set $S = \Sigma_0^\infty x^n$. Show that $S = 1 + xS$ and deduce the value of S.

30 (cont.) Set $T = \Sigma_1^\infty nx^{n-1}$. Show that $T - S = xT$ and deduce the value of T.

11-3 Expansion of Functions

We have obtained Taylor series for various functions by computing coefficients from the formula $a_n = f^{(n)}(c)/n!$. But computing successive derivatives can be extremely laborious. Try the seventh derivative of $\tan x$ or of $x^2/(1 + x^3)$ and you will soon agree. In some cases we were able to prove the validity of Taylor expansions by showing $R_n(x) \to 0$. But that also can be very hard. To avoid these difficulties, we shall discuss techniques for obtaining new valid Taylor expansions from those already known.

One basic principle underlies all the techniques we shall develop:

Uniqueness of Power Series

Suppose

$$f(x) = a_0 + \sum_{n=1}^{\infty} a_n (x - c)^n$$

in some interval $|x - c| < R$. Then

$$a_n = \frac{f^{(n)}(c)}{n!}$$

Thus if $f(x)$ is expressed as the sum of a power series, then that series must be the Taylor series of $f(x)$ at $x = c$.

This principle says that there is only one possible power series at $x = c$ for a given $f(x)$. Once you find a power series with sum $f(x)$ by any method whatever, then you have its Taylor series.

For example, we showed in Section 11-1 that

$$\frac{1}{1-x} = 1 + x + x^2 + x^3 + \cdots \quad \text{for} \quad |x| < 1$$

but we never actually verified that this series is the Taylor series of $1/(1-x)$ at $x = 0$. Now that follows automatically by the uniqueness principle.

The proof of uniqueness is based on a property of power series that will be discussed in Section 11-4: a power series can be differentiated term by term repeatedly within its interval of convergence. Assuming this, the rest is easy. If $f(x) = \Sigma\, a_k(x-c)^k$, we differentiate n times, then set $x = c$:

$$f^{(n)}(x) = n!\, a_n + \frac{(n+1)!}{1!} a_{n+1}(x-c) + \frac{(n+2)!}{2!} a_{n+2}(x-c)^2 + \cdots$$

so

$$f^{(n)}(c) = n!\, a_n \qquad \text{and} \qquad a_n = \frac{f^{(n)}(c)}{n!}$$

Addition and Subtraction

Our first technique is straightforward.

Addition of Power Series

Suppose

$$f(x) = \sum_{n=0}^{\infty} a_n(x-c)^n \qquad \text{and} \qquad g(x) = \sum_{n=0}^{\infty} b_n(x-c)^n$$

in an interval $|x - c| < R$. Then

$$f(x) \pm g(x) = \sum_{n=0}^{\infty} (a_n \pm b_n)(x-c)^n \quad \text{for} \quad |x-c| < R$$

Thus two power series may be added or subtracted term by term within their common interval of convergence.

Proof Let $s_n(x)$ and $t_n(x)$ denote the partial sums of the two series. Then $s_n(x) \to f(x)$ and $t_n(x) \to g(x)$ provided $|x - c| < R$. By a basic property of limits,

$$\sum_{k=0}^{n} (a_k \pm b_k)(x-c)^k = s_n(x) \pm t_n(x) \to f(x) \pm g(x)$$

In other words

$$\sum_{n=0}^{\infty} (a_n \pm b_n)(x - c)^n = f(x) \pm g(x) \quad \text{for} \quad |x - c| < R$$

●

Example 1

Express as a Taylor series at $x = 0$

a $\cosh x$ **b** $\sinh x$

Solution We know that

$$e^x = 1 + \frac{x}{1!} - \frac{x^2}{2!} + \frac{x^3}{3!} + \frac{x^4}{4!} + \cdots \qquad \text{on } \mathbf{R}$$

Hence

$$e^{-x} = 1 - \frac{x}{1!} + \frac{x^2}{2!} - \frac{x^3}{3!} + \frac{x^4}{4!} - \cdots \qquad \text{on } \mathbf{R}$$

To expand $\cosh x$, add these series term by term, then divide by 2:

$$\cosh x = \tfrac{1}{2}(e^x + e^{-x}) = 1 + \frac{x^2}{2!} + \frac{x^4}{4!} + \frac{x^6}{6!} + \cdots$$

To expand $\sinh x$, subtract the two exponential series and divide by 2:

$$\sinh x = \tfrac{1}{2}(e^x - e^{-x}) = x + \frac{x^3}{3!} + \frac{x^5}{5!} + \frac{x^7}{7!} + \cdots$$

Both expansions are valid on $\mathbf{R}$.

●

A polynomial is itself a power series with infinite radius of convergence. Hence it may be added to or subtracted from any power series. For example,

$$\begin{aligned} e^x + (4 + 3x + x^2) &= \left(1 + \frac{x}{1!} + \frac{x^2}{2!} + \frac{x^3}{3!} + \cdots\right) \\ &\quad + (4 + 3x + x^2) \\ &= 5 + 4x + \tfrac{3}{2}x^2 + \frac{x^3}{3!} + \frac{x^4}{4!} + \cdots \end{aligned}$$

Remark Sums and differences of power series are special cases of *linear combinations:* If

$$f(x) = \sum_{n=0}^{\infty} a_n(x - c)^n \quad \text{and} \quad g(x) = \sum_{n=0}^{\infty} b_n(x - c)^n$$

and A and B are constants, then

$$Af(x) + Bg(x) = \sum_{n=0}^{\infty} (Aa_n + Bb_n)(x - c)^n$$

Figure 11-3-1

If $f(x) = \sum_{m=0}^{\infty} a_m(x - c)^m$

and $g(x) = \sum_{n=0}^{\infty} b_n(x - c)^n$

then the n-th coefficient of $f(x)g(x)$ is obtained by adding all the products on the n-th northeast diagonal.

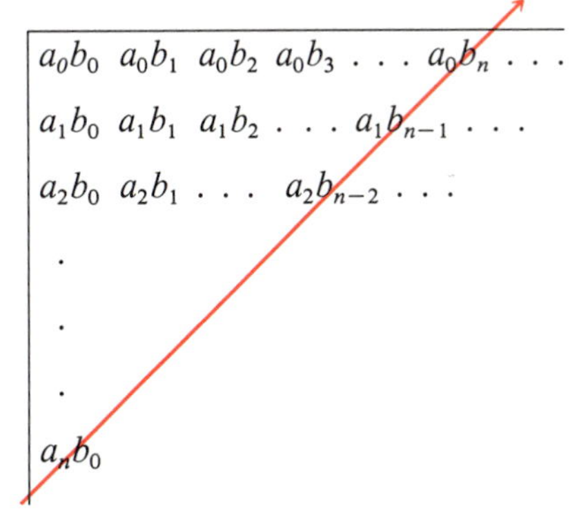

Multiplication

Our next technique is formal multiplication of power series. To simplify notation, let us take $c = 0$. Formal multiplication of two power series $\Sigma\, a_nx^n$ and $\Sigma\, b_nx^n$ is the operation of multiplying each term a_jx^j of the first by each term b_kx^k of the second and collecting terms, just as in multiplication of polynomials. Start with the lower terms and work up:

$$\begin{aligned}(a_0 + a_1x + a_2x^2 + \cdots)(b_0 + b_1x + b_2x^2 + \cdots)\\ = a_0b_0 + (a_0b_1 + a_1b_0)x + (a_0b_2 + a_1b_1 + a_2b_0)x^2 + \cdots\\ + (a_0b_n + a_1b_{n-1} + \cdots + a_kb_{n-k} + \cdots + a_nb_0)x^n\\ + \cdots\end{aligned}$$

In Figure 11-3-1, we show all products of a's by b's. They are added together along the diagonals to form the coefficients of the product function's Taylor series. We summarize the result in sigma notation.

Multiplication of Power Series

Suppose

$$f(x) = \sum_{n=0}^{\infty} a_nx^n \qquad \text{and} \qquad g(x) = \sum_{n=0}^{\infty} b_nx^n$$

both converge for $|x| < R$. Then

$$f(x)g(x) = \sum_{n=0}^{\infty}\left(\sum_{k=0}^{n} a_kb_{n-k}\right)x^n \qquad \text{for } |x| < R$$

Thus two power series may be formally multiplied within their common interval of convergence.

The proof is difficult and best postponed to a more advanced course.

Since a polynomial is also a power series, formal multiplication allows us to deduce by inspection expansions such as

$$x^3\cos x = x^3 - \frac{x^5}{2!} + \frac{x^7}{4!} - \frac{x^9}{6!} + \cdots \qquad \text{for } -\infty < x < \infty$$

$$x^4e^x = x^4 + \frac{x^5}{1!} + \frac{x^6}{2!} + \frac{x^7}{3!} + \cdots \qquad \text{for } -\infty < x < \infty$$

Example 2

Find the Taylor series at $x = 0$ of

$$\left(\frac{1}{1 - x}\right)\ln\left(\frac{1}{1 - x}\right)$$

Solution In Example 1a, page 583, we found the Taylor series for $\ln(1 + x)$ at $x = 0$. Replacing x by $-x$, we have

$$\ln\frac{1}{1 - x} = -\ln(1 - x) = x + \frac{x^2}{2} + \frac{x^3}{3} + \cdots$$

for $|x| < 1$. Hence by formal multiplication,

$$\frac{1}{1-x}\ln\frac{1}{1-x}$$
$$= (1 + x + x^2 + \cdots)\left(x + \frac{x^2}{2} + \frac{x^3}{3} + \cdots\right)$$
$$= x + \left(1 + \frac{1}{2}\right)x^2 + \left(1 + \frac{1}{2} + \frac{1}{3}\right)x^3 + \cdots$$
$$= \sum_{n=1}^{\infty}\left(1 + \frac{1}{2} + \frac{1}{3} + \cdots + \frac{1}{n}\right)x^n \qquad \text{for } |x| < 1$$

Substitution

This is a technique we have used already: if a power series $\Sigma\, a_n x^n$ converges for $|x| < R$ and if $|g(x)| < R$, then we may substitute $g(x)$ for x.

Example 3

a $$\frac{1}{1+x^2} = 1 + (-x^2) + (-x^2)^2 + (-x^2)^3 + \cdots$$
$$= 1 - x^2 + x^4 - x^6 + \cdots \qquad \text{for } |x| < 1$$

b $$\frac{1}{1-2x^3} = 1 + 2x^3 + 4x^6 + 8x^9 + \cdots \qquad \text{for } |x| < \sqrt[3]{\tfrac{1}{2}}$$

c $$e^{-x^2/2} = 1 + \left(-\frac{x^2}{2}\right) + \frac{1}{2!}\left(-\frac{x^2}{2}\right)^2 + \frac{1}{3!}\left(-\frac{x^2}{2}\right)^3 + \cdots$$
$$= 1 - \frac{x^2}{2} + \frac{x^4}{2^2 \cdot 2!} - \frac{x^6}{2^3 \cdot 3!} + \cdots \qquad \text{for } |x| < \infty$$

Odd and Even Functions

The power series at $x = 0$ of the odd function $\sin x$ involves only odd powers of x; the power series of the even function $\cos x$ involves only even powers of x. These examples illustrate a general principle:

Power Series of Odd and Even Functions

If $f(x)$ is an odd function, $f(-x) = -f(x)$, then its Taylor series at $x = 0$ has the form

$$a_1x + a_3x^3 + a_5x^5 + a_7x^7 + \cdots$$

If $f(x)$ is an even function, $f(-x) = f(x)$, then its Taylor series at $x = 0$ has the form

$$a_0 + a_2x^2 + a_4x^4 + a_6x^6 + \cdots$$

Thus the Taylor series at $x = 0$ of an odd (even) function contains only odd (even) powers of x.

The proof follows easily from the basic fact that the derivative of an odd function is even and the derivative of an even function is odd. See Exercise 31, and see Exercises 33 and 34 for another proof.

Not all functions are even or odd. However, each function defined in an interval $|x| < R$ can be expressed as the sum of an even function and an odd function. If the function $f(x)$ is defined in an interval $|x| < R$, then we can write

$$\begin{aligned} f(x) &= \tfrac{1}{2}[f(x) + f(-x)] + \tfrac{1}{2}[f(x) - f(-x)] \\ &= g(x) + h(x) \end{aligned}$$

where $g(x)$ is an even function and $h(x)$ is an odd function. If the Taylor series of $f(x)$ is

$$f(x) = \sum_{n=0}^{\infty} a_n x^n$$

then

$$g(x) = \sum_{n=0}^{\infty} a_{2n} x^{2n} \quad \text{and} \quad h(x) = \sum_{n=0}^{\infty} a_{2n+1} x^{2n+1}$$

Thus the Taylor series at $x = 0$ of $f(x)$ splits into two power series, one corresponding to the even part of $f(x)$ and the other corresponding to the odd part of $f(x)$.

Example 4

$$\begin{aligned} f(x) = e^x &= 1 + \frac{x}{1!} + \frac{x^2}{2!} + \frac{x^3}{3!} + \frac{x^4}{4!} + \cdots \\ &= g(x) + h(x) \end{aligned}$$

The even part of e^x is

$$\begin{aligned} g(x) &= \tfrac{1}{2}[f(x) + f(-x)] \\ &= \tfrac{1}{2}[e^x + e^{-x}] = \cosh x \\ &= 1 + \frac{x^2}{2!} + \frac{x^4}{4!} + \cdots \end{aligned}$$

The odd part of e^x is

$$\begin{aligned} h(x) &= \tfrac{1}{2}[f(x) - f(-x)] \\ &= \tfrac{1}{2}[e^x - e^{-x}] = \sinh x \\ &= \frac{x}{1!} + \frac{x^3}{3!} + \frac{x^5}{5!} + \cdots \end{aligned}$$

Clearly $e^x = \cosh x + \sinh x$. ●

Example 5

Find the sum of the series

$$h(x) = x + \tfrac{1}{3}x^3 + \tfrac{1}{5}x^5 + \tfrac{1}{7}x^7 + \cdots$$

Solution This is the odd part of the series

$$f(x) = -\ln(1 - x) = x + \tfrac{1}{2}x^2 + \tfrac{1}{3}x^3 + \tfrac{1}{4}x^4 + \cdots$$

according to Example 2. Therefore

$$\begin{aligned} h(x) &= \tfrac{1}{2}[f(x) - f(-x)] = \tfrac{1}{2}[-\ln(1 - x) + \ln(1 + x)] \\ &= \tfrac{1}{2}\ln\left(\frac{1 + x}{1 - x}\right) \end{aligned}$$

●

Our final example uses another technique, sometimes called the method of undetermined coefficients.

Example 6

Compute the terms up to x^7 in the Taylor series for $\tan x$ at $x = 0$.

Solution In theory, the problem requires long division of the series for $\sin x$ by the series for $\cos x$. In practice, the long division is carried out by assuming an expansion

$$\tan x = a_1x + a_3x^3 + a_5x^5 + a_7x^7 + \cdots$$

(Only odd powers are necessary since $\tan x$ is an odd function.) Now write the identity $\tan x \cos x = \sin x$ in terms of power series:

$$\begin{aligned} (a_1x + a_3x^3 + a_5x^5 + a_7x^7 + \cdots)\left(1 - \frac{x^2}{2} + \frac{x^4}{24} - \frac{x^6}{720} + \cdots\right) \\ = x - \frac{x^3}{6} + \frac{x^5}{120} - \frac{x^7}{5040} + \cdots \end{aligned}$$

Multiply the two series on the left, then equate coefficients:

$$\begin{aligned} &x \text{ terms:} && a_1 = 1 \\ &x^3 \text{ terms:} && a_3 - \tfrac{1}{2}a_1 = -\tfrac{1}{6} \\ &x^5 \text{ terms:} && a_5 - \tfrac{1}{2}a_3 + \tfrac{1}{24}a_1 = \tfrac{1}{120} \\ &x^7 \text{ terms:} && a_7 - \tfrac{1}{2}a_5 + \tfrac{1}{24}a_3 - \tfrac{1}{720}a_1 = -\tfrac{1}{5040} \end{aligned}$$

Solve these equations successively for $a_1,\ a_3,\ a_5,\ a_7$:

$$a_1 = 1 \qquad a_3 = \tfrac{1}{3} \qquad a_5 = \tfrac{2}{15} \qquad a_7 = \tfrac{17}{315}$$

Hence

$$\tan x = x + \tfrac{1}{3}x^3 + \tfrac{2}{15}x^5 + \tfrac{17}{315}x^7 + \cdots$$

●

Remark It can be shown that the Taylor series for $\tan x$ at $x = 0$ converges to $\tan x$ for $|x| < \frac{1}{2}\pi$. This is the largest interval about $x = 0$ in which the denominator $\cos x$ is non-zero.

There is a list of Taylor series for familiar functions inside the front cover. Use it as needed for the exercises.

Exercises

Find the Taylor series of the given function at $x = 0$

1 $\dfrac{1}{1 - 5x^2}$

2 $\dfrac{x}{1 - x^3}$

3 $\dfrac{1 - x}{1 + x}$

4 $\dfrac{1 + x^2}{1 + x^4}$

5 $x(\sin x + \sin 3x)$

6 $\sinh x + \cosh x$

7 $\dfrac{1 - \cos x}{x^2}$

8 $\dfrac{\sin x - x}{x^3}$

9 $\dfrac{2 - 2x + 3x^2}{1 - x}$

10 $(x^2 - 1)\cosh x$

11 $\sin^2 x$

12 $\dfrac{1}{1 + x + x^2 + x^3}$

Compute the terms up to x^6 in the Taylor series at $x = 0$

13 $\dfrac{1}{(1 - 2x^2)(1 - 3x^2)}$

14 $e^x \sin(x^2)$

15 $\dfrac{1}{1 - x^2 e^x}$

16 $\dfrac{1}{1 + x + x^3}$

17 $\sin^3 x$

18 $\ln \cos x$

19 $e^{x^3} \cos x$

20 $x^2 \cot x$

Compute $f^{(7)}(0)$

21 $\dfrac{2x + 3}{1 - x^3}$

22 $x \cos \frac{1}{2}x$

23 $\tan x$

24 $\dfrac{x^2}{1 + x^3}$

25 If $f(x) = \Sigma_0^\infty a_n x^n$, show that

$$\frac{1}{1 - x} f(x) = \sum_{n=0}^{\infty} (a_0 + a_1 + \cdots + a_n) x^n$$

26 (cont.) Find the sum of the series $\Sigma_0^\infty (n + 1)x^n$

Find the sum of the series.

27 $x + x^2 - x^3 + x^4 + x^5 - x^6 + + - \cdots$

28 $\dfrac{1}{2!} + \dfrac{x}{3!} + \dfrac{x^2}{4!} + \dfrac{x^3}{5!} + \cdots$

***29** $1 + \dfrac{x^4}{4!} + \dfrac{x^8}{8!} + \dfrac{x^{12}}{12!} + \cdots$

30 $\dfrac{x^2}{2} + \dfrac{x^4}{4} + \dfrac{x^6}{6} + \dfrac{x^8}{8} + \cdots$

31 Prove that the derivative of an odd (even) function is even (odd). Conclude that the Taylor series at $x = 0$ of an even (odd) function contains only even (odd) powers of x.

32 Prove that a function defined for $|x| < R$ can be written in *only* one way as the sum of an even function and an odd function.

33 Suppose $f(x) = \Sigma_0^\infty a_n x^n$ is even. Use *uniqueness* to prove that $a_{2n-1} = 0$.

34 (cont.) Give the corresponding proof for odd functions.

35 Set $f(x) = \Sigma_0^\infty n x^n$. Compute $(1 - x)f(x)$ and use your result to find $f(x)$.

36 (cont.) Set $g(x) = \Sigma_0^\infty n(n + 1)x^n$ and use the same technique to find $g(x)$.

37 Use $\sec x = 1/[1 - (1 - \cos x)]$ to find the terms up to x^6 in the Taylor series of $\sec x$ at $x = 0$.

38 Suppose the coefficients of $f(x) = \Sigma_0^\infty a_n x^n$ are **periodic** of period p. That is, p is a positive integer and $a_{n+p} = a_n$ for all n. Prove that $f(x)$ is a rational function.

39 Find the terms up to x^5 of the expansion at $x = 0$ of $f(x) = \sin x + \tan(\sin x)$.

40 Show that

$$f(x) = \begin{cases} \dfrac{x}{e^x - 1} & \text{for } x \neq 0 \\ 1 & \text{for } x = 0 \end{cases}$$

has a Taylor series

$$f(x) = \sum_{n=0}^{\infty} \frac{B_n}{n!} x^n$$

This relation defines the **Bernoulli numbers.** Compute B_0 through B_3.

11-4 Calculus of Power Series

Differentiation and integration are important techniques for deriving new power series from known ones. We state the main results, but omit their proof.

Term-by-Term Differentiation and Integration

Suppose

$$f(x) = a_0 + a_1 x + a_2 x^2 + \cdots = \sum_{n=0}^{\infty} a_n x^n$$

for $|x| < R$. Then

$$f'(x) = a_1 + 2a_2 x + 3a_3 x^2 + \cdots = \sum_{n=1}^{\infty} n a_n x^{n-1}$$

and

$$\int_0^x f(t)\,dt = a_0 x + \frac{a_1}{2} x^2 + \frac{a_2}{3} x^3 + \cdots = \sum_{n=0}^{\infty} \frac{a_n}{n+1} x^{n+1}$$

hold for $|x| < R$. In words, a power series may be differentiated or integrated term by term within its interval of convergence.

Example 1

Find the Taylor series at $x = 0$ for $1/(1-x)^2$ and $1/(1-x)^3$.

Solution First observe that

$$\frac{1}{(1-x)^2} = \frac{d}{dx}\left(\frac{1}{1-x}\right)$$

Then expand $1/(1-x)$ in a Taylor series and differentiate term by term. For $|x| < 1$,

$$\begin{aligned}
\frac{1}{(1-x)^2} &= \frac{d}{dx}(1 + x + x^2 + \cdots + x^n + \cdots) \\
&= \frac{d}{dx}(1) + \frac{d}{dx}(x) + \frac{d}{dx}(x^2) + \cdots \\
&= 1 + 2x + 3x^2 + \cdots + nx^{n-1} + \cdots \\
&= \sum_{n=1}^{\infty} n x^{n-1} = \sum_{n=0}^{\infty} (n+1) x^n
\end{aligned}$$

Differentiate again:

$$\begin{aligned}
\frac{2}{(1-x)^3} &= \frac{d}{dx}\left[\frac{1}{(1-x)^2}\right] \\
&= \frac{d}{dx}(1 + 2x + 3x^2 + \cdots + nx^{n-1} + \cdots)
\end{aligned}$$

Hence

$$\frac{1}{(1-x)^3}$$

$$= \tfrac{1}{2}[2 + 6x + 12x^2 + \cdots + n(n-1)x^{n-2} + \cdots]$$

$$= \sum_{n=2}^{\infty} \frac{n(n-1)}{2} x^{n-2} = \sum_{n=0}^{\infty} \frac{(n+2)(n+1)}{2} x^n$$

Example 2

Find the sum of the series

$$x + \frac{x^3}{3} + \frac{x^5}{5} + \cdots + \frac{x^{2n-1}}{2n-1} + \cdots$$

Solution By the ratio test, the series converges for $|x| < 1$ to some function

$$f(x) = x + \frac{x^3}{3} + \frac{x^5}{5} + \cdots + \frac{x^{2n-1}}{2n-1} + \cdots$$

Each term is the integral of a power of x. This suggests that $f(x)$ is the integral of some simple function. Differentiate term by term:

$$f'(x) = 1 + x^2 + x^4 + \cdots + x^{2n-2} + \cdots = \frac{1}{1-x^2}$$

Therefore, $f(x)$ is an antiderivative of $1/(1-x^2)$. Since $f(0) = 0$, it follows that

$$f(x) = \int_0^x \frac{dt}{1-t^2} = \tfrac{1}{2}\ln \frac{1+t}{1-t}\bigg|_0^x = \tfrac{1}{2}\ln \frac{1+x}{1-x}$$

for $|x| < 1$.

Example 3

Find the Taylor series at $x = 0$ of $\ln(1+x)$.

Solution

$$\frac{d}{dx}\ln(1+x) = \frac{1}{1+x} = 1 - x + x^2 - x^3 + \cdots$$

for $|x| < 1$. Integrate term by term:

$$\ln(1+x) = \int_0^x \frac{dt}{1+t} = \int_0^x [1 - t + t^2 - + \cdots + (-1)^n t^n + \cdots]\, dt$$

$$= \int_0^x dt - \int_0^x t\, dt + \int_0^x t^2\, dt - + \cdots + (-1)^n \int_0^x t^n\, dt + \cdots$$

Therefore for $|x| < 1$,

$$\ln(1 + x) = x - \frac{x^2}{2} + \frac{x^3}{3} - \cdots + (-1)^n \frac{x^{n+1}}{n + 1} + \cdots$$

$$= \sum_{n=1}^{\infty} (-1)^{n-1} \frac{x^n}{n}$$

●

Example 4

Find the Taylor series at $x = 0$ of arc tan x.

Solution

$$\frac{d}{dx} \text{arc tan } x = \frac{1}{1 + x^2} = 1 - x^2 + x^4 - + \cdots$$

for $|x| < 1$. Integrate term by term:

$$\text{arc tan } x = \int_0^x \frac{dt}{1 + t^2} = \int_0^x (1 - t^2 + t^4 - + \cdots)\, dt$$

$$= x - \frac{x^3}{3} + \frac{x^5}{5} - + \cdots = \sum_{n=0}^{\infty} (-1)^n \frac{x^{2n+1}}{2n + 1}$$

for $|x| < 1$. ●

Example 5

Find the sum of the series

$$x + 4x^2 + 9x^3 + \cdots = \sum_{n=1}^{\infty} n^2 x^n$$

Solution By the ratio test, the series converges for $|x| < 1$ to a function $f(x)$. Write

$$f(x) = x + 4x^2 + 9x^3 + \cdots + n^2 x^n + \cdots = xg(x)$$

where

$$g(x) = 1 + 2^2 x + 3^2 x^2 + \cdots + n^2 x^{n-1} + \cdots$$

Integrate term by term:

$$\int_0^x g(t)\, dt = x + 2x^2 + 3x^3 + \cdots$$

$$= x(1 + 2x + 3x^2 + \cdots) = \frac{x}{(1 - x)^2}$$

by Example 1. Now differentiate to recover $g(x)$:

$$g(x) = \frac{d}{dx}\left[\frac{x}{(1 - x)^2}\right] = \frac{1 + x}{(1 - x)^3}$$

Therefore

$$f(x) = xg(x) = \frac{x + x^2}{(1 - x)^3} \quad \text{for} \quad |x| < 1$$

●

An Application to Probability [Optional]

Imagine an experiment in which the outcome is one of the integers 0, 1, 2, · · · , such as counting the number of cars that pass a certain spot in a given hour, or the number of tails that precede the first head in a sequence of tosses of a coin. Suppose the probability of the outcome n is a number p_n, where $0 \leq p_n \leq 1$ and $\Sigma_0^\infty p_n = 1$. We say that $\{p_n\}$ is a **probability distribution** on the set $\{0, 1, 2, \cdots\}$.

The **expected value** (also called the **mean**) of this distribution is defined to be

$$E = \sum_{n=0}^{\infty} np_n$$

if the series converges; otherwise $E = +\infty$. Since $\Sigma p_n = 1$, we can think of E as a weighted average: each integer is weighted by the probability that it will occur. If E is finite, then the **variance** of the distribution is defined to be

$$\sigma^2 = \sum_{n=0}^{\infty} (n - E)^2 p_n$$

It is a measure of the spread of the various outcomes about their expected value.

Since $\Sigma p_n = 1$ and $\Sigma np_n = E$, we have

$$\begin{aligned}\sigma^2 &= \sum_{n=0}^{\infty} (n^2 - 2nE + E^2) p_n \\ &= \sum_{n=0}^{\infty} n^2 p_n - 2E \sum_{n=0}^{\infty} np_n + E^2 \sum_{n=0}^{\infty} p_n \\ &= \sum_{n=0}^{\infty} n^2 p_n - 2E \cdot E^2 \cdot 1 = \sum_{n=0}^{\infty} n^2 p_n - E^2\end{aligned}$$

The quantity $\Sigma n^2 p_n$ is called the **second moment** of the distribution. Sometimes the mean, Σnp_n, is called the **first moment.**

Example 6

In a sequence of throws of a die, what is the expected value of the number of the throw at which the first 4 occurs?

Solution The probability of a 4 turning up is $\frac{1}{6}$; the probability of any other number turning up is $\frac{5}{6}$. The first 4 occurs at throw n provided that the first $n - 1$ throws are not 4, but the n-th throw is 4. The probability of this event is

$$p_n = \tfrac{5}{6} \cdot \tfrac{5}{6} \cdots \tfrac{5}{6} \cdot \tfrac{1}{6} = \tfrac{1}{6}(\tfrac{5}{6})^{n-1} \qquad \text{for} \quad n = 1, 2, 3, \cdots$$

(You should confirm that $\Sigma_1^\infty p_n = 1$.) By definition, the expected value of the number of the throw at which a 4 first shows up is

$$E = \sum_{n=1}^{\infty} np_n = \sum_{n=1}^{\infty} n \cdot \tfrac{1}{6}(\tfrac{5}{6})^{n-1} = \tfrac{1}{6}\sum_{n=1}^{\infty} n(\tfrac{5}{6})^{n-1}$$

We can find this sum explicitly. According to Example 1,

$$\sum_{n=1}^{\infty} nx^{n-1} = \frac{1}{(1-x)^2} \qquad \text{hence} \qquad E = \tfrac{1}{6}\cdot\frac{1}{(1-\frac{5}{6})^2} = 6$$

The expected number of throws is 6. ●

Example 7

Compute the variance σ^2 of the probability distribution in Example 6.

Solution Compute the second moment:

$$\sum_{n=1}^{\infty} n^2 p_n = \tfrac{1}{6}\sum_{n=1}^{\infty} n^2(\tfrac{5}{6})^{n-1} = \tfrac{1}{6}\cdot\tfrac{6}{5}\sum_{n=1}^{\infty} n^2(\tfrac{5}{6})^n$$

By Example 5,

$$\sum_{n=1}^{\infty} n^2(\tfrac{5}{6})^n = \frac{\frac{5}{6}(1+\frac{5}{6})}{(1-\frac{5}{6})^3} = 5\cdot 36(1+\tfrac{5}{6})$$

Since $E = 6$,

$$\sigma^2 = \sum_{n=1}^{\infty} n^2 p_n - E^2 = \tfrac{1}{5}\cdot 5\cdot 36(1+\tfrac{5}{6}) - 36$$
$$= 36\cdot\tfrac{5}{6} = 30$$ ●

Exercises

Verify by expressing both sides as power series

1 $\dfrac{d}{dx}(\sin x) = \cos x$

2 $\dfrac{d}{dx}(e^{kx}) = ke^{kx}$

3 $\dfrac{d^2}{dx^2}(\cosh kx) = k^2\cosh kx$

4 $\dfrac{d^2}{dx^2}(xe^x) = (x+2)e^x$

5 $\displaystyle\int_0^x \frac{2t\,dt}{1-t^2} = -\ln(1-x^2)$ for $|x| < 1$

6 $\displaystyle\int_0^x \arctan t\,dt = x\arctan x - \tfrac{1}{2}\ln(1+x^2)$ for $|x| < 1$

7 $\displaystyle\int_0^x \frac{dt}{(1+t^2)^2} = \frac{1}{2}\left[\frac{x}{1+x^2} + \arctan x\right]$

8 $\displaystyle\int_0^x t^2\sin t\,dt = 2x\sin x + (2-x^2)\cos x - 2$

Find the sum of the power series by using differentiation or integration

9 $4 + 5x + 6x^2 + 7x^3 + \cdots$

10 $1 + 4x + 9x^2 + 16x^3 + 25x^4 + \cdots$

11 $\dfrac{x^4}{4} + \dfrac{x^8}{8} + \dfrac{x^{12}}{12} + \dfrac{x^{16}}{16} + \cdots$

12 $\dfrac{x^2}{1\cdot 2} - \dfrac{x^3}{2\cdot 3} + \dfrac{x^4}{3\cdot 4} - \dfrac{x^5}{4\cdot 5} + \cdots$

13 $x - \dfrac{x^3}{3^2} + \dfrac{x^5}{5^2} - \dfrac{x^7}{7^2} + \cdots$
[Hint Express as an integral.]

14 $x + 2^3x^2 + 3^3x^3 + 4^3x^4 + \cdots$
[Hint Use Exercise 10.]

15 $\sum_{n=0}^{\infty} (2n+1)x^n$

16 $\sum_{n=0}^{\infty} (n+2)(n+1)x^n$

17 $\sum_{n=0}^{\infty} \dfrac{nx^{n+1}}{n+1}$

*18 $\sum_{n=0}^{\infty} \dfrac{nx^{n+1}}{n+2}$

Find the sum of the numerical series

19 $\dfrac{3}{2 \cdot 1!} - \dfrac{5}{2^2 \cdot 2!} + \dfrac{7}{2^3 \cdot 3!} - \dfrac{9}{2^4 \cdot 4!} + \cdots$
[Hint Start with the power series for $e^{-x^2/2}$.]

20 $\dfrac{1}{2!} + \dfrac{2}{3!} + \dfrac{3}{4!} + \dfrac{4}{5!} + \cdots$

21 A fair coin is tossed repeatedly until a head first appears. Find the expected value of the number of tosses.

22 A coin is tossed repeatedly until either two consecutive heads or two consecutive tails occur. What is the expected value of the number of tosses?

23 A coin is tossed repeatedly until both a head and a tail have appeared. Find the expected value of the number of tosses.

24 (cont.) Find the variance of the probability distribution that you found in Exercise 23.

25 Let's play this game. We toss a coin. If heads, I pay you \$1 and the game ends. If tails, you pay me \$1 and we toss again. This time, if it's heads, I pay you \$2 and the game ends. If it's tails, you pay me \$2 and we play again for \$3, and so on. Compute your expected gain or loss, $\Sigma_1^\infty a_n p_n$, where a_n is your payoff (gain or loss) at throw n, and p_n is the probability that the game ends at throw n.

26 A certain dime slot machine has three identical wheels, each with ten different fruits. You continue to play without gain until you hit three lemons, when the payoff is the jackpot, \$$J$. Find J so the game is fair, assuming all outcomes of the spinning wheels are equally likely.

11-5 Binomial Series [Optional]

The binomial theorem asserts that for each positive integer p,

$$(1+x)^p = 1 + px + \frac{p(p-1)}{2!}x^2 + \frac{p(p-1)(p-2)}{3!}x^3 + \cdots + \frac{p(p-1)(p-2)\cdots(p-n+1)}{n!}x^n + \cdots + \frac{p!}{p!}x^p$$

Standard notation for the binomial coefficients is

$$\binom{p}{0} = 1 \qquad \binom{p}{n} = \frac{p!}{n!(p-n)!} = \frac{p(p-1)(p-2)\cdots(p-n+1)}{n!}$$

where $1 \le n \le p$. With this notation the expansion of $(1+x)^p$ can be abbreviated:

$$(1+x)^p = \sum_{n=0}^{p} \binom{p}{n} x^n$$

A generalization of the binomial theorem is the **binomial series** for $(1+x)^p$, where p is not necessarily a positive integer.

Binomial Series

For any number p, and $|x| < 1$,

$$(1 + x)^p = \sum_{n=0}^{\infty} \binom{p}{n} x^n$$

where the coefficients are

$$\binom{p}{0} = 1$$

$$\binom{p}{n} = \frac{p(p-1)(p-2)\cdots(p-n+1)}{n!} \qquad n \geq 1$$

Remark In case p happens to be a positive integer and $n > p$, then the coefficient $\binom{p}{n}$ equals 0 because it has a factor $(p - p)$. In this case, the series breaks off after the term in x^p. The resulting formula,

$$(1 + x)^p = \sum_{n=0}^{p} \binom{p}{n} x^n$$

is the old binomial theorem again. But if p is not a positive integer or zero, then each coefficient is non-zero, so the series has infinitely many terms.

The binomial series is just the Taylor series for $y(x) = (1 + x)^p$ at $x = 0$. Indeed,

$$y'(x) = p(1 + x)^{p-1}$$

$$y''(x) = p(p-1)(1 + x)^{p-2}$$

$$\cdots\cdots\cdots\cdots\cdots\cdots$$

$$y^{(n)}(x) = p(p-1)(p-2)\cdots(p-n+1)(1 + x)^{p-n}$$

Therefore the coefficient of x^n in the Taylor series is

$$\frac{y^{(n)}(0)}{n!} = \frac{p(p-1)(p-2)\cdots(p-n+1)}{n!} = \binom{p}{n}$$

The binomial series converges for $|x| < 1$. When p is a non-negative integer this is obvious because the series terminates. When p is not a non-negative integer the ratio test applies:

$$\left|\frac{a_n}{a_{n+1}}\right| = \left|\binom{p}{n} \Big/ \binom{p}{n+1}\right| = \left|\frac{n+1}{p-n}\right| \to 1$$

so the radius of convergence is 1. This, however, does not prove that the sum of the series is $(1 + x)^p$. That requires a delicate piece of analysis beyond the scope of this course.

Example 1

Find the Taylor series for $1/(1 + x)^2$ at $x = 0$.

Solution Use the binomial series with $p = -2$. The coefficient of x^n is

$$\binom{-2}{n} = \frac{(-2)(-3)(-4)\cdots(-2-n+1)}{n!}$$
$$= (-1)^n \frac{2\cdot 3\cdot 4\cdots n\cdot(n+1)}{n!}$$
$$= (-1)^n(n+1)$$

Hence for $|x| < 1$,

$$\frac{1}{(1+x)^2} = \sum_{n=0}^{\infty}\binom{-2}{n}x^n$$
$$= \sum_{n=0}^{\infty}(-1)^n(n+1)x^n$$
$$= 1 - 2x + 3x^2 - 4x^3 + \cdots$$

Check:

$$\frac{1}{(1+x)^2} = -\frac{d}{dx}\left(\frac{1}{1+x}\right)$$
$$= -\frac{d}{dx}(1 - x + x^2 - x^3 + - \cdots)$$
$$= -(-1 + 2x - 3x^2 + - \cdots)$$

●

The following formulas will be useful in the next examples. The first is valid for $n \geq 0$, the second for $n > 0$:

$$\binom{-\frac{1}{2}}{n} = \frac{(-1)^n(2n)!}{4^n(n!)^2}$$
$$\binom{\frac{1}{2}}{n} = \frac{(-1)^{n-1}(2n)!}{4^n(2n-1)(n!)^2}$$

To derive these, we start with the formula

$$1\cdot 3\cdot 5\cdots(2n-1) = \frac{(2n)!}{2^n n!}$$

which we derived on page 585. Now we have

$$\binom{-\frac{1}{2}}{n} = \frac{\left(-\frac{1}{2}\right)\left(-\frac{3}{2}\right)\left(-\frac{5}{2}\right)\cdots\left(-\frac{2n-1}{2}\right)}{n!}$$
$$= (-1)^n\frac{1\cdot 3\cdot 5\cdots(2n-1)}{2^n n!} = (-1)^n\frac{(2n)!}{2^{2n}(n!)^2}$$
$$= \frac{(-1)^n}{2^n n!}\cdot\frac{(2n)!}{2^n n!} = \frac{(-1)^n(2n)!}{4^n(n!)^2}$$

The formula for $\binom{\frac{1}{2}}{n}$ is derived similarly, or by substituting $p = \frac{1}{2}$ in the identity

$$\binom{p}{n} = \frac{p}{n}\binom{p-1}{n-1}$$

Example 2

Find the Taylor series for $\sqrt{1-x}$ at $x = 0$.

Solution Use the binomial series with $p = \frac{1}{2}$ and x replaced by $-x$. The constant term in the series is

$$\binom{\frac{1}{2}}{0} = 1$$

and the term in x^n is

$$\binom{\frac{1}{2}}{n}(-x)^n = \frac{-(2n)!}{4^n(2n-1)(n!)^2}x^n$$

Therefore

$$\sqrt{1-x} = 1 - \sum_{n=1}^{\infty} \frac{(2n)!}{(2^n \cdot n!)^2(2n-1)}x^n \quad \text{for} \quad |x| < 1$$

●

Example 3

Find the Taylor series for $1/\sqrt{x}$ at $x = 9$.

Solution Write $\sqrt{x} = \sqrt{9 + (x-9)}$. Then

$$\begin{aligned}
\frac{1}{\sqrt{x}} &= \frac{1}{3\sqrt{1 + \frac{1}{9}(x-9)}} \\
&= \frac{1}{3}\left(1 + \frac{x-9}{9}\right)^{-1/2} \\
&= \frac{1}{3}\sum_{n=0}^{\infty}\binom{-\frac{1}{2}}{n}\frac{(x-9)^n}{9^n}
\end{aligned}$$

By our previous formula for the binomial coefficient,

$$\frac{1}{\sqrt{x}} = \sum_{n=0}^{\infty} \frac{(-1)^n(2n)!}{3 \cdot 2^{2n} \cdot 9^n(n!)^2}(x-9)^n$$

This answer checks with Example 2, page 584. ●

Example 4

Find the Taylor series for $\arcsin x$ at $x = 0$.

Solution Expand its derivative $1/\sqrt{1-x^2}$ in a power series and integrate term by term. For $|x| < 1$,

$$\begin{aligned}\text{arc sin } x &= \int_0^x \frac{dt}{\sqrt{1-t^2}} \\ &= \int_0^x \sum_{n=0}^{\infty} \binom{-\frac{1}{2}}{n} (-t^2)^n \, dt \\ &= \sum_{n=0}^{\infty} (-1)^n \binom{-\frac{1}{2}}{n} \int_0^x t^{2n} \, dt \\ &= \sum_{n=0}^{\infty} (-1)^n \binom{-\frac{1}{2}}{n} \frac{x^{2n+1}}{2n+1} \\ &= \sum_{n=0}^{\infty} \frac{(2n)!}{2^{2n}(n!)^2(2n+1)} x^{2n+1}\end{aligned}$$

Remark The formula can also be written

$$\text{arc sin } x = x + \frac{x^3}{2 \cdot 3} + \frac{1 \cdot 3}{2 \cdot 4 \cdot 5} x^5 + \frac{1 \cdot 3 \cdot 5}{2 \cdot 4 \cdot 6 \cdot 7} x^7 + \cdots$$

Exercises

Expand in a power series at $x = 0$

1 $\dfrac{1}{(1+x)^3}$

2 $\dfrac{1}{(1-x)^{1/3}}$

3 $\dfrac{1}{(1-4x^2)^2}$

4 $\dfrac{1}{(3-4x^2)^2}$

5 $(1+2x^3)^{1/4}$

6 $\sqrt{2-x}$

7 $\dfrac{x^2}{\sqrt{1+2x}}$

8 $\left(\dfrac{x}{1-x}\right)^{10}$

9 $\sinh^{-1} x$

10 $\sqrt{1+2x^2}$

11 $\sqrt{1-\frac{1}{2}x^2}$

12 $(1-4x)^{-1/2}$

Expand in a power series at $x = 1$

13 $\sqrt{1+x}$

14 $1/(3+x)^2$

Compute to four-place accuracy, using the binomial series

15 $\sqrt{16.1}$

16 $\sqrt[4]{82}$

17 $1/(1.03)^5$

18 $1/\sqrt[3]{970}$

19 Starting with $2 \cdot 71^2 = 10^4 + 82$, show that $\sqrt{2} = \frac{100}{71}(1+\varepsilon)^{1/2}$, where $\varepsilon = 0.0082$. Estimate $\sqrt{2}$ by expanding up to ε^2. You may use a calculator without $\sqrt{\ }$.

20 (cont.) Estimate $\sqrt{3}$ similarly.
[Hint Find an integer k for which $3k^2$ is near 10^4.]

21 Let $p_n = 2n \sin(\pi/n)$ be the perimeter of the regular polygon of n sides inscribed in a unit circle. By expanding $\sin(\pi/n)$ into a Taylor series in π/n, show that there is a relation of the form

$$\pi = \tfrac{1}{2}p_n + \frac{a_1}{n^2} + \frac{a_2}{n^4} + \frac{a_3}{n^6} + \cdots$$

22 (cont.) Conclude from this formula that $3\pi \approx 2p_{2n} - \frac{1}{2}p_n$. Estimate π to four places using $n = 3$ and $n = 6$. You may use $\sin(\frac{1}{3}\pi) = \frac{1}{2}\sqrt{3}$, $\sin(\frac{1}{6}\pi) = \frac{1}{2}$, and $\sin(\frac{1}{12}\pi) = \frac{1}{2}\sqrt{2-\sqrt{3}}$. You may use a calculator with $\sqrt{\ }$ but without trig keys.

11-6 Review Exercises

Find the radius of convergence

1 $\sum e^{-n}x^n$

2 $\sum \dfrac{x^{2n+1}}{n}$

Expand in a Taylor series

3 $1 + x + x^2 + x^3$ at $x = 1$

4 xe^x at $x = -1$

5 $x\cos x - \sin x$ at $x = 0$

6 $[\ln(1 - x)]/x$ at $x = 0$

7 $\displaystyle\int_0^x \frac{\sin t}{t}\,dt$ at $x = 0$

8 $\displaystyle\int_0^x \frac{1 - \cos t}{t^2}\,dt$ at $x = 0$

9 $\displaystyle\int_0^x \frac{t}{1 + t^4}\,dt$ at $x = 0$

10 $\displaystyle\int_0^x \sin t^2\,dt$ at $x = 0$

11 $\dfrac{d}{dx}\left(\dfrac{1}{1 + x^2}\right)$ at $x = 0$

12 $\dfrac{d}{dx}\ln(1 - x^3)$ at $x = 0$

Sum the series

13 $\frac{1}{4}x + \frac{3}{16}x^2 + \frac{5}{64}x^3 + \cdots + \dfrac{2n - 1}{4^n}x^n + \cdots$

14 $\frac{1}{4} + \frac{5}{64}x^4 + \frac{9}{1024}x^8 + \cdots + \dfrac{4n + 1}{2^{2+4n}}x^{4n} + \cdots$

15 $x + \dfrac{x^5}{5!} + \dfrac{x^9}{9!} + \dfrac{x^{13}}{13!} + \cdots$

16 $2^2x^2 + 4^2x^4 + 6^2x^6 + 8^2x^8 + \cdots$

17 $1 + \frac{2}{3} + \frac{3}{9} + \frac{4}{27} + \frac{5}{8} + \cdots$

18 $1 + \frac{1}{300} + \frac{1}{50{,}000} + \frac{1}{7{,}000{,}000} + \cdots$

Evaluate

19 $\left.\dfrac{d^{20}}{dx^{20}}(x^6\cos 2x)\right|_{x=0}$

20 $\left.\dfrac{d^5}{dx^5}(e^x\sin x)\right|_{x=0}$

21 How many terms of $\frac{1}{4}\pi = 1 - \frac{1}{3} + \frac{1}{5} - \frac{1}{7} + \cdots$ are needed to estimate π to four decimal places?

22 (cont.) Use the formula

$$\tfrac{1}{4}\pi = 4\arctan\tfrac{1}{5} - \arctan\tfrac{1}{239}$$

and the series for $\arctan x$ to estimate π to four places. (See Exercise 58, page 269.)

12 Space Geometry

In this chapter we present the analytic geometry of three-dimensional euclidean space, denoted $\mathbf{R}^3$. This subject is traditionally called solid analytic geometry; our treatment will emphasize vectors. Three-dimensional geometry is important and interesting in its own right, and it is the key to our future study of geometric applications of calculus and of the calculus of functions of two or more variables.

12-1 Rectangular Coordinates

Plane analytic geometry begins with the introduction of an origin and two perpendicular coordinate axes in the Euclidean plane $\mathbf{R}^2$. First, one axis is directed (oriented) arbitrarily and called the x-axis. Then the other, called the y-axis, is directed so that the pair in the order x, y is a *right-handed* system, that is, so a positive (counter-clockwise) rotation through 90° from the positive x-axis brings us to the positive y-axis (Figure 12-1-1a, next page). Of course, this choice of a right-handed system is quite arbitrary. In a world in which most people were left-handed, clockwise might be preferred for the positive sense of rotation. Take a plain sheet of paper; you arbitrarily label one surface the top and the other the bottom. Draw a right-handed system of axes on the top. Now inspect the bottom against a light; the system appears left-handed.

Having this in mind, let us pass to space. First we choose a fixed point $\mathbf{0}$, called the **origin.** Then we choose three mutually perpendicular coordinate axes through $\mathbf{0}$. Two of these axes are directed and labeled the x-axis

Figure 12-1-1
Placing coordinate axes

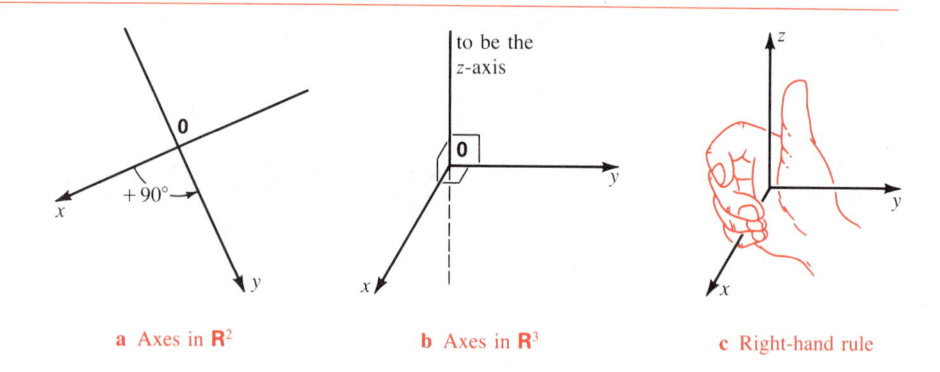

a Axes in $\mathbf{R}^2$ **b** Axes in $\mathbf{R}^3$ **c** Right-hand rule

and the y-axis respectively (Figure 12-1-1b). The third axis is labeled the z-axis, and its direction is determined by the **right-hand rule** (Figure 12-1-1c): when a right hand curls around the z-axis with its fingers curling from the positive x-axis through a right angle to the positive y-axis, its thumb points in the positive z-direction. The choice of a right-handed system is called *orienting* space. Again, there are two choices for an orientation of space, left- and right-handed, and we quite arbitrarily choose right-handed.

In case you have trouble "seeing" space figures from plane drawings, a few more words and figures are in order. In Figure 12-1-2a, drawn in perspective, we are facing the far wall of a rectangular room, and the origin of the coordinate system is the far left corner of the floor. The x-axis is chosen running toward us at the intersection of the floor and left side wall. The y-axis runs along the back of the floor to the right. Therefore either the z-axis runs straight up or it runs straight down; by the right-hand rule it runs *up*.

In Figure 12-1-2b, imagine the right-hand page of the open book flat on a desk. The y- and z-axes are drawn on the page in their usual position. The problem is to place a pointed pencil representing the x-axis, eraser end at $\mathbf{0}$. The solution, in agreement with the right-hand rule, is shown in Figure 12-1-2c.

Figure 12-1-2

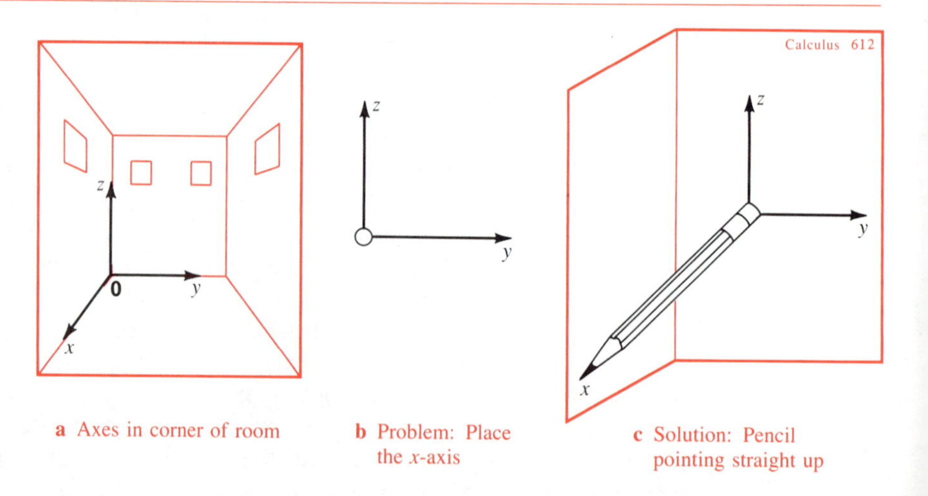

a Axes in corner of room **b** Problem: Place the x-axis **c** Solution: Pencil pointing straight up

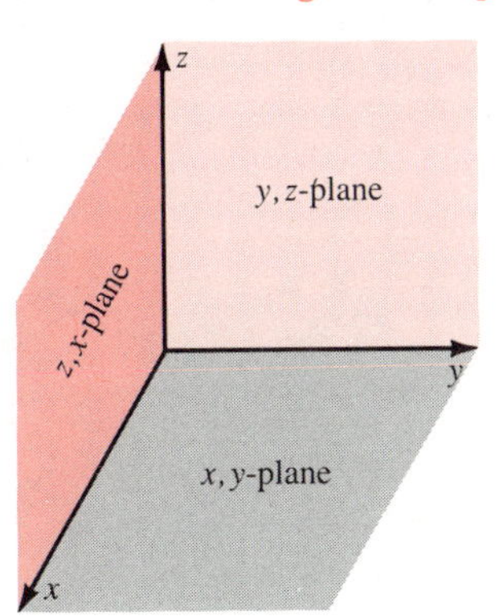

Figure 12-1-3
The coordinate planes. The y, z-plane is the plane of this page. Try to visualize the x-axis as coming out of the page at you.

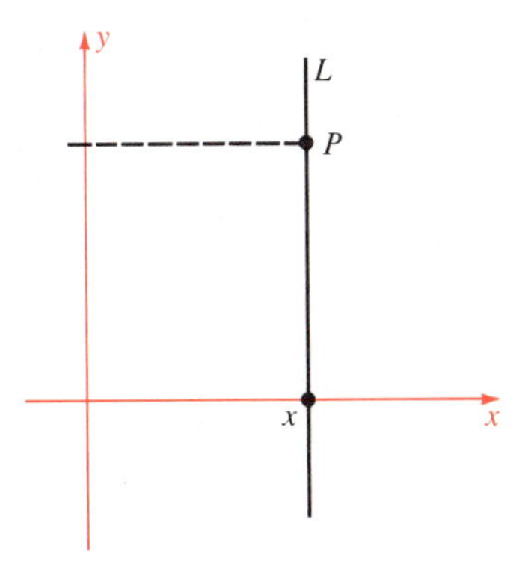

Figure 12-1-4
To find the x-coordinate of P, draw L through P perpendicular to the x-axis.

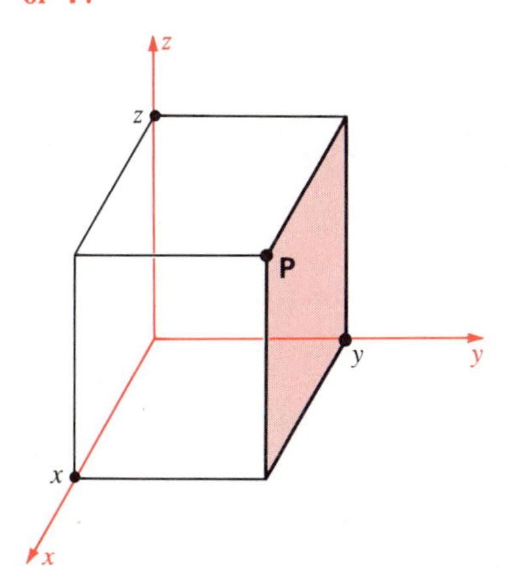

Figure 12-1-5
The plane through **P** perpendicular to the y-axis determines the y-coordinate of **P**.

The three axes determine three **coordinate planes.** For instance, the plane of the y- and z-axes is called the ***y, z*-plane,** and the other two planes are named similarly (Figure 12-1-3).

Recall how the coordinates of a point P in the plane are determined (Figure 12-1-4). To find the x-coordinate of P, we draw a line through P perpendicular to the x-axis. Where the line meets the x-axis is the desired x-coordinate. Similarly, a line through P perpendicular to the y-axis meets the y-axis in the y-coordinate of P.

Now take any point **P** in space. Through **P** pass planes perpendicular to the three coordinate axes. Their intersections with the coordinate axes determine three numbers $x, y, z,$ called the **coordinates** of **P**.

For example, in Figure 12-1-5, a plane is drawn through **P** perpendicular to the y-axis. The plane meets the y-axis at the y-coordinate of **P**. Note incidentally that a plane is perpendicular to the y-axis if and only if it is parallel to the z, x-coordinate plane.

Conversely, each triple (x, y, z) of real numbers determines a unique point **P** in space, and we write

$$\mathbf{P} = (x, y, z)$$

A point (x, y, z) is located by marking its projection $(x, y, 0)$ in the x, y-plane and going up the corresponding amount z (down if $z < 0$). See Figure 12-1-6 for examples.

The three coordinate planes divide space into eight **octants.** The octant where x, y, and z are positive is called the first octant. (No one numbers the other seven octants.)

In our figures so far, we have projected into the y, z-plane; that is, we have taken the y, z-plane in the plane of the page. Then the angle at which we draw the x-axis is arbitrary. We always try to choose it so our drawings are as uncluttered as possible. We could just as well project into one of the other two coordinate planes, or into some other plane altogether. But no matter how we set up our axes, points are located in basically the same way (Figure 12-1-7, next page). A little care is needed to make sure the drawing is right-handed.

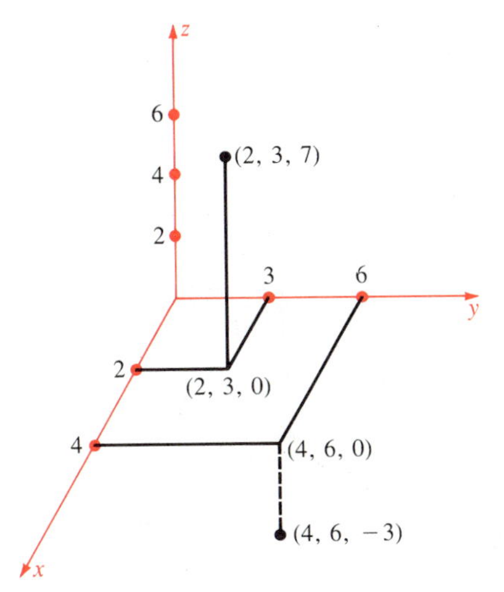

Figure 12-1-6
Locating points in space by their coordinates

Figure 12-1-7
Various projections

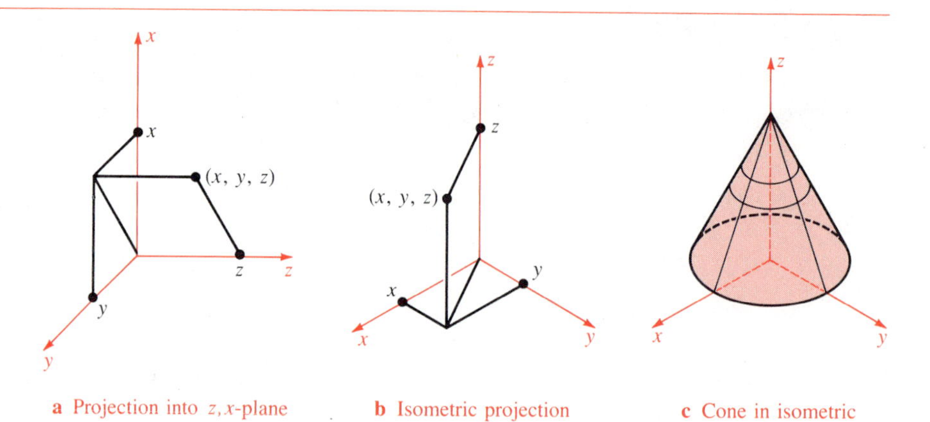

a Projection into z, x-plane b Isometric projection c Cone in isometric projection

Exercises

In Figure 12-1-2a, draw the coordinate axes from the given origin along edges of the room

1	origin rear lower left	y-axis forward
2	origin rear lower right	x-axis to left
3	origin rear upper right	z-axis down
4	origin rear upper left	y-axis down
5	origin front lower left	x-axis to right
6	origin front upper left	y-axis down

Locate the points in Figure 12-1-8

7 $(1, 2, 3)$ $(1, 3, 4)$
8 $(2, 4, 3)$ $(2, -3, 3)$
9 $(1, -2, 1)$ $(2, -3, -1)$
10 $(1, -3, -2)$ $(3, 2, -2)$

Locate the points in Figure 12-1-9

11 $(3, 4, -1)$ $(-3, -3, 1)$
12 $(0, 0, -3)$ $(-2, -2, 3)$
13 $(3, -2, 2)$ $(-2, 4, 4)$
14 $(0, -3, 2)$ $(3, 3, 3)$

Locate the points in Figure 12-1-10

15 $(1, 2, 3)$ $(0, 1, 4)$
16 $(3, 4, 2)$ $(2, -3, 3)$
17 $(3, -2, 3)$ $(4, 4, 4)$
18 $(2, 4, -1)$ $(-3, 2, 4)$

Figure 12-1-8

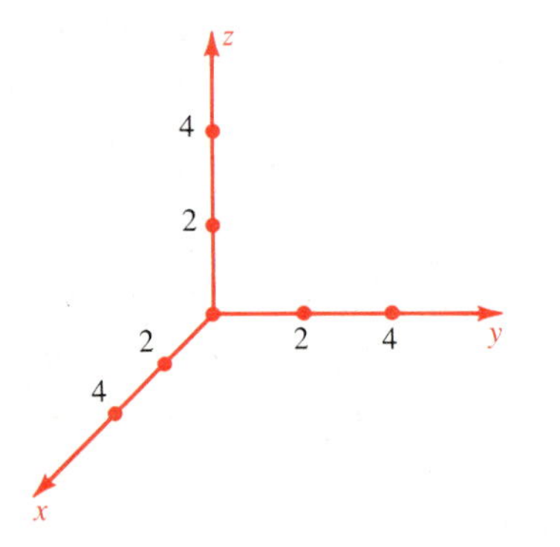

Figure 12-1-9

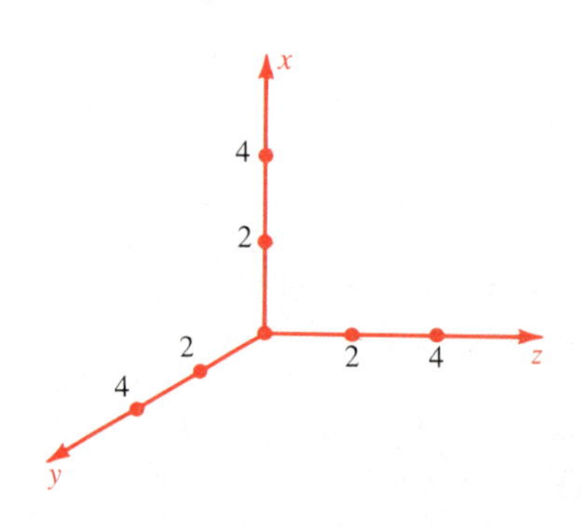

Figure 12-1-10

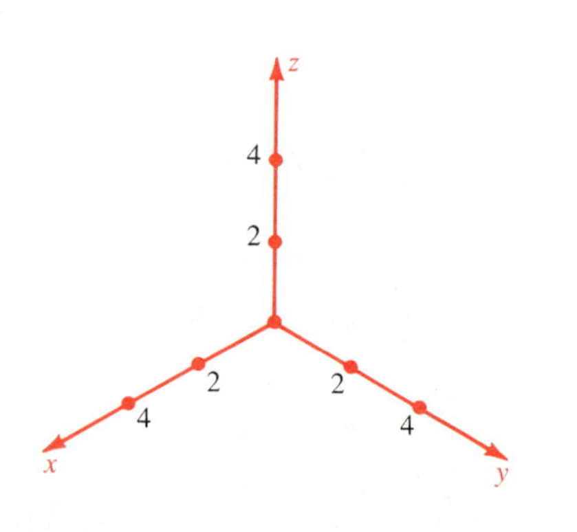

12-2 Vector Algebra

We now introduce the concept of a vector. Vectors are useful for handling problems in space because (1) equations in vector form are independent of the choice of coordinate axes, hence are well suited for describing physical situations, (2) each vector equation replaces three ordinary equations, and (3) several frequently used procedures can be summarized neatly in vector form.

A vector is determined by two quantities, its *length* (or *magnitude*) and its *direction.* Many physical quantities are vectors: directed distance, force, velocity, acceleration, electric field intensity, and others. A good example to have in mind is a force pulling an object (Figure 12-2-1). The force is represented by a vector **F** whose direction is the direction of the pull and whose length is the magnitude of the pull. (Note that to describe this situation completely, we need not only the force vector **F** but also its point of application P.) In order to do computations with a vector **F**, we always model it by a directed segment that starts at the origin.

Figure 12-2-1
Force pulling object

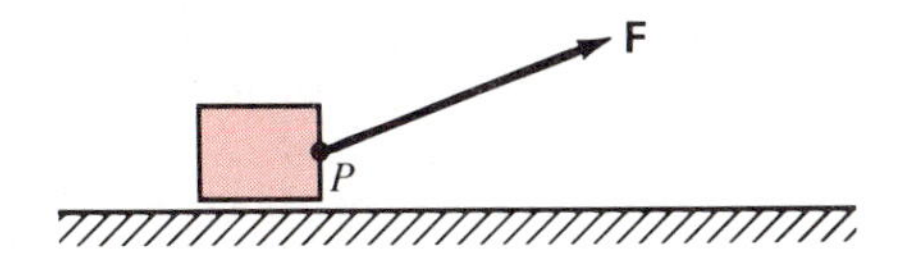

Vectors in the Plane

The plane is easier to visualize than space, so we start our work with plane vectors.

Let the origin **0** of the plane be fixed once and for all. A **vector** in the plane is a directed line segment that begins at **0**. It is completely determined by its terminal point. We shall denote vectors by bold-face letters, such as **x**, **v**, **F**, and **r**. (In your written work you may use $\underline{x}$ or $\vec{x}$.) A point (x, y) in the plane is often identified with the vector **x** from the origin to the point. The **zero vector** (origin) will be written **0** $= (0, 0)$. For this vector only, direction is undefined. When we write **x** $= (x, y)$, the numbers x and y are called the **components,** or **coordinates,** of **x**.

Remember that the origin **0** is fixed, and that each vector starts at **0**. We often draw vectors starting at other points, but in computations they all originate at **0**. For example, if a force **F** is applied at a point **x**, we may draw Figure 12-2-2a because it is suggestive. But for computations the drawing to use is Figure 12-2-2b. To describe fully the physical situation, one must specify both the force vector **F** (magnitude and direction) and its point of application **x**.

Figure 12-2-2
Drawing vectors

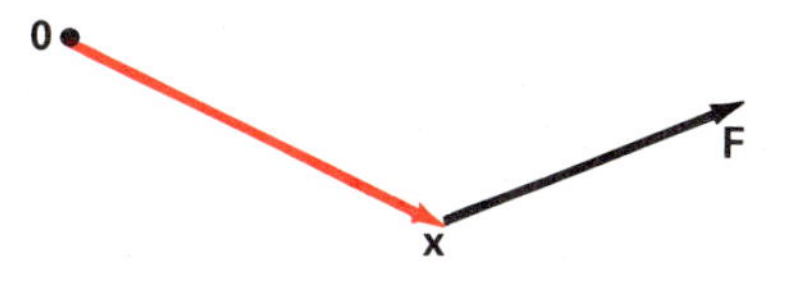

a Descriptive: **F** applied at **x**

b Analytic

Notation In working with vectors, we frequently name the axes the x_1-axis and the x_2-axis instead of the x- and y-axes. This has the advantage that as soon as we give a name to a vector, we automatically have names for its components. For instance,

$$\mathbf{x} = (x_1, x_2)$$
$$\mathbf{v} = (v_1, v_2)$$
$$\mathbf{a} = (a_1, a_2)$$

Addition of Vectors

The **sum** **u** + **v** of two vectors is defined geometrically by the parallelogram law (Figure 12-2-3). The points **0**, **u**, **v**, **u** + **v** are the vertices of a parallelogram, with **u** + **v** opposite to **0**.

Vectors are added numerically by adding their components:

Figure 12-2-3
Parallelogram law for vector addition

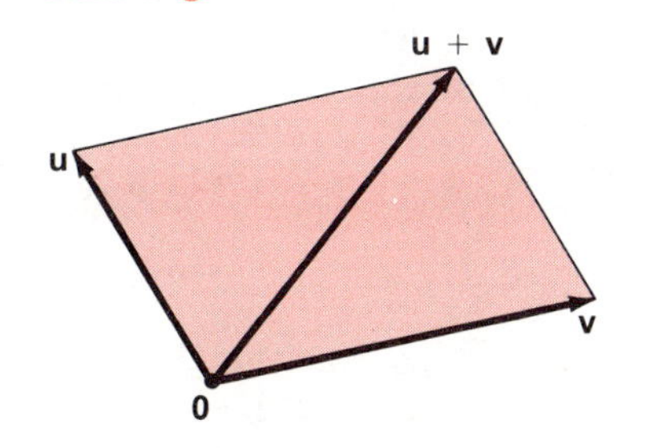

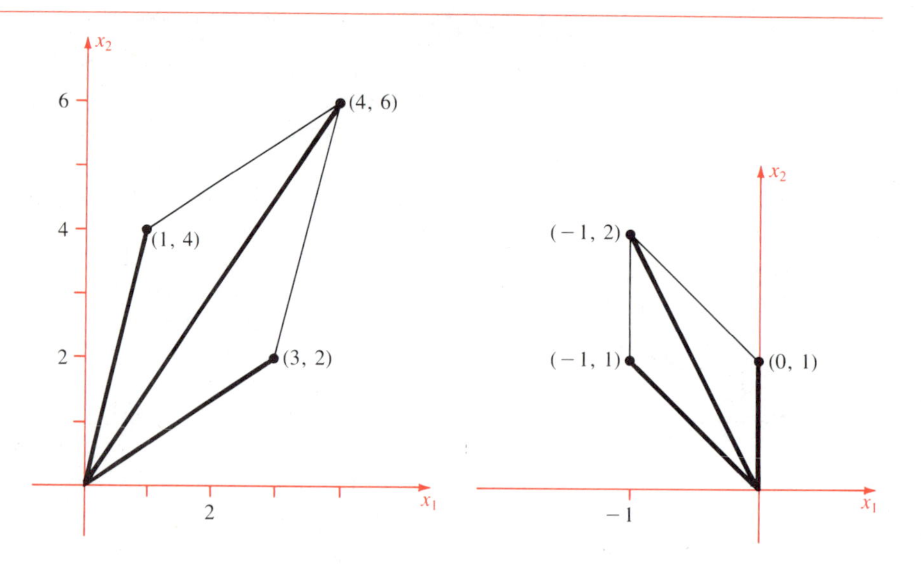

Figure 12-2-4
Numerical addition of vectors

Componentwise Addition of Vectors

$$(u_1, u_2) + (v_1, v_2) = (u_1 + v_1, u_2 + v_2)$$

For example (Figure 12-2-4)

$$(3, 2) + (1, 4) = (4, 6) \qquad (0, 1) + (-1, 1) = (-1, 2)$$

Let us prove that the sum of vectors, defined *geometrically* by the parallelogram law, can be computed *algebraically* by adding corresponding components. We pass lines L, M, and N through $\mathbf{u}$, $\mathbf{v}$, and $\mathbf{w} = \mathbf{u} + \mathbf{v}$ respectively, perpendicular to the x_1-axis (Figure 12-2-5). They meet the x_1-axis at u_1, v_1, and w_1. Because the line segments $\mathbf{vw}$ and $\mathbf{Ou}$ are parallel translates of each other, the directed distance from M to N equals the directed distance from the x_1-axis to L. Hence $w_1 - v_1 = u_1$; that is, $w_1 = u_1 + v_1$. Similarly, $w_2 = u_2 + v_2$.

Figure 12-2-5
The parallelogram law is equivalent to the numerical rule for addition of vectors.

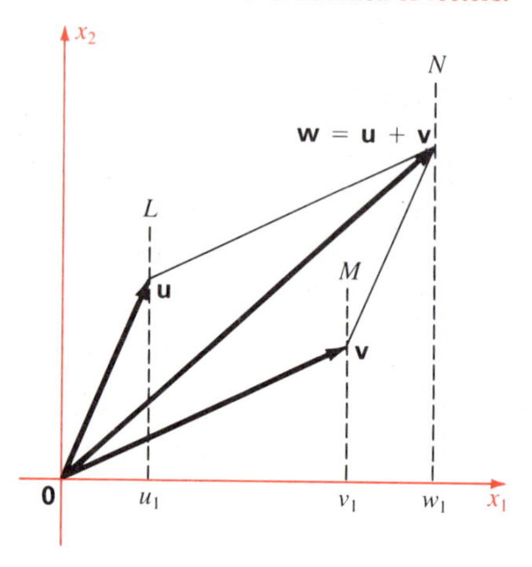

Multiplication by Scalars

Let $\mathbf{v}$ be a vector in the plane and let a be a number (called a *scalar*). We define the product $a\mathbf{v}$ to be the vector whose length is $|a|$ times the length of $\mathbf{v}$ and that points in the same direction as $\mathbf{v}$ if $a > 0$, in the opposite direction if $a < 0$. If $a = 0$, then $a\mathbf{v} = \mathbf{0}$.

There is a simple physical idea behind this definition. If a particle moving in a certain direction doubles its speed, its velocity vector is doubled; if a horse pulling a cart in a certain direction triples its pull, the force vector triples. Figure 12-2-6 illustrates multiples of a vector.

Scalar multiples are computed in components by the following rule.

Figure 12-2-6
Scalar multiples of a vector.
Note that $-\mathbf{v} = (-1)\mathbf{v}$.

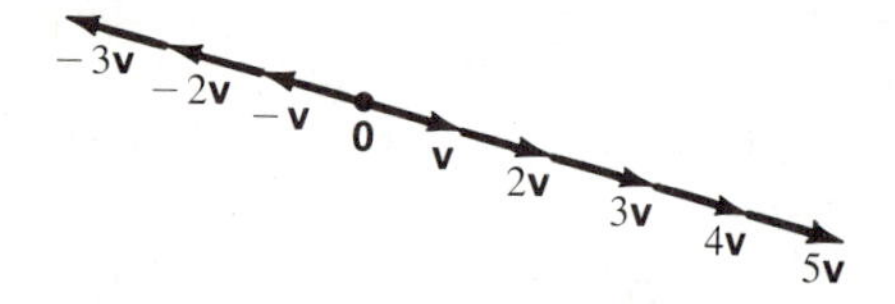

Multiples by a Scalar

$$a(v_1, v_2) = (av_1, av_2)$$

This rule is proved by similar triangles (Figure 12-2-7). The triangle $\mathbf{O}v_1\mathbf{v}$ is similar to $\mathbf{O}w_1\mathbf{w}$; hence $w_1 = av_1$ and so on. For example

Figure 12-2-7
Proof of the scalar-multiple rule

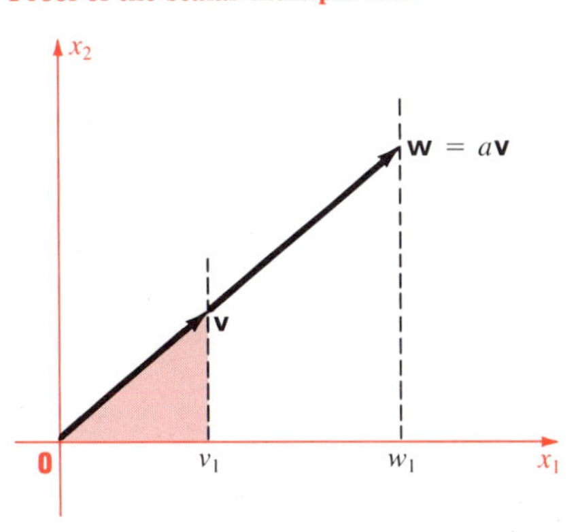

$$7(3, -1) = (21, -7) \qquad -7(3, -1) = (-21, 7)$$

The **difference** $\mathbf{v} - \mathbf{w}$ of two vectors is defined by

$$\mathbf{v} - \mathbf{w} = \mathbf{v} + (-\mathbf{w})$$

See Figure 12-2-8. (The vector $-\mathbf{w}$ has the same length as $\mathbf{w}$ but points in the opposite direction.) The segment from the tip of $\mathbf{w}$ to the tip of $\mathbf{v}$ (the dashed line in Figure 12-2-8b) has the same length and direction as $\mathbf{v} - \mathbf{w}$. Hence if two points are represented by vectors $\mathbf{v}$ and $\mathbf{w}$, the distance between them is the length of $\mathbf{v} - \mathbf{w}$.

Figure 12-2-8
Difference of vectors

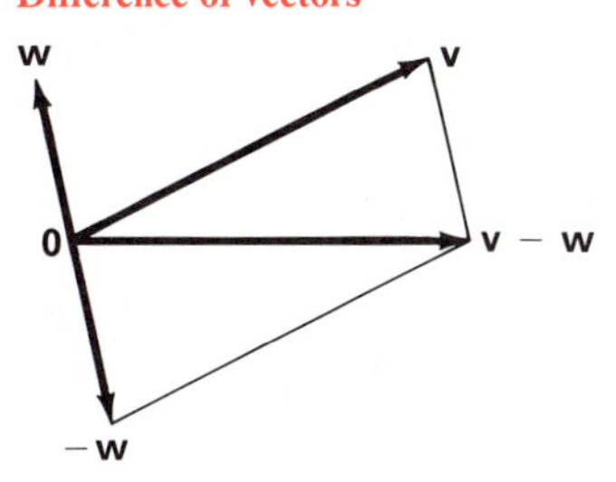

a

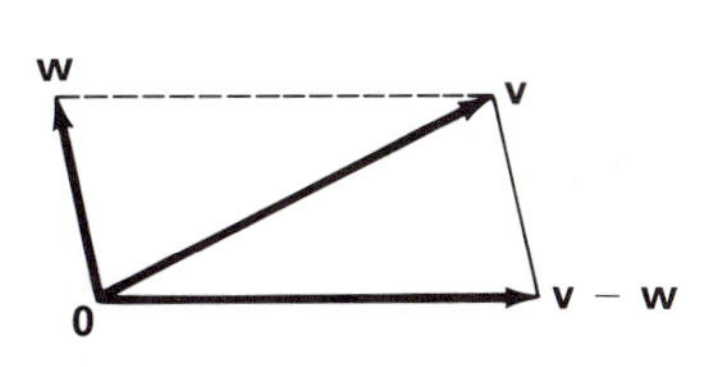

b

Vectors in Space

Now we define addition and multiplication by a scalar for vectors in $\mathbf{R}^3$. As in the plane, it is convenient to name the three axes the x_1-axis, the x_2-axis, and the x_3-axis instead of the x-, y-, and z-axes. Again this has the advantage that as soon as we name a vector, we automatically have names for its components. For instance,

$$\mathbf{x} = (x_1, x_2, x_3) \qquad \mathbf{v} = (v_1, v_2, v_3) \qquad \mathbf{a} = (a_1, a_2, a_3)$$

Also as in the plane, a point (x_1, x_2, x_3) is often identified with the vector $\mathbf{x}$ from the origin to the point.

The **sum** $\mathbf{u} + \mathbf{v}$ of two vectors is again defined by the parallelogram law (Figure 12-2-9). The points $\mathbf{0}$, $\mathbf{u}$, $\mathbf{v}$, $\mathbf{u} + \mathbf{v}$ are the vertices of a parallelogram, with $\mathbf{u} + \mathbf{v}$ opposite to $\mathbf{0}$.

Vectors are added numerically by adding their components:

$$(u_1, u_2, u_3) + (v_1, v_2, v_3) = (u_1 + v_1, u_2 + v_2, u_3 + v_3)$$

For example (Figure 12-2-10):

$$(0, 1, 3) + (1, 3, 2) = (1, 4, 5)$$

We shall omit a formal proof that the geometric parallelogram definition of addition corresponds to the arithmetic addition of components; however, the essential idea is shown in Figure 12-2-11.

Figure 12-2-9
Sum of vectors in space: parallelogram rule

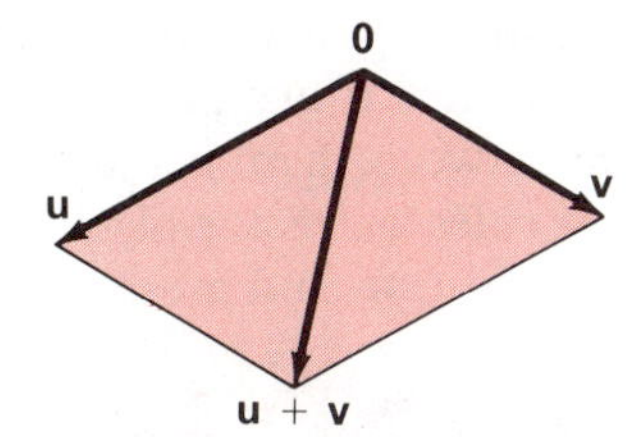

Figure 12-2-10
Geometry of
(0, 1, 3) + (1, 3, 2) = (1, 4, 5)

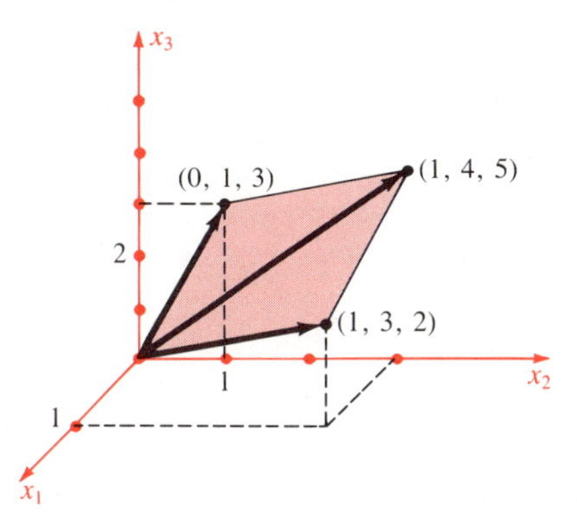

Figure 12-2-11
Proof that vectors are added componentwise: Planes P, Q, and R are perpendicular to the x_2-axis.

$\overline{v_2 w_2} = \overline{\mathbf{0}\mathbf{u}_2}$

Hence $w_2 - v_2 = u_2$ so $w_2 = u_2 + v_2$.

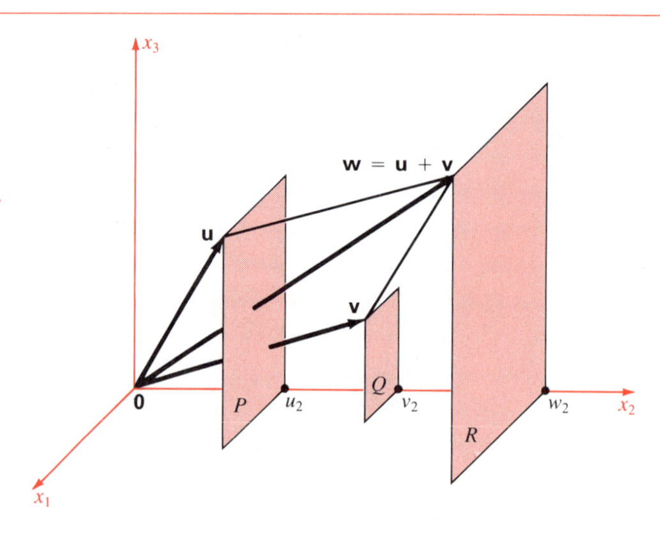

Figure 12-2-12
Proof that scalar multiplication goes componentwise: Triangles $\mathbf{0}v_2\mathbf{v}$ and $\mathbf{0}w_2\mathbf{w}$ are similar; hence $w_2 = av_2$ since $\overline{\mathbf{0w}} = a\overline{\mathbf{0v}}$.

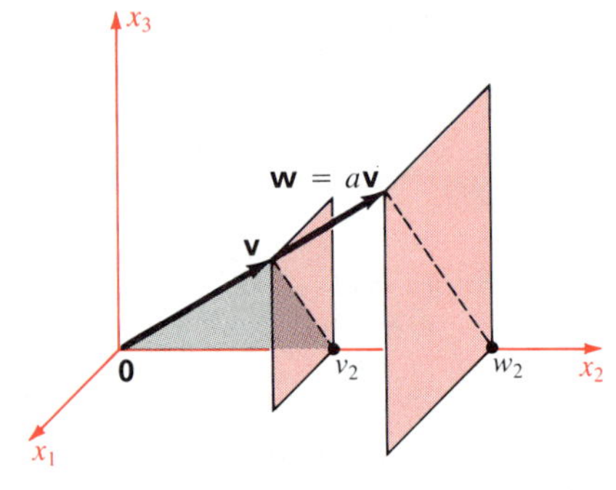

Let $\mathbf{v}$ be a vector and let a be a number (scalar). As in the plane, we define the product $a\mathbf{v}$ to be the vector whose length is $|a|$ times the length of $\mathbf{v}$ and which points in the same direction as $\mathbf{v}$ if $a > 0$, in the opposite direction if $a < 0$. If $a = 0$, then $a\mathbf{v} = \mathbf{0}$. See Figure 12-2-6 again.

- **Scalar multiples** are computed in components by the following rule:

$$a(v_1, v_2, v_3) = (av_1, av_2, av_3)$$

(This rule may be proved by similar triangles as in Figure 12-2-12.) For example

$$7(3, 2, -1) = (21, 14, -7)$$
$$-7(3, 2, -1) = (-21, -14, 7)$$

- The **difference** $\mathbf{v} - \mathbf{w}$ of two vectors is defined by

$$\mathbf{v} - \mathbf{w} = \mathbf{v} + (-\mathbf{w})$$

See Figure 12-2-8 again. (The vector $-\mathbf{w}$ has the same length as $\mathbf{w}$ but points in the opposite direction.) The segment from the tip of $\mathbf{w}$ to the tip of $\mathbf{v}$ (the dashed line in Figure 12-2-8b) has the same length and direction as $\mathbf{v} - \mathbf{w}$. Hence if two points are represented by vectors $\mathbf{v}$ and $\mathbf{w}$, the distance between them is the length of $\mathbf{v} - \mathbf{w}$.

The basic rules of vector algebra follow directly from the coordinate formulas for addition and multiplication by a scalar.

Rules of Vector Algebra

1 $\mathbf{v} + \mathbf{0} = \mathbf{0} + \mathbf{v} = \mathbf{v}$ (identity law for addition)
2 $\mathbf{v} + (-\mathbf{v}) = (-\mathbf{v}) + \mathbf{v} = \mathbf{0}$ (existence of an additive inverse)
3 $\mathbf{u} + \mathbf{v} = \mathbf{v} + \mathbf{u}$ (commutative law for addition)
4 $\mathbf{u} + (\mathbf{v} + \mathbf{w}) = (\mathbf{u} + \mathbf{v}) + \mathbf{w}$ (associative law for addition)

Figure 12-2-13
The midpoint of **uv** is
$\mathbf{m} = \frac{1}{2}(\mathbf{u} + \mathbf{v})$

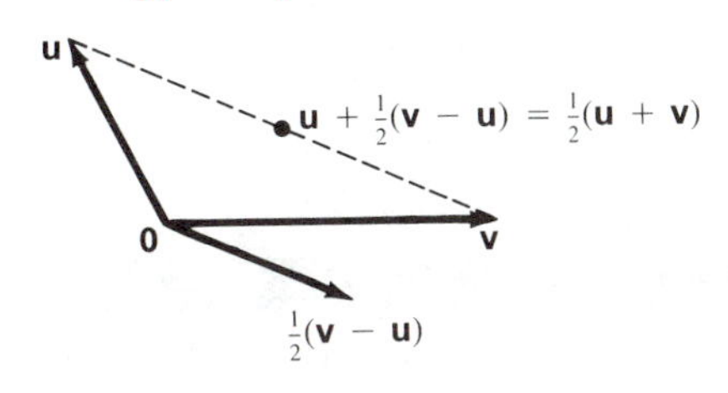

5 $0\mathbf{v} = \mathbf{0}$ and $1\mathbf{v} = \mathbf{v}$ (special rules for 0 and 1)

6 $a(b\mathbf{v}) = (ab)\mathbf{v}$ (associative law for multiplication by a scalar)

7 $\left.\begin{aligned}(a + b)\mathbf{v} &= a\mathbf{v} + b\mathbf{v}\\ a(\mathbf{v} + \mathbf{w}) &= a\mathbf{v} + a\mathbf{w}\end{aligned}\right\}$ (distributive laws)

For an example of the convenience of vector algebra, let us find the midpoint **m** of a given segment **uv**. Now to go from **u** to **v**, we add $\mathbf{v} - \mathbf{u}$ to **u**. But to go from **u** only to the midpoint **m**, we go halfway from **u** to **v**; that is, we add $\frac{1}{2}(\mathbf{v} - \mathbf{u})$ to **u**. Thus $\mathbf{m} = \mathbf{u} + \frac{1}{2}(\mathbf{v} - \mathbf{u}) = \frac{1}{2}(\mathbf{u} + \mathbf{v})$. We may think of **m** as the *average* of **u** and **v**. See Figure 12-2-13.

Figure 12-2-14
The midpoint of (3, 8, 5) and (1, 2, 7)

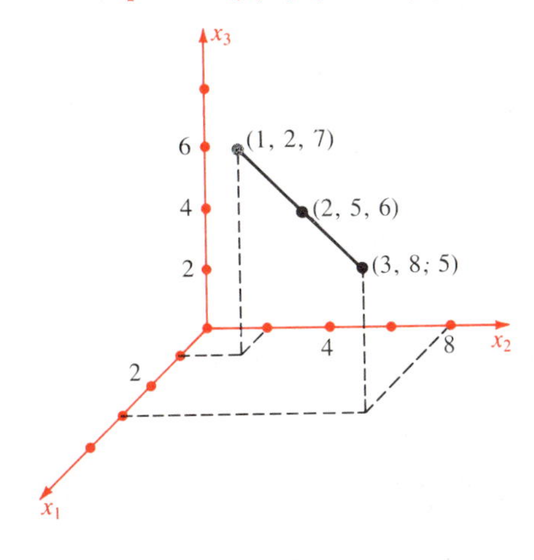

Midpoint Formula

If **u** and **v** are any two points in $\mathbf{R}^3$, then their midpoint is

$$\mathbf{m} = \tfrac{1}{2}(\mathbf{u} + \mathbf{v}) = (\tfrac{1}{2}(u_1 + v_1), \tfrac{1}{2}(u_2 + v_2), \tfrac{1}{2}(u_3 + v_3))$$

For example (Figure 12-2-14), the midpoint of (3, 8, 5) and (1, 2, 7) is

$$(\tfrac{1}{2}(3 + 1), \tfrac{1}{2}(8 + 2), \tfrac{1}{2}(5 + 7)) = (2, 5, 6)$$

Exercises

Compute

1 $(1, 2, -3) + (4, 0, 7)$

2 $(-1, -1, 0) + (3, 5, 2)$

3 $(4, 0, 7) - (1, 2, -3)$

4 $(2, 1, 1) - (3, -1, -2)$

5 $(1, 2, 3) - 6(0, 3, -1)$

6 $4[(1, -2, -7) - (1, 1, 1)]$

7 $3(1, 4, 2) - 2(2, 1, 1)$

8 $4(1, -1, 2) - 3(1, -1, 2)$

9 $3(1, 1, 0) - 2(0, 1, 1) + (1, 0, 1)$

10 $-5[3(1, 1, 1) - (2, 1, 4)] + 4(-2, -2, 1)$

Find the midpoint of the two vectors

11 $(4, 0, 0)$ $(0, 0, 2)$

12 $(1, 1, 1)$ $(3, -1, 3)$

13 $(8, 0, 6)$ $(-2, 8, -4)$

14 $(3, 2, 1)$ $(1, 2, 3)$

15 $(3, 4, -2)$ $(0, 0, 1)$

16 $(4, 4, 2)$ $(-1, -3, -5)$

Find a vector parallel to the line through

17 $(1, 1, 2)$ and $(1, -1, -1)$

18 $(-3, 4, 1)$ and $(7, -2, -2)$

Let $\mathbf{a} = (1, 1, 3)$, $\mathbf{b} = (-2, -1, 0)$, **u**, and **v** be the vertices of a parallelogram, with **a** opposite to **b**. Find **v** provided

19 $\mathbf{u} = (0, 0, 1)$

20 $\mathbf{u} = (1, 1, 1)$

Use coordinates to show

21 $\mathbf{u} + \mathbf{v} = \mathbf{v} + \mathbf{u}$

22 $\mathbf{u} + (\mathbf{v} + \mathbf{w}) = (\mathbf{u} + \mathbf{v}) + \mathbf{w}$

23 $(a + b)\mathbf{u} = a\mathbf{u} + b\mathbf{u}$

24 $a(\mathbf{v} + \mathbf{w}) = a\mathbf{v} + a\mathbf{w}$

25 $(ab)\mathbf{u} = a(b\mathbf{u}) = b(a\mathbf{u})$

26 Show that the segments joining the midpoints of opposite sides of a (skew) quadrilateral bisect each other.

27 (cont.) Find the point of intersection of these segments when the vertices of the quadrilateral, in order, are

$\mathbf{v}_1 = \mathbf{0}$ $\mathbf{v}_2 = (1, 0, 0)$ $\mathbf{v}_3 = (0, 1, 0)$ $\mathbf{v}_4 = (0, 0, 1)$

28 Find the vector that is the intersection of the medians of the triangle with vertices $\mathbf{a}$, $\mathbf{b}$, $\mathbf{c}$.

29 (cont.) In a tetrahedron, prove that the four lines joining each vertex to the centroid (intersection of the medians) of the opposite face are concurrent.

***30** Space billiards—no gravity. An astronaut cues a ball toward the corner of a rectangular room, with velocity $\mathbf{v}$. The ball misses the corner, but rebounds perfectly off each of the three adjacent walls. Find its returning velocity vector.

12-3 Length and Inner Product

The **length** of a vector $\mathbf{v}$, denoted $|\mathbf{v}|$, is the distance of its terminal point from $\mathbf{0}$. Suppose first that we work in the plane $\mathbf{R}^2$. Then $|\mathbf{v}|$ is the length of the diagonal of a certain rectangle (Figure 12-3-1). However, in space $\mathbf{R}^3$, the number $|\mathbf{v}|$ is the length of the diagonal of a certain rectangular solid (Figure 12-3-2). From these figures we have the following formulas:

Figure 12-3-1
Length of a plane vector:
$|\mathbf{v}|^2 = |v_1|^2 + |v_2|^2 = v_1{}^2 + v_2{}^2$

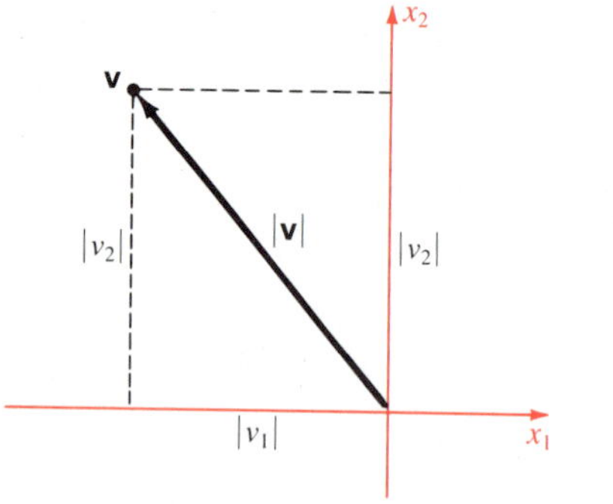

Length of a Vector

If $\mathbf{v} = (v_1, v_2)$ is a plane vector, then

$$|\mathbf{v}| = \sqrt{v_1{}^2 + v_2{}^2}$$

If $\mathbf{v}$ is a space vector, then

$$|\mathbf{v}| = \sqrt{v_1{}^2 + v_2{}^2 + v_3{}^2}$$

We now state some basic properties of length:

Properties of Length

1 $|\mathbf{0}| = 0$
2 $|\mathbf{v}| > 0$ if $\mathbf{v} \neq 0$
3 $|a\mathbf{v}| = |a| \cdot |\mathbf{v}|$
4 $|\mathbf{v} + \mathbf{w}| \leq |\mathbf{v}| + |\mathbf{w}|$ (triangle inequality)

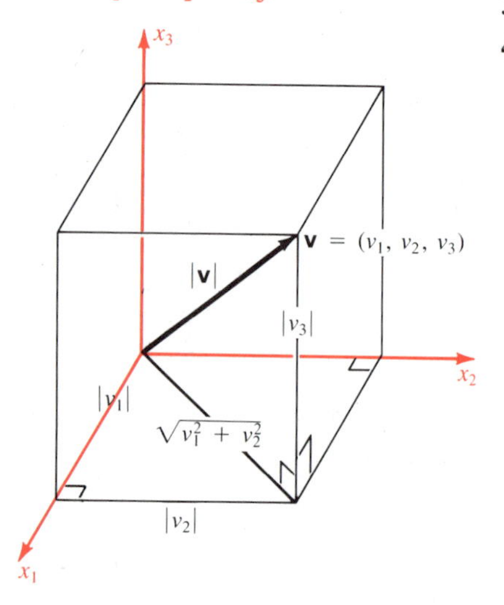

Figure 12-3-2
Length of a space vector:
$|\mathbf{v}|^2 = (\sqrt{v_1{}^2 + v_2{}^2})^2 + v_3{}^2$
$= v_1{}^2 + v_2{}^2 + v_3{}^2$

Properties 1–3 are fairly obvious geometrically, and follow easily from the length formulas. The geometric reason for the triangle inequality is shown in Figure 12-3-3. See Exercises 29 and 30 for a proof based on the algebraic formula for $|\mathbf{v}|$.

The distance between two points can be expressed in terms of vector length. In fact, by Figure 12-3-4, the distance between the points $\mathbf{v}$ and $\mathbf{w}$ equals the length of the vector $\mathbf{v} - \mathbf{w}$.

Distance Formulas

The distance between two points $\mathbf{v} = (v_1, v_2)$ and $\mathbf{w} = (w_1, w_2)$ in the plane $\mathbf{R}^2$ is

$$d(\mathbf{v}, \mathbf{w}) = |\mathbf{v} - \mathbf{w}| = \sqrt{(v_1 - w_1)^2 + (v_2 - w_2)^2}$$

Similarly, the distance between two points $\mathbf{v} = (v_1, v_2, v_3)$ and $\mathbf{w} = (w_1, w_2, w_3)$ in space $\mathbf{R}^3$ is

$$d(\mathbf{v}, \mathbf{w}) = |\mathbf{v} - \mathbf{w}| = \sqrt{(v_1 - w_1)^2 + (v_2 - w_2)^2 + (v_3 - w_3)^2}$$

Figure 12-3-3
In a triangle, the length of any side is less than the sum of the lengths of the other two sides:
$|\mathbf{v} + \mathbf{w}| < |\mathbf{v}| + |\mathbf{w}|$

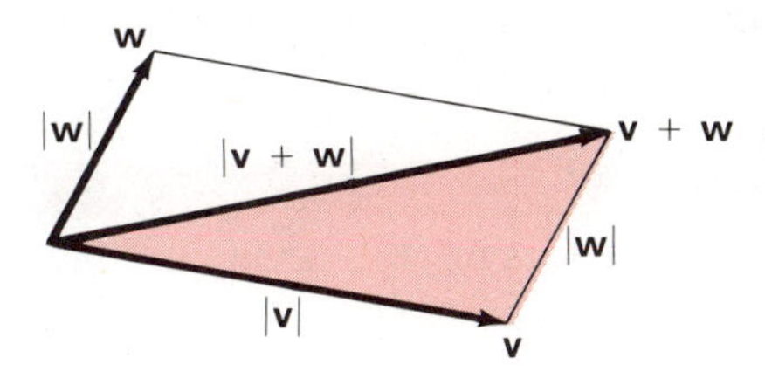

Figure 12-3-4
Distance from **v** to **w**:
$d(\mathbf{v}, \mathbf{w}) = |\mathbf{v} - \mathbf{w}|$
because opposite sides of a parallelogram have equal lengths.

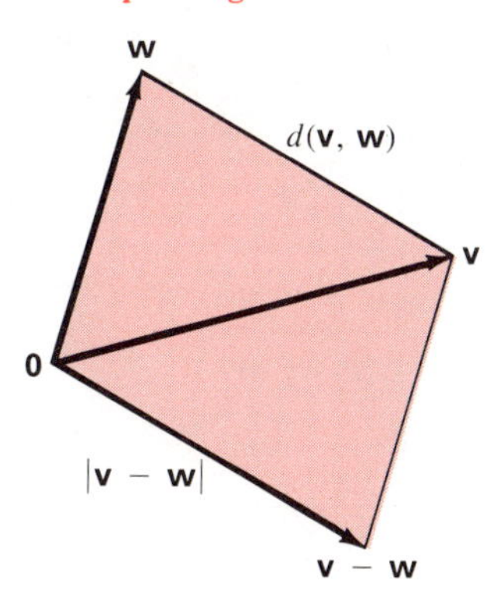

Example 1

a The distance from $(1, 2, 4)$ to $(3, 0, -1)$ is

$$|(1, 2, 4) - (3, 0, -1)| = \sqrt{(1 - 3)^2 + (2 - 0)^2 + (4 + 1)^2}$$
$$= \sqrt{4 + 4 + 25} = \sqrt{33}$$

b The set of all points (v_1, v_2, v_3) satisfying

$$v_1^2 + v_2^2 + v_3^2 = 1$$

is the sphere with center $\mathbf{0}$ and radius 1.

c A vector equation for the sphere with center $\mathbf{x}_0$ and radius $\mathbf{r}$ is

$$|\mathbf{x} - \mathbf{x}_0| = r$$

Another important vector operation is the **inner product** of two vectors, also called the **dot product.**

Inner Product

Let $\mathbf{v}$ and $\mathbf{w}$ be vectors and θ the angle between them. Their **inner product** is

$$\mathbf{v} \cdot \mathbf{w} = |\mathbf{v}| \cdot |\mathbf{w}| \cos\theta$$

If $\mathbf{v} = \mathbf{0}$ or $\mathbf{w} = \mathbf{0}$, we define $\mathbf{v} \cdot \mathbf{w} = 0$ (even though $\cos\theta$ is undefined).

See Figure 12-3-5. The angle θ between the vectors can be measured either from $\mathbf{v}$ to $\mathbf{w}$ or from $\mathbf{w}$ to $\mathbf{v}$. We shall always take $0 \le \theta \le \pi$. We see from Figure 12-3-6 that $|\mathbf{w}| \cos\theta$ is the (signed) projection of $\mathbf{w}$ on $\mathbf{v}$, hence $\mathbf{v} \cdot \mathbf{w}$ is $|\mathbf{v}|$ times the projection of $\mathbf{w}$ on $\mathbf{v}$.

Figure 12-3-5
Inner product: $\mathbf{v} \cdot \mathbf{w} = |\mathbf{v}| \cdot |\mathbf{w}| \cos\theta$

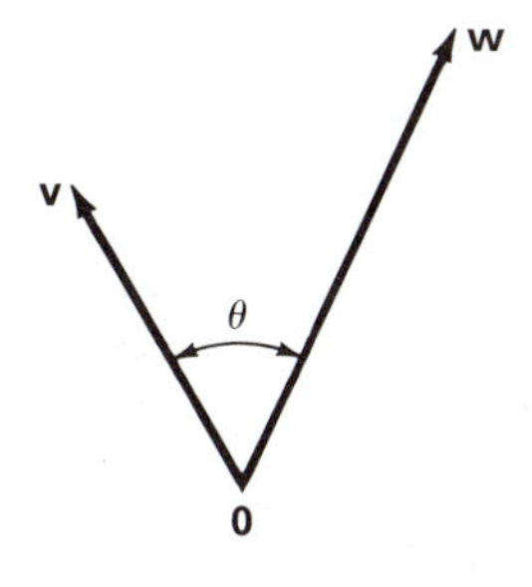

Figure 12-3-6
$\mathbf{v} \cdot \mathbf{w} = (\text{projection of } \mathbf{w} \text{ on } \mathbf{v}) \cdot |\mathbf{v}|$

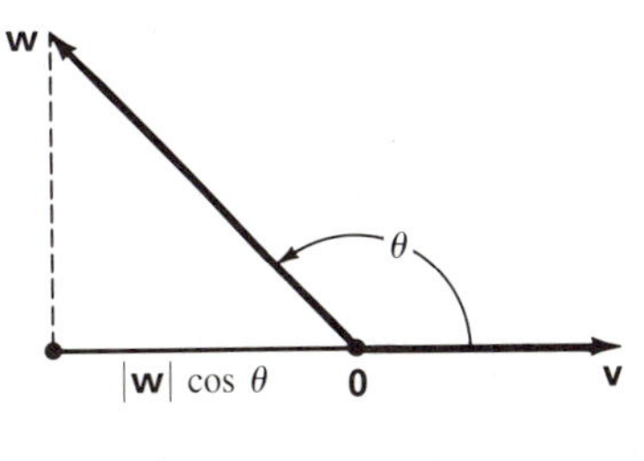

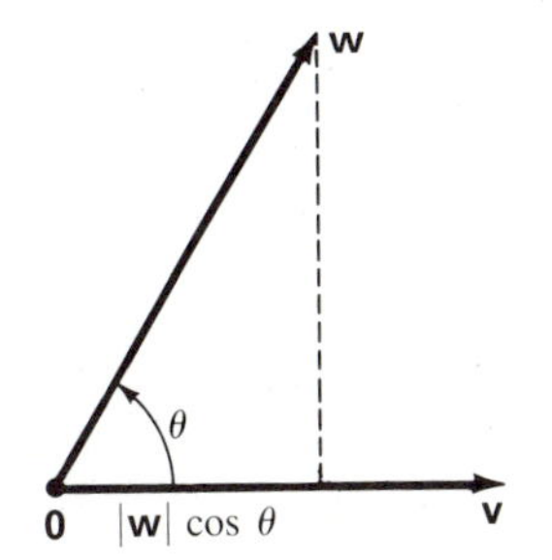

Warning The inner product of two vectors is *not* a vector; it is a *scalar* (number).

To work with inner products effectively, we need formulas for computing $\mathbf{v} \cdot \mathbf{w}$ in terms of the components of $\mathbf{v}$ and $\mathbf{w}$. Fortunately there are remarkably simple rules:

Formulas for Inner Products

If $\mathbf{v} = (v_1, v_2)$ and $\mathbf{w} = (w_1, w_2)$ are vectors in $\mathbf{R}^2$, then

$$\mathbf{v} \cdot \mathbf{w} = v_1 w_1 + v_2 w_2$$

If $\mathbf{v} = (v_1, v_2, v_3)$ and $\mathbf{w} = (w_1, w_2, w_3)$ are vectors in $\mathbf{R}^3$, then

$$\mathbf{v} \cdot \mathbf{w} = v_1 w_1 + v_2 w_2 + v_3 w_3$$

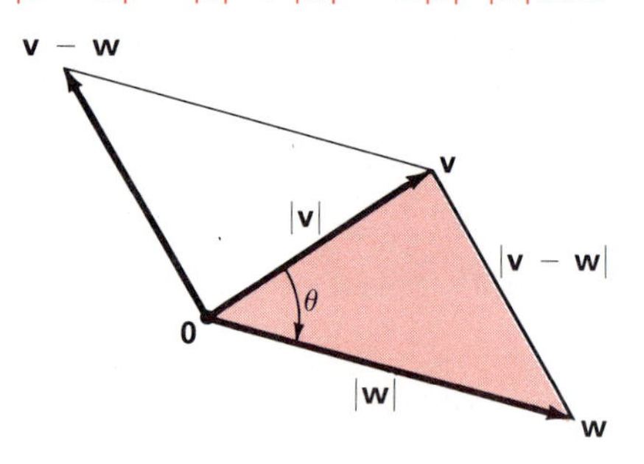

Figure 12-3-7
$|\mathbf{v} - \mathbf{w}|^2 = |\mathbf{v}|^2 + |\mathbf{w}|^2 - 2|\mathbf{v}| \cdot |\mathbf{w}| \cos\theta$

Proof By Figure 12-3-7 and the law of cosines,

$$|\mathbf{v} - \mathbf{w}|^2 = |\mathbf{v}|^2 + |\mathbf{w}|^2 - 2|\mathbf{v}| \cdot |\mathbf{w}| \cos\theta = |\mathbf{v}|^2 + |\mathbf{w}|^2 - 2\mathbf{v} \cdot \mathbf{w}$$

Hence

$$2\mathbf{v} \cdot \mathbf{w} = |\mathbf{v}|^2 + |\mathbf{w}|^2 - |\mathbf{v} - \mathbf{w}|^2$$

In the plane we have

$$\begin{aligned} 2\mathbf{v} \cdot \mathbf{w} &= |(v_1, v_2)|^2 + |(w_1, w_2)|^2 - |(v_1 - w_1, v_2 - w_2)|^2 \\ &= (v_1^2 + v_2^2) + (w_1^2 + w_2^2) \\ &\quad - (v_1 - w_1)^2 - (v_2 - w_2)^2 \\ &= (v_1^2 + v_2^2) + (w_1^2 + w_2^2) \\ &\quad - (v_1^2 - 2v_1 w_1 + w_1^2) - (v_2^2 - 2v_2 w_2 + w_2^2) \\ &= 2(v_1 w_1 + v_2 w_2) \end{aligned}$$

Therefore $\mathbf{v} \cdot \mathbf{w} = v_1 w_1 + v_2 w_2$ as asserted. The calculation for space is similar.

Example 2

a $(7, -4) \cdot (3, 2) = 7 \cdot 3 + (-4) \cdot 2 = 21 - 8 = 13$
b $(5, -1, 3) \cdot (2, 7, 2) = 5 \cdot 2 + (-1) \cdot 7 + 3 \cdot 2 = 9$
c $(3, 4, 0) \cdot (1, -5, 6) = 3 \cdot 1 + 4(-5) + 0 \cdot 6 = -17$

We now state the main algebraic properties of the inner product. Their proofs follow easily from the formulas for inner products in terms of coordinates and are left as exercises:

Properties of the Inner Product

1 $\mathbf{v} \cdot \mathbf{w} = \mathbf{w} \cdot \mathbf{v}$ (commutative law)
2 $(a\mathbf{v}) \cdot \mathbf{w} = \mathbf{v} \cdot (a\mathbf{w}) = a(\mathbf{v} \cdot \mathbf{w})$
3 $(\mathbf{u} + \mathbf{v}) \cdot \mathbf{w} = \mathbf{u} \cdot \mathbf{w} + \mathbf{v} \cdot \mathbf{w}$ (distributive law)

Figure 12-3-8
The vectors are perpendicular because
$(-1, -1, 1) \cdot (1, 2, 3) = 0$

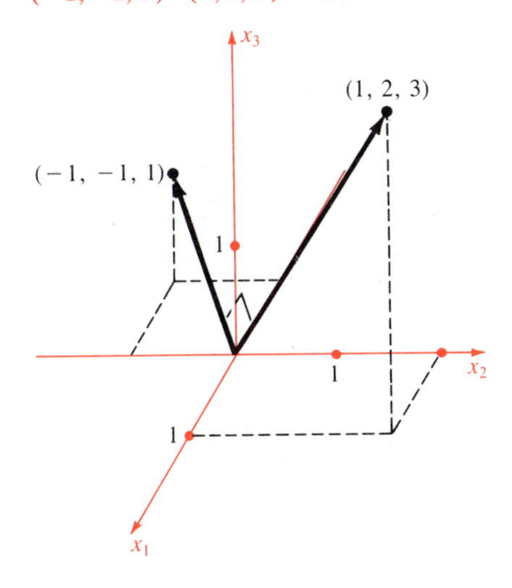

Two non-zero vectors $\mathbf{v}$ and $\mathbf{w}$ are perpendicular when the angle between them is $\theta = \frac{1}{2}\pi$. Then $\mathbf{v} \cdot \mathbf{w} = |\mathbf{v}| \cdot |\mathbf{w}| \cos\theta = 0$. Conversely, if $\mathbf{v} \cdot \mathbf{w} = 0$, either $\mathbf{v} = \mathbf{0}$ or $\mathbf{w} = \mathbf{0}$ or $\cos\theta = 0$. In the latter case, $\theta = \frac{1}{2}\pi$. We shall consider the zero vector $\mathbf{0}$ as perpendicular to all vectors.

Perpendicular Vectors

Vectors $\mathbf{v}$ and $\mathbf{w}$ are perpendicular if and only if

$$\mathbf{v} \cdot \mathbf{w} = 0$$

For example, $(1, 2, 3) \cdot (-1, -1, 1) = -1 - 2 + 3 = 0$, so $(1, 2, 3)$ and $(-1, -1, 1)$ are perpendicular (Figure 12-3-8).

Suppose $\mathbf{v} = (v_1, v_2)$ and $\mathbf{w} = (w_1, w_2)$, are non-zero vectors in $\mathbf{R}^2$. Then $\mathbf{v}$ and $\mathbf{w}$ are perpendicular if $\mathbf{v} \cdot \mathbf{w} = 0$, that is, if

$$v_1w_1 + v_2w_2 = 0 \qquad \text{so} \qquad w_2/w_1 = \frac{-1}{v_2/v_1}$$

(provided the divisions are permissible). This is just the familiar condition that the slopes of perpendicular lines are negative reciprocals of each other. [The slope of a line in the direction of $\mathbf{v} = (v_1, v_2)$ is v_2/v_1.] See Figure 12-3-9.

Figure 12-3-9
Condition for perpendicularity of plane vectors: $v_1w_1 + v_2w_2 = 0$
that is, $\mathbf{v} \cdot \mathbf{w} = 0$

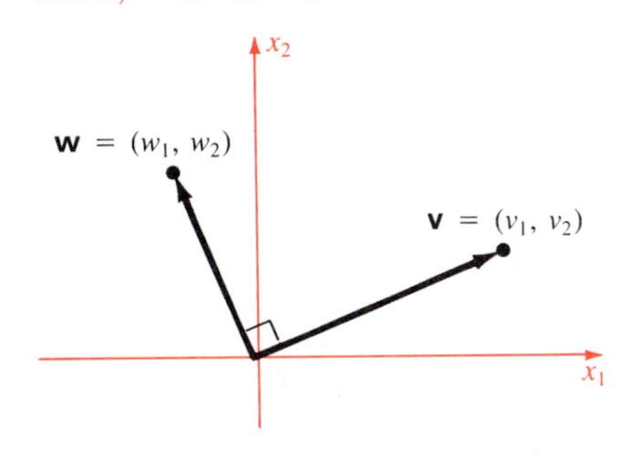

Terminology When speaking of vectors, **orthogonal** is a common synonym for perpendicular. Thus $(1, 2, 3)$ and $(-1, -1, 1)$ are *orthogonal* vectors.

Relations Involving Inner Products

There are several useful relations between length, inner product, and the angle between vectors. First, length can be expressed in terms of the inner product:

$$|\mathbf{v}|^2 = \mathbf{v} \cdot \mathbf{v}$$

This follows from the formulas in terms of coordinates, or from $\mathbf{v} \cdot \mathbf{v} = |\mathbf{v}|^2 \cos\theta$ with $\theta = 0$. Conversely, the inner product can be expressed in terms of length by means of a short calculation:

$$|\mathbf{v} + \mathbf{w}|^2 = (\mathbf{v} + \mathbf{w}) \cdot (\mathbf{v} + \mathbf{w}) = |\mathbf{v}|^2 + 2\mathbf{v} \cdot \mathbf{w} + |\mathbf{w}|^2$$
$$|\mathbf{v} - \mathbf{w}|^2 = (\mathbf{v} - \mathbf{w}) \cdot (\mathbf{v} - \mathbf{w}) = |\mathbf{v}|^2 - 2\mathbf{v} \cdot \mathbf{w} + |\mathbf{w}|^2$$

Therefore $|\mathbf{v} + \mathbf{w}|^2 - |\mathbf{v} - \mathbf{w}|^2 = 4\mathbf{v} \cdot \mathbf{w}$ so we have

$$\mathbf{v} \cdot \mathbf{w} = \tfrac{1}{4}(|\mathbf{v} + \mathbf{w}|^2 - |\mathbf{v} - \mathbf{w}|^2)$$

The angle θ between $\mathbf{v}$ and $\mathbf{w}$ can be found from

$$\mathbf{v} \cdot \mathbf{w} = |\mathbf{v}| \cdot |\mathbf{w}| \cos\theta$$

by solving for $\cos\theta$.

Relations between Length and Inner Product

Let $\mathbf{v}$ and $\mathbf{w}$ be two vectors and let θ be the angle between them. Then

$$|\mathbf{v}|^2 = \mathbf{v} \cdot \mathbf{v}$$

$$\mathbf{v} \cdot \mathbf{w} = \tfrac{1}{4}(|\mathbf{v} + \mathbf{w}|^2 - |\mathbf{v} - \mathbf{w}|^2)$$

$$\cos\theta = \frac{\mathbf{v} \cdot \mathbf{w}}{|\mathbf{v}| \cdot |\mathbf{w}|} \qquad (\mathbf{v} \neq \mathbf{0},\ \mathbf{w} \neq \mathbf{0})$$

Example 3

Find the angle between $\mathbf{v} = (1, 2, 1)$ and $\mathbf{w} = (3, -1, 1)$.

Solution

$$\mathbf{v} \cdot \mathbf{w} = 3 - 2 + 1 = 2$$

$$|\mathbf{v}|^2 = 1 + 4 + 1 = 6$$

$$|\mathbf{w}|^2 = 9 + 1 + 1 = 11$$

Hence

$$\cos\theta = \frac{\mathbf{v} \cdot \mathbf{w}}{|\mathbf{v}| \cdot |\mathbf{w}|} = \frac{2}{\sqrt{6}\sqrt{11}}$$

$$\theta = \arccos(2/\sqrt{66}) \approx 1.3221 \text{ rad}$$

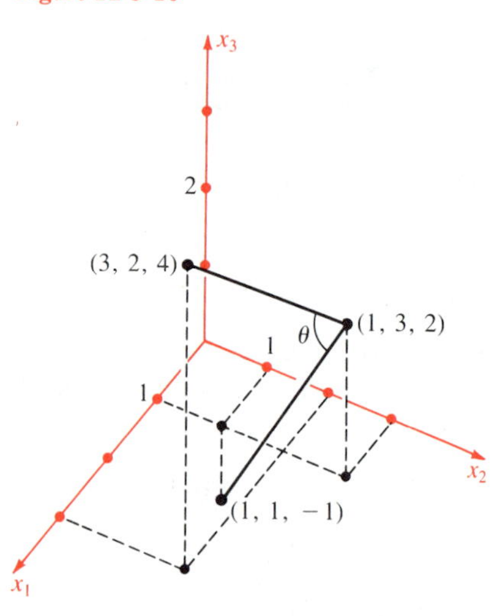

Figure 12-3-10

Example 4

The point $(1, 3, 2)$ is joined to the points $(1, 1, -1)$ and $(3, 2, 4)$ by lines L_1 and L_2. Find the angle θ between these lines.

Solution Plot the data (Figure 12-3-10). The vector

$$\mathbf{v} = (1, 1, -1) - (1, 3, 2) = (0, -2, -3)$$

is parallel to L_1 (but starts at $\mathbf{0}$). Likewise

$$\mathbf{w} = (3, 2, 4) - (1, 3, 2) = (2, -1, 2)$$

is parallel to L_2. Hence

$$\cos\theta = \frac{\mathbf{v} \cdot \mathbf{w}}{|\mathbf{v}| \cdot |\mathbf{w}|} = \frac{0 + 2 - 6}{\sqrt{0 + 4 + 9}\sqrt{4 + 1 + 4}}$$

$$= \frac{-4}{\sqrt{13}\sqrt{9}}$$

$$\theta = \arccos[-4/(3\sqrt{13})] \approx 1.9496 \text{ rad}$$

Note When we find $\cos\theta < 0$ for an angle between two lines, then θ is an angle in the second quadrant. Hence θ is not the smaller angle between the lines, but its supplement. The basic fact to remember here is that $\cos(\pi - \theta) = -\cos\theta$.

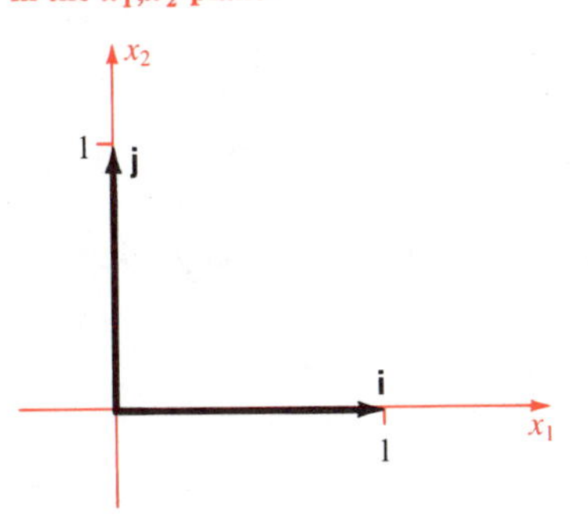

Figure 12-3-11
The basic unit vectors in the x_1,x_2-plane

The Basic Unit Vectors

In $\mathbf{R}^2$ the unit vectors along the positive coordinate axes (Figure 12-3-11) are

$$\mathbf{i} = (1, 0) \quad \text{and} \quad \mathbf{j} = (0, 1)$$

If $\mathbf{v}$ is any plane vector, then

$$\mathbf{v} = (v_1, v_2) = v_1\mathbf{i} + v_2\mathbf{j}$$

Thus $\mathbf{v}$ is the sum of two vectors, $v_1\mathbf{i}$ and $v_2\mathbf{j}$, that lie along the coordinate axes (Figure 12-3-12). If $\mathbf{u}$ is any unit vector, then

$$\mathbf{u} = (\cos\alpha)\mathbf{i} + (\cos\beta)\mathbf{j} = (\cos\alpha)\mathbf{i} + (\sin\alpha)\mathbf{j}$$

See Figure 12-3-13. Note that $\cos^2\alpha + \cos^2\beta = \cos^2\alpha + \sin^2\alpha = 1$.

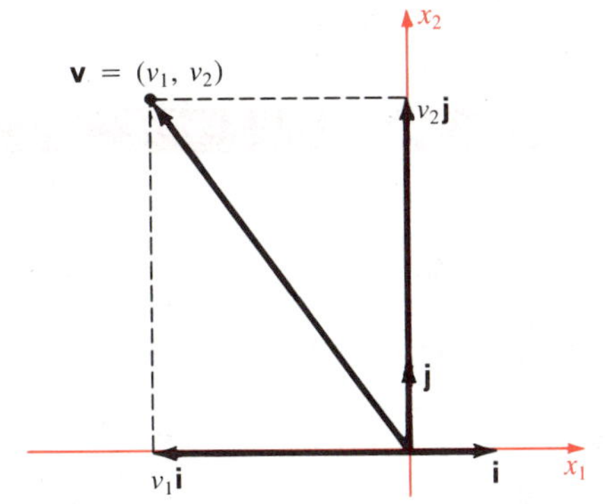

Figure 12-3-12
Plane vector $\mathbf{v}$

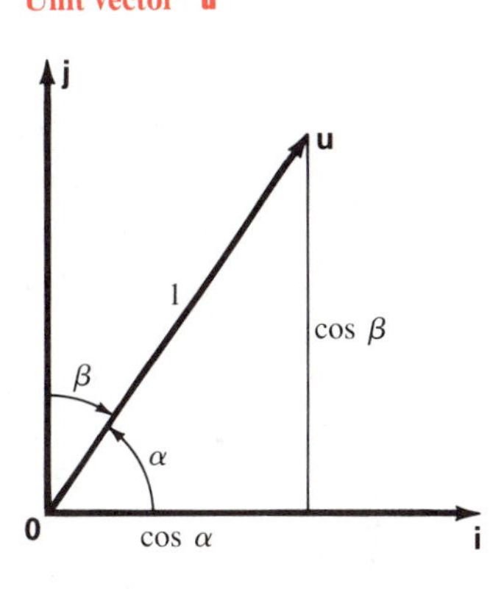

Figure 12-3-13
Unit vector $\mathbf{u}$

In space, the situation is this. The three unit-length vectors along the positive coordinate axes (Figure 12-3-14) are

$$\mathbf{i} = (1, 0, 0) \qquad \mathbf{j} = (0, 1, 0) \qquad \mathbf{k} = (0, 0, 1)$$

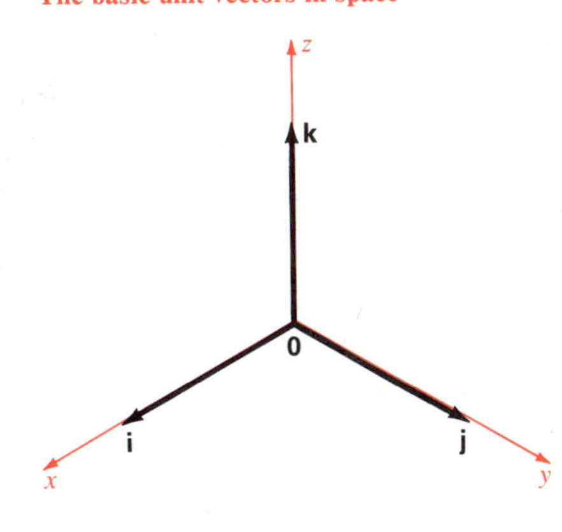

Figure 12-3-14
The basic unit vectors in space

If $\mathbf{v}$ is any vector, then

$$\begin{aligned}\mathbf{v} &= (v_1, v_2, v_3) = v_1(1, 0, 0) + v_2(0, 1, 0) + v_3(0, 0, 1)\\ &= v_1\mathbf{i} + v_2\mathbf{j} + v_3\mathbf{k}\end{aligned}$$

Thus $\mathbf{v}$ is the sum of three vectors $v_1\mathbf{i}$, $v_2\mathbf{j}$, $v_3\mathbf{k}$ that lie along the three coordinate axes, and its components can be expressed as dot products:

$$\mathbf{v}\cdot\mathbf{i} = (v_1, v_2, v_3)\cdot(1, 0, 0) = v_1 \qquad \mathbf{v}\cdot\mathbf{j} = v_2 \qquad \mathbf{v}\cdot\mathbf{k} = v_3$$

Components of a Vector

Each vector $\mathbf{v} = (v_1, v_2, v_3)$ can be expressed as

$$\mathbf{v} = v_1\mathbf{i} + v_2\mathbf{j} + v_3\mathbf{k}$$

where $v_1 = \mathbf{v}\cdot\mathbf{i}$, $v_2 = \mathbf{v}\cdot\mathbf{j}$, $v_3 = \mathbf{v}\cdot\mathbf{k}$.

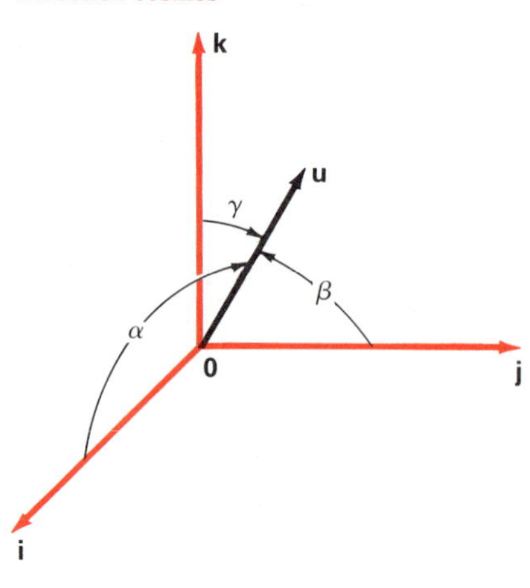

Figure 12-3-15
Direction cosines

For example, if $\mathbf{v} = (-3, 2, -2)$, then $\mathbf{v} \cdot \mathbf{i} = -3$, $\mathbf{v} \cdot \mathbf{j} = 2$, and $\mathbf{v} \cdot \mathbf{k} = -2$. Thus

$$\mathbf{v} = (\mathbf{v} \cdot \mathbf{i})\mathbf{i} + (\mathbf{v} \cdot \mathbf{j})\mathbf{j} + (\mathbf{v} \cdot \mathbf{k})\mathbf{k} = -3\mathbf{i} + 2\mathbf{j} - 2\mathbf{k}$$

Now suppose $\mathbf{u}$ is a **unit vector,** that is, a vector of length 1 (Figure 12-3-15). Let α be the angle from $\mathbf{i}$ to $\mathbf{u}$. Define β and γ similarly. Then $\mathbf{u} \cdot \mathbf{i} = \cos\alpha$, $\mathbf{u} \cdot \mathbf{j} = \cos\beta$, and $\mathbf{u} \cdot \mathbf{k} = \cos\gamma$. Hence

$$\mathbf{u} = (\cos\alpha)\mathbf{i} + (\cos\beta)\mathbf{j} + (\cos\gamma)\mathbf{k} = (\cos\alpha, \cos\beta, \cos\gamma)$$

Since $|\mathbf{u}| = 1$, we have

$$\cos^2\alpha + \cos^2\beta + \cos^2\gamma = 1$$

Unit vectors are direction indicators. Any non-zero vector $\mathbf{v}$ is a positive multiple of a unit vector $\mathbf{u}$ in the same direction as $\mathbf{v}$. In fact $\mathbf{v} = |\mathbf{v}|\mathbf{u}$, so

$$\mathbf{u} = \frac{1}{|\mathbf{v}|}\mathbf{v} \qquad (\mathbf{v} \neq \mathbf{0})$$

Direction Cosines

Each non-zero vector $\mathbf{v}$ can be expressed as

$$\mathbf{v} = |\mathbf{v}|\mathbf{u}$$

where $\mathbf{u}$ is a unit vector or as

$$\mathbf{v} = |\mathbf{v}|(\cos\alpha, \cos\beta, \cos\gamma)$$

where α, β, and γ are the angles between $\mathbf{v}$ and the x-, y-, and z-axes respectively, as shown in Figure 12-3-15.

The numbers $\cos\alpha$, $\cos\beta$, $\cos\gamma$ are called the **direction cosines** of $\mathbf{v}$. They satisfy

$$\cos^2\alpha + \cos^2\beta + \cos^2\gamma = 1$$

Exercises

Compute

1 $(8, 2, 1) \cdot (3, 0, 5)$
2 $(-1, -1, -1) \cdot (1, 2, 3)$
3 $(1, 0, 2) \cdot [(1, 4, 1) + (2, 0, -3)]$
4 $|(2, -4, 7)|$
5 $|3\mathbf{i} - \mathbf{j} + \mathbf{k}|$
6 $|(-1, -1, 0) - (3, 5, 2)|$
7 $|\frac{1}{3}\sqrt{3}(-1, 1, 1)|$
8 $[3\mathbf{j} - (1, 1, 2)] \cdot (4\mathbf{j} - \mathbf{k})$

Find the angle between the vectors

9 $(4, 3, 0)$ $(-3, 0, 4)$
10 $(1, 2, 2)$ $(-2, 1, -2)$
11 $(6, 1, 5)$ $(-2, -3, 3)$
12 $(-5, 6, 1)$ $(2, 3, -8)$
13 $(1, 1, -1)$ $(2, 0, 4)$
14 $(2, 2, 2)$ $(-2, 2, -2)$

Compute the distance between the points

15 $(0, 1, 2)$ $(5, -3, 1)$
16 $(1, 1, 1)$ $(1, -1, 2)$
17 $(7, 0, 0)$ $(2, 3, 4)$
18 $(8, 5, -1)$ $(7, 9, 3)$

Find the direction cosines

19 $(1, 0, 1)$

20 $(-1, -1, -1)$

21 $(2, 1, -3)$

22 $(4, -7, -4)$

23 Find two vectors perpendicular to $(1, -1, 2)$, neither a multiple of the other (linearly independent).

24 Find the angle between the line joining $(0, 0, 0)$ to $(1, 1, 1)$ and the line joining $(1, 0, 0)$ to $(0, 1, 0)$. [Hint Translate the angle to the origin.]

Use coordinates to show that

25 $\mathbf{v} \cdot \mathbf{w} = \mathbf{w} \cdot \mathbf{v}$

26 $(a\mathbf{v}) \cdot \mathbf{w} = \mathbf{v} \cdot (a\mathbf{w}) = a(\mathbf{v} \cdot \mathbf{w})$

27 $(\mathbf{u} + \mathbf{v}) \cdot \mathbf{w} = \mathbf{u} \cdot \mathbf{w} + \mathbf{v} \cdot \mathbf{w}$

28 Let $\mathbf{u}$ be a unit vector. Show that the formula $\mathbf{v} = (\mathbf{v} \cdot \mathbf{u})\mathbf{u} + [\mathbf{v} - (\mathbf{v} \cdot \mathbf{u})\mathbf{u}]$ expresses $\mathbf{v}$ as the sum of two vectors, one parallel to $\mathbf{u}$, the other perpendicular to $\mathbf{u}$, and is the only such expression.

29 Prove the **Cauchy-Schwarz inequality:**

$$|\mathbf{v} \cdot \mathbf{w}| \leq |\mathbf{v}| \cdot |\mathbf{w}|$$

30 (cont.) Now prove the triangle inequality: $|\mathbf{v} + \mathbf{w}| \leq |\mathbf{v}| + |\mathbf{w}|$. Hint:

$$\begin{aligned} |\mathbf{v} + \mathbf{w}|^2 &= |(\mathbf{v} + \mathbf{w}) \cdot (\mathbf{v} + \mathbf{w})| \\ &= |(\mathbf{v} + \mathbf{w}) \cdot \mathbf{v} + (\mathbf{v} + \mathbf{w}) \cdot \mathbf{w}| \\ &\leq |\mathbf{v} + \mathbf{w}| \cdot |\mathbf{v}| + |\mathbf{v} + \mathbf{w}| \cdot |\mathbf{w}| \end{aligned}$$

31 Let $\mathbf{a}$ and $\mathbf{b}$ be points of $\mathbf{R}^3$ and $c > 0$. Show that $(\mathbf{x} - \mathbf{a}) \cdot (\mathbf{x} - \mathbf{b}) = c^2$ is the equation of a sphere. Find its center and radius.

32 Show that $\Sigma_1^n |\mathbf{x} - \mathbf{a}_i|^2 = k^2$ is the equation of a sphere provided k is sufficiently large.

33 Suppose $\mathbf{u}$, $\mathbf{v}$, $\mathbf{w}$ are mutually orthogonal and $\mathbf{x} = a\mathbf{u} + b\mathbf{v} + c\mathbf{w}$. Find an expression for $|\mathbf{x}|$.

34 Use vectors to prove that the diagonals of a rhombus are perpendicular.

35 Interpret geometrically the conditions

$$|\mathbf{a}|^2 = |\mathbf{b}|^2 = 2\mathbf{a} \cdot \mathbf{b} \neq 0$$

36 Let $\mathbf{u} = (\cos\alpha_1, \cos\alpha_2, \cos\alpha_3)$ and $\mathbf{v} = (\cos\beta_1, \cos\beta_2, \cos\beta_3)$ be two unit vectors. Show that the angle θ between them satisfies

$$\cos\theta = \cos\alpha_1 \cos\beta_1 + \cos\alpha_2 \cos\beta_2 + \cos\alpha_3 \cos\beta_3$$

Interpret the formula when $\alpha_3 = \beta_3 = \frac{1}{2}\pi$.

12-4 Lines and Planes

In this section we learn how to describe lines and planes in $\mathbf{R}^3$ by equations. Let us begin with lines. A line in space can be given geometrically in three ways: as the line through two points, as the intersection of two planes, or as the line through a point in a specified direction. Let us start with the third way.

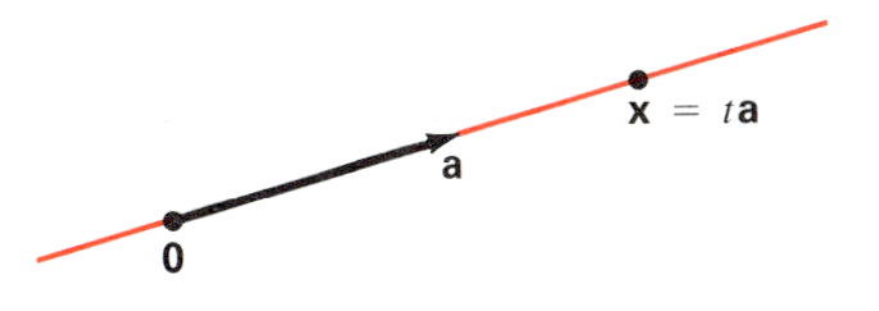

Figure 12-4-1
The line of multiples of $\mathbf{a}$

Parametric Form of a Line

A direction in space is described by a non-zero vector $\mathbf{a}$ in that direction. Let L be the line through $\mathbf{0}$ in the direction of $\mathbf{a}$. Then, as Figure 12-4-1 shows, each point $\mathbf{x}$ of L is a multiple $t\mathbf{a}$ of $\mathbf{a}$. Here $t > 0$ if $\mathbf{x}$ is on the same side of $\mathbf{0}$ as the terminal point of $\mathbf{a}$, and $t < 0$ if $\mathbf{x}$ is on the opposite side of $\mathbf{0}$. Therefore L is the set of *all* multiples $\mathbf{x} = t\mathbf{a}$ of $\mathbf{a}$, including $\mathbf{0} = 0 \cdot \mathbf{a}$.

Now suppose we are given $\mathbf{a} \neq 0$ and a point $\mathbf{b}$. We want the line through $\mathbf{b}$ in the direction determined by $\mathbf{a}$, that is, the line through $\mathbf{b}$ parallel to $\mathbf{a}$. Take any point $\mathbf{x}$ of this line. The line through $\mathbf{x}$ parallel to the vector $\mathbf{b}$ meets the line through $\mathbf{a}$, determining a parallelogram (Figure 12-4-2, next page) with vertices $\mathbf{0}$, $\mathbf{b}$, $\mathbf{x}$, and $t\mathbf{a}$. By the parallelogram rule for vector addition, $\mathbf{x} = t\mathbf{a} + \mathbf{b}$.

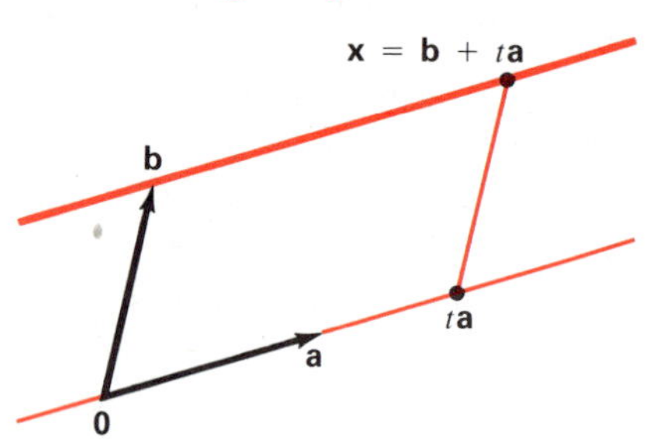

Figure 12-4-2
The line through **b** parallel to **a**

Conversely, each point of the form $\mathbf{a} = t\mathbf{a} + \mathbf{b}$ is on the line through $\mathbf{b}$ parallel to $\mathbf{a}$, again by the parallelogram rule for vector addition.

Parametric Equation of a Line

The line through a point $\mathbf{b}$ parallel to a non-zero vector $\mathbf{a}$ consists of all points

$$\mathbf{x} = t\mathbf{a} + \mathbf{b} \qquad \textbf{(parametric vector equation)}$$

where the **parameter** t varies over all real numbers. In terms of coordinates,

$$\begin{cases} x_1 = a_1 t + b_1 \\ x_2 = a_2 t + b_2 \\ x_3 = a_3 t + b_3 \end{cases} \qquad \textbf{(parametric scalar equations)}$$

Note The word *parameter* is another word for variable, used for instance when a curve is described by a function $\mathbf{x} = \mathbf{x}(t)$ on a real interval into the plane (or space). The variable t is then referred to by the fancier word "parameter."

Example 1

Find all points (x, y, z) on the line through the point $(-2, 1, 2)$ parallel to the vector $(2, -1, 3)$. Are the points $(0, 0, 4)$ or $(-6, 3, -4)$ on this line?

Solution The typical point on the line in question is

$$\begin{aligned}(x, y, z) &= t(2, -1, 3) + (-2, 1, 2) \\ &= (2t - 2, -t + 1, 3t + 2)\end{aligned}$$

In scalar form, the line consists of all (x, y, z) such that

$$x = 2t - 2 \qquad y = -t + 1 \qquad z = 3t + 2 \qquad (-\infty < t < \infty)$$

Clearly $x = 0$ only if $2t - 2 = 0$, that is, $t = 1$. But then $y = 0$ and $z = 5 \neq 4$, so $(0, 0, 4)$ is *not* on the line. However, $x = -6$ implies $2t - 2 = -6$, $t = -2$; then $y = 3$ and $z = -4$. Hence $(-6, 3, -4)$ *is* a point of the line.

Suppose we want the line through two distinct points $\mathbf{a}$ and $\mathbf{b}$. Its direction is determined by the vector $\mathbf{b} - \mathbf{a}$. Hence the line we want is the line that passes through $\mathbf{a}$ (or $\mathbf{b}$) and is parallel to the vector $\mathbf{b} - \mathbf{a}$. We know this line is

$$\mathbf{x} = t(\mathbf{b} - \mathbf{a}) + \mathbf{a} = (1 - t)\mathbf{a} + t\mathbf{b} \qquad (-\infty < t < \infty)$$

Line through Two Points

Given $\mathbf{a} \neq \mathbf{b}$, the line through $\mathbf{a}$ and $\mathbf{b}$ consists of all points

$$\mathbf{x} = (1 - t)\mathbf{a} + t\mathbf{b} \qquad \text{where} \quad -\infty < t < \infty$$

In scalar form,

$$x_1 = (1 - t)a_1 + tb_1$$
$$x_2 = (1 - t)a_2 + tb_2$$
$$x_3 = (1 - t)a_3 + tb_3$$

Example 2

Find the line through $(3, -1, 2)$ and $(4, 1, 1)$. Where does it meet the x_1, x_2-plane?

Solution The line is

$$\mathbf{x} = (1 - t)(3, -1, 2) + t(4, 1, 1) = (t + 3, 2t - 1, -t + 2)$$

This line meets the x_1, x_2-plane where $x_3 = 0$. This happens where $-t + 2 = 0$, that is, $t = 2$. Then $x_1 = 5$ and $x_2 = 3$, hence $\mathbf{x} = (5, 3, 0)$.

Equation of a Plane

A plane can be described in several ways, for example as the plane through three non-collinear points, as the plane through a line and a point not on the line, or as the plane through two intersecting (or parallel) lines.

We begin our study of planes with yet another possibility: the plane P through a point $\mathbf{c}$ and perpendicular to the direction determined by a non-zero vector $\mathbf{a}$. See Figure 12-4-3. Now a point $\mathbf{x}$ lies on P if and only if the segment $\mathbf{xc}$ is perpendicular to $\mathbf{a}$, that is, the vector $\mathbf{x} - \mathbf{c}$ is perpendicular to $\mathbf{a}$. But two vectors are perpendicular if and only if their inner product is 0.

Figure 12-4-3
The plane through $\mathbf{c}$ perpendicular to $\mathbf{a}$

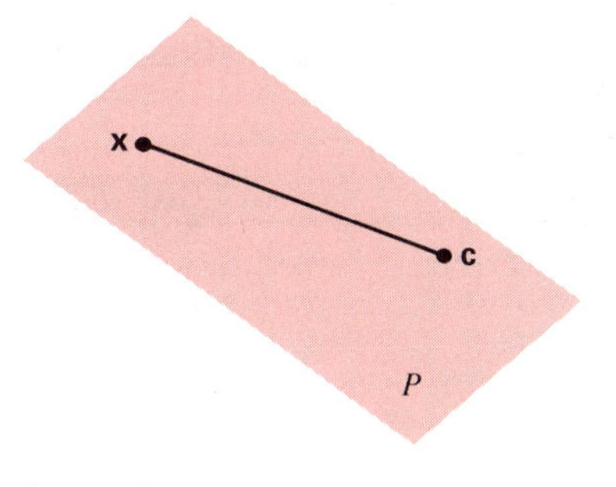

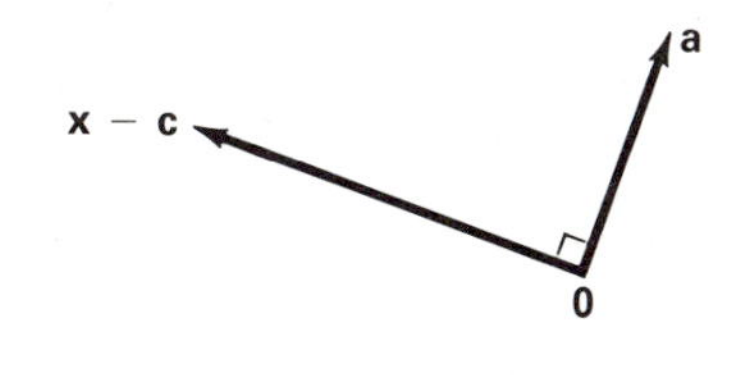

Equations of Planes

The plane that passes through a point $\mathbf{c}$ and is perpendicular to a non-zero vector $\mathbf{a}$ consists of all points $\mathbf{x}$ such that

$$(\mathbf{x} - \mathbf{c}) \cdot \mathbf{a} = 0 \qquad \text{that is} \qquad \mathbf{x} \cdot \mathbf{a} = \mathbf{c} \cdot \mathbf{a}$$

The corresponding scalar equation is

$$a_1(x_1 - c_1) + a_2(x_2 - c_2) + a_3(x_3 - c_3) = 0$$

that is,

$$a_1x_1 + a_2x_2 + a_3x_3 = b \quad \text{where} \quad b = a_1c_1 + a_2c_2 + a_3c_3$$

Conversely, each equation

$$a_1x_1 + a_2x_2 + a_3x_3 = b$$

where $(a_1, a_2, a_3) \neq (0, 0, 0)$, is the equation of a plane. For if we set $\mathbf{a} = (a_1, a_2, a_3)$, then the equation can be written

$$\mathbf{a} \cdot \mathbf{x} = b$$

To write this in the form $\mathbf{a} \cdot (\mathbf{x} - \mathbf{c}) = b$, we need one definite solution of this equation, one vector $\mathbf{c}$ such that $\mathbf{a} \cdot \mathbf{c} = b$. Not hard to find! For instance, if $a_3 \neq 0$, then $\mathbf{c} = (0, 0, b/a_3)$ is such a vector. More systematically $\mathbf{c} = (b/|\mathbf{a}|^2)\mathbf{a}$ does the trick in all cases because

$$\mathbf{a} \cdot \mathbf{c} = \mathbf{a} \cdot (b/|\mathbf{a}|^2)\mathbf{a} = (b/|\mathbf{a}|^2)\mathbf{a} \cdot \mathbf{a} = b$$

since $|\mathbf{a}|^2 = \mathbf{a} \cdot \mathbf{a}$.

Thus there is a vector $\mathbf{c}$ such that $\mathbf{a} \cdot \mathbf{c} = b$, so the equation $\mathbf{a} \cdot \mathbf{x} = b$ can be written

$$\mathbf{a} \cdot \mathbf{x} = \mathbf{a} \cdot \mathbf{c} \qquad \text{that is} \qquad \mathbf{a} \cdot (\mathbf{x} - \mathbf{c}) = 0$$

This is exactly the condition that $\mathbf{x}$ is on the plane through $\mathbf{c}$ perpendicular to $\mathbf{a}$.

For example, consider the plane given by the scalar equation

$$3x_1 - x_2 - 4x_3 = 2$$

Obviously, $\mathbf{c} = (0, -2, 0)$ satisfies the equation, so the equation can be written as

$$(3, -1, -4) \cdot [(x_1, x_2, x_3) - (0, -2, 0)] = 0$$

This describes the plane through the point $\mathbf{c} = (0, -2, 0)$ orthogonal to the vector $\mathbf{a} = (3, -1, -4)$.

Normal Form

We have seen that a given plane P can be described by an equation of the form

$$\mathbf{a} \cdot \mathbf{x} = b \qquad \text{where} \quad \mathbf{a} \neq 0$$

If the vector $\mathbf{a}$ is a *unit* vector (length 1), we say that the equation is in **normal form.**

Suppose $\mathbf{a} \cdot \mathbf{x} = b$ is not in normal form. Then there is an equivalent equation for P that is in normal form. Indeed, the equation

$$(c\mathbf{a}) \cdot \mathbf{x} = cb$$

describes P just as well as $\mathbf{a} \cdot \mathbf{x} = b$ does, provided $c \neq 0$. In particular, there are two choices of c that make $c\mathbf{a}$ a unit vector:

$$c = \frac{1}{|\mathbf{a}|} \quad \text{and} \quad c = \frac{-1}{|\mathbf{a}|}$$

We choose either one of these to obtain a normal form. Thus we set

$$n = \pm\frac{1}{|\mathbf{a}|}\mathbf{a} \quad \text{and} \quad p = \pm\frac{1}{|\mathbf{a}|}b$$

and obtain

$$\mathbf{n} \cdot \mathbf{x} = p \qquad \text{where} \quad |\mathbf{n}| = 1$$

Figure 12-4-4
Normal form of a plane:
$\mathbf{n} \cdot \mathbf{x} = p$ where $|\mathbf{n}| = 1$

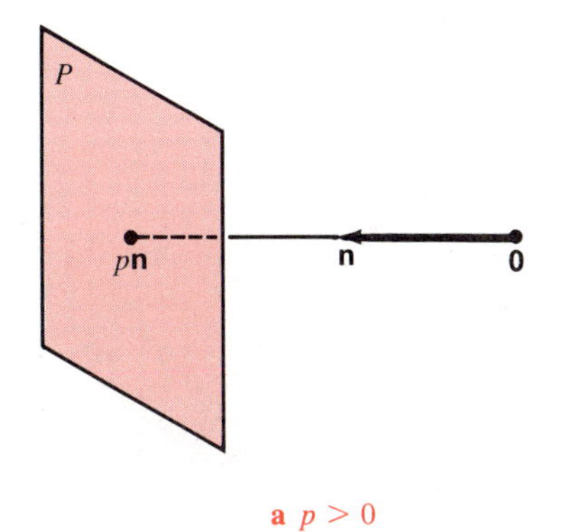

a $p > 0$

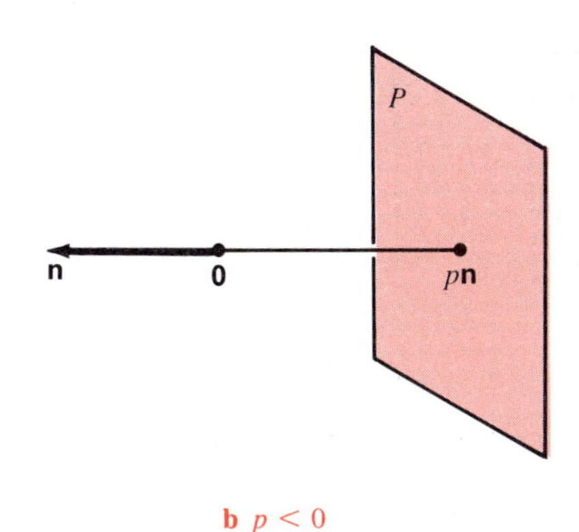

b $p < 0$

Basically, all we do is replace the vector $\mathbf{a}$, which is perpendicular (normal) to the plane P, by a unit vector $\mathbf{n}$ in the direction of $\mathbf{a}$, hence also perpendicular to P.

Let us show that the constant p in the normal form has a neat geometric interpretation (Figure 12-4-4). Since $\mathbf{n}$ is perpendicular to P, the line $\mathbf{x} = t\mathbf{n}$ of the vector $\mathbf{n}$ pierces P at the point of P closest to $\mathbf{0}$. This must be the point $p\mathbf{n}$ for the following reasons. First, $p\mathbf{n}$ obviously lies on the line. Second, $p\mathbf{n}$ lies on the plane because

$$\mathbf{n} \cdot (p\mathbf{n}) = p(\mathbf{n} \cdot \mathbf{n}) = p$$

Therefore the distance from $\mathbf{0}$ to the plane P is $|p\mathbf{n}| = |p|$. The number p itself is the *signed* distance from $\mathbf{0}$ to P. If $\mathbf{n}$ points toward P, then $p > 0$; if $\mathbf{n}$ points away from P, then $p < 0$.

Normal Form

Each plane P has a normal form

$$\mathbf{n} \cdot \mathbf{x} = p \qquad \text{where} \quad |\mathbf{n}| = 1$$

(There are two choices of the unit vector $\mathbf{n}$.) The constant p is the signed distance from $\mathbf{0}$ to P along $\mathbf{n}$.

Example 3

Give a normal form for the plane

$$2x_1 - x_2 + 2x_3 = -15$$

Also find the distance from the origin to the plane.

Solution Set $\mathbf{a} = (2, -1, 2)$ so the given equation in vector form is $\mathbf{a} \cdot \mathbf{x} = -15$. Now

$$|\mathbf{a}|^2 = 2^2 + (-1)^2 + 2^2 = 9 \qquad \text{so} \qquad |\mathbf{a}| = 3$$

Set

$$\mathbf{n} = (1/|\mathbf{a}|)\mathbf{a} = \tfrac{1}{3}\mathbf{a} = (\tfrac{2}{3}, -\tfrac{1}{3}, \tfrac{2}{3})$$

Then $\mathbf{n}$ is a unit vector. From the given equation $\mathbf{a} \cdot \mathbf{x} = -15$ we have the relation $(\tfrac{1}{3}\mathbf{a}) \cdot \mathbf{x} = \tfrac{1}{3}(-15)$, that is,

$$\mathbf{n} \cdot \mathbf{x} = -5 \qquad \text{or} \qquad \tfrac{2}{3}x_1 - \tfrac{1}{3}x_2 + \tfrac{2}{3}x_3 = -5$$

Either of these is a normal form. Here $p = -5$, so the distance from $\mathbf{0}$ to the plane is $|p| = 5$.

Figure 12-4-5
Distance, point $\mathbf{c}$ to plane P:
$D = |t| = |p - \mathbf{c} \cdot \mathbf{n}|$

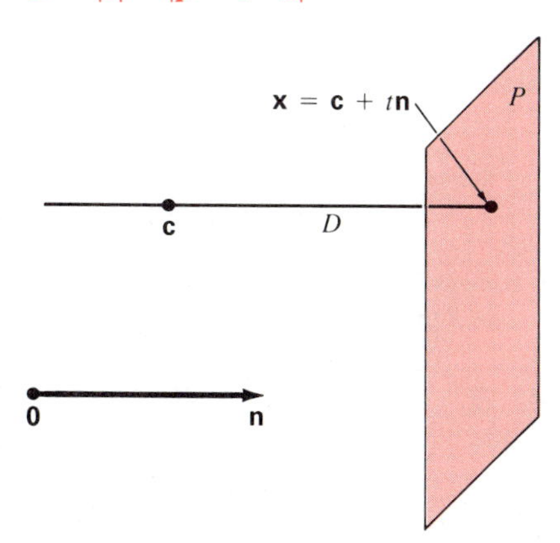

Distance from a Point to a Plane

Suppose we are given a plane P and a point $\mathbf{c}$ in space. We want the distance D from $\mathbf{c}$ to P. See Figure 12-4-5. First we express the plane in normal form $\mathbf{x} \cdot \mathbf{n} = p$. Then we drop a perpendicular from $\mathbf{c}$

onto P. It is parallel to $\mathbf{n}$, hence lies on the line $\mathbf{x} = \mathbf{c} + t\mathbf{n}$. Its foot is the point $\mathbf{x}$ on this line that satisfies the equation of the plane:

$$(\mathbf{c} + t\mathbf{n}) \cdot \mathbf{n} = p \qquad \text{that is} \qquad \mathbf{c} \cdot \mathbf{n} + t = p$$

Thus

$$t = p - \mathbf{c} \cdot \mathbf{n} \quad \text{and} \quad \mathbf{x} = \mathbf{c} + (p - \mathbf{c} \cdot \mathbf{n})\mathbf{n}$$

Therefore the required distance is

$$D = |\mathbf{x} - \mathbf{c}| = |(p - \mathbf{c} \cdot \mathbf{n})\mathbf{n}| = |p - \mathbf{c} \cdot \mathbf{n}| \cdot |\mathbf{n}| = |p - \mathbf{c} \cdot \mathbf{n}|$$

Distance from a Point to a Plane

The distance from a point $\mathbf{c}$ to a plane P with the normal form $\mathbf{n} \cdot \mathbf{x} = p$ is

$$D = |p - \mathbf{c} \cdot \mathbf{n}|$$

Example 4

Find the distance from $(2, -1, -4)$ to the plane

$$3x_1 - 6x_2 + 2x_3 = 28$$

Solution Since $3^2 + (-6)^2 + 2^2 = 49 = 7^2$, a normal form is

$$\tfrac{3}{7}x_1 - \tfrac{6}{7}x_2 + \tfrac{2}{7}x_3 = \tfrac{28}{7} = 4$$

Thus $\mathbf{n} = (\frac{3}{7}, -\frac{6}{7}, \frac{2}{7})$ and $p = 4$, so by the distance formula the required distance is

$$D = |4 - (2, -1, -4) \cdot (\tfrac{3}{7}, -\tfrac{6}{7}, \tfrac{2}{7})| = |4 - \tfrac{4}{7}| = \tfrac{24}{7}$$

Exercises

Find whether $\mathbf{c}$ is on the segment $\mathbf{ab}$, where $\mathbf{a}$, $\mathbf{b}$, $\mathbf{c}$ are

1 $(0, 3, 0)$ $(3, 0, 3)$ $(2, 2, 2)$

2 $(2, 1, -1)$ $(5, -2, 1)$ $(4, -1, 0)$

3 $(3, 3, -2)$ $(-2, 3, -2)$ $(2, 3, -2)$

4 $(2, 2, -2)$ $(4, -6, 1)$ $(3, -2, -1)$

Find an angle between line $\mathbf{ab}$ and line $\mathbf{cd}$, where $\mathbf{a}$, $\mathbf{b}$, $\mathbf{c}$, $\mathbf{d}$ are

5 $(3, 3, 1)$ $(-2, 1, -1)$ $(1, 1, 1)$ $(2, 1, -2)$

6 $(-1, -1, -3)$ $(1, 1, 1)$ $(2, 2, 2)$ $(0, 1, 1)$

The point that is r of the way from $\mathbf{a}$ to $\mathbf{b}$ is $\mathbf{a} + r(\mathbf{b} - \mathbf{a}) = (1 - r)\mathbf{a} + r\mathbf{b}$.

7 Find the point $\frac{1}{3}$ of the way from $(1, 1, 1)$ to $(0, 0, 0)$.

8 Find the point $\frac{2}{7}$ of the way from $(1, 0, 1)$ to $(-1, -1, -1)$.

Find the intersections of the line $\mathbf{ab}$ with the three coordinate planes, where $\mathbf{a}$ and $\mathbf{b}$ are

9 $(-2, 3, -4)$ $(-1, -2, 5)$

10 $(-3, 4, 4)$ $(-4, 4, -3)$

11 $(-1, 1, -1)$ $(2, -2, -1)$

12 $(1, 1, 1)$ $(-3, -3, -3)$

13 Find the distance from $\mathbf{0}$ to the line through $(1, 1, 1)$ and $(1, 0, 1)$.
[Hint If $\mathbf{x}$ is the point of the line that is closest to $\mathbf{0}$, then $\mathbf{0x}$ is orthogonal to the line.]

14 Find the distance from $(1, 1, 1)$ to the line through $(1, 2, 3)$ and $(-3, -2, -1)$.
[Hint See the previous hint.]

15 If t is time, when and where does the parametric line $\mathbf{x} = t(4, 4, 1) + (3, 2, -5)$ pierce the x, y-plane?

16 Describe the set of vectors
$\mathbf{x}(t) = t^2(4, 4, 1) + (3, 2, -5)$
as t goes from $-\infty$ to $+\infty$.

17 Give a parametric vector equation for the line through $\mathbf{a}$ parallel to the line $\mathbf{bc}$.

18 Give parametric equations for the line through $(1, 2, -2)$ perpendicular to the plane $x_1 + x_2 + x_3 = 0$.

Express in normal form

19 $x_1 - 2x_2 + 2x_3 = 1$

20 $2x_1 + 6x_2 - 3x_3 = 14$

21 $-8x_1 + x_2 - 4x_3 = 27$

22 $3x_1 - 2x_2 - 6x_3 = 4$

23 $x_1 + x_2 + x_3 = 3$

24 $x_1 - x_2 + x_3 = -12$

Find the distance from the plane to the point

25 $x_1 + x_2 + x_3 = 2 \quad (1, 1, 1)$

26 $2x_1 - x_2 - 2x_3 = 4 \quad (0, 0, 1)$

27 $-3x_1 - x_2 + 4x_3 = 8 \quad (0, 0, 2)$

28 $-3x_1 + 12x_2 + 4x_3 = 13 \quad (1, 0, -1)$

29 Find the angle θ between the line $\mathbf{x} = t\mathbf{a} + \mathbf{b}$ and the plane in normal form $\mathbf{n} \cdot \mathbf{x} = p$.

30 Find the distance between the parallel planes $4x - y - 3z = 1$ and $4x - y - 3z = 6$.

Let $\mathbf{x} = t\mathbf{a} + \mathbf{b}$ be a line and $\mathbf{n} \cdot \mathbf{x} = p$ be a plane in normal form

31 Prove that the line and plane are parallel if and only if $\mathbf{a} \cdot \mathbf{n} = 0$.

32 (cont.) Prove that the line is on the plane if and only if $\mathbf{a} \cdot \mathbf{n} = 0$ and $\mathbf{b} \cdot \mathbf{n} = p$.

33 (cont.) Suppose $\mathbf{a} \cdot \mathbf{n} \neq 0$. Prove that the point of intersection of the line and the plane is

$$\mathbf{z} = [(p - \mathbf{n} \cdot \mathbf{b})/(\mathbf{a} \cdot \mathbf{n})]\mathbf{a} + \mathbf{b}.$$

34 Let $\mathbf{m} \cdot \mathbf{x} = p$ and $\mathbf{n} \cdot \mathbf{x} = q$ be two non-parallel planes in normal form. Let θ be one of their (dihedral) angles of intersection. Find $\cos\theta$.

35 Let L be the line of intersection of two non-parallel planes $\mathbf{a} \cdot \mathbf{x} = c$ and $\mathbf{b} \cdot \mathbf{x} = d$. Show that the most general plane containing L is

$$(s\mathbf{a} + t\mathbf{b}) \cdot \mathbf{x} = sc + td$$

where $s^2 + t^2 = 1$.

36 Describe the set of points $\mathbf{x} = (1 - t)\mathbf{a} + t\mathbf{b}$ for

a $0 \leq t < +\infty$ **b** $-\infty < t \leq 1$.

37 Let $\mathbf{a}_1, \cdots, \mathbf{a}_r$ be points of $\mathbf{R}^3$. Consider all planes $\mathbf{n} \cdot \mathbf{x} = p$ in normal form such that

$$(\mathbf{a}_1 \cdot \mathbf{n} - p) + \cdots + (\mathbf{a}_r \cdot \mathbf{n} - p) = 0$$

Show that these planes pass through a common point.

38 Let $\mathbf{x} \cdot \mathbf{m} = p$ and $\mathbf{x} \cdot \mathbf{n} = q$ be non-parallel planes in normal form. Show that $\mathbf{x} \cdot \mathbf{m} - p = \pm(\mathbf{x} \cdot \mathbf{n} - q)$ are the two planes through their intersection that bisect their dihedral angles.

Let $\mathbf{x} = t\mathbf{u} + \mathbf{b}$ be a parametric line, where $\mathbf{u}$ is a unit vector, and let $\mathbf{c}$ be a point.

***39** Find the point on the line closest to $\mathbf{c}$.

***40** Find the distance D from $\mathbf{c}$ to the line (in terms of $\mathbf{u}$, $\mathbf{b}$, and $\mathbf{c}$).

12-5 Linear Systems

Suppose we are given three planes

$$\mathbf{a}_1 \cdot \mathbf{x} = d_1 \qquad \mathbf{a}_2 \cdot \mathbf{x} = d_2 \qquad \mathbf{a}_3 \cdot \mathbf{x} = d_3$$

How can we find their intersection? We must find an $\mathbf{x} = (x, y, z)$ that satisfies all three equations. In coordinates, the problem is to solve the system of three linear equations

$$\begin{cases} a_1x + b_1y + c_1z = d_1 \\ a_2x + b_2y + c_2z = d_2 \\ a_3x + b_3y + c_3z = d_3 \end{cases}$$

for x, y, z where $a_1, \cdots, d_3$ are given constants.

Generally, there is a single common point (Figure 12-5-1a, next page). However, if the planes are parallel or if one is parallel to the intersection of the other two, then there is no common point (Figure 12-5-1b). In this

Figure 12-5-1
Possible intersection of three planes

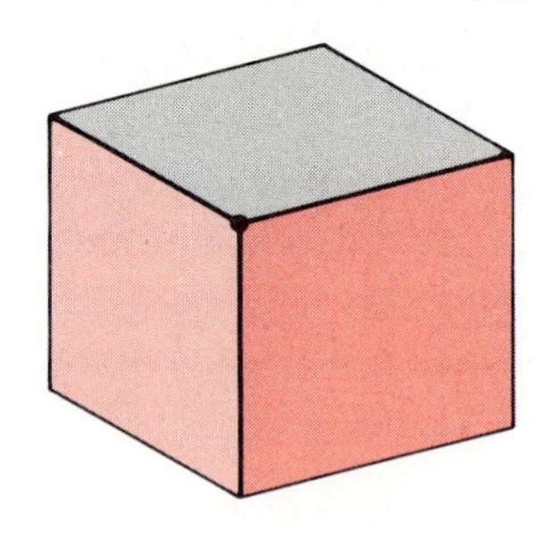

a One common point

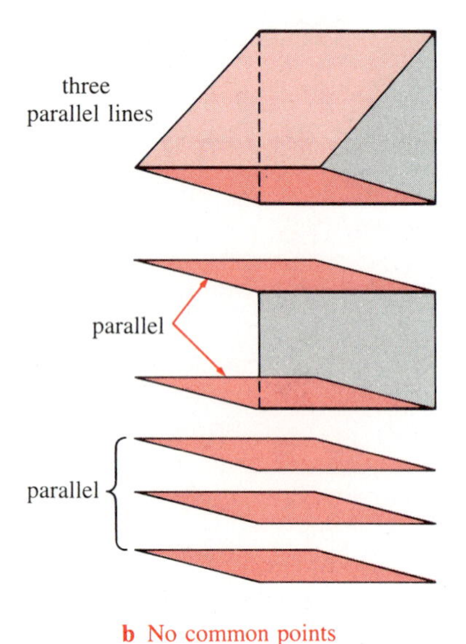

b No common points

case, the corresponding system of equations is called **inconsistent.** For example, the system

$$\begin{cases} x + y + z = 1 \\ x + y + z = 2 \\ 3x - 2y + 4z = 7 \end{cases}$$

is obviously inconsistent; the first two equations cannot both be satisfied. Geometrically, the first two planes are parallel. [From their equations, we see that both planes are perpendicular to the vector $(1, 1, 1)$.]

Three planes have more than one common point if they pass through a common line or if all of them coincide (Figure 12-5-2). In these cases, the corresponding system of equations is called **underdetermined.** For example the system

$$\begin{cases} x - 2y + 3z = 5 \\ 8x + 7y + z = 2 \\ 2x - 4y + 6z = 10 \end{cases}$$

is underdetermined; the third equation is twice the first. Geometrically, the first and third planes coincide. The system represents two distinct planes that have a line in common.

Elimination

Geometric reasoning shows that the set of solutions of a linear system of three equations in three unknowns is either (a) empty, (b) a single point, (c) a line, (d) a plane, or (e) the whole space $\mathbf{R}^3$. Let us recall briefly two methods of solving linear systems. The first, elimination, is practically self-explanatory.

Example 1

Solve the system

$$\begin{cases} 2x - y + z = 4 \\ 2x + 2y + 3z = 3 \\ 6x - 9y - 2z = 17 \end{cases}$$

Solution Eliminate x from the second and third equations as follows: subtract the first equation from the second, and subtract 3 times the first equation from the third. The result is an equivalent system of three equations (the first the same as before):

$$\begin{cases} 2x - y + z = 4 \\ 3y + 2z = -1 \\ -6y - 5z = 5 \end{cases}$$

Now eliminate y from the third equation. Add twice the second equation to the third, but keep the first two equations:

Figure 12-5-2
Three planes with more than one common point

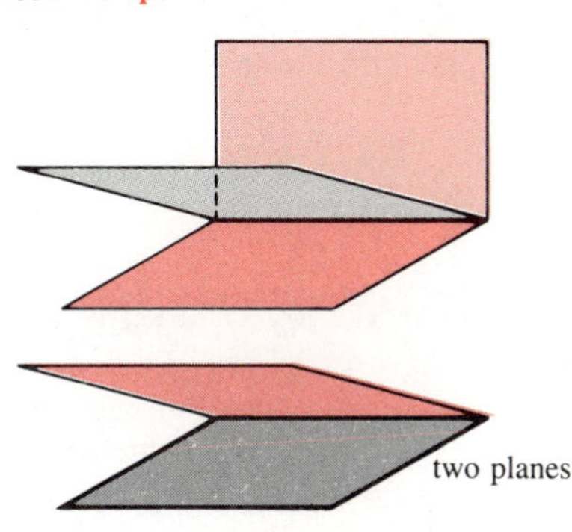

a Line of intersection

b Plane of intersection

$$\begin{cases} 2x - y + z = 4 \\ 3y + 2z = -1 \\ -z = 3 \end{cases}$$

By the third equation, $z = -3$. Substitute this into the first two equations. The result is a new system of two equations for x and y:

$$\begin{cases} 2x - y = 4 - (-3) = 7 \\ 3y = -1 - 2(-3) = 5 \end{cases}$$

By the second equation, $y = \frac{5}{3}$. Substitute this into the first equation; the result is a single equation for x:

$$2x = 7 + \tfrac{5}{3} = \tfrac{26}{3} \qquad x = \tfrac{13}{3}$$

The solution is $\mathbf{x} = (\frac{13}{3}, \frac{5}{3}, -3)$. ●

Example 2

Solve

$$\begin{cases} x + y + z = 1 \\ x - 2y + 2z = 4 \\ 2x - y + 3z = 5 \end{cases}$$

Solution Eliminate x from the second and third equations:

$$\begin{cases} x + y + z = 1 \\ -3y + z = 3 \\ -3y + z = 3 \end{cases}$$

The last two equations both say the same thing. Therefore the system is equivalent to the system

$$\begin{cases} x + y + z = 1 \\ -3y + z = 3 \end{cases}$$

and that is as far as the elimination method will go. Geometrically, this reduced system represents the intersection of *two* planes. These planes are not parallel since their normal vectors $(1, 1, 1)$ and $(0, -3, 1)$ point in different directions. Hence it is a safe bet that the solutions form a line. To get a parametric solution, set $y = t$. Then $z = 3 + 3y = 3 + 3t$ and

$$x = 1 - y - z = 1 - t - (3 + 3t) = -2 - 4t$$

Therefore the most general solution is

$$(x, y, z) = (-2 - 4t, t, 3 + 3t)$$

where $-\infty < t < \infty$. The set of solutions is the line

$$\mathbf{x} = t(-4, 1, 3) + (-2, 0, 3)$$

As a geometric check, note that this line is in the direction of the vector $(-4, 1, 3)$. Since the line supposedly lies in *both* planes, it should be perpendicular to both normals, that is, $(-4, 1, 3)$ should be perpendicular to both $(1, 1, 1)$ and $(0, -3, 1)$. It is perpendicular:

$$(-4, 1, 3) \cdot (1, 1, 1) = -4 + 1 + 3 = 0$$

$$(-4, 1, 3) \cdot (0, -3, 1) = 0 - 3 + 3 = 0$$

Review of Determinants

In elementary algebra, **determinants** of orders 2 and 3 are defined:

$$\begin{vmatrix} a_1 & b_1 \\ a_2 & b_2 \end{vmatrix} = a_1b_2 - a_2b_1$$

$$\begin{vmatrix} a_1 & b_1 & c_1 \\ a_2 & b_2 & c_2 \\ a_3 & b_3 & c_3 \end{vmatrix} = a_1b_2c_3 + a_2b_3c_1 + a_3b_1c_2 - a_1b_3c_2 - a_2b_1c_3 - a_3b_2c_1$$

The following properties of determinants follow from the definitions.

Properties of Determinants

- If two rows (columns) are equal, the determinant is zero.
- If two rows (columns) are transposed, the determinant changes sign.
- If a multiple of one row (column) is added to another row (column), the determinant is unchanged.
- If all the terms if one row (column) are multiplied by a scalar, the determinant is multiplied by the same scalar.

We can evaluate a determinant of order 3 by the technique of **expansion by minors** of a row or column. For instance,

$$\begin{vmatrix} a_1 & b_1 & c_1 \\ a_2 & b_2 & c_2 \\ a_3 & b_3 & c_3 \end{vmatrix} = a_1\begin{vmatrix} b_2 & c_2 \\ b_3 & c_3 \end{vmatrix} - b_1\begin{vmatrix} a_2 & c_2 \\ a_3 & c_3 \end{vmatrix} + c_1\begin{vmatrix} a_2 & b_2 \\ a_3 & b_3 \end{vmatrix}$$

is the expansion by minors of the first row. Here,

$$\begin{vmatrix} a_2 & c_2 \\ a_3 & c_3 \end{vmatrix}$$

is the **minor** of b_1. It is the 2×2 determinant remaining after the row and the column containing b_1 are crossed off. Such expansions by minors follow directly from the definitions of 2×2 and 3×3 determinants.

Example 3

An expansion by minors of the second row:

$$\begin{vmatrix} 1 & -2 & -3 \\ 2 & 4 & 1 \\ 6 & -5 & 3 \end{vmatrix} = -2\begin{vmatrix} -2 & -3 \\ -5 & 3 \end{vmatrix} + 4\begin{vmatrix} 1 & -3 \\ 6 & 3 \end{vmatrix} - 1\begin{vmatrix} 1 & -2 \\ 6 & -5 \end{vmatrix}$$

A system of equations

$$\begin{cases} a_1x + b_1y + c_1z = d_1 \\ a_2x + b_2y + c_2z = d_2 \\ a_3x + b_3y + c_3z = d_3 \end{cases}$$

is both **consistent** (not inconsistent) and **determined** (not underdetermined) if and only if the **system determinant** $D \neq 0$, where

$$D = \begin{vmatrix} a_1 & b_1 & c_1 \\ a_2 & b_2 & c_2 \\ a_3 & b_3 & c_3 \end{vmatrix}$$

When this is so, the system has a unique solution, given explicitly by **Cramer's rule:**

$$x = \frac{1}{D}\begin{vmatrix} d_1 & b_1 & c_1 \\ d_2 & b_2 & c_2 \\ d_3 & b_3 & c_3 \end{vmatrix}$$

$$y = \frac{1}{D}\begin{vmatrix} a_1 & d_1 & c_1 \\ a_2 & d_2 & c_2 \\ a_3 & d_3 & c_3 \end{vmatrix}$$

$$z = \frac{1}{D}\begin{vmatrix} a_1 & b_1 & d_1 \\ a_2 & b_2 & d_2 \\ a_3 & b_3 & d_3 \end{vmatrix}$$

Cramer's rule is our second method for solving linear systems. It is derived in linear algebra courses. Note how the numerators are formed. For instance, the numerator in the expression for x, the *first* unknown, is the system determinant with its *first* column replaced by the column of d's.

Example 4

Find the intersection of the three planes

$$\begin{cases} 2x - y + z = 1 \\ 2x - 3y - 2z = 4 \\ 6x + 2y + 9z = -5 \end{cases}$$

Solution Let us try Cramer's rule. The system determinant, expanded by its first row, is

$$D = \begin{vmatrix} 2 & -1 & 1 \\ 2 & -3 & -2 \\ 6 & 2 & 9 \end{vmatrix} = 2\begin{vmatrix} -3 & -2 \\ 2 & 9 \end{vmatrix} + \begin{vmatrix} 2 & -2 \\ 6 & 9 \end{vmatrix} + \begin{vmatrix} 2 & -3 \\ 6 & 2 \end{vmatrix}$$

$$= (2)(-23) + 30 + 22 = 6 \neq 0$$

The numerator of x is the system determinant with its first column replaced by the column of constant terms. Hence

$$x = \frac{1}{6}\begin{vmatrix} 1 & -1 & 1 \\ 4 & -3 & -2 \\ -5 & 2 & 9 \end{vmatrix} = -\frac{2}{3}$$

Similarly $y = -2$ and $z = \frac{1}{3}$. The planes intersect in one point: $\mathbf{x} = (-\frac{2}{3}, -2, \frac{1}{3})$. ●

Plane through Three Points

Suppose $\mathbf{p}_1$, $\mathbf{p}_2$, and $\mathbf{p}_3$ are three non-collinear points in space. How can we find the unique plane that they determine? Suppose that $\mathbf{p}_i = (x_i, y_i, z_i)$ for $i = 1, 2, 3$. We take the plane in the form

$$ax + by + cz = d$$

The constants a, b, c, d must be found to satisfy the system

$$ax_i + by_i + cz_i = d \qquad (i = 1, 2, 3)$$

This seems to be a system of three equations in *four* unknowns! Actually not, because if $\lambda \neq 0$, then (a, b, c, d) is a solution if and only if $(\lambda a, \lambda b, \lambda c, \lambda d)$ is a solution. Thus by our choice of the scale factor λ we can reduce the problem to the case in which one of the unknowns is 1; then there are three equations in three unknowns.

Simplest is first to try $d = 1$. If the resulting system has a unique solution (a, b, c), done. If, however, the original system forces $d = 0$, then this approach leads to an inconsistent system. When that happens, we set $d = 0$ and find a, b, c anyhow. (Geometrically, the plane passes through the origin when $d = 0$.)

Example 5

Find the plane through

$$(1, -1, -3) \quad (2, 2, 4) \quad \text{and} \quad (2, 1, 1)$$

Solution We assume the plane can be written as $ax + by + cz = 1$. Then (a, b, c) must satisfy the system

$$\begin{cases} a - b - 3c = 1 \\ 2a + 2b + 4c = 1 \\ 2a + b + c = 1 \end{cases}$$

Subtract the second equation from the third, then twice the first equation from the second:

$$\begin{cases} a - b - 3c = 1 \\ \quad 4b + 10c = -1 \\ \quad -b - 3c = 0 \end{cases}$$

From the resulting second and third equations, $b = -\frac{3}{2}$, $c = \frac{1}{2}$. From the first equation, $a = 1$. The equation of the plane is

$$x - \tfrac{3}{2}y + \tfrac{1}{2}z = 1 \qquad \text{that is} \qquad 2x - 3y + z = 2$$

Alternative Solution Another possibility is to find a normal vector to the plane. Clearly the vectors

$$\mathbf{u} = (2, 2, 4) - (1, -1, -3) = (1, 3, 7)$$
$$\mathbf{v} = (2, 1, 1) - (1, -1, -3) = (1, 2, 4)$$

are parallel to the plane. Any normal vector $\mathbf{a}$ must satisfy $\mathbf{u} \cdot \mathbf{a} = 0$ and $\mathbf{v} \cdot \mathbf{a} = 0$. If $\mathbf{a} = (a, b, c)$, we must have

$$\begin{cases} a + 3b + 7c = 0 \\ a + 2b + 4c = 0 \end{cases}$$

Subtract:

$$\begin{cases} a + 3b + 7c = 0 \\ \phantom{a + {}} b + 3c = 0 \end{cases}$$

It follows that $b = -3c$ and $a = 2c$. Thus the system has a whole line of solutions $\mathbf{a} = (2c, -3c, c) = c(2, -3, 1)$, consisting of vectors perpendicular to the plane. Choose for instance $c = 1$, that is, $\mathbf{a} = (2, -3, 1)$. Then the plane has the form $\mathbf{a} \cdot \mathbf{x} = k$. Since we already know that $\mathbf{x} = (2, 2, 4)$ is on the plane, we obtain the value $k = \mathbf{a} \cdot \mathbf{x} = (2, -3, 1) \cdot (2, 2, 4) = 2$. Thus the equation of the plane is

$$2x - 3y + z = 2$$ ●

Exercises

Solve by elimination

1 $\begin{cases} x + 2y = 1 \\ 3y = 2 \end{cases}$

2 $\begin{cases} 2x = 3 \\ -x + y = 0 \end{cases}$

3 $\begin{cases} x + 2y = 1 \\ x + 3y = 2 \end{cases}$

4 $\begin{cases} x + y = a \\ x - y = b \end{cases}$

5 $\begin{cases} 2x - 3y = -1 \\ 3x + 5y = 2 \end{cases}$

6 $\begin{cases} 2x - 3y = -1 \\ -3x + 5y = 2 \end{cases}$

7 $\begin{cases} x + y + z = 0 \\ 2y - 3z = -1 \\ 3y + 5z = 2 \end{cases}$

8 $\begin{cases} 2x - y - z = 1 \\ 2y - 3z = -1 \\ -3y + 5z = 2 \end{cases}$

9 $\begin{cases} x + y - z = 0 \\ x - y + z = 0 \\ -x + y + z = 0 \end{cases}$

10 $\begin{cases} 2x + y + 3z = 1 \\ -x + 4y + 2z = 0 \\ 3x + y + z = -1 \end{cases}$

11 $\begin{cases} 2x - y - 3z = 1 \\ -x - 4y - 2z = 1 \\ 3x - y - z = 1 \end{cases}$

12 $\begin{cases} 4x + 2y - z = 0 \\ x + 3y + 2z = 0 \\ x + y + 3z = 4 \end{cases}$

Show that the system is inconsistent and interpret geometrically

13 $\begin{cases} x - y = 1 \\ -x + y = 1 \end{cases}$

14 $\begin{cases} x = 2 \\ x = 3 \end{cases}$

15 $\begin{cases} x + y + 2z = 1 \\ 3x + 5y + 7z = 2 \\ -x - y - 2z = 0 \end{cases}$

16 $\begin{cases} x + y + 2z = 1 \\ -x + 2y + z = 3 \\ y + z = 1 \end{cases}$

Find all solutions of each underdetermined system

17 $\begin{cases} 2x - 3y = 1 \\ -4x + 6y = -2 \end{cases}$

18 $\begin{cases} x + y = 0 \\ x + y = 0 \end{cases}$

19 $\begin{cases} 2x - y + z = 1 \\ 3x + y + z = 0 \\ 7x - y + 3z = 2 \end{cases}$

20 $\begin{cases} 11x + 10y + 9z = 5 \\ x + 2y + 3z = 1 \\ 3x + 2y + z = 1 \end{cases}$

Solve by Cramer's rule

21 $\begin{cases} 2x + y + 2z = 1 \\ -x + 4y + 2z = 0 \\ 3x + y + z = -1 \end{cases}$

22 $\begin{cases} 2x - y - 3z = 1 \\ -x + 4y - 2z = 1 \\ 3x - y - z = 1 \end{cases}$

23 $\begin{cases} 3x - y - z = 6 \\ -x - 4y + 2z = 0 \\ 2x - y - 3z = -1 \end{cases}$

24 $\begin{cases} x + y + 3z = 1 \\ 3x + y + z = -1 \\ -x + 4y + 2z = 0 \end{cases}$

25 Show that

$$\begin{vmatrix} x & y & 1 \\ a_1 & b_1 & 1 \\ a_2 & b_2 & 1 \end{vmatrix} = 0$$

is an equation for the line in $\mathbf{R}^2$ through the two distinct points (a_1, b_1) and (a_2, b_2).

Evaluate

26 $\begin{vmatrix} a^2 & ab & ac \\ ba & b^2 & bc \\ ca & cb & c^2 \end{vmatrix}$

27 $\begin{vmatrix} 1 & 1 & 1 \\ a & b & c \\ a^2 & b^2 & c^2 \end{vmatrix}$

28 (cont.) Suppose $a \neq b$, $b \neq c$, $c \neq a$. Show that the system

$$\begin{cases} x + y + z = d_1 \\ ax + by + cz = d_2 \\ a^2x + b^2y + c^2z = d_3 \end{cases}$$

has a unique solution.

Find the intersection of the three planes

29 $\begin{cases} 2x - y + 3z = 6 \\ -x - 4y - z = 2 \\ 3x + 2y + z = 2 \end{cases}$

30 $\begin{cases} 4x + y + z = 10 \\ 2x + 3y + z = 8 \\ -x + 2y + 3z = 3 \end{cases}$

31 $\begin{cases} x + 2y + 3z = 2 \\ -x + 8y + 7z = -2 \\ 2x - y + z = 4 \end{cases}$

32 $\begin{cases} -x - y + 2z = 4 \\ 2x + y + 6z = -1 \\ 3x + 2y + 4z = -3 \end{cases}$

Find the plane through the three points

33 $(1, 0, 1)$ $(2, 2, -2)$ $(-3, 3, 2)$

34 $(1, -1, 1)$ $(4, -1, -2)$ $(-3, 2, 8)$

35 $(1, 2, 2)$ $(-1, 1, 7)$ $(3, 5, 1)$

36 $(2, 1, -1)$ $(1, 2, 3)$ $(-4, 1, 8)$

12-6 Cross Product

Given a pair of vectors $\mathbf{v}$ and $\mathbf{w}$ in $\mathbf{R}^3$, we define a new vector $\mathbf{v} \times \mathbf{w}$.

Cross Product—Geometric Definition

The **cross product** of $\mathbf{v}$ and $\mathbf{w}$, written

$$\mathbf{v} \times \mathbf{w}$$

is the vector perpendicular to $\mathbf{v}$ and $\mathbf{w}$ whose direction is determined by the right-hand rule from the pair $\mathbf{v}$, $\mathbf{w}$, and whose magnitude is the area of the parallelogram based on $\mathbf{v}$ and $\mathbf{w}$. See Figure 12-6-1.

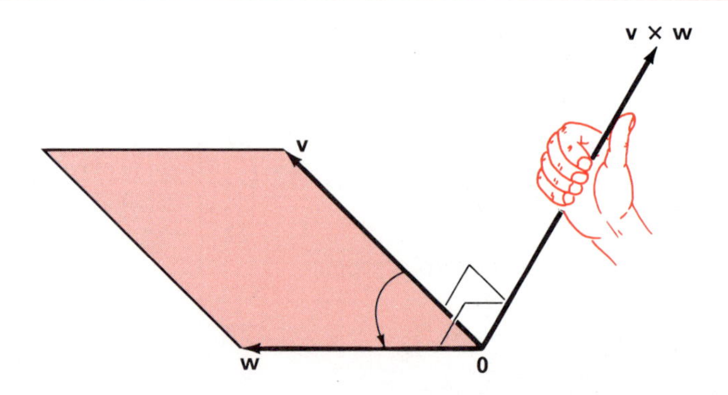

Figure 12-6-1
Cross product, geometric definition:
$|\mathbf{v} \times \mathbf{w}| = $ **(shaded area)**

Note first that the definition determines a *unique* vector $\mathbf{v} \times \mathbf{w}$, given $\mathbf{v}$ and $\mathbf{w}$. Next, if the vectors $\mathbf{v}$ and $\mathbf{w}$ are parallel, then the parallelogram collapses, so $\mathbf{v} \times \mathbf{w} = \mathbf{0}$. In particular $\mathbf{v} \times \mathbf{v} = \mathbf{0}$. Next, if $\mathbf{v}$ and $\mathbf{w}$ are interchanged, then the thumb reverses direction; hence we have $\mathbf{w} \times \mathbf{v} = -\mathbf{v} \times \mathbf{w}$. Finally, for pairs of the basic unit vectors

$$\mathbf{i} = (1, 0, 0) \qquad \mathbf{j} = (0, 1, 0) \qquad \mathbf{k} = (0, 0, 1)$$

cross products are obvious, for instance $\mathbf{i} \times \mathbf{j} = \mathbf{k}$. Let us summarize:

- $\mathbf{v} \times \mathbf{v} = \mathbf{0} \qquad \mathbf{w} \times \mathbf{v} = -\mathbf{v} \times \mathbf{w}$
- $\mathbf{i} \times \mathbf{j} = \mathbf{k} \qquad \mathbf{j} \times \mathbf{k} = \mathbf{i} \qquad \mathbf{k} \times \mathbf{i} = \mathbf{j}$

Motivated by this geometric discussion, we can *define* analytically all possible cross products of pairs of basic unit vectors:

$$\mathbf{i} \times \mathbf{i} = \mathbf{j} \times \mathbf{j} = \mathbf{k} \times \mathbf{k} = \mathbf{0}$$
$$\mathbf{i} \times \mathbf{j} = \mathbf{k} \qquad \mathbf{j} \times \mathbf{k} = \mathbf{i} \qquad \mathbf{k} \times \mathbf{i} = \mathbf{j}$$
$$\mathbf{j} \times \mathbf{i} = -\mathbf{k} \qquad \mathbf{k} \times \mathbf{j} = -\mathbf{i} \qquad \mathbf{i} \times \mathbf{k} = -\mathbf{j}$$

We use this multiplication table and the distributive law to extend the definition to $\mathbf{v} = v_1\mathbf{i} + v_2\mathbf{j} + v_3\mathbf{k}$ and $\mathbf{w} = w_1\mathbf{i} + w_2\mathbf{j} + w_3\mathbf{k}$:

$$\begin{aligned} \mathbf{v} \times \mathbf{w} &= (v_1\mathbf{i} + v_2\mathbf{j} + v_3\mathbf{k}) \times (w_1\mathbf{i} + w_2\mathbf{j} + w_3\mathbf{k}) \\ &= v_1w_2\mathbf{i} \times \mathbf{j} + v_2w_1\mathbf{j} \times \mathbf{i} + v_1w_3\mathbf{i} \times \mathbf{k} + v_3w_1\mathbf{k} \times \mathbf{i} \\ &\qquad + v_2w_3\mathbf{j} \times \mathbf{k} + v_3w_2\mathbf{k} \times \mathbf{j} \\ &= (v_2w_3 - v_3w_2)\mathbf{i} + (v_3w_1 - v_1w_3)\mathbf{j} + (v_1w_2 - v_2w_1)\mathbf{k} \end{aligned}$$

Cross Product — Analytic Definition

Let $\mathbf{v} = (v_1, v_2, v_3)$ and $\mathbf{w} = (w_1, w_2, w_3)$. Then

$$\begin{aligned} \mathbf{v} \times \mathbf{w} &= (v_2w_3 - v_3w_2, v_3w_1 - v_1w_3, v_1w_2 - v_2w_1) \\ &= \left(\begin{vmatrix} v_2 & v_3 \\ w_2 & w_3 \end{vmatrix}, \begin{vmatrix} v_3 & v_1 \\ w_3 & w_1 \end{vmatrix}, \begin{vmatrix} v_1 & v_2 \\ w_1 & w_2 \end{vmatrix} \right) \end{aligned}$$

We shall prove that the geometric and analytic definitions of the cross product are equivalent at the end of this section.

Example 1

a $$(4, 3, -1) \times (-2, 2, 1) = \left(\begin{vmatrix} 3 & -1 \\ 2 & 1 \end{vmatrix}, \begin{vmatrix} -1 & 4 \\ 1 & -2 \end{vmatrix}, \begin{vmatrix} 4 & 3 \\ -2 & 2 \end{vmatrix} \right)$$
$$= (3 + 2, 2 - 4, 8 + 6) = (5, -2, 14)$$

b $$(1, 0, 1) \times (0, 1, 1) = \left(\begin{vmatrix} 0 & 1 \\ 1 & 1 \end{vmatrix}, \begin{vmatrix} 1 & 1 \\ 1 & 0 \end{vmatrix}, \begin{vmatrix} 1 & 0 \\ 0 & 1 \end{vmatrix} \right) = (-1, -1, 1)$$

- A device for remembering the cross product is a symbolic determinant—expanded by the first row:

$$(v_1, v_2, v_3) \times (w_1, w_2, w_3) = \begin{vmatrix} \mathbf{i} & \mathbf{j} & \mathbf{k} \\ v_1 & v_2 & v_3 \\ w_1 & w_2 & w_3 \end{vmatrix}$$

$$= \begin{vmatrix} v_2 & v_3 \\ w_3 & w_3 \end{vmatrix} \mathbf{i} - \begin{vmatrix} v_1 & v_3 \\ w_1 & w_3 \end{vmatrix} \mathbf{j} + \begin{vmatrix} v_1 & v_2 \\ w_1 & w_2 \end{vmatrix} \mathbf{k}$$

From this determinant form, we see again that the cross product is anti-commutative: $\mathbf{w} \times \mathbf{v} = -\mathbf{v} \times \mathbf{w}$. For interchanging $\mathbf{v}$ and $\mathbf{w}$ switches two rows of the determinant, hence reverses its sign.

Example 2

Find a vector equation for the line through the point $(1, 1, 1)$ that is perpendicular to the plane through the three points $(0, 0, 0)$, $(1, 0, 1)$, and $(0, 1, 1)$.

Solution The vector

$$(1, 0, 1) \times (0, 1, 1) = \begin{vmatrix} \mathbf{i} & \mathbf{j} & \mathbf{k} \\ 1 & 0 & 1 \\ 0 & 1 & 1 \end{vmatrix}$$

$$= -\mathbf{i} - \mathbf{j} + \mathbf{k} = (-1, -1, 1)$$

is perpendicular to the given plane, hence in the direction of the required line. Therefore that line is

$$\mathbf{x} = t(-1, -1, 1) + (1, 1, 1) = (-t + 1, -t + 1, t + 1)$$

Scalar Triple Product

We have two definitions of the cross product. We connect them by using the following formula. It expresses as a determinant the inner product of a vector with the cross product of two other vectors:

$$\mathbf{u} \cdot (\mathbf{v} \times \mathbf{w}) = \begin{vmatrix} u_1 & u_2 & u_3 \\ v_1 & v_2 & v_3 \\ w_1 & w_2 & w_3 \end{vmatrix}$$

To prove the formula, we expand the determinant by its first row:

$$\begin{vmatrix} u_1 & u_2 & u_3 \\ v_1 & v_2 & v_3 \\ w_1 & w_2 & w_3 \end{vmatrix} = u_1 \begin{vmatrix} v_2 & v_3 \\ w_2 & w_3 \end{vmatrix} - u_2 \begin{vmatrix} v_1 & v_3 \\ w_1 & w_3 \end{vmatrix} + u_3 \begin{vmatrix} v_1 & v_2 \\ w_1 & w_2 \end{vmatrix}$$

$$= (u_1, u_2, u_3) \cdot \left(\begin{vmatrix} v_2 & v_3 \\ w_2 & w_3 \end{vmatrix}, \begin{vmatrix} v_3 & v_1 \\ w_3 & w_1 \end{vmatrix}, \begin{vmatrix} v_1 & v_2 \\ w_1 & w_2 \end{vmatrix} \right)$$

$$= \mathbf{u} \cdot (\mathbf{v} \times \mathbf{w})$$

Example 3

a $\mathbf{u} = (u_1, u_2, u_3) \qquad \mathbf{v} = (0, v_2, v_3) \qquad \mathbf{w} = (0, 0, w_3)$

$$\mathbf{u} \cdot (\mathbf{v} \times \mathbf{w}) = \begin{vmatrix} u_1 & u_2 & u_3 \\ 0 & v_2 & v_3 \\ 0 & 0 & w_3 \end{vmatrix} = u_1 v_2 w_3$$

b $\mathbf{u} = (2, 3, 4) \qquad \mathbf{v} = (1, 1, -2) \qquad \mathbf{w} = (5, 3, 1)$

$$\mathbf{u} \cdot (\mathbf{v} \times \mathbf{w}) = \begin{vmatrix} 2 & 3 & 4 \\ 1 & 1 & -2 \\ 5 & 3 & 1 \end{vmatrix}$$

$$= 2\begin{vmatrix} 1 & -2 \\ 3 & 1 \end{vmatrix} - 3\begin{vmatrix} 1 & -2 \\ 5 & 1 \end{vmatrix} + 4\begin{vmatrix} 1 & 1 \\ 5 & 3 \end{vmatrix}$$

$$= 2 \cdot 7 - 3 \cdot 11 + 4(-2)$$

$$= -27$$

●

The quantity $\mathbf{u} \cdot (\mathbf{v} \times \mathbf{w})$ occurs frequently in applications; it is called the **scalar triple product** and written $[\mathbf{u}, \mathbf{v}, \mathbf{w}]$.

Scalar Triple Product

$$[\mathbf{u}, \mathbf{v}, \mathbf{w}] = \mathbf{u} \cdot (\mathbf{v} \times \mathbf{w}) = \begin{vmatrix} u_1 & u_2 & u_3 \\ v_1 & v_2 & v_3 \\ w_1 & w_2 & w_3 \end{vmatrix}$$

●

Because the scalar triple product has an expression as a determinant, properties of determinants imply properties of the scalar triple product. We list some important ones:

Properties of the Scalar Triple Product

- If any two of the vectors $\mathbf{u}$, $\mathbf{v}$, $\mathbf{w}$ are equal, then $[\mathbf{u}, \mathbf{v}, \mathbf{w}] = 0$.
- If any two of the vectors $\mathbf{u}$, $\mathbf{v}$, $\mathbf{w}$ are interchanged, then $[\mathbf{u}, \mathbf{v}, \mathbf{w}]$ changes sign. For instance, $[\mathbf{w}, \mathbf{v}, \mathbf{u}] = -[\mathbf{u}, \mathbf{v}, \mathbf{w}]$.
- $[\mathbf{u}, \mathbf{v}, \mathbf{w}]$ is a homogeneous linear function of each of $\mathbf{u}$, $\mathbf{v}$, and $\mathbf{w}$. For example,

$$[\mathbf{u}, \mathbf{v}, \mathbf{w}_1 + \mathbf{w}_2] = [\mathbf{u}, \mathbf{v}, \mathbf{w}_1] + [\mathbf{u}, \mathbf{v}, \mathbf{w}_2]$$

$$[\mathbf{u}, \mathbf{v}, c\mathbf{w}] = c[\mathbf{u}, \mathbf{v}, \mathbf{w}]$$

$$[\mathbf{u} + 3\mathbf{v}, \mathbf{v}, \mathbf{w}] = [\mathbf{u}, \mathbf{v}, \mathbf{w}] + 3[\mathbf{v}, \mathbf{v}, \mathbf{w}] = [\mathbf{u}, \mathbf{v}, \mathbf{w}]$$

●

Warning Remember, the cross product of two vectors is a *vector.* The inner product of two vectors is a *scalar* (number). The scalar triple product of three vectors is also a *scalar* (because it is an inner product of two vectors).

We summarize the main algebraic properties of the cross product. They follow readily from the definition.

Properties of the Cross Product

- $\mathbf{v} \times \mathbf{v} = \mathbf{0} \qquad \mathbf{w} \times \mathbf{v} = -\mathbf{v} \times \mathbf{w}$
- $(a\mathbf{u} + b\mathbf{v}) \times \mathbf{w} = a(\mathbf{u} \times \mathbf{w}) + b(\mathbf{v} \times \mathbf{w})$ (linearity)
- $\mathbf{u} \times (a\mathbf{v} + b\mathbf{w}) = a(\mathbf{u} \times \mathbf{v}) + b(\mathbf{u} \times \mathbf{w})$ (linearity)
- $\mathbf{u} \cdot (\mathbf{v} \times \mathbf{w}) = \mathbf{v} \cdot (\mathbf{w} \times \mathbf{u}) = \mathbf{w} \cdot (\mathbf{u} \times \mathbf{v}) = [\mathbf{u}, \mathbf{v}, \mathbf{w}]$
- $\mathbf{v} \times \mathbf{w} = \mathbf{0}$ if and only if $\mathbf{v}$ and $\mathbf{w}$ are parallel.

Remark The associative law and the commutative law are not true in general for the cross product. For instance

$$\mathbf{i} \times \mathbf{j} = \mathbf{k} \quad \text{and} \quad \mathbf{j} \times \mathbf{i} = -\mathbf{k} \quad \text{so} \quad \mathbf{i} \times \mathbf{j} \neq \mathbf{j} \times \mathbf{i}$$

$$\mathbf{i} \times (\mathbf{j} \times \mathbf{j}) = \mathbf{i} \times \mathbf{0} = \mathbf{0} \quad \text{and} \quad (\mathbf{i} \times \mathbf{j}) \times \mathbf{j} = \mathbf{k} \times \mathbf{j} = -\mathbf{i}$$

$$\text{so} \quad \mathbf{i} \times (\mathbf{j} \times \mathbf{j}) \neq (\mathbf{i} \times \mathbf{j}) \times \mathbf{j}$$

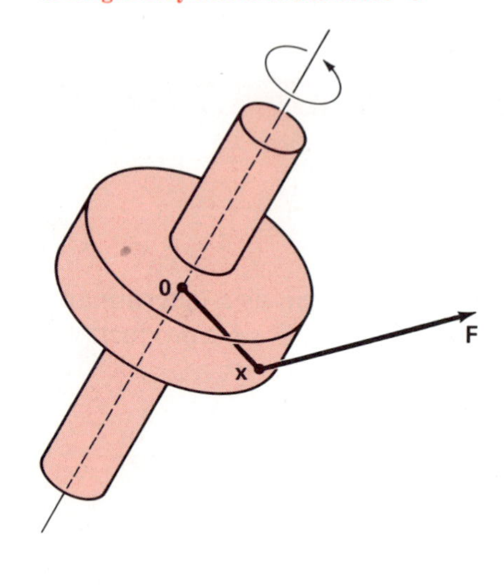

Figure 12-6-2
A force $\mathbf{F}$ applied at a point $\mathbf{x}$ of a rigid body free to rotate about $\mathbf{0}$

Torque [Optional]

The original motivation for the cross product of vectors came from physics.

Suppose a rigid body is free to turn about the origin. A force $\mathbf{F}$ acts at a point $\mathbf{x}$ of the body. As a result the body wants to rotate about an axis through $\mathbf{0}$ and perpendicular to the plane of $\mathbf{x}$ and $\mathbf{F}$ (unless $\mathbf{x}$ and $\mathbf{F}$ are collinear; then there is no turning). See Figure 12-6-2. As usual, the force vector $\mathbf{F}$ is *drawn* at its point of application $\mathbf{x}$. But analytically it starts at $\mathbf{0}$. The positive axis of rotation is determined by the right-hand rule as applied to the pair $\mathbf{x}, \mathbf{F}$ in that order: $\mathbf{x}$ first, $\mathbf{F}$ second.

The **torque** resulting from the force $\mathbf{F}$ applied at $\mathbf{x}$ is a vector that measures the tendency of a body to rotate under the action of the force. The obvious choice for the direction of the torque vector is that of the rotation axis. What should we choose for its magnitude?

By experiment, if $\mathbf{F}$ is tripled in magnitude, the torque is tripled; if $\mathbf{x}$ is moved out twice as far along the same line and the same $\mathbf{F}$ is applied there, the torque is doubled. Hence the torque is proportional to the length of $\mathbf{x}$ and to the length of $\mathbf{F}$. An educated guess is that the torque might just be proportional to the area of the parallelogram determined by $\mathbf{x}$ and $\mathbf{F}$. See Figure 12-6-3.

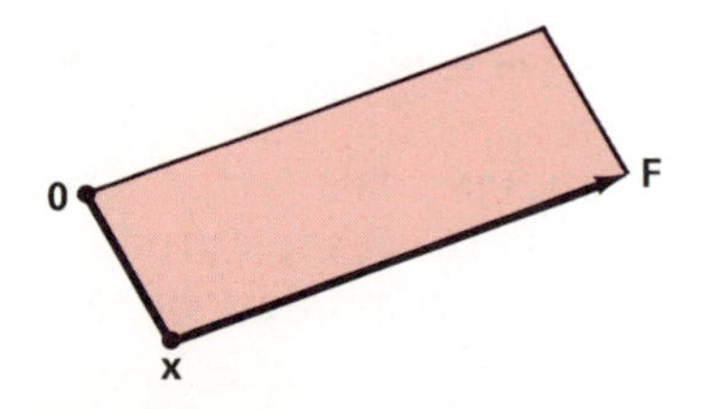

Figure 12-6-3
The torque vector is probably proportional to the area of the rectangle determined by $\mathbf{x}$ and $\mathbf{F}$.

Let us resolve $\mathbf{F}$ into a component $\mathbf{G}$ parallel to $\mathbf{x}$ and a component $\mathbf{H}$ perpendicular to $\mathbf{x}$. See Figure 12-6-4. The component $\mathbf{G}$, being on line with $\mathbf{x}$, produces no tendency to turn; only $\mathbf{H}$ produces torque. By the argument above, the amount of torque is proportional to $|\mathbf{x}| \cdot |\mathbf{H}|$, the length of the lever arm times the length of $\mathbf{H}$. But this product is the area of the parallelogram (Figure 12-6-5) determined by $\mathbf{x}$ and $\mathbf{F}$.

Therefore the torque (about the origin) is completely described by the vector $\mathbf{x} \times \mathbf{F}$. The length of $\mathbf{x} \times \mathbf{F}$ is proportional to the magnitude of the torque. The direction of $\mathbf{x} \times \mathbf{F}$ is the positive axis of rotation: with your right thumb along $\mathbf{x} \times \mathbf{F}$, your fingers curl in the direction of turning. In physics, torque about the origin is *defined* to be precisely the vector $\mathbf{x} \times \mathbf{F}$.

Figure 12-6-4
Resolve **F** into **G** parallel to **x** and **H** perpendicular to **x**:
F = **G** + **H**

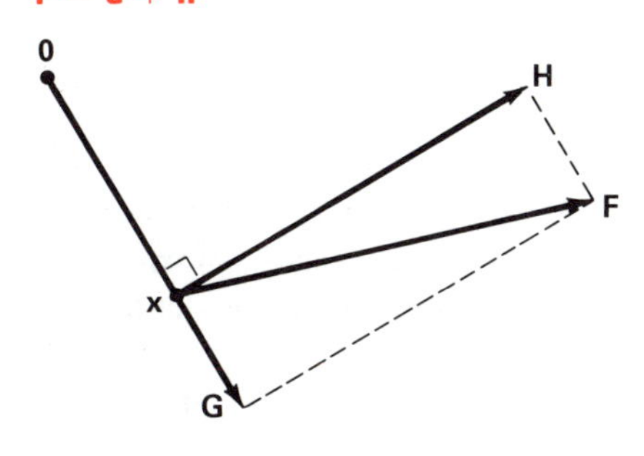

Figure 12-6-5
area = |**x**| · |**H**| = |**x** × **F**|

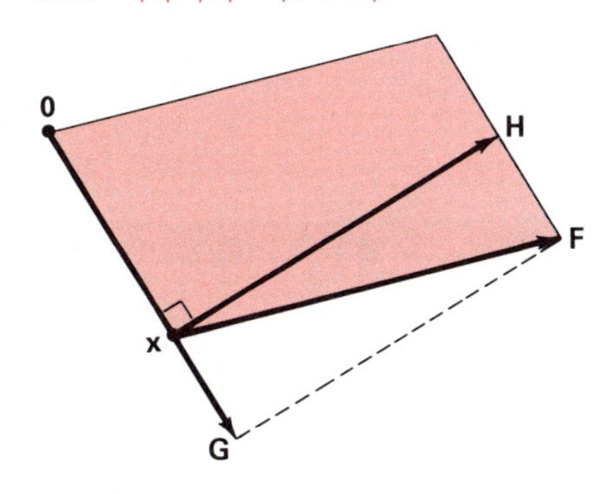

Figure 12-6-6
$|\mathbf{v} \times \mathbf{w}| = |\mathbf{v}| \cdot |\mathbf{w}| \sin\theta$ = area

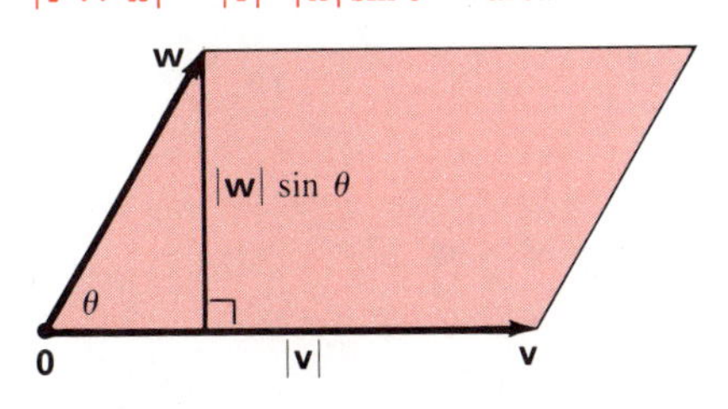

Equivalence of the Definitions [Optional]

Let us show that the analytic definition of the cross product satisfies the three conditions of the geometric definition:

- $\mathbf{v} \times \mathbf{w}$ is orthogonal to $\mathbf{v}$ and to $\mathbf{w}$.
- $\mathbf{v}$, $\mathbf{w}$, $\mathbf{v} \times \mathbf{w}$ is a right-handed system (provided $\mathbf{v}$ and $\mathbf{w}$ are not parallel).
- $|\mathbf{v} \times \mathbf{w}|$ is the area of the parallelogram based on $\mathbf{v}$ and $\mathbf{w}$ (that is, the parallelogram with vertices $\mathbf{0}$, $\mathbf{v}$, $\mathbf{w}$, and $\mathbf{v} + \mathbf{w}$, with $\mathbf{v}$ and $\mathbf{w}$ opposite).

The first geometric condition is a consequence of a property of the scalar triple product:

$$\mathbf{v} \cdot (\mathbf{v} \times \mathbf{w}) = [\mathbf{v}, \mathbf{v}, \mathbf{w}] = 0 \qquad \mathbf{w} \cdot (\mathbf{v} \times \mathbf{w}) = [\mathbf{w}, \mathbf{v}, \mathbf{w}] = 0$$

hence $\mathbf{v}$ and $\mathbf{w}$ are both orthogonal to $\mathbf{v} \times \mathbf{w}$.

The third geometric condition requires a formula for the length of $\mathbf{v} \times \mathbf{w}$, where $\mathbf{v} \times \mathbf{w}$ is taken in its analytic sense:

$$|\mathbf{v} \times \mathbf{w}|^2 = |\mathbf{v}|^2|\mathbf{w}|^2 - (\mathbf{v} \cdot \mathbf{w})^2$$

By direct computation:

$$\begin{aligned} |\mathbf{v} \times \mathbf{w}|^2 &= (v_2w_3 - v_3w_2)^2 + (v_3w_1 - v_1w_3)^2 + (v_1w_2 - v_2w_1)^2 \\ &= \sum_{6\text{ terms}} v_2^2w_3^2 - 2\sum_{3\text{ terms}} v_2v_3w_2w_3 \end{aligned}$$

The sum notation here means sum over all similar terms. For instance, the terms similar to $v_2^2w_3^2$ are

$$v_3^2w_2^2 \qquad v_1^2w_2^2 \qquad v_2^2w_1^2 \qquad v_1^2w_3^2 \qquad v_3^2w_1^2$$

Similarly,

$$\begin{aligned} |\mathbf{v}|^2|\mathbf{w}|^2 - (\mathbf{v} \cdot \mathbf{w})^2 &= (v_1^2 + v_2^2 + v_3^2)(w_1^2 + w_2^2 + w_3^2) \\ &\quad - (v_1w_1 + v_2w_2 + v_3w_3)^2 \\ &= \left(\sum_{3\text{ terms}} v_1^2w_1^2 + \sum_{6\text{ terms}} v_2^2w_3^2\right) \\ &\quad - \left(\sum_{3\text{ terms}} v_1^2w_1^2 + 2\sum_{3\text{ terms}} v_2v_3w_2w_3\right) \end{aligned}$$

All terms like $v_1^2w_1^2$ cancel; the formula follows.

Now we let θ be the angle between $\mathbf{v}$ and $\mathbf{w}$. Then we have the formula $\mathbf{v} \cdot \mathbf{w} = |\mathbf{v}| \cdot |\mathbf{w}| \cos\theta$. Hence

$$\begin{aligned} |\mathbf{v} \times \mathbf{w}|^2 &= |\mathbf{v}|^2|\mathbf{w}|^2 - (\mathbf{v} \cdot \mathbf{w})^2 = |\mathbf{v}|^2|\mathbf{w}|^2(1 - \cos^2\theta) \\ &= |\mathbf{v}|^2|\mathbf{w}|^2 \sin^2\theta \end{aligned}$$

and $|\mathbf{v} \times \mathbf{w}| = |\mathbf{v}| \cdot |\mathbf{w}| \sin\theta$ follows. This last expression is precisely the required area (Figure 12-6-6).

It remains to verify the second geometric condition. To do so, we need some analytic way of deciding whether a given triple $\mathbf{u}, \mathbf{v}, \mathbf{w}$ is a right-handed system or not.

Observe that $\mathbf{u}, \mathbf{v}, \mathbf{w}$ and $\mathbf{v}, \mathbf{u}, \mathbf{w}$ have opposite *orientations,* that is, one is a right-handed system and the other is left-handed. By analogy, the determinants $[\mathbf{u}, \mathbf{v}, \mathbf{w}]$ and $[\mathbf{v}, \mathbf{u}, \mathbf{w}]$ have opposite signs. This suggests that the sign of $[\mathbf{u}, \mathbf{v}, \mathbf{w}]$ corresponds to the orientation of $\mathbf{u}, \mathbf{v}, \mathbf{w}$. Since $\mathbf{i}, \mathbf{j}, \mathbf{k}$ is right-handed and $[\mathbf{i}, \mathbf{j}, \mathbf{k}] = 1$, we suspect that $\mathbf{u}, \mathbf{v}, \mathbf{w}$ is right-handed if $[\mathbf{u}, \mathbf{v}, \mathbf{w}] > 0$. This indeed is the case, but instead of proving it, we shall simply take the determinant criterion as the definition of right-handedness.

In view of this definition, we must prove that $[\mathbf{v}, \mathbf{w}, \mathbf{v} \times \mathbf{w}] > 0$. But

$$[\mathbf{v}, \mathbf{w}, \mathbf{v} \times \mathbf{w}] = -[\mathbf{v}, \mathbf{v} \times \mathbf{w}, \mathbf{w}] = [\mathbf{v} \times \mathbf{w}, \mathbf{v}, \mathbf{w}]$$
$$= (\mathbf{v} \times \mathbf{w}) \cdot (\mathbf{v} \times \mathbf{w}) = |\mathbf{v} \times \mathbf{w}|^2 > 0$$

This completes our proof that the vector $\mathbf{v} \times \mathbf{w}$, defined analytically, satisfies the three properties that define $\mathbf{v} \times \mathbf{w}$ geometrically. Hence the definitions are equivalent.

Exercises

Compute the cross product

1 $(-2, 2, 1) \times (4, 3, -1)$
2 $(1, 0, 1) \times (1, 1, 0)$
3 $(1, 2, 3) \times (3, 2, 1)$
4 $(3, 1, -1) \times (3, -1, -1)$
5 $(-2, -2, -2) \times (1, 1, 0)$
6 $(-1, 2, 2) \times (3, -1, 2)$
7 $(\mathbf{i} + \mathbf{j}) \times (\mathbf{i} + \mathbf{j} + \mathbf{k})$
8 $(\cos\theta\,\mathbf{i} + \sin\theta\,\mathbf{j}) \times (-\sin\theta\,\mathbf{i} + \cos\theta\,\mathbf{j})$
9 $(2\mathbf{i} + \mathbf{j} + 3\mathbf{k}) \times (2\mathbf{i} + 2\mathbf{j} - \mathbf{k})$
10 $(\mathbf{i} + 2\mathbf{j} + 3\mathbf{k}) \times (4\mathbf{i} + 5\mathbf{j} + 6\mathbf{k})$

Find a unit vector perpendicular to both

11 $(-3, 0, 1)$ and $(2, -1, -1)$
12 $(1, 2, 3)$ and $(3, 2, 1)$

13 Find a parametric vector equation for the line through $(6, 1, -2)$ and perpendicular to the plane through $\mathbf{0}$, $(-1, 1, 2)$, and $(2, 3, 4)$.
14 Find a parametric vector equation for the line through $(1, 2, 1)$ and perpendicular to the plane through $(1, 0, 0)$, $(0, 1, 0)$, and $(0, 0, 1)$.

Compute the area of a parallelogram, three of whose vertices are

15 $(0, 0, 0)$ $(1, 1, 1)$ and $(2, 3, 5)$
16 $(1, 0, 0)$ $(2, 1, 0)$ and $(3, 2, 1)$

Compute the scalar triple product

17 $[\mathbf{i}, \mathbf{i} + \mathbf{j}, \mathbf{i} + \mathbf{j} + \mathbf{k}]$
18 $[3\mathbf{i} - \mathbf{j}, 2\mathbf{j} + \mathbf{k}, \mathbf{i} - 4\mathbf{j} - 5\mathbf{k}]$
19 $(4, 1, 1) \cdot (3, 6, 0) \times (2, 5, 4)$
20 $(1, 1, 1) \cdot (1, 2, 3) \times (1, 4, 9)$

A force $\mathbf{F}$ is applied at point $\mathbf{x}$. Find its torque about the origin.

21 $\mathbf{F} = (-1, 1, 1)$ $\quad \mathbf{x} = (10, 0, 0)$
22 $\mathbf{F} = (3, 0, 0)$ $\quad \mathbf{x} = (0, 0, 1)$
23 $\mathbf{F} = (-1, 1, 1)$ $\quad \mathbf{x} = (2, 2, -1)$
24 $\mathbf{F} = (2, -1, 5)$ $\quad \mathbf{x} = (-7, 1, 0)$

25 Find all vectors $\mathbf{v}$ for which $[\mathbf{i}, \mathbf{j}, \mathbf{v}] = 3$.
26 Let $\mathbf{u}$, $\mathbf{v}$, $\mathbf{w}$ be a right-handed system of mutually orthogonal unit vectors. Find $[\mathbf{u}, \mathbf{v}, \mathbf{w}]$.
27 Given $\mathbf{a}$, find all vectors $\mathbf{x}$ such that $\mathbf{a} \times \mathbf{x} = \mathbf{x}$.
28 Given $\mathbf{a}$, find all vectors $\mathbf{x}$ such that $\mathbf{a} \times \mathbf{x} = \mathbf{a} + \mathbf{x}$.

Prove analytically

29 $\mathbf{u} \cdot (\mathbf{v} \times \mathbf{w}) = \mathbf{v} \cdot (\mathbf{w} \times \mathbf{u})$
30 $(\mathbf{u} + \mathbf{v}) \times \mathbf{w} = \mathbf{u} \times \mathbf{w} + \mathbf{v} \times \mathbf{w}$
31 $(a\mathbf{v}) \times \mathbf{w} = a(\mathbf{v} \times \mathbf{w})$
32 $\mathbf{v} \times (b\mathbf{w}) = b(\mathbf{v} \times \mathbf{w})$

33 Show analytically that

$$\begin{vmatrix} \mathbf{a}\cdot\mathbf{u} & \mathbf{a}\cdot\mathbf{v} \\ \mathbf{b}\cdot\mathbf{u} & \mathbf{b}\cdot\mathbf{v} \end{vmatrix} = (\mathbf{a}\times\mathbf{b})\cdot(\mathbf{u}\times\mathbf{v})$$

provided $\mathbf{a}$ and $\mathbf{b}$ are chosen from $\mathbf{i}$, $\mathbf{j}$, $\mathbf{k}$.

34 (cont.) Now prove the formula in general, for any four vectors.

35 Suppose $\mathbf{a}$ and $\mathbf{b}$ are vectors such that $\mathbf{a}\cdot\mathbf{v} = \mathbf{b}\cdot\mathbf{v}$ for *all* vectors $\mathbf{v}$. Prove that $\mathbf{a} = \mathbf{b}$.

***36** (cont.) Use this and the result of Exercise 34 to prove the identity

$$\mathbf{u}\times(\mathbf{v}\times\mathbf{w}) = (\mathbf{u}\cdot\mathbf{w})\mathbf{v} - (\mathbf{u}\cdot\mathbf{v})\mathbf{w}$$

37 Use the result of Exercise 34 for a proof of $|\mathbf{v}\times\mathbf{w}|^2 = |\mathbf{v}|^2|\mathbf{w}|^2 - (\mathbf{v}\cdot\mathbf{w})^2$ different from the one given in the text.

38 Prove that $\mathbf{v} = (\mathbf{a}\times\mathbf{b})\times(\mathbf{a}\times\mathbf{c})$ is parallel to $\mathbf{a}$.

39 Suppose $a > 0$, $b > 0$, $c > 0$. Find the area of the triangle with vertices $(a, 0, 0)$, $(0, b, 0)$, $(0, 0, c)$.

***40** Prove the **Jacobi identity**

$$\mathbf{u}\times(\mathbf{v}\times\mathbf{w}) + \mathbf{v}\times(\mathbf{w}\times\mathbf{u}) + \mathbf{w}\times(\mathbf{u}\times\mathbf{v}) = \mathbf{0}$$

12-7 Applications of the Cross Product

Two non-collinear (non-parallel) vectors $\mathbf{u}$ and $\mathbf{v}$ (based at $\mathbf{0}$ as usual) determine a parallelogram, whose area is $|\mathbf{u}\times\mathbf{v}|$. Similarly, three non-coplanar vectors determine a parallelepiped (Figure 12-7-1) whose volume is given by a formula involving a scalar triple product.

Figure 12-7-1
The parallelepiped determined by three non-coplanar vectors

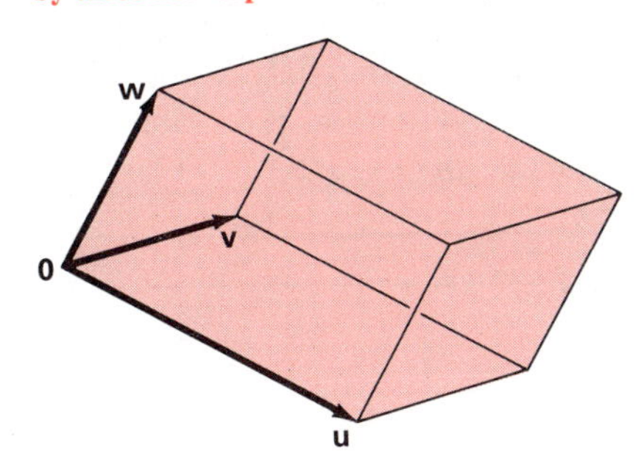

Volume

The volume of the parallelepiped determined by three non-coplanar vectors $\mathbf{u}$, $\mathbf{v}$, $\mathbf{w}$ is

$$V = |[\mathbf{u}, \mathbf{v}, \mathbf{w}]| = |\mathbf{u}\cdot(\mathbf{v}\times\mathbf{w})|$$

Proof The volume is

$$V = (\text{area of base})(\text{height})$$

For the base, take the parallelogram determined by $\mathbf{u}$ and $\mathbf{v}$. Its area is $|\mathbf{u}\times\mathbf{v}|$; furthermore, the vector $\mathbf{u}\times\mathbf{v}$ is perpendicular to the base.

Suppose first that $\mathbf{w}$ lies on the same side of the base as $\mathbf{u}\times\mathbf{v}$. By definition, the height of the parallelepiped is the projection (Figure 12-7-2) of $\mathbf{w}$ onto $\mathbf{u}\times\mathbf{v}$:

$$(\text{height}) = |\mathbf{w}|\cos\theta$$

Figure 12-7-2
Volume of parallepiped = (base area) × (height)

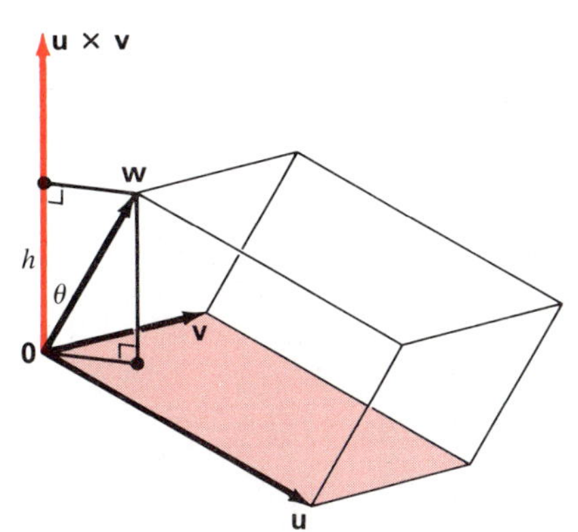

Therefore

$$V = |\mathbf{u}\times\mathbf{v}|\cdot|\mathbf{w}|\cos\theta = (\mathbf{u}\times\mathbf{v})\cdot\mathbf{w}$$
$$= [\mathbf{u}, \mathbf{v}, \mathbf{w}] = |[\mathbf{u}, \mathbf{v}, \mathbf{w}]|$$

If $\mathbf{w}$ and $\mathbf{u}\times\mathbf{v}$ lie on opposite sides of the base, then

$$V = -[\mathbf{u}, \mathbf{v}, \mathbf{w}] = |[\mathbf{u}, \mathbf{v}, \mathbf{w}]|$$

As a corollary, we derive a test for three vectors to lie on a plane through $\mathbf{0}$. This happens if and only if the parallelepiped they determine collapses, that is, has volume 0. But its volume is $|[\mathbf{u}, \mathbf{v}, \mathbf{w}]|$, so we have the following test:

- Three vectors **u**, **v**, and **w** lie on a plane through **0** if and only if $[\mathbf{u}, \mathbf{v}, \mathbf{w}] = 0$.

Example 1

Show that $\mathbf{w} = (1, -3, 3)$ lies in the plane of $\mathbf{u} = (5, 6, 1)$ and $\mathbf{v} = (2, 3, 0)$.

Solution Compute $[\mathbf{u}, \mathbf{v}, \mathbf{w}]$ by minors of the second row (since there is a zero there—it shortens the computation):

$$[\mathbf{u}, \mathbf{v}, \mathbf{w}] = \begin{vmatrix} 5 & 6 & 1 \\ 2 & 3 & 0 \\ 1 & -3 & 3 \end{vmatrix} = -2\begin{vmatrix} 6 & 1 \\ -3 & 3 \end{vmatrix} + 3\begin{vmatrix} 5 & 1 \\ 1 & 3 \end{vmatrix}$$

$$= (-2)(21) + (3)(14) = 0$$

Hence **u**, **v**, and **w** are coplanar. Obviously **u** and **v** are not parallel, so they determine a plane through **0**, and **w** must lie in this plane.

Intersection of Two Planes

Given two planes $\mathbf{x} \cdot \mathbf{m} = p$ and $\mathbf{x} \cdot \mathbf{n} = q$, how can we find their line of intersection? We must assume the planes are not parallel, that is, their normal vectors **m** and **n** are not parallel. Then $\mathbf{a} = \mathbf{m} \times \mathbf{n} \neq \mathbf{0}$. Note that **a** is parallel to the line of intersection (Figure 12-7-3).

Example 2

Find the line of intersection of the planes

$$x + y + z = -1 \quad \text{and} \quad 2x + y - z = 3$$

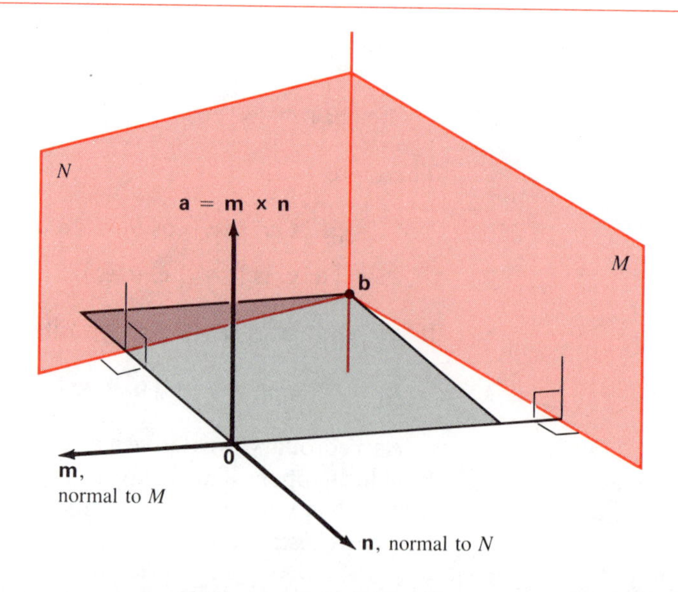

Figure 12-7-3
The line of intersection of planes M and N is given parametrically by $\mathbf{x} = t\mathbf{a} + \mathbf{b}$.

Solution The equations of the planes are $\mathbf{x}\cdot\mathbf{m} = p$ and $\mathbf{x}\cdot\mathbf{n} = q$, where

$$\mathbf{m} = (1, 1, 1) \qquad \mathbf{n} = (2, 1, -1)$$
$$p = -1 \qquad q = 3$$

Treat the two planes as a linear system:

$$\begin{cases} x + y + z = -1 \\ 2x + y - z = 3 \end{cases}$$

Keep the first equation, and add -2 times the first to the second to eliminate x:

$$\begin{cases} x + y + z = -1 \\ \quad -y - 3z = 5 \end{cases}$$

Set $z = t$, then solve for y and x:

$$y = -3z - 5 = -3t - 5$$
$$x = -1 - y - z = -1 - (-3t - 5) - t$$
$$= 2t + 4$$

Hence

$$\mathbf{x} = (2t + 4, -3t - 5, t)$$

This is a parametric form of the line of intersection. Note that if we set

$$\mathbf{a} = \mathbf{m} \times \mathbf{n} = (-2, 3, -1) \neq \mathbf{0}$$

then this line is parallel to $\mathbf{a}$ and passes through $(4, -5, 0)$, a point on both planes. ●

Figure 12-7-4
Skew lines in $\mathbf{R}^3$

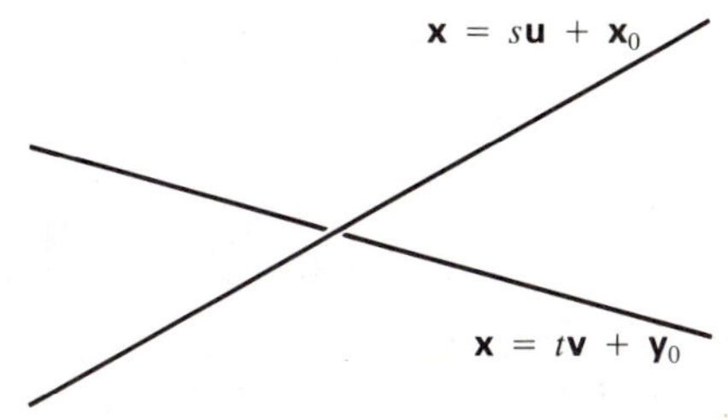

Figure 12-7-5
The skew lines lie on parallel planes, at distance $|(\mathbf{x}_0 - \mathbf{y}_0)\cdot\mathbf{n}|$.

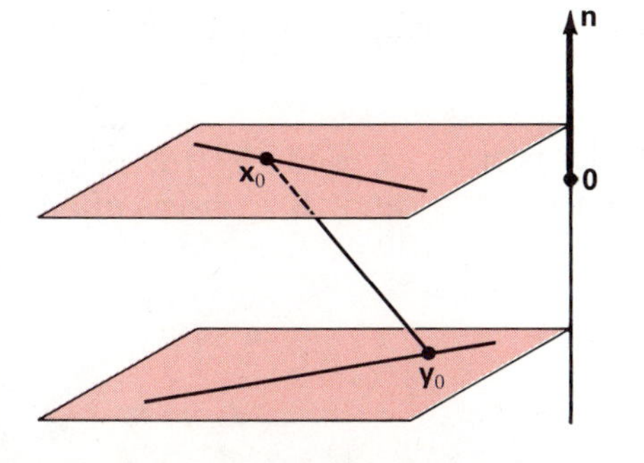

Skew Lines

Let $\mathbf{x} = s\mathbf{u} + \mathbf{x}_0$ and $\mathbf{x} = t\mathbf{v} + \mathbf{y}_0$ be two lines in $\mathbf{R}^3$ that do not intersect and are not parallel, that is, two skew lines. We ask how far apart they are (Figure 12-7-4).

The cross product $\mathbf{u} \times \mathbf{v}$ is perpendicular to both lines, so the vector $\mathbf{n} = (\mathbf{u} \times \mathbf{v})/|\mathbf{u} \times \mathbf{v}|$ is a unit vector perpendicular to both lines. Each of the lines lies on a plane that is perpendicular to $\mathbf{n}$, as shown in Figure 12-7-5. Thus the two lines lie on two parallel planes. Clearly, the closest distance between the lines equals the distance between the two planes. To find the distance, we simply project any segment with its end points on the two planes onto $\mathbf{n}$. But the segment (vector) $\mathbf{x}_0 - \mathbf{y}_0$ is such a segment, readily available, so the required distance equals $|(\mathbf{x}_0 - \mathbf{y}_0)\cdot\mathbf{n}|$.

Example 3

Find the distance between the lines

$$\mathbf{x} = (s - 1, s, 2s + 2) \quad \text{and} \quad \mathbf{x} = (-t + 1, -t + 1, -t + 1)$$

Solution The two lines are $\mathbf{x} = s\mathbf{u} + \mathbf{x}_0$ and $\mathbf{x} = t\mathbf{v} + \mathbf{y}_0$, where

$$\mathbf{u} = (1, 1, 2) \qquad \mathbf{v} = (-1, -1, -1)$$
$$\mathbf{x}_0 = (-1, 0, 2) \qquad \mathbf{y}_0 = (1, 1, 1)$$

Therefore $\mathbf{u} \times \mathbf{v} = (1, -1, 0)$ and

$$\mathbf{n} = \frac{\mathbf{u} \times \mathbf{v}}{|\mathbf{u} \times \mathbf{v}|} = \tfrac{1}{2}\sqrt{2}\,(1, -1, 0)$$

Finally, the distance between the lines is

$$|(\mathbf{x}_0 - \mathbf{y}_0) \cdot \mathbf{n}| = |(-2, -1, 1) \cdot \tfrac{1}{2}\sqrt{2}\,(1, -1, 0)|$$
$$= |-\tfrac{1}{2}\sqrt{2}\,| = \tfrac{1}{2}\sqrt{2}$$

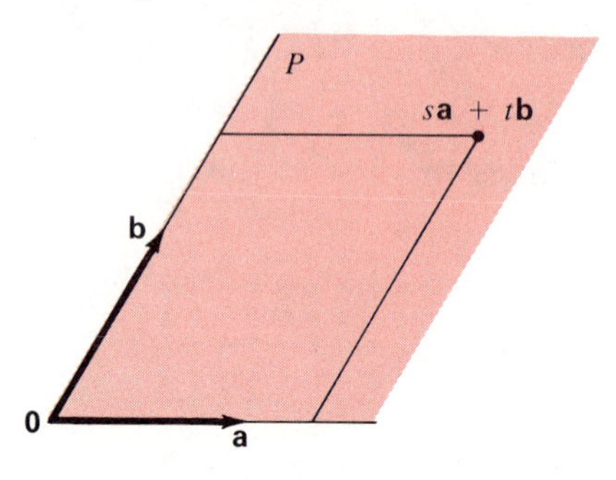

Figure 12-7-6
The plane determined by two non-parallel vectors

Parametric Form of a Plane

Suppose the *vectors,* $\mathbf{a}$ and $\mathbf{b}$ are not parallel; that is, the three *points* $\mathbf{0}$, $\mathbf{a}$, $\mathbf{b}$ are non-collinear. Then $\mathbf{a}$ and $\mathbf{b}$ lie on a unique plane P. See Figure 12-7-6. All scalar multiples $s\mathbf{a}$ and $t\mathbf{b}$ of $\mathbf{a}$ and $\mathbf{b}$ also lie on P; by the parallelogram law, so do all sums $s\mathbf{a} + t\mathbf{b}$.

Conversely each vector $\mathbf{x}$ in the plane P can be expressed uniquely in the form $s\mathbf{a} + t\mathbf{b}$. For we can construct a parallelogram with $\mathbf{x}$ as a diagonal as shown in Figure 12-7-7. This displays $\mathbf{x}$ as a sum of two vectors, a multiple of $\mathbf{a}$ and a multiple of $\mathbf{b}$. We say that $\mathbf{a}$ and $\mathbf{b}$ **span** the plane P.

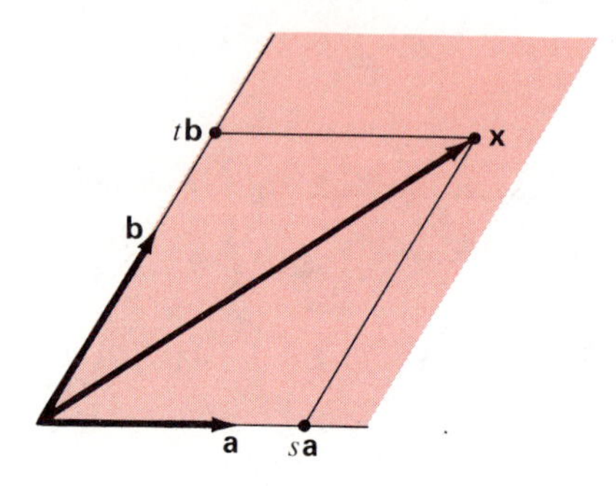

Figure 12-7-7
Each point $\mathbf{x}$ of the plane determined by $\mathbf{a}$ and $\mathbf{b}$ has the form $\mathbf{x} = s\mathbf{a} + t\mathbf{b}$.

- Two non-parallel vectors $\mathbf{a}$ and $\mathbf{b}$ span the plane consisting of all vectors

$$\mathbf{x} = s\mathbf{a} + t\mathbf{b}$$

where $-\infty < s < +\infty$ and $-\infty < t < +\infty$.

Remark In the subject called *linear algebra,* vectors $\mathbf{a}$ and $\mathbf{b}$ that are non-parallel are called **linearly independent.** It means in effect that neither is a scalar multiple of the other. Alternatively, if $s\mathbf{a} + t\mathbf{b} = \mathbf{0}$, then $s = t = 0$.

Next, suppose two non-parallel vectors $\mathbf{a}$ and $\mathbf{b}$ span a plane Q, and suppose $\mathbf{c}$ is a point not on Q. We seek the plane P through $\mathbf{c}$ parallel to Q. In fact, P consists of all $\mathbf{x} = \mathbf{z} + \mathbf{c}$, where $\mathbf{z}$ belongs to Q. See Figure 12-7-8. But we know all such vectors $\mathbf{z}$, namely $\mathbf{z} = s\mathbf{a} + t\mathbf{b}$.

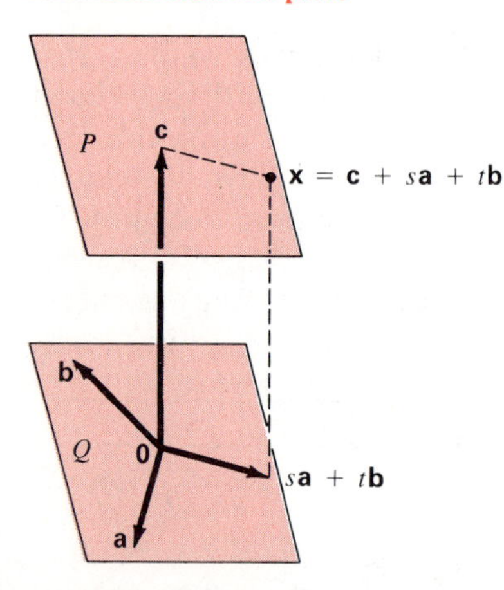

Figure 12-7-8
Parametric form of a plane

- Given a point $\mathbf{c}$ and two non-parallel vectors $\mathbf{a}$ and $\mathbf{b}$, the plane through $\mathbf{c}$ parallel to the plane spanned by $\mathbf{a}$ and $\mathbf{b}$ consists of all points

$$\mathbf{x} = s\mathbf{a} + t\mathbf{b} + \mathbf{c}$$

where $-\infty < s < +\infty$ and $-\infty < t < +\infty$.

The variables s and t are called **parameters,** and a plane presented in this fashion is said to be in **parametric form.**

Given a plane in parametric form $\mathbf{x} = s\mathbf{a} + t\mathbf{b} + \mathbf{c}$, how do we find a non-parametric equation for the plane, that is, an equation of the form $\mathbf{n} \cdot \mathbf{x} = p$ or $ax + by + cz = d$? First we find a vector $\mathbf{n}$ normal to the plane. That is easy: we take $\mathbf{n} = \mathbf{a} \times \mathbf{b}$, which is orthogonal to both $\mathbf{a}$ and $\mathbf{b}$, hence to the plane. Because $\mathbf{a}$ and $\mathbf{b}$ are non-parallel, the vector $\mathbf{n}$ is guaranteed to be non-zero. Then we note that $\mathbf{n} \cdot (\mathbf{x} - \mathbf{c}) = 0$ because $\mathbf{x} - \mathbf{c}$ is in the plane of $\mathbf{a}$ and $\mathbf{b}$, hence is perpendicular to $\mathbf{n}$. Therefore $\mathbf{n} \cdot \mathbf{x} = \mathbf{n} \cdot \mathbf{c}$ is a non-parametric equation of the plane, where $\mathbf{n} = \mathbf{a} \times \mathbf{b}$.

Example 4

Set $\quad \mathbf{a} = (1, 0, 1) \qquad \mathbf{b} = (1, 1, -1) \qquad \mathbf{c} = (-1, 1, 2)$

a Find a parametric form for the plane through $\mathbf{c}$ parallel to the plane spanned by $\mathbf{a}$ and $\mathbf{b}$.

b Find a non-parametric equation of the plane.

Solution

a Neither $\mathbf{a}$ nor $\mathbf{b}$ is a multiple of the other, so they are non-parallel. The plane through $\mathbf{c}$ parallel to the plane spanned by $\mathbf{a}$ and $\mathbf{b}$ consists of all $\mathbf{x}$ such that

$$\mathbf{x} = s\mathbf{a} + t\mathbf{b} + \mathbf{c} = s(1, 0, 1) + t(1, 1, -1) + (-1, 1, 2)$$

In coordinates:

$$x = s + t - 1 \qquad y = t + 1 \qquad z = s - t + 2$$

b The vector

$$\mathbf{n} = \mathbf{a} \times \mathbf{b} = (1, 0, 1) \times (1, 1, -1) = (-1, 2, 1)$$

is normal to the plane. Also

$$\mathbf{n} \cdot \mathbf{c} = (-1, 2, 1) \cdot (-1, 1, 2) = 5$$

so an equation of the plane is $\mathbf{n} \cdot \mathbf{x} = 5$, that is

$$-x + 2y + z = 5$$

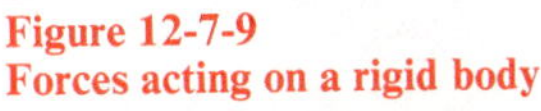

Figure 12-7-9
Forces acting on a rigid body

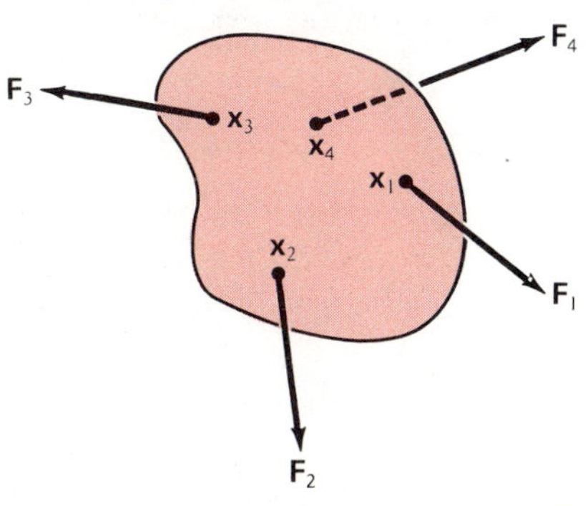

Equilibrium of Forces [Optional] dropout

Suppose forces $\mathbf{F}_1, \cdots, \mathbf{F}_n$ are applied at points $\mathbf{x}_1, \cdots, \mathbf{x}_n$ of a rigid body (Figure 12-7-9). It is shown in physics that the rigid body is in equilibrium when both the sum of the forces vanishes and the sum of the turning moments (torques) of the forces about $\mathbf{0}$ vanishes. Thus the conditions for equilibrium are the two vector equations:

$$\begin{cases} \mathbf{F}_1 + \mathbf{F}_2 + \cdots + \mathbf{F}_n = \mathbf{0} \\ \mathbf{x}_1 \times \mathbf{F}_1 + \mathbf{x}_2 \times \mathbf{F}_2 + \cdots + \mathbf{x}_n \times \mathbf{F}_n = \mathbf{0} \end{cases}$$

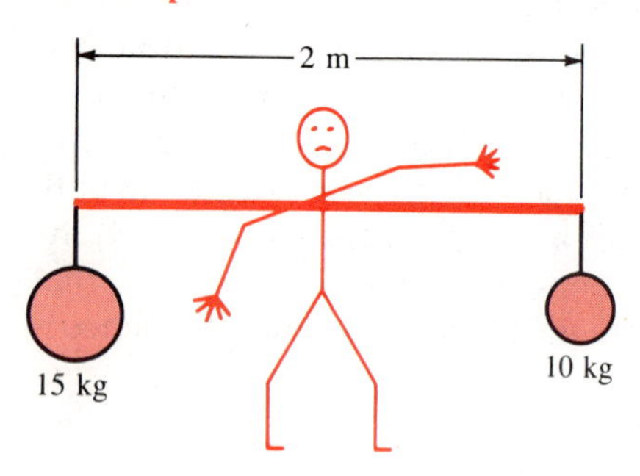

Figure 12-7-10
See Example 5

Example 5

A man has a light-weight 2-meter pole (Figure 12-7-10) balanced on his shoulder. Weights of 10 kg and 15 kg are suspended from the ends as indicated. Find the distance from the 15-kilogram end to the man's shoulder.

Solution This is a plane problem. Place axes as indicated in Figure 12-7-11, and let $\mathbf{F} = (0, F, 0)$ denote the force that the man's shoulder exerts against the pole. The pole is in equilibrium; the first conclusion is that the sum of the forces is $\mathbf{0}$:

$$(0, -15, 0) + (0, F, 0) + (0, -10, 0) = (0, 0)$$

that is, $F = 25$ kg.

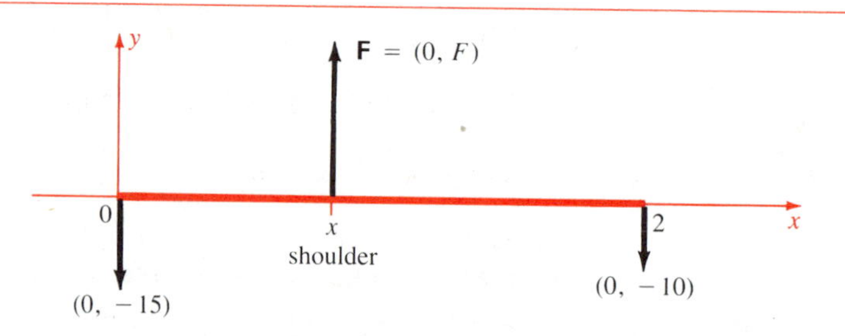

Figure 12-7-11
See Example 5

Each force tends to rotate the pole about an axis perpendicular to the plane. The total torque (tendency to rotate) must also be zero, hence

$$(0, 0, 0) \times (0, -15, 0) + (x, 0, 0) \times (0, 25, 0) + (2, 0, 0) \times (0, -10, 0) = (0, 0, 0)$$

That is,

$$(0, 0, 0) + (0, 0, 25x) + (0, 0, -20) = (0, 0, 0)$$

so $25x - 20 = 0$ and $x = \frac{20}{25} = 0.8$ meter. ●

Exercises

Find the volume of the parallelepiped determined by the origin and the three vectors

1 $(1, 1, 0)$ $(0, 1, 1)$ $(1, 0, 1)$

2 $(4, -1, 0)$ $(3, 0, 2)$ $(1, 1, 1)$

Find the line of intersection (parametric form) of the planes

3 $\begin{cases} x + 2y + 3z = 0 \\ y - z = 1 \end{cases}$

4 $\begin{cases} x - y + z = 0 \\ x + y + z = 3 \end{cases}$

5 $\begin{cases} x + 2y + z = 3 \\ 2x - y + z = 4 \end{cases}$

6 $\begin{cases} x + y = 1 \\ y + z = -1 \end{cases}$

Find a non-parametric equation for the parametric plane

7 $\mathbf{x} = (1, s, t)$

8 $\mathbf{x} = (s, s + t, t - 1)$

9 $\mathbf{x} = (s + 2, s + t + 1, s - t)$

10 $\mathbf{x} = (3s, 2s - t, 2t + 1)$

11 $\mathbf{x} = (-s + t + 1, s - t + 2, -2s - t + 3)$

12 $\mathbf{x} = (s - 3t + 2, 2s + t + 1, 2s - t + 3)$

Find the distance between the lines

13 $\mathbf{x} = s(1, 1, 1) + (0, 0, -1)$ and $\mathbf{x} = t(-1, 2, 2) + (2, 3, 4)$

14 $\mathbf{x} = s(-1, 1, 0) + (1, 0, 0)$ and $\mathbf{x} = t(-1, -1, 0) + (1, 1, 1)$

15 $\mathbf{ab}$ and $\mathbf{cd}$, where

$$\mathbf{a} = (0, 0, 1) \quad \mathbf{b} = (1, 2, 3)$$
$$\mathbf{c} = (1, 1, 0) \quad \mathbf{d} = (-1, -1, -1)$$

16 $\mathbf{ac}$ and $\mathbf{bd}$ with the same $\mathbf{a}$, $\mathbf{b}$, $\mathbf{c}$, $\mathbf{d}$.

Fnd the nearest points on the two lines

17 the lines of Exercise 13

18 the lines of Exercise 14

19 Find a parametric form for the plane through $(1, 1, 0)$, $(1, 2, 1)$, and $(-1, -1, -1)$.

20 Find a non-parametric equation for the same plane.

21 Prove $[\mathbf{a}, \mathbf{b}, \mathbf{c}]^2 \leq |\mathbf{a}|^2|\mathbf{b}|^2|\mathbf{c}|^2$.

22 Prove that the points $\mathbf{a}$, $\mathbf{b}$, $\mathbf{c}$ are collinear if and only if $\mathbf{a} \times \mathbf{b} + \mathbf{b} \times \mathbf{c} + \mathbf{c} \times \mathbf{a} = \mathbf{0}$.

23 Suppose $\mathbf{a}$, $\mathbf{b}$, $\mathbf{c}$ are non-collinear points. Describe the set of points $\mathbf{x} = r\mathbf{a} + s\mathbf{b} + t\mathbf{c}$ where $r \geq 0$, $s \geq 0$, $t \geq 0$, and $r + s + t = 1$.

24 (cont.) Suppose in addition that $\mathbf{0}$ does not lie on the plane of $\mathbf{a}$, $\mathbf{b}$, and $\mathbf{c}$, and replace the last condition by $r + s + t \leq 1$. Now what is the set?

*25 Vectors $\mathbf{a}$, $\mathbf{b}$, $\mathbf{c}$ are called **linearly independent** if the only solution of $r\mathbf{a} + s\mathbf{b} + t\mathbf{c} = \mathbf{0}$ is $r = s = t = 0$. Show that $\mathbf{a}$, $\mathbf{b}$, and $\mathbf{c}$ are linearly independent if and only if $[\mathbf{a}, \mathbf{b}, \mathbf{c}] \neq 0$.

*26 Suppose the points $\mathbf{a}$, $\mathbf{b}$, $\mathbf{c}$ are non-collinear. Prove that

$$[\mathbf{x} - \mathbf{a}, \mathbf{x} - \mathbf{b}, \mathbf{x} - \mathbf{c}] = 0$$

is an equation of the plane through $\mathbf{a}$, $\mathbf{b}$, and $\mathbf{c}$.

27 A seesaw with unequal arms of lengths a and b is in horizontal equilibrium. Find the relations between weights A and B at the ends and the upward force C at the fulcrum.

28 Unit vertical forces act downward at the points $\mathbf{p}_1, \cdots, \mathbf{p}_n$ of the horizontal x,y-plane. Suppose a single force $\mathbf{F}$ acts at another point $\mathbf{p}$ of the plane so that the rigid system is in equilibrium. Find $\mathbf{F}$ and $\mathbf{p}$.

29 A force $\mathbf{F}$ is applied at a point $\mathbf{x}$. Its **torque about a point** $\mathbf{p}$ is $(\mathbf{x} - \mathbf{p}) \times \mathbf{F}$. Suppose $\mathbf{F}_1, \cdots, \mathbf{F}_n$ are applied at points $\mathbf{x}_1, \cdots, \mathbf{x}_n$ of a rigid body and the body is in equilibrium. Show that the sum of the torques about $\mathbf{p}$ equals zero. (Here $\mathbf{p}$ is any point of space, not just $\mathbf{0}$.)

30 A **couple** consists of a pair of opposite forces $\mathbf{F}$ and $-\mathbf{F}$ applied at two different points $\mathbf{p}$ and $\mathbf{q}$. Show that the total torque is unchanged if $\mathbf{p}$ and $\mathbf{q}$ are displaced the same amount, that is, replaced by $\mathbf{p} + \mathbf{c}$ and $\mathbf{q} + \mathbf{c}$. (Note that the torque of a couple is thus independent of the origin.)

12-8 Review Exercises

1 Give a parametric vector equation for the line that passes through $(1, -2, -1)$ and is parallel to the line through $(1, 0, 0)$ and $(0, 0, 1)$.

2 Find the point that is halfway from $(1, 1, 1)$ to $(-1, 2, 2)$.

3 Find the distance between the points $(-4, 3, 0)$ and $(-1, -2, 3)$.

4 Find the angle between the vectors $(-4, 3, 0)$ and $(-1, -2, 3)$.

5 Find a unit vector in the direction of the line joining the point $(-4, 3, 0)$ to the point $(-1, -2, 3)$.

6 Find $(\mathbf{i} - 2\mathbf{j} - 3\mathbf{k}) \cdot (-1, -3, 4)$.

7 Find a normal form for the plane $3x_1 - x_2 + 4x_3 = 7$.

8 (cont.) Find the distance from the point $(1, 1, -1)$ to the plane in Exercise 7.

Solve

9 $\begin{cases} 2x + 3y = -4 \\ 5x + 7y = 2 \end{cases}$

10 $\begin{cases} x + y + 2z = 6 \\ x + 3y + 4z = 11 \\ 2x - y + 2z = 4 \end{cases}$

11 Find $(2\mathbf{i} + \mathbf{j} - \mathbf{k}) \times (\mathbf{i} + 2\mathbf{j} - 3\mathbf{k})$.

12 Find $[\mathbf{i} + \mathbf{j}, \mathbf{j} + \mathbf{k}, \mathbf{k} + \mathbf{i}]$.

13 Points $(1, 1, 1)$ and $(1, -1, 1)$ are opposite vertices of a parallelogram, a third vertex of which is $(0, 0, 0)$. Find the fourth vertex and the area.

14 Find the line of intersection of the planes

$$x - y - 2z = 3 \quad \text{and} \quad 3x + y - 3z = -1$$

15 Find a scalar equation for the parametric plane

$$\mathbf{x} = (2s + t, s - t + 1, 3s + t - 1)$$

16 Find all solutions

$$\begin{cases} x + y + z = 1 \\ x + 2y + 3z = -1 \\ x + 3y + 5z = -3 \end{cases}$$

17 Show that the line through the midpoints of two sides of a triangle is parallel to the third side.

18 Let $\mathbf{a}$, $\mathbf{b}$, $\mathbf{c}$, $\mathbf{d}$ be the vertices in order of a skew quadrilateral. Show that the midpoints of its sides are the vertices of a parallelogram.

13 Vector Functions

13-1 Differentiation

In this chapter we study functions whose values are vectors. For example, the position $\mathbf{x}$ of a moving particle at time t and the gravitational force $\mathbf{F}$ on an orbiting satellite at time t are vector functions. To indicate that $\mathbf{x}$ is a function of time, we write

$$\mathbf{x} = \mathbf{x}(t)$$

in components,

$$\mathbf{x}(t) = (x_1(t), x_2(t), x_3(t))$$

Thus a vector function is a single expression for three ordinary (scalar) functions

$$x_1 = x_1(t) \qquad x_2 = x_2(t) \qquad x_3 = x_3(t)$$

We often think of a vector function as describing a curve in space, the trajectory of a moving particle. For example, if $\mathbf{b}$ and $\mathbf{c}$ are fixed vectors, then the function

$$\mathbf{x}(t) = t\mathbf{b} + \mathbf{c}$$

describes a line traversed by a moving point that is at $\mathbf{c}$ when $t = 0$, at $\mathbf{b} + \mathbf{c}$ when $t = 1$, at $2\mathbf{b} + \mathbf{c}$ when $t = 2$, and so on.

Note that $\mathbf{y}(t) = 2t\mathbf{b} + \mathbf{c}$ describes the same line but traversed at twice the speed, and $\mathbf{z}(t) = t^3\mathbf{b} + \mathbf{c}$ is again the same line, but traversed by an accelerating particle. We shall deal with the measurement of speed and acceleration shortly.

Limits

In order to do calculus in space, we need derivatives of vector functions. We want to define the derivative of a vector function as the *limit* of a difference quotient. First, we had better know what we mean by the limit of a vector function. Our problem is to define $\lim_{t\to a}\mathbf{x}(t) = \mathbf{c}$. Intuitively, we want to say that $\mathbf{x}(t)$ approaches closer and closer to $\mathbf{c}$ as t approaches $\mathbf{a}$.

The distance $|\mathbf{x}(t) - \mathbf{c}|$ measures how close $\mathbf{x}(t)$ is to $\mathbf{c}$, and this distance is a scalar function for which we know limits.

Definition Limit

Let $\mathbf{x}(t)$ be a vector function whose domain is an interval that includes $t = a$. Let $\mathbf{c}$ be a constant vector. Then we define

$$\lim_{t\to a}\mathbf{x}(t) = \mathbf{c}$$

to mean $\lim_{t\to a}|\mathbf{x}(t) - \mathbf{c}| = 0$.

When we actually come to computing a limit, we frequently use an alternative definition in terms of coordinates:

Alternative Definition of Limit

Let

$$\mathbf{x}(t) = (x_1(t), x_2(t), x_3(t)) \quad \text{and} \quad \mathbf{c} = (c_1, c_2, c_3)$$

Then

$$\lim_{t\to a}\mathbf{x}(t) = \mathbf{c}$$

if and only if $\lim_{t\to a} x_j(t) = c_j$ for $j = 1, 2, 3$.

It seems reasonable from the geometry that the definitions are equivalent. See Figure 13-1-1 for the plane case.

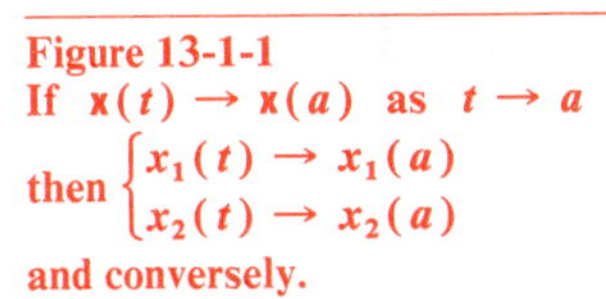
Figure 13-1-1
If $\mathbf{x}(t) \to \mathbf{x}(a)$ as $t \to a$
then $\begin{cases} x_1(t) \to x_1(a) \\ x_2(t) \to x_2(a) \end{cases}$
and conversely.

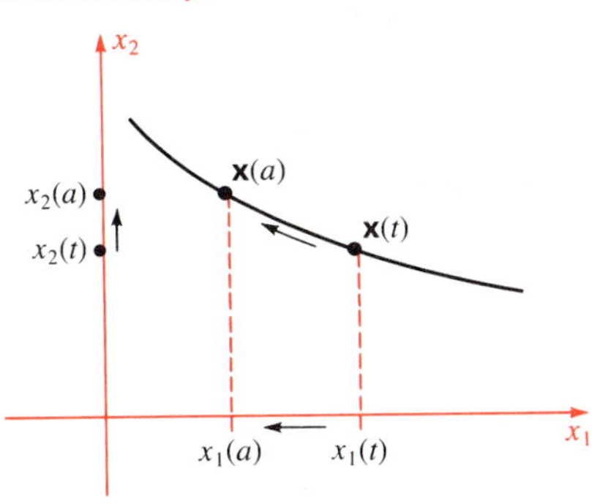

Proof Equivalence of the Definitions The relation

$$|\mathbf{x}(t) - \mathbf{c}|^2 = \sum_{j=1}^{3}|x_j(t) - c_j|^2$$

implies $|\mathbf{x}(t) - \mathbf{c}| \ge |x_j(t) - c_j|$ for each j. Therefore if

$$|\mathbf{x}(t) - \mathbf{c}| \to 0 \qquad \text{then} \qquad |x_j(t) - c_j| \to 0$$

for each j as $t \to a$. Conversely, if $|x_j(t) - c_j| \to 0$ for each j, then

$$|\mathbf{x}(t) - \mathbf{c}|^2 = \sum_{1}^{3}|x_j(t) - c_j|^2 \to 0$$

Hence $|\mathbf{x}(t) - \mathbf{c}| \to 0$ as $t \to a$.

Example 1

$$\lim_{t\to\pi}(\cos t, \sin t, t) = (-1, 0, \pi)$$ ●

The Derivative

Think of $\mathbf{x} = \mathbf{x}(t)$ as tracing a path in space (Figure 13-1-2). For h small, the difference vector

$$\mathbf{x}(t+h) - \mathbf{x}(t)$$

represents the chord from $\mathbf{x}(t)$ to $\mathbf{x}(t+h)$. Hence the difference quotient

$$\frac{\mathbf{x}(t+h) - \mathbf{x}(t)}{h}$$

represents this (short) chord divided by the small number h. Its limit as $h \to 0$, if it exists, is called the **derivative** of the vector function $\mathbf{x}$:

Figure 13-1-2

Definition of $\dfrac{d\mathbf{x}}{dt}$

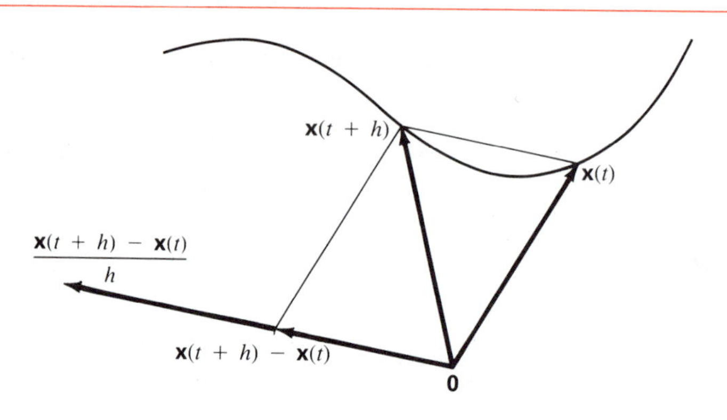

Derivative of a Vector Function

$$\frac{d\mathbf{x}}{dt} = \lim_{h\to 0}\frac{\mathbf{x}(t+h) - \mathbf{x}(t)}{h}$$ ●

Because the direction of $d\mathbf{x}/dt$ is the limiting direction of the chord, $d\mathbf{x}/dt$ is tangent to the curve.

Example 2

Let $\mathbf{x}(t) = t\mathbf{b} + \mathbf{c}$. Then

$$\frac{d\mathbf{x}}{dt} = \lim_{h\to 0}\frac{\mathbf{x}(t+h) - \mathbf{x}(t)}{h}$$

$$= \lim_{h\to 0}\frac{[(t+h)\mathbf{b} + \mathbf{c}] - [t\mathbf{b} + \mathbf{c}]}{h} = \lim_{h\to 0}\frac{h\mathbf{b}}{h} = \mathbf{b}$$

Hence $d\mathbf{x}/dt$ is the constant vector $\mathbf{b}$. This is reasonable because $\mathbf{x}(t)$ traces a line parallel to $\mathbf{b}$ at constant speed. ●

Usually, we compute the derivative $d\mathbf{x}/dt$ in coordinates. We suppose $\mathbf{x}(t) = (x_1(t), x_2(t), x_3(t))$. From the definition of $d\mathbf{x}/dt$ as the limit of a difference quotient it follows, for example, that the second coordinate of $d\mathbf{x}/dt$ is

$$\lim_{h\to 0} \frac{x_2(t+h) - x_2(t)}{h} = \frac{dx_2}{dt}$$

The same holds for the other coordinates of $d\mathbf{x}/dt$. Thus we may compute $d\mathbf{x}/dt$ by computing three ordinary derivatives.

- The derivative of a vector function $\mathbf{x}(t) = (x_1(t), x_2(t), x_3(t))$ is the vector function

$$\frac{d\mathbf{x}}{dt} = \left(\frac{dx_1}{dt}, \frac{dx_2}{dt}, \frac{dx_3}{dt}\right)$$ ●

Example 3

Suppose $\mathbf{x}(t) = t^3\mathbf{b} + \mathbf{c}$. Then

$$\mathbf{x}(t) = t^3\mathbf{b} + \mathbf{c} = (b_1t^3 + c_1, b_2t^3 + c_2, b_3t^3 + c_3)$$

and

$$\begin{aligned}\frac{d\mathbf{x}}{dt} &= (3b_1t^2, 3b_2t^2, 3b_3t^2)\\ &= 3t^2(b_1, b_2, b_3) = 3t^2\mathbf{b}\end{aligned}$$ ●

Although the words *velocity* and *speed* are used interchangeably in everyday life, there is an important distinction between them in science. Velocity is a *vector* function; speed is a *scalar* (real-valued) function.

Definition Velocity and Speed

Let the vector function $\mathbf{x} = \mathbf{x}(t)$ be the position of a particle at time t. Its **velocity** is defined as

$$\mathbf{v} = \mathbf{v}(t) = \frac{d\mathbf{x}}{dt}$$

Its **speed** is defined as the magnitude (length) of its velocity:

$$\text{speed} = |\mathbf{v}(t)| = |d\mathbf{x}/dt|$$ ●

Note carefully the distinction between velocity and speed. Velocity has direction as well as size. A car traveling at constant speed does not have constant velocity unless the road is absolutely straight (in all three dimensions)!

Example 4

For the curve $\mathbf{x}(t) = (t, t^2, t^3)$ we have

$$\text{velocity} = \mathbf{v}(t) = \frac{d\mathbf{x}}{dt} = (1, 2t, 3t^2)$$

and

$$\text{speed} = |\mathbf{v}(t)| = \sqrt{1 + 4t^2 + 9t^4}$$

For each t, the velocity vector $\mathbf{v}(t)$ has its initial point at the origin, as does any vector. However, it is helpful to think of $\mathbf{v}(t)$ as attached to the point $\mathbf{x}(t)$. See Figure 13-1-3. Then $\mathbf{v}(t)$ is pictured as tangential to the curve; its direction at each point is the direction of the motion and its magnitude is the speed. Double the speed and you double the velocity vector. (It still points in the same direction, but is twice as long.)

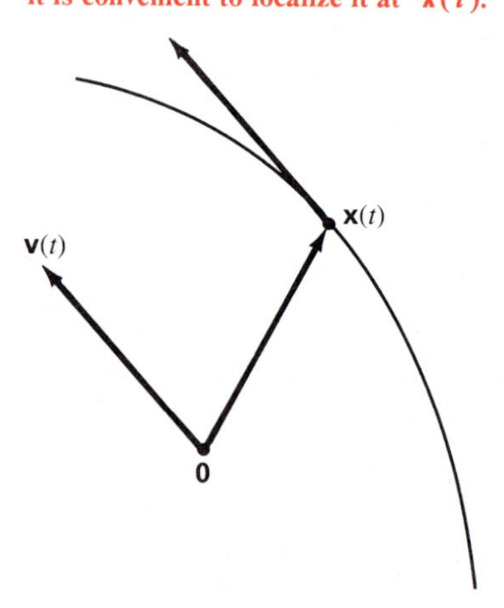

Figure 13-1-3
Even though $\mathbf{v}(t)$ starts at $\mathbf{0}$, it is convenient to localize it at $\mathbf{x}(t)$.

Zero Speed The function $\mathbf{x}(t)$ has zero speed if and only if it has zero velocity. When this is so for all t on an interval, then

$$\frac{d\mathbf{x}}{dt} = \left(\frac{dx_1}{dt}, \frac{dx_2}{dt}, \frac{dx_3}{dt}\right) = (0, 0, 0)$$

Therefore $dx_i/dt = 0$, so $x_i = c_i$ (constant) for $i = 1, 2, 3$. Hence

$$\mathbf{x}(t) = (c_1, c_2, c_3) = \mathbf{c}$$

Physically, this simply says that an object with zero speed (or velocity) is standing still.

The following formulas are essential for differentiating vector functions.

Rules for Vector Derivatives

Let $\mathbf{x}(t)$ and $\mathbf{y}(t)$ be vector functions and let $f(t)$ and $u(t)$ be scalar functions. Then

1 $$\frac{d}{dt}[\mathbf{x}(t) + \mathbf{y}(t)] = \frac{d\mathbf{x}}{dt} + \frac{d\mathbf{y}}{dt}$$

2 $$\frac{d}{dt}[f(t)\mathbf{x}(t)] = \frac{df}{dt}\mathbf{x} + f\frac{d\mathbf{x}}{dt}$$

3 $$\frac{d}{dt}[\mathbf{x}(t) \cdot \mathbf{y}(t)] = \frac{d\mathbf{x}}{dt} \cdot \mathbf{y} + \mathbf{x} \cdot \frac{d\mathbf{y}}{dt}$$

4 $$\frac{d}{dt}[\mathbf{x}(t) \times \mathbf{y}(t)] = \frac{d\mathbf{x}}{dt} \times \mathbf{y} + \mathbf{x} \times \frac{d\mathbf{y}}{dt}$$

5 $$\frac{d}{dt}\mathbf{x}[u(t)] = \frac{du}{dt} \cdot \frac{d\mathbf{x}}{du} \qquad \text{(chain rule)}$$

Note the similarity to ordinary differentiation formulas. In particular, the second, third, and fourth formulas, each involving a kind of product, all resemble the product rule for derivatives.

To prove the second formula, for example, note that the first coordinate of $f\mathbf{x}$ is fx_1, and

$$\frac{d}{dt}[f(t)x_1(t)] = \frac{df}{dt}x_1 + f\frac{dx_1}{dt}$$

which is the first component of $f'\mathbf{x} + f\mathbf{x}'$. The other formulas can be verified similarly. See Exercises 13–16.

Example 5

Use of rule 2:

$$\begin{aligned}\frac{d}{dt}(t^3, t^4, t^5) &= \frac{d}{dt}[t^3(1, t, t^2)] \\ &= \left(\frac{d}{dt}t^3\right)(1, t, t^2) + t^3\frac{d}{dt}(1, t, t^2) \\ &= 3t^2(1, t, t^2) + t^3(0, 1, 2t) = (3t^2, 4t^3, 5t^4)\end{aligned}$$

An example of rule 4: we start with

$$(1, t, -t^2) \times (1, t, t^2) = (2t^3, -2t^2, 0)$$

On the one hand, by componentwise differentiation,

$$\frac{d}{dt}(2t^3, -2t^2, 0) = (6t^2, -4t, 0)$$

On the other hand, by rule 4,

$$\begin{aligned}&\left[\frac{d}{dt}(1, t, -t^2)\right] \times (1, t, t^2) + (1, t, -t^2) \times \frac{d}{dt}(1, t, t^2) \\ &\qquad = (0, 1, -2t) \times (1, t, t^2) + (1, t, -t^2) \times (0, 1, 2t) \\ &\qquad = (3t^2, -2t, -1) + (3t^2, -2t, 1) = (6t^2, -4t, 0)\end{aligned}$$

The answers agree. ●

Example 6

Let $\mathbf{x} = \mathbf{x}(t)$ be the path of a point moving on the sphere $|\mathbf{x}| = r$. Show that the velocity $\mathbf{v}(t)$ at each instant is perpendicular to $\mathbf{x}(t)$.

Solution

$$\mathbf{x}(t) \cdot \mathbf{x}(t) = |\mathbf{x}(t)|^2 = r^2 \quad \text{so} \quad \frac{d}{dt}[\mathbf{x}(t) \cdot \mathbf{x}(t)] = \frac{d}{dt}r^2 = 0$$

But

$$\frac{d}{dt}(\mathbf{x} \cdot \mathbf{x}) = \frac{d\mathbf{x}}{dt} \cdot \mathbf{x} + \mathbf{x} \cdot \frac{d\mathbf{x}}{dt} = 2\mathbf{x} \cdot \frac{d\mathbf{x}}{dt} = 2\mathbf{x} \cdot \mathbf{v}$$

Therefore $\mathbf{x} \cdot \mathbf{v} = 0$. ●

We shall use the result of Example 6 several times in the following sections. Let us restate it in slightly different terms:

Vector of Constant Magnitude

If $\mathbf{x}(t)$ has constant magnitude, then $\mathbf{x}(t)$ is perpendicular to its derivative.

This statement makes good sense geometrically. For the curve $\mathbf{x}(t)$ lies on the surface of a sphere $|\mathbf{x}| = a$. The derivative vector $d\mathbf{x}/dt$, imagined attached at each point of the curve, is tangent to the curve, hence is tangent to the sphere. Therefore the derivative is perpendicular to the radius vector, which is $\mathbf{x}(t)$. See Figure 13-1-4.

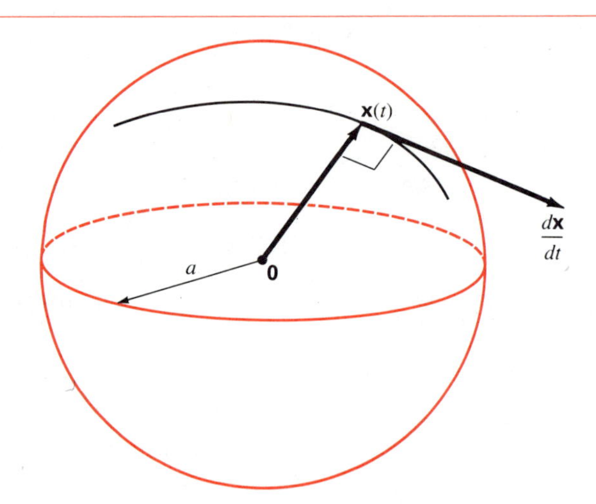

Figure 13-1-4
The tangent to a curve on the sphere is tangent to the sphere, hence perpendicular to the radius vector.

Exercises

Differentiate $\mathbf{x}(t)$

1 (e^t, e^{2t}, e^{3t})

2 (t^4, t^5, t^6)

3 $(t + 1, 3t - 1, 4t)$

4 $(t^2, 0, t^3)$

5 $(t, \cos t, \sin t)$

6 $(t^2, \arctan t, e^{-t})$

Find the speed at t of $\mathbf{x}(t) =$

7 $(t^2, t^3 + t^4, 1)$

8 $(2t - 1, 3t + 1, -2t + 1)$

9 $(A \cos \omega t, A \sin \omega t, B)$

10 $(A \cos \omega t, A \sin \omega t, Bt)$

11 $(1, t^4, t^6)$

12 $(t, \cosh t, 3)$.

Prove

13 $\dfrac{d}{dt}[\mathbf{x}(t) + \mathbf{y}(t)] = \dfrac{d\mathbf{x}}{dt} + \dfrac{d\mathbf{y}}{dt}$

14 $\dfrac{d}{dt}[\mathbf{x}(t) \cdot \mathbf{y}(t)] = \dfrac{d\mathbf{x}}{dt} \cdot \mathbf{y} + \mathbf{x} \cdot \dfrac{d\mathbf{y}}{dt}$

15 $\dfrac{d}{dt}(\mathbf{x} \times \mathbf{y}) = \dfrac{d\mathbf{x}}{dt} \times \mathbf{y} + \mathbf{x} \times \dfrac{d\mathbf{y}}{dt}$

16 $\dfrac{d}{dt}\mathbf{x}[u(t)] = \left(\dfrac{d\mathbf{x}}{du}\bigg|_{u(t)}\right)\left(\dfrac{du}{dt}\right)$

17 Suppose $d\mathbf{x}/dt = \mathbf{b}$. Find $\mathbf{x}(t)$.

18 Suppose $d\mathbf{x}/dt = \mathbf{b}t + \mathbf{c}$. Find $\mathbf{x}(t)$.

19 Suppose $d\mathbf{x}/dt = k\mathbf{x}(t)$. Find $\mathbf{x}(t)$.

20 Suppose $d^2\mathbf{x}/dt^2 = \mathbf{0}$. Find $\mathbf{x}(t)$.

21 Suppose that $\mathbf{x} = \mathbf{x}(t)$ is a moving point such that $d\mathbf{x}/dt$ is always perpendicular to $\mathbf{x}(t)$. Show that $\mathbf{x}(t)$ moves on a sphere with center at $\mathbf{0}$. [Hint Differentiate $|\mathbf{x}|^2$.]

22 Suppose $\mathbf{x}(t) \neq \mathbf{0}$. Show that $d|\mathbf{x}|/dt = |\mathbf{x}|^{-1}\mathbf{x} \cdot (d\mathbf{x}/dt)$.

23 Prove that

$$\frac{d}{dt}[\mathbf{u}(t), \mathbf{v}(t), \mathbf{w}(t)]$$
$$= [d\mathbf{u}/dt, \mathbf{v}, \mathbf{w},] + [\mathbf{u}, d\mathbf{v}/dt, \mathbf{w}] + [\mathbf{u}, \mathbf{v}, d\mathbf{w}/dt]$$

24 (cont.) Find $\dfrac{d}{dt}\left[\mathbf{u}, \dfrac{d\mathbf{u}}{dt}, \dfrac{d^2\mathbf{u}}{dt^2}\right]$

25 A particle oscillates on the line segment $\mathbf{ab}$. It starts at $\mathbf{a}$, then moves halfway to $\mathbf{b}$, then halfway back towards $\mathbf{a}$, then half again as far towards $\mathbf{b}$, and so on. Express its position $\mathbf{x}_n$ after n steps in terms of $\mathbf{a}$, $\mathbf{b}$, and $\mathbf{n}$.

26 (cont.) Find its limit as $n \to +\infty$ in terms of $\mathbf{a}$ and $\mathbf{b}$.

Let $\mathbf{x} = \mathbf{x}(t) = (x_1(t), x_2(t), x_3(t))$ for $a \le t \le b$. Define

$$\int_a^b \mathbf{x}(t)\,dt = \left(\int_a^b x_1(t)\,dt, \int_a^b x_2(t)\,dt, \int_a^b x_3(t)\,dt\right)$$

Prove

27 $\displaystyle\int_a^b c\mathbf{x}\,dt = c\int_a^b \mathbf{x}\,dt$

28 $\displaystyle\int_a^b (\mathbf{x} + \mathbf{y})\,dt = \int_a^b \mathbf{x}\,dt + \int_a^b \mathbf{y}\,dt$

29 $\displaystyle\int_a^b \mathbf{c}\cdot\mathbf{x}\,dt = \mathbf{c}\cdot\int_a^b \mathbf{x}\,dt$

30 $\displaystyle\int_a^b \mathbf{c}\times\mathbf{x}\,dt = \mathbf{c}\times\int_a^b \mathbf{x}\,dt$

31 $\displaystyle\left|\mathbf{c}\cdot\int_a^b \mathbf{x}\,dt\right| \le |\mathbf{c}|\int_a^b |\mathbf{x}|\,dt$

32 $\displaystyle\left|\int_a^b \mathbf{x}\,dt\right| \le \int_a^b |\mathbf{x}|\,dt$

[Hint Use Exercise 31.]

13-2 Arc Length

In this section we revert to the conventional notation $(x(t), y(t), z(t))$ for the components of a vector function $\mathbf{x}(t)$ instead of the more cumbersome (x_1, x_2, x_3).

Let $\mathbf{x} = \mathbf{x}(t)$ be a space curve defined for $a \le t \le b$. We want to assign a length to the curve. Actually, we do more: we define a function $s = s(t)$ that gives the length of the arc from $\mathbf{x}(a)$ to $\mathbf{x}(t)$ for any t.

We give two definitions for $s(t)$. The first is motivated by the physical idea of speed. The second, discussed at the end of this section, is based on the lengths of polygons. Both definitions involve integration in a natural way.

Arc Length from Velocity

Let $\mathbf{x} = \mathbf{x}(t)$ be a space curve defined for $a \le t \le b$ by a smooth (continuously differentiable) vector function. Suppose there is a function $s = s(t)$ that measures the arc length of the curve from time a to time t. What can we say about it? At each point of the curve, the velocity vector $\mathbf{v}(t)$ is tangential (Figure 13-2-1). Its magnitude $|\mathbf{v}(t)|$ is the speed, intuitively the rate at which the arc length $s(t)$ of the curve increases with respect to time. Therefore we *define* the arc length $s = s(t)$ by $s(a) = 0$ and

$$\frac{ds}{dt} = |\mathbf{v}(t)| = \left|\frac{d\mathbf{x}}{dt}\right|$$

Figure 13-2-1 $\dfrac{ds}{dt} = |\mathbf{v}(t)|$

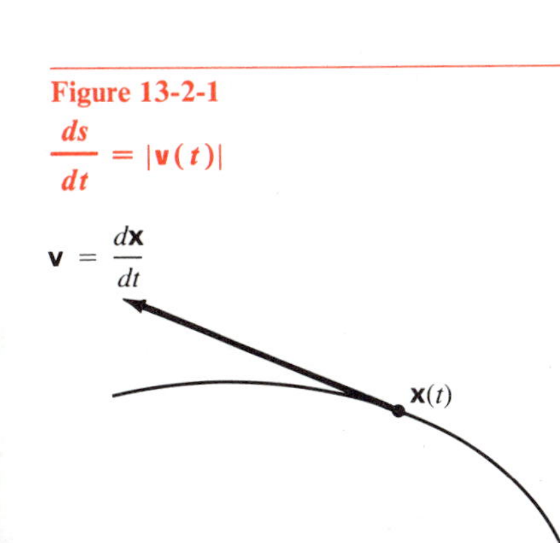

Now

$$|\mathbf{v}(t)|^2 = \left|\frac{dx}{dt}, \frac{dy}{dt}, \frac{dz}{dt}\right|^2 = \left(\frac{dx}{dt}\right)^2 + \left(\frac{dy}{dt}\right)^2 + \left(\frac{dz}{dt}\right)^2$$

Consequently

$$\frac{ds}{dt} = \sqrt{\left(\frac{dx}{dt}\right)^2 + \left(\frac{dy}{dt}\right)^2 + \left(\frac{dz}{dt}\right)^2}$$

To obtain $s(t)$ itself we integrate, starting at $t = a$ so that $s(a) = 0$:

Definition Arc Length

Let $\mathbf{x} = \mathbf{x}(t)$ describe a space curve. Then its **arc length** from $\mathbf{x}(a)$ to $\mathbf{x}(t)$ is

$$s(t) = \int_a^t \sqrt{\left(\frac{dx}{dt}\right)^2 + \left(\frac{dy}{dt}\right)^2 + \left(\frac{dz}{dt}\right)^2}\, dt$$

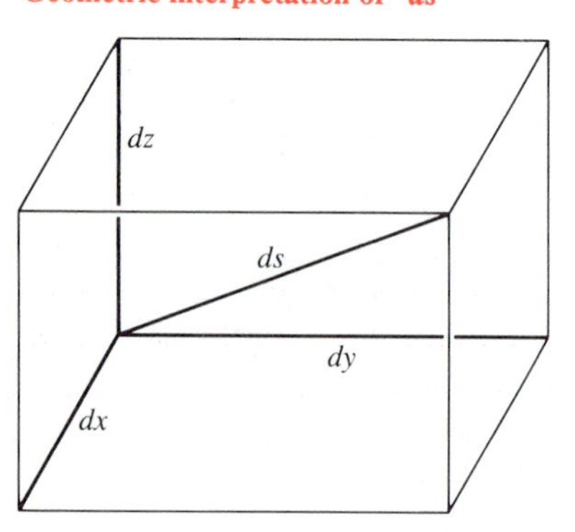

Figure 13-2-2
Geometric interpretation of *ds*

Remark The arc-length formula has a direct geometric interpretation (Figure 13-2-2). The tiny bit of arc length ds corresponds to three "displacements" dx, dy, and dz along the coordinate axes. By the distance formula,

$$(ds)^2 = (dx)^2 + (dy)^2 + (dz)^2$$

Divide formally by $(dt)^2$ and take square roots. The result is again

$$\frac{ds}{dt} = \sqrt{\left(\frac{dx}{dt}\right)^2 + \left(\frac{dy}{dt}\right)^2 + \left(\frac{dz}{dt}\right)^2}$$

Let us write the arc-length formula for the special case of a curve in the x, y-plane:

Plane Curves

Let $\mathbf{x} = (x(t), y(t))$ be a plane curve for $a \le t \le b$. Then its length is

$$L = \int_a^b \sqrt{\left(\frac{dx}{dt}\right)^2 + \left(\frac{dy}{dt}\right)^2}\, dt$$

If the curve is the graph of a function $y = f(x)$ for $a \le x \le b$, then its length is

$$L = \int_a^b \sqrt{1 + [f'(x)]^2}\, dx$$

The second formula is a special case of the first formula. Just set $x = t$, $y = f(t)$, where $a \le t \le b$. Then $dx/dt = 1$ and $dy/dt = f'(t) = f'(x)$, so

$$\frac{ds}{dt} = \sqrt{1 + (f')^2}$$

and the formula for L follows.

Example 1

Find the length of the curve

a $\mathbf{x}(t) = (\frac{1}{3}t^3, \frac{1}{2}t^2) \qquad (1 \le t \le 2)$

b $y = \sin x \qquad (0 \le x \le \pi)$

Solution

a Use the first formula for arc length of a plane curve:

$$L = \int_1^2 \sqrt{\left(\frac{dx}{dt}\right)^2 + \left(\frac{dy}{dt}\right)^2}\, dt = \int_1^2 \sqrt{(t^2)^2 + (t)^2}\, dt$$

$$= \int_1^2 t\sqrt{t^2+1}\, dt = \tfrac{1}{3}(t^2+1)^{3/2}\Big|_1^2 = \tfrac{1}{3}(5\sqrt{5} - 2\sqrt{2})$$

b $$L = \int_0^\pi \sqrt{1 + \left(\frac{dy}{dx}\right)^2}\, dx = \int_0^\pi \sqrt{1 + \cos^2 x}\, dx \approx 3.820$$

The exact integral (an elliptic integral) is impossible to express in terms of elementary functions. It can be approximated, for instance, by Simpson's rule. ●

Figure 13-2-3
Graph of $\mathbf{x}(t) = (t\cos t, t\sin t, 2t)$

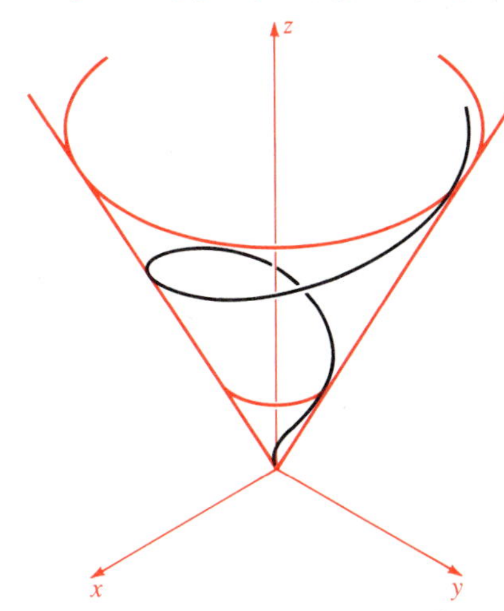

a The curve lies on the cone $x^2 + y^2 = \frac{1}{4}z^2$

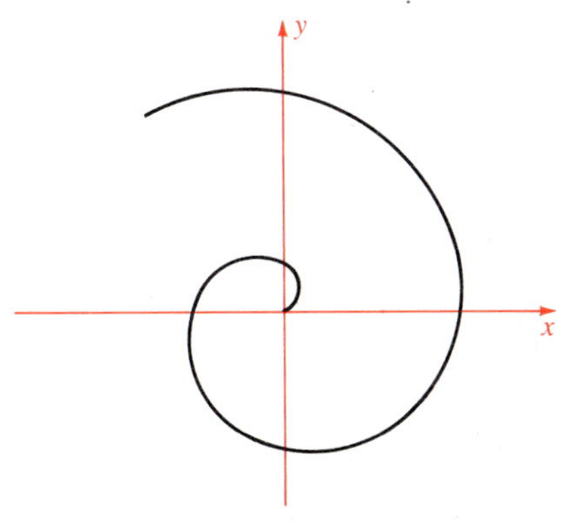

b Projection of the curve on the x,y-plane $(x,y) = t(\cos t, \sin t)$

Example 2

Find the length of the curve

$$\mathbf{x}(t) = (t\cos t, t\sin t, 2t) \quad \text{for} \quad 0 \le t \le 4\pi$$

Solution

$$\left(\frac{ds}{dt}\right)^2 = \left(\frac{dx}{dt}\right)^2 + \left(\frac{dy}{dt}\right)^2 + \left(\frac{dz}{dt}\right)^2$$

$$= (\cos t - t\sin t)^2 + (\sin t + t\cos t)^2 + 2^2 = 5 + t^2$$

By the arc-length formula and an integral table,

$$L = \int_0^{4\pi} \frac{ds}{dt}\, dt = \int_0^{4\pi} \sqrt{5+t^2}\, dt$$

$$= \tfrac{1}{2}[t\sqrt{5+t^2} + 5\ln(t + \sqrt{5+t^2})]\Big|_0^{4\pi}$$

$$= \tfrac{1}{2}[4\pi a + 5\ln(4\pi + a) - \tfrac{5}{2}\ln 5]$$

where $a = (5 + 16\pi^2)^{1/2}$. Approximately, $L \approx 86.3$. ●

The curve in Example 2 is a spiral (Figure 13-2-3). As t increases, z increases at a steady rate, while the projection of $\mathbf{x}(t)$ on the x, y-plane traces the spiral $(t\cos t, t\sin t) = t(\cos t, \sin t)$. Actually, the curve lies on a right circular cone because

$$x^2 + y^2 = (t\cos t)^2 + (t\sin t)^2 = t^2 = \tfrac{1}{4}z^2$$

is an equation describing a cone, as will be shown in the next chapter.

Arc Length Is Independent of the Parameter

Suppose a high-flying jet at unknown altitude passes across the sky, leaving a contrail. An (ideal) movie of the flight will give a dynamic record of the plane's position $\mathbf{x}(t)$ at each instant of time t. But for the *length* of the trajectory, a snapshot should suffice. But we have no idea at all from a snapshot of how the contrail was produced, how fast the plane flew or whether its speed was steady. So the arc length of a curve should not depend on the vector function $\mathbf{x}(t)$ that produced the curve, only the trajectory, a certain set of points in space.

Suppose that a curve has two different parametrizations. For instance, it may be given by $\mathbf{x} = \mathbf{x}(t)$, where $a \leq t \leq b$, also by $\mathbf{x} = \mathbf{x}(u)$, where $c \leq u \leq d$. How do we know that the arc-length formula yields the same length in each case? We suppose that either parametrization can be obtained from the other by a smooth change of variable. For example, let us take $t = t(u)$ for the change of variable and assume that $a = t(c)$, $b = t(d)$, and $dt/du > 0$. The t-length and the u-length of the curve are

$$L_t = \int_a^b \left| \frac{d\mathbf{x}}{dt} \right| dt \qquad \text{and} \qquad L_u = \int_c^d \left| \frac{d\mathbf{x}}{du} \right| du$$

By the chain rule, $d\mathbf{x}/du = (d\mathbf{x}/dt)(dt/du)$. The formula for change of variable in a definite integral implies

$$L_u = \int_c^d \left| \frac{d\mathbf{x}}{dt} \right| \frac{dt}{du}\, du = \int_a^b \left| \frac{d\mathbf{x}}{dt} \right| dt = L_t$$

Thus the formula yields the same length in each case. This proves that the length of a curve is a geometric quantity, independent of the analytic presentation of the curve.

Arc Length as the Parameter

The arc length of a space curve is a built-in attribute, independent of how the curve is parametrized. Therefore a natural parameter for a curve is its own arc length.

Example 3

$$\mathbf{x}(t) = (a \cos 2\pi t, a \sin 2\pi t)$$

This describes a circle of radius a traced counterclockwise at constant speed. The complete circle (of length $2\pi a$) is traced for $0 \leq t \leq 1$. Take $s = 0$ at $t = 0$ and $s > 0$ for $t > 0$. Then s is proportional to t; that is, $s = ct$. But $s = 2\pi a$ when $t = 1$, so $c = 2\pi a$. Thus $s = 2\pi a t$, so $2\pi t = s/a$ and the motion can be described by

$$\mathbf{x} = \left(a \cos \frac{s}{a}, a \sin \frac{s}{a} \right)$$

This is a formula for $\mathbf{x}$ as a function of s. In other words, it is a parametrization of the circle with the arc length itself as the parameter. ●

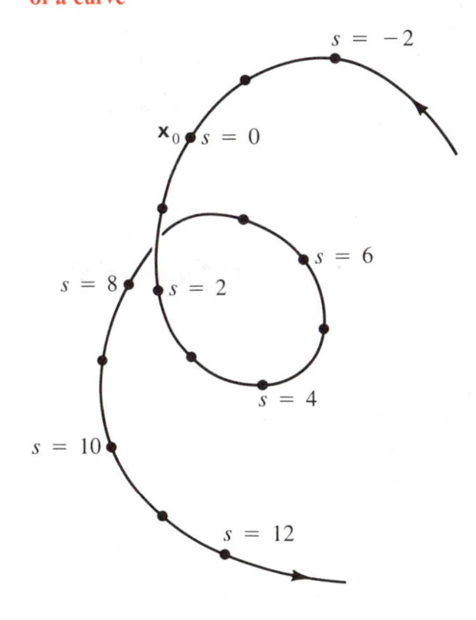

Figure 13-2-4
Arc length as the parameter of a curve

In principle, a similar parametrization is possible for every reasonable space curve. Given $\mathbf{x} = \mathbf{x}(t)$, we fix a point $\mathbf{x}_0 = \mathbf{x}(a)$ and measure s from $\mathbf{x}_0$, positive in the direction of increasing t, negative in the opposite direction (Figure 13-2-4). Let us assume $\mathbf{v}(t) = d\mathbf{x}/dt$ is never $\mathbf{0}$. Then

$$s(t) = \int_a^t |\mathbf{v}(u)|\,du$$

is a *strictly increasing* function of t because

$$\frac{ds}{dt} = |\mathbf{v}(t)| > 0$$

It follows that $s = s(t)$ has an inverse function $t = t(s)$. Therefore we may write $\mathbf{x} = \mathbf{x}(t)$ in the form

$$\mathbf{x} = \mathbf{x}[t(s)]$$

presenting $\mathbf{x}$ as a function of s. This is the desired parametrization of the curve in terms of its own arc length.

Remark 1 (bad news) For most curves it is difficult or impossible to carry out these computations explicitly. Usually the integral defining $s(t)$ is hard to evaluate because of the square root in the integrand,

$$|\mathbf{x}(t)| = \sqrt{(dx/dt)^2 + \cdot\cdot\cdot + (dz/dt)^2}$$

Even if $s(t)$ can be computed explicitly, it may be hard to find its inverse function.

The above example of a circle is one of the few cases in which we can find $s(t)$ explicitly. There the speed is *constant,* so s is just a constant multiple of t.

Remark 2 (good news) We seldom actually need $s = s(t)$ itself or $\mathbf{x}$ expressed in terms of s explicitly. The *idea* of using arc length as the parameter is important for understanding curves, but for calculations we can usually manage with ds/dt (and d^2s/dt^2), which we know.

Arc Length as a Limit

Let $\mathbf{x} = \mathbf{x}(t)$ be a space curve, where $a \le t \le b$. One way to approximate its arc length is to inscribe a polygonal arc in the given curve and take the length of the polygonal arc as an estimate of the arc length.

To this end, we partition $[a, b]$ by

$$a = t_0 < t_1 < \cdot\cdot\cdot < t_n = b$$

We denote the corresponding points on the curve by $\mathbf{x}_0, \mathbf{x}_1, \cdot\cdot\cdot, \mathbf{x}_n$. We connect these points to form a polygonal arc (Figure 13-2-5) consisting of the n straight segments

$$\mathbf{x}_{i-1}\mathbf{x}_i \qquad (i = 1, \cdot\cdot\cdot, n)$$

Figure 13-2-5
Inscribed polygonal arc corresponding to a partition of the t-interval $[a, b]$

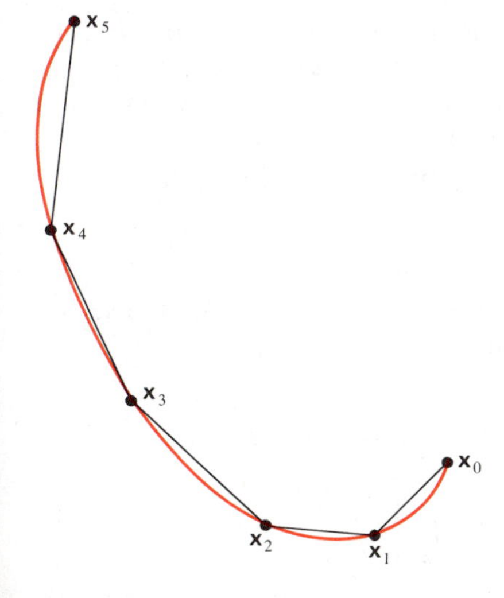

Thus $\mathbf{x}_i = (x_i, y_i, z_i)$, where $x_i = x(t_i)$, $y_i = y(t_i)$, and $z_i = z(t_i)$. The length of the i-th segment is

$$|\mathbf{x}_i - \mathbf{x}_{i-1}| = \sqrt{(x_i - x_{i-1})^2 + (y_i - y_{i-1})^2 + (z_i - z_{i-1})^2}$$

and the length of the polygonal arc—it depends on the partition—is

$$L = L(\text{part.}) = \sum_{i=1}^{n} |\mathbf{x}_i - \mathbf{x}_{i-1}|$$

Now we seek the limit of L over finer and finer partitions, that is, partitions such that

$$\max_{i=1, \cdots, n} (t_i - t_{i-1}) \to 0$$

We use the mean value theorem to approximate $|\mathbf{x}_i - \mathbf{x}_{i-1}|$ and use the result to find $\lim L$. By one application of the MVT,

$$x_i - x_{i-1} = x(t_i) - x(t_{i-1}) = (t_i - t_{i-1}) \frac{dx}{dt}(u_i)$$

where $t_{i-1} < u_i < t_i$. Similarly

$$y_i - y_{i-1} = (t_i - t_{i-1}) \frac{dy}{dt}(v_i)$$

$$z_i - z_{i-1} = (t_i - t_{i-1}) \frac{dz}{dt}(w_i)$$

where $t_{i-1} < v_i < t_i$ and $t_{i-1} < w_i < t_i$. Consequently

$$|\mathbf{x}_i - \mathbf{x}_{i-1}| = (t_i - t_{i-1}) \sqrt{x'(u_i)^2 + y'(v_i)^2 + z'(w_i)^2}$$

so

$$L(\text{part.}) = \sum_{i=1}^{n} [\sqrt{x'(u_i)^2 + y'(v_i)^2 + z'(w_i)^2}\,](t_i - t_{i-1})$$

This sum looks very much like a Riemann sum for definite integral. In order to produce an integral, let us assume that $\mathbf{x}(t)$ is *continuously differentiable,* that is, each of its coordinate functions $x(t)$, $y(t)$, $z(t)$ is a continuously differentiable function of t on $[a, b]$. Then it can be shown by a technical argument involving uniform continuity that

$$\left|\sqrt{x'(u_i)^2 + \cdots + z'(w_i)^2} - \sqrt{x'(t_i)^2 + \cdots + z'(t_i)^2}\right|$$

can be made as small as we please for all i provided we choose the partition sufficiently fine. It follows that

$$L \approx \sum_{i=1}^{n} [\sqrt{x'(t_i)^2 + y'(t_i)^2 + z'(t_i)^2}\,](t_i - t_{i-1})$$

$$\to \int_a^b \sqrt{x'^2 + y'^2 + z'^2}\, dt = \int_a^b \left|\frac{d\mathbf{x}}{dt}\right| dt$$

Thus polygonal approximation leads, in the limit, to the same formula for arc length that we arrived at earlier by a physical argument. Note that the method works for piecewise smooth curves, that is, curves that are smooth (continuously differentiable) except for a finite number of exceptional points (corners).

Exercises

Find the length of $\mathbf{x}(t) =$

1 $(a_1t + b_1, a_2t + b_2, a_3t + b_3) \quad 0 \le t \le 1$

2 $(t^2, t^3) \quad 0 \le t \le 2$

3 $(t^4, t^5) \quad 0 \le t \le 1$

4 $(t^2, t^3, t^2) \quad 0 \le t \le 1$

Set up each arc length as an integral, but do not evaluate

5 $y = x^3 \quad 0 \le x \le b$

6 $y = e^x \quad a \le x \le b$

7 $\mathbf{x}(t) = (t^m, t^n, t^r) \quad a \le t \le b$

8 $\mathbf{x}(t) = (\cos t, \sin t, \cos t + \sin t) \quad 0 \le t \le 2\pi$

Find the arc length

9 $y = \ln x \quad 1 \le x \le 2$

10 $y = 2 \sec x \quad -\frac{1}{4}\pi \le x \le \frac{1}{4}\pi$

11 $\mathbf{x}(t) = (t, t^2, \frac{4}{3}t^{3/2}) \quad 0 \le t \le b$

12 $\mathbf{x}(t) = (t, \sqrt{2}\cos t, \frac{1}{2}t - \frac{1}{4}\sin 2t) \quad 0 \le t \le \pi$

13 Show that the curve $\mathbf{x} = (\sin^2 t, \sin t \cos t, \cos t)$ lies on the unit sphere, and verify the relation $\mathbf{x} \cdot d\mathbf{x}/dt = 0$.

14 (cont.) Express its length for $a \le t \le b$ as an integral.

15 For the curve of Exercise 11, express t in terms of s; take $s = 0$ at $t = 0$.

16 For which functions $z(t)$ is the curve $\mathbf{x}(t) = (\cos t, \sin t, z(t))$ a plane curve?

17 Let $\mathbf{x}(t)$ be a curve in the x, y-plane joining $(a, 0)$ to $(b, 0)$, where $a < b$. Use the formula for arc length to prove that the length $L \ge b - a$.

18 (cont.) Suppose instead that $\mathbf{x}(t)$ joins (a, A) to (b, B). Show that

$$L \ge \sqrt{(b-a)^2 + (B-A)^2}$$

(Hence the shortest curve joining two points is a line segment!) [Hint Rotate and shift coordinates.]

13-3 Plane Curves

In this section we study plane curves in the parametric form

$$\mathbf{x} = \mathbf{x}(t) = (x(t), y(t))$$

Sometimes such a curve may be considered as the graph of a function $y = f(x)$. Suppose, for instance, that $dx/dt > 0$ for $t_0 < t < t_1$. Then $x = x(t)$ is strictly increasing on this interval, hence has an inverse function $t = t(x)$. The substitution $y = y(t) = y[t(x)]$ eliminates the parameter t and expresses y as a function of x. The given curve then is the graph of $y = y[t(x)]$ on the x-interval $x(t_0) \le x \le x(t_1)$.

How do we compute the derivatives

$$\frac{dy}{dx} \quad \text{and} \quad \frac{d^2y}{dx^2}$$

when the function $y(x)$ is so presented parametrically? Of course, if the relation $x = x(t)$ can be solved (inverted) for t explicitly as a function of x, namely $t = t(x)$, then $y = y[t(x)]$ can be differentiated directly. However, finding $t(x)$ usually is difficult or impossible, so we need another method. The key is the chain rule. First we have

$$\frac{dy}{dt} = \frac{dy}{dx} \cdot \frac{dx}{dt} \quad \text{so} \quad y' = \frac{dy}{dx} = \frac{dy/dt}{dx/dt}$$

To express the second derivative d^2y/dx^2 in terms of derivatives with respect to t, we apply the chain rule again:

$$\frac{d^2y}{dx^2} = \frac{dy'}{dx} = \frac{dy'/dt}{dx/dt}$$

Since y' is already expressed in terms of t, the numerator can be calculated.

Derivative Formulas

Suppose $\mathbf{x} = \mathbf{x}(t)$ is a parametric plane curve defined on an open interval (t_0, t_1) and suppose that dx/dt is never 0. Then the curve is the graph of a function $y = f(x)$ and

$$\frac{dy}{dx} = \frac{dy/dt}{dx/dt} \qquad \text{and} \qquad \frac{d^2y}{dx^2} = \frac{\dfrac{d}{dt}\left(\dfrac{dy}{dx}\right)}{\dfrac{dx}{dt}}$$

Example 1

Show that the equations $x(t) = t^3 + t$ and $y(t) = t^4 + 1$ define y as a function of x. Compute

$$\frac{dy}{dx} \quad \text{and} \quad \frac{d^2y}{dx^2}$$

Solution Since $dx/dt = 3t^2 + 1 > 0$, the equations determine y as a function of x. It would be quite difficult to solve $x = t^3 + t$ for t as a function of x and so find the derivative directly. Therefore our best bet is to use the derivative formulas. We have

$$\frac{dx}{dt} = 3t^2 + 1 \qquad \text{and} \qquad \frac{dy}{dt} = 4t^3$$

so

$$\frac{dy}{dx} = \frac{dy/dt}{dx/dt} = \frac{4t^3}{3t^2 + 1}$$

To compute d^2y/dx^2 we need the t-derivative of dy/dx also:

$$\frac{d}{dt}\left(\frac{dy}{dx}\right) = \frac{d}{dt}\left(\frac{4t^3}{3t^2 + 1}\right) = \frac{12t^2(3t^2 + 1) - 6t(4t^3)}{(3t^2 + 1)^2}$$

$$= \frac{12t^2(t^2 + 1)}{(3t^2 + 1)^2}$$

Consequently

$$\frac{d^2y}{dx^2} = \frac{\dfrac{d}{dt}\left(\dfrac{dy}{dx}\right)}{\dfrac{dx}{dt}} = \frac{12t^2(t^2+1)}{(3t^2+1)^3}$$

Area

Suppose $\mathbf{x} = \mathbf{x}(t)$ is a parametric plane curve defined on a closed interval $[a, b]$ of the t-axis. Suppose also that $dx/dt > 0$ for $a < t < b$ so that the curve may be considered as the graph of a function $y = f(x)$. Our problem is to express the area under this graph (Figure 13-3-1) in terms of $x(t)$, $y(t)$, and their derivatives. Note, incidentally, that x increases as t increases because we are assuming $dx/dt > 0$.

We use the change-of-variable formula for definite integrals to derive the desired expression. The enclosed area is

$$A = \int_{x(a)}^{x(b)} y[t(x)]\,dx = \int_a^b y(t)\frac{dx}{dt}\,dt$$

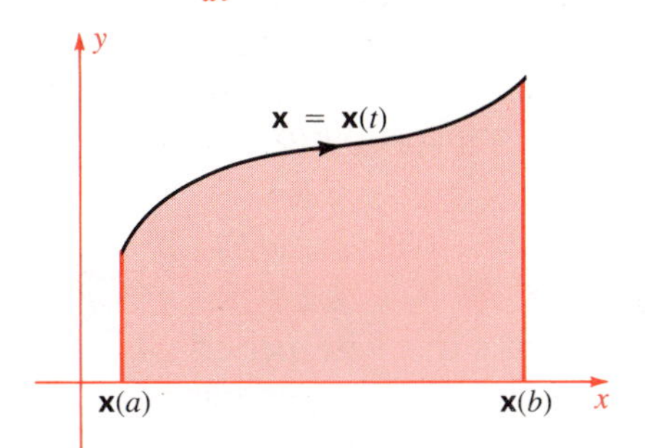

Figure 13-3-1
Area under $\mathbf{x} = \mathbf{x}(t)$
Assumed: $\dfrac{dx}{dt} > 0$ for $a \le t \le b$

Area under a Curve

If $dx/dt > 0$ for $a < t < b$, then the area under the plane curve $\mathbf{x} = \mathbf{x}(t) = (x(t), y(t))$ is

$$A = \int_a^b y\frac{dx}{dt}\,dt$$

Example 2

Find the area under the portion $y \ge 0$ of the ellipse

$$\frac{x^2}{a^2} + \frac{y^2}{b^2} = 1$$

using the usual parametrization ($x = a\cos\theta$ and $y = b\sin\theta$, suitably adjusted) of the ellipse.

Solution The upper half is traced for $0 \le \theta \le \pi$. However, we have $dx/d\theta = -a\sin\theta < 0$ for $0 < \theta < \pi$ so our formula does not apply directly. (It actually yields the negative of the area.) A related parametrization (Figure 13-3-2) is

$$x = -a\cos\theta \qquad y = b\sin\theta \qquad 0 \le \theta \le \pi$$

and now $dx/d\theta = a\sin\theta > 0$ for $0 < \theta < \pi$. We have

$$A = \int_0^\pi y\frac{dx}{d\theta}\,d\theta = \int_0^\pi (b\sin\theta)(a\sin\theta)\,d\theta$$

$$= ab\int_0^\pi \sin^2\theta\,d\theta = \tfrac{1}{2}\pi ab$$

Compare Exercise 20 in Section 5-4.

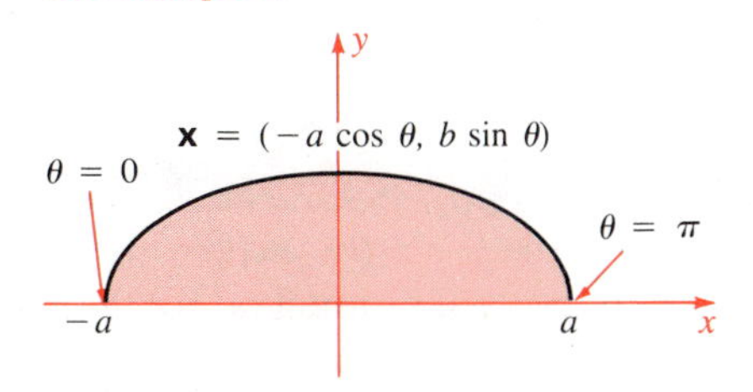

Figure 13-3-2
See Example 2.

For the area surrounded by a *closed* curve, there is a formula worth knowing at this point, although we postpone its proof until Section 18-6. A closed curve is given by a periodic vector function of period p:

$$\mathbf{x}(t+p) = \mathbf{x}(t)$$

Thus $\mathbf{x}(p) = \mathbf{x}(0)$, so a loop is formed as t runs from 0 to p. We assume the curve is **simple,** that is, never crosses itself. Precisely,

$$\mathbf{x}(t_1) \neq \mathbf{x}(t_2) \quad \text{for} \quad 0 \leq t_1 < t_2 < p$$

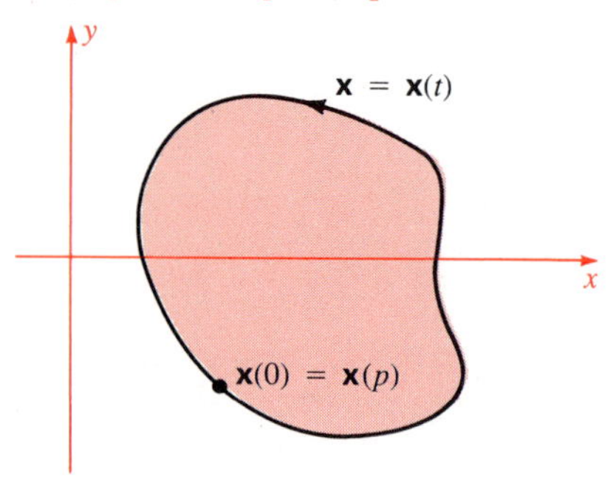

Figure 13-3-3
Simple closed curve:
$\mathbf{x}(t)$ periodic of period p

We also assume that the curve is traversed in the counterclockwise direction, that is, if you stand on the curve facing in the direction of increasing t, then the region enclosed by the curve is on your immediate left side. (This isn't very precise, but we only intend application to simple examples.) See Figure 13-3-3.

Closed Curves

Let $\mathbf{x} = \mathbf{x}(t)$ parametrize a simple closed curve with period p, traversed counterclockwise. Then the area enclosed by the curve is

$$A = \frac{1}{2}\int_0^p \left(x\frac{dy}{dt} - y\frac{dx}{dt}\right) dt$$

Example 3

Find the area enclosed by the ellipse

$$\frac{x^2}{a^2} + \frac{y^2}{b^2} = 1$$

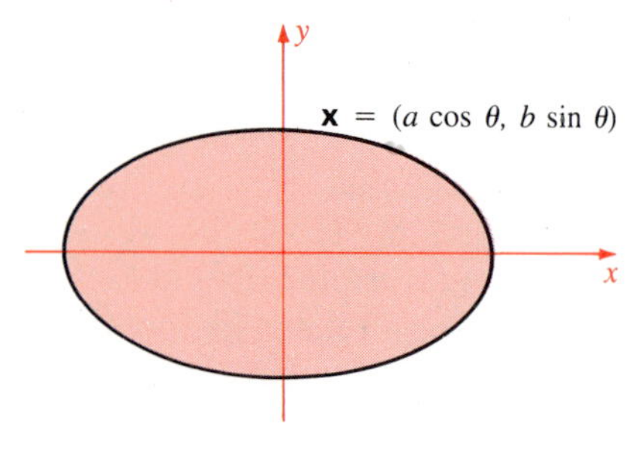

Figure 13-3-4
See Example 3.

Solution The curve (Figure 13-3-4) is parametrized by the vector function

$$\mathbf{x}(\theta) = (a\cos\theta, b\sin\theta)$$

periodic of period 2π. Clearly the ellipse does not cross itself, and is traversed counterclockwise, hence

$$\begin{aligned} A &= \frac{1}{2}\int_0^{2\pi} \left(x\frac{dy}{d\theta} - y\frac{dx}{d\theta}\right) d\theta \\ &= \frac{1}{2}\int_0^{2\pi} [(a\cos\theta)(b\cos\theta) - (b\sin\theta)(-a\sin\theta)]\, d\theta \\ &= \frac{1}{2}\int_0^{2\pi} ab\, d\theta = \pi ab \end{aligned}$$

Parametrization of the Circle

The unit circle has the parametrization $\mathbf{x} = (\cos\theta, \sin\theta)$, where θ is the central angle. Another parametrization in terms of rational rather than trigonometric functions is sometimes useful.

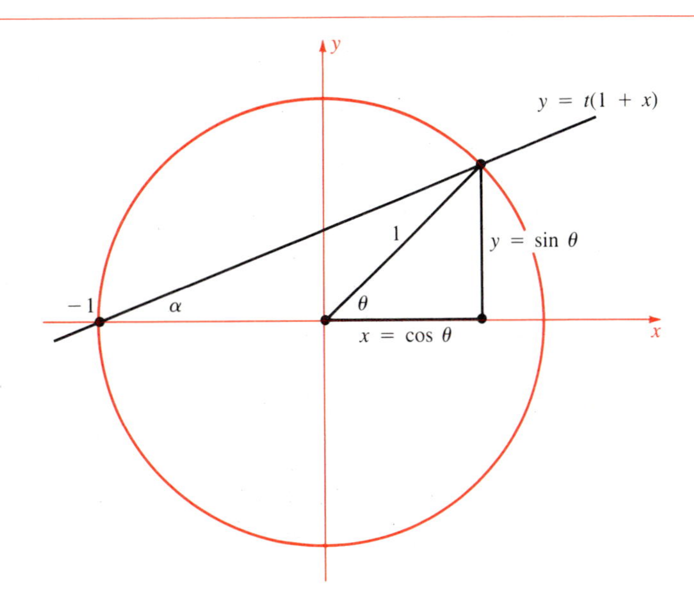

Figure 13-3-5
Rational parametrization of the unit circle

Consider a variable line through $(-1, 0)$ with slope t. See Figure 13-3-5. Its equation is

$$y = t(1 + x)$$

The line meets the circle in two points; one is $(-1, 0)$. To find the other we eliminate y from the system

$$x^2 + y^2 = 1 \qquad y = t(1 + x)$$

and solve for x:

$$x^2 + t^2(1 + x)^2 = 1 \quad \text{so} \quad x^2 + t^2(x^2 + 2x + 1) = 1$$

that is, $(t^2 + 1)x^2 + 2t^2x + (t^2 - 1) = 0$. Finally,

$$x = \frac{-t^2 \pm \sqrt{t^4 - (t^2 + 1)(t^2 - 1)}}{t^2 + 1}$$

$$= \frac{-t^2 \pm \sqrt{1}}{t^2 + 1} = \frac{-t^2 \pm 1}{t^2 + 1}$$

The minus sign leads to $x = -1$, the known solution, so we choose $+$:

$$x = \frac{1 - t^2}{1 + t^2} \qquad y = t(1 + x) = \frac{2t}{1 + t^2}$$

This is the desired rational parametrization of the circle. Note incidentally that the parameter t can be expressed rationally in terms of x and y:

$$y = t(1 + x) \quad \text{so} \quad t = \frac{y}{1 + x}$$

For $-\infty < t < +\infty$, we get every point of the circle with one exception, the point $(-1, 0)$.

Remark In a certain sense of algebraic justice, the point $(-1, 0)$ corresponds to $t = \pm\infty$. For if the slope is $\pm\infty$, the variable line is vertical, tangent to the circle at $(-1, 0)$, so it meets the circle *twice* at $(-1, 0)$.

Rational Parametrization of the Circle

The unit circle $x^2 + y^2 = 1$ is parametrized by

$$x = \frac{1 - t^2}{1 + t^2} \qquad y = \frac{2t}{1 + t^2}$$

Also

$$t = \frac{y}{1 + x}$$

Have a second look at Figure 13-3-5. The slope t equals $\tan\alpha$. But the inscribed angle α is half the corresponding central angle θ, so that $t = \tan\frac{1}{2}\theta$. Also $x = \cos\theta$ and $y = \sin\theta$. We may substitute

$$t = \tan\tfrac{1}{2}\theta \qquad x = \cos\theta \qquad y = \sin\theta$$

in the formulas above. Result:

Half-Angle Formulas

$$\cos\theta = \frac{1 - \tan^2\frac{1}{2}\theta}{1 + \tan^2\frac{1}{2}\theta} \qquad \sin\theta = \frac{2\tan\frac{1}{2}\theta}{1 + \tan^2\frac{1}{2}\theta}$$

$$\tan\tfrac{1}{2}\theta = \frac{\sin\theta}{1 + \cos\theta}$$

The Hyperbola

The hyperbola

$$\frac{x^2}{a^2} - \frac{y^2}{b^2} = 1$$

can be parametrized by hyperbolic functions:

$$x = a\cosh t \qquad y = b\sinh t$$

For a geometric interpretation of the parameter t, we compute the shaded area A in Figure 13-3-6. This area is the difference between the area bounded by the curve and the y-axis (for $0 \le y \le b\sinh t$) and the area of the shaded triangle. Therefore, with dummy variable u,

$$\begin{aligned} A &= \left(\int_0^t x(u)\frac{dy}{dt}(u)\,du\right) - \tfrac{1}{2}(a\cosh t)(b\sinh t) \\ &= \left(\int_0^t (a\cosh u)(b\cosh u)\,du\right) - \tfrac{1}{2}ab\cosh t\sinh t \\ &= ab\left(\int_0^t \cosh^2 u\,du\right) - \tfrac{1}{4}ab\sinh 2t \end{aligned}$$

Figure 13-3-6
Hyperbola with a geometric interpretation of the parameter t

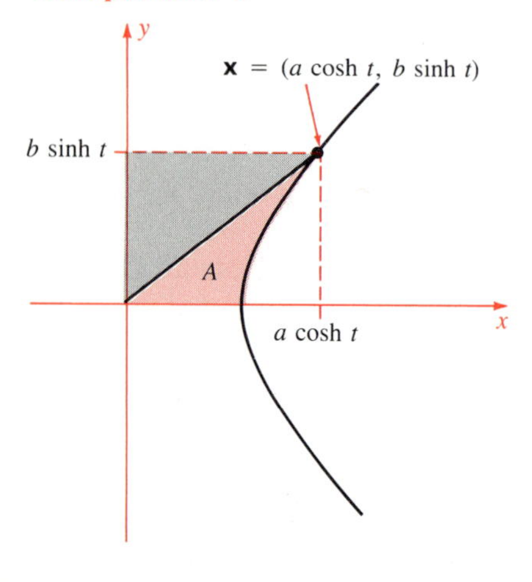

But

$$\int_0^t \cosh^2 u\, du = \left(\tfrac{1}{4}\sinh 2u + \tfrac{1}{2}u\right)\Big|_0^t = \tfrac{1}{4}\sinh 2t + \tfrac{1}{2}t$$

Hence the $\sinh 2t$ terms cancel and $A = \frac{1}{2}abt$. This implies the expression

$$t = \frac{2A}{ab}$$

for t in terms of geometric quantities.

Exercises

Find dy/dx and d^2y/dx^2 for the parametric curve $\mathbf{x}(t) =$

1 $(t^2, t^3) \quad t > 0$

2 $(t^3, t^4) \quad t \neq 0$

3 $(e^t + t, t^2)$

4 $(t - 1/t, t^3 - t) \quad t \neq 0$

5 $(\cos t, \sin t) \quad 0 < t < \pi$

6 $(3\cos t, 4\sin t) \quad 0 < t < \pi$

7 $(\cosh t, \sinh t)$

8 $(\tan t, \sec t)$

Consider the parametric curve $\mathbf{x} = \mathbf{x}(t)$ as the graph of a function $y = f(x)$. Find the area under the graph, where $\mathbf{x}(t) =$

9 $(t^2, t^3) \quad 0 \le t \le 1$

10 $(t^3, t^2) \quad 0 \le t \le 1$

11 $(\sin t, \tan t) \quad 0 \le t \le \frac{1}{4}\pi$

12 $(e^t + t, t^2) \quad 1 \le t \le 2$

13 $(-\cos^2 t, \sin^2 t) \quad 0 \le t \le \frac{1}{2}\pi$

14 $(-\cos^2 t, \sin^6 t) \quad 0 \le t \le \frac{1}{4}\pi$

15 Show that $\mathbf{x}(t) = (t^2 - 1, t^3 - t)$ parametrizes $y^2 = x^3 + x^2$.

16 (cont.) Sketch the curve and identify t geometrically.

17 Show that the **witch of Agnesi** $y = a^3/(a^2 + x^2)$ is parametrized by $\mathbf{x}(\theta) = a(\cot\theta, \sin^2\theta)$.

18 (cont.) Sketch the curve. Try to identify the geometric angle θ in your figure.

19 Show that the **serpentine** $(a^2 + x^2)y = abx$ is parametrized by $\mathbf{x}(\theta) = (a\cot\theta, b\sin\theta\cos\theta)$.

20 (cont.) Sketch the curve.

21 Show that the **folium of Descartes** $x^3 + y^3 = 3axy$ is parametrized by $\mathbf{x}(t) = 3a(t/(1 + t^3), t^2/(1 + t^3))$.

***22** (cont.) Sketch the curve.

23 Show that the **cissoid of Diocles** $y^2(a - x) = x^3$ can be parametrized by $\mathbf{x}(t) = a(t^2/(1 + t^2), t^3/(1 + t^2))$.

24 (cont.) Sketch the curve.

13-4 Curvature

Let $\mathbf{x} = \mathbf{x}(t)$ be a plane curve. Thus $\mathbf{x}(t) = (x(t), y(t))$. Let us assume that $d\mathbf{x}/dt$ is never $\mathbf{0}$. For convenience we write $\mathbf{v} = d\mathbf{x}/dt \neq \mathbf{0}$. Then $\mathbf{v}$ is a tangent vector to the curve, so

$$\mathbf{T} = \frac{1}{|\mathbf{v}|}\mathbf{v}$$

is a unit vector in the direction of $\mathbf{v}$, called the **unit tangent vector** of the parametric curve $\mathbf{x} = \mathbf{x}(t)$.

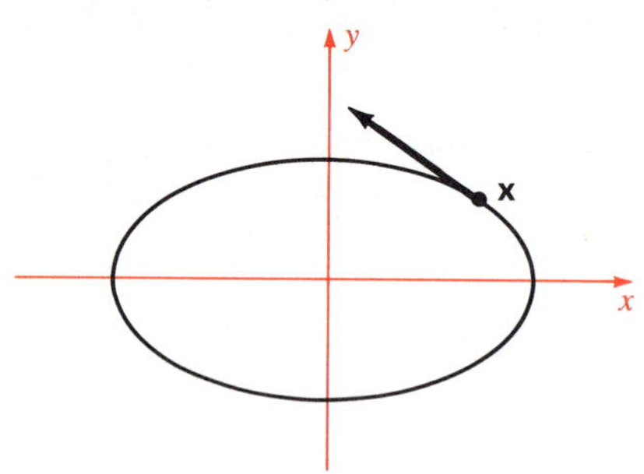

Figure 13-4-1
The unit tangent to the ellipse $\mathbf{x} = (a \cos \theta, b \sin \theta)$

Example 1

Find the unit tangent to the ellipse (Figure 13-4-1)

$$\mathbf{x}(\theta) = (a \cos \theta, b \sin \theta)$$

Solution

$$\mathbf{v} = d\mathbf{x}/d\theta = (-a \sin \theta, b \cos \theta)$$

$$|\mathbf{v}|^2 = a^2 \sin^2\theta + b^2 \cos^2\theta$$

Hence

$$\mathbf{T} = \frac{1}{|\mathbf{v}|}\mathbf{v} = \frac{1}{\sqrt{a^2 \sin^2\theta + b^2 \cos^2\theta}}(-a \sin \theta, b \cos \theta)$$

The assumption that $\mathbf{v} = d\mathbf{x}/dt \neq \mathbf{0}$ implies that the curve can be parametrized in terms of its arc length. Now the arc length s was defined by

$$\frac{ds}{dt} = \left|\frac{d\mathbf{x}}{dt}\right| = |\mathbf{v}|$$

If we compare this with the equations

$$\mathbf{v} = \frac{d\mathbf{x}}{dt} = \frac{ds}{dt} \cdot \frac{d\mathbf{x}}{ds} \quad \text{and} \quad \mathbf{v} = |\mathbf{v}|\mathbf{T}$$

we conclude that $\mathbf{T} = d\mathbf{x}/ds$.

Unit Tangent Vector

If $\mathbf{x} = \mathbf{x}(t)$ is a plane curve with $\mathbf{v} = d\mathbf{x}/dt \neq \mathbf{0}$, then its unit tangent vector is

$$\mathbf{T} = \frac{1}{|\mathbf{v}|}\mathbf{v} = \frac{d\mathbf{x}}{ds}$$

where s denotes arc length.

Cusps

It is worth exploring what can happen at a point where $\mathbf{v}(t) = \mathbf{0}$. For instance, consider $\mathbf{x}(t) = (t^3, 0)$. Then $\mathbf{v}(t) = (3t^2, 0)$, so $\mathbf{v}(0) = \mathbf{0}$. The curve is the x-axis, traced more and more slowly as t approaches 0. It is instantaneously stationary at the origin.

Another possibility is $\mathbf{x}(t) = (x(t), y(t))$, where $x(t) = 0$ for $t < 0$ but $x(t) > 0$ for $t > 0$, and $y(t) = 0$ for $t > 0$ but $y(t) > 0$ for $t < 0$. Then the curve slithers along the positive y-axis into the origin as $t \to 0-$ and then slithers out along the positive x-axis as t increases from 0. Thus the curve manages to turn a sharp corner at the origin. Such functions $x(t)$ and $y(t)$ are rather exotic, but they do exist, just as differentiable as we please.

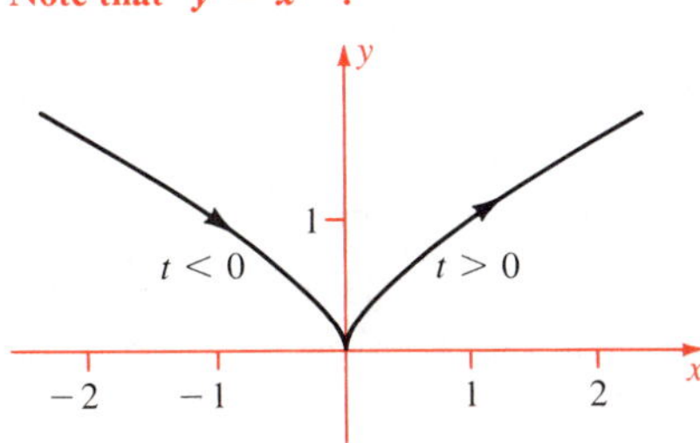

Figure 13-4-2
Graph of $\mathbf{x} = (t^3, t^2)$.
Note that $y = x^{2/3}$.

More natural is a plane curve like $\mathbf{x}(t) = (t^3, t^2)$. Then

$$\mathbf{v} = d\mathbf{x}/dt = (3t^2, 2t)$$

so $\mathbf{v}(t) \neq \mathbf{0}$ if $t \neq 0$, but $\mathbf{v}(0) = \mathbf{0}$. Plotting the curve (Figure 13-4-2), we see that there is a sharp point at $\mathbf{0}$, called a **cusp.** At the cusp, the curve changes direction abruptly. If $t < 0$ and is very small, the tangent points nearly in the direction of the negative y-axis. But if $t > 0$ and is very small, the tangent points nearly in the direction of the positive y-axis. Thus the curve does an almost instantaneous about-face. A particle moving on the curve slows down as it comes towards $\mathbf{0}$, stops instantaneously at $\mathbf{0}$, then speeds up as it leaves $\mathbf{0}$ in the opposite direction.

For the remainder of this section we shall assume that $\mathbf{v} \neq \mathbf{0}$, so that $\mathbf{T}$ is defined; hence stationary points, corners, and cusps are ruled out.

Curvature and Normal

The unit tangent vector $\mathbf{T}$ at each point of $\mathbf{x} = \mathbf{x}(s)$ shows us the direction of the curve at that point. As we move along the curve, its direction changes, rapidly if the curve bends sharply, less rapidly if the curve is fairly straight. We shall define the curvature of the curve as the rate of change of its direction with respect to arc length. The simplest example is a circle of radius a, traversed clockwise. As we move along an arc of length $a\theta$, the direction of the tangent increases by angle θ; hence the curvature is $\theta/(a\theta) = 1/a$.

On a general curve, we must examine the derivative $d\mathbf{T}/ds$. Because $\mathbf{T}$ is a unit vector, its derivative is perpendicular to it. Indeed, from

$$\mathbf{T} \cdot \mathbf{T} = 1 \qquad \text{follows} \qquad \mathbf{T} \cdot \frac{d\mathbf{T}}{ds} = 0$$

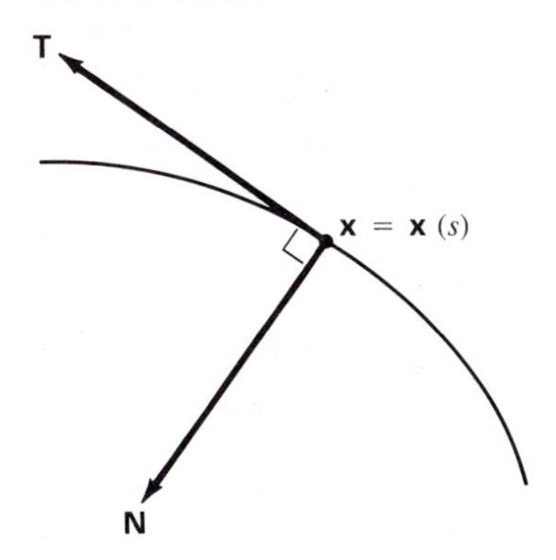

Figure 13-4-3
The unit normal

We want the curvature to be a scalar function $k(s)$. Therefore we write $d\mathbf{T}/ds$ as k times a unit vector $\mathbf{N}$. Since $d\mathbf{T}/ds$ is perpendicular to $\mathbf{T}$, we choose the unit vector $\mathbf{N}$ to be perpendicular to $\mathbf{T}$ and such that the pair $\mathbf{T}, \mathbf{N}$ is a right-handed system. In other words, we rotate $\mathbf{T}$ through angle $+\frac{1}{2}\pi$ to obtain $\mathbf{N}$. See Figure 13-4-3. This vector $\mathbf{N}$ is called the **unit normal** to the curve. (The choice of $\mathbf{N}$ rather than $-\mathbf{N}$ for the unit normal is merely a convention. We have to choose one or the other, and we might as well opt for a right-handed system.)

Curvature and Normal

Let $\mathbf{x} = \mathbf{x}(s)$ be a plane curve parametrized by arc length s and with unit tangent $\mathbf{T} = \mathbf{T}(s)$. The **unit normal** to $\mathbf{x}$ is the vector $\mathbf{N}$ obtained by rotating $\mathbf{T}$ counterclockwise one quarter rotation. Then

$$\frac{d\mathbf{T}}{ds} = k\mathbf{N}$$

defines the **curvature** $k = k(s)$ of the curve.

Note that the sign of k depends on the direction of the curve in the given parametrization. If the direction is reversed, then k is changed to $-k$ at each point of the curve.

Example 2

Find the unit tangent, unit normal, and curvature of the circle

$$\mathbf{x} = \mathbf{c} + a(\cos\theta, \sin\theta)$$

of center $\mathbf{c}$ and radius $a > 0$.

Solution If $a \neq 1$, then the parameter θ is not the arc length. We have

$$\mathbf{v} = \frac{d\mathbf{x}}{d\theta} = a(-\sin\theta, \cos\theta)$$

But $a > 0$ and $(-\sin\theta, \cos\theta)$ is a unit vector. Hence

$$\mathbf{T} = (-\sin\theta, \cos\theta) \qquad \text{and} \qquad \frac{ds}{d\theta} = \left|\frac{d\mathbf{x}}{d\theta}\right| = a$$

We may take $s = a\theta$ so $\theta = s/a$ and

$$\mathbf{T} = (-\sin(s/a), \cos(s/a))$$

expresses the unit tangent $\mathbf{T}$ in terms of s. Now the effect of rotating any vector $\mathbf{u} = (u_1, u_2)$ forward $\frac{1}{2}\pi$ is $(-u_2, u_1)$. Hence

$$\begin{aligned}\mathbf{N} &= (-\cos(s/a), -\sin(s/a)) \\ &= -(\cos(s/a), \sin(s/a))\end{aligned}$$

Therefore $\mathbf{N}$ is directed from $\mathbf{x}$ towards the center $\mathbf{c}$. See Figure 13-4-4. It remains to find the curvature k. Now

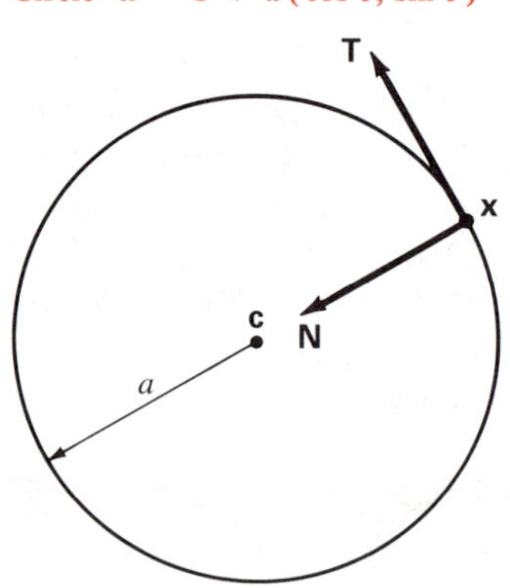

Figure 13-4-4
Circle $\mathbf{x} = \mathbf{c} + a(\cos\theta, \sin\theta)$

$$\frac{d\mathbf{T}}{ds} = \frac{1}{a}(-\cos(s/a), -\sin(s/a)) = \frac{1}{a}\mathbf{N}$$

Therefore $k = 1/a$, which we have already anticipated geometrically. Note that a circle of large radius a changes direction slowly relative to its arc length, so $k = 1/a$ is small. A circle of small radius changes direction rapidly relative to its arc length, so $k = 1/a$ is large. ●

Computation of Curvature

Let $\mathbf{x} = \mathbf{x}(t) = (x(t), y(t))$ be a plane curve. For convenience we write

$$\mathbf{v} = \frac{d\mathbf{x}}{dt} = \left(\frac{dx}{dt}, \frac{dy}{dt}\right) \qquad \mathbf{a} = \frac{d^2\mathbf{x}}{dt^2} = \left(\frac{d^2x}{dt^2}, \frac{d^2y}{dt^2}\right)$$

(Later we shall see that these vectors represent velocity and acceleration.) We assume that $\mathbf{v} \neq \mathbf{0}$ as usual. We propose to develop a formula for k in terms of the four quantities

$$\frac{dx}{dt} \qquad \frac{dy}{dt} \qquad \frac{d^2x}{dt^2} \qquad \frac{d^2y}{dt^2}$$

These quantities are directly computable, so the formula will avoid the need for finding $\mathbf{T}$, $\mathbf{N}$, and s.

We begin with one application of the chain rule:

$$\mathbf{v} = \frac{d\mathbf{x}}{dt} = \frac{ds}{dt} \cdot \frac{d\mathbf{x}}{ds} = \frac{ds}{dt}\mathbf{T}$$

We differentiate again:

$$\mathbf{a} = \frac{d^2\mathbf{x}}{dt^2} = \frac{d}{dt}\left(\frac{d\mathbf{x}}{dt}\right)$$

$$= \frac{d\mathbf{v}}{dt} = \frac{d}{dt}\left(\frac{ds}{dt}\mathbf{T}\right)$$

$$= \frac{d^2s}{dt^2}\mathbf{T} + \frac{ds}{dt} \cdot \frac{d}{dt}\mathbf{T}$$

By the chain rule again,

$$\frac{d}{dt}\mathbf{T} = \frac{ds}{dt} \cdot \frac{d\mathbf{T}}{ds} = \frac{ds}{dt} k\mathbf{N}$$

Therefore

$$\mathbf{a} = \frac{d^2s}{dt^2}\mathbf{T} + \left(\frac{ds}{dt}\right)^2 k\mathbf{N}$$

Our problem is to extract k from the pair of formulas

$$\mathbf{v} = \frac{ds}{dt}\mathbf{T} \qquad \text{and} \qquad \mathbf{a} = \frac{d^2s}{dt^2}\mathbf{T} + \left(\frac{ds}{dt}\right)^2 k\mathbf{N}$$

We do so by rotating $\mathbf{v}$ forward $\frac{1}{2}\pi$ and taking the dot product with $\mathbf{a}$. First

$$\mathbf{v} = \left(\frac{dx}{dt}, \frac{dy}{dt}\right) = \frac{ds}{dt}\mathbf{T} \qquad \text{rotates to} \qquad \left(-\frac{dy}{dt}, \frac{dx}{dt}\right) = \frac{ds}{dt}\mathbf{N}$$

Next we take the dot product with

$$\mathbf{a} = \left(\frac{d^2x}{dt^2}, \frac{d^2y}{dt^2}\right) = \frac{d^2s}{dt^2}\mathbf{T} + \left(\frac{ds}{dt}\right)^2 k\mathbf{N}$$

using $\mathbf{N} \cdot \mathbf{T} = 0$ and $\mathbf{N} \cdot \mathbf{N} = 1$:

$$\frac{dx}{dt} \cdot \frac{d^2y}{dt^2} - \frac{dy}{dt} \cdot \frac{d^2x}{dt^2} = \left(\frac{ds}{dt}\right)^3 k$$

But

$$\frac{ds}{dt} = |\mathbf{v}| = \left[\left(\frac{dx}{dt}\right)^2 + \left(\frac{dy}{dt}\right)^2\right]^{1/2}$$

so we reach the following conclusion:

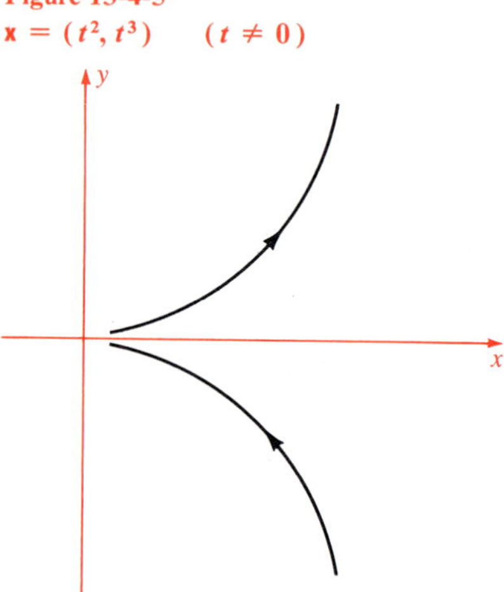

Figure 13-4-5
$\mathbf{x} = (t^2, t^3) \quad (t \neq 0)$

Formula for Curvature

If $\mathbf{x} = \mathbf{x}(t)$ is a plane curve with $d\mathbf{x}/dt \neq \mathbf{0}$ then its curvature is given by

$$k = \frac{\dfrac{dx}{dt} \cdot \dfrac{d^2y}{dt^2} - \dfrac{dy}{dt} \cdot \dfrac{d^2x}{dt^2}}{\left[\left(\dfrac{dx}{dt}\right)^2 + \left(\dfrac{dy}{dt}\right)^2\right]^{3/2}}$$

Example 3

Find the curvature of $\mathbf{x} = (t^2, t^3)$ for $t \neq 0$.

Solution The curve is shown in Figure 13-4-5. We have

$$\mathbf{v} = \frac{d\mathbf{x}}{dt} = (2t, 3t^2) \qquad \text{and} \qquad \mathbf{a} = \frac{d^2\mathbf{x}}{dt^2} = (2, 6t)$$

By the formula,

$$k = \frac{(2t)(6t) - (3t^2)(2)}{[(2t)^2 + (3t^2)^2]^{3/2}} = \frac{6t^2}{(4t^2 + 9t^4)^{3/2}}$$

The denominator is positive. If you are tempted to simplify it by taking out t, be careful because $t < 0$ on the fourth-quadrant portion of the curve. Instead, take $|t|^3$ out of the denominator and write the answer as $k = 6/|t|(4 + 9t^2)^{3/2}$.

Graph of a Function

The graph of $y = f(x)$ is a special case. Then we parametrize the curve by $\mathbf{x} = (t, f(t))$. Therefore

$$\mathbf{v} = (1, f'(t)) = (1, f'(x)) \quad \text{and} \quad \mathbf{a} = (0, f''(t)) = (0, f''(x))$$

so the formula yields

$$k = \frac{f''(x)}{[1 + f'(x)^2]^{3/2}}$$

Curvature of a Graph

The curvature of the graph of $y = f(x)$ is given by

$$k = \frac{f''(x)}{[1 + f'(x)^2]^{3/2}}$$

Example 4

Find the curvature of $y = x^3$.

Solution We have $f'(x) = 3x^2$ and $f''(x) = 6x$, so

$$k = \frac{6x}{(1 + 9x^4)^{3/2}}$$

Note that $k > 0$ for $x > 0$ and $k < 0$ for $x < 0$. That is because the curve turns towards its unit normal where $x > 0$ and turns away from its unit normal where $x < 0$. The curve stops turning instantaneously at $x = 0$; there $k = 0$. See Figure 13-4-6. ●

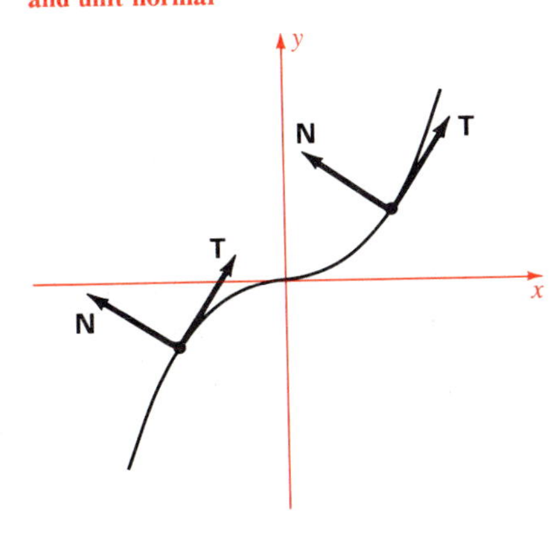

Figure 13-4-6
$y = x^3$ with its unit tangent and unit normal

The Cycloid [Optional]

The cycloid is the curve traced by an outermost point on a bicycle tire.

Example 5

A circle of radius a rolls along the x-axis in the upper half-plane. The point on the circle initially at the origin traces a **cycloid.**

a Parametrize the cycloid by the central angle θ in Figure 13-4-7

b Find the length L of one arch of the cycloid.

c Find the area A under one arch of the cycloid.

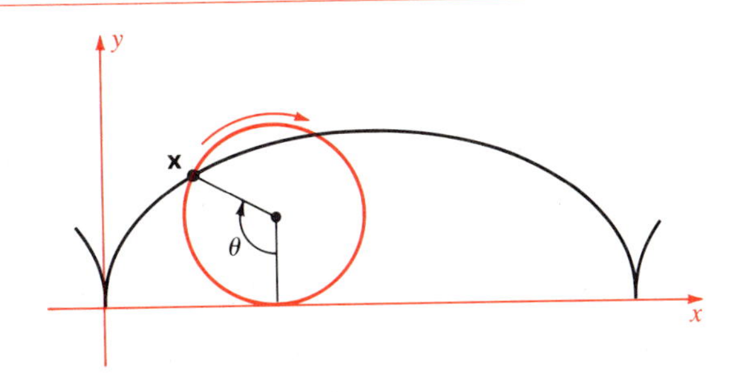

Figure 13-4-7
Cycloid generated by rolling circle

Solution

a By carefully marking various lengths (Figure 13-4-8) we can read the relations

$$x + a\sin\theta = a\theta \quad \text{and} \quad y + a\cos\theta = a$$

(Note that the corresponding circular arc has length $a\theta$, which equals the distance from 0 to the point of contact because the circle *rolls.*) Hence

$$\mathbf{x} = \mathbf{x}(\theta) = a(\theta - \sin\theta, 1 - \cos\theta)$$

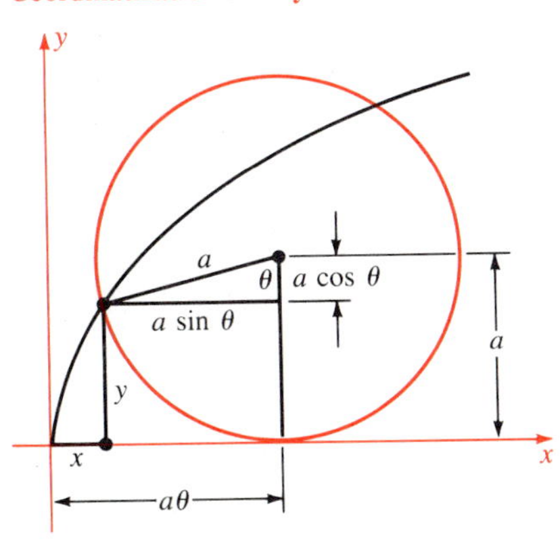

Figure 13-4-8
Coordinatization of a cycloid

b $$\frac{d\mathbf{x}}{d\theta} = (x'(\theta), y'(\theta)) = a(1 - \cos\theta, \sin\theta)$$

$$x'(\theta)^2 + y'(\theta)^2 = [a(1 - \cos\theta)]^2 + (a\sin\theta)^2$$
$$= a^2(2 - 2\cos\theta) = 4a^2\sin^2\tfrac{1}{2}\theta$$

Therefore

$$L = \int_0^{2\pi} \sqrt{(x')^2 + (y')^2}\, d\theta = \int_0^{2\pi} 2a\sin\tfrac{1}{2}\theta\, d\theta$$
$$= -4a\cos\tfrac{1}{2}\theta \Big|_0^{2\pi} = 8a$$

c $A = \int_0^{2\pi} y \frac{dx}{d\theta}\, d\theta = \int_0^{2\pi} [a(1 - \cos\theta)][a(1 - \cos\theta)]\, d\theta$

$$= a^2 \int_0^{2\pi} (1 - \cos\theta)^2\, d\theta$$

$$= a^2 \int_0^{2\pi} (1 - 2\cos\theta + \cos^2\theta)\, d\theta$$

$$= a^2 \int_0^{2\pi} (1 + \cos^2\theta)\, d\theta = a^2(2\pi + \pi) = 3\pi a^2$$

These interesting results are shown in Figure 13-4-9. ●

Figure 13-4-9
Geometry of the cycloid. Note that $2\pi a \approx 6.28a < 8a$, so the rolling circle is a little shorter than one arch of the cycloid.

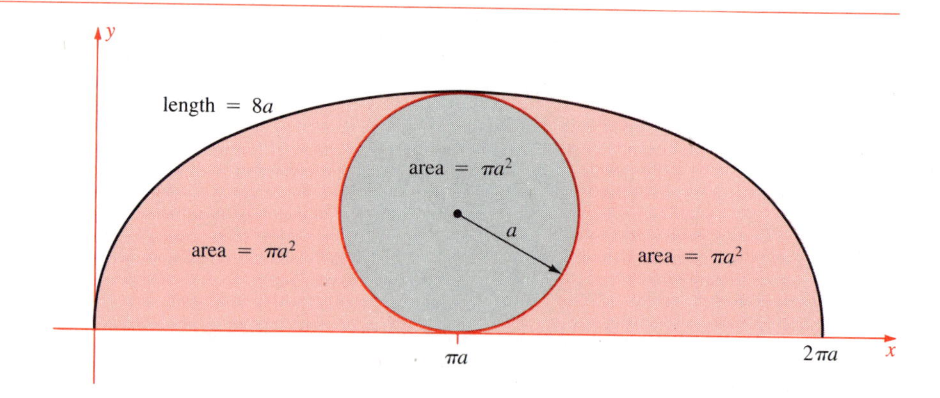

Figure 13-4-10
Hanging cable

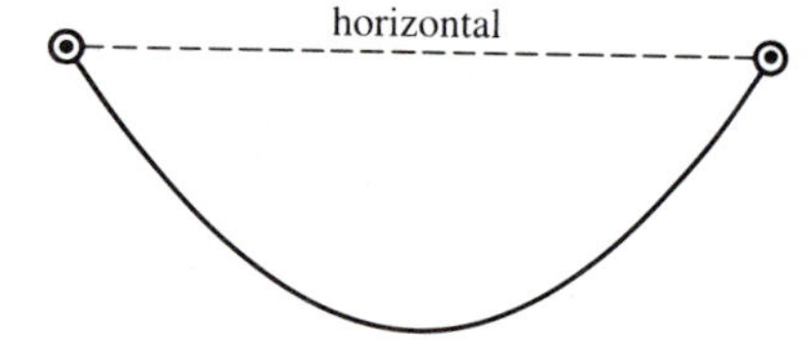

a Flexible, heavy, uniform cable

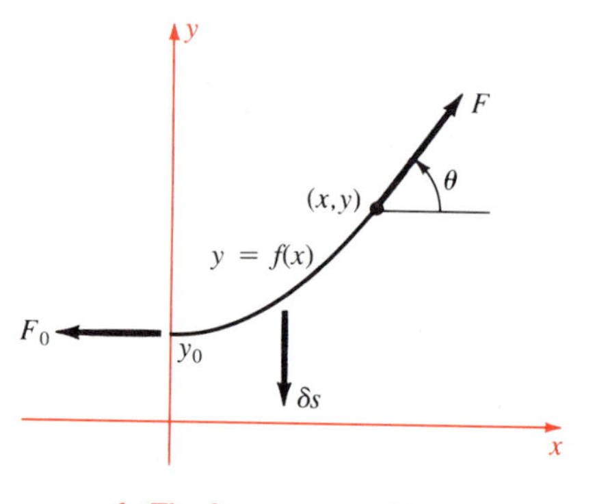

b The forces on an arbitrary segment of the cable

The Hanging Cable [Optional]

We shall find the shape of a uniform, heavy, flexible cable suspended between two points at the same level (Figure 13-4-10a). Here "flexible" means that the only internal force in the cable is tension acting in the tangential direction. "Heavy" means that gravity must be taken into account. "Uniform" means that the density of the cable, in weight per unit length, is a constant, δ.

We choose axes as in Figure 13-4-10b, so the lowest point of the cable is on the y-axis. The shape of the hanging cable is then the graph of some function $y = f(x)$, to be found.

We examine the portion of cable of length s to the right of the y-axis. It goes from $(0, y_0)$ to (x, y), where y_0 is a constant and $x = x(s)$ and $y = y(s)$. Three forces act on this portion: the horizontal tension of magnitude F_0 at $(0, y_0)$, the tangential tension of magnitude $F = F(s)$ at (x, y), and the downward gravitational force of magnitude δs, where δ is the density. When the chain hangs in equilibrium, these forces balance; in particular the horizontal components balance, and similarly the vertical components.

Let $\theta = \theta(s)$ denote the angle at (x, y) between the cable and the positive x-axis. Then the tension at (x, y) resolves into a horizontal component $F\cos\theta$ and a vertical component $F\sin\theta$. The forces balance:

$$F\cos\theta = F_0 \qquad \text{and} \qquad F\sin\theta = \delta s$$

By dividing we obtain

$$\tan\theta = \frac{1}{k}s \qquad \text{where} \quad k = \frac{F_0}{\delta}$$

However, $\tan\theta = dy/dx$ is the slope of the cable at (x, y). Hence

$$\frac{dy}{dx} = \frac{1}{k}s$$

We also know another relation between x, y, and s; the relation defining arc length:

$$dx^2 + dy^2 = ds^2 \qquad \text{that is} \qquad 1 + \left(\frac{dy}{dx}\right)^2 = \left(\frac{ds}{dx}\right)^2$$

Therefore

$$\left(\frac{ds}{dx}\right)^2 = 1 + \frac{s^2}{k^2} = \frac{k^2 + s^2}{k^2}$$

which we rewrite as

$$\frac{dx}{ds} = \frac{k}{\sqrt{k^2 + s^2}}$$

By our choice of axes, $x = 0$ at $s = 0$. Hence

$$x = \int_0^s \frac{k\,dt}{\sqrt{k^2 + t^2}} = k\sinh^{-1}(s/k)$$

We solve for s:

$$\frac{x}{k} = \sinh^{-1}\frac{s}{k} \qquad \text{so} \qquad s = k\sinh\frac{x}{k}$$

From this,

$$\frac{dy}{dx} = \frac{1}{k}s = \sinh\frac{x}{k}$$

Therefore

$$y = y_0 + \int_0^x \sinh\frac{u}{k}\,du$$

$$= y_0 + k\cosh\frac{u}{k}\bigg|_0^x = (y_0 - k) + k\cosh\frac{x}{k}$$

Clearly, it is convenient to set $y_0 = k$; then

$$y = k\cosh\frac{x}{k} \qquad \text{where} \quad k = \frac{F_0}{\delta}$$

is the equation of the hanging cable. The shape of a hanging cable is called a **catenary** after the Latin for *chain.* We now see that a catenary is simply a hyperbolic cosine curve.

If the loading on the cable is uniform in x rather than s, then we have a model of a suspension bridge. See page 408.

Exercises

Find the unit tangent

1 $\mathbf{x} = t\mathbf{c} + \mathbf{b} \quad (\mathbf{c} \neq \mathbf{0})$

2 $\mathbf{x} = (t^3, t^2) \quad (t \neq 0)$

3 $\mathbf{x} = (\cosh t, \sinh t)$

4 $y = x^3$

Plot carefully near $t = 0$ to show the cusp

5 $\mathbf{x} = (t^2, t^5)$

6 $\mathbf{x} = (t^2, 2t^4 + t^5)$

Find the curvature

7 $y = x^2$

8 $xy = 1$

9 $\mathbf{x} = (t^3, t^2) \quad (t \neq 0)$

10 $\mathbf{x} = (a\cos t, b\sin t)$

11 $\mathbf{x} = (t\cos t, t\sin t)$

12 $y = e^x$

Find the maximum absolute value of the curvature of

13 $y = \sin x$

14 $y = \ln x$

15 $y = x^2$

16 $y = x^4$

17 $xy = 1$

18 $x^2/a^2 + y^2/b^2 = 1 \quad (a > b > 0)$

19 Find the unit tangent and unit normal of the cycloid (Example 5).

20 (cont.) Find the curvature of the cycloid.

21 As a wheel of radius a rolls along the x-axis, a point *inside* the wheel at distance b from the center, $b < a$, traces a **curtate cycloid** (Figure 13-4-11). Apply the method of Example 5 to parametrize the curve.

Figure 13-4-11
Curtate cycloid

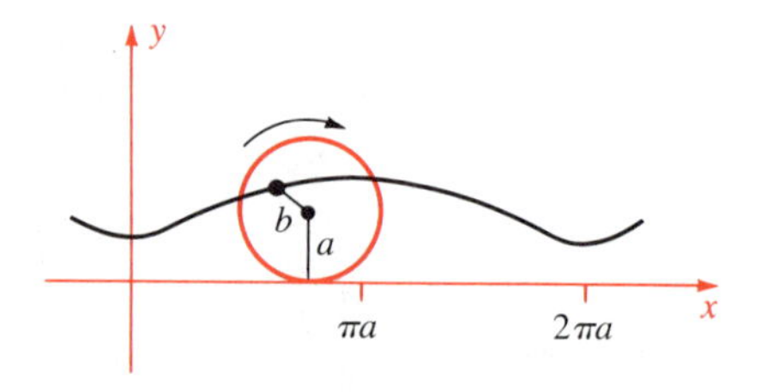

22 (cont.) Find the area under one arch of the curtate cycloid.

23 (cont.) Suppose $b > a$, so the point lies on a flange outside the wheel. (Think of the rim of a railroad wheel.) Then the curve is a **prolate cycloid** (Figure 13-4-12). Parametrize this curve.

***24** (cont.) Find the shaded area in Figure 13-4-12.

Figure 13-4-12
Prolate cycloid

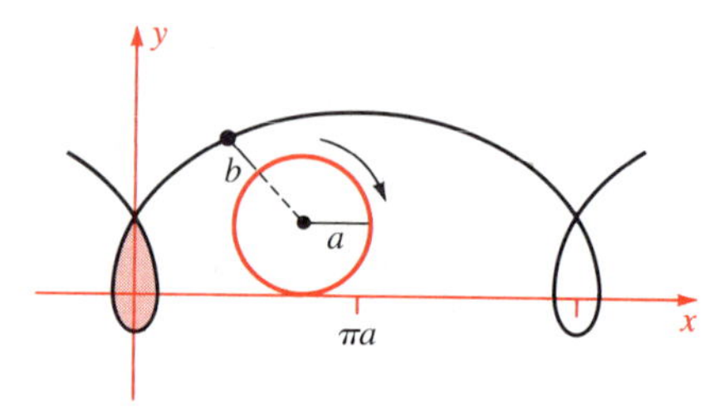

25 A (spool of) thread is wound clockwise around the unit circle so its outer end is at $(1, 0)$. Now it is unwound, always kept taut. The end traces a curve called the **involute** of the circle. Parametrize the curve, using the central angle θ in Figure 13-4-13 as the parameter.

26 (cont.) Parametrize in terms of the arc length s, measured from $(1, 0)$.

27 Express θ in terms of s for the hanging cable.

28 (cont.) Express the tension F in terms of s.

Figure 13-4-13
Involute of circle

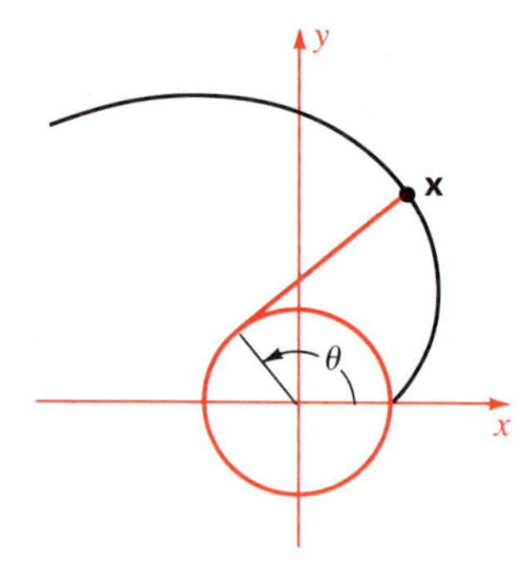

The following exercises deal with hypocycloids. A circle of radius b rolls on the inside of the circle with center $\mathbf{0}$ and radius $a > b$. The locus of a point on the boundary of the moving circle is called a **hypocycloid.** We take $(a, 0)$ for the initial position of the moving point (Figure 13-4-14).

29 Parametrize the curve by the angle θ.

30 Describe the curve if $a = 2b$.

Figure 13-4-14
Hypocycloid

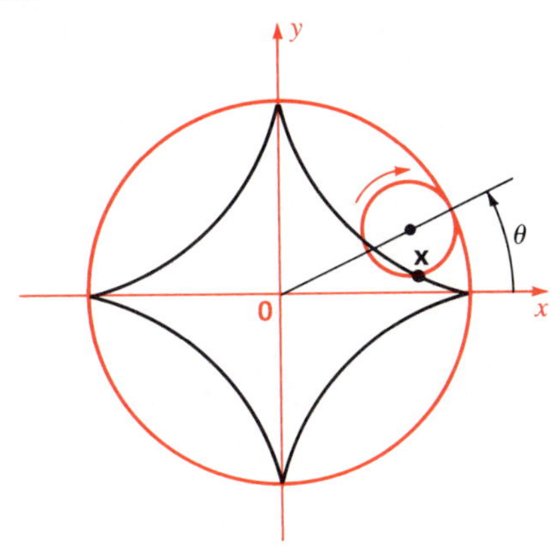

Figure 13-4-15
Epicycloid

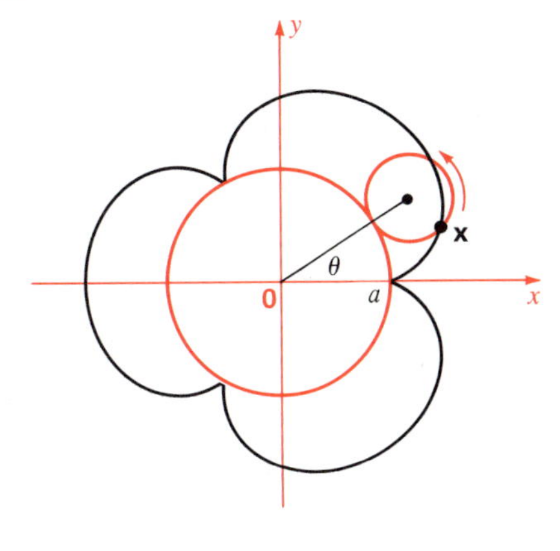

31 Suppose $a = 3b$. The curve is a hypocycloid of three cusps, or **deltoid.** Give the parametric equation, and find the area enclosed by the curve.

32 (cont.) Sketch the curve and find its length.

33 Suppose $a = 4b$, so the curve is a hypocycloid of four cusps (Figure 13-4-14). Give the parametric equation in as simple a form as possible.

34 (cont.) Find the length of the curve.

35 (cont.) Find the area enclosed by the curve.

***36** Fix a and let $b = a/n$. Find the length L_n of the hypocycloid of n cusps inscribed in a circle of radius a.

***37** (cont.) Show that $\{L_n\}$ is an increasing sequence and find $\lim L_n$.

***38** (cont.) Find the area A_n enclosed by the curve and find $\lim A_n$.

Suppose a circle of radius b rolls on the *outside* of the fixed circle with center **0** and radius a. A point on its rim traces an **epicycloid.** (These curves are important in the design of gear teeth.) Take the point initially at $(a, 0)$. See Figure 13-4-15.

39 Parametrize the curve by the angle θ.

***40** Fix a and let $b = a/n$. Find the area A_n enclosed by the resulting epicycloid of n cusps, and find $\lim A_n$.

41 Let $a = 2b$. Find the length of the corresponding **nephroid** (epicycloid of two cusps).

***42** Fix a and let $b = a/n$. Find the length L_n of the resulting epicycloid of n cusps, and find $\lim L_n$.

43 The point P in Figure 13-4-16 moves steadily around the circle of radius a. Also P is the center of a circle of radius $b < a$ that moves with it. Point **x** on the boundary of the small circle rotates steadily *backwards,* completing one revolution in one revolution of P. Find the locus of **x**, assuming it starts at $\mathbf{x}_0 = (a + b, 0)$. (Ptolemy proposed this eccentric circle as the orbit of the planet **x** around the Earth, at **0**. He called it an **epicycle-deferent.**)

Figure 13-4-16
Epicycle-deferent (see Exercise 43)

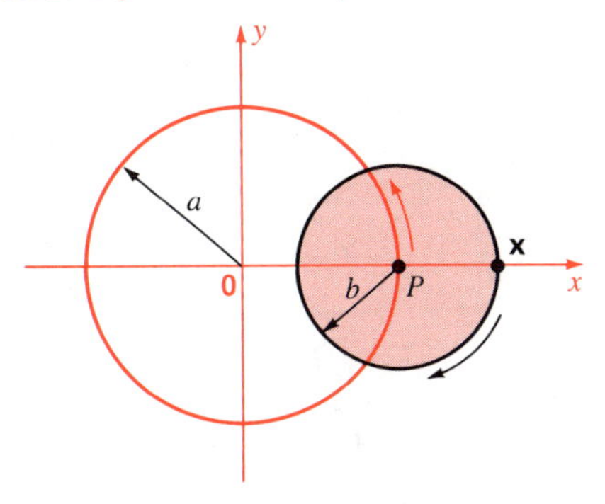

***44** A curve (Figure 13-4-17) separates two media. In the colored region the speed of light is c_1; in the gray region it is c_2. A beam of light goes from **a** to **b** as indicated. Assuming Fermat's principle of least time for the path, prove Snell's law: $(\sin \alpha_1)/c_1 = (\sin \alpha_2)/c_2$.

Figure 13-4-17
See Exercise 44

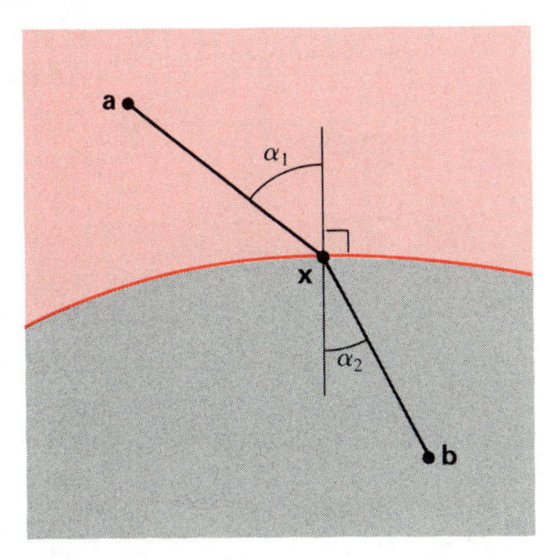

13-5 Space Curves

Let $\mathbf{x} = \mathbf{x}(t)$ be a space curve. We shall always assume that its tangent vector $\mathbf{v} = d\mathbf{x}/dt \neq \mathbf{0}$. As we know, we can then parametrize the curve in terms of its arc length s. We have

$$\mathbf{v} = \frac{d\mathbf{x}}{dt} = \frac{ds}{dt} \cdot \frac{d\mathbf{x}}{ds} \quad \text{and} \quad |\mathbf{v}| = \frac{ds}{dt}$$

so $d\mathbf{x}/ds$ is a unit vector.

Unit Tangent Vector

Let $\mathbf{x} = \mathbf{x}(t)$ be a space curve with $d\mathbf{x}/dt = \mathbf{v}(t) \neq \mathbf{0}$. Its **unit tangent vector** is

$$\mathbf{T} = \frac{\mathbf{v}}{|\mathbf{v}|} = \frac{d\mathbf{x}}{ds}$$

where s is arc length.

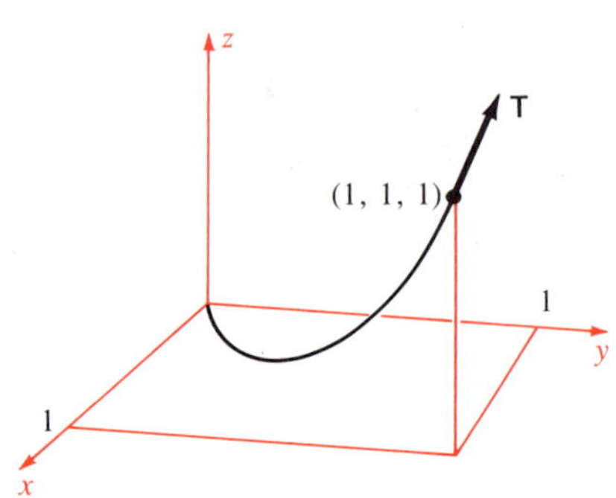

Figure 13-5-1
See Example 1

Example 1

Find the unit tangent vector to the curve $\mathbf{x}(t) = (t, t^2, t^3)$ at the point $\mathbf{x}(1) = (1, 1, 1)$. See Figure 13-5-1.

Solution

$$\mathbf{v}(t) = \frac{d\mathbf{x}}{dt} = (1, 2t, 3t^2)$$

so $\mathbf{v}(1) = (1, 2, 3)$. Therefore

$$|\mathbf{v}(1)| = \sqrt{1 + 4 + 9} = \sqrt{14}$$

so

$$\mathbf{T}(1) = \frac{\mathbf{v}(1)}{|\mathbf{v}(1)|} = \frac{1}{\sqrt{14}}(1, 2, 3)$$

Curvature

As for plane curves, the curvature of a space curve measures how fast the curve is turning with respect to its arc length. There is a basic difference, however, between plane and space curves. A plane curve has a unique unit normal $\mathbf{N}$, determined by rotating its unit tangent $+\frac{1}{2}\pi$. Then the curvature k of the plane curve is given by $d\mathbf{T}/ds = k\mathbf{N}$, and k has a definite sign. Given a space curve for which $d\mathbf{T}/ds \neq \mathbf{0}$ at each point, we can define a unique unit normal vector $\mathbf{N}$ as the unit vector in the direction of $d\mathbf{T}/ds$. But if it happens that $d\mathbf{T}/ds = \mathbf{0}$ at some points, then it may be impossible to define $\mathbf{N}$ so that it is continuous as a vector function of s, and is parallel to $d\mathbf{T}/ds$ whenever that vector is nonzero. To define curvature for a space curve, simply forget about $\mathbf{N}$ and take k to be the magnitude of $d\mathbf{T}/ds$.

Curvature

If $\mathbf{x} = \mathbf{x}(s)$ is a space curve, where s denotes arc length, its **curvature** is

$$k = \left|\frac{d\mathbf{T}}{ds}\right|$$

Example 2

Find all curves with curvature identically zero.

Solution A straight line has curvature zero, so a natural guess is that any curve with identically zero curvature must be a straight line. Let us prove this is so. We are given $k = 0$. Therefore

$$\left|\frac{d\mathbf{T}}{ds}\right| = 0 \qquad \text{hence} \qquad \frac{d\mathbf{T}}{ds} = \mathbf{0}$$

It follows that $\mathbf{T} = \mathbf{a}$, a constant unit vector. Consequently

$$\frac{d\mathbf{x}}{ds} = \mathbf{T} = \mathbf{a} = \frac{d}{ds}(s\mathbf{a}) \qquad \text{so} \qquad \mathbf{x}(s) = s\mathbf{a} + \mathbf{b}$$

where $\mathbf{b}$ is constant. This is the parametric vector equation of a straight line through $\mathbf{b}$ parallel to $\mathbf{a}$, so our guess was correct.

We shall next give two formulas for the curvature of a space curve $\mathbf{x} = \mathbf{x}(t)$ in terms of $\mathbf{x}$ and its t-derivatives. These formulas avoid passing through s en route to k.

Formulas for Curvature

Let $\mathbf{x} = \mathbf{x}(t)$ be a space curve and set

$$\mathbf{v} = \frac{d\mathbf{x}}{dt} \qquad \text{and} \qquad \mathbf{a} = \frac{d\mathbf{v}}{dt} = \frac{d^2\mathbf{x}}{dt^2}$$

Assume $\mathbf{v} \neq \mathbf{0}$. Then the curvature is given by

$$k = \frac{|\mathbf{v} \times \mathbf{a}|}{|\mathbf{v}|^3}$$

An alternative expression is

$$k = \frac{[|\mathbf{v}|^2|\mathbf{a}|^2 - (\mathbf{v} \cdot \mathbf{a})^2]^{1/2}}{|\mathbf{v}|^3}$$

Proof To prove the first formula, we start with

$$\mathbf{v} = \frac{d\mathbf{x}}{dt} = \frac{ds}{dt} \cdot \frac{d\mathbf{x}}{ds} = \frac{ds}{dt}\mathbf{T}$$

$$\mathbf{a} = \frac{d}{dt}\mathbf{v} = \frac{ds}{dt} \cdot \frac{d\mathbf{v}}{ds} = \frac{ds}{dt}\left(\frac{d^2s}{dt^2}\mathbf{T} + \frac{ds}{dt} \cdot \frac{d\mathbf{T}}{ds}\right)$$

Since $\mathbf{T} \times \mathbf{T} = \mathbf{0}$, computation of $\mathbf{v} \times \mathbf{a}$ yields

$$\mathbf{v} \times \mathbf{a} = \left(\frac{ds}{dt}\right)^3 \mathbf{T} \times \frac{d\mathbf{T}}{ds}$$

Now $\mathbf{T}$ is a unit vector, so $\mathbf{T}$ is perpendicular to its derivative $d\mathbf{T}/ds$. Therefore $|\mathbf{T} \times d\mathbf{T}/ds| = |\mathbf{T}| \cdot |d\mathbf{T}/ds| = 1 \cdot k = k$. We conclude that

$$|\mathbf{v} \times \mathbf{a}| = \left|\frac{ds}{dt}\right|^3 k = |\mathbf{v}|^3 k$$

which proves the first formula. The second formula is a consequence of the first formula and a vector identity derived on page 641:

$$|\mathbf{v} \times \mathbf{a}|^2 = |\mathbf{v}|^2|\mathbf{a}|^2 - (\mathbf{v} \cdot \mathbf{a})^2$$

●

Example 3

Find the curvature of $\mathbf{x} = (t, t^2, t^3)$.

Solution We compute $\mathbf{v} = (1, 2t, 3t^2)$ and $\mathbf{a} = (0, 2, 6t)$. Therefore

$$|\mathbf{v}|^2 = 1 + 4t^2 + 9t^4$$

$$\mathbf{v} \times \mathbf{a} = \begin{vmatrix} \mathbf{i} & \mathbf{j} & \mathbf{k} \\ 1 & 2t & 3t^2 \\ 0 & 2 & 6t \end{vmatrix} = (6t^2, -6t, 2)$$

$$|\mathbf{v} \times \mathbf{a}|^2 = 36t^4 + 36t^2 + 4$$

Finally

$$k = \frac{|\mathbf{v} \times \mathbf{a}|}{|\mathbf{v}|^3} = \frac{(36t^4 + 36t^2 + 4)^{1/2}}{(9t^4 + 4t^2 + 1)^{3/2}}$$

●

The Unit Normal

Suppose $k = |d\mathbf{T}/ds| > 0$. Then

$$\mathbf{N} = \frac{d\mathbf{T}/ds}{|d\mathbf{T}/ds|} = \frac{d\mathbf{T}/ds}{k}$$

is a unit vector in the direction of $d\mathbf{T}/ds$. Since $\mathbf{T}$ is a unit vector, $\mathbf{T}$ and $d\mathbf{T}/ds$ are perpendicular, that is, $\mathbf{T}$ and $\mathbf{N}$ are perpendicular (Figure 13-5-2). The vector $\mathbf{N}$ is called the unit normal to the curve.

Figure 13-5-2
Unit tangent and normal

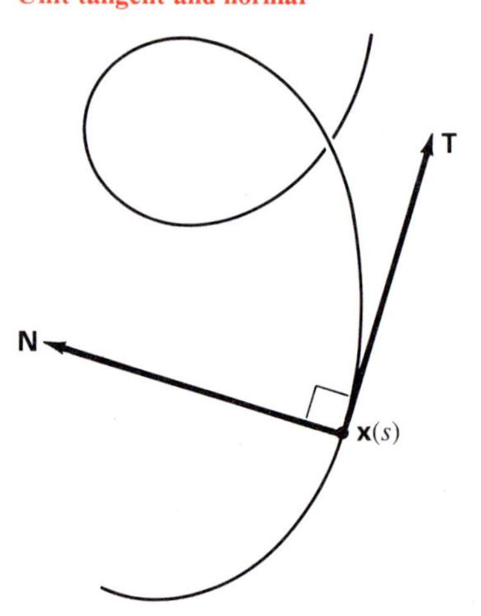

Unit Normal

Let $\mathbf{x} = \mathbf{x}(s)$ be a space curve with unit tangent $\mathbf{T}$ and curvature $k(s) > 0$. Its **unit normal** $\mathbf{N}$ is defined by $\mathbf{N} = (d\mathbf{T}/ds)/k$. It satisfies

$$|\mathbf{T}| = |\mathbf{N}| = 1 \quad \text{and} \quad \mathbf{T} \cdot \mathbf{N} = 0$$

●

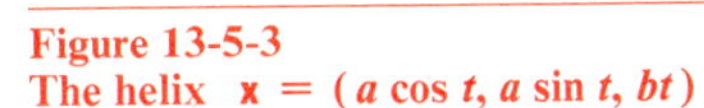
Figure 13-5-3
The helix $\mathbf{x} = (a \cos t, a \sin t, bt)$

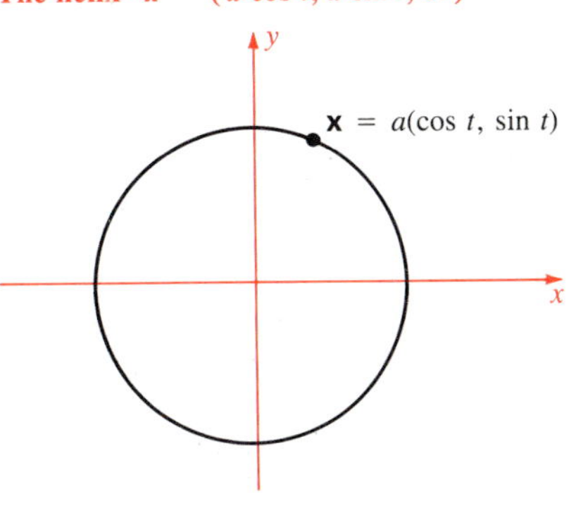

a Projection on the x,y-plane

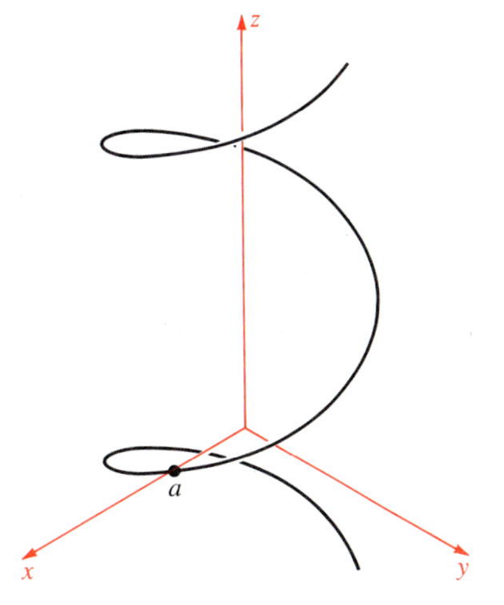

b The curve in space

Example 4

Compute $\mathbf{T}$, $\mathbf{N}$, and k for the helix (circular spiral)

$$\mathbf{x}(t) = (a \cos t, a \sin t, bt)$$

where $a > 0$ and $b > 0$.

Solution The projection of the point $\mathbf{x}(t)$ on the x, y-plane is $(a \cos t, a \sin t, 0)$. As a particle describes the curve $\mathbf{x}(t)$, its projection (Figure 13-5-3a) describes a circle of radius a. The third component of $\mathbf{x}(t)$ is bt; the particle moves upward at a steady rate. Thus, the curve is a steadily rising, circular spiral. (Figure 13-5-3b).

To find $\mathbf{T}$, differentiate, setting $c^2 = a^2 + b^2$ for convenience:

$$\mathbf{v} = d\mathbf{x}/dt = (-a \sin t, a \cos t, b)$$

so

$$|\mathbf{v}|^2 = a^2 + b^2 = c^2$$

Therefore $c = |\mathbf{v}| = ds/dt$ and

$$\mathbf{T} = \frac{1}{|\mathbf{v}|}\mathbf{v} = \frac{1}{c}(-a \sin t, a \cos t, b)$$

Next,

$$\frac{d\mathbf{T}}{ds} = \frac{dt}{ds} \cdot \frac{d\mathbf{T}}{dt} = \frac{1}{c} \cdot \frac{d\mathbf{T}}{dt} = \frac{a}{c^2}(-\cos t, -\sin t, 0)$$

Since $a/c^2 > 0$ and $(-\cos t, -\sin t, 0)$ is a unit vector, we conclude that

$$k = \left|\frac{d\mathbf{T}}{ds}\right| = \frac{a}{c^2} \quad \text{and} \quad \mathbf{N} = (-\cos t, -\sin t, 0)$$

Thus $k = a/(a^2 + b^2)$ is constant for the helix. (Note that the helix is a circle if $b = 0$.) ●

The Binormal and Torsion

It is very convenient to study a curve by attaching a coordinate system to each of its points. (Think of a particle moving along the curve. Attached to the particle is the origin of a coordinate system that moves with the particle.) The vectors $\mathbf{T}$ and $\mathbf{N}$ are two of the three required mutually perpendicular unit vectors. The third is defined so as to make a right-handed system with $\mathbf{T}$ and $\mathbf{N}$. It is called the **binormal** to the curve, and is defined by $\mathbf{B} = \mathbf{T} \times \mathbf{N}$. It can be proved that $d\mathbf{B}/ds = -\tau\mathbf{N}$, where $\tau = \tau(s)$ is a scalar function called the **torsion** of the curve. In courses on differential geometry it is shown that the two scalar functions $k(s)$ and $\tau(s)$ completely determine a curve, up to a rigid motion of space. Remarkably enough, the formula $d\mathbf{N}/ds = -k\mathbf{T} + \tau\mathbf{B}$ holds also.

Exercises

Find the unit tangent and curvature

1 $\mathbf{x} = t\mathbf{c} + \mathbf{b} \qquad (\mathbf{c} \neq \mathbf{0})$

2 $\mathbf{x} = \frac{1}{2}t^2\mathbf{a} + t\mathbf{b} + \mathbf{c}$

3 $\mathbf{x} = (t^2, t^3, t^4) \qquad (t \neq 0)$

4 $\mathbf{x} = (t\cos t, t\sin t, 2t)$

5 $\mathbf{x} = (\cos t, \sin t, \sin 2t)$

6 $\mathbf{x} = (e^{-t}, t, e^t)$

Find the unit normal and curvature

7 $\mathbf{x} = (3t, \sin t, \cos t)$

8 $\mathbf{x} = (\cos 2t, t, \sin 2t)$

9 Let $\mathbf{x}(t) = (x(t), y(t), z(t))$ for $a \leq t \leq b$ be a space curve with length L. Suppose its projection $\mathbf{x}_1(t) = (x(t), y(t))$ on the x,y-plane has length L_1. Prove that

$$L_1 \leq L \quad \text{and} \quad L \leq L_1 + \int_a^b |dz/dt|\, dt$$

***10** Let $\mathbf{x} = \mathbf{x}(s)$ be a curve on the unit sphere $|\mathbf{x}| = 1$. Prove that $k(s) \geq 1$.

13-6 Velocity and Acceleration

Suppose $\mathbf{x} = \mathbf{x}(t)$ is the path of a moving particle. The **velocity** of the particle is the vector

$$\mathbf{v} = \mathbf{v}(t) = \frac{d\mathbf{x}}{dt}$$

This velocity vector is tangential to the curve; its length is the **speed** of the particle. The **acceleration** of the particle is the vector

$$\mathbf{a} = \mathbf{a}(t) = \frac{d\mathbf{v}}{dt} = \frac{d^2\mathbf{x}}{dt^2}$$

Its length and direction are generally not as apparent as those of the velocity vector.

Example 1

Suppose the position of a moving particle is given as a function of time by

$$\mathbf{x}(t) = (b\cos\omega t, b\sin\omega t)$$

where $b > 0$ and $\omega > 0$. Since $|\mathbf{x}| = b$, the particle moves on the circle of radius b with center at the origin. Its velocity and acceleration are

$$\mathbf{v} = \frac{d\mathbf{x}}{dt} = b\omega(-\sin\omega t, \cos\omega t)$$

$$\mathbf{a} = \frac{d\mathbf{v}}{dt} = b\omega^2(-\cos\omega t, -\sin\omega t) = -\omega^2\mathbf{x}$$

The speed is $|\mathbf{v}| = b\omega$, a constant, so the motion is uniform circular motion. The velocity vector $\mathbf{v}(t)$ is perpendicular to the position vector since $\mathbf{x}(t) \cdot \mathbf{v}(t) = 0$. This is not surprising since the circle is centered at the origin and each tangent to a circle is perpendicular to the corresponding radius. The acceleration is $\mathbf{a}(t) = -\omega^2\mathbf{x}(t)$, so the acceleration vector $\mathbf{a}(t)$ is directed opposite to the position vector $\mathbf{x}(t)$. See Figure 13-6-1a. Why should that be?

The acceleration $\mathbf{a}(t)$ measures the rate of change of the velocity. Observe the velocity vectors at t and at an instant later, $t+h$. See Figure 13-6-1b. The difference $\mathbf{v}(t+h) - \mathbf{v}(t)$ is nearly parallel to $\mathbf{x}(t)$, but oppositely directed. Therefore the instantaneous rate of change of the velocity is in a direction opposite to that of $\mathbf{x}(t)$. ●

Figure 13-6-1
Uniform circular motion:
$\mathbf{x}(t) = b(\cos\omega t, \sin\omega t)$

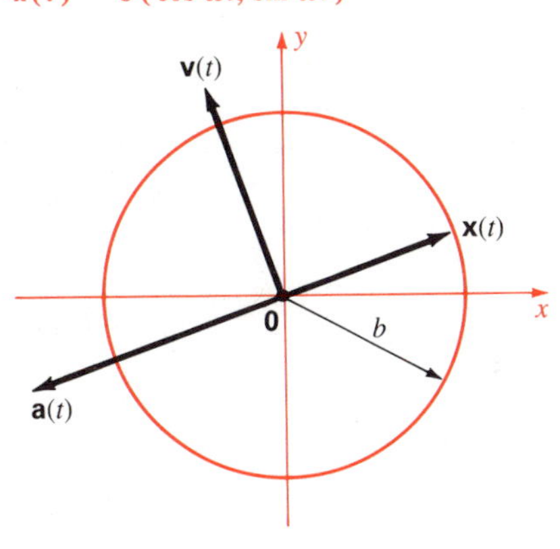

a Relative positions of **x**, **v**, and **a**

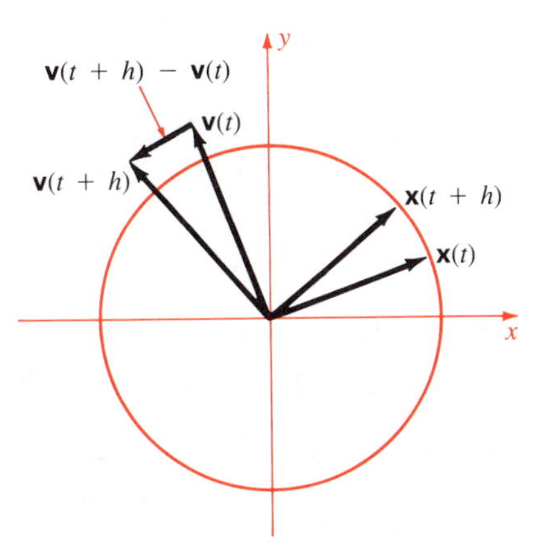

b Why **a** and **x** are oppositely directed

Newton's Law of Motion

This famous principle for the motion of a particle states that

$$\text{force} = \text{mass} \times \text{acceleration}$$

But force and acceleration are vectors, both having magnitude and direction. Thus Newton's law for the position $\mathbf{x} = \mathbf{x}(t)$ of a particle of mass m is a vector equation:

$$\mathbf{F} = m\mathbf{a} = m\frac{d^2\mathbf{x}}{dt^2}$$

Example 2

A particle of mass m is subject to zero force. Find its trajectory.

Solution By Newton's law,

$$m\frac{d^2\mathbf{x}}{dt^2} = \mathbf{0} \qquad \text{so} \qquad \frac{d^2\mathbf{x}}{dt^2} = \mathbf{0} \qquad \text{that is} \qquad \frac{d\mathbf{v}}{dt} = \mathbf{0}$$

Therefore $\mathbf{v} = \mathbf{v}_0$ is a constant, and

$$\frac{d\mathbf{x}}{dt} = \mathbf{v}_0$$

Integrate:

$$\mathbf{x} = t\mathbf{v}_0 + \mathbf{x}_0$$

Therefore the trajectory is a straight line, traversed at constant speed. ●

Example 3

A shell is fired at an angle α with the ground and initial speed v_0. What is its path? Neglect air resistance.

Solution Draw a figure, taking the axes as indicated (Figure 13-6-2). Let $\mathbf{v}_0$ be the initial velocity vector, so $\mathbf{v}_0 = v_0(\cos\alpha, \sin\alpha)$. Let m denote the mass of the shell. The force of gravity at each point is constant, $\mathbf{F} = (0, -mg)$. The equation of motion is

$$m\mathbf{a} = \mathbf{F} \qquad \text{that is} \qquad \frac{d^2\mathbf{x}}{dt^2} = (0, -g)$$

Figure 13-6-2
See Example 3

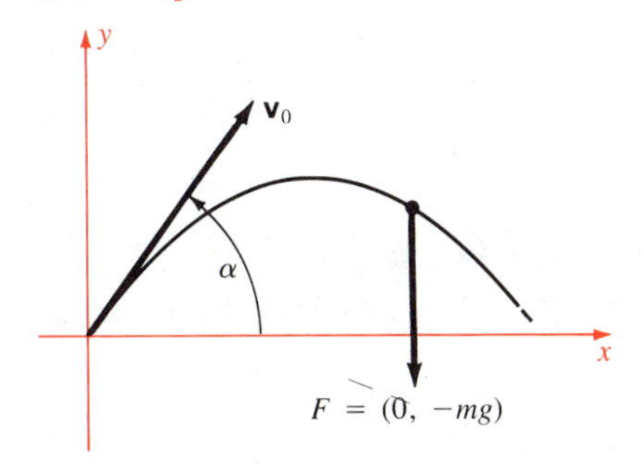

Integrate:

$$\frac{d\mathbf{x}}{dt} = (0, -gt) + \mathbf{v}_0$$

Integrate again, noting that $\mathbf{x}_0 = \mathbf{0}$ by the choice of axes:

$$\mathbf{x} = (0, -\tfrac{1}{2}gt^2) + \mathbf{v}_0 t$$

Hence

$$\begin{aligned}(x(t), y(t)) &= (0, -\tfrac{1}{2}gt^2) + tv_0(\cos\alpha, \sin\alpha)\\ &= (v_0 t\cos\alpha, v_0 t\sin\alpha - \tfrac{1}{2}gt^2)\end{aligned}$$

To describe the path, eliminate t by using

$$x = v_0 t\cos\alpha \qquad \text{that is} \qquad t = \frac{x}{v_0\cos\alpha}$$

to obtain

$$\begin{aligned}y &= v_0 t\sin\alpha - \tfrac{1}{2}gt^2\\ &= x\tan\alpha - \frac{g}{2v_0^2\cos^2\alpha}x^2 = -ax^2 + bx\end{aligned}$$

where $a = g/2v_0^2\cos^2\alpha$ and $b = \tan\alpha$. The graph of this quadratic is a parabola. ●

Example 4

In Example 3, what is the maximum range (ground distance) for fixed v_0?

Solution The shell hits ground when $y = 0$:

$$(v_0\sin\alpha - \tfrac{1}{2}gt)t = 0$$

This equation has two roots. The root $t = 0$ indicates the initial point. We want the other root, $t = 2v_0(\sin\alpha)/g$. The range is the value of x at this time:

$$x = (v_0\cos\alpha)\left(\frac{2v_0\sin\alpha}{g}\right) = \frac{v_0^2}{g}\sin 2\alpha$$

Clearly x is maximum when $\sin 2\alpha = 1$, or $\alpha = \tfrac{1}{4}\pi$. Therefore the maximum range is v_0^2/g, and it is achieved by firing at elevation $45°$. ●

Angular Momentum

The **angular momentum** (with respect to $\mathbf{0}$) of a particle of mass m at position $\mathbf{x} = \mathbf{x}(t)$ is defined as

$$\mathbf{L} = m\mathbf{x} \times \mathbf{v} \qquad \text{where} \qquad \mathbf{v} = d\mathbf{x}/dt$$

Example 5

Suppose a particle of mass m moves in a trajectory $\mathbf{x} = \mathbf{x}(t)$ under the influence of a *central force* $\mathbf{F}$, that is, a force directed towards (or against) $\mathbf{x}$. Then $\mathbf{F} = mf(\mathbf{x})\mathbf{x}$. Show that the angular momentum of the particle is constant.

Solution We have

$$\frac{d\mathbf{L}}{dt} = m\frac{d\mathbf{x}}{dt} \times \mathbf{v} + m\mathbf{x} \times \frac{d\mathbf{v}}{dt}$$

$$= m\mathbf{v} \times \mathbf{v} + m\mathbf{x} \times \mathbf{a} = m\mathbf{x} \times \mathbf{a}$$

where $\mathbf{a}$ is the acceleration of the particle. By Newton's law of motion, $\mathbf{F} = m\mathbf{a}$. Hence

$$\frac{d\mathbf{L}}{dt} = \mathbf{x} \times (m\mathbf{a}) = \mathbf{x} \times \mathbf{F} = \mathbf{0}$$

because $\mathbf{F}$ is parallel to $\mathbf{x}$. Thus $d\mathbf{L}/dt = \mathbf{0}$, so $\mathbf{L}$ is a constant. ●

Components of Acceleration

If a particle moves on a space curve, it is useful to express its velocity and acceleration in terms of $\mathbf{T}$, $\mathbf{N}$, and k since these are quantities built into the curve (independent of parametrization). We already know

$$\mathbf{v} = \frac{ds}{dt}\mathbf{T}$$

which simply says again that the motion is directed along the tangent with speed ds/dt.

For further information, we differentiate $\mathbf{v}$ with respect to time, using the chain rule carefully:

$$\mathbf{a} = \frac{d\mathbf{v}}{dt} = \frac{d^2s}{dt^2}\mathbf{T} + \frac{ds}{dt} \cdot \frac{d\mathbf{T}}{dt}$$

But

$$\frac{d\mathbf{T}}{dt} = \frac{ds}{dt} \cdot \frac{d\mathbf{T}}{ds} = \frac{ds}{dt}k\mathbf{N}$$

where k is the curvature. Therefore we have the following result:

Tangential and Normal Components of Acceleration

$$\mathbf{a} = \frac{d^2s}{dt^2}\mathbf{T} + k\left(\frac{ds}{dt}\right)^2\mathbf{N}$$ ●

This is an important equation in mechanics. It resolves the acceleration into two vectors, one tangential to the direction of motion, the other normal (perpendicular) to the direction of motion. The normal component $k(ds/dt)^2\mathbf{N}$ is called the **centripetal acceleration.**

Normal and tangential components of acceleration have a natural interpretation. Remember that acceleration is the rate of change of the velocity vector $\mathbf{v}$. Now a vector can change for two reasons: either its length changes or its direction changes. Since $|\mathbf{v}| = ds/dt$, a change in $|\mathbf{v}|$ is measured by the second derivative d^2s/dt^2. Hence the tangential component of $\mathbf{a}$ corresponds to the change in *length* of $|\mathbf{v}|$, that is, the changing speed. A change in direction of $\mathbf{v}$ is measured by the curvature k. Hence the normal component of $\mathbf{a}$ corresponds to the changing *direction* of $\mathbf{v}$.

Example 6

A particle moves counterclockwise on a circle of radius b. Resolve its acceleration into tangential and normal components.

Solution Place the circle in the x, y-plane with center at $\mathbf{0}$. Denote the central angle at time t by $\theta = \theta(t)$. Then the path is given by

$$\mathbf{x}(t) = b(\cos\theta, \sin\theta)$$

Differentiate:

$$\mathbf{v} = \frac{d\mathbf{x}}{dt} = b\frac{d\theta}{dt}(-\sin\theta, \cos\theta)$$

However $\mathbf{v} = |\mathbf{v}|\mathbf{T}$. Since $b(d\theta/dt) > 0$ and $(-\sin\theta, \cos\theta)$ is a unit vector, we conclude that

$$|\mathbf{v}| = b\frac{d\theta}{dt} \quad \text{and} \quad \mathbf{T} = (-\sin\theta, \cos\theta)$$

The **angular speed** $d\theta/dt$ is usually denoted by $\omega = \omega(t)$. Therefore $\mathbf{v} = b\omega\mathbf{T}$. By the chain rule, we have

$$\mathbf{a} = \frac{d\mathbf{v}}{dt} = b\frac{d\omega}{dt}\mathbf{T} + b\omega\frac{d\mathbf{T}}{dt}$$

But

$$\frac{d\mathbf{T}}{dt} = \frac{d\theta}{dt}\cdot\frac{d\mathbf{T}}{d\theta} = \omega(-\cos\theta, -\sin\theta)$$

Since $(-\cos\theta, -\sin\theta)$ is perpendicular to $\mathbf{T}$, it must equal $\pm\mathbf{N}$. The correct sign is plus because the normal component of $\mathbf{a}$ is a *positive* multiple of $\mathbf{N}$ and $b\omega^2 > 0$. Therefore $d\mathbf{T}/dt = \omega\mathbf{N}$, so the normal component of $\mathbf{a}$ is $b\omega^2\mathbf{N}$. The tangential component is $b(d\omega/dt)\mathbf{T}$, so we have the required resolution of $\mathbf{a}$:

$$\mathbf{a} = \mathbf{a}_{\text{tan}} + \mathbf{a}_{\text{nor}} \qquad \text{where} \quad \mathbf{a}_{\text{tan}} = b\frac{d\omega}{dt}\mathbf{T} \quad \text{and} \quad \mathbf{a}_{\text{nor}} = b\omega^2\mathbf{N}$$ ●

Remark When the motion is uniform (ω constant), then we have $\mathbf{a} = b\omega^2\mathbf{N}$, so the acceleration is all centripetal, perpendicular to the direction of motion. This agrees with Example 1.

Example 7

The position of a moving particle is given by $\mathbf{x}(t) = (5t + 1, t^2, 3t^2)$. Resolve its acceleration into tangential and normal components.

Solution First we have

$$\mathbf{v} = \frac{d\mathbf{x}}{dt} = (5, 2t, 6t) \qquad \text{so} \qquad \frac{ds}{dt} = |\mathbf{v}| = (25 + 40t^2)^{1/2}$$

We know that $\mathbf{a}_{\tan} = \dfrac{d^2s}{dt^2}\mathbf{T}$. Now $\mathbf{T} = \dfrac{1}{|\mathbf{v}|}\mathbf{v}$ and

$$\frac{d^2s}{dt^2} = \frac{d}{dt}(25 + 40t^2)^{1/2} = \frac{40t}{(25 + 40t^2)^{1/2}} = \frac{40t}{|\mathbf{v}|}$$

Hence

$$\mathbf{a}_{\tan} = \left(\frac{40t}{|\mathbf{v}|}\right)\left(\frac{1}{|\mathbf{v}|}\mathbf{v}\right) = \frac{40t}{25 + 40t^2}\mathbf{v} = \frac{8t}{5 + 8t^2}(5, 2t, 6t)$$

Now $\mathbf{a} = d\mathbf{v}/dt = (0, 2, 6)$ and $\mathbf{a} = \mathbf{a}_{\tan} + \mathbf{a}_{\text{nor}}$. Hence

$$\begin{aligned}\mathbf{a}_{\text{nor}} = \mathbf{a} - \mathbf{a}_{\tan} &= (0, 2, 6) - \frac{8t}{5 + 8t^2}(5, 2t, 6t)\\ &= \frac{5 + 8t^2}{5 + 8t^2}(0, 2, 6) - \frac{8t}{5 + 8t^2}(5, 2t, 6t)\\ &= \frac{1}{5 + 8t^2}(-40t, 10, 30) = \frac{10}{5 + 8t^2}(-4t, 1, 3)\end{aligned}$$

●

Figure 13-6-3
Angular velocity

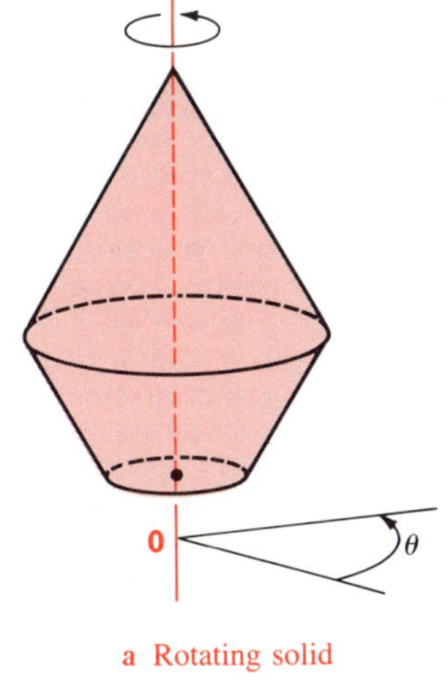

a Rotating solid

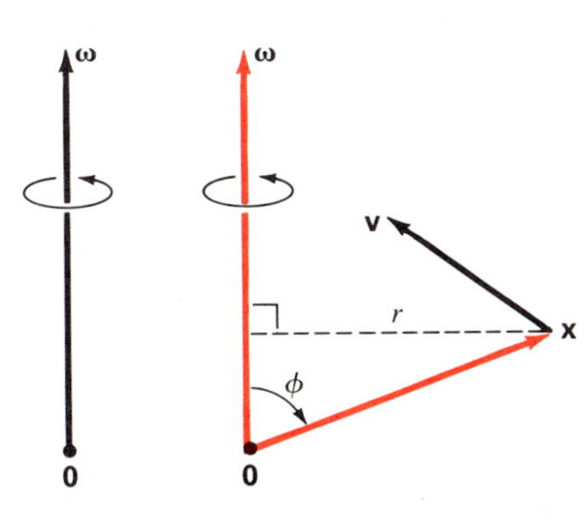

b

c Velocity $\mathbf{v} = \boldsymbol{\omega} \times \mathbf{x}$

Angular Velocity

Suppose that a rigid body rotates about an axis through $\mathbf{0}$. See Figure 13-6-3a. The central angle is $\theta = \theta(t)$, so $\omega = d\theta/dt$ is its **angular speed,** the rate of rotation in radians per second. Its **angular velocity** is defined to be the vector $\boldsymbol{\omega}$ having magnitude $d\theta/dt$ and pointing along the (positive) axis of rotation according to the right-hand rule (Figure 13-6-3b).

Once the angular velocity vector $\boldsymbol{\omega}$ is known, it is easy to find the velocity $\mathbf{v}$ of any point $\mathbf{x}$ in the rigid body. See Figure 13-6-3c.

Since the point $\mathbf{x}$ is rotating about the axis of $\boldsymbol{\omega}$, its velocity vector $\mathbf{v}$ is perpendicular to the plane of $\boldsymbol{\omega}$ and $\mathbf{x}$. By the right-hand rule, $\mathbf{v}$ points in the direction of $\boldsymbol{\omega} \times \mathbf{x}$. The speed $|\mathbf{v}|$ is the product of the angular speed $\omega = |\boldsymbol{\omega}|$ and the distance r of $\mathbf{x}$ from the axis of rotation. But $r = |\mathbf{x}|\sin\phi$, hence

$$|\mathbf{v}| = |\boldsymbol{\omega}| \cdot |\mathbf{x}|\sin\phi = |\boldsymbol{\omega} \times \mathbf{x}|$$

Therefore:

- Suppose a rigid body rotates with angular velocity $\boldsymbol{\omega}$ about an axis through $\mathbf{0}$. The velocity of a point $\mathbf{x}$ in the body is $\mathbf{v} = \boldsymbol{\omega} \times \mathbf{x}$. ●

Exercises

1 A hill makes angle β with the horizontal (Figure 13-6-4). A shell is fired with initial speed v_0 from the base of the hill at angle α with the horizontal. Show that the x-coordinate of the position where the shell strikes the hill is

$$x = (2v_0{}^2/g)(\sin\alpha\cos\alpha - \tan\beta\cos^2\alpha)$$

2 (cont.) For what value of α is x maximized, and what is its maximum value?

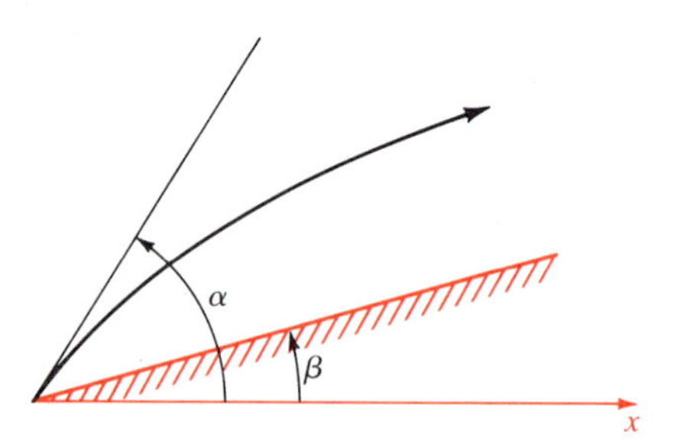

Figure 13-6-4
See Exercises 1 and 2

Find the tangential and normal components of the acceleration vector

3 $\mathbf{x} = (t, t^2)$

4 $\mathbf{x} = (t^2, t^3) \qquad (t > 0)$

5 $\mathbf{x} = (t, \sin t)$

6 $\mathbf{x} = (e^t, t)$

7 $\mathbf{x} = (\cos^2 t, \sin^2 t) \qquad (0 < t < \frac{1}{2}\pi)$

8 $\mathbf{x} = (2\cos t, 3\sin t)$

9 $\mathbf{x} = (b\cos\omega t, b\sin\omega t, ct) \qquad (\omega \text{ constant})$

10 $\mathbf{x} = (t, t^2, t^3)$

11 $\mathbf{x} = (\sin t, \cos t, \sin t)$

12 $\mathbf{x} = (t, t, t^2)$

13 A particle moves with constant speed 1 on the surface of the unit sphere $|\mathbf{x}| = 1$. Show that the normal component of the acceleration has magnitude at least 1.

14 A particle moves on the surface $z = x^2 + y^2$ with constant speed 1. At a certain instant t_0 it passes through $\mathbf{0}$. Show that the tangential component of $\mathbf{a}$ at that instant is $\mathbf{0}$ and the normal component is $(x''(t_0), y''(t_0), 2)$. Show also that $x'x'' + y'y'' = 0$ at t_0.

13-7 Plane Curves in Polar Coordinates

We study plane curves whose polar coordinates are given as functions of time: $r = r(t)$ and $\theta = \theta(t)$.

Certain problems require finding the area swept out by the segment joining $\mathbf{0}$ to a moving point on a curve (Figure 13-7-1a). Suppose the curve is given in parametric polar coordinates by

$$r = r(t) \quad \text{and} \quad \theta = \theta(t) \qquad (a \le t \le b)$$

In a small interval of time the segment sweeps out a thin triangle of base $r\,d\theta$ and height r [ignoring negligible errors (Figure 13-7-1b)]. Hence

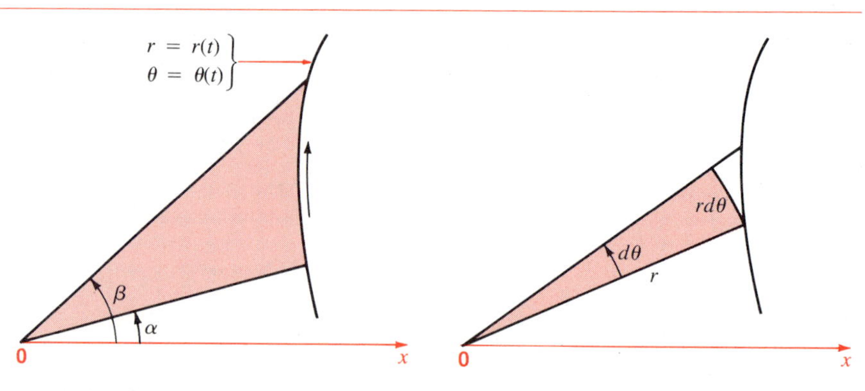

Figure 13-7-1
Area swept out by the radius vector

a Area between two positions of the radius vector

b Element of area

$$dA = \tfrac{1}{2}r^2\,d\theta = \tfrac{1}{2}r^2\frac{d\theta}{dt}\,dt$$

If the curve is given as the polar graph of a function $r = r(\theta)$ for $\alpha \le \theta \le \beta$, then we choose $t = \theta$. The area formula specializes to $dA = \frac{1}{2}r^2\,d\theta$.

Area Formula in Polar Coordinates

Suppose a curve is given by $r = r(t)$ and $\theta = \theta(t)$ for $a \le t \le b$. Then the area swept out by the segment joining the origin to a moving point on the curve is

$$A = \tfrac{1}{2}\int_a^b r^2\frac{d\theta}{dt}\,dt$$

In particular, a polar graph $r = r(\theta)$ sweeps out the area

$$A = \tfrac{1}{2}\int_\alpha^\beta [r(\theta)]^2\,d\theta$$

(The rough argument given here can be made rigorous by using Riemann sums.)

Figure 13-7-2
Four-petal rose curve

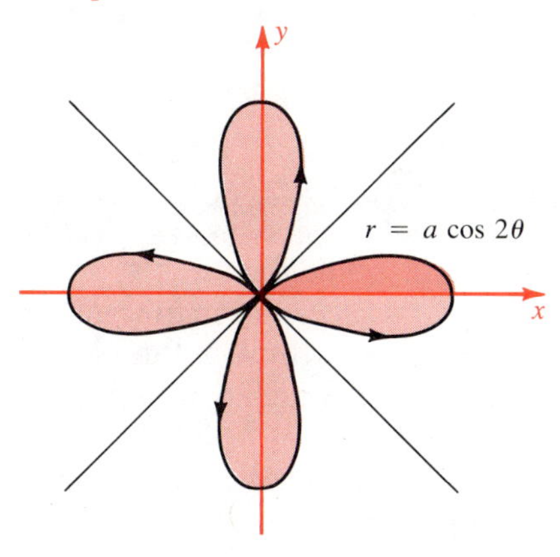

Example 1

Compute the area of the four-petal rose $r = a\cos 2\theta$.

Solution Figure 13-7-2 shows the graph, emphasizing the part where $0 \le \theta \le \frac{1}{4}\pi$. Because of symmetry it suffices to compute the area of half of one petal. Thus

$$A = 8\int_0^{\pi/4}\tfrac{1}{2}(a\cos 2\theta)^2\,d\theta = 4a^2\int_0^{\pi/4}\cos^2 2\theta\,d\theta$$

$$= 2a^2\int_0^{\pi/4}(1+\cos 4\theta)\,d\theta = \tfrac{1}{2}\pi a^2$$

The Natural Frame

Figure 13-7-3
The natural frame (a pair of orthogonal unit vectors) attached to $\mathbf{x} = [r, \theta]$

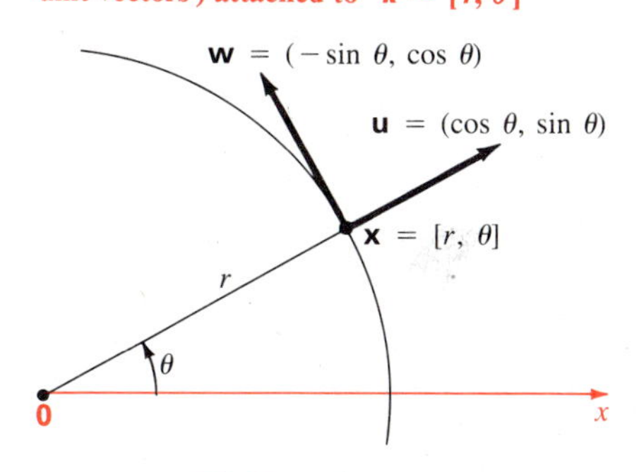

a The frame drawn at **x**

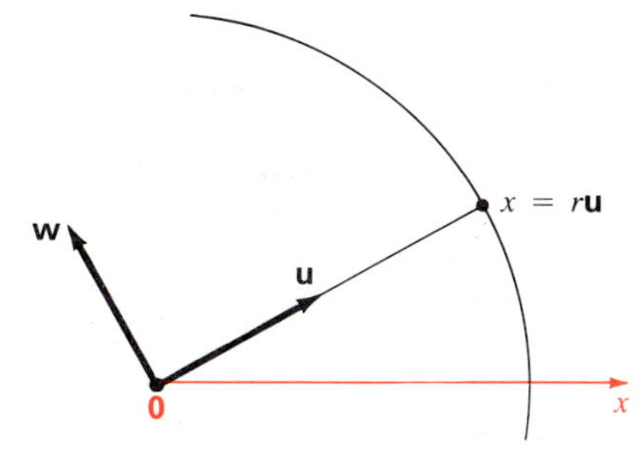

b The frame correctly drawn at **0**

When dealing with curves given in parametric polar form, we use a pair of unit vectors that do for polar coordinates what $\mathbf{i}$ and $\mathbf{j}$ do for rectangular coordinates. We associate to $[r, \theta]$, where $r \ne 0$, a unit vector $\mathbf{u}$ in the direction of increasing r and a unit vector $\mathbf{w}$ in the direction of increasing θ. See Figure 13-7-3a. From the figure we find that

$$\mathbf{u} = (\cos\theta, \sin\theta) \qquad \text{and} \qquad \mathbf{w} = (-\sin\theta, \cos\theta)$$

Like $\mathbf{i}$ and $\mathbf{j}$ the vectors $\mathbf{u}$ and $\mathbf{w}$ are perpendicular. But unlike $\mathbf{i}$ and $\mathbf{j}$, the vectors $\mathbf{u}$ and $\mathbf{w}$ associated with $\mathbf{x} = [r, \theta]$ vary with $\mathbf{x}$. Actually, they depend on θ alone. For future reference we note their derivatives:

$$d\mathbf{u}/d\theta = \mathbf{w} \qquad \text{and} \qquad d\mathbf{w}/d\theta = -\mathbf{u}$$

Natural Frame (Polar Coordinates)

The unit vectors $\mathbf{u} = (\cos\theta, \sin\theta)$ and $\mathbf{w} = (-\sin\theta, \cos\theta)$ indicate respectively the directions of increasing r and increasing θ at each point $[r, \theta]$ where $r \neq 0$. Their derivatives satisfy the relations

$$\frac{d\mathbf{u}}{d\theta} = \mathbf{w} \quad \text{and} \quad \frac{d\mathbf{w}}{d\theta} = -\mathbf{u}$$

In applications, $\mathbf{u}$ and $\mathbf{w}$ associate to each point a natural coordinate system (frame) in which computations are often simpler than in rectangular coordinates. For example, a polar curve given by $r = r(t)$ and $\theta = \theta(t)$ has a convenient vector form (Figure 13-7-3b).

$$\mathbf{x} = (r\cos\theta, r\sin\theta) = r\mathbf{u}$$

Velocity and Acceleration

Suppose the path of a particle is described in polar form, $r = r(t)$, $\theta = \theta(t)$. It is natural to express its velocity and acceleration in terms of the perpendicular unit vectors $\mathbf{u}$ and $\mathbf{w}$. (You can think of this pair of vectors as a frame moving along the path, providing at each point a convenient coordinate system.)

We need the time derivatives $d\mathbf{u}/dt$ and $d\mathbf{w}/dt$. We already have the derivatives $d\mathbf{u}/d\theta = \mathbf{w}$ and $d\mathbf{w}/d\theta = -\mathbf{u}$. From this information we find the derivatives with respect to t by applying the chain rule:

$$\frac{d\mathbf{u}}{dt} = \frac{d\theta}{dt} \cdot \frac{d\mathbf{u}}{d\theta} = \frac{d\theta}{dt}\mathbf{w}$$

$$\frac{d\mathbf{w}}{dt} = \frac{d\theta}{dt} \cdot \frac{d\mathbf{w}}{d\theta} = -\frac{d\theta}{dt}\mathbf{u}$$

To find the velocity and acceleration vectors, we first write the given curve in vector form, $\mathbf{x} = r\mathbf{u}$. Then we differentiate twice using the product rule carefully:

$$\mathbf{v} = \frac{d\mathbf{v}}{dt} = \frac{dr}{dt}\mathbf{u} + r\frac{d\mathbf{u}}{dt} = \frac{dr}{dt}\mathbf{u} + r\frac{d\theta}{dt}\mathbf{w}$$

$$\begin{aligned}
\mathbf{a} = \frac{d\mathbf{v}}{dt} &= \left(\frac{d^2r}{dt^2}\mathbf{u} + \frac{dr}{dt} \cdot \frac{d\mathbf{u}}{dt}\right) \\
&\quad + \left(\frac{dr}{dt} \cdot \frac{d\theta}{dt}\mathbf{w} + r\frac{d^2\theta}{dt^2}\mathbf{w} + r\frac{d\theta}{dt} \cdot \frac{d\mathbf{w}}{dt}\right) \\
&= \left(\frac{d^2r}{dt^2}\mathbf{u} + \frac{dr}{dt} \cdot \frac{d\theta}{dt}\mathbf{w}\right) \\
&\quad + \left(\frac{dr}{dt} \cdot \frac{d\theta}{dt}\mathbf{w} + r\frac{d^2\theta}{dt^2}\mathbf{w} - r\frac{d\theta}{dt} \cdot \frac{d\theta}{dt}\mathbf{u}\right) \\
&= \left[\frac{d^2r}{dt^2} - r\left(\frac{d\theta}{dt}\right)^2\right]\mathbf{u} + \left[r\frac{d^2\theta}{dt^2} + 2\frac{dr}{dt} \cdot \frac{d\theta}{dt}\right]\mathbf{w}
\end{aligned}$$

Velocity and Acceleration in Polar Coordinates

If the motion of a particle is given by $r = r(t), \quad \theta = \theta(t),$ then its velocity and acceleration are given by

$$\mathbf{v} = \frac{dr}{dt}\mathbf{u} + r\frac{d\theta}{dt}\mathbf{w}$$

$$\mathbf{a} = \left[\frac{d^2r}{dt^2} - r\left(\frac{d\theta}{dt}\right)^2\right]\mathbf{u} + \left[r\frac{d^2\theta}{dt^2} + 2\frac{dr}{dt}\cdot\frac{d\theta}{dt}\right]\mathbf{w}$$

Example 2

The spiral $r = t, \quad \theta = t.$ We have

$$\frac{dr}{dt} = 1 \qquad \frac{d\theta}{dt} = 1 \qquad \frac{d^2r}{dt^2} = 0 \qquad \frac{d^2\theta}{dt^2} = 0$$

Therefore

$$\mathbf{v} = \mathbf{u} + t\mathbf{w} \qquad \mathbf{a} = -r\mathbf{u} + 2\mathbf{w}$$

Central Force and Kepler's Laws

Suppose a particle moves under the influence of a **central force**

$$\mathbf{F} = mf(\mathbf{x})\mathbf{x}$$

At each instant, the force is directed toward or away from the origin. Since $m\mathbf{a} = \mathbf{F},$ the component of $\mathbf{a}$ in the direction of $\mathbf{w}$ is zero:

$$r\frac{d^2\theta}{dt^2} + 2\frac{dr}{dt}\cdot\frac{d\theta}{dt} = 0$$

that is,

$$\tfrac{1}{2}r^2\frac{d^2\theta}{dt^2} + r\frac{dr}{dt}\cdot\frac{d\theta}{dt} = 0$$

This is equivalent to

$$\frac{d}{dt}\left(\tfrac{1}{2}r^2\frac{d\theta}{dt}\right) = 0 \qquad \text{that is} \qquad \tfrac{1}{2}r^2\frac{d\theta}{dt} = \tfrac{1}{2}J$$

where J is a constant. Now the area swept out by the segment joining the origin to the curve between times a and b is

$$A = \tfrac{1}{2}\int_a^b r^2\frac{d\theta}{dt}\,dt = \tfrac{1}{2}\int_a^b J\,dt = \tfrac{1}{2}J(b - a)$$

Therefore the same area is swept out in equal time anywhere along the curve. This remarkable property of plane motion under a central force is **Kepler's second law of planetary motion.**

Kepler's first law of planetary motion asserts that a planet moves on an ellipse, and his third law gives a formula for the period (length of its year) of a planet. We shall now sketch derivations of these laws.

Suppose that the (fixed) sun has mass M and the planet has mass m. Then the central force is given by the inverse square law:

$$\mathbf{F} = -G\frac{Mm}{r^2}\mathbf{u} \qquad \text{where } G \text{ is the "constant of gravity"}$$

From $\mathbf{F} = m\mathbf{a}$ and our determination above of the $\mathbf{u}$ component of $\mathbf{a}$, we deduce that

$$\frac{d^2r}{dt^2} - r\left(\frac{d\theta}{dt}\right)^2 = -\frac{GM}{r^2}$$

Thus our problem is to solve the system of differential equations

$$\frac{d\theta}{dt} = \frac{J}{r^2} \qquad \text{and} \qquad \frac{d^2r}{dt^2} - r\left(\frac{d\theta}{dt}\right)^2 = -\frac{GM}{r^2}$$

We want the solution in the form of a relation between r and θ (which we will recognize as the equation of a conic section).

It is convenient to work with $u = 1/r$ instead of r. Then we have $d\theta/dt = Ju^2$, and the second differential equation can be written

$$\frac{d^2r}{dt^2} - J^2u^3 = -GMu^2 \tag{*}$$

By the chain rule,

$$\frac{du}{d\theta} = \frac{du/dt}{d\theta/dt} = \frac{(-1/r^2)(dr/dt)}{Ju^2} = -\frac{1}{J}\cdot\frac{dr}{dt}$$

Therefore

$$\frac{dr}{dt} = -J\frac{du}{d\theta} \qquad \text{so} \qquad \frac{d^2r}{dt^2} = -J\frac{d^2u}{d\theta^2}\cdot\frac{d\theta}{dt} = -J^2u^2\frac{d^2u}{d\theta^2}$$

The differential equation (*) now becomes

$$-J^2u^2\frac{d^2u}{d\theta^2} - J^2u^3 = -GMu^2$$

that is,

$$\frac{d^2u}{d\theta^2} + u = \frac{GM}{J^2}$$

This differential equation is a standard type, and it is known that each solution has the form

$$u = -A\cos(\theta - \theta_0) + GM/J^2$$

where A and θ_0 are constants. (This will be discussed in Chapter 19.)

Now we multiply by $r = 1/u$ to obtain

$$r[GM/J^2 - A\cos(\theta - \theta_0)] = 1$$

By choosing the x-axis suitably we can make $\theta_0 = 0$ and $A > 0$, so this equation can be written in the standard form (page 464)

$$r(1 - e\cos\theta) = 2pe$$

where $e = AJ^2/GM$ and $2pe = J^2/GM$. As we know, this is the polar form of a conic section of eccentricity e. Let us assume that the trajectory is periodic, that is, a closed curve. Then the conic section must be an ellipse, that is, $0 < e < 1$. We have proved **Kepler's first law:** the planets in a solar system move in ellipses with the sun at one focus.

The ends of the major axis occur at $\theta = 0$ and $\theta = \pi$, so in standard ellipse notation (page 460)

$$2a = \frac{2pe}{1-e} + \frac{2pe}{1+e} = \frac{4pe}{1-e^2} \qquad \text{so} \qquad a = \frac{2pe}{1-e^2}$$

We also have $c = ae$ and $b^2 = 2pc = 2pae$. The area of the ellipse is $A = \pi ab$. Hence

$$A^2 = \pi^2a^2b^2 = 2\pi^2a^3pe$$

Kepler's third law relates a^3 to the period T, the time to complete one orbit. By Kepler's second law, $A = \frac{1}{2}JT$, so that

$$\tfrac{1}{4}J^2T^2 = A^2 = 2\pi^2a^3pe$$

But $2pe = J^2/GM$. Hence

$$T^2 = \frac{4\pi^2}{GM}a^3$$

Note that the constant $4\pi^2/GM$ does not depend on the planet, so for every planet, asteroid, comet, what have you in the solar system, T^2/a^3 equals the same constant. This is **Kepler's third law.**

Arc Length

Finally, we derive a formula for the arc length of a curve in parametric polar form $r = r(t)$, $\theta = \theta(t)$, where $a \le t \le b$. We use the formulas $ds/dt = |\mathbf{v}| = |d\mathbf{x}/dt|$ and $\mathbf{v} = (dr/dt)\mathbf{u} + r(d\theta/dt)\mathbf{w}$ found earlier. Since $\mathbf{u}$ and $\mathbf{w}$ are orthogonal unit vectors,

$$\left(\frac{ds}{dt}\right)^2 = |\mathbf{v}|^2 = \left(\frac{dr}{dt}\right)^2 + r^2\left(\frac{d\theta}{dt}\right)^2$$

Arc Length

The length of a polar curve $r = r(t)$, $\theta = \theta(t)$ for $a \le t \le b$ is

$$L = \int_a^b \sqrt{\left(\frac{dr}{dt}\right)^2 + r^2\left(\frac{d\theta}{dt}\right)^2}\, dt$$

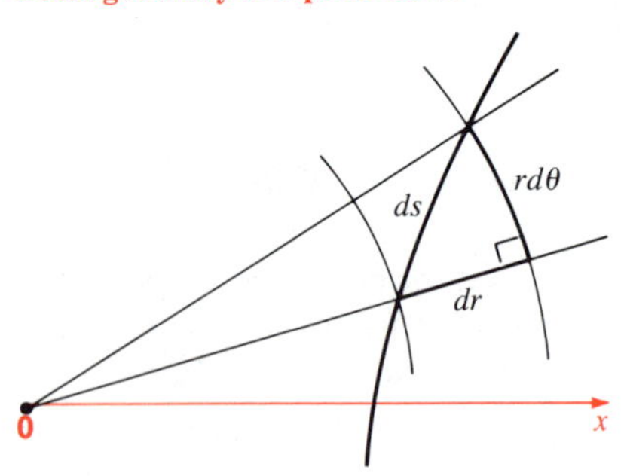

Figure 13-7-4
Local geometry of a polar curve

The length of a curve given as a polar graph $r = r(\theta)$ for $\alpha \le \theta \le \beta$ is

$$L = \int_\alpha^\beta \sqrt{\left(\frac{dr}{d\theta}\right)^2 + r^2}$$

The second formula follows from the first by setting $t = \theta$. Figure 13-7-4 provides an aid to memory. The "right triangle" has legs dr and $r\,d\theta$ and hypotenuse ds, so the Pythagorean theorem suggests

$$(ds)^2 = (dr)^2 + r^2(d\theta)^2$$

Example 3

Find the length of the spiral $r = \theta^2$ for $0 \le \theta \le 2\pi$.

Solution

$$\begin{aligned} L &= \int_0^{2\pi} \sqrt{r^2 + \left(\frac{dr}{d\theta}\right)^2}\, d\theta \\ &= \int_0^{2\pi} \sqrt{\theta^4 + (2\theta)^2}\, d\theta \\ &= \int_0^{2\pi} \theta\sqrt{\theta^2 + 4}\, d\theta = \tfrac{1}{3}(\theta^2 + 4)^{3/2}\Big|_0^{2\pi} \\ &= \tfrac{8}{3}(\pi^2 + 1)^{3/2} - \tfrac{8}{3} \approx 92.90 \end{aligned}$$

Exercises

Find the area enclosed by

1 $r = a\sin\theta$

2 $r = a\cos 3\theta$ (rose)

3 $r = a\cos(2n+1)\theta$ (rose)

4 $r = a\cos 2n\theta$ (rose)

5 $r = a\cos^2 2n\theta$

6 $r = a(1 - \cos\theta)$ (cardioid)

7 $r^2 = a^2\cos 2\theta$ (lemniscate)

8 the closed loop of the strophoid $r = a\cos 2\theta\sec\theta$

9 Find the area outside the circle $r = 1$ and inside the rose $r = 2\cos 2\theta$.

10 Find the area swept out by the segment from **0** to the spiral $r = \theta$ for $0 \le \theta \le 2\pi$.

11 Find the area common to $r = a\cos\theta$ and $r = a\sin\theta$.

***12** Find the area between the two loops of the limaçon $r = b + a\cos\theta$, $0 < b < a$.

Compute **v** and **a** in polar and rectangular form for

13 $r = t \quad \theta = 2t$

14 $r = t \quad \theta = t^2$

15 $r = \cos t \quad \theta = t$

16 $r = \sin 2t \quad \theta = t$

17 Compute **v** and **a** for $r = e^t$, $\theta = t$.

18 (cont.) Find the angle between **v** and **a**.

Set up, but do not evaluate, an integral for the arc length

19 $r = a\theta \quad 0 \le \theta \le 2\pi$

20 $r = a\cos 2\theta$

21 $r = a\cos 3\theta$

22 $r = a(1 - \cos\theta)$

Set up the arc-length integral of the closed curve; evaluate it if you can

23 $r = a(1 + \cos\theta)$ (cardioid)

24 $r^2 = a^2\cos 2\theta$ (lemniscate)

13-8 Review Exercises

Find the velocity and speed for $\mathbf{x}(t) =$

1 $(\tan t, \sec t)$

2 $(e^{-t}\cos t, e^{-t}\sin t, e^{-t})$

3 Find the length of the catenary $y = a\cosh(x/a)$ for $-b \leq x \leq b$.

4 Find the length of $\mathbf{x}(t) = (e^t, 2e^t, 3e^t)$ for $0 \leq t \leq 1$.

5 A point moves with constant unit speed along a curve C. Show that at each point of C, the curvature is the same as the length of the acceleration vector.

6 Find the quadratic $y = a + bx + cx^2$ that passes through $(0, 1)$ and agrees with the curve $y = e^x$ in slope and curvature at $(0, 1)$.

7 Find the maximum curvature of $x = 3y^2$.

8 The ellipse with foci $\mathbf{p}$ and $\mathbf{q}$ and length sum $2a$ is defined by $|\mathbf{x} - \mathbf{p}| + |\mathbf{x} - \mathbf{q}| = 2a$. Prove the reflection property of the ellipse by differentiating this relation with respect to arc length.
[Hint $|\mathbf{x} - \mathbf{p}|^2 = (\mathbf{x} - \mathbf{p}) \cdot (\mathbf{x} - \mathbf{p})$.]

9 (cont.) Do the same for the parabola.

10 Describe the curve $\mathbf{x} = (a\sec t, b\tan t)$.

11 Show that the curve $\mathbf{x} = (\sin t\cos t, \sin^2 t, \cos t)$ lies on a sphere.

12 (cont.) Find its curvature.

***13** Describe the locus $\mathbf{x} = (a\tan(t + \alpha), b\tan(t + \beta))$, where a, b, α, and β are constants and $\alpha - \beta$ is not a multiple of π.

14 Find the tangential and normal components of the acceleration vector of $\mathbf{x} = (t^3, 6t)$.

15 Find all curves $r = r(\theta)$ such that the angle γ between the tangent and the radius vector is constant.

***16** A plane curve is given by an equation $r = 1/f(\theta)$. Express the curvature of the curve in terms of $f(\theta)$ and its derivatives. Assume $f > 0$ and $f + f'' > 0$.

14 Functions of Several Variables

Up to now we have been concerned with functions such as $y = f(x)$, where y depends on one real variable x. In all sorts of situations, however, a quantity may depend on several real variables. Here are some examples:

1 The speed v of sound in an ideal gas is

$$v = \sqrt{\gamma \frac{p}{d}}$$

where d is the density of the gas, p is the pressure, and γ is a constant characteristic of the gas. Then v depends on (is a function of) the two variables p and d. We may write

$$v = f(p, d) \quad \text{or} \quad v = v(p, d)$$

The function v is defined for a certain set of pairs (p, d), which we can think of as a subset of the p, d-plane.

2 In physiology it is shown that the surface area A of a human body is related to the person's height H and weight W by

$$A = cH^a W^b$$

where a, b, and c are constants. Thus A is a function of the two variables H and W, defined for H and W on certain intervals.

3 In manufacturing a certain item, three types of components are used. The cost of the first component is p per unit, and x units are used. Similarly the second and third components cost q and r per unit and y and z units respectively are used. It follows that the cost of materials for manufacturing the item is

$$C = px + qy + rz$$

This C is a function of the six variables p, q, r, x, y, z defined for positive values of these variables.

14-1 Functions and Graphs

In general, a real-valued function f of two variables is an assignment of a real number to each point of subset $\mathbf{D}$ of the plane $\mathbf{R}^2$. A real-valued function of three variables is an assignment of a real number to each point of a subset $\mathbf{D}$ of space $\mathbf{R}^3$. The set $\mathbf{D}$ is the **domain** of f.

Suppose f is a function of two variables and (x, y) is a point of its domain. We denote the real number assigned to (x, y) by $f(x, y)$. We often use vector notation, writing $\mathbf{x} = (x, y)$ and $f(\mathbf{x}) = f(x, y)$. There is similar notation for functions of three variables.

A common notation is

$$f\colon \mathbf{D} \rightarrow \mathbf{R}$$

suggesting that f carries, or maps, the set $\mathbf{D}$ into the set $\mathbf{R}$ of real numbers.

Figure 14-1-1
$f(x, y) = \dfrac{x^2 + y^2}{2xy}$
Domain: $\mathbf{R}^2$ with the x- and y-axes deleted

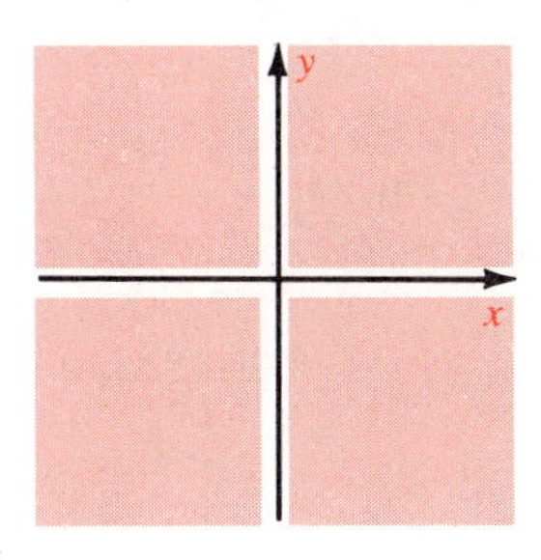

We want to extend the concepts of one-variable calculus to functions of several variables, concepts such as continuity, derivative, and integral. For these concepts to be meaningful in the one-variable situation, the domain of the function has to be a reasonably nice set, generally an interval or the union of several intervals. The same is so in the plane and in space. The functions we usually study in the plane have as their domains plane regions bounded by arcs of a few simple curves. In space, the domains are usually bounded by portions of familiar surfaces such as planes and spheres. Part of the boundary may be excluded from the domain. Let us look at some examples.

Polynomials such as

$$3x^5 - 2xy^2 + x^3y^3 - 4y^7 \qquad \text{and} \qquad (x - y)^{10} - 10x^5y^5$$

are defined on the whole plane. Rational functions such as

$$\frac{x^2 + y^2}{2xy} \quad \text{and} \quad \frac{x + y}{x - y}$$

Figure 14-1-2
$f(x, y) = \dfrac{x + y}{x - y}$
Domain: $\mathbf{R}^2$ with the line $x = y$ deleted

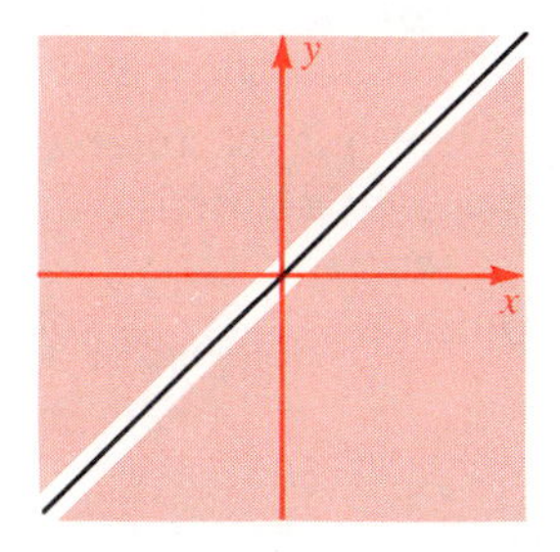

are defined wherever their denominators are non-zero. The first function is defined on the whole x, y-plane except for the x-axis and the y-axis; its domain consists of the four quadrants without their boundaries (Figure 14-1-1). The second is defined everywhere except on the line $x = y$; its domain consists of two half-planes without their common boundary (Figure 14-1-2).

The function

$$\sqrt{x + y}$$

is defined wherever $x + y \geq 0$; its domain consists of a half-plane including its boundary (Figure 14-1-3).

Figure 14-1-3
$f(x, y) = \sqrt{x + y}$
Domain: $\{(x, y) \mid x + y \geq 0\}$

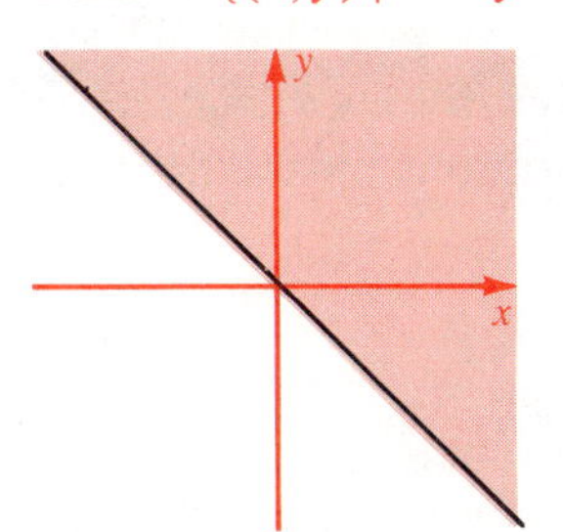

Continuous Functions

For a function of several variables to be useful, it must have some reasonable properties. The most basic such property is continuity. Here is the formal definition, a direct generalization of the definition of continuity for a function of one variable:

Definition Continuity

Let f be a real-valued function defined on $\mathbf{D}$, a subset of $\mathbf{R}^2$ or $\mathbf{R}^3$. Let $\mathbf{a}$ be a point of $\mathbf{D}$. We say f is **continuous** at $\mathbf{a}$ if

$$\lim_{\mathbf{x}\to\mathbf{a}} f(\mathbf{x}) = f(\mathbf{a})$$

Precisely, for each $\varepsilon > 0$ there exists $\delta > 0$ such that $|f(\mathbf{x}) - f(\mathbf{a})| < \varepsilon$ whenever $\mathbf{x}$ is in $\mathbf{D}$ and $|\mathbf{x} - \mathbf{a}| < \delta$.

We say f is **continuous on** $\mathbf{D}$ if f is continuous at each point of $\mathbf{D}$.

Notation We frequently write

$$f(\mathbf{x}) \to f(\mathbf{a}) \quad \text{as} \quad \mathbf{x} \to \mathbf{a}$$

instead of $\lim_{\mathbf{x}\to\mathbf{a}} f(\mathbf{x}) = f(\mathbf{a})$.

We know from the last chapter that $\mathbf{x} \to \mathbf{a}$ if and only if $x_1 \to a_1$, $x_2 \to a_2$, and $x_3 \to a_3$. That is because

$$|\mathbf{x} - \mathbf{a}|^2 = |x_1 - a_1|^2 + |x_2 - a|^2 + |x_3 - a_3|^2$$

The elementary properties of continuous functions of one variable carry over easily to functions of more than one variable. In particular, *sums, products, and quotients of continuous functions are continuous.* (For quotients, we assume non-zero denominators.)

Constant functions and the two functions defined by $f(x, y) = x$ and $g(x, y) = y$ are continuous. By forming products and sums starting with these building blocks, we deduce that *each polynomial is a continuous function on* $\mathbf{R}^2$ (*on* $\mathbf{R}^3$). From this we deduce that *each rational function is continuous wherever its denominator is not zero.* (Recall that a rational function is a quotient of polynomials.)

If $f(x)$ is a continuous function of *one* variable, then it is reasonable (and easy to prove) that $F(x, y) = f(x)$ is continuous as a function of *two* variables. Similarly, if $g(y)$ is continuous in y, then $G(x, y) = g(y)$ is continuous in x and y. Therefore the product

$$h(x, y) = f(x)g(y)$$

is continuous. By adding such products, we deduce that functions like

$$f(x, y) = e^x \sin y + 3e^y \cos x$$

are continuous "by inspection."

Composite Functions

Suppose we want to prove $f(x, y) = y^x$ is continuous on the domain $x > 0$, $y > 0$. We could write

$$f(x, y) = e^{x \ln y} = e^{g(x, y)}$$

Thus $f(x, y)$ is the composite of the continuous functions $h(t) = e^t$ and $t = x \ln y$. It is reasonable that $f(x, y)$ is continuous also.

Here is a more complicated example. Suppose we could prove that

$$P(x, y, z) = \int_0^x (y^3 + t^4)\sin(zt^2)\,dt$$

is continuous in x, y, z-space. We want to conclude that $P(u, v^u, uv)$ is continuous on the domain $u > 0,\ v > 0$. What we need is a theorem stating (roughly) that if $P(x, y, z)$ is continuous, and if x, y, and z are continuous functions of u and v, then P is continuous as a function of u and v.

The following is typical of this sort of situation. Suppose $x = x(u, v)$ and $y = y(u, v)$ are continuous functions on a domain **D** of the u, v-plane. Suppose that $F(x, y)$ is a continuous function on a domain **E** of the x, y-plane. Suppose that whenever the point (u, v) is in **D**, then the corresponding point (x, y) is in **E**. See Figure 14-1-4. Then the **composite function**

$$f(u, v) = F[x(u, v), y(u, v)]$$

is defined on domain **D**. The following theorem states that in this and similar situations, the composite function is also continuous.

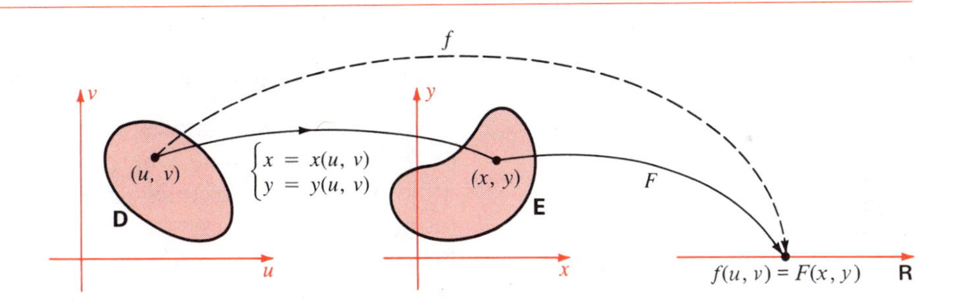

Figure 14-1-4
f: **D** → **R** is the composite function $f(u, v) = F(x(u, v), y(u, v))$

Theorem Continuity of Composite Functions

Let $P(x, y, z)$ be continuous on a domain **D** in x, y, z-space. Let f, g, and h be continuous on a domain **E** of the u, v-plane, and suppose that $(f(u, v), g(u, v), h(u, v))$ is a point of **D** whenever (u, v) is a point of **E**. Then the composite function

$$p(u, v) = P[f(u, v), g(u, v), h(u, v)]$$

is continuous on **E**.

Proof See Figure 14-1-5 (next page). Let $(u, v) \to (u_0, v_0)$. Then

$$f(u, v) \to f(u_0, v_0) \qquad g(u, v) \to g(u_0, v_0)$$
$$h(u, v) \to h(u_0, v_0).$$

Therefore

$$(f(u, v), g(u, v), h(u, v)) \to (f(u_0, v_0), g(u_0, v_0), h(u_0, v_0))$$

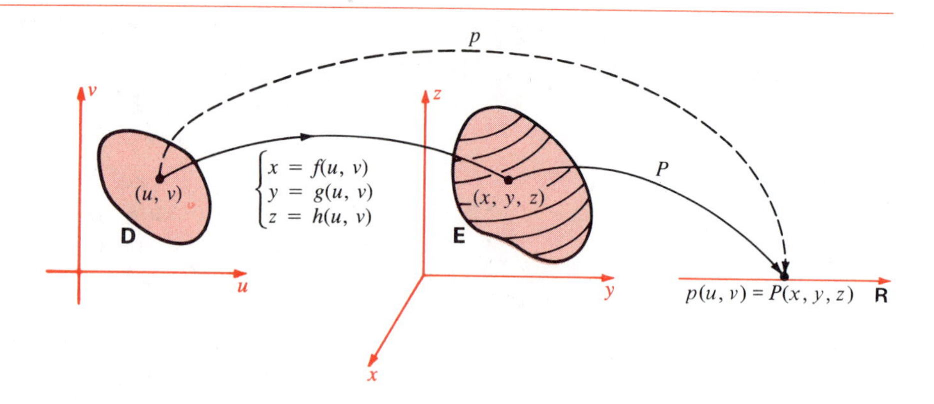

Figure 14-1-5
If f, g, and h are continuous, and if P is continuous, then the composite function p: **D** → **R** is continuous.

But P is continuous, so

$$p(u, v) = P[f(u, v), g(u, v), h(u, v)]$$
$$\rightarrow P[f(u_0, v_0), g(u_0, v_0), h(u_0, v_0)] = p(u_0, v_0)$$

Therefore p is continuous. ●

The theorem is stated for a function of three variables, where each variable is replaced by a function of two variables. However, there is nothing special about three and two, and the result may be modified as needed.

Graphs

Given a function of one variable $y = f(x)$, its graph is the set of points $(x, f(x))$ in the plane, where x is in the domain of f. Similarly, given a function of two variables $z = f(x, y)$, its **graph** is the set of points $(x, y, f(x, y))$ where (x, y) is in the domain of f. See Figure 14-1-6. For a function of one variable, the graph is a curve in $\mathbf{R}^2$; for a function of two variables, the graph is a surface in $\mathbf{R}^3$. In either case, we picture the graph as lying above (or below) the domain.

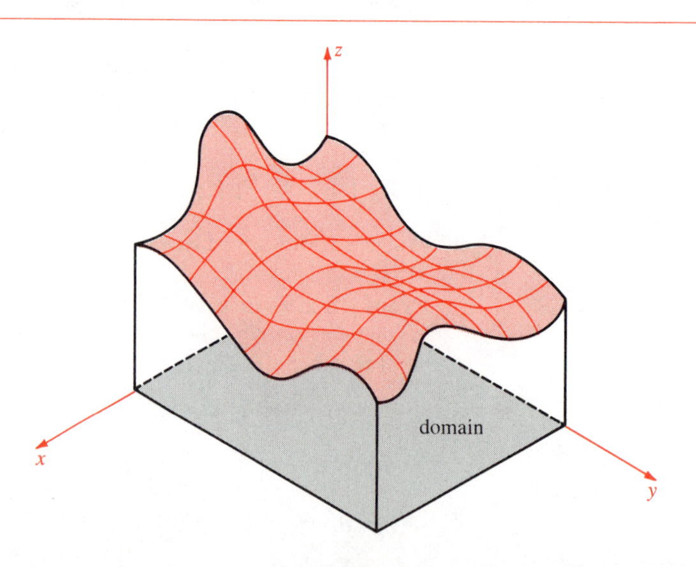

Figure 14-1-6
Graph of $z = f(x, y)$

Since a surface can be difficult to visualize, we use various techniques for picturing the graph of a function of two variables. One technique is to slice the surface by various planes and examine the cross sections.

Example 1

Graph the function $z = x^2 + y^2$.

Figure 14-1-7
Graph of $z = x^2 + y^2$

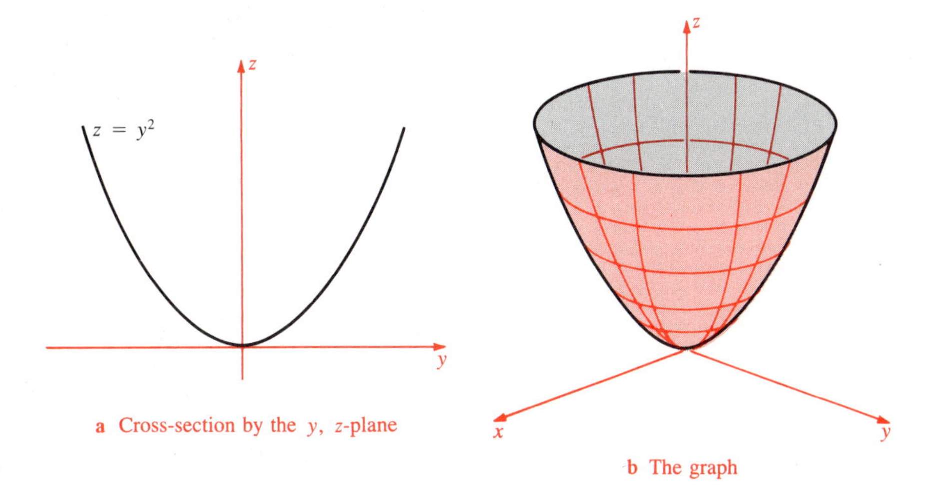

a Cross-section by the y, z-plane

b The graph

Solution Each cross section by a plane $z = c > 0$ parallel to the x, y-plane is a circle. Therefore the graph is a surface of revolution. It intersects the y, z-plane in the curve $z = y^2$, a parabola (Figure 14-1-7a). This is enough information for a sketch (Figure 14-1-7b). The surface is called a **paraboloid of revolution.** ●

Figure 14-1-8
Graph of $z = 1 - x^2$

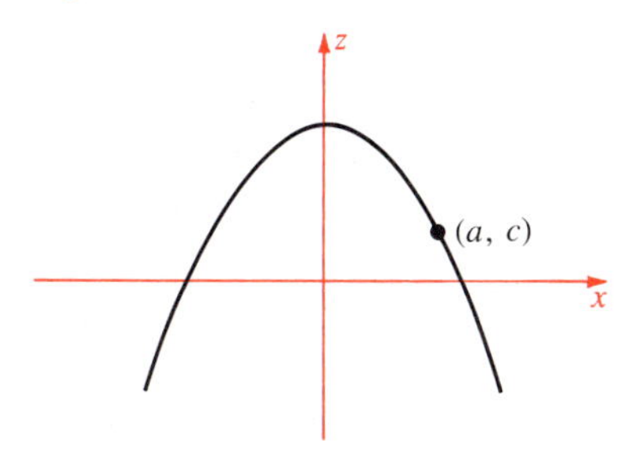

a Cross-section by the z, x-plane

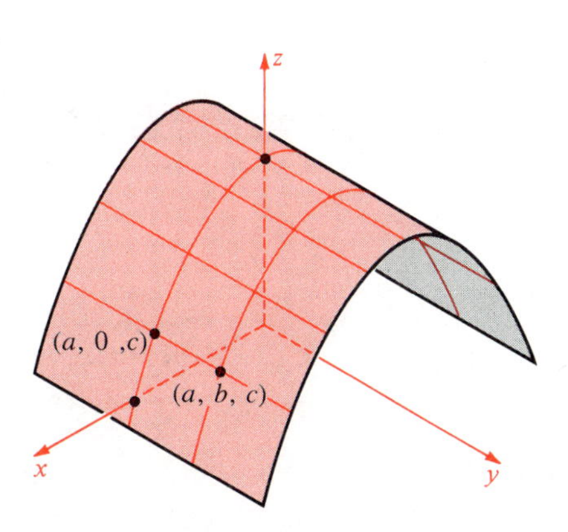

b The graph

Example 2

Graph the function $z = f(x, y) = 1 - x^2$.

Solution The function $f(x, y)$ is independent of y. Its graph is a cylinder with generators parallel to the y-axis. To see this, first graph the parabola $z = 1 - x^2$ in the x, z-plane (Figure 14-1-8a). If (a, c) is any point on this parabola and b is any value of y whatsoever, then (a, b, c) is on the graph of $z = f(x, y)$. Therefore the graph is a parabolic cylinder with generators parallel to the y-axis (Figure 14-1-8b). ●

Level Curves

A systematic way of visualizing a surface $z = f(x, y)$ is drawing its contour map. We slice the surface by planes $z = c$ at various equally spaced levels (Figure 14-1-9, next page).

Each plane $z = c$ intersects the surface in a curve. The projection of this curve onto the x, y-plane is the **level curve** or **contour line** at level c. It is the locus of $f(x, y) = c$ and indicates where the surface has "height" c. A set of level curves with equally spaced c's form a **contour map** of the surface. Where level curves are close together the surface is steep;

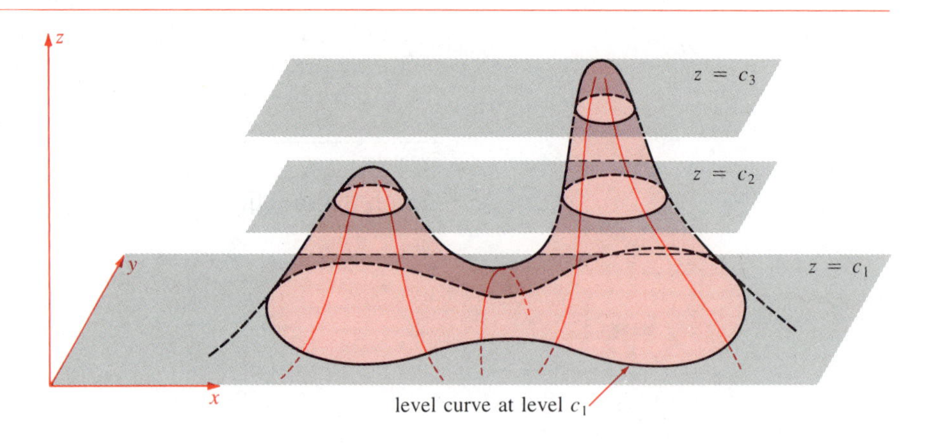

Figure 14-1-9
Level curves

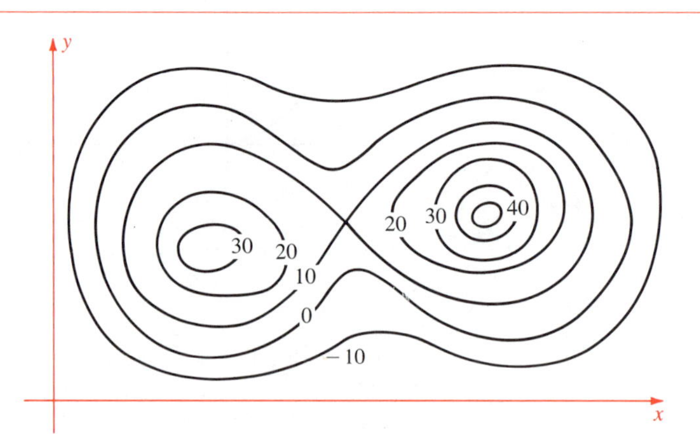

Figure 14-1-10
Contour map of a function

where they are farther apart it is flatter. Figure 14-1-10 shows a contour map of the mountain pass in Figure 14-1-9.

A contour map is the next best thing to a good drawing of a surface. Consider the surface $z = xy$ for example. A drawing is difficult (though not impossible). Still, a reasonable idea of the graph is given by a contour map, which is easy to draw; each level curve $xy = c$ is a hyperbola $(c \neq 0)$ or degenerate hyperbola $(c = 0)$. See Figure 14-1-11.

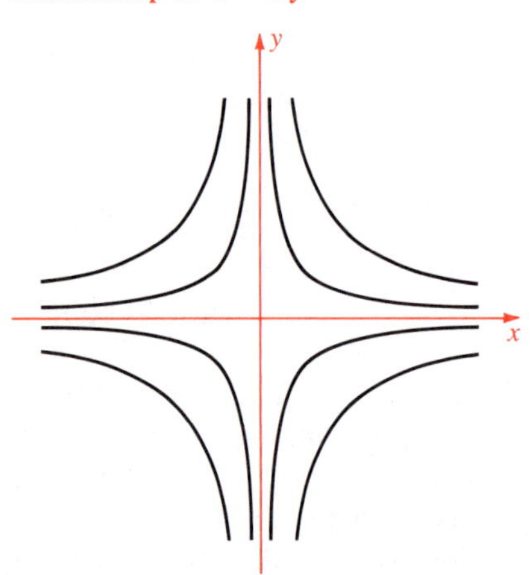

Figure 14-1-11
Contour map of $z = xy$

Level Surfaces of Functions of Three Variables

We cannot graph a function of three variables

$$w = f(x, y, z)$$

because the graph would be four-dimensional. We can, however, learn a good deal about the function by plotting in three-space the **level surfaces**

$$f(x, y, z) = k$$

For example, suppose

$$w = f(x, y, z) = x + 2y + 3z$$

Then the level surfaces of f are planes, all parallel to each other. In particular, $f(x, y, z) = 6$ is one of these planes, shown in Figure 14-1-12.

Again, consider

$$w = f(x, y, z) = x^2 + y^2 + z^2$$

Then the level surfaces of f are the spheres with center **0** (and the point **0** itself—a degenerate level surface). In particular, the level surface $f(x, y, z) = 1$ is the unit sphere (Figure 14-1-13).

Figure 14-1-12
A level surface of
$f(x, y, z) = x + 2y + 3z$

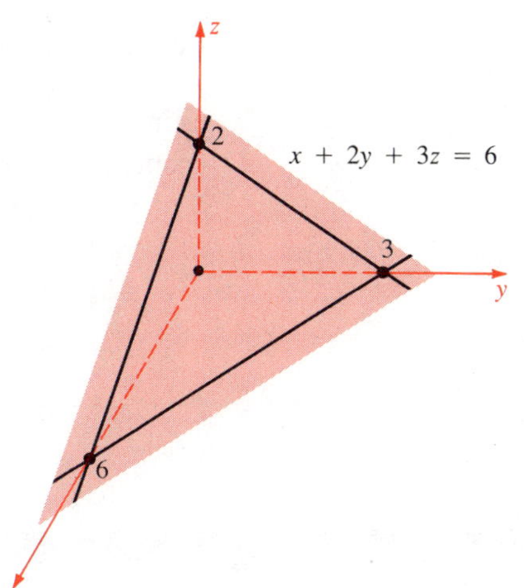

Figure 14-1-13
The sphere $x^2 + y^2 + z^2 = 1$,
a level surface of
$f(x, y, z) = x^2 + y^2 + z^2$

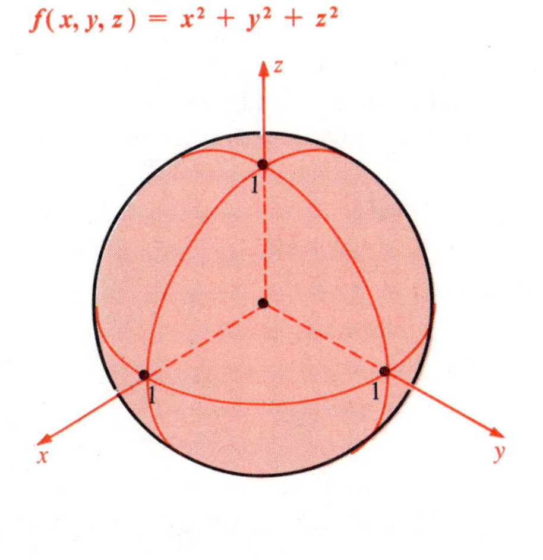

Exercises

Express as a function of two variables

1 the volume V of a cone of radius r and height h

2 the total area A of a tin can (including top and bottom) of radius r and height h

3 the distance D from $(2, 4, 3)$ to a point on the plane $x + y + z = 1$, variables x and y

4 the maximum M of $x^p e^{-qx}$ for $x \geq 0$, variables $p > 0$ and $q > 0$

Express as a function of three variables

5 the side c of a triangle, given the sides a and b and their included angle θ, in terms of a, b, and θ

6 the distance d from **0** to the plane $ax + by + cz = 1$, in terms of a, b, and c

7 the magnitude of the gravitational attraction F between masses m_1 and m_2 at distance d apart, in terms of m_1, m_2, and d

8 the value V of a deposit P after n years, interest at the annual rate $r\%$, compounded quarterly, in terms of P, n, and r

Give the domain of the function

9 $4x - 11y + 2$

10 $(x - 3y)^2 e^{xy}$

11 $\dfrac{1 + x^2y^3}{x^2 - y^2}$

12 $\ln(y - 2x)$

13 $\sqrt{x^2 - y}$

14 $\sqrt{4x^2 + 9y^2 - 36}$

15 $x \sec y$

16 $\tan(x - y)$

17 $\ln(x + 2y + 3z - 4)$

18 $1/xyz$

19 $\arcsin(x^2 + y^2 + z^2)$

20 $\dfrac{x + 2y}{x^4 - (y + 3z)^4}$

Make a contour map of the function

21 $f(x, y) = x - 3y$

22 $f(x, y) = x^2 - y^2$

23 $f(x, y) = x^2 + 4y^2$

24 $f(x, y) = |x| + |y|$

25 $f(x, y) = \ln(y - x^2)$

26 $f(x, y) = \dfrac{1}{x + y}$

Sketch the graph

27 $z = 1 - 2x$

28 $z = x^2$

29 $z = \sqrt{y}$

30 $z = x + y^2$

31 $z = x + \frac{1}{2}y$

32 $z = 1 - x^2 - y^2$

33 The gravitational potential at (x, y, z) due to a point mass at the origin is $k/\sqrt{x^2 + y^2 + z^2}$. What are the equipotential surfaces?

34 The atmospheric pressure p at sea level around the center of an anticyclone is given by $p = a^2 - (x - bt)^2 - y^2$. Plot the isobars (level curves of p). Show that the weather system is moving with constant velocity.

Suppose $\mathbf{x}_n \to \mathbf{a}$ and $\mathbf{y}_n \to \mathbf{b}$. Prove

35 $\mathbf{x}_n + \mathbf{y}_n \to \mathbf{a} + \mathbf{b}$

36 $\mathbf{x}_n \cdot \mathbf{y}_n \to \mathbf{a} \cdot \mathbf{b}$

37 If $f(x, y)$ and $g(x, y)$ are continuous, prove that their sum is continuous.

*38 If $f(x, y)$ and $g(x, y)$ are continuous, prove that $h(x, y) = f(x, y)g(x, y)$ is continuous.

39 Can $f(x, y) = \dfrac{\sin(x^2 + y^2)}{x^2 + y^2}$ be defined at $(0, 0)$ so as to be continuous?

40 Answer the same question for $f(x, y) = \dfrac{xy}{x^2 + y^2}$.

41 The line $x = -1$ is parametrized by $\mathbf{x} = (-1, \ln u)$. Similarly $x = 0$ and $x = 1$ are parametrized by $\mathbf{x} = (0, \frac{1}{2}\ln w)$ and $\mathbf{x} = (1, \ln v)$. Now a ruler is placed connecting any point $(-1, \ln u)$ of $x = -1$ with any point $(1, \ln v)$ of $x = 1$. The ruler crosses the y-axis at $(0, \frac{1}{2}\ln w)$. Express w as a function of u and v. (Write in the u, v, and w scales; the result is an example of an **alignment chart,** a useful graphical device for quick estimates of functions of two or more variables.)

42 (cont.) Construct an alignment chart for $w = \sqrt{u^2 + 2v^2}$.

14-2 Partial Derivatives

Let $z = f(x, y)$ be a function of two variables and let $\mathbf{a} = (a, b)$ be a point of its domain, not on the boundary. Suppose we set $y = b$ and allow only x to vary. Then $f(x, b)$ is a function of the single variable x, defined at least in some open interval including a. See Figure 14-2-1 for the geometry of this situation.

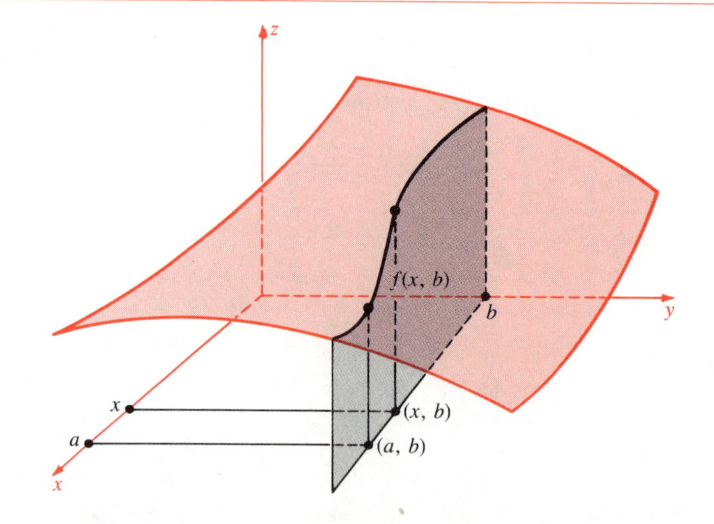

Figure 14-2-1
If $y = b$ is fixed, then $f(x, y) = f(x, b)$ is a function of x alone. We can differentiate this function of one variable.

We define

$$\frac{\partial f}{\partial x}(a, b) = \frac{d}{dx} f(x, b)\Big|_{x=a}$$

This quantity is called the **partial derivative** (or simply **partial**) of f with respect to x. (The ∂ is a curly d.) It measures the rate of change of f with respect to x while y is held constant.

We similarly define the partial derivative of f with respect to y:

$$\frac{\partial f}{\partial y}(a, b) = \frac{d}{dy} f(a, y)\Big|_{y=b}$$

In like manner, given a function $w = f(x, y, z)$ of three variables, we define the three partial derivatives $\partial w/\partial x$, $\partial w/\partial y$, and $\partial w/\partial z$. For instance,

$$\frac{\partial w}{\partial y}(a, b, c) = \frac{d}{dy} f(a, y, c)\Big|_{y=b}$$

Each of the partials is the derivative of w with respect to the variable in question, taken while all other variables are held fixed.

Example 1

Let $z = f(x, y) = xy^2$. Find

$$\frac{\partial z}{\partial x}(1, 3) \qquad \frac{\partial z}{\partial y}(-4, 2) \qquad \frac{\partial z}{\partial x} \quad \text{and} \quad \frac{\partial z}{\partial y} \quad \text{in general}$$

Solution Set $y = 3$. Then $z = f(x, 3) = x(3^2) = 9x$, so that

$$\frac{\partial z}{\partial x}(1, 3) = \frac{d}{dx}(9x)\Big|_{x=1} = 9$$

Set $x = -4$. Then $z = f(-4, y) = -4y^2$, so that

$$\frac{\partial z}{\partial y}(-4, 2) = \frac{d}{dy}(-4y^2)\Big|_{y=2} = -16$$

To compute $\partial z/\partial x$ in general, differentiate as usual, treating y as a constant:

$$\frac{\partial z}{\partial x} = \frac{\partial(xy^2)}{\partial x} = y^2 \frac{d}{dx}(x) = y^2$$

To compute $\partial z/\partial y$, differentiate while treating x as a constant:

$$\frac{\partial z}{\partial y} = \frac{\partial(xy^2)}{\partial y} = x \frac{d}{dy}(y^2) = 2xy$$

●

Example 2

a The gas law for a fixed mass of n moles of an ideal gas is $P = nRT/V$, where R is the universal gas constant. Thus P is a function of the two variables T and V. Its partial derivatives are

$$\frac{\partial P}{\partial T} = nR\frac{1}{V} \qquad \text{and} \qquad \frac{\partial P}{\partial V} = -nR\frac{T}{V^2}$$

b The area A of a parallelogram of base b, slant height s, and angle α is

$$A = f(b, s, \alpha) = bs \sin \alpha$$

a function of b, s, and α. Its partial derivatives are

$$\frac{\partial A}{\partial b} = s \sin \alpha \qquad \frac{\partial A}{\partial s} = b \sin \alpha \qquad \frac{\partial A}{\partial \alpha} = bs \cos \alpha$$

Notation There are several different notations for partial derivatives in common use. Become familiar with them; they come up again and again in applications. Suppose $w = f(x, y, z)$. Common notations for $\partial w/\partial x$ are

$$f_x \qquad f_x(x, y, z) \qquad w_x \qquad w_x(x, y, z) \qquad D_x f$$

For example, if $w = f(x, y, z) = x^3y^2\sin z$ then

$$f_x = 3x^2y^2\sin z \qquad w_y = 2x^3y \sin z \qquad D_z f = x^3y^2\cos z$$

Geometric Interpretation

The graph of $z = f(x, y)$ is a surface in three dimensions. A plane $x = a$ cuts the graph in a plane curve $x = a$, $z = f(a, y)$. See Figure 14-2-2a. The perpendicular projection of this curve onto the y, z-plane

Figure 14-2-2
Geometric interpretation of $\partial f/\partial y$

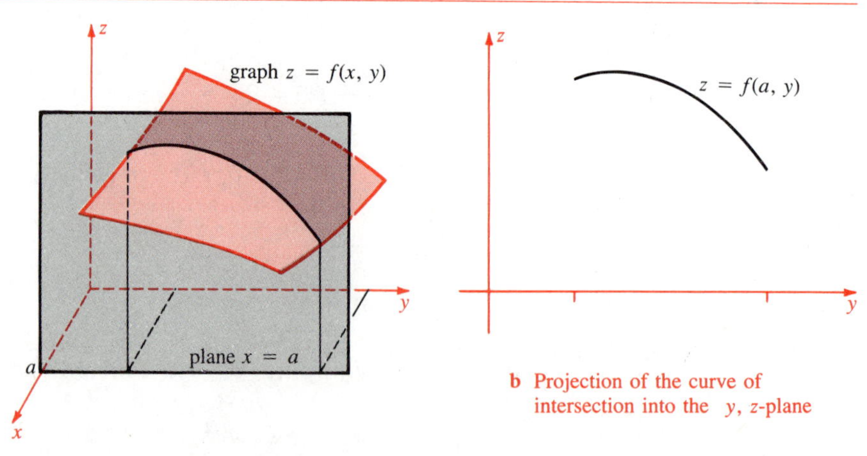

a Cross-section by plane $x = a$

b Projection of the curve of intersection into the y, z-plane

is the graph of the function $z = f(a, y)$. See Figure 14-2-2b. The partial derivative

$$\frac{\partial f}{\partial y}(a, y)$$

is the slope of this graph.

For example, suppose the graph of the function $z = x^2 + y^2$ is sliced by the plane $x = a$. The resulting plane curve is the parabola $x = a$, $z = y^2 + a^2$. Its projection onto the y, z-plane is the parabola $z = y^2 + a^2$, with slope $\partial z/\partial y = 2y$. See Figure 14-2-3.

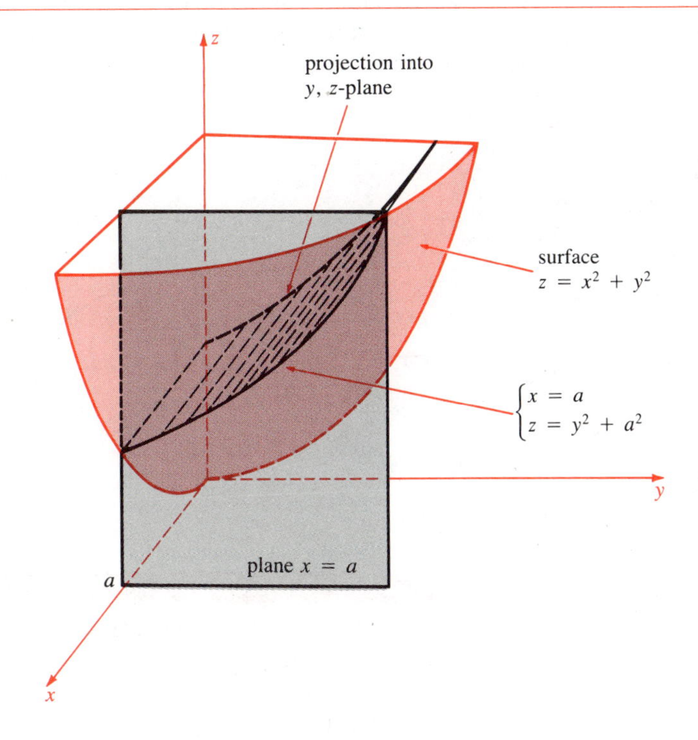

Figure 14-2-3
The projection into the y, z-plane is the plane curve $z = y^2 + a^2$.

The Chain Rule

The chain rule for functions of one variable first states that the composite of differentiable functions is itself differentiable and then gives a formula for the derivative of a composite function: If $y = f(x)$ where $x = x(t)$, then

$$\frac{dy}{dt} = \frac{dy}{dx} \cdot \frac{dx}{dt}$$

The chain rule for functions of several variables also has two parts: an assertion that a composite function built out of differentiable functions is itself differentiable, and a practical formula for computing partial derivatives of a composite function.

We consider the following special situation, which, as we shall see, is typical of the chain rule. Suppose $z = f(x, y)$ where $x = x(t)$ and $y = y(t)$. Then z is a composite function of t. In this situation, the chain rule asserts that

$$\frac{dz}{dt} = \frac{\partial z}{\partial x} \cdot \frac{dx}{dt} + \frac{\partial z}{\partial y} \cdot \frac{dy}{dt}$$

It helps to interpret this formula geometrically (Figure 14-2-4). We write $z = f(x, y)$ as $z = f(\mathbf{x})$ and write the composite function $z(t) = f[x(t), y(t)]$ as $z(t) = f[\mathbf{x}(t)]$. We think of t as time and $\mathbf{x}(t)$ as the path of a moving point in the plane. Then the composite function $f[\mathbf{x}(t)]$ assigns a number z to each instant of the motion. The chain rule is a formula for dz/dt. For instance if $f(\mathbf{x})$ is the temperature at position $\mathbf{x}$, then the chain rule tells how fast the temperature is changing as the point moves along the curve $\mathbf{x}(t)$.

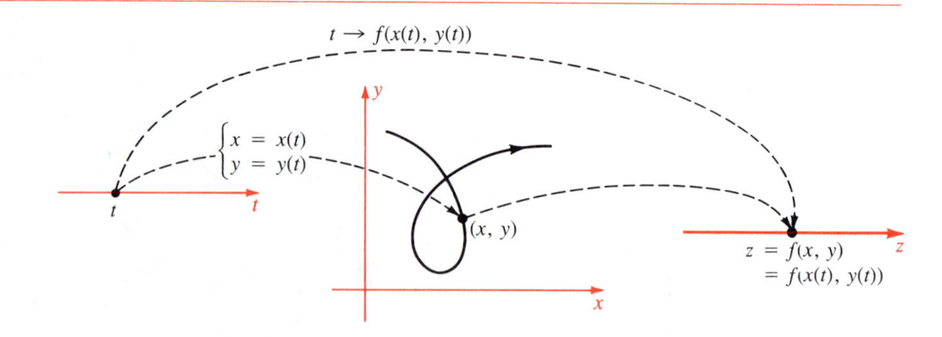

Figure 14-2-4
Chain rule. The rate of change dz/dt of z with respect to t is the sum of two terms:
$$\left(\frac{\partial f}{\partial x}\right)\left(\frac{dx}{dt}\right) \text{ and } \left(\frac{\partial f}{\partial y}\right)\left(\frac{dy}{dt}\right)$$

Chain Rule (Special Case)

Let $z = f(x, y)$ be a continuously differentiable function. That is, f is continuous and its partials $\partial f/\partial x$ and $\partial f/\partial y$ exist and are also continuous functions. Let $x = x(t)$ and $y = y(t)$ be differentiable functions of t. Then

$$\frac{dz}{dt} = \frac{\partial z}{\partial x} \cdot \frac{dx}{dt} + \frac{\partial z}{\partial y} \cdot \frac{dy}{dt}$$

where $\partial z/\partial x$ and $\partial z/\partial y$ are evaluated at $(x(t), y(t))$.

Similar rules hold for functions of more than two variables. For instance, if $w = f(x, y, z)$ is continuously differentiable, and if $x = x(t)$, $y = y(t)$, and $z = z(t)$ are differentiable, then

$$\frac{dw}{dt} = \frac{\partial w}{\partial x} \cdot \frac{dx}{dt} + \frac{\partial w}{\partial y} \cdot \frac{dy}{dt} + \frac{\partial w}{\partial z} \cdot \frac{dz}{dt}$$

The chain rule also is valid under weaker restrictions on $f(x, y)$, but we postpone a proof until Section 14-7. Until then, whenever we refer to a "continuously differentiable" function, we mean that its (first) partials exist and are continuous. Merely "differentiable" will simply mean that the function's (first) partials exist. (The word "differentiable" will be given another meaning in Section 14-7.)

Example 3

Let $z = f(x, y) = x^3y^2$, where $x = \cos t$ and $y = e^t$. Compute dz/dt at $t = 0$ by the chain rule.

Solution

$$\left.\frac{dz}{dt}\right|_0 = \frac{\partial z}{\partial x} \cdot \left.\frac{dx}{dt}\right|_0 + \frac{\partial z}{\partial y} \cdot \left.\frac{dy}{dt}\right|_0$$

where the partials are evaluated at $\mathbf{a} = (x(0), y(0)) = (1, 1)$. From $z = x^3y^2$ follows

$$\left.\frac{\partial z}{\partial x}\right|_{\mathbf{a}} = 3x^2y^2\Big|_{(1,1)} = 3 \qquad \text{and} \qquad \left.\frac{\partial z}{\partial y}\right|_{\mathbf{a}} = 2x^3y\Big|_{(1,1)} = 2$$

Therefore

$$\left.\frac{dz}{dt}\right|_0 = 3\left.\frac{dx}{dt}\right|_0 + 2\left.\frac{dy}{dt}\right|_0$$

But

$$\left.\frac{dx}{dt}\right|_0 = -\sin t\Big|_0 = 0 \qquad \text{and} \qquad \left.\frac{dy}{dt}\right|_0 = e^t\Big|_0 = 1$$

Hence

$$\left.\frac{dz}{dt}\right|_0 = 3 \cdot 0 + 2 \cdot 1 = 2$$

●

Example 4

Let $w = x^2y^3z^4$ and $x = t^4$, $y = t^3$, and $z = t^2$. Find dw/dt by the chain rule and check the answer.

Solution

$$\begin{aligned}\frac{dw}{dt} &= \frac{\partial w}{\partial x} \cdot \frac{dx}{dt} + \frac{\partial w}{\partial y} \cdot \frac{dy}{dt} + \frac{\partial w}{\partial z} \cdot \frac{dz}{dt} \\ &= (2xy^3z^4)(4t^3) + (3x^2y^2z^4)(3t^2) + (4x^2y^3z^3)(2t) \\ &= (2t^{21})(4t^3) + (3t^{22})(3t^2) + (4t^{23})(2t) = 25t^{24}\end{aligned}$$

To find dw/dt directly, we express w in terms of t and differentiate:

$$w = x^2y^3z^4 = (t^4)^2(t^3)^3(t^2)^4 = t^{25}$$

$$\frac{dw}{dt} = 25t^{24}$$

●

Suppose $z = f(x, y)$, where this time x and y are functions of *two* variables, $x = x(s, t)$ and $y = y(s, t)$. Then indirectly, z is a function of the variables s and t. There is a chain rule for computing $\partial z/\partial s$ and $\partial z/\partial t$:

Chain Rule

If $z = f(x, y)$ is a continuously differentiable function of two variables x and y, and $x = x(s, t)$ and $y(s, t)$ are differentiable, then

$$\frac{\partial z}{\partial s} = \frac{\partial z}{\partial x} \cdot \frac{\partial x}{\partial s} + \frac{\partial z}{\partial y} \cdot \frac{\partial y}{\partial s}$$

$$\frac{\partial z}{\partial t} = \frac{\partial z}{\partial x} \cdot \frac{\partial x}{\partial t} + \frac{\partial z}{\partial y} \cdot \frac{\partial y}{\partial t}$$

where $\partial z/\partial x$ and $\partial z/\partial y$ are evaluated at $(x(s, t), y(s, t))$. ●

This version of the chain rule is a consequence of the previous special case. For instance, to compute $\partial z/\partial s$, hold t fixed, making $x(s, t)$ and $y(s, t)$ effectively functions of the one variable s. Then apply the previous chain rule.

Example 5

Let $w = x^2 y$, where $x = s^2 + t^2$ and $y = \cos st$. Compute $\partial w/\partial s$.

Solution

$$\frac{\partial w}{\partial s} = \frac{\partial w}{\partial x} \cdot \frac{\partial x}{\partial s} + \frac{\partial w}{\partial y} \cdot \frac{\partial y}{\partial s} = (2xy)(2s) + (x^2)(-t \sin st)$$

$$= 2(s^2 + t^2)(\cos st)(2s) + (s^2 + t^2)^2(-t \sin st)$$

$$= (s^2 + t^2)[4s \cos st - t(s^2 + t^2) \sin st]$$ ●

The next example is important in physical applications.

Example 6

If $w = f(x, y)$, where $x = r \cos \theta$ and $y = r \sin \theta$, show that

$$\left(\frac{\partial w}{\partial x}\right)^2 + \left(\frac{\partial w}{\partial y}\right)^2 = \left(\frac{\partial w}{\partial r}\right)^2 + \frac{1}{r^2}\left(\frac{\partial w}{\partial \theta}\right)^2$$

Solution Use the chain rule to compute $\partial w/\partial r$ and $\partial w/\partial \theta$:

$$\frac{\partial w}{\partial r} = \frac{\partial w}{\partial x} \cdot \frac{\partial x}{\partial r} + \frac{\partial w}{\partial y} \cdot \frac{\partial y}{\partial r} = \frac{\partial w}{\partial x} \cos \theta + \frac{\partial w}{\partial y} \sin \theta$$

$$\frac{\partial w}{\partial \theta} = \frac{\partial w}{\partial x} \cdot \frac{\partial x}{\partial \theta} + \frac{\partial w}{\partial y} \cdot \frac{\partial y}{\partial \theta} = \frac{\partial w}{\partial x}(-r \sin \theta) + \frac{\partial w}{\partial y} r \cos \theta$$

$$= r\left(-\frac{\partial w}{\partial x} \sin \theta + \frac{\partial w}{\partial y} \cos \theta\right)$$

From these formulas follow

$$\left(\frac{\partial w}{\partial r}\right)^2 = \left(\frac{\partial w}{\partial x}\right)^2 \cos^2\theta + 2\frac{\partial w}{\partial x}\cdot\frac{\partial w}{\partial y}\sin\theta\cos\theta + \left(\frac{\partial w}{\partial y}\right)^2 \sin^2\theta$$

$$\frac{1}{r^2}\left(\frac{\partial w}{\partial \theta}\right)^2 = \left(\frac{\partial w}{\partial x}\right)^2 \sin^2\theta - 2\frac{\partial w}{\partial x}\cdot\frac{\partial w}{\partial y}\sin\theta\cos\theta + \left(\frac{\partial w}{\partial y}\right)^2 \cos^2\theta$$

Add:

$$\left(\frac{\partial w}{\partial r}\right)^2 + \frac{1}{r^2}\left(\frac{\partial w}{\partial \theta}\right)^2 = \left[\left(\frac{\partial w}{\partial x}\right)^2 + \left(\frac{\partial w}{\partial y}\right)^2\right](\cos^2\theta + \sin^2\theta) = \left(\frac{\partial w}{\partial x}\right)^2 + \left(\frac{\partial w}{\partial y}\right)^2$$

●

The most general form of the chain rule is this. Let z be a function of m variables: $z = F(y_1, \cdots, y_m)$, and let each y_i be a function of the n variables $x_1, \cdots, x_n$: namely, $y_i = y_i(x_1, \cdots, x_n)$. Then z is a composite function of the x's: $z = f(x_1, \cdots, x_n)$. With suitable assumptions about differentiability and domains, f is a differentiable function and

$$\frac{\partial z}{\partial x_j} = \sum_{i=1}^{m} \frac{\partial z}{\partial y_i}\cdot\frac{\partial y_i}{\partial x_j} \qquad \text{for } j = 1, 2, \cdots, n$$

Exercises

Find $\partial z/\partial x$ and $\partial z/\partial y$ at the point (a, b)

1 $z = x + 2y$ $\quad (0, 0)$
2 $z = 3x + 4y$ $\quad (-1, 1)$
3 $z = 3xy$ $\quad (2, -1)$
4 $z = x^2y$ $\quad (3, 1)$
5 $z = 2x^2/(y + 1)$ $\quad (1, 1)$
6 $z = x^3y^2 - 2xy^4$ $\quad (-1, -1)$
7 $z = x\sin y$ $\quad (2, \frac{1}{2}\pi)$
8 $z = y^2\cos x$ $\quad (\frac{1}{4}\pi, 1)$
9 $z = \tan 2x + \cot 3y$ $\quad (0, \frac{1}{6}\pi)$
10 $z = x\tan y + y\tan x$ $\quad (\frac{1}{4}\pi, \frac{1}{4}\pi)$

Find $\partial f/\partial x$ and $\partial f/\partial y$ for $f(x, y) =$

11 $\sin 2xy$
12 $\cos(2x + y)$
13 $\dfrac{x}{x^2 + y^2}$
14 $\dfrac{1}{x + 2y + 5}$
15 $\ln(x^2 + 3y)$
16 $\sqrt{x^2 + 3y}$
17 $e^{2x}\sin(x - y)$
18 $(1 + 2x^2 - 3y^3)^5$
19 $\sqrt{\dfrac{x - y}{x + y}}$
20 $\dfrac{(x + y)(2x + y)}{(x - y)(2x - y)}$ [Hint Use logarithmic differentiation.]

21 Let $z = x^2y$. Find $\partial z/\partial x$ for $y = 2$, and $\partial z/\partial y$ for $x = -1$.

22 Let $z = y^2/x$. Find z_x for $y = 3$.

23 Let $w = xy^2z^3$. Find w_x for $y = 2$ and $z = 2$, find w_y for $x = 1$ and $z = 0$, and find w_z for $x = y$.

24 Let $w = xy - xz - yz$. Find

$$\frac{\partial w}{\partial x} + \frac{\partial w}{\partial y} + \frac{\partial w}{\partial z}$$

Show that

25 $z = (3x - y)^2$ satisfies $\partial z/\partial x + 3\,\partial z/\partial y = 0$

26 $z = f(x) + y^2$ satisfies $\partial z/\partial y = 2y$

27 $z = x^2 - y^2$ satisfies $(\partial z/\partial x)^2 - (\partial z/\partial y)^2 = 4z$

28 $z = x^6 - x^5y + 7x^3y^3$ satisfies $x(\partial z/\partial x) + y(\partial z/\partial y) = 6z$

29 $w = \dfrac{xyz}{x^4 + y^4 + z^4}$ satisfies $xw_x + yw_y + zw_z = -w$

30 $w = e^{nr}\cos n\theta$ satisfies $(w_r^2 + w_\theta^2)\cos^2 n\theta = n^2w^2$

Find dz/dt by the chain rule

31 $z = e^{xy} \quad x = 3t + 1 \quad y = t^2$

32 $z = x/y \quad x = t + 1 \quad y = t - 1$

33 $z = x^2\cos y - x \quad x = t^2 \quad y = 1/t$

34 $z = x/y \quad x = \cos t \quad y = 1 + t^2$

Find dw/dt by the chain rule

35 $w = xyz \quad x = t^2 \quad y = t^3 \quad z = t^4$

36 $w = e^x\cos(y + z) \quad x = 1/t \quad y = t^2 \quad z = -t$

37 $w = e^{-x}y^2\sin z \quad x = t \quad y = 2t \quad z = 4t$

38 $w = (e^{-x}\sec z)/y^2 \quad x = t^2 \quad y = 1 + t \quad z = t^3$

Find $\partial z/\partial s$ and $\partial z/\partial t$ by the chain rule

39 $z = x^3/y^2 \quad x = s^2 - t \quad y = 2st$

40 $z = (x + y^2)^4 \quad x = se^t \quad y = se^{-t}$

41 $z = \sqrt{1 + x^2 + y^2} \quad x = st^2 \quad y = 1 + st$

42 $z = \exp(x^2y) \quad x = s/\sqrt{1 + t^2} \quad y = st$

43 The radius r and height h of a conical tank increase at rates $dr/dt = 0.025$ ft/hr and $dh/dt = 0.050$ ft/hr. Find the rate of increase of the volume V when $r = 6$ ft and $h = 30$ ft.

44 (cont.) Also find the rate of increase dS/dt of the lateral area at the same instant.

45 The surface area of a man of weight W kg and height H cm is

$$A = (71.84)\,W^{0.425}H^{0.725}\ \text{cm}^2$$

Find A, $\partial A/\partial W$, and $\partial A/\partial H$ for $W = 80$ and $H = 180$.

46 The wingspan W of a flying vertebrate is related to the length L of its humerus bone and variables Y and α by $W = YL^\alpha$. Find the partials of W with respect to Y, L, and α.

47 The marginal revenue M for sales of a certain good is related to the demand x for the good and to the elasticity η of the demand by $M = x(1 + \eta)$. Find the partials of M with respect to x and η.

48 Van der Waals's equation relates the pressure P, volume V, and temperature T of a fixed quantity of a gas:

$$\left(P + \frac{a}{V^2}\right)(V - b) = kT$$

where a, b, and k are constants. Assuming that this relation defines P as a function of V and T, find the partials of P with respect to V and T.

49 A metal bar of length 120 cm lies on the x-axis from $x = 0$ to $x = 120$. At each point x of the bar, the temperature at time t is

$$u(x, t) = 100e^{-kt}\sin(\pi x/120)$$

where $k > 0$. Show that the bar is cooling at all points except its ends, and that at each instant the rate of cooling at the midpoint is twice the rate at the points $\frac{1}{6}$ of the way across.

50 (cont.) Show that $u(x, t)$ satisfies $\partial u/\partial t = -ku$. Now find *all* solutions $u(x, t)$ of this partial differential equation.

51 Given $F(x, y)$, show that

$$\frac{\partial}{\partial u}F(u + v, u - v) + \frac{\partial}{\partial v}F(u + v, u - v) = 2\frac{\partial}{\partial x}F(u + v, u - v)$$

Verify the formula for $F = y\sin(xy)$.

52 Given $f(x)$, show that

$$b\frac{\partial}{\partial u}f(au + bv) = a\frac{\partial}{\partial v}f(au + bv)$$

Verify the formula for $f(x) = x^3$.

53 Let $z = \arctan(y/x)$. Given that $x = r\cos\theta$ and $y = r\sin\theta$, compute $\partial z/\partial r$ and $\partial z/\partial\theta$ by the chain rule. Check your answers by expressing z in terms of r and θ first, then differentiating.

***54** Prove that

$$\frac{d}{dx}\int_{g(x)}^{h(x)} F(t)\,dt = F[h(x)]h'(x) - F[g(x)]g'(x)$$

14-3 Gradients and Directional Derivatives

The chain rule can be expressed in a concise vector notation. Suppose $w = f(x, y, z) = f(\mathbf{x})$ and $\mathbf{x} = \mathbf{x}(t)$. According to the chain rule, the t-derivative of the composite function $w = f(\mathbf{x}(t))$ equals

$$\frac{dw}{dt} = f_x \frac{dx}{dt} + f_y \frac{dy}{dt} + f_z \frac{dz}{dt}$$

Notice that the right side is the inner product of two vectors, (f_x, f_y, f_z) and $d\mathbf{x}/dt = (dx/dt, dy/dt, dz/dt)$, so we have

$$\frac{dw}{dt} = \left(\frac{\partial f}{\partial x}, \frac{\partial f}{\partial y}, \frac{\partial f}{\partial z}\right) \cdot \left(\frac{dx}{dt}, \frac{dy}{dt}, \frac{dz}{dt}\right)$$

The first vector in the inner product depends only on the function $f(x, y, z)$. This first vector will be quite important for our work, and it has a special name, given in the following definition, which includes functions of two and three variables.

Gradient

Suppose $f(x, y)$ is differentiable on a domain $\mathbf{D}$ in the x, y-plane. Let $\mathbf{a}$ be a point of $\mathbf{D}$, not on the boundary. The **gradient** of f at $\mathbf{a}$ is the vector

$$\operatorname{grad} f(\mathbf{a}) = (f_x, f_y)\Big|_{\mathbf{a}}$$

Suppose $f(x, y, z)$ is differentiable on a domain $\mathbf{D}$ in space, and $\mathbf{a}$ is a point inside $\mathbf{D}$. Then the **gradient** of f at $\mathbf{a}$ is the vector

$$\operatorname{grad} f(\mathbf{a}) = (f_x, f_y, f_z)\Big|_{\mathbf{a}}$$

The **gradient field** of f is the assignment of the vector $\operatorname{grad} f(\mathbf{a})$ to each point $\mathbf{a}$ inside $\mathbf{D}$.

We were led to this definition from the chain rule. Let us now state the relation explicitly.

Relation of Gradient to Chain Rule

Suppose $f(\mathbf{x})$ is continuously differentiable and $\mathbf{x}(t)$ is differentiable. Then the t-derivative of the composite function $w = f(\mathbf{x}(t))$ equals

$$\frac{dw}{dt} = (\operatorname{grad} f) \cdot \frac{d\mathbf{x}}{dt}$$

Example 1

a $f(x, y) = x^2 + y^3 \qquad \operatorname{grad} f = (2x, 3y^2)$

b $f(x, y, z) = |\mathbf{x}|^2 = x^2 + y^2 + z^2 \qquad \operatorname{grad} f = (2x, 2y, 2z) = 2\mathbf{x}$

c $f(x, y, z) = xyz \qquad \operatorname{grad} f = (yz, zx, xy)$

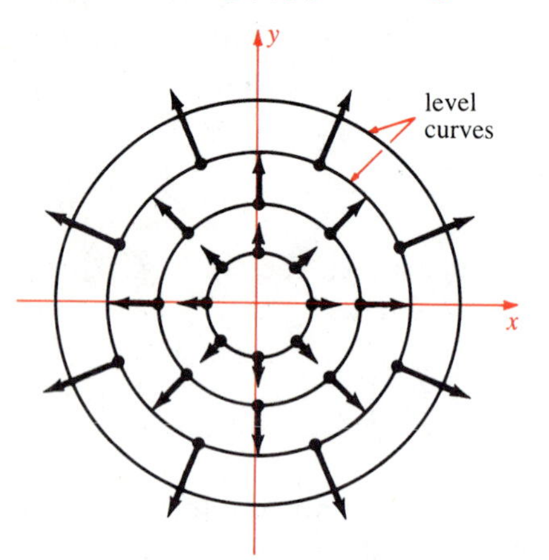

Figure 14-3-1.
Gradient field of $f(x, y) = x^2 + y^2$

We can visualize the gradient field of f by drawing the vector $\operatorname{grad} f(\mathbf{a})$ at each point $\mathbf{a}$. For example, the field $\operatorname{grad}(x^2 + y^2)$ is shown in Figure 14-3-1.

Notation A common notation for $\operatorname{grad} f$ is ∇f. The upside-down delta is called "del" or "nabla."

Gradients and Level Curves

Recall that the level curves of a function $f(x, y)$ are the curves $f(x, y) = c$ in the x, y-plane. There is an important relation between the level curves and the gradient field: at each point of a level curve, the gradient vector is perpendicular to the curve.

Theorem

The field $\operatorname{grad} f$ of a continuously differentiable function f is orthogonal (perpendicular) to the level curves of f.

Before giving a proof, let us note that Figure 14-3-1 illustrates this statement. There $f(x, y) = x^2 + y^2$, so the level curves are the circles $x^2 + y^2 = a^2$, while the gradient field is $\operatorname{grad} f(x, y) = 2\mathbf{x}$. At each $\mathbf{x}$, the vector $\operatorname{grad} f(\mathbf{x})$ points along the radius, hence is orthogonal to the circle.

Proof Suppose a point moves along a level curve $f(x, y) = c$. Let its position at time t be $\mathbf{x} = \mathbf{x}(t)$. Then the composite function $f[\mathbf{x}(t)] = c$ is constant, so its time derivative is 0. By the chain rule

$$\frac{df}{dt} = [\operatorname{grad} f(\mathbf{x})] \cdot \frac{d\mathbf{x}}{dt}$$

Since $df/dt = 0$, we see that $\operatorname{grad} f(\mathbf{x})$ is perpendicular to $d\mathbf{x}/dt$, the vector tangent to the level curve at the point in question.

Here is a practical application of the orthogonality of gradients and level curves. Suppose we want an equation for the tangent line to a level curve $f(x, y) = c$ at a point $\mathbf{a}$. This tangent line passes through $\mathbf{a}$ and is perpendicular to $\operatorname{grad} f(\mathbf{a})$. Hence an equation is

$$[\operatorname{grad} f(\mathbf{a})] \cdot (\mathbf{x} - \mathbf{a}) = 0$$

Tangent Line

Let $\mathbf{a} = (a, b)$ be a point of the level curve $f(x, y) = c$. Then the tangent line to the level curve at $\mathbf{a}$ is

$$[\operatorname{grad} f(\mathbf{a})] \cdot (\mathbf{x} - \mathbf{a}) = 0$$

In coordinates,

$$\frac{\partial f}{\partial x} \cdot (x - a) + \frac{\partial f}{\partial y} \cdot (y - b) = 0$$

where the partials $\partial f/\partial x$ and $\partial f/\partial y$ are evaluated at (a, b).

Example 2

Find the tangent to the cubic $x^2 = y^3$ at $(-1, 1)$.

Solution Set $f(x, y) = x^2 - y^3$, so the cubic is the level curve $f = 0$. Then $\mathbf{a} = (-1, 1)$ and

$$\operatorname{grad} f(\mathbf{a}) = (2x, -3y^2)\Big|_{(-1,1)} = (-2, -3)$$

Hence the tangent is

$$(-2, -3) \cdot [(x, y) - (-1, 1)] = 0$$
$$(-2, -3) \cdot (x + 1, y - 1) = 0$$
$$-2(x + 1) - 3(y - 1) = 0 \qquad \text{that is} \qquad 2x + 3y = 1$$

See Figure 14-3-2.

Figure 14-3-2
$f(x, y) = x^2 - y^3$. See Example 2.

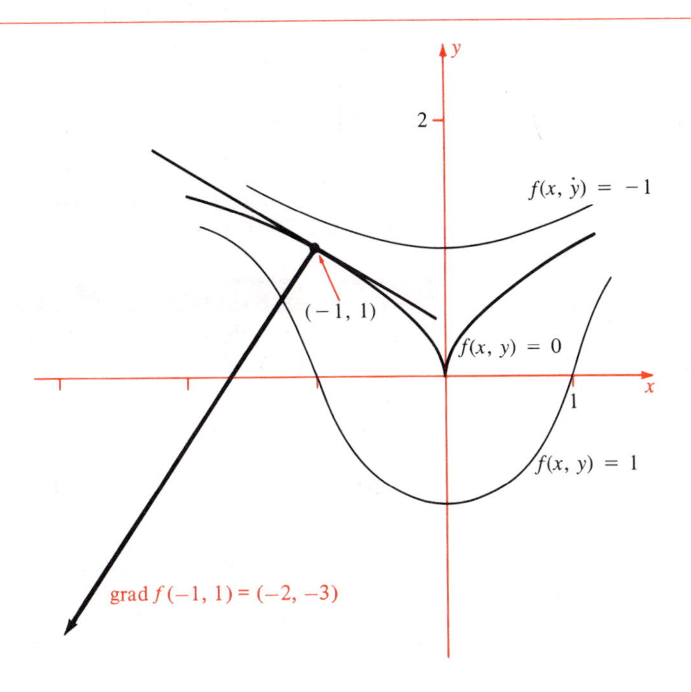

Directional Derivatives

The partial derivatives of a function give the rates of change of the function in directions parallel to the axes. Sometimes, however, we need the rate of change of a function in an arbitrary direction, that is, a *directional* derivative.

A direction is given by a unit vector $\mathbf{u}$. Suppose $f(x, y, z)$ is a function and $\mathbf{a}$ is a point of its domain. Imagine a point moving with constant velocity $\mathbf{u}$ along a straight line, passing through $\mathbf{a}$ when

Figure 14-3-3
Directional derivative of $f(x, y, z)$ at $\mathbf{a}$ in the direction of the unit vector $\mathbf{u}$:

$$D_{\mathbf{u}}f(\mathbf{a}) = \frac{d}{dt}f(\mathbf{a} + t\mathbf{u})\Big|_0$$

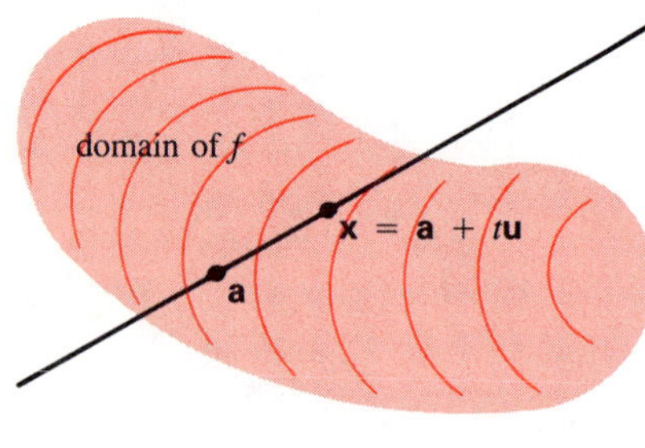

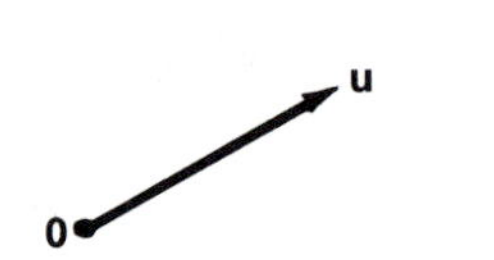

$t = 0$. See Figure 14-3-3. To each point $\mathbf{a} + t\mathbf{u}$ of the path is assigned the number $w(t) = f(\mathbf{a} + t\mathbf{u})$. We define the directional derivative of f at $\mathbf{a}$ in the direction $\mathbf{u}$ to be dw/dt at $t = 0$.

Definition Directional Derivative

The **directional derivative** of a continuously differentiable function $f(x, y, z)$ at a point $\mathbf{a}$ in the direction $\mathbf{u}$ is

$$D_{\mathbf{u}}f(\mathbf{a}) = \frac{d}{dt}f(\mathbf{a} + t\mathbf{u})\Big|_{t=0}$$

For example, suppose $f(x, y, z)$ is the steady temperature at each point (x, y, z) of a fluid. Suppose a particle moves with unit speed through a point $\mathbf{a}$ in the direction $\mathbf{u}$. Then $D_{\mathbf{u}}f(\mathbf{a})$ measures the time rate of change of the particle's temperature.

To compute the directional derivative, write

$$\mathbf{x}(t) = \mathbf{a} + t\mathbf{u}$$

so that $w(t) = f[\mathbf{x}(t)]$. By definition, $D_{\mathbf{u}}f(\mathbf{a}) = dw/dt$ at $t = 0$. By the chain rule,

$$\frac{dw}{dt}\Big|_0 = [\operatorname{grad} f(\mathbf{a})] \cdot \frac{d\mathbf{x}}{dt}\Big|_0 = [\operatorname{grad} f(\mathbf{a})] \cdot \mathbf{u}$$

Since $\mathbf{u}$ is a unit vector, $(\operatorname{grad} f) \cdot \mathbf{u}$ equals the projection of the vector $\operatorname{grad} f$ on $\mathbf{u}$.

Theorem

The directional derivative of a continuously differentiable function f in the direction $\mathbf{u}$ is the projection of $\operatorname{grad} f$ on $\mathbf{u}$:

$$D_{\mathbf{u}}f(\mathbf{a}) = [\operatorname{grad} f(\mathbf{a})] \cdot \mathbf{u}$$

In particular if $\mathbf{u} = \mathbf{i} = (1, 0, 0)$, then

$$D_{\mathbf{i}}f(\mathbf{a}) = (\operatorname{grad} f) \cdot (1, 0, 0) = \left(\frac{\partial f}{\partial x}, \frac{\partial f}{\partial y}, \frac{\partial f}{\partial z}\right) \cdot (1, 0, 0)$$

$$= \frac{\partial f}{\partial x}\Big|_{\mathbf{a}}$$

A similar situation holds for $\mathbf{j} = (0, 1, 0)$ and $\mathbf{k} = (0, 0, 1)$. Therefore we have the following result:

Corollary The directional derivatives of $f(x, y, z)$ in the directions $\mathbf{i}$, $\mathbf{j}$, and $\mathbf{k}$ are the partial derivatives:

$$D_{\mathbf{i}}f = \frac{\partial f}{\partial x} \qquad D_{\mathbf{j}}f = \frac{\partial f}{\partial y} \qquad D_{\mathbf{k}}f = \frac{\partial f}{\partial z}$$

Completely analogous results hold for functions of two variables.

Example 3

Compute the directional derivatives of $f(x, y, z) = xy^2z^3$ at $(3, 2, 1)$, in the direction of the vectors

a $(-2, -1, 0)$ **b** $(5, 4, 1)$

Solution

$$D_{\mathbf{u}}f(\mathbf{a}) = [\operatorname{grad} f(\mathbf{a})] \cdot \mathbf{u}$$

where $\mathbf{u}$ is a unit vector in the desired direction and $\mathbf{a} = (3, 2, 1)$. Now

$$\operatorname{grad} f(\mathbf{a}) = (y^2z^3, 2xyz^3, 3xy^2z^2)\Big|_{(3,2,1)} = (4, 12, 36)$$

Hence

$$D_{\mathbf{u}}f(\mathbf{x}) = (4, 12, 36) \cdot \mathbf{u}$$

Cases **a** and **b** are

a $\mathbf{u} = \dfrac{1}{\sqrt{5}}(-2, -1, 0)$

$$D_{\mathbf{u}}f(\mathbf{x}) = (4, 12, 36) \cdot \frac{1}{\sqrt{5}}(-2, -1, 0) = \frac{-20}{\sqrt{5}}$$

b $\mathbf{u} = \dfrac{1}{\sqrt{42}}(5, 4, 1)$

$$D_{\mathbf{u}}f(\mathbf{x}) = (4, 12, 36) \cdot \frac{1}{\sqrt{42}}(5, 4, 1) = \frac{104}{\sqrt{42}}$$

●

Directions of Most Rapid Increase and Decrease

In which direction is a continuously differentiable function $f(\mathbf{x})$ increasing fastest? decreasing fastest? In other words, given a point $\mathbf{a}$ in space, for which unit vector $\mathbf{u}$ is $D_{\mathbf{u}}f(\mathbf{a})$ largest? smallest? The answers follow immediately from the formula

$$D_{\mathbf{u}}f(\mathbf{a}) = [\operatorname{grad} f(\mathbf{a})] \cdot \mathbf{u} = |\operatorname{grad} f(\mathbf{a})| \cos \theta$$

where θ is the angle between $\operatorname{grad} f(\mathbf{a})$ and $\mathbf{u}$. Therefore the largest value of $D_{\mathbf{u}}f(\mathbf{a})$ is $|\operatorname{grad} f(\mathbf{a})|$, taken where $\cos \theta = 1$, that is, for $\theta = 0$. The smallest value is $-|\operatorname{grad} f(\mathbf{a})|$, taken in exactly the opposite direction, that is, for $\theta = \pi$.

Theorem

The direction of most rapid increase of a continuously differentiable function $f(x, y, z)$ at a point $\mathbf{a}$ is the direction of the gradient. The directional derivative in that direction is $|\operatorname{grad} f(\mathbf{a})|$.

The direction of most rapid decrease is opposite to the direction of the gradient. The directional derivative in that direction is $-|\operatorname{grad} f(\mathbf{a})|$. ●

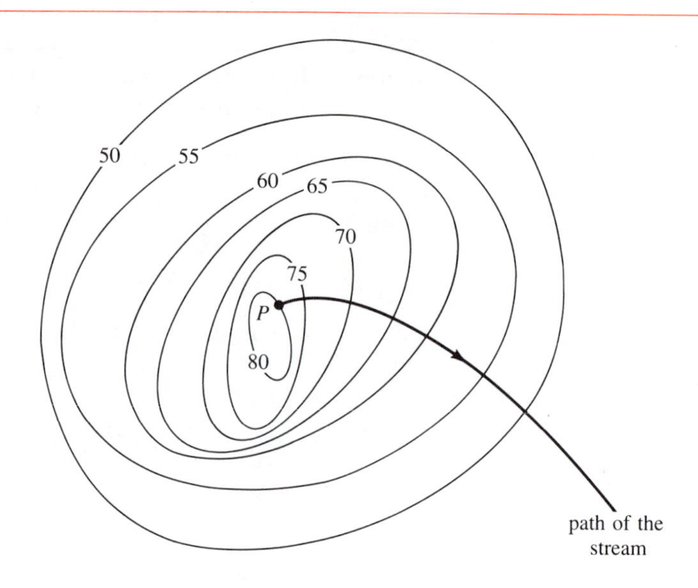

Figure 14-3-4
Contour map; path of quickest descent from spring at P

Here is an application of the direction of most rapid decrease. Suppose on a contour map (Figure 14-3-4) you want to plot the path of water flowing downhill from a spring at P. A physical principle says that gravity will cause the water to flow in such a way that its potential energy decreases as rapidly as possible. But the potential energy of a water particle equals its mass (constant) times its height. Therefore the particle "chooses" the direction of steepest descent (most rapid change of altitude). If the hill is represented by the surface $z = f(x, y)$, then water will flow in the direction of $-\text{grad}\, f$, that is, *perpendicular to the level curves.*

Example 4

Find the direction of most rapid increase of $f(x, y, z) = x^2 + yz$ at $(1, 1, 1)$ and give the rate of increase in this direction.

Solution

$$\text{grad}\, f\Big|_{(1,1,1)} = (2x, z, y)\Big|_{(1,1,1)} = (2, 1, 1)$$

The most rapid increase is

$$D_{\mathbf{u}} f = |\text{grad}\, f| = \sqrt{2^2 + 1^2 + 1^2} = \sqrt{6}$$

where $\mathbf{u}$ is the direction of most rapid increase, the direction of $\text{grad}\, f$:

$$\mathbf{u} = \frac{\text{grad}\, f}{|\text{grad}\, f|} = \frac{1}{\sqrt{6}}(2, 1, 1)$$

●

It is customary to define $D_{\mathbf{v}} f$ for *any* vector $\mathbf{v}$, not just unit vectors. Formally, the definition is the same as before for a continuously differentiable function f:

General Directional Derivative

$$D_{\mathbf{v}}f(\mathbf{a}) = \frac{d}{dt}f(\mathbf{a} + t\mathbf{v})\bigg|_{t=0}$$

We shall refer to this derivative as $D_{\mathbf{v}}f(\mathbf{a})$ and reserve the name *directional derivative* for $D_{\mathbf{u}}f(\mathbf{a})$, where $\mathbf{u}$ is a unit vector. By the chain rule,

$$D_{\mathbf{v}}f(\mathbf{a}) = [\operatorname{grad} f(\mathbf{a})] \cdot \mathbf{v}$$

Suppose $\mathbf{v} = k\mathbf{u}$, where $\mathbf{u}$ is a unit vector and $k = |\mathbf{v}|$. Then

$$\begin{aligned} D_{\mathbf{v}}f(\mathbf{a}) &= D_{k\mathbf{u}}f(\mathbf{a}) = [\operatorname{grad} f(\mathbf{a})] \cdot (k\mathbf{u}) \\ &= k[\operatorname{grad} f(\mathbf{a})] \cdot \mathbf{u} = kD_{\mathbf{u}}f(\mathbf{a}) \end{aligned}$$

Therefore

$$D_{\mathbf{v}}f(\mathbf{x}) = |\mathbf{v}|D_{\mathbf{u}}f(\mathbf{x})$$

Hence, for a fixed vector $\mathbf{v}$ the derivative $D_{\mathbf{v}}f(\mathbf{x})$ is nothing terribly new: it is simply proportional to the directional derivative $D_{\mathbf{u}}f(\mathbf{x})$ in the direction $\mathbf{u}$ of $\mathbf{v}$.

Example 5

$$f(\mathbf{x}) = xyz \qquad \mathbf{a} = (2, 1, -1) \qquad \mathbf{v} = (1, 1, 2)$$

$$\begin{aligned} D_{\mathbf{v}}f(\mathbf{a}) &= \left[\operatorname{grad} f(\mathbf{x})\bigg|_{(2,1,-1)}\right] \cdot (1, 1, 2) \\ &= \left[(yz, zx, xy)\bigg|_{(2,1,-1)}\right] \cdot (1, 1, 2) \\ &= (-1, -2, 2) \cdot (1, 1, 2) = 1 \end{aligned}$$

Let us look again at the definition

$$D_{\mathbf{v}}f(\mathbf{a}) = \frac{d}{dt}f(\mathbf{a} + t\mathbf{v})\bigg|_{t=0}$$

It gives special importance to the *line*

$$\mathbf{x} = \mathbf{x}(t) = \mathbf{a} + t\mathbf{v}$$

which passes through $\mathbf{a}$ with velocity $\mathbf{v}$. Consider instead *any curve* $\mathbf{x} = \mathbf{x}(t)$ that passes through $\mathbf{x} = \mathbf{a}$ at time $t = 0$ with velocity $\mathbf{v}$:

$$\mathbf{x}(0) = \mathbf{a} \qquad \text{and} \qquad \frac{d\mathbf{x}}{dt}\bigg|_{t=0} = \mathbf{v}$$

Now differentiate the composite function $f[\mathbf{x}(t)]$ at $t = 0$:

$$\begin{aligned} \frac{d}{dt}f[\mathbf{x}(t)]\bigg|_{t=0} &= \left[\operatorname{grad} f\bigg|_{\mathbf{x}(0)}\right] \cdot \frac{d\mathbf{x}}{dt}\bigg|_{t=0} \\ &= [\operatorname{grad} f(\mathbf{a})] \cdot \mathbf{v} = D_{\mathbf{v}}f(\mathbf{a}) \end{aligned}$$

The result is completely independent of what curve $\mathbf{x}(t)$ you take; it depends only on $\mathbf{x}(0) = \mathbf{a}$ and $(d\mathbf{x}/dt)_0 = \mathbf{v}$.

- Let $\mathbf{x}(t)$ be any differentiable curve such that $\mathbf{x}(0) = \mathbf{a}$ and $(d\mathbf{x}/dt)_0 = \mathbf{v}$. Then

$$\frac{d}{dt} f[\mathbf{x}(t)]\bigg|_{t=0} = D_{\mathbf{v}} f(\mathbf{a})$$

Example 6

$$f(x, y) = x^2 y \qquad \mathbf{a} = (1, 0) \qquad \mathbf{v} = (0, 1)$$

$$\mathbf{x}(t) = (\cos t, \sin t)$$

On the one hand

$$D_{\mathbf{v}} f(\mathbf{a}) = [\operatorname{grad} f(\mathbf{a})] \cdot \mathbf{v} = \left[(2xy, x^2)\Big|_{(1,0)} \right] \cdot (0, 1)$$

$$= (0, 1) \cdot (0, 1) = 1$$

On the other hand, the curve $\mathbf{x}(t)$ satisfies

$$\mathbf{x}(0) = (1, 0) = \mathbf{a} \qquad d\mathbf{x}/dt\Big|_0 = (0, 1) = \mathbf{v}$$

Now calculate the derivative of $f[\mathbf{x}(t)] = \cos^2 t \sin t$ directly:

$$\frac{d}{dt} f[\mathbf{x}(t)]\bigg|_{t=0} = (-2\cos t \sin^2 t + \cos^3 t)\Big|_{t=0} = \cos^3 0 = 1$$

Exercises

Compute $\operatorname{grad} f$

1 $f = x^2 y + 3xy^3$

2 $f = y^2 e^{xy}$

3 $f = \dfrac{ax + by}{cx + dy}$

4 $f = \sqrt{x^2 + 3y^2}$

5 $f = (x^2 + y^2)e^z$

6 $f = \log(3x - y - 4z)$

7 $f = \sqrt{1 + x^2 y^2 z^4}$

8 $f = x^2 \cos(yz)$

9 Let $f(x, y, z) = x^2 + y^2 + z^2 - 2xy + 3yz + 6zx$. Find all points where $\operatorname{grad} f$ is parallel to the x, y-plane.

10 Let $z = 1/(x^2 + y^2 + 10)$. Show that $\operatorname{grad} z$ points in the direction of $-\mathbf{x}$ at all points $\mathbf{x} \neq \mathbf{0}$.

11 Suppose $z = f(r, \theta)$ is given in terms of polar coordinates. Show that

$$\operatorname{grad} z = f_r \mathbf{u} + r^{-1} f_\theta \mathbf{w}$$

where $\mathbf{u} = (\cos\theta, \sin\theta)$ and $\mathbf{w} = (-\sin\theta, \cos\theta)$.

12 (cont.) Find $\operatorname{grad}(r^{-2}\cos 2\theta)$.

13 (cont.) Use Exercise 11 to do Exercise 10.

14 (cont.) Let $f(x, y) = \arctan(y/x)$. Compute $\operatorname{grad} f$ in both rectangular and polar coordinates. Show that your results agree.

15 Suppose $\operatorname{grad} f(a, b) \cdot \operatorname{grad} g(a, b) = 0$ and the two gradients are non-zero. Translate this into a statement about level curves.

16 (cont.) What can you say if the condition on the gradients is replaced by $\operatorname{grad} f(a, b) = \lambda \operatorname{grad} g(a, b)$, where λ is a scalar?

Find the tangent line to

17 $x^2 - 3xy + y^2 = -1$ at $(1, 2)$

18 $x + x^3 y^4 - y = 0$ at $(0, 0)$

19 $y + \sin xy = 1$ at $(0, 1)$

20 $x\sqrt{3x + y} = 6$ at $(2, 3)$

Compute the directional derivatives at $\mathbf{a}$ in the directions of $\mathbf{v}_1$ and $\mathbf{v}_2$

21 $f(x, y) = e^x \cos y$
$\mathbf{a} = \mathbf{0} \quad \mathbf{v}_1 = (1, 0) \quad \mathbf{v}_2 = (0, 1)$

22 $f(x, y) = \ln(x + 2y)$
$\mathbf{a} = (0, 1) \quad \mathbf{v}_1 = (1, 1) \quad \mathbf{v}_2 = (3, 4)$

23 $f(x, y, z) = xyz$
$\mathbf{a} = (1, -1, 2) \quad \mathbf{v}_1 = (1, 1, 0) \quad \mathbf{v}_2 = (1, 0, 1)$

24 $f(x, y, z) = \sqrt{x + 3y + 5z}$
$\mathbf{a} = (1, 1, 1) \quad \mathbf{v}_1 = (1, 1, -1) \quad \mathbf{v}_2 = (1, 2, 3)$

Find the largest directional derivative of $f(x, y, z)$ at $\mathbf{a}$:

25 $f(x, y, z) = x^3 + y^2 + z \quad \mathbf{a} = \mathbf{0}$

26 $f(x, y, z) = x^2 - y^2 + 4z^2 \quad \mathbf{a} = (-1, -1, 1)$

Find all unit vectors $\mathbf{u}$ for which $D_{\mathbf{u}}f(\mathbf{a}) = 0$:

27 $f(x, y) = (2 + xy^2)^5 \quad \mathbf{a} = (0, 1)$

28 $f(x, y, z) = x^2 + xy + yz \quad \mathbf{a} = (-1, 1, 1)$

29 Given $f(x, y)$ and a point $\mathbf{a}$, show that the value of $[D_{\mathbf{u}}f(\mathbf{a})]^2 + [D_{\mathbf{v}}f(\mathbf{a})]^2$ is constant for all pairs of perpendicular unit vectors $\mathbf{u}$ and $\mathbf{v}$. What is the constant value?

*30 (cont.) Given k, show that there exist perpendicular unit vectors $\mathbf{u}$ and $\mathbf{v}$ such that $D_{\mathbf{v}}f(\mathbf{a}) = kD_{\mathbf{u}}f(\mathbf{a})$.

Compute the general directional derivative $D_{\mathbf{v}}f(\mathbf{a})$

31 $f(x, y) = 1/(2x + 5y)$
$\mathbf{a} = (2, 1) \quad \mathbf{v} = (-3, 4)$

32 $f(x, y, z) = xye^{yz} \quad \mathbf{a} = (1, 1, 1) \quad \mathbf{v} = (2, 2, 5)$

33 Through each point $\mathbf{a} \neq \mathbf{0}$ of the plane pass a level curve of xy and a level curve of $x^2 - y^2$. Find their angle of intersection.

34 Show that the curves of steepest ascent for the function $f(x, y) = \frac{1}{2}x^2 + y^2$ are the parabolas $y = kx^2$. Draw a contour map of f to see if this looks reasonable.

Prove

35 $D_{\mathbf{v}+\mathbf{w}}f(\mathbf{a}) = D_{\mathbf{v}}f(\mathbf{a}) + D_{\mathbf{w}}f(\mathbf{a})$

36 $D_{\mathbf{v}}(fg)(\mathbf{a}) = f(\mathbf{a})D_{\mathbf{v}}g(\mathbf{a}) + g(\mathbf{a})D_{\mathbf{v}}f(\mathbf{a})$

A function $w = f(x, y, z)$ is **homogeneous of degree n** if $f(tx, ty, tz) = t^n f(x, y, z)$ for all $t > 0$. The condition of homogeneity can be written vectorially:

$$f(t\mathbf{x}) = t^n f(\mathbf{x})$$

Show that the function is homogeneous: What degree?

37 $x^2 + yz$

38 $x - y + 2z$

39 $x^3 + y^3 + z^3 - 3xyz$

40 $x^2 e^{-y/z}$

41 $\dfrac{xyz}{x^4 + y^4 + z^4}$

42 $\dfrac{1}{x + y}$

43 Suppose f and g are homogeneous of degree m and n respectively. Show that fg is homogeneous of degree mn.

44 Let $f(x, y, z)$ be homogeneous of degree n. Show that f_x is homogeneous of degree $n - 1$. (Exception: $n = 0$ and f constant.)

45 Let $f(x, y, z)$ be homogeneous of degree n. Prove **Euler's relation:** $xf_x + yf_y + zf_z = nf$. [Hint Differentiate $f(tx, ty, tz) = t^n f(x, y, z)$ with respect to t, using the chain rule; then set $t = 1$.]

*46 (cont.) (Converse of Euler's relation) Let $f(\mathbf{x})$ be differentiable for $\mathbf{x} \neq \mathbf{0}$, and suppose $\mathbf{x} \cdot \operatorname{grad} f = nf$. Prove f is homogeneous of degree n. [Hint Show that $\partial[t^{-n}f(t\mathbf{x})]/\partial t = 0$.]

14-4 Surfaces and Tangent Planes

In this section we shall study surfaces. The simplest way for a surface to be presented is as the graph of a function $z = f(x, y)$ of two variables. A second way for a surface to be presented is as a level surface $F(x, y, z) = k$ of a continuously differentiable function of three variables. A level surface is the graph of a *relation* rather than the graph of a *function.* Actually, the graph of a function is a special case of the graph of a relation. Indeed, the graph of the *function* $z = f(x, y)$ is the same as the graph of the *relation*

$$f(x, y) - z = 0$$

which is the level surface at level 0 of the function

$$F(x, y, z) = f(x, y) - z$$

of three variables. Therefore we concentrate first on surfaces of the form $F(x, y, z) = k$ and later apply our findings to the special case of the graph of a function.

Examples show that points on such a surface where $\operatorname{grad} F = \mathbf{0}$ are generally singularities like peaks, corners, places where the surface intersects itself, and ridges. See Figures 14-4-1, 14-4-2, and 14-4-3. The geometry of surfaces near such points is difficult to study and beyond the scope of this course. We simply rule such points out of bounds, so we only consider points $\mathbf{a}$ such that $F(\mathbf{a}) = k$ and $\operatorname{grad} F(\mathbf{a}) \neq \mathbf{0}$.

Figure 14-4-1
$F(x, y, z) = x^2 + y^2 - (z - 2)^2$
Level surface: $F(x, y, z) = 0$
$\operatorname{grad} F(0, 0, 2) = \mathbf{0}$
Peak at $(0, 0, 2)$

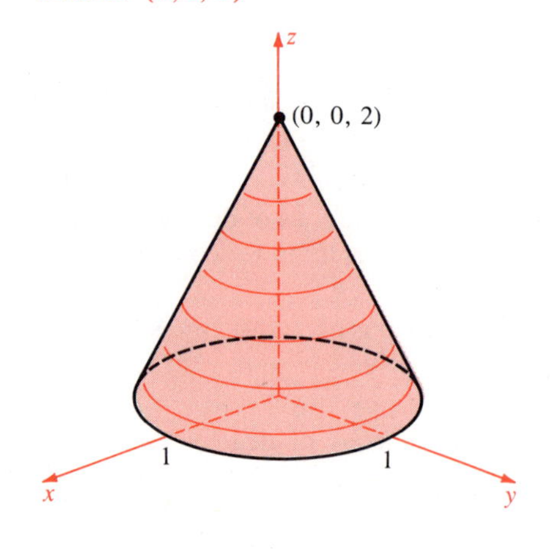

Figure 14-4-2
$F(x, y, z) = xy$
Level surface: $F(x, y, z) = 0$
$\operatorname{grad} F(0, 0, z) = \mathbf{0}$
Self-intersection along z-axis

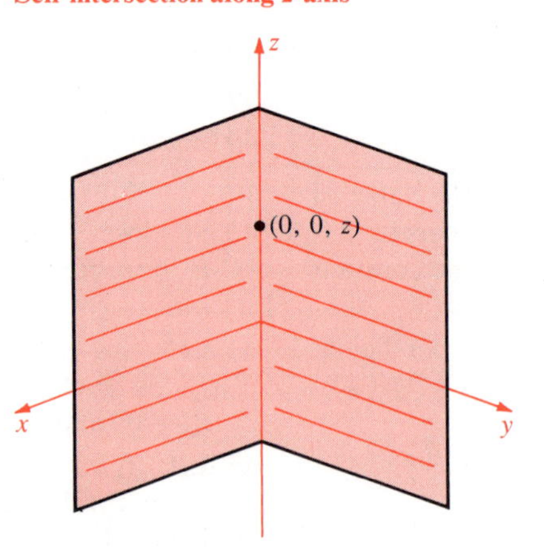

Figure 14-4-3
$F(x, y, z) = x^3 - y^2$
Level surface: $F(x, y, z) = 0$
$\operatorname{grad} f(0, 0, z) = \mathbf{0}$
Ridge along z-axis

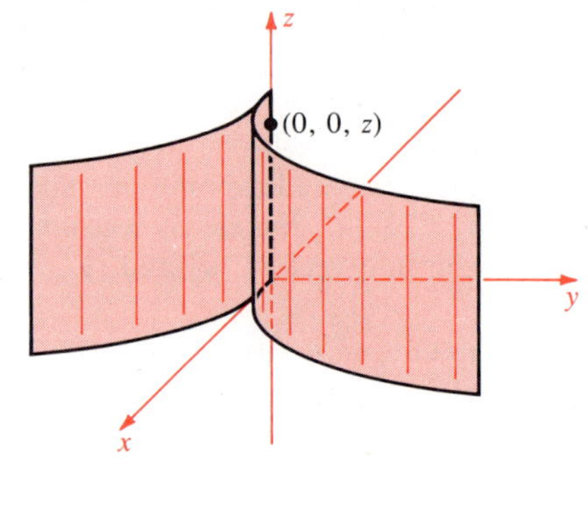

Example 1

a $F(x, y, z) = x^2 + y^2 + z^2 \qquad k = a^2 > 0$

The surface $F(x, y, z) = a^2$ is the sphere (Figure 14-4-4)

$$x^2 + y^2 + z^2 = a^2$$

with center $\mathbf{0}$ and radius a. In vector notation

$$F(\mathbf{x}) = |\mathbf{x}|^2 \qquad \operatorname{grad} F(\mathbf{x}) = (2x, 2y, 2z) = 2\mathbf{x}$$

and $F(\mathbf{x}) = a^2$ is equivalent to $|\mathbf{x}| = a$. For each $\mathbf{x}$ on the sphere,

$$\operatorname{grad} F(\mathbf{x}) = 2\mathbf{x} \neq \mathbf{0}$$

since $|\mathbf{x}| = a > 0$.

b $F(x, y, z) = z - x^2 - y^2$.

The surface $F(x, y, z) = 0$ is

$$z = x^2 + y^2 \qquad \text{and} \qquad \operatorname{grad} F(\mathbf{x}) = (-2x, -2y, 1) \neq \mathbf{0}$$

In Example 1 in Section 14-1 we showed that this surface (Figure 14-4-5) is a paraboloid of revolution. ●

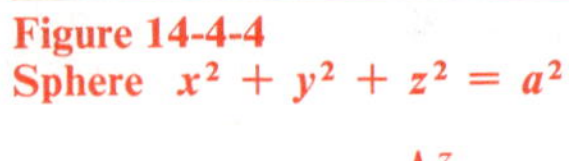
Figure 14-4-4
Sphere $x^2 + y^2 + z^2 = a^2$

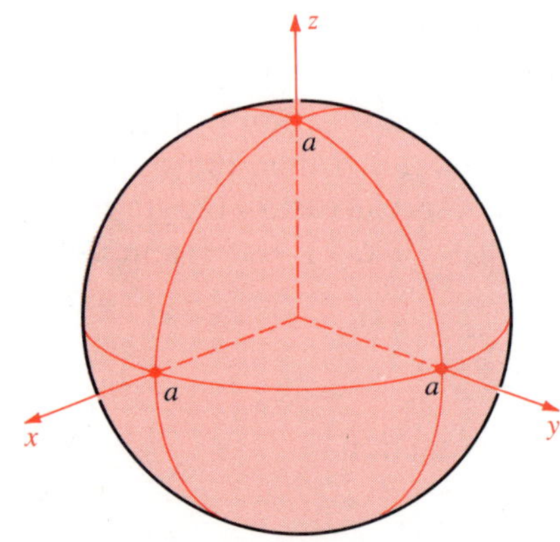

Figure 14-4-5
$z - x^2 - y^2 = 0$
Paraboloid of revolution

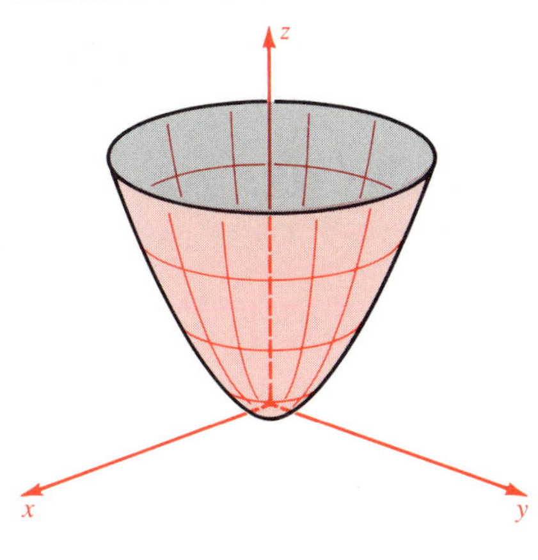

The Tangent Plane

When we studied curves, the tangent line was useful. Among other things, it shows the direction of a curve and it provides the closest linear approximation to the curve. The corresponding object for a surface is the *tangent plane.*

Given a surface $F(x, y, z) = k$ and one of its points $\mathbf{a}$, we want to define the plane tangent to the surface at $\mathbf{a}$ and find an equation for this plane. In the previous section, we found the tangent *line* to a level *curve* $f(x, y) = c$. Now we want the tangent *plane* to a level *surface* $F(x, y, z) = k$. It is not surprising, therefore, that both the discussion and the resulting equation will be similar.

We begin by proving a fundamental geometric fact: Given a point $\mathbf{a}$ on the surface $F(\mathbf{x}) = k$, consider all possible smooth curves that lie entirely on the surface and pass through $\mathbf{a}$. Then all of the velocity vectors at $\mathbf{a}$ of all of these curves are coplanar. In fact, they are all perpendicular to the non-zero vector $\operatorname{grad} F(\mathbf{a})$. See Figure 14-4-6.

This fact is remarkable because $\operatorname{grad} F(\mathbf{a})$ is a fixed vector, whereas the assertion holds for any curve whatsoever lying on the surface and passing through $\mathbf{a}$.

Figure 14-4-6
Each curve through $\mathbf{a}$ on the surface $F(\mathbf{x}) = k$ is orthogonal to $\operatorname{grad} F(\mathbf{a})$ at $\mathbf{a}$.

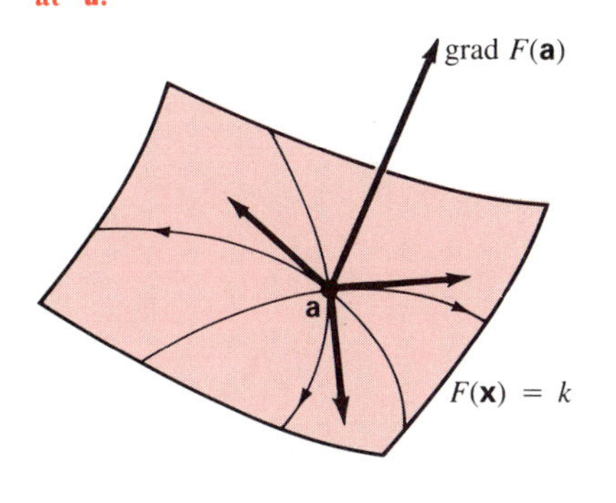

Theorem

Let F be continuously differentiable, and let $\mathbf{a}$ be a point of a level surface $F(\mathbf{x}) = k$. Then the velocity vector $\mathbf{v}$ at $\mathbf{a}$ of each curve on this surface that passes through $\mathbf{a}$ satisfies

$$\mathbf{v} \cdot \operatorname{grad} F(\mathbf{a}) = 0$$

Proof Let $\mathbf{x} = \mathbf{x}(t)$ be any differentiable curve lying on $F(\mathbf{x}) = k$ and passing through $\mathbf{a}$ at $t = 0$. Thus $F[\mathbf{x}(t)] = k$ and $\mathbf{x}(0) = \mathbf{a}$. Because the composite function $F[\mathbf{x}(t)]$ is constant, its t-derivative is zero. By the vector form of the chain rule,

$$\frac{d}{dt} F[\mathbf{x}(t)] = \frac{d\mathbf{x}}{dt} \cdot \operatorname{grad} F[\mathbf{x}(t)] = 0$$

We set $t = 0$ in this relation. Since $\mathbf{x}(0) = \mathbf{a}$ and $d\mathbf{x}(0)/dt = \mathbf{v}$, we obtain

$$\mathbf{v} \cdot \operatorname{grad} F(\mathbf{a}) = 0$$

as claimed.

Figure 14-4-7
Tangent plane to $F(\mathbf{x}) = k$ at $\mathbf{x} = \mathbf{a}$. It is the plane through $\mathbf{a}$ orthogonal to $\operatorname{grad} F(\mathbf{a})$.

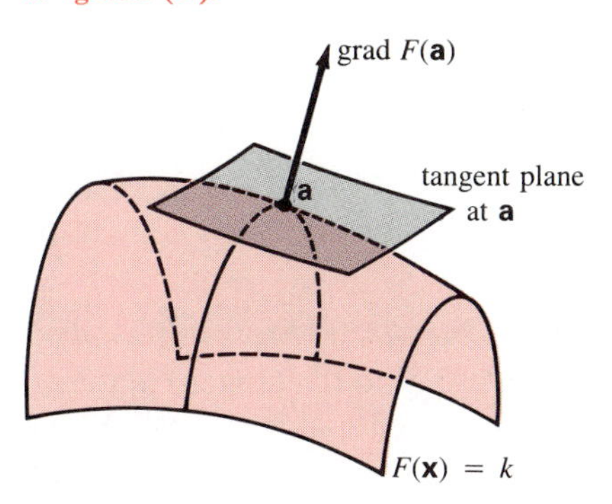

We therefore define the tangent plane at $\mathbf{a}$ as the plane through $\mathbf{a}$ perpendicular to $\operatorname{grad} F(\mathbf{a})$. See Figure 14-4-7. This definition is reasonable geometrically. The plane passes through $\mathbf{a}$ and is parallel to the velocity vector at $\mathbf{a}$ of any curve through $\mathbf{a}$ that lies on the surface. To obtain an equation of this plane, let $\mathbf{x}$ be any point on it. Then $\mathbf{x} - \mathbf{a}$ is perpendicular to $\operatorname{grad} F(\mathbf{a})$, hence

$$(\mathbf{x} - \mathbf{a}) \cdot \operatorname{grad} F(\mathbf{a}) = 0$$

Tangent Plane

Let $\mathbf{a} = (a, b, c)$ be a point of the surface $F(\mathbf{x}) = k$. The **tangent plane** to the surface at $\mathbf{x} = \mathbf{a}$ is the plane

$$[\operatorname{grad} F(\mathbf{a})] \cdot (\mathbf{x} - \mathbf{a}) = 0$$

In coordinate notation, this equation is

$$\frac{\partial F}{\partial x} \cdot (x - a) + \frac{\partial F}{\partial y} \cdot (y - b) + \frac{\partial F}{\partial z} \cdot (z - c) = 0$$

where the partial derivatives are evaluated at (a, b, c).

Note the similarity to the corresponding statement about the tangent line to a level curve (page 716).

Example 2

Find the tangent plane to

a $x + 2y + 3z = 6$ at $(1, 1, 1)$

b $x^2 + y^2 + z^2 = 14$ at $(3, 2, 1)$

Solution

a Set $F(x, y, z) = x + 2y + 3z$. Then the given surface (plane) is $F(\mathbf{x}) = 6$, and the point $\mathbf{a} = (1, 1, 1)$ is on the surface since $F(\mathbf{a}) = 6$. We have

$$\operatorname{grad} F(\mathbf{a}) = \operatorname{grad} F(\mathbf{x})\Big|_{\mathbf{a}} = (1, 2, 3)\Big|_{\mathbf{a}} = (1, 2, 3)$$

Hence the tangent plane at $\mathbf{a}$ is

$$(1, 2, 3) \cdot (\mathbf{x} - \mathbf{a}) = 0 \qquad (1, 2, 3) \cdot (x - 1, y - 1, z - 1) = 0$$

$$x + 2y + 3z = 1 + 2 + 3 = 6$$

Not surprising! The tangent plane to a plane at any of its points is the plane itself.

b Set $F(\mathbf{x}) = x^2 + y^2 + z^2 = |\mathbf{x}|^2$. Then the given surface (sphere) is $F(\mathbf{x}) = 14$, and the point $\mathbf{a} = (3, 2, 1)$ is on the surface since $F(\mathbf{a}) = 14$. We have

$$\operatorname{grad} F(\mathbf{a}) = \operatorname{grad} F(\mathbf{x})\Big|_{\mathbf{a}} = 2(x, y, z)\Big|_{\mathbf{a}} = 2\mathbf{a}$$

Hence the tangent plane at $\mathbf{a}$ is

$$2\mathbf{a} \cdot (\mathbf{x} - \mathbf{a}) = 0 \qquad \mathbf{a} \cdot \mathbf{x} = \mathbf{a} \cdot \mathbf{a} = 14 \qquad 3x + 2y + z = 14$$

The Unit Normal

The vector $\operatorname{grad} F(\mathbf{a})$ is normal (perpendicular) to the tangent plane at $\mathbf{x} = \mathbf{a}$. Therefore a *unit* vector normal to the tangent plane is $\operatorname{grad} F(\mathbf{a})$ divided by its own length.

Unit Normal

Let F be continuously differentiable, and let $\mathbf{a}$ be a point of a level surface $F(\mathbf{x}) = k$. The **unit normal** to this surface at $\mathbf{x} = \mathbf{a}$ is the vector

$$\mathbf{n} = \frac{1}{|\text{grad } F(\mathbf{a})|} \text{grad } F(\mathbf{a})$$

Example 3

Find the unit normal to $xyz = 3$ at $(-1, -3, 1)$.

Solution Set $F(\mathbf{x}) = xyz$ and $\mathbf{a} = (-1, -3, 1)$. Then $\mathbf{a}$ is on the surface because $F(\mathbf{a}) = 3$, and

$$\text{grad } F(\mathbf{a}) = \text{grad } F(\mathbf{x})\Big|_{\mathbf{a}} = (yz, zx, xy)\Big|_{\mathbf{a}} = (-3, -1, 3)$$

Since $|\text{grad } F(\mathbf{a})|^2 = 19$, we have

$$\mathbf{n} = \frac{1}{|\text{grad } F(\mathbf{a})|} \text{grad } F(\mathbf{a}) = \frac{1}{\sqrt{19}}(-3, -1, 3)$$

Remark "The" unit normal is not quite fair, because really $\mathbf{n}$ is only determined up to sign. If we define the same surface by $-F(\mathbf{x}) = -k$, then $\mathbf{n}$ is changed to $-\mathbf{n}$. Nonetheless, the inaccurate terminology "*the* unit normal" is common, and either $+\mathbf{n}$ or $-\mathbf{n}$ can be chosen.

Figure 14-4-8
Angle between a line through $\mathbf{a}$ and the unit normal $\mathbf{n}$ to $F = k$ at $\mathbf{a}$. We have
$\mathbf{n} \cdot \mathbf{u} = \cos\theta$
where $\mathbf{u}$ is a unit vector along the line.

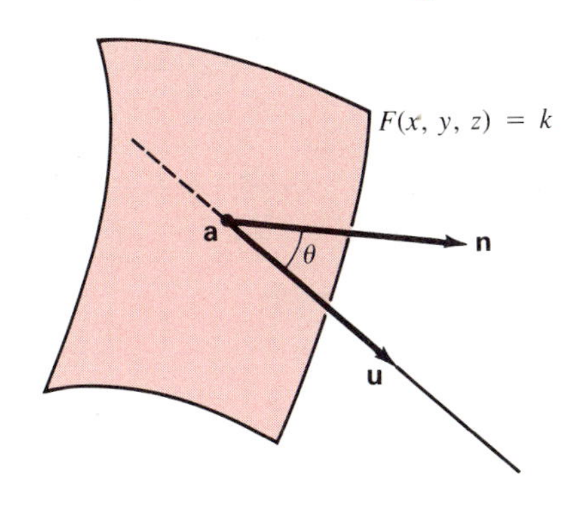

The tangent plane at each point of a level surface $F(x, y, z) = k$ is orthogonal to grad $F(\mathbf{x})$ at that point. As an application we can compute the angle at which a line intersects the surface $F(x, y, z) = k$. We define that angle as the complement of the (acute) angle between the line and the normal to the surface at the point of intersection. See Figure 14-4-8. If θ is the angle between the line and the normal then $\cos\theta = \mathbf{u} \cdot \mathbf{n}$, where $\mathbf{u}$ is a unit vector parallel to the line and $\mathbf{n}$ is the unit normal at the point of intersection.

Example 4

Find the angle at which the line $\mathbf{x}(t) = (t, t, t)$ intersects the surface $x^2 + y^2 + 2z^2 = 1$ in the first octant.

Solution A point of intersection corresponds to a value of t satisfying

$$t^2 + t^2 + 2t^2 = 1$$

$$4t^2 = 1$$

$$t = \pm\tfrac{1}{2}$$

Figure 14-4-9
The ellipsoid $x^2 + y^2 + 2z^2 = 1$ meets the line at $\mathbf{a} = (\frac{1}{2}, \frac{1}{2}, \frac{1}{2})$ in the first octant.

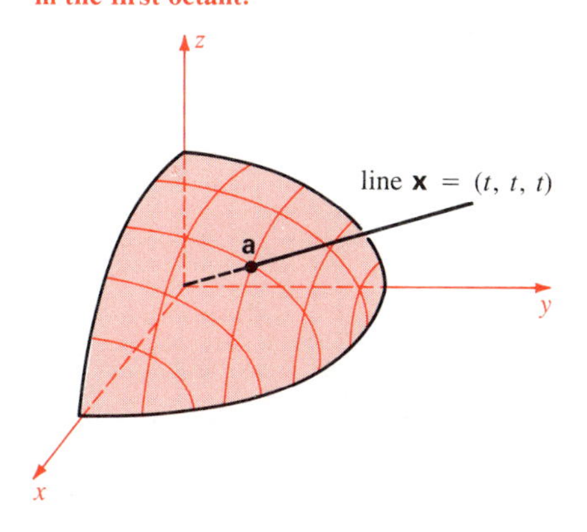

There are two points of intersection. The point in the first octant is $\mathbf{a} = (\frac{1}{2}, \frac{1}{2}, \frac{1}{2})$. See Figure 14-4-9.

Write the surface as $F(x, y, z) = x^2 + y^2 + 2z^2 = 1$. Then

$$\text{grad } F(\mathbf{a}) = (2x, 2y, 4z)\Big|_{\mathbf{a}} = (1, 1, 2)$$

and the unit normal at $\mathbf{a}$ is

$$\mathbf{n} = \frac{1}{|(1, 1, 2)|}(1, 1, 2) = \frac{1}{\sqrt{6}}(1, 1, 2)$$

Obviously $(1, 1, 1)$ is a vector parallel to the given line. A unit vector in the same direction is

$$\mathbf{u} = \frac{1}{|(1, 1, 1)|}(1, 1, 1) = \frac{1}{\sqrt{3}}(1, 1, 1)$$

Therefore, if θ is the angle between the line and the normal to the surface,

$$\cos\theta = \mathbf{u}\cdot\mathbf{n} = \frac{1}{\sqrt{3}}(1, 1, 1)\cdot\frac{1}{\sqrt{6}}(1, 1, 2) = \frac{4}{\sqrt{18}} = \tfrac{2}{3}\sqrt{2}$$

The desired angle is the complement, $\arcsin(\tfrac{2}{3}\sqrt{2}) \approx 70.53°$. ●

Graph of a Function

Let us now consider the special case of the graph $z = f(x, y)$. Here $f(x, y)$ is a function of two variables defined over some plane domain $\mathbf{D}$ and assumed to have continuous partial derivatives (Figure 14-4-10). The graph is a surface; to fit it into our previous pattern, we set

$$F(\mathbf{x}) = F(x, y, z) = z - f(x, y)$$

Then the graph $z = f(x, y)$ is the level surface $F(\mathbf{x}) = 0$. Now

$$\operatorname{grad} F(\mathbf{x}) = (-f_x, -f_y, 1)$$

Since the third component is 1, this gradient is never $\mathbf{0}$, so our basic assumption is satisfied.

If (a, b) is a point of $\mathbf{D}$, then $(a, b, f(a, b))$ is the corresponding point on the surface. The gradient there is

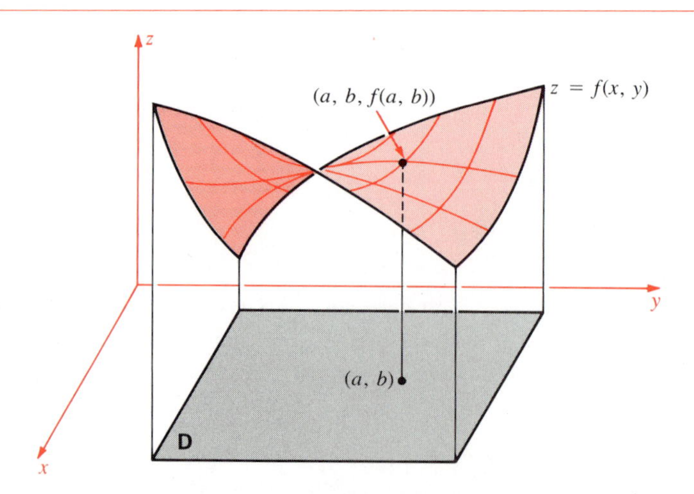

Figure 14-4-10
A surface defined as the graph of a function $z = f(x, y)$

$$\text{grad } F\bigg|_{(a,b,f(a,b))} = (-f_x(a, b), -f_y(a, b), 1)$$

and the tangent plane is

$$(\text{grad } F) \cdot (x - a, y - b, z - f(a, b)) = 0$$

which simplifies to

$$z = f(a, b) + f_x(a, b)(x - a) + f_y(a, b)(y - b)$$

The unit normal is

$$\mathbf{n} = \text{grad } F/|\text{grad } F|$$

(Note that this one of the two possible choices has a positive z-coordinate.) Clearly $|\text{grad } F|^2 = 1 + f_x^2 + f_y^2$.

Tangent Plane and Unit Normal to the Graph of a Function

Let $f(x, y)$ be continuously differentiable, and let (a, b) be a (non-boundary) point of the domain of $f(x, y)$. Then the tangent plane to the graph $z = f(x, y)$ at $(a, b, f(a, b))$ is

$$z = f(a, b) + f_x(a, b)(x - a) + f_y(a, b)(y - b)$$

and the unit normal is

$$\mathbf{n} = \frac{1}{\sqrt{1 + f_x^2(a, b) + f_y^2(a, b)}}(-f_x(a, b), -f_y(a, b), 1)$$

Example 5

Find the tangent plane and unit normal to $z = x^2y$ for $(x, y) = (3, 2)$.

Solution Set $f(x, y) = x^2y$. Then $f(3, 2) = 18$, and

$$(-f_x, -f_y, 1)\bigg|_{(3,2)} = (-2xy, -x^2, 1)\bigg|_{(3,2)} = (-12, -9, 1)$$

The tangent plane is

$$\begin{aligned} z &= 18 + f_x \cdot (x - 3) + f_y \cdot (y - 2) \\ &= 18 + 12(x - 3) + 9(y - 2) \end{aligned}$$

that is,

$$12x + 9y - z = 36$$

The unit normal is

$$\mathbf{n} = \frac{1}{|(-12, -9, 1)|}(-12, -9, 1) = \frac{1}{\sqrt{226}}(-12, -9, 1)$$

Exercises

Find the tangent plane and the unit normal to the given surface at the given point

1 $x^3 + y^3 + z^3 = 3 \quad (4, 4, -5)$

2 $x^2 + y^3 + z^4 = 18 \quad (3, 2, 1)$

3 $x^2 + yz = 7 \quad (1, 2, 3)$

4 $x^3 + y^3 + z^3 + 3xyz = 16 \quad (1, 1, 2)$

5 $xy^2z^3 + yz^5 = 14 \quad (3, 2, 1)$

6 $z^5 - xz^4 + yz^3 - 1 = 0 \quad (1, 1, 1)$

7 $xy + yz + zx = 0 \quad (2, 2, -1)$

8 $z\cos(xy) = y^2 \quad (0, 1, 1)$

9 $e^{xy} + y = e^z + 1 \quad (1, 1, 1)$

10 $x\sin y + y\sin z + xyz = 0 \quad (1, 0, 0)$

11 $x^3 + xe^{yz} = 10 \quad (2, 0, 1)$

12 $(x + 2y)/\cosh z = 1 \quad (5, -2, 0)$

13 Find the angle of intersection of the line through $\mathbf{0}$ and $(1, 1, 1)$ and the surface $z = x^2 + y^2$ at each of their intersections.

14 Compute the angle at which a line through the north pole $\mathbf{k}$ of the unit sphere $|\mathbf{x}| = 1$ intersects the sphere at a typical point $\mathbf{a} = (a, b, c) \neq (0, 0, 1)$.

15 Find all points on the surface $xyz = 1$ where the normal line intersects the z-axis. (The **normal line** is the line through the point with the direction of the unit normal.)

16 Suppose $F(\mathbf{x})$ is an **even** function, that is, it satisfies $F(-\mathbf{x}) = F(\mathbf{x})$. Show that the tangent planes at $\mathbf{a}$ and $-\mathbf{a}$ to the surface $F(\mathbf{x}) = k$ are parallel.

Find the tangent plane and unit normal to the graph $z = f(x, y)$ at the point with the given (x, y)

17 $z = x^2 - y^2 \quad (0, 0)$

18 $z = x^2 - y^2 \quad (1, -1)$

19 $z = x^2 + 4y^2 \quad (2, 1)$

20 $z = x^2e^y \quad (-1, 2)$

21 $z = x^2y + y^3 \quad (-1, 2)$

22 $z = x\cos y + y\cos x \quad (0, 0)$

23 $z = x + x^2y^3 + y \quad (0, 0)$

24 $z = x^3 + y^3 \quad (1, -1)$

25 $z = \ln(1 + x^2 + 2y^2) \quad (1, 1)$

26 $z = \sqrt{1 - x^2 - y^2} \quad (\frac{1}{2}, \frac{1}{2})$

14-5 Parametric Surfaces and Surfaces of Revolution

A curve can be parametrized by a vector function $\mathbf{x}(t)$ of one real variable. Physically speaking, a curve is a one-dimensional set of points; you can move on a curve with one degree of freedom (forwards or backwards). In slightly different terms, the points of a curve can be specified by one real number (parameter), for instance, by the directed distance along the curve from a fixed point.

Now we want to parametrize a surface. In contrast to a curve, a surface is a two-dimensional set of points; you can move on a surface with two degrees of freedom. Thus it requires *two* real numbers to specify a point, for instance, latitude and longitude on a sphere.

This suggests presenting a surface by a vector function

$$\mathbf{x} = \mathbf{x}(u, v)$$

of *two* real variables u and v, the parameters. In coordinates,

$$\mathbf{x} = (x, y, z) = (x(u, v), y(u, v), z(u, v))$$

Here (u, v) varies over a domain $\mathbf{D}$ in the u, v-plane. Each point of $\mathbf{D}$ is identified by a pair of real numbers (u, v); to this pair the function assigns a point $\mathbf{x}(u, v)$ on the surface (Figure 14-5-1). In other words, $\mathbf{x}(u, v)$ maps the plane domain $\mathbf{D}$ onto the surface. We assume $\mathbf{x}(u, v)$ is continuously differentiable, that is, the six partials

Figure 14-5-1
Surface presented in parametric form

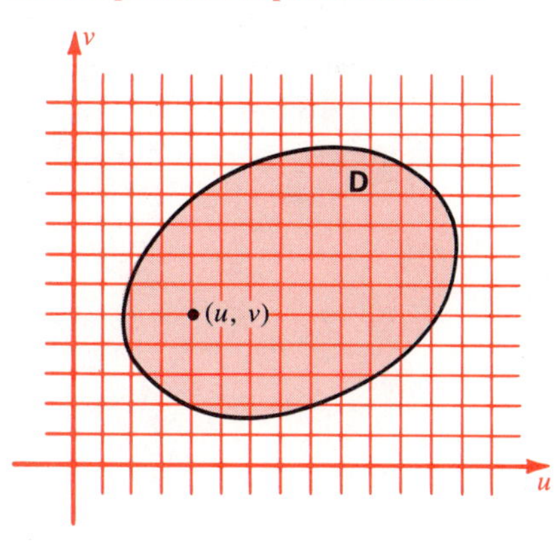

a Domain **D**

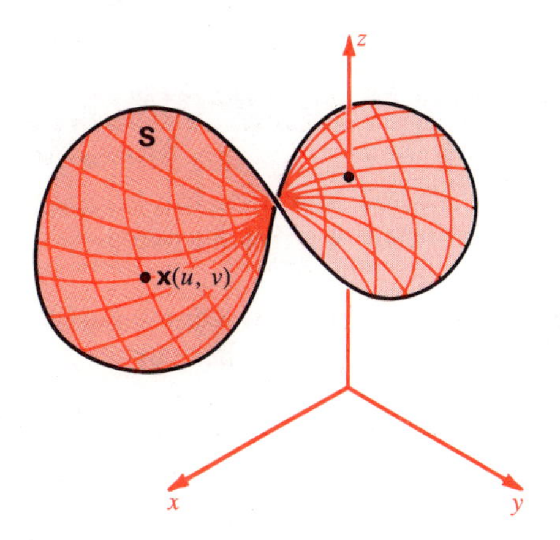

b Surface **S**

$$\frac{\partial x}{\partial u} \quad \frac{\partial x}{\partial v} \quad \frac{\partial y}{\partial u} \quad \frac{\partial y}{\partial v} \quad \frac{\partial z}{\partial u} \quad \frac{\partial z}{\partial v}$$

exist and are continuous. [We have previously studied the special case in which x and y are the parameters and $\mathbf{x} = (x, y, z(x, y))$, where $z(x, y)$ is a function of x and y.]

Example 1

Describe the surface given parametrically by

$$\mathbf{x} = (x, y, z) = (u + v, u - v, u)$$
$$-\infty < u < +\infty$$
$$-\infty < v < +\infty$$

Solution

$$x + y = (u + v) + (u - v)$$
$$= 2u = 2z$$

so the function $x(u, v)$ maps the u,v-plane into the plane $x + y = 2z$. Conversely, if $x + y = 2z$, then there are unique u and v such that

$$x = u + v \quad \text{and} \quad z = u$$

namely $u = z$ and $v = x - u = x - z$. But now

$$y = 2z - x$$
$$= 2u - (u + v) = u - v$$

Hence

$$\mathbf{x} = (x, y, z) = (u + v, u - v, u)$$

Therefore the relation $\mathbf{x} = (u + v, u - v, u)$ is a parametrization of the plane $x + y = 2z$. ●

Example 2

$$\mathbf{x} = (u \cos v, u \sin v, u) \qquad (u \neq 0)$$

In this case

$$x^2 + y^2 = (u \cos v)^2 + (u \sin v)^2$$
$$= u^2 = z^2$$

Conversely, given that $x^2 + y^2 = z^2$ and $z \neq 0$, set $u = z$. Then $x^2 + y^2 = u^2$, so there is a v, unique up to a multiple of 2π, such that $x = u \cos v$ and $y = u \sin v$. Thus the given form of $\mathbf{x}$ parametrizes $x^2 + y^2 = z^2$. We shall see later that this surface is a right circular cone. ●

Consider the example

$$\mathbf{x} = (u + v, u + v, u + v)$$

There is something wrong here because this is just another version of the *line* $\mathbf{x} = (t, t, t)$. To assure that a surface is genuinely two-dimensional we shall assume that

$$\left(\frac{\partial x}{\partial u}, \frac{\partial y}{\partial u}, \frac{\partial z}{\partial u}\right) \times \left(\frac{\partial x}{\partial v}, \frac{\partial y}{\partial v}, \frac{\partial z}{\partial v}\right) \neq \mathbf{0}$$

In vector notation

$$\mathbf{x}_u \times \mathbf{x}_v \neq \mathbf{0} \qquad \text{that is} \qquad \frac{\partial \mathbf{x}}{\partial u} \times \frac{\partial \mathbf{x}}{\partial v} \neq \mathbf{0}$$

This assumption plays a similar role to that of $\operatorname{grad} F(\mathbf{x}) \neq \mathbf{0}$ for surfaces $F(\mathbf{x}) = k$. It guarantees the existence of a well-defined tangent plane, as we are about to show.

Figure 14-5-2
The two special curves through $\mathbf{c}$
$\begin{cases} C_u: & \mathbf{x} = \mathbf{x}(u, b) \\ C_v: & \mathbf{x} = \mathbf{x}(a, v) \end{cases}$
correspond to the lines $v = b$ and $u = a$ respectively in the u, v-plane. Their tangent vectors at $\mathbf{c}$ are $\mathbf{x}_u$ and $\mathbf{x}_v$ respectively.

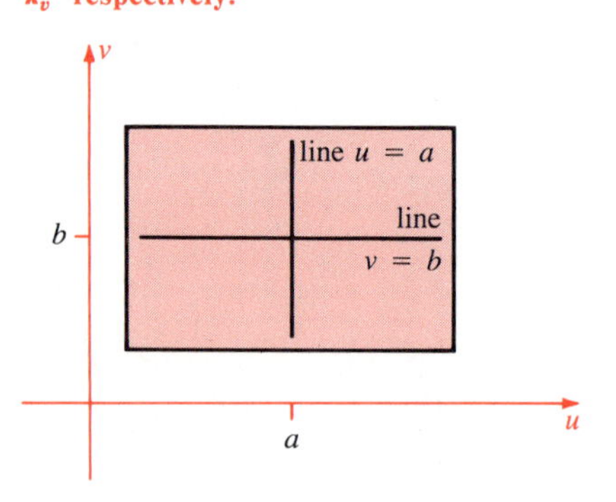

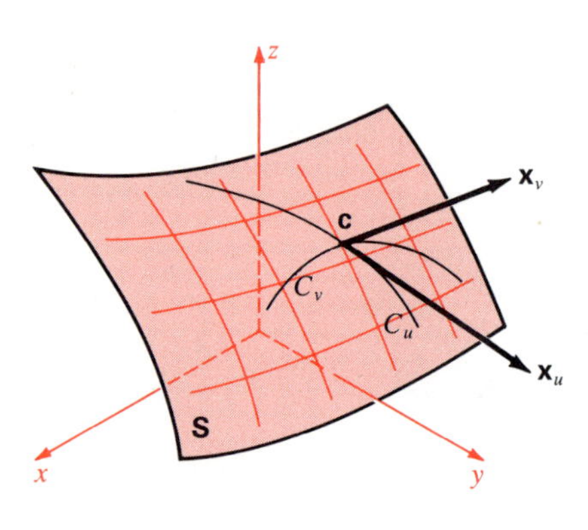

Tangent Plane and Unit Normal

Let us find the tangent plane at the point $\mathbf{c} = \mathbf{x}(a, b)$ to a surface $\mathbf{S}$ presented parametrically by $\mathbf{x} = \mathbf{x}(u, v)$. We show that the velocity vectors at $\mathbf{c}$ of all curves on $\mathbf{S}$ that pass through $\mathbf{c}$ fill out a plane. The parallel plane through $\mathbf{c}$ is the tangent plane.

To start, let us look at the two special curves on the surface, the curves obtained by holding one parameter fixed and letting the other vary:

$$\mathbf{x}(u) = (x(u, b), y(u, b), z(u, b))$$

$$\mathbf{x}(v) = (x(a, v), y(a, v), z(a, v))$$

Both pass through $\mathbf{c}$, the first for $u = a$, the second for $v = b$. See Figure 14-5-2. Their velocity vectors at $\mathbf{c}$ are

$$\left.\left(\frac{\partial x}{\partial u}, \frac{\partial y}{\partial u}, \frac{\partial z}{\partial u}\right)\right|_{(a,b)} = \mathbf{x}_u(a, b)$$

and

$$\left.\left(\frac{\partial x}{\partial v}, \frac{\partial y}{\partial v}, \frac{\partial z}{\partial v}\right)\right|_{(a,b)} = \mathbf{x}_v(a, b)$$

Our basic assumption $\mathbf{x}_u \times \mathbf{x}_v \neq \mathbf{0}$ guarantees that $\mathbf{x}_u$ and $\mathbf{x}_v$ are non-parallel. Hence they span a plane $\mathbf{P}$ consisting of all vectors

$$h\mathbf{x}_u(a, b) + k\mathbf{x}_v(a, b)$$

Figure 14-5-3
The set of all velocity vectors of all curves on $\mathbf{S}$ through $\mathbf{c}$ fill a plane $\mathbf{P}$.

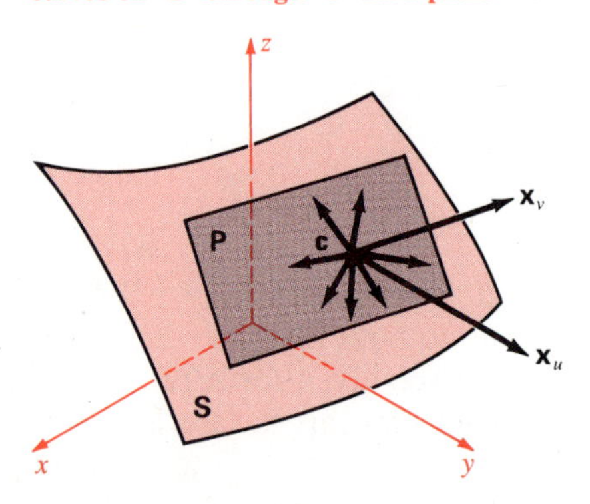

It turns out that the velocity vectors at $\mathbf{c}$ of all curves on $\mathbf{S}$ that pass through $\mathbf{c}$ fill out the plane $\mathbf{P}$. The proof, which we omit, is like the corresponding proof on page 725 for level surfaces, but longer. See Figure 14-5-3.

The vector $\mathbf{x}_u \times \mathbf{x}_v$ is perpendicular to $\mathbf{P}$, hence normal to the surface at $\mathbf{c}$. Therefore the unit normal is $(\mathbf{x}_u \times \mathbf{x}_v)/|\mathbf{x}_u \times \mathbf{x}_v|$.

Tangent Plane and Unit Normal — Parametric Surfaces

Let $\mathbf{x} = \mathbf{x}(u, v)$ be a surface in parametric form and let $\mathbf{c} = \mathbf{x}(a, b)$. The tangent plane at $\mathbf{c}$, in parametric form, is

$$\mathbf{x} = \mathbf{c} + h\mathbf{x}_u(a, b) + k\mathbf{x}_v(a, b)$$

where h and k are arbitrary real numbers. The unit normal at $\mathbf{c}$ is

$$\mathbf{n} = \frac{\mathbf{x}_u \times \mathbf{x}_v}{|\mathbf{x}_u \times \mathbf{x}_v|}$$

Example 3

Find the tangent plane at $\mathbf{c} = (0, 1, \frac{1}{2}\pi)$, in both parametric and non-parametric form, and the unit normal for the surface

$$\mathbf{x} = (u \cos v, u \sin v, v) \qquad (u > 0)$$

Solution The point $\mathbf{c}$ corresponds to $(u, v) = (1, \frac{1}{2}\pi)$. We have

$$\mathbf{x}_u(1, \tfrac{1}{2}\pi) = (\cos v, \sin v, 0)|_{(1,\pi/2)} = (0, 1, 0)$$
$$\mathbf{x}_v(1, \tfrac{1}{2}\pi) = (-u \sin v, u \cos v, 1)|_{(1,\pi/2)} = (-1, 0, 1)$$

The parametric form of the tangent plane is easily computed:

$$\begin{aligned}\mathbf{x} &= \mathbf{c} + h\mathbf{x}_u(1, \tfrac{1}{2}\pi) + k\mathbf{x}_v(1, \tfrac{1}{2}\pi)\\ &= (0, 1, \tfrac{1}{2}\pi) + h(0, 1, 0) + k(-1, 0, 1)\\ &= (-k, 1 + h, \tfrac{1}{2}\pi + k)\end{aligned}$$

Thus

$$\mathbf{x} = (-k, 1 + h, \tfrac{1}{2}\pi + k)$$

The cross product of the two tangent vectors $\mathbf{x}_u$ and $\mathbf{x}_v$ is a normal vector; call it $\mathbf{w}$:

$$\mathbf{w} = \mathbf{x}_u \times \mathbf{x}_v = (0, 1, 0) \times (-1, 0, 1) = (1, 0, 1)$$

Consequently

$$\mathbf{n} = \frac{\mathbf{w}}{|\mathbf{w}|} = \frac{(1, 0, 1)}{|(1, 0, 1)|} = \frac{1}{\sqrt{2}}(1, 0, 1)$$

We easily find the non-parametric form:

$$(\mathbf{x} - \mathbf{c}) \cdot \mathbf{w} = 0 \qquad (\mathbf{x} - \mathbf{c}) \cdot (1, 0, 1) = 0$$
$$(x, y - 1, z - \tfrac{1}{2}\pi) \cdot (1, 0, 1) = 0$$

Therefore the non-parametric form is

$$x + z = \tfrac{1}{2}\pi$$

The surface is a spiral ramp (Figure 14-5-4).

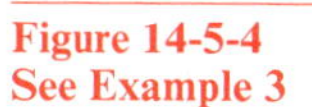
Figure 14-5-4
See Example 3

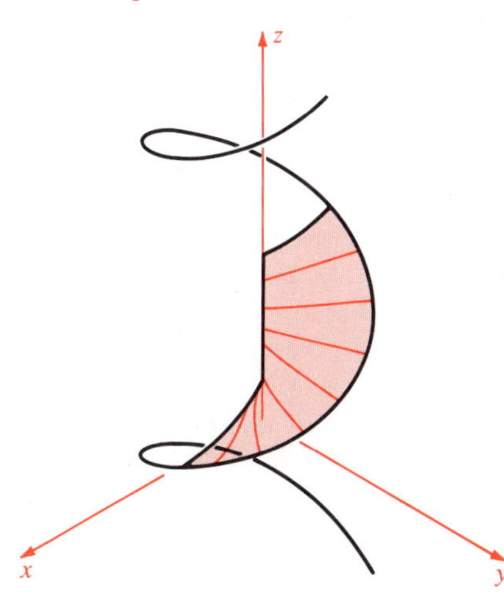

Surfaces of Revolution

Let $f(x, z) = 0$ be the equation of a curve in the part $x \geq 0$ of the x, z-plane (Figure 14-5-5). If the curve is revolved about the z-axis, a surface, called a **surface of revolution,** is swept out (Figure 14-5-6).

To find its equation, let $\mathbf{x} = (x, y, z)$ be any point of the surface. The distance from $\mathbf{x}$ to the z-axis is $\sqrt{x^2 + y^2}$. See Figure 14-5-6. Hence $\mathbf{x}$ is swept out by the point $(\sqrt{x^2 + y^2}, z)$ on the original curve. This point satisfies the equation of the curve so

$$f(\sqrt{x^2 + y^2}, z) = 0$$

Thus $\mathbf{x} = (x, y, z)$ satisfies this equation. Hence it is the equation of the surface of revolution.

Figure 14-5-5
Curve in x, z-plane $(x \geq 0)$

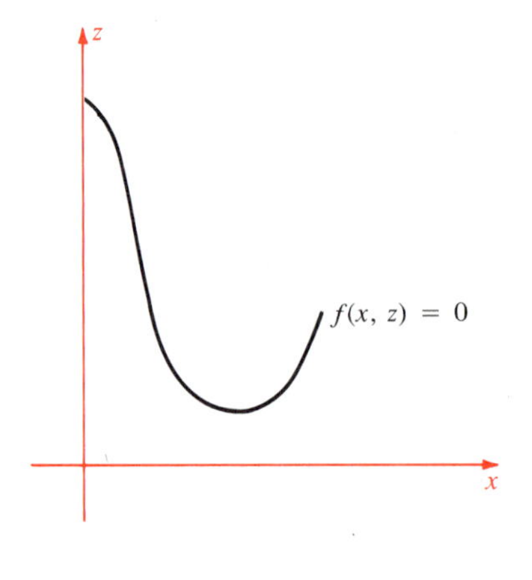

Figure 14-5-6
The corresponding surface of revolution about the z-axis

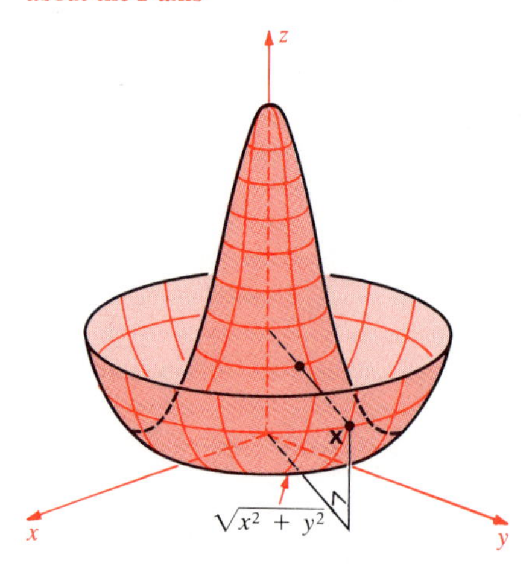

Example 4

a A paraboloid of revolution is swept out when the curve $x^2 - z = 0$, $x \geq 0$, is revolved about the z-axis. Its equation is

$$(\sqrt{x^2 + y^2})^2 - z = 0 \qquad \text{that is} \qquad z = x^2 + y^2$$

b A right circular cone is swept out when the line $x - z = 0$, $x \geq 0$ is revolved about the z-axis. Its equation is

$$\sqrt{x^2 + y^2} - z = 0 \qquad \text{that is} \qquad z^2 = x^2 + y^2 \qquad (z \geq 0)$$ ●

Revolution about other coordinate axes than the z-axis is handled similarly. For example if $f(x, y) = 0$, $y \geq 0$, is revolved about the x-axis, the equation of the resulting surface is

$$f(x, \sqrt{y^2 + z^2}) = 0$$

Exercises

Express the parametric surface in the form $F(x, y, z) = k$ and describe the surface

1 $\mathbf{x} = (u, v, u + v)$

2 $\mathbf{x} = (2u - v + 3, -u + 4v + 1, -u - 2v - 6)$

3 $\mathbf{x} = (v \cos u, v \sin u, v)$

4 $\mathbf{x} = (a \cos u, a \sin u, v)$

5 $\mathbf{x} = (a \sin u, v, a \cos u)$

6 $\mathbf{x} = (u^3, u^3, v)$

Find all points (a, b) in the u,v-domain where the given parametrization of the surface is degenerate, that is, where $\mathbf{x}_u$ and $\mathbf{x}_v$ are parallel $(\mathbf{x}_u \times \mathbf{x}_v = \mathbf{0})$

7 $\mathbf{x} = (u + v, u - v, uv)$

8 $\mathbf{x} = (u, u^2, v^3)$

9 $\mathbf{x} = (u \cos v, u \sin v, v)$

10 $\mathbf{x} = (u + v, uv, u^2 + v^2)$

11 $\mathbf{x} = (u^2, v, u^3)$

12 $\mathbf{x} = (1 - u^2, u - u^3, v)$

Give the tangent plane in parametric form and the unit normal for the surface at the given parametric values (u, v)

13 $\mathbf{x} = (u - v, u + v, u + 3v)$ $\quad (1, 0)$

14 $\mathbf{x} = (u \sin v, v, u \cos v)$ $\quad (2, \frac{1}{6}\pi)$

15 $\mathbf{x} = (\sin u \cos v, \sin u \sin v, 2 + \cos u)$ $\quad (\frac{1}{2}\pi, \frac{1}{4}\pi)$

16 $\mathbf{x} = (u, v/u, 1/v)$ $\quad (1, -2)$

17 $\mathbf{x} = (u^2 - v^2, 2uv, u^3 + v^3)$ $\quad (0, -2)$

18 $\mathbf{x} = (u^2v^2, u^3 + v^3, u^2v + uv^2)$ $\quad (-1, -1)$

19 Let $0 < a < b$. Revolve about the z-axis the circle in the x,z-plane with center $(b, 0)$ and radius a. Give an equation for the resulting torus.

***20** (cont.) Parametrize the surface in terms of two angles.

21 What is a simple way of parametrizing the graph $z = f(x, y)$ of a function?

***22** Suppose a curve in the x,z-plane is revolved about the z-axis. Show that the normal line at each point of the resulting surface of revolution either intersects the z-axis or is parallel to it. (The **normal line** is the line through the point in the direction of the unit normal.)

***23** Let $\mathbf{x} = \mathbf{x}(s)$ be a curve on the unit sphere $|\mathbf{x}| = 1$, where s is arc length. Consider the parametric cone $\mathbf{x} = \mathbf{x}(u, v) = v\mathbf{x}(u)$, $v > 0$. Show that this surface has a tangent plane at each of its points.

***24** (cont.) Show that the unit normal to the surface is constant along each generator, and the tangent plane is also the same at each point of a generator.

25 Let $\mathbf{x} = \mathbf{x}(s)$ be a space curve with arc length s and curvature $k = k(s) > 0$. The set of (positive) tangent lines to the curve sweep out the parametric surface $\mathbf{x} = \mathbf{x}(u, v) = \mathbf{x}(u) + v\mathbf{t}(u)$, $v > 0$, where $\mathbf{t}(s)$ is the unit tangent at $\mathbf{x}(s)$. Find $\mathbf{x}_u$ and $\mathbf{x}_v$ and show that these vectors are never parallel.

26 (cont.) Find an equation for the tangent plane at $\mathbf{x}(u, v)$, and show that it is independent of v.

14-6 Differentials and Approximation

If $f(x)$ is differentiable function of one variable, then its first order approximation at $x = a$ is

$$f(a + t) \approx f(a) + f'(a)t$$

This is highly accurate provided t is sufficiently small. In fact, we know by Taylor's formula that if $f''(x)$ is continuous, then

$$|f(a + t) - f(a) - f'(a)t| < Kt^2$$

for t sufficiently small, where K is a constant.

There is a similar result for functions of several variables.

First Order Approximation

Let $f(\mathbf{x})$ be continuously differentiable near $\mathbf{x} = \mathbf{a}$. Then

$$f(\mathbf{a} + \mathbf{v}) \approx f(\mathbf{a}) + D_{\mathbf{v}}f(\mathbf{a}) = f(\mathbf{a}) + [\operatorname{grad} f(\mathbf{a})] \cdot \mathbf{v}$$

for $|\mathbf{v}|$ small.

To see why, write $\mathbf{v} = t\mathbf{u}$, where $\mathbf{u}$ is a unit vector and t is small, and set

$$g(t) = f(\mathbf{a} + \mathbf{v}) = f(\mathbf{a} + t\mathbf{u})$$

Then, since t is small,

$$\begin{aligned} g(t) &\approx g(0) + g'(0)t \\ &= f(\mathbf{a}) + g'(0)t \end{aligned}$$

But

$$g'(0) = D_{\mathbf{u}}f(\mathbf{a}) = \mathbf{u} \cdot \operatorname{grad} f(\mathbf{a})$$

by definition of directional derivative. Hence

$$\begin{aligned} g(t) &\approx f(\mathbf{a}) + t\mathbf{u} \cdot \operatorname{grad} f(\mathbf{a}) \\ &= f(\mathbf{a}) + [\operatorname{grad} f(\mathbf{a})] \cdot \mathbf{v} \end{aligned}$$

Example 1

Given $f(5, 7) = 10$ and $\operatorname{grad} f(5, 7) = (2, 3)$, estimate $f(5.01, 6.98)$.

Solution Set $\mathbf{a} = (5, 7)$ and $\mathbf{v} = (0.01, -0.02)$. Then

$$\begin{aligned} f(5.01, 6.98) &= f(\mathbf{a} + \mathbf{v}) \\ &\approx f(\mathbf{a}) + [\operatorname{grad} f(\mathbf{a})] \cdot \mathbf{v} \\ &= 10 + (2, 3) \cdot (0.01, -0.02) \\ &= 10 + 0.02 - 0.06 \\ &= 9.96 \end{aligned}$$

●

Later we shall make the meaning of the symbol $\approx$ in first order approximation more precise.

Differentials

In the first order approximation

$$f(\mathbf{a} + \mathbf{v}) \approx f(\mathbf{a}) + [\operatorname{grad} f(\mathbf{a})] \cdot \mathbf{v}$$

suppose we allow *both* $\mathbf{a}$ and $\mathbf{v}$ to vary. It is customary to replace $\mathbf{a}$ by $\mathbf{x}$ and $\mathbf{v}$ by a new symbol

$$d\mathbf{x} = (dx, dy, dz)$$

In this notation,

$$f(\mathbf{x} + d\mathbf{x}) \approx f(\mathbf{x}) + [\operatorname{grad} f(\mathbf{x})] \cdot d\mathbf{x}$$

The second term on the right is especially important. We call it the **differential** of f, and denote it by df:

Differential

Suppose $f(x, y, z)$ is continuously differentiable on its domain. Its **differential** is

$$df = [\operatorname{grad} f(\mathbf{x})] \cdot d\mathbf{x}$$
$$= \frac{\partial f}{\partial x}\,dx + \frac{\partial f}{\partial y}\,dy + \frac{\partial f}{\partial z}\,dz$$

Example 2

a $f(x, y, z) = xy^2z^3 \qquad df = y^2z^3\,dx + 2xyz^3\,dy + 3xy^2z^2\,dz$

b $f(x, y) = \dfrac{x}{y} \qquad df = \dfrac{1}{y}\,dx - \dfrac{x}{y^2}\,dy$

For a function $f(x, y, z)$ of three variables, the differential df is a function of six variables x, y, z, dx, dy, dz. Technically we should use a notation such as $df = df(\mathbf{x}, d\mathbf{x})$, but that would be cumbersome.

For each fixed $\mathbf{x}$ in the domain of f, the differential is a *linear* function of the variables dx, dy, dz, whose domains are unrestricted. For instance, if $f(x, y, z) = xy^2z^3$ as in Example 1a, then

$$df = dx + 2\,dy + 3\,dz \qquad \text{at } (1, 1, 1)$$
$$df = -27\,dx - 108\,dy + 54\,dz \qquad \text{at } (2, 1, -3)$$

Our discussion of the first order approximation shows that in general df is the best linear approximation to $f(\mathbf{x}) - f(\mathbf{a})$ near the point $\mathbf{a}$.

The differential has elementary algebraic properties, which correspond to analogous results for derivatives:

Rules for Differentials

$$d(f + g) = df + dg \qquad d(af) = a\,df$$
$$d(fg) = (df)g + f\,dg \qquad d(f/g) = \frac{(df)g - f\,dg}{g^2} \quad (g \neq 0)$$

For example,

$$d(f + g) = \frac{\partial}{\partial x}(f + g)\,dx + \frac{\partial}{\partial y}(f + g)\,dy + \frac{\partial}{\partial z}(f + g)\,dz$$
$$= \left(\frac{\partial f}{\partial x} + \frac{\partial g}{\partial x}\right)dx + \left(\frac{\partial f}{\partial y} + \frac{\partial g}{\partial y}\right)dy + \left(\frac{\partial f}{\partial z} + \frac{\partial g}{\partial z}\right)dz$$
$$= \left(\frac{\partial f}{\partial x}\,dx + \frac{\partial f}{\partial y}\,dy + \frac{\partial f}{\partial z}\,dz\right) + \left(\frac{\partial g}{\partial x}\,dx + \frac{\partial g}{\partial y}\,dy + \frac{\partial g}{\partial z}\,dz\right)$$
$$= df + dg$$

Numerical Approximations

The variables dx, dy, dz that appear in the differential are free to take all real values. However, in practice, we usually think of them as small changes in x, y, and z respectively. A typical application is the approximation

$$f(\mathbf{x} + d\mathbf{x}) \approx f(\mathbf{x}) + df$$

Example 3

Given $f(x, y, z) = \dfrac{xy^2}{1 + z^4}$, estimate

a $f(1.96, 3.02, 0.99)$

b $f(2.06, 3.00, 1.01)$

Solution Both parts ask for values of $f(\mathbf{x})$ near $\mathbf{x} = (2, 3, 1)$. Use the approximation $f(\mathbf{x} + d\mathbf{x}) \approx f(\mathbf{x}) + df$ for $\mathbf{x} = (2, 3, 1)$. Now $f(2, 3, 1) = 9$, and at $(2, 3, 1)$ we have

$$\begin{aligned} df &= f_x\,dx + f_y\,dy + f_z\,dz \\ &= \frac{y^2}{1 + z^4}\,dx + \frac{2xy}{1 + z^4}\,dy - \frac{4xy^2z^3}{(1 + z^4)^2}\,dz \\ &= \tfrac{9}{2}\,dx + 6\,dy - 18\,dz \end{aligned}$$

a Set $d\mathbf{x} = (-0.04, 0.02, -0.01)$. Then

$$\begin{aligned} f(1.96, 3.02, 0.99) &\approx f(2, 3, 1) + df \\ &= 9 + \tfrac{9}{2}(-0.04) + 6(0.02) - 18(-0.01) \\ &= 9.12 \end{aligned}$$

b Set $d\mathbf{x} = (0.06, 0.00, 0.01)$. Then

$$\begin{aligned} f(2.06, 3.00, 1.01) &\approx 9 + \tfrac{9}{2}(0.06) + 6(0) - 18(0.01) \\ &= 9.09 \end{aligned}$$

●

Differentials and the Chain Rule

Suppose $f = f(x, y, z)$ it continuously differentiable so that

$$df = f_x\,dx + f_y\,dy + fz\,dz$$

Now suppose

$$x = x(u, v) \qquad y = y(u, v) \qquad \text{and} \qquad z = z(u, v)$$

are also continuously differentiable. We can look at the composite function $f[\mathbf{x}(u, v)]$ and compute its differential, another "df":

$$df = f_u\,du + f_v\,dv$$

Is this different? No, because by the chain rule

$$\begin{aligned} &f_u\,du + f_v\,dv \\ &\quad = (f_x x_u + f_y y_u + f_z z_u)\,du + (f_x x_v + f_y y_v + f_z z_v)\,dv \\ &\quad = f_x(x_u du + x_v dv) + f_y(y_u du + y_v dv) + f_z(z_u du + z_v dv) \\ &\quad = f_x dx + f_y dy + f_z dz \end{aligned}$$

This may appear to be no more than a consequence of sloppy notation for composite functions, but there is something more to it.

Suppose we have a function f of independent variables u and v, not given directly, but possibly as a composite function of other functions or as an implicit function. Suppose that after some computation we arrive at a relation

$$df = M\,du + N\,dv$$

Then we know automatically that $M = f_u$ and $N = f_v$ no matter how we obtained the relation.

Example 4

Let $z = z(x, y)$ be defined implicitly by $x^2 + y^2 + z^2 = 1$. Compute $\partial z/\partial x$ and $\partial z/\partial y$.

Solution Since the differential of a constant function is 0,

$$x\,dx + y\,dy + z\,dz = 0$$

Hence

$$dz = -\frac{x}{z}\,dx - \frac{y}{z}\,dy$$

This is a relation of the form $dz = M\,dx + N\,dy$, from which it follows that $\partial z/\partial x = M$ and $\partial z/\partial y = N$. Hence

$$\frac{\partial z}{\partial x} = -\frac{x}{z} = -\frac{\pm x}{\sqrt{1 - x^2 - y^2}}$$

$$\frac{\partial z}{\partial y} = -\frac{y}{z} = -\frac{\pm y}{\sqrt{1 - x^2 - y^2}}$$

We can check these answers by directly differentiating

$$z = \pm\sqrt{1 - x^2 - y^2}$$

●

Implicit Functions

The method used in Example 4 applies in general. First let us back up one dimension and state a result on implicit differentiation of functions of one variable. Although we have used implicit differentiation since Chapter 2, we never made a precise statement about its scope up to now. That is because it requires the material of this chapter to do so.

Implicit Differentiation—One Variable

Suppose $y = y(x)$ is a function of x that satisfies a relation

$$F(x, y) = 0$$

where F is a continuously differentiable function. Then

$$\frac{dy}{dx} = -\frac{F_x(x, y(x))}{F_y(x, y(x))}$$

at each point $(x, y(x))$ where $F_y \neq 0$.

Proof We have

$$F_x\,dx + F_y\,dy = d0 = 0$$

Hence

$$dy = -\frac{F_x}{F_y}\,dx$$

It follows that $dy/dx = -F_x/F_y$.

Remark The minus sign in the formula seems to contradict common sense. In the ordinary chain rule, differentials "cancel":

$$\frac{dy}{dx} = \frac{dy}{du} \cdot \frac{du}{dx} \quad \text{implies} \quad \frac{dy/dx}{dy/du} = \frac{du}{dx}$$

Thus the dy appears to have "canceled." But here we are writing

$$\frac{\partial F/\partial x}{\partial F/\partial y} = -\frac{dy}{dx}$$

and ∂F "cancels" with a mysterious sign change.

The reason for the sign change is that when we write $F(x, y) = 0$, we "take y to the other side." The equation $y = f(x)$ is equivalent to $F(x, y) = y - f(x) = 0$. Now

$$\frac{F_x}{F_y} = \frac{-f'(x)}{1} = -f'(x)$$

There is the minus sign!

Example 5

Find dy/dx for $y(x) = (1 - x^{2/3})^{3/2}$ where $0 < x < 1$.

Solution Set $F(x, y) = x^{2/3} + y^{2/3} - 1$. Then the relation satisfied by $y(x)$ is $F(x, y) = 0$. Therefore

$$\frac{dy}{dx} = -\frac{F_x}{F_y} = -\frac{\frac{2}{3}x^{-1/3}}{\frac{2}{3}y^{-1/3}} = -\left(\frac{y}{x}\right)^{1/3}$$

A similar implicit differentiation formula holds for functions of several variables. We state it in the special case of two variables.

Implicit Differentiation — Several Variables

Let $z(x, y)$ be a function that satisfies a relation

$$F(x, y, z) = 0$$

where F is a continuously differentiable function. Then

$$\frac{\partial z}{\partial x} = -\frac{F_x}{F_z} \quad \text{and} \quad \frac{\partial z}{\partial y} = -\frac{F_y}{F_z}$$

at each point $(x, y, z(x, y))$ where $F_z = 0$.

Indeed, from $F(x, y, z) = 0$ follows

$$F_x\,dx + F_y\,dy + F_z\,dz = 0$$

which implies

$$dz = -\frac{F_x}{F_z}\,dx - \frac{F_y}{F_z}\,dy$$

whenever $F_z \neq 0$ at $(x, y, z(x, y))$. Therefore

$$\partial z/\partial x = -F_x/F_z \quad \text{and} \quad \partial z/\partial y = -F_y/F_z$$

Existence of Implicit Functions

An advanced theorem called the *implicit function theorem* states that a relation $F(x, y, z) = 0$ determines a function $z = f(x, y)$ in the neighborhood of any point where $F_z \neq 0$.

More precisely, suppose $F(x, y, z)$ is continuously differentiable, $F(a, b, c) = 0$, and $F_z(a, b, c) \neq 0$. Then there exists a unique function $z = f(x, y)$, defined and differentiable in a neighborhood of (a, b), such that $f(a, b) = c$ and $F(x, y, f(x, y)) = 0$. The function $z = f(x, y)$ is differentiable and its partials can be computed implicitly as indicated above. See Figure 14-6-1.

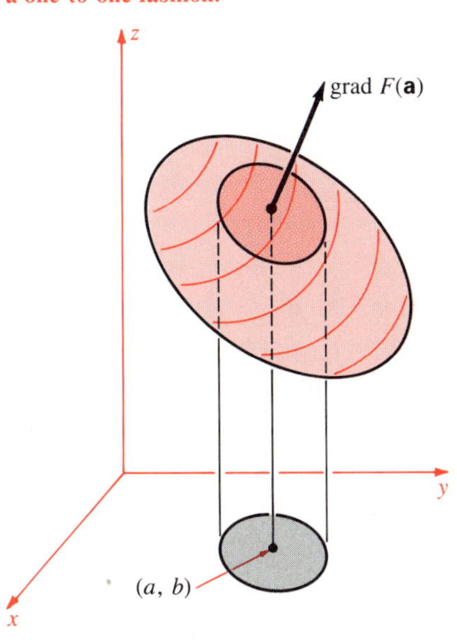

Figure 14-6-1
Assume $F_z(\mathbf{a}) \neq 0$, that is, grad $F(\mathbf{a})$ has a non-zero z-component. Then a neighborhood of $\mathbf{a}$ on $F(\mathbf{x}) = k$ projects into the x, y-plane in a one-to-one fashion.

Example 6

Find the partials at $(1, 1)$ of the function $z = f(x, y)$ defined by

$$x^3 + xy^2z + z^5 = 3 \quad \text{and} \quad f(1, 1) = 1$$

Solution Set

$$F(x, y, z) = x^3 + xy^2z + z^5 - 3$$

Then $F(1, 1, 1) = 0$ and

$$\left.\frac{\partial F}{\partial z}\right|_{(1,1,1)} = xy^2 + 5z^4\Big|_{(1,1,1)} = 6 \neq 0$$

Therefore there is a unique differentiable function $z = f(x, y)$ defined near $(x, y) = (1, 1)$ and such that $F(x, y, f(x, y)) = 0$ and $f(1, 1) = 1$. Its partials are given by

$$\frac{\partial f}{\partial x} = -\frac{F_x}{F_z} = -\frac{3x^2 + y^2 z}{xy^2 + 5z^4}$$

$$\frac{\partial f}{\partial y} = -\frac{F_y}{F_z} = -\frac{2xyz}{xy^2 + 5z^4}$$

Therefore

$$\left.\frac{\partial f}{\partial x}\right|_{(1,1)} = -\frac{3 + 1}{1 + 5} = -\tfrac{2}{3}$$

and

$$\left.\frac{\partial f}{\partial y}\right|_{(1,1)} = -\frac{2}{1 + 5} = -\tfrac{1}{3}$$

●

Tangent Plane

Suppose that we want the tangent plane at a point $\mathbf{a}$ of a level surface $F(\mathbf{x}) = k$. The following is a snappy way to write down its equation. First, we compute

$$\left. dF \right|_{\mathbf{a}} = L\,dx + M\,dy + N\,dz$$

Along the level surface F is constant, so $dF = 0$. Now we substitute $\mathbf{x} - \mathbf{a}$ for (dx, dy, dz), and presto! the tangent plane appears:

$$L(x - a) + M(y - b) + N(z - c) = 0$$

This is correct because

$$(L, M, N) = \text{grad}\, F(\mathbf{a})$$

For instance, to compute the tangent plane at $(1, 2, 3)$ to the sphere $x^2 + y^2 + z^2 = 14$, write

$$\begin{aligned} \left. dF \right|_{(1,2,3)} &= \left.(2x\,dx + 2y\,dy + 2z\,dz)\right|_{(1,2,3)} \\ &= 2\,dx + 4\,dy + 6\,dz \end{aligned}$$

and the tangent plane is

$$2(x - 1) + 4(y - 2) + 6(z - 3) = 0$$

that is, $x + 2y + 3z = 14$.

Exercises

Compute df at $(x, y) = (2, 1)$ and at $(x, y) = (1, -3)$

1 $f(x, y) = x^3y$ **2** $f(x, y) = \sqrt{x^2 + y^2}$

Compute df in general

3 $f(x, y) = 3x^2y - xy^2$

4 $f(x, y) = xe^{xy}$

5 $f(x, y) = \ln(x^2 + 3y^2)$

6 $f(x, y) = \sin(x + 2y)$

7 $f(x, y, z) = \dfrac{x}{y^2z}$

8 $f(x, y z) = \dfrac{xyz}{1 + y^2}$

9 $f(x, y, z) = e^x \cos y + e^y \cos z$

10 $f(x, y, z) = x^2 - y^2 - 5z^2 + xy + yz + 3x$

Prove

11 $d(fg) = (df)g + f\,dg$

12 $d(f/g) = \dfrac{(df)g - f\,dg}{g^2} \quad (g \neq 0)$

13 If $x = r\cos\theta$ and $y = r\sin\theta$, prove that $x\,dx + y\,dy = r\,dr$ and $x\,dy - y\,dx = r^2\,d\theta$.

14 Do Exercise 13 by computing $d(r^2)$ and $d\theta$ from $r^2 = x^2 + y^2$ and $\theta = \arctan(y/x)$.

Estimate by using differentials

15 $5.1 \times 7.1 \times 9.9$

16 $\sqrt{(5.99)^2 + (8.03)^2}$

17 $(2.01)^{0.98}$

18 $e^{-0.1}\tan(0.24\pi)$

19 the distance from the origin to $(3.05, 4.02, 11.96)$

20 the percentage increase in the volume V of a rectangular box caused by a 1% increase in each of its dimensions. [Hint Look at dV/V.]

21 At a point of space, the intensity I of illumination due to a point source of light is proportional to the power of the source and inversely proportional to the square of the distance from the source. Estimate the percentage change in I caused by 1% increases in both the power of the source and in the distance from the source. [Hint Look at dI/I.]

***22** The sides of a triangle are 20, 30, and 40 cm. Estimate the change in the largest angle if each side is shortened by 1 cm.

23 The range of a shell fired at angle α with the ground is $x = (v^2 \sin 2\alpha)/g$, where v is the muzzle velocity and g is the gravitational constant. If $v = 300$ m/sec and $g = 10$ m/sec^2, the shell will hit a target at distance $4500\sqrt{3} \approx 7794$ m if fired at a 30° angle. Suppose, in an actual test, that $v = 297.0$ m/sec, $g = 10.1$ m/sec^2, and $\alpha = 30.2°$. Estimate how far the shell will miss the target.

24 (cont.) Show that

$$\frac{dx}{x} = 2\frac{dv}{v} + (2\alpha\cot 2\alpha)\frac{d\alpha}{\alpha} - \frac{dg}{g}$$

Discuss the relative importance of a 1% error in α as compared to a 1% error in v for $\alpha = 30°$ and 45°.

Compute dz implicitly.

25 $x^2yz = 1$

26 $a^2x^2 + b^2y^2 + c^2z^2 = 10$

27 $z^5 + 6xz^2 + 8y^2z = 3$

28 $z^3 = e^{-xyz}$

29 $xy^2z^3 = 1 + y^2 + z^2$

30 $x = yz + z^4$

31 Let $z = z(x, y)$ be the larger solution of the equation $z^2 + 2xz + y = 0$, where $x^2 - y > 0$. Find dz in terms of x, y, dx, dy only.

32 Let $y = y(x)$ be a curve in the x, y-plane with $y'' \neq 0$. Then $(u, v) = (dy/dx, x\,dy/dx - y)$ is a curve in the u, v-plane. Express its slope simply in terms of x and y.

33 Suppose $f(x, y, z) = F[g(x, y, z)]$. Give a formula for df.

***34** An investment fund keeps equal amounts of capital in three types of account, A, B, and C, returning respectively 6%, 7%, and 10% annually compounded continuously. Because of a recession, these interest rates are cut by $\frac{1}{3}$%, $\frac{1}{2}$%, and $\frac{1}{4}$%. If the fund has $\$10^7$ in each account before the recession, estimate by differentials how much more total capital it must invest to maintain the same total earnings. Also find the *exact* additional capital.

***35** Given $f(x, y, z)$ and $\mathbf{a} = (a, b, c)$ such that $f(\mathbf{a}) = 0$, $f_x(\mathbf{a}) \neq 0$, $f_y(\mathbf{a}) \neq 0$, $f_z(\mathbf{a}) \neq 0$. Then $f(x, y, z) = 0$ can be solved for either x, y, or z as a function of the other two variables. Thus $x = P(y, z)$, where $a = P(b, c)$ and $f[P(y, z), y, z] = 0$. Let $y = Q(z, x)$ and $z = R(x, y)$ be the other two implicit functions. Find the product $P_yQ_zR_x$.

***36** Show that the function $z = z(x, y)$ defined implicitly by eliminating t from the two relations

$$\begin{cases} z\phi'(t) = [y - \phi(t)]^2 \\ (x + t)\phi'(t) = y - \phi(t) \end{cases}$$

satisfies $z = (\partial z/\partial x)(\partial z/\partial y)$.
[Hint Use differentials and eliminate dt.]

14-7 Differentiable Functions

Differential calculus is the study of functions that have linear approximations at each point. We now extend this point of view to functions of several variables, but first let us review the one-variable situation.

Suppose $f(x)$ has derivative $f'(a)$ at $x = a$. Then for x near a, the approximation $f(x) \approx f(a) + f'(a)(x - a)$ is accurate. The term "accurate" is not well defined; let us express what we mean geometrically. We know that $y = f(a) + f'(a)(x - a)$ is the equation of the tangent line to the graph of $y = f(x)$ at $(a, f(a))$. Geometrically, the tangent "hugs" the curve near the point of tangency (Figure 14-7-1). Therefore if $E(x)$ is the vertical distance between the curve and the tangent line, $E(x)$ ought to be small compared to $|x - a|$. In fact, the closer x is to a, the smaller the ratio $E(x)/|x - a|$ should be.

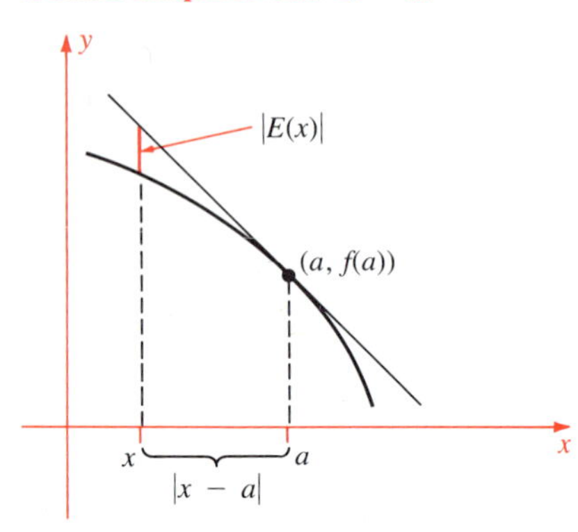

Figure 14-7-1
$E(x) = f(x) - f(a) - f'(a)(x - a)$ is small compared with $x - a$.

Let us now state these ideas precisely. It is convenient to introduce $u = x - a$.

- Suppose $f(x)$ is differentiable at $x = a$. Set $f(a + u) = f(a) + f'(a)u + E(u)$. Then

$$\lim_{u \to 0} \frac{E(u)}{u} = 0$$

For example, suppose $f(x) = x^2$ and $a = 3$. Then

$$f(3 + u) = (3 + u)^2 = 9 + 6u + u^2 = f(3) + f'(3)u + u^2$$

Therefore $E(u) = u^2$ and $E(u)/u = u \to 0$ as $u \to 0$.

The relation $E(u)/u \to 0$ characterizes differentiability. For suppose $f(x)$ satisfies

$$f(a + u) = f(a) + Bu + E(u)$$

where B is a constant and $E(u)/u \to 0$ as $u \to 0$. Then

$$\frac{f(a + u) - f(a)}{u} = B + \frac{E(u)}{u} \to B \qquad \text{as} \qquad u \to 0$$

This proves the following:

- Suppose $f(x)$ is defined near $x = a$ and satisfies

$$f(a + u) = f(a) + Bu + E(u)$$

where B is a constant and $\lim_{u \to 0} E(u)/u = 0$. Then $f(x)$ is differentiable at $x = a$ and $f'(a) = B$.

The relation $\lim_{u \to 0} E(u)/u = 0$ means that the linear function $f(a) + Bu$ closely approximates $f(a + u)$, and as u approaches 0, in fact, the error $E(u) = f(a + u) - [f(a) + Bu]$ approaches 0 *faster* than u approaches 0.

Thus for a function of *one* variable, the relation between derivative and linear approximation is clear: if the derivative exists, the linear approximation exists; if the linear approximation exists, the derivative exists.

Differentiable Functions of Two Variables

For a function of *several* variables, the situation is more complicated. It can happen that $\partial f/\partial x$ and $\partial f/\partial y$ both exist at a point (a, b), but no linear approximation $c + hx + ky$ exists!

Example 1

Let $f(0,0) = 0$, and for $(x, y) \neq (0,0)$, let

$$f(x, y) = \frac{xy}{\sqrt{x^2 + y^2}}$$

Show that

a f is continuous at $(0,0)$
b $f_x(0,0)$ and $f_y(0,0)$ exists
c f does not have a linear approximation at $(0,0)$.

Solution

a To handle points near $(0,0)$, write $f(x, y)$ in terms of polar coordinates:

$$f(x, y) = \frac{(r\cos\theta)(r\sin\theta)}{r} = r\cos\theta\sin\theta$$

Clearly $|f(x, y)| \le r$, so $f(x, y) \to 0$ as $(x, y) \to (0,0)$. This says that f is continuous at $(0,0)$.

b $$\frac{f(x,0) - f(0,0)}{x} = 0 \to 0 \quad \text{as} \quad x \to 0$$

Hence $f_x(0,0)$ exists and $f_x(0,0) = 0$. Similarly, $f_y(0,0) = 0$.

c Since $f(x, y) = 0$ on the x-axis and on the y-axis, the only possible linear approximation is 0. If 0 is truly a linear approximation to $f(x, y)$, then as (x, y) approaches $(0,0)$ the quantity $f(x, y) - 0$ must approach 0 *faster* than (x, y) approaches $(0,0)$. But along the line $x = y$,

$$f(x, x) - 0 = f(x, x) = \frac{x^2}{|x|} = |x|$$

which obviously approaches 0 at the same rate as (x, x) approaches $(0,0)$. Hence, $f(x, y)$ does *not* have a linear approximation at $(0,0)$. ●

Example 1 shows that the existence of both partials of $f(x, y)$ at a point does not guarantee that $f(x, y)$ is a reasonably behaved function. (Note that the two partials are not continuous in this case.) Now we shall examine the most useful functions of two variables, those that have good linear approximations. We shall call such functions *differentiable*. (This is a change from our previous usage, in which "differentiable" merely meant that first partials exist.)

Differentiable Functions

A function $f(x, y)$ is **differentiable** at (a, b) if there exists a linear function $Hu + Kv$ such that

$$f(a + u, b + v) = f(a, b) + Hu + Kv + E(u, v)$$

where

$$\lim_{(u,v)\to(0,0)} \frac{E(u, v)}{|(u, v)|} = 0$$

(Recall that $|(u, v)| = \sqrt{u^2 + v^2}$.)

Our first job is to show that differentiability implies both continuity and the existence of partials.

Theorem

Suppose $f(x, y)$ is differentiable at a point (a, b) of its domain:

$$f(a + u, b + v) = f(a, b) + Hu + Kv + E(u, v)$$

where $E(u, v)/|(u, v)| \to 0$ as $(u, v) \to (0, 0)$. Then

1 The function f is continuous at (a, b).

2 If (a, b) is an *interior* point of the domain, then the first partials of $f(x, y)$ at (a, b) exist, and

$$\frac{\partial f}{\partial x}(a, b) = H \qquad \text{and} \qquad \frac{\partial f}{\partial y}(a, b) = K$$

Proof If $(u, v) \to (0, 0)$ then $E(u, v) \to 0$. Consequently $f(a + u, b + v) \to f(a, b)$. This proves continuity.

To say that (a, b) is an **interior** point of the domain of f means that $(a + u, b + v)$ is in the domain of f whenever u and v are both small enough. If we set $v = 0$ in the formula above for $f(a + u, b + v)$, then we obtain

$$f(a + u, b) = f(a, b) + Hu + E(u, 0)$$

Consequently we have

$$\frac{\partial f}{\partial u}(a, b) = \lim_{u\to 0} \frac{f(a + u, b) - f(a, b)}{u}$$

$$= \lim_{u\to 0} \frac{Hu + E(u, 0)}{u} = H + \lim_{u\to 0} \frac{E(u, 0)}{u}$$

But

$$\left|\frac{E(u, 0)}{u}\right| = \left|\frac{E(u, 0)}{|(u, 0)|}\right| \to 0$$

so we have proved that $f_x(a, b) = H$. Similarly, $f_y(a, b) = K$.

The existence of partials $\partial f/\partial x$ and $\partial f/\partial y$ does not necessarily mean that f is differentiable. However, if the partials are *continuous*, then f is indeed differentiable.

Test for Differentiability

Suppose that $f(x, y)$ has partials $\partial f/\partial x$ and $\partial f/\partial y$ at each point of its domain. Suppose also that (a, b) is an interior point of the domain and that $\partial f/\partial x$ and $\partial f/\partial y$ are both continuous at (a, b). Then $f(x, y)$ is differentiable at (a, b).

Proof [Optional] The proof depends on the device of subtracting and adding a term. The idea is to go from (a, b) to a nearby point $(a + u, b + v)$ in two steps: vertical to $(a, b + v)$, then horizontal to $(a + u, b + v)$. We write

$$f(a + u, b + v) - f(a, b) = [f(a + u, b + v) - f(a, b + v)] + [f(a, b + v) - f(a, b)]$$

Now we apply the MVT twice:

$$\begin{cases} f(a + u, b + v) - f(a, b + v) = uf_x(a + \theta u, b + v) \\ f(a, b + v) - f(a, b) = vf_y(a, b + \gamma v) \end{cases}$$

where θ and γ depend on u and v, and also $0 < \theta < 1$ and $0 < \gamma < 1$. Therefore

$$\begin{aligned} &f(a + u, b + v) \\ &\quad = f(a, b) + uf_x(a + \theta u, b + v) + vf_y(a, b + \gamma v) \end{aligned}$$

Next we *define* $E(u, v)$ by

$$f(a + u, b + v) = f(a, b) + uf_x(a, b) + vf_y(a, b) + E(u, v)$$

Then

$$\begin{aligned} E(u, v) &= ug(u, v) + vh(u, v) \\ &= (u, v) \cdot (g(u, v), h(u, v)) \end{aligned}$$

where

$$\begin{cases} g(u, v) = f_x(a + \theta u, b + v) - f_x(a, b) \\ h(u, v) = f_y(a, b + \gamma v) - f_y(a, b) \end{cases}$$

Clearly $|E(u, v)| \le |(u, v)| \cdot |(g, h)|$. Therefore

$$\left| \frac{E(u, v)}{|(u, v)|} \right| \le |(g, h)|$$

Now we let $(u, v) \to (0, 0)$. Then

$$(a + \theta u, b + v) \to (a, b) \quad \text{and} \quad (a, b + \gamma v) \to (a, b)$$

But f_x and f_y are continuous at (a, b). Hence

$$g(u, v) \to 0 \quad \text{and} \quad h(u, v) \to 0$$

It follows that

$$\frac{E(u, v)}{|(u, v)|} \to 0 \qquad \text{as} \qquad (u, v) \to (0, 0)$$

This completes the proof. ●

The Chain Rule

Now we are in a position to give a precise statement and proof of the chain rule. This is important because the chain rule is the backbone of several variable differential calculus. For simplicity we limit the discussion to two variables.

Chain Rule

Let $f(x, y)$ be differentiable at the point $(x, y) = (a, b)$. Suppose that $x = x(t)$ and $y = y(t)$ are both differentiable at $t = c$ and $(x(c), y(c)) = (a, b)$. Then the function

$$z(t) = f[x(t), y(t)]$$

is differentiable at $t = c$ and

$$\left.\frac{dz}{dt}\right|_c = f_x(a, b)\left.\frac{dx}{dt}\right|_c + f_y(a, b)\left.\frac{dy}{dt}\right|_c$$

$$= \operatorname{grad} f(a, b) \cdot \left.\left(\frac{dx}{dt}, \frac{dy}{dt}\right)\right|_c$$

Proof [Optional] Since $f(x, y)$ is differentiable at (a, b),

$$f(a + u, b + v) = f(a, b) + Hu + Kv + E(u, v)$$

where

$$H = f_x(a, b) \qquad K = f_y(a, b) \qquad \text{and} \qquad \frac{E(u, v)}{|(u, v)|} \to 0$$

as $(u, v) \to (0, 0)$. We substitute

$$u = x(c + t) - a \qquad \text{and} \qquad v = y(c + t) - b$$

into this relation:

$$\begin{aligned} & f[x(c + t), y(c + t)] \\ & \quad = f(a, b) + H \cdot [x(c + t) - a] \\ & \qquad\qquad + K \cdot [y(c + t) - b] + E(u, v) \end{aligned}$$

Now we form the difference quotient

$$\frac{f[x(c+t), y(c+t)] - f(a,b)}{t}$$
$$= H\frac{x(c+t)-a}{t} + K\frac{y(c+t)-b}{t} + \frac{E(u,v)}{t}$$

and see what happens as $t \to 0$. Certainly $u \to 0$, $v \to 0$, and

$$\frac{x(c+t)-a}{t} \to \left.\frac{dx}{dt}\right|_c \quad\text{and}\quad \frac{y(c+t)-b}{t} \to \left.\frac{dy}{dt}\right|_c$$

since $a = x(c)$ and $b = y(c)$. Therefore

$$\lim_{t\to 0}\frac{z(c+t)-z(c)}{t} = \lim_{t\to 0}\frac{f[x(c+t), y(c+t)] - f(a,b)}{t}$$
$$= H\left.\frac{dx}{dt}\right|_c + K\left.\frac{dy}{dt}\right|_c + \lim_{t\to 0}\frac{E(u,v)}{t}$$

It remains to show that $E(u,v)/t \to 0$ as $t \to 0$. Now

$$\left|\frac{E(u,v)}{t}\right| = \frac{|E(u,v)|}{|(u,v)|}\cdot\frac{|(u,v)|}{|t|}$$

The first factor approaches 0 because $f(x,y)$ is differentiable at (a,b). The second factor approaches a limit because

$$\frac{(u,v)}{t} = \left(\frac{x(c+t)-a}{t}, \frac{y(c+t)-b}{t}\right) \to \left.\left(\frac{dx}{dt}, \frac{dy}{dt}\right)\right|_c$$

Therefore $E(u,v)/t \to 0$, and the proof is finished. ●

Exercises

Show that $f(x,y)$ is differentiable by finding a linear function $Hu + Kv$ and a function $E(u,v)$ that satisfy the definition of differentiability given on page 746. [Hint Set $\mathbf{u} = (u,v)$ and use the estimates $|u| \le |\mathbf{u}|$, $|v| \le |\mathbf{u}|$ as needed to prove $E(\mathbf{u})/|\mathbf{u}| \to 0$.]

1 $3x - 7y + 4$ at (a,b)

2 $x^2 + y^2$ at $(-2,1)$

3 xy^2 at $(0,0)$

4 xy^2 at $(3,2)$

5 y/x at $(1,1)$

6 $1/xy$ at $(-1,2)$

7 Prove that the sum of two differentiable functions is differentiable.

***8** Prove that the product of two differentiable functions is differentiable.

Given $f(0,0) = 0$. **a** Determine whether $f(x,y)$ is continuous at $(0,0)$. **b** Determine whether $f_x(0,0)$ and $f_y(0,0)$ exist and find their values if they do. **c** Determine whether $f(x,y)$ is differentiable at $(0,0)$.

9 $f(x,y) = xy/(x^2+y^2)$ if $(x,y) \ne (0,0)$

10 $f(x,y) = x^3/(x^2+y^2)$ if $(x,y) \ne (0,0)$

11 $f(x,y) = (x^6+y^6)/(x^4+y^4)$ if $(x,y) \ne (0,0)$

12 $f(x,y) = (x^5+y^6)/(x^4+y^4)$ if $(x,y) \ne (0,0)$

The analogue of **Newton's method** for solving a system $f(x,y) = 0$, $g(x,y) = 0$ is an iterative scheme that replaces an approximation $\mathbf{x}$ to a solution by $\phi(\mathbf{x})$, a closer approximation. The formula is

$$\phi(\mathbf{x}) = \mathbf{x} - \frac{1}{\begin{vmatrix} f_x & f_y \\ g_x & g_y \end{vmatrix}}\left(\begin{vmatrix} f & f_y \\ g & g_y \end{vmatrix}, \begin{vmatrix} f_x & f \\ g_x & g \end{vmatrix}\right)$$

with all functions evaluated at $\mathbf{x}$. (The denominator must be non-zero near the solution.) Given an initial "guess" $\mathbf{x}_0$, we construct

$$\mathbf{x}_1 = \phi(\mathbf{x}_0), \quad \mathbf{x}_2 = \phi(\mathbf{x}_1), \quad \mathbf{x}_3 = \phi(\mathbf{x}_2), \quad \cdots$$

Under reasonable conditions, the sequence $\{\mathbf{x}_n\}$ converges to a solution of the system $f(\mathbf{x}) = 0$, $g(\mathbf{x}) = 0$. Test this: in each case compute $\mathbf{x}_1$ and $\mathbf{x}_2$

13 $f(x, y) = x + y^2 \qquad g(x, y) = y + x^2$
$\mathbf{x}_0 = (\frac{1}{2}, 0)$

14 $f(x, y) = x + y^2 \qquad g(x, y) = x + y$
$\mathbf{x}_0 = (1, \frac{1}{4})$

15 $f(x, y) = x^2 + y^2 - 1 \qquad g(x, y) = x + 4y - 2$
$\mathbf{x}_0 = (1, 1)$

***16** Use the definition of differentiability, with the error ignored, to derive Newton's method.

14-8 Review Exercises

Find $\partial f/\partial x$ and $\partial f/\partial y$ for $f(x, y) =$

1 $e^x \cos y$

2 $\sin(x^2 y)$

3 $\exp(x - y)$

4 $x^8 y^{10} - x^{10} y^8$

Find grad f

5 $f = x^2 y$

6 $f = x^3 y^2 - 2xy^4$

7 $f = xy^2 z^3$

8 $f = x^2 + y^3 + z^4$

9 Find the largest directional derivative at $(1, 1, 1)$ of $f = x^4 + y^5 + z^6$.

10 Find the tangent plane and unit normal at $(1, -2, 1)$ to $x^4 + y^4 + z^4 = 18$.

11 Find the tangent plane and unit normal to $\mathbf{x} = (u^2, uv, v^2)$ at the point corresponding to $(u, v) = (1, 2)$.

12 Compute the differential of $f = x^3 y^4 z^5$.

Find df/dt, where

13 $f(t) = x^3 + y^3 \qquad x = \cos t \qquad y = \sin t$

14 $f(t) = \sin(uv) \qquad u = t^2 \qquad v = e^t$

Use differentials to estimate

15 $(1.01)^3 (0.98)^4 (1.02)^5$

16 $e^{-0.02} \sin 0.01$

Find $\partial z/\partial x$ and $\partial z/\partial y$ where

17 $xyz + z^2 = x^2 + y^2$

18 $xz + ye^z = 1$

15 Surfaces and Optimization

We begin this chapter with surfaces defined by quadratic relations; these surfaces arise frequently in applications. The main part of the chapter covers optimization (finding maxima and minima). Sections 15-2 and 15-3 deal with finding the maxima or the minima of functions of two or more variables. This is a very satisfactory extension of our work on optimizing functions of one variable. Sections 15-4 and 15-5 deal with a new type of optimization problem in which the independent variables are not completely independent, but are subject to side conditions (constraints).

15-1 Quadric Surfaces

These special surfaces occur frequently in calculus examples. A **quadric surface** is a special type of level surface, defined by an equation $f(x, y, z) = 0$, where $f(x, y, z)$ is a quadratic polynomial. The most general quadratic polynomial is

$$f(x, y, z) = Ax^2 + By^2 + Cz^2 + Dxy + Eyz + Fzx + px + qy + rz + k$$

We shall study quadric surfaces only for the following special types of quadratic polynomials:

a $Ax^2 + By^2 + Cz^2 + k$

b $Ax^2 + By^2 + rz$ **b′** $Ax^2 + By^2 + k$

c $Ax^2 + qy$ **c′** $Ax^2 + k$

Later we shall observe that every quadric surface is one of these types with respect to a suitably chosen rectangular coordinate system. We assume throughout that *some* second degree terms are present; otherwise the surface $f(x, y, z) = 0$ is a plane, which we studied in Chapter 12.

We begin with type **a**, assuming $ABC \neq 0$ and $k \neq 0$. With a bit of juggling of constants, the equation $f(x, y, z) = 0$ can be written in the form

$$\pm\frac{x^2}{a^2} \pm \frac{y^2}{b^2} \pm \frac{z^2}{c^2} = 1$$

The nature of this quadric surface depends on how many of the signs are plus and how many are minus. If all three signs are minus, then no points satisfy the equation, so the surface is the empty set. Otherwise it is symmetric in each coordinate plane, because if (x, y, z) is on the surface, then so are all eight points $(\pm x, \pm y, \pm z)$. Therefore the whole surface is determined by symmetry from its part in the first octant.

Ellipsoids

Consider the quadric surface

$$\frac{x^2}{a^2} + \frac{y^2}{b^2} + \frac{z^2}{c^2} = 1 \qquad (a, b, c > 0)$$

Since squares are non-negative, each of its points satisfies

$$\frac{x^2}{a^2} \le 1 \qquad \frac{y^2}{b^2} \le 1 \qquad \frac{z^2}{c^2} \le 1$$

This means the surface is confined to the box

$$-a \le x \le a$$
$$-b \le y \le b$$
$$-c \le z \le c$$

Suppose $-c \le z_0 \le c$. The horizontal plane $z = z_0$ intersects the surface in the curve whose x, y-equation is

$$\frac{x^2}{a^2} + \frac{y^2}{b^2} = 1 - \frac{{z_0}^2}{c^2}$$

This curve is an ellipse. It is as large as possible when $z_0 = 0$, and it becomes smaller and smaller as $z_0 \to c$ or $z_0 \to -c$. Thus each such cross section by a horizontal plane is an ellipse, except at the extremes $z_0 = \pm c$, where it is a single point.

The same argument applies to plane sections parallel to the other coordinate planes. This gives us enough information for a sketch. The surface is called an **ellipsoid** (Figure 15-1-1). In the special case $a = b = c$, it is a sphere.

There is a nice formula for the tangent plane to an ellipsoid, which we derive by applying our method for level surfaces (pages 725–726).

Figure 15-1-1

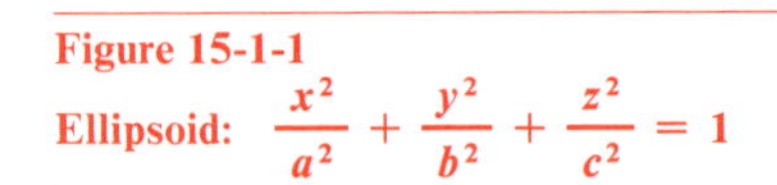

Ellipsoid: $\dfrac{x^2}{a^2} + \dfrac{y^2}{b^2} + \dfrac{z^2}{c^2} = 1$

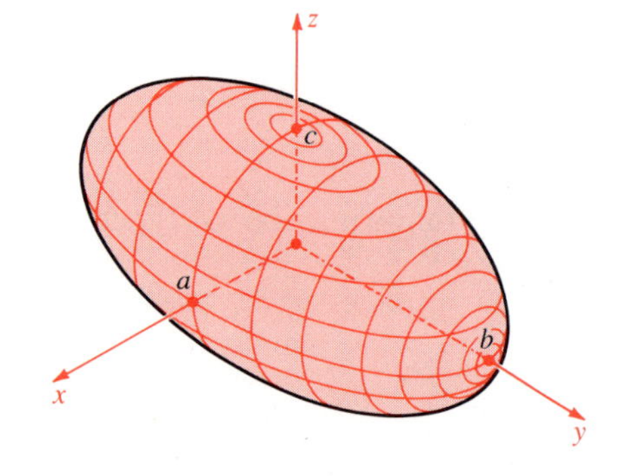

Tangent Plane

Let $\mathbf{m} = (\ell, m, n)$ be a point of the ellipsoid

$$\frac{x^2}{a^2} + \frac{y^2}{b^2} + \frac{z^2}{c^2} = 1$$

Then the tangent plane at $\mathbf{m}$ is

$$\frac{\ell x}{a^2} + \frac{my}{b^2} + \frac{nz}{c^2} = 1$$

To derive the formula, we first compute the gradient:

$$\operatorname{grad}\left(\frac{x^2}{a^2} + \frac{y^2}{b^2} + \frac{z^2}{c^2}\right)\bigg|_{\mathbf{m}} = 2\left(\frac{x}{a^2}, \frac{y}{b^2}, \frac{z}{c^2}\right)\bigg|_{\mathbf{m}} = 2\left(\frac{\ell}{a^2}, \frac{m}{b^2}, \frac{n}{c^2}\right)$$

Therefore $\mathbf{x}$ is on the tangent plane at $\mathbf{m}$ if and only if

$$2\left(\frac{\ell}{a^2}, \frac{m}{b^2}, \frac{n}{c^2}\right) \cdot (\mathbf{x} - \mathbf{m}) = 0$$

This condition is equivalent to

$$\frac{\ell x}{a^2} + \frac{my}{b^2} + \frac{nz}{c^2} = \frac{\ell^2}{a^2} + \frac{m^2}{b^2} + \frac{n^2}{c^2} = 1$$

The right-hand side equals 1 because $\mathbf{m}$ is on the ellipsoid.

Hyperboloids of One Sheet

Consider the quadric surface

$$\frac{x^2}{a^2} + \frac{y^2}{b^2} - \frac{z^2}{c^2} = 1 \qquad (a, b, c > 0)$$

Each horizontal cross section cut out by a plane $z = z_0$ is an ellipse with x, y-equation

$$\frac{x^2}{a^2} + \frac{y^2}{b^2} = 1 + \frac{{z_0}^2}{c^2}$$

no matter what value z_0 is. The smallest ellipse occurs for $z_0 = 0$; as $z_0 \to \infty$ or $z_0 \to -\infty$, the ellipses get larger and larger.

The surface meets the y,z-plane in the hyperbola

$$\frac{y^2}{b^2} - \frac{z^2}{c^2} = 1$$

and it meets the z,x-plane in the hyperbola

$$\frac{x^2}{a^2} - \frac{z^2}{c^2} = 1$$

Figure 15-1-2
Hyperboloid of one sheet:
$$\frac{x^2}{a^2} + \frac{y^2}{b^2} - \frac{z^2}{c^2} = 1$$

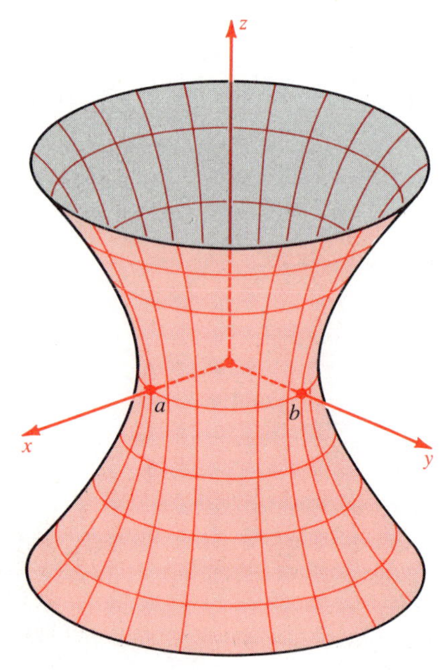

Figure 15-1-3
Hyperboloid of two sheets:
$$-\frac{x^2}{a^2} - \frac{y^2}{b^2} + \frac{z^2}{c^2} = 1$$

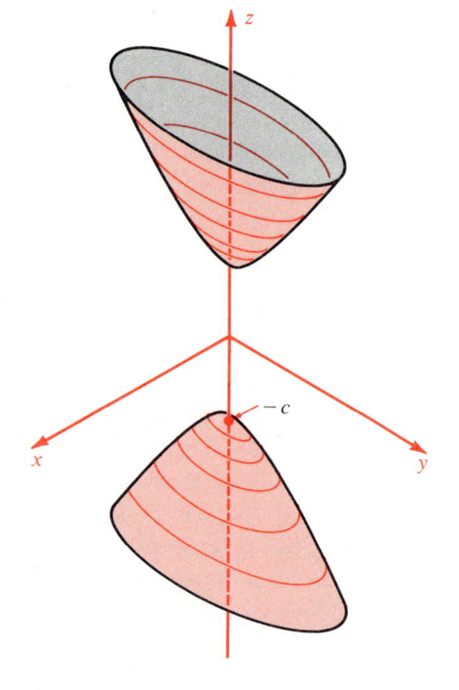

This information is enough to sketch the surface, called a **hyperboloid of one sheet** (Figure 15-1-2). Exactly as for the ellipsoid, we can show that the tangent plane at (ℓ, m, n) to $x^2/a^2 + y^2/b^2 - z^2/c^2 = 1$ is

$$\frac{\ell x}{a^2} + \frac{my}{b^2} - \frac{nz}{c^2} = 1$$

Hyperboloids of Two Sheets

Consider the quadric surface

$$-\frac{x^2}{a^2} - \frac{y^2}{b^2} + \frac{z^2}{c^2} = 1 \qquad (a, b, c > 0)$$

If (x, y, z) is a point on the surface, then

$$\frac{z^2}{c^2} = 1 + \frac{x^2}{a^2} + \frac{y^2}{b^2} \geq 1$$

Hence $z^2 \geq c^2$. This means either $z \geq c$ or $z \leq -c$; that is, there are no points of the surface between the horizontal planes $z = c$ and $z = -c$.

If $z_0{}^2 > c^2$, the horizontal plane $z = z_0$ meets the surface in the curve whose x, y-equation is

$$\frac{x^2}{a^2} + \frac{y^2}{b^2} = \frac{z_0{}^2}{c^2} - 1 > 0$$

an ellipse. Also the surface meets the y, z-plane and the z, x-plane in the hyperbolas

$$-\frac{y^2}{b^2} + \frac{z^2}{c^2} = 1 \qquad \text{and} \qquad -\frac{x^2}{a^2} + \frac{z^2}{c^2} = 1$$

respectively. The surface breaks into two parts, and it is called a **hyperboloid of two sheets** (Figure 15-1-3). Its tangent plane at (ℓ, m, n) is

$$-\frac{\ell x}{a^2} - \frac{my}{b^2} + \frac{nz}{c^2} = 1$$

Quadratic Cones

To complete the classification of quadric surfaces of the form

$$Ax^2 + By^2 + Cz^2 = k \qquad (ABC \neq 0)$$

we must consider the case $k = 0$. Then the equation can be written

$$\pm\frac{x^2}{a^2} \pm \frac{y^2}{b^2} \pm \frac{z^2}{c^2} = 0 \qquad (a, b, c > 0)$$

If the signs are all the same, then $(0, 0, 0)$ is the only point on the graph; not interesting. If the three signs are not all the same, then two of the signs are the same and the third is opposite. Changing signs if necessary, and renaming constants, we have

$$z^2 = \frac{x^2}{a^2} + \frac{y^2}{b^2} \qquad (a, b > 0)$$

For each point $\mathbf{x}_0 = (x_0, y_0, z_0)$ on the surface, the entire line $\mathbf{x} = t\mathbf{x}_0$ lies on the surface. A surface with this property is called a **cone,** and the lines $\mathbf{x} = t\mathbf{x}_0$ are called **generators** of the cone.

To show that

$$z^2 = \frac{x^2}{a^2} + \frac{y^2}{b^2}$$

really has the cone property, we take any point $\mathbf{x}_0 = (x_0, y_0, z_0)$ on the surface and check that $t\mathbf{x}_0 = (tx_0, ty_0, tz_0)$ is also on the surface. We have

$$(tz_0)^2 = t^2 z_0{}^2 = t^2\left(\frac{x_0{}^2}{a^2} + \frac{y_0{}^2}{b^2}\right) = \frac{(tx_0)^2}{a^2} + \frac{(ty_0)^2}{b^2}$$

Thus $t\mathbf{x}_0$ is indeed on the surface.

To sketch the cone, we note that it meets the horizontal plane $z = 1$ in the ellipse

$$\frac{x^2}{a^2} + \frac{y^2}{b^2} = 1$$

For each point on this ellipse, we draw the line through the point and $\mathbf{0}$. See Figure 15-1-4.

The origin is a singular point on the cone (no tangent plane). The tangent plane at any other point (ℓ, m, n) is

$$nz = \frac{\ell x}{a^2} + \frac{my}{b^2}$$

Figure 15-1-4

Quadratic cone: $z^2 = \dfrac{x^2}{a^2} + \dfrac{y^2}{b^2}$

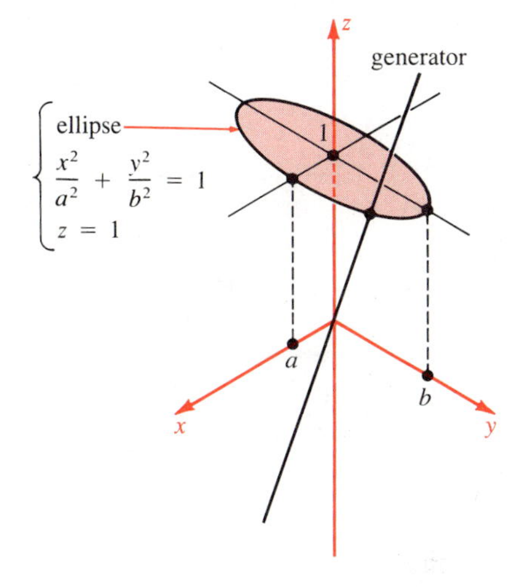

a A generator

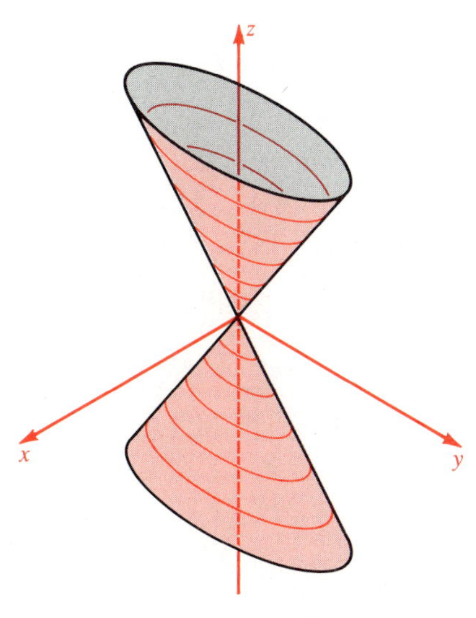

b The cone

Paraboloids

The next type of quadric surface we consider is given by

$$f(x, y, z) = Ax^2 + By^2 + rz \qquad (AB \neq 0, \quad r \neq 0)$$

We may write the equation $f(x, y, z) = 0$ in the form

$$z = \pm\frac{x^2}{a^2} \pm \frac{y^2}{b^2} \qquad (a, b > 0)$$

Changing z to $-z$ merely turns the surface upside down, so there are essentially two distinct cases: $++$ and $-+$.

The first case is the surface

$$z = \frac{x^2}{a^2} + \frac{y^2}{b^2} \qquad (a, b > 0)$$

Figure 15-1-5
Elliptic paraboloid:
$z = \dfrac{x^2}{a^2} + \dfrac{y^2}{b^2}$

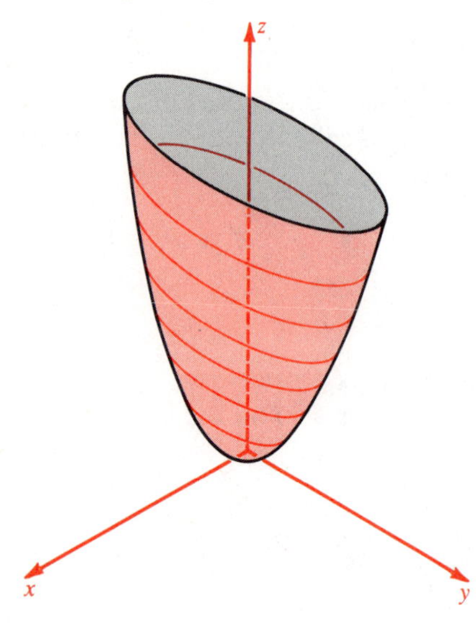

called an **elliptic paraboloid.** It lies above the x,y-plane and is symmetric in the y,z- and z,x-planes. Each horizontal cross-section

$$\frac{x^2}{a^2} + \frac{y^2}{b^2} = z \qquad z = z_0 > 0$$

is an ellipse. As z_0 increases, the ellipse grows larger. The surface is called a paraboloid because it meets the y,z-plane and the z,x-plane in the parabolas $z = y^2/b^2$ and $z = x^2/a^2$ respectively (Figure 15-1-5). Indeed, the surface meets each plane through the z-axis in a parabola. The tangent plane to the elliptic paraboloid at any point (ℓ, m, n) is

$$z = 2\left(\frac{\ell x}{a^2} + \frac{my}{b^2}\right) - n$$

The second case is the **hyperbolic paraboloid,** the graph of

$$z = -\frac{x^2}{a^2} + \frac{y^2}{b^2} \qquad (a, b > 0)$$

This surface is symmetric in the y,z- and z,x-planes. The horizontal planes $z = z_0 > 0$ meet it in hyperbolas whose branches open out in the y-direction. The horizontal planes $z = z_0 < 0$ meet it in hyperbolas that open out in the x-direction. The y,z-plane meets the surface in the parabola $z = y^2/b^2$, which opens upward; and the z,x-plane meets it in the parabola $z = -x^2/a^2$, which opens downward. The best description is "saddle-shaped." See Figure 15-1-6. Note that the special horizontal plane $z = 0$ meets the surface in $y^2/b^2 = x^2/a^2$, that is, in the pair of lines $y/b = \pm x/a$.

Figure 15-1-6
Hyperbolic paraboloid:
$z = -\dfrac{x^2}{a^2} + \dfrac{y^2}{b^2}$

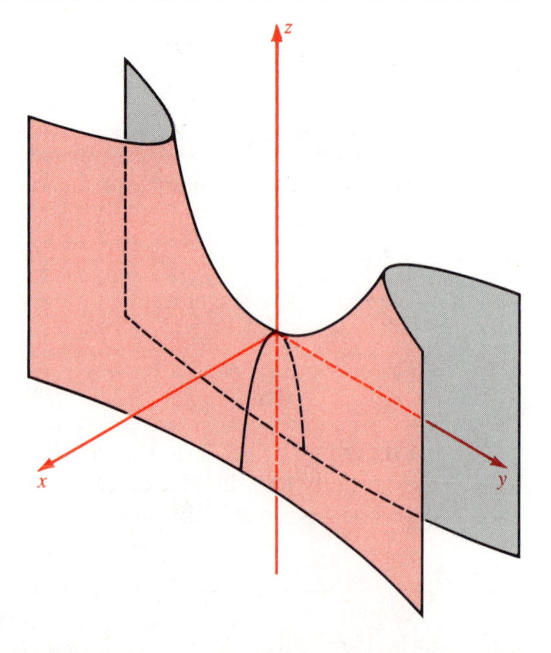

The tangent plane to the hyperbolic paraboloid at (ℓ, m, n) is

$$z = 2\left(-\frac{\ell x}{a^2} + \frac{my}{b^2}\right) - n$$

Quadratic Cylinders

Now we consider quadric surfaces of the types

$$Ax^2 + By^2 + k = 0 \quad \text{and} \quad Ax^2 + qy = 0$$

In both, the variable z is missing. Generally, when one variable is missing the surface is a cylinder over a base curve in the plane of the missing variable. Take for example $Ax^2 + By^2 + k = 0$, where $A > 0$, $B > 0$, and $k < 0$. This can be written in the form

$$\frac{x^2}{a^2} + \frac{y^2}{b^2} = 1 \qquad (a, b > 0)$$

Figure 15-1-7
Quadratic cylinders

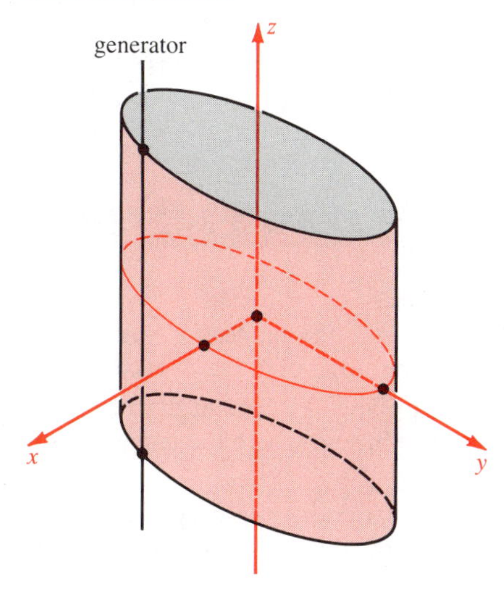

a Elliptic cylinder: $\frac{x^2}{a^2} + \frac{y^2}{b^2} = 1$

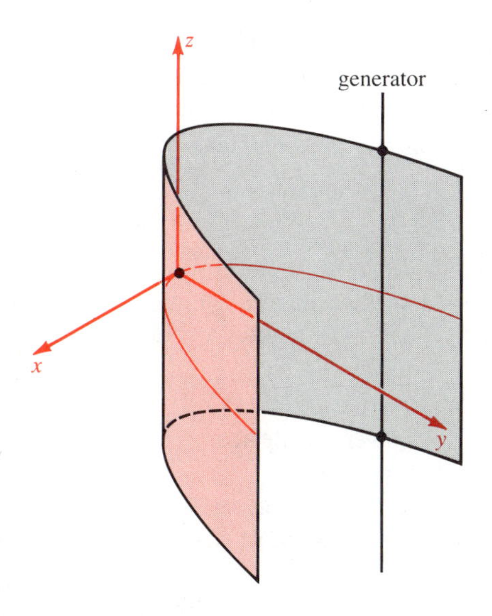

b Parabolic cylinder: $Ax^2 + qy = 0$

The surface meets each horizontal plane $z = z_0$ in the same-size ellipse. If (x_0, y_0, z_0) is any point of the surface, the whole vertical line (x_0, y_0, z) for $-\infty < z < \infty$ lies on the surface. The surface is an **elliptic cylinder** and these vertical lines that lie on the surface are called **generators** of the cylinder (Figure 15-1-7a).

Any curve $f(x, y) = 0$ in the x, y-plane generates a cylinder in space consisting of all points (x_0, y_0, z) for which $f(x_0, y_0) = 0$. In particular, a circle generates a (right) circular cylinder, a hyperbola generates a hyperbolic cylinder, and so on.

The equation $Ax^2 + qy = 0$ $(A \neq 0, q \neq 0)$ is a parabola in the x, y-plane, so in space it is **parabolic cylinder** (Figure 15-1-7b).

This completes our discussion of quadrics except for surfaces of the type $Ax^2 + k = 0$ where both y and z are missing. Depending on the signs of A and k, the surface $Ax^2 + k = 0$ is empty or consists of one plane or two planes parallel to the y, z-plane. In general, the surface $f(x) = 0$ in $\mathbf{R}^3$ is a set of planes parallel to the y, z-plane. For each zero x_0 of $f(x)$, the plane $x = x_0$ is included in the surface.

Reduction of Quadratic Polynomials [Optional]

Given a general quadratic polynomial

$$f(x, y, z) = Ax^2 + By^2 + Cz^2 + Dxy + Eyz + Fzx + px + qy + rz + k$$

a fairly deep theorem in linear algebra says that the mixed terms, xy, yz, zx, can be eliminated by a suitable rotation of the coordinate system. A rotation does not affect the nature of the quadric surface $f(x, y, z) = 0$, so in studying quadrics we may assume that the mixed terms have been eliminated:

$$f(x, y, z) = Ax^2 + By^2 + Cz^2 + px + qy + rz + k$$

If $A \neq 0$, the translation $x = x' - p/2A$ eliminates the term px. Similarly if $B \neq 0$ then qy can be eliminated, and if $C \neq 0$ then

rz can be eliminated. Thus by a rotation and a translation we reduce the study of quadratic polynomials $f(x, y, z)$ to one of the forms:

i $Ax^2 + By^2 + Cz^2 + k$

ii $Ax^2 + By^2 + rz + k$

iii $Ax^2 + qy + rz + k$

These include all possibilities, provided we are willing to permute the variables. For instance, the polynomial $Cz^2 + px + qy + k$ becomes type **iii** when x and z are interchanged.

In type **ii**, if $r \neq 0$, then a translation in the z-direction eliminates k. In type **iii**, a rotation in the y, z-plane, taken so that $qy + rz = 0$ is the new z-axis, changes the function to the form $ax^2 + qy + k$. Again, k can be eliminated by translation if $q \neq 0$.

This discussion shows that we may study all quadric surfaces by concentrating on the five types of quadratic polynomials listed at the beginning of this section.

Exercises

Sketch the first octant portion

1 $\frac{1}{4}x^2 + y^2 + \frac{1}{4}z^2 = 1$

2 $x^2 + \frac{1}{9}y^2 + \frac{1}{4}z^2 = 1$

3 $x^2 + y^2 - z^2 = 1$

4 $-x^2 - y^2 + z^2 = 1$

5 $x^2 - y^2 + z^2 = 1$

6 $-x^2 + y^2 - z^2 = 1$

7 $z = x^2 + y^2$

8 $z = \frac{1}{4}x^2 + y^2$

9 $z = -x^2 + y^2$

10 $z = x^2 - y^2$

Identify the quadric surface

11 $z = x^2 + 2x + y^2$
[Hint Complete the square.]

12 $x^2 + 2y^2 + 3z^2 - 2x - 8y + 6z = 0$

13 $x^2 + y^2 - a^2(z - 1)^2 = 0$

14 $z = xy$
[Hint Rotate 45° about the z-axis.]

Sketch the paraboloids

15 $x = y^2 + z^2$

16 $y = x^2 - z^2$

Sketch the surface in x, y, z-space

17 $x - z = 1$

18 $y = x^2$

19 $xy = 1$

20 $-x^2 + y^2 = 1$

21 $x = z^2$

22 $y^2 + 4z^2 = 1$

23 $z = x^2 - x$

24 $x^2 + 4z^2 = 1$

25 $y^2 = z^2 + 4x^2$

26 $x^2 = y^2$

27 Suppose $f(x, y) = 0$ and $z = 1$ describe a curve on the plane $z = 1$. Find an equation for the cone obtained by taking all points on all lines through $\mathbf{0}$ and points of the curve.

28 (cont.) Test your result on $x^2 + y^2 = 1,\ z = 1$.

29 Show that all cross sections of the ellipsoid $x^2/a^2 + y^2/b^2 + z^2/c^2 = 1$ by planes $x = k$, where $-a < x < a$, are ellipses of the same eccentricity.

30 Given an ellipsoid and a plane P, prove that there are exactly two points on the ellipsoid where the tangent plane is parallel to P.

31 Let $\mathbf{n}$ be a fixed unit vector and α a fixed acute angle. Find a vector equation for the cone with vertex $\mathbf{0}$ whose generators all make angle α with $\mathbf{n}$.

32 (cont.) Find an equation for the cone with vertex $\mathbf{0}$ whose generators make angle 45° with the line through $\mathbf{0}$ and $(1, 1, 0)$.

33 Show that the tangent planes at all points (u, v, k), where $k > 0$ is fixed, of the elliptic paraboloid $z = x^2/a^2 + y^2/b^2$ have a common point. What point?

34 Find the intersection of the tangent plane at $\mathbf{0}$ to the hyperbolic paraboloid $x = y^2/a^2 - z^2/b^2$ with the quadric itself.

The following exercises assume some knowledge of matrices. Let A be a 3×3 non-zero symmetric matrix and $\mathbf{x} = (x, y, z)$. Then $\mathbf{x}'$ denotes a column vector, the *transpose* of $\mathbf{x}$.

***35** Set $f(\mathbf{x}) = \mathbf{x}A\mathbf{x}'$. Show that $\operatorname{grad} f(\mathbf{x}) = 2\mathbf{x}A$.

36 (cont.) Show that the tangent plane at a point $\mathbf{m}$ of the quadric $f(\mathbf{x}) = 1$ is $\mathbf{m}A\mathbf{x}' = 1$.

37 (cont.) Let $\mathbf{b}$ be a fixed vector and set $f(\mathbf{x}) = \mathbf{x}A\mathbf{x}' + 2\mathbf{b}\mathbf{x}' + k$, the most general quadratic polynomial. Find $\operatorname{grad} f(\mathbf{x})$.

***38** (cont.) Let $\mathbf{m}$ be a point of the quadric $f(\mathbf{x}) = 0$. Find the tangent plane at $\mathbf{m}$.

39 Let $\mathbf{n}$ be a unit vector. Show that the set of points on the ellipsoid $x^2/a^2 + y^2/b^2 + z^2/c^2 = 1$ where the tangent plane is parallel to $\mathbf{n}$ is an ellipse. (You may assume that any plane section of an ellipsoid is an ellipse.)

40 (cont.) What does this say about the shadow of an ellipsoid when the sun is directly overhead?

41 Let $\mathbf{a} = (a, b, c)$ be a point outside of the paraboloid $z = x^2 + y^2$, that is, such that $c < a^2 + b^2$. Show that the set of all points $\mathbf{x}$ on the paraboloid such that tangent plane at $\mathbf{x}$ passes through $\mathbf{a}$ lies on a plane.

42 (cont.) Show that the intersection of the plane found in Exercise 41 and the paraboloid is also the intersection of the paraboloid with a right circular cylinder.

43 Let $\mathbf{a}$ be a point of the hyperbolic paraboloid $z = x^2 - y^2$. Show that the tangent plane at $\mathbf{a}$ intersects the surface in two straight lines.
[Hint Eliminate z from the equations of the quadric and the tangent plane.]

44 (cont.) Show the same thing for the hyperboloid of one sheet $z^2 = x^2 + y^2 - 1$.

45 Consider the cone $x^2 + y^2 = z^2 \tan^2\alpha$ and the cylinder $(z - a\csc\alpha)^2 + y^2 = a^2$, where $a > 0$ and $0 < \alpha < \frac{1}{2}\pi$. Find the two points of tangency of the two surfaces.

***46** (cont.) Show that the intersection of the two surfaces consists of a pair of ellipses.

47 The straight line that passes through $(1, 0, 1)$ and $(0, 1, 0)$ is revolved about the z-axis. What is the corresponding surface of revolution?

48 Each point of the line $(1, t, t)$ is connected to the corresponding point of the line $(-1, t, -t)$ by a straight line. What surface does this line sweep out? [Hint See Exercise 14.]

15-2 Optimization: Two Variables

The word "optimization" means choosing the most favorable value—for us, the maximum or minimum of a function of several variables. Before we plunge in, let us recall some facts about extrema of functions of one variable. Let $f(x)$ be a differentiable function. Suppose $f(x)$ has a *local* maximum or a *local* minimum at a point $x = c$ that is not a boundary point of its domain. Then $y = f(x)$ has a horizontal tangent at $x = c$. See Figures 15-2-1a and 15-2-1b. But as Figure 15-2-1c shows, a horizontal tangent does not guarantee a local maximum or a local minimum.

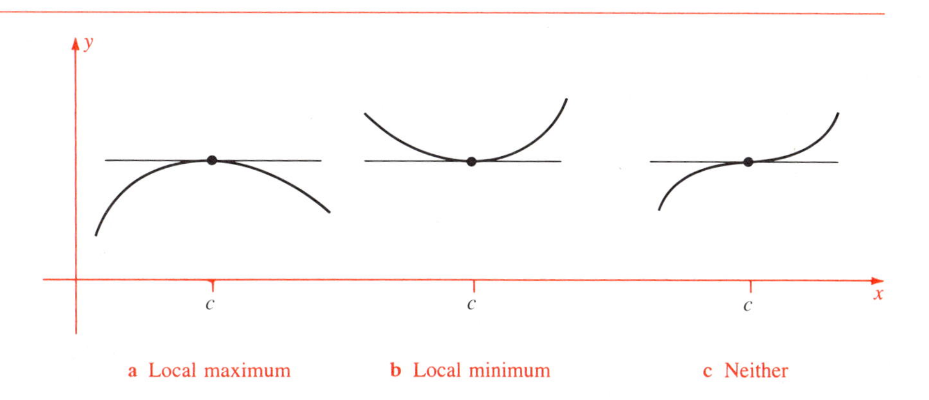

Figure 15-2-1
Horizontal tangents

Now let $f(x, y)$ be a continuously differentiable function on a domain $\mathbf{D}$ in the x,y-plane. Suppose $f(x, y)$ has a local maximum at $(x, y) = (a, b)$, an interior point of $\mathbf{D}$ (not a boundary point). This means that

$$f(x, y) \le f(a, b)$$

Figure 15-2-2
Horizontal tangent plane at a local maximum

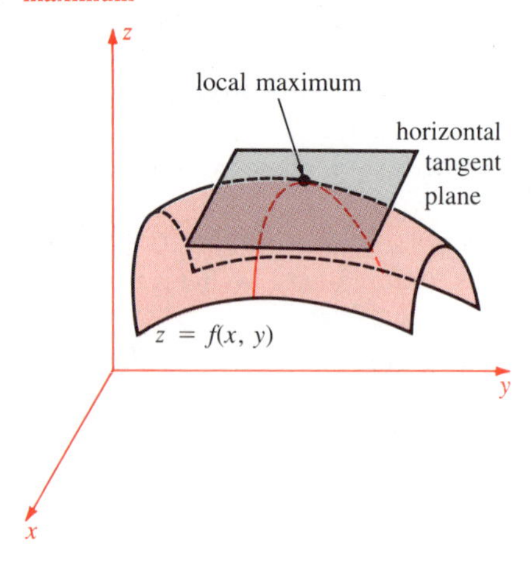

Figure 15-2-3
The cross section $z = f(x, b)$ of $z = f(x, y)$ at $y = b$ has a local maximum at $x = a$.

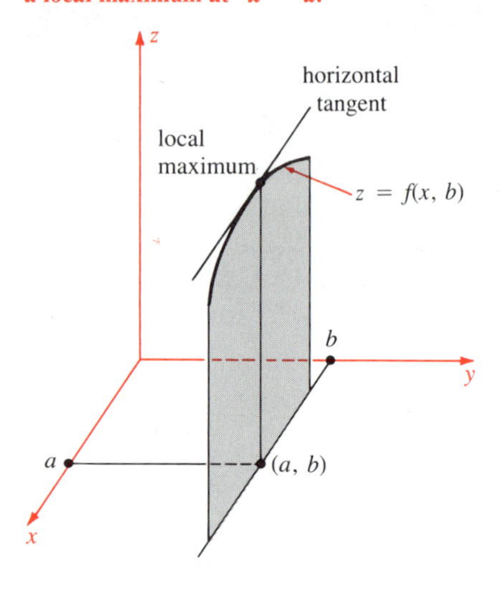

for all points (x, y) of **D** *sufficiently near* (a, b). We shall show that the tangent plane to the surface $z = f(x, y)$ at (a, b) is horizontal (Figure 15-2-2).

We set $g(x) = f(x, b)$. Then $g(x)$, a function of one variable, has a local maximum at $x = a$. See Figure 15-2-3. Therefore $g'(a) = 0$. But by definition, $g'(a) = f_x(a, b) = 0$, and similarly, $f_y(a, b) = 0$. Consequently the equation of the tangent plane is

$$\begin{aligned} z &= f(a, b) + f_x(a, b)(x - a) + f_y(a, b)(y - b) \\ &= f(a, b) \end{aligned}$$

Hence the tangent plane is horizontal. Conversely, if the plane is horizontal, then $f_x(a, b) = f_y(a, b) = 0$.

First Derivative Test

Let $f(x, y)$ be a continuously differentiable function and let (a, b) be an interior point of its domain. Suppose $f(x, y)$ has either a local maximum or a local minimum at (a, b). Then the tangent plane to the surface $z = f(x, y)$ at $(a, b, f(a, b))$ is horizontal. Equivalently,

$$\frac{\partial f}{\partial x}(a, b) = 0 \qquad \text{and} \qquad \frac{\partial f}{\partial y}(a, b) = 0$$

Intuitively it is almost obvious that the tangent plane is horizontal at a local maximum or local minimum. Suppose, for example, that $f(x, y) \le f(a, b)$ for all (x, y) near (a, b). Then the surface $z = f(x, y)$ is on or below the horizontal plane $z = f(a, b)$ but touches it at $(x, y, z) = (a, b, f(a, b))$. So the horizontal plane $z = f(a, b)$ must be the tangent plane. See Figure 15-2-4.

The conditions

$$f_x(a, b) = 0 \qquad \text{and} \qquad f_y(a, b) = 0$$

are like the first derivative test for functions of one variable. As in the one-variable case, this test has limitations that you should be aware of. First, the conditions $f_x = f_y = 0$ are *necessary* for a local maximum or a local minimum, but they are *not sufficient.* The graph $z = f(x, y)$ may have a saddle point (Figure 15-2-4c), where both partials are 0 yet there is neither a maximum nor a minimum. A second derivative test that distinguishes these cases is developed in the next chapter.

Second, the test does not apply at the boundary of the domain. For example, take $f(x, y) = x^2 + y^2$ on the disk $x^2 + y^2 \le 1$. Obviously the maximum value is 1, taken on the boundary. But $f_x = f_y = 0$ only at $(0, 0)$. See Figure 15-2-5.

Applications

In practice, we look for the maximum or minimum of a given function by solving the system

$$\frac{\partial f}{\partial x} = 0 \qquad \frac{\partial f}{\partial y} = 0$$

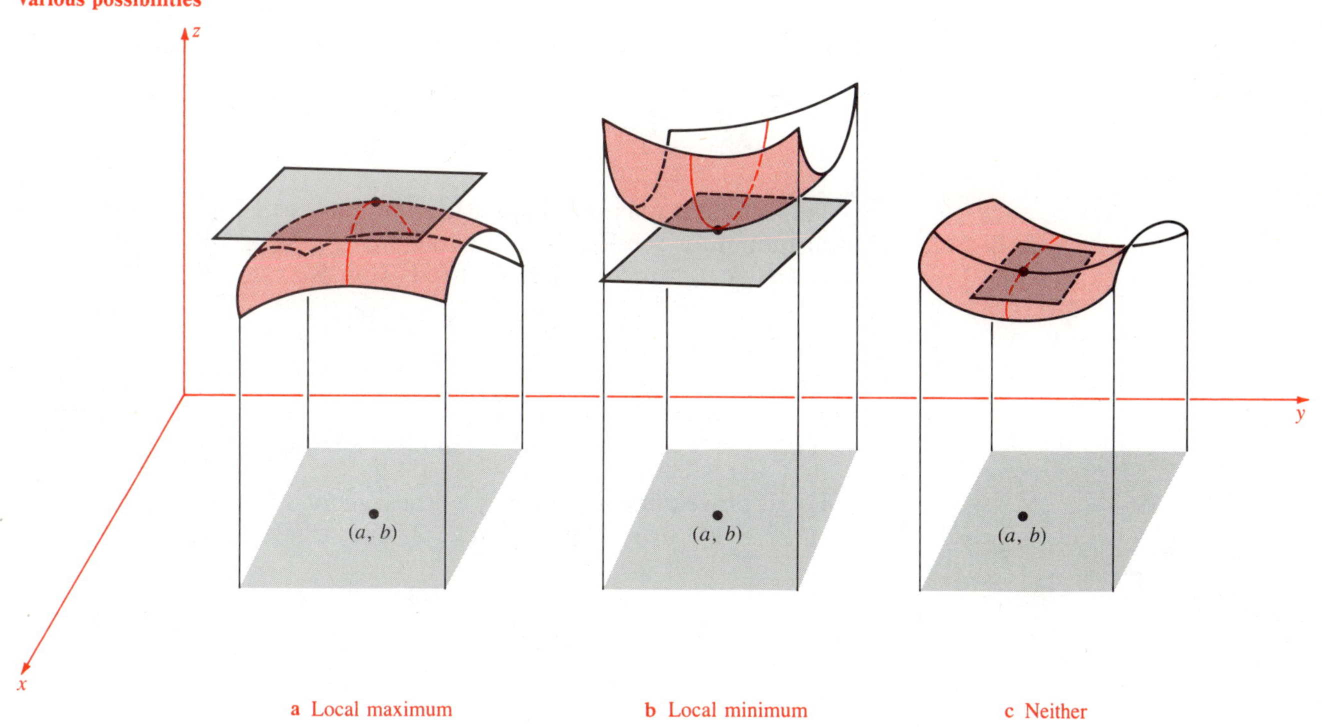

Figure 15-2-4
Horizontal tangent plane ($\partial f/\partial x = 0$, $\partial f/\partial y = 0$): various possibilities

a Local maximum b Local minimum c Neither

of two equations in two unknowns. This may give a number of possibilities. Then we try to sort out those that give maxima, minima, or neither. We must also give special attention to the boundary of the domain. However, we can often rule out the boundary, using additional information, physical properties, or if all else fails, common sense.

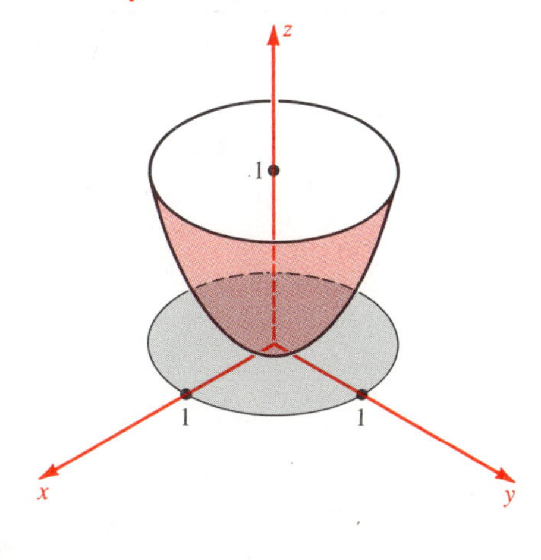

Figure 15-2-5
$f(x, y) = x^2 + y^2$ on domain $x^2 + y^2 \le 1$. Maximum on the boundary.

Example 1

Find the local extrema of

$$f(x, y) = x^2 - xy + y^2 + 3x$$

Solution The function is defined for all values of x and y; there is no boundary. Begin by finding all points (x, y) at which both

$$\frac{\partial f}{\partial x} = 0 \quad \text{and} \quad \frac{\partial f}{\partial y} = 0$$

Now

$$\begin{cases} \dfrac{\partial f}{\partial x} = \dfrac{\partial}{\partial x}(x^2 - xy + y^2 + 3x) = 2x - y + 3 \\[2ex] \dfrac{\partial f}{\partial y} = \dfrac{\partial}{\partial y}(x^2 - xy + y^2 + 3x) = -x + 2y \end{cases}$$

The conditions $\partial f/\partial x = 0$ and $\partial f/\partial y = 0$ are

$$2x - y + 3 = 0 \qquad \text{and} \qquad -x + 2y = 0$$

Solve: $x = -2$ and $y = -1$. The corresponding value of $f(x, y)$ is

$$\begin{aligned} f(-2, -1) &= (-2)^2 - (-2)(-1) + (-1)^2 + 3(-2) \\ &= 4 - 2 + 1 - 6 = -3 \end{aligned}$$

Is this a local maximum, a local minimum, or neither? We suspect a local minimum because the value of $f(x, y)$ seem to increase as $|x|$ increases or $|y|$ increases. For a fixed $y = b$,

$$f(x, b) = x^2 + (3 - b)x + b^2 \to +\infty \quad \text{as} \quad x \to \pm\infty$$

Similarly,

$$f(a, y) = y^2 - ay + (a^2 + 3a) \to +\infty \quad \text{as} \quad y \to \pm\infty$$

Let us prove our conjecture by a little algebra. First we move the origin to $(-2, -1)$ by setting $x = u - 2$ and $y = v - 1$. Then

$$\begin{aligned} f(x, y) &= (u - 2)^2 - (u - 2)(v - 1) \\ &\quad + (v - 1)^2 + 3(u - 2) \\ &= u^2 - uv + v^2 - 3 \end{aligned}$$

Next, we complete the square:

$$\begin{aligned} f(x, y) &= (u - \tfrac{1}{2}v)^2 - \tfrac{1}{4}v^2 + v^2 - 3 \\ &= (u - \tfrac{1}{2}v)^2 + \tfrac{3}{4}v^2 - 3 \ge -3 \end{aligned}$$

Conclusion: the absolute minimum of $f(x, y)$ is

$$f_{\min} = f(-2, -1) = -3$$

There is no other local maximum or local minimum. ●

Figure 15-2-6
See Example 2

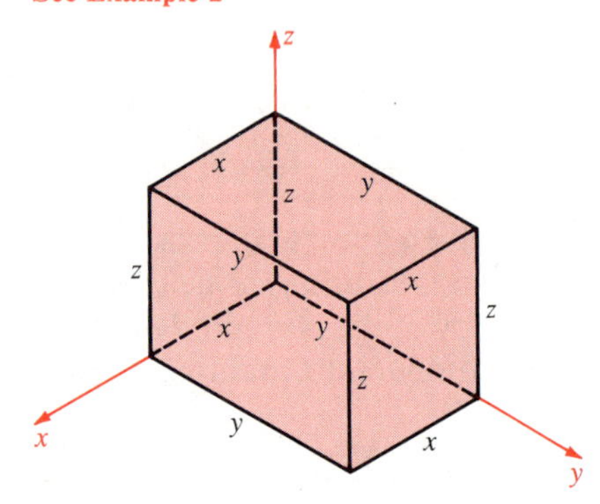

a Total edge length: $4x + 4y + 4z$

b The domain is a triangle, and we seek a solution in its interior.

Example 2

Find the rectangular solid of maximum volume whose total edge length is a given constant.

Solution As drawn in Figure 15-2-6a, the total length of the 12 edges is $4x + 4y + 4z$. Thus $4x + 4y + 4z = 4k$, a constant. That is,

$$x + y + z = k$$

The volume is

$$V = xyz = xy(k - x - y) = kxy - x^2y - xy^2$$

so write

$$V = V(x, y) = kxy - x^2y - xy^2$$

By the nature of the problem, $x > 0$ and $y > 0$; also $x + y < k$ because $z > 0$. These conditions describe the interior of the domain of $V(x, y)$, a triangle (Figure 15-2-6b). Clearly $V > 0$ on this interior and $V = 0$ on its boundary (since either $x = 0$, $y = 0$, or $z = 0$ at each point of the boundary). Hence the system V has a positive absolute maximum on the domain. To locate it, solve

$$\frac{\partial V}{\partial x} = 0 \qquad \frac{\partial V}{\partial y} = 0$$

that is,

$$\begin{cases} ky - 2xy - y^2 = 0 \\ kx - x^2 - 2xy = 0 \end{cases}$$

Since $x > 0$ and $y > 0$, cancel y from the first equation and x from the second:

$$\begin{cases} 2x + y = k \\ x + 2y = k \end{cases}$$

This pair of simultaneous linear equations has the unique solution

$$x = \tfrac{1}{3}k \qquad y = \tfrac{1}{3}k$$

Hence $z = k - x - y = \tfrac{1}{3}k$; the solid is a cube. ●

Remark We have used an important principle: suppose $f(x, y) > 0$ on a domain **D** and $f(x, y) = 0$ on the boundary of **D**. If $f_x = f_y = 0$ at only one interior point (a, b) of **D**, then $f(x, y)$ has its maximum at (a, b). For (assuming the maximum exists) it certainly does not occur on the boundary. But (a, b) is the *only* interior point that satisfies the necessary conditions for a local maximum, so $f(a, b)$ is the absolute maximum.

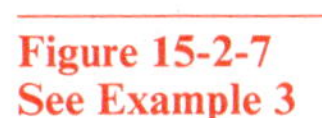
Figure 15-2-7
See Example 3

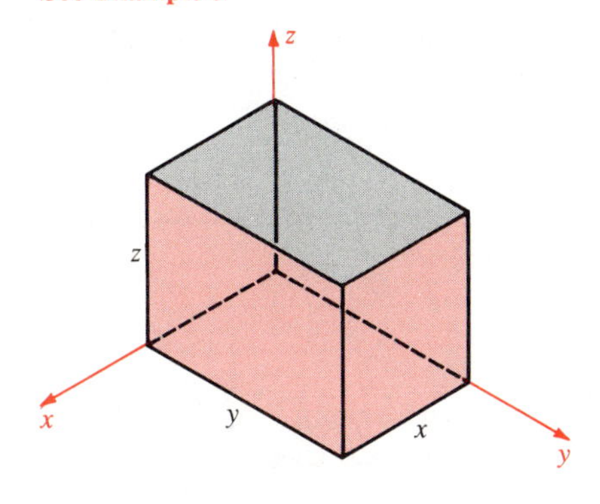

a Volume = xyz
Surface area = $xy + 2yz + 2zx$

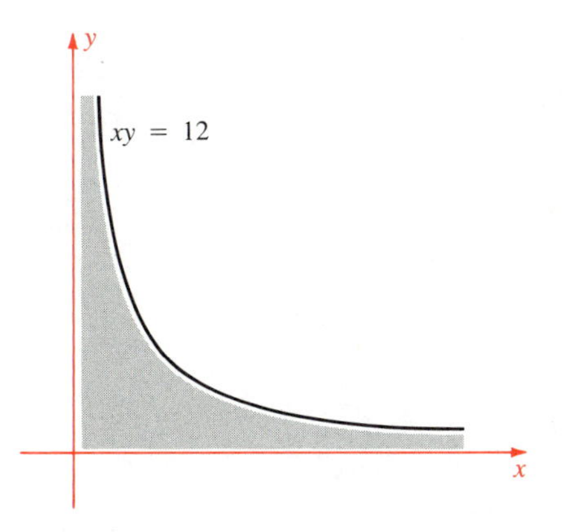

b The domain is defined by $x \geq 0$, $y \geq 0$, and $xy \leq 12$. We seek a solution in its interior.

Example 3

What is the largest possible volume, and what are the dimensions of an open rectangular aquarium constructed from $12\ \text{ft}^2$ of Plexiglas? Ignore the thickness of the plastic.

Solution See Figure 15-2-7a. The volume is $V = xyz$. The total surface area of the bottom and four sides is

$$xy + 2yz + 2zx = 12$$

Solve for z, then substitute into the formula for V:

$$z = \frac{12 - xy}{2(x + y)} \qquad V = \frac{(12 - xy)xy}{2(x + y)}$$

By the nature of the problem, $x > 0$, $y > 0$, and $z > 0$; hence $xy < 12$. Therefore the domain of V is the region of the first quadrant bounded by the axes and the curve $xy = 12$, and we seek a solution in its interior. See Figure 15-2-7b. Since $V > 0$ in the interior and $V = 0$ on the boundary, there is probably a maximum in the interior.

The domain extends to $+\infty$ in the x- and y-directions. What happens to V if, say, x is taken very large? Since $0 < xy < 12$, it follows that $(12 - xy)xy < 12 \cdot 12 = 144$, so

$$0 < V = \frac{(12 - xy)xy}{2(x + y)} < \frac{144}{2(x + y)} < \frac{72}{x + y} < \frac{72}{x}$$

Therefore V is small if x is large; similarly, V is small if y is large.

It follows that the maximum of V occurs for values of x and y neither too near 0 nor too large. To find the maximum, compute the partial derivatives of V (only one computation is needed because of the symmetry in x and y):

$$\frac{\partial V}{\partial x} = \frac{y^2(-x^2 - 2xy + 12)}{2(x + y)^2} \qquad \frac{\partial V}{\partial y} = \frac{x^2(-y^2 - 2xy + 12)}{2(x + y)^2}$$

Now find all points (x, y) where both partials are zero. Such points must satisfy the system

$$\begin{cases} y^2(-x^2 - 2xy + 12) = 0 \\ x^2(-y^2 - 2xy + 12) = 0 \end{cases}$$

Both x and y are positive, so y^2 and x^2 may safely be canceled:

$$\begin{cases} -x^2 - 2xy + 12 = 0 \\ -y^2 - 2xy + 12 = 0 \end{cases}$$

Solve these equations for x and y. Subtract to obtain $x^2 - y^2 = 0$. Hence $y = \pm x$. Since both x and y are positive, only $y = x$ applies. Now substitute $y = x$ into the first equation:

$$-x^2 - 2x^2 + 12 = 0 \qquad \text{hence} \qquad 3x^2 = 12 \quad \text{so} \quad x = 2$$

Therefore $(x, y) = (2, 2)$ is the *only* point where $V_x = V_y = 0$, so it must yield $V_{\max}$. Now when $x = 2$ ft and $y = 2$ ft, then

$$z = \frac{12 - xy}{2(x + y)} = \frac{12 - 4}{2 \cdot 4} = 1 \text{ ft}$$

Hence

$$V_{\max} = xyz \Big|_{(2,2,1)} = 4 \text{ ft}^3$$

Existence of Extrema [Optional]

Recall one of the basic facts about continuous functions of one variable:

- If f is continuous on a closed interval $[a, b]$, then there exist points x_0 and x_1 in the interval such that $f(x_0) \le f(x) \le f(x_1)$ for all x in $[a, b]$.

This result says that

$$f(x_0) = \min_{a\le x\le b} f(x)$$

$$f(x_1) = \max_{a\le x\le b} f(x)$$

If the interval is not closed, then f need not have a maximum or a minimum. For example, $f(x) = x$ has neither a maximum nor a minimum on the *open* interval $a < x < b$. The same holds for any continuous increasing or decreasing function.

Furthermore, the result is not true on an unbounded domain, a domain that contains points arbitrarily far from the origin. For example, on the domain $0 \le x < +\infty$, the function $f(x) = e^{-x}$ has a maximum but no minimum; on $0 < x < +\infty$ it has neither. The correct generalization of the preceding theorem requires a domain that is both closed and bounded.

Definition

A set $\mathbf{D}$ in the plane (in space) is **bounded** if there is a number B such that $|\mathbf{x}| \le B$ for all $\mathbf{x}$ in $\mathbf{D}$. In other words $\mathbf{D}$ is contained in some circle (sphere) centered at the origin.

It is not obvious how to define a closed domain in the plane or in space so as to generalize a closed interval on the line. The following definition turns out to be operational.

Definition

A domain $\mathbf{D}$ in the plane (in space) is **closed** if whenever the points $\mathbf{x}_n$ are in $\mathbf{D}$ and $\mathbf{x}_n \to \mathbf{x}$, then $\mathbf{x}$ is in $\mathbf{D}$.

For example a disk

$$\{\mathbf{x} \mid |\mathbf{x} - \mathbf{a}| \le r\}$$

with center $\mathbf{a}$ and radius r is a closed set in the plane. The unit cube

$$\{\mathbf{x} \mid 0 \le x_1 \le 1,\ 0 \le x_2 \le 1,\ 0 \le x_3 \le 1\}$$

is a closed set in space.

Intuitively, then, closed means *includes its boundary points.* Now we can state the basic existence property of maxima and minima.

Existence of Maxima and Minima

Let $f(\mathbf{x})$ be a continuous function on a bounded, closed domain $\mathbf{D}$. Then there exist points $\mathbf{x}_0$ and $\mathbf{x}_1$ in $\mathbf{D}$ such that

$$f(\mathbf{x}_0) \le f(\mathbf{x}) \le f(\mathbf{x}_1) \quad \text{for all } \mathbf{x} \text{ in } \mathbf{D}$$

This result justifies all of our work on maxima and minima. We leave its proof to more advanced courses.

Exercises

Find the maximum and minimum of $f(x, y) =$

1 $4 - 2x^2 - y^2$

2 $x^2 + y^2 - 1$

3 $(x-2)^2 + (y+3)^2$

4 $(x-1)^2 + y^2 + 3$

5 $x^2 - 2xy + 2y^2 + 4$

6 $xy - x^2 - 2y^2 + x + 2y$

7 $xy\exp(-x^2 - y^2)$

8 $\exp(-x^2 - y^2 + y)$

*9 $\sin x + \sin y + \sin(x+y)$

*10 $\cos x + \cos y + \cos(x+y)$

Find the minimum for $x \geq 0$, $y \geq 0$ of $f(x, y) =$

11 $\dfrac{1}{xy} + x + y^2$

12 $x^3 + y^3 - 3axy \quad (a > 0)$

13 $xy + \dfrac{a}{x} + \dfrac{b}{y} \quad (a > 0, b > 0)$

*14 $x^5 + y^3 - 15ax^3y \quad (a > 0)$

15 $xy^3(2x + y - 10)$

16 $xy^2(x + y - 1)^3$

Find the maximum and minimum of $f(x, y)$ in the given domain. Be sure to check the boundary!

17 $xy + 1/xy \quad (x > 0, y > 0, x + y \leq 1)$

18 $3(x-1)^2 + (y-2)^2 \quad (0 \leq x \leq 2, 0 \leq y \leq 3)$

19 $x^2(x^2 + y^2) \quad (x^2 + y^2 \leq r^2)$

20 $x^3 - y^2 + 6xy \quad (0 \leq x \leq 1, 0 \leq y \leq 1)$

21 Find the largest possible volume of a rectangular solid inscribed in the unit sphere.

22 Compute the largest possible volume of a rectangular box, edges parallel to the axes, that is inscribed in the solid bounded below by the x,y-plane and above by the paraboloid $z = 1 - x^2/a^2 - y^2/b^2$.

23 Find the dimensions of an open-top rectangular box of minimal surface area whose volume V is given.

24 Find the dimensions of the cheapest (open-top) aquarium of given volume V whose slate base costs 3 times as much per unit area as its glass sides.

*25 The base of a prism is a right triangle, its lateral sides are perpendicular to its base, and its volume V is given. Find its smallest possible total surface area.

26 Do Example 3, page 763, using the face areas as variables.

*27 Find the plane farthest from $\mathbf{0}$ among all planes tangent to the surface $xyz^2 = 1$ at a point of the first octant.

*28 A rectangle of dimension $a \times b$ is cut into four subrectangles by two lines. If $p > 1$, find the maximum and the minimum of $A_1^p + A_2^p + A_3^p + A_4^p$, where A_i is the area of the i-th piece.

15-3 Optimization: Three Variables

Suppose $f(a, b, c)$ is a local maximum or local minimum of a differentiable function $f(x, y, z)$ of three variables and (a, b, c) is an interior point of its domain. Then the following conditions must hold:

$$\frac{\partial f}{\partial x}(a, b, c) = 0 \qquad \frac{\partial f}{\partial y}(a, b, c) = 0 \qquad \frac{\partial f}{\partial z}(a, b, c) = 0$$

The proof is practically the same as for two variables.

The practical procedure for locating extrema for functions of three (or more) variables is quite similar to that for two, except that we start by solving the system

$$f_x = 0 \qquad f_y = 0 \qquad f_z = 0$$

of three equations in three unknowns.

Example 1

Find the maximum of

$$f(x, y, z) = \frac{x + 2y + 3z}{1 + x^2 + y^2 + z^2}$$

Solution The function is defined on the whole of space. However, there is no sense in looking for a maximum far from the origin because $|f|$ is very small if $|(x, y, z)|$ is large. To see why, set $\rho^2 = x^2 + y^2 + z^2$. Then $|x| \le \rho$, $|y| \le \rho$, and $|z| \le \rho$, so

$$|f| = \left|\frac{x + 2y + 3z}{1 + x^2 + y^2 + z^2}\right| \le \frac{\rho + 2\rho + 3\rho}{1 + \rho^2} = \frac{6\rho}{1 + \rho^2} < \frac{6}{\rho}$$

Clearly $f \to 0$ as $\rho \to +\infty$. For example, $|f| < 0.1$ if $|\mathbf{x}| \ge 60$. But $f(0, 0, \pm 1) = \pm\frac{3}{2}$. Therefore, $f(x, y, z)$ certainly has a positive maximum (and a negative minimum) inside the sphere $|\mathbf{x}| = 60$.

To find the maximum, we solve the equations $f_x = 0$, $f_y = 0$, $f_z = 0$. This computation gets messy, so it is best to organize it. For simplicity, we write

$$f = \frac{u}{v}$$

where $u = x + 2y + 3z$ and $v = 1 + x^2 + y^2 + z^2$. Then

$$f_x = \frac{vu_x - uv_x}{v^2} \qquad f_y = \frac{vu_y - uv_y}{v^2} \qquad f_z = \frac{vu_z - uv_z}{v^2}$$

The system of equations to be solved becomes

$$vu_x = uv_x \qquad vu_y = uv_y \qquad vu_z = uv_z$$

that is,

$$v = 2xu \qquad 2v = 2yu \qquad 3v = 2zu$$

Now $v > 0$ for any (x, y, z), so $2zu = 3v > 0$; hence $u \ne 0$. Divide each equation by u:

$$2x = y = \tfrac{2}{3}z = \frac{v}{u}$$

Therefore $y = 2x$ and $z = 3x$, and from here on it's smooth sailing. Substitute these expressions for y and z:

$$u = x + 2y + 3z = x + 4x + 9x = 14x$$
$$v = 1 + x^2 + y^2 + z^2 = 1 + x^2 + 4x^2 + 9x^2 = 1 + 14x^2$$

Now substitute these values of u and v into the equation $v = 2xu$ and solve for x:

$$1 + 14x^2 = 2x(14x) = 28x^2 \qquad 14x^2 = 1 \qquad x = \pm\tfrac{1}{14}\sqrt{14}$$

Also $y = 2x$, and $z = 3x$, so the conditions $f_x = f_y = f_z = 0$ hold only at

$$(x, y, z) = \pm\tfrac{1}{14}\sqrt{14}\,(1, 2, 3)$$

At these points, $v = 1 + 14x^2 = 2$ and $u = 14x = \pm\sqrt{14}$, so $f = u/v = \pm\frac{1}{2}\sqrt{14}$. Therefore

$$f_{\max} = f(\tfrac{1}{14}\sqrt{14}, \tfrac{2}{14}\sqrt{14}, \tfrac{3}{14}\sqrt{14}) = \tfrac{1}{2}\sqrt{14}$$

The solution shows also that $f_{\min} = -\frac{1}{2}\sqrt{14}$. ●

Figure 15-3-1

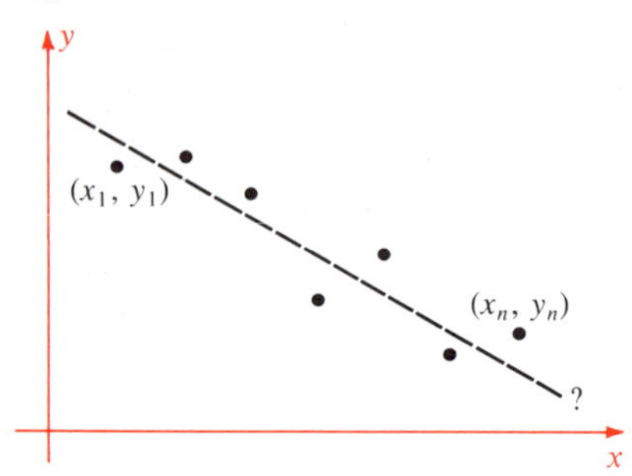

a Find the line that best "fits" the data.

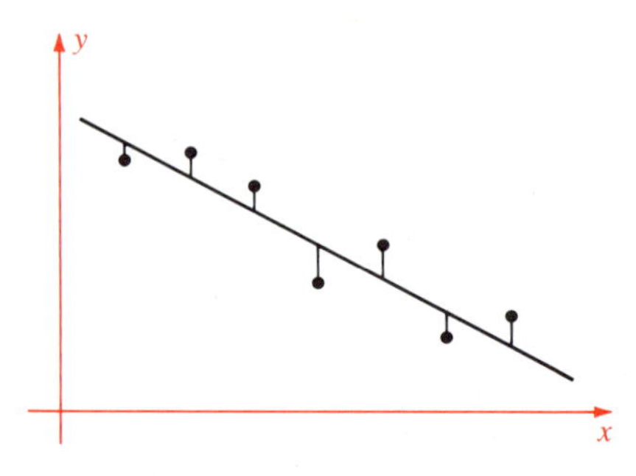

b Least-squares fit: The sum of the squares of the vertical deviations is minimal.

Least Squares and Linear Regression

Often an experiment produces a sequence of readings:

$$(x_1, y_1) \quad (x_2, y_2) \quad \cdots \quad (x_n, y_n)$$

When plotted (Figure 15-3-1a) the points may cluster in roughly the form of a straight line, suggesting that y is a linear function of x (which may be just what the experimenter would like to establish). A practical question arises: which straight line $y = Ax + B$ *most closely fits* the data?

The answer depends on what is meant by "fit." For an experimenter interested in how y varies with x, the most popular measure of fit is by **least squares:** the line is chosen to minimize the sum of the squares of the *vertical* deviations from the line (Figure 15-3-1b). The resulting line is called the line of regression of y with respect to x.

Example 2

Find the line $y = Ax + B$ that is the least-squares fit to the points $(0, 2)$, $(1, 3)$, $(2, 3)$.

Solution Write $y(x) = Ax + B$, and choose A and B to minimize

$$\begin{aligned} f(A, B) &= [y(0) - 2]^2 + [y(1) - 3]^2 + [y(2) - 3]^2 \\ &= (B - 2)^2 + (A + B - 3)^2 + (2A + B - 3)^2 \end{aligned}$$

Necessary conditions for a minimum are

$$\frac{\partial f}{\partial A} = 0 \qquad \text{and} \qquad \frac{\partial f}{\partial B} = 0$$

Differentiate:

$$\frac{\partial f}{\partial A} = 2(A + B - 3) + 4(2A + B - 3) = 2(5A + 3B - 9)$$

$$\begin{aligned} \frac{\partial f}{\partial B} &= 2(B - 2) + 2(A + B - 3) + 2(2A + B - 3) \\ &= 2(3A + 3B - 8) \end{aligned}$$

Set these partial derivatives equal to zero:

$$\begin{cases} 5A + 3B - 9 = 0 \\ 3A + 3B - 8 = 0 \end{cases}$$

The system has a unique solution: $A = \frac{1}{2}$ and $B = \frac{13}{6}$.

Now $f(A, B)$ must have a minimum at $(\frac{1}{2}, \frac{13}{6})$; here is why. If either A or B is large, then $f(A, B)$ is large (by inspection). On the circle $A^2 + B^2 = (1000)^2$, for example, $f(A, B)$ is very large. The minimum of $f(A, B)$ in the region bounded by this circle occurs either on the boundary or at a point where $\partial f/\partial A = \partial f/\partial B = 0$. But the boundary is ruled out; hence the minimum occurs at $(\frac{1}{2}, \frac{13}{6})$, the only point where $\partial f/\partial A = \partial f/\partial B = 0$.

Therefore the answer (Figure 15-3-2) is $y = \frac{1}{2}x + \frac{13}{6}$. ●

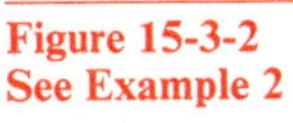
Figure 15-3-2
See Example 2

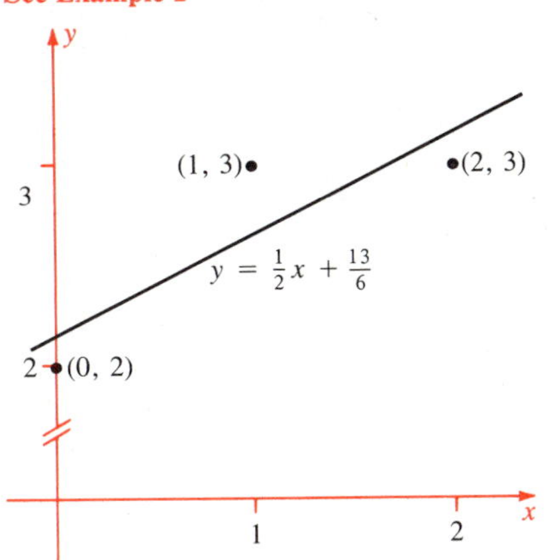

To solve the general problem of least-squares fit, we imitate the method used in the example. Given readings

$$(x_1, y_1) \quad (x_2, y_2) \quad \cdots \quad (x_n, y_n)$$

for n distinct values of (x, y) we seek a linear function $y = Ax + B$ that minimizes

$$f(A, B) = \sum_{i=1}^{n} [(Ax_i + B) - y_i]^2$$

Now

$$\begin{aligned}\frac{\partial f}{\partial A} &= 2 \sum x_i(Ax_i + B - y_i) \\ &= 2\left[A\left(\sum x_i^2\right) + B\left(\sum x_i\right) - \left(\sum x_i y_i\right)\right] \\ \frac{\partial f}{\partial B} &= 2 \sum (Ax_i + B - y_i) = 2\left[A\left(\sum x_i\right) + nB - \left(\sum y_i\right)\right]\end{aligned}$$

We set these partial derivatives equal to zero and obtain two equations for the two unknowns A and B:

$$\left(\sum x_i^2\right) A + \left(\sum x_i\right) B = \sum x_i y_i \qquad \left(\sum x_i\right) A + nB = \sum y_i$$

All coefficients in this system of equations are computable from the data. We shall show that there is a unique solution (A, B), if the x_i are not all equal. It can be shown that this solution does minimize $f(A, B)$.

Terminology

To express the solution in a concise form, we introduce some standard statistical terminology:

$$\bar{x} = \frac{1}{n} \sum x_i \qquad \bar{y} = \frac{1}{n} \sum y_i \qquad \sigma_x^2 = \frac{1}{n} \sum (x_i - \bar{x})^2$$

$$s_{xy} = \frac{1}{n} \sum (x_i - \bar{x})(y_i - \bar{y})$$

Then $\bar{x}$ is called the **mean** (average) of $x_1, \cdots, x_n$ and σ_x^2 the **variance** of $x_1, \cdots, x_n$. Likewise $\bar{y}$ is the mean of the y_i. Finally, s_{xy} is called the **covariance** of the x's and y's. ●

It is an easy exercise to verify the relations

$$\sigma_x^2 = \frac{1}{n}\sum x_i^2 - \bar{x}^2 \quad \text{and} \quad s_{xy} = \frac{1}{n}\sum x_i y_i - \bar{x}\bar{y}$$

Let us return to the linear system of equations for A and B. We divide both equations by n. Then in terms of this new notation, the system becomes

$$\begin{cases} (\sigma_x^2 + \bar{x}^2)A + \bar{x}B = s_{xy} + \bar{x}\bar{y} \\ \qquad\qquad \bar{x}A + \ B = \bar{y} \end{cases}$$

We multiply the second equation by $\bar{x}$, subtract it from the first equation, and divide by σ_x^2, which is non-zero since the x_i are not all the same. This gives us

$$\sigma_x^2 A = s_{xy} \qquad \text{so that} \qquad A = \frac{s_{xy}}{\sigma_x^2}$$

Since $B = \bar{y} - \bar{x}A$, the desired linear fit $y = Ax + B$ can be written

$$y = Ax + \bar{y} - \bar{x}A \qquad \text{that is} \qquad y - \bar{y} = A(x - \bar{x})$$

and finally

$$y - \bar{y} = \frac{s_{xy}}{\sigma_x^2}(x - \bar{x})$$

This is the required least-squares fit to the data. In statistics it is called the **regression line** and $A = s_{xy}/\sigma_x^2$ is called the **regression coefficient.**

Regression Line of y with Respect to x

The best linear fit to the data

$$(x_1, y_1) \quad (x_2, y_2) \quad \cdots \quad (x_n, y_n)$$

in the sense of least squares is the line

$$y - \bar{y} = \frac{s_{xy}}{\sigma_x^2}(x - \bar{x})$$

Remark We should mention that σ_x is the **standard deviation** of the x's, and $\rho_{xy} = s_{xy}/\sigma_x\sigma_y$ is the **correlation coefficient.** Note that the regression line passes through $(\bar{x}, \bar{y})$. Does that seem reasonable?

Example 3

Find the regression line for the data

x	1	2	3	4	5	6	7	8
y	10	9	7	7	6	5	3	2

Solution Here $n = 8$ and

$$\bar{x} = \tfrac{1}{8}\sum x_i = \tfrac{1}{8}(1 + 2 + \cdots + 8) = \tfrac{9}{2}$$

$$\bar{y} = \tfrac{1}{8}\sum y_i = \tfrac{1}{8}(10 + 9 + 7 + \cdots + 2) = \tfrac{49}{8}$$

$$\sigma_x^{\,2} = \tfrac{1}{8}\sum x_i^2 - \bar{x}^2 = \tfrac{1}{8}(1^2 + 2^2 + \cdots + 8^2) - (\tfrac{9}{2})^2$$

$$= \tfrac{1}{8}(204) - \tfrac{81}{4} = \tfrac{21}{4}$$

$$s_{xy} = \tfrac{1}{8}\sum x_i y_i - \bar{x}\bar{y}$$

$$= \tfrac{1}{8}(1 \cdot 10 + 2 \cdot 9 + \cdots + 8 \cdot 2) - (\tfrac{9}{2})(\tfrac{49}{8})$$

$$= \tfrac{1}{8}(174) - \tfrac{441}{16} = -\tfrac{93}{16}$$

The regression coefficient is

$$A = \frac{s_{xy}}{\sigma_x^{\,2}} = -\tfrac{31}{28}$$

and the least-square fit is

$$y - \tfrac{49}{8} = -\tfrac{31}{28}(x - \tfrac{9}{2})$$

that is,

$$y = -\tfrac{31}{28}x + \tfrac{311}{28} \approx -1.107x + 11.12$$

See Figure 15-3-3.

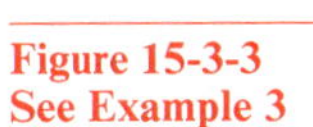
Figure 15-3-3
See Example 3

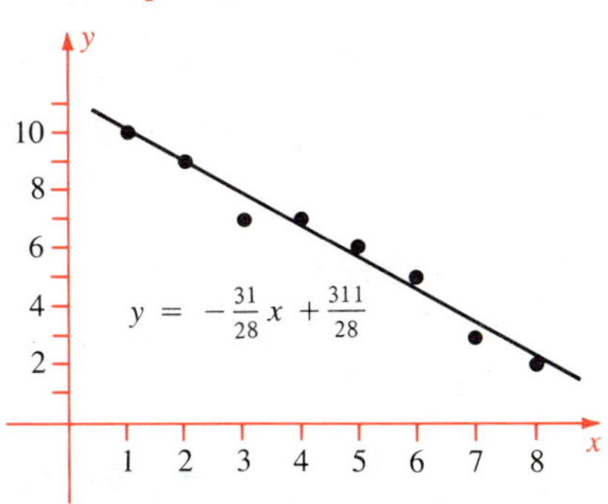

The least-squares idea can be used in more complicated situations. For example, one might seek $y = Ax^2 + Bx + C$, the *quadratic* regression curve that most closely fits the data

$$(x_1, y_1) \quad (x_2, y_2) \quad \cdots \quad (x_n, y_n)$$

Then A, B, and C must be found so that

$$F(A, B, C) = \sum_{i=1}^{n} [(Ax_i^2 + Bx_i + C) - y_i]^2$$

is minimized.

Still more complicated is the problem of approximating a *function* rather than discrete data. For example, suppose $y = x^2$ is to be approximated on the interval $0 \le x \le 1$ by a linear function $y = Ax + B$ in the sense of least squares. Then

$$F(A, B) = \int_0^1 [(Ax + B) - x^2]^2\, dx$$

must be minimized.

Exercises

1 Find the minimum of $\dfrac{9}{x} + \dfrac{4}{y} + \dfrac{1}{z} + xyz$ for $x > 0,\ y > 0,\ z > 0$.

2 Find the minimum of $x^5 + y^5 + z^5 - 5xyz$ for $x > 0,\ y > 0,\ z > 0$.

3 Find the maximum of $xyz(10 - x^2 - 2y^2 - 3z^2)$ for $x > 0,\ y > 0,\ z > 0$.

4 Find the maximum of $(x + 2y + 4z)\exp(-x^2 - y^2 - z^2)$.

5 Find the maximum of $\dfrac{ax + by + cz}{1 + x^2 + y^2 + z^2}$ $(abc \neq 0)$.

6 Find the maximum and minimum of

$$\frac{xyz}{(1 + x^2 + y^2 + z)^2}$$

7 Find the maximum and minimum of

$$\frac{xyz}{(1 + x^2 + 2y^2 + 3z^2)^2}$$

***8** Find the maximum and minimum of $\cos x + \cos y + \cos z - \cos(x + y + z)$.

9 A segment of length L is cut into four pieces. What is the largest possible product of their four lengths?

10 Find the minimum of

$$Q(x, y, z) = x^2 + 3y^2 + 14z^2 - 2xy + 2yz - 6zx$$

Complete squares to prove you really have the minimum.

Find the least-squares straight-line fit to the data.

11 $(0, 0)$ $(1, 1)$ $(2, 3)$

12 $(1, 0)$ $(2, -1)$ $(3, -4)$

13 $(0, 0)$ $(\frac{1}{2}, \frac{1}{4})$ $(1, 1)$ $(\frac{3}{2}, \frac{9}{4})$ $(2, 4)$

14 $(1, 1)$ $(2, \frac{1}{2})$ $(3, \frac{1}{3})$ $(4, \frac{1}{4})$

15 $(1, 5.0)$ $(2, 5.3)$ $(3, 5.4)$ $(4, 5.6)$

16 $(-2, 3.0)$ $(-1, 1.8)$ $(0, 1.1)$ $(1, 0.6)$

17 $(0, 0)$ $(1, -1)$ $(2, -2)$ $(3, -1)$ $(4, 0)$

18 $(-3, 0)$ $(-2, 0)$ $(-1, 0)$ $(0, 1)$ $(1, 2)$ $(2, 2)$ $(3, 2)$

19 $(-1, 6.2)$ $(0, 5.4)$ $(1, 5.5)$ $(2, 5.0)$

20 $(1, -1)$ $(2, 1)$ $(3, -1)$ $(4, 1)$ $\cdots$ $(9, -1)$

Approximate $y = f(x)$ on the interval $[0, 1]$ by a linear function $y = Ax + B$ in the sense of least squares, that is, minimize

$$F(A, B) = \int_0^1 [(Ax + B) - f(x)]^2\, dx$$

21 $f(x) = x^2$ **22** $f(x) = x^3$ **23** $f(x) = e^x$

24 (cont.) Prove in general that $A = 6(2\beta - \alpha)$ and $B = 2(2\alpha - 3\beta)$, where

$$\alpha = \int_0^1 f(x)\, dx \quad \text{and} \quad \beta = \int_0^1 xf(x)\, dx$$

25 Find the best least-squares fit to the data $(1, 7)$, $(2, 4)$, $(3, 3)$ by $y = ax + b/x$.

26 The population of a city (in thousands) was

year	1960	1965	1970	1975	1980
population	100	104	108	113	120

A demographer assumes an exponential curve $y = ae^{bt}$, with $t = 0$ at year 1960. He obtains the curve by fitting a straight line by least squares to $\ln y$ versus t. What growth curve does he find?

15-4 Lagrange Multipliers: Two Variables

Consider several problems that have a common feature:

a Of all rectangles with perimeter 1, which has the shortest diagonal? That is, minimize $(x^2 + y^2)^{1/2}$ subject to $2x + 2y = 1$.

b Of all right triangles with perimeter 1, which has largest area? That is, maximize $\frac{1}{2}xy$ subject to $x + y + (x^2 + y^2)^{1/2} = 1$.

c Find the largest value of $x + 2y + 3z$ for points (x, y, z) on the unit sphere. That is, maximize the function $x + 2y + 3z$ subject to $x^2 + y^2 + z^2 = 1$.

d Of all closed rectangular boxes with fixed surface area, which has greatest volume? That is, maximize xyz subject to $xy + yz + zx = c$.

Each of these problems asks for the maximum (or minimum) of a function of several variables, where the variables must satisfy a certain relation (constraint).

Figure 15-4-1
Maximize $f(x, y)$ subject to $g(x, y) = c$.

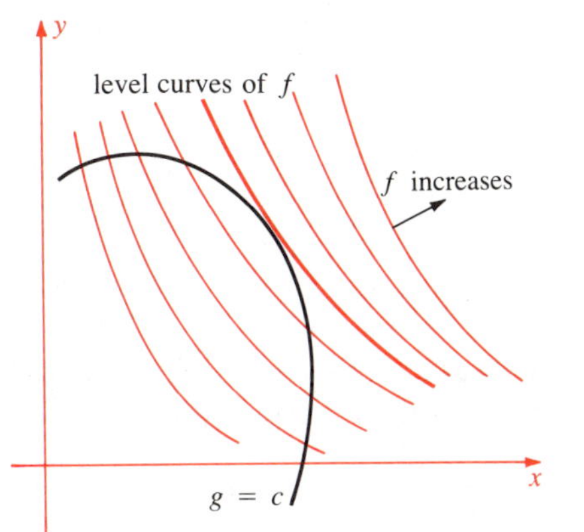

a Find the highest level of $f(x, y)$ that meets $g(x, y) = c$.

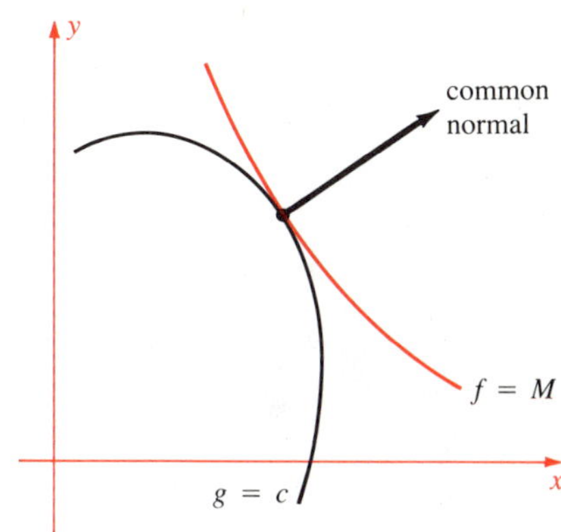

b Equality of the normals at the maximum

Such problems can be analyzed geometrically. Suppose you are asked to maximize a function $f(x, y)$, subject to a constraint $g(x, y) = c$. On the same graph plot $g(x, y) = c$ and several level curves of $f(x, y)$, noting the direction of increase of the level (Figure 15-4-1a). To find the largest value of $f(x, y)$ on the curve $g(x, y) = c$, find the highest level curve that intersects $g = c$. If there is a highest one and the intersection does not take place at an end point, this level curve and the graph $g = c$ are tangent.

Suppose $f(x, y) = M$ defines a level curve tangent to $g(x, y) = c$ at a point (x, y). See Figure 15-4-1b. Since the two graphs are tangent at (x, y), their normals at (x, y) are parallel. But the vectors

$$\operatorname{grad} f(x, y) \quad \text{and} \quad \operatorname{grad} g(x, y)$$

point in the respective normal directions; hence one is a multiple of the other:

$$\operatorname{grad} f(x, y) = \lambda \operatorname{grad} g(x, y)$$

for some number λ. (The argument presupposes that $\operatorname{grad} g \neq \mathbf{0}$ at the point in question.)

This geometric argument yields a practical rule for locating points on $g(x, y) = c$ where $f(x, y)$ may have a maximum or minimum. [However, note that where the condition of tangency is satisfied, there may be a maximum, a minimum, or neither (Figure 15-4-2).]

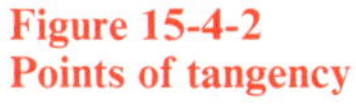
Figure 15-4-2
Points of tangency

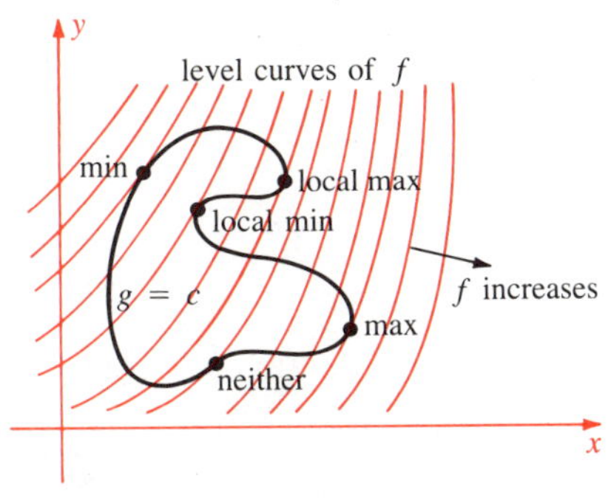

Lagrange Multiplier Rule

To maximize or minimize a function $f(x, y)$ subject to a constraint $g(x, y) = c$, solve the system of equations

$$(f_x, f_y) = \lambda(g_x, g_y) \qquad g(x, y) = c$$

in the three unknowns x, y, λ. Each resulting point (x, y) is a candidate. The number λ is called a **Lagrange multiplier,** or simply a **multiplier.**

This rule requires three simultaneous equations

$$\begin{cases} f_x(x, y) = \lambda g_x(x, y) \\ f_y(x, y) = \lambda g_y(x, y) \\ g(x, y) = c \end{cases}$$

to be solved for three unknowns $x, \; y, \; \lambda$.

Remark There is another way to interpret the multiplier rule that is quite interesting. Consider the function

$$F(x, y, \lambda) = f(x, y) - \lambda[g(x, y) - c]$$

of *three* independent (unconstrained) variables $x, \; y, \; \lambda$. This function has a local extremum where

$$\frac{\partial F}{\partial x} = 0 \qquad \frac{\partial F}{\partial y} = 0 \qquad \frac{\partial F}{\partial \lambda} = 0$$

that is, where

$$f_x - \lambda g_x = 0 \qquad f_y - \lambda g_y = 0 \qquad g(x, y) - c = 0$$

These three conditions are those of the multiplier rule.

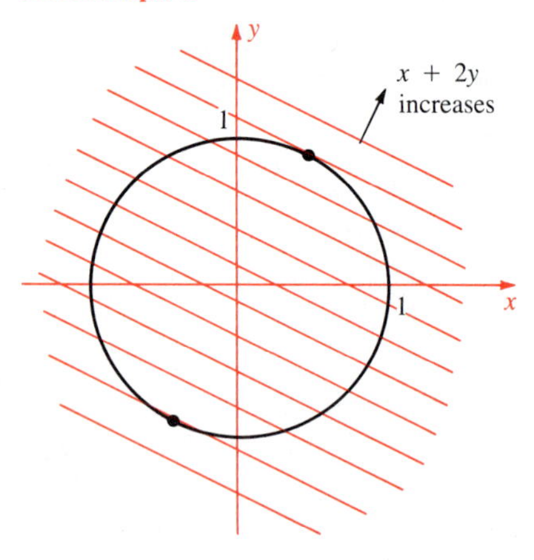

Figure 15-4-3
See Example 1

Example 1

Find the largest and smallest values of $f(x, y) = x + 2y$ on the circle $x^2 + y^2 = 1$.

Solution Draw a figure (Figure 15-4-3). As seen from the figure, $f(x, y)$ takes its maximum at a point in the first quadrant, and its minimum at a point in the third quadrant. Apply the method of Lagrange multipliers with

$$f(x, y) = x + 2y \qquad \text{and} \qquad g(x, y) = x^2 + y^2$$

so that

$$\text{grad} f = (1, 2) \qquad \text{and} \qquad \text{grad}\, g = (2x, 2y)$$

The conditions

$$(f_x, f_y) = \lambda(g_x, g_y) \qquad \text{and} \qquad g(x, y) = 1$$

become

$$(1, 2) = \lambda(2x, 2y) \qquad \text{and} \qquad x^2 + y^2 = 1$$

Rewrite the conditions in the form

$$x = \frac{1}{2\lambda} \qquad y = \frac{1}{\lambda} \qquad \left(\frac{1}{2\lambda}\right)^2 + \left(\frac{1}{\lambda}\right)^2 = 1$$

By the third equation, $\lambda^2 = \frac{5}{4}$ so $\lambda = \pm\frac{1}{2}\sqrt{5}$. The value $\lambda = \frac{1}{2}\sqrt{5}$ yields

$$x = \frac{1}{\sqrt{5}} \quad \text{and} \quad y = \frac{2}{\sqrt{5}} \quad \text{so} \quad f(x, y) = \frac{5}{\sqrt{5}} = \sqrt{5}$$

The value $\lambda = -\frac{1}{2}\sqrt{5}$ yields

$$x = -\frac{1}{\sqrt{5}} \quad \text{and} \quad y = -\frac{2}{\sqrt{5}} \quad \text{so} \quad f(x, y) = -\frac{5}{\sqrt{5}} = -\sqrt{5}$$

The largest value of $f(x)$ is $\sqrt{5}$, the smallest, $-\sqrt{5}$. ●

Figure 15-4-4
See Example 2

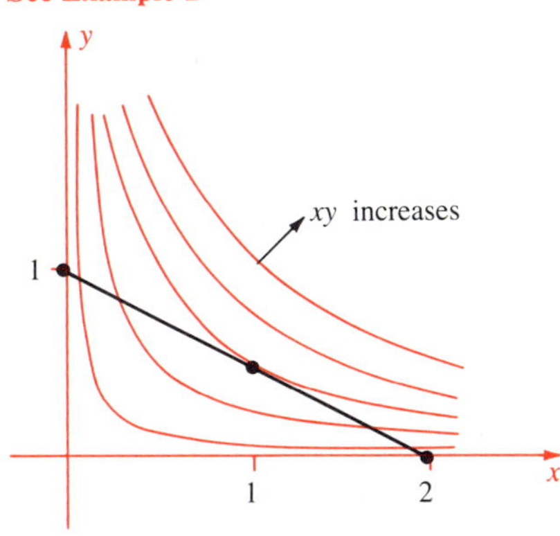

Example 2

Find the largest and smallest values of xy on the segment $x + 2y = 2$, $x \geq 0$, $y \geq 0$.

Solution Draw a graph (Figure 15-4-4). Evidently the smallest value of xy is 0, taken at either end point. To find the largest value, use the multiplier technique with

$$f(x, y) = xy \quad \text{and} \quad g(x, y) = x + 2y$$

The resulting system of Lagrange multiplier equations is

$$(y, x) = \lambda(1, 2) \quad \text{and} \quad x + 2y = 2$$

Consequently $x = 2\lambda$ and $y = \lambda$. Substitute into $x + 2y = 2$:

$$2\lambda + 2\lambda = 2 \quad \text{that is} \quad \lambda = \tfrac{1}{2}$$

Therefore

$$(x, y) = (1, \tfrac{1}{2}) \quad \text{so} \quad f_{\max} = f(1, \tfrac{1}{2}) = \tfrac{1}{2}$$

As already noted, $f_{\min} = f(2, 0) = f(0, 1) = 0$. ●

Figure 15-4-5
See Example 3

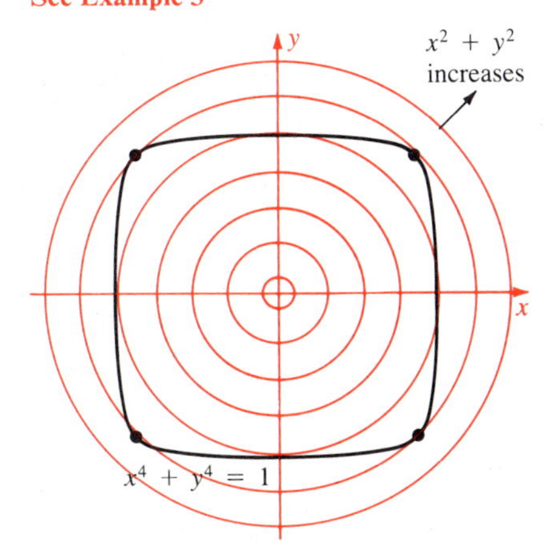

Example 3

Find the largest and smallest values of $x^2 + y^2$, subject to the constraint $x^4 + y^4 = 1$.

Solution Graph the curve $x^4 + y^4 = 1$ and the level curves of $f(x, y) = x^2 + y^2$. See Figure 15-4-5. By drawing $x^4 + y^4 = 1$ *accurately,* you see that the graph is quite flat where it crosses the axes and most sharply curved where it crosses the $45°$ lines $y = \pm x$. It is closest to the origin ($x^2 + y^2$ is smallest) at $(\pm 1, 0)$ and $(0, \pm 1)$ and farthest where $y = \pm x$.

The analysis confirms this. Use the multiplier technique with

$$f(x, y) = x^2 + y^2 \quad \text{and} \quad g(x, y) = x^4 + y^4$$

The multiplier equations are

$$(2x, 2y) = \lambda(4x^3, 4y^3) \quad \text{and} \quad x^4 + y^4 = 1$$

Four obvious solutions are

$$(x, y, \lambda) = (0, \pm 1, \tfrac{1}{2}) \quad \text{and} \quad (\pm 1, 0, \tfrac{1}{2})$$

Thus the points $(0, \pm 1)$ and $(\pm 1, 0)$ are candidates for the maximum or minimum. At each of these points $f(x, y) = 1$.

Suppose both $x \neq 0$ and $y \neq 0$. From

$$2x = 4\lambda x^3 \quad \text{and} \quad 2y = 4\lambda y^3 \qquad \text{follows} \qquad x^2 = y^2 = \frac{1}{2\lambda}$$

Hence $\lambda = 1/(2x^2) > 0$. From $x^4 + y^4 = 1$ follow

$$\left(\frac{1}{2\lambda}\right)^2 + \left(\frac{1}{2\lambda}\right)^2 = 1 \qquad \text{so} \qquad \lambda^2 = \tfrac{1}{2} \qquad \text{and} \qquad \lambda = \frac{1}{\sqrt{2}}$$

Hence

$$x^2 = y^2 = \frac{1}{2\lambda} = \frac{\sqrt{2}}{2} = \frac{1}{\sqrt{2}}$$

Consequently, the four points

$$\left(\pm \frac{1}{\sqrt[4]{2}}, \pm \frac{1}{\sqrt[4]{2}}\right)$$

are candidates for the maximum or minimum. At each of these points

$$f(x, y) = x^2 + y^2 = 2/\sqrt{2} = \sqrt{2}$$

Therefore the largest value is $\sqrt{2}$, the smallest, 1. ●

Example 4

Find the proportions of a right circular cone with fixed lateral area and maximal volume (Figure 15-4-6).

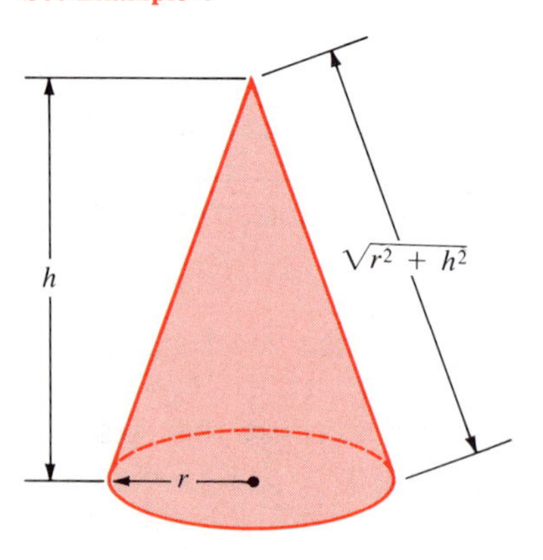

Figure 15-4-6
See Example 4

Solution Denote the radius, height, fixed lateral area, and volume by r, h, A, and V. Then

$$V = \tfrac{1}{3}\pi r^2 h \qquad \text{and} \qquad A = \tfrac{1}{2}(2\pi r)\sqrt{r^2 + h^2} = \pi r\sqrt{r^2 + h^2}$$

because $\sqrt{r^2 + h^2}$ is the slant height. Since A is fixed, so is A^2/π^2. Therefore the problem is equivalent to maximizing

$$f(r, h) = r^2 h \quad \text{subject to} \quad g(r, h) = r^2(r^2 + h^2) = A^2/\pi^2$$

The three Lagrange multiplier equations are $f_r = \lambda g_r$, $f_h = \lambda g_h$, and $g = A^2/\pi^2$, that is,

$$2rh = \lambda(4r^3 + 2rh^2) \qquad r^2 = 2\lambda r^2 h \qquad r^2(r^2 + h^2) = A^2/\pi^2$$

Since $r \neq 0$, the equations are equivalent to

$$h = (2r^2 + h^2)\lambda \qquad 2\lambda h = 1 \qquad r^2(r^2 + h^2) = A^2/\pi^2$$

Multiply the first equation by $2h$ and use the second equation to eliminate λ:

$$2h^2 = (2r^2 + h^2)(2\lambda h) = 2r^2 + h^2$$

Hence

$$h^2 = 2r^2 \qquad \text{so} \qquad h = r\sqrt{2}$$

Let us show that this proportion yields maximal volume. By the nature of the problem, $r > 0$. Also $h > 0$ and the relation

$$\pi^2 r^2 (r^2 + h^2) = A^2 = \text{constant}$$

implies that $r \to \sqrt{A/\pi}\,-$ as $h \to 0+$.

Conversely, $h^2 = A^2/\pi^2 r^2 - r^2$, so $h \to 0+$ as $r \to \sqrt{A/\pi}\,-$. Thus, as a function of r, the volume V is defined on the interval $0 < r < \sqrt{A/\pi}$. Furthermore, $h < \sqrt{r^2 + h^2}$, so as $r \to 0+$,

$$V = \tfrac{1}{3}\pi r^2 h < \tfrac{1}{3}\pi r^2 \sqrt{r^2 + h^2} = \tfrac{1}{3} rA \to 0$$

As $r \to \sqrt{A/\pi}$ from below, $h \to 0+$ and

$$V = \tfrac{1}{3}\pi r^2 h \to \tfrac{1}{3}\pi (A/\pi) \cdot 0 = 0$$

Thus V is a positive function of r on the interval $0 < r < \sqrt{A/\pi}$, and V approaches 0 towards the end points of the interval. Therefore V certainly has a maximum in the interval, but not necessarily a minimum. Since $h = r\sqrt{2}$ represents the only candidate for either a maximum or a minimum, this condition must yield the maximal volume. ●

Exercises

Use the Lagrange multiplier method to find the maximum and minimum of the function, subject to the constraint, and state where they occur

1 xy on $x^2 + y^2 = 1$

2 xy on $\frac{1}{4}x^2 + \frac{1}{9}y^2 = 1$

3 $x + y$ on $\frac{1}{4}x^2 + \frac{1}{9}y^2 = 1$

4 xy on $x^2 + xy + 4y^2 = 1$

5 xy^2 on $x^2 + y^2 = 1$

6 $x + y^2$ on $x^2 + y^2 = 1$

7 $x - y$ on the branch $x > 0$ of $\frac{1}{9}x^2 - \frac{1}{4}y^2 = 1$

8 (cont.) $x - y$ on $\frac{1}{9}x^2 - \frac{1}{4}y^2 = 1$

9 $x - y$ on the branch $x > 0$ of $\frac{1}{4}x^2 - \frac{1}{9}y^2 = 1$

10 $x - y$ on the branch $x > 0$ of $x^2 - y^2 = 1$

***11** $8y - (x - 3)^2$ on $y^2 = x$

***12** $y - (x - 1)^2$ on $4y^2 - 3x = 0$

13 $x^3 y$ on $\sqrt{x} + \sqrt{y} = 1$

14 y/x on $(x - 3)^2 + (y - 3)^2 = 6$

15 $x + y$ on $y^2 = x^2 + x^3 \quad (x \le 0)$

16 $x^2 y^3$ on $\frac{1}{2}x + \frac{1}{6}y = 1 \quad (-\frac{1}{4} \le x \le 2)$

17 $y - 2x$ on $x^3 = y^2 \quad (y \ge 0)$

18 $x^2 y^3$ on $x^2 + y^2 = 15$

19 Of all rectangles with perimeter 1, which has the shortest diagonal? That is, minimize $(x^2 + y^2)^{1/2}$ subject to $2x + 2y = 1$.

20 Of all right triangles with perimeter 1, which has largest area? That is, maximize $\frac{1}{2}xy$ subject to $x + y + (x^2 + y^2)^{1/2} = 1$.

21 Find the dimensions of the right circular cylinder of fixed total surface area A, including top and bottom, with maximum volume.

22 Find the dimensions of the right circular cylinder of fixed lateral area A with maximum volume.

23 Find the maximal area of a triangle of perimeter $2 + \sqrt{2 - \sqrt{2}}$ with a $45°$ angle. [Hint Use the law of cosines.]

24 Find the dimensions of the right circular cone of fixed total surface area A, including the base, with maximum volume.

25 Maximize $x + y - \sqrt{x^2 + y^2}$, where x and y are the legs of a right triangle of area 1.

26 Maximize $x + y - \sqrt{x^2 + y^2}$, where x and y are the legs of a right triangle of perimeter $3 + 2\sqrt{2}$.

27 Maximize $(x^2 + y^2)(2x + y)$ on $x^2 + y^2 \le 9$.

28 Let $0 < p < q$ and $b > 0$. Find the maximum and minimum of $x^p + y^p$ on $x^q + y^q = b^q$ for $x \ge 0$ and $y \ge 0$.

29 (cont.) Let $0 < p < q$ and $x \ge 0$, $y \ge 0$. Prove

$$\frac{1}{2^{1/p-1/q}}\left(\frac{x^q + y^q}{2}\right)^{1/q} \le \left(\frac{x^p + y^p}{2}\right)^{1/p} \le \left(\frac{x^q + y^q}{2}\right)^{1/q}$$

[Hint Set $x^q + y^q = b^q$.]

30 Let (x_0, y_0, λ) be a solution of the Lagrange multiplier problem for the extrema of $f(x, y)$ subject to the constraint $g(x, y) = c$. Show that $df = \lambda\, dg$ at (x_0, y_0) so that λ represents the ratio of the differentials df and dg for a change in any direction at (x_0, y_0).

15-5 Lagrange Multipliers: Three Variables

The first constraint problem in three variables that we consider is to maximize (or minimize) a function $f(x, y, z)$ subject to a single constraint $g(x, y, z) = c$. Geometrically, this problem is completely analogous to the two-variable problems treated in the preceding section. We try to find the level surface of $f(x, y, z)$ of highest level that intersects the surface $g(x, y, z) = c$. See Figure 15-5-1a. Each point of tangency is a candidate for a maximum or a minimum. At such points, the normals to the two surfaces are parallel (Figure 15-5-1b). But the vectors

$$\operatorname{grad} f(\mathbf{x}) \quad \text{and} \quad \operatorname{grad} g(\mathbf{x})$$

point in the respective normal directions, so one is a multiple of the other:

$$\operatorname{grad} f(\mathbf{x}) = \lambda \operatorname{grad} g(\mathbf{x})$$

for some number λ, provided $\operatorname{grad} g(\mathbf{x}) \ne \mathbf{0}$. This observation leads to a practical method for locating possible maxima and minima.

Figure 15-5-1
Lagrange multiplier rule

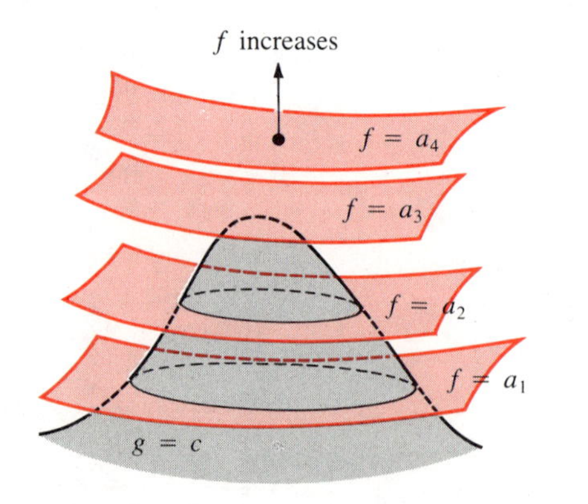

a To find the highest level surface of f that meets $g = c$

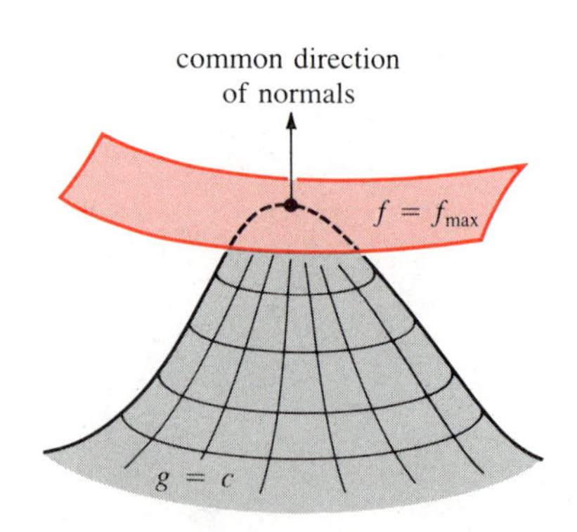

b The highest level $f = f_{max}$ and $g = c$ are tangent at their common contact point.

Lagrange Multiplier Rule (One Constraint)

To maximize (minimize) a function $f(x, y, z)$ subject to a constraint $g(x, y, z) = c$, solve the system of equations

$$(f_x, f_y, f_z) = \lambda(g_x, g_y, g_z) \qquad g(x, y, z) = c$$

in the four unknowns x, y, z, λ. Each resulting point (x, y, z) is a candidate.

In applications, the usual precautions concerning the boundary must be observed.

Figure 15-5-2

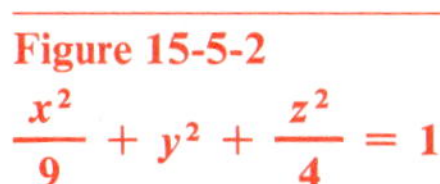

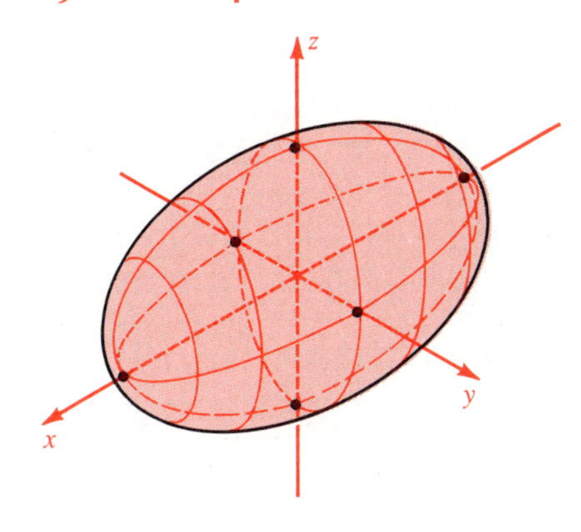

Example 1

Find the points on the ellipsoid $\frac{1}{9}x^2 + y^2 + \frac{1}{4}z^2 = 1$ nearest to and farthest from the origin (Figure 15-5-2).

Solution The problem calls for the extrema of

$$f(x, y, z) = x^2 + y^2 + z^2$$

subject to

$$g(x, y, z) = \tfrac{1}{9}x^2 + y^2 + \tfrac{1}{4}z^2 = 1$$

Now

$$\operatorname{grad} f = (2x, 2y, 2z) \qquad \text{and} \qquad \operatorname{grad} g = (\tfrac{2}{9}x, 2y, \tfrac{1}{2}z)$$

According to the Lagrange multiplier rule, we must solve the equations

$$(2x, 2y, 2z) = \lambda(\tfrac{2}{9}x, 2y, \tfrac{1}{2}z) \qquad \text{and} \qquad \tfrac{1}{9}x^2 + y^2 + \tfrac{1}{4}z^2 = 1$$

that is,

$$2x = \tfrac{2}{9}\lambda x \qquad 2y = 2\lambda y \qquad 2z = \tfrac{1}{2}\lambda z \quad \text{and} \quad \tfrac{1}{9}x^2 + y^2 + \tfrac{1}{4}z^2 = 1$$

At least one of x, y, z is non-zero. If $x \neq 0$, then the first equation implies $\lambda = 9$. Then the second and third equations imply that $y = z = 0$. Now by the last equation, $x^2 = 9$ so $x = \pm 3$. Thus the points $(\pm 3, 0, 0)$ are candidates for extrema.

If $y \neq 0$, then the second equation implies $\lambda = 1$, which leads in the same way to candidates $(0, \pm 1, 0)$. If $z \neq 0$, then $\lambda = 4$ and the candidates are $(0, 0, \pm 2)$. Since

$$f(\pm 3, 0, 0) = 9 \qquad f(0, \pm 1, 0) = 1 \qquad f(0, 0, \pm 2) = 4$$

the maximum distance is $\sqrt{9} = 3$, the minimum is 1.

Example 2

Find the volume of the largest rectangular solid with sides parallel to the coordinate axes that can be inscribed in the ellipsoid $x^2 + \frac{1}{9}y^2 + \frac{1}{4}z^2 = 1$.

Figure 15-5-3
See Example 2

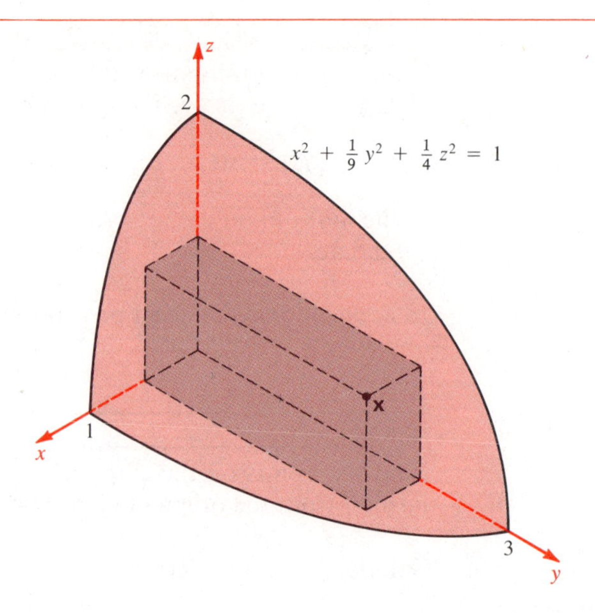

Solution As Figure 15-5-3 shows, one-eighth of the volume of the solid is xyz, where $x > 0$, $y > 0$, $z > 0$. Hence it suffices to maximize

$$f(x, y, z) = xyz$$

subject to the constraint $g(x, y, z) = x^2 + \frac{1}{9}y^2 + \frac{1}{4}z^2 = 1$. Apply the Lagrange multiplier rule by setting $\operatorname{grad} f = \lambda \operatorname{grad} g$ and $g = 1$ to obtain

$$(yz, zx, xy) = \lambda(2x, \tfrac{2}{9}y, \tfrac{1}{2}z) \qquad \text{and} \qquad x^2 + \tfrac{1}{9}y^2 + \tfrac{1}{4}z^2 = 1$$

that is

$$yz = 2\lambda x \qquad zx = \tfrac{2}{9}\lambda y \qquad xy = \tfrac{1}{2}\lambda z \quad \text{and} \quad x^2 + \tfrac{1}{9}y^2 + \tfrac{1}{4}z^2 = 1$$

To solve these equations, multiply the first two and cancel xy:

$$z^2 = \tfrac{4}{9}\lambda^2$$

Likewise $x^2 = \frac{1}{9}\lambda^2$ and $y^2 = \lambda^2$. (This is valid because if x or y is zero, then the volume is 0; not for us.) Substitute into the fourth equation:

$$\tfrac{1}{9}\lambda^2 + \tfrac{1}{9}\lambda^2 + \tfrac{1}{9}\lambda^2 = 1 \qquad \text{so that} \qquad \lambda^2 = 3$$

Hence

$$x^2 = \tfrac{1}{3} \qquad y^2 = 3 \qquad z^2 = \tfrac{4}{3}$$

Therefore

$$f(x, y, z)^2 = x^2y^2z^2 = \tfrac{4}{3} \qquad f_{\max} = \tfrac{2}{3}\sqrt{3}$$

The maximal volume is 8 times this: $V_{\max} = \frac{16}{3}\sqrt{3}$ ●

Figure 15-5-4
Lagrange multiplier rule for two constraints

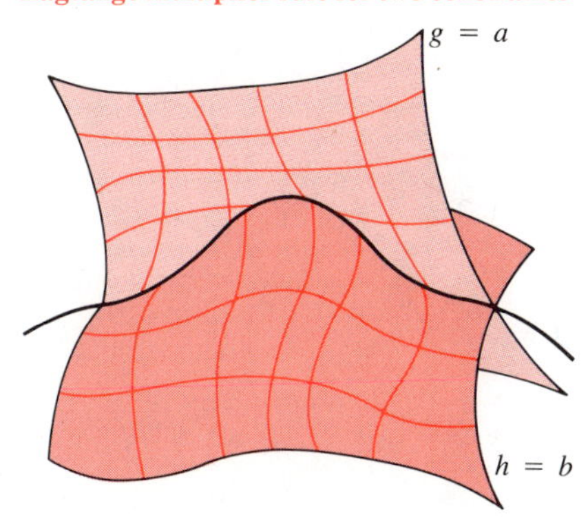

a Two constraints determine a curve.

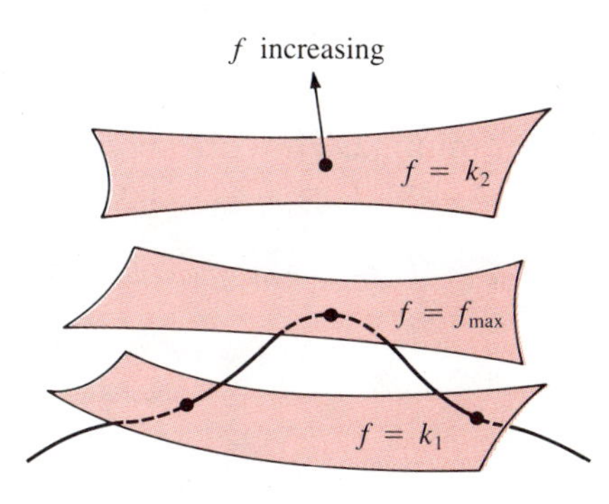

b The highest level of f that meets the curve is tangent to it.

Remark Another way of solving for x, y, z, λ is to solve the first three equations for λ and equate the results:

$$\lambda = \frac{yz}{2x} = \frac{9zx}{2y} = \frac{2xy}{z}$$

It follows easily that $y^2 = 9x^2$ and $z^2 = 4x^2$. Substituting these values into the fourth equation, we find $3x^2 = 1$, $x^2 = \frac{1}{3}$, and the rest follows.

Two Constraints

Suppose the problem is to maximize (minimize) $f(x, y, z)$ where (x, y, z) is subject to *two* constraints, $g(x, y, z) = a$ and $h(x, y, z) = b$. Each constraint defines a surface and these two surfaces in general have a curve of intersection (Figure 15-5-4a). A candidate for a maximum or minimum of $f(\mathbf{x})$ is a point $\mathbf{x}$ where a level surface of f is tangent to this curve of intersection (Figure 15-5-4b). The vector grad $f(\mathbf{x})$ is normal to the level surface at $\mathbf{x}$, hence normal to the curve. But the vectors grad $g(\mathbf{x})$ and grad $h(\mathbf{x})$ *determine* the normal plane to the curve at $\mathbf{x}$. Hence for some constants λ and μ,

$$\operatorname{grad} f(\mathbf{x}) = \lambda \operatorname{grad} g(\mathbf{x}) + \mu \operatorname{grad} h(\mathbf{x})$$

Remark The existence of such **multipliers** λ and μ presupposes that grad $g \neq \mathbf{0}$, grad $h \neq \mathbf{0}$, and that neither is a multiple of the other. Concisely, grad g and grad h must be *linearly independent* vectors at $\mathbf{x}$.

Lagrange Multiplier Rule (Two Constraints)

To maximize (minimize) a function $f(x, y, z)$ subject to two constraints $g(x, y, z) = a$ and $h(x, y, z) = b$, solve the system of five equations

$$\begin{cases} (f_x, f_y, f_z) = \lambda(g_x, g_y, g_z) + \mu(h_x, h_y, h_z) \\ g(x, y, z) = a \qquad h(x, y, z) = b \end{cases}$$

in five unknowns x, y, z, λ, μ. Each resulting point (x, y, z) is a candidate.

Figure 15-5-5
See Example 3

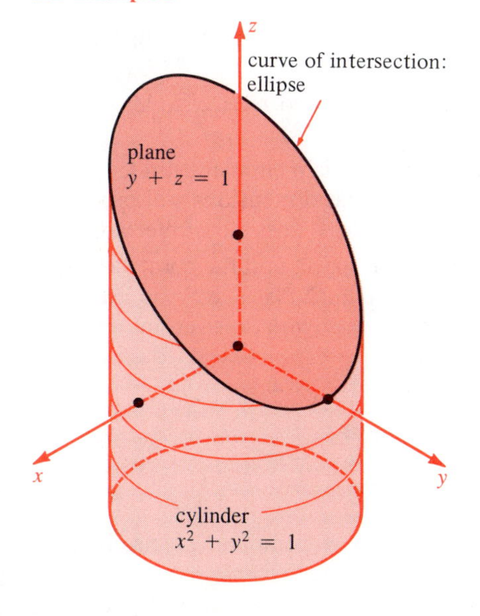

Example 3

Find the maximum and minimum of $f(x, y, z) = x + 2y + z$ on the ellipse $x^2 + y^2 = 1$, $y + z = 1$.

Solution The ellipse is the intersection of the cylinder $x^2 + y^2 = 1$ and the plane $y + z = 1$. See Figure 15-5-5. Maximizing $f(x, y, z)$ on the ellipse is equivalent to maximizing $f(x, y, z)$ subject to the constraints $g(x, y, z) = 1$ and $h(x, y, z) = 1$, where

$$g(x, y, z) = x^2 + y^2 \qquad \text{and} \qquad h(x, y, z) = y + z$$

According to the Lagrange multiplier rule, the equations to be solved are

$$\begin{cases}(1, 2, 1) = \lambda(2x, 2y, 0) + \mu(0, 1, 1) \\ x^2 + y^2 = 1 \quad \text{and} \quad y + z = 1\end{cases}$$

From the first equation,

$$1 = 2\lambda x \qquad 2 = 2\lambda y + \mu \qquad 1 = \mu$$

Hence $x = \frac{1}{2\lambda}$ and $y = \frac{1}{2\lambda}$. Therefore

$$\left(\frac{1}{2\lambda}\right)^2 + \left(\frac{1}{2\lambda}\right)^2 = 1 \qquad \text{so } \lambda^2 = \tfrac{1}{2} \qquad \text{and} \qquad \lambda = \pm\tfrac{1}{2}\sqrt{2}$$

The solutions $\lambda = \frac{1}{2}\sqrt{2}$ and $\lambda = -\frac{1}{2}\sqrt{2}$ lead respectively to

$$x_1 = (\tfrac{1}{2}\sqrt{2}, \tfrac{1}{2}\sqrt{2}, 1 - \tfrac{1}{2}\sqrt{2})$$
$$x_2 = (-\tfrac{1}{2}\sqrt{2}, -\tfrac{1}{2}\sqrt{2}, 1 + \tfrac{1}{2}\sqrt{2})$$

These are the only candidates for maxima or minima. The corresponding values of f are $f(\mathbf{x}_1) = 1 + \sqrt{2}$ and $f(\mathbf{x}_2) = 1 - \sqrt{2}$. Therefore

$$f_{\max} = f(\mathbf{x}_1) = 1 + \sqrt{2} \qquad \text{and} \qquad f_{\min} = f(\mathbf{x}_2) = 1 - \sqrt{2}$$ ●

Exercises

Find the maximum and the minimum and the points where they are taken

1 $2x + y - 5z$ on $x^2 + y^2 + z^2 = 1$

2 $x + y + z$ on $\dfrac{x^2}{a^2} + \dfrac{y^2}{b^2} + \dfrac{z^2}{c^2} = 1$ $\quad (a, b, c > 0)$

3 $x + y + z$ on $\dfrac{1}{x} + \dfrac{4}{y} + \dfrac{9}{z} = 1$ $\quad (x > 0, y > 0, z > 0)$

4 xy^2z^3 on $x + y + z = 3$ $(x \ge 0, y \ge 0, z \ge 0)$

5 xyz on $x^2 + y^2 + 3z^2 = 5$

6 $e^x yz$ on $4x + 2y + z = 12$ $(x \ge 0, y \ge 0, z \ge 0)$

7 xyz on $\dfrac{x}{a} + \dfrac{y}{b} + \dfrac{z}{c} = 1$ $\quad (x \ge 0, y \ge 0, z \ge 0, a, b, c > 0)$

8 $x^4 + y^4 + z^4$ on $x^2 + y^2 + z^2 = 1$

9 Show that the maximum of $ax + by + cz$ subject to $x^2 + y^2 + z^2 = r^2$ is $r\sqrt{a^2 + b^2 + c^2}$. Use Lagrange multipliers. Assume $(a, b, c) \ne \mathbf{0}$.

10 (cont.) Obtain the same conclusion by means of vectors and the Cauchy-Schwarz inequality.

Use the result of Exercise 9 to maximize

11 $\dfrac{2x - 2y + z}{1 + x^2 + y^2 + z^2}$ **12** $\dfrac{ax + by + cz}{\exp(x^2 + y^2 + z^2)}$

13 Find the rectangular solid of fixed volume with minimal surface area.

14 Find the rectangular solid of fixed total edge length with maximal surface area.

***15** A silo is built in the form of a right circular cylinder topped by a right circular cone. What dimensions will produce a given volume V with the least possible surface area?

***16** Given $\mathbf{a}_1, \cdots, \mathbf{a}_n$ in space, find the point $\mathbf{x}$ of the plane $\mathbf{x} \cdot \mathbf{c} = p$ such that $\Sigma_1^n |\mathbf{x} - \mathbf{a}_j|^2$ is least. Assume $|\mathbf{c}| = 1$, and express your answer in terms of $\mathbf{a} = n^{-1} \Sigma \mathbf{a}_j$.

17 Maximize $x^2y^2z^2$ on $x^2 + y^2 + z^2 = r^2$.

18 (cont.) Prove the inequality

$$\sqrt[3]{xyz} \le \frac{x + y + z}{3}$$

for $x > 0$, $y > 0$, $z > 0$. (Compare Exercise 16, page 222.)

19 Prove the inequalities

$$-\tfrac{1}{2}(x^2 + y^2 + z^2) \le xy + yz + zx$$
$$\le x^2 + y^2 + z^2$$

20 Let $\mathbf{a}_1, \cdots, \mathbf{a}_n$ be points of space. Set $\mathbf{a} = n^{-1}\sum_1^n \mathbf{a}_j$ and assume $\mathbf{a} \ne \mathbf{0}$. Find the points $\mathbf{x}$ of the unit sphere $|\mathbf{x}| = 1$ that yield the maxima and minima of $\sum_1^n |\mathbf{x} - \mathbf{a}_j|^2$.

21 Use Heron's formula and Lagrange multipliers to maximize the area of a triangle with fixed semiperimeter s.

22 Find the maximal area in Figure 15-5-6 if the perimeter is $4 + 2\sqrt{3}$.

Figure 15-5-6
See Exercise 22

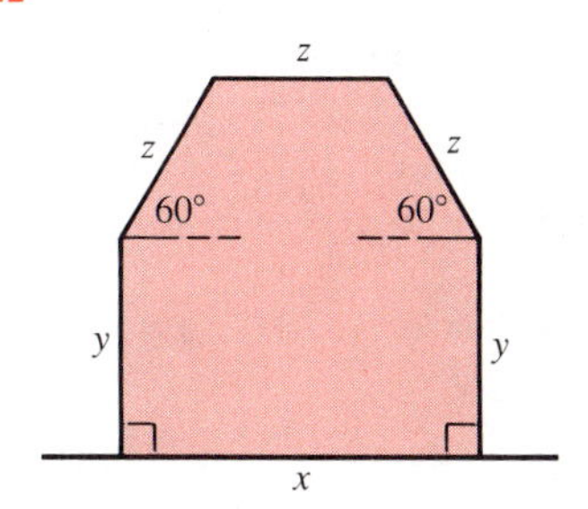

23 Prove that $\cot x - 1/x$ is strictly decreasing for $0 < x < \pi$.

24 (cont.) Let x, y, z be the angles of a triangle. Find the maximum of

$$\frac{\sin x \sin y \sin z}{xyz}$$

and where this maximum is taken.

Find the point(s) of the first octant nearest the origin and on the surface

25 $xyz = 1$

26 $xyz^2 = 1$

27 $x^2 + yz = 1$

28 $x^3 + 3yz = 3$

29 $x(y^2 + z^2) = 1$

30 $xy^2z^3 = 6\sqrt{3}$

31 Let $0 < p < q$. Find the maximum and minimum of $x^p + y^p + z^p$ on the surface $x^q + y^q + z^q = b^q$ where $x \ge 0$, $y \ge 0$, and $z \ge 0$. (Use Exercise 28, page 778.)

32 (cont.) Let $0 < p < q$ and $x \ge 0$, $y \ge 0$, $z \ge 0$. Show that

$$\left(\frac{x^p + y^p + z^p}{3}\right)^{1/p} \le \left(\frac{x^q + y^q + z^q}{3}\right)^{1/q}$$

33 Find the maximum and minimum of $\sqrt{x^2 + y^2 + z^2}$ on $x^3 + y^2z = 62$, $x \ge 0$, $y \ge 0$, $z \ge 0$, and where they occur.

34 Find the maximum of xyz on $x^2 + y^3 + z^4 = \frac{13}{3}$, $x \ge 0$, $y \ge 0$, $z \ge 0$, and where it occurs.

***35** Find the maximal area in Figure 15-5-7 provided that $x + y + z = 6k$. [Hint Introduce the altitude and treat this as a problem in four variables with one constraint.]

Figure 15-5-7
See Exercise 35

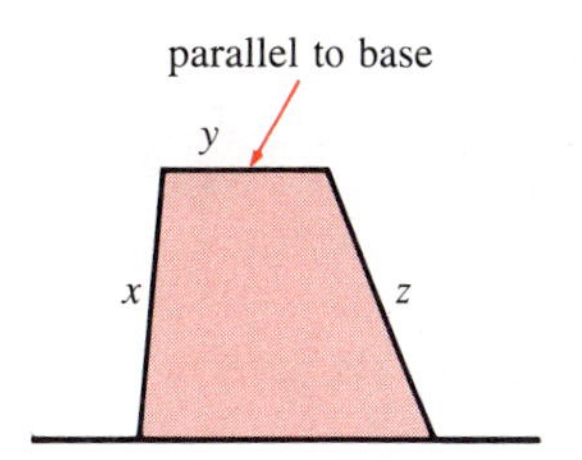

***36** Suppose $p > 1$, $q > 1$, $1/p + 1/q = 1$, $a > 0$, $b > 0$, $c > 0$, and $a^q + b^q + c^q = d^q$. Find the maximum of $ax + by + cz$ on $x^p + y^p + z^p = r^p$, $x \ge 0$, $y \ge 0$, $z \ge 0$.

Find the closest point(s) to the origin

37 on the line $x + y + z = 1$, $x + 2y + 3z = 1$

***38** on the ellipse $x + y + z = 0$, $x^2 + \frac{1}{4}y^2 + \frac{1}{4}z^2 = 2$

39 Find the maximum and minimum volumes for a rectangular solid whose total edge length is 24 ft and whose surface area is 22 ft^2.

40 Find the maximum of x on the circle $x^2 + y^2 + z^2 = 182$, $x + 2y + 3z = 0$.

41 Find the minimum of $2x^2 + y^2 + z^2$ on the line $2x - y + z = 1$, $x - 2y - z = 2$.

***42** Find the maximum and minimum of $x + y + z$ on the curve $xyz = 1$, $x^2 + y^2 + z^2 = 5$, $x > 0$, $y > 0$, $z > 0$.

43 A rectangular solid has volume 30. The sum of the area of its base and the square of its height is 19. Find the maximum and minimum of the perimeter of its base.

44 Find the maximal area of a right triangle of perimeter 1 by treating its three edge lengths as variables subject to 2 constraints.

45 A production process uses input materials $X_1, \cdots, X_n$, and the output is a product Y. Let p_i be the (fixed) price per unit amount of X_i and x_i the amount of X_i used. Then the cost of producing Y is $C = \sum p_i x_i$. The amount of Y produced is $y = f(x_1, \cdots, x_n)$. We assume the **production function** f is homogeneous of degree 1. (In economics, this property of f is called *con-*

stant returns to scale.) Suppose $(x_1, \cdots, x_n)$ minimizes the cost C subject to the single constraint $f(x_1, \cdots, x_n) = k$, and all $x_i > 0$. Prove the **law of marginal productivity:** For each i,

$$\frac{p_i}{\partial f/\partial x_i} = \frac{C}{k} \quad \text{at} \quad (x_1, \cdots, x_n)$$

(In a competitive market, C/k will be the unit price of Y.) [Hint Use Euler's relation, Exercise 45 in Section 14-3.]

*46 (cont.) A special case is the **Cobb-Douglas** production function

$$f = x_1^{\alpha_1} x_2^{\alpha_2} \cdots x_n^{\alpha_n}$$

where $\alpha_j > 0$ and $\alpha_1 + \alpha_2 + \cdots + \alpha_n = 1$. Show that

$$\frac{p_i x_i}{\alpha_i} = \frac{C}{k}$$

at a minimum. In this case λ is called the **equimarginal productivity.**

15-6 Review Exercises

Identify and sketch the quadric surface

1 $x^2 + 4y^2 + 9z^2 = 25$

2 $x^2 + 4y^2 - z^2 = 25$

3 $-x^2 - y^2 + z^2 = 9$

4 $z = x^2 + 4y^2$

5 Find the tangent plane to $x^2 + y^2 + z^2 = 17$ at $(2, -2, 3)$.

6 Find the tangent plane to $z = x^2 - 9y^2$ at $(1, 1, -8)$.

7 Find the largest rectangular area inscribed in a semicircle of radius a.

8 Find the maximum and minimum in the first quadrant of $f(x, y) = xy^2 \exp(-x^3 - y^3)$.

9 A right circular cone has volume 1. How small can its lateral area be? How large?

10 Given $\mathbf{a}_1, \cdots, \mathbf{a}_n$ in the plane, find the $\mathbf{x}$ on the circle $|\mathbf{x}| = 1$ for which $\Sigma_1^n |\mathbf{x} - \mathbf{a}_j|^2$ is least.

11 Find the polynomial $p(x) = x^2 + ax + b$ that minimizes $\int_{-1}^{1} p(x)^2\,dx$.

12 (cont.) The same for $p(x) = x^3 + ax^2 + bx + c$.

13 Find the maximum of $x^3y^2(12 - x - y)$ in the first quadrant.

14 Solve Exercise 13 by maximizing x^3y^2z subject to a suitable condition on x, y, and z.

15 Find the largest possible area of a triangle of perimeter 1, if one of its angles is a given α. [Hint Law of cosines]

16 Assume $a, b, c > 0$. Find the largest possible volume of a rectangular solid (with sides parallel to the coordinate planes) inscribed in the ellipsoid

$$\frac{x^2}{a^2} + \frac{y^2}{b^2} + \frac{z^2}{c^2} = 1$$

17 Compute the maximum of xy^2z^3 on the sphere $x^2 + y^2 + z^2 = 1$.

18 Find the greatest distance from the x-axis to the ellipse $4x^2 - 2xy + y^2 = 1$.

19 Find the minimum of $a/x + b/y + c/z + xyz$ for $x > 0$, $y > 0$, $z > 0$. Assume $a > 0$, $b > 0$, $c > 0$.

20 Let $f(x, y) = x^{3/2} + 2y^{3/2}$. Maximize $|\text{grad} f|$ on $x^2 + (y - 1)^2 \le 17$.

21 Find the largest value of $x + 2y + 3z$ for points (x, y, z) on the unit sphere $x^2 + y^2 + z^2 = 1$.

22 Of all closed rectangular boxes with fixed surface area, which has greatest volume? That is, maximize xyz subject to $xy + yz + zx = 3c^2$.

***23** Find the maximum of $f(x, y, z) = x + y - z$ on the region

$$\begin{cases} x + y + z = 1 \\ 0 \le x \le y \le z \end{cases}$$

***24** Consider an ammonia synthesis reaction

$$N_2 + 3H_2 \rightleftharpoons 2NH_3$$

at fixed pressure P and also at fixed temperature. Then $P = x + y + z$, where x, y, z are the partial pressures of N_2, H_2, and NH_3 respectively. In this situation it is known that $z^2/xy^3 = 3k^2$, where k is a constant. Also, the concentration of each of the three ingredients is proportional to its partial pressure. Maximize z, so as much ammonia as possible is produced.

16 Higher Partial Derivatives

After we studied first derivatives of functions of one variable, we went on to second and higher derivatives. These were quite useful for graphing, approximating, and finding extrema. Now we shall extend our work on the partial derivatives of functions of several variables to second and higher derivatives and their applications. We shall see in Section 16-1 that there is a phenomenon for several variables that has no precedent in one variable: the mixed second partials. In Section 16-2 we study Taylor approximations for functions of several variables. Sections 16-3 and 16-4 cover second-derivative tests for extrema, and Section 16-5 covers some basic theory.

16-1 Mixed Partials

A function of two variables $f(x, y)$ has two first partial derivatives,

$$f_x(x, y) \quad \text{and} \quad f_y(x, y)$$

each itself a function of two variables. Each in turn has two first partial derivatives; these four new functions are the second derivatives of $f(x, y)$. Figure 16-1-1 (next page) shows their evolution:

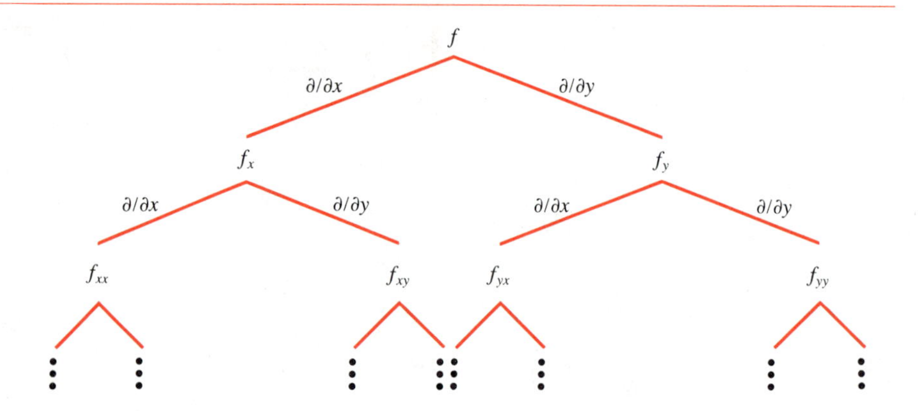

Figure 16-1-1
Partial derivative tree

The **pure second partials** f_{xx} and f_{yy} represent nothing really new. Each is found by holding one variable constant and differentiating twice with respect to the other variable. Alternative notation:

$$f_{xx} = \frac{\partial^2 f}{\partial x^2} \qquad f_{yy} = \frac{\partial^2 f}{\partial y^2}$$

For example, if $f(x, y) = x^3y^4 + \cos 5y$, then

$$f_x = 3x^2y^4 \qquad f_y = 4x^3y^3 - 5 \sin 5y$$
$$f_{xx} = 6xy^4 \qquad f_{yy} = 12x^3y^2 - 25 \cos 5y$$

The **mixed second partials**

$$f_{xy} = \frac{\partial}{\partial y}\left(\frac{\partial f}{\partial x}\right) = \frac{\partial^2 f}{\partial y \partial x} \quad \text{and} \quad f_{yx} = \frac{\partial}{\partial x}\left(\frac{\partial f}{\partial y}\right) = \frac{\partial^2 f}{\partial x \partial y}$$

are new. The mixed partial f_{xy} measures the rate of change in the y-direction of the rate of change of f in the x-direction. The mixed partial f_{yx} measures the rate of change in the x-direction of the rate of change of f in the y-direction. It is not easy to see how, if at all, these two mixed partials are related to each other.

Let us compute the mixed partials of $f(x, y) = x^3y^4 + \cos 5y$:

$$f_x = 3x^2y^4 \qquad f_{xy} = 3 \cdot 4x^2y^3$$
$$f_y = 4x^3y^3 - 5 \sin 5y \qquad f_{yx} = 4 \cdot 3x^2y^3 + 0$$

The mixed partials are equal! This is not an accident but a special case of an equality that holds for functions normally encountered in applications.

Equality of Mixed Partials

Let $f(x, y)$ be defined on a domain **D**. If f_{xy} and f_{yx} exist at each point of **D** and are continuous at (a, b), then

$$\frac{\partial^2 f}{\partial x \partial y} = \frac{\partial^2 f}{\partial y \partial x} \quad \text{at} \quad (a, b)$$

We postpone the proof until Section 16-5.

Higher Partials

The following discussion assumes the existence and continuity of all partials involved.

A function $z = f(x, y)$ has two distinct first partials,

$$\frac{\partial z}{\partial x} \quad \text{and} \quad \frac{\partial z}{\partial y}$$

and three distinct second partials,

$$\frac{\partial^2 z}{\partial x^2} \qquad \frac{\partial^2 z}{\partial x \partial y} = \frac{\partial^2 z}{\partial y \partial x} \qquad \frac{\partial^2 z}{\partial y^2}$$

Because the mixed second partials are equal, so are certain mixed third partials. For example,

$$\frac{\partial^2}{\partial x^2}\left(\frac{\partial z}{\partial y}\right) = \frac{\partial}{\partial y}\left(\frac{\partial^2 z}{\partial x^2}\right)$$

because

$$\frac{\partial^2}{\partial x^2}\left(\frac{\partial z}{\partial y}\right) = \frac{\partial}{\partial x}\left[\frac{\partial}{\partial x}\left(\frac{\partial z}{\partial y}\right)\right] = \frac{\partial}{\partial x}\left[\frac{\partial}{\partial y}\left(\frac{\partial z}{\partial x}\right)\right]$$
$$= \frac{\partial}{\partial y}\left[\frac{\partial}{\partial x}\left(\frac{\partial z}{\partial x}\right)\right] = \frac{\partial}{\partial y}\left(\frac{\partial^2 z}{\partial x^2}\right)$$

Thus there are precisely four distinct third partials:

$$\frac{\partial^3 z}{\partial x^3} \qquad \frac{\partial^3 z}{\partial x^2 \partial y} \qquad \frac{\partial^3 z}{\partial x \partial y^2} \qquad \frac{\partial^3 z}{\partial y^3}$$

Provided all partials up to order n exist and are continuous, there are only $n + 1$ distinct partials of order n:

$$\frac{\partial^n z}{\partial x^k \partial y^{n-k}} \qquad (k = 0, 1, 2, \cdots, n)$$

Each is completely determined by k, the number of times $\partial/\partial x$ is taken (so $\partial/\partial y$ is taken $n - k$ times). The order in which the $\partial/\partial x$'s and $\partial/\partial y$'s are taken is irrelevant.

Example 1

Find all first and second partials of $z = f(x, y) = e^{x-y^2}$.

Solution First partials:

$$\frac{\partial z}{\partial x} = e^{x-y^2} = z$$

$$\frac{\partial z}{\partial y} = -2ye^{x-y^2} = -2yz$$

Second partials:

$$\frac{\partial^2 z}{\partial x^2} = \frac{\partial}{\partial x}\left(\frac{\partial z}{\partial x}\right) = \frac{\partial}{\partial x}(z) = z$$

$$\frac{\partial^2 z}{\partial x \partial y} = \frac{\partial}{\partial y}\left(\frac{\partial z}{\partial x}\right) = \frac{\partial}{\partial y}(z) = -2yz$$

$$\frac{\partial^2 z}{\partial y^2} = \frac{\partial}{\partial y}\left(\frac{\partial z}{\partial y}\right) = \frac{\partial}{\partial y}(-2yz) = -2\frac{\partial}{\partial y}(yz)$$

$$= -2[z + y(-2yz)] = -2z(1 - 2y^2)$$

●

More Variables

Everything said applies to functions of three or more variables. For example, suppose $w = f(x, y, z)$. Then w has three first partials:

$$\frac{\partial w}{\partial x} \qquad \frac{\partial w}{\partial y} \qquad \frac{\partial w}{\partial z}$$

The nine possible second partials may be written in matrix form:

$$\begin{bmatrix} \dfrac{\partial^2 w}{\partial x^2} & \dfrac{\partial^2 w}{\partial x \partial y} & \dfrac{\partial^2 w}{\partial x \partial z} \\ \dfrac{\partial^2 w}{\partial y \partial x} & \dfrac{\partial^2 w}{\partial y^2} & \dfrac{\partial^2 w}{\partial y \partial z} \\ \dfrac{\partial^2 w}{\partial z \partial x} & \dfrac{\partial^2 w}{\partial z \partial y} & \dfrac{\partial^2 w}{\partial z^2} \end{bmatrix}$$

This matrix, called the **Hessian matrix** of the function $f(x, y, z)$ is symmetric since the mixed second partials are equal in pairs:

$$\frac{\partial^2 w}{\partial y \partial x} = \frac{\partial^2 w}{\partial x \partial y} \qquad \frac{\partial^2 w}{\partial x \partial z} = \frac{\partial^2 w}{\partial z \partial x} \qquad \frac{\partial^2 w}{\partial z \partial y} = \frac{\partial^2 w}{\partial y \partial z}$$

Recall that an $n \times n$ square matrix $A = [a_{ij}]$ is symmetric if $a_{ij} = a_{ji}$ for all i and j from 1 to n.

Some Partial Differential Equations

In the following examples we shall find all functions whose partials satisfy given relations. We assume that all necessary partials exist and, for simplicity, that the functions are defined on the whole plane. All the solutions are based on one fact: if

$$\frac{\partial z}{\partial x} = 0 \qquad \text{and} \qquad \frac{\partial z}{\partial y} = 0$$

on the whole plane, then z is a constant. For $\partial z/\partial x = 0$ means $z(x, y)$ is constant in x, that is, $z = z(y)$ does not depend on x. But $\partial z/\partial y = 0$ implies this function does not depend on y either, so z is a constant.

Example 2

Find all functions $z = f(x, y)$ that satisfy the system of partial differential equations

$$\frac{\partial^2 z}{\partial x^2} = 0 \qquad \frac{\partial^2 z}{\partial x \partial y} = 0 \qquad \frac{\partial^2 z}{\partial y^2} = 0$$

Solution First, look at the partials of $\partial z/\partial x$:

$$\frac{\partial}{\partial x}\left(\frac{\partial z}{\partial x}\right) = \frac{\partial^2 z}{\partial x^2} = 0$$

$$\frac{\partial}{\partial y}\left(\frac{\partial z}{\partial x}\right) = \frac{\partial^2 z}{\partial y \partial x} = \frac{\partial^2 z}{\partial x \partial y} = 0$$

It follows that $\partial z/\partial x = A,$ a constant. Consequently

$$\frac{\partial}{\partial x}[z(x, y) - Ax] = \frac{\partial z}{\partial x} - A = 0$$

so $z - Ax$ is a function of y alone:

$$z - Ax = g(y) \qquad \text{that is} \qquad z = Ax + g(y)$$

But $\partial^2 z/\partial y^2 = 0;$ hence $g''(y) = 0$ so $g(y) = By + C.$ Therefore

$$z(x, y) = Ax + By + C$$

a linear polynomial. ●

Remark In Example 2, we spelled out in detail the passage from $\partial z/\partial x = A$ to $z = Ax + g(y).$ After this we shall just refer to such a step as "integrating (with respect to x)."

Example 3

Find all functions $z = f(x, y)$ whose third partials are all $0.$

Solution The second partials of $\partial z/\partial x$ are all $0.$ By the last example,

$$\frac{\partial z}{\partial x} = Ax + By + C$$

Integrate:

$$z = \tfrac{1}{2}Ax^2 + Bxy + Cx + g(y)$$

But $0 = \partial^3 z/\partial y^3 = d^3 g/dy^3,$ so $g(y)$ is a quadratic polynomial in $y.$ Therefore z is a quadratic in x and y:

$$z(x, y) = ax^2 + bxy + cy^2 + dx + ey + f$$

●

Example 4

Find all functions $z = f(x, y)$ that satisfy $\dfrac{\partial^2 z}{\partial x^2} = 0.$

Solution Write the condition in the form

$$\frac{\partial}{\partial x}\left(\frac{\partial z}{\partial x}\right) = 0 \qquad \text{then integrate:} \qquad \frac{\partial z}{\partial x} = g(y)$$

where $g(y)$ is an arbitrary differentiable function of y alone. Integrate again:

$$z = g(y)x + h(y)$$

where $h(y)$ is an arbitrary differentiable function of y alone. Check:

$$\frac{\partial^2 z}{\partial x^2} = \frac{\partial}{\partial x}\left[\frac{\partial}{\partial x}(g(y)x + h(y))\right] = \frac{\partial}{\partial x} g(y) = 0$$

●

Example 5

Find all functions $z = f(x, y)$ that satisfy $\dfrac{\partial^2 z}{\partial x \partial y} = 0.$

Solution Write the condition in the form

$$\frac{\partial}{\partial y}\left(\frac{\partial z}{\partial x}\right) = 0 \qquad \text{then integrate:} \qquad \frac{\partial z}{\partial x} = p(x)$$

where $p(x)$ is an arbitrary function of x. Integrate again, with respect to x:

$$z = g(x) + h(y)$$

where $g(x)$ is an antiderivative of $p(x)$ and $h(y)$ is an arbitrary function of y. Note that $g(x)$ is an arbitrary differentiable function of x because $p(x)$ is. Check:

$$\frac{\partial^2}{\partial x \partial y}[g(x) + h(y)] = \frac{\partial}{\partial x}\left\{\frac{\partial}{\partial y}[g(x) + h(y)]\right\}$$

$$= \frac{\partial}{\partial x}[h'(y)] = 0$$

●

Example 6

Find all functions $z = f(x, y)$ that satisfy the system of partial differential equations

$$\frac{\partial z}{\partial x} = y \qquad \frac{\partial z}{\partial y} = 1$$

Solution Integrate the first equation:

$$z = xy + g(y)$$

Substitute this into the second equation:

$$\frac{\partial}{\partial y}[xy + g(y)] = 1 \qquad x + g'(y) = 1$$

$$g'(y) = 1 - x$$

This is impossible since $g'(y)$ is a function of y alone. Therefore the problem has no solution. ●

Remark Example 6 illustrates an important point. A system of partial differential equations may have no solution at all! Could we have foreseen this catastrophe for the system above? Yes; for suppose there *were* a function $f(x, y)$ satisfying

$$\frac{\partial f}{\partial x} = y \qquad \text{and} \qquad \frac{\partial f}{\partial y} = 1$$

Then

$$\frac{\partial^2 f}{\partial y \partial x} = \frac{\partial}{\partial y}(y) = 1 \qquad \text{and} \qquad \frac{\partial^2 f}{\partial x \partial y} = \frac{\partial}{\partial x}(1) = 0$$

so the mixed partials would be unequal, a contradiction. Thus we have the following result:

• If the system of equations

$$\frac{\partial z}{\partial x} = p(x, y) \qquad \frac{\partial z}{\partial y} = q(x, y)$$

has a solution, then $\partial p/\partial y = \partial q/\partial x$. ●

Indeed,

$$\frac{\partial p}{\partial y} = \frac{\partial}{\partial y}\left(\frac{\partial z}{\partial x}\right) = \frac{\partial}{\partial x}\left(\frac{\partial z}{\partial y}\right) = \frac{\partial q}{\partial x}$$

This result will be useful later in the study of line integrals.

Exercises

Compute $\dfrac{\partial^2 f}{\partial x^2}$ $\dfrac{\partial^2 f}{\partial x \partial y}$ and $\dfrac{\partial^2 f}{\partial y^2}$ for $f =$

1 $\sin(x - 3y)$
2 xy^6
3 $x^2 \arcsin y$
4 $e^{2x} \cosh y$
5 $ax^2 + 2bxy + cy^2 + dx + ey$
6 $\ln(1 + x - 2y)$

Verify that $\dfrac{\partial^2 f}{\partial x \partial y} = \dfrac{\partial^2 f}{\partial y \partial x}$ for $f =$

7 x/y^2
8 $x + x^3y + y^4$
9 $x^m y^n$
10 $g(x)h(y)$
11 $\dfrac{x + y}{x - y}$
12 $(x - y)(x - 2y) \times (x - 3y)$

Compute $\dfrac{\partial^3 f}{\partial x^2 \partial y}$ and $\dfrac{\partial^3 f}{\partial x \partial y^2}$ for $f =$

13 x^3y^4

14 $x^4y^5 - xy^2$

15 $\cos(xy)$

16 $\sin(x^2y)$

17 $e^{2y}\sin(x + y)$

18 x^y

Write the matrix of nine second partials

19 $xy + yz + zx$

20 $x^m y^n z^p$

21 $\sin(x + 2y + 3z)$

22 $x^2 e^{yz}$

23 There is a function $z = f(x, y)$ such that $z^2 + x^2 + y^2 = 1$ and $f(0, 0) = -1$. Use implicit differentiation to find $\partial^2 f/\partial x \partial y$ at $(0, 0)$. Now check your answer by explicit differentiation.

24 There is a function $z = f(x, y)$ such that $z^5 - xz^4 + yz^3 - 1 = 0$ and $f(1, 1) = 1$. Find $\partial^2 f/\partial x \partial y$ at $(1, 1)$.

25 How many distinct second partials does $f(x, y, z, w)$ have? How many distinct third partials?

26 Show that if f and g are sufficiently differentiable, then each function of the form $f(x, t) = g(x + ct) + h(x - ct)$ satisfies the **wave equation**

$$c^2 \frac{\partial^2 f}{\partial x^2} - \frac{\partial^2 f}{\partial t^2} = 0$$

27 How many possible distinct third partials does $f(x, y, z)$ have?

28 (cont.) Find a function for which these partials are actually distinct.

Find all functions $f(x, y)$ satisfying

29 $\dfrac{\partial^3 f}{\partial x^2 \partial y} = 0$

[Hint Use Example 5.]

30 (cont.) $\dfrac{\partial^3 f}{\partial x^2 \partial y} = 0$ and $\dfrac{\partial^3 f}{\partial x \partial y^2} = 0$

31 all fourth partial derivatives equal 0

32 $\dfrac{\partial^4 f}{\partial x^2 \partial y^2} = 0$

33 $\dfrac{\partial f}{\partial x} = a \qquad \dfrac{\partial f}{\partial y} = b$

34 $\dfrac{\partial f}{\partial x} = y \qquad \dfrac{\partial f}{\partial y} = x$

35 $\dfrac{\partial f}{\partial x} = y^2 \qquad \dfrac{\partial f}{\partial y} = x^2$

36 $\dfrac{\partial^2 f}{\partial x^2} = 2y^3 \qquad \dfrac{\partial f}{\partial y} = 3x^2y^2$

37 Find all functions $f(x, y, z)$ that satisfy $\partial f/\partial x = 0$.

38 Find all functions $f(x, y, z)$ that satisfy $\partial^2 f/\partial x \partial y = 0$.

39 Given a function $f(x, y)$, let $u = x + y$ and $v = x - y$. Show that

$$\frac{\partial^2 f}{\partial x^2} - \frac{\partial^2 f}{\partial y^2} = 4\frac{\partial^2 f}{\partial u \partial v}$$

40 (cont.) Find all functions that satisfy the partial differential equation

$$\frac{\partial^2 f}{\partial x^2} - \frac{\partial^2 f}{\partial y^2} = 0$$

41 Show that Laplace's equation in two variables, $\partial^2 f/\partial x^2 + \partial^2 f/\partial y^2 = 0$, expressed in polar coordinates, becomes

$$\frac{\partial^2 f}{\partial r^2} + \frac{1}{r}\cdot\frac{\partial f}{\partial r} + \frac{1}{r^2}\cdot\frac{\partial^2 f}{\partial \theta^2} = 0$$

42 (cont.) Show that the **cylindrical harmonics** $g_k(x, y) = r^k \cos k\theta$ and $h_k(x, y) = r^k \sin k\theta$ satisfy Laplace's equation for all integers k.

***43** Show that each solution of the partial differential equation

$$\frac{\partial f}{\partial x} + 2\frac{\partial f}{\partial y} = 0$$

is of the form $f(x, y) = g(y - 2x)$. [Hint Set $x = u$ and $y = 2u + v$, then solve a partial differential equation for $h(u, v) = f(u, 2u + v)$.]

***44** (cont.) Apply the same technique to solve

$$\frac{\partial^2 f}{\partial x^2} + 4\frac{\partial^2 f}{\partial x \partial y} + 4\frac{\partial^2 f}{\partial y^2} = 0$$

[Hint Use Example 4.]

45 A uniform thin metal rod (surrounded by insulation) lies on the x-axis from $x = 0$ to $x = L$. Its temperature $u(x, t)$ at $(x, 0)$ at time t satisfies the **heat equation**

$$\frac{\partial^2 u}{\partial x^2} = \frac{1}{k}\cdot\frac{\partial u}{\partial t}$$

Here k is a physical constant. Suppose the ends of the bar are kept at temperature 0. Show that the functions

$$u_n(x, t) = c_n \exp(-kn^2\pi^2 t/L^2)\sin\frac{n\pi x}{L}$$

are solutions for integers n.

46 A taut guitar or piano string is fixed at $(0, 0)$ and $(L, 0)$ and plucked or struck when $t = 0$ so as to vibrate in the x, y-plane. Its displacement $y(x, t)$ from the x-axis (if small) satisfies the **wave equation**

$$\frac{\partial^2 y}{\partial x^2} = \frac{1}{c^2}\cdot\frac{\partial^2 y}{\partial t^2}$$

where c^2 depends on physical properties of the string.

Verify that the functions

$$y_n(x, t) = \left(\sin \frac{n\pi x}{L}\right)\left(a_n \cos \frac{n\pi ct}{L} + b_n \sin \frac{n\pi ct}{L}\right)$$

are solutions for integers n.

47 If the bar in Exercise 45 is moving with constant speed v, then there is some heat loss through its side, and the corresponding heat equation is

$$k\frac{\partial^2 u}{\partial x^2} = v\frac{\partial u}{\partial x} + \frac{\partial u}{\partial t}$$

(The ends are no longer kept at temperature 0.) Show that for each integer n there is an a such that $u(x, t) = \exp(-at + vx/2k)\sin(n\pi x/L)$ is a solution.

48 The current $I(x, t)$ along a transmission line satisfies

$$LC\frac{\partial^2 I}{\partial t^2} + RC\frac{\partial I}{\partial t} = \frac{\partial^2 I}{\partial x^2} \qquad (R, L, C \text{ constant})$$

Find the system of ordinary differential equations that $Y(x)$ and $Z(x)$ must satisfy so that $I = Y(x)\cos nt + Z(x)\sin nt$ is a solution.

49 Show that $V(x, y) = e^{-ax}\cos a(y - y_0)$ satisfies Laplace's equation

$$\partial^2 V/\partial x^2 + \partial^2 V/\partial y^2 = 0$$

50 Find all quadratic forms in x, y, z that satisfy Laplace's equation

$$\partial^2 V/\partial x^2 + \partial^2 V/\partial y^2 + \partial^2 V/\partial z^2 = 0$$

51 Find conditions under which

$$z(x, y, t) = \sin\frac{m\pi x}{a}\sin\frac{n\pi y}{b}\quad(A\cos pt + B\sin pt)$$

(where $p \neq 0$, $A^2 + B^2 \neq 0$, $m \neq 0$, and $n \neq 0$) satisfies the **wave equation** in two dimensions:

$$\frac{\partial^2 z}{\partial x^2} + \frac{\partial^2 z}{\partial y^2} = \frac{1}{c^2}\cdot\frac{\partial^2 z}{\partial t^2}$$

52 The temperature u at distance r from the z-axis at time t after a flash of heat along the z-axis is given by

$$u(r, t) = \frac{a}{t}\exp(-r^2/4kt)$$

Show that

$$\frac{\partial^2 u}{\partial r^2} + \frac{1}{r}\cdot\frac{\partial u}{\partial r} = \frac{1}{k}\cdot\frac{\partial u}{\partial t}$$

16-2 Taylor Approximations

Let us recall some facts about Taylor approximations of functions of one variable. If $y = f(x)$, then

$$f(x) = f(a) + f'(a)(x - a) + r_1(x)$$

and

$$f(x) = f(a) + f'(a)(x - a) + \tfrac{1}{2}f''(a)(x - a)^2 + r_2(x)$$

where

$$|r_1(x)| \leq \frac{M_2}{2!}(x - a)^2 \qquad \text{and} \qquad |r_2(x)| \leq \frac{M_3}{3!}|x - a|^3$$

and where M_2 and M_3 are bounds for $|f''(x)|$ and $|f'''(x)|$ respectively. The Taylor polynomial

$$p_1(x) = f(a) + f'(a)(x - a)$$

is constructed so that

$$p_1(a) = f(a) \qquad \text{and} \qquad p_1'(a) = f'(a)$$

The Taylor polynomial

$$p_2(x) = f(a) + f'(a)(x - a) + \tfrac{1}{2}f''(a)(x - a)^2$$

is constructed so that

$$p_2(a) = f(a) \qquad p_2'(a) = f'(a) \qquad \text{and} \qquad p_2''(a) = f''(a)$$

In a similar way, one cannot construct linear and quadratic polynomials in two variables approximating a given function of two variables by matching its partial derivatives.

Taylor Polynomials in Two Variables

Let $f(x, y)$ have continuous first and second partials on a domain **D**. The **first degree** and **second degree Taylor polynomials** of f at (a, b) are

$$p_1(x, y) = f(a, b) + f_x \cdot (x - a) + f_y \cdot (y - b)$$

$$p_2(x, y) = p_1(x, y) + \tfrac{1}{2}[f_{xx} \cdot (x - a)^2 + 2f_{xy} \cdot (x - a)(y - b) + f_{yy} \cdot (y - b)^2]$$

where all the partials are evaluated at (a, b).

It is easy to check that $p_1(a, b) = f(a, b)$ and that the first partials of p_1 agree with those of f at (a, b). Similarly, $p_2(x, b) = f(a, b)$ and all first and second partials of p_2 agree with the corresponding partials of f at (a, b).

Example 1

Compute the Taylor polynomials $p_1(x, y)$ and $p_2(x, y)$ of the function $f(x, y) = \sqrt{x^2 + y^2}$ at $(3, 4)$.

Solution The first partials of $f(x, y)$ are

$$\frac{\partial f}{\partial x} = \frac{x}{\sqrt{x^2 + y^2}} \qquad \frac{\partial f}{\partial y} = \frac{y}{\sqrt{x^2 + y^2}}$$

The second partials are

$$\frac{\partial^2 f}{\partial x^2} = \frac{y^2}{(x^2 + y^2)^{3/2}}$$

$$\frac{\partial^2 f}{\partial x \partial y} = \frac{-xy}{(x^2 + y^2)^{3/2}}$$

$$\frac{\partial^2 f}{\partial y^2} = \frac{x^2}{(x^2 + y^2)^{3/2}}$$

Their values at $(3, 4)$ are

$$\frac{\partial f}{\partial x} = \tfrac{3}{5} \qquad \frac{\partial f}{\partial y} = \tfrac{4}{5}$$

$$\frac{\partial^2 f}{\partial x^2} = \tfrac{16}{125} \qquad \frac{\partial^2 f}{\partial x \partial y} = -\tfrac{12}{125} \qquad \frac{\partial^2 f}{\partial y^2} = \tfrac{9}{125}$$

Therefore

$$p_1(x, y) = 5 + \tfrac{3}{5}(x - 3) + \tfrac{4}{5}(y - 4)$$

and

$$p_2(x, y) = p_1(x, y) + \tfrac{1}{2}\left[\tfrac{16}{125}(x - 3)^2 - \tfrac{24}{125}(x - 3)(y - 4) + \tfrac{9}{125}(y - 4)^2\right]$$

●

Example 2

Use the Taylor approximations of Example 1 to estimate $\sqrt{(3.1)^2 + (4.02)^2}$

a by $p_1(x, y)$ **b** by $p_2(x, y)$

Solution Set $f(x, y) = \sqrt{x^2 + y^2}$. Near $(3, 4)$

$$f(x, y) \approx p_1(x, y) = \tfrac{1}{5}[25 + 3(x - 3) + 4(y - 4)]$$

Hence

$$f(3.1, 4.02) \approx \tfrac{1}{5}[25 + 3(0.1) + 4(0.02)] = 5.076$$

Next,

$$f(x, y) \approx p_2(x, y) = p_1(x, y) + \tfrac{1}{250}[16(x - 3)^2 - 24(x - 3)(y - 4) + 9(y - 4)^2]$$

Hence

$$\begin{aligned} f(3.1, 4.02) &\approx p_1(3.1, 4.02) + \tfrac{1}{250}[16(0.1)^2 - 24(0.1)(0.02) + 9(0.02)^2] \\ &= 5.076 + \frac{0.1156}{250} \\ &= 5.076 + 0.0004624 \\ &= 5.07646\ 24 \end{aligned}$$

(Actual value to seven places: 5.07645 55.) ●

Now we ask how closely these Taylor polynomials approximate $f(x, y)$ for (x, y) near (a, b). In other words, we want estimates for the errors in $f(x, y) \approx p_1(x, y)$ and $f(x, y) \approx p_2(x, y)$.

The first of these approximations is familiar. For

$$z = f(a, b) + f_x \cdot (x - a) + f_y \cdot (y - b)$$

is the equation of the tangent plane to the graph of $z = f(x, y)$ at the point $(a, b, f(a, b))$. So we are approximating a surface by its tangent plane.

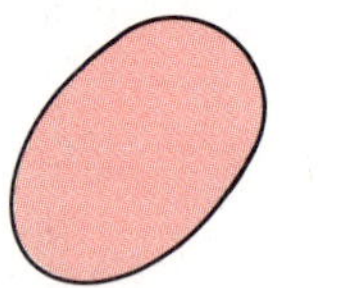
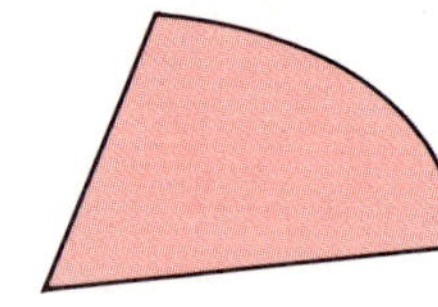

Figure 16-2-1
Convexity in the plane

a Convex domains

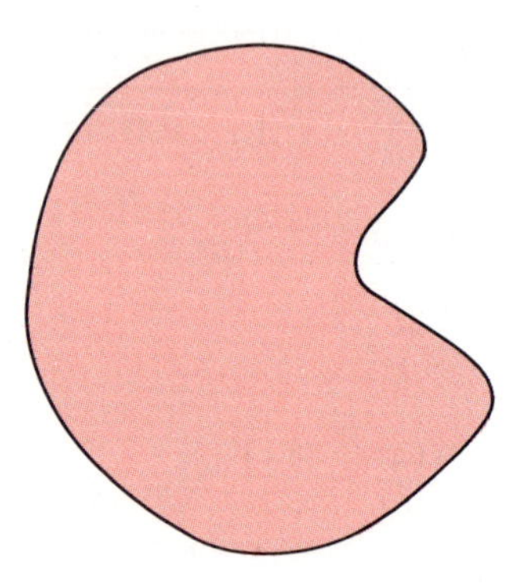

b A *non-convex* domain

Let us state some error estimates subject to the mild restriction that f is defined on a convex domain. A domain $\mathbf{D}$ in the plane or space is called **convex** if it contains the whole segment joining any two of its points. For example, a plane domain bounded by a triangle, square, or ellipse is convex; a space domain bounded by a tetrahedron, cube, or ellipsoid is convex. See Figure 16-2-1.

Error in Taylor Approximation

Let $f(x, y)$ have a convex domain $\mathbf{D}$ and let $\mathbf{a}$ be a point of $\mathbf{D}$.

1 Suppose f has continuous first and second derivatives on $\mathbf{D}$ and the second derivatives satisfy

$$|f_{xx}| \le M_2 \qquad |f_{xy}| \le M_2 \qquad |f_{yy}| \le M_2 \qquad \text{for all } \mathbf{x} \text{ in } \mathbf{D}$$

Let p_1 be the first degree Taylor polynomial of f at $\mathbf{a}$. Then

$$f(\mathbf{x}) = p_1(\mathbf{x}) + r_1(\mathbf{x})$$

where $|r_1(\mathbf{x})| \le M_2|\mathbf{x} - \mathbf{a}|^2$

2 Suppose f has continuous first, second, and third derivatives on $\mathbf{D}$ and the third derivatives satisfy

$$|f_{xxx}| \le M_3 \qquad |f_{xxy}| \le M_3 \qquad |f_{xyy}| \le M_3 \qquad |f_{yyy}| \le M_3$$

Let p_2 be the second degree Taylor polynomial of f at $\mathbf{a}$. Then

$$f(\mathbf{x}) = p_2(\mathbf{x}) + r_2(\mathbf{x})$$

where $|r_2(\mathbf{x})| \le \frac{1}{3}\sqrt{2}\, M_3|\mathbf{x} - \mathbf{a}|^3$.

We postpone the proof until Section 16-5.

Remark There are Taylor polynomials of higher degree and corresponding error estimates. The notation for these polynomials is complicated, and since we shall not need them, we leave their study to an advanced calculus course.

Example 3

If $|x| < 0.1$ and $|y| < 0.1$, prove that

$$|e^x \sin(x + y) - (x + y)| < 0.05$$

Solution Set $f(x, y) = e^x \sin(x + y)$. Then $f(0, 0) = 0$ and

$$f_x(0, 0) = e^x \sin(x + y) + e^x \cos(x + y)\Big|_{(0,0)} = 1$$

$$f_y(0, 0) = e^x \cos(x + y)\Big|_{(0,0)} = 1$$

Hence at $(0, 0)$

$$p_1(x, y) = x + y$$

Therefore the remaining problem is to show that $|r_1(\mathbf{x})| < 0.05$ for points $\mathbf{x} = (x, y)$ with $|x| < 0.1$ and $|y| < 0.1$. Such points satisfy $|\mathbf{x}|^2 < (0.1)^2 + (0.1)^2$, so we restrict the domain of f to the disk $|\mathbf{x}|^2 < 0.02$.

According to the error estimate,

$$|r_1(\mathbf{x})| \le M_2|\mathbf{x}|^2 < (0.02)M_2$$

where M_2 is bound for $|f_{xx}|$, $|f_{xy}|$, and $|f_{yy}|$. To find a suitable value for M_2, we compute the second partials:

$$f_{xx} = 2e^x \cos(x + y) \qquad f_{xy} = e^x[\cos(x + y) - \sin(x + y)]$$
$$f_{yy} = -e^x \sin(x + y)$$

Since $|\sin(x + y)| \le 1$ and $|\cos(x + y)| \le 1$, we have

$$|f_{xx}| \le 2e^x \qquad |f_{xy}| \le 2e^x \qquad |f_{yy}| \le e^x$$

Furthermore, $|x| < 0.1$, so $e^x < e^{0.1} \approx 1.1052 < 1.11$. Therefore $M_2 = 2(1.11) = 2.22$ is a suitable bound. It follows that

$$|r_1(\mathbf{x})| < (0.02)(2.22) = 0.0444 < 0.05$$ ●

Exercises

Compute the Taylor polynomials $p_1(x, y)$ and $p_2(x, y)$ of $f(x + y) =$

1 x^2y^2 at $(1, 1)$
2 x^4y^3 at $(2, -1)$
3 $\sin(xy)$ at $(0, 0)$
4 e^{xy} at $(0, 0)$
5 x^y at $(1, 0)$
6 x^y at $(1, 1)$
7 $\cos(x + y)$ at $(0, \frac{1}{2}\pi)$
8 $1 + xy$ at $(1, 1)$
9 $\ln(x + 2y)$ at $(\frac{1}{2}, \frac{1}{4})$
10 x^2e^y at $(1, 0)$

Estimate by using the second degree Taylor polynomial; carry your work to five places

11 $(1, 1)^{1.2}$
12 $[(1.2)^2 + 7.2]^{1/3}$
13 $f(1.01, 2.01)$ where $f(x, y) = x^3y^2 - 2xy^4 + y^5$
14 $f(2.01, 0.98)$ where $f(x, y) = x^7y^{10}/2^7$

15 If $p_1(x)$, $p_2(x)$ and $q_1(x)$, $q_2(x)$ are first and second degree Taylor polynomials of $f(x)$ and $g(x)$ at a and b respectively, find the first and second degree Taylor polynomials of $h(x, y) = f(x)g(y)$ at (a, b).

16 Let $p_1(x, y)$ be the first degree Taylor polynomial of $f(x, y)$ at $(a, b) = \mathbf{a}$. Show that $p_1(\mathbf{x}) = f(\mathbf{a}) + D_{\mathbf{x}-\mathbf{a}}f(\mathbf{a})$.

Prove the inequality, given $|x| < 0.1$ and $|y| < 0.1$

17 $|\sqrt{1 + x + 2y} - (1 + \frac{1}{2}x + y)| < 0.04$
18 $|e^x \sin(x + y) - (1 + x)(x + y)| < 0.01$ (even < 0.005 with more careful estimates).

19 Given a function $f(x, y, z)$ of three variables, define its first and second degree Taylor polynomials $p_1(x, y, z)$ and $p_2(x, y, z)$ at (a, b, c). Check that their first partial derivatives agree with those of $f(x, y, z)$ at (a, b, c). Check that this is true also of the second partials in p_2.

20 (cont.) Assuming $\operatorname{grad} f(a, b, c) \ne \mathbf{0}$, identify the graph of $p_1(x, y, z) = f(a, b, c)$.

*21 Compute $p_2(x, y)$ at $(0, 0)$ for $f(x, y) = (1 - x - 2y + 3xy)^{-1}$. Verify that you get the same result by expanding $1/(1 - z)$ in power series, where $z = x + 2y - 3xy$, then collecting all terms of degree 2 or less.

*22 (cont.) Use a similar technique to find $p_2(x, y)$ for $f(x) = \sqrt{1 + 3x - 4y + x^2}$.

16-3 Second Derivative Test: Two Variables

Suppose (a, b) is an interior point of the domain of $f(x, y)$ and

$$f_x(a, b) = 0 \quad \text{and} \quad f_y(a, b) = 0$$

Then (a, b) is called a **critical point** of $f(x, y)$. We now consider the behavior of a function near one of its critical points. We shall assume in this section that $f(x, y)$ has continuous first, second, and third partial derivatives. With this assumption, the Taylor approximations of the previous section apply.

Let us briefly review the one-variable case. We consider a function $g(t)$ and a critical point c, that is, a point where $g'(c) = 0$. Suppose $g''(c) > 0$; then $g(c)$ is a local minimum of g. Suppose $g''(c) < 0$; then $g(c)$ is a local maximum of $g(t)$. However, if $g''(c) = 0$, no conclusion can be drawn.

Now we let (a, b) be a critical point of a function $f(x, y)$ of two variables. We want a second derivative condition to guarantee that, say, $f(a, b)$ is a local minimum. But there are *three* second derivatives to reckon with: $f_{xx}(a, b)$, $f_{xy}(a, b)$, and $f_{yy}(a, b)$. What magic combination of signs will give us the desired local minimum?

Second Derivative Test for a Minimum

Let (a, b) be a critical point of $f(x, y)$. Suppose

$$f_{xx}(a, b) > 0 \qquad \text{and} \qquad (f_{xx}f_{yy} - f_{xy}^2)\Big|_{(a,b)} > 0$$

Then $f(a, b)$ is a *strong* local minimum of $f(x, y)$ in the following sense: there exists a $\delta > 0$ such that

$$f(x, y) > f(a, b) \qquad \text{whenever} \quad 0 < |(x, y) - (a, b)| < \delta$$

(Note that $f_{xx}f_{yy} - f_{xy}^2 = \begin{vmatrix} f_{xx} & f_{xy} \\ f_{xy} & f_{yy} \end{vmatrix}$, a determinant.)

The test is easy to use, but its proof involves a delicate technique and is postponed until Section 16-5. The proof uses the Taylor approximation

$$f(a + x, b + y) = f(a, b) + \tfrac{1}{2}[Ax^2 + 2Bxy + Cy^2] + r_2(x, y)$$

where $A = f_{xx}(a, b)$, $B = f_{xy}(a, b)$, and $C = f_{yy}(a, b)$. The point is to prove that the second degree term is positive and dominates the remainder for (x, y) small, hence that

$$f(a + x, b + y) > f(a, b)$$

for (x, y) small.

Example 1

Show that

$$f(x, y) = x^2y - 3xy + xy^2$$

has a strong local minimum at $(1, 1)$.

Solution We have

$$f_x = 2xy - 3y + y^2 \quad \text{and} \quad f_x(1, 1) = 0$$
$$f_y = x^2 - 3x + 2xy \quad \text{and} \quad f_y(1, 1) = 0$$

so $(1, 1)$ is a critical point of $f(x, y)$. Next,

$$f_{xx} = 2y \qquad f_{xy} = 2x - 3 + 2y \qquad \text{and} \qquad f_{yy} = 2x$$

so that

$$f_{xx}(1, 1) = 2 > 0$$
$$(f_{xx}f_{yy} - f_{xy}^2)\Big|_{(1,1)} = 2 \cdot 2 - 1^2 = 3 > 0$$

By the second derivative test, $f(1, 1)$ is a strong local minimum of $f(x, y)$. ●

The second derivative test for a minimum, applied to $-f(x, y)$, implies a test for a maximum.

Second Derivative Test for a Maximum

Let (a, b) be a critical point of $f(x, y)$. Suppose

$$f_{xx}(a, b) < 0 \qquad \text{and} \qquad (f_{xx}f_{yy} - f_{xy}^2)\Big|_{(a,b)} > 0$$

Then $f(a, b)$ is a *strong* local maximum of $f(x, y)$ in the following sense: there exists a $\delta > 0$ such that

$$f(x, y) < f(a, b) \qquad \text{whenever} \quad |(x, y) - (a, b)| < \delta$$ ●

Example 2

Of all triangles with fixed perimeter $2S$, which has the largest area?

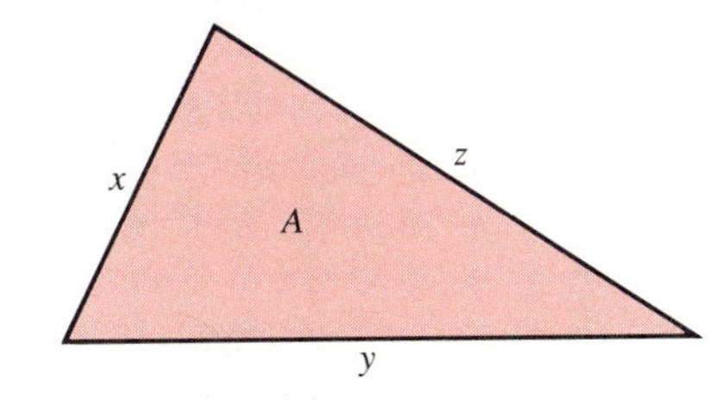

Figure 16-3-1
See Example 2

Solution Let the sides be x, y, z. See Figure 16-3-1. By Heron's formula, the area A satisfies

$$A^2 = S(S - x)(S - y)(S - z)$$

Since $2S = x + y + z$, we can eliminate z:

$$S - z = x + y - S$$
$$A^2 = S(S - x)(S - y)(x + y - S)$$

To maximize A, it suffices to maximize

$$f(x, y) = (S - x)(S - y)(x + y - S)$$

for $x < S$, $y < S$, $x + y > S$. Now

$$f_x = (S-y)[-(x+y-S)+(S-x)]$$
$$= (S-y)(2S-2x-y)$$
$$f_y = (S-x)[-(x+y-S)+(S-y)]$$
$$= (S-x)(2S-x-2y)$$

Hence $f_x = 0$ and $f_y = 0$ imply

$$2S-2x-y=0 \quad \text{and} \quad 2S-x-2y=0$$

The only solution is $x = \frac{2}{3}S$, $y = \frac{2}{3}S$, Hence $f(x, y)$ has the unique critical point $(\frac{2}{3}S, \frac{2}{3}S)$. At this point,

$$f_{xx} = -2(S-y) = -\tfrac{2}{3}S < 0$$
$$f_{yy} = -2(S-x) = -\tfrac{2}{3}S$$
$$f_{xy} = -(2S-2x-y)-(S-y)$$
$$= -3S+2x+2y = -\tfrac{1}{3}s$$

Consequently

$$f_{xx}f_{yy} - f_{xy}^2 = \tfrac{4}{9}S^2 - \tfrac{1}{9}S^2 = \tfrac{1}{3}S^2 > 0$$

Therefore $f(x, y)$ has its only local maximum at $(\frac{2}{3}S, \frac{2}{3}S)$. Since $f(x, y) = 0$ on the boundary of the domain of f, the point just found yields an absolute maximum. Now the third side is

$$z = 2S - x - y = 2S - \tfrac{2}{3}S - \tfrac{2}{3}S = \tfrac{2}{3}S$$

Therefore the three sides are equal for maximal area; the triangle is equilateral. ●

Saddle Points

Let us review our method for finding maxima and minima. We find a critical point (a, b) of $f(x, y)$. Then we compute the **(Hessian)** matrix

$$H = \begin{bmatrix} f_{xx} & f_{xy} \\ f_{yx} & f_{yy} \end{bmatrix}_{(a,b)}$$

There are three possibilities for the determinant

$$\det(H) = (f_{xx}f_{yy} - f_{xy}^2)\Big|_{(a,b)}$$

Case 1 $\det(H) > 0$. Then

$$f_{xx}f_{yy} > f_{xy}^2 \geq 0$$

so $f_{xx}(a, b) \neq 0$. Either $f_{xx} > 0$ and $f(a, b)$ is a strong local minimum, or $f_{xx} < 0$ and $f(a, b)$ is a strong local maximum.

Case 2 $\det(H) = 0$. No conclusion can be drawn. Look at these examples; in both cases $(a, b) = (0, 0)$ and all first and second partials equal 0 at $(0, 0)$:

- $f(x, y) = x^4y^4$, local minimum.
- $f(x, y) = x^3y^3$, neither local maximum nor local minimum.

Case 3 $\det(H) < 0$. In this case we are *sure* there is neither a local maximum nor a local minimum. Here is the exact statement:

Saddle Point Test

Let (a, b) be a critical point of $f(x, y)$. Suppose

$$(f_{xx}f_{yy} - f_{xy}^2)\Big|_{(a,b)} < 0$$

Then $f(a, b)$ is neither a local maximum nor a local minimum of $f(x, y)$. Precisely, there exist points (x, y) arbitrarily close to (a, b) where $f(x, y) < f(a, b)$ and there exist points (x, y) arbitrarily close to (a, b) where $f(x, y) > f(a, b)$. ●

Figure 16-3-2
A function with a saddle point:
$z = -x^2 + y^2$

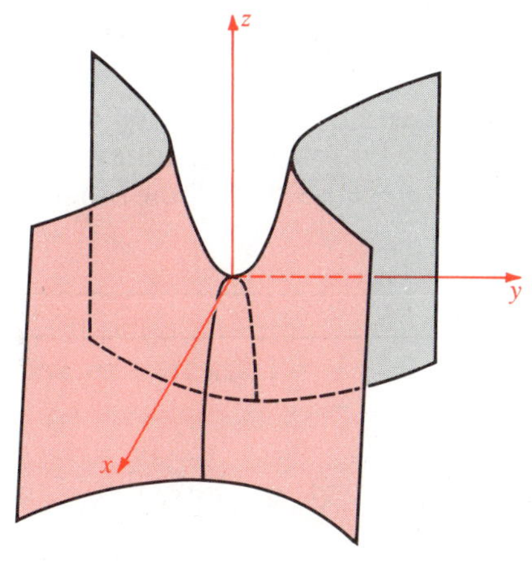

a Saddle point at (0, 0, 0)

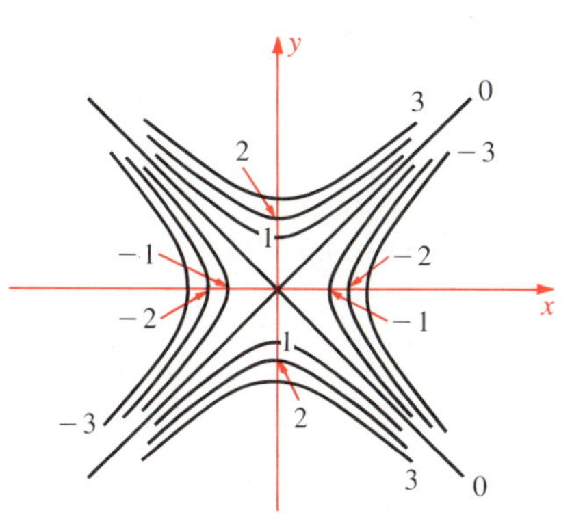

b Level curves

If (a, b) is a critical point of $f(x, y)$ and if $f(a, b)$ is neither a local maximum nor a local minimum of $f(x, y)$, then $f(x, y)$ is said to have a **saddle point** at (a, b); the surface $z = f(x, y)$ is generally shaped like a saddle or a mountain pass near $(a, b, f(a, b))$. The tangent plane is horizontal; the surface rises in some directions, falls in others, so that it crosses its tangent plane. See Figure 16-3-2 for an example. We shall postpone a proof of the saddle point test until Section 16-5. Let us summarize our results:

Summary Let (a, b) be a critical point of $f(x, y)$. That is, (a, b) is an interior point of the domain of $f(x, y)$ and

$$f_x(a, b) = 0 \quad \text{and} \quad f_y(a, b) = 0$$

Let f_{xx}, f_{xy}, and f_{yy} be evaluated at (a, b).

1. If $f_{xx} > 0$ and $f_{xx}f_{yy} - f_{xy}^2 > 0$, then $f(a, b)$ is a strong local minimum.
2. If $f_{xx} < 0$ and $f_{xx}f_{yy} - f_{xy}^2 > 0$, then $f(a, b)$ is a strong local maximum.
3. If $f_{xx}f_{yy} - f_{xy}^2 < 0$, then $f(a, b)$ is a saddle point, neither a local maximum nor local minimum.
4. If $f_{xx}f_{yy} - f_{xy}^2 = 0$, no conclusion can be drawn from this information alone. ●

Imagine a particle constrained to move on the surface $z = f(x, y)$. (This is analogous to having a bead constrained to slide on a wire, although the surface constraint is harder to realize physically.) Suppose the particle is subjected to the downward force of gravity. Finally, suppose the particle is at rest (in equilibrium) at a point $(a, b, f(a, b))$, that is, its height z is stationary: $f_x(a, b) = 0$ and $f_y(a, b) = 0$.

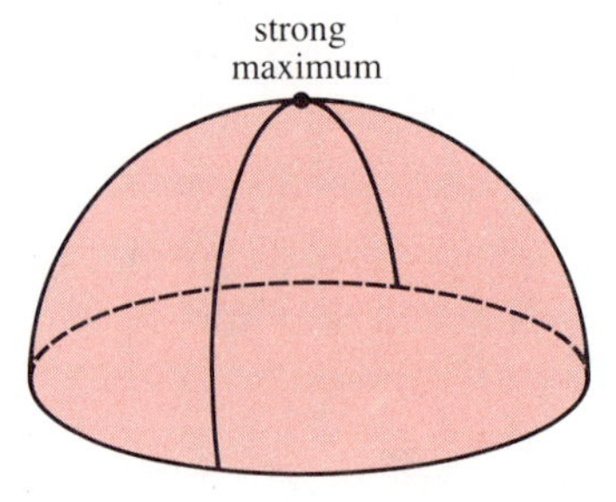

Figure 16-3-3
A point of unstable equilibrium

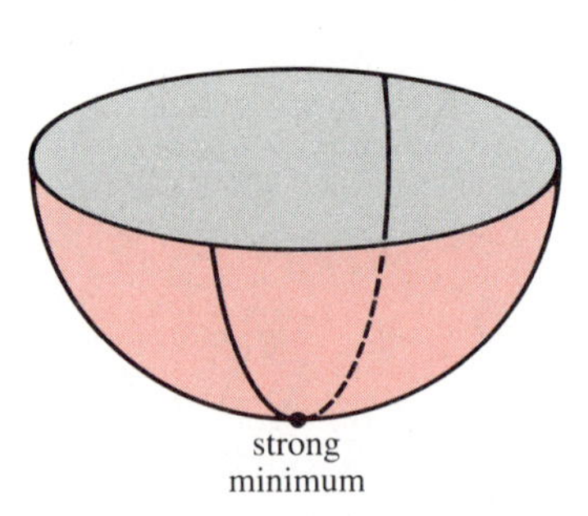

Figure 16-3-4
A point of stable equilibrium

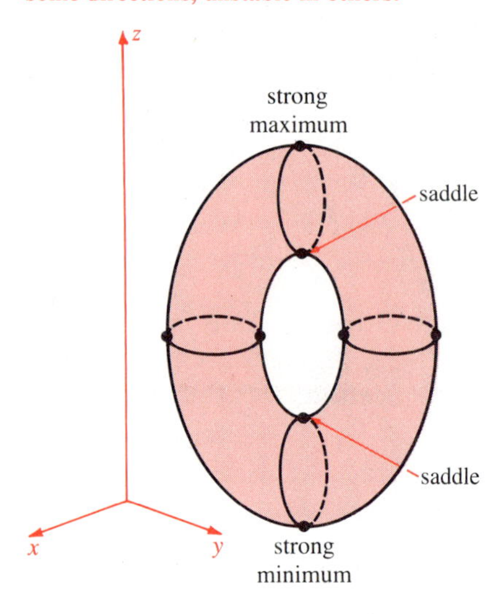

Figure 16-3-5
Equilibrium at saddle points is stable in some directions, unstable in others.

Now suppose the particle is displaced slightly—given a little shove. What happens? If $f(a, b)$ is a strong local maximum of z, then the particle tumbles downward—its equilibrium was **unstable** (Figure 16-3-3). If $f(a, b)$ is a strong local minimum of z, then the particle returns to its equilibrium point—its equilibrium was **stable** (Figure 16-3-4). The saddle point case is smack in between. For some displacement directions the particle tumbles downward, and for others it returns to equilibrium.

As a concrete example of such a surface, imagine an inflated inner tube hanging from a string (Figure 16-3-5). Near each of the marked points, z may be considered locally as a function of x and y. At the top of the inner tube, z has a local maximum, and at the bottom a local minimum. There are two saddle points of z, one at the top of the "hole," the other at its bottom.

Example 3

Find the points on the ellipsoid $\frac{1}{9}x^2 + y^2 + \frac{1}{4}z^2 = 1$ nearest to and farthest from the origin (Figure 16-3-6).

Solution The square of the distance from (x, y, z) on the ellipsoid to $(0, 0, 0)$ is

$$f(x, y) = D^2 = x^2 + y^2 + z^2 = x^2 + y^2 + 4(1 - \tfrac{1}{9}x^2 - y^2)$$
$$= 4 + \tfrac{5}{9}x^2 - 3y^2$$

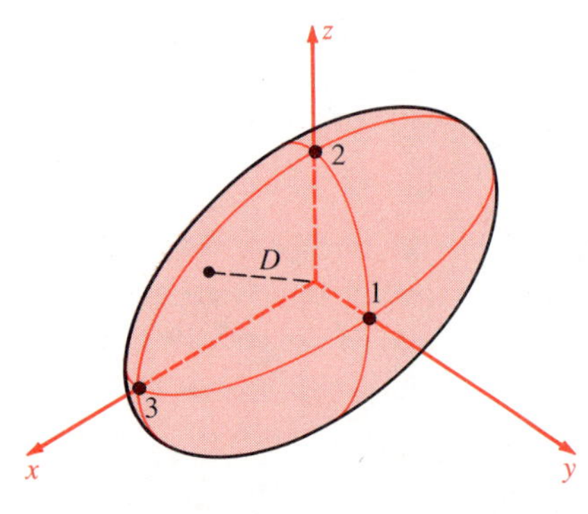

Figure 16-3-6
See Example 3

If (x, y, z) is on the ellipsoid, then

$$\tfrac{1}{9}x^2 + y^2 \le \tfrac{1}{9}x^2 + y^2 + \tfrac{1}{4}z^2 = 1$$

Conversely, if $\frac{1}{9}x^2 + y^2 \le 1$, then (x, y, z) is on the ellipsoid provided that $z = 2\sqrt{1 - \frac{1}{9}x^2 - y^2}$. Therefore the domain of $f(x, y)$ is the elliptical region

$$\tfrac{1}{9}x^2 + y^2 \le 1$$

The first partials of $f(x, y)$ are

$$f_x = \tfrac{10}{9}x \quad \text{and} \quad f_y = -6y$$

which are simultaneously 0 only for $(x, y) = (0, 0)$. However, at $(0, 0)$

$$f_{xx}f_{yy} - f_{xy}^2 = \tfrac{10}{9}(-6) - 0 < 0$$

Therefore $f(x, y)$ has neither a maximum nor a minimum at $(0, 0)$, but a saddle point. There seems to be no possible maximum or minimum.

We have forgotten the boundary! Of course, the first derivative test isn't going to help us on the boundary. The continuous function $f(x, y)$ has both a maximum and a minimum in its *bounded closed* domain, and since they do not occur inside the domain, they must occur on the boundary curve (Figure 16-3-7), an ellipse. On this ellipse, $z = 0$ so $f(x, y) = x^2 + y^2$. By inspection, we observe $f_{\min} = f(0, \pm 1) = 1$ and $f_{\max} = f(\pm 3, 0) = 9$.

Therefore the points of the original ellipsoid nearest to the origin are $(0, \pm 1, 0)$ and the points farthest from the origin are $(\pm 3, 0, 0)$. [The points $(0, 0, \pm 2)$ are saddle points for the distance function.] ●

Figure 16-3-7
See Example 3

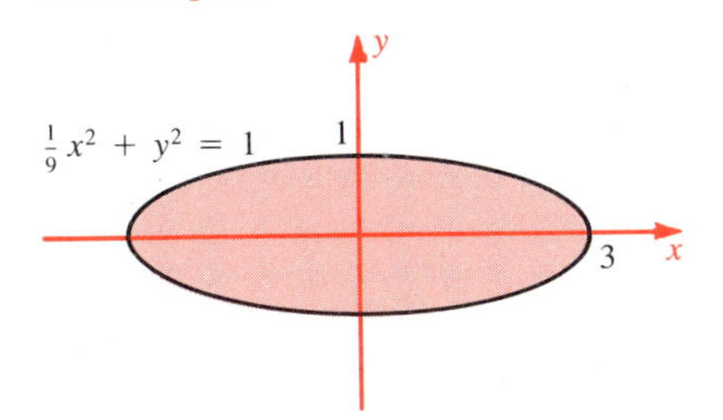

Remark The problem can also be solved by means of the parametrization

$$(x, y, z) = (3 \sin \phi \cos \theta, \sin \phi \sin \theta, 2 \cos \phi)$$

of the ellipsoid.

Exercises

Does the function $f(x, y)$ have a local maximum, a local minimum, or a saddle point at the origin? (Use the first and second derivative tests; if all else fails, sketch a few level curves near the origin.)

1 $x^2 + 3y^2$
2 xy
3 $x^2 - 4y^2$
4 $x^2 + 2xy + y^2$
5 $-x^2 + 2xy - y^2$
6 $x^4 + y^2$
7 x^4
8 $-x^2y^2$
9 $x^3 + y^3$
10 $x^2 + y^5$
11 $x^2y^2 - 3x^4y^4$
12 $x^6 + y^6 - 6x^2y^2$

Find all local maxima, local minima, and saddle points

13 $x^4 + y^4 - 2(x - y)^2$
14 $2x^2y - y^2 + 4x$
15 $(x^2 + y^2)^2 - 4(x^2 + 2y^2) + 1$
16 $x^3 + y^3 - 3axy \quad (a > 0)$
17 $x^3 - 4x^2 - xy - y^2$
18 $e^{4y}(x^2 + x + y)$
19 $(y - x)(x^2 + y^2 - 1)$
20 $[(x + 1)^2 + y^2 - 1][(x - 1)^2 + y^2 - 1]$
21 $(y - x^2)(2 - x - y)$
22 $(y - x^2)(x^2 + y^2 - 1)$

23 Let $f(x)$ and $g(y)$ be differentiable. Suppose $f(x)$ has a strong local maximum at $x = a$ and a strong local minimum at $x = b$, and $g(y)$ has a strong local maximum at $y = c$ and a strong local minimum at $y = d$. What can you say about $h(x, y) = f(x) + g(y)$ at (a, c), (a, d), (b, c), and (b, d)?

24 Explain geometrically the remark about saddle points in the very last [bracketed] sentence of the solution of Example 3.

25 A manufacturer produces two lines of a product, at a cost of \$2 per unit for the regular model and \$3 per unit for the special model. If he fixes their prices at $x > 2$ dollars and $y > 3$ dollars respectively, then the demand for the regular model is $y - x$ and the demand for the special model is $14 + x - 2y$, in thousand units per week. What prices maximize his profits?

26 Suppose $\mathbf{a}$ does not lie on a closed surface $F(\mathbf{x}) = 0$ and $\mathbf{b}$ is a point of the surface that maximizes or minimizes the distance $|F(\mathbf{x}) - \mathbf{a}|$. Show that $\mathbf{a} - \mathbf{b}$ is normal to the surface at $\mathbf{b}$. [Hint Parametrize.]

16-4 More on Stability

A straightforward extension of the second-derivative test applies to functions of three (or more) variables. It hinges on the signs of certain minors of the Hessian matrix (matrix of second partial derivatives).

Second Derivative Test

Let **a** be a critical point of $f(x, y, z)$. Set

$$D_1 = f_{xx} \qquad D_2 = \begin{vmatrix} f_{xx} & f_{xy} \\ f_{yx} & f_{yy} \end{vmatrix} \qquad \text{and} \qquad D_3 = \begin{vmatrix} f_{xx} & f_{xy} & f_{xz} \\ f_{yx} & f_{yy} & f_{yz} \\ f_{zx} & f_{zy} & f_{zz} \end{vmatrix}$$

all evaluated at $\mathbf{x} = \mathbf{a}$. Assume $D_1D_2D_3 \neq 0$.

1 If

$$D_1 > 0 \qquad D_2 > 0 \qquad D_3 > 0$$

then $f(\mathbf{a})$ is a strong local minimum of $f(\mathbf{x})$.

2 If

$$D_1 < 0 \qquad D_2 > 0 \qquad D_3 < 0$$

then $f(\mathbf{a})$ is a strong local maximum of $f(x)$.

3 In all other cases, **a** is a saddle point of $f(\mathbf{x})$.

Remark 1 If $D_1D_2D_3 = 0$, then nothing can be concluded. However, it may be possible to permute the variables x, y, and z in such a way that after the permutation, $D_1D_2D_3 \neq 0$, so the second derivative test can be applied after all. For instance, suppose that the Hessian matrix at a critical point **a** is

$$\begin{bmatrix} 1 & 1 & 0 \\ 1 & 1 & 1 \\ 0 & 1 & 1 \end{bmatrix}$$

Then $D_1 = 1$, $D_2 = 0$, and $D_3 = -1$, so the test doesn't apply. But if y and z are interchanged, then the new Hessian matrix is

$$\begin{bmatrix} 1 & 0 & 1 \\ 0 & 1 & 1 \\ 1 & 1 & 1 \end{bmatrix}$$

with $D_1 = 1$, $D_2 = 1$, and $D_3 = -1$, so the second derivative test implies that **a** is a saddle point of f.

Remark 2 The quantity D_2 is the **principal 2 × 2 minor** of the Hessian matrix. For a function of four variables $f(x, y, z, w)$ at a stationary point **a** we look at the principal minors D_1, D_2, and D_3 as above and the principal 4 × 4 minor (the determinant) D_4 of the Hessian matrix, all evaluated at **a**. Assuming none of these is zero, if they are all positive, then there is a strong local minimum at **a**. If

$$D_1 < 0 \qquad D_2 > 0 \qquad D_3 < 0 \qquad D_4 > 0$$

then there is a strong local maximum at **a**. Otherwise there is a saddle point at **a**. The generalization to five or more variables should be clear.

Example 1

Discuss the behavior at $(0, 0, 0)$ of

a $f = x^2 + y^2 + z^2$
b $f = x^2 - y^2 + z^2$
c $f = x^2 - y^2 - z^2$
d $f = -x^2 - y^2 - z^2$

Solution In all cases

$$\operatorname{grad} f\Big|_{\mathbf{0}} = (\pm 2x, \pm 2y, \pm 2z)\Big|_{\mathbf{0}} = \mathbf{0}$$

so $\mathbf{0}$ is a critical point. The four matrices of second derivatives, evaluated at $\mathbf{x} = \mathbf{0}$, are

a $\begin{bmatrix} 2 & 0 & 0 \\ 0 & 2 & 0 \\ 0 & 0 & 2 \end{bmatrix}$ **c** $\begin{bmatrix} 2 & 0 & 0 \\ 0 & -2 & 0 \\ 0 & 0 & -2 \end{bmatrix}$

b $\begin{bmatrix} 2 & 0 & 0 \\ 0 & -2 & 0 \\ 0 & 0 & 2 \end{bmatrix}$ **d** $\begin{bmatrix} -2 & 0 & 0 \\ 0 & -2 & 0 \\ 0 & 0 & -2 \end{bmatrix}$

a $D_1 = 2 > 0,\quad D_2 = 4 > 0,\quad D_3 = 8 > 0$: local minimum
b $D_1 = 2 > 0,\quad D_2 = -4 < 0,\quad D_3 = -8 < 0$: saddle point
c $D_1 = 2 > 0,\quad D_2 = -4 < 0,\quad D_3 = 8 > 0$: saddle point
d $D_1 = -2 < 0,\quad D_2 = 4 > 0,\quad D_3 = -8 < 0$: local maximum ●

Example 2

Find all local minima of

$$f(x, y, z) = x^4 + y^4 + z^4 - 108x + 4y - 4z$$

Solution

$$\operatorname{grad} f = (4x^3 - 108, 4y^3 + 4, 4z^3 - 4)$$

so $\operatorname{grad} f = \mathbf{0}$ only for $(x, y, z) = (3, -1, 1)$. All the mixed second partials equal zero. The pure second partials are $f_{xx} = 12x^2$, $f_{yy} = 12y^2$, and $f_{zz} = 12z^2$, so

$$f_{xx}(3, -1, 1) = 108$$
$$f_{yy}(3, -1, 1) = 12$$
$$f_{zz}(3, -1, 1) = 12$$

At $(3, -1, 1)$,

$$f_{xx} = 108 > 0 \qquad \begin{vmatrix} f_{xx} & f_{xy} \\ f_{yx} & f_{yy} \end{vmatrix} = \begin{vmatrix} 108 & 0 \\ 0 & 12 \end{vmatrix} = 108 \cdot 12 > 0$$

$$\begin{vmatrix} f_{xx} & f_{xy} & f_{xz} \\ f_{yx} & f_{yy} & f_{yz} \\ f_{zx} & f_{zy} & f_{zz} \end{vmatrix} = \begin{vmatrix} 108 & 0 & 0 \\ 0 & 12 & 0 \\ 0 & 0 & 12 \end{vmatrix} = 108 \cdot 12 \cdot 12 > 0$$

Therefore the only local minimum of $f(x, y, z)$ is

$$\begin{aligned} f(3, -1, 1) &= 81 + 1 + 1 - 324 - 4 - 4 \\ &= -249 \end{aligned}$$

(It is fairly easy to show that $f(\mathbf{x}) \to +\infty$ as $|\mathbf{x}| \to \infty$. Hence we actually have an absolute minimum.) ●

Lagrange Multipliers

The method of Lagrange multipliers (Section 15-4) is like the first derivative test for an unconstrained function. There is also a Lagrange multiplier test that is like the second derivative test for an unconstrained maximum or minimum.

Second Derivative Test with Constraints

Suppose $f(x, y)$ and $g(x, y)$ have a common domain $\mathbf{D}$. We seek extrema of $f(x, y)$ along a level curve $g(x, y) = c$. Suppose (a, b) is a candidate for an extremum found by the Lagrange multiplier (first derivative) test. Thus (a, b) is a fixed interior point of $\mathbf{D}$ that is on the curve, $\operatorname{grad} g(a, b) \neq \mathbf{0}$, and $\operatorname{grad} f(a, b)$ is proportional to $\operatorname{grad} g(a, b)$. Consequently for a fixed multiplier λ,

$$\operatorname{grad} f(a, b) = \lambda \operatorname{grad} g(a, b) \quad \text{and} \quad g(a, b) = c$$

Set

$$D_1 = f_{xx} - \lambda g_{xx} \qquad \text{and} \qquad D_2 = \begin{vmatrix} f_{xx} - \lambda g_{xx} & f_{xy} - \lambda g_{xy} \\ f_{yx} - \lambda g_{yx} & f_{yy} - \lambda g_{yy} \end{vmatrix}$$

evaluated at (a, b). [Thus D_2 is the determinant of the Hessian matrix of $f(x, y) - \lambda g(x, y)$ with λ fixed.]

1. If $D_1 > 0$ and $D_2 > 0$, then $f(a, b)$ is a strong local minimum of $f(x, y)$ along the level curve $g(x, y) = c$.
2. If $D_1 < 0$ and $D_2 > 0$, then $f(a, b)$ is a strong local maximum of $f(x, y)$ along the level curve $g(x, y) = c$.
3. No further conclusions are possible on the basis of the signs D_1 and of D_2. ●

Example 3

Discuss the extrema of $f(x, y) = x^3 + y^3$ on the circle $x^2 + y^2 = 1$.

Solution Set $g(x, y) = x^2 + y^2$. Then

$$\operatorname{grad} f = (3x^2, 3y^2) \quad \text{and} \quad \operatorname{grad} g = (2x, 2y)$$

The Lagrange multiplier equations are

$$3x^2 = 2x\lambda$$
$$3y^2 = 2y\lambda$$
$$x^2 + y^2 = 1$$

The solutions with $x = 0$ or $y = 0$ are

$$x = 0 \qquad y = \pm 1 \qquad \lambda = \pm\tfrac{3}{2} \qquad f(x, y) = \pm 1$$
$$x = \pm 1 \qquad y = 0 \qquad \lambda = \pm\tfrac{3}{2} \qquad f(x, y) = \pm 1$$

Otherwise $x \neq 0$ and $y \neq 0$, so

$$3x = 2\lambda \qquad 3y = 2\lambda \qquad x = y = \tfrac{2}{3}\lambda$$

The corresponding solutions are

$$x = \tfrac{1}{2}\sqrt{2} \qquad y = \tfrac{1}{2}\sqrt{2} \qquad \lambda = \tfrac{3}{4}\sqrt{2} \qquad f(x, y) = \tfrac{1}{2}\sqrt{2}$$
$$x = -\tfrac{1}{2}\sqrt{2} \qquad y = -\tfrac{1}{2}\sqrt{2} \qquad \lambda = -\tfrac{3}{4}\sqrt{2} \qquad f(x, y) = -\tfrac{1}{2}\sqrt{2}$$

In any case, the matrix of second derivatives is

$$\begin{bmatrix} f_{xx} - \lambda g_{xx} & f_{xy} - \lambda g_{xy} \\ f_{yx} - \lambda g_{yx} & f_{yy} - \lambda g_{yy} \end{bmatrix} = \begin{bmatrix} 6x - 2\lambda & 0 \\ 0 & 6y - 2\lambda \end{bmatrix}$$

What can we learn from the second derivative test? At $\mathbf{x} = (0, \pm 1)$, where $\lambda = \pm\frac{3}{2}$ the matrix equals

$$\begin{bmatrix} \mp 3 & 0 \\ 0 & \pm 3 \end{bmatrix}$$

The determinant is negative, so we draw no conclusion. Similarly, at $\mathbf{x} = (\pm 1, 0)$ the second derivative test gives no conclusion.

At $\mathbf{x} = (\frac{1}{2}\sqrt{2}, \frac{1}{2}\sqrt{2})$, where $\lambda = \frac{3}{4}\sqrt{2}$, we have

$$\begin{bmatrix} \frac{3}{2}\sqrt{2} & 0 \\ 0 & \frac{3}{2}\sqrt{2} \end{bmatrix}$$

so $f(\frac{1}{2}\sqrt{2}, \frac{1}{2}\sqrt{2}) = \frac{1}{2}\sqrt{2}$ is a local minimum. Similarly $f(-\frac{1}{2}\sqrt{2}, -\frac{1}{2}\sqrt{2}) = -\frac{1}{2}\sqrt{2}$ is a local maximum of $f(x, y)$ on the circle $g(x, y) = 1$. See Figure 16-4-1. ●

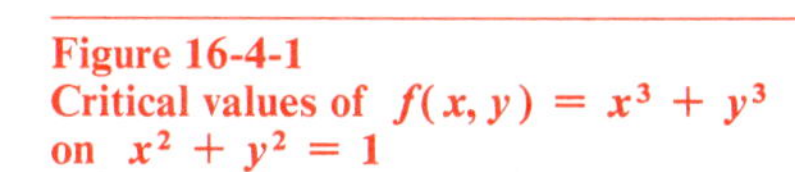
Figure 16-4-1
Critical values of $f(x, y) = x^3 + y^3$ on $x^2 + y^2 = 1$

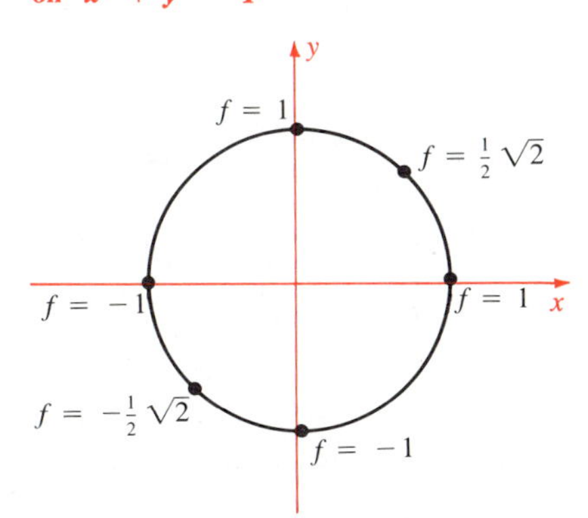

Remark This example is a striking illustration of the limitation of the second derivative test for a problem with constraints. For although the test identified a local maximum and a local minimum, it failed to locate the absolute maxima $f(1, 0) = f(0, 1) = 1$ and the absolute minima $f(-1, 0) = f(0, -1) = -1$.

The corresponding second derivative test for a function $f(x, y, z)$ subject to a constraint $g(x, y, z) = c$ depends on the three quantities

$$D_1 = f_{xx} - \lambda g_{xx} \qquad D_2 = \begin{vmatrix} f_{xx} - \lambda g_{xx} & f_{xy} - \lambda g_{xy} \\ f_{yx} - \lambda g_{yx} & f_{yy} - \lambda g_{yy} \end{vmatrix}$$

and

$$D_3 = \begin{vmatrix} f_{xx} - \lambda g_{xx} & f_{xy} - \lambda g_{xy} & f_{xz} - \lambda g_{xz} \\ f_{yx} - \lambda g_{yx} & f_{yy} - \lambda g_{yy} & f_{yz} - \lambda g_{yz} \\ f_{zx} - \lambda g_{zx} & f_{zy} - \lambda g_{zy} & f_{zz} - \lambda g_{zz} \end{vmatrix}$$

Briefly, assuming $\operatorname{grad} f = \lambda \operatorname{grad} g$ and $\operatorname{grad} g \neq 0$, we have:

1 If $D_1 > 0$, $D_2 > 0$, $D_3 > 0$, then a local minimum.
2 If $D_1 < 0$, $D_2 > 0$, $D_3 < 0$, then a local maximum.
3 Otherwise inconclusive.

Exercises

Show that the second derivative test is inconclusive at $(x, y, z) = (0, 0, 0)$. Determine nonetheless whether the function f has a local maximum, local minimum, or neither at the origin, where $f =$

1 $x^2 + y^2 + z^4$
2 $x^2 + y^2z^2$
3 $x^2 + y^2$
4 $x^4 + y^2 - z^6$
5 $x^2 + y^4 + z^6$
6 $x^3y^3z^3$
7 $x^4 + y^3z^3$
8 $x^4y^4 - z^5$
9 $x^4 + y^2z^2$
10 $x^3 + y^3 + z^3$
11 $x^4y^6z^3$
12 $x^2y^2z^2$

Use the first and second derivative tests to find all local maxima and minima of $f =$

13 $-2x^2 - y^2 - 3z^2 + 2xy - 2xz$
14 $x^2 + 2y^2 + z^2 + 2xy - 4yz$
15 $2x^2 + y^2 + 2z^2 + 2xy + 2yz + 2zx + x - 3z$
16 $x^2 + 3xy + y^2 - z^2 - x - 2y + z + 3$

Does the function have a local maximum, local minimum, or neither at $(0, 0, 0)$?

17 $f = x^2 + y^2 + z^2 + xy + yz + zx$
18 $f = x^2 + 4y^2 + 9z^2 - xy - 2yz$
19 $f = -x^2 - 2y^2 - z^2 + yz$
20 $f = x^2 + y^2 + 2z^2 - 10yz$
21 $f = x^2 - y^2 + 3z^2 + 12xy$
22 $f = 3x^2 + y^2 + 4z^2 - xy - yz - zx$

23 Use the first and second derivative tests on

$$f(x, y, z) = x^3 + x^2 + 4y^2 + 9z^2 + 4xy - 12zx$$

to locate the point where f has a local minimum.

***24** Derive the second derivative test for constrained maxima and minima from the second derivative test for unconstrained maxima and minima.

25 Discuss the extrema of $f(x, y) = xy$ on $x^2 + y^2 = 1$.

26 Discuss the extrema of $f(x, y) = x^2y$ on $x^2 + y^2 = 1$.

16-5 Theory [Optional]

We first prove the equality of mixed second partials, as stated in Section 16-1. We next prove the error estimate in Taylor approximations for functions of several variables. This is the first step in establishing the second derivative tests of Section 16-3. The Taylor approximation allows us to describe the behavior of a function near a critical point by a quadratic form. Thus we are led to study quadratic forms and to establish an algebraic test for their maxima, minima, and saddle points. Finally, we apply this test to prove the second derivative tests.

Equality of Mixed Partials

We must prove that $f_{xy}(a, b) = f_{yx}(a, b)$ under the hypotheses that f_{xy} and f_{yx} exist at all points near (a, b) and are continuous at (a, b).

We consider the mixed second *difference*

$$\begin{aligned}\Delta &= [f(a + h, b + k) - f(a + h, b)] \\ &\quad - [f(a, b + k) - f(a, b)]\end{aligned}$$

and its alternative form

$$\begin{aligned}\Delta &= [f(a + h, b + k) - f(a, b + k)] \\ &\quad - [f(a + h, b) - f(a, b)]\end{aligned}$$

We shall apply the mean value theorem (twice) to the first expression for Δ. To do this, we set

$$g(x) = f(x, b + k) - f(x, b)$$

Then

$$\begin{aligned}\Delta &= g(a + h) - g(a) \\ &= hg'(x_1)\end{aligned}$$

where x_1 is between a and $a + h$. Next,

$$\begin{aligned}g'(x_1) &= f_x(x_1, b + k) - f_x(x_1, b) \\ &= kf_{xy}(x_1, y_1)\end{aligned}$$

where y_1 is between b and $b + k$. Hence

$$\Delta = hkf_{xy}(x_1, y_1)$$

We apply similar reasoning to the second expression for Δ to obtain

$$\Delta = hkf_{yx}(x_2, y_2)$$

where x_2 is between a and $a + h$ and y_2 is between b and $b + k$.

Now we take $h = k \neq 0$. By equating the expressions for Δ we obtain

$$f_{xy}(x_1, y_1) = f_{yx}(x_2, y_2)$$

Let $h \to 0$. Then $(x_1, y_1) \to (a, b)$ and $(x_2, y_2) \to (a, b)$. Since f_{xy} and f_{yx} are continuous at (a, b), we deduce that

$$f_{xy}(a, b) = f_{yx}(a, b)$$

●

Error in Taylor Approximation

We now prove the error estimates stated on page 796. The idea is to interpret the two-variable situation in such a way that we can use the error estimates we developed in Section 9-5 for a function of one variable.

Assume at first that $\mathbf{a} = \mathbf{0}$; this will simplify the notation considerably. Now fix a point $\mathbf{x} = (x, y)$ in $\mathbf{D}$. By convexity, $\mathbf{D}$ contains the entire line segment connecting $\mathbf{0}$ and $\mathbf{x}$; that is, $\mathbf{D}$ contains all points $t\mathbf{x}$ for $0 \le t \le 1$.

Set $g(t) = f(t\mathbf{x})$. Then $g(t)$ is a function of one variable defined for $0 \le t \le 1$ and

$$g(0) = f(\mathbf{0}) \qquad \text{and} \qquad g(1) = f(\mathbf{x})$$

Let us compute the first and second degree Taylor polynomials of g at $t = 0$. First we need derivatives. By the chain rule,

$$g'(t) = f_x(t\mathbf{x})x + f_y(t\mathbf{x})y \qquad \text{and} \qquad g'(0) = f_x(\mathbf{0})x + f_y(\mathbf{0})y$$

Again,

$$g''(t) = f_{xx}(t\mathbf{x})x^2 + 2f_{xy}(t\mathbf{x})xy + f_{yy}(t\mathbf{x})y^2$$

$$g''(0) = f_{xx}(\mathbf{0})x^2 + 2f_{xy}(\mathbf{0})xy + f_{yy}(\mathbf{0})y^2$$

Finally,

$$g^{(3)}(t) = f_{xxx}(t\mathbf{x})x^3 + 3f_{xxy}(t\mathbf{x})x^2y + 3f_{xyy}(t\mathbf{x})xy^2 + f_{yyy}(t\mathbf{x})y^3$$

Therefore the required Taylor polynomials of $g(t)$ are

$$P_1(t) = f(\mathbf{0}) + [f_x(\mathbf{0})x + f_y(\mathbf{0})y]t$$

$$P_2(t) = P_1(t) + \tfrac{1}{2}[f_{xx}(\mathbf{0})x^2 + 2f_{xy}(\mathbf{0})xy + f_{yy}(\mathbf{0})y^2]t^2$$

The corresponding Taylor approximations of $g(t)$ are

$$g(t) = P_1(t) + R_1(t) \qquad \text{and} \qquad g(t) = P_2(t) + R_2(t)$$

The error terms $R_1(t)$ and $R_2(t)$ satisfy

$$|R_1(t)| \le \tfrac{1}{2}K_2|t|^2 \qquad \text{and} \qquad |R_2(t)| \le \tfrac{1}{6}K_3|t|^3$$

where

$$|g''(t)| \le K_2 \qquad \text{and} \qquad |g^{(3)}(t)| \le K_3$$

We now make the important observation that

$$g(1) = f(x, y) \qquad P_1(1) = p_1(x, y) \qquad P_2(1) = p_2(x, y)$$

Consequently

$$f(x, y) = p_1(x, y) + r_1(x, y) \qquad f(x, y) = p_2(x, y) + r_2(x, y)$$

are the first and second degree Taylor expansions of $f(x, y)$, with

$$r_1(x, y) = R_1(1) \quad \text{and} \quad r_2(x, y) = R_2(1)$$

It remains to estimate K_2 and K_3.

As on page 796, we let M_2 denote a common upper bound for the absolute values of three second partials of $f(x, y)$; M_3 is the same for the four third partials. Then from the previous expressions for g'' and $g^{(3)}$ we have

$$|g''(t)| \le M_2|x|^2 + 2M_2|x| \cdot |y| + M_2|y|^2 = M_2(|x| + |y|)^2$$

and

$$|g^{(3)}(t)| \le M_3|x|^3 + 3M_3|x|^2|y| + 3M_3|x| \cdot |y|^2 + M^3|y|^3$$
$$= M_3(|x| + |y|)^3$$

Now we modify these estimates slightly as follows. From $(|x| - |y|)^2 \ge 0$ we have $2|x| \cdot |y| \le |x|^2 + |y|^2$, hence

$$(|x| + |y|)^2 = |x|^2 + 2|x| \cdot |y| + |y|^2 \le 2(|x|^2 + |y|^2) = 2|\mathbf{x}|^2$$

We take the $\frac{3}{2}$ power:

$$(|x| + |y|)^3 \le 2^{3/2}|\mathbf{x}|^3$$

Therefore

$$|g''(t)| \le 2M_2|\mathbf{x}|^2 = K_2 \qquad \text{and} \qquad |g^{(3)}(t)| \le 2\sqrt{2}\,M_3|\mathbf{x}|^3 = K_3$$

The estimates

$$|r_1(x, y)| \le M_2|\mathbf{x}|^2 \qquad \text{and} \qquad |r_2(x, y)| \le \tfrac{1}{3}\sqrt{2}\,M_3|\mathbf{x}|^3$$

follow. This completes the proof, assuming $\mathbf{a} = \mathbf{0}$. In the general case, we define $g(t) = f[\mathbf{a} + t(\mathbf{x} - \mathbf{a})]$. The proof proceeds as before, except that (x, y) is replaced by $(x - a, y - b)$ and the partials of f are all evaluated at $\mathbf{a}$.

Quadratic Forms

A **quadratic form** is a homogeneous quadratic polynomial

$$Q(x, y) = ax^2 + 2bxy + cy^2$$

Of particular importance are those quadratic forms that have an absolute maximum or minimum at $(0, 0)$. Now the extrema of $Q(x, y)$ are found from solutions of the system $Q_x = 0$, $Q_y = 0$, that is,

$$\begin{cases} ax + by = 0 \\ bx + cy = 0 \end{cases}$$

Certainly $(0, 0)$ is a solution; hence $(0, 0)$ is a candidate for a maximum or a minimum. Since $Q(0, 0) = 0$, obviously Q has an absolute minimum at $(0, 0)$ if

$$Q(x, y) > 0 \qquad \text{for} \quad (x, y) \ne (0, 0)$$

When this condition is satisfied, we say that $Q(x, y)$ is **positive definite.** It is helpful to have a criterion for positive definiteness:

Test for Positive Definiteness

Let $Q(x, y) = ax^2 + 2bxy + cy^2$. Then $Q(x, y)$ is positive definite if and only if

$$a > 0 \quad \text{and} \quad \begin{vmatrix} a & b \\ b & c \end{vmatrix} > 0$$

Proof Suppose $Q(x, y)$ is positive definite, that is,

$$Q(x, y) = ax^2 + 2bxy + cy^2 > 0$$

whenever $(x, y) \neq (0, 0)$. In particular $a = Q(1, 0) > 0$. Now complete the square:

$$Q(x, y) = a\left(x + \frac{b}{a}y\right)^2 + \left(\frac{ac - b^2}{a}\right)y^2$$

Then $Q(-b/a, 1) > 0$; hence

$$\frac{ac - b^2}{a} > 0 \qquad \text{so that} \qquad ac - b^2 = a\frac{ac - b^2}{a} > 0$$

Therefore $ac - b^2 > 0$.

Conversely, suppose $a > 0$ and $ac - b^2 > 0$. Then certainly

$$Q(x, y) = a\left(x + \frac{b}{a}y\right)^2 + \left(\frac{ac - b^2}{a}\right)y^2 \geq 0$$

for any (x, y). Furthermore, $Q(x, y) = 0$ only if each of the squared quantities is zero:

$$x + \frac{b}{a}y = 0 \qquad y = 0$$

hence only for $(x, y) = (0, 0)$. This completes the proof. ●

Example 1

Positive definite:

a $3x^2 - 2xy + y^2$ $\quad a = 3 > 0 \quad ac - b^2 = 2 > 0$

b $5x^2 + 6xy + 2y^2$ $\quad a = 5 > 0 \quad ac - b^2 = 1 > 0$

Not positive definite:

c $x^2 + 4xy + y^2$ $\quad a = 1 > 0 \quad ac - b^2 = -3 \leq 0$

d $-2x^2 + 5xy + y^2$ $\quad a = -2 \leq 0 \quad ac - b^2 = -\frac{33}{4} \leq 0$

e $-2x^2 + 2xy - y^2$ $\quad a = -2 \leq 0 \quad ac - b^2 = 1 > 0$

f $x^2 + 6xy + 9y^2$ $\quad a = 1 > 0 \quad ac - b^2 = 0 \leq 0$ ●

We define $Q(x, y)$ to be **negative definite** if $Q(x, y) < 0$ whenever $(x, y) \neq (0, 0)$. This is the same as $-Q(x, y)$ positive definite, so the conditions are

$$-a > 0 \qquad \begin{vmatrix} -a & -b \\ -b & -c \end{vmatrix} = \begin{vmatrix} a & b \\ b & c \end{vmatrix} > 0$$

Test for Negative Definiteness

Let $Q(x, y) = ax^2 + 2bxy + cy^2$. Then $Q(x, y)$ is negative definite if and only if

$$a < 0 \qquad \text{and} \qquad \begin{vmatrix} a & b \\ b & c \end{vmatrix} > 0$$

The Second Derivative Test

The second derivative test (for a minimum) was stated on page 798 and its proof postponed until now. Let us begin by reviewing the one-variable case. We consider a function $g(t)$ and a critical point c, that is, a point where $g'(c) = 0$. Suppose $g''(c) > 0$. We want to conclude that $g(c)$ is a local minimum of g. For this purpose, the second degree Taylor approximation of g at c is an excellent tool:

$$g(t) = g(c) + \tfrac{1}{2}g''(c)(t - c)^2 + r_2(t)$$

$$|r_2(t)| \leq k|t - c|^3 \qquad k > 0$$

It follows that

$$\begin{aligned} g(t) - g(c) &= \tfrac{1}{2}g''(c)(t - c)^2 + r_2(t) \\ &\geq \tfrac{1}{2}g''(c)(t - c)^2 - k|t - c|^3 \\ &= (t - c)^2[\tfrac{1}{2}g''(c) - k|t - c|] \end{aligned}$$

Since $g''(c) > 0$, the quantity on the right is positive provided that $0 < |t - c| < \frac{1}{2}g''(c)/k$. Consequently, there is a positive number $\delta = \frac{1}{2}g''(c)/k$ such that the inequality $g(t) - g(c) > 0$ holds whenever $0 < |t - c| < \delta$. In other words, $g(c)$ is smaller than any other value of g in an interval of radius δ and center c. Hence $g(c)$ is a local minimum of g.

Similarly if $t = c$ is a critical point of $g(t)$ and $g''(c) < 0$, then $g(c)$ is a local maximum of $g(t)$. However, if $g''(c) = 0$, no conclusion can be drawn.

To prove the second derivative test in the two-variable case, we require a preliminary lemma about positive definite quadratic forms.

Lemma Let $Ax^2 + 2Bxy + Cy^2$ be a positive definite quadratic form. Then there exists a constant $k > 0$ such that

$$Ax^2 + 2Bxy + Cy^2 > k(x^2 + y^2) \qquad \text{for all} \quad (x, y) \neq (0, 0)$$

Proof We are given $A > 0$ and $AC - B^2 > 0$. We choose $k > 0$ so small that

$$A - k > 0 \quad \text{and} \quad (A - k)(C - k) - B^2 > 0$$

[Since $h(t) = (A - t)(C - t) - B^2$ is a continuous function and $h(0) = AC - B^2 > 0$, this is certainly possible.] These inequalities imply that

$$(A - k)x^2 + 2Bxy + (C - k)y^2$$

is positive definite, that is, takes positive values for all $(x, y) \neq (0, 0)$. This means

$$Ax^2 + 2Bxy + Cy^2 > k(x^2 + y^2)$$

whenever $(x, y) \neq (0, 0)$. ●

Now we can prove the second derivative test. To make the notation simple, let us take $(a, b) = (0, 0)$. Then the second degree Taylor approximation of f at $(0, 0)$ is

$$f(x, y) = f(0, 0) + \tfrac{1}{2}[Ax^2 + 2Bxy + Cy^2] + r(x, y)$$
$$|r(x, y)| < h|\mathbf{x}|^3$$

where $A = f_{xx}(0, 0)$, $B = f_{xy}(0, 0)$, $C = f_{yy}(0, 0)$, and h is a positive constant. Since $Ax^2 + 2Bxy + Cy^2$ is positive definite (by hypothesis), the lemma provides a constant $k > 0$ such that

$$Ax^2 + 2Bxy + Cy^2 \geq k|\mathbf{x}|^2$$

Consequently

$$f(x, y) - f(0, 0) \geq \tfrac{1}{2}k|\mathbf{x}|^2 - h|\mathbf{x}|^3 = |\mathbf{x}|^2(\tfrac{1}{2}k - h|\mathbf{x}|)$$

This implies $f(x, y) - f(0, 0) > 0$, that is $f(x, y) > f(0, 0)$, provided $0 < |\mathbf{x}| < \tfrac{1}{2}k/h$. Finally, for δ we may choose any number such that $0 < \delta < \tfrac{1}{2}k/h$ and so small that the disk $|\mathbf{x}| < \delta$ is contained in the domain of $f(x, y)$. ●

Remark The essential point in the proof is that for $|\mathbf{x}|$ small, the positive definite quadratic form $Ax^2 + 2Bxy + Cy^2$ is much larger than the remainder $r(x, y)$ because r is of third order: $|r(x, y)| \leq h|\mathbf{x}|^3$.

The Saddle Point Test

Before we prove the saddle point test (stated on page 801), we need another lemma about quadratic forms.

Lemma Let $Q(x, y) = Ax^2 + 2Bxy + Cy^2$, and suppose that $AC - B^2 < 0$. Then there are points (x_1, y_1) and (x_2, y_2) such that

$$Q(x_1, y_1) < 0 \quad \text{and} \quad Q(x_2, y_2) > 0$$ ●

Proof We assume $A \neq 0$ and complete the square:

$$\begin{aligned} Q(x, y) &= Ax^2 + 2Bxy + Cy^2 \\ &= A\left(x + \frac{B}{A}y\right)^2 + \frac{AC - B^2}{A}y^2 \\ &= A\left(x + \frac{B}{A}y\right)^2 + C_1y^2 \end{aligned}$$

where

$$AC_1 = A\left(\frac{AC - B^2}{A}\right) = AC - B^2 < 0$$

Clearly

$$Q(1, 0) = A \quad \text{and} \quad Q(-B/A, 1) = C_1$$

have opposite signs.

The same argument applies if $A = 0$ but $C \neq 0$. If both $A = 0$ and $C = 0$, then $B \neq 0$ since $-B^2 = AC - B^2 < 0$. In that case $Q(x, y) = 2Bxy$, and $Q(1, 1)$ and $Q(1, -1)$ have opposite signs. ●

Now we can prove the saddle point test. We take $(a, b) = (0, 0)$ for simplicity and consider the Taylor approximation

$$f(x, y) = f(0, 0) + \tfrac{1}{2}Q(x, y) + r(x, y) \qquad |r(x, y)| < h(|\mathbf{x}|^3$$

where

$$Q(x, y) = Ax^2 + 2Bxy + Cy^2$$

with $A = f_{xx}(0, 0)$, $B = f_{xy}(0, 0)$, and $C = f_{yy}(0, 0)$. By hypothesis, $AC - B^2 < 0$, so the lemma gives us points (x_1, y_1) and (x_2, y_2) such that $Q(x_1, y_1) < 0$ and $Q(x_2, y_2) > 0$.

Consider $(x, y) = (tx_1, ty_1)$, where $t > 0$ and small. Then

$$\tfrac{1}{2}Q(tx_1, ty_1) = \tfrac{1}{2}Q(x_1, y_1)t^2 = kt^2$$

with $k < 0$, and

$$\begin{aligned} |r(tx_1, ty_1)| &< h(x_1^2 + y_1^2)^{3/2}t^3 \\ &= h_1 t^3 \end{aligned}$$

Hence

$$\begin{aligned} f(tx_1, ty_1) - f(0, 0) &< kt^2 + h_1 t^3 \\ &= t^2(k + h_1 t) \end{aligned}$$

For t sufficiently small, $t^2(k + h_1 t) < 0$ because $k < 0$. Consequently

$$f(tx_1, ty_1) < f(0, 0)$$

for points (tx_1, ty_1) as close to $(0, 0)$ as we please. Similarly

$$f(tx_2, ty_2) > f(0, 0)$$

for points (tx_2, ty_2) as close to $(0, 0)$ as we please. Therefore $f(0, 0)$ is neither a maximum nor a minimum of f.

Exercises

The following four exercises show that f_{xy} and f_{yx} are not necessarily equal if they fail to be continuous

1 Set $f(0, 0) = 0$ and

$$f(x, y) = \frac{(x - y)(x^3 + y^3)}{x^2 + y^2}$$

for $(x, y) \neq (0, 0)$. Show that $f(x, y)$ is continuous everywhere.

2 (cont.) Compute f_x and f_y at $(x, y) \neq (0, 0)$.

3 (cont.) Compute $f_x(0, 0)$ and $f_y(0, 0)$. Conclude that f_x and f_y are continuous everywhere.

4 (cont.) Compute $f_{xy}(0, 0)$ and $f_{yx}(0, 0)$.

Suppose $f(x, y)$ has continuous second partials

5 Assume $f(x, y) = f(y, x)$.
Prove $f_{xx}(c, c) = f_{yy}(c, c)$.

6 Assume $f(x, y) = -f(y, x)$. Prove $f_{xy}(c, c) = 0$.

7 Suppose $f(x, y)$ is a function of two variables and $x = x(t)$ and $y = y(t)$ are functions of time. Form the composite function $g(t) = f[x(t), y(t)]$. By the chain rule,

$$g'(t) = f_x[x(t), y(t)]x'(t) + f_y[x(t), y(t)]y'(t)$$

where $'$ denotes d/dt. Show that

$$g'' = f_{xx}x'^2 + 2f_{xy}x'y' + f_{yy}y'^2 + f_x x'' + f_y y''$$

8 (cont.) Suppose also that $f_x(0, 0) = f_y(0, 0) = 0$, and that only curves $\mathbf{x}(t) = (x(t), y(t))$ are allowed that pass through $(0, 0)$ with speed 1 at $t = 0$. Suppose for each such curve $g''(0) > 0$. Show that $g'(0) = 0$ and $f_{xx}(0, 0) > 0$ and $f_{yy}(0, 0) > 0$.

***9** (cont.) Show also that
$f_{xx}(0, 0)f_{yy}(0, 0) - f_{xy}(0, 0)^2 > 0$.

10 Suppose $g(x, y) = Ax^2 + 2Bxy + Cy^2$ and $AC - B^2 = 0$. Conclude that
$g(x, y) = \pm(ax + by)^2$.

Find whether the quadratic form is positive definite, negative definite, or neither

11 $x^2 + 4xy + 2y^2$

12 $9x^2 - 12xy + y^2$

13 $2x^2 - xy + 3y^2$

14 $7x^2 + 5xy + 4y^2$

15 $-x^2 + 3xy - 3y^2$

16 $-5x^2 + 2xy + y^2$

17 $4xy$

18 $x^2 + 2xy$

19 Suppose that
$Q = ax^2 + 2bxy + cy^2 = (Ax + By)(Cx + Dy)$ where $AD - BC \neq 0$. Show that $ac - b^2 < 0$.

***20** Suppose $ac - b^2 < 0$. Show that
$Q = ax^2 + 2bxy + cy^2$ is **indefinite,** that is, takes on both positive and negative values. [Hint Complete the square.]

21 Prove that $f(x, y) = x^2 - 6xy + 10y^2$ has a positive minimum value p on the circle $x^2 + y^2 = 1$.

22 If $f(x, y) = ax^2 + bxy + cy^2$ has a positive minimum p on $x^2 + y^2 = 1$, prove that $f(x, y) > 0$ for all (x, y) except $(0, 0)$.

23 The form $x^2 + 2xy + 2y^2$ is positive definite so there is a $k > 0$ such that $x^2 + 2xy + 2y^2 \geq k(x^2 + y^2)$ for all (x, y). Find the largest such k.

***24** (cont.) Solve this problem in general for a positive definite $f(\mathbf{x}) = ax^2 + 2bxy + cy^2$.

25 Let $Q(x, y) = ax^2 + 2bxy + cy^2$ be an indefinite quadratic form, that is, $ac - b^2 < 0$. Show that Q can be expressed as the product $(a_1x + b_1y)(a_2x + b_2y)$ of two linear forms.
[Hint Solve $at^2 + 2bt + c = 0$.]

***26** Let $f(x, y) = ax^2 + 2bxy + cy^2 + px + qy + r$, where $ax^2 + 2bxy + cy^2$ is positive definite. Prove that $f(x, y) \to +\infty$ as $|x| \to +\infty$.
[Hint First prove that $ax^2 + 2bxy + cy^2 \geq m|\mathbf{x}|^2$, where $m > 0$.]

16-6 Review Exercises

A function is called **harmonic** if it satisfies **Laplace's equation**

$$\frac{\partial^2 f}{\partial x^2} + \frac{\partial^2 f}{\partial y^2} + \frac{\partial^2 f}{\partial z^2} = 0$$

Verify that the function is harmonic

1 $\arctan(y/x)$

2 $x \sin x \cosh y - y \cos x \sinh y$

3 $x^5 - 10x^3y^2 + 5xy^4$

4 $e^{13x} \sin 5y \cos 12z$

5 $\ln(x^2 + y^2)$

6 $1/\sqrt{x^2 + y^2 + z^2}$

Find the second degree Taylor polynomial

7 $\dfrac{x}{1 - y^2}$ at $(0, 0)$

8 $\dfrac{1}{1 + x + y^2}$ at $(0, 0)$

9 $\arctan \dfrac{x - y}{1 + xy}$ at $(0, 0)$

10 $\dfrac{x}{1 - x + y}$ at $(3, 1)$

11 Estimate $e^{1.01}(0.98)^{1/2}$ numerically by using the second degree Taylor approximation of $e^x(1 - y)^{1/2}$.

12 The period of a pendulum is given approximately by $T = 2\pi\sqrt{L/g}$ sec, where L is its length in meters and g is the acceleration of gravity in m/sec^2. Suppose *relative* errors of h and k are made in measuring L and g. Show that the relative error in T, up to second order, is

$$\tfrac{1}{2}(h - k) - \tfrac{1}{8}[h^2 + 2hk - 3k^2]$$

Does the function have a local maximum, local minimum, or saddle point at the origin? Use first and second derivative tests

13 $x^2 - 4xy + 7y^2$

14 $x^2 - 4xy + y^2$

15 $x^2 + 2y^2 + 2z^2 - 2xy - 2yz$

16 $x^2 + 2y^2 - 2xy - 2yz$

Find all local maxima, minima, and saddle points

17 $x^2 + y^2 - 6x + 10y$

18 $x^2 - y^2 + 6x - 10y$

19 $x^2 - 4xy + y^2 - 8x - 4y$

20 $x^2 + 2y^2 + z^2 - 2xy - 2yz - 14x - 12z$

Find all stationary points for the distance from the origin to a variable point on the given surface, and identify these points as local maxima, local minima, or saddle points

21 $xy - z^2 = 1$

22 $xyz + 1 = 0$

17 The Double Integral

In single-variable calculus we discussed the problem of finding the area under a curve $y = f(x)$ defined on a closed interval $[a, b]$. Its solution led to the definite integral of a one-variable function, which we now call the *simple integral.* Now we take up the problem of finding the volume under a surface $z = f(x, y)$ defined on a closed domain $\mathbf{D}$ of the x, y-plane (Figure 17-1-1). This leads to a new kind of integral, called the *double integral,* which we develop in this chapter.

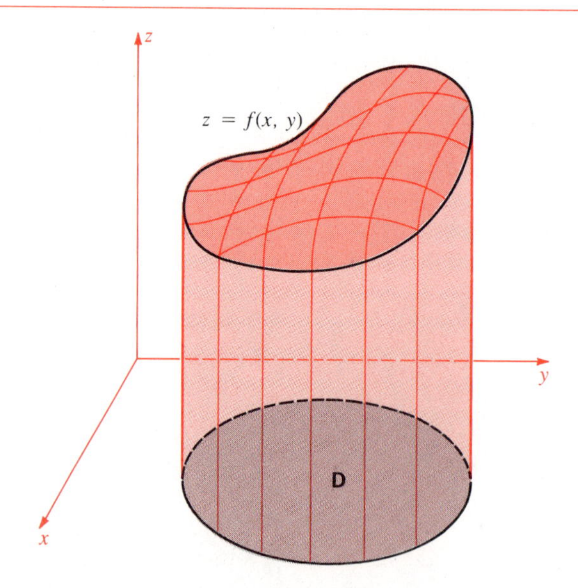

Figure 17-1-1
Problem: To define and evaluate the volume of the space region $0 \le z \le f(x, y)$, (x, y) in $\mathbf{D}$.

17-1 The Volume Problem

We attack the volume problem in basically the same way as we did the area problem. Assuming the volume of the region in Figure 17-1-1 exists, we make reasonable approximation to it, then we find the exact value by taking a limit.

Figure 17-1-2

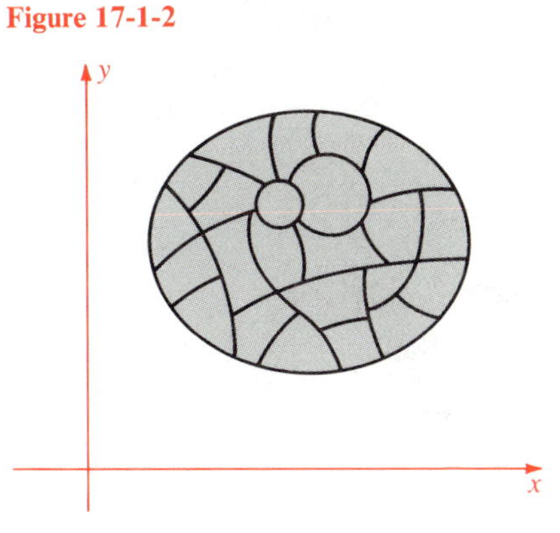

a Partition of the domain into many small subdomains

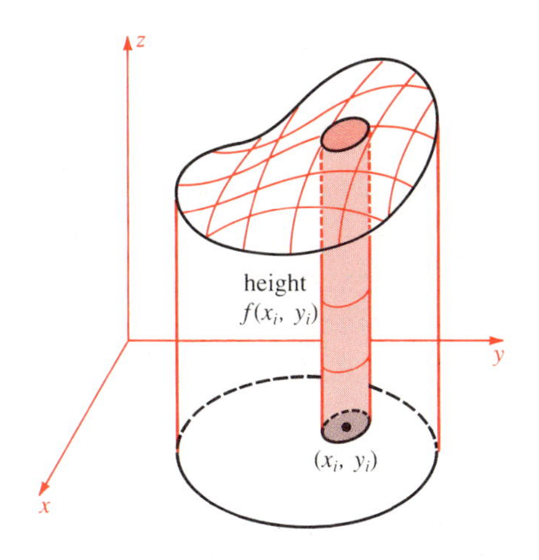

b Long thin cylinder approximates part of the volume under the surface

First, we partition the domain **D** into many small domains $\mathbf{D}_1$, $\mathbf{D}_2, \cdots, \mathbf{D}_n$. See Figure 17-1-2a. In each $\mathbf{D}_i$, we choose a point (x_i, y_i). Then we approximate that part of the region above $\mathbf{D}_i$ by a thin solid of height $f(x_i, y_i)$. Its volume is

$$(\text{height})(\text{area of base}) = f(x_i, y_i) \cdot \text{area}(\mathbf{D}_i)$$

See Figure 17-1-2b. The sum of these volumes,

$$\sum_{i=1}^{n} f(x_i, y_i) \cdot \text{area}(\mathbf{D}_i)$$

ought to be close to the desired volume provided the $\mathbf{D}_i$'s are small enough. Therefore, we look for a limit as the $\mathbf{D}_i$'s are made smaller and smaller. If the limit exists, we define the volume to be the limit and denote it by

$$\iint_{\mathbf{D}} f(x, y)\, dx\, dy$$

Then we say that $f(x, y)$ is **integrable** on **D**.

So far everything looks practically the same as in a discussion of simple integrals. However, in two dimensions there are difficulties not found in one dimension. For one thing, plane domains **D** can be much more complicated than intervals $[a, b]$. Consequently, partitions of **D** into smaller domains can be nasty. Also there is the question of which functions are integrable.

These matters require attention, but we do not attempt a full treatment here. We do give a discussion of double integrals that conveys the flavor of the subject and teaches techniques needed to solve problems. We shall justify the properties of double integrals intuitively and focus on how to evaluate and to apply double integrals.

Domains

Previously we integrated a function $f(x)$ over a *closed interval* $[a, b]$. All closed intervals on the line look alike; if you've seen one, you've seen them all. In contrast, plane domains can be complicated. We avoid difficulties by limiting attention to a restricted type of domain, which we shall call a *domain of integration.* Such domains are general enough for practical purposes. They are bounded (stay in a finite part of the plane—do not go off to infinity in any direction) and have reasonable boundaries.

- A **domain of integration** in the plane is a bounded domain **D** that includes all points of its boundary. Its boundary must consist of a finite number of graphs of convex or concave functions.

Figure 17-1-3
Typical domains of integration

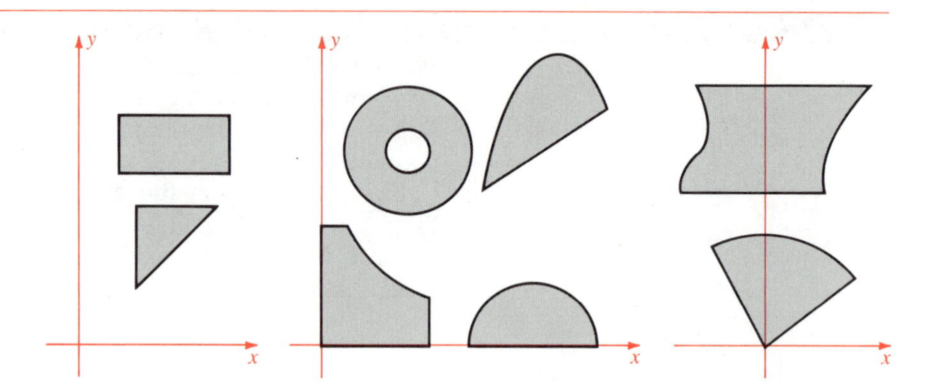

Here "graph of a convex function" means either

$$y = f(x) \qquad a \le x \le b \qquad f''(x) \ge 0$$

or

$$x = g(y) \qquad c \le y \le d \qquad g''(y) \ge 0$$

with a similar interpretation for "concave." Figure 17-1-3 shows some examples of domains of integration.

Partitions

We partition a domain of integration into smaller domains. Exactly what does it mean to say a domain is small? We want small to mean that any two points of the domain are close together.

As a measure of smallness, we define the **radius** of a domain $\mathbf{D}$, written $\operatorname{rad}(\mathbf{D})$, as the radius of the smallest circle that includes $\mathbf{D}$. See Figure 17-1-4. Clearly

Figure 17-1-4
Radius of a domain: $\operatorname{rad}(\mathbf{D}) = r$

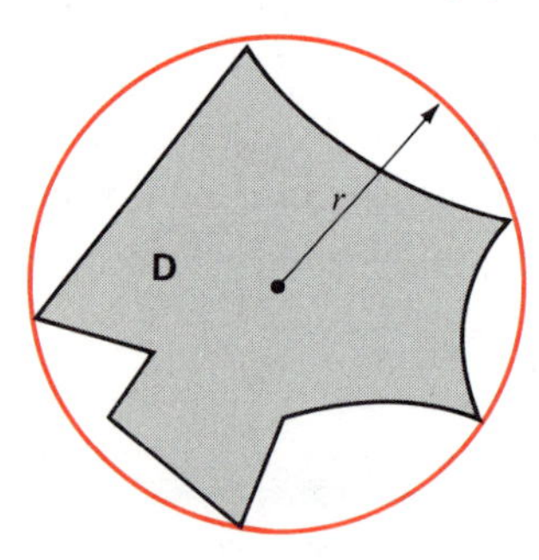

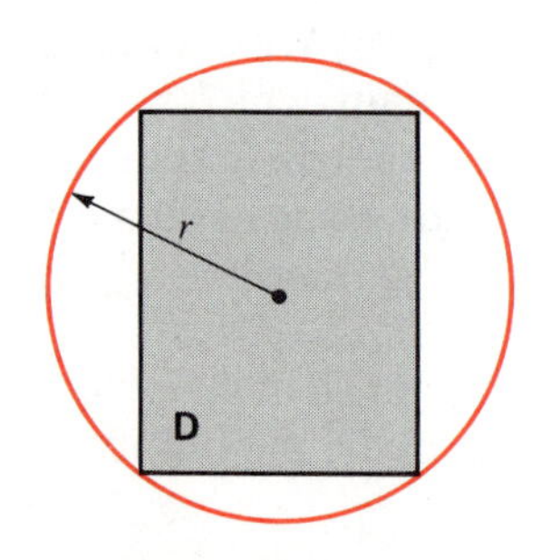

$$\operatorname{area}(\mathbf{D}) \le \pi[\operatorname{rad}(\mathbf{D})]^2$$

so $\operatorname{area}(\mathbf{D})$ is small if $\operatorname{rad}(\mathbf{D})$ is small.

We say that

$$P = \{\mathbf{D}_1, \mathbf{D}_2, \cdots, \mathbf{D}_n\}$$

is a **partition** of $\mathbf{D}$ provided

1 Each $\mathbf{D}_i$ is a domain of integration.

2 The $\mathbf{D}_i$ are subsets of $\mathbf{D}$ and fill $\mathbf{D}$, that is,

$$\mathbf{D} = \mathbf{D}_1 \cup \mathbf{D}_2 \cup \cdots \cup \mathbf{D}_n$$

3 If $i \ne j$, then $\mathbf{D}_i$ and $\mathbf{D}_j$ have no interior points in common. Thus $\mathbf{D}_i \cap \mathbf{D}_j$ is either empty or part of the boundaries of $\mathbf{D}_i$ and $\mathbf{D}_j$.

Given a partition P of $\mathbf{D}$, each subdomain $\mathbf{D}_i$ of the partition has a radius $\operatorname{rad}(\mathbf{D}_i)$. The largest of these radii is called the **mesh** of the partition, and is written

$$\operatorname{mesh}(P) = \max_{1 \le i \le n} \operatorname{rad}(\mathbf{D}_i)$$

The Double Integral

We are ready to define the double integral of a bounded function $f(x, y)$ on a domain of integration $\mathbf{D}$. Let P be a partition of $\mathbf{D}$ into subdomains $\mathbf{D}_1, \mathbf{D}_2, \cdots, \mathbf{D}_n$. Choose a single point (x_i, y_i) in each $\mathbf{D}_i$. To each such choice corresponds a **Riemann sum**

$$R = \sum_{i=1}^{n} f(x_i, y_i) \cdot \text{area}(\mathbf{D}_i)$$

The function $f(x, y)$ will be called integrable on $\mathbf{D}$ if these Riemann sums approach a limit as $\text{mesh}(P) \to 0$. We make this limit concept precise with the following definition:

The Double Integral

A bounded function $f(x, y)$ on a domain of integration $\mathbf{D}$ is **integrable** on $\mathbf{D}$ if there is a number I such that for each $\varepsilon > 0$ there is a number $\delta > 0$ with the following property. For each partition P with $\text{mesh}(P) < \delta$ and each Riemann sum R corresponding to P we have

$$|R - I| < \varepsilon$$

Then I is called the **double integral** of $f(x, y)$ on $\mathbf{D}$ and we write

$$I = \iint_{\mathbf{D}} f(x, y)\, dx\, dy$$

Briefly we can write

$$\iint_{\mathbf{D}} f(x, y)\, dx\, dy = \lim_{\text{mesh}(P) \to 0} R$$

where R denotes any Riemann sum corresponding to the partition P.

This theoretical definition is fine, but immediately two practical questions arise: (1) For what functions, if any, does the limit exist? (2) If the limit does exist, how do we evaluate it?

The answer to the second question is the business of Sections 17-2 and 17-3. The complete answer to the first question is too technical to include here, but the following basic assertion covers most situations that arise in practice.

Integrability of Continuous Functions

If $f(x, y)$ is continuous on a domain of integration $\mathbf{D}$, then $f(x, y)$ is integrable on $\mathbf{D}$; that is, the double integral

$$\iint_{\mathbf{D}} f(x, y)\, dx\, dy$$

exists.

We omit the proof. It is similar to the proof of the corresponding theorem for functions of one variable, and depends on a property of continuous functions called uniform continuity.

Remark 1 The theorem solves (at least theoretically) the volume problem for continuous functions $f(x, y) > 0$. However, the theorem does not require that f be positive. If f takes both positive and negative values, the double integral represents an *algebraic volume* rather than a geometric volume. The volume between the surface $z = f(x, y)$ and the x, y-plane counts positively where $f > 0$ and negatively where $f < 0$.

Remark 2 It is often helpful to write $dA = dx\,dy$ and to refer to the quantity dA as the **element of area.** Thus

$$\iint_{\mathbf{D}} f(x, y)\,dx\,dy = \iint_{\mathbf{D}} f(x, y)\,dA$$

Properties of the Double Integral

Several basic properties of double integrals follow from the corresponding properties of Riemann sums, just as they do for simple integrals. For instance, the formulas

$$\sum_{i=1}^{n} kf(x_i, y_i) \cdot \text{area}(\mathbf{D}_i) = k\sum_{i=1}^{n} f(x_i, y_i) \cdot \text{area}(\mathbf{D}_i)$$

$$\sum_{i=1}^{n} [f(x_i, y_i) + g(x_i, y_i)] \cdot \text{area}(\mathbf{D}_i)$$
$$= \sum_{i=1}^{n} f(x_i, y_i) \cdot \text{area}(\mathbf{D}_i) + \sum_{i=1}^{n} g(x_i, y_i) \cdot \text{area}(\mathbf{D}_i)$$

imply the relations

- $$\iint_{\mathbf{D}} kf(x, y)\,dx\,dy = k\iint_{\mathbf{D}} f(x, y)\,dx\,dy$$
- $$\iint_{\mathbf{D}} [f(x, y) + g(x, y)]\,dx\,dy = \iint_{\mathbf{D}} f(x, y)\,dx\,dy + \iint_{\mathbf{D}} g(x, y)\,dx\,dy$$

linear properties

Next, suppose $f(x, y) \le g(x, y)$ on $\mathbf{D}$. The inequality carries over to Riemann sums:

$$\sum_{1}^{n} f(x_i, y_i) \cdot \text{area}(\mathbf{D}_i) \le \sum_{1}^{n} g(x_i, y_i) \cdot \text{area}(\mathbf{D}_i)$$

The relation carries over to the limit. If $f(x, y) \le g(x, y)$ on $\mathbf{D}$, then

- $$\iint_{\mathbf{D}} f(x, y)\, dx\, dy \le \iint_{\mathbf{D}} g(x, y)\, dx\, dy$$

This inequality is useful, just as the corresponding one-dimensional inequality is. We next mention two inequalities that are most useful in making estimates:

- $$\left| \iint_{\mathbf{D}} f(x, y)\, dx\, dy \right| \le \iint_{\mathbf{D}} |f(x, y)|\, dx\, dy$$

- If $|f(x, y)| \le M$, then
$$\left| \iint_{\mathbf{D}} f(x, y) g(x, y)\, dx\, dy \right| \le M \iint_{\mathbf{D}} |g(x, y)|\, dx\, dy$$

In the next two sections, we shall evaluate double integrals. Meanwhile, we observe that one double integral is obvious, that of a constant function.

- $$\iint_{\mathbf{D}} k\, dx\, dy = k \cdot \operatorname{area}(\mathbf{D})$$

This is clear because all the Riemann sums have the same value:

$$\begin{aligned} \sum_1^n f(x_i, y_i) \cdot \operatorname{area}(\mathbf{D}_i) &= \sum_1^n k \cdot \operatorname{area}(\mathbf{D}_i) \\ &= k \sum_1^n \operatorname{area}(\mathbf{D}_i) \\ &= k \cdot \operatorname{area}(\mathbf{D}) \end{aligned}$$

17-2 Rectangular Domains

How do we evaluate a double integral

$$\iint_{\mathbf{D}} f(x, y)\, dA = \iint_{\mathbf{D}} f(x, y)\, dx\, dy?$$

In most cases, direct application of the definition is practically hopeless. (It is tough enough for simple integrals on intervals.) Fortunately, there is a way of reducing the problem to the evaluation of simple integrals. This method makes the double integral a practical tool as well as a theoretical one.

Let us avoid difficulties with domains by dealing throughout this section only with rectangular domains (Figure 17-2-1, next page), that is, domains of the form

$$\mathbf{D} = \{(x, y) \mid \text{where } a \le x \le b \text{ and } c \le y \le d\}$$

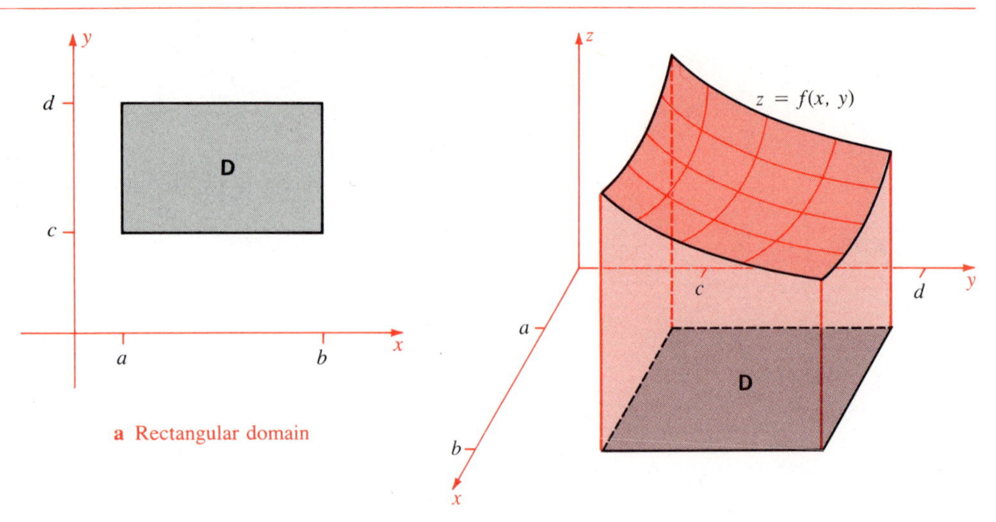

Figure 17-2-1

a Rectangular domain

b The corresponding volume problem

The simplest way to partition a rectangular domain is to partition into subrectangles. Take arbitrary partitions

$$a = x_0 < x_1 < x_2 < \cdots < x_m = b$$
$$c = y_0 < y_1 < y_2 < \cdots < y_n = d$$

of the intervals $[a, b]$ and $[c, d]$ and put together the corresponding parallel rulings of the plane (Figure 17-2-2). The rectangle $\mathbf{D}$ is partitioned into mn subrectangles $\mathbf{D}_{11}, \cdots, \mathbf{D}_{mn}$, where

$$\mathbf{D}_{ij} = \{(x, y) \mid \text{where } x_{i-1} \le x \le x_i \text{ and } y_{j-1} \le y \le y_j\}$$

and

$$\text{area}(\mathbf{D}_{ij}) = (x_i - x_{i-1})(y_j - y_{j-1})$$

Figure 17-2-2
Partition of $\mathbf{D}$ into subintervals $\mathbf{D}_{ij}$

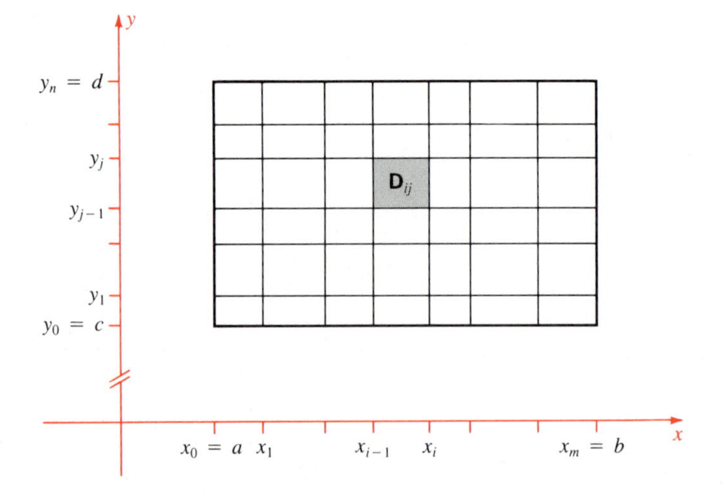

Clearly $\text{rad}(\mathbf{D}_{ij})$ equals half the diagonal of $\mathbf{D}_{ij}$. It follows that $\text{rad}(\mathbf{D}_{ij})$ is small if $x_i - x_{i-1}$ and $y_j - y_{j-1}$ are both small.

The problem is to evaluate $\iint_{\mathbf{D}} f(x, y)\,dx\,dy$. Choose any real numbers $\bar{x}_i$ and $\bar{y}_j$ satisfying

$$x_{i-1} \le \bar{x}_i \le x_i \quad \text{and} \quad y_{j-1} \le \bar{y}_j \le y_j$$

Then $(\bar{x}_i, \bar{y}_j)$ is a point of $\mathbf{D}_{ij}$. Form the corresponding Riemann sum over all the subrectangles $\mathbf{D}_{ij}$:

$$S_{mn} = \sum_{i=1}^{m} \sum_{j=1}^{n} f(\bar{x}_i, \bar{y}_j)(x_i - x_{i-1})(y_j - y_{j-1})$$

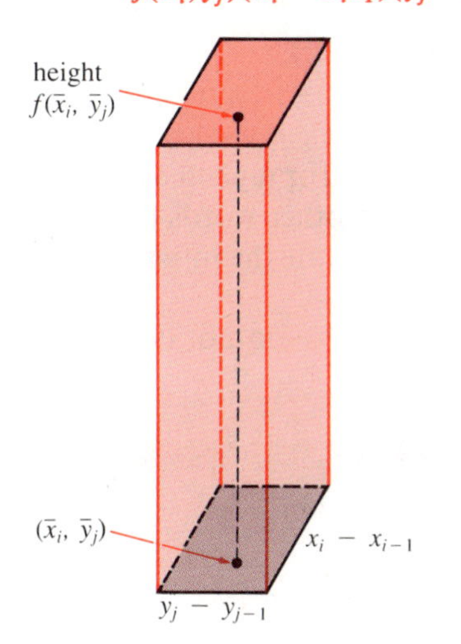

Figure 17-2-3
Typical term in the Riemann sum: volume = $f(\bar{x}_i, \bar{y}_j)(x_i - x_{i-1})(y_j - y_{j-1})$

Each term in this sum is the (algebraic) volume of a thin rectangular solid (Figure 17-2-3).

The terms in the Riemann sum can be added together in any order. Think of the mn terms as arranged in a rectangle with n rows and m columns. First add the i-th column (i fixed, and j running), then sum the column totals:

$$S_{mn} = \sum_{i=1}^{m} \left(\sum_{j=1}^{n} f(\bar{x}_i, \bar{y}_j)(y_j - y_{j-1}) \right)(x_i - x_{i-1})$$

Now here is the crucial step. Look closely at the inner sum

$$\sum_{j=1}^{n} f(\bar{x}_i, \bar{y}_j)(y_j - y_{j-1})$$

It looks familiar. In fact, since $\bar{x}_i$ is constant in each term, this inner sum is a Riemann sum for the integral

$$\int_c^d f(\bar{x}_i, y)\,dy$$

Note that one variable is held fixed and the other is integrated out. Thus we have

$$\sum_{j=1}^{n} f(\bar{x}_i, \bar{y}_j)(y_j - y_{j-1}) \approx \int_c^d f(\bar{x}_i, y)\,dy$$

Substitute this integral into the double Riemann sum:

$$S_{mn} \approx \sum_{i=1}^{m} \left(\int_c^d f(\bar{x}_i, y)\,dy \right)(x_i - x_{i-1})$$

Theory (omitted) says this is a good approximation if all $x_i - x_{i-1}$ and $y_j - y_{j-1}$ are small.

Again the expression looks like a Riemann sum for a simple integral. To see this clearly, we set

$$g(x) = \int_c^d f(x, y)\,dy$$

Then

$$S_{mn} \approx \sum_{i=1}^{m} g(\bar{x}_i)(x_i - x_{i-1}) \approx \int_a^b g(x)\,dx$$

Putting everything together, we have

$$S_{mn} \approx \int_a^b \left(\int_c^d f(x, y)\,dy \right) dx$$

But S_{mn} is a Riemann sum for

$$\iint_{\mathbf{D}} f(x, y)\,dx\,dy$$

so we have strong reason to believe that

$$\iint_{\mathbf{D}} f(x, y)\,dx\,dy = \int_a^b \left(\int_c^d f(x, y)\,dy \right) dx$$

The right-hand expression is called an **iterated integral** or **repeated integral.** The value of the inner integral depends on x. In other words the inner integral is a function of x, so it makes sense to integrate the inner integral with respect to x.

This informal argument can be made rigorous. We do not do so, but rather we summarize the results:

Iteration Formulas

Let $f(x, y)$ be continuous on the rectangular domain

$$\mathbf{D} = \{(x, y) | \text{ where } a \le x \le b \text{ and } c \le y \le d\}$$

Then

$$g(x) = \int_c^d f(x, y)\,dy$$

is continuous on $[a, b]$ and

$$\iint_{\mathbf{D}} f(x, y)\,dx\,dy = \int_a^b \left(\int_c^d f(x, y)\,dy \right) dx$$

Similarly

$$\iint_{\mathbf{D}} f(x, y)\,dx\,dy = \int_c^d \left(\int_a^b f(x, y)\,dx \right) dy$$

This result is a real breakthrough, for it reduces the calculation of double integrals to the calculation of simple integrals, something we are pretty good at by now. True, a *rectangular* domain is rather special, but we shall overcome that restriction in the next section.

Example 1

Find $\iint x^2y^3\,dx\,dy$ over $0 \le x \le 2,\ 1 \le y \le 3.$

Solution

$$\iint_{D} x^2y^3\,dx\,dy = \int_0^2 \left(\int_1^3 x^2y^3\,dy\right) dx$$

In the inner integration, x^2 is a constant, so

$$\int_1^3 x^2y^3\,dy = x^2 \int_1^3 y^3\,dy = x^2\left(\tfrac{1}{4}y^4\Big|_1^3\right) = 20x^2$$

Therefore

$$\iint_{D} x^2y^3\,dx\,dy = \int_0^2 20x^2\,dx = \tfrac{20}{3}x^3\Big|_0^2 = \tfrac{160}{3}$$

●

Example 2

Find $\iint e^x \cos y\,dx\,dy$ over $0 \le x \le 1,\ \frac{1}{2}\pi \le y \le \pi.$

Solution

$$\iint_{D} e^x \cos y\,dx\,dy = \int_{\pi/2}^{\pi} \left(\int_0^1 e^x \cos y\,dx\right) dy$$

But

$$\int_0^1 e^x \cos y\,dx = (\cos y) \int_0^1 e^x\,dx = (e - 1)\cos y$$

Hence

$$\iint_{D} e^x \cos y\,dx\,dy = \int_{\pi/2}^{\pi} (e-1)\cos y\,dy$$

$$= (e-1)\sin y\Big|_{\pi/2}^{\pi} = (e-1)(-1) = 1 - e$$

●

Examples 1 and 2 are special in that the integrand is a product

$$f(x, y) = g(x)h(y)$$

of a function of x alone by a function of y alone. When this is the case, the iteration formula says

$$\iint_{\mathbf{D}} f(x, y)\, dx\, dy = \iint_{\mathbf{D}} g(x)h(y)\, dx\, dy$$

$$= \int_a^b \left(\int_c^d g(x)h(y)\, dy \right) dx$$

But $g(x)$ is constant in the inner integration, so

$$\int_c^d g(x)h(y)\, dy = g(x) \int_c^d h(y)\, dy$$

Consequently

$$\iint_{\mathbf{D}} f(x, y)\, dx\, dy = \int_a^b \left(g(x) \int_c^d h(y)\, dy \right) dx$$

However,

$$\int_c^d h(y)\, dy$$

is a constant and can be factored out. Hence

$$\iint_{\mathbf{D}} f(x, y)\, dx\, dy = \left(\int_c^d h(y)\, dy \right) \left(\int_a^b g(x)\, dx \right)$$

Iteration Formula (Factored Integrand)

$$\iint_{\mathbf{D}} g(x)h(y)\, dx\, dy = \left(\int_a^b g(x)\, dx \right) \left(\int_c^d h(y)\, dy \right)$$

where $\mathbf{D} = \{(x, y) \mid \text{where } a \leq x \leq b \text{ and } c \leq y \leq d\}$ ●

Example 3

Find $\iint e^{x-y}\, dx\, dy$ over $0 \leq x \leq 1$, $-2 \leq y \leq -1$.

Solution

$$\iint_{\mathbf{D}} e^{x-y}\, dx\, dy = \iint_{\mathbf{D}} e^x e^{-y}\, dx\, dy = \left(\int_0^1 e^x\, dx \right) \left(\int_{-2}^{-1} e^{-y}\, dy \right)$$

$$= \left(e^x \Big|_0^1 \right) \left(-e^{-y} \Big|_{-2}^{-1} \right) = (e - 1)(e^2 - e) = e(e - 1)^2$$ ●

Example 4

Find $V = \iint (x^2 y - 3xy^2)\, dx\, dy$

over $1 \leq x \leq 2$, $-1 \leq y \leq 1$.

Solution Use the linear property of the double integral:

$$\iint_D (x^2y - 3xy^2)\,dx\,dy = \iint_D x^2y\,dx\,dy - 3\iint_D xy^2\,dx\,dy$$

Evaluate these two integrals separately:

$$\iint_D x^2y\,dx\,dy = \left(\int_1^2 x^2\,dx\right)\left(\int_{-1}^1 y\,dy\right) = 0$$

$$\iint_D xy^2\,dx\,dy = \left(\int_1^2 x\,dx\right)\left(\int_{-1}^1 y^2\,dy\right) = \tfrac{3}{2}\cdot\tfrac{2}{3} = 1$$

Therefore $V = 0 - 3\cdot 1 = -3$. ●

If the integrand is not of the form $g(x)h(y)$ or a sum of such functions, then we must use the iteration formulas in their general form.

Example 5

Find the volume of the solid between the surface $z = 1/(x+y)$ and the x, y-domain $0 \le x \le 1,\ 1 \le y \le 2$.

Solution The volume is

$$V = \iint_D z\,dA = \iint_D \frac{1}{x+y}\,dx\,dy$$

Iterate: $$V = \iint_D \frac{dx\,dy}{x+y} = \int_0^1\left(\int_1^2 \frac{dy}{x+y}\right)dx$$

For fixed x,

$$\int_1^2 \frac{dy}{x+y} = \ln(x+y)\Big|_{y=1}^{y=2} = \ln(2+x) - \ln(1+x)$$

Hence

$$V = \int_0^1 [\ln(2+x) - \ln(1+x)]\,dx$$

But $\int \ln u\,du = u\ln u - u + C$. Hence

$$\begin{aligned} V &= \int_0^1 [\ln(2+x) - \ln(1+x)]\,dx \\ &= [(2+x)\ln(2+x) - (2+x) \\ &\qquad - (1+x)\ln(1+x) + (1+x)]\Big|_0^1 \\ &= 3\ln 3 - 2\ln 2 - 2\ln 2 = 3\ln 3 - 4\ln 2 = \ln\tfrac{27}{16} \end{aligned}$$ ●

Remark Note carefully the expression

$$\ln(x+y)\Big|_{y=1}^{y=2}$$

We wrote $y = 1$ and $y = 2$ because there were two variables. If we had written only 1 and 2, it would not have been clear how to evaluate $\ln(x + y)$.

Iteration may be done in either order. Sometimes the computation is difficult in one order but relatively easy in the opposite order.

Example 6

Find $\displaystyle\iint y\cos(xy)\,dx\,dy$ over $0 \le x \le 1,\ 0 \le y \le \pi$.

Solution Try integrating first on y:

$$\iint_{\mathbf{D}} = \int_0^1 \left(\int_0^{\pi} y\cos(xy)\,dy \right) dx$$

The inner integral can be done by parts:

$$\int y\cos(xy)\,dy = \frac{y\sin(xy)}{x} + \frac{1}{x^2}\cos(xy) + C$$

When the definite (inner) integral is evaluated, it leads to

$$\iint_{\mathbf{D}} = \int_0^1 \left[\frac{\pi\sin(\pi x)}{x} - \frac{1}{x^2} + \frac{\cos(\pi x)}{x^2} \right] dx$$

This seems to be getting in too deep, so try iteration in the opposite order:

$$\iint_{\mathbf{D}} = \int_0^{\pi} \left(\int_0^1 y\cos(xy)\,dx \right) dy$$

$$\int_0^1 \cos(xy)\,dx = \frac{1}{y}\sin(xy)\Big|_{x=0}^{x=1} = \frac{\sin y}{y}$$

Hence $\displaystyle\iint_{\mathbf{D}} = \int_0^{\pi} y\frac{\sin y}{y}\,dy = \int_0^{\pi} \sin y\,dy = 2.$

Another way of writing the solution:

$$\iint_{\mathbf{D}} y\cos(xy)\,dx\,dy = \int_0^{\pi} y\,dy \int_0^1 \cos(xy)\,dx$$

$$= \int_0^{\pi} y\left(\frac{\sin(xy)}{y}\Big|_0^1 \right) dy = \int_0^{\pi} \sin y\,dy = 2$$

Exercises

Compute $\int_0^1 f(x, y)\, dx$ and $\int_1^3 f(x, y)\, dy$

1 $f(x, y) = xy^3$

2 $f(x, y) = e^{xy}$

3 $f(x, y) = (2x - y - 1)^4$

4 $f(x, y) = \sqrt{x + y + 2}$

5 $f(x, y) = \dfrac{1 + y^2}{1 + x^2}$

6 $f(x, y) = x^3y^2 + 5x - 7y$

Evaluate

7 $\iint (3x - 1)\, dx\, dy$ $\quad -1 \le x \le 2,\ 0 \le y \le 5$

8 $\iint e^y\, dx\, dy$ $\quad -1 \le x \le 1,\ 0 \le y \le \ln 2$

9 $\iint x^2y^2\, dx\, dy$ $\quad -1 \le x \le 1,\ -1 \le y \le 1$

10 $\iint x^3y^3\, dx\, dy$ $\quad -1 \le x \le 1,\ -1 \le y \le 1$

11 $\iint (x^5 - y^5)\, dx\, dy$ $\quad 0 \le x \le 1,\ 0 \le y \le 1$

12 $\iint [\exp(x^2) - \exp(y^2)]\, dx\, dy$ $\quad 0 \le x \le 1,\ 0 \le y \le 1$

13 $\iint (1 + x - 2y)^3\, dx\, dy$ $\quad 0 \le x \le 1,\ 1 \le y \le 2$

14 $\iint \sqrt{x + y + 2}\, dx\, dy$ $\quad 0 \le x \le 1,\ 1 \le y \le 3$

15 $\iint \dfrac{dx\, dy}{(1 + x + y)^2}$ $\quad 0 \le x \le 1,\ 0 \le y \le 1$

16 $\iint \sin(x + y)\, dx\, dy$ $\quad 0 \le x \le \frac{1}{2}\pi,\ 0 \le y \le \frac{1}{2}\pi$

17 $\iint (1 - 2x)\sin(y^2)\, dx\, dy$ $\quad 0 \le x \le 1,\ 0 \le y \le 1$

18 $\iint \dfrac{x^2}{y^3}\, dx\, dy$ $\quad 1 \le x \le 2,\ 1 \le y \le 4$

19 $\iint \dfrac{x}{1 + y^2}\, dx\, dy$ $\quad 0 \le x \le 2,\ 0 \le y \le 1$

20 $\iint xy \ln x\, dx\, dy$ $\quad 1 \le x \le 4,\ -1 \le y \le 2$

21 $\iint x \ln(xy)\, dx\, dy$ $\quad 2 \le x \le 3,\ 1 \le y \le 2$

22 $\iint e^{x+y} \cos 2x\, dx\, dy$ $\quad 0 \le x \le \pi,\ 1 \le y \le 2$

Find the volume between the surface determined by z and the indicated portion of the x, y-plane

23 $z = 2 - (x^2 + y^2)$ $\quad -1 \le x \le 1 \quad -1 \le y \le 1$

24 $z = 1 - xy$ $\quad 0 \le x \le 1 \quad 0 \le y \le 1$

25 $z = x^2 + 4y^2$ $\quad 0 \le x \le 2 \quad 0 \le y \le 1$

26 $z = \sin x \sin y$ $\quad 0 \le x \le \pi \quad 0 \le y \le \pi$

27 $z = x^2y + y^2x$ $\quad 1 \le x \le 2 \quad 2 \le y \le 3$

28 $z = (1 + x^3)y^2$ $\quad -1 \le x \le 1 \quad -1 \le y \le 1$

29 Suppose $f(x, y)$ is continuous and $f(x, -y) = -f(x, y)$. Prove that $\iint f(x, y)\, dx\, dy = 0$ on each rectangle of the form $a \le x \le b,\ -c \le y \le c$. Verify this for $f(x, y) = x^2y^3$.

30 Suppose $f(x, y)$ is continuous and $f(-x, -y) = -f(x, y)$. Prove that $\iint f(x, y)\, dx\, dy = 0$ on each rectangle of the form $-a \le x \le a,\ -b \le y \le b$. Verify this for $f(x, y) = (3x - 2y)^5$.

31 Find the constant A that best approximates $f(x, y)$ on the domain $0 \le x \le 1,\ 0 \le y \le 1$ in the **least-squares** sense. In other words, minimize

$$\iint [f(x, y) - A]^2\, dx\, dy$$

32 (cont.) Show that the coefficients for the least-squares linear approximation $A + Bx + Cy$ to $f(x, y)$ on the domain $0 \le x \le 1,\ 0 \le y \le 1$ satisfy

$$A + \tfrac{1}{2}B + \tfrac{1}{2}C = \iint f\, dx\, dy$$

$$\tfrac{1}{2}A + \tfrac{1}{3}B + \tfrac{1}{4}C = \iint xf\, dx\, dy$$

$$\tfrac{1}{2}A + \tfrac{1}{4}B + \tfrac{1}{3}C = \iint yf\, dx\, dy$$

17-3 Domains between Graphs

Let us venture away from the security of rectangles and consider double integrals over some curved domains. We shall discuss an important class of domains, more general than rectangles but not too general. They are "four-sided" domains, bounded on two sides by smooth curves and on the other two sides by parallel lines. To be precise, they are described either as

$$\mathbf{D} = \{(x, y) \mid \text{where } g(x) \leq y \leq h(x) \text{ for } a \leq x \leq b\}$$

or as

$$\mathbf{D} = \{(x, y) \mid \text{where } g(y) \leq x \leq h(y) \text{ for } c \leq y \leq d\}$$

where g and h are smooth functions. See Figure 17-3-1. Sometimes we describe the domains in Figure 17-3-1a by saying that for each fixed x from a to b, the variable y runs from $g(x)$ to $h(x)$.

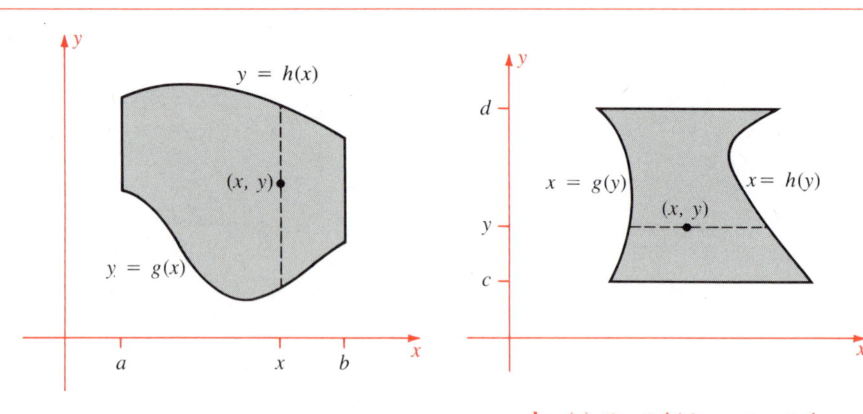

Figure 17-3-1
Domains between graphs

a $g(x) \leq y \leq h(x), \quad a \leq x \leq b$

b $g(y) \leq x \leq h(y), \quad c \leq y \leq d$

Let us examine the volume problem for such a domain. Suppose $f(x, y) \geq 0$ is a continuous function on the domain

$$\mathbf{D} = \{(x, y) \mid \text{where } g(y) \leq x \leq h(y) \text{ for } c \leq y \leq d\}$$

The solid under the graph consists of all points (x, y, z) in space such that (x, y) is in $\mathbf{D}$ and $0 \leq z \leq f(x, y)$. See Figure 17-3-2. Its volume is

$$V = \iint_{\mathbf{D}} f(x, y)\, dx\, dy$$

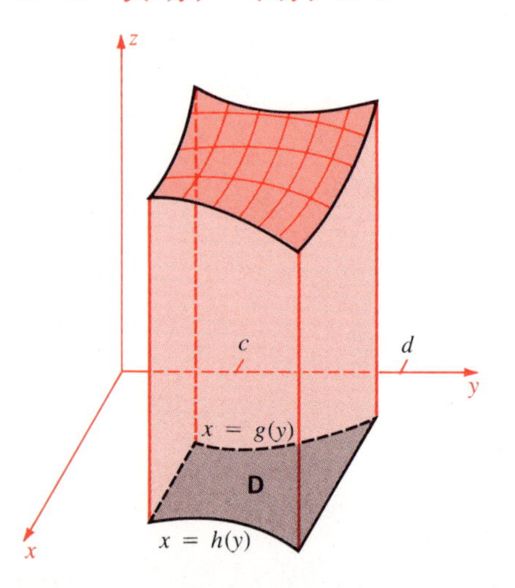

Figure 17-3-2
Volume problem for the solid
$0 \leq z \leq f(x, y) \quad (x, y)$ in D

We compute V by slicing. We slice the solid by a vertical plane $y =$ constant and let $V(y)$ denote the volume to the left of the plane (Figure 17-3-3a). The area of the cross section (Figure 17-3-3b) is

$$A(y) = \int_{g(y)}^{h(y)} f(x, y)\, dx$$

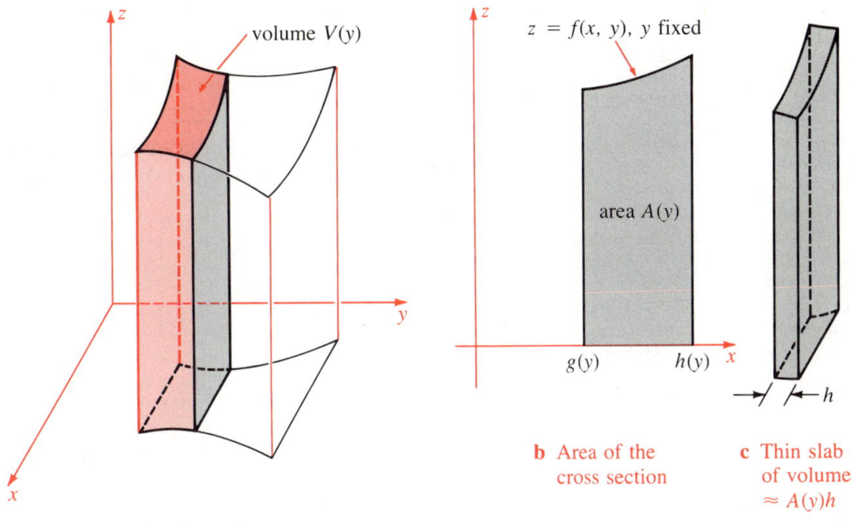

Figure 17-3-3
Solution of the volume problem by sectioning

a Cross-section by a plane y = constant

b Area of the cross section

c Thin slab of volume $\approx A(y)h$

We argue intuitively that

$$\frac{dV}{dy} = A(y)$$

Indeed, the derivative is the limit as $h \to 0$ of

$$\frac{V(y+h) - V(y)}{h}$$

For h very small, the numerator is the volume of a thin slab (Figure 17-3-3c) of width h and cross-sectional area approximately $A(y)$. Hence the quotient is approximately $A(y)$. It follows that

$$V = \int_c^d \frac{dV}{dy}\,dy = \int_c^d A(y)\,dy = \int_c^d \left(\int_{g(y)}^{h(y)} f(x, y)\,dx\right) dy$$

We are thus led intuitively to the following rules:

Iteration Formulas

Let $f(x, y)$ be continuous on **D**.

- If $\mathbf{D} = \{(x, y) \mid \text{where } g(y) \leq x \leq h(y) \text{ for } c \leq y \leq d\}$, then

$$\iint_{\mathbf{D}} f(x, y)\,dx\,dy = \int_c^d \left(\int_{g(y)}^{h(y)} f(x, y)\,dx\right) dy$$

- If $\mathbf{D} = \{(x, y) \mid \text{where } g(x) \leq y \leq h(x) \text{ for } a \leq x \leq b\}$, then

$$\iint_{\mathbf{D}} f(x, y)\,dx\,dy = \int_a^b \left(\int_{g(x)}^{h(x)} f(x, y)\,dy\right) dx$$

Remark In advanced courses, these formulas are proved by the use of Riemann sums.

Example 1

Find the volume between the surface $z = x + y$ and the domain of the x, y-plane bounded by the y-axis, the parabola $x = y^2$, and the lines $y = 1$ and $y = 2$.

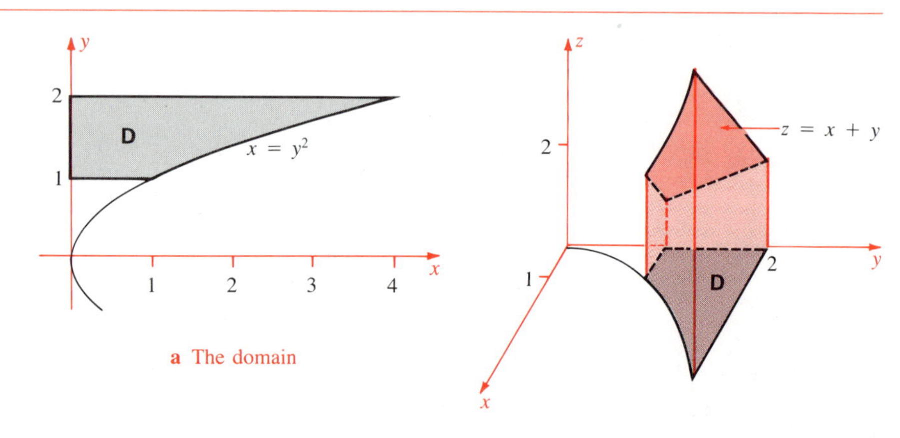

Figure 17-3-4
A volume problem over the domain
$\mathbf{D} = \{(x, y) \mid 0 \le x \le y^2,\ 1 \le y \le 2\}$

a The domain

b The solid

Solution First draw the domain (Figure 17-3-4a). It doesn't hurt to draw the solid also (Figure 17-3-4b). The domain is the region between the graphs $x = 0$ and $x = y^2$ for $1 \le y \le 2$. Therefore

$$V = \iint_{\mathbf{D}} z\,dA = \iint_{\mathbf{D}} (x + y)\,dx\,dy$$

$$= \int_1^2 \left(\int_0^{y^2} (x + y)\,dx \right) dy$$

(This is the crucial step. Study the set-up carefully and be sure you understand it.) Now

$$\int_0^{y^2} (x + y)\,dx = \left(\tfrac{1}{2}x^2 + xy\right)\Big|_{x=0}^{x=y^2}$$

$$= \tfrac{1}{2}y^4 + y^3$$

so

$$V = \int_1^2 \left(\tfrac{1}{2}y^4 + y^3\right) dy = \tfrac{1}{10}y^5 + \tfrac{1}{4}y^4 \Big|_1^2$$

$$= \tfrac{1}{10}(32 - 1) + \tfrac{1}{4}(16 - 1)$$

$$= \tfrac{31}{10} + \tfrac{15}{4} = \tfrac{137}{20}$$

●

Figure 17-3-5
See Example 2

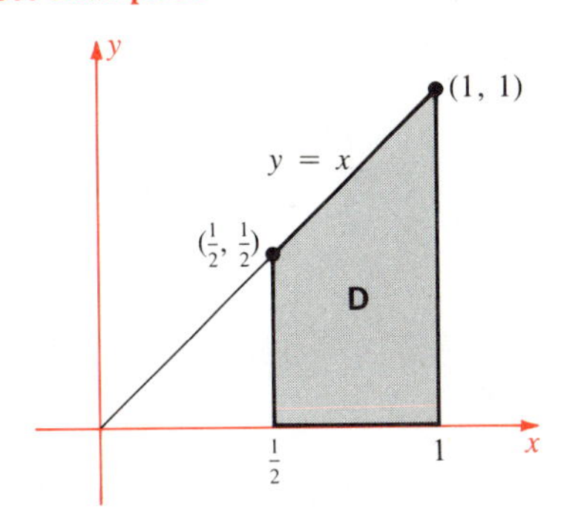

a D: $0 \le y \le x$, $\frac{1}{2} \le x \le 1$

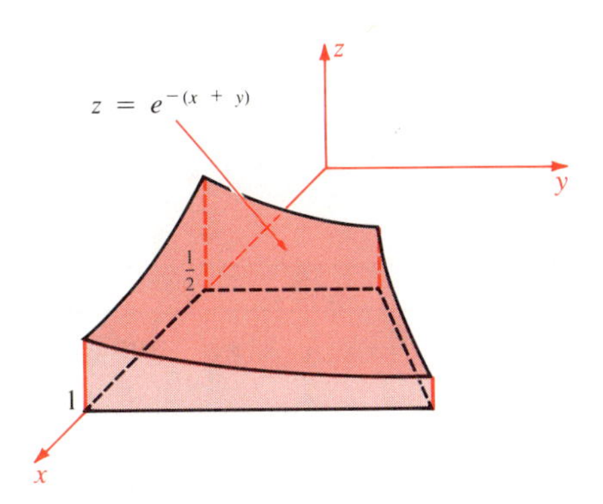

b The solid

Example 2

Find the volume between the surface $z = e^{-(x+y)}$ and the domain of the x, y-plane bounded by the x-axis, the line $y = x$, and the lines $x = \frac{1}{2}$ and $x = 1$.

Solution First draw the domain and the solid (Figure 17-3-5). By iteration,

$$V = \iint_{\mathbf{D}} z\,dA = \iint_{\mathbf{D}} e^{-(x+y)}\,dx\,dy = \int_{1/2}^{1}\left(\int_0^x e^{-(x+y)}\,dy\right)dx$$

Now

$$\int_0^x e^{-(x+y)}\,dy = \int_0^x e^{-x}e^{-y}\,dy = e^{-x}\int_0^x e^{-y}\,dy$$

$$= -e^{-x}(e^{-y})\Big|_{y=0}^{y=x} = e^{-x} - e^{-2x}$$

so

$$V = \int_{1/2}^{1}(e^{-x} - e^{-2x})\,dx = (\tfrac{1}{2}e^{-2x} - e^{-x})\Big|_{1/2}^{1}$$

$$= \tfrac{1}{2}e^{-2} - \tfrac{3}{2}e^{-1} + e^{-1/2}$$

$$= \tfrac{1}{2}e^{-2}(1 - 3e + 2e^{3/2})$$

●

Example 3

Find the volume between the surface $z = 1 - x^2 - y^2$, and the square in the x, y-plane with vertices $(\pm 1, 0)$ and $(0, \pm 1)$.

Solution First draw the square (Figure 17-3-6a). Observe that, by symmetry, it suffices to find the volume over the triangular portion in the

Figure 17-3-6
See Example 3

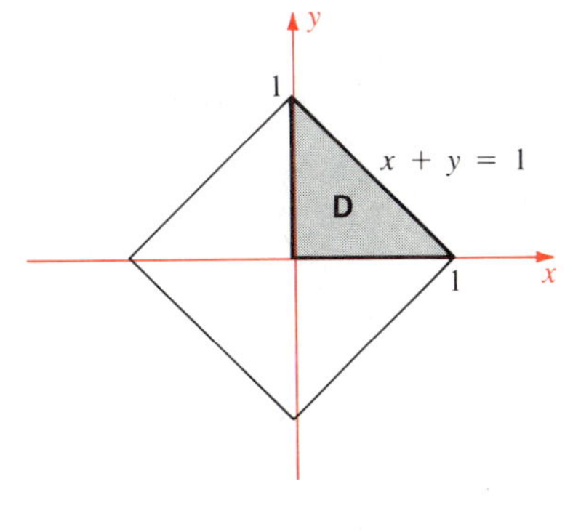

a D: $0 \le y \le 1 - x$, $0 \le x \le 1$

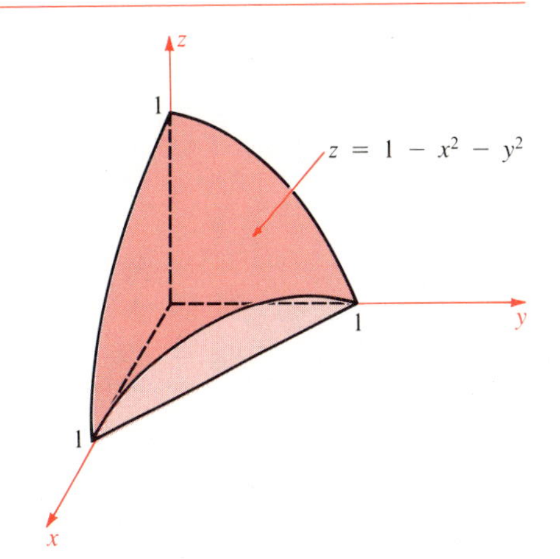

b The first octant part of the solid

first quadrant, and then quadruple it. The corresponding quarter of the solid is shown in Figure 17-3-6b. By the iteration formula,

$$\begin{aligned}
V &= 4\iint_{\mathbf{D}} (1 - x^2 - y^2)\,dx\,dy \\
&= 4\int_0^1 \left(\int_0^{1-x} (1 - x^2 - y^2)\,dy\right) dx \\
&= 4\int_0^1 \left[(y - x^2y - \tfrac{1}{3}y^3)\Big|_{y=0}^{y=1-x}\right] dx \\
&= 4\int_0^1 [(1 - x) - x^2(1 - x) - \tfrac{1}{3}(1 - x)^3]\,dx \\
&= 4\int_0^1 [1 - x - x^2 + x^3 - \tfrac{1}{3}(1 - x)^3]\,dx \\
&= 4[1 - \tfrac{1}{2} - \tfrac{1}{3} + \tfrac{1}{4} - \tfrac{1}{12}] \\
&= 4 \cdot \tfrac{4}{12} = \tfrac{4}{3}
\end{aligned}$$

Example 4

Find the volume between the plane $z = 1 + x + y$, and the x,y-domain bounded by the lines $x = \frac{1}{2}$ and $x = 1$ and the curves $y = x^2$ and $y = 2x^2$.

Solution The domain and the solid are sketched in Figure 17-3-7. Since the domain is between the graphs of two functions, the iteration formula applies. First integrate on y:

Figure 17-3-7
See Example 4

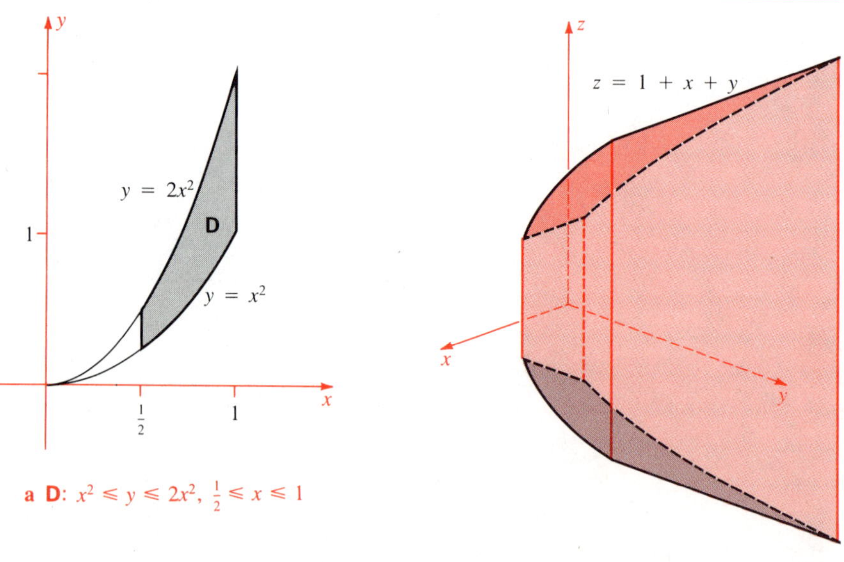

a D: $x^2 \leq y \leq 2x^2$, $\frac{1}{2} \leq x \leq 1$

b The solid

$$V = \iint_{\mathbf{D}} (1 + x + y)\, dx\, dy = \int_{1/2}^{1} \left(\int_{x^2}^{2x^2} (1 + x + y)\, dy \right) dx$$

The inside integral equals

$$\left(y + xy + \tfrac{1}{2}y^2\right)\Big|_{y=x^2}^{y=2x^2} = (2x^2 + 2x^3 + 2x^4) - (x^2 + x^3 + \tfrac{1}{2}x^4) = x^2 + x^3 + \tfrac{3}{2}x^4$$

so

$$V = \int_{1/2}^{1} (x^2 + x^3 + \tfrac{3}{2}x^4)\, dx = \left(\tfrac{1}{3}x^3 + \tfrac{1}{4}x^4 + \tfrac{3}{10}x^5\right)\Big|_{1/2}^{1}$$

$$= \left(\tfrac{1}{3} + \tfrac{1}{4} + \tfrac{3}{10}\right) - \tfrac{1}{8}\left(\tfrac{1}{3} + \tfrac{1}{8} + \tfrac{3}{40}\right) = \tfrac{49}{60}$$ ●

Example 5

Find

$$\iint xy\, dx\, dy$$

over the domain **D** bounded by $y = x$ and $y = x^2$.

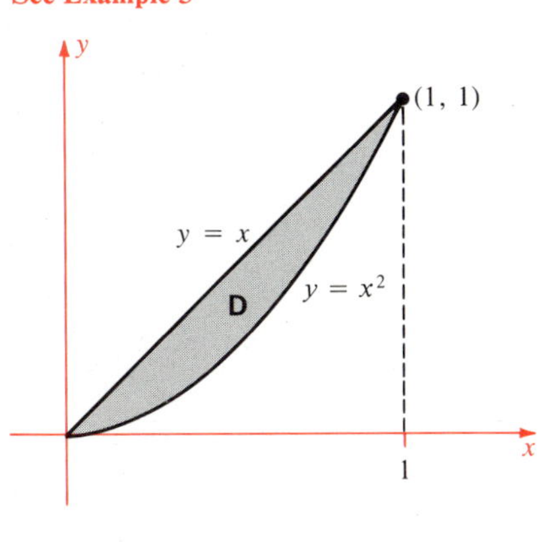

Figure 17-3-8
See Example 5

a **D**: $x^2 \le y \le x$, $0 \le x \le 1$

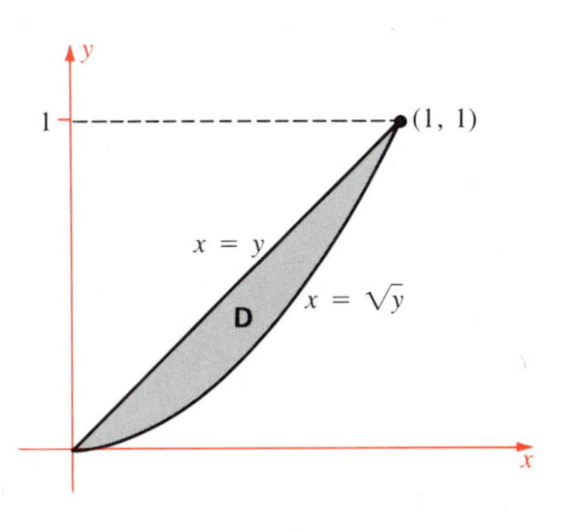

b **D**: $y \le x \le \sqrt{y}$, $0 \le y \le 1$

Solution The first problem is to describe **D** in a way that shows the limits of integration. The line $y = x$ and the parabola $y = x^2$ intersect at $(0, 0)$ and at $(1, 1)$. This information and a drawing (Figure 17-3-8a) suggest that **D** is the domain between the graphs $y = x^2$ and $y = x$ for $0 \le x \le 1$. Therefore

$$\iint_{\mathbf{D}} xy\, dx\, dy = \int_0^1 \left(\int_{x^2}^{x} xy\, dy \right) dx$$

$$= \int_0^1 \left(\tfrac{1}{2}xy^2 \Big|_{y=x^2}^{y=x} \right) dx = \int_0^1 \tfrac{1}{2}(x^3 - x^5)\, dx$$

$$= \tfrac{1}{2}\left(\tfrac{1}{4} - \tfrac{1}{6}\right) = \tfrac{1}{24}$$

Alternative Solution The domain **D** may be thought of as bounded by $x = y$ (below) and $x = \sqrt{y}$ (above), where $0 \le y \le 1$. See Figure 17-3-8b. This approach allows us to integrate with respect to x first, which is possibly easier to do. For each y, the range of x is $y \le x \le \sqrt{y}$. Therefore the set-up for the iteration is

$$\iint_{\mathbf{D}} xy\, dx\, dy = \int_0^1 \left(\int_{y}^{\sqrt{y}} xy\, dx \right) dy$$

$$= \int_0^1 \left(\tfrac{1}{2}x^2 y \Big|_{x=y}^{x=\sqrt{y}} \right) dy = \int_0^1 \tfrac{1}{2}(y^2 - y^3)\, dy$$

$$= \tfrac{1}{2}\left(\tfrac{1}{3} - \tfrac{1}{4}\right) = \tfrac{1}{24}$$ ●

Exercises

The table of *definite* integrals inside the front cover will prove useful for this and subsequent exercise sets.

Compute the volume between the surface $z = f(x, y)$ and the indicated domain of the x, y-plane

1 $z = 1$

2 $z = y$

3 $z = x$

4 $z = y^2 \sin \pi x$

5 $z = x^2$

6 $z = xy$

7 $z = \sqrt{x}$

8 $z = 1/(1 + x^2)$

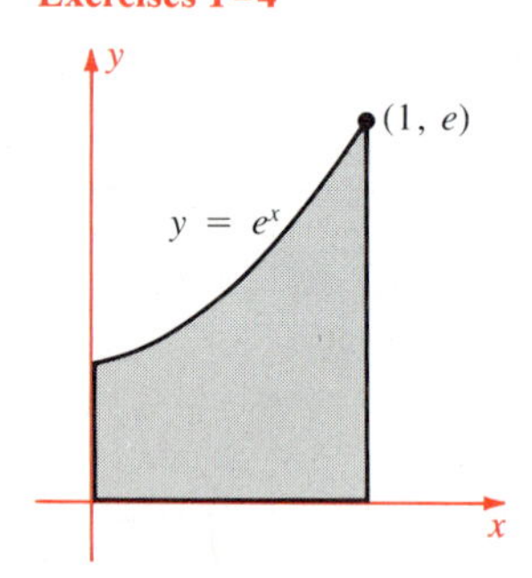

Figure 17-3-9
Exercises 1–4

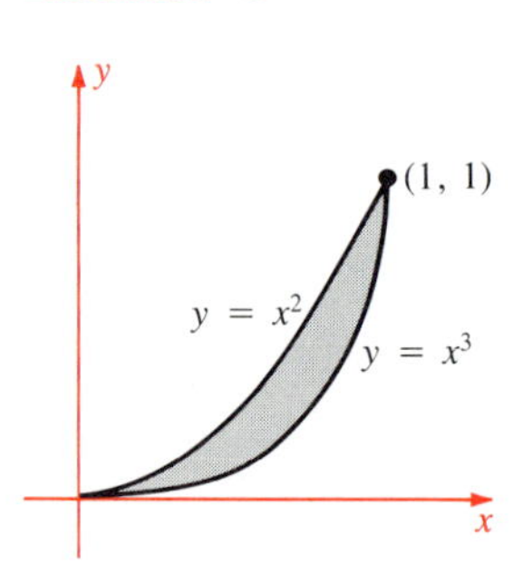

Figure 17-3-10
Exercises 5–8

9 $z = x/y^2$

10 $z = x + 2y$

11 $z = 1/x^3y^3$

12 $z = x^2e^{xy}$

13 $z = ye^x$

14 $z = x^2y$

15 $z = y^3$

16 $z = x\sqrt{x^2 + y^4}$

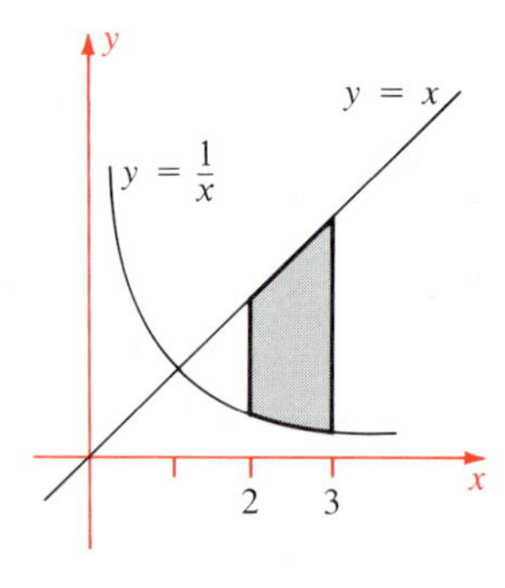

Figure 17-3-11
Exercises 9–12

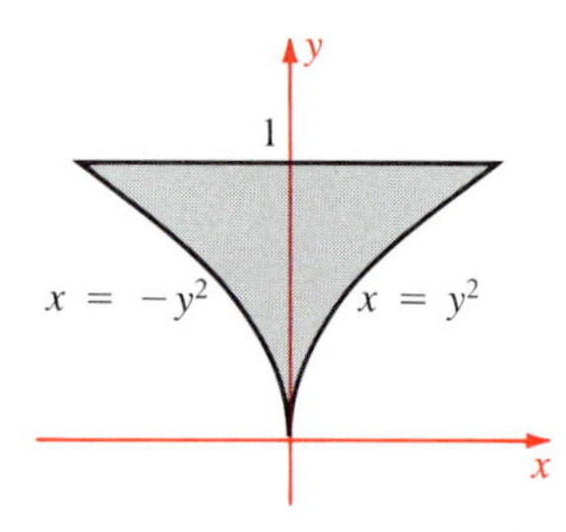

Figure 17-3-12
Exercises 13–16

Express $\iint f(x, y)\,dx\,dy$ as one iterated integral (not a sum of several) over the domain indicated, showing suitable limits of integration. Do it in two ways if possible

17 triangle vertices $(0, 0)$, $(0, 2)$, $(3, 3)$

18 triangle vertices $(-2, 5)$, $(-1, 1)$, $(5, 5)$

19 parallelogram vertices $(-4, 1)$, $(0, 1)$, $(-3, 3)$, $(1, 3)$

20 trapezoid vertices $(0, 2)$, $(1, 4)$, $(4, 4)$, $(8, 2)$

21 the domain bounded by the x-axis and the upper half of the circle with center $(5, 0)$ and radius 5.

22 the domain bounded by the lines $x = 3$ and $y = 1$ and the hyperbola $xy = 1$

23 the domain in the first quadrant bounded by the ellipse $x^2 + \frac{1}{9}y^2 = 1$ and the line $y = 3 - 3x$

24 the domain bounded by the parabola $y = x^2$ and the line $5x + y = 14$

Compute the double integral over the domain bounded by the given curves

25 $\iint xe^{xy}\,dx\,dy$ $\quad x = 1 \quad x = 3 \quad xy = 1 \quad xy = 2$

26 $\iint x^2y\,dx\,dy$ $\quad y = 0 \quad x = 0 \quad x = (y - 1)^2$

27 $\iint (x^3 + y^3)\,dx\,dy$ $\quad x^2 + y^2 = 1$

28 $\iint (x + y)^2\,dx\,dy$ $\quad x + y = 0 \quad y = x^2 + x$

29 $\iint (1 + xy)\,dx\,dy$ $\quad y = 0 \quad y = x \quad y = 1 - x$

30 $\iint \exp(x^2)\,dx\,dy$ $\quad x = 1 \quad x = 2 \quad y = x \quad y = x^3$

31 $\iint x^2y\,dx\,dy$ $\quad y = x^4 \quad y = x^2$

32 $\iint x^4y^2\,dx\,dy$ $\quad |x + y| = 1 \quad |x - y| = 1$

33 $\iint (1 + x)\,dx\,dy$ $\quad x + y = 0 \quad x^2 + y = 1$

34 $\iint x\,dx\,dy$ $\quad y = x \quad x = 3 \quad y = x^2 - x$

Justify the formulas

35 $$\int_0^1 \left(\int_0^x f(x, y)\,dy \right) dx = \int_0^1 \left(\int_y^1 f(x, y)\,dx \right) dy$$

36 $$\int_{-a}^{a} \left(\int_0^{\sqrt{a^2-x^2}} f(x, y)\,dy \right) dx = \int_0^a \left(\int_{-\sqrt{a^2-y^2}}^{\sqrt{a^2-y^2}} f(x, y)\,dx \right) dy$$

Interchange the order of integration

37 $$\int_4^9 \left(\int_{\sqrt{x}}^3 f(x, y)\,dy \right) dx$$

38 $$\int_0^2 \left(\int_0^{3\sqrt{1-y^2/4}} f(x, y)\,dx \right) dy$$

***39** Suppose

$$\int_0^1 f(x)\,dx = 0$$

Prove that

$$\iint\limits_{\mathbf{D}} f(x)f(y)\,dx\,dy = 0$$

where $\mathbf{D}$ is the triangle $0 \le x \le y \le 1$.

***40** Suppose $f(x)$ is a strictly decreasing function for $0 \le x \le a$, that $f(0) = b > 0$, and $f(a) = 0$. Let $g(y)$ be the inverse function of $f(x)$. Compute in two ways area$(\mathbf{D})$, where $\mathbf{D}$ is the region under the graph of $f(x)$, to conclude that

$$\int_0^b g(y)\,dy = \int_0^a f(x)\,dx$$

17-4 Arbitrary Domains

The domains between graphs studied in Section 17-3 are fundamental in double integration. They are simple enough that the iteration formulas hold; you can compute double integrals on them. But at the same time, domains between graphs are complicated enough that any domain of integration can be partitioned into a finite number of such:

- Each domain of integration $\mathbf{D}$ can be partitioned into a finite number of domains of integration

$$\mathbf{D}_1,\ \mathbf{D}_2,\ \cdots,\ \mathbf{D}_n$$

each a domain between the graphs of two functions.

Figure 17-4-1
Arbitrary domains divided into domains of integration

a Two subdomains

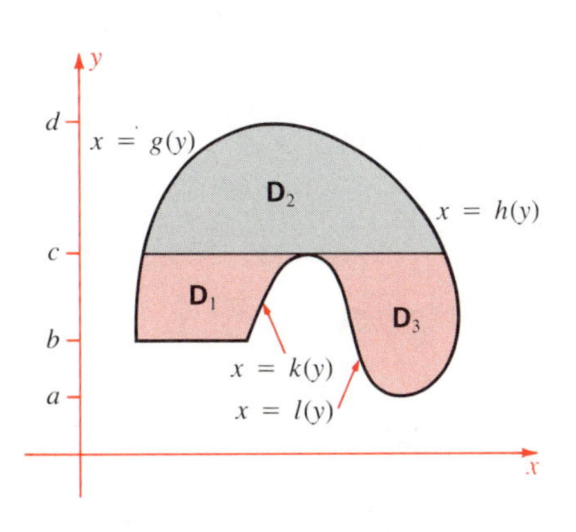

b Three subdomains

Some examples are shown in Figure 17-4-1.

This is the key to integration over arbitrary domains. Split the domain, preferably using horizontal or vertical line segments, into subdomains, each between the graphs of two functions. Then apply iteration on each subdomain and add the results. In Figure 17-4-1a the domain splits into two nice domains $\mathbf{D}_1$ and $\mathbf{D}_2$. The integration set-up is

$$\iint\limits_{\mathbf{D}} f(x,y)\,dx\,dy = \iint\limits_{\mathbf{D}_1} f(x,y)\,dx\,dy + \iint\limits_{\mathbf{D}_2} f(x,y)\,dx\,dy$$

$$= \int_a^c \left(\int_{g(x)}^{h(x)} f(x,y)\,dy\right)dx + \int_c^b \left(\int_{g(x)}^{k(x)} f(x,y)\,dy\right)dx$$

The set-up for Figure 17-4-1b is

$$\iint\limits_{\mathbf{D}} f(x,y)\,dx\,dy = \left(\iint\limits_{\mathbf{D}_1} + \iint\limits_{\mathbf{D}_2} + \iint\limits_{\mathbf{D}_3}\right) f(x,y)\,dx\,dy$$

$$= \int_b^c \left(\int_{g(y)}^{k(y)} f(x,y)\,dx\right)dy + \int_c^d \left(\int_{g(y)}^{h(y)} f(x,y)\,dx\right)dy$$

$$+ \int_a^c \left(\int_{\ell(y)}^{h(y)} f(x\ y)\,dx\right)dy$$

Figure 17-4-2
See Example 1

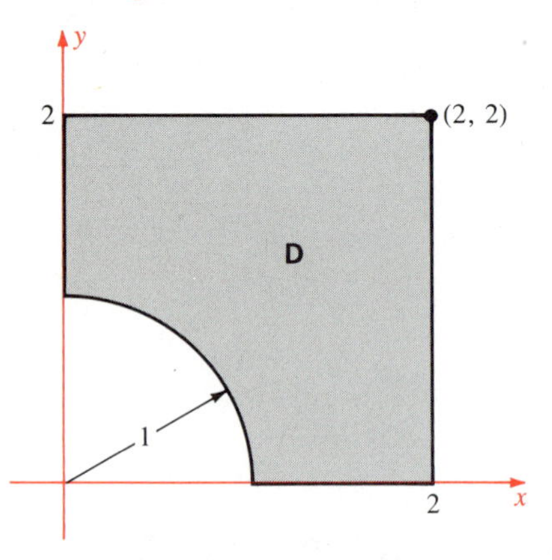

a The domain

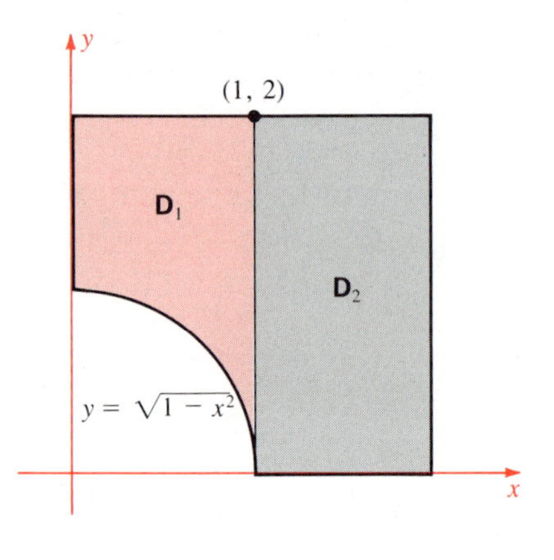

b The partition

Example 1

Set up $\iint f(x, y)\,dx\,dy$ as an iterated integral over the domain in Figure 17-4-2a.

Solution Partition $\mathbf{D}$ by the vertical segment through $(1, 0)$. See Figure 17-4-2b. Domain $\mathbf{D}_1$ is the domain between $y = \sqrt{1 - x^2}$ and $y = 2$ for $0 \le x \le 1$, and $\mathbf{D}_2$ is the (rectangular) domain between $y = 0$ and $y = 2$ for $1 \le x \le 2$. Therefore

$$\iint_{\mathbf{D}} f(x, y)\,dx\,dy = \left(\iint_{\mathbf{D}_1} + \iint_{\mathbf{D}_2}\right) f(x, y)\,dx\,dy$$

$$= \int_0^1 \left(\int_{\sqrt{1-x^2}}^2 f(x, y)\,dy\right) dx + \int_1^2 \left(\int_0^2 f(x, y)\,dy\right) dx$$

Alternative Solution Partition with the horizontal segment through $(0, 1)$:

$$\iint_{\mathbf{D}} f(x, y)\,dx\,dy$$

$$= \int_0^1 \left(\int_{\sqrt{1-y^2}}^2 f(x, y)\,dx\right) dy + \int_1^2 \left(\int_0^2 f(x, y)\,dx\right) dy$$ ●

Domains and Inequalities

A domain in the plane (or space) is frequently specified by a system of inequalities. A single inequality $f(x, y) \le 0$ determines a domain whose boundary is $f(x, y) = 0$. To find the domain described by several such inequalities, draw the domain each describes, then form their intersection.

Example 2

Sketch $\mathbf{D} = \{(x, y) \mid \text{where } x + y \le 0 \text{ and } y \ge x^2 + 2x\}$ and set up a double integral on $\mathbf{D}$.

Solution The first inequality determines the domain below (and on) the line $x + y = 0$. See Figure 17-4-3a. The second inequality determines the domain above (and on) the parabola $y = x^2 + 2x$. See Figure 17-4-3b. The line and parabola intersect at $(0, 0)$ and $(-3, 3)$. The domain $\mathbf{D}$, satisfying both inequalities, is shown in Figure 17-4-3c. Clearly $\mathbf{D}$ is a domain beween two functions of x. Hence

$$\iint_{\mathbf{D}} f(x, y)\,dx\,dy = \int_{-3}^0 \left(\int_{x^2+2x}^{-x} f(x, y)\,dy\right) dx$$ ●

Example 3

Sketch the domain $\mathbf{D}$ specified by $0 \le x \le y \le b$. Equate the two corresponding iterated integrals.

Figure 17-4-3
See Example 2

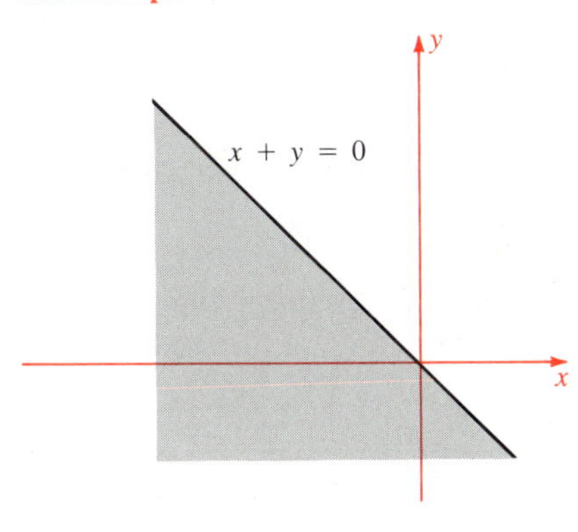

a $x + y \leqslant 0$

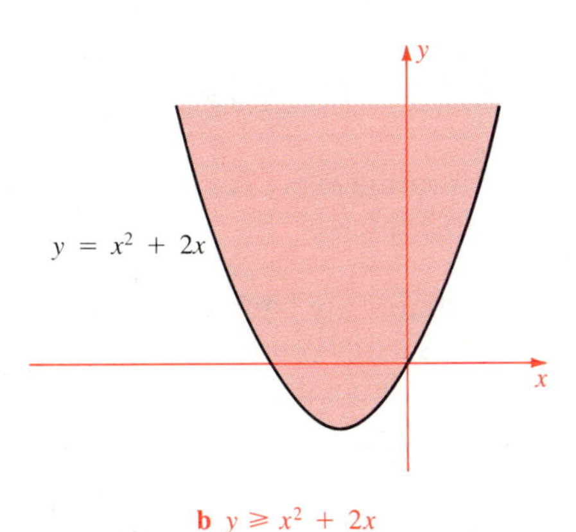

b $y \geqslant x^2 + 2x$

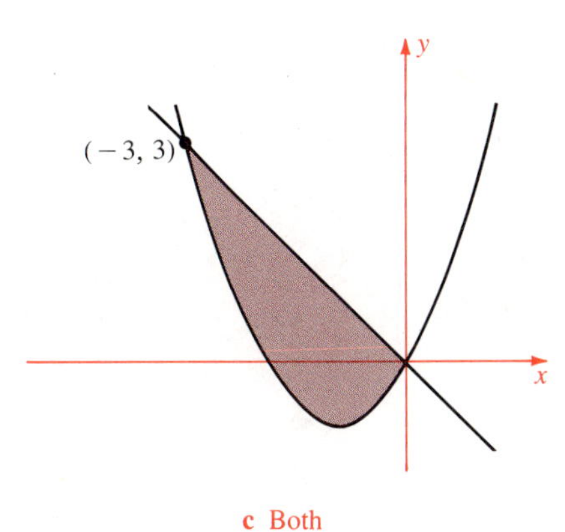

c Both

Solution The region is described by the three inequalities

$$x \geq 0 \qquad y \geq x \qquad y \leq b$$

Figure 17-4-4
See Example 3

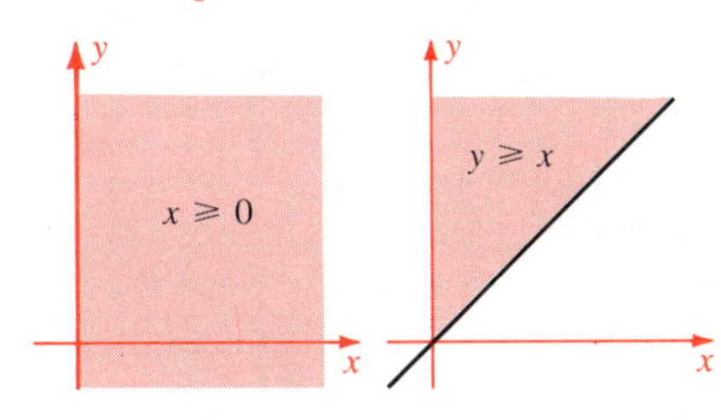

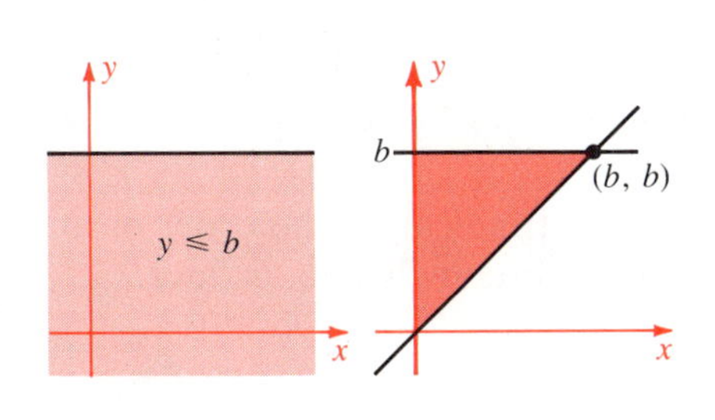

Draw the corresponding domains and take their intersection, a triangle (Figure 17-4-4). The domain can be considered in two ways as the domain between two graphs. Either as

$$\mathbf{D} = \{(x, y) \mid \text{where } \; x \leq y \leq b \; \text{ and } \; 0 \leq x \leq b\}$$

or as

$$\mathbf{D} = \{(x, y) \mid \text{where } \; 0 \leq x \leq y \; \text{ and } \; 0 \leq y \leq b\}$$

Correspondingly

$$\iint_{\mathbf{D}} f(x, y)\, dx\, dy = \int_0^b \left(\int_x^b f(x, y)\, dy \right) dx$$

and

$$\iint_{\mathbf{D}} f(x, y)\, dx\, dy = \int_0^b \left(\int_0^y f(x, y)\, dx \right) dy$$

Therefore

$$\int_0^b \left(\int_0^y f(x, y)\, dx \right) dy = \int_0^b \left(\int_x^b f(x, y)\, dy \right) dx$$

●

Remark A special case is interesting. Suppose $f(x, y) = g(x)$, a function of x alone. Then the right side of the preceding equation is

$$\int_0^b \left(\int_x^b g(x)\, dy \right) dx = \int_0^b g(x) \left(\int_x^b dy \right) dx$$

$$= \int_0^b (b - x) g(x)\, dx$$

Figure 17-4-5
See Example 4

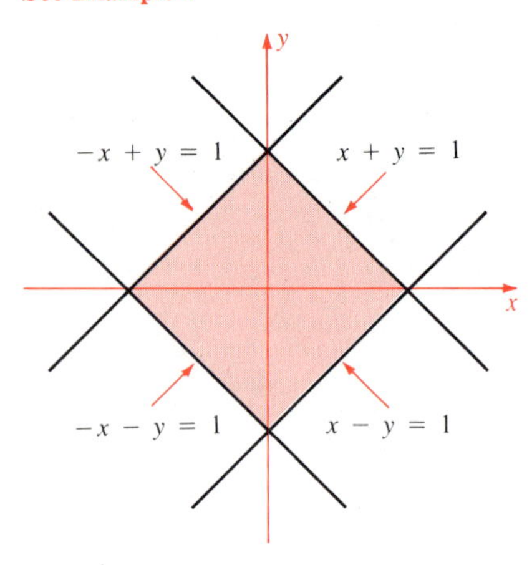

a $|x| + |y| \le 1$

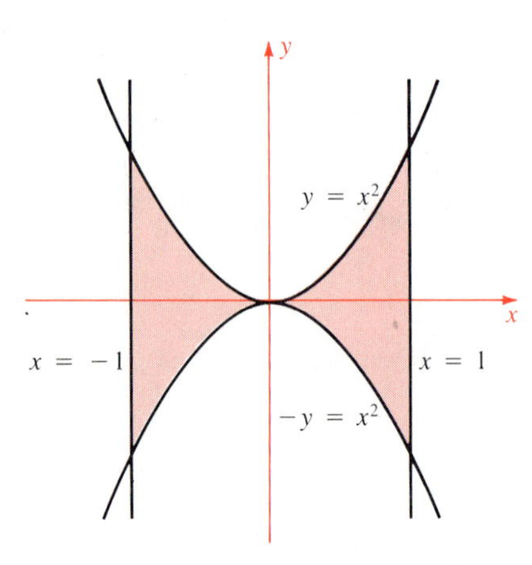

b $|y| \le x^2$ and $|x| \le 1$

Therefore

$$\int_0^b \left(\int_0^y g(x)\, dx \right) dy = \int_0^b (b - x) g(x)\, dx$$

This formula expresses the iterated integral of a function of one variable as a simple integral.

Domains are sometimes described by inequalities involving absolute values. To draw such a domain, find its boundary, and be sure to take into account all possible signs.

Example 4

Sketch the domain

a $|x| + |y| \le 1$ **b** $|y| \le x^2$ and $|x| \le 1$

Solution

a The boundary satisfies $|x| + |y| = 1$. In each quadrant this has a different expression. For instance, in the fourth quadrant it becomes $x - y = 1$. Thus four lines bound the region:

$$x + y = 1 \qquad -x + y = 1 \qquad -x - y = 1 \qquad x - y = 1$$

These lines cut the plane into nine pieces (Figure 17-4-5a). Which is the right one? Obviously the central square because $(0, 0)$ satisfies $|x| + |y| \le 1$; alternatively because $|x| + |y| \le 1$ certainly describes a bounded region, and the square is the only one of the nine pieces that is bounded.

b The boundary consists of $|y| = x^2$ (that is, the two parabolas $y = x^2$ and $-y = x^2$) and $|x| = 1$ (that is, the two vertical lines $x = -1$ and $x = 1$). This is enough for a sketch (Figure 17-4-5b). ●

Remark There are several accepted ways of indicating the order of an iterated integral. The following string of equalities will give the idea:

$$\int_1^3 \left(\int_x^{x^2} f(x, y)\, dy \right) dx = \int_1^3 dx \int_x^{x^2} f(x, y)\, dy$$
$$= \int_1^3 \int_x^{x^2} f(x, y)\, dy\, dx$$

Exercises

Sketch the domain satisfying

1 $x^2 + y^2 \le 1$ and $y + x^2 \ge 0$

2 $x^2 + y^2 \le 1$ and $-x^2 \le y \le x^2$

3 $x^2 + y^2 \ge 1$ and $(x - 2)^2 + y^2 \le 9$

4 $x \ge 3 \quad y \le -5$ and $y - x \ge -10$

5 $1 \le x \le y \le 4$

6 $\frac{1}{2} \le y \le \sqrt{1 - x^2}$

7 $(x + y)^2 \le 1$ and $(x - y)^2 \le 1$

8 $x + y \le 0 \quad xy \le 1$ and $(x - y)^2 \le 1$

Express the double integral of $f(x, y)$ over the specified domain as a sum of one or more iterated integrals in which y is the first variable integrated

9 $x^2 + y^2 \le 1 \quad x^2 + (y - 1)^2 \le 1$

10 $y \ge (x + 1)^2 \quad y + 2x \le 3$

11 $x \ge 0 \quad 0 \le y \le \pi \quad x \le \sin y$

12 $x \ge 0 \quad x^2 - y^2 \ge 1 \quad x^2 + y^2 \le 9$

13 the triangle with vertices $(0, 0)$, $(-1, 4)$, $(2, 3)$

14 the parallelogram with vertices $(0, 0)$, $(1, 5)$, $(6, 7)$, $(5, 2)$

Compute $\iint f(x, y)\,dx\,dy$ over the domain **D**

15 $f(x, y) = \dfrac{1}{x + y}$ — **D** in Figure 17-4-6

16 $f(x, y) = xy$ — **D** in Figure 17-4-6

17 $f(x, y) = xy$ — **D** in Figure 17-4-7

18 $f(x, y) = \sqrt{xy^3}$ — **D** in Figure 17-4-7

19 $f(x, y) = x^2$ — **D** in Figure 17-4-8

20 $f(x, y) = xy$ — **D** in Figure 17-4-8

21 $f(x, y) = 2x + y$ — **D** in Figure 17-4-9

22 $f(x, y) = x^2y^2$ — **D** in Figure 17-4-9

23 $f(x, y) = e^y$ — **D** bounded by $y = \pm x$ and $y = \frac{1}{2}x + 3$

24 $f(x, y) = 1 + xy^2$ — **D** bounded by $x = -y^2$ and the segments from $(2, 0)$ to $(-1, \pm 1)$

25 $f(x, y) = x$ — **D** determined by $x^2 + y^2 \le 1$ and $y \le 2x + 1$

26 $f(x, y) = y/x^2$ — **D** the quadrilateral with vertices $(1, 1)$, $(2, 0)$, $(4, 0)$, $(7, 3)$

27 $f(x, y) = x$ — **D** is determined by $x^2 + y^2 \le 1$ and $x + y \ge 0$

28 $f(x, y) = x$ — **D** bounded by $y = 0$, $y = 2x$, and $5y - 3x = 21$

Express as an iterated integral in which x is the first variable integrated

29 $\displaystyle\int_0^1 \left(\int_0^x f(x, y)\,dy \right) dx + \int_1^e \left(\int_{\ln x}^1 f(x, y)\,dy \right) dx$

30 $\displaystyle\int_1^4 \left(\int_1^{\sqrt{x}} f(x, y)\,dy \right) dx + \int_4^5 \left(\int_1^{6-x} f(x, y)\,dy \right) dx.$

Use double integrals to find the area of **D** in

31 Figure 17-4-6

32 Figure 17-4-7

33 Figure 17-4-8

34 Figure 17-4-9

35 Exercise 24

36 Exercise 28

Figure 17-4-6
See Exercises 15, 16, 31

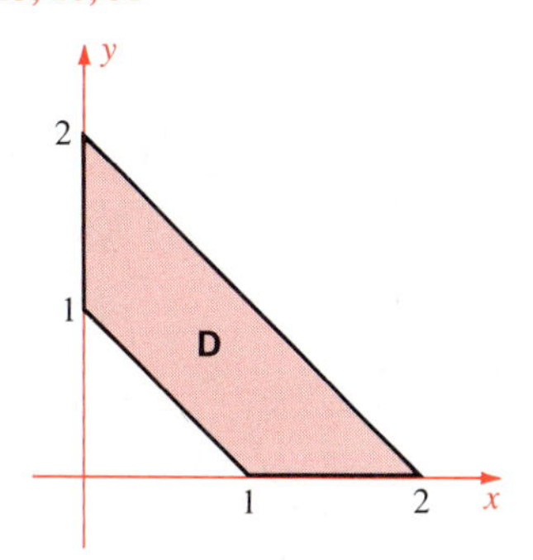

Figure 17-4-7
See Exercises 17, 18, 32

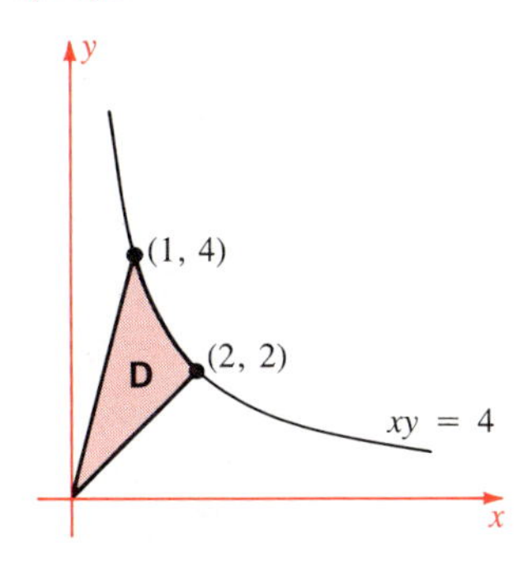

Figure 17-4-8
See Exercises 19, 20, 33

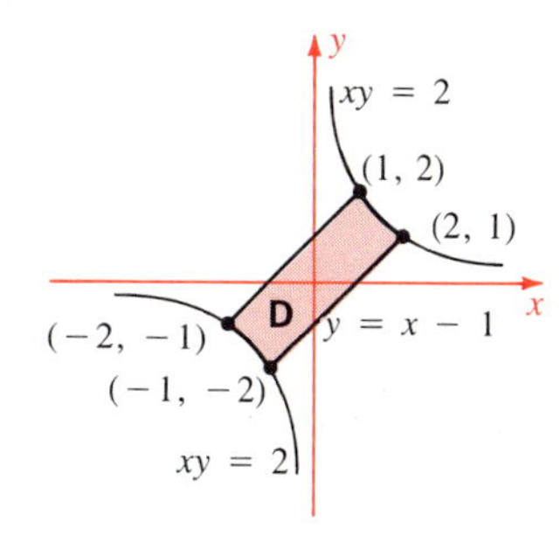

Figure 17-4-9
See Exercises 21, 22, 34

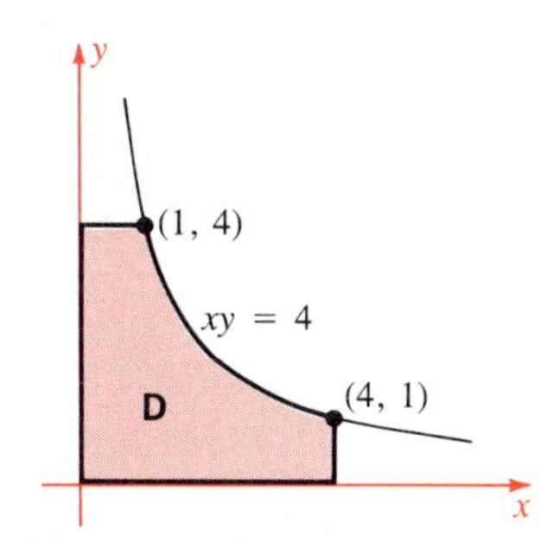

17-5 Polar Coordinates

Sometimes it is convenient to compute a double integral $\iint f(x, y)\,dx\,dy$ in polar coordinates rather than rectangular coordinates. This is frequently the case if the domain has rays $\theta = \theta_0$ and $\theta = \theta_1$ as part of its boundary. Certainly polar coordinates are more suitable for domains of the type shown in Figure 17-5-1.

Figure 17-5-1
Domains for which polar coordinates are preferred

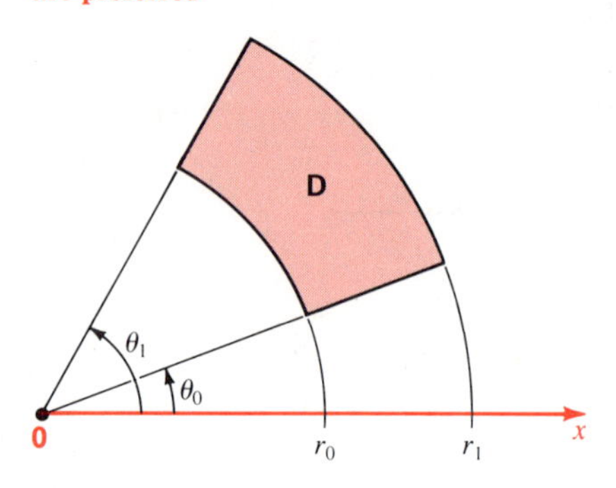

a **D** consists of all points $[r, \theta]$ such that $r_0 \leq r \leq r_1$ and $\theta_0 \leq \theta \leq \theta_1$.

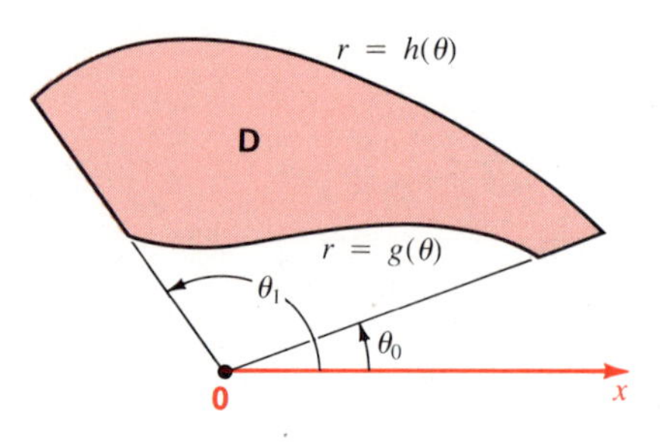

b **D** consists of all points $[r, \theta]$ such that $g(\theta) \leq r \leq h(\theta)$ and $\theta_0 \leq \theta \leq \theta_1$.

Given such a domain **D**, the problem is to express

$$\iint_{\mathbf{D}} f(x, y)\,dx\,dy$$

as an integral involving the variables r and θ. Since $x = r\cos\theta$ and $y = r\sin\theta$,

$$f(x, y) = f(r\cos\theta, r\sin\theta)$$

so there is no trouble with the integrand. But what do we do with $dx\,dy$, just replace it with $dr\,d\theta$? No, this won't do because $dx\,dy$ represents an area (with dimension *length squared*) whereas $dr\,d\theta$ has dimension *length* since the angle θ is dimensionless.

Element of Area

The "element of area" $dA = dx\,dy$ in rectangular coordinates is the area of the rectangle swept out by an increase dx in x and an increase dy in y. See Figure 17-5-2a. Now suppose in polar coordinates r increases from r to $r + dr$ and θ from θ to $\theta + d\theta$. Then a small region (Figure 17-5-2b) is swept out. It is almost a rectangle of sides dr and $r\,d\theta$ (the arc of a circle of radius r with central angle $d\theta$). A natural guess, therefore, is that the corresponding polar "element of area" is $dA = r\,dr\,d\theta$.

If our guess is right, we should have

$$\iint_{\mathbf{D}} r\,dr\,d\theta = \text{area}(\mathbf{D})$$

Figure 17-5-2
Element of area

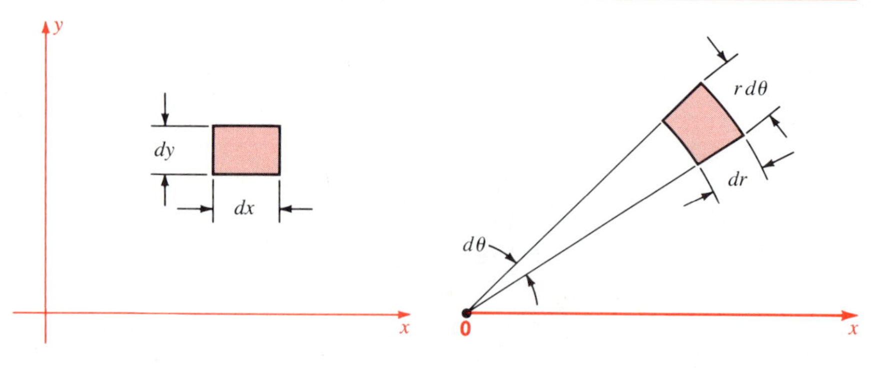

a $dA = dx\,dy$

b $dA = r\,dr\,d\theta$

Let us confirm this formula in the case that **D** is the disk $0 \le r \le a$, $0 \le \theta \le 2\pi$. The formula yields

$$\iint_{\mathbf{D}} r\,dr\,d\theta = \left(\int_0^{2\pi} d\theta\right)\left(\int_0^a r\,dr\right) = 2\pi \cdot \tfrac{1}{2}a^2 = \pi a^2$$

which is the correct area. This is further evidence that $r\,dr\,d\theta$ is the correct element of area in polar coordinates.

Element of Area (Polar Coordinates)

$$dA = r\,dr\,d\theta$$

Now we can write general double integrals in polar coordinates.

Double Integral in Polar Coordinates

$$\iint_{\mathbf{D}} f(x, y)\,dA = \iint f(r\cos\theta, r\sin\theta)\,r\,dr\,d\theta$$

Example 1

Find **a** $\displaystyle\iint_{r\le 1} x^2\,dx\,dy$ **b** $\displaystyle\iint_{\substack{r\le a\\ 0\le\theta\le\pi/2}} y\,dx\,dy$

Solution

a The domain is a full circle, so $0 \le \theta \le 2\pi$:

$$\iint_{r\le 1} x^2\,dx\,dy = \iint_{r\le 1} (r\cos\theta)^2(r\,dr\,d\theta) = \iint_{r\le 1} r^3\cos^2\theta\,dr\,d\theta$$

$$= \left(\int_0^{2\pi} \cos^2\theta\,d\theta\right)\left(\int_0^1 r^3\,dr\right) = \tfrac{1}{4}\pi$$

b The domain is a quarter-circle of radius a:

$$\iint_{\mathbf{D}} y\,dx\,dy = \iint_{\mathbf{D}} r\sin\theta\,r\,dr\,d\theta$$

$$= \left(\int_0^{\pi/2} \sin\theta\,d\theta\right)\left(\int_0^a r^2\,dr\right) = \tfrac{1}{3}a^3$$

Example 2

Find the volume between the cone $z = 3\sqrt{x^2 + y^2}$ and the circle $x^2 + y^2 \le a^2$.

Solution Use polar coordinates. The integrand is $z = 3r$ and the domain is $r \le a$. Hence

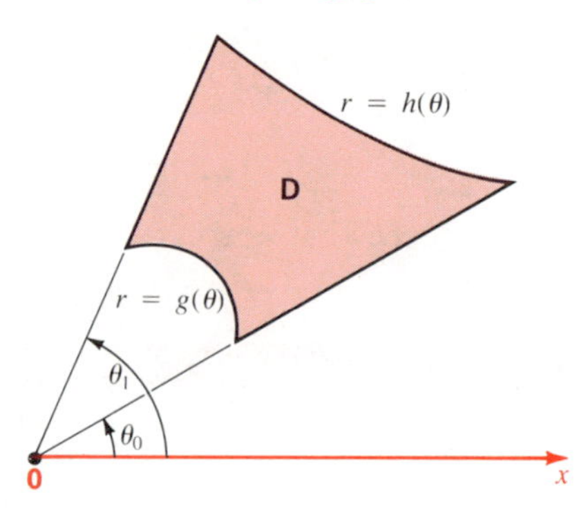

Figure 17-5-3
Domain between polar graphs

$$V = \iint z\,dx\,dy = \iint 3r\,r\,dr\,d\theta = 3\iint r^2\,dr\,d\theta$$

$$= 3\left(\int_0^{2\pi} d\theta\right)\left(\int_0^a r^2\,dr\right) = 3(2\pi)(\tfrac{1}{3}a^3) = 2\pi a^3$$

Now we look at some examples where the domain is like that of Figure 17-5-3. Assume the integrand is already expressed as a function of r and θ. Then the set-up is

$$\iint_{\mathbf{D}} F(r,\theta)\,r\,dr\,d\theta = \int_{\theta_0}^{\theta_1}\left(\int_{g(\theta)}^{h(\theta)} r\,F(r,\theta)\,dr\right)d\theta$$

Example 3

Find $\iint_{\mathbf{D}} \cos^2\theta\,dx\,dy$ over the domain $\mathbf{D}$ of Figure 17-5-4.

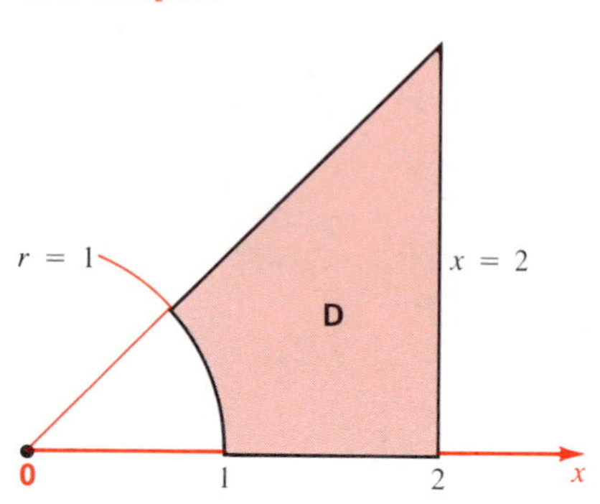

Figure 17-5-4
See Example 3

Solution The domain is bounded by $\theta = 0$, $\theta = \frac{1}{4}\pi$, $r = 1$, and $x = 2$. Since $x = r\cos\theta$, the line $x = 2$ has the polar equation $r\cos\theta = 2$, that is, $r = 2\sec\theta$. Hence

$$\mathbf{D} = \{[r,\theta] \mid \text{where} \quad 1 \le r \le 2\sec\theta \quad \text{and} \quad 0 \le \theta \le \tfrac{1}{4}\pi\}$$

Therefore

$$\iint_{\mathbf{D}} \cos^2\theta\,dx\,dy = \iint_{\mathbf{D}} \cos^2\theta\,r\,dr\,d\theta$$

$$= \int_0^{\pi/4} \cos^2\theta\,d\theta\left(\int_1^{2\sec\theta} r\,dr\right)$$

$$= \int_0^{\pi/4} \cos^2\theta(2\sec^2\theta - \tfrac{1}{2})\,d\theta$$

$$= \int_0^{\pi/4} (2 - \tfrac{1}{2}\cos^2\theta)\,d\theta$$

$$= \tfrac{1}{2}\pi - (\tfrac{1}{16}\pi + \tfrac{1}{8}) = \tfrac{7}{16}\pi - \tfrac{1}{8}$$

Example 4

Compute $\iint_{\mathbf{D}} xy\,dx\,dy$ where $\mathbf{D}$ is the domain bounded by $r = \sin 2\theta$ for $0 \le \theta \le \frac{1}{2}\pi$.

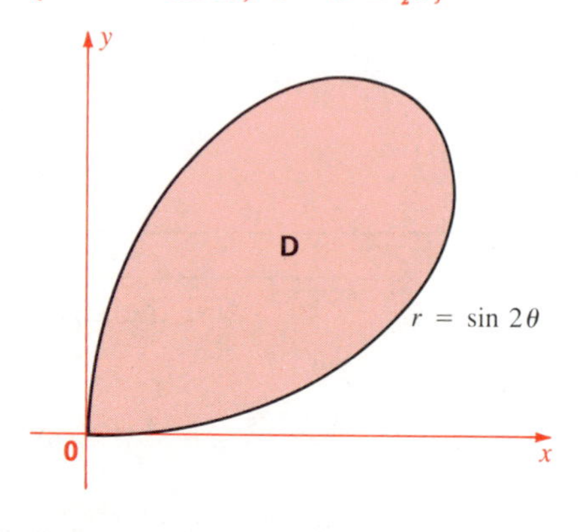

Figure 17-5-5
Domain $\mathbf{D} = \{0 \le r \le \sin 2\theta,\ 0 \le \theta \le \frac{1}{2}\pi\}$

Solution The domain is bounded by one petal of a four-petal rose curve (Figure 17-5-5). We have

$$\mathbf{D} = \{[r,\theta] \mid \text{where} \quad 0 \le r \le \sin 2\theta \quad \text{and} \quad 0 \le \theta \le \tfrac{1}{2}\pi\}$$

Therefore

$$\iint_{\mathbf{D}} xy\,dx\,dy = \iint_{\mathbf{D}} (r\cos\theta)(r\sin\theta)(r\,dr\,d\theta)$$

$$= \iint_{\mathbf{D}} r^3\cdot\tfrac{1}{2}\sin 2\theta\,dr\,d\theta = \tfrac{1}{2}\int_0^{\pi/2}(\sin 2\theta)\left(\int_0^{\sin 2\theta} r^3\,dr\right)d\theta$$

$$= \tfrac{1}{2}\int_0^{\pi/2}(\sin 2\theta)(\tfrac{1}{4}\sin^4 2\theta)\,d\theta = \tfrac{1}{8}\int_0^{\pi/2}\sin^5 2\theta\,d\theta$$

$$= \tfrac{1}{16}\int_0^{\pi}\sin^5\alpha\,d\alpha = \tfrac{1}{8}\int_0^{\pi/2}\sin^5\alpha\,d\alpha$$

$$= \tfrac{1}{8}\cdot\frac{2\cdot 4}{1\cdot 3\cdot 5} = \tfrac{1}{15} \qquad \text{(by tables)}$$

●

Example 5

Two solid right-circular cylinders of radius 1 intersect at a right angle on center. Find their common volume.

Solution Choose coordinates (Figure 17-5-6a) so the solid cylinders are

$$x^2 + y^2 \le 1 \qquad \text{and} \qquad x^2 + z^2 \le 1$$

Their intersection is the solid $\mathbf{S}$ consisting of all points (x, y, z) that satisfy both inequalities. From the first inequality, $\mathbf{S}$ lies above and below the circle $r \le 1$ in the x, y-plane. From the second, $z^2 \le 1 - x^2$ so that $\mathbf{S}$ lies between the graphs

$$z = +\sqrt{1 - x^2} \qquad \text{and} \qquad z = -\sqrt{1 - x^2}$$

Set up a double integral in polar coordinates for the volume V. By symmetry, V equals 8 times the volume of the part in the first octant. Hence use the domain

$$\mathbf{D} = \{(r, \theta] \mid \text{where } 0 \le r \le 1 \text{ and } 0 \le \theta \le \tfrac{1}{2}\pi\}$$

The volume is

$$V = 8\iint_{\mathbf{D}}\sqrt{1 - x^2}\,dx\,dy = 8\iint_{\mathbf{D}}\sqrt{1 - r^2\cos^2\theta}\;r\,dr\,d\theta$$

$$= 8\int_0^{\pi/2}\left[\int_0^1 r\sqrt{1 - r^2\cos^2\theta}\,dr\right]d\theta$$

$$= 8\int_0^{\pi/2}\left[\left(\frac{-1}{3\cos^2\theta}\right)(1 - r^2\cos^2\theta)^{3/2}\Bigg|_{r=0}^{r=1}\right]d\theta$$

$$= \tfrac{8}{3}\int_0^{\pi/2}\frac{1 - \sin^3\theta}{\cos^2\theta}\,d\theta$$

Figure 17-5-6
See Example 5

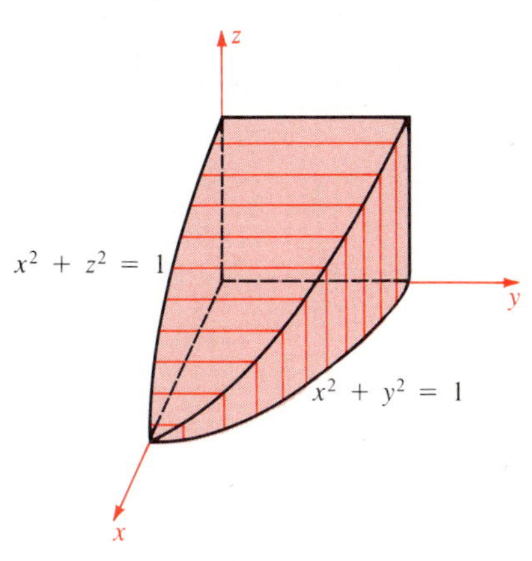

a First octant portion of the intersection

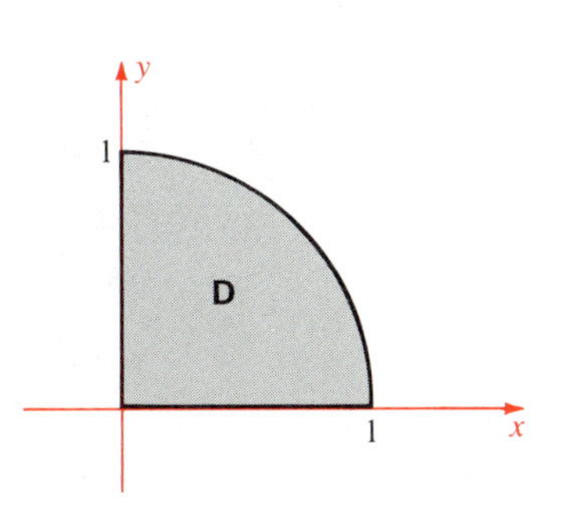

b The domain of integration

Now

$$\frac{1-\sin^3\theta}{\cos^2\theta}=\frac{1-\sin^3\theta}{1-\sin^2\theta}=\frac{1+\sin\theta+\sin^2\theta}{1+\sin\theta}$$

$$=\sin\theta+\frac{1}{1+\sin\theta}$$

From tables,

$$\int\frac{d\theta}{1+\sin\theta}=-\tan\left(\tfrac{1}{4}\pi-\tfrac{1}{2}\theta\right)+C$$

Hence

$$V=\tfrac{8}{3}\left[-\cos\theta-\tan\left(\tfrac{1}{4}\pi-\tfrac{1}{2}\theta\right)\right]\Big|_0^{\pi/2}=\tfrac{8}{3}(1+1)=\tfrac{16}{3}$$

●

Example 6

A cylindrical hole of radius a is bored through a sphere of radius $2a$. The surface of the hole passes through the center of the sphere. How much material is removed?

Solution Choose coordinates (Figure 17-5-7a) so **0** is the center of the sphere, the axis of the cylinder is parallel to the z-axis, and the cylinder intersects the x, y-plane in a circular disk **D** centered on the positive x-axis. By Figure 17-5-7b, the disk **D** of the cylinder is described by

$$\mathbf{D}=\{[r,\theta]\mid\text{where}\quad 0\le r\le 2a\cos\theta\quad\text{and}\quad -\tfrac{1}{2}\pi\le\theta\le\tfrac{1}{2}\pi\}$$

The upper surface of the hole is

$$z=\sqrt{(2a)^2-x^2-y^2}=\sqrt{4a^2-r^2}$$

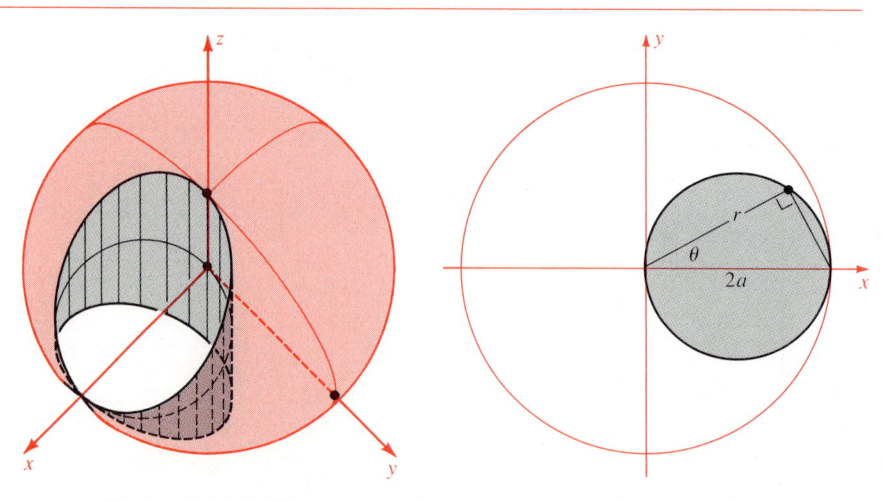

Figure 17-5-7
Cylindrical hole through sphere. See Example 6

a The choice of coordinates

b The domain; on its boundary, $r=2a\cos\theta$

and similarly the lower surface is $z = -\sqrt{4a^2 - r^2}$. Therefore the volume of the hole is

$$V = \iint_D 2\sqrt{4a^2 - r^2}\, r\, dr\, d\theta$$
$$= 2\int_{-\pi/2}^{\pi/2} d\theta \int_0^{2a\cos\theta} r\sqrt{4a^2 - r^2}\, dr$$
$$= 4\int_0^{\pi/2} d\theta \int_0^{2a\cos\theta} r\sqrt{4a^2 - r^2}\, dr$$
$$= \tfrac{4}{3}\int_0^{\pi/2} - (4a^2 - r^2)^{3/2}\Big|_0^{2a\cos\theta} d\theta$$
$$= \tfrac{4}{3}\int_0^{\pi/2} [(4a^2)^{3/2} - (4a^2\sin^2\theta)^{3/2}]\, d\theta$$
$$= \frac{32a^3}{3}\int_0^{\pi/2} (1 - \sin^3\theta)\, d\theta$$
$$= \frac{32a^3}{3}\left(\frac{\pi}{2} - \frac{2}{3}\right) = \frac{16\pi a^3}{3} - \frac{64a^3}{9}$$

(See definite integral 5 inside the front cover.) ●

Remark An instructive mistake is possible in this example. By symmetry we equated the integral from $-\frac{1}{2}\pi$ to $\frac{1}{2}\pi$ with twice the integral from 0 to $\frac{1}{2}\pi$. Suppose we had not done this. Then we would have reached

$$V = \tfrac{2}{3}\int_{-\pi/2}^{\pi/2} [(4a^2)^{3/2} - (4a^2\sin^2\theta)^{3/2}]\, d\theta$$
$$= \frac{16a^3}{3}\int_{-\pi/2}^{\pi/2} (1 - \sin^3\theta)\, d\theta = \frac{16a^3}{3}\int_{-\pi/2}^{\pi/2} d\theta = \frac{16\pi a^3}{3}$$

since $\sin^3\theta$ is an odd function. This is certainly a different answer! Where is the goof?

It is a subtle application of the false argument

$$3 = \sqrt{3^2} = \sqrt{(-3)^2} = [(-3)^2]^{1/2} = -3$$

The point is that $\sin\theta$ is *negative* in the fourth quadrant; consequently for $-\frac{1}{2}\pi < \theta < 0$,

$$(\sin^2\theta)^{3/2} = -\sin^3\theta \qquad \text{not} \qquad \sin^3\theta$$

Therefore, the correct argument is

$$V = \frac{16a^3}{3}\int_{-\pi/2}^{\pi/2} [1 - (\sin^2\theta)^{3/2}]\, d\theta$$
$$= \frac{16a^3}{3}\left[\int_{-\pi/2}^{0} (1 + \sin^3\theta)\, d\theta + \int_0^{\pi/2} (1 - \sin^3\theta)\, d\theta\right]$$

Avoid this blunder!

Figure 17-5-8
Correspondence between domains **D** and **E**

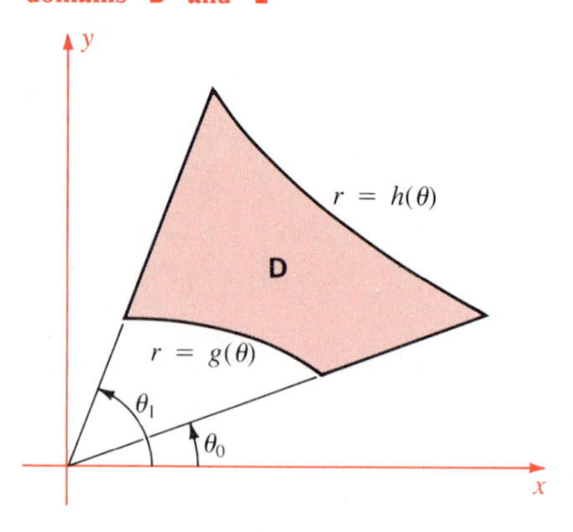

a **D** = {[r, θ], where $g(\theta) \le r \le h(\theta)$, $\theta_0 \le \theta \le \theta_1$}
D is in the x, y-plane.

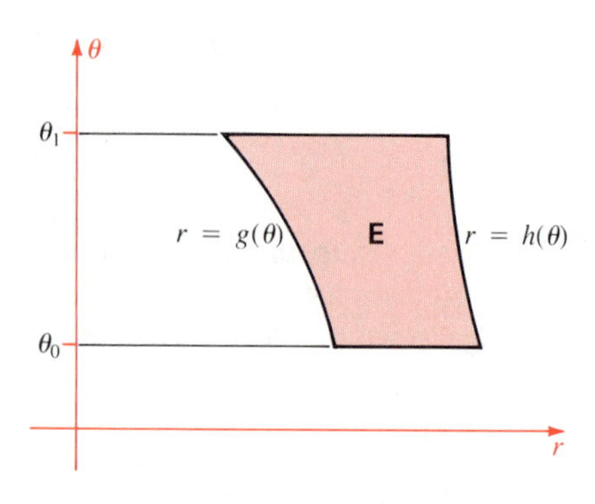

b **E** = {(r, θ), where $g(\theta) \le r \le h(\theta)$, $\theta_0 \le \theta \le \theta_1$}
E is in the r, θ-plane.

Change of Variables

The polar-coordinate transformation

$$x = r\cos\theta \qquad y = r\sin\theta$$

can be interpreted as a correspondence between domains. To a domain **D** in the x, y-plane corresponds another domain **E** in the (rectangular) r, θ-plane. For instance, the domain (Figure 17-5-8a)

$$\mathbf{D} = \{[r, \theta] \mid \text{where} \quad g(\theta) \le r \le h(\theta) \quad \text{and} \quad \theta_0 \le \theta \le \theta_1\}$$

in the x, y-plane corresponds to the domain (Figure 17-5-8b)

$$\mathbf{E} = \{(r, \theta) \mid \text{where} \quad g(\theta) \le r \le h(\theta) \quad \text{and} \quad \theta_0 \le \theta \le \theta_1\}$$

in the r, θ-plane.

The formula for integrating in polar coordinates can be expressed as

$$\iint_{\mathbf{D}} f(x, y)\, dx\, dy = \iint_{\mathbf{E}} f(r\cos\theta, r\sin\theta)\, r\, dr\, d\theta$$

Thus you may substitute $x = r\cos\theta$ and $y = r\sin\theta$ in the integral on the left, but then you must replace $dx\,dy$ by $r\,dr\,d\theta$ and **D** by the corresponding domain **E** in the r, θ-plane. (The latter corresponds to changing limits in a simple integral.)

The formula is a special case of a very general formula for changing variables in double integrals. Its proof is beyond the scope of this course, but we shall state the result. First we need a definition.

Jacobian

Suppose

$$\begin{cases} x = x(u, v) \\ y = y(u, v) \end{cases}$$

is a pair of differentiable functions of two variables. Their **Jacobian** is the determinant

$$\frac{\partial(x, y)}{\partial(u, v)} = \begin{vmatrix} x_u & x_v \\ y_u & y_v \end{vmatrix} = x_u y_v - x_v y_u$$

(The expression on the left is the usual notation.) For example, the change to polar coordinates

$$x = r\cos\theta \qquad y = r\sin\theta$$

has the Jacobian

$$\frac{\partial(x, y)}{\partial(r, \theta)} = \begin{vmatrix} x_r & x_\theta \\ y_r & y_\theta \end{vmatrix} = \begin{vmatrix} \cos\theta & -r\sin\theta \\ \sin\theta & r\cos\theta \end{vmatrix}$$
$$= r\cos^2\theta + r\sin^2\theta = r$$

Now we can state the rule for changing variables.

Change of Variables

Suppose that

$$\begin{cases} x = x(u, v) \\ y = y(u, v) \end{cases}$$

maps a domain **E** of the u, v-plane in a one-to-one manner onto a domain **D** of the x, y-plane as shown in Figure 17-5-9. Suppose also that

$$\frac{\partial(x, y)}{\partial(u, v)} > 0$$

on **E**. Then

$$\iint_{\mathbf{D}} f(x, y)\, dx\, dy = \iint_{\mathbf{E}} f[x(u, v), y(u, v)] \frac{\partial(x, y)}{\partial(u, v)}\, du\, dv$$

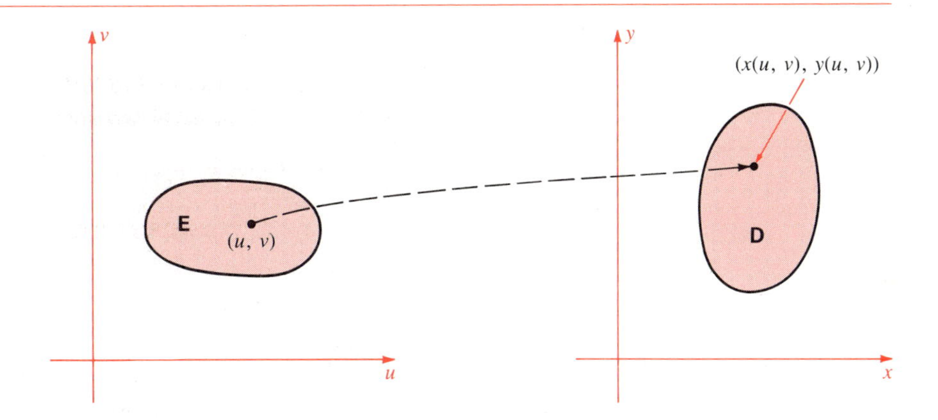

Figure 17-5-9
Transformation on a domain **E** in the u, v-plane onto a domain **D** in the x, y-plane

The formula is usually valid even if the Jacobian equals zero at some points of the boundary of **E**. Note that the formula is analogous to the formula

$$\int_{x(a)}^{x(b)} f(x)\, dx = \int_a^b f[x(u)] \frac{dx}{du}\, du$$

for a change of variable in a simple integral.

Example 7

Find the area enclosed by the ellipse

$$\frac{x^2}{a^2} + \frac{y^2}{b^2} = 1 \qquad (a > 0, b > 0)$$

Solution An ellipse is an elongated circle. The simple transformation

$$x = au \qquad y = bv$$

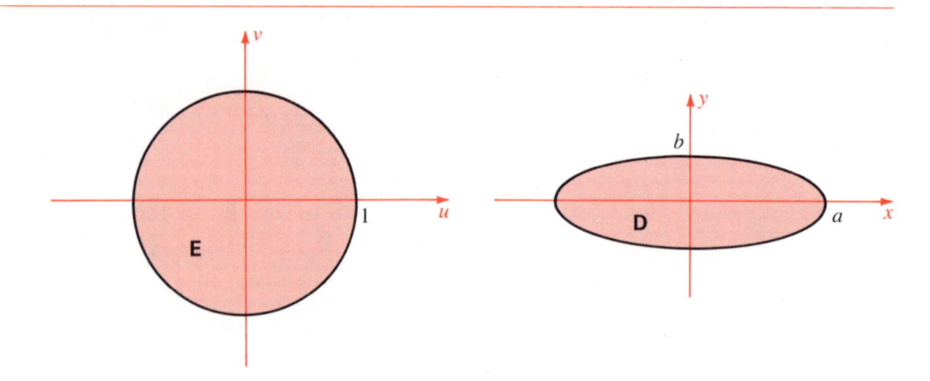

Figure 17-5-10
$\left.\begin{array}{l} x = au \\ y = bv \end{array}\right\}$ circle **E** → ellipse **D**

which scales each variable by a factor, takes the circle **E**: $u^2 + v^2 \le 1$ onto the domain **D** enclosed by the ellipse (Figure 17-5-10). The Jacobian is

$$\frac{\partial(x, y)}{\partial(u, v)} = \begin{vmatrix} a & 0 \\ 0 & b \end{vmatrix} = ab > 0$$

Therefore

$$\text{area}(\mathbf{D}) = \iint_{\mathbf{D}} dx\, dy = \iint_{\mathbf{E}} ab\, du\, dv = ab \iint_{\mathbf{E}} du\, dv$$

$$= ab \cdot \text{area}(\mathbf{E}) = \pi ab$$

●

Exercises

Evaluate $\iint f(x, y)\, dx\, dy$ over the disk $x^2 + y^2 \le a^2$

1 $f = xy^2$

2 $f = \exp(x^2 + y^2)$

3 $f = \ln(a^2 + x^2 + y^2)$

4 $f = \dfrac{1}{a + \sqrt{x^2 + y^2}}$

5 $f = x^4$

6 $f = x^2y^2$

7 $f = \dfrac{1}{a^2 + x^2 + y^2}$

8 $f = \sin(xy)$

Use polar coordinates to compute the volume of

9 $0 \le z \le r^2 \quad 1 \le r \le 2 \quad 0 \le \theta \le \frac{1}{2}\pi$

10 $0 \le z \le x \quad 0 \le r \le 1 \quad -\frac{1}{2}\pi \le \theta \le \frac{1}{2}\pi$

11 $0 \le z \le xy \quad 1 \le r \le 2 \quad \frac{1}{4}\pi \le \theta \le \frac{1}{2}\pi$

12 $0 \le z \le r^6 \quad 0 \le r \le 1 \quad 0 \le \theta \le \frac{1}{2}\pi$

13 the hemisphere of radius a

14 the region bounded by the two paraboloids $z = x^2 + y^2$ and $z = 4 - 3(x^2 + y^2)$

15 the lens-shaped region common to the sphere of radius 1 centered at $(0, 0, 0)$ and the sphere of radius 1 centered at $(0, 0, 1)$

16 the material removed when a drill of radius b bores on center through a sphere of radius a, where $b < a$

17 the region under the cone $z = 5r$ and over one petal of the rose $r = \sin 3\theta$

18 the region under the paraboloid $z = x^2 + y^2$ and over the circle $x^2 + y^2 = 2x$

19 $0 \le z \le x^4y^4 \quad x^2 + y^2 \le 1$

20 $0 \le z \le r^3 \quad 0 \le r \le \theta \quad 0 \le \theta \le \pi$

21 $0 \le z \le x^2y^2 \quad x^2 + y^2 \le 1$

***22** the region common to three right-circular cylinders of radius 1 intersecting at mutual right angles on center

Compute the Jacobian $\dfrac{\partial(x, y)}{\partial(u, v)}$

23 $x = Au + Bv \quad y = Cu + Dv$

24 $x = u + v \quad y = uv$

25 $x = u^2 + v^2 \quad y = u - v$

26 $x = ue^v \quad y = u^2 - v^2$

27 Assume $0 < a < b$. Show that $x = u - v$, $y = v$ takes the domain

$$\mathbf{E} = \{(u, v) \mid a \le u \le b \text{ and } 0 \le v \le u\}$$

onto

$$\mathbf{D} = \{(x, y) \mid x \ge 0, \; y \ge 0, \text{ and } a \le x + y \le b\}$$

and compute the Jacobian. (See Figure 17-5-11.)

Figure 17-5-11
See Exercises 27–30

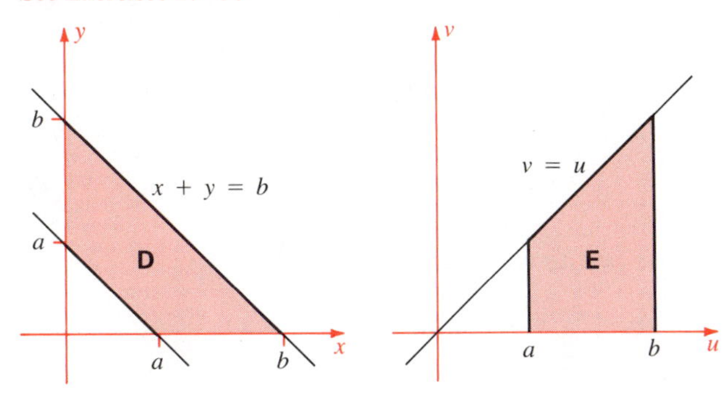

28 (cont.) Compute $\iint xy \, dx \, dy$ over **D** by changing variables.

29 (cont.) Express $\iint f(x + y) \, dx \, dy$ over **D** as a simple integral.

30 (cont.) Evaluate $\iint \exp(x + y)^2 \, dx \, dy$ over **D**.

31 Compute the Jacobian of $x = u/v$, $y = v$.

32 (cont.) Suppose $0 < a < b$ and let

$$\mathbf{D} = \{(x, y) \mid ax \le by, \; ay \le bx, \text{ and } xy \le ab\}$$

To what domain **E** in the u, v-plane does **D** correspond? (See Figure 17-5-12.)

33 (cont.) Use the change of variables in Exercise 31 to compute $\iint \sqrt{xy^3} \, dx \, dy$ over **D**.

34 Use the transformation of Example 7 and polar coordinates to evaluate $\iint x^2 y \, dx \, dy$ over the domain $x^2/a^2 + y^2/b^2 \le 1$, $y \ge 0$.

Figure 17-5-12
See Exercises 31–33

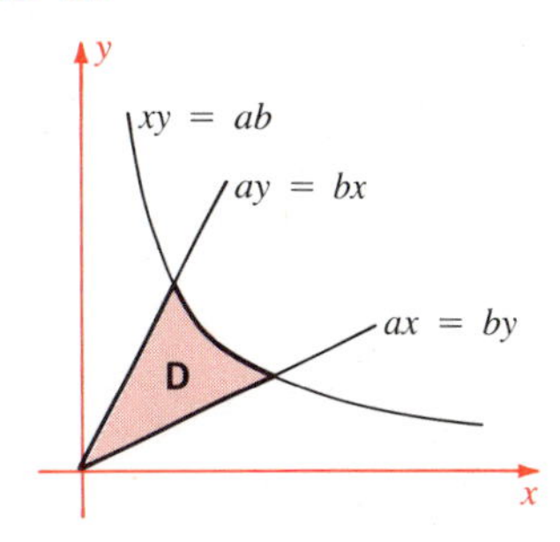

35 Let $0 < a < b$ and consider the domain

$$\mathbf{D} = \{(x, y) \mid 0 \le y - x \le b - a \text{ and } xy \le ab\}$$

To what domain **E** in the u, v-plane does **D** correspond under the coordinate transformation $x = \frac{1}{2}(u - v)$, $y = \frac{1}{2}(u + v)$? (See the part of Figure 17-5-13 where $y \ge x$.)

36 (cont.) Use the transformation to compute $\iint (y - x) \, dx \, dy$ over **D**.

Figure 17-5-13
See Exercises 35–36

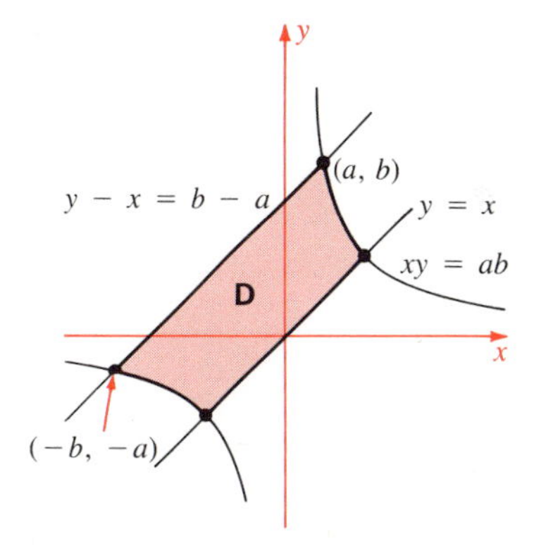

17-6 Applications

In this section we take up several important applications of double integrals. The first concerns differentiation under the integral sign. Consider a function defined by a definite integral

$$F(t) = \int_a^b f(x, t) \, dx$$

When the variable x is "integrated out," there remains a function of t. Problem: find the derivative $F'(t)$. The answer is called the Leibniz rule, or the rule for differentiating under the integral sign.

Leibniz Rule

Suppose $f(x, t)$ and the partial derivative $f_t(x, t)$ are continuous on a rectangle

$$a \le x \le b \qquad c \le t \le d$$

Then

$$\frac{d}{dt}\int_a^b f(x, t)\, dx = \int_a^b f_t(x, t)\, dx$$

for $c \le t \le d$.

Proof Fix t and let $\mathbf{D}$ be the rectangle

$$\mathbf{D} = \{(x, s) \mid \text{where } a \le x \le b \text{ and } c \le s \le t\}$$

in the x, s-plane. Let

$$G(t) = \iint_{\mathbf{D}} f_s(x, s)\, dx\, ds$$

The idea is to iterate the double integral both ways, then to compute $G'(t)$. On the one hand, by the fundamental theorem of calculus (note that x is fixed)

$$\int_c^t f_s(x, s)\, ds = f(x, t) - f(x, c)$$

Consequently

$$G(t) = \int_a^b \left(\int_c^t f_s(x, s)\, ds\right) dx$$
$$= \int_a^b [f(x, t) - f(x, c)]\, dx$$

so that

$$\frac{d}{dt} G(t) = \frac{d}{dt}\int_a^b f(x, t)\, dx - \frac{d}{dt}\int_a^b f(x, c)\, dx$$
$$= \frac{d}{dt}\int_a^b f(x, t)\, dx$$

On the other hand,

$$G(t) = \int_c^t \left(\int_a^b f_s(x, s)\, dx\right) ds$$

By the fundamental theorem of calculus again,

$$\frac{d}{dt}G(t) = \frac{d}{dt}\int_c^t \left(\int_a^b f_s(x, s)\,dx\right) ds$$
$$= \int_a^b f_t(x, t)\,dx$$

The Leibniz rule follows upon equating these expressions for $G'(t)$. ●

Example 1

Find $\displaystyle \frac{d}{dt}\int_0^\pi \frac{\sin tx}{x}\,dx$

Solution

$$\frac{d}{dt}\int_0^\pi \frac{\sin tx}{x}\,dx = \int_0^\pi \frac{\partial}{\partial t}\left(\frac{\sin tx}{x}\right) dx$$
$$= \int_0^\pi \cos tx\,dx = \frac{\sin \pi t}{t}$$ ●

Remark It is known that $F(t) = \int [(\sin tx)/x]\,dx$ cannot be expressed in terms of (a finite number of) the usual functions of calculus; you won't find it in a table of integrals, except as an infinite series. Nevertheless, $F(t)$ is a perfectly good differentiable function. But to compute its derivative, you need the Leibniz rule.

Surface Area

We are given a piece of surface in $\mathbf{R}^3$. What is its area; that is, how much paint is needed to cover one side? Suppose the surface is given in parametric form:

$$\mathbf{x} = \mathbf{x}(u, v)$$

where the point (u, v) varies over a domain $\mathbf{D}$ of the u, v-plane and, as usual,

$$\mathbf{x}_u \times \mathbf{x}_v \neq \mathbf{0}$$

Recall that the unit normal to the surface is

$$\mathbf{n} = \frac{\mathbf{x}_u \times \mathbf{x}_v}{|\mathbf{x}_u \times \mathbf{x}_v|}$$

A small rectangle in $\mathbf{D}$ with sides du and dv maps to a small region in the surface. By the definition of differential, $d\mathbf{x} = \mathbf{x}_u\,du + \mathbf{x}_v\,dv$. Hence the small surface region is closely approximated by the parallelogram (Figure 17-6-1) in the tangent plane with sides $\mathbf{x}_u\,du$ and $\mathbf{x}_v\,dv$, whose area is

$$dA = |(\mathbf{x}_u\,du) \times (\mathbf{x}_v\,dv)| = |\mathbf{x}_u \times \mathbf{x}_v|\,du\,dv$$

Figure 17-6-1
Approximation to area by a small "parallelogram"

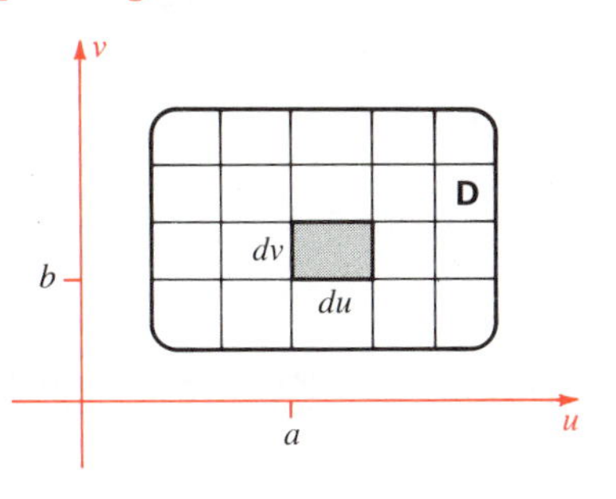

a Domain that maps into the surface

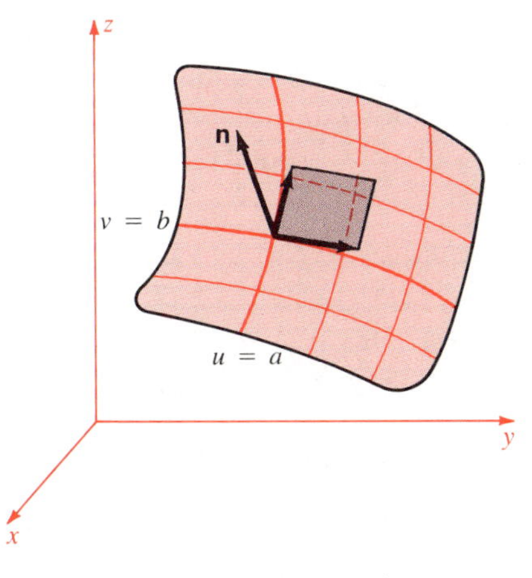

b The parallelogram has sides $\mathbf{x}_u\,du$ and $\mathbf{x}_v\,dv$

Recall that

$$\mathbf{x}_u \times \mathbf{x}_v = \begin{vmatrix} \mathbf{i} & \mathbf{j} & \mathbf{k} \\ x_u & y_u & z_u \\ x_v & y_v & z_v \end{vmatrix} = \begin{vmatrix} y_u & z_u \\ y_v & z_v \end{vmatrix} \mathbf{i} + \begin{vmatrix} z_u & x_u \\ z_v & x_v \end{vmatrix} \mathbf{j} + \begin{vmatrix} x_u & y_u \\ x_v & y_v \end{vmatrix} \mathbf{k}$$

$$= \frac{\partial(y, z)}{\partial(u, v)} \mathbf{i} + \frac{\partial(z, x)}{\partial(u, v)} \mathbf{j} + \frac{\partial(x, y)}{\partial(u, v)} \mathbf{k}$$

Consequently

$$|\mathbf{x}_u \times \mathbf{x}_v|^2 = \left(\frac{\partial(y, z)}{\partial(u, v)}\right)^2 + \left(\frac{\partial(z, x)}{\partial(u, v)}\right)^2 + \left(\frac{\partial(x, y)}{\partial(u, v)}\right)^2$$

Therefore the area of the small shaded region on the surface is approximately the **element of surface area**

$$d\sigma = |\mathbf{x}_u \times \mathbf{x}_v| \, du \, dv$$

$$= \sqrt{\left(\frac{\partial(y, z)}{\partial(u, v)}\right)^2 + \left(\frac{\partial(z, x)}{\partial(u, v)}\right)^2 + \left(\frac{\partial(x, y)}{\partial(u, v)}\right)^2} \, du \, dv$$

If we add up these elements of surface area, the sum approximates the surface area. If we divide $\mathbf{D}$ into smaller and smaller rectangles, these sums approach the double integral

$$\iint |\mathbf{x}_u \times \mathbf{x}_v| \, du \, dv$$

This discussion leads us to the following definition:

Surface Area

Let $\mathbf{x} = \mathbf{x}(u, v)$ define a parametric surface with domain $\mathbf{D}$. Its **surface area** is

$$S = \iint_{\mathbf{D}} |\mathbf{x}_u \times \mathbf{x}_v| \, du \, dv$$

$$= \iint_{\mathbf{D}} \sqrt{\left(\frac{\partial(y, z)}{\partial(u, v)}\right)^2 + \left(\frac{\partial(z, x)}{\partial(u, v)}\right)^2 + \left(\frac{\partial(x, y)}{\partial(u, v)}\right)^2} \, du \, dv$$

Figure 17-6-2
Spiral ramp. See Example 2

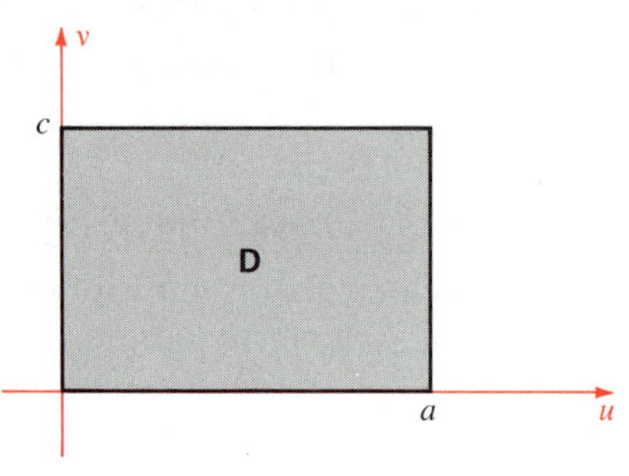

a The domain

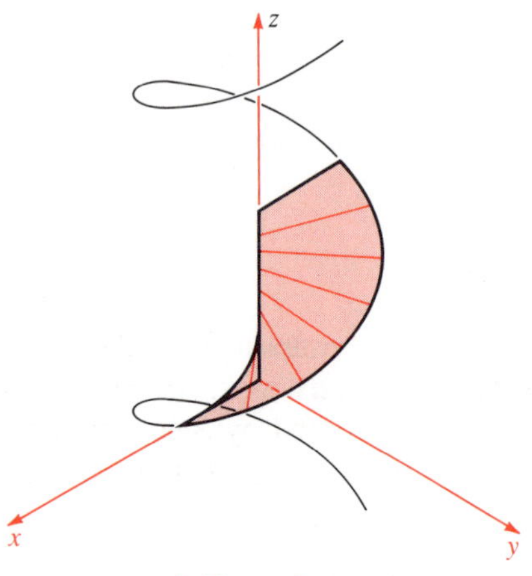

b The surface

Example 2

Find the area of the spiral ramp $\mathbf{x} = (u \cos v, u \sin v, bv)$ corresponding to the rectangle $\mathbf{D}$: $0 \le u \le a, \; 0 \le v \le c$.

Solution Although not necessary, it is nice to sketch the surface (Figure 17-6-2). Since

$$\mathbf{x}_u = (\cos v, \sin v, 0) \quad \text{and} \quad \mathbf{x}_v = (-u \sin v, u \cos v, b)$$

the element of surface area is

$$d\sigma = \sqrt{\begin{vmatrix} \sin v & 0 \\ u\cos v & b \end{vmatrix}^2 + \begin{vmatrix} 0 & \cos v \\ b & -u\sin v \end{vmatrix}^2 + \begin{vmatrix} \cos v & \sin v \\ -u\sin v & u\cos v \end{vmatrix}^2}\, du\, dv$$

$$= \sqrt{b^2\sin^2 v + b^2\cos^2 v + u^2}\, du\, dv = \sqrt{b^2 + u^2}\, du\, dv$$

As (u, v) ranges over the rectangle, the point $\mathbf{x}(u, v)$ runs over the spiral ramp. Hence

$$S = \iint_{\mathbf{D}} \sqrt{b^2 + u^2}\, du\, dv$$

$$= \left(\int_0^a \sqrt{b^2 + u^2}\, du\right)\left(\int_0^c dv\right)$$

$$= \frac{c}{2}\left[a\sqrt{a^2 + b^2} + b^2 \ln\left(\frac{a + \sqrt{a^2 + b^2}}{b}\right)\right]$$

●

Area of a Graph

Suppose a surface is given as the graph of a function $z = f(x, y)$, where (x, y) varies over a domain $\mathbf{D}$ in the x, y-plane. This is a special case of a parametric surface, where the parameters are x and y and the surface is defined by

$$\mathbf{x} = (x, y, f(x, y))$$

Now $d\sigma = |\mathbf{x}_x \times \mathbf{x}_y|\, dx\, dy$. But

$$\frac{\partial \mathbf{x}}{\partial x} = (1, 0, f_x) \qquad \text{and} \qquad \frac{\partial \mathbf{x}}{\partial y} = (0, 1, f_y)$$

so we have

$$\frac{\partial \mathbf{x}}{\partial x} \times \frac{\partial \mathbf{x}}{\partial y} = (1, 0, f_x) \times (0, 1, f_y) = (-f_x, -f_y, 1)$$

Hence the resulting formula for the element of surface area is $d\sigma = \sqrt{1 + f_x^2 + f_y^2}\, dx\, dy$.

Figure 17-6-3
$(\cos\gamma)\, dA = dx\, dy$

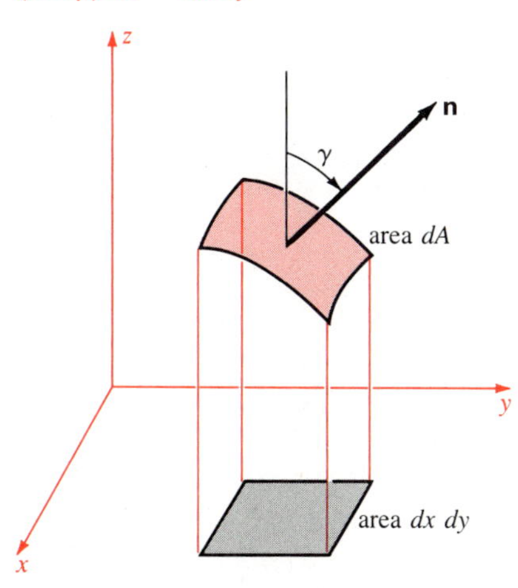

Area of a Graph

The area of the graph of a continuously differentiable function $z = f(x, y)$ with domain $\mathbf{D}$ is

$$S = \iint_{\mathbf{D}} \sqrt{1 + f_x^2 + f_y^2}\, dx\, dy$$

●

The formula for $d\sigma$ has a geometric interpretation (Figure 17-6-3). The unit normal to the surface is

$$\mathbf{n} = \frac{1}{\sqrt{1 + f_x^2 + f_y^2}}(-f_x, -f_y, 1)$$

Its third component (direction cosine) is

$$\cos\gamma = \frac{1}{\sqrt{1 + f_x^2 + f_y^2}}$$

where γ is the angle between the normal and the z-axis. Thus

$$(\cos\gamma)\,d\sigma = dx\,dy$$

which means that the element of surface area $d\sigma$ projects directly down onto a small portion of the x, y-plane of area $dx\,dy$.

Example 3

Find the area of the graph of $z = 2 - x^2 + y^2$ over the circle $x^2 + y^2 \le a^2$.

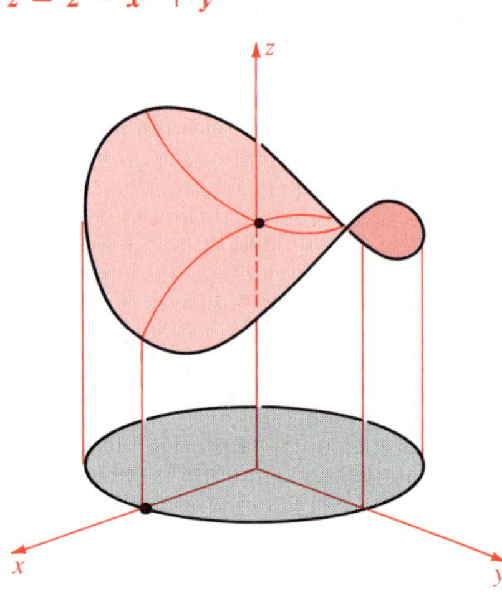

Figure 17-6-4
$z = 2 - x^2 + y^2$

Solution The saddle-shaped surface is part of a hyperbolic paraboloid (Figure 17-6-4). We have

$$S = \iint_D \sqrt{1 + z_x^2 + z_y^2}\,dx\,dy$$

$$= \iint_D \sqrt{1 + (-2x)^2 + (2y)^2}\,dx\,dy$$

$$= \iint_D \sqrt{1 + 4x^2 + 4y^2}\,dx\,dy$$

It pays to use polar coordinates here. Then $x^2 + y^2 = r^2$ and $dx\,dy = r\,dr\,d\theta$, so

$$S = \iint_D \sqrt{1 + 4r^2}\,r\,dr\,d\theta$$

$$= \int_0^{2\pi} d\theta \int_0^a r\sqrt{1 + 4r^2}\,dr$$

$$= (2\pi)\cdot\tfrac{1}{12}[(1 + 4a^2)^{3/2} - 1]$$

$$= \tfrac{1}{6}\pi[(1 + 4a^2)^{3/2} - 1]$$

●

A Probability Integral

The improper integral

$$I = \int_{-\infty}^{\infty} e^{-x^2}\,dx$$

is important in probability. Its exact value can be found by using polar coordinates and a clever trick. Write

$$I = \int_{-\infty}^{\infty} e^{-y^2}\,dy$$

The trick is to write I^2 as a double integral:

$$I^2 = \left(\int_{-\infty}^{\infty} e^{-x^2}\,dx\right)\left(\int_{-\infty}^{\infty} e^{-y^2}\,dy\right) = \iint_{\mathbf{R}^2} e^{-(x^2+y^2)}\,dx\,dy$$

Here the plane $\mathbf{R}^2$ can be thought of as the limit of expanding squares centered at the origin. But it also can be thought of as the limit of expanding circles. This suggests that we switch from rectangular to polar coordinates. Then $x^2 + y^2 = r^2$ and $dx\,dy = r\,dr\,d\theta$. Hence

$$I^2 = \int_0^{2\pi}\left(\int_0^{\infty} e^{-r^2}\,r\,dr\right)d\theta = 2\pi(-\tfrac{1}{2}e^{-r^2})\Big|_0^{\infty} = 2\pi(\tfrac{1}{2}) = \pi$$

Since I is positive,

$$\int_{-\infty}^{\infty} e^{-x^2}\,dx = \sqrt{\pi}$$

The function

$$\phi(x) = \frac{1}{\sqrt{2\pi}}\,e^{-x^2/2}$$

is known as the density function of the **normal distribution.** Its graph is the familiar bell-shaped curve (Figure 17-6-5) and encloses area 1 (by a simple modification of the definite integral above).

Remark 1 Expressing an antiderivative of e^{-x^2} in terms of elementary functions is known to be impossible. Thus it is quite remarkable that the integral of this function over the whole x-axis can be evaluated exactly.

Remark 2 The derivation above is not quite complete because we have not discussed improper double integrals, a subject we leave for a later course. A basic result of this subject is that for non-negative integrands, any method that gets an answer yields the correct answer. You may take this for granted in the exercises.

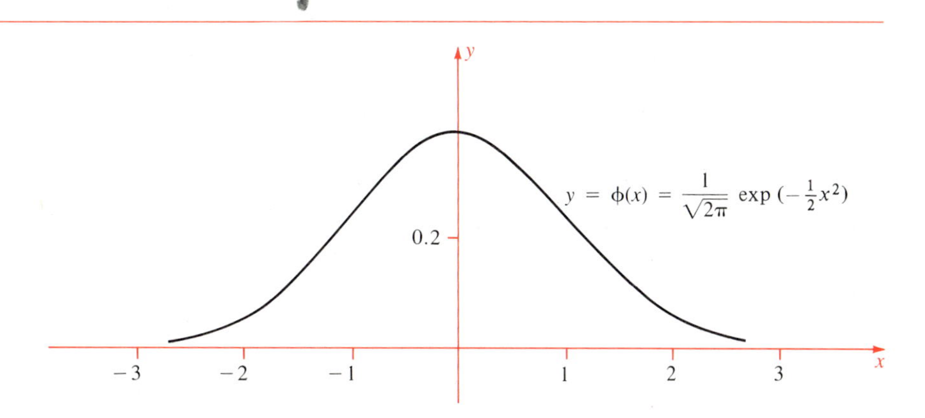

Figure 17-6-5
Normal distribution

Exercises

Evaluate $F(t)$ for $t > 0$ and then compute $F'(t)$. Now compute $F'(t)$ by the Leibniz rule and compare your answers

1 $F(t) = \int_0^1 e^{-tx}\,dx$

2 $F(t) = \int_0^1 \left(\arc\tan \dfrac{x}{t}\right) dx \qquad t > \dfrac{2}{\pi}$

3 $F(t) = \int_0^1 (t + x)^n\,dx$

4 $F(t) = \int_0^1 x^t\,dx, \quad t > -1$

Parametrize the surface and set up an integral for its surface area; evaluate the integral if you can

5 triangle with vertices $\mathbf{a}$, $\mathbf{b}$, $\mathbf{c}$

6 (cont.) triangle with vertices $(a, 0, 0)$, $(0, b, 0)$, $(0, 0, c)$

7 lateral surface of a right circular cylinder of radius a and height h

8 lateral surface of a right circular cone of radius a and lateral height L

9 right circular torus obtained by revolving a circle of radius a about an axis in its plane at distance b from its center (assume that $a < b$)

10 sphere $|\mathbf{x}| = a$
[Hint Write $\mathbf{x} = a(\sin u \cos v, \sin u \sin v, \cos u)$.]

***11** ellipsoid $\dfrac{x^2}{a^2} + \dfrac{y^2}{b^2} + \dfrac{z^2}{c^2} = 1$
[Hint Write
$\mathbf{x} = (a \sin u \cos v, b \sin u \sin v, c \cos u)$.]

***12** (cont.) Reduce the double integral to a simple integral in the special case $a = b$.

Set up an integral for the area of the given graph, and evaluate it if you can

13 $z = ax + by \qquad (x, y)$ in a domain $\mathbf{D}$

14 $z = x^2 + y^2 \qquad x^2 + y^2 \le 1$

15 $z = \sqrt{1 - x^2 - y^2} \qquad x^2 + y^2 \le 1$

16 $bz = xy \qquad x^2 + y^2 \le a^2$
where $a > 0$ and $b > 0$

17 Find the area of the top piece of a spherical surface that is inside the hole in Figure 17-5-7.

18 (cont.) Find the lateral surface area of the hole.

***19** A curve in the part $x > 0$ of the z, x-plane is given parametrically: $z = z(s)$, $x = x(s)$, where s is arc length and $a \le s \le b$. The curve is rotated about the z-axis, generating a surface. Express the area of this surface as a simple integral.

***20** Suppose a graph in $\mathbf{R}^3$ is given in the form $z = g(r, \theta)$, where (r, θ) varies over a domain $\mathbf{E}$ in the r, θ-plane. Express the area of the graph as an integral over $\mathbf{E}$.

Evaluate

21 $\displaystyle\lim_{a \to 0+} \iint_{a \le r \le 1} \ln r\,dx\,dy$

22 $\displaystyle\lim_{a \to 0+} \iint_{a \le r \le 1} (1 - r)^p\,dx\,dy \qquad (p > -1)$

23 $\displaystyle\lim_{a \to +\infty} \iint_{r \le a} \frac{dx\,dy}{1 + x^2 + y^2}$

24 $\displaystyle\lim_{a \to +\infty} \iint_{r \le a} \frac{dx\,dy}{(1 + x^2 + y^2)^p} \qquad (p > 1)$

25 $\displaystyle\lim_{a \to 0+} \iint_{a \le r \le 1} \frac{dx\,dy}{r^{2p}} \qquad (p < 1)$

26 $\displaystyle\lim_{a \to +\infty} \iint_{1 \le r \le a} \frac{dx\,dy}{r^{2p}} \qquad (p > 1)$

27 $\displaystyle\int_0^\infty \exp(-ax^2)\,dx \qquad (a > 0)$

***28** $\displaystyle\int_0^\infty x^2 \exp(-x^2)\,dx$

29 Set $f_n(t) = \displaystyle\int_0^\infty x^{2n+1} \exp(-tx^2)\,dx$ for $t > 0$ and $n \ge 0$. Assume the Leibniz rule applies, and use it to derive a relation between $f_n(t)$ and $f_{n+1}(t)$.

30 (cont.) Evaluate $f_0(t)$ and then $f_n(t)$ in general.

31 Find $\iint e^{-xy}\,dx\,dy$ over the first quadrant.

32 Use the formula

$$\frac{\sin x}{x} = \int_0^1 \cos(tx)\,dt$$

and the Leibniz rule to prove the inequality

$$\left| \frac{d^n}{dx^n} \frac{\sin x}{x} \right| \le \frac{1}{n + 1}$$

17-7 Physical Applications

Figure 17-7-1
Mass and density

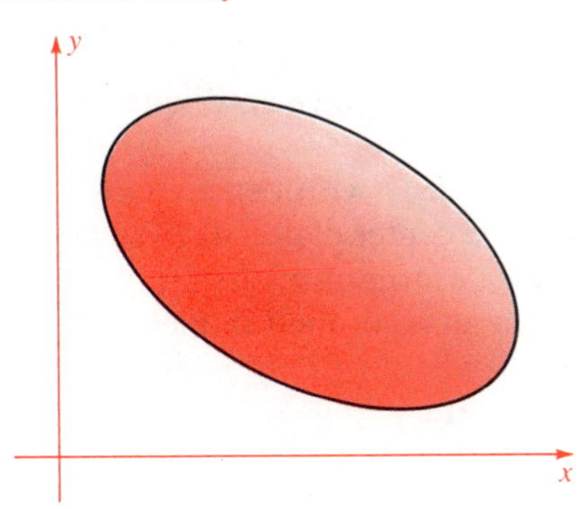

a Non-homogeneous material

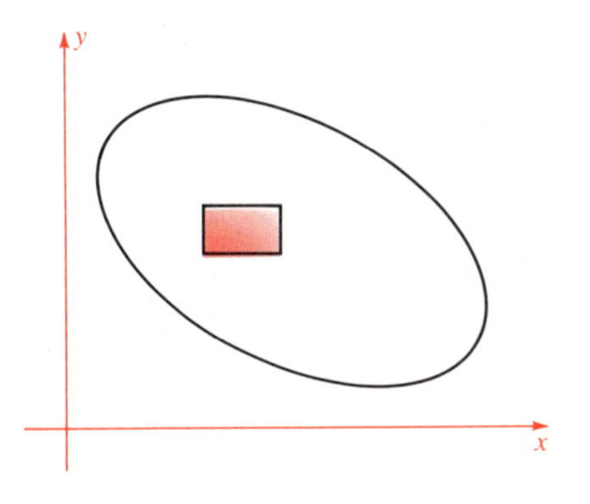

b Element of mass $dM = \rho(x, y)\,dx\,dy$

Suppose a sheet of non-homogeneous material covers a plane domain **D**. See Figure 17-7-1a. At each point (x, y), let $\rho(x, y)$ denote the **density** of the material, that is, the mass per unit area. Dimensionally, planar density is mass divided by length squared. Common units are gm/cm^2 and lb/ft^2. The mass of a small rectangular portion of the sheet (figure 17-7-1b) is the **element of mass**

$$dM \approx \rho(x, y)\,dA = \rho(x, y)\,dx\,dy$$

Therefore the total mass of the sheet is

$$M = \iint_{\mathbf{D}} \rho(x, y)\,dA = \iint_{\mathbf{D}} \rho(x, y)\,dx\,dy$$

Example 1

The density in lb/ft^2 at each point of a one-foot square of plastic is the product of the four distances of the point from the sides of the square. Find the total mass.

Solution Take the square in the position $0 \le x \le 1$, $0 \le y \le 1$. Then the density is

$$\rho(x, y) = x(1 - x)y(1 - y)$$

The element of mass is $dM = \rho(x, y)\,dx\,dy$, so the mass is

$$M = \iint \rho(x, y)\,dx\,dy = \left(\int_0^1 x(1 - x)\,dx\right)\left(\int_0^1 y(1 - y)\,dy\right)$$

$$= \tfrac{1}{6} \cdot \tfrac{1}{6} = \tfrac{1}{36}\text{ lb}$$

●

Moment and Center of Gravity

Suppose gravity (perpendicular to the plane of the figure) acts on the sheet of Figure 17-7-1a. The sheet is to be suspended by a single point so that it will balance. This point of balance is the **center of gravity** of the sheet and is denoted $\bar{\mathbf{x}} = (\bar{x}, \bar{y})$.

To motivate the formulas that follow, let us recall what happens when, instead of a sheet, we have a finite system of point masses glued to the weightless plane, mass M_1 at $\mathbf{x}_1$, mass M_2 at $\mathbf{x}_2, \cdots,$ mass M_n at $\mathbf{x}_n$. Then the center of gravity $\bar{\mathbf{x}}$ of the system is the weighted average of the $\mathbf{x}_i$, weighted with the M_i:

$$\mathbf{x} = \frac{M_1\mathbf{x}_1 + M_2\mathbf{x}_2 + \cdots + M_n\mathbf{x}_n}{M_1 + M_2 + \cdots + M_n} = \frac{1}{\sum M_i} \sum M_i\mathbf{x}_i$$

We can write

$$\bar{\mathbf{x}} = (\bar{x}, \bar{y}) = \frac{1}{M}\mathbf{m} = \frac{1}{M}(m_x, m_y)$$

where $M = \Sigma M_i$, the total mass, and the **moments** of the system of masses are

$$m_x = \sum M_i x_i$$

$$m_y = \sum M_i y_i$$

Note that $\bar{x} = m_x / M$ is the weighted average of the x-coordinates of the individual masses, and similarly for $\bar{y}$.

Now suppose mass is distributed continuously with density $\rho(x, y)$ over a plane sheet $\mathbf{D}$. How do we define the center of gravity? We argue very roughly as follows. We partition $\mathbf{D}$ into a large number of tiny pieces and pretend each piece is a point mass. The mass of the i-th piece $\mathbf{D}_i$ is approximately $M_i = \rho(x_i, y_i)\,\text{area}(\mathbf{D}_i)$ where (x_i, y_i) is a point in $\mathbf{D}_i$. For this finite system, the moments are

$$m_x = \sum x_i M_i = \sum x_i \rho(x_i, y_i)\,\text{area}(\mathbf{D}_i)$$

$$m_y = \sum y_i M_i = \sum y_i \rho(x_i, y_i)\,\text{area}(\mathbf{D}_i)$$

The center of gravity

$$\frac{1}{\sum M_i}\left(\sum x_i M_i, \sum y_i M_i\right)$$

approximates that of the sheet. The approximation improves as the partitions become finer and finer. But when that happens, $\Sigma x_i M_i$, $\Sigma y_i M_i$, and ΣM_i approach double integrals. This suggests the following definitions.

- The **moment** of a domain $\mathbf{D}$ with density distribution $\rho(x, y)$ is

$$\mathbf{m} = (m_x, m_y) = \left(\iint_{\mathbf{D}} x\, dM, \iint_{\mathbf{D}} y\, dM\right) = \iint_{\mathbf{D}} \rho(\mathbf{x})\mathbf{x}\, dx\, dy$$

where $dM = \rho(x, y)\, dx\, dy$ is the element of mass.

The **center of gravity** of the domain is

$$\bar{\mathbf{x}} = (\bar{x}, \bar{y}) = \frac{1}{M}\mathbf{m} = \frac{1}{M}(m_x, m_y)$$

where the total **mass** is

$$M = \iint_{\mathbf{D}} dM = \iint_{\mathbf{D}} \rho(x, y)\, dx\, dy$$

We can still think of $\bar{x}$ as the weighted average of x over the sheet:

$$\bar{x} = \iint_{\mathbf{D}} x\rho(x, y)\,dx\,dy \Big/ \iint_{\mathbf{D}} \rho(x, y)\,dx\,dy$$

and similarly for $\bar{y}$.

After two examples, we shall prove that the sheet balances if suspended from $\bar{\mathbf{x}}$.

Example 2

Find the center of gravity of a homogeneous rectangluar sheet.

Solution "Homogeneous" means the density ρ is constant. Take the sheet in the position $0 \le x \le a$ and $0 \le y \le b$. Its mass is then $M = \rho ab$, and its moment is

$$\begin{aligned}\mathbf{m} &= \iint \rho\mathbf{x}\,dx\,dy = \rho \iint \mathbf{x}\,dx\,dy \\ &= \rho \cdot \left(\iint x\,dx\,dy, \iint y\,dx\,dy\right) \\ &= \rho \cdot \left(\int_0^a x\,dx \int_0^b dy, \int_0^a dx \int_0^b y\,dy\right) = \rho \cdot (\tfrac{1}{2}a^2 b, \tfrac{1}{2}ab^2)\end{aligned}$$

Therefore its center of gravity is

$$\bar{\mathbf{x}} = \frac{1}{M}\mathbf{m} = \frac{1}{\rho ab}\rho \cdot (\tfrac{1}{2}a^2 b, \tfrac{1}{2}ab^2) = \tfrac{1}{2}(a, b)$$

This is the midpoint (intersection of the diagonals) of the rectangle. (Of course the rectangle balances on its midpoint; no one needs calculus for this, but it is reassuring that the analytic method gives the intuitive answer.) ●

Remark We sometimes speak of the "center of gravity of a domain $\mathbf{D}$," without reference to a density. Then it is understood that $\rho = 1$, so the center of gravity is a purely geometric quantity associated with $\mathbf{D}$. It is also known as the **centroid** of $\mathbf{D}$.

Example 3

The triangular sheet $x \ge 0$, $y \ge 0$, $x + y \le 1$ has density $\rho = xy$. Find its center of gravity.

Solution

$$\begin{aligned}M &= \iint_{\mathbf{D}} xy\,dx\,dy = \int_0^1 y\,dy \int_0^{1-y} x\,dx = \tfrac{1}{2}\int_0^1 y(1-y)^2\,dy \\ &= \tfrac{1}{2}\int_0^1 (y - 2y^2 + y^3)\,dy = \tfrac{1}{2}(\tfrac{1}{2} - \tfrac{2}{3} + \tfrac{1}{4}) = \tfrac{1}{24}\end{aligned}$$

By symmetry, $m_x = m_y$ and

$$m_x = \iint_D x \cdot xy\,dx\,dy = \int_0^1 y\,dy \int_0^{1-y} x^2\,dx$$

$$= \tfrac{1}{3}\int_0^1 y(1-y)^3\,dy = \tfrac{1}{3}\int_0^1 (1-u)\,u^3\,du$$

$$= \tfrac{1}{3}\int_0^1 (u^3 - u^4)\,du = \tfrac{1}{3}\left(\tfrac{1}{4} - \tfrac{1}{5}\right) = \tfrac{1}{60}$$

Therefore

$$\bar{x} = \bar{y} = \tfrac{1}{60}/\tfrac{1}{24} = \tfrac{2}{5} \quad \text{and} \quad \bar{\mathbf{x}} = \left(\tfrac{2}{5}, \tfrac{2}{5}\right)$$

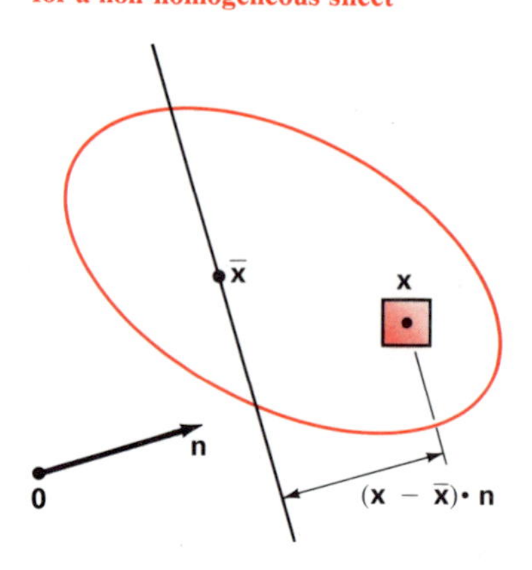

Figure 17-7-2
Proof of the center of gravity formula for a non-homogeneous sheet

Proof That a Sheet Balances

Suppose any knife edge passes through $\bar{\mathbf{x}}$, To prove that the sheet balances about the knife edge (Figure 17-7-2), divide the sheet into many small rectangles. The turning moments of these pieces about the knife edge must add up to zero.

Let $\mathbf{n}$ be a unit vector in the plane of the rectangle that is perpendicular to the knife edge. A small rectangle with sides dx and dy located at $\mathbf{x}$ has (signed) distance $(\mathbf{x} - \bar{\mathbf{x}}) \cdot \mathbf{n}$ from the knife edge and has mass $dM = \rho(x, y)\,dx\,dy$. Hence its turning moment is $(\mathbf{x} - \bar{\mathbf{x}}) \cdot \mathbf{n}\,dM$. The sum of all such turning moments must be zero:

$$\iint_D (\mathbf{x} - \bar{\mathbf{x}}) \cdot \mathbf{n}\,dM = 0$$

Since $\bar{\mathbf{x}}$ and $\mathbf{n}$ are constant, this relation may be written

$$\mathbf{n} \cdot \iint_D \mathbf{x}\,dM = (\mathbf{n} \cdot \bar{\mathbf{x}}) \iint_D dM$$

that is, $\mathbf{n} \cdot \mathbf{m} = M\mathbf{n} \cdot \bar{\mathbf{x}}$. But $\mathbf{m} = M\bar{\mathbf{x}}$, so $\mathbf{n} \cdot \mathbf{m} = M\mathbf{n} \cdot \bar{\mathbf{x}}$ as required. This completes the proof.

Mass and Center of Gravity in Polar Coordinates

Suppose a non-homogeneous sheet covers a domain $\mathbf{D}$ described by polar coordinates. The element of area is $dA = r\,dr\,d\theta$, so the element of mass is

$$dM = \rho r\,dr\,d\theta$$

where the density $\rho = \rho(r, \theta)$ is expressed as a function of r and θ. Therefore the mass of the sheet is expressed as an integral.

Mass in Polar Coordinates

$$M = \iint_D \rho r\,dr\,d\theta$$

To find the moment of the sheet express $\mathbf{x}$ in terms of its polar coordinates, $\mathbf{x} = (x, y) = (r\cos\theta, r\sin\theta)$, and integrate:

Moment in Polar Coordinates

$$\mathbf{m} = \iint_{\mathbf{D}} \mathbf{x}\, dM = \iint \mathbf{x}\rho r\, dr\, d\theta = \iint_{\mathbf{D}} \rho r^2(\cos\theta, \sin\theta)\, dr\, d\theta$$

$$= \left(\iint_{\mathbf{D}} \rho r^2 \cos\theta\, dr\, d\theta, \iint_{\mathbf{D}} \rho r^2 \sin\theta\, dr\, d\theta\right)$$

Once M and $\mathbf{m}$ are computed, we know $\bar{\mathbf{x}} = \mathbf{m}/M$.

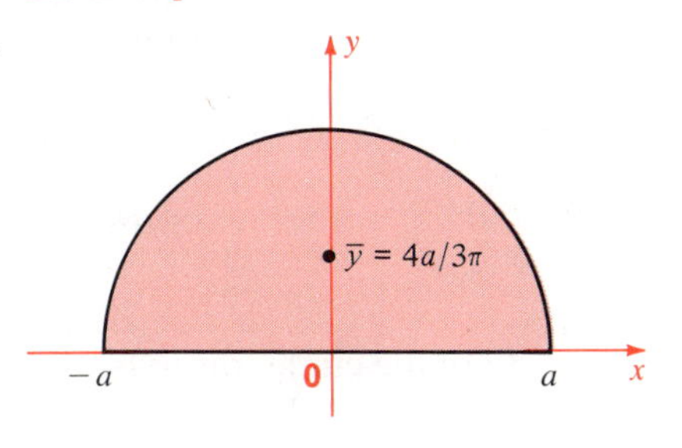

Figure 17-7-3
See Example 4

Example 4

Find the center of gravity of a uniform semicircular sheet of radius a. See Figure 17-7-3.

Solution Take the center at $\mathbf{0}$ and the diameter along the x-axis. Then the domain $\mathbf{D}$ covered by the sheet is $0 \le r \le a$, $0 \le \theta \le \pi$. Since ρ is constant, $M = \frac{1}{2}\pi a^2 \rho$ and

$$\mathbf{m} = \rho \cdot \left(\iint_{\mathbf{D}} r^2 \cos\theta\, dr\, d\theta, \iint_{\mathbf{D}} r^2 \sin\theta\, dr\, d\theta\right)$$

$$= \rho \cdot \left(\int_0^{\pi} \cos\theta\, d\theta \int_0^a r^2\, dr, \int_0^{\pi} \sin\theta\, d\theta \int_0^a r^2\, dr\right) = \rho \cdot (0, \tfrac{2}{3}a^3)$$

Therefore

$$\bar{\mathbf{x}} = \frac{1}{M}\mathbf{m} = \frac{2}{\pi a^2 \rho}\rho \cdot (0, \tfrac{2}{3}a^3) = \left(0, \frac{4a}{3\pi}\right)$$

Example 5

Find the center of gravity of a sheet in the shape of a quarter circle, whose density is proportional to the distance from the center of the circle.

Figure 17-7-4
$\bar{\mathbf{x}} = \left(\frac{3a}{2\pi}, \frac{3a}{2\pi}\right)$

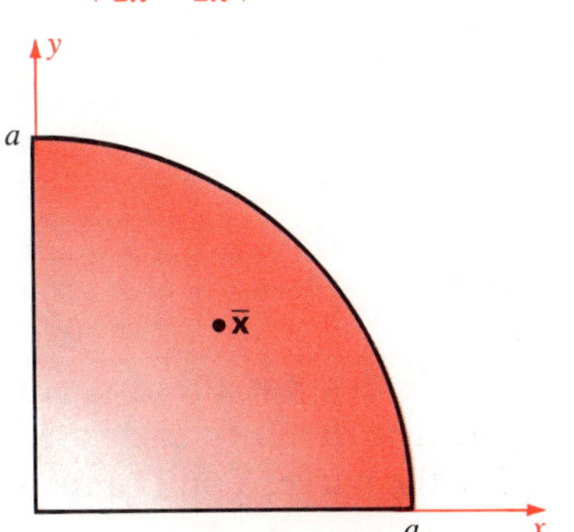

Solution Place the sheet as in Figure 17-7-4. Then $\rho = kr$, where k is a constant. We know that k will cancel in the end, so we may take $k = 1$, that is, $\rho = r$. The mass of the sheet is

$$M = \iint \rho r\, dr\, d\theta = \iint r^2\, dr\, d\theta = \int_0^a r^2\, dr \int_0^{\pi/2} d\theta = \frac{\pi a^3}{6}$$

and its moment is

$$\mathbf{m} = \iint \mathbf{x}\rho r\, dr\, d\theta = \iint (r\cos\theta, r\sin\theta) r^2\, dr\, d\theta$$

$$= \int_0^a r^3\, dr \int_0^{\pi/2} (\cos\theta, \sin\theta)\, d\theta = \frac{a^4}{4}(1, 1)$$

Therefore

$$\bar{\mathbf{x}} = \frac{1}{M}\mathbf{m} = \frac{6}{\pi a^3} \cdot \frac{a^4}{4}(1, 1)$$
$$= \left(\frac{3a}{2\pi}, \frac{3a}{2\pi}\right)$$

Could you have predicted that $\bar{\mathbf{x}}$ would lie on the line $y = x$ and that $|\bar{\mathbf{x}}| > \frac{1}{2}a$? ●

There is a useful connection between the centers of gravity and volumes of revolution.

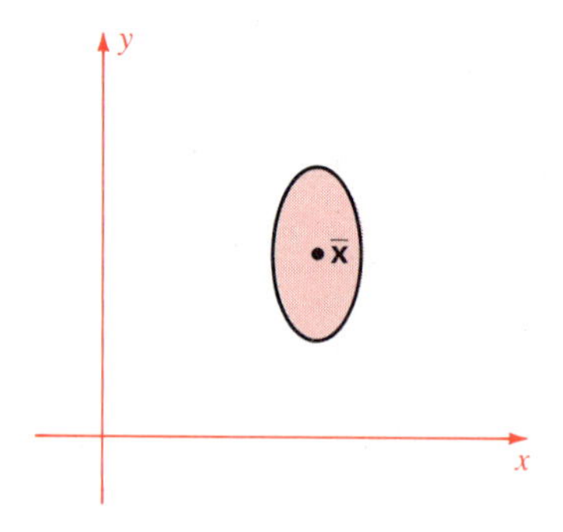

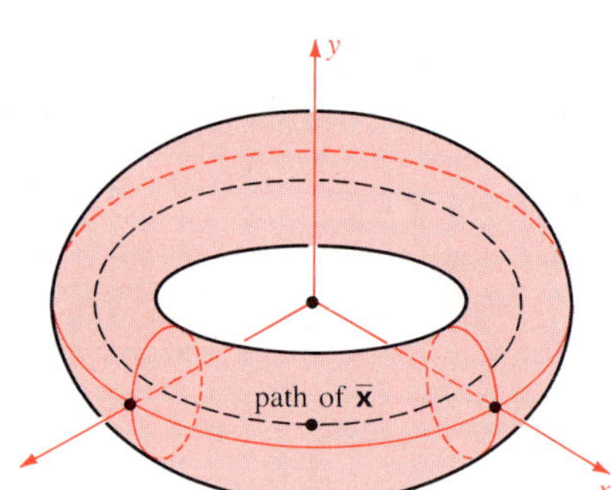

Figure 17-7-5
First Pappus theorem

First Pappus Theorem

Suppose a region $\mathbf{D}$ in the x, y-plane, to the right of the y-axis, is revolved about the y-axis. Then the volume of the resulting solid is

$$V = 2\pi\bar{x}A$$

where A is the area of the plane region $\mathbf{D}$ and $\bar{x}$ is the x-coordinate of its center of gravity (Figure 17-7-5). ●

In words, the volume is the area times the length of the circle traced by the center of gravity.

Proof A small portion $dx\,dy$ at $\mathbf{x}$ revolves into a thin ring of volume

$$dV = 2\pi x\,dx\,dy$$

Consequently

$$V = \iint_{\mathbf{D}} 2\pi x\,dx\,dy = 2\pi m_x$$

But $m_x = \bar{x}A$; hence $V = 2\pi\bar{x}A$. ●

Wires

A non-homogeneous wire is described by its position—a space curve $\mathbf{x} = \mathbf{x}(s)$ where $a \leq s \leq b$—and its density $\delta = \delta(s)$. (Here s denotes arc length.) Its mass, moment, and center of gravity are

$$M = \int_a^b \delta(s)\,ds$$

$$\mathbf{m} = \int_a^b \mathbf{x}(s)\delta(s)\,ds \qquad \text{and} \qquad \bar{\mathbf{x}} = \frac{1}{M}\mathbf{m}$$

If the wire is uniform, then $\delta(s)$ is a constant. In this case, the center of gravity is independent of δ, hence it is a geometric quantity associated with the curve and also known as the **centroid** of the curve. You can then take $\delta = 1$ and replace M by L, the length.

Figure 17-7-6
See Example 6

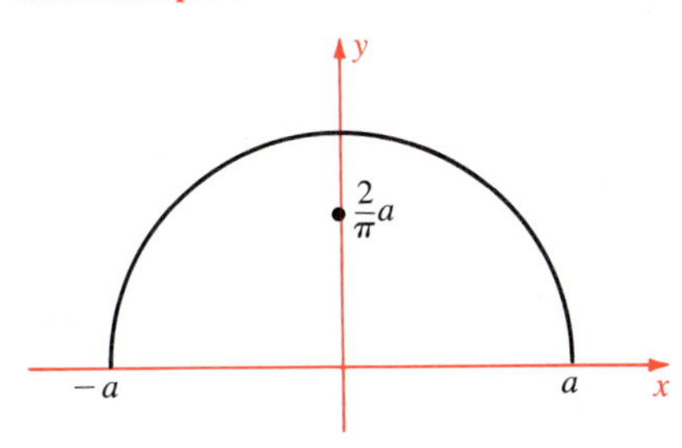

Example 6

Find the center of gravity of the uniform semicircle $r = a, \quad y \geq 0.$

Solution The length is $L = \pi a.$ The moment is

$$\mathbf{m} = \int \mathbf{x}\, ds$$

$$= \int_0^\pi (a\cos\theta, a\sin\theta)\, a\, d\theta$$

$$= a^2 \int_0^\pi (\cos\theta, \sin\theta)\, d\theta = a^2(0, 2)$$

The centroid (Figure 17-7-6) is

$$\bar{\mathbf{x}} = \frac{1}{L}\mathbf{m} = \frac{1}{\pi a}a^2(0, 2) = \frac{a}{\pi}(0, 2) = \left(0, \frac{2}{\pi}a\right)$$

Suppose a plane curve is revolved about an axis in its plane, generating a surface of revolution. There is a useful relation between the center of gravity of the curve and the area of the surface.

Second Pappus Theorem

Suppose a curve in the x, y-plane to the right of the y-axis is revolved about the y-axis. Then the area of the resulting surface is

$$A = 2\pi\bar{x}L$$

where L is the length of the curve and $\bar{x}$ is the x-coordinate of the center of gravity (Figure 17-7-7).

In words, the area is the length of the curve times the length of the circle traced by the center of gravity.

Figure 17-7-7
Second Pappus theorem

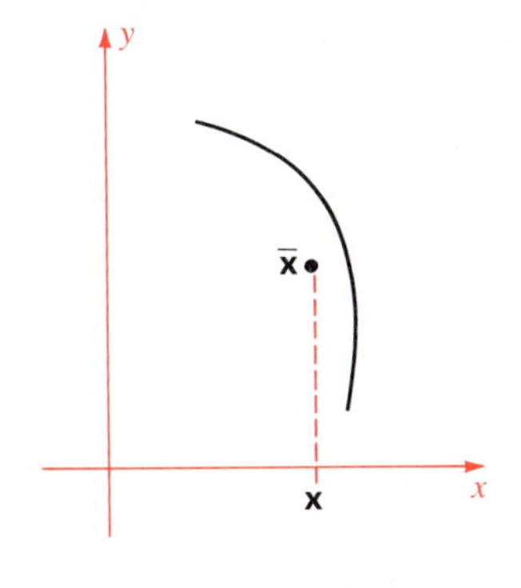

a Curve in x, y-plane

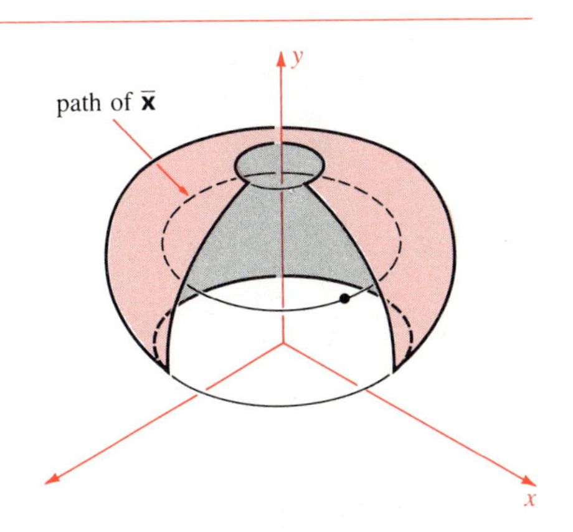

b Corresponding surface of revolution (cut-away view)

Figure 17-7-8
Element of rotated area

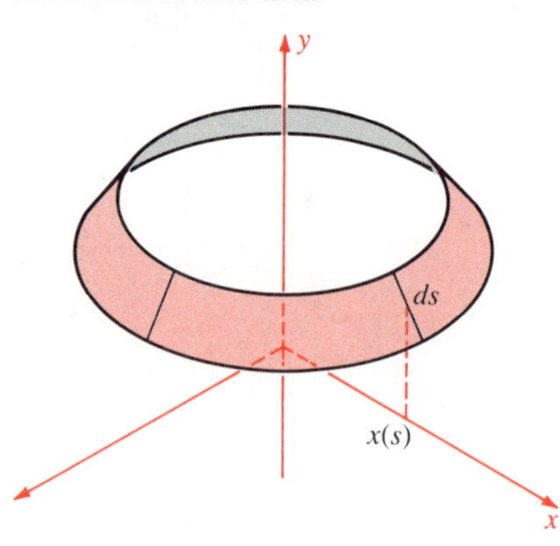

Proof A short segment of length ds of the curve at the point $\mathbf{x}(s)$ revolves into the frustum of a cone with lateral area $dA = 2\pi x\, ds$. See Figure 17-7-8. Hence

$$A = \int_a^b 2\pi x\, ds = 2\pi \int_a^b x\, ds = 2\pi m_x = 2\pi \bar{x} L$$

Moment of Inertia

Suppose a plane sheet over domain $\mathbf{D}$ rotates about $\mathbf{0}$ with angular speed ω rad/sec (Figure 17-7-9). An element of mass dM at distance r from $\mathbf{0}$ has speed $r\omega$, so its element of kinetic energy is

$$\begin{aligned} dE &= \tfrac{1}{2}(dM)(\text{speed})^2 = \tfrac{1}{2}(\rho\, dx\, dy)(r\omega)^2 \\ &= \tfrac{1}{2}\omega^2 r^2 \rho\, dx\, dy = \tfrac{1}{2}\omega^2(x^2 + y^2)\rho\, dx\, dy \end{aligned}$$

Therefore the total kinetic energy of the rotating sheet is

Figure 17-7-9
Rotating sheet, angular speed ω rad/sec

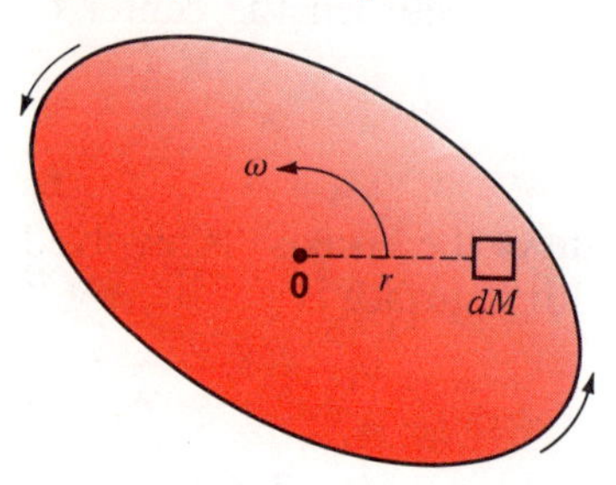

$$\begin{aligned} E &= \tfrac{1}{2}\omega^2 \iint_{\mathbf{D}} (x^2 + y^2)\rho\, dx\, dy = \tfrac{1}{2} I\omega^2 \\ &= \tfrac{1}{2}\left[\iint_{\mathbf{D}} (x^2 + y^2)\rho(x, y)\, dx\, dy\right]\omega^2 \end{aligned}$$

The bracketed quantity I is called the moment of inertia of the sheet (with respect to the origin). Thus $E = \tfrac{1}{2}I\omega^2$, which is the analogue for rotational motion of $E = \tfrac{1}{2}mv^2$ for linear motion.

Moment of Inertia

Suppose a sheet (lamina) covers a plane domain $\mathbf{D}$ and its density function is $\rho(x, y)$. Then its **moment of inertia** is

$$I = \iint_{\mathbf{D}} (x^2 + y^2)\, dM = \iint_{\mathbf{D}} (x^2 + y^2)\rho(x, y)\, dx\, dy$$

Example 7

Find the moment of inertia of a uniform square of side a about its center.

Solution Place the square in the standard position $-\tfrac{1}{2}a \le x \le \tfrac{1}{2}a$, $-\tfrac{1}{2}a \le y \le \tfrac{1}{2}a$. The density ρ is constant so the moment of inertia is

$$\begin{aligned} I &= \iint_{\mathbf{D}} (x^2 + y^2)\rho\, dx\, dy = \rho \iint_{\mathbf{D}} x^2\, dx\, dy + \rho \iint_{\mathbf{D}} y^2\, dx\, dy \\ &= \rho\left(\int_{-a/2}^{a/2} x^2\, dx \int_{-a/2}^{a/2} dy + \int_{-a/2}^{a/2} dx \int_{-a/2}^{a/2} y^2\, dy\right) \\ &= \rho[\tfrac{2}{3}(\tfrac{1}{2}a)^3 a + a(\tfrac{2}{3})(\tfrac{1}{2}a)^3] = \tfrac{1}{6}\rho a^4 \end{aligned}$$

Since the mass of the square is $M = \rho a^2$, the answer can be written $I = \tfrac{1}{6}Ma^2$.

Exercises

Find the mass and the center of gravity of each region where, as usual, ρ denotes the density

1 $\rho = (1 + x)(1 + y)$
$0 \le x \le 1 \quad 0 \le y \le 1$

2 $\rho = xy \quad 1 \le x \le 2 \quad 1 \le y < 3$

3 $\rho = 2 - x \quad 0 \le x \le 1 \quad -1 \le y \le 1$

4 $\rho = 1 + x \quad 0 \le x \le 3 \quad 0 \le y \le 2$

5 $\rho = (1 - x)(1 - y) + 1$
$0 \le x \le 1 \quad 0 \le y \le 1$

6 $\rho = \sin x \quad 0 \le x \le \pi \quad 0 \le y \le 1$

7 $\rho = 1 + x^2 + y^2 \quad -1 \le x \le 1 \quad 1 \le y \le 4$

8 $\rho = 2 + x^2y^2 \quad -1 \le x \le 1 \quad 0 \le y \le 1$

9 $\rho = 1 \quad 0 \le y \le 1 - x^2$

10 $\rho = y \quad 0 \le y \le 1 - x^2$

11 $\rho = 1 \quad 0 \le x \le 1 \quad 0 \le y \le x^2$

12 $\rho = x \quad 0 \le x \le 1 \quad 0 \le y \le x^2$

13 $\rho = 1 \quad x^2/a^2 + y^2/b^2 \le 1 \quad y \ge 0$

14 $\rho = 2 + x/a \quad x^2/a^2 + y^2/b^2 \le 1$

15 $\rho = 1 \quad r \le a \quad 0 \le \theta \le \frac{1}{2}\pi$

16 $\rho = 1 - r^2/a^2 \quad r \le a \quad 0 \le \theta \le \frac{1}{2}\pi$

17 $\rho = 1 \quad r \le a \quad 0 \le \theta \le \beta$

18 (cont.) Let $\beta \to 0$ in Exercise 17. Find the limiting position of $\bar{\mathbf{x}}$.

Find the mass and the center of gravity of the indicated wire with density ρ

19 $\rho = 1 \quad r = a \quad 0 \le \theta \le \frac{1}{2}\pi$

20 $\rho = \sin\theta \quad r = a \quad 0 \le \theta \le \pi$

21 $\rho = k\theta \quad r = a \quad 0 \le \theta \le \pi$

22 $\rho = 1 + k\theta \quad r = a \quad 0 \le \theta \le \pi$

23 $\rho = 1$ triangle: vertices $(0, 0)$, $(4, 0)$, $(0, 3)$

24 $\rho = 1$ triangle: vertices $(0, 0)$, $(1, 0)$, $(0, 1)$

25 Verify the first Pappus theorem for a semicircle revolved about its diameter.

26 Use the first Pappus theorem to find the volume of a right circular torus.

27 Verify the first Pappus theorem for a right triangle revolved about a leg.

28 Use the second Pappus theorem to find the surface area of a right circular torus. (See Exercise 9, page 860.)

29 Use the second Pappus theorem to obtain another solution of Example 6.

30 Use the second Pappus theorem to obtain another solution of Exercise 23.

Find the moment of inertia with respect to the origin; give the answer in the form $I = M \cdot (?)$

31 $\rho = 1$ circle $r \le a$

32 $\rho = r^n$ circle $r \le a$

33 $\rho = 1$ circle $(x - a)^2 + y^2 \le a^2$

34 $\rho = 1$ rectangle $|x| \le a, \quad |y| \le b$

35 $\rho = 1 + x$
triangle: vertices $(0, 0)$, $(1, 0)$, $(0, 1)$

36 $\rho = xy$ square $0 \le x \le b, \quad 0 \le y \le b$

17-8 Approximate Integration [Optional]

In this section we discuss an extension of Simpson's rule to approximation of double integrals. For simplicity, we shall allow only rectangular domains.

Let us recall Simpson's rule. To approximate the integral

$$\int_a^b f(x)\,dx$$

we divide the interval $a \le x \le b$ into $2m$ equal parts of length h:

$$a = x_0 < x_1 < x_2 < \cdots < x_{2m} = b$$

$$h = \frac{b - a}{2m}$$

Then

$$\int_a^b f(x)\,dx \approx \frac{h}{3}\sum_{i=0}^{2m} B_i f(x_i)$$

where the coefficients B_i are $1, 4, 2, 4, 2, 4, 2, \cdots, 2, 4, 1$.

We extend Simpson's rule to double integrals in the following way. To approximate

$$\iint_{\mathbf{D}} f(x, y)\,dx\,dy$$

where **D** denotes the rectangle $a \le x \le b$ and $c \le y \le d$, we divide the x-interval into $2m$ parts as before and likewise divide the y-interval into $2n$ equal parts of length k:

$$c = y_0 < y_1 < y_2 < \cdots < y_{2n} = d \qquad k = \frac{d - c}{2n}$$

We obtain $(2m + 1)(2n + 1)$ points of the rectangle (Figure 17-8-1). The approximate integration rule is

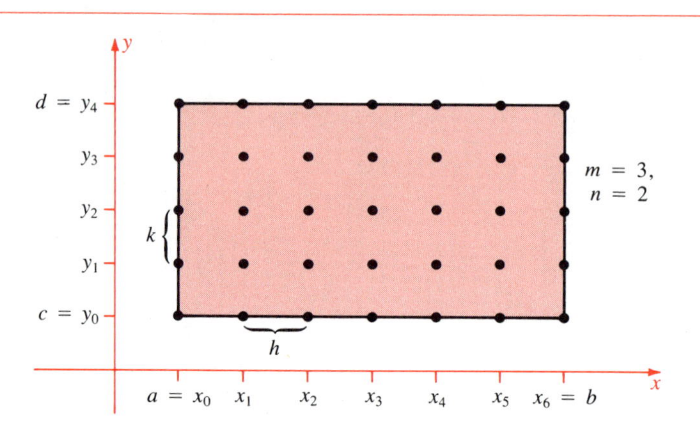

Figure 17-8-1
Division of rectangular domain

$$\iint_{\mathbf{D}} f(x, y)\,dx\,dy \approx \frac{hk}{9}\sum_{i=0}^{2m}\sum_{j=0}^{2n} A_{ij} f(x_i, y_j)$$

where the coefficients A_{ij} are certain products of the coefficients in the ordinary Simpson's rule. Precisely,

$$A_{ij} = B_i C_j$$

where $B_0, B_1, \cdots, B_{2m}$ are the coefficients in the ordinary Simpson's rule:

$$1 \quad 4 \quad 2 \quad 4 \quad 2 \quad 4 \quad \cdots \quad 2 \quad 4 \quad 1$$

and similarly $C_0, C_1, \cdots, C_{2n}$ are the coefficients in the ordinary Simpson's rule.

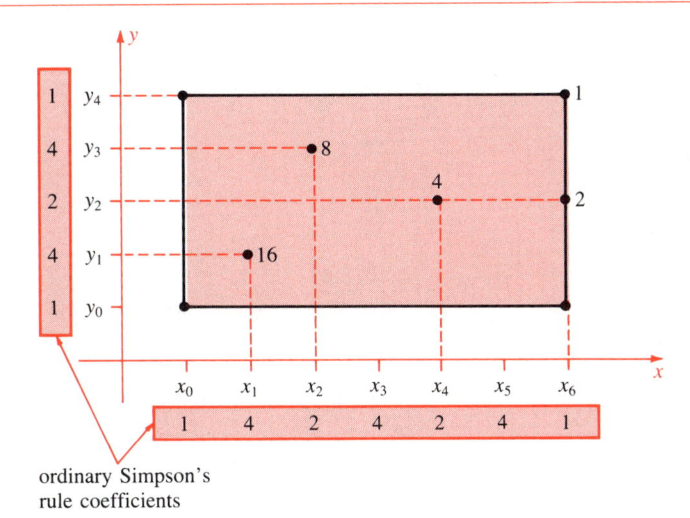

Figure 17-8-2
Coefficients for double Simpson's rule

In Figure 17-8-2, several of these products are formed. Since B_i and C_j take values 1, 2, and 4, the coefficients A_{ij} take values 1, 2, 4, 8, and 16. The A_{ij} can be written in a matrix corresponding to the points (x_i, y_j) as in Figure 17-8-1. For example, if $m = 3$ and $n = 2$, the matrix is

$$\begin{bmatrix} 1 & 4 & 2 & 4 & 2 & 4 & 1 \\ 4 & 16 & 8 & 16 & 8 & 16 & 4 \\ 2 & 8 & 4 & 8 & 4 & 8 & 2 \\ 4 & 16 & 8 & 16 & 8 & 16 & 4 \\ 1 & 4 & 2 & 4 & 2 & 4 & 1 \end{bmatrix}$$

Example 1

Estimate

$$I = \iint_{\mathbf{D}} (x + y)^3 \, dx \, dy$$

by using Simpson's rule with $m = n = 1$, where $\mathbf{D}$ is the rectangle $0 \le x \le 1$, $0 \le y \le 1$. Compare the result with the exact answer.

Solution Here $h = k = \frac{1}{2}$. The coefficient matrix is

$$[A_{ij}] = \begin{bmatrix} 1 & 4 & 1 \\ 4 & 16 & 4 \\ 1 & 4 & 1 \end{bmatrix}$$

Write the values $(x_i + y_j)^3$ in a matrix:

$$[(x_i + y_j)^3] = \begin{bmatrix} (1+0)^3 & (1+\frac{1}{2})^3 & (1+1)^3 \\ (\frac{1}{2}+0)^3 & (\frac{1}{2}+\frac{1}{2})^3 & (\frac{1}{2}+1)^3 \\ (0+0)^3 & (0+\frac{1}{2})^3 & (0+1)^3 \end{bmatrix} = \begin{bmatrix} 1 & \frac{27}{8} & 8 \\ \frac{1}{8} & 1 & \frac{27}{8} \\ 0 & \frac{1}{8} & 1 \end{bmatrix}$$

Now estimate the integral by

$$I \approx \frac{hk}{9} \sum_{i=0}^{2} \sum_{j=0}^{2} A_{ij}(x_i + y_j)^3$$

To evaluate the sum, multiply corresponding terms of the two matrices and add the nine products:

$$\begin{aligned} I \approx (\tfrac{1}{9})(\tfrac{1}{2})(\tfrac{1}{2})[1 \cdot 1 &+ 4 \cdot \tfrac{27}{8} + 1 \cdot 8 \\ &+ 4 \cdot \tfrac{1}{8} + 16 \cdot 1 + 4 \cdot \tfrac{27}{8} \\ &+ 1 \cdot 0 + 4 \cdot \tfrac{1}{8} + 1 \cdot 1] \\ = \tfrac{1}{36}[1 + \tfrac{27}{2} + 8 + \tfrac{1}{2} &+ 16 + \tfrac{27}{2} + \tfrac{1}{2} + 1] = \tfrac{54}{36} = \tfrac{3}{2} \end{aligned}$$

The exact value is

$$\begin{aligned} I &= \iint (x + y)^3 \, dx \, dy \\ &= \int_0^1 \left(\int_0^1 (x + y)^3 \, dx \right) dy \\ &= \tfrac{1}{4} \int_0^1 [(y + 1)^4 - y^4] \, dy \\ &= \tfrac{1}{20}[(y + 1)^5 - y^5] \Big|_0^1 = \tfrac{30}{20} = \tfrac{3}{2} \end{aligned}$$

so the estimate is exact in this case. ●

Remark 1 Because Simpson's rule is exact for cubics, the double integral rule is exact for cubics in two variables. (See Exercises 19–20).

Remark 2 The matrix of values $[f(x_i, y_j)]$ is arranged to conform to the layout of points (x_i, y_j) in the plane (Figure 17-8-1, page 870).

The next example is an integral that cannot be evaluated exactly, only approximated.

Example 2

Estimate

$$I = \iint_{\mathbf{D}} \sin(xy) \, dx \, dy$$

using $m = n = 1$, where $\mathbf{D}$ is the square $0 \le x \le \frac{1}{2}\pi$, $0 \le y \le \frac{1}{2}\pi$.

Solution Here $h = k = \frac{1}{4}\pi$, and the coefficient matrix is

$$[A_{ij}] = \begin{bmatrix} 1 & 4 & 1 \\ 4 & 16 & 4 \\ 1 & 4 & 1 \end{bmatrix}$$

The matrix of values of $\sin xy$ is

$$[\sin x_i y_j] = \begin{bmatrix} \sin 0 & \sin \frac{1}{8}\pi^2 & \sin \frac{1}{4}\pi^2 \\ \sin 0 & \sin \frac{1}{16}\pi^2 & \sin \frac{1}{8}\pi^2 \\ \sin 0 & \sin 0 & \sin 0 \end{bmatrix}$$

Therefore

$$I = \iint_{\mathbf{D}} \sin xy \, dx \, dy$$

$$\approx \frac{\pi^2}{144}\left(16 \sin \frac{\pi^2}{16} + 8 \sin \frac{\pi^2}{8} + \sin \frac{\pi^2}{4}\right) \approx 1.195$$

Error Estimate

The error estimate for Simpson's rule in two variables is analogous to that in one variable:

$$|\text{error}| \le \frac{(b-a)(d-c)}{180}(h^4M + k^4N)$$

where

$$\left|\frac{\partial^4 f}{\partial x^4}\right| \le M \quad \text{and} \quad \left|\frac{\partial^4 f}{\partial y^4}\right| \le N$$

We omit the proof.

Example 3

Estimate the error in Example 2.

Solution

$$\frac{\partial^4}{\partial x^4}(\sin xy) = y^4 \sin xy$$

$$\frac{\partial^4}{\partial y^4}(\sin xy) = x^4 \sin xy$$

But $|\sin xy| \le 1$. Hence in the square $0 \le x \le \frac{1}{2}\pi$, $0 \le y \le \frac{1}{2}\pi$, the inequalities

$$\left|\frac{\partial^4 f}{\partial x^4}\right| = |y^4 \sin xy| \le \left(\frac{\pi}{2}\right)^4 \qquad \left|\frac{\partial^4 f}{\partial y^4}\right| = |x^4 \sin xy| \le \left(\frac{\pi}{2}\right)^4$$

hold. Apply the error estimate with $m = n = 1$, $h = k = \frac{1}{4}\pi$, and $M = N = (\frac{1}{2}\pi)^4$:

$$|\text{error}| \le \frac{1}{180}\left(\frac{\pi}{2}\right)^2\left[2\left(\frac{\pi}{4}\right)^4\left(\frac{\pi}{2}\right)^4\right] = \frac{1}{45} \cdot \frac{\pi^{10}}{2^{15}} < 0.064$$

Exercises

Estimate to four significant figures; take $m = n = 1$

1 $\displaystyle\iint\limits_{\substack{0\le x\le\frac12\pi\\0\le y\le\frac12\pi}} \cos(xy)\,dx\,dy$

2 $\displaystyle\iint\limits_{\substack{0\le x\le1\\0\le y\le1}} \frac{dx\,dy}{1+x+y}$

3 $\displaystyle\iint\limits_{\substack{0\le x\le1\\0\le y\le1}} \frac{dx\,dy}{1+x^2+y^2}$

4 $\displaystyle\iint\limits_{\substack{0\le x\le1\\1\le y\le2}} \exp(x^2/y)\,dx\,dy$

5 $\displaystyle\iint\limits_{\substack{0\le x\le2\\0\le y\le2}} e^{-x^2-y^2}\,dx\,dy$

6 $\displaystyle\iint\limits_{\substack{0\le x\le1\\0\le y\le1}} e^{-x^2y^2}\,dx\,dy$

7 $\displaystyle\iint\limits_{\substack{0\le x\le1\\0\le y\le1}} x^4y^3\,dx\,dy$

8 $\displaystyle\iint\limits_{\substack{0\le x\le\frac12\pi\\0\le y\le1}} \tan(xy)\,dx\,dy$

Estimate to five significant figures; take $m = n = 2$

9 $\displaystyle\iint\limits_{\substack{0\le x\le\frac12\pi\\0\le y\le\frac12\pi}} \sin(xy)\,dx\,dy$

10 $\displaystyle\iint\limits_{\substack{0\le x\le\frac12\pi\\0\le y\le\frac12\pi}} \cos(xy)\,dx\,dy$

11 $\displaystyle\iint\limits_{\substack{0\le x\le1\\1\le y\le2}} \frac{xy^2}{x+y}\,dx\,dy$

12 $\displaystyle\iint\limits_{\substack{0\le x\le2\\0\le y\le2}} e^{-x^2-y^2}\,dx\,dy$

13 $\displaystyle\iint\limits_{\substack{0\le x\le1\\0\le y\le\pi}} \sin(xy)\,dx\,dy$

14 $\displaystyle\iint\limits_{\substack{0\le x\le1\\0\le y\le2}} \frac{dx\,dy}{1+x^3+y^4}$

Give an upper bound for the error in

15 Exercise 1

16 Exercise 2

17 Exercise 9

18 Exercise 13

19 Suppose $f(x, y) = p(x)q(y)$. Show that the double integral Simpson's rule estimate is just the product of the Simpson's rule estimate for $\int p(x)\,dx$ by that for $\int q(y)\,dy$.

20 (cont.) Conclude that the rule is exact for polynomials involving only x^3y^3, x^3y^2, x^2y^3, x^3y, x^2y^2, xy^3, and lower degree terms.

21 For double integrals, the analogue of the trapezoidal rule for simple integrals is

$$\iint\limits_{\substack{0\le x\le1\\0\le y\le1}} f(x, y)\,dx\,dy \approx \frac{1}{4}[f(0,0) + f(0,1) + f(1,1) + f(1,0)]$$

Show that this rule is exact for polynomials $f(x, y) = A + Bx + Cy + Dxy$.

22 (cont.) Find the corresponding rule for a rectangle $a \le x \le b$, $c \le x \le d$, divided into rectangles of size h by k with $h = (b - a)/m$ and $k = (d - c)/n$.

23 (cont.) Test the resulting rule on

$$\iint\limits_{\substack{0\le x\le1\\0\le y\le1}} x^4y^4\,dx\,dy$$

with $m = n = 4$.

***24** Let $\mathbf{I}$ denote the unit square $0 \le x \le 1$, $0 \le y \le 1$. Suppose $f(x, y) = 0$ at its four vertices. Prove that

$$\iint\limits_{\mathbf{I}} f(x, y)\,dx\,dy = -\tfrac12 \iint\limits_{\mathbf{I}} y(1-y)f_{yy}(x, y)\,dx\,dy - \tfrac14 \int_0^1 x(1-x)[f_{xx}(x,1) + f_{xx}(x,0)]\,dx$$

25 (cont.) Suppose also that $|f_{xx}| \le M$ and $|f_{yy}| \le N$ on $\mathbf{I}$. Prove that

$$\left|\iint\limits_{\mathbf{I}} f(x, y)\,dx\,dy\right| \le \tfrac{1}{12}M + \tfrac{1}{12}N$$

26 (cont.) Conclude that for any function on the square $0 \le x \le 1$, $0 \le y \le 1$, the error in the trapezoidal estimate (Exercise 21) is at a most $\frac{1}{12}(M + N)$. [Hint Use the result of Exercise 21 and interpolation.]

27 (cont.) Suppose $f(x, y)$ has domain $a \le x \le b$, $c \le y \le d$ and satisfies $|f_{xx}| \le M$, $|f_{yy}| \le N$. Show that the error in the trapezoidal approximation (Exercise 22) with $m = n = 1$ is at most $\frac{1}{12}hk(h^2M + k^2N)$, where $h = b - a$ and $k = d - c$.

28 (cont.) Deduce the corresponding error estimate for arbitrary m and n in Exercise 22.

29 (cont.) What does this result give for Exercise 23?

30 (cont.) Suppose the trapezoidal approximation is used to estimate the integral of Exercise 9, but with any m and n. Give an upper bound for the error.

17-9 Review Exercises

1 Let C denote the volume of the cone with base $x^2/a^2 + y^2/b^2 = 1$ and apex $(0, 0, c)$. Let P denote the volume of the inverted paraboloid given by $0 \le z \le c(1 - x^2/a^2 - y^2/b^2)$. Assume a, b, c, are all positive. Find the relation between P and C **(Archimedes' relation).**

2 Find the volume of the solid $0 \le z \le cy/b$, $x^2/a^2 + y^2/b^2 \le 1$, where $a > 0$, $b > 0$, $c > 0$.

3 Integrate xy over the domain $0 \le x \le a$, $0 \le y \le a$, $x^2 + y^2 \ge a^2$.

4 Evaluate $\iint dx\,dy$ over $a \le x \le y \le b$.

5 An exponential horn loudspeaker is bounded by the six surfaces $x = e^{az}$, $x = -e^{az}$, $y = e^{az}$, $y = -e^{az}$, $z = 0$, $z = b$. Find its volume.

6 Evaluate $\iint dx\,dy/r^4$ over $1 \le r \le 2$, $r^2 \ge 2x$.

7 A certain solid lies between the planes $z = a$ and $z = b$, where $a < b$. Let $A(z)$ be the cross-sectional area of the solid at height z. Give a formula for the volume of the solid **(Cavalieri's principle).**

8 (cont.) Let $a > 0$, $b > 0$, $c > 0$. Join each point $(x, y, 0)$ in the elliptic domain $x^2/a^2 + y^2/b^2 \le 1$, $z = 0$ by a segment to $(0, y, c)$. These segments then sweep out a tentlike solid. Find its volume.

9 Let $\mathbf{x} = \mathbf{x}(s)$ be a curve of length L on the sphere $|\mathbf{x}| = a$. The segments $\overline{\mathbf{0x}(s)}$ sweep out a (conical) surface. Find its area.

***10** Find the volume of the portion of a cone cut off by a plane: Let $0 < a < b$ and $c > 0$. Find the volume of the solid $cy/b \le z \le c(1 - r/a)$, where $r^2 = x^2 + y^2$ as usual.

***11** Find the area and the moment m_x for **D** in Figure 17-9-1, where $\rho = 1$. [Hint You may use Exercise 17 in Section 17-7.]

Figure 17-9-1
See Exercises 11 and 12

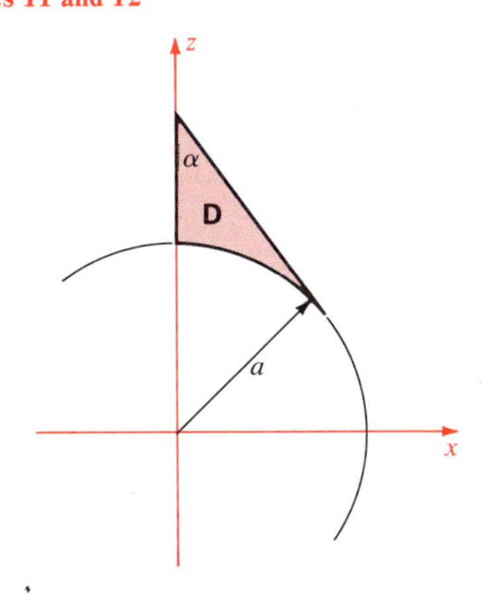

12 (cont.) A sphere of radius a is inscribed in one nappe of a right circular cone of apex angle 2α. Find the volume of the portion of the cone between the sphere and the apex.

13 Find the moment m_x for **D** in Figure 17-9-2; $\rho = 1$. [Hint Use Example 4 in Section 17-7.]

Figure 17-9-2
See Exercises 13 and 14

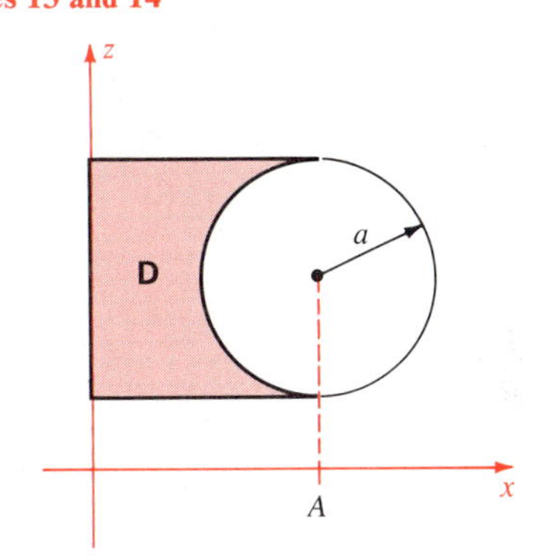

14 (cont.) Find the volume of the hole in a donut, **D** rotated about the z-axis.

15 Find the mass and center of gravity of the spiral wire $r = \theta$, $0 \le \theta \le 2\pi$, where the density is $\rho = \sqrt{1 + \theta^2}$.

***16** Suppose $f(a) = 0$. Express

$$2\int_a^b \left(\int_a^y f(x)f(y)[1 - f'(x)]\,dx\right) dy$$

in terms of $\int_a^b f(x)\,dx$ and $\int_a^b f(x)^3\,dx$.

17 (cont.) Suppose $0 = f(a) \le f(x)$ and $f'(x) \le 1$ for $a \le x \le b$. Prove

$$\int_a^b [f(x)]^3\,dx \le \left(\int_a^b f(x)\,dx\right)^2$$

18 Evaluate $\displaystyle\int_0^1 \frac{dx}{(1 + x^2)^2}$ by differentiating

$$\int_0^1 \frac{dx}{x^2 + t^2} = \frac{1}{t}\arctan\frac{1}{t}$$

Suppose $f(x)$ and $g(x)$ are continuous for $x \ge 0$. Their **convolution** is the function $h = f * g$ defined by

$$h(x) = \int_0^x f(t)g(x - t)\,dt$$

19 Prove $g * f = f * g$.

20 (cont.) Prove $f * (g * h) = (f * g) * h$.

***21** (cont.) Recall that the **Laplace transform** of $f(x)$ is the function

$$L(f)(s) = \int_0^\infty e^{-sx} f(x)\,dx$$

where convergent. Assuming everything in sight converges and changing the order of integration is valid, prove $L(f * g)(s) = L(f)(s) \cdot L(g)(s)$.

22 Criticize:

$$I = \iint\limits_{\substack{0 \le x \le 1 \\ 0 \le y \le 1}} \frac{x^2 - y^2}{(x^2 + y^2)^2}\,dx\,dy$$

$$= \int_0^1 \left(\int_0^1 \frac{x^2 - y^2}{(x^2 + y^2)^2}\,dx \right) dy$$

$$= \int_0^1 \left(\frac{-x}{x^2 + y^2} \right) \bigg|_{x=0}^{x=1} dy$$

$$= \int_0^1 \frac{-1}{1 + y^2}\,dy = -\tfrac{1}{4}\pi$$

In a snow pile, let $p = p(\mathbf{x})$ denote the pressure and $\rho = \rho(\mathbf{x})$ the density. Assume that p and ρ depend only on the depth z of the snow above $\mathbf{x}$. Thus in a vertical column of unit cross-sectional area, we have

$$p(z) = \int_0^z \rho(u)\,du \qquad \text{at depth } z$$

Assume also that the density at any point depends only on the pressure there, so $\rho = f(p)$. If the snow is not too deep (so it doesn't pack into ice), a reasonable assumption is $\rho = \rho_0 + kp$, where $\rho_0 > 0$ and $k > 0$.

23 Prove $p(z) = (\rho_0/k)(e^{kz} - 1)$ and $\rho(z) = \rho_0 e^{kz}$.

24 (cont.) Suppose the snow pile has the shape of a right circular cone of radius a and height h. Find the weight of the snow pile.

25 (cont.) Suppose the snow pile has the shape of a hemisphere of radius a. Find the weight of the snow pile.

26 (cont.) Suppose the snow pile has the shape of an inverted paraboloid of revolution of height h and base radius a. Find its weight.

27 Suppose $f(x)$ is increasing on $a \le x \le b$. Prove

$$(b + a) \int_a^b f(x)\,dx \le 2 \int_a^b x f(x)\,dx$$

[Hint Consider $(x - y)[f(x) - f(y)]$.]

***28** Suppose $f(x)$ is increasing on $a \le x \le b$ and $f(x) > 0$. Prove

$$\frac{\displaystyle\int_a^b x[f(x)]^2\,dx}{\displaystyle\int_a^b [f(x)]^2\,dx} \ge \frac{\displaystyle\int_a^b x f(x)\,dx}{\displaystyle\int_a^b f(x)\,dx}$$

29 Let a line segment in $\mathbf{R}^2$ have length L, and let L_1 and L_2 be the lengths of its projections on the axes. Prove $L^2 = L_1{}^2 + L_2{}^2$.

30 (cont.) Find the analogous relation for a plane region of area A in $\mathbf{R}^3$.

18 Multiple Integrals

18-1 Triple Integrals

Triple integrals are used to solve problems of the following type. Suppose we have a bounded domain $\mathbf{D}$ in space filled by a non-homogeneous solid. At each point $\mathbf{x}$ the density of the solid is $\delta(\mathbf{x})$ gm/cm^3. What is the total mass?

Our work with double integrals suggests how to approach this problem. We decompose $\mathbf{D}$ into many small subdomains $\mathbf{D}_i$ and choose a point $\mathbf{x}_i$ in each $\mathbf{D}_i$. Then the mass in $\mathbf{D}_i$ is approximately $\delta(\mathbf{x}_i)|\mathbf{D}_i|$, where $|\mathbf{D}_i|$ is the volume. The total mass is approximately

$$\sum \delta(\mathbf{x}_i)|\mathbf{D}_i|$$

Finally, we take the limit of these Riemann sums as the subdivisions of $\mathbf{D}$ become finer and finer. The limit, if it exists, is the triple integral

$$\iiint_{\mathbf{D}} \delta(\mathbf{x})\, dx\, dy\, dz = \iiint_{\mathbf{D}} \delta(\mathbf{x})\, dV$$

where $dV = dx\, dy\, dz$ is called the **element of volume** (in rectangular coordinates).

The theory of triple integrals is quite similar to the theory of double integrals and presents no really new difficulties, so we shall omit most of it and pass quickly to applications. The main theoretical fact is that the triple integral exists if $\delta(\mathbf{x})$ is continuous and the boundary of $\mathbf{D}$ is not too complicated. The main practical fact is that the integral can be evaluated by iteration.

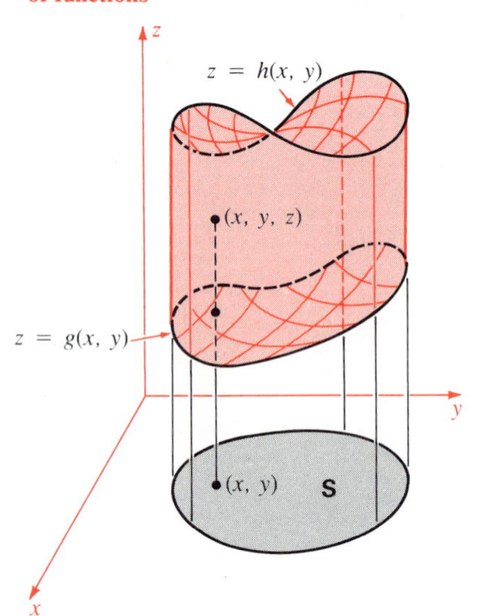

Figure 18-1-1
Domain in $\mathbf{R}^3$ between two graphs of functions

Iteration

A convenient domain for triple integration is the part of a cylinder bounded between two surfaces, each the graph of a function. Precisely, suppose two surfaces $z = g(x, y)$ and $z = h(x, y)$ are defined over a domain **S** in the x, y-plane, and that $g(x, y) < h(x, y)$. See Figure 18-1-1. These surfaces can be considered as the top and bottom of a domain **D** in the cylinder over **S**. Thus **D** consists of all points (x, y, z) where (x, y) is in **S** and

$$g(x, y) \le z \le h(x, y)$$

We then have the following formula, reducing the triple integral to a single integral followed by a double integral.

Iteration Formula

$$\iiint_{\mathbf{D}} \delta(\mathbf{x})\, dV = \iint_{\mathbf{S}} \left(\int_{z=g(x,y)}^{z=h(x,y)} \delta(x, y, z)\, dz \right) dA$$

Some prefer the notation

$$\iint_{\mathbf{S}} dx\, dy \int_{g(x,y)}^{h(x,y)} \delta(x, y, z)\, dz$$

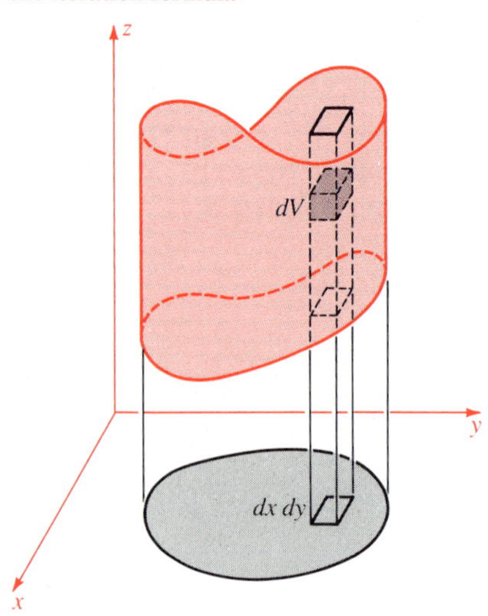

Figure 18-1-2
Intuitive reason for the iteration formula

The intuitive logic behind the iteration formula is shown in Figure 18-1-2. First, the elements of mass $dM = \delta(x, y, z)\, dV$ in one column are added up by an integral in the vertical direction (x and y fixed, z variable). Intuitively, $dV = dz\, dA$, where $dA = dx\, dy$ is the element of area of **S**. Consequently, the mass of an individual column is

$$\left(\int_{g(x,y)}^{h(x,y)} \delta(x, y, z)\, dz \right) dA$$

Then these masses are totaled by a double integral over **S**.

Example 1

Find $\iiint (x^2 + y) z\, dx\, dy\, dz$, taken over the block $1 \le x \le 2$, $0 \le y \le 1$, $3 \le z \le 5$.

Solution The upper and lower boundaries are the planes $z = 5$ and $z = 3$. By the iteration formula,

$$\iiint (x^2 + y) z\, dV = \iint_{\mathbf{S}} \left(\int_3^5 (x^2 + y) z\, dz \right) dA$$

where **S** is the rectangle $1 \le x \le 2$ and $0 \le y \le 1$. Now x and y are constant in the inner integral, so

$$\int_3^5 (x^2 + y) z\, dz = (x^2 + y) \int_3^5 z\, dz = 8(x^2 + y)$$

we have reduced the triple integral to a double integral, which we evaluate by iteration:

$$\iiint (x^2 + y)z\,dV = \iint_{\mathbf{S}} 8(x^2 + y)\,dA$$

$$= \iint_{\mathbf{S}} 8(x^2 + y)\,dx\,dy$$

$$= \int_1^2 \left(\int_0^1 8(x^2 + y)\,dy \right) dx$$

$$= 8 \int_1^2 (x^2 + \tfrac{1}{2})\,dx$$

$$= 8(\tfrac{7}{3} + \tfrac{1}{2}) = \tfrac{68}{3}$$

●

Remark The solution of Example 1 can be set up as a triply iterated integral

$$\iiint (x^2 + y)z\,dx\,dy\,dz = \int_1^2 \left[\int_0^1 \left(\int_3^5 (x^2 + y)z\,dz \right) dy \right] dx$$

Figure 18-1-3
See Example 2

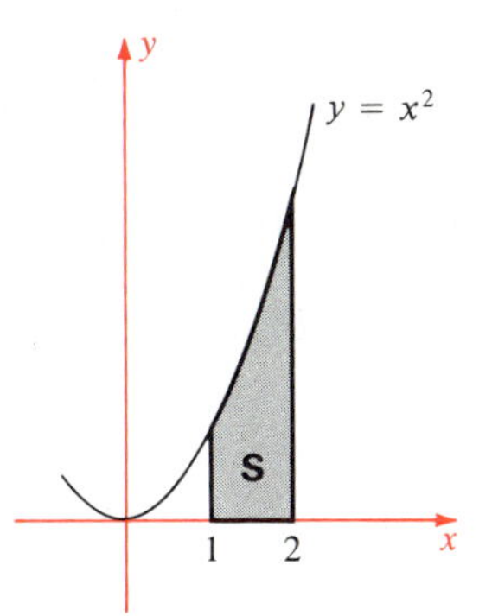

a The domain **S**, the base of domain **D**

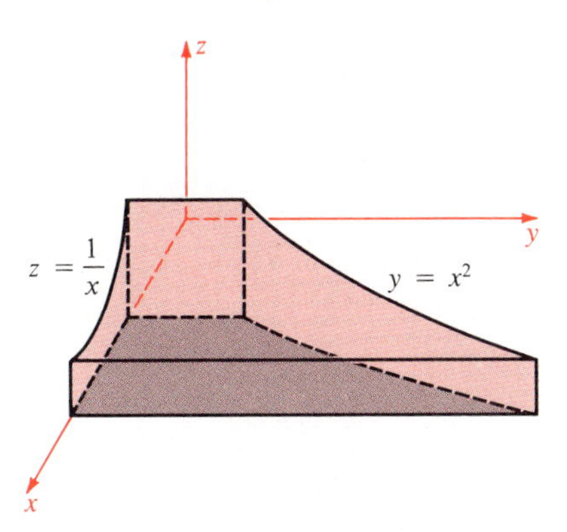

b Domain **D** of integration

Example 2

Compute $\iiint x^3y^2z\,dx\,dy\,dz$ over the domain **D** bounded by $x = 1$, $x = 2$; $y = 0$, $y = x^2$; and $z = 0$, $z = 1/x$.

Solution The domain **D** is the portion between the surfaces $z = 0$ and $z = 1/x$ of a solid cylinder parallel to the z-axis. The cylinder has base **S** in the x, y-plane, where **S** is as shown in Figure 18-1-3a. The solid **D** itself is sketched in Figure 18-1-3b. (A rough sketch showing the general shape is satisfactory.) The iterated integral is

$$\iiint_{\mathbf{D}} x^3y^2z\,dx\,dy\,dz = \iint_{\mathbf{S}} \left(\int_0^{1/x} x^3y^2z\,dz \right) dx\,dy$$

$$= \iint_{\mathbf{S}} x^3y^2 \left(\tfrac{1}{2}z^2 \Big|_0^{1/x} \right) dx\,dy$$

$$= \tfrac{1}{2} \iint_{\mathbf{S}} xy^2\,dx\,dy = \tfrac{1}{2} \int_1^2 x \left(\int_0^{x^2} y^2\,dy \right) dx$$

$$= \tfrac{1}{6} \int_1^2 x^7\,dx = \tfrac{1}{48}(2^8 - 1) = \tfrac{255}{48} = \tfrac{85}{16}$$

Figure 18-1-4
See Alternative Solution of Example 2

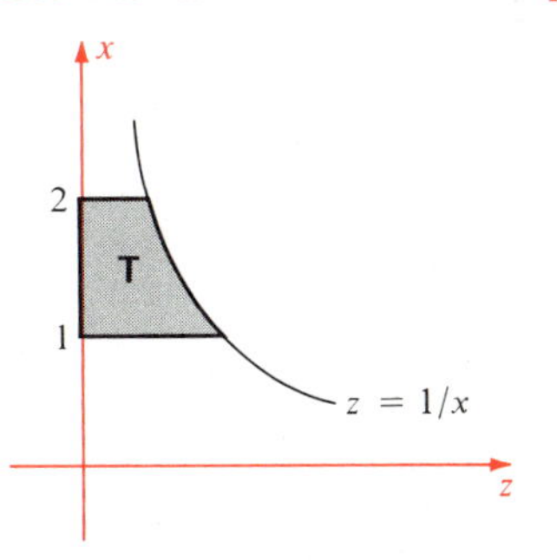

Alternative Solution The domain also may be considered as the portion between the surfaces $y = 0$ and $y = x^2$ of a solid cylinder parallel to the y-axis. The cylinder has base **T** in the z, x-plane (Figure 18-1-4). From this viewpoint, the first integration is with respect to y; the iterated integral is

$$\iiint\limits_{\mathbf{D}} x^3y^2z\,dx\,dy\,dz = \iint\limits_{\mathbf{T}} x^3z\left(\int_0^{x^2} y^2\,dy\right)dx\,dz$$
$$= \iint\limits_{\mathbf{T}} \tfrac{1}{3}x^9z\,dx\,dz$$
$$= \tfrac{1}{3}\int_1^2 x^9\left(\int_0^{1/x} z\,dz\right)dx$$
$$= \tfrac{1}{6}\int_1^2 x^7\,dx = \tfrac{1}{48}(2^8 - 1) = \tfrac{85}{16}$$

Remark It is poor technique to consider the region as a solid cylinder parallel to the x-axis because you must break the projection of the solid into the y, z-plane into four parts in order to integrate. Therefore the solid **D** itself must be decomposed into four parts, and the triple integral correspondingly expressed as a sum of four triple integrals (Figure 18-1-5). The resulting computation is much longer than that in either of the previous solutions.

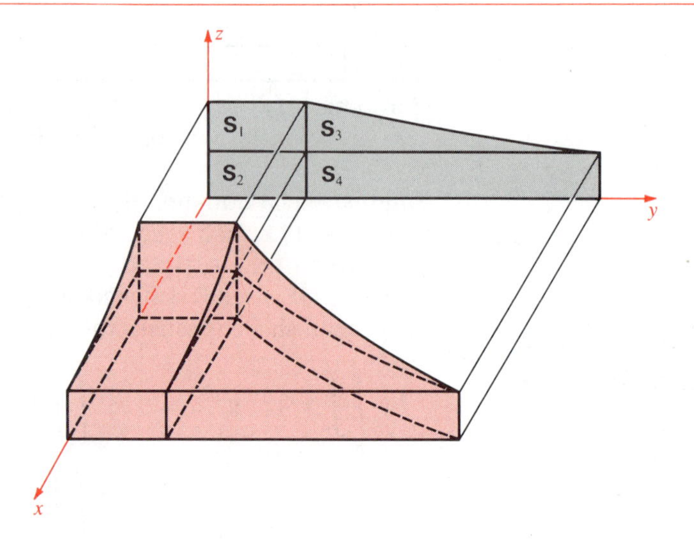

Figure 18-1-5
Projection of **D** onto the y, z-plane forces you to decompose the integral into four pieces.

In general, try to pick an order of iteration that expresses the required triple integral as a sum of as few terms as possible, at best only one. The typical summand has the form

$$\int_a^b\left[\int_{k(x)}^{h(x)}\left(\int_{g(x,y)}^{f(x,y)} \delta(x, y, z)\,dz\right)dy\right]dx$$

(The variables may be in some other order.) Once the integral

$$\int_{g(x,y)}^{f(x,y)} \delta(x, y, z)\,dz$$

is evaluated, the result is a function of x and y alone; z does not appear. Likewise, once the integral

$$\int_{k(x)}^{h(x)} \left(\int_{g(x,y)}^{f(x,y)} \delta(x, y, z)\, dz \right) dy$$

is evaluated, the result is a function of x alone; y does not appear.

Keep in mind that there are six possible orders of iteration for triple integrals. If trying one of them leads to an integrand you cannot work out or find in tables, try a different order of iteration.

Domains and Inequalities

If a domain **D** is specified by inequalities, it may be possible to arrange the inequalities so that limits of integration can be set up easily. For example, suppose the inequalities can be arranged in this form:

$$a \le x \le b \qquad h(x) \le y \le k(x) \qquad g(x, y) \le z \le f(x, y)$$

Then

$$\iiint_{\mathbf{D}} \delta(x, y, z)\, dx\, dy\, dz$$

$$= \int_a^b \left[\int_{h(x)}^{k(x)} \left(\int_{g(x,y)}^{f(x,y)} \delta(x, y, z)\, dz \right) dy \right] dx$$

Tetrahedral domains can be expressed by such inequalities, and they occur frequently enough that it is useful to practice setting up integrals over them.

Example 3

A tetrahedron **T** has vertices at $(0, 0, 0)$, $(a, 0, 0)$, $(0, b, 0)$, $(0, 0, c)$, where a, b, $c > 0$. Set up $\iiint_{\mathbf{T}} \delta(x, y, z)\, dx\, dy\, dz$ as an interated integral.

Solution The slanted surface (Figure 18-1-6, next page) has equation

$$\frac{x}{a} + \frac{y}{b} + \frac{z}{c} = 1$$

The domain is defined by the inequalities

$$0 \le x \qquad 0 \le y \qquad 0 \le z \qquad \frac{x}{a} + \frac{y}{b} + \frac{z}{c} \le 1$$

Any order of iteration is satisfactory; for instance, choose the order of integration

$$\int \left[\int \left(\int \delta(x, y, z)\, dz \right) dy \right] dx$$

Figure 18-1-6
Tetrahedral domain

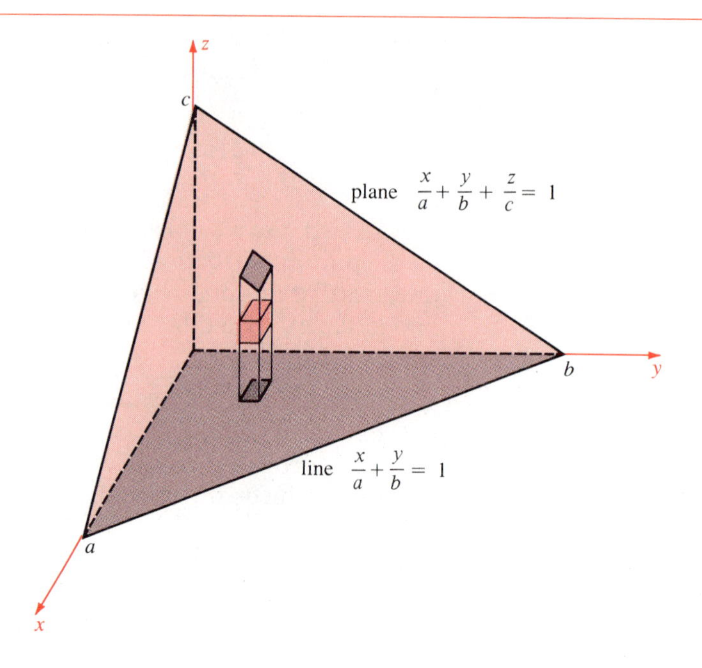

To find the limits of integration, we must replace the system of inequalities defining **D** by an equivalent system of the form

$$a \le x \le b \qquad h(x) \le y \le k(x) \qquad g(x, y) \le z \le f(x, y)$$

The original inequalities imply $0 \le x \le a$. Once we choose such an x, then

$$0 \le y \le b\left(1 - \frac{x}{a} - \frac{z}{c}\right) \le b\left(1 - \frac{x}{a}\right)$$

since $z \ge 0$. Once we choose x and y then

$$0 \le z \le c\left(1 - \frac{x}{a} - \frac{y}{b}\right)$$

Thus we obtain the equivalent system of inequalities:

$$0 \le x \le a \qquad 0 \le y \le b\left(1 - \frac{x}{a}\right)$$

$$0 \le z \le c\left(1 - \frac{x}{a} - \frac{y}{b}\right)$$

The corresponding iteration is

$$\iiint_{\mathbf{T}} \delta(x, y, z)\, dx\, dy\, dz$$

$$= \int_0^a \left[\int_0^{b\left(1-\frac{x}{a}\right)} \left(\int_0^{c\left(1-\frac{x}{a}-\frac{y}{b}\right)} \delta(x, y, z)\, dz\right) dy\right] dx$$

Example 4

Set up an evaluation of the triple integral

$$I = \iiint_{\mathbf{D}} f(x, y, z)\, dx\, dy\, dz$$

where the domain of integration **D** is specified by the inequalities

$$0 \le x \le 2 \qquad 0 \le y \le 2 \qquad 0 \le z \le 2$$
$$y + z \le 3 \qquad x + y \le 3$$

Solution The first order of business in solving such a problem is drawing **D**. We start with the cube $0 \le x \le 2$, $0 \le y \le 2$, $0 \le z \le 2$. The plane $y + z = 3$, shown in Figure 18-1-7a, cuts the cube into two pieces. Clearly $y + z \le 3$ is the lower one, so we chop off the prism $y + z \ge 3$. The result is the solid shown in Figure 18-1-7b. The plane $x + y = 3$, shown in Figure 18-1-7b, cuts the solid into two pieces, and we want the rear one, the one that includes the origin. We chop off the front piece and are left with **D** itself, shown in Figure 18-1-7c.

Figure 18-1-7
See Example 4

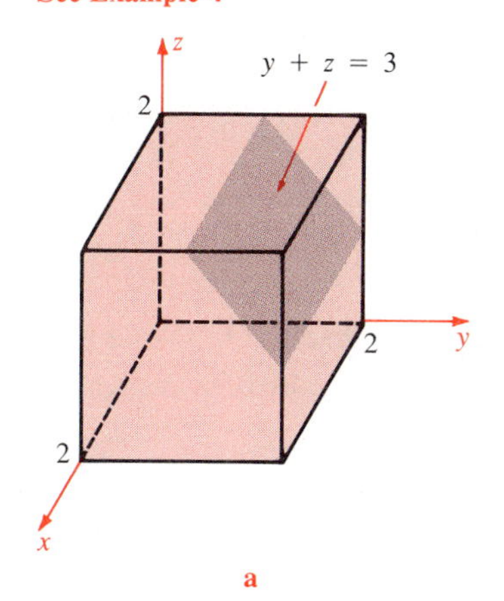

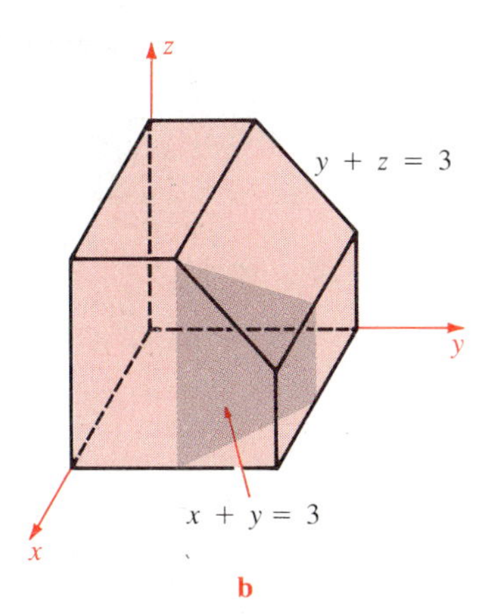

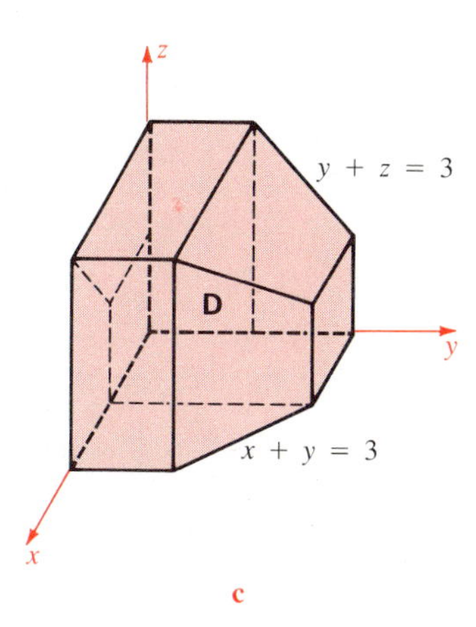

Clearly we have to split the integral into pieces. If for instance we project **D** into the z, x-plane, then the resulting plane domain splits naturally into three subdomains (outlined in Figure 18-1-7c). Better is to project **D** into the x, y-plane or into the y, z-plane; in either case the resulting plane domain splits into only two pieces (again refer to Figure 18-1-7c).

We choose the projection into the y, z-plane. The shadow of **D** is a plane domain $\mathbf{S} = \mathbf{S}_1 + \mathbf{S}_2$, where

$$\mathbf{S}_1 = \{0 \le y \le 1,\ 0 \le z \le 2\}$$
$$\mathbf{S}_2 = \{1 \le y \le 2,\ 0 \le z \le 3 - y\}$$

The part $\mathbf{D}_1$ of $\mathbf{D}$ that projects onto $\mathbf{S}_1$ is specified by $0 \le x \le 2$. The part $\mathbf{D}_2$ that projects onto $\mathbf{S}_2$ is specified by $0 \le x \le 3 - y$. Accordingly,

$$\begin{aligned} I &= \iiint_{\mathbf{D}_1} f\,dx\,dy\,dz + \iiint_{\mathbf{D}_2} f\,dx\,dy\,dz \\ &= \iint_{\mathbf{S}_1} \left(\int_0^2 f\,dx\right) dy\,dz + \iint_{\mathbf{S}_2} \left(\int_0^{3-y} f\,dx\right) dy\,dz \\ &= \int_0^1 \left[\int_0^2 \left(\int_0^2 f\,dx\right) dz\right] dy + \int_1^2 \left[\int_0^{3-y} \left(\int_0^{3-y} f\,dx\right) dz\right] dy \end{aligned}$$

Exercises

In working these and other exercises in this chapter, be sure to use the table of definite integrals inside the front cover.

Evaluate the triple integral over the indicated domain

1 $\iiint xy^2z\,dx\,dy\,dz$

$0 \le x \le 1 \quad 0 \le y \le 1 \quad 0 \le z \le 2$

2 $\iiint \dfrac{xy}{z}\,dx\,dy\,dz$

$0 \le x \le 1 \quad 1 \le y \le 2 \quad 1 \le z \le 3$

3 $\iiint \dfrac{x}{y+z}\,dx\,dy\,dz$

$-1 \le x \le 2 \quad 0 \le y \le 1 \quad 1 \le z \le 3$

4 $\iiint xy^2 \sin(xyz)\,dx\,dy\,dz$

$0 \le x \le 1 \quad 0 \le y \le \pi \quad 0 \le z \le 1$

5 $\iiint (x-y)(y-z)(z-x)\,dx\,dy\,dz$

$0 \le x \le 1 \le y \le 2 \le z \le 3$

6 $\iiint (x+y)(y+z)(z+x)\,dx\,dy\,dz$

$0 \le x \le 1 \quad 0 \le y \le 1 \quad 0 \le z \le 1$

7 $\iiint 120(x+y+z)^3\,dx\,dy\,dz$

$0 \le x \le 1 \quad 0 \le y \le 1 \quad 0 \le z \le 1$

8 $\iiint \dfrac{x+y}{y+z}\,dx\,dy\,dz$

$1 \le x \le 3 \quad 1 \le y \le 2 \quad 0 \le z \le 1$

9 $\iiint z\,dx\,dy\,dz$

$0 \le x \quad 0 \le y \quad x + y \le 1 \quad 0 \le z \le 1 - x^2$

10 $\iiint y\,dx\,dy\,dz$

$0 \le x \quad 0 \le y \quad x + y \le 1$
$0 \le z \le x^2 + 2y^2$

11 $\iiint \dfrac{xz}{(1+y)^2}\,dx\,dy\,dz$

$0 \le x \quad 0 \le z \le 1 - y^2 \quad y \ge x^2$

12 $\iiint (3x^2 - z^2)\,dx\,dy\,dz$

$y \le 1 \quad -y \le x \le y \quad -y^2 \le z \le y^2$

13 $\iiint z^3\,dx\,dy\,dz$

pyramid with apex $(0, 0, 1)$, and with base the square with vertices $(\pm 1, \pm 1, 0)$

14 the same as Exercise 13, except the square base has vertices $(\pm 1, 0, 0)$, $(0, \pm 1, 0)$

15 $\iiint xyz\,dx\,dy\,dz$

tetrahedron with vertices $(0, 0, 0)$, $(1, 0, 0)$, $(0, 1, 0)$, $(0, 0, 1)$

16 $\iiint y\,dx\,dy\,dz$

tetrahedron with vertices $(0, 0, 0)$, $(0, 0, 1)$, $(1, 1, 0)$, $(-1, 1, 0)$

17 $\iiint x\,dx\,dy\,dz$

tetrahedron with vertices $(0, 0, 0)$, $(0, 0, 1)$, $(0, 1, 0)$, $(1, 1, 1)$

18 $\iiint (y+z)\,dx\,dy\,dz$

tetrahedron with vertices $(1, 0, 0)$, $(0, 0, 2)$, $(1, 0, 1)$, $(1, 1, 1)$

*19 $\iiint x\,dx\,dy\,dz$

tetrahedron with vertices
$(1, 0, 0), \quad (-1, 1, 0), \quad (1, 1, 1), \quad (2, 2, 0)$

*20 $\iiint y\,dx\,dy\,dz$

tetrahedron with vertices
$(0, 0, 0), \quad (1, 1, 0), \quad (2, -2, 0), \quad (3, 0, 2)$

21 $\iiint (x + y + z)^2\,dx\,dy\,dz$

$0 \le x \le 1 \quad 0 \le y \le 1$
$0 \le z \le 1 \quad x + y + z \le 2$

22 $\iiint xy\,dx\,dy\,dz$

$0 \le x \le 2 \quad 0 \le y \le 2$
$0 \le z \quad x + y + 3z \le 3$

23 $\iiint (x + 2y + 3z)\,dx\,dy\,dz$

$0 \le x \le 2y \le 3z \le 6$

24 $\iiint z^2\,dx\,dy\,dz$

$(x - 1)^2 + y^2 \le 4 \quad (x + 1)^2 + y^2 \le 4$
$0 \le z \le y$

25 A solid cube has side a. Its density at each point is k times the product of the six distances of the point to the faces of the cube, where k is constant. Find the mass.

26 Electric charge is distributed over the tetrahedron with vertices **0**, **i**, **j**, **k**. The charge density at each point is a constant k times the product of the four distances from the point to the faces of the tetrahedron. Find the total charge.

27 Express $\int_0^a \left[\int_0^z \left(\int_0^y g(x)\,dx\right) dy\right] dz$ as a simple integral.

*28 Refer to the table of definite integrals inside the front cover. Assuming formula 7 for $\int_0^1 x^m(1 - x)^n\,dx$, derive formula 9 for $\iiint x^p y^q z^r (1 - x - y - z)^s\,dx\,dy\,dz$ over the tetrahedron $x \ge 0$, $y \ge 0$, $z \ge 0$, $x + y + z \le 1$. Assume p, q, r, and s are non-negative integers.

29 Take four vertices of a unit cube, no two adjacent. Find the volume of the tetrahedron with these points as vertices.

30 (cont.) Now take the tetrahedron whose vertices are the remaining four vertices of the cube. The two tetrahedra intersect in a certain polyhedron. Describe it and find its volume.

Here are a few more integrals on which to sharpen your skills.

31 $\iiint \dfrac{z}{x + y}\,dx\,dy\,dz$

$x \ge 0 \quad y \ge 0 \quad 1 \le x + y \le 2 \quad y \le z \le x$

32 $\iiint z\,dx\,dy\,dz$

$0 \le x \le 2 \quad 0 \le y \le 2 \quad 0 \le z \le xy$
$y \le (x - 2)^2 \quad x \le (y - 2)^2$

33 $\iiint x^2y^2z^2\,dx\,dy\,dz$

regular octahedron with vertices
$(\pm 1, 0, 0), \quad (0, \pm 1, 0), \quad (0, 0, \pm 1)$

34 $\iiint (x^2 + 2xy)\,dx\,dy\,dz$

$|2xy| \le z \le 1 - x^2 - y^2$

18-2 Cylindrical Coordinates

Cylindrical coordinates are designed to fit situations with rotational symmetry about an axis. (Such symmetry is called **axial symmetry.**) The axis is usually taken to be the z-axis.

The **cylindrical coordinates** of a point $\mathbf{x} = (x, y, z)$ are $[r, \theta, z]$, where $[r, \theta]$ are the polar coordinates of (x, y), and z is the third rectangular coordinate of $\mathbf{x}$ (Figure 18-2-1a, next page). Each surface $r =$ constant is a right circular cylinder; hence the name *cylindrical coordinates* (Figure 18-2-1b, next page).

Through each point $\mathbf{x}$ (not on the z-axis) pass the three surfaces $r =$ constant, $\theta =$ constant, $z =$ constant (Figure 18-2-2, next page). Each is orthogonal (perpendicular) to the other two at their common intersection $\mathbf{x}$.

The relations between the rectangular coordinates (x, y, z) and the cylindrical coordinates $[r, \theta, z]$ of a point are

Figure 18-2-1
Cylindrical coordinates

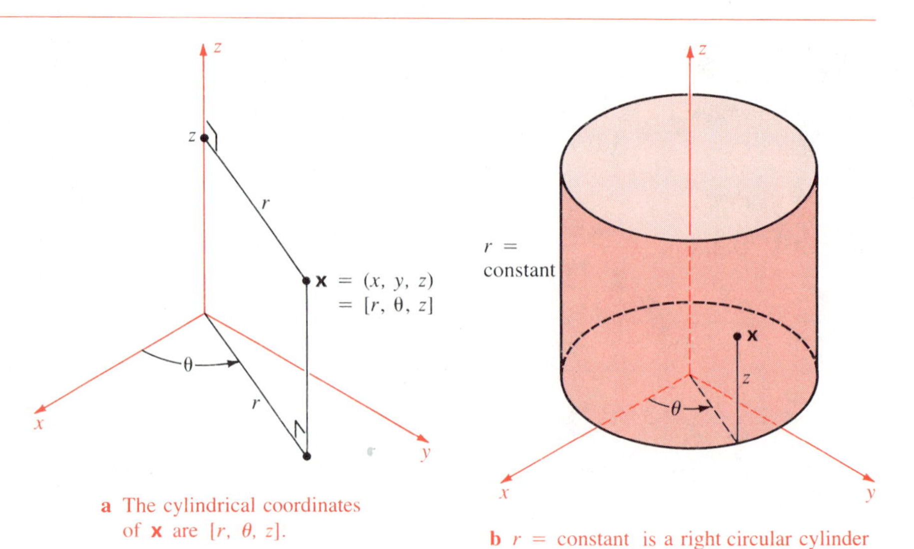

a The cylindrical coordinates of **x** are $[r, \theta, z]$.

b $r =$ constant is a right circular cylinder

Figure 18-2-2
The (mutually orthogonal) level surfaces of the cylindrical coordinates

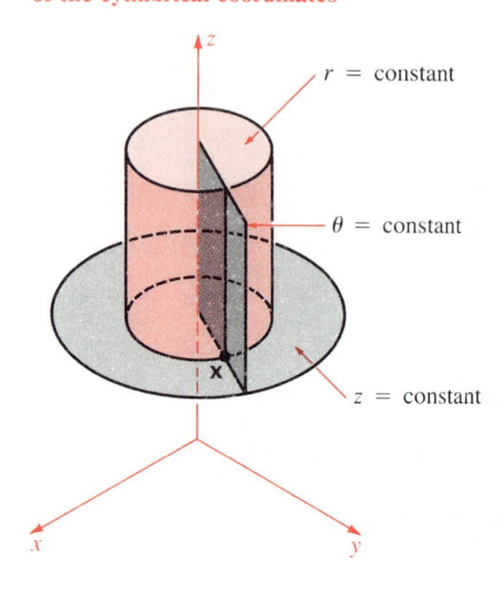

Rectangular and Cylindrical Coordinates

For any point **x** not on the z-axis,

$$\begin{cases} x = r\cos\theta \\ y = r\sin\theta \\ z = z \end{cases} \qquad \begin{cases} r^2 = x^2 + y^2 \\ \cos\theta = x/r \qquad \sin\theta = y/r \\ z = z \end{cases}$$

The origin in the plane is given in polar coordinates by $r = 0$; the angle θ is undefined. Similarly, a point on the z-axis is given in cylindrical coordinates by $r = 0$, $z =$ constant; θ is undefined.

Example 1

Graph the surfaces **a** $z = 2r$ **b** $z = r^2$.

Solution Both are surfaces of revolution about the z-axis, as is any surface $z = f(r)$. Since z depends only on r, not on θ, the height of the surface is constant above each circle $r = c$ in the x, y-plane. Thus the level curves are circles in the x, y-plane, centered at the origin.

a The surface meets the first quadrant of the y, z-plane in the line $z = 2y$. (Note that in the first quadrant of the y, z-plane, $x = 0$ and $y \geq 0$. Since $r^2 = x^2 + y^2 = y^2$, it follows that $r = y$.) Rotated about the z-axis, this line spans a cone with apex at **0**. See Figure 18-2-3a.

b The surface meets the y, z-plane in the parabola $z = y^2$. Rotated about the z-axis, this parabola generates a paraboloid of revolution (Figure 18-2-3b).

Integrals

If a solid has axial symmetry, it is often convenient to place the z-axis on the axis of symmetry and use cylindrical coordinates $[r, \theta, z]$ for the computation of integrals.

Figure 18-2-3
See Example 1

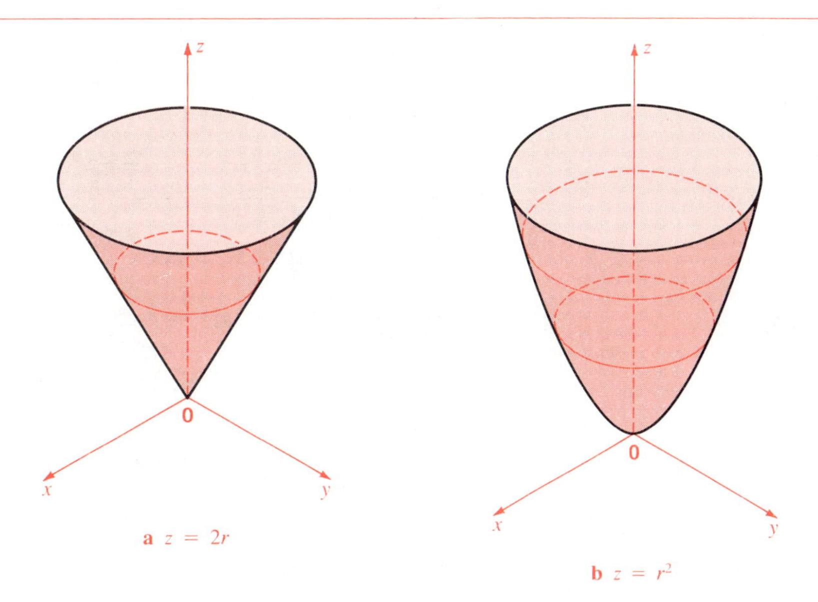

In polar coordinates $[r, \theta]$, the element of area is $dA = r\,dr\,d\theta$. Correspondingly, the element of volume in cylindrical coordinates $[r, \theta, z]$ is:

Element of Volume in Cylindrical Coordinates

$$dV = r\,dr\,d\theta\,dz$$

Let us justify this formula intuitively. We start at a point $\mathbf{x} = [r, \theta, z]$ and give small displacements dr, $d\theta$, dz to its cylindrical coordinates. According to Figure 18-2-4, the displacement of $\mathbf{x}$ in the r-direction has length dr, that in the θ-direction has length $r\,d\theta$, and that in the

Figure 18-2-4
Intuitive proof that
$dV = (dr)(r\,d\theta)(dz)$

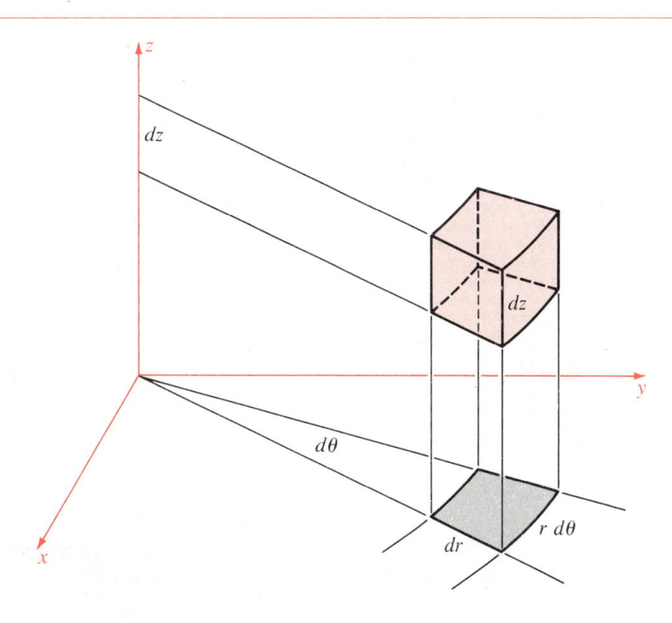

z-direction has length dz. These three displacements are mutually orthogonal, so they span a "rectangular" box of approximate volume $(dr)(r\,d\theta)(dz)$; hence the formula $dV = r\,dr\,d\theta\,dz$.

Example 2

Evaluate $\iiint z\sqrt{x^2 + y^2}\,dx\,dy\,dz$ taken over the first-octant portion of the solid cone with apex $(0, 0, 2)$ and base $x^2 + y^2 \le 1$.

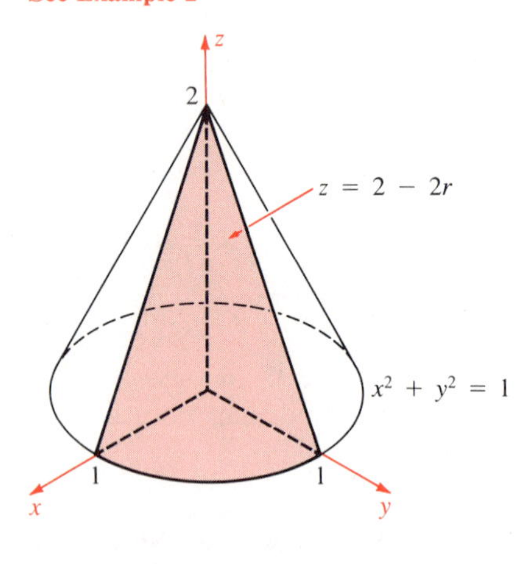

Figure 18-2-5
See Example 2

Solution The axial symmetry of the cone (Figure 18-2-5) plus the expression $\sqrt{x^2 + y^2} = r$ in the integrand make this problem a natural for cylindrical coordinates. The integral becomes

$$I = \iiint (rz)r\,dr\,d\theta\,dz = \iiint r^2 z\,dr\,d\theta\,dz$$

The surface of the cone must be of the form $z = f(r)$. Since $f(r)$ is obviously linear, and $f(0) = 2$ and $f(1) = 0$, the surface must be $z = 2 - 2r$. Therefore the solid domain of integration is described by the inequalities

$$0 \le \theta \le \tfrac{1}{2}\pi \qquad 0 \le r \le 1 \qquad 0 \le z \le 2 - 2r$$

By iteration,

$$\begin{aligned} I &= \left(\int_0^{\pi/2} d\theta\right)\left(\int_0^1 r^2\,dr \int_0^{2-2r} z\,dz\right) \\ &= \tfrac{1}{2}\pi \int_0^1 \tfrac{1}{2}r^2(2 - 2r)^2\,dr \\ &= \pi \int_0^1 r^2(1 - r)^2\,dr \\ &= \pi \int_0^1 (r^2 - 2r^3 + r^4)\,dr = \tfrac{1}{30}\pi \end{aligned}$$

●

Example 3

A region **D** in space is generated by revolving the plane region bounded by $z = 2x^2$, the x-axis, and $x = 1$ about the z-axis. Mass is distributed in **D** so that the density at each point in **D** equals $\delta = k(x^2 + y^2)(z + 1)$, where k is constant. Compute the total mass.

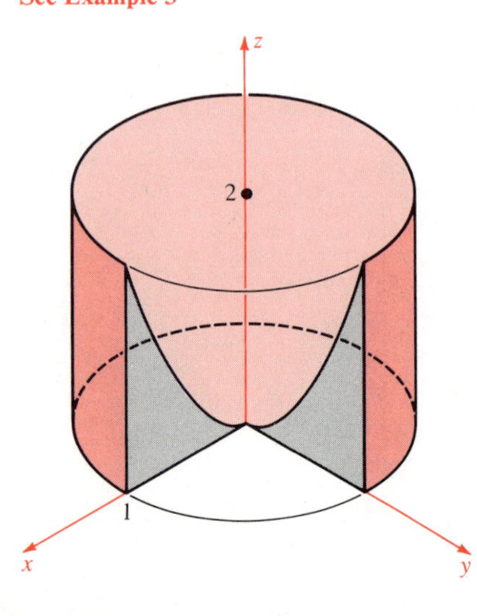

Figure 18-2-6
See Example 3

Solution We may write $\delta = kr^2(z + 1)$. A cut-away view of the solid is shown in Figure 18-2-6. In cylindrical coordinates, the solid is described by the inequalities

$$0 \le \theta \le 2\pi \qquad 0 \le r \le 1 \qquad 0 \le z \le 2r^2$$

For as Figure 18-2-6 shows, $0 \le r \le 1$. And fixing a value of r in this range determines the surface of a vertical cylinder on which z runs from the level $z = 0$ to the level $z = 2r^2$.

The total mass of the solid is

$$\iiint \delta(x, y, z)\, dx\, dy\, dz = \iiint kr^2(z + 1)r\, dr\, d\theta\, dz$$
$$= k\left(\int_0^{2\pi} d\theta\right)\int_0^1 r^3\left[\int_0^{2r^2} (z + 1)\, dz\right] dr$$
$$= 2\pi k \int_0^1 r^3[\tfrac{1}{2}(2r^2)^2 + (2r^2)]\, dr$$
$$= 4\pi k \int_0^1 (r^7 + r^5)\, dr = \tfrac{7}{6}\pi k$$

The Natural Frame

It is convenient to fit a frame of three mutually perpendicular vectors to cylindrical coordinates just as the frame **i**, **j**, **k** fits rectangular coordinates. At each point $[r, \theta, z]$ of space, attach three mutually perpendicular unit vectors **u**, **w**, **k** chosen so that

$$\left.\begin{matrix}\mathbf{u}\\ \mathbf{w}\\ \mathbf{k}\end{matrix}\right\} \text{ points in the direction of increasing } \left\{\begin{matrix} r\\ \theta\\ z\end{matrix}\right.$$

(Points on the z-axis must be excluded because angle θ is not defined there.) The vectors **u**, **w**, **k** at a point **x** are shown in Figure 18-2-7, and we can read their rectangular coordinates from the figure.

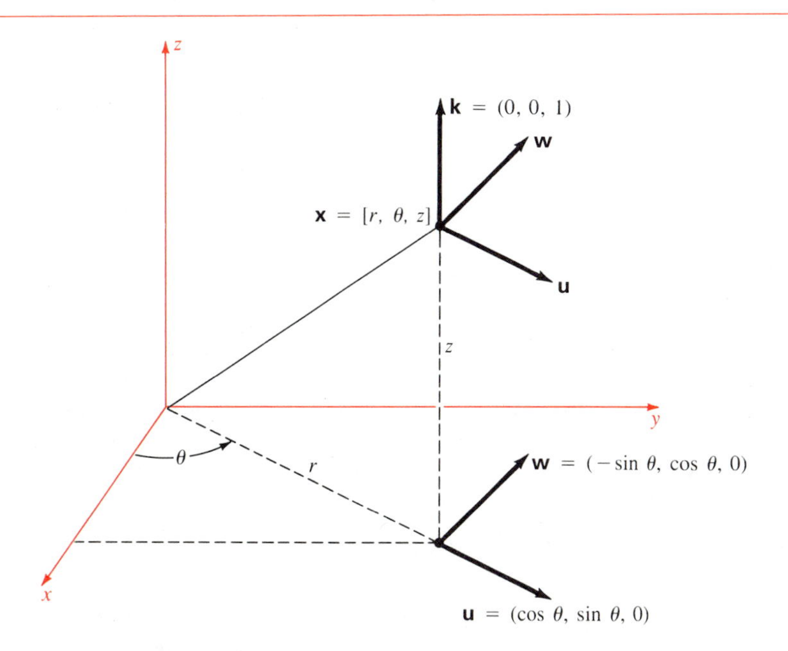

Figure 18-2-7
The natural frame for cylindrical coordinates

Natural Frame (Cylindrical Coordinates)

At each point $\mathbf{x} = [r, \theta, z]$ not on the z-axis, the triple of vectors

$$\mathbf{u} = (\cos\theta, \sin\theta, 0) \qquad \mathbf{w} = (-\sin\theta, \cos\theta, 0) \qquad \mathbf{k} = (0, 0, 1)$$

is the **natural frame** for cylindrical coordinates.

Note also the relation $\mathbf{x} = (x, y, z) = r\mathbf{u} + z\mathbf{k}$. Here we are viewing $\mathbf{x}$ as a vector function of r, θ, z.

At each point $\mathbf{x}$, the vectors $\mathbf{u}$, $\mathbf{w}$, $\mathbf{k}$ form a right-hand system:

$$\mathbf{u} \times \mathbf{w} = \mathbf{k} \qquad \mathbf{w} \times \mathbf{k} = \mathbf{u} \qquad \mathbf{k} \times \mathbf{u} = \mathbf{w}$$

Note that $\mathbf{u}$ and $\mathbf{w}$ vary with position but depend on θ alone, but that $\mathbf{k}$ is a constant vector, our old friend from the trio $\mathbf{i}$, $\mathbf{j}$, $\mathbf{k}$. Note also that

$$\frac{\partial\mathbf{u}}{\partial\theta} = \mathbf{w} \qquad \frac{\partial\mathbf{w}}{\partial\theta} = -\mathbf{u}$$

In situations with axial symmetry, it is frequently better to express vectors in terms of $\mathbf{u}$, $\mathbf{w}$, $\mathbf{k}$ rather than $\mathbf{i}$, $\mathbf{j}$, $\mathbf{k}$.

Let us express $d\mathbf{x}$ in terms of dr, $d\theta$, and dz. Intuitively (Figure 18-2-8), if r, θ, z are given small increments dr, $d\theta$, dz, then the displacement of $\mathbf{x}$ in the $\mathbf{u}$-direction is $dr\,\mathbf{u}$, in the $\mathbf{w}$-direction is $r\,d\theta\,\mathbf{w}$, and in the $\mathbf{k}$-direction is $dz\,\mathbf{k}$. Accordingly

Expansion of $d\mathbf{x}$ in the Natural Frame

$$d\mathbf{x} = dr\,\mathbf{u} + r\,d\theta\,\mathbf{w} + dz\,\mathbf{k}$$

The formula can be derived most directly by using differentials. We have $d\mathbf{u} = (\partial\mathbf{u}/\partial\theta)\,d\theta = \mathbf{w}\,d\theta$. Hence

$$\begin{aligned} d\mathbf{x} &= d(r\mathbf{u} + z\mathbf{k}) = dr\,\mathbf{u} + r\,d\mathbf{u} + dz\,\mathbf{k} \\ &= dr\,\mathbf{u} + r\,d\theta\,\mathbf{w} + dz\,\mathbf{k} \end{aligned}$$

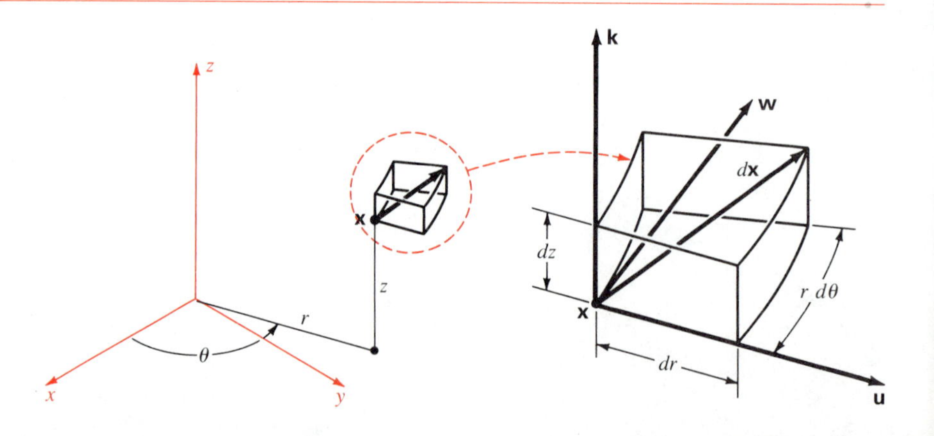

Figure 18-2-8
Geometric proof that
$d\mathbf{x} = dr\,\mathbf{u} + r\,d\theta\,\mathbf{w} + dz\,\mathbf{k}$

Exercises

Give an equation for the surface in cylindrical coordinates

1 $\dfrac{x}{a} + \dfrac{y}{b} + \dfrac{z}{c} = 1$

2 sphere, center **0**, radius a

3 cylinder parallel to z-axis, base the circle in the x, y-plane with center $(0, a)$ and radius a

4 hyperboloid $z = 2xy$

Use cylindrical coordinates to evaluate the integral over the indicated domain

5 $\displaystyle\iiint xyz\,dx\,dy\,dz$

$x \geq 0 \quad y \geq 0 \quad 0 \leq z \leq b \quad x^2 + y^2 \leq a^2$

6 $\displaystyle\iiint (x^2 + y^2 + z^2)\,dx\,dy\,dz$

$x^2 + y^2 \leq a^2 \quad |z| \leq b$

7 $\displaystyle\iiint yz\,dx\,dy\,dz$

$0 \leq z \leq y \quad x^2 + y^2 \leq a^2$

8 $\displaystyle\iiint z^2\,dx\,dy\,dz$

$x^2 + y^2 + z^2 \leq a^2 \quad x^2 + y^2 \leq b^2$
$(0 < b < a)$

9 $\displaystyle\iiint e^z\,dx\,dy\,dz$

$x^2 + y^2 \leq z \leq 2(x^2 + y^2) \leq 2$

10 $\displaystyle\iiint z^4\,dx\,dy\,dz$

$-(x^2 + y^2) \leq z \leq 0$
$x \geq 0 \quad y \leq 0 \quad x^2 + y^2 \leq a^2$

11 $\displaystyle\iiint z\,dx\,dy\,dz$

$1 \leq x^2 + y^2 + z^2 \quad 1 \leq x^2 + y^2 + (z-2)^2$
$x^2 + y^2 \leq 1 \quad 0 \leq z \leq 2$

12 $\displaystyle\iiint e^z\,dx\,dy\,dz$

$b \leq z \leq 2b + \sqrt{a^2 - x^2 - y^2}$
$x^2 + y^2 \leq a^2 \quad (b > 0)$

13 $\displaystyle\iiint z\,dx\,dy\,dz$

$(x - a)^2 + y^2 \leq a^2 \quad 2x \leq z \leq 3x$

14 $\displaystyle\iiint z\,dx\,dy\,dz$

$0 \leq r \leq \cos 2\theta \quad -\tfrac{1}{4}\pi \leq \theta \leq \tfrac{1}{4}\pi$
$0 \leq z \leq 1 - r^2$

15 $\displaystyle\iiint (y - 5z)\,dx\,dy\,dz \qquad \begin{array}{l} x^2 + z^2 \leq 4 \\ 0 \leq y \leq 1 \end{array}$

16 $\displaystyle\iiint (x^3 + y^3)\,dx\,dy\,dz \qquad \begin{array}{l} 0 \leq y \leq 2z \\ z^2 + x^2 \leq a^2 \end{array}$

17 $\displaystyle\iiint xy\,dx\,dy\,dz$

$y^2 + z^2 \leq a^2 \quad y \geq 0 \quad z \leq x \leq z + b$

18 $\displaystyle\iiint (y - x)\,dx\,dy\,dz$

$0 \leq ax \leq y^2 + z^2 \leq a^2 \quad y \leq 0$

19 $\displaystyle\iiint z^2\,dx\,dy\,dz$

$\tfrac{1}{4}x^2 + \tfrac{1}{9}y^2 \leq 1 \quad 0 \leq z \leq 2$

***20** $\displaystyle\iiint z\,dx\,dy\,dz$

$\tfrac{1}{4}x^2 + \tfrac{1}{9}y^2 \leq 1 \quad x \geq 0 \quad y \geq 0$
$x + y + z \leq \sqrt{13} \quad z \geq 0$

A space curve is given in parametric form:

$$r = r(t) \qquad \theta = \theta(t) \qquad z = z(t)$$

21 Express its velocity in terms of the natural frame **u**, **w**, **k**.

22 Express its acceleration in terms of the natural frame.

23 Use Exercise 21 to express its arc length for $a \leq t \leq b$ in terms of r, θ, z and their time derivatives.

24 Find the length of the spiral $r = A$, $\theta = Bt$, $z = Ct$, $a \leq t \leq b$.

A surface is given in the parametric form

$$r = r(s, t) \qquad \theta = \theta(s, t) \qquad z = z(s, t)$$

where (s, t) varies over a domain **D**

25 Express $\partial\mathbf{x}/\partial s$ and $\partial\mathbf{x}/\partial t$ in terms of the natural frame **u**, **w**, **k**.

26 Express $\partial^2\mathbf{x}/\partial s\,\partial t$ in terms of the natural frame.

27 Use Exercise 25 to express the surface area in terms of r, θ, z and their derivatives.

Use Exercise 27 to find the area

28 lateral surface of a right circular cone of radius a and height h

29 hemisphere of radius a

30 paraboloid $bz = x^2 + y^2$, $x^2 + y^2 \leq a^2$

31 Given a function f on a domain in $\mathbf{R}^3$, express df in terms of dr, $d\theta$, and dz.

32 (cont.) Express grad f in terms of the natural frame **u**, **w**, **k**. [Hint $df = (\text{grad} f) \cdot d\mathbf{x}$.]

18-3 Spherical Coordinates

Spherical coordinates are designed to fit situations with central symmetry. The **spherical coordinates** $\langle \rho, \phi, \theta \rangle$ of a point $\mathbf{x}$ are its distance $\rho = |\mathbf{x}|$ from the origin, its elevation angle ϕ, and it azimuth angle θ. (Observe that θ is the same polar coordinate as in cylindrical coordinates. Often θ is called the longitude and ϕ the co-latitude.) Note that θ is not determined on the z-axis, so points of this axis are usually avoided. In general, θ is determined up to a multiple of 2π, and $0 < \phi < \pi$. See Figure 18-3-1a.

Figure 18-3-1
Spherical coordinates

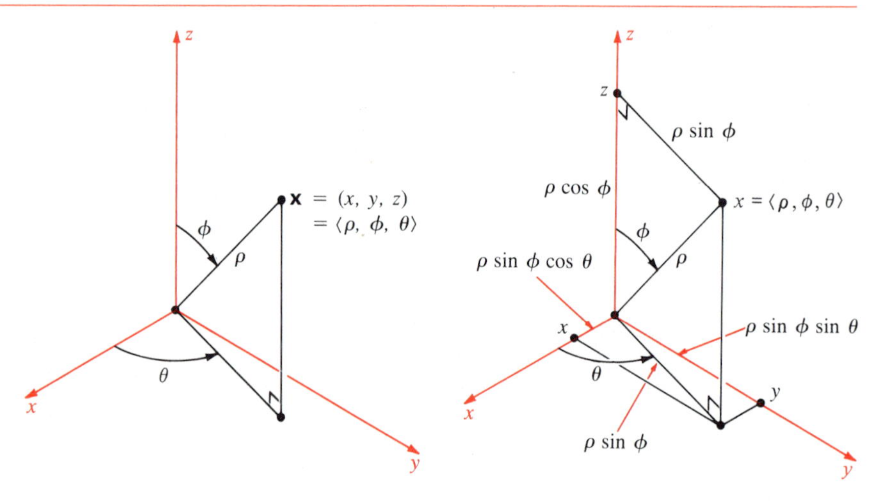

a The spherical coordinates of $\mathbf{x}$ are $\langle \rho, \phi, \theta \rangle$.

b The relation between (x, y, z) and $\langle \rho, \phi, \theta \rangle$

Figure 18-3-2
The (mutually orthogonal) level surfaces of the spherical coordinates

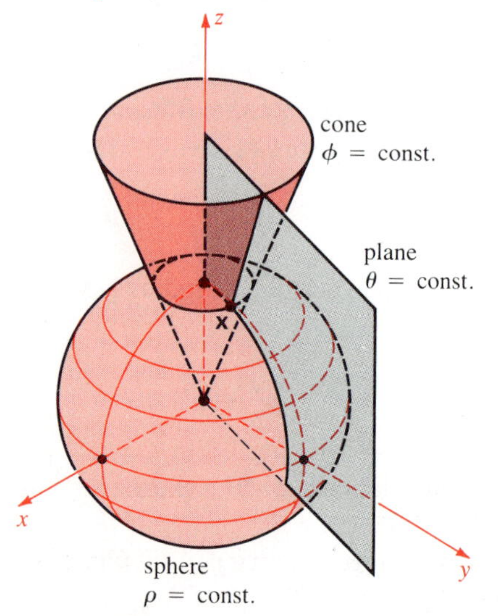

Relations between the rectangular coordinates (x, y, z) of a point and its spherical coordinates $\langle \rho, \theta, \phi \rangle$ may be read from Figure 18-3-1b. They are

Spherical and Rectangular Coordinates

For any point not on the z-axis

$$\begin{cases} x = \rho \sin\phi \cos\theta \\ y = \rho \sin\phi \sin\theta \\ z = \rho \cos\phi \end{cases} \qquad \begin{cases} \rho^2 = x^2 + y^2 + z^2 \\ \cos\phi = z/\rho \\ \tan\theta = y/x \end{cases}$$

The level surfaces

$$\left.\begin{matrix} \rho = \text{constant} \\ \phi = \text{constant} \\ \theta = \text{constant} \end{matrix}\right\} \quad \text{are} \quad \begin{cases} \text{concentric spheres about } \mathbf{0} \\ \text{right circular cones, apex } \mathbf{0} \\ \text{planes through the } z\text{-axis} \end{cases}$$

The three level surfaces through each point $\mathbf{x}$ intersect orthogonally (Figure 18-3-2).

Integrals

If a solid has central symmetry, it is convenient to place the origin at the center of symmetry and use spherical coordinates $\langle \rho, \phi, \theta \rangle$ for the computation of integrals.

Let us find a formula for the element of volume dV in terms of ρ, ϕ, and θ. We start at a point $\mathbf{x}$ and give small increments $d\rho$, $d\phi$, $d\theta$ to its spherical coordinates. According to Figure 18-3-3 the displacement of $\mathbf{x}$ in the ρ-direction has length $d\rho$, that in the ϕ-direction has length $\rho\, d\phi$, and that in the θ-direction has length $\rho \sin\phi\, d\theta$. The three displacements are mutually orthogonal, so they span a "rectangular" box of approximate volume $(d\rho)(\rho\, d\phi)(\rho \sin\phi\, d\theta)$, hence the formula

Element of Volume in Spherical Coordinates

$$dV = \rho^2 \sin\phi\, d\rho\, d\phi\, d\theta$$

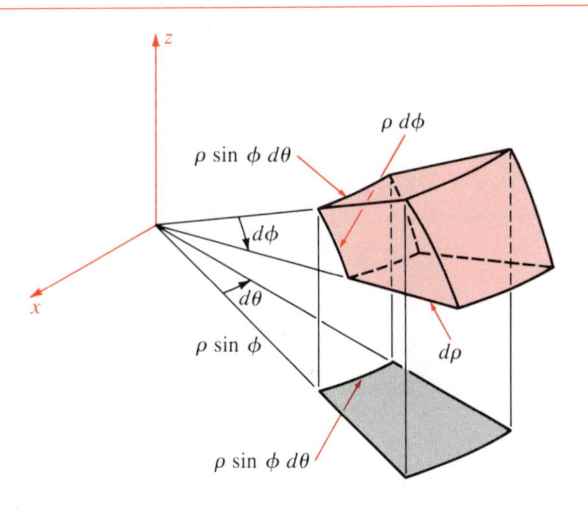

Figure 18-3-3
Intuitive proof that $dV = (d\rho)(\rho\, d\phi)(\rho \sin\phi\, d\theta)$

Example 1

Use spherical coordinates to find the volume of a sphere of radius a.

Solution The set-up and iteration is

$$V = \iiint dV = \iiint \rho^2 \sin\phi\, d\rho\, d\phi\, d\theta$$

$$= \left(\int_0^a \rho^2\, d\rho\right)\left(\int_0^\pi \sin\phi\, d\phi\right)\left(\int_0^{2\pi} d\theta\right)$$

$$= (\tfrac{1}{3}a^3)(2)(2\pi) = \tfrac{4}{3}\pi a^3$$

Example 2

Find the volume of the portion of the unit sphere that lies in the right circular cone having its apex at the origin and making angle α with the positive z-axis.

Figure 18-3-4
See Example 2

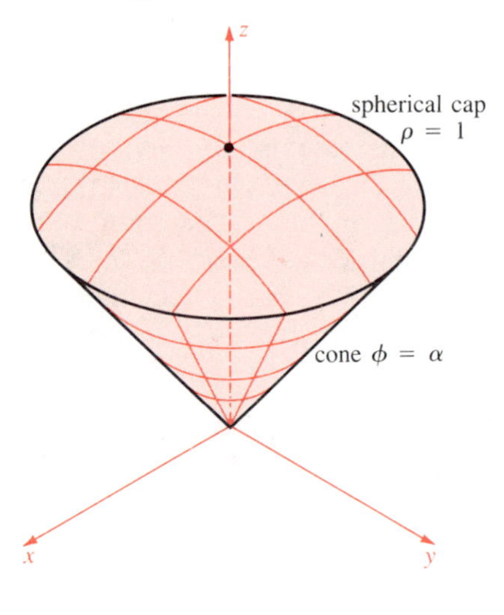

Solution The cone is specified by $0 \le \phi \le \alpha$, so the portion of the sphere is determined by $0 \le \theta \le 2\pi$, $0 \le \phi \le \alpha$, and $0 \le \rho \le 1$. See Figure 18-3-4. Hence the volume is

$$V = \left(\int_0^{2\pi} d\theta\right)\left(\int_0^{\alpha} \sin\phi\, d\phi\right)\left(\int_0^{1} \rho^2\, d\rho\right)$$
$$= (2\pi)(1 - \cos\alpha)(\tfrac{1}{3}) = \tfrac{2}{3}\pi(1 - \cos\alpha)$$

●

Remark As a check, let $\alpha \to \pi$. Then the volume should approach the volume of a sphere of radius 1. Does it?

Example 3

A solid fills the region between concentric spheres of radii a and b, where $0 < a < b$. The density at each point is inversely proportional to its distance from the center. Find the total mass.

Solution The solid is specified by $a \le \rho \le b$; the density is $\delta = k/\rho$. Hence

$$M = \iiint \delta(\mathbf{x})\, dV = \iiint \frac{k}{\rho}\rho^2 \sin\phi\, d\rho\, d\phi\, d\theta$$
$$= k\left(\int_0^{2\pi} d\theta\right)\left(\int_0^{\pi} \sin\phi\, d\phi\right)\left(\int_a^{b} \rho\, d\rho\right)$$
$$= (2\pi k)(2)\left(\frac{b^2 - a^2}{2}\right) = 2\pi k(b^2 - a^2)$$

●

Remark As $a \to 0$, the solid tends to the whole sphere, with infinite density at the center. But $M \to 2\pi k b^2$, which is finite.

Spherical Area

Suppose a domain $\mathbf{S}$ lies on the surface of the sphere $\rho = a$. We should be able to use spherical coordinates to find its area $|\mathbf{S}|$. What we need is a formula for the element of spherical area dA. Now ρ is constant. Small displacements $d\phi$ and $d\theta$ result in a small rectangular region (Figure 18-3-5) whose sides are $a\, d\phi$ and $a \sin\phi\, d\theta$. The element of area is the product of the sides:

Figure 18-3-5
Element of spherical area
$dA = (a\, d\phi)(a \sin\phi)\, d\theta$

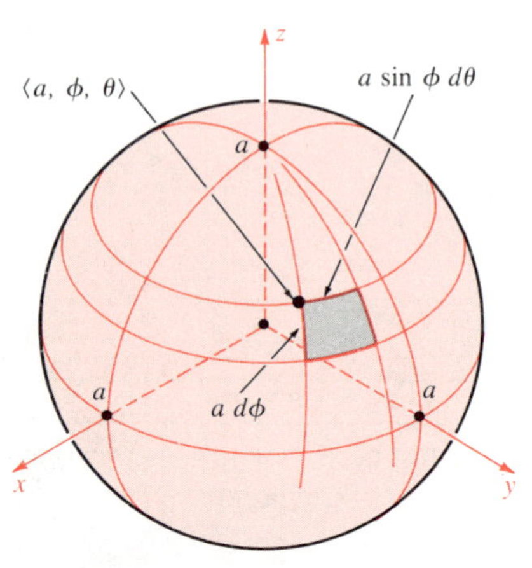

Element of Spherical Area

$$dA = a^2 \sin\phi\, d\phi\, d\theta$$

●

Remark This formula also follows from the method used in Section 17-6 to find the element of surface area in rectangular coordinates.

Example 4

Find the area of the polar cap, all points of co-latitude α or less on the unit sphere. Use the result to find the area of the whole sphere.

Solution Refer back to Figure 18-3-4. The region is defined on the sphere $\rho = 1$ by $0 \le \phi \le \alpha$, hence

$$A = \int_0^{2\pi}\left(\int_0^{\alpha} \sin\phi\, d\phi\right) d\theta = 2\pi(1 - \cos\alpha)$$

For the whole sphere, $\alpha = \pi$; hence $A = 2\pi[1 - (-1)] = 4\pi$. ●

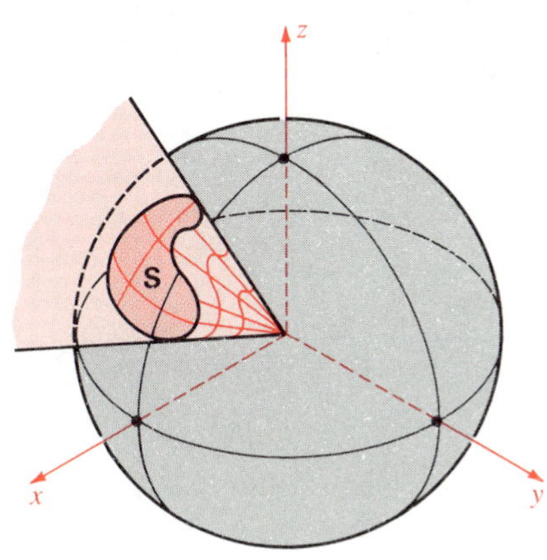

Figure 18-3-6
Solid angle

Remark Suppose **S** is a region on the *unit* sphere. The totality of infinite rays starting at **0** and passing through points of **S** is a cone, which is called a **solid angle** (Figure 18-3-6). A solid angle is measured by the area of the base region **S**. The unit for solid angles is the **steradian** (sr). The solid angle determined by the whole sphere equals 4π sr. The solid angle determined by the first octant equals $\frac{1}{2}\pi$ sr. The solid angle determined by the polar cap in Example 4 equals $2\pi(1 - \cos\alpha)$ sr.

The Natural Frame

As for cylindrical coordinates, there is a natural frame of unit vectors suited to spherical coordinates. At each point $\langle \rho, \phi, \theta \rangle$, we select unit vectors λ, μ, ν:

$$\left.\begin{matrix}\lambda\\ \mu\\ \nu\end{matrix}\right\} \text{ points in the direction of increasing } \left\{\begin{matrix}\rho\\ \phi\\ \theta\end{matrix}\right.$$

Points on the z-axis must be excluded because θ is not defined there. See Figure 18-3-7.

Figure 18-3-7
The frame λ, μ, ν

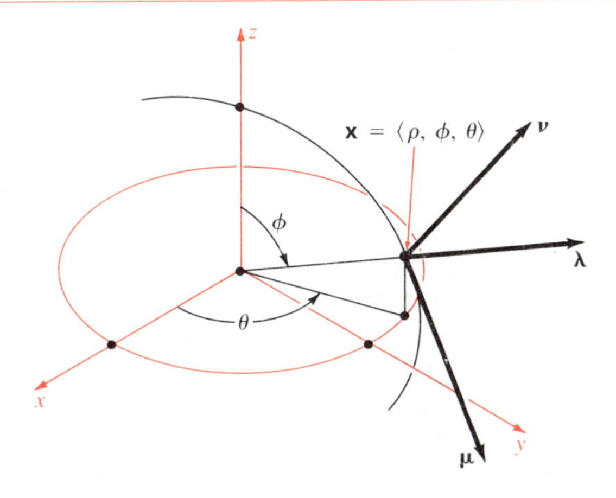

To express λ, μ, ν in terms of ρ, ϕ, θ, we use a short cut, starting with the formulas on page 892 for x, y, and z in terms of ρ, ϕ, and θ. We first rewrite these formulas as a vector relation:

$$\mathbf{x} = \rho(\sin\phi\cos\theta, \sin\phi\sin\theta, \cos\phi)$$

We take differentials:

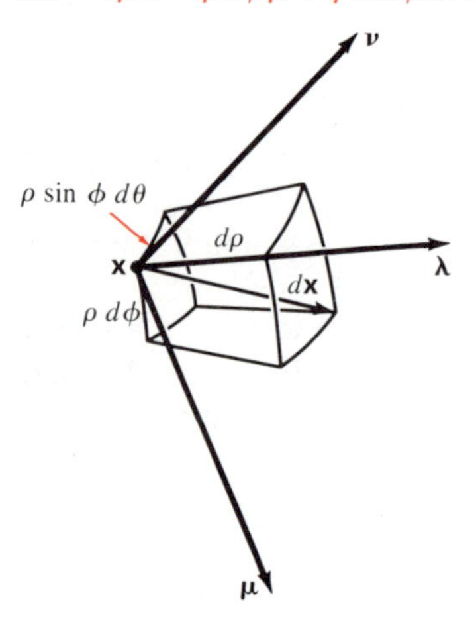

Figure 18-3-8
Geometric proof that
$d\mathbf{x} = d\rho\,\boldsymbol{\lambda} + \rho\,d\phi\,\boldsymbol{\mu} + \rho\sin\phi\,d\theta\,\boldsymbol{\nu}$

$$\begin{aligned} d\mathbf{x} &= (\sin\phi\cos\theta, \sin\phi\sin\theta, \cos\phi)\,d\rho \\ &\quad + \rho(\cos\phi\cos\theta, \cos\phi\sin\theta, -\sin\phi)\,d\phi \\ &\quad + \rho(-\sin\phi\sin\theta, \sin\phi\cos\theta, 0)\,d\theta \\ &= \boldsymbol{\lambda}\,d\rho + \boldsymbol{\mu}\rho\,d\phi + \boldsymbol{\nu}\rho\sin\phi\,d\theta \end{aligned}$$

(A geometric approach to this formula is shown in Figure 18-3-8.)
From the formula for $d\mathbf{x}$ follow:

Natural Frame (Spherical Coordinates)

At each point $\mathbf{x} = \langle \rho, \phi, \theta \rangle$ not on the z-axis,

$$\begin{cases} \boldsymbol{\lambda} = (\sin\phi\cos\theta, \sin\phi\sin\theta, \cos\phi) \\ \boldsymbol{\mu} = (\cos\phi\cos\theta, \cos\phi\sin\theta, -\sin\phi) \\ \boldsymbol{\nu} = (-\sin\theta, \cos\theta, 0) \end{cases}$$

Also $\mathbf{x} = \rho\boldsymbol{\lambda}$, and the displacement vector $d\mathbf{x}$ has the expansion

$$d\mathbf{x} = d\rho\,\boldsymbol{\lambda} + \rho\,d\phi\,\boldsymbol{\mu} + \rho\sin\phi\,d\theta\,\boldsymbol{\nu}$$

As is easily verified, the vectors $\boldsymbol{\lambda}$, $\boldsymbol{\mu}$, $\boldsymbol{\nu}$ given by these formulas are mutually orthogonal unit vectors. Furthermore, they point in the directions of increasing ρ, ϕ, θ respectively: if only ρ increases, then $d\phi = d\theta = 0$; hence $d\mathbf{x} = \boldsymbol{\lambda}\,d\rho$. Similarly, if only ϕ increases, then $d\mathbf{x} = \boldsymbol{\mu}\,\rho\,d\phi$; and if only θ increases, then $d\mathbf{x} = \boldsymbol{\nu}\,\rho\sin\phi\,d\theta$.

We are now in a position to prove analytically the formula for dV, which was derived intuitively at the beginning of this section. We have proved that $d\mathbf{x}$ is the sum of three mutually orthogonal displacements $d\rho\,\boldsymbol{\lambda}$, $\rho\,d\phi\,\boldsymbol{\mu}$, and $\rho\sin\phi\,d\theta\,\boldsymbol{\nu}$. These vectors span a box whose volume dV is a scalar triple product:

$$\begin{aligned} dV &= [d\rho\,\boldsymbol{\lambda}, \rho\,d\phi\,\boldsymbol{\mu}, \rho\sin\phi\,d\theta\,\boldsymbol{\nu}] \\ &= (d\rho)(\rho\,d\phi)(\rho\sin\phi\,d\theta)[\boldsymbol{\lambda}, \boldsymbol{\mu}, \boldsymbol{\nu}] \\ &= \rho^2\sin\phi\,d\rho\,d\phi\,d\theta \end{aligned}$$

The scalar triple product $[\boldsymbol{\lambda}, \boldsymbol{\mu}, \boldsymbol{\nu}]$ equals 1 because $\boldsymbol{\lambda}$, $\boldsymbol{\mu}$, $\boldsymbol{\nu}$ are mutually perpendicular unit vectors and form a right-handed system. (You should check that $\boldsymbol{\mu} \times \boldsymbol{\nu} = \boldsymbol{\lambda}$.)

Change of Variables

We now mention very briefly the general rule for changing variables in triple integrals. It includes the cylindrical and spherical coordinate changes as special cases. This discussion is a continuation of that in the previous chapter, p. 850.

First we define the **Jacobian** of a change of variables.

If $\begin{cases} x = x(u, v, w) \\ y = y(u, v, w) \\ z = z(u, v, w) \end{cases}$ then $\dfrac{\partial(x, y, z)}{\partial(u, v, w)} = \begin{vmatrix} x_u & x_v & x_w \\ y_u & y_v & y_w \\ z_u & z_v & z_w \end{vmatrix}$

Change of Variables

Suppose

$$\begin{cases} x = x(u, v, w) \\ y = y(u, v, w) \\ z = z(u, v, w) \end{cases}$$

is a one-one transformation of a domain **E** in **u**-space onto a domain **D** in **x**-space. Suppose the functions $x(u, v, w), \cdots$ are continuously differentiable and that

$$\frac{\partial(x, y, z)}{\partial(u, v, w)} > 0$$

at all points of **E**. Then

$$\iiint_{\mathbf{D}} f(x, y, z)\, dx\, dy\, dz$$
$$= \iiint_{\mathbf{E}} f[x(u, v, w), y(u, v, w), z(u, v, w)] \frac{\partial(x, y, z)}{\partial(u, v, w)}\, du\, dv\, dw$$

The main thing to remember is that

$$dx\, dy\, dz$$

is replaced by

$$\frac{\partial(x, y, z)}{\partial(u, v, w)}\, du\, dv\, dw$$

For spherical coordinates $u = \rho,\ v = \phi,\ w = \theta$ and

$$\begin{cases} x = \rho \sin\phi \cos\theta \\ y = \rho \sin\phi \sin\theta \\ z = \rho \cos\phi \end{cases}$$

The Jacobian is

$$\frac{\partial(x, y, z)}{\partial(\rho, \phi, \theta)} = \begin{vmatrix} \sin\phi\cos\theta & \rho\cos\phi\cos\theta & -\rho\sin\phi\sin\theta \\ \sin\phi\sin\theta & \rho\cos\phi\sin\theta & \rho\sin\phi\cos\theta \\ \cos\phi & -\rho\sin\phi & 0 \end{vmatrix}$$

$$= \rho^2\sin\phi \begin{vmatrix} \sin\phi\cos\theta & \cos\phi\cos\theta & -\sin\theta \\ \sin\phi\sin\theta & \cos\phi\sin\theta & \cos\theta \\ \cos\phi & -\sin\phi & 0 \end{vmatrix}$$

$$= \rho^2 \sin\phi$$

(The determinant can be expanded easily by minors of the third row.) Therefore $dx\, dy\, dz$ is replaced by $\rho^2 \sin\phi\, d\rho\, d\phi\, d\theta$.

Example 5

Find the volume enclosed by the ellipsoid

$$\frac{x^2}{a^2} + \frac{y^2}{b^2} + \frac{z^2}{c^2} = 1 \qquad (a, b, c > 0)$$

Solution We want

$$\text{vol}(\mathbf{D}) = \iiint_{\mathbf{D}} dx\,dy\,dz$$

where

$$\mathbf{D} = \left\{(x, y, z) \;\middle|\; \frac{x^2}{a^2} + \frac{y^2}{b^2} + \frac{z^2}{c^2} \le 1\right\}$$

Let $\mathbf{E} = \{(u, v, w) \mid u^2 + v^2 + w^2 \le 1\}$ be the unit sphere in $\mathbf{u}$-space and define the transformation

$$x = au \qquad y = bv \qquad z = cw$$

which takes $\mathbf{E}$ onto $\mathbf{D}$ in a one-one manner. Also

$$\frac{\partial(x, y, z)}{\partial(u, v, w)} = \begin{vmatrix} a & 0 & 0 \\ 0 & b & 0 \\ 0 & 0 & c \end{vmatrix} = abc > 0$$

Therefore

$$\text{vol}(\mathbf{D}) = \iiint_{\mathbf{D}} dx\,dy\,dz = \iiint_{\mathbf{E}} abc\,du\,dv\,dw$$

$$= abc \iiint_{\mathbf{E}} du\,dv\,dw$$

$$= abc\,\text{vol}(\mathbf{E}) = \tfrac{4}{3}\pi abc$$

●

Exercises

Given an equation for the surface in spherical coordinates

1 sphere, center $(0, 0, a)$, radius a

2 the cylinder of all points at distance a from the z-axis

3 paraboloid $z = x^2 + y^2$

4 hyperbolic paraboloid $z = x^2 - y^2$

5 right circular cylinder, axis through $(a, 0, 0)$ and parallel to the z-axis, radius a.

6 (cont.) Find the intersection of this cylinder with the sphere of radius $2a$ and center $\mathbf{0}$. Give your answer in the form of two relations between the spherical coordinates.

Use spherical coordinates to evaluate the integral over the indicated domain

7 $\displaystyle\iiint z\,dx\,dy\,dz$

$\rho \le a \qquad x \ge 0 \qquad y \ge 0 \qquad z \ge 0$

8 $\displaystyle\iiint x^2\,dx\,dy\,dz \qquad \rho \le a \qquad x \ge 0 \qquad y \ge 0$

9 $\displaystyle\iiint \rho^n\,dx\,dy\,dz \qquad n \ge -3 \qquad \rho \le a$

10 $\iiint (a-\rho)^n\,dx\,dy\,dz \qquad n \ge 0 \qquad \rho \le a$

11 $\iiint z\,dx\,dy\,dz \qquad 1 \le z \qquad \rho \le \frac{2}{3}\sqrt{3}$

12 $\iiint \rho^{-2}\,dx\,dy\,dz \qquad a \le \rho \le b \qquad (0 < a < b)$

13 $\iiint x^4\,dx\,dy\,dz \qquad \rho \le a$

14 $\iiint x^2y^2\,dx\,dy\,dz \qquad \rho \le a$

15 $\iiint (x^2+y^2)^2\,dx\,dy\,dz \qquad \rho \le a$

16 $\iiint (x^2+z^2)^2\,dx\,dy\,dz \qquad \rho \le a$

17 $\iiint z^2\,dx\,dy\,dz \qquad \rho \le a \qquad \frac{1}{4}\pi \le \phi \le \frac{3}{4}\pi$

18 $\iiint x^2\,dx\,dy\,dz \qquad \rho \le a \qquad \frac{5}{12}\pi \le \theta \le \frac{7}{12}\pi$

19 $\lim_{\varepsilon\to 0+} \iiint (\ln \rho)\,dx\,dy\,dz \qquad \varepsilon \le \rho \le 1$

20 $\lim_{\varepsilon\to 0+} \iiint \dfrac{dx\,dy\,dz}{\rho^{5/2}} \qquad \varepsilon \le \rho \le 1$

21 $\iiint z\,dx\,dy\,dz \qquad \rho \ge 1 \quad x^2+y^2 \le 1 \quad 0 \le z \le 1$

22 $\iiint \rho^2\,dx\,dy\,dz \qquad \rho \ge 1 \quad 0 \le x \le 1 \quad 0 \le y \le 1 \quad 0 \le z \le 1$

23 Given a function f on a domain in $\mathbf{R}^3$, express grad f in terms of the natural frame $\boldsymbol{\lambda}, \boldsymbol{\mu}, \mathbf{v}$. [Hint $df = (\text{grad} f)\cdot d\mathbf{x}$.]

24 Derive the three formulas

$$d\boldsymbol{\lambda} = d\phi\,\boldsymbol{\mu} + \sin\phi\,d\theta\,\mathbf{v}$$
$$d\boldsymbol{\mu} = -d\phi\,\boldsymbol{\lambda} + \cos\phi\,d\theta\,\mathbf{v}$$
$$d\mathbf{v} = -\sin\phi\,d\theta\,\boldsymbol{\lambda} - \cos\phi\,d\theta\,\boldsymbol{\mu}$$

by taking differentials of the formulas for $\boldsymbol{\lambda}, \boldsymbol{\mu}, \mathbf{v}$ in coordinates (page 896).

A space curve is given in the parametric form

$$\rho = \rho(t) \qquad \phi = \phi(t) \qquad \theta = \theta(t) \qquad a \le t \le b$$

25 Express its velocity $\mathbf{v}$ in terms of the natural frame $\boldsymbol{\lambda}$, $\boldsymbol{\mu}$, $\mathbf{v}$.

26 Do the same for its acceleration $\mathbf{a}$. [Hint Use Exercise 24.]

27 Use Exercise 25 to express its arc length in terms of ρ, ϕ, θ and their time derivatives.

28 A **rhumb line** on a sphere of radius a is a curve that intersects each meridian at the same angle α. (Follow a constant compass setting.) Find the length of a rhumb line of angle α from the equator to the north pole.

29 Set up a definite integral for the length of the conical spiral $\phi = \alpha$, $\rho = \theta$, $a \le \theta \le b$. Here α is a constant. [Use Exercise 27.]

30 Set up a definite integral for the length of the upper part of the curve of intersection in Exercise 6. [Use Exercise 27.]

A surface is given in the parametric form

$$\rho = \rho(u, v) \qquad \phi = \phi(u, v) \qquad \theta = \theta(u, v)$$

where (u, v) varies over a domain **D**

31 Express $\partial\mathbf{x}/\partial u$ and $\partial\mathbf{x}/\partial v$ in terms of the natural frame $\boldsymbol{\lambda}, \boldsymbol{\mu}, \mathbf{v}$.

32 Express $\partial^2\mathbf{x}/\partial u\,\partial v$ in terms of the natural frame [Hint Use the result of Exercise 24.]

33 Use Exercise 31 to express the surface area in terms of ρ, ϕ, θ and their derivatives.

34 Give a simple expression for the element of area on the cone $\phi = \alpha$, a constant.

35 (cont.) Find the area of the region $0 \le \rho \le b + a\sin n\theta$ on the cone, $b > a > 0$.

36 Use the result of Exercise 33 to work out in spherical coordinates the area of the cylindrical surface $x^2+y^2=a^2$, $0 \le z \le h$.

Use one or more changes of variables to evaluate

37 $\iiint xyz\,dx\,dy\,dz$

$\dfrac{x^2}{a^2} + \dfrac{y^2}{b^2} + \dfrac{z^2}{c^2} \le 1 \qquad x \ge 0 \qquad y \ge 0 \qquad z \ge 0$

38 $\iiint \dfrac{dx\,dy\,dz}{(x+y+z)^2}$

$a \le x+y+z \le b \qquad (0 < a < b)$

$x \ge 0 \qquad y \ge 0 \qquad z \ge 0$

39 $\iiint z^3\,dx\,dy\,dz$

$\dfrac{x^2}{a^2} + \dfrac{y^2}{b^2} \le 1 \qquad 0 \le \dfrac{z}{c} \le 1 - \dfrac{x^2}{a^2} - \dfrac{y^2}{b^2}$

***40** $\iiint (ax+by+cz)^{2n}\,dx\,dy\,dz$

$x^2+y^2+z^2 \le k^2$

41 The tetrahedron **T** has vertices **a**, **b**, **c**, **d**. Set up $I = \iiint f(\mathbf{x})\,dx\,dy\,dz$ over **T**, using the change of variables $\mathbf{x} = (1-u-v-w)\mathbf{a} + u\mathbf{b} + v\mathbf{c} + w\mathbf{d}$.

42 (cont.) Evaluate $\iiint y\,dx\,dy\,dz$ over the tetrahedron with vertices $(0, 0, 0)$, $(1, 1, 0)$, $(2, -2, 0)$, $(3, 0, 2)$.

18-4 Center of Gravity and Moments of Inertia

We have studied the center of gravity for non-homogeneous plane sheets and for wires in the plane. Now we extend this concept to solids.

Suppose a solid $\mathbf{D}$ has density $\delta(\mathbf{x})$ at each point $\mathbf{x}$. Its element of mass is $dM = \delta(\mathbf{x})\,dV$, and its (total) mass is

$$M = \iiint_{\mathbf{D}} dM = \iiint_{\mathbf{D}} \delta(\mathbf{x})\,dV = \iiint_{\mathbf{D}} \delta(\mathbf{x})\,dx\,dy\,dz$$

The **moment** of $\mathbf{D}$ is

$$\mathbf{m} = (m_x, m_y, m_z) = \iiint_{\mathbf{D}} \mathbf{x}\,dM = \iiint_{\mathbf{D}} \delta(\mathbf{x})\mathbf{x}\,dx\,dy\,dz$$

The **center of gravity** of $\mathbf{D}$ is

$$\bar{\mathbf{x}} = (\bar{x}, \bar{y}, \bar{z}) = \frac{1}{M}\mathbf{m}$$

Thus, for instance, the x-component of the center of gravity is

$$\bar{x} = \iiint_{\mathbf{D}} x\,dM \bigg/ \iiint_{\mathbf{D}} dM$$

So $\bar{x}$ is the weighted average of x over $\mathbf{D}$. Similar statements apply to $\bar{y}$ and $\bar{z}$. The center of gravity is the weighted average of the points of the solid. Recall in this connection that the center of gravity of a system of point-masses $M_1, \cdots, M_n$ located at $\mathbf{x}_1, \cdots, \mathbf{x}_n$ is also a weighted average, namely,

$$\bar{\mathbf{x}} = \frac{1}{M}(M_1\mathbf{x}_1 + M_2\mathbf{x}_2 + \cdots + M_n\mathbf{x}_n)$$

where $M = M_1 + \cdots + M_n$.

Symmetry

If a solid and its density function are symmetric in a coordinate plane, then the center of gravity lies on that coordinate plane. For example, suppose $\mathbf{D}$ is **symmetric** about the x, y-plane. This means that whenever a point (x, y, z) is in the solid, then $(x, y, -z)$ is in the solid, *and* $\delta(x, y, z) = \delta(x, y, -z)$. The contribution to m_z at (x, y, z) is

$$\delta(x, y, z)\,z\,dx\,dy\,dz$$

and it is cancelled by the contribution

$$\delta(x, y, -z)(-z)\,dx\,dy\,dz = -\delta(x, y, z)\,z\,dx\,dy\,dz$$

at $(x, y, -z)$. Hence $m_z = 0$ and $\bar{z} = 0$.

Similarly, if $\mathbf{D}$ and its density function are symmetric in a coordinate axis, then the center of gravity lies on that axis. Finally, if $\mathbf{D}$ and its density function are symmetric about the origin, then $\bar{\mathbf{x}} = \mathbf{0}$.

Similar conclusions apply to symmetries in arbitrary planes, lines, or points. Another physically intuitive fact is that the center of gravity of a solid $\mathbf{D}$ depends only on the solid and its density function, not on how the rectangular coordinate system is chosen.

Remark Suppose $\mathbf{D}$ is a domain in space. When we refer to the **center of gravity** (or **centroid**) of $\mathbf{D}$ without mentioning a density function, then it is understood that $\mathbf{D}$ is a uniform solid with $\delta = 1$. For such a solid, $M = V$ and $\bar{\mathbf{x}}$ depends only on the shape of the solid.

To compute the center of gravity of a solid, exploit any symmetry it has by choosing an appropriate coordinate system and expressing the element of volume dV in that system.

Example 1

Find the center of gravity of a hemisphere of radius a.

Solution Here $\delta = 1$ so

$$M = V = \tfrac{1}{2}(\tfrac{4}{3}\pi a^3) = \tfrac{2}{3}\pi a^3$$

To exploit symmetry, choose spherical coordinates. The hemisphere is symmetric in the z-axis; hence $\bar{\mathbf{x}}$ lies on the z-axis, $\bar{x} = \bar{y} = 0$. Therefore we need only compute $\bar{z}$. The z-moment is

$$\begin{aligned} m_z &= \iiint_{\mathbf{D}} z\,dV = \iiint_{\mathbf{D}} \rho\cos\phi\,\rho^2\sin\phi\,d\rho\,d\phi\,d\theta \\ &= \int_0^{2\pi} d\theta \int_0^{\pi/2} \cos\phi\sin\phi\,d\phi \int_0^a \rho^3\,d\rho \\ &= (2\pi)(\tfrac{1}{2})(\tfrac{1}{4}a^4) = \tfrac{1}{4}\pi a^4 \end{aligned}$$

Therefore

$$\bar{z} = \frac{m_z}{M} = \frac{\tfrac{1}{4}\pi a^4}{\tfrac{2}{3}\pi a^3} = \tfrac{3}{8}a$$

It follows that the center of gravity lies on the axis of the hemisphere, $\tfrac{3}{8}$ of the distance from the center to the pole (Figure 18-4-1). ●

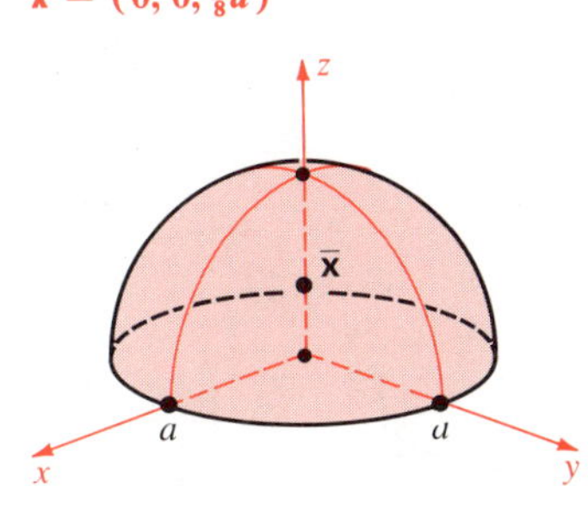

Figure 18-4-1
Uniform hemisphere:
$\bar{\mathbf{x}} = (0, 0, \tfrac{3}{8}a)$

Example 2

Find the center of gravity of a right circular cone of radius a and height h.

Solution Here $\delta = 1$, so

$$M = V = \tfrac{1}{3}\pi a^2 h$$

Use cylindrical coordinates with axes as in Figure 18-4-2a (next page). By symmetry, $\bar{x} = \bar{y} = 0$, so we need only compute m_z and $\bar{z}$:

Figure 18-4-2
Center of gravity of uniform right circular cone

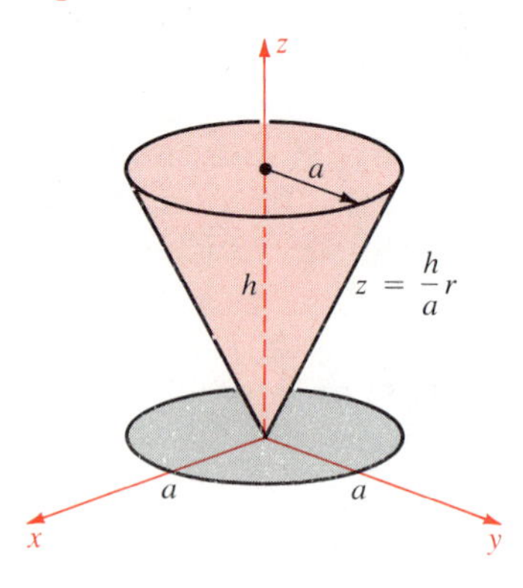

a $0 \le \theta \le 2\pi,\ 0 \le r \le a,\ \frac{h}{a}r \le z \le h$

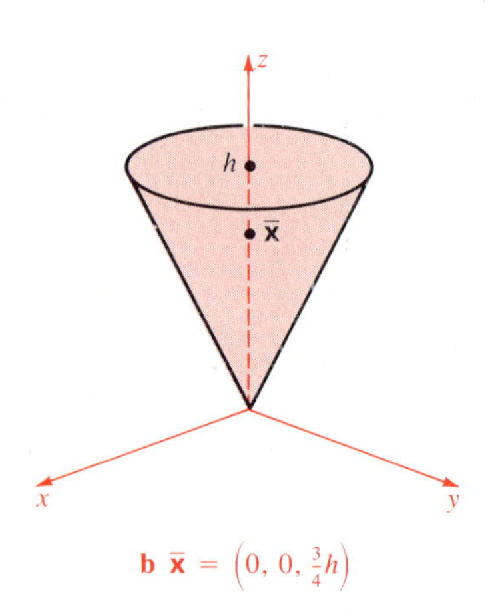

b $\bar{\mathbf{x}} = \left(0, 0, \frac{3}{4}h\right)$

$$m_z = \iiint_{\mathbf{D}} z\, dV = \iiint_{\mathbf{D}} z\, r\, dr\, d\theta\, dz$$

$$= \int_0^{2\pi} d\theta \int_0^a r\, dr \int_{hr/a}^h z\, dz$$

$$= (2\pi)\left(\frac{h^2}{2}\right)\int_0^a r\left(1 - \frac{r^2}{a^2}\right) dr$$

$$= \pi h^2\left(\frac{a^2}{2} - \frac{a^4}{4a^2}\right) = \tfrac{1}{4}\pi a^2 h^2$$

Therefore

$$\bar{z} = \frac{m_z}{M} = \frac{\frac{1}{4}\pi a^2 h^2}{\frac{1}{3}\pi a^2 h} = \tfrac{3}{4}h$$

It follows that the center of gravity is on the axis of the cone, $\frac{1}{4}$ of the distance from its base to its apex (Figure 18-4-2b). ●

Example 3

The solid $0 \le x \le 1$ m, $0 \le y \le 2$ m, $0 \le z \le 3$ m has density xyz kg/m³. Find its center of gravity.

Solution The mass is

$$M = \iiint xyz\, dx\, dy\, dz = \int_0^1 x\, dx \int_0^2 y\, dy \int_0^3 z\, dz$$

$$= \tfrac{1}{2}\cdot\tfrac{4}{2}\cdot\tfrac{9}{2} = \tfrac{9}{2}\text{ kg}$$

The moment is

$$\mathbf{m} = \iiint (x, y, z)\, xyz\, dx\, dy\, dz$$

$$= \left(\iiint x^2yz\, dx\, dy\, dz, \iiint xy^2z\, dx\, dy\, dz, \iiint xyz^2\, dx\, dy\, dz\right)$$

$$= \left(\int_0^1 x^2\, dx \int_0^2 y\, dy \int_0^3 z\, dz, \int_0^1 x\, dx \int_0^2 y^2\, dy \int_0^3 z\, dz, \int_0^1 x\, dx \int_0^2 y\, dy \int_0^3 z^2\, dz\right)$$

$$= \left(\tfrac{1}{3}\cdot\tfrac{4}{2}\cdot\tfrac{9}{2},\ \tfrac{1}{2}\cdot\tfrac{8}{3}\cdot\tfrac{9}{2},\ \tfrac{1}{2}\cdot\tfrac{4}{2}\cdot\tfrac{27}{3}\right)$$

$$= \frac{4\cdot 9}{3\cdot 2\cdot 2}(1, 2, 3) = 3(1, 2, 3)\text{ kg-m}$$

Hence, $\bar{\mathbf{x}} = \dfrac{\mathbf{m}}{M} = \tfrac{2}{9}\mathbf{m} = \tfrac{2}{9}(3)(1, 2, 3) = \left(\tfrac{2}{3}, \tfrac{4}{3}, 2\right)$ m. ●

Sometimes it is convenient to decompose a solid $\mathbf{D}$ into two or more pieces. Then the center of gravity of $\mathbf{D}$ can be expressed in terms of the centers of gravity of the pieces.

Addition Law

Suppose a solid $\mathbf{D}$ of mass M and center of gravity $\bar{\mathbf{x}}$ is made up of two pieces $\mathbf{D}_0$ and $\mathbf{D}_1$, of masses M_0 and M_1, and centers of gravity $\bar{\mathbf{x}}_0$ and $\bar{\mathbf{x}}_1$. Then

$$M = M_0 + M_1 \qquad \bar{\mathbf{x}} = \frac{1}{M}(M_0\bar{\mathbf{x}}_0 + M_1\bar{\mathbf{x}}_1)$$

The first formula is obvious. The second is just a decomposition of the moment integral.

$$M\bar{\mathbf{x}} = \iiint_{\mathbf{D}} \delta(\mathbf{x})\mathbf{x}\,dV = \iiint_{\mathbf{D}_0} + \iiint_{\mathbf{D}_1} = M_0\bar{\mathbf{x}}_0 + M_1\bar{\mathbf{x}}_1$$

A similar principle applies to plane sheets and wires.

Example 4

A solid consists of a cylindrical can of radius a and height h capped by a hemisphere at one end (Figure 18-4-3a). Locate its center of gravity.

Figure 18-4-3
Center of gravity of uniform cylinder with hemispherical cap

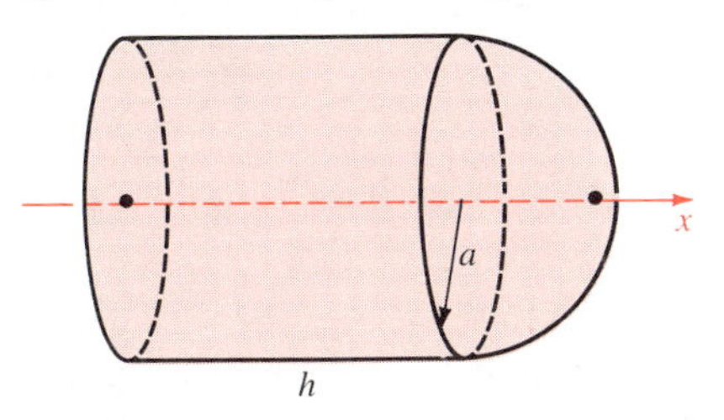

a Three-dimensional view

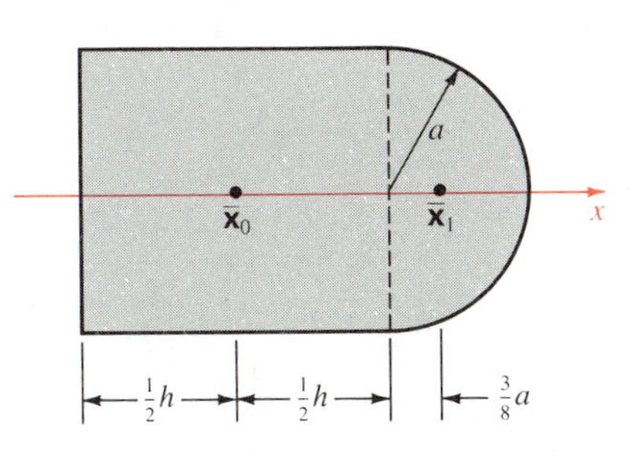

b Cross section

Solution Since no density is given, we may assume that $\delta = 1$ and work with volumes instead of masses. We choose the x-axis along the axis of symmetry of the solid, with 0 at its base. The cylinder has volume

$$V_0 = \pi a^2 h \qquad \text{and centroid} \qquad \bar{x}_0 = \tfrac{1}{2}h$$

By Example 1, the hemisphere has volume

$$V_1 = \tfrac{2}{3}\pi a^3 \qquad \text{and centroid} \qquad \bar{x}_1 = h + \tfrac{3}{8}a$$

See Figure 18-4-3b. Therefore, the volume of the whole solid is

$$V = V_0 + V_1 = \pi a^2 h + \tfrac{2}{3}\pi a^3 = \tfrac{1}{3}\pi a^2(3h + 2a)$$

and its center of gravity is on the axis at $(\bar{x}, 0, 0)$, where

$$\begin{aligned}\bar{x} &= \frac{1}{V}(V_0\bar{x}_0 + V_1\bar{x}_1) = \frac{1}{V}[(\pi a^2 h)(\tfrac{1}{2}h) + (\tfrac{2}{3}\pi a^3)(h + \tfrac{3}{8}a)]\\ &= \frac{1}{V}(\pi a^2)[\tfrac{1}{2}h^2 + (\tfrac{2}{3}a)(h + \tfrac{3}{8}a)]\\ &= \frac{1}{V}(\tfrac{1}{12}\pi a^2)(6h^2 + 8ah + 3a^2) = \frac{6h^2 + 8ah + 3a^2}{4(3h + 2a)}\end{aligned}$$

To check, note that $\bar{x} \to \frac{3}{8}a$ as $h \to 0$, and $\bar{x} \to \frac{1}{2}h$ as $a \to 0$.

Moments of Inertia: Arbitrary Axis

Let **D** be a solid with density $\delta(\mathbf{x})$ and let α be any straight line (axis) in space. The **moment of inertia** of **D** about α is

$$I_\alpha = \iiint_{\mathbf{D}} p(\mathbf{x})^2 \, \delta(\mathbf{x}) \, dV = \iiint_{\mathbf{D}} p(\mathbf{x})^2 \, dM$$

where $p(\mathbf{x})$ is the distance from $\mathbf{x}$ to the axis α. See Figure 18-4-4a. [Moments of inertia are sometimes called *second moments,* to distinguish them from the first moments used to define center of gravity. The *products of inertia* (Exercises 37–42) are also second moments.]

Figure 18-4-4
Moment of inertia and kinetic energy

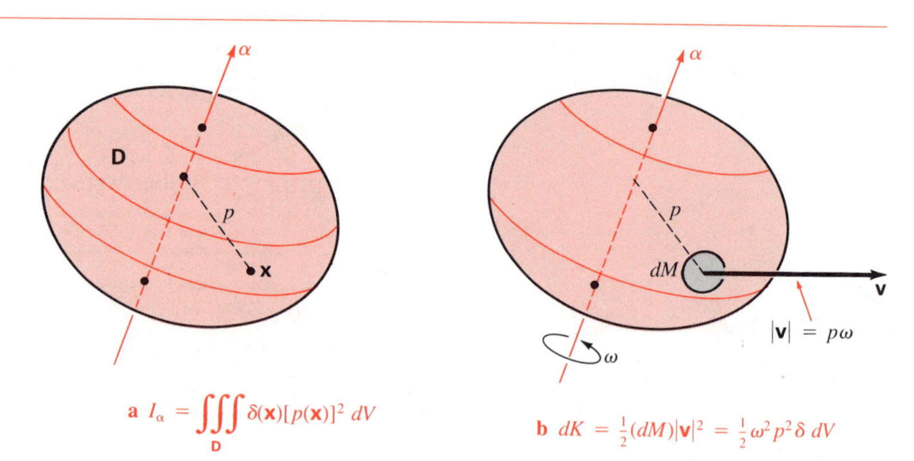

The quantity I_α is used in computing kinetic energies of rotating bodies. Suppose **D** rotates about α with angular speed ω. See Figure 18-4-4b. A point $\mathbf{x}$ in **D** moves with speed $p(\mathbf{x})\omega$. Since kinetic energy equals one-half the mass times the speed squared, an element of mass dM at $\mathbf{x}$ has kinetic energy $dK = \frac{1}{2}[p(\mathbf{x})\omega]^2 \, dM$. Hence the total kinetic energy of **D** is

$$K = \tfrac{1}{2}\omega^2 \iiint_{\mathbf{D}} p(\mathbf{x})^2 \, dM = \tfrac{1}{2} I_\alpha \omega^2$$

We denote the moments of inertia of **D** with respect to the x-axis, y-axis, and z-axis, respectively, by I_x, I_y, and I_z. For these quantities, $p(\mathbf{x})^2$ is particularly simple. For instance, if $\mathbf{x} = (x, y, z)$ and $p(\mathbf{x})$ is the distance to the x-axis, then $p(\mathbf{x})^2 = y^2 + z^2$. Similar formulas hold relative to the y-axis and the z-axis.

Moments of Inertia: Coordinate Axes

$$I_x = \iiint_{\mathbf{D}} (y^2 + z^2) \, \delta(x, y, z) \, dV$$

$$I_y = \iiint\limits_{\mathbf{D}} (z^2 + x^2)\delta(x, y, z)\, dV$$

$$I_z = \iiint\limits_{\mathbf{D}} (x^2 + y^2)\delta(x, y, z)\, dV$$

Example 5

Compute the moments of inertia I_x, I_y, I_z of a uniform sphere of center **0**, radius a, and mass M.

Solution If δ is the constant density, then $M = \frac{4}{3}\pi a^3 \delta$. By symmetry $I_x = I_y = I_z$. It seems most natural to use spherical coordinates to compute I_z:

$$\begin{aligned} I_z &= \delta \iiint (x^2 + y^2)\, dV \\ &= \delta \iiint (\rho^2 \sin^2\phi \cos^2\theta + \rho^2 \sin^2\phi \sin^2\theta)\, dV \\ &= \delta \iiint (\rho^2 \sin^2\phi)\rho^2 \sin\phi\, d\rho\, d\phi\, d\theta \\ &= \delta \int_0^{2\pi} d\theta \int_0^{\pi} \sin^3\phi\, d\phi \int_0^a \rho^4\, d\rho \\ &= \delta(2\pi)(\tfrac{4}{3})(\tfrac{1}{5}a^5) = \tfrac{2}{5}(\tfrac{4}{3}\pi a^3 \delta)a^2 = \tfrac{2}{5}Ma^2 \end{aligned}$$

Units If M is measured in kilograms and a in meters, then I_α is measured in kg-m^2. In the formula for kinetic energy $K = \frac{1}{2}I_\alpha \omega^2$, if I_α is in kg-m^2 and ω in rad/sec, then K is in joules = newton-meters. A newton, the metric unit of force, equals one kg-m/sec^2.

Suppose α is an axis through the center of gravity of a solid **D**, and β is an axis parallel to α. There is a formula that expresses the moment of inertia I_β in terms of I_α.

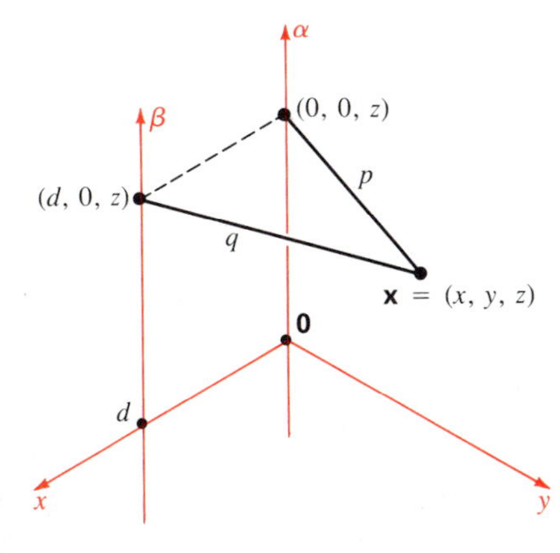

Figure 18-4-5
Parallel axis theorem

Parallel Axis Theorem

If α is an axis through the center of gravity of a solid **D**, and β is an axis parallel to α, then

$$I_\beta = I_\alpha + Md^2$$

where M is the mass of **D** and d is the distance between the axes.

Proof Choose the coordinate system with $\bar{\mathbf{x}} = \mathbf{0}$, and with α the z-axis and β the line $x = d$, $y = 0$. See Figure 18-4-5. For any point $\mathbf{x} = (x, y, z)$ of **D**, let $p(\mathbf{x})$ be the distance from $\mathbf{x}$ to the axis α and $q(\mathbf{x})$ the distance from $\mathbf{x}$ to the axis β. Then

$$p(\mathbf{x})^2 = x^2 + y^2$$
$$q(\mathbf{x})^2 = (x - d)^2 + y^2 = x^2 + y^2 - 2\,dx + d^2$$
$$= p(\mathbf{x})^2 - 2\,dx + d^2$$

Therefore

$$I_\beta = \iiint_{\mathbf{D}} q(\mathbf{x})^2\,dM = \iiint_{\mathbf{D}} [p(\mathbf{x})^2 - 2\,dx + d^2]\,dM$$
$$= I_\alpha - 2d \iiint_{\mathbf{D}} x\,dM + d^2 \iiint_{\mathbf{D}} dM$$

The second integral is the moment m_x, and the third integral is M. But $m_x = 0$ because the center of gravity is at $\mathbf{0}$. Hence

$$I_\beta = I_\alpha + Md^2$$ ●

Example 6

Find the moment of inertia of a uniform sphere of radius a and mass M about an axis tangent to the sphere.

Solution From Example 5, the moment of inertia about any axis through the center (of gravity) is $\frac{2}{5}Ma^2$. The distance from a tangent axis β to the center is a, so the parallel axis theorem implies

$$I_\beta = \tfrac{2}{5}Ma^2 + Ma^2 = \tfrac{7}{5}Ma^2$$ ●

Exercises

Find the center of gravity of

1 the first-octant portion of the uniform sphere $\rho \le a$
2 the hemisphere $\rho \le a$, $z \ge 0$, density $\delta = a - \rho$
3 the uniform spherical cone $\rho \le a$, $0 \le \phi \le \alpha$
4 the uniform hemispherical shell $a \le \rho \le b$, $z \ge 0$
5 the uniform solid $0 < a \le r \le b$, $0 \le z \le r$
6 the uniform solid $x^2 + y^2 \le z \le 1$
7 the uniform spherical cap (surface) $\rho = a$, $0 \le \phi \le \alpha$
*8 the uniform solid spherical cap $\rho \le a$, $a - h \le z$
9 the uniform sheet $x^2 + y^2 \le 4a^2$, $(x - a)^2 + y^2 \ge a^2$
10 the uniform solid $x^2 + y^2 + z^2 \le 4a^2$, $(x - a)^2 + y^2 + z^2 \ge a^2$
11 the lateral surface of the cone $\phi = \alpha$, $0 \le \rho \le a$
12 the uniform solid $r \le a$, $0 \le az \le r^2$
13 the uniform spherical triangle $\rho = a$, $x \ge 0$, $y \ge 0$, $z \ge 0$
14 the uniform wedge $\rho \le a$, $-\alpha \le \theta \le \alpha$
15 the uniform lune $\rho = a$, $-\alpha \le \theta \le \alpha$
16 the sphere $\rho \le a$ with density $\delta = a + z$
17 the uniform frustum of a right circular cone of height h and base radii $a < b$
*18 the uniform cone with apex $(0, 0, c)$ and base a domain $\mathbf{D}$ in the x, y-plane
19 the octant of a uniform ellipsoid $x \ge 0$, $y \ge 0$, $z \ge 0$, $x^2/a^2 + y^2/b^2 + z^2/c^2 \le 1$
*20 the uniform tetrahedron with vertices $\mathbf{a}$, $\mathbf{b}$, $\mathbf{c}$, $\mathbf{d}$ [Hint Use Exercise 41 in Section 18-3.]

Find the moments of inertia; give your answer in the form $I_x = M \cdot (\;)$, etc.

21 uniform solid $|x| \le a$, $|y| \le b$, $|z| \le c$
22 uniform hemisphere $\rho \le a$, $z \ge 0$
23 uniform cylinder $r \le a$, $|z| \le h$
24 uniform cone $hr \le az \le ah$
25 uniform cylinder $(x - a)^2 + y^2 \le a^2$, $|z| \le h$

6 uniform solid torus $(r-b)^2+z^2 \le a^2,\ 0<a<b$
7 uniform double cone $a|z| \le h(a-r)$
8 sphere $\rho \le a$, density $\delta = 1/\rho$
9 uniform solid paraboloid $0 \le z \le h(1-r^2/a^2)$
0 uniform solid $\rho \ge a,\ |x| \le a,\ |y| \le a,\ |z| \le a$
1 uniform spherical shell $\rho = a$
2 uniform lateral surface of cone $r \le a,\ hr = az$
3 uniform toroidal shell
$(r-b)^2+z^2 = a^2,\ 0<a<b$
[Hint Use the parametrization
$x=(b+a\cos\phi)\cos\theta,\ y=(b+a\cos\phi)\sin\theta,$
$z = a\sin\phi,\ 0 \le \phi \le 2\pi,\ 0 \le \theta \le 2\pi.$]
4 uniform ellipsoid $x^2/a^2+y^2/b^2+z^2/c^2 \le 1$

5 In Example 5, use symmetry to prove, without integrating, that $I_z = \frac{2}{3}\delta \iiint \rho^2\,dV$. Then evaluate the integral.

6 A domain $\mathbf{D}$ in the x, y-plane has density $\delta = 1$, area A, and moment of inertia $I = Ad^2$ (with respect to the z-axis). Find I_z for the cone with base $\mathbf{D}$, apex $(0, 0, c)$ and density 1.

The **products of inertia** of a domain $\mathbf{D}$ with density function δ and element of mass $dM = \delta\,dV$ are defined by

$$I_{yz} = -\iiint_{\mathbf{D}} yz\,dM \qquad I_{zx} = -\iiint_{\mathbf{D}} zx\,dM$$

$$I_{xy} = -\iiint_{\mathbf{D}} xy\,dM$$

Find the products of inertia for the

37 uniform box $0 \le x \le a,\ 0 \le y \le b,\ 0 \le z \le c$
38 uniform tetrahedron, vertices
$(0,0,0),\ (a,0,0),\ (0,b,0),\ (0,0,c)$
39 uniform quarter cylinder
$r \le a,\ 0 \le \theta \le \frac{1}{2}\pi,\ 0 \le z \le h$
40 uniform hemisphere
$(x-a)^2+y^2+z^2 \le a^2,\ z \ge 0$
41 cube $|x| \le a,\ |y| \le a,\ |z| \le a$,
density $\delta = (x+a)(y+a)(z+a)$
42 prism $0 \le x,\ 0 \le y,\ |z| \le h,\ x+y \le a$,
density $\delta = xy(z+h)$

18-5 Line Integrals

The remainder of this chapter deals with certain new kinds of integrals and their relations to double and triple integrals. The discussion starts here with integrals over curves in space and in the plane. The next section studies an important relation between integrals over closed plane curves and certain double integrals.

A **vector field** is the assignment of a vector $\mathbf{F}(\mathbf{x})$ to each point $\mathbf{x}$ of a region $\mathbf{D}$ in space. Familiar examples of vector fields are the gradient of a function, the wind velocity field on a weather map, the gravitational field in space, and the magnetic field near a magnet.

Let $\mathbf{F}(\mathbf{x})$ be a vector field on $\mathbf{D}$ and suppose $\mathbf{c}$ is directed curve in $\mathbf{D}$. See Figure 18-5-1. Written in components,

Figure 18-5-1
A vector field $\mathbf{F}$ and directed curve $\mathbf{c}$ in domain $\mathbf{D}$

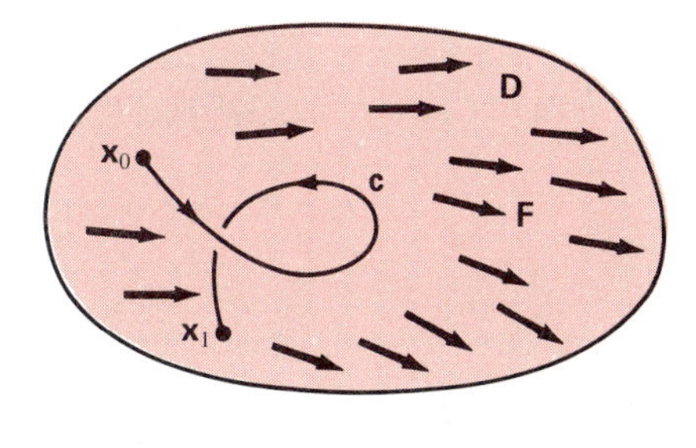

$$\mathbf{F}(\mathbf{x}) = (P(\mathbf{x}), Q(\mathbf{x}), R(\mathbf{x}))$$

By restricting our attention to $\mathbf{F}$ on $\mathbf{c}$ only, we define a new kind of integral denoted by

$$\int_{\mathbf{c}} \mathbf{F}\cdot d\mathbf{x} \qquad \text{or} \qquad \int_{\mathbf{c}} P\,dx + Q\,dy + R\,dz$$

Line Integral

Let $\mathbf{F} = \mathbf{F}(\mathbf{x})$ be a continuous vector field on $\mathbf{D}$ and let $\mathbf{c}$ be a directed curve in $\mathbf{D}$. Suppose $\mathbf{c}$ is described by $\mathbf{x} = \mathbf{x}(t)$, where $a \le t \le b$. Define the **line integral**

$$\int_{\mathbf{c}} \mathbf{F}\cdot d\mathbf{x} = \int_a^b \mathbf{F}[\mathbf{x}(t)]\cdot \frac{d\mathbf{x}}{dt}\,dt$$

It follows from the change of variable rule for simple integrals that the line integral depends on $\mathbf{c}$, not on how $\mathbf{c}$ is parametrized. Precisely, if τ is another parameter for the curve, running from α to β as t runs from a to b, then

$$\int_a^b \mathbf{F}\cdot\frac{d\mathbf{x}}{dt}\,dt = \int_\alpha^\beta \mathbf{F}\cdot\frac{d\mathbf{x}}{dt}\cdot\frac{dt}{d\tau}\,d\tau = \int_\alpha^\beta \mathbf{F}\cdot\frac{d\mathbf{x}}{d\tau}\,d\tau$$

The definition is similar for a plane curve $\mathbf{c}$ and a plane vector field $\mathbf{F}(\mathbf{x}) = (P(\mathbf{x}), Q(\mathbf{x}))$. In this case, the line integral can be written

$$\int_{\mathbf{c}} P\,dx + Q\,dy$$

Example 1

Let $\mathbf{F}(\mathbf{x}) = (x, y, x + y + z)$. Compute the line integral

$$\int_{\mathbf{c}} \mathbf{F}\cdot d\mathbf{x}$$

over the path $\mathbf{c}$ given by

a $\mathbf{x}(t) = (t, t, t),\quad 0 \le t \le 1$
b $\mathbf{x}(t) = (t, t^2, t^3),\quad 0 \le t \le 1$

Solution

a We are given $\mathbf{x}(t) = (t, t, t)$ so $d\mathbf{x}/dt = (1, 1, 1)$. Along the curve, $\mathbf{F}(\mathbf{x}) = \mathbf{F}[\mathbf{x}(t)] = (t, t, 3t)$. Therefore

$$\int_{\mathbf{c}} \mathbf{F}\cdot d\mathbf{x} = \int_0^1 (t, t, 3t)\cdot(1, 1, 1)\,dt = \int_0^1 5t\,dt = \tfrac{5}{2}$$

b We are given $\mathbf{x}(t) = (t, t^2, t^3)$ so $d\mathbf{x}/dt = (1, 2t, 3t^2)$. Also $\mathbf{F}[\mathbf{x}(t)] = (t, t^2, t + t^2 + t^3)$. Therefore

$$\int_{\mathbf{c}} \mathbf{F}\cdot d\mathbf{x} = \int_0^1 (t, t^2, t + t^2 + t^3)\cdot(1, 2t, 3t^2)\,dt$$

$$= \int_0^1 [t + 2t^3 + 3(t^3 + t^4 + t^5)]\,dt$$

$$= \int_0^1 (t + 5t^3 + 3t^4 + 3t^5) = \tfrac{1}{2} + \tfrac{5}{4} + \tfrac{3}{5} + \tfrac{1}{2} = \tfrac{57}{20}$$ ●

For an interpretation of line integrals, suppose $\mathbf{x} = \mathbf{x}(s)$ is a parametrization of $\mathbf{c}$ with arc length as the parameter, and $a \le s \le b$. Then $d\mathbf{x}/ds = \mathbf{t}$, the unit tangent. Therefore the definition of the line integral in this case is

$$\int_{\mathbf{c}} \mathbf{F} \cdot d\mathbf{x} = \int_a^b \mathbf{F}[\mathbf{x}(s)] \cdot \mathbf{t}(s)\, ds$$

The quantity $\mathbf{F} \cdot \mathbf{t}$ is the tangential component of $\mathbf{F}$. See Figure 18-5-2. Hence

$$\int_{\mathbf{c}} \mathbf{F} \cdot d\mathbf{x} = \int_a^b [\text{tangential component of } \mathbf{F}(s)]\, ds$$

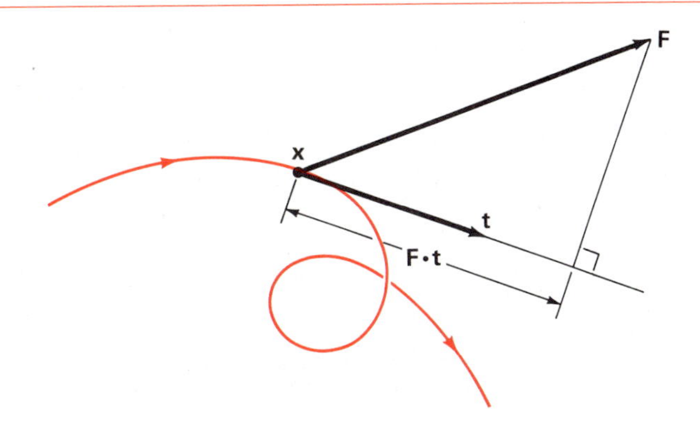

Figure 18-5-2
Interpretation of $\mathbf{F} \cdot d\mathbf{x} = (\mathbf{F} \cdot \mathbf{t})\, ds$ as the element of work

This formula gives an immediate physical interpretation of the line integral as *work*. Recall that work equals force times distance. When a particle moves on a curve in a force field $\mathbf{F}(\mathbf{x})$, it is the component of motion tangent to the force that does work. Thus the element of work is

$$dW = (\mathbf{F} \cdot \mathbf{t})\, ds$$

The line integral adds up these small bits of work and yields the total work done by the field in moving the particle along its path.

Independence of Path

In Example 1, the same vector field is integrated over two different curves. Both curves have exactly the same end points, $\mathbf{x}(0) = (0, 0, 0)$ and $\mathbf{x}(1) = (1, 1, 1)$, yet the line integrals are unequal. Therefore, in general, a line integral depends on the curve, not just on its end points.

In an important special case, however, the line integral does *not* depend on the curve $\mathbf{c}$, but only on its initial and terminal points:

- Suppose the vector field $\mathbf{F}(\mathbf{x})$ is the gradient of some function $f(\mathbf{x})$:

  $$\mathbf{F} = \operatorname{grad} f$$

  Then for each curve $\mathbf{c}$ going from $\mathbf{x}_0$ to $\mathbf{x}_1$.

  $$\int_{\mathbf{c}} \mathbf{F} \cdot d\mathbf{x} = f(\mathbf{x}_1) - f(\mathbf{x}_0)$$

Therefore the line integral depends only on the end points, not on what path is taken between them. In particular, as we shall see, if $\mathbf{c}$ is a *closed* path, then

$$\int_{\mathbf{c}} \mathbf{F} \cdot d\mathbf{x} = 0$$

Proof By the chain rule (read backwards),

$$\mathbf{F} \cdot \frac{d\mathbf{x}}{dt} = [\operatorname{grad} f(\mathbf{x})] \cdot \frac{d\mathbf{x}}{dt} = \frac{d}{dt} f[\mathbf{x}(t)]$$

Therefore

$$\int_{\mathbf{c}} \mathbf{F} \cdot d\mathbf{x} = \int_a^b \mathbf{F} \cdot \dot{\mathbf{x}}\, dt = \int_a^b \frac{d}{dt} f[\mathbf{x}(t)]\, dt$$

$$= f[\mathbf{x}(t)]\Big|_a^b = f(\mathbf{x}_1) - f(\mathbf{x}_0)$$

If $\mathbf{c}$ is closed, then $\mathbf{x}_1 = \mathbf{x}_0$ so $f(\mathbf{x}_1) - f(\mathbf{x}_0) = 0$.

Conservation of Energy

Suppose $\mathbf{F} = \mathbf{F}(\mathbf{x})$ is a force field, and a particle of mass m moves under the influence of this force along a path $\mathbf{c}$ from $\mathbf{x}_0$ to $\mathbf{x}_1$. Say the path is parametrized by time: $\mathbf{x} = \mathbf{x}(t)$, where $t_0 \le t \le t_1$.

The work done by $\mathbf{F}$ in moving the particle is

$$W = \int_{\mathbf{c}} \mathbf{F} \cdot d\mathbf{x} = \int_{t_0}^{t_1} \mathbf{F}[\mathbf{x}(t)] \cdot \frac{d\mathbf{x}}{dt}\, dt$$

Now the motion of the particle is determined by Newton's law

$$\mathbf{F} = m\frac{d^2\mathbf{x}}{dt^2}$$

It follows that

$$\mathbf{F} \cdot \frac{d\mathbf{x}}{dt} = m\frac{d\mathbf{x}}{dt} \cdot \frac{d^2\mathbf{x}}{dt^2}$$

$$= \tfrac{1}{2}m\frac{d}{dt}\left(\frac{d\mathbf{x}}{dt} \cdot \frac{d\mathbf{x}}{dt}\right) = \tfrac{1}{2}m\frac{d}{dt}|\mathbf{v}|^2$$

where $\mathbf{v} = d\mathbf{x}/dt$ is the velocity. Therefore

$$W = \int_{t_0}^{t_1} \tfrac{1}{2}m\frac{d}{dt}|\mathbf{v}|^2 = \tfrac{1}{2}m|\mathbf{v}_1|^2 - \tfrac{1}{2}m|\mathbf{v}_0|^2$$

The quantity $K = \tfrac{1}{2}m|\mathbf{v}|^2$ is the kinetic energy of the particle. The result of this discussion is the *law of conservation of energy:* work done equals change in kinetic energy.

Inverse Square Law

If $\mathbf{F} = \text{grad}\, f$ is a force, the net work done by this force in moving a particle from $\mathbf{x}_0$ to $\mathbf{x}_1$ is $f(\mathbf{x}_1) - f(\mathbf{x}_0)$, independent of the path. The function f, which is unique up to an additive constant, is called the **potential** of the force.

Figure 18-5-3

Inverse square law: $E = \dfrac{\mathbf{x}}{\rho^3}$

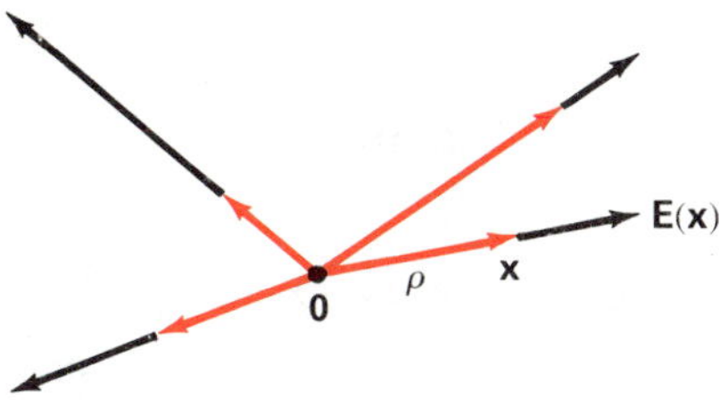

An important example is a central force subject to the inverse square law, for instance, the electric force $\mathbf{E}$ on a unit charge at $\mathbf{x}$ due to a unit charge of the same sign at the origin. The magnitude of the vector $\mathbf{E}$ is inversely proportional to $|\mathbf{x}|^2$. Its direction is the same as that of $\mathbf{x}$. See Figure 18-5-3. The unit vector in the direction of $\mathbf{x}$ is

$$\frac{\mathbf{x}}{|\mathbf{x}|} = \frac{\mathbf{x}}{\rho} \qquad \rho = |\mathbf{x}|$$

Therefore, expressed in suitable units,

$$\mathbf{E} = \frac{1}{\rho^2} \cdot \frac{\mathbf{x}}{\rho} = \frac{\mathbf{x}}{\rho^3}$$

The force field $\mathbf{E}$ is defined at all points of space except the origin. We shall prove that $\mathbf{E}$ is the gradient of a function. In fact, we shall show that

$$\mathbf{E} = \text{grad}\, f \qquad \text{where} \quad f(x, y, z) = -\frac{1}{\rho} = \frac{-1}{\sqrt{x^2 + y^2 + z^2}}$$

Let us compute the gradient of $f = -1/\rho$:

$$\frac{\partial f}{\partial x} = \frac{x}{(x^2 + y^2 + z^2)^{3/2}} = \frac{x}{\rho^3}$$

Similarly,

$$\partial f/\partial y = y/\rho^3 \quad \text{and} \quad \partial f/\partial z = z/\rho^3$$

Therefore

$$\text{grad}\, f = \left(\frac{x}{\rho^3}, \frac{y}{\rho^3}, \frac{z}{\rho^3}\right) = \frac{1}{\rho^3}(x, y, z) = \frac{\mathbf{x}}{\rho^3} = \mathbf{E}$$

It follows that

$$\int_{\mathbf{x}_0}^{\mathbf{x}_1} \mathbf{E} \cdot d\mathbf{x} = f(\mathbf{x}_1) - f(\mathbf{x}_0) = \frac{1}{|\mathbf{x}_0|} - \frac{1}{|\mathbf{x}_1|}$$

The right-hand side is the **potential difference** or **voltage.** It represents the work done by the electric force when a unit charge moves from $\mathbf{x}_0$ to $\mathbf{x}_1$ *along any path.*

If $\mathbf{x}_1$ is far out, then $1/|\mathbf{x}_1|$ is small, so

$$\int_{\mathbf{x}_0}^{\mathbf{x}_1} \mathbf{E} \cdot d\mathbf{x} \approx \frac{1}{|\mathbf{x}_0|}$$

As $\mathbf{x}_1$ moves farther out, the approximation improves:

$$\int_{\mathbf{x}_0}^{\mathbf{x}_1} \mathbf{E} \cdot d\mathbf{x} \rightarrow \frac{1}{|\mathbf{x}_0|} \quad \text{as} \quad |\mathbf{x}_1| \rightarrow \infty$$

that is,

$$\int_{\mathbf{x}_0}^{\infty} \mathbf{E} \cdot d\mathbf{x} = \frac{1}{|\mathbf{x}_0|} = \frac{1}{\rho_0}$$

The final result, $1/|\mathbf{x}_0|$, is called the **potential** at $\mathbf{x}_0$. It is the work done by the force in moving a unit charge from $\mathbf{x}_0$ to infinity, along any path.

Conservation of Momentum

Recall that the integral (simple or multiple) of a vector-valued function is defined componentwise. For instance, if

$$\mathbf{u}(t) = (u(t), v(t), w(t))$$

then

$$\int_a^b \mathbf{u}(t)\, dt = \left(\int_a^b u(t)\, dt, \int_a^b v(t)\, dt, \int_a^b w(t)\, dt\right)$$

The following useful formula can be easily proved componentwise:

$$\int_a^b \frac{d\mathbf{w}}{dt}\, dt = \mathbf{w}(b) - \mathbf{w}(a)$$

For example,

$$\int_0^2 (2t, 3t^2, 4t^3)\, dt = \int_0^2 \frac{d}{dt}(t^2, t^3, t^4)\, dt$$
$$= (t^2, t^3, t^4)\Big|_0^2 = (4, 8, 16)$$

Now suppose a particle of mass m moves on a path $\mathbf{c}$ under the influence of the force field $\mathbf{F} = \mathbf{F}(\mathbf{x})$. According to Sir Isaac Newton,

$$\mathbf{F} = m\frac{d^2\mathbf{x}}{dt^2} = m\frac{d}{dt}\left(\frac{d\mathbf{x}}{dt}\right) = m\frac{d\mathbf{v}}{dt}$$

Say the path is described by $\mathbf{x} = \mathbf{x}(t)$, where $t_0 \le t \le t_1$. Then

$$\int_{t_0}^{t_1} \mathbf{F}[\mathbf{x}(t)]\, dt = \int_{t_0}^{t_1} m\frac{d\mathbf{v}}{dt}\, dt = m\mathbf{v}_1 - m\mathbf{v}_0$$

The quantity $m\mathbf{v}$ is the **momentum** of the particle. The quantity

$$\int_{t_0}^{t_1} \mathbf{F}\, dt$$

is the **impulse** of the force during the time interval $[t_0, t_1]$. The equation

$$\int_{t_0}^{t_1} \mathbf{F}\,dt = m\mathbf{v}\Big|_{t_0}^{t_1}$$

is the law of conservation of momentum: impulse equals change in momentum.

The **angular momentum** of the particle with respect to the origin $\mathbf{0}$ is defined as

$$m\mathbf{x} \times \mathbf{v} = m\mathbf{x} \times \frac{d\mathbf{x}}{dt}$$

Now

$$\frac{d}{dt}(m\mathbf{x} \times \mathbf{v}) = \frac{d}{dt}\left(m\mathbf{x} \times \frac{d\mathbf{x}}{dt}\right)$$

$$= m\frac{d\mathbf{x}}{dt} \times \frac{d\mathbf{x}}{dt} + m\mathbf{x} \times \frac{d^2\mathbf{x}}{dt^2} = m\mathbf{x} \times \frac{d^2\mathbf{x}}{dt^2}$$

since $d\mathbf{x}/dt \times d\mathbf{x}/dt = \mathbf{0}$. But $md^2\mathbf{x}/dt^2 = \mathbf{F}$, hence

$$m\mathbf{x} \times \frac{d^2\mathbf{x}}{dt^2} = \mathbf{x} \times m\frac{d^2\mathbf{x}}{dt^2} = \mathbf{x} \times \mathbf{F}$$

which is the torque of $\mathbf{F}$ at $\mathbf{x}$. Therefore

$$\frac{d}{dt}(m\mathbf{x} \times \mathbf{v}) = \mathbf{x} \times \mathbf{F}$$

Integrate

$$\int_{t_0}^{t_1} \mathbf{x} \times \mathbf{F}\,dt = m\mathbf{x} \times \mathbf{v}\Big|_{t_0}^{t_1}$$

This result, called the *law of conservation of angular momentum,* asserts that the time integral of the torque equals the change in angular momentum.

Exercises

Evaluate the line integral over the indicated path

1 $\displaystyle\int y\,dx - x\,dy$

$\mathbf{x} = (\cos t, \sin t) \qquad 0 \le t \le \frac{1}{2}\pi$.

2 $\displaystyle\int x\,dx + y\,dy$

$\mathbf{x} = (t^3, t^2) \qquad 0 \le t \le 1$

3 $\displaystyle\int z\,dx + x\,dy + y\,dz$

straight path from $(0, 0, 0)$ to (a, b, c)

4 $\displaystyle\int z\,dx + x\,dy + y\,dz$

straight path from $(1, 1, 2)$ to $(-1, 0, 4)$

5 $\displaystyle\int xy\,dx + (1 + 2y)\,dy$

straight path from $(0, 1)$ to $(0, -1)$

6 $\displaystyle\int xy\,dx + (1 + 2y)\,dy$

semi-circle $r = 1 \qquad \frac{1}{2}\pi \le \theta \le \frac{3}{2}\pi$

7 $\int -dy + x\,dz$

$\mathbf{x} = (t^2, t^3, t^4) \qquad 1 \le t \le 2$

8 $\int z\,dx + x\,dz$

$\mathbf{x} = (\sin t, \cos t, t^2) \qquad 0 \le t \le \frac{1}{2}\pi$

9 $\int yz\,dx + zx\,dy + xy\,dz$

$\mathbf{x} = (t^2, t^3, t^{-4}) \qquad a \le t \le b \qquad (0 < a)$

10 $\int \sin y \cos z\,dx + x \cos y \cos z\,dy - x \sin y \sin z\,dz$

$\mathbf{x} = (\cos t, \sin t, \cos t) \qquad 0 \le t \le 2\pi$

11 Let $\mathbf{F} = (3x^2y^2z, 2x^3yz, x^3y^2)$. Show that $\int_{(0,0,0)}^{(1,1,1)} \mathbf{F} \cdot d\mathbf{x}$ is independent of the path, and evaluate it.

12 Let $\mathbf{F} = (x^2 + yz, y^2 + zx, z^2 + xy)$. Show that $\int_{(0,0,0)}^{(a,bc)} \mathbf{F} \cdot d\mathbf{x}$ is independent of the path, and evaluate it.

13 Let θ denote the polar angle in the plane. Show that

$$\text{grad}\,\theta = \left(\frac{-y}{x^2+y^2}, \frac{x}{x^2+y^2}\right)$$

14 (cont.) Find

$$\int \frac{-y\,dx + x\,dy}{x^2+y^2}$$

over the circle $|\mathbf{x}| = a$.

15 Find the work done by the central force field

$$\mathbf{F} = -\frac{1}{|\mathbf{x}|^3}\mathbf{x}$$

in moving a particle from $(1, 8, 4)$ to $(2, 1, 2)$ along a straight path.

16 Find the work done by the uniform gravitational field $\mathbf{F} = (0, 0, -g)$ in moving a particle from $(0, 0, 1)$ to $(1, 1, 0)$ along a straight path.

17 Let $\mathbf{F} = \mathbf{x}/|\mathbf{x}|^5$ and suppose $\mathbf{a} \ne \mathbf{0}$. Show that $\int_{\mathbf{a}}^{\infty} \mathbf{F} \cdot d\mathbf{x}$, taken along any path from $\mathbf{a}$ not passing through $\mathbf{0}$ and going out indefinitely, depends only on $a = |\mathbf{a}|$. Evaluate the integral.

18 (cont.) Do the same for $\mathbf{F} = \mathbf{x}/|\mathbf{x}|^n$, for any $n > 2$.

19 (cont.) Show that $\mathbf{F} = \mathbf{x}/|\mathbf{x}|^2$ is a gradient.

20 (cont.) What value should be assigned to $\int_{\mathbf{a}}^{\infty} \mathbf{F} \cdot d\mathbf{x}$, where $\mathbf{F} = \mathbf{x}/|\mathbf{x}|^2$, taken along a path from $\mathbf{a}$ not passing through $\mathbf{0}$ and going out indefinitely?

Evaluate

21 $\int_0^1 (1 + t, 1 + 2t, 1 + 3t)\,dt$

22 $\int_0^{2\pi} (\cos t, \sin t, 1)\,dt$

23 $\int_{-1}^1 (t^3, t^4, t^5)\,dt$

24 $\int_1^4 \left(\frac{1}{t}, \frac{1}{t^2}, \frac{1}{t^3}\right) dt$

25 The force $\mathbf{F}(t) = (1 - t, 1 - t^2, 1 - t^3)$ acts from $t = 0$ to $t = 2$. Find its impulse.

26 The force $\mathbf{F}(t) = (e^t, e^{2t}, e^{3t})$ acts from $t = -1$ to $t = 0$. Find its impulse.

27 An electron of mass m in a uniform magnetic field follows the spiral path

$$\mathbf{x}(t) = (a \cos t, a \sin t, bt)$$

Find its angular momentum with respect to $\mathbf{0}$.

28 A particle of unit mass moves on the unit sphere $|\mathbf{x}| = 1$ with unit speed. Show that its angular momentum with respect to $\mathbf{0}$ is a unit vector.

18-6 Green's Theorem

In this section we discuss some properties of line integrals in the plane, that is, integrals of the form

$$\int_{\mathbf{c}} P\,dx + Q\,dy$$

where $P = P(x, y)$, $Q = Q(x, y)$, and $\mathbf{c}$ is a directed plane curve.

Before coming to the main business of this section, let us make some practical remarks about the evaluation of such integrals. By definition, we parametrize $\mathbf{c}$ by $x = x(t)$, $y = y(t)$ and then evaluate

$$\int P[x(t), y(t)]\frac{dx}{dt}\,dt + Q[x(t), y(t)]\frac{dy}{dt}\,dt$$

Now suppose $\mathbf{c}$ happens to be part of the graph of a function $y = f(x)$, say for x running from a to b. Then it is most natural to parametrize $\mathbf{c}$ by x itself. Accordingly we write $x = x$, $y = f(x)$. For example, suppose $\mathbf{c}$ is a horizontal segment. Then we write $x = x$ and $y = c$, so $dy/dx = 0$ and

$$\int_{\mathbf{c}} P\,dx + Q\,dy = \int_{\mathbf{c}} P(x, c)\,dx$$

The $Q\,dy$ part of the integral drops out. We describe this situation by saying that on a horizontal segment $dy = 0$. Similarly, on a vertical segment $dx = 0$, and the $P\,dx$ part of the integral drops out.

Figure 18-6-1
Orientation of the boundary $\partial\mathbf{D}$ of a plane domain $\mathbf{D}$

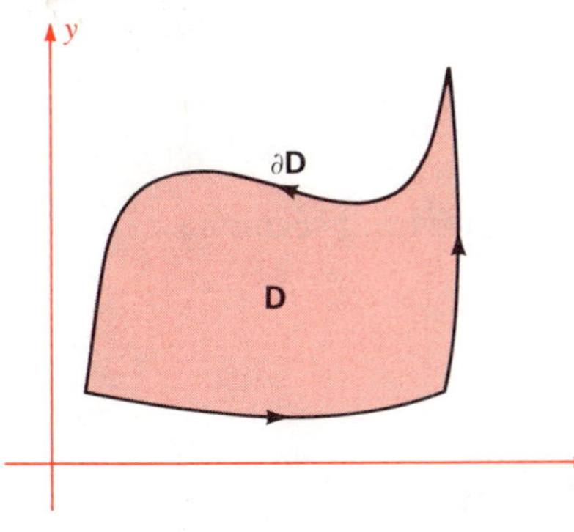

a Boundary consists of one closed curve

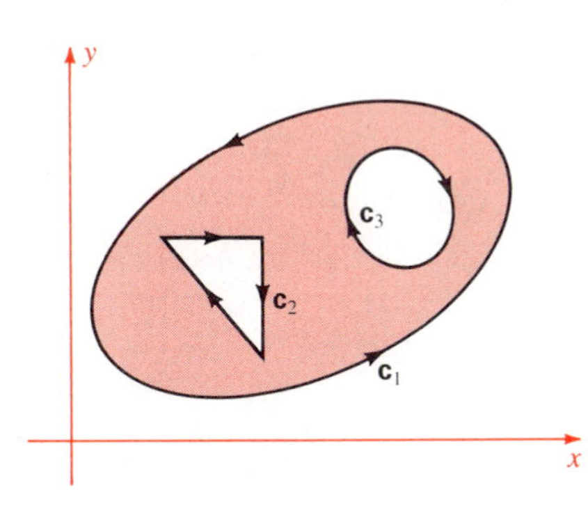

b Boundary consists of three closed curves: $\partial\mathbf{D} = \mathbf{c}_1 + \mathbf{c}_2 + \mathbf{c}_3$

There is a useful connection between certain line integrals and double integrals. Suppose $\mathbf{D}$ is a domain in the x, y-plane bounded by one or more closed curves, each composed of arcs with continuously turning tangent vectors. We assign a direction to each of the boundary curves (Figure 18-6-1) such that if we walk around any boundary curve in this direction, the region $\mathbf{D}$ is always on our left.

The symbol $\partial\mathbf{D}$ will denote the whole directed boundary of $\mathbf{D}$. We are interested in the line integral

$$\int_{\partial\mathbf{D}} P\,dx + Q\,dy$$

If $\partial\mathbf{D}$ consists of the directed closed paths, $\mathbf{c}_1, \mathbf{c}_2, \cdots, \mathbf{c}_n$, we write

$$\partial\mathbf{D} = \mathbf{c}_1 + \mathbf{c}_2 + \cdots + \mathbf{c}_n$$

and define

$$\int_{\partial\mathbf{D}} P\,dx + Q\,dy = \int_{\mathbf{c}_1} (P\,dx + Q\,dy) + \cdots + \int_{\mathbf{c}_n} (P\,dx + Q\,dy)$$

Notation There is a time-honored tradition of writing

$$\oint P\,dx + Q\,dy$$

for a line integral over a *closed* path.

An important theorem says that the *line* integral over the boundary $\partial\mathbf{D}$ is equal to a certain *double* integral over the region $\mathbf{D}$.

Green's Theorem

Suppose $P(x, y)$ and $Q(x, y)$ are continuously differentiable functions on a plane domain $\mathbf{D}$. Then

$$\oint_{\partial\mathbf{D}} P\,dx + Q\,dy = \iint_{\mathbf{D}} \left(\frac{\partial Q}{\partial x} - \frac{\partial P}{\partial y}\right) dx\,dy$$

The theorem may be viewed as two independent formulas,

$$\oint_{\partial\mathbf{D}} P\,dx = -\iint_{\mathbf{D}} \frac{\partial P}{\partial y}\,dx\,dy \qquad \oint_{\partial\mathbf{D}} Q\,dy = \iint_{\mathbf{D}} \frac{\partial Q}{\partial x}\,dx\,dy$$

It is fairly easy to see that the second follows from the first by interchanging x and y. (This changes the sense of turning in the plane, which accounts for the sign change.) We shall prove the first formula, not in the most general case, but when $\mathbf{D}$ can be decomposed by line segments into subdomains, each of which is the domain between the graphs of two functions of x.

If $\mathbf{D}$ is so decomposed $(\mathbf{D} = \mathbf{D}_1 + \cdots + \mathbf{D}_n)$, then

$$\oint_{\partial\mathbf{D}} P\,dx = \oint_{\partial\mathbf{D}_1} P\,dx + \cdots + \oint_{\partial\mathbf{D}_n} P\,dx$$

because the contributions over the common division segments cancel in pairs (Figure 18-6-2a).

Figure 18-6-2
Proof of Green's theorem

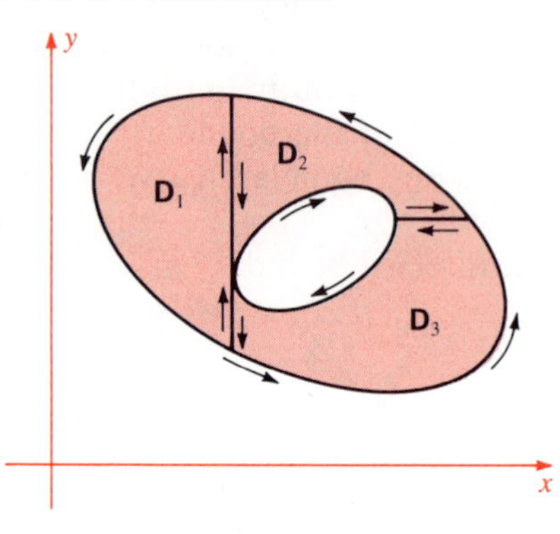

a Cancellation of the common boundary contribution

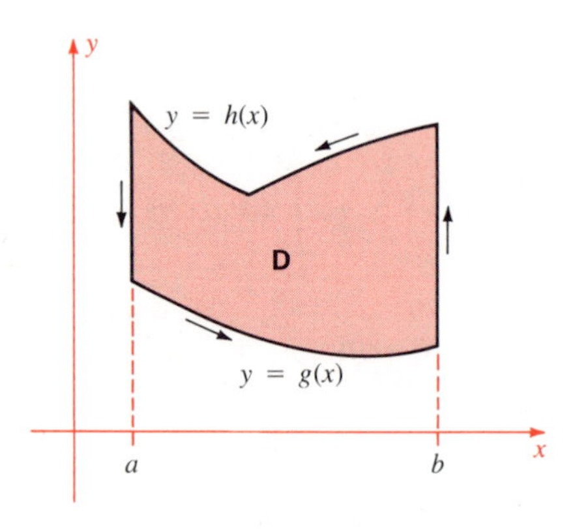

b Domain between two graphs

We shall prove

$$\oint_{\partial\mathbf{D}} P\,dx = -\iint_{\mathbf{D}} \frac{\partial P}{\partial y}\,dx\,dy$$

for the domain $\mathbf{D}$ of Figure 18-6-2b. Its boundary $\partial\mathbf{D}$ consists of four pieces; accordingly the line integral decomposes into four summands. On the two vertical sides x is constant; hence $dx = 0$, no contribution. Therefore

$$\begin{aligned}\oint_{\partial\mathbf{D}} P\,dx &= \int_b^a P[x, h(x)]\,dx + \int_a^b P[x, g(x)]\,dx\\ &= \int_a^b \{-P[x, h(x)] + P[x, g(x)]\}\,dx\\ &= -\int_a^b \left(\int_{g(x)}^{h(x)} \frac{\partial P}{\partial y}\,dy\right) dx = -\iint_{\mathbf{D}} \frac{\partial P}{\partial y}\,dx\,dy\end{aligned}$$

This completes the proof.

Remark Green's theorem is a two-dimensional generalization of the fundamental theorem of calculus,

$$\int_a^b f'(x)\,dx = f(b) - f(a)$$

which equates the integral of a derivative $f'(x)$ over an interval $[a, b]$ to something computed at the *boundary* (the two points a and b) of the interval. Green's theorem has important applications in physical situations with two-dimensional models, like fluid flow on a plane region. We shall give some details of the analogous three-dimensional case in Section 18-8.

Example 1

Compute the line integral

$$\oint (x^2 + y^2)\,dx + (x + 2)\,dy$$

taken over the boundary of the triangle T with vertices at the points $(0, 0)$, $(0, 1)$, and $(1, 0)$. Use two methods:

a direct computation and **b** Green's theorem

Solution

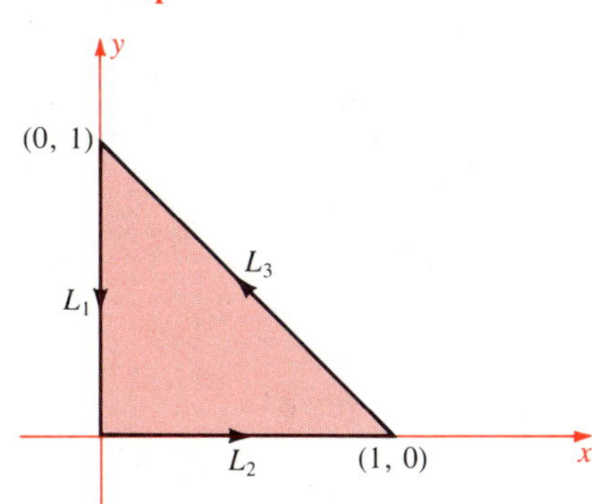

Figure 18-6-3
See Example 1

a The integral breaks into integrals over line segments L_1, L_2, L_3 as shown in Figure 18-6-3. Compute each separately.

On L_1, we have $x = 0$ and $dx = 0$ since x is constant. Hence

$$\int_{L_1} (x^2 + y^2)\,dx + (x + 2)\,dy = \int_1^0 2\,dy = -2$$

On L_2, we have $y = 0$ and $dy = 0$. Hence

$$\int_{L_2} (x^2 + y^2)\,dx + (x + 2)\,dy = \int_0^1 x^2\,dx = \tfrac{1}{3}$$

On L_3 we use x as the parameter, running from 1 to 0. Then $y = 1 - x$ and $dy = -dx$. Hence

$$\int_{L_3} (x^2 + y^2)\,dx + (x + 2)\,dy$$

$$= \int_1^0 [x^2 + (1 - x)^2 - (x + 2)]\,dx$$

$$= \int_1^0 (-1 - 3x + 2x^2)\,dx$$

$$= \int_0^1 (1 + 3x - 2x^2)\,dx = \tfrac{11}{6}$$

Adding the results, we have

$$\oint (x^2 + y^2)\,dx + (x + 2)\,dy = -2 + \tfrac{1}{3} + \tfrac{11}{6} = \tfrac{1}{6}$$

b Let $P(x, y) = x^2 + y^2$ and $Q(x, y) = x + 2$. By Green's theorem,

$$\oint P\,dx + Q\,dy = \iint_{\mathbf{T}} \left(\frac{\partial Q}{\partial x} - \frac{\partial P}{\partial y}\right) dx\,dy$$

$$= \iint_{\mathbf{T}} (1 - 2y)\,dx\,dy = \int_0^1 (1 - 2y)\left(\int_0^{1-y} dx\right) dy$$

$$= \int_0^1 (1 - 2y)(1 - y)\,dy = \int_0^1 (1 - 3y + 2y^2)\,dy = \tfrac{1}{6}$$ ●

A useful application of Green's theorem is a formula for the area of a plane domain in terms of an integral over its boundary.

Area Formula

If $\mathbf{D}$ is a plane domain with a smooth boundary, then the area of $\mathbf{D}$ is given by

$$\text{area}(\mathbf{D}) = \tfrac{1}{2} \oint_{\partial \mathbf{D}} -y\,dx + x\,dy$$

Proof Apply Green's theorem with $P = -y$ and $Q = x$. Then $Q_x - P_y = 2$, so

$$\oint_{\partial \mathbf{D}} -y\,dx + x\,dy = \iint_{\mathbf{D}} 2\,dx\,dy = 2\,\text{area}(\mathbf{D})$$ ●

Remark Since $Q_x = 1$ and $P_y = -1$, we also have

$$\text{area}(\mathbf{D}) = \oint_{\partial \mathbf{D}} x\,dy = -\oint_{\partial \mathbf{D}} y\,dx$$

However, the formula above with $-y\,dx + x\,dy$ is often more convenient than either of these one-term formulas because it has a certain amount of symmetry.

Example 2

Find the area enclosed by the ellipse $x^2/a^2 + y^2/b^2 = 1$.

Solution Let $\mathbf{D}$ denote the domain bounded by the ellipse. Parametrize the ellipse as usual by

$$x = a\cos\theta \qquad y = b\sin\theta \qquad \text{where} \quad 0 \le \theta \le 2\pi$$

Then

$$\text{area}(\mathbf{D}) = \tfrac{1}{2}\oint_{\partial\mathbf{D}} -y\,dx + x\,dy$$

$$= \tfrac{1}{2}\int_0^{2\pi} -(b\sin\theta)(-a\sin\theta\,d\theta) + (a\cos\theta)(b\cos\theta\,d\theta)$$

$$= \tfrac{1}{2}\int_0^{2\pi} (ab\sin^2\theta + ab\cos^2\theta)\,d\theta = \tfrac{1}{2}\int_0^{2\pi} ab\,d\theta = \pi ab$$

●

Exercises

Evaluate the line integral over the indicated closed curve

1 $\oint 2y\,dx + 4x\,dy$ Figure 18-6-4

2 $\oint x^2\,dx + xy\,dy$ Figure 18-6-5

3 $\oint y^2e^x\,dx + 2ye^x\,dy$ Figure 18-6-5

4 $\oint -y^2\,dx + x^2\,dy$ Figure 18-6-6

5 $\oint (y - x^2)\,dx + (2x + y^2)\,dy$ Figure 18-6-6

6 $\oint \cos x \sin y\,dx + \sin x \cos y\,dy$ Figure 18-6-6

7 $\oint xy\,dx + xy\,dy$ Figure 18-6-5

8 $\oint xy\,dx + xy\,dy$ Figure 18-6-7

9 $\oint -y^3\,dx + x^3\,dy$ Figure 18-6-7

10 $\oint -y^3\,dx + x^3\,dy$ Figure 18-6-8

11 $\oint x^2y^2\,dx - x^3y\,dy$ Figure 18-6-8

12 $\oint \dfrac{x\,dx}{(x^2 + y^2)^3} + \dfrac{y\,dy}{(x^2 + y^2)^3}$ Figure 18-6-9

Figure 18-6-4

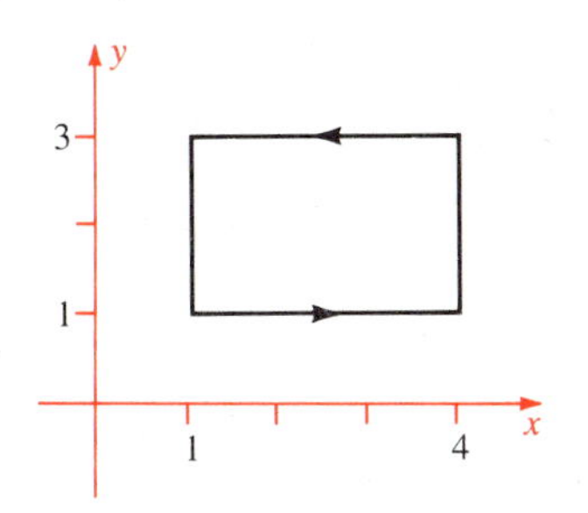

Figure 18-6-5

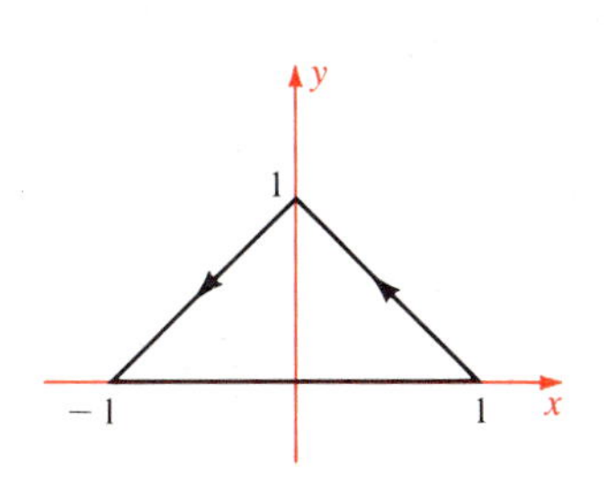

Figure 18-6-6

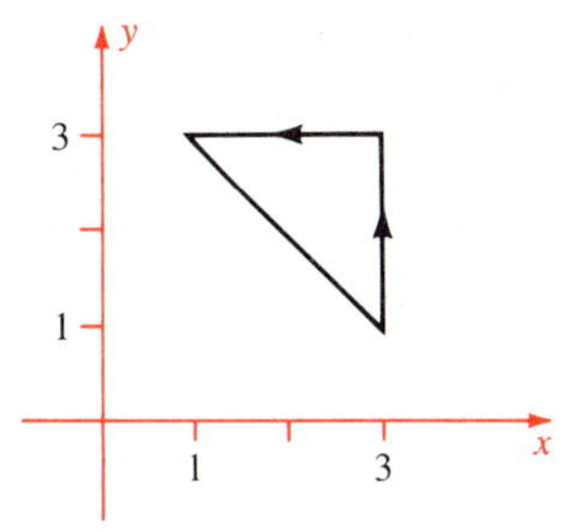

Figure 18-6-7
Ellipse

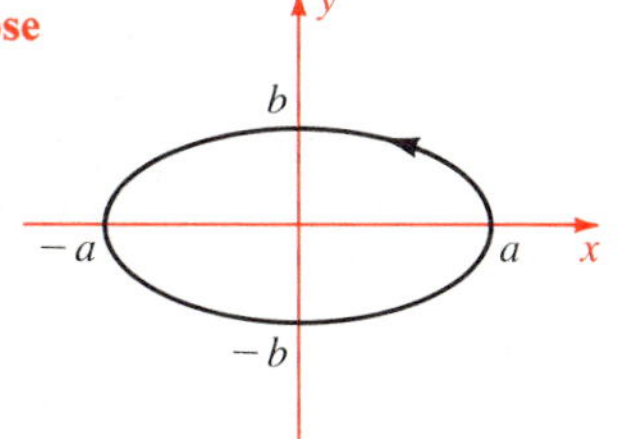

Figure 18-6-8
Semicircle

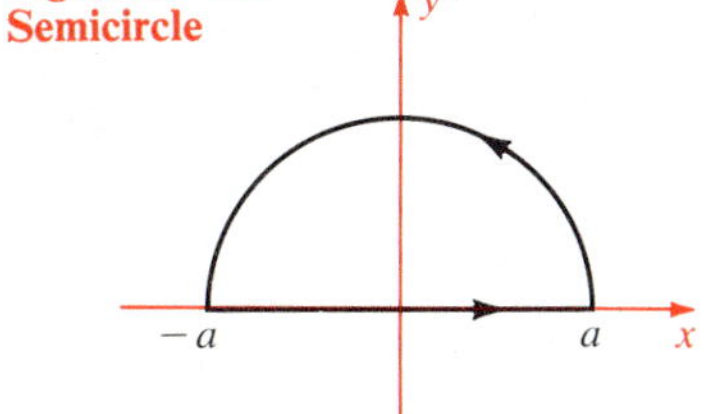

Figure 18-6-9
Circle

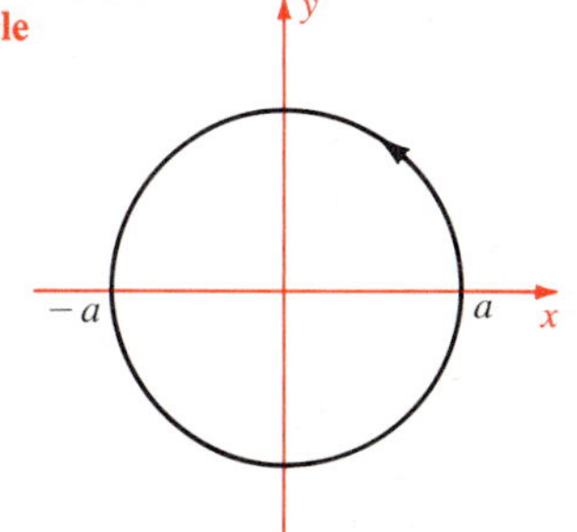

13 Find

$$\oint \frac{-y\,dx + x\,dy}{x^2 + y^2}$$

over the circle in Figure 18-6-9. [Hint Express the integral in terms of polar coordinates. The answer is *not* 0.]

14 (cont.) Evaluate the integral over the closed curves in
a Figure 18-6-4 **b** Figure 18-6-6
c Figure 18-6-10.

15 (cont.) Evaluate

$$\oint \frac{-y\,dx + (x-1)\,dy}{(x-1)^2 + y^2} + \frac{y\,dx - (x+1)\,dy}{(x+1)^2 + y^2}$$

over the contour in Figure 18-6-11.

16 (cont.) Evaluate the same integral over the contour in Figure 18-6-12.

17 Evaluate

$$\oint \frac{(-3x^2y + y^3)\,dx + (x^3 - 3xy^2)\,dy}{(x^2 + y^2)^3}$$

over the rectangle in Figure 18-6-4.

***18** (cont.) Evaluate the same integral over the contour in Figure 18-6-10.

19 Prove **Green's formula** under suitable hypotheses:

$$\oint_{\partial\mathbf{D}} (-uv_y\,dx + uv_x\,dy) - \oint_{\partial\mathbf{D}} (-vu_y\,dx + vu_x\,dy)$$

$$= \iint_{\mathbf{D}} [u(v_{xx} + v_{yy}) - v(u_{xx} + u_{yy})]\,dx\,dy$$

20 (cont.) Test this formula when $\mathbf{D}$ is the unit disk, $u = 1$, and $v = \ln r$. Explain the result.

21 Suppose P and Q are continuously differentiable on a rectangle $a \le x \le b$, $c \le y \le d$. For each point (x, y) of the rectangle define

$$F(x, y) = \int_a^x P(u, c)\,du + \int_c^y Q(x, v)\,dv$$

and

$$G(x, y) = \int_a^x P(u, y)\,dy + \int_c^y Q(a, v)\,dv$$

Find $\partial F/\partial y$ and $\partial G/\partial x$.

22 (cont.) Assume also that $\partial P/\partial y = \partial Q/\partial x$. Prove that $F = G$ and deduce that $(P, Q) = \operatorname{grad} F$. [Hint Use Green's theorem.]

Figure 18-6-10

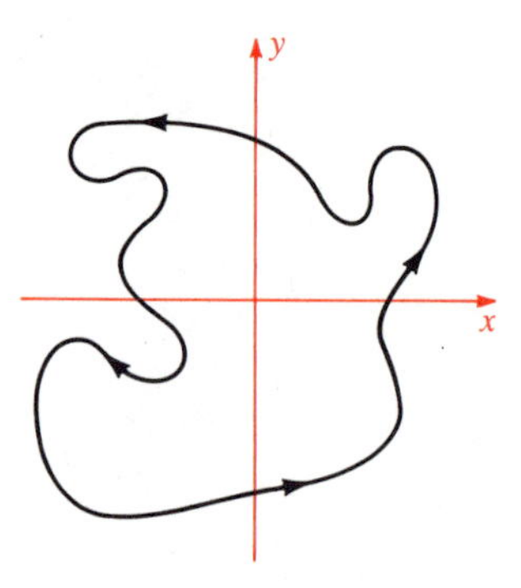

Figure 18-6-11

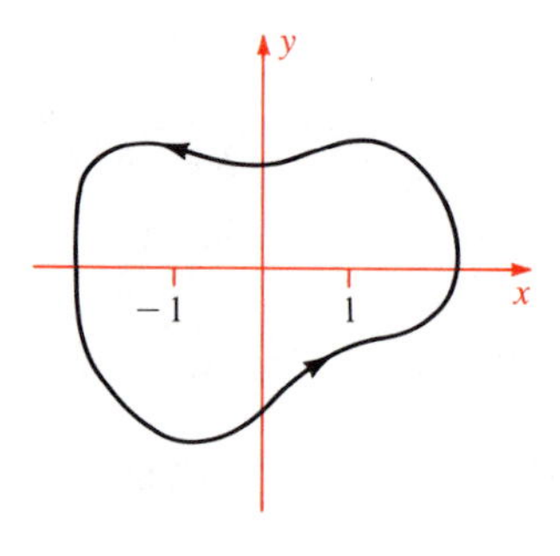

Figure 18-6-12

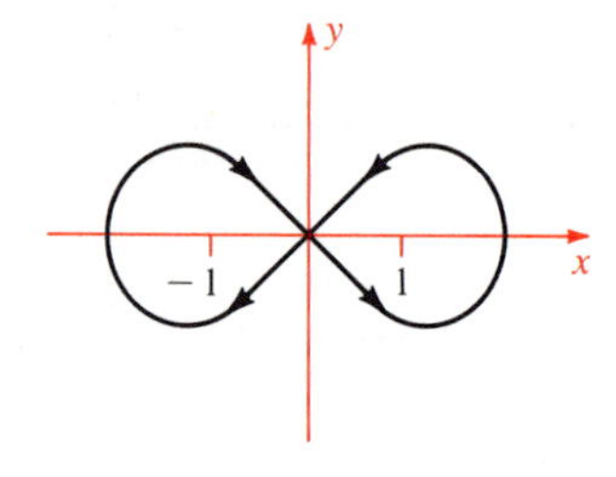

23 Express the area formula

$$A = \tfrac{1}{2}\oint -y\,dx + x\,dy$$

in terms of polar coordinates.

24 (cont.) Apply your answer to find the total area enclosed by the five-petal rose curve $r = a\cos 5\theta$.

18-7 Surface Integrals

In this section we shall study integrals of the form

$$\iint_{\mathbf{S}} A\,dy\,dz + B\,dz\,dx + C\,dx\,dy$$

where $\mathbf{S}$ is an oriented surface in $\mathbf{R}^3$. By oriented surface we mean a two-sided surface with one side designated as the top. (There do exist one-sided surfaces, such as the Möbius strip, but we won't look for trouble.) We choose the unit normal vector $\mathbf{n}$ to point out of the top of the surface.

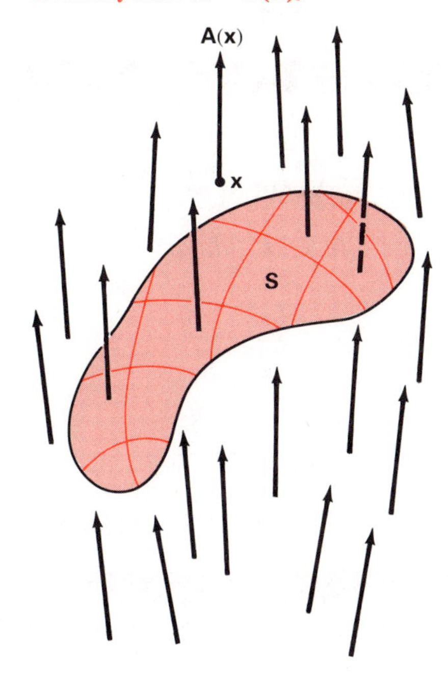

Figure 18-7-1
Find the flux through surface $\mathbf{S}$ of velocity field $\mathbf{A} = \mathbf{A}(\mathbf{x})$.

Such **surface integrals** arise in problems of the following type. Suppose fluid is flowing in a region of space with velocity field $\mathbf{A} = \mathbf{A}(\mathbf{x})$. Given an oriented surface $\mathbf{S}$ fixed somewhere in space, the problem is to find how fast the fluid flows through $\mathbf{S}$, through the bottom and out the top (Figure 18-7-1). In other words, to compute how much fluid crosses $\mathbf{S}$ per unit time. (The rate of flow across $\mathbf{S}$ is called the **flux** of $\mathbf{A}$ through $\mathbf{S}$.)

To find the flux Φ, we decompose $\mathbf{S}$ into small "elements" of surface, compute the flux through each, and sum the results (integrate). We consider an element of surface (Figure 18-7-2a) and two associated quantities, its (element of) area $d\sigma$ and its unit normal $\mathbf{n}$, pointing out of the *top* of the surface. The fluid velocity $\mathbf{A}$ at the surface element decomposes into a tangential component and a normal component. The tangential component doesn't cross the surface; its contribution to the flux is 0. The normal component (Figure 18-7-2b), which flows straight through the surface, equals $\mathbf{A}\cdot\mathbf{n}$; its contribution to the flux is

$$d\Phi = (\mathbf{A}\cdot\mathbf{n})\,d\sigma = \mathbf{A}\cdot(\mathbf{n}\,d\sigma)$$

Therefore

$$\Phi = \iint_{\mathbf{S}} (\mathbf{A}\cdot\mathbf{n})\,d\sigma$$

Figure 18-7-2
Element of flux

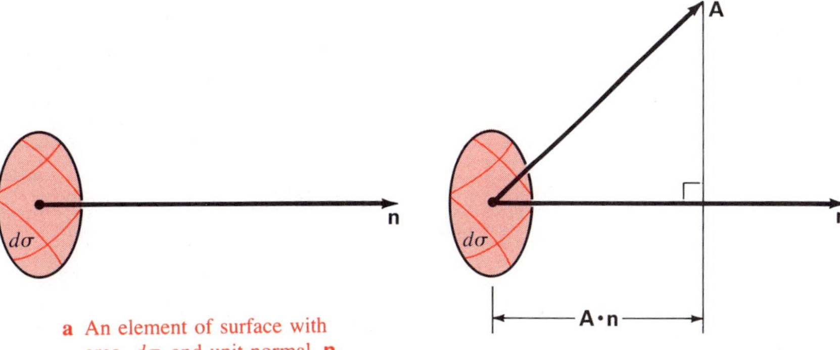

a An element of surface with area $d\sigma$ and unit normal $\mathbf{n}$

b The flux across the element of surface is $d\Phi = (\mathbf{A}\cdot\mathbf{n})\,d\sigma$.

Figure 18-7-3
Parametrization of $\mathbf{S}$ by $\mathbf{x} = \mathbf{x}(u, v)$, where (u, v) is in $\mathbf{D}$

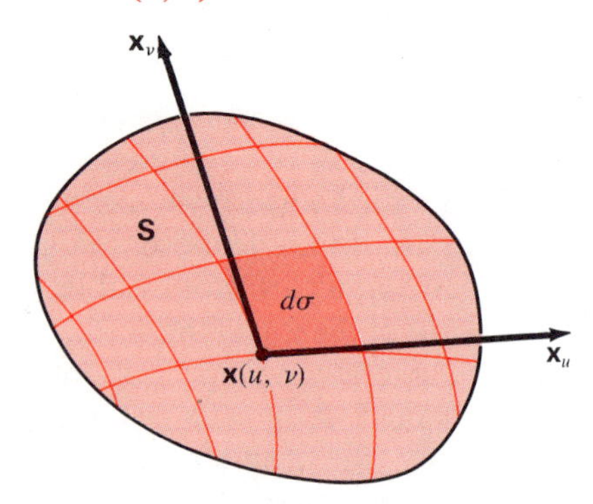

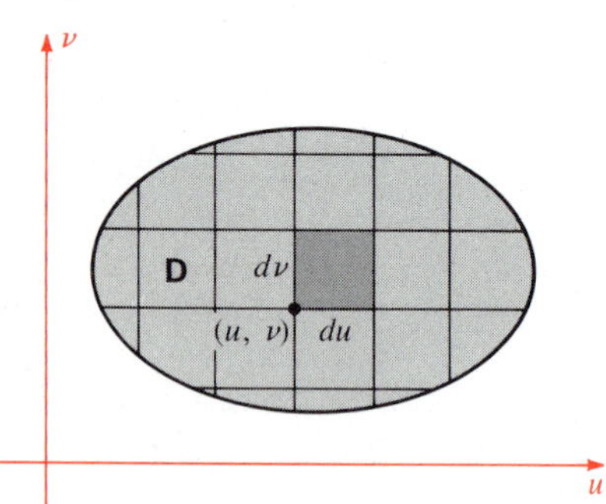

To evaluate the integral, we suppose that $\mathbf{S}$ is parametrized by $\mathbf{x} = \mathbf{x}(u, v)$, where (u, v) runs over a domain $\mathbf{D}$ in the u, v-plane (Figure 18-7-3). We recall the formulas (pp. 733 and 856) for the unit normal and the element of area;

$$\mathbf{n} = \frac{\mathbf{x}_u \times \mathbf{x}_v}{|\mathbf{x}_u \times \mathbf{x}_v|} \qquad \text{and} \qquad d\sigma = |\mathbf{x}_u \times \mathbf{x}_v|\, du\, dv$$

It follows that $\mathbf{n}\, d\sigma = \mathbf{x}_u \times \mathbf{x}_v\, du\, dv$. This quantity is important in the study of surface integrals and is given a special name:

Element of Vectorial Area

$$d\boldsymbol{\sigma} = \mathbf{n}\, d\sigma = \mathbf{x}_u \times \mathbf{x}_v\, du\, dv$$

In terms of u and v the element of flux works out to be

$$d\Phi = \mathbf{A} \cdot d\boldsymbol{\sigma} = \mathbf{A} \cdot (\mathbf{x}_u \times \mathbf{x}_v)\, du\, dv = [\mathbf{A}, \mathbf{x}_u, \mathbf{x}_v]\, du\, dv$$

$$= \begin{vmatrix} A & B & C \\ x_u & y_u & z_u \\ x_v & y_v & z_v \end{vmatrix} du\, dv$$

$$= \left[A \frac{\partial(y, z)}{\partial(u, v)} + B \frac{\partial(z, x)}{\partial(u, v)} + C \frac{\partial(x, y)}{\partial(u, v)} \right] du\, dv$$

This completes our set-up for the formal definition of surface integrals.

Definition Surface Integral

Let $\mathbf{A} = \mathbf{A}(\mathbf{x})$ be a continuous vector field in a region of $\mathbf{R}^3$ and let $\mathbf{S}$ be an oriented surface in that region, given by $\mathbf{x} = \mathbf{x}(u, v)$, where (u, v) varies over a domain $\mathbf{D}$. Assume $\mathbf{x}(u, v)$ is continuously differentiable. Then we define

$$\iint_{\mathbf{S}} A\, dy\, dz + B\, dz\, dx + C\, dx\, dy = \iint_{\mathbf{S}} \mathbf{A} \cdot d\boldsymbol{\sigma}$$

$$= \iint_{\mathbf{D}} \left[A \frac{\partial(y, z)}{\partial(u, v)} + B \frac{\partial(z, x)}{\partial(u, v)} + C \frac{\partial(x, y)}{\partial(u, v)} \right] du\, dv$$

It is understood that A is the composite function $A[\mathbf{x}(u, v)]$, and so on, in the last integral.

Since the integrand $A\, dy\, dz + B\, dz\, dx + C\, dx\, dy$ can be written

$$(A, B, C) \cdot (dy\, dz, dz\, dx, dx\, dy) = \mathbf{A} \cdot (dy\, dz, dz\, dx, dx\, dy)$$

where

$$d\boldsymbol{\sigma} = (dy\, dz, dz\, dx, dx\, dy)$$

This formula is important because it gives the vectorial area element in a form independent of the parametrization of the surface.

Remark 1 The physical interpretation of the surface integral as flux makes it clear that the integral depends only on the vector field $\mathbf{A}$ and the oriented surface $\mathbf{S}$, not on how the surface is parametrized. An analytic proof of this fact is possible using the change of variables rule.

Remark 2 If a complete oriented surface cannot be parametrized (because it is closed or has handles, for example) cut it into pieces and parametrize each piece; then sum the corresponding integrals to obtain the surface integral over the complete surface.

Remark 3 Choosing a parametrization $\mathbf{x} = \mathbf{x}(u, v)$ for a surface $\mathbf{S}$ fixes a normal vector

$$\mathbf{n} = \mathbf{x}_u \times \mathbf{x}_v = \left(\frac{\partial(y, z)}{\partial(u, v)}, \frac{\partial(z, x)}{\partial(u, v)}, \frac{\partial(x, y)}{\partial(u, v)} \right)$$

This normal either agrees with the normal determined by the orientation of $\mathbf{S}$ or is the negative of it. In the latter case, the surface integral will give $-\Phi$, not Φ.

Example 1

Let $\mathbf{T}$ be the triangle with vertices $(1, 0, 0)$, $(0, 1, 0)$, $(0, 0, 1)$ and normal pointing away from $\mathbf{0}$. Let $\mathbf{A}(\mathbf{x}) = (y, z, 0)$. Find

$$\iint_{\mathbf{T}} \mathbf{A} \cdot d\sigma = \iint_{\mathbf{T}} y\, dy\, dz + z\, dz\, dx$$

Solution The triangle is given by inequalities $x \geq 0$, $y \geq 0$, $z \geq 0$, and $x + y + z = 1$. Parametrize it by

$$x = u \qquad y = v \qquad z = 1 - u - v$$

where $u \geq 0$, $u + v \leq 1$. See Figure 18-7-4. Then

$$\frac{\partial(y, z)}{\partial(u, v)} = \begin{vmatrix} y_u & y_v \\ z_u & z_v \end{vmatrix} = \begin{vmatrix} 0 & 1 \\ -1 & -1 \end{vmatrix} = 1$$

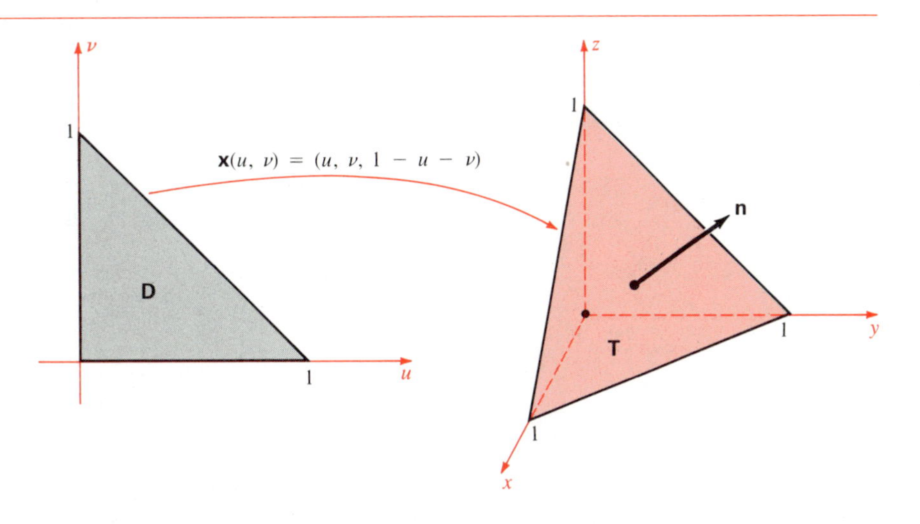

Figure 18-7-4
See Example 1

Similarly,

$$\frac{\partial(z, x)}{\partial(u, v)} = \begin{vmatrix} -1 & -1 \\ 1 & 0 \end{vmatrix} = 1 \quad \text{and} \quad \frac{\partial(x, y)}{\partial(u, v)} = \begin{vmatrix} 1 & 0 \\ 0 & 1 \end{vmatrix} = 1$$

Hence $d\boldsymbol{\sigma} = (1, 1, 1)\, du\, dv,$ so we have

$$\begin{aligned} \mathbf{A} \cdot d\boldsymbol{\sigma} &= (y, z, 0) \cdot (1, 1, 1)\, du\, dv = (y + z)\, du\, dv \\ &= (1 - u)\, du\, dv \end{aligned}$$

Therefore

$$\begin{aligned} \iint_{\mathbf{T}} \mathbf{A} \cdot d\boldsymbol{\sigma} &= \iint_{\mathbf{D}} (1 - u)\, du\, dv = \int_0^1 dv \int_0^{1-v} (1 - u)\, du \\ &= \int_0^1 [(1 - v) - \tfrac{1}{2}(1 - v)^2]\, dv \\ &= \int_0^1 (\tfrac{1}{2} - \tfrac{1}{2}v^2)\, dv = \tfrac{1}{2} - \tfrac{1}{6} = \tfrac{1}{3} \end{aligned}$$

Note that the parametrization chosen does indeed yield the normal directed away from $\mathbf{0}$, because all three Jacobians are positive. ●

Example 2

Find

$$\Phi = \iiint_{\mathbf{H}} dx\, dy + 2\, dy\, dz + 3\, dz\, dx$$

where $\mathbf{H}$ is the hemisphere $\rho = a,\ z \geq 0,$ taken with its normal *inward.*

Solution It is easiest to treat the three terms separately. For the $dx\, dy$ term, parametrize by projection onto the circle $\mathbf{D}$: $x^2 + y^2 \leq a^2$. That is, let x and y be the parameters, with (x, y) running over $\mathbf{D}$, and parametrize $\mathbf{H}$ by

$$\mathbf{x}(x, y) = (x, y, \sqrt{a^2 - x^2 - y^2})$$

This parametrization determines the outward (upward) normal $\mathbf{n}$, because the z-component of $\mathbf{n}$ is

$$\frac{\partial(x, y)}{\partial(x, y)} = \begin{vmatrix} 1 & 0 \\ 0 & 1 \end{vmatrix} = 1 > 0$$

Since this is *opposite* to the given orientation,

$$\iint_{\mathbf{H}} dx\, dy = -\iint_{\mathbf{D}} dx\, dy = -\text{area}(\mathbf{D}) = -\pi a^2$$

For the $dy\,dz$ term, break the hemisphere into two quarter spheres, $x \ge 0$ and $x \le 0$. Parametrize each by projection onto the semi-circle **E** in the y, z-plane: $y^2 + z^2 \le a^2$, $z \ge 0$. For the quarter sphere $x \ge 0$, this parametrization defines the outward normal; for the quarter sphere $x \le 0$, it defines the inward normal. (The key fact is that the x-component of **n** is 1, proved by the same reasoning as above.) Therefore

$$\iint_{\mathbf{H}} dy\,dz = -\iint_{\mathbf{E}} dy\,dz + \iint_{\mathbf{E}} dy\,dz = 0$$

Likewise

$$\iint_{\mathbf{H}} dz\,dx = 0 \qquad \text{so} \qquad \Phi = \iiint_{\mathbf{H}} dx\,dy = -\pi a^2$$

Alternative Solution It makes sense to exploit radial symmetry in this problem. That suggests using spherical instead of rectangular coordinates. Thus parametrize **H** by spherical coordinates ϕ, θ:

$$x = a \sin\phi \cos\theta \qquad y = a \sin\phi \sin\theta \qquad z = a\cos\phi$$

where $0 \le \theta \le 2\pi$, $0 \le \phi \le \frac{1}{2}\pi$. Then the three Jacobians are

$$\frac{\partial(y,z)}{\partial(\phi,\theta)} = \begin{vmatrix} a\cos\phi\sin\theta & a\sin\phi\cos\theta \\ -a\sin\phi & 0 \end{vmatrix} = a^2\sin^2\phi\cos\theta$$

$$\frac{\partial(z,x)}{\partial(\phi,\theta)} = \begin{vmatrix} -a\sin\phi & 0 \\ a\cos\phi\cos\theta & -a\sin\phi\sin\theta \end{vmatrix} = a^2\sin^2\phi\sin\theta$$

$$\frac{\partial(x,y)}{\partial(\phi,\theta)} = \begin{vmatrix} a\cos\phi\cos\theta & -a\sin\phi\sin\theta \\ a\cos\phi\sin\theta & a\sin\phi\cos\theta \end{vmatrix} = a^2\sin\phi\cos\phi$$

Therefore

$$\begin{aligned} d\boldsymbol{\sigma} &= (dy\,dz,\, dz\,dx,\, dx\,dy) \\ &= (a^2\sin^2\phi\cos\theta,\, a^2\sin^2\phi\sin\theta,\, a^2\sin\phi\cos\phi)\, d\phi\, d\theta \end{aligned}$$

In the first octant $(0 < \theta < \frac{1}{2}\pi,\ 0 < \phi < \frac{1}{2}\pi)$, all components are positive; hence the parametrization yields the *outward* normal, the wrong one, and we must change sign. We have

$$\begin{aligned} \Phi &= \iint_{\mathbf{H}} 2\,dy\,dz + 3\,dz\,dx + dx\,dy \\ &= -\iint_{\mathbf{D}} (2a^2\sin^2\phi\cos\theta + 3a^2\sin^2\phi\sin\theta \\ &\qquad\qquad + a^2\sin\phi\cos\phi)\, d\phi\, d\theta \end{aligned}$$

with **D** the rectangle $0 \le \theta \le 2\pi$, $0 \le \phi \le \frac{1}{2}\pi$ in the ϕ, θ-plane.

Clearly

$$\iint_{\mathbf{D}} \sin^2\phi \cos\theta \, d\phi \, d\theta = \int_0^{2\pi} \cos\theta \, d\theta \int_0^{\pi/2} \sin^2\phi \, d\phi = 0$$

Similarly

$$\iint_{\mathbf{D}} \sin^2\phi \sin\theta \, d\phi \, d\theta = 0$$

Hence

$$\begin{aligned}\Phi &= -\iint_{\mathbf{D}} a^2 \sin\phi \cos\phi \, d\phi \, d\theta \\ &= -2\pi a^2 \int_0^{\pi/2} \sin\phi \cos\phi \, d\phi \\ &= -2\pi a^2 (\tfrac{1}{2}) = -\pi a^2\end{aligned}$$

●

Exercises

Evaluate the surface integral over the indicated portion of the surface; always take the normal to have a non-negative z-component unless you are instructed otherwise

1 $\iint x\,dy\,dz + y\,dz\,dx + z\,dx\,dy$

plane $2x - 2y + z = 3$, $0 \le x \le 1 \quad 0 \le y \le 1$

2 $\iint x^2\,dy\,dz + xy\,dz\,dx$

plane $x + 2y + 3z = 1$ $\quad x \ge 0 \quad y \ge 0 \quad x + y \le 1$

3 $\iint xz\,dy\,dz$ $\quad$ sphere $\rho = a \quad z \ge 0$

4 $\iint 3x\,dy\,dz + 2z^2\,dx\,dy$

paraboloid $z = x^2 + 2y^2 \quad x^2 + y^2 \le 1$

5 $\iint z\,dy\,dz + z\,dz\,dx + dx\,dy$

cone $z = \sqrt{x^2 + y^2} \quad x^2 + y^2 \le 4 \quad x \ge 0$

6 $\iint z\,dy\,dz + z\,dz\,dx + dx\,dy$

cone $z = 2 - \sqrt{x^2 + y^2} \quad x^2 + y^2 \le 4$

7 $\iint 4\,dy\,dz - 3\,dz\,dx$

cylinder $z = \sqrt{1 - x^2} \quad 0 \le x \le 1 \quad 0 \le y \le 2$

8 $\iint x^2\,dy\,dz$

cylinder $z = -\sqrt{1 - x^2}$ $\quad -1 \le x \le 0 \quad -1 \le y \le 1$

9 $\iint x^6\,dy\,dz + z^5\,dx\,dy$

cylinder $x^2 + y^2 = 1 \quad 0 \le z \le 1$ (outward normal)

10 $\iint x\,dy\,dz + y\,dz\,dx + z\,dx\,dy$

cylinder $x^2 + y^2 = a^2 \quad 0 \le z \le h$ (inward normal)

11 $\iint x^n\,dy\,dz$ $\quad$ sphere $\rho = a$ (outward normal)

12 $\iint y^n\,dy\,dz$ $\quad$ sphere $\rho = a$ (outward normal)

18-8 Theorems of Gauss and Stokes

We now pave the way towards an important theorem that relates surface integrals over closed surfaces to certain triple integrals. It is a generalization of Green's theorem for line integrals. First we define a scalar quantity associated with a vector field and called its divergence. We shall see shortly that the divergence of the velocity vector of a moving gas can be interpreted in terms of expansion and contraction of the gas.

Divergence

Let $\mathbf{A} = \mathbf{A}(\mathbf{x}) = (A, B, C)$ be a differentiable vector field on a domain in $\mathbf{R}^3$. The **divergence** of $\mathbf{A}$ is the real-valued function

$$\operatorname{div} \mathbf{A}(\mathbf{x}) = \frac{\partial A}{\partial x} + \frac{\partial B}{\partial y} + \frac{\partial C}{\partial z}$$

Example 1

a $\operatorname{div} \mathbf{x} = \operatorname{div}(x, y, z) = 1 + 1 + 1 = 3$

b $$\operatorname{div}(x^3, xyz, xz^2) = \frac{\partial}{\partial x}(x^3) + \frac{\partial}{\partial y}(xyz) + \frac{\partial}{\partial z}(xz^2)$$
$$= 3x^2 + xz + 2xz = 3(x^2 + xz)$$

The divergence operator satisfies an important relation that is analogous to the product rule for differentiation.

- Let $\mathbf{A} = \mathbf{A}(\mathbf{x})$ be a vector field and $f = f(\mathbf{x})$ a scalar. Then

$$\operatorname{div}(f\mathbf{A}) = (\operatorname{grad} f) \cdot \mathbf{A} + f \operatorname{div} \mathbf{A}$$

Its proof is left for Exercise 1.

Now we are ready for the first result of this section, a formula that has numerous applications in mathematics and in the physical sciences.

Divergence Theorem (Gauss's Theorem)

Let $\mathbf{A} = \mathbf{A}(\mathbf{x})$ be a continuously differentiable vector field on a domain $\mathbf{D}$ in $\mathbf{R}^3$. Orient the boundary $\partial\mathbf{D}$ by the outward normal. Then

$$\iint_{\partial\mathbf{D}} \mathbf{A} \cdot d\sigma = \iiint_{\mathbf{D}} (\operatorname{div} \mathbf{A})\, dV$$

In coordinate notation,

$$\iint_{\partial\mathbf{D}} A\, dy\, dz + B\, dz\, dx + C\, dx\, dy$$
$$= \iiint_{\mathbf{D}} \left(\frac{\partial A}{\partial x} + \frac{\partial B}{\partial y} + \frac{\partial C}{\partial z}\right) dx\, dy\, dz$$

As with Green's theorem, we shall give only a partial proof. We shall prove

$$\iint_{\partial \mathbf{D}} C\,dx\,dy = \iiint_{\mathbf{D}} \frac{\partial C}{\partial z}\,dx\,dy\,dz$$

when $\mathbf{D}$ is the domain between the graphs of two functions of (x, y):

$$\mathbf{D} = \{(x, y, z) \mid (x, y) \text{ is in } \mathbf{E} \text{ and } g(x, y) \le z \le h(x, y)\}$$

Here $\mathbf{E}$ is a domain in the x, y-plane (Figure 18-8-1a).

Figure 18-8-1
Proof of the divergence theorem

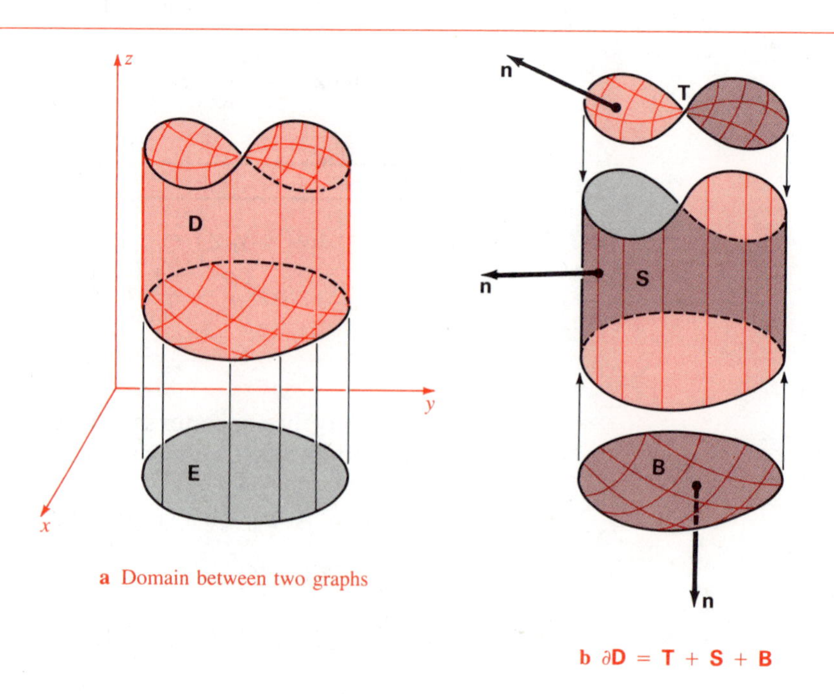

a Domain between two graphs

b $\partial\mathbf{D} = \mathbf{T} + \mathbf{S} + \mathbf{B}$

The boundary of $\mathbf{D}$ splits naturally into top, bottom, and lateral side (Figure 18-8-1b):

$$\partial\mathbf{D} = \mathbf{T} + \mathbf{B} + \mathbf{S}$$

We observe first that

$$\iint_{\mathbf{S}} C\,dx\,dy = 0$$

For, by definition,

$$\iint_{\mathbf{S}} C\,dx\,dy = \iint_{\mathbf{S}} (0, 0, C) \cdot \mathbf{n}\,d\sigma$$

But on the *vertical* side $\mathbf{S}$, the normal vector $\mathbf{n}$ is parallel to the x, y-plane, so its z-component is 0. Therefore, $(0, 0, C) \cdot \mathbf{n} = 0$.

Next we parametrize both the top **T** and the bottom **B** by projection onto **E**. This parametrization gives the outward normal on **T** and the inward normal on **B**. Therefore, on the one hand,

$$\iint_{\partial \mathbf{D}} C\,dx\,dy = \left(\iint_{\mathbf{T}} + \iint_{\mathbf{B}} + \iint_{\mathbf{S}}\right) C\,dx\,dy$$
$$= \iint_{\mathbf{T}} C\,dx\,dy + \iint_{\mathbf{B}} C\,dx\,dy$$
$$= \iint_{\mathbf{E}} C[x, y, h(x, y)]\,dx\,dy - \iint_{\mathbf{E}} C[x, y, g(x, y)]\,dx\,dy$$

On the other hand, we iterate the triple integral:

$$\iiint_{\mathbf{D}} \frac{\partial C}{\partial z}\,dx\,dy\,dz = \iint_{\mathbf{E}} \left(\int_{g(x,y)}^{h(x,y)} \frac{\partial C}{\partial z}\,dz\right) dx\,dy$$
$$= \iint_{\mathbf{E}} \left(C(x, y, z)\Big|_{z=g(x,y)}^{z=h(x,y)} \right) dx\,dy$$
$$= \iint_{\mathbf{E}} \{C[x, y, h(x, y)] - C[x, y, g(x, y)]\}\,dx\,dy$$

The results are equal so the proof is complete ●

Just as Green's theorem implies the area formula, so does the divergence theorem imply a formula for the volume of a domain in terms of an integral over its boundary.

Volume Formula

Let **D** be a domain in $\mathbf{R}^3$. Then

$$\operatorname{vol}(\mathbf{D}) = \tfrac{1}{3}\iint_{\partial \mathbf{D}} x\,dy\,dz + y\,dz\,dx + z\,dx\,dy = \tfrac{1}{3}\iint_{\partial \mathbf{D}} \mathbf{x} \cdot d\boldsymbol{\sigma}$$ ●

Proof By the divergence theorem,

$$\iint_{\partial \mathbf{D}} x\,dy\,dz + y\,dz\,dx + z\,dx\,dy = \iiint_{\mathbf{D}} \operatorname{div}(x, y, z)\,dV$$
$$= \iiint_{\mathbf{D}} 3\,dV$$
$$= 3\operatorname{vol}(\mathbf{D})$$ ●

Example 2

Find the volume V of the ellipsoid

$$\frac{x^2}{a^2}+\frac{y^2}{b^2}+\frac{z^2}{c^2}\le 1$$

Solution Parametrize the boundary of the ellipsoid by

$$x=a\sin\phi\cos\theta \qquad y=b\sin\phi\sin\theta \qquad z=c\cos\phi$$

where $0\le\theta\le 2\pi$, $0\le\phi\le\pi$. Then

$$x\,dy\,dz+y\,dz\,dx+z\,dx\,dy=\begin{vmatrix} x & y & z\\ x_\phi & y_\phi & z_\phi\\ x_\theta & y_\theta & z_\theta\end{vmatrix}d\phi\,d\theta$$

$$=abc\begin{vmatrix} \sin\phi\cos\theta & \sin\phi\sin\theta & \cos\phi\\ \cos\phi\cos\theta & \cos\phi\sin\theta & -\sin\phi\\ -\sin\phi\sin\theta & \sin\phi\cos\theta & 0\end{vmatrix}d\phi\,d\theta$$

$$=abc\sin\phi\,d\phi\,d\theta$$

By the volume formula,

$$V=\tfrac{1}{3}\iint x\,dy\,dz+y\,dz\,dx+z\,dx\,dy=\tfrac{1}{3}abc\iint\sin\phi\,d\phi\,d\theta$$

$$=\tfrac{1}{3}abc\int_0^{2\pi}d\theta\int_0^{\pi}\sin\phi\,d\phi=\tfrac{1}{3}abc(2\pi)(2)=\tfrac{4}{3}\pi abc$$

●

Figure 18-8-2
Flux through a closed surface S of a field that obeys the inverse square law

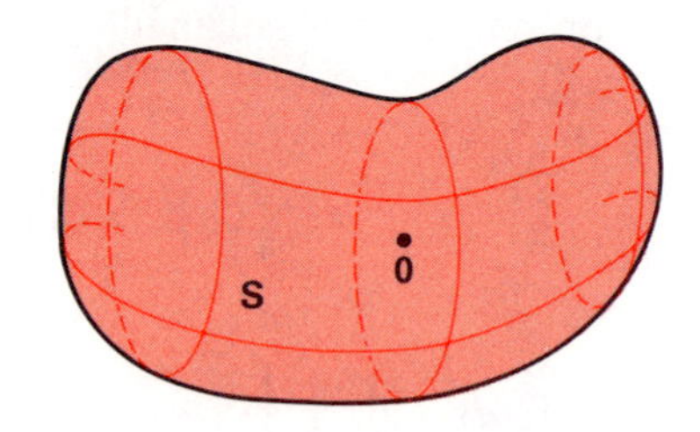

a Surface **S** surrounds **0**

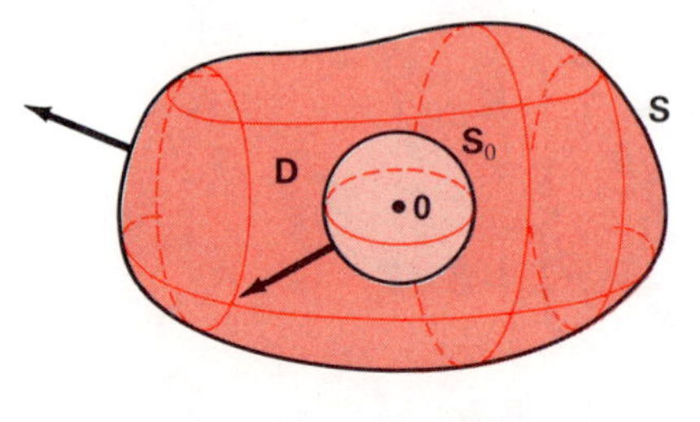

b Domain with a cavity $\partial\mathbf{D}=\mathbf{S}-\mathbf{S}_0$

Inverse Square Law

Suppose a closed surface **S** bounds a domain that includes the origin (Figure 18-8-2a). We want to compute the flux

$$\Phi=\iint_{\mathbf{S}}\mathbf{F}\cdot d\sigma$$

where the force field **F** obeys the inverse square law:

$$\mathbf{F}=\frac{1}{\rho^3}\mathbf{x} \qquad \text{with} \qquad \rho=|\mathbf{x}|$$

First suppose **S** is the sphere $\rho=a$. Then

$$\iint_{\mathbf{S}}\mathbf{F}\cdot d\sigma=\frac{1}{a^3}\iint_{\mathbf{S}}\mathbf{x}\cdot d\sigma=\frac{1}{a^3}(3V)$$

according to the volume formula, where V is the volume of the ball $\rho\le a$, that is, $V=\tfrac{4}{3}\pi a^3$. Hence

$$\iint_{\rho=a}\mathbf{F}\cdot d\sigma=4\pi$$

Note that $\Phi = 4\pi$ no matter what the radius a is.

Let us pass to the general case. We shall prove a remarkable fact: $\Phi = 4\pi$ for every closed surface $\mathbf{S}$ that includes the origin in its interior. We let $\mathbf{D}$ be the domain inside of $\mathbf{S}$ and outside of a small sphere $\mathbf{S}_0$ with center $\mathbf{0}$. See Figure 18-8-2b. We take the *outward* normal on $\mathbf{S}_0$. Then the oriented boundary of $\mathbf{D}$ is

$$\partial\mathbf{D} = \mathbf{S} - \mathbf{S}_0$$

Hence by the divergence theorem

$$\iint_{\mathbf{S}} \mathbf{F}\cdot d\sigma - \iint_{\mathbf{S}_0} \mathbf{F}\cdot d\sigma = \iint_{\partial\mathbf{D}} \mathbf{F}\cdot d\sigma = \iiint_{\mathbf{D}} (\operatorname{div}\mathbf{F})\, dV$$

We must compute $\operatorname{div}\mathbf{F}$. Now $\rho^2 = x^2 + y^2 + z^2$, so $\rho\rho_x = x$, $\rho\rho_y = y$, $\rho\rho_z = z$, and

$$\begin{aligned}\operatorname{div}\mathbf{F} &= \frac{\partial}{\partial x}\left(\frac{x}{\rho^3}\right) + \frac{\partial}{\partial y}\left(\frac{y}{\rho^3}\right) + \frac{\partial}{\partial z}\left(\frac{z}{\rho^3}\right)\\ &= \frac{\rho^3 - 3\rho x^2}{\rho^6} + \frac{\rho^3 - 3\rho y^2}{\rho^6} + \frac{\rho^3 - 3\rho z^2}{\rho^3}\\ &= \frac{3\rho^3 - 3\rho(x^2 + y^2 + z^2)}{\rho^6} = \frac{3\rho^3 - 3\rho^3}{\rho^6} = 0\end{aligned}$$

It follows that

$$\iint_{\mathbf{S}} \mathbf{F}\cdot d\sigma = \iint_{\mathbf{S}_0} \mathbf{F}\cdot d\sigma = 4\pi$$

The flux is constant, independent of the surface $\mathbf{S}$.

- Let $\mathbf{S}$ be a closed surface in $\mathbf{R}^3$ that surrounds $\mathbf{0}$, and take its normal outward. Then

$$\iint_{\mathbf{S}} \frac{x\,dy\,dz + y\,dz\,dx + z\,dx\,dy}{(x^2 + y^2 + z^2)^{3/2}} = 4\pi$$

Gas (or Fluid) Flow

Suppose gas flows throughout a region of space. At each point $\mathbf{x}$ and time t the gas has density $\delta = \delta(\mathbf{x}, t)$, and its velocity of flow is a time-dependent vector field $\mathbf{v} = \mathbf{v}(\mathbf{x}, t)$.

Let $\mathbf{D}$ denote a fixed domain in the region of flow. Assume there are no inlets (sources) or outlets (sinks) within $\mathbf{D}$ so that δ and $\mathbf{v}$ are continuously differentiable in $\mathbf{D}$. The total mass of gas inside $\mathbf{D}$ at any instant is

$$M = \iiint_{\mathbf{D}} \delta(\mathbf{x}, t)\, dV$$

The time rate of change of M is dM/dt. Please accept the fact, analogous to the Leibniz rule (page 854) that we can differentiate under the integral sign. Then we have

$$\frac{dM}{dt} = \iiint_{\mathbf{D}} \frac{\partial \delta}{\partial t}\, dV$$

There is another way to compute dM/dt, the change in the mass of gas inside $\mathbf{D}$ per unit time. The mass inside $\mathbf{D}$ can change *only* as the result of gas flow across the boundary of $\mathbf{D}$ from outside $\mathbf{D}$ into $\mathbf{D}$ (or from inside out). This is a consequence at the *principle of conservation of matter:* matter can neither be created nor be destroyed.

Remember that $d\boldsymbol{\sigma}$ is the vectorial element of area of $\mathbf{D}$ pointing *outward.* Therefore the rate of flow of gas *into* $\mathbf{D}$ through an element $d\sigma$ of surface area of $\partial\mathbf{D}$ equals $-\delta\mathbf{v} \cdot d\boldsymbol{\sigma}$. Consequently we have a second formula for dM/dt:

$$\frac{dM}{dt} = -\iint_{\partial\mathbf{D}} \delta\mathbf{v} \cdot d\boldsymbol{\sigma}$$

By Gauss's theorem, applied to the right side,

$$\frac{dM}{dt} = -\iiint_{\mathbf{D}} [\operatorname{div}(\delta\mathbf{v})]\, dV$$

The two formulas for dM/dt yield

$$\iiint_{\mathbf{D}} \left[\frac{\partial \delta}{\partial t} + \operatorname{div}(\delta\mathbf{v})\right] dV = 0$$

This is true for *every* domain $\mathbf{D}$ in the region of the gas, no matter how small the domain. It follows that

$$\frac{\partial \delta}{\partial t} + \operatorname{div}(\delta\mathbf{v}) = 0$$

This relation, called the **continuity equation,** is one of the cornerstones of fluid mechanics. For an incompressible fluid, rather than a gas, δ is constant. Then the continuity equation reduces to $\operatorname{div}\mathbf{v} = 0$.

Stokes's Theorem

We close with a brief mention of another important result relating integrals. It may also be considered as a generalization of Green's theorem. We first require the definition of the curl (or rotation) of a vector field.

Curl

Let $\mathbf{F} = (P, Q, R)$ be a differentiable vector field on a domain in $\mathbf{R}^3$. The **curl** of $\mathbf{F}$ is the *vector field*

$$\operatorname{curl}\mathbf{F}(\mathbf{x}) = (R_y - Q_z, P_z - R_x, Q_x - P_y)$$

Example 3

$$\operatorname{curl}(y^2 + z^2, z^2 + x^2, x^2 + y^2) = (2y - 2z, 2z - 2x, 2x - 2y)$$

Stokes's Theorem

Let $\mathbf{F} = \mathbf{F}(\mathbf{x})$ be a continuously differentiable vector field in a region of space, and let $\mathbf{S}$ be an oriented surface in that region. Orient the boundary $\partial\mathbf{S}$ of $\mathbf{S}$ to be consistent with the orientation of $\mathbf{S}$ according to the right-hand rule. Then

$$\oint_{\partial\mathbf{S}} \mathbf{F} \cdot d\mathbf{x} = \iint_{\mathbf{S}} (\operatorname{curl} \mathbf{F}) \cdot d\sigma$$

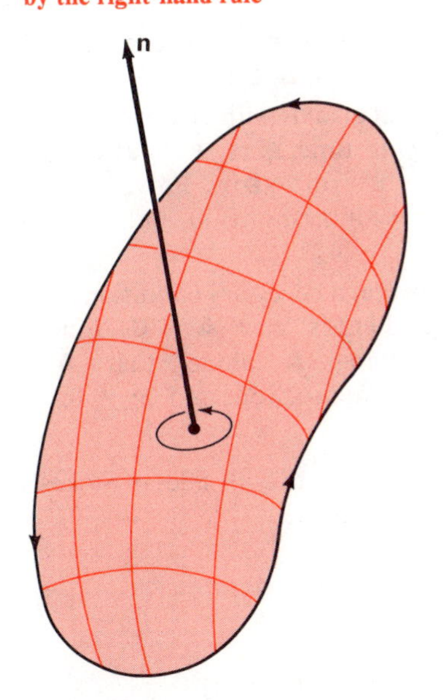

Figure 18-8-3
Orientation of the boundary $\partial\mathbf{S}$ of a directed surface $\mathbf{S}$ by the right-hand rule

A proof of this is best left to a later course. See Figure 18-8-3 for the orientation of the boundary.

Stokes's theorem can be explained in terms of our fluid flow model above. We assume the (imaginary) surface $\mathbf{S}$ lies entirely inside the region of flow and, as above, that $\mathbf{v}$ is the fluid velocity vector. Then $(\operatorname{curl} \mathbf{v}) \cdot \mathbf{n}$ measures the amount the fluid tends to rotate about the unit normal to the surface and

$$\iint_{\mathbf{S}} (\operatorname{curl} \mathbf{v}) \cdot \mathbf{n}\, d\sigma = \iint_{\mathbf{S}} (\operatorname{curl} \mathbf{v}) \cdot d\sigma$$

sums the "elements of rotation" of the fluid over $\mathbf{S}$. Stoke's theorem says that this total amount of rotation of the fluid over $\mathbf{S}$ can be found computing

$$\oint_{\partial\mathbf{S}} \mathbf{v} \cdot d\mathbf{x} = \oint_{\partial\mathbf{S}} \mathbf{v} \cdot \mathbf{T}\, ds$$

which is the amount $\mathbf{v}$ rotates in one traversal of the boundary of $\mathbf{S}$. Here $\mathbf{T}$ is the unit tangent and s the arc length of $\partial\mathbf{S}$.

Exercises

1 Prove the formula $\operatorname{div}(f\mathbf{A}) = (\operatorname{grad} f) \cdot \mathbf{A} + f \operatorname{div} \mathbf{A}$.

2 Prove the following vector relation, a consequence of the divergence theorem:

$$\iint_{\partial\mathbf{D}} f\, d\sigma = \iiint_{\mathbf{D}} (\operatorname{grad} f)\, dV$$

Note that both sides are vectors. [Hint Dot an arbitrary constant vector into both sides.]

3 (Archimedes' principle) A body $\mathbf{D}$ is completely submerged in a fluid of constant density δ. Show that the buoyant force on $\mathbf{D}$ due to the surrounding fluid pressure equals the weight of the fluid that $\mathbf{D}$ displaces. Recall that the pressure at depth z equals the weight of a vertical column of fluid of unit cross-sectional area over z, and use Exercise 2.

4 (cont.) Complete the proof of Archimedes' principle by deriving the same result for a partly submerged body.

Evaluate over a closed surface

5 $\displaystyle\iint_{\mathbf{S}} d\sigma$ [Hint Use Exercise 2.]

*6 $\displaystyle\iint_{\mathbf{S}} \mathbf{x} \times d\sigma$ [Hint Use the hint in Exercise 2.]

The gravitational force of a mass m_1 at $\mathbf{x}_1$ on a mass m_2 at $\mathbf{x}_2$ is

$$\mathbf{F} = Gm_1m_2\frac{\mathbf{x}_1 - \mathbf{x}_2}{|\mathbf{x}_1 - \mathbf{x}_2|^3}$$

The gravitational attraction of one rigid body on another is obtained from this basic inverse square law of Newton by integration.

7 Find the gravitational attraction $\mathbf{F}$ exerted on a unit mass at $(0, 0, c)$ by the uniform disk $x^2 + y^2 \leq a^2$, $z = 0$ of density δ.

8 (cont.) Find $\lim_{a\to\infty} \mathbf{F}$.

9 (cont.) Express $\mathbf{F}$ in terms of the mass M of the disk instead of δ. Find $\lim_{a\to 0} \mathbf{F}$, assuming M constant.

10 Hold an "ideally" transparent, closed surface near a point source of light. Light rays pass in at some places and out at others; none remains inside. Explain that fact by mathematics in terms of the light flux Φ.

11 The solid sphere $\rho \leq a$ has density $\delta = \delta(\rho)$ with radial symmetry. Using symmetry only, what can you say about the gravitational force field $\mathbf{F} = \mathbf{F}(\mathbf{x})$ of the sphere on a unit mass at a typical point $\mathbf{x}$ of space. Your answer should involve $\mathbf{x}$ and $\rho = |\mathbf{x}|$.

***12** (cont.) Let $\mathbf{S}$ be *any* closed surface that includes the sphere in its interior. Find the flux of $\mathbf{F}$ over $\mathbf{S}$.

***13** (cont.) Now choose $\mathbf{S}$ carefully to prove the famous result of Newton, $\mathbf{F} = -GM\mathbf{x}/|\mathbf{x}|^3$ for a unit mass at $\mathbf{x}$ outside of the sphere. It says that the gravitational attraction of the sphere on an external unit mass at $\mathbf{x}$ is the same as if the whole mass of the sphere were concentrated at its center.

14 (cont.) Find the gravitational attraction of the sphere on a unit mass at $\mathbf{x}$ *inside* the sphere. Assume constant density δ.

15 (cont.) Suppose that a spherical planet of radius a consists of a homogeneous fluid of constant density δ. Find the pressure $p = p(\rho)$ in the planet at distance ρ from its center. Then find the pressure at the center.

Prove

16 $\operatorname{div}[f(\rho)\mathbf{x}] = [\rho^3 f(\rho)]'/\rho^2 \qquad \rho = |\mathbf{x}|$

17 $\operatorname{curl}(f\mathbf{A}) = (\operatorname{grad} f) \times \mathbf{A} + f \operatorname{curl} \mathbf{A}$

18 $\operatorname{curl}(\operatorname{grad} f) = \mathbf{0}$

19 $\operatorname{curl}[f(\rho)\mathbf{x}] = \mathbf{0}$

20 $\operatorname{div}(\operatorname{curl} \mathbf{A}) = 0$

21 $\operatorname{curl}(\mathbf{a} \times \mathbf{x}) = 2\mathbf{a}, \qquad \mathbf{a}$ constant

22 $\operatorname{div}[\operatorname{grad} f(\rho)] = [\rho f(\rho)]''/\rho$

The **Laplacian** Δf of a function f is defined by

$$\Delta f = \frac{\partial^2 f}{\partial x^2} + \frac{\partial^2 f}{\partial y^2} + \frac{\partial^2 f}{\partial z^2}$$

23 Express Δ in terms of grad and div.

24 Express $\operatorname{div}(f \operatorname{grad} g - g \operatorname{grad} f)$ in terms of Laplacians.

25 Compute $\Delta(\rho^n)$.

26 Suppose $\operatorname{div} \mathbf{A} = 0$ and $\operatorname{curl} \mathbf{A} = \mathbf{0}$. Prove $\Delta A = \Delta B = \Delta C = 0$, where $\mathbf{A} = (A, B, C)$.

27 Suppose $\mathbf{D}$ is rotating about an axis with angular velocity ω. (Possibly ω varies with time.) Let $\mathbf{v} = \mathbf{v}(\mathbf{x})$ denote the velocity of $\mathbf{x}$ in $\mathbf{D}$. Prove $\omega = \frac{1}{2}\operatorname{curl} \mathbf{v}$. [Hint Express $\mathbf{v}$ in terms of ω and $\mathbf{x}$ and use Exercise 21.]

***28** From Exercise 18 we know that a necessary condition for a vector field $\mathbf{A}$ to be a gradient is $\operatorname{curl} \mathbf{A} = \mathbf{0}$. Find a necessary condition (involving $\mathbf{A}$ alone) that $\mathbf{A} = g \operatorname{grad} f$, where f and g are functions and $g \neq 0$.

Set $\mathbf{F} = (y^2z, z^2x, x^2y)$. Compute both sides of Stokes's theorem and verify their equality

29 $\mathbf{S}$ = spherical octant
$\rho = 1 \qquad x \geq 0 \qquad y \geq 0 \qquad z \geq 0$

30 $\mathbf{S}$ = cone $\quad z = r \qquad 0 \leq r \leq 1$

31 $\mathbf{S}$ = cylinder $\quad r = 1 \qquad 0 \leq z \leq 1$

32 $\mathbf{S}$ = cone
$(1 - z)^2 = x^2 + y^2 \qquad z \geq 0 \qquad x^2 + y^2 \leq 1$

18-9 Review Exercises

1 Two parallel planes at distance h from each other intersect a sphere of radius a. Find the surface area of the spherical zone between the planes.

2 Find the center of gravity of the uniform spiral $\mathbf{x}(t) = (a\cos t, a\sin t, bt)$, $0 \le t \le T$.

3 Suppose $f\langle \rho, \phi, \theta\rangle$ is homogeneous of degree n with respect to rectangular coordinates, that is, $f(t\mathbf{x}) = t^n f(\mathbf{x})$ for all $t > 0$. Prove $f\langle \rho, \phi, \theta\rangle = \rho^n g(\phi, \theta)$, where $g(\phi, \theta) = f\langle 1, \phi, \theta\rangle$.

4 Evaluate

$$\oint \frac{3x^2y^2}{z}\,dx + \frac{2x^3y}{z}\,dy - \frac{x^3y^2}{z^2}\,dz$$

over the curve $\mathbf{x} = (e^{\sin t}\cos t, e^{\cos t}\sin t, 2 + \sin 7t)$, $0 \le t \le 2\pi$.

5 Evaluate $\displaystyle\iiint (x^n + y^n + z^n)\,dx\,dy\,dz$ over $\rho \le a$.

6 Satellite observations determine that the moment of inertia of Earth about its axis of rotation is $I = \frac{1}{3}Ma^2$, where M is the mass and a the radius of Earth. Assume Earth is a sphere whose density δ at any point $\mathbf{x}$ depends only on $\rho = |\mathbf{x}|$ and is linear: $\delta = ab - c\rho$, where b and c are constants. Prove that the density at the center is five times the surface density.

Toroidal coordinates ρ, ϕ, θ of $\mathbf{x}$ in $\mathbf{R}^3$ are defined by the change of variables

$$x = (b + \rho\sin\phi)\cos\theta$$

$$y = (b + \rho\sin\phi)\sin\theta \qquad z = \rho\cos\phi$$

Here $0 \le \rho \le a$, where $a \le b$, $0 \le \phi \le 2\pi$, $0 \le \theta \le 2\pi$

7 Evaluate the Jacobian $\partial(x, y, z)/\partial(\rho, \phi, \theta)$.

8 (cont.) What are the level surfaces $\rho =$ constant, $\phi =$ constant, $\theta =$ constant? Find their mutual angles of intersection.

9 Describe the domain $0 \le x \le y \le 1$, $0 \le z \le y$ and find its volume.

10 The z-axis is an infinite wire with uniform density δ. Find its gravitational attraction on a unit mass at $(b, 0, 0)$.

11 (cont.) The infinite cylinder $x^2 + y^2 = a^2$, $-\infty < z < +\infty$ has uniform density δ. Find its gravitational attraction on a unit mass at $(b, 0, 0)$, where $b > a$.

$$\left[\text{Hint} \quad \int_0^{2\pi} \frac{d\theta}{A + B\cos\theta} = \frac{2\pi}{\sqrt{A^2 - B^2}}\right]$$

12 (cont.) What is the gravitational force if $0 < b < a$?

13 By a long series of steps involving reduction formulas one arrives at the formulas

$$\int_0^{\pi/2} \sin^{2n}\theta\,d\theta = \frac{(2n)!}{2^{2n}(n!)^2}\frac{\pi}{2}$$

and

$$\int_0^{\pi/2} \sin^{2n+1}\theta\,d\theta = \frac{2^{2n}(n!)^2}{(2n+1)!}$$

A consequence is the simple-looking formula

$$\left(\int_0^{\pi/2} \sin^{2n}\theta\,d\theta\right)\left(\int_0^{\pi/2} \sin^{2n+1}\theta\,d\theta\right) = \frac{\pi}{2(2n+1)}$$

Use multiple integrals over the unit sphere to give a simple proof of the latter formula.

14 If $\mathbf{v} = (v_1, v_2, v_3)$ is a vector field, its divergence is $\operatorname{div}\mathbf{v} = \sum \partial v_i/\partial x_i$. Suppose we take another rectangular coordinate system (with the same origin). Then its coordinates are $\bar{x}_i = \sum a_{ij}x_j$, where the a_{ij} are constants. Express $\mathbf{v}$ in the new coordinate system, and show that $\overline{\operatorname{div}}\,\mathbf{v} = \operatorname{div}\mathbf{v}$, that is, that $\operatorname{div}\mathbf{v}$ is a geometric quantity, independent of the coordinate system.

***15** Find in any way the volume of the solid $x^2 + y^2 \le a^2$, $|y + z| \le b$, $|y - z| \le b$, where $0 < b \le a$. Note that the solid may be interpreted as a square hole on center through a cylinder, the sides of the hole at 45° to the axis of the cylinder.

16 A cylindrical hold of radius a is bored through a solid cylinder of radius $2a$; the hole is perpendicular to the solid cylinder and just touches a generator. Find in any way the volume removed.

17 From each point of the space curve $\mathbf{x} = \mathbf{x}(s)$ draw a segment of length 1 in the direction of the unit tangent. These segments sweep out a surface. Show that its area is $\frac{1}{2}\int k(s)\,ds$, where $k(s)$ is the curvature and the integral is taken over the length of the curve.

18 Find the volume of the four-dimensional sphere $x^2 + y^2 + z^2 + w^2 \le a^2$.

19 A regular icosahedron has its center at the origin. Find the solid angle subtended by each of its faces.

***20** A rigid body $\mathbf{D}$ is rotating about an axis through $\mathbf{0}$ with angular velocity $\boldsymbol{\omega}$. If $\mathbf{v} = \mathbf{v}(\mathbf{x})$ is the velocity at $\mathbf{x}$, the **angular momentum** of the rotating body is

$$\mathbf{J} = \iiint_{\mathbf{D}} \mathbf{x} \times \mathbf{v}\,dM.$$

Prove $\mathbf{J} = \boldsymbol{\omega}\begin{bmatrix} I_x & I_{xy} & I_{xz} \\ I_{yx} & I_y & I_{yz} \\ I_{zx} & I_{zy} & I_z \end{bmatrix}$

Thus $\mathbf{J}$ equals the product of the row vector $\boldsymbol{\omega}$ with the inertia matrix. (See the instructions before Exercise 37, page 907.)

21 A vector field $\mathbf{A}$ and a function f are defined on a region in space. Suppose for each $\mathbf{x}$ inside the region and for each $\varepsilon > 0$ there is a domain $\mathbf{D}$ containing $\mathbf{x}$ such that $\operatorname{rad}(\mathbf{D}) < \varepsilon$, such that $\mathbf{D}$ lies in the region, and such that

$$\iint_{\partial \mathbf{D}} \mathbf{A} \cdot d\sigma = \iiint_{\mathbf{D}} f\,dV$$

Prove $f = \operatorname{div} \mathbf{A}$. Assume f continuous and $\mathbf{A}$ continuously differentiable.

In the next seven exercises we shall obtain the Laplacian $\Delta f = \operatorname{div}(\operatorname{grad} f)$ in spherical coordinates. (See Exercise 23, page 899.) We need the result of Exercise 21 and the following formula:

$$\operatorname{grad} f = f_\rho \boldsymbol{\lambda} + \frac{1}{\rho} f_\phi \boldsymbol{\mu} + \frac{1}{\rho \sin \phi} f_\theta \boldsymbol{\nu}$$

Let $\mathbf{D}$ denote the domain $\rho_0 \le \rho \le \rho_1$, $\phi_0 \le \phi \le \phi_1$, $\theta_0 \le \theta \le \theta_1$, with the outward normal on $\partial \mathbf{D}$.

22 Compute $d\sigma$ on all six faces of $\partial \mathbf{D}$.

23 Prove $\displaystyle\iint_{\partial \mathbf{D}} (A\boldsymbol{\lambda}) \cdot d\sigma = \iiint_{\mathbf{D}} \frac{1}{\rho^2} \cdot \frac{\partial}{\partial \rho}(\rho^2 A)\,dV$

24 Obtain a similar formula for $\displaystyle\iint_{\partial \mathbf{D}} (B\boldsymbol{\mu}) \cdot d\sigma$

25 Obtain a similar formula for $\displaystyle\iiint_{\partial \mathbf{D}} (C\boldsymbol{\nu}) \cdot d\sigma$

26 Assemble the last three steps into one; let $\mathbf{A} = A\boldsymbol{\lambda} + B\boldsymbol{\mu} + C\boldsymbol{\nu}$ and express $\displaystyle\iint_{\partial \mathbf{D}} \mathbf{A} \cdot d\sigma$ as an integral over $\mathbf{D}$.

27 Use Exercise 21 and the result of Exercise 26 to express $\operatorname{div} \mathbf{A}$ in terms of spherical coordinates.

28 Apply the result to $\mathbf{A} = \operatorname{grad} f$ to obtain Δf in terms of spherical coordinates.

29 (cont.) The **Yukawa potential** is the function $f(\rho) = e^{-k\rho}/\rho$ with k constant. Show that $\Delta f = k^2 f$.

***30** Use the result of Exercise 28 to show that if $f\langle \rho, \phi, \theta \rangle$ satisfies $\Delta f = 0$, then its **Kelvin transform**

$$g\langle \rho, \phi, \theta \rangle = \frac{1}{\rho} f\left\langle \frac{1}{\rho}, \phi, \theta \right\rangle$$

satisfies $\Delta g = 0$.

***31** Show that

$$\int_{-1}^{1} \cdots \int_{-1}^{1} (x_1 + x_2 + \cdots + x_n)^4\,dx_1\,dx_2 \cdots dx_n = \frac{2^n n}{15}(5n - 2)$$

***32** Show that

$$\int_{-1}^{1} \cdots \int_{-1}^{1} (x_1 + \cdots + x_n)^6\,dx_1 \cdots dx_n = \frac{2^n n}{63}(35n^2 - 42n + 16)$$

19 Differential Equations

A differential equation is a relation between a variable x, a function $y(x)$ of that variable, and several derivatives $y'(x)$, $y''(x)$, $\cdots$. In this brief introduction to the study of differential equations we shall study first order equations of the form

$$\frac{dy}{dx} = F(x, y)$$

and later a particular type of second order differential equation, one involving x, y, dy/dx, and d^2y/dx^2.

To solve a differential equation is to find all functions $y(x)$ that satisfy it. With some exceptions, that is the main problem of this subject. Many of the laws of physics, chemistry, economics, and other fields are expressed as differential equations, so this subject is one of the really important parts of applicable mathematics. Indeed, some say that the purpose of calculus is to prepare for differential equations.

19-1 Separation of Variables

We begin our work with equations of the form

$$\frac{dy}{dx} = F(x, y)$$

Such an equation is called a **first order differential equation.** It is called a "differential equation" because it involves a derivative of the unknown function $y(x)$. It is called "first order" because it involves only the first derivative dy/dx.

The equation poses a problem: find all functions $y(x)$ such that

$$\frac{dy(x)}{dx} = F(x, y(x))$$

is satisfied. For example, the most general solution of the familiar natural growth equation

$$\frac{dy}{dx} = y$$

is $y(x) = Ce^x$, where C is any constant. A more complicated example is

$$\frac{dy}{dx} = y - x$$

It is easy to verify that if C is any constant, then $y(x) = Ce^x + x + 1$ is a solution. We shall soon see that every solution at this particular differential equation is so given.

Example 1

Show that the differential equation

$$\frac{dy}{dx} = 2\frac{y}{x} \qquad (x \neq 0)$$

is satisfied by each function $y = Cx^2$, where C is a constant.

Solution Differentiate $y = Cx^2$. The result is $y' = 2Cx$. Therefore $xy' = 2Cx^2 = 2y$ so $y' = 2y/x$ as asserted. ●

Geometric Interpretation

Let us look at a differential equation of the form

$$\frac{dy}{dx} = F(x, y)$$

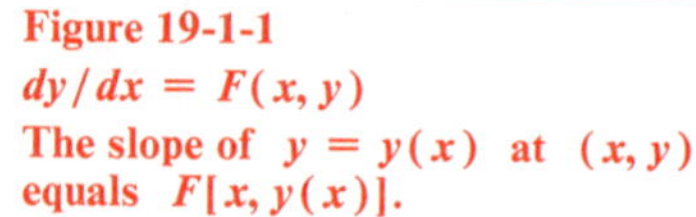
Figure 19-1-1
$dy/dx = F(x, y)$
The slope of $y = y(x)$ at (x, y) equals $F[x, y(x)]$.

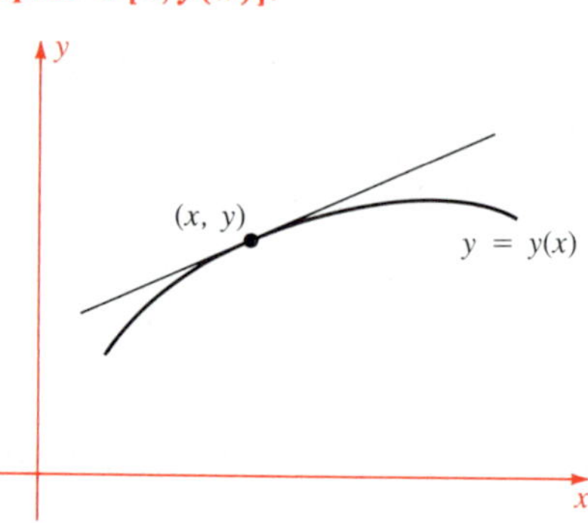

where $F(x, y)$ is a smooth function on some domain **D** in the x, y-plane. Suppose $y = y(x)$ is a solution of the differential equation, and let (x, y) be a point of **D** on the graph of $y = y(x)$. See Figure 19-1-1. Then the slope of the graph of $y = y(x)$ at (x, y) is

$$\left.\frac{dy}{dx}\right|_x = F(x, y(x))$$

This suggests a geometric way of seeking solutions.

At each point (x, y) of the domain of $F(x, y)$ we draw a short segment with slope $F(x, y)$. The result is called the **direction field** of the differential equation. Each solution curve of the equation must be tangent to the direction field at each point of the curve. Thus the direction field gives us a feeling for the way solutions flow. This picture also suggests that

Figure 19-1-2
$dy/dx = y - x$

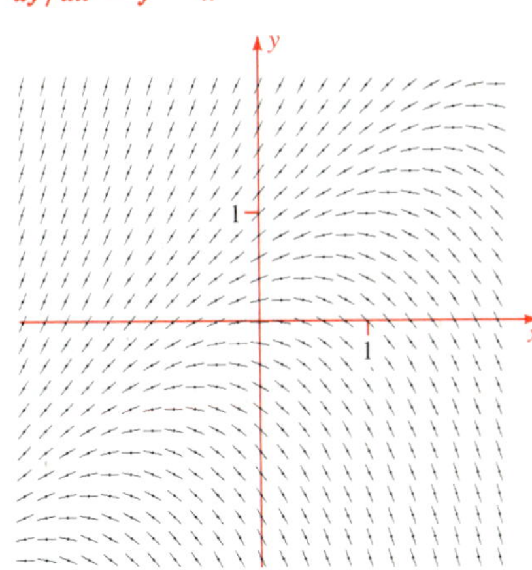

Figure 19-1-3
$dy/dx = x + y$

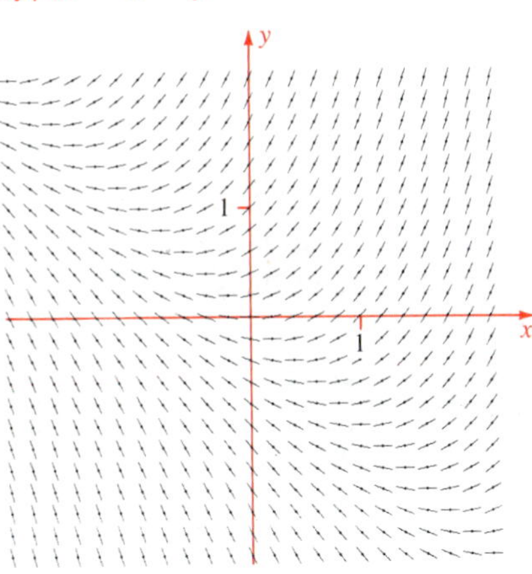

Figure 19-1-4
$dy/dx = x^2/y$

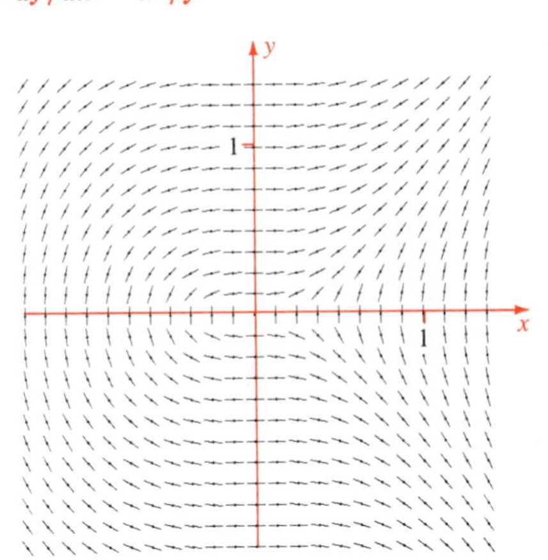

through any point (x_0, y_0) there is exactly one solution curve. Figures 19-1-2 through 19-1-4 show examples of direction fields.

We now consider a special type of differential equation, in which the function $F(x, y)$ factors into a function of x alone and a function of y alone: $F(x, y) = g(x)h(y)$. Examples are

$$\frac{dy}{dx} = x^2y^3 \qquad \frac{dy}{dx} = e^x \cos y \qquad \frac{dy}{dx} = \frac{y}{\sqrt{1 + x^2}}$$

We solve

$$\frac{dy}{dx} = g(x)h(y)$$

by **separating variables.** This means we put all the terms involving y on one side and all the terms involving x on the other. That is, we write

$$\frac{dy}{h(y)} = g(x)\,dx$$

Then we integrate:

$$\int \frac{dy}{h(y)} = \int g(x)\,dx + C$$

where C is a constant. If we carry out the integrations, we have a relation of the form

$$H(y) = G(x) + C$$

We solve this relation for y in terms of x and the constant C. The result, for each value of C, is a solution of the original differential equation.

This method is strictly a formal one, and we have ignored domains completely. One thing we should point out, however. When we pass from

$$\frac{dy}{dx} = g(x)h(y) \qquad \text{to} \qquad \frac{dy}{h(y)} = g(x)\,dx$$

we must assume $h(y) \neq 0$ on the domain of h. Otherwise, if $h(y_0) = 0$, then the constant function $y = y_0$ is a solution of the original differential equation.

Example 2

Solve $\quad \dfrac{dy}{dx} = xy$

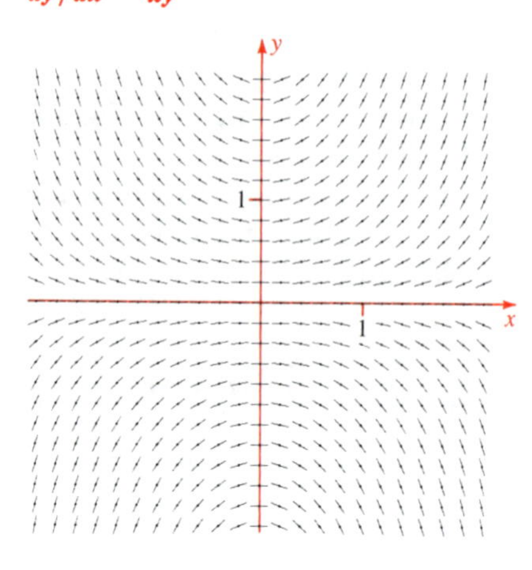

Figure 19-1-5
$dy/dx = xy$

Solution The direction field is sketched in Figure 19-1-5. Obviously $y = y(x) = 0$ is a solution. Suppose $y \neq 0$ and separate variables:

$$\frac{dy}{y} = x\,dx$$

Now integrate:

$$\int \frac{dy}{y} = \int x\,dx + C \qquad \text{that is} \qquad \ln|y| = \tfrac{1}{2}x^2 + C$$

To solve for y, take exponentials:

$$|y| = K\exp(\tfrac{1}{2}x^2) \qquad \text{where} \qquad K = e^C > 0$$

Since $|y| = \pm y$, the result is $y = \pm K\exp(\frac{1}{2}x^2)$. We simply rename $\pm K$ a new constant to obtain the most general solution

$$y = K\exp(\tfrac{1}{2}x^2) \qquad (K \text{ any constant})$$

This includes the solution $y = 0$ for the choice $K = 0$.

Check Let us check that $y = K\exp(\frac{1}{2}x^2)$ really satisfies the given differential equation:

$$\frac{dy}{dx} = \frac{d}{dx}[K\exp(\tfrac{1}{2}x^2)] = Kx\exp(\tfrac{1}{2}x^2) = xy$$

●

Note that we can always test a proposed solution to a given differential equation to see if it indeed satisfies the equation.

Example 3

Solve

$$\frac{dy}{dx} = \frac{2x(y-1)}{x^2+1}$$

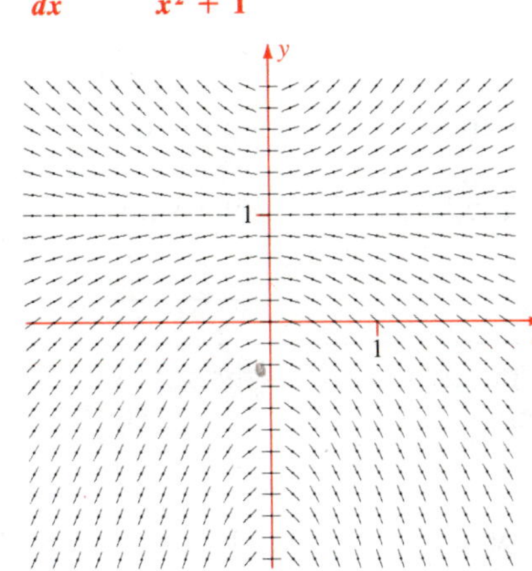

Figure 19-1-6
$\dfrac{dy}{dx} = \dfrac{2x(y-1)}{x^2+1}$

Solution The direction field of the equation is sketched in Figure 19-1-6. The equation has the form

$$\frac{dy}{dx} = g(x)h(y) \qquad g(x) = \frac{2x}{x^2+1} \quad \text{and} \quad h(y) = y - 1$$

Since $h(1) = 0$, the function $y = 1$ is a solution. Now assume $y \neq 1$ and separate variables:

$$\frac{dy}{y-1} = \frac{2x\,dx}{x^2+1} \qquad \text{so} \qquad \int \frac{dy}{y-1} = \int \frac{2x\,dx}{x^2+1} + C$$

Integrate, then take exponentials:

$$\ln|y-1| = \ln(x^2+1) + C$$
$$|y-1| = K(x^2+1) \qquad \text{where} \quad K = e^C > 0$$

That is, $y - 1 = \pm K(x^2+1)$. Therefore the answer is

$$y = 1 + K(x^2+1)$$

where K is any constant. Now check that this really is a solution:

$$\frac{dy}{dx} = \frac{d}{dx}[1 + K(x^2+1)] = 2Kx$$
$$\frac{2x(y-1)}{x^2+1} = \frac{2x[K(x^2+1)]}{x^2+1} = 2Kx$$

●

Example 4

Solve $\quad \dfrac{dy}{dx} = e^{x-y}$

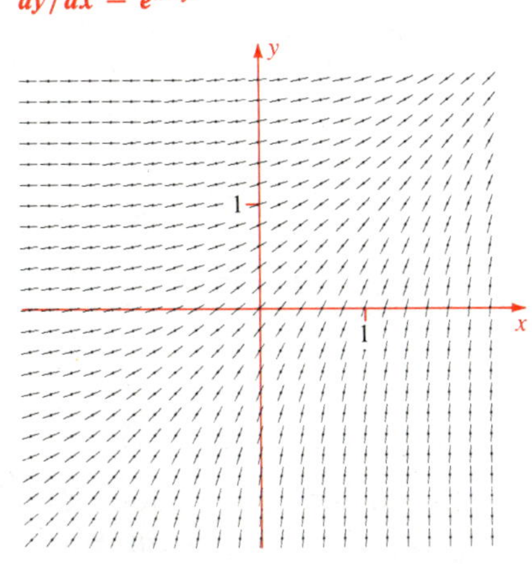

Figure 19-1-7
$dy/dx = e^{x-y}$

Solution The direction field is sketched in Figure 19-1-7. We can write the right side of the equation as the product $e^x e^{-y}$, so we can separate variables:

$$e^y\,dy = e^x\,dx \qquad \text{hence} \qquad \int e^y\,dy = \int e^x\,dx + C$$

that is, $e^y = e^x + C$. Now solve for y:

$$y = \ln(e^x + C)$$

Check:

$$\frac{dy}{dx} = \frac{1}{e^x + C} \cdot \frac{d}{dx}(e^x + c)$$
$$= \frac{e^x}{e^x + C} = \frac{e^x}{e^y} = e^{x-y}$$

●

Remark We may be unable to solve for y explicitly in the relation

$$\int \frac{dy}{h(y)} = \int g(x)\,dx + C$$

Still, just having the relation may be sufficient for solving a practical problem.

We are now ready to try our solution method on an application.

Example 5

Find a curve such that the slope at each point (x, y) of the curve is the reciprocal of the slope of the line through $(0, 0)$ and (x, y). In fact, find all such curves.

Figure 19-1-8
$dy/dx = x/y$

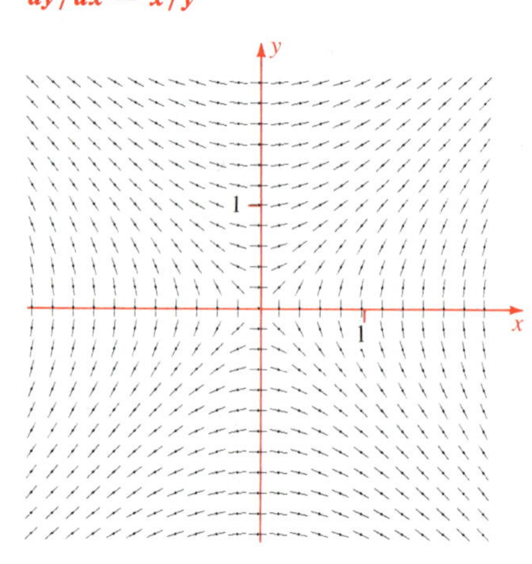

Solution The geometric nature of the problem is shown in Figure 19-1-8. Along each radial line, the direction field has constant slope. The steeper the radial line, the flatter the slope of the direction field. Now the line through $(0, 0)$ and (x, y) has slope y/x, so we are asking for solutions of the differential equation

$$\frac{dy}{dx} = \frac{x}{y}$$

Separate variables and integrate:

$$y\,dy = x\,dx \qquad \text{hence} \qquad \tfrac{1}{2}y^2 = \tfrac{1}{2}x^2 + C$$

Replace C by $\frac{1}{2}C$; the answer is

$$y^2 - x^2 = C$$

This is a family of rectangular hyperbolas (Figure 19-1-9). ●

Remark In the future we shall replace one constant by another without comment if this simplifies an answer. So don't be confused if $2C$ becomes C or $K + 3$ becomes K.

Figure 19-1-9
The general solution of $dy/dx = x/y$ is all hyperbolas $y^2 - x^2 = C$.

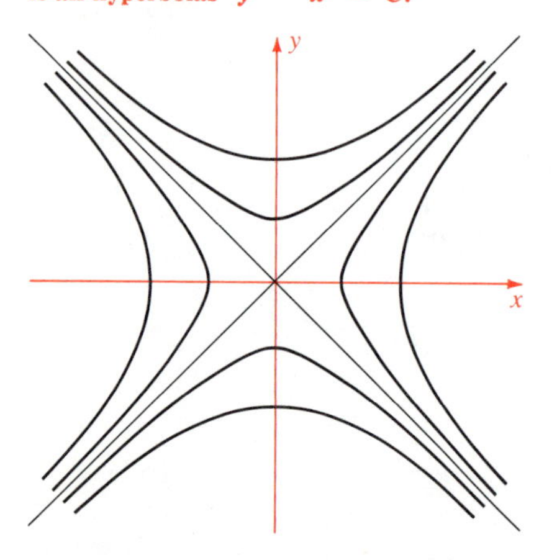

Initial Value Problems

Given a differential equation

$$\frac{dy}{dx} = F(x, y)$$

and a point (a, b) in the domain $\mathbf{D}$ of $F(x, y)$, the **initial value problem** is to find a function $y = y(x)$ such that

$$\frac{dy}{dx} = F(x, y(x)) \qquad \text{and} \qquad y(a) = b$$

In other words, $y(x)$ must be a solution of the differential equation and its graph must pass through the (initial) point (a, b). Much numerical

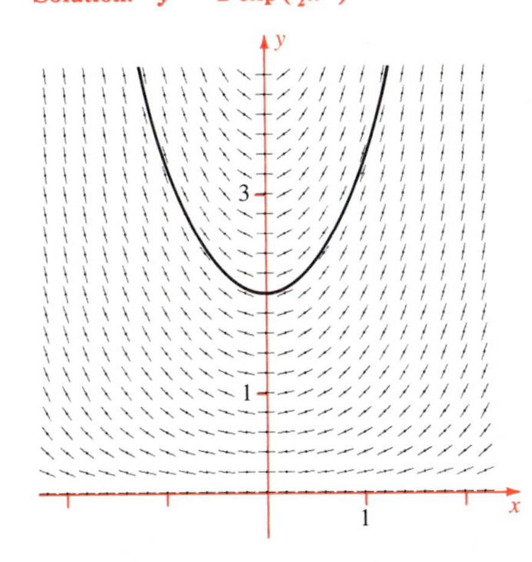

Figure 19-1-10
Initial value problem:
$dy/dx = xy \qquad y(0) = 2$
Solution: $y = 2\exp(\frac{1}{2}x^2)$

work in differential equations concentrates on this problem. Also, many applications lead to such initial value problems.

In each of our previous examples, the general solution depends on a constant. When this is so, we can usually choose the constant to solve the initial value problem.

Example 6

Solve the initial value problem

$$\frac{dy}{dx} = xy \qquad y(0) = 2$$

Solution The general solution $y = K\exp(\frac{1}{2}x^2)$ was worked out earlier, in Example 2. Substitute $(x, y) = (0, 2)$:

$$2 = K\exp(0) = K \qquad \text{so} \qquad y = 2\exp(\tfrac{1}{2}x^2).$$

The solution is sketched in Figure 19-1-10.

Exercises

Show that the function satisfies the differential equation (C is a constant)

1 $y = C/x \qquad dy/dx = -y/x$

2 $y = C(x^2+1) \qquad dy/dx = 2xy/(x^2+1)$

3 $y = Cx + 1 \qquad dy/dx = (y-1)/x$

4 $y = C\sin x \qquad dy/dx = y\cot x$

5 $y = Ce^x + xe^x \qquad dy/dx = y + e^x$

6 $y = 1/(x+C) \qquad dy/dx = -y^2$

Solve

7 $\dfrac{dy}{dx} = \dfrac{x^2}{y}$

8 $\dfrac{dy}{dx} = \dfrac{1}{xy}$

9 $\dfrac{dy}{dx} = \sqrt{\dfrac{y}{x}}$

10 $\dfrac{dy}{dx} = \dfrac{x+1}{xy}$

11 $\dfrac{dy}{dx} = xe^y$

12 $\dfrac{dy}{dx} = \left(\dfrac{1+y}{1+x}\right)^2$

13 $\dfrac{dy}{dx} = -\dfrac{y}{x}$

14 $\dfrac{dy}{dx} = -\dfrac{x}{y}$

15 $\dfrac{dy}{dx} = \dfrac{y}{x}$

16 $\dfrac{dy}{dx} = -\dfrac{y}{2x}$

17 $\dfrac{dy}{dx} = 2y\cot x$

18 $\dfrac{dy}{dx} = x(y^2+1)$

19 $\dfrac{dy}{dx} = y^2\cos x$

20 $\dfrac{dy}{dx} = xy(y-1)$

21 $\dfrac{dy}{dx} = \dfrac{x^2}{y^2}$

22 $\dfrac{dy}{dx} = x^2y^2$

23 $\dfrac{dy}{dx} = \sqrt{\dfrac{1-y^2}{1-x^2}}$

24 $\dfrac{dy}{dx} = \dfrac{1+y^2}{1+x^2}$

Sketch the direction field

25 $\dfrac{dy}{dx} = x^2y$

26 $\dfrac{dy}{dx} = x - y^2$

27 $\dfrac{dy}{dx} = \dfrac{3x^2}{y}$

28 $\dfrac{dy}{dx} = \dfrac{3x^2+1}{5y^4-1}$

29 The slope at each point (x, y) of a curve is twice the slope of the line through $(0, 0)$ and (x, y). Find all such curves.

30 The slope at each point (x, y) of a curve is twice the slope of the circle through (x, y) with center $(0, 0)$. Find all such curves.

Show that the substitution $y = ux$ changes the given differential equation into an equation for $u = u(x)$ whose variables separate. Then solve for u and obtain y.

31 $\dfrac{dy}{dx} = \dfrac{x^2+y^2}{2x^2}$

32 $\dfrac{dy}{dx} = \dfrac{x+y}{x-y}$

Solve the initial value problem

33 $\dfrac{dy}{dx} = (x+2)y \qquad y(0) = 0$

34 $\dfrac{dy}{dx} = x^2y^3 \qquad y(-1) = 2$

35 $\dfrac{dy}{dx} = \dfrac{x^2}{y} \qquad y(0) = -1$

36 $\dfrac{dy}{dx} = \dfrac{e^x}{y^2} \qquad y(0) = -3$

37 $\dfrac{dy}{dx} = -\dfrac{y}{2x}$ $\qquad y(1) = 3$

38 $\dfrac{dy}{dx} = 2y \cot x$ $\qquad y(\frac{1}{2}\pi) = 5$

39 $\dfrac{dy}{dx} = \dfrac{(y-1)(y-2)}{x}$ $\qquad y(4) = 0$

40 $\dfrac{dy}{dx} = y^3 \sin x$ $\qquad y(0) = \frac{1}{2}$

41 Sketch the direction field for $dy/dx = F(x, y)$ where

$$F(x, y) = \begin{cases} 2\sqrt{y} & \text{if } y \geq 0 \\ -2\sqrt{-y} & \text{if } y < 0 \end{cases}$$

***42** (cont.) Solve the initial value problem with the initial point $(0, 0)$. You should find many solutions!

43 Solve $xy' + y = 2x$ [Hint Compute $(xy)'$.]

44 Solve $y' + y = e^{-x} \sin x$
[Hint Compute $(e^x y)'$.]

45 Solve $x + yy' = x^2 + y^2$
[Hint Compute the derivative of $\ln(x^2 + y^2)$.]

46 Solve $xy' - y = x^2 \sin x$
[Find the trick yourself this time.]

19-2 First Order Linear Equations

A first order **linear** differential equation is a first order differential equation that can be written in the form

$$\frac{dy}{dx} + p(x)y = q(x)$$

Differential equations of this type arise in many applications, so they are important. What is more, if we can integrate $p(x)$ and another function related to $p(x)$ and $q(x)$, then we can solve the equation explicitly.

The given equation can be written in the form

$$\frac{dy}{dx} = F(x, y) \qquad \text{where} \quad F(x, y) = -p(x)y + q(x)$$

Thus $F(x, y)$ is defined for all y, so its domain is an infinite strip (Figure 19-2-1):

$$\mathbf{D} = \{(x, y) \mid x_0 < x < x_1, \ y \text{ arbitrary}\}$$

Figure 19-2-1
The domain of $F(x, y) = -p(x)y - q(x)$ is an infinite strip. It is assumed that $p(x)$ and $q(x)$ have domain (x_0, x_1).

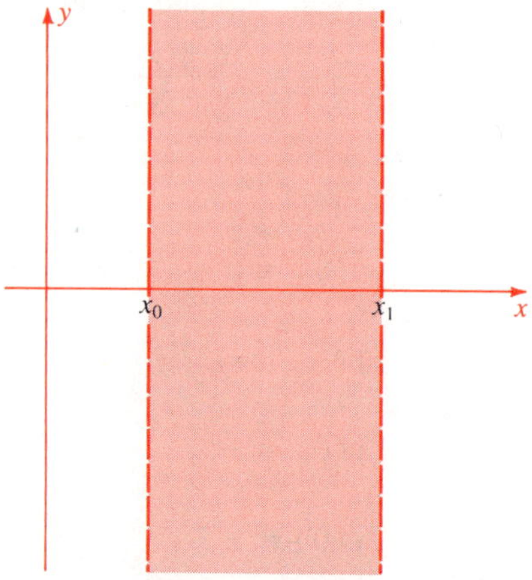

The Homogeneous Case

When $q(x) = 0$, the first order linear equation is

$$\frac{dy}{dx} + p(x)y = 0$$

This is called a **homogeneous** first order linear differential equation. We can separate variables and solve that way. However, we prefer another approach based on finding an *integrating factor.* This approach applies to many types of differential equations.

To solve

$$\frac{dy}{dx} + p(x)y = 0$$

we seek a function $m(x)$, called an **integrating factor,** such that

$$m(x)\left[\frac{dy}{dx} + p(x)y\right] = \frac{d}{dx}[m(x)y]$$

This gives us a way of rewriting the equation so that it is easily solved. If we can find such an $m(x) \neq 0$, then the given differential equation is equivalent to

$$\frac{d}{dx}[m(x)y] = 0$$

so $m(x)y = K$; that is, $y(x) = K/m(x)$.

Now

$$\frac{d}{dx}[m(x)y] = m(x)\frac{dy}{dx} + \frac{dm}{dx}y$$

so the definition of an integrating factor requires $m(x)$ to satisfy

$$\frac{dm}{dx} = p(x)m$$

This is a new differential equation, an equation for $m(x)$. We solve it by separating variables:

$$\frac{dm}{m} = p\,dx \qquad \text{so} \qquad \ln|m| = \int_a^x p(t)\,dt + C$$

$$m(x) = Ke^{P(x)} \qquad \text{where} \quad P(x) = \int_a^x p(t)\,dt$$

The choice of the initial point a is irrelevant. We simply require $P(x)$ to be *any* antiderivative of $p(x)$. Furthermore, we can choose $C = 0$, so $m(x) = e^{P(x)}$ is an integrating factor, and the solution of the original differential equation is

$$y(x) = K/m(x) = Ke^{-P(x)}$$

We have obtained the following result.

Homogeneous First Order Equations

Suppose $p(x)$ is continuous on an open interval (a, b). Let $P(x)$ be any antiderivative of $p(x)$ on this interval. Then the solutions of the differential equation

$$\frac{dy}{dx} + p(x)y = 0$$

are precisely the functions

$$y(x) = Ke^{-P(x)} \qquad \text{where} \quad K \text{ is any constant}$$

Example 1

Solve the initial value problem

$$\frac{dy}{dx} + 2y = 0$$

$$y(1) = 0.4$$

Solution The direction field is sketched in Figure 19-2-2. Here $p(x) = 2$ so $P(x) = 2x$ is an antiderivative. Therefore the general solution is

$$y(x) = Ke^{-2x}$$

To satisfy the initial condition, set $(x, y) = (1, 0.4)$:

$$0.4 = Ke^{-2}$$

$$K = 0.4e^2$$

Therefore

$$y = y(x) = 0.4e^2e^{-2x} = 0.4e^{2-2x}$$

Note that this solution is indicated in Figure 19-2-2. ●

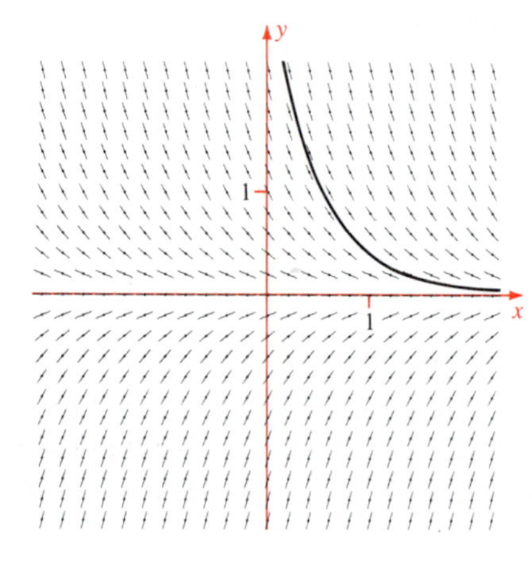

Figure 19-2-2
$dy/dx + 2y = 0$
(The solution with initial value $y(1) = 0.4$ is shown.)

Example 2

Solve the initial value problem

$$\frac{dy}{dx} + \frac{1}{x}y = 0 \qquad y(1) = 2$$

Solution The direction field is sketched in Figure 19-2-3. We stick to the interval $x > 0$ since it contains the x-coordinate of the initial point $(1, 2)$ and $p(x) = 1/x$ is continuous there. An obvious antiderivative of $p(x)$ is $P(x) = \ln x$. Hence the general solution of the differential equation is

$$y = Ke^{-\ln x} = K/x$$

The initial condition $y(1) = 2$ is satisfied for $K = 2$. Therefore the solution (Figure 19-2-3) is

$$y = 2/x \qquad (x > 0)$$

This function is sketched in the figure. ●

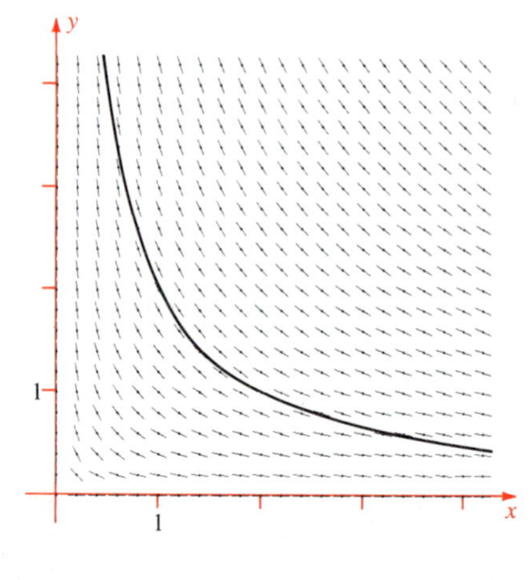

Figure 19-2-3
$dy/dx + y/x = 0$
(The solution with initial value $y(1) = 2$ is shown.)

The General Case

We return to the general, or **non-homogeneous,** linear first order differential equation

$$\frac{dy}{dx} + p(x)y = q(x)$$

To solve this equation, we use the integrating factor technique. Suppose $P(x)$ is any antiderivative of $p(x)$ and $m(x) = \exp[P(x)]$. Then as before

$$\frac{dm}{dx} = p(x)m(x)$$

so that

$$\begin{aligned} \frac{d}{dx}[m(x)y] &= m(x)\frac{dy}{dx} + p(x)m(x)y \\ &= m(x)\left[\frac{dy}{dx} + p(x)y\right] \\ &= m(x)q(x) \end{aligned}$$

Thus we need only solve the simpler equation

$$\frac{d}{dx}[m(x)y] = m(x)q(x)$$

We solve it by one integration:

$$m(x)y(x) = \int_a^x m(t)(t)\,dt + C$$

To obtain $y(x)$, we simply divide by $m(x)$.

Example 3

Find the general solution of the differential equation

$$\frac{dy}{dx} + 2y = e^x$$

Figure 19-2-4

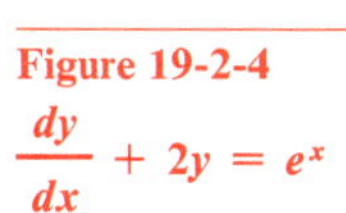

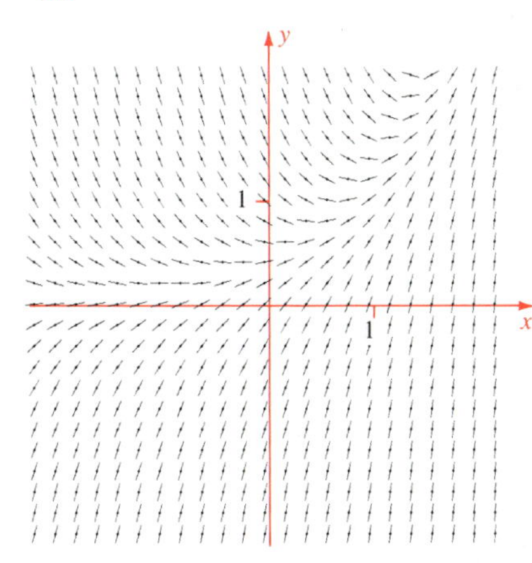

Solution The direction field is sketched in Figure 19-2-4. In this example, $p(x) = 2$, with one antiderivative $P(x) = 2x$. Therefore $m(x) = e^{P(x)} = e^{2x}$ is an integrating factor. We multiply both sides of the given equation by $m(x)$ and rewrite the left side as a derivative:

$$e^{2x}\frac{dy}{dx} + 2e^{2x}y = e^{3x} \qquad \text{so} \qquad \frac{d}{dx}(e^{2x}y) = e^{3x}$$

Next, we integrate:

$$e^{2x}y(x) = \tfrac{1}{3}e^{3x} + C$$

Finally we divide by e^{2x}, that is, multiply by e^{-2x}:

$$y(x) = \tfrac{1}{3}e^x + Ce^{-2x}$$

This is the general solution. ●

Let us summarize what we have just found by using integrating factors.

First Order Linear Differential Equation

To solve

$$\frac{dy}{dx} + p(x)y = q(x)$$

let $P(x)$ be any antiderivative of $p(x)$. Then

$$y(x) = e^{-P(x)}Q(x) + Ce^{-P(x)}$$

where $Q(x)$ is any antiderivative of $e^{P(x)}q(x)$.

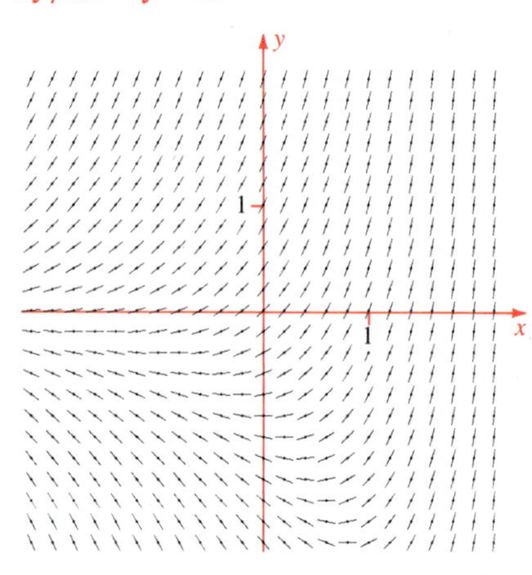

Figure 19-2-5
$dy/dx - y = e^x$

Example 4

Solve $dy/dx - y = e^x$.

Solution The direction field is sketched in Figure 19-2-5. In this case $p(x) = -1$, so take $P(x) = -x$. Then

$$Q(x) = \int e^{P(x)}q(x)\,dx = \int e^{-x}e^x\,dx$$

$$= \int dx = x + C$$

and

$$y = y(x) = e^{-P(x)}(x + C) = e^x(x + C)$$

$$= xe^x + Ce^x$$

Example 5

Solve $\quad \dfrac{dy}{dx} - \dfrac{y}{x} = x^3 + 1 \qquad (x > 0)$

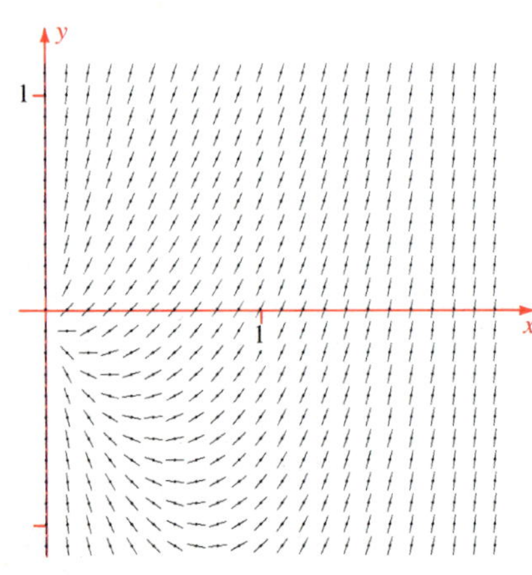

Figure 19-2-6
$dy/dx - y/x = 1 + x^3$

Solution The direction field is sketched in Figure 19-2-6. For this equation $p(x) = -1/x$ and $q(x) = x^3 + 1$. Clearly $P(x) = -\ln x$ is an antiderivative of $p(x)$, so $m(x) = e^{P(x)} = 1/x$ is an integrating factor. The given differential equation is equivalent to

$$\frac{d}{dx}(my) = mq = x^2 + 1/x$$

Therefore

$$my = \tfrac{1}{3}x^3 + \ln x + C$$

Since $m(x) = 1/x$ it follows that

$$y = y(x) = \tfrac{1}{3}x^4 + x\ln x + Cx$$

Figure 19-2-7
$dy/dx + 2xy = x$

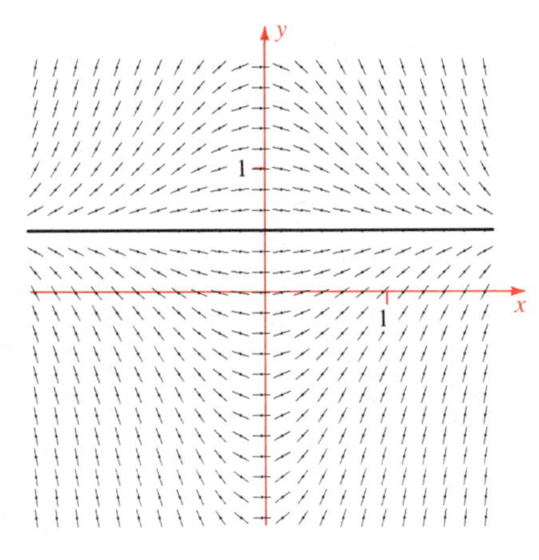

Example 6

Solve $$\frac{dy}{dx} + 2xy = x.$$

Solution The direction field is sketched in Figure 19-2-7. For this equation, $p(x) = 2x$ and $q(x) = x$. Thus $P(x) = x^2$; consequently $m(x) = e^{P(x)} = \exp(x^2)$ is an integrating factor. Hence

$$Q(x) = \int mq\,dx = \int x\exp(x^2)\,dx = \tfrac{1}{2}\exp(x^2) + C$$

so that

$$y = y(x) = Q(x)/m(x) = \tfrac{1}{2} + C\exp(-x^2)$$

In particular $y = \frac{1}{2}$ is a solution, as is clear in the figure. ●

Example 7

Solve $$\frac{dy}{dx} - \frac{y}{x} = \frac{1}{1-x} \qquad (0 < x < 1)$$

Figure 19-2-8
$dy/dx - y/x = 1/(1-x)$
$(0 < x < 1)$

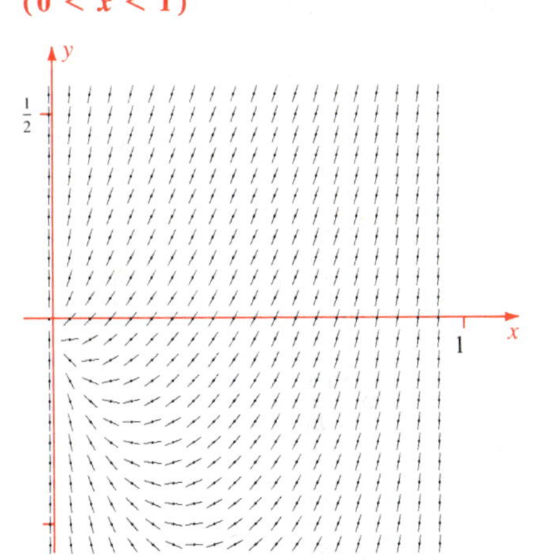

Solution The direction field is sketched in Figure 19-2-8. Note how it appears impossible for a solution that starts in the strip $0 < x < 1$ to ever escape from that domain. Note also that $p(x) = -1/x$ and $q(x) = 1/(1-x)$ are continuous on the strip.

Clearly $P(x) = -\ln x$ is an indefinite integral of $p(x)$, so $m(x) = e^{P(x)} = 1/x$ is an integrating factor. Thus

$$my = \int mq\,dx + C = \int \frac{dx}{x(1-x)} + C$$

But

$$\int \frac{dx}{x(1-x)} = \int \left(\frac{1}{x} + \frac{1}{1-x}\right)dx$$
$$= \ln x - \ln(1-x) + C$$
$$= \ln\frac{x}{1-x} + C$$

so we finally have

$$y(x) = \frac{1}{m}\left(\ln\frac{x}{1-x} + C\right) = x\ln\frac{x}{1-x} + Cx$$ ●

Example 8

Solve the initial value problem

$$\frac{dy}{dx} - y = -x \qquad y(1) = 0$$

Figure 19-2-9
Initial value problem:
$dy/dx - y = -x \quad y(1) = 0$
Solution: $y = -2e^{x-1} + x + 1$

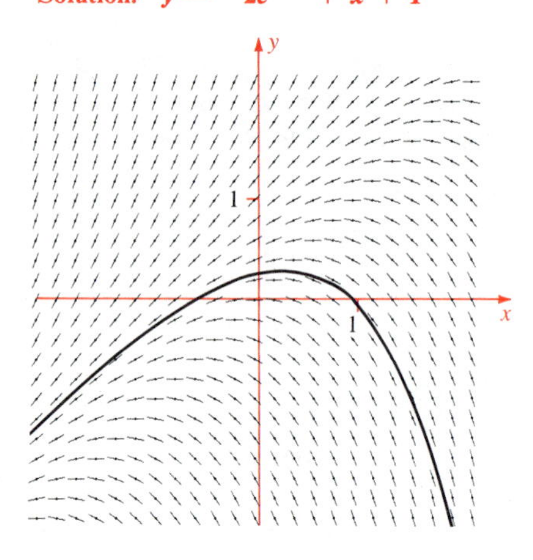

Solution In this case, $p(x) = -1$, so $P(x) = -x$ and

$$Q(x) = \int q(x)e^{P(x)} = -\int xe^{-x}\,dx = xe^{-x} + e^{-x} + C$$

Therefore the general solution is

$$y = y(x) = e^{-P(x)}Q(x) = e^{x}(xe^{-x} + e^{-x} + C)$$
$$= x + 1 + Ce^{x}$$

To satisfy the initial condition, substitute $(x, y) = (1, 0)$:

$$0 = 1 + 1 + Ce \qquad Ce = -2 \qquad C = -2e^{-1}$$

Consequently the solution to the problem is

$$y = x + 1 - 2e^{x-1}$$

The direction field of the differential equation is sketched in Figure 19-2-9, and the solution to the initial value problem is shown. ●

Exercises

Solve

1 $\dfrac{dy}{dx} + 3\dfrac{y}{x} = 0$

2 $L\dfrac{di}{dt} + Ri = 0$

3 $\dfrac{dy}{d\theta} + y\tan\theta = 0$

4 $\dfrac{1}{2x}\cdot\dfrac{dy}{dx} + \dfrac{y}{x^2+1} = 0$

5 $x\dfrac{dy}{dx} + (1 + x)y = 0$

6 $\dfrac{dy}{dx} + y\sin x = 0$

Solve the initial value problem

7 $\dfrac{dy}{dx} + xy = 0 \qquad y(0) = -1$

8 $(1 - x^2)\dfrac{dy}{dx} + xy = 0 \qquad y(0) = 3$

9 $\dfrac{dy}{dx} + y\sqrt{x} = 0 \qquad y(0) = 2$

10 $\dfrac{dy}{dx} + \dfrac{2y}{x} = 0 \qquad y(2) = \frac{1}{5}$

Solve

11 $\dfrac{dy}{dx} + 2y = x$

12 $\dfrac{dy}{dx} - y = 3x - 2$

13 $\dfrac{dy}{dx} + 4y = e^{x}$

14 $\dfrac{dy}{dx} - y = xe^{x}$

15 $\dfrac{dy}{dx} + y = x^2e^{-x}$

16 $\dfrac{dy}{dx} + 2y = \cos x$

17 $2\dfrac{dy}{dx} + 5y = 3\sin x$

18 $6\dfrac{dy}{dx} + y = e^{3x} - x - 1$

19 $\dfrac{dy}{dx} + 2y = e^{-x}\cos x$

20 $L\dfrac{di}{dt} + Ri = e^{-kt}$

21 $L\dfrac{di}{dt} + Ri = E\cos t$

22 $\dfrac{dy}{d\theta} + ay = \cos 2\theta$

23 $\dfrac{dy}{dx} - y = x^5$

24 $\dfrac{dy}{dx} + 2y = e^{-2x}$

25 $\dfrac{dy}{dx} - y = 2e^{x}$

26 $\dfrac{dy}{dx} + y = xe^{3x}$

27 $\dfrac{dy}{dx} - 3y = x^2$

28 $\dfrac{dy}{dx} - y = \sin x$

29 $\dfrac{dy}{dx} + xy = x$

30 $\dfrac{dy}{dx} - \dfrac{2y}{x} = x^2 + 1$

31 $\dfrac{dy}{dx} - \dfrac{y}{x} = \dfrac{1}{x^2}$

32 $\dfrac{dy}{dx} - \dfrac{2xy}{x^2+1} = x$

Solve the initial value problem

33 $\dfrac{dy}{dx} + 3y = 1 \qquad y(0) = 0$

34 $\dfrac{dy}{dx} - 2y = 4 \qquad y(-1) = 0$

35 $\dfrac{dy}{dx} - 2y = 2x + 1 \qquad y(0) = 0$

36 $\dfrac{dy}{dx} + y = xe^{2x} \qquad y(0) = 0$

37 $\dfrac{dy}{dx} - 2y = x^2e^x \qquad y(1) = 0$

38 $\dfrac{dy}{dx} + 2xy = 2x \qquad y(1) = 1$

39 $\dfrac{dy}{dx} - 3x^2y = \exp(x^3) \qquad y(0) = -1$

40 $\dfrac{dy}{dx} + \dfrac{2xy}{1 + x^2} = 2x \qquad y(1) = 2$

19-3 Applications

We now apply the material of the previous sections to problems taken from various subjects. The first application is to find the curves that are orthogonal (perpendicular) to a given family of curves. This application is important in fluid flow, electricity, meteorology, and other areas. For example, a weather map usually shows the family of **isobars,** curves of constant barometric pressure. The curves that are orthogonal to the isobars are the paths of the wind.

In general, given a family of plane curves, the family of curves orthogonal to these are called their **orthogonal trajectories.** For example, the orthogonal trajectories of the family of all circles centered at the origin are the radial lines.

Given a family of curves that depend on a constant C, we can usually find a differential equation for the family in the form $dy/dx = f(x, y)$. Then at each point (x, y), the value $f(x, y)$ is the slope of the curve through (x, y). The slope of the orthogonal trajectory through (x, y) is the negative reciprocal $-1/f(x, y)$. Thus

$$\frac{dy}{dx} = \frac{-1}{f(x, y)}$$

is the differential equation for the orthogonal trajectories. Solve this equation and we have the orthogonal trajectories.

Example 1

Find the orthogonal trajectories of the family of rectangular hyperbolas $xy = C$.

Solution The family of hyperbolas is indicated in Figure 19-3-1 (next page). Each has the x- and y-axes as asymptotes. Differentiate $xy = C$ to find the corresponding slopes (hence the direction field):

$$y + x\frac{dy}{dx} = 0 \qquad \text{that is} \qquad \frac{dy}{dx} = -\frac{y}{x}$$

The orthogonal trajectory through (x, y) has slope the negative reciprocal of this slope. Therefore the differential equation of the orthogonal trajectories is

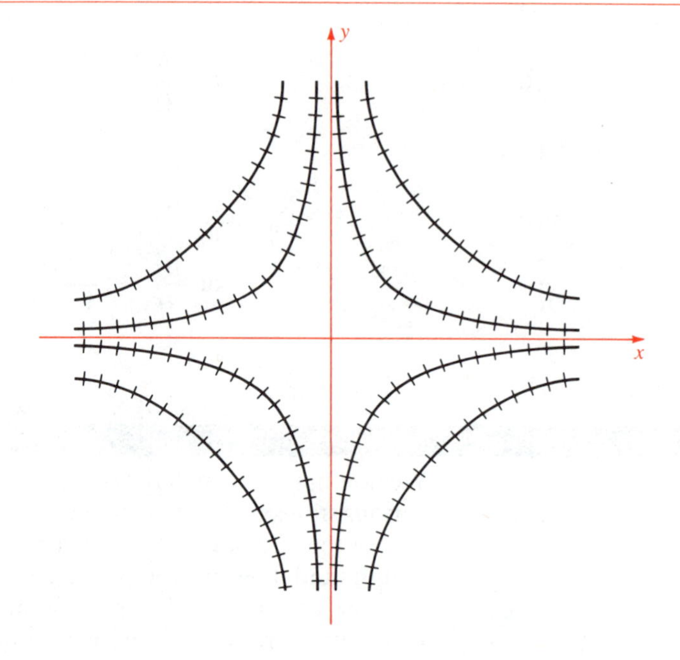

Figure 19-3-1
Family of curves $xy = C$ (hyperbolas)
Orthogonal directions marked

$$\frac{dy}{dx} = \frac{-1}{-y/x} = \frac{x}{y}$$

We considered this equation in Example 5 of Section 19-1 and sketched its direction field in Figure 19-1-8. By separating variables we found the general solution

$$y^2 - x^2 = C$$

This is the family of rectangular hyperbolas with axes the coordinate axes. See Figure 19-3-2. ●

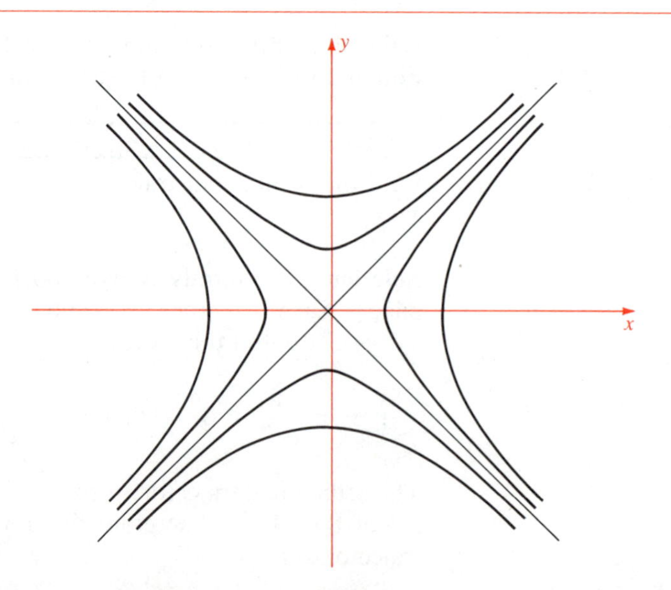

Figure 19-3-2
Orthogonal trajectories of the family of curves in Figure 19-3-1: hyperbolas $y^2 - x^2 = C$

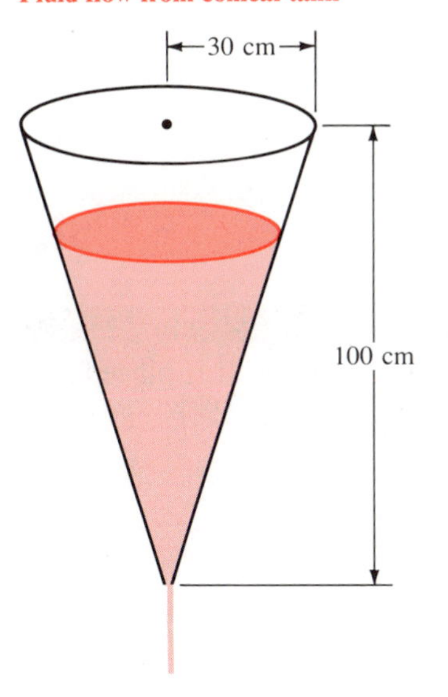

Figure 19-3-3
Fluid flow from conical tank

Flow through an Orifice

When you drain a tank, the water flows out more slowly as the level drops. How long will it take to empty the tank? Now when water leaves a tank through a small hole at the bottom, the fluid speed at the exit is

$$v = \sqrt{2gy}$$

where y is the fluid depth and g is the gravitational acceleration. This is called Torricelli's law, and we can use it to solve problems such as the following one.

Example 2

An open conical tank has the dimensions shown in Figure 19-3-3. At the bottom is a hole of area 4 cm^2. Initially the tank is filled with water. How long will it take to empty? Take $g = 980 \text{ cm/sec}^2$.

Solution At time t, the depth is $y = y(t)$ and the volume of water is $V = V(t)$. The rate the volume decreases is the velocity v times the area of the hole. Hence

$$\frac{dV}{dt} = -4v = -4\sqrt{2gy}$$

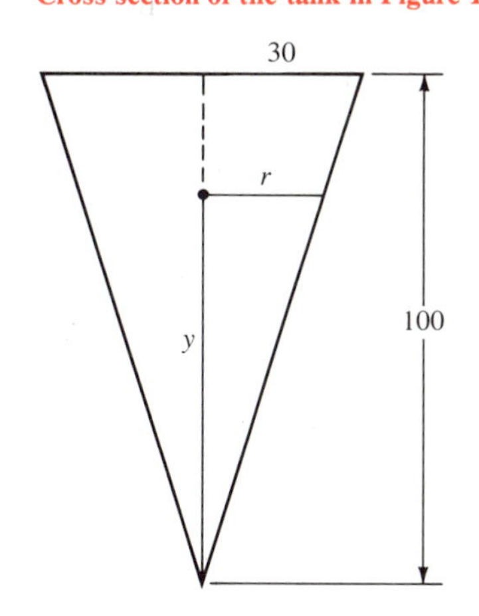

Figure 19-3-4
Cross section of the tank in Figure 19-3-3

It is readily shown by similar triangles in Figure 19-3-4 that the radius of the surface of the fluid at this time is

$$r = \frac{30}{100}y = \tfrac{3}{10}y$$

Therefore the volume of the fluid is

$$V = \tfrac{1}{3}\pi r^2 y = \tfrac{1}{3}\pi \cdot \tfrac{9}{100}y^3 = \tfrac{3}{100}\pi y^3$$

Differentiate with respect to time:

$$\frac{dV}{dt} = \tfrac{9}{100}\pi y^2 \frac{dy}{dt}$$

Consequently

$$\tfrac{9}{100}\pi y^2 \frac{dy}{dt} = -4\sqrt{2gy}$$

This differential equation may be written

$$\frac{dy}{dt} = -Ay^{-3/2} \qquad \text{where} \quad A = \frac{400\sqrt{2g}}{9\pi} \approx 626.32$$

This is the set-up; it remains to solve the initial value problem with $y(0) = 100$ cm. Separate variables:

$$y^{3/2}dy = -A\,dt \qquad \text{hence} \qquad \tfrac{2}{5}y^{5/2} = -At + C$$

Now set $t = 0$ and $y = 100$:

$$C = \tfrac{2}{5} \cdot 10^5 = 4 \times 10^4$$

The tank is empty when $y = 0$. At that instant,

$$At = C \qquad \text{hence} \qquad t = \frac{C}{A} \approx 64 \text{ sec}$$

●

Newton's Law of Cooling

Suppose $T(t)$ represents the temperature of an object. It is placed in a cooling bath of constant temperature T_0. Then the rate of change of T is proportional to $T - T_0$. That is,

$$\frac{dT}{dt} = -k(T - T_0)$$

where k is a constant. This is Newton's law of cooling. It applies to objects heated or cooled by conduction of heat.

Example 3

A ball bearing at $500\ ^\circ\mathrm{C}$ is placed in a cooling bath at fixed temperature $50\ ^\circ\mathrm{C}$. In 10 sec it cools to $350\ ^\circ\mathrm{C}$. How long will it take to cool from $350\ ^\circ\mathrm{C}$ to $100\ ^\circ\mathrm{C}$? (The figures are realistic if the ball bearing is fairly small.)

Solution The law of cooling is the differential equation

$$\frac{dT}{dt} = -k(T - 50) = -kT + 50k$$

that is

$$\frac{dT}{dt} + kT = 50k$$

This is a first order linear equation of the form

$$\frac{dT}{dt} + p(t)T = q(t)$$

with $p(t) = k$ and $q(t) = 50k$. As usual, $m(t) = e^{kt}$ is an integrating factor, so

$$\frac{d}{dt}(e^{kt}T) = 50ke^{kt}$$

Hence

$$e^{kt}T = 50e^{kt} + C \qquad \text{that is} \qquad T = 50 + Ce^{-kt}$$

Since $T(0) = 500$ we have $C = 450$ so

$$T = 50 + 450e^{-kt}$$

We also know that $T(10) = 350$. Hence

$$350 = 50 + 450e^{-10k} \qquad \text{that is} \qquad e^{-10k} = \frac{300}{450} = \tfrac{2}{3}$$

Therefore $e^{10k} = 1.5$ so that $e^k = 1.5^{0.1}$. Next we want to know t when $T = 100$ °C. We set

$$100 = 50 + 450e^{-kt} \qquad \text{so that} \qquad 450e^{-kt} = 50$$

Then

$$e^{kt} = \frac{450}{50} = 9 \qquad \text{that is} \qquad (1.5)^{0.1t} = 9$$

We take logs to find t:

$$t = \frac{\ln 9}{0.1 \ln 1.5} \approx 54.2 \text{ sec}$$

This is the time for cooling from its original temperature of 500 °C to 100 °C. Hence the metal cools down from 350 °C to 100 °C in about $54.2 - 10 = 44.2$ sec. ●

Population Growth

When a population is small so that its food supply and space is virtually unlimited, it grows at a rate proportional to its own size. But as it gets larger, its members compete with each other for food and living space. Studies indicate that the population's growth rate must be corrected by a factor proportional to the square of the population. This leads to a basic model in ecology, the **Verhulst logistic equation** for population growth:

$$\frac{dP}{dt} = kP(M - P)$$

where $P = P(t)$ is the population at time t, and k and M are positive constants. The initial population is $P_0 = P(0)$, and we assume $P_0 < M$. We can solve the equation systematically by separating variables and using a decomposition into partial fractions. First

$$\frac{dP}{P(M - P)} = k\,dt \qquad \text{that is} \qquad \left(\frac{1}{P} + \frac{1}{M - P}\right) dP = kM\,dt$$

Therefore

$$\ln \frac{P}{M - P} = kMt + C$$

By applying exp to both sides we obtain

$$\frac{P}{M - P} = Ae^{kMt}$$

By setting $t = 0$ we find $A = P_0/(M - P_0)$. Next we solve for P:

$$P = (M - P)Ae^{kMt} \qquad \text{hence} \qquad (1 + Ae^{kMt})P = MAe^{kMt}$$

so that

$$P = \frac{MAe^{kMt}}{1 + Ae^{kMt}} = \frac{MA}{A + e^{-kMt}}$$

Now we substitute $A = P_0/(M - P_0)$ to obtain the solution of the logistic equation:

$$P(t) = \frac{MP_0}{P_0 + (M - P_0)e^{-kMt}}$$

We can see from this solution that $P(t)$ is a strictly increasing function for $t \geq 0$ and

$$\lim_{t \to +\infty} P(t) = M$$

Consequently, M represents the maximum population—never quite achieved. See Figure 19-3-5 for a graph of P.

Figure 19-3-5
Population growth: the Verhulst model

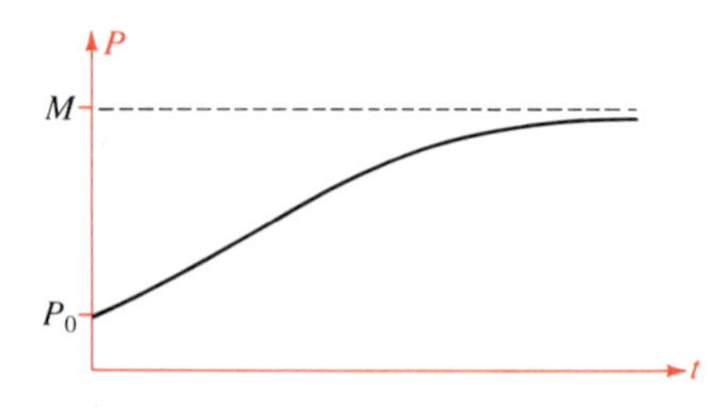

Telescope Mirrors

We know that any light ray from the focus of a parabola is reflected by the parabola into a ray parallel to the axis of the parabola. Parabolic reflectors are used in this way in automobile headlights. Conversely, reflecting telescopes gather light rays coming in parallel to the axis of the telescope mirror and bring them together to the focus. But let us suppose we don't know this property of parabolas and try to discover it. We put the light source at the origin and seek a curve that reflects any ray from the origin into a ray parallel to the y-axis (Figure 19-3-6).

Take any point $(x, y) \neq (0, 0)$. The curve we seek through (x, y) has unit tangent

$$\mathbf{t} = (1, y')/\sqrt{1 + y'^2}$$

The unit vector from the origin towards (x, y) is

$$\mathbf{u} = (x, y)/\sqrt{x^2 + y^2}$$

and the unit vector in the positive y-direction is $\mathbf{j} = (0, 1)$. The reflection property is simply $\mathbf{u} \cdot \mathbf{t} = \mathbf{j} \cdot \mathbf{t}$, that is,

$$\frac{(x, y)}{\sqrt{x^2 + y^2}} \cdot \frac{(1, y')}{\sqrt{1 + (y')^2}} = (0, 1) \cdot \frac{(1, y')}{\sqrt{1 + (y')^2}}$$

Figure 19-3-6
Set-up for telescope mirror:
$\mathbf{j} = (0, 1)$
$\mathbf{u} = (x^2 + y^2)^{-1/2}(x, y)$
$\mathbf{t} = [1 + (y')^2]^{-1/2}(1, y')$

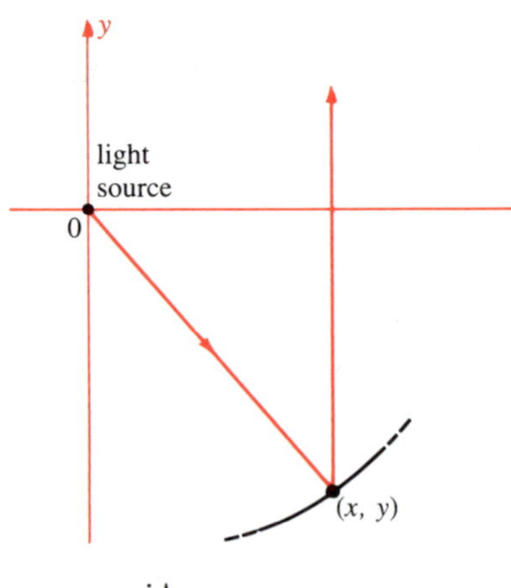

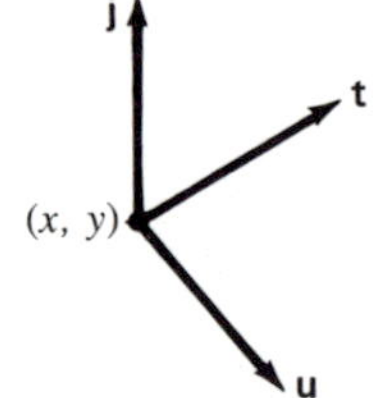

We can drop $\sqrt{1 + (y')^2}$, and this relation becomes

$$\frac{x + yy'}{\sqrt{x^2 + y^2}} = y' \qquad \text{that is} \qquad x + yy' = y'\sqrt{x^2 + y^2}$$

This looks like a terribly hard differential equation to solve; it is not even linear. However, the expression $x^2 + y^2$ suggests that we try the variable change

$$r^2 = x^2 + y^2 \qquad \text{so} \qquad rr' = x + yy'$$

Then the equation is simply

$$rr' = y'r \qquad \text{that is} \qquad r' = y'$$

This integrates immediately: $r = y + C$. Therefore

$$x^2 + y^2 = r^2 = (y + C)^2$$
$$x^2 = 2Cy + C^2$$
$$x^2 = 2C(y + \tfrac{1}{2}C)$$

Set $2C = 4p$ and this becomes the standard form

$$x^2 = 4p(y + p)$$

of a parabola with focus at $(0, 0)$ and vertex at $(0, -p)$. See Figure 19-3-7. (Note, incidentally, that the relation $r = y + C$ is already the geometric relation defining the parabola with focus the origin and directrix $y = -c$.

Figure 19-3-7
The mirror is a parabola.

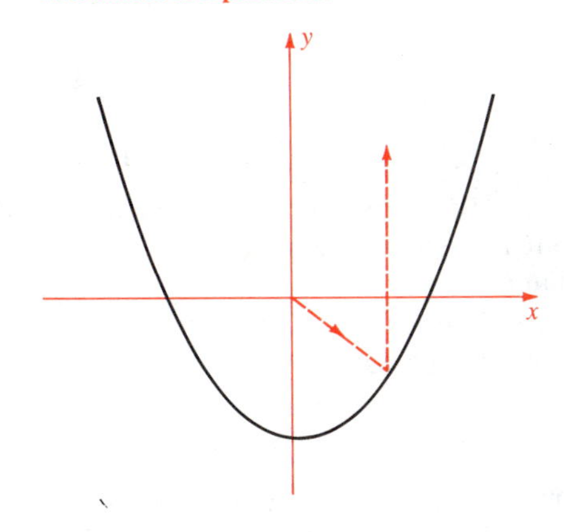

Exercises

Find the orthogonal trajectories of

1 the family of ellipses $x^2 + 3y^2 = C$

2 the family of curves $y = C\exp(-x^2)$

3 the family of parabolas $y = Cx^2$

4 the family of half parabolas $y = (x - C)^2$, $x > C$

5 How long will it take to empty an upright cylindrical tank of water if the radius is 50 cm and the height 200 cm? The orifice at the bottom has area 2 cm².

6 How long will it take to empty a spherical tank of water of radius 100 cm through a 3-cm² hole at the bottom. Assume there is an air vent at the top so the air pressure inside equals atmospheric pressure.

7 When quenched in oil at 50 °C, a piece of hot metal cools from 250 °C to 150 °C in 30 sec. At what constant oil temperature will the metal cool from 250 °C to 150 °C in 45 sec?

8 (cont.) The same piece of metal is quenched in oil whose temperature is 50 °C, but rising 0.5 °C/sec. How long will it take for the metal to cool from 250 °C to 150 °C?

9 When a cool, thermally conducting object is placed in hot gas it heats up at a rate proportional to the difference in temperature between the gas and the object. A cold metal bar at 0 °C is placed in a tank of gas at constant temperature 200 °C. In 40 sec its temperature rises to 50 °C. How long would it take to heat up from 20 °C to 100 °C?

10 (cont.) How hot should the gas be so that the metal bar heats up from 0 °C to 50 °C in 20 sec?

11 Two substances, U and V, combine chemically to form a substance X; one gram of U combines with 2 gm of V to form 3 gm of X. Suppose 50 gm of U are allowed to react with 100 gm of V. According to the law of mass action, the chemicals combine at a rate proportional to the product of the untransformed masses. Denote

by $x(t)$ the mass of X at time t. (The reaction starts at $t = 0$.) Show that $x(t)$ satisfies the initial-value problem

$$\frac{dx}{dt} = k(50 - \tfrac{1}{3}x)(100 - \tfrac{2}{3}x) \qquad x(0) = 0$$

If 75 gm of X are formed in the first second, find a formula for $x(t)$.

12 (cont.) Suppose that 50 gm of U are allowed to react with 150 gm of V. (Not all of V will be transformed.) Set up an initial value problem for $x(t)$ and solve. Verify that $x(t) \to 150$ as $t \to +\infty$.

13 A quantity of gas undergoes ionization at the constant rate r. Let $p = p(t)$ denote the number of positive ions at time t. At the same time, the ions tend to recombine with electrons at the rate αp^2, where α is the **coefficient of recombination.** Thus

$$\frac{dp}{dt} = r - \alpha p^2$$

Solve for p in terms of t, given that initially $p(0) = 0$.

***14** Suppose a quantity of hot fluid is stirred so that at any time t, its temperature $u(t)$ is uniform throughout the fluid. Suppose it loses heat to the outside only by *radiation.* Then according to the **Stefan-Boltzmann law,**

$$\frac{du}{dt} = -k(u^4 - a^4)$$

where $k > 0$ and a is the (constant) outside temperature. Express $t = t(u)$ in terms of u.

15 The US population in 1790 was 3.93×10^6. Assume that $M = 2.569 \times 10^8$ and $k = 1.189 \times 10^{-10}$. What does the Verhulst equation predict for 1850, 1900, 1950, and 2000?

16 For world population, some ecologists estimate the constants $M = 9.86 \times 10^9$ and $k = 2.94 \times 10^{-12}$. The UN Statistical Office estimates world population in the middle of 1970 as 3.63×10^9. Estimate world population by the Verhulst equation for 2000, 2050, and 2100.

17 Another model for population growth is the Gompertz equation

$$\frac{dP}{dt} = kP \ln\left(\frac{M}{P}\right) \qquad P(0) = P_0$$

Solve the equation. [Hint Set $u = \ln(M/P)$.]

18 A bacterial colony grows in the presence of a toxin. Let $N = N(t)$ be the quantity of bacteria and $X = X(t)$ be that of the toxin. Then the toxin inhibits the growth of the colony by an amount proportional to the product of the amounts of toxin and bacteria. Hence

$$\frac{dN}{dt} = kN - aNX \qquad (a > 0,\ k > 0)$$

Suppose $X = ct$, where $c > 0$. Solve the initial value problem with $N(0) = N_0$. Find $\lim_{t\to+\infty} N(t)$.

19 Here is a model for the growth of a concentrated colony of bacteria with limited space and limited food supply. Under such conditions, the colony tends to die out, so its concentration decreases at a rate proportional to itself. If nutrient is added, however, the concentration also increases at a rate proportional to the rate of nutrient addition. Let $x(t)$ be the concentration of cells, and assume that nutrient is added at a constant rate. Then

$$\frac{dx}{dt} = -ax + b \qquad (a > 0,\ b > 0)$$

Solve for $x(t)$ and find the steady-state concentration $x_\infty = \lim_{t\to+\infty} x(t)$.

20 Under certain conditions, the air density $\rho(y)$ at height y meters satisfies an equation of the form

$$\rho(y) = k\int_y^{y_1} \rho(u)\, du$$

where k and y_1 are constants, $k \approx 1.25 \times 10^{-4}\ \text{m}^{-1}$ and $y_1 \approx 10^5$ m. Differentiate to obtain a first order linear differential equation for $\rho(t)$, and solve that equation under the initial condition $\rho(0) = \rho_0 \approx 1.03 \times 10^4\ \text{kg/m}^3$. About how many atmospheres pressure is found on the summit of Mt. Everest (8848 m)?

21 A projectile is shot straight up. If we assume air resistance is proportional to its velocity, then the equation of motion is

$$\frac{dv}{dt} = -kv - g \qquad \text{where} \quad v = \frac{dx}{dt}$$

and k and g are positive constants. Show that

$$\frac{dx}{dv} = -\frac{1}{k} + \frac{g}{k^2}\left(\frac{1}{v + g/k}\right)$$

22 (cont.) Find the relation between x and v, taking x_0 and v_0 for their initial values.

Newton's law of motion, "force equals mass times acceleration," does not hold in this form for a variable mass. The correct law in that case is "force equals derivative of momentum":

$$F = \frac{d}{dt}(mv) \qquad \text{where} \quad v = \frac{dx}{dt}$$

Exercises 23–27 depend on this relation.

23 One model for a falling raindrop says that its mass $m = m(t)$ grows at a rate proportional to itself. Newton's law of motion for the falling raindrop is

$$\frac{d}{dt}(mv) = mg$$

where v is its downward speed and g is the gravitational constant. Express v in terms of t and find the **terminal velocity,** $\lim_{t\to+\infty} v$.

24 A rocket is traveling in a space region of negligible gravity. Let $m = m(t)$ denote its mass, $v = v(t)$ its speed, and c the constant speed (relative to the rocket) of its rocket exhaust. The rocket is propelled forward because hot gases are propelled backward (action equals reaction). The mass of the rocket is decreasing because it loses its expelled gas, so its law of motion is

$$\frac{d}{dt}(mv) = (v - c)\frac{dm}{dt}$$

Show that

$$m\frac{dv}{dt} = -c\frac{dm}{dt}$$

25 (cont.) Now use the chain rule to show that

$$\frac{dm}{dv} = -\frac{1}{c}m$$

Deduce that $m = m_0 e^{-v/c}$, assuming that $v(0) = 0$ and $m(0) = m_0$.

26 (cont.) Let m_1 be the mass of the initial fuel supply and V the maximum speed. Show that

$$e^{-V/c} = \frac{m_0 - m_1}{m_0}$$

***27** (cont.) Suppose in addition that gravity is acting against the motion of the rocket. Then the equation of motion is

$$\frac{d}{dt}(mv) = (v - c)\frac{dm}{dt} - mg$$

where g is the gravitational constant. Show that $m = m_0 e^{-(v+gt)/c}$. [Hint Set $w = v + gt$ and find dm/dw.]

28 Make the substitution $P = 1/Q$ in the Verhulst logistic equation (page 955). Show that the result is a linear differential equation for Q. Solve it and eventually get back to P.

19-4 Second Order Equations

A **second order** differential equation is a differential equation that includes first and second derivatives and that can be put into the form

$$\frac{d^2y}{dt^2} = F(t, y, dy/dt)$$

Examples are

$$\frac{d^2y}{dt^2} = t^2y + e^t\frac{dy}{dt} \quad \text{and} \quad \frac{d^2y}{dt^2} = (\sin t)y\frac{dy}{dt}$$

Any, all, or none of the quantities t, y, dy/dt may be present in the function F. Thus

$$\frac{d^2y}{dt^2} = y^3 \qquad \frac{d^2y}{dt^2} = t\frac{dy}{dt} \quad \text{and} \quad \frac{d^2y}{dt^2} = t + y$$

are also examples.

In this section we work mainly with **linear** second order equations, differential equations that can be written in the form

$$\frac{d^2y}{dt^2} + p(t)\frac{dy}{dt} + q(t)y = r(t)$$

Such an equation is called **homogeneous** if $r(t) = 0$.

Constant Coefficients

We shall concentrate now on the special, but important, case of a *second order linear homogeneous differential equation with constant coefficients:*

$$\frac{d^2y}{dt^2} + p\frac{dy}{dt} + qy = 0$$

where p and q are constants.

Our method of solving the equation is to *assume* a solution of the form $y = e^{\lambda t}$ where λ is a constant. If we substitute this assumed solution into the differential equation, we find that λ must satisfy an ordinary quadratic equation in order for $y = e^{\lambda t}$ to be a solution. By suitably interpreting the roots of that quadratic, we find the actual general solution of the differential equation.

If $y = e^{\lambda t}$, then

$$\frac{dy}{dt} = \lambda e^{\lambda t} \quad \text{and} \quad \frac{d^2y}{dt^2} = \lambda^2 e^{\lambda t}$$

Therefore

$$\begin{aligned}\frac{d^2y}{dt^2} + p\frac{dy}{dt} + qy &= \lambda^2 e^{\lambda t} + p\lambda e^{\lambda t} + qe^{\lambda t} \\ &= (\lambda^2 + p\lambda + q)e^{\lambda t}\end{aligned}$$

The original differential equation (left side equals 0) is satisfied provided

$$\lambda^2 + p\lambda + q = 0$$

This ordinary quadratic equation is called the **characteristic equation** or the **auxiliary equation** of the given differential equation. It has either two distinct real roots (discriminant positive) or a double real root (discriminant zero) or two non-real conjugate complex roots (discriminant negative).

Case 1 $D = p^2 - 4q > 0.$

Then the roots are real and are

$$\lambda = \tfrac{1}{2}(-p + \sqrt{D}) \quad \text{and} \quad \mu = \tfrac{1}{2}(-p - \sqrt{D})$$

In this case the general solution of the differential equation is

$$y = y(t) = Ae^{\lambda t} + Be^{\mu t}$$

where A and B are arbitrary constants.

For example, consider

$$\frac{d^2y}{dt^2} - 3\frac{dy}{dt} + 2y = 0$$

The characteristic equation is

$$\lambda^2 - 3\lambda + 2 = 0$$

with discriminant $D = 1 > 0$ and roots $\lambda = 2$ and $\mu = 1$. The general solution of the differential equation is

$$y(t) = Ae^{2t} + Be^{t}$$

where A and B are arbitrary constants.

Case 2 $D = p^2 - 4q = 0$

Then $\lambda = -\frac{1}{2}p$ is a double root of the characteristic equation. It can be easily checked that not only is $y = e^{\lambda t}$ a solution of the differential equation, but so is $y = te^{\lambda t}$ also. In this case the general solution is

$$y = y(t) = Ae^{\lambda t} + Bte^{\lambda t}$$

where A and B are constants.

For example, consider

$$\frac{d^2y}{dt^2} + 6\frac{dy}{dt} + 9y = 0$$

The characteristic equation is

$$\lambda^2 + 6\lambda + 9 = 0$$

with discriminant $D = 0$ and only the (double) root $\lambda = -3$. The general solution of the differential equation is

$$y(t) = Ae^{-3t} + Bte^{-3t}$$

where A and B are arbitrary constants. (Please check that this really is a solution.)

Case 3 $D = p^2 - 4q < 0$.

It is convenient to write

$$D = -4k^2 \qquad \text{where} \qquad k > 0$$

The roots of the characteristic equation are the conjugate complex numbers

$$\lambda = -\tfrac{1}{2}p + ki \quad \text{and} \quad \mu = -\tfrac{1}{2}p - ki$$

where i denotes the complex unit such that $i^2 = -1$. So now we have the *formal* solutions

$$y = e^{\lambda t} \quad \text{and} \quad y = e^{\mu t}$$

of the differential equation. How shall we make any sense out of these? It really is a matter for more advanced courses so, to cut a rather long story short, we simply give the conclusion. For $\lambda = -\frac{1}{2}p + ki$, the correct interpretation of $e^{\lambda t}$ is

$$e^{\lambda t} = e^{-pt/2}(\cos kt + i \sin kt)$$

Similarly, $e^{\mu t} = e^{-pt/2}(\cos kt - i \sin kt)$. For our purpose, this interpretation leads to the general solution

$$y(t) = Ae^{-pt/2} \cos kt + Be^{-pt/2} \sin kt$$

where A and B are arbitrary real constants.

For example, consider

$$\frac{d^2y}{dt^2} + 2\frac{dy}{dt} + 5y = 0$$

The characteristic equation is

$$\lambda^2 + 2\lambda + 5 = 0$$

with discriminant $D = -16 = -4 \cdot 2^2$. The roots are

$$\lambda = -1 + 2i$$
$$\mu = -1 - 2i$$

The general solution of the differential equation is

$$y(t) = Ae^{-t}\cos 2t + Be^{-t}\sin 2t$$

where A and B are arbitrary constants. (Please check that this really is a solution.)

General Solution of $d^2y/dt^2 + p(dy/dt) + qy = 0$

Let $D = p^2 - 4q$ be the discriminant of the characteristic equation

$$\lambda^2 + p\lambda + q = 0$$

- If $D > 0$ then

$$y(t) = Ae^{\lambda t} + Be^{\mu t}$$

where $\lambda = \frac{1}{2}(-p + \sqrt{D})$ and $\mu = \frac{1}{2}(-p - \sqrt{D})$.

- If $D = 0$ then

$$y(t) = Ae^{\lambda t} + Bte^{\lambda t}$$

where $\lambda = -\frac{1}{2}p$.

- If $D = -4k^2 < 0$ then

$$y(t) = Ae^{-pt/2}\cos kt + Be^{-pt/2}\sin kt$$

In all cases, A and B denote arbitrary constants.

The following example illustrates all cases with $p = 0$. It is worth remembering.

Example 1

Solve

a $\dfrac{d^2y}{dt^2} - k^2y = 0$ **b** $\dfrac{d^2y}{dt^2} = 0$ **c** $\dfrac{d^2y}{dt^2} + k^2y = 0$

where $k > 0$.

Solution

a In this case the characteristic equation is $\lambda^2 - k^2 = 0$ with roots $\lambda = k$ and $\mu = -k$. Hence

$$y(t) = Ae^{kt} + Be^{-kt}$$

b Clearly $D = 0$ and the double root is $\lambda = 0$. Hence $e^{\lambda t} = 1$ so that

$$y(t) = A + Bt$$

c The characteristic equation is $\lambda^2 + k^2 = 0$ and $D = -4k^2$. Since $p = 0$ we have

$$y(t) = A \cos kt + B \sin kt$$

●

Initial Value Problem

The solution of a second order differential equation

$$\frac{d^2y}{dt^2} = F\left(t, y, \frac{dy}{dt}\right)$$

generally depends on two constants. Thus if two conditions can be specified, they generally force a unique solution. Usually, **initial data** of the form

$$y(t_0) = y_0 \qquad \frac{dy}{dt}(t_0) = y_0'$$

is given. That is, the value of the function y and the value of its first derivative y' are specified at one instant of time t_0.

Example 2

Solve the initial value problem

$$\frac{d^2y}{dt^2} + 2\frac{dy}{dt} + 5y = 0 \qquad y(0) = 1 \qquad y'(0) = 0$$

Solution We solved this differential equation earlier (page 962). We found the general solution

$$y(t) = Ae^{-t}\cos 2t + Be^{-t}\sin 2t$$

Now we differentiate:

$$y'(t) = (-A + 2B)e^{-t}\cos 2t + (-2A - B)e^{-t}\sin 2t$$

The initial conditions $y(0) = 1$ and $y'(0) = 0$ become

$$A = 1 \qquad \text{and} \qquad -A + 2B = 0$$

with solution $A = 1$, $B = \frac{1}{2}$. Therefore

$$y(t) = e^{-t}\cos 2t + \tfrac{1}{2}e^{-t}\sin 2t$$

is the unique solution of the problem.

●

Instead of two initial conditions, the problem might specify only one. Then the resulting solution will generally depend on one constant.

Example 3

Find all solutions of the initial value problem

$$\frac{d^2y}{dt^2} - 3\frac{dy}{dt} + 2y = 0$$

$$3y(0) - y'(0) = 3$$

Solution Earlier (page 960) we found the general solution of this differential equation:

$$y(t) = Ae^{2t} + Be^{t}$$

Differentiate:

$$y'(t) = 2Ae^{2t} + Be^{t}$$

Substitute the initial condition $3y(0) - y'(0) = 3$:

$$3(A + B) - (2A + B) = 3$$

$$A + 2B = 3 \qquad \text{that is} \qquad A = 3 - 2B$$

Therefore

$$y(t) = (3 - 2B)e^{2t} + Be^{t}$$

is the most general solution of the problem. It depends on one constant, B. ●

The General Second Order Linear Equation

We return now to the non-homogeneous second order linear differential equation

$$\frac{d^2y}{dt^2} + p(t)\frac{dy}{dt} + q(t)y = r(t)$$

We are mostly interested in the case of constant coefficients, that is, equations in which $p(t) = p$ and $q(t) = q$ are constants. If we can find *any* solution at all of such an equation, then we can obtain the *general* solution of the equation by adding the one solution we know to the general solution of the **associated homogeneous equation**

$$\frac{d^2y}{dt^2} + p\frac{dy}{dt} + qy = 0$$

Often we can guess by inspection one solution to the given differential equation [the one with $r(t)$] and for many cases we can systematize the guessing process. We refer to any single solution (without arbitrary constants) of the given differential equation as a **particular solution.**

Example 4

Find a particular solution of

$$\frac{d^2y}{dt^2} + 3\frac{dy}{dt} - y = t^2 - 1$$

Solution Try a quadratic polynomial $y(t) = At^2 + Bt + C$. This guess is motivated by the observation that we want a quadratic $t^2 - 1$ on the right side, and the second order differential operator

$$y \to \frac{d^2y}{dt^2} + 3\frac{dy}{dt} - y$$

will take a quadratic into a quadratic. Then

$$\frac{d^2}{dt^2}(At^2 + Bt + C) + 3\frac{d}{dt}(At^2 + Bt + C) - (At^2 + Bt + C)$$
$$= -At^2 + (6A - B)t + (2A + 3B - C)$$

This expression equals $t^2 - 1$ provided

$$-A = 1 \qquad 6A - B = 0 \qquad \text{and} \qquad 2A + 3B - C = -1$$

The solution of this linear system is

$$A = -1 \qquad B = -6 \qquad C = -19$$

so $y(t) = -t^2 - 6t - 19$ is a particular solution. ●

Example 5

Find the general solution of

$$\frac{d^2y}{dt^2} + 3\frac{dy}{dt} + 2y = 60e^{4t}$$

Solution Try $y = Ce^{4t}$. Then

$$16Ce^{4t} + 12Ce^{4t} + 2Ce^{4t} = 60e^{4t}$$

Hence $30C = 60$ so $C = 2$. Thus $y = 2e^{4t}$ is a particular solution. The associated homogeneous equation

$$\frac{d^2y}{dt^2} + 3\frac{dy}{dt} + 2y = 0$$

has characteristic equation $\lambda^2 + 3\lambda + 2 = 0$ with roots $\lambda = -1$ and $\mu = -2$. Consequently the general solution of the *homogeneous* equation is $y = Ae^{-t} + Be^{-2t}$. The general solution of the given (nonhomogeneous) equation is the sum of this and the particular solution:

$$y(t) = Ae^{-t} + Be^{-2t} + 2e^{4t}$$

where A and B are arbitrary constants. ●

Sometimes the obvious guess fails because it is a solution of the associated homogeneous equation. The usual fix is to toss in a factor t.

Example 6

Find a particular solution of

$$\frac{d^2y}{dt^2} - 9y = e^{-3t}$$

Solution Our first guess is $y(t) = Ae^{-3t}$. For this function

$$\frac{d^2y}{dt^2} - 9y = 9Ae^{-3t} - 9Ae^{-3t} = 0$$

so no value of a will work. Therefore guess again: $y(t) = Ate^{-3t}$. Now

$$\frac{d^2y}{dt^2} - 9y = (-6Ae^{-3t} + 9Ae^{-3t}) - 9Ate^{-3t} = -6Ae^{-3t}$$

so $y(t)$ is a solution provided

$$-6Ae^{-t} = e^{-3t} \qquad \text{that is} \qquad A = -\tfrac{1}{6}$$

Hence $y(t) = -\frac{1}{6}te^{-3t}$ is a particular solution. ●

Example 7

Find a particular solution of

$$\frac{d^2y}{dt^2} + y = \cos t - \sin t$$

Solution The natural first guess is $y(t) = A\cos t + B\sin t$ because the functions $\cos t$ and $\sin t$ reproduce each other under differentiation. This guess fails because $\cos t$ and $\sin t$ are precisely the solutions of the associated homogeneous equation. Hence we guess

$$y(t) = At\cos t + Bt\sin t$$

Substitute into $y'' + y = \cos t - \sin t$:

$$\begin{aligned} A(-2\sin t - t\cos t) + B(2\cos t - t\sin t) \\ + (At\cos t + Bt\sin t) \\ = \cos t - \sin t \end{aligned}$$

Collect terms:

$$-2A\sin t + 2B\cos t = \cos t - \sin t$$

Therefore $A = \frac{1}{2}$ and $B = \frac{1}{2}$, so

$$y(t) = \tfrac{1}{2}t(\cos t + \sin t)$$

is a particular solution. ●

Example 8

Solve the initial value problem

$$\frac{d^2y}{dt^2} - y = 3t - 4 \qquad y(0) = 0 \qquad y'(0) = -1$$

Solution We guess a solution $y = At + B$ of the given equation, and find $A = -3$ and $B = 4$, so $y(t) = -3t + 4$ is a particular solution.

From Example 1a, with $k = 1$, we know that the associated homogeneous equation $d^2y/dt^2 - y = 0$ has the general solution

$$y(t) = Ae^t + Be^{-t}$$

Therefore the general solution of the given (non-homogeneous) equation is

$$y(t) = Ae^t + Be^{-t} - 3t + 4$$

Its derivative is

$$\frac{dy}{dt} = Ae^t - Be^{-t} - 3$$

We substitute the initial conditions $t = 0$, $y = 0$, $dy/dt = -1$:

$$\begin{cases} A + B + 4 = 0 \\ A - B - 3 = -1 \end{cases} \qquad \text{that is} \qquad \begin{cases} A + B = -4 \\ A - B = 2 \end{cases}$$

The solution of this linear system is $A = -1$ and $B = -3$, so

$$y(t) = -e^t - 3e^{-t} - 3t + 4$$

is the (unique) solution of the problem. ●

Exercises

Solve the homogeneous equation

1 $\dfrac{d^2y}{dt^2} - 6\dfrac{dy}{dt} + 5y = 0$

2 $\dfrac{d^2y}{dt^2} + 7\dfrac{dy}{dt} + 6y = 0$

3 $\dfrac{d^2r}{d\theta^2} + 4r = 0$

4 $\dfrac{d^2y}{dx^2} + 6\dfrac{dy}{dx} + 13y = 0$

5 $2\dfrac{d^2y}{dt^2} - \dfrac{dy}{dt} - y = 0$

6 $\dfrac{d^2y}{dt^2} - 8\dfrac{dy}{dt} + 16y = 0$

7 $\dfrac{d^2y}{dt^2} + 6\dfrac{dy}{dt} = 0$

8 $4\dfrac{d^2y}{dt^2} + 4\dfrac{dy}{dt} + y = 0$

9 $\dfrac{d^2y}{dt^2} + 5\dfrac{dy}{dt} + y = 0$

10 $2\dfrac{d^2y}{dx^2} - 5\dfrac{dy}{dx} - 3y = 0$

11 $\dfrac{1}{L^3} \cdot \dfrac{d^2x}{dt^2} - \dfrac{4}{L} \cdot \dfrac{dx}{dt} + 4Lx = 0$

12 $\dfrac{d^2y}{dt^2} + \dfrac{dy}{dt} + y = 0$

Solve the initial value problem

13 $\frac{d^2y}{dt^2} + 9y = 0$ $\quad y(0) = 0$, $y'(0) = 4$

14 $\frac{d^2y}{dt^2} + 4y = 0$ $\quad y(0) = -1$, $y'(0) = 0$

15 $\frac{d^2y}{dt^2} - 3\frac{dy}{dt} + 2y = 0$ $\quad y(0) = 1$, $y'(0) = -1$

16 $\frac{d^2y}{dt^2} - 10\frac{dy}{dt} + 25y = 0$ $\quad y(0) = 0$, $y'(0) = 5$

17 $\frac{d^2y}{dt^2} - 16y = 0$ $\quad y(0) = 3$, $y'(0) = -3$

18 $\frac{d^2y}{dt^2} + 6\frac{dy}{dt} + 8y = 0$ $\quad y(0) = 2$, $y'(0) = -3$

19 $\frac{d^2y}{dt^2} - 8\frac{dy}{dt} + 25y = 0$ $\quad y(0) = 3$, $y'(0) = 0$

20 $\frac{d^2y}{dt^2} + \frac{dy}{dt} + 2y = 0$ $\quad y(0) = 0$, $y'(0) = 0$

Find all solutions

21 $\frac{d^2r}{d\theta^2} - 2\frac{dr}{d\theta} + 2r = 0$ $\quad r(\pi) = 1$

22 $\frac{d^2x}{dt^2} - 16x = 0$ $\quad x(0) = 0$

23 $\frac{d^2y}{dt^2} + 6y = 0$ $\quad y'(0) = 1$

24 $\frac{d^2y}{dt^2} - 2\frac{dy}{dt} + y = 0$ $\quad y(0) = 0$

25 $\frac{d^2y}{dt^2} + 6\frac{dy}{dt} + 8y = 0$ $\quad y(0) + y'(0) = 3$

26 $\frac{d^2y}{dt^2} - 5\frac{dy}{dt} + 6y = 0$ $\quad y'(0) = y(0)$

Find a particular solution by guessing

27 $\frac{d^2x}{dt^2} + 3\frac{dx}{dt} = 1$

28 $2\frac{d^2x}{dt^2} + 3\frac{dx}{dt} = 5$

29 $\frac{d^2x}{dt^2} - 4x = 2t + 1$

30 $\frac{d^2x}{dt^2} + x = t^2$

31 $\frac{d^2x}{dt^2} + 3\frac{dx}{dt} - x = t^2 + 4t + 6$

32 $\frac{d^2x}{dt^2} + \frac{dx}{dt} + x = t^3$

33 $\frac{d^2x}{dt^2} + \frac{dx}{dt} + 2x = 2e^{3t}$

34 $\frac{d^2x}{dt^2} + x = \cosh t$

35 $3\frac{d^2y}{dx^2} + \frac{dy}{dx} - y = e^{-2x} + x$

36 $2\frac{d^2y}{dx^2} - 3y = \sin x$

37 $\frac{d^2y}{dx^2} + \frac{dy}{dx} - 3y = 4\sin 2x$

38 $\frac{d^2y}{dx^2} + 2\frac{dy}{dx} + y = 1 + x + \cos 3x$

39 $\frac{d^2y}{dx^2} + \frac{dy}{dx} + 2y = e^x \cos x$

40 $2\frac{d^2y}{dx^2} - y = x\cos x$

41 $\frac{d^2y}{dx^2} + y = e^x + e^{2x} + e^{3x} + e^{4x} + e^{5x}$

42 $\frac{d^2y}{dx^2} - \frac{dy}{dx} - y = xe^{3x}$

43 $\frac{d^2x}{dt^2} - 4x = e^{2t} + e^{-2t}$

44 $\frac{d^2x}{dt^2} + 4x = 3\cos 2t$

Find the general solution

45 $\frac{d^2x}{dt^2} + x = t^2$

46 $\frac{d^2x}{dt^2} - 7\frac{dx}{dt} + 10x = 3e^t$

47 $\frac{d^2x}{dt^2} + \frac{dx}{dt} - 6x = te^{-t}$

48 $\frac{d^2x}{dt^2} + \frac{dx}{dt} - 5x = 2t + 3$

49 $\frac{d^2x}{dt^2} + 3\frac{dx}{dt} = \cosh 2t$

50 $3\frac{d^2y}{dx^2} - \frac{dy}{dx} + y = x^2 + 5$

51 $3\frac{d^2i}{dt^2} + 4\frac{di}{dt} + 2i = 10\cos t$

52 $\frac{d^2r}{d\theta^2} - 2r = \sin\theta + \cos\theta$

Solve the initial value problem:

53 $\dfrac{d^2x}{dt^2} + x = 2t - 5$ $\quad x(0) = 0 \quad x'(0) = 1$

54 $\dfrac{d^2x}{dt^2} + 3x = e^t$ $\quad x(0) = 0 \quad x'(0) = 1$

55 $\dfrac{d^2x}{dt^2} - 4x = \sin 2t$ $\quad x(0) = 1 \quad x'(0) = 0$

56 $\dfrac{d^2x}{dt^2} - 2\dfrac{dx}{dt} - 15x = 1$ $\quad x(1) = 0 \quad x'(1) = 0$

57 $4\dfrac{d^2x}{dt^2} - 7\dfrac{dx}{dt} + 3x = e^{2t}$ $\quad x(0) = -1 \quad x'(0) = 2$

58 $\dfrac{d^2x}{dt^2} + 2\dfrac{dx}{dt} + x = t^2$ $\quad x(0) = 0 \quad x'(0) = 5$

The following two exercises show how some of the theory of differential equations might be established.

59 Suppose $k > 0$ and $y = y(x)$ is a solution of

$$\frac{d^2y}{dt^2} + k^2y = 0$$

Set

$$\begin{cases} u(t) = ky\cos kt - y'\sin kt \\ v(t) = ky\sin kt + y'\cos kt \end{cases}$$

Find u' and v'

60 (cont.) Conclude that $y = A\cos kt + B\sin bt$, where A and B are constants.

61 Find all functions $x = x(t)$ such that for $t > 0$

$$\frac{dx}{dt}\cdot\frac{d^2x}{dt^2} = \tfrac{9}{8}A^2 \quad \text{and} \quad \frac{dx}{dt}(0) = 0$$

[Hint Differentiate $(dx/dt)^2$.]

62 (cont.) Solve for $t > 0$

$$\left(\frac{dx}{dt}\right)^2\frac{d^2x}{dt^2} = \tfrac{1}{3}A^3 \qquad \frac{dx}{dt}(0) = 0$$

19-5 Applications to Vibration

Our applications of the material of Section 19-4 concern various kinds of vibrations. If there is no friction and no external force, a particle vibrates in *simple harmonic motion.* If there is friction then we have *damped harmonic motion.* Finally if there is a periodic external force acting on the particle we have what is called *forced vibration.* These concepts will be explained as we work through the section.

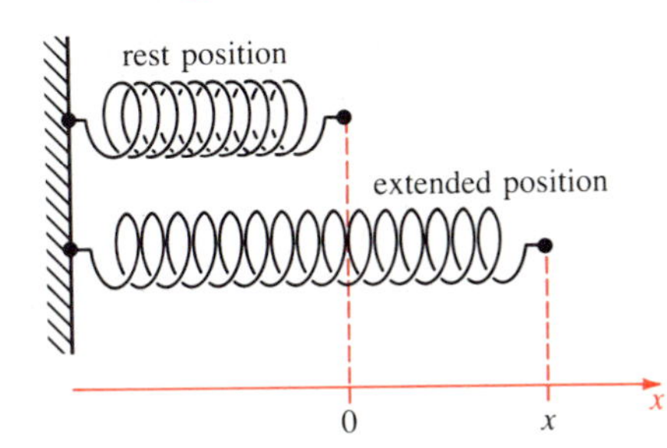

Figure 19-5-1
See Example 1

Example 1

A particle moves along a line. The only force acting on the particle is a force directed from the particle towards a fixed point of the line and proportional in magnitude to the displacement of the particle from the fixed point. Describe the motion.

Solution Set up the x-axis as in Figure 19-5-1, so the rest position of the particle is $x = 0$. We may think of the force as caused by a coiled spring attached somewhere to the left of $x = 0$ and whose free end is at $x = 0$ when the spring is neither stretched nor compressed. Call the mass of the particle m. The spring force can be written $-mk^2x$, where $k > 0$. According to Newton's law of motion $F = ma$ we have

$$m\frac{d^2x}{dt^2} = -mk^2x \qquad \text{that is} \qquad \frac{d^2x}{dt^2} + k^2x = 0$$

From Example 1c in Section 19-4, we know the general solution of this homogeneous second order linear differential equation to be

$$x = A\cos kt + B\sin kt$$

where A and B are constants. It is convenient to write this solution as a single cosine (or sine). Define $C = (A^2 + B^2)^{1/2}$. We may assume $C > 0$, for otherwise $x = 0$ and the particle is at rest forever. Then

$$(A/C)^2 + (B/C)^2 = 1$$

so there is an angle ϕ_0, determined up to a multiple of 2π, such that $\cos\phi_0 = A/C$ and $\sin\phi_0 = B/C$. We have

$$x = C(\cos kt \cos\phi_0 + \sin kt \sin\phi_0) = C\cos(kt - \phi_0)$$

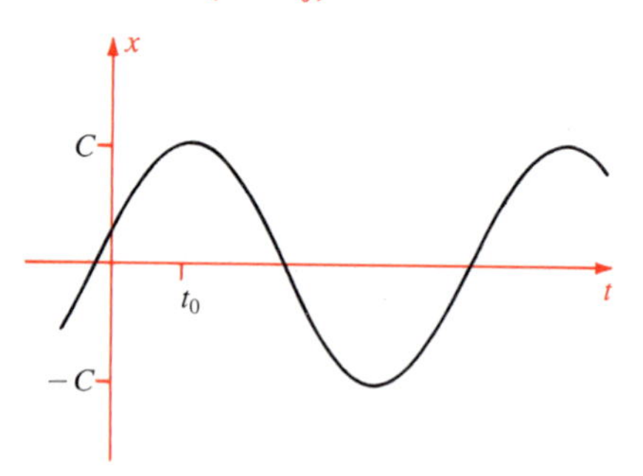

Figure 19-5-2
Simple harmonic motion:
$x = C\cos k(t - t_0)$

The periodic motion of $x = x(t)$ is called **simple harmonic** motion. The **period** of the motion is $p = 2\pi/k$, and its reciprocal $f = k/2\pi$ is called the **frequency** of the motion. Frequency is measured in cycles per second, or **hertz** (abbreviated cps or Hz). The angle ϕ_0 is called the **phase angle** of the motion. Finally, C is called the **amplitude** of the motion. A graph is shown in Figure 19-5-2. It is convenient to write $\phi_0 = kt_0$ so that

$$x = C\cos k(t - t_0) = C\cos(kt - \phi_0)$$

●

Example 2

Solve the initial value problem for Example 1 with initial conditions

a $x(0) = 0 \qquad \dfrac{dx}{dt}(0) = v_0 > 0$

b $x(0) = x_0 > 0 \qquad \dfrac{dx}{dt}(0) = 0$

Solution

a The initial conditions mean that the particle is moving through its rest position at time $t = 0$. From

$$x = C\cos(kt - \phi_0) \qquad \text{follows} \qquad \frac{dx}{dt} = -kC\sin(kt - \phi_0)$$

Set $t = 0$:

$$C\cos\phi_0 = 0 \qquad \text{and} \qquad kC\sin\phi_0 = v_0$$

We may take $\phi_0 = \frac{1}{2}\pi$ and have $kC = v_0$. Hence the solution is

$$x = x(t) = \frac{v_0}{k}\cos(kt - \tfrac{1}{2}\pi) = \frac{v_0}{k}\sin kt$$

b The initial conditions mean that the particle is pulled out to x_0, stretching the spring, held still there, then released. Again we set $t = 0$ in the formulas for x and dx/dt:

$$x_0 = C\cos\phi_0 \quad \text{and} \quad 0 = kC\sin\phi_0$$

Therefore we may take $\phi_0 = 0$, so that $C = x_0$. Thus the solution is $x = x(t) = x_0 \cos kt$.

●

We next consider the effect of a frictional force in addition to the spring force. The resulting motion is known as **damped harmonic motion.**

Example 3

Suppose in Example 1 that the mass slides along a track (Figure 19-5-3). In addition to the restoring force due to the spring, the mass is subjected to a retarding frictional force proportional to its speed. Describe the motion of the object, assuming that it starts at the origin, where the spring force is 0, with initial speed $v_0 > 0$.

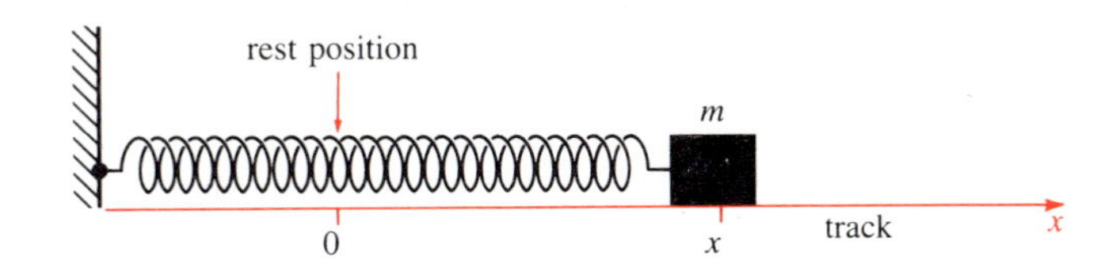

Figure 19-5-3
Vibration with friction

Solution When the object is at x, the forces acting on it are the spring force $-mqx$ and the friction force $-mpx'$, where p and q are positive constants. The friction force is $-mpx'$ (not $+mpx'$) because friction opposes motion. The equation of motion is

$$m\frac{d^2x}{dt^2} = -mqx - mpx'$$

that is,

$$\frac{d^2x}{dt^2} + p\frac{dx}{dt} + qx = 0$$

with the initial conditions

$$x(0) = 0 \qquad x'(0) = v_0$$

Before solving, think about the problem. If there is no friction (that is, if p is zero), then the motion is simple harmonic. If the friction is small, the motion should be nearly simple harmonic motion, except for a gradual slowing down due to friction. If the friction is large, the motion should be considerably inhibited, in fact there might not be oscillations. From this physical reasoning, it seems clear that the relative size of the constants p and q is crucial. (You might look ahead at the next two figures for these cases.)

As usual we consider the characteristic equation

$$\lambda^2 + p\lambda + q = 0$$

and its discriminant $D = p^2 - 4q$.

Case 1 $D < 0$. Then $D = -4k^2$ with $k > 0$ and the general solution of the differential equation is

$$x = e^{-pt/2}(A\cos kt + B\sin kt) = Ce^{-pt/2}\cos(kt - \phi_0)$$

The initial condition $x(0) = 0$ is satisfied for $\phi_0 = \frac{1}{2}\pi$. Consequently $x = Ce^{-pt/2} \sin kt$. The initial condition $x'(0) = v_0$ implies that $Ck = v_0$ so we finally have the solution for Case 1:

$$x = \frac{v_0}{k} e^{-pt/2} \sin kt$$

The resulting motion is a damped harmonic motion. The particle oscillates with constant period $2\pi/k$, but the amplitude gets smaller and smaller as $t \to +\infty$. The number $k/2\pi$ is the **natural frequency** of the motion. See Figure 19-5-4.

Case 1 is called the **underdamped** case: friction is small compared to spring force.

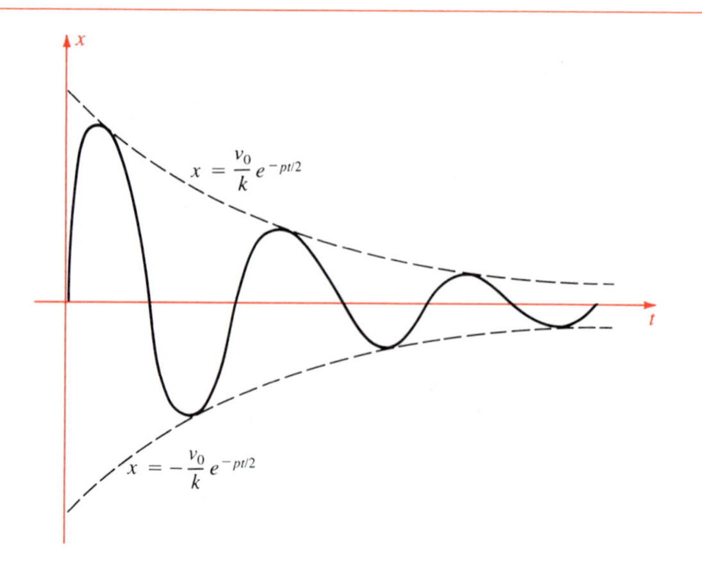

Figure 19-5-4
Underdamped harmonic motion:
$x = \frac{v_0}{k} e^{-pt/2} \sin kt$

Case 2 $D > 0$. Then $D = k^2$ where $k > 0$, and the roots of the characteristic equation are

$$\lambda = \tfrac{1}{2}(-p + k) \quad \text{and} \quad \mu = \tfrac{1}{2}(-p - k)$$

Both λ and μ are negative. That is because $p > 0$ and also $q > 0$, so that

$$k^2 = p^2 - 4q < p^2 \qquad \text{hence} \qquad k < p$$

It follows easily that $-p + k < 0$ so $\lambda < 0$ and $\mu < 0$. It is convenient to write $\lambda = -r$ and $\mu = -s$. Then $0 < r < s$ and the general solution of the differential equation is

$$x = Ae^{-rt} + Be^{-st}$$

The initial condition $x(0) = 0$ implies $B = -A$ so that

$$x = A(e^{-rt} - e^{-st})$$

Figure 19-5-5
Overdamped harmonic motion:
$x = \frac{v_0}{s-r}(e^{-rt} - e^{-st})$
$(0 < r < s)$

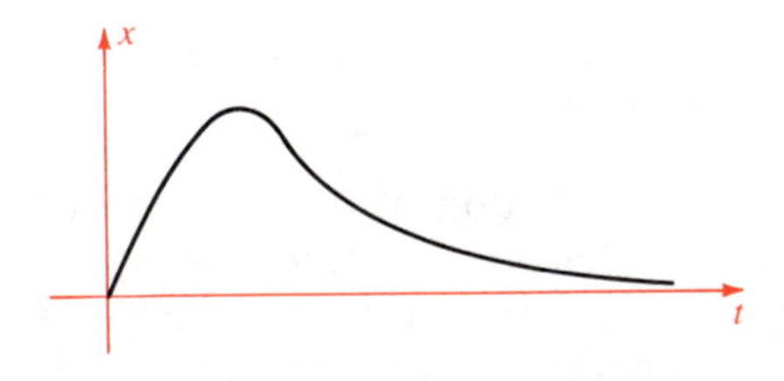

The other initial condition $x'(0) = v_0$ implies $v_0 = A(s - r)$ so that $A = v_0/(s - r)$ and we finally have the solution for Case 2:

$$x = \frac{v_0}{s - r}(e^{-rt} - e^{-st}) \qquad (0 < r < s)$$

The graph of x as a function of t is shown in Figure 19-5-5. The particle moves away from the origin at first, then reverses direction and approaches the origin, never quite reaching it again.

Case 2 is called the **overdamped** case: friction is large compared to spring force.

Case 3 $D = 0$. Then the general solution of the differential equation is

$$x = (A + Bt)e^{-pt/2}$$

Figure 19-5-6
Critically damped harmonic motion:
$x = v_0 te^{-pt/2}$

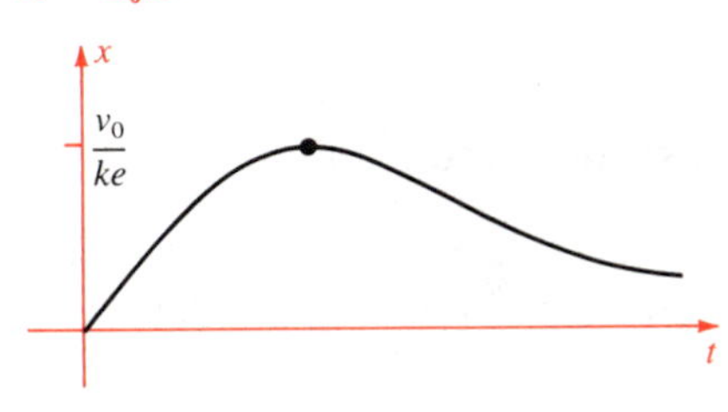

The initial condition $x(0) = 0$ implies $A = 0$, and the condition $x'(0) = v_0$ implies $B = v_0$, so we finally have the solution for Case 3:

$$x = v_0 te^{-pt/2}$$

The graph of $x = x(t)$ is shown in Figure 19-5-6. The motion is similar to that of Case 2.

Case 3 is called the **critically damped** case. ●

Now we consider the effect of an additional external periodic force. This is the situation of **forced vibrations.**

Example 4

Suppose in Example 3 that the friction force is relatively small (underdamped) and that in addition to the spring force and the frictional force, an external periodic force of the form $F(t) = Am \sin 2\pi\omega t$ acts on the particle. Describe the nature of the motion for t large.

Solution The differential equation of motion is

$$\frac{d^2x}{dt^2} + p\frac{dx}{dt} + qx = A \sin 2\pi\omega t \qquad x(0) = 0 \qquad x'(0) = v_0$$

where $p > 0$ and $q > 0$, and also $D = p^2 - 4q < 0$. If, as usual, we write $D = -4k^2$, then the associated homogeneous equation

$$x'' + px' + qx = 0$$

has the general solution

$$x = x_1(t) = Ce^{-pt/2}\cos(kt - \phi_0)$$

as we saw in the solution to Case 1 of Example 3. We need to add on a particular solution of the non-homogeneous equation. We assume a solution of the form

$$x = B_1 \cos 2\pi\omega t + B_2 \sin 2\pi\omega t$$

By a rather tedious calculation we eventually find the B_1 and B_2 that work. If we then use the method in the solution of Example 1 to write this linear combination of $\cos 2\pi\omega t$ and $\sin 2\pi\omega t$ as a sine, we end up with the required particular solution:

$$x = x_2(t) = \frac{A}{\sqrt{(q - 4\pi^2\omega^2)^2 + 4\pi^2\omega^2 p^2}} \sin(2\pi\omega t - \theta_0)$$

The function $x = x_1(t) + x_2(t)$ is the solution of the equation of motion. The constant θ_0 is completely determined in terms of p, q, and ω, and the constants C and ϕ_0 can be determined from the initial conditions. But their values are unimportant if we are only interested in large t because the exponential damping factor $e^{-pt/2}$ approaches 0 rapidly. The function $x_1(t)$ is called a **transient** solution. It represents a damped vibration that becomes negligible after sufficient time has passed. For t sufficiently large,

$$x(t) \approx x_2(t) = \frac{A}{\sqrt{\cdots}} \sin(2\pi\omega t - \theta_0)$$

is a simple harmonic motion called the **steady-state** solution. ●

Remark The amplitude of the steady-state solution is

$$F(\omega) = \frac{A}{\sqrt{(q - 4\pi^2\omega^2)^2 + 4\pi^2\omega^2 p^2}}$$

The square of the denominator is

$$H(\omega) = (q - 4\pi^2\omega^2)^2 + 4\pi^2\omega^2 p^2$$

By expanding and then completing the square, we obtain

$$H(\omega) = [4\pi^2\omega^2 - \tfrac{1}{2}(2q - p^2)]^2 + \tfrac{1}{4}p^2(4q - p^2)$$

We know that p^2 is small compared to q because we have assumed $D = p^2 - 4q < 0$. If we make the assumption that p is even smaller, so small that $p^2 - 2q < 0$, then $H(\omega)$ takes its minimum provided

$$8\pi^2\omega^2 = 2q - p^2$$

[Otherwise $H(\omega)$ takes its minimum at $\omega = 0$.] The corresponding ω, called ω_r, is the **resonant frequency** of this mechanical system. Thus

$$\omega_r = \frac{1}{2\pi}\sqrt{\frac{2q - p^2}{2}}$$

so that

$$\begin{aligned} H_{\min} &= H(\omega_r) = \tfrac{1}{4}p^2(4q - p^2) \\ &= \tfrac{1}{4}p^2[2(2q - p^2) + p^2] \end{aligned}$$

Since $2q - p^2 = 8\pi^2\omega_r^2$, we can write $H_{\min}$ in the form

$$H_{\min} = \tfrac{1}{4}p^2(16\pi^2\omega_r^2 + p^2) \approx 4\pi^2p^2\omega_r^2$$

Clearly $F(\omega)$ is maximal if $H(\omega)$ is minimal, so

$$F_{\max} = \frac{A}{\sqrt{H_{\min}}} \approx \frac{A}{2\pi p\omega_r}$$

When p is very small, $F_{\max}$, the amplitude of the steady-state solution, is very large. This is the phenomenon of **resonance,** whereby a relatively small harmonic force of suitable frequence can build up enormous vibrations that tear a piece of machinery to pieces.

Exercises

1 Suppose a spherical planet of mass M and radius R has uniform density (mass per unit volume). A narrow tunnel is bored straight through the planet's center C from one side to the other. It is known that the gravitational attraction of the whole planet on a particle of mass m in the tunnel at distance x from C is the same as that due to a particle at C whose mass is that of the part of the planet within the sphere of radius x and center C. Express this force in terms of M, R, and x. Take the origin at C and the x-axis in the tunnel.

2 (cont.) At time 0 a particle of mass m is dropped into the tunnel from the planet's surface. Set up its differential equation of motion and initial conditions, and show that the solution is simple harmonic motion.

3 A rocket sled is subjected to $6g$ acceleration for 5 sec. After the engine shuts down, the sled undergoes a deceleration (in ft/sec^2) equal to 0.05 times its velocity (in ft/sec). What is the speed of the sled 10 sec after engine shutdown? How far has it traveled?
Assume $g = 32.2$ ft/sec^2.

4 A cylindrical buoy floats vertically in the water. Its weight is 100 lb and its diameter is 2 ft. When depressed slightly and released, it oscillates with simple harmonic motion. Find the period of the oscillation. [Hint This is just a spring problem in disguise. Use Archimedes' law: A body in water is subjected to an upward buoyant force equal to the weight of the water displaced. Take the density of water to be 62.4 lb/ft^3 and $g = 32.2$ ft/sec^2.]

5 A falling body of mass m is subjected to a downward force mg, where g is the gravitational constant. Due to air resistance, it is subjected also to a retarding (upward) force mkv proportional to its velocity, where $k > 0$. Set up the initial value problem.

6 (cont.) Solve, and find $\lim_{t\to+\infty} v(t)$.

7 Consider the overdamped case of Example 3 as graphed in Figure 19-5-5. Show that $x(t)$ increases from 0 up to its maximum, then decreases towards 0.

8 (cont.) Find the maximum of $x(t)$ and where it occurs.

9 Consider the critically damped case of Example 3 as graphed in Figure 19-5-6. Show that $x(t)$ increases from 0 to its maximum, then decreases towards 0.

10 (cont.) Find the maximum of $x(t)$ and where it occurs.

11 Consider the simple electric circuit of Figure 19-5-7. If $Q = Q(t)$ is the charge on the capacitor, then the current is $I = I(t) = dQ/dt$. It is known that

$$L\frac{d^2Q}{dt^2} + R\frac{dQ}{dt} + \frac{1}{C}Q = -\frac{A}{2\pi\omega}\cos 2\pi\omega t$$

where A, L, C, R are constants. Find the steady-state solution for I (yes I, not Q). Show that resonance occurs for $\omega^2 = 1/LC$ and the resonant steady-state solution is

$$I(t) = \frac{A}{R}\sin(\omega t - \tfrac{1}{2}\pi)$$

Figure 19-5-7
See Exercise 11

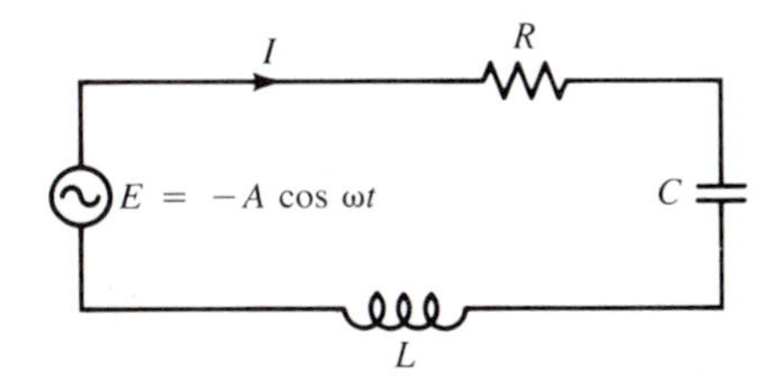

12 A mass m is towed, starting from rest, through a viscous fluid by a constant force F. Assume the opposing friction force of the fluid is kv, where $k > 0$. Set up the differential equation of motion and initial conditions, and solve. Find the terminal speed. (This is a crude model of a tugboat towing a barge through an oil spill.)

13 A pendulum is made of a weight at the end of a long wire. Its motion is described by the differential equation

$$\frac{d^2\theta}{dt^2} + \frac{g}{L}\sin\theta = 0$$

where θ is the angle between the wire and the vertical. If the pendulum swings only through a small arc, then $\sin\theta$ can be approximated by θ, thus simplifying the differential equation. Do so and find the approximate period of the pendulum.

14 A model for the thermal breakdown of dielectrics leads to the initial value problem

$$\frac{d^2u}{dx^2} + \beta e^u = 0 \qquad u(0) = u_0 \qquad u'(0) = 0$$

Find a solution of the form $u = a + b\ln\cosh cx$.

The external forces acting on a projectile are gravity and air resistance. At low altitude and low speed, it may be assumed that air resistance is proportional to speed.

If a projectile is shot straight up, it rises to its maximum height and then falls to the ground. Whether it takes longer to rise or longer to fall, or equal times, is not obvious. However, as we shall see, in falling there is a terminal speed. Hence a projectile shot up with initial speed faster than the terminal speed necessarily takes longer in falling than in rising. The next five examples show this is *always* so.

15 A projectile is shot straight up with initial velocity v_0. Show that its height satisfies the initial value problem $y'' + ky' = -g$, $y(0) = v_0$. Derive the solution.

16 (cont.) Show that v approaches a terminal velocity as $t \to +\infty$. Find it.

17 (cont.) Show that the projectile reaches its maximum height at time

$$t_1 = \frac{1}{k}\ln\left(\frac{g + kv_0}{g}\right)$$

Show that the projectile returns to ground at time $t_2 > 0$, where

$$\left(\frac{g + kv_0}{g}\right)(1 - e^{-kt_2}) = kt_2$$

18 (cont.) Find $\lim_{k\to 0} t_1$ and $\lim_{k\to 0} t_2$.

***19** (cont.) Prove $(t_2 - t_1) > t_1$. Begin by showing that $y'(t_1 + t) + y'(t_1 - t) > 0$ for $t > 0$. Interpret physically. Then integrate the inequality over the interval $0 \le t \le t_1$ to obtain $y(t_1 + t) > y(t_1 - t)$ for $t > 0$. Deduce that $2t_1 < t_2$.

***20** (cont.) Now try a concrete problem. Suppose the initial velocity is 200 m/sec and the maximum height is 1000 m. Assume $g = 9.80$ m/sec^2. Estimate k to three significant digits. Then estimate t_1 and t_2.

19-6 Numerical Methods

Consider an initial value problem

$$\frac{dy}{dx} = F(x, y) \qquad y(x_0) = y_0$$

If $F(x, y)$ is sufficiently smooth, say differentiable, then the theory of differential equations says that the problem has a unique solution valid in some open interval $(x_0 - \delta,\ x_0 + \delta)$. But theory is one thing and actually finding a solution is quite another. Although in some cases we can find an explicit solution, these cases are extremely rare, and usually we must use methods that approximate a solution rather than find an exact one.

We shall discuss two numerical methods in this section, power series and Heun's method. We shall use the following test examples:

$$\begin{cases} \dfrac{dy}{dx} = y \\ y(0) = 1 \end{cases} \qquad \begin{cases} \dfrac{dy}{dx} = x^2y - 1 \\ y(0) = 2 \end{cases} \qquad \begin{cases} \dfrac{dy}{dx} = x + y^2 \\ y(0) = 0 \end{cases}$$

The first one has the solution $y(x) = e^x$.

The second test example is a linear equation. By the usual methods we find

$$y(x) = 2e^{x^3/3} - e^{x^3/3} \int_0^x e^{-t^3/3}\, dt$$

The integration cannot be carried out exactly. However, we can use the Taylor expansions

$$e^{x^3/3} = 1 + \tfrac{1}{3}x^3 + \tfrac{1}{18}x^6 + \tfrac{1}{162}x^9 + \cdots$$

$$e^{-t^3/3} = 1 - \tfrac{1}{3}t^3 + \tfrac{1}{18}t^6 - \tfrac{1}{162}t^9 + \cdots$$

If we carry out the integration term by term and combine the series, we find

$$y(x) = 2 - x + \tfrac{2}{3}x^3 - \tfrac{1}{4}x^4 + \tfrac{1}{9}x^6 - \tfrac{1}{28}x^7 + \tfrac{1}{81}x^9 + \cdots$$

This will be useful for comparison later.

The third test example,

$$\frac{dy}{dx} = x + y^2 \qquad y(0) = 0$$

cannot be solved exactly, nor can the answer be expressed in terms of an integral (as far as I know).

Power Series Method

To solve the initial value problem

$$\frac{dy}{dx} = F(x, y) \qquad y(0) = a_0$$

we can assume that $y(x)$ has a convergent Taylor series

$$y = a_0 + a_1x + a_2x^2 + \cdots$$

We substitute into the differential equation and equate coefficients of the various powers of x:

$$a_1 + 2a_2x + 3a_3x^2 + \cdots = F(x, a_0 + a_1x + a_2x^2 + \cdots)$$

If we can effectively express the right-hand side of the equation as a Taylor series, then the result will be a sequence of relations from which we can successively compute a_1, a_2, $\cdots$.

Example 1

Solve by the power series method

a $\dfrac{dy}{dx} = y \qquad y(0) = 1$

b $\dfrac{dy}{dx} = x^2y - 1 \qquad y(0) = 2$

c $\dfrac{dy}{dx} = x + y^2 \qquad y(0) = 0$

Solution

a Set $y = 1 + a_1x + a_2x^2 + a_3x^3 + \cdots$ and substitute into the differential equation $y' = y$:

$$a_1 + 2a_2x + 3a_3x^2 + \cdots = 1 + a_1x + a_2x^2 + a_3x^3 + \cdots$$

Equate coefficients

$$a_1 = 1 \qquad 2a_2 = a_1 \qquad 3a_3 = a_2 \qquad 4a_4 = a_3 \qquad \cdots$$

It follows that

$$a_2 = \tfrac{1}{2}a_1 = \frac{1}{2!} \qquad a_3 = \tfrac{1}{3}a_2 = \frac{1}{3!} \qquad \cdots \qquad a_n = \frac{1}{n!}$$

Therefore

$$y = 1 + \sum_{n=1}^{\infty} \frac{x^n}{n!} = e^x$$

b Substitute $y = 2 + a_1x + a_2x^2 + a_3x^3 + \cdots$ into the differential equation $y' = x^2y - 1$:

$$\begin{aligned} a_1 + 2a_2x + 3a_3x^2 + \cdots \\ = -1 + 2x^2 + a_1x^3 + a_2x^4 + a_3x^5 + \cdots \end{aligned}$$

Equate coefficients:

$$\begin{array}{lll} a_1 = -1 & 2a_2 = 0 & 3a_3 = 2 \\ 4a_4 = a_1 & 5a_5 = a_2 & 6a_6 = a_3 \quad \cdots \end{array}$$

It follows that

$$\begin{array}{lll} a_1 = -1 & a_2 = 0 & a_3 = \tfrac{2}{3} \\ a_4 = -\tfrac{1}{4} & a_5 = 0 & a_6 = \dfrac{2}{3 \cdot 6} \\ a_7 = \dfrac{-1}{4 \cdot 7} & a_8 = 0 & a_9 = \dfrac{2}{3 \cdot 6 \cdot 9} \\ a_{10} = \dfrac{-1}{4 \cdot 7 \cdot 10} & a_{11} = 0 & a_{12} = \dfrac{2}{3 \cdot 6 \cdot 9 \cdot 12} \quad \cdots \end{array}$$

Thus

$$a_{3n} = \frac{2}{3^n n!} \qquad a_{3n+1} = \frac{-1}{4 \cdot 7 \cdot 10 \cdots (3n+1)} \qquad a_{3n+2} = 0$$

so we can write a formula for the whole Taylor series, and compute as many terms as we please:

$$y = 2 - x + \tfrac{2}{3}x^3 - \tfrac{1}{4}x^4 + \tfrac{1}{9}x^6 - \tfrac{1}{28}x^7 + \tfrac{1}{81}x^9 + \cdots$$

c Substitute $y = a_1x + a_2x^2 + a_3x^3 + \cdots$ into $y' = x + y^2$:

$$a_1 + 2a_2x + 3a_3x^2 + 4a_4x^3 + \cdots$$
$$= x + (a_1x + a_2x^2 + a_3x^3 + \cdots)^2$$

The term of lowest degree in the square is $a_1{}^2x^2$. Therefore we conclude without further computation that

$$a_1 = 0 \qquad a_2 = \tfrac{1}{2} \qquad a_3 = 0$$

Now the term of lowest degree in the square is $a_2{}^2x^4$, so we also conclude that

$$a_4 = 0 \qquad 5a_5 = a_2{}^2 = \tfrac{1}{4} \qquad a_5 = \tfrac{1}{20}$$

Now let's rewrite the equation and square:

$$x + \tfrac{1}{4}x^4 + 6a_6x^5 + 7a_7x^6 + 8a_8x^7 + \cdots$$
$$= x + (\tfrac{1}{2}x^2 + \tfrac{1}{20}x^5 + a_6x^6 + \cdots)^2$$
$$= x + \tfrac{1}{4}x^4 + \tfrac{1}{20}x^7 + a_6x^8 + \cdots$$

We conclude that

$$a_6 = 0 \qquad a_7 = 0 \qquad 8a_8 = \tfrac{1}{20} \qquad a_8 = \tfrac{1}{160}$$

So far we have

$$y = \tfrac{1}{2}x^2 + \tfrac{1}{20}x^5 + \tfrac{1}{160}x^8 + \cdots$$

From this much, an educated guess is that y has the form

$$y = x^2(b_0 + b_1x^3 + b_2x^6 + b_3x^9 + \cdots)$$

If we start from scratch with this expression for y, the differential equation becomes

$$2b_0x + 5b_1x^4 + 8b_2x^7 + 11b_3x^{10} + \cdots$$
$$= x + x^4(b_0 + b_1x^3 + b_2x^6 + \cdots)^2$$
$$= x + b_0{}^2x^4 + 2b_0b_1x^7 + (2b_0b_2 + b_1{}^2)x^{10} + \cdots$$

Equating coefficients, we find

$$2b_0 = 1 \qquad 5b_1 = b_0{}^2 \qquad 8b_2 = 2b_0b_1 \qquad 11b_3 = 2b_0b_2 + b_1{}^2 \qquad \cdots$$

from which

$$b_0 = \tfrac{1}{2} \qquad b_1 = \tfrac{1}{20} \qquad b_2 = \tfrac{1}{160} \qquad b_3 = \tfrac{7}{8800} \qquad \cdots$$

Therefore

$$y = \tfrac{1}{2}x^2 + \tfrac{1}{20}x^5 + \tfrac{1}{160}x^8 + \tfrac{7}{8800}x^{11} + b_4x^{14} + \cdots$$

●

Next we try the power series method on a second order equation.

Example 2

Solve by the power series method

$$\frac{d^2y}{dx^2} = xy \qquad y(0) = a \qquad y'(0) = b$$

Solution Set

$$y = a + bx + \sum_{n=2}^{\infty} a_n x^n$$

so that

$$\frac{d^2y}{dx^2} = \sum_{n=2}^{\infty} n(n-1)a_n x^{n-2}$$

The initial conditions are satisfied, and the differential equation becomes

$$\sum_{n=2}^{\infty} n(n-1)a_n x^{n-2} = ax + bx^2 + \sum_{n=2}^{\infty} a_n x^{n+1}$$

that is

$$\sum_{n=2}^{\infty} n(n-1)a_n x^{n-2} = ax + bx^2 + \sum_{n=5}^{\infty} a_{n-3} x^{n-2}$$

Equate coefficients:

$$2 \cdot 1 \cdot a_2 = 0 \qquad 3 \cdot 2 \cdot a_3 = a \qquad 4 \cdot 3 \cdot a_4 = b$$

$$5 \cdot 4 \cdot a_5 = a_2 \qquad 6 \cdot 5 \cdot a_6 = a_3 \qquad \cdots$$

$$n(n-1)a_n = a_{n-3} \qquad \cdots$$

Solve first for the a_{3n+2}:

$$a_2 = 0 \qquad a_5 = 0 \qquad \cdots \qquad a_{3n+2} = 0$$

Next solve for the a_{3n}:

$$a_3 = \frac{a}{2 \cdot 3} \qquad a_6 = \frac{a_3}{5 \cdot 6} = \frac{a}{2 \cdot 3 \cdot 5 \cdot 6} = \frac{1 \cdot 4}{6!} a$$

$$a_9 = \frac{a_6}{8 \cdot 9} = \frac{1 \cdot 4 \cdot 7}{9!} a \qquad \cdots$$

Finally solve for the a_{3n+1}:

$$a_4 = \frac{b}{3 \cdot 4} = \frac{2}{4!} b$$

$$a_7 = \frac{a_4}{6 \cdot 7} = \frac{2 \cdot 5}{7!} b \qquad \cdots$$

Therefore

$$y = a\left(1 + \frac{1}{3!}x^3 + \frac{1 \cdot 4}{6!}x^6 + \frac{1 \cdot 4 \cdot 7}{9!}x^9 + \cdots\right)$$
$$+ b\left(x + \frac{2}{4!}x^4 + \frac{2 \cdot 5}{7!}x^7 + \frac{2 \cdot 5 \cdot 8}{10!}x^{10} + \cdots\right)$$

The function y is sometimes written $A(a, b; x)$ and called an **Airy function.** ●

The Heun Method

Suppose we have an initial value problem

$$\frac{dy}{dx} = F(x, y) \qquad y(a) = y_0$$

and we cannot find an exact solution. Suppose that what we really want is only the value $y(b)$ of the solution function at a specific point $x = b$. This is quite different from approximating the whole solution *function* $y(x)$; now we want only to approximate one single solution *value:* $y(b)$.

The geometric basis of the Heun method is shown in Figure 19-6-1. We introduce the notation $h = b - a$, so that $b = a + h$. The first step in approximating $y(b)$ is to compute

$$y_0' = F(a, y_0) \qquad \text{and} \qquad y_1 = y_0 + hy_0'$$

so y_1 is a first approximation to $y(b)$. We reason that the slope of the direction field at (b, y_1) must be close to the slope of the direction field at $(b, y(b))$, so we compute

$$y_1' = F(b, y_1)$$

Now we go back to (a, y_0) and move forward again with the *average* of the two computed slopes. The corrected approximation is

$$y(b) \approx y_0 + \tfrac{1}{2}h(y_0' + y_1')$$

The analytic basis of this approximation is a proof that it agrees with the second order Taylor approximation of $y(x)$ at $x = a$. See Exercises 27 and 28.

For reasonable accuracy, we divide the interval $[a, b]$ into n parts and proceed by steps from a to b. Note that $2n$ evaluations of $F(x, y)$ are required, and that is a measure of the computing cost. A flowchart is given in Figure 19-6-2.

Figure 19-6-1
Heun method:
slope$(L) = y_0' = F(a, y_0)$
$y_1 = y_0 + hy_0' \quad (h = b - a)$
slope$(M) = y_1' = F(b, y_1)$
slope$(N) =$ average $= \frac{1}{2}(y_0' + y_1')$
new $y_1 = y_0 + h \cdot$ slope(N)

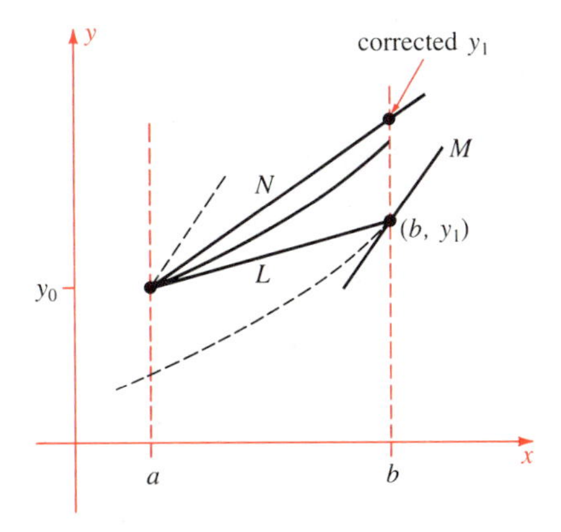

Figure 19-6-2
Flowchart for Heun method
(← denotes assignment)

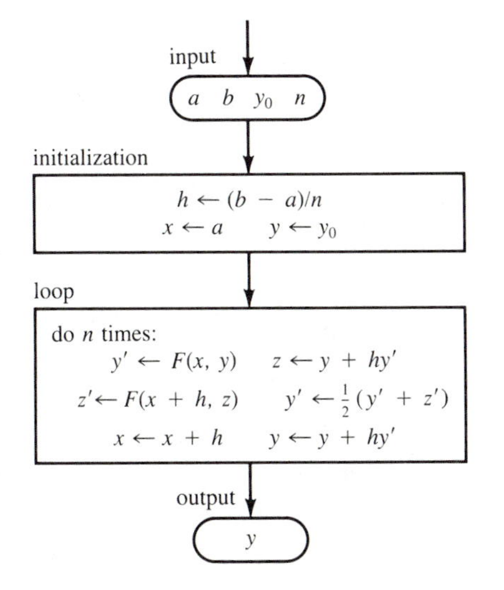

Example 3

Estimate $y(1)$ for the solution of

$$\frac{dy}{dx} = y \qquad y(0) = 1$$

Solution Because $f(x, y) = y$ is such a simple function, we can find an exact value for the answer. Let us not fix n yet, but remember that $h = (b - a)/n = 1/n$. We proceed from y_j to y_{j+1} as follows. First we compute

$$y'_j = F(x_j, y_j) = y_j \qquad \text{and} \qquad y_{j+1} = y_j + hy'_j = (1 + h)y_j$$

Then we compute

$$y'_{j+1} = F(x_{j+1}, y_{j+1}) = y_{j+1} = (1 + h)y_j$$

and average y'_j and y'_{j+1} for the new value of y'_j:

$$y'_j = \tfrac{1}{2}[y_j + (1 + h)y_j] = (1 + \tfrac{1}{2}h)y_j$$

Finally the corrected value of y_{j+1} is

$$y_{j+1} = y_j + hy'_j = [1 + h(1 + \tfrac{1}{2}h)]y_j = (1 + h + \tfrac{1}{2}h^2)y_j$$

We conclude easily that

$$y_n = (1 + h + \tfrac{1}{2}h^2)^n y_0 = (1 + h + \tfrac{1}{2}h^2)^n$$

First take $n = 10$, so $h = 0.1$. Then $1 + h + \frac{1}{2}h^2 = 1.105$ so that

$$y(1) \approx (1.105)^{10} \approx 2.714$$

Next take $n = 20$ so $h = 0.05$. Then $1 + h + \frac{1}{2}h^2 = 1.05125$ so that

$$y(1) \approx (1.05125)^{20} \approx 2.7172$$

which is reasonably close to $e \approx 2.7183$. ●

Example 4

Estimate $y(1)$ for the solution of

$$\frac{dy}{dx} = x^2y - 1$$

$$y(0) = 2$$

Solution This time we need a computer. We show partial results in Table 19-6-1 for $n = 10, 20,$ and 100 ($20, 40,$ and 200 function evaluations respectively).

The following comparison is interesting. In Example 1b, we found

$$y(x) \approx 2 - x + \tfrac{2}{3}x^3 - \tfrac{1}{4}x^4 + \tfrac{1}{9}x^6 - \tfrac{1}{28}x^7 + \tfrac{1}{81}x^9$$

This approximation yields

$$y(0.5) \approx 1.5692$$

$$y(1.0) \approx 1.5044$$ ●

Table 19-6-1
Heun method
for $y' = x^2y - 1 \quad y'(0) = 2$

n	10	20	100
h	0.1	0.05	0.01
x	$y(x)$	$y(x)$	$y(x)$
0.0	2.0000	2.0000	2.0000
0.1	1.9009	1.9007	1.9006
0.2	1.8055	1.8051	1.8049
0.3	1.7168	1.7162	1.7161
0.4	1.6376	1.6369	1.6367
0.5	1.5703	1.5695	1.5692
0.6	1.5171	1.5162	1.5159
0.7	1.4804	1.4795	1.4792
0.8	1.4632	1.4622	1.4619
0.9	1.4690	1.4681	1.4677
1.0	1.5029	1.5020	1.5017

Exercises

Find the power series solution through the terms in x^3 of

1 $y' = x + y$ $\quad y(0) = 2$

2 $y' = x^2 + y$ $\quad y(0) = -1$

3 $y' = 1 - x^2 - y^2$ $\quad y(0) = 0$

4 $y' = x^2 + y^2$ $\quad y(0) = 1$

5 $y' = 1 + x^2e^{-y}$ $\quad y(0) = 1$

6 $y' = e^x + xy^2$ $\quad y(0) = -1$

7 $y' = \cos(xy) - 1$ $\quad y(0) = -1$

8 $y' = y \sin x^2$ $\quad y(0) = 1$

9 $y' = (1 + x^2)(1 + y^2)$ $\quad y(0) = 2$

10 $y' = \dfrac{x}{x + y + 1}$ $\quad y(0) = 0$

11 $y'' = -4y$ $\quad y(0) = 0$, $y'(0) = 1$

12 $y'' = x + y^2$ $\quad y(0) = 0$, $y'(0) = 0$

13 $y'' = x^2 + y^2$ $\quad y(0) = 0$, $y'(0) = 1$

14 $y'' = x^2y^2$ $\quad y(0) = 0$, $y'(0) = -1$

15 $y'' = xy + (y')^2$ $\quad y(0) = 1$, $y'(0) = 2$

16 $y'' = y - xy'$ $\quad y(0) = 0$, $y'(0) = -2$

Use the Heun method: calculator computation with $n = 2$ and $n = 5$ or computer (or programmable calculator) computation with $n = 20, 50$, and 100

17 $y' = xy + x$ $\quad y(0) = 0$ Estimate $y(1)$

18 $y' = xy + y$ $\quad y(0) = 1$ Estimate $y(1)$

19 $y' = xy + 1$ $\quad y(0) = 2$ Estimate $y(1)$

20 $y' = y + \sin x$ $\quad y(0) = 1$ Estimate $y(1)$

21 $y' = x^3y - x$ $\quad y(0) = 1$ Estimate $y(1)$

22 $y' = 1 + x^2y$ $\quad y(0) = 1$ Estimate $y(1)$

23 $y' = (1 + y)\cos x$ $\quad y(0) = 0$ Estimate $y(\frac{1}{4}\pi)$

24 $y' = (1 - x)(1 - y)$ $\quad y(0) = -1$ Estimate $y(0.5)$

25 $y' = 2xy^2$ $\quad y(1) = 1$ Estimate $y(0)$

26 $y' = x^2 + y^2$ $\quad y(0) = -1$ Estimate $y(-0.5)$

The next two exercises will show that the Heun method yields the second order Taylor polynomial of the solution. Start with

$$\frac{dy}{dx} = F(x, y) \qquad y(a) = y_0 \qquad y_0' = F(a, y_0)$$

$$y_1' = F(a + h, y_0 + hy_0') \qquad y_1 = y_0 + \tfrac{1}{2}h(y_0' + y_1')$$

27 Set $G_0 = \partial F(a, y_0)/\partial x$ and $H_0 = \partial F(a, y_0)/\partial y$. Find the second order Taylor polynomial of y_1 in powers of h.

28 (cont.) Let $y = y(x)$ be the exact solution. Find the second order Taylor polynomial of $y(a + h)$ in powers of h. It should be the same as the answer to Exercise 27.

19-7 Review Exercises

Solve

1 $\dfrac{dy}{dx} = xy^4$

2 $\dfrac{dy}{dx} = \dfrac{1 + x^2}{y^3}$

3 $\dfrac{dy}{dx} = \dfrac{e^x}{y}$ $\quad y(0) = -1$

4 $\dfrac{dy}{dx} = 2x(1 + y^2)$ $\quad y(\frac{1}{2}\sqrt{\pi}) = -1$

5 $\dfrac{dy}{dx} - 16y = e^{4x}$

6 $\dfrac{dy}{dx} + 16y = e^x \cos x$ $\quad y(0) = 0$

7 The tank in Figure 19-7-1 (next page) is a paraboloid of revolution with vertical axis. How long will it take the tankful of water to empty through a hole of area 3 cm^2 at the bottom? (See Example 2 in Section 19-3.) Take $g = 980\text{ cm/sec}^2$.

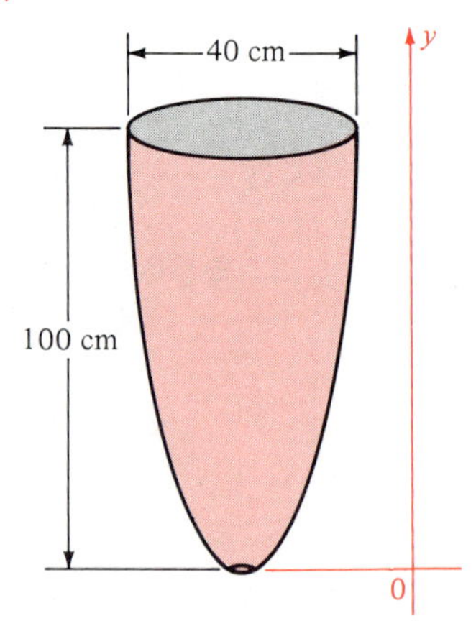

Figure 19-7-1
See Exercise 7

8 A thermometer is plunged into a hot liquid. The thermometer initially reads 50 °C. Five seconds later it reads 70 °C, and five more seconds later it reads 78 °C. Find the temperature of the liquid. (Assume that the reading T changes at a rate proportional to the difference between the constant liquid temperature and T.)

Solve

9 $y'' - 2y = 0$

10 $y'' - 2y' = 0$

11 $y'' - 2y = e^x$

12 $y'' - 2y' = 1$

13 Show that the equation $x'' + k^2x = r(t)$ has the particular solution

$$x(t) = \frac{1}{k}\int_0^t \sin k(t-s)r(s)\,ds$$

You may presuppose the formula

$$\frac{d}{dt}\left(\int_0^t F(s,t)\,ds\right) = F(t,t) + \int_0^t \frac{\partial F}{\partial t}\,ds$$

14 (cont.) Apply this to $x'' + x = \cos t$.

Solve by power series up to the x^5 term

15 $(x^2 - 1)y'' = 6y$ $\quad y(0) = a$, $y'(0) = b$

16 $(1 - x^2)y'' - xy' + 49y = 0$ $\quad y(0) = 0$, $y'(0) = 1$

17 Show that the **Bernoulli Equation** $y' + p(x)y = q(x)y^n$ can be reduced to a linear equation by the substitution $z = y^{1-n}$.

18 (cont.) Solve $y' + xy = \sqrt{y}$; find a formula for the answer, but do not try to evaluate it.

19 Suppose $x = x(t)$ satisfies the differential equation $dx/dt = c\sin x$. If $y = 2x$, show that $d^2y/dt^2 = c^2 \sin y$.

20 Suppose $q(x)$ satisfies $|q(x)| \le M$ and that $y(x)$ is a solution of the initial value problem $y' + y = q(x)$, $y(0) = 0$. Show that $|y(x)| \le M$ for $x \ge 0$.

21 Solve by the Heun method for $y(2)$

$$\frac{dy}{dx} = x^2 - y \qquad y(0) = -1$$

with $n = 10$ and $n = 50$.

22 Solve by the Heun method for $y(1)$

$$\frac{dy}{dx} = x + y^2 \qquad y(0) = 0$$

with $n = 10$ and $n = 100$.

Answers to Odd-Numbered Exercises

Chapter 1 Functions and Graphs

Section 1-1 The Real Number System

Page 12

1 7

3 $\frac{7}{5}$

5 21

7 First $y > 0$, so $y^{-1} > 0$. Second, $0 < x < y$ and $y^{-1} > 0$, so $0 \cdot y^{-1} < x \cdot y^{-1} < y \cdot y^{-1}$, that is, $0 < x/y < 1$.

9 Either $|a| = a$ or $|a| = -a$. In either case, $|a| \le b$.

11 Suppose $a > 0$ and $b = -c < 0$. Then $|a + b| = |a - c| = a - c$ or $c - a$, and $|a| + |b| = a + c$. Since $2c > 0$ and $2a > 0$, we have $a - c < a - c + 2c = a + c$ and $c - a < c - a + 2a = a + c$. In either case of $|a + b|$ we have $|a + b| < |a| + |b|$.

13 $x < 4$

15 $x > -\frac{2}{3}$

17 $x \ge 30$

19 $x > \frac{12}{5}$

21 $9 \le x \le 17$

23 $x \le -21$

25 $-5 < x < 3$

27 $(-\infty, -\frac{1}{4}) \cup (0, +\infty)$

29 $(0, 3)$

31 $|x| = 2$

33 $|x - a| \le |x - b|$

35 $|x - 17| < 1$

37 $-1, 2$

39 If there were such an x, then 1 and 12 would be within $2 + 3 = 5$ of each other; absurd. Detail:
$11 = |12 - 1| = |(12 - x) + (x - 1)|$
$\le |x - 12| + |x - 1| < 3 + 2$

41 $[-4, 4]$

43 $[-0.01, 0.01]$

45 $[2, 4]$

47 $(-2, 2)$

49 $(-1, 9)$

51 $(-\infty, -\frac{2}{3}) \cup (0, +\infty)$

53 $(-3, 3)$

55 $(-2.1, -2) \cup (-2, -1.9)$

57 $[0, 4]$

59 $|7x - 7a| = 7 \cdot |x - a| < 7 \times 10^{-6} < 10^{-5}$

61 $|(x - y) + 7| = |(x + 3) + (4 - y)|$
$\le |x + 3| + |y - 4| < 10^{-5} + 10^{-5} = 2 \times 10^{-5}$

63 $|xy - 35| = |(x - 5)y + 5(y - 7)|$
$\le |x - 5| \cdot |y| + 5 \cdot |y - 7| < |y| \cdot 10^{-6} + 5 \cdot 10^{-6}$
Clearly $|y| < 8$, so $|xy - 35| < 13 \times 10^{-6}$
$= 1.3 \times 10^{-5} < 2 \times 10^{-5}$.

65 $|xy - 6| = |(x + 2)y - 2(y + 3)|$
$\le |x + 2| \cdot |y| + 2 \cdot |y + 3| < 10^{-6} \cdot |y| + 2 \times 10^{-6}$
Clearly $|y| \le 4$; hence $|xy - 6| < 6 \times 10^{-6}$.

67 $x^3 - 27 = (x - 3)(x^2 + 3x + 9)$. Since $|x| < 4$, we have $|x^2 + 3x + 9| \le |x|^2 + 3 \cdot |x| + 9 < 37$, so
$|x^3 - 27| \le 37 \cdot |x - 3| < 37 \times 10^{-6}$
$= 3.7 \times 10^{-5} < 5 \times 10^{-5}$.

Section 1-2 Functions and Their Graphs

Page 20

1 5, 9, 6, $2/x + 5$, $2x - 1$

3 **R**, **R**

5 **R**, **R**

7 $\mathbf{R} - \{\frac{3}{2}\}$, $\mathbf{R} - \{0\}$

9 $\mathbf{R} - \{\frac{5}{3}\}$, $\mathbf{R} - \{\frac{1}{3}\}$

11 $[6, +\infty)$, $[0, +\infty)$

13 $[-\frac{2}{3}, \frac{2}{3}]$, $[0, 2]$

15 $[\frac{3}{2}, +\infty)$, $[0, +\infty)$

17 $[-\frac{1}{2}, \frac{1}{2}]$, $[0, \frac{1}{2}]$

19 $(-\infty, 1] \cup [4, +\infty)$, $[0, +\infty)$

21 $3x - 1$, $-6x - 2$

23 $(x - 1)^2$, $-2x^3 + x^2$

25 $3x - 5$, $3x - 1$

27 $2x^2 + 4x + 2$, $-2x^2 - 1$

29 $-4x$, $-4x$

31 9, 3

33 $x = -3$

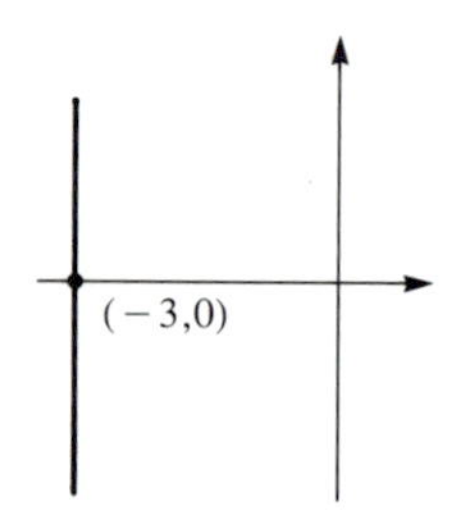

35 $x > 0$ and $y > 0$

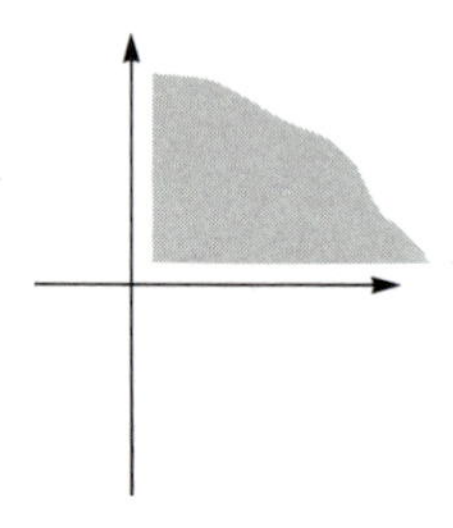

37 $1 \le x \le 3$

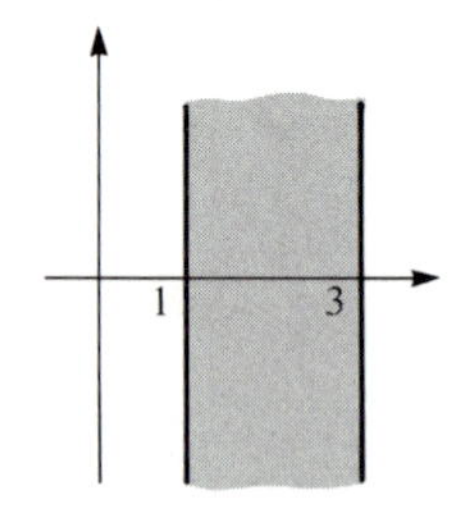

39 $-2 \le x \le 2$ and $-2 \le y \le 2$

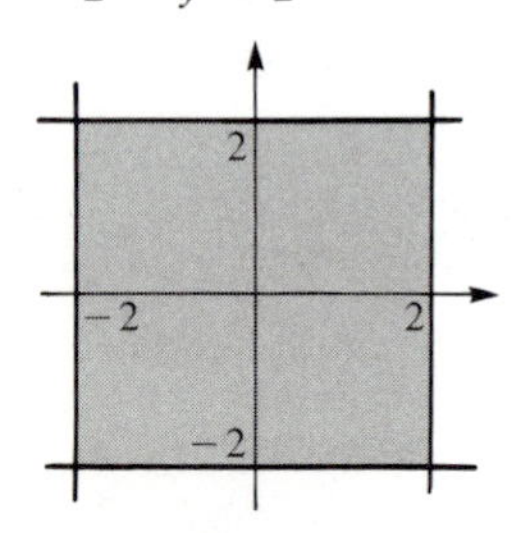

41 x and y integers

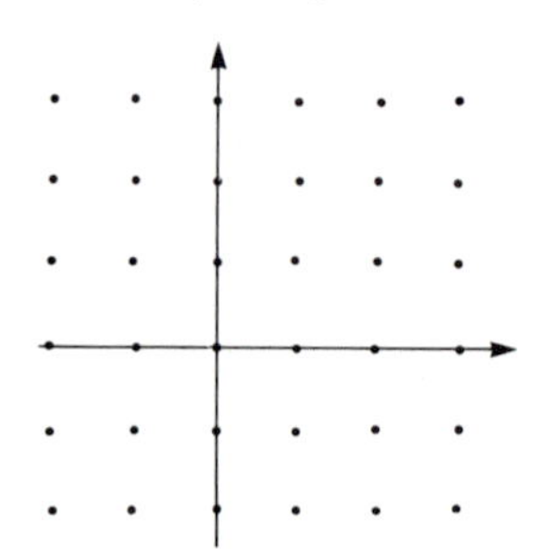

43 $|x| \ge 1$ and $|y| \le 2$

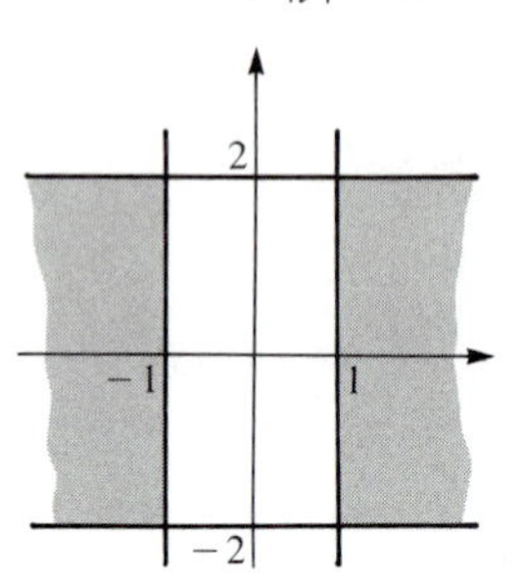

45 $xy > 0$ and $|x| \le 3$

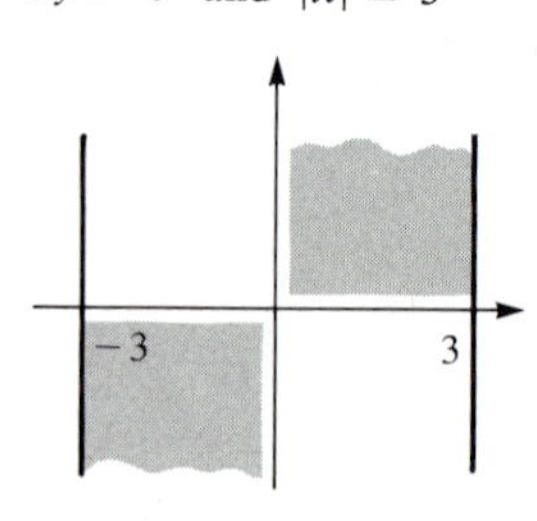

47 $y = x + 2$

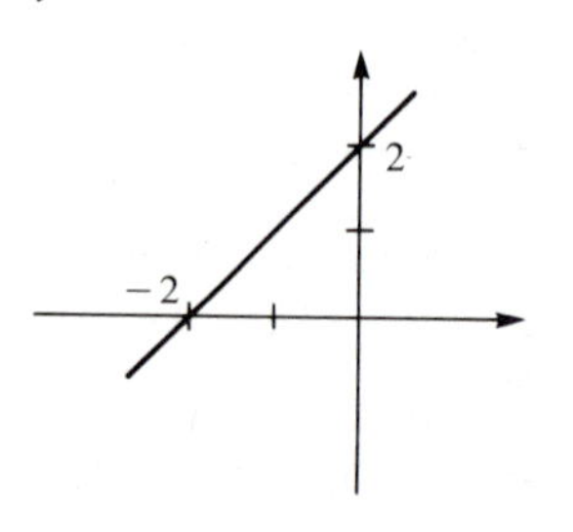

49 $y = -x$

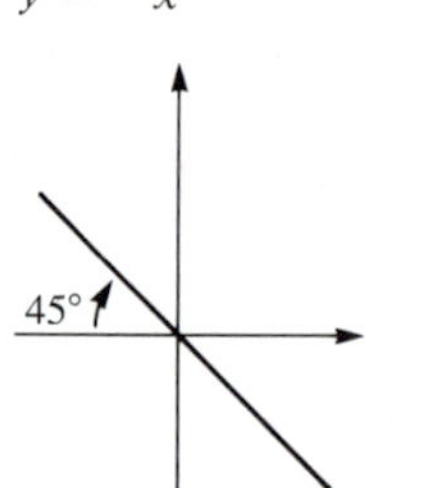

51 $y = -17$

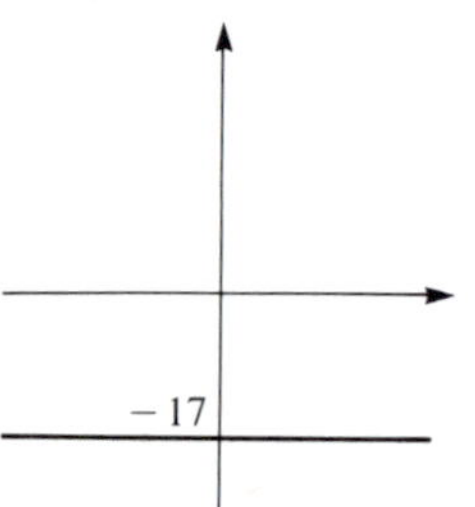

53 $y = x + 0.01$

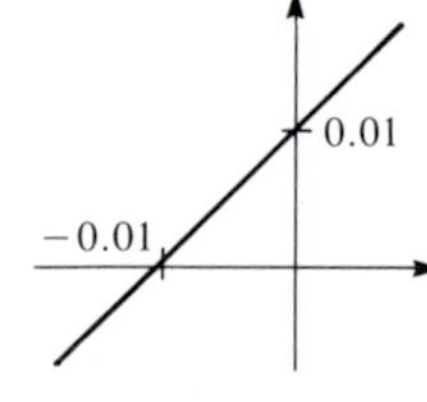

55 $y = |x|$

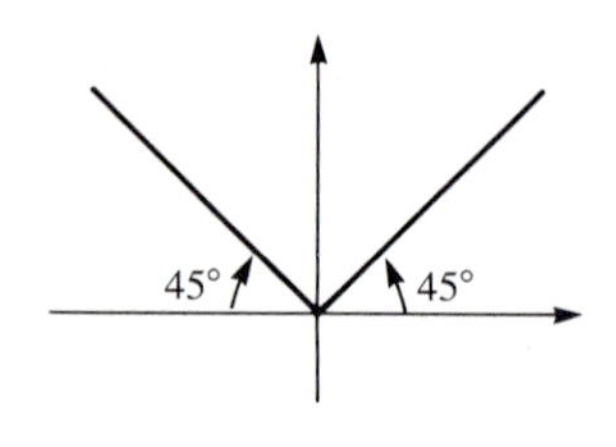

57 $y = 0$ for $x \le 0$
$y = 2x$ for $x > 0$

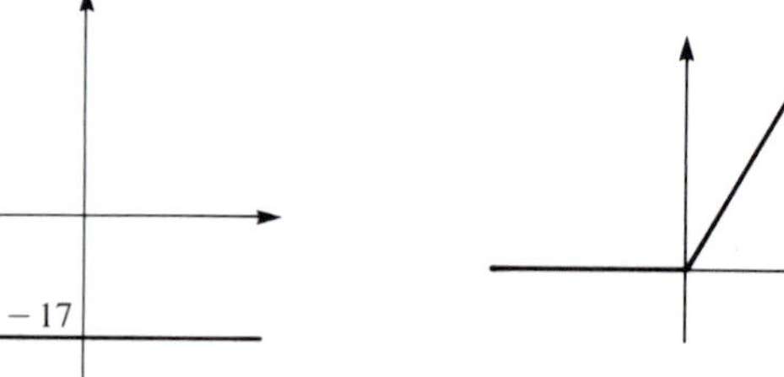

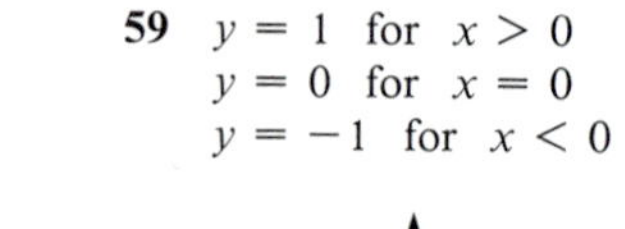

59 $y = 1$ for $x > 0$
$y = 0$ for $x = 0$
$y = -1$ for $x < 0$

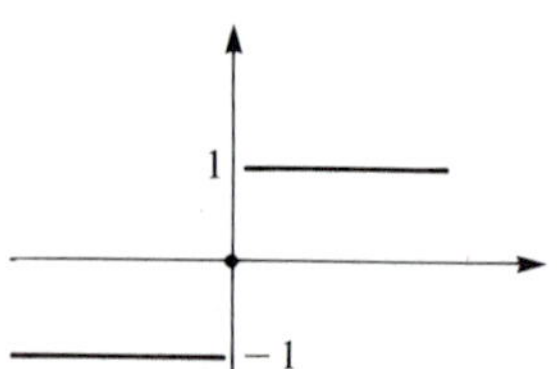

61 No. The domains $(-\infty, 1]$ and $[2, +\infty)$ have empty intersection.

63 $f \circ g = g$

65 $(f \circ f)(x) = x$

67 $f[\frac{1}{2}(x_0 + x_1)] = 3[\frac{1}{2}(x_0 + x_1)] - 5$
$= \frac{1}{2}(3x_0 - 5) + \frac{1}{2}(3x_1 - 5) = \frac{1}{2}[f(x_0) + f(x_1)]$

69 $f(x_0x_1) = 1/(x_0x_1)^2 = 1/(x_0^2x_1^2) = (1/x_0^2)(1/x_1^2)$
$= f(x_0)f(x_1)$

71 $f(x) = 1 - x$, $f[f(x)] = x$

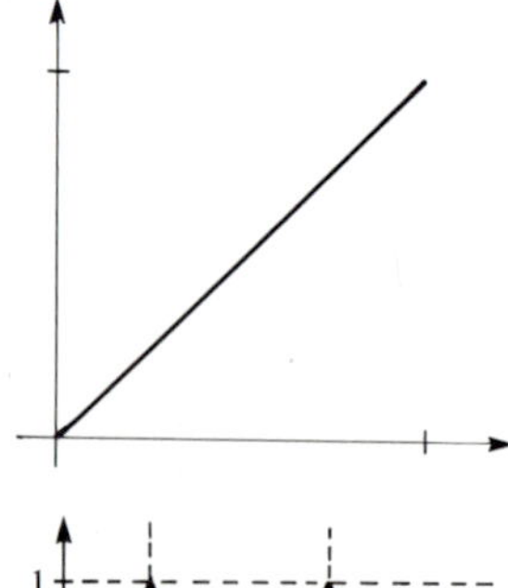

73

1

$\frac{1}{2}$ 1

Section 1-3 Linear and Quadratic Functions

Page 31

1 $y = 2x - 3,\ [0, 4]$

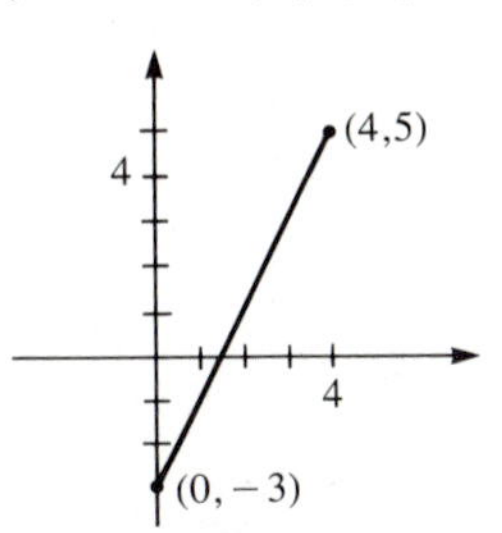

3 $y = 2x + 9,\ [1, 2]$

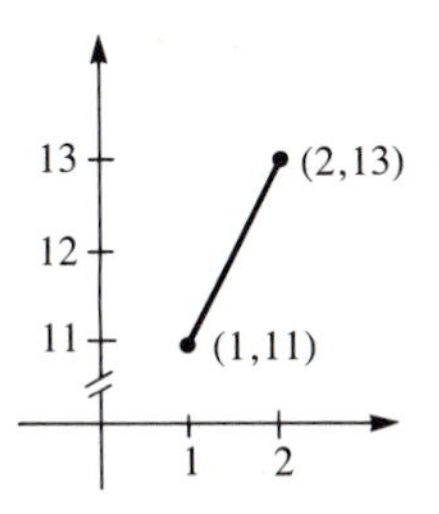

5 $y = -3x + 1,\ [-5, 5]$

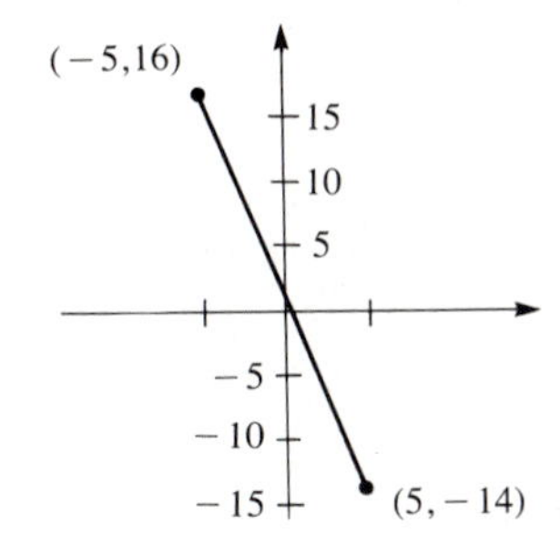

7 $y = 3x + 40,\ [25, 50]$

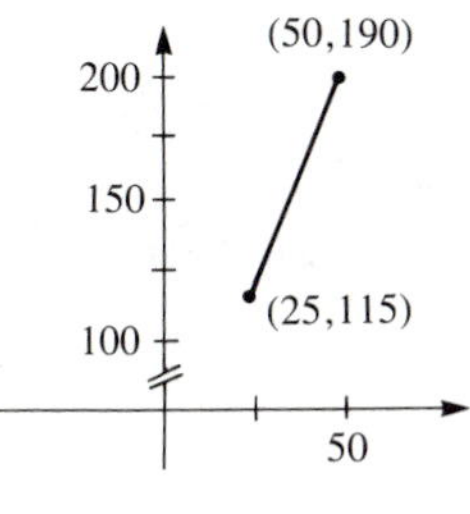

9 $y = 0.1x + 1.5,\ [2, 3]$

(different scales)

11 $x = 0.2t - 1,\ [0, 5]$

13 $x = 9t - 9,\ [1, 2]$

15 $x = -t + 10,\ [25, 50]$

17 $\frac{4}{3}$

19 0

21 1

23 $\frac{3}{2}$

25 1

27 $y = x + 1$

29 $y = 3$

31 $y = \frac{1}{2}x - 3$

33 $y = 2x$

35 $y = \frac{4}{3}x + \frac{4}{3}$

37 $y = x + \frac{1}{2}$

39 $y = -5.0x + 3.5$

41 $3,\ -7$

43 $-1,\ 7$

45 x: 2, y: 3

47 x: $\frac{1}{2}$, y: $\frac{1}{3}$

49 $y = -5x$

51 $y = x - 3$

53 $y = -3x + 17$

55 $y = -\frac{3}{5}x$

57 $y = 0.1x^2,\ [0, 100]$

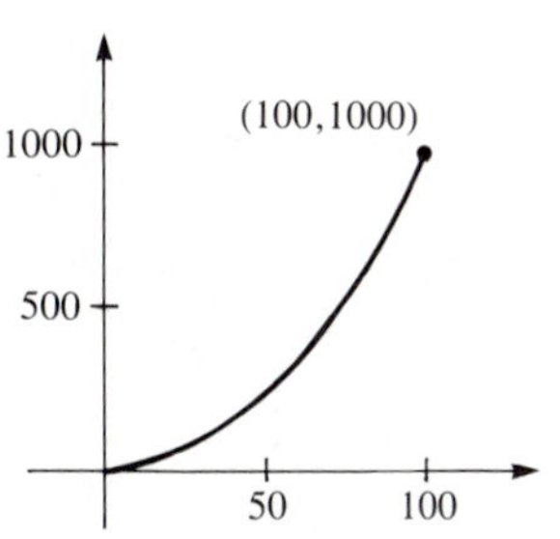

59 $y = -\frac{1}{2}x^2$

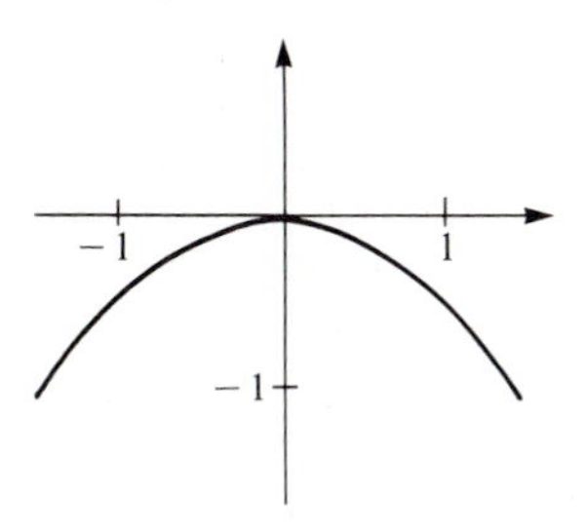

61 $y = x^2 + 3$

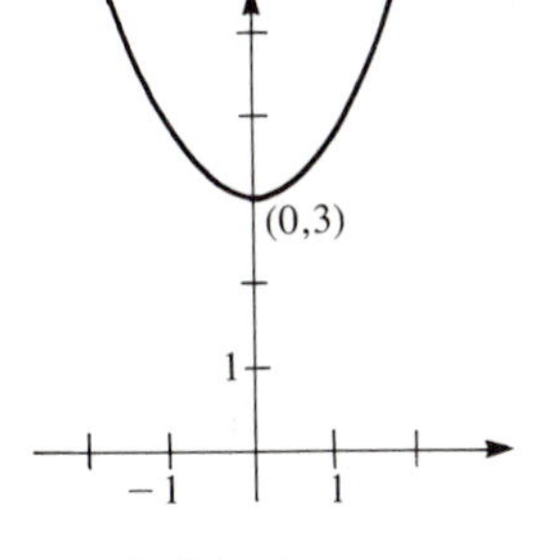

63 $y = 2x^2 - 1$

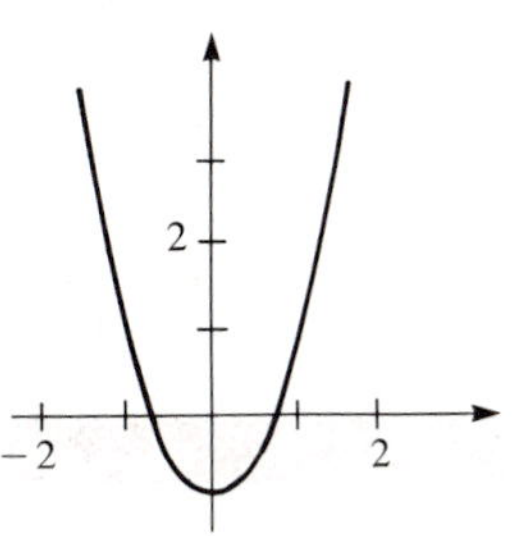

65 $y = -\frac{1}{4}x^2 + 2$

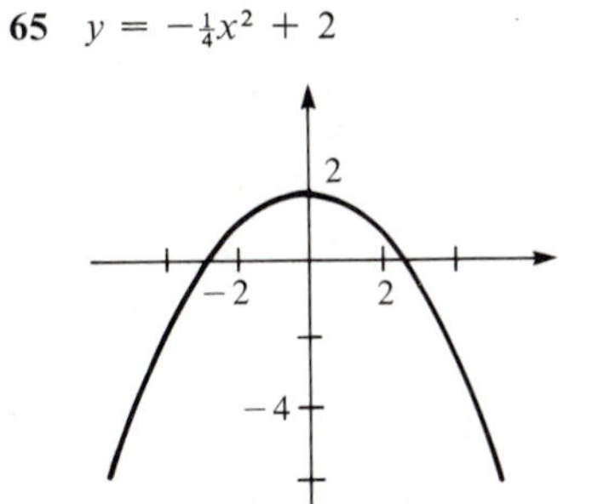

67 $y = x^2 - 4x + 1$
$= (x - 2)^2 - 3$

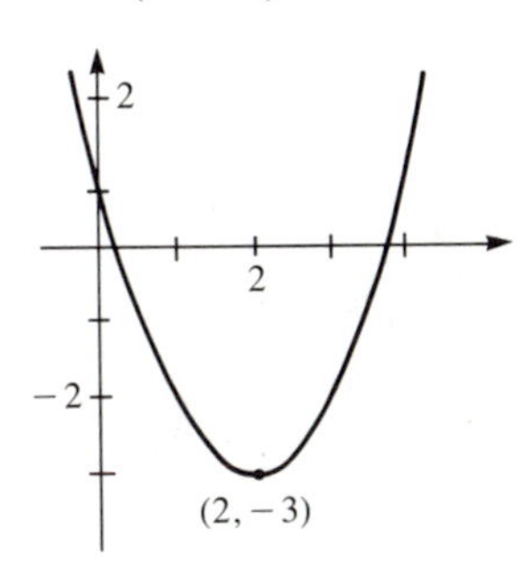

69 $y = x^2 - x + 1$
$= (x - \frac{1}{2})^2 + \frac{3}{4}$

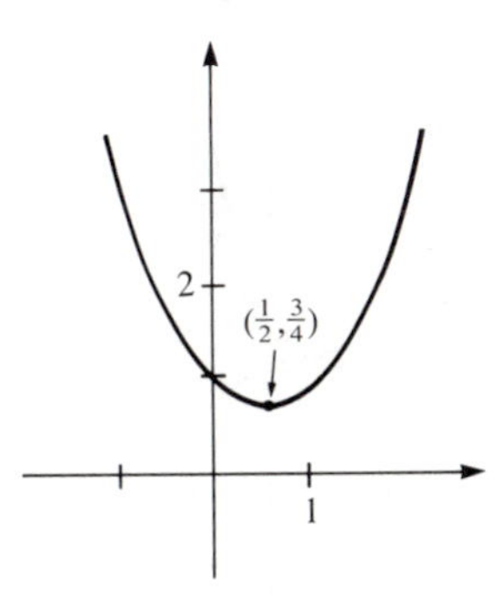

71 $y = -x^2 - 4x - 3$
$= -(x + 2)^2 + 1$

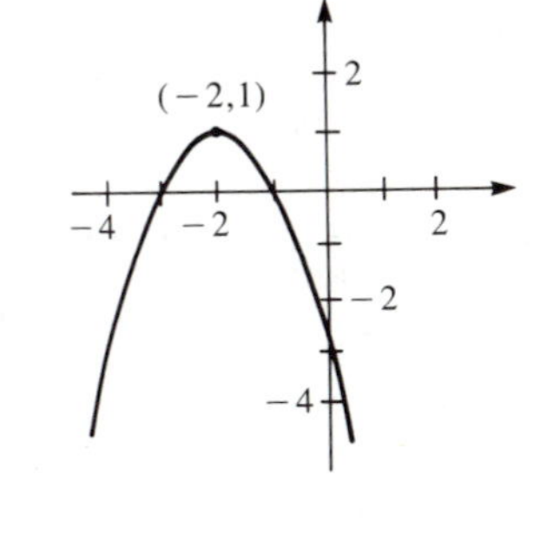

73 $y = 2x^2 + 4x$
$= 2(x + 1)^2 - 2$

(−1, −2)

75 $y = -2x^2 + 8x - 10$
$= -2(x-2)^2 - 2$

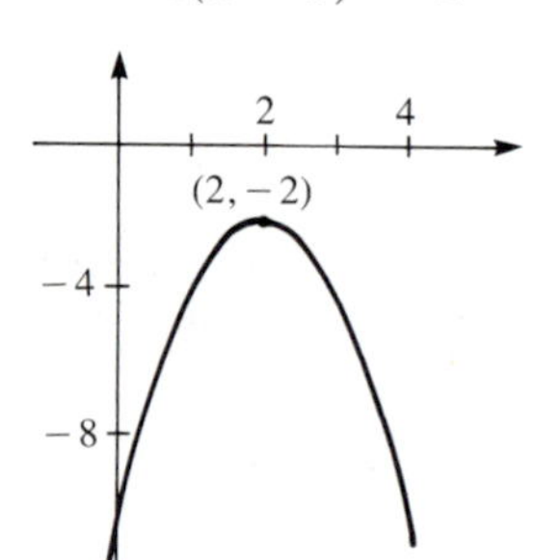

79 $y = x^2 + x - 4$
$= (x + \frac{1}{2})^2 - \frac{17}{4}$

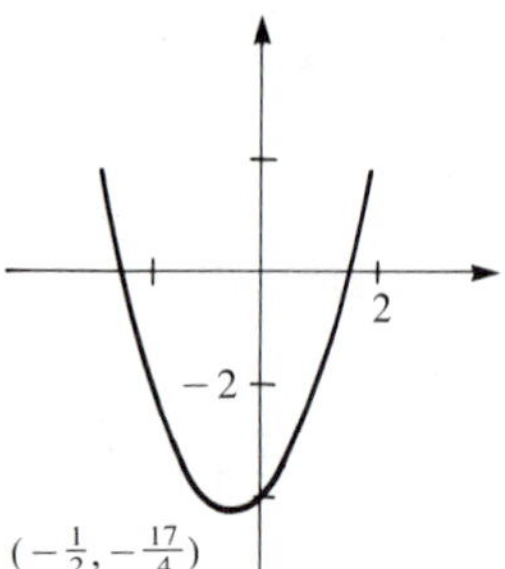

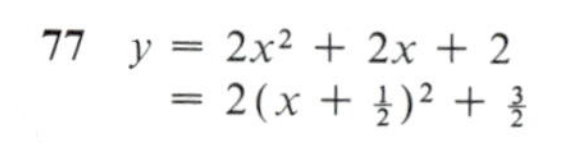

77 $y = 2x^2 + 2x + 2$
$= 2(x + \frac{1}{2})^2 + \frac{3}{2}$

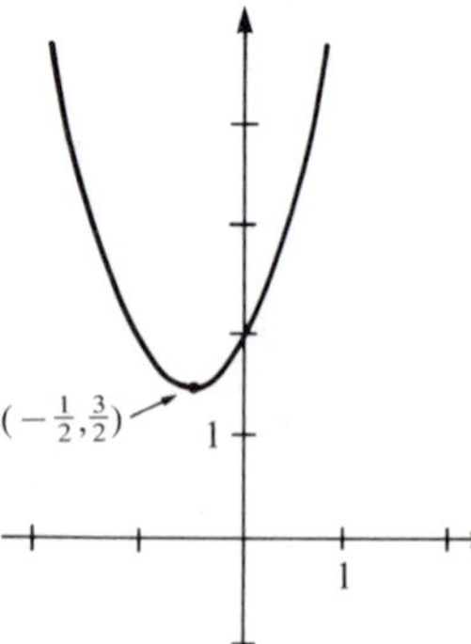

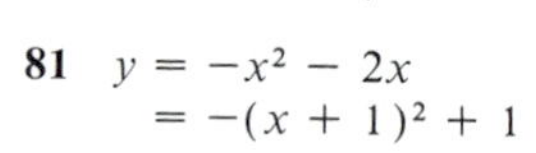

81 $y = -x^2 - 2x$
$= -(x+1)^2 + 1$

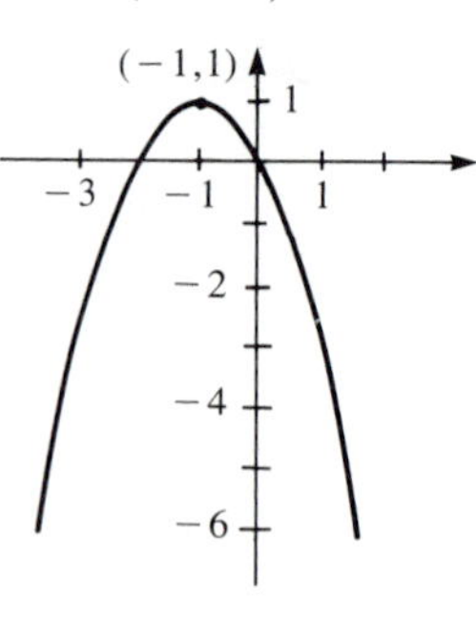

83 If $x = 0$, then
$y = a \cdot 0^2 + b \cdot 0 = 0.$

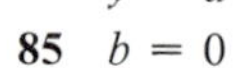

85 $b = 0$

87 $A^2 = \frac{1}{4}x^2(16 - x^2) = -\frac{1}{4}(x^4 - 16x^2)$
$= -\frac{1}{4}(x^2 - 8)^2 + 16$
$(A_{\max})^2 = 16$ at $x^2 = 8$, $A_{\max} = 4\text{ ft}^2$ at $x = \sqrt{8}$ ft

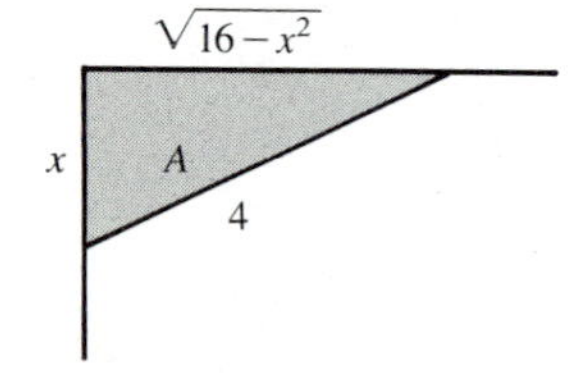

Section 1-4 Graphing Techniques

Page 43

1 x^2, x^4, $1/(x^2+1)$

3 $y = \frac{1}{4}x^4$

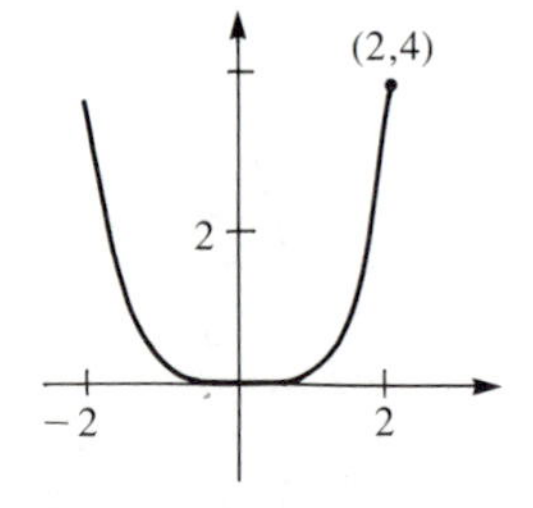

5 $y = \frac{1}{8}x^3$

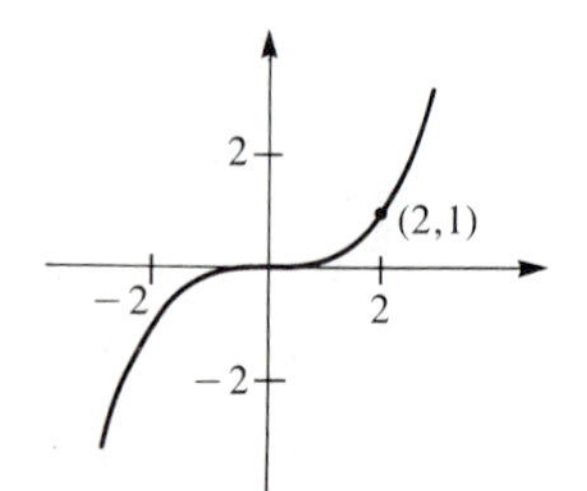

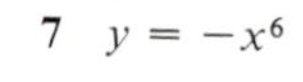

7 $y = -x^6$

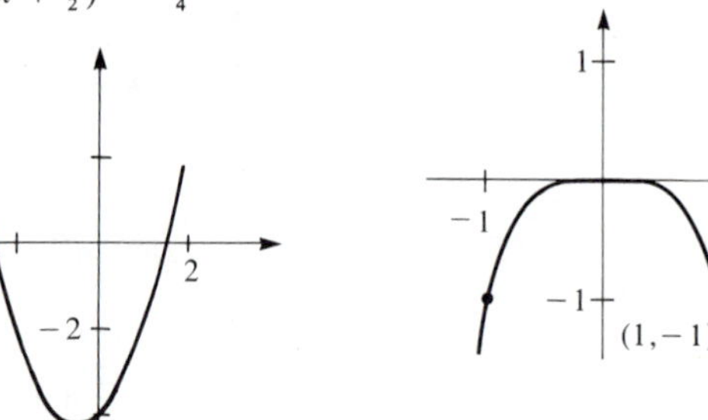

17 $y = (x-1)^3 + 1$

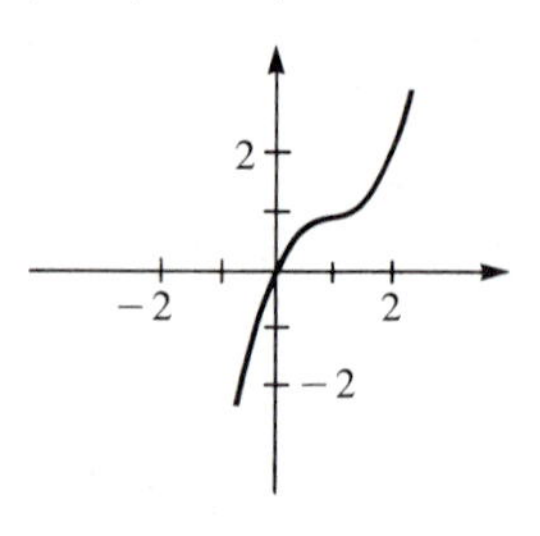

9 $y = x^4 - 1$

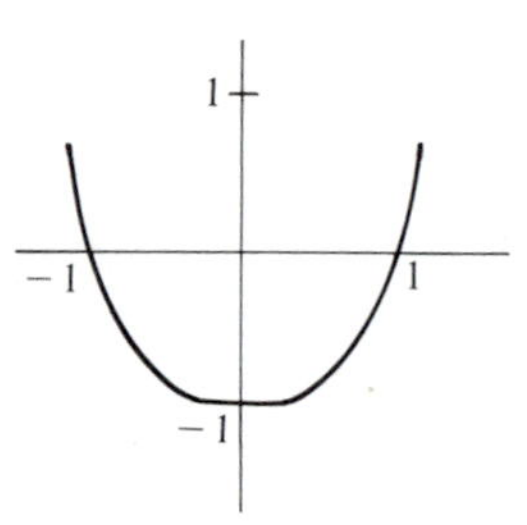

19 $y = \frac{1}{3}(x+2)^4$

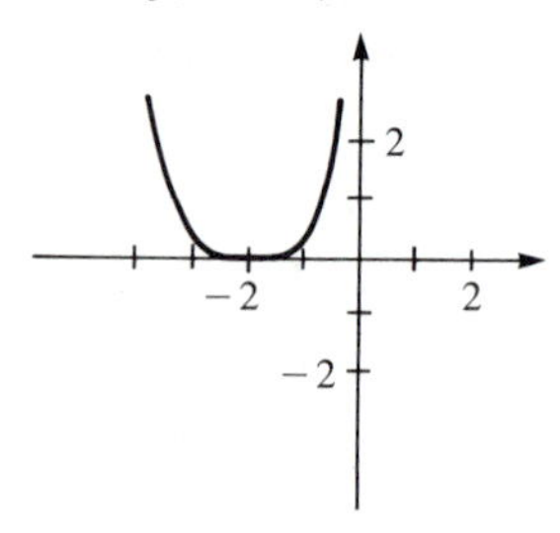

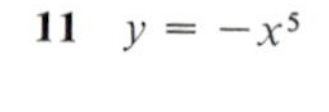

11 $y = -x^5$

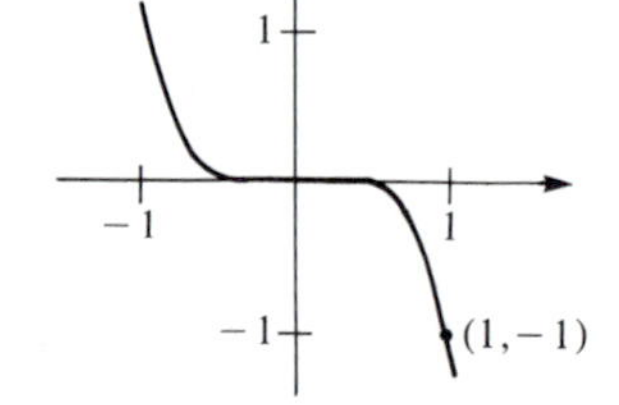

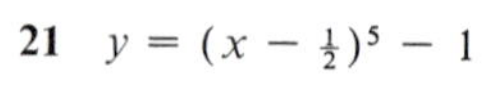

21 $y = (x - \frac{1}{2})^5 - 1$

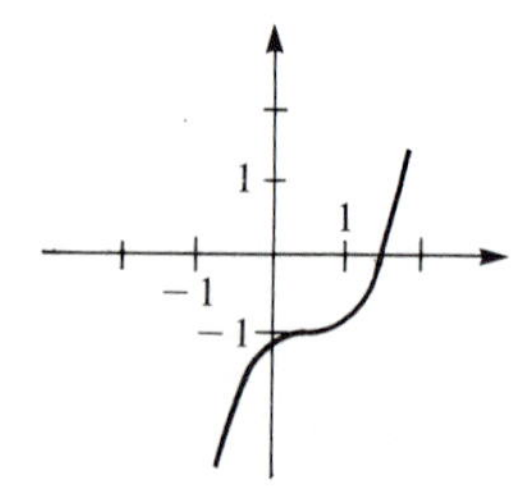

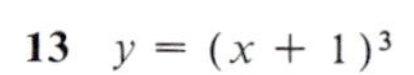

13 $y = (x+1)^3$

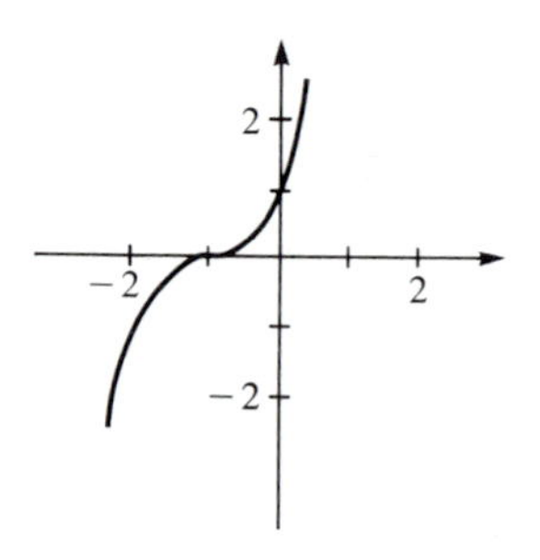

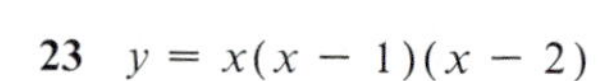

23 $y = x(x-1)(x-2)$

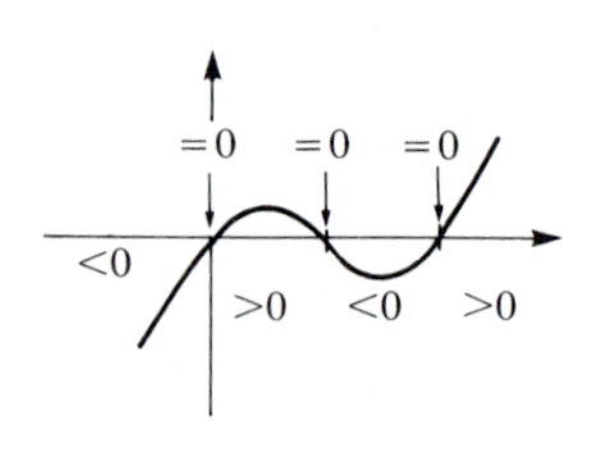

15 $y = -(x+1)^3$

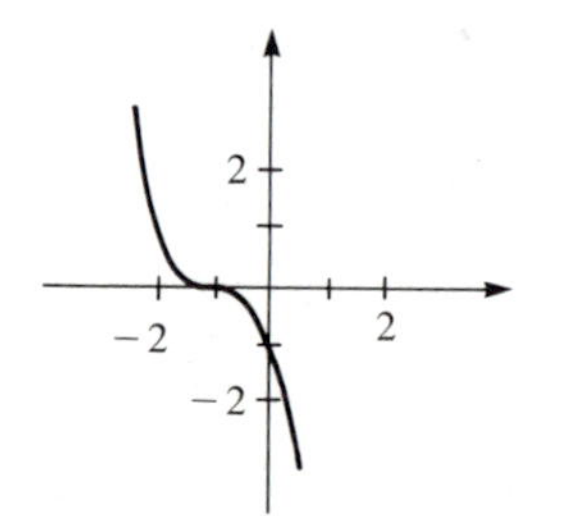

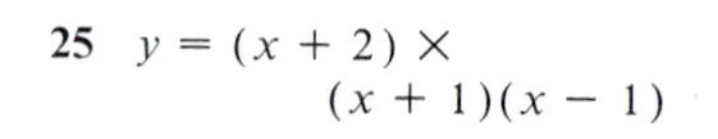

25 $y = (x+2) \times$
$(x+1)(x-1)$

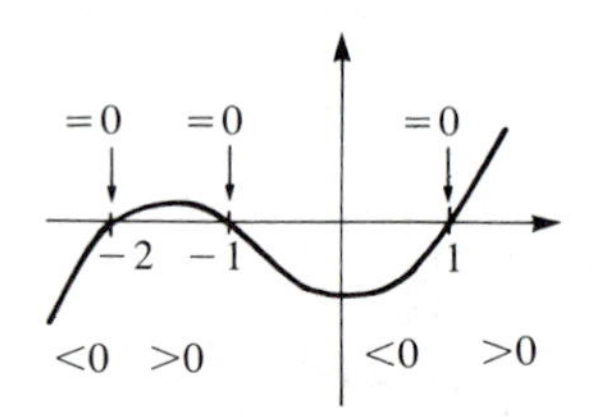

27 $y = x^2(x - 1)$

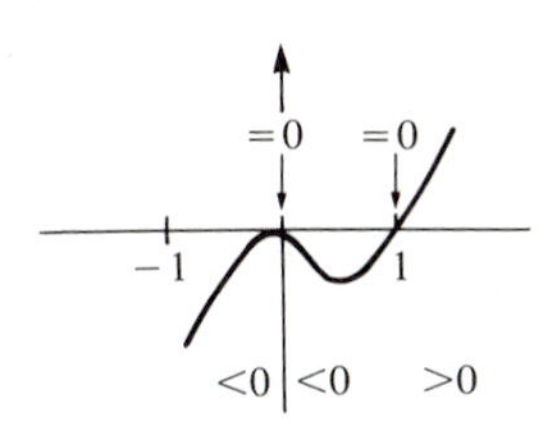

29 $y = -(x + 1)(x - 1)^2$

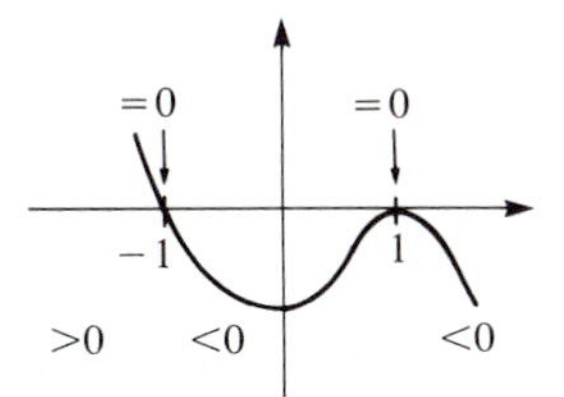

31 $y = \frac{1}{6}(x - 1)(x - 2) \times (x - 3)(x - 4)$

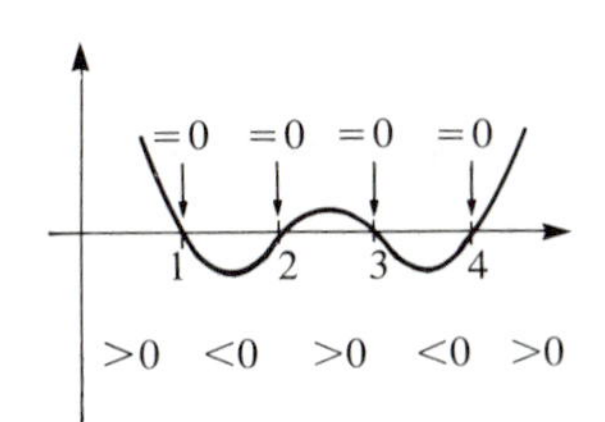

33 $y = x^2(x^2 - 1)$

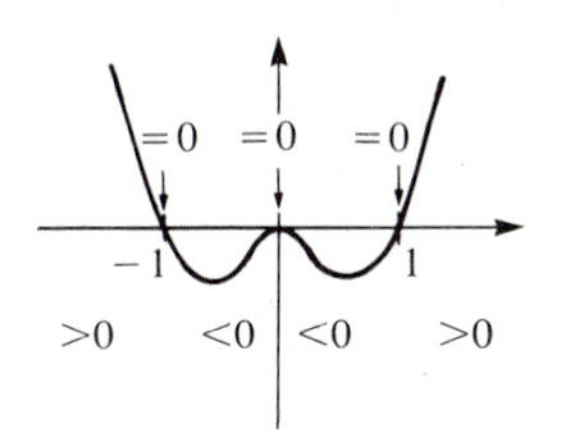

35 $y = -\frac{1}{4}(x + 1)^2(x^2 - 4)$

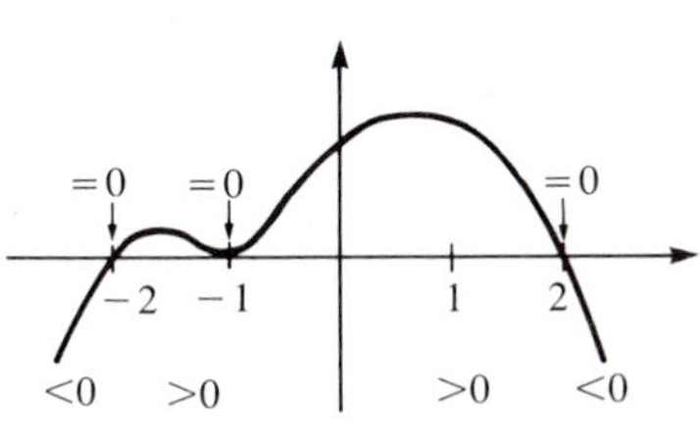

37 $y = x(x - 1)^3$

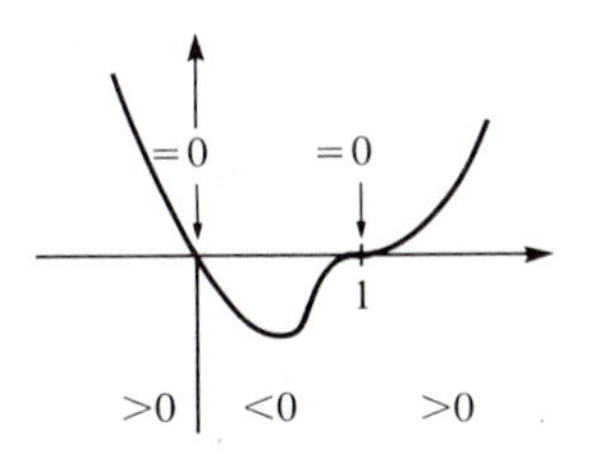

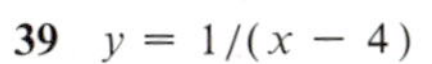
39 $y = 1/(x - 4)$

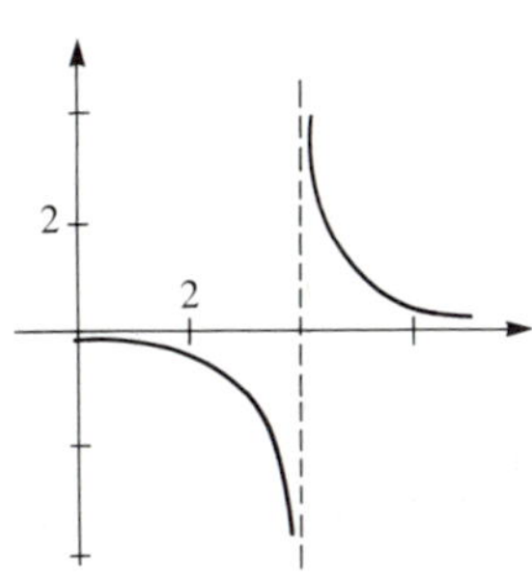

41 $y = 1/(x + 1)^2 - 2$

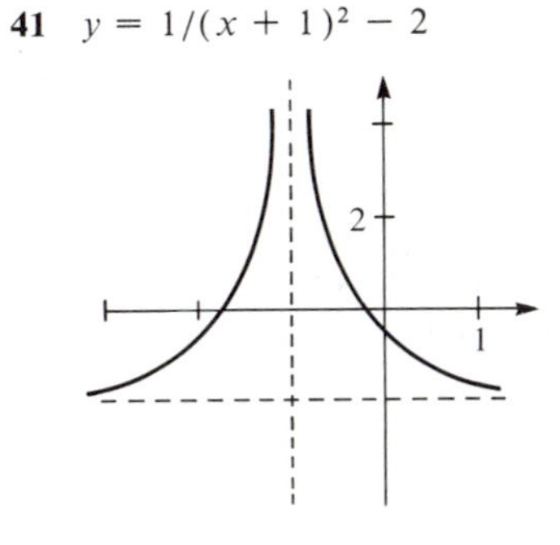

43 $y = 1/(1 + x^2)$

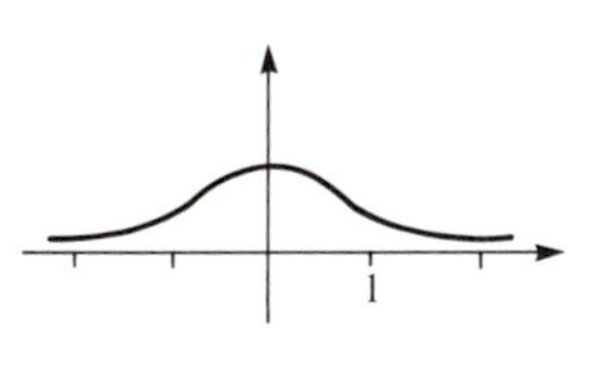

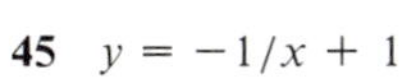
45 $y = -1/x + 1$

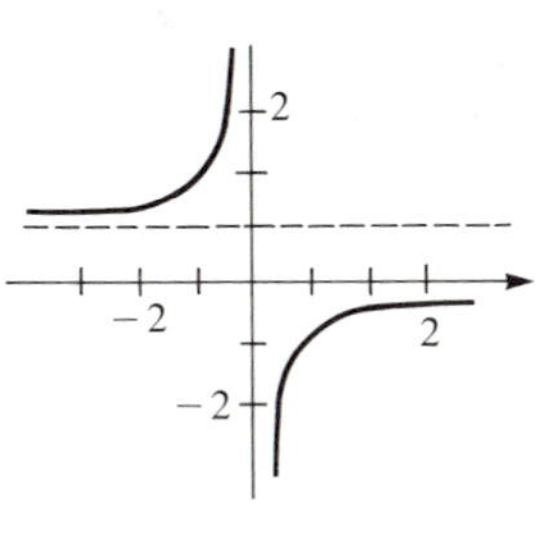

47 $y = x/(5x - 3)$

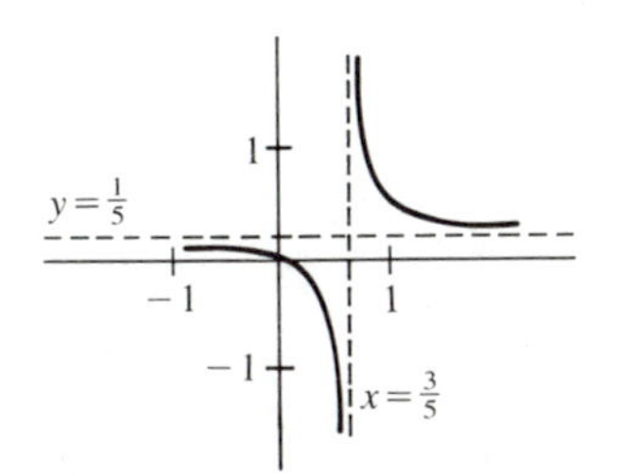

49 $y = -1/(4 + x^2)$

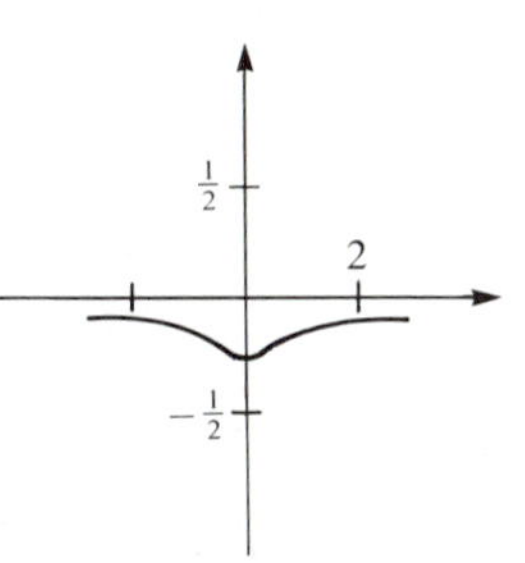

51 $y = x^2/(x + 1)$

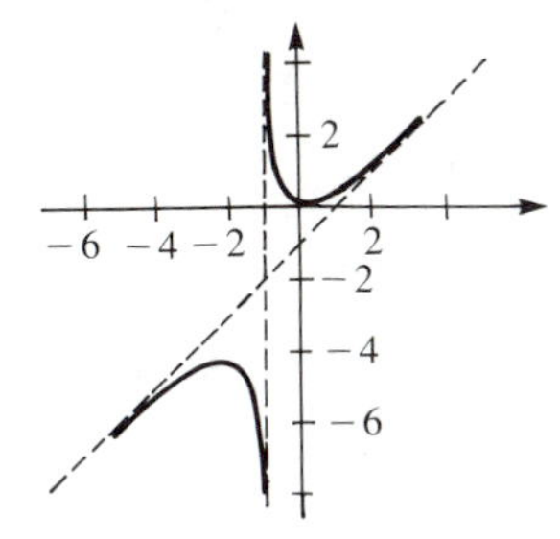

53 $r(x) \to 0-$

55 $y = (x + 1)(x - 1)/x^3$

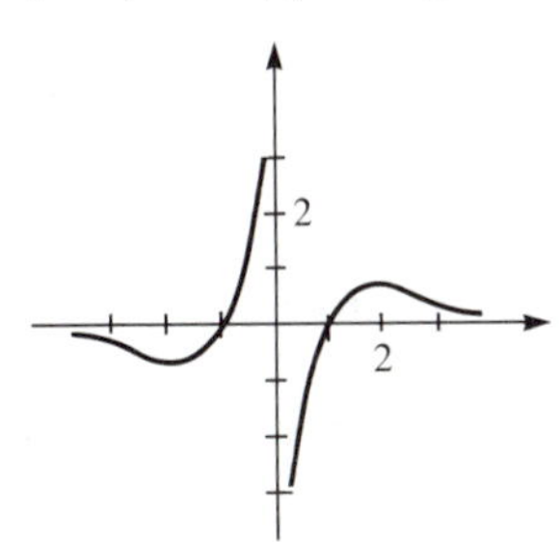

57 $y = (x + 2)^2/x^3$

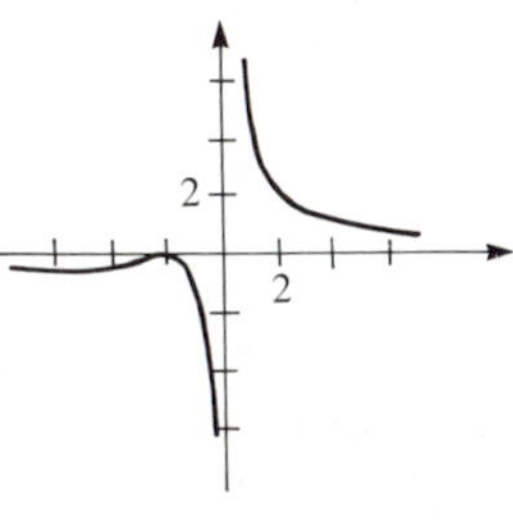

59 $y = -x^2/(x + 1)^2$

61

x	f	g	f/g
10	2.2439×10^2	3.0000×10^2	0.74797
10^2	2.9294×10^4	3.0000×10^4	0.97647
10^3	2.9930×10^6	3.0000×10^6	0.99766
10^6	3.0000×10^{12}	3.0000×10^{12}	1.0000
-10	3.6453×10^2	3.0000×10^2	1.2151
-10^2	3.0694×10^4	3.0000×10^4	1.0231
-10^3	3.0070×10^6	3.0000×10^6	1.0023
-10^6	3.0000×10^{12}	3.0000×10^{12}	1.0000

63

x	f	g	f/g
10	1.0310×10^{-1}	10^{-1}	1.0310
10^2	1.0004×10^{-2}	10^{-2}	1.0004
10^3	1.0000×10^{-3}	10^{-3}	1.0000
10^6	1.0000×10^{-6}	10^{-6}	1.0000
-10	-1.0499×10^{-1}	-10^{-1}	1.0499
-10^2	-1.0004×10^{-2}	-10^{-2}	1.0004
-10^3	-1.0000×10^{-3}	-10^{-3}	1.0000
-10^6	-1.0000×10^{-6}	-10^{-6}	1.0000

65

x	f	g	f/g
10	0.64513	0.66667	0.96770
10^2	0.66445	·	0.99668
10^3	0.66644	·	0.99967
10^6	0.66667	·	1.0000
-10	0.68962	·	1.0344
-10^2	0.66890	·	1.0033
-10^3	0.66689	·	1.0003
-10^6	0.66667	·	1.0000

Section 1-5 The Distance Formula

Page 49

1 $(x-1)^2+(y-3)^2=36$

3 $(x+4)^2+(y-3)^2=25$

5 $(x-1)^2+(y-5)^2=26$

7 $(x+5)^2+(y-2)^2=25$

9 $(x-\frac{3}{2})^2+(y-2)^2=\frac{13}{4}$

11 $(x-a)^2+(y\pm 3)^2=9$

13 $(x-a)^2+(y-b)^2=a^2+b^2>0$

15 No

17 To prove: $\sqrt{x^2+y^2}=2\sqrt{(x-3)^2+y^2}$.
Square and expand:
$x^2+y^2=4[(x-3)^2+y^2]$,
$3x^2+3y^2-24x+36=0$, $x^2+y^2-8x+12=0$,
$(x-4)^2+y^2=4$. Now read backwards.

19 Outside

21 Inside

23 $(1,2)$

25 $(4,0)$

27 $y=\frac{3}{5}x+\frac{17}{5}$

29 $y=\frac{3}{7}x-\frac{3}{7}$

31 Focus $(\frac{1}{4},0)$, directrix $x=-\frac{1}{4}$

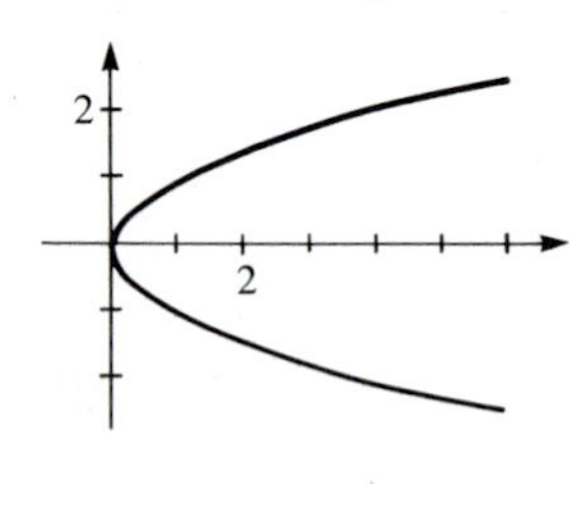

33 Focus $(2,-\frac{7}{4})$, directrix $y=-\frac{9}{4}$

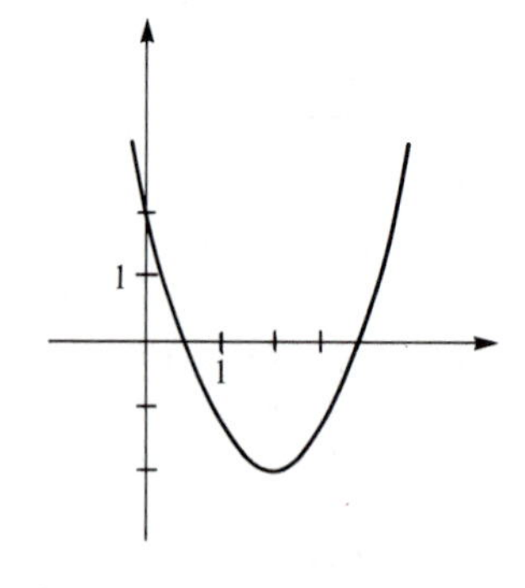

35 Parabola $y=\frac{1}{4}x^2+1$, focus $(2,0)$, directrix x-axis

37 Their midpoint, $(\frac{1}{2}(5x+3x+4y),\frac{1}{2}(5y+4x-3y))$ $=(4x+2y,2x+y)$, lies on the line $2y=x$, and the slope of the segment joining them

$$m=\frac{(4x-3y)-5y}{(3x+4y)-5x}=\frac{4x-8y}{-2x+4y}=-2$$

is the negative reciprocal of the slope of the line $2y=x$. Hence segment and line are perpendicular.

Section 1-6 Sine and Cosine Functions

Page 55

1 $\frac{1}{3}\pi$ $\frac{5}{6}\pi$ $-\frac{4}{3}\pi$
$\frac{13}{6}\pi$ $-\frac{5}{2}\pi$ 5π

3 $\frac{1}{4}\pi$ $-\pi$ $\frac{3}{2}\pi$
$\frac{7}{2}\pi$ $-\frac{3}{4}\pi$ $\frac{11}{4}\pi$

5 $90°$ $-120°$ $300°$
$540°$ $-480°$ $3000°$

7 $45°$ $24°$ $75°$
$5°$ $-165°$ $-12°$

9 $(\frac{1}{2}\sqrt{2},\frac{1}{2}\sqrt{2})$
$(-\frac{1}{2}\sqrt{2},-\frac{1}{2}\sqrt{2})$
$(-\frac{1}{2}\sqrt{2},\frac{1}{2}\sqrt{2})$

11 $(-1,0)$ $(-\frac{1}{2},-\frac{1}{2}\sqrt{3})$
$(\frac{1}{2}\sqrt{3},\frac{1}{2})$

13 $\frac{1}{4}\pi$ $\frac{3}{4}\pi$

15 $\frac{5}{6}\pi$ $\frac{7}{6}\pi$

17 $\frac{3}{2}\pi$

19 $\frac{1}{4}\pi$ $\frac{5}{4}\pi$ because the unit circle meets $y=x$ in two points only.

21 $p=1$

23 $p=\pi$

25 $p=\pi$

27 Replace θ by $-\theta$ in $\cos(\theta+\pi)=-\cos\theta$ and $\sin(\theta+\pi)=-\sin\theta$, and use parity.

29 $\sin(\alpha+\frac{1}{4}\pi)=\frac{1}{2}\sqrt{2}(\sin\alpha+\cos\alpha)$
$\cos(\alpha+\frac{1}{4}\pi)=\frac{1}{2}\sqrt{2}(\cos\alpha-\sin\alpha)$

31 $\frac{1}{2}\cos\theta+\frac{1}{2}\sqrt{3}\sin\theta$

33 $3\sin\theta\cos^2\theta-\sin^3\theta=3\sin\theta-4\sin^3\theta$

35 $4\sin\theta\cos\theta(\cos^2\theta-\sin^2\theta)$

37 $\cos(\alpha+\beta)+\cos(\alpha-\beta)$
$=(\cos\alpha\cos\beta-\sin\alpha\sin\beta)+(\cos\alpha\cos\beta+\sin\alpha\sin\beta)$
$=2\cos\alpha\cos\beta$

39 Use Exercise 37 seven times:

$$\begin{aligned}(\cos x+\cos 3x)&+(\cos 5x+\cos 7x)+\cdots+(\cos 13x+\cos 15x)\\ &=2\cos x\cos 2x+2\cos x\cos 6x+2\cos x\cos 10x+2\cos x\cos 14x\\ &=2\cos x[(\cos 2x+\cos 6x)+(\cos 10x+\cos 14x)]\\ &=2\cos x[2\cos 2x\cos 4x+2\cos 2x\cos 12x]\\ &=4\cos x\cos 2x(\cos 4x+\cos 12x)\\ &=4\cos x\cos 2x(2\cos 4x\cos 8x)\end{aligned}$$

41 $\cos 2x=1-2\sin^2 x$, so $\cos\theta=1-2\sin^2\frac{1}{2}\theta$,
$\sin^2\frac{1}{2}\theta=\frac{1}{2}(1-\cos\theta)$, $\sin\frac{1}{2}\theta=\pm\sqrt{\frac{1}{2}(1-\cos\theta)}$
with $+$ if $0\le\theta\le 2\pi$ and $-$ if $-2\pi\le\theta<0$.

43 $y=\cos 3\theta$

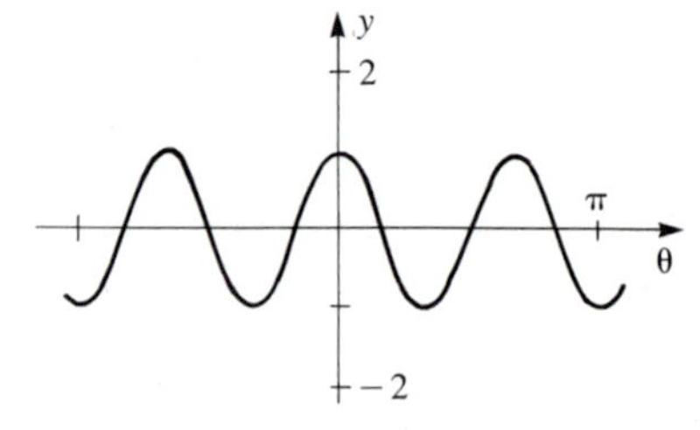

45 $y=2-\sin\theta$

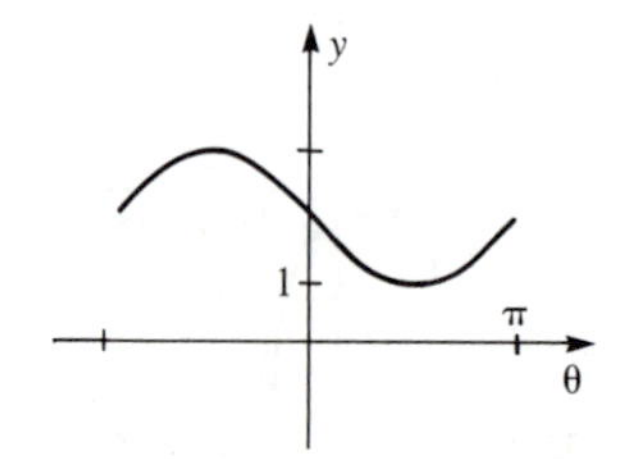

47 $y = \cos 2\pi t$

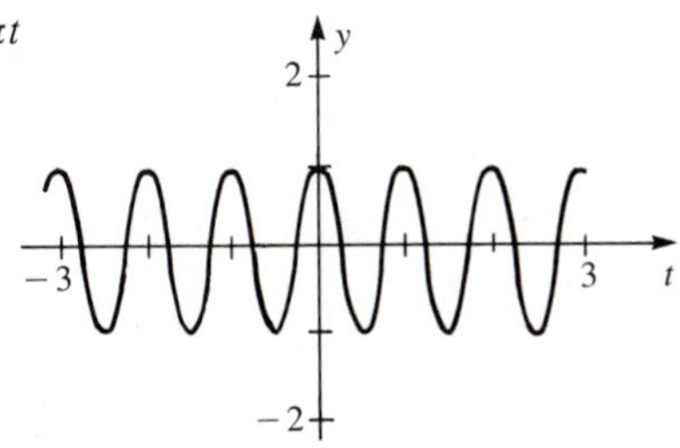

49 0.6678
51 −0.6250
53 16.40°
55 −13.42°
57 0.36390
59 0.20944
61 0.95409

Section 1-7 Additional Trigonometric Functions

Page 61

1 $\sin^2 x = 1/\csc^2 x = 1/(1 + \cot^2 x)$, so $\sin x = \pm\sqrt{1/(1 + \cot^2 x)}$ with $+$ in quadrants 1 and 2, $-$ in 3 and 4.

3 $\cot^2 x = \cos^2 x/\sin^2 x = \cos^2 x/(1 - \cos^2 x)$

5
$$\cot(\alpha - \beta) = \frac{1}{\tan(\alpha - \beta)} = \frac{1 - \tan\alpha\tan(-\beta)}{\tan\alpha + \tan(-\beta)} = \frac{1 + \tan\alpha\tan\beta}{\tan\alpha - \tan\beta} = \frac{\cot\alpha\cot\beta + 1}{\cot\beta - \cot\alpha}$$

(Last step: divide numerator and denominator by $\tan\alpha\tan\beta$.)

7
$$\frac{\sin 2\theta}{1 + \cos 2\theta} = \frac{2\sin\theta\cos\theta}{1 + (2\cos^2\theta - 1)} = \frac{\sin\theta}{\cos\theta} = \tan\theta$$

9 $p = \frac{1}{2}\pi$

11 $y = \cot x$

13 $y = \tan^2 x$

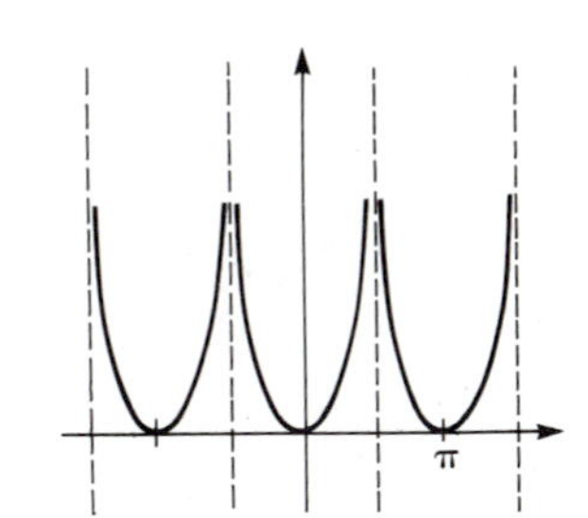

15 $y = \tan(x - \frac{1}{4}\pi)$

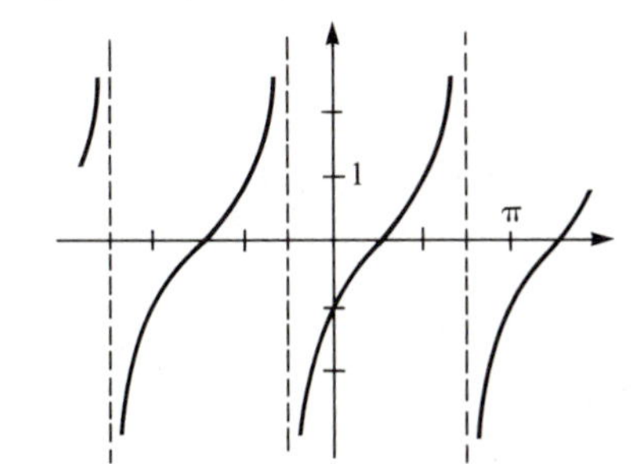

17 $\cos\theta = \cos 2(\frac{1}{2}\theta) = 1 - 2\sin^2\frac{1}{2}\theta$, $\sin^2\frac{1}{2}\theta = \frac{1}{2}(1 - \cos\theta)$, etc.

19
$$\begin{aligned}\sin\theta/(1 + \cos\theta) &= (2\sin\tfrac{1}{2}\theta\cos\tfrac{1}{2}\theta)/[1 + (2\cos^2\tfrac{1}{2}\theta - 1)] \\ &= (2\sin\tfrac{1}{2}\theta\cos\tfrac{1}{2}\theta)/(2\cos^2\tfrac{1}{2}\theta) \\ &= (\sin\tfrac{1}{2}\theta)/(\cos\tfrac{1}{2}\theta) = \tan\tfrac{1}{2}\theta\end{aligned}$$

21 $\alpha = x + y$ and $\beta = x - y$, where $x = \frac{1}{2}(\alpha + \beta)$ and $y = \frac{1}{2}(\alpha = \beta)$. Thus
$$\begin{aligned}\sin\alpha + \sin\beta &= \sin(x + y) + \sin(x - y) \\ &= (\sin x\cos y + \cos x\sin y) + (\sin x\cos y - \cos x\sin y) \\ &= 2\sin x\cos y.\end{aligned}$$

23 $\alpha \approx 55.04°$ $\beta \approx 83.09°$ $\gamma \approx 41.87°$

25 $c \approx 3.540$ $\alpha \approx 1.6420$ $\beta \approx 0.6966$

27 $\alpha \approx 1.0886$ $b \approx 4.511$ $c \approx 1.927$

Section 1-8 Exponents and Exponential Functions

Page 65

1 $y = 5^x$

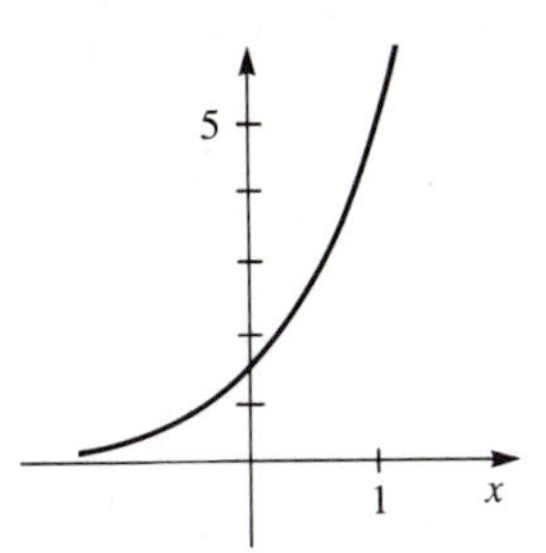

3 $y = (0.8)^x$

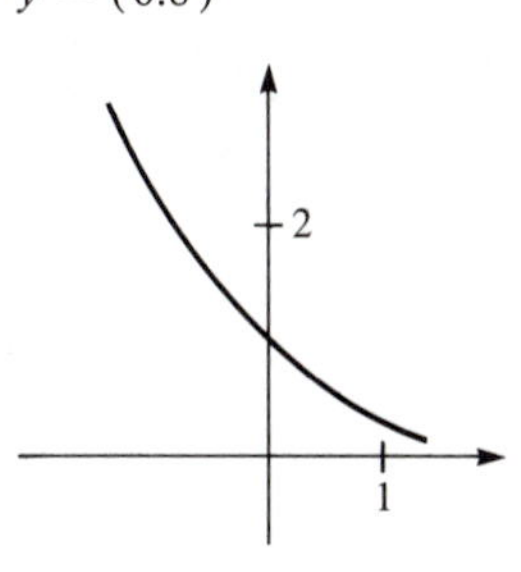

5 $y = 2^{x+1}$

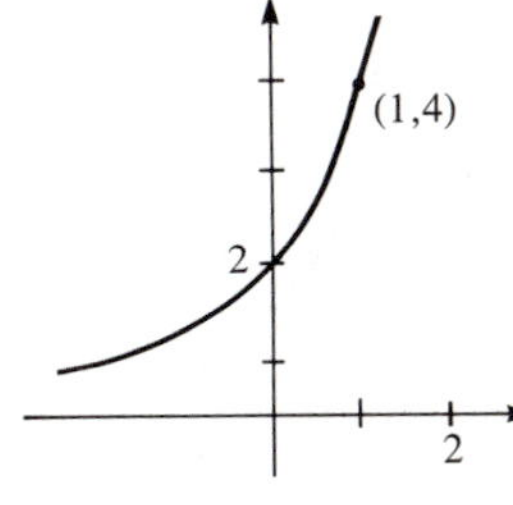

7 $y = 3^{|x|}$

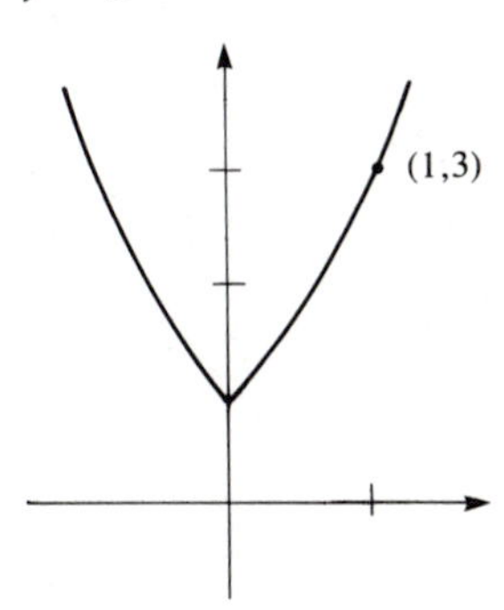

9 $y = 1 - 2^x$

11 $b/a > 1$, $(b/a)^x > 1$, $b^x/a^x > 1$, $b^x > a^x$

13 $y = 2^x/x$, $(0, +\infty)$

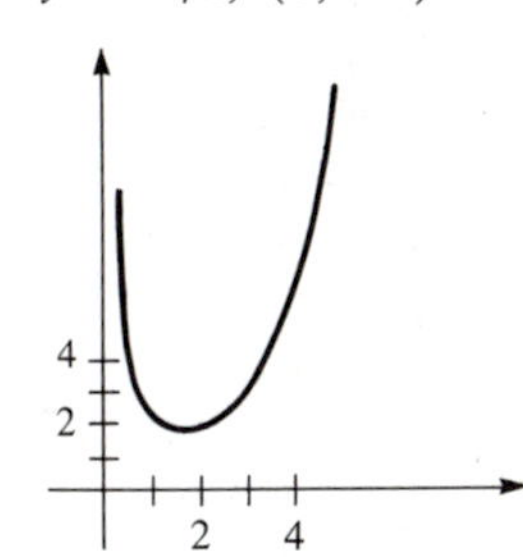

15 $y = 3^x - 100 \cdot 2^x$

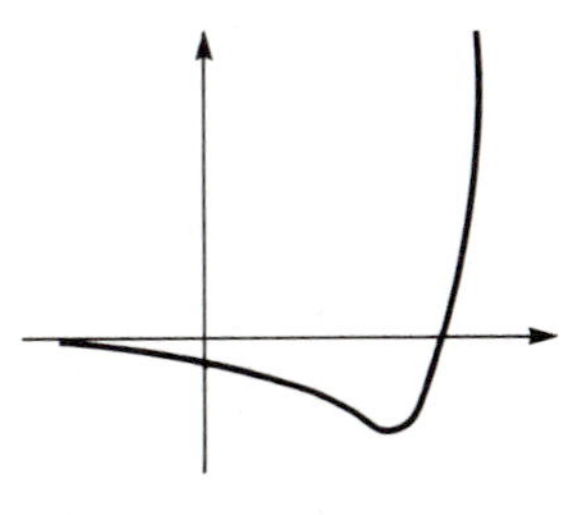

Section 1-9 Review Exercises

Page 66

1 **a** $1, \ -5, \ 9x + 4$

b $f(a+b)$
$= 3(a+b) + 1$
$= 3a + 3b + 1$
$= (3a + 1)$
$\quad + (3b + 1) - 1$
$= f(a) + f(b) - 1$

3 $y = 2x^2 - 12x + 14$

(3, −4)

5 $a = -1, \ b = 6,$
$f(x) = -x + 6$

7

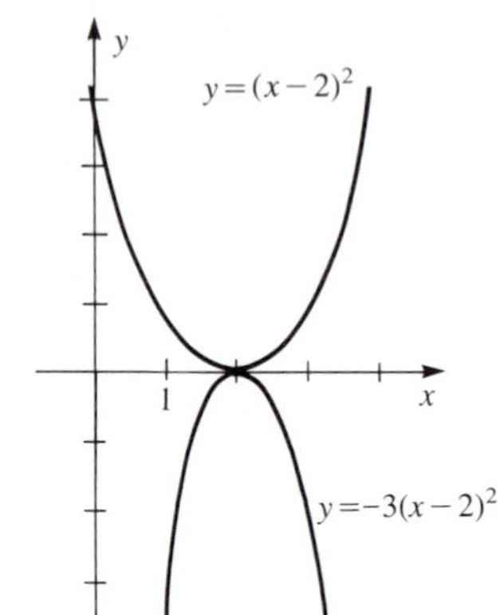

9 $r(x) = \dfrac{3x^2 + 1}{x(x-4)}$

11 $y = x^3 - 3x$

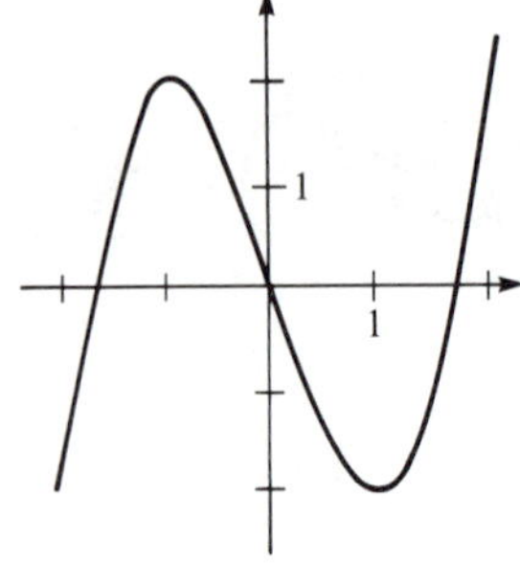

13 $y = x^3 - 2x^2$

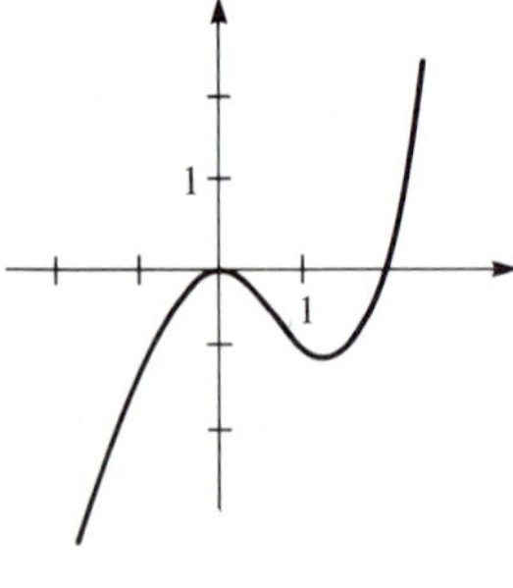

15 $y = x + 1/x$

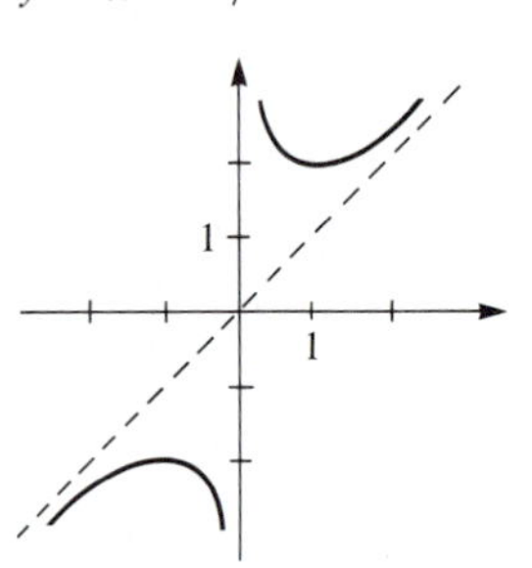

17 $y = x(x-1)/(x^2 - 4)$

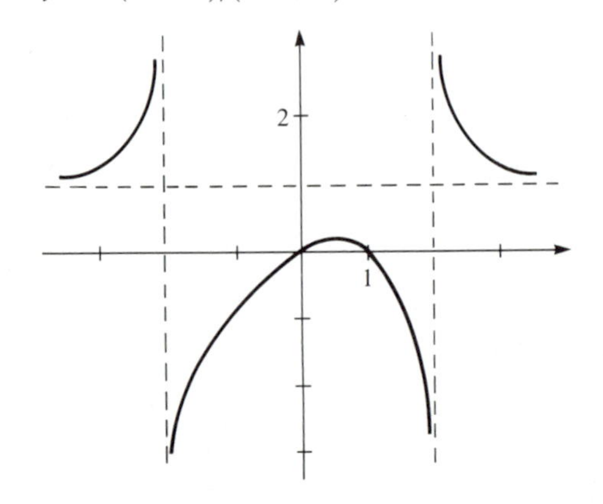

19 $\dfrac{1 - \tan^2 x}{1 + \tan^2 x} = \dfrac{\cos^2 x - \sin^2 x}{\cos^2 x + \sin^2 x} = \dfrac{\cos 2x}{1}$

Alternative: $\dfrac{1 - \tan^2 x}{1 + \tan^2 x} = \dfrac{1 - \tan^2 x}{\sec^2 x}$
$= (\cos^2 x)(1 - \tan^2 x) = \cos^2 x - \sin^2 x = \cos 2x$

21 Set $x = \frac{1}{2}(\alpha + \beta)$ and $y = \frac{1}{2}(\alpha - \beta)$ so $x + y = \alpha$ and $x - y = \beta$. Then subtract the relations

$$\begin{cases} \sin\alpha = \sin(x+y) = \sin x \cos y + \sin y \cos x \\ \sin\beta = \sin(x-y) = \sin x \cos y - \sin y \cos x \end{cases}$$

23 $y = \cos(x - \frac{1}{4}\pi)$

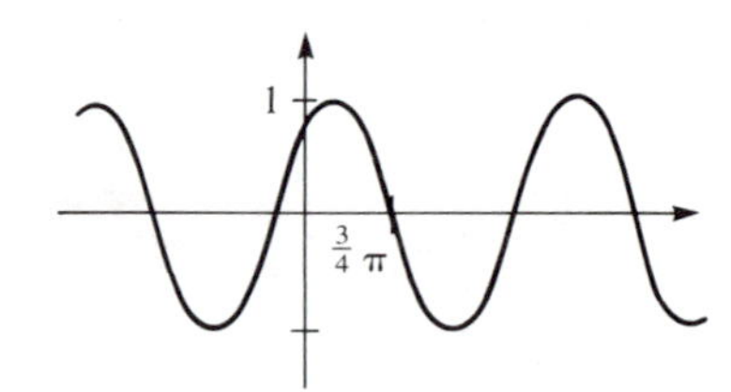

25 $y = \sin(2\pi x^2)$

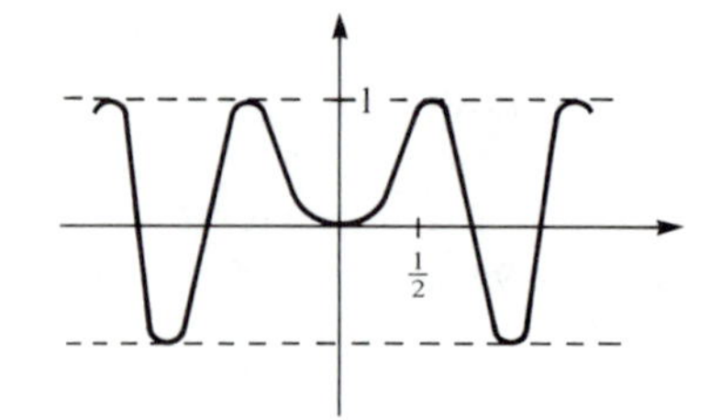

27 $y = 2^{-1.5x}$

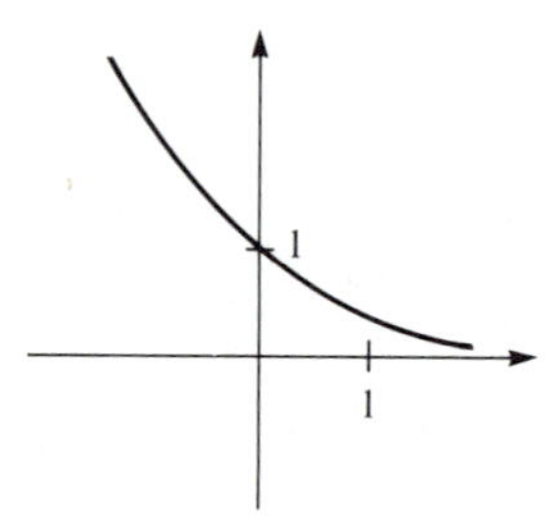

29 $y = 3^{1/(1+x^2)}$

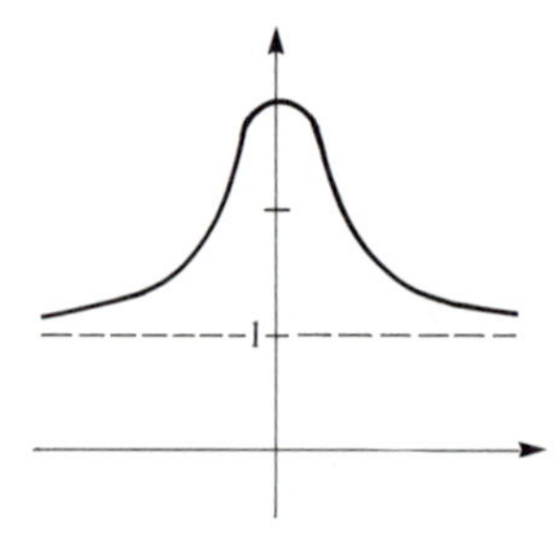

Chapter 2 The Derivative

Section 2-1 The Slope Problem

Page 70

	0.01	−0.01	0.001	−0.001	0.0001	−0.0001	slope
1	12.03	11.97	12.003	11.997	12.000	12.000	12
	$P = (2, 12)$						
3	1.01	0.99	1.001	0.999	1.0001	0.9999	1
	$P = (1, 0)$						
5	7.01	6.99	7.001	6.999	7.0001	6.9999	7
	$P = (2, 10)$						
7	55.180	54.820	55.018	54.982	55.002	54.998	55
	$P = (3, 57)$						
9	32.900	33.100	32.990	33.010	32.999	33.001	33
	$P = (-3, -36)$						
11	−2.0299	−1.9699	−2.0030	−1.9970	−2.0003	−1.9997	−2
	$P = (-1, 8)$						

13 $[3(a+h)^2 - 3a^2]/h = 6a + 3h \approx 6a$

15 $[(a+h)^2 + 3(a+h) - (a^2 + 3a)]/h$
$= (2a + 3) + h \approx 2a + 3$

17 $[(a+h)^3 - (a+h)^2 - (a^3 - a^2)]/h$
$= (3a^2 - 2a) + (3a - 1)h + h^2 \approx 3a^2 - 2a$

19 $[1/(0.5 + h) - 1/0.5]/h = -2/(0.5 + h)$

0.01	−0.01	0.00001	−0.00001	slope
−3.9216	−4.0816	−3.9999	−4.0001	−4

21

0.01	−0.01	0.001	−0.001	slope
0.99998	0.99998	1.00000	1.00000	1

Section 2-2 Limits of Functions

Page 79

1 $|4x - (-4)| = 4|x + 1| < \varepsilon$ provided $|x - (-1)| < \delta = \frac{1}{4}\varepsilon$. $L = -4$.

3 $|(3x - 1) - 5| = 3|x - 2| < \varepsilon$ provided $|x - 2| < \delta = \frac{1}{3}\varepsilon$. $L = 5$.

5 $|3x^2 - 12| = 3|x + 2| \cdot |x - 2| < \varepsilon$ provided $|x - 2| < \delta$, the smaller of 1 and $\frac{1}{15}\varepsilon$. $L = 12$.

7 $|(x^2 + 2x) - 0| = |x + 2| \cdot |x| < \varepsilon$ provided $|x| < \delta$, the smaller of 1 and $\frac{1}{3}\varepsilon$. $L = 0$.

9 $|1/x - 1| = |x - 1|/x < \varepsilon$ provided $|x - 1| < \delta$, the smaller of $\frac{1}{2}$ and $\frac{1}{2}\varepsilon$. $L = 1$.

11 $|-1/(2x) - (-\frac{1}{4})| = [1/(4x)]|x - 2| < \varepsilon$ provided $|x - 2| < \delta$, the smaller of 1 and 4ε. $L = -\frac{1}{4}$.

13 $|1/(1 + x) - \frac{1}{4}| = [1/4(1 + x)]|x - 3| < \varepsilon$ provided $|x - 3| < \delta$, the smaller of 3 and 4ε. $L = \frac{1}{4}$.

15 $x^2 - x - 2 = (x + 1)(x - 2)$, hence $f(x) = (x + 1)/(x^2 - x - 2) = 1/(x - 2)$ for $x \neq -1$. $|1/(x - 2) - (-\frac{1}{3})| = [1/3(2 - x)]|x - (-1)| < \varepsilon$ provided $|x - (-1)| < \delta$, the smaller of 1 and 6ε. $L = -\frac{1}{3}$.

17 $|(x + 2)/(x + 1) - 2| = |1/(x + 1)| \cdot |x| < \varepsilon$ provided $|x| < \delta$, the smaller of $\frac{1}{2}$ and $\frac{1}{2}\varepsilon$. $L = 2$.

19 $(x^2 + 5x + 6)/(x^2 + 3x + 2) = (x + 3)/(x + 1)$ for $x \neq -2$. $|(x + 3)/(x + 1) - (-1)| = |2/(x + 1)| \cdot |x - (-2)| < \varepsilon$ provided $|x - (-2)| < \delta$, the smaller of $\frac{1}{2}$ and $\frac{1}{4}\varepsilon$. $L = -1$.

	−0.01	−0.0001	0.0001	0.01	limit
21	1.01010	1.00010	0.99990	0.99010	1
23	0.98998	0.99990	1.00010	1.00998	1
25	2.73200	2.71842	2.71815	2.70481	2.718 · · ·

Section 2-3 Rules for Limits

Page 84

1 $\lim_{x\to a} x^3 = (\lim_{x\to a} x)^3 = a^3$

3 $\lim_{x\to 0} 1/(1 + x) = 1/\lim_{x\to 0}(1 + x)$
$= 1/(1 + \lim_{x\to 0} x) = 1/(1 + 0) = 1$

(Some steps will be omitted in the remaining solutions.)

5 $\lim_{x\to 0} 1/(x - 1)^3 = 1/[\lim_{x\to 0}(x - 1)]^3 = 1/(-1)^3 = -1$

7 $\lim_{t\to 2}(t^2 - 1) = \lim t^2 - 1 = 4 - 1 = 3$

9 $\lim_{h\to 1}(1 - 2h)/(1 + h) = [\lim(1 - 2h)]/[\lim(1 + h)]$
$= (1 - 2)/(1 + 1) = -\frac{1}{2}$

11 $\lim_{y\to 0} y/[(1 + 2y)^2 - 1] = \lim y/(4y + 4y^2)$
$= \lim 1/(4 + 4y) = 1/\lim(4 + 4y) = 1/(4 + 0) = \frac{1}{4}$

13 $\lim_{z\to 1}(2 - 1)/(2^4 - 1)$
$= \lim(z - 1)/[(z - 1)(z^3 + z^2 + z + 1)]$
$= \lim 1/(z^3 + z^2 + z + 1) = \frac{1}{4}$

15 $\sqrt{x + 3}/(x + 1) \to \sqrt{-2 + 3}/(-2 + 1) = -1$ as $x \to -2$

17 $[1 + (1 + x)^{-1}]^{-1} \to [1 + (1 + 1)^{-1}]^{-1}$
$= (1 + \frac{1}{2})^{-1} = (\frac{3}{2})^{-1} = \frac{2}{3}$ as $x \to 1$

19 0 **21** 1 **23** 0 **25** 0

Section 2-4 More on Limits [Optional]

Page 90

1 $x \to 2$ so $1/x \to 1/2 = \frac{1}{2}$ by rule 4.

3 $x^2 + 1 \to (-2)^2 + 1 = 5$ by rules 3 and 2; $1/(x^2 + 1) \to 1/5 = \frac{1}{5}$ by rule 4.

5 $x^2/x^2 = 1$ for $x \neq 0$, so $x^2/x^2 \to 1$.

7 $x^2/x = x$ for $x \neq 0$, so $x^2/x \to 0$.

9 $1 + x^3 \to 1 + 0 = 1$ by rules 3 and 2; $\sqrt{1 + x^3} \to \sqrt{1} = 1$ by rule 7.

11 $3 + h + h^2 \to 3$ and $5 - 4h + h^3 \to 5$ by rules 3 and 2. By rule 4, quotient $\to 3/5 = \frac{3}{5}$.

13 $\lim_{x\to 0+} f(x) = \lim_{x\to 0+} 1 = 1$ and $\lim_{x\to 0-} f(x) = \lim_{x\to 0-} 0 = 0$. Suppose $\lim_{x\to 0} f(x) = L$. Choose $\varepsilon = \frac{1}{2}$. Then there exists a $\delta > 0$ such that $|f(x) - L| < \frac{1}{2}$ for all x such that $0 < |x| < \delta$. By choosing $0 < x < \delta$ we have $|1 - L| < \frac{1}{2}$, and by choosing $-\delta < x < 0$ we have $|L| < \frac{1}{2}$. By the triangle inequality,

$$1 = |1| = |(1 - L) + L| \leq |1 - L| + |L| < \tfrac{1}{2} + \tfrac{1}{2} = 1,$$

which is impossible. Therefore the assumption $\lim_{x\to 0} f(x) = L$ is false.

15 Set $L = \lim_{x\to a+} f(x) = \lim_{x\to a-} f(x)$. Let $\varepsilon > 0$. Then there exist $\delta_1 > 0$ and $\delta_2 > 0$ such that if $a < x < a + \delta_1$, then $|f(x) - L| < \varepsilon$ and if $a - \delta_2 < x < a$, then $|f(x) - L| < \varepsilon$. Let δ be the smaller of δ_1 and δ_2. If $0 < |x - a| < \delta$, then either $a - \delta_2 \le a - \delta < x < a$ or $a < x < a + \delta \le a + \delta_1$, so $|f(x) - L| < \varepsilon$. Therefore $f(x) \to L$ as $x \to a$.

17 If $x \ne 0$, then $|\sin(1/x)| \le 1$; hence $|x \sin(1/x)| \le |x| \to 0$.

19 If $|h| < 1$, then $|3 + h| > 2$, so $|h/(3 + h)| < \frac{1}{2}|h|$. We want $\frac{1}{2}\delta \le 10^{-5}$, so we choose $\delta = 2 \times 10^{-5}$.

21 $\lim_{x\to a} [f(x) - 10] = [\lim f(x)] - 10 > 0$. By rule 5, $f(x) - 10 > 0$ for $0 < |x - a| < \delta$; hence $f(x) > 10$.

Section 2-5 Continuous Functions

Page 98

1 $g(x) = x^2$ is the square of the identity function, hence continuous by Theorem 1, part 3. Also, $h(x) = 4x$ is a constant function times the identity function, hence continuous. Finally, $k(x) = 6$ is a constant function, hence continuous. Now $f(x) = g(x) - h(x) + k(x)$ is continuous by Theorem 1, part 2.

3 $g(x) = x =$ (identity function) is continuous, and $h(x) = x^2 + 3 =$ (identity function)^2 + (constant function) is continuous by Theorem 1, parts 3 and 2. $f(x) = g(x)/h(x)$ is continuous by Theorem 1, part 4. Note that $g(x) \ge 3 > 0$.

5 $g(x) = 1 + x^2 =$ (constant function) + (identity function)^2 is continuous by Theorem 1, parts 3 and 2, and $g(x) \ge 1 > 0$. $f(x) = \sqrt{g(x)}$ is continuous by Theorem 4.

7 $|x| = \sqrt{x^2} = \sqrt{f(x)}$ where $f(x) = x^2$ is a polynomial, hence continuous by Theorem 2. By Theorem 4, $\sqrt{f}$ is continuous.

9 Choose $\delta > 0$ so that $|f(x) - f(a)| < f(a)$ provided $|x - a| < \delta$. Then $0 = f(a) - f(a) < f(x) < f(a) + f(a)$; in particular, $f(x) > 0$ for $|x - a| < \delta$. [By choosing $\varepsilon = \frac{1}{2}f(a)$ we even assure that $f(x)$ is bounded away from 0 for $|x - a| < \delta$.]

11 $f(x) = \frac{1}{2}(-x + \sqrt{x^2 + 12})$. The function $g(x) = x^2 + 12$ is a polynomial, hence continuous. Also $g(x) > 0$, so $\sqrt{g(x)}$ is continuous on **R**. It follows by Theorem 1, parts 2 and 1, that $f(x)$ is continuous on **R**.

13 If $x > 0$, then $\sqrt[3]{x^2} + \sqrt[3]{a}\sqrt[3]{x} + \sqrt[3]{a^2} > \sqrt[3]{a^2}$,

$$|\sqrt[3]{x} - \sqrt[3]{a}| = |x - a|/|\sqrt[3]{x^2} + \cdots| < |x - a|/\sqrt[3]{a^2}$$

Given $\varepsilon > 0$, let δ be the smaller of a and $a^{2/3}\varepsilon$. If $|x - a| < \delta$, then $x > 0$ and

$$|\sqrt[3]{x} - \sqrt[3]{a}| < (a^{2/3}\varepsilon)/a^{2/3} = \varepsilon$$

15

h	0.005	0.001	0.0005	0.0001
$(3 + h)^2$	9.03002	9.00600	9.00300	9.00060
$(3 - h)^2$	8.97002	8.99400	8.99700	8.99940

17

h	0.005	0.001	0.0005	0.0001
$f(4 + h)$	2.00125	2.00025	2.00012	2.00003
$f(4 - h)$	1.99875	1.99975	1.99987	1.99998

19

h	0.005	0.001	0.0005	0.0001
$f(\frac{1}{6}\pi + h)$	0.50432	0.50087	0.50043	0.50009
$f(\frac{1}{6}\pi - h)$	0.49566	0.49913	0.49957	0.49991

The evidence indicates $|\sin(\frac{1}{6}\pi + h) - \frac{1}{2}| < 0.9|h|$. [$\sin(\frac{1}{6}\pi + h) = (\sin\frac{1}{6}\pi)\cos h + (\cos\frac{1}{6}\pi)\sin h$ $= \frac{1}{2}\cos h + \frac{1}{2}\sqrt{3}\sin h \approx \frac{1}{2}(1 - \frac{1}{2}h^2) + \frac{1}{2}\sqrt{3}(h - \frac{1}{6}h^3)$ $\approx \frac{1}{2} + \frac{1}{2}\sqrt{3}h$ and $\frac{1}{2}\sqrt{3} \approx 0.866$.]

Section 2-6 Derivative of a Function

Page 105

1 $2x$

3 -4

5 1

7 3

9 8

11 $3y^2$

13 0

15 $3x^2$, 0, 27, 27

17 108

19 48, $3a^2$

21 -12, 24, 2

23 -1, -1, $-1/a^2$, $-1/a^2$

25 $-1/a^2$, $-a^2$

27 $-1/b^2$

29 6, 14, 22

31 13

33 6

35 -32

37 $(3, 9)$

39 $(2, 8)$, $(-2, -8)$

41 $(0, 0)$

43 $0 < x < \frac{2}{3}$

45 $\lim_{x\to 1}(x^n - 1)/(x - 1) = d(x^n)/dx|_1 = n$, $(1 + x + \cdots + x^{n-1})|_1 = n$.

Section 2-7 Sums, Products, and Quotients

Page 114

1 $2x + 3$

3 $-4x^3 + 4x$

5 $1 - 1/x^2$

7 0

9 $\frac{27}{2}$

11 $3x^2 + 2x + 1$

13 $9x^2 - 4x - 17$

15 $8x^7 + 6x^5 - 15x^4 - 6x^2 - 2$

17 $3x^2 + 12x + 11$

19 $18x^2 - 14x - 29$

21 $2x$; $(x - 1) + (x + 1) = 2x$

23 $4x^3$; $2x(x^2 - 1) + (x^2 + 1)(2x)$ $= (2x^3 - 2x) + (2x^3 + 2x) = 4x^3$

25 $5x^4$; $(x^4 + x^3 + x^2 + x + 1) + (x - 1)(4x^3 + 3x^2 + 2x + 1)$ $= (x^4 + x^3 + x^2 + x + 1) + (4x^4 - x^3 - x^2 - x - 1) = 5x^4$

27 $(u/v)' = 1' = 0$, $u'/v' = 1/1 = 1$

29 $(u/v)' = (1/x)' = -1/x^2$, $u'/v' = 1/(2x)$

31 $1/(x + 1)^2$

33 $14/(x + 5)^2$

35 $(x^2 + 8x + 1)/(x + 4)^2$

37 $-4x/(x^2 - 1)^2$

39 $(-2x^2 + 2)/(x^2 + x + 1)^2$

41 $(-2x^2 - 14x + 6)/(x^2 + 3)^2$

43 $(-x^4 - 2x^3 + 3x^2 + 4x + 2)/(x^3 + 2)^2$

45 $(4x^2 + 20x + 22)/[(x + 3)^2(x + 4)^2]$

47 $2x + 2x^{-3}$

49 $-x^{-4} + x^{-8}$

51 $-3x^{-7} + \frac{1}{27}x^{-8/9}$

53 $$\left(\frac{fg}{h}\right)' = \left[\frac{fgh}{h^2}\right]' = \frac{h^2(fgh)' - (fgh)(2hh')}{h^4} = \frac{(fgh)' - 2fgh'}{h^2}$$

Section 2-8 The Chain Rule

Page 123

1 $6x^2(x^3 + 1)[2(x^3 + 1)^2 - 3]$

3 $(x^2 - 1)(x^2 + 1)^2(5x^4 + 7x^2 + 5)/x^6$

5 $-30(2x - 1/x)^9(2 + 1/x^2)$

7 $-10x(x^2 - 1)^4$

9 $6(x - 1)^5$

11 $2(x^2 + 1)^2(x - 1)(4x^2 - 3x + 1)$

13 $(x^3 + 6)(5x^3 - 6)/x^2$

15 $-9/x^{10}$

17 $-4(x^2 + x)^{-5}(2x + 1)$

19 $4(x + 1)/(x + 3)^3$

21 $-12x/(x^2 + 1)^7$

23 $-10x/(2x^2 - 1)^2$

25 $\frac{1}{2}(x + 3)^{-1/2}$

27 $\frac{1}{2}x^{-1/2}(x + 1)^{-3/2}$

29 $(1 - x^2)^{-3/2}$

31 $x(3x^2 - 2a^2)(x^2 - a^2)^{-1/2}$

33 $-16(8x - x^2)^{-3/2}$

35 $3x^2(\frac{1}{2}\sqrt{x} + 1)/(\sqrt{x} + 1)^4$

37 $-(1 + 3x)(1 + 2x + 3x^2)^{-3/2}$

39 $\frac{12}{5}x(3x^2 + 1)[(3x^2 + 1)^2 + 1]^{-4/5}$

41 $\frac{1}{2}[(x^2 + 2x)^{1/3} - 3x]^{1/2}[\frac{1}{3}(x^2 + 2x)^{-2/3}(2x + 2) - 3]$

43 $[3(x^3 + 2x)^2 + 2](3x^2 + 2)$

45 $-\frac{1}{3}x^{-2}(5x^2 + 3a^2)(x^2 + a^2)^{-4/3}$

47 1

49 $8x(x^2 + 1)[(x^2 + 1)^2 + 1]$

51 $\frac{1}{3}x^{-1/3}(x^{2/3} + 2)^{-1/2}$

53 $\frac{1}{3}(2x + 1)(x^2 + x)^{-2/3}[2(x^2 + x)^{1/3} + 1]$

Section 2-9 Trigonometric Functions

Page 130

1 $4 \sin^3 x \cos x$

3 $-x \sin x$

5 $-\frac{1}{2} \sin x(\cos x)^{-1/2}$

7 $(x \cos x - \sin x)/x^2$

9 $\tan x + x \sec^2 x$

11 $2 \tan x \sec^2 x$

13 $-2(\sin x - \cos x)^{-2}$

15 $\sec x \tan x + \csc x \cot x$

17 $\sec^2 x - \csc^2 x$

19 $3 \sec^2 x \tan x(\sec x - 2)$

21 $1 + 2 \tan^2 x$

23 $\sec^2 x \tan x(1 + \sec^2 x)^{-1/2}$

Section 2-10 Implicit and Inverse Functions

Page 137

1 $y = 2x - 6$

3 $y = x - \sqrt{x^2 + 4}$

5 $y' = -x^2/y^2$

7 $y' = -(y + 2x)/(3y^2 + x)$

9 $y' = (y - 2x)/(3y^2 - x)$

11 $y' = 1/(\cos y - 1)$

13 $y' = (1 - 4x^3)/(12y^3 - 1)$

15 $y' = (4x^3 + 3x^2y^5)/(1 - 5x^3y^4)$

17 $y = \frac{1}{3}(x + 7)$

19 $y = -1/x$

21 $y = (4x + 7)/(2 - x)$

23 $y = (3x - 2)/(1 - x)$

25 $y = [(2 - 3x)/(x - 1)]^{1/3}$

27 $y = \frac{1}{2}(x^2 + 8)$

29 $y = 9/(x + 7) - 7$

31 $\frac{1}{28}$

33 $\frac{1}{2}$

35 $\frac{25}{28}$

37 $-\frac{1}{12}$

39 $2/\sqrt{3}$

41 $\frac{1}{3}$

43 $\frac{1}{2}$

45 $12/(6 + \pi\sqrt{3})$

Section 2-11 Higher Derivatives

Page 142

1 6

3 $2 - 6x$

5 $30x^4 + 40x^3 + 12x^2$

7 $(6x^2 - 2)/(1 + x^2)^3$

9 $4/(x - 2)^3$

11 $4a(3x^2 + a)/(x^2 - a)^3$

13 $\frac{27}{4}(1 - 3x)^{-5/2}$

15 $9(x^2 + 9)^{-3/2}$

17 $-\frac{2}{9}x^{-5/3} + \frac{10}{9}(1 + x)^{-5/3}$

19 $2 \cot x \csc^2 x$

21 $\csc x(2 \csc^2 x - 1)$

23 $6 \tan x \sec^2 x(\tan^2 x + \sec^2 x)$

25 16π

27 $\frac{5}{864}$

29 18

31 -2π

33 $8(2 - \pi)/\pi^3$

35 $y''(0) = -1$

37 $y^{(n)} = (-1)^{n+1}2^{-n}[1 \cdot 3 \cdot 5 \cdots (2n - 3)]x^{-n+1/2}$

39 $y^{(n)} = (-1)^{n-1}3^{-n}[2 \cdot 5 \cdot 8 \cdots (3n - 4)]x^{1/3-n}$

41 $12x^2(2x + 1)^4(30x^2 + 12x + 1)$

43 $8 \cdot 7 \cdot 6 \cdot 5 \cdot (9x - 15)(x - 3)^3$

45 $3 \cdot 5 \cdot 7 \cdots 15 \cdot 2^{-10}x^{-19/2}(3x - 17)$

47 0

49 $ff''/(f')^2$

Section 2-12 Review

Page 145

1 $-\frac{1}{2}$

3 $-\frac{1}{2}$

5 0

7 $\frac{1}{4}\sqrt{2}$

9 See Theorem 2, part 5, page 82. Note that $10^{-6} > 0$.

11 Continuous on **R**, differentiable on **R**. $f'(0) = \lim_{x\to 0}(x|x|)/x = \lim_{x\to 0}|x| = 0$.

13 Multiply both sides by $a^2(a + h)$:

$$a^2 = a(a + h) - h(a + h) + h^2$$

which checks out. Next,

$$\lim_{h\to 0}[(a + h)^{-1} - a^{-1}]/h$$
$$= \lim_{h\to 0}[-(1/a^2) + h/a^2(a + h)]$$
$$= -1/a^2$$

By definition, this is the derivative of $1/x$ at $x = a$.

15 $9x^8 \sin^2 x + 2x^9 \sin x \cos x$

17 $\frac{15}{2}(2 + \sqrt{3x})^4(3x)^{-1/2}$

19 $(2x + 1)(3x + 1)^2(36x^2 + 48x + 14)$

21 $2x \tan x + x^2 \sec^2 x$

23 $4x^3 + 3y^2y' = 0$, $y' = -\frac{4}{3}x^3y^{-2}$

25 $y' = (5y^4 + 1)^{-1}$

27 $2/(x + 2)^3$

29 $-6 \sin x - x \cos x$

31 a

33 $y' = 1/(2x - 1)^2$

Chapter 3 Applications of Differentiation

Section 3-1 Tangents, Graphs, and Approximation

Page 151

1 $y = 4x - 4$

3 $y - \frac{1}{2} = -\frac{1}{2}\sqrt{3}(x - \frac{1}{3}\pi)$

5 $y = 2(x + 1)$

7 $y = \frac{2}{27}x + \frac{1}{3}$

9 $y = 7x - 5$, $(0, -5)$, $(\frac{5}{7}, 0)$

11 $y = -81x + 18$, $(0, 18)$, $(\frac{2}{9}, 0)$; $y = -81x - 18$, $(0, -18)$, $(-\frac{2}{9}, 0)$

13 $(5, 25)$, $y = 10x - 25$

15 $(4, 16)$, $y = 8x - 16$; $(-4, 16)$, $y = -8x - 16$

17 $y = x^3 - x$

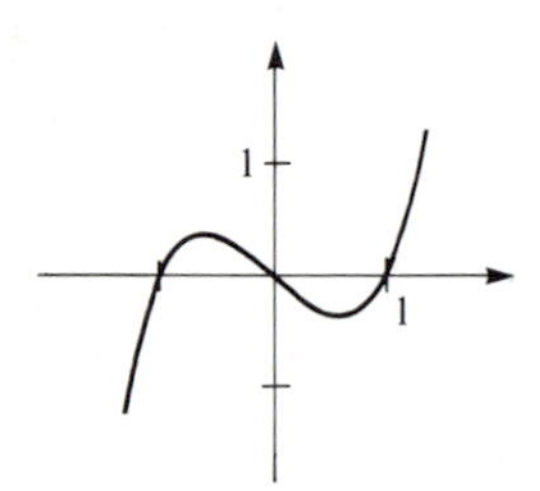

21 $y = 4x^3 - 2x^2$

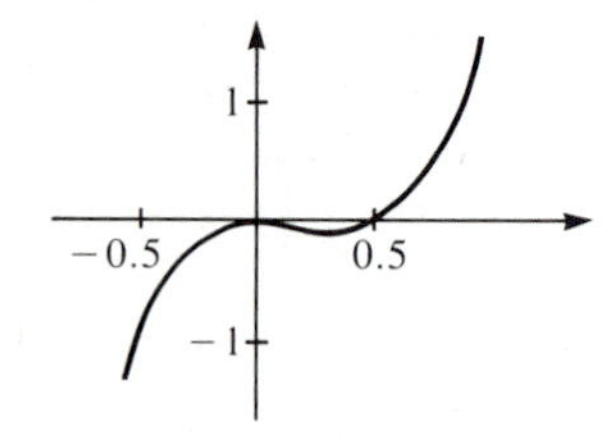

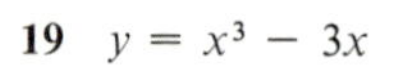

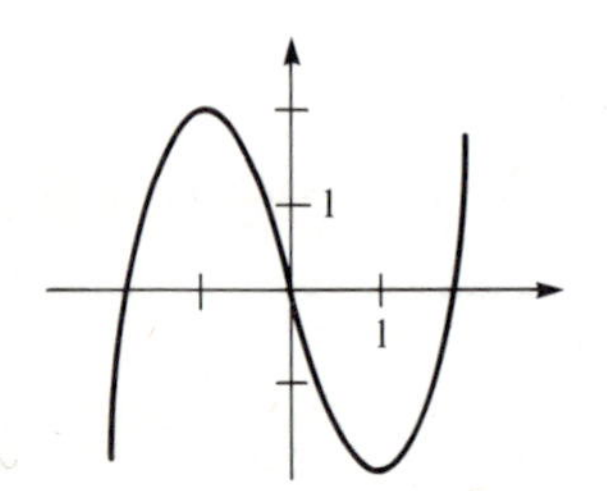

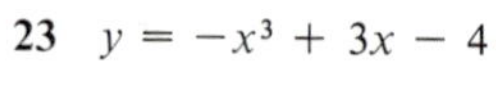

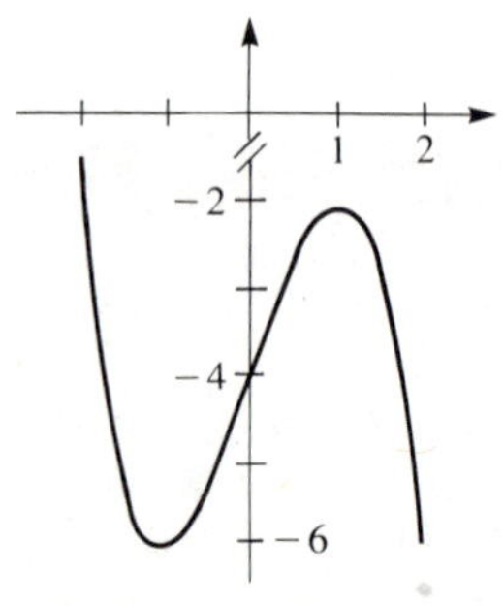

25 $y = \frac{1}{3}x - x^3$

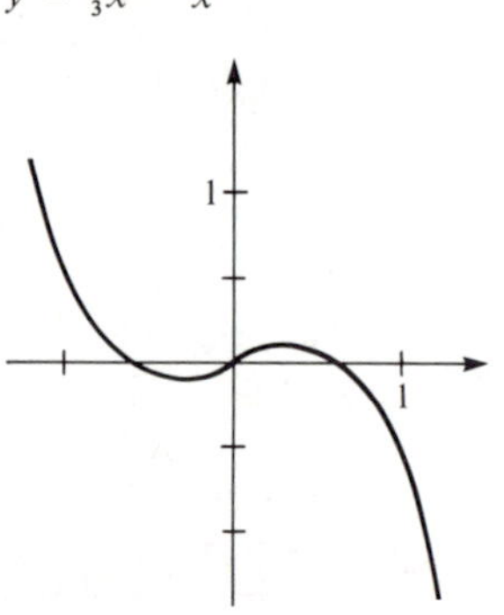

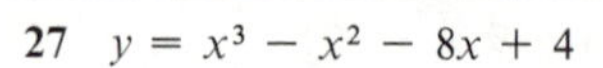

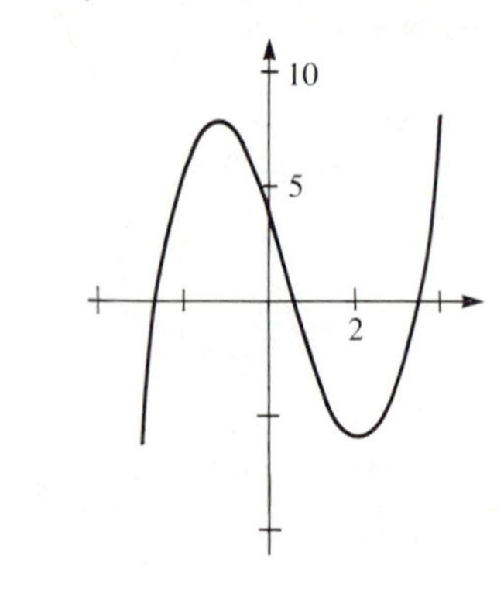

29 $y = \sin x + \cos 2x$

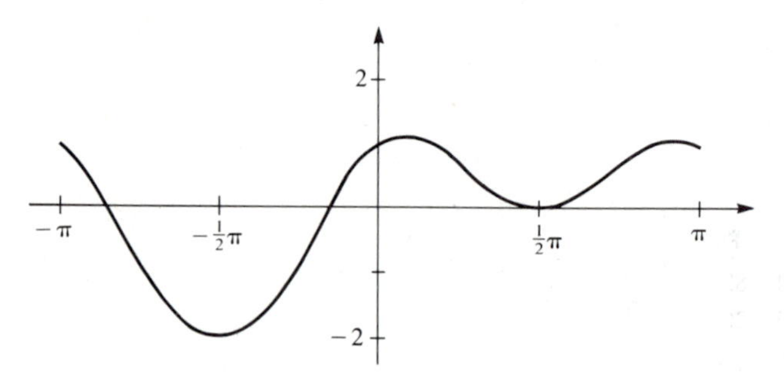

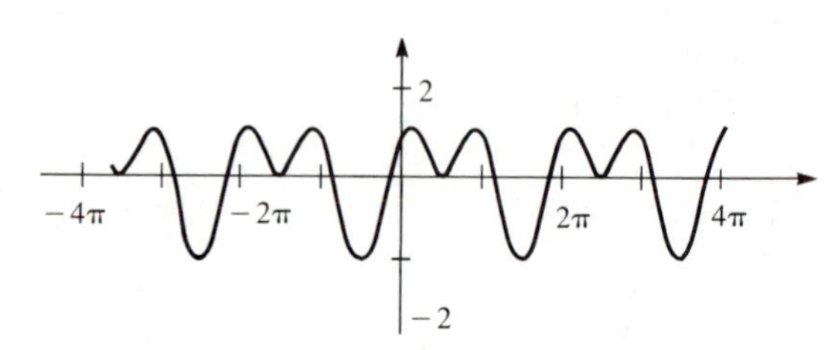

31 $y = \tan x + \cot x$

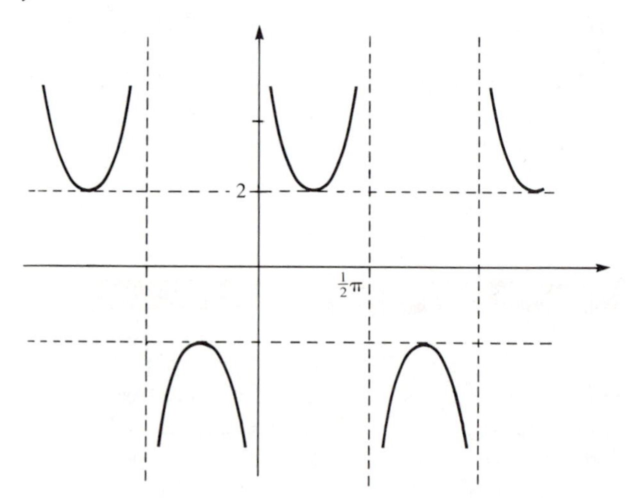

33 $y = 1$, $E = -x^2$

35 $y = 27(x - 2)$, $E = (x - 3)^2(x + 6)$

37 $y = -3x - 2$, $E = 9(x + 1)^2/(3x + 4)$

39 $f(-1) = -1$, $f'(-1) = 3$. Tangent: $y + 1 = 3(x + 1)$. $E(x) = x^3 - [-1 + 3(x + 1)] = (x - 2)(x + 1)^2$. If $|x + 1| < 1$, then $-2 < x < 0$, so $-4 < x - 2 < -2$, $|x - 2| < 4$; hence $|E(x)| \le 4|x + 1|^2$.

41 $g(x) = \frac{1}{2}(x + 1)$, $E(x) = \sqrt{x} - \frac{1}{2}(x + 1)$

x	0.98	0.99	1.02	1.01
$\sqrt{x}$	0.98995	0.99499	1.00995	1.00499
$g(x)$	0.99	0.995	1.01	1.005
$E(x)$	-0.00005	-0.00001	-0.00005	-0.00001

43 Tangent: $y - \frac{1}{2} = -\frac{1}{4}(x - 2)$
Intercepts: $(0, 1)$, $(4, 0)$
Area: $\frac{1}{2} \cdot 1 \cdot 4 = 2$

45 Tangents: $y - 9 = \pm 6(x \mp 3)$, $y = \pm 6x - 9$
y-intercepts: $(0, -9)$

47 $g(x) = 8 + 12(x - 2) = 12x - 16$,
$x^3 - g(x) = (x - 2)^2(x + 4)$
If $|x - 2| < 1$, then $5 < x + 4 < 7$; hence $|x^3 - g(x)| < 7|x - 2|^2$.

Section 3-2 Rectilinear Motion

Page 158

1 820 ft/sec, 834.4 ft/sec, 836 ft/sec

3 2000 m, -50 m/sec, -200 m/sec

5 $\frac{3}{4}$ sec, 16 ft/sec, 16 ft/sec, up speed = down speed

7 **a** Increasing: $t < 2$, $t > 4$; decreasing: $2 < t < 4$
b Increasing: $t > 3$; decreasing: $t < 3$ **c** 44 ft

9 40 ft

11 6 ft/sec^2

13 480 m, 16 sec

15 $1366\frac{2}{3}$ m

17 $y = -8x^2 + 12$

19 $y = -16t^2 + 64t$

21 $y = \frac{1}{3}t^3 - \frac{1}{2}t^2 + 3t + 5$

Section 3-3 Related Rates

Page 164

1 $-\frac{10}{27}$

3 -1

5 $\frac{5}{9}$

7 24π ft^2/sec

9 $h' = 0.2/\pi$
≈ 0.0637 m/min

11 $21/\sqrt{5}$ cm/sec

13 0.024 cm/sec

15 620π ft^3/sec

17 $4\sqrt{3}$ ft/sec

19 $dP/dt = -P^2(dV/dt)/k$

21 $a\omega \cos \frac{1}{2}\theta$

23 -4π in./min

25 200π ft/sec

27 $dI/d\theta = k \sec^2\theta$,
$S = [(k \tan\theta)/\theta](\cos^2\theta)/k = (\sin\theta \cos\theta)/\theta$.
As $\theta \to 0$, $(\sin\theta)/\theta \to 1$ and $\cos\theta \to 1$, hence $S \to 1$.

Section 3-4 Continuous and Differentiable Functions

Page 174

1 0

3 $\frac{3}{2}$

5 $\frac{1}{2}\pi$, $\frac{3}{2}\pi$, $\frac{5}{2}\pi$, $\frac{7}{2}\pi$

7 $\frac{1}{8}(9 - \sqrt{17})$

9 0

11 $y = x^2 \quad (0 \le x \le 1)$

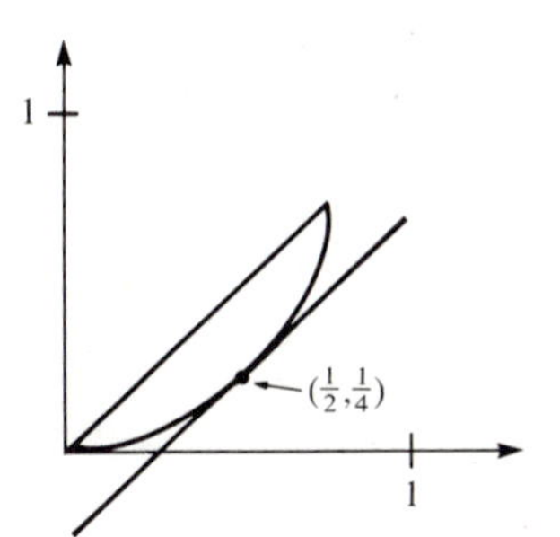

13 $y = x^3 \quad (0 \le x \le 1)$

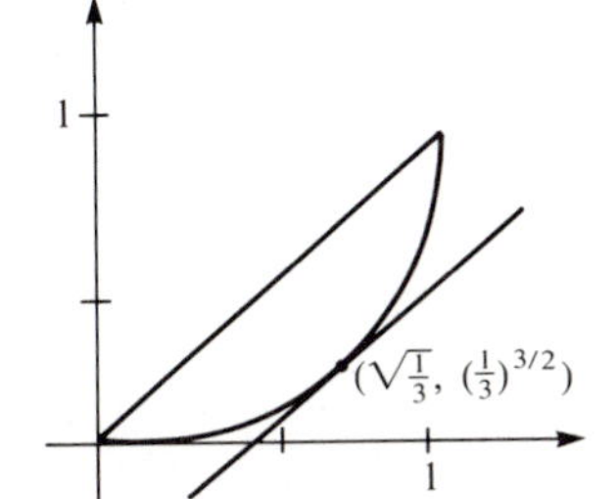

15 $y = x^3 \quad (-1 \le x \le 2)$

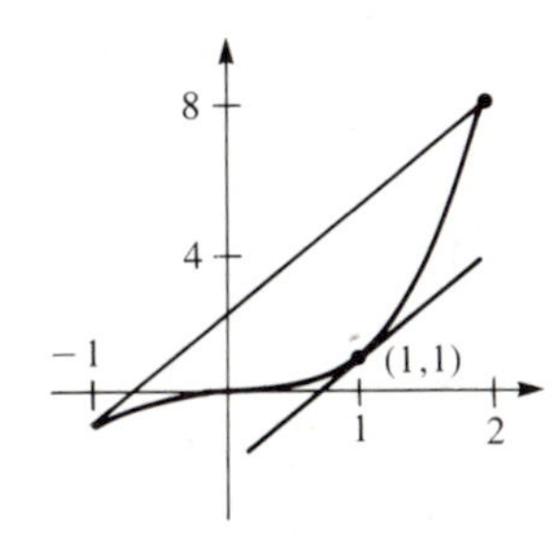

17 $y = \sqrt{1 - x^2}$
$(-1 \le x \le 0)$

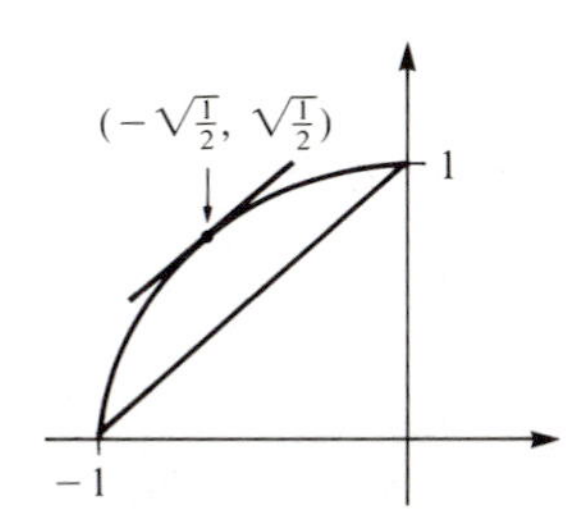

19 $f(x)$ has a maximum on the closed interval $[0, 1]$ because it is continuous there. But $f(\frac{1}{2}) > 0$, so the maximum is greater than 0.

21 No

Section 3-5 Curve Sketching

Page 184

1 C. up everywhere

3 C. up: $x > 2$. C. down: $x < 2$. Inflection at $(2, 4)$.

5 C. up: $x < 0$. C. down: $x > 0$. Inflection at $(0, -5)$.

7 C. up: $|x| > \sqrt{\frac{5}{2}}$. C. down: $|x| < \sqrt{\frac{5}{2}}$.
Inflection at $(\pm\sqrt{\frac{5}{2}}, \sqrt{\frac{2}{15}})$.

9 C. up: $(2n + 1)\pi < x < (2n + 2)\pi$.
C. down: $2n\pi < x < (2n + 1)\pi$. Inflections at $(n\pi, 0)$.

11 C. up: $n\pi < x < (n + \frac{1}{2})\pi$.
C. down: $(n - \frac{1}{2})\pi < x < n\pi$. Inflections at $(n\pi, 0)$.

13 C. up: $(2n - \frac{1}{2})\pi < x < (2n + \frac{1}{2})\pi$.
C. down: $(2n + \frac{1}{2})\pi < x < (2n + \frac{3}{2})\pi$. No inflections.

15 $y = 3x^3 + x$

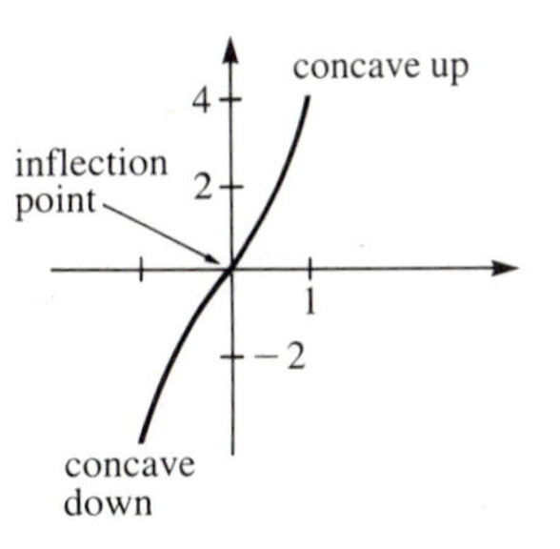

17 $y = x^2(x - 3)$
Inflection point $(1, -2)$

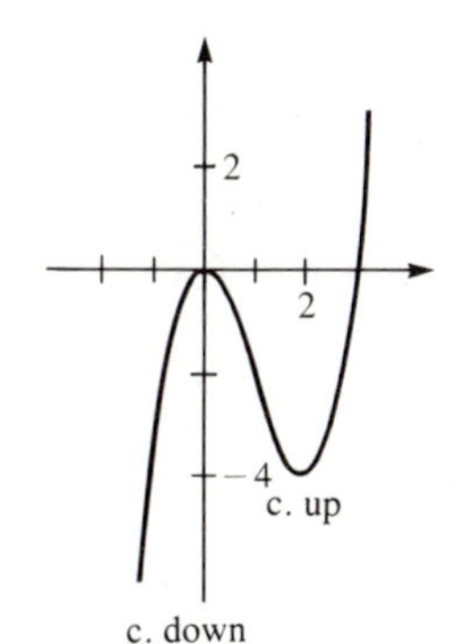

19 $y = x^3(x + 2)$
Inflection points $(0, 0)$, $(-1, -1)$

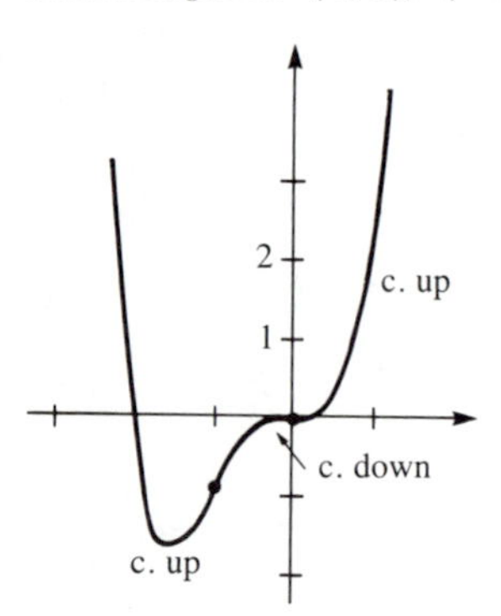

21 $y = 1/(x^2 + 1)$
Inflection points $(\pm\sqrt{\frac{1}{3}}, \frac{3}{4})$

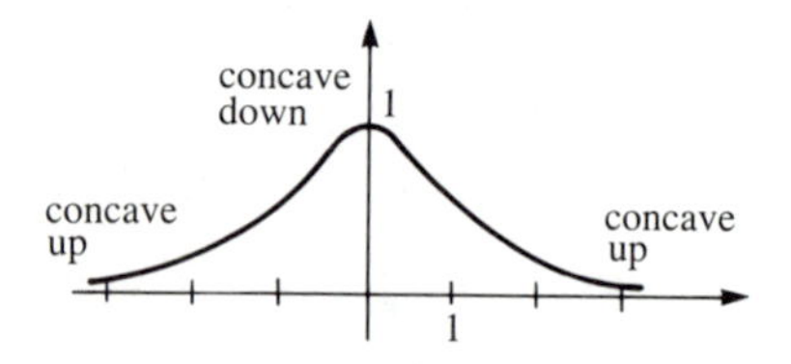

23 $y = x/(x^2 - 1)$
Inflection point $(0, 0)$

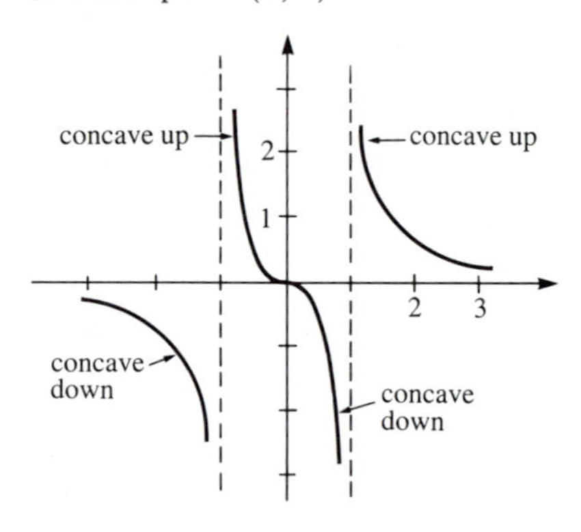

25 $y = x^2 + x^{1/2}$
$(x \geq 0)$

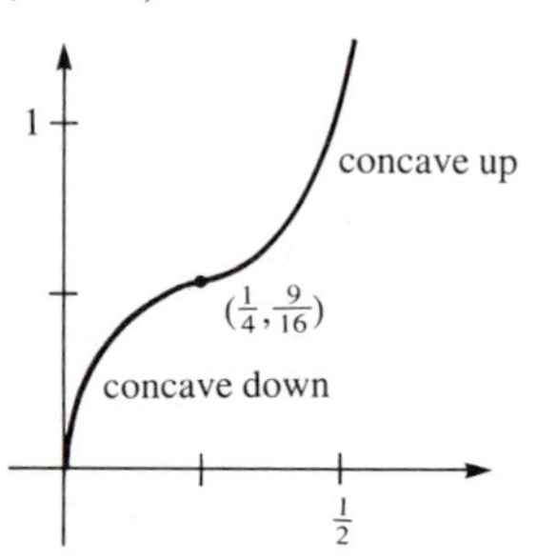

27 $y = x^2/(1 + x^3) \qquad (x \neq -1)$
$$y'' = \frac{2(x^6 - 7x^3 + 1)}{(1 + x^3)^3}$$
Inflection points $((\frac{1}{2}(7 \pm \sqrt{45}))^{1/3}, \frac{1}{3}(\frac{1}{2}(3 \pm \sqrt{5}))^{1/3})$

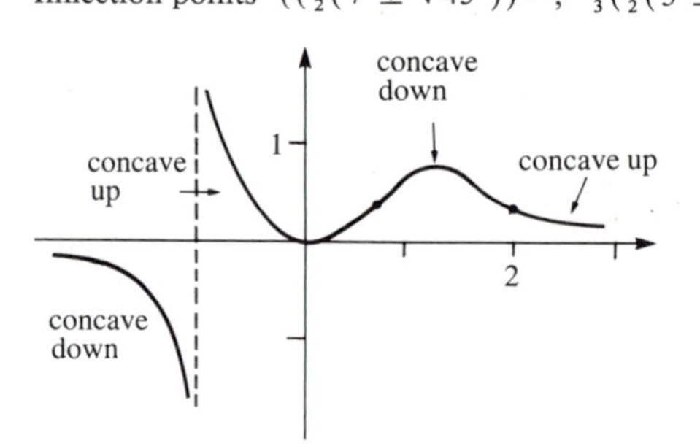

29 $y = 4 \sin x + \sin 2x$
$(-\frac{1}{2}\pi \leq x \leq \frac{1}{2}\pi)$
$y'' = -4(1 + 2 \cos x) \sin x$
Inflection point $(0, 0)$

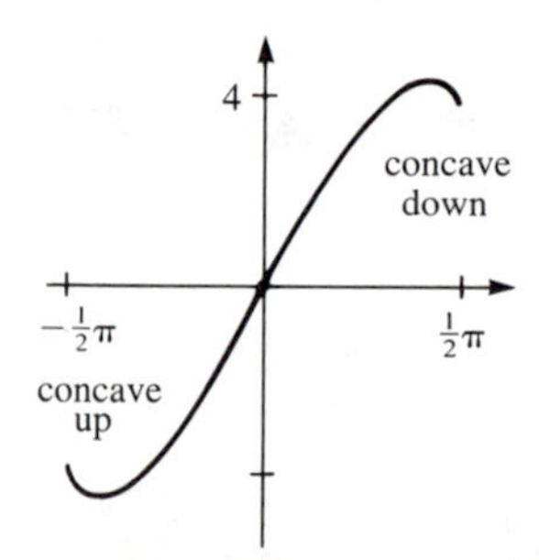

Section 3-6 Optimization

Page 195

1 $y = x^2 - 4x + 6$

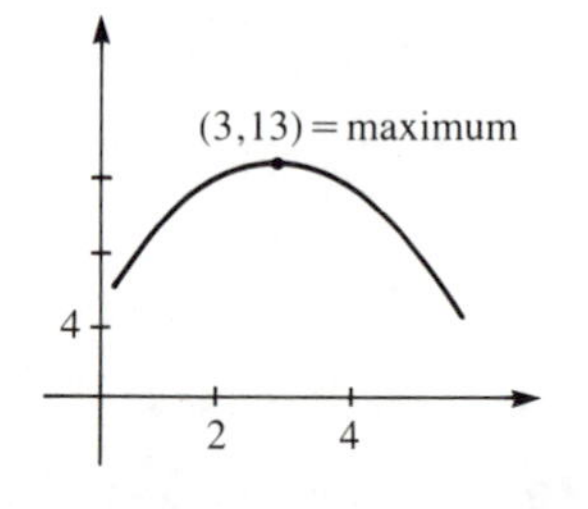

3 $y = -x^2 + 6x + 4$

(3,13) = maximum
4
2
4

5 $y = 12x + 1/(3x)$
$(x > 0)$

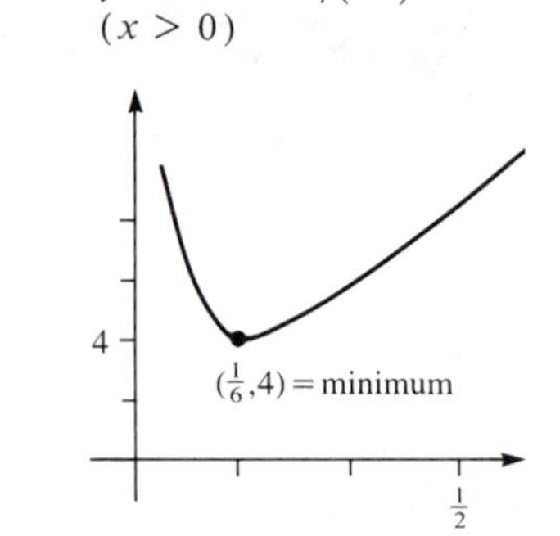

7 $y = 2x^3 - 3x^2 - 12x + 1$
$(0 \leq x \leq 3)$

(0,1) = maximum
3
−10
−20
(2, −19) = minimum

9 $y = 1/[(x-1)(2-x)] \quad (1 < x < 2)$

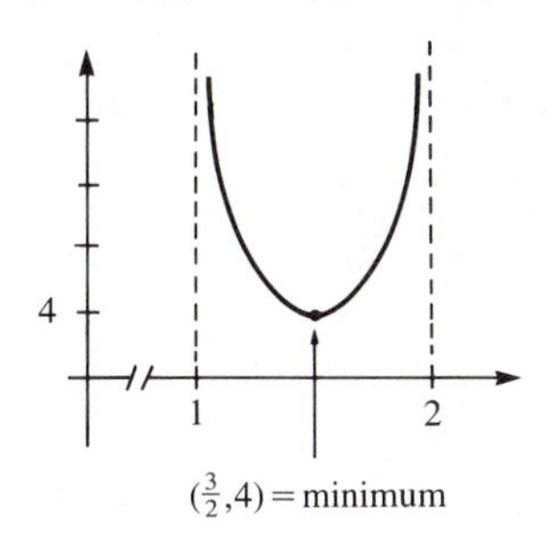

11 $y = \frac{1}{3}x + 1/x \quad (1 \le x \le 3)$

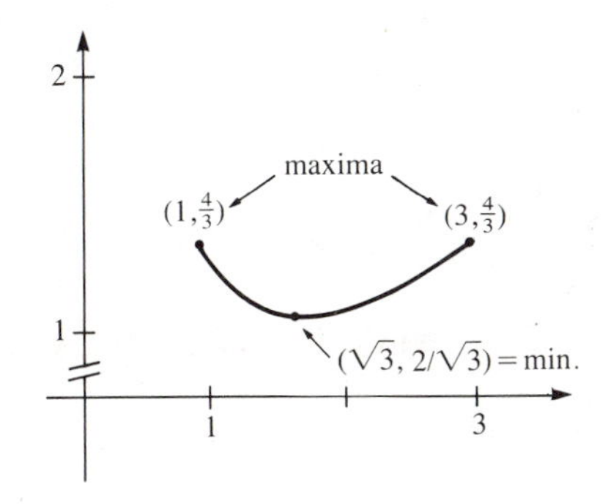

13 $y = 1/x - 1/x^2 - 1/x^3 \quad (-3 \le x \le -1)$

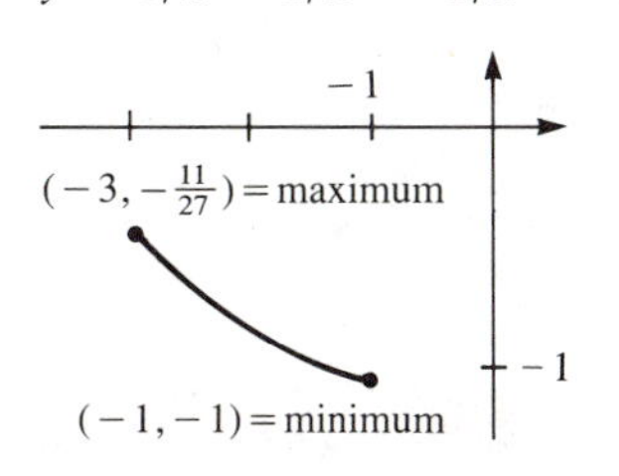

15 $y = 4x^2 + 1/x \quad (\frac{1}{4} \le x \le 1)$

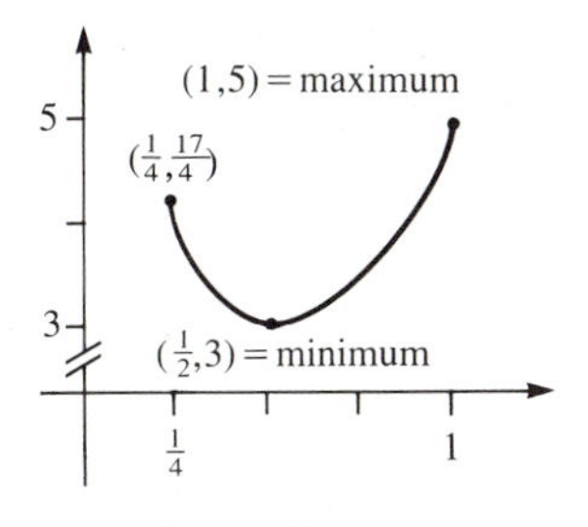

17 $y = x^4 - 2x^2 + 1 \quad (-1 \le x \le 2)$
Note that $y = (x^2-1)^2$

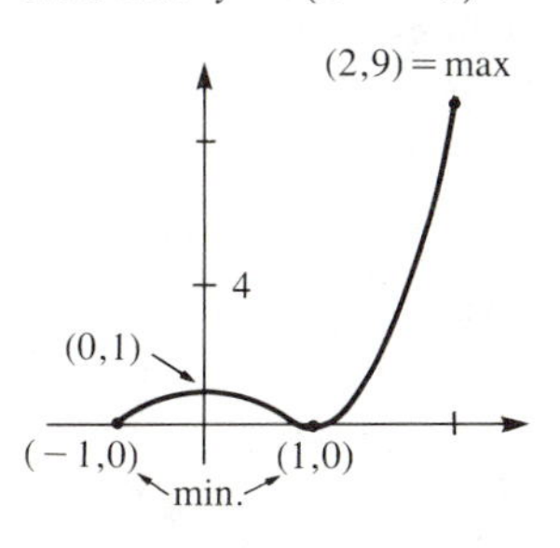

19 $y = -3x^4 - 16x^3 - 18x^2 - 12$
$(-3 \le x \le 0)$

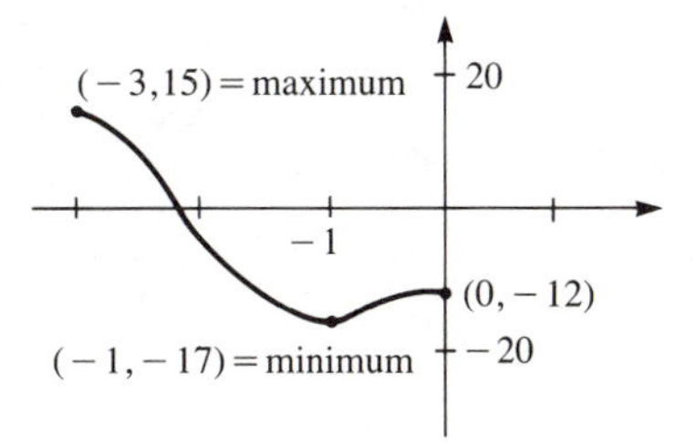

21 $y = \sqrt{1 - x^4}$
$(-1 \le x \le 1)$

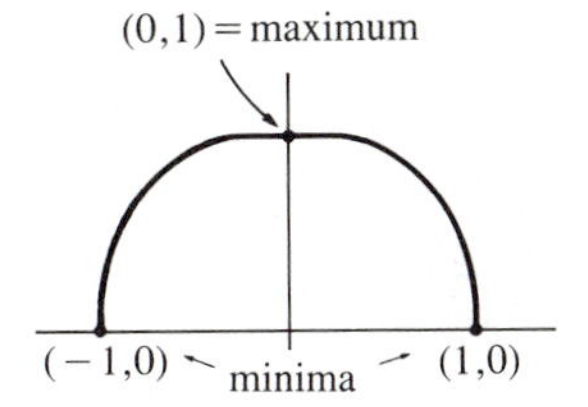

23 $y = x\sqrt{1 - x^2}$
$(-1 \le x \le 1)$

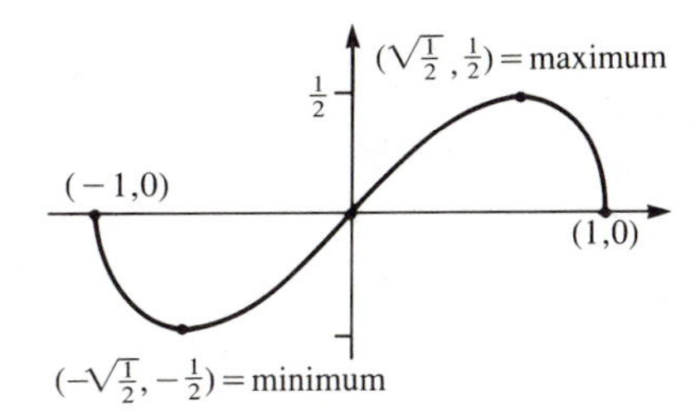

25 $y = x/(1 + x^2)$

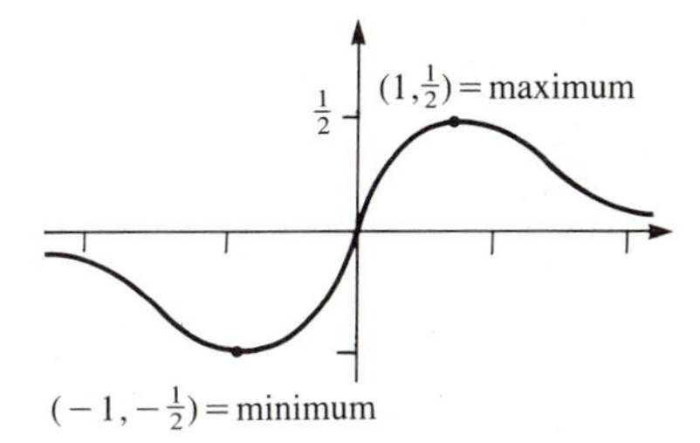

27 $y = x^2/(1 + x^3)$
$(x > -1)$

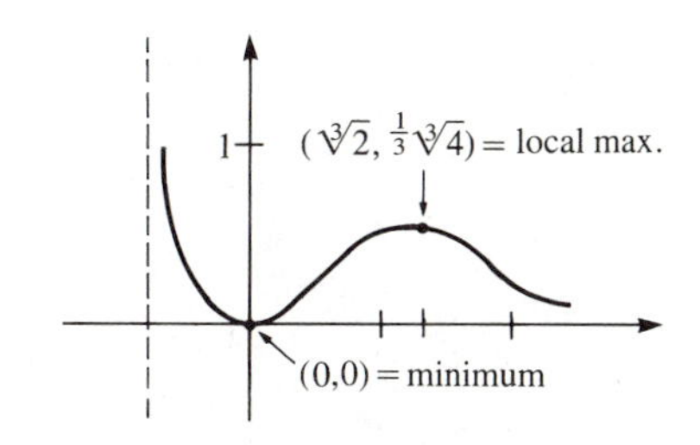

29 $y = (\cos^2 x)/(1 + \cos^2 x)$
maxima: $(n\pi, \frac{1}{2})$
minima: $((n + \frac{1}{2})\pi, 0)$

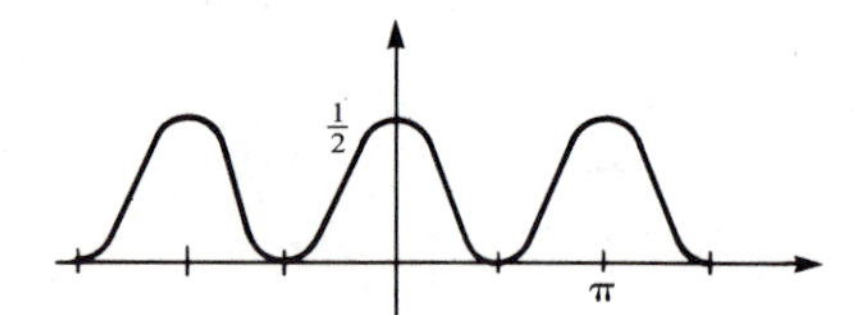

31 $y = \sin x - \cos 2x$
$y' = (\cos x)(1 + 4 \sin x)$
maxima: $((2n + \frac{1}{2})\pi, 2)$
local maxima: $((2n - \frac{1}{2})\pi, 0)$
minima: $(2n\pi - \theta, -\frac{9}{8})$ and $((2n + 1)\pi + \theta, -\frac{9}{8})$,
where $\theta = \arcsin \frac{1}{4}$

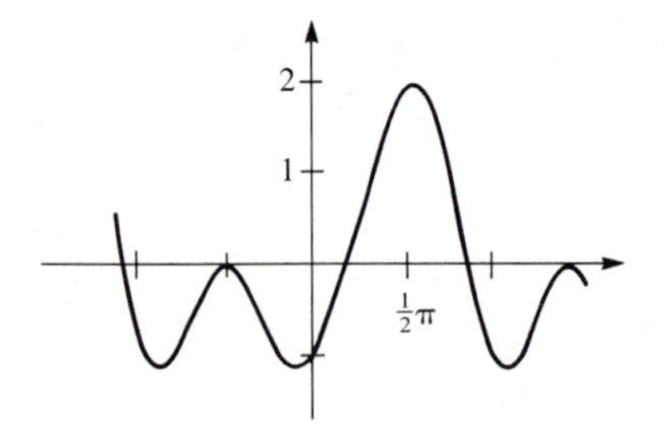

33 $y = x^{2/3} - x$
$y' = \frac{2}{3}x^{-1/3} - 1$

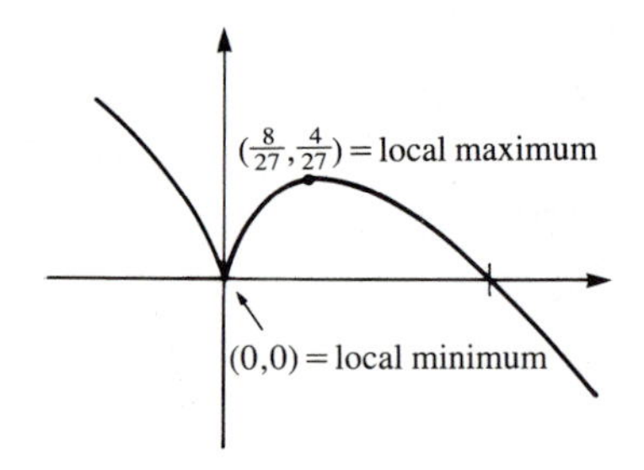

35 $y = (\sin x)^{2/3}$
$y' = \frac{2}{3}(\sin x)^{-1/3} \cos x$
maxima: $((n + \frac{1}{2})\pi, 1)$
minima: $(n\pi, 0)$

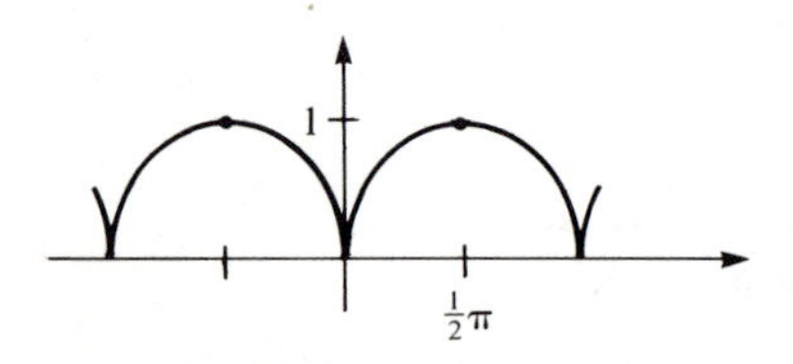

37 $y = x^{2/3} + (1 - x)^{2/3}$
$y' = \frac{2}{3}[x^{-1/3} - (1 - x)^{-1/3}]$

1.5
$(\frac{1}{2}, \sqrt[3]{2})$ = local maximum
(0,1)
(1,1)
1

39 $y = |x^2 - 1|$
$y' = 2x$ for $|x| > 1$,
$y' = -2x$ for $|x| < 1$

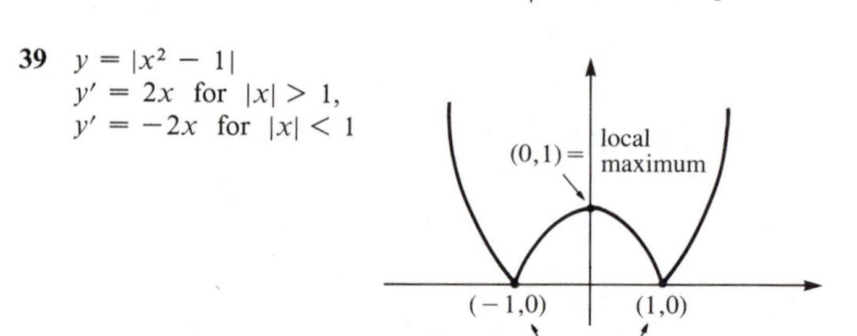

Section 3-7 On Problem Solving

Page 203

1 28 ft
3 25 m, 75 m
5 $L = 3\sqrt[3]{10}$, $H = \frac{1}{2}\sqrt[3]{10}$, $W = \sqrt[3]{10}$ ft
7 $\frac{98}{5}$ ft^2
9 $(2^{-1/3}, \pm 2^{1/6})$
11 1
13 12
15 $\frac{1}{9}a^3\sqrt{3}$
17 At noon
19 80 ft
21 \$2725
23 $1000/\sqrt{3}$ mph
25 2×10^{-7} moles
27 $\frac{1}{4}b^2/a$

Section 3-8 Further Applications

Page 212

1 $\frac{1}{2}r^2$
3 $\frac{1}{8}L^2$
5 width $= 200/\pi$, straight length $= 100$ m
7 $\frac{1}{2}b$ by $\frac{1}{2}a$
9 equilateral, side $= 2r\sqrt{3}$
11 $\frac{1}{16}\sqrt{2}(3b + c)(4a^2 - b^2 + bc)^{1/2}$ where $c = (b^2 + 8a^2)^{1/2}$
13 $16/(\pi + 4) \approx 2.24$ ft
15 $\overline{CX} = (\sqrt{2} - 1)a$
17 Width $= 2r\sqrt{\frac{1}{3}}$, Depth $= 2r\sqrt{\frac{2}{3}}$, $D/W = \sqrt{2}$
19 $8(1 - \frac{1}{2}\sqrt[3]{2})^{3/2} \approx 1.8008$ ft
21 Row straight to a point $5 - \sqrt{3}$ mi from the destination, then walk.
23 $\frac{2}{75}a^3/\sqrt{5}$
25 $I_{\max} = \frac{2}{9}ka^{-2}\sqrt{3}$ for $z = a\sqrt{\frac{1}{2}}$
27 Radius $= s\sqrt{\frac{2}{3}}$, height $= s\sqrt{\frac{1}{3}}$
29 Height $=$ radius $= (V/\pi)^{1/3}$ decimeters
31 Height $= 2r\sqrt{\frac{1}{3}}$, radius $= r\sqrt{\frac{2}{3}}$
33 $A_{\max} = \frac{1}{2}\pi h^2 r/(h - r)$
35 $A_{\min} = 3r^2\sqrt{3}$

Section 3-9 Second Derivative Test

Page 217

1 Local maximum: $x = 0$. Local minima: $x = \pm 1$.

3 Local maximum: $x = 0$. Local minima: $x = -1, 2$.

5 Local maximum: $x = -3$.

7 Local maximum: $x = \frac{3}{2}$.

9 Local maxima: $x = \pm 1$. Local minimum: $x = 0$.

11 Local maximum: $x = 2$. Local minimum: $x = -2$.

13 $y_{max} = (n-1)^{n-1}/(2^{n-1}n^n)$

15 $y' = 2Q(x)/\sqrt{b^2 + (a-x)^2}$, where $Q(x) = x^2 - ax + \frac{1}{2}b^2$. **Case 1**: $a^2 < 2b^2$. Then $Q(x) > 0$ for all x; no maximum or minimum. **Case 2**: $a^2 = 2b^2$. Then $Q(x) = (x - \frac{1}{2}a)^2$; horizontal inflection at $x = \frac{1}{2}a$, but no maximum or minimum. **Case 3**: $a^2 > 2b^2$. Then $Q(x) = (x - x_1)(x - x_2)$, where $x_1 = \frac{1}{2}(a - \sqrt{a^2 - 2b^2})$ and $x_2 = \frac{1}{2}(a + \sqrt{a^2 - 2b^2})$; maximum at x_1, minimum at x_2.

17 There is an inflection point at $(c, f(c))$. Just to the left, the graph is concave down; just to the right, concave up. Typical example: $y = x^3 + ax + b$, $c = 0$.

19 $y_{max} = \frac{1}{4}$

21 $f = \Sigma(x - a_i)^2$, $f' = 2\Sigma(x - a_i) = 2(nx - \Sigma a_i)$. $f' = 0$ for $x = (\Sigma a_i)/n = \bar{a}$. $f \to +\infty$ as $x \to \pm\infty$; hence f has its minimum at $x = \bar{a}$.

Section 3-10 Basic Theory and Proofs [Optional]

Page 222

1 This is the chord theorem with $a = x$, $b = y$, and $c = tx + (1-t)y$, so $a < c < b$.

3 If f' and g' are strictly increasing, then so is $(f+g)' = f' + g'$.

5 $g(x) = x^{3/2}$ is concave up, $f(x) = \sqrt{g(x)} = x^{3/4}$ is concave down.

7 $\{f[g(x)]\}'' = f''[g(x)][g'(x)]^2 + f'[g(x)]g''(x) > 0$, so $f \circ g$ is concave up.

9 Being a cubic, y' has at most three distinct zeros; hence y has at most three local maxima and minima, one at $x = c$. Since $y''(c) < 0$, $y(x)$ decreases immediately to the right of $x = c$ by the tangent theorem. But $y \to +\infty$ as $x \to +\infty$, so y has at least one local minimum to the right of $x = c$. Similarly, y has at least one local minimum to the left of $x = c$. Hence, y has exactly two local minima.

11 $y = e^x$ is concave up. It is above the tangent $y = 1 + x$ at $(0, 1)$. For $0 < x < 1$, it is below the chord $y = 1 + (e-1)x$ joining $(0, 1)$ and $(1, e)$.

13 $\theta(x)$ is the angle the tangent at $(x, f(x))$ makes with the positive x-axis. But $\theta(x) = \arctan f'(x)$ is the composite of differentiable functions, hence differentiable by the chain rule.

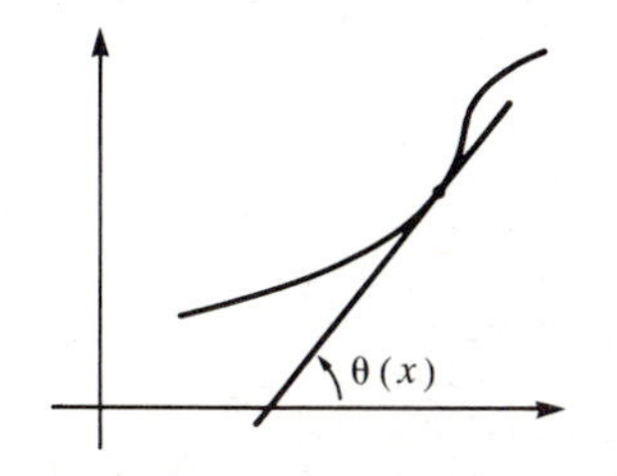

15 The non-negative function $f(x)$ satisfies $f(0) = 0$ and $\lim_{x\to+\infty} f(x) = 0$. Also $f'(x) = 0$ only for one positive x, namely $x = A_n$. Hence A_n yields a maximum, so $f(x) \le f(A_n)$, with "=" only if $x = A_n$. In particular, $f(a_{n+1}) \le f(A_n)$, equality holding only if $a_{n+1} = A_n$. This implies $G_{n+1}/A_{n+1} \le (G_n/A_n)^{n/(n+1)}$.

Section 3-11 Review Exercises

Page 222

1 $y = (2x+1)/(x^2+4)$

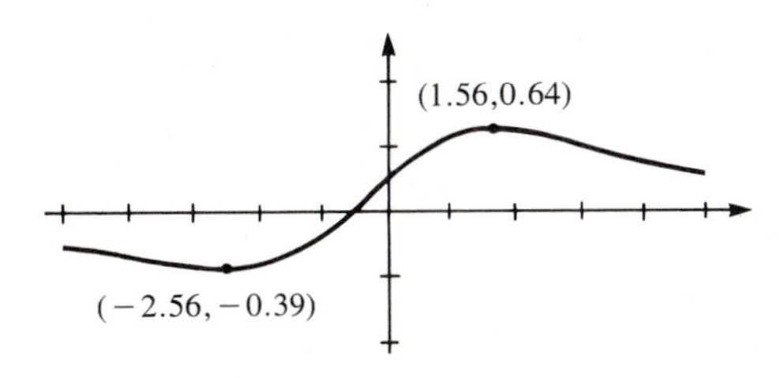

3 $\sqrt{2}$ ft/sec

5 15 sec, 675 ft

7 $\frac{1}{40}$ rad/sec

9 Starts at $x = 0$ and oscillates with period 1 between $x = -A$ and $x = A$; also $v = 2\pi A \cos 2\pi t$ and $a = -4\pi^2 A \sin 2\pi t$. If $v = 0$, then $x = A$ and $t = \frac{1}{4}$ + integer, or $x = -A$ and $t = -\frac{1}{4}$ + integer. This is an example of **simple harmonic motion.**

11 width 6 in, height 18 in.

13 $\frac{2}{27}\pi c^3\sqrt{3}$

15 $\sqrt{\frac{1}{3}}$

17 625 m^2

19 $2\sqrt{2}$

21 $\frac{11}{6}\pi + \frac{1}{2}\sqrt{3} \approx 6.63$

23 500

25 circumference $= 30\pi/(\pi + 4)$ in.

27 Set $\alpha = \arccos\sqrt{\frac{1}{3}} \approx 0.9553$. $f_{max} = \frac{4}{9}\sqrt{3}$ at $x = \alpha$ and $x = 2\pi - \alpha$ $f_{min} = -\frac{4}{9}\sqrt{3}$ at $x = \pi - \alpha$ and $x = \pi + \alpha$ local maximum: $(\pi, 0)$; local minima: $(0, 0)$ and $(2\pi, 0)$

29 $(b/a)^{1/4}$

31 $4f$

33 $x_{max} = (v_0^2/g)(1 - \sin\alpha)/\cos\alpha$ for $\theta = \frac{1}{2}\alpha + \frac{1}{4}\pi$, halfway between the slope of the hill and the vertical.

35 $dB/dt = B' = \frac{1}{2}k^{1/2}(PP')^{-1/2}[PP'' + (P')^2]$

37 $\frac{1}{18}a^2\sqrt{3}$

39 Set $y = (x^4+1)^3/(x^3+1)^4$. $y' = 12x^2(x^4+1)^2(x-1)/(x^3+1)^5$, which changes sign from $-$ to $+$ at $x = 1$. Hence $y_{min} = y(1) = \frac{1}{2}$. Therefore $y \ge \frac{1}{2}$ for all $x \ge 0$; that is, the given inequality holds.

Chapter 4 Special Functions

Section 4-1 Exponential Functions

Page 237

1 $y = e^{x-1}$

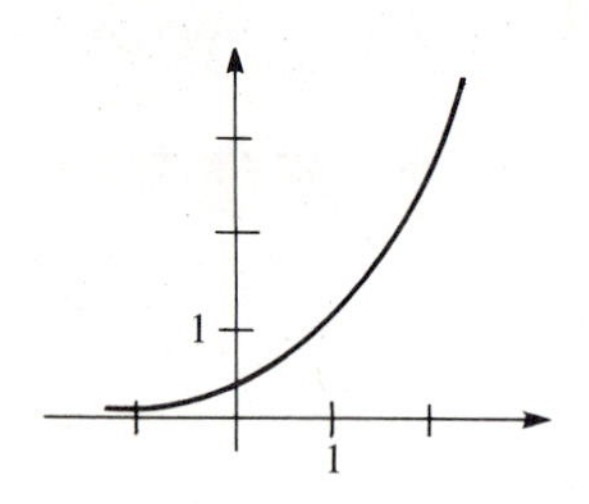

3 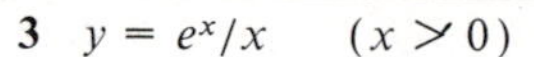 $y = e^x/x \quad (x > 0)$

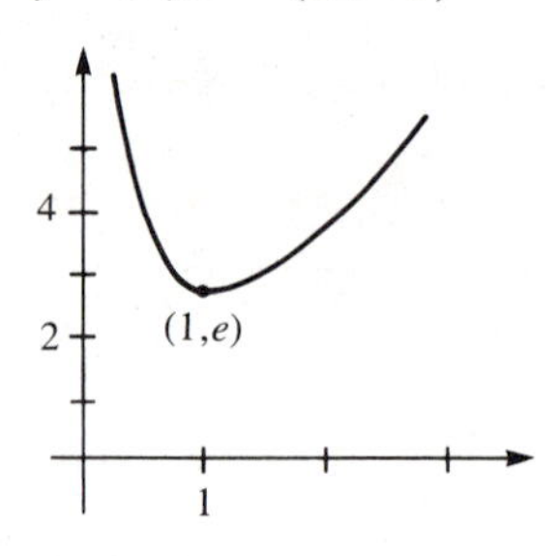

5 $y = \exp(\sin x)$

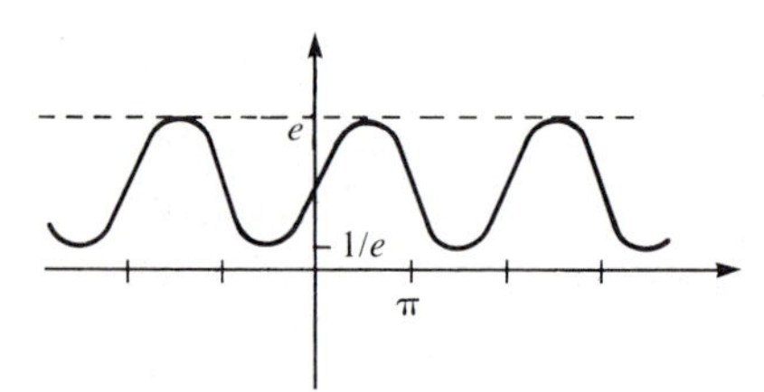

7 $y = e^{-x} \sin 10x \quad (x > 0)$

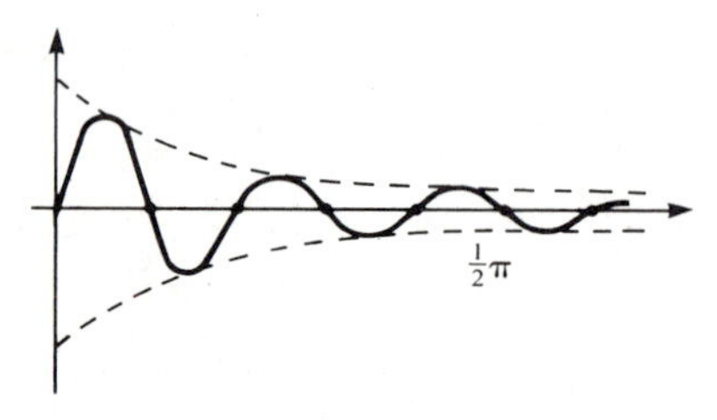

9 $y = \exp(-x^2)$

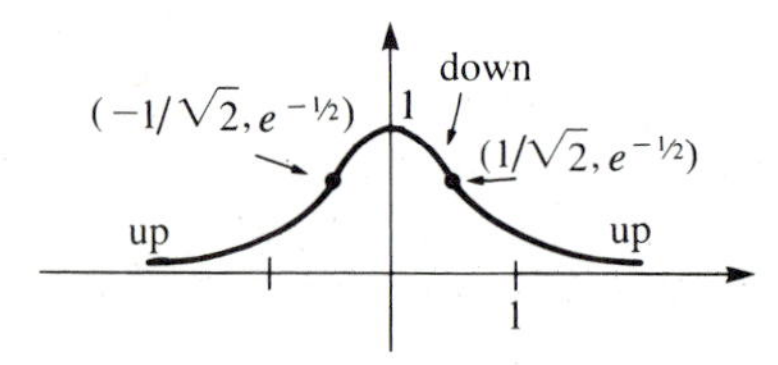

11 $y = e^{1/x} \quad (x > 0)$

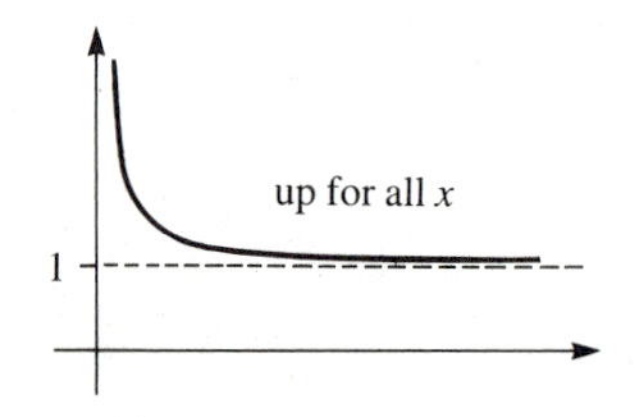

13 $6x \exp(3x^2)$

15 $4x^3 \exp(x^4)$

17 $(x - 1)e^x/x^2$

19 $e^x/(1 + e^x)^2$

21 $\dfrac{3x^2(1 + e^{2x} - 2xe^{2x})}{(e^{2x} + 1)^4}$

23 $-4/(e^x - e^{-x})^2$

25 $(\cos x)\exp(\sin x)$

27 $ab \exp[bx + u(x)]$

29 $64e^{-2x}$

31 $f_{\max} = f(1) = e^{-1}$

33 $f_{\max} = f(0) = 2$

35

h	$(5^h - 1)/h$	$(5^{-h} - 1)/(-h)$
0.0009	1.61060	1.60827
0.0008	1.61047	1.60840
0.0007	1.61034	1.60853
0.0006	1.61022	1.60866
0.0005	1.61009	1.60879
0.0004	1.60996	1.60892
0.0003	1.60983	1.60905
0.0002	1.60970	1.60918
0.0001	1.60957	1.60931

37 $(2^x - 1)/x \approx d(2^x)/dx|_0 \approx 0.6931$

39 $(e^x - 1)/x \approx d(e^x)/dx|_0 = 1$

Section 4-2 Applications of Exponential Functions

Page 242

1 1.26 hr

3 2000: 2.88×10^8, 2025: 3.87×10^8, 2050: 5.21×10^8

5 $m = m_0 2^{-t/3.64}$, 5.77 days

7 After ≈ 2.187 days

9 1458 students

11 69.3%

13 $u(t) = a + ce^{-kt} \to a + 0 = a$ as $t \to +\infty$

15 k

17 Let $m(t) = m_0 e^{-\lambda t}$ be the mass of ^{14}C per gram of wood killed at the time of Hammurabi. Here t is measured in years from that time, m_0 is the mass of ^{14}C per gram of living wood, and $e^{\lambda} = 2^{1/5568}$. The decay rate is $dm/dt = -\lambda m$, so the ratio of the decay rate initially to that in 1950 is $m_0/m(1950) = e^{\lambda t}$. Hence $e^{\lambda t} = 6.68/4.09$,

$$t = [\log(6.68/4.09)]/\log(2^{1/5568}) \approx 3941 \text{ yr}$$

This is the time from the date we want to 1950 A.D., so Hammurabi's reign was about 3941 – 1950 ≈ 1990 B.C. (Some give his dates as 1955 – 1913 B.C. Of course this particular tree could have been cut earlier.)

Section 4-3 The Logarithm Function

Page 249

1 $y = -\ln x \quad (x > 0)$

3 $y = e^{1/x} \quad (x > 0)$

5 $y = e^x - 5$

7 $a + 2$

9 $1/x$

11 $\frac{1}{2}$

13 $1/x$

15 $4/x$

17 $-1/x$

19 $\cot x$

21 $-2/(x^2 - 1)$

23 $1/(2x\sqrt{\ln x})$

25 $-1/[x(\ln x)^2]$

27 $1/\sqrt{x^2 + 1}$

29 $1/[x\sqrt{x^2 + 4}]$

31 $(\ln x)^{n-1}/x$

33 $y = x - 1$

35 $y'' = -1/x^2 < 0$

37 $y_{\min} = y(1/e) = -1/e$

39 $y_{\min} = y(e^{-1/3}) = -\frac{1}{3}e^{-1}$

41 $y = x \ln x$

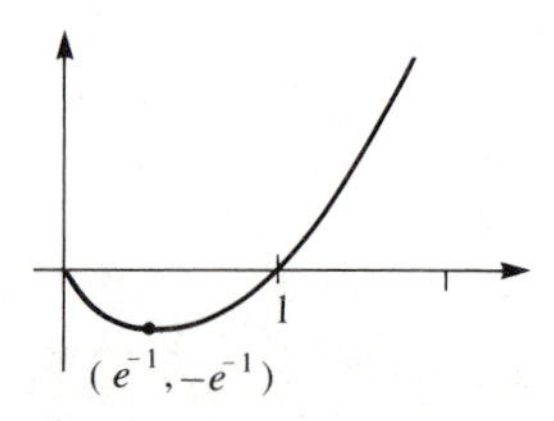

43 $+\infty$

45 0

47 $+\infty$

49 0

51 0

53 $dx/dy = (1 - \ln y)/y^2 < 0$ for $y > e$, so this strictly decreasing function has an inverse. In fact, $dy/dx = y^2/(1 - \ln y)$, $g(3/e^3) = e^3$, and $g'(3/e^3) = e^6/(1 - \ln e^3) = -\frac{1}{2}e^6$.

55 6

57 The graph of $y = \ln x$ is concave down because $y'' = -1/x^2 < 0$. Its tangent at $x = 1$ is $y = x - 1$. By the tangent theorem, $\ln x \le x - 1$ with equality only at $x = 1$.

59 $y/x > 1$, so

$$(y/x - 1)/(y/x) < \ln(y/x < y/x - 1,$$
$$(y - x)/y < \ln y - \ln x < (y - x)/x, \text{ etc.}$$

61 Replace x by $e^x > 1$ in 60:

$$x < n(e^{x/n} - 1),\quad x/n < e^{x/n} - 1,$$
$$1 + x/n < e^{x/n},\quad (1 + x/n)^n < e^x$$

63 Replace x by $e^x > 1$ in 62:

$$x > n(1 - e^{-x/n}),\quad e^{-x/n} > 1 - x/n,$$
$$e^{x/n} < (1 - x/n)^{-1},\quad e^x < (1 - x/n)^{-n}$$

65 $f' = (x - 1)^2/(2x^2) > 0$ for $x > 1$, $f(x) > f(1) = 0$, $\ln x < \frac{1}{2}(x - 1/x)$.

67

x	$\ln x$	$x - 1$	$2(\sqrt{x} - 1)$	$\sqrt{x} - 1/\sqrt{x}$
1.2	0.18232	0.20000	0.19089	0.18257
1.5	0.40547	0.50000	0.44949	0.40825
2.0	0.69315	1.00000	0.82843	0.70711
e	1.00000	1.71828	1.29744	1.04219
3.0	1.09861	2.00000	1.46410	1.15470

69 $f' = (x - 2)^2/4x > 0$ for $1 < x < 2$; hence f is strictly increasing. $\ln 2 = f(2) > f(1) = \frac{5}{8}$.

71 $\ln 3 > \ln 2 + \frac{7}{18} > \frac{5}{8} + \frac{7}{18} = \frac{73}{72} > 1$, $3 > e$.

73 $y = \ln x$ is concave down because $y'' < 0$. Apply the tangent theorem at $(a, \ln a)$:

$$\ln x \le \ln a + (x - a)/a$$

with equality only if $x = a$.

75 Set $x = n$ and $a = n + 1$ in 73.

Section 4-4 Applications of Logarithms

Page 258

1 1

3 1

5 0

7 1

9 25 ln 5

11 $\frac{1}{2}(\ln 3)\sqrt{2} \cdot 3^{1/\sqrt{2}}$

13 2.09590

15 1.58496

17 $8.55 \times 10^{13{,}394}$

19 $y_{\max} = y(e^{1/2}) = 1/(2e)$

21 $y_{\max} = y(\frac{1}{3}e) = 3/e$

23 $y = x^x$

minimum $(1/e, (1/e)^{1/e})$

25 $(x \ln x + x - 1)x^{x-2}$

27 $(1 - \ln x)x^{1/x-2}$

29 $(\ln 3)x^{(\ln 3)-1}$

31 $(2 \ln 10)xy$

33 $(\ln \ln x + 1/\ln x) \cdot y$

35 $-y/(x \ln x)$

37 $2(\ln x)y/x$

39 $\frac{3}{2}x^2/(x^3 + 2)^{1/2}$

41 $6y \cdot [2x/(x^2 + 4) - 1/(x + 7)]$

43 $y \cdot [\frac{2}{3}/(2x + 3) - 2x]$

45 $y \cdot [1 + 3x^2/(x^3 - 1) - 1/(2x + 1)]$

47 $\frac{5}{6}/x$

49 $1/(x - 3) + 1/(x - 5)$

51 $1/(x + 2) + 3/(x + 7)$

53
$$\frac{E(uv)}{Ex} = \frac{x}{uv}\frac{d(uv)}{dx} = \frac{x}{uv}\left(v\frac{du}{dx} + u\frac{dv}{dx}\right) = \frac{x}{u}\frac{du}{dx} + \frac{x}{v}\frac{dv}{dx} = \frac{Eu}{Ex} + \frac{Ev}{Ex}$$

55
$$\frac{Ey}{Ex} = \frac{x}{y}\frac{dy}{dx} = \frac{x}{y}\frac{dy}{du}\frac{du}{dx} = \left(\frac{u}{y}\frac{dy}{du}\right)\left(\frac{x}{u}\frac{du}{dx}\right) = \frac{Ey}{Eu}\frac{Eu}{Ex}$$

57 p

59 kx

61 The domain is $(0, +\infty)$. As x increases, e^{-x} decreases, $1 - e^{-x}$ increases, $y = \ln(1 - e^{-x})$ increases.

63 $y = ck^x$, $c > 0$, $k > 0$

65 By Example 6a, $(\ln x)/x$ is strictly decreasing for $x > e$; hence $(\ln a)/a > (\ln b)/b$, $b \ln a > a \ln b$, $a^b > b^a$.

67 Take logs: for large x,

$$(\ln x)^4 < \sqrt{x} \ln x < x \ln 2 < \tfrac{1}{2}x \ln x < x^2$$

Hence $y_2 < y_4 < y_1 < y_3 < y_5$.

69 $m(S_1) - m(S_2) = \frac{5}{2}\log[f(S_2)/f(S_1)]$

Section 4-5 Inverse Trigonometric Functions

Page 268

1 $\frac{1}{4}\pi$

3 $\frac{1}{3}\pi$

5 $\frac{1}{3}\pi$

7 $\frac{1}{2}\pi$

9 $\frac{1}{7}\pi$

11 $\frac{4}{5}$

13 $\frac{1}{2}\pi$

15 $\theta = \text{arc sin } e^x$, $x \le 0$

17 $1/\sqrt{9 - x^2}$

19 $2x/(1 + x^4)$

21 $6(\text{arc sin } 3x)/\sqrt{1 - 9x^2}$

23 $1/(1 + x^2)$

25 $1/(1 + x^2)$

27 2 arc tan $2x$

29 arc sin $\frac{1}{4}x$

31 $d\theta/dx = -a/(x^2 + a^2) - b/(x^2 + b^2)$

33 $d\theta/dx = -bx/[(x^2 + a^2 + b^2)\sqrt{x^2 + a^2}\,]$

35 $d\theta/dx = -a(a + x)/[(a^2 + b^2 + 2ax)\sqrt{b^2 - x^2}\,]$

37 $\theta_{\max} = \text{arc cot}\sqrt{a/b} - \text{arc cot}\sqrt{b/a}$

39 $\theta = \alpha + \beta = \text{arc cot } x/a + \text{arc cot}(c - x)/b$, $d\theta/dx = b/[b^2 + (c - x)^2] - a/(a^2 + x^2)$, $d\theta/dx = 0$ provided $b/[b^2 + (c - x)^2] = a/(a^2 + x^2)$. But $\sin\alpha = a/\sqrt{a^2 + x^2}$ and $\sin\beta = b/\sqrt{b^2 + (c - x)^2}$, so $d\theta/dx = 0$ provided $b \sin^2\alpha = a \sin^2\beta$.

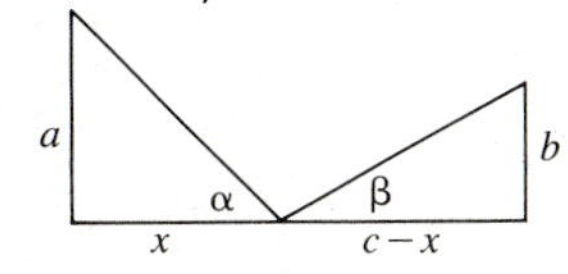

41 2 rad/sec

43 $\theta = \arc\tan y$, $\cot(\frac{1}{2}\pi - \theta) = y/1 = y$,
$\frac{1}{2}\pi - \theta = \text{arc cot } y$,
$\text{arc tan } y + \text{arc cot } y = \theta + (\frac{1}{2}\pi - \theta) = \frac{1}{2}\pi$

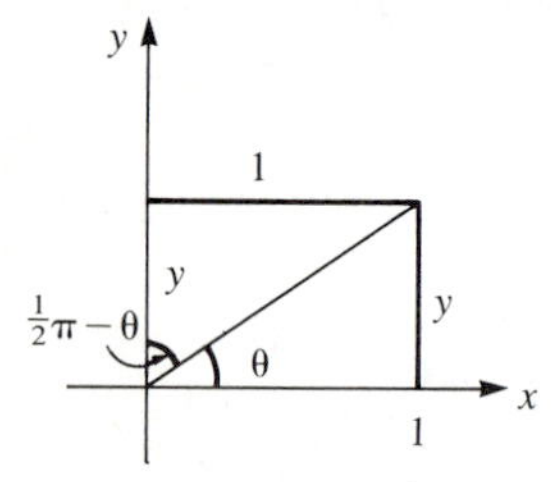

45 Set $\theta = \text{arc csc } x$ so $-\frac{1}{2}\pi \le \theta \le \frac{1}{2}\pi$.
Then $\sec(\frac{1}{2}\pi - \theta) = 1/\cos(\frac{1}{2}\pi - \theta) = 1/\sin\theta$
$= \csc\theta = x$ and $0 \le \frac{1}{2}\pi - \theta \le \pi$,
so $\frac{1}{2}\pi - \theta = \text{arc sec } x$.

47 Set $\theta = \text{arc tan}(1/x)$, so $0 < \theta < \frac{1}{2}\pi$. Then
$\cot\theta = 1/\tan\theta = 1/(1/x) = x$ so $\theta = \text{arc cot } x$.

49 Set $\theta = \text{arc cos } x$ so $x = \cos\theta$, $1 - x^2 = \sin^2\theta$. Since
$0 \le \theta \le \pi$, we have $\sin\theta \ge 0$, so $\sin\theta = \sqrt{1 - x^2}$.

51 Set $\theta = \text{arc cos } x$ so $0 \le \theta \le \frac{1}{2}\pi$.
Then $\cos 2\theta = 2\cos^2\theta - 1 = 2x^2 - 1$ and $0 \le 2\theta \le \pi$;
hence $2\theta = \text{arc cos}(2x^2 - 1)$.

53 Set $\alpha = \text{arc tan } x$ and $\beta = \text{arc tan } y$ so
$-\frac{1}{2}\pi < \alpha + \beta < \frac{1}{2}\pi$. Then

$$\tan(\alpha + \beta) = (\tan\alpha + \tan\beta)/(1 - \tan\alpha\tan\beta) = (x + y)/(1 - xy)$$

Hence $\alpha + \beta = \text{arc tan}[(x + y)/(1 - xy)]$

In Exercises **55–59**, set $f(x) = \text{arc tan } x$

55 $(\frac{1}{3} + \frac{1}{7})/(1 - \frac{1}{3}\cdot\frac{1}{7}) = \frac{1}{2}$ and $(\frac{1}{2} + \frac{1}{3})/(1 - \frac{1}{2}\cdot\frac{1}{3}) = 1$
so $2f(\frac{1}{3}) + f(\frac{1}{7}) = f(1) = \frac{1}{4}\pi$

57 $(\frac{1}{7} + \frac{3}{79})/(1 - \frac{1}{7}\cdot\frac{3}{79}) = \frac{2}{11}$ and $(\frac{1}{7} + \frac{2}{11})/(1 - \frac{1}{7}\cdot\frac{2}{11}) = \frac{1}{3}$
so $5f(\frac{1}{7}) + 2f(\frac{3}{79}) = f(\frac{1}{7}) + 2[2f(\frac{1}{7}) + f(\frac{3}{79})]$
$= f(\frac{1}{7}) + 2f(\frac{1}{3}) = \frac{1}{4}\pi$ by Exercise **55**.

59 $(\frac{1}{8} + \frac{1}{57})/(1 - \frac{1}{8}\cdot\frac{1}{57}) = \frac{1}{7}$
$(\frac{1}{8} + \frac{1}{7})/(1 - \frac{1}{8}\cdot\frac{1}{7}) = \frac{3}{11}$
$(\frac{1}{8} + \frac{3}{11})/(1 - \frac{1}{8}\cdot\frac{3}{11}) = \frac{7}{17}$
Therefore $3f(\frac{1}{8}) + f(\frac{1}{57}) = f(\frac{7}{17})$,
$(\frac{7}{17} + \frac{1}{239})/(1 - \frac{7}{17}\cdot\frac{1}{239}) = \frac{5}{12}$,
$(\frac{7}{17} + \frac{5}{12})/(1 - \frac{7}{17}\cdot\frac{5}{12}) = 1$.
Therefore $6f(\frac{1}{8}) + 2f(\frac{1}{57}) + f(\frac{1}{239}) = \frac{1}{4}\pi$.

Section 4-6 Hyperbolic Functions

Page 273

1 $\text{RHS} = \frac{1}{4}(e^u - e^{-u})(e^v + e^{-v}) + \frac{1}{4}(e^u + e^{-u})(e^v - e^{-v})$
$= \frac{1}{2}(e^ue^v - e^{-u}e^{-v}) = \text{LHS}$

3 Set $u = v = x$ in the text formula for $\cosh(u + v)$.

5 $\sinh 3x = \sinh(2x + x) = \sinh 2x\cosh x + \cosh 2x \sinh x$
$= 2\sinh x\cosh^2 x + (\cosh^2 x + \sinh^2 x)\sinh x$
$= 3\sinh x\cosh^2 x + \sinh^3 x$
$= 3\sinh x(\sinh^2 x + 1) + \sinh^3 x$, etc.

7 $5\cosh 5x$

9 $x(x^2 + 1)^{-/2}\,\text{sech}^2(x^2 + 1)^{1/2}$

11 $e^{2x}\cosh x$

13 $\tanh x$

15 $x^2\cosh x$

17 $\frac{1}{2}$

19 2

21 0

23 $$\text{LHS} = \frac{d}{dx}\left(\frac{\sinh x}{\cosh x}\right) = \frac{\cosh^2 x - \sinh^2 x}{\cosh^2 x} = \frac{1}{\cosh^2 x} = \text{sech}^2 x$$

25 $y = \cosh^{-1}x$ implies $x = \cosh y = \frac{1}{2}(e^y + e^{-y})$ so
$e^{2y} - 2xe^y + 1 = 0$. Solve for e^y: $e^y = x \pm \sqrt{x^2 - 1}$.
Since $y > 0$ for $x > 1$, we have $e^y > 1$, so we must take $+$. Now $y = \ln(x + \sqrt{x^2 - 1})$.

27 $x = \cosh y$ implies $1 = (\sinh y)(dy/dx)$; hence
$dy/dx = 1/\sinh y = 1/\sqrt{\cosh^2 y - 1} = 1/\sqrt{x^2 - 1}$.

29 $\text{LHS} = \ln(|\tan\theta| + \sqrt{\tan^2\theta + 1}) = \ln(|\tan\theta| + \sec\theta)$,
$\text{RHS} = \ln(\sec\theta + \sqrt{\sec^2\theta - 1}) = \ln(\sec\theta + |\tan\theta|)$

Section 4-7 Review Exercises

Page 274

1 $(2x + 3)e^{-3/x}$

3 $(1 - x)/e^x$

5 $\frac{4}{3}(x\ln x)^{1/3}(1 + \ln x)$

7 $\frac{1}{2}(\ln 2)2^{\sqrt{x}}/\sqrt{x}$

9 $\dfrac{1 + x\,\text{arc tan } x}{\sqrt{1 + x^2}}$

11 $1/(1 - x^3)$

13 -1

15 1

17 $\frac{5}{2}$

19 $e^{-1/e}$

21 Apply the tangent theorem to $y = e^x$, which is concave up, at $(0, 1)$.

23 1.84×10^5

25 $I = (E/R)(1 - e^{-Rt/L})$

27 5.28×10^8

29 $(2e/(e^2 - 1), 2/(e^2 - 1))$

31 $t = 2026.87 - (1.79 \times 10^{11}/N)^{1/0.99}$

33 $z'' = [ff'' - (f')^2]/f^2 < 0$

35 $\text{LHS} = (\cosh^2 x - \sinh^2 x)(\cosh^2 x + \sinh^2 x)$
$= \cosh^2 x + \sinh^2 x = \cosh 2x$

37 $$\text{RHS} = \frac{\sinh 2x + \cosh 2x - 1}{\sinh 2x + \cosh 2x + 1} = \frac{e^{2x} - 1}{e^{2x} + 1} = \frac{e^x - e^{-x}}{e^x + e^{-x}} = \tanh x$$

39 Set $f(x) = \frac{1}{2}\pi + \text{arc tan } x - 2\,\text{arc tan } 2x$. Then
$f'(x) = -3/(1 + x^2)(1 + 4x^2) < 0$ so $f(x)$
is strictly decreasing. But $\lim_{x\to\infty} f(x) = \frac{1}{2}\pi + \frac{1}{2}\pi - 2(\frac{1}{2}\pi) = 0$.
Therefore $f(x) > 0$ for all x.

41 $\text{arc cos } x = \text{arc tan}[(\sqrt{1 - x^2})/x]$ for $0 < x \le 1$

43 As $x \to 0+$, $(e^x - 1)/x \to de^x/dx|_0 = 1$ and $e^x \to 1$,
so $f(x) \to 1$. For $x \to +\infty$,

$$f(x) = f(x)\frac{e^{2x}}{e^{2x}} = \frac{x^2e^x}{e^{2x}}\left(\frac{e^x}{e^x - 1}\right)^2 = \left(\frac{x^2}{e^x}\right)\left(\frac{1}{1 - e^{-x}}\right)^2 \to 0\cdot 1^2 = 0$$

Chapter 5 Integration

Section 5-1 The Area Problems

Page 286

1 $\int = 6$

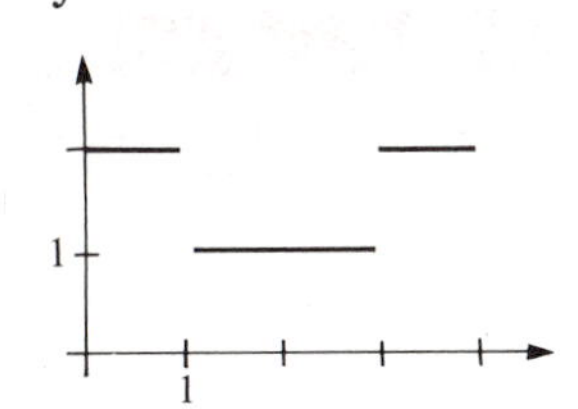

3 $\int = \frac{3}{4}$

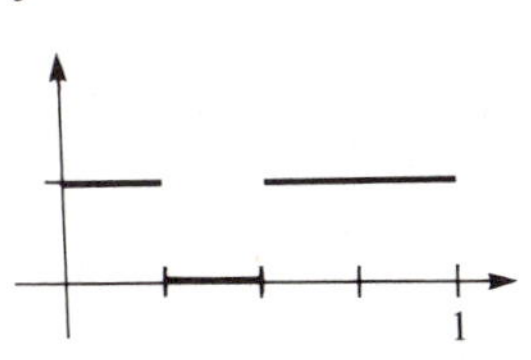

5 $\int = 4$

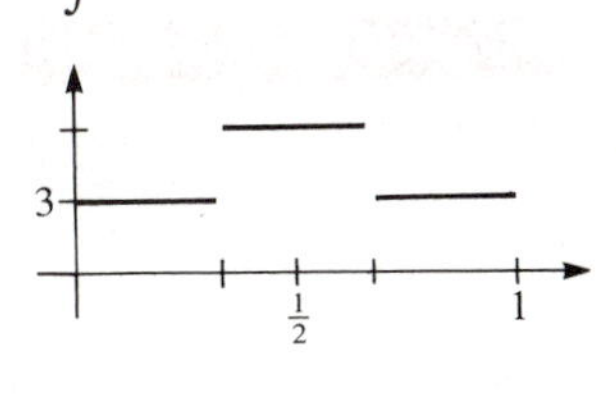

7 $\int = 3$

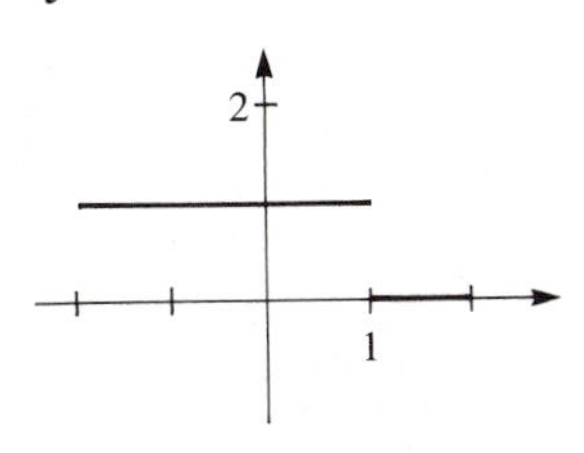

9 0.33

Section 5-2 Sums and Examples

Page 295

1 0

3 101

5 5050

7 385

9 Set $j = n - i$. As i goes from 0 to n, the j goes from n to 0.

11 Set $k = j - 10$. As j goes from 11 to $n + 10$, then k goes from 1 to n. Hence

$$\sum_{k=1}^{n} a_k = \sum_{j=11}^{n+10} a_k = \sum_{j=11}^{n+10} a_{j-10}$$

Change the dummy index k to j on the left side.

13 $S_n - S_{n-1} = \frac{1}{2}n(n+1) - \frac{1}{2}(n-1)(n)$
$= \frac{1}{2}n[(n+1) - (n-1)] = n$ and $S_0 = 0$,
so $1 + 2 + \cdots + n$
$= (S_1 - S_0) + (S_2 - S_1) + \cdots + (S_n - S_{n-1}) = S_n$

15 $\int_0^b x^2\,dx = \int_0^a x^2\,dx + \int_a^b x^2\,dx$. Hence
$\int_a^b x^2\,dx = \frac{1}{3}b^3 - \frac{1}{3}a^3 = \frac{1}{3}(b^3 - a^3)$.

17 $x^3 \ge s(x)$ on $[0, b]$ so $\int_0^b x^3\,dx \ge \int_0^b s(x)\,dx$.

$$\text{Now } \int_0^b s(x)\,dx = h\sum_{j=0}^{n-1} x_j^3 = h^4 \sum_{j=0}^{n-1} j^3$$
$$= \tfrac{1}{4}h^4 n^2 (n-1)^2$$
$$= \tfrac{1}{4}(b/n)^4 n^2 (n-1)^2$$
$$= \tfrac{1}{4}b^4(1 - 1/n)^2 \to \tfrac{1}{4}b^4$$

Therefore $\int_0^b x^3\,dx \ge \frac{1}{4}b^4$

19 $n(b^{1/n} - 1)$

21 Define $s(x)$ by $s(1) = 1$ and $s(x) = r^{-j-1}$ on $(r^j, r^{j+1}]$ so that $1/x \ge s(x)$ on $[1, b]$. Then

$$\int_1^b s(x)\,dx = \sum_{j=0}^{n-1} r^{-j-1}(r^{j+1} - r^j) = \sum_{j=0}^{n-1}(1 - 1/r)$$
$$= n(1 - 1/r) = n(b^{1/n} - 1)/b^{1/n}.$$

But $b^{1/n} \to 1$ as $n \to +\infty$ and

$$\lim_{n\to+\infty} n(b^{1/n} - 1) = \lim_{x\to 0+} \frac{b^x - 1}{x} = \left.\frac{db^x}{dx}\right|_{x=0} = \ln b$$

Therefore $\int_1^b \frac{dx}{x} \ge \ln b$.

Section 5-3 The Fundamental Theorem

Page 305

1 2

3 $\frac{3}{2}$

5 -3

7 0

9 16

11 45

13 -12

15 $\frac{1}{2}$

17 $\frac{8}{3}$

19 $-\frac{16}{3}$

21 $\frac{8}{3}a^3$

23 $1 - e^{-1}$

25 $4e^{-2}$

27 1

29 0

31 2

33 0

35 1

37 -16

39 -6

41 $\sec x$

43 $\sqrt{\sin x}\,(\cos x)$

45 $3\ \text{cm}^2/\text{sec}$

Section 5-4 Applications

Page 312

1 $y = (x - 1)^2 \quad A = 9$

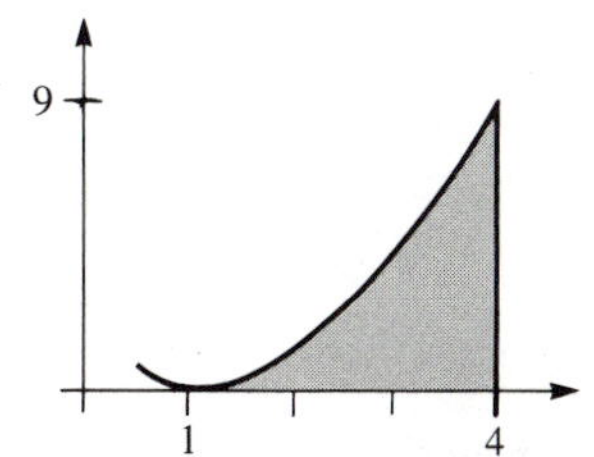

3 $y = 4(x - 3)^2 \quad A = 72$

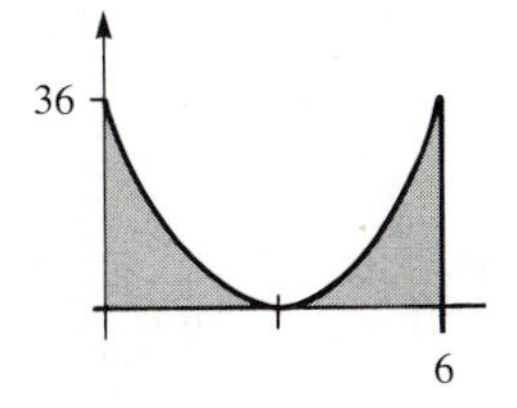

5 $y = \cos x \quad A = 2$

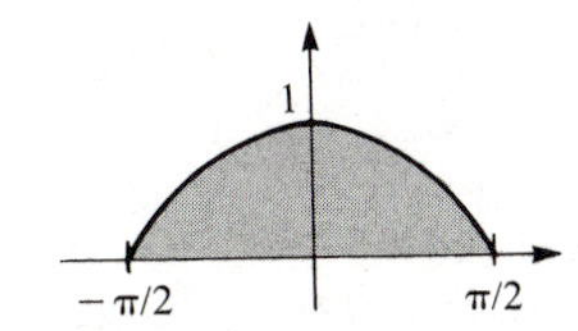

7 $y = \sin x + 4\cos x$
$A = 5$

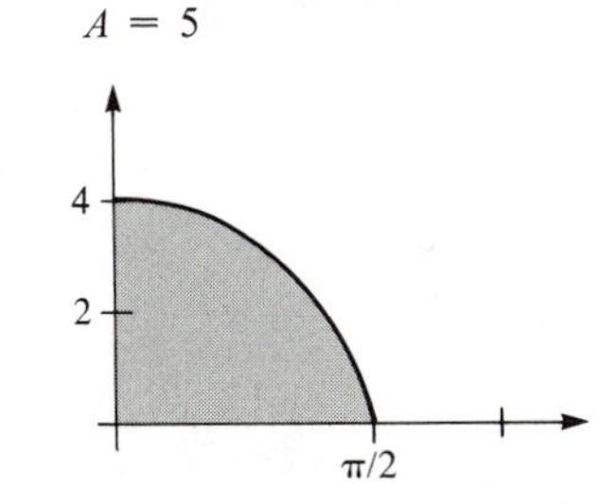

9 $y = (x^2 - 1)^2 \quad A = \frac{16}{15}$

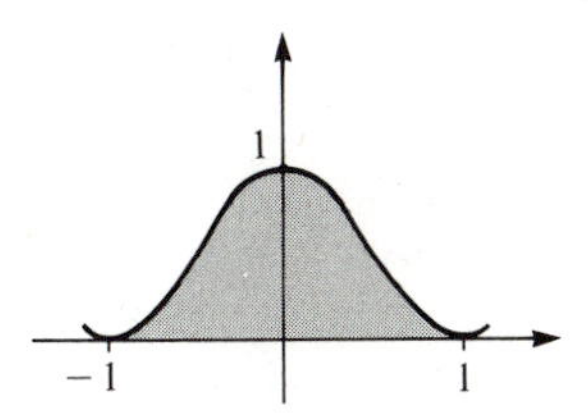

11 $A = 12$

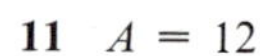

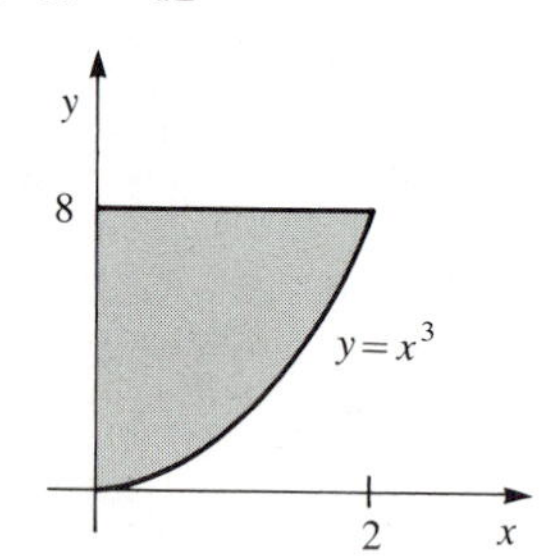

15 $A = \frac{29}{6}$

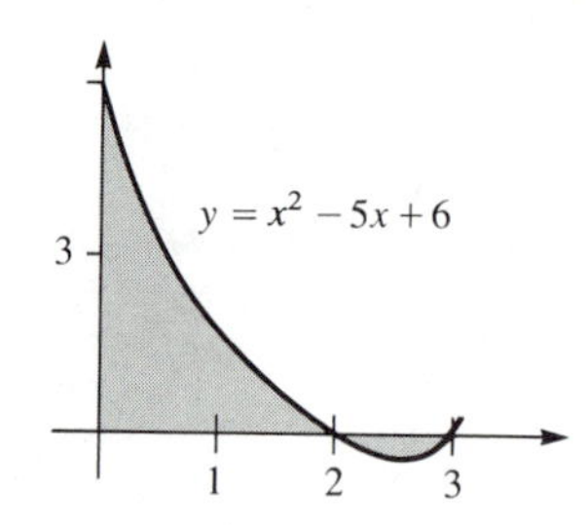

13 $A = 2$

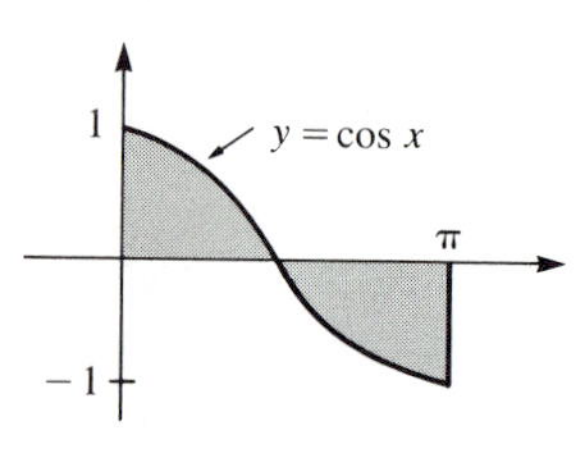

17 $A + B = b \cdot b^2 = b^3$

$$A = \int_0^b x^2\,dx = \tfrac{1}{3}b^3$$

$$B = \int_0^{b^2} \sqrt{y}\,dy = \tfrac{2}{3}y^{3/2}\Big|_0^{b^2} = \tfrac{2}{3}b^3$$

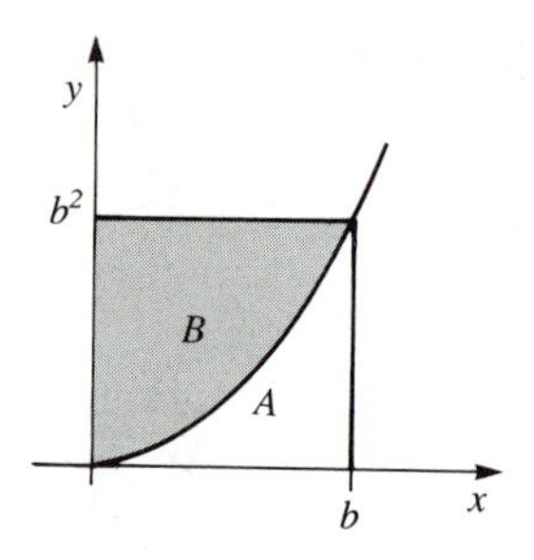

19 $\displaystyle\int_{-a}^{a} \sqrt{a^2 - x^2}\,dx = \tfrac{1}{2}$ area (circle of radius a) $= \tfrac{1}{2}\pi a^2$

21 0

23 $\frac{81}{5}$

25 $\frac{1}{2}(e - e^{-1})$

27 \$39

29 ≈ 1.095 cm/day

31 $y = t(x) = -\frac{1}{4}x + 1$. $f''(x) > 0$, so $y = f(x) = 1/x$ is concave up. By the tangent theorem $t(x) \le 1/x$ with "=" only for $x = 2$.

33 $f'(x) = 1/x > 0$; hence $f(x)$ is strictly increasing. Since it is continuous on an interval, its range is an interval. A continuous, strictly increasing function on an interval has an inverse function, also strictly increasing.

35 Suppose $F(x) = x_1$ and $F(y) = y_1$. Then $x + y = f(x_1) + f(y_1) = f(x_1y_1)$; hence $x + y$ is in the range of f, that is, in the domain of F. Indeed, $F(x + y) = x_1y_1 = F(x)F(y)$. Next, replace x_1 in $f(x_1y_1) = f(x_1) + f(y_1)$ by x_1/y_1 to obtain $f(x_1) = f(x_1/y_1) + f(y_1)$, that is, $f(x_1/y_1) = f(x_1) - f(y_1) = x - y$. Therefore $x - y$ is in the range of f, that is, in the domain of F. Indeed, $F(x - y) = x_1/y_1 = F(x)/F(y)$.

Section 5-5 Inequalities and Estimates

Page 317

1 For $0 < x < 1$ we have $0 < x^2 < x$, hence $1 = e^0 < \exp(x^2) < e^x$, so

$$\int_0^1 dx < \int_0^1 \exp(x^2) < \int_0^1 e^x\,dx$$

3 $3 < \sqrt{3 + 2x} < \sqrt{13} < 4$ for $3 < x < 5$, hence

$$\frac{3}{x} < \frac{\sqrt{3 + 2x}}{x} < \frac{4}{x} \qquad \text{etc.}$$

5 $0 \le \sin^2 x \le 1$ with equality only if $x = n\pi$ or $\frac{1}{2}\pi + n\pi$. Thus $e^{-x}\sin^2 x < e^{-x}$ except at a few points. But

$$\int_0^{100} e^{-x}\,dx = 1 - \varepsilon < 1.$$

7 $25 < 5 + 4x < 49$ for $5 < x < 11$, hence $5 < \sqrt{5 + 4x} < 7$ and

$$\frac{x}{7} < \frac{x}{\sqrt{5 + 4x}} < \frac{x}{5} \qquad \text{etc.}$$

9 $1 - k^2 < 1 - k^2\sin^2\theta < 1$ for $0 < \theta < \frac{1}{2}\pi$, hence

$$1 < \frac{1}{\sqrt{1 - k^2\sin^2\theta}} < \frac{1}{\sqrt{1 - k^2}} \qquad \text{etc.}$$

and $0 < \sin 2\theta$.

11 $\frac{1}{4} < 2^{-x} < \frac{1}{2}$ for $1 < x < 2$, hence $\frac{1}{4}x^3 < x^3 2^{-x} < \frac{1}{2}x^3$, etc.

13 $\sin x + \cos x = \sqrt{2}\cos(x - \frac{1}{4}\pi) < \sqrt{2}$ for $0 < x < 1$, hence

$$\frac{\sin x + \cos x}{(1 + x)^2} < \frac{\sqrt{2}}{(1 + x)^2} \qquad \text{etc.}$$

15 $\frac{1}{4}\pi < \arctan x < \frac{1}{2}\pi$ for $1 < x < 100$, hence

$$\frac{\pi}{4x^2} < \frac{\arctan x}{x^2} < \frac{\pi}{2x^2} \qquad \text{etc.}$$

17 $\displaystyle \text{Error} = \int_2^3 \frac{dx}{(x^2 + 1)^2} < \int_2^3 \tfrac{1}{25}\,dx = 0.04$

19 Suppose $f(c) = 2\varepsilon > 0$. By continuity there is a $\delta > 0$ so that $|f(x) - 2\varepsilon| < \varepsilon$ whenever x is in $[a, b]$ and $|x - c| < \delta$. By decreasing δ, and even taking $\frac{1}{2}\delta$ for δ in case c is an end point, we can find an interval $[a_1, b_1]$ in $[a, b]$ of positive length δ such that the inequality $|f(x) - 2\varepsilon| < \varepsilon$ holds on $[a_1, b_1]$. But this inequality implies $f(x) > \varepsilon$. Define $s(x) = \varepsilon$ on $[a_1, b_1]$ and $s(x) = 0$ elsewhere on $[a, b]$. Then $s(x) \le f(x)$ on $[a, b]$ and

$$\int_a^b s(x)\,dx = \int_{a_1}^{b_1} s(x)\,dx = \delta\varepsilon > 0$$

Therefore $\displaystyle\int_a^b f(x)\,dx > 0.$

Section 5-6 Approximate Integration

Page 324

	$n = 4$	$n = 10$	$n = 50$	exact	6 sig. figs.
1	0.697	0.6938	0.693172	$\ln 2$	0.693147
3	0.606	0.6125	0.613655	$2 - 2\ln 2$	0.613706
5	0.458	0.4580	0.458141	$\frac{1}{2}\ln\frac{5}{2}$	0.458145
7	9.37	9.307	9.29408	$\frac{1}{2}[4\sqrt{17} + \ln(4 + \sqrt{17})]$	9.29357
9	0.608	0.9201	0.996173	$1 - 11e^{-10}$	0.999501
11	0.504	0.5046	0.504785	unknown	0.504792
13	0.167	0.1676	0.167758	$(2\ln 2 - 1)/\ln 10$	0.167766
15	0.350	0.3471	0.346594	$\ln\sqrt{2}$	0.346574
17	1.18	1.148	1.14185	$\pi - 2$	1.14159
19	0.785	0.7854	0.785398	$\frac{1}{4}\pi$	0.785398
21	0.691	0.6928	0.693135		
23	0.618	0.6143	0.613730		
25	0.458	0.4582	0.458147		
27	9.25	9.287	9.29331		
29	1.149	1.0377	1.001162		
31	0.505	0.5049	0.504796		
33	0.168	0.1679	0.167769		
35	0.345	0.3463	0.346563		
37	1.12	1.138	1.14146		
39	0.785	0.7854	0.785398		

41 592.8 ft

	exact	trap.	midpoint
43	$\frac{8}{3} \approx 2.667$	3	$\frac{5}{2} = 2.5$
45	$\ln 3 \approx 1.099$	$\frac{7}{6} \approx 1.667$	$\frac{16}{15} \approx 1.067$

47 $|f''| = |-3e^{-x}\sin 2x - 4e^{-x}\cos 2x| < 7e^{-x} < \frac{7}{50}$ for $4 \le x \le 7$. Hence

$$|\text{error}| < \frac{(\frac{7}{50})(100)(\frac{3}{100})^3}{12} < (\frac{7}{50})(\frac{27}{12})(10^{-4}) < 4 \times 10^{-5}$$

49 993

51 $y = \ln x$ is concave down, so it lies *above* the trapezoidal approximation graph by the chord theorem. Therefore the trapezoidal rule yields too small an answer.

Section 5-7 Review Exercises

Page 325

1 $\int_0^{10} = 5$

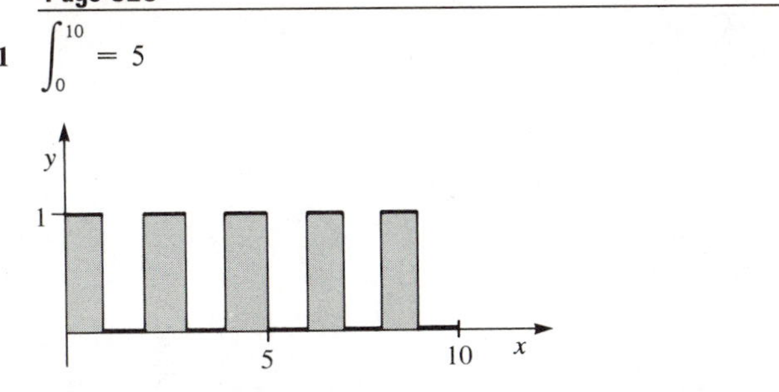

3 $\frac{512}{15}$

5 1

7 $\frac{1}{2}$

9 $-\exp(-x^2)$

11 $\frac{15}{2}$

13 $2/\pi$

15 $\frac{5}{4}$

17 $\frac{1}{2}x|x|$

19 For $0 \le x \le 1$,

$$0 \le f(x) = \int_0^x \le \int_0^x dt = x \le 1 < \tfrac{4}{3}$$

For $x > 1$,

$$0 \le f(x) = \int_0^1 + \int_1^x \le 1 + \int_1^x \frac{dt}{t^4} = 1 - \tfrac{1}{3}t^{-3}\Big|_1^x < \tfrac{4}{3}$$

For $x < 0$, $|f(x)| = |-f(-x)| < \frac{4}{3}$.

21 3.116

Chapter 6 Techniques of Integration*

Section 6-2 Differentials and Change of Variable

Page 335

* Constants of integration are omitted.

1 $\frac{1}{2}\sin^2 x$

3 $-\frac{1}{3}\cos^3 x$

5 e^{5x}

7 $-\exp(\cos x)$

9 $\frac{1}{4}(1 + \sin x)^4$

11 $\frac{1}{3}\sin 3x$

13 $\frac{1}{3}\tan x^3$

15 $\ln(e^x + x^2 + 1)$

17 $-\sqrt{1 - x^2}$

19 $-\frac{1}{4}(1 + x^2)^{-2}$

21 $\frac{1}{15}(3x + 1)^5$

23 $\arctan e^x$

25 $\frac{1}{3}(5 - 3x)^{-1}$

27 $\ln(e^x + e^{-x})$

29 $\frac{1}{4}\ln(1 + x^4)$

31 $\ln|\ln x|$

33 $\frac{1}{15}(5x^2 + 3)^{3/2}$

35 $\tan^3 \frac{1}{3}x$

37 $\exp(e^x)$

39 $\frac{1}{4}\ln|x^4 - 2x^2|$

41 $\frac{255}{16}$

43 $\frac{1}{2}(e^9 - e)$

45 1

47 $\frac{15}{4}$

49 $\frac{3}{16}$

51 2

53 $\frac{7}{576}$

55 $\frac{1}{4}(\ln 2)^4$

57 $\frac{1}{8}$

59 $\frac{9}{2}$

61 ≈ 2.308 m

Section 6-3 Examples of Substitutions

Page 339

1 $\frac{2}{5}(x-2)(x+3)^{3/2}$

3 $\frac{1}{3}(x-5)\sqrt{2x+5}$

5 $\ln|x-1| - \frac{1}{2}(4x-3)/(x-1)^2$

7 $2\sqrt{x} - 2\ln(1+\sqrt{x})$

9 $\frac{4}{3}\ln(x^{3/4}+1)$

11 $\frac{1}{2}\arctan e^{2x}$

13 $\frac{1}{5}\arctan(5x+2)$

15 $-\frac{1}{3}(x^2+2)\sqrt{1-x^2}$

17 $-\frac{1}{2}\ln(3+\cos 2x)$

19 $\frac{3}{112}(8x-3)(2x+1)^{4/3}$

21 $\frac{1}{2}x^2 - \frac{1}{2}\ln(x^2+1)$

23 $\frac{1}{3}\ln(3x+\sqrt{9x^2+1})$

25 $\frac{49}{3}$

27 $\frac{11}{36}$

29 $\frac{1}{14}$

31 $2 - 2\ln\frac{3}{2}$

33 $\frac{16}{3} - 3\sqrt{3}$

35 0

37 $c + x = u$

$$\int_a^b f(c+x)\,dx = \int_{c+a}^{c+b} f(u)\,du = \int_{c+a}^{c+b} f(x)\,dx$$

39 $t = cu \qquad \int_{ca}^{cb} dt/t = \int_a^b (c\,du)/(cu) = \int_a^b du/u = \int_a^b dt/t$

That is, $\ln cb - \ln ca = \ln b - \ln a$

Section 6-4 Integration by Parts

Page 346

1 $-\frac{1}{2}x\cos 2x + \frac{1}{4}\sin 2x$

3 $\frac{1}{4}(2x-1)e^{2x}$

5 $\frac{2}{3}x^{3/2}\ln x - \frac{4}{9}x^{3/2}$

7 $x\arctan x - \frac{1}{2}\ln(1+x^2)$

9 $\frac{1}{2}(x^2+1)\arctan x - \frac{1}{2}x$

11 $\frac{1}{13}e^{2x}(2\sin 3x - 3\cos 3x)$

13 $x\sinh x - \cosh x$

15 $a^{-3}(a^2x^2-2)\sin ax + 2a^{-2}x\cos ax$

17 $x(\ln x)^2 - 2x\ln x + 2x$

19 $x\ln\ln x$

21 $\frac{1}{2}x(\sin\ln x - \cos\ln x)$

23 2

25 $\frac{2}{9}e^3 + \frac{1}{9}$

27 $e - 2$

29 $\frac{1}{4}\pi - \frac{1}{2}$

31 $10e - 5$

33 $\frac{1}{24}\pi\sqrt{3} - \frac{5}{72}$

35 $$\text{LHS} = \frac{1}{2}\int_{-1}^{1} x\,d(\exp x^2) = \frac{1}{2}x\exp x^2\Big|_{-1}^{1} - \frac{1}{2}\int_{-1}^{1}\exp x^2\,dx = \text{RHS}$$

37 $$\text{LHS} = \int_0^{2\pi} f(x)\,d(\sin x) = f(x)\sin x\Big|_0^{2\pi} - \int_0^{2\pi} f'(x)\sin x\,dx = \text{RHS}$$

39 $$I = \int_0^{2\pi}\cos x\cos 2x\,dx = \frac{1}{2}\int_0^{2\pi}\sin x\sin 2x\,dx = \frac{1}{4}\int_0^{2\pi}\cos x\cos 2x\,dx = \frac{1}{4}I;\ \text{hence}\ \frac{3}{4}I = 0\ \text{and}\ I = 0.$$

41 $$\int_0^x \frac{\sin t}{t}\,dt = \int_0^x \frac{1}{t}\,d(1-\cos t) = \frac{1-\cos t}{t}\Big|_0^x + \int_0^x \frac{1-\cos t}{t}\,dt = \frac{1-\cos x}{x} + \int_0^x \frac{1-\cos t}{t^2}\,dt > 0$$

Section 6-5 Use of Algebra

Page 353

1 $\frac{1}{3}x^3 - 9x + \frac{82}{3}\arctan\frac{1}{3}x$

3 $2x + 9\ln|x-4|$

5 $\frac{1}{2}x^2 + 3x + 3\ln|x| - 1/x$

7 $a^{-1}(\arcsin ax - \sqrt{1-a^2x^2})$

9 $\ln|\tan x|$

11 $\frac{1}{5}\sin^5 x - \frac{1}{7}\sin^7 x$

13 $\frac{1}{5}\cos^5 x - \frac{1}{3}\cos^3 x$

15 $\frac{1}{3}\sin 3x - \frac{2}{9}\sin^3 3x + \frac{1}{15}\sin^5 3x$

17 $\frac{3}{8}x + \frac{1}{4}\sin 2x + \frac{1}{32}\sin 4x$

19 $\frac{1}{3}\tan^3 x - \tan x + x$

21 $-\frac{1}{2}\ln|\cos x^2|$

23 $\tan x + \sec x$

25 $-\ln\sqrt{2}$

27 1

29 0

31 $\frac{1}{2}\arctan\frac{1}{2}(x+1)$

33 $\arcsin\frac{1}{3}(x-3)$

35 $2\arcsin\frac{1}{2}(x-2) - \sqrt{4x-x^2}$

37 $(2\sqrt{3})^{-1}\ln|3x^2 - 2 + \sqrt{3}\sqrt{3x^4-4x^2+1}|$

39 $b^{-1}\ln|x/(ax-b)|$

Section 6-6 Change of Independent Variable

Page 358

1 $\ln|x + \sqrt{1+x^2}|$

3 $-\frac{1}{3}(x^2+8)\sqrt{4-x^2}$

5 $-\frac{1}{16}\sqrt{16-x^2}/x$

7 $-\sqrt{x^2+a^2}/x + \ln(x+\sqrt{x^2+a^2})$

9 $\frac{1}{2}\arctan x - \frac{1}{2}x/(1+x^2)$

11 $\ln(x+\sqrt{x^2-a^2})$

13 $\ln(x+\sqrt{x^2+a^2})$

15 $\frac{1}{2}x\sqrt{x^2+a^2} - \frac{1}{2}a^2\ln(x+\sqrt{x^2+a^2})$

17 $a^{-1}\tanh^{-1}(x/a) = (2a)^{-1}\ln[(a+x)/(a-x)]$

19 $\ln(1+\sqrt{2})$

21 $\frac{1}{2}\ln 3$

23 $\dfrac{dx}{xy} = \dfrac{x\,dx}{x^2y} = \dfrac{y\,dy}{x^2y} = \dfrac{dy}{x^2} = \dfrac{dy}{y^2 \mp a^2}$

$\displaystyle\int \frac{dx}{xy} = \int \frac{dy}{y^2 - a^2}$

$\displaystyle = \frac{1}{2a} \ln\left(\frac{y-a}{y+a}\right) + C$ if $y^2 = x^2 + a^2$

$\displaystyle\int \frac{dx}{xy} = \int \frac{dy}{y^2 + a^2}$

$\displaystyle = \frac{1}{a} \text{arc tan} \frac{y}{a} + C$ if $y^2 = x^2 - a^2$

25 $\dfrac{y\,dx}{x^2} = d\left(-\dfrac{y}{x}\right) + \dfrac{dy}{x}$

$\displaystyle\int \frac{y\,dx}{x^2} = -\frac{y}{x} + \ln(x+y) + C$ by Example 7.

27 $\dfrac{dx}{xy} = \dfrac{x\,dx}{x^2y} = \dfrac{-y\,dy}{x^2y} = -\dfrac{dy}{x^2} = -\dfrac{dy}{a^2 - y^2}$

$\displaystyle\int \frac{dx}{xy} = -\frac{1}{2a} \ln\left(\frac{a+y}{a-y}\right) + C$

29 $d\left(\dfrac{y}{x}\right) = \dfrac{x\,dy - y\,dx}{x^2} = \dfrac{xy\,dy - y^2\,dx}{x^2y}$

$\displaystyle = \frac{-x^2\,dx - y^2\,dx}{x^2\,y} = -a^2 \frac{dx}{x^2y}$

$\displaystyle\int \frac{dx}{x^2y} = \frac{-y}{a^2x} + C$

Section 6-7 Rational Functions

Page 366

1 $\frac{1}{2}\ln|(x-1)/(x+1)|$

3 $x + \frac{1}{3}\ln|(x-2)^4/(x+1)|$

5 $\frac{1}{2}\ln|(x+2)^4/[(x+1)(x+3)^3]|$

7 $x - \frac{3}{2}\text{arc tan}\, x + \frac{1}{2}x/(x^2+1)$

9 $\frac{2}{5}\ln|x-1| - \frac{1}{5}\ln(x^2+4) + \frac{3}{10}\text{arc tan}\,\frac{1}{2}x$

11 $\ln|(x-2)/(x-1)|$

13 $\ln|x^3/(x+1)^2|$

15 $\frac{3}{2}\ln|x^2/(x^2+1)| + 2\,\text{arc tan}\, x$

17 $\frac{1}{169}\{\ln[(x^2+9)^2/(x-2)^4] - 13/(x-2) - \frac{5}{3}\text{arc tan}\,\frac{1}{3}x\}$

19 $\frac{1}{2}x^2 + \frac{1}{6}\ln[(x-1)/(x^2+x+1)]$
$\quad + (1/\sqrt{3})\,\text{arc tan}[(2x+1)/\sqrt{3}]$

21 $\frac{1}{2}x^2 - 3x + \ln|(x+2)^8/(x+1)|$

23 $\frac{13}{17}\ln|x-3| + \frac{2}{17}\ln(x^2+2x+2) - \frac{1}{17}\text{arc tan}(x+1)$

25 $\ln\frac{4}{3}$

27 $\frac{1}{36}\pi\sqrt{3}$

29 $\frac{1}{3}\ln\frac{5}{4}$

31 $\frac{1}{2}\sqrt{2}\,\ln|(\tan\frac{1}{2}\theta - 1 + \sqrt{2})/(\tan\frac{1}{2}\theta - 1 - \sqrt{2})|$

Section 6-8 Integral Tables and Loose Ends

Page 373

1 $-\frac{1}{29}e^{-2x}(2\sin 5x + 5\cos 5x)$

3 $\frac{1}{32}x(8x^2-1)\sqrt{1-4x^2} + \frac{1}{64}\text{arc sin}\, 2x$

5 $\frac{1}{5}x - \frac{1}{25}\sqrt{10}\,\text{arc tan}(\frac{1}{2}x\sqrt{10})$

7 $12(x^2-8)\sin\frac{1}{2}x - 2x(x^2-24)\cos\frac{1}{2}x$

9 $\frac{1}{10}\sqrt{10x^2+7} - \frac{3}{5}\sqrt{10}\,\ln(x\sqrt{10} + \sqrt{10x^2+7})$
$\quad + \frac{2}{7}\sqrt{7}\,\ln|(\sqrt{7} + \sqrt{10x^2+7})/x|$

11 $-\frac{1}{2}\pi$

13 $\frac{77}{325}(e^{3\pi} - 1)$

15 $\frac{3}{4} + 5\ln\frac{7}{8}$

17 $\displaystyle\int_1^2 (x^3 - 5x)\,dx$

19 $\displaystyle\int_2^3 \frac{x\,dx}{x^4 + x^2 + 1}$

21 $\displaystyle 2\int_0^{\pi/2} \sin x\,dx$

23 $\displaystyle 8\int_0^{\pi/2} \sin^2 x\,dx$

25 $\displaystyle -2(\sin 4)\int_0^{4\pi} \cos\tfrac{1}{12}x\,dx$

27 $\displaystyle\int_{c-a}^{c} f(x)\,dx = \int_{-a}^{0} f(c+t)\,dt$

$\displaystyle = \int_{-a}^{0} f(c-t)\,dt$

$\displaystyle = -\int_{c+a}^{c} f(x)\,dx$

$\displaystyle = \int_{c}^{c+a} f(x)\,dx$

$\displaystyle\int_{c-a}^{c+a} f(x)\,dx = \int_{c-a}^{c} + \int_{c}^{c+a} = 2\int_{c}^{c+a} f(x)\,dx$

29 Line: $x = 0$

31 Point: $(-1, 0)$

33 Lines: $x = (n + \frac{1}{4})\pi$ Points: $((n - \frac{1}{4})\pi, 0)$

35 Line: $x = -1$

37 $f(x) = 0$ because $f(x) = f(-x) = -f(x)$, so $2f(x) = 0$.

39 $f(a) = b$

41 $\displaystyle\int_a^{a+p} f'(x)\,dx = f(x)\Big|_a^{a+p} = f(a+p) - f(a) = 0$

43 $f(x) = \dfrac{\sqrt{2}\sin x}{\sin x + \cos x}$ and $f(\frac{1}{2}\pi - x) = \dfrac{\sqrt{2}\cos x}{\cos x + \sin x}$

so $f(x) + f(\frac{1}{2}\pi - x) = \sqrt{2}$ and $\displaystyle\int_0^{\pi/2} f(x)\,dx = \tfrac{1}{4}\pi\sqrt{2}$

Section 6-9 Iteration Formulas

Page 377

1 $J_n = x\ln^n x - nJ_{n-1}$

3 $J_n = \frac{1}{3}x^3\ln^n x - \frac{1}{3}nJ_{n-1}$

5 $J_n = \dfrac{-1}{2(n-1)a^2}\left[\dfrac{x}{(x^2-a^2)^{n-1}} + (2n-3)J_{n-1}\right]$

7 $J_n = (n-1)^{-1}\tan^{n-1}x - J_{n-2}$

9 $J_n = (n-1)^{-1}[\sec^{n-2}x\tan x + (n-2)J_{n-2}]$

11 $J_n = n^{-1}[-\sin^{n-1}x\cos x + (n-1)J_{n-2}]$

13 $\frac{16}{35}$

15 $\frac{263}{315} - \frac{1}{4}\pi$

17 $2a^4 - 8a^3 + 24a^2 - 48a + 24,\ a = \ln 2$

19 $J_n = (n-1)^{-1}(e^{ax}\tan^{n-1}x - aJ_{n-1}) - J_{n-2}$

Section 6-10 Review Exercises

Page 378

1 $(a-b)^{-1}\ln|(x-a)/(x-b)|$
3 $-\sqrt{5+\cos 2x}$
5 $\frac{1}{12}\tan^4 3x + \frac{1}{18}\tan^6 3x$
7 $\frac{1}{7}\sin^7 x - \frac{1}{9}\sin^9 x$
9 $\frac{1}{105}(15x^4 - 12x^2 + 8)(1 + x^2)^{3/2}$
11 $2(\sin\sqrt{x} - \sqrt{x}\cos\sqrt{x})$
13 $\arcsin(x-1) - \sqrt{2x - x^2}$
15 $\frac{1}{3}x^3 \arctan x - \frac{1}{6}x^2 + \frac{1}{6}\ln(1 + x^2)$
17 $a^{-4}[-1/x - (b/a^2)\arctan(bx/a^2)]$
19 $\frac{1}{3}\ln[(e^x + 1)/(e^x + 4)]$
21 $\ln|\ln \ln x|$
23 $(2 - x^2)\cos x + 2x \sin x$
25 $-\ln|(1 + \sqrt{1 + x^2})/x|$
27 $\frac{1}{2}\pi a^2$
29 $b^{n+2}/[(n+1)(n+2)]$

7 $A = \frac{8}{3}$

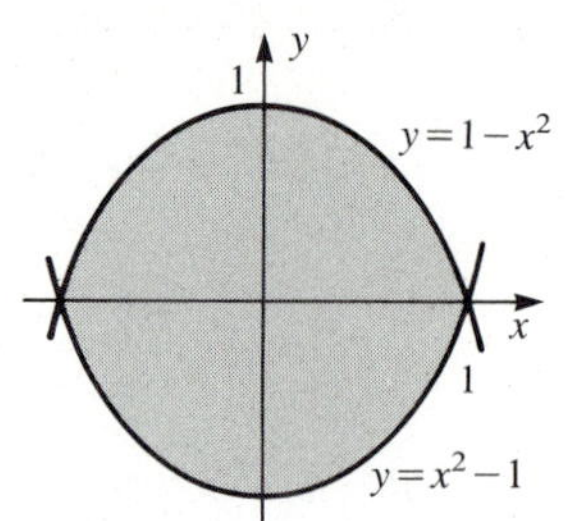

9 $A = 2$

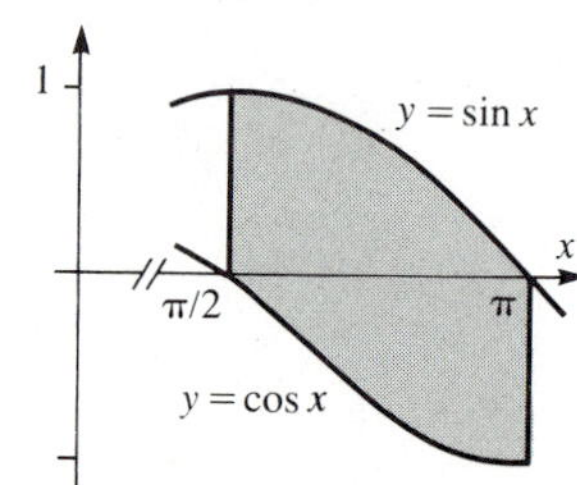

11 36
13 4
15 $\frac{36}{25}$
17 $2\sqrt{2}$
19 $2e^2 + 2$
21 $\frac{3}{2}$
23 $\frac{128}{3}$
25 $a = \frac{3}{2}$
27 $\frac{1}{6}k(b-a)^3$
29

Chapter 7 Applications of Integration

Section 7-2 Area

Page 386

1 $A = e - \frac{5}{2}$

3 $A = \frac{3}{4}$

5 $A = \ln 3 - \frac{2}{3}$

Section 7-3 Volume

Page 394

1 $\frac{1}{6}\pi r^3$
3 $\frac{1}{15}\pi b^{5/2}$
5 117π
7 $\frac{4}{5}\pi$
9 $\frac{26}{81}\pi$
11 $\frac{1}{2}\pi(e^{2b} - e^{2a})$
13 72π
15 $\pi(\frac{1}{4}\sinh 2a - \frac{1}{2}a)$
17 102π
19 $\frac{1}{4}\pi(e^4 + 1)(e^4 + 3)$
21 $\frac{1}{3}\pi h(a^2 + ab + b^2)$
23 $\frac{1}{3}\pi(r-b)^2(2r+b)$
25 $\frac{1}{6}\pi h^3$
27 8
29 $(3.75 \times 10^6)\pi/\ln 2 \approx 17.0 \times 10^6 \text{ ft}^3$
31 $\frac{1}{3}Sr$

Section 7-4 Work

Page 399

1 84 J
3 2500 ft-lb
5 1, 4, 16 ft-lb
7 4500 ft-lb
9 $\approx 2.28 \times 10^9$, $\approx 4.25 \times 10^9$ J
11 $\frac{4}{3}\pi\delta a^3(a+b)$
13 $\frac{113}{1024}wh$
15 $mga(1 - \cos\phi)$
17 $\approx$27,582 J

Section 7-5 Fluid Pressure

Page 403

1 $\frac{1}{6}\delta gbh^2$
3 $\frac{1}{6}\delta g(a + 2b)h^2$
5 $\frac{16}{3}\delta ga^3$
7 $(\frac{1}{2}\pi - \frac{2}{3})a^3\delta g$
9 $\frac{1}{6}a^4\delta g$
11 $\approx 4.79 \times 10^7$ kg
13 $c = -3$
15 $\frac{1}{2}\delta ghA$

Section 7-6 Miscellaneous Applications

Page 412

1 $(c/r)(1 - e^{-rT})$

3 $(b/r^2)[1 - (rT + 1)e^{-rT}]$

5 $\dfrac{c}{r}\left(\dfrac{1 - e^{-r}}{1 + e^{-r}}\right)(1 - e^{-2Nr})$

7 $\dfrac{c\pi}{\pi^2 + r^2}(1 - e^{-2Nr})$

9 $f(t) = ce^{rt}$

11 For the given $A(t)$,

$$A'(t) = A_0e^{\phi(t)}\phi'(t) = A_0e^{\phi(t)}r(t) = A(t)r(t),$$

and $A(0) = A_0$. Thus the natural growth law for the growth of A_0 at interest rate $r(t)$ is satisfied.

13 $|t| \geq 0$ and $\displaystyle\int_{-1}^{1} |t|\,dt = 1$. $E = 0$, $\text{var} = \frac{1}{2}$.

15 $3t^2 \geq 0$ and $\displaystyle\int_0^1 3t^2\,dt = 1$. $E = \frac{3}{4}$, $\text{var} = \frac{3}{80}$.

17 $1 - \cos t \geq 0$ and $\displaystyle\frac{1}{2\pi}\int_{-\pi}^{\pi}(1 - \cos t)\,dt = 1$
$E = 0$, $\text{var} = \frac{1}{3}\pi^2 + 2$

19 $f(t) \geq 0$ and $\displaystyle\int_0^1 f(t)\,dt = \left(\int_0^{1/3} + \int_{2/3}^{1}\right)(\tfrac{3}{2}\,dt) = 1$
$E = \frac{1}{2}$, $\text{var} = \frac{13}{108}$

21 $F = \frac{1}{2} + \frac{1}{2}t|t|$

23 $F'(t) = f(t)$

25 $T(x) = \sqrt{T_0^2 + \delta^2x^2}$

27 $\displaystyle\bar{x} = \frac{1}{M}\int_a^b x\delta(x)\,dx$ where $\displaystyle M = \int_a^b \delta(x)\,dx$

29 $T_0 \approx 2.37 \times 10^3\,\text{K}$

Section 7-7 Review Exercises

Page 713

1 $\frac{4}{3}$

3 $x = \frac{20}{11}$

5 $\frac{4}{3}$

7 $\frac{873}{128}$

9 $\frac{1}{7}\pi$

11 $\frac{56}{5}\pi\,\text{ft}^3$

13 $abh + \frac{1}{2}(a\tan\beta + b\tan\alpha)h^2 + \frac{1}{3}h^3\tan\alpha\tan\beta$

15 $\frac{1}{6}\,\delta g\,h^2(3b + h\tan\beta)$

17 $\frac{2}{3}(\sqrt{13} - 1)$

19 $\displaystyle V = \int_{-a}^{a} 2\sqrt{a^2 - y^2}\,(h - y\tan\alpha)\,dy$

21 $\displaystyle W = \int_{-a}^{a} \delta g(h - y\tan\alpha)^2\sqrt{a^2 - y^2}\,dy$

23 $p(x) = p_0x_0/x$

25 102,500 gm-cm

27 $V \approx 6.372 \times 10^6\,\text{ft}^3$, $A \approx 4.828 \times 10^5\,\text{ft}^2$

29 $k = \frac{1}{4}$, $E = \frac{8}{5}$

Chapter 8 Analytic Geometry

Section 8-1 Translation and Circles

Page 420

1 $(8, 3)$

3 $(7, -3)$

5 $(2, -9)$

7 $(-6, 6)$

9 $(-13, -4)$

11 $(-14, 6)$

13 $x = \bar{x} + h$, $y = \bar{y} + k$, where $3h - 2k = 1$

15 $\bar{x} = x + 1$, $\bar{y} = y$

17 $\bar{x} = x - \frac{1}{6}\pi$, $\bar{y} = y + 1$

19 $(3, -3)$

21 $x = x_1 + r(x_2 - x_1)$ and $y = y_1 + r(y_2 - y_1)$. Point (x, y) is on the *ray* from (x_1, y_1) through (x_2, y_2), at distance from (x_1, y_1) equal to r times the length of the segment from (x_1, y_1) to (x_2, y_2).

23 Circle: center $(2, 2)$, radius $\sqrt{8}$

25 Circle: center $(-1, -3)$, radius 6

27 Circle: center $(\frac{1}{2}, -1)$, radius $\frac{1}{2}\sqrt{5}$

29 Circle: center $(\frac{3}{4}, \frac{5}{4})$, radius $\sqrt{\frac{13}{8}}$

31 The equation simplifies to

$$(a_2 - a_1)x + (b_2 - b_1)y = k$$

where k is a constant. This is a line of slope $m = -(a_2 - a_1)/(b_2 - b_1)$, so it is perpendicular to the line of centers, which has slope the negative reciprocal of m. (The usual horizontal and vertical special cases obtain.)

Section 8-2 Intersections of Lines and Circles

Page 426

1 $(-\frac{1}{2} \pm \frac{1}{2}\sqrt{17}, \frac{1}{2} \pm \frac{1}{2}\sqrt{17})$

3 $(1, 2)$: point of tangency

5 Empty

7 $(\frac{6}{5} \pm \frac{4}{5}\sqrt{31}, \frac{7}{5} \mp \frac{2}{5}\sqrt{31})$

9 $(-\frac{21}{8}, \pm\frac{3}{8}\sqrt{15})$

11 Empty

13 $x = -1$

15 $x - y = \sqrt{2}$

17 $\frac{1}{2}\sqrt{3}\,x + \frac{1}{2}y = 1$

19 $x = 0$ and $y = \frac{3}{4}x$

21 $-2x + 3y = 13$ and $3x + 2y = -13$

23 Distance between centers $= \sqrt{(3 - 1)^2 + (6 - 2)^2} = 2\sqrt{5}$
Sum of radii $= \sqrt{\frac{45}{4}} + \sqrt{\frac{5}{4}} = 2\sqrt{5}$

25 The second is $(x - \frac{4}{25})^2 + (y - \frac{3}{25})^2 = (\frac{4}{5})^2$
Distance between centers $= \sqrt{(\frac{4}{25} - 0)^2 + (\frac{3}{25} - 0)^2} = \frac{1}{5}$
Difference of radii $= 1 - \frac{4}{5} = \frac{1}{5}$
Circles are tangent, second inside first.

27 $3x - 4y = 25$ and $-3x + 4y = 25$

29 $(x - 5)^2 + (y - 5)^2 = 25$ and
$(x - 13)^2 + (y - 13)^2 = 169$

31 $(x - 4)^2 + (y + 1)^2 = 9$ and $(x - 1)^2 + (y + 5)^2 = 4$
Centers: $(4, -1)$, $(1, -5)$ Radii: 3, 2
Distance between centers: $\sqrt{(4 - 1)^2 + (-1 + 5)^2} = 5$
Sum of radii: $3 + 2 = 5$. The circles are tangent, and their point of tangency is $\frac{3}{5}$ of the way from $(4, -1)$ to $(1, -5)$, at $(\frac{2}{5}\cdot 4 + \frac{3}{5}\cdot 1, \frac{2}{5}(-1) + \frac{3}{5}(-5)) = (\frac{11}{5}, -\frac{17}{5})$

33 $x^2 + (mx)^2 + 2ax + 2b(mx) + c = 0$,
$(1 + m^2)x^2 + 2(a + mb)x + c = 0$,
$D = 4[(a + mb)^2 - c(1 + m^2)]$.
Tangent if and only if unique intersection if and only if $D = 0$.

Section 8-3 Curves Defined Geometrically

Page 429

1 The line parallel to L, halfway from A to L

3 Concentric circle of radius 4

5 Circle: center $(\frac{25}{8}, \frac{25}{6})$, radius $\frac{25}{24}$

7 Circle of radius $\frac{1}{2}r$, tangent internally to the given circle at A

9 Line, perpendicular to the ray from the origin through the center of C, at distance $(2r)^{-1}$ from the origin

11 Line if $a = 1$, circle if $a \neq 1$

13 (x, y) where $x = (ac + b^2u - abv)/(a^2 + b^2)$ and $y = (bc + a^2v - abu)/(a^2 + b^2)$

Section 8-4 Conics — The Parabola

Page 435

1 $(\frac{37}{12}, 0)$, $x = \frac{35}{12}$, $y = 0$, $(3, 0)$

3 $(\frac{1}{8}, -1)$, $x = -\frac{1}{8}$, $y = -1$, $(0, -1)$

5 $(-2, \frac{5}{6})$, $y = -\frac{13}{6}$, $x = -2$, $(-2, -\frac{2}{3})$

7 $y = 3x^2$

9 $x - 1 = -(y - 2)^2$

11 $y + 3 = \frac{1}{16}(x - 2)^2$

13 x-axis and the parabola $x = \frac{1}{4}y^2$

15 Parabolas $x + \frac{1}{2} = \frac{1}{6}y^2$ and $x - \frac{3}{2} = -\frac{1}{2}y^2$

17 Ray, parallel to the axis of P, end point on P, inside P.

19 $P = (0, 1)$. Let $X = (u, v)$ so $4v = u^2$. The center of the circle is $(\frac{1}{2}u, \frac{1}{2}(1 + v))$ and its radius r satisfies

$$(2r)^2 = \overline{PX}^2 = u^2 + (v - 1)^2$$
$$= 4v + (v - 1)^2 = (v + 1)^2$$
$$r = \tfrac{1}{2}(v + 1) = y\text{-coordinate of center}$$

The result follows.

21 $y = -x - 1$

23 $y = 4x - 8$

25 $y = \frac{1}{2}(-1 \pm \sqrt{5})x + \frac{1}{2}(-3 \pm \sqrt{5})$

27 $y = -2x + 1$, $y = -6x - 3$

29 $v > au^2$ and $y + v = 2aux$, so

$$y = 2aux - v < 2aux - au^2$$
$$= ax^2 - a(u - x)^2 \leq ax^2$$

Hence (x, y) is outside.

31 $4p(y + p) = x^2 \quad (p \neq 0)$

33 $\arctan(\frac{4}{3}\sqrt{3})$

35 Take $A = (u, u^2)$ and $B = (v, v^2)$. Then OA and OB have slopes u and v respectively; hence $uv = -1$, $v = -1/u$, $B = (-1/u, 1/u^2)$. The slope of AB is

$$\frac{u^2 - 1/u^2}{u + 1/u} = \frac{u^2 - 1}{u}$$
$$= \frac{u^2 - 1}{u - 0}$$

This equals the slope of $A(0, 1)$, so $(0, 1)$ is on AB.

37 Parabola $y = \frac{1}{2}x^2 + 1$

Section 8-5 Conics — The Ellipse

Page 444

1 $(0, 0)$, 5 and 3, $(\pm 5, 0)$ and $(0, \pm 3)$, $(\pm 4, 0)$

3 $(-1, 2)$, $\sqrt{2}$ and 1, $(-1, 2 \pm \sqrt{2})$ and $(0, 2)$ and $(-2, 2)$, $(-1, 3)$ and $(-1, 1)$

5 $(3, 2)$, 1 and $\frac{1}{2}\sqrt{2}$, $(3, 3)$ and $(3, 1)$ and $(3 \pm \frac{1}{2}\sqrt{2}, 2)$, $(3, 2 \pm \frac{1}{2}\sqrt{2})$

7 $\frac{1}{81}(x - 1)^2 + \frac{1}{4}(y - 4)^2 = 1$

9 $\frac{1}{25}(x - 5)^2 + \frac{1}{16}y^2 = 1$

11 $\frac{1}{16}(x - 1)^2 + \frac{1}{12}y^2 = 1$

13 $x/a^2 + y(dy/dx)/b^2 = 0$, so $dy/dx = 0$ at $(0, \pm b)$. Similarly $x(dx/dy)/a^2 + y/b^2 = 0$, so $dx/dy = 0$ at $(\pm a, 0)$.

15 Take $x^2/a^2 + y^2/b^2 = 1$, with $a > b > 0$. If (x, y) is on the ellipse, then

$$1 = \frac{x^2}{a^2} + \frac{y^2}{b^2} \geq \frac{x^2}{a^2} + \frac{y^2}{a^2}$$

hence $\quad x^2 + y^2 \leq a^2$

that is, $\sqrt{x^2 + y^2} \leq a$. Clearly there is equality if and only if $y = 0$.

17 Ellipse, center the origin, axes the x- and y-axes.

19 Ellipse, major axis D, semi-axes r and $\frac{1}{2}r$

21 Ellipse, foci $(a, 0)$ and $(b, 0)$, length sum $a + b$

23 Let $x = a\cos\theta$ and $y = b\sin\theta$. The arc from $(a, 0)$ to (x, y) is given by $x = a\cos t$, $y = b\sin t$, where $0 \leq t \leq \theta$.

$$A = \int_{t=\theta}^{t=0} y\,dx = -ab\int_{\theta}^{0} \sin^2 t\,dt = \tfrac{1}{2}ab(\theta - \tfrac{1}{2}\sin 2\theta)$$
$$= \tfrac{1}{2}ab\theta - \tfrac{1}{2}ab\sin\theta\cos\theta = \tfrac{1}{2}ab\theta - \tfrac{1}{2}xy$$

25 $x = 1$

27 $2x + y = 10$

29 $x = 1$ and $-17x + 12y = 19$

31 $x = -1$ and $-x + 4y = 5$

Section 8-6 Conics — The Hyperbola

Page 450

1 x-axis, $(0, 0)$, $(\pm\sqrt{13}, 0)$, $y = \pm\frac{3}{2}x$

3 y-axis, $(0, 0)$, $(0, \pm\sqrt{13})$, $y = \pm\frac{2}{3}x$

5 $y = 1$, $(-1, 1)$, $(-1 \pm \sqrt{2}, 1)$, $y = x + 2$ and $y = -x$

7 $x = -2$, $(-2, -2)$, $(-2, -2 \pm \sqrt{\frac{96}{5}})$, $x + 2 = \pm\sqrt{5}(y + 2)$

9 $y = -1$, $(3, -1)$, $(3 \pm \sqrt{5}, -1)$, $2x - y = 7$ and $2x + y = 5$

11 $-\frac{1}{9}x^2 + \frac{1}{16}y^2 = 1$

13 $\frac{1}{4}x^2 - \frac{1}{16}y^2 = 1$

15 $\frac{1}{3}(x - 1)^2 - \frac{1}{3}y^2 = 1$

17 $(x + a)^2 - (y - b)^2 = a^2 - b^2 - c \neq 0$. Hyperbola with asymptotes $y - b = \pm(x + a)$ of slopes ± 1, hence perpendicular.

19 $(x/a^2)(dx/dy) - y/b^2 = 0$, $dx/dy = (a^2/b^2)y/x$. Hence $dx/dy = 0$ at $(\pm a, 0)$.

21 Right branch of the hyperbola with foci P, Q and distance difference $r - s$

23 The set of centers of circles that are tangent to the given circles, to one externally and to the other internally, is a hyperbola.

25 Take $A = (-5, 0)$, $B = (5, 0)$ and $C = (5, -8)$. The explosion is at $(\sqrt{\frac{5}{3}}, -4)$.

27 $x = a\cosh t \geq a$ and $x^2/a^2 - y^2/b^2 = \cosh^2 t - \sinh^2 t = 1$. Conversely, if $x^2/a^2 - y^2/b^2 = 1$ and $x \geq a$, set $t = \operatorname{arg\,sinh} y/b$. Then $x^2/a^2 = 1 + y^2/b^2 = 1 + \sinh^2 t = \cosh^2 t$, so $x = a\cosh t$.

29 The unshaded portion of the triangle.

31 Differentiate: $x/a^2 - (y/b^2)(dy/dx) = 0$. Hence

$$\left.\frac{dy}{dx}\right|_{(u,v)} = \frac{b^2}{a^2}\cdot\frac{u}{v}$$

The tangent at (u, v) is $y - v = (b^2/a^2)(u/v)(x - u)$, which simplifies to the stated equation. (Remember that $u^2/a^2 - v^2/b^2 = 1$.)

33 Let the conics be $\dfrac{x^2}{c+\lambda} + \dfrac{y^2}{\lambda} = 1$, $\dfrac{x^2}{c-\mu} - \dfrac{y^2}{\mu} = 1$, where $0 < \lambda$ and $0 < \mu < c$. Subtract, simplify, and cancel $\lambda + \mu$:

$$\frac{x^2}{(c+\lambda)(c-\mu)} = \frac{y^2}{\lambda\mu}, \quad \left[\frac{-\lambda x}{(c+\lambda)y}\right]\left[\frac{\mu x}{(c-\mu)y}\right] = -1$$

This is exactly the condition that the respective slopes at an intersection point (x, y) are negative reciprocals of each other.

Section 8-7 Polar Coordinates

Page 457

1 $(0, 1)$

3 $(\frac{1}{2}\sqrt{3}, -\frac{1}{2})$

5 $(-\sqrt{2}, -\sqrt{2})$

7 $[\sqrt{2}, \frac{1}{4}\pi]$

9 $[\sqrt{2}, \frac{3}{4}\pi]$

11 $[2, -\frac{1}{6}\pi]$

13 $\theta = \frac{1}{4}\pi$

15 $r(\cos\theta + \sin\theta) = 1$

17 $r = -2a\cos\theta$

19 $r = 2\cos(\theta + \frac{3}{4}\pi)$

21 $r(\cos\theta + 2\sin\theta) = 5$

23 $r^2 - 10r\cos(\theta - \frac{1}{6}\pi) + 9 = 0$

25 $\frac{3}{5}x - \frac{4}{5}y = 1$

27 $-\frac{7}{25}x - \frac{24}{25}y = 2$

29 $\frac{1}{2}\sqrt{2}\,x + \frac{1}{2}\sqrt{2}\,y = \frac{3}{2}\sqrt{2}$

31 $x = \bar{x} + h,\ y = \bar{y} + k,$

$\bar{x}\cos\alpha + \bar{y}\sin\alpha = p - (h\cos\alpha + k\sin\alpha) = \bar{p}$

If $\bar{p} \geq 0$ this is it; otherwise

$\bar{x}\cos(\alpha + \pi) + \bar{y}\sin(\alpha + \pi) = -\bar{p}$

Section 8-8 Polar Graphs

Page 464

1 $r = 2\sin 2\theta$

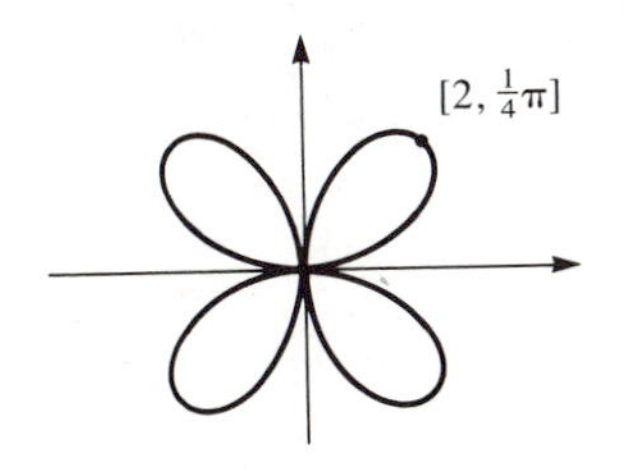

3 $r = \cos 3\theta$

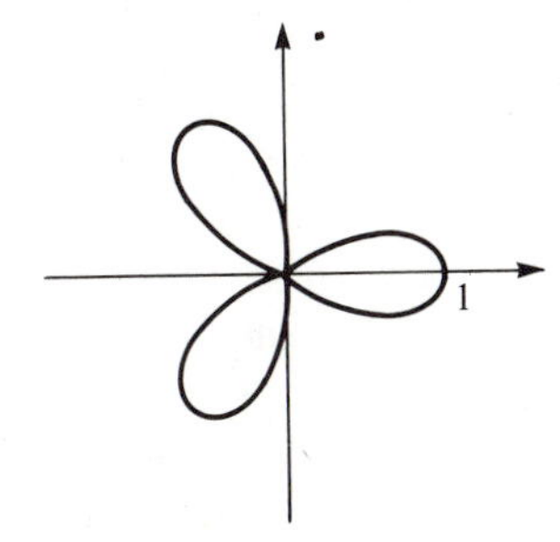

5 $r = \theta^2$

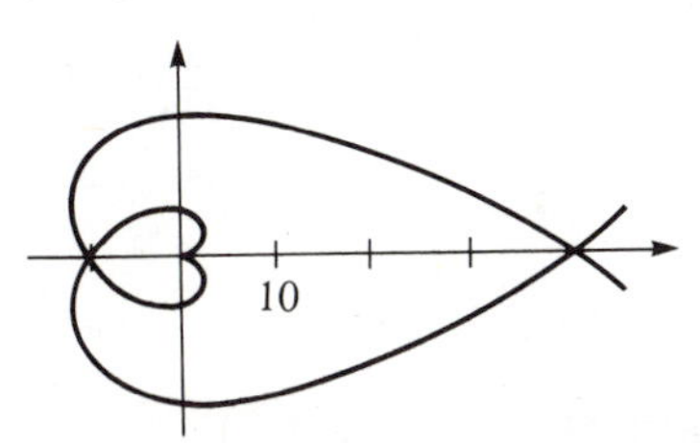

7 $r = \sec\theta - \cos\theta$

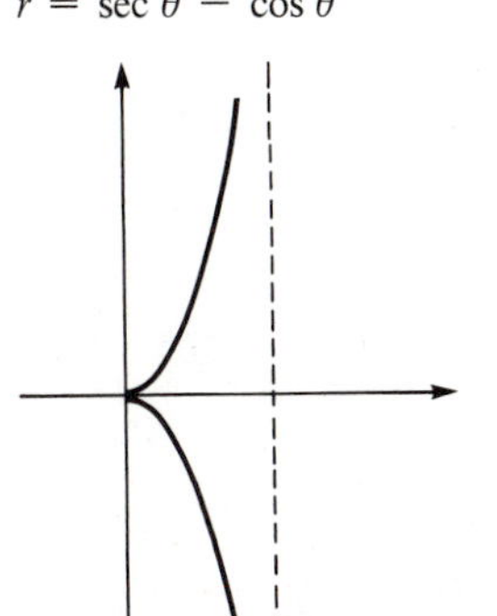

11 $r = 1 + 2\cos\theta$

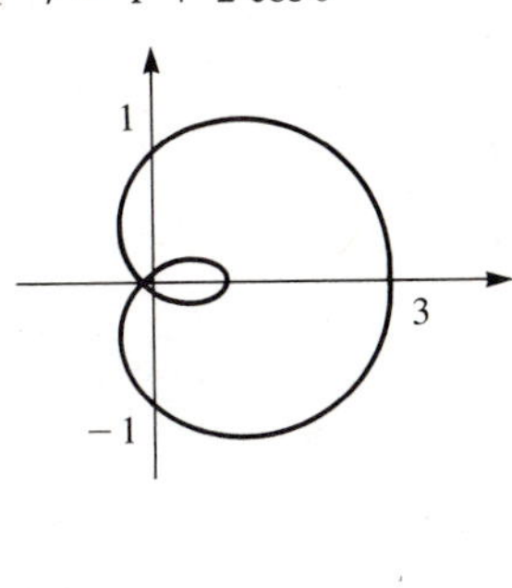

9 $r = 1 - \cos\theta$

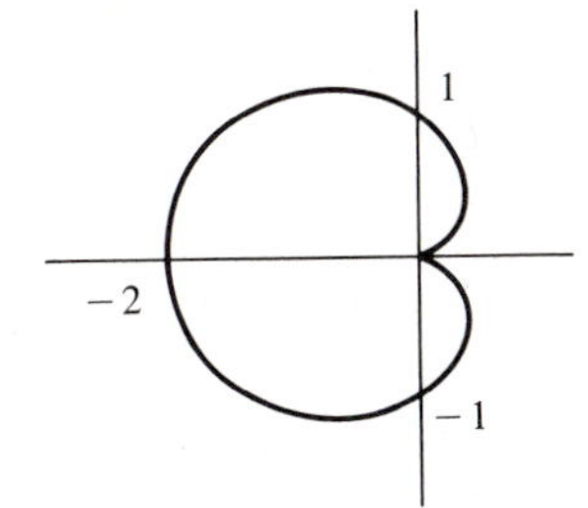

13 $r = \csc\theta - 2$

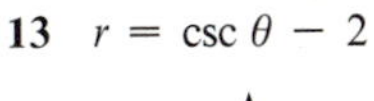

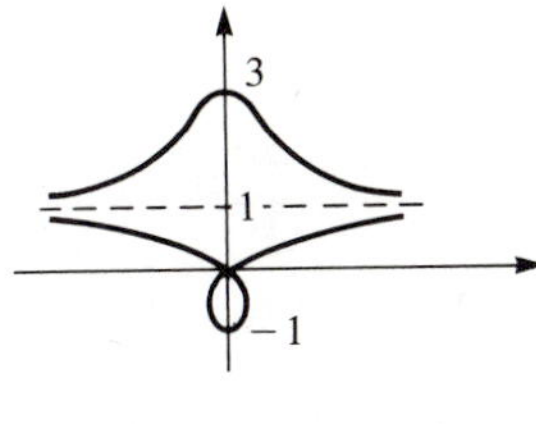

15

$\frac{b}{a} = \sin\alpha$

17 $e \approx 0.206$

19 $\sqrt{2}$

21 Take $P = (0, 0)$ and the ellipse $r(1 - e\cos\theta) = 2ep$.
$X = [2ep/(1 - e\cos\theta), \theta]$,
$Y = [2ep/(1 + e\cos\theta), \theta + \pi]$,
$1/\overline{XP} + 1/\overline{YP} = (1 - e\cos\theta)/2ep + (1 + e\cos\theta)/2ep = 1/ep$

23 $r(1 - \cos\theta) = 2p$

Section 8-9 Rotation of Axes

Page 470

1 $\bar{x} = x\cos\alpha + y\sin\alpha,\ \bar{y} = -x\sin\alpha + y\cos\alpha$
$(x, y) \to (\bar{x}, \bar{y})$ by a rotation through $-\alpha$.

3 $x_1x_2 + y_1y_2 = \bar{x}_1\bar{x}_2 + \bar{y}_1\bar{y}_2$

5 The polar angle of the point with respect to the rotated axes is $\theta - \alpha$.

7 $\alpha = \frac{3}{8}\pi,\ -\frac{1}{2}(\sqrt{2} - 1)\bar{x}^2 + \frac{1}{2}(\sqrt{2} + 1)\bar{y}^2 = 1$, hyperbola

9 $\alpha = \frac{3}{8}\pi,\ \frac{1}{2}(\sqrt{2} + 1)\bar{x}^2 - \frac{1}{2}(\sqrt{2} - 1)\bar{y}^2 = 1$, hyperbola

11 $\alpha = \frac{1}{4}\pi,\ \bar{x}^2/\frac{2}{3} + \bar{y}^2/2 = 1$, ellipse, major axis at angle $\frac{3}{4}\pi$

13 $\frac{1}{2}\sqrt{5}\,\bar{x}^2 - \frac{1}{2}\sqrt{5}\,\bar{y}^2 = 1$, hyperbola, principal axis at angle $\frac{1}{2}\text{arc tan }\frac{1}{2}$

15 $\alpha = \frac{3}{8}\pi$, $\frac{1}{2}(3 + \sqrt{2})\bar{x}^2 + \frac{1}{2}(3 - \sqrt{2})\bar{y}^2 = 1$, ellipse, major axis at angle $\frac{7}{8}\pi$

17 $\alpha = \frac{1}{4}\pi$, $\bar{x} + \frac{1}{8}\sqrt{2} = \sqrt{2}(y - \frac{1}{4}\sqrt{2})^2$, parabola, axis at angle $\frac{1}{4}\pi$

19 $-\frac{1}{2}(\sqrt{37} - 3)\bar{x}^2 + \frac{1}{2}(\sqrt{37} + 3)\bar{y}^2 = 1$, hyperbola, principal axis at angle $\pi - \frac{1}{2}\text{arc tan }6$

21
$$\bar{a} + \bar{c} = (a\cos^2\alpha + b\cos\alpha\sin\alpha + c\sin^2\alpha) + (a\sin^2\alpha - b\sin\alpha\cos\alpha + c\cos^2\alpha) = a(\cos^2\alpha + \sin^2\alpha) + c(\sin^2 + \cos^2\alpha) = a + c$$

Section 8-10 Review Exercises

Page 470

1 $(x - 4)^2 + (y + 6)^2 = 9$ Circle. Center: $(4, -6)$. Radius: 3.

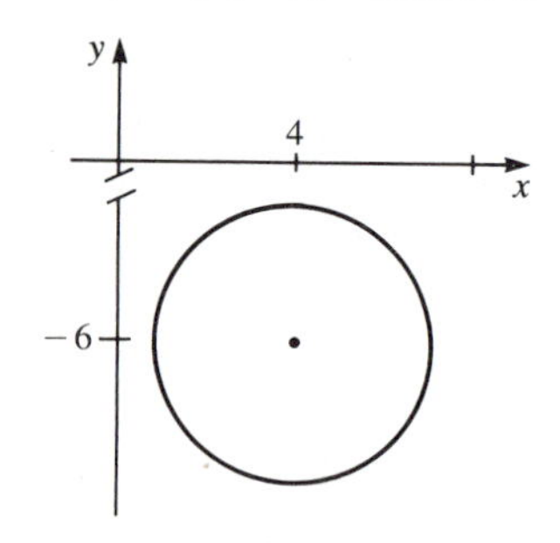

3 $\frac{1}{100}(x - 3)^2 + \frac{1}{4}(y + 1)^2 = 1$ Ellipse. Center: $(3, -1)$. Major axis: $x = 3$. $a = 10$, $b = 2$, $c = 4\sqrt{6}$. Foci: $(3 \pm 4\sqrt{6}, -1)$. Vertices: $(3 \pm 10, -1)$, $(3, -1 \pm 2)$

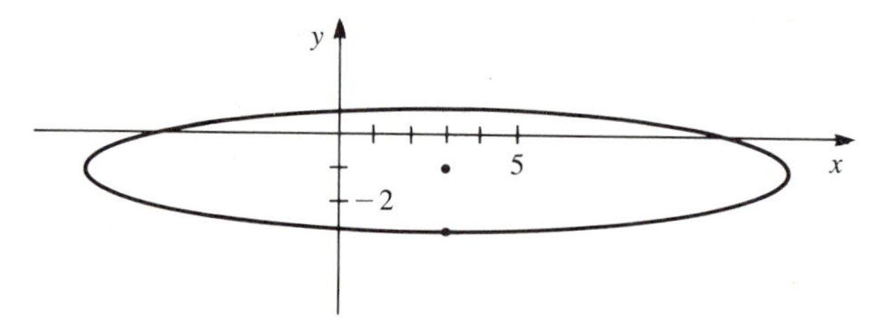

5 Line: $r(\cos\theta - 2\sin\theta) = 3$

7 Parabola, axis parallel to axis of original

9 $r\theta = 1$

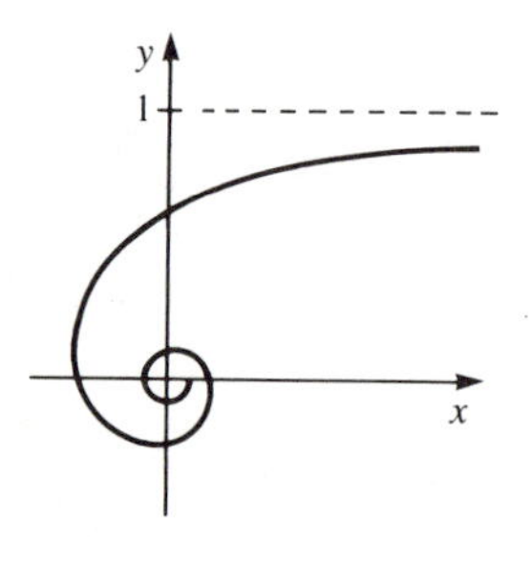

11 $\ln r = \theta$

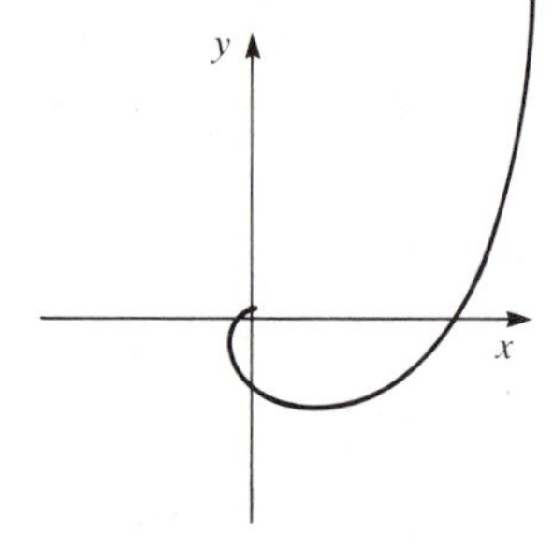

13 The normal has slope $-\frac{1}{2}ax$, and meets the y-axis at $\eta = y + \frac{1}{2}a$. Hence $d = \frac{1}{2}a$.

15 Let the ellipse be $x^2/a^2 + y^2/b^2 = 1$. Then
$$\frac{uv}{w^2} = \frac{(a + x)(a - x)}{y^2} = \frac{a^2 - x^2}{y^2} = \frac{a^2(y^2/b^2)}{y^2} = \frac{a^2}{b^2}$$

17 $r^2 = 2a^2\cos 2\theta$

19 Parabola with focus P and vertical directrix

Chapter 9 Numerical Calculus

Section 9-1 Mean Value Theorem

Page 476

1 $f'(c) = e^{-kc}(\cos c - k\sin c) = 0$ for some c in $(0, \pi)$. Hence $\cot c = k$, so the range of $\cot x$ is $\mathbf{R}$.

3 $\ln 51 - \ln 50 = (51 - 50)/c = 1/c$ for some c in $(50, 51)$. Hence, $\ln 51 - \ln 50 < \frac{1}{50} = 0.02$.

5 $f(x) = \text{arc tan }x$: $\text{arc tan }6 - \text{arc tan }5 = 1/(1 + c^2)$ for some c in $(5, 6)$. Thus
$\text{arc tan }6 - \text{arc tan }5 < 1/(1 + 5^2) = \frac{1}{26} < 0.04$.

7 $f(x) = \text{arc sin }x$: $\text{arc sin }\frac{3}{5} - \text{arc sin }\frac{1}{2} = (\frac{3}{5} - \frac{1}{2})/\sqrt{1 - c^2}$ where c is in $(\frac{1}{2}, \frac{3}{5})$. Thus
$$\text{arc sin }\tfrac{3}{5} = \tfrac{1}{6}\pi + \tfrac{1}{10}/\sqrt{1 - c^2}$$
But for $\frac{1}{2} < c < \frac{3}{5}$,
$$\tfrac{2}{3}\sqrt{3} < 1/\sqrt{1 - c^2} < \tfrac{5}{4}$$
and the asserted inequalities follow.

9 $f(x)$ and $g(x)$ are differentiable on $[0, 1]$, $g'(x) = 2x \neq 0$ on $(0, 1)$. $[f(1) - f(0)]/[g(1) - g(0)] = 1$, $f'(c)/g'(c) = (3c^2)/(2c) = \frac{3}{2}c = 1$ for $c = \frac{2}{3}$.

11 $f(x)$ and $g(x)$ are differentiable on $[0, 1]$, $g'(x) = 2(x - 1) \neq 0$ on $(0, 1)$. $[f(1) - f(0)]/[g(1) - g(0)] = -1$, $f'(c)/g'(c) = (3c^2)/2(c - 1) = -1$ for $3c^2 + 2c - 2 = 0$ with root $c = \frac{1}{3}(-1 + \sqrt{7})$ in $(0, 1)$.

13 $\displaystyle\int_0^1 x^3e^x\,dx = e^c\int_0^1 x^3\,dx$ for some c in $[0, 1]$

15 $f(x) = x^{10} - 2x^8 + 2x^3 - x$ so $f(0) = f(1) = 0$. By Rolle's theorem, $f'(c) = 0$ for some c in $(0, 1)$. But $f'(x) = g(x)$.

17 Set $g(x) = xf(x)$. Then $g(a) = g(b) = 0$ and $g'(x) = f(x) + xf'(x)$. By Rolle's theorem, $f(c) + cf'(c) = 0$ for some c in (a, b).

19 $f(-1) = f(1) = 0$, so
$$c^2 - 1 + \sin\pi c + c(2c + \pi\cos\pi c) = 0$$
for some c in $(-1, 1)$.

21 $\tan c = c$ for some c in $(\pi, \frac{3}{2}\pi)$.

23 Suppose $g(x)$ has no zero in (a, b). Clearly $g(a) \neq 0$ and $g(b) \neq 0$ because, for instance, $(f'g - fg')(a) = f'(a)g(a) \neq 0$. Thus $h(x) = f(x)/g(x)$ is differentiable on $[a, b]$ and $h(a) = h(b) = 0$. By Rolle's theorem, $h'(c) = 0$ for some c in (a, b), that is, $f'(c)g(c) - f(c)g'(c) = 0$, a contradiction.

25 $g(a) = g(b) = 0$ so $g'(c) = -(c - a)f'(c) + f(b) - f(c) = 0$ for some c in (a, b), that is, $f'(c) = [f(b) - f(c)]/(c - a)$.

27 By the MVT, $[f(a+h)-f(a-h)]/2h = 2hf'(x)/2h = f'(x)$, where x is between $a-h$ and $a+h$. As $h \to 0$, $x \to a$, so $f'(x) \to f'(a)$.

29 The c that works for f is not necessarily the same as the c that works for g.

Section 9-2 L'Hospital's Rule

Page 483

1 1

3 $\frac{1}{6}$

5 $\frac{1}{2}$

7 -1

9 -2

11 0

13 1

15 1

17 0

19 1

21 0

23 0

25 $+\infty$

27 0

29 $a^{a/(a-1)}$

31 $f'(c)$

33 $f(x+1)-f(x) \to L-L=0$ and $f(x+1)-f(x) = f'[c(x)]$ where $x < c(x) < x+1$, so $f'[c(x)] \to 0$ as $x \to +\infty$. But $c(x) \to +\infty$ also so $M = 0$.

Section 9-3 Polynomial Approximation

Page 489

	$f(a)$	$f'(a)$	$P_1(x)$
1	1	0	1
3	-1	3	$-1+3(x+1)$
5	2	-4	$2-4(x-\frac{1}{2})$
7	1	4	$1+4(x-1)$
9	1	0	1
11	0	1	x

13 $P_1 = 1+2(x+1)$ $E = (x+1)^2[(1-2x)/x^2]$

15 $P_2 = 1+x$ $E = 0$

17 $P_2 = -\frac{1}{2}-\frac{1}{4}(x+2)-\frac{1}{8}(x+2)^2$ $E = (x+2)^3/(8x)$

19 $P_2 = x^2$ $E = x^3[-1/(x+1)]$

21 $P_2 = 1+2x+2x^2$

23 $P_2 = x$

25

x	xe^x	$P_1(x)$	$P_2(x)$
0.1	0.1105	0.1000	0.1100
0.2	0.2443	0.2000	0.2400
0.3	0.4050	0.3000	0.3900
0.4	0.5967	0.4000	0.5600
0.5	0.8244	0.5000	0.7500

27 $\ln x$ and $2(x-1)/(x+1)$ have the same $P_2(x) = (x-1)-\frac{1}{2}(x-1)^2$ at $a=1$.

x	0.5	0.8	1.2	1.5	2.0
$\ln x$	-0.6931	-0.2231	0.1823	0.4055	0.6931
$\frac{2(x-1)}{(x+1)}$	-0.6667	-0.2222	0.1818	0.4000	0.6667
$P_2(x)$	-0.6250	-0.2200	0.1800	0.3750	0.5000

Section 9-4 Taylor Approximations

Page 497

1 $p(x) = 8+7u+u^2$

3 $p(x) = 9u-u^2+2u^3$

5 $p(x) = 18u^2-11u^3+2u^4$

7 $p(x) = 5u-19u^2+31u^3-21u^4+5u^5$

9 $P_5 = x-x^2+\frac{1}{2}x^3-\frac{1}{6}x^4+\frac{1}{24}x^5$

11 $P_5 = x-\frac{1}{6}x^3+x^4+\frac{1}{120}x^5$

13 $P_5(x) = \frac{1}{2}\sqrt{2}\,(1+u-\frac{1}{2}u^2-\frac{1}{6}u^3+\frac{1}{24}u^4+\frac{1}{120}u^5)$ where $u = x-\frac{1}{4}\pi$

15 $P_5(x) = -1-3u-6u^2-10u^3-15u^4-21u^5$ where $u = x+1$

17 $P_5(x) = 1-\frac{1}{2}u^2+\frac{1}{24}u^4$ where $u = x-\frac{1}{2}\pi$

19 $P_5(x) = \ln 2+\frac{1}{2}u-\frac{1}{8}u^2+\frac{1}{24}u^3-\frac{1}{64}u^4+\frac{1}{160}u^5$ where $u = x-2$

21 $P_5(x) = e(1+u+\frac{1}{2}u^2+\frac{1}{6}u^3+\frac{1}{24}u^4+\frac{1}{120}u^5)$ where $u = x-1$

23 $P_5 = 2+x-x^2-\frac{1}{6}x^3+\frac{1}{12}x^4+\frac{1}{120}x^5$

25 $P_5 = 3x-\frac{9}{2}x^3+\frac{81}{40}x^5$

27 $f(x) = xe^x$
$P_4(x) = x+x^2+\frac{1}{2}x^3+\frac{1}{6}x^4$

x	$f(x)$	$P_4(x)$
-0.5	-0.30327	-0.30208
-0.4	-0.26813	-0.26773
-0.3	-0.22225	-0.22215
-0.2	-0.16375	-0.16373
-0.1	-0.09048	-0.09048
0.0	0.00000	0.00000
0.1	0.11052	0.11052
0.2	0.24428	0.24427
0.3	0.40496	0.40485
0.4	0.59673	0.59627
0.5	0.82436	0.82292

29 $f(x) = 1/(1-x)$
$P_4(x) = 1+x+x^2+x^3+x^4$

x	$f(x)$	$P_4(x)$
-0.5	0.66667	0.68750
-0.4	0.71429	0.72160
-0.3	0.76923	0.77110
-0.2	0.83333	0.83360
-0.1	0.90909	0.90910
0.0	1.00000	1.00000
0.1	1.11111	1.11110
0.2	1.25000	1.24960
0.3	1.42857	1.42510
0.4	1.66667	1.64960
0.5	2.00000	1.93750

Section 9-5 Taylor's Formula

Page 503

1 $P_{2m}(x) = P_{2m-1}(x) = \sum_{j=1}^{m} (-1)^{j-1} \frac{2^{2j-1}}{(2j-1)!} x^{2j-1}$

$|R_{2m-1}(x)| = |R_{2m}(x)| \le \frac{2^{2m+1}}{(2m+1)!} |x|^{2m+1}$

3 $P_n(x) = \sum_{j=1}^{n} \frac{x^j}{(j-1)!}$

$|R_n(x)| \le \frac{(x+n+1)}{(n+1)!} e^x x^{n+1}$ if $x \ge 0$

$|R_n(x)| \le \frac{|x|^{n+1}}{n!}$ if $x \le 0$

5 $P_n(x) = (x-1) + \frac{3}{2}(x-1)^2 + 2\sum_{j=3}^{n} (-1)^{j-1} \frac{(x-1)^j}{j(j-1)(j-2)}$

Assuming $n \ge 4$,

$|R_n(x)| \le \frac{2}{n(n^2-1)}(x-1)^{n+1}$ if $x \ge 1$

$|R_n(x)| \le \frac{2(1-x)^{n+1}}{n(n^2-1)x^{n-1}}$ if $0 < x \le 1$

7 $P_n(x) = \sum_{j=2}^{n} (-1)^j \frac{x^j}{(j-2)!}$

$|R_n(x)| \le \frac{x^{n+1}}{(n-1)!}$ for $x \ge 0$

$|R_n(x)| \le \frac{[x^2 - 2(n+1)x + n(n+1)]|x|^{n+1}}{e^x(n+1)!}$

for $x \le 0$

9 $P_{2m}(x) = P_{2m+1}(x) = \sum_{j=1}^{m} (-1)^{j-1} \frac{x^{2j}}{(2j-1)!}$

$|R_{2m+1}(x)| = |R_{2m}(x)| \le \frac{|x| + 2m + 2}{(2m+2)!} |x|^{2m+2}$

11 $P_n(x) = \sum_{j=1}^{n} \frac{\pm x^j}{j!}$ where the signs are

$+ + - - + + - - + + - - \cdots$

$|R_n(x)| \le \frac{1}{2}\sqrt{2}\, \frac{|x|^{n+1}}{(n+1)!}$

13 $P_{2m-1}(x) = P_{2m}(x) = \sum_{j=1}^{m} \frac{x^{2j-1}}{(2j-1)!}$

$|R_{2m}(x)| = |R_{2m-1}(x)| \le \frac{\cosh x}{(2m+1)!} |x|^{2m+1}$

15 $P_n(x) = \frac{1}{2} \sum_{j=0}^{n} \frac{(-1)^j}{2^j} (x-1)^j$

$|R_n(x)| \le \frac{(1-x)^{n+1}}{2^{n+1}(1+x)^{n+2}}$ for $-1 < x \le 1$

$|R_n(x)| \le \frac{(x-1)^{n+1}}{2^{n+1}}$ for $x \ge 1$

17 $|R_{10}(x)| < 5 \times 10^{-5}$

19 $\sin^2 x \approx P_4(x) = x^2 - \frac{1}{3}x^4$

$|R_4(x)| = |R_5(x)| < 4.4 \times 10^{-8}$

21 $|R_3(x)| = |R_4(x)| < (0.22)^5/5! < 4.3 \times 10^{-6}$

23 $|R_7(x)| = |R_8(x)| < (1.06)^9/9! < 4.7 \times 10^{-6}$

25 $k = 0.0033$

27 6

29 $|R_3(x)| \le |x - \frac{1}{4}\pi|^4/4! < (0.1)^4/4! < 4.2 \times 10^{-6}$

31 $P_3(x) = 1 - \frac{1}{2}x - \frac{1}{8}x^2 - \frac{1}{16}x^3$

$|R_4(x)| \le \frac{5}{128}x^4$ for $x \le 0$

$|R_4(x)| \le \frac{5}{128} \frac{x^4}{(1-x)^{7/2}}$ for $0 \le x < 1$

33 Compute $P_1(I)$ at $I = 50$ for $f(I) = \ln(1 + I/1200)$.

$\ln(1 + I/1200) \approx \ln(1 + 50/1200) + \frac{1}{1250}(I - 50)$

But $\ln(1 + 50/1200) \approx 50/1200$,
so $\ln(1 + I/1200) \approx I/1250$. Therefore

$n = \frac{\frac{1}{2}\ln 2}{\ln(1 + I/1200)} \approx \frac{\frac{1250}{12}\ln 2}{I} \approx \frac{72}{I}$

35 $f/g = [\frac{1}{6}x^3 + R_3(x)]/x^3 = \frac{1}{6} + x^{-3}R_3(x) \to \frac{1}{6}$

37 $f/g = [\frac{1}{120}x^5 + R_5(x)]/x^5 \to \frac{1}{120}$

39 $f(x) = R_0(x) = f'(c_1)$ and $g(x) = g'(c_2)$ where c_1 and c_2 are between 0 and x.

$\lim_{x \to 0} \frac{f(x)}{g(x)} = \lim_{x \to 0} \frac{f'(c_1)}{g'(c_2)} = \frac{f'(0)}{g'(0)}$

41 $f(x) = f(a) + R_n(x)$
$= f(a) + f^{(n+1)}(c)(x-a)^{n+1}/(n+1)!$
where c is between a and x. If $|x - a|$ is sufficiently small, then $f^{(n+1)}(c)$ has the sign of $f^{(n+1)}(a)$. But $(x-a)^{n+1} > 0$ for $0 < |x - c| < \delta$ because $n + 1$ is even. Thus if $f^{(n+1)}(a) < 0$, then $f(x) < f(a)$ for $0 < |x - c| < \delta$, that is, $f(a)$ is a local maximum, and so on.

43 $R_{n-1}(x) = \frac{1}{(n-1)!} \int_a^x (x-t)^{n-1} f^{(n)}(t)\, dt$

$(x-t)^{n-1} = [(x-a)(1-u)]^{n-1}$ and $dt = (x-a)\, du$

$R_{n-1}(x) = \frac{(x-a)^n}{(n-1)!} \int_0^1 (1-u)^{n-1} f^{(n)}[a + u(x-a)]\, du$

Section 9-6 Approximate Integration

Page 512

1 1.0023 0.0023
3 1.4174 0.0032
5 4.6723 0.0015
7 0.5046 0.0046
9 1.71832 1.71828
11 0.10869 0.10065
13 2.98209 2.98200
15 0.61036 0.61989
(Exploiting symmetry:
0.61989 0.62050)
17 4.87757 4.85802
19 6.03354 6.03343
21 1.71828 19
23 8

25 $A(x) = \pi x \quad V = \frac{1}{6}a(\pi \cdot 0 + 4\pi \cdot \frac{1}{2}a + \pi a) = \frac{1}{2}\pi a^2$

27 $A(x) = (\pi b^2/a^2)(a^2 - x^2)$

$V = \frac{1}{6}(2a)(\pi b^2/a^2)[(a^2 - (-a)^2) + 4a^2 + (a^2 - a^2)]$
$= \frac{4}{3}\pi a b^2$

29 By slices,

$$V = \int_a^{a+h} A(x)\, dx = \frac{1}{6}h[A(0) + 4A(a + \frac{1}{2}h) + A(a+h)] = \frac{1}{6}h(A_0 + 4M + A_1)$$

since Simpson's rule is exact for cubics.

31

$n = 2$	$n = 4$	$n = 10$	$n = 20$	exact
−0.2452	−0.2488	−0.24981	−0.24995	−0.25

Section 9-7 Root Finding; Newton's Method

Page 521

1 $x_8 = x_9 = 0.7035$

3 $x_6 = x_7 = 2.3028$

5 $x_8 = x_9 = 2.3311$

7 $x_8 = x_9 = 1.000$

9 $x_{16} = x_{17} = 0.4999$

11 $x_3 = x_4 = 1.02986\ 653$

13 ± 3.16228
$\pm 3.16227\ 76602$

15 ± 0.66874
$\pm 0.66874\ 03050$

17 -1.32472
$-1.32471\ 79572$

19 1.31460 1.31459 62123

21 -0.56714
$-0.56714\ 32904$

23 -1.27178
$-1.27177\ 97887$
0.47178 0.47177 97887

25 -1.14790
$-1.14789\ 90357$

27 -1.45367
$-1.45367\ 36665$
0.53979 0.53978 51608

29 ± 1.89579
$\pm 1.89549\ 42670$

31 -1.54165
$-1.54165\ 16841$
0.20006 0.20006 41026
1.44050 1.44050 03973

33 ± 0.72705
$\pm 0.72704\ 72898$
± 3.90336
$\pm 3.90335\ 68641$

35 0; ± 1.52861
$\pm 1.52861\ 47266$

37 0.33651

39 1.03159

41 $|x_0 - \sqrt{2}| < |1.42 - 1| = 0.42$, so the estimate becomes $|x_6 - \sqrt{2}| < 2^{-63}(0.42)^{64} \approx 8.4 \times 10^{-44}$

43 $|\phi'(x)| = |\frac{1}{4}\sqrt{2}\,(1 + x)^{-1/2}| < \frac{1}{4}\sqrt{2} < 1$, hence $x_n \to \phi(1) = 1$.

Section 9-8 Review Exercises

Page 521

1 $\sqrt[3]{28} - 3 = \sqrt[3]{28} - \sqrt[3]{27} = \frac{1}{3}c^{-2/3} < \frac{1}{3}(27)^{-2/3} = \frac{1}{27}$

3 $+\infty$

5 1

7 $P_5(x) = \frac{1}{2} - \frac{1}{2}\sqrt{3}\,t - \frac{1}{4}t^2 + \frac{1}{12}\sqrt{3}\,t^3 + \frac{1}{48}t^4 - \frac{1}{240}\sqrt{3}\,t^5$ where $t = x - \frac{1}{3}\pi$

9 For $x \approx 0$ and $x \neq 0$,

$$\sqrt{1 - x^2} \approx 1 - \tfrac{1}{2}x^2 - \tfrac{1}{8}x^4 < \cos x \approx 1 - \tfrac{1}{2}x^2 + \tfrac{1}{24}x^4$$

so $y = \cos x$ is outside the circle.

11 8.0387

13 0.4091

15 3.4076

17 -1.18841

19 0.15411

21 $\frac{2}{3}$

23 $f(x) - f(a) = (x - a)f'(c) \geq m(x - a) \to +\infty$ so $f(x) \to +\infty$.

25 It is readily checked the LHS and RHS have the same Taylor polynomial $P_{14}(x)$ at $x = 0$. Hence if

$$R(x) = \text{LHS} - \text{RHS}, \text{ then } |R(x)| \leq \frac{M}{15!}(\tfrac{1}{2}\pi)^{15}$$

for $|x| < \frac{1}{2}\pi$, where $M = \max|r^{(15)}(x)|$ for $|x| < \frac{1}{2}\pi$. Now $r^{(15)}(x) = -\cos x - \frac{1}{2}(e^x + e^{-x})$, so by calculation, $M \leq 1 + \frac{1}{2}(e^{\pi/2} + e^{-\pi/2}) < 3.51$,

$$|R(x)| < \frac{3.51}{(15)!}(\tfrac{1}{2}\pi)^{15} \approx 2.35 \times 10^{-9}.$$

Chapter 10 Sequences and Series

Section 10-1 Sequences and Limits

Page 532

1 $0\ \ \frac{1}{2}\ \ \frac{2}{3}\ \ \frac{3}{4}\ \ \frac{4}{5}$ $L = 1$

3 $0\ \ \frac{1}{3}\ \ \frac{2}{4}\ \ \frac{3}{5}\ \ \frac{4}{6}$ $L = 1$

5 $\frac{1}{3}\ \ \frac{2}{7}\ \ \frac{3}{13}\ \ \frac{4}{21}\ \ \frac{5}{31}$ $L = 0$

7 $-\frac{1}{2}\ \ \frac{1}{4}\ \ -\frac{1}{8}\ \ \frac{1}{16}\ \ -\frac{1}{32}$ $L = 0$

9 1.01 1.0201 1.030301 1.04060 401 1.05101 00501
$L = +\infty$

11 687

13 Let $N > 1/\varepsilon^2$. Then for $n \geq N$, $|1/\sqrt{n} - 0| = 1/\sqrt{n} \leq 1/\sqrt{N} < \varepsilon$.

15 $\frac{1}{2}$

17 0

19 1

21 $\frac{1}{2}$

23 1

25 0

27 3

29 0

31 $\frac{2}{3}$

33 Let $L = \lim a_n$. If $\varepsilon > 0$, then there exists an N so $|a_n - L| < \varepsilon$ for all $n \geq N$. Hence $a_n - L < \varepsilon$, $L > a_n - \varepsilon \geq -\varepsilon$. Therefore $L \geq -\varepsilon$ for all $\varepsilon > 0$, so $L \geq 0$.

35 $a_n = b_n = n$

37 $a_n = b_n = n$

39 No; $a_n = (-1)^n$.

41 Let $\{b_n\}$ be the new sequence. Then there are integers K and M such that $b_n = a_{K+n}$ for all $n \geq M$. Given $\varepsilon > 0$, there is an N such that $|a_n - L| < \varepsilon$ for all $n \geq N$. Hence $|a_{K+n} - L| < \varepsilon$ for all n such that $K + n \geq N$. Hence $|b_n - L| < \varepsilon$ for all $n \geq \max(M, N - K)$.

43 Only if it is eventually constant

45 If $\varepsilon > 0$, choose N so that $|a_n| < \varepsilon/B$ for all $n \geq N$. Then

$$|a_n b_n| = |a_n||b_n| < (\varepsilon/B)B = \varepsilon$$

for all $n \geq N$. Hence $a_n b_n \to 0$.

47 Let $\{b_n\}$ be a subsequence of a_n. Then for each n, there is a $k = k(n) \geq n$ such that $b_n = a_k$. Now let $\varepsilon > 0$ and choose N so that $|a_n - L| < \varepsilon$ for all $n \geq N$. Then if $n \geq N$,

$$|b_n - L| = |a_{k(n)} - L| < \varepsilon$$

because $k(n) \geq n \geq N$. Therefore $b_n \to L$.

49 $a_n = (-1)^n \dfrac{1 \cdot 3 \cdot 5 \cdots (2n - 1)}{2n}$ for $n \geq 1$

Section 10-2 Properties of Limits

Page 540

1 1

3 $\frac{1}{4}\pi$

5 $a_n > 0$ and $a_{n+1} = a_n[(3n - 2)/(3n - 1)] < a_n$ so $a_1 > a_2 > a_3 > \cdots > 0$, $\lim a_n = L \geq 0$.

7 $a_{n+1} = a_n + (\ ) > a_n$ and

$$a_n \leq 1 + \frac{1}{2^2} + \frac{1}{(2^2)^2} + \cdots + \frac{1}{(2^{n-1})^2}$$

$$= 1 + \tfrac{1}{4} + \frac{1}{4^2} + \cdots + \frac{1}{4^{n-1}} < \tfrac{4}{3}$$

so $a_1 < a_2 < a_3 < \cdots < \frac{4}{3}$.

9 $a_{n+1} = ca_n < a_n$ and $a_n > 0$ so $a_1 > a_2 > a_3 > \cdots > 0$

11 Induction: $a_0 < 2$. If $a_n < 2$ then $a_{n+1} = 1 + \frac{1}{2}a_n < 1 + 1 = 2$. Next

$$a_{n+1} = 1 + \tfrac{1}{2}a_n > \tfrac{1}{2}a_n + \tfrac{1}{2}a_n = a_n$$

Thus $a_1 < a_2 < a_3 < \cdots < 2$.

13 5

15 $|x_{n+1} - x_{n+2}| = |x_{n+1} - \frac{1}{2}(x_n + x_{n-1})| = \frac{1}{2}|x_n - x_{n+1}|$, so the second comparison test applies.

17 $a_1 < 2$. Suppose $a_n < 2$. Then $a_{n+1} < \sqrt{2 + 2} = 2$. Hence $a_n < 2$ for all n. Therefore $a_{n+1} = \sqrt{2 + a_n} > \sqrt{a_n + a_n} = \sqrt{2a_n} > \sqrt{a_n \cdot a_n} = a_n$. Thus $a_1 < a_2 < a_3 < \cdots < 2$; the sequence converges to a positive limit $L \le 2$.

19 **Case 1:** $0 < x \le 2$. Then $a_1 = \sqrt{x} < 2$. If $a_n < 2$, then $a_{n+1} = \sqrt{x + a_n} < \sqrt{2 + 2} = 2$. Therefore $a_n < 2$ for all n. **Case 2:** $2 < x$. Then $a_1 = \sqrt{x} < x$. If $a_n < x$, then $a_{n+1} = \sqrt{x + a_n} < \sqrt{x + x} = \sqrt{2x} < \sqrt{x \cdot x} = x$. Therefore $a_n < x$ for all n.

21 Suppose $a_n < b_n$. Then b_{n+1} is the average of a_n and b_n so $a_n < b_{n+1} < b_n$. Likewise $1/a_{n+1}$ is the average of $1/a_n$ and $1/b_n$, so $1/b_n < 1/a_{n+1} < 1/a_n$, that is, $a_n < a_{n+1} < b_n$. Next, $4a_nb_n < (a_n + b_n)^2$, so $2a_nb_n/(a_n + b_n) < (a_n + b_n)/2$, that is, $a_{n+1} < b_{n+1}$.

23 $|b_0| < 1$ and, by induction, $b_n = b_0{}^{2^n}$. The sequence $\{b_0{}^n\}$ converges to 0; so does its subsequence $\{b_n\}$.

25 Clearly $|b_0| < 1$. Suppose $|b_n| < |b_0|^{2^n}$. Then $|b_n| < 1$ and

$$|b_{n+1}| < \tfrac{1}{4}|b_0|^{2^{n+1}}(3 + b_n) < \tfrac{4}{4}|b_0|^{2^{n+1}} = |b_0|^{2^{n+1}}$$

Hence $|b_n| < |b_0|^{2^n}$ for all n. It follows as in Ex. 23 that $b_n \to 0$.

27 0

29 $|a_n - a_{n+1}| = 1/(n+1)! < (\frac{1}{2})^n$ so $\{a_n\}$ converges

31 $x + 1/x = (\sqrt{x} - 1/\sqrt{x})^2 + 2 \ge 2$, hence $b_0 \ge 1$. If $b_n \ge 1$, then $b_{n+1}^2 = \frac{1}{2}(1 + b_n) \ge \frac{1}{2}(1 + 1) = 1$ so $b_{n+1} \ge 1$. By induction, $b_n \ge 1$ for all n. Next, $1 \le b_n \le b_n{}^2$, so $1 + b_n \le 2b_n{}^2$, $b_{n+1}^2 = \frac{1}{2}(1 + b_n) \le b_n{}^2$, $b_{n+1} \le b_n$. Thus $\{b_n\}$ is decreasing and bounded below, so $L = \lim b_n$ exists. But $2b_{n+1}^2 = 1 + b_n$ implies $2L^2 = 1 + L$, so $L = 1$.

33 In either case the quantity $c_n + (\log_2 a_n)/2^n$ doesn't change when n increases to $n + 1$. But for $n = 0$ it equals $\log_2 x$.

35 $\dfrac{y^{n+1} - x^{n+1}}{y - x} = (n+1)c^n$ where $x < c < y$. Clearly $x^n < c^n < y^n$.

37 Take $x = 1 + 1/(n+1)$ and $y = 1 + 1/n$ in the left inequality:

$$\left(1 + \frac{1}{n+1}\right)^n\left(\frac{(n+1)^2}{n} - \frac{n(n+2)}{n+1}\right) < b_n$$

It suffices to prove

$$\left(1 + \frac{1}{n+1}\right)^2 < \frac{(n+1)^2}{n} - \frac{n(n+2)}{n+1}$$

When fractions are cleared, this boils down to

$$n^3 + 4n^2 + 4n < n^3 + 4n^2 + 4n + 1$$

Section 10-3 Infinite Series

Page 545

1 $\frac{3}{2}(1 - 3^{-10})$

3 $\frac{255}{256}$

5 $3(x^{n+1} - 3^{n+1})/[x^n(x - 3)]$ if $x \ne 3$; $3(n+1)$ if $x = 3$.

7 $r^{1/2}(1 - r^4)/(1 - r^{1/2})$ if $0 < r$ and $r \ne 1$; 8 if $r = 1$.

9 $\frac{5}{7}$

11 2^{-9}

13 $\frac{3}{2}$

15 $1/(1 + x^2)$

17 9 feet

19 The trains meet in 1 hour; all that time, the fly is flying at 60 mph.

21 $\frac{1}{9}$

23 $\frac{43}{99}$

25 $S = \frac{1}{2}(1 + \frac{1}{2} + \frac{1}{3} + \cdots) = +\infty$ (harmonic series)

27 1600 (743 just works)

29 $\dfrac{1}{n(n+1)} = \dfrac{1}{n} - \dfrac{1}{n+1}$; 1.

31 $a_n = 0$ for $n \ge N$

33 Write b_{N+j} as in the hint. The first summand on the right approaches 0 as $j \to +\infty$ because N is fixed. The second summand has absolute value less than $[j/(N+j)](\frac{1}{2}\varepsilon) < \frac{1}{2}\varepsilon$. Hence for j sufficiently large, $|b_{N+j}| < \frac{1}{2}\varepsilon + \frac{1}{2}\varepsilon = \varepsilon$.

Section 10-4 Series of Positive Terms

Page 552

1 Converges

3 Diverges

5 Converges

7 Converges

9 Converges

11 $\displaystyle\sum_{j=1}^{n}(a_j + b_j) = \sum_{j=1}^{n} a_j + \sum_{j=1}^{n} b_j \to \sum_{j=1}^{\infty} a_j + \sum_{j=1}^{\infty} b_j$

13 $a_n \to 0$, so $0 < a_n < 1$ for $n \ge N$. Then $0 < a_n{}^2 < a_n$ so $\sum a_n{}^2$ converges by comparison with $\sum a_n$.

15 Converges

17 Diverges

19 Converges

21 Converges

23 Converges

25 Converges

27 If $\sqrt[n]{a_n} \le r < 1$ for all $n \ge N$, then $a_n \le r^n$, so $\sum a_n$ converges by comparison with $\sum r^n$.

29 $p_1 = 1 + a_1 \ge 1 + a_1$. If $p_n \ge 1 + a_1 + \cdots + a_n$, then

$$\begin{aligned} p_{n+1} &\ge (1 + a_1 + \cdots + a_n)(1 + a_{n+1}) \\ &= 1 + a_1 + \cdots + a_{n+1} + a_{n+1}(a_1 + \cdots + a_n) \\ &\ge 1 + a_1 + \cdots + a_{n+1} \end{aligned}$$

31 By the tangent theorem $\ln(1 + x) \le x$. Hence $\ln p_n = \sum_1^n \ln(1 + a_j) \le \sum_1^n a_j = s_n$

33 Let $a_1 = kb_1$. Suppose $a_n \le kb_n$. Then $a_{n+1} \le a_n(b_{n+1}/b_n) \le (kb_n)(b_{n+1}/b_n) = kb_{n+1}$. Hence $a_n \le kb_n$ for all n, so $\sum a_n$ converges by comparison with $\sum kb_n$.

35 $\sum \frac{1}{n^2} = \sum \frac{1}{(2n)^2} + \sum \frac{1}{(2n+1)^2}$
$= \frac{1}{4}\sum \frac{1}{n^2} + \sum \frac{1}{(2n+1)^2},$
hence $\sum \frac{1}{(2n+1)^2} = \frac{3}{4}\sum \frac{1}{n^2} = \frac{3}{4} \cdot \frac{\pi^2}{6} = \frac{\pi^2}{8}.$

Section 10-5 Series with Positive and Negative Terms

Page 558

1 Converges; diverges absolutely
3 Diverges
5 Converges absolutely
7 Converges; diverges absolutely
9 Converges absolutely
11 Converges; diverges absolutely
13 Diverges
15 Converges; diverges absolutely
17 Diverges
19 Converges; diverges absolutely
21 ≈ 0.6065
23 $2|xy| \le x^2 + y^2$, hence $\sum |a_n b_n|$ converges by comparison with $\sum \frac{1}{2}(a_n^2 + b_n^2)$
25 Converges absolutely for all x
27 Converges (and converges absolutely) for $|x| < \frac{1}{3}$

Section 10-6 Improper Integrals

Page 563

1 $\frac{1}{8}$
3 1
5 $\frac{1}{4}\pi$
7 2
9 $\frac{1}{3}\ln 2$
11 $\frac{1}{4}\pi$
13 π
15 $1/s^2$
17 $n!/s^{n+1}$
19 $1/(s^2+1)$
21 $1/(s-a)^2$
23 Infinite
25 Infinite
27 Infinite
29 0
31 L/a
33 $\int = -\frac{\exp(-x^2/2)}{x}\Big|_a^{\infty} = [\exp(-a^2/2)]/a$
35 True for $k = 0$. By induction and parts,
$$I_k = -x^{2k-1}e^{-x^2/2}\Big|_{-\infty}^{\infty} + (2k-1)\int_{-\infty}^{\infty} x^{2k-2}e^{-x^2/2}\,dx$$
$$= (2k-1)I_{k-1}$$
so I_k exists if I_{k-1} does.

Section 10-7 Convergence and Divergence Tests

Page 568

1 Diverges
3 Converges
5 Diverges
7 Converges
9 Converges
11 Diverges
13 Converges
15 $\int = \int_{\ln 2}^{\infty} \frac{du}{u^p}$, converges if $p > 1$, diverges if $p \le 1$
17 If $p = 1$, $\int = \frac{1}{2}(\ln x)^2\Big|_1^{\infty} = +\infty.$
If $p < 1$, then $(\ln x)/x^p \ge (\ln x)/x$ for $x \ge 1$, so diverges by comparison. If $p > 1$, choose p' so $p > p' > 1$. Then $0 < (\ln x)/x^p < 1/x^{p'}$ for x sufficiently large; hence converges by comparison with $1/x^{p'}$.
19 $s > 4$
21 $s > \frac{1}{2}$
23 $s > 1$
25 All s
27 Integrate by parts with $u = e^{-t}$ and $v = \ln t$; then a second time with $u = e^{-t}$ and $v = t\ln t - t$.
29 Integrate K_{n+2} by parts with $u = x^{n+1}$ and $v = -\frac{1}{2}e^{-x^2}$. This gives $K_{n+2} = \frac{1}{2}(n+1)K_n$. Now use induction (or just multiply and telescope).
31 $\int_0^{\infty} \frac{f(ax)}{x}\,dx = \int_{t_0}^{\infty} \frac{f(t)}{t}\,dt$
33 $\beta > -1$ and $1/(p+2) = \frac{1}{2}(1+\beta)$

Section 10-8 Relation Between Integrals and Series

Page 572

1 Diverges
3 Converges
5 Diverges
7 Converges
9 Converges
11 $S < 1 + \int_1^{\infty} dx/x^2 = 2$
13 Somewhere between 7.247×10^{433} and 1.971×10^{434}
15 Since $\ln n$ increases,
$$\ln(n!) = \sum_{k=2}^{n} \ln k > \sum_{k=2}^{n} \int_{k-1}^{k} \ln x\,dx$$
$$= \int_1^n \ln x\,dx = x\ln x - x\Big|_1^n = n\ln n - n + 1$$
Hence $n! > \exp(n\ln n - n + 1) = e(n/e)^n$.
17 By parts with $u = 1 - \cos x$ and $v = -1/x$:
$$\int_0^b \frac{1-\cos x}{x^2}\,dx = -\frac{1-\cos x}{x}\Big|_0^b + \int_0^b \frac{\sin x}{x}\,dx$$
$$= -\frac{1-\cos b}{b} + \int_0^b \frac{\sin x}{x}\,dx$$
Let $b \to \infty$.
19 By parts with $u = 1/\sqrt{x}$ and $v = -\cos x$:
$$\int_1^b \frac{\sin x}{\sqrt{x}}\,dx = \cos 1 - \frac{\cos b}{\sqrt{b}} - \frac{1}{2}\int_1^b \frac{\cos x}{x^{3/2}}\,dx$$
$$\to \cos 1 - \frac{1}{2}\int_1^{\infty} \frac{\cos x}{x^{3/2}}\,dx$$
Since $|(\cos x)/x^{3/2}| \le 1/x^{3/2}$, the RHS exists by a comparison test.
21 Diverges

Section 10-9 Review Exercises

Page 573

1 1
3 $a_{2n} \to L$, $b_{2n+1} \to M$ (subsequences), $a_{2n} - b_{2n+1} \to L - M$
5 Call the new sequence $\{b_n\}$. Suppose $|a_n - L| < \varepsilon$ for $n \ge N$. Then $|b_m - L| < \varepsilon$ for $m \ge \frac{1}{2}N(N-1) = 1 + 2 + 3 + \cdots + (N-1)$.

7 Diverges
9 Diverges
11 Converges
13 Converges
15 Converges; diverges absolutely
17 Converges; diverges absolutely
19 Converges
21 Diverges
23 $a_n = a_{n-1}(1+n^2)/(2+n^2) < a_{n-1}$ so $a_1 > a_2 > a_3 > \cdots > 0$
25 $\sum_1^N a_{2n} < \sum_1^{2N} a_n < \sum_1^\infty a_n$, so the partial sums are bounded, also increasing.

Chapter 11 Power Series

Section 11-1 Convergence and Divergence

Page 581

1 $R = 1$
3 $R = 1$
5 $R = 3$
7 $R = 1$
9 $R = 1/e$
11 $R = +\infty$
13 $R = 1$
15 $R = \sqrt[4]{\frac{1}{2}}$
17 $R = 1$
19 $S(x) = 1/(4-x)$, $|x-3| < 1$
21 $S(x) = 1/(1+e^x)$, $x < 0$
23 $S(x) = 2\ln x$, $x > 0$
25 $[-1, 1]$
27 $(-1, 1)$
29 $R = \frac{1}{2}$
31 $\sum x^n/\pi^n$
33 $S(1) = \sum a_n$ diverges since $a_n \not\to 0$. Hence $R \le 1$.
35 All x
37 $x < 0$

Section 11-2 Taylor Series

Page 587

1 $\sum_{n=0}^\infty \frac{3^n}{n!} x^n$

3 $\sum_{n=0}^\infty \frac{1}{2}\sqrt{2}\left(\frac{(-1)^n}{(2n)!} t^{2n} + \frac{(-1)^{n+1}}{(2n+1)!} t^{2n+1}\right)$, $t = x - \frac{1}{4}\pi$

5 $\sum_{n=0}^\infty \frac{x^{2n+1}}{(2n+1)!}$

7 $\sum_{n=0}^\infty (\frac{3}{4})^{n+1}(x - \frac{1}{3})^n$

9 $\ln a + \sum_{n=1}^\infty \frac{(-1)^{n-1}}{na^n} x^n$

11 $\sum_{n=0}^\infty \frac{n+1}{n!} x^n$

13 $2 + \sum_{n=1}^\infty \frac{(-1)^{n-1}(2n-2)!}{2^{4n-2}n!(n-1)!}(x-4)^n$

15 $-7 + 12x + 5x^3$

17 $x + \frac{1}{3}x^3 + \frac{2}{15}x^5$

19 $\exp[-(x-1)^2]$

21 $\sum_1^\infty \frac{a_{n-1}}{n} x^n$

23 $\sum_0^\infty x^{2n}/(2n)!$. Since $f^{(n+1)}(x) = \sinh x$ or $\cosh x$, we have $|f^{(n+1)}(x)| < e^x$. Apply Exercise 22 with $a = e^B$ and $K = 1$ or use the argument given in the text for the Taylor series of e^x.

25 $f'(0) = \lim_{x\to 0} f(x)/x = \lim_{t\to\pm\infty} tf(1/t)$
$= \lim_{t\to\pm\infty} t\exp(-t^2) = 0$

27 $f'(0) = 0$ by Exercise 25. Assume $f^{(n)}(0) = 0$. Then $f^{(n+1)}(0) = \lim_{x\to 0}[f^{(n)}(x) - f^{(n)}(0)]/x$. By Exercise 26 and the induction hypothesis, $f^{(n+1)}(0)$ is the sum of limits of the form

$$\lim_{x\to 0} ae^{-1/x^2}/x^{k+1} = a \lim_{y\to+\infty} y^{k+1}e^{-y^2} = 0$$

Hence $f^{(n+1)}(0)$ exists and equals 0.

29 $xS = \sum_0^\infty x^{n+1} = S - 1$, $S = 1 + xS$, $(1-x)S = 1$, $S = 1/(1-x)$

Section 11-3 Expansion of Functions

Page 595

1 $\sum_{n=0}^\infty 5^n x^{2n}$

3 $1 + 2\sum_{n=1}^\infty (-1)^n x^n$

5 $\sum_{n=1}^\infty \frac{(-1)^{n-1}(1+3^{2n+1})}{(2n-1)!} x^{2n}$

7 $\sum_{n=0}^\infty \frac{(-1)^n}{(2n+2)!} x^{2n}$

9 $2 + \sum_{n=2}^\infty 3x^n$

11 $\sum_{n=1}^\infty \frac{(-1)^{n-1}2^{2n-1}}{(2n)!} x^{2n}$

13 $1 + 5x^2 + 19x^4 + 65x^6 + \cdots$

15 $1 + x^2 + x^3 + \frac{3}{2}x^4 + \frac{13}{6}x^5 + \frac{73}{24}x^6 + \cdots$

17 $x^3 - \frac{1}{2}x^5 + \cdots$

19 $1 - \frac{1}{2}x^2 + x^3 + \frac{1}{24}x^4 - \frac{1}{2}x^5 + \frac{359}{720}x^6 + \cdots$

21 10080

23 272

25 $\frac{1}{1-x} f(x)$
$= (1 + x + x^2 + \cdots)(a_0 + a_1 x + a_2 x^2 + \cdots)$
$= a_0 + (a_0 + a_1)x + (a_0 + a_1 + a_2)x^2 + \cdots$

27 $(x + x^2 - x^3)/(1 - x^3)$ for $|x| < 1$

29 $\frac{1}{2}(\cos x + \cosh x)$

31 Suppose $f(x)$ is odd, $f(-x) = -f(x)$. Differentiate using the chain rule: $-f'(-x) = -f'(x)$, which says $f'(x)$ is even. Similarly the derivative of an even function is odd. If $f(x)$ is even, then $f'(x)$, $f'''(x)$, $f^{(5)}(x)$, $\cdots$ are odd. Hence $f'(0) = f'''(0) = \cdots = 0$, because an odd function is 0 at $x = 0$. Therefore the odd Taylor coefficients are 0. Similarly for an even function.

33 From $f(x) = f(-x)$ we have $\sum a_n x^n = \sum (-1)^n a_n x^n$. By uniqueness, $a_n = (-1)^n a_n$ for all n; hence $a_{2n-1} = -a_{2n-1}$ so that $a_{2n-1} = 0$.

35 $(1-x)f(x) = \sum_1^\infty nx^n - \sum_1^\infty nx^{n+1}$
$= \sum_1^\infty nx^n - \sum_2^\infty (n-1)x^n = \sum_1^\infty x^n = x/(1-x);$
hence $f(x) = x/(1-x)^2$.

37 $1 + \frac{1}{2}x^2 + \frac{5}{24}x^4 + \frac{61}{720}x^6 + \cdots$

39 $2x - \frac{1}{60}x^5 + \cdots$

Section 11-4 Calculus of Power Series

Page 600

1 $\dfrac{d}{dx}\left(x - \dfrac{x^3}{3!} + \dfrac{x^5}{5!} - \cdots\right) = 1 - \dfrac{x^2}{2!} + \dfrac{x^4}{4!} - \cdots = \cos x$

3 $\dfrac{d^2}{dx^2}\left(1 + \dfrac{k^2x^2}{2!} + \dfrac{k^4x^4}{4!} + \cdots\right) = \dfrac{1\cdot 2k^2}{2!} + \dfrac{3\cdot 4k^4x^2}{4!} + \cdots = k^2\left(1 + \dfrac{x^2}{2!} + \dfrac{x^4}{4!} + \cdots\right) = k^2 \cosh kx$

5 $\int_0^x 2(t + t^3 + t^5 + \cdots)\,dt = 2\left(\dfrac{x^2}{2} + \dfrac{x^4}{4} + \dfrac{x^6}{6} + \cdots\right) = x^2 + \dfrac{(x^2)^2}{2} + \dfrac{(x^2)^3}{3} + \cdots = -\ln(1 - x^2)$

7 Both sides equal $x - \frac{2}{3}x^3 + \frac{3}{5}x^5 - \frac{4}{7}x^7 + \cdots$.

9 $\dfrac{4 - 3x}{(1-x)^2}$ for $|x| < 1$

11 $-\frac{1}{4}\ln(1 - x^4)$ for $|x| < 1$

13 $\displaystyle\int_0^x \frac{\arctan t}{t}\,dt$

15 $\dfrac{1 + x^2}{(1 - x^2)^2}$ for $|x| < 1$

17 $\ln(1 - x) + \dfrac{x}{1 - x}$ for $|x| < 1$

19 1

21 2

23 3

25 0; the game is fair.

Section 11-5 Binomial Series

Page 605

1 $\sum_0^\infty (-1)^n \dfrac{(n+1)(n+2)}{2} x^n$ for $|x| < \frac{1}{2}$

3 $\sum_0^\infty 4^n(n+1)x^{2n}$ for $|x| < \frac{1}{2}$

5 $1 + \frac{1}{2}x^3 + \sum_2^\infty (-1)^{n-1} \dfrac{3\cdot 7 \cdots (4n-5)}{2^n n!} x^{3n}$ for $|x| < 1/\sqrt[3]{2}$

7 $\sum_0^\infty (-1)^n \dfrac{(2n)!}{2^n(n!)^2} x^{n+2}$ for $|x| < \frac{1}{2}$

9 $\sum_0^\infty (-1)^n \dfrac{(2n)!}{4^n(n!)^2(2n+1)} x^{2n+1}$ for $|x| < 1$

11 $1 - \sum_1^\infty \dfrac{(2n)!}{8^n(2n-1)(n!)^2} x^{2n}$ for $|x| < \sqrt{2}$

13 $\sqrt{2} + \sum_1^\infty (-1)^{n-1} \dfrac{\sqrt{2}\,(2n)!}{8^n(n!)^2(2n-1)} (x-1)^n$

15 4.0125

17 0.8626

19 1.41421 3514

21 $p_n = 2n \sin \dfrac{\pi}{n} = 2n\left(\dfrac{\pi}{n} - \dfrac{\pi^3}{6n^3} + \dfrac{\pi^5}{120n^5} - \dfrac{\pi^7}{7!n^7} + \cdots\right)$

hence $\pi = \frac{1}{2}p_n + \dfrac{\pi^3}{6n^2} - \dfrac{\pi^5}{120n^4} + \dfrac{\pi^7}{7!n^6} - \cdots$.

Section 11-6 Review Exercises

Page 606

1 $R = e$

3 $4 + 6(x-1) + 4(x-1)^2 + (x-1)^3$

5 $\sum_0^\infty (-1)^n \dfrac{2n}{(2n+1)!} x^{2n+1}$

7 $\sum_1^\infty (-1)^{n-1} \dfrac{x^{2n-1}}{(2n-1)!(2n-1)}$

9 $\sum_0^\infty \dfrac{(-1)^n}{4n+2} x^{4n+2}$

11 $\sum_1^\infty (-1)^n(2n)x^{2n-1}$

13 $\dfrac{x^2 + 4x}{(4 - x)^2}$

15 $\frac{1}{2}(\sin x + \sinh x)$

17 $\frac{9}{4}$

19 $-2^{14}(20!)/(14!)$

21 To have $4/(2n-1) < 5 \times 10^{-5}$ requires $n \geq 40{,}001$.

Chapter 12 Space Geometry

Section 12-1 Rectangular Coordinates

Page 610

1 x-axis up

3 x-axis forward

5 z-axis up

7 and 9

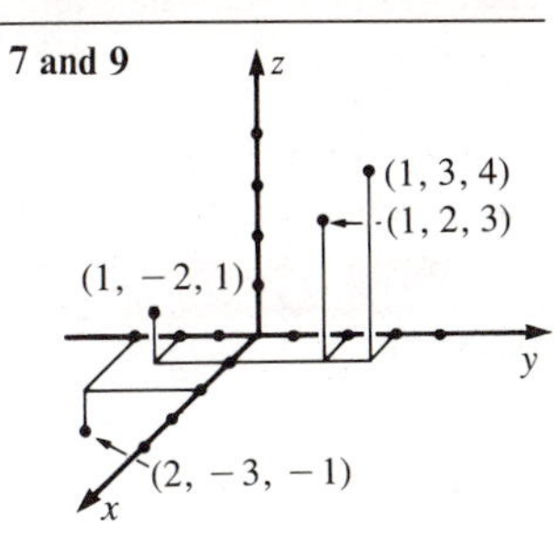

11 and 13

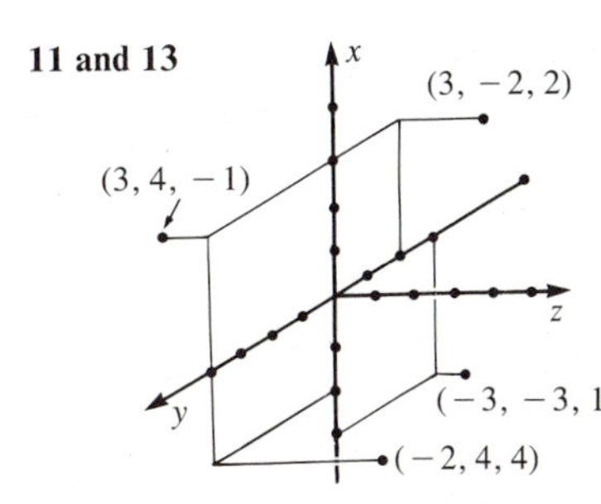

15 and 17

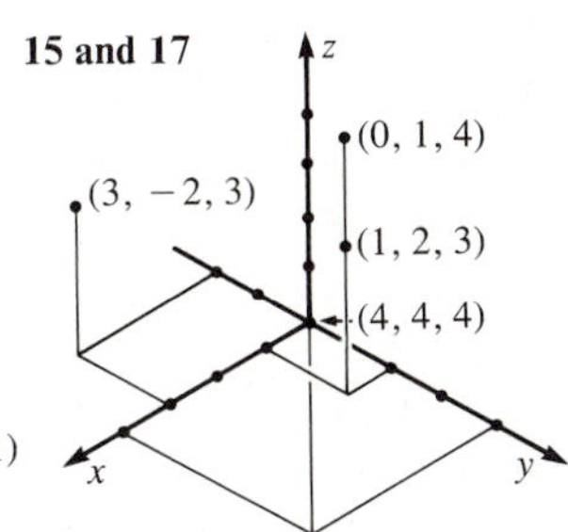

Section 12-2 Vector Algebra

Page 615

1 $(5, 2, 4)$

3 $(3, -2, 10)$

5 $(1, -16, 9)$

7 $(-1, 10, 4)$

9 $(4, 1, -1)$

11 $(2, 0, 1)$

13 $(3, 4, 1)$

15 $(\frac{3}{2}, 2, -\frac{1}{2})$

17 $(0, 2, 3)$

19 $(-1, 0, 2)$

21 $u_1 + v_1 = v_1 + u_1$, etc.

23 $(a + b)u_1 = au_1 + bu_1$, etc.

25 $(ab)u_1 = a(bu_1)$, etc.

27 $(\frac{1}{4}, \frac{1}{4}, \frac{1}{4})$

29 Let the vertices be $\mathbf{u}$, $\mathbf{v}$, $\mathbf{w}$, $\mathbf{z}$. The centroid of $\mathbf{uvw}$ is $\frac{1}{3}(\mathbf{u} + \mathbf{v} + \mathbf{w})$. The point $\mathbf{c} = \frac{3}{4}[\frac{1}{3}(\mathbf{u} + \mathbf{v} + \mathbf{w})] + \frac{1}{4}\mathbf{z}$ is on the segment joining this centroid to $\mathbf{z}$; it is $\frac{3}{4}$ of the way from $\mathbf{z}$ to the centroid. But $\mathbf{c} = \frac{1}{4}(\mathbf{u} + \mathbf{v} + \mathbf{w} + \mathbf{z})$ is *symmetric* in the four vertices, so the same construction starting with any face and its opposite vertex leads to the same $\mathbf{c}$.

Section 12-3 Length and Inner Product

Page 622

1 29

3 -1

5 $\sqrt{11}$

7 1

9 $\arccos(-\frac{12}{25})$

11 $\frac{1}{2}\pi$

13 $\arccos(-1/\sqrt{15})$

15 $\sqrt{42}$

17 $\sqrt{50}$

19 $\frac{1}{2}\sqrt{2}$, 0, $\frac{1}{2}\sqrt{2}$

21 $\frac{1}{7}\sqrt{14}$, $\frac{1}{14}\sqrt{14}$, $-\frac{3}{14}\sqrt{14}$

23 $(1, 1, 0)$ and $(0, 2, 1)$, for instance

25 $\mathbf{v} \cdot \mathbf{w} = v_1w_1 + \cdots = w_1v_1 + \cdots = \mathbf{w} \cdot \mathbf{v}$

27
$$\begin{aligned}(\mathbf{u} + \mathbf{v}) \cdot \mathbf{w} &= (u_1 + v_1, \cdots) \cdot (w_1, \cdots)\\ &= (u_1 + v_1)w_1 + \cdots\\ &= (u_1w_1 + \cdots) + (v_1w_1 + \cdots)\\ &= \mathbf{u} \cdot \mathbf{w} + \mathbf{v} \cdot \mathbf{w}\end{aligned}$$

29 $|\mathbf{v} \cdot \mathbf{w}| = |\mathbf{v}| \cdot |\mathbf{w}| \cdot |\cos\theta| \le |\mathbf{v}| \cdot |\mathbf{w}|$. Equality if $\mathbf{v} = \mathbf{0}$, $\mathbf{w} = \mathbf{0}$, or $\theta = 0, \pi$.

31 The equation is equivalent to $|\mathbf{x} - \frac{1}{2}(\mathbf{a} + \mathbf{b})|^2 = c^2 + \frac{1}{4}|\mathbf{a} + \mathbf{b}|^2$; center $\frac{1}{2}(\mathbf{a} + \mathbf{b})$, radius r. $r^2 = c^2 + \frac{1}{4}|\mathbf{a} + \mathbf{b}|^2$

33 $|\mathbf{x}|^2 = a^2|\mathbf{u}|^2 + b^2|\mathbf{v}|^2 + c^2|\mathbf{w}|^2$

35 $\mathbf{0ab}$ is equilateral.

Section 12-4 Lines and Planes

Page 628

1 No

3 Yes

5 $\arccos\sqrt{\frac{1}{330}}$

7 $(\frac{2}{3}, \frac{2}{3}, \frac{2}{3})$

9 $(0, -7, 14)$, $(-\frac{7}{5}, 0, \frac{7}{5})$, $(-\frac{14}{9}, \frac{7}{9}, 0)$

11 $(0, 0, -1)$, $(0, 0, -1)$, no intersection

13 $\sqrt{2}$

15 $(23, 22, 0)$

17 $\mathbf{x} = t(\mathbf{b} - \mathbf{c}) + \mathbf{a}$

19 $\frac{1}{3}x_1 - \frac{2}{3}x_2 + \frac{2}{3}x_3 = \frac{1}{3}$

21 $-\frac{8}{9}x_1 + \frac{1}{9}x_2 - \frac{4}{9}x_3 = 3$

23 $\frac{1}{3}\sqrt{3}\,x_1 + \frac{1}{3}\sqrt{3}\,x_2 + \frac{1}{3}\sqrt{3}\,x_3 = \sqrt{3}$

25 $\frac{1}{3}\sqrt{3}$

27 0

29 $\theta = \arcsin(\mathbf{a} \cdot \mathbf{n}/|\mathbf{a}|)$

31 The line is parallel to the plane if and only if it is perpendicular to the normal $\mathbf{n}$ to the plane—if and only if $\mathbf{a} \cdot \mathbf{n} = 0$.

33 We want $\mathbf{z} = t\mathbf{a} + \mathbf{b}$ and $\mathbf{z} \cdot \mathbf{n} = p$; hence $(t\mathbf{a} + \mathbf{b}) \cdot \mathbf{n} = p$ so $t = (p - \mathbf{b} \cdot \mathbf{n})/(\mathbf{a} \cdot \mathbf{n})$, so $\mathbf{z}$ is as stated.

35 Let $\mathbf{u} \cdot \mathbf{x} = k$ be any plane containing L. Its normal $\mathbf{u}$ must lie in the plane spanned by the normals $\mathbf{a}$ and $\mathbf{b}$, hence $\mathbf{u} = s\mathbf{a} + t\mathbf{b}$. Since $\mathbf{u} \ne \mathbf{0}$, we can adjust a factor so $s^2 + t^2 = 1$. The plane is $s(\mathbf{a} \cdot \mathbf{x}) + t(\mathbf{b} \cdot \mathbf{x}) = k$. Choose $\mathbf{x}$ on L. Then $\mathbf{a} \cdot \mathbf{x} = C$ and $\mathbf{b} \cdot \mathbf{x} = d$; hence $k = sc + td$.

37 $\sum \mathbf{a}_j \cdot \mathbf{n} - rp = 0$, so the planes all pass through $\mathbf{c} = (\sum \mathbf{a}_j)/r$, the centroid of the $\mathbf{a}_j$.

39 $\mathbf{x} = [(\mathbf{c} - \mathbf{b}) \cdot \mathbf{u}]\mathbf{u} + \mathbf{b}$

Section 12-5 Linear Systems

Page 635

1 $(-\frac{1}{3}, \frac{2}{3})$

3 $(-1, 1)$

5 $(\frac{1}{19}, \frac{7}{19})$

7 $(-\frac{8}{19}, \frac{1}{19}, \frac{7}{19})$

9 $(0, 0, 0)$

11 $(\frac{3}{14}, -\frac{1}{4}, -\frac{3}{28})$

13 Add $0 = 2$; parallel lines

15 Add first and third: $0 = 1$; the first and third planes are parallel.

17 $(t, \frac{2}{3}t - \frac{1}{3})$

19 $(t, -\frac{1}{2}t - \frac{1}{2}, -\frac{5}{2}t + \frac{1}{2})$

21 $-\frac{1}{15}(8, 13, -22)$

23 $\frac{1}{32}(90, 11, 67)$

25 Expanded by the first row, it is a linear equation, hence a line. If (a_i, b_i) is substituted for (x, y), a determinant with two equal rows results, value 0. Hence the line passes through the given points. Note that the coefficients of x and y are not both 0.

27 $(b - a)(c - a)(c - b)$

29 $(1, -1, 1)$

31 Line $\mathbf{x} = (t + 2, t, -t)$

33 $x + y + z = 2$

35 $7x - 4y + 2z = 3$

Section 12-6 Cross Product

Page 642

1 $(-5, 2, -14)$

3 $(-4, 8, -4)$

5 $(2, -2, 0)$

7 $\mathbf{i} - \mathbf{j}$

9 $-7\mathbf{i} + 8\mathbf{j} + 2\mathbf{k}$

11 $\frac{1}{11}\sqrt{11}\,(1, -1, 3)$

13 $\mathbf{x} = (-2t + 6, 8t + 1, -5t - 2)$

15 $\sqrt{14}$

17 1

19 87

21 $(0, -10, 10)$

23 $(3, -1, 4)$

25 $\mathbf{v} = (v_1, v_2, 3)$

27 $\mathbf{x} = \mathbf{0}$

29 A determinant changes sign when two rows are interchanged, hence $\mathbf{u} \cdot (\mathbf{v} \times \mathbf{w}) = -\mathbf{v} \cdot (\mathbf{u} \times \mathbf{w}) = +\mathbf{v} \cdot (\mathbf{w} \times \mathbf{u})$.

31 $(av_2)w_3 - (av_3)w_2 = a(v_2w_3 - v_3w_2)$, etc.

33 Clearly the formula is true if $\mathbf{a} = \mathbf{b}$ because both sides equal $\mathbf{0}$. Now for example if $\mathbf{a} = \mathbf{i}$ and $\mathbf{b} = \mathbf{j}$, then
$$\begin{aligned}\text{LHS} &= (\mathbf{i} \cdot \mathbf{u})(\mathbf{j} \cdot \mathbf{v}) - (\mathbf{i} \cdot \mathbf{v})(\mathbf{j} \cdot \mathbf{u}) = u_1v_2 - u_2v_1\\ &= \mathbf{k} \cdot (\mathbf{u} \times \mathbf{v}) = (\mathbf{a} \times \mathbf{b}) \cdot (\mathbf{u} \times \mathbf{v}) = \text{RHS, etc.}\end{aligned}$$

35 Write $(\mathbf{a} - \mathbf{b}) \cdot \mathbf{v} = 0$ and choose $\mathbf{v} = \mathbf{a} - \mathbf{b}$. Then $|\mathbf{a} - \mathbf{b}|^2 = (\mathbf{a} - \mathbf{b}) \cdot (\mathbf{a} - \mathbf{b}) = 0$. Hence $\mathbf{a} - \mathbf{b} = \mathbf{0}$, $\mathbf{a} = \mathbf{b}$.

37 Take $\mathbf{a} = \mathbf{u} = \mathbf{w}$ and $\mathbf{b} = \mathbf{v}$.

39 $\frac{1}{2}\sqrt{a^2b^2 + b^2c^2 + c^2a^2}$

Section 12-7 Applications of the Cross Product

Page 648

1 2

3 $\mathbf{x} = t(-5, 1, 1) + (-2, 1, 0)$

5 $\mathbf{x} = t(3, 1, -5) + (1, 0, 2)$

7 $x = 1$

9 $2x - y - z = 3$

11 $x + y = 3$

13 $\sqrt{2}$

15 $\frac{3}{17}\sqrt{17}$

17 $\mathbf{x} = (\frac{8}{3}, \frac{8}{3}, \frac{5}{3}), \quad y = (\frac{8}{3}, \frac{5}{3}, \frac{8}{3})$

19 $\mathbf{x} = (-2t + 1, s - 2t + 1, s - t)$

21 The left side is the square of the volume of the parallelepiped determined by $\mathbf{a}$, $\mathbf{b}$, $\mathbf{c}$, clearly at most the square of the product of the edges.

23 It is the solid triangle with vertices $\mathbf{a}$, $\mathbf{b}$, $\mathbf{c}$.

25 We show that the following three conditions are equivalent: (i) $[\mathbf{a}, \mathbf{b}, \mathbf{c}] \neq 0$, (ii) no one of the vectors $\mathbf{a}$, $\mathbf{b}$, $\mathbf{c}$ lies in the plane spanned by the other two, and (iii) $\mathbf{a}$, $\mathbf{b}$, $\mathbf{c}$ are linearly independent. Suppose (i). Then the parallelepiped determined by $\mathbf{a}$, $\mathbf{b}$, and $\mathbf{c}$ has positive volume; hence (ii). Conversely, if (ii), then the parallelepiped has positive volume; hence (i). Now assume (ii) and $r\mathbf{a} + s\mathbf{b} + t\mathbf{c} = \mathbf{0}$. If say $t \neq 0$, then $\mathbf{c} = (-r/t)\mathbf{a} + (-s/t)\mathbf{b}$; impossible. Hence $t = 0$ and similarly $r = s = 0$, so (iii) follows. Conversely, if (iii) and say $\mathbf{c} = r\mathbf{a} + s\mathbf{b}$, then $r\mathbf{a} + s\mathbf{b} + (-1)\mathbf{c} = \mathbf{0}$, impossible; hence (ii). This proves (i) if and only if (ii) if and only if (iii).

27 $C = A + B, \ aA = bB$

29 $\sum(\mathbf{x}_j - \mathbf{p}) \times \mathbf{F}_j = \sum \mathbf{x}_j \times \mathbf{F}_j - \mathbf{p} \times \sum \mathbf{F}_j = \mathbf{0} - \mathbf{0} = \mathbf{0}$

Section 12-8 Review Exercises

Page 649

1 $\mathbf{x} = t(1, 0, 1) + (1, -2, -1)$

3 $\sqrt{43}$

5 $\frac{1}{43}\sqrt{43}\,(3, -5, 3)$

7 $\frac{1}{26}\sqrt{26}\,(3, -1, 4)\cdot\mathbf{x} = \frac{7}{26}\sqrt{26}$

9 $(34, -24)$

11 $-\mathbf{i} + 5\mathbf{j} + 3\mathbf{k}$

13 $(2, 0, 2), \ \sqrt{8}$

15 $4x + y - 3z = 4$

17 Let $\mathbf{a}$, $\mathbf{b}$, $\mathbf{c}$ be the vertices. The midpoints of $\mathbf{ab}$ and $\mathbf{ac}$ are $\mathbf{m} = \frac{1}{2}(\mathbf{a} + \mathbf{b})$ and $\mathbf{n} = \frac{1}{2}(\mathbf{a} + \mathbf{c})$. The vector $\mathbf{n} - \mathbf{m} = \frac{1}{2}(\mathbf{c} - \mathbf{b})$ is clearly parallel to $\mathbf{bc}$.

Chapter 13 Vector Functions

Section 13-1 Differentiation

Page 656

1 $(e^t, 2e^{2t}, 3e^{3t})$

3 $(1, 3, 4)$

5 $(1, -\sin t, \cos t)$

7 $\sqrt{(2t)^2 + (3t^2 + 4t^3)^2}$

9 $|A\omega|$

11 $\sqrt{16t^6 + 36t^{10}}$

13 $(x_1 + y_1)' = x_1' + y_1'$, etc.

15 $(x_2y_3 - x_3y_2)' = (x_2'y_3 - x_3'y_2) + (x_2y_3' - x_3y_2')$, etc.

17 $\mathbf{x} = t\mathbf{b} + \mathbf{c}$

19 $\mathbf{x} = e^{kt}\mathbf{c}$

21 $(|\mathbf{x}|^2)' = (\mathbf{x}\cdot\mathbf{x})' = 2\mathbf{x}\cdot\mathbf{x}' = 0, \ |\mathbf{x}|^2 = r^2$

23
$$\begin{aligned}[\mathbf{u}, \mathbf{v}, \mathbf{w}]' &= [\mathbf{u}\cdot(\mathbf{v}\times\mathbf{w})]' = \mathbf{u}'\cdot(\mathbf{v}\times\mathbf{w}) + \mathbf{u}\cdot(\mathbf{v}\times\mathbf{w})' \\ &= [\mathbf{u}', \mathbf{v}, \mathbf{w}] + \mathbf{u}\cdot(\mathbf{v}'\times\mathbf{w} + \mathbf{v}\times\mathbf{w}') \\ &= [\mathbf{u}', \mathbf{v}, \mathbf{w}] + [\mathbf{u}, \mathbf{v}', \mathbf{w}] + [\mathbf{u}, \mathbf{v}, \mathbf{w}']\end{aligned}$$

25 $\mathbf{x}_n = \mathbf{a} + \frac{1}{3}[1 - (-\frac{1}{2})^n](\mathbf{b} - \mathbf{a})$

27 The i-th component is $\int_a^b cx_i\,dt = c\int_a^b x_i\,dt$

29
$$\begin{aligned}\int_a^b \mathbf{c}\cdot\mathbf{x}\,dt &= \int_a^b \sum c_ix_i\,dt = \sum c_i\left(\int_a^b x_i\,dt\right) \\ &= \mathbf{c}\cdot\int_a^b \mathbf{x}\,dt\end{aligned}$$

31
$$\begin{aligned}\left|\mathbf{c}\cdot\int_a^b \mathbf{x}\,dt\right| &= \left|\int_a^b \mathbf{c}\cdot\mathbf{x}\,dt\right| \le \int_a^b |\mathbf{c}\cdot\mathbf{x}|\,dt \\ &\le \int_a^b |\mathbf{c}|\cdot|\mathbf{x}|\,dt = |\mathbf{c}|\int_a^b |\mathbf{x}|\,dt\end{aligned}$$

Section 13-2 Arc Length

Page 663

1 $\sqrt{a_1^2 + a_2^2 + a_3^2}$

3 $\frac{1}{9375}(43\cdot 41^{3/2} + 2048)$

5 $\int_0^b \sqrt{1 + 9x^4}\,dx$

7 $\int_a^b [m^2t^{2m-2} + n^2t^{2n-2} + r^2t^{2r-2}]^{1/2}\,dt$

9 $\sqrt{5} - \sqrt{2} + \ln\dfrac{\sqrt{5} - 1}{2(\sqrt{2} - 1)}$

11 $b^2 + b$

13 $|\mathbf{x}|^2 = \sin^4 t + \sin^2 t\cos^2 t + \cos^2 t = 1$

$\mathbf{x}\cdot\mathbf{x}' = (\sin^2 t)(2\sin t\cos t) + (\sin t\cos t)(\cos^2 t - \sin^2 t) + (\cos t)(-\sin t) = 0$

15 $t = \frac{1}{2}(-1 + \sqrt{1 + 4s})$

17
$$\begin{aligned}L &= \int_\alpha^\beta \sqrt{x'^2 + y'^2}\,dt \ge \int_\alpha^\beta |x'|\,dt \\ &\ge \left|\int_\alpha^\beta x'\,dt\right| = |x(\beta) - x(\alpha)| = b - a\end{aligned}$$

Section 13-3 Plane Curves

Page 669

1 $\frac{3}{2}t, \ 3/4t$

3 $2t/(e^t + 1), \ \dfrac{2[(1 - t)e^t + 1]}{(e^t + 1)^3}$

5 $-\cot t, \ -\csc^3 t$

7 $\coth t, \ -\operatorname{csch}^3 t$

9 $\frac{2}{5}$

11 $1 - \frac{1}{2}\sqrt{2}$

13 $\frac{1}{2}$

15 $y = tx$; hence $y^2 = t^2x^2 = (x + 1)x^2 = x^3 + x^2$. Conversely, if $y^2 = x^3 + x^2$, set $y = tx$. Then $t^2 = x + 1, \ x = t^2 - 1, \ y = t^3 - t$.

17 $a^2 + x^2 = a^2 + a^2\cot^2\theta = a^2\csc^2\theta$; hence $a^3/(a^2 + x^2) = a/\csc^2\theta = a\sin^2\theta = y$. Conversely, if $(a^2 + x^2)y = a^3$, set $x = a\cot\theta$. Then $y = a\sin\theta$.

19 $(a^2+x^2)y = (a^2+a^2\cot^2\theta)(b\sin\theta\cos\theta)$
$= a^2 b\csc^2\theta\sin\theta\cos\theta = a^2 b\csc\theta\cos\theta$
$= a^2 b\cot\theta = abx$

Conversely, if $(a^2+x^2)y = abx$, set $x = a\cot\theta$. Then $a^2\csc^2\theta\, y = a^2 b\cot\theta$, $y = b\sin\theta\cos\theta$.

21 $x^3+y^3 = 27a^3\dfrac{t^3+t^6}{(1+t^3)^3} = 27a^3\dfrac{t^3}{(1+t^3)^2} = 3axy$

Conversely, if $x^3+y^3 = 3axy$, set $t = y/x$. Then $x^3(1+t^3) = 3atx^2$, $x = 3at/(1+t^3)$, $y = 3at^2/(1+t^3)$.

23 $y^2(a-x) = \left[\dfrac{a^2t^6}{(1+t^2)^2}\right]\left[\dfrac{a}{1+t^2}\right] = \dfrac{a^3t^6}{(1+t^2)^3} = x^3$

Conversely, if $y^2(a-x) = x^3$, set $y = tx$. Then $t^2(a-x) = x$, $x = at^2/(1+t^2)$, $y = at^3/(1+t^2)$.

Section 13-4 Curvature

Page 678

1 $\mathbf{T} = \mathbf{c}/|\mathbf{c}|$

3 $\mathbf{T} = \dfrac{1}{\sqrt{9t^4+4t^2}}(3t^2, 2t)$

5 $\mathbf{x} = (t^2, t^5)$

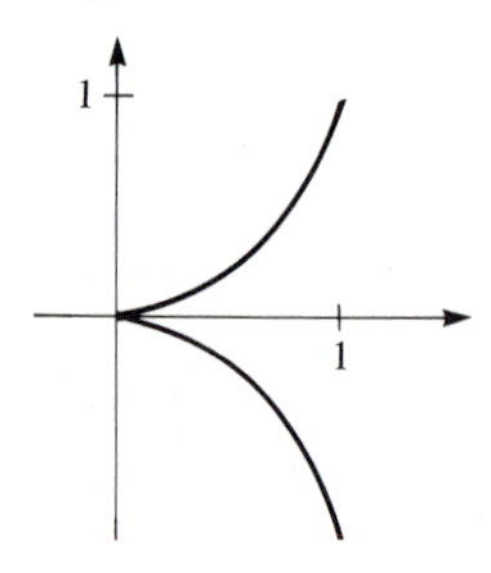

7 $2/(1+4x^2)^{3/2}$

9 $-6t^2/(9t^4+4t^2)^{3/2}$

11 $(2+t^2)/(1+t^2)^{3/2}$

13 1

15 2

17 $\frac{1}{2}\sqrt{2}$

19 $(2-2\cos\theta)^{-1/2}(1-\cos\theta, \sin\theta)$,
$(2-2\cos\theta)^{-1/2}(-\sin\theta, 1-\cos\theta)$

21 $\mathbf{x} = (a\theta - b\sin\theta, a - b\cos\theta)$

23 $\mathbf{x} = (a\theta - b\sin\theta, a - b\cos\theta)$

25 $\mathbf{x} = (\cos\theta + \theta\sin\theta, \sin\theta - \theta\cos\theta)$

27 $\theta = \arctan(s/k)$

29 $\mathbf{x} = ((a-b)\cos\theta + b\cos\alpha, (a-b)\sin\theta - b\sin\alpha)$, where $\alpha = (a-b)\theta/b$

31 $\mathbf{x} = b(2\cos\theta + \cos 2\theta, 2\sin\theta - \sin 2\theta)$, $A = \frac{2}{9}\pi a^2$

33 $\mathbf{x} = 4b(\cos^3\theta, \sin^3\theta)$

35 $\frac{3}{8}\pi a^2$

37 $L_n = 8a(1-1/n)$ increases because $1/n$ decreases. $L_n \to 8a$

39 $\mathbf{x} = ((a+b)\cos\theta - b\cos\alpha, (a+b)\sin\theta - b\sin\alpha)$, where $\alpha = (a+b)\theta/b$

41 $12a$

43 Ellipse $x^2/(a+b)^2 + y^2/(a-b)^2 = 1$

Section 13-5 Space Curves

Page 684

1 $\mathbf{c}/|\mathbf{c}|$, 0

3 $(4t^2+9t^4+16t^6)^{-1/2}(2t, 3t^2, 4t^3)$,
$2t^2(36t^4+64t^2+9)^{1/2}/(16t^6+9t^4+4t^2)^{3/2}$

5 $(1+4\cos^2 2t)^{-1/2}(-\sin t, \cos t, 2\cos 2t)$,
$(17-12\cos^2 2t)^{1/2}/(1+4\cos^2 2t)^{3/2}$

7 $(0, -\sin t, -\cos t)$, $\frac{1}{10}$

9 Use $\sqrt{a^2+v^2} \le u+v$ for $u \ge 0$ and $v \ge 0$. Choose $u = \sqrt{x'^2+y'^2}$, $v = |z'|$:

$$L = \int_a^b \sqrt{x'^2+y'^2+z'^2}\,dt \le \int_a^b (\sqrt{x'^2+y'^2} + |z'|)\,dt$$
$$\le L_1 + \int_a^b |z'|\,dt$$

Section 13-6 Velocity and Acceleration

Page 690

1 The shell strikes the hill when $y = x\tan\beta$. Substitute this into the solution $y = -ax^2 + bx$ of Example 3, cancel x, solve for x, etc.

3 $[4t/(1+4t^2)](1, 2t)$, $[2/(1+4t^2)](-2t, 1)$

5 $[-\sin t\cos t/(1+\cos^2 t)](1, \cos t)$,
$[\sin t/(1+\cos^2 t)](\cos t, -1)$

7 $2\cos 2t(-1, 1)$, $\mathbf{0}$

9 $\mathbf{0}$, $b\omega^2(-\cos\omega t, -\sin\omega t, 0)$

11 $-[(\sin t\cos t)/(1+\cos^2 t)](\cos t, -\sin t, \cos t)$,
$[-1/(1+\cos^2 t)](\sin t, 2\cos t, \sin t)$

13 $\mathbf{x}\cdot\mathbf{x} = 1$; hence $\mathbf{x}\cdot\mathbf{v} = 0$, $\mathbf{x}\cdot\mathbf{a} + \mathbf{v}\cdot\mathbf{v} = 0$, $\mathbf{x}\cdot\mathbf{a} = -1$. Since $\mathbf{x}$ is a unit vector perpendicular to $\mathbf{t} = \mathbf{v}$, and since the projection $\mathbf{a}\cdot\mathbf{x}$ of $\mathbf{a}$ on $\mathbf{x}$ is -1, it follows that the normal component, the component of $\mathbf{a}$ perpendicular to $\mathbf{t}$, has length at least 1.

Section 13-7 Plane Curves in Polar Coordinates

Page 696

1 $\frac{1}{4}\pi a^2$

3 $\frac{1}{4}\pi a^2$

5 $\frac{3}{8}\pi a^2$

7 a^2

9 $\frac{2}{3}\pi + \sqrt{3}$

11 $(\frac{1}{8}\pi - \frac{1}{4})a^2$

13 $\mathbf{v} = \mathbf{u} + 2t\mathbf{w} = (\cos 2t - 2t\sin 2t, \sin 2t + 2t\cos 2t)$
$\mathbf{a} = -4t\mathbf{u} + 4\mathbf{w}$
$= (-4t\cos 2t - 4\sin 2t, -4t\sin 2t + 4\cos 2t)$

15 $\mathbf{v} = -\sin t\,\mathbf{u} + \cos t\,\mathbf{w} = (-\sin 2t, \cos 2t)$
$\mathbf{a} = -2\cos t\,\mathbf{u} - 2\sin t\,\mathbf{w}$
$= -2(\cos 2t, \sin 2t)$

17 $\mathbf{v} = e^t(\mathbf{u}+\mathbf{w}) = e^t(\cos t - \sin t, \cos t + \sin t)$
$\mathbf{a} = 2e^t\mathbf{w} = 2e^t(-\sin t, \cos t)$

19 $a\displaystyle\int_0^{2\pi}\sqrt{\theta^2+1}\,d\theta$

21 $2a\displaystyle\int_0^{\pi/2}\sqrt{\cos^2\alpha + 9\sin^2\alpha}\,d\alpha$

23 $a\displaystyle\int_0^{2\pi}\sqrt{2+2\cos\theta}\,d\theta = 8a$

Section 13-8 Review Exercises

Page 697

1 $(\sec^2 t, \sec t\tan t)$, $|\sec t|\sqrt{\sec^2 t + \tan^2 t}$

3 $2a\sinh(b/a)$

5 $\mathbf{v} = \mathbf{T}$, $ds/dt = 1$, $\mathbf{a} = k\mathbf{N}$, $k = |\mathbf{a}|$

7 6

9 Let $\mathbf{p}$ be the focus, D the directrix and $\mathbf{c}$ the constant unit vector perpendicular to D directed from D towards $\mathbf{p}$. If $\mathbf{x}$ is on the parabola, let $\mathbf{z}$ denote the corresponding point on D. Then $\mathbf{x} = \mathbf{z} + |\mathbf{x} - \mathbf{p}|\mathbf{c}$. Hence

$$\mathbf{T} = \mathbf{z}' + \left(\frac{\mathbf{x} - \mathbf{p}}{|\mathbf{x} - \mathbf{p}|} \cdot \mathbf{T}\right)\mathbf{c}$$

Dot in $\mathbf{c}$: $\mathbf{T} \cdot \mathbf{c} = \mathbf{a} \cdot \mathbf{T}$, where $\mathbf{a}$ is the unit vector in the direction $\mathbf{px}$. Thus the angle between $\mathbf{a}$ and $\mathbf{T}$ equals the angle between $\mathbf{c}$ and $\mathbf{T}$.

11 $x^2 + y^2 + z^2 = (\sin^2 t)(\cos^2 t + \sin^2 t) + \cos^2 t = 1$. The curve is on the unit sphere $|\mathbf{x}| = 1$.

13 Rectangular hyperbola

15 $r = ke^{a\theta}$

Chapter 14 Functions of Several Variables

Section 14-1 Functions and Graphs

Page 705

1 $V = \frac{1}{3}\pi r^2 h$

3 $D = [(x-2)^2 + (y-4)^2 + (2 + x + y)^2]^{1/2}$

5 $c = (a^2 + b^2 - 2ab\cos\theta)^{1/2}$

7 $F = Gm_1m_2/d^2$, G constant

9 The entire x, y-plane

11 The x, y-plane excluding the two lines $y = \pm x$

13 The region on and below the parabola $y = x^2$

15 The x, y-plane excluding the horizontal lines $y = \frac{1}{2}(2n - 1)\pi$, n all integers

17 The region above the plane $x + 2y + 3z = 4$

19 The solid unit ball $|\mathbf{x}| \leq 1$

21 $f(x, y) = x - 3y$

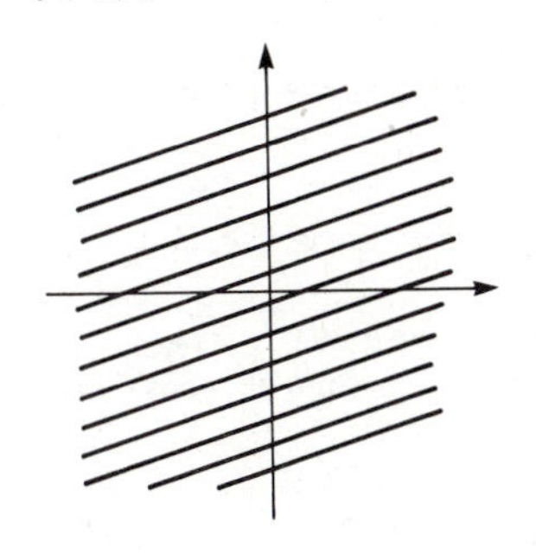

25 $f(x, y) = \ln(y - x^2)$

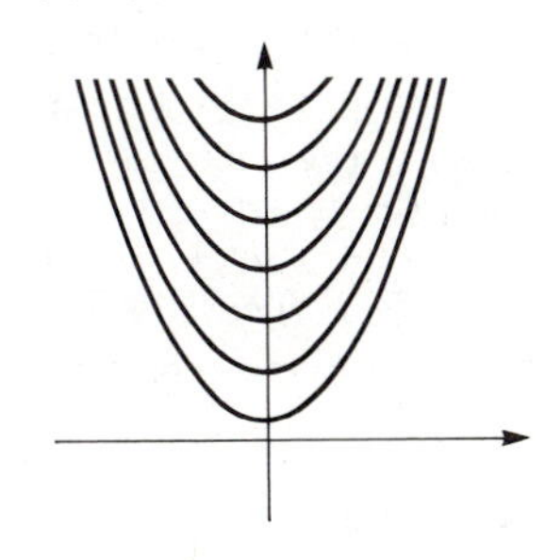

23 $f(x, y) = x^2 + 4y^2$

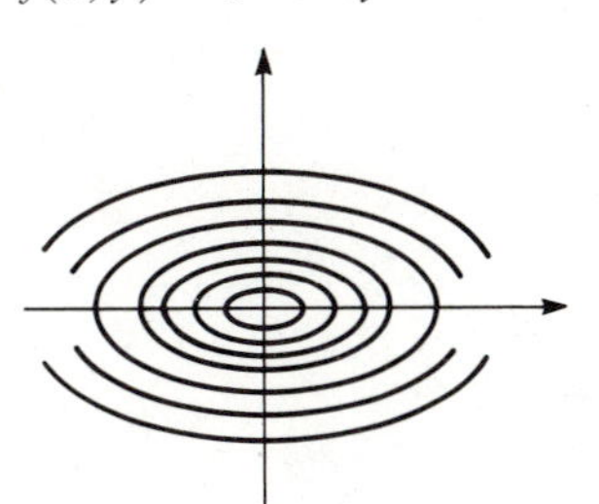

27 $z = 1 - 2x$

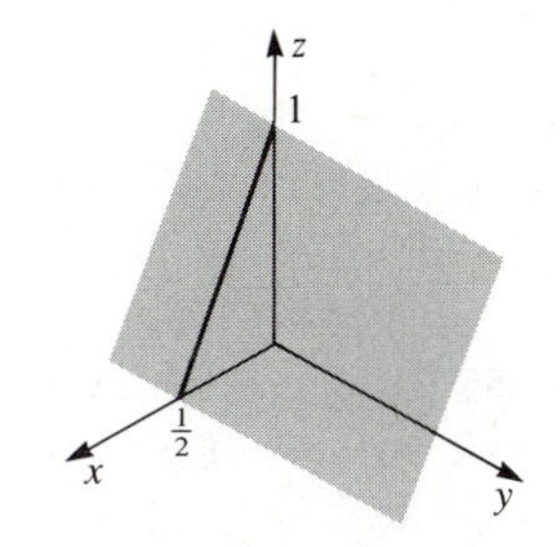

29 $z = \sqrt{y}$

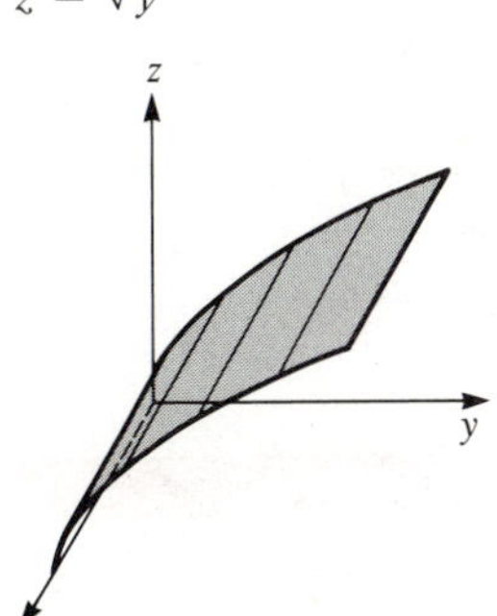

31 $z = x + \frac{1}{2}y$

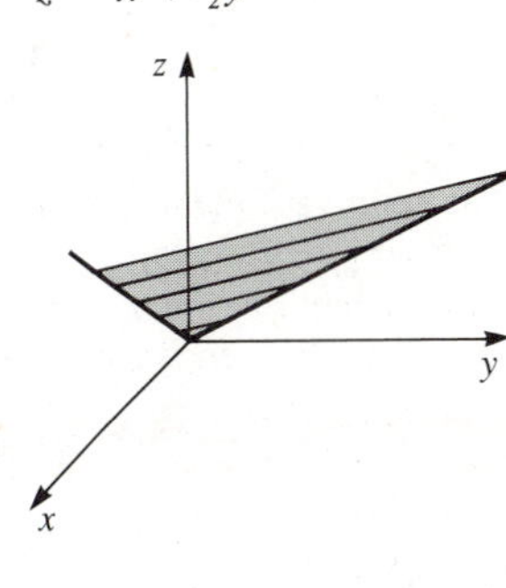

33 All the spheres with center $\mathbf{0}$

35 $|(\mathbf{x}_n + \mathbf{y}_n) - (\mathbf{a} + \mathbf{b})| \leq |\mathbf{x}_n - \mathbf{a}| + |\mathbf{y}_n - \mathbf{b}|$, etc.

37 $|[f(\mathbf{x}) + g(\mathbf{x})] - [f(\mathbf{a}) + g(\mathbf{a})]|$
$\leq |f(\mathbf{x}) - f(\mathbf{a})| + |g(\mathbf{x}) - g(\mathbf{a})|$, etc.

39 Yes. Set $f(0, 0) = 1$ and use $(\sin t)/t \to 1$ as $t \to 0$.

41 $w = uv$

Section 14-2 Partial Derivatives

Page 713

1 1, 2

3 -3, 6

5 2, $-\frac{1}{2}$

7 1, 0

9 2, -3

11 $2y\cos 2xy$, $2x\cos 2xy$

13 $(y^2 - x^2)/(x^2 + y^2)^2$, $-2xy/(x^2 + y^2)^2$

15 $2x/(x^2 + 3y)$, $3/(x^2 + 3y)$

17 $e^{2x}[2\sin(x - y) + \cos(x - y)]$, $-e^{2x}\cos(x - y)$

19 $\dfrac{yf}{(x - y)(x + y)}$, $\dfrac{-xf}{(x - y)(x + y)}$

21 $4x$, 1

23 32, 0, $3x^3z^2$

25 $z_x + 3z_y = 6(3x - y) - 3 \cdot 2(3x - y) = 0$

27 $(z_x)^2 - (z_y)^2 = (2x)^2 - (-2y)^2 = 4(x^2 - y^2) = 4z$

29 Set $D = x^4 + y^4 + z^4$. Then $xw_x = xyz/D - 4x^4(xyz)/D^2 = w - 4x^4w/D$. By symmetry, $xw_x + yw_y + zw_z = 3w - 4(x^4 + y^4 + z^4)w/D = 3w - 4w = -w$.

31 $dz/dt = (9t^2 + 2t)\exp(3t^3 + t^2)$

33 $dz/dt = 4t^3\cos(1/t) + t^2\sin(1/t) - 2t$

35 $dw/dt = 9t^8$

37 $dw/dt = 4te^{-t}[(2 - t)\sin 4t + 4t\cos 4t]$

39 $\partial z/\partial s = (s^2 - t)^2(2s^2 + t)/(2s^3t^2)$
$\partial z/\partial t = -\frac{1}{4}(s^2 - t)^2(2s^2 + t)/(s^2t^3)$

41 $\partial z/\partial s = t(st^3 + st + 1)/z$
$\partial z/\partial t = s(2st^3 + st + 1)/z$

43 3.6π ft^3/hr

45 19,960 cm^2, 106 cm^2/kg, 80.4 cm

47 $1 + \eta$, x

49 $u_t = -ku < 0$; "$=0$" if and only if $x = 0$ or 120. Also $u_t(60, t)/u_t(20, t) = (\sin\frac{1}{2}\pi)/(\sin\frac{1}{6}\pi) = 2$.

51 $(\partial/\partial u)F(u+v,u-v) = (\partial F/\partial x)\cdot 1 + (\partial F/\partial y)\cdot 1$
$= F_x(u+v,u-v) + F_y(u+v,u-v)$
$(\partial/\partial v)F(u+v,u-v)$
$= F_x(u+v,u-v) - F_y(u+v,u-v)$
Their sum $= 2F_x(u+v,u-v)$. For $F = y\sin xy$, both sides $= 2y^2\cos xy$.

53 By the chain rule, $\partial z/\partial r = 0$, $\partial z/\partial\theta = 1$. Also $z = \theta$; hence same result.

Section 14-3 Gradients and Directional Derivatives

Page 722

1 $(2xy + 3y^3, x^2 + 9xy^2)$

3 $(ad - bc)(cx + dy)^{-2}(y, -x)$

5 $e^z(2x, 2y, x^2 + y^2)$

7 $(xyz^3/f)(yz, xz, 2xy)$

9 The plane $6x + 3y + 2z = 0$

11 RHS $= (z_x\cos\theta + z_y\sin\theta)(\cos\theta, \sin\theta)$
$+ r^{-1}(-z_x r\sin\theta + z_y r\cos\theta)(-\sin\theta, \cos\theta)$
$= (\cos^2\theta + \sin^2\theta)(z_x, z_y) = \text{grad } z$

13 $z = 1/(r^2+10)$, grad $z = [-2r/(r^2+10)^2]\mathbf{u}$ has direction $-\mathbf{u}$.

15 The level curves of f and g at (a, b) are orthogonal.

17 $-4x + y = -2$

19 $x + y = 1$

21 1, 0

23 $0, -\frac{3}{2}\sqrt{2}$

25 1

27 $\pm(0, 1)$

29 grad $f(\mathbf{a}) = b\mathbf{u} + c\mathbf{v}$ so $|\text{grad } f|^2 = b^2 + c^2$. But $b = (\text{grad } f)\cdot\mathbf{u} = D_{\mathbf{u}}f(\mathbf{a})$ and $c = (\text{grad } f)\cdot\mathbf{v} = D_{\mathbf{v}}f(\mathbf{a})$. Hence $[D_{\mathbf{u}}f(\mathbf{a})]^2 + [D_{\mathbf{v}}f(\mathbf{a})]^2 = |\text{grad } f|^2$.

31 $-\frac{14}{81}$

33 $\frac{1}{2}\pi$

35 LHS $= [\text{grad } f(\mathbf{a})]\cdot(\mathbf{v} + \mathbf{w})$
$= [\text{grad } f(\mathbf{a})]\cdot\mathbf{v} + [\text{grad } f(\mathbf{a})]\cdot\mathbf{w} = $ RHS

37 $(tx)^2 + (ty)(tz) = t^2(x^2 + yz)$; degree 2

39 $(tx)^3 + (ty)^3 + (tz)^3 - 3(tx)(ty)(tz) = t^3 f$; degree 3

41 $\dfrac{(tx)(ty)(tz)}{(tx)^4 + (ty)^4 + (tz)^4} = \dfrac{f}{t}$; degree -1

43 $[fg](t\mathbf{x}) = f(t\mathbf{x})g(t\mathbf{x}) = t^m f(\mathbf{x}) t^n g(\mathbf{x}) = t^{m+n}[fg](\mathbf{x})$

45 $xf_x(t\mathbf{x}) + yf_y(t\mathbf{x}) + zf_z(t\mathbf{x}) = nt^{n-1}f(\mathbf{x})$. Set $t = 1$.

Section 14-4 Surfaces and Tangent Planes

Page 730

1 $16(x-4) + 16(y-4) + 25(z+25) = 0$
$\mathbf{n} = (1137)^{-1/2}(16, 16, 25)$

3 $2x + 3y + 2z = 14$ $\quad\mathbf{n} = (17)^{-1/2}(2, 3, 2)$

5 $4x + 13y + 46z = 84$
$\mathbf{n} = (2301)^{-1/2}(4, 13, 46)$

7 $x + y + 4z = 0$ $\quad\mathbf{n} = \frac{1}{6}\sqrt{2}(1, 1, 4)$

9 $ex + (e+1)y - ez = e + 1$
$\mathbf{n} = (3e^2 + 2e + 1)^{-1/2}(e, e+1, -e)$

11 $13x + 2y = 26$ $\quad\mathbf{n} = (173)^{-1/2}(13, 2, 0)$

13 At $(0, 0, 0)$, arc sin $\frac{1}{3}\sqrt{3}$; at $(\frac{1}{2}, \frac{1}{2}, \frac{1}{2})$, arc sin $\frac{1}{3}$

15 $(a, \pm a, \pm 1/a^2)$ where $a \neq 0$

17 $z = 0$ $\quad\mathbf{n} = (0, 0, 1)$

19 $z = 4x + 8y - 8$ $\quad\mathbf{n} = \frac{1}{9}(-4, -8, 1)$

21 $z = -4x + 13y - 20$ $\quad\mathbf{n} = (186)^{-1/2}(4, -13, 1)$

23 $z = x + y$ $\quad\mathbf{n} = \frac{1}{3}\sqrt{3}(-1, -1, 1)$

25 $z = \frac{1}{2}x + y + (\ln 4 - \frac{3}{2})$ $\quad\mathbf{n} = \frac{2}{3}(-\frac{1}{2}, -1, 1)$

Section 14-5 Parametric Surfaces and Surfaces of Revolution

Page 735

1 $z = x + y$; plane

3 $x^2 + y^2 = z^2$; cone with axis the z-axis

5 $x^2 + z^2 = a^2$; cylinder of radius a with axis the y-axis

7 None

9 None

11 All $(0, b)$

13 $\mathbf{x} = (1 + h - k, 1 + h + k, 1 + h + 3k)$
$\mathbf{n} = \frac{1}{6}\sqrt{6}(1, -2, 1)$

15 $\mathbf{x} = (\frac{1}{2}\sqrt{2} - \frac{1}{2}k\sqrt{2}, \frac{1}{2}\sqrt{2} + \frac{1}{2}k\sqrt{2}, 2 - h)$
$\mathbf{n} = \frac{1}{2}\sqrt{2}(1, 1, 0)$

17 $\mathbf{x} = (-4 + 4k, -4h, -8 + 12k)$
$\mathbf{n} = \frac{1}{10}\sqrt{10}(-3, 0, 1)$

19 $(\sqrt{x^2 + y^2} - b)^2 + z^2 = a^2$

21 $\mathbf{x} = (u, v, f(u, v))$

23 Since $|\mathbf{x}(s)| = 1$, $\mathbf{x}(s)$ and the unit tangent $\mathbf{t}(s) = d\mathbf{x}/ds$ are perpendicular unit vectors. Now $\mathbf{x}_u \times \mathbf{x}_v = v\mathbf{t} \times \mathbf{x} \neq \mathbf{0}$, so there is a tangent plane.

25 $\mathbf{x}_u(u, v) = \mathbf{t}(u) + k(u)v\mathbf{n}(u)$, where $\mathbf{n}(s)$ is the unit normal to the curve at $\mathbf{x}(s)$, and $\mathbf{x}_v(u, v) = \mathbf{t}(u)$. Also $\mathbf{x}_u \times \mathbf{x}_v = k(u)v\mathbf{t}(u) \times \mathbf{n}(u) \neq \mathbf{0}$ since $k > 0$, $v > 0$, and $\mathbf{t}$ and $\mathbf{n}$ are orthogonal unit vectors.

Section 14-6 Differentials and Approximation

Page 743

1 $12\,dx + 8\,dy$, $\ -9\,dx + dy$

3 $df = (6xy - y^2)\,dx + (3x^2 - 2xy)\,dy$

5 $df = (x^2 + 3y^2)^{-1}(2x\,dx + 6y\,dy)$

7 $df = dx/y^2z - 2x\,dy/y^3z - x\,dz/y^2z^2$

9 $df = e^x\cos y\,dx + (-e^x\sin y + e^y\cos z)\,dy - e^y\sin z\,dz$

11 $d(fg) = (fg)_x\,dx + (fg)_y\,dy + (fg)_z\,dz$
$= (f_x g + fg_x)\,dx + \cdots$
$= (f_x\,dx + f_y\,dy + f_z\,dz)g + f(g_x\,dx + g_y\,dy + g_z\,dz)$
$= (df)g + f(dg)$

13 $x\,dx + y\,dy = (r\cos\theta)(dr\cos\theta - r\sin\theta\,d\theta)$
$+ (r\sin\theta)(dr\sin\theta + r\cos\theta\,d\theta)$
$= r\,dr(\cos^2\theta + \sin^2\theta) = r\,dr$
$x\,dy - y\,dx = (r\cos\theta)(dr\sin\theta + r\cos\theta\,d\theta)$
$- (r\sin\theta)(dr\cos\theta - r\sin\theta\,d\theta)$
$= r^2(\cos^2\theta + \sin^2\theta)\,d\theta = r^2\,d\theta$

15 358.5

17 1.98227

19 12.98077

21 1% decrease

23 Short -202.4 meters

25 $-2(z/x)\,dx - (z/y)\,dy$

27 $(-6z^2\,dx - 16yz\,dy)/(5z^4 + 12xz + 8y^2)$

29 $[-y^2z^3\,dx + 2y(1 - xz^2)\,dy]/(3xy^2z^2 - 2z)$

31 $-dx + (x\,dx - \frac{1}{2}dy)/\sqrt{x^2 - y}$

33 $df = F'[g(\mathbf{x})]\, dg(\mathbf{x})$

35 -1

Section 14-7 Differentiable Functions

Page 749

1 $f(a + u, b + v) = f(a, b) + 3u - 7v, \quad E(u, v) = 0$

3 $f(u, v) = uv^2, \quad H = K = 0, \quad E = uv^2,$
$|E|/|(u, v)| \le |v|^2 \to 0$

5 $f(1 + u, 1 + v) - 1 - (v - u) = u(u - v)/(1 + u)$
$Hu + Kv = -u + v, \quad E(u, v) = u(u - v)/(1 + u),$
$|E| \le 2|\mathbf{u}|^2/(1 + u) < 4|\mathbf{u}|^2$ for $|u| < \frac{1}{2}$,
$|E|/|\mathbf{u}| < 4|\mathbf{u}| \to 0.$

7 Let $f_i(a + u, b + v) = f_i(a, b) + H_i u + K_i v + E_i(u, v)$ for $i = 1, 2$, where $|E_i(u, v)|/|\mathbf{u}| \to 0$. Set $g = f_1 + f_2$. Then $g(a + u, b + v)$
$= g(a, b) + (H_1 + H_2)u + (K_1 + K_2)v + E_1(\mathbf{u}) + E_2(\mathbf{u})$
$= g(a, b) + Hu + Kv + E(\mathbf{u})$
$|E(\mathbf{u})|/|\mathbf{u}| \le |E_1|/|\mathbf{u}| + |E_2|/|\mathbf{u}| \to 0.$

9 **a** Discontinuous
b $\partial f(0, 0)/\partial x = 0, \quad \partial f(0, 0)/\partial y = 0$
c Not differentiable

11 **a** Continuous
b $\partial f(0, 0)/\partial x = 0, \quad \partial f(0, 0)/\partial y = 0$
c Differentiable

13 $\mathbf{x}_1 = (0, \frac{1}{4}), \quad \mathbf{x}_2 = (\frac{1}{16}, 0)$

15 $\mathbf{x}_1 = (\frac{4}{3}, \frac{1}{6}), \quad \mathbf{x}_2 = (\frac{95}{93}, \frac{91}{372}) \approx (1.02, 0.24)$

Section 14-8 Review Exercises

Page 750

1 $e^x \cos y, \quad -e^x \sin y$

3 $\exp(x - y), \quad -\exp(x - y)$

5 $(2xy, x^2)$

7 $(y^2z^3, 2xyz^3, 3xy^2z^2)$

9 $\sqrt{77}$

11 $\mathbf{x} = (1 + 2h, 2 + 2h + k, \; 4 + 4k)$
$\mathbf{n} = (4, -4, 1)/\sqrt{33}$

13 $3(\sin t)(\cos t)(-\cos t + \sin t)$

15 1.05

17 $(2x - yz)/(2z + xy), \quad (2y - xz)/(2z + xy)$

Chapter 15 Surfaces and Optimization

Section 15-1 Quadric Surfaces

Page 758

1 $\frac{1}{4}x^2 + y^2 + \frac{1}{4}z^2 = 1$
ellipsoid

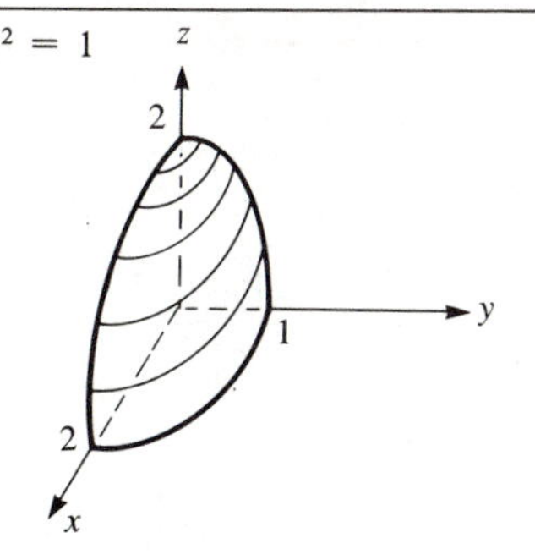

3 $x^2 + y^2 - z^2 = 1$
hyperboloid of one sheet

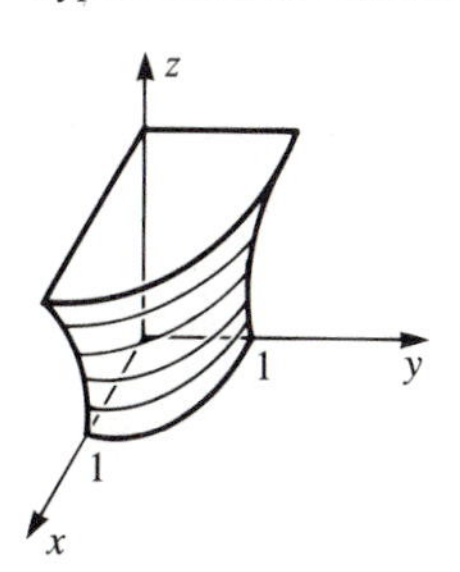

5 $x^2 - y^2 + z^2 = 1$
hyperboloid of one sheet

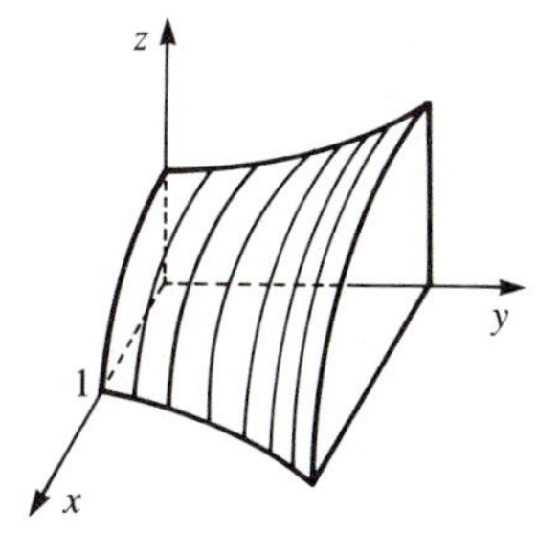

7 $z = x^2 + y^2$
paraboloid (of revolution)

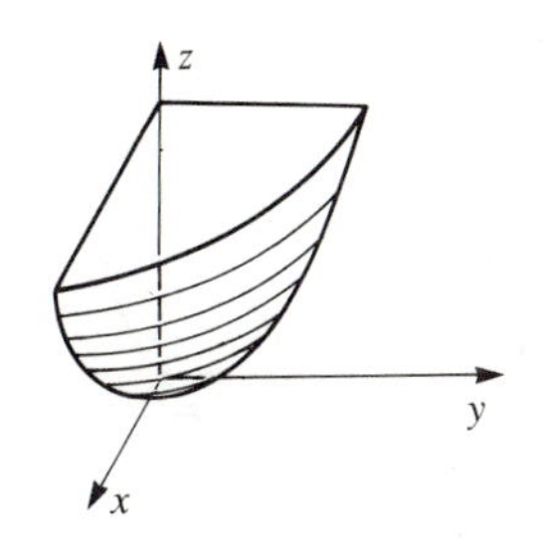

9 $z = -x^2 + y^2$
hyperbolic paraboloid

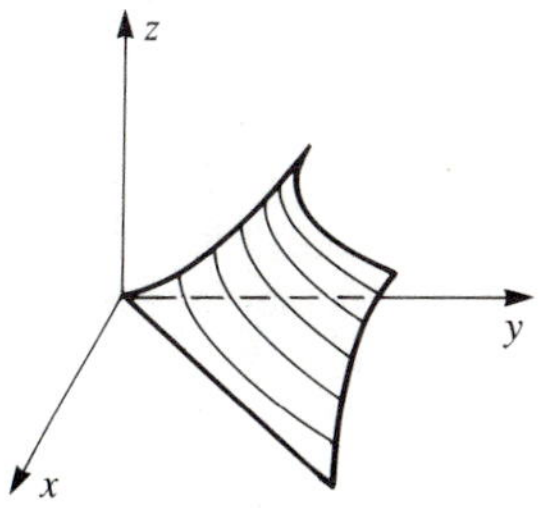

11 $(x + 1)^2 + y^2 = z + 1$
circular paraboloid, vertex $(-1, 0, -1)$, axis parallel to z-axis

13 $x^2/a^2 + y^2/a^2 = (z - 1)^2$
right circular cone generated by revolving the line $y = a(z - 1), \; x = 0$ about the z-axis

15 $x = y^2 + z^2$
paraboloid (of revolution)

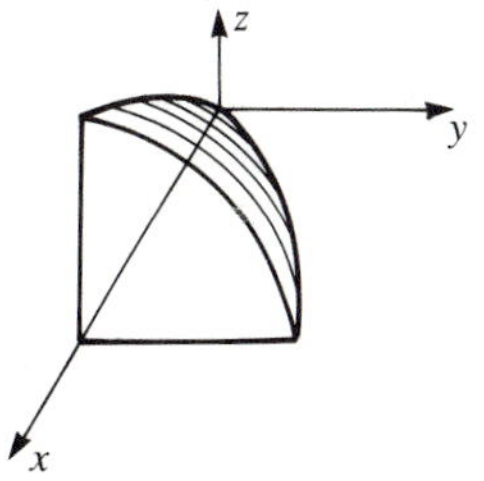

19 $xy = 1$
hyperbolic cylinder

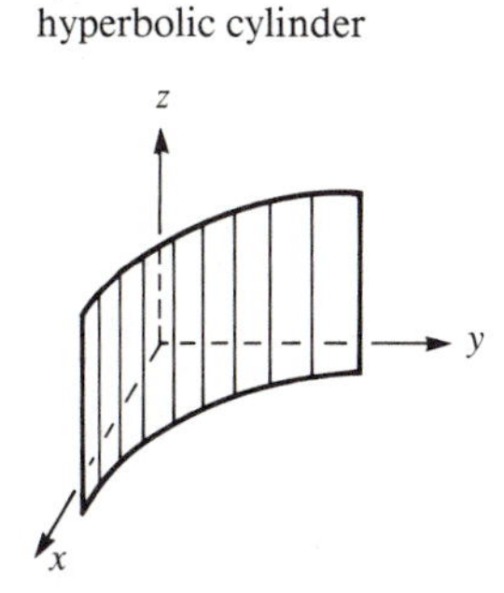

17 $x - z = 1$; plane

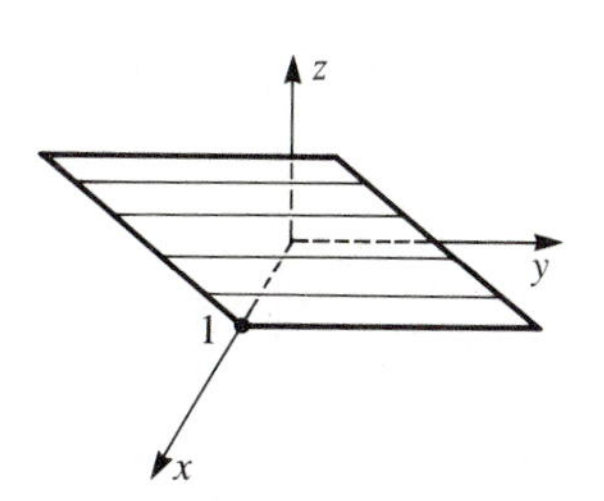

21 $x = z^2$
parabolic cylinder

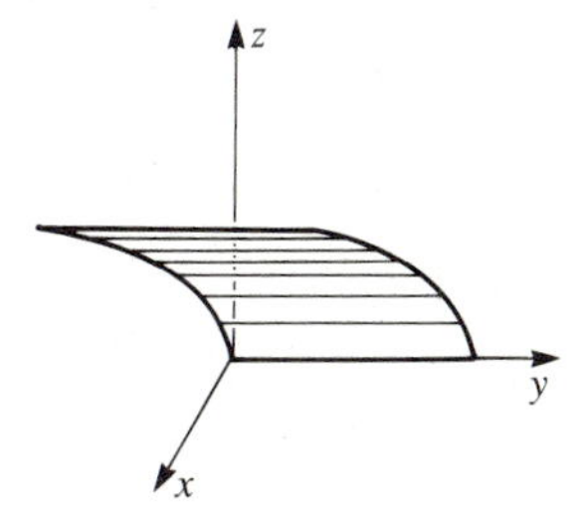

23 $z = x^2 - x$
$z + \frac{1}{4} = (x - \frac{1}{2})^2$
parabolic cylinder

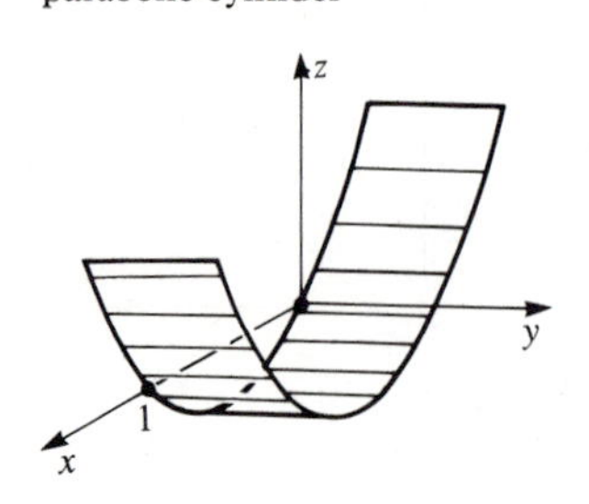

25 $y^2 = z^2 + 4x^2$
cone, axis the y-axis

27 $f(x/z, y/z) = 0$

29 Each cross section is $\dfrac{y^2}{b^2} + \dfrac{z^2}{c^2} = 1 - \dfrac{k^2}{a^2}$, which can be written in the form $\dfrac{y^2}{B^2} + \dfrac{z^2}{C^2} = 1$, where $B = \lambda b$, $C = \lambda c$. The eccentricity of this ellipse depends only on $B/C = b/c$, independent of λ, hence of k.

31 $(\mathbf{x} \cdot \mathbf{n})^2 = \mathbf{x} \cdot \mathbf{x} \cos^2\alpha$

33 The tangent planes $z = 2(ux/a^2 + vy/b^2) - k$ all pass through $(0, 0, -k)$.

35 $df = d\mathbf{x}\,A\mathbf{x}' + \mathbf{x}A\,d\mathbf{x}' = 2\mathbf{x}A\,d\mathbf{x}'$ because A is symmetric, hence $\operatorname{grad} f = 2\mathbf{x}A$.

37 By Exercise 35, $\operatorname{grad}(\mathbf{x}A\mathbf{x}') = 2\mathbf{x}A$, so $\operatorname{grad} f = 2(\mathbf{x}A + \mathbf{b})$

39 The tangent plane at (u, v, w) is $ux/a^2 + vy/b^2 + wz/c^2 = 1$. It is parallel to $\mathbf{n} = (\ell, m, n)$ provided its normal is orthogonal to $\mathbf{n}$. The condition is $(u/a^2, v/b^2, w/c^2) \cdot (\ell, m, n) = 0$. Thus (u, v, w) lies on a *plane* (through $\mathbf{0}$), so the required set of points is a plane section of the ellipsoid, hence an ellipse.

41 The tangent plane at (ℓ, m, n) is $2\ell x + 2my = n + z$, and $\mathbf{a}$ is on this plane if $2a\ell + 2bm = c + n$. Now replace (ℓ, m, n) by (x, y, z); then $\mathbf{a}$ is on the tangent plane at $\mathbf{x}$ provided $x^2 + y^2 = z$ and $2ax + 2by = z + c$. The latter is the equation of a plane.

43 Take $\mathbf{a} = (a, b, a^2 - b^2)$. Then the tangent plane is $2a(x - a) - 2b(y - b) = z - (a^2 - b^2)$. Eliminate z from this and the equation $z = x^2 - y^2$ of the surface:
$2a(x - a) - 2b(y - b) = (x^2 - a^2) - (y^2 - b^2)$,
$(x - a)^2 = (y - b)^2$, $y - b = \pm(x - a)$.
The lines are $\{y - b = \pm(x - a),$
$2a(x - a) - 2b(y - b) = z - (a^2 - b^2)\}$.

45 $(0, \pm a\cos\alpha, a\cos\alpha\cot\alpha)$

47 $2x^2 + 2y^2 - 4(z - \frac{1}{2})^2 = 1$; hyperboloid of one sheet

Section 15-2 Optimization: Two Variables

Page 766

1 $f_{\max} = f(0, 0) = 4$, no minimum

3 $f_{\min} = f(2, -3) = 0$, no maximum

5 $f_{\min} = f(0, 0) = 4$, no maximum

7 $f_{\max} = f(\pm\frac{1}{2}\sqrt{2}, \pm\frac{1}{2}\sqrt{2}) = 1/2e$
$f_{\min} = f(\pm\frac{1}{2}\sqrt{2}, \mp\frac{1}{2}\sqrt{2}) = -1/2e$

9 $f_{\max} = f(\frac{1}{3}\pi + 2m\pi, \frac{1}{3}\pi + 2n\pi) = \frac{3}{2}\sqrt{3}$
$f_{\min} = f(-\frac{1}{3}\pi + 2m\pi, -\frac{1}{3}\pi + 2n\pi) = -\frac{3}{2}\sqrt{3}$

11 $f_{\min} = f(\sqrt[5]{2}, \frac{1}{2}\sqrt[5]{8}) = \frac{5}{2}\sqrt[5]{2}$

13 $f_{\min} = f(\sqrt[3]{a^2/b}, \sqrt[3]{b^2/a}) = 3\sqrt[3]{ab}$

15 $f_{\min} = f(1, 6) = -432$

17 No maximum; $f_{\min} = f(\frac{1}{2}, \frac{1}{2}) = \frac{17}{4}$

19 $f_{\min} = f(0, y) = 0$; $f_{\max} = f(\pm r, 0) = r^4$

21 $V_{\max} = \frac{8}{9}\sqrt{3}$ for a cube

23 $\sqrt[3]{2V}$ by $\sqrt[3]{2V}$ by $\frac{1}{2}\sqrt[3]{2V}$

25 $3a^2V^{2/3}$

27 $x + y + z\sqrt{2} = 2\sqrt[4]{8}$

Section 15-3 Optimization: Three Variables

Page 772

1 $4\sqrt{6}$

3 $\frac{8}{3}\sqrt{3}$

5 $\frac{1}{2}\sqrt{a^2 + b^2 + c^2}$

7 $\frac{1}{96}\sqrt{6}$, $-\frac{1}{96}\sqrt{6}$

9 $(\frac{1}{4}L)^4$

11 $y = \frac{3}{2}x - \frac{1}{6}$

13 $y = 2x - \frac{1}{2}$

15 $y = 0.19x + 4.85$

17 $y = -\frac{4}{5}$

19 $y = -0.35x + 5.7$

21 $y = x - \frac{1}{6}$

23 $y = (18 - 6e)x + (4e - 10)$

25 $y = \frac{24}{181}(2x + 51/x)$

Section 15-4 Lagrange Multipliers: Two Variables

Page 777

1 $f_{\max} = f(\pm\frac{1}{2}\sqrt{2}, \pm\frac{1}{2}\sqrt{2}) = \frac{1}{2}$
$f_{\min} = f(\pm\frac{1}{2}\sqrt{2}, \mp\frac{1}{2}\sqrt{2}) = -\frac{1}{2}$

3 $f_{\max} = f(\mathbf{a}) = \sqrt{13}$, $f_{\min} = f(-\mathbf{a}) = -\sqrt{13}$, where $\mathbf{a} = \frac{1}{13}\sqrt{13}\,(4, 9)$

5 $f_{\max} = f(\frac{1}{3}\sqrt{3}, \pm\frac{1}{3}\sqrt{6}) = \frac{2}{9}\sqrt{3}$
$f_{\min} = f(-\frac{1}{3}\sqrt{3}, \pm\frac{1}{3}\sqrt{6}) = -\frac{2}{9}\sqrt{3}$

7 No maximum; $f_{\min} = f(\frac{2}{5}\sqrt{5}, \frac{4}{5}\sqrt{5}) = \sqrt{5}$

9 No maximum; no minimum

11 $f_{\max} = f(4, 2) = 15$; no minimum

13 $f_{\max} = f(\frac{9}{16}, \frac{1}{16}) = 9^3/16^4$
$f_{\min} = f(1, 0) = f(0, 1) = 0$

15 $f_{\max} = f(0, 0, 0) = 0$
$f_{\min} = f(-\frac{8}{9}, -\frac{8}{27}) = -\frac{32}{27}$

17 No maximum; $f_{\min} = f(\frac{16}{9}, \frac{64}{27}) = -\frac{32}{27}$

19 The square: $d_{\min} = \frac{1}{4}\sqrt{2}$

21 Radius $\sqrt{A/6\pi}$, height $2\sqrt{A/6\pi}$

23 $\frac{1}{4}\sqrt{2}$

25 $2\sqrt{2} - 2$

27 $27\sqrt{5}$

29 If $x = y = 0$, the inequalities are true. Otherwise, set $b = (x^q + y^q)^{1/q}$. Then $b^p \le x^p + y^p \le 2^{1-p/q}b^p$, that is, $b^p/2 \le (x^p + y^p)/2 \le b^p/2^{p/q}$, $b/2^{1/p} \le [(x^p + y^p)/2]^{1/p} \le b/2^{1/q}$, etc.

Section 15-5 Lagrange Multipliers: Three Variables

Page 782

1 $f_{\max} = f(\mathbf{a}) = \sqrt{30}$, $f_{\min} = f(-\mathbf{a}) = -\sqrt{30}$ where $\mathbf{a} = \frac{1}{30}\sqrt{30}\,(2, 1, -5)$

3 No maximum; $f_{\min} = f(6, 12, 18) = 36$

5 $f_{\max} = \frac{5}{9}\sqrt{5}$, $f_{\min} = -\frac{5}{9}\sqrt{5}$; the maximum is taken at the points $(\pm\sqrt{\frac{5}{3}}, \pm\sqrt{\frac{5}{3}}, \pm\frac{1}{3}\sqrt{5})$ with an odd number of $+$ signs, the minimum at the points with an even number of $+$ signs.

7 $f_{\min} = f(\text{boundary}) = 0$ $\quad f_{\max} = f(\frac{1}{3}a, \frac{1}{3}b, \frac{1}{3}c) = \frac{1}{27}abc$

9 $a = 2\lambda x,\ b = 2\lambda y,\ c = 2\lambda z$
Multiply by x, y, z and sum: $f = 2\lambda r^2$.
Multiply by a, b, c and sum: $a^2 + b^2 + c^2 = 2\lambda f$.
Hence $f^2/r^2 = a^2 + b^2 + c^2$,
$f_{max} = r\sqrt{a^2 + b^2 + c^2}$ (and $f_{min} = -f_{max}$).

11 $f \leq \dfrac{3r}{1 + r^2}$ with maximum $\frac{3}{2}$ and $r = 1$

13 Cube

15 Radius $r = (3V/\pi\sqrt{5})^{1/3}$, cone height $\frac{2}{5}\sqrt{5}\,r$, cylinder height $\frac{1}{5}\sqrt{5}\,h$

17 $\frac{1}{27}r^6$ for $x^2 = y^2 = z^2 = \frac{1}{3}r^2$

19 Maximize $f = xy + yz + zx$ on $x^2 + y^2 + z^2 = r^2$. The multiplier equations are $y + z = 2\lambda x,\ z + x = 2\lambda y,$ $x + y = 2\lambda z$. Subtract the first two: $y - x = 2\lambda(x - y)$, etc. We conclude either $x = y = z$ or $2\lambda = -1$. If $x = y = z$, then $3x^2 = r^2,\ f = x^2 + x^2 + x^2 = r^2$. If $2\lambda = -1$, then $x + y + z = 0$. Square: $r^2 + 2f = 0$, $f = -\frac{1}{2}r^2$. Therefore $-\frac{1}{2}r^2 \leq f \leq r^2$.

21 $A_{max} = \frac{1}{9}s^2\sqrt{3}$, equilateral triangle

23 $f'(x) = (\sin^2 x - x^2)/(x^2 \sin^2 x) < 0$ because $0 < \sin x < x$ for $0 < x < \pi$.

25 $(1, 1, 1)$ **27** $(1, 0, 0)$

29 $(\sqrt[3]{\frac{1}{2}}, y, z)$ where $y \geq 0,\ z \geq 0,\ y^2 + z^2 = \sqrt[3]{2}$

31 $3^{1-p/q}b^p,\ b^p$

33 No maximum; minimum $\sqrt[3]{62}$ at $(\sqrt[3]{62}, 0, 0)$

35 $3k^2\sqrt{3}$

37 $(\frac{5}{6}, \frac{1}{3}, -\frac{1}{6})$

39 $6 + \frac{2}{9}\sqrt{3}$ ft³, $6 - \frac{2}{9}\sqrt{3}$ ft³

41 $\frac{3}{4}$

43 No maximum; minimum $= 4\sqrt{10}$

45 $\partial C/\partial x_i = \lambda \partial f/\partial x_i$, that is, $p_i = \lambda \partial f/\partial x_i$
Multiply by x_i and sum, using Euler's relation $\sum x_i \partial f/\partial x_i = f,\ C = \lambda f$.
Hence $p_i/(\partial f/\partial x_i) = \lambda = C/f = C/k$.

Section 15-6 Review Exercises

Page 784

1 $x^2 = 4y^2 + 9z^2 = 25$
ellipsoid

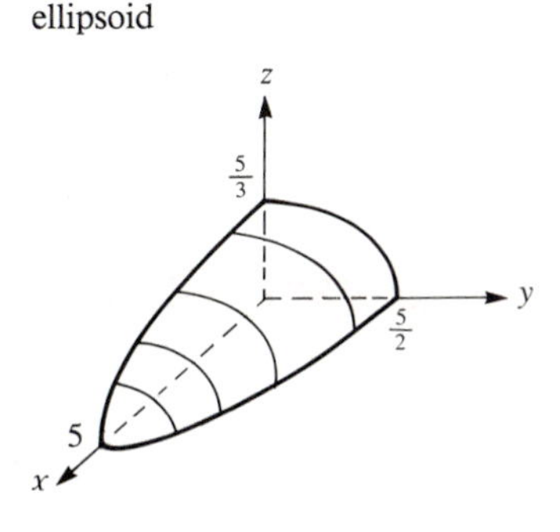

3 $-x^2 - y^2 + z^2 = 9$
hyperboloid of two sheets

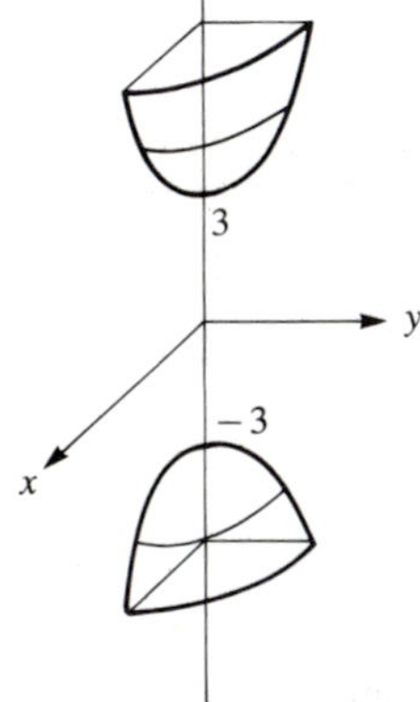

5 $2x - 2y + 3z = 17$

7 a^2

9 $3(\frac{3}{4}\pi^2)^{1/6}$, arbitrarily large

11 $x^3 - \frac{1}{3}$

13 6912

15 $\frac{1}{8}(\sin\alpha)/(1 + \sin\frac{1}{2}\alpha)^2$

17 $\frac{1}{36}\sqrt{3}$

19 $4\sqrt[4]{abc}$

21 $\sqrt{14}$

23 $\frac{1}{3}$

Chapter 16 Higher Partial Derivatives

Section 16-1 Mixed Partials

Page 791

1 $-f,\ -3f,\ -9f$

3 $2 \arcsin y,\ 2x/\sqrt{1 - y^2},\ x^2y/(1 - y^2)^{3/2}$

5 $2a,\ 2b,\ 2c$

7 Both equal $-2/y^3$

9 Both equal $mnx^{m-1}y^{n-1}$

11 Both equal $-2(x + y)/(x - y)^3$

13 $24xy^3,\ 36x^2y^2$

15 $xy^2 \sin xy - 2y \cos xy,\ x^2y \sin xy - 2x \cos xy$

17 $-2e^{2y}\sin(x + y) - e^{2y}\cos(x + y)$,
$3e^{2y}\cos(x + y) - 4e^{2y}\sin(x + y)$

19
$$\begin{bmatrix} 0 & 1 & 1 \\ 1 & 0 & 1 \\ 1 & 1 & 0 \end{bmatrix}$$

21
$$-[\sin(x + 2y + 3z)]\begin{bmatrix} 1 & 2 & 3 \\ 2 & 4 & 6 \\ 3 & 6 & 9 \end{bmatrix}$$

23 0

25 10, 20

27 10

29 $f(x, y) = g(x) + xh(y) + k(y)$

31 Any cubic polynomial in x and y

33 $f(x, y) = ax + by + c$

35 No solution

37 $f(x, y, z) = g(y, z)$

39 $f_x = f_u u_x + f_v v_x = f_u + f_v$. Similarly,
$f_{xx} = f_{uu} + 2f_{uv} + f_{vv},\ f_y = f_u - f_v,\ f_{yy} = f_{uu} - 2f_{uv} + f_{vv},$
$f_{xx} - f_{yy} = 4f_{uv}$.

41 $x = r\cos\theta,\ y = r\sin\theta,\ f_r = f_x\cos\theta + f_y\sin\theta,$
$f_{rr} = f_{xx}\cos^2\theta + 2f_{xy}\cos\theta\sin\theta + f_{yy}\sin^2\theta,$
$f_\theta = r[-f_x\sin\theta + f_y\cos\theta],$
$f_{\theta\theta} = r[-f_x\cos\theta - f_y\sin\theta]$
$+ r^2[f_{xx}\sin^2\theta - 2f_{xy}\sin\theta\cos\theta + f_{yy}\cos^2\theta].$
Now combine: $f_{rr} + f_r/r + f_{\theta\theta}/r^2 = f_{xx} + f_{yy}$.

43 By the chain rule, $h_u = f_x \cdot 1 + f_y \cdot 2 = 0$;
hence $h(u, v) = g(v)$, that is, $f(u, 2u + v) = g(v)$.
But $v = (2u + v) - 2u = y - 2x$;
hence $f(x, y) = f(u, 2u + v) = g(v) = g(y - 2x)$.

45 $k\,\partial^2 u_n/\partial x^2 = -k(n\pi/L)^2 u_n = \partial u_n/\partial t$
and $u_n(0, t) = u_n(L, t) = 0$.

47 Direct substitution yields $a = (n^2\pi^2 k/L^2) + (v^2/4k)$.

49 $V_{xx} = a^2V$ and $V_{yy} = -a^2V$, so $V_{xx} + V_{yy} = 0$.

51 $z_{xx} = -(m\pi/a)^2 z$ and $z_{yy} = -(n\pi/a)^2 z,\ z_{tt} = -p^2 z$,
so there is one condition:
$$p^2/c^2 = (m^2/a^2 + n^2/b^2)\pi^2$$

Section 16-2 Taylor Approximation

Page 797

1 $p_1(x, y) = 1 + 2(x - 1) + 2(y - 1)$
$p_2(x, y) =$
$p_1(x, y) + (x - 1)^2 + 4(x - 1)(y - 1) + (y - 1)^2$

3 $p_1(x, y) = 0, \quad p_2(x, y) = xy$

5 $p_1 = 1, \quad p_2 = 1 + (x - 1)y$

7 $p_1 = -x - (y - \frac{1}{2}\pi), \quad p_2 = p_1$

9 $p_1 = (x - \frac{1}{2}) + 2(y - \frac{1}{4})$
$p_2 = p_1 - \frac{1}{2}(x - \frac{1}{2})^2 - 2(x - \frac{1}{2})(y - \frac{1}{4}) - 2(y - \frac{1}{4})^2$

11 1.12000

13 3.99930

15 $r_1(x, y) =$ the first-degree part of $p_1(x)q_1(y)$
$= g(b)p_1(x) + f(a)q_1(y) - f(a)g(b)$
$r_2(x, y) =$ the second-degree part of $p_2(x)q_2(y)$
$= g(b)p_2(x) + f(a)q_2(x) + p_1(x)q_1(x) - f(a)g(b) - r_1(x, y)$

17 $p_1(x, y) = 1 + \frac{1}{2}x + y$ at $(0, 0)$. Also $f_{xx} = -1/4(1 + x + 2y)^{3/2}$, $f_{xy} = -1/2(\cdots)^{3/2}$, $f_{yy} = -1/(\cdots)^{3/2}$. Hence $|f_{xx}|$, $|f_{yy}|$ are bounded by

$$\frac{1}{(1 + x + 2y)^{3/2}} \le \frac{1}{(1 - 0.1 - 0.2)^{3/2}} = \frac{1}{(0.7)^{3/2}} < \frac{1}{(0.64)^{3/2}} = \frac{1}{(0.8)^3} = \frac{1}{0.512} < 2.$$

Take $M_2 = 2$: $|r_1(x, y)| < M_2|(x, y)|^2 \le 2(0.02) = 0.04$.

19 $p_1(x, y, z) =$
$f(a, b, c) + f_x \cdot (x - a) + f_y \cdot (y - b) + f_z \cdot (z - c)$
$p_2(x, y, z) = p_1(x, y, z)$
$+ \frac{1}{2}[f_{xx} \cdot (x - a)^2 + f_{yy} \cdot (y - b)^2 + f_{zz} \cdot (z - c)^2 + 2f_{xy} \cdot (x - a)(y - b) + 2f_{yz} \cdot (y - b)(z - c) + 2f_{zx} \cdot (z - c)(x - a)]$
All partials evaluated at (a, b, c)

21 $p_2 = 1 + x + 2y + x^2 + xy + 4y^2$, the quadratic part of $1 + z + z^2 + \cdots$
$= 1 + (x + 2y - 3xy) + (x + 2y - 3xy)^2 + \cdots$

Section 16-3 Second Derivative Test: Two Variables

Page 803

1 Local minimum

3 Saddle point

5 Maximum

7 Minimum

9 Saddle point

11 Local minimum

13 $(0, 0)$, saddle point; $(\pm\sqrt{2}, \mp\sqrt{2})$, local minima

15 $(0, 0)$, local maximum; $(0, \pm 2)$, local minima; $(\pm\sqrt{2}, 0)$, saddle points

17 $(0, 0)$, local maximum; $(\frac{5}{2}, -\frac{5}{4})$, saddle point

19 $(\pm\frac{1}{2}\sqrt{2}, \pm\frac{1}{2}\sqrt{2})$, saddle points; $(\frac{1}{6}\sqrt{6}, -\frac{1}{6}\sqrt{6})$, local maximum; $(-\frac{1}{6}\sqrt{6}, \frac{1}{6}\sqrt{6})$, local minimum

21 $(1, 1)$, $(-2, 4)$, saddle points; $(-\frac{1}{2}, \frac{11}{8})$, local maximum

23 (a, c), strong local maximum; (b, d), strong local minimum; (a, d), (b, c), saddle points

25 Regular \$8.00, special \$8.50

Section 16-4 More on Stability

Page 808

1 $D_3 = 0$, inconclusive; minimum

3 $D_3 = 0$, inconclusive; minimum

5 $D_3 = 0$, inconclusive; minimum

7 $D_3 = 0$, inconclusive; neither

9 $H = 0$, inconclusive; minimum

11 $H = 0$, inconclusive; neither

13 $(0, 0, 0)$, local minimum

15 $(-\frac{1}{2}, -1, \frac{3}{2})$, local minimum

17 Local minimum

19 Local maximum

21 Saddle point

23 $(\frac{8}{3}, -\frac{4}{3}, \frac{16}{9})$

25 $f_{max} = f(\pm\frac{1}{2}\sqrt{2}, \pm\frac{1}{2}\sqrt{2}) = \frac{1}{2}$
$f_{min} = f(\pm\frac{1}{2}\sqrt{2}, \mp\frac{1}{2}\sqrt{2}) = -\frac{1}{2}$

Section 16-5 Theory [Optional]

Page 816

1 From $|x| \le |\mathbf{x}|$ and $|y| \le |\mathbf{x}|$ follows $|f| \le 4|\mathbf{x}|^2$, hence $f \to 0$ as $\mathbf{x} \to \mathbf{0}$. This proves continuity at $\mathbf{0}$; elsewhere it is obvious.

3 $[f(x, 0) - f(0, 0)]/x = x \to 0$; hence $f_x(0, 0) = 0$. Similarly, $f_y(0, 0) = 0$. As in Exercise 1, $|f_x| \le 14|\mathbf{x}| \to 0$, etc.

5 $f(x, y) = f(y, x)$ implies $f_x(x, y) = f_y(y, x)$. Now apply $\partial/\partial x$ again: $f_{xx}(x, y) = f_{yy}(y, x)$. Set $x = y = c$.

7 $g'' = f_x x'' + f_y y'' + (f_{xx}x' + f_{xy}y')x' + (f_{xy}x' + f_{yy}y')y'$
$= (f_{xx}x'^2 + 2f_{xy}x'y' + f_{yy}y'^2) + (f_x x'' + f_y y'')$

9 Set $A = f_{xx}(0, 0) > 0$, $B = f_{xy}(0, 0)$, $C = f_{yy}(0, 0) > 0$, and $\mathbf{x}(t) = (Bt, -At)/(A^2 + B^2)^{1/2}$. Then $g''(0) = A(AC - B^2)/(A^2 + B^2) > 0$; hence $AC - B^2 > 0$.

11 Neither

13 Positive definite

15 Negative definite

17 Neither

19 $ac - b^2 = (AC)(BD) - \frac{1}{4}(AD + BC)^2$
$= -\frac{1}{4}(AD - BC)^2 < 0$

21 Since f is continuous on a bounded closed set, the circle, f has a minimum there. But $f(x, y) = (x - 3y)^2 + y^2 > 0$ on the circle; hence the minimum is positive.

23 $\frac{1}{2}(3 - \sqrt{5})$

25 If $a = 0$, then $Q = (2bx + y)y$. If $a \ne 0$, then $at^2 + 2bt + c = 0$ has two *real* roots λ and μ; hence $at^2 + 2bt + c = a(t - \lambda)(t - \mu)$. Replace t by x/y and multiply by y^2:

$$Q(x, y) = a(x - \lambda y)(x - \mu y) = (ax - a\lambda y)(x - \mu y)$$

Section 16-6 Review Exercises

Page 816

1 $f_{xx} = 2xy/(x^2 + y^2) = -f_{yy}$

3 $f_{xx} = 20x^3 - 60xy^2 = -f_{yy}$

5 $f_{xx} = 2(y^2 - x^2)/(x^2 + y^2)^2 = -f_{yy}$

7 x

9 $x - y$

11 $0.9999e$

13 Local minimum

15 Local minimum

17 $f_{min} = f(3, -5) = -34$

19 $(-\frac{8}{3}, -\frac{10}{3})$, saddle point

21 $(\pm 1, \pm 1, 0)$, local minimum

Chapter 17 The Double Integral

Section 17-2 Rectangular Domains

Page 831

1 $\frac{1}{3}y^3,\ 20x$

3 $\frac{1}{10}[(1+y)^5+(1-y)^5]$, $\frac{32}{5}[(x-1)^5-(x-2)^5]$

5 $\frac{1}{4}\pi(1+y^2),\ \frac{32}{3}/(1+x^2)$

7 $\frac{15}{2}$

9 $\frac{4}{9}$

11 0

13 $-\frac{21}{4}$

15 $\ln\frac{4}{3}$

17 0

19 $\frac{1}{2}\pi$

21 $\frac{9}{2}\ln 3+3\ln 2-\frac{15}{4}$

23 $\frac{16}{3}$

25 $\frac{16}{3}$

27 $\frac{46}{3}$

29 $\displaystyle\int_a^b dx\int_{-c}^{c} f(x,y)\,dy=\int_a^b 0\cdot dx=0$
because for each x, the integrand $f(x,y)$ is an odd function of y.
$\displaystyle\int_a^b x^2\,dx\int_{-c}^{c} y^3\,dy=\int_a^b x^2\cdot 0\,dx=0$

31 $A=\displaystyle\iint f(x,y)\,dx\,dy$ over the square

Section 17-3 Domains between Graphs

Page 838

1 $e-1$

3 1

5 $\frac{1}{30}$

7 $\frac{4}{63}$

9 $\frac{16}{3}$

11 $\frac{1}{2}\ln\frac{3}{2}-\frac{1}{8}(\frac{1}{16}-\frac{1}{81})$

13 $2\sinh^2\frac{1}{2}=\frac{1}{2}(e+e^{-1})-1$

15 $\frac{1}{3}$

17 $\displaystyle\int_0^3 dx\int_x^{2+x/3} f(x,y)\,dy$

19 $\displaystyle\int_1^3 dy\int_{(y-9)/2}^{(y-1)/2} f(x,y)\,dx$

21 $\displaystyle\int_0^{10} dx\int_0^{\sqrt{10x-x^2}} f(x,y)\,dy=\int_0^5 dy\int_{5-\sqrt{25-y^2}}^{5+\sqrt{25-y^2}} f(x,y)\,dx$

23 $\displaystyle\int_0^1 dx\int_{3-3x}^{3\sqrt{1-x^2}} f\,dy=\int_0^3 dy\int_{1-y/3}^{\sqrt{1-y^2/9}} f\,dx$

25 $2(e^2-e)$

27 0

29 $\frac{13}{48}$

31 $\frac{4}{77}$

33 $\frac{5}{4}\sqrt{5}$

35 Both equal the integral of $f(x,y)$ over the triangle $0\le y\le x\le 1$; vertices $(0,0)$, $(1,0)$, $(1,1)$.

37 $\displaystyle\int_2^3 dy\int_4^{y^2} f\,dx$

39 By symmetry $I=\displaystyle\iint_{\mathbf{E}} f(x)f(y)\,dx\,dy$, where
$\mathbf{E}=\{x\mid 0\le y\le x\le 1\}$. Hence
$2I=\displaystyle\iint_{\mathbf{D}}+\iint_{\mathbf{E}}=\int_0^1 f(x)\,dx\int_0^1 f(y)\,dy=0$

Section 17-4 Arbitrary Domains

Page 842

1 $x^2+y^2\le 1$,
$y+x^2\ge 0$

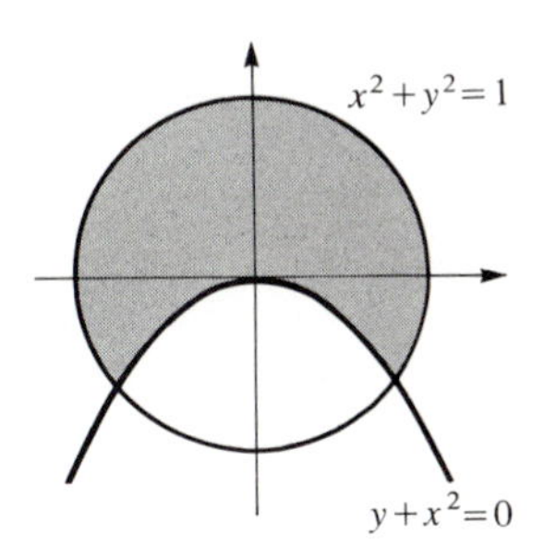

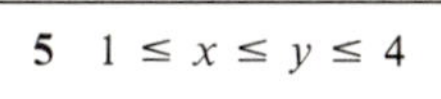
5 $1\le x\le y\le 4$

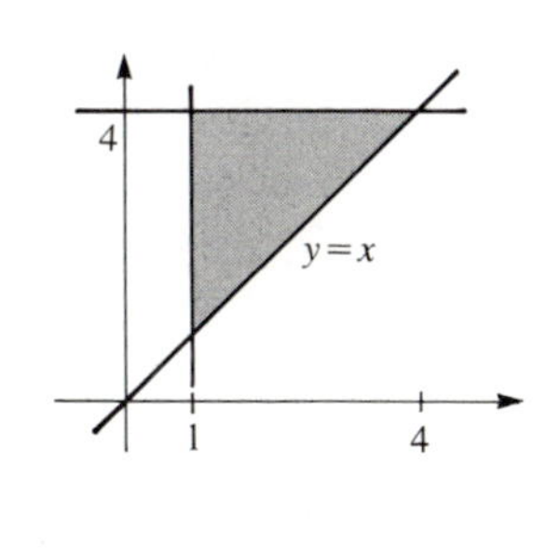

3 $x^2+y^2\ge 1$,
$(x-2)^2+y^2\le 9$

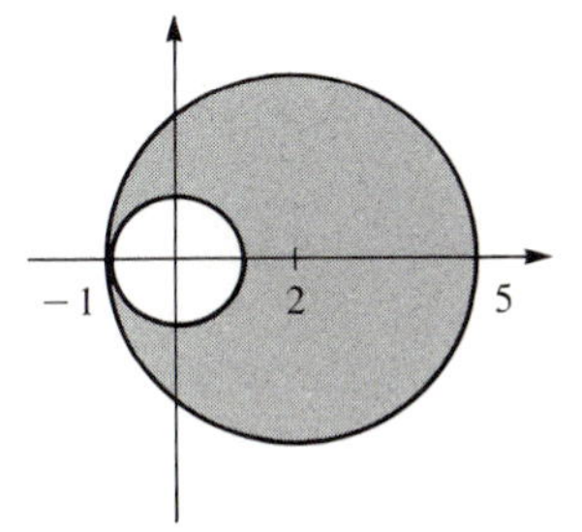

7 $(x+y)^2\le 1$,
$(x-y)^2\le 1$

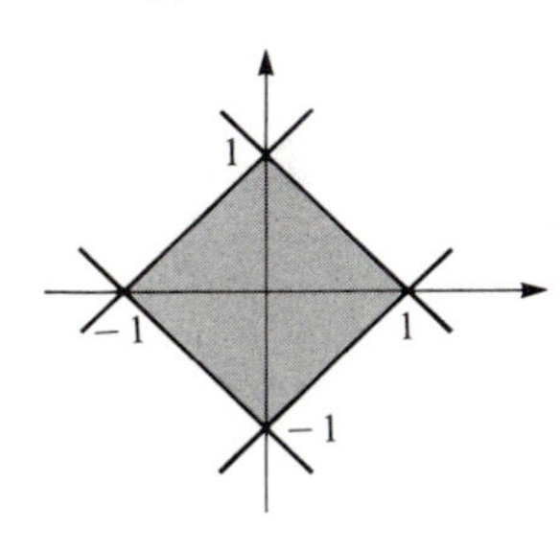

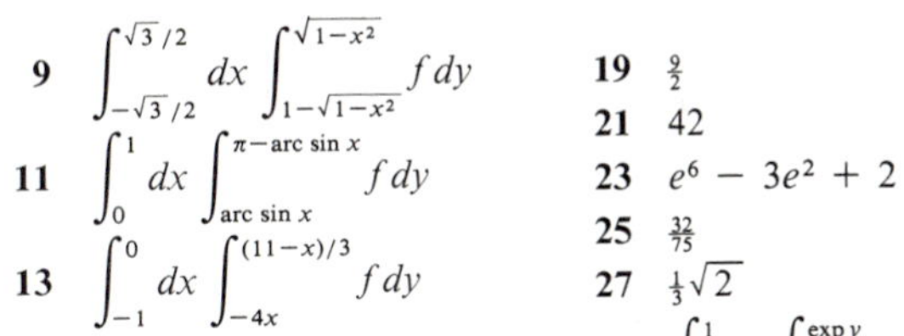

9 $\displaystyle\int_{-\sqrt{3}/2}^{\sqrt{3}/2} dx\int_{1-\sqrt{1-x^2}}^{\sqrt{1-x^2}} f\,dy$

11 $\displaystyle\int_0^1 dx\int_{\arcsin x}^{\pi-\arcsin x} f\,dy$

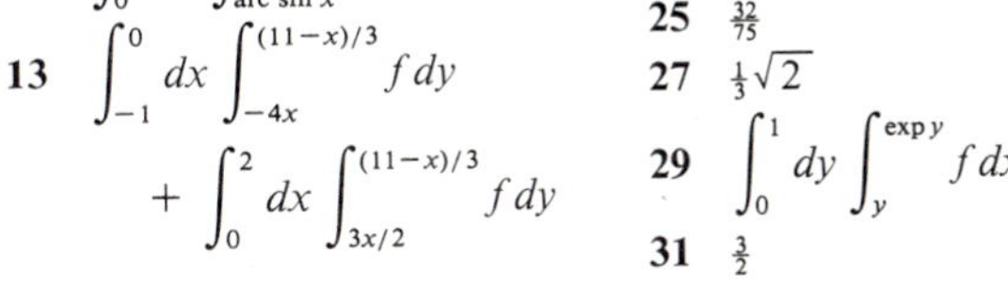

13 $\displaystyle\int_{-1}^0 dx\int_{-4x}^{(11-x)/3} f\,dy+\int_0^2 dx\int_{3x/2}^{(11-x)/3} f\,dy$

15 1

17 $8\ln 2$

19 $\frac{9}{2}$

21 42

23 e^6-3e^2+2

25 $\frac{32}{75}$

27 $\frac{1}{3}\sqrt{2}$

29 $\displaystyle\int_0^1 dy\int_y^{\exp y} f\,dx$

31 $\frac{3}{2}$

33 $3+4\ln 2$

35 $\frac{5}{3}$

Section 17-5 Polar Coordinates

Page 852

1 0

3 $\pi a^2(2\ln 2a-1)$

5 $\frac{1}{8}\pi a^6$

7 $\pi\ln 2$

9 $\frac{15}{8}\pi$

11 $\frac{15}{16}$

13 $\frac{2}{3}\pi a^3$

15 $\frac{5}{12}\pi$

17 $\frac{20}{27}$

19 $\frac{3}{640}\pi$

21 $\frac{1}{24}\pi$

23 $AD-BC$

25 $-2(u+v)$

27 Clearly line segments go to line segments, so it suffices to check where the vertices go:

(u,v)	$(a,0)$	$(b,0)$	(b,b)	(a,a)
(x,y)	$(a,0)$	$(b,0)$	$(0,b)$	$(0,a)$

Note that the counterclockwise order is preserved. Finally, $J=1$.

29 $\int_a^b uf(u)\,du$

31 $1/v$

33 $\frac{1}{2}(ab)^{3/2}(b-a)$

35 The domain $\mathbf{E}$ is bounded by the horizontal lines $v=0$, $v=b-a$ and lies between the branches of the hyperbola $u^2-v^2=4ab$.

Section 17-6 Applications

Page 860

1 $F=(1-e^{-t})/t,\quad F'=(te^{-t}-1+e^{-t})/t^2$

3 $F=[(t+1)^{n+1}-t^{n+1}]/(n+1),\quad F'=(t+1)^n-t^n$

5 $\mathbf{x}=u(\mathbf{b}-\mathbf{a})+v(\mathbf{c}-\mathbf{a}),\quad u\ge 0,\quad v\ge 0,\quad u+v\le 1.$ $A=\frac{1}{2}|\mathbf{a}\times\mathbf{b}+\mathbf{b}\times\mathbf{c}+\mathbf{c}\times\mathbf{a}|$

7 $\mathbf{x}=(a\cos u, a\sin u, v),\quad 0\le u\le 2\pi,\quad 0\le v\le h.$ $A=2\pi ah$

9 $\mathbf{x}=((b+a\cos\phi)\cos\theta,\ (b+a\cos\phi)\sin\theta,\ a\sin\phi),$ $0\le\theta\le 2\pi,\quad 0\le\phi\le 2\pi.\quad A=4\pi^2ab$

11 $$A=\int_0^{2\pi}dv\int_0^{\pi}[c^2(\sin^2u)(b^2\cos^2v+a^2\sin^2v)+a^2b^2\cos^2u]^{1/2}(\sin u)\,du$$

13 $A=\sqrt{1+a^2+b^2}\cdot\text{area}(\mathbf{D})$

15 2π

17 $4(\pi-2)a^2$

19 $2\pi\int_a^b x(s)\,ds$

21 $-\frac{1}{2}\pi$

23 $+\infty$

25 $\pi/(1-p)$

27 $\frac{1}{2}\sqrt{\pi/a}$

29 $f_n'=-f_{n+1}$

31 $+\infty$

Section 17-7 Physical Applications

Page 869

1 $\frac{9}{4},\ (\frac{5}{9},\frac{5}{9})$

3 $3,\ (\frac{4}{9},0)$

5 $\frac{5}{4},\ (\frac{7}{15},\frac{7}{15})$

7 $50,\ (0,\frac{59}{20})$

9 $\frac{4}{3},\ (0,\frac{2}{5})$

11 $\frac{1}{3},\ (\frac{3}{4},\frac{3}{10})$

13 $\frac{1}{2}\pi ab,\ (0,\frac{4}{3}b/\pi)$

15 $\frac{1}{4}\pi a^2,\ (4a/3\pi, 4a/3\pi)$

17 $\frac{1}{2}a^2\beta,\ (2a/3\beta)(\sin\beta, 1-\cos\beta)$

19 $\frac{1}{2}\pi a,\ (2a/\pi, 2a/\pi)$

21 $\frac{1}{2}ak\pi^2,\ (-4a/\pi^2, 2a/\pi)$

23 $12,\ (\frac{3}{2},1)$

25 $V=2\pi\bar{y}A=2\pi(\frac{4}{3}a/\pi)(\frac{1}{2}\pi a^2)=\frac{4}{3}\pi a^3$

27 A right triangular domain of base a revolves about its leg of length h into a right circular cone of volume $V=\frac{1}{3}\pi a^2h$. The area is $A=\frac{1}{2}ah$ and $\bar{x}=\frac{1}{3}a$, so $V=2\pi\bar{x}A$.

29 $A=4\pi a^2,\ L=\pi a,$ and $A=2\pi\bar{y}L,$ so $4\pi a^2=2\pi^2a\bar{y},\ \bar{y}=2a/\pi.$

31 $\frac{1}{2}Ma^2$

33 $\frac{3}{2}Ma^2$

35 $\frac{7}{20}M$

Section 17-8 Approximate Integration

Page 874

1 1.776

3 0.6377

5 0.6888

7 0.05208

9 1.1847

11 0.52772

13 1.6502

15 0.064

17 <0.004

19 $$\left(\tfrac{1}{3}h\sum B_ip_i\right)\left(\tfrac{1}{3}k\sum C_jq_j\right)=\tfrac{1}{9}hk\sum B_iC_jp_iq_j=\tfrac{1}{9}hk\sum A_{ij}f_{ij}$$

21 Both sides evaluate to $A+\frac{1}{2}B+\frac{1}{2}C+\frac{1}{4}D$

23 0.0487

25 $$\left|\iint\right|\le\tfrac{1}{2}N\iint y(1-y)\,dx\,dy+\tfrac{1}{2}M\int_0^1 x(1-x)\,dx=\tfrac{1}{12}N+\tfrac{1}{12}M$$

27 Apply Exercise 26 to $g(u,v)=f(a+hu, c+kv)$, where $0\le u\le 1,\ 0\le v\le 1$. Then

$$|g_{uu}|=|h^2f_{xx}|\le h^2M\quad\text{and}\quad|g_{vv}|\le k^2N$$

Thus

$$\left|\int_c^d dy\int_a^b f(x,y)\,dx-\text{trap. approx.}\right|=hk\left|\int_0^1 dv\int_0^1 g(u,v)\,du-\text{trap. approx.}\right|\le\tfrac{1}{12}hk(h^2M+k^2N)$$

29 $|\text{error}|<\frac{1}{8}$

Section 17-9 Review Exercises

Page 875

1 $2P=3C$

3 $\frac{1}{8}a^4$

5 $(2/a)(e^{2ab}-1)$

7 $V=\int_a^b A(z)\,dz$

9 $\frac{1}{2}aL$

11 $\frac{1}{4}a^2(2\cot\alpha-\pi+2\alpha),\ \frac{1}{6}a^3(1-\sin\alpha)^2/\sin\alpha$

13 $\frac{1}{6}a(6A^2-3\pi aA+4a^2)$

15 $\frac{2}{3}\pi(3+4\pi^2),\ 3(3+4\pi^2)^{-1}(6\pi, 5-4\pi^2)$

17 By Exercise 16, $\left(\int f\right)^2-\int f^3=2\iint(\text{non-negative})\ge 0.$

19 The change of variables $s=x-t$ does the trick.

21 $$\begin{aligned}L(f*g)(s)&=\int_0^\infty e^{-sx}\left(\int_0^x f(t)g(x-t)\,dt\right)dx\\&=\int_0^\infty f(t)\left(\int_t^\infty e^{-sx}g(x-t)\,dx\right)dt\\&=\int_0^\infty f(t)e^{-st}\left(\int_t^\infty e^{-s(x-t)}g(x-t)\,dx\right)dt\\&=\int_0^\infty f(t)e^{-st}\left(\int_0^\infty e^{-sy}g(y)\,dy\right)dt\\&=\left(\int_0^\infty f(t)e^{-st}\,dt\right)\left(\int_0^\infty e^{-sy}g(y)\,dy\right)\\&=L(f)(s)\cdot L(g)(s)\end{aligned}$$

23 First, $p-p_0=kp=k\int_0^z\rho(u)\,du$. Differentiate with respect to z: $dp/dz=k\rho(z)$. Therefore $\rho(z)$ is an exponential function, $\rho(z)=\rho_0e^{kz}$, so $p=k^{-1}(\rho-\rho_0)=k^{-1}\rho_0(e^{kz}-1)$.

25 $(2\pi\rho_0/k^3)[(ka-1)e^{ka}+1-\frac{1}{2}a^2k^2]$

27 $\iint_{\mathbf{D}}(x-y)[f(x)-f(y)]\,dx\,dy\ge 0$ where $\mathbf{D}$ denotes the square $a\le x\le b,\ a\le y\le b$. Expand and iterate.

29 Pythagorean theorem

Chapter 18 Multiple Integrals

Section 18-1 Triple Integrals

Page 884

1 $\frac{1}{3}$
3 $\frac{9}{2}\ln\frac{4}{3}$
5 2
7 540
9 $\frac{11}{60}$
11 $\frac{1}{48}$
13 $\frac{1}{15}$
15 $\frac{1}{720}$
17 $\frac{1}{24}$
19 $\frac{5}{8}$
21 $\frac{33}{20}$
23 54
25 $\frac{1}{216}ka^9$
27 $\frac{1}{2}\int_0^a (a-x)^2 g(x)\,dx$
29 $\frac{1}{3}$
31 $\frac{7}{24}$
33 $2^6/9!$

Section 18-2 Cylindrical Coordinates

Page 891

1 $r(a^{-1}\cos\theta + b^{-1}\sin\theta) + c^{-1}z = 1$
3 $r = 2a\sin\theta$
5 $\frac{1}{16}a^4b^2$
7 $\frac{2}{15}a^5$
9 $\frac{1}{2}\pi(e-1)^2$
11 $\frac{2}{3}\pi$
13 $\frac{25}{8}\pi a^4$
15 2π
17 $\frac{1}{3}a^3b^2$
19 16π
21 $\mathbf{x}' = r'\mathbf{u} + r\theta'\mathbf{w} + z'\mathbf{k}$
23 $L = \int_a^b [(r')^2 + r^2(\theta')^2 + (z')^2]^{1/2}\,dt$
25 $\mathbf{x}_s = r_s\mathbf{u} + r\theta_s\mathbf{w} + z_s\mathbf{k}$
$\mathbf{x}_t = r_t\mathbf{u} + r\theta_t\mathbf{w} + z_t\mathbf{k}$
27 $A =$
$$\iint_{\mathbf{D}} \sqrt{r^2\left[\frac{\partial(\theta,z)}{\partial(s,t)}\right]^2 + \left[\frac{\partial(z,r)}{\partial(s,t)}\right]^2 + r^2\left[\frac{\partial(r,\theta)}{\partial(s,t)}\right]^2}\,ds\,dt$$
29 $2\pi a^2$
31 $df = f_r\,dr + f_\theta\,d\theta + f_z\,dz$

Section 18-3 Spherical Coordinates

Page 898

1 $\rho = 2a\cos\phi$
3 $\rho\sin^2\phi = \cos\phi$
5 $\rho\sin\phi = 2a\cos\theta$
7 $\frac{1}{16}\pi a^4$
9 $\dfrac{4\pi a^{n+3}}{n+3}$
11 $\frac{1}{36}\pi$
13 $\frac{4}{35}\pi a^7$
15 $\frac{32}{105}\pi a^7$
17 $\frac{1}{15}\pi\sqrt{2}\,a^5$
19 $-\frac{4}{9}\pi$
21 $\frac{1}{4}\pi$
23 $\operatorname{grad} f = f_\rho\boldsymbol{\lambda} + \dfrac{1}{\rho}f_\phi\boldsymbol{\mu} + \dfrac{1}{\rho\sin\phi}f_\theta\boldsymbol{\nu}$
25 $\mathbf{v} = \rho'\boldsymbol{\lambda} + \rho\phi'\boldsymbol{\mu} + \rho(\sin\phi)\theta'\boldsymbol{\nu}$
27 $L = \int_a^b \sqrt{(\rho')^2 + \rho^2(\phi')^2 + \rho^2(\sin^2\phi)(\theta')^2}\,dt$
29 $\int_a^b \sqrt{1+\theta^2\sin^2\alpha}\,d\theta$
31 $\mathbf{x}_u = \rho_u\boldsymbol{\lambda} + \rho\phi_u\boldsymbol{\mu} + \rho(\sin\phi)\theta_u\boldsymbol{\nu}$
$\mathbf{x}_v = \rho_v\boldsymbol{\lambda} + \rho\phi_v\boldsymbol{\mu} + \rho\theta_v(\sin\phi)\boldsymbol{\nu}$
33
$$A = \iint_{\mathbf{D}} \rho\left\{\rho^2\sin^2\phi\left[\frac{\partial(\phi,\theta)}{\partial(u,v)}\right]^2 + \sin^2\phi\left[\frac{\partial(\theta,\rho)}{\partial(u,v)}\right]^2 + \left[\frac{\partial(\rho,\phi)}{\partial(u,v)}\right]^2\right\}^{1/2} du\,dv$$
35 $\frac{1}{2}(\sin\alpha)(2\pi b^2 + \pi a^2)$
37 $\frac{1}{48}a^2b^2c^2$
39 $\frac{1}{20}\pi abc^4$
41 $I = |[\mathbf{b}-\mathbf{a}, \mathbf{c}-\mathbf{a}, \mathbf{d}-\mathbf{a}]| \iiint f[\mathbf{x}(\mathbf{u})]\,du\,dv\,dw$
taken over $u \ge 0,\ v \ge 0,\ w \ge 0,\ u+v+w \le 1$

Section 18-4 Center of Gravity and Moments of Inertia

Page 906

[If $\delta =$ constant, we take $\delta = 1$.]

1 $(\frac{3}{8}a, \frac{3}{8}a, \frac{3}{8}a)$
3 $\frac{3}{8}(1+\cos\alpha)(0,0,1)$
5 $\bar{x} = \bar{y} = 0,\ \bar{z} = \frac{3}{8}(b^4 - a^4)/(b^3 - a^3)$
7 $\bar{x} = \bar{y} = 0,\ \bar{z} = \frac{1}{2}a(1+\cos\alpha)$
9 $\bar{x} = -\frac{1}{4}a,\ \bar{y} = 0$
11 $(0, 0, \frac{2}{3}a\cos\alpha)$
13 $\frac{1}{2}a(1,1,1)$
15 $\bar{x} = \frac{1}{4}\pi a(\sin\alpha)/\alpha,\ \bar{y} = \bar{z} = 0$
17 Let b be the bottom radius, a the top radius, $b > a$. Measure z upwards from the bottom. Then
$$\bar{z} = \frac{1}{4}h\frac{b^2 + 2ab + 3a^2}{b^2 + ab + a^2}$$
19 $\frac{3}{8}(a, b, c)$
21 $I_x = \frac{1}{3}M(b^2+c^2),\ I_y = \frac{1}{3}M(c^2+a^2),\ I_z = \frac{1}{3}M(a^2+b^2)$
23 $I_x = I_y = \frac{1}{12}M(3a^2+4h^2),\ I_z = \frac{1}{2}Ma^2$
25 $I_x = \frac{1}{12}M(3a^2+4h^2),\ I_y = \frac{1}{12}M(15a^2+4h^2),\ I_z = \frac{3}{2}Ma^2$
27 $I_x = I_y = \frac{3}{20}M(3a^2+2h^2),\ I_z = \frac{3}{10}Ma^2$
29 $I_x = I_y = \frac{1}{6}M(a^2+h^2),\ I_z = \frac{1}{3}Ma^2$
31 $I_x = I_y = I_z = \frac{2}{3}Ma^2$
33 $I_x = I_y = \frac{1}{4}M(5a^2+2b^2),\ I_z = \frac{1}{2}M(3a^2+2b^2)$
35
$$\iiint x^2\,dM = \iiint y^2\,dM = \iiint z^2\,dM = \frac{1}{3}\iiint \rho^2\,dM$$
$$I_z = \frac{2}{3}\iiint \rho^2 dM = \frac{2}{3}\delta\int_0^{2\pi} d\theta \int_0^\pi \sin\phi\,d\phi \int_0^a \rho^4\,d\rho = \frac{2}{3}\delta\cdot 2\pi\cdot 2\cdot\frac{1}{5}a^5 = \frac{2}{5}(\frac{4}{3}\pi a^3\delta)a^2 = \frac{2}{5}Ma^2$$
37 $I_{yz} = -\frac{1}{4}Mbc,\ I_{xy} = -\frac{1}{4}Mab,\ I_{zx} = -\frac{1}{4}Mca$
39 $I_{zx} = I_{yz} = -\dfrac{2ah}{3\pi}M,\ I_{xy} = -\dfrac{a^2}{2\pi}M$
41 $I_{xy} = I_{yz} = I_{zx} = -\frac{1}{9}Ma^2$

Section 18-5 Line Integrals

Page 913

1 $-\frac{1}{2}\pi$
3 $\frac{1}{2}(ab + bc + ca)$
5 -2
7 35
9 $b - a$
11 1

13 $-y\,dx + x\,dy$
$= -(r\sin\theta)(dr\cos\theta - r\,d\theta\sin\theta)$
$\quad + (r\cos\theta)(dr\sin\theta + r\,d\theta\cos\theta)$
$= r^2(\sin\theta + \cos^2\theta)\,d\theta = r^2\,d\theta,$
hence $d\theta = (-y\,dx + x\,dy)/r^2$

15 $\frac{2}{9}$

17 $\mathbf{F} = \operatorname{grad}(-\frac{1}{3}\rho^{-3})$

$$\int_{\mathbf{a}}^{\mathbf{b}} \mathbf{F}\cdot\mathbf{dx} = \tfrac{1}{3}|\mathbf{a}|^{-3} - \tfrac{1}{3}|\mathbf{b}|^{-3} \to \tfrac{1}{3}|\mathbf{a}|^{-3} \text{ as } |\mathbf{b}| \to +\infty.$$

19 $\mathbf{F} = \operatorname{grad}\ln\rho$

21 $(\frac{3}{2}, 2, \frac{5}{2})$

23 $(0, \frac{2}{5}, 0)$

25 $(0, -\frac{2}{3}, -2)$

27 $ma(b\sin t - bt\cos t, -b\cos t - bt\sin t, a)$

Section 18-6 Green's Theorem

Page 919

1 12

3 0

5 2

7 $\frac{1}{3}$

9 $\frac{3}{4}\pi ab(a^2 + b^2)$

11 $-\frac{2}{3}a^5$

13 2π

15 0

17 0

19 $\oint_{\partial\mathbf{D}} (-uv_y + vu_y)\,dx + (uv_x - vu_x)\,dy$

$$= \iint_{\mathbf{D}} [(uv_x - vu_x)_x - (-uv_y + vu_y)_y]\,dx\,dy$$

$$= \iint_{\mathbf{D}} [u(v_{xx} + v_{yy}) - v(u_{xx} + u_{yy})]\,dx\,dy.$$

The other terms cancel each other. The proof requires u and v to have continuous second partials on $\mathbf{D}$.

21 $F_y(x, y) = Q(x, y),\ \ G_x(x, y) = P(x, y)$

23 $A = \frac{1}{2}\oint r^2\,d\theta$

Section 18-7 Surface Integrals

Page 926

1 3

3 $\frac{1}{4}\pi a^4$

5 $2\pi - \frac{16}{3}$

7 8

9 0

11 $\dfrac{4\pi a^{n+2}}{n+2}$

Section 18-8 Theorems of Gauss and Stokes

Page 933

1 $\dfrac{\partial}{\partial x}(fA) + \dfrac{\partial}{\partial y}(fB) + \dfrac{\partial}{\partial z}(fC) = (f_xA + fA_x) + \cdots$
$= (f_xA + f_yB + f_zC) + f(A_x + B_y + C_z) = \text{etc.}$

3 Choose the z-axis upward, with 0 at the fluid surface (to avoid any funny stuff with orientation). Then $p = -\delta gz,$ and the buoyant force is

$$\mathbf{F} = -\iint_{\partial\mathbf{D}} p\,d\sigma = -\iiint_{\mathbf{D}} (\operatorname{grad} p)\,dV$$

$$= \delta g \iiint_{\mathbf{D}} \mathbf{k}\,dV = \delta g|\mathbf{D}|\mathbf{k}$$

Thus the force is directed upward, and its magnitude is the mass of fluid displaced by $\mathbf{D}$ times the constant of gravity, i.e., the weight of the displaced fluid. (The first sign is negative because $\mathbf{n}$ is outward, but pressure acts inward.)

5 $\mathbf{0}$

7 $\mathbf{F} = (0, 0, -F_3),\ \ F_3 = 2\pi\delta Gc\left(\dfrac{1}{c} - \dfrac{1}{\sqrt{a^2 + c^2}}\right)$

9 $F_3 = \dfrac{2GMc}{a^2}\left(\dfrac{1}{c} - \dfrac{1}{\sqrt{a^2 + c^2}}\right)$

$\mathbf{F} \to -\dfrac{GM}{c^2}(0, 0, 1)$

11 $\mathbf{F}(\mathbf{x}) = -f(\rho)\mathbf{x}$

13 Let $\mathbf{S}$ be the sphere through $\mathbf{x}$ with center $\mathbf{0}$. Then

$$-4\pi GM = \iint_{\mathbf{S}} \mathbf{F}\cdot d\sigma = -\iint_{\mathbf{S}} f(\rho)\mathbf{x}\cdot d\sigma$$

$$= -f(\rho)\iint_{\mathbf{S}} \mathbf{x}\cdot d\sigma$$

But $\mathbf{x}\cdot d\sigma = (\rho\mathbf{n})\cdot(\mathbf{n}\,d\sigma) = \rho\,d\sigma,$ so

$$\iint_{\mathbf{S}} \mathbf{x}\cdot d\sigma = \rho\iint_{\mathbf{S}} d\sigma = 4\pi\rho^3.$$

Therefore $-4\pi GM = -4\pi\rho^3 f(\rho),\ \ f(\rho) = GM/\rho^3.$

15 $p(\rho) = \frac{2}{3}\pi\delta^2G(a^2 - \rho^2),\ \ p(0) = \frac{2}{3}\pi G\delta^2a^2$

17 $\operatorname{curl}(f\mathbf{A}) = \operatorname{curl}(fA, fB, fC) = ((fC)_y - (fB)_z, \cdots)$
$= (f_yC - f_zB + f(C_y - B_z), \cdots)$
$= (f_yC - f_zB, \cdots) + f(C_y - B_z, \cdots)$
$= (f_x, f_y, f_z) \times (A, B, C) + f\operatorname{curl}\mathbf{A} = \operatorname{grad} f \times \mathbf{A} + f\operatorname{curl}\mathbf{A}$

19 By Exercise 17,
$\operatorname{curl}[f(\rho)\mathbf{x}] = [\operatorname{grad} f(\rho)] \times \mathbf{x} + f(\rho)\operatorname{curl}\mathbf{x}$
$= [(f'(\rho)/\rho)\mathbf{x}] \times \mathbf{x} + \mathbf{0} = \mathbf{0}.$

21 $\operatorname{curl}(\mathbf{a}\times\mathbf{x}) = \operatorname{curl}(bz - cy, cx - az, ay - bx)$
$= ((ay - bx)_y - (cx - az)_z, (bz - cy)_z - (ay - bx)_x,$
$\quad (cx - az)_x - (bz - cy)_y)$
$= (2a, 2b, 2c) = 2\mathbf{a}$

23 $\Delta f = \operatorname{div}(\operatorname{grad} f)$

25 $n(n + 1)\rho^{n-2}$

27 We have $\mathbf{v} = \omega \times \mathbf{x}$ (see page 689), hence $\operatorname{curl}\mathbf{v} = 2\omega$ so $\omega = \frac{1}{2}\operatorname{curl}\mathbf{v}.$

29 0

31 $-\pi$

Section 18-9 Review Exercises

Page 935

1 $2\pi ah$

3 $f(\mathbf{x}) = f[\rho(\rho^{-1}\mathbf{x})] = \rho^n f(\rho^{-1}\mathbf{x}),$ that is,
$f\langle\rho, \phi, \theta\rangle = \rho^n f\langle 1, \phi, \theta\rangle$

5 $\dfrac{12\pi a^{n+3}}{(n+1)(n+3)}$

7 $\rho(b + \rho\sin\phi)$

9 $\frac{1}{3}$

11 $\mathbf{F} = (-F, 0, 0),\ \ F = 4\pi a\delta G/b$

13 By symmetry, $\displaystyle\iiint y^{2n}\,dV = \iiint z^{2n}\,dV$ taken over $\rho \le 1.$ Iterate:

$$\int_0^{2\pi} \sin^{2n}\theta \, d\theta \int_0^{\pi} \sin^{2n+1}\phi \, d\phi = \int_0^{2\pi} d\theta \int_0^{\pi} \cos^{2n}\phi \sin\phi \, d\phi$$

Hence

$$8\int_0^{\pi/2} \sin^{2n}\theta \, d\theta \int_0^{\pi/2} \sin^{2n+1}\theta \, d\theta = \frac{4\pi}{2n+1}$$

15 $4a^2b \arcsin(b/a) + \frac{4}{3}(2a^2 + b^2)\sqrt{a^2 - b^2} - \frac{8}{3}a^3$

17 Parametrize the surface by $\mathbf{x}(s, v) = \mathbf{x}(s) + v\mathbf{t}(s)$. Then $|\mathbf{x}_s \times \mathbf{x}_v| = |(\mathbf{t} + vk\mathbf{n}) \times \mathbf{t}| = vk$; hence

$$A = \int_0^1 v\, dv \int_0^L k(s)\, ds = \tfrac{1}{2}\int_0^L k(s)\, ds$$

19 $\frac{1}{3}\pi$ steradians

21 By the divergence theorem, $\iiint_{\mathbf{D}} (f - \operatorname{div}\mathbf{A})\, dV = 0$ for such $\mathbf{D}$. If $f - \operatorname{div}\mathbf{A} > 0$ at some point $\mathbf{x}_0$, then by continuity $f - \operatorname{div}\mathbf{A} > 0$ at all $\mathbf{x}$ sufficiently near $\mathbf{x}_0$; hence if $\mathbf{D}$ contains $\mathbf{x}_0$ and is sufficiently small, $\iiint_{\mathbf{D}} (f - \operatorname{div}\mathbf{A}) > 0$, a contradiction, etc.

23 By 22,

$$\iint_{\partial\mathbf{D}} (A\lambda)\cdot d\sigma = \iint_{\substack{\phi_0\le\phi\le\phi_1\\ \theta_0\le\theta\le\theta_1}} [A(\rho_1,\phi,\theta)\rho_1^2 - A(\rho_0,\phi,\theta)\rho_0^2]\sin\phi\, d\phi\, d\theta$$

$$= \iint_{\substack{\phi_0\le\phi\le\phi_1\\ \theta_0\le\theta\le\theta_1}} \left(\int_{\rho_0}^{\rho_1} \frac{\partial}{\partial\rho}(\rho^2 A)\, d\rho\right)\sin\phi\, d\phi\, d\theta$$

$$= \iiint_{\mathbf{D}} \frac{\partial}{\partial\rho}(\rho^2 A)\sin\phi\, d\rho\, d\phi\, d\theta$$

$$= \iiint_{\mathbf{D}} \frac{1}{\rho^2}\frac{\partial}{\partial\rho}(\rho^2 A)\rho^2\sin\phi\, d\rho\, d\phi\, d\theta$$

$$= \iiint_{\mathbf{D}} \frac{1}{\rho^2}\frac{\partial}{\partial\rho}(\rho^2 A)\, dV$$

25 $\iint_{\partial\mathbf{D}} (C\nu)\cdot d\sigma = \iiint_{\mathbf{D}} \frac{1}{\rho\sin\phi}\frac{\partial C}{\partial\theta}\, dV$

27 If $\mathbf{A} = A\lambda + B\mu + C\nu$, then

$$\operatorname{div}\mathbf{A} = \frac{1}{\rho^2}(\rho^2 A)_\rho + \frac{1}{\rho\sin\phi}(B\sin\phi)_\phi + \frac{1}{\rho\sin\phi}C_\theta$$

29 $\Delta f = \rho^{-1}(\rho f)_{\rho\rho} = \rho^{-1}(e^{-k\rho})_{\rho\rho} = k^2\rho^{-1}e^{-k\rho} = k^2 f$

31 Expand $(x_1 + \cdots + x_n)^4$ and drop all terms with an odd power of any x_i:

$$\int\cdots\int = \int\cdots\int (nx_1^4 + 6\cdot\tfrac{1}{2}n(n-1)x_1^2x_2^2)\, dx_1\cdots dx_n$$
$$= n\cdot 2^n[\tfrac{1}{5} + \tfrac{1}{3}(n-1)] = \tfrac{1}{15}n2^n(5n-2)$$

Chapter 19 Differential Equations

Section 19-1 Separation of Variables

Page 943

1 $y' - C/x^2 = -xy/x^2 = -y/x$

3 $y' = C = (y-1)/x$

5 $y' = Ce^x + (e^x + xe^x) = (y - xe^x) + (e^x + xe^x) = y + e^x$

7 $y = \pm\sqrt{\frac{2}{3}x^3 + C}$

9 $y = (\sqrt{x} + C)^2$

11 $y = -\ln(-\frac{1}{2}x^2 + C)$

13 $y = C/x$

15 $y = Cx$

17 $y = C\sin^2 x$

19 $y = 1/(C - \sin x)$

21 $y = (x^3 + C)^{1/3}$

23 $y = (\cos C)x + (\sin C)\sqrt{1 - x^2}$

25 $\dfrac{dy}{dx} = x^2 y$

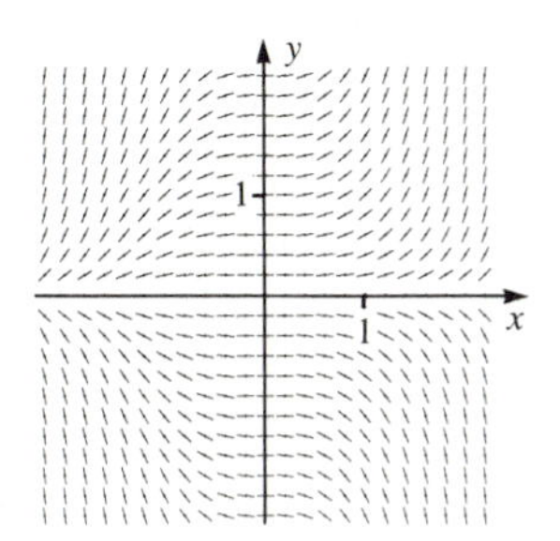

27 $\dfrac{dy}{dx} = 3\dfrac{x^2}{y}$

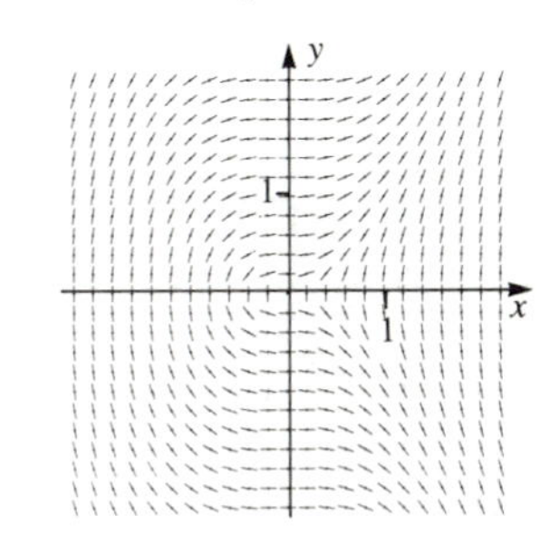

29 $y = Cx^2$; parabolas

31 $y = x + 2x/(C - \ln|x|)$ or $y = x$

33 $y = 0$

35 $y = -\sqrt{\frac{2}{3}x^3 + 1}$

37 $y = 3/\sqrt{x}$

39 $y = (4 - x)/(2 - x)$

41 $\dfrac{dy}{dx} = \begin{cases} 2\sqrt{y} & \text{for } y \ge 0 \\ -2\sqrt{-y} & \text{for } y < 0 \end{cases}$

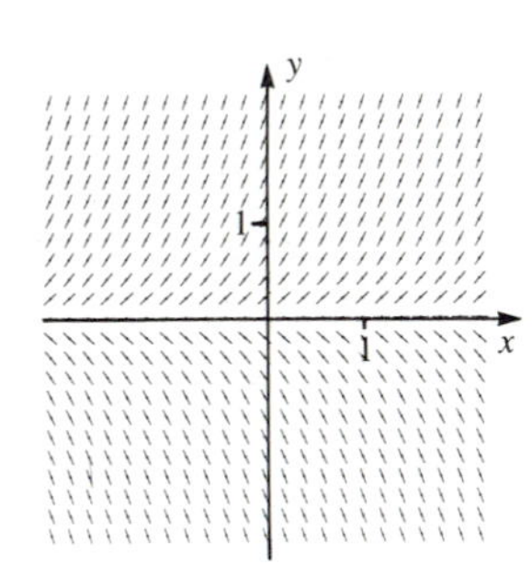

43 $y = x + C/x$

45 $y^2 = Ce^{2x} - x^2$

Section 19-2 First Order Linear Equations

Page 950

1 $y = Kx^{-3}$

3 $y = K\cos\theta$

5 $y = Ke^{-x}/x$

7 $y = -\exp(-\frac{1}{2}x^2)$

9 $y = 2\exp(-\frac{2}{3}x^{3/2})$

11 $y = Ce^{-2x} + \frac{1}{2}x - \frac{1}{4}$

13 $y = Ce^{-4x} + \frac{1}{5}e^x$

15 $y = Ce^{-x} + \frac{1}{3}x^3e^{-x}$

17 $y = Ce^{-5x/2} + \frac{15}{29}\sin x - \frac{6}{29}\cos x$

19 $y = Ce^{-2x} + \frac{1}{2}e^{-x}(\cos x + \sin x)$

21 $i = Ce^{-Rt/L} + E(R^2 + L^2)^{-1}(R\cos t + L\sin t)$

23 $y = Ce^{-x} - x^5 - 5x^4 - 20x^3 - 60x^2 - 120x - 120$
25 $y = Ce^x + 2xe^x$
27 $y = Ce^{3x} - \frac{1}{3}x^2 - \frac{2}{9}x - \frac{2}{27}$
29 $y = C\exp(-\frac{1}{2}x^2) + 1$
31 $y = Cx - \frac{1}{2}x^{-1}$
33 $y = \frac{1}{3} - \frac{1}{3}e^{-3x}$
35 $y = e^{2x} - x - 1$
37 $y = 5e^{2x-1} - (x^2 + 2x + 2)e^x$
39 $y = (x - 1)\exp(x^3)$

Section 19-3 Applications

Page 957

1 $y = Cx^3$
3 $\frac{1}{2}x^2 + y^2 = C^2$; ellipses
5 $t = (50)^2\pi\sqrt{100/980} \approx 2509$ sec
7 $T_0 = \dfrac{150 - 250 \cdot 2^{-3/2}}{1 - 2^{-3/2}} \approx 95.3°C$
9 $40(\ln\frac{9}{5})/(\ln\frac{4}{3}) \approx 81.7$ sec
11 $x = 150t/(t + 1)$
13 $p(t) = \sqrt{r/\alpha}\ \tanh(t\sqrt{r\alpha})$
15

Year	t	P
1790	0	3.93×10^6
1850	60	2.27×10^7
1900	110	7.94×10^7
1950	160	1.73×10^8
2000	210	2.32×10^8

17 $P = M(P_0/M)^{\exp(-kt)}$
19 $x = x(t) = (x_0 - b/a)e^{-at} + b/a$
$x_\infty = b/a$
21 Divide (chain rule):

$$\frac{dx}{dv} = \frac{dx/dt}{dv/dt} = \frac{v}{-kv - g} = -\frac{1}{k} + \frac{g}{k^2}\left(\frac{1}{v + g/k}\right)$$

23 $v(t) = g/k + (v_0 - g/k)e^{-kt}$
$v_\infty = g/k$
25 $m = m_0 e^{-v/c}$
27 $m = m_0\exp[-(v + gt)/c]$

Section 19-4 Second Order Equations

Page 967

1 $y = Ae^{5t} + Be^t$
3 $r = A\cos 2\theta + B\sin 2\theta$
5 $y = Ae^{-t/2} + Be^t$
7 $y = Ae^{-6t} + B$
9 $y = Ae^{\lambda t} + Be^{\mu t}$
11 $x = A\exp(2L^2t) + Bt\exp(2L^2t)$
13 $y = \frac{4}{3}\sin 3t$
15 $y = -2e^{2t} + 3e^t$
17 $y = \frac{9}{8}e^{4t} + \frac{15}{8}e^{-4t}$
19 $y = e^{4t}(3\cos 3t - 4\sin 3t)$
21 $r = e^\theta(-e^{-\pi}\cos\theta + B\sin\theta)$
23 $y = A\cos(\sqrt{6}\,t) + \frac{1}{6}\sqrt{6}\sin(\sqrt{6}\,t)$
25 $y = -3e^{-2t} + B(e^{-4t} - 3e^{-2t})$
27 $x = \frac{1}{3}t$
29 $x = -\frac{1}{2}t - \frac{1}{4}$
31 $x = -t^2 - 10t - 38$
33 $x = \frac{1}{7}e^{3t}$
35 $y = \frac{1}{9}e^{-2x} - x - 1$
37 $y = -\frac{28}{53}\sin 2x - \frac{8}{53}\cos 2x$
39 $y = \frac{1}{6}e^x(\cos x + \sin x)$
41 $y = \frac{1}{2}e^x + \frac{1}{5}e^{2x} + \frac{1}{10}e^{3x} + \frac{1}{17}e^{4x} + \frac{1}{26}e^{5x}$
43 $x = \frac{1}{4}te^{2t} - \frac{1}{4}te^{-2t}$
45 $x = A\cos t + B\sin t + t^2 - 2$
47 $x = Ae^{2t} + Be^{-3t} - \frac{1}{6}te^{-t} + \frac{1}{36}e^{-t}$
49 $x = Ae^{-3t} + B - \frac{1}{5}\cosh 2t + \frac{3}{10}\sinh 2t$
51 $i =$
$e^{-2t/3}[A\cos(\frac{1}{3}\sqrt{2}\,t) + B\sin(\frac{1}{3}\sqrt{2}\,t)] - \frac{10}{17}\cos t + \frac{40}{17}\sin t$
53 $x = 5\cos t - \sin t + 2t - 5$
55 $x = \frac{9}{16}e^{2t} + \frac{7}{16}e^{-2t} - \frac{1}{8}\sin 2t$
57 $x = -\frac{56}{5}e^{3t/4} + 10e^t + \frac{1}{5}e^{2t}$
59 $u' = 0,\ v' = 0$
61 $x = \pm At^{3/2} + C$

Section 19-5 Applications to Vibrations

Page 975

1 $F = -(GMm/R^3)x$
3 $v(10) = 30(32.2)e^{-0.5} \approx 585.9$ ft/sec
Tot. dist. $= 75g + 600g(1 - e^{-0.5}) \approx 10{,}017$ ft
5 $\dfrac{d^2x}{dt^2} + k\dfrac{dx}{dt} = -g,\ x(0) = x_0,\ v(0) = 0$
7 $x = c(e^{-rt} - e^{-st})$ where $0 < r < s,\ c > 0$. Clearly $x(0) = 0$, and $x(t) > 0$ for $t > 0$. Also $x(t) \to 0$ as $t \to +\infty$.

$$\frac{dx}{dt} = c(se^{-st} - re^{-rt})$$

so $dx/dt = 0$ only for

$$e^{(s-r)t} = s/r,\quad t = t_0 = \frac{\ln s - \ln r}{s - r}\ (>0)$$

For $0 < t < t_0$, $dx/dt > 0$; and for $t > t_0$, $dx/dt < 0$; so $x(t)$ increases to its maximum at t_0, then decreases.
9 $x = v_0te^{-pt/2}$. Clearly $x(0) = 0$, $x(t) > 0$ for $t > 0$, and $x(t) \to 0$ for $t \to +\infty$.

$$\frac{dx}{dt} = v_0e^{-pt/2}(1 - \tfrac{1}{2}pt)$$

so $dx/dt = 0$ for $t = t_0 = 2/p$. Clearly $dx/dt > 0$ for $0 < t < t_0$ and $dx/dt < 0$ for $t > t_0$.

11 Differentiate:

$$\frac{d^2I}{dt^2} + \frac{R}{L}\frac{dI}{dt} + \frac{1}{LC}I = A \sin 2\pi\omega t$$

By the text, the steady state solution is

$$I = \frac{A}{Z}\sin(2\pi\omega t - \phi_0)$$

where

$$Z = \left[\left(\frac{1}{LC} - 4\pi^2\omega^2\right)^2 + 4\pi^2\omega^2\frac{R^2}{L^2}\right]^{1/2}.$$

Finally

$$\omega_r = \frac{1}{2\pi}\sqrt{\frac{1}{LC} - \frac{R^2}{2L^2}}$$

13 $2\pi\sqrt{L/g}$

15 The (upward) forces on the projectile are $-mg$ and $-kmy'$, where $k > 0$; hence $my'' = -mg - kmy'$, that is, $y'' + ky' = -g$.
$y(t) = A(1 - e^{-kt}) - (g/k)t$ where $A = (g + kv_0)/k^2$

17 $v(t) = 0$ at $e^{-kt} = g/(k^2A) = g/(g + kv_0)$, $t = t_1 = k^{-1}\ln[(g + kv_0)/g]$. Clearly $v(t) > 0$ for $0 < t < t_1$ and $v(t) < 0$ for $t > t_1$.
Next, $y(t) = 0$ for $t = t_2 > 0$ provided $A(1 - e^{-kt_2}) - (g/k)t_2 = 0$, that is,

$$\left(\frac{g + kv_0}{g}\right)(1 - e^{-kt_2}) = kt_2$$

19
$$\begin{aligned} y'(t_1 + t) + y'(t_1 - t) &= [kAe^{-k(t_1+t)} - g/k] + [kAe^{-k(t_1-t)} - g/k] \\ &= kA[e^{-k(t_1+t)} + e^{-k(t_1-t)}] - 2g/k \\ &= kAe^{-kt_1}(e^{-kt} + e^{kt}) - 2g/k \\ &= (g/k)(e^{kt} + e^{-kt}) - 2(g/k) \\ &= 2(g/k)\cosh kt > 0 \end{aligned}$$

Thus the speed t seconds before the peak exceeds the speed t seconds after the peak. Consequently for $t > 0$,

$$\int_0^t [y'(t_1 + u) + y'(t_1 - u)]\,du > 0$$

that is, $y(t_1 + t) > y(t_1 - t)$. In particular for $t = t_1$, $y(2t_1) > y(0) = 0$, so the projectile is still in the air at time $2t_1$; hence $2t_1 < t_2 = t_1 + (\text{time down})$. Hence it takes longer coming down than going up.

Section 19-6 Numerical Methods
Page 983

1 $y = 2 + 2x + \frac{3}{2}x^2 + \frac{1}{2}x^3 + \cdots$

3 $y = x - \frac{2}{3}x^3 + \cdots$

5 $y = 1 + x + \frac{1}{3}e^{-1}x^3 + \cdots$

7 $y = -1 - \frac{1}{6}x^3 + \cdots$

9 $y = 2 + 5x + 10x^2 + \frac{70}{3}x^3 + \cdots$

11 $y = x - \frac{2}{3}x^3 + \cdots$

13 $y = x + a_4x^4 + \cdots$

15 $y = 1 + 2x + 2x^2 + \frac{17}{6}x^3 + \cdots$

	$n = 2$	$n = 5$	$n = 20$	$n = 50$	$n = 100$
17	0.61719	0.64484	0.64853	0.64869	0.64871
19	4.66797	4.70441	4.70803	4.70812	4.70813
21	0.73779	0.69818	0.69213	0.69179	0.69174
23	0.98601	1.02068	1.02763	1.02804	1.02810
25	0.43750	0.50361	0.50037	0.50006	0.50001

27 $y_1 = y_0 + y_0'h + \frac{1}{2}(G_0 + H_0y_0')h^2 + \cdots$

Section 19-7 Review Exercises
Page 983

1 $y^3 = 2/(C - 3x^2)$

3 $y = -\sqrt{2e^x - 1}$

5 $y = -\frac{1}{12}e^{4x} + Ce^{16x}$

7 $(8000\pi)/(9\sqrt{2g}) \approx 63.1$ sec

9 $y = A\exp(\sqrt{2}\,t) + B\exp(-\sqrt{2}\,t)$

11 $y = Ae^{\sqrt{2}x} + Be^{-\sqrt{2}x} - e^x$

13
$$\begin{aligned} x' &= \frac{1}{k}\cdot\frac{d}{dt}\int_0^t \sin k(t - s)r(s)\,ds \\ &= \frac{1}{k}\left[\sin k(t - s)r(s)\Big|_{s=t} + k\int_0^t \cos k(t - s)r(s)\,ds\right] \\ &= \int_0^t \cos k(t - s)r(s)\,ds \end{aligned}$$

$$\begin{aligned} x'' &= \cos k(t - s)r(s)\Big|_{s=t} - k\int_0^t \sin k(t - s)r(s)\,ds \\ &= r(t) - k^2x \end{aligned}$$

15 $y = a + bx - 3ax^2 - bx^3 + ax^4 + c_6x^6 + \cdots$

17 $z' + (1 - n)p(x)z = (1 - n)q(x)$

19 $dy/dt = 2c\sin x$, $d^2y/dt^2 = 2c(\cos x)(dx/dt) = 2c^2\sin x\cos x = c^2\sin 2x = c^2\sin y$

21 For $n = 10$, $y(2) \approx 1.6068$; for $n = 50$, $y(2) \approx 1.59448$.

Index

C

D

E

F

G

Q

R

S

T

U

V

W

Y

Z